KANGAROOS, WALLABIES and RAT-KANGAROOS

KANGAROOS, WALLABIES and RAT-KANGAROOS

Volume 1

Edited by

Gordon Grigg, Peter Jarman and Ian Hume

Published by

Surrey Beatty & Sons Pty Limited

The National Library of Australia
Cataloguing-in-Publication entry:

Kangaroos, Wallabies and Rat-kangaroos.

Bibliography.
Includes index.
ISBN 0 949324 20 5.
ISBN 0 949324 22 1 (v. 1).
ISBN 0 949324 23 X (v. 2).

1. Marsupialia — Australia. 2. Marsupialia — Papua New Guinea. 3. Kangaroos. 4. Wallabies. 5. Potoroidae. I. Grigg, Gordon C. (Gordon Clifford). II. Jarman, P. (Peter). III. Hume, Ian D. (Ian Douglas).

599.20994

Published November, 1989

PRINTED AND PUBLISHED IN AUSTRALIA BY
SURREY BEATTY & SONS PTY LIMITED
43 Rickard Road, Chipping Norton, NSW 2170

Opening Address — A chromosome phylogeny of Kangaroos

G. B. SHARMAN[1]

The grounds upon which Linnaeus would fain to have banished the whales from the waters, he states as follows: On account of their warm bilocular heart, their lungs, their moveable eyelids, their hollow ears, "penem intrantem feminam mammis lactantem" and finally, "ex lege naturae jure meritoque". I submitted all this to my friends Simeon Macey and Charley Coffin, of Nantucket, both messmates of mine in a certain voyage, and they united in the opinion that the reasons set forth were altogether insufficient. Charley profanely hinted they were humbug.

Herman Melville — "Moby Dick"

SOME fifty thousand years after the discovery of kangaroos by the first human inhabitants of Australia, three hundred and sixty years after the first written observations on their mode of generation and thirty years after the formation of the Australian Mammal Society, by a group of hopefuls on a field excursion to Portee Station, it is appropriate to hold a meeting to find what is known about kangaroos and their allies. The organizers have arranged an interesting agenda and the contributors have laboured to bring completed manuscripts to the meeting. The proceedings will make interesting and informative reading.

There are about fifty Australian species of kangaroos, wallabies and rat-kangaroos grouped in what is now called the superfamily Macropodoidea. My interest in the kangaroos predates the formation of the Society as I hopefully began construction of a chromosome phylogeny of the group thirty-eight years ago. A year previously I had seen some chromosome preparations of kangaroos made by Gordon Lyne, a founder of the Society, while he was an undergraduate student in Botany at the University of Tasmania. Chromosome data for about 12% of the extant species of Australian kangaroos was then available and it was better to travel hopefully. Now it is possible to arrive because over 90% of described species, thought to be extant today, have been studied making the kangaroos chromosomally one of the best known groups.

One group of living macropodoids has a short urogenital sinus, a long posterior vaginal sinus and a large anterior vaginal expansion (Pearson 1950). *Hypsiprymnodon* with a diploid 22 chromosomes is a member of this group as is *Potorous longipes* which has 24. Johnston, Davey and Seebeck (1984) showed that *P. tridactylus* may be derived from a 24 chromosome ancestor by the occurrence of six chromosome fusion events. *Hypsiprymnodon, Aepyprymmus* and *Bettongia* may also be derived from a 22 chromosome ancestor but one with different chromosome morphology to that of *P. longipes. Lagostrophus* has another different 2n=24 chromosome morphology but it is placed with the above potoroids on grounds of reproductive system morphology (Tyndale-Biscoe 1965). Flannery (1983) considered *Lagostrophus* to be a sthenurine which group he placed with Macropodidae rather than Potoroidae.

The remaining living macropodoids have a long urogenital sinus, short posterior vaginal sinus and lack a definite vaginal expansion (Pearson 1950). They were formerly placed in the monophyletic genus *Macropus* but numerous later attempts have been made to split *Macropus* into a number of smaller groups of generic rank. None has been satisfactory. Thus when all the species of Macropodinae listed in the accompanying figure (except *Lagostrophus*) were included in *Macropus* that was a monophyletic group. However, some of the split-off groups including *Macropus*, which became the rump when everything else of considered generic status was removed, now became polyphyletic — that is they included species, the closest living relatives of which were placed in different groups. Admittedly these were not as polyphyletic as Charley Coffin would have the "group" containing both fish and whales to be, but they were far from monophyletic. The

[1]1 Logan Road, Evandale, Tasmania 7212.

Pages v-vii in KANGAROOS, WALLABIES AND RAT-KANGAROOS ed. by G. Grigg, P. Jarman and I. Hume. Surrey Beatty & Sons Pty Limited, New South Wales, Australia, 1989.

subdivision of *Macropus* into subgenera was considered conservative by Dawson and Flannery (1985) and with that I am in agreement.

The accompanying figure represents yet another attempt to regroup the various species of Macropodidae so that each group becomes a monophyletic unit. The numerals on the figure were derived from the following sources:

- 1,2 Changed shape chromosome No. 1 (Sharman 1954; Hayman and Martin 1974).
- 3 Changed shape chromosome No. 6 (unpub.).
- 4-6 Changed shape chromosomes 2, 8, X (Hayman and Martin 1974).
- 7 Changed shape X chromosome (unpub.).
- 8,9ff Numerous changes (Sharman, Close and Maynes 1989).
- 10 Various changes in chromosomes 1, 2 mostly leading to acrocentric shape (unpub.).
- 11 Tandem fusion 2-6 (unpub. with C. E. Murtagh, *not* Sharman 1961).
- 12 Numerous changes not elucidated, derivation of *L. conspicillatus* assumes a like derivation to that of *L. hirsutus*.
- 13-17 Tandem fusion 1-10, centric fusions 2-6, 5-9, changed shape chromosome 7, centric fusion 7-8 (unpub. with P. M. Johnson).
- 18,19 Numerous changes (unpub. with C. E. Murtagh).
- 20-27 Centric fusions 6-10, 1-9, 5-8, changed shape chromosomes 4, 7, 6-10, fusions 4-6-10, 2-7 (Rofe 1978).
- 28,29 Centric fusions 1-8, 5-9 (Rofe 1978).
- 30-34 Centric fusion 1-10 (Rofe 1978; unpub. with C. E. Murtagh for *M. (O). antilopinus, bernardus, M (N.) agilis, dorsalis, rufogriseus*).
- 35 Changed shape chromosome 4 (Sharman 1961; Hayman and Martin 1974).

There are two obvious solutions if taxonomy is to reflect phylogeny. 1. Replace the split-off genera so that *Macropus*, as the sole genus of the subfamily Macropodinae, becomes monophyletic, or 2. Split the subfamily into genera which are monophyletic units as in Figure 1. Thus *Thylogale* (with *Petrogale*) is a monophyletic unit as are *Lagorchestes, Dendrolagus, Onychogalea, Wallabia* with *M. giganteus* and *M. fuliginosus* and *Osphranter* with *Notamacropus,* both as defined by Flannery and Dawson (1985).

The cladistic analysis hopefully follows the rules of Baverstock (1987). I am deeply indebted to Carolyn Murtagh who assisted with the chromosome studies and to Peter Johnson of Queensland National Parks and Wildlife Service who provided some of the material. Research was supported by Macquarie University and the Australian Research Grants Scheme.

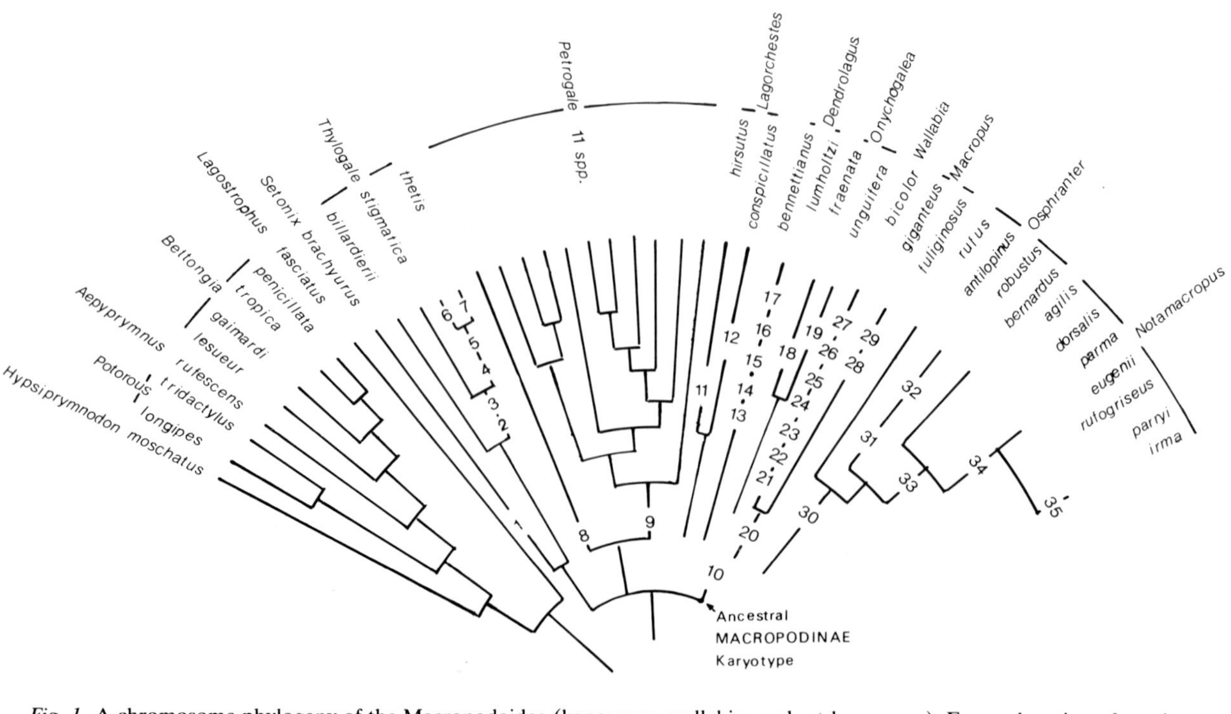

Fig. 1. A chromosome phylogeny of the Macropodoidea (kangaroos, wallabies and rat-kangaroos). For explanation of numbers on cladogram see text.

In conclusion I thank the Australian Mammal Society for inviting me to open the meeting and extend my best wishes to the contributors and participants.

REFERENCES

BAVERSTOCK, P. R., 1987. Modern taxonomic character sets. Pp. 287-93 *in* Fauna of Australia general articles ed. by D. W. Walton. Australian Government Publishing Service: Canberra.

DAWSON, L. AND FLANNERY, T., 1985. Taxonomic and phylogenetic status of living and fossil kangaroos and wallabies of the genus *Macropus* Shaw (Macropodidae: Marsupialia) with a new subgeneric name for the larger wallabies. *Aust. J. Zool.* **33:** 473-98.

FLANNERY, T. F., 1983. Revision in the macropodid subfamily Sthenurinae (Marsupialia: Macropodoidea) and the relationships of the species of *Tropsodon* and *Lagostrophus*. *Aust. Mammal.* **6:** 15-28.

HAYMAN, D. L. AND MARTIN, P. G., 1974. Animal Cytogenetics 4, Chordata 4, Mammalia I: Monotremata and Marsupialia. Gebruder Borntraeger: Berlin.

JOHNSTON, P. G., DAVEY, R. J. AND SEEBECK, J. H., 1982. Chromosome homologies in *Potorous tridactylus* and *P. longipes* (Marsupialia: Macropodidae) based on G-banding patterns. *Aust. J. Zool.* **32:** 319-24.

PEARSON, J., 1950. The relationship of the Potoroidae to the Macropodidae (Marsupialia). *Pap. Roy. Soc. Tas.* **1949:** 211-30.

ROFE, R., 1978. G-banded chromosomes and the evolution of the Macropodidae. *Aust. Mammal.* **2:** 53-63.

SHARMAN, G. B., 1954. The relationships of the quokka *(Setonix brachyurus)*. *West. Aust. Nat.* **4:** 159-68.

SHARMAN, G. B., 1961. The mitotic chromosomes of marsupials and their bearing on taxonomy and phylogeny. *Aust. J. Zool.* **9:** 38-60.

SHARMAN, G. B., CLOSE, R. L. AND MAYNES, G. M., 1989. Chromosome evolution, phylogeny and speciation of rock wallabies (*Petrogale:* Macropodidae). *Aust. J. Zool.* (in press).

TYNDALE-BISCOE, C. H., 1965. The female urogenital system and reproduction of the marsupial *Lagostrophus fasciatus*. *Aust. J. Zool.* **13:** 255-67.

CONTENTS

(VOLUME 1)

LIST OF COLOUR PLATES
(VOLUME 1)

Frontispiece. A female Eastern Grey Kangaroo *(Macropus giganteus)* adopts the correct stance for her young to re-enter the pouch. Photograph by P. J. Jarman.

INTRODUCTION AND ACKNOWLEDGEMENTS

The superfamily Macropodoidea (the macropods) is the most speciose and the most numerous and conspicuous of all marsupial groups, comprising about 50 extant species from Australia and a further 13 or so from Papua New Guinea. Two families are represented, the Macropodidae (kangaroos and wallabies) and the Potoroidae (potoroos, bettongs and rat-kangaroos). In Australia, macropod species make up nearly 40% of the marsupial fauna and about 20% of the total mammalian fauna. They occupy habitats as diverse as tropical rainforests, alpine tussock grasslands and arid inland plains, and range in size from less than 0.5 kg to more than 90 kg.

The largest macropods, Red Kangaroos, have come to be a symbol of Australia. It was appropriate, therefore, that in 1988, Australia's Bicentennial Year, there should be an occasion on which to review our knowledge of the biology of macropod marsupials. This occasion took the form of a Bicentennial Symposium of the Australian Mammal Society. The Symposium was held in May 1988 at The University of Sydney as part of the Society's annual scientific meeting. The name of the Symposium was chosen to be "Kangaroos, Wallabies and Rat-Kangaroos" to emphasise the desire of the convenors to review the biology of all members of the superfamily Macropodoidea, not just the large and conspicuous or economically important species. Indeed, one of the themes running through the Symposium was a common concern over the effects of 200 years of settlement in Australia by non-indigenous peoples on the status of the less conspicuous macropod marsupials.

Topics were therefore chosen and arranged with an eye to both broad coverage of current research areas and the conservation status of kangaroos, wallabies and rat-kangaroos. The publication you now hold is based on the Bicentennial Symposium and hopefully reflects these aims. It contains 43 review papers written by recognized or emerging experts in all facets of the biology of macropods, including their evolutionary history and biogeography, their physiology, reproduction and development, behaviour, ecology and management. For an illustrated compendium of all the Australian species of Macropodoidea, the reader is referred to the Complete Book of Australian Mammals (Edited by R. Strahan, Angus & Robertson 1983.

That the book has come to fruition is a tribute to the authors, reviewers and the publisher, who shared our faith in the project. The editors can now breathe a sigh of relief. Although we were always confident of the success of the venture, to the extent of including the cost of the book at a discounted rate in the Symposium registration fee, ultimately the appearance of the publication depended on Ivor Beatty deciding that it was a project worth doing. Fortunately he did, without any financial guarantee from the Australian Mammal Society.

Our first acknowledgement of gratitude therefore goes to Ivor Beatty and the team at Surrey Beatty and Sons. We are delighted that "Kangaroos, Wallabies and Rat-Kangaroos" has joined the string of significant Surrey Beatty publications on Australia's fauna.

The Australian Mammal Society made a grant of $2000 which was used to subsidize attendance at the Symposium by students and the preparation of the index.

Authors have been generous with original data and we thank them for that and for their participation in the Symposium and the book. Most were very prompt in responding to queries and returning revised manuscripts and corrected page proofs. Each paper was refereed by two experts and we are grateful for their assistance in this task.

Thanks go also to The University of Sydney for allowing us the use of the Stephen Roberts Lecture Theatre free of charge, and to the many post-graduate and honours students and technical staff who helped at the time of the Symposium.

We wish to record also our special thanks to Sylvia Warren, School of Biological Sciences, The University of Sydney, for her pivotal role as Symposium Secretary, and to Fay Marsack, Zoology Department, The University of Queensland, for significant help in the latter stages of the project.

Gordon Grigg
Peter Jarman
Ian Hume

June 1989

Participants in the Bicentennial Symposium on Kangaroos, Wallabies and Rat-Kangaroos, at The University of Sydney, May 12-13, 1988

1. Peter Jarman
2. Karen Higgenbottom
3. Robin Stuart-Dick
4. John Seebeck
5. Randy Rose
6. Dominic Fanning
7. Sandy Ingleby
8.
9. Leon Hughes
10. Lyn Hinds
11. Marilyn Renfree
12. Stuart Cairns
13. Ian Wallis
14. Colin Southwell
15. Chris Johnson
16. Michael Messer
17. Eva Crisp
18. Robert Close
19. Jan Bell
20. Andrew Bennett
21. Ken Richardson
22. Tony Press
23. Gerry Maynes
24. Dan Lunney
25. Geoff Sharman
26. Russ Baudinette
27. Tony Lee
28. Colin Groves
29. Eleanor Russell
30. Tim Flannery
31. Udo Ganslosser
32. Robert Gemmel
33. Jim Merchant
34. Peter Janssens
35. Ian Hume
36. Barry Fox
37. Ken Johnson
38. Brian Green
39. John Nelson
40. John Calaby
41. Ian McDonald
42. Grant Norbury
43. Graeme Coulson
44. Meredith Smith
45.
46. Hugh McCarron
47. Terry Dawson
48. Gordon Sanson
49. Hugh Tyndale-Biscoe
50. Tim Clancy
51. Glen Edwards
52. Tony Pople
53. Steve Cork
54. David Freudenberger
55. Jim Menzies
56. Mike Beal
57. Russell Jones
58. David Croft
59.
60. Joanna Clarke
61. Darren Quin
62. Gordon Grigg

FOREWORD

The Symposium on Kangaroos, Wallabies and Rat-kangaroos held at Sydney University in May 1988 and these handsome volumes arising from it are the Australian Mammal Society's contribution to the celebration of the bicentenary of English settlement in Australia. There is no more fitting subject for such a celebration than this group of marsupials.

Momentous events seemingly happened in years ending in '88'. We can pass over the defeat of the Spanish Armada in 1588 and the sacking of James II and the installation of William and Mary in 1688. Of much more importance to us was the visit in the latter years of the Englishman William Dampier to what is now the Kimberley coast. Dampier was navigator on a buccaneering voyage in the ship *Cygnet* with a quarrelsome crew under Captain Read and a master rejoicing in the name Mr Teat, a patronym that should appeal to mammalogists. Dampier had a lively curiosity about people and places and natural history which he recorded in detail in his journal, but although the *Cygnet* spent two months on the Australian coast, he "saw no sort of Animal" except the tracks of a large dog. After his return Dampier published his journal as *A New Voyage Round the World.* It was an immediate success and initiated a tradition of great travel literature that was to last for almost two centuries. The *New Voyage* and his later book had a lasting influence on English literature, playing a part in the first true English novel *Robinson Crusoe,* and directly inspiring the English literary masterpiece *Gulliver's Travels,* Jonathon Swift's scathing satire on Humankind.

The success of the *New Voyage* also led to Dampier being appointed to the Navy and being sent on an official voyage of discovery in the ship *Roebuck.* On this occasion he spent about five weeks on the western coast, and again he published a detailed narrative, entitled *A Voyage to New Holland, &c. in the year 1699.* It contains the first description of the Banded Hare-wallaby, that Dampier observed on Dirk Hartog Island. This was not the first observation of an Australian macropodoid however, as Dutch navigators had already encountered two other species on islands off the western coast. The Tammar was recorded on islands in Houtman's Abrolhos by Pelsaert in 1629 and the Quokka was observed by Volckersen on Rottnest Island in 1658, and again by de Vlamingh in 1696. For most of the 18th century only one further species was added, the Aru Island Wallaby *Thylogale brunii* which the Dutch traveller de Bruin had seen in captivity in Java. His book of travels published in 1714 contained an engraving of the animal, the first published illustration of any macropodoid.

As the 17th century observers described the animals they saw as kinds of rats, cats or raccoons little notice was taken of them. It was the observations on large kangaroos by James Cook and his scientific party at Endeavour River in 1770 that brought the kangaroo to prominence as a new and very different kind of animal. The publication of the narrative of the voyage with an engraving of the kangaroo in 1773 resulted in the animal becoming a popular conversation piece among the learned and its name, kangaroo, became the first word of Australian Aboriginal origin to be adopted into the English language. Samuel Johnson, who was asked his opinion on most things, declared the voyage a failure – "they have found very little, only one new animal, I think". The engraving, a poor representation, was based on an oil painting by the famous animal painter George Stubbs. It was endlessly copied in popular works for many years and it even appeared on such objects as pottery and medallions.

When the English colony was established at Sydney Cove in 1788 the kangaroo was already in the minds of the new settlers, not only as a natural curiosity but perhaps more importantly as a source of fresh meat. The "founding books" on the colony published between 1789 and 1802 have extensive descriptions of the kangaroo including its appearance, measurements, habits and culinary properties. Few external features escaped these chroniclers. For example, Captain of Marines Watkin Tench in his *Narrative of the Expedition to Botany Bay . . .* (1789) gave a detailed description of a large male kangaroo and rather delicately observed that "the testicles of the male are placed contrary to the usual order of nature".

The year 1788 also saw the foundation of the Linnean Society of London, the oldest society devoted to natural history that is still flourishing. Australian mammals have never figured much in its publications, however the first scientific descriptions and figures of the Thylacine and Tasmanian Devil appeared in its *Transactions* in 1808. In the previous year a description and figure were published of a large wallaby named *Macropus elegans.* This is usually considered indeterminable or at most an inadequate description of *Macropus parryi.* However, I believe that the paper contains internal evidence that it refers to the Red-necked Wallaby, as J. E. Gray believed.

Within a decade or so of the beginning of settlement the kangaroo had become *the* symbol of Australia and has remained so ever since. In 1819 the judge Barron Field published *First Fruits of Australian Poetry,* the earliest book of poems to be composed and printed in Australia. The centrepiece is entitled "The Kangaroo" and begins:

"Kangaroo, Kangaroo!
Thou Spirit of Australia"

Down the years the kangaroo's image has been used as a supporter on coats-of-arms, and it has appeared on tokens and coins and on postage stamps, not only in Australia but in other countries as far away as Cuba, Bulgaria and Hungary. It has appeared in political cartoons and in numerous company symbols and advertisements ranging from QANTAS to Billy Tea. Since Barron Field the kangaroo has continued to appear in novels and poetry. During the few months that the major English novelist D. H. Lawrence lived in Australia in 1922 he wrote his well known novel about right-wing Australian politics entitled *Kangaroo* and the only poem that he wrote in Australia had the same title. At least the poem *was* about a kangaroo. Popular culture also includes the kangaroo; who has forgotten "Tie me kangaroo down sport" or the television star Skippy? And pre-dating all this are the Aborigines' rich traditions describing the deeds of their kangaroo ancestors and their depictions of the animal in rock art.

While the kangaroo had caught the popular imagination early in its European history the naturalists were left behind. Within a few years of the beginning of European settlement it was known that there were a number of kinds of macropodoids in the Sydney area. For example, among drawings made in the early 1790s is one of a Swamp Wallaby (see Calaby 1988) and "rock kangaroos" were observed to the south-west of Sydney as early as 1798. Yet by 1800 only two species had been described, *Macropus giganteus* and *Potorous tridactylus.* For the next thirty years French expeditions visited the coast at intervals and French naturalists took over the running. By 1830 14 species had been described, 11 of them by French naturalists. In the next decade the English regained the lead with the appointment of J. E. Gray to the British Museum and the visit to Australia of John Gould and his chief collector John Gilbert. Gould was greatly interested in the macropodoids and over the years described no less than 16 species still regarded as valid. At the end of the Gouldian period the majority of species had been discovered and described. The discovery of occasional species has continued down the years. The descriptive phase is never over however, and new species continue to appear, the latest being *Potorous longipes* (1980) and *Petrogale persephone* (1982).

For most of the 19th century and the first half of the present one, observations on natural history continued to be gathered, and the major interests of biologists were comparative anatomy and taxonomy. Two important events accompanied the centenary of English settlement in 1888. The first congress of the Australasian Association for the Advancement of Science (now ANZAAS) was held in Sydney in that year. The Macropodoidea didn't rate a mention in its deliberations; the only paper on mammals however was a notable one as it was the first announcement of the discovery of the Marsupial Mole (Stirling 1889). Of much greater importance to mammalogists was the publication of Oldfield Thomas's *Catalogue of the Marsupialia and Monotremata in the Collection of the British Museum (Natural History).* This seminal work gives detailed descriptions of all species known at that date, and the systematics and classification proposed are essentially modern. The book is still regularly consulted by present-day systematists and other biologists. During his professional career Thomas proposed many more names for new taxa of Australian mammals, mostly as subspecies, than any other author. Two of the species of macropodids he described are still recognized, *Petrogale rothschildi* and *P. godmani.* He also described the genera *Caloprymnus* and *Peradorcas,* the latter being relegated back to the synonymy of *Petrogale* only recently.

As is well known to all of us the modern era of research began in the early 1950s and happily it has continued to flourish and expand, in spite of financial restraints. Many research centres and people are involved and most fields are now well represented, including evolution and systematics, ecology, behaviour, reproductive biology, physiology, endocrinology, and so on. The major stocktake of our knowledge contained in these volumes will remain the first source of reference on the kangaroo families for many years to come. We must all be grateful to Gordon Grigg, Peter Jarman and Ian Hume for the enormous call on their time in organizing and managing our bicentennial symposium, and for their careful editing of these fine volumes.

REFERENCES

Calaby, J. H., 1988. The natural history drawings. Pp. 140-97 *in* The Art of the First Fleet, ed. by B. Smith and A. Wheeler. Oxford University Press: Melbourne.

Stirling, E. C., 1889. On a new Australian mammal. Pp. 349-51 *in* Rept. 1st Meeting Australasian Assoc. Adv. Sci., Sydney, 1888.

John Calaby
Canberra
June 1989.

Phylogeny of the Macropodoidea; a study in convergence

T. F. FLANNERY[1]

Macropodoid relationships are investigated using cladistic methodology and parsimony-based computer programmes.

Two monophyletic families of the Superfamily Macropodoidea are recognized: Potoroidae, containing (in order of branching sequence) subfamilies Palaeopotoroinae, Propleopinae, Hypsiprymnodontinae, Bulungamayinae and Potoroinae; and Macropodoidea, containing (in order of branching sequence) subfamilies Balbarinae (which may be paraphyletic), Sthenurinae and Macropodinae.

Within Potoroidae, *Ekaltadeta ima* is the sister group of the species of *Propleopus*, and *P. oscillans* is the sister group of a clade containing *P. chillagoensis* and *P. wellingtonensis*. A dentary from the Ngama local fauna may represent the sister group of bulungamayines. Species of *Potorous* are the sister group of other potoroines. Species of *Bettongia* are probably monophyletic and *Aepyprymnus rufescens* and *Caloprymnus campestris* form a clade.

Within Macropodidae, Sthenurinae and Macropodinae form a clade. Balbarines are possibly ancestral to sthenurines and macropodines. Within Balbarinae, *Nambaroo saltavus* is the sister group to other species. *Balbaroo camfieldensis*, *B. gregoriensis* and an unnamed species from the Kutjumarpu local fauna form a clade. Within Sthenurinae, *Lagostrophus fasciatus* is the sister group to other species. Within *Troposodon*, *T. bluffensis* is most primitive, and *T. minor* and *T. kenti* form a clade. Within Macropodinae, interrelationships of most genera are so obscured by convergence and/or reticulate evolution that they cannot at present be determined. However, species of *Onychogalea* and *Lagorchestes* may be monophyletic, and the species of *Macropus* are apparently monophyletic.

Examination of the fossil record shows that many features previously used to support the monophyly of the Macropodoidea, Potoroidae and lower level groups have been developed independently several times in the evolution of the groups in question. Convergence is a far more common event than the development of unique features in the Macropodoidea.

INTRODUCTION

THE principal aim of this study is to elucidate the interrelationships of the macropodoids and to construct a phylogeny and classification for the group. For the past 20 years, a Miocene fossil record has been accumulating, the study of which has led to a major re-assessment of macropodoid phylogeny. In addition, over the last 15 years, many phylogenetic hypotheses have been generated by biochemical studies. These have provided ideas to test against earlier phylogenetic hypotheses based on morphology and the fossil record. Also over this period, cladistic methodology has become widely accepted, and valuable syntheses of the study of phylogenetic systematics, such as that of Wiley (1981), have provided a new and consistent framework within which to operate.

Previous interest in macropodoid systematics has been episodic. With the exploration of much of Australia and New Guinea by Europeans in the early 1800s, there was great interest in the newly discovered fauna and a flurry of species descriptions. After this initial work, new species were described only sporadically until the 1970s when biochemical methods that then became available, together with a renewed interest in systematics, sparked a contemporary but more modest burst of species descriptions. Likewise, there was much interest in Australian palaeontology in the nineteenth century following the discovery of extensive fossiliferous deposits of Pleistocene age in New South Wales and south-eastern Queensland. Interest lapsed in the first half of the present century, but was renewed in the late 1950s, resulting in a rapid increase in species descriptions. This upsurge in interest, which has continued to the present, is largely a result of the pioneering work of Stirton, Tedford and their associates in discovering major Tertiary mammal-bearing strata in central and northern Australia.

Macropodoids appear to have received more attention from systematists than all the other Australian mammal groups. This is partly because they are the dominant large terrestrial and arboreal herbivores of the Australasian region and also because throughout the latter part of the Neogene they have dominated the medium-sized terrestrial herbivore niche. Over 60 species exist today and an equal or greater number of extinct species are known. Macropodoids are also of increasing interest to biostratigraphers (e.g., Woodburne *et al.* 1985) because they have been

[1]The Australian Museum, 6-8 College Street, Sydney, New South Wales 2000.

Pages 1-46 in KANGAROOS, WALLABIES AND RAT-KANGAROOS, ed. by G. Grigg, P. Jarman and I. Hume. Surrey Beatty & Sons Pty Limited, New South Wales, Australia, 1989.

extremely dynamic throughout their history, rapidly adapting to new ecological opportunities and producing large radiations with a wide diversity of forms.

Physiologists and anatomists have also studied macropodoids extensively, being interested in their evolutionary adaptations to a wide variety of ecological niches. Indeed, the breadth of the macropodoid radiation is extraordinary: they appear to be ecological vicars of primates, cervids and lagomorphs, and to cover a wider range of adaptations than I can find in any placental family or superfamily, unless it be murids. Thus they can be found in habitats as diverse as desert dunefields and gibber plains (where some excavate complex burrow systems), and tropical rainforests (where several taxa have become arboreal). This breadth of adaptation is the more surprising when one considers the underlying similarity of structure that is unique to and typical of all macropodoids.

As a result of these studies, an image of macropodoids is emerging that is antithetical to the previously widely held view that they are primitive because they are marsupials. Dawson (1977) has shown that hopping is actually a more efficient gait than running at medium to high speeds. Some species have developed highly efficient reproductive strategies, such as delayed implantation of the embryo and staggering of the production of young so that three dependent young at different growth stages can be maintained (Russell 1974). Some species such as the Tammar Wallaby, *Macropus (Notamacropus) eugenii* have developed the ability to drink seawater (Chilcott and Hume 1981), and some desert species, given the right conditions, drink rarely or not at all (e.g., the Euro, *Macropus (Osphranter) robustus* (Ealey 1967). It is therefore hardly surprising that macropodoids can successfully compete with their placental equivalents on foreign soil. Red-Necked Wallabies *(Macropus (Notamacropus) rufogriseus)* are now found as feral animals in Britain, Germany and New Zealand, and Rock Wallabies *(Petrogale penicillata)* are feral in Hawaii (Gilmore 1977).

HISTORICAL REVIEW

Ride (1978) is a comprehensive review of the literature relating to macropodoids up to that date, and should be referred to by those interested in the earlier literature. Here, I will review the literature relating to macropodoid systematics that has been published between 1978-1987.

In 1978 "Australian Mammalogy" devoted a volume (Vol. 2) to papers dealing with the phylogeny and evolution of the Macropodoidea, which included Ride's review. Several papers are comprehensive in their approach, including those of Bartholomai (dealing with the fossil macropodoids), Sanson (occlusal cycles and dentition), Hume (the digestive system) and Rose (female reproduction).

Bartholomai (1978b) presents an overview of macropodid relationships as understood at that time, but also suggests some innovations. *Hadronomas puckridgi* is regarded as possibly being ancestral to the species of *Sthenurus* and *Procoptodon* or possibly *Protemnodon*. The origins of the other Macropodinae are not speculated upon, primarily because of the dearth of a Miocene fossil record. *Dorcopsoides* is synonomised with *Wallabia*.

Sanson (1978), studying masticatory cycles and dentition, divides the macropodids into browsing and grazing grades, with some intermediates. In the browsing group are placed the species of *Wallabia*, *Dorcopsis*, *Dorcopsulus*, *Dendrolagus*, *Thylogale*, *Setonix*, *Lagostrophus*, *Protemnodon*, *Hadronomas*, *Dorcopsoides*, *Prionotemnus* and *Sthenurus*. Among the grazers are included the species of *Macropus*, *Peradorcas*, *Onychogalea*, *Lagorchestes*, *Procoptodon*, *Megaleia* and *Fissuridon* (the last two of which are considered synonyms of *Macropus* here). In an intermediate category are placed the species of *Petrogale* and *Troposodon*. Kangaroos of the grazing grade are thought to have arisen from browsers, and the former group is indicated as possibly being polyphyletic.

Hume (1978) deals with the evolution of the digestive system within the Macropodoidea. The digestive systems of potoroids are characterized by a very large forestomach, a relatively undeveloped midstomach, lack of a ventricular groove and little sacculation: the caecum and colon are well-developed. Stomachs of the species of *Thylogale* resemble those of potoroids in some proportions. However, grazing macropodines have a small forestomach and a large midstomach which is similar to the equine colon. He speculates that browsing macropodines arose from potoroids, and that browsing macropodines in turn gave rise to grazing macropodines.

Rose (1978) concludes that oestrous cycle length is probably not a taxonomically useful tool within the Macropodoidea, with the possible exception of the species of *Bettongia*.

Subsequent to the publication of the wealth of material in the 1978 Australian Mammalogy volume, the most striking development of the decade has been the revelation of the Miocene fossil record of the Macropodoidea. Although the first Miocene macropodoids were described by Woodburne (1967), he had only two late Miocene taxa of little phylogenetic significance at his disposal. Archer (1979) described the first middle Miocene taxon; *Wabularoo naughtoni*, from "D" site on Riversleigh Station, northwestern Queensland. The relationships and significance of this species remained uncertain until the publication of Flannery, Archer and Plane (1983), who described additional taxa from Riversleigh as well as other Miocene localities in northern and central Australia. They described two new subfamilies of macropodoids: the Bulungamayinae, which were characterized as potoroids with lophodont molars, and included *W. naughtoni* and a newly described genus and species; and the Balbarinae, including a single new genus with two new species, which were classified as primitive macropodids. New genera and species of potoroines and macropodines were also described.

Case (1984) and Woodburne (1984) describe two new potoroid genera and species from middle Miocene localities in central Australia. Both are referred to the subfamily Potoroinae. Case also includes a re-definition of the Potoroinae (the equivalent of Flannery, Archer and Plane's Potoroidae), and excludes bulungamayines from the group. Flannery, Archer and Plane (1984) re-examine the relationships of the potoroids, centring on the Miocene taxa, and reconfirm their hypothesis that bulungamayines do indeed represent potoroids, albeit with lophodont molars.

Archer and Flannery (1985) review the genus *Propleopus* as well as describing a new, related middle Miocene genus and species. They propose that these gigantic rat kangaroos be placed in a separate subfamily, Propleopinae, and that they represent the most plesiomorphic of potoroids.

Flannery and Rich (1986) describe a remarkable assemblage of extremely plesiomorphic kangaroos from the Pinpa Formation, South Australia. Although represented by only a few scraps, those from Lakes Pinpa and Yanda are the most plesiomorphic macropodoids known. A diverse assemblage from Lake Tarkarooloo represents slightly more advanced, yet none the less mainly exceedingly plesiomorphic taxa. They propose yet another subfamily of potoroids, Palaeopotoroinae, for a plesiomorphic species from Lake Tarkarooloo. They note that all species from these sites are tiny, and that they may date from close to the point of divergence of the macropodoid families.

Flannery and Archer (1987a,b) describe new species of *Hypsiprymnodon* and *Bettongia* from Miocene sediments from Riversleigh, Queensland. They indicate that considerable convergence has occurred in potoroid evolution, and that the potoroid genera were clearly differentiated by Miocene times.

Along with this unfolding of the early fossil record of the macropodoids have come many other developments and schemes of macropodoid evolution, sometimes producing conflicting results.

Archer (1981) includes a review of the relationships within the Macropodoidea. It differs notably from previous works in that he places within the Sthenurinae the species of *Sthenurus*, *Procoptodon*, *Setonix*, *Hadronomas*, *Dendrolagus*, *Dorcopsis*, *Dorcopsulus* and possibly *Dorcopsoides*, relegating all other genera to the Macropodinae. Within this classification the sthenurines are clearly regarded as plesiomorphic. The species of *Troposodon* and *Lagostrophus* are regarded as close relatives.

Archer (1984) includes an overview of macropodoid systematics. The scheme adopted is similar to that presented here. However, the Sthenurinae is considered to have originated from within the Macropodinae, and thus the Macropodinae as understood here is considered to be paraphyletic.

Marshall (1981), although primarily a checklist of the genera and families of marsupials, also contains some innovations and a historical review. He includes all macropodoids within the Macropodidae. Within the Macropodinae he proposes a division into two tribes: the tribe Macropodini includes the species of *Dendrolagus, Dorcopsis, Dorcopsulus, Dorcopsoides, Fissuridon, Hadronomas*, *Lagorchestes*, *Lagostrophus*, *Macropus*, *Onychogalea*, *Peradorcas*, *Petrogale*, *Prionotemnus*, *Protemnodon*, *Setonix*, *Synaptodon*, *Thylogale*, *Troposodon*, *Wabularoo* and *Wallabia;* within the tribe Sthenurini are included the species of *Sthenurus* and *Procoptodon*. The subfamily Potoroinae is divided into the tribe Hypsiprymnodontini; which includes the species of *Hypsiprymnodon* and *Propleopus;* and the tribe Potoroini which includes the species of *Aepyprymnus*, *Bettongia*, *Caloprymnus* and *Potorous*.

Flannery (1982) is an overview of macropodoid evolution centring on adaptive changes in the hindlimb. The families Macropodidae (containing Macropodinae and Sthenurinae) and Potoroidae (containing Hypsiprymnodontinae and Potoroinae) are recognized, Flannery (1984a,b) are more comprehensive overviews of macropodoid evolution, and broadly reflect the hypotheses supported in this study.

In addition to these overviews, studies of a narrower scope have also been numerous. Studies focussing on the Potoroidae have been productive of many new ideas.

Langer (1979, 1980a,b) include studies of the anatomy of the stomach of *Potorous tridactylus*, *Bettongia lesueur* and *Aepyprymnus rufescens*. All of the above species are shown to possess a large complex sacciform forestomach not seen in other macropodoids, and *Hypsiprymnodon moschatus* and phalangeroids are shown to have simple stomachs. Two modes of evolution are noted as being possible; 1, that macropodines evolved complex stomachs independently, and 2, that potoroines and macropodines have evolved complex stomachs independently.

Archer, Bartholomai and Marshall (1978) describe a new species of *Propleopus*, *P. chilagoensis*, from the Pleistocene Chinchilla fissure fill deposits of north Queensland. It is considered to be more derived in morphology than *P. oscillans*. Pledge (1981) describes the M^1 of *P. oscillans*, and similarities between this tooth and that of the species of *Bettongia* are noted.

Johnson and Strahan (1982) re-describe *Hypsiprymnodon moschatus*, and re-examine its status and relationships. In many instances primitive traits are recognized in behaviour, reproduction and morphology. It is concluded that *H. moschatus* and other potoroids share a common ancestor and the possibility that potoroids gave rise to macropodids is discounted. However, relationships between the groups of macropodoids is postulated to be best represented by recognizing two macropodoid families, the Macropodidae and the Potoroidae. Within the Potoroidae are included the Hypsiprymnodontinae and Potoroinae.

Sharman *et al.* (1980) examine the chromosome number of *Bettongia tropica* and conclude that *B. tropica* and *B. penicillata* are probably synonyms. Seebeck and Johnson (1980) describe a new species of *Potorous*, *P. longipes*, from eastern Victoria, which is related to *P. tridactylus*. Balme (1980) records a P_3 of a species of *Potorous* from early Pleistocene sediments at Jandacot, Western Australia. The tooth cannot be assigned to a living species, and is the oldest record of the genus.

It is, however, within the Macropodidae that the most prolific advances have been made. Flannery (1983) revises the subfamily Sthenurinae on a cladistic basis, and includes within it the species of *Lagostrophus*, *Troposodon*, *Sthenurus*, *Simosthenurus* and *Procoptodon*. He concludes that Macropodinae and Sthenurinae are co-equal groupings, and that both share a common ancestor.

Flannery and Archer (1983b) is a revision of the genus *Troposodon*. Two new species (*T. bowensis* and *T. gurar*) are recognized. *Troposodon bluffensis* is shown to be the most primitive member of the genus, and *Troposodon minor* and *T. kenti* are shown to form a clade.

Wells and Murray (1979), describe a new Pleistocene species of *Simosthenurus*, Flannery and Hope (1983) a major range extension of that species, and Archer (1978b) describes a new Pleistocene species of *Procoptodon*.

Within the subfamily Macropodinae many points of evolutionary history have been clarified. Genera that have received at least partial revision during the past decade are *Petrogale* (Poole 1978; Briscoe *et al.* 1982), *Macropus* (Dawson and Flannery 1985; Flannery 1981; Richardson and McDermid 1978), and *Dendrolagus* (Groves 1982). In addition the following have described new macropodine species, and in so doing have shed some light on macropodine relationships; Maynes (1982), *Petrogale persephone*, extant; Hope (1981), *Thylogale christenseni*, subfossil; Flannery (1980), *Macropus mundjabus*, extinct; Flannery and Szalay (1982), *Bohra paulae*, extinct; Kitchener and Sanson (1978), *Petrogale burbidgei*, extant; Bartholomai (1978a), *Protemnodon snewini*, extinct; Flannery, Mountain and Aplin (1983), *Protemnodon nombe*, *Protemnodon tumbuna*, *Dendrolagus noibano*, all extinct; Flannery and Archer (1984), *Kurrabi mahoneyi*, *K. merriwaensis*, both extinct; Flannery and Hann (1984) *Baringa nelsonensis*, extinct.

Flannery and Szalay (1982), in addition to describing an extinct gigantic tree-kangaroo (*Bohra paulae*) from Australia, divide the tree-kangaroos into two groups on the basis of tibial and fibular morphology. The structurally primitive group includes *Bohra paulae*, *D. lumholtzi*, *D. bennettianus* and *D. inustus*, and the derived group contains *D. matschiei*, *D. goodfellowi*, *D. dorianus* and *D. ursinus*. Groves (1982) concurs with Flannery and Szalay on this division of the tree-kangaroos, and finds further support for it on the basis of dental, cranial and pedal characters. Groves also synonomises *D. matschiei* with *D. goodfellowi*. Flannery, Mountain and Aplin (1983) describe a new species of *Dendrolagus* (*D. noibano*) from late Pleistocene sediments from Simbu Province, Papua New Guinea. It appears to be related to *D. dorianus*.

Poole (1978) deals briefly with the species of *Petrogale*, and notes that work by Sharman and others is in progress. He divides the species of *Petrogale* into three groups. Group 1 is characterized by possessing 22 chromosomes, and contains *P. xanthopus*, *P. rothschildi* and *P. persephone* Maynes, 1982 (then known but unnamed). Group 2 contains the brush-tailed rock wallabies, *P. lateralis* and *P. penicillata*, and Group 3 contains the Kimberley and Arnhem Land species, *P. brachyotis*, *P. burbidgei* and *P. concinna* (syn. *Peradorcas*). Briscoe *et al.* (1983) provide a comprehensive examination of chromosomal and electrophoretic variation within *Petrogale*. They reduce the 23 named forms to 10 species; *P. lateralis, P. rothschildi*, *P. inornata*, *P. godmani*, *P. penicillata*, *P. xanthopus*, *P. persephone*, *P. brachyotis*, *P. burbidgei* and *P. concinna*. They also speculate that the species of *Petrogale* arose on the Australian shield as a western equivalent of the species of *Thylogale*. The karyotype of *Thylogale stigmatica* is considered to be close to that of the ancestral *Petrogale*. A picture of far greater complexity than was previously recognized is emerging. Further morphological and biochemical work is in progress that may modify the conclusions presented in the paper.

Rofe (1978) studied "G"-banding of chromosomes, mostly in the species of *Macropus*. Three different "G"-banded karyotypes are present in the 2n=16 chromosome species of *Macropus*. One is seen in the species of *Macropus (Macropus)* tested, (*M. giganteus* and *M. fuliginosus);* one in the only species of *Macropus (Osphranter)* tested, *(M. robustus)*; and one is shared between the three species — *Macropus (Notamacropus)* tested, (*M. eugenii*, *M. parryi* and *M. rufogriseus*). Thus several chromosomes in the species of the "16 chromosome kangaroos" are only analogous and chromosome number alone is probably not indicative of a close relationship.

Richardson and McDermid (1978) studied the species of *Macropus* using allozyme, cytological and immunological data. This work also speculates to a minor extent about the relationships of the species of *Aepyprymnus*, *Petrogale*, *Thylogale*, *Peradorcas* and *Setonix*. Within *Macropus* they examined the species *M. (M.) giganteus, M. (M.) fuliginosus, M. (N.) agilis, M. (N.) irma, M. (N.) rufogriseus, M. (N.) dorsalis, M. (N.) parma, M. (N.) eugenii, M. (O.) robustus, M. (O.) antilopinus,* and *M. (O.) rufus.* They also include *Wallabia bicolor*, which they believe is closely related to this group. They conclude that the species of *Macropus* and *Wallabia* are monophyletic, and cite three synapomorphic character states shared by the species tested. The two species of *Macropus (Macropus)* are shown to be very closely related, the species of *Macropus (Notamacropus)* also cluster together, as do *M. (O.) robustus* and *M. (O.) antilopinus*. *Macropus (O.) rufus* and *Wallabia bicolor* do not cluster tightly with any group. The relationships of the three groups to each other could

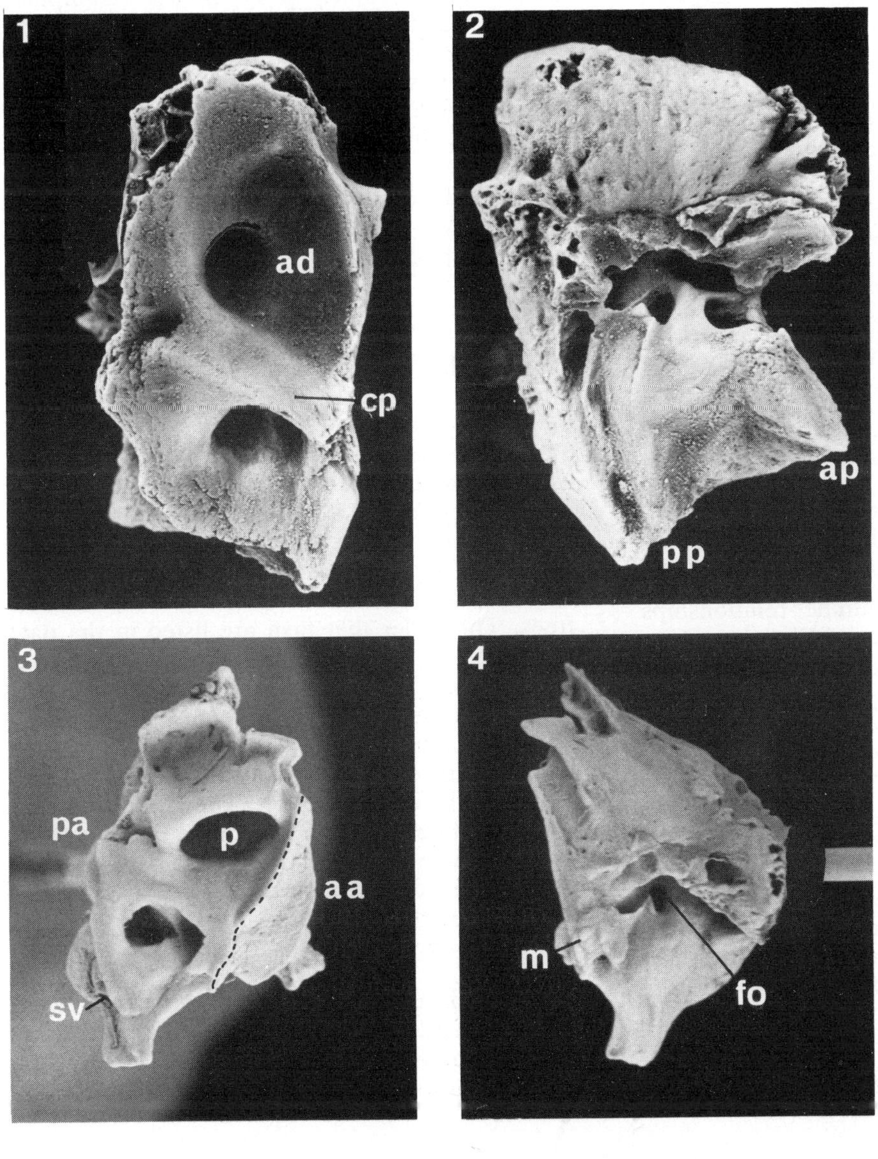

Fig. 1. 1, internal surface and 2, external surface of left periotic of *Protemnodon* sp. cf. *P. brehus*, and 3, internal surface and 4, external surface of left periotic of *Hypsiprymnodon moschatus*. ad=depression anterior to parafloccular fossa, aa=antero-internal face of periotic, pa=posterointernal face of periotic (the dotted line marks the division between these two entities, ap=anteroventral process, cp=crista petrosa, fo=fenesta ovale, m=mastoid process, p=parafloccular fossa, pp=posteroventral process, sv=groove for inferior petrosal sinus vein.

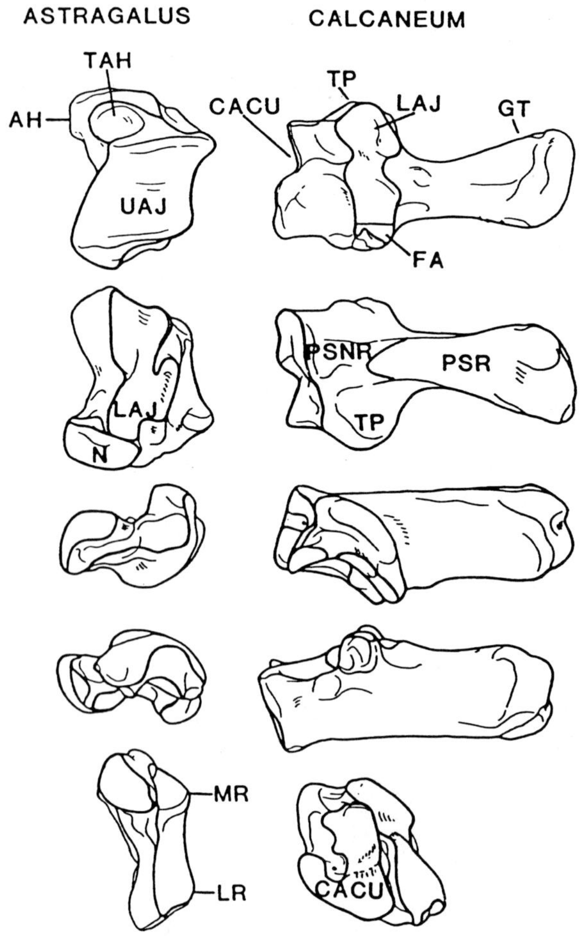

Fig. 2. Astragalus and calcaneum of a potoroine illustrating the pedal terminology used here. AH=astragalar head, CACU=calcaneal-cuboid articulation, FA=calcaneal-fibular articulation, GT=greater tuberosity, LAJ=lower ankle joint, LR=lateral ridge of UAJ, MR=median ridge of UAJ, N=astralalar-navicular articulation, PSR=rugose area of plantar surface of calcaneum, PSNR=non-rugose area of plantar surface of calcanuem, TAH=tibial articulation of astragalar head, TP=trochlear process, UAJ=upper ankle joint.

not be determined. They recommend that all species be included within the genus *Macropus*, with the possible exception of *Wallabia bicolor*.

Peacock *et al.* (1981) examine relationships between species of *Macropus* and *Wallabia*, using highly repeated DNA sequences. *Macropus (Osphranter) rufus*, *M. (O.) antilopinus*, *Macropus (Macropus) giganteus*, *M. (M.) fuliginosus*, *Macropus (Notamacropus) rufogriseus*, *M. (N.) parryi*, *M. (N.) eugenii*, *M. (N.) agilis* and *Wallabia bicolor* are included in the study. Results support the contention that *Macropus* is monophyletic, but also indicates that *M. (N.) rufogriseus* is related to *M. (O.) rufus*. Relationships are also detected between *M. (N.) eugenii* and *M. (N.) agilis*, and a very close relationship was detected between *M. (M.) giganteus* and *M. (M.) fuliginosus*.

Dawson and Flannery (1985) is a revision of the taxonomic status of all species of *Macropus*. They exclude *Prionotemnus palankarinnicus* from the genus, and propose a new subgeneric name, *Notamacropus*, for the wallaby species previously placed in *Prionotemnus*.

METHODS

Terminology and homology. Dental terminology follows Flannery and Rich (1986) and Archer and Flannery (1985). Cheektooth enumeration follows Archer (1978a). No suitable terminology for the topography of the marsupial periotic exists. The terminology utilized here is modified from Macintyre (1972), and conforms with studies by Aplin (in prep.) for the marsupial periotic. The terminology is illustrated in Fig. 1. Pedal terminology follows Szalay (1982) and Flannery and Szalay (1982) with some modifications (see Fig. 2). Use of the terms fauna and local fauna follows Rich *et al.* (1982). Phalangerids have been used as the outgroup to the Macropodoidea (see Flannery 1987).

Use of computer programmes. Computer analysis was used to evaluate the large numbers of characters and taxa examined. The programme package chosen was PHYLIP, a phylogenetic inference package (version 2.2), produced by J. Felsenstein. Four of the seven programmes provided for analysis of discrete state data (the Dollo, Wagner, Sokal and Polym programmes) were chosen to investigate relationships within the Macropodoidea.

Characters with only two states, primitive (coded as "0") and derived (coded as "1") were used. Multistate characters were broken down into a series of two state characters utilizing the method outlined by Sokal and Sneath (1963). States that were unknown for a taxon were coded as "?", and characters that were polymorphic for a taxon were coded as "p". No other options provided in the package were used. Several runs of each data set, with taxon order shuffled, were usually made in an attempt to find the shortest trees. The order that taxa are listed in the data set can affect tree construction since the trees are constructed in an accretionary manner, the first three taxa being taken, relationships assessed, then the fourth taxa listed being added and relationships of the whole group reassessed, and so on for the entire data set. The number of runs considered necessary to determine that a reasonable estimate of the shortest possible tree had been obtained depended on the consistency and structure of the trees produced by a given data set.

Although their assumptions were somewhat different, all four programmes used are based on the parsimony method (Felsenstein 1983). The assumptions of the methods are as follows.

Sokal:

(1) That the characters are evolving independently.

(2) That different lineages are evolving independently.

(3) That we know that state 0 is the ancestral state.

(4) That the rates of evolution are sufficiently slow over the lengths of evolutionary time involved, that a change 0-1 is *a priori* an improbable event.

(5) That reversion (1-0) or retention of polymorphism for both states (0 and 1) are far less probable events than (0-1).

(6) That inequality of lengths of branches of the true tree is not so great that having two changes on a long branch is more probable than one change on a short one.

For Wagner:

(1) As above

(2) As above

(3) We do not know which state is ancestral; however, by including a completely primitive ancestor the tree can be polarized.

(4) Over the evolutionary times which are relevant for the particular group, changes 0-1 and 1-0 are *a priori* improbable.

(5) Retention of polymorphism for both states is far less probable than are changes of state.

(6) Inequality of lengths of segments of the tree is not so extreme that two changes of state along a long segment are more probable than one change along a short segment.

Because the Wagner programme regards two evolutionary events (independent derivation and reversion) as equally likely when interpreting homoplasious characters, it is impossible to determine which of these events was the one utilized in the construction of the tree. For this reason character state changes involving homoplasious characters cannot be placed on the Wagner trees as they can for trees generated by the other programmes.

For Dollo:

(1) As above

(2) As above

(3) As for Sokal

(4) The probability of a forward change (0-1) is small over the evolutionary time involved.

(5) The probability of a reversion (1-0) is also small, but still far larger than the probability of a forward change, so that many reversions are easier to envisage than even one extra forward change.

(6) Retention of polymorphism for both states (0 and 1) is highly improbable.

(7) The lengths of the segments of the true tree are not so unequal that two changes in a long segment are as probable as one in a short segment.

For Polym:

(1) As above

(2) As above

(3) As above

(4) Forward change (0-1) is highly improbable over the length of time involved in the evolution of the group.

(5) Retention of polymorphism is also improbable, but far more probable than forward change, so that we can more easily envisage much polymorphism than even one additional forward change.

(6) Once state 1 is reached, re-occurrence of state 0 is very improbable, much less probable than multiple retentions of polymorphism.

(7) The lengths of segments in the true tree are not so unequal that we can more easily envisage retention events occurring in both of two long segments than one retention event in a short segment.

These programmes were chosen for the following reasons. Felsenstein (1979) presents a probablistic model of evolution in which there are three evolutionary events: 1) origination of a synapomorphic state; 2) reversion to a primitive state; and 3) loss of polymorphism. There is also an event corresponding to total character misinterpretation by the taxonomist. When the probabilities of these four events are taken to the extreme, four different phylogenetic inference methods emerge as maximum-likelihood models. These are the Sokal, Dollo, Polym and the Estabrook, Johnson, McMorris compatibility methods, all available in PHYLIP. Of these, only the Estabrook *et al.* method as not utilized here as it could not accept incomplete data sets. In addition to these models, where the branching sequence is constructed to reflect only a single evolutionary event after the initial gain of a synapomorphy, the Wagner programmme, which takes two evolutionary events (reversion to a primitive state and independent derivation) as being equally likely was utilized for comparison. However, trees produced by this programme are at a disadvantage as they cannot be analysed fully for character state placement (see above).

To ascertain when a stable branching sequence had been generated, the trees were evaluated according to consistency (between different runs of the same method and different methods). The stable regions of the dendrograms are usually supported by many characters and no alternative arrangement is likely to be supported using a different subset of characters from the data set. When such configurations were found, they were taken to represent a close approximation of the true phylogenetic relationships. Taxa in areas of instability in the original trees were sometimes run again using a different data set (see below). This method resulted in a nested set of trees.

In many cases not all species of a genus, subfamily or plesion were included in an analysis, but instead an hypothetical or "ur-species" was used. This word, coined by Ernst Haeckel (1834-1919) was used by him to connote a primitive hypothetical ancestor: although most of his ur-species have proved illusory, the term has proved to be useful in a cladistic sense. It is constructed by taking all of the most plesiomorphic states present in the members of the clade being considered, and combining these to produce the most plesiomorphic hypothetical taxon. This concept is similar to that of the HTU (hypothetical taxonomic unit) constructed by computer programmes such as Wagner. This approach was chosen as it allows one to sidestep problems of taxonomic instability at the species level. This allows for a reduction in the number of taxa in a data set, without at the same time suffering much loss of information. An ur-species member of a clade can be more confidently placed in a phylogeny than a randomly chosen member of that clade (which may possess many autapomorphies, some of which may be homoplasious).

A problem that could not be resolved in these programmes was the assumption, present in all programmes, that characters are evolving independently. In clear cut cases (e.g., two interdependent sets of measurements), highly correlated characters could be identified and discarded. However, characters can be highly correlated in species groups for several reasons. They may be elements of character complexes that are truly interdependent (i.e., expressions of a single character), or they may be a suite of characters inherited from a common ancestor – what could be called phylogenetically correlated characters. Since the programmes used are all based on the parsimony method, blanket removal of highly correlated characters (e.g., by using a Spearman Rank Coefficient), including the phylogenetically correlated kind, would seriously affect the analysis. Because of this, correlated characters were not removed. Fortunately, Felsenstein (1983) has noted that the programmes used here are robust with respect to highly correlated characters.

Felsenstein (1978) has dealt with the problems caused by homoplaisy in computer-generated phylogenies based on the parsimony method. As one would intuitively expect, when homoplasious characters outnumber non-homoplasious ones these programmes are almost guaranteed to produce an "incorrect" phylogeny. Unfortunately, a high degree of homoplaisy appears to be almost inevitable when dealing with any aspect of macropodoid phylogenetics, and this has been a major factor in limiting the usefulness of these computer programmes in this study.

Overall, the four programmes utilized produced most parsimonious trees of generally similar topography. This was a surprising result given the different assumptions of the likelihood of various evolutionary events made by the programmes.

RESULTS

Relationships within the Macropodoidea

I will here consider the interrelationships of the seven major groups (subfamilies) of macropodoids: the Hypsiprymnodontinae; Propleopinae; Potoroinae; Bulungamayinae; Macropodinae; Sthenurinae; and Balbarinae. Six of these are thought to represent monophyletic groups, but one (Balbarinae) may be paraphyletic. The composition of and evidence for monophyly of these subfamilies are discussed in the following sections. One subfamily, the Palaeopotoroinae, is excluded from the computer analysis because it contains a single very poorly-known taxon that is phenetically similar to *Hypsiprymnodon moschatus*. Members of the seven subfamilies have been divided into two families, the Potoroidae and Macropodidae, in a number of previous works (e.g., Archer and Bartholomai 1978; Johnson and Strahan 1982; Strahan 1983a; Flannery, Archer and Plane 1983; Archer 1984). Within the Macropodidae are placed the Macropodinae, Sthenurinae and (Flannery, Archer and Plane 1983 and Archer 1984 only) Balbarinae. Within the Potoroidae are placed the Hypsiprymnodontinae, Potoroinae and (Flannery, Archer and Plane 1983; Archer 1984 only), Bulungamayinae, and (Archer and Flannery 1985 only), Propleopinae. However, this is not the only reasonable phylogenetic classification. Given a dichotomously branching sequence, a large number of phylogenetic hypotheses can be generated using seven taxa. Using the seven subfamilies of macropodoids, however, only a few hypotheses are defensible. The five hypotheses that I regard to be the most probable are considered below. The data used in construction of these hypotheses is run through four algorithms for interpreting phylogenetic relationships which are provided in PHYLIP. The differences between the algorithm generated hypotheses and those produced by me are discussed. The five hypotheses considered by me to be most likely (given macropodoid monophyly) are:

(1) That potoroids are monophyletic and that macropodids are monophyletic (Fig. 3a).

(2) That macropodids are descended from potoroids and are the sister group of the bulungamayines (Fig. 3b).

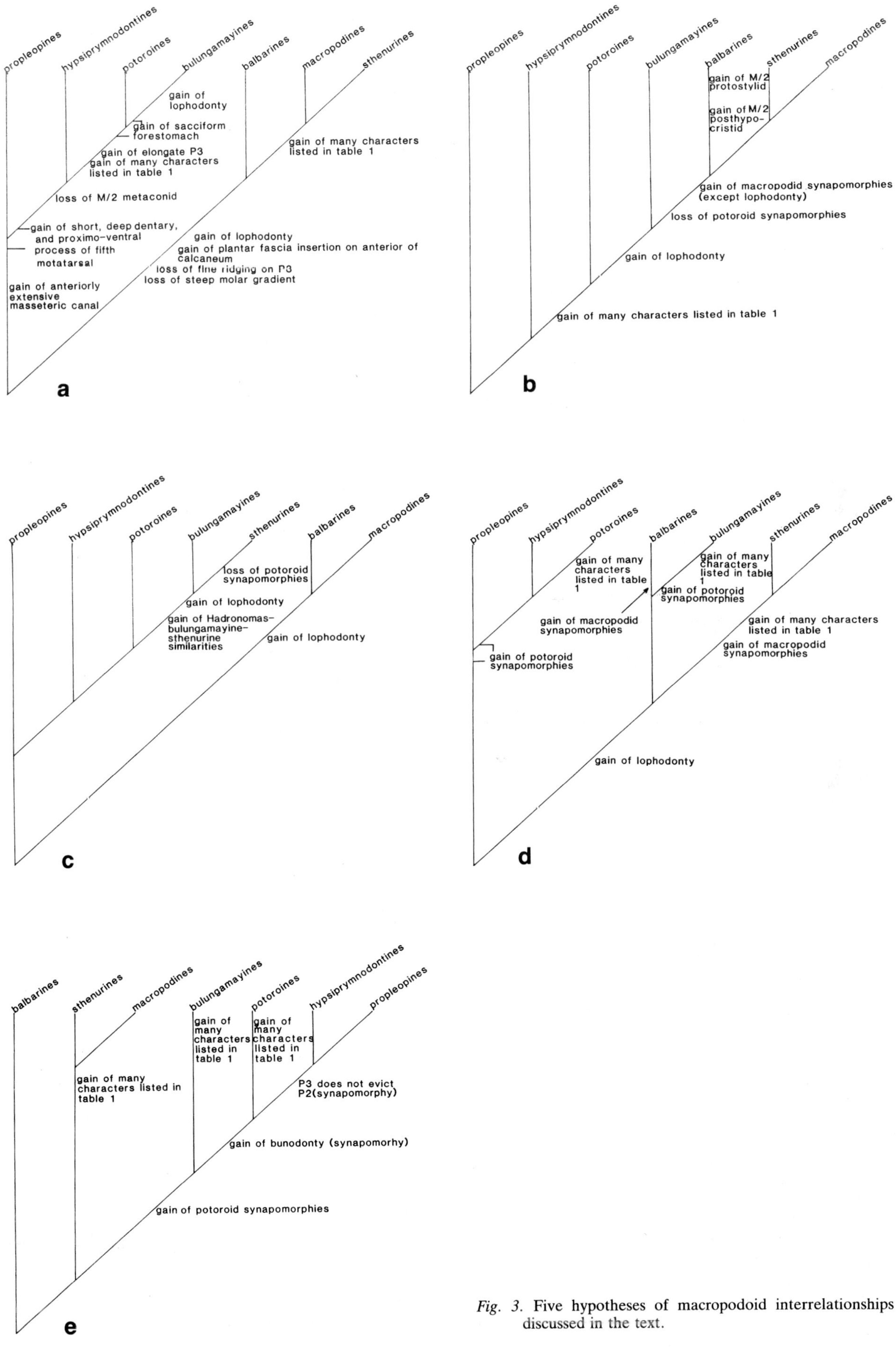

Fig. 3. Five hypotheses of macropodoid interrelationships discussed in the text.

(3) That sthenurines are derived from potoroids (they are the sister group of bulungamayines) but that macropodines are derived from balbarines (Fig. 3c).

(4) That bulungamayines are derived from macropodids but that all other potoroids are monophyletic (see Fig. 3d).

(5) That all potoroids are derived from macropodids (Fig. 3e).

1: *Macropodids are monophyletic and potoroids are monophyletic.*

In this hypothesis, propleopines, hypsiprymnodontines, potoroines and bulungamayines form a monophyletic grouping (Potoroidae), as do balbarines, sthenurines and macropodines (Macropodidae). Propleopines and hypsiprymnodontines are best regarded as the most primitive potoroids, and balbarines as the most primitive macropodids. Evidence supporting this phylogeny is drawn from several areas. Immunological studies by Kirsch (1977) using whole sera, and by Baverstock (pers. comm.), using albumin indicate that the extant potoroids tested are monophyletic, as are the extant macropodids. Baverstock's study further indicates that *Lagostrophus fasciatus*, the only living sthenurine, is more closely related to macropodines than potoroids (Baverstock, pers. comm.). Kirsch's study placed *Lagostrophus fasciatus* within the Macropodinae. The discovery of apparent morphological synapomorphies uniting these groups also lends strong support. For potoroids, these include dental, cranial, stomach and pedal synapomorphies (see below). The evidence of monophyly is less strong for macropodids but includes dental and pedal synapomorphies. If this phylogenetic hypothesis is correct, however, then a large number of characters must have been subject to parallelism in advanced potoroids and macropodids. Twenty-three such characters are listed in Table 1. A few of these features can be shown to be almost certainly homoplasious in potoroids and macropodids because of (18) the distribution of other synapomorphies. One can be shown to be only a superficial similarity because the structures

Table 1. The following characters and their derived states (1) are presumed to have been independently developed in some macropodid-potoroid lineages under hypothesis 1 (see Fig. 3) in some cases derived states in some lineages can be almost certainly shown to be independently derived. (S) indicates presence of the advanced state in only some members of the subfamily.

Derived states of characters	Balbarines	Macropodines	Sthenurines	Hypsiprymnodontines	Propleopines	Potoroines	Bulungamayines
1. P3 replaces M1 and P2	1	1	1		(S)1	1	1
2. Hopping developed	?	1	1		?	(S)1	?
3. Loss of I_2		1	1			1	1
4. P3 elongate		(S)1	1			1	1
5. P3 has postero-internal cusp	1	1	1			(S)1	1
6. Postlink lost on upper molars		(S)1			1	1	
7. Loss of hallux	?	1	1		?	1	?
8. Loss of striated digital pads	?	1	1		?	1	?
9. Angle of internal face of periotic acute	1	1	1		?	1	
10. Infra-orbital canal opens above P^3	?	1	1			1	1
11. Foot elongate	?	(S)1	1		?	(S)1	?
12. Forelimbs much shorter than hindlimbs	?	1	1		?	1	?
13. Lophodonty developed	1	1	1				1
14. P3 aligned with molar row		1	1			(S)1	1
15. M^1 has shortened parastyle	?	(S)1	1		1	(S)1	
16. Loss of fine ridging on P3	1	1	1			(S)1	
17. Loss of steep molar gradient	1	1	1		(S)1	(S)1	1
18. M_2 anterior loph broadened		1	1		(S)1	1	1
19. Development of masseteric process of maxillary	1	(S)1	(S)1			(S)1	
20. Plate lacks vacuities	?	(S)1	(S)1			(S)1	
21. Canines reduced	?	(S)1	1		?	(S)1	
22. 2 roots on buccal side of upper molars	(S)1	(S)1	1			1	1
23. Squamosal-frontal contact				(S)1		(S)1	

involved have been found to be non-homologous. However, the great majority can be shown to be homoplasious only by validation of this hypothesis.

The following list of potoroid synapomorphies support the hypothesis that the potoroids form a monophyletic group. Some characteristics are synapomorphies for all potoroids, others for a large group of potoroids.

1. In all potoroids the masseteric canal extends into the body of the dentary to below the posterior edge of P_3. Dissection has shown that the masseter muscle also inserts this far anteriorly at least in *Aepyprymnus rufescens*. In addition, the masseteric and dental canals are completely confluent in the body of the dentary in all potoroids except *Ekaltadeta ima*. A masseteric canal is not present in phalangerids and in macropodids the masseteric canal does not extend as far anteriorly. Also the masseteric canal is usually partially separated from the dental canal by a bony septum in macropodids. Even in macropodids where the P3 is extremely elongate relative to molar size, such as the species of *Dorcopsis* and *Protemnodon*, the masseteric canal does not reach as far anteriorly as in potoroids. On this basis it seems logical to assume that the more extreme development seen in potoroids is the derived one.

2. The dentary is short and deep, being deepest below the middle portion of the cheektooth row. The ventral prominence and the area lingual to it are often rugose, or a ridge is present. This area presumably serves as a point of insertion for the digastric muscle. Where visible, this rugose muscle attachment in macropodids and phalangerids is usually more posteriorly placed, well posterior to the root of the I_1, not immediately behind it as in potoroids. Also the dentary is not so sharply convex ventrally nor so deep in macropodines. The anterior attachment of the digastric in potoroids parallels the anterior extension of the masseteric canal. These features may be functionally interrelated. However, they can occur independently, as in the species of *Propleopus* and *Potorous*, where anterior masseter insertion is present but the dentary is shallow, without the marked ventral convexity and anterior digastric muscle attachment. The lack of this synapomorphy in these taxa could be interpreted as a reversion to the primitive state, or a retained symplesiomorphy.

3. The P3 is elongate. Apart from propleopines and hypsiprymnodontines, all potoroids possess an elongate P3. The ancestral condition is presumed to have been similar to that seen in hypsiprymnodontines because these species have a P3 that is most similar in shape to the P3 in the outgroup (phalangerids). Subsequent elongation of the finely ridged blade occurred in potoroines and bulungamayines. Elongation of P3 is probably convergent in some advanced macropodids and potoroids.

4. In all potoroids except propleopines and (to a lesser extent) palaeopotoroines the metaconid of M_2 is reduced, posteriorly displaced or absent. The homology of this cuspid is discussed in Flannery and Rich (1986) and Archer and Flannery (1985).

5. In all potoroids except *Palaeopotorous priscus*, a cristid running anteriorly from the protostylid of M_2 is present. This feature is intimately related to feature 5 (above) because both involve changes in the structure of the trigonid of M_2. The derived states of both 4 and 5 appear together in all potoroid taxa except propleopines, where feature 5 but not 4 is present (Archer and Flannery 1985).

6. The fifth metatarsal possesses a proximoventral process. This process is not present on the fifth metatarsal of phalangeroids, nor in macropodids, with the exception of the species of *Protemnodon* where it is probably convergent. Even within *Protemnodon* it is poorly developed in some species (e.g., *P. otibandus*). Significantly, the process is absent in *Hadronomas puckridgi*, a form possibly ancestral to the species of *Protemnodon*. Species of *Protemnodon* are unusual within the macropodids in having extremely short, robust feet. Presence of a proximoventral process on the fifth metatarsal may be related to the development of this feature in species of *Protemnodon*. The process probably serves as a point of attachment for the plantar fascia, thereby more securely binding the bones of the foot. In potoroids, the proximoventral process is proportionately much larger than in the species of *Protemnodon*. Despite these arguments it is just possible that this feature represents a macropodoid symplesiomorphy retained in potoroids and the species of *Protemnodon*.

7. With the exception of *Hypsiprymnodon moschatus*, living potoroids possess a sacciform forestomach. Extreme hypertrophy of this structure is unique to the Potoroidae within the Macropodoidea, where it is so large that it completely overshadows the tubiform midsection of the stomach (Langer 1979, 1980a, 1980b) and presumably harbours microbes used in the production of volatile fatty acids. While probably not a synapomorphy for all potoroids, this is a striking synapomorphy for all living species except *H. moschatus*.

It should be noted here that possession of a frontal-squamosal contact, long regarded as a potoroid synapomorphy, is not accepted as such here. Flannery and Archer (1987a, b) show that this feature has probably been developed at least four times independently within the Potoroidae (in *Hypsiprymnodon moschatus*, *Potorous*, most

Bettongia and *Aepyprymnus/Caloprymnus*). The presence of a frontal-squamosal contact in potoroids may be dictated by rates of ossification in the frontal, squamosal, parietal and alisphenoid, and may be due to the need for early, firm attachment points for the temporal muscles. Alternatively, it may be that a suture parallel to the muscle fibres is preferable to one running at 90 degrees to it particularly when the musculature is powerful.

It is necessary to clarify here why the features discussed above are thought to be potoroid synapomorphies and not symplesiomorphies. A scenario could be developed that envisaged stem macropodoids evolving highly derived characteristics related to premolar shear (i.e., features here considered potoroid synapomorphies), and that macropodids, derived from this group, lost these characteristics due to secondary emphasis on molar attrition. It is clear that primitive macropodoids did indeed develop a complex functional system to enhance premolar shear (Flannery 1984b). However, it does not necessarily follow that the unique potoroid adaptations discussed above, some of which presumably further advance this character complex, are also primitive. It is more parsimonious to surmize that macropodids and potoroids parted ways before these further adaptations developed in potoroids. After this initial dichotomy, potoroids could have gone on to further enhance premolar shear, while most macropodids subsequently de-emphasized this aspect of dental mechanics, but developed more efficient molar attrition.

While potoroid monophyly seems to be reasonably well-supported and as such is strong support for the hypothesis presented above, macropodid monophyly is less clearly indicated by the morphological evidence. The following morphological characters, however, seem to support it.

1. All macropodids possess well-developed lophs on molars and (where present) the posthypocristid is lingually shifted. With the exception of bulungamayines, potoroids possess less well-developed molar lophs and a posthypocristid that originates at the hypoconid. On the basis of the distribution of other synapomorphies, the development of fully lophodont molars in bulungamayines and macropodids is believed to be convergent. The development of lophodont molars may be related to functional changes that led to de-emphasis of premolar shear in macropodids and the development of more efficient molar attrition.

2. With the exception of some primitive balbarines, all macropodids have apparently lost the protostylid of M_2 (Flannery and Rich 1986). This structure is retained in phalangerids and potoroids.

3. Macropodids have molars that increase in size backwards, with M5 not markedly smaller than M4. A much steeper molar gradient (with posterior molars smaller) is usually seen in phalangerids and potoroids. In potoroids that lack a steep molar gradient (i.e., in *Aepyprymnus rufescens*), this is clearly due to convergence as these species share many synapomorphies with other potoroids.

4. In macropodids, P3 is almost never as finely-ridged as the P3 of most potoroids. This loss of fine ridging is thought to be derived because finely-ridged premolars are present in phalangerids and most potoroids. Because the presence of fine ridging on premolars seems to enhance premolar shear, it may be that loss of this ridging is related to the shift of emphasis from premolar shear to molar attrition that has taken place in macropodids. Even in primitive macropodids such as balbarines from the Kutjumarpu local fauna, where the P3 is similar in shape to that of *Hypsiprymnodon moschatus*, it is only irregularly and coarsely ridged. However, a P3 tentatively referred to *Nambaroo saltavus* from the Tarkarooloo local fauna, one of the most primitive of macropodids, is finely ridged.

5. The plantar fascia inserts for nearly the whole length of the PSR. This adaptation presumably provides further stability for the tarsal joints. That this feature is not purely size-related is shown by the fact that large potoroids (e.g., *Aepyprymnus rufescens* with a maximum adult body weight of about 3.5 kg, Johnson 1983) lack this feature while small macropodids (e.g., *Lagorchesteshirsutus*, *Peradorcas concinna* and *Lagostrophus fasciatus*, with adult body weights between 1 and 3 kg, Burbidge and Johnson 1983; Prince 1983; Sanson 1983) possess it.

Synapomorphies 1 to 4 are probably features related to the increased emphasis on molar attrition that seems to be a major trend in macropodid evolution. This change in emphasis in macropodids from premolar shear to molar attrition undoubtably led to the development of the browsing masticatory cycle described in Sanson (1982), which is postulated to be primitive within the Macropodidae. The trend away from emphasis on premolar shear has been carried furthest in the most derived grazing macropodids, where P3 has been so reduced that it is lost early in life, opening the way for a serial use of the molars and also to the production of supernumary molars as in *Petrogale concinna*.

In addition to the morphological evidence presented here, immunological studies also support macropodid monophyly, indicating that sthenurines *(Lagostrophus fasciatus)* and macropodines are monophyletic (Kirsch 1977). Further investigation, particularly of the soft anatomy of *Lagostrophus fasciatus* and primitive macropodines may well result in the identification of more synapomorphies for this clade.

Finally, it is worth examining closely the characters listed in Table 1 that are thought to be subject to homoplaisy if this hypothesis is correct. These characters can be divided into several groups. First, there are some with derived character states that are found only in two or more highly divergent taxa. These include characters 19-21. *Aepyprymnus rufescens* lacks palatal vacuities, as do several macropodids (20). *Caloprymnus campestris* possesses a masseteric process (19) and has highly reduced canines (21), features also seen in many macropodids. Because these taxa possess many synapomorphies shared with other potoroids, these macropodid-like developments are almost certainly convergent.

A second group of homoplasious characters includes many in which the derived states are widespread throughout more advanced macropodids and potoroids but are absent in primitive species. Characters 1-18 fall into this category. Characters 1-2 have been discussed above because they are so widespread that they have previously been considered as probable macropodoid synapomorphies. In most cases, there is nothing intrinsic in the structure of these derived states to suggest that they are non-homologous. However, the broadening of the anterior loph of M_2 (18) appears to have been achieved independently three times using different structures. Potoroids (except propleopines) have lost the metaconid, macropodids the protostylid, and propleopines none of these structures. Thus the anterior loph of M_2 in most potoroids is a protoconid-protostylid crest, in macropodids a metaconid-protoconid crest; and in propleopines a protostylid-protoconid-metaconid crest (Flannery and Rich 1986; Archer and Flannery 1985).

2: *Macropodids are descended from potoroids and are the sister group of bulungamayines.*

If macropodids are descended from potoroids (particularly such apparently derived species as bulungamayines), then it is necessary to postulate that features previously regarded as potoroid synapomorphies (deep insertion of the masseters, anterior attachment of the digastric musculature to the dentary and presence of a deep, short dentary) are macropodoid synapomorphies that have been lost in macropodids. Alternatively, these features could have been independently derived in hypsiprymnodontines, propleopines and some advanced macropodoids (including potoroines and bulungamayines). These features are probably functionally correlated to enhance premolar shear, and are partially developed even in such primitive potoroids as hypsiprymnodontines. The loss of these features in the earliest macropodids would represent a massive reversion to the primitive condition.

The main advantage of this hypothesis is that it allows a large number of features which are considered under some other hypotheses to be subject to convergence to be developed only once. These include the derived states of characters 3-17 of Table 1. This is because *Hypsiprymnodon moschatus* is regarded as near basal to the macropodoid radiation rather than as part of the potoroid clade, and thus is interpreted as retaining the primitive macropodoid condition for many features. However, if balbarines are also considered to have developed from bulungamayines, then massive reversion to the primitive condition would be required to explain, for instance, the presence of I_2, a posthypocristid on lower molars and a protostylid on M_2 in many of these species. However, if lophodonty is envisaged as developing independently in balbarines and in macropodines-sthenurines-bulungamayines, then balbarines could be interpreted as a separate lineage derived from primitive macropodoids (e.g., hypsiprymnodontines).

3: *Sthenurines are derived from potoroids (bulungamayines) via* Hadronomas puckridgi, *and macropodines are derived from bulungamayines.*

In this hypothesis, similarities between *Wabularoo naughtoni* and *Hadronomas puckridgi* may be viewed as synapomorphies. These features could conceivably include similarity in premolar shape and the presence of low-crowned lophodont molars that lack posterior cingula. However, there is no cladistic evidence which suggests a relationship between *H. puckridgi* and sthenurines.

Data contradictory to this hypothesis are provided by immunological studies (Kirsch 1977), suggesting a close relationship between *Lagostrophus fasciatus* and macropodines (but not potoroids), and probable synapomorphies shared between sthenurines and macropodines (see below). Thus this hypothesis assumes considerable convergence between macropodines and sthenurines. In addition, apparent potoroid synapomorphies would have had to have been lost in sthenurines (see Fig. 3c).

4: *Bulungamayines are derived from macropodids.*

This hypothesis is attractive because lophodonty need only have arisen once (Fig. 3d). Balbarines thus would be envisaged as the most primitive lophodont macropodoids and bulungamayines as having arisen from them. However, if this interpretation is correct, the potoroid synapomorphies presented in hypothesis 1 would have to be viewed as symplesiomorphic for macropodoids, or to have been developed independently in bulungamayines and other potoroids. While either of these alternatives is possible, it seems more likely that a single feature, lophodonty, has been independently derived in bulungamayines and macropodids.

5: *All potoroids are derived from macropodids.*

Under this hypothesis, lophodonty is regarded to be symplesiomorphic within macropodoids, and bulungamayines are seen as basal potoroids. Bunodonty could be considered a synapomorphy for all other potoroids (see Fig. 3e). This hypothesis would allow the potoroid synapomorphies listed in hypothesis 1 to stand, but the macropodid synapomorphies listed there to be regarded as symplesiomorphies independently lost in several potoroid lineages. Hypsiprymnodontines and propleopines are extremely difficult to deal with in this hypothesis because much of their morphology previously treated as plesiomorphic would have to be interpreted as derived (from a bulungamayine-like ancestor). Aspects of this morphology, such as cheektooth displacement, M_2 trigonid morphology and P3 morphology are difficult to reconcile with this interpretation and this hypothesis is regarded as the least likely of those outlined here.

The computer generated trees

In an attempt to shed further light on the hypotheses discussed above, the characters used in constructing these hypotheses were analysed using programmes of the PHYLIP package. Character state polarities were determined on the basis of hypothesis 1. No attempt was made to deal with convergence, except where there was some indication from morphology that characters were non-homologous. In addition, the Balbarinae was split into three taxonomic units because the monophyly of this subfamily is in doubt.

An initial run of this data set (Tables 2-3) through the Sokal, Dollo and Polym programmes produced the trees shown in Fig. 4. All of the trees are similar in that advanced potoroids and macropodids are monophyletic and hypsiprymnodontines and propleopines are primitive relative to all other groups. The trees generated by the Polym and Sokal programmes are similar in recognizing a bulungamayine-potoroine clade and a *Nambaroo novus* (and *N. saltavus* in Polym)-*Balbaroo*-sthenurine-macropodine clade. The Dollo tree differs from these in recognizing a *Nambaroo saltavus*-potoroine group and a bulungamayine-*Nambaroo novus*-*Balbaroo*-macropodine-sthenurinae group. The Dollo tree is clearly closest to hypothesis 2 of those discussed above. The Sokal and Polym trees are also close to hypothesis 2, except that the bulungamayines are not the sister group of the macropodids. In all the computer generated hypotheses, advanced potoroids and macropodids are monophyletic. Therefore, depending on the programme, lophodonty is interpreted as having been independently

Table 2. Characters used in computer analysis of macropodoid interrelationships. The first of the two described states is the primitive one.

(1)	I^1 occlusal edge narrow, broad.
(2)	I^2 larger, smaller than I^1 or I^3.
(3)	I^3 does not possess, possesses a groove on the buccal face.
(4)	I_1 possesses, lacks a dorsal enamel crest.
(5)	I_2 is retained, lost.
(6)	Lower incisors lie on top of upper incisor crowns, lie inside upper incisor arcade.
(7)	P1 retained, lost.
(8)	P2 small, hypertrophied.
(9)	P3 short, elongate.
(10)	P3 finely, coarsely grooved.
(11)	P3 lacks, has a posterointernal cusp.
(12)	P3 is, is not flexed buccally out of allignment with the molar row.
(13)	Postlink is not, is hypertrophied.
(14)	A neomorphic posterior cingulum is not, is developed.
(15)	The posthypocristid is not, is lost.
(16)	The cristid obliqua runs, does not run to the protostylid on M_2.
(17)	The protostylid of M_2 is present, lost.
(18)	The postglenoid process is absent, present.
(19)	A frontal-squamosal contact is absent, present.
(20)	The molar gradient is steep, not steep.
(21)	The dentary is not, is convex below the middle of the molar row.
(22)	The masseteric canal does not extend to below P_3, extends to below P_3.
(23)	The hallux is present, lost.
(24)	Molar enamel is coarsely crenulate, not coarsely crenulate.
(25)	The calcaneal cuboid facet is not, is stepped.
(26)	The tibia-fibular contact is very small, extensive.
(27)	The dental foramen opens anterior to P^3, above P^3.
(28)	The masseteric and dental canals are not, are confluent anteriorly.
(29)	The metaconid of M_2 is present, lost.

Table 3. The data matrix used in the computer programmes for the analysis of the Macropodoidea; 0 = plesiomorphic, 1 = derived, ? = state not known. The characters are listed in the same order as in Table 2.

ur-Phalanger	0 0 0 0 0 0 0 0 0 0 0 0 0 0 0 0 0 0 0 0 0 0 0 0 0 1 0 0 0
ur-hypsiprymnodontine	0 0 0 0 0 0 1 1 0 0 0 0 0 0 0 0 0 0 1 1 0 0 1 0 0 0 0 1 1
ur-bulungamayine	? ? ? 0 1 0 ? 1 1 0 1 0 0 ? 1 0 ? ? 1 1 ? 1 ? 0 0 ? 1 1 ?
ur-potoroine	0 0 0 0 1 0 1 1 1 0 0 0 0 0 0 0 0 0 1 1 0 0 0 0 1 0 1 1 1
Nambaroo saltavus	? ? ? 0 ? ? ? ? 1 ? ? ? ? 0 0 0 0 ? ? ? ? 1 ? ? ? ? ? ? 0
ur-Balbaroo	? ? ? 0 0 ? ? 1 0 1 1 0 0 1 1 1 1 ? 0 0 ? 1 ? ? 1 ? ? 1 0
ur-macropodine	1 1 1 0 1 1 1 1 1 1 1 1 0 1 1 1 1 1 0 0 1 1 1 1 1 0 1 1 0
ur-propleopine	? ? ? 1 0 ? ? 1 0 1 0 0 0 0 0 0 0 ? 0 1 ? 1 ? ? ? ? 0 0 0
ur-sthenurine	0 0 1 0 1 0 1 1 1 1 1 1 1 1 1 1 1 1 0 0 1 1 1 1 ? 0 1 1 0

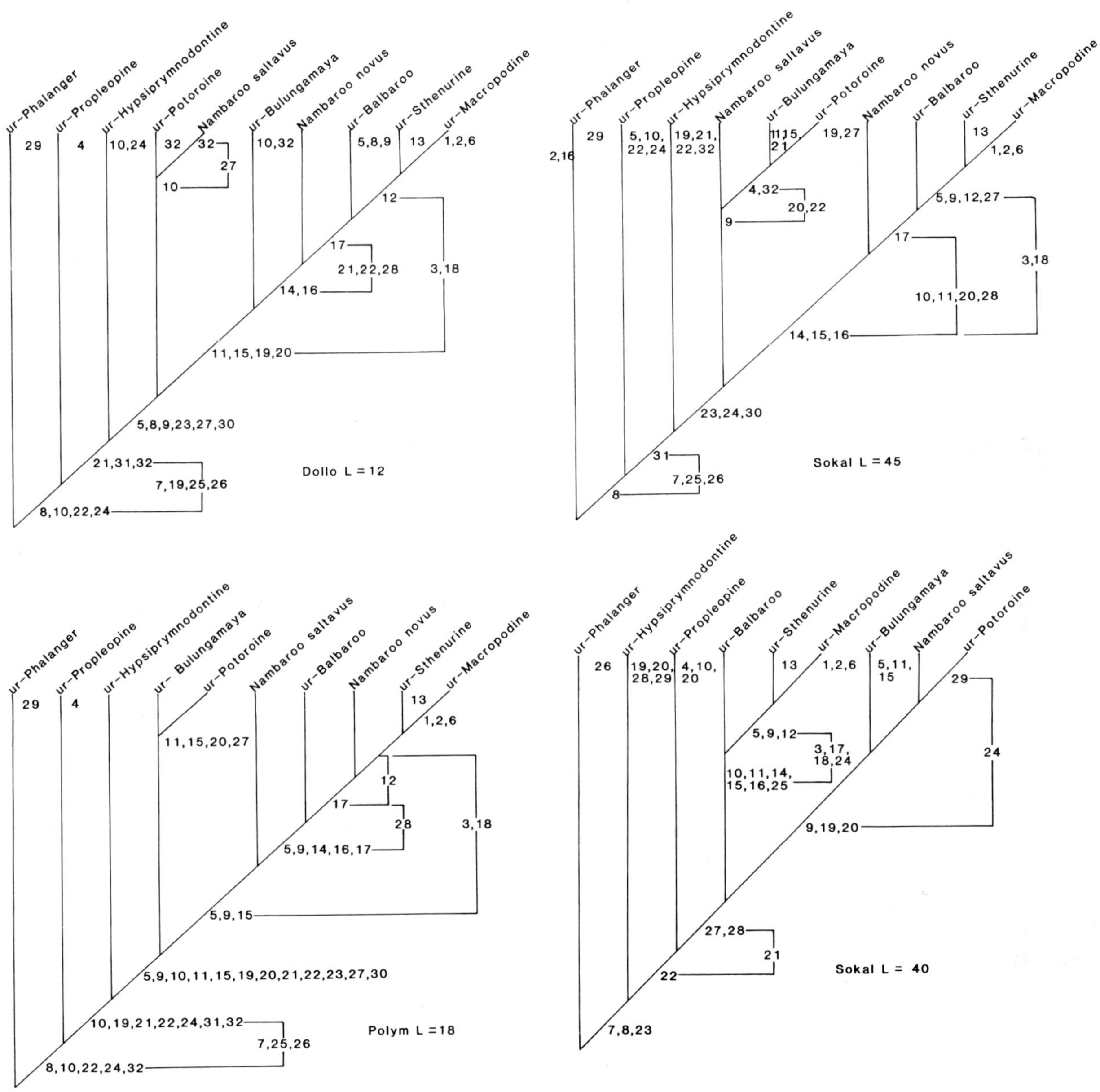

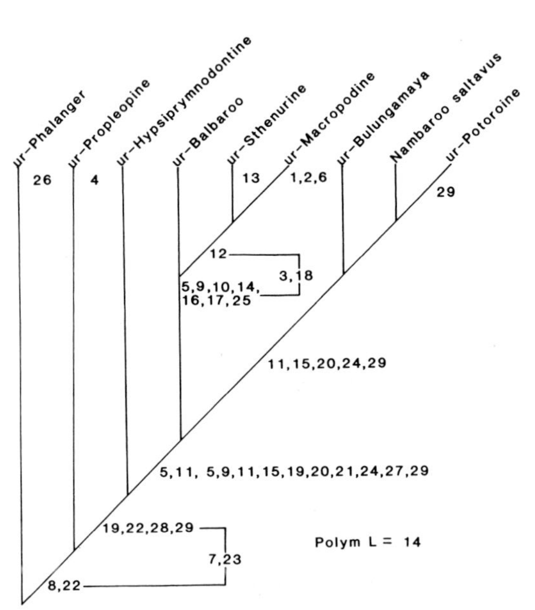

Fig. 4. Cladograms depicting hypotheses of macropodoid interrelationships generated by: (using data set 1), (upper left) the Dollo programme, (upper right) the Sokal programme, (middle left) the Polym programme; and (using data set 2), (middle right) the Sokal programme and (lower left) the Polym programme.

derived in bulungamayines and macropodids (Sokal), or present in the common ancestor of the advanced potoroids and macropodids (Dollo), and there has been a reversion to bunodonty in potoroines. The placement of *Nambaroo saltavus* with potoroines in the Dollo programme is the reason for need of this reversion. However, the placement of *N. saltavus* is weakly supported because it is a very poorly-known species. Lophodonty could just as easily have arisen in the common ancestor of macropodids and bulungamayines.

The most important similarity between these trees and hypothesis 2 appears to be that potoroid synapomorphies proposed in hypothesis 1 are not recognized as such. At least some are seen as having been independently derived in hypsiprymnodontines and other potoroids or as having been lost in macropodids (plesiomorphic for the Macropodoidea).

In light of these results, it seemed reasonable to conclude that the large number of characters suspected of being homoplasious in Table 1 were biasing the results of the computer programmes. Accordingly, three characters (23, 24, 30) that supported the "advanced potoroine/macropodid" clade (they were acquired at the HTU basal to the advanced potoroid-macropodid clade in Sokal) were removed from the matrix. *Nambaroo novus* was also removed from the matrix because it is a poorly-known taxon and because it is always grouped with *Balbaroo*. The new data set was run twice through the Sokal and once through the Polym programmes. These were chosen because of their concordance in the previous run and because of the clustered macropodids and potoroids as monophyletic units. Surprisingly, the new data set produced trees similar to those produced by the previous data set. It is clear that the large number of suspected homoplasies presented in Table 1 bias the programmes so that only trees of this sort are produced. This is because the clustering programmes rely solely on parsimony, and because the features that support the derived macropodid-potoroid clade are more numerous than those supporting hypotheses necessitating other clades.

Of course, it would be possible to remove all characters presented in Table 1 that are used in this analysis. However, this, along with the polarity of the characters given, would virtually guarantee that trees concordant with hypothesis 1 would be produced. The aim of the computer analysis, was, after all, to determine, given as few apriorisms as possible, which hypothesis is the most parsimonious. Given extreme manipulation of the input data, use of a computer becomes unnecessary because the trees generated have already been predicted by the operator.

It appears then, that use of computer programmes in this way give us an estimate of the most parsimonious interpretation of the input data given the programme assumptions. Without weighting of characters and a consideration of functional interrelationships I feel that these computer programmes produce rather simplistic solutions to complex problems. With the use of such biasses, however, the operator "predicts" the desired phylogeny and does away with simple parsimony.

Computer programmes such as these do, of course, have great value in some systematic studies. Where large number of taxa and characters are involved and where polarity cannot be easily determined, these programmes are useful in generating hypotheses that can then be tested. They are also useful as they organize data in ways that make it easier to interpret because features are divided into groups that support one hypothesis or another.

Felsenstein (1978) discusses the effect of homoplaisy on computer programmes based on maximum parsimony and shows that when more than a given number of characters are homoplasious, the trees generated become statistically inconsistent; ". . . The conditions for this failure . . . are essentially that parallel changes exceed informative, nonparallel changes . . . " (Felsenstein 1978, p401). Here, suspected homoplasious characters far outweigh suspected "true" synapomorphies. Thus the trees produced consistently give the "wrong" answer.

Summary

The approaches taken and the hypotheses presented above would seem to cover the most important of the myriad possible ways of interpreting higher-level macropodoid systematics. Preference for one hypothesis above another depends upon interpretation of polarity, likelihood of homoplaisy in characters and reliance on parsimony. For me, hypothesis 1 is by far the most preferable because the synapomorphies that support it seem well-understood in terms of function, and are integrated parts of character complexes that, because of their intricate nature are less likely to be subject to homoplaisy. Also, the evolution of these character complexes seem to signify fundamental shifts in adaptation, leading for instance to the development of specialized browsers. The large number of characters that would have to be homoplasious if this hypothesis is correct (see Table 1) are not particularly worrisome. On the basis of structure, one can be inferred to be non-homologous; others involve loss of teeth already reduced in primitive macropodoids (e.g., C^1, I_2), and yet others, on the basis of the distribution of many other synapomorphies can be shown to be almost certainly homoplasious (e.g., characters 19-21). In addition, there are a number of reasons to suspect that convergence has been a major feature of macropodoid evolution. Macropodoids display changes in functional complexes that seem to have pre-adapted them for further developments. For example, hindlimb adaptations in the *ur*-macropodoid involved structural changes to cope with terrestriality. If the lineages split at an early stage, is it surprising that potoroids and macropodids have independently gone on to lose the hallux and striated plantar pads, to elongate the foot and develop true hopping? After all, these adaptations may be essential in developing a more efficient terrestrial gait for macropodoids and they are not complex adaptations. Also, members of most macropodoid lineages have given rise to species adapted to life in arid regions following the development of these areas in the late Tertiary. The development of aridity is an all pervasive environmental change. Given the similar genetic base of all macropodoids perhaps the responses of the group to such change are very limited. Thus the same structural solutions to the "problem" of aridity have arisen many times in different lineages.

The Potoroidae

Introduction

Potoroids are small to large-sized macropodoids which, although few in species, display a wide range of morphological variation and exploit (now or in the past) diverse ecological niches. Although potoroids were a conspicuous element of many Miocene faunas, they had lapsed into relative insignificance by the end of that period, eclipsed in importance by the vast radiation of macropodids. The 11 living species vary in adult body weight from 0.4 to 3.5 kilograms and are omnivores or herbivores. The 15 extinct taxa recognized here were considerably more varied in size and diet, some weighing up to an estimated 70 kilograms. These large species may have been partly or wholly carnivorous (Archer and Flannery 1985). Several smaller Miocene taxa (the bulungamayines) exhibit fully lophodont molars and were presumably more specialized browsers.

Sharman *et al.* (1980) suggest that *Bettongia tropica* may be synonymous with *Bettongia penicillata* because of chromosomal similarities. Unfortunately, Sharman *et al.* do not illustrate the morphology of the karyotyped specimen of *B. tropica* reported by them, so it cannot be determined if it conforms to the morphology of the type series. Because of their distinct morphology these taxa are maintained as separate species here.

Phylogenetic relationships

Having argued a case for the monophyly of the Potoroidae it is my intention here to examine the interrelationships of the members of this clade. A primary assessment of potoroid interrelationships has been obtained by running a data set containing taxa representing the main kinds of potoroids through four programmes of the PHYLIP package (Dollo, Wagner, Polym and Sokal). The data set includes a hypothetical ancestor, *Palaeopotorous priscus*, *Hypsiprymnodon moschatus*, an *ur-Propleopus*, an ur-bulungamayine, an ur-Potorous, *Aepyprymnus rufescens*, *Caloprymnus campestris*, *Bettongia penicillata*, *Bettongia lesueur*, *Gumardee pascuali*, *Wakiewakie lawsoni* and an undescribed genus and species of potoroine from Pliocene sediments at Hamilton. Some important taxa are excluded as they were either unknown at the time that this work was undertaken (e.g., *Ekaltadata ima* and *Bettongia moyesi*), or were unavailable to the author at that time (e.g., *Purtia mosaicus*).

The data set (Tables 4-5) was run three times (with taxon order shuffled) in all four programmes. The most parsimonious tree produced by each programme was analysed for character placement. These four trees (and most of the others), were consistent in their placement of the following taxa in the following branch order; hypothetical ancestor, *Palaeopotorous priscus*, *Hypsiprymnodon moschatus*, *ur-Propleopus*, *ur*-bulungamayine and *ur-Potorous* (see Fig. 5). Because of the consistent relative placement of these taxa and the nature of the characters supporting their placement, this section of the cladograms is considered to be stable. Synapomorphies that are not subject to convergence, reversion or retained polymorphism in these phylogenies and which are identified within this section in all four programmes are:

(1) For all taxa other than *Palaeopotorous priscus:* 13, presence of an anterior crista emanating from the protostylid of M_2. This structure is known in all potoroid species (including propleopines), except *P. priscus*, and is unique in the Macropodoidea to these species. At present there is no reason to doubt that this structure is homologous in all the potoroids that possess it.

(2) For taxa other than *Palaeopotorous priscus* and *Hypsiprymnodon moschatus:* 12, the parastyle of M^1 is reduced in size; and 14, the protolophid of M_2 is broadened. Unfortunately, at the time that this analysis was undertaken, investigation into the homology of cuspids of the M_2 trigonid of potoroids was not complete. The discovery and study, subsequent to completion of this analysis, of *Ekaltadeta ima* from Riversleigh in Queensland, has shown that the anterior lophid of M_2 in propleopines and hypsiprymnodontines/potoroines is not homologous. In propleopines the protoconid, metaconid and protostylid are all incorporated into the protolophid. In hypsiprymnodontines/potoroines, however, the metaconid is displaced posteriorly and reduced or lost. Thus

Table 4. A list of characters used in the computer analysis of potoroid interrelationships. The characters are listed in the same order as in Table 5. The first described of the two states is the primitive one. In general, polarity is determined by outgroup comparisons with the phalangerids, particularly the less specialized species of *Phalanger* (see Flannery 1987). Unfortunately in some cases such comparisons were not useful and other comparisons had to be made (e.g., see periotic characters). The characters are for the most part self explanatory. However, some merit further discussion.

Character 16 — A bunodont macropodoid molar is defined here as one that possesses a posthypocristid and very well-developed cristid obliqua, paracristid and prehypocristid (lower molars), and very well-developed pre and posthypocrista and pre and postprotocrista (upper molars). These longitudinally oriented structures are generally better-developed in bunodont molars than the transverse lophids. Lophodont molars are defined as those which lack a posthypocristid, or at least have it markedly shifted lingually, and where the lophids (transverse structures) are better-developed than the longitudinal structures.

Character 17 — A strongly skewed posthypocristid is present on all lower molars of *Aepyprymnus rufescens* but may only be present on the more posterior molars of *Caloprymnus campestris*.

Characters 27, 29-35 — The periotic characters included here have presented some problems in analysis. Comparisons with the phalangerids were not generally successful, primarily because of the highly specialized nature of the phalangerid basicranium. Also, little comparative material in the way of isolated phalangerid periotics is available. Therefore, the primitive potoroid condition was determined largely by comparison with primitive macropodids and *Hypsiprymnodon moschatus*. An isolated periotic from the Kutjumarpu local fauna is here assigned to a species of bulungamayine for the following reasons. Bulungamayines and balbarines are the most abundant mammals in the Kutjumarpu local fauna. The periotic is undoubtedly of macropodoid-type morphology, and is too small to belong to a balbarine. The only other possible candidate is *Wakiewakie lawsoni*. This species is a much rarer element in the fauna than the bulungamayine species, and is probably a potoroine. The periotic is markedly non-potorine like in morphology, but resembles that of *H. moschatus*. Thus on the basis of frequency, size and morphology, it is considered likely that the periotic belongs to one of the two closely related bulungamayine species from the Kutjumarpu local fauna.

A final problem encountered in the use of periotic characters has been the limited material available for study. Only 1-3 periotics of adults were available for the taxa studied, and sometimes the usefulness of these specimens was decreased by their being buried in skulls. Clearly, much more work needs to be done before the full use of the periotic as a taxonomic tool within macropodoids is realized. This preliminary investigation has revealed much variation within the potoroid periotic, and the bone clearly has great potential.

(1) I_1 possesses a dorsal enamel crest, does not posses the crest.
(2) I_2 is present, absent.
(3) C^1 is large, very small.
(4) P^3_3 is short, elongate.
(5) P^3_3 is elongate, very elongate with more than 10 ridgelets.
(6) P^3_3 is finely grooved, coarsely grooved.
(7) P^3 is not bulbous, is bulbous and half-moon shaped.
(8) P^3 lacks posterointernal cusp, has cusp.
(9) P^3_3 is deflected from molar row, is alligned with molar row.
(10) P^3_3 has arcuate occlusal edge, has straight occlusal edge.
(11) P^3 lacks basal buccal cingulum, has basal buccal cingulum.
(12) M^1 has a very large parastyle, parastyle is smaller.
(13) M_2 lacks anterior cristid from the protostylid, has cristid.
(14) Protolophid of M_2 is narrow, broad.
(15) Molar lophids well formed, very weakly formed (i.e., molars almost quadritubercular).
(16) Molars bunodont or sublophodont, fully lophodont (see discussion).
(17) Posterior cingulum of lower molars is formed by the posthypocristid and postentocristid, posthypocristid dominates posterior cingulum.
(18) Molar enamel crenulate, non crenulate.
(19) Molar gradient steep, not steep.
(20) Dentary not convex below molar row, convex below molar row.
(21) Postglenoid process absent, present.
(22) Frontal squamosal contact narrow, broad.
(23) Masseteric process absent, present.
(24) Hallux present, absent.
(25) Bullae not greatly inflated, greatly inflated.
(26) Infraorbital canal opens anterodorsal to P^3, opens dorsal to P^3.
(27) Flat area of endocranial exposure of the periotic gently angled from the rest of the endocranial exposure, steeply angled.
(28) Digital pads of the pes separate, fused into a single unit.
(29) Crista petrosa weak, strong.
(30) Anterior end of the crista petrosa is knoblike and low, a strut like spike.
(31) Anterior subarcuate sulcus (arterial) weakly developed, strongly developed.
(32) Mastoid does not project above promontory, does project above promontory.
(33) Tensor tympanic fossa poorly defined, well defined.
(34) Ectotympanic process of promontory small, large.
(35) Ectotympanic process is not discrete, is discrete.
(36) A single buccal root is present on upper molars, two buccal roots are present.

Table 5. The data matrix for the Potoroidae used in the computer programmes; 0 = plesiomorphic, ? = state unknown, 1 = derived state, p = polymorphic for that one character (0 and 1 present).

Hypothetical ancestor	0 0 0 0 0 0 0 0 0 0 0 0 0 0 0 0 0 0 0 0 0 0 0 0 0 0 0 0 0 0 0 0 0 0 0 0
Palaeopotorous priscus	? ? ? ? ? ? ? ? ? ? ? ? 0 0 0 0 0 1 ? ? ? ? ? ? ? ? ? ? ? ? ? ? ? ? ? ?
Hypsiprymnodon moschatus	0 0 0 0 0 0 0 0 0 0 0 0 1 0 1 0 0 0 0 1 0 0 0 0 0 0 0 0 0 0 0 0 0 0 0 0
ur-bulungamayine	0 1 ? 1 0 0 1 1 1 1 0 ? 1 1 0 1 0 1 1 1 ? ? ? ? ? ? 0 ? 0 0 0 0 0 1 0 0
ur-propleopine	1 0 ? 0 0 0 0 0 0 0 0 1 1 1 0 0 0 1 0 0 ? ? ? ? ? 0 ? ? ? ? ? ? ? ? ? 1
ur-Potorous	0 1 0 1 0 1 0 0 1 1 0 1 1 1 0 0 0 1 0 0 0 1 0 1 0 1 1 0 1 1 1 0 1 1 0 1
Bettongia penicillata	1 1 0 1 0 0 0 1 0 1 0 1 1 1 0 0 0 1 0 1 1 1 0 1 1 1 1 1 1 ? ? ? ? ? ? 1
Bettongia lesueur	1 1 0 1 0 0 0 1 1 1 0 1 1 1 0 0 0 1 0 1 1 1 0 1 1 1 1 1 1 1 1 1 1 1 1 1
Aepyprymnus rufescens	1 1 0 1 0 0 0 p 1 1 0 1 1 1 0 0 1 1 1 1 1 1 0 1 0 1 1 1 1 1 1 1 1 1 1 1
Wakiewakie lawsoni	1 ? ? 1 1 0 0 1 1 1 0 ? 1 1 0 0 0 1 1 1 ? ? ? ? ? ? ? ? ? ? ? ? ? ? ? 1
Gumardee pascuali	? ? ? 1 1 0 0 1 1 1 1 ? ? ? 0 0 ? 1 0 ? ? ? 0 ? ? 1 ? ? ? ? ? ? ? ? ? 0
Hamilton potoroid	? ? ? 1 1 0 0 0 ? 1 0 ? ? ? 0 0 0 1 1 ? ? ? ? ? ? ? ? ? ? ? ? ? ? ? ? ?
Caloprymnus campestris	1 1 1 1 0 1 0 1 1 1 0 1 1 1 0 0 1 1 0 1 1 1 1 1 0 1 1 1 1 1 ? ? ? ? 1 ? 1

posterior displacement and loss of the metaconid is a synapomorphy linking hypsiprymnodontines and potoroines. For this reason the relative positions of the *ur-Propleopus* and *H. moschatus* in the trees should be reversed. In light of this discovery, it seems sensible to regard 12, (reduction of the parastyle of M^1) to have occurred independently in propleopines and potoroines.

(3) For taxa other than *Palaeopotorous priscus*, *Hypsiprymnodon moschatus* and *ur-Propleopus:* 2, loss of the I_2; 4, elongation of the P^3_3; 10, the occlusal edge of the premolars is subhorizontal (not convex). Also included somewhere between this branching point and that separating the Potoroinae from all other potoroids are characters 22 (the gain of a broadened frontal-squamosal contact) and 24 (the loss of the hallux). However, they could not be placed any more specifically than this because the polarity of these states is unknown in many taxa (see Table 5). These characters seem convincing as indicators of monophyly as some (e.g., 4, 10, 22) form part of a character complex involving increased specialization of premolar shear that is complex and thus is unlikely to be subject to convergence. Others (e.g., 24), are adaptations to terrestriality.

(4) Characters supporting the monophyly of the Potoroinae (i.e., all taxa included except *Palaeopotorous priscus*, *Hypsiprymnodon moschatus*, *ur-Propleopus* and *ur*-bulungamayine) are: 27, the endocranial exposure of the anterior face of the periotic is at a sharp angle relative to the more posterior endocranial surface; 29, the crista petrosa of the periotic is strongly-developed; 30, the transverse process of the periotic is spine-like; 31, the anterior subarcuate sulcus of the periotic is well-defined and 33, the tensor tympanic fossa is well-defined. It is difficult to determine how constant these features are as only a small number of periotics could be examined for any taxa. They are also difficult to evaluate in terms of function. However, they probably reflect increasing specialization in the auditory apparatus.

(5) Monophyly of the Bettongini is strongly supported in all 4 cladograms by the following characters: 21, possession of a postglenoid process; 28, fusion of the pedal digital pads into a single unit; 32, the mastoid process of the periotic projects markedly and 35, the ectotympanic process of the periotic is discrete. As the polarity of characters 32 and 35 are unknown in *Caloprymnus campestris*, it is placed at the base of the Bettongini in the Dollo tree. Precise placement of these characters in the Dollo tree is not possible (see Fig. 5).

Determination of the polarity of the presence versus absence of the postglenoid process in macropodoids is difficult, but is a matter of great phylogenetic importance. Here, absence of the process is coded as primitive, because the structure is unique to the Macropodoidea and thus is absent in the outgroup (although a postglenoid process of a different kind *is* present in phalangerids). However, since this analysis was carried out, the Miocene potoroid *Hypsiprymnodon bartholomaii*, which possesses an incipient postglenoid process, was discovered. This specimen is discussed below. Overall it detracts from the certainty of the polarity of these states as discussed here. Despite this doubt, the three remaining characters cited here support the monophyly of this grouping and thus it appears to be justified in terms of the computer generated trees. The fusion of the pedal digital pads into a single unit is a constant character and seems to be related to increasing terrestriality. Again, the periotic characters are difficult to evaluate.

Relationships within the Bettongini

It is clear from even a brief examination of these trees that there are two areas of instability. These are the placement of the species of the genera *Wakiewakie*, *Gumardee*, and the Hamilton potoroine, and branching order within the Bettongini. The instability of the three first mentioned genera is partly a result of the large

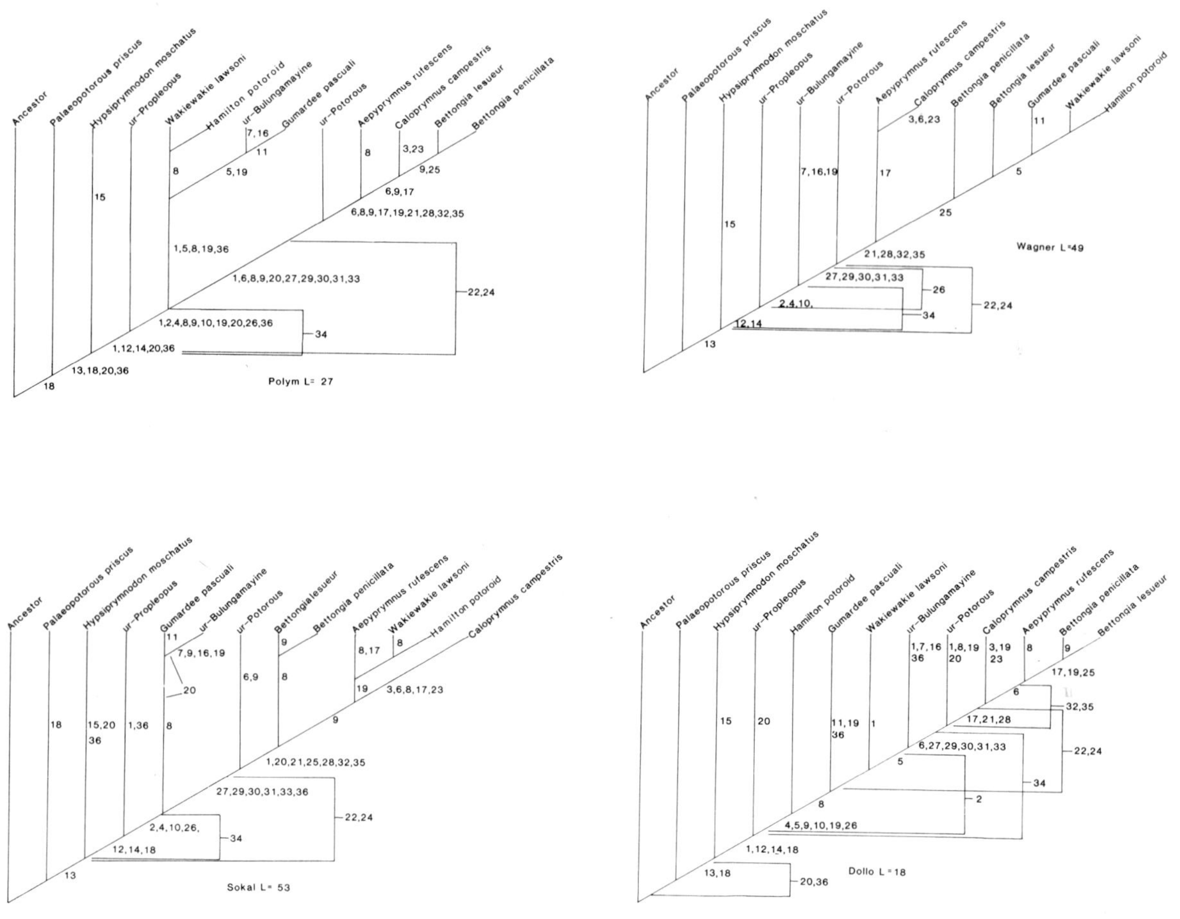

Fig. 5. Most parsimonious cladograms depicting hypotheses of potoroid interrelationships generated by: top left, the Polym programme; top right, the Wagner programme; lower left, the Sokal programme; and lower right, the Dollo programme.

numbers of homoplasious characters and characters where states are unknown in these taxa. However, in the Wagner and Sokal programmes, all or part of this "group" fell within the Bettongini. Because of a lack of resolution regarding relationships within the Bettongini and because of the placement of the above mentioned three genera within this tribe in two programmes, it was decided that further examination of these taxa was necessary. To this end a data set for these taxa was made, using characters that seemed to hold promise of elucidating relationships within the group. This data set was run through the programmes in the same manner as the larger data set.

The more detailed analysis of the Bettongini includes two taxa (*Bettongia tropica* and *B. gaimardi*) not included in the larger study and a smaller, only partially overlapping, data set (Tables 6-7). Because of the less consistent results obtained with this data set, ten runs (with taxon order shuffled) were completed for each programme. Instability within these trees is probably caused by the high number of homoplasious characters (in Sokal, Wagner and Dollo) and by the high number of retained polymorphisms necessary in Polym. In the most parsimonious Wagner tree produced, 10 of the 23 characters were homoplasious, while in the most parsimonious Sokal tree, 12 of the 23 characters were homoplasious. The most parsimonious trees produced by each programme were analysed for character placement. In some cases equally parsimonious trees were produced by the one programme. For the Wagner programme, three trees of equally short length were produced, all of which were similar in topography. However, for the Sokal programme, two trees of equally short length produced quite different phylogenies. In this case, both trees were analysed and utilized and referred to as Sokal 1 and Sokal 2 (see Fig. 6).

Despite the unstable nature of these trees, some taxa are consistently grouped together. Perhaps the most significant of these is the grouping of *Aepyprymnus rufescens* with *Caloprymnus campestris*, present in all the most parsimonious trees. The synapomorphies uniting these taxa are: 1, the I^3 occlusal crest is rotated medially; 9, the posthypocristid is well developed, crossing

Table 6. Characters used in computer analysis of the interrelationships of the Bettongini. The characters are listed in the same order as in Table 7. The first of the two described states is the primitive one.

(1)	The occlusal crest of I^3 is alligned with the occlusal rests of I^{1-2}, is flexed lingually.
(2)	The canine is large, small.
(3)	The P3 occlusal crest is flexed buccally, is parallel with the molar row.
(4)	The P3 is finely, coarsely grooved.
(5)	The P3 is elongate, very elongate.
(6)	P^3_3 lacks a basal buccal cingulum, has basal buccal cingulum.
(7)	P^3 lacks, has posterointernal cusp.
(8)	M^1 very small, is more equal in size with M^2.
(9)	Upper molars lack, possess very well-developed posthypocrista that crosses the rear face of the hypoloph diagonally.
(10)	Molars are complexly, coarsely crenulate, not crenulate.
(11)	Molar gradient steep, not steep.
(12)	Single buccal root on upper molars, two buccal roots present.
(13)	Upper molars with steep, gently sloping lingual margins.
(14)	Masseteric process of maxilla small, large.
(15)	Bullae small, large.
(16)	Bullae small or large, extremely large.
(17)	Premaxilla elongate, foreshortened.
(18)	P^3-C^1 diastema elongate, foreshortened.
(19)	Foramen ovale not hooded, hooded.
(20)	Periotic exposed on basicranium, not exposed.
(21)	Palatal vacuities present, not present.
(22)	No dorsal crest is present on the tail, a crest is present.
(23)	The jugal does not reach high on the orbit, does reach high on the orbit.

Table 7. Distribution states within the Bettongini. 0 = plesiomorphic state, 1 = apomorphic state, ? = state not known, p = polymorphic for that character (0 and 1 present).

Ancestor	00000000000000000000000
Bettongia lesueur	00100010010100110111001
Gumardee pascuali	??10111?010000?????????
Aepyprymnus rufescens	101000p1111100001101100
Caloprymnus campestris	11110010110101101110000
Hamilton potoroid	???01001011?1??????????
Bettongia penicillata	00000010000100110011011
Bettongia tropica	00100011000100110011011
Bettongia gaimardi	001000p1000100110011011
Wakiewakie lawsoni	??10101101110??????????

diagonally almost the whole width of the rear face of the hypolophid; 17, the premaxilla is foreshortened; and 18 (also present in *B. lesueur*), the P^3-canine diastema is shortened. In the Polym programme, 1 and 17 were the only non-homoplasious characters that supported this grouping because the inclusion of the species of *Gumardee, Milliyowi* and *Wakiewakie* in the group necessitate a retained polymorphism in 9. Because the derived state of 18 is also present in *B. lesueur*, this character is homoplasious in Sokal, Wagner and Dollo, and subject to retained polymorphism in Polym.

These four synapomorphies are convincing evidence of monophyly. The unusual molar pattern (relatively high-crowned molars with very well-developed posthypocristae that cross diagonally almost the whole width of the rear face of the hypolophid) is unique in the Potoroidae but, peculiarly, is mimicked in highly specialized grazing macropodines (e.g., *Macropus pearsoni*). The I^3, with its cutting crest offset from that of I^{1-2} is again unique to this clade among potoroids but a similar adaptation is seen in the grooved $I^{2-3}$of *Macropus* species, which form a series of obliquely set cutting edges. These features may thus be related to dietary specializations, particularly grazing. However, the shortening of the rostrum seen in *A. rufescens* and *C. campestris* is not seen in grazing macropodines, where the opposite trend is apparent. That shortening of the premaxilla and anterior part of the maxilla can occur independently is corroborated by *Bettongia lesueur*, where the premaxilla is foreshortened but the maxilla is not. This corroborates the decision to code these as independent characters.

A further clade, supported in all the most parsimonious trees except Dollo, where the group is considered polyphyletic, is that including the species of *Bettongia*. Non-homoplasious synapomorphic features that unite this group include characters 16 (the bullae are highly inflated) and 23 (the jugal reaches high on the orbit). Homoplasious characters 12, 15, 19 and 20 (see Table 6) also support this placement in the Sokal trees. It is surprising to find that monophyly in the genus *Bettongia* is less strongly supported by morphology than the *Aepyprymnus rufescens*/*Caloprymnus campestris* clade. This is thought to be because the species of *Bettongia* are generally plesiomorphic in morphology. However, the high jugal is particularly convincing as a synapomorphy because its form in the species of *Bettongia* is unique within the Potoroidae and it is consistently present in all of the species of *Bettongia*.

For those trees that recognized *Bettongia* as a monophyletic group, species were arranged within the genus in different ways. Both the Polym and Sokal trees recognized a terminal clade containing *B. tropica* and *B. gaimardi* on the basis of characters 7, the P^3 has a posterointernal cusp and 10, the molar enamel is not coarsely crenulate, both of which are homoplasious. Polym placed *B. penicillata* as the sister group to this clade on the basis of 22, presence of a tail crest, while Sokal 2 placed *B. lesueur* as the

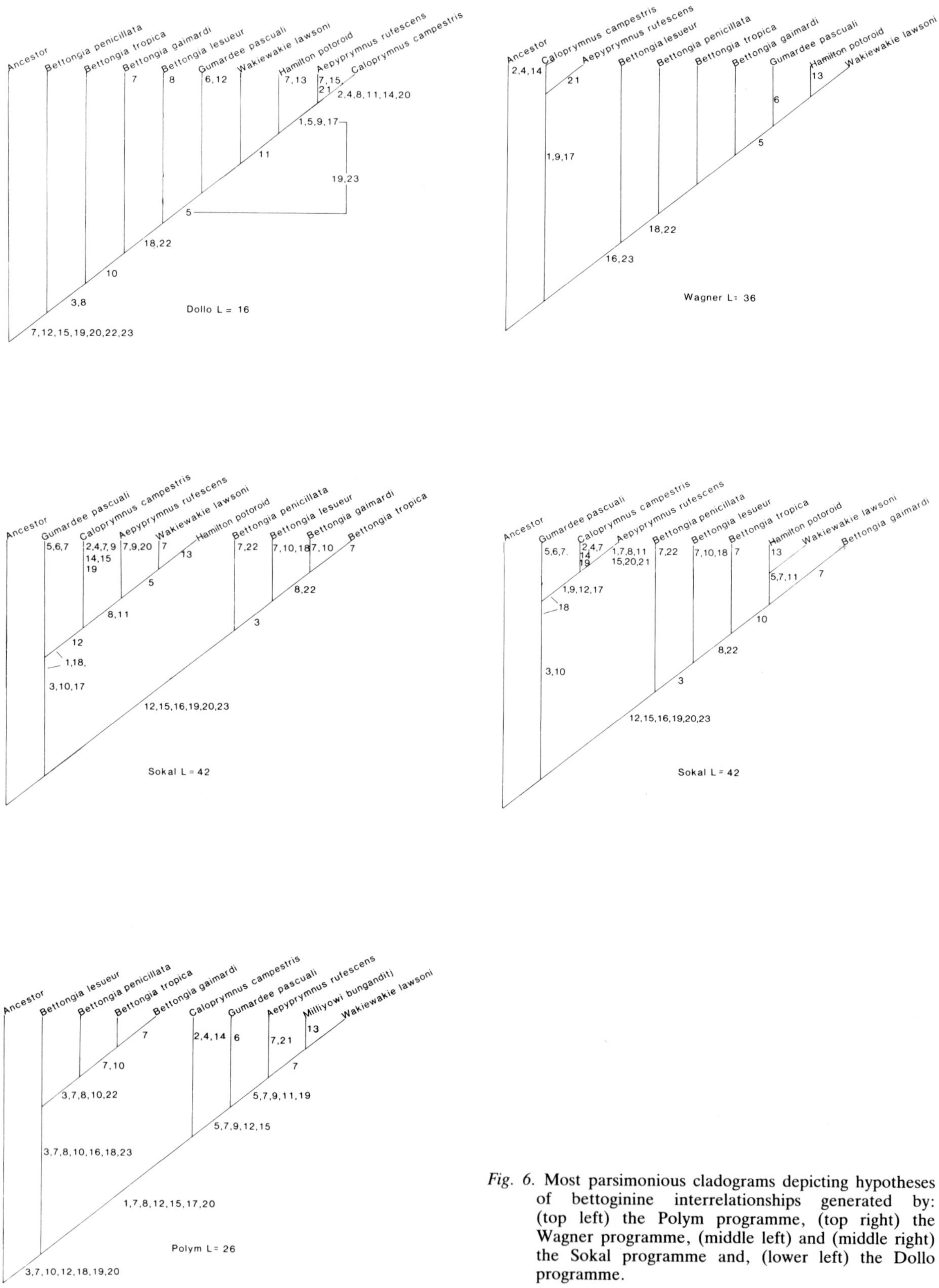

Fig. 6. Most parsimonious cladograms depicting hypotheses of bettoginine interrelationships generated by: (top left) the Polym programme, (top right) the Wagner programme, (middle left) and (middle right) the Sokal programme and, (lower left) the Dollo programme.

sister group to the clade on the basis of 3, the P3 is not deflected from the molar row (which is homoplasious). While the evidence is slim, the Polym programme placement within *Bettongia* seems more reasonable because character 22 is non-homoplasious.

In only one tree (Wagner) did the Hamilton potoroid, *Gumardee pascuali* and *Wakiewakie lawsoni* form a clade, and this grouping is supported by only a single character, 5, possession of a very elongate P^3. The grouping of the Hamilton potoroid, *G. pascuali* or *W. lawsoni* within the *Bettongia* clade or the *Aepyprymnus/Caloprymnus* clade by some programmes is weakly supported as the states of the group synapomorphies are unknown for the Hamilton potoroid *G. pascuali* and *W. lawsoni*, or else the synapomorphies are homoplasious. For example, in Sokal 2 *Aepyprymnus rufescens*, *Caloprymnus campestris*, the Hamilton potoroid, *G. pascuali* and *W. lawsoni* are united by characters 3, 10, 17, and possibly 1, 8, 15, and 18. Characters 3 and 10 are homoplasious in this tree, while the states of the remaining characters are not known for the species of *Gumardee*, the Hamilton potoroid and *Wakiewakie*. Thus, in fact, the only characters uniting the group are the two homoplasious characters (3 and 10).

Overall, the clade containing *Aepyprymnus rufescens* and *Caloprymnus campestris* is strongly supported in all of the parsimonious trees. That containing the species of *Bettongia* is also well supported in all trees except Dollo. The interrelationships of the species of *Bettongia* are not greatly illuminated by this study. However, there is some evidence that *B. lesueur* represents the most primitive branch, followed by *B. penicillata*, and that *B. tropica* and *B. gaimardi* form a terminal clade. Evidence supporting the monophyly of a group containing the Hamilton potoroid, *G. pascuali* and *W. lawsoni* is weak; a single character in a single tree. Evidence for including these taxa within the Bettongini is also tenuous. For convenience, they will be grouped at the base of the Bettongini radiation in the classification given here, but will be considered *incertae sedis*.

Relationships within the Propleopinae and Hypsiprymnodontinae

In addition to elucidating the homology of the cuspids of M_2 in propleopines, describing new taxa and discussing the relationships of the propleopines to other macropodoids, Archer and Flannery (1985) give an interpretation of the interrelationships of the propleopines. Synapomorphies uniting the propleopines include the presence of an anterior cristid emanating from both the protostylid and metaconid on M_2 and possibly the presence of a continuously growing I_1 with enamel restricted to the ventral margin. More synapomorphies may be found as the species become better known. Because the species are so poorly known, they do not lend themselves to analysis using the PHYLIP package and thus little can be added here.

With only two taxa presently grouped in the Hypsiprymnodontinae, little can be said about relationships (see Flannery and Archer 1987a).

Relationships within the Bulungamayinae

Few of the known bulungamayines have been described. The author recognizes five closely related bulungamayine species all of which possess lophodont molars (a synapomorphy within the Potoroidae). These are *Wabularoo naughtoni* and *Bulungamaya delicata*, both from the Miocene Riversleigh local fauna of northwestern Queensland, a new undescribed species of *Bulungamaya* from the Gag Site (Dwornamoor local fauna), northwestern Queensland, and two undescribed species of *Bulungamaya* from the Miocene Kutjumarpu local fauna, northern South Australia. In addition to these species, which are clearly monophyletic, there are further undescribed potoroids that may be related to this clade.

Most significant of these undescribed taxa are a dentary (referred to as "cf. *Aepyprymus rufescens*" in Rich *et al.* 1982) from the Miocene Ngama local fauna from Mammalon Hill, South Australia, and *Purtia mosaicus* Case (1984), from the Miocene Ngapakaldi local fauna from Lake Ngapakaldi, South Australia.

Flannery, Archer and Plane (1983) regard *Bulungamaya delicata* to be more primitive than *Wabularoo naughtoni* for the following reasons. A posthypocristid is apparently present on at least M_4 of *B. delicata*, but is unknown on any molars of *W. naughtoni*, and the paracristid and cristid obliqua are more lingually shifted in *W. naughtoni* than in *B. delicata*. The posthypocristid and buccally positioned longitudinal links present on *B. delicata* are reminiscent of bunodont molars and thus, on the basis of outgroup comparisons, are here considered primitive. The interrelationships of the species of *Bulungamaya* remain undetermined.

The Mammalon Hill dentary, although unnamed, merits brief note here, as it displays a combination of phylogenetically interesting features. It shows no more than a gross resemblance to *Aepyprymnus rufescens*, lacking the unique posthypocristid shape seen in that species (see above), although both taxa possess rectangular rather than squarish molars. The masseteric canal is buccally expanded and as anteriorly extensive as in other potoroids. The ventral margin of the dentary is strongly convex as in bulungamayines, most potoroines and, to a lesser extent, *Hypsiprymnodon moschatus*. The P_3 is elongate and bulbous, has an arcuate occlusal crest and is finely grooved. In general it is reminiscent of the P_3 of bulungamayines. The posthypocristid and postentocristid of the molars are short, enclosing a shallow postcingular pocket placed high on the rear face of the hypolophid. The cristid obliqua and paracristid are shifted further lingually than in most potoroines and a shallow fossette is present buccal to the paracristid. The molars are rectangular in occlusal outline and the lophids are extremely well-developed relative to those of potoroines.

These molars thus clearly display a morphology intermediate between the typical bunodont and lophodont patterns. This "sublophodont" molar pattern and the close resemblance of the P_3 to that of bulungamayines, along with a potoroid-like dentary, all combine to suggest that this taxon may represent a primitive bulungamayine.

Purtia *mosaicus* Case (1984) from the Ngapakaldi local fauna of the Miocene Etadunna Formation bears some resemblance to the Mammalon Hill dentary. The P_3 shape is similar, although the molars are more square in occlusal outline and have a cristid obliqua and paracristid that are less lingually shifted and have less well-developed lophs. However, a well-developed fossette is present buccal to the paracristid. This taxon may conceivably represent an even more primitive member of the bulungamayine lineage.

Flannery *et al.* (1984) have suggested that *Gumardee pascuali* may possibly be related to *Purtia mosaicus*. A feature supporting this hypothesis is the presence of a lingual extension of the anterior cingulum on upper molars of *G. pascuali*, which match the fossette placed buccal to the paracristid on the lower molars of the *P. mosaicus*. These features are also seen in macropodoids with lophodont molars as well as some unusual bunodont species (e.g., some propleopines).

In summary, it is difficult at present to elucidate bulumgamayine interrelationships because so little of the known material has been adequately described. However, the known undescribed material is of interest in that it demonstrates the existence of taxa which are intermediate between fully lophodont bulungamayines and somewhat primitive potoroines. A phylogeny for the potoroids is presented in Fig. 7.

Taxonomy

Using data presented above, the taxonomy of the Potoroidae can be expressed as follows. The conventions used here follow Wiley (1981).

Family Potoroidae
 Subfamily Palaeopotoroinae Flannery and Rich, 1986
 Palaeoptorous priscus Flannery and Rich, 1986

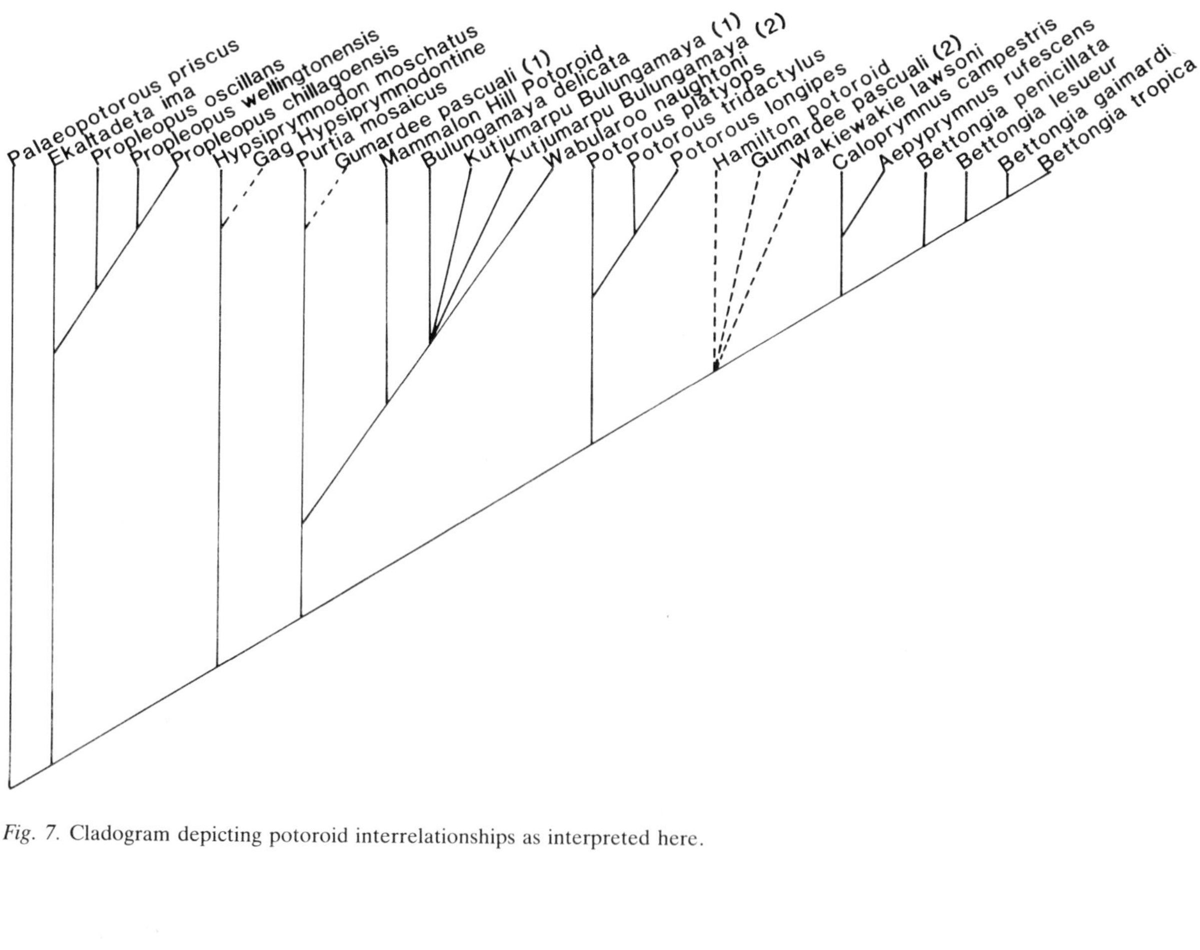

Fig. 7. Cladogram depicting potoroid interrelationships as interpreted here.

Subfamily Propelopinae Archer and Flannery, 1985

Ekaltadeta ima Archer and Flannery, 1985

Propleopus oscillans (De Vis, 1888)

Propleopus wellingtonensis Archer and Flannery, 1985

Propleopus chillagoensis Archer, Bartholomai and Marshall, 1979

Subfamily Hypsiprymnodontinae Collett, 1877

Hypsiprymnodon bartholomaii Flannery and Archer, 1987

Hypsiprymnodon moschatus Ramsay, 1876

Subfamily Bulungamayinae Flannery, Archer and Plane, 1983

Purtia mosaicus Case, 1984 *incertae sedis*
Mammalon Hill dentary

Bulungamaya delicata Flannery, Archer and Plane, 1983

Kutjumarpu *Bulungamaya* 1 *sedis mutabilis*

Kutjumarpu *Bulungamaya* 2 *sedis mutabilis*

Gag Site *Bulungamaya sedis mutabilis*

Wabularoo naughtoni Archer, 1979

Subfamily Potoroinae (Gray, 1821)

Tribe Potoroini n. tribe

Potorous platyops (Gould, 1844) *sedis mutabilis*

Potorous tridactylus (Kerr, 1792)

Potorous longipes Seebeck and Johnson, 1980

Tribe Bettongini n. tribe

Gumardee pascuali Flannery, Archer and Plane, 1983 *incertae sedis*

Hamilton potoroine *incertae sedis*

Wakiewakie lawsoni Woodburne, 1984 *incertae* sedis

Bettongia moyesi Flannery and Archer, 1987

Bettongia lesueur (Quoy and Gaimard, 1824)

Bettongia penicillata Gray, 1837

Bettongia tropica Wakefield, 1967

Bettongia gaimardi (Desmarest, 1822)

Caloprymnus campestris (Gould, 1843)

Aepyprymnus rufescens (Gray, 1837)

The Macropodidae

Introduction

Macropodids include members of the subfamilies Balbarinae, Sthenurinae and Macropodinae. Members of this morphologically diverse group include the largest macropodoids and the only truly arboreal species. Balbarines are restricted to sediments of middle Miocene age and may be ancestral to sthenurines and macropodines, which appear in the early Pliocene and late Miocene respectively. During the Pliocene and Pleistocene, macropodines and sthenurines dominated the medium-sized terrestrial herbivore niche in Australia. However, sthenurines suffered drastic extinctions through the Pleistocene leaving macropodines to dominate this niche in Holocene times.

Two probable synapomorphies suggest that the sthenurines and macropodines form a monophyletic clade. These are the possession on M_2 of a protolophid and hypolophid that are subequal in width, and the complete lack of a posthypocristid on any lower molars. However, more synapomorphies may become apparent as balbarines (the outgroup) become better known.

The Balbarinae

The Balbarinae is a subfamily of extinct macropodids described by Flannery, Archer and Plane (1983), and so far are restricted to sediments of middle to late Miocene age. The known species are estimated to have ranged in size from about 1 to 20 kg in weight. Balbarines were probably browsers and, in some deposits (e.g., the Leaf Locality of central Australia), they are the dominant medium-sized terrestrial herbivores.

Balbarine taxa include: *Balbaroo camfieldensis*, known from the holotype dentary containing $M_{2\text{-}5}$ and a referred specimen containing $M_{3\text{-}5}$, both from the Bullock Creek local fauna, Northern Territory; and *Balbaroo gregoriensis*, known from a single M_2 from the Riversleigh local fauna, Queensland. A dentary fragment containing M_4, of an unnamed species of *Balbaroo* from the Kangaroo Well local fauna, described in Flannery, Archer and Plane (1983), and *Nambaroo tarrinyeri*, *N. novus* and *N. saltavus*, all based on isolated M_2's and referred isolated teeth from the Tarkarooloo local fauna of South Australia (Flannery and Rich 1986). Further material referable to two species of *Nambaroo* has recently been recognized from the Kutjumarpu local fauna. These taxa are referred to here as Kutjumarpu *Balbaroo* 1 and 2. A brief description of these taxa will be included below.

The Kutjumarpu *Balbaroo* species are close to *B. gregoriensis* in size and morphology. The two Kutjumarpu species are immediately distinguishable from one another in that Kutjumarpu *Balbaroo* 1 possesses a protostylid on M_2 while Kutjumarpu *Balbaroo* 2 lacks this structure. Kutjumarpu *Balbaroo* 1 is by far the better known form, being represented by several near complete dentaries. Kutjumarpu *Balbaroo* 2 is represented by a single dentary containing $M_{2\text{-}4}$). An alveolus for I_2 is present in Kutjumarpu *Balbaroo* 1 specimens preserving this area of the dentary, and the P_3 is short and coarsely ridged. The character state polarities of the species used in this analysis are listed in Tables 8-9. The juvenile dentition is unknown for all balbarines.

Table 8. A list of characters used in computer analysis of interrelationships within the Balbarinae. The first of the two described states is the primitive one.

(1)	A protostylid is present, is absent on M_2.
(2)	A neomorphic posterior cingulum on lower molars is absent or ill-defined, is clearly defined.
(3)	The cristid obliqua does not connect to the protoconid on M_2, does connect to the protoconid.
(4)	A posthypocristid is clearly present, is ill-defined or absent.
(5)	A paraconid is present on M_2, absent.
(6)	The paracristid is straight and very well developed on M_2, is more weakly developed.
(7)	The protolophid of M_2 is subequal in height to the hypolophid, is lower than the hypolophid.
(8)	The protoconid is much taller than the metaconid on M_2, the conids are more equal in height.
(9)	The rear face of the hypolophid of M_2 is convex or planar, is concave.
(10)	The P3 is finely grooved, coarsely grooved.

IS THE BALBARINAE MONOPHYLETIC?

As a result of the hypothesis of macropodoid interrelationships developed earlier, balbarines are regarded as the most primitive macropodids. While they are united here within a single subfamily, there is as yet no convincing evidence of monophyly. This may be in part due to their overall plesiomorphic morphology as well as the fact that most species are extremely poorly known. However, some of the better known balbarines do possess distinctive structures that may eventually prove to be synapomorphies uniting the whole group. Included among these are the form of the I_1 in the Kutjumarpu species of *Balbaroo* and the presence of a posterobuccal cusp on some isolated P^3's from the Tarkarooloo and Kutjumarpu local faunas that are probably referable to species of *Nambaroo* and *Balbaroo*. The lower incisors of the Kutjumarpu species of *Balbaroo* are highly distinctive. They are enamelled lingually and possess a series of almost linear antero-posteriorly oriented enamel ridglets on the dorsal surface of the crown. This I_1 structure is unique within the Macropodoidea.

Alternatively, it is possible that the balbarines are in fact a paraphyletic group from which arose both sthenurines and macropodines. Evidence supporting this hypothesis includes the overall similarity of the most advanced balbarines (e.g., *Balbaroo camfieldensis*) to primitive macropodines. Thus loss of the protostylid, contact of the cristid obliqua with the protoconid, gain of the neomorphic posterior cingulum and loss of the posthypocristid could be considered synapomorphies uniting the advanced balbarines with sthenurines and macropodines. These two interpretations result in radically different taxonomies. If the group is considered paraphyletic many additional ranks would become necessary in a ranked classification. For the sake of taxonomic stability, the species discussed here are placed within a single subfamily (Balbarinae) even though it is not certainly a monophyletic taxon.

Table 9. Data matrix used in computer analysis for the Balbarinae. The characters are listed in the same order as in Table 8. 0 = primitive, 1 = derived, ? = state unknown.

Nambaroo saltavus	0	0	0	0	1	0	1	1	0	0
Nambaroo tarrinyeri	0	0	1	0	0	0	1	0	0	0
Nambaroo novus	0	1	1	1	1	0	0	1	1	?
Kutjumarpu *Balbaroo* 1	0	1	1	1	0	0	1	0	1	1
Kutjumarpu *Balbaroo* 2	1	1	1	1	1	0	1	1	1	?
Balbaroo gregoriensis	1	1	1	1	0	0	0	0	1	?
Balbaroo camfieldensis	1	1	1	1	0	1	1	1	1	?

BALBARINE INTERRELATIONSHIPS

In an attempt to determine interrelationships within the Balbarinae, a data set of 10 characters coded for a hypothetical ancestor and for *Nambaroo saltavus*, *N. tarrinyeri*, *N. novus*, *Balbaroo camfieldensis*, *B. gregoriensis*, Kutjumarpu *Balbaroo* 1 and Kutjumarpu *Balbaroo* 2 (see Tables 8-9) was run through the Dollo, Wagner, Sokal and Polym programmes. As the data set is very small, a single run through each programme was considered sufficient to obtain a reasonable approximation of the most parsimonious tree. Because many taxa are known only from M_2, a limited number of characters is available for analysis in most taxa. For this reason, the trees generated here are less likely to be close estimates of the "true" phylogenetic relationships of the taxa than trees based on more comprehensive data sets.

The trees produced are all closely similar in some respects (see Fig. 8). All include a clade containing all balbarines except *Nambaroo tarrinyeri* and *N. saltavus*. The following synapomorphies define this clade; 2, a neomorphic posterior cingulum is developed on M_2, 4, the posthypocristid is very poorly-developed or absent on M_2 and 9, the rear face of the hypolophid of M_2 is near vertical or concave. The monophyly of this clade thus seems well supported. The neomorphic development of the posterior cingulum and the loss of the well-developed posthypocristid are features seen in many advanced macropodids.

Although the Polym, Wagner and Dollo programmes recognize *N. saltavus* as the sister group of this clade, they do so only on the basis of characters subject to homoplaisy or retained polymorphism. These are characters 5 (presence/absence of a paraconid) and 8, (the protoconid is taller than or subequal in height to the metaconid) in the Dollo and Polym programmes. The Sokal programme recognizes *N. tarrinyeri* as the sister group of the clade containing most balbarines on

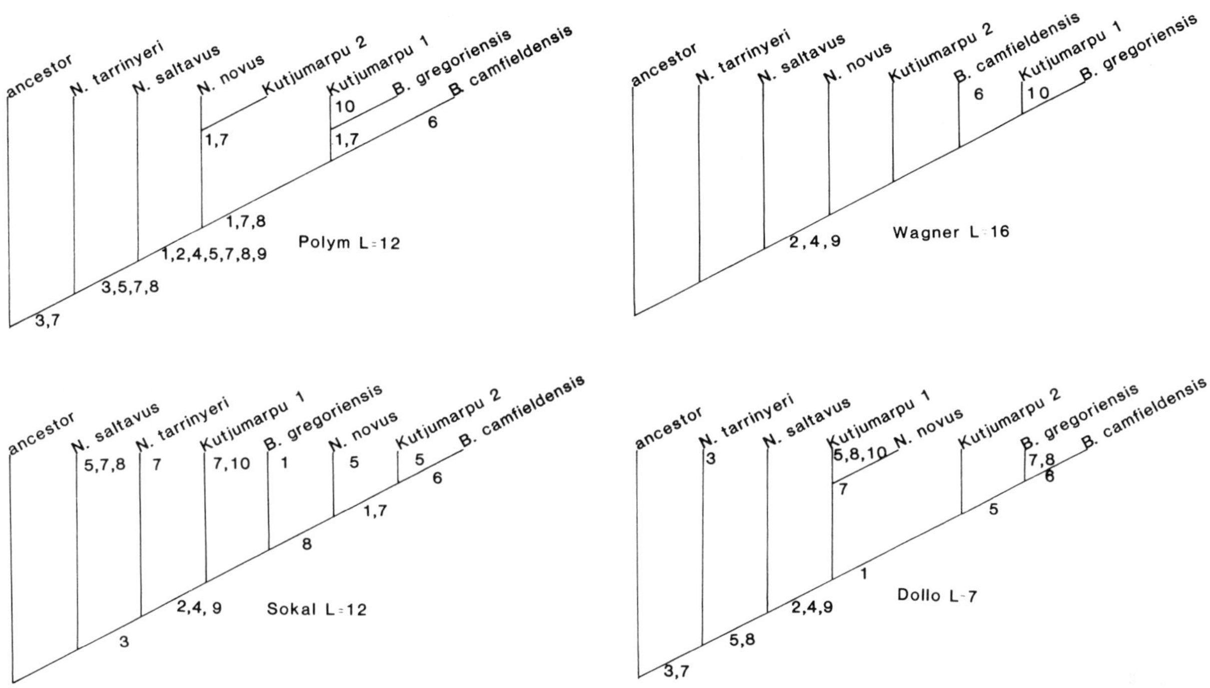

Fig. 8. Cladograms depicting hypotheses of balbarine interrelationships generated by: Top left, the Polym programme; Top right, the Wagner programme; Lower left, the Sokal programme; and Lower right, the Dollo programme.

the basis of a non-homoplasious synapomorphy, the derived state of character 3 (that the cristid obliqua contacts the base of the protoconid). *Nambaroo tarrinyeri* possesses a bifid cristid obliqua, one branch of which terminates at the base of the protolophid and one at the protostylid. Thus it was coded as derived for character 3. The presence of this state in *N. tarrinyeri* is considered to be of some importance here because establishment of contact between the cristid obliqua and the protoconid is characteristic of all macropodines and sthenurines as well as advanced balbarines. Thus the interpretation presented in the Sokal tree is thought to be the correct one.

Apart from these clades, only a single additional clade supported by non-homoplasious characters was recognized by any programme. This was the grouping of Kutjumarpu *Balbaroo* 2, *B. gregoriensis* and *B. camfieldensis* in the Dollo tree, which was supported by a single synapomorphy, 1, the absence of a protostylid on M_2. Again, protostylid loss characterizes macropodines and sthenurines and appears to be a feature of some importance. Thus this clade is accepted as probably being monophyletic.

Despite the paucity of well-supported clades, the trees did tend to place taxa in a generally similar manner (see Fig. 8). For instance, *Balbaroo gregoriensis* and/or *B. camfieldensis* were always grouped in or near a terminal clade in the tree and *Nambaroo novus* was always placed further towards the base. As a result of the consistent placement of the taxa in these trees some trends in balbarine evolution are apparent. These can be summarized in the following ways. There is a trend towards an increased obscuring of the posthypocristid. In the most primitive species it is a distinct structure, originating two-thirds of the distance lingual of the apex of the hypoconid on M_2. In the most derived species (e.g., *B. camfieldensis*), it is represented by a barely distinguishable ridge bounding the buccal edge of a fissure located on the buccal edge of the rear face of the hypolophid. There is also a trend towards loss of contact between the cristid obliqua and the protostylid on M_2 followed by loss of the protostylid itself, and towards weakening of the strength of the paracristid on M_2, and also (probably) loss of the grooving on P3. Concomitant with these changes comes an increase in size. As a result of these trends, the most derived balbarines are rather similar to what one would imagine to have been the ancestral morphology for macropodines and sthenurines. A phylogeny for the Balbarinae is given in Fig. 9.

The balbarines can be classified as follows. The conventions used here follow Wiley (1981).

Subfamily Balbarinae Flannery, Archer and Plane, 1983

Nambaroo saltavus Flannery and Rich 1986

Nambaroo tarrinyeri Flannery and Rich 1986

Nambaroo novus Flannery and Rich 1986 *sedis mutabilis*

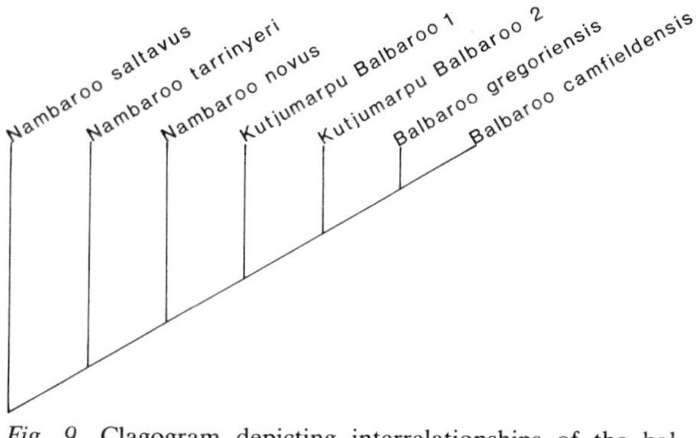

Fig. 9. Clagogram depicting interrelationships of the balbarines as interpreted here.

Kutjumarpu *Balbaroo* 1 *sedis mutabilis*

Kutjumarpu Balbaroo 2

Balbaroo gregoriensis Flannery, Archer and Plane, 1983

Balbaroo camfieldensis Flannery, Archer and Plane, 1983

The Sthenurinae

Sthenurines (including the species of *Lagostrophus*, *Troposodon*, *Simosthenurus*, *Sthenurus* and *Procoptodon*) first appear in the fossil record in the early Pliocene. At this time, the species of *Troposodon* were abundant, species of *Procoptodon* and *Sthenurus* unknown and species of *Simosthenurus* rare, being represented thus far by a few fragmentary teeth. Throughout the Pliocene and Pleistocene the species of *Troposodon* became rarer, and those of *Simosthenurus* and related species became more abundant. Before the end of the Pleistocene the species of *Troposodon* had become extinct. During the same period *Simosthenurus* was represented by about seven species, *Sthenurus* by about three and *Procoptodon* by about four species. However, nothing of this vast radiation survived the Pleistocene extinctions. Only the primitive, diminutive Banded Hare Wallaby *(Lagostrophus fasciatus)*, whose known fossil record extends only as far back as the early Pleistocene (although the lineage is undoubtably much older), survived to represent this previously diverse subfamily in the modern world.

STHENURINE MONOPHYLY

The Subfamily Sthenurinae has recently been revised (Flannery 1983). Concepts presented there regarding composition of the group differ from those of earlier studies. Synapomorphies uniting sthenurine taxa are identified by Flannery (1983) as follows: 1) presence of a well-developed postlink on upper molars; 2) presence of an overlap between the postmetacrista and posthypocrista, the posthypocrista being placed higher on the rear face of the hypoloph than the postmetacrista; 3) the premetacristid is well-developed, usually contacting the paracristid; 4) the mandibular symphysis is usually strongly ankylosed, much more so than in potoroids, balbarines and macropodines, and a "boss" is often present on the dentary; 5) the I_1 has a sinuous (shaped) occlusal edge.

The unusual sthenurine premetacristid, posthypocrista and postlink add to the complexity to the cutting edges of the molars. The premetacristid adds a more or less transversely-oriented cutting ridge to the lower molars, and the posthypocrista an opposing transverse cutting ridge on the upper molars. The function of these adaptations is unknown although it is possible that they assist in the fine comminution of leaves. Increased complexity of the molars is seen in the most derived species of the genera *Simosthenurus* and *Procoptodon*. In these, highly complex enamel ridging of the molars is present. Studies of occlusion in *Lagostrophus fasciatus* may shed further light on the functional significance of these adaptations. The distribution and nature of these synapomorphies are discussed in Flannery (1983).

INTERRELATIONSHIPS WITHIN THE STHENURINAE

A single dental and several pedal synapomorphies indicating the monophyly of the species of *Troposodon Simosthenurus*, *Sthenurus* and *Procoptodon* are identified in Flannery (1983). The dental synapomorphy is that the crown of I_1 is completely enamelled, not just enamelled buccally as in most other macropodoids. The pedal synapomorphies include extreme reduction of the fifth digit of the foot and the presence of high, rounded crests on the tibial articular surface of the astragalus. Evidence of the nature of the pedal elements in the species of *Troposodon* is presented in Flannery and Archer (1983b).

Several synapomorphies have been identified that indicate that the species of *Simosthenurus*, *Sthenurus* and *Procoptodon* are monophyletic. These include presence of a discrete posterobuccal crest on P_3, extreme development of the raised lingual cingulum of P_3, and the anteriorly shifted premetacristid which occurs on the anterior cingulum. An important exception to this last synapomorphy is noted below. The distribution and significance of these states is discussed in Flannery (1983). A pedal synapomorphy, having the LAJ divided into an anterior and posterior section, also characterizes this group (although the state is unknown in *Troposodon*). This division tends to stabilize the LAJ, further restricting movement.

Several synapomorphies indicate that the species of *Simosthenurus* and *Procoptodon* are monophyletic. These include presence of a foreshortened and deepened facial area of the skull (where known) and a contribution to the masseteric process from the jugal (except in *Simosthenurus gilli* where the masseteric process is absent). These states are discussed in Flannery (1983).

Only a single specimen adding to the phylogenetic conclusions of Flannery (1983) has come to light subsequently. This specimen is the protolophid of a species of *Simosthenurus* from the Pliocene Hamilton local fauna. On this specimen the premetacristid crosses the anterior face of the protolophid but terminates before reaching the paracristid. This morphology is similar to that seen in the most primitive species of *Troposodon* and indicates that this configuration was also present in the most primitive species of *Simosthenurus*. The extremely primitive molar morphology of the Hamilton *Simosthenurus* suggests that members of this genus may have been ancestral to the species of *Sthenurus* and/or *Procoptodon* and that a secondary simplification of the molars has occurred in the species of *Sthenurus*. If this is so then the genus *Simosthenurus* as now constituted would be paraphyletic. Alternatively, the anteriorly shifted premetacristid could have been independently developed in both *Sthenurus* and *Simosthenurus* lineages. Paraphyly of the genus *Simosthenurus* was already suspected on the basis of the close similarity of some species to the species of *Procoptodon*. Of particular interest in this respect is *Simosthenurus gilli*. Is the lack of a masseteric process primitive or autapomorphic in this species? If primitive, it may well represent a species basal to most other *Simosthenurus* and *Procoptodon*.

Flannery and Archer (1983b) have reviewed the species of *Troposodon* as well as described two new species; *T. bowensis* and *T. gurar* from the Pliocene of New South Wales and Queensland respectively. Synapomorphies for the species of *Troposodon* that are unique within the Sthenurinae include presence of a non-fenestrate palate and possession of a very narrow I_1 that develops a near horizontal wear facet and that is heavily enamelled both lingually and buccally. *Troposodon bluffensis* and *T. bowensis* appear to be the most plesiomorphic species, both having relatively low lingual cingulum and a distinct posterolingual cusp rather than a blade on P^3, and contact between the premetacristid and the metaconid (but not the paracristid) on lower molars. *Troposodon bluffensis* possesses even less ornate molars and a lower lingual cingulum on P^3 than *T. bowensis*. It also lacks a posterior cingulum on the lower molars. These first two characteristics suggest that it is even more plesiomorphic than *T. bowensis*. *Troposodon gurar* retains the primitive premetacristid structure seen in the above forms, but also possesses an "accessory paracristid" (see Flannery and Archer 1983b), a structure occasionally seen in *T. bowensis* and even more rarely in *T. minor*. *T. gurar* is highly autapomorphic in possessing very hypsodont molars with a high anterior cingulum relative to other species and in having a reduced premetacristid on the anterior molars.This taxon may have been a grazer, perhaps derived from a *T. bowensis*-like ancestor.

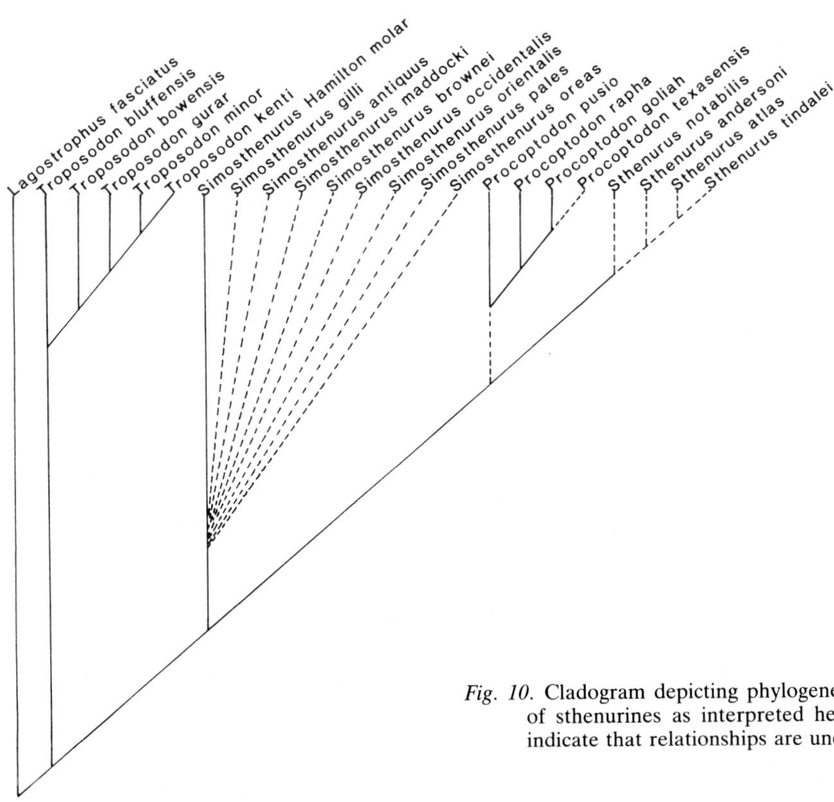

Fig. 10. Cladogram depicting phylogenetic interrelationships of sthenurines as interpreted here. The dotted lines indicate that relationships are uncertain.

Troposodon kenti and *T. minor* are similar in having a premetacristid that contacts the paracristid but not always the metaconid, in possessing squarish, low-crowned molars and in being large. The first of these features is probably a synapomorphy uniting these species. *Troposodon kenti* is highly unusual in possessing a "disjunct paracristid". This feature is seen elsewhere within the Macropodoidea only in the species of *Simosthenurus* and *Procoptodon*. Is this a synapomorphy uniting *T. kenti* with *Simosthenurus?* If so, the genus *Troposodon* must be regarded as paraphyletic because the *Simosthenurus* clade would be envisaged as arising from within it. However, this seems unlikely for a number of reasons. *Troposodon kenti* appears to be an advanced species of *Troposodon* that does not appear until late in the fossil record. Thus it cannot be directly ancestral to species of the *Simosthenurus* which appear much earlier. Until there is better evidence to the contrary, this similarity will be regarded as the result of parallelism. A phylogeny for the Sthenurinae is presented in Fig. 10.

The sthenurines can be classified as follows. The conventions used here follow Wiley (1981).

Subfamily Sthenurinae (Glauert, 1926)
  Tribe Lagostrophini n. tribe.
    Lagostrophus fasciatus Peron and Lesueur,1807
  Tribe Sthenurini (Glauert, 1926)
    Troposodon bluffensis Bartholomai, 1978
      ,, *bowensis* Flannery and Archer, 1983
      ,, *gurar* Flannery and Archer, 1983
      ,, *minor* (Owen, 1877)
      ,, *kenti* Campbell, 1973
    Simosthenurus "Hamilton molar"
      ,, *gilli* (Merrilees, 1965)
      ,, *antiquus* (Bartholomai, 1963) *sedis mutabilis*
      ,, *maddocki* (Wells and Murray, (1979) *sedis mutabilis*
      ,, *brownei* (Merrilees, 1967) *sedis mutabilis*
      ,, *occidentalis* (Glauert, 1910) *sedis mutabilis*
      ,, *orientalis* (Tedford, 1966) *sedis mutabilis*
      ,, *pales* (De Vis, 1895) *sedis mutabilis*
      ,, *oreas* (De Vis, 1895) *sedis mutabilis*
    Sthenurus andersoni Marcus, 1962 *sedis mutabilis*
    Sthenurus atlas (Owen, 1838) *sedis mutabilis*
      ,, *tindalei* Tedford, 1966 *sedis mutabilis*
      ,, *notabilis* Bartholomai, 1963 *sedis mutabilis*
    Procoptodon pusio Owen, 1874
      ,, *rapha* Owen, 1873
      ,, *goliah* (Owen, 1846)
      ,, *texasensis* Archer, 1978 *sedis mutabilis*

The Macropodines

The macropodines are the most abundant and widespread of macropodoids. Their fossil record extends back to the late Miocene and is the best of any Australian mammal group. Nevertheless, study of relationships within the subfamily poses some of the most intractable problems in marsupial systematics. The attempt at systematic analysis made here has done little to help resolve these problems. However, some hitherto unsuspected relationships have been determined, and the subfamily is defined on the basis of morphological synapomorphies for the first time.

The taxonomy of the living species of macropodines is at present for the most part stable. However, several genera need a thorough revision. Foremost among these are *Dorcopsis* (where the possible number of species is probably somewhere between one and six), *Petrogale* (although recent work, e.g., Briscoe *et al.* 1983; Maynes 1982 and Poole 1978 is beginning to clarify the situation), *Dendrolagus* (although Groves (1982) has recently revised the group there are probably several unrecognized species in New Guinea), and (to a lesser extent) *Lagorchestes* (the validity and interrelationships of *L. asomatus*, *L. leporides* and *L. hirsutus* are still in doubt). For these genera, and several others, an *ur*-species has been constructed for use in the computer programmes to avoid the problems of species level systematics.

Fossil macropodines present many more taxonomic problems than the living species. Genera that contain many extinct taxa such as *Protemnodon* and *Wallabia* are desperately in need of taxonomic revision because much new material and many species have been described since they were last reviewed (in part in Bartholomai 1973a and Bartholomai 1976 respectively). Many fossil macropodines are known but undescribed (e.g., the Gag Site macropodine and the Mammoth Cave wallaby). The undescribed taxa have been included here because they are zoogeographically and biochronologically significant and because they shed considerable light on the phylogenetic history of the group. A brief

Table 10. Characters used in computer analysis of macropodine interrelationships. The characters are listed in the same order as in Table 11. The first of the two described states is the primitive one. Although the following characters are mostly self-explanatory, the following need comment.

Character 16. The forelink is interpreted here as being homologous to the preprotocrista of bunodont molars. However, in derived grazers such as the species of *Macropus* this structure may be secondarily boosted or derived.

Character 18. *Prionotemnus palankarinnicus* is coded as 1 here. At the time that this work was undertaken only worn M_1's of this species were available for study, and it appeared that a slight protolophid was present. However, subsequent examination of an unworn M_1 shows than no protolophid is present.

(1) When unworn the I^1 crown extends much further ventally than I^{2-3}, the crowns of I^{1-3} extend a similar distance ventrally.
(2) The crowns of I^{2-3} are short, tall (hypsodont).
(3) An extensive groove is present on the buccal face of I^3, groove is primarily restricted to occlusal edge.
(4) C^1 is large, reduced or lost.
(5) A slight postlink is present on any of M^{2-5}, a postlink is lacking.
(6) P^2 lacks, has a posterointernal cusp.
(7) P^2 lacks, has a lingual cingulum.
(8) A well-developed crest is absent, present on the dentary below the masseteric foramen.
(9) P^3 has, lacks a lingual cingulum.
(10) P3 is not, is extremely elongate.
(11) P3 is not, is extremely short.
(12) A large parastyle is present, absent on M^1.
(13) The preparacristae on M^{2-5} are very well-developed, reduced.
(14) The postparacrista and premetacrista on the upper molars are extremely well-developed, less well developed.
(15) The postparacrista and premetacrista are moderately or well-developed, poorly developed.
(16) Forelink is present, absent.
(17) I_1 crown is narrow, broad.
(18) M_1 protolophid is undeveloped, at least narrow lophid is present.
(19) M_1 protolophid is markedly narrower than hypolophid, subequal in width.
(20) Molars brachydont, hypsodont.
(21) Molars brachydont or hypsodont, extremely hypsodont.
(22) Cristid obliqua does not, does originate at hypoconid apex.
(23) Rear face of the hypolophid is not, is convex.
(24) A groove is not present, is present on the rear face of the hypolophid.
(25) The anterior cingulum of the upper molars is not, is restricted lingually.
(26) The upper molars have one, two buccal roots.
(27) The cheektooth rows are parallel, bowed.
(28) A posterior cingulum is present, absent on lower molars.
(29) Posterior molars are not, are markedly larger than anterior molars.
(30) The anterior cingulum of lower molars is not, is extremely narrow and elongate.
(31) Premolars are not lost, are lost through molar progression.
(32) The anterior molars, but not premolars are not lost, are lost through molar progression.
(33) The molar alveolar margin of the dentary is not arched, is arched.
(34) The palate is not, is restricted posterior to the upper incisors.
(35) The masseteric process of the maxillary is not, is well-developed.
(36) Molars are not, are markedly narrower at the apex than the crown base.
(37) The palate has, lacks vacuities.
(38) The bullae are not, are inflated.
(39) A premetacristid is, is not present on M_{2-5}.
(40) Diastema is not elongate, is elongate.
(41) The ectotympanic does not, does form a complete bony ring.
(42) The occlusal groove does not exit, does exit onto the posterior face of I^2.
(43) Dentary is not, is markedly deeper below M_2 than M_5.
(44) Parafloccular fossa of periotic is deep, shallow.
(45) The depression anterior to the parafloccular fossa is deep and well-defined, poorly defined.
(46) The fifth metatarsal lacks, has a proximoventral process.
(47) The posteroventral process of the periotic is subequal in size with the anteroventral process, is smaller.
(48) The I^1 is not, is broader than I^2 or I^3.
(49) The crista petrosa lacks, forms a large, elongate spink-like process at its ventral end.
(50) The ilium forms a flattened blade, is triangular in cross section.
(51) Marked sexual dimorphism in size is not, is present.

description of the Mammoth Cave wallaby and the Gag Site macropodine are included below.

The Mammoth Cave wallaby is known from isolated dentaries and maxilla fragments from Pleistocene deposits in Mammoth Cave, southwestern Western Australia. Although it has been recognized for some years as a distinct form (e.g., Merrilees 1968; Merrilees and Porter 1979) it has remained undescribed. It resembles both *Wallabia bicolor* and some species of *Thylogale* in dental morphology, but is larger than any other known species of these genera. The molars are high-crowned and narrow relative to those of *W. bicolor*, and the P3s are shorter. On M_1, the protolophid is distinctly narrower than the hypolophid. The character states for this taxon used in the computer analysis are listed in Tables 10-11.

The Gag Site macropodine is known from excellently preserved material, including crushed but almost complete crania and dentaries recovered in 1983 from Miocene limestones at the Gag Site, Riversleigh, northwestern Queensland. A single apparent macropodine synapomorphy is present in the Gag Site macropodine. This is the presence of a more oval-shaped condyle of the dentary, preserved in a single juvenile specimen (AR 3651). However, isolated calcanea from the Gag Site possess the laterally bilobed UAJ that is unique to macropodines. These calcanea may belong to the Gag Site macropodine. The Gag Site macropodine also displays a number of primitive dental characteristics. The premolars are elongate and deeply grooved. The I^3 forms a simple blade as it does in sthenurines such as *Simosthenurus occidentalis*. The C^1 is as large as in the species of *Dorcopsis*. The M^1 possesses a large parastyle and a slight postlink is present on the rear face of the hypoloph in M^{1-3}. The protoloph is broader than the hypoloph on upper molars. The crown of the I_1 forms a short, deep blade with well-developed dorsal and ventral ridges. The mandibular symphysis is very weakly ankylosed. The protolophid of M_1 consists of a single anteriorly-directed crest and cuspid. The premetacristid is well-developed on the lower molars, forming the lingual margin of the anterior part of the tooth. The premolars are all strongly-grooved. The P^2 lacks a posterointernal cusp. The P^3 possesses a posterolingual cusp and an extensive lingual cingulum. The character state polarities for this taxon used in the computer analysis are listed in Tables 10-11.

Some taxa are included in this analysis that have not been convincingly demonstrated to possess any macropodine synapomorphies. However, their morphology does not exclude the possibility of their being macropodines and they share no synapomorphies with other macropodid groups. These taxa include *Galanarla tesselata* and *Hadronomas puckridgi*, both poorly-known species from the Miocene of northern Australia. Because they are not treated elsewhere it was thought best to include them here.

MACROPODINE MONOPHYLY

The following character states have been identified as synapomorphies for the Macropodinae.

(1) When the dentaries are articulated with the skull, the lower incisors lie inside the dental arcade formed by the upper incisors. In all other macropodoids where the character is known (potoroids, most sthenurines), the lower incisors lie on the crowns of the upper incisors. This latter condition is considered to be plesiomorphic because it is present in phalangerids. However, *Simosthenurus maddocki* appears to have independently developed a macropodine-like condition (Flannery 1983).

(2) The occlusal edge of I^1 is broad, forming an elongate cutting blade; it does not taper to a point as it does (when unworn) in potoroids, sthenurines and phalangerids.

(3) The mandibular condyle is oval in shape, not transversely elongate or barrel-shaped as in potoroines, sthenurines, balbarines and phalangerids.

(4) The mandibular symphysis is very loosely ankylosed in most macropodines; more so than in potoroids or phalangerids and far more so than in sthenurines, where the dentaries are extremely firmly ankylosed relative to all other macropodoid taxa.

(5) The calcaneal-astragalar joint (LAJ) in all macropodines is hourglass-shaped, being markedly constricted in its central portion; in extreme cases (e.g., within *Macropus*) the joint is completely divided into a lateral and medial section; primitively, in phalangerids, potoroids and primitive sthenurines, the LAJ is broadly continuous; in advanced sthenurines the LAJ is divided into an anterior and a posterior section; the effect of separation of the LAJ into two units is to further restrict movement at the joint (see Szalay 1982). The macropodine solution to this problem of LAJ stability is unique among the Australian marsupials and is a striking, constant synapomorphy, undoubtably produced as a response to the increased stress placed on this joint due to increased size and specialization in the saltatory gait.

It is easy to envisage how the macropodine cranial synapomorphies (1-4 above) may be functionally interrelated. The broadening of I^1 lengthens the occlusal edge of the upper incisor arcade producing a more equal lower and upper incisor occlusal edge. The broadening of the upper incisor arcade has allowed the left and right lower incisors to operate independently and to slide

Table 11. Data matrix used in the computer analysis of the Macropodinae. The characters are listed in the same order (from left to right) as in Table 10. 0 = primitive, 1 = derived, ? = state unknown.

Taxon	Characters
ur-Dorcopsis	0 0 0 0 1 0 0 0 1 1 0 0 0 0 0 0 0 0 0 0 0 0 0 0 0 1 0 1 0 0 0 0 0 0 0 0 0 0 0 0 0 0 0 0 0 0 1 0 0 0 0
ur-Dendrolagus	0 1 0 0 0 0 1 0 1 0 0 0 0 0 0 1 0 0 0 0 0 0 0 0 1 0 1 0 0 0 0 0 0 1 0 1 0 0 0 0 0 0 0 0 1 0 1 0 0 0 0
Wallabia bicolor	1 1 1 1 1 1 1 0 1 0 0 1 0 0 0 1 1 1 1 0 0 0 0 0 0 0 1 1 1 0 0 0 0 0 0 1 1 0 0 0 1 0 1 1 0 0 0 0 1 0 0 1 0
Thylogale billardieri	1 1 1 1 1 1 1 0 0 0 0 0 1 1 0 1 1 1 0 0 0 0 0 0 1 1 1 1 0 1 0 0 0 1 1 0 0 0 1 0 0 1 0 0 0 0 1 0 1 ? 0
Setonix brachyurus	1 1 1 1 1 1 1 0 1 0 0 0 0 1 0 1 1 1 0 0 0 0 0 0 0 0 1 1 1 0 0 0 0 0 1 1 0 0 0 1 0 0 1 0 0 0 ? 1 0 0 ? 0
ur-Onychogalea	1 1 1 1 1 1 0 0 0 0 1 0 0 1 0 1 0 1 1 1 0 1 1 0 0 1 1 1 0 0 1 1 0 1 1 1 0 1 1 0 0 1 0 0 1 0 1 1 0 ? 0
ur-Lagorchestes	1 1 1 1 1 0 0 0 0 0 0 0 0 1 0 1 0 1 0 1 0 0 0 0 0 1 1 1 0 0 0 ? 0 1 1 0 0 1 1 0 0 1 0 0 0 0 ? 1 0 ? 0
ur-Macropus (Macropus)	1 1 1 1 1 1 1 0 0 0 1 1 1 1 1 0 1 1 1 1 1 1 1 1 0 1 1 1 1 0 1 1 1 1 1 1 1 1 0 1 1 1 1 0 0 1 0 1 0 0 1 1
ur-Macropus (Osphranter)	1 1 1 1 1 1 1 0 1 0 0 1 1 1 1 0 1 1 1 1 1 1 1 1 0 1 1 1 1 0 1 1 1 1 1 1 1 1 0 1 1 1 1 1 0 1 0 1 0 0 1 1
Macropus agilis	1 1 1 1 1 1 1 0 1 0 0 1 1 1 1 0 1 1 1 1 0 1 1 0 0 1 1 1 0 0 0 1 1 1 1 1 0 0 1 1 1 1 0 0 0 0 1 0 0 ? 0
ur-Petrogale	1 1 1 1 1 1 1 0 1 0 0 0 0 1 0 1 1 1 1 1 0 1 1 0 0 1 1 1 0 0 0 1 0 1 1 1 0 0 1 0 1 0 0 0 0 0 1 0 1 ? 0
Thylogale christenseni	? ? ? 1 1 1 0 0 0 0 0 0 1 0 0 1 1 1 0 0 0 0 0 0 1 1 1 1 0 0 0 0 0 1 1 0 0 ? 1 0 ? ? 0 ? ? ? ? 0 ? ? ?
Dorcopsoides fossilis	? ? ? ? 1 0 0 0 1 0 0 0 1 1 0 1 ? 1 0 0 0 0 0 0 0 0 1 0 0 0 0 0 0 ? ? 0 0 0 1 0 0 ? 0 ? ? ? ? ? ? ? ?
Hadronomas puckridgi	? ? ? ? 1 ? ? 0 1 1 0 ? 1 1 0 1 1 ? ? 0 0 0 0 0 0 0 0 ? 1 0 0 0 0 0 ? 0 0 ? ? 1 0 ? ? 0 ? ? 0 ? ? ? ? ?
Baringa nelsonensis	? 1 1 ? 1 1 1 1 0 0 0 1 1 1 1 1 1 1 1 0 0 0 0 0 1 1 ? 1 0 1 0 0 0 ? 1 0 ? ? 1 ? ? ? 1 ? ? ? ? ? ? ? ?
ur-Protemnodon	1 1 1 1 0 1 1 0 1 1 0 0 0 0 0 0 1 1 0 0 0 0 0 0 0 0 0 1 0 0 0 0 0 0 1 1 0 1 0 1 0 ? 1 0 1 0 1 0 0 0 0 ?
ur-Kurrabi	? ? ? ? 1 1 1 0 1 1 0 1 1 1 1 1 1 1 1 0 0 0 0 0 1 1 1 1 0 1 0 0 0 ? 1 0 1 ? 1 0 ? ? 0 ? ? ? ? ? ? ? ?
Galanarla tesselata	? 0 1 0 ? ? ? 0 ? 0 0 ? 0 0 0 0 1 ? ? 0 0 0 0 0 0 0 1 1 0 0 0 0 0 0 0 ? ? 0 ? ? 0 0 ? ? 0 ? ? ? ? 0 ? ? ?
Gag Macropodine	? 1 0 0 0 0 0 0 1 0 0 0 0 0 0 1 1 0 0 0 0 0 0 0 0 0 0 ? 1 0 0 0 0 0 ? ? 1 ? 0 0 0 ? ? 0 0 0 ? ? ? 0 ? ?
Mammoth Cave Wallaby	? ? ? ? ? ? ? 0 1 0 0 ? ? ? ? ? 1 1 0 0 0 0 0 0 ? 1 1 1 0 0 0 0 0 ? ? 0 ? ? 1 0 ? ? 0 ? ? ? ? ? ? ? ?
Prionotemnus palankarinnicus	? 1 1 ? 1 0 1 0 1 0 0 1 1 1 0 1 1 1 0 1 0 1 1 0 0 1 1 0 0 0 0 ? 0 ? 1 0 ? ? 1 ? ? 0 0 ? ? ? ? ? ? ? ?
Macropus eugenii	1 1 1 1 1 1 1 0 0 0 1 1 1 1 1 1 1 1 1 1 0 1 1 0 0 1 1 1 0 0 1 ? 1 1 1 1 0 0 1 1 1 1 1 0 0 1 0 1 0 0 1 0
Ancestor	0 0 0 0 0 0 0 0 0 0 0 0 0 0 0 0 0 0 0 0 0 0 0 0 0 0 0 0 0 0 0 0 0 0 0 0 0 0 0 0 0 0 0 0 0 0 0 0 0 0 0

past the occlusal edge of the upper incisors in a slicing action. This movement is impossible when the lower incisors lie on top of the upper incisor crowns (i.e., as in potoroids, phalangerids and most sthenurines, see Flannery 1983). Changes in condyle shape as well as a loosening of the mandibular symphysis facilitate independent movement of the left and right dentaries as well as possibly allowing a more precise incisor occlusion.

Thus, macropodines seem to have developed a unique means of incisor occlusion that allows an independent shearing motion of the left and right lower incisor blades. More advanced macropodines have gone on to further develop this system by adding extra cutting edges, altering their shape and position, and by lengthening and shortening the relative occlusal lengths of the incisors. Further studies are clearly necessary, perhaps using X-ray cinematography, to discover how the unique macropodine morphology affects incisor use and action. Ride (1959) examined mastication in the macropodine *Macropus rufogriseus* and hypothesized that masticatory adaptations seen in the Macropodinae may be related to grazing. However, it is felt that the adaptations outlined here were not developed in response to grazing because they are seen even in primitive browsing species such as the species of *Dorcopsis*.

Undoubtably many macropodine synapomorphies are to be found in the soft anatomy. However, further research on *Lagostrophus fasciatus* for use as an outgroup is necessary before they can be identified.

MACROPODINE INTERRELATIONSHIPS

In an attempt to interpret macropodine interrelationships, a large data set including 23 taxa and 51 characters was run through the four Felsenstein programmes (Polym, Sokal, Wagner and Dollo). The characters used are presented in Tables 10-11. The taxa included were; a hypothetical ancestor, *Galanarla tesselata*, the Gag macropodine, an *ur-Dorcopsis*, an *ur-Dendrolagus*, an *ur-Protemnodon*, *Dorcopsoides fossilis*, the Mammoth Cave wallaby, *Thylogale christenseni*, *T. billardierii*, *Setonix brachyurus*, an *ur-Kurrabi*, *Baringa nelsonensis*, *Prionotemnus palankarinnicus*, an *ur-Petrogale*, an *ur-Lagorchestes*, an *ur-Onychogalea*, *Macropus (Notamacropus) agilis*, *M.* (N.) *eugenii*, an *ur-Macropus (Macropus) and an ur-Macropus (Osphranter)*. Surprisingly, given the nature of the problem and the number of taxa involved, the trees produced concurred in a number of areas (see Fig. 11).

Strongly supported in all of the programmes is a clade containing the species of *Macropus*. The non-homoplasious characters 33 (the cheektooth alveolar margin of the dentary is arched) and 40 (the diastema is elongate), unite the *Macropus* clade in all programmes. Homoplasious characters also group these taxa in the Dollo and Sokal trees. Within the *Macropus* clade all trees group *Macropus (Macropus)* and *Macropus (Osphranter)* on the basis of the non-homoplasious characters 21 (molars extremely hypsodont), 24 (a groove is present on the rear face of the hypolophid) and 29 (M5 is usually as large as or larger than M4). Again, additional homoplasious characters support this clade in the Sokal and Dollo trees. The Dollo programme groups *M.* (N.) *agilis* and *M.* (N.) *eugenii* (Fig. 11), on the basis of characters 37, 43 and 51 (see Table 10), all of which are subject to homoplaisy.

Monophyly of the species of *Macropus* is strongly supported by biochemical studies such as those of Richardson and McDermid (1978) and Kirsch (1977) and this evidence, combined with concurrence of the rather slim morphological data, makes a case that is accepted here. The grouping of the species of *Macropus (Macropus)* and *Macropus (Osphranter)* to the exclusion of *Macropus (Notamacropus)*, however, is not strongly supported. In fact, Baverstock (pers. comm.) has indicated that his current serological studies suggest that *M. (Macropus)* may be the sister group to *M. (Osphranter)* and *M. (Notamacropus)*. The case based on morphology is not strong and could be the result of unrecognized convergence. The grouping together of the two species of *M. (Notamacropus)* is very weakly supported by this study with only three homoplasious characters in a single tree supporting it (although they are grouped in Polym, it is only on the basis of retained polymorphisms in hypothetical ancestors). However, some biochemical evidence supporting the monophyly of *M. (Notamacropus)* (the presence of Pep C-5 in all species tested) is given in Richardson and McDermid (1978).

The Sokal and Wagner programmes group *ur-Petrogale* as the sister group to the *Macropus* clade on the basis of homoplasious characters only. The Wagner programme does not allow determination of which characters support this grouping. In Sokal the homoplasious characters 14, 20, 22, 23, 32 and 36 (see Table 10) support it. The Polym programme places a clade containing the *ur-Petrogale* and the *ur-Onychogalea* as the sister group of the *Macropus* clade on the basis of the non-homoplasious characters 32 (loss of anterior molars occurs with age) and 36 (the molars are markedly narrower at the crown apex than at the base). The Dollo programme places a clade containing the *ur-Onychogalea*, *ur-Lagorchestes* and the *ur-Petrogale*, as the sister group to the *Macropus* clade on the basis of the non-homoplasious character 32. An insufficient sample of *Lagorchestes* specimens was available to determine polarity in 32 and it's polarity was left unknown in the data set. Thus there is no conflict in including the *ur-Lagorchestes* in this clade (see Fig. 11). Despite the morphological support for this hypothesis presented here, there is good evidence to suggest that it is incorrect. Both the immunological study of Kirsch (1977) and the exhaustive allozyme study of Richardson and McDermid (1978) have indicated that the species of *Petrogale* are most closely related to the species of *Thylogale* and that the species of *Macropus* are most closely related to *Wallabia bicolor*. Although this last relationship is strongly supported by the above biochemical studies, the highly divergent dental and cranial morphology of these taxa must bias any traditional morphological study. Nevertheless, morphology does not convincingly support any alternative interpretation (the characters cited above are not difficult to accept as homoplasious) and as a result the conclusions of the biochemical studies are accepted here.

One other clade is recognized in the Wagner, Dollo and Sokal Programmes; that containing the *ur-Onychogalea* and *ur-Lagorchestes* (and as indicated above, in the Dollo tree this clade is placed as the sister group to the *ur-Petrògale*). This clade is supported in Wagner and Dollo by the non-homoplasious synapomorphies 38 (the bullae are inflated) and 48 (the I_1 is broader than I^2 or I^3). Only 48 supports the clade in Sokal. This grouping is interesting because a relationship between the species of *Onychogalea* and *Lagorchestes* had hitherto been unsuspected. However, one must suspect convergence when a clade is supported by so few characters, particularly when both taxa include arid-adapted browsers occupying similar habitats. The grouping of this clade with *Petrogale* and close to *Macropus* is also dubious. Serology indicates that the species of *Onychogalea* are only distantly related to *Petrogale* and *Macropus* species (Kirsch 1977) and the characters supporting such a relationship could easily be subject to homoplaisy.

One of the striking features of all of the computer-generated cladograms for the macropodines is their rigid ordering of macropodine taxa by geological age and/or habitat. In all of the cladograms produced, the taxa can easily be divided into three groups. The first (group 1) includes all species which branch below the HTU where the derived states of characters 1, 4, 18, 34 and 39 are gained in the cladograms (Fig. 11). It contains the most plesiomorphic species, all extant rainforest-dwelling species (e.g., the *ur-Dorcopsis* and *ur-Dendrolagus* or fossil taxa of Miocene age (e.g., *Galanarla tesselata*, *Hadronomas puckridgi*, the Gag macropodine and *Dorcopsoides fossilis*). Exceptions are the Pliocene-Pleistocene species represented by the *ur-Protemnodon* which falls into group I in Dollo and *H. puckridgi* and *D. fossilis* which fall into group 2 in Wagner and Polym.

The members of group 2 are intermediate in both age and habitat between those of groups 1 and 3. It includes extant species that are browsing scrub-dwellers either in xeric (e.g., species of *Lagorchestes* and *Onychogalea),* or mesic (e.g., *Wallabia* and *Thylogale* species) habitats, or extinct Pliocene and Pleistocene presumed scrub-dwelling browsers (e.g., *Baringa nelsonensis*, the species of *Protemnodon* and *Kurrabi*). Group 3 contains the species of *Macropus* and *Petrogale*, all of which are grazers (see Sanson 1978) and inhabit woodland, plains or rocky areas. Also falling into group 3 in the Polym tree is the *ur-Onychogalea* and in the Dollo tree, the *ur-Onychogalea* and *ur-Lagorchestes*.

A number of synapomorphies support the monophyly of members of groups 2 and 3 and of members of group 3. The latter have been discussed above in the section on *Macropus* and *Petrogale*. Synapomorphies not subject to homoplaisy that support the monophyly of the clade including members of groups 2 and 3 that are recognized by all programmes are 1, the cutting edges of the crowns of I^{1-3} are equal in height (i.e., I^1 does not project downwards; placement uncertain in Sokal, see Fig. 11), 4, the canine is reduced in size or is lost, 18, a protolophid is formed on the trigonid of M_1 (placement uncertain in sokal, see Fig. 11), 34, the anterior part of the palate immediately behind the upper incisors is constricted (placement uncertain in Sokal, see Fig. 11) and 39, the premetacristid is reduced or lost on lower molars (see Fig. 11).

The monophyly of *all* non-rainforest-dwelling and post-Miocene macropodines, implied in this hypothesis, is regarded sceptically here. Despite the fact that this cluster is based on the strongest support from morphology of any phylogenetic hypothesis for macropodines generated by the computer programmes, and the fact that there is no good morphological or biochemical evidence to contradict the hypothesis, the great convergences that have occurred in other macropodoid groups makes me wary of unquestioningly accepting it. For it implies that a single ancestral species of macropodine became successfully adapted to life outside the rainforest (if one discounts the possibility that Miocene forms were not rainforest-dwellers). Intuitively one would suspect that many lineages would have independently adapted to the new and fast-growing relatively xeric habitats that developed in the late Tertiary, and perhaps became markedly convergent in morphology in the process.

My treatment of the results provided by the computer programmes here is only one of several alternatives. One could accept at face value "synapomorphies" identified by programmes. Hennig suggested that homoplaisy should be taken to have occurred only when it can be proved to have occurred (Hennig 1966). However, this guideline has not been adopted here. This is because it is strongly suspected, on the basis of biochemical studies and the distribution of a few complex synapomorphies, that by far the greater number of characters examined in this study are subject to homoplaisy. Thus I suspect that macropodine evolution has not followed a parsimonious pathway, at least in the sense that parsimony is used in the computer programmes.

It may well be that the acquisition of certain adaptations is highly advantageous to all lophodont macropodoids. For example, one feature that seems to have developed independently in several macropodoid lineages is a trend towards adding transverse cutting ridges to the anterior part of the cheektooth row while at the same time obliterating or shortening longitudinal cutting ridges. As a result, various macropodoids have developed a transverse lophid on the trigonid on M_1, while at the same time reducing the size of the parastyle of M^1, and strengthening the protoloph of M^1. A posterointernal cusp or transversely oriented crest is also added to P^2 and P^3, and the posterior ends of P_2 and P_3 are twisted lingually to form more transversely oriented crests. Several of these features can be shown to have almost certainly been derived independently in at least one case within the Macropodinae. Within the tree-kangaroos *(Dendrolagus)*, the most primitive species, *D. lumholtzi* and *D. bennettianus* (see Groves 1982; Flannery and Szalay 1982), the trigonid of M_1 lacks a transverse lophid, P^2 lacks a posterointernal cusp and M^1 has a large parastyle. However, in the most advanced species (e.g., *D. dorianus*), a transverse lophid on the M_1 trigonid, a posterointernal cusp on P^2 and a small parastyle on M^1 are all present. These features are also seen in other advanced macropodines. If these structures represent synapomorphies, one would have to view the numerous apparent synapomorphies for the tree-kangaroos as homoplasious. I feel that the greatest weight of evidence indicates that the tree-kangaroos are monophyletic (see below) and that the structures noted above are homoplasious. Within other macropodine genera (e.g., *Lagorchestes* and *Thylogale*) primitive species are found that lack or possess only an incipient trigonid lophid on M_1 and posterointernal cusp on P^2, while other species of the same genera possess well-developed structures. Examples of these plesiomorphic species are *Lagorchestes hirsutus* and *Thylogale billardierii*.

Even more bizarre features can be shown to have been independently developed in various macropodine lineages. One is the development of hypsodont molars and associated shedding of worn anterior molars and premolars. While this is a striking adaptation, it would be difficult to argue that *Petrogale concinna* and the species of *Macropus (Macropus)* are monophyletic just because they are unique among the Macropodoidea (and indeed Marsupialia) in constantly possessing these features.

In light of the degree of convergence thought to have taken place in these unusual features, it seems likely that many characters of a less complex nature have been subject to homoplaisy. These would include the development of hypsodonty and associated dentary and dental structure, reduction in premolar size, canine reduction

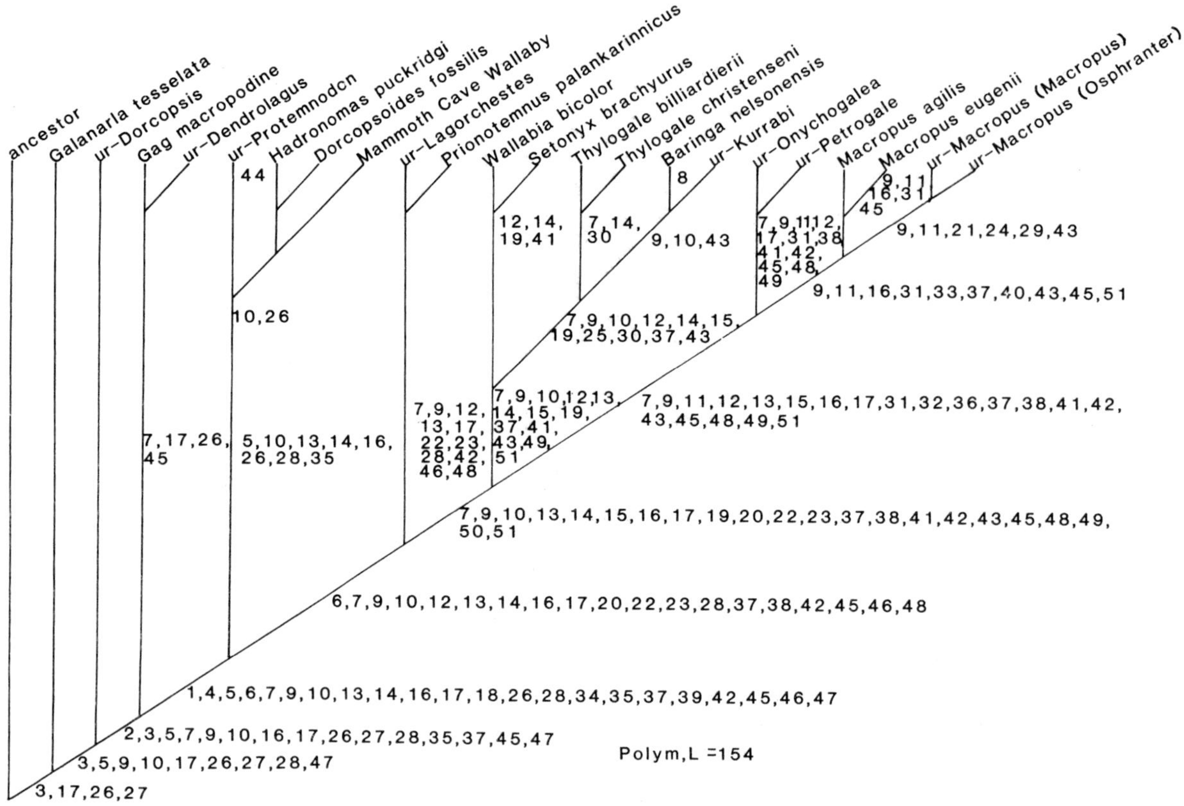

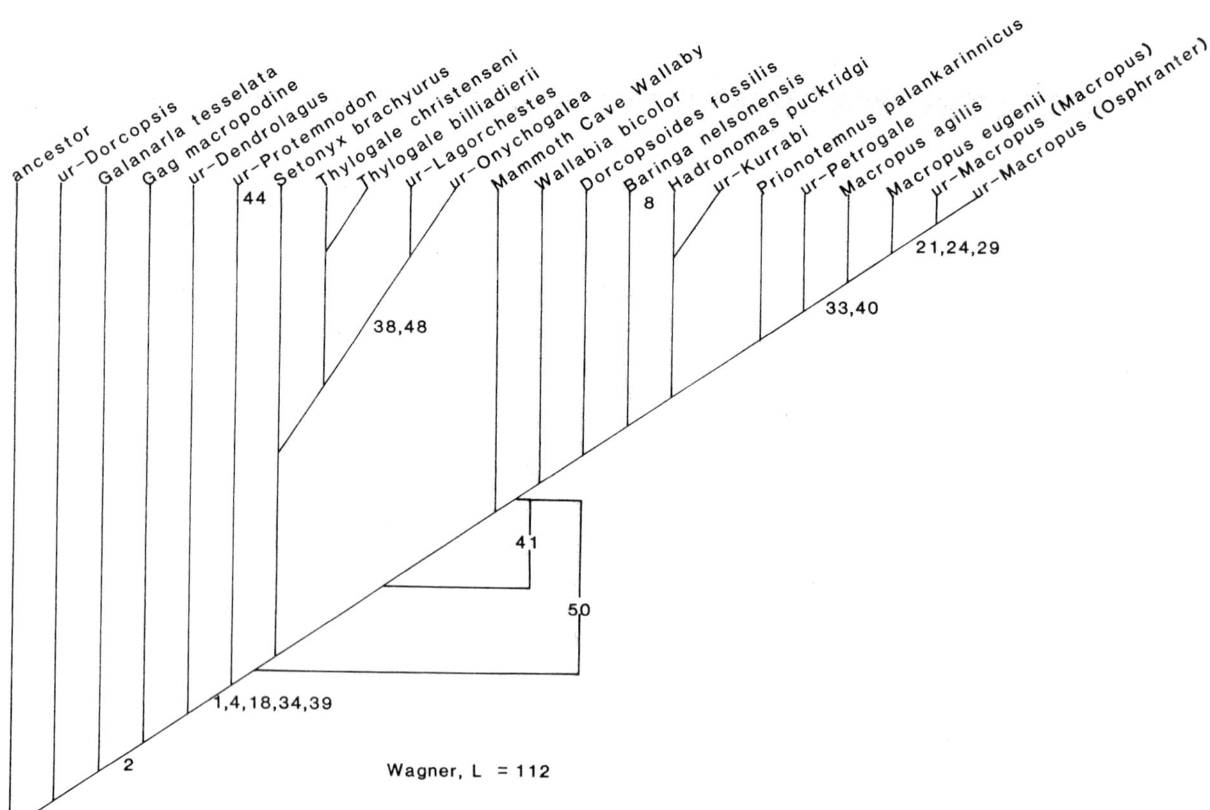

Fig. 11. Cladograms depicting hypotheses of macropodine interrelationships generated by: top left, the Polym programme; bottom left, the Wagner programme; top right, the Sokal programme; and bottom right, the Dollo programme.

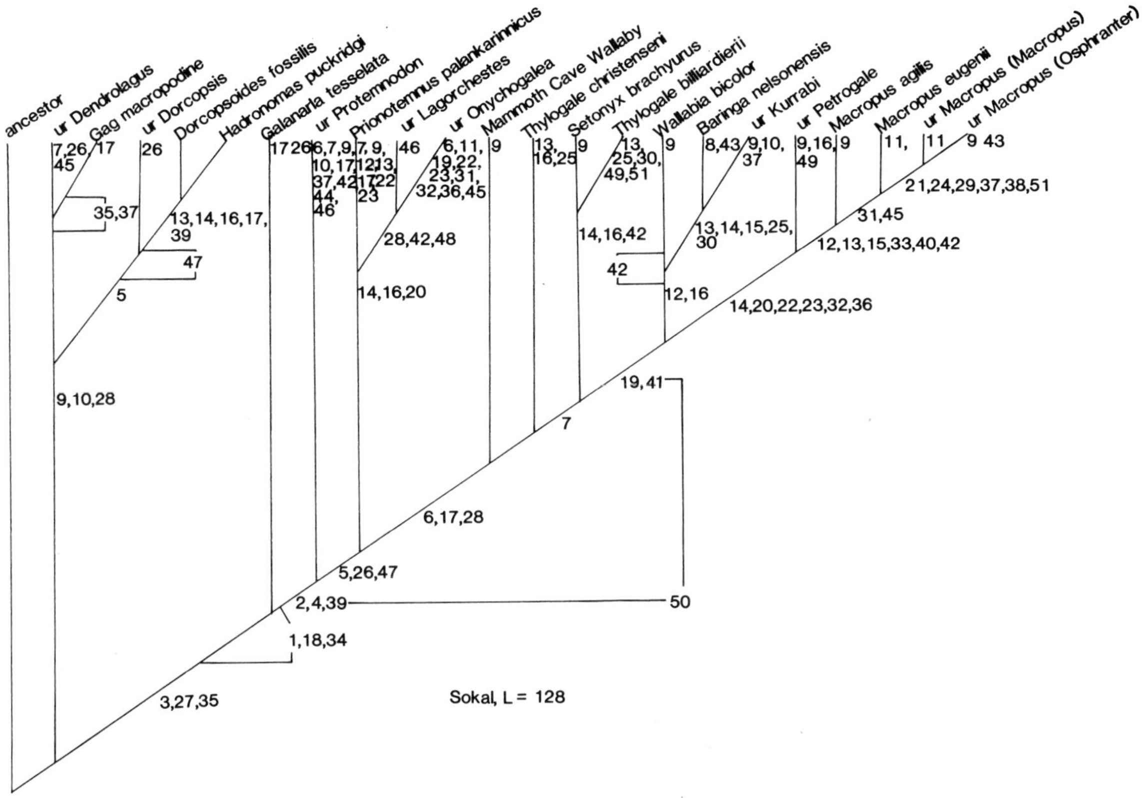
ancestor
ur Dendrolagus
Gag macropodine
ur Dorcopsis
Dorcopsoides fossilis
Hadronomas puckridgi
Galanarla tesselata
ur Protemnodon
Prionotemnus palankarinnicus
ur Lagorchestes
ur Onychogalea
Mammoth Cave Wallaby
Thylogale christenseni
Setonyx brachyurus
Thylogale billiardierii
Wallabia bicolor
Baringa nelsonensis
ur Kurrabi
ur Petrogale
Macropus agilis
Macropus eugenii
ur Macropus (Macropus)
ur Macropus (Osphranter)
Sokal, L = 128

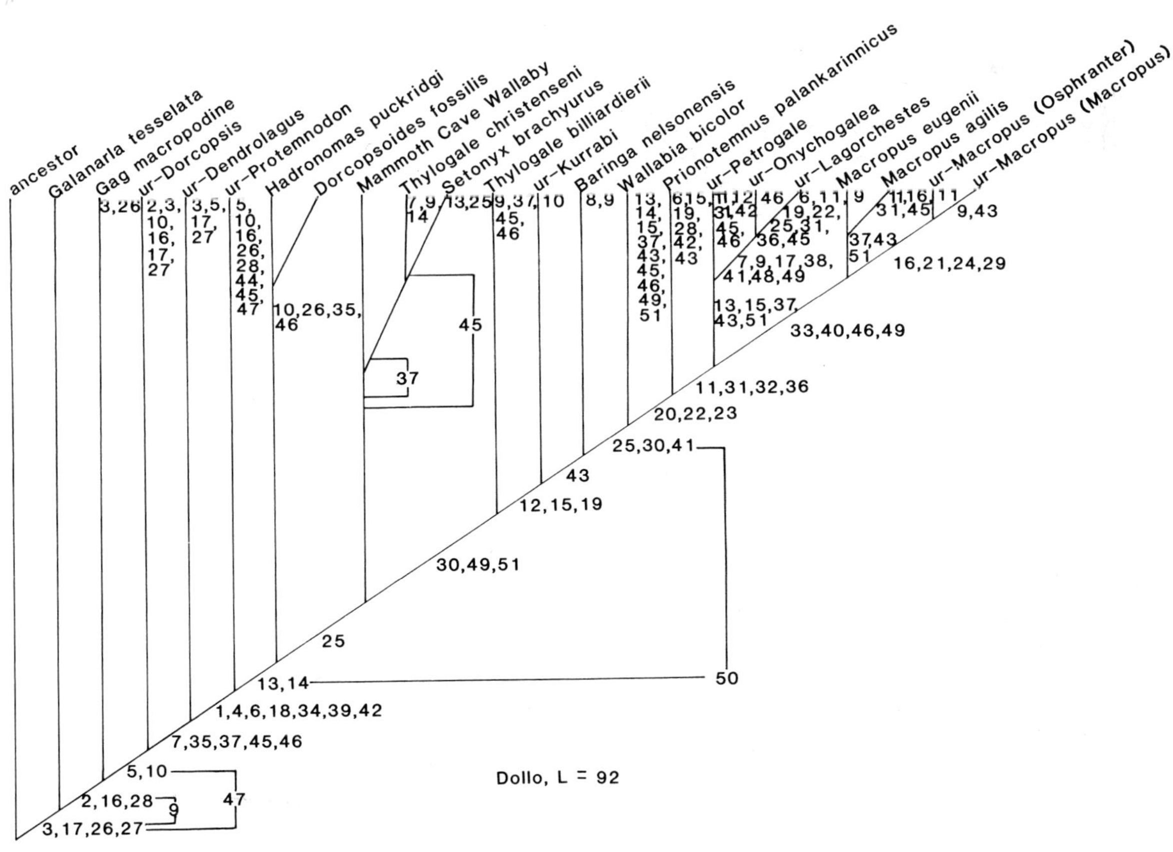
ancestor
Galanarla tesselata
Gag macropodine
ur-Dorcopsis
ur-Dendrolagus
ur-Protemnodon
Hadronomas puckridgi
Dorcopsoides fossilis
Mammoth Cave Wallaby
Thylogale christenseni
Setonyx brachyurus
Thylogale billiardierii
ur-Kurrabi
Baringa nelsonensis
Wallabia bicolor
Prionotemnus palankarinnicus
ur-Petrogale
ur-Onychogalea
ur-Lagorchestes
Macropus eugenii
Macropus agilis
ur-Macropus (Osphranter)
ur-Macropus (Macropus)
Dollo, L = 92

and loss, loss of palatal fenestration, development of a masseteric process (the latter again seen in some advanced species of *Dendrolagus* but not primitive ones) and reduction of the fifth digit of the foot; in short, almost all of the characters used in past morphological studies to determine interrelationships within the Macropodinae. Perhaps these adaptations are responses to the need to adapt to such all pervasive environmental changes as the replacement of rainforest with more open habitats.

A curious aspect of macropodine systematics is that while the interrelationships of the genera seem for the most part to be uninterpretable on the basis of morphology, the genera themselves are clearly recognizable, and most are able to be characterised on the basis of unique (within the Macropodinae) synapomorphies. That is, there are plenty of autapomorphies for macropodine genera, but almost no non-homoplasious synapomorphies. This may indicate that the period of hybridization and/or convergence occurred mainly in the past, and that now the lineages are "separate" and evolving along their own pathways. Whatever the reason, interrelationships within some macropodine genera are easier to interpret than intergeneric interrelationships, and are relatively well-understood. Interrelationships within the genera *Dendrolagus*, *Macropus* and *Thylogale* will be discussed below.

Morphological data strongly suggests that the Dendrolagini is monophyletic. It includes the following synapomorphies; presence of a calcaneal-cuboid facet that has partially "smoothed out" the primitive macropodoid stepped CACU joint; great UAJ mobility; presence of an astragalar head with its long axis less vertically oriented than in other macropodines; presence of a low broad calcaneum; and presence of a high medial ridge on the tibial articular surface of the astragalus (Flannery and Szalay 1982). Within the Dendrolagini several clades can be recognized. *Bohra paulae*, an extinct Pleistocene taxon from New South Wales, is the most plesiomorphic dendrolaginine. In the greater tuberosity length/calcaneal length ratio, the length-width ratio of the proximal epiphysis of the tibia, and larger tibial-fibular contact *B. paulae* resembles non-arboreal macropodines. Within the genus *Dendrolagus* three species are more primitive than the others. These are *D. bennettianus* and *D. lumholtzi* from Australia and *D. inustus* from the north coast of New Guinea. These species possess an anteriorly placed hair whorl, internally naked ears, a reduced but still large area of tibial-fibular contact, a relatively longer fourth toe and more elongate feet than in other species (Flannery and Szalay 1982; Groves 1982). The most derived group of tree-kangaroos includes *D. ursinus* (which may be the most primitive of this group on the basis of hair whorl position and other characters, see Groves 1982), *D. goodfellowi*, *D. matschiei* and *D. spadix*, which are all similar in morphology and coat colour and are probably closely related. It also includes a fossil species, *D. noibano* (Flannery, Mountain and Aplin 1983) known only from dental remains; and the most autapomorphic species, *Dendrolagus dorianus*. All of these taxa (where known) have a distal tibial-fibular contact restricted to the distal epiphysis, the fourth and fifth toe subequal in length, an extremely short, broad foot, a furred internal area of the ear and more posteriorly positioned hair whorls than in the three plesiomorphic taxa (Flannery and Szalay 1982; Groves 1982). Members of this clade are restricted to the New Guinean Central Cordillera, the Huon peninsula and nearby Umboi Island and Genderawasih. All of the adaptations listed above as being present in these taxa can be related to increasing arboreality, but peculiarly, *D. dorianus* is more terrestrial than other species (Ganslosser 1981). Despite the lack of resolution of species level problems within New Guinea, particularly within the *D. goodfellowi* and *D. dorianus* groups, interrelationships within this genus are clearly among the best understood for any macropodine taxa.

Members of the genus *Macropus* have probably had more time and publications devoted to their systematics than have any other macropodoids, and consensus regarding some relationships within the genus is emerging. There is good biochemical and morphological evidence which suggests that the species of *Macropus (Osphranter)*, including *M. rufus*, *M. robustus*, *M. antilopinus*, *M. barnardus* and *M. pavana* are monophyletic (see Dawson and Flannery 1985; Rofe 1978; Richardson and Sharman 1976; Richardson and McDermid 1978 and Kirsch 1977). Within this group interrelationships are difficult to determine. However, the $I^{2\text{-}3}$ of *M. bernardus* are much more distinctly grooved than in other species and the pattern of grooving more closely resembles that present in *M. (Macropus)* and *M. (Notamacropus)*. On this basis, this taxon could be considered to be the most primitive of the subgenus. In possessing more elongate hindlimbs and similar coat colour, and in spending more time away from rocky outcrops than other taxa, *M. antilopinus* and *H. rufus* are similar, perhaps suggesting that they form a clade.

On the basis of morphological synapomorphies, the species of *Macropus (Macropus)* appear to be monophyletic. However, interrelationships within the group are less clear, partly because the majority of taxa are known only as fossils. Species of *M. (Macropus)* include *M. giganteus*, *M. fuliginosus*, *M. ferragus*, *M. mundjabus*, *M. pan* and *M. pearsoni*. Synapomorphies identified for *Macropus (Macropus)* in Flannery

(1981) include presence of a thick ventrolingual enamel band on I_1, consistently well-developed forelinks on upper molars and presence of an extremely elongate I_1-P_3 diastema. *Macropus (M.) giganteus* and *M. (M.) fuliginosus* are the only extant species and are probably sibling taxa. *Macropus (M.) pan* and *M. (M.) pearsoni* share two synapomorphies; presence of a diagonally-placed, gaping hypolophid groove, and a high anterior cingulum on lower molars (Flannery and Archer 1983a).

Dawson and Flannery (1985) is a revision of the genus *Macropus*. They show that *Prionotemnus palankarinnicus* does not in fact belong within *Macropus*, but instead represents a much more plesiomorphic member of the Macropodinae. For species other than the subgeneric type grouped within *Macropus (Prionotemnus)* by Bartholomai (1975) and Richardson and McDermid (1978), a new subgeneric name, *Notamacropus*, is proposed. *Macropus (Notamacropus)* includes the species *M. agilis*, *M. irma*, *M. parryi*, *M. greyi*, *M. parma*, *M. rufogriseus*, *M. eugenii* and *M. thor*. Richardson and McDermid (1978) conclude that the presence of Pep C-5 in the extant species of this group tested by them may indicate that it is monophyletic. Some relationships can be determined within this group. These include the grouping of *M. (N.) parma* with *M. (H.) dorsalis* on the basis of pelage and cranial morphology (Ride 1957). Also, *M. (N.) greyi* and *M. (N.) irma* share similar coat colouration, short molars and premolars and narrow lower incisors and appear to be sibling species. *Macropus (N.) thor* also resembles *M. (N.) parryi* (Bartholomai 1975). *Macropus (N.) parryi* shares some intriguing pedal similarities with the species of *M. (Osphranter)*, such as having the distal end of the fourth metatarsal dorsoventrally compressed (Flannery 1981), which may indicate a relationship between these forms.

Six species of *Thylogale; T. billardierii*, *T.* "Hamilton" (an undescribed species from the early Pliocene Hamilton local fauna, Flannery, Rich, Turnbull and Lundelius in prep.), *T. christenseni*, *T. brunii*, *T. stigmatica* and *T. thetis* are recognized here. Of these *T.* "Hamilton" and *T. christenseni* are only known from fossil and subfossil cranial material. The species *T. brunii*, *T. stigmatica* and *T. thetis* are clearly monophyletic on the basis of a number of synapomorphies. These include the presence of a posteriorly positioned groove on I^3 (discussed in Ride 1957), having the lophids of M_1 subequal in width, a relatively short parastyle on M^1 and a lingual cingulum on P^3. *Thylogale billardierii* and *T.* "Hamilton" are closely related, clearly plesiomorphic species relative to the above clade. They possess a more centrally positioned I^3 groove (as is seen in most macropodine species), a protolophid on M_1 that is much narrower than the hypolophid, a large parastyle on M^1 and no lingual cingulum on P^3. The anterior cingulum of the lower molars of these species is extremely narrow and elongate (especially on the anterior molars) relative to other species and the I_1 is broader. These last two features may represent synapomorphies uniting these taxa or symplesiomorphies for the whole genus. The immunological study of Kirsch (1977) alone thus far provides evidence that *T. billardierii* is related to the members of the more advanced clade. The possession of 2n=22 chromosomes (regarded by Ride 1962 and Sharman 1961 as a feature uniting *Thylogale* species) may be a symplesiomorphy for the group, as Hayman and Martin (1974) regard this as the primitive chromosome number for macropodoids. *Thylogale christenseni* Hope, 1981 is an enigmatic species known only from subfossil cranial remains from Mt Jaya, Irian Jaya. The molars are extremely low crowned relative to other species, the parastyle of M_1 is larger and the protolophid of M_1 even narrower than in *T. billardierii*, and the I_1 is extremely narrow relative to other taxa. While *T. christenseni* may represent the most plesiomorphic species of *Thylogale* known, there is at present no compelling evidence that it belongs within that genus at all. Figure 12 is a phylogeny for the Macropodinae.

The macropodines and *Galanarla tesselata* can be classified as follows. The conventions used here follow Wiley (1981). *= species or genera tentatively accepted as valid pending further research.

Galanarla tesselata Flannery, Archer and Plane, 1983 *incertae sedis*

Subfamily Macropodinae (Gray, 1821)

Tribe indet

Genus indet. *piltonensis* (Bartholomai, 1975) *incertae sedis*

Hadronomas puckridgi Woodburne, 1967 *incertae sedis*

Gag macropodine *incertae sedis*

Dorcopsoides fossilis Woodburne, 1967 *incertae sedis*

Protemnodon nombe Flannery, Mountain and Aplin 1984 *sedis mutabilis*

,, *otibandus* Plane, 1967 *sedis mutabilis*

,, *tumbuna* Flannery, Mountain and Aplin, 1984 *sedis mutabilis*

,, *chinchillaensis* Bartholomai, 1973 *incertae sedis*

,, *snewini* Bartholomai, 1978 *sedis mutabilis*

,, *devisi* Bartholomai, 1973 *sedis mutabilis*

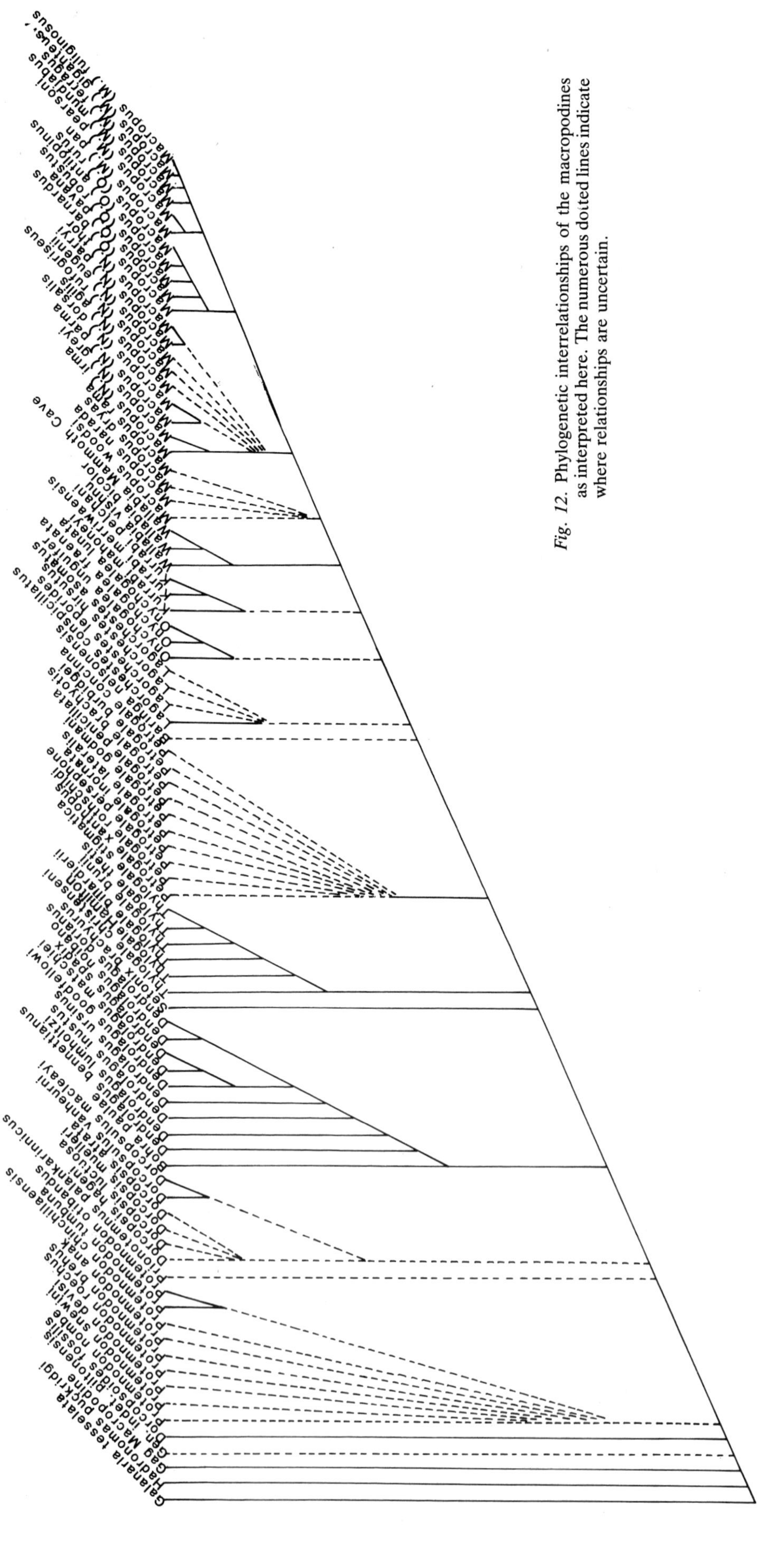

Fig. 12. Phylogenetic interrelationships of the macropodines as interpreted here. The numerous dotted lines indicate where relationships are uncertain.

Protemnodon anak Owen, 1874 *sedis mutabilis*
,, *brehus** (Owen, 1874) *sedis mutabilis*
,, *roechus** Owen, 1874 *sedis mutabilis*
Prionotemnus palankarinnicus Stirton, 1955 *incertae sedis*
Dorcopsis hageni Heller, 1896 *sedis mutabilis*
,, *luctuosa** D'Albertis, 1874 *sedis mutabilis*
,, *muelleri** Garrod, 1875 *sedis mutabilis*
,, *atrata* Van Deusen, 1957 *sedis mutabilis*
*Dorcopsulus** *vanheurni* Thomas, 1922 *sedis mutabilis*
,, *macleayi** (Mikluoho Maclay, 1885) *sedis mutabilis*

Tribe Dendrolagini n. tribe

Bohra paulae Flannery and Szalay, 1982
Dendrolagus bennettianus De Vis, 1887
,, *lumholtzi* Collett, 1884
inustus Muller, 1840
,, *ursinus* Muller, 1840
,, *goodfellowi* Thomas, 1908
,, *spadix** Troughton and Le Soeuf, 1936
,, *matschiei* Forster and Rothschild, 1907
,, *noibano* Flannery, Mountain and Aplin, 1984
,, *dorianus* Ramsay, 1883

Tribe Macropodini n. tribe

Setonix brachyurus (Quoy and Gaimard, 1830)
Thylogale christenseni Hope, 1982 *incertae sedis*
,, "Hamilton"
,, *billardierii* (Desmarest, 1822)
,, *brunii* (Schreber, 1777)
,, *thetis* (Lesson, 1827)
,, *stigmatica* Gould, 1860
Petrogale xanthopus Gray, 1854 *sedis mutabilis*
,, *rothschildi* Thomas, 1904 *sedis mutabilis*
,, *persephone* Maynes, 1982 *sedis mutabilis*
,, *inornata* Gould, 1842 *sedis mutabilis*
,, *lateralis* Gould, 1842 *sedis mutabilis*
Petrogale godmani Thomas, 1923 *sedis mutabilis*
,, *penicillata* (Griffith, 1827) *sedis mutabilis*
,, *brachyotis* (Gould, 1840) *sedis mutabilis*
,, *burbidgei* Kitchener and Sanson, 1978 *sedis mutabilis*
,, *concinna* (Gould, 1842) *sedis mutabilis*
Baringa nelsonensis Flannery and Hand, 1986
Lagorchestes asomatus Finlayson, 1943 *sedis mutabilis*
,, *leporides* (Gould, 1840) *sedis mutabilis*
,, *hirsutus* Gould, 1842 *sedis mutabilis*
,, *conspicillatus* Gould, 1841 *sedis mutabilis*
Onychogalea lunata (Gould, 1840)
,, *fraenata* (Gould, 1840)
,, *unguifer* (Gould, 1840)
Kurrabi "Hamilton"
,, *merriwaensis* Flannery and Archer, 1984
,, *mahoneyi* Flannery and Archer, 1984
Wallabia vishnu (De Vis, 1895) *sedis mutabilis*
,, *bicolor* (Desmarest, 1804)
,, "Mammoth Cave"
Macropus indet. *woodsi* Bartholomai, 1975 *sedis mutabilis*
,, ,, *narada* Bartholomai, 1978 *sedis mutabilis*
,, ,, *dryas* (De Vis, 1895) *sedis mutabilis*
,, ,, *rama* Bartholomai, 1975 *sedis mutabilis*
Macropus (Notamacropus) greyi Waterhouse, 1846 *sedis mutabilis*
,, ,, *irma* (Jourdan, 1837) *sedis mutabilis*
,, ,, *parma* Waterhouse, 1846 *sedis mutabilis*
,, ,, *dorsalis* (Gray, 1837) *sedis mutabilis*
,, ,, *agilis* (Gould, 1841) *sedis mutabilis*

Macropus (Notamacropus) parryi Bennett,
,, ,, 1834 *sedis mutabilis*
,, ,, *eugenii* (Desmarest, 1817) *sedis mutabilis*
,, ,, *rufogriseus* (Desmarest, *1817) sedis mutabilis*
,, ,, *thor* (De Vis, 1895) *sedis mutabilis*
Macropus (Macropus) pan De Vis, 1895
,, ,, *pearsoni* (Bartholomai, 1973)
,, ,, *fuliginosus* (Desmarest, 1817)
,, ,, *giganteus* Shaw, 1790
,, ,, *ferragus* (Owen, 1874)
,, ,, *mundjabus* Flannery,
,, ,, 1980
Macropus (Osphranter) barnardus Rothschild, 1903)
,, ,, *robustus* Gould, 1840
,, ,, *pavana*
,, ,, Bartholomai, 1978
,, ,, *antilopinus* (Gould, 1842)
,, ,, *rufus* (Desmarest, 1822).

OVERVIEW

Placing the major events of kangaroo evolution into a temporal and ecological framework is rather difficult because of a paucity of dated pre-Pliocene fossil sites and a lack of detailed palaeoecological studies. Despite the uncertainties, I have attempted in Figure (13) to summarize in a simple way my view of the chronology and major events in macropodoid evolution. The long period of evolutionary quiescence shown in this figure during the early Tertiary is based upon Flannery's (1987) hypothesis that the kangaroos are the sister taxon of all other Phalangeriformes. One suspects that the major possum groups have been evolving independently for a considerable time, and this implies in turn that kangaroos are an ancient group. Perplexingly, however, kangaroo assemblages from the Namba Formation (South Australia), are very primitive, and seem to indicate that they do not greatly postdate the differentiation of the macropodoid families. No estimates of the age of the fossiliferous part of the Namba Formation date it as older than about 30 million years.

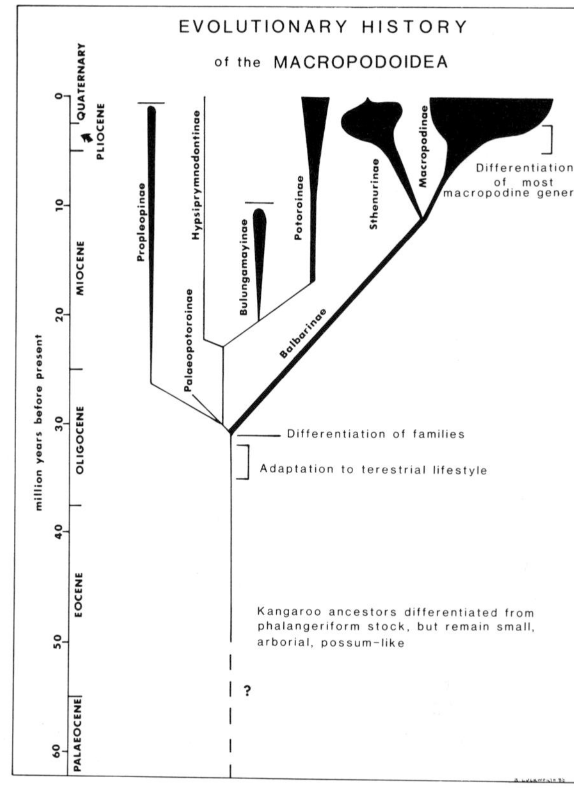

Fig. 13. Evolutionary history of the Macropodoidea.

These data strongly suggest that the kangaroos we know today shared a common ancestor perhaps as recently as 30 million years ago, but that the group itself has a much longer history. The Namba kangaroos all share (where known) a very primitive hindlimb morphology, and all are tiny. Although clearly macropodoid in form, and indicative of a terrestrial lifestyle, the kangaroo hindlimb by this time had not evolved many of the specializations seen in almost all modern lineages which allow for efficient terrestrial locomotion (Flannery and Rich 1986). This, along with their tiny size (all estimated as a kilo or less in weight) suggest to me that their common ancestor had recently adapted to a terrestrial lifestyle.

Thus, perhaps around 30 million years ago, kangaroos underwent a fundamental adaptational shift, and, if the number of taxa at the Lake Tarkarooloo site are indicative (Flannery and Rich 1986), began a phase of rapid speciation. Most of the Miocene kangaroo assemblages are characterized by a diverse radiation of potoroids. Propleopines, palaeopotoroines and hypsiprymnodontines seem to have differentiated early. All possess a plesiomorphic molar morphology, and probably had (and in the case of *H. moschatus* still have) diets that did not require the development of a complex stomach (Archer and Flannery 1985; Langer 1979, 1980a,b). Given the largely quadrupedal bound of the living

H. moschatus (Johnson and Strahan 1982), it seems probable that these kangaroos did not hop at all. By the middle Miocene, hypsiprymnodontine-like species had given rise to a lineage with more specialized molars, and a more exclusively herbivorous diet. This lineage differentiated into potoroines, which remained somewhat generalized, and bulungamayines, which developed a lophodont molar pattern and were probably specialized browsers. Both groups doubtless had complex stomachs (living potoroines are characterized by possession of a complex sacciform forestomach). Despite their rather spectacular Miocene radiation, it is clear that by the Pliocene potoroids had become a relatively insignificant part of the Australian fauna, eclipsed in importance by the macropodids.

The evolutionary development of the macropodids stands in stark contrast to that of the potoroids. When first detected in the fossil record (in the Lake Tarkarooloo kangaroo assemblage), the macropodids are represented by a single genus of primitive, tiny, but nontheless already lophodont species. This suggests that even at this stage they were adapted to browsing, and may have had a complex stomach. Although the species increase in size through the early part of their fossil record, they are never represented by more than one genus at a time, and their morphology remains primitive. All of these early species are balbarines, and it is not until the potoroid radiation is well underway that the macropodids show any sign of greatly diversifying. In the youngest of the Riversleigh sites, and at Alcoota (the youngest Miocene site known), the first macropodines are present. These are presumably rainforest-dwellers, similar in size and morphology to the species of *Dorcopsis*. The next time we encounter macropodines (in the first well-dated suite of fossil sites; 4-5 million years old), the macropodines are represented by a truly staggering radiation. Almost all the modern genera have differentiated and are already speciose. This denotes an extremely rapid, indeed explosive, macropodine radiation during the latest Miocene. The macropodines may well have crossed an adaptational threshold by this time, with perhaps their unique incisor occlusal cycle and stomach morphology opening up new adaptational niches for them. By Alcoota times, the bulungamyines (lophodont potoroids) have disappeared, perhaps displaced by the newly evolved (but still only modestly radiated) macropodines.

Cladistic analysis suggests that sthenurines evolved from balbarines, and not macropodines, so they also must have been differentiated by the end of the middle Miocene. However, they have left no fossil record dating from this period, so must have remained relatively insignificant. Even in the early Pliocene sites (e.g., Hamilton; 4.5 million years old) they remain rare. By the early Pleistocene, however, they had begun to blossom, with perhaps as many as 20 species in five genera being present. Pleistocene extinctions, probably human caused, resulted in the loss of all except one sthenurine species, all propleopines, and all members of one genus of macropodine *(Protemnodon)* as well as some species of *Macropus*.

In overview, the most striking aspect of the evolutionary history of the kangaroos has been the remarkable, indeed all pervasive, convergence that characterizes the lineage. Repeatedly, members of the various subfamilies and families have developed morphologically identical or closely similar adaptations, perhaps in response to similar environmental challenges or opportunities. Doubtless, the apparently rigid and universal macropodoid ground plan has limited the number of evolutionary responses possible for kangaroos, resulting in the same evolutionary responses being elicited time and again. Finally, it is worth meditating on the fact that without their unsurpassed fossil record, any attempt to unravel the relationships of the modern genera based upon parsimony would result in incorrect phylogenies, precisely because of this widespread convergence.

ACKNOWLEDGEMENTS

This work formed part of my doctoral thesis, so first and foremost thanks must go to my supervisor Dr M. Archer who has, more than anyone else, given me my grounding in systematics and provided numerous invaluable opportunities to undertake research and fieldwork. I also owe thanks to my wife Paula, without whose patience, understanding and aid in drafting the many figures this work would never have been completed.

Tom and Pat Rich also deserve my warmest thanks for reading with great patience drafts of my work, supplying valuable advice and for making available to me many of the important specimens that have shaped the conclusions of this work. Others who have offered valuable criticism and suggestions are Mr K. Aplin and Dr L. Dawson of the University of New South Wales, Dr H. Woodburne and Mr J. Case of the University of California at Riverside, Dr F. Szalay of the City University of New York and Dr P. Baverstock of the South Australian Museum.

The following colleagues and their institutions are thanked for their generous loans of specimens: Professor W. Clemens of the University of California at Berkeley; Dr R, Molnar of the Queensland Museum; Dr T. Rich and Ms J. Dixon of the Museum of Victoria; Mr N. Pledge and Mr P. Aitken of the South Australian Museum; Dr K. McNamara of the Western Australian Museum; and Dr A. Ritchie of the Australian Museum.

REFERENCES

ARCHER, M., 1978a. The nature of the molar-premolar boundary in marsupials and a reinterpretation of the homology of marsupial cheek teeth. *Mem. Qd Mus.* **18**: 157-64.

ARCHER, M., 1978b. Quaternary vertebrate faunas from the Texas Caves of southeastern Queensland. *Mem. Qd Mus.* **19**: 69-109.

ARCHER, M., 1979. *Wabularoo naughtoni gen. et. sp. nov.*, an enigmatic kangaroo (Marsupialia) from the middle Tertiary Cark Creek Limestone of northwestern Queensland. Results of the Ray E. Lemley Expeditions, part 4. *Mem. Qd Mus.* **19**: 299-307.

ARCHER, M., 1981. A review of the origins and radiation of Australian mammals. Pp. 1437-87 *in* Biogeography and Ecology in Australia ed. by A. Keast. Junk Press: The Hague.

ARCHER, M., 1984. The Australian mammal radiation. Pp. 633-808 *in* Vertebrate Zoogeography and Evolution in Australasia ed. by M. Archer and G. Clayton. Hesperian Press: Perth, Western Australia.

ARCHER, M. AND BARTHOLOMAI, A, A., 1978. Tertiary Mammals of Australia: a synoptic review. *Alcheringa* **2**: 1-19.

ARCHER, M., BARTHOLOMAI, A. AND MARSHALL, L. G., 1978. *Proleopus chillagoensis*, a new north Queensland species of extinct giant rat-kangaroo (Macropodidae: Potoroinae). *Mem. Nat. Mus. Vict.* **39**: 55-60.

ARCHER, M. AND FLANNERY, T. F., 1985. Revision of the extinct gigantic rat kangaroos (Potoroidae: Marsupialia) with a description of a new Miocene genus and species and a new Pleistocene species of *Propleopus*. *J. Paleo.* **59**: 1331-49.

BALME, J., 1980. An early Pleistocene macropod from Jandakot, Western Australia. *West. Aust. Nat.* **14**: 233-35.

BARTHOLOMAI, A., 1973a. The genus *Protemnodon* Owen (Marsupialia: Macropodidae) in the Upper Cainozoic deposits of Queensland. *Mem. Qd Mus.* **16**: 309-63.

BARTHOLOMAI, A., 1975. The genus *Macropus* Shaw (Macropodidae: Marsupialia) in the Upper Cainozoic deposits of Queensland. *Mem. Qd Mus.* **17**: 195-235.

BARTHOLOMAI, A., 1976. The genus *Wallabia* Trouessart (Marsupialia: Macropodidae) in the Upper Cainozoic deposits of Queensland. *Mem. Qd Mus.* **17**: 375-77.

BARTHOLOMAI, A., 1978a. The Macropodidae (Marsupialia) of the Allingham Formation, northern Queensland. Results of the Ray E. Lemley Expeditions. Part 3. *Mem. Qd Mus.* **18**: 127-42.

BARTHOLOMAI, A., 1978b. The fossil kangaroos. *Aust. Mammal.* **2**: 15-22.

BRISCOE, D. A., CALABY, R. L., MAYNES, G. M., MURTAGH, C. E. AND SHARMAN, G. B., 1982. Isolation, introgression and genetic variation in rock wallabies. Pp. 73-87 *in* Species at Risk ed. by R. H. Groves and W. D. L. Ride. Springer Verlag: Berlin.

BURBIDGE, A. A. AND JOHNSON, K. A., 1983. Rufous Hare Wallaby. P. 199 *in* The Australian Museum Complete Book of Australian Mammals ed. by R. Strahan. Angus & Robertson: Sydney.

CASE, J. A., 1984. A new genus of Potoroinae (Marsupialia: Macropodidae) from the Miocene Ngapakaldi local fauna, South Australia, and a definition of the Potoroinae. *J. Paleo.* **58**: 1074-86.

CHILCOTT, M. J. AND HUME, I. D., 1981. The effects of water restriction on the nitrogen metabolism of the Tammar Wallaby *(Macropus eugenii)* fed high and low protein diets. *Bull. Aust. Mammal. Soc.* **1981**: 29 (abst.).

DAWSON, T. J., 1977. Kangaroos. *Scientific American.* **237**: 78-89.

DAWSON, L. AND FLANNERY, T. F., 1985. Taxonomic and phylogenetic status of living and fossil kangaroos and wallabies of the genus *Macropus* Shaw (Macropodidae: Marsupialia) with a new subgeneric name for the large wallabies. *Aust. J. Zool.* **33**: 473-98.

EALEY, E. H. M., 1967. Ecology of the Euro *Macropus robustus* (Gould) in northwestern Australia. 2. Behaviour, movement and drinking patterns. *CSIRO Wildl. Res.* **12**: 27-51.

FELSENSTEIN, J., 1978. Cases in which parsimony or compatability methods will be positively misleading. *Syst. Zool.* **27**: 401-10.

FELSENSTEIN, J., 1979. Alternative methods of phylogenetic inference and their interrelationship. *Syst. Zool.* **28**: 49-62.

FELSENSTEIN, J., 1983. Parsimony in systematics: Biological and statistical issues. *Ann. Rev. Ecol. Syst.* **14**: 313-33.

FLANNERY, T. F., 1980. *Macropus mundjabus*, a new kangaroo (Marsupialia: Macropodidae) of uncertain age from Victoria, Australia. *Aust. Mammal.* **3**: 35-51.

FLANNERY, T. F., 1981. A review of the genus *Macropus*, the living Grey Kangaroos and their fossil allies. M.Sc. Thesis. Earth Sciences, Monash University, Clayton, Victoria.

FLANNERY, T. F., 1982. Hindlimb structure and evolution in the kangaroos (Marsupialia: Macropodoidea). Pp. 508-24 *in* The Vertebrate Fossil Record of Australasia ed. by P. V. Rich and E. M. Thompson. Monash University Press: Clayton, Victoria.

FLANNERY, T. F., 1983. Review of the subfamily Sthenurinae (Marsupialia) and the relationships of the species of *Troposodon* and *Lagostrophus*. *Aust. Mammal.* **6**: 15-28.

FLANNERY, T. F., 1984a. The kangaroos: 15 million years of bounders. Pp. 817-36 *in* Vertebrate Zoogeography and Evolution in Australasia ed. by M. Archer and G. Clayton. Hesperian Press: Perth, Western Australia.

FLANNERY, T. F., 1984b. Systematics, Biochronology and Zoogeography of the Macropodoidea. Ph.D. Thesis. Zoology, University of New South Wales.

FLANNERY, T. F., 1987. The relationships of the macropodoids (Marsupialia) and the polarity of some morphological features within the Phalangeriformes. Pp. 741-47 *in* Possums and Opossums: Studies in Evolution ed. by M. Archer. Surrey Beatty & Sons: New South Wales.

FLANNERY, T. F. AND SZALAY, F. S., 1982. *Bohra paulae*, a new giant fossil tree-kangaroo (Marsupialia: Macropodidae) from New South Wales, Australia. *Aust. Mammal.* **5**: 38-54.

FLANNERY, T. F. AND ARCHER, M., 1983a. The taxonomy and distribution of *Macropus pearsoni*. *Aust. Mammal.* **5**: 261-65.

FLANNERY, T. F. AND ARCHER, M., 1983b. Revision of the genus *Troposodon* Bartholomai (Macropodidae: Marsupialia). *Alcheringa* **7**: 263-79.

FLANNERY, T. F. AND HOPE, J. H., 1983. Occurrences of the extinct macropodid marsupial *Simosthenurus maddocki* in New South Wales. *Aust. Mammal.* **5**: 38-54.

FLANNERY, T. F. AND ARCHER, M., 1984. The macropodids (Marsupialia) of the early Pliocene Bow Local Fauna, central eastern New South Wales. *Aust. Zool.* **21**: 357-84.

FLANNERY, T. F. AND HANN, C., 1984. A new macropodine genus and species (Macropodidae: Marsupialia) from the early Pleistocene of southwestern Victoria. *Aust. Mammal.* **7**: 193-205.

FLANNERY, T. F. AND RICH, T. H. R., 1986. Macropodoids of the middle Miocene Namba Formation, South Australia, and the homology of some dental structures in kangaroos. *J. Paleo.* **60**: 418-47.

FLANNERY, T. F. AND ARCHER, M., 1987a. A new species of *Hypsiprymnodon* (Potoroidae: Marsupialia) from the Miocene of northwestern Queensland, and a reinterpretation of the morphology of the living *Hypsiprymnodon moschatus*. Pp. 749-58 *in* Possums and Opossums: Studies in Evolution ed. by M. Archer. Surrey Beatty & Sons: New South Wales.

FLANNERY, T. F. AND ARCHER, M., 1987b. *Bettongia moysei*, a new plesiomorphic kangaroo (Marsupialia: Potoroidae) from Miocene sediments of northwestern Queensland. Pp. 759-67 *in* Possums and Opossums: Studies in Evolution ed. by M. Archer. Surrey Beatty & Sons: New South Wales.

FLANNERY, T. F., ARCHER, M. AND PLANE, M., 1983. Miocene kangaroos (Macropodidae: Marsupialia) from three localities in northern Australia, with a description of two new subfamilies. *Bull. Bur. Miner. Res. Aust. Geol. Geophys.* **7**: 287-302.

FLANNERY, T. F., MOUNTAIN, M. J. AND APLIN, K., 1983. Quaternary kangaroos (Macropodidae: Marsupialia) from Nombe rockshelter, Papua New Guinea, with comments on the nature of megafaunal extinction in the New Guinea highlands. *Proc. Linn. Soc. NSW* **107**: 77-99.

FLANNERY, T. F., ARCHER, M. AND PLANE, M., 1984. Phylogenetic relationships and a reconsideration of higher level systematics within the Potoroidae (Marsupialia). *J. Paleo.* **58**: 1087-97.

GANSLOSSER, V. U., 1981. Vertleichende Untersuchungen zur Kletterfhahigkeit einer Baumkanguruharten (*Dendrolagus*, Muller 1839). *Zool. Anz. Jena.* **206**: 62-86.

GILMORE, D. P., 1977. The success of marsupials as introduced species. Pp. 169-78 *in* The Biology of Marsupials ed. by B. Stonehouse and D. Gilmore. Macmillan Press: London.

GROVES, C. P., 1982. The systematics of the tree kangaroos (*Dendrolagus*; Marsupialia, Macropodidae). *Aust. Mammal.* **5**: 157-86.

HAYMAN, D. L. AND MARTIN, P. G., 1974. Cytogenetics of marsupials. Pp. 91-217 *in* Comparative Mammalian Cytogenetics ed. by K. Benirschke. Springer: New York.

HENNIG, W., 1966. *Phylogenetic Systematics*. University of Illinois Press: Urbana.

HOPE, J. H., 1981. A new species of *Thylogale* (Marsupialia: Macropodidae) from Mapala rock shelter, Jaya (Carstenz) Mountains, Irian Jaya (Western New Guinea), Indonesia. *Rec. Aust. Mus.* **33**: 369-87.

HUME, I. D., 1978. Evolution of the Macropodidae digestive system. *Aust. Mammal.* **2**: 37-42.

JOHNSON, P. M., 1983. Rufous Bettong. Pp. 190-91 *in* The Australian Museum Complete Book of Australian Mammals ed. by R. Strahan. Angus & Robertson: Sydney.

JOHNSON, P. M. AND STRAHAN, R., 1982. A further description of the musky rat-kangaroo, *Hypsiprymnodon moschatus* Ramsay, 1876 (Marsupialia: Potoroidae), with notes on its biology. *Aust. Zool.* **21**: 28-46.

KIRSCH, J. A. W., 1977. A comparative serology of the Marsupialia, and a classification of the marsupials. *Aust. J. Zool.* supp. ser. no. 52.

KITCHENER, D. J. AND SANSON, G., 1978. *Petrogale burbidgei* (Marsupialia, Macropodidae), a new rock wallaby from Kimberley, Western Australia. *Rec. West. Aust. Mus.* **6**: 269-85.

LANGER, P., 1979. Phylogenetic adaptation of the stomach of the Macropodidae Owen, 1839, to food. *Z. Saugetierkunde* **44**: 321-33.

LANGER, P., 1980a. Anatomy of the stomach in three species of Potoroinae (Marsupialia: Macropodidae). *Aust. J. Zool.* **20**: 19-32.

LANGER, P., 1980b. Stomach evolution in the Macropodidae Owen, 1839 (Mammalia: Marsupialia). *Z. Zool. Syst. Evolut. Forsch.* **18**: 211-32.

MACKINTYRE, G. T., 1972. The trisulcate petrosal pattern of mammals. Pp. 275-303 *in* Evolutionary Biology V6 ed. by T. Dobzhansky, M. Hecht and W. C. Steere. Appleton, Century and Crofts: New York.

MARSHALL, L. G., 1981. The families and genera of Marsupialia. *Fieldiana*, Geol., n.s. **8**: 1-65.

MAYNES, G. M., 1982. A new species of rock wallaby *Petrogale persephone* (Marsupialia: Macropodidae) from Proserpine, central Queensland. *Aust. Mammal.* **5**: 47-58.

MERRILEES, D., 1968. Man the destroyer: late Quaternary changes in the Australian marsupial fauna. *J. Proc. R. Soc. Western Aust.* **51**: 1-24.

MERRILEES, D. AND PORTER, J. K., 1979. Guide to the Identification of Teeth and Some Bones of Native Land Mammals Occurring in the Extreme Southwest of Western Australia. Western Australian Museum, Perth.

PEACOCK, W. J., DENNIS, G. S., ELIZUR, A. AND CALABY, J. H., 1981. Repeated DNA sequences and kangaroo phylogeny. *Aust. J. Biol. Sci.* **34**: 325-40.

PLEDGE, N. S., 1981. The giant rat kangaroo, *Propleopus oscillans* (de Vis), (Potoroidae: Marsupialia) in South Australia. *Trans. R. Soc. S. Aust.* **105**: 41-47.

POOLE, W. E., 1978. The status of the Australian Macropodidae. Pp. 13-27 *in* The Status of Endangered Species ed. by M. J. Tyler. Royal Zoological Society of South Australia, Adelaide.

PRINCE, R. T. T., 1983. Banded Hare Wallaby. Pp. 201-02 *in* The Australian Museum Complete Book of Australian Mammals ed. by R. Strahan. Angus & Robertson: Sydney.

RICH, T. H., ARCHER, M., PLANE, M. D., FLANNERY, T. F., PLEDGE, N. S., HAND, S. AND RICH, P. V., 1982. Australian Tertiary mammal localities. Pp. 525-72 *in* The Fossil Vertebrate Record of Australasia ed. by P. V. Rich and E. M. Thompson. Monash University Press: Clayton, Victoria.

RICHARDSON, B. J. AND SHARMAN, G. B., 1976. Biochemical and morphological observations on the wallaroos (Macropodidae: Marsupialia) with a suggested new taxonomy. *J. Zool. Lond.* **179**: 499-513.

RICHARDSON, B. J. AND MCDERMID, E. M., 1978. A comparison of genetic relationships within the Macropodidae as determined from allozyme, cytological and immunological data. *Aust. Mammal.* **2**: 43-52.

RIDE, W. D. L., 1957. *Protemnodon parma* (Waterhouse) and the classification of related wallabies (*Protemnodon, Thylogale, Setonix*). *Proc. Zool. Soc. Lond.* **128**: 327-46.

RIDE, W. D. L., 1959. Mastication and taxonomy in the macropodine skull. Pp. 33-59 *in* Function and Taxonomic Importance ed. by J. Cain. London Systematic Association Publication No 3, London.

RIDE, W. D. L., 1962. On the evolution of Australian marsupials. Pp. 281-306 *in* The Evolution of Living Organisms ed. by G. W. Leeper. University of Melbourne Press: Melbourne.

RIDE, W. D. L., 1978. Historical introduction to studies on the evolution and phylogeny of the Macropodidae. *Aust. Mammal.* **2**: 1-14.

ROFE, R. H., 1978. G-banded chromosomes and the evolution of the Macropodidae. *Aust. Mammal.* **2**: 53-64.

ROSE, R. W., 1978. Reproduction and evolution in the female Macropodidae. *Aust. Mammal.* **2**: 65-72.

RUSSELL, E. M., 1974. The biology of kangaroos (Marsupialia – Macropodidae). *Mammal. Rev.* **4**: 1-59.

SANSON, G. D., 1978. The evolution and significance of mastication in the Macropodidae. *Aust. Mammal.* **2**: 23-28.

SANSON, G. D., 1982. Evolution and feeding adaptations in fossil and recent macropodoids. Pp. 489-506 *in* The Fossil Vertebrate Record of Australasia ed. by P. V. Rich and E. M. Thompson. Monash University Press.

SANSON, G. D., 1983. Narbalek. Pp. 223-24 *in* The Australian Museum Complete Book of Australian Mammals ed. by R. Strahan. Angus & Robertson: Australia.

SEEBECK, J. H. AND JOHNSON, P. G., 1980. *Potorous longipes* (Marsupialia: Macropodidae): a new species from Eastern Victoria. *Aust. J. Zool.* **28**: 119-34.

SHARMAN, G. B., 1961. The mitotic chromosomes of marsupials and their bearing on taxonomy and phylogeny. *Aust. J. Zool.* **9**: 38-60.

SHARMAN, G. B., MURTAGH, C., JOHNSON, P. M. AND WEAVER, C. M., 1980. The chromosomes of a rat-kangaroo attributable to *Bettongia tropica* (Marsupialia: Macropodidae). *Aust. J. Zool.* **28**: 59-63.

SOKAL, R. R. AND SNEATH, P. H. A., 1963. *The principles of numerical taxonomy*. W. H. Freeman and Co. San Francisco, California.

STRAHAN, R. (ed.), 1983. The Australian Museum Complete Book of Australian Mammals. Angus & Robertson.

SZALAY, F. S., 1982. A new appraisal of marsupial phylogeny and classification. Pp. 621-40 *in* Carnivorous Marsupials. ed. by M. Archer. Royal Zoological Society of New South Wales: Sydney.

WELLS, R. T. AND MURRAY, P., 1979. A new sthenurine kangaroo (Marsupialia: Macropodidae) from southeastern South Australia. *Trans. R. Soc. S. Aust.* **103**: 213-19.

WILEY, E. O., 1981. Phylogenetics. The Theory and Practice of Phylogenetic Systematics. John Wiley and Sons: New York.

WOODBURNE, M. O., 1967. The Alcoota fauna, central Australia. An integrated palaeontological and geological study. *Bull. Bur. Miner. Resour. Geol. Geophys. Aust.* **87**: 1-187.

WOODBURNE, M. O., 1984. *Wakiewakie lawsoni*, a new genus and species of Potoroinae (Marsupialia: Macropodidae) of medial Miocene age, South Australia. *J. Paleo.* **58**: 1074-86.

WOODBURNE, M. O., TEDFORD, R. H., ARCHER, M., TURNBULL, W. D., PLANE, M. AND LUNDELIUS, E. L. 1985. Biochronology of the continental mammal record of Australia and New Guinea. *Spec. Publ. S. Aust. Dept. Mines. and Energy* **5**: 347-63.

Zoogeography of the Macropodoidea

G. M. MAYNES[1]

The present diversity of the superfamily Macropodoidea traces its origins to the establishment of contact between Australia and New Guinea during the Miocene and subsequent climatic fluctuations, particularly during the Pliocene and Pleistocene. The relationship of the faunas of these two land masses and their distinction from faunas elsewhere in the world has been recognized by grouping them together as a single zoogeographic region (the Australian Region). This region has been further subdivided into seven subregions by some zoo-geographers on the basis of similarities and differences in the distributional ranges of individual taxa.

The Papuan Subregion includes the island of New Guinea and is the most distinctive subregion with 13 endemic species of macropodids from a total fauna of 15 species. These species are derivatives of rainforest-adapted genera first noted in the Australian fossil record during the Pliocene. The subregion is also characterized by the absence of any representatives of the family Potoroidae and rock-adapted species of macropodids.

The Torresian Subregion in northeast Queensland shares two species with the Papuan Subregion to the north. It is characterized by five endemic species, three of which are restricted to the remnant rainforest at the base of Cape York Peninsula, while the remaining two endemics are rock-adapted species in the genus *Petrogale*. The rainforest species are clearly relicts of the formerly more widespread rainforest fauna of the Pliocene. In contrast, the two species of *Petrogale* appear to be in the process of active speciation with five distinctive chromosomal races distributed parapatrically.

The presence of only rock-adapted endemic species (4) is a unique characteristic of the Timorean Subregion. The relationships of the fauna in this subregion lie to the east with the Torresian Subregion (five common species) rather than to the south with the Eyrean Subregion (only three common species).

In the south-east of the continent, the Kosciuskan Subregion covers an area of great topographic and climatic variation. Four of the six endemic species in this subregion are restricted to the highest rainfall areas and could be considered as survivors from previously more mesic environments. Another distinctive feature of this subregion is the concentration of members of the genus *Macropus*; eight species have been recorded in it. For six of these species, the subregion is central to their distributional ranges.

The Tasmanian Subregion is not distinguished by any endemic macropodoid species, but does possess a distinctive endemic subspecies of *M. rufogriseus*. The limited former distribution of *T. billardierii* along coastal Victoria, however, strongly suggests that it may have originated in Tasmania and invaded the mainland during a period of lowered sea levels.

Relationships of the three macropodoid species endemic to the Southwestern Subregion lie to the east with the Kosciuskan fauna, rather than to the adjacent Eyrean Subregion. Nevertheless, the Eyrean Subregion has a strong influence in the Southwestern Subregion through the joint sharing of eight species adapted to arid or semi-arid conditions.

Seven species were endemic to the Eyrean Subregion and two of these are rock-adapted species. Six of the eight genera of macropodoids recorded from this subregion were first recorded in the fossil record during the Pleistocene. This subregion shares a large number of species in the semi-arid interface with the surrounding peripheral subregions.

Potoroids are most diverse south of the Tropic of Capricorn (23°28′S) and are completely absent north of latitude 16°S. A relationship between latitude and diploid chromosomal number is evident in the genus *Petrogale* where eight of nine chromosomal races north of 19°30′S have a diploid number of 20 or less while only one of nine chromosomal races south of that latitude has a diploid number of 20, the remainder having $2n=22$. The date of the origin of the southern 20-chromosomal race *(P. lateralis hacketti)* can be placed reliably at between about 12 500 and 15 000 B.P. based upon its distribution on islands of the Recherche Archipelago. This corresponds with a period of transition from the most arid in the last 50 000 years to a period of higher rainfall between about 7 000 and 5 000 years B.P.

The high level of speciation within the genus *Petrogale* is contrasted with the limited differentiation within *Macropus robustus,* a rock-adapted species with a distribution almost as extensive as all species of *Petrogale* combined. Differences in size and social organization are believed to be major contributors to the high degree of speciation in *Petrogale* and its absence in *M. robustus.*

Twenty-four species of macropodoids are recorded from islands around Australia, while a further five species are recorded from islands adjacent to New Guinea. Island distributions are notable for the limited presence or absence of the five largest species of macropodoids, the high proportion of species of *Petrogale* on islands and the absence of Torresian rainforest species. The presence of macropodoids on islands off the continental shelf in the vicinity of New Guinea is attributed to human intervention.

Examination of introductions of macropodoids throughout the world indicates that feral populations may be established from minimal founding populations of two to five animals, given the right combination of circumstances.

[1]Australian National Parks and Wildlife Service, G.P.O. Box 636, Canberra City, Australian Capital Territory 2601.

Pages 47-66 in KANGAROOS, WALLABIES AND RAT-KANGAROOS, ed. by G. Grigg, P. Jarman and I. Hume. Surrey Beatty & Sons Pty Limited, New South Wales, Australia, 1989.

INTRODUCTION

FOLLOWING the breakup of Gondwanaland in the early Cretaceous, the northward moving Australian block collided with the Papua New Guinea Mobile Belt and the central part of New Guinea, including the Aure Trough, about 15 million years ago in the Mid-Miocene (Crook 1981). By the Pliocene (5-6 million years ago) most of the present-day island of New Guinea was dry land (Doutch 1972). The central cordillera of New Guinea was formed by uplifting along major fault lines which has raised the mountains by several thousand metres. This process is continuing still with the main centre of tectonic activity now occurring in the northern ranges where maximum rates of uplift are in the order of 30 cm per century (Loffler 1982).

Continental Australia is now separated from New Guinea by the 40-80 m deep Arafura Sea to the West and 13 m deep Torres Strait to the East, and from Tasmania in the South by Bass Strait (average depth 60-80 m). These three areas and associated continental islands have been united at various times of lowered sea levels in the past to form a single land mass with a total area of up to 11 million km^2. This combined area has variously been referred to as "Terra Australis" (Nix 1981), "Sahulland" (Bulmer 1982) or "Meganesia" (Filewood 1984) to emphasize that New Guinea has been an integral part of the evolving continental land mass.

During the Miocene, the amount of ice on Antarctica increased sharply to reach approximately its present volume by the late Miocene (Kemp 1981). In the latest Miocene, there was a dramatic decrease in sea-surface temperature accompanied by a rapid expansion of the ice shelves between five and six million years ago (Kemp 1981) and accompanied by a major phase of regression in Australian sedimentary basins (Quilty 1977) so that a direct connection was made between Australia and proto-New Guinea for the first time. Kemp (1981) suggested that the relatively rapid retreat of rainforest vegetation shown in pollen counts during the late Miocene in southeastern Australia could be associated with an increase in aridity resulting from expansion of the Antarctic ice.

Galloway and Kemp (1981) gave a detailed exposition of the broad scale changes in climate from the Miocene to the Recent and reviewed data on late Cainozoic environments in Australia and their potential influence on the biogeography of Australia. They reported that fluctuations of more than 200 metres in sea level have repeatedly occurred during the Pleistocene and that the sea level may have been substantially lower than at present for substantial periods of the late Tertiary. Contact with New Guinea to the north and Tasmania to the south via land bridges would have occurred for substantial periods during the last few million years, although the duration of contact with Tasmania would have been more restricted due to the greater depth of Bass Strait (Galloway and Kemp 1981). Major changes in the distribution and composition of plant communities (and presumably associated animal communities) were associated with the climatic deterioration which occurred during the late Tertiary. Rainforest became restricted to the eastern coastal areas and more arid-adapted vegetation evolved. At the same time, changes in the macropodoid communities tracked these changes in the plant communities (Flannery 1984). Kershaw (1981) reviewed data on vegetation changes during the Quaternary based upon analyses of pollen deposits and reported a continuation of the trend for a reduction in the extent and diversity of rainforest and expansion of plant communities associated with more arid conditions that emerged with climatic fluctuations during the Quaternary.

The earliest known macropodoids in the fossil record are believed to be mid-Miocene in age. All appear to be near the base of the kangaroo evolutionary tree and were small, rainforest forms (Flannery 1984). The origin and early differentiation of the Macropodoidea therefore appear to have been contemporaneous with the formation of New Guinea, while the radiation and increase in diversity of the superfamily appear to be a result of the oscillations in climate and vegetation communities since the late Miocene.

ZOOGEOGRAPHIC PATTERNS

The pattern of distribution displayed by a species is the result of the interaction of numerous factors. These include physical and chemical factors of the environment as well as biological characteristics of the species such as its evolutionary history, dispersal ability, habitat preference, interactions with other species and the genetic flexibility to respond positively to environmental perturbations. The distribution pattern of a species is dynamic and its outer limits may fluctuate in response to variation in the limiting factors over time. Despite this, biogeographers have sought to delimit areas which have a homogeneous species composition in order to make generalizations across a large geographic area.

Australia and New Guinea have traditionally been united as the Australian Region which was considered to be sharply divided from the Oriental Region along Wallace's Line. In a summary of recent data on the distribution of various taxa in the contact zone between the Oriental and Australian Regions, Keast (1981) concluded that it was best to consider the Australian Region as that area bound by the Sahul

Shelf while the Oriental Region was that area bound by the Sunda Shelf (Fig. 1). The intervening archipelagos between these boundaries represent an area of addition and subtraction of the faunal components of each region, sometimes referred to as Wallacea. The limits to distribution and degree of penetration of this region by the Oriental and Australian faunal components is a reflection of the dispersal ability across water of the individual taxa concerned and the availability of appropriate niches.

Further sub-divisions within the Australian Region have been proposed as either Subregions or Provinces and these are summarized by Keast (1981) and Heatwole (1987). New Guinea is recognised as the Papuan Subregion and an additional six subregions (Timorean, Torresian, Kosciuskan, Tasmanian, Southwest and Eyrean Subregions, Fig. 1) are identified as the most commonly accepted and generally applicable divisions within Australia (Heatwole 1987). Table 1 summarizes the information available on the distribution of endemic species of macropodoids in each of these zoogeographic subregions, compared with the total number of species having part or all of their distributional ranges within

Table 1. Matrix showing number of endemic species and shared species between subregions. Diagonal represents number of species endemic to each subregion. Endemic species restricted to rock habitats given in brackets.

	Papuan	Timorean	Torresian	Kosciuskan	Tasmanian	Southwestern	Eyrean	Total number of species
Papuan	13(0)	1	2	2	0	0	0	15
Timorean		4(4)	5	1	0	0	3	9
Torresian			5(2)	8	1	0	6	16
Kosciuskan				6(2)	5	1	10	21
Tasmanian					0(0)	1	1	5
Southwestern						3(0)	8	11
Eyrean							7(2)	26

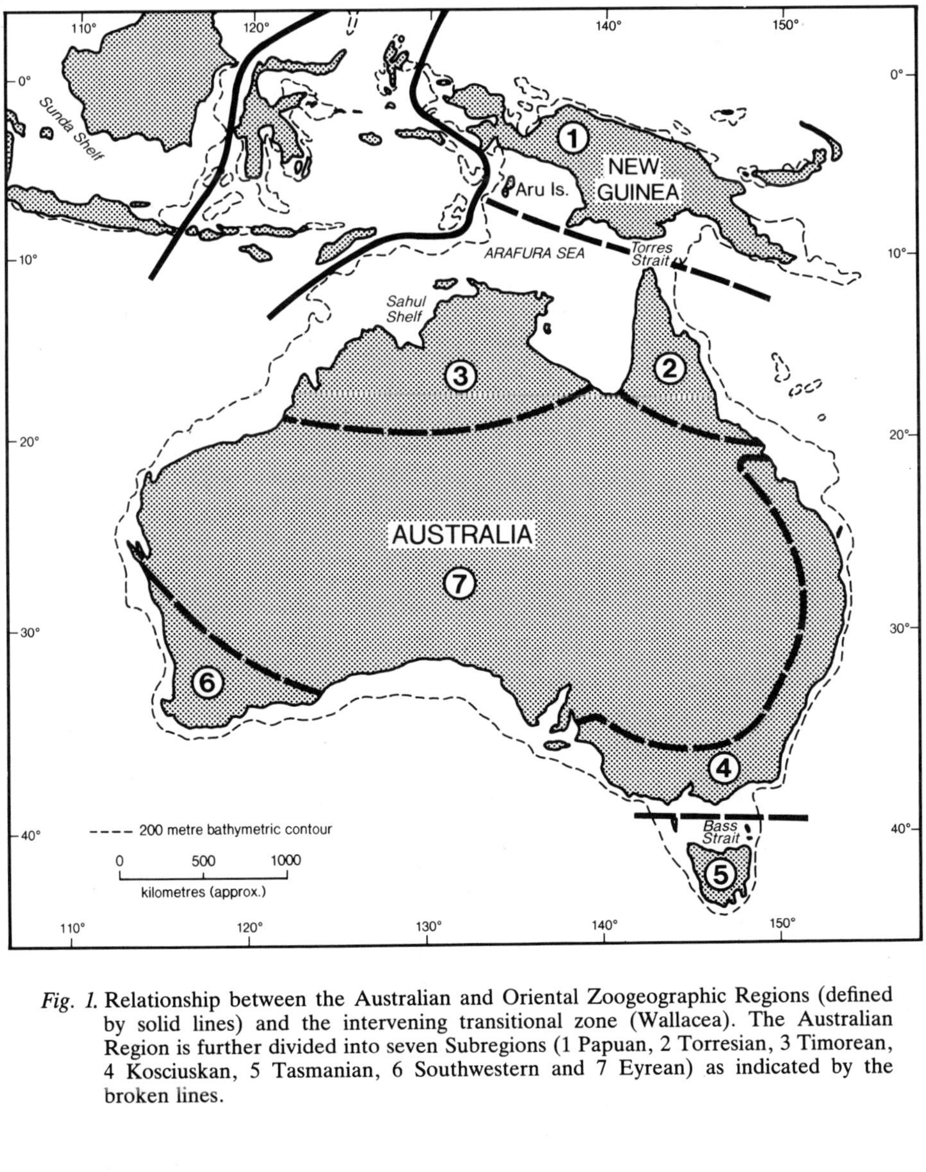

Fig. 1. Relationship between the Australian and Oriental Zoogeographic Regions (defined by solid lines) and the intervening transitional zone (Wallacea). The Australian Region is further divided into seven Subregions (1 Papuan, 2 Torresian, 3 Timorean, 4 Kosciuskan, 5 Tasmanian, 6 Southwestern and 7 Eyrean) as indicated by the broken lines.

them. The high level of specific endemism in the Papuan Subregion immediately distinguishes this area from all other subregions. Despite the total number of endemic species recorded from this subregion, four species is the maximum number likely to be recorded from any specific area while many areas have only two or three species present. Schodde and Calaby (1972) reviewed the biogeography of both the bird and mammal faunas in relation to Torres Strait and concluded that the basic stocks of birds, marsupials and monotremes in Australia and New Guinea are jointly shared and that the divergence of these stocks is due to their separate adaptation to wet environments (rainforest) and dry environments (eucalypt woodland and other relatively arid vegetation formations). This view is reinforced in the Torresian Subregion where three of the five endemic species of macropodoids are restricted to the limited rainforest habitat at the base of Cape York, while the remaining two are restricted to rock habitats in open woodlands. Additional support is provided by the fact that both the Tasmanian and Papuan Subregions experienced similar periods of connection to mainland Australia during intervals of lowered sea levels, yet no species endemic to the Tasmanian Subregion evolved (unless one counts *Thylogale billardierii*).

The Timorean Subregion is unique in that the four species endemic to this subregion (three species of *Petrogale* and *Macropus bernardus*) are all restricted to rocky habitats. It is also noteworthy that the two nonrainforest endemic species in the Torresian subregion are also rock-adapted species (*Petrogale godmani* and *P. assimilis*). An additional species (*M. antilopinus*) is restricted to both the Timorean and Torresion Subregions and *Onychogalea unguifera* is primarily restricted to these two subregions but also extends into the periphery of the Eyrean Subregion.

The Kosciuskan Subregion covers the higher rainfall eastern and southeastern coasts from Townsville to Adelaide and is topographically quite diverse. Of the species endemic to this subregion, two (*Petrogale persephone* and *Potorous longipes*) have extremely restricted distributional ranges and are regarded as relictual species. As mentioned above, none of the five species of macropodoids found in the Tasmanian Subregion are endemic to that subregion as they have all been recorded from the Kosciuskan Subregion. The formerly restricted distribution of *Thylogale billardierii* in coastal Victoria, however, strongly suggests that the species arose in Tasmania and invaded the Kosciuskan Subregion during a period of lowered sea level. The nominate Tasmanian subspecies of *Macropus rufogriseus* has a genetically programmed breeding season in contrast to the continuously breeding Kosciuskan subspecies *(M. r. banksianus)* and is morphologically distinct (Merchant and Calaby 1981). It is found on the Bass Strait islands but has not established on the mainland. This subspecies is postulated to have arisen in Tasmania, possibly at a more recent date than *Thylogale billardierii. Bettongia gaimardi* also has a distribution restricted to both the Tasmanian and Kosciuskan Subregions but there are insufficient data at present to postulate the likely original source for this species.

In the Southwestern Subregion there are three endemic species (*Setonix brachyurus, Macropus irma* and *Potorous platyops*) and the relationships of these lie to the east with the Kosciuskan Subregion. Populations of *Potorus tridactylus* were also present in both of these subregions and the relationships between *M. fuliginosus* and *M. giganteus* also reflects an earlier separation between these subregions.

The remaining subregion, the Eyrean, covers the largest area and includes both arid and semi-arid zones. The high total number of species recorded from the Eyrean Subregion (26) compared to all other subregions reflects the large number of species which have a limited proportion of their total range extending into the semi-arid perimeter of this subregion rather than an inordinate diversity of species within the subregion. This is highlighted in Table 1 which shows the number of species shared between subregions. The Eyrean Subregion shares six and 10 species respectively with the Torresian and Kosciuskan Subregions to the east and eight species with the Southwestern Subregion but only three species with the Timorean Subregion to the north. Within it the presence of eastern and western pairs of species of *Petrogale, Onychogalea* and *Lagorchestes* and the large number of species shared with the eastern coast subregions suggests a possible further sub-division into an eastern, largely semi-arid, area and a western more arid area for macropodoids.

If one excludes the peripheral species for which the distributional range is centred on surrounding subregions, then the Eyrean Subregion is reduced to a core fauna of 15 species which have the majority of their distributional ranges within the subregion, indicating adaptation primarily to arid and semi-arid conditions. Seven of these species are endemic.

There is also a large degree of overlap between the Torresian and Kosciuskan Subregions (eight species) but this is primarily due to the northward extension of species whose distribution is centred to the south or those jointly shared with the Eyrean Subregion to the west. Only two species (*T. stigmatica* and *M. agilis*) are here regarded as northern species which have extended south into the Kosciuskan Subregion. In the case of *M. agilis* there is evidence from the relictual presence on Stradbroke and Peel Islands in Moreton Bay (Martin 1975) and the present southern limit of its range on the mainland near Rockhampton, that it had a more extensive range in the past which contracted northwards in prehistoric times (probably during the last 5 000 years).

Prehistoric contractions in range have also been indicated for the Southwestern *Potorous platyops* which is recorded in fossil deposits from Dongara in Western Australia to Kangaroo Island in South Australia (Butler and Merrilees 1971; Hope *et al.* 1977); the Kosciuskan *Macropus greyi,* formerly recorded from Kangaroo Island to Tasmania (Hope 1984); and *Caloprymnus campestris* from the Eyrean Subregion which previously extended to the coastal areas of the Nullarbor Plain (Lundelius 1963).

NEW GUINEA

Macropodids probably first entered New Guinea from Australia across the Torres Strait region at the end of the Miocene or beginning of the Pliocene. Species of *Dendrolagus, Dorcopsis* and *Thylogale* have been recorded from early Pliocene deposits at Hamilton in Victoria that have been dated at 4.46 million years (Archer and Hand 1984). This locality is also the earliest record of the extinct genus *Protemnodon.* Plane (1967) has reported *Protemnodon* (two species) and *Dorcopsis* (one species) from a late Pliocene New Guinean deposit estimated to be three to five million years old (Archer and Hand 1984). At present, this is the earliest, and only known Tertiary, mammal deposit in New Guinea.

Radiation of both *Dendrolagus* and *Dorcopsis* has occurred within New Guinea to produce six endemic species in each genus (Ziegler 1982; Groves and Flannery 1989). Each of these two genera has produced both lowland and highland species. The moderate radiation of both *Dorcopsis* and *Dendrolagus* in New Guinea suggests a long history in the island and it is significant that *Dendrolagus* has evolved species which have been found to be sympatric at the broad distribution level. However, whereas the lowland species of *Dorcopsis, (D. hageni, D. luctuosa* and *D. muelleri)* are widely and parapatrically distributed over the coastal plains of both the north and the south coasts of New Guinea, the low altitude species of *Dendrolagus* are restricted to the foothills and mountains of the Vogelkop and North-west coastal ranges *(Dendrolagus inustus, Dendrolagus ursinus)* and to a limited area on the southern foothills of the central cordillera between the headwaters of the Fly and Purari Rivers *(Dendrolagus spadix).* The absence of a species of *Dendrolagus* from the extensive lowland forest of the southern trans-Fly area could be due to the lack of topographic relief and extensive swamps and wetlands in this area or may reflect the history of incursions of open forest into this area during periods of lower sea level (Nix and Kalma 1972), the most recent being between about 17 000 and 5 000 years ago. Nevertheless, a period of 5 000 years would appear to be sufficient time for *Dendrolagus spadix* to expand its range into the southern lowland forests if these were a suitable habitat for the species.

Speciation within the New Guinean *Thylogale* has been less marked, with only two endemic species currently recognized (*T. brunii* and the presumably extinct *T. christenseni*), although Tate (1948) recognized morphological differences between southern New Guinea *T. brunii* and two northern New Guinea races (one lowland and one highland race). The pattern of distribution of *Thylogale* is somewhat enigmatic and a review of the taxonomy of *T. brunii* is required to determine whether more than one species has been subsumed under this name. Nevertheless, it would appear from the distribution of this taxon that it may have had a more recent origin in New Guinea than the species of *Dendrolagus* or *Dorcopsis,* perhaps at some time in the Pleistocene.

The remaining species of *Thylogale* in New Guinea *(T. stigmatica)* has a restricted distribution in southern New Guinea opposite Cape York. It is almost certainly the most recent invader from Australia and probably entered New Guinea just prior to the most recent inundation of Torres Strait at about 5 000 years ago.

Macropus agilis, the only other macropodid recorded from New Guinea, is also an immigrant from Australia. Its preferred habitat is along streams and rivers in open woodland and the adjacent grasslands. Nix and Kalma (1972) postulated three different climatic regimes and associated vegetation patterns related to three different lowered sea levels between Australia and New Guinea. Under these postulated regimes, woodland and open forest would have been the predominant vegetation in the areas joining Australia to southern New Guinea and it is reasonable to assume that *M. agilis* would have been widespread in such areas.

Currently *M. agilis* is common in the savannah of the trans-Fly. It is also found in the Port Moresby savannah, along the southeast coast and around to about Popondetta on the north coast. Island populations occur on Kiriwina in the Trobriand Islands and Goodenough, Ferguson and Normanby Islands of the D'Entrecasteaux Group (Brass 1956, 1959). These island populations could be due to introductions by Melanesian man as they were all connected by a traditional exchange system (the Kula Ring). Alternatively, the populations may have become established during a period of lower sea levels when the islands were joined together and through East Cape to the mainland, the most recent connection being about 17 000 to 14 000 years ago, a time which corresponds with the arid extreme of the palaeoclimates postulated by Nix and Kalma (1972).

No member of the family Potoroidae has ever been reported from New Guinea. In fact no recent species of this family has been recorded north of 16°S. This suggests that distribution of species in this family may have been restricted by climatic conditions in the past to areas south of Torres Strait such that no species was able to benefit from the land bridge across Torres Strait when it existed. The one species which would appear to be preadapted to the New Guinea rainforests *(Hypsiprymnodon moschatus)* is unlikely to have successfully crossed the extensive areas of open woodland between rainforest pockets which are postulated to have existed across Torres Strait during periods of lower sea levels (Nix and Kalma 1972), due to its small size and presumed limited ability to disperse through unsuitable habitats.

AUSTRALIA

Within Australia there are 14 genera of macropodoids, six of which are monotypic, with a total of 50 species. In contrast, New Guinea contains four genera, none of which are monotypic, and a total of 15 extant species. The New Guinea species are primarily adapted to rainforest while the majority of Australian species (about 80%) are adapted to habitats with lower rainfall.

All of the genera currently recorded in New Guinea have been recorded in Australian fossil deposits since the Pliocene, and it is reasonable to assume that they evolved in Australia before invading New Guinea to undergo further speciation there. In that sense, New Guinea can be regarded as a refuge for rainforest genera which subsequently were severely restricted or extinguished in Australia. The Australian macropodoid fauna meanwhile proceeded to evolve adaptations to the increasing aridity of the continent. Keast (1961) suggested a variety of refuge areas in Australia which he related to the speciation of birds (Fig. 2). These areas appear to be largely applicable to macropodoids as well.

The extant genera of the Potoroidae, except for *Hypsiprymnodon,* first appear in the fossil record during the Pleistocene (Flannery 1984). Of the ten recent species of this family, the distributional ranges of six are restricted to south of the Tropic of Capricorn (28°28′S) while two (*B. tropica* and *H. moschatus*) are found north of the Tropic and the remaining two species (*A. rufescens* and *B. lesueur)* have ranges which straddle the Tropic of Capricorn. However, none of these species have been recorded north of 16°S. *Potorous* and *Hypsiprymnodon* are both

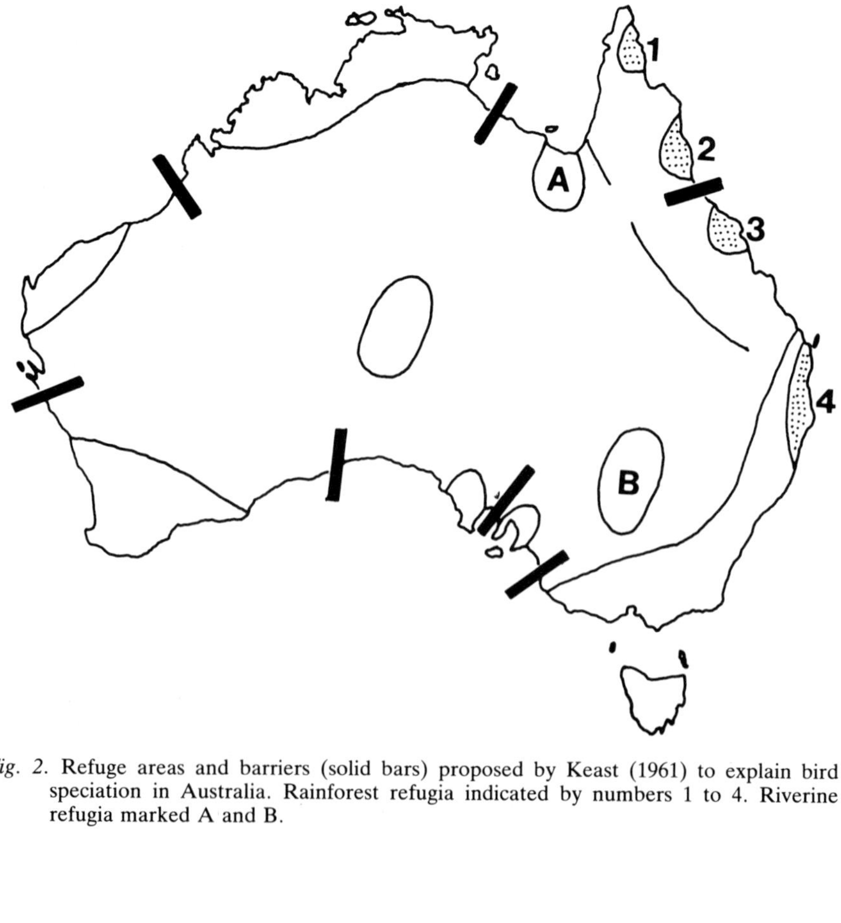

Fig. 2. Refuge areas and barriers (solid bars) proposed by Keast (1961) to explain bird speciation in Australia. Rainforest refugia indicated by numbers 1 to 4. Riverine refugia marked A and B.

restricted to the forested higher rainfall areas of coastal Australia whereas species of other genera of Potoroidae prefer more open habitats, such as woodland, shrubland or grassland (Strahan 1983; Bridgewater 1987).

Within the Australian genera of macropodoids, there is only limited overlap in the ranges of individual congeneric species. In most cases where this ocours, there is a clear separation of the species on size into a smaller and larger species e.g., *Lagorchestes hirsutus* and *L. conspicillatus, Potorous tridactylus* and *Potorous longipes, Potorous platyops* and *Potorous tridactylus, Petrogale concinna* and *Petrogale brachyotis, Petrogale burbidgei* and *Petrogale brachyotis, Dendrolagus lumholtzi* and *D. bennettianus, Macropus bernadus* and *M. robustus* (Strahan 1983). In those cases where congeneric species are similar in size and sympatric on the broad scale, separation occurs at the local level by habitat preference (e.g., *M. parryi* and *M. rufogriseus*) or possibly by diet (e.g., *Petrogale burbidgei* and *P. concinna.)*

The genus *Petrogale* is distributed over the entire continent of Australia but has not penetrated to the Papuan or Tasmanian Subregions despite the connection of both of these subregions to continental Australia by land bridges at various times during the Quaternary. A similar pattern of distribution is also shown by *Macropus robustus* which is adapted to rocky habitats as well. These animals are primarily adapted to open habitats over rocky hills and gorges and their absence from the Papuan Subregion is readily explained by the absence of suitable habitats. Although both *Macropus robustus* and *Petrogale penicillata* are found in the southern Kosciuskan Subregion, both species apparently occurred only as scattered localized populations during historic times and it is likely that similar environmental factors operated to prevent the establishment of populations in the Tasmanian Subregion. These factors could have been either climatic or unsuitable habitats across the Bass Strait Bridge when it existed, or a combination of both.

Earliest records of *Petrogale* are in the Pliocene (Flannery 1984), a period of increasingly drier, more open vegetation compared to the preceding Miocene. In this regard it can be postulated that the earliest *Petrogale* was probably a species of the wet/dry forest margins, perhaps similar in habitat to the present *Petrogale persephone* (Maynes 1982). Briscoe *et al.* (1982) assumed *Petrogale* to have had its origin on the Australian shield as the western equivalent of the genus *Thylogale.* It is possible that *Petrogale* evolved from a rainforest ancestor of *Thylogale* as the rainforest vegetation retreated during the late Miocene. Evolution of the proto-*Petrogale* then opened a new niche for exploitation by macropodids: areas of rocky cliffs and gorges and the associated vegetation, using the rock caves and crevices as an environmental ameliorant and as protection from predators.

Although 11 species of *Petrogale* are currently recognized (Walton 1988), and there are several additional chromosomal races, some of which may warrant separate taxonomic recognition, these are primarily distributed on a parapatric or allopatric basis. In two cases only have species pairs been recorded sympatrically in the same habitat, *P. brachyotis* and *P. burbidgei* (Kitchener and Sanson 1978) and *P. brachyotis* and *P. concinna* (Sanson *et al.* 1985), and these both occur within the Timorean Subregion. Both *P. burbidgei* and *P. concinna* are the same size and have not been recorded from the same island or mainland locality. The closest recorded populations of each species were 15 km apart (Kitchener and Sanson 1978).

Three broad groupings of species can be discerned within the extant species of *Petrogale*. The three species restricted to the Timorean Subregion (*P. brachyotis, P. burbidgei* and *P. concinna*) are more closely related to one another than to other *Petrogale* and are quite distinct chromosomally from other rock wallabies (Briscoe *et al.* 1982). It is presumed that these three species have evolved from a single ancestral form which penetrated the extensive rock habitats of the Timorean Subregion early in the evolutionary history of the genus and subsequently speciated in isolation from other members of the genus. McKenzie *et al.* (1978) reported that both *P. burbidgei* and *P. concinna* have been recorded only from high rainfall areas while *P. brachyotis* occurs in areas with low to medium rainfall and is less common than *P. concinna* in the higher rainfall areas of the Prince Regent River Nature Reserve where the two species are sympatric. Early divergence of the stocks which gave rise to *P. brachyotis* and the *P. burbidgei*/*P. concinna* lineage may therefore have been related to adaptations to different climatic conditions. Subsequent separation of *P. concinna* from *P. burbidgei* by the evolution of continuously erupting molars in the former has permitted *P. concinna* to exploit a highly abrasive diet (Sanson *et al.* 1985) which could only have limited availability to *P. burbidgei.*

The second group of *Petrogale* species includes *P. rothschildi, P. persephone* and *P. xanthopus.* These species have each retained chromosomal complements which are little changed from the presumed ancestral chromosomal type (Briscoe *et al.* 1982; G. B. Sharman and R. L. Close, pers. comm.). *P. persephone* has the most restricted distribution, being found in areas of rocky outcrops covered by rainforest associated with nearby open forest in the vicinity of Proserpine, central Queensland, (i.e., in the Kosciuskan Subregion). The other two species are found in the

Eyrean Subregion. The eastern *P. xanthopus* shows many similarities to *P. persephone* and these two species are considered to be xeric and mesic representatives of a common lineage. This group is considered with the first group to represent the products of an early radiation within the genus.

The remaining taxa forming the *P. lateralis-penicillata* complex constitute the third group of *Petrogale* species. Of these species, the nominate race of *P. lateralis* retains a near ancestral chromosomal complement and may also represent the end-product of an early radiation. This group has representatives in four subregions (Eyrean, Southwestern, Kosciuskan and Torresian) and has undergone considerable evolutionary change within the Kosciuskan and Torresian Subregions where a total of eight recognizable major chromosomal races have been identified (Briscoe *et al.* 1982), ranging from Cape York to Western Victoria. Entry of the *P. lateralis-pencillata* group to the eastern coast subregions is assumed to have occurred via the Euroka Ridge, an elevated area separating the drainage basins of rivers flowing to the Gulf of Carpentaria and those flowing to Lake Eyre (Briscoe *et al.* 1982).

Populations of *P. lateralis* with the presumed near ancestral chromosomal complement have a peripheral distribution from Barrow Island in the west through south-west Western Australia to Pearson Island in South Australia. A reduction in the diploid number from 2n=22 to 2n=20 occurred between about 15 000 and 12 500 years ago in the common land area which contained the present Combe, Mondrain and Wilson Islands of the Recherche Archipelago, to produce *P. lateralis hacketti.* The nominate subspecies with 2n= 22 was retained on Salisbury Island which was isolated at about 15 000 years B.P. Kershaw (1981) considered the period 15 000 to 10 000 years B.P. to represent a transition from the drier conditions of the precedings 10 000 years (which was the driest period in the last 50 000 years) to conditions of higher rainfall between about 7 000 to 5 000 years B.P., followed by a reduction in rainfall in the last few thousand years. Merrilees (1979) reported that *Petrogale lateralis* may have entered the Cape Leeuwin-Cape Naturaliste region of Western Australia at some time prior to 30 000 years B.P. and disappeared from the region at some time after 5 000 years B.P. but well before historic time. He suggested that the disappearance of *Petrogale* from the region may have been due to the replacement of more open vegetation by dense forest in the early Holocene, consequent on a change in climate. A similar change of climate in the area of the Recherche Archipelago may have selected for the fixation of the *hacketti* chromosomal rearrangement.

During a previous period of expansion, *P. lateralis* penetrated to the hills fringing the northern border of the Great Sandy Desert where another distinctive 20-chromosomal race evolved (the West Kimberley race of Briscoe *et al.* 1982). The populations present in Central Australia and northwestern Queensland represent another variation on the 22-chromosomal complement. The wide distribution of this species throughout arid and semi-arid Australia and its presence in historic times on many isolated rock outcrops is evidence of the species' dispersal ability, at least during periods of favourable climatic conditions. An intriguing aspect of this dispersal ability is the presence of *P. lateralis* populations on Depuch and Barrow Islands off the Pilbara coast and on North-west Cape, while *P. rothschildi* is recorded from mainland localities in the Pilbara and islands of the Dampier Archipelago. Further populations of *P. lateralis* occur to the east of the Pilbara in suitable habitat, but those in which the karyotype has been examined are chromosomally distinct from the peripheral populations with near ancestral chromosomal complements.

On the eastern coast, a major centre of speciation in *Petrogale* is centred in the Torresian Subregion of northern Queensland beyond Townsville where five major chromosomal races have been identified (Briscoe *et al.* 1982). Kershaw (1981) reported considerable variation in the vegetation of the Atherton Tablelands over the last 140 000 years based on pollen analysis from Lynch's Crater. He attributed these changes to climatic fluctuations affecting the distribution and nature of forest cover in the area. It is likely that such climatic changes could also have had a significant effect on the distribution of *Petrogale* in northern Queensland and acted as a strong selective force upon novel adaptive chromosomal rearrangements as they arose. Although the strongest evidence of climatic fluctuations is related to the last 140 000 years, Kershaw (1981) reported that deep sea cores through at least the last 800 000 years showed cyclical changes similar to those which were related to changes recorded in the pollen analyses for the more recent time interval. The Quaternary is regarded as a time characterized by frequent large climatic oscillations which must have greatly affected the stability of vegetation communities (Kershaw 1981) and consequently the mammalian communities associated with them.

The advantages of reducing the diploid chromosome number for members of the genus *Petrogale* is evidently greatest in tropical Australia north of 19°30′S. Thus of nine chromosomal races (species or subspecies) north of that latitude, only one has a diploid number of 22, while the remainder have a diploid number of 20 or less. In contrast, eight of nine chromosomal races south of latitude 19°30′S have a diploid number of 22 and only one has a diploid number of 20, that being the recently evolved (12 500-15 000 years B.P.) *P. lateralis hacketti* (Briscoe *et al.* 1982). The large number of chromosomal races found within *Petrogale* and their primarily allopatric or parapatric distribution may be a consequence of the adaptation of populations to localized climatic conditions. The relatively small size of the animals (adult size three to eight kg for all species except *P. burbidgei* and *P. concinna* which are about 1.3 kg) and social organization of the species may be important factors in contributing to the pattern of speciation observed.

Macropus robustus is another species adapted to rock habitats throughout Australia. It has a total distributional range almost as extensive as the total ranges of all species of *Petrogale* combined. In contrast to the considerable speciation which has occurred in *Petrogale, M. robustus* has maintained its coherence as a single species. Variation does occur within the species with up to four subspecies recognized (Richardson and Sharman 1976), but the evidence for two subspecies in Queensland is equivocal and more consistent with clinal variation related to increasing aridity (Maynes pers. obs.; Kirkpatrick pers. comm.). The principal reasons for the lack of speciation in *M. robustus* compared with *Petrogale* may be its larger size (adults up to 46.5 kg male, 25 kg female, Strahan 1983) which facilitates mobility and dispersion, and its more dispersed social organization (Croft 1981). The only other macropodid to occupy rocky habitats, *M. bernardus*, is intermediate in size between *M. robustus* and the majority of *Petrogale* species and has only a limited distribution in Arnhem Land. Insufficient is known of the biology of this species to hypothesize on the reasons for its extremely limited distribution or absence from the Kimberley within the Timorean Subregion.

The two species of macropodoids with the largest individual distributional ranges are members of the genus *Macropus. M. rufus* occurs throughout the entire Eyrean Subregion, its distribution reflecting the interaction of mean annual precipation and mean annual temperature (Caughley *et al.* 1987). An even more extensive distribution is found in *M. robustus* which occurs in all subregions except the Southwestern and Tasmanian. The two other largest macropodoids (*M. giganteus* and *M. fuliginosus*) also have substantial ranges. Areas where rainfall shows little seasonal trend or where summer rainfall exceeds the winter rainfall define the distributional range of *M. giganteus,* whereas *M. fuliginosus* occupies areas of uniform or winter rainfall (Caughley *et al.* 1987). Remaining members of this genus all have distributional ranges centred upon the periphery of the continent with the greatest concentration of species (eight) occurring within the Kosciuskan Subregion. Species in this genus are primarily grazing animals occupying forest, woodland and grassland habitats and the concentration within the Kosciuskan Subregion is a reflection of the variation in topography, vegetation and climate found in this subregion. The genus *Macropus* provides evidence for speciation of eastern and western species from each other following isolation of populations and subsequent climatic changes related to fluctuating sea levels e.g., *M. fuliginosus* and *M. giganteus, M. greyi* and *M. irma,* and *M. dorsalis, M. parma* and *M. eugenii.* There is also evidence of a northern centre of speciation for *M. bernardus, M. antilopinus* and *M. agilis,* based on their individual distributional ranges.

Along the eastern coast, *Wallabia bicolor* occurs throughout both the Torresian and Kosciuskan Subregions as well as extending into the periphery of the Eyrean Subregion. The genus *Thylogale* however, has separated and evolved into three species (*stigmatica, thetis* and *billardierii*) while *M. rufogriseus* has separated at the subspecies level into a mainland subspecies (*M. r. banksianus*) and a Tasmanian and Bass Strait islands subspecies (*M. r. rufogriseus*)

Bass Strait also has acted apparently as an ecological filter as only five species have been recorded from Tasmania while eight species have been reported from the mainland in the Gippsland Lakes region. The species which apparently did not enter Tasmania are *Potorous longipes, Wallabia bicolor* and *Petrogale penicillata* (Norris *et al.* 1983).

ISLAND DISTRIBUTIONS

The distribution of macropodoids on islands was first considered by Main (1961) who examined data on their known distribution on the western and southern coasts of Western Australia and the ecological and climatic implications of those data. Subsequently Main and Yadav (1971) examined a more complete data set for the same region in terms of its implications for conservation and the minimum size for reserves. Hope (1969, 1973) considered the distribution of mammals on the Bass Strait Islands and later Abbott (1980) reviewed the distribution of macropodoids on islands around Australia, hypothesizing that the lower number of islands on the northwestern, northern and eastern coastlines with species recorded on them was due to the activities of Aboriginal man.

To date 24 species of macropodids and potoroids have been recorded from continental islands around Australia and an additional five species of macropodids have been recorded from islands adjacent to New Guinea (Table 2). Two of the five largest species of kangaroos (*Macropus rufus* and *M. antilopinus*) have not been recorded from any islands while *M. fuliginosus* is found

Table 2. Number of species of each genus recorded on islands around Australia and New Guinea (including introductions).

Genus	Number of Species	Number recorded on islands
Potoroidae		
Aepyprymnus	1	0
Bettongia	4	3
Caloprymnus	1	0
Hypsiprymnodon	1	0
Potorous	3	1
	10	4
Macropodidae		
Dendrolagus	8	2
Dorcopsis	6	2
Lagorchestes	4	2
Lagostrophus	1	1
Macropus	14	6
Onychogalea	3	0
Petrogale	11	8
Setonix	1	1
Thylogale	5	2
Wallabia	1	1
	54	25

only on 445 100 ha Kangaroo Island (S.A.) and 240 ha Woody Island (W.A.), the latter no doubt an unrecorded recent introduction. *Macropus giganteus* has been recorded from Tasmania (6 792 785 ha), Stradbroke Island (27 520 ha), Fraser Island (162 262 ha), and Little Snake Island (600 ha). These last three islands are all separated from the mainland by narrow channels and have probably only recently separated from the mainland. It is perhaps significant that, although *M. giganteus* is recorded on both sides of Bass Strait, this species has been found in fossil deposits greater than 8 000 years B.P. only on Flinders Island (132 800 ha), the largest of the Bass Strait Islands (Hope 1973). It is also notable that the known island populations of *M. robustus* appear to be dwarf versions of the mainland populations.

The inland distributional ranges of seven Australian species and six New Guinean species precludes their presence on islands but notably absent from islands are the four rainforest species of the Torresian Subregion.

The minimum size of an island needed to maintain a population of a small macropod species appears to be between 100 and 200 hectares (Table 3). In almost all cases where more than one species occurs on an island there is a clear separation in size between the species involved (Table 4). Habitat preference separates the open forest *M. agilis* from *W. bicolor* on Stradbroke Island where the two species are found together with *M. giganteus*. Differing individual responses to habitat variation are also indicated for the *B. lesueur, L. hirsutus, L. fasciatus* community on Bernier and Dorre Islands (Ride *et al.* 1962).

ISLAND DWARFING

Reduction in size of animals in island populations has occurred in several species of macropodoids. Richardson and Sharman (1976) report

Table 3. Presence of macropodoids in island groups related to size of island.

Furneaux Group Tasmania

Size (ha)	0-50	51-100	101-200	201-400	401-800	801-1 600	>1 601
Absent	27	8	5	1	0	0	0
Present	0	0	2	4	4	2	3

Recherche Archipelago WA 121°3′E to 124°10′E

Size (ha)	0-50	51-100	101-200	201-400	401-800	801-1 600	1 601-3 200
Absent	60	18	10	3	0	0	0
Present	0	1	1	3	0	2	0

Dampier Archipelago WA

Size (ha)	0-50	51-100	101-200	201-400	401-800	801-1 600	1 600-3 200
Absent	15	5	1	3	0	4	1
Present	0	0	0	0	0	1	3

Kimberley Islands WA 123°E to 127°E

Size (ha)	0-50	51-100	101-200	201-400	401-800	801-1 600	1 600-3 200	3 200-6 400	6 401-12 800	>12 801
Absent	146	19	29	35	13	9	5	1	0	0
Present	0	0	0	0	1	1	1	3	0	2
Number Surveyed	4	1	0	5	3	7	3	4	0	2

Wessel Islands NT

Size (ha)	0-50	51-100	101-200	201-400	401-800	801-1 600	1 600-3 200	3 200-6 400	6 401-12 800
Absent	0	0	0	1	0	0	0	0	0
Present	0	0	1	2	2	1	0	1	1

Species present:			
	Furneaux Group	*T. billardierii* (7.0,3.9 kg)	*M. rufogriseus* (19.7,14.0 kg)
	Recherche Archipelago	*P. lateralis* (4.5, 3.4 kg)	*M. eugenii* (7.5,5.5 kg)
	Dampier Archipelago	*P. rothschildi* (3.8, 2.6 kg)	*M. robustus* (>15 kg)
	Kimberley Islands	*P. burbidgei* (1.3 kg)	*P. concinna* (1.3 kg)
	Wessel Islands	*P. brachyotis* (<4.5 kg)	

Table 4. Natural communities of macropodoid species found on islands showing mean adult weights (male, female). Based on records held in ANPWS Compisle Database of Macropodoids on Islands, weights primarily from Strahan (1983).

2. Species		
T. billardierii (7, 3.9 kg)	*M. rufogriseus* (19.7, 14 kg)	
P. tridactylus (1.2, 1.0 kg)	*T. billardierii* (7, 3.9 kg)	
B. lesueur (1.2 kg)	*L. conspicillatus* (3.5 kg)	
P. tridactylus (1.2, 1.0 kg)	*W. bicolor* (17, 13 kg)	
P. brachyotis (3.6, 2.6 kg)	*M. agilis* (19, 11 kg)	
P. rothschildi (3.8, 2.6 kg)	*M. robustus* (>15 kg)	
M. eugenii (7.5, 5.5 kg)	*M. fuliginosus* (50, 25 kg)	
3. Species		
P. tridactylus (1.2, 1.0 kg)	*T. billardierii* (7, 3.9 kg)	*M. rufogriseus* (19.7, 14 kg)
M. giganteus (50, 25 kg)	*M. agilis* (19, 11 kg)	*W. bicolor* (17, 13 kg)
B. lesueur (1.2 kg)	*L. fasciatus* (1.75 kg)	*L. hirsutus* (1.6, 1.7 kg)
4. Species		
B. lesueur (1.2 kg); *L. conspicillatus* (3.5 kg); *P. lateralis* (4.5 kg); and *M. robustus* (>15 kg)		
5 Species		
P. tridactylus (1.2, 1.0 kg); *B. gaimardi* (1.7 kg); *T. billardierii* (7, 3.9 kg); *M. rufogriseus* (19.7, 14 kg); and *M. giganteus* (50, 25 kg).		

the Barrow Island population of *M. robustus* to be a miniature version of the adjacent mainland type. Main and Yadav (1971) similarly reported *M. robustus* from Barrow Island to be considerably reduced in size and suggested from sight records that the Dolphin Island population may be similarly reduced in size.

Rosemary and Enderby Island populations of *Petrogale rothschildi* are both reduced in size when compared to mainland populations of this species (Maynes, unpublished data) but data are not available on the size of the animals on Dolphin or Lewis Islands. Based on preliminary studies of *Thylogale billardierii,* Hope (1969) suggested that animals in the extinct Victorian population were larger than those in populations in Tasmania or on the Bass Strait islands and that the smallest-sized animals were those found on the smallest islands of the Furneaux Group. Hope (1969) also suggested that *M. rufogriseus* varied in size on the islands of Bass Strait.

Burbidge (1983) reports considerable variation in the size of different populations of *Bettongia lesueur* with the smallest individuals being recorded from Barrow Island. Populations of *Potorous tridactylus* from Flinders Island and Clarke Island were smaller than those found in the adjacent northeastern Tasmania but similar in size to King Island and northwestern Tasmanian populations (Hope 1969; Johnston and Sharman 1976).

The population of *M. parma* on Kawau Island where it was introduced about 1870 (Wodzicki and Flux 1967a), occurs at a much higher density than any population encountered in Australia. Females on Kawau Island are significantly smaller than those in Australia although males from Australia and Kawau Island are similar in size (Maynes 1977). This difference in size between females presumably arose during the 100 years since founding of the Kawau Island population and may reflect selection due to the effects of nutritional stress. It is possible that this situation reflects an early stage in the process of island dwarfing.

Detailed studies of the relative sizes of animals from island and mainland populations of other species of macropodoids have yet to be made but it is possible that such studies may find other evidence for island dwarfing in this taxon.

INTRODUCTIONS OF MACROPODOIDS

Feral populations of macropodids have been established by European man in Great Britain (Yalden and Hosey 1971), New Zealand (Wodzicki and Flux 1967a, 1967b), Hawaii (Tomich 1969; Kramer 1971), and Germany (Grzimek 1967; Boettger 1943). It is likely that Melanesians have introduced *Thylogale brunii* to the Bismarck Archipelago (Umboi, Bagabag and New Britain Islands), *Dendrolagus matschiei* to Umboi Island, and possibly *Thylogale brunii* to the Kei Islands.

In Great Britain two separate feral populations of *Macropus rufogriseus rufogriseus* have been reported (Taylor-Page 1970; Yalden and Hosey 1971). The population in the Peak District is reported to be derived from five animals which escaped during the early years of World War II (Yalden and Hosey 1971).

An Hawaiian population of *Petrogale penicillata* has become successfully established on the island of Oahu from a single pair which escaped on 19 August, 1916 (Tinker 1938). These animals were purchased from a collection of animals which was on its way to the United States mainland from Taronga Zoological Park in Sydney. The population is currently restricted to the lower Ko'olau Range on rocky slopes between 90 and 425 m (Gilmore 1977). In 1981 the population was estimated to be about 250 animals but this was reduced, as a result of drought conditions in 1983 and 1984, to about 100 animals by 1987 (Lazell 1987). Details of the founding of this population are given in Tomich (1969) and Kramer (1971). It is fortuitous that the founding animals reached suitable habitat and shelter close to the original site from which they escaped.

In New Zealand five species of macropodids (*Macropus eugenii, M. parma, M. rufogriseus, Petrogale penicillata* and *Wallabia bicolor*) have been introduced and successfully established feral populations. The claims for the presence of a sixth species *(Macropus dorsalis)* have not been substantiated and appear to be based primarily on early misidentification of specimens of *Macropus parma* from Kawau Island (Maynes 1977). Wodzicki and Flux (1967b) published measurements of a reputed *M. dorsalis* female killed on Kawau Island in 1954 during poisoning operations. These measurements are clearly outside the values obtained for *M. dorsalis* females in Australia but do correspond with the values of *M. rufogriseus*. Additionally, Wodzicki and Flux (1967b) reported that local residents on Kawau Island did not recognize a description of *M. dorsalis* nor did they find skulls of *M. dorsalis* in a collection from that island. It is concluded that the claim for the presence of *M. dorsalis* in New Zealand is not supported by the evidence but rather suggests that the fifth species of macropodid on Kawau Island was *M. rufogriseus*. The origins of the species on Kawau Island are not known but probably are South Australia for *M. eugenii* and New South Wales for *M. parma, W. bicolor* and *Petrogale penicillata,* with the most likely area for the latter three species being either the Illawarra district or the Wattagan Mountains near Gosford.

Populations of *M. eugenii* near Rotorua on the North Island of New Zealand were apparently introduced about 1912 (Wodzicki and Flux 1967b) most likely derived from Kawau Island animals, although this is not known with certainty. The remaining species in New Zealand *(M. rufogriseus)* was introduced to the Waimate area of the South Island in 1874 and quickly spread into the Hunter Hills which is now the stronghold of the species. A further introduction occurred about 1948 on Mount Burke Station between Lakes Hawea and Wanaka but attempts have been made to exterminate the colony (Wodzicki and Flux 1967b).

Petrogale penicillata was also introduced to Motutapu Island (1 200 ha) at the entrance to Waitemata Harbour, Auckland from which it crossed to the adjoining Rangitoto Island (1 600 ha) by a narrow neck of land (Wodzicki and Flux 1967b). Gilmore (1977) reported that development on Motutapu has now restricted the species to about 20 ha of bluffs on the east of the island while the animals on Rangitoto are kept in check by periodic government control operations and persistent trapping by government caretakers.

Grzimek (1967) reported that small populations of wallabies which were established on the Channel Islands of Great Britain and in Rhineland and Silesia (Germany) during the 19th Century were all hunted to extinction by the end of the First World War (1914-1918). Boettger (1943) similarly reported the extermination of a population of wallabies which were released near Hamburg in 1940.

In the New Guinea region wallabies have been recorded off the Sahul Shelf only on Kei Island (*Thylogale brunii brunii,* Thomas 1888; Tate 1948), the Bismarck Archipelago (*Thylogale brunii brunii,* Thomas 1888; Tate 1948; Koopman 1979) and Umboi Island (*Dendrolagus matschiei,* Koopman 1979). Although Umboi Island and New Britain would have joined at times of minimal sea levels (−200 m), the conjoined area would have been separated from mainland Sahulland by the Vitiaz Strait which would have been about 40 km wide at the time. Similarly the Kei Islands would have been separated from Sahulland by about 80 km of deep water. Given that Melanesian man is known to have transported pigs and dogs into New Guinea and that young wallabies are occasionally kept as pets or offered for sale in markets in Papua New Guinea, it is not unreasonable to postulate that these records of wallabies on islands off the Sahul Shelf may be the result of introduction by Melanesian man. *Macropus agilis* is known to have been introduced to Baniara Island in Milne Bay Province, Papua New Guinea in recent times (Seri, pers. comm.) and the presence of this species on Kiriwina Island and Goodenough and Ferguson islands of the D'Entrecasteaux Group has been postulated by Brass (1956, 1959) as due to the action of man. These latter three islands would have been joined to each other and to the mainland via East Cape and Normanby Island about 17 000 to 14 000 years B.P., although Nix and Kalma (1972) postulate that the area would have been covered by closed forest. It is nevertheless possible that the presence of *M. agilis* on these islands may date from around this period although the more recent action of man cannot be ruled out.

Within Australia, a number of introductions and re-introductions of macropodoids to islands has been attempted by fishermen and wildlife authorities with varying degrees of success. Unfortunately, the opportunities which these introductions have provided to investigate the conditions operating during the initial stages of establishment of a population in the absence of mammalian predators and to identify potential factors limiting population numbers or to investigate these factors experimentally, have largely been lost due to the shortage of funds and manpower committed to the detailed studies required.

Delroy *et al.* (1986) give the only published account of a study which attempts to address some of these considerations and proposes continued monitoring of the populations involved.

Niven (1970, 1971) used population models based upon biological data for the quokka *(Setonix brachyurus)* to investigate the effects of stochastic events on the establishment of populations of "pseudo-quokkas". His studies suggested that the successful establishment of a population of animals may require a founding population of up to 200 animals and that it may take 20 or more years before a population becomes successfully established.

Nevertheless, the successful introduction of some of the above populations is known to have resulted from minimal numbers. The population of *M. rufogriseus* established in the Waimate area of New Zealand is attributed to the release of one male and two females in 1874 (Wodzicki and Flux 1967b). The *M. rufogriseus* established in the Peak District of Great Britain is known to be derived from a single captive pair which had increased to a total of five animals in 1939 immediately prior to their escape (Yalden and Hosey 1971). The Hawaiian population of *Petrogale penicillata* was founded by a single pair (Tinker 1938; Tomich 1969; Kramer 1971). Following an examination of genetic variation in Kawau Island and Australian populations of *Macropus parma,* I have hypothesized that the founding population in New Zealand may have been less than five animals (Maynes 1976).

Despite the restricted sample of genetic variability inherent in such small founding populations it is evident that some species of macropodoids can successfully establish populations in alien environments and maintain these in spite of likely predation by dogs and cats (Kawau Island) and foxes (Great Britain) and control operations by man (New Zealand).

CONCLUSIONS

The concept of zoogeographic subregions within the Australian region has been criticized by some authors as having limited value in understanding speciation events (e.g., Keast 1961) but considered by others (e.g., Nix 1986) to have heuristic value. The generally recognized Australian subregions are useful for an initial assessment of the distribution of macropodoids but it is evident that the boundaries of most subregions are not defined by the concurrence of the individual distributional ranges of species endemic to the subregion. Considerable interdigitation occurs between all subregions except the Papuan (Table 1), which is characterized by high endemism (13 species from a total fauna of 15 macropodids). The striking difference between the Papuan Subregion and the remaining Australian Subregions is due to the adaptation of the Papuan macropodids primarily to rainforest environments compared with the extremely limited extent of rainforest in Australia. When considered at the generic level, all of the Papuan genera have been recorded from the Australian fossil record and it is presumed that the modern radiation of macropodids within New Guinea is derived from an early invasion of *Dendrolagus* and *Dorcopsis,* probably in the late Miocene or early Pliocene, followed by a later intrusion of *Thylogale.* Of the two species shared by the Papuan Subregion with the Torresian Subregion (*M. agilis* and *T. stigmatica*), the former species probably had extensive contact during the Pleistocene but the latter is presumed to have entered New Guinea only just prior to the most recent inundation of Torres Strait at about 5 000 years B.P.

The high number of Papuan endemics is a reflection of the dramatic topographic variation in New Guinea. The island is divided into northern and southern components by a central cordillera running from east to west where many of the mountains have peaks at over 4 000 m. However, at any individual location within New Guinea, the maximum number of macropodoid species likely to be encountered is only four, while two or three species is the more usual number. This level of species diversity is comparable to that found in the remnant tropical rainforests at the base of Cape York Peninsula in northern Australia.

The Timorean Subregion is unique in that all four species endemic to this subregion are adapted to rock habitats. This may be a reflection of the extensive amount of such habitats in the subregion and their isolation from similar habitats in adjacent subregions. A high degree of habitat partitioning is suggested by the division of the endemic species into three distinct size categories with mean adult body weights of about 1.3 kg, 4.5 kg and 17 kg, and further separation of the two smallest species by the evolution of a unique process of continuous molar eruption in *P. concinna.* An additional size category is provided by the ubiquitous *M. robustus* which is also adapted to rock habitats.

Freeland *et al.* (1988) have argued that the high diversity of rock-adapted mammals in the seasonally dry tropics is a consequence of patterns of moisture availabiliby in the tropical environment that result in water accumulation near rocks and a more diverse and productive flora. They considered the diversity and high level of endemicity in the rock faunas of northern and northwestern Australia to be a consequence of the seasonally dry climate, rather than any unusual opportunity for speciation. Comparison of the rock-adapted macropodid fauna in the two subregions of the seasonally dry tropics (Torresian and Timorean Subregions) tends not to support this view. In the Torresian Subregion, one is likely to encounter

Macropus robustus and one of two species of *Petrogale* at any specific rock site. Two of the other size categories of rock-adapted macropodids (1.3 kg and 17 kg) are not represented in the Torresian Subregion or elsewhere in Australia. Given the large number of islands off the Kimberley coast (71 greater than 200 ha or 100 greater than 100 ha in size, Table 3), the restriction of both species of rock-adapted macropodids in the smallest size category (1.3 kg) to the peripheral, higher rainfall areas of the Timorean Subregion (including some islands of the Kimberley), and the fact that island dwarfing is known for some macropodids, it may be postulated that the *P. burbidgei-concinna* lineage arose as a dwarf population of *Petrogale* isolated on one of the Kimberley Islands. Separation of *P. burbidgei* and *P. concinna* may have occurred during one of the climatic fluctuations of the Quaternary. The restricted distribution of *Macropus bernardus* [the intermediate size category (17 kg) of rock-adapted macropodid] to a limited area of western Arnhem Land and the greater proportion of rock-adapted macropodoids as a proportion of the total macropodoid fauna in the Timorean Subregion compared with other subregions, suggests that some factor other than greater plant diversity has resulted in the increased diversity of rock-adapted macropodoids in localized areas within the Timorean Subregion. A detailed study of habitat partitioning of the rock-adapted macropodoids of the north Kimberley and western Arnhem Land and a comparison with the Torresian rock-adapted species would clarify this issue.

Species shared between the Timorean and Torresian Subregions are those primarily adapted to the open habitats (woodlands and grasslands) of the plains and riverine edges. These species (*M. antilopinus, M. agilis* and *Onychogalea unguifera*) also show a clear division into three separate size categories with mean adult body weights of about 30 kg, 15 kg and 6.5 kg respectively.

Endemic species in the Torresian Subregion clearly separate into rock-adapted species (two species of *Petrogale*) and rainforest-adapted species (two species of *Dendrolagus* and *Hypsiprymnodon moschatus*). The Torresian Subregion has been a centre of speciation for the genus *Petrogale* since the two recognized species in this subregion include five distinctive chromosomal races which are distributed parapatrically. This concentration of localized forms is attributed to the selection for localized adaptive chromosomal rearrangements during the rapid climatic and vegetation fluctuations in the subregion during the Quaternary, especially the last 140 000 years. The rainforest endemics are considered to be relictual remnants of the rainforest-adapted communities from the Miocene and Pliocene.

High rainfall and topographic relief characterize the Kosciuskan Subregion and result in a diversity of habitats ranging from grassland through shrublands and open forest to rainforest (Bridgewater 1987). This complex of environments is reflected in the diversity of macropodoid species occurring in the subregion. The highest diversity of macropodoids from a restricted area (11 species) was recorded in this subregion from the Upper Richmond and Clarence River study area of Calaby (1966). Eight species of *Macropus* are recorded from this subregion and for six of these species (two of which are endemic), the subregion is central to their distributional ranges. Sequential replacement of three parapatric chromosomal races of the *Petrogale lateralis-penicillata* complex occurs from north to south in the subregion, while one other endemic and relictual *Petrogale (P. persephone)* is restricted to the interface between rainforest and open forest in limited areas near Proserpine. The relictual *Potorous longipes* also occurs in the Kosciuskan Subregion and both it and *Petrogale persephone* may represent the end products of earlier radiations during more mesic times. Both *Macropus parma* and *Thylogale thetis,* two of the other endemic species, are also restricted to the higher rainfall areas of the subregion. Another feature of this subregion is that seven of the 10 species common to both the Kosciuskan and Eyrean Subregions extend into only the eastern periphery of the Eyrean Subregion.

The relationship of the Southwestern Subregion fauna to that of the Kosciuskan Subregion suggests that contact between the two subregions across the Nullarbor Plain and Great Australian Bight may have occurred on a number of occasions during the past. *Setonix* is considered to be an early isolate from a *Thylogale* type ancestor. Later contacts are suggested by the *Macropus irma-Macropus greyii* pair of species and the presence of *Potorous tridactylus* in both subregions. *Macropus fuliginosus* is believed to have evolved from a common ancestor to *Macropus giganteus* when isolated in the southwest, then to have expanded its range across the Nullarbor Plain eventually to become sympatric with *M. giganteus* in a broad arc through northern Victoria, western New South Wales and south-western Queensland. *Potorous platyops* is clearly related to the other *Potorous* species of the Kosciuskan Subregion and presumably speciated in the south-west prior to *Potorous tridactylus* reaching the subregion. Insufficient is known of the biology of *P. platyops* to speculate on the possible interactions of the two species of *Potorous* in the south-west.

The Tasmanian Subregion is basically a depauperate extension of the Kosciuskan Subregion with regard to macropodoids. However, the presence of an endemic subspecies of

M. rufogriseus in the Tasmanian Subregion and the former limited distribution of *Thylogale billardierii* in the Kosciuskan Subregion suggest that there is some validity to considering it as a separate subregion.

The Eyrean Subregion covers the remainder of Australia and contains a fauna which has adapted to the arid and semi-arid habitats which expanded concurrently with the decline in rainforest habitats from the Miocene. The subregion formerly contained seven endemics (two of which are *Petrogale* species) from a total fauna of 26 species, 15 of which were adapted primarily to arid or semi-arid conditions.

Examination of the distribution patterns for the monospecific genera of macropodoids indicates that five genera are located in the peripheral eastern and southwestern higher-rainfall areas, while only one *(Caloprymnus)* occurs in the arid interior. Greatest species richness occurs in those genera which were first recorded in the fossil record during the Pliocene (*Dendrolagus, Dorcopsis, Macropus, Petrogale* and *Thylogale,* Flannery 1984).

Potoroids are absent from the tropics north of 16°S and the majority of species (eight) have part or all of their distribution south of the Tropic of Capricorn (23°28′S). There is also a relationship between latitude and diploid chromosome number in the genus *Petrogale* where eight of nine chromosomal races north of 19°S have a diploid number of 20 or less, while only one of nine chromosomal races south of that latitude has a diploid number of 20, the remaining eight having a diploid number of 22.

Congeric species of the smaller macropodoids generally have an allopatric or parapatric distribution pattern. Where species do occur sympatrically, there is usually a marked difference in size between the two species. This pattern is demonstrated most notably in the genus *Petrogale* but also is found in *Onychogalea, Thylogale, Lagorchestes, Dorcopsis, Dendrolagus, Potorous* and to a lesser extent in *Bettongia.* These genera primarily reflect differing herbivore specializations while the individual species within each genus are most likely a reflection of adaptations to specific climatic regimes.

Macropus robustus, a large rock-adapted macropodid, occupies a total distributional range almost as extensive as the combined ranges of the 11 species in the genus *Petrogale,* a group of small (1.3-8 kg) rock-adapted macropodids. Numerous localized chromosomal races are found amongst the species of *Petrogale,* yet *M. robustus* shows only limited intraspecific variation. This striking difference in response to essentially the same climatic conditions is attributed to the difference in size (and consequently mobility) and the highly social organization of *Petrogale* colonies compared with the more dispersed social organization of *M. robustus.* These differences would facilitate fixation of locally adaptive gene combinations in *Petrogale* populations whereas the dispersal of less adaptive gene combinations from other populations of *M. robustus* would tend to prevent the development of a comparable degree of local adaptation in *M. robustus.*

Island distributions of macropodoids are marked by the absence or limited presence of the five largest macropodoids on islands, the high incidence of *Petrogale* species and the absence of Torresian rainforest species. Where more than one species of macropodoid occurs on an island, the partitioning of resources in the community appears to be correlated primarily with the size of the species concerned. The minimum size of an island necessary to support a population of a small macropodoid of sufficient size to withstand stochastic environmental changes in the absence of mammalian predators and interspecific competition over a time interval of 8 000 to 15 000 years appears to be between 100 and 200 ha. In contrast, an area of as much as 445 000 ha may be necessary to ensure the survival of the largest macropodoids over the same time period.

The limits to distribution of macropodoids are set at the broadest scale by climatic factors and at the local level by ecological factors such as habitat preference, predation and competition with other species. A more detailed examination of the broad-scale limits to distribution and biogeographic patterns awaits the application of the BIOCLIM package (Nix 1986) to all species of macropodoids. Caughley *et al.* (1987) have demonstrated the value of BIOCLIM in analysing the distributions of the three largest macropodids to determine their climatic characteristics. They found that the area of sympatry and allopatry between *M. rufus, M. giganteus* and *M. fuliginosus* appeared to be determined by the independent reaction of each species to specific and differing climatic stimuli rather than by biological interaction between the species. Nix (1986) applied a BIOCLIM analysis to distributional data of elapid snakes and determined various biogeographic patterns for species with similar climatic requirements. The advantage of this technique for analysing distributional data lies in its ability to predict distributions of species in relation to postulated palaeoclimates, thereby helping to clarify the influence of historic factors on present distributional pattern.

REFERENCES

ABBOTT, I., 1980. Aboriginal man as an exterminator of wallaby and kangaroo populations on islands around Australia. *Oecologia.* **44:** 347-54.

ALLEN, H., 1983. Nineteenth century faunal change in Western New South Wales and North-West Victoria. *In* Dept. Anthropology University of Auckland Working Papers in Anthropology, Archaelogy Linguistics, Maori Studies. No. **64**: 1-69.

ARCHER, M. AND HAND, S., 1984. Background to the search for Australia's oldest mammals. Pp. 517-66 *in* Vertebrate Zoogeography and Evolution in Australasia ed. by M. Archer and G. Clayton. Hesperian Press: Western Australia.

ATKINSON, E. D., 1890. Notes of a short trip to the islands of Western Bass Straits. *Vic. Nat.* **6**: 156-64.

BARRY, D. H. AND CAMPBELL, P. R., 1977. A survey of the mammals and herptiles of Fraser Island, with comments on the Cooloola Peninsula, North Stradbroke, Moreton and Bribie Islands. *Occas. Pap. Anthropology Univ. Qld.* No. **8**: 147-78.

BOETTGER, C. R., 1943. Kanguruhs in Deutschland in Freier Wildbahn. *Natur. Volk.* **73**: 331-36.

BRASS, L. J., 1956. Results of the Archbold Expeditions. No. 75. Summary of the Fourth Archbold Expedition to New Guinea (1953). *Bull. Amer. Mus. Nat. Hist.* **102**: 135-206.

BRASS, L. J., 1959. Results of the Archbold Expeditions. No. 76. Summary of the Fifth Archbold Expedition to New Guinea (1956-1957). *Bull. Amer. Mus. Nat. Hist.* **118**: 1-70.

BRIDGEWATER, P. B., 1987. The present Australian environment — terrestrial and freshwater. Pp. 69-100 *in* Fauna of Australia ed. by G. R. Dyne and D. W. Walton. Vol 1A General Articles AGPS, Canberra.

BRISCOE, D. L., CALABY, J. H., CLOSE, R. L., MAYNES, G. M., MURTAGH, C. E. AND SHARMAN, G. B., 1982. Isolation, introgression and genetic variation in Rock-wallabies. Pp. 73-87 *in* Species at risk: research in Australia ed. by R. H. Groves and W. D. L. Ride. Australian Academy of Science, Canberra.

BROWNRIGG, M. B., 1872. The cruise of the Freak — Launceston.

BULMER, S., 1982. Human ecology and cultural variation in prehistoric New Guinea. Vol 1. Pp. 169-206 *in* Biogeography and Ecology of Papua New Guinea ed. by J. L. Gressitt. Dr W. Junk: The Hague.

BURBIDGE, A. A., 1983. Burrowing Bettong. Pp. 187-89 *in* The Complete Book of Australian Mammals ed. by R. Strahan. Angus & Robertson: Sydney.

BURBIDGE, A. A. AND GEORGE, A. S., 1978. The flora and fauna of Dirk Hartog Island, Western Australia. *J. Roy. Soc. West. Aust.* **60**: 71-90.

BURBIDGE, A. A. AND PRINCE, R. I. T., 1971. The fauna, flora and planned useage of the Dampier Archipelago. *Dept. Fish. Fauna West. Aust.* Report No. 11.

BUTLER, W. H. AND MERRILEES, D., 1971. Remains of *Potorous platyops* (Marsupialia, Macropodidae) and other mammals from Bremer Bay, Western Australia. *J. Roy. Soc. W.A.* **54**: 53-58.

CALABY, J. H., 1966. Mammals of the Upper Richmond and Clarence Rivers, New South Wales. *C.S.I.R.O. Div. Wildl. Res.* Tech. Paper No. 10.

CALABY, J. H., 1971. Man, fauna and climate in Aboriginal Australia. Pp. 80-93 *in* Aboriginal man and environment in Australia ed. by D. J. Mulvaney and J. Golson. ANU Press: Canberra.

CAUGHLEY, G., SHORT, J., GRIGG, G. C. AND NIX, H., 1987. Kangaroos and climate: an analysis of distribution. *J. Anim. Ecol.* **56**: 751-61.

CHAPELL, J. M. A. AND THOM, B. G., 1977. Sea levels and coasts. Pp.279-91 *in* Sunda and Sahul. Prehistoric Studies in South East Asia, Melanesia and Australia ed. by J. Allen *et al.* Academic Press: London.

COLLETT, R., 1897. On a collection of mammals from North and North-west Australia. *Proc. Zool. Soc. Lond.* **1897**: 317-36.

CROFT, D. B., 1981. Social behaviour of the Euro *Macropus robustus* (Gould) in the Australian arid zone. *Aust. Wildl. Res.* **8**(1): 13-49.

CROOK, K. A. W., 1981. The break up of the Australian Antarctic segment of Gondwanaland. *In* Ecological Biogeography of Australia ed. by A Keast. Dr W. Junk: The Hague.

DAHL, K., 1926. In Savage Australia. An account of a hunting and collecting expedition to Arnhem Land and Dampier Land. Philip Allen: London.

DELROY, L. B., EARL, J., RADBONE, I., ROBINSON, A. C. AND HAWETT, M., 1972. The breeding and re-establishment of the Brush-tailed Bettong, *Bettongia penicillata*, in South Australia. *Aust. Wildl. Res.* **13**: 387-96.

DIXON, J. M. AND HUXLEY, L. (eds), 1985. Donald Thompson's Mammals and Fishes of Northern Australia. Nelson: Melbourne Pp. i-xi, 1-210.

DOUTCH, H. F., 1972. The palaeogeography of northern Australia and New Guinea and its relevance to the Torres Strait area. Pp. 1-10 *in* Bridge and Barrier: the natural and cultural history of Torres Strait ed. by D. Walker. Dept. Biogeog. and Geomorph. Publ. BG/3 ANU, Canberra.

EMISON, W. B., PORTER, J. W., NORRIS, K. C. AND APPS, G. J., 1978. Survey of the vertebrate fauna in the Grampians — Edenhope area of south-western Victoria. *Mem. Nat. Mus. Vic.* **39**: 281-363.

FILEWOOD, W., 1984. The Torres Connection: zoogeography of New Guinea. Pp. 1121-32 *in* Vertebrate Zoogeography and Evolution in Australasia ed. by M. Archer and G. Clayton. Hesperian Press: Western Australia.

FINLAYSON, H. H., 1930. Observations on the South Australian species of the subgenus *Wallabia* Part II. *Trans. Proc. Roy. Soc. S.A.* **54**: 47-56.

FINLAYSON, H. H., 1932. *Caloprymnus campestris*. Its recurrence and characters. *Trans. Roy. Soc. S.A.* **56**: 148-67. Plates vii, viii, ix.

FINLAYSON, H. H., 1935. The Red Centre. Man and beast in the heart of Australia. Angus & Robertson: Sydney.

FINLAYSON, H. H., 1961. On central Australian mammals. Part IV — The distribution and status of central Australian species. *Rec. S.A. Mus.* **14**: 141-96.

FINLAYSON, H. H., 1938. On a new species of *Potorous* (Marsupialia) from a cave deposit on Kangaroo Island, South Australia. *Trans. Roy. Soc. S.A.* **62**: 132-40.

FLANNERY, T. F., 1984. Kangaroos: 15 million years of Australian bounders. Pp. 817-35 *in* Vertebrate Zoogeography and Evolution in Australasia ed. by M. Archer and G. Clayton. Hesperian Press: Western Australia.

FREELAND, W. J., WINTER, J. W. AND RASKIN, S., 1988. Australian Rock-Mammals: A phenomenon of the seasonally dry tropics. *Biotropica* **20**(1): 70-79.

FRITH, H. J. AND CALABY, J. H., 1974. Fauna survey of the Port Essington District, Cobourg Peninsula, Northern Territory of Australia. *C.S.I.R.O. Div. Wildl. Res.* Tech. Paper No. **28**: 179-208.

GALLOWAY, R. W. AND KEMP, E. M., 1981. Late Cainozoic environments in Australia. Vol 1. Pp. 53-80 *in* Ecological Biogeography of Australia ed. by A. Keast. Dr W. Junk: The Hague.

GILMORE, D. P., 1977. The success of marsupials as introduced species. Pp. 169-78 *in* The Biology of Marsupials ed. by B. Stonehouse and D. P. Gilmore. Macmillen Press Ltd.

GLAUERT, L., 1933. The distribution of the marsupials in Western Australia. *J. Roy. Soc. W.A.* **19**: 17-32.

GOODSELL, J., TINGAY, A. AND TINGAY, S. R., 1976. A resource survey of Woody Island, Archipelago of the Recherche. *Dept. Fish. Wildl. W.A. Rept. No. 21.* Pp. 1-32.

GROVES, C. AND FLANNERY, T. F., This volume. Revision of the genus *Dorcopsis*.

GRZIMEK, B., 1967. Four-legged Australians. Collins: London.

HARRISON, J. L., 1962a. Mammals of Innisfail I. Species and distribution. *Aust. J. Zool.* **10**: 45-83.

HARRISON, J. L., 1962b. The distribution of feeding habits among animals in tropical rainforest. *J. Anim. Ecol.* **31**: 53-63.

HARRISON, J. L., 1963. The food of some Innisfail mammals. *Proc. Roy. Soc. Qld.* **73(3)**: 37-43.

HAYMAN, D. L. AND MARTIN, P. G., 1974. Animal Cytogenetics Vol. 4: Chordata 4. Mammalia 1: Monotremata and Marsupialia ed. by B. John. Gebruder Borntraeger: Berlin, Stuttgart.

HEATWOLE, H., 1987. Major components and distributions of the terrestrial fauna. Pp. 101-35 *in* Fauna of Australia ed. by G. R. Dyne and D. W. Walton. Vol. 1A General Articles, AGPS, Canberra.

HOPE, J. H., 1969. Biogeography of the mammals of the islands of Bass Strait; with an account of variation in the genus *Potorous*. Ph.D. Thesis, Monash University.

HOPE, J. H., 1973. Mammals of the Bass Strait Islands. *Proc. Roy. Soc. Vic.* **85**: 163-95.

HOPE, J. H., 1977. The effect of prehistoric man on the fauna of New Guinea. Pp. 21-27 *in* The Melanesian environment ed. by J. Winslow. ANU Press: Canberra.

HOPE, J. H., 1984. The Australian Quaternary. Pp. 69-81 *in* Vertebrate Zoogeography and Evolution in Australasia ed. by M. Archer and G. Clayton. Hesperian Press: Western Australia.

HOPE, J. F., LAMBERT, R. J., EDMONDSON, E., SMITH, M. J. AND VAN TETS, G. F., 1977. Pleistocene fauna remains from Seton rock shelter, Kangaroo Is. South Australia. *J. Biogeo.* **4**: 363-85.

JENNINGS, J. N., 1971. Sea level changes and land links. Pp. 1-13 *in* Aboriginal man and environment in Australia ed. by D. J. Mulvaney and J. Golson. ANU Press: Canberra.

JOHNSON, D. H., 1964. Mammals of the Arnhem Land Expedition. Vol. 4 (Zoology): 427-515 *in* Records of the American-Australian Scientific Expedition to Arnhem Land ed. by R. L. Specht. Melbourne University Press: Melbourne.

JOHNSTON, P. G. AND SHARMAN, G. B., 1976. Studies on populations of *Potorous* Desmarest (Marsupialia) I. Morphological variation. *Aust. J. Zool.* **23**: 573-88.

JONES, F. W., 1923. The flora and fauna of Nuyts Archipelago and the Investigator Group No. 6. The Didelphian Mammals. *Trans. Roy. Soc. S.A.* **47**: 82-94.

JONES, F. W., 1924a. The flora and fauna of Nuyts Archipelago and the Investigator Group No. 15. The Pearson Island Rat and Flinders Island Wallaby. *Trans. Roy. Soc. S.A.* **48**: 10-14.

JONES, F. W., 1924b. The Mammals of South Australia. Part II. The Bandicoots and Herbivorous Marsupials. Government Printer: Adelaide.

KEAST, A., 1961. Bird speciation on the Australian continent. *Bull. Mus. Comp. Zool.* **123**(8): 305-495.

KEAST, A., 1968. Evolution of mammals on southern continents IV. Australian mammals: zoogeography and evolution. *Q. Rev. Bio.* **43**(4): 373-408.

KEAST, A., 1969. Evolution of mammals on southern continents VII. Comparisons of the contemporary mammalian faunas of the southern continents. *Q. Rev. Biol.* **44**(2): 121-67.

KEAST, A., 1981. Origins and relationships of the Australian biota. Vol. 3: 2000-50 *in* Ecological Biogeography of Australia ed. by A. Keast. W. Junk: The Hague.

KEITH, K., 1968. Mammals of the Pellew Islands. Mimag (Mount Isa Mines Magazine) June 1968.

KEMP, E. M., 1981. Tertiary palaeogeography and the evolution of Australian climate. Vol. 1: 31-49 *in* Ecological Biogeography of Australia ed. by A. Keast. W. Junk: The Hague.

KERSHAW, A. P., 1981. Quaternary vegetation and environments. Vol. 1: 81-101 *in* Ecological Biogeography of Australia ed. by A. Keast. W. Junk: The Hague.

KIKKAWA, J., MONTIETH, G. B. AND INGRAM, G., 1981. Cape York Peninsula: major region of faunal interchange. Vol. 3: 1695-742 *in* Ecological Biogeography of Australia ed. by A. Keast. W. Junk: The Hague.

KITCHENER, D. J., 1978. Mammals of the Ord River Area, Kimberley, Western Australia. *Rec. W.A. Mus.* **6**(2): 189-219.

KITCHENER, D. J., CHAPMAN, A. AND DELL, J., 1975. A biological survey of Cape Le Grand National Park. *Rec. W.A. Mus. Suppl.* No. 1. Pp. 48.

KITCHENER, D. J., CHAPMAN, A., MUIR, B. G. AND PALMER, M., 1980. The conservation value for mammals of reserves in the Western Australian wheatbelt. *Biol. Conserv.* **18**: 179-207.

KITCHENER, D. J., KELLER, L. E., CHAPMAN, A., MCKENZIE, N. L., START, A. N. AND KENEALLY, K. F., 1981. Observations on mammals of the Mitchell Plateau Area, Kimberley, Western Australia. *Rec. W.A. Mus. Suppl.* 7.

KITCHENER, D. J. AND SANSON, G. D., 1978. *Petrogale burbidgei* (Marsupialia: Macropodidae), a new Rock-wallaby from Kimberley, Western Australia. *Rec. W.A. Mus.* **6**(2): 269-85.

KOOPMAN, K. F., 1979. Zoogeography of mammals from islands off the north-eastern coast of New Guinea. *Amer. Mus. Novit.* **2690**: 1-17.

KRAMER, R. J., 1971. Hawaiian Land Mammals. Charles E. Tuttle Co.: Rutland, Vermont and Tokyo.

KREFFT, G., 1864. Catalogue of Mammalia in the Collection of the Australian Museum. Australian Museum, Sydney.

LAZELL, J. D., 1987. Evolution on the hop. BBC Wildlife. Dec. 1987: 666-68.

LE SOUEF, A. S., 1929. Notes on some mammals from the Bass Strait Islands. *Aust. Zool.* **5**: 329-32.

LEWIS, F., 1931. Rock Wallaby in Victoria. *Vic. Nat.* **58**: 120.

LOFFLER, E., 1982. Landforms and landform development. Vol. 1: 57-72 *in* Biogeography and Ecology of Papua New Guinea ed. by J. L. Gressitt. W. Junk: The Hague.

LONNBERG, E. AND MJOBERG, E., 1916. Results of Dr E. Mjoberg's Swedish scientific expeditions to Australia 1910-1913. II Mammals of Queensland. *K. Svenska Vetensk. Acad. Handl.* **52**(2): 1-20.

LUNDELIUS, E. L., 1963. Vertebrate remains from the Nullarbor Caves Western Australia. *J. Roy. Soc. W.A.* **46**: 75-80.

MCKENZIE, N. L., 1981. Mammals of the Phanerozoic Southwest Kimberley. *J. Biogeo.* **8**: 263-80.

MCKENZIE, N. L., BURBIDGE, A. A., CHAPMAN, A. AND YOUNGSON, W. K., 1978. Mammals. *In* The islands of the North-West Kimberley Western Australia ed. by A. A. Burbidge and N. L. McKenzie. *Wildl. Res. Bull.* **7**: 22-28.

MAIN, A. R., 1961. The occurrence of Macropodidae on islands and its climatic and ecological implications. *J. Roy. Soc. W.A.* **44**: 84-89.

MAIN, A. R. AND YADAV, M., 1971. Conservation of macropods in reserves in Western Australia. *Biol. Cons.* **3**: 123-33.

MARLOW, B. J., 1958. Survey of the marsupials of New South Wales. *C.S.I.R.O. Wildl. Res.* **3**: 71-114.

MARSHALL, L. G. AND CORRUCCINI, R. S., 1978. Variability, evolutionary rates, and allometry in dwarfing lineages. *Paleobiology* **4**(2): 101-19.

MARTIN, J. H. D., 1975. A list of mammals from Stradbroke Island. *Proc. Roy. Soc. Qld.* **86**: 73-76.

MAYNES, G. M., 1976. Biology of the Parma Wallaby, *Macropus parma* Waterhouse. Ph.D. Thesis, Macquarie University.

MAYNES, G. M., 1977. Breeding and age structure of the population of *Macropus parma* on Kawau Island, New Zealand. *Aust. J. Ecol.* **2**: 207-14.

MAYNES, G. M., 1982. A new species of Rock-wallaby, *Petrogale persephone* (Marsupialia: Macropodidae), from Proserpine Central Queensland. *Aust. Mammal.* **5**: 47-58.

MENKHORST, P. W. AND GILMORE, A. M., 1979. Mammals and reptiles of North-central Victoria. *Mem. Nat. Mus. Vict.* **40**: 1-34.

MERCHANT, J. C. AND CALABY, J. H., 1981. Reproductive biology of the Rednecked Wallaby *(Macropus rufogriseus banksianus)* and Bennett's Wallaby *(M. r. rufogriseus)* in captivity. *J. Zool. Lond.* **194**: 203-17.

MERRILEES, D., 1979. Prehistoric rock-wallabies (Marsupialia: Macropodidae, *Petrogale*) in the far south-west of Western Australia. *J. Roy. Soc. W.A.* **61**(3): 73-97.

MITCHELL, F. J. AND BEHRNDT, A. C., 1949. Fauna and flora of the Greenly Islands, Pt 1. Introductory narrative and vertebrate fauna. *Rec. S. Aust. Mus.* **9**: 169-79.

NIVEN, B. S., 1970. Mathematics of populations of the Quokka, *Setonix brachyurus* (Marsupialia). I. A simple deterministic model for Quokka populations. *Aust. J. Zool.* **18**: 209-14.

NIVEN, B. S., 1971. Mathematics of populations of the Quokka, *Setonix brachyurus* (Marsupialia). II. A stochastic model for Quokka populations. *Aust. J. Zool.* **19**: 393-99.

NIX, H. A., 1981. The environment of Terra Australis. Vol. 1: 733-95 *in* Ecological Biography of Australia ed. by A. Keast. W. Junk: The Hague.

NIX, H. A., 1986. A biogeographic analysis of Australian Elapid snakes. *In* Atlas of Elapid Snakes of Australia ed. by R. Longmore. Australian Flora and Fauna Series No. 7 AGPS, Canberra.

NIX, H. A. AND KALMA, J. D., 1972. Climate as a dominant control in the biogeography of Northern Australia and New Guinea. Pp. 61-91 *in* Bridge and Barrier: the natural and cultural history of Torres Strait ed. by D. Walker. ANU Press: Canberra.

NORRIS, K. C., MANSERGH, I. M., AHERN, L. D., BELCHER, C. A., TEMBY, I. D. AND WALSH, N. G., 1983. Vertebrate fauna of the Gippsland Lakes catchment, Victoria. Fisheries and Wildlife Division, *Vic. Occ.* Paper Series Number 1. Pp. 1-58.

PARKER, S., 1973. An annotated checklist of the native land mammals of the Northern Territory. *Rec. S.A. Mus.* **16**(1): 1-57.

PLANE, M. D., 1967. Stratigraphy and vertebrate fauna of the Otibanda Formation, New Guinea. *Bull. Bur. Min. Res. Geol. Geophys. Aust.* **86**: 1-64.

QUILTY, P. G., 1977. Cenozoic sedimentation cycles in Western Australia. Geology **5**: 336-40.

RICHARDSON, B. J. AND SHARMAN, G. B., 1976. Biochemical and morphological observations on Wallaroos (Macropodidae: Marsupialia) with a suggested new taxonomy. *J. Zool. Lond.* **179**: 499-513.

RIDE, W. D. L., 1957. *Protemnodon parma* (Waterhouse) and the classification of related wallabies (*Protemnodon, Thylogale* and *Setonix*). *Proc. Zool. Soc. Lond.* **128**: 327-46.

RIDE, W. D. L., CRAWFORD, I. M., STORR, G. M., BERNDT, R. M. AND ROYCE, R. D., 1964. Report on the Aboriginal engravings and flora and fauna of Depuch Island, Western Australia. *W.A. Mus. Spec. Publ. No. 2.*

RIDE, W. D. L., MEES, G. F., DOUGLAS, A. M. AND ROYCE, R. D., 1962. The results of an expedition to Bernier and Dorre Islands, Shark Bay, Western Australia. *W.A. Fisheries Dept.* Fauna Bulletin No. **2**: 1-131.

ROBINSON, A. C., 1980. Notes on the mammals and reptiles of Pearson, Dorothee and Greenly Islands, South Australia. *Trans. Roy. Soc. S.A.* **104**: 93-100.

ROBINSON, A. C. AND SMITH, M. E. B., 1976. The vertebrate fauna of Nuyts Archipelago, South Australia. *Trans. Roy. Soc. S.A.* **100**: 171-76.

SANSON, G. D., NELSON, J. E. AND FELL, P., 1985. Ecology of *Peradorcas concinna* in Arnhem Land in a Wet and Dry season. *Proc. Ecol. Soc. Aust.* **13**: 65-72.

SCHODDE, R. AND CALABY, J. H., 1972. The biogeography of the Australo-papuan bird and mammal faunas in relation to Torres Strait. *In* Bridge and barrier: the natural and cultural history of Torres Strait ed. by D. Walker. Dept. Biogeogr. and Geomorph. Publ. No. BG/3 ANU, Canberra.

SCOTT, J., 1828. Account of the group of islands situated in Bass Strait generally known in the charts by the name of Furneaux Islands (taken from information by James Campbell, boatman). Hobart Town.

SEEBECK, J. H., FRANKENBERG, J. AND HAMPTON, J. W. F., 1968. The mammal fauna of Darlimurla. *Vic. Nat.* **85**: 184-93.

SERVENTY, D. L. AND MARSHALL, A. J., 1964. A natural history reconnaissance of Barrow and Montebellow Islands, 1958. *C.S.I.R.O. Div. Wildl. Res.* Tech. Paper No. **6**: 1-23.

SHORTRIDGE, G. C., 1909. An account of the geographical distribution of the marsupials and monotremes of South-west Australia having special reference to the specimens collected during the Balston Expedition of 1904-1907. *Proc. Zool. Soc. Lond.* **55**: 803-48.

SMART, J., 1977. Late Quaternary sea-level changes. Gulf of Carpentaria, Australia. Geology **5**: 755-59.

SMITH, M. J. AND ROGERS, P. A. W., 1981. Skulls of *Bettongia lesueur* (Mammalia: Macropodidae) from a cave in the Flinders Ranges, South Australia. *Trans. Roy. Soc. S.A.* **105**(4): 217.

SA NPWS, 1984. Kangaroo conservation in South Australia. South Australian National Parks and Wildlife Service, Dept. of Environment and Planning.

SPENCER, W. B., 1896a. Mammalia. Pp. 1-52 *in* Report on the work of the Horn Scientific Expedition to Central Australia. Pt 2 (Zoology).

SPENCER, W. B., 1896b. Narrative. Pp.1-136 *in* Report on the work of the Horn Scientific Expedition to Central Australia Pt 1 ed. by W. B. Spencer. Dulau: London.

STORR, G. M., 1960. The physiography, vegetation and vertebrate fauna of North Island, Houtman Abrolhos. *J. Roy. Soc. W.A.* **43**: 59-62.

STORR, G. M., 1965. The physiography, vegetation and vertebrate fauna of the Wallabi Group, Houtman Abrolhos. *J. Roy. Soc. W.A.* **48**: 1-14.

STRAHAN, R., 1983. Complete Book of Australian Mammals. Angus & Robertson: Sydney.

TATE, G. H. H., 1948. Results of the Archbold Expeditions No. 59. Studies on the anatomy and phylogeny of the Macropodidae (Marsupialia). *Bull. Am. Mus. Nat. Hist.* **91**: 233-351.

TATE, G. H. H., 1952. Mammals of Cape York Peninsula, with notes on the occurrence of rainforest in Queensland. Results of the Archbold Expeditions No. 66. *Bull. Am. Mus. Nat. Hist.* **98**: 563-616.

TAYLOR-PAGE, F. J., 1970. The Sussex mammal report 1969. Sussex Naturalists Trust Ltd: Henfield.

THOMAS, I. M. AND DELROY, L. B., 1971. Pearson Island expedition 1969 4. The Pearson Island Wallaby. *Trans. Roy. Soc. S.A.* **95**: 143-45.

THOMAS, O., 1888. Catalogue of the Marsupialia and Monotremata in the Collection of the British Museum (Natural History). British Museum (Natural History), London.

THOMAS, O., 1897. The mammals collected in British New Guinea by Dr Lambert Loria. *Ann. Mus. Stor. Nat. Genova, Ser. 2,* **18**: 606-22.

THOMAS, O., 1904. On a collection of mammals made by Mr J. T. Tunney in Arnhem Land, Northern Territory of South Australia. *Novit. Zool.* **11**: 222-29.

THOMAS, O., 1906a. On mammals collected in south-west Australia by Mr W. E. Balston. *Proc. Zool. Soc. Lond.* **1906**: 486-78.

THOMAS, O., 1906b. On mammals from northern Australia, presented to the National Museum by Sir Wm. Ingram, Bt, and the Hon John Forrest. *Proc. Zool. Soc. Lond.* **1906**: 536-43.

THOMAS, O., 1909. Some mammals from N. E. Kimberley, Northern Australia. *Ann. Mag. Nat. Hist. Ser. 8,* **3**: 149-52.

THOMAS, O., 1926. On various mammals obtained during Capt. Wilkins expedition in Australia. *Ann. Mag. Nat. Hist. (Series 9)* **17**: 625-35.

THOMPSON, G. M., 1922. The naturalization of animals and plants in New Zealand. Cambridge University Press. 607 Pp.

TINKER, S. W., 1938. Animals of Hawaii. Nippu Jiju Co. Ltd: Honolulu 188 Pp.

TOMICH, P. Q., 1969. Mammals in Hawaii. Bishop Museum Press: Honolulu.

TORGENSEN, T., HUTCHINSON, M. F., SEARLE, D. E. AND NIX, H. A., 1983. General bathymetry of the Gulf of Carpentaria and the Quaternary physiography of Lake Carpentaria. *Palaeog. Palaeoclim. Palaeoecol.* **41**: 207-25.

TURNBULL, W. D. AND LUNDELIUS, E. L., 1970. The Hamilton fauna, a late Pliocene mammalian fauna from the Grange Burn, Victoria Australia. Fieldiana: Geology **19**: 1-163.

WAKEFIELD, N. A., 1954. The rediscovery of the Rock-wallaby in Victoria. *Vic. Nat.* **70**: 202-06.

WAKEFIELD, N. A., 1961. Victoria's Rock-wallabies. *Vic. Nat.* **77**: 322-32.

WAKEFIELD, N. A., 1963. Notes on Rock-wallabies. *Vic. Nat.* **80**(6): 169-76.

WAKEFIELD, N. A., 1963. Mammal remains from the Grampians, Victoria. *Vic. Nat.* **80**(5): 130-33.

WAKEFIELD, N. A., 1964a. Mammal sub-fossils from Basalt Caves in south-western Victoria. *Vic. Nat.* **80**(9): 274-78.

WAKEFIELD, N. A., 1964b. Recent mammalian sub-fossils of the Basalt Plains of Victoria. *Proc. Roy. Soc. Vic.* **77**(2): 419-25.

WAKEFIELD, N. A., 1966. Mammals of the Blandowski expedition to northwestern Victoria, 1856-1857. *Proc. Roy. Soc. Vic.* **79**: 371-91.

WAKEFIELD, N. A., 1967. Some taxonomic revision in the Australian marsupial genus *Bettongia* (Macropodidae) with description of a new species. *Vic. Nat.* **84**(1): 8-22.

WAKEFIELD, N. A., 1971. The brush-tailed Rock-wallaby *(Petrogale penicillata)* in Western Victoria. *Vic. Nat.* **88**: 92-102.

WALTON, D. W. (ed.), 1988. Zoological Catalogue of Australia. Volume 5. Mammalia AGPS, Canberra.

WEBB, L. J. AND TRACEY, J. G., 1981. Australian rainforests: patterns and change. Pp. 605-94 *in* Ecological Biogeography of Australia ed. by A. Keast. W. Junk: The Hague.

WEBSTER, P. J. AND STRETEN, N. A., 1972. Aspects of Late Quaternary climate in tropical Australia. Pp. 39-60 *in* Bridge and Barrier: the natural and cultural history of Torres Strait ed. by D. Walker. Dept. Biogeog. and Geomorph. Publ. B.G/3 ANU: Canberra.

WEBSTER, P. J. AND STRETEN, N. A., 1978. Late Quaternary ice age climates of Tropical Australasia: Interpretation and reconstruction. *Quaternary Res.* **10**: 279-309.

WINTER, J. W. AND ALLISON, F. R., 1980. The native mammals of Cape York Peninsula. Changes in status since the 1948 Archbold Expedition. Pp. 31-44 *in* Contemporary Cape York Peninsula ed. by N. C. Stevens and A. Bailey. The Royal Society of Queensland: St Lucia.

WODZICKI, K., 1950. Introduced mammals of New Zealand. DSIR Bulletin 98.

WODZICKI, K. AND FLUX, J. E. C., 1967a. Re-discovery of the Whitethroated Wallaby *Macropus parma* Waterhouse 1846, on Kawau Island, New Zealand. *Aust. J. Sci.* **29**(11): 429-30.

WODZICKI, K. AND FLUX, J. E. C., 1967b. Guide to introduced wallabies in New Zealand. Tuatara **15**: 47-59.

YALDEN, D. W. AND HOSEY, G. R., 1971. Feral wallabies in the Peak District. *J. Zool. Lond.* **165**(4): 513-20.

ZIEGLER, A. C., 1977. Evolution of New Guinea's marsupial fauna in response to a forested environment. Pp.117-38 *in* The Biology of Marsupials ed. by B. Stonehouse and D. P. Gilmore. Macmillen Press Ltd.

ZIEGLER, A. C., 1982. An ecological checklist of the New Guinea recent mammals. Pp. 867-93 *in* Biogeography and Ecology of Papua New Guinea ed. by J. L. Gressitt. W. Junk: The Hague.

Ecology of the Potoroidae — A Review

J. H. SEEBECK[1], A. F. BENNETT[1] and DAVID J. SCOTTS[1]

"In the modern wreck of this remarkable group of mammals the plan of its unfolding into the vast territories, which it formerly occupied, is but dimly to be seen"

H.H. Finlayson, 1958

Members of the family Potoroidae (rat kangaroos), formerly widespread across Australia, have been dramatically affected by European settlement and only three of 10 modern species remain relatively common. Habitats ranging from arid hummock grassland to tropical rainforest are occupied, but woodland or forest in the southern half of the continent contained the greatest species richness. Although omnivorous, many species are now known to be primarily mycophagous, and the distribution of some species may be determined by the presence of fungal resources. One species is diurnal but most are mainly nocturnal. Stable populations, continuous reproduction and dispersal of young all serve to maintain populations, although predation by canids appears to be a controlling influence for some species. Home ranges of open-country species are large compared with those of species of dense habitat. Most species mate promiscuously and are not territorial away from the nest-site, but one species (*Potorous longipes*) may be territorially monogamous.

INTRODUCTION

AMONG the first marsupials to be encountered by the First Fleet settlers was the Long-nosed Potoroo, *Potorous tridactylus*. Figured and described by both Phillip (1789) and White (1790), the generic name was based on the Aboriginal "poto roo", and has subsequently formed the basis for the family name, the Potoroidae. Commonly known as Rat-kangaroos, the family contains 10 modern species in two sub-families: the Hypsiprymnodontinae (one species) and the Potoroinae (nine species in four genera) (Table 1). They are small, compact animals, which are superficially similar in external appearance. They range in size from *Hypsiprymnodon moschatus* (mean adult body weight 530 g) to *Aepyprymnus rufescens* (3 250 g), but most species are between 1 000 and 2 000 g in weight (Table 5). In this review we describe the distributional pattern for each species, and address comparative aspects of habitat use, diet, activity patterns and population ecology. *(See Table 1 overleaf.)*

DISTRIBUTION

Rat-kangaroos were widely distributed throughout the southern half of the continent, mostly below about Latitude 20°S, before European settlement. The arid and semi-arid parts of the continent supported four species including *Bettongia lesueur* and *B. penicillata*, which had the widest continental range. The temperate/tropical east and south-east supported six species, and the south-west four species (Fig. 1a — *overleaf*).

The available historical evidence suggests that five species *(H. moschatus, Potorous platyops, P. longipes, B. tropica, Caloprymnus campestris)* have always had restricted distributions, and only five species (*P. tridactylus, B. lesueur, B. penicillata, B. gaimardi* and *Aepyprymnus rufescens*) were widespread.

However, changes caused by European activity, such as the introduction of ruminant stock, clearing, agriculture, introduction of predators/competitors and hunting, caused a very rapid decline in the range of many species, especially in the more arid parts of Australia. Two species (*P. platyops, C. campestris*) are probably extinct, and three *(B. lesueur, B. penicillata, B. tropica*) are restricted to remnants of their former range. Only two species (*P. tridactylus* and *A. rufescens*) remain relatively widespread and common within their former ranges, and a third (*B. gaimardi*) is common in Tasmania, although now extinct on the mainland.

The present and former distribution of each species is shown in Figs 1b-d *(overleaf)* (after Seebeck and Rose 1988), and is discussed below.

Hypsiprymnodon moschatus

H. moschatus is restricted to areas of tropical rainforest in north Queensland from just south of Cooktown to Mt Lee south-west of Ingham (J. W. Winter, pers. comm. 1988). Changes in distribution since European settlement have been essentially due to clearing of rainforest for agriculture, but where it remains *H. moschatus* seems relatively common (Johnson and Strahan

---

[1]Arthur Rylah Institute for Environmental Research, National Parks and Wildlife Division, Department of Conservation, Forests and Lands 123 Brown Street, Heidelberg, Victoria 3084.

Pages 67-88 in KANGAROOS, WALLABIES AND RAT-KANGAROOS, ed. by G. Grigg, P. Jarman and I. Hume. Surrey Beatty & Sons Pty Limited, New South Wales, Australia 1989.

Table 1. Modern species of Potoroidae and their present status. Taxonomy follows Seebeck and Rose 1988.

Species	Common Name	Present status
Sub-family Hypsiprymnodontinae		
Hypsiprymnodon moschatus	Musky Rat-kangaroo	Uncommon
Sub-family Potoroinae		
Potorous platyops	Broad-faced Potoroo	Extinct
P. tridactylus	Long-nosed Potoroo	Common
P. longipes	Long-footed Potoroo	Rare
Bettongia lesueur	Burrowing Bettong	Rare
B. penicillata	Brush-tailed Bettong	Rare
B. tropica	Queensland Bettong	Unknown
B. gaimardi	Tasmanian Bettong	Common
Caloprymnus campestris	Desert Rat-kangaroo	Extinct
Aepyprymnus rufescens	Rufous Bettong	Common

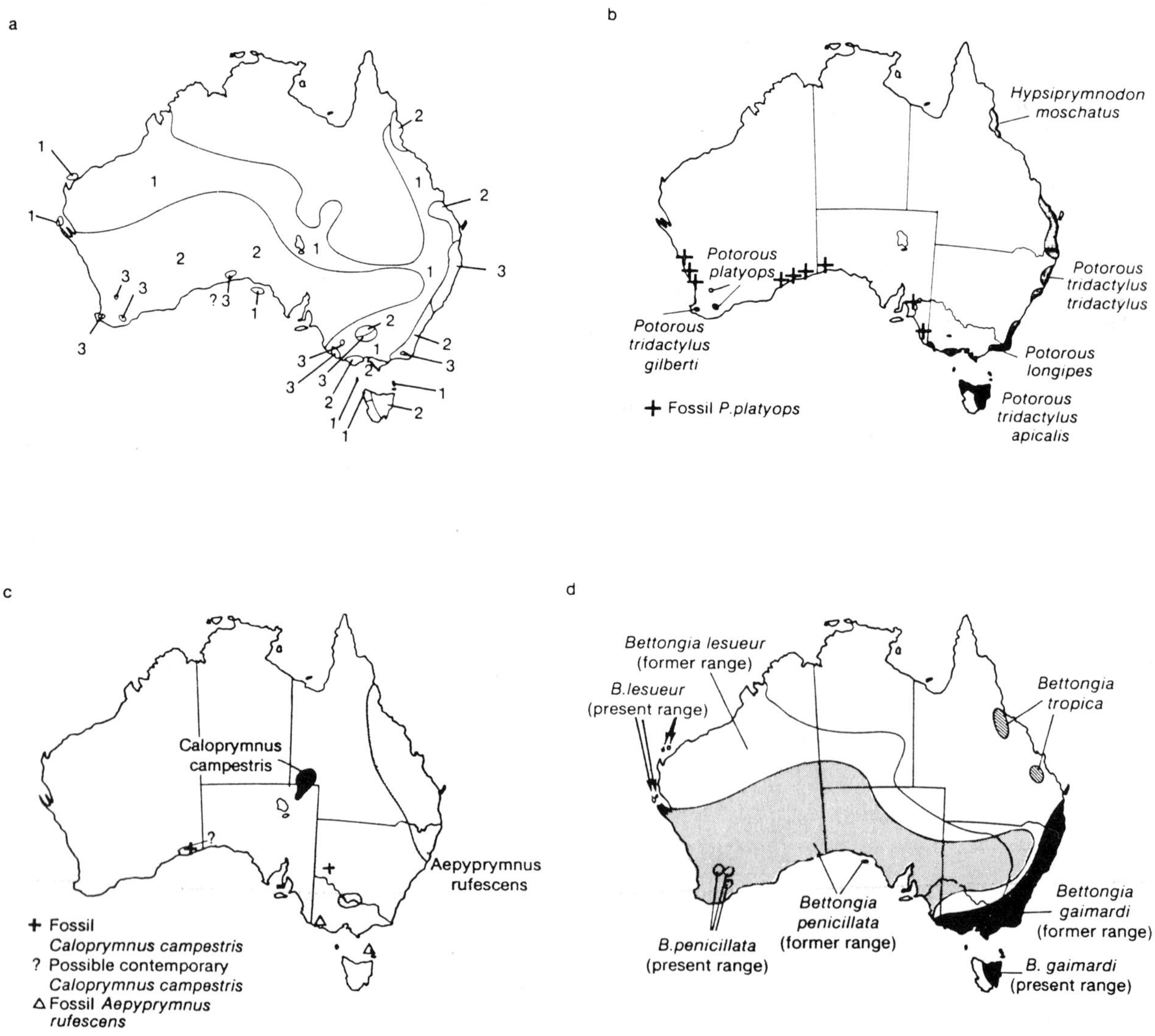

Fig. 1. Distribution of modern potoroids: (a) isoclines of potoroid species in Australia; (b) distribution of *Potorous* and *Hypsiprymnodon* species; (c) distribution of *Aepyprymnus* and *Caloprymnus* species; (d) distribution of *Betongia* species. (From Seebeck and Rose 1988, with permission from Bureau of Flora and Fauna).

1982). The altitudinal range encompassed by this species is from sea-level to about 1500 m.

Potorous platyops

P. platyops was known as a modern species only from a small area of southwestern Western Australia and may have been rare before European settlement (Kitchener 1983). Few localities are known precisely, and little habitat detail was recorded. Sub-fossil remains indicate that the species formerly occurred from southeastern South Australia (including Kangaroo Island), across the Nullarbor Plain and possibly as far north as Northwest Cape in Western Australia (Butler and Merrilees 1971).

The altitudinal range of *P. platyops* was up to about 300 m.

Potorous tridactylus

This species occurs along the eastern seaboard from southeastern Queensland (about Gladstone) (Amos 1982) to the south-west of Victoria, some Bass Strait islands and northern and eastern Tasmania. It occurs on French Island and Phillip Island (Victoria), on King Island and Flinders Island (Tasmania); and it formerly occurred on Cape Barren Island and Clarke Island (Tasmania) (Hope 1974; Seebeck 1981). An isolated population formerly occurred near Augusta in southwestern Western Australia but this is now presumed to be extinct. Before European settlement the species probably also occurred in the extreme south-east of South Australia. Most populations occur within 50 km of the sea, but in Victoria an isolated population in The Grampians is some 130 km from the coast (Seebeck 1976).

Subfossil remains indicate that the species was only slightly more widely distributed in prehistoric times, mainly in the south-east of South Australia, further inland in southeastern New South Wales and on certain Bass Strait islands from which it is not now recorded (Hope 1974).

The distribution of *P. tridactylus* has been made disjunct by land-use changes in many parts of its range, and in all States where it now occurs its distribution is patchy (Seebeck 1981; Amos 1982; Johnston 1983). For Victoria, Seebeck (1981) has shown that this is due in part to physiographic disjunctions in habitat rather than the influence of settlement; and perhaps the same may be true in other parts of its range.

P. tridactylus is found mainly from sea-level to about 800 m (Seebeck 1981), although it has been recorded at almost twice that altitude in New England National Park in northeastern New South Wales (Schlager 1981).

Potorous longipes

P. longipes has been found only in a very small area of east Gippsland in Victoria (Seebeck and Johnston 1980; Drinnan and Hill 1984; Hill and Triggs 1985; Scotts and Seebeck 1989). Within this area of some 80 000 ha it is sparsely and disjunctly distributed, at altitudes ranging from 140 m to 800 m.

A fossil skull from Yarrangobilly Caves (Seebeck, unpubl.) indicates that the species formerly occurred in southeastern New South Wales. Recently, remains of *P. longipes* were identified in fox scats from near Bombala, New South Wales (Dovey 1987), indicating that a population may be present in that region.

Bettongia lesueur

Formerly the most widespread of all potoroids, its range extended generally south-east from near Broome, Western Australia, through the Northern Territory, South Australia, western New South Wales to northwestern Victoria. It is now extinct on the mainland, but still occurs on Bernier, Dorre, Barrow and Boodie Islands off the Western Australian coast. *B. lesueur* digs burrows, and extensive warren systems may still be seen in areas from which the species has long since vanished.

Fossil remains indicate little extension of its reported historic range.

The altitudinal range of *B. lesueur* was from sea-level to about 300 m.

Bettongia penicillata

Formerly widely distributed in southern Australia, mostly south of about 30°S, this species was known from northeastern New South Wales (Namoi River area), across the southern Northern Territory, South Australia and the Nullarbor to southwestern Western Australia, perhaps as far north as Shark Bay. One island population (St Francis, in Nuyts Archipelago, South Australia) was known (Delroy *et al.* 1986).

Based on extensive interviews with Aborigines, Finlayson (1958) believed, *B. penicillata* to have a much wider distribution, particularly in the Northern Territory and Queensland, but this cannot be substantiated. He also included *B. tropica* with *B. penicillata* and indicated that the range extended across central and southern Queensland.

B. penicillata is extinct over most of its former range except for three small areas in the southwest of Western Australia at Tutanning, Dryandra and Tone/Perup River (Christensen 1983). It has been recently reintroduced to St Francis Island and introduced to several other islands off South Australia, where breeding populations have been established on Island A, Venus Bay; Bairds Bay Island; and Wedge Island (Delroy *et al.* 1986).

B. penicillata occurred in an altitudinal range from sea-level (St Francis Island) to about 300 m.

Bettongia tropica

Except for a single animal taken in 1884 in southeastern Queensland, and a possible record by John Gilbert in the same area in 1844 (Crossman and Reimer 1986), *B. tropica* is restricted to tropical forest in north Queensland, inland of the Cairns/Townsville area (Wakefield 1967). The most northerly records are on the Mt Windsor Tableland, and the most southerly at "Red Road" west of Koombooloomba Dam near Tully (J. W. Winter, pers. comm. 1988).

The altitudinal range of *B. tropica* is from 300 m at Coomooboolaroo to 1 300 m at Mt Spurgeon. All recent records are above 450 m.

Bettongia gaimardi

This species is now confined to Tasmania, where it remains common and widespread in the eastern part of the state (Rose 1986). It is now extinct on the mainland, where it formerly occurred from southeastern Queensland to southeastern South Australia (Wakefield 1967). It has not been recorded from any of the Bass Strait islands (Hope 1974), but subfossil remains indicate a formerly wider range in the southeastern area of South Australia (Smith 1971).

The species has been introduced to Maria Island, Tasmania (Rose 1986).

In Tasmania, the species occurs over an altitudinal range from sea level to 1 000 m.

Caloprymnus campestris

This species was recorded as a living animal only from the Lake Eyre Basin of South Australia and southwestern Queensland (Finlayson 1932, 1958). It has not been encountered since 1935 and is presumed extinct. Tate (1879) reported the species to have occurred in southwestern South Australia at the head of the Bight, but this has never been confirmed, and his description tallies more with *B. lesueur* than *Caloprymnus*. Lundelius (1957) identified fossils from caves near Eucla (Western Australia) as *Caloprymnus*, but Baynes (1987) has recently indicated that this material belongs to an as yet undescribed potoroid. Fossils in archaeological sites near Lake Menindee and Lake Willaura, in southwestern New South Wales, indicate a former much more widespread distribution (Tedford 1967).

C. campestris occurred at an altitude of about 150 m (but also near sea-level if Tates' record is included).

Aepyprymnus rufescens

A. rufescens is widely distributed from about Newcastle, New South Wales (Schlager 1981), to Cooktown, Queensland. However, there is a wide gap between Barrington Tops and the nearest northerly record in New South Wales. Although mainly a coastal species, it is present on both sides of the Dividing Range in Queensland, especially in the northern part of its range, where it occurs almost 300 km west of Townsville (A. Dennis, James Cook University, pers. comm. 1988). It may also occur well inland near Mitchell, in southern Queensland (P. McRae, Queensland NPWS, pers. comm. 1988). It has vanished from inland areas of New South Wales and from former collection areas near the junction of the Murray and Murrumbidgee Rivers and at Violet Town in Victoria. It is not known whether this southern population was isolated from that of the major range. The altitudinal range is from about 100 m to about 700 m.

Fossil remains are known from several sites in southwestern Victoria (Gill 1953; Wakefield 1963a,b, 1964a,b), from near Bacchus Marsh, Victoria (L. Conole, Victoria Archaeological Survey, pers. comm. 1988) and from Flinders Island (Calaby 1971; Hope 1974).

HABITAT AND HABITAT USE

Hypsiprymnodon moschatus

This species occurs in tall closed forest (rainforest) at all altitudes (Johnson and Strahan 1982). Ramsay (1876) found it in the "dense and damp portions of the scrubs which fringe the rivers", and Johnson and Strahan (1982) indicated that within the rainforest this species prefers the most mesic areas close to watercourses or springs. At Mission Beach (Lat. 17°53′S, Long. 146°06′E) *H. moschatus* was recorded living in rainforest within a few metres of the beach. Nests of leaves, lichen and ferns, which serve as nocturnal shelter for *H. moschatus*, are built in clumps of lawyer vine or between the plank buttresses of trees (Johnson and Strahan 1982).

Potorous platyops

The only information available on the habitat occupied by the now extinct *P. platyops* is "thicket surrounding one of the salt lagoons" (Kitchener 1983). Typically these lagoons are fringed by low shrubs of *Arthrocnemum* grading into heaths dominated by *Melaleuca* spp. The effective shelter provided by these species-associations may be quite dense, with up to 70% canopy cover (Muir 1979).

Potorous tridactylus

In southern Australia, the habitats in which *P. tridactylus* is mainly found range from woodland with dry heathy understorey, to wet heaths, dense coastal scrubs and mixed species open-forest.

Near Hobart, Tasmania, Kitchener (1967) found *P. tridactylus* to prefer the moister and denser components of a dry eucalypt woodland. He reported a significant preference for moderately dense *E. globulus* forest, with a thick understorey of sclerophyllous shrubs (on the western slopes of his study area), but the species also frequented areas of thick understorey beneath the drier *E. viminalis-E. linearis* association developed on the eastern and southern aspects. On the sandy coastal plains of northern Tasmania, Heinsohn (1968) found *P. tridactylus* in dense tall wet heath of *Melaleuca* and *Leptospermum*, with emergent *E. ovata* and *E. obliqua*, and thick tussocks of *Lepidosperma laterale* and *Lomandra longifolia*, throughout which runways were common.

Tall dense wet heath was also found to be a habitat favoured by *P. tridactylus* in parts of the Otway Ranges of southern Victoria (Emison *et al.* 1975). They considered that the optimum habitat was moist gullies in *E. ovata* open-forest, where there was an understorey of myrtaceous plants and tall sedges. The species was less abundant on drier, less densely-vegetated slopes. *P. tridactylus* has been only rarely reported from wet tall open-forest in the Otway Ranges (Seebeck 1981; K. A. Cherry, Arthur Rylah Institute, pers. comm. 1988). The habitat at those sites comprises a canopy of *E. regnans-E. obliqua*, with a tall dense mesic understorey of *Acacia*, *Bedfordia* and tree ferns, and a very dense field layer of ferns and shrubs. In southwestern Victoria, Menkhorst and Beardsell (1982) encountered *P. tridactylus* in open-forest of *E. obliqua* or *E. baxteri* and in *Banksia marginata* closed-heath with emergent eucalypts. Seebeck (1981) described the habitat of *P. tridactylus* in several discrete regions of southern Victoria and found it to occur in a wide variety of open-forests. In most regions drier sites were characterized by *E. obliqua, E. baxteri* or *E. aromophloia,* and the wetter sites by *E. ovata* or *E. viminalis*; but in east Gippsland by a more mixed array including *E. sieberi, E. globoidea, E muellerana, E. baxteri,* and in wetter areas, *E. cypellocarpa*. Favoured locations of *P. tridactylus* in southern Victoria were often ecotonal between forest and heath.

Its habitat in southern coastal New South Wales is very similar to that occupied in east Gippsland (Johnston 1973; Newsome *et al.* 1974), but in northern New South Wales most records are from rainforest and associated wet sclerophyll forests (Schlager 1981). Schlager (1981) believed that there were two distinct components to the distribution (and habitat) of *P. tridactylus* in northern New South Wales and southern Queensland: an inland component in the moist woodlands and forests of the eastern slopes and highlands; and a coastal component in sandy wet heaths.

He described the rainforest (sub-tropical rainforest of *Dysoxylem* or *Sloana*; warm temperate rainforest of *Ceratopetalum-Schizomeria*; and temperate rainforest of *Nothofagus-Ceratopetalum* associations) as the "backbone" of the distribution, with the "meat" being the surrounding wet sclerophyll forests of *Tristania, E. dunnii* or *E. microcorys-E. saligna* associations. *P. tridactylus* avoided virgin rainforest but utilized its edge in disturbed areas, especially along verges of roads (Schlager 1981). The wet sclerophyll forests supported either an understorey of dense mesophyll shrubs or an equally dense field layer of grasses *(Poa, Imperata)*, ferns *(Pteridium)*, or the sedge-like *Lomandra.* Amos (1982) reported *P. tridactylus* from notophyll vine forest in Queensland.

Wet coastal heaths in northern New South Wales and southeastern Queensland are typically very dense, 1-3 m tall, with a few small emergent trees — *Banksia, Melaleuca, Casuarina, Leptospermum, Acacia, Eucalyptus* and *Angophora.* Despite the extensive coastal plains presently or formerly covered by this vegetation, *P. tridactylus* is considered rare in that environment (Schlager 1981).

P. tridactylus shelter during the day in a shallow squat, usually excavated at the base of a tussock or under dense shrubs, and do not construct complex nests.

Although generalized habitat preferences of *P. tridactylus* have been well documented throughout its range, less is known about microhabitat use. Animals require dense understorey cover, and do not venture far from it during feeding or exploratory behaviour, despite the significant number of road-kills reported in Tasmania (TASPAWS) and to a lesser extent in Victoria (Atlas of Victorian Wildlife) (Guiler 1958; Kitchener 1967; Heinsohn 1968; Seebeck 1981).

An examination of microhabitat use by this species in southwestern Victoria (Bennett 1987a) found no simple correlations between captures of *P. tridactylus* and a particular vegetation type. Rather, animals used a range of both floristic and structural vegetation classes within their home ranges. It was proposed (Bennett 1987a) that microhabitat selection involved the use of dense, structurally-complex vegetation for diurnal shelter and predator avoidance, and the adjacent more open and floristically more diverse areas for foraging.

Potorous longipes

Seebeck and Johnston (1980) described the habitat of *P. longipes* at two sites in eastern Victoria, which lay at the opposite ends of the then known range. One, in the hinterland, was in wet forest; the other, near the coast, was in dry forest. They were markedly different in climate and topography, and shared few floristic similarities. The hinterland site, and other inland sites subsequently described (Chesterfield *et al.* 1983; Horrocks *et al.* 1984; Opie *et al.* 1984) have been in tall wet forest. The eucalypts present at Seebeck and Johnston's site were *E. cypellocarpa* with some *E. obliqua*, above a dense understorey of *Acacia, Bedfordia* and other mesic tall shrubs and tree-ferns. The dense field layer included *Gahnia, Pteridium,* and other ferns, as well as mesic shrubs. The locality in which Opie *et al.* (1984) found *P. longipes* was floristically and structurally similar; and that at which Horroccks *et al.* (1984) encountered the species was slightly drier, with a less dense understorey and ground layer, except along a drainage line which was a very dense fern gully.

Most of the occupied lowland sites are dry sclerophyll forest, dominated by *E. sieberi, E. obliqua, E. globoidea, E. baxteri* and *E. muellerana.* Superficially, this habitat is similar to that occupied by *P. tridactylus* in east Gippsland, and, indeed, both species occur sympatrically at one location, Bellbird Creek-Manorina. The understorey at these lowland sites includes tall shrubs of species of *Acacia, Pultenaea, Platylobium* and *Pomaderris,* tangles of *Tetrarrhena juncea,* and a field layer of ferns and *Lomandra.* At Bellbird Creek, the riparian vegetation included a tall overstorey of *E. cypellocarpa, Acacia melanoxylon* and *Acmena smithii,* overtopping a dense "jungle" of temperate rainforest (Seebeck and Johnston 1980).

P. longipes shelters during the day in shallow squats in a fashion similar to *P. tridactylus.*

Scotts and Seebeck (1989) have recently investigated microhabitat use at one lowland site, Bellbird Track, in east Gippsland. The site, at about 300 m ASL ranged from a ridge supporting a dry sclerophyll forest characterized by *E. sieberi* and *Pultenaea juniperina,* through mixed *Eucalyptus* wet and damp sclerophyll forest, to a gully supporting a warm-temperate rainforest dominated by *Acmena smithii, Bedfordia arborescens* and *Cyathea australis.* The dispersion of animal captures was analysed in relation to six floristic and five structural vegetation groups and revealed that *P. longipes* had showed a clear preference for the wetter floristic communities and the structural groups that were closely allied to them (Scotts and Seebeck 1989). The driest floristic and structural groups, on the ridges, were avoided by the animals. Habibat use was also analysed on a seasonal basis, and indicated that during the autumn-winter months most activity was concentrated in the less-damp forest upslope from the gully; while during spring-summer the animals remained mostly in the gully and adjacent vegetation. Both of these latter habitats were characterized by a tall dense treefern and tall-shrub understorey. The gully was avoided during the wettest parts of the year.

Like its congener *P. tridactylus,* microhabitat use is based on the need for shelter (both from climatic extremes and predators) and food supply and is further discussed below.

Whereas *Potorous* spp. are usually associated with very dense cover, members of the genus *Bettongia* are found in more open habitats.

Bettongia lesueur

B. lesueur formerly occurred over a wide range of arid and semi-arid Australia in habitats encompassing "most types of country: grassy and herbaceous loam flats within the major ranges, open mulga and ironwood parks skirting the ranges, and deep into the true sandridge areas . . ." but avoided rocky hills and dense thickets (Finlayson 1958). Dahl (1926) reported that, in northwestern Western Australia, *B. lesueur* inhabited wooded regions, especially those with undulating sandhills. He described one such area as being clad with "spinifex . . . A few green bushes . . . (and) two or three stunted trees. . . ." Near Adelaide *B. lesueur* was commonly found on the coastal plains area, which supported woodland and open-forest with an herbaceous understorey (Specht 1972).

B. lesueur is now confined to the Western Australian islands of Bernier Island and Dorre Island in Shark Bay, and Barrow Island and Boodie Island off Onslow. On these it inhabits hummock grassland (Open Steppe) dominated by species of *Triodia* with small shrubs (on Barrow, Boodie and Dorre Islands), or Tall Scrub of *Diplolaena, Olearia* and *Hibiscus* (on Bernier Island) (Ride and Tyndale-Biscoe 1962; Royce 1962; Main and Yadav 1971). Royce (1962) has provided a detailed description of habitats on Bernier and Dorre Islands, and Serventy and Marshall (1964), Butler (1970), Burbidge and Main (1971) and Main and Yadav (1971) have described habitats on Barrow Island.

Habitat selection by *B. lesueur* is also influenced by its burrowing habits. *B. lesueur* frequently burrows in firm loamy soils, rather than in sandridges (Burbidge and Main 1972; Burbidge 1983), but also in outcrops of friable rock (Finlayson 1958), or under capping layers of stone (Ride and Tyndale-Biscoe 1962). Giles (1889) found "burrowing wallabies" (undoubtedly

B. lesueur) to be abundant near Lake Christopher, north-west of the Rawlinson Ranges, South Australia, where the sparsely-vegetated limestone rises were "completely honeycombed" by burrow systems. In the north-west of its range, in Western Australia, where the species burrowed in the sandhills, heavy rain caused the collapse of the burrow systems and consequent death of many animals (Dahl 1926).

The burrows, which are used by *B. lesueur* to avoid extremes of climate (and predators), are often complex interconnecting warrens, which support many animals (Gould 1863; Dahl 1926). Ward (1907) encountered colonies of 20-70 animals in southwestern Western Australia. Warren size on Barrow Island may be of the same order (R. Prince, Conservation and Land Management, Western Australia, pers. comm. 1987). Finlayson (1958) averred that, in Central Australia, the burrow systems were less complicated. Ride and Tyndale-Biscoe (1962) found both simple and complex types of burrow on Dorre Island, and related these to differing topography and soils, the more complex systems being excavated in elevated areas and the simpler ones among the loams of the Open Steppe. Finlayson (1958) described "tchungoo roads", the conspicuous radiating pathways formed by *B. lesueur* from warren entrances to feeding grounds.

Bettongia penicillata

This species also occurred throughout a wide range of vegetation types in arid and semi-arid Australia, but was also present in higher rainfall areas. Finlayson (1958) reported its occurrence in South Australia both in mallee and in *E. obliqua* forest in the ranges. Favoured habitat on the Fleurieu Peninsula was *E. leucoxylon* open-forest with a ground layer of sedges, especially *Lepidosperma.* St Francis Island (South Australia) supports a low shrubland of *Atriplex*, and, on sand dunes, a tall open shrubland of *Olearia* with a variety of understorey shrubs. Higher areas on the island also support a low shrubland of legumes and other shrubs and patches of *Triodia* (Delroy *et al.* 1986). [Morton 1861] referred, on a number of occasions, to "kangaroo rats" which built nests being abundant in mallee scrub in northwestern Victoria. These were probably *B. penicillata.* Krefft (1866) found *B. penicillata* to be common among *Muehlenbeckia* scrub under *E. largiflorens* along the Murray River and its tributaries. Presumably it utilized extensively this habitat along watercourses. Gould (1863) was familiar with the species in inland river systems of New South Wales.

B. penicillata is now confined to Western Australia, where its original habitat ranged from sandy spinifex country (Burbidge and Fuller 1984) to woodland and open-forests of the south-west. It is now found only in the south-west, in a few small areas of multi-layered open-forest of *E. marginata* and *E. wandoo* which occur on low flat ridges, separated by wide shallow valleys in which the understorey vegetation (especially *Gastrolobium* sp.) may be "quite tall" and "relatively dense" with few emergent trees (Christensen and Leftwich 1980). The species seems to prefer well-drained sandier soils and requires moderately dense cover for shelter (Sampson 1971; Christensen 1980a).

Christensen and Leftwich (1980) believed that the distribution of *B. penicillata* in these open-forests may be influenced by the distribution of the distinctive vegetation types used in constructing the nest. They have described in detail the nest-building behaviour of *B. penicillata.* Nests are built at the base of shrubs and tussocks, and are often entirely concealed. The complex nests are used for daytime shelter from predators and temperature extremes. Several nests may be used at random within an individual's home range.

Bettongia tropica

J. W. Winter (pers. comm. 1988) describes the habitat of *B. tropica* as upland grassy open forest and the immediately adjacent rainforest, on the western edge of the rainforest region in north Queensland. Most records have been from forest developed on poorer soils but it has also been observed in rainforest on basalt soils.

Bettongia gaimardi

B. gaimardi was formerly present on mainland Australia, from southeastern South Australia to southeastern Queensland (Wakefield 1967; Smith 1971), but no details of its habitat were recorded. It is now found only in eastern Tasmania, where its preferred habitat is open-forest of *E. obliqua, E. viminalis* or *E. amygdalina* developed on infertile dolerite soils (Rose 1986). Taylor (1986b) also found *B. gaimardi* in grassy sub-alpine *E. rodwayi* woodland and *E. pauciflora* forest. He determined that forests with dense undergrowth did not support *B. gaimardi.* The understorey is usually low (0-1m) and is composed of grasses (*Poa, Danthonia* and *Agrostis*) and the sedge *Lomandra longifolia.* The relatively sparse shrub layer usually comprises *Leptospermum, Melaleuca* or *Banksia* (Rose 1986). This species also constructs a complex nest, usually of grass and bark, which is located under the shelter of fallen limbs, shrubs or tussocks (Johnson 1979; Rose 1986).

Caloprymnus campestris

C. campestris inhabited the most arid environment in which potoroids are found. Gould (1863) stated "The stony and sandy plains . . . partially clothed with scrub are its native habitat . . .", and this was proved to be the case when Finlayson (1932) rediscovered the species near Lake Eyre. There he found it on stony flats between sandhills, which he described as transitional between true gibber plains and loamy flats. Vegetation (present on a "comparatively lavish scale" due to good rains) was sparse Saltbush and Emu Bush with scattered *Hakea* (Corkwood). Finlayson (1932) graphically illustrated the uncompromising habitat.

Nests were constructed of leaves and grass, lining a shallow excavation, and were situated either under a bush or out in the open. Finlayson (1932) observed that there were radiating paths from many of the nests he saw; this observation recalls the "tchungoo roads" that he described for *B. lesueur.*

Aepyprymnus rufescens

Gould's (1863) original observations of the habitat in which *A. rufescens* occurred ("the stony sterile ridges bordering the grassy flats") were refined by Finlayson (1931) who reported that, in the Dawson and Fitzroy River valleys of Queensland, the species had a preference for "grassy lands free from bush growth", especially in riparian situations; although he found the species in both open and forested country. Recent studies by Southgate (1980), Schlager (1981) and Southwell (1987) have provided a more detailed understanding of habitat use among populations in northern New South Wales. Generally, *A. rufescens* is found in dry open woodland, mostly in areas dominated by *E. maculata*, but with other eucalypts, *Casuarina* and *Angophora* often present. Wetter forests are little-used, and the species has not been recorded from rainforest, although fringe areas may be used (Johnson and Bradshaw 1977). Favoured eucalypt associations range in structure from open woodland to tall open-forest, with a maximum projective foliage cover of 40-50%. This latter value reflects the logging history of the forests in northeastern New South Wales.

A conspicuous characteristic of all habitats examined by Schlager (1981) in which *A. rufescens* occurred was the virtual absence of a shrub layer. He found, as did Southgate (1980), that a dense cover of tall native grasses was preferred — usually tussocks of *Poa* or areas of *Imperata.* Southgate (1980) found that the animals' "activity area" incorporated both woodland and pasture, and access to nesting areas with sufficient tussocks and logs. Southwell (1987) reported that at his study site, *A. rufescens* favoured a mosaic of grassland and open forest vegetation on the valley floor. Sightings were positively associated with ground vegetation diversity, tussock grass greenness and cattle density, and negatively associated with canopy cover and evenness of other grasses. Although earlier authors had implied the use of more rugged terrain, both Southgate (1980) and Schlager (1981) found that *A. rufescens* selected flat or gently sloping areas. Nevertheless, Jarman (pers. comm. 1988) has observed the species nesting in quite steep country, and moving to flatter country to feed. Johnson and Bradshaw (1977) reported *A. rufescens* in hummock grassland in Queensland.

At Black Rock (265 km west of Townsville) *A. rufescens* is common in open woodland of either *E. crebra* or *E. dichromophloia*, with *Melaleuca, Brachychiton, Terminalia* and *Petaloshagma* present in low-flying areas (A. Dennis, pers. comm. 1988).

Nests used for diurnal shelter by *A. rufescens* are constructed of grass or bark, lining and covering a shallow excavation. A number of nests are used, probably in a random manner, by individuals (Schlager 1981).

In Table 2 we provide a summary of the general habitat of modern potoroids.

DIET

All modern potoroids are omnivores, although Flannery (1984) has suggested that the extinct giant potoroid *Propleopus*, from the Pleistocene, was a carnivore. Much of the early literature about the diet of potoroids has been anecdotal, and this is still all that is known for certain species. Recent studies have demonstrated that some species, notably those from temperate habitats with predictable climatic conditions, are, to a large degree, mycophagous. Species of *Potorous* and *Bettongia* in particular depend significantly on fungi. All species except *H. moschatus* obtain most of their food by excavating pits in the soil with their well-adapted forepaws.

Hypsiprymnodon moschatus appears to be primarily a frugivore, consuming fruit from rainforest trees such as *Archontophoenix, Eugenia, Diploglotis* and *Sarcotoechia.* Tuberous roots, and soil or litter-dwelling invertebrates are also eaten (Ramsay 1876; Johnson and Strahan 1982). Johnson and Strahan (1982) described in graphic detail the dissection and consumption of grasshoppers by *H. moschatus.* Hume (1982) believed that fungi were also eaten. Considering the manner in which *H. moschatus* obtains its food this seems highly likely, but there is no published evidence that the species excavates hypogeal fungi.

Conversely, the two extant species of *Potorous* have been found to rely considerably on fungi, particularly the sporocarps of hypogeal species. Kitchener (1967) and Guiler (1971) examined the diet of *P. tridactylus* in Tasmania,

Table 2. Summary of the habitats of modern potoroids.

	Rainforest		Sclerophyll Forest						
	Trop.	Temp.	Wet	Dry	Woodland	Shrubland	Heath	Grassland	Sources
Hypsiprymnodon moschatus	×								a
Potorous platyops						×	×		b
P. tridactylus		×	×	×	×	×	×		c
P. longipes		×	×	×					d
Bettongia lesueur				×	×	×		×	e
B. penicillata				×	×	×		×	f
B. tropica	×			×					g
B. gaimardi				×	×				h
Caloprymnus campestris						×			i
Aepyprymnus rufescens				×	×				j

Sources: a) Johnson and Strahan 1982; *b)* Kitchener 1983, Muir 1979; *c)* Schlager 1981, Seebeck 1981, Amos 1982; *d)* Seebeck and Johnston 1980, Scotts and Seebeck 1989; *e)* Finlayson 1958, Ride and Tyndale-Biscoe 1962, Burbidge 1983. *f)* Sampson 1971, Christensen 1980, 1983; *g)* Winter pers. comm. 1988; *h)* Rose 1986, Taylor 1986b; *i)* Finlayson 1932; *j)* Southgate 1980, Schlager 1981.

Table 3. Summary of the diets of modern potoroids.

			Vascular Plants					Fungi				
	Flower	Fruit	Seed	Leaf Stem	Root Tuber	Bulb	Exudate	Hypogeal	Epigeal	Inverts	Other	Sources
Hypsiprymnodon moschatus		(×)			×			?	?	×		a
Potorous platyops												
P. tridactylus	×	×	×	×	×			(×)	×	×		b
P. longipes		×	×	×				(×)	×	×		c
Bettongia lesueur			×	×	(×)	×					carrion	d
B. penicillata			×	×	×		×	(×)		×		e
B. tropica												
B. gaimardi			×		×	×	×	(×)		?		f
Caloprymnus campestris				?	?					×		g
Aepyprymnus rufescens	×	×	×	×	(×)		×	×	×	×		h

(×): Major food

Sources: a) Johnson and Strahan 1982; *b)* Kitchener 1967, Guiler 1971, Bennett 1987a; *c)* Scotts and Seebeck 1989; *d)* Finlayson 1958, Ride and Tyndale-Biscoe 1962; *e)* Sampson 1971, Christensen 1980; *f)* Rose 1982; *g)* Dixon 1988; *h)* Finlayson 1931, Southgate 1980, Schlager 1981.

where they found that fungi (although not identified in detail) comprised an average of nearly 80% of the diet throughout the late autumn to early summer, reaching a peak of over 90% in winter. Other dietary items reported included insects (mostly Lepidoptera, in spring-summer), grasses, sedges, herbaceous dicots, bryophytes, and seeds and fruit, particularly of *Astroloma*. However, only small quantities of any of these foods were taken, with the exception of insects which made up about 15% of the diet in summer.

Bennett (1987a) and Seebeck (unpubl.) investigated the diet of *P. tridactylus* on the mainland and also found fungi to be of major importance throughout the year. Bennett (1987a) reported a maximum occurrence of fungi in faecal samples during autumn and winter, when it exceeded 50% of the total diet, and Seebeck found even higher levels (up to 77%) in June. The consumption of fungi declined during spring and summer but always exceeded 25% of the diet. Most of the fungi were hypogeal gasteromycetes, and more than 50 species (based on spore morphology) were consumed (Bennett 1987a). Other dietary items were arthropods, fleshy fruits, seeds of monocots and vascular plant tissue (Bennett 1987a). These foods were consumed in seasonally varying amounts, generally in proportion to their seasonal availability.

Scotts and Seebeck (1989) studied the diet of *P. longipes* over 13 months, during which period the species was found to be primarily mycophagous. During most of the year the occurrence of fungi in faecal samples exceeded 90% of dietary volume, and was never less than 80%. Other items identified were insects, vascular plant tissue and seeds. Many taxa of fungi were identified (by spore form) from *P. longipes* faeces. Most were hypogeal or sub-hypogeal gasteromycetes, although a few epigeal species were also eaten.

The diet of *B. lesueur* has not been studied in detail. However, numerous observations suggest that roots and tubers form an important part of the diet. Finlayson (1958) recorded that *B. lesueur* ate leaves and stems also. Species of *Boerhaavia, Cyperus* and *Crinum,* and a yam-like plant were favoured sources of bulbs and roots in Central Australia, while in the south-west quandongs *(Santalum acuminatum)* were eaten, and even taken down into the burrow. It has been suggested (Lyne 1974) that *Bettongia* may have influenced the spread of quandong and sandalwood *(Santalum spicatum)* by burying nuts. Dahl (1926) reported that a ground-nut or tuber, locally called "nalgoa" was excavated by *B. lesueur* in sandhills in northwestern Western Australia. Ride and Tyndale-Biscoe (1962) speculated that on the Shark Bay Islands, roots, possibly from *Carpobrotus,* formed the major part of the diet. Shortridge (1909) stated that, on Bernier Island, *B. lesueur* fed to "a great extent on marine refuse and dead matter, even dead sheep . . .". Whatever its natural diet, *B. lesueur* soon adapted to the ready availability of food supplied by settlers' crops. Gould (1863) stated that it was one of the most destructive pests, attacking almost all crops, especially peas and beans. Finlayson (1958) reported that it was a major pest in cereal-growing areas north of Adelaide, damaging both standing and stored crops. Many thousands were destroyed by embattled farmers (Seebeck and Rose 1988).

Main and Yadav (1971) suggested that the feeding behaviour of *B. lesueur,* namely removal of the entire standing crop of tubers and roots, meant that the species requires large areas of home range in which to find sufficient food. However, this "destructive" feeding process is common to all *Bettongia* and to *Potorous. P. tridactylus* has relatively small home ranges, and, for *P. longipes,* Scotts and Seebeck (1989) have shown that there is a temporal change in habitat use which is apparently related to the seasonal productivity of the fungi on which it feeds.

Main and Yadav (1971) reported a very low water intake by *B. lesueur* (3% of body weight). Dahl (1926) did not observe the species to drink at all. Main and Yadav (1971) also showed that *B. lesueur* has water-conserving renal adaptations, and considered that, because of these and its burrowing behaviour, it was able to occupy a wide latitudinal range without the need for free water.

Christensen (1980a) studied the diet of *B. penicillata* at Boyicup, Western Australia, and found that the major food item was fungi. Most were hypogeal species and *Mesophellia* sp. was the most frequently identified. Cellular plant tissue was also found, and was at a maximum in winter-spring. Plant tissue was found more frequently in faecal samples from Tutanning, 160 km NNE of Boyicup, where Sampson (1971) believed that roots and tubers formed the bulk of the diet, with some leaf tissue and arthropods as minor items. He observed animals extracting only the internal tissue from both seeds and insects, and thought that this behaviour would bias estimates of intake based on either stomach or scat analyses. Both Christensen (1980a) and Sampson (1971) reported the burying of seeds by *B. penicillata,* a habit which had also been reported for *B. lesueur*. Kinnear *et al.* (1979) recorded that gum exudate was also eaten in addition to fungi. Sampson (1971) found that *B. penicillata* did not drink free water, even at the height of the annual summer drought.

Green and Nelson (1988) found that *B. penicillata*, which had been introduced to a South

Australian offshore island, were primarily feeding on dicotyledon plant material, including seeds and fruit. Fungi were not a major food, although endomycorrhizal fungi were dominant in faeces at the end of winter. Green and Nelson (1988) suggested that this may have been ingested as gut content of scarab larvae which were also a major part of the *Bettongia* diet at that time.

The study island was not previously known to support *B. penicillata* so the animals have had to adjust to available food. The average adult body weight is significantly lower than that of adults in Western Australia, and may result from their lack of access to high-quality food (Green and Nelson 1988).

Like its congener, *B. penicillata* was also an agricultural pest during early settlement, and the only known island population (St Francis Island, South Australia) was exterminated because of the animal's depredations in the gardens (Jones 1924). They were destroyed in huge numbers by mainland farmers (Seebeck and Rose 1988).

B. gaimardi is also dependent on hypogeal fungi as its main food source (Rose 1982, 1985). The fungi were found by Taylor (1986a,b) to be more abundant in poorer soils, and he speculated that the population density might be influenced by the relative abundance of food. Seeds, roots and bulbs are also eaten, and *Acacia* exudate has been found in some stomachs. Insects do not appear to be important in the diet (Johnson and Rose 1983).

A recent study by Dixon (1988) of a single preserved gut of *C. campestris* showed the presence of coleopteran remains, possibly of weevils. Finlayson (1932) considered that *C. campestris* ate fibrous plant tissue but was unable to be more precise. He reported that the species was independent of free water, but wondered why sandhills, with an abundance of succulent plants, were not used by the animals.

Finlayson's (1931) observation that *A. rufescens* was largely rhizophagous was confirmed by Southgate (1980) and Schlager (1981), who found that roots were the main food items selected by this species in northern New South Wales. The species most commonly eaten were introduced perennial pasture flatweeds — *Hypochoeris, Centaurea* and *Taraxacum* — all of which have long tapering tap-roots. A number of native annual and perennial herbs were also taken, particularly in native forest areas. Hypogeal and epigeal fungi were also eaten, and more than 30 different fungal spore-types were found in faecal samples. *A. rufescens* was observed to shift its home range to take advantage of a seasonal flush of fungi. Insects were eaten during spring and early summer, flower parts were sometimes eaten, and grass seeds were harvested. As with the bettongs, *Acacia* exudate was occasionally eaten. Although Finlayson (1931) averred that *A. rufescens* was dependent on free water Schlager (1981) believed that this was not the case.

In Table 3 we present a summary of the diet of modern potoroids, indicating the major food of each species.

ACTIVITY PATTERNS

H. moschatus is wholly diurnal, and thus differs from other potoroids which are primarily nocturnal. Johnson and Strahan (1982) reported maximum activity in the early morning and late afternoon. During the night and in the middle of the day it sleeps in a nest.

The two extant species of *Potorous* differ in their degree of nocturnality. Guiler (1958) suggested that *P. tridactylus* was strictly nocturnal, but Kitchener (1967) found that activity began about one hour before complete darkness. Seebeck (unpubl.) found *P. tridactylus* to be frequently active during the day. Between June 1973 and November 1974 traps were examined at four times during each 24-hour period (0800, 1200, 1800, 2000 hrs) on 41 days. During that period 671 captures were made of *P. tridactylus*; 239 of these (36%) were made during daylight hours (i.e., animals were in traps at either 1200 or 1800 hrs). There seemed to be little, if any, correlation with season, although the highest percentages of diurnal captures were recorded during summer months. By contrast, *P. longipes* is completely nocturnal. By the use of radiotelemetry, Scotts and Seebeck (1989) found that activity was restricted to the period between dusk and early morning, ceasing some two hours before sunrise when animals returned to and stayed at their nest site. No daylight activity was recorded.

All species of *Bettongia* which have been studied appear to be completely nocturnal. *B. lesueur* remains in the burrow until well after nightfall (Krefft 1866; Dahl 1926). Finlayson (1958) reported that during summer emergence was not until "long after dark". Stodart (1966) found that captive *B. lesueur* did not emerge until more than 30 minutes after nightfall. *B. penicillata* is somewhat less rigidly nocturnal in its activity, leaving the nest at dusk, although returning before dawn (Christensen and Leftwich 1980). Similarly, *B. gaimardi* leaves the nest at dusk, although Kershaw (1971) reported observing two animals active during the day.

C. campestris is presumed to be nocturnal. Finlayson (1932) stated that they fed at night, and animals encountered by his party were all disturbed from daytime shelter in nests.

A. rufescens is also strongly nocturnal. Schlager (1981) reported that emergence was always after sunset, and that return to the nest was prior to sunrise. Males retired earlier than females.

In Table 4 we summarize activity patterns of modern potoroids.

Table 4. Summary of activity patterns of modern potoroids.

	Diurnal	Partly Diurnal	Nocturnal
Hypsiprymnodon moschatus	×		
*Potorous platyops**			
P. tridactylus		×	
P. longipes			×
Bettongia lesueur			×
B. penicillata			×
*B. tropica**			
B. gaimardi			×
Caloprymnus campestris			×
Aepyprymnus rufescens			×

*Not known

POPULATION ECOLOGY

There are large gaps in our knowledge of the population ecology of the potoroids. Nothing is known concerning the extinct species *C. campestris* and *P. platyops*, and only limited information is available for the extant species *B. lesueur, B. tropica* and *H. moschatus* (Table 5). Those species for which data are available are discussed below.

Potorous

BODY SIZE

Bennett (1987a) described a seasonal variation in weight in *P. tridactylus*, with maxima in autumn and winter. The mean body weight of males (but not females) at this time (May-August) was significantly greater than that for the rest of the year (Bennett, unpubl.). He suggested that this was linked to changes in food availability. Weight gains were correlated with an increase in fungi in the diet. However, Kitchener (1967) found that seasonal weight changes observed were not significantly different, although the trends that he observed paralleled those of Bennett. Scotts and Seebeck (1989), although able to sample only a small number of animals, found no seasonal variation in mean body weight of *P. longipes*. They believed that *P. longipes* altered habitat use rather than diet, and thus did not undergo the dietary transition described by Bennett (1987a).

Bennett (1987a) found no sexual dimorphism in body weight in his study population of *P. tridactylus*, although males are generally larger than females (Seebeck, unpubl.). Johnson and Sharman (1976) found evidence of sexual dimorphism in some cranial and external measurements in *P. tridactylus* but their results were not consistent within or between populations. Jarman (1983) considered that potoroids were homomorphic. Male *P. longipes* are some 20% heavier than females (Scotts and Seebeck 1989).

POPULATION SIZE, DENSITY AND COMPOSITION

P. tridactylus (Bennett 1987a,b) and *P. longipes* (Scotts and Seebeck 1989) both maintained relatively stable populations over two years and 13 months respectively, with low turnover of individuals. Both species are relatively long-lived; up to 7+ years for wild *P. tridactylus* (Guiler and Kitchener 1967) and over 10 years for captive *P. longipes* (Seebeck, unpubl.) so this is not unexpected. Further evidence of population stability is found in the low level of transient animals encountered. Bennett (1987a,b) reported a very high mean density for *P. tridactylus* of 2.55 individuals per ha; this contrasts strongly with the density reported by Kitchener (1967, 1973) which was an order of magnitude lower, at 0.2 animals per ha. The density of *P. longipes* was about 0.6 animals per ha, although absolute numbers were small (Scotts and Seebeck 1989).

The composition of the population in both species was relatively constant, and was largely composed of reproductively mature animals (Bennett 1987a,b; Scotts and Seebeck 1989). Bennett (1987a) found that females of *P. tridactylus* always outnumbered males, so that the apparent sex ratio was less than parity. Nevertheless, the sex ratio of pouch young was not significantly different from parity. Scotts and Seebeck (1989) found a ratio of about 1:1 for *P. longipes*.

HOME RANGE AND SOCIAL ORGANIZATION

That home range may be dependent on habitat quality is suggested by the disparity in home-range sizes calculated for *P. tridactylus* in different localities. Kitchener (1967, 1973) reported ranges of 19.4 ha for males and 5.2 ha for females; and Bennett (1987a,b) and Seebeck (unpubl.) both found home ranges of around 2.0 ha for males and 1.5 ha for females in western Victoria. Scotts and Seebeck (1989) reported a trap-revealed home range of about 5 ha for *P. longipes*, but believed this to be an underestimate, based on additional data gathered using radiotelemetry. Hill and Triggs (1985), using radiotelemetry, gave a preliminary estimate of about 10 ha for a male *P. longipes*. The mean range probably falls between five and 10 ha.

Bennett (1987a,b) described a high level of range overlap both within and between sexes of *P. tridactylus*, with few trapping sites occupied exclusively by a single individual. On average, the range of each individual overlapped with the ranges of at least four other individuals of each sex. Male *P. tridactylus* overlapped with females at a significantly greater proportion of sites than they did with other males. Females also shared their home range with others of each sex but there was no significant difference in the extent of overlap with males or females. Kitchener (1967) reported a dominance hierarchy among *P. tridactylus* at an experimental feeding site, but thought that this was because the dominant animal had the site in her home range.

Home ranges of *P. longipes*, on the other hand, were found to overlap only with their single respective mate (or sub-adult offspring) (Scotts and Seebeck 1989). Thus this species appears to be territorial and monogamous — quite different to its congener. Lee and Cockburn (1985) have described this social system spatially as "territorial monogamy" and temporally as "permanent monogamy". *P. tridactylus* appears to fall into the "overlap promiscuity" spatial classification of Lee and Cockburn (1985).

REPRODUCTION

P. tridactylus has a gestation period of 38 days, which is one of the longest reported for marsupials. The gestation period for *P. longipes* is unknown. Both species produce a single young. Pouch life is longer (140-150 days) in *P. longipes* than in *P. tridactylus* (120-130 days), and reproductive maturity is probably reached later in *P. longipes* than in *P. tridactylus* (Seebeck, unpubl.). However, both species probably breed continuously in the wild, and have similar reproductive potential of between 2.5 and three young per year. Bennett (1987a,b) found that at least 70%, but frequently 90%, of female *P. tridactylus* were carrying or suckling young throughout the year.

PREDATION

Heinsohn (1968) believed that the absence of Foxes *Vulpes vulpes* was a major factor in the continuing high population levels of *P. tridactylus* in Tasmania. However, his assertion that the Fox is a more efficient predator than the native carnivores *Dasyurus maculatus* and *D. viverrinus* remains unproven. Both of the latter species are still common in Tasmania, but are either rare or extinct on the mainland. Seebeck (1978) suggested that, in western Victoria, Foxes do not appear to be a controlling influence on *P. tridactylus* populations. However, Dingoes and dogs *(Canis familiaris)* are significant predators on *P. longipes* (Scotts and Seebeck 1989) and no doubt would have preyed on *P. tridactylus* throughout its range in the past as they still do in east Gippsland. Foxes are known to take *P. longipes*, and, in fact, the only modern record of that species from New South Wales is from remains in Fox scats (Dovey 1987).

In summary, the two species of *Potorous* demonstrate rather different population strategies. Although population stability and composition are similar, use of home range and mating systems are quite different. Bennett (1987a) concluded that small body-size, stable population structure, restricted and overlapping home ranges, continuous reproduction and dispersal of both sexes enable *P. tridactylus* to maintain populations in habitats which are naturally patchy in distribution, or have become fragmented. Scotts and Seebeck (1989) believed that *P. longipes* responded to a naturally patchy distribution by maintaining population stability, territorial monogamy, continuous reproduction and dispersal of young and were able to make efficient use of a specific food resource by a low population density and by harvesting the fungal food resource within their home range on a rotation basis.

Bettongia

Bettongia penicillata

BODY SIZE

B. penicillata weighs between 1.1 and 1.6 kg, with a mean of 1.3 kg (Christensen 1983). No significant weight differences between sexes were found by Christensen (1980a) but he demonstrated seasonal variation in weight. Males were heaviest in December and lightest in August. Females showed less well-defined changes, but the greatest weights were recorded during summer.

POPULATION SIZE, DENSITY AND COMPOSITION

Sampson (1971) and Christensen (1980a) studied *B. penicillata* in Western Australia at sites some 160 km apart, with differing thermal and rainfall intensities. Sampson's study site was in the more arid area.

In both study areas *B. penicillata* was relatively common, and sample sizes were large. Sampson (1971) found a density of about 0.07 animals per ha, but Christensen (1980a) found density to be much higher, at 0.45 animals per ha in favoured habitat. Christensen (1980a) found no significant departure from parity in the sex ratio, although males always outnumbered females. Adults always formed the bulk of the population, but sub-adults were present in large numbers. Sampson (1971) found the sex ratio of captured animals to be heavily biased towards males (2.5:1).

HOME RANGE AND SOCIAL ORGANIZATION

Sampson (1971) described trap-revealed home ranges of 28-43 ha (mean 35 ha) for males and 15-28 ha for females. One female which was radio-tracked had a range of about 22 ha. Sampson (1971) believed that, within the larger home range, each animal had a core feeding range of 4-6 ha. He suggested that only males defended territory, and only in the vicinity of their nests. This view was supported by Christensen (1980a) who found that the nest areas (2-3 ha) of males did not overlap with those of other males at all although the larger (9 ha) feeding areas did. His estimates of size of home ranges, also from trapping, were similar to, although slightly smaller than, Sampson's (1971) estimates. Male *B. penicillata* averaged 27 ha, and females about 20 ha in size. He found considerable overlap between adjacent males and females. Animals were solitary, except during courtship and mating and when females had young at foot (Christensen 1980a).

REPRODUCTION AND DISPERSAL

B. penicillata breeds throughout the year. Sampson (1971) found no seasonal variation in testis size in male *B. penicillata*, and thus concluded that there was no evidence for seasonal breeding. Christensen (1980a) found that females first bred at 170-180 days and then continued to give birth to a single young at about 100-day intervals. Significantly more males than females were born in summer. The annual reproductive potential of *B. penicillata* is three young per female.

Christensen (1980a) found very high mortality of young, particularly following pouch vacation, with only 11-15% surviving to adulthood. Young did not disperse large distances but filled nearby vacated areas upon the disappearance of the residents.

PREDATION

Christensen (1980b) convincingly demonstrated the severe impact that predation by Foxes can have on *B. penicillata* in southwestern Western Australia. A complex interrelationship between Fox numbers, rabbits *Oryctolagus cuniculus*, 1080 poison and vegetation influenced the levels of predation on native mammals in areas of forest with an open understorey. When rabbit numbers were controlled by heavy use of 1080 poison, Fox numbers also remained low.

However, with a decline in use of 1080 in the 1970s, and an increased reliance on other rabbit control measures, the abundance of Foxes has greatly increased in southwestern Western Australia and native mammals such as *B. penicillata* have come under greater threat. In open areas lacking patches of dense ground cover *B. penicillata* (among other species) was more easily caught. King *et al.* (1981) examined the effects of 1080 on *B. gaimardi, B. lesueur* and *B. penicillata* and found markedly different tolerances to fluoroacetate, with *B. penicillata* being exceptionally highly tolerant. They also examined the relationships between the decline in rabbit numbers, the increase in Fox numbers and the consequent apparent increase in native mammal predation by Foxes, and recommended that, for *B. penicillata* at least, control of Foxes by 1080 was a potentially valuable management tool.

B. lesueur

BODY SIZE

B. lesueur from Bernier and Dorre Islands seem to be slightly larger than those on Barrow Island. Males are heavier than females, and may reach more than 1.5 kg. Females as small as 750 g may be sexually mature, but the average weight of adult females is about 1.1-1.3 kg. Weight changes of between 10% and 15% may occur with changes in physical condition (R. Prince, pers. comm., 1986; Western Australian Museum records).

POPULATION SIZE, DENSITY AND COMPOSITION

The smallest island on which *B. lesueur* occurs, Boodie Island probably supports only about 50 animals, primarily because only a small part of the island is suitable for burrows. Barrow Island has extensive large warrens that may have 40-50 occupants. The island exceeds 20 000 ha and the total number of *B. lesueur* may be quite large (R. Prince, pers. comm.). Burbidge and Main (1971) found fewer animals in the south (disturbed) of the island than the north (undisturbed), but were unable to account satisfactorily for this phenomenon. Population size is not known for Bernier and Dorre Islands, but Ride and Tyndale-Biscoe (1962) found *B. lesueur* to be "relatively common" and in some parts "extremely abundant", and Burbidge (1983) considered it to be abundant there.

The sex ratio of trapped animals on Dorre Island was 1:1 (R. Prince, pers. comm.).

HOME RANGE AND SOCIAL ORGANIZATION

On Barrow Island animals move up to 5 km from their warrens (R. Prince and W. H. Butler, pers. comm.).

B. lesueur is the only routinely burrowing macropodoid and the only communal nester, occupying large complex warrens (Gould 1863; Krefft 1866; Finlayson 1958). Dahl (1926) regarded *B. lesueur* as a "sociable animal" but noted that "the sociable disposition does not appear to be very prominent in each individual". Stodart (1966) found that captive males were aggressive towards one another and that females were "sociable and peaceful" but showed that there was some evidence of a hierarchy.

Males defended females rather than territory, i.e., controlled a harem. Butler (1970) observed apparently unprovoked attacks by *B. lesueur* on other species, particularly *Trichosurus*, and he believed those attacks to be evidence of territoriality.

REPRODUCTION

The gestation length is 21 days, after which the single young remains in the pouch for 115 days and becomes sexually mature at about 9-10 months. The annual reproductive potential is, like other potoroids, 2-3 young per female. On Bernier and Dorre Islands there is a peak of breeding in April and May, and possibly another in November/December. In captivity, breeding occurred regularly from February to September, after which time the oestrus cycle of the females ceased for several months (Tyndale-Biscoe 1968). This seasonality in breeding makes *B. lesueur* unique among its continuously breeding congeners and other potoroids.

PREDATION

Jones (1924) remarked that Foxes were a major predator on the remaining South Australian population of *B. lesueur* in the 1920s, and Finlayson (1958) also asserted that Foxes were the major factor in the dramatic range reductions he had observed for this species since the 1930s. Other known predators include goannas (Varanidae) (Finlayson 1958), pythons (Boidae) (Dahl 1926), Wedge-tailed Eagle *(Aquila audax)* and Ospreys *(Pandion haliaetus)* (Ride and Tyndale-Biscoe 1962).

B. gaimardi

BODY SIZE

Rose (1982) reported an adult weight of 1.8 kg, although Johnson and Rose (1983) considered that the average adult weight was about 1.7 kg, with a range of 1.2-2.3 kg. No sexual dimorphism was found in weight, but Taylor (1986b) found that males were larger than females in some body dimensions. He attributed the lack of weight difference to the poorer condition of males he encountered.

POPULATION SIZE, DENSITY AND COMPOSITION

Taylor (1986b) trapped *B. gaimardi* at 15 sites in Tasmania, with a trapping success rate ranging from 0.4-21.1 animals per 100 trap-nights. At most sites the rate was between three and eight animals per 100 trap-nights. The sex ratio of trapped animals (and their pouch young) was about parity. More immature males than immature females were caught. Taylor (pers. comm. 1988) reported a density of about 0.19 animals per ha at one site.

HOME RANGE AND SOCIAL ORGANIZATION

The home range reported by Johnson and Rose (1983) of between 65 and 135 ha is very large for members of the Potoroidae. Taylor (pers. comm. 1988) reported the mean monthly home range of males to be 65 ha (range 54-76) and of females to be 45 ha (range 32-54). The species is capable of long movements, of as much as 1.5 km between nest and feeding area. *B. gaimardi* is normally solitary except when mating or when females have young-at-heel.

REPRODUCTION AND DISPERSAL

The gestation period in *B. gaimardi* is 21-22 days, after which one young is born. Pouch life is about 106 days. Breeding is continuous throughout the year, with a slight trend towards more births in winter, and the annual fecundity is about three young per female. Both sexes are reproductively mature at between nine and 12 months (Rose 1987).

PREDATION

Rose (1986) considered *D. maculatus* to prey upon *B. gaimardi* but regarded *D. viverrinus* as unimportant. Probably *Thylacinus* was also a predator. The effects of Feral Cats *Felis catus* are not known.

Johnson and Rose (1983) suggested that, rather than the absence of the Fox from Tasmania being the major factor in the continued survival of bettongs in that state, it is the relatively low density of rabbits and the concomitant retention of suitable cover which has been important. Taylor (1986b) has suggested that land clearing, and an increase in undergrowth density in forests due to an altered fire regime, may also have contributed to the species' decline on the mainland.

Aepyprymnus rufescens

BODY SIZE

A. rufescens is the largest living potoroid, with weights of up to 3.5 kg being reported. Johnson and Bradshaw (1977) stated that females were significantly larger than males, but Schlager (1981) provided evidence that the reverse was true. He found that males weighed about 2.7 kg (range 2.5-3.0) and females 2.5 kg (range 2.2-2.8).

POPULATION SIZE, DENSITY AND COMPOSITION

The species is relatively abundant in suitable habitat within its present range. Southgate (1980) found that population numbers were highest in January and September, and estimated a mean density of about 0.4 animals per ha. Changes in seasonal abundance were considered to be related to the availability of food. Although more females than males were captured by Southgate (1980) the sex ratio was not significantly different from parity.

HOME RANGE AND SOCIAL ORGANIZATION

Southgate (1980) estimated the home range of *A. rufescens* to be about 20 ha, but defined a much smaller (1-2 ha) nightly activity area, in which there were sufficient food resources. Activity areas overlapped between sexes, but there was regular spacing between animals of the same sex. The maximum nightly movement recorded was 625 m. *A. rufescens* individuals may feed in close association. Southgate (1980) found that 12.5% of groups seen were in pairs, the remainder being single individuals; and hence he calculated an overall typical group size of 1.22. Surprisingly few of these "pairs" were of females with young-at-foot. *A. rufescens* may not be as solitary as other potoroids.

Males are aggressive towards other males in captivity but Southgate (1980) found no evidence of aggression in the wild.

REPRODUCTION

Breeding occurs throughout the year but Johnson (1978) found that fewer young were born in summer. The gestation period is 22-24 days, and the single young remains in the pouch for about 114 days. Females may become sexually mature at eight to nine months (P. Jarman, University of New England, pers. comm. 1988). Annual fecundity is about three young per female (Johnson 1978).

PREDATION

Despite the absence of firm evidence of predation on *A. rufescens* by Dingoes during his study, Schlager (1981) considered that they had always been a major population control mechanism. He reported that numbers of *A. rufescens* at Cheviot Hills in northeastern New South Wales rose considerably following the commencement of aerial baiting of Dingoes with 1080 poison in 1968. Foxes were also considered to be a significant predator, as they were observed on several occasions with *A. rufescens* as prey. It was suggested by one observer that Foxes killed many animals but did not eat them (Schlager 1981). Cats are also predators of *A. rufescens* (Marlow 1957), as are Wedge-tailed Eagles (Schlager 1981).

In Table 5 we present a summary of biological and population parameters of modern potoroids discussed in this Section.

DISCUSSION

As Strahan (1983) has suggested, the modern family Potoroidae comprises three groups. There is a single extant species, *H. moschatus*, in the sub-family Hypsiprymnodontinae while the nine modern members of the sub-family Potoroinae have been divided into two tribes, the Potoroini (containing *Potorous*) and the Bettongini (containing *Bettongia, Caloprymnus* and *Aepyprymnus*) (Flannery and Archer 1987). *H. moschatus* differs from other members of the family in many aspects of its morphology, physiology and behaviour: for example, it possesses a second pair of lower incisors (although this feature is also found in the extinct *Bettongia moysei*; Flannery and Archer 1987), and a single lachrymal foramen; it has a first pedal digit, a naked, scaly tail, and a simple stomach; it normally bears two young at a time; and it has a non-saltatory gait. Its taxonomic separation in a separate sub-family is certainly warranted.

These taxonomic divisions are, in general, supported by the ecological evidence that is available. *H. moschatus* is endemic to Wet Tropical Rainforest, and its diet and behaviour (e.g., diurnal activity pattern) differ from other members of the family. Unfortunately, virtually nothing is known as yet of its population biology. *Potorous* spp. (apart from *P. platyops*, for which few ecological data are available) display ecological differences from other family members, primarily by their occurrence in denser habitats in temperate forest environments, their relatively small home ranges and that they do not construct complex nests. *Bettongia* spp. and *Aepyprymnus* occur mainly in more open habitats, and have larger home ranges or are capable of regular, long distance movements. Both genera plus *Caloprymnus* construct complex nests for shelter.

A thorough understanding of distributional patterns of the Potoroidae is hampered by the rapid decline in status and reduction in range of many species following European settlement, combined with a paucity of accurate historical records. However, it is clear that the geographical ranges of the modern species of Potoroidae together encompassed most of the Australian continent; and spanned a wide variety of environments, ranging from tropical rainforest to shrublands of the arid interior. It is also evident that while several species had a wide geographical distribution (e.g., *B. penicillata, B. lesueur*) others (e.g., *P. longipes, H. moschatus*) have had an apparently restricted distribution. For several other species (e.g., *P. platyops, C. campestris*), sub-fossil remains distant from the known occurrence suggest either that the known historical distribution is incomplete, or that a contraction in range was occurring prior to European settlement.

The greatest species-density of potoroids was in the coastal and near-coastal regions of southern Australia, where up to three species were sympatric (Fig. 1). The temperate climate experienced in these forest and woodland habitats is among the most reliable and predictable in the continent. In contrast, the lowest species-density, frequently only a single species, was in the arid

Table 5. Summary of biological and population parameters of modern potoroids.

	Weight (g)[a] Range-mean	Pop^{n} density (anim./ha)	Home Range (ha)		Reproduction[e]			Shelter	Contemporary predators
			M	F	Gest. period (d)	Pouch life (d)	Fecund. (yg/yr)		
Hypsiprymnodon moschatus	360-680(530)					147		nest	dog
Potorous platyops	ca. 800								
P. tridactylus	660-1 640(1 100)	0.2[2]-2.55[d]	19.4[c] 2.0[d]	5.2[c] 1.5[d]	38	120-130	2.5-3	squat	dog, fox
P. longipes	1 600-2 200(1 900)	0.6[f]	5-10[f]			140-150	2.5-3	squat	dog, fox
Bettongia lesueur	970-1 530(1 250)				21	115	2-3	nest in burrow	fox, goanna python, osprey
B. penicillata	1 100-1 600(1 300)	0.07[g]-0.45[h]	28-43[g] 27[h]	15-28[g] 20[h]	21	100	3	nest	fox
B. tropica	n.a.							nest	
B. gaimardi	1 200-2 250(1 660)	0.19[i]	54-76[i]	32-54[i]	21	109	3	nest	dog
Caloprymnus campestris	637-1 060(889)							nest	
Aepyprymnus rufescens	3 000-3 500(3 250)		ca. 20[j]		21-30	114	3	nest	dog, fox, cat, eagle

Sources: a) Seebeck and Rose 1988; *b)* Johnson and Strahan 1982; *c)* Kitchener 1967; *d)* Bennett 1987a; *e)* Tyndale-Biscoe and Renfree 1987; *f)* Scotts and Seebeck 1989; *g)* Sampson 1971; *h)* Christensen 1980; *i)* Taylor 1986b; *j)* Schlager 1981.

and unpredictable environment of inland Australia, and the wet tropical environment of the eastern coast. The absence of potoroids from the tropical woodlands and forests of northern Australia (Fig. 1) is surprising in view of the successful radiation of this group into habitats as forbidding as the arid interior. Knowledge of the availability of potential food items such as hypogeal fungi, tuberous roots and fruits, and their reliability as a food source throughout the sharp seasonal fluctuations in climate in the tropical environment may provide some insights.

In areas of sympatry between two or more species, ecological overlap is usually minimized by a marked separation in the habitats occupied. For example, *P. tridactylus* is sympatric with *B. gaimardi* in eastern Tasmania, and with *A. rufescens* in parts of eastern Australia (Fig. 1). In each region, however, *P. tridactylus* is restricted to vegetation types with a dense understorey cover (Heinsohn 1968; Schlager 1981) whereas *B. gaimardi* and *A. rufescens*, respectively, occur in adjacent vegetation with a more open understorey (Schlager 1981; Rose 1986; Southwell 1987). However, little is known of ecological overlap between potoroids in semi-arid and arid environments. *B. lesueur* and *B. penicillata* were generally sympatric over a large area of southern Australia, but the available information on habitat use and diet of each species in these areas is not sufficient to resolve the basis for any resource partitioning.

Habitats occupied by potoroids are rarely uniform in distribution, but are generally dispersed in a mosaic pattern in response to topographic or edaphic variation in the landscape, or to seral successional stages in vegetation communities. Studies of micro-habitat use for several species *(P. tridactylus, P. longipes, B. pencillata, A. rufescens)* have indicated the importance of fine-scale patchiness in the habitat: patches of denser cover are usually required for shelter, whereas more open vegetation may be favoured for foraging (Christensen 1980; Bennett 1987a; Scotts and Seebeck 1989). Christensen (1980) has discussed the role of fire in creating and maintaining patches of vegetation of suitable seral stage to provide favoured habitat for *B. pencillata*. Wildfire and anthropogenic fire regimes have probably had an important influence on the availability and quality of habitat for other potoroids, also. Kitchener *et al.* (1980) linked the decline and loss of medium-sized mammals (including *B. lesueur, B. penicillata,* and *P. platyops*) in semi-arid Western Australia to changes in the scale of environmental patchiness due to alteration to pre-European fire regimes.

Hume (1982) reviewed the digestive physiology of the potoroids, and categorized the group as herbivorous; but recent evidence led Lee and Cockburn (1985) to classify most potoroids as fungivores/omnivores, with the exception of *H. moschatus* which is a frugivore/omnivore. Detailed observations of diet are available only for those species of temperate forests and woodlands, *A. rufescens, B. gaimardi, B. penicillata, P. longipes,* and *P. tridactylus* (Table 3). These studies have revealed a broad range of food items, and have indicated a dependence on non-fibrous foods, particularly the sporocarps of hypogeal fungi.

Little is known of the relative abundance, environmental distribution, fruiting pattern, or nutritional status of hypogeal fungi, and so the adaptive value of mycophagy as a feeding strategy is poorly understood. Kinnear *et al.* (1979) reported an inbalance of amino acids in fungal sporocarps consumed by *B. penicillata* in Western Australia, but demonstrated that a nutritional balance was achieved by the microbial fermentation of fungi in the large sacciform forestomach. Frappel and Rose (1986) provided evidence for this process in *P. tridactylus*, and have also demonstrated that cellulose plant materials are mostly digested in the caecum and proximal colon. A large forestomach is characteristic of most potoroids, suggesting that other species may also employ microbial fermentation to obtain a nutritionally balanced diet. Hume (1982) suggested that a large forestomach may also serve as a storage organ, facilitating rapid feeding as a means of limiting the risk of predation.

Hypogeal fungi form mycorrhizal relationships with forest trees, enhancing their nutrient uptake and growth. The fungi lack active spore-dispersal mechanisms and rely on mycophagous animals, such as potoroids, for this process. This complex inter-relationship between mycophagous mammals, hypogeal fungi and the host vegetation has only recently begun to be investigated in Australia (Bennett 1987a; Malajczuk *et. al.* 1987; Scotts and Seebeck 1989) and has important ecological implications for both the conservation of the mammals and the health and vigour of their forest habitat.

Further information on the diet and on the seasonal availability of food items of potoroids inhabiting arid and tropical environments is required before the dietary ecology of this group can be fully understood. For example, was *B. penicillata* mycophagous throughout its geographic range (which included some arid areas)? The general dependence of hypogeal fungi on adequate soil moisture (Beaton *et al.* 1984) suggests that fungi, although periodically present, would be an unreliable resource in arid areas. It is likely that potoroids in drier environments depended to a large extent on other non-fibrous foods such as subterranean bulbs, tubers and roots. If the recent examination by Dixon (1988) of preserved material of extinct mammals could be extended to include any museum specimens of *Bettongia* it might assist in solving this problem.

A general synthesis of the population ecology of the Potoroidae is hampered by the lack of basic information for at least four species *(C. campestris, B. tropica, H. moschatus, P. platyops)* and the incomplete nature of the data for those species that have received attention. However, some trends may be recognized. Life-history attributes which appear to be generally common to potoroids include: the limitation of a single young per litter (with the exception of *H. moschatus*, the only macropodoid which rears two young per litter); a continuous pattern of reproduction; embryonic diapause of developing young during pouch-life of the previous young; and an annual fecundity of about three young. There is no marked seasonal fluctuation in abundance, and, with an estimated average longevity of about five years, this contributes to a relatively stable population structure.

Differences between species in aspects of their population ecology generally reflect the different habitats that they occupy. *Bettongia* spp., which generally occur in dry environments and in habitats which are relatively open, have a shorter gestation period, a shorter period of pouch life and hence a slightly higher fecundity than do extant species of *Potorous* which favour densely vegetated habitats in mesic and reliable environments. Similarly, species inhabiting drier environments have relatively larger home ranges and a lower population density than do species in mesic environments. These differences probably reflect variation in climatic predictability and reliability, and the relative abundance of food resources. Nests are constructed for diurnal shelter by all species of potoroids, with the exception of *Potorous* spp., both of which extant species occur in habitats where dense ground cover is readily available. The function of nests is not clear, but appears more likely to be a means of thermoregulatory control, particularly in hot weather, than as a means of reducing predation (Rübsamen *et al.* 1983).

Quantitative evidence of the impact of predation on the population dynamics of potoroids is lacking, but circumstantial evidence from studies on *Aepyprymnus, Bettongia* and *Potorous* suggests that predation by canids may have a controlling influence on populations. In comparison with some other terrestrial small mammals with which they co-exist (e.g., peramelids, dasyurids, rodents), potoroids have a relatively low fecundity; and thus even a low level of predation may influence population size and growth.

Potoroids display little, or no evidence, of sexual dimorphism in body size. This is in marked contrast with the larger macropodids, in which the degree of dimorphism is positively correlated with increasing body size (Jarman 1983). Jarman (1983) has discussed the adaptiveness of homomorphism in *A. rufescens* in relation to its social organization and mating system. Animals live in solitary home ranges which overlap with a few individuals of both sexes, in habitats in which visibility is low. The probability for a male of chance encounters with females in oestrus is low; and so, to gain matings, males invest in knowledge of the status and locations of females, rather than in increased body-size. Females may need to match the size of males to defend access to nest sites and food resources. Jarman (1983) proposes that homomorphism in this and similar species arises from a balance between these adaptive pressures.

The Potoroidae are among those native mammals which have experienced the most dramatic decline in status since European settlement in Australia, and they are now amongst those which are most in need of conservation measures. Unfortunately, it is too late for *C. campestris* and *P. platyops*. Ecological studies, at least of a preliminary nature, of all extant species have recently been completed or are underway, and our knowledge of the biology of this group is steadily growing. Nevertheless, there are many gaps in our knowledge, and it is important that these be addressed so that conservation and management strategies have a sound ecological basis, particularly when population manipulation, such as re-introductions and introductions (Delroy *et al.* 1986), are to be attempted.

The feeding ecology of potoroids requires further investigation, in particular to quantify the availability and seasonal abundance of major foods such as hypogeal fungi, and to relate these parameters to trends in the size and use of home range. Further quantitative descriptions of habitat use are required as are investigations of factors, such as fire, which influence the quality and both the temporal and spatial availability of the habitat. There are few demographic data on age-specific mortality in potoroids. Quantitative data on the relative impact of various sources of mortality, particularly of predation by canids, are urgently required.

The value that ecological knowledge can have in the conservation of small mammals has been recently addressed by Bennett (1987b), who considered that autecological studies formed a sound basis for conservation strategies, especially in disturbed environments. With few exceptions, disturbed environments are the norm for surviving potoroids, and the urgent need for further ecological studies on these uncommonly interesting marsupials is clear, if Finlayson's (1958) gloomy prophesy is not to be their epilogue.

ACKNOWLEDGEMENTS

We wish to thank the many colleagues from the Department of Conservation, Forests and Lands, Victoria, who have helped during field and laboratory investigations on Potoroos. For providing us with unpublished material which has greatly expanded our meagre knowledge of certain species we warmly thank Bob Prince and Tony Friend (Conservation and Land Management, Western Australia), Harry Butler (Western Australia), John Winter (Ravenshoe, Queensland), Andrew Dennis (James Cook University), Peter Jarman (University of New England) and Robert Taylor (Forests Commission, Tasmania). The Australian National Parks and Wildlife Service, Canberra, permitted us access to several unpublished reports. Judy Firman ably typed the manuscript.

REFERENCES

AMOS, P. J, 1982. The potoroo in Queensland. *Queensland Agric. J.* **108**(1): 5-6.

BAYNES, A., 1987. The original mammal fauna of the Nullarbor and Southern Peripheral Regions: evidence from skeletal remains in superficial cave deposits. Pp. 137-52 *in* A Biological Survey of the Nullarbor Region South and Western Australia in 1984 ed. by N. L. McKenzie and A. C. Robinson. Government Printer: Adelaide.

BEATON, G., PEGLER, D. N. AND YOUNG, T. W., 1984. Gasteroid basidiomyceta of Victoria, Australia: 8-9. *Kew Bull.* **40**: 827-42.

BENNETT, A. F., 1987a. Biogeography and conservation of mammals within a fragmented forest environment in southwestern Victoria. Ph.D. Thesis, University of Melbourne.

BENNETT, A. F., 1987b. Conservation of mammals within a fragmented forest environment: the contributions of insular biogeography and autecology. Pp. 41-52 *in* Nature Conservation: The Role of Remnants of Native Vegetation ed. by D. A. Saunders, G. W. Arnold, A. A. Burbidge and A. J. M. Hopkins. Surrey Beatty & Sons: New South Wales.

BURBIDGE, A. A., 1983. Burrowing Bettong, *Bettongia lesueur.* Pp. 187-89 *in* The Complete Book of Australian Mammals ed. by R. Strahan. Angus & Robertson: Sydney.

BURBIDGE, A. AND FULLER, P., 1984. Finding out about desert mammals. SWANS **14**(1): 9-13.

BURBIDGE, A. A. AND MAIN, A. R., 1971. Report on a visit of inspection to Barrow Island November, 1969. Report VIII, Dept. of Fisheries and Fauna, Western Australia.

BUTLER, W. H., 1970. A summary of the vertebrate fauna of Barrow Island, W.A. *West. Aust. Nat.* **11**: 149-60.

BUTLER, W. H. AND MERRILEES, D., 1971. Remains of *Potorous platyops* (Marsupialia, Macropodidae) and other mammals from Bremer Bay, Western Australia. *J. R. Soc. West Aust.* **54**: 53-58.

CALABY, J. H., 1971. The current status of Australian Macropodidae. *Aust. Zool.* **16**: 17-31.

CHESTERFIELD, E. A., MACFARLANE, M. A., ALLEN, D., HUTCHINSON, M. N., TRIGGS, B. AND BARLEY, R., 1983. The flora and fauna of the Rodger forest block, east Gippsland, Victoria. Department of Conservation, Forests and Lands, Victoria. *Ecol. Surv. Rep.* No. **1**.

CHRISTENSEN, P. E. S., 1980a. The biology of *Bettongia penicillata* Gray 1837, and *Macropus eugenii* (Desmarest, 1817) in relation to fire. *For. Dept. West. Aust. Bull.* No. **91**: ix+90 pp.

CHRISTENSEN, P. E. S., 1980b. A sad day for native fauna. *Forest Focus* No. **23**: 1-12.

CHRISTENSEN, P., 1983. Brush-tailed Bettong, *Bettongia penicillata.* Pp. 184-85 *in* The Complete Book of Australian Mammals ed. by R. Strahan. Angus & Robertson: Sydney.

CHRISTENSEN, P. AND LEFTWICH, T., 1980. Observations on the nest-building habits of the Brush-tailed Rat-kangaroo or Woylie *(Bettongia penicillata). J. R. Soc. West. Aust.* **63**(2): 33-38.

CROSSMAN, D. G. AND REIMER, D. S., 1986. Mammals, birds, reptiles and amphibians of the Taroom Shire, central Queensland. *Qld. J. Agric. Anim. Sci.* **43**: 55-72.

DAHL, K., 1926. In Savage Australia. Phillip Allan & Co.: London.

DELROY, L. B., EARL, J., RADBONE, I., ROBINSON, A. C. AND HEWETT, M., 1986. The breeding and re-establishment of the Brush-tailed Bettong *(Bettongia penicillata)* in South Australia. *Aust. Wildl. Res.* **13**: 387-96.

DIXON, J. M., 1988. Notes on the diet of three mammals presumed to be extinct: the Pig Footed Bandicoot, the Lesser Bilby and the Desert Rat Kangaroo. *Victorian Nat.* **105**: 208-11.

DOVEY, L., 1987. Use of databases, indirect survey and predictive techniques in the discovery of the Long-footed Potoroo in New South Wales. *Abstr. 33rd Sci. Mtg. Aust. Mammal Soc.* p. 16.

DRINNAN, A. N. AND HILL, F. A. R., 1984. A study of the distribution and habitat of *Potorous longipes* in East Gippsland, Victoria. *For. Comm. Victoria Res. Br. Rep.* **248**: 16 pp. and App. (unpubl.).

EMISON, W. B., PORTER, J. W., NORRIS, K. C. AND APPS, G. J., 1975. Ecological distribution of the vertebrate animals of the Volcanic Plains-Otway Range area of Victoria. *Fish. Wildl. Pap. Victoria* **6**: 1-93.

FINLAYSON, H. H., 1931. On mammals from the Dawson Valley, Queensland. Part 1. *Trans. Proc. R. Soc. S. Aust.* **50**: 67-89.

FINLAYSON, H. H., 1932. *Caloprymnus campestris.* Its recurrence and characters. *Trans. Proc. R. Soc. S. Aust.* **56**: 148-67.

FINLAYSON, H. H., 1958. On Central Australian mammals (with notice of related species from adjacent tracts). Part III, the Potoroinae. *Rec. S. Aust. Mus.* **3**(2): 235-302.

FLANNERY, T. F., 1984. Systematics, zoogeography and biochronology of the Macropodoidea (Marsupialia). Ph.D. Thesis, University of New South Wales.

FLANNERY, T. F. AND ARCHER, M., 1987. *Bettongia moyesi,* a new and plesiomorphic kangaroo (Marsupialia: Potoroidae) from Miocene sediments of northwestern Queensland. Pp. 759-67 *in* Possums and Opossums: Studies in Evolution ed. by M. Archer. Surrey Beatty & Sons: New South Wales.

FRAPPELL, P. B. AND ROSE, R. W., 1986. A radiographic study of the gastrointestinal tract of *Potorous tridactylus,* with a suggestion as to the role of the foregut and hindgut in potoroine marsupials. *Aust. J. Zool.* **34**: 463-71.

GILES, E., 1889. Australia Twice Traversed: The Romance of Exploration, being a Narrative Compiled from the Journals of Five Exploring Expeditions into and through

Central South Australia, and Western Australia, from 1872 to 1876. Sampson Low, Marston, Searle and Rivington: London.

GILL, E. D., 1953. Geological evidence in Western Victoria relative to the antiquity of the Australian aborigines. *Mem. Nat. Mus. Vic.* **18:** 25-92.

GLOUERT, L., 1929. Eagles and wallabies. *Victorian Nat.* **45:** 243.

GOULD, J., 1863. The Mammals of Australia. 3 Vols. The Author: London.

GREEN, B. AND NELSON, L., 1988. The field biology of the Woylie, *Bettongia penicillata,* on Island 'A', Venus Bay, South Australia. Unpubl. rept. to Australian National Parks and Wildlife Service, 41 pp.

GUILER, E. R., 1958. Observations on a population of small marsupials in Tasmania. *J. Mammal.* **39:** 44-58.

GUILER, E. R., 1971. Food of the potoroo (Marsupialia: Macropodidae). *J. Mammal.* **52**(1): 232-34.

GUILER, E. R. AND KITCHENER, D. A., 1967. Further observations on longevity in the wild potoroo, *Potorous tridactylus. Aust. J. Sci.* **30:** 105-06.

HEINSOHN, G. E., 1968. Habitat requirements and reproductive potential of the macropod marsupial *Potorous tridactylus* in Tasmania. *Mammalia* **32**(1): 30-43.

HILL, F. A. R. AND TRIGGS, B. E., 1985. Ecology and distribution of the Long-footed Potoroo *(Potorous longipes)* — a second preliminary examination. *State For. Lands Serv. Victoria. Res. Br. Rep.* No. **310** (unpubl.).

HOPE, J. H., 1974. The biogeography of the mammals of the islands of Bass Strait. Pp. 397-415 *in* Biogeography and Ecology in Tasmania ed. by W. D. Williams. Dr W. Junk: The Hague.

HORROCKS, G. F. B., OPIE, A. M., CARR, G. W., CHERRY, K. A., BROWN, G. W., CRAIG, S. A. AND TRIGGS, B. E., 1984. Flora and fauna of the Ellery forest block, east Gippsland, Victoria. Department of Conservation, Forests and Lands, Victoria. *Ecol. Surv. Rep.* No. 3.

HUME, I. D., 1982. Digestive Physiology and Nutrition of Marsupials. Cambridge University Press: Melbourne.

JARMAN, P., 1983. Mating system and sexual dimorphism in large, terrestrial, mammalian herbivores. *Biol. Rev.* **58:** 485-520.

JOHNSON, K. A., 1979. Construction and site selection of nests by Gaimard's rat kangaroo, *Bettongia gaimardi,* in Tasmania. *Bull. Aust. Mammal Soc.* **6:** 29.

JOHNSON, K. A. AND ROSE, R., 1983. Tasmanian Bettong, *Bettongia gaimardi.* P. 186 *in* The Complete Book of Australian Mammals ed. by R. Strahan. Angus & Robertson: Sydney.

JOHNSON, P. M., 1978. Studies of macropodidae in Queensland. Reproduction of the Rufous Rat-kangaroo (*Aepyprymnus rufescens* (Gray)) in captivity with age estimation of pouch young. *Qld. J. Agric. Anim. Sci.* **35**(1): 69-72.

JOHNSON, P. M. AND BRADSHAW, I. M., 1977. Rufous Rat-Kangaroo in Queensland. *Qld. Agric. J.* **100:** 181-83.

JOHNSON, P. M. AND STRAHAN, R., 1982. A further description of the Musky Rat-kangaroo, *Hypsiprymnodon moschatus* Ramsay, 1876 (Marsupialia, Potoroidae), with notes on its biology. *Aust. Zool.* **21:** 27-46.

JOHNSTON, P. G., 1973. Variation in island and mainland populations of *Potorous tridactylus* and *Macropus rufogriseus* (Marsupialia). Ph.D. Thesis, University of New South Wales.

JOHNSTON, P. G., 1983. Long-nosed Potoroo, *Potorous tridactylus.* P. 181 *in* The Complete Book of Australian Mammals ed by R. Strahan. Angus & Robertson: Sydney.

JOHNSTON, P. G. AND SHARMAN, G. B., 1976. Studies on populations of *Potorous* Desmarest (Marsupialia). I. Morphological Variation. *Aust. J. Zool.* **24:** 573-88.

JONES, F. W., 1924. The Mammals of South Australia. Volume II. Government Printer: Adelaide.

KERSHAW, R. C., 1971. Notes on Tasmanian Rat Kangaroos. *Victorian Nat.* **88:** 4-10.

KING, D. R., OLIVER, A. J. AND MEAD, R. J., 1981. *Bettongia* and fluoroacetate: a role for 1080 in fauna management. *Aust. Wildl. Res.* **8:** 529-36.

KINNEAR, J. E., COCKSON, A., CHRISTENSEN, P. AND MAIN, A. R., 1979. The nutritional biology of the ruminants and ruminant-like mammals — a new approach. *Comp. Biochem. Physiol.* **64a:** 357-65.

KITCHENER, D. J., 1967. Some aspects of the biology of the potoroo — *Potorous tridactylus* (Kerr). B.Sc. (Hons.) Thesis, University of Tasmania.

KITCHENER, D. J., 1973. Notes on home range and movement in two small macropods, the Potoroo *(Potorous apicalis)* and the Quokka *(Setonix brachyurus). Mammalia* **37:** 231-40.

KITCHENER, D. J., 1983. Broad-faced Potoroo, *Potorous platyops.* P. 182 *in* The Complete Book of Australian Mammals ed. by R. Strahan. Angus & Robertson: Sydney.

KREFFT, G., 1866. On the vertebrated animals of the lower Murray and Darling, their habits, economy and geographical distribution. *Trans. Phil. Soc. New South Wales* **1866:** 1-38.

LEE, A. K. AND COCKBURN, A., 1985. Evolutionary Ecology of Marsupials. Cambridge University Press: Melbourne.

LUNDELIUS, E. L. Jnr., 1957. Additions to knowledge of the ranges of Western Australian mammals. *West. Aust. Nat.* **5**(7): 173-82.

LYNE, A. G., 1974. Bettongs. *Aust. Wildl. Her.* **3**(41): 1310-12.

MAIN, A. R. AND YADAV, M., 1971. Conservation of macropods in reserves in Western Australia. *Biol. Conserv.* **3**(2): 123-33.

MALAJCZUK, N., TRAPPE, J. M. AND MOLINA, R., 1987. Interrelationships among some extomycorrhizal trees, hypogeous fungi and small mammals: Western Australian and northwestern American parallels. *Aust. J. Ecol.* **12:** 53-55.

MARLOW, B. J., 1957. A recent record from New South Wales of the Rufous Rat-Kangaroo, *Aepyprymnus rufescens* (Gray) (Macropodidae). *C.S.I.R.O. Wildl. Res.* **2:** 166-76.

MENKHORST, P. W. AND BEARDSELL, C. M., 1982. Mammals of southwestern Victoria from the Little Desert to the coast. *Proc. R. Soc. Victoria* **94:** 221-47.

[MORTON, W. L.] AN OLD BUSHMAN, 1861. Notes of a tour in the Wimmera District. *The Yeoman, and Australian Acclimatizer* Vol. 1.

MUIR, B. G., 1979. Vegetation of Marchagee Nature Reserve. *Rec. West. Aust. Mus.* Suppl. No. **8:** 15-28.

NEWSOME, A. E., MCILROY, J. AND CATLING, O., 1974. The effects of an extensive wildfire on populations of twenty ground vertebrates in south-west Australia. Pp. 107-23 *in* Managing Terrestrial Ecosystems (Vol. 8 of *Proc. Ecol. Soc. Aust.*) ed by J. Kikkawa and H. A. Nix.

OPIE, A. M., CHERRY, K. A., HORROCKS, G. F. B., CARR, G. W., SCHULZ, M. AND TRIGGS, B. E., 1984. Flora and fauna of the Yalmy forest block, east Gippsland, Victoria. Department of Conservation, Forests and Lands. Victoria. *Ecol. Surv. Rep.* No. **2**.

[PHILLIP, A.,] 1789. The Voyage of Governor Phillip to Botany Bay: with an account of the establishment of the colonies of Port Jackson & Norfolk Island. lxxiv+298 pp. John Stockdale: London.

RAMSAY, E. P., 1876. Description of a new genus and species of rat kangaroo, allied to the genus *Hypsiprymnus*, proposed to be called *Hypsiprymnodon moschatus*. *Proc. Linn. Soc. New South Wales* **1**: 33-35.

RIDE, W. D. L. AND TYNDALE-BISCOE, C. H., 1962. Mammals. Pp. 54-97 *in* The Results of an expedition to Bernier and Dorre Islands, Shark Bay, Western Australia, in July 1959 ed. by A. J. Fraser. Fauna Bulletin No. **2**, Western Australian Fisheries Department.

ROSE, R. W., 1982. Tasmanian Bettong *Bettongia gaimardi*: maintenance and breeding in captivity. Pp. 108-10 *in* The Management of Australian Mammals in Captivity ed. by D. D. Evans. Australian Mammal Society and Zoological Board of Victoria: Melbourne.

ROSE, R. W., 1985. The reproductive biology of the Tasmanian Bettong, *Bettongia gaimardi*. Ph.D. Thesis, University of Tasmania.

ROSE, R. W., 1986. The habitat, distribution and conservation status of the Tasmanian Bettong, *Bettongia gaimardi* (Desmarest). *Aust. Wildl Res.* **13**: 1-6.

ROSE, R. W., 1987. Reproductive biology of the Tasmanian Bettong *(Bettongia gaimardi:* Macropodidae). *J. Zool., Lond.* **212**: 59-67.

ROYCE, R. D., 1962. Botany. Pp. 31-53 *in* The Results of an Expedition to Bernier and Dorre Islands, Shark Bay, Western Australia, in July, 1959 ed. by A. J. Fraser. Fauna Bulletin No. **2**. Western Australian Fisheries Department.

RÜBSAMEN, U., HUME, I. D. AND RÜBSAMEN, K., 1983. Effect of ambient temperature on autonomic thermoregulation and activity patterns in the rufous rat-kangaroo *(Aepyrymnus rufescens:* Marsupialia). *J. Comp. Physiol.* **153**: 175-79.

SAMPSON, J. C., 1971. The biology of *Bettongia penicillata* Gray 1837. Ph.D. Thesis, University of Western Australia.

SCHLAGER, F. E., 1981. The distribution and status of the Rufous Rat-kangaroo, *Aepyprymnus rufescens* and the Long-nosed Potoroo, *Potorous tridactylus* in northern New South Wales. Report to the New South Wales National Parks & Wildlife Foundation. Department of Ecosystem Management, University of New England. Report No. **18**.

SCOTTS, D. J. AND SEEBECK, J. H., 1989. Ecology of *Potorous longipes* (Marsupialia: Potoroidae); with preliminary recommendations for its conservation in Victoria. *Arthur Rylah Inst. Env. Res. Tech. Rept. Ser.* **62**. National Parks and Wildlife Division, Victoria.

SEEBECK, J. H., 1976. Mammals in the Pomonal district, the Grampians. *Victorian Nat.* **93**: 138-47.

SEEBECK, J. H., 1978. Diet of the fox *Vulpes vulpes* in a western Victorian forest. *Aust. J. Ecol.* **3**: 105-08.

SEEBECK, J. H., 1981. *Potorous tridactylus* (Kerr), (Marsupialia: Macropodidae): its distribution, status and habitat preferences in Victoria. *Aust. Wildl. Res.* **8**: 285-306.

SEEBECK, J. H. AND JOHNSON, P. G., 1980. *Potorous longipes* (Marsupialia: Macropodidae): a new species from eastern Victoria. *Aust. J. Zool.* **28**: 119-34.

SEEBECK, J. H. AND ROSE, R. W., 1988. Potoroidae. Ch. 30 *in* Fauna of Australia. Vol. 1B ed. by G. R. Dyne and D. W. Walton. Australian Government Publishing Service: Canberra.

SERVENTY, D. L. AND MARSHALL, A. J., 1964. A natural history reconnaissance of Barrow and Montebello Islands, 1958. *C.S.I.R.O. Div. Wildl. Res. Tech. Pap.* **6**.

SHORTRIDGE, G. C., 1909. An account of the geographical distribution of the marsupials and monotremes of south-west Australia, having special reference to the specimens collected during the Balston Expedition of 1904-1907. *Proc. Zool. Soc. Lond.* **55**: 803-48.

SMITH, M. J., 1971. Small fossil vertebrates from Victoria Cave, Naracoorte, South Australia. 1. Potoroinae (Macropodidae), Petauridae & Burramyidae (Marsupialia). *Trans. R. Soc. S. Aust.* **95**: 185-98.

SOUTHGATE, R., 1980. Factors affecting the spatial and temporal use of habitat by the rufous rat-kangaroo *Aepyprymnus rufescens*. B. Nat. Res. Thesis, University of New England.

SPECHT, R. L., 1972. The Vegetation of South Australia. Government Printer: Adelaide.

STODART, E., 1966. Observations on the behaviour of the marsupial *Bettongia lesueuri* (Quoy & Gaimard) in an enclosure. *C.S.I.R.O. Wildl. Res.* **11**: 91-101.

STRAHAN, R. (ed.), 1983. The Complete Book of Australian Mammals. Angus & Robertson: Sydney.

TATE, R., 1879. The natural history of the country around the head of the Great Australian Bight. *Trans. Phil. Soc. S. Aust.* **2**: 94-128.

TAYLOR, R. J., 1986a. Habitat requirements of the Tasmanian Bettong, *Bettongia gaimardi*. *Abstr. 32nd Sci. Mtg., Aust. Mammal Soc.* P. 19.

TAYLOR, R. J., 1986b. Habitat requirements and conservation of the Tasmanian Bettong, *Bettongia gaimardi*. Unpubl. rept. to Australian National Parks and Wildlife Service, 35 pp.

TEDFORD, R. H., 1967. The fossil Macropodidae from Lake Menindee, New South Wales. *Univ. Calif. Publ. Geol. Sci.* **64**: 1-156.

TYNDALE-BISCOE, C. H., 1968. Reproduction and postnatal development in the marsupial *Bettongia lesueur* (Quoy & Gaimard). *Aust. J. Zool.* **16**: 577-692.

TYNDALE-BISCOE, H. AND RENFREE, M., 1987. Reproductive Physiology of Marsupials. Cambridge University Press: Cambridge.

WAKEFIELD, N. A., 1963a. Sub-fossils from Mt Hamilton, Victoria. *Victorian Nat.* **79**: 323-30.

WAKEFIELD, N. A., 1963b. Mammal sub-fossils from near Portland, Victoria. *Victorian. Nat.* **80**: 39-45.

WAKEFIELD, N. A., 1964. Mammal sub-fossils from basalt caves in southwestern Victoria. *Victorian Nat.* **80**: 275-80.

WAKEFIELD, N. A., 1964b. Mammal sub-fossils of the basalt plains of Victoria. *Proc. R. Soc. Victoria* **77**: 419-25.

WAKEFIELD, N. A., 1967. Some taxonomic revision in the Australian marsupial genus *Bettongia* (Macropodidae) with description of a new species. *Victorian Nat.* **84**: 8-22.

[WARD, T.] written by Fountain, P., 1907. Rambles of an Australian Naturalist. John Murray: London.

WHITE, J., 1790. Journal of a Voyage to New South Wales with Sixty-five Plates of Non descript Animals, Birds, Lizards, Serpents, Curious Cones of Trees and other Natural productions. J. Debrett: London.

Community ecology of macropodoids

BARRY J. FOX[1]

Macropod communities in Australia are not generally very species-rich (with some exceptions). Almost half of the continent now supports only two species although the maximum richness reaches 12 on the east coast and 10 or 11 in the southwestern and northeastern. Macropods can be grouped on body size. Large kangaroos occupy the entire continent, exhibiting a maximum richness of four species in semi-arid western New South Wales. Wallabies occupy the forested margins of the continent reaching a peak richness of five species in southeastern Queensland and northeastern New South Wales. Small wallabies are found in denser moist forests and the richest communities comprise three species. A fourth group, rock wallabies, should be considered separately because of their specialized habitat requirements, rock piles or cliff faces. A fifth group, the now almost extinct hare-wallabies, are also treated separately, because of the massive range decreases suffered since the European invasion; they formerly reached their maximum richness of six species in the arid zone which they shared with the Red Kangaroo and the Euro.

Available information indicates strong ecological separation between species, with partitioning of resources, along habitat and diet selection axes. However, there are no controlled manipulation experiments to demonstrate competitive interactions between these species although agonistic behaviour has been recorded when resources are in short supply. All available evidence suggests that habitat selection plays an important role in the organization of macropod communities, but very few studies have attempted to obtain direct evidence of current interspecific interaction. However, there is strong indirect evidence indicating the importance of competition between species and future research should provide a better understanding of the relative roles of these two important factors.

INTRODUCTION

By far the majority of studies on the ecology of macropods have been essentially autecological (e.g., Ealey 1962, 1967a,b,c and Newsome 1965a,b). However, some studies have had as their primary aim the comparison of pairs of sympatric species (e.g., Caughley 1964; Newsome 1975; Taylor 1985). The first attempt to study the ecology of a community of macropods was made by Kaufmann (1974). However, multivariate methods of analysis were not applied until later (Christensen 1980). These are now routinely applied in such studies (see, for example, Algar 1986; Southwell 1987).

In this review I have taken the broad view of what constitutes community ecology and include any studies which report interactions between macropods. The ecology of potoroids is described elsewhere in this volume (Seebeck, Bennett and Scotts) and they will not be considered here apart from their incidental inclusion with studies of macropodids and their inclusion in maps of the numbers of species present over the continent. As a framework for analysis I use body size (male adult body weight) as a convenient factor to classify the macropods and I then consider their geographical distributions within these body-size groups to examine patterns in community organization. These body-size groupings show distinct differences in the types of macrohabitat used, so influencing the geographic distribution of macropods. Areas with high or low species-richness and the trends observed along latitudinal and longitudinal gradients can be identified readily; and this further highlights the need to consider present and past distributions separately because of the drastic reductions in the range of many species following the European invasion. However, the lack of detailed survey information on past distributions in the species-rich east of the continent severely limits such analyses. Macropods demonstrate strong community organization with species showing marked ecological separation. This is analysed by a detailed examination of case studies from the literature that have addressed the question: how do sympatric species of macropods interact and partition the resources available to them? Macropod diets and species interactions are then reviewed as separate topics. I conclude that macropod communities separate most strongly along an axis related to habitat selection. Further partitioning may occur in diet preference, but this is of less importance except where it is complementary for species with high overlap in habitat use.

THE DISTRIBUTION OF SPECIES RICHNESS

Distribution maps are based on Strahan (1983) and Archer, Flannery and Grigg (1984) with additional updating from more recent sources (A.N.P.W.S. 1988; Burbidge *et al.* 1989). Because the maps of species richness have been constructed from distribution maps drawn by a number of different people, some small anomalies with the actual distributions on the ground may

[1]School of Biological Science, University of New South Wales, Kensington, New South Wales 2033.

Pages 89-104 in KANGAROOS, WALLABIES AND RAT-KANGAROOS ed. by G. Grigg, P. Jarman and I. Hume. Surrey Beatty & Sons, Pty Limited, New South Wales, Australia, 1989.

result but do not affect the broad patterns considered here.

Present distributions

I have used previously published boundaries for size classes (A.N.P.W.S. 1988). A boundary at about 25 kg usefully separates the six large species of kangaroo: the Black Wallaroo, *Macropus bernardus*; the Wallaroo and Euro sub-species of *M. robustus*; the Antilopine Wallaroo, *M. antilopinus*; the Red Kangaroo, *M. rufus*; the Eastern Grey Kangaroo, *M. giganteus*; and the Western Grey Kangaroo, *M. fuliginosus*. The distribution of the numbers of species in this group found at any place is illustrated in Fig. 1A. The greater part of Australia, the arid zone and the eastern coast, has only two species of large kangaroo, while areas in the north of the continent, and the southwestern and southeastern extremities of Australia have only a single species. Three species are found in a swathe along the semi-arid zone, with one area in western New South Wales, and extending into Queensland, supporting four species. The ways in which the species of large kangaroos are able to coexist will be examined in detail later in this paper.

Large wallabies and small wallabies can be conveniently separated at 10 kg with the large wallaby group (10-25 kg) comprising *Macropus agilis, M. dorsalis, M. irma, M. parryi, M. rufogriseus, Wallabia bicolor* and the two tree Kangaroos *Dendrolagus bennettianus* and *D. lumholtzi.* The distribution of species richness in this group is shown in Fig. 1B which clearly demonstrates how much of Australia does not support any macropods from this group. All members of this group are forest or woodland species and this is reflected very markedly in their distributions which do not extend beyond the coastal border of the continent where the rainfall is sufficient to support forests or woodlands. The maximum species richness is five, in a narrow coastal strip in southeastern Queensland and extending across the New South Wales border, with an orderly decrease in richness as one moves away from this area.

With the exception of the Northern Nailtail Wallaby *(Onychogalea unguifera)* across the north of the continent, and perhaps *O. fraenata* in southeastern Queensland, all of the species of small wallaby (3-10 kg) considered here are essentially forest species *(M. eugenii, M. parma, Setonix*

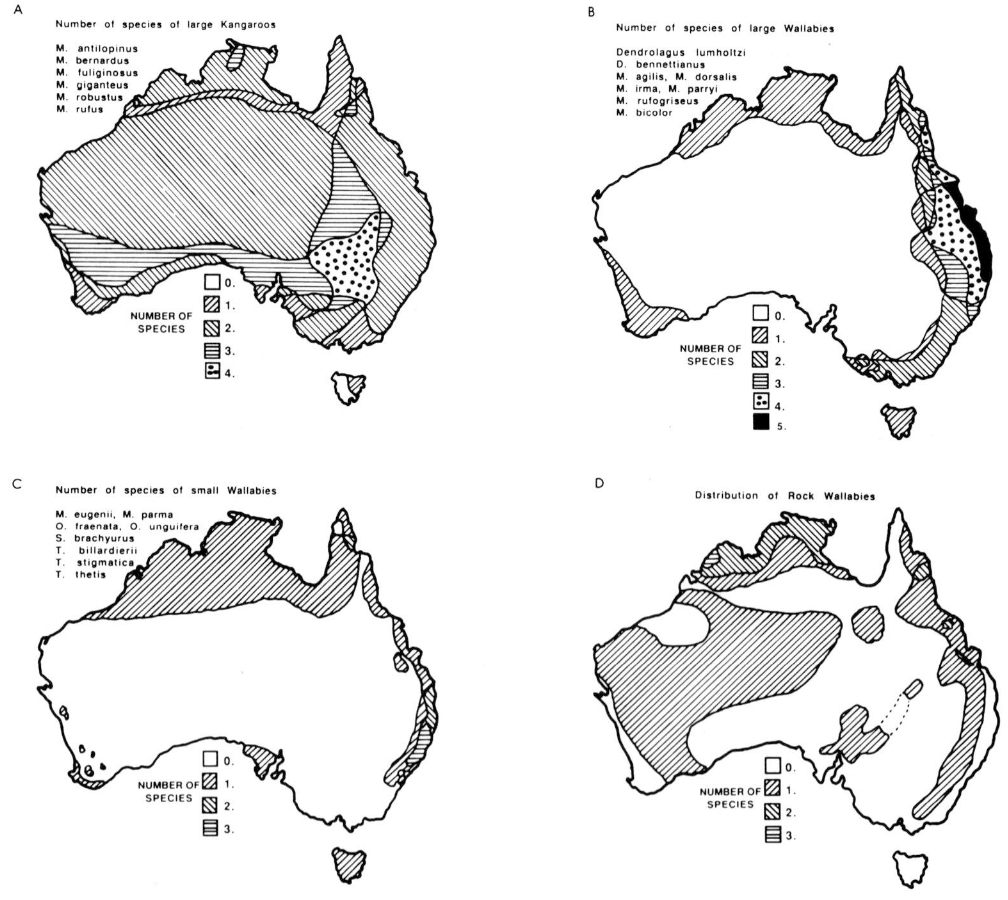

Fig. 1. Distributions of the number of species for (A) large kangaroos (>25 kg), (B) large wallabies (10-25 kg), (C) small wallabies (<10 kg), (D) rock-wallabies. The species included in each figure are listed in the text and the key shows the number of species present.

brachyurus, Thylogale billardierii, T. stigmatica and *T. thetis*). The extreme case is represented by *T. stigmatica* which is restricted almost entirely to dense, closed forest habitat for feeding as well as resting or sheltering. The maximum richness for this group is three and occurs in northeastern New South Wales, leaving a large area with but a single species and an even larger area with no species from this group (Fig. 1C). The formerly more extensive distribution of *O. fraenata* through inland Queensland and New South Wales would merely have expanded the area carrying one species from this group.

The rock-wallabies' distribution of species is almost entirely devoid of overlap (Fig. 1D). The only exceptions are three small patches in Queensland, and much of the Top End which carry two species, and a small area of the Kimberleys with three species. This, combined with their specialized habitat requirements, rock piles or cliff faces, means that their contribution to the structure of macropod communities is limited to those areas with appropriate habitat within the distributions indicated. For these reasons they have been separated for special attention.

Past distributions

The other group deserving special attention is the hare-wallabies, not because of their present distribution, rather, for their past distribution. With these I have included the past distributions of small wallaby species that have suffered substantial range contractions: *O. fraenata* and *O. lunata* and the potoroids (Fig. 2A).

Here we see this group of diminutive macropods (all except *Onychogalea* are below 3 kg) as the only ones sharing the arid zone with the large kangaroos. This is an interesting comment on the physiological limitations existing in the arid zone. There seems to be no niche available to medium-sized species. Macropods in the arid zone need to be small enough to have low absolute resource requirements for both energy and nitrogen, and most especially for shade or shelter. Otherwise they need to be large enough to endure the rigours of low-nitrogen diets and the stressful heatloads encountered. The Red Kangaroo (*M. rufus*) and the Euro (*M. robustus*) represent two different solutions to this latter problem while the hare-wallabies (*Lagostrophus* and *Lagorchestes*) represent solutions to the former.

The map of past distribution of these diminutive species (Fig. 2A) should be interpreted with care because of the different levels of reliability in the data presented. Distributions in the western half of the continent have had the benefit of an up-to-date survey to locate positive sighting records within living memory of Aboriginal

A

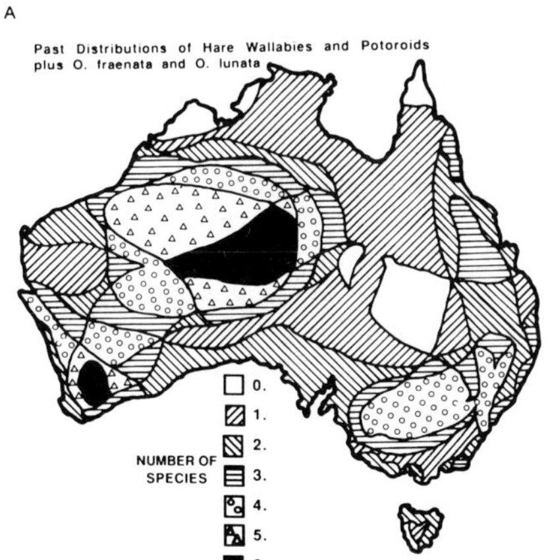

B

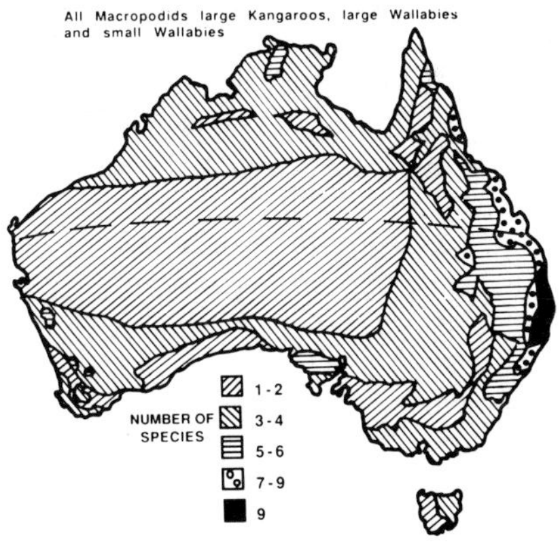

C

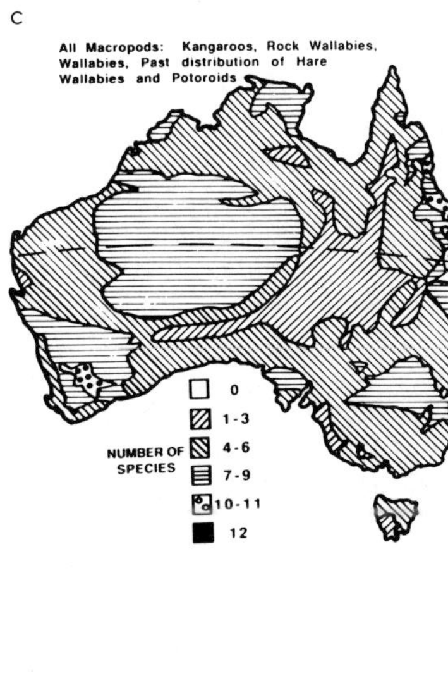

Fig. 2. (A) The past distributions of the number of species of hare-wallabies and potoroids, with the addition of *Onychogalea fraenata* and *O. lunata*. (B) The present distribution for the total number of species of macropodids. (C) The past and present distributions of all species of macropods. The longitudinal transects, in Fig. 3A, are taken along the Tropic of Capricorn shown on the figure.

peoples (Burbidge *et al.* 1989). Such information is not yet available for the eastern states, hence the two areas with zero species-richness may be an artifact of the lack of published data. More detailed surveys of past distributions for eastern Australia could produce centres of species richness as in the south-west and central Australia. The largest area of high species-richness (five and six species) covers much of the arid zone and complements the ranges of the two species of large kangaroo which are the only other macropods in this area.

Unfortunately, being small became a less viable solution after the European invasion and its attendant disturbance to both the habitat and the custodial activities of the Aboriginal peoples. This conclusion is inescapable when one considers the extinction and decline of macropods over the 200 years of European influence (see Johnson and Burbidge, this volume). All endangered or extinct species except *M. greyii* come from what Johnson and Burbidge call the critical weight range (less than 5.5 kg mean adult body weight). A comparison of the distributions of species richness shown in Figs 1 and 2 clearly emphasizes the markedly different patterns for each group of macropods.

Total number of species of macropods

(i) PRESENT MACROPODS

The present day distribution of all macropods (Fig. 2B) has a roughly concentric pattern that can be related to present patterns of total annual rainfall, with increasing species-richness as one moves from west to east across the southern half of the continent. However, the pattern of species-richness is not related to rainfall alone because the increase in the tropics is less than one might expect. The pattern more likely also relates to factors such as the density of canopy cover. Another factor likely to be important in maintaining species richness is the patchiness of the habitat mosaic, where mixtures of rainforest, wet sclerophyll and dry sclerophyll forest, woodland and grassland may be found in the same local area. The ecotone effect, where areas of grassland or grassy understorey abut forests with denser lateral and canopy cover, is particularly important for the maintenance of locally high species-richness. The area with the highest species-richness of nine species occurs in the region of the Macpherson-Macleay overlap where there is a substantial overlap of floristic elements from the tropical and temperate zones (see Burbidge 1960). It also lies in an area that is not strongly seasonal with respect to rainfall, so that plants there benefit from both summer and winter rainfall. This aseasonality of rainfall is also a feature of the area of maximum richness (four species) for the large kangaroos, as it is for the whole superfamily (Fig. 2C).

(ii) PAST AND PRESENT MACROPODS

The distribution of total species richness for all macropods (past and present distributions) is shown in Fig. 2C. Three areas of high species-richness are identified. The richest area, with 12 species, lies in a narrow band along the east coast from 27°S to 31°S latitude, straddling the New South Wales-Queensland border. This area is bordered by a band with 10 or 11 species. Detailed studies of the macropod communities from this area (Kaufmann 1974; Southwell 1987) will be reviewed later. The second rich area (10 or 11 species) lies in the central coastal area of Queensland (17°S to 25°S), while the third rich area (10-11 species) is in the south-west of Western Australia (31°S to 33°S) near the site of another study to be reviewed later (Algar 1986). Areas with seven to nine species border each of these rich areas and another very extensive area is centred on the arid zone while small areas occur in the Top End and the Kimberleys in northern Australia and on the Eyre Peninsula in South Australia.

Trends in species-richness

Latitudinal trends are most commonly considered; however, because of the concentric patterns observed for species-richness, I consider here the longitudinal trends as well.

(i) LONGITUDINAL TRENDS IN PAST AND PRESENT SPECIES-RICHNESS

The considerable difference between past and present distributions is very evident if one considers a longitudinal transect along the Tropic of Capricorn. While such a transect does not pass through the richest area, the transect is readily identifiable and is representative of the longitudinal changes. The distribution of present richness, as a slice through Fig. 2B demonstrates, is an ordered increase in species-richness as one moves from west to east which correlates with rainfall (Fig. 3A). When the past distributions are included, and one takes a slice through Fig. 2C, two peaks in species-richness are demonstrated, one on the east coast and one centred on the arid zone, with lower species-richness for the semi-arid zone and the edge of the arid zone (Fig. 3A). A comparison of the two profiles graphically illustrates the loss of species. The greatest difference between the two occurs in the arid zone where there has been the greatest loss of small species, below the critical weight limit (Johnson and Burbidge, this volume). However, east of longitude 135° the difference should be regarded as a minimum because the past distributions have not benefitted from the detailed survey information available west of that point (Burbidge *et al.* 1989).

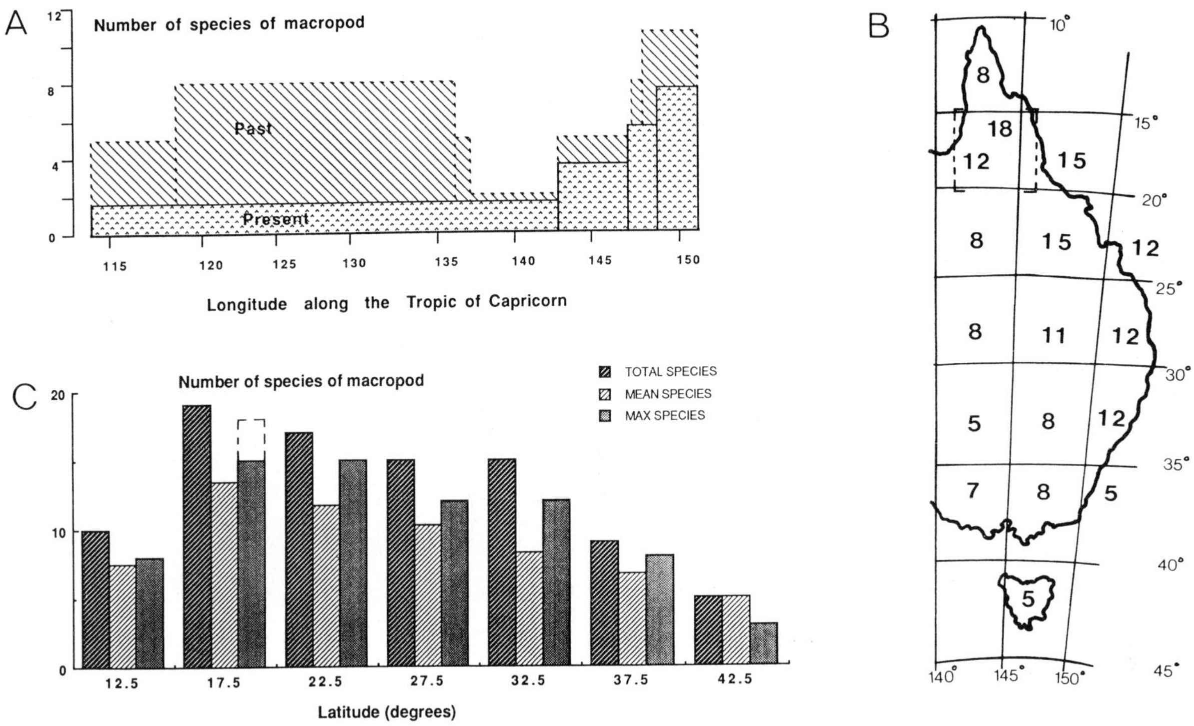

Fig. 3. (A) A profile of the number of species of macropods for a longitudinal transect along the Tropic of Capricorn (see Fig. 2B,C). Past distribution is shown as a dashed line and present distribution is shown as a solid line. (B) A map showing the distribution of the present number of species of macropods in five-degree squares east of longitude 140°E. (C) The latitudinal trends for the total number of species, the mean number of species and the maximum number of species in five-degree bands of latitude (see text).

(ii) LATITUDINAL TRENDS IN PRESENT SPECIES-RICHNESS

Using a grid of five-degree squares one can examine latitudinal trends in species-richness along the east coast (Fig. 3B). Three measures can be used: the number of species in each of the five-degree bands of latitude; the mean number of species per five-degree square for each of the latitude bands; or the five degree square with the maximum number of species for each of the latitude bands. The trends in these three measures are shown in Fig. 3C. The shape of the Australian east coast creates an inequality of land areas available in the coastal squares, with substantial reductions in area at both the southern and northern ends of the transects. This can be corrected to some extent by considering the square with the maximum value for each band. A further correction for the 15°-20° band can be made by centring the square on the land area available at the base of Cape York (shown dashed in Fig. 3B). This results in an increase in richness from 15 to 18, while a similar correction (eight to nine species) can be made by combining the land area for the two squares in Victoria. The two extreme latitudinal bands fall below values that might be expected, reflecting the much reduced land area available on the peninsula and island respectively. There is no simple way to correct for the reduced land area in these squares and for this reason I have omitted them from the analysis and consider five bands only, 15°-40°. These data can be tested against the one-tailed prediction that species-richness will increase as one moves toward the equator. The linear regressions for all three variables are significant, strongly supporting the expected latitudinal trend: total number of species [$y=27.1-0.44x$, $n=5$, $r=-0.93$, $p<0.025$]; mean number of species [$y=19.45-0.34x$, $r=-0.999$, $p<0.0025$]; maximum number of species [$y=24.75-0.42x$, $r=-0.97$, $p<0.005$]. Even without the corrections the maximum number of species is significantly related to latitude [$r=0.93$, $p<0.025$]. The values predicted by the extrapolated regression line would be: Cape York 17.4 and Tasmania 9.0. Latitudinal trends usually reflect increasing temperature and, with the exception of the extreme bands, these data would seem to support that.

THE ECOLOGICAL SEPARATION OF MACROPODS WITHIN COMMUNITIES

A large proportion of the studies on macropods have been directed at the large kangaroos and much of this work has been autoecological. Questions about the organization of species in large kangaroo communities have not been asked specifically until recently. However, it is possible to piece together some information from comparisons made between pairs of sympatric species.

Studies of pairs of kangaroo species

The Red Kangaroo *(M. rufus)* and the Euro *(M. robustus)* show the greatest area of overlap and much work has been carried out on each species (e.g., Newsome 1965b; Ealey 1967b and related papers). A number of studies have identified the physiological constraints within which these two species must operate in the arid zone (e.g., Dawson and Denny 1969 and related papers), and the review by Russell (1974) was followed by the first direct ecological comparison of these two species (Newsome 1975).

Newsome (1975) comments on the similarity of these two species, in terms of physical parameters and the physiological processes they have in common, but notes their marked ecological separation. The Euro leads a sedentary existence in the ranges and hills, feeding on a low-nitrogen diet while the Red Kangaroo leads a nomadic existence on the plains, feeding on nitrogenous herbage (Newsome 1975).

Another early study compared the Red Kangaroo with the Eastern Grey Kangaroo *(M. giganteus)* in southwestern Queensland (Caughley 1964). Caughley demonstrated a correlation between the density of *M. giganteus* in five different habitats and a measure of horizontal or lateral cover in those habitats, but found no relationship for *M. rufus*. The five habitats are shown ranked from the least dense (Mitchell Grass) to the most dense (Spinifex) in Fig. 4A. The spinifex habitat showed the least, and the thick gidgee habitat the greatest proportion of use by the Red Kangaroo. When the habitats are ranked in terms of canopy cover (Fig. 4B) the Spearman rank correlation with habitat use for the Red Kangaroo is $r_s=0.943$ (when for two-tailed tests $r_{s(0.1)}=0.900$ and $r_{s(0.05)}=1.000$). The Grey Kangaroo has its greatest proportion of use in the low-canopy-cover spinifex habitat and demonstrates a clear partitioning of habitat along an axis for canopy cover. I have interpreted this axis as one of heat-load avoidance and the horizontal-cover axis as one of predator avoidance, as Caughley had concluded. The trends in habitat use by the two species would seem to indicate that predator avoidance is more important to *M. giganteus*, while *M. rufus* places greater importance on the shade to be found under denser canopy cover.

The Western Grey Kangaroo *M. fuliginosus* has also been shown to partition resources when sympatric with *M. rufus*, with much of the work detailing this reported by Caughley, Shepherd and Short (1987). Priddel (1987) demonstrated that at Kinchega National Park the sand-plain landform was used much more than the floodplain landform by both *M. rufus*, (0.76±0.04) and *M. fuliginosus* (0.70±0.07) during 1980 (an average rainfall year, Wellard 1987). The figures represent

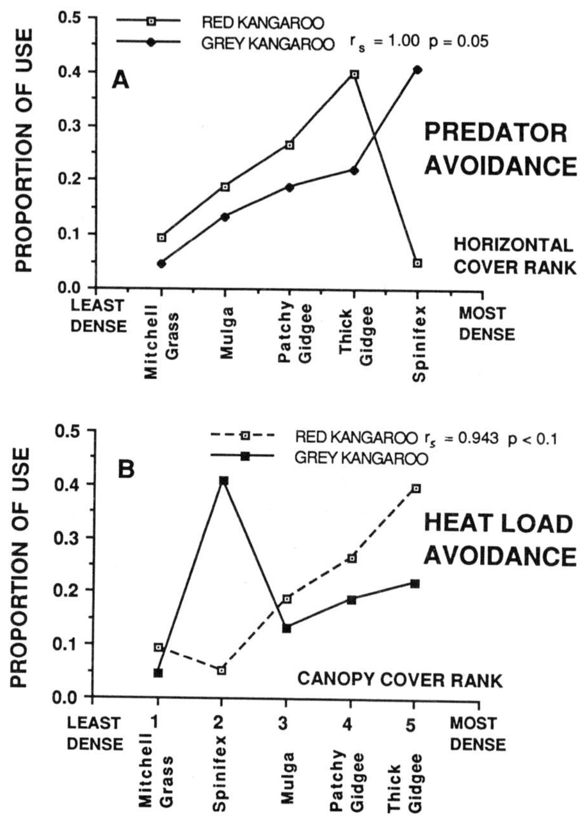

Fig. 4. The proportional use of habitats by the Red Kangaroo *Macropus rufus* (open symbols) and the Eastern Grey Kangaroo *M. giganteus* (solid symbols) at Cunnamulla in south-west Queensland, measured as the density of each species in the five habitats from the density of each species' faeces. (A) The relation between density and horizontal cover rank (Grey Kangaroo: Spearman rank correlation $r_s=1.00, p<0.05$). (B) The relation between density and canopy cover rank (Red Kangaroo: $r_s=0.943, <0.1$). Data taken from Caughley (1964).

the mean and standard error for the proportion of use in each landform determined from the number of animals counted in two replicate surveys flown during each quarter of the year. There is no significant difference between each species' proportional use of these landforms. However, for 1981 (a below-average rainfall year, Wellard 1987) there is a significant difference ($t=2.86$, $p<0.05$) with continuing high proportional use of sand plains by the Red Kangaroo (0.68±0.03) while the Western Grey (0.46±0.07) shows stronger preference for the floodplain. Johnson and Bayliss (1981) demonstrate that *M. fuliginosus* showed stronger selection for grey-black soil types (floodplain) which supported black box woodlands or cane grass, usually dry, swamplands (38%) than for the red soil shrublands (29%). Transitional soils between these two were most favoured by *M. rufus* (43%).

In good years there is little difference in the diets of these two species but during periods of low rainfall *M. rufus* concentrates more on chenopod shrubs such as *Maireana pyramidata* while *M. fuliginosus* concentrates on other shrubs such as *Acacia victoriae* rather than the grasses and forbs both species favour when they are readily available (Barker 1987). The patterns of grazing for both species are similar although *M. fuliginosus* does show a 22% drop in time spent grazing during summer (Priddel 1986). Another important point to be considered here is the ability of each species to utilize the food available. Western Grey Kangaroos are unable to maintain any food intake below a vegetation biomass of 180 kg ha^{-1} while Red Kangaroos show negligible reduction in food intake down to 100 kg ha^{-1} and are able to maintain some food intake while any vegetation remains (Short 1985). In addition, in contrast to the Red Kangaroo's very steep rise in food intake with increasing vegetation biomass, the Western Grey Kangaroo's intake rises slowly and does not reach that of the Red Kangaroo until almost 600 kg ha^{-1}, and continues to increase at high levels of vegetation biomass (Short 1987). The effect of these notable differences is clearly reflected in the mortality rates of these two species in the arid zone; Western Grey Kangaroos suffer higher mortalities, and die sooner in drought (Robertson 1986).

Data on the daily activity patterns of the Eastern and Western Grey Kangaroos indicate that they are active in the day (sunrise to sunset) for 19% and 39% of the time, respectively (Coulson 1978). The same study showed that in this community the Red-necked Wallaby *(M. rufogriseus)* was much more diurnal with 60% of its sightings during daylight hours.

Research has also been conducted to examine the ecological separation between *M. robustus* and other species of large kangaroos; the most extensive study was designed as an examination of ecological differences between sympatric

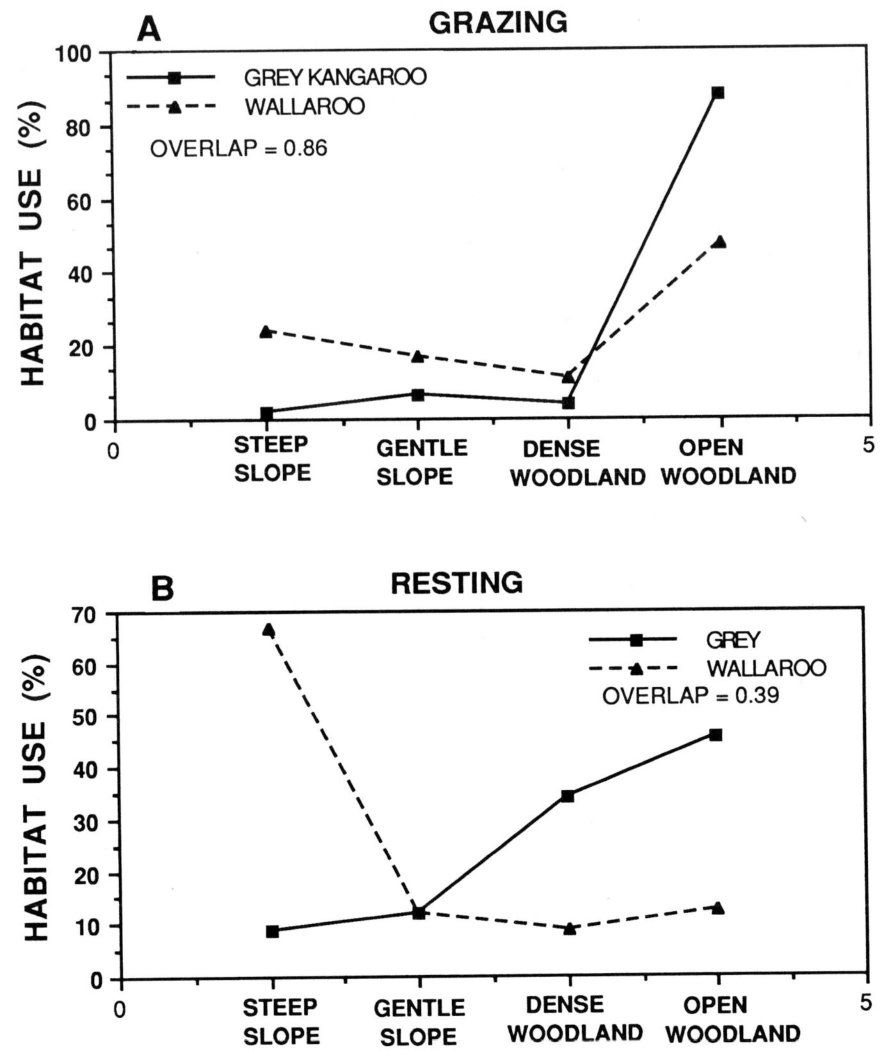

Fig. 5. The percentage use of habitats by the Eastern Grey Kangaroo *Macropus giganteus* (square symbols) and the Wallaroo *M. robustus* (triangular symbols) near Armidale, New South Wales, measured as the density of each species from direct sightings along transects. (A) the strong overlap (0.86) for grazing habitat. (B) The reduced overlap (0.39) for resting habitat. Data taken from Taylor (1985).

M. robustus and *M. giganteus* in northeastern New South Wales (Taylor 1983, 1984, 1985). A simplification of the habitats recognized by Taylor (1985) into four categories (steep slope, gentle slope, dense woodland and open woodland) demonstrates that, although these two species show a high overlap in their use of these habitats for grazing (0.86, Fig. 5A), there is substantial ecological separation in their use of these habitats for resting (Fig. 5B). The Wallaroo clearly prefers the steep slope while the Grey Kangaroo prefers the woodland habitats, demonstrating a low spatial overlap (0.39).

Marked habitat separation was also reported between the Common Wallaroo and the Antilopine Wallaroo in northern Australia (Russell and Richardson 1971). They reported *M. robustus* to be confined to rugged country with caves and rock shelters, a very specific habitat preference; *M. antilopinus* uses more open savannah woodland and open forest but is also found in the rocky areas occupied by *M. robustus*.

A study of sympatric *M. robustus* and the Black Wallaroo *M. bernardus* found the latter to be three times as abundant and revealed differing dispersion for these two species (Press, this volume). *Macropus robustus* showed no significant differences in density among any sites while *M. bernardus* showed significantly different densities among monsoon forest sites and also among spinifex sites, but no differences among woodland sites. However, the data are insufficient to allow direct comparisons of habitat preference between these species (Press, this volume). Further studies including the large kangaroos are considered below in the wider context of studies of macropod communities.

Studies of more species-rich macropod communities

A descriptive study from northern Queensland (Bell 1973) shows a gross partitioning of habitats, ranging from open forest on rugged hilly slopes to open scrub on alluvial plains, between four species: *M. giganteus, M. agilis, M. parryi* and *Petrogale* sp. Each selects a different range of habitats from those available (Table 1).

Marked spatial separation between three species, recorded during a study of the socio-ecology of *M. antilopinus* south of Darwin (Croft 1987), is shown in Fig. 6 which demonstrates the lack of spatial overlap between these species. Where limited overlap does occur it represents the peripheral parts of each species' local distribution. The spatial separation is related to habitat selection; *M. antilopinus* occupied ridge top, slope and valley but was not found on the steep-sided rocky ridges with thick scrub, most favoured by *M. robustus. Macropus agilis* was seldom seen on ridges and was most often observed in narrow valleys. Croft (1987) was able to link the habitat occupied, and the tendency to aggregate, to the size of groups and an hypothesized difference in predation pressure experienced by these species in the different habitats and to their body size as well.

Four species of macropods were studied by Algar (1986) on a 250 ha reserve near Perth. Three species, *M. fuliginosus, M. irma* and *Setonix brachyurus*, would normally occur sympatrically in the area, although the latter usually showed marked habitat segregation into swampy, densely vegetated sites with plenty of lateral cover. The fourth species, *M. robustus*, was purposely introduced to the area; the southern limits to its distribution lie not too far north of the site of the reserve. The four species partitioned the habitat available on the reserve despite the fact that it was much less varied than they might encounter in the wild. This study has been one of the few to apply the power of multivariate analysis to help understand complexity in macropod communities. Fig. 7A shows a plot of the first two discriminant function axes based on structural components of the feeding

Table 1. The results of a descriptive study of the habitat use of four species of macropods, *Macropus giganteus, M. agilis, M. parryi,* and *Petrogale* sp. Data adapted from Bell (1973).

TOPOGRAPHY	RUGGED HILLY SLOPES	UPLAND UNDULATING	FLAT UNDULATING	ALLUVIAL PLAINS		
VEGETATION	OPEN FOREST	WOODLAND	SAVANNNAH WOODLAND	OPEN FOREST	OPEN FOREST	OPEN SCRUB
TREE SPECIES	IRONBARK/ RED BLOODWOOD	IRONBARK/ GHOST GUM		POPLAR GUM/ GREY BLOODWOOD	POPLAR GUM	TEA-TREE
MACROPODIDS						
PETROGALE SP.	XXXXXXXXXX					
MACROPUS PARRYI	XXXXXXXXXX	XXXXXXXXXX				
M. GIGANTEUS		XXXXXXXXXX	XXXXXXXXXX			
M. AGILIS		XXXXXXXXXX	XXXXXXXXXX	XXXXXXXXXX	XXXXXXXXXX	XXXXXXXXXX

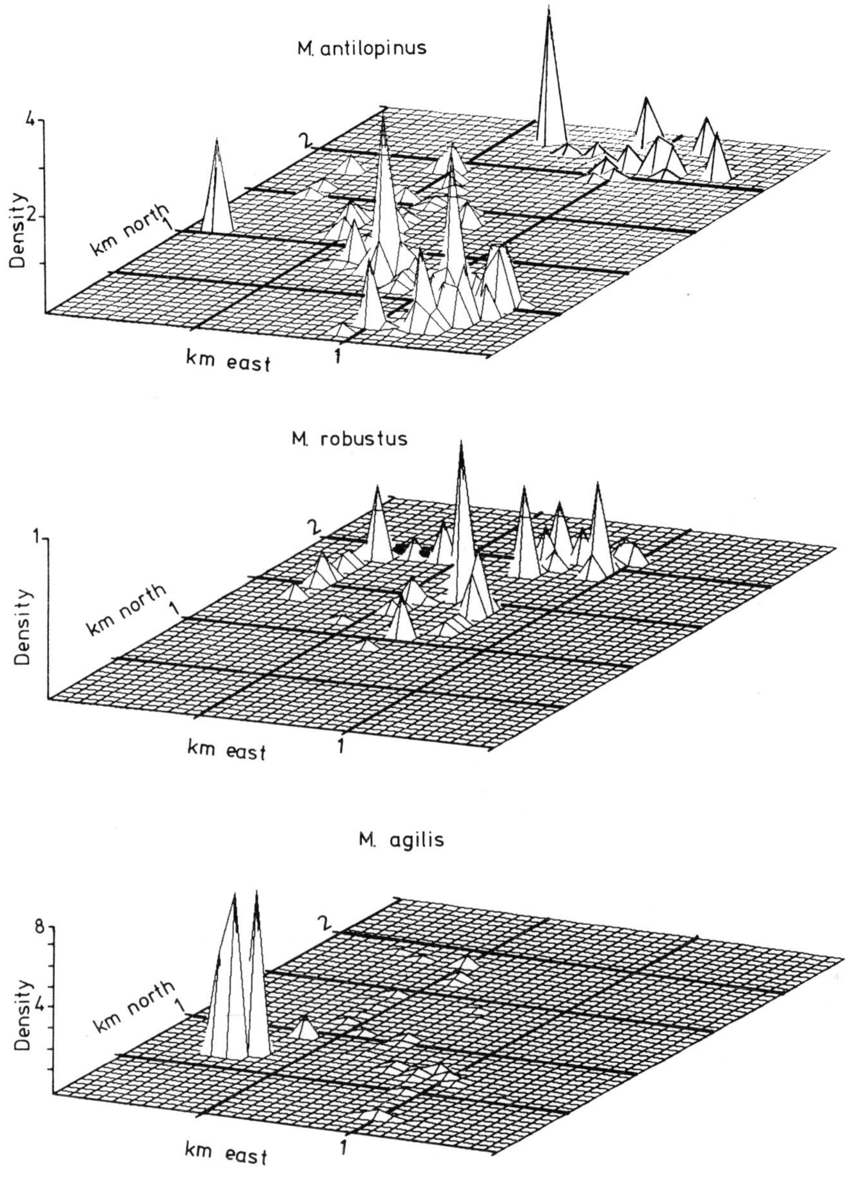

Fig. 6. Spatial separation of three northern species of macropods, *Macropus antilopinus, M. robustus* and *M. agilis*, near Darwin, measured as the density of each species from direct sightings along transects (figure adapted from Croft 1987).

habitat, with the centroid and 95% confidence ellipse indicated for each species. The first axis can be simplified to relate to dense lateral cover and sedge thicket to the positive end, and high shrub abundance to the negative end, with the axis being a site-specific step-function rather than a gradient. The positive direction of the second axis also relates to high shrub abundance, but associated with woodland, while the negative direction relates to sites with low shrub abundance that are grass dominated. The Brush Wallaby *M. irma* and the Western Grey Kangaroo show considerable overlap while the Euro separates on the second axis and the Quokka *S. brachyurus* is well separated along the first axis which is a good representation of the swamp and lake margins it occupies. A seasonal cycle in species' centroid separation was recorded also. Algar (1986) then examined the diets for these species using a similar discriminant function analysis. The results are shown in Fig. 7B using a greatly simplified version from Algar (1986) to show how the axes related to diet components. The first axis was dominated by grasses and sedges at the positive end with perennial herbaceous and non-herbaceous dicots at the negative end. The second axis is all dicots with species from the wetter swamp area at the positive end and from the drier woodland area at the negative end. The two species that overlapped on habitat use show marked separation in their diet, the Western Grey Kangaroo favouring monocots in the diet while Brush Wallabies' diet oontained a high proportion of dicots, although monocots were also eaten. This is a good example of niche complementarity, where species that overlap on one niche axis most often show clear ecological separation on a second axis. This is often taken to imply the existence of interspecific interaction but the evidence is indirect.

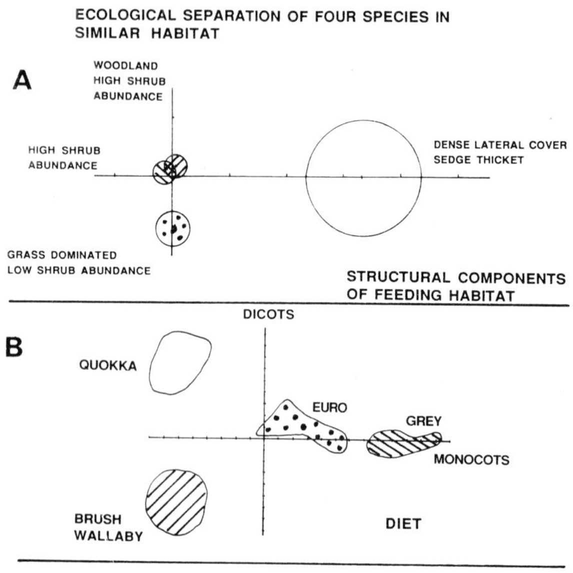

Fig. 7. Ecological separation of four species of macropod in similar habitat, near Perth, Western Australia, illustrated on discriminant function axes as labelled. (A) Structural components of feeding habitats. (B) Diet components. Figures adapted from Algar (1986).

A similar pattern of separation is shown by the Tammar Wallaby *(M. eugenii)* and the Woylie *(Bettongia penicillata)*, two smaller macropods found in forests south of Perth. Christensen (1980) was the first to use multivariate methods to study the structure of macropod communities. Figure 8 shows a principal component analysis of the plant species encountered in quadrats at all sites within the study area. The heavy outline defines the two-dimensional projection of all quadrats sampled in the forest. The first axis relates to wet or seasonally wet sites at the positive end while conversely at the negative end, one will find well drained sites. The second axis is more geographic and probably relates to rainfall and the potential productivity associated with that rainfall. Overlaid on to this figure is an identification of those sites which had Bettong nests nearby (shown hatched). When a similar exercise is carried out for the Tammar it is very notable that, with a few small exceptions where overlap occurs, the Tammar distribution falls in areas where the Bettong does not occur. This marked spatial separation reflects the ecological separation of the two species as they partition the habitats in these forests.

Another study that has made very successful use of multivariate techniques is the major project being conducted at Wallaby Creek (Jarman *et al.* 1987). Southwell (1987) reports the densities and distributions of seven species of macropods on that 370-ha study area in northeastern New South Wales. Four species were sufficiently abundant for their association with habitat factors to be analysed using canonical correlation analysis. An abbreviated and greatly simplified summary of these results is presented in Table 2, where the six most important variables were selected from the 19 reported by Southwell (1987). The Rufous Bettong *Aepyprymnus rufescens* shows the strongest correlation with Altitude and Canopy Cover (both negative) while the same two variables show important positive correlations with the Whiptail Wallaby *M. parryi* which has its strongest correlation with slope. The Red-necked Wallaby shows its strongest correlation with the distance from the edge of the forest (−) and also with the diversity of the ground vegetation (+). The Eastern Grey Kangaroo showed weaker correlations than the other species but its strongest correlation was a negative relation with canopy cover. A potentially more useful analysis for considering the community is shown in Southwell (1987: Table 6) and relates the species that occur together and the environmental parameters they relate to. Eastern Grey Kangaroos and Rufous Bettongs both use open areas close to the valley floor and distant from the forest edge where tussock grass was green and cattle density high, while Whiptail Wallabies show a negative association with these variables. A second grouping of species was identified as the three large species without the Rufous Bettong and was found close to the forest edge of high elevation with abundant other grass and patchy Blady Grass. The third grouping of species was a high

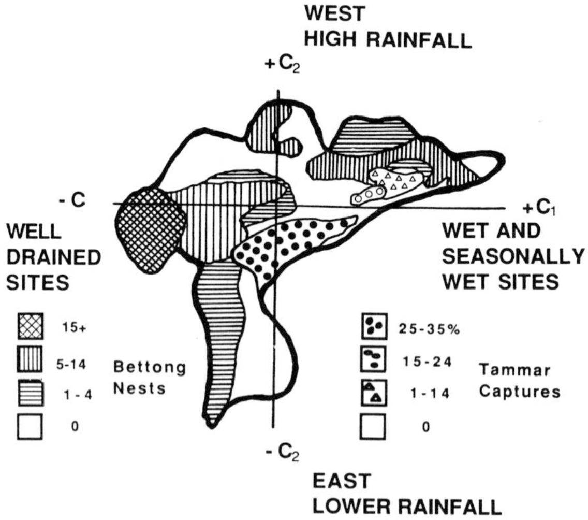

Fig. 8. Ecological separation of Tammar Wallabies and Brush-tailed Bettongs shown on the two-dimensional component space, measured from a principal component analysis of the plant species present in each quadrat. The locations of all quadrats are defined by the heavy outline and overlain by density classes for Bettong nests and Tammar captures, demonstrating the clear separation between species. Adapted from Christensen (1980).

Table 2. Spatial separation among four species of macropods, *Macropus giganteus, M. parrryi, M. rufogriseus* and *Aepyprymnus rufescens,* using a canonical correlation analysis based on densities measured from direct observations along transects. Each species is associated with a component, and the loading of physical habitat parameters on to these components is shown (adapted from Southwell 1987).

	MACROPOD DENSITY			
	A.rufescens	*M. rufogriseus*	*M. parryi*	*M. giganteus*
Altitude	-0.86	0.24	0.37	0.27
Canopy Cover	-0.48	-0.44	0.42	-0.44
Distance from forest edge	0.21	-0.50	-0.25	0.21
Ground Vegetation Diversity	0.03	0.45	-0.05	0.03
Slope	0.09	-0.28	0.49	0.03
Tussock grass greenness	0.34	-0.19	-0.14	0.25

density of Red-necked Wallabies with low densities of the Grey Kangaroo and the Whiptail Wallaby; this group relates to flat, open areas close to forest edges with diverse ground vegetation and sparse but uniform distribution of other grass.

The first attempt to analyse the structure, behaviour and other elements of a complete macropod community was undertaken by Kaufmann (1974). Although he did not use multivariate techniques it is possible to take the data presented and re-analyse them using a principal component analysis. Kaufmann (1974: Table 1) presented values for the percent of total observations in pastures on ridge tops and lower slopes, and in open forest on hill or ridgeface for the five most common species. In the text he provided information on the remaining species and also some limited data on observations in the area of closed forest. Additional information was available from the literature. A detailed study of the habitat requirements of *Petrogale penicillata* provided a predictive equation incorporating five variables: the aspect of the cliff face and four variables relating to properties of ledges on the cliff faces (Short 1982). Hence I added an additional habitat of rocky cliffs for which *Petrogale penicillata* alone was scored as dependent. Detailed information is also available for use of habitat by *Thylogale thetis* (Johnson 1980), indicating that activity is partitioned into a large diurnal home range within the rainforest and a much smaller nocturnal home range in pasture abutting this rainforest. Little information is available for *M. dorsalis*, but it also appears to require dense cover (Kirkpatrick 1983). Hence I have scored both species highly on an additional habitat type of closed forest with dense cover which is consistent with Kaufmann's (1974) text. The results of this analysis are presented in Fig. 9. The X-axis correlates strongly and positively with the open, cleared habitats: ridge top, with scattered eucalypts and hill, with some densely wooded patches remaining. There was a weaker negative correlation with the closed forest habitat with dense canopy-cover. The Y-axis correlates positively with the ridge-face habitat which is relatively undisturbed eucalypt open-forest, while there is a negative correlation with the lower slopes habitat which is largely cleared for pasture. A third axis correlates positively with rocky cliffs and negatively with closed-forests and if it were placed on the figure would separate *Petrogale penicillata* from *M. dorsalis* and *Thylogale* sp. The ecological separation between species illustrated in Fig. 9 is largely due to habitat selection and relates well to groups of species that may be encountered together in the same habitat.

An alternative way to demonstrate this is to use a very simplistic approach that considers a hierarchy of dichotomous choices that represent habitat selection (Fig. 10). One additional parameter is included — body size. The nine species can be split into five that prefer cover to be reasonably dense and are usually associated with forest, while the four remaining species prefer sparser, more open cover for foraging although this is usually not far from a forest edge. A second habitat division is shown dividing the open cover group into: species preferring ridge top and hill habitats, which are one large species *M. giganteus* and one medium species *M. rufogriseus*; and those preferring the even more open slopes and pastures, which are the medium sized *M. parryii* and the small *A. rufescens.* The five species in the habitats with better cover divide into: those preferring hillsides and ridge face with rocks and caves, which are the large *M. robustus* and the small *P. penicillata*; and the species preferring denser cover habitats, which are the medium sized *W. bicolor* and the small wallabies *M. dorsalis* and *Thylogale* sp. which are not able to be separated on Kaufmann's data.

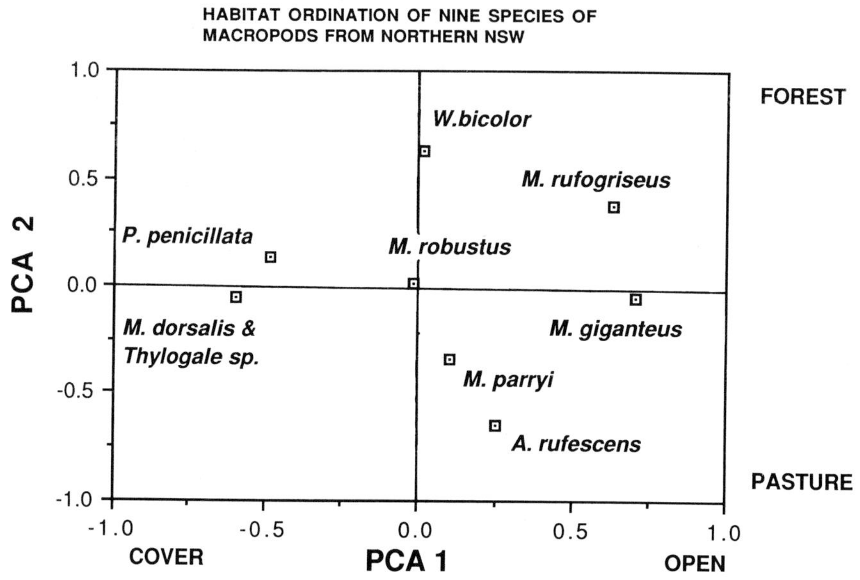

Fig. 9. The habitat ordination of nine species of macropods from northeastern New South Wales, from a principal component analysis of densities from sightings along transects and additional information (see text). Data taken from Kaufmann (1974).

Macropod diets

The study of macropod diets has provided much useful information on the species' ecological separation. Studies on single species, such as *Wallabia bicolor* (Hollis, Robertshaw and Harden 1986), or pairs of species, such as *M. giganteus* and *M. robustus* (Taylor 1983), can be incorporated into studies of larger communities of macropods (e.g., Jarman and Phillips, this volume).

Jarman and Phillips (this volume) have analysed the diets of the 10 species occurring at Wallaby Creek, six from analyses performed on samples from Wallaby Creek and the other four from studies on other comparable areas within the region. I will very briefly summarize these results as they nicely complement the habitat separation observed at Wallaby Creek (Southwell 1987). The larger species, the Eastern Grey Kangaroo, Wallaroo and Whiptail Wallaby feed on the abundant grasses with only 1% to 4% of dicot leaf material. The medium-sized Red-necked and Black-striped Wallabies have larger proportions of dicots (16% to 19%) and consequent reductions in grass intake. The Swamp Wallaby is, as in many other characteristics, an exception, taking only 14% grasses and 43% dicot leaf but 25% of fern and 18% above-ground fungi. Of the small species, the Red-Necked Pademelon is more similar to larger species, taking 67% grass and 33% dicot leaf but the Red-Legged Pademelon takes 88% dicot leaf and 10% dicot fruit. The two potoroids take very different items that they dig for and also include invertebrates. These diets complement the habitat selection exhibited by these species. In the case of the Eastern Grey Kangaroo and the Whiptail Wallaby, which overlap in habitat and also in the proportions of grasses selected, closer analysis demonstrates that they select different species of grass. These differences emphasize the ecological separation between the macropods in this community.

A long-term study of diets of macropods, in the arid zone of northwestern New South Wales, has clearly demonstrated marked separation between the macropods (Dawson *et al.* 1975; Ellis *et al.* 1977; Dawson and Ellis 1979). This is particularly apparent during times of low resource availability, although considerable overlap may be observed in times of plenty, emphasizing the partitioning of the available resources among the macropod species and the introduced species, both domestic and feral, that inhabit the area. This aspect is reviewed in detail (Dawson, this volume) and I will consider only aspects that throw light on potential interactions between these arid zone species.

Barker (1987) has reviewed studies on the diet of the Red Kangaroo, and both the Eastern and Western Grey Kangaroos in western New South Wales and southern Queensland. In times of plenty there is considerable overlap between these species. However, in dry times there is more separation, with Western Greys moving entirely away from grasses and concentrating on shrubs in a very marked shift in diet (Barker 1987:81, Fig. 5.6). Whether or not this shift in diet results from the shift in habitat use also exhibited by this species (Priddel 1987), the result is a marked reduction in the potential for detrimental effects on the species involved. Such evidence, although

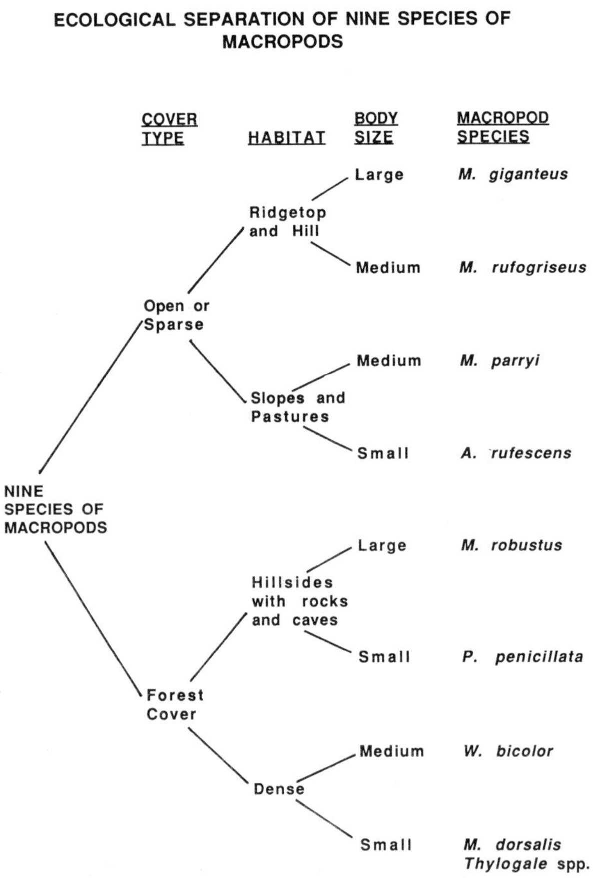

Fig. 10. An alternative simple analysis of the data from Kaufmann (1974), demonstrating the ecological separation of the nine species of macropod, on cover type, habitat and body size.

indirect, is consistent with competitive interactions between these species.

SPECIES INTERACTIONS

Surprisingly little information is available to shed any light on the presence or nature of interactions between species of macropods. No removal experiments have been conducted to investigate whether the necessary conditions to demonstrate competition are in fact met, although many authors have implied or inferred the existence of interspecific competition; for example, "the species are almost ecological antithesis and apparent examples of Gaussian exclusion" (Newsome 1975:419, referring to *M. robustus* and *M. rufus*).

There is strong circumstantial evidence to indicate that the shift of habitat and diet, by the Western Grey Kangaroo, during dry times is a direct response to its functional response on vegetation whose biomass is declining. The fitness of the Western Grey Kangaroo becomes increasingly inferior to that of the Red Kangaroo below 600 kg ha^{-1}, they suffer much higher mortalities, and die earlier as drought conditions prevail (Robertson 1987; Short 1987). However, experiments to provide the direct evidence to link such effects to interspecific competition have not yet been carried out.

The existence of the strong ecological separation between species in all of the studies reviewed here is consistent with interspecific competition being an important component in shaping the structure of macropod communities. However, without conclusive evidence from manipulation experiments alternative explanations must also be considered, the most obvious of which is that the separate species each have their own habitat requirements which are expressed independently of the presence of other species. That such behaviour may have resulted from past competition, "the ghost of competition past" (Schroder and Rosenzweig 1975), must be considered, but equally well it may have resulted from genetically fixed patterns of habitat selection that owed nothing to species interactions. One review of competition in small marsupials (<5 kg) considered the guild of small herbivore/omnivores to include bettongs, potoroos, rat kangaroos and *Setonix* (Dickman 1984). Dickman reviewed 118 studies of this guild involving 11 species. Four studies (3.4%) were suggestive of competition between marsupial species while a further four (3.4%) involved probable competition with eutherians. Hence there was a total of 6.8% of studies suggestive of competition for this guild compared with totals of 13.8% for insectivores, 10% for insectivore/omnivores, 7.1% for folivores, but only 0.9% for nectarivore/insectivores. The total of 356 studies reviewed dealt with "field research on the ecology, geographical distribution, behaviour, taxonomy, biogeography and reproduction of one or more species of small marsupials" (Dickman 1984:28). Because of this breadth these results are equivocal; "it is not clear whether small marsupials rarely compete, whether appropriate studies have not been conducted, or whether the interpretation of competition is seldom applied" (Dickman 1984:28).

Of the studies I have reviewed only two have the potential to provide direct evidence of interspecific competition. Algar (1986) did in fact report on a manipulation experiment; *M. robustus* was introduced into a large reserve, outside its geographical distribution, containing a relatively narrow range of habitat that did not include "typical" Euro habitat. The Euro coexisted with the other three species and it seems that all species may have adjusted their habitat requirements. However, as no data were collected on the habitat selection of the other species in the absence of *M. robustus*, the experiment remains uncontrolled and hence cannot provide unequivocal evidence of interspecific competition. The niche dimension complementarity exhibited in Figs 7A and B did meet the criteria Dickman (1984) used for indirect evidence of past competition.

One other study unequivocally demonstrated the existence of both intra and interspecific competition for a limiting resource, water (Croft 1985). Observations were made of Red Kangaroos and Euros drinking from a water trough, paying particular attention to supplanting encounters where one animal attempts to displace another already drinking. Animals from each species were divided into size-sex classes: large, medium and small for males, medium and small for females. In general Euros were smaller than Red Kangaroos from the same size-sex class (e.g., Large male: Red Kangaroo = 60-90 kg; Euro = 40-60 kg). Within species the priority of access for water acquisition depended on the size of the animal. "Between species the smaller euro tends to win more contests for a drinking place with a Red Kangaroo from an equivalent size-sex class" (Croft 1985:345). Croft concludes that conflict at restricted water sources has probably always been part of the interactions within and between these species with the Euro being the more successful in interactions with Red Kangaroos. Rather than trying to relate these outcomes to the less gregarious social organization of Euros or to a higher aggressiveness and lower tolerance of other individuals, Croft concluded that these outcomes may be influenced by the physiological condition in which the two species reached the water source. Euros moved from the hills onto the plains to reach the water source; they moved the greater distance and did so less frequently than Red Kangaroos, when the water demand was higher.

These conclusions reflect the different evolutionary strategies for survival followed by these two species in the Australian arid zone. Euros avoid the daily radiant heat load by restricting themselves to shade in resting sites usually found in the hilly habitats. This means they are sedentary with the consequence that they must be able to survive on the lower nutrient forage found nearby. With this comes a necessity to be able to retain their resting site successfully. The Red Kangaroo on the other hand requires a higher nutrient diet and retains the mobility to obtain this with adaptations, such as their dense fur, which allow them to survive the higher heat loads encountered in the plains environment (see Dawson 1972; Denny 1973, 1982).

CONCLUSION

In all of the reviewed studies, marked ecological separation has been observed. While of itself this does not demonstrate interaction between species, it is consistent with such an interpretation and is remarkably widespread. Controlled, replicated manipulation experiments have not been conducted and few studies have attempted to obtain the direct evidence needed to provide an unequivocal answer. However, these studies give limited support to the occurrence of interspecific competition. There is considerable indirect and circumstantial evidence indicating that interaction between species is an important factor affecting the structure of macropod communities. Evidence also exists which lends some support to both predation and habitat selection as likely contributing factors. Habitat selection does appear to be the single most important factor and should be accepted as such until experimental manipulations show whether or not species interactions contribute to the habitat selection. The weight of indirect evidence tends to indicate that future research will show this to be so.

ACKNOWLEDGEMENTS

I would like to thank Elizabeth Archer for her patience in drawing and redrawing the distribution maps after checking what seemed to be constantly shifting distribution boundaries. Many people offered suggestions or advice at various stages but I would like particularly to thank Andy Burbidge for the maps he sent, and Ken Johnson for bringing them to my attention, and Peter Jarman for access to his diet information and his constructive comments on the manuscript. Dave Croft, Gerry Maynes, Tony Press and Ian Rowley also provided useful suggestions to me. As with any review of a topic the result can only be as good as the material that is available in the literature; the studies of Algar, Christensen, Croft, Kaufmann and Southwell were of particular importance to the study of macropod communities. I gladly acknowledge the contribution of all the authors whose work I have been able to include in this paper; without their efforts I would have had very little to say.

REFERENCES

ALGAR, D., 1986. An ecological study of macropodid marsupial species on a reserve. Ph.D. Thesis, University of Western Australia.

A.N.P.W.S., 1988. Kangaroos in Australia: Conservation Stautus and Management. Occasional Paper 14, Australian National Parks and Wildlife Service, Canberra.

ARCHER, M., FLANNERY, T. AND GRIGG, G., 1984. The Kangaroo. Weldons: Sydney.

BARKER, R. D., 1987. The diet of herbivores in the sheep rangelands. Pp. 69-83 *in* Kangaroos: their ecology and management in sheep rangelands ed. by G. Caughley, N. Shepherd and J. Short. Cambridge University Press: Cambridge.

BELL, H. M., 1973. The ecology of three macropod marsupial species in an area of open forest and savannah woodland in north Queensland, Australia. *Mammalia* **37**: 527-44.

BURBIDGE, N. T., 1960. The phytogeography of the Australian region. *Aust. J. Bot.* **8**: 75-211.

BURBIDGE, A. A., JOHNSON, K. A., FULLER, P. J. AND SOUTHGATE, R. I., 1989. Aboriginal knowledge of the mammals of the central deserts of Australia. *Aust. Wildl. Res.*

CAUGHLEY, G. J., 1964. Density and dispersion of two species of kangaroo in relation to habitat. *Aust. J. Zool.* **12:** 238-49.

CAUGHLEY, G., SHEPHERD, N. AND SHORT, J., 1987. Kangaroos: their ecology and management in sheep rangelands. Cambridge University Press: Cambridge.

CHRISTENSEN, P., 1980. The biology of *Bettongia penicillata* (Gray, 1837) and *Macropus eugenii* (Desmarest, 1817) in relation to fire. Forests Department Western Australia Bulletin 91.

COULSON, G. M., 1978. Daily activity patterns of three sympatric macropodid species, and the effects of a solar eclipse. *Search* **9:** 153-56.

CROFT, D. B., 1985. Inter- and intraspecific conflict between arid-zone kangaroos at watering points. *Aust. Wildl. Res.* **12:** 337-48.

CROFT, D. B., 1987. Socio-ecology of the Antilopine Walleroo, *Macropus antilopinus*, in the Northern Territory, with observations on sympatric *M. robustus woodwardii* and *M. agilis*. *Aust. Wildl. Res.* **14:** 243-55.

DAWSON, T. J., 1972. Thermoregulation in Australian desert Kangaroos. *Symp. Zool. Soc.* Lond. **31:** 133-46.

DAWSON, T. J. (this volume). Diets of macropodoid marsupials: general patterns and environmental influences.

DAWSON, T. J. AND DENNY, M. J. S., 1969. A bioclimatological comparison of the summer day microenvironment of two species of arid-zone kangaroos. *Ecology* **50:** 328-32.

DAWSON, T. J. AND ELLIS, B. A., 1979. Comparison of the diets of yellow-footed rock wallabies and sympatric herbivores in western New South Wales. *Aust. Wildl. Res.* **6:** 245-54.

DAWSON, T. J., DENNY, M. J. S., RUSSELL, E. M. AND ELLIS, B. A., 1975. Water usage and diet preference of free ranging kangaroos, sheep and feral goats in the Australian arid zone during summer. *J. Zool. (Lond.)* **177:** 1-23.

DENNY, M. J. S., 1973. Water relations in arid zone macropodids. Unpublished Ph.D. Thesis, University of New South Wales.

DENNY, M. J. S., 1982. Adaptations of the red kangaroo and euro (Macropodidae) to aridity. Pp. 179-83 *in* Evolution of the Flora and Fauna of Arid Australia ed. by W. R. Barker and P. J. M. Greenslade. Peacock Publication: Frewville, South Australia.

DICKMAN, C. R., 1984. Competition and coexistence among the small marsupials of Australia and New Guinea. *Acta. Zool. Fennica* **172:** 27-31.

EALEY, E. M. H., 1962. Biology of the Euro, *Macropus robustus cervinus* Thomas, with emphasis on its interaction with sheep in the Pilbara district of Western Australia. Unpublished Ph.D. Thesis, University of Western Australia.

EALEY, E. M. H., 1967a. Ecology of the euro *Macropus robustus* (Gould), in northwestern Australia. I. *CSIRO Wildl. Res.* **12:** 9-25.

EALEY, E. M. H., 1967b. Ecology of the euro *Macropus robustus* (Gould), in northwestern Australia. II. Behaviour, movements and drinking patterns. *CSIRO Wildl. Res.* **12:** 27-51.

EALEY, E. M. H., 1967c. Ecology of the euro *Macropus robustus* (Gould), in northwestern Australia. III. *CSIRO Wildl. Res.* **12:** 53-65.

ELLIS, B. A., RUSSELL, E. M., DAWSON, T. J. AND HARROP, C. J. F., 1977. Seasonal changes in diet preferences of free-ranging red kangaroos, euros and sheep in western New South Wales. *Aust. Wildl. Res.* **4:** 127-44.

HOLLIS, C. J., ROBERTSHAW, J. D. AND HARDEN, R. H., 1986. Ecology of the Swamp Wallaby *(Wallabia bicolor)* in northeastern New South Wales. I. Diet. *Aust. Wildl. Res.* **13:** 355-65.

JARMAN, P. J., JOHNSON, C. N., SOUTHWELL, C. J. AND STUART-DICK, R., 1987. Macropod studies at Wallaby Creek. I. The area and animals. *Aust. Wildl. Res.* **14:** 1-14.

JARMAN, P. J. AND PHILLIPS, C. M. (this volume). Diets in a community of macropodoid species.

JOHNSON, C. N. AND BAYLISS, P. G., 1981. Habitat selection by sex, age and reproductive class in the red kangaroo, *Macropus rufus*, in western New South Wales. *Aust. Wildl. Res.* **8:** 465-74.

JOHNSON, K. A., 1980. Spatial and temporal use of habitat by the red-necked pademelon, *Thylogale thetis* (Marsupialia: Macropodidae). *Aust. Wildl. Res.* **7:** 157-66.

JOHNSON, K. A. AND BURBIDGE, A. A. (this volume). Australian macropods: status, causes of decline, future research and management.

KAUFMANN, J. H., 1974. Habitat use and social organization of nine sympatric species of macropod marsupials. *J. Mammal.* **55:** 66-80.

KIRKPATRICK, T. H., 1983. Black-striped Wallaby. P. 238 *in* The complete book of Australian mammals ed. by R. Strahan. Angus & Robertson: Sydney.

NEWSOME, A. E., 1965a. The abundance of red kangaroos, *Megaleia rufa* (Desmarest), in central Australia. *Aust. J. Zool.* **13:** 269-87.

NEWSOME, A. E., 1965b. The distribution of red kangaroos, *Megaleia rufa* (Desmarest), about sources of persistent food and water in central Australia. *Aust. J. Zool.* **13:** 289-99.

NEWSOME, A. E., 1975. An ecological comparison of the two arid-zone kangaroos of Australia, and their anomalous prosperity since the introduction of ruminant stock to their environment. *The Quarterly Review of Biology.* **50:** 389-424.

PRESS, A. J. (this volume). The abundance and distribution of Black Wallaroos and Common Wallaroos on the Arnhem Land Plateau, Australia.

PRIDDEL, D., 1986. The diurnal and seasonal patterns of grazing of the Red Kangaroo, *Macropus rufus*, and the Western Grey Kangaroo, *M. fuliginosus*. *Aust. Wildl. Res.* **13:** 113-20.

PRIDDEL, D., 1987. The mobility and habitat utilization of kangaroos. Pp. 100-18 *in* Kangaroos: their ecology and management in sheep rangelands ed. by G. Caughley, N. Shepherd and J. Short. Cambridge University Press: Cambridge.

ROBERTSON, G., 1986. The mortality of kangaroos in drought. *Aust. Wildl. Res.* **13:** 349-54.

ROBERTSON, G., 1987. Plant Dynamics. Pp. 50-68 *in* Kangaroos: their ecology and management in sheep rangelands ed. by G. Caughley, N. Shepherd and J. Short. Cambridge University Press: Cambridge.

RUSSELL, E. M., 1974. The biology of Kangaroos (Marsupialia-Macropodidae). *Mammal Rev.* **4:** 1-59.

RUSSELL, E. M. AND RICHARDSON, B. J., 1971. Some observations on the breeding, age structure, dispersion and habitat of populations of *Macropus robustus* and *Macropus antilopinus* (Marsupialia). *J. Zool. Lond.* **165:** 131-42.

SCHRODER, G. D. AND ROSENZWEIG, M. L., 1975. Perturbation analysis of competition and overlap in habitat utilization between *Dipodomys ordii* and *Dipodomys merriami*. *Oecologia (Berl.).* **19:** 9-28.

SEEBECK, J. H., BENNETT, A. F. AND SCOTTS, D. J. (this volume). Ecology of the Potoroidae — a review.

SHORT, J., 1982. Habitat requirements of the Brush-tailed Rock-Wallaby *Petrogale penicillata* in New South Wales. *Aust. Wildl. Res.* **9:** 239-46.

SHORT, J., 1985. The functional response of kangaroos, sheep and rabbits in an arid grazing system. *J. Appl. Ecol.* **22:** 435-47.

SHORT, J., 1987. Factors affecting food intake of rangelands herbivores. Pp. 84-99 *in* Kangaroos: their ecology and management in sheep rangelands ed. by G. Caughley, N. Shepherd and J. Short. Cambridge University Press: Cambridge.

SOUTHWELL, C., 1987. Macropod studies at Wallaby Creek. II. Density and distribution of macropod species in relation to environmental variables. *Aust. Wildl. Res.* **14:** 15-34.

STRAHAN, R., 1983. The complete book of Australian mammals. Angus & Robertson: Sydney.

TAYLOR, R. J., 1983. The diet of the eastern grey kangaroo and wallaroo in areas of improved and native pasture in the New England tablelands. *Aust. Wildl. Res.* **10:** 203-12.

TAYLOR, R. J., 1984. Foraging in the eastern grey kangaroo and the wallaroo. *J. Anim. Ecol.* **53:** 65-74.

TAYLOR, R. J., 1985. Habitat use by the eastern grey kangaroo and wallaroo in an area of sympatry. *Mammalia* **49:** 173-86.

WELLARD, G., 1987. The effect of weather on soil moisture and plant growth in the arid zone. Pp. 35-49 *in* Kangaroos: their ecology and management in sheep rangelands ed. by G. Caughley, N. Shepherd and J. Short. Cambridge University Press: Cambridge.

Life histories of macropodoid marsupials

ANTHONY K. LEE[1] and SIMON J. WARD[1]

Much of the variation in life history traits which sets the Macropodoidea aside from other diprotodonts and the Dasyuroidea can be attributed to differences in body size. The relationships of mass of the litter at birth and at weaning, and the ages of females and males at maturity, to maternal body mass are similar in these taxa. Differences occur in relationships of mass of the litter at pouch exit, and age at pouch exit and at weaning, to maternal body mass. The young of large macropodoids are heavier than the litters of the other taxa at pouch exit; macropodoids and large diprotodonts tend to leave the pouch later than large dasyurids; and potoroids are weaned earlier than similar sized other diprotodonts. The Macropodoidea is unusual among marsupial taxa in the number of species capable of reproducing at the time of weaning.

Within the Macropodoidea, macropodids tend to be less fecund and have a greater age at pouch exit and weaning, and age at maturity in females, than the open-habitat potoroids. The closed-habitat potoroids more closely resemble the macropodids in these traits. Within the genus *Macropus*, species in the subgenus *Macropus* tend to be less fecund and have longer developmental times than those in *Osphranter* and some *Notomacropus* species. This difference is reflected in distribution of these subgenera.

INTRODUCTION

Of the major taxa of the Marsupialia, the Macropodoidea should furnish the most satisfying analysis of life history strategies. It is a relatively large taxon of 52 extant species encompassing a body mass range of two orders of magnitude, and is represented in all of the major terrestrial systems in Australia and New Guinea. At present this potential cannot be fully realized because of the limited knowledge of the life history traits of many macropodoid species. Nevertheless we see benefit in reviewing this knowledge both for its heuristic value, and to draw attention to gaps in our data set.

In this review we will attempt to identify those life history traits of the Macropodoidea which set this taxon apart from other diprotodont marsupials and from the Dasyuroidea. We have restricted our comparisons to these "taxa" because they alone among the higher taxa of the Marsupialia provide a data set which permits statistical comparison. We will also examine variation in life history traits within the Macropodoidea.

MATERIALS AND METHODS

Our data set for the Macropodoidea is confined to 29 species (Table 1). It is confined to traits which define reproductive rates, rates of development and generation times, reflecting our dependence upon studies of reproduction as sources of data. Most of our information comes from Tyndale-Biscoe and Renfree (1987, Tables 2.1, 2.2), but some is drawn from Lee and Cockburn (1985) and Tyndale-Biscoe (1984).

Much of the variation in life history traits can be accounted for by their scaling with body mass (maternal body mass or MBM). We have determined allometric relationships for these traits by the commonly adopted procedure of linear regression analysis of logarithm-transformed data. To test whether there are differences between taxa which are independent of the influence of body mass, the relationships for individual life history traits against MBM for the three superfamilies were then compared by ANCOVA. To compare the life history traits of the Potoroidae and the Macropodidae, a common regression line for all macropodoids was calculated. The residuals (deviations from the regression line) were then subjected to ANOVA.

Comparisons at the superfamily level were restricted to generic and subgeneric averages (Table 2) rather than species values to reduce interdependence of data points (see Harvey and Mace 1982). We have followed Dawson and Flannery (1985) in recognizing *Macropus, Osphranter* and *Notomacropus* as subgenera within the genus *Macropus,* and split the *Antechinus* spp. into two subgeneric groupings, one containing *A. swainsonii-A. minimus,* and the other, the remaining *Antechinus* spp.

LIFE HISTORY TRAITS OF THE MACROPODOIDEA

Litter size and fecundity. With the exception of *Hypsiprymnodon moschatus,* which produces twins (Johnson and Strahan 1982), macropodoids have a litter size of one. In this they resemble other large diprotodonts, but differ from other marsupials which usually have litters of more than one young. As a result, variation in fecundity in the macropodoids depends upon variation in the interval between litters, whereas in some other marsupials it results from the interplay between

[1]Department of Zoology, Monash University, Clayton, Victoria 3168.

Pages 105-15 in KANGAROOS, WALLABIES AND RAT-KANGAROOS ed. by G. Grigg, P. Jarman and I. Hume. Surrey Beatty & Sons, Pty Limited, New South Wales, Australia, 1989.

Table 1. Values for life-history traits used in allometric comparisons of macropodoids. B, at birth; P.E., at permanent pouch exit; W, at weaning; MF, at maturity (females); MM, at maturity (males).

Taxon	Maternal body mass (g)	Mass of young (g)			Age (d) at		MF(m)	MM(m)	Gest. length
		B	P.E.	W	P.E.	W			
Aepyprymnus									
rufescens	3 000				114	155	10	12	22-24
Bettongia									
gaimardi	1 800	0.300			109	160	8-11		21.1
lesueur	1 100	0.317	230	500	115	165	7	14	21.3
penicillata	1 350	0.290			100	130	6		21
Hypsiprymnodon									
moschatus	510				147		>12		
Potorus									
tridactylus	810	0.330	200		130	147	12		38
Dendrolagus									
matschiei	15 000				400				
Lagorchestes									
conspicullatus	2 000				150		12		
Lagostrophus									
fasciatus	1 770		330		180	270			
Macropus									
agilis	11 000	0.630	2 286		219	328	12	14	29.4
antilopinus	17 500	0.664			269	360			33.9
dorsalis	6 500				210		14	20	33-35
eugenii	5 000	0.370	1 000	3 048	250	270	8	24	29.3
fuliginosus	27 600	0.828			310	540	14	31	30.6
giganteus	27 600	0.740	5 000		319	540	18	48	36.4
parma	3 500	0.510	754	1 800	212	300	16	22	34.5
parryi	14 000				275	420	24	18-24	36.3
r. robustus	16 000			7 000	231-270	380	15	24	32.7
r. erubescens	16 000	0.703			243	360			33.4
rufogriseus	14 000	0.450	2 300	4 400	270	360	11-21	13	29.4
rufus	27 300	0.817	4 600	12 000	235	360	14-20	24-36	33.4
Peradorcas									
concinna	1 300				180	300-360	12-24		30-32
Petrogale									
inornata	4 000				204	286	18		31
penicillata	(6 500)				204	290	18	20	31
xanthopus	6 500				194				31-32
Setonix									
brachyurus	2 750	0.350	500		190	240	8-9	13	27
Thylogale									
billardierii	3 900	0.400	1 000	3 000	200	300	14		30
thetis	3 900		800		181	210	17	17	
Wallabia									
bicolor	11 500	0.610			256	450	15		35.5

Sources; Lee and Cockburn (1985), Poole and Merchant (1987), Russell (1982), Strahan (1983), Tyndale-Biscoe and Renfree (1987).

this interval and litter size. MBM accounts for very little (8 to 37%) of the variation in the interval between litters in the superfamilies we have chosen for comparison, largely due to the number of species which breed once annually. If the analysis is restricted to those species which breed continuously or at least twice during a breeding season, and applied to the combined data for the three superfamilies, we find that MBM accounts for 64% of variation, and the slope of the regression line is positive (0.155). Inspection of the data (Fig. 1a) suggests that macropodoids have no advantage in this trait, and being larger, have long intervals between litters. Consequently fecundity rates are generally low in macropodoids when compared to other marsupials and, with the exception of certain continuously breeding potoroids which produce up to three young annually, usually lie between one and two young a year (Table 3). In this they resemble the similar-sized koala and wombats.

Size of young. Most of the variation in the mass of individual neonates at birth can be accounted for by MBM (macropodoids 91%, "other diprotodonts" 81%, dasyuroids 67%). The regression lines for macropodoids and "other diprotodonts" have similar slopes, but differ in elevation (Tables 4 and 5). Inspection of the data shows that macropodoid neonates tend to be larger than those of the other taxa for a given MBM (Fig. 2a). Among multiparous diprotodonts, those above 100 g (*Petaurus* and *Pseudocheirus*) have neonate masses close to those predicted for uniparous diprotodonts, including

Table 2. Average values for life-history traits for genera of marsupials used in the comparison of macropodoids, diprotodonts other than macropodoids, and dasyuroids. Ages at pouch exit, weaning and maturity are averages of minimum ages at these stages.

Genus	Female body mass (g)	Litter size	Mass (g) at birth	Mass (g) at pouch exit	Mass (g) at weaning	Gest. length (d)	Age (d) at pouch exit	Age (d) at weaning	Age (d) at maturity female	Age (d) at maturity male
MACROPODOIDEA										
Aepyprymnus	3 000	1				22	114	155	10	12
Bettongia	1 417	1	0.302	230.0	500.0	21.1	108	152	7	14
Hypsiprymnodon	510	2					147			
Potorous	810	1	0.330	200.0		38.0	130	147	12	
Dendrolagus	15 000	1					400			
Lagorchestes	2 000	1					150		12	
Lagostrophus	1 770	1		330.0			180	270		
Peradorcas	1 300	1				31.0	180	300	12	
Petrogale	5 667	1				31.0	201	288	18	20
Setonix	2 750	1	0.350	500.0		27.0	190	240	8	13
Thylogale	3 900	1	0.400	900.0	3 000.0	30.0	190	255	15.5	17.7
M. Notomacropus	9 000	1.0	0.490	1 585.0	3 085.0	32.2	239	336	13.5	18.5
M. Macropus	27 600	1.0	0.784	5 000.0		33.5	310	540	16	38.5
M. Osphranter	20 267	1.0	0.728	4 600.0	9 500.0	33.4	247	363	14.5	24
"OTHER DIPROTODONTS"										
Trichosurus	2 078	1	0.200	375.0	923.0	16.8	169	216	17	30
Petaurus	307.5	1.5	0.194	30.0	180.0	16.0	83	160	16	18
Petauroides	1 200	1.0	0.230	150.0	550.0		150	240	24	24
Pseudocheirus	835	2	0.300	85.0	260.0		120	200	12	12
Acrobates	13	3.5	0.018	2.0	8.5		65	100	8	8
Burramys	40	4.0			22.0		35	73	12	12
Cercartetus	19	3.5	0.034	4.3	9.3		35	64	4.5	4.5
Phascolarctos	7 900	1.0	0.360	700.0	1 200.0	35.0	255	360	24	24
Lasiorhinus	25 000	1.0	0.500	2 400.0	6 500.0		300	400		
Vombatus	26 000	1.0	0.500	3 000.0	6 000.0		300	400	24	24
Tarsipes	11	2.6	0.004	2.5	5.0		63	90	6	
Gymnobelideus	130	1.5			42.0		87	120		
DASYUROIDEA										
Antechinus[1]	38.3	7.0	0.016	2.2	13.8	27.0	40	95	9.25	10
Dasyurus	1 300	5.7	0.013	21.0	217.0	19.0	87	133	12	12
Sminthopsis	15.3	8.3	0.010		6.3	13.0	36.5	68	4	6
Sarcophilus	6 000	3.0	0.180	200.0	2 500.0		105	240	24	
Ningaui	8.0	7.0	0.005	1.1	5.0		46	79		
Antechinomys	20	5.0		12.6?			48	80		
Antechinus[2]	41	7.5				38	98	9	10	
Phascogale	145	7.5		15.0	58.5		54	120	7.5	
Dasycercus	85	6.0		5.5	25.0		88	120	12	12
Dasyuroides	100	6.0		6.0	45.0		74	110	12	12
Parantechinus	46	6.5	4.0		13.5			90	10	
Planigale	6.6	7.7		1.0	4.4		40	75	10	

macropodoids, while those less than 40 g (*Tarsipes, Acrobates* and *Cercartetus*) more closely resemble dasyuroids. Differences in litter size do not account for this dichotomy, since *Pseudocheirus* has a similar litter size to *Tarsipes*.

At birth, MBM accounts for 91, 61 and 54% of the variation in mass of the litters of macropodoids, "other diprotodonts" and dasyuroids, respectively. There are no differences in the slopes or elevations of the regression lines describing the relationship between weight of the litter at birth and MBM for these taxa (Tables 4 and 5, Fig. 2b).

Almost all of the variation in the mass of the litter at pouch exit is accounted for by MBM (macropodoids and "other diprotodonts" 98%, dasyuroids 94%). The slopes of the regression lines for this relationship differ between taxa (Tables 4 and 5). The mass of the litter at pouch exit in small macropodoids is approximately the same as in similar sized "other diprotodonts" and dasyuroids, but is greater in large macropodoids (Fig. 2c).

MBM also accounts for most of the variation in mass of the litter at weaning (macropodoids 86%, "other diprotodonts" 96%, dasyuroids 98%). The regression lines for this relationship (Fig. 2d) have similar slopes and elevations (Table 4). Unfortunately there are few values for this trait in macropodoids; inspection of the data suggests that large macropodoids may wean young at a greater mass than similar sized "other diprotodonts".

The information available for gestation length in "other diprotodonts" and dasyuroids precludes satisfactory comparison with macropodoids. We have therefore combined the data for all taxa, and in the combined data, MBM accounts for only 37% of the variation in gestation length

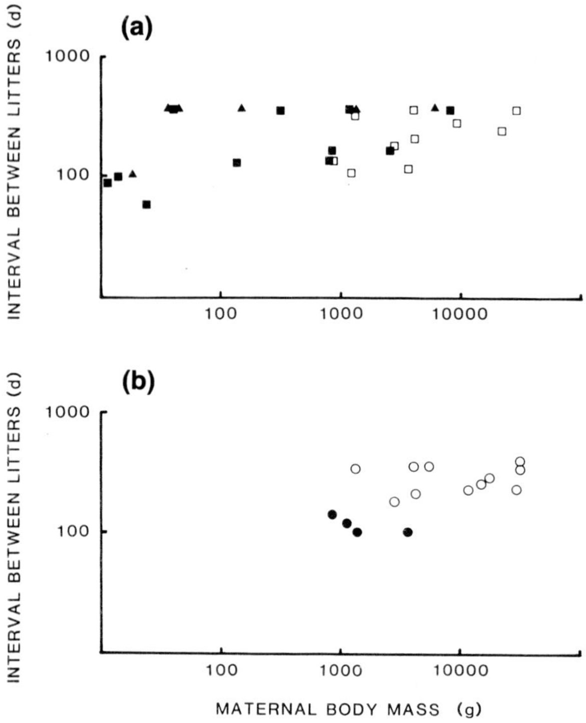

Fig. 1. Minimal interval between litters in (a) genera of macropodoids (open squares), "other diprotodonts" (closed squares) and dasyuroids (triangles); (b) potoroids (closed circles) and macropodids (open circles). Data from Table 2.

(Fig. 3a, Table 4). Macropodoids, particularly macropodids, appear to have relatively long gestation times among marsupials.

MBM accounts for 64, 86 and 52% of the variation in age at pouch exit of macropodoids, "other diprotodonts" and dasyuroids, respectively. The slopes of the regression lines describing the relationship between this trait and MBM are similar, but they differ in elevation (Table 5). Large dasyuroids leave the pouch at an earlier age than similar sized members of the other taxa (Fig. 3b). Since large macropodoids tend to leave the pouch at a greater weight than the litters of "other diprotodonts" and dasyuroids, they must invest more in young during pouch life.

Most of the variation in age at weaning can be attributed to MBM in all taxa (macropodoids 63%, "other diprotodonts" 92%, dasyuroids 89%). The slopes of the regression lines for the relationship between age at weaning and MBM are similar, but again there is a difference in elevations (Table 5). Inspection of the data suggests that this may be due to a low age at weaning in some potoroids (Fig. 3c). The ages at weaning in large macropodoids and large "other diprotodonts" appear similar.

Growth. We have used the growth constant from the Gompertz equation to compare the rates of growth in different marsupials. Cockburn and Johnson (1988) provide five values for macropodoids, all but one of which fall below, but close to the regression line relating the growth constant to MBM for the 15 marsupials for which we have adequate data (Fig. 4).

Age at maturity. MBM accounts for 55, 63 and 66% of variation in the age at maturity in female, and 80, 65 and 72% of the variation in age at maturity in male macropodoids, "other diprotodonts" and dasyuroids, respectively. The slopes and elevations of the regression lines for each of these traits and MBM are similar (Table 5, Fig. 3d,e).

It has been previously noted that females of a number of macropodoid species may have their first fertile mating at or slightly before weaning (Tyndale-Biscoe and Renfree 1987). Published information on the timing of first fertile mating and on the minimum age at maturity suggest that this phenomenon, which is unusual, if not unique

Table 3. Minimum intervals between litters and maximum annual fecundities of macropodoid species and genera of macropodoids*, "other diprotodonts" and dasyuroids.

Taxa	Minimum interval between litters	Maximum annual fecundity (young per year)
MACROPODOIDAE		
Aepyprymnus rufescens	119	3
Bettongia lesueur	122	3
B. penicillata	100	3
Potorous tridactylus	140	2.5
Macropus agilis	217	1.8
M. eugenii	365	1
M. fuliginosus	372	1
M. giganteus	362	1
M. robustus	255	1.4
M. rufogriseus	260	1.4
M. rufus	236	1.5
Peradorcas concinna	335	1.1
Petrogale inornata	210	1.7
Setonix brachyurus	185	1.8
Thylogale billardierii	365	1
Bettongia	111	3
M. Notomacropus	245	1.5
M. Macropus	367	1
M. Osphranter	280	1.3
"OTHER DIPROTODONTS"		
Petaurus	365	1.5
Petauroides	365	1
Pseudocheirus	135	6
Acrobates	100	8
Burramys	365	4
Cercartetus	60	12
Phascolarctos	365	1
Tarsipes	90	12
Gymnobelideus	130	4
DASYUROIDEA		
Antechinus[1]	365	12
Dasyurus	365	8
Sminthopsis	107	10
Sarcophilus	365	4
Antechinus[2]	365	8
Phascogale	365	8

*Where data were available for one species in a genus, those data were used for generic comparisons.
[1]*Antechinus stuartii* species group
[2]*Antechinus swainsonii–A. minimus.*

among marsupials, may be common among macropodoids (Fig. 5).

VARIATION IN LIFE HISTORY TRAITS WITHIN THE MACROPODOIDEA

Litter size and annual fecundity. We have already commented upon the lack of variation in the litter size of macropodoids (*H. moschatus* excepted), and the influence of the interval between litters on fecundity rates. MBM accounts for 31% of the variation in the minimum interval between litters among macropodoid species. The continuously breeding potoroids have shorter minimum intervals than the macropodids of similar MBM (Fig. 1b), and there is a significant difference in residuals ($F_{1,13}=6.57$, $p<0.05$). They have higher maximum fecundity rates of between 2.5 and 3 young per year, in contrast to 1 to 1.8 young per year for the macropodids (Table 3). Among the macropodids, relatively short intervals between litters and high fecundity rates are found in *M. rufus* and *M. agilis,* species in which juvenile mortality is often high (Newsome 1975; Bolton, Newsome and Merchant 1982).

Size of young. Most of the variation in neonate mass at birth (85%), and in mass of the young at pouch exit (99%) and weaning (92%), can be accounted for by variation in MBM (Fig. 6a,b,c). Deviations from the relationship of neonate mass at birth and MBM are greatest in *M. parma,* which has a relatively high, and *M. eugenii* and *M. rufogriseus,* which have relatively low neonate masses (Fig. 6a). The importance of this variation is not obvious.

MBM accounts for 23, 68 and 65% of the variation in length of gestation, age at pouch exit and age at weaning respectively (Fig. 6e,f,g). Comparisons of residuals show that potoroids differ from macropodids in age at pouch exit ($F_{1,26}=6.18$, $p<0.05$), and at weaning ($F_{1,21}=9.15$, $p<0.05$), but not in gestation length ($F_{1,21}=2.96$, $p>0.05$). Potoroids clearly leave the pouch and are weaned at earlier ages than similar sized macropodids. The lack of difference in gestation length is hardly surprising as the Potoroidae includes species with the shortest (*Bettongia* spp.) and longest *(P. tridactylus)* gestation lengths among macropodoids (Table 1).

There is also variation in the age at pouch exit within the potoroids. The *Bettongia* spp. and *A. rufescens* leave the pouch earlier than *P. tridactylus* and *H. moschatus,* which depart at an age similar to that predicted for macropodids of comparable size (Fig. 6f). Among the macropodids, *D. matschiei* has an unusually long

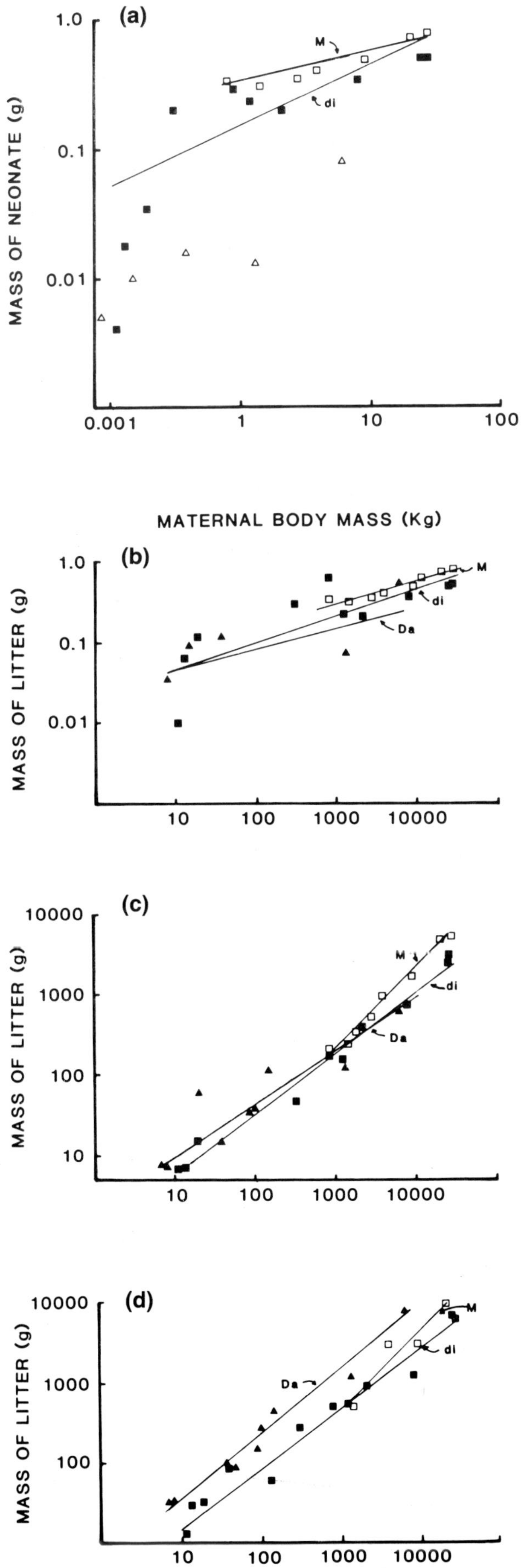

Fig. 2. Mass of the neonate (a), and mass of litters of three taxa of marsupials at birth (b), pouch exit (c), and weaning (d). Regression lines for macropodoids (M), "other diprotodonts" (di) and dasyuroids (Da). Statistics for regression lines are presented in Table 4. Symbols as in Figure 1; data from Table 2.

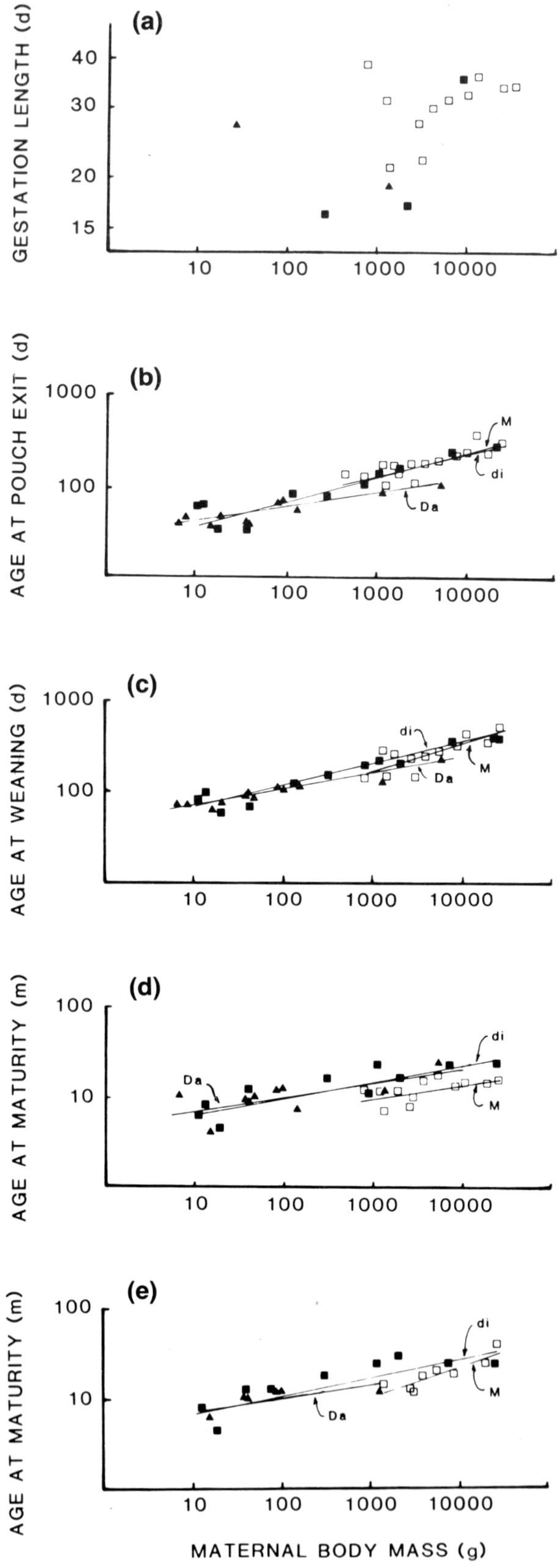

Fig. 3. Gestation lengths (a), and age of young at pouch exit (b), weaning (c) and maturity for females (d) and males (e) of three marsupial taxa. Statistics for regression lines are presented in Table 4. Symbols as for Figure 1; data from Table 2.

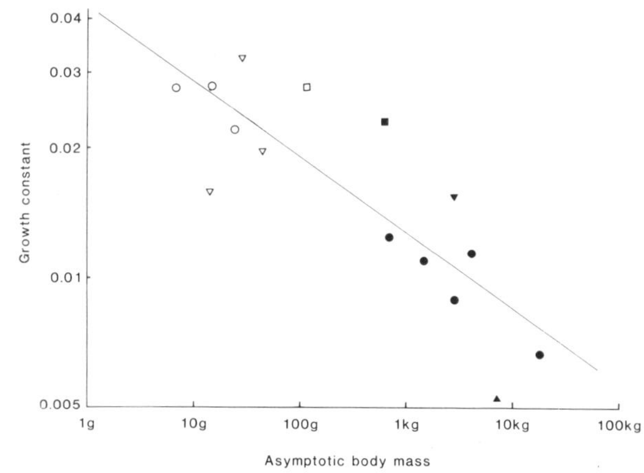

Fig. 4. The Gompertz growth constant as a function of asymptotic body mass in marsupials. In order of increasing body mass; *Ningaui ridei, Sminthopsis murina, Antechinus stuartii* (open circles); *Acrobates pygmaeus, Cercartetus nanus, C. caudatus* (inverted open triangles); *Petaurus breviceps* (open squares); *Perameles gunnii* (filled squares); *Peradorcas concinna, Bettongia lesueur, macropus parma, Thylogale billardierii, Macropus giganteus* (filled circles); *Trichosurus vulpecula* (inverted filled triangles); *Phascolarctos cinereus* (filled triangles). The regression line has the equation: Growth constant=MBM$^{-0.18}$-1.37, r^2=0.71, F=31.1, $p<0.05$. (from Ward 1988).

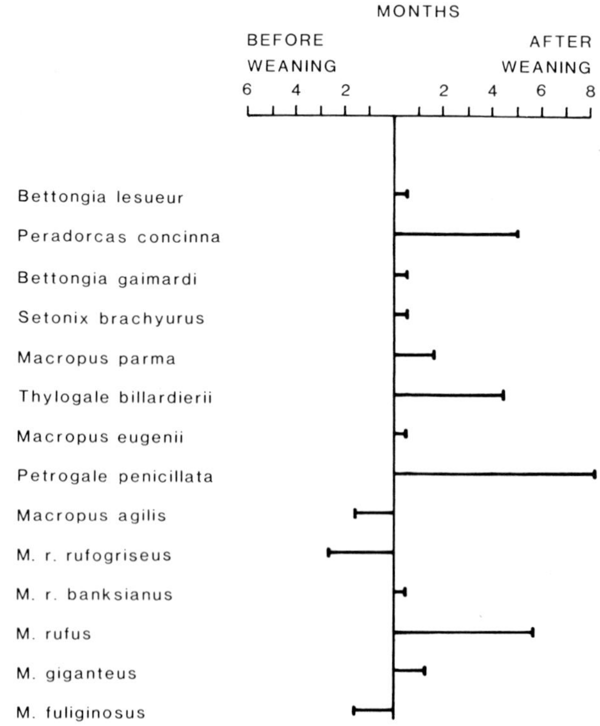

Fig. 5. The time of first fertile mating or age at maturity of female macropodoids in relation to age at weaning.

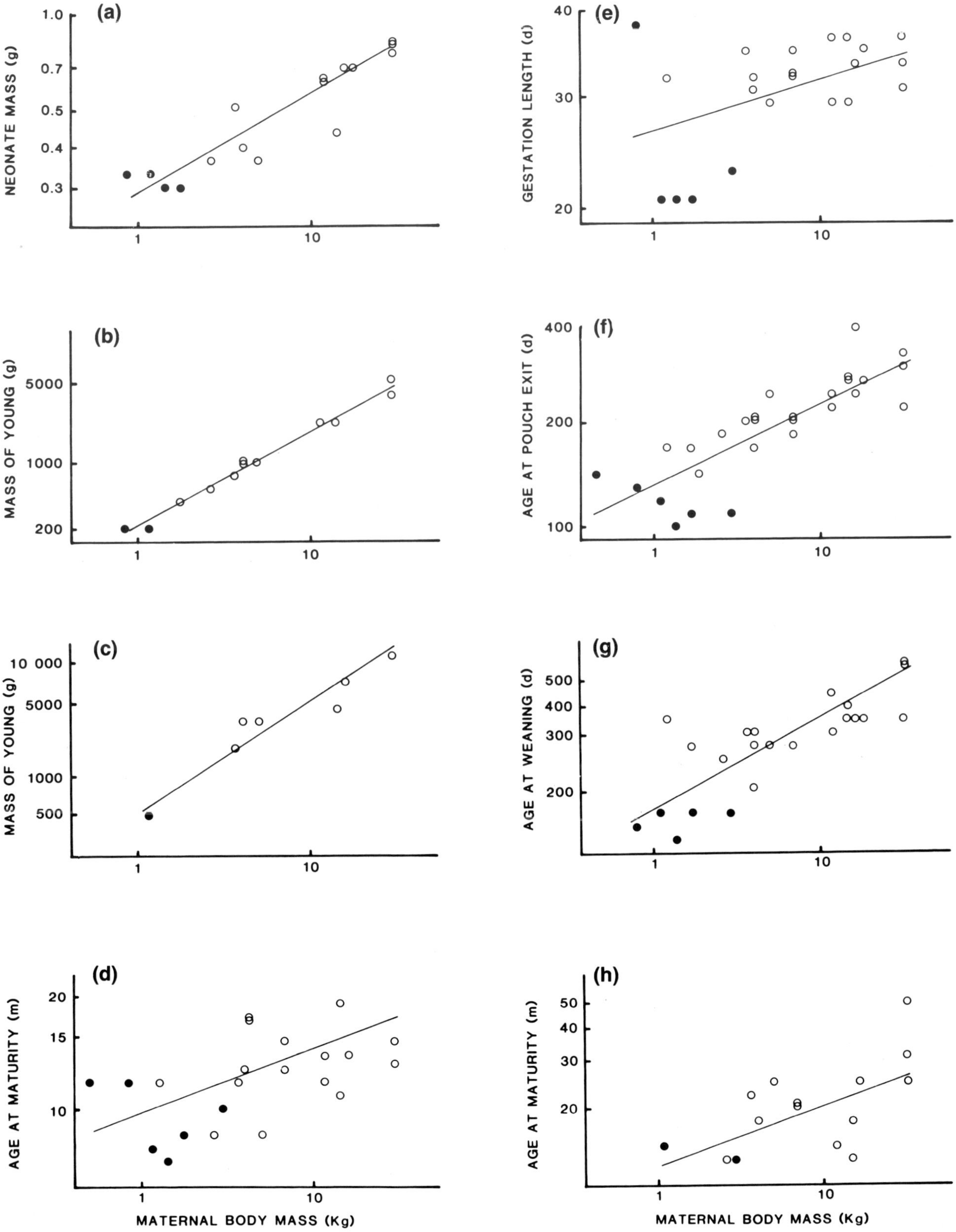

Fig. 6. Life history traits of macropodoid species as functions of maternal body mass. Mass of neonates (a); mass of young at pouch exit (b); mass of young at weaning (c); age at maturity in females (d); length of gestation (e); age at pouch exit (f); age at weaning (g); age at maturity in males (h). Potoroids (solid circles), macropodids (open circles). Data from Table 1. Statistics for regression lines are presented in Table 4.

Table 4. Regression constants and statistics for allometric equations relating life-history traits to MBM(g). Mg, regression for genera of macropodoids; Di, regression for "other diprotodonts"; Da, regression for dasyuroids; Ms, regression for species of macropodoids; ***, p<0.001; **, p<0.01; *, p<0.05; N.S., p>0.05.

Trait	Taxon	Slope	Intercept	n	r^2	F	Sig
Neonate mass (g)	Mg	0.281	-1.38	8	0.91	73.1	***
	Di	0.498	-2.32	10	0.81	40.3	***
	Da	0.421	-2.61	5	0.67	9.01	N.S.
	Ms	0.287	-1.40	16	0.85	84.1	**
Litter mass (g)							
at birth	Mg	0.281	-1.38	8	0.91	73.1	***
	Di	0.334	-1.66	10	0.61	14.8	**
	Da	0.291	-1.58	5	0.54	5.72	N.S.
	Ms	0.287	-1.40	16	0.85	84.1	***
at pouch exit	Mg	0.990	-0.67	8	0.98	414.2	***
	Di	0.749	-0.02	10	0.98	431.7	***
	Da	0.653	0.31	8	0.94	113.2	***
	Ms	0.929	-0.44	12	0.99	959.9	***
at weaning	Mg	1.020	-0.40	4	0.86	19.1	*
	Di	0.712	0.57	12	0.96	297.5	***
	Da	0.817	0.75	11	0.98	522.5	***
	Ms	0.893	0.10	7	0.92	73.5	***
Age (d) at							
pouch exit	Mg	0.245	1.40	15	0.64	25.4	***
	Di	0.264	1.32	13	0.86	72.5	***
	Da	0.154	1.48	12	0.52	13.1	**
	Ms	0.253	1.36	28	0.68	57.6	***
at weaning	Mg	0.301	1.34	12	0.63	19.5	**
	Di	0.230	1.61	13	0.92	137.1	***
	Da	0.161	1.72	11	0.89	89.1	***
	Ms	0.299	1.34	23	0.65	41.6	***
at maturity							
female	Mg	0.150	0.55	12	0.29	5.39	*
	Di	0.190	0.63	10	0.75	28.4	**
	Da	0.170	0.66	9	0.47	8.12	*
	Ms	0.158	0.51	23	0.26	8.52	**
at maturity							
male	Mg	0.331	0.80	8	0.77	24.2	**
	Di	0.196	0.65	9	0.66	16.3	**
	Da	0.151	0.72	6	0.43	4.82	N.S.
	Ms	0.238	0.36	15	0.33	8.00	*
Gestation (d)	Mg	0.085	1.17	18	0.37	10.96	**
	Di						
	Da						
	Ms	0.082	1.17	22	0.22	6.80	*

pouch life and this may be related to its scansorial habit where it may be advantageous to extend care within the pouch. Among the large kangaroos, *M. rufus* has a shorter pouch life than the similar sized grey kangaroos.

Unfortunately our data for the mass of the young at weaning do not allow us to distinguish whether the early age at weaning in potoroids reflects a lower investment at weaning or a faster rate of investment. The single value, for *B. lesueur,* suggests that the latter may apply (Fig. 6c). Among the macropodids, *P. concinna* and the grey kangaroos are weaned relatively late, and *M. eugenii* and *M. rufus,* relatively early.

Variation in MBM accounts for only 26 and 33% of the variation in age at maturity of female and male macropodoids, respectively. Among potoroids, *P. tridactylus* and *H. moschatus* mature later than the *Bettongia* spp. and *A. rufescens,* while among the macropodids, late maturation occurs in *P. inornata, T. thetis* and *M. parryi,* and relatively early maturation in *S. brachyurus, M. eugenii, M. rufogriseus,* and

Table 5. Results of ANCOVA on the regression lines for the relationships between life-history traits and MBM for macropodoids, "other diprotodonts", and dasyuroids.

Trait	Difference in slopes F	Difference in slopes p	Difference in elevations F	Difference in elevations p
Neonate mass (g)	0.859	0.441	7.969	0.003
Litter mass (g)				
at birth	0.159	0.854	2.491	0.109
pouch exit	5.905	0.010	–	–
weaning	1.875	0.178	2.296	0.123
Age (d) at				
pouch exit	2.568	0.092	16.429	0
weaning	3.223	0.054	9.372	0.001
♀ maturity	0.203	0.817	0.438	0.650
♂ maturity	1.266	0.307	0.953	0.403

among the large kangaroos, *M. rufus* and *M. fuliginosus*. In most macropodoids, the minimum age at maturity in males is two to 12 months later than females (Table 1). We have only two values for age at maturation in male potoroids, both of which fall within the macropodid range (Fig. 6h). In the Macropodidae, males of *M. agilis* and *M. rufogriseus* mature at a relatively early age, and maturity is appreciably delayed in males of *M. giganteus*.

DISCUSSION

In comparison to other marsupial taxa, the Macropodoidea is dominated by species of relatively large body mass, and this is thought to convey advantages in gut capacity and energy requirements essential for their subsistence as herbivores. However, large body size carries with it constraints on those life history traits, such as litter size, the interval between litters and age at maturity, which determine fecundity and generation time. Of the three superfamilies we have compared, macropodoids generally have the smallest litters (Table 2) and tend to have a long minimum interval between litters (Fig. 1a, Table 3), so that their fecundity rates are low and indeed, among the lowest for marsupials (Table 3). Macropodoids also tend to have the greatest ages at maturity and therefore longest generation times. Large body size, however, confers the advantage of greater longevity, and increased opportunity for iteroparity, both of which are probably typical of macropodoids.

Our analysis showed that MBM accounted for more than 50% of the variation of most of the life history traits we examined. However, there were superfamily differences in several traits in their relationship to MBM, and examination of these relationships allows us to define what is distinct about macropodoids.

Recently Hughes and Hall (1988; Hughes *et al.*, this volume) have shown that the neonates of macropodoids are developmentally more advanced than the neonates of other marsupials. Earlier Russell (1982) observed that the neonates of marsupials with litters of more than one young were sometimes much smaller than the neonates of marsupials with single young. Inspection of Figure 2a shows that this is true for very small diprotodonts. The difference in the elevations of the regression lines for the relationship between neonate mass and MBM for macropodoids and "other diprotodonts" is probably due to inclusion of values for very small diprotodonts. However, the data in Figure 2a suggest that the neonates of macropodoids may also be consistently heavier than neonates from the other marsupial taxa, including "other diprotodonts" which give birth to single young, when MBM is taken into account (Fig. 2a).

Although gestation lengths in macropodoids, particularly macropodids, are among the longest for marsupials (Tyndale-Biscoe and Renfree 1987), there does not appear to be any direct relationship between gestation length and mass of the neonate (cf. Fig. 2a and Fig. 3a). This is not surprising as most of the mass of the embryo is acquired during organogenesis which occupies a highly variable fraction of about a third or less of gestation (Lee and Cockburn 1985; Tyndale-Biscoe and Renfree 1987).

At birth, maternal investment in litters is similar in the three taxa. By the time of pouch exit, however, investment in the litter of large macropodoids is clearly greater than in the other taxa (Fig. 2c). Large macropodoids are generally found in open habitats where retention of young within the pouch may reduce predation. In other marsupials, young at this stage are secreted in a nest or burrow or are carried on their mother's back. Since there is no difference in the age at pouch exit in large macropodoids and large "other diprotodonts" (Fig. 3b), the rate of investment to this stage must be greater in large macropodoids. The same pattern to investment in the litter may occur at weaning (cf. Fig. 2c and Fig. 3b), but here we have only four values for macropodoids. These differences in investment might be expected to be reflected in differences in overall growth rates. Unfortunately we have growth curves for only two species in this size range (*Phascolarctos cinereus* and *M. giganteus*), although these conform with our prediction (Fig. 4).

Another trait in which macropodoids are distinctive is the number of species in which females first reproduce at the time of weaning (Fig. 5). Since we were unable to demonstrate differences in the relationships between female age at maturity and MBM between the three taxa, this coincidence could result from a trend towards prolongation of lactation relative to the age of maturity with increasing MBM in marsupials. Alternatively, the indeterminancy of growth prevalent in macropodoids may have resulted in a disassociation of the age of maturity and growth and development (A. Cockburn, pers. comm.).

One reproductive trait which sets most macropodoids apart from dasyuroids and many "other diprotodonts" is embryonic diapause. Embryonic diapause has been found to occur in all macropodoids so far studied, except *M. fuliginosus* (Tyndale-Biscoe and Renfree 1987; Tyndale-Biscoe this volume), and Thompson (1987) has argued that the major ecological advantage of embryonic diapause to the macropodoids is a reduction in the interval between litters. In embryonic diapause, quiescent embryos are carried *in utero* during lactation, and hence the interval between successive births may be reduced by up to the period of "normal" gestation. This is in the order of one month in macropodoids. One advantage of a reduced interval between litters is a greater fecundity. Since the period of gestation in the macropodoids is independent of body size, so is the reduction in the interval due to embryonic diapause. Hence, the relative influence of embryonic diapause on the fecundity is least in the large macropodids, which have a longer absolute duration of lactation. In these species, the interval may be reduced by only about one in nine to twelve months. However, in populations in unpredictable environments, this may result in the production of more than one young during occasional extended periods of plant growth.

Within the Macropodoidea, there is a dichotomy in a number of the traits we have examined between certain potoroids and the macropodids. In comparison to similar sized macropodids, potoroids have shorter minimal intervals between litters and, consequently, higher maximum fecundity rates (Table 3). Potoroids also emerge from the pouch and may be weaned earlier than macropodids. Since the masses of the pouch young at each of these stages are similar when differences in MBM are taken into account, the rate of investment in potoroid young must be greater. Surprisingly this was not reflected in the growth rate of *B. lesueur* (Fig. 4). This capacity of potoroids to invest in young at a faster rate may be a benefit flowing from their selection of diets of higher quality than those fed on by macropodids (Norbury, Sanson and Lee, this volume).

There is also a dichotomy within the Potoroidae between the closed-habitat species, *P. tridactylus* and *H. moschatus,* and the relatively open-habitat species of *Bettongia* and *A. rufescens.* The former more closely resemble macropodids in both age at pouch exit and in age at maturity in females. Within the genus *Macropus,* (Macropodidae), species in the *Osphranter* group, which includes *M. rufus* and *M. robustus,* are weaned and mature earlier, and have a shorter interval between litters and a higher maximum fecundity rate than species in the *Macropus* group, *M. giganteus* and *M. fuliginosus.* Some species in the *Notomacropus* group resemble the former (e.g., *M. agilis*), while others are more conservative (e.g., *M. eugenii*). Of these groups, only the *Osphranter* group and the less conservative species in the *Notomacropus* group occur where there is a low probability of occurrence of resources sufficient to raise young (see Newsome 1977).

In this review we have not discussed intraspecific variability in life history traits, but this topic is dealt with elsewhere (see Lee and Cockburn 1985; Tyndale-Biscoe, this volume). The tractability of some macropodids is considerable, with variability in the interval between litters (*M. r. robustus* 245 days, *M. r. erubescens* 265 days, Poole and Merchant 1987), in the frequency of litters (continuously breeding and seasonal breeding *S. brachyurus,* see Tyndale-Biscoe, this volume), and in age at sexual maturity (mean age at maturity of *M. agilis* in woodland 26 months, grassland 20 months, Bolton, Newsome and Merchant 1982), among other traits.

A weakness of this review is the quality of the data set. Our information for macropodoids is strongly biased in favour of species from temperate Australia. Twenty-two macropodoids have ranges confined to the tropics, and we have been able to obtain some information for only six of these. This weakness is also common to the data sets for the "other diprotodonts" and the dasyuroids. There is also the problem that the data were obtained under different circumstances; some were derived from natural populations, some from captive populations which were provided with food and water *ad libitum.* Finally, as with other reviews of marsupial life histories (i.e., Russell 1982; Lee and Cockburn 1985), we were largely restricted to consideration of reproductive traits and growth. This reflects our present dependence on information derived from studies of reproduction, and emphasises the paucity of information on age-specific reproduction and mortality of marsupials.

REFERENCES

BOLTON, B. L., NEWSOME, A. E. AND MERCHANT, J. C., 1982. Reproduction in the agile wallaby *Macropus agilis* (Gould) in the tropical lowlands of the Northern Territory: opportunism in a seasonal environment. *Aust. J. Ecol.* **7**: 261-77.

COCKBURN, A. AND JOHNSON, C. N., 1988. Patterns of growth. Pp. 28-40 *in* The Developing Marsupial: Models for Biomedical Research ed. by C. H. Tyndale-Biscoe and P. A. Janssens. Springer Verlag: Berlin.

DAWSON, L. AND FLANNERY, T., 1985. Taxonomic and phylogenetic status of living and fossil kangaroos and wallabies of the genus *Macropus* Shaw (Macropodidae: Marsupialia), with a new subgeneric name for the larger wallabies. *Aust. J. Zool.* **33**: 473-98

HARVEY, P. H. AND MACE, G. M., 1982. Comparisons between taxa and current adaptive trends: problems of methodology. Pp. 343-61 *in* Current Problems in Sociobiology. Kings College Sociobiology Group. Cambridge University Press: Cambridge.

HUGHES, R. L. AND HALL, L. S., 1988. Structural adaptations of the newborn marsupial. Pp. 8-27 *in* The Developing Marsupial. Models for Biomedical Research ed. by C. H. Tyndale-Biscoe and P. A. Janssens. Springer-Verlag: Berlin.

HUGHES, R. L., HALL, L. S., TYNDALE-BISCOE, C. H. AND HINES, L. A. (this volume). Evolutionary implications of macropodid organogenesis.

JOHNSON, P. M. AND STRAHAN, R., 1982. A further description of the Dusty Rat-kangaroo, *Hypsiprymnodon moschatus* Ramsay, 1876 (Marsupialia, Potoroidae) with notes on its biology. *Aust. Zool.* **21**: 27-46.

LEE, A. K. AND COCKBURN, A., 1985. Evolutionary Ecology of Marsupials. Cambridge University Press: Cambridge.

NEWSOME, A. E., 1975. An ecological comparison of two arid zone kangaroos of Australia, and their anomalous prosperity since the introduction of ruminant stock to their environment. *Quart. Rev. Biol.* **50**: 389-424.

NEWSOME, A. E., 1977. Imbalance in the sex ratio and age structure of the Red Kangaroo, *Macropus rufus*, in central Australia. Pp. 221-33 *in* The Biology of Marsupials ed. by B. Stonehouse and D. Gilmore. Macmillan: London.

NORBURY, G., SANSON, G. D. AND LEE, A. K. (this volume). Feeding ecology of the Macropodoidea.

POOLE, W. E. AND MERCHANT, J. C., 1987. Reproduction in captive wallaroos; the eastern wallaroo, *Macropus robustus robustus*, the euro, *M. r. erubescens*, and the antilopine wallaroo, *M. antilopinus*. *Aust. Wildl. Res.* **14**: 225-42.

RUSSELL, E. M., 1982. Patterns of parental care and parental investment in marsupials. *Biol. Rev.* **57**: 423-86.

STRAHAN, R., 1983. The Australian Museum Complete Book of Australian Mammals. Angus & Robertson: Sydney.

THOMPSON, S. D., 1987. Body size, duration of parental care, and the intrinsic rate of natural increase in eutherian and metatherian mammals. *Oecologia* **71**: 201-9.

TYNDALE-BISCOE, C. H., 1984. Mammals: Marsupials. Pp. 386-454 *in* Marshall's Physiology of Reproduction Vol 1. ed. by G. E. Lamming. Churchill Livingstone: London.

TYNDALE-BISCOE, C. H. (this volume). The adaptiveness of reproductive processes.

TYNDALE-BISCOE, C. H. AND RENFREE, M. B., 1987. Reproductive Physiology in Marsupials. Cambridge University Press: Cambridge.

WARD, S. J., 1988. Life Histories of the Small Diprotodont Marsupials. Ph.D. Thesis, Monash University, Clayton.

Revision of the Genus *Dorcopsis* (Macropodidae: Marsupialia)

COLIN P. GROVES[1] and T. F. FLANNERY[2]

Multivariate analysis and morphological studies indicate that there are four species of *Dorcopsis*. *Dorcopsis atrata* and *D. hageni* are monotypic, while *D. luctuosa* includes two distinct subspecies (one newly described here), and *D. muelleri* four (also one newly described here). The three mainland New Guinean species are probably parapatric, but only in one instance are taxa recorded as occurring in close proximity. *Dorcopsis luctuosa* is probably the sister taxon to the remaining taxa, while the autapomorphic *D. atrata* is probably convergent upon the species of *Dorcopsulus*.

INTRODUCTION

THE forest wallabies of the genus *Dorcopsis* are among the most plesiomorphic of living macropodids (Flannery 1984). Along with the closely related *Dorcopsulus,* they are the only genus of macropodoid presently restricted to the island of New Guinea and its sattelite islands. However, fossils of *Dorcopsulus* and the closely related genus *Dorcopsoides* are known from Tertiary sediments in Australia.

A species of *Dorcopsis* was first made known by Quoy and Gaimard (1830); but they described their material under the name *Didelphis brunii,* thinking that is was the species seen by De Bruijn and named *Didelphis brunii* Schreber; which was however, a species of *Thylogale.* Quoy and Gaimard's animal was probably the western species of *Dorcopsis,* to which Owen (1839) was referring under the name *Hypsiprymnus dorsocephalus,* and Gray (1843) as *Halmaturus asiaticus*. This is made more likely by the fact that the latter specifically mentioned Lobo Bay as a locality, and that the British Museum had received a specimen from there via the Leiden Museum some time previously. It was not until 1866 that the first available name was awarded to a New Guinea forest Wallaby: *Macropus muelleri* Schlegel.

Husson (1955) argued that the earliest name for a species of this group is actually *Kangurus veterum* Lesson and Garnot, 1826. George and Schurer (1975), however, maintain that this name actually refers not to a member of this group but to *Dendrolagus inustus.* Lesson and Garnot's description reads as follows (translated by CPG): " . . . This animal we provisionally name kangaroo of Aru *(Kangurus veterum* N.), is called *Podin* by the Papuans of the port of Dorey, New Guinea. Differs from the Australian kangaroos by the dimensions of its limbs. Size is that of a common Hare: its ears are proportionately shorter than other species. Head rounded, muzzle more conical, less shortened than the "oualabat". Neck less slender. Forelimbs more elongated, stronger, more robust; hindlimbs less long, bigger. Tail is one-third shorter. Pelage is uniformly brown on upper parts, passing to grey on underside . . . " (Lesson and Garnot 1826, p.164).

Perhaps the most significant part of the vague description (which to judge from the size given must have been of a juvenile) is the limb proportions, which would certainly fit best with a tree-kangaroo as George and Schurer proposed. The short tail and colour, however, do not fit well with any of the Vogelkop tree-kangaroos, and are indeed reminiscent of a species of *Dorcopsis.* Given the vagueness of the description we suggest that it be regarded as a *nomen dubium* and that it be allowed to lapse.

Schlegel and Muller (1845) gave *Didelphis brunii* Quoy and Gaimard, 1830 (*nec* Schreber 1778) generic rank as *Dorcopsis.* Troughton (1937) substituted the name *Conoyces* Lesson, 1842, for the genus, but as shown by Tate (1945) the type of Lesson's name is *Didelphis brunii* Gmelin, 1788, who gives as his primary reference Schreber (1778): not *Didelphis brunii* Quoy and Gaimard, 1830. *Conoyces* is hence a junior synonym of *Thylogale* Gray, 1837.

A second species of *Dorcopsis* was described by d'Albertis (1874) as *Halmaturus luctuosus;* Miklouho-Maclay described (1884) *Dorcopsis chalmersii* and (1885) *D. beccarii;* Heller (1897) described *D. hageni;* Rothschild and Rothschild (1898), *D. rufolateralis;* Jentink (1908), *D. lorentzii;* Thomas (1913), *D. mysoliae.* Thomas (1922) and Troughton (1937) described new taxa as subspecies of *D. hageni* Heller, 1897. Finally in 1957 Van Deusen made known a further new species, *D. atrata.* Of these taxa, *D. mysoliae* was from Misool Island, west of the Cenderawasih

[1]Prehistory and Anthropology, ANU, P.O. Box 4, Canberra, A.C.T. 2601.
[2]The Australian Museum, 6-8 College Street, Sydney, New South Wales 2000.

Pages 117-28 in KANGAROOS, WALLABIES AND RAT-KANGAROOS ed. by G. Grigg, P. Jarman and I. Hume. Surrey Beatty & Sons Pty Limited, New South Wales, Australia, 1989.

Peninsula, Irian Jaya; *D. atrata* from Goodenough Island, east of mainland Papua New Guinea; all other putative taxa were from the New Guinea mainland itself.

In the meantime a series of related, smaller species were being made known, beginning with *Dorcopsis macleayi* Miklouho-Maclay (1885). Matschie (1916) placed this species in a new genus, *Dorcopsulus,* about whose validity opinions have differed since then, some putting it into *Dorcopsis,* others leaving it as distinct. The question of the relationships between *Dorcopsis* and *Dorcopsulus* we leave to a future publication, and at this time we deal only with the larger-sized species, *Dorcopsis* in the strict sense.

MATERIAL AND METHODS

Between us we have studied material in the following collections: Australian Museum, Sydney (AM M); Macleay Museum, Sydney (MM); Museum of Victoria, Melbourne (NMV); National Wildlife Collection, Canberra (CM); Western Australian Museum (WAM); British Museum (Natural History), London (BM); Museum National d'Historie Naturelle, Paris (MNP); Rijksmuseum van Natuurlijke Historie, Leiden (RML); Institut voor Taxonomische Zoologie, Amsterdam (ZMA); Zoologisches Museum A. Humboldt, Berlin (ZMB); Landesmuseum fur Naturkunde, Karlsruhe (LNK); American Museum of Natural History, New York (AMNH). Some of the specimens in the Australian Museum were collected by one of us (TFF).

We made notes on external and craniodental characters, and took a series of measurements on the skulls and teeth of each specimen. The measurements are listed and described in the Appendix. These measurements were utilized in a series of multivariate analyses (SPSS "Direct" method, Discriminant Functions). Three analyses were performed: Skulls, adults only; Skulls, adults and older juveniles only; Teeth, all ages. Skull and tooth measurements were treated separately because they are different kinds of data: skulls grow, teeth do not, hence skulls are subject to ontogenetic allometry, teeth only to static allometry, reducing correlations between them to a point where it loses biological validity to treat them together except perhaps as a bivariate diagnostic tool.

RESULTS

The *Dorcopsis* specimens we have studied fall very neatly into four groups, which are well-defined geographically and differ by colour and colour pattern, pattern of dorsal hair-tracts, body proportions, and skull characters. We rate then as four distinct, probably allopatric, species. The earliest available names are as follows: for the western brown to reddish species, *D. muelleri* (Schlegel 1866); for the north coastal fawn to blackish brown species with a light dorsal stripe, *D. hageni* Heller, 1897; for the southeastern dark grey species, *D. luctuosa* (D'Albertis 1874); and for the black species from Goodenough Island, *D. atrata* Van Deusen, 1955.

Within two of these four species (*D. muelleri* and *D. luctuosa*) there are quite sharply distinct geographic forms, distinguished from each other by size and other features and are sharply differentiated by multivariate analysis (Figs 1-2). These we rate as subspecies; some are already named, others not; in two cases the evidence is good enough to warrant the description of two new subspecies, but in some other cases we merely point to the probable existence of new subspecies for the evidence is not yet enough to describe and define them.

Within *D. hageni* there is some polymorphism, which appears to have a geographic component, but not sharply enough to warrant the recognition of subspecies. At least one of the subspecies of *D. muelleri* is also polymorphic in an interesting manner.

TAXONOMY AND SYSTEMATICS

(1) *DORCOPSIS MUELLERI* (Schlegel 1866). Brown Dorcopsis.

Diagnosis. Colour of dorsal surface some shade of brown (as opposed to grey in *D. luctuosa* and black in *D. atrata*), and lacking the white dorsal stripe of *D. hageni.* Underside grey to white or

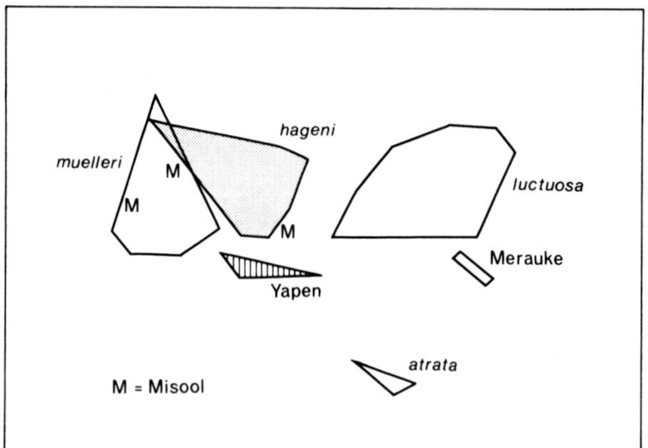

Fig. 1. Graph of First and Second Discriminant Functions for 18 cranio-mandibular measurements of *Dorcopsis* species. As in Fig. 2, individual specimens are not plotted, only distribution polygons. Discriminant Function 1 (61% of total variance), on the horizontal axis, denotes small size with posteriorly broadened nasals, long palatal vacuities and long incisive foramina. Discriminant Function 2 (14% of total variance), on the vertical axis, denotes an elongated skull but short incisive foramina, combined again with posteriorly broad nasals. Note that, as for the dental graph (Fig. 2), *D. l. phyllis* n. ssp. ("Merauke") and *D. m. yapeni* n. ssp. ("Yapen") are separated from conspecifics, but that *D. m. mysoliae* ("Misool") is not.

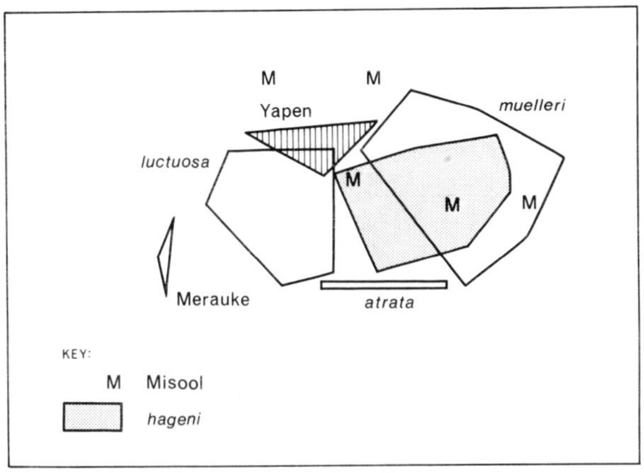

Fig. 2. Graph of First and Second Discriminant Functions for seven dental measurements of the species of *Dorcopsis*. Polygons are constructed by joining up the most outlying specimens from each sample. Discriminant Function 1 (accounting for 72% of total variance) is on the horizontal axis. It is strongly dependent on absolute size, especially total toothrow length. Discriminant Function 2 (19% of the total variance) is on the vertical axis; high scores on this Function denote large-sized secator, broad M^2 and short incisor row. Note separation of "Merauke" (*D. l. phyllis* n. sp.) from other *D. luctuosa*, and of "Yapen" (*D. m. yapeni* n. sp.), but not "Misool" (*D. m. mysoliae*) from other *D. muelleri*.

creamy (as opposed to blackish brown in *D. atrata*), this tone sharply set off from that of upper side (not sharply set off in *D. luctuosa*); throat whiter than rest of underside; limbs paler than rest of upper side; dorsal hair tracts divided by either a whorl, or pair of whorls, level with scapulae, or by a pair of transverse partings from axillae running towards middle of back (as opposed to the "V" pattern of *D. luctuosa*). Dorsal outline of facial skeleton straight in lateral view; braincase rounded (except for sagittal crest, usual in adults), falling away abruptly behind, in front of nuchal crest; nasals broad, posteriorly rounded; rostrum broad, rounded; palatine vacuities tend to be short and broad; posterior margin of palate somewhat concave. Toothrows straight, parallel; secators (P3) long (shorter in *D. luctuosa*), equal to two and a quarter to two and a half molar lengths, narrow, mesially pointed.

Distribution: Lowlands of Irian Jaya, as far east as Danau (=Lake) Biru in the north, and the Setakwa/Mimika Rivers district in the south; the offshore islands of Misool, Salawatti, and Yapen (Fig. 3).

(a) *Dorcopsis muelleri muelleri* (Schlegel 1866)

1830 *Didelphis brunii* Quoy and Gaimard. Dorey (=Manokwari), in the Cenderawasih (formerly Vogelkop) district of Irian Jaya. Not of Schreber, 1778 (a *Thylogale*).

?1839 *Hypsiprymnus dorsocephalus* Owen. New Guinea. *Nomen nudum.*

1843 *Halmaturus asiaticus* Gray. Lobo Bay (see Laurie and Hill 1954, p. 24). *Nomen nudum.*

1866 *Macropus muelleri* Schlegel. Lobo Bay.

1898 *Dorcopsis rufolateralis* Rothschild and Rothschild. "North New Guinea".

Diagnosis: Dorsal colour overall reddish brown; hair shafts medium red-brown to clay-brown, bases only slightly paler. Arms creamy-brown; legs also contrastingly lighter than body. Underside creamy with a red tinge, extending up posterior flanks in a broad wedge in front of stifle. One or two scapular whorls. Claws dark. Tail naked for one-sixth of its length. Size large, at

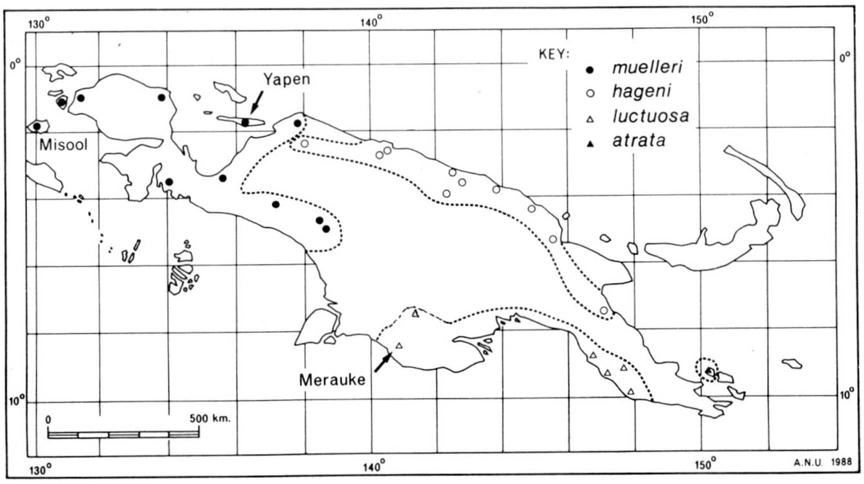

Fig. 3 Distribution of the species of *Dorcopsis*. The labelled localities are as follows: *D. muelleri* – Misool, type locality (and entire distribution) of *D. m. mysoliae;* Yapen, type locality (and entire distribution) of *D. m. yapeni* n. ssp. *Dorcopsis luctuosa* – Merauke, type locality of *D. l. phyllis* n. ssp.; Fly delta localities are in same general area. Other localities for *D. luctosa* represent nominotypical *D. l. luctuosa.* Other localities for *D. muelleri* represent nominotypical *D. muelleri* or allied (undescribed) taxa, on the Cenderawasih Peninsula; more easterly localities represent *D. m. lorentzii.*

Table 1. Skull measurements of adult *Dorcopsis.*

	Specimen	Total Skull Length	Condylo-Basal Length	Bizygo-matic Breadth	Nasal Length	Posterior Nasal Breadth	Rostrum Breadth	Rostrum Height	Maxillary Toothrow Length	Height Mandibular Ramus	Length of Palatine Vacuities
(A) MALES											
Adult											
D. muelleri											
D. m. muelleri	RML13475	ca.140	—	58	—	16	—	—	40	14	14.5
D. m. lorentzii	RML24235	149	140	60	56	15	20.5	22	41	15	—
D. m. mysoliae	BM61.12.11.22	121	115.3	55	46	13	16.5	19	39	13.5	14
D. m. yapeni	BM46.691	120	114	55	47	12.2	17	17.2	37.2	14	ca.17
D. hageni	RML24256	—	—	—	52	14.5	19	21	40	—	17
	AM M6211	—	—	59	55	16	20	22	39	—	16
	LNK535	134	129	57	52	14.5	19	22.5	41.5	14	15.5
	AM M13636	133.5	130	60	50	15	21	21	38	—	16.6
D. luctuosa											
D. l. luctuosa	AM M7385	135	125	56	54	19	20	22	36	15.5	13
	AM M9139	134.9	126.6	59.9	58.4	16.7	17.3	21.8	38.8	15.1	16
	AMNH108112	—	—	56.4	50.2	13.3	23.2	21.2	37.4	12.9	—
D. l. phyllis	AMNH105988	—	—	51.4	39.9	14.3	22.0	20.4	32.7	11.9	—
	AMNH105981	—	—	52.8	40.8	14.8	22.4	19.5	35.0	11.7	—
	AMNH105979	—	—	52.6	43.7	16.6	22.9	19.7	34.9	12.3	—
	NMV C7096	ca.111	104	50.0	46.0	12.5	17.0	17.8	37.0	—	—
D. atrata	AMNH157277	112.9	109.3	53.6	42.5	11.1	18.8	17.8	38.5	—	17.0
	AMNH157276	117.9	111.9	51.0	45.2	10.2	16.8	18.1	39.0	13.0	14.8
	AM M19461	121	116	55.5	48.0	11.5	19.0	20.0	41.0	13.5	145
Young Adult											
D. muelleri											
D. m. muelleri	RML13481	128	126	54.5	51	14	19	22	40.5	13.5	14
	RML13473	129	122	54	51	12	19	20	39	12	12
	RML cat.ost.m	—	—	54	50	16.5	19.5	21.5	42	14	16
D. m. lorentzii	RML24254	131	124	57	50	13.5	18	21	41	13	10
	RML74268	127	119	54	50	13	18	21	40	13.5	12
	ZMA17991	123	117	55	46	13	17	19	40	13	12
	BM13.6.18.97	116.5	112.5	56	45	14	17	18	40.5	13	11.5
	ZMB91751	117	113	54	44	12	16	18	40	12	ca.14
D. m. mysoliae	RML cat.ost.f	—	—	60	55	16	18	22.5	43	14	17
D. hageni	ZMB22369	134	128	57	56	14	18	23.5	42	14	15
	AM M13634	129.5	122	55	51	16	18	20	40	—	15.4
	AM M6098	131.7	125.5	61.7	51.2	13.2	17.2	19.8	39.3	13.8	15.8
	BM22.2.2.63	144	136	61.0	54	16	21.5	22	42.2	15	14
D. luctuosa											
D. l. luctuosa	ZBM22717	120	114	52	48	13	18	19.5	38	12.2	15.2
	NMV C10784	119	111.7	53.8	42.4	14.6	17.2	18.6	37.9	—	—
	RML cat.ost.a	121	114	54	49	14	18	20	39	12	15
	AM M7052	120	111	53	47	16	17	21	37	13	15
	AM M19464	120	111	52.7	47.7	13	14.8	19.5	35.5	12.7	13.5
	BM8.10.8.37	114	109.5	55	45.5	14	17.2	20	36	12	16
D. atrata	AM M19458	107.5	104	51	41	10.5	16	16.2	37	12	11
(B) FEMALES											
Adult											
D. muelleri											
D. m. lorentzii	RML24253	121	114	52	—	11.5	16	17	39	12	12
	ZBM91753	120	114	54	46	13	17	19	42	13	14
	ZMB91752	121	114	50	45	13.2	17	17	38	12.5	14.5
	BM11.11.11.109	118.5	112.5	55	45	14.5	16.5	17.5	38	13.5	10.5
D. m. mysoliae	BM61.12.11.23	118	110.5	51.5	40+	14	16.5	18	37	13	10
	AMB22402	—	—	51	41	12	12	14.5	38	13	11
D. m. yapeni	BM46.693	115	108.5	52.3	43	12.5	15	15.2	37	13	ca.17
D. hageni	AM13633	117.2	109.8	53.6	45.9	13.3	15.0	16.5	37.8	12.7	11.5
	AM M13632	119.2	112.9	51.5	45.6	11.9	15.1	16.8	38.3	12.1	ca.16.5
D. luctuosa											
D. l. luctuosa	ZMB6035	123	114	54	51	17	20	22.5	35	16	14
	ZMB11521	124	117	58	53	16.5	19	20	36	14.5	ca.12
	AM M790	ca.108+	—	50	45	14	16	18	36	11	12
	AM M3663	108	101	52	45	15	16	17.5	35	12	14
D. l. phyllis	AMNH105998	—	—	46.4	37.0	12.0	20.0	17.5	32.8	10.5	—
	AMNH105982	—	—	48.5	40.9	9.0	20.3	19.2	33.7	11.6	—
	RML17239	107	101	51	41.0	16.0	18.0	18	33.0	12.0	12.5
	WAM M2808	—	—	48.3	41.0	13.0	14.5	15.5	32.8	10.4	13.2

Table 1 continued

	Specimen	Total Skull Length	Condylo-Basal Length	Bizygo-matic Breadth	Nasal Length	Posterior Nasal Breadth	Rostrum Breadth	Rostrum Height	Maxillary Toothrow Length	Height Mandibular Ramus	Length of Palatine Vacuities
	(B) FEMALES										
	Adult										
D. atrata	AMNH157275	104.0	98.3	50.4	40.7	10.2	17.1	15.2	36.1	—	16.3
	AM M19457	109	105.6	52.4	44.2	11.3	13.3	16.2	34.0	13.7	16.6
	Young adult										
D. muelleri											
D. m. lorentzii	RML24264	121	114	51	47	11	17	18	39	12	—
	BM13.6.18.99	118.5	111.5	52	46	13	16	16.5	39	12	12
	BM13.6.18.98	116	112	53.3	45	15	17	17	40	12	7.5
D. m. mysoliae	BM13.3.6.23	107	99.5	48	41	13	14.5	15	35	10	6.5
D. m. yapeni	BM46.692	109	104	52.5	42	12	15	15	35	12	13.5
D. hageni	NMV7577	121	117	53	47	14	18	20	40	12	13
	ZMB A.228.10	115	106	52	44	14.5	16	18	38	12	ca.9
	ZMB A101.10	120	113	54	49	15	18	19	39	12	13
	ZMB VII.56	116	109	52	42	14	17	17	40	11.5	11.2
	RML12494	115	108	56	46	13	19	17	37	13	15
	ZMB VII41	109	—	—	42	11	16	17	36	11	ca.11
	ZMB[unreg.]	116	109	52	47	12.5	17	19	37	12	11
	ZMB11361	—	—	51	44	12	17	19	38.5	12	14
	RML16436	113	107	51	45	14	16	18	40	12	15
	RML16430	114	108	53	45	14	18	17	38	13	15
D. luctuosa											
D. l. luctuosa	BM61.8.23	114	106	52	42+	14.5	17	17	35	12	13
	NMW2981	112	105	52	45	14	18	18.5	37	11	ca.12
	AM M7449	108	102	51	46	14	17	17	36	12	15
D. l. phyllis	ZMA5243	100	94	49.5	46	13	16	16	32	11	6.5
D. atrata	AM M19460	119	115	54.5	48.5	11	16.5	19	37	13.3	14.6
	AM M19456	100.6	95.5	49.3	39.8	10	12.3	14.5	35.5	10.3	13.8

least in males; the single available fully adult male skull is considerably larger than the three young adults (i.e., with all molars erupted but spheno-occipital synchondrosis ["basilar suture"] unfused) (Table 1). Nasals somewhat broadened posteriorly, less rounded into frontals than other subspecies. Zygomatic arches bowed outward, Teeth (incisors, canines, cheekteeth) large (Table 2).

Distribution: Apparently the southern peninsula of Cenderawasih.

Specimens seen: RML 13473 (Jentink's cat.b, cat.ost.a), skin, skull and skeleton, from Lobo Bay 3.51°S, 134.04°E, lectotype of *muelleri* Schlegel; RML 13474 (cat.a, cat.ost.c), skin and skull of a juvenile, from Lobo Bay, paralectotype of *muelleri;* RML 13475 (cat.ost.b), skull and skeleton, from Lobo Bay; RML cat.d (cat.ost.n), skin and skull, juvenile, from Andai (0.51°S, 134.01°E); RML cat.c (cat.ost.m), skin and skull, juvenile, from Andai; RML 13481 (cat.q, cat.ost.o), head-skin and skull, from Sorong (0.54°S, 131.15°E); RML 13482 (cat.ost.p), juvenile skull, from Sorong; BM unregistered, skin only, type of *rufolateralis* Rothschild and Rothschild; BM 81.2581, -2, -3, skins only, collected by de Bruijn, so probably from Lobo

Table 2. Dental measurements of *Dorcopsis.*

	I^{1-3} Male $\bar{x}$ s.d. n	I^{1-3} Female $\bar{x}$ s.d. n	P^3 length Male $\bar{x}$ s.d. n	P^3 length Female $\bar{x}$ s.d. n	P^3 as percent of P^3-M^5 Both sexes $\bar{x}$ O.R. n
D. muelleri					
D. m. muelleri	10.8±0.96(4)	10.3±0.35(2)	14.6±0.44(5)	14.5 (1)	56.0(55.0-57.7)(5)
D. m. lorentzii	10.3±1.06(12)	9.7±0.53(12)	14.6±0.48(12)	14.3±0.71(10)	56.2(53.8-60.4)(16)
D. m. mysoliae	11.0 (2)	9.0±1.00(3)	14.8 (2)	13.2±0.93(3)	56.0(51.9-59.2)(5)
D. m. yapeni	9.0 (1)	8.5 (2)	14.0 (1)	12.8 (2)	55.4(53.4-58.3)(3)
D. hageni	11.4±0.68(8)	9.4±0.45(7)	14.2±0.74(16)	13.6±0.50(7)	54.9(50.0-59.6)(23)
D. luctuosa					
D. l. luctuosa	10.8±1.16(7)	9.1±1.07(7)	13.3±0.79(9)	12.6±0.52(10)	53.5(50.0-57.7)(18)
D. l. phyllis	8.4±0.06(3)	8.7±0.41(4)	12.6 (1)	11.3 (2)	52.4(50.0-54.8)(2)
D. atrata	10.7±1.30(4)	9.2±0.94(4)	13.5±0.37(4)	12.2±0.58(4)	♂52.7(51.2-54.3)(4) ♀46.4(42.2-49.6)(4)

Bay, and probably syntypes of *asiaticus* Gray; BM 41.1109, skin, no locality.

Notes: There are possibly two separate subspecies among these specimens. The two Lobo Bay skins have the hair-shafts more reddish with the bases barely detectably paler, while the two from Andai have darker, more clay-brown shafts with the bases a clearly (but not sharply) lighter version of the same. Without larger series from each locality, preferably backed up by cranial characters, it is impossible to confirm the consistency of these differences.

(b) *Dorcopsis muelleri lorentzii* Jentink, 1908

1908 *Dorcopsis lorentzii* Jentink. Alkmaar, Noord River, Irian Jaya.

Diagnosis: Dorsal colour medium fawn to deep chocolate brown, the dark skins predominating; hairs absolutely unicoloured from base to tip. Arms fawn; legs not contrasting as much with body. Underside grey-white to creamy yellow, forming a wedge in front of stifle. Throat whiter than rest of underside; muzzle white. A single scapular whorl, inconspicuously expressed; often showing a slight tendency to double (Fig. 4a, b). Claws dark, to black. Tail naked for one-tenth to one-sixth of its length, slightly depigmented at tip. Skull apparently as in nominotypical *muellerii*, but palatine vacuities tend to be shorter, more oval, often divided by struts into two or more on one or both sides.

Distribution: Mainland of Irian Jaya, at least south of the main cordillera.

Specimens seen: RML 24254, subadult skull, from Alkmaar (4.40°S, 138.43°E), lectotype of *lorentzii;* RML 24253 and -5, skulls, from Alkmaar, paratypes of *lorentzii*; RML 24235, 24265, -, -7, -9, -70, skins and skulls, from Alkmaar; RML 24268, skull only, Alkmaar; RML unregistered, skin only, Alkmaar; RML 24264, subadult, skull only, Van Weel's Camp (4.51°S, 138.45°E); ZMB 91751, -2, -3, skins and skulls, from Lower Menoo, Weyland Range (3.43°S, 135.24°E); BM 1939.2912, skin and skull, Setakwa River (4.18°S, 137.17°E); BM 13.6.18.96., -7, -8, -9, skins and skulls, Setakwa River; BM 11.11.11.105, -6, -7, -8, -9, -10, skins and skulls (mostly subadult), from Mimika River (4.43°S, 138.28°E); BM 22.3.23.10, skin and skull from Kloof Bivak, Noord River (4.39°S, 138.42°E); ZMA 17.990, -1, skins and skulls, Kloof Bivak.

Notes: This is the best-represented of the subspecies of *D. muelleri*, well distinguished from all others, especially by the unparallelled unicoloured nature of the hairs. The three Weyland Range specimens are rather darker than most others, their undersides more creamy yellow. The adult male skulls are all much larger

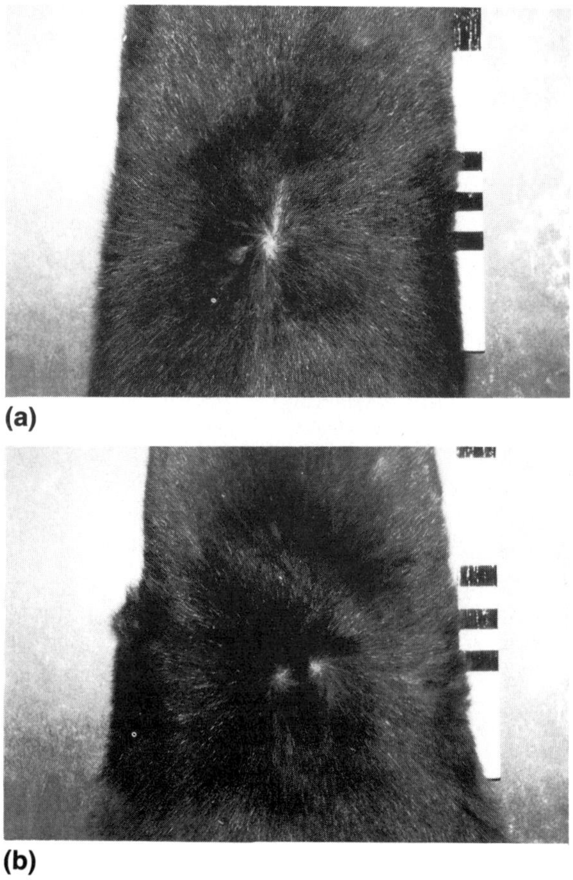

Fig. 4. Hairwhorl patterns in *D. m. lorentzii* (a) BM 22.3.23.10 Kloof-bivak; (b) BM 11.11.11.109 Mimika River.

than those of adult females or subadults of either sex; evidently sexual dimorphism is established largely by male hypermorphosis. The teeth of the two sexes, with the possible exception of the incisors, are the same size (Table 2).

(c) *Dorcopsis muelleri mysoliae* Thomas, 1913

1913 *Dorcopsis mysoliae* Thomas. Mysol Island (=Pulau Misool).

Diagnosis: Overall tone dull brown; shafts dark brown, bases very pale grey. The limbs, especially the legs, contrast very considerably with the body in their lighter, greyer colour. Underside grey to white, sharply set-off from upper side, without a lateral wedge in front of stifle; throat white; cheeks light, grading into dark colour of head. Claws light in colour. Hair plush; most skins have long conspicuous partings from armpits up to shoulders (Fig. 5), but a few have paired whorls as in nominotypical *muelleri.* Nearly one-fifth of tail is naked, this region being very sharply set-off from haired part; tail-tip is dark. Posterior ends of nasal evenly convex, less rounded than in mainland races; premaxillary/maxillary suture straight, not sinuous like mainland forms; zygomata straight-sided rather than outbowed; palatine vacuities broad, rounded. Size small; skulls of both sexes equivalent in size, in both

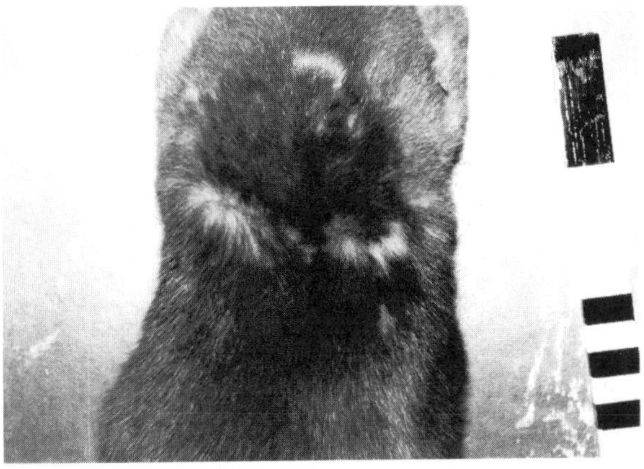

Fig. 5. Hairwhorl of *D. m. mysoliae*. BM 61.12.11.23 Misool.

adults and (apparently) young adults — i.e. no hypermorphosis in male.

Distribution: P. Misool.

Specimens seen: BM 61.12.11.22, skin and skull, type of *mysoliae;* BM 61.12.11.23-4, skins and skulls; BM 13.3.6.23, skin and skull; RML *muelleri* cat.e, f, g, h, i, j, k, l, m, skins, and cat.ost.i, j, k, l, m, skulls; ZMA, unregistered, skin only; ZMB 22402, skull only; ZMZ, unregistered, skin only. (All are from Misool).

Notes: The difference between the skins with the parting and those with the whorls is so striking that we initially considered whether they might represent two species; but no other differences are apparent, so we ascribe it to polymorphism. (There are no intermediate states). RML cat.f, k and m have whorls; all the rest, in all collections, have partings.

(d) *Dorcopsis muelleri yapeni* n.subsp.

Type: BM 46.691, adult male skin and skull, Pulau Yapen (=Japen Island), Irian Jaya, Indonesia.

Diagnosis: Most closely resembles *D. m. mysoliae*, but with the following differences: overall colour browner; hair bases pure white; arms more contrasting in tone. Transverse partings present as in most *mysoliae,* but fail to reach midline of shoulders (Fig. 6). Claws pale brown. Fur still thicker, more plush. Compared to mainland subspecies (but *mysoliae* is unknown in this respect), the extremities (tail, hindfoot, ears) are relatively shorter (Table 3); this is not a purely paedomorphic (progenetic) phenomenon, as extremities cease growth earlier than the body as a whole, so that in mainland taxa, and other species of the genus, the ratios of various extremities to head plus body length are higher in young than in adults. Premaxillary-maxillary suture strongly angled in middle. Posterior nasal margin straight, less convex than in any other subspecies. Size apparently as in *mysoliae,* with perhaps greater increase between young and full adult stages, in both sexes. Midlinks on molars more strongly developed than other subspecies. The incisor row is shorter than in other subspecies, and the secator averages smaller especially compared to the molar row (Table 2).

Distribution: Pulau Yapen (formerly spelt Japen).

Specimens seen: BM 46.691 (the type), -2, -3, skins and skulls, all from P. Yapen.

Notes: Although there are only three specimens in collections, this subspecies seems so distinctive in so many characters that we have no qualms about naming and describing it. Other characters, less clearly discernable on univariate analysis (Table 1), are shown by the Discriminant Function analysis, on both cranial and dental characters (Figs 1-2). All three skulls show a feature which is very rare, though not unknown, in other taxa: the palatal vacuities are open behind, reflecting failure of ossification of the post-palatine bars.

A photograph of a living *D. muelleri* from Danau Biru, in the Mamberamo River district of Irian Jaya, kindly sent to us by Dr Ron Petocz, shows an animal rsembling *D. m. yapeni* in its short, straight transverse partings, but seemingly much greyer in tone. Museum specimens from this region are entirely lacking.

Table 3. External measurements of *Dorcopsis*.

	(Adults and subadults only)		
	Tail as % head plus body	Hindfoot as % head plus body	Ear as % head plus body
D. muelleri			
D. m. lorentzii	60.2-70.6(11)	20.7-23.6(10)	7.1- 9.7(8)
D. m. salawatti	71.3 (1)	23.0 (1)	9.4 (1)
D. m. yapeni	57.8-59.2(2)	18.3-18.9(2)	5.5- 6.1(2)
D. hageni	68.8-75.0(4)	22.3-26.7(3)	8.3- 8.7(3)
D. luctuosa			
D. l. luctuosa	59.0-67.9(5)	21.1-24.5(4)	8.0-10.9(4)
D. l. phyllis	57.0-58.8(2)	20.9-21.4(2)	10.1-10.3(2)
D. atrata	51.3-56.9(6)	16.3-18.2(5)	4.0- 6.0(5)

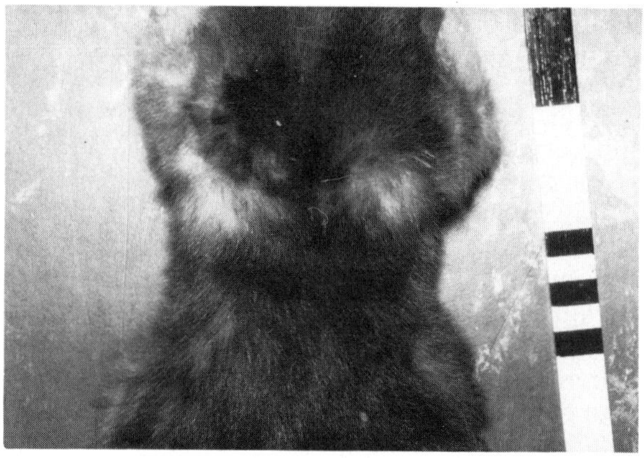

Fig. 6. Hairwhorl of *D. m. yapeni.* BM 46.692 Yapen.

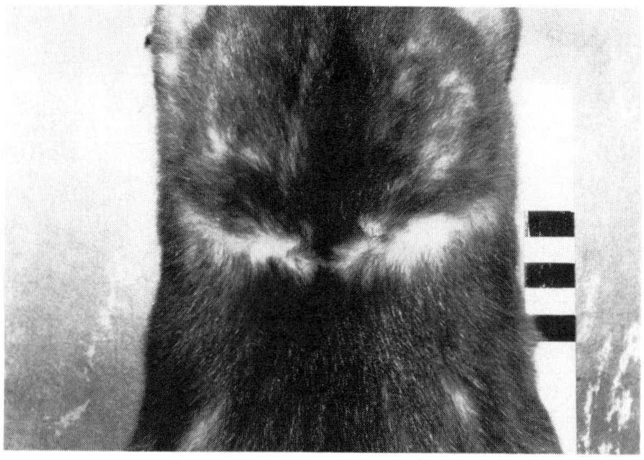

Fig. 7. Hairwhorl of *D. muelleri.* BM 33.11.4.4 Salawatti.

(e) *Dorcopsis muelleri* subsp. from Salawatti.

Specimens seen: BM 33.11.4.4, skin and skull; RML 13477 (cat.n.), skin only; RML 13478 (cat.o, cat.ost.i), 13479 (cat.p, cat.ost.j), skins and skulls; RML 13480 (cat.ost.e), skull only. All these specimens are juveniles.

We are fairly sure that the P. Salawatti *Dorcopsis* will prove to be a distinct subspecies of *D. muelleri,* but it cannot be adequately defined at present because all museum specimens are juvenile. The skins all have transverse partings, as long as those of (most skins of) *mysoliae* (Fig. 7). They are medium to dark brown, the hair-bases nearly white (not as pure white as in *yapeni*); there is a white wedge from the underside in front of the stifle, the limbs are contrastingly paler than the body; the tail-tip is depigmented; the claws are an off-white colour. The secator is curved inward anteriorly (one specimen only). Other characters resemble nominotypical *muelleri,* as far as can be made out.

(2) *DORCOPSIS LUCTUOSA* (D'Albertis, 1874). Grey Dorcopsis.

Diagnosis: Colour very dark, greyish ("olive-silvery-brown" is an apt description of many specimens) (as opposed to brown as in *D. hageni* and *D. meulleri* and black in *D. atrata*); hair-bases pale brown or grey. Limbs not contrastingly lighter than body; hands and feet blackish. Underside drab greyish to creamy orange, not sharply set-off from upperside (sharply set-off in *D. muelleri*, and blackish brown in *D. atrata*); chest nearly black; throat pale grey-brown, contrasting with head which is nearly black. Claws dark. A whorl just behind each axilla is prolonged into a back-curving hair-part; the hair-stream anterior to this does not run forward, but merely somewhat more medially than that posterior, so that the whole effect is of a V-pattern running back from axillae to just behind scapula, which is unique in the genus (Fig. 8). Cranially, the rostrum is very narrow (broader in *D. muelleri* and *D. hageni*); the premaxillary-maxillary suture is angled; the nasals are straight-sided; the naso-frontal suture is flared but does not wedge far backward into the frontals; the postorbital processes are less produced, more shelflike, than in *D. muelleri;* the palatine vacuities are long, with very thin (sometimes unossified) bars behind; the basicranium is constricted in the middle. Toothrows are curved in at the back, not straight and parallel (as in *D. muelleri* and *D. hageni*). Secators are short, and more sexually dimorphic in relative size than in *D. muelleri.* The midlinks on the molars are poorly developed.

Distribution: The south-east coastal lowlands of New Guinea, from Merauke to Milne Bay (Fig. 3).

Notes: Except for George and Schurer (1978), George (1979) and Archer, Flannery and Grigg (1985), this taxon has not been distinguished, in recent years, even subspecifically, from *D. muelleri*. We find it to be a highly distinctive species, cranially and dentally perhaps the most distinctive of all in the genus.

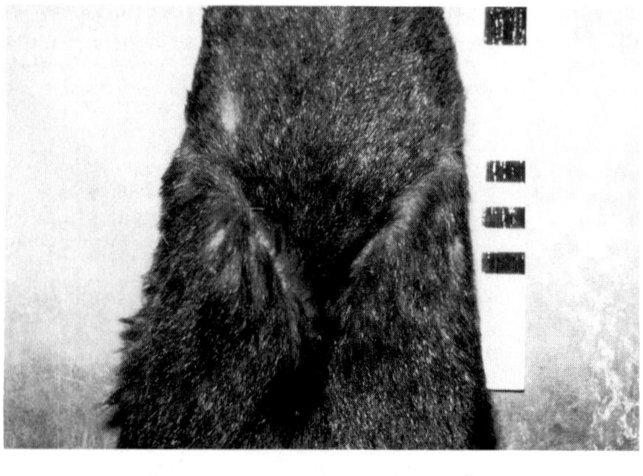

Fig. 8. Hairwhorl of *Dorcopsis luctuosa* holotype BM75.4.16.1 south-east New Guinea.

(a) *Dorcopsis luctuosa luctuosa* (D'Albertis 1874).

1874 *Halmaturus luctuosus* D'Albertis. "South-east New Guinea".

1884 *Dorcopsis chalmersii* Miklouho-Maclay. South coast of New Guinea, opposite Samarai Island.

1885 *Dorcopsis beccarii* Miklouho-Maclay. Hills around Port Moresby.

Diagnosis: Size large; strongly sexually dimorphic; teeth relatively large; tail, hindfoot, ears relatively long; colour medium to dark grey or fawn-grey.

Specimens seen: BM 75.4.16.1, juvenile skin and skull, "south-east New Guinea", type of *luctuosa*; RML cat.ost.a, skull only; BM 8.10.8.3, skin and skull, Epa (8.44°S, 146.43°E); ZMB 22717, skull only, Epa; BM 76.10.28.3, -4, -5, juvenile skins and skulls, Aleya; AM 7385 and 7449, skins and skulls, Laloki district, Port Moresby; AM 8665, 7052, 3663, skins and skulls, Port Moresby; AM 15288, skin and skull, Taronga Zoo, of stock from Brown River area, Port Moresby; AM M19464, skin and skull, Vaimuri logging area, near Port Moresby; NMV 2981, skull only, Port Moresby; Macleay, M383, skin only, Port Moresby, type of *beccarii;* Macleay, M399, skin only, Port Moresby; ZMB 6035, skull only, Laloki; AM 790, skin and skull, Astrolabe Range (9.33°S, 147.26°E); ZMB 11521, Astrolabe Range; BM 6.1.8.23, skull only, Upper Gira River (7.59°S, 147.59°E); AM S1863, skull only, Mt. Lamington; NMV C26610, 23027, 16188, 18536, 18513, 18599, 18600, skins and skulls, mostly juvenile, all from Healesville Sanctuary, ex Port Moresby; NMV C10784, 11300, skins and skulls, Melbourne Zoo, ex Port Moresby.

Notes: All these specimens are remarkably homogenous, varying only in the tone of the body colour. AM 790, however, has an elongated whorl in the midline of the back as well as the typical V pattern.

(b) *Dorcopsis luctuosa phyllis* n.subsp.

Type: ZMA 5243, adult female skin and skull, Merauke (8.27°S, 140.22°E) (Fig. 10).

Diagnosis: Much smaller than nominotypical *D. l. luctuosa,* with little sexual dimorphism; teeth relatively smaller; extremities shorter (Table 3); colour darker, nearly black.

Notes: This highly distinctive subspecies is, in some respects, analogous to *D. muelleri yapeni;* in particular the shortened tail, feet and ear, in combination with the smaller size, are the same in both. The type has an anomalous hair-tract pattern: the right side of the V is present, but on the left this is replaced by two small, partly connected, whorls near the midline of the back.

Specimens seen: the type; RML 17239, skin and skull, Kali Koembe, Kaisa village, west of Merauke; NMV C7096, coll. A. Goldie, 1881 (and so probably from Fly Delta); WAM M2808 (received in exchange from American Museum, formerly AMNH 108012), skull only, L. Daviumbu, Middle Fly River. Tate and Archbold (1936) give additional data on specimens in the AMNH from the Fly River.

Etymology: For Phyllis R. Groves, in recognition of her invaluable assistance to us in studying specimens, in discussion and (in the case of CPG) in general support.

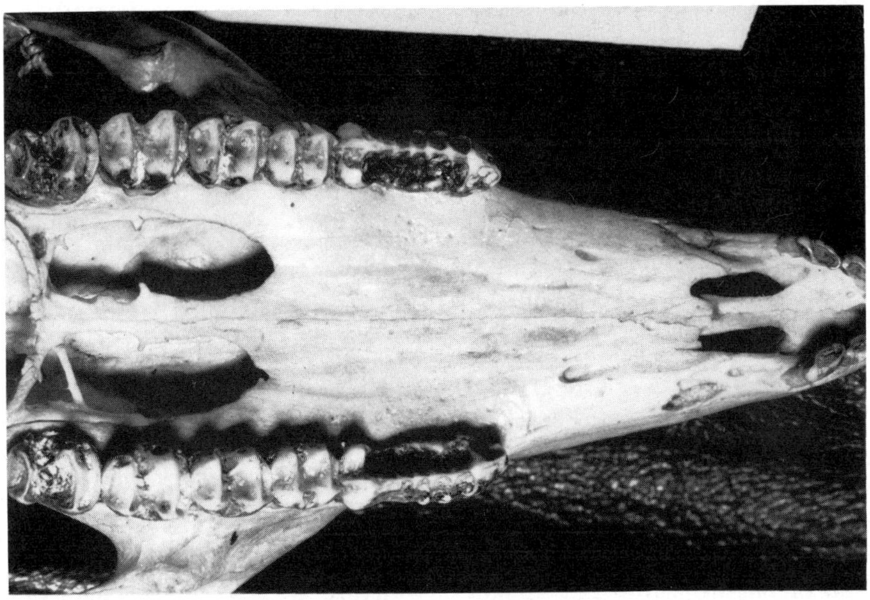

Fig. 9. Dorcopsis luctuosa luctuosa RML cat.ost. a, palatal view.

Fig. 10. Dorsal views of study skins of (from left to right) ZMA 5243, holotype of *Dorcopsis luctuosa phyllis* n. ssp.; ZMA 17991 and 17990, *Dorcopsis muelleri lorentzi,* Kloof Bivak, Irian Jaya.

(3) *DORCOPSIS HAGENI* Heller, 1897. White-striped Dorcopsis.

1897 *Dorcopsis hageni* Heller. Stefansort, Astrolabe Bay.

1922 *Dorcopsis hageni caurina* Thomas. Pionierbivak, Mamberamo River.

1937 *Conoyces hageni eitape* Troughton. Eitape (=Aitape).

Diagnosis: Fawn-brown, sometimes with a greyish cast, to nearly blackish-brown; a white or very pale fawn dorsal stripe from nape to root of tail (which is unique in the genus). Hair bases white. Underside grey-white. Claws light brown. Limbs contrastingly pale, fawn. Tail fawn, with a brown crest along dorsal midline of its proximal third; often a hairless callus on dorsum of root. Whorls behind axillae, but not extended into partings like *D. luctuosa.* Tail averaging longer than other species; hindfoot long. Size very large. Naso-frontal suture flared laterally, not wedging much into frontals; premaxillary-maxillary suture straight; secator (P3) broad, mesially blunt not pointed as in other taxa, and short, not more than two molar lengths (see also Table 2). Palatine vacuities tend to be short; posterior margin of palate V-shaped.

Distribution: Lowlands of northern New Guinea, from the region of the Mamberamo River in the west to the Morobe Coast (Fig. 3).

Specimens seen: LNK 535, skin and skull, Stephansort, type of *hageni;* ZMB 22369, skin and skull, Stephansort; AM M6211, skin and skull, Aitape, type of *eitape*; AM M6 098, juvenile, skull only, Aitape; BM 1939.2911, skin and skull, Hol; RML 16430, skin and skull, Humboldt's Bay; BM 22.2.2.63, skin and skull, young adult, Pionierbivak, type of *caurina*; ZMB VII.41 and an unregistered specimen, skins and skulls, Momofolag, Lower Sepik River; ZMB A.101.10, skin and skull, Berlinhafen; ZMB VII.56, skin and skull, Malu, Lower Sepik; RML 24256, skin and skull, Tami River (2.41°S, 140.51°E); ZMB A.228.10, skin and skull; RML 12494, skin and skull, Tami River; RML 16436, skin and skull, Lake Sentani; NMV 7576, skin and skull, juvenile, Ramu River (Arim); ZMB 11861, skin and skull, Ramu River; NMV 7577, skin and skull, Potsdamhafen; AM13633-4, 17202-3 skin and skull, AM13636, skull only, Yapsiei area, Lower August River; AM19062, skin and skull, Wara Bruk, Madang Province.

Notes: Despite its strongly divergent external appearance, this species is cranially and dentally very close to *D. muelleri*. There is considerable individual variation, which to some extent is geographically circumscribed. A blackish morph, in which the dorsal stripe is relatively inconspicuous, occurs at Pionierbivak, Malu and Monofolag; a grey-fawn-brown type with a white dorsal stripe occurs at Hol, Berlinhafen, Queen Augusta River, Tami River, and Aitape; while a dull brown form with fawn-toned stripe is from Astrolabe Bay. The series from Yapsiei in the Australian Museum range from fawn to dark brown. The type of *caurina,* though far from mature, is the largest specimen of the whole genus seen in the present study.

(4) *DORCOPSIS ATRATA* Van Deusen, 1955. Black Dorcopsis.

Diagnosis: Colour black dorsally (unique in the genus), venter blackish brown (also unique), both with white underfur. Ears and tail relatively shorter than in other *Dorcopsis*. Pelage long and luxuriant relative to all other *Dorcopsis* species. Dorsal hair tracts divided by a single whorl at the level of the scapulae. Skull and particularly nasals narrower than in other *Dorcopsis* species. Premolars hypsodont and short. On average smaller than any mainland *Dorcopsis* species.

Distribution: Montane rainforests between 1 000 and 1 800 m on Goodenough Island, D'Entrecasteaux Group, Papua New Guinea (Fig. 3).

Specimens seen: AM M19456 female skin and skull; AM M19260 male in spirit; AM M19459 female skin and skull; AM M19461 male skin, skull, skeleton; AM M19458 male skin and skull; AM M19460 female skin and skull; AM M19457 female skin and skull; AM M19462 skeleton.

DISCUSSION

Distribution: All *Dorcopsis* species except *D. atrata* are confined to lowlands (below 400 m). *D. atrata* differs markedly in habitat needs from the others in that it inhabits mid-montane rainforest on very steep terrain (pers. obs. TFF). In the experience of TFF in West Sepik Province is that *Dorcopsis hageni* is never found in even the lowest foothill forest, but can be abundant on the river floodplains. Present understanding of the distribution of the species of *Dorcopsis* suggest that the three mainland taxa recognized by us are parapatric. Unfortunately, collections of *Dorcopsis* from regions where the species are likely to abut are few (Fig. 3). Only in northern Irian Jaya have we records of two taxa in close proximity, where *hageni* and *muelleri* have been recorded on the lower Mamberamo River. No intergradation is evident in this case, and the Mamberamo may form a boundary between these two taxa.

Relationships: A number of features suggest that *Dorcopsis luctuosa* may be the sister taxon to all other species of *Dorcopsis*. Apparently derived features shared by all taxa except *D. luctuosa* include: posterior nasal narrowing, straight toothrows, concave post palatal margins, relatively short nasals, and a relatively broad basicranium. Within the clade containing the three remaining species, *D. atrata* is the most autapomorphic, and may be the sister taxon to a *muelleri-hageni* clade. Its short ears and tail are reminiscent of *Dorcopsulus* spp. from montane New Guinea, but this may be convergence due to habitat similarities.

SUMMARY

There are four species of *Dorcopsis*. *Dorcopsis atrata* and *D. hageni* and monotypic, while *D. luctuosa* includes two subspecies and *D. muelleri* four. *Dorcopsis luctuosa* may be the sister taxon to the remaining species.

ACKNOWLEDGEMENTS

Special thanks go to Paula Flannery, who has sacrificed much in allowing one of us (TFF) to conduct fieldwork in New Guinea that resulted in the collection of many of the specimens listed here, and to Phyllis Groves, who has aided us in our work. A grant from ARGS to Drs Gardner and Attenborough, on which Dr Flannery was an associate, funded a major expedition during which much material was collected. We collectively thank those curators who have helped us in our effort to locate *Dorcopsis* material, and particularly Paula Jenkins of the British Museum, who has provided invaluable data regarding the type of *D. rufolateralis*.

REFERENCES

Archer, M., Flannery, T. F. and Grigg, G. C., 1985. The Kangaroo. Wheldon's Pty Ltd: Sydney.

D'Albertis, L. M., 1874. "Letter from Mr D'Albertis". *Proc. Zool. Soc. Lond.* **1874:** 110.

Flannery, T. F., 1984. The kangaroos: 15 million years of bounders. Pp. 817-36 *in* Vertebrate Zoogeography and Evolution in Australasia. Hesperian Press: Perth.

George, G. G. and Schurer, U., 1978. Some notes on macropods commonly misidentified in zoos. Pp. 152-56 *in Int. Zoo. Yearb.* 18 ed. by P. J. S. Olney.

Gray, J. E., 1843. List of the Specimens of Mammalia in the Collection of the British Museum. British Museum.

Heller, K. M., 1897. Zwei neue beutelthiere aus Deutsch New Guinea nebst einer aufzahlung der bekannten papuanischen saugthiere. *Abhand. und Bericthe Konig. Zool. Anthrop. Mus. Dresden.* **6:** 1-7.

Husson, A. M. 1955. Notes on the mammals collected by the Swedish New Guinea Expedition 1948-1949. *Nova Guinea* (n.s.) **6:** 283-306.

Jentkin, 1908. Mammals collected by the Dutch New Guinea Expedition. *Nova Guinea* **9 (zool.):** 165-83.

Lesson, R. P., 1827. Description des mammiferes. Pp. 137-76 *in* Voyage Autor de Monde, sur la corvette . . . La Coquille, pendant les annees 1822, 1823, 1824 et 1825. Zoologie 1 ed. by L. I. Duperrey.

Matschie, P., 1916. Die verbreitung der beuteltiere auf New-guinea mit einegen Bemerkungen uber ihre einteilung in untergattungen. *Mitt. Zool. Mus. Berlin* **8:** 142-82.

Miklouho-Maclay, N. de., 1884. On a new species of Kangaroo *(Dorcopsis chalmersii)* from the south-east end of New Guinea. *Proc. Linn. Soc. NSW.* **9:** 569-70.

Miklouho-Maclay, N. de., 1885. On two new species of *Dorcopsis* from the south coast of New Guinea. *Proc. Linn. Soc. NSW.* **10:** 145-50.

Owen, R., 1839. Outlines of a classification of the Marsupialia. *Proc. Zool. Soc. Lond.* **7:** 3-19.

QUOY AND GAIMARD, 1830. Zoologie. Pp. 5-268 *in* Voyage ed Decouvertes de l'Astrolabe execute par ordre du Roi, pendant les annees 1826-1827-1828-1829. Vol. 1 ed. by J. Dumont D'Urville.

ROTHSCHILD, LORD AND ROTHSCHILD, N. C., 1898. Descriptions of three new kangaroos, and notes on the skull of *Dendrolagus bennettianus* De Vis. *Novit. Zool.* **5**: 511-13.

SCHLEGEL, H., 1866. Observations zoologiques 11. *Ned. Tijdschr. Dierk.* **3**: 249-58.

SCHLEGEL, H. AND MUELLER, S., 1845. Over drie buideldieren uit de familie der kangoeroes. Pp. 129-48 *in Verh. Nat. Gesch. Overz. Bezitt. Zoologie* ed. by C. J. Temminck.

SCHREBER, J. C. D. VON, 1778. Die Saugethiere in Abbildungen nach der Natur mit Beschreibungen. 283-590.

TATE, G. H. H., 1948. Results of the Archbold Expeditions. No. 59. Studies on the anatomy and phylogeny of the Macropodidae. *Bull. Amer. Mus. Nat. Hist.* **91**: 237-351.

THOMAS, M. O., 1913. On some mammals from Bali and Mysol, collected by Mr E. Streseman during the "Freiburger Molukken-Expedition". *Ann. Mag. N.H.* **11**: 504-07.

THOMAS, M. O., 1922. New mammals from New Guinea and neighbouring islands. *Ann. Mag. N.H.* **9**: 261-65.

TROUGHTON, E. LeG., 1937. Descriptions of some New Guinea mammals. *Rec. Aust. Mus.* **20**: 116-27.

VAN DEUSEN, H. N., 1955. Results of the Archbold Expeditions. No. 76. A new species of wallaby (genus *Dorcopsis*) from Goodenough Island, Papua. *Amer. Mus. Novit.* 1826.

APPENDIX

Measurements used in discriminate function analysis: greatest skull length; condylobasal length; basilar length; bimastoid breadth; braincase breadth; interorbital breadth; nasal length; least nasal breadth; greatest nasal breadth; snout breadth; snout height; upper cheektoothrow length; upper incisor row length; total upper toothrow length; molar row length; palate breadth; incisive foramen length; P^3 length; P^3 breadth; M^2 length; M^2 breadth; Dentary height; palatal fenestra maximum length.

Diets of macropodoid marsupials: General patterns and environmental influences

TERENCE J. DAWSON[1]

Information available on the diets of the Australian macropodoids has been reviewed. The characteristics of the vegetation have also been considered in terms of digestibility by enzymic or fermentative means. Rough or fibrous plant material requires extended fermentative digestion and is consequently not a reasonable diet option for most small species.

Diet variation shows some basic patterns, with body size being the most important. The small species, such as rat-kangaroos, tend to be specialized, feeding on high quality but scattered foods. Intake of underground fungi is especially significant to some species. The "medium" size macropodoids tend to be more generalist feeders and a wide variety of vegetation appears in their diet, the more common components varying with species and environment. Characteristic of their intakes is a relatively low fibre content. Even when browse is taken, on close observation, it is apparent that the animals are selecting the lower fibre parts of the plants. The large macropodoids, especially the kangaroos, are specialist grass eaters. The Red Kangaroo may be less selective than the other species, perhaps by necessity.

When communities of macropodoids are examined, resource partitioning appears evident. However, the basis of this is unclear and we cannot yet assess the role of competition in diet selection or community structure. A similar picture is seen when the interaction between macropodoids and domestic stock is considered. Diet overlap can occur between the kangaroos and sheep and cattle, notably with grass, but the fine detail of such overlap has yet to be examined. Competition among these species is a very complex issue. While it is suspected to occur in some situations, firm evidence is sparse.

INTRODUCTION

THE macropodoids are the dominant native mammalian herbivores of the Australian continent. Numerous herbivorous, quadrupedal marsupials also existed until the late Pleistocene megafaunal extinctions some 20 000-40 000 years ago (Archer 1984), but whether it was Aboriginal man, climatic change or both that led to these extinctions, the macropodoids survived with considerable diversity. Currently, there is concern for the survival of these unusual herbivores, but also worry about competition with domestic herbivores. My interest in the diets of the macropodoids is related to both of these concerns. However, this review will examine only the general patterns in the diets of macropodoids because other papers in this volume will deal with specific aspects e.g., kangaroo-sheep competition (Edwards, this volume).

Insights gained from this examination of diets may form a basis to help answer a series of important questions. What habitats do the macropodoids require for continued food and shelter? What vegetation do they consume that might have been available for animal production? Does the combined impact of wild and domestic herbivores result in pasture degradation to the disadvantage of all? Pasture degradation seems to have occurred in the early days of the grazing industry in our arid and semi-arid rangelands (Williams and Oxley 1979) and associated with this there was a marked decline in the species richness of native mammalian herbivores.

The earliest of the macropodoids probably arose in the mid-Tertiary, 15-20 MYBP (Archer 1984). These were inhabitants of the floor of the temperate forests that then covered Australia (Truswell and Harris 1982). The Musky Rat-kangaroo, *Hypsiprymnodon moschatus,* of the northern Queensland rainforests is considered a modern example of this body-form and life-style. It feeds on invertebrates, fruit, seeds and leaves. The stomach is relatively simple but, as pointed out by Heighway (1939, cited in Hume 1982a), it is partly compartmentalised (Fig. 1) and therefore is intermediate between the simple stomachs of the possums and the complex forms of the wallabies and kangaroos.

Complex stomachs are the hallmark of the macropodoids and the basis of their success as herbivores (Hume 1982a). The expanded forestomach is basically a large vat in which plant material is partially digested by micro-organisms. The role of these micro-organisms is very complex. They provide a macropodoid with many necessary products but it is their part in the extraction of energy from complex plant carbohydrates that is most important. These carbohydrates are largely the structural components of cell walls, often referred to as fibre, and they can be broken down only by fermentation.

The fermentation by micro-organisms, mainly bacteria, is carried out in an oxygen-free (anaerobic) environment. Because oxygen is not available the micro-organisms are able to extract only a small amount of energy from the

[1]School of Biological Science, University of New South Wales, Kensington, New South Wales 2033.

Pages 129-42 in KANGAROOS, WALLABIES AND RAT-KANGAROOS ed. by G. Grigg, P. Jarman and I. Hume. Surrey Beatty & Sons Pty Limited, New South Wales, Australia, 1989.

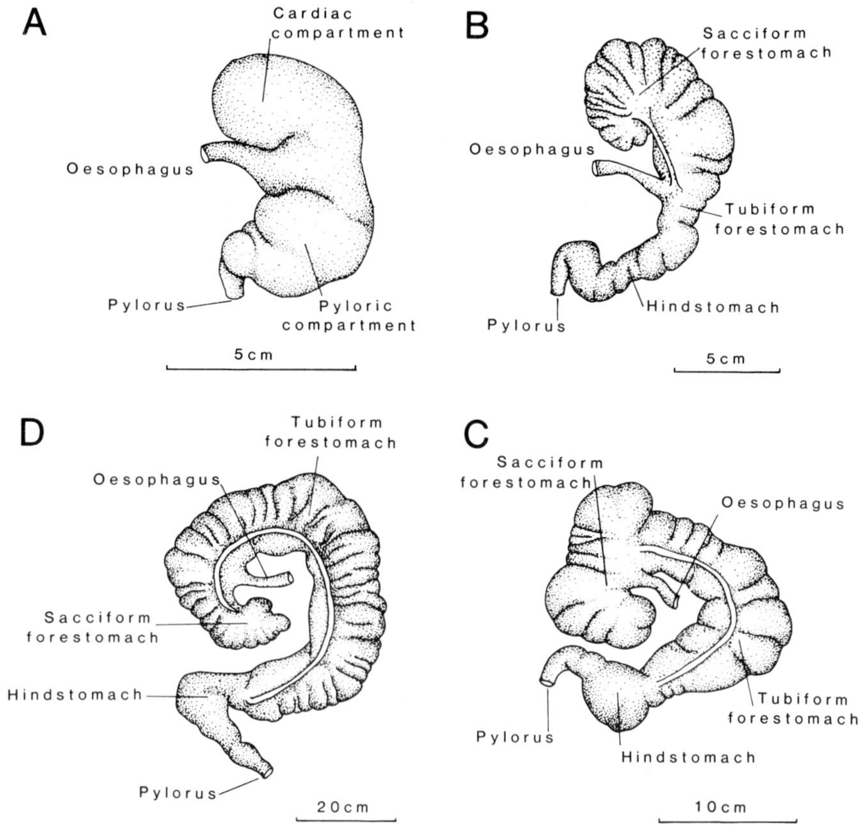

Fig. 1. Variation in the structure of the forestomachs of macropodoid marsupials with body size and diet. A. *Hypsiprymnodon moschatus*, body weight, ca 0.5 kg; omnivorous diet. B. *Bettongia gaimardi* body weight 1.2-2.3 kg; variety of plant material especially fungus in diet. C. *Thylogale thetis*, average body weight 7.0 kg (males), 3.8 kg (females); short grasses and soft browse in diet. D. *Macropus giganteus*, average body weight ca 55 kg (males), ca 25 kg (females); grass principal diet component. (After Hume 1982a, 1982b).

carbohydrates, and the end products of their fermentation are short-chain fatty acids such as acetic, butyric and propionic. These steam-volatile fatty acids (VFAs) are then available to a kangaroo or wallaby as its principal energy source. The success of the ruminants, among the placentals, is based on a similar digestive strategy. It is this strategy that enables the kangaroos, as well as sheep and cattle, to utilize the coarse vegetation that is the primary production of the extended rangelands of Australia.

When we consider the factors that influence diet selection by mammalian herbivores, characteristics of the vegetation, such as chemical composition, the presence of secondary compounds, or palatability, generally come to mind. These factors are secondary, however, to the basic characteristics of the herbivore itself. Such characteristics are encompassed by the fundamental physiology and ethology of an animal, but the most important of these is its size (Hanley 1982). The consequences of size are nicely seen in the successful mammalian herbivore groups such as the ruminants and in the group now under discussion, the macropodoid marsupials.

INFLUENCE OF BODY SIZE ON DIET SELECTION

Why is the size of the animal so important in its diet choice? Firstly, size is one of the primary determinants of metabolic rate or metabolic intensity (Kleiber 1961; Dawson and Hulbert 1970). Smaller animals have a proportionally greater metabolic intensity. Secondly, fermentative digestion of plant fibre is a relatively slow process that does not vary much in its rate. The microbial activity in the forestomach of macropodoids is similar in animals of different body size and similar to that in ruminants (Dellow *et al.* 1983). The relatively lower fibre digestion that is seen in some wallabies and kangaroos is due to the generally faster rates of passage of material through their guts, not a less efficient microbial fermentation.

Small mammals need to process more feed relative to their size than do larger ones. Short (1985) demonstrated this when estimating the intake of pasture by rabbits, Red Kangaroos *Macropus rufus* and sheep grazing in small yards with similar vegetation. On the basis of their metabolic body

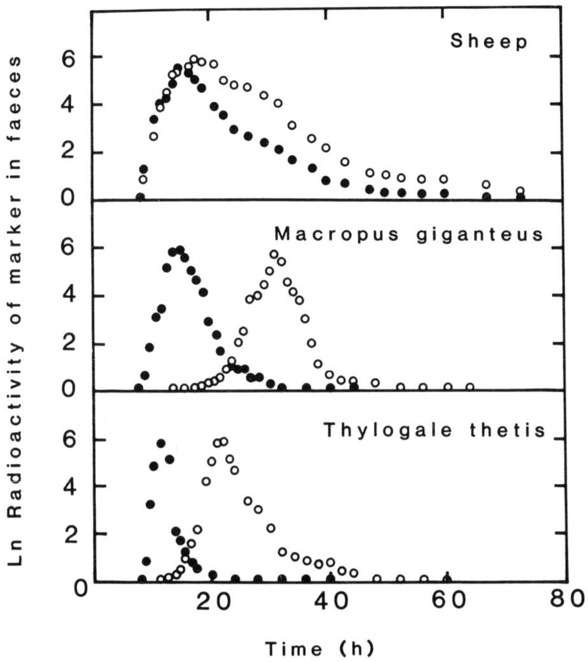

Fig. 2. Influence of body size and gut morphology on pattern of appearance in faeces of digesta markers. ^{51}Cr EDTA (●) marks fluid fraction and ^{103}RuP (○) marks particulate fraction; radioactivity measured in $10^3 \times$ counts min^{-1} per gram dry matter. NOTE: the fluid and particulate fractions became more separated in the tubiform forestomachs of the marsupials, *M. giganteus* and *T. thetis*. (After Dellow 1982).

size, i.e., body weight raised to the 0.75 power, the three mammals took in roughly the same amount. The estimated maximum dry matter intakes of rabbits, kangaroos and sheep were 68, 62 and 61 g $kg^{-0.75}day^{-1}$, respectively. The average weights of the animals were 1.25, 25.6 and 57.5 kg for rabbits, kangaroos and sheep, respectively. From this information, and the assumption that the pasture was 30% dry matter, the rabbits would be daily eating pasture weighing 21% of their body weight. Sheep would take in pasture equivalent to only 7% and kangaroos 9% of body weight.

Obviously, smaller wallabies need a faster through-put than large kangaroos if eating the same vegetation. This was noted by Dellow (1982) (Fig. 2). However, the digestibility of a fibrous diet decreases as rate of passage increases. Prince (1976) demonstrated this well (Fig. 3), as did McIntosh (1966) in his study of the digestion of Red Kangaroos and sheep. Therefore, the holding of fibrous diets longer to facilitate digestion is not a reasonable option for smaller mammals and their only broad option is to select plant material that is low in cell wall constituents (fibre) and high in cell contents. The cell contents, largely proteins, soluble carbohydrates and fats, are much more rapidly digested, even by fermentation (Dellow and Hume 1982).

The size effect is further compounded in the field. Field metabolic rates (FMR) have been accumulated for marsupials by Nagy (1987) and these are shown in Fig. 4 together with the standard (or basal) metabolic rates (SMR). The FMRs are higher than the SMRs but the FMRs for smaller animals are more elevated than those of the larger species, such as the kangaroos. This

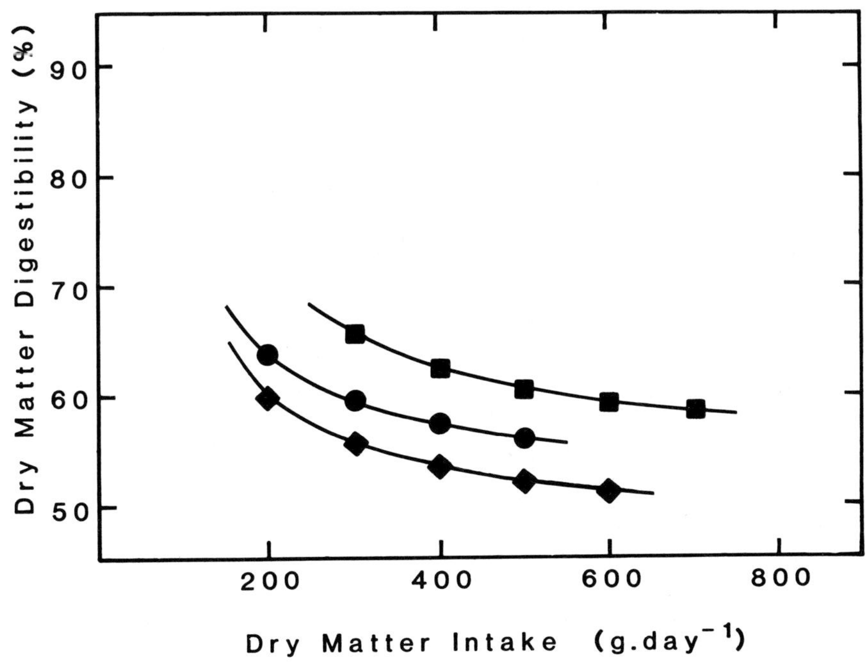

Fig. 3. Influence of rate of dry matter intake on the digestibility of high fibre, high nitrogen diet by *M. fuliginosus* (■), *M. robustus erubescens* (●), and *M. rufus* (♦). Derived from Prince (1976) p. 95. Note: McIntosh (1966) demonstrated that dry matter retention time decreased as dry matter intake increased.

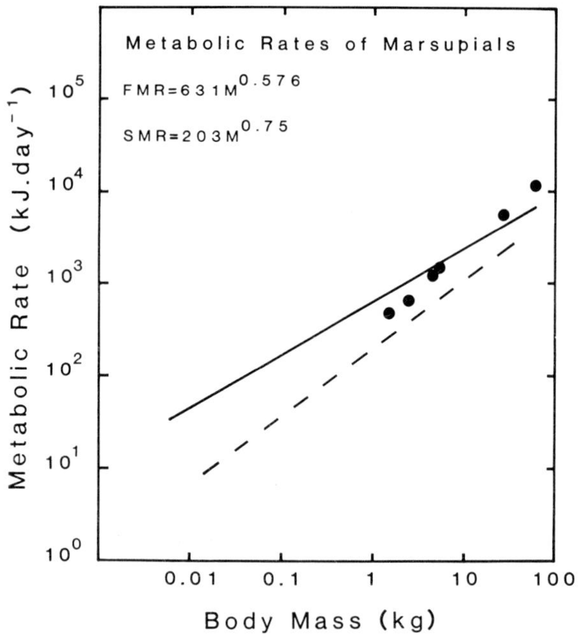

Fig. 4. Comparison of the scaling of field metabolic rate, FMR (–), and the standard metabolic rate, SMR, (– –) of marsupials. The points (●) are FMR values for four species of macropodids; the conditions under which these values were obtained are not available. (After Nagy 1987).

effect is partially related to the smaller species being further out of their thermoneutral zones. In such circumstances extra energy must be used for thermoregulation and this means extra food has to be obtained and processed. As we will see, soft (low fibre) high quality (high in cell contents) vegetation is generally selected by the smaller species of rat-kangaroo and wallabies. On the other hand the larger kangaroos can cope with tough, poor quality vegetation but this may not be their preference if sufficient high-quality feed is available.

The effect of size on diet choice is usually considered as an interspecific effect. However, intraspecific effects will occur. Firstly, there is the effect of sexual dimorphism. In kangaroos the males may be double (or more) the size of mature females. The larger males should be able to process high fibre feed better. Forbes and Tribe (1970) note such an effect; a large Eastern Grey Kangaroo *Macropus giganteus* was better able to digest poor quality oaten straw than were smaller individuals, but digestibilities were only slightly different on good lucerne hay. Newsome (1980) noted differences in the diets of male and female Red Kangaroos in central Australia. Females tended to select more forbs but the differences were small, being most obvious when conditions became dry. The other case concerns young animals. There are potentially severe feeding disadvantages for the newly weaned sub-adults. Their ability to digest roughage would be relatively poor and consequently for growth and survival they should require a relatively good, low fibre diet. The high mortality of young kangaroos during droughts is well known (Newsome 1977; Robertson 1986).

DIET CATEGORIES

The diets of macropodoids are diverse. While we generally consider that they eat vegetation, it must be remembered that smaller species can take invertebrates, tubers, bulbs and seeds, and also fungi. Vegetation or herbage is of two general types: monocotyledons, mainly grasses and sedges; and dicotyledons, forbs, shrubs and trees. Ferns also may be used by a few species at times. There are structural differences both between and within these groups in the amounts and the structural arrangements of the cell wall components, fibre (Short *et al.* 1974). These differences have marked effects on total digestibility, as well as the rate of digestion (Table 1). Obviously, such factors as these operate together with availability to determine intake.

Table 1. Rates for fermentative digestion of principal types of forages. Samples suspended in nylon bags inside goat rumen. (After Short *et al.* 1974).

Forage type	DMD, %, maximum	DMD, %, at 4 h	Time to maximum digestibility, h
Browse, new leaves*	89	87	8
Mature forbs	49	39	72
Young grass	94	75	32
Mature grass	57	15	168
Twigs of woody shrubs	33	24	96

DMD is Dry Matter digestibility.
*Values for young forbs are similar.

We cannot examine all the vegetation intake of macropodoids in detail because large numbers of plant species may be involved. Algar (1986) identified 98 plant species in the diets of the Quokkas *Setonix brachyurus,* Western Brush Wallabies *M. irma,* Euros *M. robustus* and Western Grey Kangaroos *M. fuliginosus.* Generally, the following broad categories will be used:

Forbs — small herbaceous dicots, often annuals or ephemerals. Generally of good guality but water content may be high.

Shrubs — small bushes, often perennials e.g., saltbushes. The leaves may be of good quality but the stems and twigs take time to digest.

Browse — woody bushes and trees. The leaves are generally heavily lignified, i.e., fibre content is high but the nutrient levels may also be high. Young leaves are generally of good quality.

Grass — young grass is nutritious, while hayed-off grass can be of very poor quality and low digestibility. Old grass is often avoided by all but the largest species.

I have not commented on the methods used in diet analysis. These have been discussed in many of the papers cited and it is apparent to most workers in this field that precise estimation of diets is difficult and laborious. In the discussion of diets that follows I have used the best available data but for some species these are very modest.

DIETS OF GROUPS OF MACROPODOIDS

Rat-kangaroos

Rat-kangaroos are the small macropodoids of the genera Hypsiprymnodon, Bettongia, Aepyprymnus and Potorous. Hume (1982a) has reviewed their gut structure and, except for Hypsiprymnodon (Fig. 1), their forestomachs are fully sacciform. This implies that they retain food for fermentation. The few studies of the diets of the rat-kangaroos are summarized by Seebeck, Bennett and Scotts (this volume). The studies of Guiler (1971) on the potoroo *Potorous tridactylus* and Christensen (1977) on the Brush-tailed Bettong *Bettongia penicillata* both point to underground fungal fruiting bodies as the principal food source. Other foods are eaten such as bulbs, tubers, seeds, insects and *Acacia* gums as well as small amounts of young grass and forbs. Similar diets have been reported for other rat-kangaroos: *Aepyprymnus rufescens* (Schlager cited in Hume 1982a); *B. gaimardi* (Johnson and Rose 1983); *B. lesueur* (Burbidge 1983). These items, except perhaps the fungi, are all largely digestible without recourse to fermentation. Fungal spores are very indigestible and the cellular supporting structures are high in complex carbohydrates (Martin 1978). It seems to me that the large forestomachs of these rat-kangaroos are involved in the fermentation of the fungal carbohydrates; Kinnear *et al.* (1979) demonstrated a high VFA concentration in the forestomach of *B. penicillata*.

Hare-wallabies

Hare-wallabies are small wallabies of the open forests and plains. They have largely vanished from southern Australia and only *Lagorchestes conspicillatus* is secure. No detailed studies of their diets are available but they seem to feed on forbs, especially succulents like *Portulaca,* some green grass and grass seed heads (Main and Yadav 1971; Burbidge and Johnson 1983). The remnant population of *L. hirsutus* in central Australia appears similar in diet (Bolton and Latz 1978). Given the size of these species (1.8-4 kg) they should not eat high fibre material such as dried off grass or woody shrubs.

Rock-wallabies

The diets of this widely scattered group of species have not been examined in detail except for *Petrogale xanthopus*. This attractive dry country rock-wallaby has been studied because of concerns about its survival. Dawson and Ellis (1979) worked in western New South Wales and Copley and Robinson (1983) in the Flinders Ranges of South Australia. During good seasons in western New South Wales herbaceous forbs and young grass made up the bulk of the diet. Leaves of trees and woody shrubs were present in small amounts but became the biggest single component with the onset of drought. Forbs were preferentially selected at all times. When the data are compared from the two sites, the basic components of the diets are seen to be similar but the actual plant species selected were different.

As far as can be assessed, a similar pattern of mixed herbage and green grass, with some browse applies to the diets of other rock-wallabies (Dawson and Ellis 1979). An exception is the Nabarlek, *Peradorcas concinna,* from northern Australia. This small rock-wallaby (1-1.5 kg) is unique among all marsupials in producing a continuous supply of molars. This leads one to suspect an unusual diet. The diet in the wet season is largely sedges and grasses that grow along the edges of the seasonal swamps. However, in the dry season *P. concinna* forages on the fern *Marsilea crenata* on the edges of the billabongs. While these diets have high digestibilities, they have high levels of silica, up to 26% of dry weight in the case of the fern (Sanson 1983). It is probable that this highly abrasive diet is the reason for the continual tooth replacement. Arid-zone kangaroos face similar problems but, being much larger, their relative through-put of feed is lower.

Pademelons

The three pademelons of the genus *Thylogale* tend to inhabit rainforest and tall open forest with a dense understorey. General observations (Johnson 1980; Johnson and Rose 1983; Johnson 1983) suggest that they take succulent short grasses and herbage. Other foods noted as part of their diets are fruits, ferns, and browse, the latter sometimes in the form of fallen leaves. Of the three species, the rainforest inhabiting *T. stigmatica* is more distinct in that it has a large intake of fallen dicot fruits, together with other dicot leaves (Jarman 1984).

Nailtail Wallabies

A recent study by Ellis, Tierney and Dawson (in prep.) on the Bridled Nailtail Wallaby, *Onychogalea fraenata,* has indicated that it also follows the expected pattern for a small wallaby. In good seasons it takes mainly herbaceous vegetation (Table 2). It seems particularly to like *Portulaca;* seeds are also eaten. Some green grass is taken if forbs are not abundant. In dry times browse becomes important. Leaf fall from the tree *Eremophila mitchellii* was a major component in the driest sampling time. Presumably,

Table 2. Diets of the Bridled Nailtail Wallaby *(Onychogalea fraenata)* and the Black-striped Wallaby *(Macropus dorsalis)* before and after summer rains at Dingo, central Queensland.

	Percentage of major forage types in diet		
	Grass	Forbs*	Browse
NAILTAIL WALLABY			
Late spring			
Nov 1982	36	17	47
Nov 1984	22	56	22
Autumn			
April 1983	38	45	17
May 1984	12	77	11
BLACK-STRIPED WALLABY			
Late spring			
Nov 1982	81	3	16
Nov 1984	93	4	3
Autumn			
April 1983	96	2	2
May 1984	94	1	5

*Forbs in this case includes ephemeral herbs and small sub-shrubs e.g., of the family Chenopodiaceae, Malvaceae and Solanaceae. Number of samples in each category, 20.

the other nailtail wallabies had or have similar diets.

Small Wallabies

This group of wallabies is like small kangaroos in body form. Their habitat is often open forest or woodland and in agricultural regions they may be seen emerging from shelter in the late afternoon to feed on pasture. While observations on their feeding generally report that they graze on short grasses and herbage, there are few detailed studies on their dietary preferences. Some of these studies are as yet not generally available.

Macropus dorsalis, the Black-striped Wallaby, presents an interesting picture. This largely Queensland species (Table 2) feeds almost exclusively on grass and sedges. Only in the driest times did the intake of browse become of significance, up to 16%. On first examination it seemed that cattle in the area had a similar diet. A closer examination revealed that *M. dorsalis* took mainly leaf material from short grasses, whereas cattle ate more rank species and took in much more stem. In the New England area of New South Wales the species was also noted to be principally a grass eater (Jarman and Phillips, this volume).

Macropus irma, the Western Brush Wallaby, was examined on a seasonal basis over three years (Algar 1986). *M. irma* is principally a browser (Table 3), feeding on low shrubs. These non-herbaceous dicotyledons represented 80 to 95% the total diet at any time. The species involved are generally not woody and the selected parts would be readily available.

Table 3. Seasonal variation in the diet of the Western Brush Wallaby *(Macropus irma)* in southwestern Western Australia. (After Algar 1986).

	Percentage of major forage types in diet		
	Grass*	Forbs	Browse
1982			
Summer	4	5	90
Autumn	3	1	97
Winter	3	1	97
Spring	9	1	90
1983			
Summer	2	2	97
Autumn	2	1	98
Winter	3	2	95
Spring	16	7	79

*Grass includes sedges. Number of samples for each season, 9-10.

M. rufogriseus (Southwell 1987), *M. agilis* and *M. parryi* (Bell 1973) have a mixed intake, but with short grasses being preferred. While these patterns are probably correct, assessments of intakes and preferences derived from simple observation have not been always particularly accurate.

Setonix brachyurus, the Quokka, is often considered along with the Swamp Wallaby *Wallabia bicolor* as taxonomically unusual. This pattern does also apply in some measure to the diet. On Rottnest Island the Quokka takes mainly dicotyledonous succulents but this does vary with availability. In one section of the island grasses were almost absent from the diet; yet, elsewhere, grasses made up 31% of the diet in winter when availability was high (Storr 1964). In a reserve south of Perth, the Quokkas were mainly browsers (Algar 1986). The leaves of larger shrubs accounted for 52%-88% of the total diet in any season across all years. The availability of grass was not high in this site but sedge intake increased in late summer and autumn. Again this small wallaby had an intake of plant material that was not high in fibre. However, there is no indication of the effect of secondary compounds, such as tannins, on digestibility of such diets.

Wallabia bicolor is regarded as the classic "browser" among the macropodoids (Sanson 1978). Again there is the problem of definition, What is browse? The studies of Harrington (1976) and Hollis *et al.* (1986) in the New England region of New South Wales showed that the Swamp Wallaby was a generalist feeder with a preference for dicotyledons. These dicots were both forbs and woody shrubs and trees. Harrington noted a preference for forbs in all seasons, while that for shrubs was seen in winter. While the proportion of grass in the diet was less than that available it was still significant, especially in the summer (Table 4). Harrington comments that *W. bicolor* liked mushrooms. The liking for fungi was again reported by Hollis *et al.* (1986); it was about 20%

of the diet except in the spring. Hollis *et al.* also noted a significant intake of ferns. That trees are readily eaten is confirmed by the examination by Horne (1975) of the effect of wallaby browsing in eucalypt plantations. *W. bicolor* was found to be the culprit. It ate the apical shoot of any of the plantation eucalypts that were below 60 cm. These depradations were most noticeable in winter. The point can be made that the material ingested was low in fibre and of good quality.

Table 4. Seasonal variation in the diet of the Swamp Wallaby, *Wallabia bicolor,* in the New England Region of New South Wales. (After Harrington 1976).

Forage type	Percentage of major forage types in diet* December n=5	March n=6	July n=5	October n=6
Grasses	32	19	20	23
Sedges	4	1	—	2
Forbs	37	21	24	38
Shrubs	14	20	44	22
Vines	3	1	5	1
Ferns	1	6	4	11
Fungus	—	29	2	1
Seeds	9	3	1	2

*Stems and veins omitted, unidentified monocots assigned to grass, unidentified dicots assigned to forbs and shrubs.

Kangaroos

There is more information on the diets of the various species of kangaroos because of their presumed competition with domestic stock. On the basis of their teeth, they are considered to be grass eaters, grazers (Sanson 1978). This may be largely true but not invariate.

Macropus rufus. Chippendale (1962) gave the earliest quantitative report of the diet of the Red Kangaroos in central Australia. This work in the Alice Springs area was subsequently extended to cover a range of seasonal conditions, including drought (Chippendale 1968). At all times, grass was over 75% of intake, Neverfail, *Eragrostis setifolia,* being by far the most common species. *E. setifolia* is a hardy perennial grass that is found in the numerous small depressions, gilgais, in dry plains. Neverfail produces some growth from its heavily grazed butt in all but the driest times. Five-minute grass, *Tripogon loliiformis,* was also selected immediately after rain and for as long as green leaf shoots were available. Chippendale (1968) did not make detailed assessments of preferences but noted selection for green grass. Stem material was eaten, however, in the driest times. Forbs were not preferred and were only eaten in significant amounts in late winter and early spring. An exception was *Portulaca oleracea,* which was selected when it was available, but this succulent was relatively rare. A later study in the same broad area (Low *et al.* 1973) generally confirmed the findings of Chippendale but found that in good seasons a wider variety of grasses and forbs were eaten.

In the dry sheep country of Western Australia, near Port Hedland, Storr (1968) found an essentially similar intake by Red Kangaroos to that noted by Chippendale (1962; 1968). "Soft" grasses (different species from those at Alice Springs) were selected as compared with the more sclerophyllous species such as the spinifexes. Forbs, up to 25%, featured in the autumn and through until the coming of the summer rains again. Similar results were seen also in the detailed investigations of Griffiths and Barker (1966) and Griffiths *et al.* (1974) at Cunnamulla, southwestern Queensland. Intake of grass was below 50% during good summer conditions (Fig. 6) but this was due to the presence of *Portulaca oleracea,* which again was highly selected. The perennial bluebush *Maireana tomentosa* was also selected by the Red Kangaroo when available. Ellis *et al.* (1977) reported a high grass intake under most conditions, but with forbs being occasionally important. They worked at Fowlers Gap in western New South Wales and forbs become abundant after good winter-spring rains.

The most detailed account of the influence of drought on the vegetation eaten by Red Kangaroos is that of Bailey et al. (1971), at Tero Creek, also in western New South Wales. The grasses *Eragrostis* spp. and *Enneapogon avenaceus* were preferred when they were available but as the drought progressed and they became rare, chenopods (principally the grey copperbur *Sclerolaena diacantha*) were selected instead of other grasses. The perennial bluebushes such as *M. pyramidata* were generally avoided. Barker (1987) obtained similar results in a qualitative study at Menindee, New South Wales, except that during the depths of a drought the perennial bluebushes were eaten. It seems that little else was available and animals were dying of starvation.

Macropus robustus, Euros and Common Wallaroos. Work on the diets of Euros was amongst the earliest diet investigations. Studies were undertaken in the Pilbara region of northwestern Western Australia to examine the interaction between *M. robustus erubescens* and domestic sheep (Ealey and Main 1967). In this semi-arid region of summer rainfall, the vegetation is now dominated by the sclerophyll grasses of the genus *Triodia* (spinifex). In October, at the end of the dry season, the diet of the Euros was made up of about 80% spinifex, mainly the soft spinifex *T. pungens.* Other grasses contributed 10% to the diet; forbs were only a very minor component. *Triodia pungens* was the most available of the grasses. After the summer rains, the Euros generally switched to other grasses and these made up the bulk of the intake. Presumably, these other grasses were in their growth stages and supplied more readily available nutrients. Although forbs were present at this time they were largely avoided.

Also near Port Hedland, Western Australia, Storr (1968) noted a similar pattern of grass preference by the Euro. Of note, however, was the high intake of *T. pungens* after good summer rain. This might be unexpected but Storr reported that this was due to the intake of the large seed heads of the spinifex. The above average summer rains provided good growth into the autumn and the Euros made use of the annual grasses and the growing parts of spinifex and other perennial grasses. Some forbs (18% of the diet) were eaten. By August the perennial grasses, *Cenchrus* spp. and *Aristida* spp. accounted for most of the feed. These were preferred over the more abundant spinifex, which still contributed significantly to the diet. This pattern continued into November in spite of a general decrease in availability of feed.

In western New South Wales, at the other end of its range, the Euro also has a preference for grass. In this chenopod shrub dominated community, grass comprised the bulk of the diet even when it was present at very low levels (Ellis *et al.* 1977; Dawson and Ellis 1979). Forbs and shrubs did show up in the diet at times. The forbs were eaten after winter rains and the shrubs in dry times, when the amount of grass declined.

M. robustus robustus, the Common Wallaroo, from the eastern highlands of Australia has had its diet examined on the New England Tablelands of New South Wales (Taylor 1983). Interestingly, the diet preferences of this sub-species are very similar in basic form to those of the Euro. Short, low fibre grasses were selected unless they were in very limited supply. Coarse grasses were eaten in preference to shrubs but some specific forbs (legumes) were taken in the winter and spring.

Macropus robustus woodwardi, the subspecies of Common Wallaroo from the far north of Australia, also seems to follow the basic pattern of grass eating. Croft (1987) found that grass made up almost all of the diet at the end of the summer wet season and into the start of the dry season. Although the grass senesced during this period it was still abundant. What happens at the end of the dry season is not known.

Macropus antilopinus, the Antilopine Wallaroo, was also examined by Croft (1987) in his study 80 km south of Darwin, Northern Territory. He found only grass in his extensive sampling of the diets of this kangaroo, but noted that they moved to areas of short green grasses as the majority of the grasses dried off.

Macropus giganteus, the Eastern Grey Kangaroo, is probably the most common of kangaroos. There have been few studies on their feeding strategies but Kirkpatrick (1965) examined the mouth contents of Eastern Grey Kangaroos shot while feeding and concluded that they were non-selective feeders. The major species of grasses eaten were in the pasture in similar ratios. The region of the study was southeastern Queensland and the area was dominated by grasses.

In the New England Tablelands of northern New South Wales, Eastern Grey Kangaroos again took almost entirely grass (Taylor 1983). The animals appeared very selective, however, in that they preferred the leaf of low-fibre (soft) grasses and generally avoided forbs. The grasses eaten varied with season and area, and in areas of unimproved pasture, tussock grasses such as *Poa* spp. were eaten in autumn and winter in proportion to their occurrence in the pasture. During these conditions of poor growth, more stem and inflorescences also were eaten Taylor (1983).

Evidence of selectivity by the Eastern Grey Kangaroo should have been seen in studies in the more arid areas of their range but vegetation sampling was not undertaken (Griffiths and Barker 1966; Griffiths *et al.* 1974). These studies did show that Eastern Grey Kangaroos ate more grass than Red Kangaroos and sheep (Fig. 6) but whether they were more selective is uncertain. Comments by Griffiths *et al.* (1974) on the apparent selection of forbs by *M. rufus* and sheep indicate a lower level of selection by *M. giganteus.* During severe drought, Grey Kangaroos ate the more abundant coarser grasses, soft spinifex *(Triodia mitchelli),* kangaroo grass *(Themeda australis)* and spear grasses (*Aristida* spp.).

Other studies on *M. giganteus* also indicate that it is not averse to coarse native perennial grasses in its woodland and forest habitats but that it appears to select shorter low fibre grasses if these are available (Bell 1973; Taylor 1980). At Fowlers Gap, which is at the far western edge of the range of *M. giganteus,* it selected for grass during drought (Dawson, Ellis and Mazanov in prep.). The selection index was the same as that obtained for Red Kangaroos. In summary, the Eastern Grey Kangaroo is primarily a grass eater and the indications are that it is selective. The level of selectivity has yet to be adequately determined.

Macropus fuliginosus, the Western Grey Kangaroo, should perhaps be called the "Southern Grey Kangaroo" because it ranges through the woodlands and forests of the winter rainfall belt of southern Australia. It overlaps with *M. giganteus* in the plains of southern and western New South Wales. Details of its diet can be derived from only four studies, two from the wetter regions of Western Australia and two from the other side of its range in New South Wales and Victoria.

Studies in Western Australia indicate that *M. fuliginosus* concentrated on sedges. In a mixed pasture and woodland habitat 90 km north-east of Perth, the diet and its variation with season was rather different to that described for Eastern Grey Kangaroos (Halford *et al.* 1984). Before the onset of the winter rains in April and May, introduced pasture grasses made up some 30% of the species taken, the sedge *Conostylis setigera* slightly more, with a variety of perennial forbs and shrubs making up the rest. After the rains, in June, the pasture grasses became more important but the intake of sedge was maintained, as was a smaller proportion of the forbs and shrubs. The sedge was generally eaten in proportion to its abundance.

On the coastal plain near Perth, Algar (1986) also noted the high proportion of sedges in the diet of *M. fuliginosus*. In all seasons except spring this group of non-grass monocotyledons made up more than 82% of the diet. In spring, grasses became important but the sedges still formed the majority of the feed intake. At all times the monocotyledons made up more than 90% of the diet. The implication was that the sedges were eaten because of their abundance while grasses were preferred if they were available in sufficient amounts. The Western Grey Kangaroo, like its eastern relative, can graze on rather fibrous feed, if necessary.

Under very different conditions, Barker (1987) qualitatively assessed the diet of *M. fuliginosus* in southwestern New South Wales during a severe drought. The study basically suggested that the Western Grey Kangaroos preferred grasses when they were available. Next in order of preference were forbs and the chenopod sub-shrubs (*Sclerolaena* spp.). Through the hard drought of summer 1982-83 large amounts of woody shrubs were eaten, *Acacia victoriae, Cassia* spp., *Eremophila*. spp. and *Dodonaea attenuata*. Apparently, these were the only plants available, except for *Maireana pyramidata,* the black bluebush, which is relatively unpalatable. Large numbers of kangaroos died at this time. After the drought *M. fuliginosus* returned to grasses and forbs.

A detailed examination of the diets of this kangaroo in semi-arid areas is still required before we understand its place in the ecosystem. A start to this process is the preliminary report of Norbury (1987) in which he noted the dynamics of the changing selection for grass by *M. fuliginosus*. In a mixed pasture, the kangaroos generally ate in excess of 75% grass. As pasture biomass declined, *M. fuliginosus* maintained this level of grass intake and as a consequence the selection of grass increased. However, below a threshold level of grass availability, forbs and shrubs became more preferred. Further work investigating the optimal foraging strategies of kangaroos is required.

Summary of diet patterns

The diets outlined clearly show the basic effect of size. The small species tend to be more specialized, feeding on high quality but scattered foods. The "medium" sized macropodoids tend to be more generalist feeders and a wide variety of vegetation appears in their diet, the more common components varying with species and environment. The characteristic of the intake is that it is generally of relatively low fibre content. Even when browse is taken, on close observation, it is apparent that the animals are selecting the lower fibre parts of the plants. The large macropodoids, especially the kangaroos, are specialist grass eaters. The Red Kangaroo may be less selective than the other species, perhaps by necessity.

OVERLAP IN DIETS BETWEEN MAOROPODOIDS

Potential competition has been one of the reasons for investigating overlap in the diets of various species of macropodoids. Unfortunately, most studies have not been aware of the complexity of the issue and have not provided evidence to resolve this issue. Simple diet overlap is not sufficient to indicate competition. For competition to be demonstrated one species must have a deleterious effect on another and this has yet to be demonstrated in the Australian context. Indeed, competition between native and domestic mammals has yet to be proven in Australia, although the data are suggestive (Newsome 1971; Edwards this volume).

While comparative diet studies have reported some overlap, resource partitioning appears generally evident. This is suggested when the animal species within a community are using different plant species or growth phases within a group of plants and/or different parts of the habitat. The controversy about the underlying mechanisms which result in resource partitioning are beyond the scope of this review.

Suggested examples of this are described by Jarman (1984) and Jarman and Phillips (this volume) for the Wallaby Creek macropodoid community, by Taylor (1983) for *Macropus giganteus* and *M. robustus* in the New England region, and by Croft (1987) for *Macropus antilopinus, M. robustus* and *M. agilis* in the Darwin area (Fig. 5). Of note, the Euros and Wallaroos separate out from the other macropodoids because of their association with rough and hilly country (see also Storr 1968; Ellis *et al.* 1977). However, where they do overlap with other species in habitat, the Euros appear to separate on diet (Dawson and Ellis 1979).

The two Grey Kangaroos and the Red Kangaroo have a broad zone of sympatry in semi-arid southeastern Australia. Their reported diets exhibit some degree of overlap but adequate data are not available to assess any level of competition. It is reported that the species form mixed "mobs" but these may occur as the result of

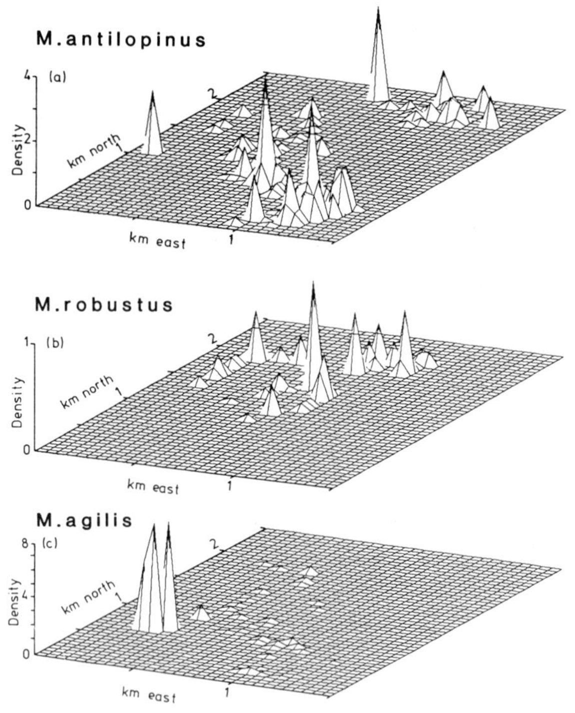

Fig. 5. Distribution and densities (animals per km^2) of three species of macropodids on a study site 80 km south of Darwin. *M. agilis* generally occupied the narrow valleys; *M. antilopinus* were found on ridge tops, slopes and valleys; *M. robustus woodwardii* occupied steep-sided rocky ridges with thick scrub. (After Croft 1987).

disturbance. In regions of sympatry, Eastern Grey Kangaroos (Caughley 1964) and Western Grey Kangaroos (Priddel 1987) appear to have differing habitat preferences to those of the Red Kangaroos. This is the case on Fowlers Gap Arid Zone Research Station in far western New South Wales. More work could be fruitfully done on the comparative feeding ecology of these species in western New South Wales.

DIET OVERLAP BETWEEN KANGAROOS AND SHEEP AND CATTLE

There is little firm evidence for competition between the kangaroos and domestic stock, although it is often considered to occur. Some very recent work (see Edwards this volume) provides evidence that actual competition between sheep and kangaroos takes place, but even in these cases there are questions to be resolved.

In the north of Australia the ground vegetation is dominated by grasses and for kangaroos and cattle it comprises the bulk of their diets. Work in this region (Chippendale 1968; Low and Low 1976; Newsome 1975) has been reviewed by Dudzinski *et al.* (1982). Cattle and Red Kangaroos tend to eat different grasses or parts of grasses and most importantly, there seems to be little spatial overlap between the species except during severe drought. No evidence was found that one species repels or attracts the other in spatial terms and it was considered that there was little actual diet overlap and that the two species could coexist if cattle numbers were reasonably controlled (Dudzinski *et al.* 1982).

The story of the interaction between sheep and kangaroos is more complex and has been dealt with by Edwards (this volume). The pattern of diets generally follows that indicated by Griffiths and Barker (1968), with sheep and Red Kangaroos having broadly similar diets (Fig. 6). However, detailed examination of the data indicate that the actual level of competiton is uncertain. In the study of Griffiths and Barker (1968), although the basic categories of plants eaten were similar, the species eaten were in many cases different. Also, plant distributions and abundances were not determined, nor were the relative numbers of animals. Storr (1968) found a similar pattern of feed intake in the Pilbara region of Western Australia, with the diet of sheep being closer to that of Red Kangaroos than to that of Euros. Potential for competition was considered to be greatest at the end of the dry season when the availability of forbs declined markedly.

The difficulties in assessing competition are considerable because for competition to be established it has to be demonstrated that the presence of one species has an adverse effect on another. The problem with this is the variability of environmental conditions in our semi-arid lands. One

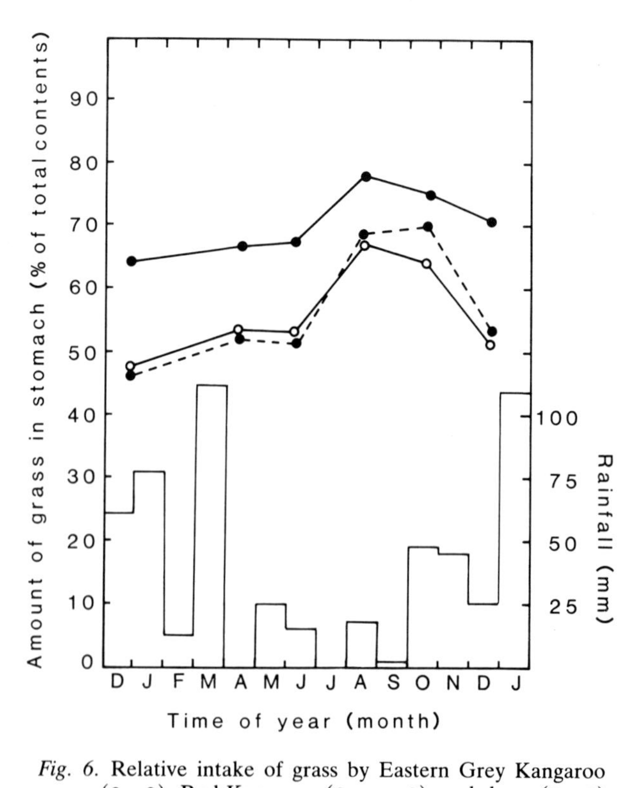

Fig. 6. Relative intake of grass by Eastern Grey Kangaroo (●—●), Red Kangaroo (● – – – ●), and sheep (○—○) in southwestern Queensland. Circles indicate sampling times from January 1963 to January 1964. Monthly rainfall is indicated by histogram. (After Griffiths and Barker 1966).

season or year may be very different from the previous one and vegetation patterns may change markedly with soil type and land form over a few metres. Such effects were seen in the diets noted in a long running (12 years) study of the mammalian herbivores on Fowlers Gap Station (Dawson, Ellis and Mazanov, in prep.). Data from three of the nine sampling periods, the wettest winter and the two driest summers, illustrate some of this variation.

The complete data set, after arc-sine transformation, was subject to a Factor analysis (Apple MacIntosh Statview 512+, orthogonal factor analysis) and most of the variance was accounted for by the first three factors; note the correlation coefficients (Table 5). Factor 1 is largely determined by saltbush and grass (anti); factor 2 is largely determined by forbs and grass (anti), while factor 3 is largely determined by browse and round chenopods, and grass (anti). This being so, the factor axes may be labelled to reflect this. In Fig. 7, the factor scores of the different animals are plotted as 95% confidence ellipses. The data for the wet winter of 1974 are shown in Fig. 7a,b and those for the two dry summers of 1978 and 1983 in Fig. 7c,d, and Fig. 7e,f.

Table 5. Correlation coefficients from factor analysis of the diets of the mammalian herbivores at Fowlers Gap, western New South Wales. (After Dawson, Ellis and Mazanov, in prep.).

Diet factors	Factor 1	Factor 2	Factor 3
Grass	−0.84	−0.44	−0.25
Saltbush	0.95	−0.26	0.01
Bluebush and Copperburrs	0.10	0.42	0.54
Sub-Shrubs*	−0.06	0.56	0.19
Forbs	−0.06	0.86	−0.24
Browse	0.08	−0.13	0.87

*Sub-shrubs mainly consisting of the family Malvaceae, plus *Solanum* spp. and *Ptilotus obovatus*.

Of note was the relative consistency of the diet mix of individual species at a particular sampling time. The pattern of intakes did vary with time, however. Additionally, while Fig. 7 indicates overlap in diets there was apparent separation of the diets of species feeding in the same area.

In the winter of 1974 most of the species were taking a broad mixture of forbs and grasses, with the forbs generally being most important. Little saltbush was eaten except by goats in the hilly country. Goats also were the only ones to have a significant intake of browse. Even with the concentration on forbs there was only a small amount of overlap in the diets of species on the same landform. The Euros and Red Kangaroos had the highest intakes of grass but these species were spatially separated. Vegetation was in abundance and it would be difficult to envisage competition.

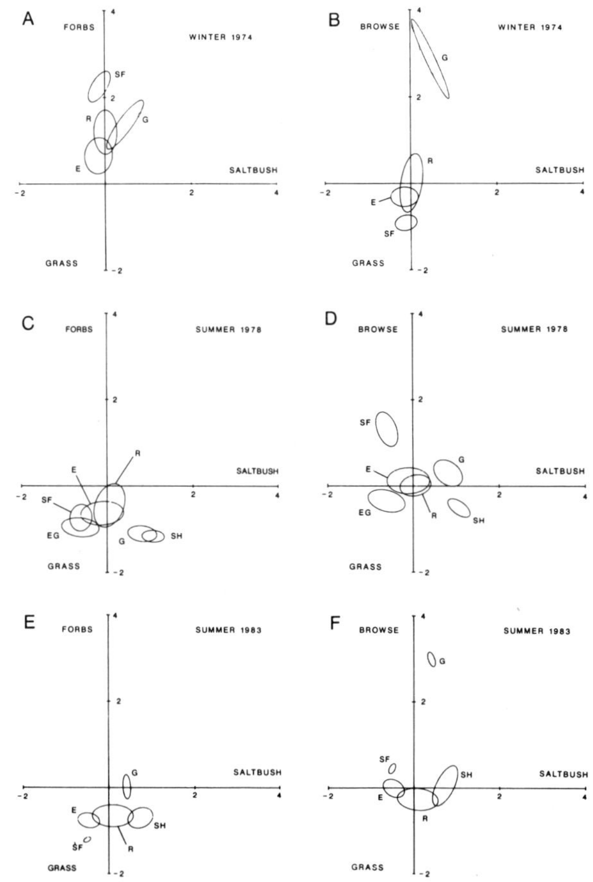

Fig. 7. Influence of seasonal conditions on the diet selection by the mammalian herbivores at Fowlers Gap Arid Zone Research Station, western New South Wales. The axes are from the factor analysis (Table 5). Factor 1 is represented by the axis saltbush-grass; factor 2 is represented by the axis forbs-grass; factor 3 is represented by the axis browse-grass. Winter 1974 (7a,b) was a very good season; summer 1978 (7c,d) was a moderate drought; summer 1983 (7e,f) was a severe drought. The ellipses are the 95% confidence ellipses for the diets. Species are indicated as follows: R, Red Kangaroos; E, Euros; EG, Eastern Grey Kangaroos; G, feral goats; SF, sheep on flat country; SH, sheep in hills. (After Dawson, Ellis and Mazanov, in prep).

The summer of 1978 represents a minor drought. Forbs were rare. The kangaroos were concentrating on grass, which was in low abundance and consequently highly selected. Their diets largely overlapped but the Euros and Red Kangaroos were separated spatially, with the Reds on the flat country and the Euros in the hills. There were a few Eastern Grey Kangaroos on the flat country and these had the highest intake of grass. The patterns of intake by sheep were different from the various kangaroo species and, interestingly, the two groups of sheep did not have similar diets. The sheep on the flats consistently ate a mix of grass and browse, while those in the hills took in a mixture of saltbush and grass. Vegetation was sparse and all species had major components that had high selection indices.

A major drought occurred in the summer of 1982/83 and deaths of sheep and kangaroos were common. These conditions again produced a separation of the patterns of intake between species and some reduction of the variation of diets within each species. The Red Kangaroos and the Euros were still highly selective for grass in their separate environments. The sheep in their two habitats were still distinct in their non-grass components and the goats were eating mainly browse. Eastern Grey Kangaroos were not available in sufficient numbers in the study area to be included in the sample.

These data suggest that there is some potential for competition between the native and introduced species. This, however, varies with conditions and the presence of other species. There are also marked differences in the general diet "niche" of the various species. Overlap, when competition is a possibility, such as during drought, appears related to the high intake of grass by the kangaroos. The reasons for the diet patterns require answers. Do the kangaroos concentrate on the rare but patchily distributed grasses because of competition with sheep or because their morphology and diet physiology makes this their optimum choice in the circumstances? The work of Belovsky (1986) is most interesting in this regard. How does the presence of one species influence the diet of another? During drought sheep in the hills concentrated on saltbush while the Euros largely ate grass and goats ate browse. However, on the flat country no goats were present and the major non-grass component of the diet of sheep was browse. The answers to these questions are not yet available. Short (1985, 1986) attempted to address some of these problems but his experimental design was restricted and much removed from natural circumstances.

ACKNOWLEDGEMENTS

Beverely Ellis's enthusiasm and skills in diet analysis have been central to my interest in the diets of macropodoids; without these this review could not have been written.

REFERENCES

ALGAR, D., 1986. An ecological study of macropodid marsupial species on a Reserve. Ph.D. Thesis, University of Western Australia, Perth.

ARCHER, M., 1984. The Australian marsupial radiation. Pp. 633-808 *in* Vertebrate Zoogeography and Evolution in Australia ed. by M. Archer and G. Clayton. Hesperian Press: Perth.

BAILEY, P. T. AND BARKER, R., 1971. The red kangaroo, *Megaleia rufa* (Desmarest), in northwestern New South Wales. II. Food. *C.S.I.R.O. Wildl. Res.* **16:** 29-39.

BARKER, R. D., 1987. The diets of herbivores in the sheep rangelands *in* Kangaroos: their Ecology and Management in Sheep Rangelands ed. by G. Caughley, N. Shepherd and J. Short. Cambridge University Press: Cambridge.

BELL, H., 1973. The ecology of three macropod marsupial species in an area of open forest and savannah woodland in north Queensland, Australia. *Mammalia* **37:** 527-43.

BELOVSKY, G. E., 1986. Optimal foraging and community structure: Implications for a guild of generalist grassland herbivores. *Oecologia* **70:** 35-52.

BOLTON, B. L. AND LATZ, P. K., 1978. The western hare-wallaby *Lagorchestes hirsutus* (Gould) (Macropodidae) in the Tanami Desert. *Aust. Wildl. Res.* **5:** 285-93.

BURBIDGE, A. A., 1983. Burrowing Bettong, *Bettongia lesueur*. Pp. 187-88 *in* The Australian Museum Complete Book of Australian Mammals ed. by R. Strahan. Angus & Robertson: Sydney.

BURBIDGE, A. A. AND JOHNSON, P. M., 1983. Spectacled Hare-wallaby *Lagorchestes conspicillatus*. Pp. 197-98 *in* The Australian Museum Complete Book of Australian Mammals ed. by R. Strahan. Angus & Robertson: Sydney.

CAUGHLEY, G. J., 1964. Density and dispersion of two species of kangaroo in relation to habitat. *Aust. J. Zool.* **12:** 238-49.

CHIPPENDALE, G. M., 1962. Botanical examination of kangaroo stomach contents and cattle rumen contents. *Aust. J. Sci.* **25:** 21-22.

CHIPPENDALE, G. M., 1968. The plants grazed by red kangaroos, *Megaleia rufa* (Desmarest) in central Australia. *Proc. Linn. Soc. New South Wales.* **93:** 98-110.

CHRISTENSEN, P. E. S., 1977. The Biology of *Bettongia penicillata* and *Macropus eugenii* in Relation to Fire. Ph.D. Thesis, University of Western Australia, Perth.

COPLEY, P. B. AND ROBERTSON, A. C., 1983. Studies on the Yellow-footed Rock-wallaby, *Petrogale xanthopus* Gray (Marsupialia: Macropodidae) II. Diet. *Aust. Wildl. Res.* **10:** 63-76.

CROFT, D. B., 1987. Socio-ecology of the Antilopine Wallaroo, *Macropus antilopinus,* in the Northern Territory, with observations on sympatric *M. robustus woodwardii* and *M. agilis*. *Aust. Wildl. Res.* **14:** 243-55.

DAWSON, T. J. AND HULBERT, A. J., 1970. Standard metabolism, body temperature, and surface areas of Australian marsupials. *Amer. J. Physiol.* **218:** 1233-238.

DAWSON, T. J. AND ELLIS, B. A., 1979. Comparison of diets of the yellow-footed rock-wallabies and sympatric herbivores in western New South Wales. *Aust. Wildl. Res.* **6:** 245-54.

DELLOW, D. W., 1982. Studies on the nutrition of macropodine marsupials. III. The flow of digesta through the stomach and intestine of macropodines and sheep. *Aust. J. Zool.* **30:** 751-65.

DELLOW, D. W. AND HUME, I. D., 1982. Studies on the nutrition of macropodine marsupials. IV. Digestion in the stomach and the intestine of *Macropus giganteus, Thylogale thetis* and *Macropus eugenii*. *Aust. J. Zool.* **30:** 767-77.

DELLOW, D. W., NOLAN, J. V. AND HUME, I. D., 1983. Studies on the nutrition of macropodine marsupials V. Microbial fermentation in the forestomach of *Thylogale thetis* and *Macropus eugenii*. *Aust. J. Zool.* **31:** 433-43.

DUDZINSKI, M. L., LOWE, W. A., MILLER, W J. AND LOW, B. S., 1982. Joint use of habitat by red kangaroos and shorthorn cattle in Central Australia. *Aust. J. Ecol.* **7:** 69-74.

EALEY, E. H. M. AND MAIN, A. R., 1967. Ecology of the Euro, *Macropus robustus,* in northwestern Australia. 3. Seasonal changes in nutrition. *C.S.I.R.O. Wildl. Res.* **12:** 53-65.

EDWARDS, G. P. (this volume). The interaction between macropods and sheep: a review.

ELLIS, B. A., RUSSELL, E. M., DAWSON, T. J. AND HARROP, C. J. F., 1977. Seasonal changes in diet preferences of free-ranging red kangaroos, euros and sheep in western New South Wales. *Aust. Wildl. Res.* **4:** 127-44.

FORBES, D. K. AND TRIBE, D. E., 1970. The utilization of roughages by sheep and kangaroos. *Aust. J. Zool.* **18:** 247-56.

GRIFFITHS, M. AND BARKER, R., 1966. The plants eaten by sheep and by kangaroos grazing together in a paddock in southwestern Queensland. *C.S.I.R.O. Wildl. Res.* **11:** 145-67.

GRIFFITHS, M., BARKER, R. AND MACLEAN, L., 1974. Further observations on the plants eaten by kangaroos and sheep grazing together in a paddock in southwestern Queensland. *Aust. Wildl. Res.* **1:** 27-43.

GUILER, E. R., 1971. Food of the potoroo (Marsupialia, macropodidae). *J. Mammal.* **52:** 232-34.

HALFORD, D. A., BELL, D. T. AND LONERAGAN, W. A., 1984. Diet of the Western Grey Kangaroo (*Macropus fuliginosus* Desm.) in a mixed pasture-woodland habitat of Western Australia. *J. Roy. Soc. West. Aust.* **66:** 119-28.

HANLEY, T. A., 1982. The nutritional basis for food selection by ungulates. *J. Range Manag.* **35:** 146-51.

HARRINGTON, J., 1976. The Diet of the Swamp Wallaby, *Wallabia bicolor,* at Diamond Flat, New South Wales. Unpublished M.Nat.Res. Thesis, University of New England, Armidale.

HEIGHWAY, F. R., 1939. The anatomy of *Hypsiprymnodon moschatus.* Unpublished Thesis, University of Sydney, Sydney, cited in Hume 1982a.

HOLLIS, C. J., ROBERTSHAW, J. D. AND HARDEN, R. H., 1986. Ecology of the Swamp Wallaby *(Wallabia bicolor)* in northeastern New South Wales. 1. Diet. *Aust. Wildl. Res.* **13:** 355-65.

HORNE, R. R., 1975. Wallaby browsing in eucalypt plantations. Pp. 80-100 *in* Agriculture, Forestry and Wildlife: Conflict or Coexistence ed. by P. J. Jarman. University of New England: Armidale.

HUME, I. D., 1982a. Digestive Physiology and Nutrition of Marsupials. Cambridge University Press: Cambridge.

HUME, I. D., 1982b. The digestive physiology of marsupials. *Comp. Biochem. Physiol.* **71A:** 1-10.

JARMAN, P. J., 1984. The dietary ecology of macropod marsupials. *Proc. Nutiriton Soc. Aust.* **9:** 82-87.

JARMAN, P. J. AND PHILLIPS, C. M. (this volume). Diets in a community of macropodoid species.

JOHNSON, K. A. AND ROSE, R., 1983. Tasmanian Bettong, *Bettongia gaimardi.* Pp. 186 *in* The Australian Museum Complete Book of Australian Mammals ed. by R. Strahan. Angus & Robertson: Sydney.

JOHNSON, K. A., 1980. Spatial and temporal use of habitat by Red-necked Pademelon, *Thylogale thetis. Aust. Wildl. Res.* **7:** 157-66.

JOHNSON, K. A., 1983. Red-necked Pademelon, *Thylogale thetis.* Pp. 228 *in* Australian Museum Complete Book of Australian Mammals ed. by R. Strahan. Angus & Robertson: Sydney.

KINNEAR, J. E., COCKSON, A., CRISTENSEN, P. AND MAIN, A. R., 1979. The nutritional biology of the ruminants and ruminant-like mammals — a new approach. *Comp. Biochem. Physiol.* **64A:** 357-65.

KIRKPATRICK, T. H., 1965. Studies of the Macropodidae in Queensland I. Food preferences of the Grey Kangaroo (*Macropus major* Shaw). *Qld. J. Agric. Animal Sci.* **22:** 89-93.

KLEIBER, M., 1961. The Fire of Life. John Wiley: New York.

LOW, B. S., BIRK, E., LONDON, C. AND LOW, W. A., 1973. Community utilization by cattle and kangaroos in Mulga near Alice Springs, Northern Territory. *Tropical Grasslands* **7:** 149-56.

LOW, B. S. AND LOW, W. A., 1975. Feeding interactions of red kangaroos and cattle in an arid ecosystem. Pp. 87-94 *in* Proceedings, III World Conference on Animal Production ed. by R. L. Reid. University of Sydney Press: Sydney.

MAIN, A. R. AND YADOV, M., 1971. Conservation of macropods in reserves in Western Australia. *Biol. Conservation* **3:** 123-33.

MARTIN, M. M., 1979. Biochemical implications of insect mycophagy. *Biol. Rev.* **54:** 1-21.

MCINTOSH, D. L., 1966. The digestibility of two roughages and the rates of passage of their residues by the red kangaroo, *Megaleia rufa* (Desmarest) and Merino sheep. *C.S.I.R.O. Wildl Res.* **11:** 125-35.

NAGY, K. A., 1987. Field metabolic rate and food requirement scaling in mammals and birds. *Ecological Monographs* **57:** 111-28.

NEWSOME, A. E., 1971. Competition between wildlife and domestic stock. *Aust. Vet. J.* **47:** 577-86.

NEWSOME, A. E., 1975. An ecological comparison of the two arid-zone kangaroos of Australia, and the anomalous prosperity since the introduction of ruminant stock to their environment. *Quart. Rev. Biol.* **50:** 389-424.

NEWSOME, A. E., 1977. Imbalance in the sex-ratio and age structure of the red kangaroo in central Australia. Pp. 221-36 *in* The Biology of Marsupials ed. by B. Stonehouse and D. Gilmore. MacMillen: London.

NEWSOME, A. E., 1980. Differences in the diets of male and female red kangaroos in central Australia. *Afr. J. Ecol.* **18:** 27-31.

NORBURY, G. L., 1987. Diet selection by Western Grey Kangaroos in relation to declining food availability. Pp. 75-76 *in* Herbivore Nutrition Research Papers presented at the Second International Symposium of the Nutrition of Herbivores ed. by M. Rose. *Australian Soc. Animal Prod:* Brisbane.

PRIDDEL, D., 1987. The mobility and habitat utilization of kangaroos. Pp. 100-118 *in* Kangaroos, their Ecology and Management in the sheep rangelands of Australia ed. by G. Caughley, N. Shepherd and J. Short. Cambridge University Press: Cambridge.

PRINCE, R. I. T., 1976. Comparative Studies of Aspects of Nutritional and Related Physiology in Macropod Marsupials. Ph.D. Thesis, University of Western Australia, Perth.

ROBERTSON, G. G., 1986. The mortality of kangaroos in drought. *Aust. Wildl. Res.* **13:** 349-54.

SANSON, G. D., 1978. The evolution and significance of mastication in the Macropodidae. *Aust. Mammal.* **2:** 23-28.

SANSON, G. D., 1983. Nabarlek, *Peradorcas concinna*. P. 223 *in* Australian Museum Complete Book of Australian Mammals ed. by R. Strahan. Angus & Robertson: Sydney.

SEEBECK, J., BENNETT, A. F. AND SCOTTS, D. J. (this volume). Ecology of the Potoroidae — A Review.

SHORT, H. L., BLAIR, R. M. AND SEGELQUIST, C. A., 1974. Fiber composition and forage digestibility by small ruminants. *J. Wildl. Manage.* **38:** 197-209.

SHORT, J., 1985. The functional response of kangaroos, sheep and rabbits in an arid grazing system. *J. Appl. Ecol.* **22:** 435-47.

SHORT, J., 1986. The effect of pasture availability on food intake, species selection and grazing behaviour of kangaroos. *J. Appl.* **23:** 559-71.

SOUTHWELL, C., 1987. Macropod studies at Wallaby Creek II. Density and distribution of macropod species in relation to environmental variables. *Aust. Wildl. Res.* **14:** 15-33.

STORR, G. M., 1964. Studies on marsupial nutrition. IV. Diet of the quokka, *Setonix brachyurus* (Quoy and Gaimard), on Rottnest Island, Western Australia *Aust. J. Biol. Sci.* **17:** 469-81.

STORR, G. M., 1968. Diet of kangaroos (*Megaleia rufa* and *Macropus robustus*) and Merino sheep near Port Hedland, Western Australia. *J. Roy. Soc. West. Aust.* **51:** 25-32.

TAYLOR, R. J., 1980. Distribution of feeding activity of the Eastern Grey Kangaroo, *Macropus giganteus*, in coastal lowland of south-east Queensland. *Aust. Wildl. Res.* **7:** 317-25.

TAYLOR, R. J., 1983. The diet of the Eastern Grey Kangaroo and Wallaroo in areas of improved and native pasture in the New England Tablelands. *Aust. Wildl. Res.* **10:** 203-11.

TRUSEWELL, E. M. AND HARRIS, W. K., 1982. The Cainozoic palneobotanical record in arid Australia: fossil evidence for the origins of an arid-adapted flora. Pp. 67-76 *in* Evolution of the Flora and Fauna of Arid Australia ed. by W. R. Barker and P. J. M. Greenslade. Peacock Publ.: Adelaide.

WILLIAMS, O. B. AND OXLEY, R. E., 1979. Historical aspects of the use of chenopod shrublands. Pp. 5-16 *in* Studies of the Australian Arid Zone. IV. Chenopod Shrublands ed. by R. D. Graetz and K. M. W. Howes. C.S.I.R.O.: Melbourne.

Diets in a community of macropod species

PETER J. JARMAN[1] AND CATHERINE M. PHILLIPS[1]

Diets of the eleven species of macropods living at or near Wallaby Creek, northern New South Wales, are discussed. Diets of seven of the most abundant species are derived from material collected at or near Wallaby Creek; diets of the other four are inferred from studies made elsewhere. The three largest species, *Macropus giganteus, M. parryi* and *M. robustus*, feed in open woodland or pasture and eat 96-100% grass. *M. giganteus* and *M. parryi* differ in the species of grasses that they eat. One species, the Brush-tailed Rock-wallaby *Petrogale penicillata*, grazes in open, grassy, eucalypt woodland adjacent to the cliffs where it finds refuge in the day. Four medium-sized species, *M. rufogriseus, M. dorsalis, Wallabia bicolor* and *Thylogale thetis*, occupy the boundary between dense cover (forest) and pasture, and mix varying proportions of monocot and dicot leaves in their diet. The last two also eat fungi and ferns. *T. stigmatica*, living entirely in rainforest, eats predominantly dicot leaves and fruit. The two potoroids *Potorous tridactylus* (living in wet eucalypt forest) and *Aepyprymnus rufescens* (living in open forest and tussocky pasture) eat mainly subterranean storage organs and some fruiting bodies of angiosperms and fungi. There is thus considerable separation between these species in their diets which, in conjunction with their choices of distinct habitats, appears to produce ecological separation within this community.

INTRODUCTION

FEW studies have been made of ecological separation within communities of macropods (see Fox: this volume). This paper reports what we know of the diets of the ten species of macropods which occur sympatrically or parapatrically in 16 km^2 of rough pasture, eucalypt and casuarina forest, and rainforest in northern New South Wales. An eleventh species, the Brush-tailed Rock-wallaby, also occurs in the district but not in the Wallaby Creek valley, which lacks appropriate cliffs. The paper complements earlier ones reporting separation by choice of habitat within this community (Jarman *et al.* 1987; Southwell 1987). The diets of seven of the species are here described from analyses of faeces or stomach contents collected at the study site; those of the other four (which are the rarer species in the area) are inferred from studies made in similar habitats elsewhere. The methods used to describe diets are sometimes poorly comparable; some sample sizes are small. Nevertheless, for the sake of being able to investigate the role that dietary separation can play in ecological separation in this species-rich macropod community, we feel that the data are worth reporting.

The paper also allows us to test the general assumption (e.g., Jarman 1984) that larger species of macropods eat more abundant, widely distributed food-items than do the smaller species.

The study area

The study area, at Wallaby Creek in north-eastern New South Wales, described by Jarman *et al.* (1987), is in the northern headwaters of the Clarence River catchment. This paper refers mainly to macropods in about 4 km of a valley that is about 4 km wide. Vegetation communities include subtropical rainforest types, wet and dry sclerophyll forests, some of which have extensive areas of narrow-leaved forest oak *Casuarina torulosa*, and pasture formed by clearing forest from one side of the valley. The pasture is improved in places and includes exotic grasses such as Kikuyu *Pennisetum clandestinum*, paspalum *Paspalum dilatatum*, and Rhodes grass *Chloris gayana*, and some introduced legumes. Much of the pasture contains tussock-forming grasses such as snow grasses *Poa* spp. and swamp foxtail *Pennisetum alopecuroides*, and beds of blady grass *Imperata cylindrica*. Between the tussocks are short swards of narrow-leaved carpet grass *Axonopus affinis*, Kikuyu and *P. dilatatum*. More open parts of the dry sclerophyll forest carry a grass community dominated by kangaroo grass *Themeda triandra*. Fuel-reduction burns are carried out in the drier areas of forest intermittently, and at least some parts of the pasture are burned in any year. The forest, creekside, and banks of gulleys in the pasture carry diverse communities of forbs and shrubs.

The animals

The macropod species in the study area and their habitat preferences and relative abundances are given in Table 1.

Methods for description of diets

The Long-nosed Potoroo, Common Wallaroo and Swamp Wallaby are too uncommon in the study area for either faecal samples or stomach contents to have been collected. Consequently their diets at the study site are assumed to be similar to those determined by studies elsewhere. For Long-nosed Potoroo, the studies consulted

[1]Department of Ecosystem Management, University of New England, Armidale, New South Wales 2351.

Pages 143-49 in KANGAROOS, WALLABIES AND RAT-KANGAROOS ed. by G. Grigg, P. Jarman and I. Hume. Surrey Beatty & Sons Pty Limited, New South Wales, Australia, 1989.

Table 1. The species of macropods which occur at (or near*) Wallaby Creek, their preferred habitats there, and their status and distribution. See also Jarman *et al.* (1987) and Southwell (1987).

Species	Preferred Habitats	Status and Distribution within Habitats
Eastern Grey Kangaroo *Macropus giganteus*	Pasture and open forest	Very abundant;widespread
Whiptail wallaby *M. parryi*	Open forest and pasture on hillsides	Uncommon; widespread within habitat
Common Wallaroo *M. robustus*	Confined to woodland on one rocky hilltop	Rare; one small area only
Red-necked Wallaby *M. rufogriseus*	Wet and dry sclerophyll forest and adjacent pasture	Very abundant; very widespread
Black-striped Wallaby *M. dorsalis*	Forest with dense under-storey and adjacent pasture	Uncommon; patchy
Swamp Wallaby *Wallabia bicolor*	Gulleys with ground cover in sclerophyll forest	Rare; very few occurrences
Red-necked Pademelon *Thylogale thetis*	Wet sclerophyll forest and edge of adjacent pasture	Common; rather restricted
Red-legged Pademelon *Thylogale stigmatica*	Rainforest	Common; widespread within habitat
Brush-tailed Rock-Wallaby* *Petrogale penicillata*	Dry sclerophyll forest adjacent to cliffs	Absent from Wallaby Creek; colonies exist within 20 km
Rufous Bettong *Aepyprymnus rufescens*	Pasture and dry sclerophyll forest with tussock grasses	Common; widespread within habitat
Long-nosed Potoroo *Potorous tridactylus*	Wet sclerophyll forest with dense ground cover	Uncommon; distribution not fully known

were those by Bennett (1987), Guiler (1971) and Kitchener (1967) (see also Seebeck *et al.* this volume), all of which were conducted in the south of the species' range. For Common Wallaroo, we have used the diet described by Taylor (1983) for wallaroos on a relatively unimproved sheep property on the New England Tablelands, comparable in many respects to the small area where the species occurs at our study site. For Swamp Wallaby we have consulted studies by Harrington (1976) and Hollis *et al.* (1986) of wallabies living on the eastern scarp of the New England Tablelands, in wet sclerophyll forests with access to patches of tussock grassland, similar to the vegetation communities in which the species has been seen at Wallaby Creek. Although the Red-necked Pademelon is fairly common in a restricted part of the study area, no analyses have yet been made of its diet there. Consequently, we have relied on analyses of their diets on the Dorrigo Plateau (Redenbach 1982), where they fed in rainforest and wet sclerophyll forest and the pastures adjacent to them, just as they do at Wallaby Creek. The diet of Brush-tailed Rock-wallaby has been described from a single collection of fresh faeces from Joel's Gulley, near Wallaby Creek.

For the remaining species, diets are based on material collected at Wallaby Creek. This usually consisted of faecal pellets, seen to be dropped by individuals of a recorded species, and collected while fresh. These were usually preserved in FAA in the field and stored thus until analysis. Except for comparisons of pasture use before and after fire, samples from all seasons have been combined. Faeces from Black-striped Wallaby and Red-legged Pademelon were also collected from sites where those species alone were known to be active. Stomach contents from four Red-legged Pademelons (two road-kills; two shot) were also analysed (Redenbach 1982). Dietary items taken by Rufous Bettong were described from field observations of the animals feeding, from remains of food plants at the feeding sites, and from faecal analysis (Schlager 1981; pers. obs.).

The methods used for analyses of faeces or stomach contents were as follows. Plant species in the diet were recognized by characteristics of their epidermis as found in a reference collection. The relative proportions of species or their parts in the diet were taken to be represented by the relative amounts of epidermal area which they contributed to the total area of epidermis recognized in an inspected sample of faecal material or stomach contents. Thus bulky items such as fruit or fungi will be underestimated, as will species whose epidermis disintegrates during digestion.

Reference material from grasses was first divided into leaf, sheath, stem and seedhead. These were cut into 5 mm lengths and put into a conical flask containing 10% chromic/nitric acid. The flask and its contents were heated intermittently over a bunsen for about five minutes or until the epidermis separated from underlying tissue. When the solution boiled it was removed from the heat, swirled around, and reheated. When the epidermis separated, water was added to the flask to prevent further disintegration, and a few drops of ammonia were added to neutralize the acid, the aim being to halt disintegration quickly. The contents of the flask were transferred to a beaker, where the epidermal fragments usually floated to the surface. Epidermal fragments were picked up with a small brush and put into a small quantity of saffranin stain in a watch glass. After a few minutes the fragments were removed to a dilute solution of glycerine which removed excess stain. The fragments were then mounted in glycerine jelly. Reference material for dicotyledonous leaves was treated in much the same way, but dicot stems were not prepared. The reference slides were kept and were also photographed.

Faecal material, after storage in FAA, was agitated with a glass rod to separate particles, and transferred to 70% alcohol. Saffranin stain was added to a subsample. After a day the particles were washed in water to remove excess stain, then stored in 70% alcohol. For analysis, the particles were agitated and samples were lifted onto a microscope slide using a Pasteur pipette. A coverslip was put on and the sample examined with a binocular, light-transmission microscope under ×100 to ×400 magnification. Fragments were compared with reference slides and photographs and recorded by species and plant part, if possible. In general a fragment of grass epidermis with over 100 cells would be recognizable if it was represented in the reference collection. Fragments which were smaller than that size were ignored. Fragments which were that large but could not be recognized were recorded as "unknown" and usually photographed in the hope that they might later become recognizable. With two years' practice, well over 90% of the grass epidermal area (consisting of fragments of more than 100 cells) in a sample could be recognized.

Each slide was "read" from one side to the other. Fragments are not of uniform size, and were recorded as being a multiple of a standard unit area (which was initially set as being the minimal size which we would confidently expect to be able to identify). The first 100 fragments on any slide, regardless of size, were recorded as from dicot or monocot, a distinction that is relatively easy to make on even very small areas of epidermis. Any fragments greater than the minimal standard size were also recorded in the next step, which was to record fragments of minimal standard size or larger, identified to species and part if possible, with the relative size of each being recorded. The results for the analysis of each sample could then be expressed as the proportion of the fragments that were of monocot or dicot plants, and the proportion of the potentially identifiable epidermal area that was formed by each species (or parts of each species, etc.).

RESULTS

The numbers of samples (i.e., separate defaecations) analysed and used in this paper were: Eastern Grey Kangaroo, 32 (16 before and 16 after the annual burning of the pasture); Whiptail Wallaby, 7; Red-necked Wallaby, 22; Black-striped Wallaby, 20; and Brush-tailed Rock-wallaby, 13. Redenbach (1982) described the diets of Red-legged Pademelon from four stomach contents and five faecal samples, and of Red-necked Pademelon from 21 stomach samples. The diets of Common Wallaroo and Swamp Wallaby are taken from publications describing larger samples (Taylor 1983; Hollis *et al.* 1986).

Table 2 compares the roles of monocots (mainly grasses), various parts of dicots (the few records of flowers are included in leaf and stem), ferns, two kinds of fungi, and invertebrates in the

Table 2. Proportions of plant types and parts in the recorded diets of macropod species which occur at Wallaby Creek. Diets of the two potoroid species are described qualitatively only. Data from our own analyses, except: 1. from Taylor 1983, not studied at Wallaby Creek; 2. from Hollis *et al.* 1986, not studied at Wallaby Creek; 3. from Redenbach 1982, not studied at Wallaby Creek; 4, from Redenbach 1982; 5. from Schlager 1981 and our observations; 6. from Guiler 1971; Kitchener 1967 and Bennett 1987, not studied at Wallaby Creek.

MACROPODIDAE	Monocot parts %	Dicot leaf and stem %	Dicot root	Dicot fruit %	Fern %	Above-ground fungi %	Below-ground fungi %	Invertebrates
Eastern Grey Kangaroo	99	1						
Whiptail Wallaby	99	1						
Common Wallaroo[1]	96	4						
Brush-tailed Rock-wallaby	95	5						
Red-necked Wallaby	84	16						
Black-striped Wallaby	81	19						
Swamp Wallaby[2]	14	43			25	18		
Red-necked Pademelon[3]	66	31		1	1	1		
Red-legged Pademelon[4]		88		10	1	1		
POTOROIDAE								
Rufous Bettong[5]	Little	V. Little	Much				Much	Regularly
Long-nosed Potoroo[6]	V. Little	V. Little	?				V. Much	Common seasonally

diets of the 11 macropod species. Diets of the two potoroids are described only qualitatively, because of the difficulties of quantifying their diversely composed diets. Within this community there is a major separation between the diets of the two potoroids, both of which eat subterranean foods and some invertebrate material, and the macropodids, none of whom do. Among the macropodids, the Red-legged Pademelon is distinctly different from the rest in eating no grass and taking a relatively high proportion of dicot fruit. The Swamp Wallaby eats little grass, and more fern and fungi than any of the other macropodids. Red-necked Pademelons (if this reported diet reflects their diet at Wallaby Creek) eat much grass, but also a large proportion of dicot material and traces of fruit, fern and fungi. Black-striped Wallabies and Red-necked Wallabies eat very similar proportions of grasses and dicots, and this pair of species is distinct from the three largest species, Eastern Grey Kangaroos, Whiptail Wallabies and Common Wallaroos, which are almost pure grazers, as were the Brush-tailed Rock-wallabies in this sample.

There is thus considerable potential for the five *Macropus* species to compete in their use of grass. Table 3 compares the use by the four least rare species of the most commonly eaten grasses (Wallaroos are rare and restricted in distribution at Wallaby Creek).

The most abundantly used grasses are narrow-leaved carpet grass *Axonopus affinis*, paspalum *Paspalum dilatatum*, swamp foxtail *Pennisetum alopecuroides*, blady grass *Imperata cylindrica*, and kangaroo grass *Themeda triandra*. Eastern Grey Kangaroos and Red-necked Wallabies took very similar proportions of these grasses, although the wallabies took more of the carpet grass and less of the foxtail and kangaroo grass than did the kangaroos. Whiptail Wallabies were outstanding in eating much kangaroo grass and more blady grass, and taking less carpet grass, than did any of the other species. Black-striped Wallabies distributed their use rather more evenly, and over a greater range of species, than the others did. Although carpet grass was the single most abundant species in their diet, they used it less heavily than did Eastern Grey Kangaroos and Red-necked Wallabies. They were the only species to make much use of snow grass *Poa labillardieri*, eating both its leaves and its seedheads.

Brush-tailed Rock-wallabies, although not sympatric with these species at Wallaby Creek, differed from them in utilizing kangaroo grass exceptionally heavily (87% of the identified grass component of their diet), paspalum (3%), swamp foxtail (3%), and blady grass (2%) being the other common components. Fourteen species of grasses were identified in their faeces.

Black-striped Wallabies exceeded the others in the species richness of their diet (21 grass species; Table 3) and the equitability of distribution of use between species. Calculated Simpson-Yule Indices of Diversity (D) (Southwood 1978) for the species were: Eastern Grey Kangaroo pre-burn 4.4, post-burn 4.2; Whiptail Wallaby 4.1; Red-necked Wallaby 3.0; Black-striped Wallaby 7.5. The higher the index value, the greater the equitability, demonstrating the reliance of Red-necked Wallaby on relatively few, heavily used species, but the equitable spread of use across many species by Black-striped Wallaby.

The degree of overlap between these species in the grass component of their diets was calculated by summing (over all grass species) the lesser of the percentages that each grass species formed of the diets of two compared macropod species, and is shown in Table 4. Average overlap was greatest (62.3%) between Eastern Grey Kangaroo and other species, and least (44.5%) between Whiptail Wallaby and other species. Points of interest

Table 3. The proportions (of all grass in their diets) formed by the grasses most commonly eaten by Eastern Grey Kangaroos, Whiptail Wallabies, Red-necked Wallabies and Black-striped Wallabies at Wallaby Creek.

	Eastern Grey Kangaroo		Whiptail Wallaby	Red-necked Wallaby	Black-striped Wallaby
	Pre Burn %	Post Burn %	%	%	%
Axonopus affinis	33.3	41.5	5.4	47.3	24.8
Capillipedium spicigerum	1.5	2.1	7.4	0.6	1.9
Cynodon dactylon	2.7	0.9	2.0	2.1	1.9
Bothriochloa decipiens	2.9	0.7	1.7	1.3	1.0
Imperata cylindrica	2.5	16.7	20.3	1.1	10.9
Paspalum dilatatum	30.6	10.9	10.8	32.0	10.9
Pennisetum alopecuroides	8.2	9.5	5.0	3.6	5.9
Pennisetum clandestinum	5.1	6.3	3.6	6.4	8.2
Poa labillardieri	0.2	0.1	1.2	0	10.2
Themeda triandra	9.4	9.9	42.2	1.7	15.9
Number of other identified grass species in diets	8	4	2	8	11

Table 4. The matrix of percentage overlap in the grass components of their diets between paris of *Macropus* species at Wallaby Creek.

Overlap Matrix:	Whiptail Wallaby	Red-necked Wallaby	Black-striped Wallaby
Eastern Grey Kangaroo	49.5%	73.9%	63.6%
Whiptail Wallaby		26.5%	57.6%
Red-necked Wallaby			52.0%

are the high degree of similarity in the diets of the two most abundant species, Eastern Grey Kangaroo and Red-necked Wallaby, and the low overlap between Whiptail and Red-necked Wallabies.

The use of substantial quantities of browse (dicot leaf material) by Red-necked and Black-striped Wallabies would reduce their overlap with the two pure grazers (and perhaps with each other) to even lower levels.

Had Brush-tailed Rock-wallabies been sympatric with these four and all had retained their diets, rock-wallabies would have overlapped most with Whiptail Wallabies (54.3%) and Black-striped Wallabies (37.0%), and least with Eastern Grey Kangaroos (20.2%) and Red-necked Wallabies (11.9%).

No allowance has been made in these crude analyses for seasonal variation in diets: but Eastern Grey Kangaroo diets have been recorded from shortly before and shortly after the annual burning of pasture. The results (Table 3) illustrate that the general form of the diet (species-richness, equitability, and species eaten) remains similar, although the extent of use of some species changes importantly. In particular, after the fire there was an increase in use of blady grass *Imperata cylindrica*. This was evident in field observations (J. L. Clarke, pers. comm.), since kangaroos would move into the burned area within hours of the fire, avidly seeking the now-exposed green shoots of blady grass. Other species did not do this.

All four *Macropus* species, whose diets were analysed at Wallaby Creek, took predominantly the leaves of grass, avoiding much sheath or stem in their intake (Table 5). The extent to which sheath and stem were avoided differed between grass species, but also between macropod species. Thus Red-necked Wallabies were particularly successful at avoiding sheath or stem, while Whiptail Wallabies and Eastern Grey Kangaroos grazing unburned pasture sometimes ate much non-leaf material. Within each *Macropus* species' diet there was a weak trend towards eating less leaf and more sheath or stem of the most commonly used grasses. This trend was most obvious for Whiptail Wallabies. Carpet grass *Axonopus affinis*, a prostrate grass forming a dense sward, was usually eaten as almost pure leaf, while kangaroo grass *Themeda triandra*, with a taller growth form, contributed sheath and stem to the macropods' diets. Blady grass *Imperata cylindrica*, which has a tall, dense growth form but has a high proportion of dead matter among its leaves, was eaten almost entirely as leaf except by Whiptail Wallabies. Swamp foxtail *Pennisetum alopecuroides*, a grass which grows as tussocks, was always eaten as leaf, perhaps reflecting the lack of stem in its growth form. Snow grass *Poa labillardieri*, which was eaten in only small quantities, is also a tussock-forming grass; it was eaten as leaf, except when Black-striped Wallabies were harvesting its seedheads. They then ingested a proportion of stem, but no sheath. Brush-tailed Rock-wallabies ate 93% of kangaroo grass, the commonest species by far in their diet, as leaf. Like the four *Macropus* species, they ate only leaf of swamp foxtail. They ate 94% leaf of paspalum *P. dilatatum*, but a mean of only 63% leaf of all the less common grasses in their diet.

DISCUSSION

Although the analyses of diets were conducted on small samples, and although we have had to use studies from other sites for our descriptions of the probable diets of some of the rare species at Wallaby Creek, the elements of dietary separation in this rich community are quite apparent.

The potoroids are distinguished by their reliance on foods which must be dug up. Although they and Swamp Wallabies all eat high proportions of fungi, those taken by Swamp Wallabies bear fruiting bodies above ground,

Table 5. The proportions of leaf (out of the total of leaf, sheath and stem area for that grass species) for five common grass species in the diets of four *Macropus* species, and the rank (R) of that grass in the species' diet.

	Eastern Grey Kangaroo				Whiptail Wallaby		Red-necked Wallaby		Black-striped Wallaby	
	Pre Burn		Post Burn							
	%	R	%	R	%	R	%	R	%	R
Axonopus affinis	97	1	99	1	100	5	100	1	99	1
Imperata cylindrica	100	7	95	2	74	2	100	9	100	4
Paspalum dilatatum	54	2	99	3	96	3	95	2	87	3
Pennisetum alopecuroides	100	4	100	5	100	6	100	4	100	7
Themeda triandra	91	3	90	4	77	1	99	6	84	2

those used by the potoroids do so below ground. Potential competition between potoroids and macropodids would arise only for the few dicots whose leaves, flowers or fruits the potoroids might occasionally eat.

The diets described here are in several cases appropriate to the habitat occupied by the species, not unexpectedly. Thus, Red-legged Pademelons can be found aggregating beneath some currently fruiting trees. The most purely grazing species, Eastern Grey Kangaroos and Whiptail Wallabies, are found in pasture and open forest habitats.

These dietary correlations with habitat must contribute to ecological separation of species that are already separated by selection for habitat. Some species are restricted to only parts of the total area of a habitat. Red-necked Pademelons, for example, emerge from wet sclerophyll forest to feed in pasture no more than 100 m from the forest edge. Their potential competition with other grazing species using the pasture habitat is thus spatially limited. Whiptail Wallabies preferentially occupy pasture on steep slopes, which restricts them to only part of the range of Eastern Grey Kangaroos. It also means that they tend to occur among grasses with tall or tussock growth forms, a tendency that is reflected in the frequencies of such species in their diets. By contrast, Eastern Grey Kangaroos feed preferentially (except immediately after fire) where there is a dense sward of short, green, leafy grass; and this again is reflected in their diet.

This separation by habitat has to be borne in mind when considering the degree of overlap in the use of grass species by the four most abundant *Macropus* species at Wallaby Creek. The values for dietary overlap between pairs of these macropod species is generally much more than those between pairs of large herbivores in an African mammal community. For example, in a study of mammals in the flooded Zambezi valley, Jarman (1971) found overlaps in diet ranging from 12 to 39%.

The spatial reaction of these macropods to burning of the pasture has been described by Southwell and Jarman (1987). These dietary findings support their account of Eastern Grey Kangaroos responding immediately by moving to the burned area, and J. L. Clarke's (pers. comm.) observations of the exposed green shoots of blady grass being eaten then.

This study confirms that macropodids, when eating grass, select for leaf, generally taking it in preference to sheath or stem. Presumably they do so to avoid the higher structural components (fibre) of the latter parts. There is little difference among the major grazing species in their ability to find a diet composed largely of leaf, although Whiptail Wallabies were slightly less successful than the other *Macropus* species.

If these grazing macropods are compared with grazing ungulates in Africa, it is clear that the macropods' diets consist far more purely of grass leaf. Even such pure grazers as Blue Wildebeest *Connochaetes taurinus* and Common Zebra *Equus burchelli* may not achieve better than 61 and 37% leaf in their diets, respectively, at the best time of year (Gwynne and Bell 1968). Smaller, more selective species, such as the 40-55 kg impala *Aepyceros melampus*, range seasonally from 24 to 54% leaf in the grass component of their diet (Jarman and Sinclair 1979). The macropods' narrower incisor arcade and the advantage gained by crouching low and being able to rotate the head may contribute to this far greater selectivity for the parts of grass which they ingest, although differences in the growth forms of swards should not be discounted.

The macropods at Wallaby Creek vary in size from the Long-nosed Potoroo (1 kg) and Rufous Bettong (2-3 kg) to the Eastern Grey Kangaroo (females 35 kg; males growing to over 90 kg). The smallest species feed on scattered food-items, rich in proteins or digestible carbohydrates: underground storage organs such as roots and rhizomes, or fruiting bodies of fungi, and invertebrates. The next smallest species are the two pademelons (4-9 kg), which select dicot leaves and fruits (Red-legged Pademelon) or grass and dicot leaves (Red-necked Pademelon). While these food items are not as individually valuable, nor take as much seeking, as those of the potoroids, they, and those of the Swamp Wallaby, nevertheless are probably lower in structural carbohydrates and require more selection overall than the diets of the next largest abundant species, Red-necked Wallaby (10-23 kg) and Black-striped Wallaby (7-20 kg). These two macropods had diets of over 80% grass, the rest being dicot leaves (mainly). Red-necked Wallaby selected most strongly of all the macropods for grass leaf. Black-striped Wallabies, which have to feed within a limited distance of the edge of the forest in which they spend the day, ate a wide range of grass species, distributing their use among them most equitably. Both Black-striped and Red-necked Wallabies included a proportion of grass seedheads in their diets (as did Brush-tailed Rock-wallabies), valuable items which were far less common in the diets of Eastern Grey Kangaroos (35-90 kg) and Whiptail Wallabies (11-26 kg). These two large species ate diets composed of food items which were abundant, well distributed, and required least seeking and selection. Thus this community illustrates the expected trend in value and selection for food items, against body size, in macropods. Nevertheless, there remains a need for the trend to be confirmed with chemical analyses of diets.

Our findings show that the macropods at Wallaby Creek are ecologically well separated by their diets. The commonest species are those using the most abundant resources, the sward-forming grasses in cattle-grazed pasture.

ACKNOWLEDGEMENTS

We would like to thank those who helped collect the samples from which these diets are described, especially Joanna Clarke, Menna Jones and Colin Southwell. Particular thanks are due to Jack and Iris Hayes, the late Ned and Margaret Hayes. and Victor Mulcahy, for access to their private land; and to the Forestry Commission, and Foresters at Urbenville, for permission to work in Beaury State Forest. We thank Joanna Clarke, Barry Fox and Robert Taylor for helpful critism of a draft of the manuscript. The work was supported by A.R.G.S. grant D18115308, and a University of New England research grant, to P. J. Jarman.

REFERENCES

BENNETT, A. F., 1987. Biogeography and conservation of mammals within a fragmented forest environment in southwestern Victoria. Ph.D. Thesis, University of Melbourne.

FOX, B. J. (this volume). Community ecology of macropodids.

GUILER, E. R., 1971. Food of the potoroo (Marsupialia, Macropodidae). *J. Mammal.* **51**(1): 232-34.

GWYNNE, M. D. AND BELL, R. H. V., 1968. Selection of vegetation components by grazing ungulates in the Serengeti National Park. *Nature* (London). **220**: 390-93.

HARRINGTON, J. R., 1976. The diet of the swamp wallaby *Wallabia bicolor* at Diamond Flat, New South Wales. Dip. Nat. Res. dissertation, University of New England.

HOLLIS, C. J., ROBERTSHAW, J. D. AND HARDEN, R. H., 1986. Ecology of the Swamp Wallaby *(Wallabia bicolor)* in northeastern New South Wales. I. Diet. *Australian Wildlife Research* **13**: 355-65.

JARMAN, P. J., 1971. Diets of large mammals in the woodlands around Lake Kariba, Rhodesia. *Oecologia* **8**: 157-78.

JARMAN, P. J., 1984. The dietary ecology of macropod marsupials. Proceedings of the Nutrition Society of Australia **9**: 82-87.

JARMAN, P. J. AND SINCLAIR, A. R. E., 1979. Feeding strategy and the patterns of resource partitioning in ungulates. Pp. 130-63 *in* Serengeti: Dynamics of an Ecosystem ed. by A. R. E. Sinclair and M. Norton-Griffiths. The University of Chicago Press: Chicago and London.

JARMAN, P. J., JOHNSON, C. N., SOUTHWELL, C. J. AND STUART-DICK, R., 1987. Macropod studies at Wallaby Creek. I. The area and animals. *Australian Wildlife Research* **14**: 1-14.

KITCHENER, D. J., 1967. Some aspects of the biology of the potoroo, *Potorous tridactylus* (Kerr). B.Sc. (Hons.) Thesis, University of Tasmania.

REDENBACH, C. J., 1982. A dietary comparison of three sympatric forest-dwelling macropodids. B. Nat. Res. Thesis, University of New England.

SCHLAGER, F. E., 1981. The distribution, status and ecology of the rufous rat-kangaroo, *Aepyprymnus rufescens,* in northern New South Wales. M. Nat. Res. Thesis, University of New England.

SEEBECK, J. H., BENNETT, A. F. AND SCOTTS, D. J. (this volume). Ecology of the Potoroidae — A Review.

SOUTHWELL, C. J., 1987. Macropod studies at Wallaby Creek. II. Density and distribution of macropod species in relation to environmental variables. *Australian Wildlife Research* **14**: 15-33.

SOUTHWELL, C. J. AND JARMAN, P. J., 1987. Macropod studies at Wallaby Creek. III. The effect of fire on pasture utilization by macropods and cattle. *Australian Wildlife Research* **14**: 117-24.

SOUTHWOOD, T. R. E., 1978. Ecological Methods, second edition. Chapman and Hall: London and New York.

TAYLOR, R. J., 1983. The diet of the eastern grey kangaroo and wallaroo in areas of improved and native pasture in the New England Tablelands. *Australian Wildlife Research* **10**: 203-11.

Morphological adaptations of teeth to diets and feeding in the Macropodoidea

G. D. SANSON[1]

Teeth are primarily concerned with mechanically reducing food to a size and shape that will optimize residence time in the gut and produce sites suitable for endogenous or bacterial enzymatic hydrolysis. Diets in the Macropodoidea range from a heterogeneous assemblage of insects, fruit, and fungi (mainly Potoroidae), through dicotyledonous browse to structurally homogeneous monocotyledonous graze (Macropodinae). This spectrum represents an increase in the content of structural plant cell wall and there is an associated suite of dental adaptations to process it. While there is a spectrum of feeding adaptations they can be usefully grouped into grades of organization which reflect the physical properties of the diet. Potoroidae are characterized by specialized premolars and simple tuberculate molars with little or no shearing capacity. Browsers have a reduced premolar and molars with shearing and crushing capability. Grazers have vestigial premolars and molars with enhanced shearing capacity. Grazing macropods do not seem to have evolved the obvious adaptations to resist the heavy wear imposed by grasses that are seen in other groups. As an indirect consequence of premolar reduction, molar progression has possibly been induced.

INTRODUCTION

THIS paper attempts to review the function of macropodoid dentitions in relation to the physical properties of their diet. This will be treated in the context of a presumed shift from mixed feeding on invertebrates, roots, tubers and fungi, to browse, and then to graze. These patterns will be approached in terms of adaptive grades of organization of the dental apparatus and the various changes involved in shifting or evolving from one grade to another. It is argued that omnivorous feeding is functionally ancestral to browsing and in turn to grazing and in this regard essentially follows the arguments of Bensley (1903) and Raven and Gregory (1946). The apparent problems of the notion of evolution through intermediate adaptive troughs will be explored. Since the associated dental changes relate to, and are a consequence of, the evolution of the Macropodoidea, the sequence of dental change from one form to another, and the patterns of tooth morphologies that can be recognized, is of some interest. However, this paper does not address the question of the phylogeny of the Macropodoidea, whether the grades are monophyletic, or which groups are lineally ancestral. In particular it does not suggest that the Macropodinae, which are seen as derived grades of functional dental organization, evolved from the Potoroidae, the extant members of which are in a more ancestral grade.

The history of the recognition of grades of organization within the Macropodoidea is also explored. The salient parts of the dental morphology, which cannot be separated from its function, are then discussed. This is done in sections on the incisors, premolars and molars and a section on molar progression, which is an important part of this character complex. The information is then drawn together in sections on each grade. Inevitably there is some repetition of arguments but this is unavoidable in describing such a complex set of interrelationships.

Taxonomy and dental grades of organization of the Macropodoidea

Initially, recognition of subgroups among the kangaroos had a conventional morphological rather than ecological or dietary base. Much discussion has centred around the taxonomic legitimacy of those groups. Some ecological and dietary information was available, however, and Waterhouse (1846) was among the first to note that some kangaroos are fitted for one kind of scrub and some for another, for example *Wallabia bicolor* and *Macropus rufogriseus.* While he did not say so, these differences in habitat preferences reflect a difference in feeding niche, i.e., browsing and grazing respectively. Owen (1859) was aware of grazing and browsing kangaroos but does not seem to have recognized the associated dental and cranial morphologies. Krefft (1875) recognized a group of large kangaroos having molar progression, a feature noted by Waterhouse (1841), broad upper third incisors and small premolars which are lost. A second group of the smaller kangaroos, which included the rat-kangaroos, had narrower third upper incisors, permanent premolars and no tooth movement.

Garrod (1875), in a paper which showed considerable insight, divided the family Macropodidae (now the superfamily Macropodoidea) into two subfamilies, the Macropodinae and the Hypsiprymninae. He further divided his Macropodinae into two sections. The first section, probably following Owen, contained the genera *Macropus, Halmaturus, Petrogale* and *Lagorchestes.* This group was characterized by premolars never much larger than the molars, which had a characteristic though undefined pattern.

[1]Department of Zoology, Monash University, Clayton, Victoria 3168.

Pages 151-68 in KANGAROOS, WALLABIES AND RAT-KANGAROOS ed. by G. Grigg, P. Jarman and I. Hume. Surrey Beatty & Sons Pty Limited, New South Wales, Australia, 1989.

The second section contained *Dorcopsis, Dendrolagus, Protemnodon* and *Sthenurus,* and had strikingly large premolars combined with characteristic molars "slightly modified upon that of Macropus". He apparently ignored relative amounts of tooth movement. Garrod's second subfamily, the Hypsiprymninae, was distinguished among other things by also having a characteristic molar pattern. The premolar structure helped differentiate the included genera. Garrod incorporated structures of the stomach in his classification and noted differences which are now thought to have considerable significance. While he linked tooth and gut structure he did not relate this to diet. Garrod evidently included some of the smaller wallabies in *Halmaturus* and separated others into *Protemnodon* which he placed in the *Dorcopsis* section. This was an interesting grouping because most authors before, and many since, have not recognized and maintained this distinction. Garrod may have been the first to recognize some difference between the predominantly browsing wallabies and those which include a lot of grass in their diet, yet he did not appear to be aware of the signficance of this.

Bensley (1903), in his classic paper, argued that the potoroine and bettongine groups with their low-crowned quadritubercular molars represented more insectivorous forms than the macropodines. However, he thought *Aepyprymnus* and *Calopyrmnus* were tending towards a bilophodont condition, and that these teeth were derived from a bunodont phalangerid type. He particularly noted the Potoroidae for their strongly developed and specialized premolars, though no particular function was ascribed to them. The macropodine lophodont tooth he in turn derived from the more primitive bunodont potoroid form. The Macropodinae he broke into two groups, the brachyodont (*Dendrolagus, Dorcopsis* and *Setonix*) and the hypsodont (*Lagorchestes, Lagostrophus, Onychogalea, Petrogale* and *Macropus*) series. He suggested the brachyodonts, while being truly lophodont with small longitudinal links, represented a transitional phase between the cuspidate potoroids and the hypsilophodont wallabies and kangaroos. However, they were also seen as being aberrant in that their evolution had been affected by a return, for at least some of them, to an arboreal life. In the hypsodonts the development of the longitudinal links supplemented the transverse lophs. Bensley noted that an assumption of herbivory involved a wearing down of the molar crowns and that this resulted in the evolution of high crowned or hypsodont teeth and the replacement of worn teeth by molar progression.

Bensley's general thesis was refined by Raven and Gregory (1946) who related the morphological patterns to diet. They included *Thylogale* in the brachyodont group and noted the intermediate nature of dental morphology and diet of many of the wallabies. This intermediate nature of many forms was recognized by some later workers, but a possible primary browsing condition within the Macropodinae tended to be overlooked. Troughton (1947) suggested that the Macropodinae were primarily grazers. This, and his own observations which included a comparison of a phalanger and a grazing macropodine (*Macropus rufogriseus*), led Ride (1959) to conclude that the Macropodinae are adapted for grazing. Later, in suggesting that "the Macropodinae are primarily specialized for grazing" (Ride 1962 p.340), he argued that there are some forms (e.g., *Dendrolagus*) where grasses have been secondarily rejected for leaves and fruits.

Ride, like Bensley, noted the anteroposterior "links" which connect the lophs in *M. rufogriseus,* and argued that they increase the "abrasive effect on harsh siliceous grasses . . . in comparison with the attritional effect of the typical phalangerine molars" (Ride 1959 p.52). In this he followed Simpson's observation "that the transition from browsing to grazing in herbivores is only possible if structures are developed to allow the grazer to deal with the harsher, more abrasive nature of grass" (Ride 1959 p.52). It seems that Ride envisaged a basal or even pre-macropodine browsing condition, but basal macropodine dental adaptations were not identified while phalangerine ones were. Ride did note that it was difficult to interpret the structure of the premolar as a grazing adaptation and that little was known of its function. It is not clear what Simpson and Ride meant by the use of the word "harsh" in describing grasses. Grasses certainly abrade teeth and in that sense are harsh, but they also have higher levels of plant fibre which are distributed differently to dicotyledonous leaves. As discussed below, the adaptations to resist wear, and to comminute grasses effectively, are different. This difference has not always been clear and has led to confusing interpretations of dental structure and function. The suggestion that molar progression is an adaptation to compensate for tooth wear is an example of this, as will be discussed below.

In an important and influential series of papers Ride (1957, 1959, 1962, 1964, 1971, 1978) certainly recognized a relationship between reduction in the size of the premolar and increased specialization for grazing, and that there were many variations and degrees of development of this pattern. He further noted that the sectorial premolars are elaborated for cutting through objects more substantial than grass and that the forest- and thicket-dwelling wallabies and tree-kangaroos have simpler molars than their grazing-adapted relatives. Ride (1964) argued that the stem forms of the macropodid radiation are the

Macropodinae, and that the potoroine and sthenurine phyletic lines are successive offshoots. The Potoroinae retain a number of primitive characters such as relatively simple molars but combine them with specialized premolars. Ride (1971, 1978) developed the argument that the macropodid hypothetical common ancestor would have had a molar not very different from that of *Dorcopsoides* and that the bunodont potoroine molar is derived by reduction. The Sthenurinae represent a secondary return to a browsing habit from the main grazing line. Ride essentially suggested that the potoroine condition, contrary to Bensley's view, is not necessarily ancestral to the macropodid condition partly because potoroids in his view are derived from the basal macropodids. This may be true, though newer work suggests otherwise (Flannery 1983), but it is still possible that the basal macropodid, which it is generally argued evolved from a phalanger-like form, was unlike a potoroid. New Miocene finds considered to be potoroid and basal macropodid forms (Flannery *et al.* 1983; Flannery and Rich 1986) have important implications for the debate about the phylogenetic relationships of the Potoroidae and Macropodidae, but I do not believe they negate the arguments presented here; if anything they support it.

The taxonomy of the medium to small macropodine wallabies has been particularly vexed. Its history up to 1962 is well described by Ride (1962), and the debate still continues (Bartholomai 1978; Sanson 1978; Dawson and Flannery 1985). The kinds of arguments put forward have generally either not supported, adequately recognized or explicitly stated the view that, in terms of functional morphology and diet, there is a basal herbivorous grade within the Macropodinae, the browser. Its adaptations are qualitatively different to those of the presumed invertebrate-, fruit- and seed-eaters on the one hand, and the grazers on the other (Sanson 1978). Whether Ride's phylogenetic arguments are supported by other evidence is not important here. What is important though, is whether evolutionary reversal from grazing to browsing is easily achieved, and whether the primitive macropodine dentition evolved through stages now retained in various potoroids. Butler (1982) has argued that "when considered in isolation, the evolution of dental characters may be reversed, but such reversals are secondary consequences of progressive evolution of a wider functional system to which the characters are subordinate. Evolution of the dentition, like that of other complex structures, seems to be irreversible" (p. 242). I argue that this is likely to be the case in the evolution of the Macropodoidea.

The concept of grades has been particularly developed in dental systems by Butler (e.g., Butler 1971) who recognized grades of dental and masticatory adaptations in mammals associated with different broad dietary types. He noted that the adaptations involved in advancing from one grade to the next may be independently acquired. Grades may reflect general improvements that evolve in a mosaic fashion in a number of different related phyletic lines (Simpson 1959). Schaeffer (1965) noted that functional demands can have a marked influence on the nature and degree of parallelism even when the common ancestry is relatively remote. As will be argued here, the limited ways different food types can be prepared is likely to have played an important part in the evolution of grades within the Macropodoidea. Since the group is probably monophyletic similar solutions to food preparation should not be a surprise. There is ample evidence that evolutionary continua frequently involve a sequence of adaptive or organizational levels (Schaeffer and Hecht 1965). The recognition of such grades may not merely be the result of deficient evidence in lineages. It may also reflect the rapid acquisition of a new and stable set of adaptive characters.

Sanson (1978, 1980) described two basic patterns of premolar and molar morphology in the Macropodinae which he argued represented grades of organization. However, the distinction is not as clear as this. Examination of the available data indicates that there is a spectrum of dietary preferences ranging from strict browsing to almost exclusive grazing. There are intermediates which take increasing amounts of grass (Sanson 1982). The simple original distinction may not be adequate and may merit more detailed analysis. The problem is that a dentition adapted to comminute browse might be expected to be poorly adapted for a grass diet. This assumption follows from the observation that dentitions of grazers have different characteristics which can be related to the organization of the higher fibre level. If the evolution of the grazing dentition is driven by selection, then it follows that it is better at chewing grass than a browsing dentition. What then is the situation in the intermediate feeders in terms of their teeth operating on a mixed diet of browse and grass? In the Macropodoidea I believe it can be argued that in general the intermediate browser/grazers are at no obvious selective disadvantage, with the possible exception of the rock-wallabies. The potential problems of shifting from one grade (or adaptive peak) to another by going through an adaptive trough, can be explained in this case as a continuous shift of the adaptive peak. No adaptive trough is envisaged. In spite of that, the recognition of grades is useful in organizing the variety of morphological patterns and dietary preferences for description. The living genera of the Macropodoidea can be assigned to the broad grades identified on the basis of their character states (Fig. 1).

POTOROID AND BASAL MACROPODOID GRADE

mixed invertebrate, seed and low plant fibre diet

Hypsiprymnodon, Potorous, Bettongia, Aepyprymnus

BROWSER GRADE

low fibre browse

Dorcopsis, Dorcopsulus, Dendrolagus, Thylogale, Setonix, Wallabia

Petrogale

Petrogale xanthopus

Peradorcas

INTERMEDIATE BROWSER / GRAZER GRADE

mixed browse and grass diet

Macropus (*Notamacropus*), *Lagorchestes*

Lagostrophus

GRAZER GRADE

grass diet

Macropus (*Macropus*), *M.* (*Osphranter*), *Onychogalea*

Fig. 1. The grades, separated by solid lines, of dental organization of the Potoroidae and the Macropodinae recognized in this paper. Dotted lines subdivide important variations. Broad characteristics of the diet for each grade are indicated. Extant examples of each grade are given. Except for the genera *Macropus* and *Petrogale,* only genera are indicated as all included species are members of the same grade. Members of the genus *Macropus* occur in two grades but at the subgeneric (Dawson and Flannery 1985) level. *Lagostrophus,* which Flannery (1983) considers to be a sthenurine, is indicated separately within the Intermediate Browser/ Grazer Grade because it has some peculiar dental features. The diet of most rock-wallaby *(Petrogale)* species is unknown in detail, if at all. Most have the dental characteristics of the Browser Grade. *Petrogale xanthopus* is known to take grass and some others probably do as well. It has some of the features of the Intermediate Browser/Grazer Grade. *Peradorcas* does eat browse and grass, but present evidence suggests it does so on a seasonal basis, rather than as a mixed diet. It has unique dental features.

It is not suggested here that macropodines evolved from potoroids, only that the evolution of the macropodoid dentition has passed through three successive basic grades. These are a mixed invertebrate and very-low-fibre plant diet (seen in most if not all living Potoroidae), a low-fibre leaf diet (browsing Macropodinae) and a high-fibre grass diet (grazing Macropodinae).

Many of the taxonomic proposals for the Macropodoidea have not reflected dietary grades of organization within the group, and there is no obvious reason why they should have. Dietary preferences and associated adaptive morphologies could be convergent or parallel and such morphological character states are not necessarily better taxonomic characters than any other character. However, little attempt has been made to understand the function and complexity of these character states, and if it had, maybe the polarity of the changes would have been better understood and perhaps differently interpreted.

The significance of tooth function and principles of tooth design

Digestion in mammals involves two major processes which influence each other. The first is the mechanical preparation of food items. This reduces material to smaller particle sizes thereby increasing total surface area. The second process chemically operates on these particles, reducing them to molecules small enough to cross the gut wall. The rate of enzymatic attack will be proportional to the surface area of the available substrate. Consequently, the initial mechanical preparation of a food item by the teeth affects the capacity of the gut for enzymatic digestion and hence digestibility (Spalinger *et al.* 1986). Also there is good evidence that the kind of plant affects the preparation (Lees *et al.* 1981a, 1981b; Akin *et al.* 1983). Particle size affects gut passage rate in ruminants (Campling and Freer 1962) and *in vivo* (Pearce and Moir 1964; Crampton and Bell 1946; Greenhalgh and Reid 1973) and *in vitro* (McLeod and Minson 1969) digestibility. Cows may vary in their ability to reduce particle size during chewing and there are important interactions between chewing and bacterial degradation in the reticulo-rumen (Balch and Campling 1965, Baker and Harriss 1947). Certain rumen bacteria preferentially adhere to damaged cell wall (Latham *et al.* 1978). Such damage is caused by teeth. Alteration of the dentition artificially, by grinding, or naturally, by wear, alters the nature and texture of particles produced by mastication and this affects *in vivo* digestibility of ring-tail possums *Pseudocheirus* spp. (Gipps and Sanson 1984).

The works outlined above are examples of empirical studies which show that food preparation affects digestibility. The manner of preparation depends on the morphology of the teeth and their interaction with the physical properties of the diet. It would appear that there are similar morphologies which disparate mammalian groups have evolved to treat particular food items (Butler 1983) and this suggests that there may be limited ways of chewing different diets.

Butler (1983) argues that the traditional way of interpreting functional molar morphology has been to infer the mechanical actions of teeth from a study of their occlusal relations and so determine which types of food could be masticated most efficiently. However, it would seem that the decision of which foods could be masticated most effectively with a given morphology usually rests mainly on a knowledge of what kinds of foods animals with such teeth actually chew. The approach is largely a comparative one. Some work has been very sophisticated and has taken the analysis to a fine level. Rensberger (1973, 1975, 1978) and Rensberger and Koenigswald (1980) in particular have explored occlusal events in great detail and have been able to relate many processes to the ultrastructure of the enamel. Other workers, reviewed by Butler (1983), have been interested in the particle sizes, and rate of particle breakdown, produced by dentitions with only slightly different morphologies. Empirically it has been shown that small differences in tooth morphology, even in related species, have different effects. These effects have been explained in terms of, for example, the relative shearing or crushing capacity of the tooth and the amount of plant fibre it normally is required to masticate.

Butler (1983) also identified a different approach, such as that taken by Lucas (1979, 1982). Lucas considered the problem from the other direction and, as Butler points out, one of engineering design. That is, given the physical properties of the food, what form of tooth is best suited to reduce it? Lucas predicted that hard, brittle foods are reduced by the propagation of cracks through the material. The force required to maintain crack propagation is not related to the shape of the occluding cusps. However, a sharp cusp chewing hard, brittle food might be more susceptible to damage than a blunt cusp. Soft, ductile foods do not fracture in the same way and need to be penetrated. Sharp cusps are less likely to be damaged by this material and would achieve penetration with less force than a blunt cusp. Hill (1985) tested the force required by different cusp shapes to break or penetrate insect cuticle, with different moduli of elasticity. She empirically confirmed Lucas' predictions. In addition, she was able to relate these findings to the different molar form of insectivores which tended to eat insects with cuticles of different properties.

There is, however, little equivalent work on plant matter. Vincent (1982) explored the mechanical "design" of grass and concluded that it could be treated as an orientated fibrous composite. He examined its resistance to tensile load and fracture and concluded that the stiffness, or strength, of the grass was predominantly due to the sclerenchyma fibre content. Small changes in the volume-fraction of sclerenchyma will produce large changes in the stiffness and strength of the leaf. He suggested that large grazers are strong enough to break the grass by tension while smaller forms would have to use their "scissor-like" teeth to nibble at individual leaves. Vincent seemed to be mainly concerned with the stiffness of the plant in the field and its resistance to trampling and ingestion. But his results and the arguments of Lucas (1982), Lucas and Luke (1984) and Gordon (1968) are important in understanding constraints which operate in dentitions and the likely resultant shape of the teeth.

Rensberger (1973) describes cutting as a process in which a relatively sharp edge is pressed against another surface. Shearing is a type of cutting in that two sharp edges appress food, narrowly missing each other, examples being carnassials and scissors. In crushing, blunt surfaces oppose each other with no translational motion. Grinding is the motion of flat or irregular surfaces across each other. Rensberger argues that grinding differs from shearing because it involves multiple repetitive contacts which at any one time are a fraction of the total area of contact made during a stroke and the pressure is proportionately high. These definitions need refining. I agree with Rensberger that grinding should be distinguished from shearing. There is no difficulty with his definition of flat or smooth grinding surfaces, such as a mortar and pestle. The problem is in distinguishing grinding between surfaces with irregular faces, from shearing. If the irregular ridges, which generate the repetitive contacts, are not aligned with any specific orientation with respect to the direction of translational motion, surface area contact will be reduced. Each ridge may repeatedly cross other ridges, and with different orientations. Surface area contact can be further reduced if the ridges are specifically oriented so that opposing ridges approach obliquely. Each set of such ridges will then shear, with the production of continuous, point contacts. An example of this would be the artiodactyl selenodont system. It is possible that one ridge may shear across a number of opposing ridges, but this seems to be different to the kind of grinding envisaged by Rensberger. Butler (1983) takes a similar view to Rensberger, emphasizing that shearing is a vertical stroke while grinding is semi-horizontal. I would argue that this difference is misleading and obscures the more important distinction between shearing and grinding.

It may help, as Lucas tried to do, to look at the problem of food preparation from the point of view of the material and not the tooth. Grinding may be more usefully seen as crushing with a translational force. Crushing between broad surfaces will only be effective if the material between the surfaces is brittle and hence fractures by crack propagation. The effect of the introduction of translational motion in such systems, i.e., grinding, is not clear. Its major impact may be on materials like plant cells which are turgid and will burst under such a load, releasing their contents. Grinding in Rensberger and Butler's sense can really be many shearing actions, but usually over a smaller amplitude than the shearing stroke of carnassial teeth. The important point is that it passes edges across each other, usually by sequential point contacts, thereby increasing the pressure.

Shearing mechanisms are effective in breaking ductile or elastic materials which are crack resistant. The material must be separated by what probably amounts to some special case of tensile failure. Materials with a composite structure such as fibre-glass are particularly resistant to crack propagation because of the crack stopping nature of the fibres (Gordon 1968) which must be shorn. Grass, as Vincent (1982) pointed out, is a composite material with fibres running through it. In addition, individual cell walls are constructed of a composite mixture of cellulose microfibrils in a hemicellulose matrix. The cellulose fibres are laid down with such an orientation (Esau 1953) that they would further resist crack formation and propagation. This presumably provides plant fibre with much of its structural strength. Plants such as grasses, with high levels of fibre, are quite resistant to breakage and consequently require high shearing forces.

Browse is a heterogeneous material with different thicknesses and levels of resistance to breakage. Stems, which are often lignified, require large, or coarse, shearing blades. These are provided by the premolars. The vascular bundles of the leaf lamina may be secondarily thickened or even lignified, and are concentrated in a reticulated system of veins, separating regions of thinner walled and more easily damaged cells. Large vascular bundles require shearing blades with sufficient relief or height, and hence potential amplitude of movement, to be effectively shorn through. However, this amplitude is much less than that of the premolars, and may be categorized as fine shearing. Relatively high ridges concommitantly means there are large faces between the ridges. If these faces occlude in a grinding action, tissue, including turgid mesophyll cells, caught between them is presumably burst, exposing both the contents and the walls to attack. Consequently, effective mastication of browse requires the teeth to perform several different actions.

Grass lamina width is generally less, and thickness is generally more uniform in size, than browse. Vascular bundles are evenly distributed and generally narrow. Molar ridges with low relief and smaller amplitude of movement provide adequate shear to comminute such material. The presence of grinding faces seems to be less important in grass feeders. This probably relates to the different proportion and distribution of cells which can be usefully prepared by grinding. These suggestions need to be confirmed, but there is an implication that grinding faces are neither necessary nor useful to grass feeders. This may be important because, if so, it means that more small ridges, or cutting edges, could be packed into a tooth, but only at the expense of the loss of grinding capacity. Large herbivores, such as the ungulates, that live on grasses require large amounts of material. Having big teeth, these animals can support many fine shearing edges, but they lose grinding capacity.

High shearing forces are best achieved by reducing the radius of curvature of the edges, i.e., increasing the sharpness, and reducing the number of edges in contact. The optimum number of edges will be a compromise between the amount of food that has to be processed and the available muscle force. I suggest that these factors are important selective pressures in the development of teeth adapted to grazing and represent quite different solutions to those occurring in browsing-adapted teeth.

Teeth, then, are an important component of the digestive system, but their efficiency, in terms of the sustained production of small particles, has been demonstrated to be affected by wear at least in Koalas *Phascolarctos cinereus* (Lanyon and Sanson 1986). Tooth wear and its implications for longevity have been explored in different populations of kangaroos (McArthur and Sanson 1988) and the phenomenon may have other implications for natural populations (Sanson and McArthur 1987). There is no reason to expect that wear does not have some influence in other species, particularly those without continuous tooth growth. Tooth wear is generated by the contact of the occluding molars and the abrasives in the diet. Grasses tend to have high levels of abrasive silica. Consequently, teeth adapted to masticating grass must cope with two different problems. The first is the mechanical treatment of the plant tissue, and the second is the wear generated by that tissue.

The problem of silica in grass arises because silica is harder than tooth enamel and consequently wears down the enamel edges (Baker *et al.* 1959). The effect of this is to increase the radius of curvature of the occluding edges, increasing the surface area contact and, because the same muscles are operating the teeth, lowering the available pressure for food comminution. Continued wear eventually breaches the enamel, exposing the dentine. Because the dentine is softer than enamel it erodes faster (Greaves 1973), leaving enamel ridges on either side. Thus initially, the length of enamel ridges or cutting edges is increased. It is not known if this is compensated for by increased muscle force. Because of the scale of the enamel/dentine boundary the enamel ridges tend to have smaller radii of curvature, reducing the surface area contact (Dea 1987). Therefore wear may initially increase the masticatory capacity of the tooth. However, eventually these enamel ridges wear to such an extent that their radius of curvature, hence bluntness, and therefore their surface area contact, is increased. With further wear the enamel separating exposed dentine areas disappears and the length of the cutting edges begins to reduce.

Many grazers evolve high-crowned or hypsodont teeth as a response to wear, and this has the effect of increasing the amount of enamel that is available for being worn away. Another adaptation is the evolution of open-rooted or hypselodont teeth. This also increases the amount of available enamel, in some cases for the life of the animal, and has the effect of maintaining the length of the cutting edges. However, both these adaptations are only really effective if the enamel and dentine are arranged in such a way that wear maintains the same spatial relationship of the exposed tissues. This is usually achieved by the development of prismatic teeth with a relatively flat surface only relieved by the differential wear of the enamel and dentine. Perhaps because of their history kangaroos have not, or were unable to, evolve a surface topography which does not change with wear. The evolution of a hypsodont or hypselodont tooth would, therefore, provide no advantage. Hypsodont teeth may be recognized as such if the height exceeds their length. While this is relatively arbitrary it does indicate a measure. On this basis no macropod tooth is hypsodont and, as discussed above, they certainly are not functionally so.

One response to higher abrasive levels is to increase the hardness of the enamel if possible. Within the Macropodinae there is evidence that the enamel of *Macropus giganteus,* a grazer, is harder than the enamel of the browsing *Wallabia bicolor* (Palamara *et al.*1984).

The capacity and function of the gut in relation to different diets of the Macropodoidea is addressed by Freudenberger *et al.* (this volume). Earlier work indicates that there are patterns of gut structure and function within the Macropodoidea (e.g., Langer 1979a, 1979b, 1979c, 1980 and the work of Hume cited in Hume 1982) and these patterns do relate to published work on tooth function and diet (Hume 1982).

Tooth form and function in Macropodoidea

Teeth are conventionally divided into four groups: incisors, canines, premolars and molars. Canines, while fairly well developed in the Potoroidae are rarely found in the Macropodinae, and are probably not functionally important in food preparation. The incisors are certainly concerned with food ingestion and may be aided in this by the premolars. The premolars may also contribute to molar function, which is primarily mastication of the food prior to being swallowed. Tooth morphology has been well described in a large body of literature (e.g., Bensley 1903; Tate 1948; and papers cited in reviews e.g., Ride 1964, 1971; Bartholomai 1978; Archer 1981; Archer and Bartholomai 1978; Flannery this volume; Dawson and Flannery 1985) and will not be repeated here.

Detailed studies of tooth function in the Macropodoidea, other than casual observation on broad dietary regimes, are less common. Sanson (1978, 1980) described in detail the occlusal relationships of the cheek teeth of a predominantly browsing wallaby, *Wallabia bicolor,* and a grazing kangaroo, *Macropus giganteus.* Implications about the evolution of these structures, particularly molar progression, were discussed in Sanson (1982). The significance of tooth wear in a grazing macropodine has been explored (McArthur and Sanson 1988). The Potoroidae have been subject to less investigation (Morton 1981). In particular, most studies have concentrated on the cheek teeth and very little comparative work has been done on the incisors

INCISORS

Anderson (1927) examined previous suggestions that the lower incisors acted like a pair of scissors and severed grass between them. He pointed out that the true function was for the lower incisors to act against the upper incisal arcade. This was achieved by rotating the mandibles around the flexible symphysis, separating the lower incisors and aligning them with the uppers. This same action brings the lower molar rows closer together, separating the upper and lower molars and preventing contact between them when the incisors were being opposed. Anderson noted that in some species the lower incisors were completely surrounded by enamel while in others there was enamel only on one side. While it has been used as a taxonomic character, its functional significance is not known.

Upper incisor development in the grades shows development in two main ways. In potoroids and browsing macropodines the first upper incisor is large and its occlusal surface extends below the surface generated by the two posterior smaller incisors (Fig. 2). This condition is also found in phalangerids. It has the effect of constraining the lower procumbent incisors from occluding on all upper incisors at once and therefore reduces the amount of upper and lower incisor contact compared with grazing macropodines (Fig. 9). The small second and third upper incisors do not constrain the lower incisors to lie within the upper arcade. What this means to the animals ingestive capacity is not known, but it would appear to be associated with a more omnivorous diet. It possibly reflects emphasis on manipulation of individual items between particular upper and lower incisors.

The second characteristic relates to the relative development of the length of the third upper incisor and the position of a labial groove which is variably developed (Ride 1957). Ride noted that, in the evolution of the grazing macropodids from the browsing phalangerids, there has been an increase in the length of the incisor row and particularly the third incisor. Probably related to this is a decrease in the height of the first upper incisors, so that all incisors tend to have the same occlusal plane, and an increased procumbency of the lower incisors in grazing macropodines. Unlike the potoroids (Fig. 2) and browsing macropodines (Fig. 4), in grazers (Fig. 9) the lower incisors occlude over the entire upper arcade, of which the third upper incisor is the largest.

PREMOLARS

Much has been written about the structure of the premolars in the Macropodoidea. It has been important in terms of understanding tooth formulae and replacement patterns. The premolar is also frequently a large and obvious tooth and is particularly well developed in the Potoroidae. No functional analysis of the plagiaulacoid premolar of *Hypsiprymnodon* has been made. However, Stirrat (1981) examined the forces required to crack seeds with different premolars, including the plagiaulacoid premolar of *Burramys.* He concluded that it required significantly less static force to penetrate *Hovea* seeds with a *Burramys* premolar compared to a similar sized *Isoodon* or *Petauroides* premolar. Interestingly, an unworn *Burramys* premolar with the ridges still present took less force to penetrate a seed than a worn premolar where the ridges had been reduced. It would appear that the *Hypsiprymnodon* premolar is adapted to penetrating materials with a resistant skin or cuticle such as seeds or insects. Molar preparation is less important since the contents are easily digestible.

Potoroids, and to a lesser extent macropodines, have an unusual feature in the insertion of the deep masseter into the masseteric canal (Abbie 1939). The effect of this has been discussed (Abbie 1939; Friant 1954; Ride 1957; Sanson 1980) but it is still not clear. Ride (1959) argued that it may serve to improve the mechanical advantage by inserting the muscle as close to the shearing premolar as possible. In addition it may provide a means to control the plane of occlusion of the sectorial premolar by opposing the rotational movement of the ramus, which displaces the tooth. A rotational movement is accommodated by the flexible symphysis. However, the orientation of the deep masseter suggests that, if anything, it will cause medial rotation of the mandible lingually, which will tend to disengage the premolars. The large pterygoid system opposes this and will tend to rotate the lower premolar buccally. By so doing the angle at which the lower premolar meets its upper counterpart, the relief angle, can be adjusted, and the animal may be able to compensate for wear on the premolar. It is significant that species with the largest premolars tend to have the best developed deep masseter and presumably the hypertrophy of the muscle is involved in enhancing its function. Vertical axes through the molars in the potoroids are not aligned. Moving from the anterior molars to the posterior ones, the vertical axis becomes rotated lingually, so that the occlusal plane twists, giving the tooth row a "twisted" appearance. The term "twisted" in this paper refers to this morphology.

Within the remaining Potoroidae the premolars are well developed longitudinal shearing blades (Fig. 2) and species of the three genera, *Potorous, Bettongia* and *Aepyprymnus,* have all been observed to use them, rather than the incisors, to initially prepare particularly resistant items (Morton 1981). The size and configuration of these teeth create potential difficulties in their efficient occlusion so as not to engage the molars and incisors at the same time. Morton (1981) showed that in the Potoroidae molar occlusion involves a medial rotation of the mandible along its anteroposterior axis which sequentially brings the molars into occlusion from front to back. This capacity for medial rotation allows the efficient occlusion of the premolars to be maintained, in spite of wear, by increasingly rotating the tooth and adjusting the relief angle. This could be seen as a consequence of having a medial rotation capacity to allow worn premolars to maintain their efficiency, but it is more likely to be the capitalization of a feature already present in the immediate ancestors, namely a twisted tooth row, which required medial rotation during sequential occlusion of the molars. The medial direction of rotation used to occlude the premolars is only an emphasis of the direction of rotation required for the sequential use of the molars in the twisted row. This system enables the molars and premolars to be brought into effective occlusion without a bodily lateral movement of the mandible. Lateral mandibular movement would not be possible during molar occlusion because the contralateral lower premolar would be blocked by its upper counterpart. Therefore a grinding movement can only be achieved in an anterior or propalinal direction.

The browsing macropodine genera, *Dorcopsis, Dorcopsulus* and *Dendrolagus,* like the Potoroidae, have permanent premolars which exceed the length of the first molars (Fig. 4) and form well developed shearing blades which cross each other with a larger amplitude than the molar shearing edges. It is not clear why these forms have such large premolars or what their precise function is. Browsing macropodids consume a food resource which in terms of its physical structure is relatively heterogeneous when compared to a diet of grass. *Dorcopsis* has been reported to use its sectorial premolar to deal with hard food items in captivity (Menzies 1988).

Other browsing genera, *Wallabia, Setonix,* and *Thylogale,* and some species of *Petrogale,* have relatively smaller premolars, but these species generally include some grass in their diet. Personal observation indicates that *Wallabia bicolor,* like *Dorcopsis*, uses its premolars in much the same fashion that a dog uses its carnassials. Large resistant food items such as twigs or small branches are placed between the premolars and severed. *Petrogale brachyotis* has been observed to use its premolars deliberately to break into *Terminalia* seeds before the fragments were processed by the cheek teeth.

The smaller intermediate grazing *Macropus* species have, relatively, even smaller premolars (Fig. 6) which are variably developed and are sometimes lost. The nailtail wallabies, *Onychogalea,* have also reduced the premolar to a particularly vestigial condition, but dietary information for them and the hare-wallabies, *Lagorchestes* and *Lagostrophus,* is minimal, though what is known indicates a mixed diet of grasses and shrubs for the hare-wallabies and probably grass for the nailtails (Strahan 1983). The premolar reaches its minimum size in the predominantly grazing macropodines (Fig. 9). This reduction has probably occurred several times independently. It has certainly taken place in the genus *Macropus* where, in the large strictly grazing species, it is particularly reduced and is rapidly lost by molar progression. In some cases the permanent premolar is pushed out of occlusion before any wear facets have been developed, indicating that it probably has limited or no function in these forms.

In contrast to the browsers, grazing macropodids are faced with a substrate which is relatively uniform in its physical properties, and there is no obvious advantage in the retention of large sectorial premolars. Whether this is sufficient explanation for the phylogenetic reduction in the size of the premolar to a vestigial condition is doubtful. As will be argued, a small premolar has important consequences in terms of molar progression, but it is not easy to envisage how reduction in the size of the premolar could have been selected for initially. Yet it seems to be smaller in all groups which eat substantial portions of grass.

A possible explanation lies in the interaction between premolar and molar function, and a shift in emphasis from the premolar in potoroids and browsers to the molars in the grazers. Lophodont macropodines with large premolars (e.g., *Dorcopsis* and *Dendrolagus*) have flat molar tooth rows where all the molars occlude at once and, unlike the potoroids, occlude in a fairly precise way (Sanson 1980). In ancestral basal macropodines (Fig. 3) medial rotation would be a disadvantage because it would probably lead to asymmetrical and inefficient wear. Presumably the evolutionary sequence would require the development of a lophodont molar as a selective response to higher levels of structural plant matter. It would most likely occur in animals with a large premolar and a twisted tooth row because the premolar would still be an important part of the ingestive system. Any other scenario requires the reduction of the premolar before selection of the molar adaptation to chew structural plant material and consequently lowers the very emphasis on the premolar. It is interesting that *Dorcopsis* and the fossil *Dorcopsoides* have very large premolars, a flat tooth row and relatively larger posterior molars. I suggest that medial rotation of much magnitude in such forms cannot be accommodated. The insertion of the deep masseter in these forms is reduced, perhaps correlating with a reduced emphasis on premolar function. Effective use of the flat tooth row therefore mitigates against a retained capacity to compensate for wear in the premolars. Selection for a smaller premolar, and its reduced functional importance, accommodates the increasing demands of more precise molar occlusion along the entire molar series at once. The argument is that the evolution of lophodonty shifted the masticatory emphasis, and hence the selective pressure, from premolar to molar function and the development of precise occlusion of the whole molar row; and that this was driven by the incorporation of structural plant material in the diet.

Grazing macropodines require very precise occlusion of their molars (Sanson 1980) and have various possible adaptations to achieve this. *Macropus giganteus* have two layers of enamel in their molars and the outer one, surprisingly, is softer. It is possible that this serves to allow the newly occluding molars to wear in rapidly and produce accurate occlusal facets (Palamara *et al.* 1984). Grazing, therefore, may increase the demand for precise occlusion further and so exert positive selection pressure to reduce the size of the premolar. A synchronous shift to a more homogeneous diet, grass, in which premolar function is no longer necessary as part of the ingestive system and, therefore, provides no competing selective pressure, facilitates this selective change. This hypothesis requires further study, in particular the significance of whether the large premolars in the browsing macropodines need to occlude with the same degree of precision. While it was only a preliminary study, interestingly, Morton (1981) noted that the molars of *Potorous tridactylus* were more variable in size than those of more specialist herbivores, supporting the idea that specialists require greater precision in their occlusion.

MOLARS

Within the modern Potoroidae there is a progressive morphocline of incipient lophodonty from the tubercular *Potorous* molar through *Bettongia* to *Aepyprymnus*. Morton (1981) relates this to the increasing incorporation of cellulose and other plant cell wall components in the diet of the species on the morphocline. Cellulose (76 kg/cm^2) has a higher impact-strength than insect cuticle (22 kg/cm^2) (Rosenberger and Kinsey 1976). Because of the high impact-strength of these materials, compared particularly to fungi and the contents of insects, which are a significant component of the diet of *Potorous,* there is an increased development of shearing edges of *Bettongia* and *Aepyprymnus*, though this does not compare to the macropodines. A similar increasing development of molar shearing edges with a change from an omnivorous diet, including insects and fruit, to a diet containing more structural plant material, is seen in the primates (Kay and Hylander 1978).

Browsing kangaroos are lophodont (Fig. 4). The lophs form transverse ridges, which is unusual. The presence of lophs distinguishes the Macropodidae from the Potoroidae. Presumably the basal macropodines were also lophodont (Fig. 3) or at least tending to that condition. It is argued that because translational movement was confined to a propalinal direction, transverse linking of the basal macropodoid cusps was inevitable. It is interesting that *Aepyprymnus*, which is incipiently lophodont, and presumably on an independent clade, also developed its lophodonty transversely.

The lophs are high in unworn browsers with sharp crests, and the surfaces over which individual molars occlude are large. Sanson (1980) argues that browsing kangaroos occlude their molars in two phases. The first involves the shearing of the major enamel ridges past each other with an anterior motion, presumably dividing the leaf fibre bundles and associated tissue into small fragments. The second phase generates occlusal pressures over large surface areas on the faces of the lophs continuing the anterior motion, but with some lingual deflection. This is obviously adequate, and indeed appropriate, because leaves have their resistant fibre bundles unevenly distributed, and the thin cell-walled mesophyll and epidermal cells can be more effectively ruptured in the second phase of grinding. The molars, therefore, have a fine shearing action and a grinding action. A very similar division of function and process is seen in folivorous primates and other marsupials (Kay and Hylander 1978).

Grasses have their fibre bundles more evenly distributed and in higher levels than dicotyledonous browse. This requires more energy to separate the grass into smaller fragments and this is most economically achieved by shearing, rather than crushing or grinding. It would appear that grazing kangaroos have evolved to concentrate the available occlusal force over a smaller surface area in the second phase of occlusion, thereby increasing the pressure. This has been achieved by the development of extra enamel ridges in the upper and lower molars which prevent the degree of occusal contact found in the browsers. However, these ridges are longitudinal and "link" the lophs. Lateral movement of the mandible in the second phase causes these ridges to shear past each other. Links seem to be either well developed, as in *Macropus,* or hardly developed at all, as in the browsing macropodines. There has been no detailed work quantifying link development, though this might be difficult to do.

The force applied between the teeth, in macropodines with links, is increased by curving the lower tooth row in a reversed Curve of Spee. This curvature is in the vertical plane and should not be confused with the "twisting" of the molar row in potoroids, which is rotational. Intermediate browser/grazers have a moderately curved tooth row (Fig. 6) while the large grazers have quite extreme curvature. Because the upper tooth row is curved in the opposite direction, the upper and lower tooth rows meet at a tangent, and occlusion is restricted to the anterior molars, lowering the overall surface area contact even further than in browsers, and thus increasing possible occlusal pressure.

The grazing kangaroos have enhanced the shearing component and lost the crushing component in molar occlusion. As a result molars are only involved in shearing which is appropriate for a higher fibre diet. A similar pattern is seen in two marsupials, *Pseudocheirus peregrinus* and *Phascolarctos cinerus,* which consume *Eucalyptus* foliage. This foliage is more like grass than most other browse in terms of its fibre level and therefore requires more shearing than grinding. *Pseusocheirus* selects young leaves and it is suggested that, by virtue of its small size, it can avoid fibre bundles. It may therefore reduce the lignified fibre intake compared to the Koala, and consequently has a shearing and grinding component in its occlusal cycle. *Phascolarctos*, which consumes the whole leaf, must deal with the fibre bundles and its teeth have adapted accordingly, with an enhanced shearing component and no grinding component (Lanyon and Sanson 1986).

In grazing macropodids, with wear, the enamel cap of each loph and link is eventually breached exposing the softer dentine. This generates two shearing edges in place of the original one (Sanson 1980; McArthur and Sanson 1988). The links are involved in only the second phase of occlusion, while the lophs are involved in both, and the links only occlude on part of the lophs in the second phase. Therefore, parts of the lophs wear at different relative rates to the links. The relative height of the links to the lophs is critical for the maintenance of effective occlusion. If the height of the link increases, the loph must become relatively higher. The maximum usable height of the crown, at the lophs, is, therefore, constrained by the relative size and geometry of the tooth. There is no advantage in having a particularly high-crowned tooth, like most eutherian grazers, because it operates in a very different manner. Most, if not all, hypsodont grazing molars are prismatic, with the height greater than the length, and with all the enamel shearing ridges at the same level and subject to equal wear. In this case these constraints do not apply. I contend that the use of the word hypsodont to describe the molars of grazing macropods, in which the height is not greater than the length, is misleading and inappropriate.

Evolution of molar progression

The phenomenon of molar progression in kangaroos is well known and has been described in a number of species. For reviews of the literature concerning particular species and its mechanism see Sanson (1980, 1982). Briefly, molar progression seems to be driven by propalinal occlusal pressure and is subject to the amount the animal chews, rather than correlating with chronological age. The amount of chewing affects the rate of wear on the molars. An animal on high quality food may not need, and hence have to chew, as much food as another on a low quality diet. In addition the level of abrasive in the food will independently affect the rate of wear on the teeth. These effects can be discriminated by the use of appropriate wear and molar progression indices. However, the degree of wear will affect the efficiency of a tooth and an animal with worn teeth might have to respond by chewing its food more, thereby increasing the rate of molar progression. The rate of wear and the degree of molar progression, therefore, have a complex relationship as has been shown by McArthur and Sanson (1988).

Extraction of occlusal counterparts stops molar progression in a molar row; remaining molars, which were not yet in occlusion, over-erupt into occlusion, re-initiating molar progression; differential mastication on one side leads to increased molar progression on that side, so older animals have an increased incidence of molar assymetry; and progressive movements are transmitted from occluding teeth to posterior molars not yet in occlusion by the transseptal fibres of the periodontal ligament. However, browsing macropodines with large premolars also occlude their molars propalinally. It is considered that the premolar acts as a buttress and prevents progression. The molars are continually driven forwards against the premolars and, as noted above, the molars are sometimes squeezed out behind the premolars, indicating the latent capacity for progression in these forms. Extraction of the premolar allows the molars to drift forward.

As the premolar reduces in size its root surface area also presumably reduces. It is suggested that, below some threshold size of premolar, which has not been determined, the occlusal forces cause some drift of the whole cheek-tooth row. With further reduction in size the degree of progression is proportional to the size of the premolar which, when very small, can be pushed right out by molar progression. After some point it is possible to envisage that with an increasingly curved tooth row there is considerable selective advantage in reducing the premolar size further so that it provides no impediment to rapid molar progression. This would allow full advantage to be taken of a curved tooth row which, by reducing the number of teeth in occlusion, increases the pressure (see above), and eventually brings all molars into occlusion over the life of the animal. However, it is difficult to see how selection could initially operate to reduce the size of the premolar in order to take advantage of an increasingly curved tooth row. It would seem that molar progression evolved as an inevitable consequence of a gradually reducing premolar. Only after molar progression was taking place could selection operate to increase its rate.

Sanson (1980) points out that molar progression in itself is unlikely to be a device to counteract wear in the grass feeders, because a grazer still wears out only four molars in its life, just as a browser does. The grazer merely wears them out two at a time rather than all at once. Only *Peradorcas* could really be seen to use molar progression as an anti-wear device, and that is because it has extra molars.

Macropodoid grades of dental organization and diet

Dietary preference is very much species specific and drawing any generalization has its dangers. However, the food items consumed by macropodoids can be grouped into six major categories: invertebrates, fruits and seeds, roots and tubers, fungi, shrubs and other dicotyledonous leaves, and grass. Members of the Potoroidae predominantly consume the first three categories of food in different proportions, while the Macropodidae are mainly restricted to shrubs, grass, or a mixture of both (Strahan 1983). Consequently, the potoroids are ingesting and masticating a variety of materials with quite different physical properties than the macropodids, which consume, by comparison, more homogeneous substrates. Within the macropodids, however, the browsers exploit a more physically heterogeneous resource than the grazers. These differences are reflected in both the ingestion apparatus, the incisors and premolars, and the masticating system, the premolars and molars. These changing relationships with diet are described below and are schematically represented and summarized in Figs 2 to 9.

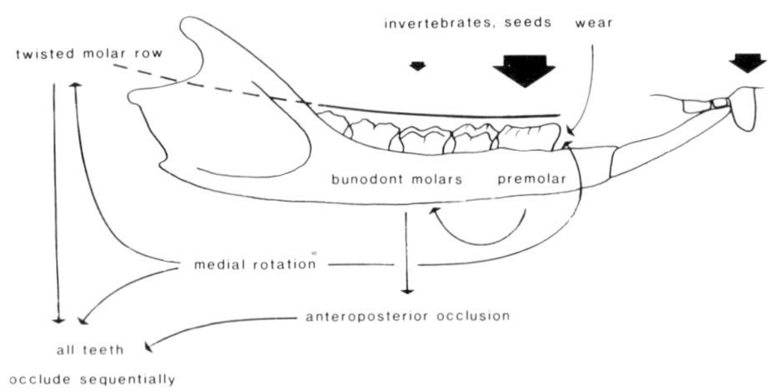

Fig. 2. Schematic representation of a lateral view of the mandible and upper incisors of a hypothetical basal macropodoid and extant potoroids (e.g., *Potorous, Bettongia,* and *Aepyprymnus*) and the interacting factors influencing its morphology. The relative size of the broad arrows indicates relative occlusal emphases within the tooth row. The continuous line above the cheek teeth (molars and premolars) represents the region over which the teeth are in occlusion. The diet of hard individual items is manipulated by the large first incisors and penetrated by the premolars. Wear of the sectorial premolar is compensated for by medial rotation of the mandible, which also sequentially occludes the molars. The large premolar constrains the bunodont molars to anteroposterior occlusion. The combined effect is that all teeth occlude.

BASAL MACROPODOIDS AND POTOROIDS (Fig. 2)

I consider the basal macropodoid grade to be invertebrate-, fruit- and seed-eating. These foods tend to be large items with respect to the size of a tooth, particularly a small potoroid tooth. It is also a material which represents a fairly concentrated package of easily digested nutrients, provided access is achieved. Such items tend to be dealt with individually and precisely. Ingestion is mainly achieved by the precise manipulation of the procumbent lower incisors against the relatively large first upper incisors. Gape must be of sufficient amplitude and the premolars of sufficient relative size to occlude successfully on the dietary item. Consequently there is a long shearing stroke and the whole process is relatively coarse. The process, like any set of shearing blades, such as a carnassial system or a pair of scissors, requires precise co-ordination involving specialized muscles or articulations. In the potoroids and presumably the basal macropodoids, lateral movement of one set of premolars, and hence that mandible, is prevented by the contralateral lower premolar butting up against its counterpart.

It is suggested here that selection for such large and significant premolars constrained molar occlusion to a propalinal direction. Consequently any future development of lophs, which are normally developed at right angles to the occlusal direction for maximum effect, would have to be in a transverse direction. This is unusual in herbivores which mostly have longitudinal lophs and lateral occlusion.

The emphasis in the cheek-tooth row would seem to be on the premolars. Since the outer covering of the dietary items is the major physical barrier, to which the premolars are adapted, the molars are merely involved with crushing and grinding the contents once they have been liberated. The molar row is twisted but the medial rotation capacity permits all molar teeth to occlude sequentially and the premolars to rotate in response to wear.

BASAL MACROPODINES (Fig. 3)

Incorporation of structural plant material as distinct from storage material, probably mainly in the form of roots, involved the development of lophodonty in the molars. In fully developed lophodont forms, probably in the ancestral macropodines, which incorporated even more structural plant matter in the diet, medial rotation of the mandible would have become a disadvantage (see above) and the teeth might have been used inefficiently. It is suggested that straightening of the tooth row reflected increasing selective emphasis on the molars and reduced importance of the premolars. The reduction of the insertion of the deep masseter may be associated with this change.

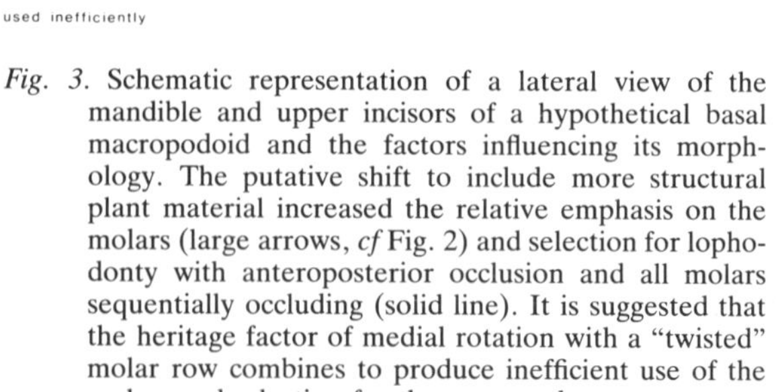

Fig. 3. Schematic representation of a lateral view of the mandible and upper incisors of a hypothetical basal macropodoid and the factors influencing its morphology. The putative shift to include more structural plant material increased the relative emphasis on the molars (large arrows, *cf* Fig. 2) and selection for lophodonty with anteroposterior occlusion and all molars sequentially occluding (solid line). It is suggested that the heritage factor of medial rotation with a "twisted" molar row combines to produce inefficient use of the molars and selection for the next grade.

BROWSERS (Fig. 4)

Browsers consume more, or only, structural plant material, but it is still heterogeneous in nature. The incisors will need to be precise at

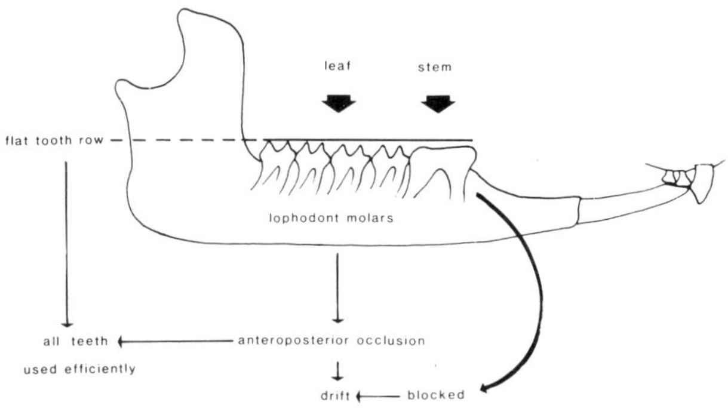

Fig. 4. Schematic representation of a lateral view of the mandible and upper incisors of a macropodine browser (e.g., *Dorcopsis, Dorcopsulus,* and *Dendrolagus*) and the interacting factors influencing its morphology. *Thylogale, Setonix* and *Wallabia* have similar features operating within the mandible but the incisors tend towards the condition seen in Fig. 6. Lophodont molars, all occluding at once (solid line), combined with anteroposterior occlusion, tend to induce anterior drift of the cheek teeth. This is blocked by the large premolar. The broad arrows indicate the regions where components of the diet are prepared, and indicate the relative (and almost equivalent) importance of these regions.

selecting particular items in an array of vegetation. The item is either easily plucked by holding it with the incisors and pulling, or directly severing it. If the leaf is supported by a tough, lignified stem, of a relatively large diameter, a shearing premolar is still advantageous and, being large, will block any tendency for the teeth to drift forward. Unlike an invertebrate or a seed, a dicotyledonous leaf is made up of a series of fairly discrete, thick cell-walled fibre bundles, which must be cut, interspersed with thin cell-walled mesophyll cells, which can be crushed. Browser molars consequently have some shearing capacity, but with a smaller or finer shear amplitude than the large sectorial premolars. In addition they generally have some grinding capacity. Emphasis shifts from the fine control of the premolar to the manipulation of the molars. The tooth row is flat for precise molar occlusion and all teeth occlude at once.

Within the *Petrogale* group (Fig. 5), *P. brachyotis* does not take much or any grass and has a large and permanent premolar (Sanson *et al.* 1985). Some rock wallabies, particularly the ones living in more arid areas, such as *P. xanthopus* which take some grass (Copley and Robinson 1983), have a curved tooth row. However, they also have a large premolar which seems to prevent any molar progression. This means that the posterior teeth often do not occlude. Sometimes the first molar, and more rarely the second molar, is squeezed out from behind the premolar. This allows the posterior molars to drift into occlusion. Such loss does not always occur, and skulls are known of animals that have died in the field in which the first two molars are worn down to the gum. These are presumably very inefficient, while the posterior molars cannot occlude. This suggests that the evolution of the curved tooth row is not necessarily linked to the evolution of molar progression, though after progression is initiated the two might reinforce each other. Some rock wallabies would seem to be poorly adapted to their present diet. In some cases it may even reduce their longevity.

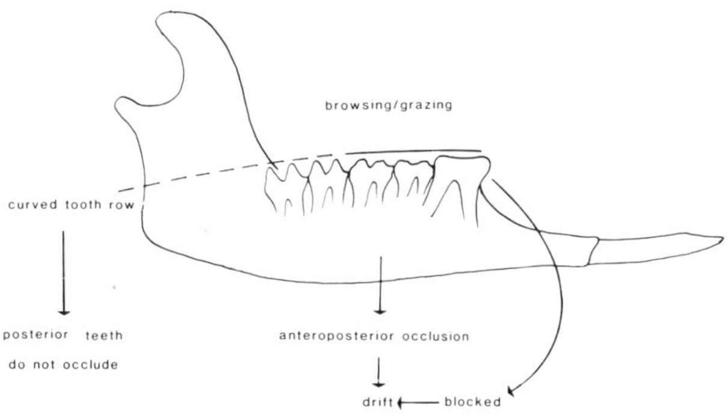

Fig. 5. Schematic representation of a lateral view of the mandible of a macropodine rock-wallaby *(Petrogale)* and the factors influencing its morphology. Grass is included as part of the diet of at least some species. The tooth row is slightly curved, reducing the number of molars in contact (continuous line). The large premolar prevents any forward drift, so not all teeth are used.

INTERMEDIATE BROWSER/GRAZERS (Fig. 6)

In *Lagorchestes, Lagostrophus,* and some species of *Macropus,* grass may constitute a major proportion of the diet at different times and the premolar is reduced further in size. The capacity of intermediate forms to take a mixture of grass and browse successfully requires some explanation otherwise the development of fully adapted grazers would have had to pass through an adaptive trough.

Bensley (1903) originally observed the curvature of the tooth row in the vertical plane in the macropodines. He noted that in the brachyodont forms all the teeth occluded at once. In the hypsodont forms, because of this curvature, not all teeth occluded at once but, because there was tooth progression, eventually all teeth were used. Sanson (1980) pointed out that one effect of this was to reduce the surface area contact in grazing forms. A consequence of this was that the efficiency of the shearing components of the cycle was probably increased. It is also likely that with a reduced surface area contact the forces in the second grinding phase would also be higher. It is therefore significant that forms which include a proportion of grass in their diet, but have not specialized their molars into shearing in both phases, have reduced the number of teeth in contact. Curvature of the tooth row means that only the anterior teeth occlude initially. However,

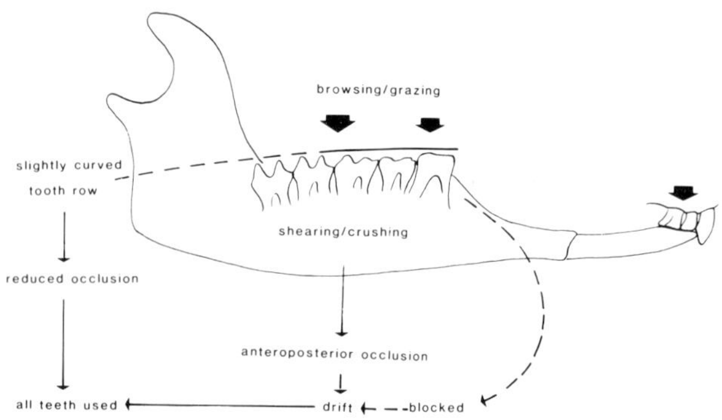

Fig. 6. Schematic representation of a lateral view of the mandible and upper incisors of a macropodine intermediate browser/grazer (e.g., *Macropus (Notamacropus)* and *Lagorchestes*) and the factors influencing its morphology. The broad arrows indicate a greater relative emphasis on the molars compared to the premolars. The molars have a grinding and shearing component. The tooth row is slightly curved, reducing the number of molars in occlusion (continuous line). The premolar is smaller and is less effective at blocking drift, so all molars are used. The third upper incisor is broader.

without some degree of molar progression the posterior teeth cannot be brought into occlusion.

I suggest that with more grass being consumed by animals with browsing-adapted teeth, the slightly curved tooth row compensated for the now inefficent teeth by concentrating the pressure on the anterior molars, thereby emphasizing molar function even more and perhaps restoring some, if not all, of the impaired function. If the premolar reduced to a size that allowed some drift, all the teeth would still be used in a way that enabled some exploitation of an abundant resource. Associated with this, the third incisor becomes broader, increasing the length of contact between the upper and lower incisors and increasing efficiency of cropping, while still retaining some of the manipulative capacity afforded by the first incisors.

Lagostrophus (Fig. 7) is unusual in having a reduced premolar which, therefore, does not block drift, combined with a flat tooth row. In other forms with molar progression the curvature of the tooth row forms an anterior ledge. Teeth approaching the edge of this ledge fall out. *Lagostrophus* appears to lack this ledge and, as might be predicted the whole tooth row drifts forward reducing the length of the diastema (Hughes 1965). This presumably has consequences for the efficiency of the muscle systems as the occlusal load migrates anteriorly.

Peradorcas (Fig. 8), which is sympatric with *P. brachyotis,* consumes grass and browse seasonally (Sanson *et al.* 1985). However, the mixed seasonal feeding regime must impose inefficiencies on the teeth through conflicting demands on the dentition. The curved tooth row may compensate for this to some extent. In such

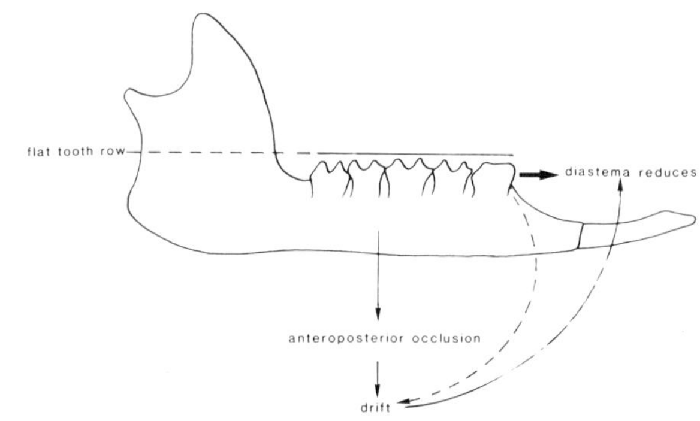

Fig. 7. Drawing of a lateral view of the mandible of *Lagostrophus* and the factors influencing its morphology. All molars occlude (continuous line) but there is a reduced premolar allowing drift, which reduces the diastema.

a condition it would seem that there would be considerable advantage in reducing the size of the premolar allowing the posterior teeth to drift forward. Only *Peradorcas* seems to have done this within the *Petrogale* group. Molar progression is not prevented in *Peradorcas* and all the teeth are used. Since the rate of molar progression is dependent on the amount of chewing, and presumably an inefficient molar means it has to be used more, it might be predicted that *Peradorcas* has rapid molar progression, as is the case. *Peradorcas* is unique among marsupials in having unlimited horizontal tooth replacement. Manatees *Trichechus* spp. also have this capacity. Supernumary molars might seem to be an advantage as a means of coping with extreme wear. However, studies of the enamel ultrastructure indicates that *Peradorcas* enamel is poorly developed with voids in the small crystalline matrix. Not surprisingly the enamel is softer than that in *Macropus* or *Wallabia* (Palamara *et al.* 1984). Soft enamel presumably further reduces the efficiency of the molars, inducing more rapid molar progression.

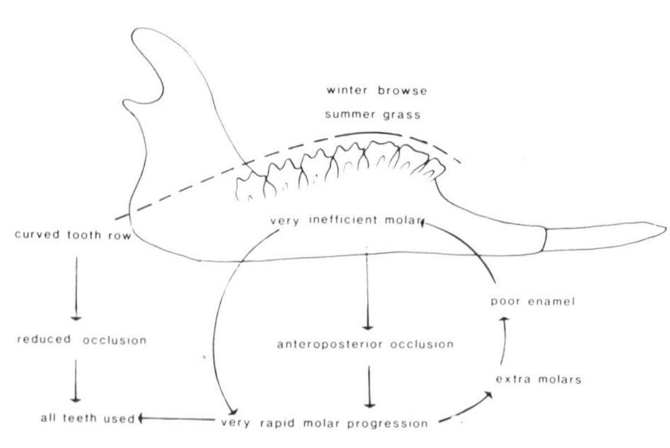

Fig. 8. Drawing of a lateral view of the mandible of the macropodine *Peradorcas,* and the factors influencing its morphology. The tooth row is curved, but the reduced premolar allows molar progression so that all teeth can come into occlusion. The molars have the shearing and grinding capacity of the browser, which is presumably inefficient on grass. The diet is very abrasive, but there are supernumery, though poor quality, molars.

GRAZERS (Fig. 9)

Grazers, in total contrast to the invertebrate- and seed-eaters, consume dietary items which tend to be smaller than the teeth and which are distributed more abundantly and evenly. Ingestion involves the cropping by the incisors of many blades at once. Recently Janis and Ehrhardt (1988) noted that in eutherian herbivores a change from browsing to grazing was also associated with a broadening of muzzle and incisor width. This may relate to the problems of ingesting small grass blades, though Janis and Ehrhardt also discussed the differences in selectivity of browsers and grazers and the need for grazers to consume more material as the substrate is less nutritious. In kangaroos the third incisor width is increased markedly and a longer occlusal contact between the uppers and lowers is produced.

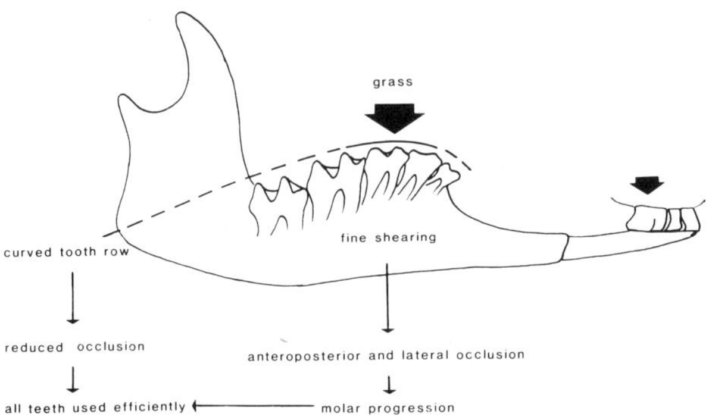

Fig. 9. Schematic representation of a lateral view of the mandible and upper incisors of a macropodine grazer (e.g., *Macropus (Macropus), M. (Osphranter)* and *Onychogalea*) and the factors influencing its morphology. Primary emphasis is on the third upper incisor and the molars, which have an enhanced shearing capacity. The tooth row is very curved, reducing the number of molars in occlusion (continuous line), but the vestigial premolar allows molar progression, so all molars are used.

The premolar plays no part in the acquisition process and becomes useless, losing its effect of retarding molar progression. Many grazers either lose the premolars or molarise them (Butler 1952).

Large quantities of cellulosic material are dealt with simultaneously, unlike the particular treatment of single dietary items in potoroids. Therefore blades which shear in a vertical plane are limited. The general shift in all grazers is through the addition of extra ridges between the existing ones and enhancement of the fine shearing component of molar occlusion. Utilization of these new shearing edges can then be achieved by translational movement in the already existing horizontal plane. However, the anterior motion already present in the macropodine browser has certain disadvantages. As the occlusal stroke progresses from back to front, the occlusal load moves further away from the fulcrum and the insertions of the muscles that provide the force. Other groups with anteroposterior occlusion all have special modifications to deal with the problem of potential loss of mechanical advantage. For example, the rodents have moved the masseter insertions onto the face of the maxilla. In other grazers (and most other browsers) which chew with a lateral motion, the occlusal load moves in an arc around the fulcrum and stays at the same relative distance. The premolars, being reduced in grazing kangaroos, no longer provide an impediment to lateral mandibular movement and so it is interesting that the new enamel ridges developed are longitudinal links between the lophs. These are most effectively used by lateral movements of the mandible, after the initial anterior movement which shears the major lophs past each other. Grazers, which have reduced the premolars to such an extent that they no longer prevent lateral movement, incorporate longitudinal links and lateral movement.

Curvature of the upper and lower tooth rows is increased, reducing the number of teeth in occlusion, and lowering the surface area contact even more. This adaptation is only effective if there is some mechanism for gradually bringing all the teeth into occlusion at some stage in the animal's life. Molar progression does do this and the evolution of this character trait is seen as a mechanism that achieves this.

CONCLUSION

This necessarily brief account of macropodoid dental form and function seeks to establish an argument that incisors, premolars and molars all evolved in response to the physical properties of the diet and occlusal demands. The initial selective response of the premolars in particular had profound consequences on the eventual molar form and, in turn, the development of molar progression. These evolutionary relationships are summarized and presented in Figs 2 to 9. I contend that the complexity of these interactions mitigate against the concepts of a reversal of evolution from grazing to browsing because of the morphological and functional ramifications involved. The argument of an independent origin of the macropodid and potoroid ingestive and masticatory apparatus, with the potoroid molar evolving by reduction, also has difficulties which have to be explained. Taxonomic arguments prior to these have tended to discuss character states as if they were relatively independent and with little effect on each other, although sometimes correlated structures were recognized. This paper seeks to emphasize the complex interactions that may have operated in the evolution of the macropodoid dentition, that parts should not be treated in isolation, and that this has implications for the recognition of polarity and synapomorphies.

It also illustrates the potential of mosaic evolution in this system.

ACKNOWLEDGEMENTS

The hypotheses generated and the conclusions drawn in this paper are the result of many years discussion and development. Over that time I have become indebted to many people whom I cannot all adequately thank. I must, however, mention my wife, Dr. Ann Sanson, Professors J. W. Warren and W. A. Miller, Ass. Professor A. K. Lee, Drs. J. E. Nelson, T. O'Brien, and P. Crabb, and Peter Fell and Graeme Farrington. I would like especially to acknowledge a group whose interest and enthusiasm helped establish an intellectual environment where many of these ideas could be explored and developed; these are Clare McArthur, Dr Grant Norbury, Janet Lanyon, Suzanne Moore, Jan Gipps, Tessa Morton, Simon Stirrat and Diana Hill. The arguments expressed here of course are not necessarily held by them. I would like to thank Clare McArthur, Suzanne Moore, and the referees for their comments on this paper.

REFERENCES

Abbie, A. A., 1939. A masticatory adaptation peculiar to some diprotodont marsupials. *Proc. Zool. Soc. Lond.* **109:** 261-79.

Akin, D. E., Wilson, J. R. and Windham, W. R., 1983. Site and rate of tissue digestion in leaves of C3, C4, and C3/C4 intermediate Panicum species. *Crop Sci.* **23:** 147-55.

Anderson, C., 1927. The incisor teeth of the Macropodinae. *Aust. Zool.* **5:** 105-12.

Archer, A., 1981. A review of the origins and radiation of Australian marsupials. Pp. 1437-488 *in* Ecological Biogeography of Australia ed. by A. Keast. Junk: The Hague.

Archer, A. and Bartholomai, A., 1978. Tertiary mammals of Australia: a synoptic review. *Alcheringa* **2:** 1-19.

Baker, F. and Harris, S. T., 1947. Microbial digestion in the rumen (and caecum), with special reference to the decomposition of structural cellulose. *Nutrition Abstracts and Reviews* **17:** 3-12.

Baker, G., Jones, L. H. P. and Wardrop, I. D., 1959. Cause of wear in sheep's teeth. *Nature* **184:** 1583-84.

Balch, C. C. and Campling, R. C., 1965. Rate of passage of digesta through the ruminant digestive tract. Pp. 108-23 *in* Physiology of Digestion in the Ruminant ed. by R. W. Dougherty. Butterworth: Washington.

Bartholomai, A., 1978. The fossil kangaroos. *Aust. Mammal.* **2:** 15-22.

Bensley, B. A., 1903. On the evolution of the Australian Marsupialia; with remarks on the evolution of the Marsupials in general. *Trans. Linn. Soc. Lond. Zool.* **9:** 83-217.

Butler, P. M., 1952. Molarization of the premolars in the Perissodactyla. *Proc. Zool. Soc. Lond.* **121:** 819-43.

Butler, P. M., 1971. Some functional aspects of molar evolution. *Evolution* **26:** 474-83.

Butler, P. M., 1982. Direction of Evolution in the Mammalian Dentition. Pp. 235-44 *in* Problems of Phylogenetic Reconstruction ed. by K. A. Joysey and A. E. Friday. Systematics Association Special Volume No. 21. Academic Press: London.

Butler, P. M., 1983. Evolution and mammalian dental morphology. *Jour. Biol. Buccale.* **11:** 285-302.

Campling, R. C. and Freer, M., 1962. The effect of specific gravity and size on the mean time of retention of inert particles in the alimentary tract of the cow. *Brit. J. Nutrit.* **16:** 607-18.

Crampton, E. W. and Bell, J. M., 1946. The effect of fineness of grinding on the utilization of oats by market hogs. *J. Anim. Sci.* **5:** 200-10.

Dawson, L. and Flannery, T., 1985. Taxonomic and phylogenetic status of living and fossil kangaroos and wallabies of the genus *Macropus* Shaw (Macropodidae: Marsupialia), with a new subgeneric name for the larger wallabies. *Aust. J. Zool.* **33:** 473-98.

Dea, C., 1987. Molar wear and masticatory efficiency in some Macropodinae. Honours Thesis, Monash University, Melbourne.

Esau, K., 1953. Plant Anatomy. Wiley: New York.

Flannery, T. F., 1983. Revision in the macropodid subfamily Sthenurinae (Marsupialia: Macropodoidea) and the relationships of the species of *Troposodon* and *Lagostrophus. Aust. Mammal.* **6:** 15-28.

Flannery, T. F., Archer, M. and Plane, M., 1983. Middle Miocene kangaroos (Macropodoidea: Marsupialia) from three localities in northern Australia, with a description of two new subfamilies. *BMR J. Aust. Geol. Geophysics* **7:** 287-302.

Flannery, T. F. and Rich, T. H. V., 1986. Macropodoids from the Middle Miocene Namba Formation, South Australia, and the homology of some dental structures in kangaroos. *J. Paleo.* **60:** 418-47.

Friant, M., 1954. Sur les affinites du *Plagiaulax,* Mammifere mesozoique. *Proc. Zool. Soc. Lond.* **124:** 501-07.

Garrod, A. H., 1875. On the kangaroo called *Halmaturus luctuosus* by D'Albertis, and its affinities. *Proc. Zool. Soc.* 48-59.

Gipps, M. and Sanson, G. D., 1984. Mastication and digestion in *Pseudocheirus*. Pp. 237-46 *in* Possums and Gliders ed. by A. Smith and I. D. Hume. Surrey Beatty & Sons: New South Wales.

Gordon, J. E., 1968. The New Science of Strong Materials. Penguin: London.

Greaves, W. S., 1973. The inference of jaw motion from tooth wear facets. *J. Paleontol.* **47:** 1000-01.

Greenhalgh, J. F. D. and Reid, G. W., 1973. The effects of pelleting various diets on intake and digestability in sheep and cattle. *Anim. Prod.* **16:** 223-33.

Hill, D., 1985. Functional Dental Morphology in some small Dasyurids. Honours Thesis, Monash University, Melbourne.

Hughes, R. D., 1965. On the age composition of a small sample of individuals from a population of the Banded Hare Wallaby, *Lagostrophus fasciatus* (Peron and Lesuer). *Aust. J. Zool.* **13:** 75-95.

Hume, I. D., 1982. Digestive physiology and nutrition of marsupials. Cambridge University Press: Great Britain.

Janis, C. M. and Ehrhardt, D., 1988. Correlation of relative muzzle width and relative incisor width with dietary preference in ungulates. *Zool. J. Linn. Soc.* **92:** 267-84.

KAY, R. F. AND HYLANDER, W. L., 1978. The dental structure of mammalian folivores with special reference to Primates and Phalangeroidea (Marsupialia). Pp. 173-91 *in* Biology of Arboreal Folivores ed. by G. G. Montgomery. Smithsonian Institution: Washington.

KREFFT, G., 1875. Remarks on Professor Owen's arrangement of the fossil kangaroos. *Ann. Mag. Nat. Hist.* **15:** 204-09.

LANYON, J. M. AND SANSON, G. D., 1986. Koala *(Phascolarctos cinereus)* dentition and nutrition. II. Implications of tooth wear in nutrition. *J. Zool. Lond.* **209:** 169-81.

LANGER, P., 1979a. Phylogenetic adaptation of the stomach of the Macropodidae Owen, 1839, to food. *Sonderdruck aus Zeitschrift fur Saugertierkunde* **44:** 321-33.

LANGER, P. M., 1979b. Functional gastric anatomy of macropod marsupials. *Ann. Rech. Vet.* **10:** 476-79.

LANGER, P., 1979c. Functional anatomy and ontogenetic development of the stomach in the macropodine species *Thylogale stigmatica* and *Thylogale thetis* (Mammalia: Marsupialia). *Zoomorphologie* **93:** 137-51.

LANGER, P., 1980. Stomach evolution in the Macropodidae Owen, 1839 (Mammalia: Marsupialia). *Sonderdruck aus Z. f. zool. Systematik u. Evolutionsforschung* **18:** 211-32.

LATHAM, M. J., BROOKER, B. E., PETTIPHER, G. L. AND HARRIS, P. J., 1978. Adhesion of *Bacteroides succinogenes* in pure culture and in the presence of *Ruminococcus flavefaciens* to cell walls in leaves of perennial ryegrass *(Lolium perenne)*. *Appl. Environ. Microbiol.* **35:** 1166-73.

LEES, G. L. HOWARTH, R. E. AND GOPLEN, B. P., 1981a. Morphological characteristics of leaves from some legume forages: relation to digestion and mechanical strength. *Can. J. Bot.* **60:** 2126-32.

LEES, G. L., HOWARTH, R. E., GOPLEN, B. P. AND FESSER, A. C., 1981b. Mechanical disruption of leaf tissues and cells in some bloat-causing and bloat-safe forage legumes. *Crop Sci.* **21:** 444-48.

LUCAS, P. W., 1979. The dental-dietary adaptation of mammals. *N. Jb. Geol. Palaont. Mh.* **8:** 486-516.

LUCAS, P. W., 1982. Basic principles of tooth design. Pp. 154-62 *in* Teeth: Form, Function and Evolution ed. by B. Kurten. Columbia University Press: New York.

LUCAS, P. W. AND LUKE, D. A., 1984. Chewing it over: basic principles in food breakdown. Pp. 283-301 *in* Food Acquisition and Processing in Primates ed. by D. J. Chivers, B. A. Wood and A. Bilsborough. Plunum: New York.

MCARTHUR, C. AND SANSON, G. D., 1988. Tooth wear in Eastern Grey Kangaroos *(Macropus giganteus)* and Western Grey Kangaroos *(Macropus fuliginosus)*, and its potential influence on diet selection, digestion and population parameters. *J. Zool. Lond.* **215:** 491-5041.

MCLEOD, M. N. AND MINSON, D. J., 1969. Sources of variation in the *in vitro* digestibility of tropical grasses. *J. Brit. Grassland Soc.* **24:** 244-49.

MENZIES, J. I., 1988. Observations on a captive forest wallaby *(Dorcopsis muelleri)* colony. *The Australian Mammal Society Newsletter,* Winter 1988 p. 56.

MORTON, T., 1981. Mastication in the Potoroinae. Honours Thesis, Monash University, Melbourne.

OWEN, R., 1859. On the classification and geographical distribution of the Mammalia. London.

PALAMARA, J., PHAKEY, P. P., RACHINGER, W. A., SANSON, G. D. AND ORAMS, H. J., 1984. On the nature of the opaque and translucent enamel regions of some Macropodinae (*Macropus giganteus, Wallabia bicolor* and *Peradorcas concinna*). *Cell Tiss. Res.* **238:** 329-37.

PEARCE, G. R. AND MOIR, R. J., 1964. Rumination in sheep. I. The influence of rumination and grinding upon the passage and digestion of food. *Aust. J. Agric. Res.* **15:** 635-42.

RAVEN, H. C. AND GREGORY, W. K., 1946. Adaptive branching of the kangaroo family in relation to habitat. *Am. Mus. Novitates* **1309:** 1-15.

RENSBERGER, J. M., 1973. An occlusion model for mastication and dental wear in herbivorous mammals. *J. Paleont.* **47:** 51-68.

RENSBERGER, J. M., 1975. Function in the cheek tooth evolution of some hypsodont geomyid rodents. *J. Paleont.* **49:** 10-22.

RENSBERGER, J. M., 1978. Scanning electron microscopy of wear and occlusal events in some small herbivores. Pp. 415-38 *in* Development, Function and Evolution of Teeth ed. by P. M. Butler and K. A. Joysey. Academic Press: New York.

RENSBERGER, J. M. AND KOENIGSWALD, W. V., 1980. Functional and phylogenetic interpretation of enamel microstructure in rhinoceroses. *Paleobiology* **6:** 477-95.

RIDE, W. D. L., 1957. *Protemnodon parma* (Waterhouse) and the classification of related wallabies (*Protemnodon, Thylogale,* and *Setonix*). *Proc. Zool. Soc. Lond.* **128:** 327-46.

RIDE, W. D. L., 1959. Mastication and Taxonomy in the macropodine skull. Pp. 33-59 *in* Function and Taxonomic Inportance ed. by A. J. Cain. Systematics Association Publication No. 3.

RIDE, W. D. L., 1962. On the use of generic names for kangaroos and wallabies (Subfamily Macropodinae). *Aust. J. Sci.* **24:** 367-72.

RIDE, W. D. L., 1964. A review of Australian fossil marsupials. *Proc. Roy. Soc. West. Aust.* **47:** 97-131.

RIDE, W. D. L., 1971. On the fossil evidence of the evolution of the Macropodidae. *Aust. Zool.* **16:** 6-16.

RIDE, W. D. L., 1978. An historical introduction to studies on the evolution and phylogeny of the Macropodidae. *Aust. Mammal.* **2:** 1-14.

ROSENBERGER, A. L. AND KINSEY, W. G., 1976. Functional patterns of molar occlusion in the platyrrhine primates. *Am. J. Phys. Anthrop.* **45:** 281-98.

SANSON, G. D., 1978. The evolution and significance of mastication in the Macropodidae. *Aust. Mammal.* **2:** 23-28.

SANSON, G. D., 1980. The morphology and occlusion of the molariform cheek teeth in some Macropodinae (Marsupialia: Macropodidae). *Aust. J. Zool.* **28:** 341-65.

SANSON, G. D., 1982. Evolution of feeding in fossil and recent macropodoids. Pp. 489-506 *in* The Fossil Vertebrate Record of Australasia ed. by P. V. Rich and E. M. Thompson. Monash University: Melbourne.

SANSON, G. D. AND MCARTHUR, C., 1987. Tooth wear and food processing in herbivorous mammals. *Occasional Publication of Aust. Soc. Anim. Prod.* Pp. 61-62.

SANSON, G. D., NELSON, J. E. AND FELL, P., 1985. Ecology of *Peradorcas concinna* in Arnhemland in a wet and a dry season. *Proc. Ecol. Soc. Aust.* **13:** 6-72.

SCHAEFFER, B., 1965. The role of experimentation in the origin of higher levels of organization. *Syst. Zool.* **14:** 317-36.

SCHAEFFER, B. AND HECHT, M. K., 1965. The origin of higher levels of organization: Introduction and historical resume. *Syst. Zool* **14:** 245-48.

SPALINGER, D. E., ROBBINS, C. T. AND HANLEY, T. A., 1986. The assessment of handling time in ruminants: the effects of plant chemical and physical structure on the rate of breakdown of plant particles in the rumen of mule, deer and elk. *Can. J. Zool.* **64**: 312-31.

SIMPSON, G. G., 1959. The nature and origin of supraspecific taxa. *Cold Spring Harbour Symposia on Quantitative Biol.* **24**: 255-71.

STIRRAT, S., 1981. The functional morphology of the plagiaulacoid premolar of *Burramys parvus* Broom. Honours Thesis, Monash University, Melbourne.

STRAHAN, R. (ed.), 1983. Complete Book of Australian Mammals. Angus & Robertson: Sydney

TATE, G. H. H., 1948. Results of the Archbold Expeditions. No. 59. Studies on the anatomy and phylogeny of the Macropodidae (Marsupialia). *Bull. Am. Mus. Nat. Hist.* **91**: 233-351.

TROUGHTON, E., 1947. Furred Animals of Australia. Charles Scribnor: New York.

VINCENT, J. F. V., 1982. The mechanical design of grass. *J. Mat. Sci.* **17**: 856-60.

WATERHOUSE, G. R., 1841. Marsupialia or Pouched Animals. Mammalia Vol. 11. The Naturalists Library: Edinburgh.

WATERHOUSE, G. R., 1846. Natural History of the Mammalia. Bailliere: London.

Feeding ecology of the Macropodoidea

G. L. NORBURY[1,2] G. D. SANSON[1] and A. K. LEE[1]

The preferred habitats of macropodoid species can be broadly categorized as desert plains, open forest, rocky shelters, closed forest, and low, dense vegetation, in order of apparent increasing cover to the animal. Preferred dietary types can be classified as fruit and invertebrates; fungi, roots, tubers and invertebrates; browse (and sometimes fruit); and grasses and herbs. This order is likely to reflect increasing field abundance of food items and their decreasing nutritional quality. Large species are restricted to habitats with low cover, where their preferred foods are highly abundant and of moderately low quality. In contrast, small species occupy all habitats and consume all food types. A highly selected diet may constrain large body size because of limited abundance of preferred foods. Interspecific differences in body size are positively correlated with home-range size. Group size increases as habitat cover decreases. Therefore, food and habitat choice may ultimately determine group size and home-range size. Jarman (1974) argued that, for African antelope, these behavioural characteristics affect the probability of a female's location at a given time and therefore the kind of mating system. This link has yet to be demonstrated for the Macropodoidea.

INTRODUCTION

In a seminal study, Jarman (1974) established relationships between interspecific differences in food and habitat choice, body size, group size and the size of the home range in African antelope. Later he suggested that similar relationships occur in the macropod marsupials, which may be viewed as ecological equivalents of the African antelope (Jarman 1984); however, no attempt has been made to confirm this in equivalent detail to his earlier analysis. Kaufmann (1974a, 1974b) related the social organization of the macropodoids to what little was then known of their feeding ecology. Since then, there have been a number of pertinent studies (e.g., K. A. Johnson 1980; Hollis *et al.* 1986), but it is still not possible to review the feeding ecology of all members of the Macropodoidea. This review summarizes and relates data available for the extant Australian species. From the little that is known of recently extinct forms we conclude that their inclusion would not alter the patterns observed here. A portion of the data comes from studies conducted in habitats that are partly the result of European settlement; for example, the increased interface between forest and improved pastures. Therefore, a note of caution is sounded regarding the representativeness of some of the data to the pristine situation. The sources of information for each species can be found in Table 1, and the data summarized in Table 2.

Analyses using inferential statistics were not attempted because much of the information from the literature was unsuitable for closer scrutiny. Hence, the derived relationships are intended only to indicate broad patterns. We recognize that there are exceptions to many of the generalizations that we will draw. However, we believe that there is heuristic value in identifying these broad patterns.

Habitat choice

We have broadly categorized macropodoid habitats as desert plains, open forest, rocky shelters, closed forest and low, dense vegetation, in order of apparent increasing overhead and lateral cover. As cover declines, the body size of the largest species using the habitat increases (Fig. 1). However, each habitat contains one or more of the smallest macropodoid species (0.5-2.5 kg). In all habitats except the desert plains, the habitat with the lowest cover, there is a more or less continuous range of body sizes.

We see two anomalies in this trend. First, only the small species, *Lagorchestes hirsutus* (1.7 kg), and the large species, *Macropus rufus* (43.2 kg), are found in the desert plains. We believe the absence of medium-sized macropodoids from this habitat may be due to the advantages of either being small enough to shelter under grass tussocks, or being very big and having advantages of temperature regulation and water conservation (Schmidt-Nielson 1979). It could be argued that sub-adult *M. rufus* are medium sized, and so there are medium-sized macropodoids in that habitat. However, these may be above the threshold for tolerance to hot, arid conditions. A smaller species would produce smaller young that may be below this threshold. Newsome (1977) argued that the Red Kangaroo relies on sporadic flush years to sustain its population numbers, pointing out that peaks in the population structure reflected survival of births that had occurred in unusually wet years. While shorter moist periods are vital to the survival of adults in these populations, it is the rarer rainy excesses which regenerate populations sufficiently to tide them over long periods of intermittent drought. Clearly, a larger animal having a greater longevity is better able to survive such conditions than a smaller one. Of course, other macropodoid species occupied

[1]Department of Zoology, Monash University, Clayton, Victoria 3168.

[2]Present Address: Agriculture Protection Board, P.O. Box 522, Carnarvon, Western Australia 6701.

Pages 169-78 in KANGAROOS, WALLABIES AND RAT-KANGAROOS ed. by G. Grigg, P. Jarman and I. Hume. Surrey Beatty & Sons Pty Limited, New South Wales, Australia, 1989.

Table 1. Sources of information for each species used in this study. Strahan (1983) was used for each; however, additional information was obtained from the following sources:

No.	Species	Reference
1	*Hypsiprymnodon moschatus*	Johnson and Strahan (1982)
2	*Potorous tridactylus*	Guiler (1971); Kitchener (1973); Seebeck (1981)
3	*P. longipes*	Hill and Triggs (1985)
4	*Bettongia penicillata*	Kinnear *et al.* (1979); Christensen (1977, 1980)
5	*B. gaimardi*	Johnson and Rose (1983); Rose (1986)
6	*B. lesueur*	Strahan (1983)
7	*Aepyprymnus rufescens*	Home range from Southgate in Schlager (1981); Strahan (1983)
8	*Lagorchestes leporides*	Strahan (1983)
9	*L. conspicillatus*	Johnson and Lyon (1985)
10	*L. hirsutus*	Bolton and Latz (1978)
11	*Lagostrophus fasciatus*	Strahan (1983)
12	*Onychogalea unguifera*	Strahan (1983)
13	*O. fraenata*	Strahan (1983)
14	*Petrogale lateralis*	Strahan (1983)
15	*P. penicillata*	Strahan (1983)
16	*P. inornata*	Strahan (1983)
17	*P. godmani*	Strahan (1983)
18	*P. rothschildi*	Strahan (1983)
18	*P. xanthopus*	Wilson *et al.* (1976); Copley and Robinson (1983)
20	*P. persephone*	Strahan (1983)
21	*P. brachyotis*	Sanson *et al.* (1985)
22	*P. burbidgei*	Strahan (1983)
23	*Peradorcas concinna*	Sanson *et al.* (1985)
24	*Thylogale stigmatica*	Strahan (1983)
25	*T. billardierii*	Strahan (1983)
26	*T. thetis*	Johnson K. A. (1980)
27	*Macropus parma*	Strahan (1983)
28	*Macropus eugenii*	Inns (1980)
29	*M. irma*	Strahan (1983)
30	*M. parryi*	Kaufmann (1974c)
31	*M. dorsalis*	McCann (1976)
32	*M. rufogriseus*	Johnson (1987)
33	*M. agilis*	Johnson P. M. (1980)
34	*M. giganteus*	Taylor (1982); Jarman and Taylor (1983); Jaremovic and Croft (1987)
35	*M. fuliginosus*	Johnson (1983); Priddle (1987); Coulson and Norbury (1988)
36	*M. robustus*	Taylor (1982)
37	*M. antilopinus*	Croft (1982)
38	*M. bernardus*	Strahan (1983)
39	*M. rufus*	Johnson (1983); Oliver (1986); Priddel (1987)
40	*Setonix brachyurus*	Holsworth (1967); Nicholls (1971); Kitchener (1973)
41	*Wallabia bicolor*	Edwards and Ealey (1972); Lee and Cockburn (1985); Hollis *et al.* (1986)
42	*Dendrolagus lumholtzi*	Strahan (1983)
43	*D. bennettianus*	Strahan (1983)

desert plains before European settlement, for example, *Bettongia lesueur* and *Onychogalea fraenata*; but these are only of small body size (maximum 6 kg).

The concept of cover as categorized here is fairly arbitrary and does not adequately take into account the cover that is actually seen by the animal. Spinifex probably provides considerable cover for an animal the size of a hare-wallaby. Cover has not been adjusted to account for differences in animal size because of the extra assumptions involved. In any case, it would appear that the trends described in this paper would not be severely altered if, for a given habitat type, smaller animals were afforded greater cover than larger animals. We do not classify the medium-sized *M. robustus erubescens* as an inhabitant of desert plains because of its use of rocky shelters for cover.

The second apparent anomaly is the absence of large herbivores from closed habitat. Since they exist on other continents (e.g., African Elephant, *Loxodonta africana*; Black Rhinoceros, *Diceros bicornis*; Okapi, *Okapia johnstoni*; and Tapir, *Tapirus* sp.), and probably occurred in Australia in the past (e.g., sthenurines and diprotodonts), their absence cannot be considered a consequence of just size, and the cause of their extinction is of considerable interest.

Food choice, abundance and quality

We have grouped the diets consumed by macropodoids into four major categories and have placed these in order of increasing local abundance in the field: fruit and invertebrates; fungi, roots, tubers and invertebrates; shrub or tree leaves and fruit; and grasses and herbs. As the local abundance of these food items increases, the body size of the largest species increases (Fig. 2). However, each food type is consumed by one or more of the smallest macropodoids. For all food types, except fruit and invertebrates, there is a more or less continuous range of body sizes of macropodoids that eat them. Fruit and invertebrates are eaten only by *Hypsiprymnodon moschatus*, which is remarkable in being the only macropodoid with a simple stomach (Johnson and Strahan 1982) and presumably no fermentative capacity. The largest species, *M. rufus, M. antilopinus* and *M. giganteus*, consume only the most abundant food types. It is likely that none of the large macropodoids eat sparsely distributed foods such as fruit, invertebrates, fungi and roots because large animals require large food intakes. Grasses, herbs and to a lesser extent, browse, support the greatest diversity of species, ranging in size from *Peradorcas concinna* (1.4 kg) to *M. rufus* (43.2 kg). Grasses and herbs probably also support the greatest biomass of macropodoids. Jarman (1974) observed similar relationships between food abundance and species richness for African antelope.

These food categories reflect different levels of structural plant material. In terms of increasing nutritional value, it is suggested here that the order of diet categories reflecting these different levels is: shrubs; grasses and herbs; tree leaves and fruit; fungi, roots, tubers and invertebrates;

Table 2. Results of a literature survey of the feeding ecology of macropodoids. Key to ranks of quality of preferred foods: 1, shrubs; 2, grasses and herbs; 3, tree leaves and fruit; 4, fungi, roots, tubers and invertebrates; 5, fruit and invertebrates. Key to ranks of field abundance of preferred foods: 1, fruit and invertebrates; 2, fungi, roots, tubers and invertebrates; 3, shrubs or tree leaves and fruit; 4, grasses and herbs. Key to ranks of resting habitat cover and cover of feeding habitat: 1, desert plains; 2, open forest; 3, rock shelters; 4, closed forest; 5, low, dense vegetation. Missing data indicated by *.

Species	Weight kg	Quality of preferred foods ranked 1-5	Field abundance of preferred foods ranked 1-5	Home range (ha)	Cover of resting habitat ranked 1-5	Cover of feeding habitat ranked 1-5	Maximum group size whilst feeding	Typical group size
1	0.5	5.0	1.0	*	4.0	4.0	3.0	1.0
2	1.1	4.0	2.0	12.0	5.0	5.0	*	1.0
3	2.0	4.0	2.0	10.0	5.0	5.0	*	*
4	1.3	4.0	2.0	16.0	2.0	2.0	*	*
5	1.7	4.0	2.5	100.0	2.0	*	*	*
6	*	4.0	2.0	*	1.0	*	*	*
7	3.3	2.5	4.0	20.0	2.0	*	3.0	*
8	*	2.0	4.0	*	1.0	*	*	*
9	3.0	1.0	3.0	9.0	1.5	*	3.0	*
10	1.7	2.0	3.5	*	1.0	*	*	*
11	2.0	2.0	3.5	*	5.0	*	*	*
12	6.7	2.0	4.0	*	2.0	*	4.0	*
13	5.0	*	*	*	2.0	1.5	3.0	*
14	4.9	*	*	*	3.0	*	*	*
15	6.6	2.0	4.0	*	3.0	*	*	*
16	4.2	*	*	*	*	*	*	*
17	5.0	*	*	*	*	*	*	*
18	5.3	*	*	*	3.0	*	*	*
19	6.5	1.5	3.0	*	*	*	*	*
20	6.5	2.0	4.0	*	3.0	*	*	*
21	4.5	1.0	3.0	*	3.0	*	*	*
22	1.3	*	*	*	3.0	*	*	*
23	1.4	1.5	3.5	*	3.0	*	2.0	*
24	4.6	2.5	3.5	*	4.0	2.0	4.0	*
25	5.5	1.5	3.5	170.0	4.5	2.0	10.0	*
26	5.4	2.0	3.5	14.0	4.0	2.0	3.0	1.0
27	4.5	2.0	3.5	*	4.5	2.0	3.0	*
28	6.5	2.0	3.5	30.0	4.5	*	3.0	*
29	8.0	2.0	4.0	*	2.0	*	*	*
30	13.5	2.0	4.0	90.0	2.0	1.0	50.0	*
31	11.3	2.0	4.0	*	4.0	*	*	*
32	16.5	2.0	4.0	15.0	2.0	1.0	30.0	*
33	15.0	2.0	4.0	*	2.0	1.0	10.0	1.0
34	26.4	2.0	4.0	317.0	2.0	1.0	*	6.6
35	22.1	2.0	4.0	502.0	2.0	1.0	*	3.5
36	21.3	1.5	3.5	*	3.0	1.0	*	3.1
37	27.2	2.0	4.0	*	2.0	1.0	30.0	2.0
38	18.0	2.0	3.5	*	3.0	*	*	*
39	43.2	2.0	4.0	1 472.0	1.0	1.0	*	3.5
40	3.3	1.0	3.0	4.1	5.0	*	3.0	*
41	15.0	2.0	3.0	5.9	3.5	3.5	3.0	1.0
42	6.7	3.0	3.0	*	4.0	4.0	4.0	*
43	11.0	3.0	3.0	*	4.0	4.0	*	*

*Missing data

and fruit and invertebrates. Shrubs may well have a lower digestibility than grasses because of their secondary plant compounds and higher levels of lignin. Most macropodoid species consume a moderately low quality diet of grasses and herbs (Fig. 3). Only a few small species consume the lowest quality food, shrubs (e.g., *Setonix brachyurus*), and the highest quality food, fruit and invertebrates *(H. moschatus)*. There may be certain costs in selecting food at either end of the food-quality spectrum. On a weight for weight basis, high quality foods such as fruit and invertebrates provide maximum nutritional value but are relatively scarce. Abundant foods like browse are readily accessible but are of low quality. We have already suggested why no large macropodoids can survive on high quality but sparsely distributed foods. However, it is anomalous that very small species such as *Lagorchestes hirsutus* and *S. brachyurus* consume such low quality foods when these animals would have relatively high metabolic requirements. It is possible that as a result of their small size, these species are capable of selecting the most nutritious component of these food types. Certainly *L. hirsutus* has been shown to select the very tips of grasses (Bolton and Latz 1978). Therefore, classifying animals as browsers or grazers may be misleading and an

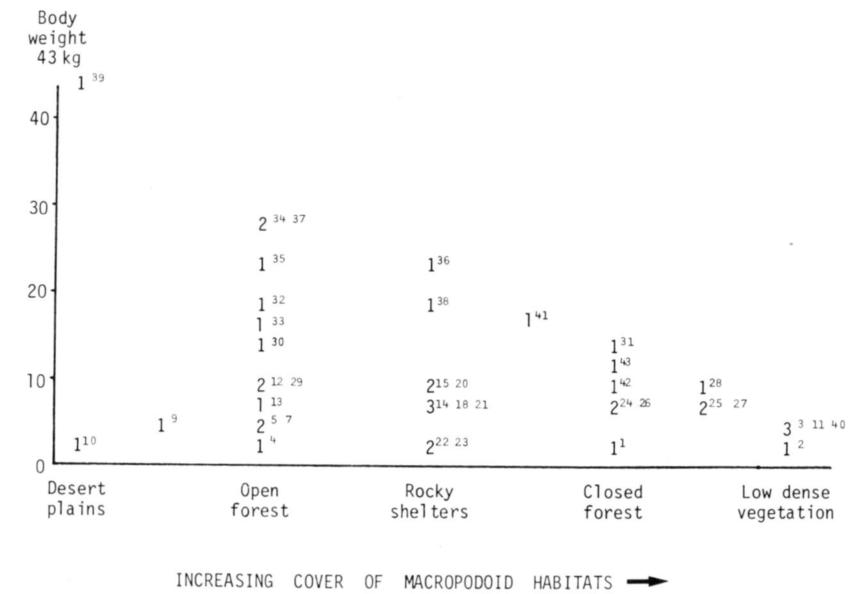

Fig. 1. The relationship between cover of preferred resting habitat and body mass of macropodoids.

overgeneralization in terms of the quality of the consumed food.

There may be constraints on the maximum body size of macropodoids in dense vegetation imposed by increasing overhead and lateral cover. Although the mobility of a large macropodoid would be restricted in dense vegetation, an alternative explanation is that dense vegetation supports smaller animals only because of the limited abundance of preferred food resources (see Fig. 6). By virtue of its size, a small animal is able to survive on relatively scarce, patchily distributed, high quality food. Larger animals require more food and so they consume abundant, evenly distributed, low quality foods. Bell (1971) argued that large body size may be a prerequisite for a low quality diet because larger animals have a relatively lower metabolic rate and a larger gut capacity.

As already pointed out there are no large browsing macropodoids, even though they were present in the Pleistocene (e.g., *Sthenurus*, Sanson 1982). The conventionally held reasons for their extinction are a decline in the abundance of

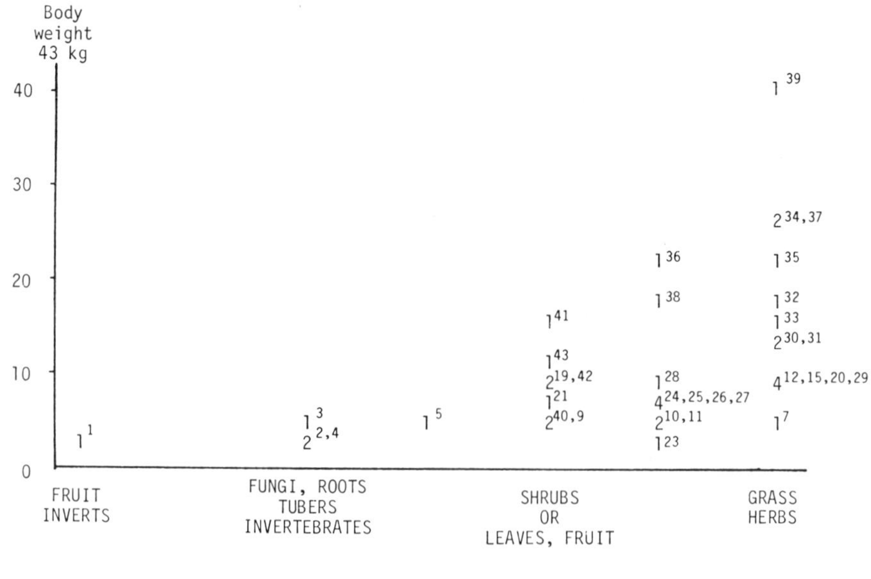

Fig. 2. The relationship between increasing abundance of preferred foods and body mass of macropodoids.

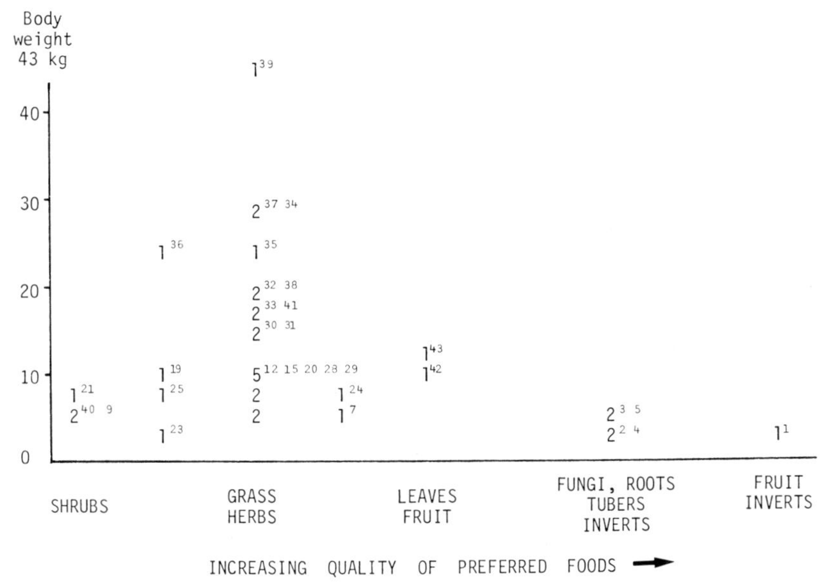

Fig. 3. The relationship between increasing quality of preferred foods and body mass of macropodoids.

browse as the Australian continent became more arid, overkill by Aborigines, or alteration of the flora by different fire regimes introduced by Aborigines (Tindale 1981). The eucalypt forests today are extensive over southeastern Australia. The disappearance of the browsing megafauna from areas that are now arid or open grassland is explicable in terms of a reduction of an appropriate food supply. Their extinction in the rest of Australia, where there is still extensive areas of browse, suggests that the food supply is not appropriate, or that causes other than nutritional ones operated in their extinction. We would question whether browse is so scarce today and suggest that a change in the quality of the browse may have been as important as a change in absolute quantity. The extinct, browsing megafauna may have eaten a higher quality browse, compared to the sclerophyllous browse found in eucalypt forests today. *Wallabia bicolor*, is the largest browser in such areas and its diet is not restricted to sclerophyllous vegetation. Eucalypt foliage, while abundant, is low in protein and rich in plant secondary compounds. Only the hindgut-fermenting *Petauroides, Pseudocheirus* and *Phascolarctos* species successfully utilize such foliage. Bell (1971) pointed out that foregut fermenters tend to consume lower quantities than hindgut fermenters. Is it possible that the extinction of the foregut-fermenting macropodoid megafauna is related to a decline of food quantity of the appropriate quality? The diprotodontoids such as *Diprotodon*, which were presumably hindgut fermenters, while they might have had some physiological advantage in digesting material rich in secondary compounds and tannins compared to their macropodoid competitors, would have required higher intakes, and the quantity may have been limiting.

Jarman (1974) in his review of the feeding ecology of African antelope and more recently in that of the macropodoids (Jarman 1984) indicated essentially similar trends to those just shown. However, the data presented here indicate some important differences that are worth further consideration.

Food abundance, group size and home-range size

Jarman (1974) also related the feeding ecology of antelope to their group size and home-range size. He argued that these behavioural characteristics determined the probability of a female's location at a given place and therefore the kind of strategy a male employs to maximize its breeding success. For the macropodoids there is a negative relationship between maximum group size whilst feeding and the cover of their feeding habitat (Fig. 4). This trend supports the hypothesis which predicts that large groups should form when cover is low in order to share and increase the total vigilance against predators. Jarman (1984) provides data which indicate that in larger groups of Grey Kangaroos each feeding individual spends less time in vigilance. However, this trend does not apply to all macropodoids in that some species, such as *Onychogalea fraenata*, occur in relatively small groups despite apparent low cover.

However, large groups can only form in areas where their food is abundant and evenly distributed. This is particularly true for large species and interestingly there is a relationship between

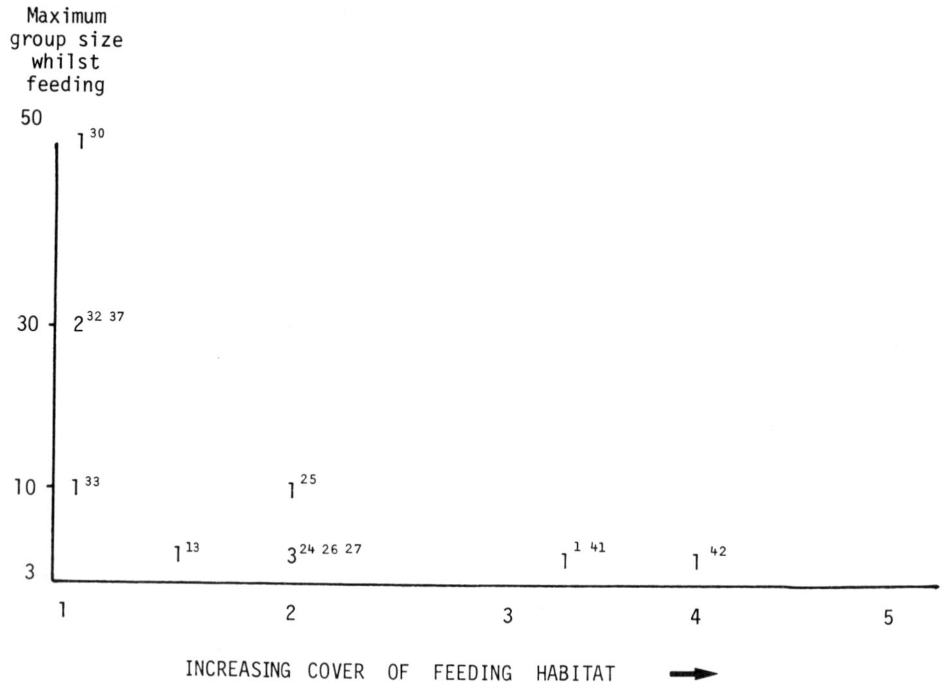

Fig. 4. The relationship between increasing cover of feeding habitat of macropodoids and maximum group size whilst feeding.

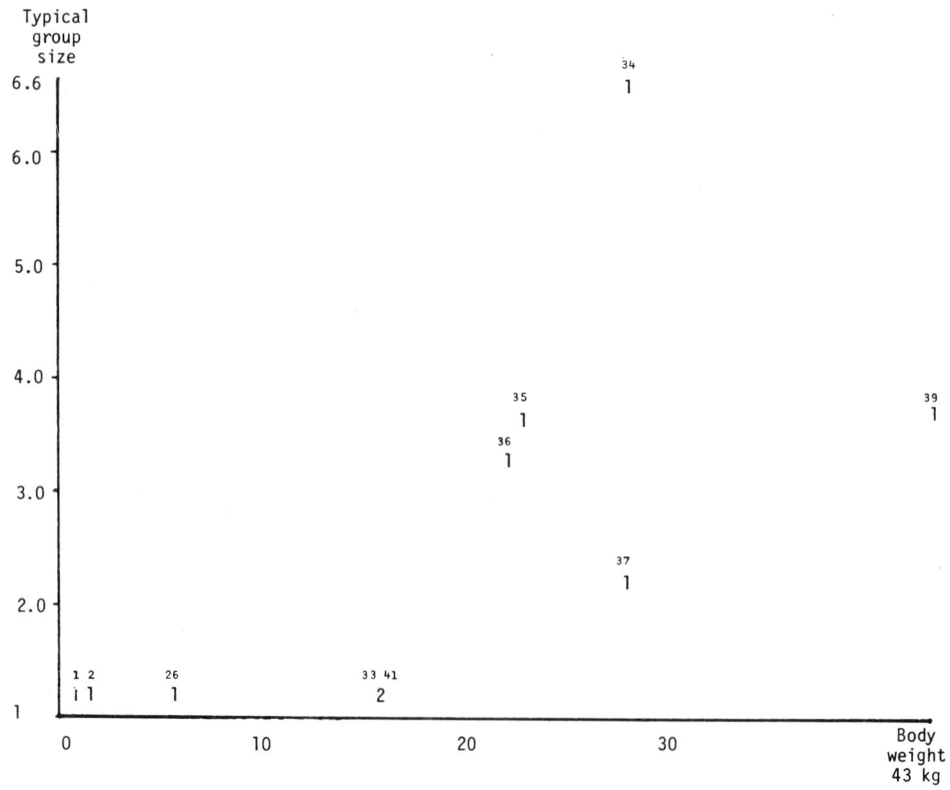

Fig. 5. The relationship between body mass of macropodoids and typical group size.

group size and body mass (Fig. 5). In addition, abundant food supply is associated with low cover (Fig. 6). It is not surprising therefore that typical group size is positively correlated with the abundance of preferred food (Fig. 7) though the data available for establishing the relationship are very limited. What is known suggests that some macropodoid species (e.g., *M. agilis* and *Potorous tridactylus*) occur in small groups independent of food abundance, whereas, as pointed out, larger groups are found, and perhaps can only be found, where food is abundant and evenly distributed. Large groups may form to increase the probability of detecting predators and/or to render each individual a less distinguishable target for attack, provided their preferred food is distributed appropriately. Consequently, large feeding groups might be expected to form only when the cover of their feeding areas is low. Caughley (1964), Heinsohn (1968), Kaufmann (1974b), Johnson, K. A. (1980) and Short (1982) argued that grouping and seeking shelter are anti-predatory behaviours in response to predation by extinct and extant carnivores. However, these arguments, which are causative, are based in part on the observed correlations of group size and cover. We point out that there are other correlations operating and that it may be impossible to distinguish cause and effect. Forming a large group might provide anti-predator advantages; it does not follow that that is why large groups formed. The significance of predation in the evolution of macropodoids is not obvious. Against this, and in spite of the often-made assertion that this continent lacks large predators, Jarman and Coulson (this volume) argue that we need to distinguish between a predator's effect and impact and that there are effective predators in Australia. Further, there were large predators in the Pleistocene (e.g., Archer 1981).

The solitary nature of small macropodoids with highly selected diets (e.g., most potoroids) may be due to intraspecific competition for food or an inability to maintain cohesiveness between individuals in the presence of sparsely distributed food items. Southwell (1984) showed that smaller groups of *M. giganteus* occurred in habitats where food abundance was lower. Therefore, diet may influence group size.

Taylor (1982), Johnson (1983) and Southwell (1984) found a positive relationship between intraspecific differences in group size and local population density of *M. rufus, M. giganteus*, and to some extent, of *M. fuliginosus*. These groups were defined as the number of individuals within 50 m of one another (e.g., Taylor 1982). Therefore, they should not be considered as groups of socially interacting individuals when inter-individual distance may be imposed on animals simply because more have to fit within a certain area. To what extent interspecific differences in local population density of macropodoid species account for group size is unknown.

There is a relationship between body size and home-range size (Fig. 8). McNab (1963) related the size of a mammal's home range to the combined effects of energetic requirements and the abundance of preferred foods. Because larger animals require more energy than smaller animals, they tend to have larger home ranges in which to search for food. It might also be expected that an animal that feeds on sparsely distributed, high-quality food will have a larger home range than an equivalent-sized animal that relies on evenly distributed, low-quality food. Kitchener (1973) attributed a 150-fold increase in the home range of *Potorous tridactylus* compared to that of *S. brachyurus* to the lower abundance

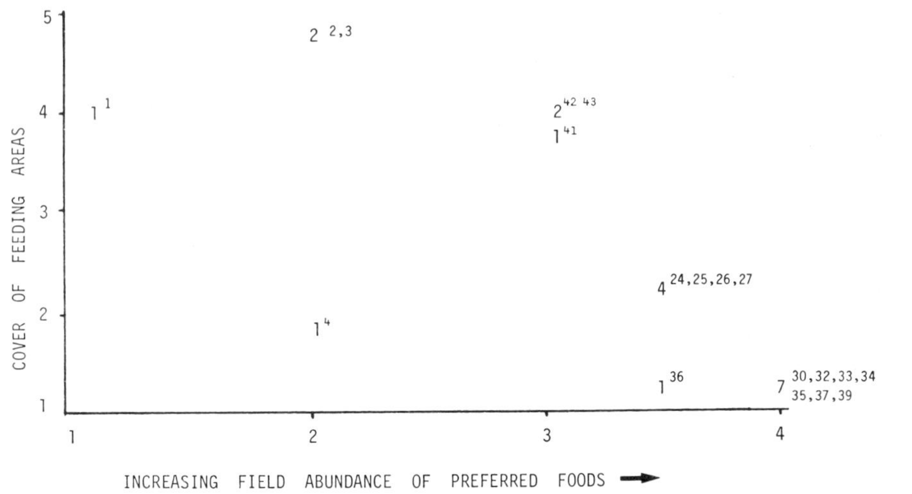

Fig. 6. The relationship between increasing field abundance of preferred foods of macropodoids and the cover of their feeding habitat.

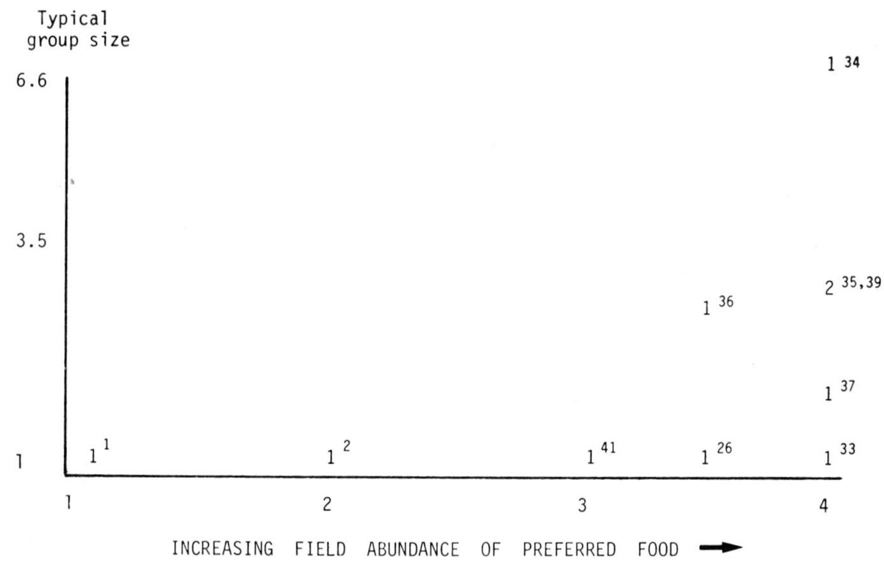

Fig. 7. The relationship between increasing field abundance of preferred food and typical group size of macropodoids.

of preferred food in their habitat. The same can be said for intraspecific differences in home-range size. Inns (1980) suggested that differences in food abundance may account for intraspecific variations in the size of the home range of *M. eugenii*. The larger home-range sizes of *S. brachyurus*, of four and eight hectares, estimated by Holsworth (1967) and Nicholls (1971), respectively, compared to the estimate of 0.13 hectares obtained by Kitchener (1973), may have been due to differences in site. The former two studies were conducted in coastal scrub, whereas the latter was located in a moist swamp area where the vegetation was presumably more digestible. Therefore, home-range size appears to be related to food abundance.

It is clear that group size and home-range size of macropodoids are related to some aspects of their feeding ecology, such as habitat cover and preferred food abundance.

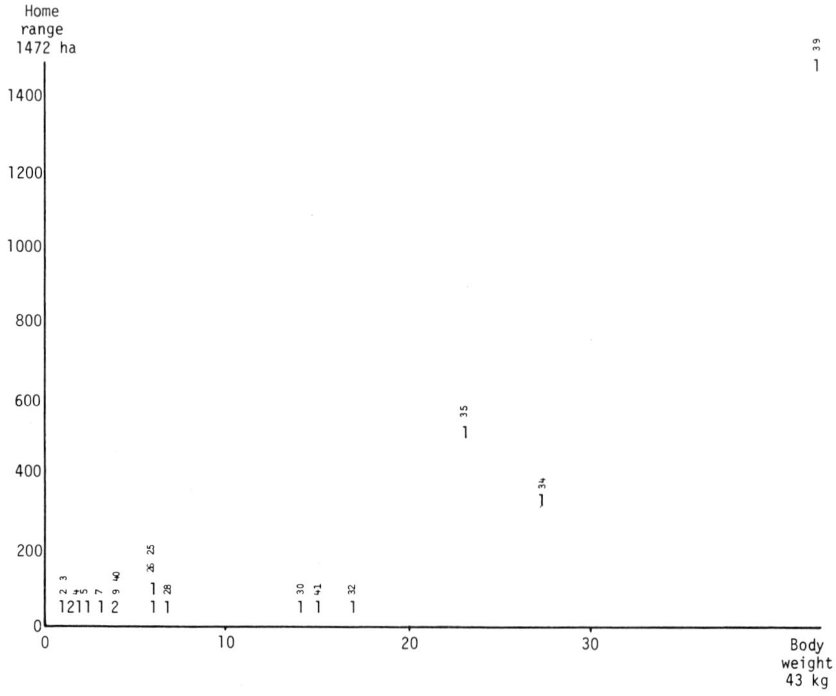

Fig. 8. The relationship between body mass of macropodoids and home range size.

CONCLUSION

Jarman (1974) demonstrated the effect of the type and dispersion of preferred food types on the social organization of African antelope. Interspecific differences in diet selection were proportional to group size, body size, habitat choice and the pattern of movement over the home range. Jarman (1974) argued that these behavioural characteristics determined the probability of a female's location at a given place and therefore the kind of strategy a male employed to maximize its breeding success.

As Jarman (1984) suggested, we can demonstrate similar relationships for the macropodoids. However, the relationships between these factors are complex and their cause and effect are unknown. Whether group size and home-range size are important behavioural characteristics that determine the kind of strategy a male macropodoid employs to maximize its breeding success, and therefore the social organization of macropodoids, remains to be seen.

ACKNOWLEDGEMENTS

We thank Dr. W. Foley, Clare McArthur, Peter Fell and Suzanne Moore for their comments and discussion.

REFERENCES

ARCHER, M., 1981. A review of the origins and radiations of Australian mammals. Pp. 1435-488 *in* Ecological Biogeography of Australia ed. by A. Keast. W. Junk: The Hague.

BELL, R. H. V., 1971. A grazing ecosystem in the Serengeti. *Scient. Amer.* **225:** 86-93.

BOLTON, B. L. AND LATZ, P. K., 1978. The Western Hare-wallaby *Lagorchestes hirsutus* (Gould) (Macropodidae) in the Tanami Desert. *Aust. Wildl. Res.* **5:** 285-93.

CAUGHLEY, G., 1964. Social organization and daily activity of the red kangaroo and the grey kangaroo. *J. Mammal.* **45:** 429-36.

CHRISTENSEN, P. E. S., 1977. The biology of *Bettongia penicillata* and *Macropus eugenii* in relation to fire. Ph.D. Thesis, University of Western Australia, Perth.

CHRISTENSEN, P. E. S., 1980. The biology of *Bettongia penicillata* Grey, 1837, and *Macropus eugenii* (Desmarest, 1817) in relation to fire. Forests Dept. of Western Australia. Bull. 91.

COPLEY, P. B. AND ROBINSON, A. C., 1983. Studies on the Yellow-footed Rock Wallaby, *Petrogale xanthopus* Gray (Marsupialia: Macropodidae). II. Diet. *Aust. Wildl. Res.* **10:** 63-76.

COULSON, G. AND NORBURY, G., 1988. Ecology and management of Western Grey Kangaroos, *Macropus fuliginosus melanops* (Desmarest), at Hattah-Kulkyne National Park. Arthur Rylah Institute for Environmental Research Technical Report Series No. 72.

CROFT, D. B., 1982. Some observations on the behaviour of the Antilopine Walleroo *Macropus antilopinus* (Marsupialia: Macropodidae). *Aust. Mammal.* **5:** 5-13.

EDWARDS, G. P. AND EALEY, E. H. M., 1972. Aspects of the ecology of the Swamp Wallaby, *Wallabia bicolor* (Marsupialia: Macropodidae). *Aust. Mammal.* **1:** 307-17.

GUILER, E. R., 1971. Food of the potoroo (Marsupialia: Macropodidae). *J. Mamm.* **52:** 232-34.

HEINSOHN, G. E., 1968. Habitat requirements of the macropod marsupial *Potorous tridactylus* in Tasmania. *Mammalia* **32:** 30-43.

HILL, F. A. R. AND TRIGGS, B. E., 1985. Ecology and distribution of the long-footed potoroo *(Potorous longipes)* — a second preliminary examination. Department Conservation Forests and Lands. State Forests and Land Service. Research Branch Report No. 310.

HOLLIS, C. J., ROBERTSON, J. D. AND HARDEN, R. H., 1986. Ecology of the swamp wallaby *(Wallabia bicolor)* in north-eastern New South Wales. I. Diet. *Aust. Wildl. Res.* **13:** 355-65.

HOLSWORTH, W. N., 1967. Population dynamics of the Quokka, *Setonix brachyurus*, on the west end of Rottnest Island, Western Australia. *Aust. J. Zool.* **15:** 29-46.

INNS, R. W., 1980. Ecology of the kangaroo Island Wallaby, *Macropus eugenii* (Desmarest), in Flinders Chase National Park, Kangaroo Island. Ph.D. Thesis, University of Adelaide, Adelaide.

JAREMOVIC, R. V. AND CROFT, D. B., 1987. Comparison of techniques to determine eastern grey kangaroo home range. *J. Wildl. Manage.* **51:** 921-30.

JARMAN, P. J., 1974. The social organization of antelope in relation to their ecology. *Behaviour* **48:** 215-67.

JARMAN, P. J., 1984. The dietary ecology of macropod marsupials. *Proc. Nutr. Soc. Aust.* **9:** 82-87.

JARMAN, P. J. AND TAYLOR, R. J., 1983. Ranging of eastern grey kangaroos and wallaroos on a New England pastoral property. *Aust. Wildl . Res.* **10:** 33-38.

JOHNSON, C. N., 1983. Variations in group size and composition in red and western grey kangaroos, *Macropus rufus* (Desmarest) and *M. fuliginosus* (Desmarest). *Aust. Wildl. Res.* **10:** 25-31.

JOHNSON, C. N., 1987. Macropod studies at Wallaby Creek. IV. Home range and movements of the red-necked wallaby. *Aust. Wildl. Res.* **14:** 125-32.

JOHNSON, K. A., 1980. Spatial and temporal use of habitat by the Red-necked Pademelon *Thylogale thetis* (Marsupialia: Macropodidae). *Aust. Wildl. Res.* **7:** 157-66.

JOHNSON, K. A. AND ROSE, R. W., 1983. Tasmanian bettong, *Bettongia gaimardi.* P. 186 *in* Complete Book of Australian Mammals ed. by R. Strahan. Angus & Robertson: Sydney.

JOHNSON, P. M., 1980. Field observations on group compositions in the Agile Wallaby, *Macropus agilis* (Gould) (Marsupialia: Macropodidae). *Aust. J. Wildl. Res.* **7:** 157-66.

JOHNSON, P. M. AND LYON, B. J., 1985. The spectacled hare-wallaby in Queensland. *Qld. Agric. J.* III p. 116.

JOHNSON, P. M. AND STRAHAN, R., 1982. A further description of the Musky Rat-Kangaroo, *Hypsiprymnodon moschatus.* Ramsay, 1876. (Marsupialia, Potoridae), with notes on its biology. *Aust. Zool.* **21:** 27-46.

KAUFMANN, J. H., 1974a. Habitat use and social organization of nine sympatric species of macropodoid marsupials. *J. Mammal.* **55:** 66-80.

KAUFMANN, J. H., 1974b. The ecology and evolution of social organization in the kangaroo family (Macropodidae). *Amer. Zool.* **14:** 51-62.

KAUFMANN, J. H., 1974c. Social ethology of the whiptail wallaby, *Macropus parryi,* in northeastern New South Wales. *Anim. Behav.* **22:** 281-369.

KINNEAR, J. E., COCKSON, A., CHRISTENSEN, R. AND MAIN, A. R., 1979. The nutritional biology of the ruminants and ruminant-like mammals — a new approach. *Comp. Biochem. Physiol.* **64a:** 357-65.

KITCHENER, D. J., 1973. Notes on home range and movement in two small macropods, the Potoroo *(Potorous apicalis)* and the Quokka *(Setonix brachyurus). Mammalia* **37:** 231-40.

LEE, A. K. AND COCKBURN, A., 1985. Evolutionary ecology of marsupials. Monographs on Marsupial Biology. Cambridge University Press: Cambridge.

MCCANN, J., 1976. Distribution and habitat of *Wallabia dorsalis* in the Upper Clarence Region. *Bull. Aust. Mammal. Soc.* (Abstract). **3**(1): p. 44.

MCNAB, B. K., 1963. Bioenergetics and the determination of home range size. *Amer. Nat.* XCVII: 133-40.

NEWSOME, A. E., 1977. Imbalance in the sex ration and age structure of the Red Kangaroo, *Macropus rufus,* in central Australia. Pp. 221-33 *in* The Biology of Marsupials ed. by B. Stonehouse and D. Gilmore. MacMillan: London.

NICHOLLS, D. G., 1971. Daily and seasonal movements of the quokka, *Setonix brachyurus* (Marsupialia), on Rottnest Island. *Aust. J. Zool.* **19:** 215-26.

OLIVER, A. J., 1986. Social organization and dispersal in the red kangaroo. Ph.D. Thesis, Murdoch University, Perth.

PRIDDEL, D., 1987. The mobility and habitat utilization of kangaroos. Pp. 100-118 *in* Kangaroos, their ecology and management in the sheep rangelands of Australia ed. by G. Caughley, N. Shepherd and J. Short. Cambridge Studies in applied ecology and resource management. Cambridge University Press: Cambridge.

ROSE, R. W., 1986. The habitat, distribution and conservation status of the Tasmanian bettong, *Bettongia gaimardi* (Desmarest). *Aust. Wildl. Res.* **13:** 1-6.

SANSON, G. D., 1982. Evolution of feeding in fossil and recent macropodoids. Pp. 489-506 *in* The Fossil Vertebrate Record of Australasia ed. by P. V. Rich and E. M. Thompson. Monash University: Melbourne.

SANSON, G. D., NELSON, J. E. AND FELL, P., 1985. Ecology of *Peradorcas concinna* in Arnhemland in a wet and a dry season. *Proc. Ecol. Soc. Aust.* **13:** 65-72.

SCHLAGER, F. E., 1981. The distribution, status and ecology of the Rufous rat-kangaroo, *Aepyprymnus rufescens* in northern New South Wales. Masters of Natural Resources Thesis, University of New England, Armidale.

SCHMIDT-NIELSON, K. S., 1979. Animal Physiology: adaptation and environment. Cambridge University Press: Cambridge.

SEEBECK, J. H., 1981. *Potorous tridactylus* (Kerr) (Marsupialia: Macropodidae): its distribution, status and habitat preferences in Victoria. *Aust. Wildl. Res.* **8:** 285-300.

SHORT, J., 1982. Habitat requirements of the Brush-tailed Rock-wallaby, *Petrogale penicillata,* in New South Wales. *Aust. Wildl. Res.* **9:** 239-46.

SOUTHWELL, C. J., 1984. Variability in grouping in the eastern grey kangaroo, *Macropus giganteus.* I. Group density and group size. *Aust. Wildl. Res.* **11:** 423-35.

STRAHAN, R. (ed.), 1981. Complete Book of Australian Mammals. Angus & Robertson: Sydney.

TAYLOR, R. J., 1982. Group size in the eastern grey kangaroo, *Macropus giganteus,* and the walleroo, *Macropus robustus. Aust. Wildl. Res.* **9:** 229-37.

TINDALE, N. B., 1981. Prehistory of the Aborigines: some interesting considerations. Pp. 1 761-789 *in* Ecological Biogeograph of Australia ed. by A. Keast. W. Junk: The Hague.

WILSON, G. R., GERRITSEN, J. AND MILTHORPE, P. L., 1976. The yellow-footed rock-wallaby, *Petrogale xanthopus* (Macropodidae), in western New South Wales. *Aust. Wildl. Res.* **3:** 73-78.

Digestive adaptations of kangaroos, wallabies and rat-kangaroos

D. O. FREUDENBERGER[1], I. R. WALLIS[1] and I. D. HUME[2]

The wide range of nutritional niches occupied by macropodoid marsupials provides an opportunity to study the relationships among natural diets, body size and digestive strategies of the dominant herbivores of the Australian vertebrate fauna. The voluminous sacciform forestomach of forest-dwelling potoroine marsupials probably has both a storage and a fermentative function, but in the Macropodidae there is a trend for the sacciform forestomach to decrease in relative size and for the tubiform forestomach to increase. This trend reaches its extreme in the largest grazing kangaroos, and is considered a necessary adaptation to allow for passage through the stomach of less digestible but abundant grasses. Other adaptations that allowed macropodid marsupials to exploit more arid habitats include relatively low metabolic rates and thus low maintenance requirements for energy, nitrogen and water, extensive recycling of endogenous urea to the digestive tract, and in some species, efficient conservation of water.

INTRODUCTION

THE ancestors of today's macropodoid marsupials are thought to have come down out of the forest canopy by the mid Miocene (Flannery 1989). The earliest ancestral forms may have been little different from the omnivorous 500 g Musky Rat-Kangaroo (*Hypsiprymnodon moschatus*) which has a simple dentition and gut morphology (Heighway 1939). Some limited fermentation probably occurred in the hindgut, particularly the caecum, as it does in the present-day omnivorous bandicoots (the Peramelidae) (Hume 1982). At some later stage, the ancestral Macropodoidea developed a fore-stomach in which a microbial symbiosis capable of degrading plant cell walls became established. Except for *H. moschatus*, all extant potoroid and macropodid species have retained a forestomach fermentation chamber. As the Australian climate became increasingly arid in the late Miocene and early Pliocene new nutritional niches appeared and a diversity of morphological and physiological adaptations occurred to exploit these niches. This diversity provides an opportunity to study the relationships between natural diets, body size, and digestive tract structure and function among the super-family Macropodoidea.

DIET, BODY SIZE, METABOLIC REQUIREMENTS AND GUT CAPACITY

Plant matter can be divided on a functional basis into structural (cell walls) and non-structural (cell contents) components. Plant cell walls are mainly cellulose and hemicellulose, digestion of which is dependent on microbial fermentation. On the other hand cell contents are digestible by the herbivore's own enzymes. As a plant matures or is stressed (e.g., by a lack of water) the ratio of cell walls to cell contents increases and the nutritive value of the plant material declines.

There is a relative paucity of information on the food preferences and feeding behaviour of many macropodoid marsupials. However, based on Hume's (1982) review, potoroid and small macropodid species generally ingest diets of higher nutritive value than do the larger (greater than 10 kg) macropodids. A relationship between body size and nutritive value of preferred diets has also been reported for eutherian herbivores (Demment and Van Soest 1985). The explanation for this relationship lies in the different ways in which metabolic requirements and gut capacity change with changes in body size. The total energy requirements of animals increase with increasing body mass, but at a progressively slower rate (Fig. 1). Thus, although large animals require more total energy (kJ/day), small animals require more energy relative to body mass (kJ/kg/day). In contrast, gut capacity tends to be a constant proportion of body mass (Fig. 1).

The non-linear increase in total metabolic rate and the linear increase in gut capacity result in higher ratios of metabolic rate (and thus total nutrient requirements) to plant processing capacity as body size decreases (Demment and Van Soest 1985; Fig. 2). Table 1 illustrates this relationship for macropodids. This means that larger animals are more likely to be able to use plant material of low nutritive value than are small animals. We propose that these relationships govern the form and function of the macropodoid marsupial gut.

POTOROID MARSUPIALS

The consequences of a small body size are pronounced in potoroid marsupials, whose body masses (0.5–3.0 kg) fall below those of most macropodids. At such low body sizes, absolute digestive tract capacity in relation to metabolic requirements is low, and this tends to prohibit the

[1]Department of Biochemistry, Microbiology and Nutrition, University of New England, Armidale, New South Wales 2351.
[2]School of Biological Sciences, University of Sydney, New South Wales 2006.

Pages 179-87 in KANGAROOS, WALLABIES AND RAT-KANGAROOS ed. by G. Grigg, P. Jarman and I. Hume. Surrey Beatty & Sons Pty Limited, New South Wales, Australia, 1989.

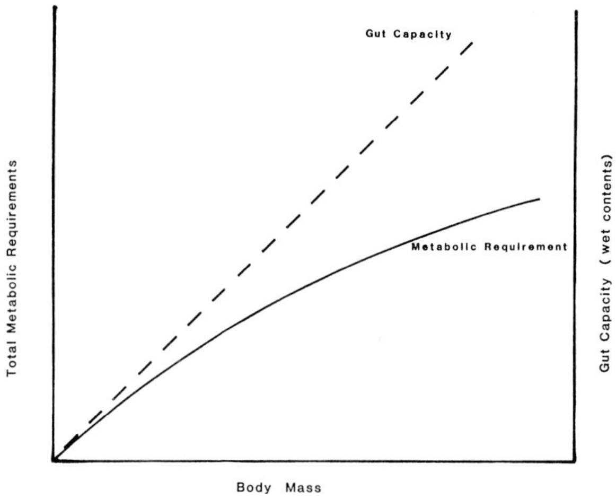

Fig. 1. Conceptual model of the relationship between total metabolic requirement (MR, kJ/d), gut capacity (GC, kg wet contents) and body mass (W, kg) of mammalian herbivores, where MR = 289 $W^{0.75}$ (Kleiber 1975) and GC = 0.116 $W^{1.032}$ (Demment and Van Soest 1985).

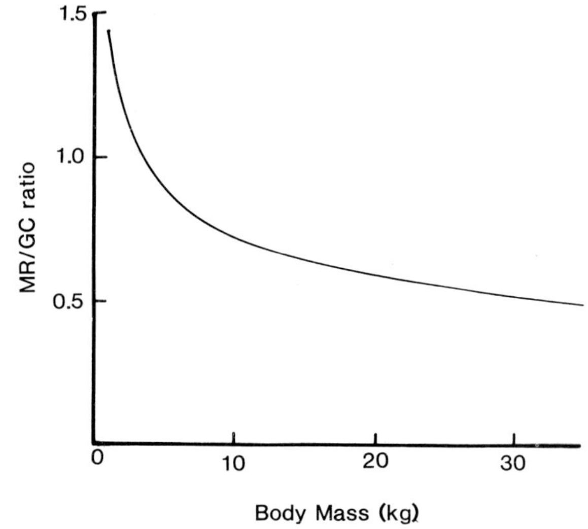

Fig. 2. The basal metabolic rate to gut capacity ratio (MR/GC) as a function of body mass (adapted from Demment 1983).

strategy of plant cell wall fermentation as a sole means of energy acquisition. We suggest that this is one of the reasons why potoroids do not rely solely on a forage diet (i.e., the aerial structural parts of plants). Instead, evidence suggests that they seek diets in which nutrients are more concentrated and as a consequence fibre intake is minimized. Such food items include grass seeds, underground plant storage organs such as rhizomes, tubers and swollen tap roots, hypogeous fungi, plant exudates and invertebrate animal material (Hume 1982; Seebeck 1989).

The dentition of potoroids indicates the nature of their diet (Morton 1981). They lack the molar progression and hypsodonty (i.e., a relative increase in the height of the cheek teeth) seen in those macropodids that are adapted to abrasive diets (Sanson 1989). The hypogeous diet of potoroids is sometimes abrasive, but these animals are observed to be fastidious feeders, manipulating food items with their forepaws and teeth and discarding unwanted components, thereby improving the nutritive value of materials before ingestion (Schlager 1982; Wallis, pers. obs.).

Except for *Hypsiprymnodon,* the stomach of potoroid marsupials can be divided into a voluminous sacciform forestomach, a short tubiform forestomach, and an acid-secreting region, the hindstomach (Fig. 3a). The large sacciform region functions as a primary mixing pool in a manner somewhat like the ruminant forestomach. Thus, when fluid and particulate markers are given orally as a pulse dose, the difference between the mean retention times of the fluid and particulate markers in rat-kangaroos is only 2 hr (Wallis, unpubl. data), similar to that in sheep (Dellow 1982). Mean retention times are also longer than might be expected in such small herbivores. This may explain why fibre digestibility can be high in the potoroid system (Table 2).

Rat-kangaroos are strongly nocturnal, sheltering in grass nests during the day (Wallis *et al.* 1989). This limits the time available for feeding, so that short feeding bouts of high intensity and ingesta storage are probably important features of the feeding ecology of these small marsupials. Radiographic evidence shows that the potoroine

Table 1. Dry matter intake[1], digestive tract capacity[2] and their ratio in four species of macropodid marsupials fed chopped hay.

Species	Ref.	Body mass (kg)	Dry matter intake (g/d)	Dry matter intake (g/kg/d)[3]	Gut capacity (kg)[4]	Ratio of 3 to 4
Macropus eugenii	a	4.8	94	19.6	0.367	53
Thylogale thetis	a	5.0	174	35.8	0.589	61
Thylogale thetis	b	5.5	250	45.5	0.663	69
M. rufogriseus	b	11.3	333	29.5	1.202	25
M. giganteus	a	20.8	549	26.4	2.405	11

[1]In adult animals of species with similar digestibilities the voluntary intake of dry matter reflects their maintenance energy requirements.
[2]In Refererence b this does not include the distal colon.
[3]The mass-specific intake.
[4]Measured as the total weight of digesta in the tract of animals fed *ad libitum*.
Reference a, Dellow and Hume (1982c); b, Hume (1977b).

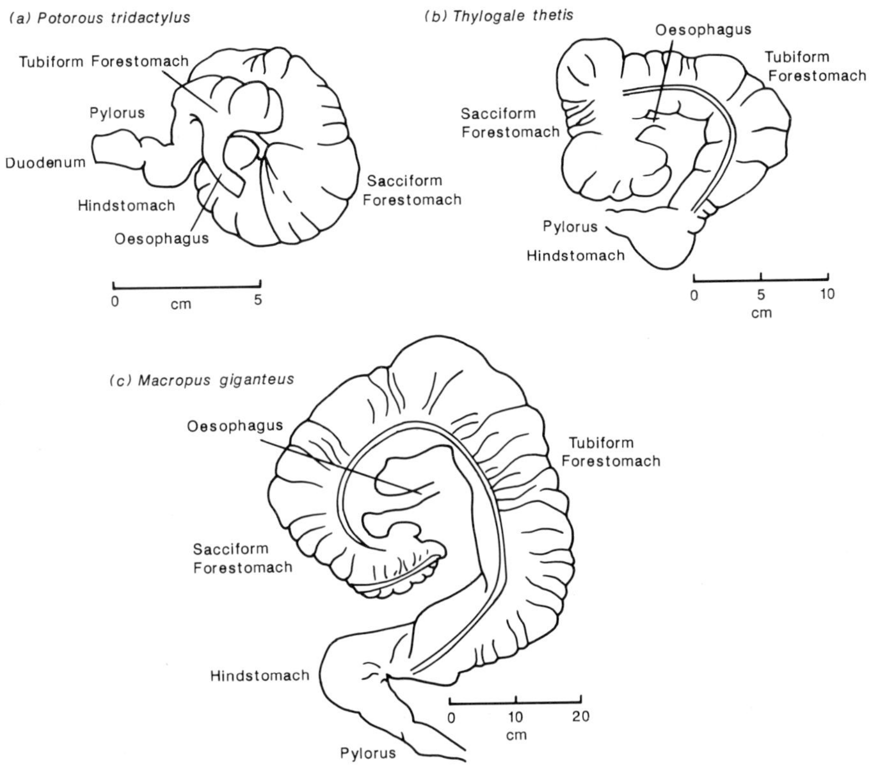

Fig. 3. Stomach morphology of (a) the potoroid marsupial *Potorous tridactylus,* (b) the forest browsing wallaby *Thylogale thetis,* and (c) the large grazing kangaroo *Macropus giganteus.* After Hume and Carlisle (1985) and Hume (1982).

forestomach is used for storage (Hume and Carlisle 1985). At dusk, just before feeding, sacciform forestomach fill is minimal. When *Aepyprymnus* begins to feed, ingesta tend to flow through the tubiform forestomach to the hindstomach and then the small intestine. As the hindstomach fills, ingesta accumulate in the tubiform forestomach and then overflow into the sacciform forestomach. At dawn sacciform forestomach fill is maximal, but it declines during the day through fermentation and through passage of ingesta from the sacciform forestomach through the tubiform forestomach and hindstomach into the small intestine.

In addition to ingesta storage and microbial digestion, the potoroine forestomach fermentation may further benefit the host by detoxifying plant allelochemicals and producing microbial amino acids and B-vitamins that may be deficient in the diet (Kinnear *et al.* 1979).

MACROPODID MARSUPIALS

Unlike the potoroids, all extant macropodids rely primarily on the fermentation of aerial plant parts to supply energy and nutrients. This dietary expansion is associated with several morphological differences between potoroid and macropodid marsupials: increases in body size,

Table 2. Intake and digestibilities of a pelleted diet of 75% chopped lucerne hay and 25% crushed maize, and containing 50% neutral-detergent fibre (NDF). (Freudenberger and Wallis, unpubl.).

	Macropus robustus	*Aepyprymnus rufescens*	*Bettongia penicillata*	*Potorous tridactylus*
n	8	3	2	5
Body mass (kg)	17.6	2.7	1.1	0.9
DM Intake ($g/kg^{0.75}/d$)	56.8	36.9	43.4	35.4
Apparent DM Digestibility (%)	63.5	58.8	56.8	57.8
Digestible DM Intake ($g/kg^{0.75}/d$)	36.1	21.7	24.5	20.6
NDF Digestibility (%)	39.3	59.6	55.5	57.9
ADF Digestibility (%)	29.3	36.0	36.0	37.8

DM = dry matter; NDF = neutral-detergent fibre (cell-wall constituents); ADF = acid-detergent fibre (cellulose plus lignin).

Table 3. Relative capacities of stomach regions of the potoroid marsupial *Aepyprymnus rufescens* and three macropodid marsupials.

	A. rufescens Ref. a	*T. thetis* Ref. b	*M. eugenii* Ref. b	*M. giganteus* Ref. b
Body mass (kg)	3.2	5.8	4.5	19.0
SFS contents				
Stomach contents %	75	51	30	23
TFS contents				
Stomach contents %	14	40	56	70
HS contents				
Stomach contents %	11	9	14	7

SFS = sacciform forestomach; TFS = tubiform forestomach; HS = hindstomach. References: a, Wallis, unpubl. data; b, Dellow and Hume (1982c).

dental specialisation and a larger tubiform forestomach.

General foregut features

The striking feature of the macropodid foregut, compared with the potoroid foregut, is the smaller sacciform forestomach and larger tubiform forestomach. There is also a trend within the Macropodidae toward increasing relative size of the tubiform forestomach with increasing body size (Hume 1984; Table 3, Fig. 3).

Microbial fermentation occurs throughout the sacciform and tubiform regions (Dellow *et al.* 1983). The musculature of the forestomach wall is organized into three longitudinal bands (taeniae). Semi-lunar folds between the taeniae form the haustrations that give the stomach its colon-like appearance (Langer *et al.* 1980). Contractions of the haustra propel digesta caudally in such a way that fluid is expressed through the particulate ingesta. As a result, the 50% excretion time of fluid in the digestive tract is much less than that of particles (12 hr versus 23 hr in *T. thetis* fed chopped lucerne hay) (Dellow 1982).

Chemical reactor theory identifies the ideal reactor vessel configuration for converting reactants to products in a minimum volume and time (Levenspiel 1972). If animal digestive tracts are assumed to function as chemical reactors, the theory can be used to predict functional differences between sacciform and tubiform forestomachs. Penry and Jumars (1987) used two reactor models, the continuous-flow, stirred tank reactor (CSTR) and the plug-flow reactor (PFR) to explain the function of herbivore digestive tracts. A CSTR is characterized by perfect mixing within and continuous flow through a spherical vessel. A PFR features incomplete mixing and orderly flow through a usually tubular vessel. Digestion within a CSTR maximizes plant fibre fermentation within a minimum volume, but may not maximize throughput (the minimum processing time of one gut volume of food). Within a PFR fibre digestion is less, but throughput per unit volume is greater, than in a CSTR. We propose that the potoroid sacciform forestomach is like a CSTR, whereas the predominantly tubiform fore-stomach of macropodids is more like a PFR.

The nature of a CSTR may account for the greater fibre digestibility of a high quality diet in *Aepyprymnus rufescens, Bettongia penicillata* and *Potorous tridactylus* compared with *Macropus robustus* recorded in Table 2. In this comparison the pelleted dietary formulations for the macropodid and potoroid diets were the same, but the chopped lucerne hay incorporated in the potoroid diet was of lower quality (50% neutral-detergent fibre) than that fed to the macropodids (36% NDF). This explains the lower dry matter digestibility in the potoroids. The higher fibre digestibilities in the potoroids do not mean that rat-kangaroos select and digest natural forages of such high fibre content. In this experiment pelleting the diet minimized dietary selection and considerably reduced feed particle size. It is thus difficult to extrapolate to field conditions. However, these results indicate that potoroids have considerable potential for fibre digestion, but the degree to which they rely on this for energy in their natural habitat is not known.

The largest macropodid forestomach compartment is generally the tubiform region (Table 3). The tubiform forestomach can be thought of as a PFR modified by semi-lunar folds between the taeniae (Langer 1988). These folds reduce material flow and achieve some axial mixing, which is not typical of a true PFR. These modifications may enhance fermentation without the slow throughput of a CSTR. We suggest that the prominent tubiform region of the forestomach of large macropodids (Table 3, Fig. 3) allows for greater flow of poorly digested forage than could be expected in a CSTR-type forestomach.

Forest herbivory

The early macropodids were inhabitants of wet forest (Flannery 1984). Some of their digestive features may have been retained by extant genera such as the pademelons (*Thylogale* spp.). Like the ancestral macropodids, pademelons are bigger than the potoroids and have different teeth adaptations. The pademelons are considered to have a lophodont, browing grade dentition which maximizes occlusal surface area and results in a crushing action suitable for soft, non-abrasive forage (Sanson 1989).

The pademelon forestomach is mostly sacciform (Table 3, Fig. 3) which was probably true of ancestral macropodids. However, about 40% is tubiform and we propose that this is a derived feature to facilitate the passage of less digestible plant material.

Tropical browse contains a wide range of potentially toxic plant allelochemicals (Waterman *et al.* 1980). Some rumen microbes are capable of metabolizing toxic plant compounds (Barry and Blaney 1987). It follows that a symbiosis with detoxifying microbial populations may have been a critical adaptation for the exploitation of browse. However, the metabolism of plant allelochemicals by browsing macropodids has not been studied, although there is evidence that Swamp Wallabies consume plants toxic to domestic stock (Edwards and Ealey 1975).

Forest ecotone herbivory

As the Australian climate began to dry out, the forest receded, creating ecotones between forests and grasslands. These ecotones are nutritionally diverse and currently support a number of medium-sized macropodids such as the Swamp Wallaby (*Wallabia bicolor*), rock-wallabies (*Petrogale* spp.) and other wallabies (e.g., *M. rufogriseus*). Using the terminology that Hofmann (1973) applied to East African ruminants, some of the forest edge macropodids may be termed intermediate feeders of both browse and grasses. Within the forest edge group, Swamp Wallabies have the ancestral browsing dentition. The other forest-edge animals have a derived grazing-grade dentition, which has a reduced area of contact, resulting in an effective cutting action (Sanson 1989). The unifying feature of the forest-edge macropodids is an enlarged tubiform forestomach and the presence of a gastric sulcus (Dellow 1979). The enlarged tubiform forestomach may allow for increased flow of less digestible but more abundant grasses (Hume 1984).

Grassland herbivory

Macropodid body size increased dramatically during the Pliocene and Pleistocene epochs (Flannery 1984). This size increase probably corresponded with the expansion of grasslands of low nutritive value, and may indicate an evolutionary pressure to minimize the ratio of total metabolic rate to gut capacity so that this new resource could be exploited. An increase in body size as a response to reduced feed quality is seen in other groups of mammalian herbivores such as African ungulates (Hofmann 1973) and baboons (Demment 1983).

The grazing macropodids tend to have the most reduced sacciform but most expanded tubiform forestomach (Table 3, Fig. 3). This, in combination with their greater body size, may have allowed them to exploit seasonally poor quality but abundant grasslands. Preliminary evidence supports this view. Hollis (1984) found that when changed from a high-quality chopped lucerne hay diet to poor-quality chopped oaten straw, the dry matter intake of 6 kg Tammar Wallabies (*M. eugenii*) fell by 50% whereas that of 20 kg Wallaroos (*M. robustus robustus*) fell by only 17%.

The reduction of the sacciform forestomach in grazing macropodids has not necessarily led to reduced fibre digestibility. The greater absolute gut size in the larger grazing macropodids allows for digesta retention times as long as those in smaller macropodids with a relatively large sacciform forestomach. For example, 50% excretion times and fibre digestibilities were similar between 5 kg wallabies and 20 kg kangaroos (Dellow 1982; Dellow and Hume 1982a) (Table 4).

Mature grasses are often low in protein and this could have been a major hindrance to the exploitation of grasslands. However, in macropodids urea, an end-product of protein catabolism, is recycled to the gastro-intestinal tract, including the forestomach, where microbes can utilize urea-nitrogen for protein synthesis (Hume 1982). Microbial cells are a well-balanced source of amino acids, and are digested and absorbed in the host's small intestine. Thus recycling of urea can reduce the need for dietary protein. Kennedy and Hume (1978) showed that incorporation of urea into non-ammonia nitrogen in the gut of

Table 4. Excretion times and fibre digestibility in kangaroos and wallabies fed chopped lucerne hay.

Species	Body mass (kg)	50% excretion time (hr) Fluid	Particles	Fibre[1] Digestibility (%)
T. thetis	4.9	12	23	38
M. eugenii	4.8	15	24	39
M. robustus robustus	16.6	19	33	40
M. giganteus	20.8	14	30	39

From Dellow (1982) and Dellow and Hume (1982a). [1]Acid-detergent fibre.

Tammar Wallabies fed a low-nitrogen diet was 103-112% of dietary nitrogen intake.

Browsing and intermediate (forest-edge) macropodids also benefit from urea recycling (Dellow and Hume 1982b). However, this is of less nutritional importance in the presence of abundant green forage in which nitrogen is not usually limiting.

The maintenance requirement for truly digestible nitrogen is substantially less in Eastern Grey Kangaroos (*M. giganteus*) than in Red-necked Pademelons (*Thylogale thetis*) (270 vs 530 mg N/kg$^{0.75}$/d, Foley *et al.* 1980; Hume 1977a). However, the basis for this difference remains unexplained since it was shown by Dellow and Hume (1982b) that the extent of urea recycling in these different macropodid species was similar, at least on diets of adequate digestible energy content.

Salivary secretions in grazing kangaroos contain significant sulphur and phosphate (Forbes and Tribe 1969; Beal 1987) which, like nitrogen, tend to be deficient in mature grasses. Saliva may be an important source of sulphur and phosphate for growth of forestomach microbes, and thus may have a strong influence on fibre digestion. Beal (1989) suggests that the higher digestive efficiency of the Grey Kangaroo over the Euro and the Red Kangaroo (Prince 1976) may be partially due to higher salivary flows and concentrations of urea and phosphate in parotid saliva from the former species.

The hind-gut (the caecum and colon) may have undergone considerable changes from the ancestral browsers to extant large grazing species. It appears that in the large grazers the length of the caecum and particularly the colon has increased relative to body weight (Osawa 1987) (Fig. 4), providing additional capacity for microbial digestion of plant cell walls.

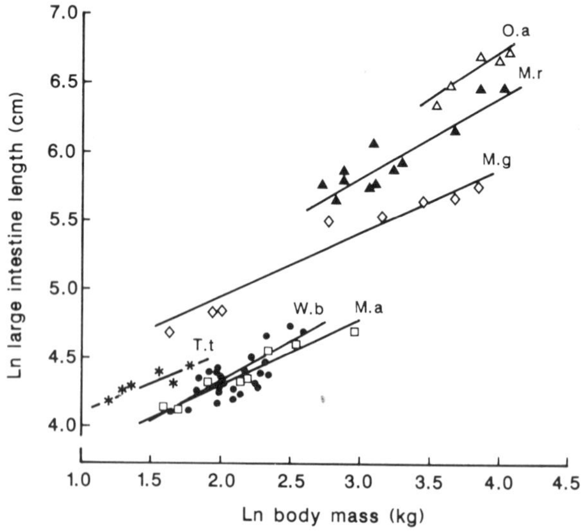

Fig. 4. Relationship between large intestine length and body mass in *Thylogale thetis* (T.t), *Wallabia bicolor* (W.b), *Macropus agilis* (M.a.), *Macropus giganteus* (M.g), *Macropus robustus* (M.r) and *Ovis aries* (O.a). From Osawa (1987).

Arid zone herbivory

Additional digestive tract changes were probably required for exploitation of arid-zone grasslands by macropodids. Water is essential for the maintenance of homeostasis as well as for microbial function. Failure to maintain water intake results in an immediate decline in feed intake (Freudenberger and Hume 1987). In laboratory studies, the sedentary arid-zone Euro (*M. robustus erubescens*) needed less water per gram of dry matter intake than the Eastern Wallaroo (*M. robustus robustus*) during both *ad libitum* (1.68 vs 2.52 ml/gm DM/kg$^{0.80}$/d) and water-restricted conditions (1.09 vs 1.30). This suggests that the Euro uses water more efficiently than the Wallaroo.

These animals lose most water from their faeces. The Euro, when given limited water, produced drier faeces (43% DM) than the Wallaroo (30% DM) — a saving of 5 ml/kg$^{0.8}$/d. This is probably achieved in the Euro by greater water resorption in a longer colon (3.3 vs 2.0 cm/kg$^{0.75}$) (Freudenberger, unpubl. data). Euros may also be able to preferentially maintain gut water during water restriction at the expense of interstitial water, as suggested by Denny and Dawson (1975).

The Euro also appears to have lower maintenance requirements for dietary nitrogen than other macropods. The truly digestible N requirement for Euros is 33% less than for Eastern Wallaroos and 41% less than for Eastern Grey Kangaroos (Brown and Main 1967; Foley *et al.* 1980). This may be a further adaptation to arid-zone herbivory.

RUMINANT COMPARISONS

The radiation of the artiodactyls, including the ruminants, began at about the same time as that of the macropodids, and was a response to similar environmental changes (Hume and Warner 1980). Thus the structure and function of the digestive tract in extant species of ruminants and macropodids is of comparative evolutionary interest and important to the understanding of the impact of introduced ruminants on the Australian flora and fauna.

One fundamental difference between the ruminants and macropodids is that, as eutherians, many ruminants have maintenance energy requirements that exceed those of similar sized macropodids by 20-30% (Hume 1982). Consequently, ruminants must either eat more, or digest more of what they eat. In general ruminants do both. When compared with various macropodids, sheep consistently consumed more chopped lucerne hay (Hume 1982). Similarly, goats consumed more lucerne than did Euros (Freudenberger, unpubl. data).

Fibre digestion depends on physical attachment of cellulolytic bacteria and fungi to plant particles (Bauchop 1979; Cheng *et al.* 1983/84). Fibre digestion is enhanced through increasing the surface area of ingested forage particles by mastication. Mechanical reduction of plant particles is greater in ruminants than macropodids. There is limited reduction in particle size between the abomasum and rectum in ruminants (Poppi and Norton 1980) and between the hindstomach and rectum in large macropodids (Freudenberger, unpubl. data). Therefore the distribution of particle sizes in faeces is indicative of the range of particles exposed to microbial attachment in the foregut. In goats 99% of wet-sieved faecal particles passed through a 1 200 μm screen, whereas in Wallaroos fed the same diet only 82% passed through the same screen (Fig. 5). This greater reduction of particle size in goats is due in part to the ruminant's ability to selectively retain and remasticate large particles (rumination) in the presence of micro-organisms (Sutherland 1988). We suggest that these factors help to explain the greater fibre digestibility usually observed in ruminants.

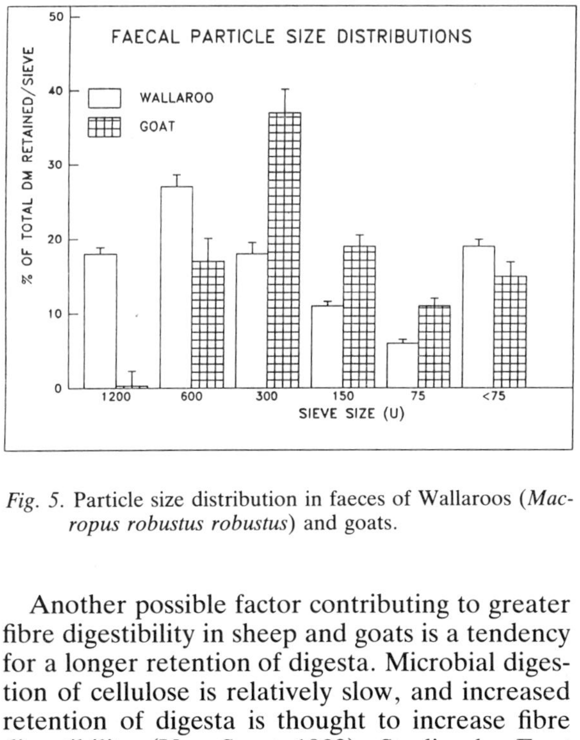

Fig. 5. Particle size distribution in faeces of Wallaroos (*Macropus robustus robustus*) and goats.

Another possible factor contributing to greater fibre digestibility in sheep and goats is a tendency for a longer retention of digesta. Microbial digestion of cellulose is relatively slow, and increased retention of digesta is thought to increase fibre digestibility (Van Soest 1982). Studies by Foot and Romberg (1965), Forbes and Tribe (1970), and Dellow (1982) have shown, with few exceptions, that digesta retention is significantly longer in sheep than in kangaroos.

Differences in retention time are probably due to differences in digesta flow. In large macropodids, there is tubular flow with only local mixing of digesta. In ruminants, retention of digesta is prolonged and fibre digestion is maximized by more complete mixing within a primary pool resembling a continuously-stirred tank reactor (Penry and Jumars 1987). There is a clear difference between macropodids and sheep in the excretion of particulate and fluid digesta. There is a 17 h difference between 90% excretion times for fluid and particles in Grey Kangaroos fed chopped lucerne hay, but only a 6 h difference in sheep fed the same diet. In each case, fluid is excreted more rapidly than particles (Dellow 1982). The effect of rapid flow of fluid on fibre digestion and on the possible flow of soluble carbohydrates to the small intestine is unknown.

Fibre intake and digestion are closely linked with supply of nitrogen to gut microbes. A deficiency in nitrogen supply results in poor microbial growth and consequently reduced fibre digestion and intake (Van Soest 1982). The maintenance nitrogen requirements of many macropodids are less than those of sheep (Hume 1982). However, the interactions among nitrogen requirements, fibre digestion and feed intake are unclear and the subject of current research (Freudenberger, unpubl.). Field evidence suggests that the Euro's lower nitrogen requirement allows it to survive in arid regions of the Pilbara which, because of rangeland degradation, can no longer support sheep (Ealey 1967).

It has been suggested that a tubular forestomach allows macropodids to eat more poor quality forage than can ruminants of similar size (Hume 1984). Dry matter intakes of chopped oaten straw, compared to intakes of chopped lucerne hay, were 58% lower in sheep but only 17% lower in Wallaroos (Hollis 1984). Foot and Romberg (1965) reported a similar difference between sheep and young Red Kangaroos. However, in studies by McIntosh (1966), Forbes and Tribe (1970) and Hume (1974), the intakes of sheep and kangaroos declined to the same degree as fibre content was increased. In more recent studies Freudenberger (unpubl.) found that dry matter intakes in adult Wallaroos, Euros and feral goats all declined to the same extent when offered either a diet containing 55% chopped barley straw or one containing 85% straw (Table 5). Both diets were offered with or without a supplement of protein-rich soybean meal.

Why grazing macropodids can better tolerate abundant low quality (high fibre, low nitrogen) forage is thus still an open question. Complex interactions between palatability, forage nitrogen content, mineral and vitamin availability, resistance to mastication, and gut structure and function make it difficult to interpret comparisons for species on poor-quality diets.

Table 5. Digestion and intake of chopped barley straw diets by Euros (*Macropus robustus erubescens*), Wallaroos (*M. r. robustus*) and goats.

	55% straw		85% straw		% decline between 55% and 85% straw
	+N	−N	+N	−N	
Dry matter intake (g/kg$^{0.75}$/d)					
Euros	51.4	42.9	39.0	41.2	20
Wallaroos	54.4	48.4	42.1	40.9	25
Goats	64.2	58.0	51.7	51.6	20
Apparent digestibility of dry matter (%)					
Euros	52.9	53.3	41.5	41.7	21
Wallaroos	57.2	43.7	36.8	40.8	29
Goats	58.3	52.1	45.9	45.9	21
Intake of apparently digestible dry matter (g/kg$^{0.75}$/d)					
Euros	27.9	23.7	16.5	17.4	38
Wallaroos	31.8	25.4	17.3	16.7	47
Goats	37.8	30.4	24.4	24.4	36

Water availability can have an important effect on intake and digestion. Freudenberger and Hume (1987) found that the western feral goat was less affected by a 50% reduction in water intake and required less water per gram dry matter intake (0.95 ml/g DM/kg$^{0.80}$/d) than did Euros (1.09). Calculated respiratory water loss was higher in goats, but their faecal water loss was lower. Relative colon length appears similar in goats and Euros (Freudenberger, unpubl. data). Urinary water loss and nitrogen balance were similar between the species, but restriction improved the goat's ability to digest dry matter.

CONCLUSIONS

We propose that within the Macropodoidea there are consistent broad relationships between natural diet, body size, gut structure and digestive function.

The small potoroids must ingest nutrient-rich food because of their high metabolic requirements relative to gut capacity. Their sacciform forestomach probably acts as both a storage organ and fermentation chamber. The synthetic capabilities of the forestomach microbes may be an important source of amino acids and vitamins that may otherwise be deficient in a hypogeous diet.

The macropodid radiation was a probable response to the evolution of extensive grasslands, with larger body size being a probable requirement for the exploitation of grasslands. Large animals have lower metabolic requirements relative to gut capacity, a significant advantage when ingesting slowly digestible herbage. In conjunction with larger body size came a reduction of the sacciform forestomach and enlargement of the tubiform forestomach which probably allows for greater flow of poorly digestible material. The enlargement of the hindgut in larger kangaroos may be a further adaptation to grassland and arid-zone herbivory.

Little is known about digestive tract adaptations in the small specialist wallabies such as the rock-wallabies (*Petrogale* spp.) and hare-wallabies (*Lagorchestes* spp.). Also, the potential and importance of microbial detoxification of plant allelochemicals in forest wallabies is unclear. Detoxification may have been an important selective pressure for the early evolution of the macropodid and potoroid forestomachs.

REFERENCES

Barry, T. N. and Blaney, B. J., 1987. Secondary compounds of forages. Pp. 91-119 *in* The Nutrition of Herbivores ed. by J. B. Hacker and J. A. Ternouth. Academic Press: Sydney.

Bauchop, T., 1979. The rumen anaerobic fungi: Colonizers of plant fibre. *Ann. Rech. Vet.* **10:** 246-48.

Beal, A. M., 1987. Effect of diet and mineralocorticoid administration on the concentration of anions in parotid saliva from the red kangaroo, *Macropus rufus*. *Aust. J. Zool.* **35:** 133-45.

Beal, A. M. (this volume). Differences in salivary flow and composition among kangaroo species: Implications for digestive efficiency.

Brown, G. D. and Main, A. R., 1967. Studies on marsupial nutrition. V. The nitrogen requirements of the Euro, *Macropus robustus*. *Aust. J. Zool.* **15:** 7-27.

Cheng, K. J., Stewart, C. S., Dinsdale, D. and Costerton, J. W., 1983-84. Electron microscopy of bacteria involved in the digestion of plant cell walls. *Anim. Feed Sci. Technol.* **10:** 93-120.

Dellow, D. W., 1979. Physiology of Digestion in the Macropodine Marsupials. Ph.D. Thesis: University of New England, Armidale.

Dellow, D. W., 1982. Studies on the nutrition of macropodine marsupials. III. The flow of digesta through the stomach and intestine of macropodines and sheep. *Aust. J. Zool.* **30:** 751-65.

Dellow, D. W. and Hume, I. D., 1982a. Studies on the nutrition of macropodine marsupials. I. Intake and digestion of lucerne hay and fresh grass, *Phalaris aquatica*. *Aust. J. Zool.* **30:** 391-98.

Dellow, D. W. and Hume, I. D., 1982b. Studies on the nutrition of macropodine marsupials. II. Urea and water metabolism in *Thylogale thetis* and *Macropus eugenii*; two wallabies from divergent habitats. *Aust. J. Zool.* **30:** 399-406.

Dellow, D. W. and Hume, I. D., 1982c. Studies on the nutrition of macropodine marsupials. IV. Digestion in the stomach and the intestine of *Macropus giganteus, Thylogale thetis* and *Macropus eugenii*. *Aust. J. Zool.* **30:** 767-77.

Dellow, D. W., Nolan, J. V. and Hume, I. D., 1983. Studies on the nutrition of macropodine marsupials. V. Microbial fermentation in the forestomach of *Thylogale thetis* and *Macropus eugenii*. *Aust. J. Zool.* **31:** 433-43.

Demment, M. W., 1983. Feeding ecology and the evolution of body size of baboons. *Afr. J. Ecol.* **21:** 219-33

Demment, M. W. and Van Soest, P. J., 1985. A nutritional explanation for body size patterns of ruminant and non-ruminant herbivores. *Am. Nat.* **125:** 641-72.

DENNY, M. J. S. AND DAWSON, T. J., 1975. Effects of dehydration on body-water distribution in desert kangaroos. *Am. J. Physiol.* **229**: 251-4.

EALEY, E. H. M, 1967. Ecology of the euro, *Macropus robustus* (Gould) in north-western Australia; I. The environment and changes in euro and sheep populations. *CSIRO Wildlife Res.* **12**: 9-25.

EDWARDS, G. P. AND EALEY, E. H. M., 1975. Aspects of the ecology of the swamp wallaby *Wallabia bicolor. Aust. Mammal.* **1**: 307-17.

FLANNERY, T. F., 1984. The kangaroos: 15 million years of bounders. Pp. 817-36 *in* Vertebrate Zoogeography and Evolution in Australasia ed. by M. Archer and G. Clayton. Hesperian Press: Perth.

FLANNERY, T. F. (this volume). Phylogeny of the Macropodoidea; a study in convergence.

FOLEY, W. J., HUME, I. D. AND TAYLOR, R., 1980. Protein intake and requirements of the eastern wallaroo and eastern grey kangaroo. *Bull. Aust. Mammal Soc.* **6**: 34-35 (abstract).

FOOT, J. Z. AND ROMBERG, B., 1975. The utilization of roughage by sheep and the red kangaroo, *Macropus rufus. Aust. J. Agric. Res.* **16**: 429-35.

FORBES, D. K. AND TRIBE, D. E., 1969. Salivary glands of kangaroos. *Aust. J. Zool.* **17**: 765-75.

FORBES, D. K. AND TRIBE, D. E., 1970. The utilization of roughages by sheep and kangaroos. *Aust. J. Zool.* **18**: 247-56.

FREUDENBERGER, D. O. AND HUME, I. D., 1987. Effects of water restriction on digestive function in the euro (*Macropus robustus erubescens*), wallaroo (*M. r. robustus*), and feral goat. Pp. 127-28 *in* Herbivore Nutrition Research ed. by M. Rose. Australian Society of Animal Production.

HEIGHWAY, F. R., 1939. The Anatomy of *Hypsiprymnodon moschatus*. Unpublished Thesis, University of Sydney, Sydney.

HOFMANN, R. R., 1973. The Ruminant Stomach: Stomach Structure and Feeding Habits of East African Game Ruminants. East African Literature Bureau: Nairobi.

HOLLIS, C. J., 1984. The Ability of Two Macropods (*Macropus eugenii* and *M. robustus robustus*) of Different Body Size to Utilize Diets of Different Fibre Content Compared with Sheep. Unpublished Thesis, University of New England, Armidale.

HUME, I. D., 1974. Nitrogen and sulphur retention and fibre digestion by euros, red kangaroos and sheep. *Aust. J. Zool.* **22**: 12-23.

HUME, I. D., 1977a. Maintenance nitrogen requirements of the macropodid marsupials *Thylogale thetis,* red-necked pademelon, and *Macropus eugenii,* tammar wallaby. *Aust. J. Zool.* **25**: 407-17.

HUMÉ, I. D., 1977b. Production of volatile fatty acids in two species of wallaby and in sheep. *Comp. Biochem. Physiol.* **56A**: 299-304.

HUME, I. D., 1982. Digestive Physiology and Nutrition of Marsupials. Cambridge University Press: Cambridge.

HUME, I. D., 1984. Principal features of digestion in kangaroos. *Proc. Nutr. Soc. Aust.* **9**: 76-81.

HUME, I. D. AND CARLISLE, C. H., 1985. Radiographic studies on the structure and function of the gastrointestinal tract of two species of potoroine marsupials. *Aust. J. Zool.* **33**: 641-54.

HUME, I. D. AND WARNER, A. C. I., 1980. Evolution of microbial digestion in mammals. Pp. 615-34 *in* Digestive Physiology and Metabolism in Ruminants ed. by Y. Ruckebush and P. Thivend. MTP Press: Lancaster.

KENNEDY, P. M. AND HUME, I. D., 1978. Recycling of urea nitrogen to the gut of the tammar wallaby (*Macropus eugenii*). *Comp. Biochem. Physiol.* **61A**: 117-21.

KINNEAR, J. E., COCKSON, A., CHRISTENSEN, P. AND MAIN, A. R., 1979. The nutritional biology of the ruminants and ruminant-like mammals — a new approach. *Comp. Biochem. Physiol.* **54A**: 357-65.

KLEIBER, M., 1975. The Fire of Life: An Introduction to Animal Energetics. Krieger: Huntington, N.Y.

LANGER, P., 1988. The Mammalian Herbivore Stomach: Comparative Anatomy, Function and Evolution. Fischer: Stuttgart.

LANGER, P., DELLOW, D. W. AND HUME, I. D., 1980. Stomach structure and function in three species of macropodine marsupials. *Aust. J. Zool.* **28**: 1-18.

LEVENSPIEL, O., 1972. Chemical Reaction Engineering. 2nd Ed. Wiley: New York.

MCINTOSH, D. L., 1966. The digestibility of two roughages and the rates of passage of their residues by the red kangaroo, *Megaleia rufa* and the merino sheep. *CSIRO Wildlife Res.* **11**: 123-35

MORTON, T., 1981. Mastication in the Potoroinae. Unpublished Thesis, Monash University, Melbourne.

OSAWA, R., 1987. Aspects of Digestive Strategies in the Macropodidae. Ph.D. Thesis, University of Queensland, Brisbane.

PENRY, D. L. AND JUMARS, P. A., 1987. Modelling animal guts as chemical reactors. *Am. Nat.* **129**: 69-96.

POPPI, D. P. AND NORTON, B. W., 1980. The validity of the critical size theory for particles leaving the rumen. *J. Agric. Sci., Camb.* **94**: 275-80.

PRINCE, R. I. T., 1976. Comparative Studies of Aspects of Nutritional and Related Physiology in Macropod Marsupials. Ph.D. Thesis, University of Western Australia, Perth.

SANSON, G. D. (this volume). Morphological adaptations of teeth to diets and feeding.

SEEBECK, J. H. (this volume). Ecology of the Potoroidae.

SUTHERLAND, T. M., 1988. Particle separation in the forestomach of sheep. *In* Aspects of Ruminant Physiology ed. by A. and M. Dobson. Compstock Publishing Co.: Ithaca.

VAN SOEST, P. J., 1982. Nutritional Ecology of the Ruminant. O. and B. Books: Corvallis.

WALLIS, I. R., JARMAN, P. J., JOHNSON, B. E. AND LIDDLE, R. W. (this volume). Rufous Bettongs: Nest sites and use of nests.

WATERMAN, P. G., MBI, C. N., MCKEY, D. B. AND GARTLAN, J. S., 1980. African rainforest vegetation and rumen microbes: phenolic compounds and nutrients as correlates of digestibility. *Oecologia.* **47**: 22-33

Differences in salivary flow and composition among kangaroo species: Implications for digestive efficiency

A. M. BEAL[1]

Salivary flow rates and the secretion of amylase activity, buffer anions and nitrogenous compounds were investigated in anaesthetized Eastern Grey Kangaroos, Red Kangaroos and Euros to ascertain whether reported differences in digestive efficiency among kangaroo species could be attributed to differences in rates of secretion of salivary constituents. Mandibular glands of all species had similar rates of secretion of nitrogen as protein plus urea and low rates of secretion of amylase activity, bicarbonate and phosphate. Thus, mandibular salivation was unlikely to differentially influence digestive efficiency of the kangaroos. Grey Kangaroos had higher maximal rates of parotid saliva flow than the other species, indicating a larger gland relative to body mass. At equivalent flow rates, Grey Kangaroos had higher rates of parotid urea clearance and lower rates of protein secretion than the other kangaroos but the rate of nitrogen secretion as urea plus protein was similar in all species. Parotid saliva of Grey Kangaroos and Euros had higher phosphate concentrations than that of Red Kangaroos, and Grey Kangaroos had higher rates of phosphate secretion than the other species. Euros had higher levels and rates of parotid secretion of amylase activity due to having higher amylase activity/protein ratios in this saliva. These differences in parotid salivary composition and secretion rates provide an explanation for the observed differences in digestive efficiency among kangaroo species.

INTRODUCTION

DIGESTIVE efficiency, defined as the ratio of dry matter digestion/dry matter intake, is reported to be greater in Western Grey Kangaroos *(Macropus fuliginosus)* than in Red Kangaroos *(M. rufus)* with Euros *(M. robustus erubescens)* being intermediate (Prince 1976). In the same study, the levels of total phosphorus and inorganic phosphorus in the digesta were found to be higher in Euros and Grey Kangaroos than in Red Kangaroos leading to the suggestion that a functional relationship exists between phosphorus concentration in the digesta and digestive efficiency. How the sources of this phosphorus were partitioned was not ascertained, but differences in dietary intake and in glandular secretion into the gut were proposed. In particular, Prince (1976) suggested that phosphorus levels in the digesta might be correlated with salivary phosphorus concentration and/or flow rates and the mandibular gland was nominated as the likely culprit because this saliva was reportedly high in phosphorus (Forbes and Tribe 1969). Subsequently, the parotid rather than the mandibular gland was shown to secrete saliva of high phosphate concentration (Beal 1984, 1986). Prince (1976) also argued that higher salivary flow rates would have the additional benefit of facilitating the recycling of endogenous nitrogen to the foregut with additional nutritional benefits. This paper reports comparisons of salivary flow rates and secretion of amylase, bicarbonate, phosphate, protein and urea by the parotid and mandibular glands of Red Kangaroos, Eastern Grey Kangaroos *(M. giganteus)* and Euros/Wallaroos *(M. robustus)* which had been maintained under standardized conditions of diet and housing.

METHODS

Experimental procedures

The experiments were performed using four males and four females of each species. Bodyweights were 41.5-43.5 and 18.5-23.5 kg, 35.0-43.0 and 23.5-28.5 kg, and 34.0-41.0 and 22.5-25.5 kg for Euros, Grey Kangaroos and Red Kangaroos respectively. Each animal had a carotid artery loop and was maintained on a diet of lucerne chaff and concentrate pellets, and drinking water containing 25 mmol.l^{-1} NaCl + 25 mmol.l^{-1} $NaHCO_3$. Two days before each experiment, one lateral tail vein was cannulated as described by Beal (1987). Food was removed 15-16 h before the commencement of each experiment, while the saline drinking solution was available until the experiment began.

At the beginning of each experiment, the animals were anaesthetized with 5% sodium pentobarbitone in saline given at rates of 20-40 mg kg^{-1} by i.v. injection through the tail vein cannula. Anaesthesia was maintained with sodium pentobarbitone throughout the experiment using the corneal reflex as a guide to level. The animals were positioned on one side (carotid loop side up) with an electrically-heated pad under the thorax to maintain normal body temperature and with an air cushion under the hind quarters to prevent pressure damage to the hip and thigh region. The trachea was intubated with a cuffed endotracheal tube which was shortened so that the dead space of the respiratory tract was not increased. NaCl:KCl (150:5 mmol.l^{-1}) solution was infused intravenously at 1.6-2.0 ml.min^{-1} for the duration of each experiment to minimize changes in body fluid volume and composition

[1]School of Biological Science, University of New South Wales, P.O. Box 1, Kensington, New South Wales, Australia 2033.

Pp. 189-95 in KANGAROOS, WALLABIES AND RAT-KANGAROOS ed. by G. Grigg, P. Jarman and I. Hume. Surrey Beatty & Sons Pty Limited, New South Wales, Australia, 1989.

resulting from transpiration and salivary loss. The carotid artery loop was cannulated with polyethylene cannulae (0.5-0.58 mm.i.d., 0.8-0.96 mm.o.d.; Dural Plastics, NSW) inserted 10 cm in the direction of the heart using the technique of Seldinger (1953). The duct of the parotid gland ipsilateral to the carotid loop was catheterized with a vinyl tube (1.57 mm.i.d, 2.08 mm.o.d; Dural Plastics, NSW) inserted 3 cm into the duct through its orifice in the mouth. Where possible, the mandibular duct was also catheterized (N = 5, 5 and 8 for Euros, Grey Kangaroos and Red Kangaroos respectively) using polyethylene tubing (0.5-0.8 mm.i.d, 0.8-1.2mm.o.d; Dural Plastics, NSW). The distal ends of the catheters were positioned about 10 cm below the duct orifice. Saliva was collected into tared polypropylene tubes which were closed except for a 20 wire gauge air-bleed.

Salivary secretion was stimulated by intracarotid infusion of acetylcholine chloride (Sigma Chemical Co., USA) at rates ranging from 2.6-475ηmol.min^{-1} using a variable speed syringe pump. Initially, a low flow rate ($< 4\mu$l.kg^{-1} bodyweight. min^{-1}) was established, with secretion being increased subsequently to maximum through a series of predetermined flow rate ranges by increasing the rate of acetylocholine infusion. The flow rate ranges used were <4, 4-8, 8-16, 16-32, 32-64, 64-128 and $>128\mu$l.kg^{-1}.min^{-1}. During any period of stimulation, salivary collection was not commenced until the salivary flow was well established and reasonably constant. Two or more timed samples were taken during every flow interval (except $<4\mu$l.kg^{-1}.min^{-1} interval) and the results obtained from their analysis were averaged to provide mean values for each flow interval of each experiment. Rest periods of 10-60 min were allowed between periods of gland stimulation. The shortest rest periods followed the period of lowest stimulation and the longest rest periods followed maximal stimulation. Blood samples (6-8 ml) were taken from the carotid artery before the first acetylcholine infusion and, thereafter, at the end of each period of gland stimulation.

At the end of each period of anaesthesia, the kangaroos were kept on the heated pad and air cushion until 15-30 min after the endotracheal tube had been removed. The endotracheal tube was removed when the swallowing reflex had returned and the animals were monitored over the following 15-30 min to ensure that respiration was unimpaired. Removal of cannulae and catheters was done while the kangaroos were anaesthetized. At the end of each experiment, the anaesthetized kangaroos were given a single intramuscular injection of procaine penicillin and dihydrostreptomycin (Strypen; May and Baker, Australia) at a rate of 0.1ml kg^{-1} as a safeguard against infection.

Analytical procedures

Blood plasma and saliva were analysed for amylase activity, protein, urea, bicarbonate and phosphate. Blood samples were taken into plastic syringes, heparinized with one drop of heparin (5000 U.ml^{-1}) and centrifuged at 2200 *g* for 10 min to obtain plasma for analysis. Duplicate estimations of total protein were made by the Folin-Lowry method (Daughaday, Lowry, Rosebrough and Fields 1952) using standards composed of a 50:50 mixture of bovine albumen and human globulin. Urea concentrations in the samples were estimated by enzymatic cleavage of urea and colorimetric determination of ammonia (Fawcett and Scott 1960). No correction for preformed ammonia was made to the urea results since the amount of ammonia in saliva always approximated zero and the ammonia in plasma caused an overestimate of plasma urea not exceeding 2% in any experiment. The amylase activity was obtained using a procedure based on the blue starch method of Ceska, Birath and Brown (1969) as described by Beal (1987). Plasma and saliva were analysed for total inorganic phosphate using the method of Baginski, Foa and Zak (1967). pH and pCO_2 were measured anaerobically at 36°C with Radiometer electrodes. The bicarbonate concentration of saliva was determined by the titration procedure of Gyory and Edwards (1967) modified for 0.2ml aliquots of sample.

Statistical procedures

Salivary flow and secretion rate data were expressed on a per kilogram bodyweight basis to allow equitable comparison of salivary parameters between species groups in which the animals were of varying sizes. Concentrations and secretion rates for each salivary constituent were then calculated for nominated percentages of the maximum flow rate (viz. 2, 5, 10, 20, 50, 100%). The values for each parameter at these percentage points were subjected to analysis of variance and when statistically significant variance ratios were obtained, differences between species were found using Student-Newman-Keuls' test (Steel and Torrie 1960).

RESULTS

Plasma

No statistically significant differences between the three species were found in amylase activity or in the concentrations of protein, urea, phosphate or bicarbonate in the plasma. The values for these parameters were similar to those reported for Red Kangaroos held under similar conditions (Beal 1984, 1987).

Parotid Saliva

The maximum flow rates for parotid saliva were 3.6±0.30, 4.9±0.35 and 4.5±0.09 ml/min for

Euros, Grey Kangaroos and Red Kangaroos respectively. When expressed as flow rate per kg bodyweight, Grey Kangaroos (157±12.5 μl/kg/min) had higher rates of secretion ($P< 0.05$) than either Euros (120±6.5 μl/kg/min) or Red Kangaroos (125±7.4 μl/kg/min). The most probable explanation for the above is that Grey Kangaroos have larger parotid glands than the other two species. Consequently, for the sake of comparison, salivary flow rates have been expressed as % of maximum flow rate.

The protein concentrations and the rates of protein secretion were similar in the parotid salivas of Euros and Red Kangaroos and were

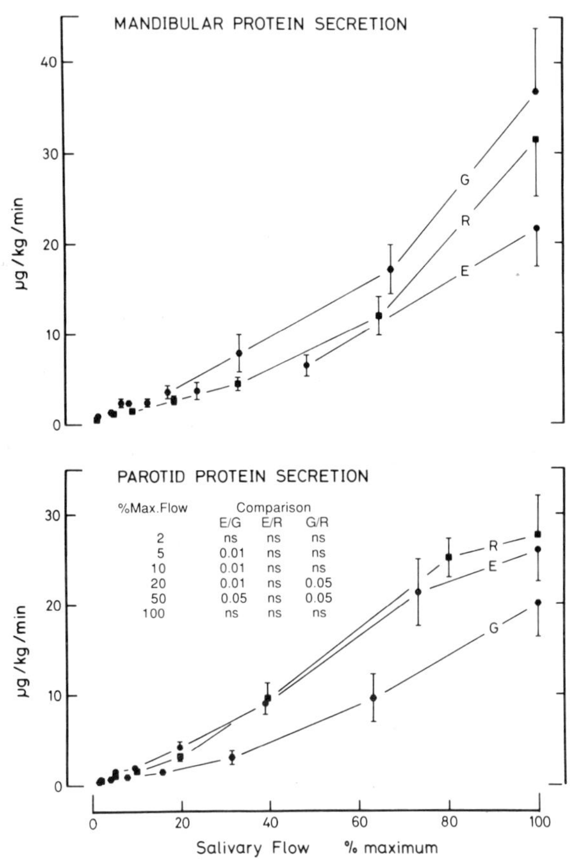

Fig. 1. Mandibular and parotid secretion of protein in Euros (E, ●), Grey Kangaroos (G, ⬢) and Red Kangaroos (R, ■). Levels of statistical significance are given for comparisons of secretion rates at equivalent flow rates. Omission of comparisons means none were significant at the flow rates analysed.

significantly higher than those of Grey Kangaroos (Fig. 1). Amylase activity as a ratio of protein concentration in Red Kangaroo saliva was lower ($P<0.01$) than that in the other two species, but due to differences in protein concentration and flow rate, amylase activities and rates of amylase secretion were similar in Red and Grey Kangaroo salivas. In contrast, amylase activity and secretion were higher in parotid saliva of Euros (Fig. 2).

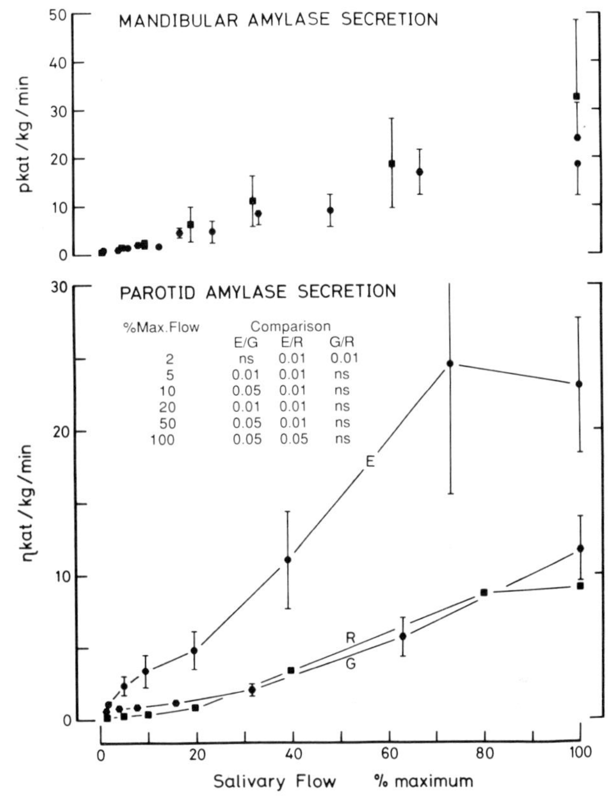

Fig. 2. Mandibular and parotid secretion of amylase in Euros (E, ●), Grey Kangaroos (G, ⬢) and Red Kangaroos (R, ■). Levels of statistical significance are given for comparisons of secretion rates at equivalent flow rates. Omission of comparisons means none were significant at the flow rates analysed.

Because it is generally accepted that incorporation of urea into saliva is a passive process (Thaysen 1960; Burgen and Emmelin 1961), urea concentrations were expressed as saliva/plasma (S/P) ratios and urea secretion rates as urea clearances. The S/P urea ratio of Red Kangaroo saliva was lower than that of the other species and the higher flow rates of the Grey Kangaroo caused the urea clearance of this species to exceed the clearances of Red Kangaroos or Euros (Fig. 3). In terms of total nitrogen (protein N + urea N) delivered into the gut via parotid saliva, there were no significant differences among the three species, the lower N concentration of the Grey Kangaroo saliva being offset by the higher rates of salivary flow in this species (Table 1).

Red Kangaroo saliva had lower phosphate concentrations ($P<0.01$) than that of either Grey Kangaroos or Euros whereas phosphate secretion was highest in the Grey Kangaroo with the Euro exceeding the Red Kangaroo at flow rates in excess of 20% maximum flow (Fig. 4). Salivary bicarbonate concentrations reflected these differences in that Red Kangaroo saliva had the highest bicarbonate levels, Euros were intermediate and Grey Kangaroos had the lowest concentrations. However, because of the differences in flow rate, bicarbonate secretion by the parotid gland was

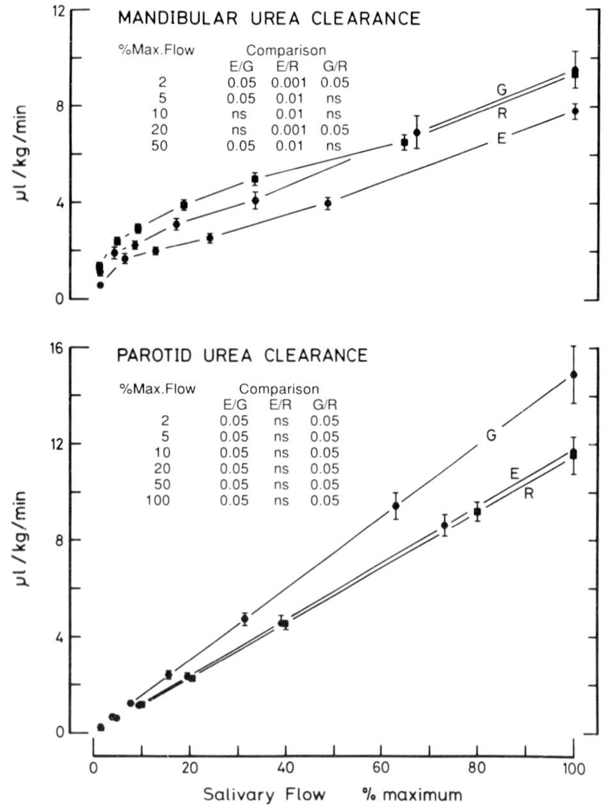

Fig. 3. Mandibular and parotid clearance of urea in Euros (E, •), Grey Kangaroos (G, ●) and Red Kangaroos (R, ■). Levels of statistical significance are given for comparisons of secretion rates at equivalent flow rates. Omission of comparisons means none were significant at the flow rates analysed.

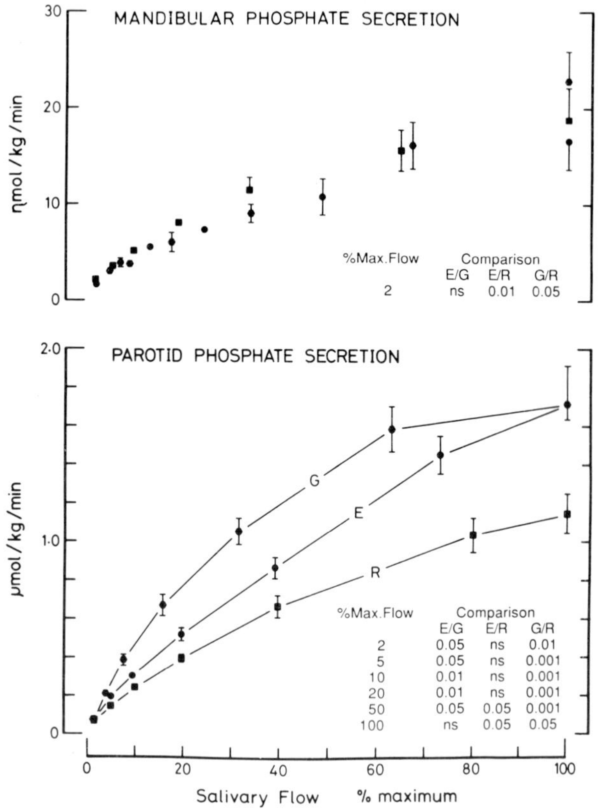

Fig. 4. Mandibular and parotid secretion of phosphate in Euros (E, •), Grey Kangaroos (G, ●) and Red Kangaroos (R, ■). Levels of statistical significance are given for comparisons of secretion rates at equivalent flow rates. Omission of comparisons means none were significant at the flow rates analysed.

Table 1. Mean secretion rates (SEM) for protein nitrogen (N), urea nitrogen and total nitrogen in mandibular and parotid salivas of Euros, Grey Kangaroos and Red Kangaroos. Means with different superscripts are significantly at P<0.05 or better.

		MANDIBULAR SALIVA			PAROTID SALIVA		
Flow %Max	Spec.	Protein N µg/kg/min	Urea N µg/kg/min	Total N µg/kg/min	Protein N µg/kg/min	Urea N µg/kg/min	Total N µg/kg/min
	Euro	0.17 (0.02)	0.06 (0.01) [a]	0.23 (0.02)	0.10 (0.01)	0.22 (0.01)[ab]	0.33 (0.02)
2	Grey	0.15 (0.02)	0.09 (0.01)[b]	0.25 (0.01)	0.09 (0.01)	0.24 (0.02)[a]	0.33 (0.03)
	Red	0.13 (0.01)	0.13 (0.01)[c]	0.27 (0.01)	0.12 (0.01)	0.18 (0.01)[b]	0.31 (0.03)
	Euro	0.32 (0.05)	0.12 (0.01)[a]	0.44 (0.05)	0.23 (0.03)[a]	0.52 (0.03)[ab]	0.75 (0.06)
5	Grey	0.24 (0.03)	0.15 (0.02)[ab]	0.38 (0.03)	0.11 (0.01)[b]	0.57 (0.05)[a]	0.68 (0.06)
	Red	0.19 (0.02)	0.18 (0.01)[b]	0.37 (0.03)	0.17 (0.02)[ab]	0.43 (0.03)[b]	0.60 (0.04)
	Euro	0.41 (0.14)	0.16 (0.01)[a]	0.57 (0.05)	0.33 (0.05)[a]	1.02 (0.06)[ab]	1.36 (0.08)
10	Grey	0.40 (0.07)	0.18 (0.02)[ab]	0.58 (0.06)	0.15 (0.03)[b]	1.11 (0.09)[a]	1.27 (0.11)
	Red	0.26 (0.04)	0.23 (0.02)[b]	0.49 (0.05)	0.25 (0.02)[ab]	0.85 (0.05)[b]	1.10 (0.06)
	Euro	0.60 (0.18)	0.21 (0.02)[a]	0.81 (0.17)	0.67 (0.08)	2.02 (0.11)[ab]	2.69 (0.17)
20	Grey	0.70 (0.17)	0.23 (0.03)[a]	0.93 (0.16)	0.29 (0.06)[a]	2.19 (0.18)[a]	2.47 (0.23)
	Red	0.42 (0.06)	0.30 (0.01)[b]	0.72 (0.06)	0.52 (0.06)	1.68 (0.10)[b]	2.19 (0.13)
	Euro	1.22 (0.21)	0.35 (0.02)[a]	1.57 (0.23)	2.11 (0.18)	5.00 (0.28)[ab]	7.11 (0.55)
50	Grey	1.89 (0.28)	0.40 (0.03)[ab]	2.29 (0.28)	1.01 (0.26)[a]	5.38 (0.43)[a]	6.39 (0.62)
	Red	1.21 (0.15)	0.45 (0.02)[b]	1.67 (0.15)	2.10 (0.21)	4.18 (0.26)[b]	6.28 (0.41)
	Euro	3.46 (0.67)	0.66 (0.03)	4.12 (0.67)	4.17 (0.57)	9.89 (0.54)	14.06 (0.75)
100	Grey	5.89 (1.12)	0.68 (0.06)	6.57 (1.15)	3.20 (0.60)	10.69 (0.87)	13.75 (1.15)
	Red	5.08 (0.97)	0.70 (0.05)	5.78 (0.96)	4.40 (0.73)	8.31 (0.56)	12.72 (0.65)

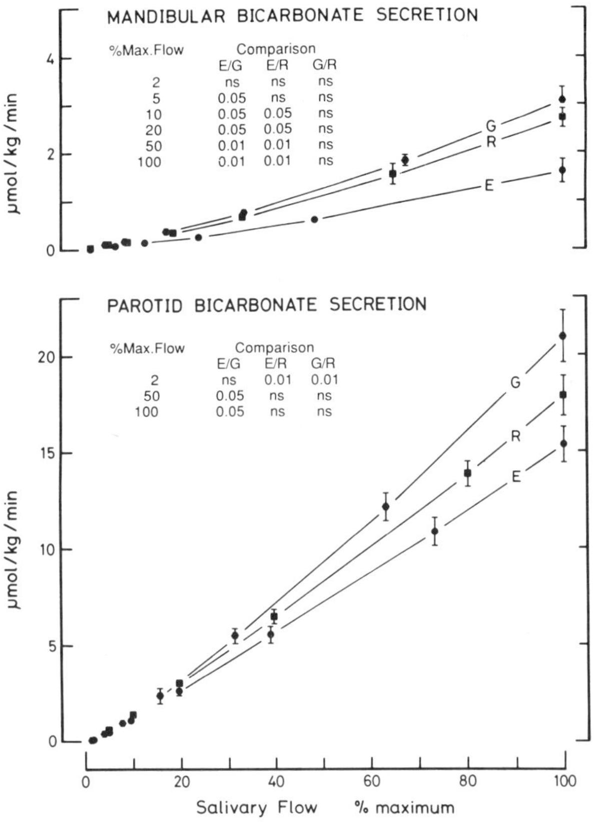

Fig. 5. Mandibular and parotid secretion of bicarbonate in Euros (E, ●), Grey Kangaroos (G, ⬢) and Red Kangaroos (R, ■). Levels of statistical significance are given for comparisons of secretion rates at equivalent flow rates. Omission of comparisons means none were significant at the flow rates analysed.

similar in all species except at the extremes of the flow range (Fig. 5).

Mandibular saliva

The maximum flow rates for the mandibular glands of Euros, Grey Kangaroos and Red Kangaroos were 3.1±0.13, 4.2±0.23 and 4.0±0.11 ml/min respectively. When expressed as flow rate per kg bodyweight, Grey Kangaroos (118±4.6 µl/kg/min) and Red Kangaroos (128±9.3 µl/kg/min) had higher maximum rates of fluid secretion than the Euros (84±5.3 µl/kg/min; Euro/Grey, $P<0.001$ and Euro/Red, $P<0.05$).

Protein concentrations of Euro saliva were higher ($P<0.05$) at low flow rates (up to 10% of maximum flow) than the saliva of other species but were similar at higher flows. The rates of protein secretion by all three species were similar for equivalent flow rates across the entire flow range (Fig. 1). Amylase activities in the three salivas were low, generally less than 1 µkat/1, resulting in trivial rates of secretion in this saliva (Fig. 2). There were no significant differences in amylase activities or secretion rates between the mandibular salivas of the three species. At flow rates above 10%, S/P urea ratios of Grey Kangaroo saliva were lower ($P<0.05$) than those of Euro or Red Kangaroo saliva whereas, due to the lower flow rates, urea clearances were lower ($P<0.05$) for the saliva of Euros than for the other two species (Fig. 3). In terms of total nitrogen (protein N + urea N) delivered into the gut via mandibular saliva, there were no differences between the three species, the lower N concentration of the Grey Kangaroo saliva being offset by the higher rates of salivary flow in this species (Table 1).

Phosphate concentrations in mandibular saliva were low, exceeding 1 mmol/1 only at the lowest flow rates. Except at maximum flow, phosphate concentrations in Grey Kangaroo saliva were lower than those of the other species. Relative to the parotid gland, rates of phosphate secretion were lower and there were no differences between species at equivalent rates of flow (Fig. 4). Bicarbonate concentrations were low and similar in level in all species. As a consequence of the lower rates of flow produced by the glands of Euros bicarbonate secretion was lowest ($P<0.05$) in this species (Fig. 5).

DISCUSSION

Flow rates: After standardization for body weight, maximum rates of parotid salivary flow in Grey Kangaroos were higher than the maxima in Red Kangaroos or Euros whereas the maximum rates of mandibular flow in Euros were lower than those of Grey Kangaroos or Red Kangaroos. These results could be explained by differences between the species in capacity per gram tissue to secrete fluid. However, given the close relationship between the three species and the report that Grey Kangaroo parotid glands are larger than Red Kangaroo glands (Forbes and Tribe 1969), the species differences in flow maxima probably result from differences in gland mass per unit body mass. Consequently, to facilitate equitable comparison of the concentrations and secretion rates for salivary constituents, maximum flow was taken as 100% and values for nominated percentage points between 2-100% were calculated for each salivary parameter. Kangaroo mandibular glands secrete only during mastication and stress/thermoregulatory licking, whereas secretion by the parotid gland is essentially continuous. This is presumably because reflex stimulation of the parotid occurs in response to conditions in the foregut fermentation as well as to masticatory and licking activities (Beal, unpubl. obs.). Thus, parotid secretion would be expected to influence digesta composition to a far greater extent than mandibular secretion and the higher rates of parotid secretion in Grey Kangaroos could contribute, totally or in part, to the lower levels of dry matter reported for Western Grey Kangaroo digesta (Prince 1976). In interpreting the parotid secretion data, it is important to remember that, except during mastication and licking, salivary flow rates will be at the low end of the flow range throughout most of the day (i.e., probably under 5% of maximum).

Nitrogen Secretion: Urea recycled to the foregut makes a significant contribution to microbial protein production (Kennedy and Hume 1978) and, consequently, could influence digestive efficiency. As urea incorporation into saliva is a passive process, the rate of delivery of urea to the foregut in the saliva will depend mainly on the plasma concentration and salivary flow rate. The data for saliva/plasma urea ratios indicate that, at any given plasma concentration, the Red Kangaroo parotid will secrete less urea than the parotid glands of the other two species, whereas the mandibular glands of all species were alike. Although mean plasma urea concentrations in Euros (13.0±1.04mmol/1) were approximately 2mmol/1 higher than those of the other kangaroos, these differences were not statistically significant. The overwhelming influence on urea secretion was salivary flow rate; the Grey Kangaroo parotid gland had the highest flow and urea clearance and the Euro mandibular gland had the lowest flow and urea clearance.

As dietary supplements of urea and casein are utilized equally well by macropods (Brown 1969; Lintern-Moore 1973), salivary protein secretion also makes a contribution to the amount of nitrogen delivered into the forestomach in the saliva. Protein concentrations in mandibular saliva of Euros were higher than for the other kangaroos, but due to the lower flow rate of the Euro gland, there were no differences in protein secretion among the three species. Protein levels in parotid saliva of Grey Kangaroos were lower than in the other species. but this was more than compensated for by its high flow rate which gave this species the highest rates of parotid protein secretion. For either saliva, there were no differences in total nitrogen delivered into the gut at equivalent flow rates although the proportion of protein-nitrogen to urea-nitrogen differed. In mandibular saliva, protein-nitrogen predominated whereas, in parotid saliva, urea provided most nitrogen. However, in the absence of evidence to the contrary, these two constituents must be considered to be equally available sources of nitrogen for the micro-organisms in the kangaroo foregut.

Amylase Secretion: The levels of amylase activity in parotid saliva of Red Kangaroos and of Grey Kangaroos and in mandibular saliva of all three species were similar to those previously reported for the Red Kangaroo (Beal 1987). The amylase activity of Euro parotid saliva was higher (3-6 times) than that of the other species and, consequently, the rate of secretion of amylase activity into the gut was about twice that of the other kangaroos. As no definite function has been attributed to salivary amylase and ruminant salivas lack the enzyme, it is not clear what advantage its presence gives to a foregut fermenter. At macropodine forestomach pH values of 6.4-7.4 (Dellow and Hume 1982), the amylase would have 60-90% of maximum activity (Beal 1987) provided that the amylase molecule was not rapidly fermented. Some α-amylase molecules can resist proteolytic enzymes for 4 h or more (Stein and Fischer 1958) which is similar to the transit time for fluid in the macropodine forestomach (Dellow and Hume 1982). One possible function may be to encourage fermentation to start in fresh ingesta by hydrolysis of starch to more readily useable maltose.

Phosphate and Bicarbonate Secretion: The optimal level of inorganic phosphorus in the forestomach digesta of kangaroos has not been determined. In ruminants, the optimal phosphorus concentration in rumen fluid is between 1.3 and 11.5mmol/1 (Anderson, Cheng and Burroughs 1966; Evans and Davis 1966; Matturi 1978); the width of the range presumably depends on the conditions of investigation. As inorganic phosphorus levels in kangaroo forestomach digesta lie within a similar range (Red Kangaroo, 4.2-6.8 mmol/1; Western Grey Kangaroo and Euro, 9.7-11.0 mmol/1; Prince 1976), digestive efficiency in the various kangaroo species may be influenced by the level of phosphorus in the digesta. The source of the phosphorus is clearly not the mandibular glands because phosphate concentrations in mandibular saliva are very low and flow is sporadic. The data for the parotid salivas indicate that this saliva would be one of the major contributors to foregut phosphate levels with Grey Kangaroos having the highest rate of secretion, Red Kangaroos the lowest and Euros intermediate. This pattern is in accord with the observed hierarchy of digestive efficiencies (Prince 1976) if Eastern and Western Grey Kangaroos are assumed to be similar.

In addition to being necessary for molecular synthesis by micro-organisms in the foregut, phosphate with bicarbonate provides buffering of pH changes in the ferment. Although the differences in rates of salivary buffer secretion were not statistically significant, the trend was for Grey Kangaroos to have the highest rate of delivery of buffer to the gut via the saliva.

Overall, there appear to be species differences in salivary composition and secretion rates for the parotid glands but not for the mandibular glands of kangaroos. These differences in parotid secretion may, at least, partially explain the reported hierarchy of digestive efficiencies among the kangaroo species.

ACKNOWLEDGEMENTS

I am indebted to Mr. David Hair for skilful technical assistance and to Mr. David Armstrong for his husbandry of the kangaroos. I would also like to thank Abbott Australia Pty Ltd for the gift of the sodium pentobarbitone and the Australian Research Grants Scheme for financial support of the project.

REFERENCES

ANDERSON, R., CHENG, E. AND BURROUGHS, W., 1956. A laboratory technique for measuring phosphorus availability of feed supplements fed to ruminants. *J. An. Sci.* **15**: 489-95.

BAGINSKI, E. S., FOA, P. P. AND ZAK, B., 1967. Microdetermination of inorganic phosphate, phospholipids and total phosphate in biological materials. *Clin. Chem.* **13**: 326-32.

BEAL, A. M., 1984. Electrolyte composition of parotid saliva from sodium-replete red kangaroos *(Macropus rufus)*. *J. exp. Biol.* **111**: 225-37.

BEAL, A. M., 1986. Effects of flow rate, duration of stimulation, and mineralocorticoids on the electrolyte concentrations of mandibular saliva from the red kangaroos *(Macropus rufus)*. *J. exp. Biol.* **126**: 315-39.

BEAL, A. M., 1987. Amylase activity, protein and urea in saliva of red kangaroos *(Macropus rufus)*. *Archs. oral Biol.* **32**: 825-32.

BROWN, G. D., 1969. Studies on marsupial nutrition. VI. The utilization of dietary urea by the euro or hill kangaroo, *Macropus robustus* (Gould) *Aust. J. Zool.* **17**: 187-94.

BURGEN, A. S. V. AND EMMELIN, N. G., 1961. *Physiology of the salivary glands*. Pp. 169-94. Arnold: London.

CESKA, M., BIRATH, K. AND BROWN, B., 1969. A new and rapid method for the clinical determination of α-amylase activities in human serum and urine. Optimal conditions. *Clin. Chim. Acta.* **26**: 437-44.

DAUGHADAY, W. H., LOWRY, O. H., ROSEBROUGH, N. J. AND FIELDS, W. S., 1952. Determination of cerebrospinal fluid protein with the Folin phenol reagent. *J. lab. clin. Med.* **39**: 663-65.

DELLOW, D. W., 1982. Studies on the nutrition of macropodine marsupials. III. The flow of digesta through the stomach and intestine of macropodines and sheep. *Aust. J. Zool.* **30**: 751-65.

DELLOW, D. W. AND HUME, I. D., 1982. Studies on the nutrition of macropodine marsupials. IV. Digestion in the stomach and the intestine of *Macropus giganteus, Thylogale thetis* and *Macropus eugenii*. *Aust. J. Zool.* **30**: 767-77.

EVANS, J. L. AND DAVIS, G. K., 1966. Influence of sulfur, molybdenum, phosphorus and copper interrelationships in cattle upon cellulose digestion *in vivo* and *in vitro*. *J. An. Sci.* **25**: 1014-018.

FAWCETT, J. K. AND SCOTT, J. E., 1960. A rapid and precise method for determination of urea. *J. clin. Path.* **13**: 156-59.

FORBES, D. K. AND TRIBE, D. E., 1969. Salivary glands of kangaroos. *Aust. J. Zool.* **17**: 765-75.

GYORY, A. Z. AND EDWARDS, K. D. G., 1967. Simultaneous titrimetric determination of bicarbonate and titratable acid in urine. *Aust. J. exp. Biol. med. Sci.* **45**: 141-47.

KENNEDY, P. M. AND HUME, I. D., 1978. Recycling of urea nitrogen to the gut of the tammar wallaby *(Macropus eugenii)*. *Comp. Biochem. Physiol.* **61**A: 117-21.

LINTERN-MOORE, S., 1973. Utilization of dietary urea by the Kangaroo Island wallaby *Protemnodon eugenii* (Desmarest). *Comp. Biochem. Physiol.* **46**A: 345-51.

MATTURI, A. S., 1978. The mineral and nitrogen requirements for the digestion of cellulose by ruminants. Pp. 98-105. Ph.D. Thesis, University of Western Australia.

PRINCE, R. I. T., 1976. Comparative studies of aspects of nutritional and related physiology in macropod marsupials. Ph.D. Thesis, University of Western Australia.

SELDINGER, S. I., 1953. Catheter replacement of the needle in precutaneous arteriography. *Acta radiol.* **39**: 368-76.

STEEL, R. D. G. AND TORRIE, J. H., 1960. Principles and procedures in statistics. Pp. 110-11. McGraw-Hill: New York.

STEIN, E. A. AND FISCHER, E. H., 1958. The resistance of α-amylase toward proteolytic attack. *J. Biol. Chem.* **232**: 867-79.

THAYSEN, J. H., 1960. Handling of alkali metals by exocrine glands other than the kidney. Pp. 432-33 *in* Handbuch der experimentallen Pharmakalogie, Vol. 13. Springer-Verlag: Berlin.

Motility patterns of the stomach and intestine of small Macropodids

R. S. WYBURN[1] and K. C. RICHARDSON[1]

Using chronically implanted bipolar electrodes on the stomach and small intestine serosal surfaces the myoelectrical activity was recorded from the Tammar Wallaby *(Macropus eugenii)* and from the Quokka *(Setonix brachyurus)*. The slow wave frequency recorded from the Tammar stomach was 5.5±0.1 per minute (x̄±SE) and 5.3±0.13 from the Quokka. Action potentials were recorded from all parts of the Quokka stomach but except for the pyloric region were absent from the Tammar stomach.

There was virtually no slow wave gradient along the length of the small intestine of the two species examined. Duodenal slow wave frequency was 26±0.15 per minute (x̄±SE) in the Tammar and 17.8±0.4 in the Quokka. Ileal slow wave frequency was 25±0.15 per minute (x̄±SE) in the Tammar and 17.5±0.4 in the Quokka. Migrating myoelectric complex activity was recorded from the small intestine of both species.

INTRODUCTION

FOR mammals to survive on a diet of herbage they must have an alimentary tract which provides a site for the microbial breakdown of cellulose. In herbivores such as the horse, rabbit and koala the main site is provided by adaptions of the caecum and colon. In others, such as the macropods and ruminants, the main site is a modified stomach. Macropods have an elongated, partially haustrated, tubular stomach in which microbial fermentation occurs in the oral four-fifths and peptidergic digestion in the aboral fifth (Hume 1982). For the stomach to provide a suitable environment for the microbial breakdown of ingested herbage it must slow the rate of passage, thoroughly mix the contents then sort the digesta particles according to size. These functions can only be achieved by the muscular action of the stomach wall.

Like other herbivorous mammals that are stomach fermenters, the macropods have a long small intestine. It would therefore be reasonable to anticipate that it would function in much the same way as in ruminants.

A number of techniques have been used to study the muscular activity of the alimentary tract wall. Some early investigations were carried out by direct observation through an abdominal incision (Bayliss and Starling 1899). The use of x-rays provided a noninvasive method of demonstrating motility of the alimentary tract and was first used by Cannon in 1898. Other methods of observation made use of various designs of exteriorized sections of the tract (Bieble 1930; Titchen 1958; Hiatt *et al.* 1966). Graphic recordings of contractions of the alimentary tract wall were achieved by a number of techniques. Changes in intraluminal pressure have been recorded using balloons either singly or in series (Hightower 1952; Smith 1959), open tipped catheters (Coombe 1966) and pressure sensitive radio pills (Ramorino and Colagrande 1964). Information on the strength and direction of contractions has been obtained using small strain gauges (Grivel and Ruckebusch 1972). However, the method that has become the most widely used to investigate alimentary tract motility is the recording of the electrical activity of the muscles in the tract wall (Bortoff 1961; Coombe 1966; Ruckebusch and Bueno 1973; Wyburn 1981; Richardson and Wyburn 1983). It was demonstrated by Ruckebusch (1970) that the electrical activity is an indication of the mechanical activity and that certain types of electrical activity are related to certain types of mechanical activity.

It has been established that there are two basic components to the electrical activity. One component, a slow change in potential which migrates aborad, is referred to as the "slow wave", "pacesetter potential", or "basic electrical rhythm". Its frequency and rate of propagation have been found to be the same at any given site within a species (Grivel and Ruckebusch 1972) and to decrease from duodenum to ileum (Bortoff 1961). The slow wave acts as a pacesetter for contractions of the gut wall (Bozler 1946; Code *et al.* 1968) and the presence of a decreasing slow wave frequency gradient from duodenum to the ileocaecal valve was thought to be essential for the aborad movement of digesta even though there is no direct mechanical activity associated with the slow wave. Richardson and Wyburn (1983) investigated the myoelectrical activity of a small macropod, the Tammar Wallaby, and reported that there was no slow wave frequency gradient over the length of the small intenstine. A similar finding was later reported in the guinea pig (Galligan *et al.* 1985). The second component

[1]School of Veterinary Studies, Murdoch University, Murdoch, Western Australia 6150.

Pages 197-203 in KANGAROOS, WALLABIES AND RAT-KANGAROOS ed. by G. Grigg, P. Jarman and I. Hume. Surrey Beatty & Sons Pty Limited, New South Wales, Australia, 1989.

of electrical activity recorded from the alimentary tract is a rapid change in potential called a "spike or action potential" which is superimposed on the slow wave. This is associated with both a visible contraction of the gut wall and a rise in intraluminal pressure (Ruckebusch *et al.* 1968). Action potentials of the small intestine occur in three basic patterns: (1) when they are totally absent, a quiescent phase; (2) when they occur in irregular bursts, an irregular phase; (3) when they occur between each slow wave, a regular phase. Together these three phases make up the migrating myoelectric complex. The form of the migrating myoelectric complex varies greatly between species and within a species can vary considerably depending on the feeding regimen imposed (Szurszewski 1969; Ruckebusch and Bueno 1976, 1977a,b; Wyburn 1981). In those herbivores which are stomach fermenters the migrating myoelectric complex has been shown to recur at regular intervals in the jejunum. It recurs at a mean time interval of 88.4 min in the sheep, 78.6 min in the goat, 57.3 min in the cow and 95.5 min in the Tammar Wallaby. This study reports the ongoing study of the myoelectrical events of the stomach and small intestine of the Tammar Wallaby (*Macropus eugenii*) and of the Quokka (*Setonix brachyurus*).

METHOD

Twenty Tammar Wallabies (*Macropus eugenii*) and six Quokkas (*Setonix brachyurus*) were held on the Murdoch University Native Fauna Research Area. While in this area they fed on the natural forage, irrigated pasture and were supplemented with oat grain. The animals were in good condition, the Tammars weighing betwen 3 and 6 kg and the Quokkas between 2 and 3 kg. One week prior to an experiment the animals were transferred to individual cages in air conditioned rooms. They had *ad libitum* access to pelleted kangaroo mix (Milne) and were supplemented twice weekly with fresh cabbage. Water was supplied from gravity feed bottles.

Bipolar electrodes made from stainless steel 28 swg orthopaedic wire (Zimmer, USA) were used. The two electrode poles were set 1 mm apart in self-curing acrylic (DUZ ALL, Coralite Dental Products) into which was incorporated a 1 mm diameter piece of lead shot to make the assembly radiographically visible. The electrodes were soldered to 40 cm of multistranded stainless steel wire coated with teflon (0.61 mm, OD., Beldon, USA). The exposed poles were 0.75 mm to 1 mm long with flat ends. Two indifferent electrodes were made by forming a 3 mm diameter loop of exposed wire at the end of a teflon coated lead. The electrode assemblies were exposed to ultraviolet light for twelve hours to complete polymerisation of the acrylic and to sterilize them. Thirty minutes before being surgically implanted they were soaked in 1% chlorhexidine gluconate (Hibitane I.G.I. Aust.) and then washed in sterile water.

All surgery was carried out under general anaesthesia using aseptic techniques. Following induction of anaesthesia by an intravenous injection of thiopentone (Pentathal R Abbot) an endotracheal tube was passed and the anaesthesia maintained on a 2% halothane (Fluothane I.G.I. Aust.) and oxygen mixture (Richardson and Cullen 1981). After surgery each animal was observed closely until it had fully recovered from the anaesthetic.

Ten Tammars and three Quokkas had seven electrodes placed on the stomach and ten Tammars and three Quokkas had seven electrodes placed on the hind stomach and small intestine (Fig. 1). The surgical approach to the stomach was through a left sided paracostal incision. The electrodes were sutured to the serosal surface 1 cm lateral to the greater curvature with their poles aligned at right angles to the long axis of the stomach. They were arranged in sequence 2-3 cm apart over the caudal gastric flexure. The surgical approach to the hind stomach and small intestine was through a right sided paracostal incision. The first electrode was sutured to the pyloric antrum, the second to the duodenal bulb, the third 1-2 cm aborally on the duodenum and the fourth a further 1-2 cm aborally. The fifth, sixth and seventh electrodes were placed on the ileum 12 cm, 7 cm and 1 cm orad of the ileocaecal valve respectively. In both surgical approaches two indifferent electrodes were sutured to the peritoneum of the abdominal wall close to the site of the incision. All electrode leads were then exteriorised in the paralumbar region and were soldered to a 15 pin plug assembly (Radio Spares, Sub-Miniature Multipole Plug).

Electrical recordings were made on a seven channel recorder with a rectilinear ink writing system (Mingograph E.E.G. junior Siemens Elema) with a time constant of 0.15 sec, a low pass filter of 70 Hz and a paper speed of 2.5 mm/sec. All channels were calibrated against a 50 μV input signal. Recording sessions commenced ten days after surgery. The animals had access to kangaroo pellets and water until the commencement of a recording session at 8.30 a.m. Throughout any session the animals were restrained by placing them in a hessian sac. As far as could be determined this form of restraint caused no stress as there was no struggling and when monitored continuously the heart and respiration rates remained within normal limits. Each recording session lasted between five and eight hours with recordings being made at weekly intervals from each Tammar on ten occasions and each Quokka on four occasions. For four recording sessions on four animals, each animal was given 10 ml of

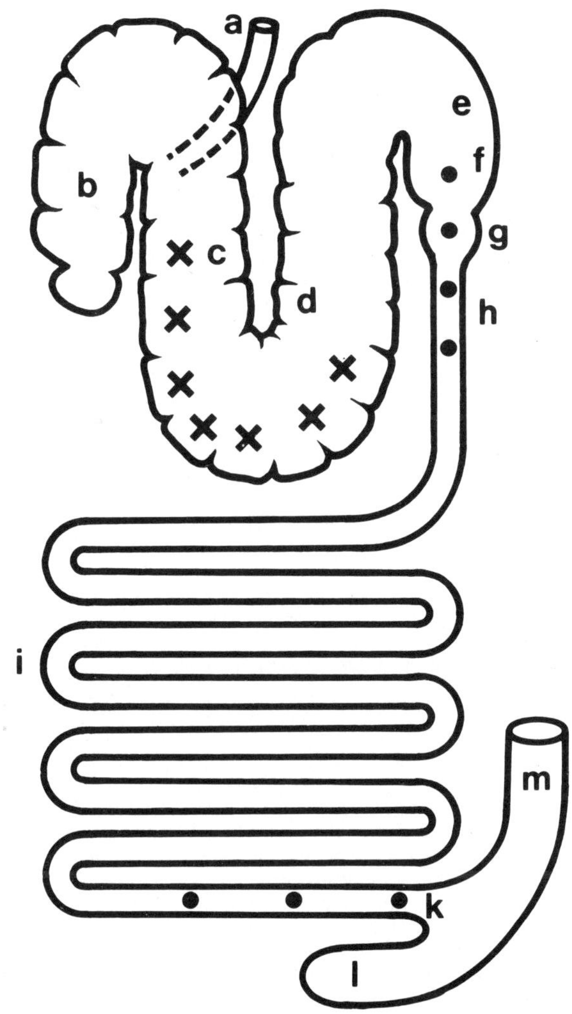

Fig. 1. Schematic representation of the macropodid alimentary tract showing the electrode positions on the stomach (x) and small intestine (•). a, oesophagus; b, blind sac; c, proximal compartment; d. middle compartment; e, gastric gland region; f, pyloric region; g, duodenal bulb; h, duodenum; i, small intestine; k, ileocaecal valve; l, caecum; m, colon.

65% w/v suspension of barium sulphate (Medibar Medicon) orally and the movement of the electrodes and the barium sulphate observed radiographically. The apparatus used was an image intensifier linked to a closed circuit television chain (Dual Field 25/15 Sircon 2, Siemens). The image was recorded on videotape and time linked through its audio channel to the seven channel recordings.

The electrical recordings were analysed in the following manner. The frequency, amplitude and propagation rate of the slow wave were measured. The frequency and amplitude of the action potentials were measured and the pattern of their occurrence noted and timed. If no action potentials were recorded for a period of 30 secs or more, that segment of the alimentary tract was considered to be in a quiescent phase. When numbers of action potentials were interspersed with periods of up to 20 secs with no action potentials the segment was considered to be in an irregular phase. When action potentials occurred on every slow wave for 4-10 min, the segment was considered to be in a regular phase. The time between the regular phases, the time a single phase took to pass one electrode and the time it took to pass between electrodes was measured. The time over which the regular pattern was recorded at one electrode was designated the duration and the velocity at which it migrated was calculated from the time it took to travel the known distance, at placement, between electrodes. The length of alimentary tract wall covered by the regular phase was determined from its velocity and duration (velocity×duration = length).

RESULTS

Slow waves of similar form were recorded from all stomach electrodes of both the Tammar and the Quokka. Each slow wave had a complex of potential changes lasting approximately two secs followed by an isopotential period lasting approximately nine secs. The only major difference in the slow wave between the two species was in the amplitude range. In the Tammar this was 50-240 μv while in the Quokka it was 50-450 μv. The frequency had a mean of 5.5 with a range of 4.5-6.0 in the Tammar and a mean of 5.3 and a range of 4.6-6.1 in the Quokka. The aborad propagation was approximately 3 mm/sec in both species. The striking difference between the two species was in the occurrence of action potentials (Fig. 2). No action potentials were recorded from the haustrated proximal and middle compartments of the Tammar stomach while in the Quokka action potentials were recorded from all the stomach electrodes. Radiographic observation invariably showed that the haustra contracted at the same time and with the same frequency as the slow wave.

The pattern of slow waves and action potentials over the pyloric antrum, duodenal bulb, duodenum and ileum was very similar in both species (Table 1). Over the pylorus bursts of action potentials lasting 2.0-2.3 secs consisting of 6-14 spikes of varying amplitude and duration commonly followed about 0.5 sec after a slow wave. The amplitude of slow waves as well as of action potentials waxed and waned depending on the electrical activity on the duodenum. When the duodenum was in regular activity or was quiescent the pylorus showed no action potentials and remained quiescent for up to 18 min. Once the duodenum went into a period of irregular activity the action potentials returned to the pyloric antrum over a period of 70-80 min. The cycle was then repeated (Fig. 3).

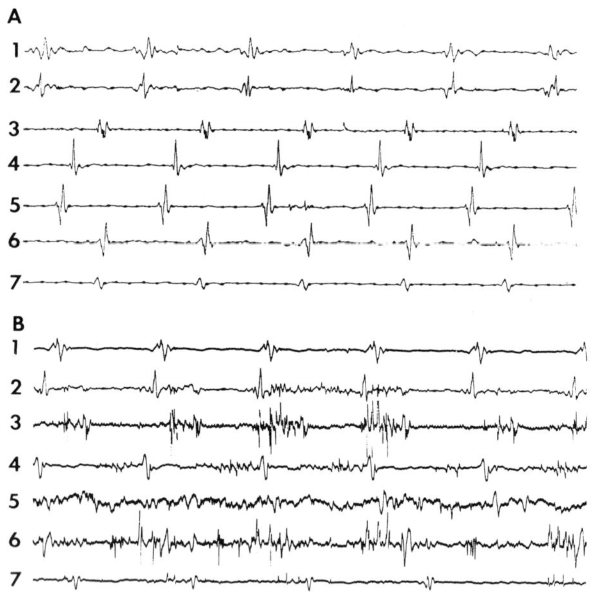

Fig. 2. Electrical recording from a Tammar Wallaby (A) and a Quokka (B). Records are for one minute with the maximal recorded amplitude being 180 μv for the Tammar Wallaby and 240 μv for the Quokka. Numbers 1-7 represent recordings from the stomach electrode sites depicted in Figure 1 with 1 being the most orad. Note the regularity of the slow wave from any single channel. Note the presence of action potential complexes in the isoelectrical period between successive slow waves on channels 2 to 6 in the Quokka.

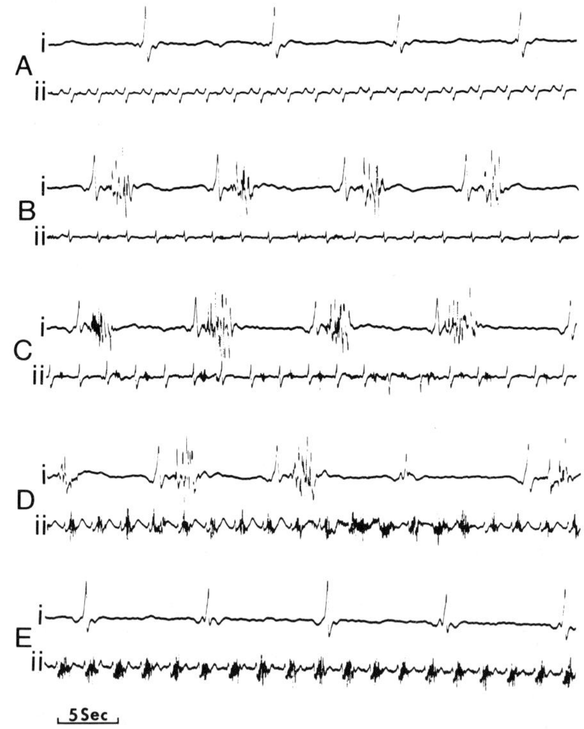

Fig. 3. A sequence of electrical signals recorded simultaneously from the pylorus (i) and the adjacent duodenum (ii) of a Tammar Wallaby. Record A, slow wave only on the pylorus and duodenum; B, action potentials occurring on the pylorus with small action potentials on the duodenum showing irregular activity; C and D show increasing irregular activity of the duodenum still with action potentials on the pylorus; E the pylorus is now quiescent while the duodenum shows regular activity.

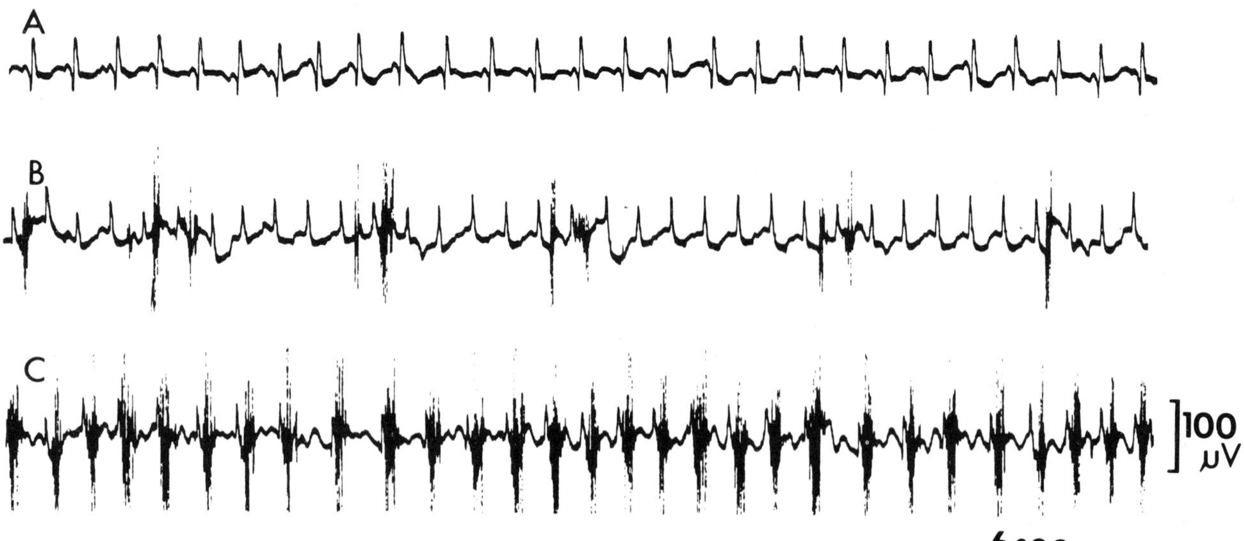

Fig. 4. Electrical recordings from a Quokka ileum. A, quiescent phase showing only slow wave activity; B, irregular phase where the slow wave activity has occasional bursts of action potentials interposed; C, regular phase where strong bursts of action potential activity occur between every sequential slow wave.

Table 1. Myoelectrical activity of the stomach and small intestine in Tammar Wallabies and Quokkas.

		Forestomach	Pylorus	Duodenal Bulb	Duodenum	Ileum
Slow wave frequency min^{-1}, x±SE (n=100)	Tammar	5.5±0.1	5.4±0.14	5.5±0.1	26±0.15	25±0.14
	Quokka	5.3±0.13	5.0±0.3	5.3±0.3	17.8±0.4	17.5±0.4
Slow wave amplitude range μV (n=100)	Tammar	50-240	80-400	50-70	38-276	25-200
	Quokka	50-450	120-300	50-220	70-130	30-180
Action potential range μV (n=100)	Tammar	—	20-300	30-150	25-360	50-250
	Quokka	—	20-420	45-400	45-280	15-360

The amplitude of the slow wave and action potentials as well as the frequency of the slow wave over the duodenum and ileum in both species is given in Table 1. Neither the Tammar nor the Quokka show any evidence of a slow wave frequency gradient over the length of the small intestine.

The sequence of electrical events most commonly recorded from any small intestinal electrode occurred in three distinct phases. A quiescent phase when only slow waves were recorded lasted for approximately 30 min. This was followed by a period of about 40 min when there was an irregular pattern of action potentials. Finally there was an approximate 5 min period of regular activity when action potentials occurred on every slow wave (Fig. 4). There was considerable variation in the duration of these phases particularly in quiescent and irregular phases. The transition from irregular to regular phase was gradual so that the precise time when the transition occurred was difficult to determine but the transition from regular to quiescent phase was abrupt. These phases migrated aborally over the duodenum at an estimated velocity of 5.5 cm/min and over the ileum at 1 cm/min. The regular phase took a mean of 5.1 min to pass an electrode on the duodenum and a mean of 5.8 min to pass one on the ileum. It was calculated that the regular activity involved an average length of 27.5 cm of the duodenum at any one time and that this reduced to 5.8 cm by the time it reached the ileum. Of those periods of regular activity that were recorded from the duodenal electrodes 39% did not appear at the ileal electrodes. The intervals between successive periods of regular activity in both species ranged between 42 and 172 min at the duodenum and 74 and 258 min at the ileum.

DISCUSSION

The electrical events recorded from these two species of marsupials are similar in many ways to those that have been reported from a number of eutherians (Grivel and Ruckebusch 1972; Szurszewski 1969; Galligan, 1985). The slow wave which is generally not associated with any mechanical activity appears to be all important in the regulation of contractions of some regions of the alimentary tract. It has been postulated that the wave length of the slow wave forms the physiological unit of the tract (Code *et al.* 1968). This view of the function of the slow wave is supported by the finding that action potentials will only occur on the slow wave at the height of depolarisation (Bortoff 1976). Therefore there is a strong indication that the frequency and rate of propagation of action potentials is controlled by the slow wave and, since action potentials are directly related to mechanical events (Ruckebusch 1970), the slow wave is controlling the frequency and rate of propagation of contractions.

In the simple-stomached animals that have been studied, no slow wave has been found over the fundic region of the stomach where contractions are controlled from higher centers via a vago-vagal reflex (Code and Carlson 1968). A slow wave is initiated in the more aboral regions of the stomach from a location that is species dependent. From this region peristaltic contractions move towards the pylorus at the same rate and frequency as the slow wave. The ruminants with their multicompartmental stomach only have slow waves over the more aboral regions of the abomasum. The regular cycle of contractions of the orad part of the abomasum, the omasum, rumen and reticulum is under central neural control mediated through the vagus nerve (Titchen 1968).

Finding a slow wave over the entire stomach in the Tammar and Quokka is a marked difference from the eutherians studied to date. It suggests that the muscular activity of the macropod stomach is basically an intrinsic property of the smooth muscle syncytium in the wall and that there is little or no control from higher nerve centres. The failure to find any evidence of action potentials over the more orad regions of the Tammar's stomach was surprising. The radiographic observations showed contractions of the haustra occurring at the same frequency as the slow wave and at the electrode sites so there is no doubt that contractions were taking place. Although it has been reported that small contractions can occur with the slow wave (Weisbrodt 1981) large forceful contractions such as observed over the haustrated region of the stomach have always been reported to be accompanied by action potentials. It is unlikely that this phenomenon was an artefact due to faulty technique as action potentials were recorded from the Quokka stomach and from the intestine of both species using the same methods. This aspect of the function of the Tammar's stomach warrants further investigation. The results of research on the slow wave of the eutherian small intestine have been reviewed by Weisbrodt (1981). In all eutherian species studied to date the slow wave has a decreasing frequency gradient from duodenum to ileum with the exception of the guinea pig which has no gradient (Galligan *et al.* 1985). The gradient is not a simple one but there is a series of plateaux: for example, in the dog the oral 30-50% of the intestine has the same slow wave frequency then it drops over a series of short plateaux. It has been suggested that this frequency pattern is the result of a series of relaxation oscillators at a number of locations in the alimentary tract wall (Sarna *et al.* 1971). In the Guinea Pig, Tammar and Quokka which do not demonstrate a slow wave gradient there may be only one oscillator, located close to the pylorus which would provide a simplified model for further studies on the control of muscular activity of the small intestine.

The three phases of small intestine action potential activity, the quiescent, the irregular and the regular phase which make up the migrating myoelectric complex occur in a sequence similar to that found in sheep (Bueno *et al.* 1975). Weisbrodt (1981) suggested that the velocity at which the migrating myoelectric complex migrates along the intestine is directly related to the length of the intestine so that in all species the time taken to traverse it is the same. As the small intestine in these two species approximates that of the dog it could be anticipated that the velocities would be similar. The velocities in the dog reported by Szurszewski (1969) were 5.3 cm/min in the duodenum and 1.5 cm/min in the ileum compared with 5.5 and 1 in both the Tammar and Quokka.

REFERENCES

Bayliss, W. M. and Starling, E. H., 1899. The movements and innervation of the small intestine. *J. Physiol. Lond.* **24**: 99-143.

Bieble, M., 1930. Graphische Darstellung der Darmbewegungen an der in einem Hautschlauch exventerierten Darmschlugl. *Klin. Wchnschr.* **9**: 1674-89.

Bortoff, A., 1961. Slow potential variations of the small intestine. *Amer. J. Physiol.* **201**: 203-08.

Bortoff, A., 1976. Myogenic control of intestinal motility. *Physiol. Rev.* **56**: 418-34.

Bozler, E., 1946. The relation of action potentials to mechanical activity in intestinal muscle. *Amer. J. Physiol.* **146**: 496-501.

Bueno, L., Fioramonti, J. and Ruckebusch, Y., 1975. The rate of flow of digesta and the electrical activity of the small intestine in dogs and sheep. *J. Physiol. Lond.* **249**: 69-85.

Cannon, W. B., 1898. The movements of the intestines studied by means of the roentgen rays. *Amer. J. Physiol.* **6**: 359-82.

Code, C. F. and Carlson, H., 1968. Motor activity of the stomach. *In* Handbook of Physiology ed. by C. F. Code and W. Heidel. American Physiology Society: Washington DC.

Code, C. F., Szurszewski, J. H., Kelly, K. A. and Smith, I. B., 1968. A concept of control of gastrointestinal motility. Pp. 2881-896 *in* Handbook of Physiology ed. by C. F. Code and W. Heidel. American Physiology Society: Washington DC.

Coombe, J. B., 1966. Observations on the motility of the small intestine in sheep. *Aust. Soc. Anim. Produc.* **6**: 407-12.

Galligan, J. J., Costa, M. and Furness, J. B., 1985. Gastrointestinal myoelectric activity in conscious guinea pigs. *Amer. J. Physiol.* **294**: G92-G99.

Grivel, M. L. and Ruckebusch, Y., 1972. Propagation of segmental contractions along the small intestine. *J. Physiol. Lond.* **227**: 611-25.

Hightower, N. C., 1952. Comparison of intraluminal pressures and motility patterns in the gastrointestinal tract of human beings when recorded by balloons and direct electrical transducer pressure system. *Fed. Proceed.* **11**: 69-81.

Hiatt, R. B., Goodman, I. and Bircher, R., 1966. Control of motility in Thiry-Vella ileal segments in dogs. *Amer. J. Physiol.* **210**: 373-78.

Hume, I. D., 1982. The digestive physiology of marsupials. *Comp. Biochem. Physiol.* **71A**: 1-10.

Ramorino, M. L. and Colagrande, C., 1964. Intestinal motility: preliminary studies with telemetering capsules and synchronized fluorocinematography. *Amer. J. Digest. Di.* **9**: 64-71.

Richardson, K. C. and Wyburn, R. S., 1983. The electromyographic events in the stomach and small intestine of a small kangaroo, the Tammar wallaby *(Macropus eugenii)*. *J. Physiol. Lond.* **342**: 453-63.

Ruckebusch, Y., Grivel, M. L. and Santini, R., 1968. Technique de derivation chronique de l'activite electrique et mechanique de la paroi intestinale chez l'animal eveille. *B.S.Vet. Med. Comp.* **70**: 337-46.

Ruchebusch, Y., 1970. The electrical activity of the digestive tract of sheep as an indication of the mechanical events in various regions. *J. Physiol. Lond.* **210**: 857-82.

Ruckebusch, Y. and Bueno, L., 1976. The effect of feeding on the motility of the stomach and small intestine of pigs. *Brit. J. Nutr.* **30**: 491-99.

Ruckebusch, Y. and Bueno, L., 1977a. Origin of migrating myoelectrical complex in sheep. *Amer. J. Physiol.* **233**: E483-87.

RUCKEBUSCH, Y. AND BUENO, L., 1977b. Migrating myo-electrical complex of the small intestine. An intrinsic activity mediated by the vagus. *Gastroenter.* **73**: 1309-314.

SARNA, S. K., DANIEL, E. E. AND KINGMA, Y. J., 1971. Simulation of slow wave electrical activity of small intestine. *Amer. J. Physiol.* **221**: 166-73.

SMITH, I. B., 1959. A Study of Small Intestinal Motility in Dogs. Thesis, Cambridge University.

SZURSZEWSKI, J., 1969. A migrating electrical complex of the canine small intestine. *Amer. J. Physiol.* **217**: 1757-763.

TITCHEN, D. A., 1958. Partial exteriorisation of the reticulum in sheep. *J. Physiol. Lond.* **143**: 35P.

TICHEN, D. A., 1968. Nervous control of motility of the fore-stomachs of ruminants. *In* Handbook of Physiology ed. by C. F. Code and W. Heidel. American Physiological Society: Washington DC.

WEISBRODT, N. W., 1981. Motility of the small intestine. *In* Physiology of the Gastrointestinal Tract ed. by L. R. Johnson. Raven Press: New York.

WYBURN, R. S., 1981. A Radiological Study of the Patterns of Contradiction and Digesta Movement in the Alimentary Tract of Sheep. Thesis, Massey University.

Radiographic studies on the form and function of the gastrointestinal tract of the Woylie (*Bettongia penicillata*)

K. C. RICHARDSON[1]

The structure and motility of the alimentary tract of the Woylie, *Bettongia penicillata,* was examined by dissection and radiographic techniques. Contrast agents (barium sulphate and radio-opaque pellets) took a long time to be cleared from the large saccular part of the stomach. Small pellets (1×1 mm) passed through the pylorus faster than did large pellets (1×2-3 mm) ($p<0.001$). Small pellets were completely cleared from the stomach and from the entire intestine faster than the large ones ($p<0.001$). The ratio of the length of the small intestine to that of the large intestine is 0.8 compared with 3.0 for the Tammar Wallaby *(Macropus eugenii)* which supports the hypothesis that the large intestine is a site for fermentation in the Potoroinae.

INTRODUCTION

THE sigmoidally arranged tubular stomach of the kangaroo was first described by Home (1814) when he made the pertinent observation that "The stomach of the kanguroo differs in many particulars from that of any known animal and bears a greater resemblance to the human caecum and colon than to any stomach". Unlike the main groups of large terrestrial herbivorous eutherians, kangaroos do not have a simple stomach as is found in the Perissodactyla nor do they have a multi-chambered stomach as is the case in the Artiodactyla (Dyce *et al.* 1987) They have a tubular stomach where microbial breakdown of digesta occurs in the orad four fifths of its length (Hume 1982). This form of stomach functions successfully if one gauges success on; the number of macropodid species, the number of individual animals, the number of different habitats in which they continue to be found, and the different dietary options that they have adopted (Hume 1982; Strahan 1983).

The superfamily Macropodoidea is divided into two distinct families, the Macropodidae comprising the kangaroos and wallabies, and the Potoroidae comprising the bettongs, potoroos and the rat-kangaroos (Strahan I983). The Potoroidae has two subfamilies, the Hypsiprymnodontinae with only one extant species *Hypsiprymnodon moschatus* which are omnivorous and have a simple stomach quite unlike that of the rest of the Macropodoidea (Hume 1982; Strahan 1983). The second subfamily, the Potoroinae, probably has six extant species (Strahan 1983) about which little is known. Some dig for food and have been reported to forage extensively for fungi, particularly after fire (Kitchener 1967; Guiler 1971; Kinnear *et al.* 1979; Christensen 1980). Such an unusual diet may explain why the potoroine tubular stomach is different from that of the Macropodidae (Shultz 1976; Kinnear *et al.* 1979; Langer 1980; Hume 1982; Hume and Carlisle 1985; Frappel and Rose 1986). This study aims to describe the form and motility of the alimentary tract of the Woylie *(Bettongia penicillata)* one of the rare potoroines of southern Australia.

MATERIALS AND METHODS

Animals

A small breeding group of five Woylies *(Bettongia penicillata)* was held in captivity in completely enclosed outside pens. These pens had a concrete pad 3×4 m at one end adjoining a run of 6×4 m with a deep sand floor. Animals were fed *ad libitum*, a diet of mixed vegetables with moistened dry dog food. Water was always available. A further five adult animals were donated for gross anatomy study after having been found dead or been euthanased on humanitarian grounds.

Gross anatomy

The five dead Woylies were dissected from the flank after removal of either the left (3) or the right (2) abdominal wall. The lengths of the saccular and tubular parts of the stomach were measured along their greater curvature. The lengths and circumferences of the small intestine, large intestine, and caecum were measured. Sequential photographs were taken throughout the dissections. Measurements of mesenteric lengths were taken to the nearest body wall or organ attachment at various sites along the alimentary tract.

Earlier terminology of potoroine stomachs is descriptive in nature and slightly misleading. The term sacciform as used by Langer 1980; Hume

[1]School of Veterinary Studies, Murdoch University, Murdoch, Western Australia 6150.

Pages 205-15 in KANGAROOS, WALLABIES AND RAT-KANGAROOS ed. by G. Grigg, P. Jarman and I. Hume. Surrey Beatty & Sons Pty Limited, New South Wales, Australia, 1989.

and Carlisle 1985 and Frappell and Rose 1986 is avoided, instead sacculated referring to the numerous haustra is used. Sacculated with its antonym non-sacculated are appropriate terms whilst saccular and tubular are simple, readily understood and nonambiguous. This report uses the terms saccular part, tubular part and pyloric part to distinguish distinct parts of the stomach. The nomenclature used is close to that recommended by the International Committee in Veterinary Anatomical Nomenclature (Habel *et al.* 1983).

Radiography

Each animal was maintained on a fixed diet of vegetables and moistened dried dog food. At 8 a.m. the animal to be examined was placed in a hessian sack where it remained for the next eight hours. Previous studies of Tammar Wallabies *(Macropus eugenii)* restrained within hessian sacks showed that both respiratory and cardiac rates remained within the normal range (Richardson and Wyburn 1983). Animals were radiographed using a condenser-discharge mobile X-ray apparatus (Shimadzu model M.C.I00 L-10) using high speed film (Ilford) and intensifier screens (Dupont) at a film focus distance of 1 meter. Exposure was at 4 mAs and 54 kV (lateral view) and 57 kV (dorsoventral view). Animals were held on the X-ray table in left lateral or sternal recumbency by gentle physical restraint. No chemical tranquillization was used.

(1) BARIUM SULPHATE STUDIES

Four animals were each dosed on four separate occasions with ten millilitres of ultrafine 65% w/w suspension of barium sulphate (Medebar by Medicon). This was given by syringe into the oropharynx. Radiographs both lateral and dorsoventral, were taken at 5, 10, 15, 30, 45, and 60 minutes. Then hourly until the eighth hour. The animal was returned to its pen where it had access to its normal diet and to water. It was then radiographed at 8 a.m. on each day until no contrast agent was visible. Times at which the contrast agent was first seen to pass through the pylorus, enter the colon, be incorporated into the faecal pellets were noted. So were the times that it took for the barium sulphate to be lost from the stomach or the entire gastrointestinal tract.

(2) MOVEMENT OF RADIO-OPAQUE PARTICLES

Each animal was radiographed in eight separate sessions following the oral administration of either (1×1 mm) or (1×2-3 mm) neutral buoyancy radio-opaque pellets. Between 40 and 120 pellets were used in any one experiment. Radiographs were taken in the same manner and sequence as for the barium sulphate studies.

(3) FLUOROSCOPY

An x-ray apparatus (Siemens) with automatic dose control of 10 microSieverts in any half hour period was used to radiograph continuously four individual animals immediately after they had been given ten millilitres of 65% w/w suspension of barium sulphate (Medebar by Medicon) while they stood in a perspex crush. Individual animals were viewed on three separate occasions, each separated by one week intervals. Images were transmitted through an image intensifier (Dualfield 25/15 Sirecon 2, Siemens) viewed by closed circuit television and recorded on videotape (Sony KCA60). Patterns of movement were subsequently analysed by playing back the videotapes. Then representative drawings of the movements were made.

RESULTS

Gross anatomy

A drawing of the gastrointestinal tract of *B. penicillata* is shown in Figure 1. The intra-abdominal oesophagus is a tube (non-distended diameter 5 mm) which runs obliquely ventrocaudally from its diaphragmatic hiatus to enter the medial aspect of the aboral fourth of the stomach. The stomach is tubular in form with its blind end portion being distinctly haustrated (Fig. 2) and its open end, towards the pylorus, being devoid of haustra.

The largest segment of the stomach is the succular part, which together with the oral half of the tubular part, account for four-fifths of the stomach's total volume. The saccular part extends

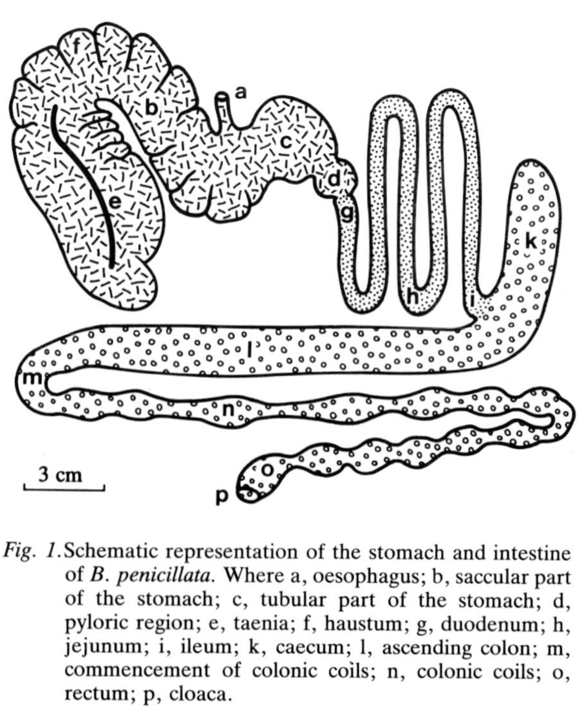

Fig. 1. Schematic representation of the stomach and intestine of *B. penicillata*. Where a, oesophagus; b, saccular part of the stomach; c, tubular part of the stomach; d, pyloric region; e, taenia; f, haustum; g, duodenum; h, jejunum; i, ileum; k, caecum; l, ascending colon; m, commencement of colonic coils; n, colonic coils; o, rectum; p, cloaca.

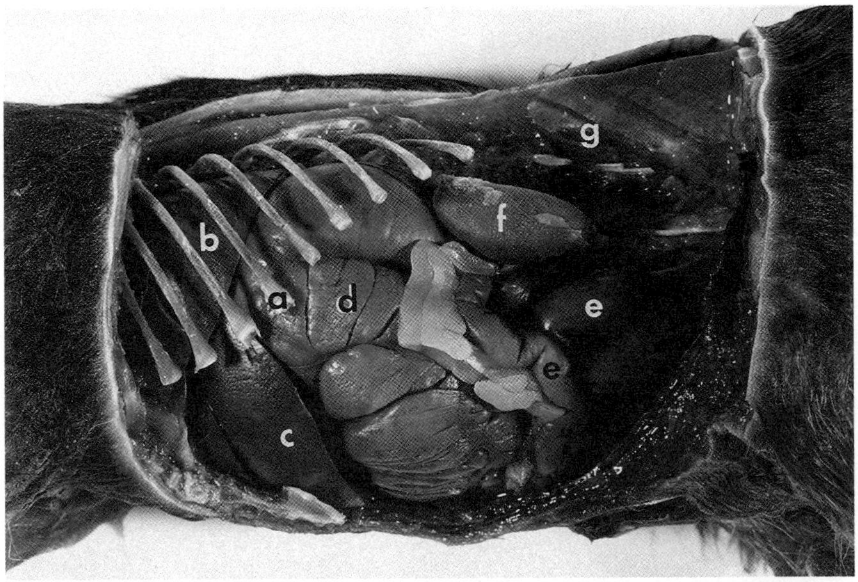

Fig. 2. Left flank exposure of the abdomen of *B. penicillata.* a, rib; b, lung; c, liver; d, haustra of saccular part of stomach; e, distal colon; f, left kidney; g, epaxial musculature.

to the dorsal left flank where it abuts the diaphragm (Fig. 2). There it runs obliquely ventral and caudally to terminate blindly (blind sac) on the floor of the abdomen. Most of this segment has distinct haustra, separated by deep sulci running transversely across the axis of the organ. A particularly prominent taenia runs over the lateral wall of the saccular part of the stomach (Fig. 2). This taenia runs from the tip of the blind sac over the greater curvature to terminate 2-3 cm orad to the cardia.

The tubular part of the stomach is only a few centimetres in length and lies transversely across the dorsocranial portion of the abdomen. Its aboral portion having the proper gastric glands unlike in the Macropodinae, is difficult to see on gross examination. A distinct sulcus, running completely around the viscus, delineates the transition from tubular part to pyloric part of the stomach. The pyloric antrum is about 1.5 cm in length and has a maximal diameter of about 1.5 cm. The pyloric part lies medial to the dorsal right lateral lobes of the liver. It is directed caudally and slightly dorsally. A shallow sulcus indicates the end of the pylorus and commencement of the duodenum.

The descending arm of the duodenum runs caudally in the dorsal right of the abdominal cavity for about 3 cm towards the pelvic cavity entrance where it turns medially at the duodenal

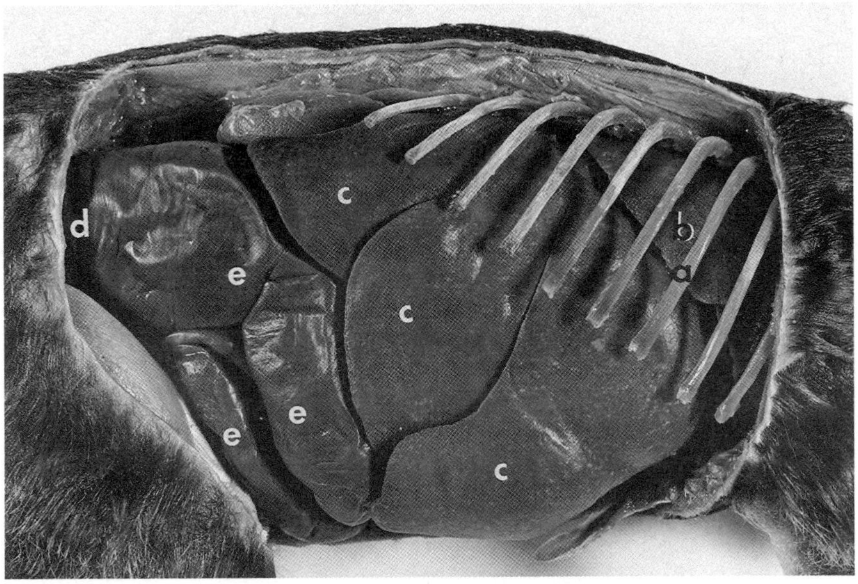

Fig. 3. Right flank exposure of the abdomen of *B. penicillata.* a, rib; b, lung; c, liver; d, caecum; e, ascending colon.

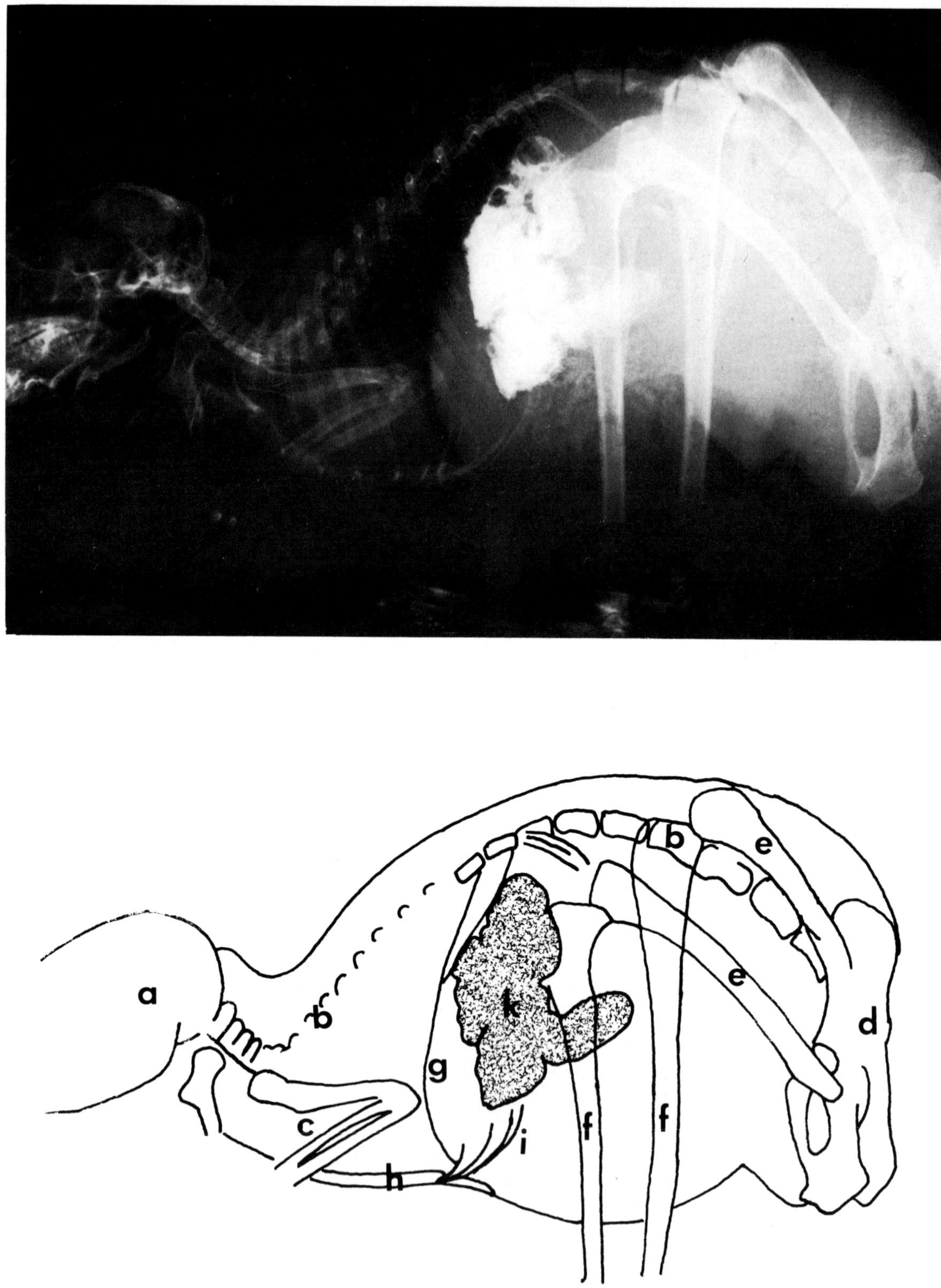

Fig. 4. Left lateral radiograph of a standing *B. penicillata.* a, head; b, vertebral column; c, forelimb; d, pelvis; e, femur; f, tibia; g, diaphragmatic silhouette; h, sternum; i, costal arch; k, barium sulphate in the stomach.

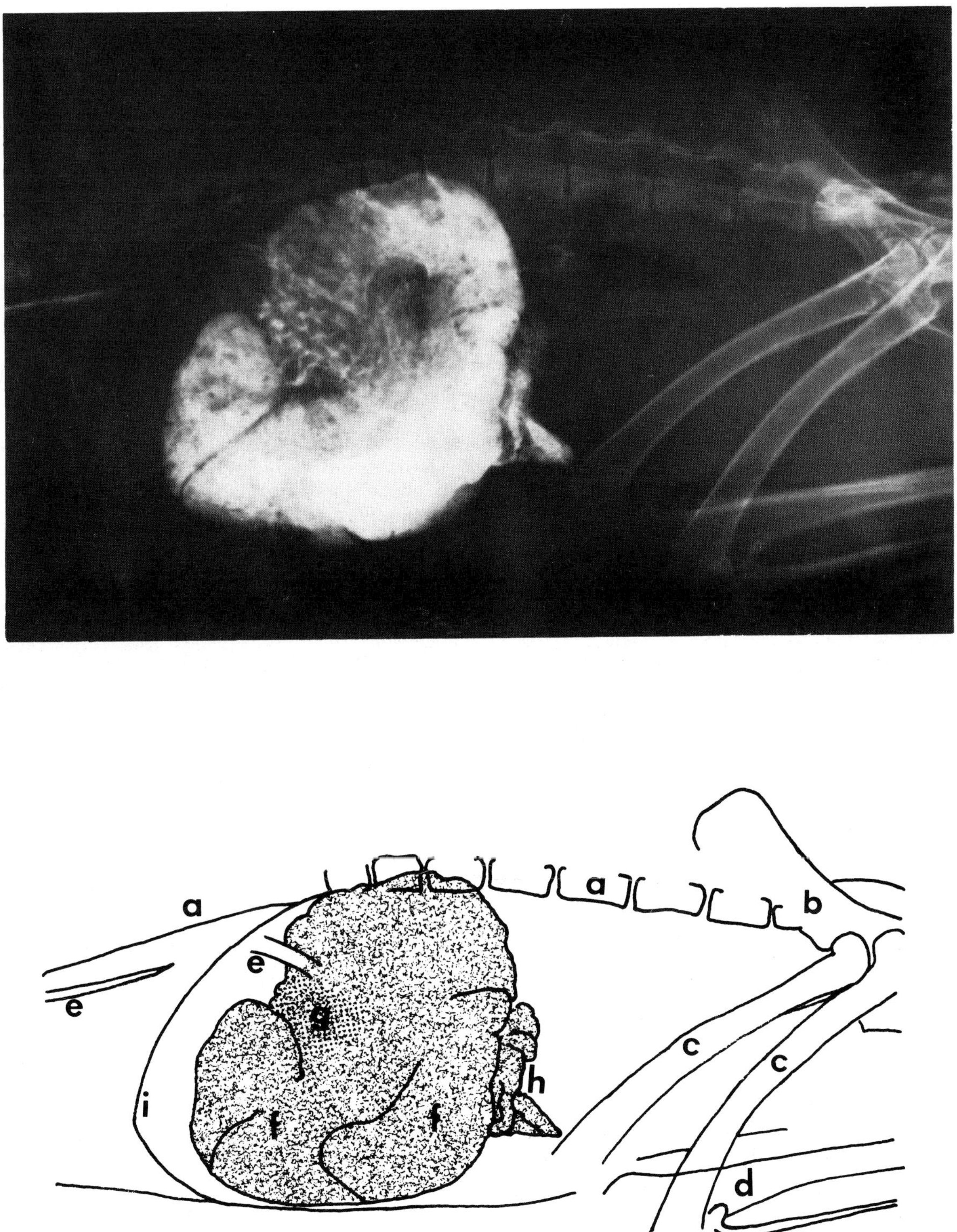

Fig. 5. Left lateral radiograph of *B. penicillata*. a, vertebral column; b, pelvis; c, femur; d, tibia; e, barium sulphate in oesophagus; f, barium sulphate in the saccular part of the stomach; g, barium sulphate outlining irregular hexagonal luminal projections; h, barium sulphate in small intestine; i, diaphragmatic silhouette.

flexure to run cranially as the ascending duodenum. The two arms of the duodenum lie close to each other being separated by the bulk of the pancreas. A short mesoduodenum (1.5-2 cm) firmly anchors the duodenum in place. The mesoduodenum has a short (2 cm) subsidiary leaf which fuses with the short gastrohepatic ligament joining the medial pyloric and tubular parts of the stomach to the liver. On the opposite side of the pylorus a gastrocolic ligament (1-1.2 cm) is prominent. The jejunum and ileum generally lie caudoventrally to the duodenum. The mesojejunoileum is long (12 cm) allowing this viscus a great deal of mobility.

The ileocaecal junction lies in the right ventrocaudal quadrant of the abdomen. The caecum is quite variable in its dimensions, being small and pyriform when fully contracted and elongate and cylindrical at other times. Usually it has its base on the right but its apex lies either on the right towards the pelvic entrance or is directed caudomedially across the abdomen to the left side. The caecum does not have haustra or taenia. The lumen of the caecum is continuous with that of the ascending colon (Fig. 3). This segment usually lies on the dorsal right and runs forwards so that its aboral portion is anchored to the adjacent tubular and pyloric parts of the stomach by the short (1-1.5 cm) gastrocolic ligament. The ascending colon has a wide lumen which narrows dramatically in the next colonic segment, the transverse colon, where the gastrocolic ligament is absent. It is here that the first faecal pellets may be seen. The subsequent colonic coils have an elongate mesentery allowing them a great deal of movement within the abdomen. The final segment, the descending colon, runs to the midline just caudal to the kidneys and then directly back to enter the pelvic cavity. Here it has a short mesentery and runs through the cavity to terminate in the cloaca (Fig. 1). The large intestine is equal to or longer than the small intestine.

Dimensions of the gastrointestinal tract are given in Table 1.

Table 1. Measurements of the gastrointestinal tract (cm).

	N	Mean+S.E.	Range
Length of saccular stomach along greater curvature	5	15.6+1.208	11-18
Length of tubular stomach along greater curvature	5	6.6+0.430	5-7.5
Length of descending duodenum	5	3.1+0.187	2.5-3.5
Length of ascending duodenum	5	3.6+0.187	3-4
Circumference of duodenum	5	2.5+0.204	2-3
Length of entire small intestine	5	46.6+5.154	30-56
Circumference of small intestine	5	2.3+0.095	2-3
Length of caecum	5	6.1+0.696	3.5-7.5
Circumference of caecum	5	5.4+0.519	3.2-6.0
Length of large intestine	5	58.8+6.689	34-71
Circumference of large intestine	5	3.81+0.512	1.5-7.5

Radiography

Plain radiographs show the stomach has a uniformly coarse granular appearance. Often gas bubbles delineate short segments of the intestine but rarely adequately to be of any use in determining the anatomy of the region. The normal anatomy could not be determined by plain radiographs alone. On the other hand contrast studies are more useful to determine the anatomy and movement patterns of the stomach and intestine.

Barium sulphate

Occasionally once through the cardia the barium sulphate passes directly into the tubular stomach and quickly into the pyloric antrum. In such cases the stomach is emptied of contrast agent within 4-5 hours. More often barium sulphate passes through the cardia and into the sacculated part of the stomach (Fig. 4) where it may take up to two hours for it to begin to pass through the pylorus (Fig. 5). The minimum time for the barium sulphate to pass completely from the stomach was six hours. Transit time along the small intestine was variable, with contrast agent usually appearing at the ileocaecal junction in about three hours after passing through the pylorus (Fig. 6).

Total elimination of the barium sulphate from the tract took a mean time of 39.4 hours (Table 2).

Movement of radio-opaque particles

Passage of radio-opaque pellets through the alimentary tract was much slower than the barium sulphate (Figs 7 and 8). The mean + SE as well as the range of transit times for large and small pellets is given in Table 2. The slower passage of the large pellets when compared to the small pellets is significant ($p<0.001$) at each of the three sites recorded.

There was a significant difference ($p<0.001$) between small and large pellets in the time it took for their passage through the alimentary tract: (a) to initially pass through the pylorus (b) until the stomach was empty (c) and until the entire alimentary tract was empty.

Fluoroscopy

Immediately following administration, barium was seen to move rapidly through the oesophagus without slowing at any site within the thorax (Fig. 4). It showed the oesophagus to be straight in its path through the abdomen before terminating on the medial aspect of the junction between

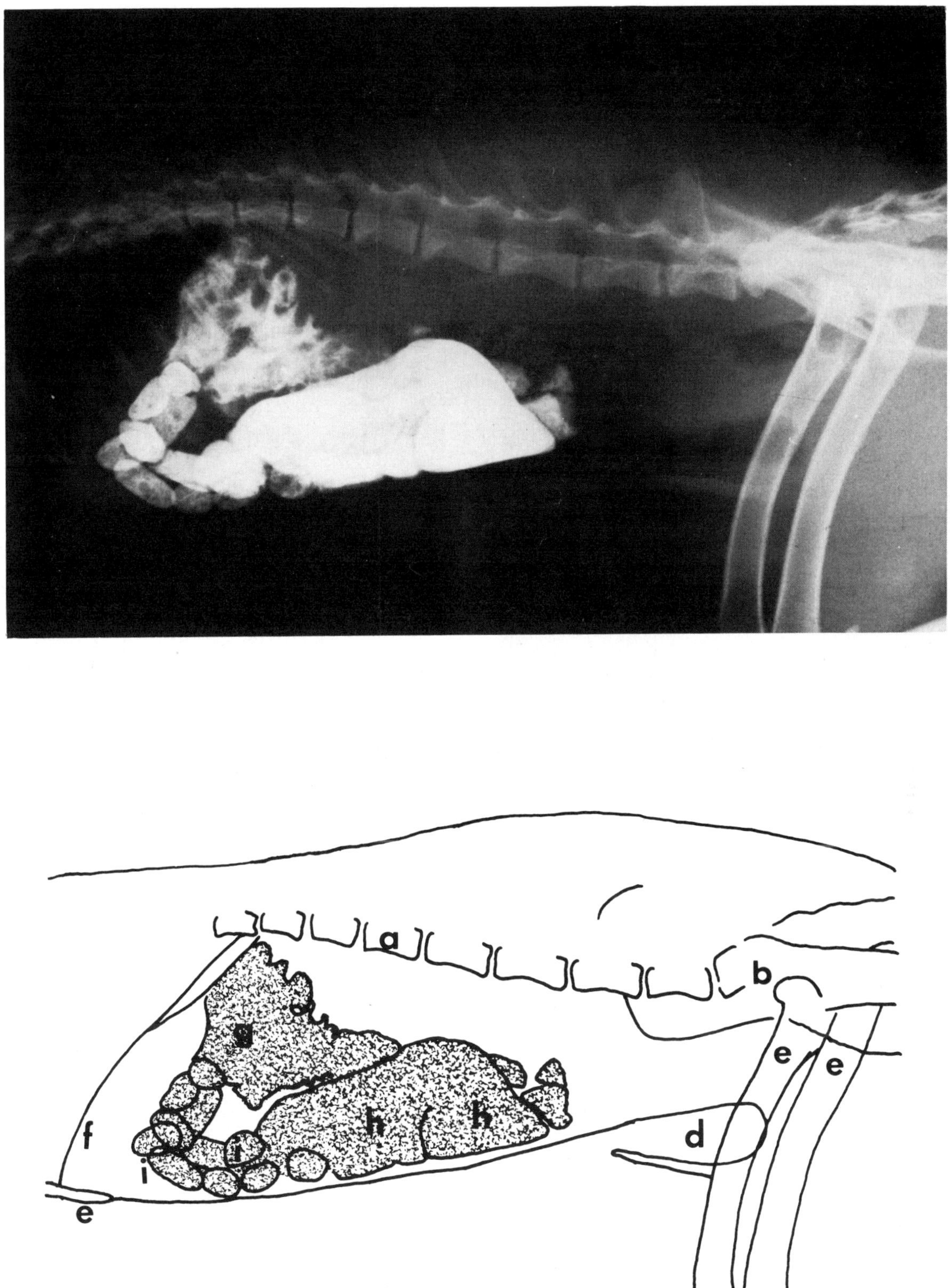

Fig. 6. Left lateral radiograph of *B. penicillata.* a, vertebral column; b, pelvis; c, femur; d, pouch; e, sternum; f, diaphragmatic silhouette; g, barium sulphate in the tubular and pyloric parts of the stomach; h, barium sulphate in the ascending colon and caecum; i, faecal pellets in the colonic coils.

Table 2. Transit times (hrs) of contrast agent through the alimentary tract.

	N	Initial passage	Empty stomach	Empty tract
Barium Sulphate				
mean+SE	16	0.80+0.110	7.94+ 0.798	38.37+1.400
(range)		(0.25-2)	(4-12)	(30-48)
Small pellets				
mean+SE	16	2.70+0.326	15+1.018	24.88+2.326
(range)		(1.5-3)	(10-18)	(16-33)
Large pellets				
mean+SE	16	5.63+0.498	22.25+1.098	35.25+1.729
(range)		(4-7)	(16-26)	(24-40)

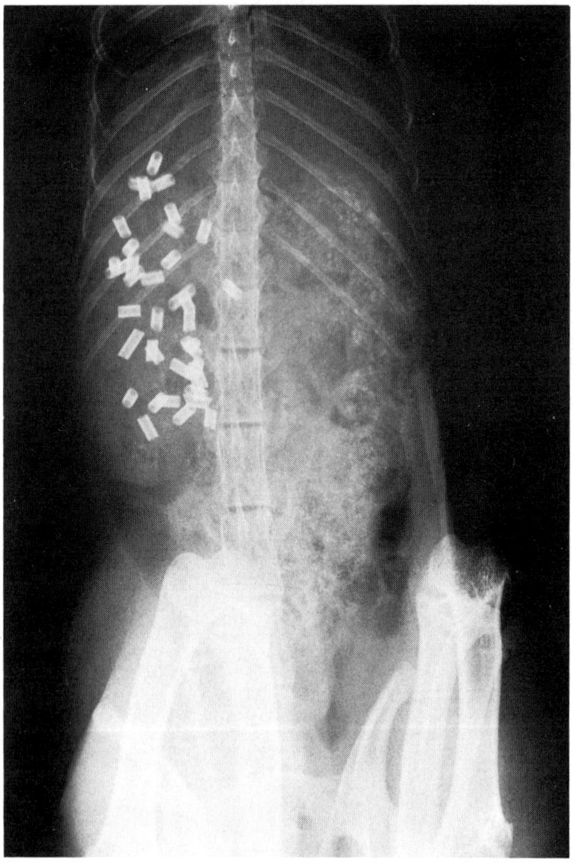

Fig. 7. Dorsoventral radiograph of *B. penicillata* showing large pellets retained in the saccular part of the stomach, (shaded); a, vertebral column; b, diaphragmatic silhouette; c, femur; d, pelvis; e, saccular stomach; f, granular appearance of digesta and faeces in the alimentary tract.

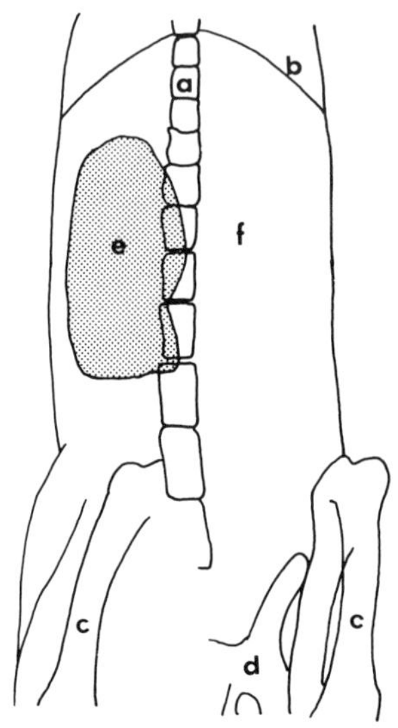

the distal fourth to fifth to the stomach. No structure was seen to direct digesta either into the saccular part or into the tubular part. However, once the barium was in the saccular part it remained there for a considerable time outlining the haustra. Three types of movement were seen: localized contractions, sequential contractions and obliterative movements. The localized contractions occurred at random intervals on different haustra. In these cases the haustrum was seen to narrow in width and to shorten in height over 1-2 seconds then to cease movement for the following 1-2 seconds before returning to its original size during the next 1-2 seconds. Sometimes a particular haustrum was seen to balloon well beyond its normal silhouette (Fig. 9).

The reoccurrence of a contraction and relaxation of any particular haustrum appeared to be random. When this cycle of contraction and relaxation was seen to move sequentially from one haustrum to the next aboral haustrum and for this to continue aborally, the pattern was categorized as a sequential contraction.

On a few occasions the two luminally projecting ends of a haustrum were seen to move towards each other (Fig. 10). In such cases the edges closer to the greater curvature appeared to move most rapidly. Ultimately the two walls merged thus obliterating their lumen. Generally, after a few seconds the walls reappeared, diverged from each other until finally re-establishing their normal form (Fig. 10). This pattern of events was classified as an obliterative movement. These events were most common on the proximal saccular part of the stomach but were virtually absent from the blind end of the saccular part as well as the tubular part of the stomach. While the three patterns of movement were the dominant feature

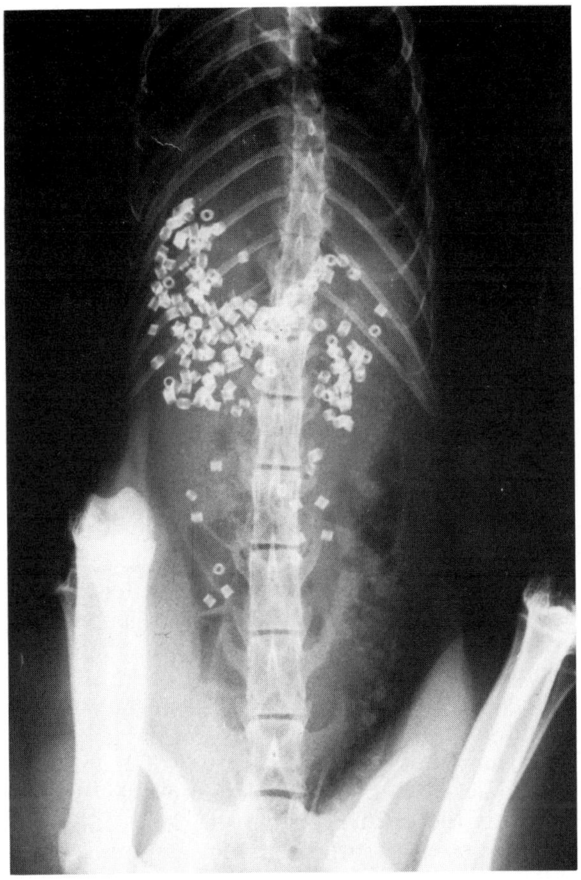

Fig. 8. Dorsoventral radiograph of *B. penicillata* showing small pellets retained primarily in the stomach, (shaded) with a few in the small intestine (scattered).
a, vertebral column;
b, diaphragmatic silhouette;
c, sternum; d, pelvis;
e, femur;
f, saccular part of the stomach;
g, tubular part of the stomach;
h, pyloric region.

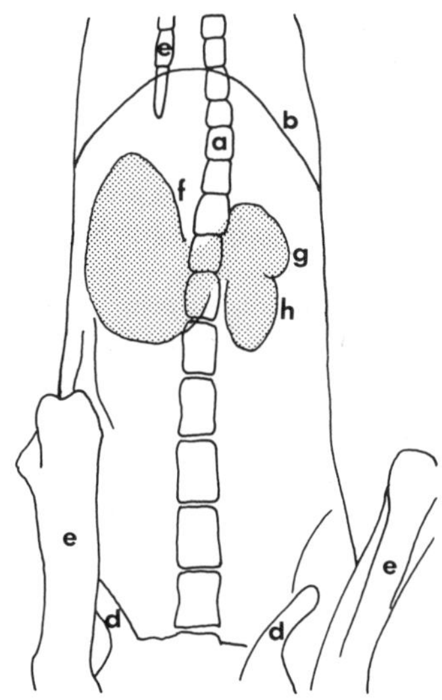

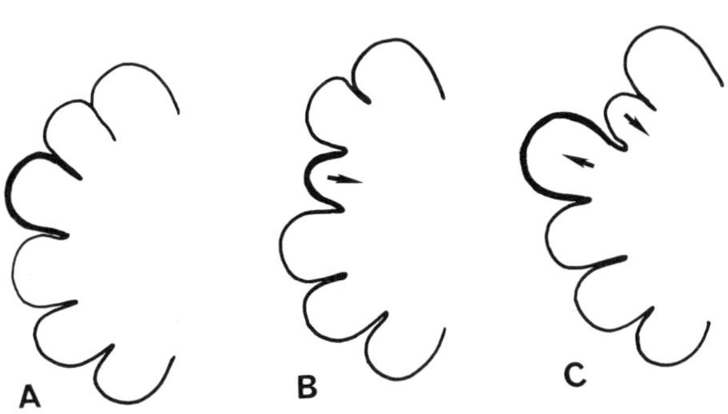

Fig. 9. Schematic representation of haustra movement over the saccular part of the stomach of *B. penicillata*. A, quiescence, the particular haustrum to be considered has a darkened outline; B, contraction of the haustrum; C, relaxation of the haustrum with contraction of the adjacent haustrum.

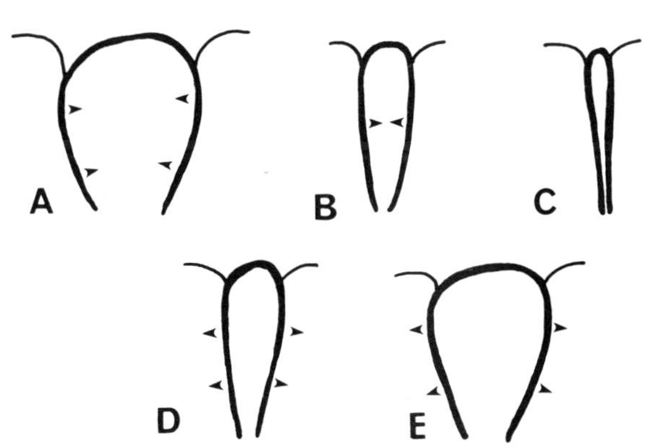

Fig. 10. Schematic representation of an obliterative cycle of haustral movement in *B. penicillata*. Movement of the semilunar folds of the haustra is indicated by arrows. A, B, C, are progressive and sequential movement of the vertical walls towards each other. D and E show their return to their original position as in A.

seen during fluoroscopy there were often periods of up to 25 min when no movement of the stomach was seen. Overall the three movement patterns appeared to be quite effective in mixing digesta although it was relatively slow. On a few occasions digesta was seen to move retrograde to normal, particularly across the outer (greater curvature) portions of the haustra. Restraints of X-ray exposure prevented determination of how digesta circulated within the blind end of the sacculated part of the stomach. This also prevented critical studies on how digesta, on occasion, was able to bypass the saccular region, enter the tubular region and rapidly pass through to the pylorus. Intestinal studies confirmed the gross morphology and topography as determined by dissection and simple contrast radiography.

DISCUSSION

This study confirms the form and function of the stomach and intestine of the Woylie *(Bettongia penicillata)* to be similar to that of the alimentary tracts of other potoroines described over the last decade (Langer 1979, 1980; Hume and Carlisle 1985; Frappell and Rose 1986). There is general agreement that the stomach of the potoroines has an elongate sacculated proximal portion and a much smaller and shorter non-sacculated but simply tubular distal portion. Whilst there has been some confusion due to different terminology there has been agreement that while the potoroine stomach resembles that of the macropodinae it is in reality quite different. Digesta, highlighted by contrast agent, passes through the entire length of the stomach in the macropodinae (Dellow 1979; Richardson 1980) but in the potoroine stomach because the cardia is in its oral fifth, digesta may on some occasions pass only through that terminal segment (Hume and Carlisle 1985; Frappell and Rose 1986). Hume and Carlisle (1985) suggest that the extent that digesta bypasses the large sacculated part of the stomach could be due to circadian factors such as the normal pattern of ingestion. They believe, quite reasonably, that the amount of food bypassing the blind ending saccular part is least in the early evening and greatest towards morning. They found that contrast agent was completely cleared from the stomach in nine hours in the Long-nosed Potoroo *(P. tridactylus)* and that it took up to 24 hours in a Rufous Rat-kangaroo *(A. rufescens)*. Frappell and Rose (1986) also studying *P. tridactylus* found that most barium sulphate given as a bolus dose bypassed the sacculated part of the stomach. No contrast agent was visible in the stomach of these animals within 10 minutes. However, when barium sulphate was given in conjunction with food it took 60 minutes for total clearance to occur. This is radically different to the retention times of up to 12 hours found in this study. The shortest time for barium sulphate to pass through the stomach was four hours when it had bypassed the sacculated part and simply moved through the aboral portion of the tubular stomach then the hind stomach into the duodenum. This happened only infrequently.

When contrast pellets were used, quite different results were obtained. The small pellets first entered the duodenum in a mean time of 2.7 hours while it took 5.6 hours for this to happen with the large pellets. This is in close agreement with Hume and Carlisle (1985) who, using a slightly different technique found radio-opaque threads passed through in 1.5-2 hours in three individuals but in greater than 8 h in a fourth. Total transit time of the pellets through the stomach took a mean of 15 h for small and 22.2 h for large pellets. There is a significant difference in the transit times for large and small pellets ($p < 0.001$) which may possibly be due to the pyloric ostium acting as a differential sorting device as suggested by Kelly (1981). He reported that in dogs, small sized particles i.e., < 2 mm diameter pass rapidly through the pyloric orifice while it took days (unspecified) for pellets having a diameter of 7 mm to pass. A similar possible differential sorting may be found in the Australian Sea-lion *(Neophoca cinerea)* where all small pellets (1 mm) were recovered from their faeces within four days but recovery of large pellets (> 3 mm) took more than six days (Richardson and Gales 1987). Morphologically the pyloric region is the most probable site for sorting the different sized particles thus causing their differential passage through the alimentary tract.

The overall form of the intestine was unremarkable. However the ratio of the length of the small intestine to the length of the large intestine is quite different from that found in macropodine species. That of the Woylie is 0.8 compared with 3.0 for the Tammar Wallaby. Assuming that small intestine length relative to body size/weight as well as factors such as metabolic rate, fermentation rate and diet are comparable in the two species, then the relatively longer large intestine of the Woylie is an important structural modification. As a consequence of a relatively longer large intestine having a similar circumference to their counterpart small macropodines such as the Quokka *(Setonix brachyurus)* and the Tammar Wallaby, Woylies must have a greater mucosal surface area. Hume and Carlisle (1985) found that the hindgut (large intestine) capacity as a proportion of total gut capacity was about 35% in *Aepyprymnus* compared to 11-17% for three macropodid species (Dellow 1979). These features all suggest that the large intestine of the potoroines is an important site for microbial fermentation.

ACKNOWLEDGEMENTS

This study has been kindly given support by the E. A. Weston and I. E. Fernie Research Fund. I would like to thank Mr and Mrs A. C. Richardson for their continued encouragement as well as for their financial assistance. I would also like to thank Associate Professors R. Wyburn and J. Grandage for their critical review of this paper. Mr J. Atherton and Mr J. Herd gave a great deal of assistance with the radiography. Mr. G. Griffiths did the photography.

REFERENCES

CHRISTENSEN, P. E. S., 1980. The biology of *Bettongia penicillata* (Gray 1837), and *Macropus eugenii* (Desmarest 1817) in relation to fire. *West. Aust. Dept. For. Bull.* No. 91 **1:** 62-72.

DELLOW, D. W., 1979. Physiology of digestion in the macropodine marsupials. Ph.D. Thesis, University of New England, Armidale.

DYCE, K. M., SACK, W. O. AND WENSING, C. J. G., 1987. Textbook of Veterinary Anatomy. Saunders: Philadelphia.

FRAPPELL, P. B. AND ROSE, R. W., 1986. A radiographic study of the gastrointestinal tract of *Potorous tridactylus*, with a suggestion as to the role of the foregut and hindgut in potoroine marsupials. *Aust. J. Zool.* **34:** 463-71.

GUILER, E. R., 1971. Food of the potoroo (Marsupialia, Macropodidae). *J. Mammal.* **52:** 232-34.

HABEL, R. E., FREWEIN, J. AND SACK, W. O., (eds), 1983. *Nomina Anatomica Veterinaria* 3rd edition. International Committee on Veterinary Anatomical Nomenclature: Ithaca, New York.

HOME, E., 1814. Lectures on Comparative Anatomy. Bulmer and Co.: London.

HUME, I. D., 1982. Digestive Physiology and Nutrition of Marsupials. Cambridge University Press: Cambridge.

HUME, I. D. AND CARLISLE, C. H., 1985. Radiographic studies on the structure and function of the gastrointestinal tract of two species of potoroine marsupials. *Aust. J. Zool.* **33:** 641-54.

KELLY, K. A., 1981. Motility of the stomach and gastroduodenal junction. Pp. 393-410 *in* Physiology of the Gastrointestinal Tract ed. by L. R. Johnson. Raven Press: New York.

KINNEAR, J. E., COCKSON, A., CHRISTENSEN, P. AND MAIN, A. R., 1979. The nutritional biology of the ruminants and ruminant-like mammals a new concept. *Comp. Biochem. Physiol. Ser. A.* **64:** 357-65.

KITCHENER, D. J., 1967. Some aspects of the biology of the potoroo – *Potorous tridactylus* (Kerr). B.Ag.Sc.Hons. Thesis, University of Tasmania.

LANGER, P., 1979. Phylogenetic adaptation of the stomach of the Macropodidae Owen, 1839, to food. *Z. Saugetierkunde* **44:** 321-33.

LANGER, P., 1980. Anatomy of the stomach in three species of Potoroine (Marsupialia: Macropodidae). *Aust. J. Zool.* **28:** 19-31.

RICHARDSON, K. C., 1980. The structure and radiographic anatomy of the alimentary tract of the Tammar wallaby, *Macropus eugenii* (Marsupialia). I. The stomach. *Aust. J. Zool.* **28:** 367-79.

RICHARDSON, K. C. AND GALES, N. J., 1987. Functional morphology of the alimentary tract of the Australian Sealion, *Neophoca cinerea. Aust. J. Zool.* **35:** 219-26.

RICHARDSON, K. C. AND WYBURN, R. S., 1980. The structure and radiographic anatomy of the alimentary tract of the Tammar wallaby, *Macropus eugenii* (Marsupialia). II The intestines. *Aust. J. Zool.* **28:** 499-509.

SCHULTZ, W., 1976. Magen-Darm-Kanal der Monotremen und Marsupialier. *Handbuch der Zoologie.* **8:** 1-177.

STRAHAN, R., 1983. Complete Book of Australian Mammals. Angus & Robertson: Sydney.

Lactose digestion in suckling macropodids

MICHAEL MESSER[1], EVA A. CRISP[1] and ROBERT CZOLIJ[1]

There is evidence suggesting that bottle-fed kangaroos frequently suffer from symptoms of lactose intolerance, such as severe diarrhoea, when they are given lactose-containing milk formulas. In eutherian mammals, diarrhoea associated with lactose intolerance is usually caused by an accumulation of unabsorbed lactose within the intestinal lumen due to a deficiency of intestinal lactase. Studies in our laboratory have shown, however, that the small intestines of suckling Tammar Wallabies have high lactase activities. In this paper we present histochemical and biochemical evidence demonstrating that these lactase activities are due, exclusively, to an acid β-galactosidase which is located intracellularly, i.e., within the enterocytes. Lactose and the galactosyl oligosaccharides which are normally found in macropodid milk are therefore digested to monosaccharides only *after* they are transported into the enterocytes; this is in distinct contrast to eutherian mammals in which lactose is digested by a neutral β-galactosidase located extracellularly on the microvillous membrane. Ultrastructural studies of enterocytes of suckling Tammars, involving light and electron microscopy, suggest that the transport of lactose and of other milk oligosaccharides from the intestinal lumen into the enterocytes may occur *via* pinocytosis. We postulate that this process is relatively slow, thus limiting the rate at which lactose- containing milk can be bottle-fed without causing diarrhoea.

INTRODUCTION

SEVERAL years ago Stephens *et al.* (1974) and Stephens (1975) suggested that bottle-fed pouch young marsupials, especially macropodids, commonly suffer from severe diarrhoea when they are fed with cows milk and that this can be prevented by using a lactose-free milk formula.

In humans diarrhoea associated with lactose intolerance is caused by an accumulation of unabsorbed lactose within the gut lumen due to a deficiency of the enzyme, intestinal lactase. Stephens (1975) therefore suggested that pouch young macropodids have low intestinal lactase activities.* This proposition was investigated by measuring the intestinal lactase (β-galactosidase; β-Gal) and other disaccharidase activities of Tammars, *Macropus eugenii*, of various ages (Walcott and Messer 1980). Although adult animals did indeed have very low levels of intestinal lactase activity, pouch young aged from six to 34 weeks had high levels (see Janssens: this volume) which were of the same order as those of suckling rats and humans. This finding could be interpreted to mean that suckling macropodids should have no difficulty in digesting the lactose of cows milk (Messer 1978). However, the Tammar intestinal lactase was found to have some rather unusual properties (Walcott and Messer 1980). Unlike the eutherian lactase activity which is due mainly to a neutral β-Gal (pH optimum $\simeq$ 6), the intestinal lactase activity of suckling Tammars had a pH optimum of 4.6 and was strongly inhibited by p-chloromercuribenzoate sulphonate (p-CMBS). These and other observations suggested that the Tammar lactase activity was due to an acid β-Gal located intracellularly within lysosomes, in contrast to the eutherian neutral β-Gal which is located extracellularly in the microvillous brush border. This acid β-Gal would be similar to a lysosomal β-Gal of uncertain function which is found in enterocytes of eutherian mammals.

An intracellular location for the macropodid lactase activity would imply that the mechanism by which lactose is digested in suckling macropodids differs fundamentally from that found in suckling eutherians and could provide an explanation for the paradox of lactose intolerance despite high intestinal lactase activities. We therefore sought to obtain more direct evidence concerning the intracellular localization of the intestinal lactase activity of pouch young macropodids, and carried out two kinds of experiments, histochemical and biochemical (Crisp *et al.* 1987).

Histochemistry. Frozen section of small intestinal mucosa from pouch young Tammars of various ages were incubated with specific indigogenic substrates for neutral and acid β-Gal at pH 6.0 and 4.5, respectively; enzyme activity was monitored by observing the development of a blue colour due to indigo, beginning at the site of highest activity (Fig. 1). In tissues of suckling rats, which were run as controls, neutral β-Gal activity was very prominent in the brush border while acid β-Gal activity of lesser intensity was observed intracellularly in the supranuclear regions of the enterocytes. In suckling Tammars, no neutral β-Gal activity whatsoever could be detected either at the brush border or intracellularly but strong acid β-Gal activity was observed in the supranuclear regions of the enterocytes (Crisp *et al.* 1987).

*Stephens *et al.* (1974, 1975) noted that bottle-fed kangaroos sometimes develop cataracts (a symptom of galactosaemia in humans) in addition to diarrhoea and therefore proposed that pouch young macropodids are deficient in the enzymes required to metabolize galactose. Studies in our laboratory showed, however, that the activities of galactose-metabolizing enzymes in the livers of suckling macropodids are similar to those of suckling rats (Vernon *et al.* 1981).

[1]Department of Biochemistry, University of Sydney, New South Wales 2006.

Pages 217-21 in KANGAROOS, WALLABIES AND RAT-KANGAROOS ed. by G. Grigg, P. Jarman and I. Hume. Surrey Beatty & Sons Pty Limited, New South Wales, Australia, 1989.

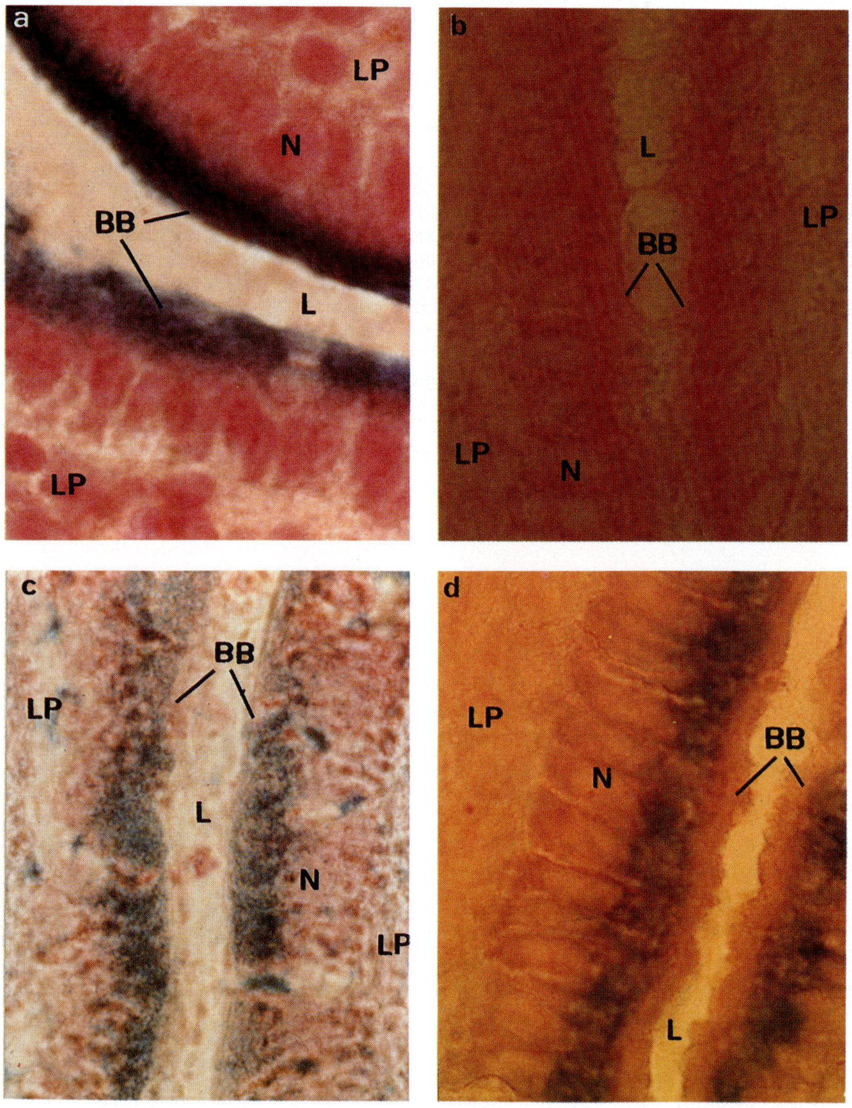

Fig. 1. Light micrographs of lingitudinally sectioned enterocytes. Frozen sections of jejunum from a 10-day-old rat and a 28-week-old Tammar were incubated at 4°C with 5-bromo-4-chloro-3-indolyl-β-D-galactoside at pH 4.5 for acid β-Gal and 5-bromo-4-chloro-3-indolyl-β-D-fucoside at pH 6.0 for neutral β-Gal (Lojda *et al.* 1979).
(a) Rat, neutral β-Gal: activity very obvious in BB after 4 hr of incubation. (b) Tammar Wallaby, neutral β-Gal, 18 hr: no activity. (c) Rat, acid β-Gal, 18 hr: activity in supranuclear region. (d) Tammar Wallaby, acid β-Gal, 4-hr: activity in supranuclear region. Brush border (BB); lumen (L); lamina propria (LP); nucleus (N). Magnification × 1 300.

Biochemical studies. Brush border membrane and soluble fractions were isolated from homogenates of mucosa from small intestines of Tammars and two other species of suckling macropodids and their β-Gal activities then measured. Almost all of the β-Gal activities were found in the soluble (non-membranous) fractions (Table 1) and were due to an acid β-Gal because they were optimal at pH 4.5 and were completely inhibited by 0.2 mM p-CMBS. Very little of the total β-Gal activities were found in the brush border fractions; moreover, the low activities present in the brush border fractions were also due to an acid β-Gal. These results were very different from those observed with suckling rats, in which most of the β-Gal activities were found in the brush border fractions and were due to a neutral β-Gal. From these biochemical data it can be concluded that the intestinal lactase of suckling macropodids must be an intracellular acid β-Gal and not a brush border neutral β-Gal.

It is important to note that, during most of lactation, lactose as such is only a minor component of the carbohydrate fraction of macropodid milk (Messer *et al.* 1984). Instead, the carbohydrate consists of a variety of unusual oligosaccharides, most of which have the structure $(Gal)_n$ Lac as

Table 1. Distribution of β-Gal activities between the brush border and soluble fractions of intestinal mucosal homogenates. The activities for macropodids were measured at pH 4.5 (optimum); those for *R. norvegicus* were determined at pH 6.0 (optimum) and represent the means of results obtained with two animals. The data do not include values obtained with a third fraction (P1) containing non-brush border membranous material; for full details, see Fig. 1. of Crisp *et al.* 1987, from which the values of Table 1 have been re-calculated.

Species	Age (weeks)	Soluble fraction	Brush border fraction
		(% of total activity)	
M. eugenii	28	98	1.9
M. giganteus	23	95	4.5
M. rufus	21	97	2.9
R. norvegicus	2	20	80

shown in Fig. 2 (Collins *et al.* 1981). We have found that the acid β-Gal of Tammar intestine readily hydrolyses these oligosaccharides to galactose and glucose. In contrast, the neutral β-Gal of rat intestine is unable to hydrolyse them; this may be due to the fact that the linkages between the galactosyl residues in the oligosaccharides are $\beta\ 1 \rightarrow 3$, and not $\beta\ 1 \rightarrow 4$ as in lactose. It is not surprising, therefore, that the macropodid intestine has an acid rather than a neutral β-Gal.

Transport of lactose and other oligosaccharides into the enterocytes

Our histochemical and biochemical studies both confirm that the intestinal lactase of macropodids is an acid β-Gal which is located inside the enterocytes. It follows that lactose and the other galactosyl oligosaccharides which are normally found in macropodid milk are digested to monosaccharides only *after* they are transported into the enterocytes. (The alternative possibility, that the acid β-Gal is secreted into the lumen of the small intestine, where it then exerts its activity, is unlikely because its pH optimum of pH 4.6 is much lower than the pH of the lumenal contents, pH 6 to 7). What is the mechanism of

$$\text{Gal}\,\beta\,1 \rightarrow 3\text{Lac}$$
$$\text{Gal}\,\beta\,1 \rightarrow 3\text{Gal}\,\beta\,1 \rightarrow 3\text{Lac}$$
$$\text{Gal}\,\beta\,1 \rightarrow 3\text{Gal}\,\beta\,1 \rightarrow 3\text{Lac}$$
$$\text{Gal}\,\beta\,1 \rightarrow 3\text{Gal}\,\beta\,1 \rightarrow 3\text{Gal}\,\beta\,1 \rightarrow 3\text{Gal}\,\beta\,1 \rightarrow 3\text{Lac}$$
$$\text{Gal}\,\beta\,1 \rightarrow 3\text{Gal}\,\beta\,1 \rightarrow 3\text{Gal}\,\beta\,1 \rightarrow 3\text{Gal}\,\beta\,1 \rightarrow 3\text{Gal}\,\beta\,1 \rightarrow 3\text{Lac}$$

Fig. 2. Structures of major oligosaccharides of Tammar milk. Lac=Lactose; Gal=galactose. In each case, the linkage between Gal and Lac is *via* carbon-3 of the galactosyl residue of Lac. Thus, Gal β 1 $\rightarrow$ 3Lac represents the trisaccharide, O-β-D-galactopyranosyl-(1 $\rightarrow$3)-O-β-D-galactopyranosyl-(1 $\rightarrow$ 4)-D-glucose (3′-galactosyllactose).

this transport? In theory, the oligosaccharides could be transported *via* a specific carrier. However, there is no precedent for this in eutherian mammals, in which the only sugars transported across membranes are monosaccharides.

A more likely possibility is that suckling macropodids transport these oligosaccharides by the same route as that which some suckling eutherians are believed to use for macromolecules, especially proteins. In suckling rats the enterocytes of the ileum exhibit an apical endocytic complex consisting of invaginations of the apical plasma membrane, a complex of anastomosing tubules, numerous small vacuoles, and a large supranuclear vacuole (Wissig and Graney 1968; Cornell and Padykula 1969). This complex is believed to be active in the absorption of both membrane-bound and fluid-phase macromolecules, which are delivered to the lysosomes and the supranuclear vacuole *via* non-selective pinocytosis and are then degraded (see Henning 1987). With repect to marsupials, Krause *et al.* (1977) have observed that the enterocytes of the entire small intestine of opposum *(Dedelphis virginiana)* pouch young show modifications for the absorption of macromolecules until shortly before weaning; the absorbing cells exhibit an apical endocytic complex like that found in ileal enterocytes of suckling rats. We have similarly examined the small intestinal mucosa of suckling Tammars of various ages by light and electron microscopy and found that cells of both jejunum and ileum exhibit an apical endocytic complex comprising invaginations of the apical plasma membrane, anastomosing tubules which extend down from the plasma membrane, numerous vesicles (including pinocytotic vesicles) below the terminal web and prominent vacuoles in a supranuclear position (Fig. 3); the supranuclear vacuoles are larger and more clearly defined in ileal than in jejunal cells. These features were not observed in enterocytes from adult Tammars.

Thus the ultrastructure of the intestinal absorbing cells of suckling Tammars is quite similar to that of suckling oppossums and rats. It therefore seems likely that in suckling marsupials, as in rats, proteins and possibly other macromolecules are transported into the enterocytes *via* pinocytosis. In this process, small water-soluble molecules such as lactose might be transported non-specifically along with macromolecules, ultimately to be digested by lysosomal enzymes such as the acid β-Gal. (For a brief discussion of how this transport might be related to the absorption of intact immunoglobulins, see Janssens and Messer 1988).

Lactose intolerance in pouch young macropodids. The above-mentioned scheme is still hypothetical since we lack direct evidence for the presence of intact milk oligosaccharides within

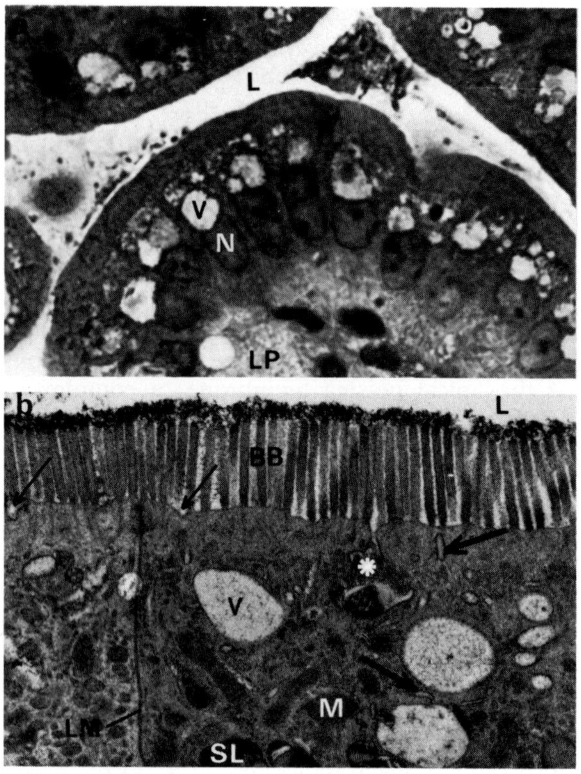

Fig. 3. (a) Light microscope micrograph of longitudinally sectioned villi tips from the ileum of an 18-week-old Tammar. Each enterocyte contains a large supranuclear vacuole (V), above which are numerous smaller vacuoles. Other symbols as in Fig. 1. Magnification × 1 000. (b) Transmission electron microscope micrograph of the apex of a longitudinally sectioned enterocyte from the jejunum of a 12-week-old Tammar. There are anastomosing tubules (thick arrows) in and beneath the terminal web and various degrees of invagination of the apical plasma membrane (thin arrows), suggesting pinocytotic activity. Some of these invaginations are extensive and contain particles (asterisk) which may be partially digested milk constituents. Lateral plasma membrane (LM); Vacuole (V); Mitochondrion (M); Presumptive secondary lysosome (SL). Other symbols as in Fig. 1. Magnification × 15 000.

the endocytic complex of suckling macropodid enterocytes. The hypothesis is, however, attractive as it could explain reduced tolerance for lactose in bottle-fed pouch young. In eutherian mammals, lactose is hydrolysed at the lumen-enterocyte membrane interface and the monosaccharide products are then transported across the membrane by a specific mechanism. In human infants the rate-limiting step in this process has been shown to be the initial hydrolysis (Gray 1983). When hydrolysis is suboptimal due to a deficiency of intestinal lactase, lactose accumulates within the intestinal lumen. The resulting osmotic effect causes water and sodium ions to be drawn into the lumen so that the volume of fluid within it increases, which in turn decreases the transit time so that even less lactose is digested. Undigested lactose enters the colon where it is converted to short-chain fatty acids (plus carbon dioxide and hydrogen) by bacterial metabolism, which increases the osmolality of the colonic contents. The net result is a watery diarrhoea, dehydration, malabsorption of other nutrients, malnutrition and retarded growth (Gray 1983; Ransome-Kuti 1985).

In our hypothetical scheme the rate-limiting step in the disposal of milk oligosaccharides is probably transport into the enterocyte since this is postulated to be a non-specific process. It seems likely that this transport *via* pinocytosis is quite slow compared with the rate at which lactose is hydrolysed in eutherians. Thus when macropodids are bottle-fed with cow's milk the capacity of the transport mechanism for lactose may be exceeded and unabsorbed lactose will accumulate within the intestinal lumen, resulting in diarrhoea and other symptoms as described above. We believe that this does not happen in mother-fed macropodids because (a) they probably suckle more slowly and thus allow the milk oligosaccharides more time to be absorbed and (b) these oligosaccharides are significantly larger than lactose (Fig. 2; Messer and Mossop 1977) and thus exert a lower osmotic effect on a weight basis (Messer and Green 1979).

The recommendations of Stephens (1975) concerning milk substitutes for bottle-feeding of marsupials have led to the almost universal use of lactose-free or low-lactose milk formulas in the hand-rearing of orphaned kangaroos, with good results (Reddacliff 1986; Smith 1986). Although these recommendations were not, initially, based on rigorous scientific documentation, Vickery and Walker (1982) and Walker and Vickery (1988) have since conducted a number of feeding trials with kangaroo pouch young in which they used milk replacers containing different amounts of either lactose or glucose. These trials showed that the animals have a markedly lower tolerance for lactose than for glucose as measured by their faecal excretions of dry matter and volatile fatty acids, and the frequency of diarrhoea (see also Walker 1986).

Walker and Vickery (1988) have suggested that there is a threshold for lactose which, if exceeded, will result in diarrhoea in a majority of pouch young kangaroos. This is in accord with our results. It would seem, therefore, that pouch young macropodids should be able to be bottle-fed either with a milk formula containing lactose below a threshold concentration or with cow's milk provided it is given sufficiently slowly.

In conclusion, it appears that, compared with eutherians, suckling macropodids have a low tolerance for lactose as suggested by Stephens (1975); however, this is not due to low levels of intestinal lactase activity but a result of the intracellular localization of this activity. The question of whether this unusual localization occurs in other suckling marsupials besides macropodids is one for further investigation*.

ACKNOWLEDGEMENTS

We thank Drs Brian Green and Des Cooper for supplying intestines from macropodids. This work was supported by the Australian Research Grants Scheme.

REFERENCES

COLLINS, J. G., BRADBURY, J. H., MESSER, M. AND TRIFONOFF, E., 1981. Structures of four new oligosaccharides from marsupial milk, determined mainly by ^{13}C-n.m.r. spectroscopy. *Carbohyd. Res.* **92:** 136-40.

CORNELL, R. AND PADYKULA, H. A., 1969. A cytological study of intestinal absorption in the suckling rat. *Am. J. Anat.* **125:** 291-316.

CRISP, E. A., CZOLIJ, R. AND MESSER, M., 1987. Absence of β-galactosidase (lactase) activity from intestinal brush borders of suckling macropods: implications for mechanism of lactose absorption. *Comp. Biochem. Physiol.* **88B:** 923-27.

GRAY, G. M., 1983. Intestinal disaccharidase deficiencies and glucose-galactose malabsorption. *In* The Metabolic Basis of Inherited Disease ed. by J. B. Stanbury, J. B. Wyngaarden, D. S. Frederickson, J. K. Goldstein and N. S. Brown. 5th edition. McGraw-Hill: New York.

HENNING, S. J., 1987. Functional development of the gastrointestinal tract. *In* Physiology of the Gastrointestinal Tract ed. by L. R. Johnston, 2nd edition. Raven Press: New York.

JANSSENS, P. A. AND MESSER, M., 1988. Changes in nutritional metabolism during weaning. *In* The Developing Marsupial. Models for Biochemical Research ed. by C. H. Tyndale-Biscoe and P. A. Janssens. Springer-Verlag: Berlin, Heidelberg.

KRAUSE, W. J., CUTTS, J. H. AND LEESON, C. R., 1977. The postnatal development of the alimentary canal in the opossum. *J. Anat.* **123:** 21-45.

LOJDA, Z., GOSSRAU, R. AND SCHIEBLER, T. H., 1979. Enzyme Histochemistry, Pp. 24-180. Springer-Verlag: Berlin, Heidelberg New York.

MESSER, M., 1978. Comments on the feeding of young marsupials. *Vict. Nat.* **95:** 61.

MESSER, M. AND GREEN, B., 1979. Milk carbohydrates of Marsupials. II. Quantitative and qualitative changes in milk carbohydrates during lactation in the Tammar Wallaby *(Macropus eugenii)*. *Aust. J. Biol. Sci.* **32:** 519-31.

MESSER, M., GRIFFITHS, M. AND GREEN, B., 1984. Changes in milk carbohydrates and electrolytes during early lactation in the Tammar Wallaby, *Macropus eugenii. Aust. J. Biol. Sci.* **37:** 1-6.

MESSER, M. AND MOSSOP, G. S., 1977. Milk carbohydrates of marsupials. I. Partial separation and characterization of neutral milk oligosaccharides of the Eastern grey kangaroo. *Aust. J. Biol. Sci.* **30:** 397-88.

RANSOME-KUTI, O., 1985. Lactose tolerance. *In* Diarrhoeal Disease and Malnutrition ed. by M. Gracey. Churchill Livingstone: Edinburgh, London, Melbourne and New York.

REDDACLIFF, G., 1986. Taronga Zoo, Sydney, New South Wales. Personal communication.

SMITH, G., 1986. National Parks and Wildlife Service of New South Wales, Sydney. Personal communication.

STEPHENS, T., 1975. Nutrition of orphan marsupials. *Aust. Vet. J.* **51:** 453-58.

STEPHENS, T., CROLLINI, C., MUTTON, M., GUPTA, J. D. AND HARLEY, J. D., 1975. Galactose metabolism in relation to cataract formation in marsupials. *Aust. J. exp. Biol. med. Sci.* **53:** 233-39.

STEPHENS, T., IRVINE, S., MUTTON, P., GUPTA, J. D. AND HARLEY, J. D., 1974. Deficiency of two enzymes of galactose metabolism in kangaroos. *Nature* **248:** 524-25.

VERNON, J. K., MESSER, M. AND GREEN, B., 1981. Enzymes of galactose metabolism in livers of suckling and adult Tammar Wallabies *(Macropus eugenii)* and other marsupials. *Aust. J. Biol. Sci.* **34:** 401-09.

VICKERY, K. AND WALKER, D. M., 1982. Threshold for glucose and lactose in the diet of young kangaroos. *Proc. Nutr. Soc. Aust.* **7:** 141.

WALCOTT, P. J. AND MESSER, M., 1980. intestinal lactase β-galactosidase) and other glycosidase activities in suckling and adult Tammar Wallabies *(Macropus eugenii)*. *Aust. J. Biol. Sci.* **33:** 521-30.

WALKER, D. M., 1986. The nutrition of pouch-young marsupials. *Proc. Nutr. Soc. Aust.* **11:** 24.

WALKER, D. M. AND VICKERY, K., 1988. Tolerance of pouch young kangaroos *(Macropodidae)* for cow's milk and milk replacers containing different amounts of glucose or lactose. *Aust. Mammal.* **11:** 125-33.

WISSIG, S. L. AND GRANEY, D. O., 1968. Membrane modifications in the apical endocytic complex of ileal epithelial cells. *J. Cell. Biol.* **39:** 564-79.

Note added in proof

*We have recently detected neutral β-Gal activity in intestinal brush border preparations of pouch young *Trichosurus vulpecula, Isoodon obesulus* and *Dasyurus viverrinus* (Crisp, Messer and Cowan, to be published), suggesting that this enzyme is absent only in macropodids.

Water and energy turnover in free-living macropodoids

B. GREEN[1]

Water influx and metabolic rates have been measured by isotopic water turnover techniques in comparatively few free-living macropodoids, however some obvious trends are apparent in the available data. Notably the rates of water influx strongly reflect the water status of the diet; high water fluxes are observed when animals have access to lush food and low rates prevail when dry food only is available.

Field water turnover rates scale to body mass $^{0.92}$ in "dry" conditions and mass $^{0.97}$ in wet conditions, but the difference between these exponents is not significant. Field metabolic rates scale to body mass $^{0.74}$

The impacts of vegetational water status and food availability on drinking water requirements, energy balance and breeding success are discussed.

INTRODUCTION

THE Macropodoidea are a group of herbivorous marsupials that predominantly graze and browse, with some of the smaller species occasionally feeding on fungi and invertebrates.

The group underwent major radiations during the late Miocene to Pliocene (Bartholomai 1972), a time when the Australian climate was becoming drier and forests were retreating, leading to the development of vast areas of arid and semi-arid rangelands (Hume and Warner 1980). This situation was successfully exploited by the macropodoids with the development of elaborate alimentary systems (Hume 1982) and masticatory systems (Sanson 1978) to enable extremely fibrous diets to be utilized.

Australia is the driest of the continents (Philips 1977) and experiences unreliable and generally limited rainfall. Consequently, the water and nitrogen status of plants can vary widely; from lush growth with high levels of available nitrogen and energy, to dry dorage with low available nitrogen and energy. The macropodoids therefore experience extremes in the water and energy status of their diets.

This paper reviews the data presently available on water and energy turnover in free-living macropodoids and addresses the major factors that influence water and energy requirements and their consequent ecological implications.

METHODS

The water fluxes and metabolic rates (CO_2 production) of free-living animals can be measured by means of isotopic water turnover. The turnover of injected tritiated or deuteriated water (^3H_2O or ^2H_2O) provides estimates of total water intake and losses, and when used in conjunction with $H_2{}^{18}O$ give estimates of CO_2 production.

The isotopes are generally injected into the peritoneal cavity and after equilibration of the isotopes in the body water pool, which usually occurs within about 3 hours in macropodoids, an initial post-injection blood sample is taken. The animals are then released into the field for several days, after which they are re-captured and a further blood sample obtained.

The initial dilution of isotopes that occurs during the equilibration period provides estimates of the total body water pool. The subsequent logarithmic dilution of ^3H_2O or ^2H_2O that occurs during the release period is determined by water influxes (food water, metabolic water, drinking, pulmo-cutaneous exchange of water) and effluxes (pulmo-cutaneous losses and exchanges, excretion, defaecation etc.). Injected $H_2{}^{18}O$ also mirrors water fluxes but additionally is lost as CO_2. When ^2H_2O or $H_2{}^{18}O$ are used it is necessary to obtain background (pre-injection) blood samples.

The necessary calculations and theoretical background to isotopic water turnover techniques have been described in detail by Lifson and McClintock (1966), with assessments of potential errors by Nagy (1980) and Nagy and Costa (1980).

The isotope levels in blood samples are determined by liquid scintillation spectrometry (^3H_2O), mass spectrometry (^2H_2O and $H_2{}^{18}O$) or the proton activation technique ($H_2{}^{18}O$; Wood *et al.* 1975).

RESULTS AND DISCUSSION

The water, energy and food requirements of free- living macropodoids have not been studied extensively, however, some clear trends are apparent in the limited data that are available.

The rates of water influx in a range of macropodoids are presented in Table 1. Most field

[1]Division of Wildlife and Ecology, C.S.I.R.O., P.O. Box 84, Lyneham, Australian Capital Territory 2602.

Pages 223-29 in KANGAROOS, WALLABIES AND RAT-KANGAROOS ed. by G. Grigg, P. Jarman and I. Hume. Surrey Beatty & Sons Pty Limited, New South Wales, Australia, 1989.

Table 1. Water and energy turnover in free-living macropodoids. (Wt=wet; D=dry; S=Summer; Au=Autumn; W=Winter; Sp=Spring; *=Values calculated from data in reference).

Species	Mass (kg)	H_2O Influx (ml/kg.day)	Metabolized Energy (kJ/day)	Metabolized Energy (kJ/kg.day)	FMR/SMR	Season	Authors
Bettongia							
penicillata	1.1	61.0±13.9	524±101	476	2.44	SD	
		90.0±13.9	590±63	536	2.49	Au	Green, Best and Turner (in prep.)
		177.8±42.5	695±206	632	3.23	W, Wt	
		*54.2±14.9				SD	
		*78.1±9.3				S+Rain	Sampson (1971)
		*192.9±25.4				W, Wt	
Setonix							
brachyurus	1.9	47.3±11.7	548±150	288	1.70	SD	Nagy, Bradley and Morris (in prep.)
		*51±11				SD	Rice (1976)
		*116±30				W, Wt	Horwitz (1980)
Lagorchestes							
conspicillatus	2.23	*40.7±10.0				Sp.D	Bradshaw and Bakker (pers. comm.)
Aepyprymnus							
rufescens	2.85	139.6±26.1	1 442±188	506	3.32	S	Green and Wallis (in prep.)
	2.90	128.7±19.9	1 491±432	514	3.44	W	
Petrogale							
inornata	3.15	80.7±6.2				Annual	Kennedy and Heinsohn (1974)
P. rothschildi	3.35	44.4±10.8				Sp.D	Bradshaw (pers. comm.)
	2.74	*133.2±21.8				S.Wt	Bradshaw (pers. comm.)
P. xanthopus	8.9	53.4±10.0	2 209±	248	2.20	SD	
		55.8±17.0				WD	Green and Robinson (in prep.)
		146.5±17.1				W.Wt	
Thylogale							
billardieri	5.98	108±30	1 630±678	273	2.17	SD	
Macropus							Nagy and Sanson (in prep.)
giganteus	43.9	60±3	8 170±3 620	186	2.51	SD	
M. eugenii	4.38	57.5±11.4	1 150±280	263	1.93	SD	Nagy, Bradley and Morris (in prep.)
		*61.3				SD	Bakker *et al.* (1982)
		40.3±3.5				SD	Inns (1980)
		182.5±38.5				W.Wt	
		*138.5				Sp.Wt	Bakker *et al.* (1982)
		188.5±20.0				W.Wt	Horwitz (1980)
M. rufus	21.8	39.5±4.2				SD	Dawson *et al.* (1975)
M. robustus	28.1	39.4±12.1				SD	

studies of water turnover in macropodoids have been carried out during the summer or under dry climatic conditions, when water influxes are low and generally reflect the reduced water status of dietary plants.

The most detailed field study of water turnover in a macropodoid is that carried out by Inns (1980) on the Tammar Wallaby *(Macropus eugenii)* at Kangaroo Island, South Australia. The seasonal variations in water influxes over a three year period are shown in Fig. 1. It can be clearly seen that water influxes are very low in summer, ranging between 40 and 89 ml/kg. day, depending on the year of study. However, following the advent of the regular winter rains the water influx rates increase three-fold to around 180 ml/kg. day.

Tammar Wallabies in Western Australia exhibit similar seasonal differences in water influxes to those recorded by Inns (Table 1), and other species that experience hot dry summers and cool wet winters show similar patterns of water influxes, e.g., the Quokka, *Setonix brachyurus* (Rice 1976; Horwitz 1980) and the Woylie, *Bettongia penicillata* (Sampson 1971; Green, Best and Turner in prep.).

The water influx rates of wild Yellow-footed Rock-wallabies *(Petrogale xanthopus)* over a two and a half year period are shown in Fig. 2. These animals inhabit areas in which winter rainfall is not guaranteed and in the winter of 1982 water influx rates were as low as in summer in response to low winter rainfall. In the winters of 1981 and 1983 there were good autumn/winter rains, and water influx rates were again about three times higher than rates measured during dry periods. Another rock-wallaby, *Petrogale rothschildi*, inhabits the wet/dry tropics of north west Australia, a region that experiences summer monsoonal rains with the driest season being Spring, immediately prior to the onset of summer rainfall. Again, the lowest water influx rates (41 ml/kg day) are associated with dry conditions, with much higher rates (133 ml/kg. day) occurring during the "wet" season (Bradshaw, pers. comm.).

The Rufous Rat-kangaroo *(Aepyprymnus rufescens)* inhabits the forests of northeastern

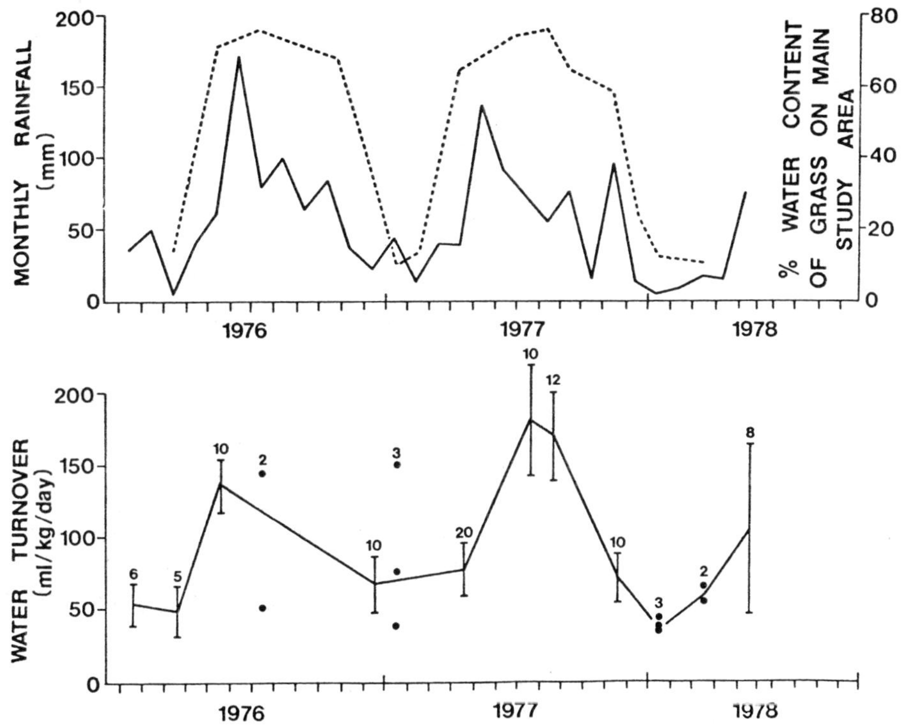

Fig. 1. Rainfall and water content of pastures and associated water influxes of wild Tammar Wallabies *(Macropus eugenii)* at Kangaroo Island, South Australia. (From Inns 1980).

Australia, areas that receive regular abundant rainfall. There was no significant difference in the water influx rates of *A. rufescens* measured in summer and winter (Green and Wallis in prep.); about 135 ml/kg. day in both seasons.

The pattern of water influxes in macropodoid marsupials is therefore dictated by rainfall patterns and its consequent influence on plant moisture content. When habitats experience dry conditions, rates of water influx are low; conversely, after extensive rainfall water influxes increase substantially.

Body mass exerts a profound influence on the physiological and ecological performance of animals (reviewed by Peters 1983) and allometric (scaling) influences on water turnover have been reported in laboratory studies of mammals. Slope exponents that have been proposed for water fluxes in mammals are; 0.8 (Richmond *et al.* 1962) and 0.82 (MacFarlane 1965) for eutherians, 0.8 for macropodoids (Denny and Dawson 1975) and 0.82 for marsupials and eutherians combined (Nicol 1978).

Field water turnover rates for macropodoids during "dry" conditions scale to body mass $^{0.98}$, however the data for *M. giganteus* and *T. billardieri* were obtained from animals in a large enclosure and are therefore not strictly field water fluxes. If these two data points are excluded

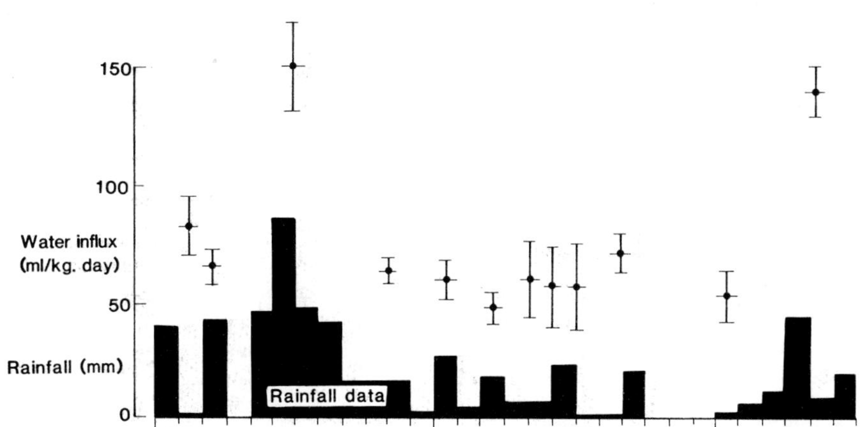

Fig. 2. Rainfall and water influxes of Yellow-footed Rock-wallabies *(Petrogale xanthopus)* in South Australia. (Green and Robinson in prep).

from the analysis it is that "dry" water influx rates scale to body mass $^{0.92}$ (Fig. 3). Water influx rates in "wet" conditions scales to mass $^{0.97}$, however an ANCOVA comparison shows that the difference between these exponents is not significant. The intercepts of both regressions, however, are significantly different (P< 0.001) and water influx rates in "wet" conditions are about three times those measured in "dry" conditions.

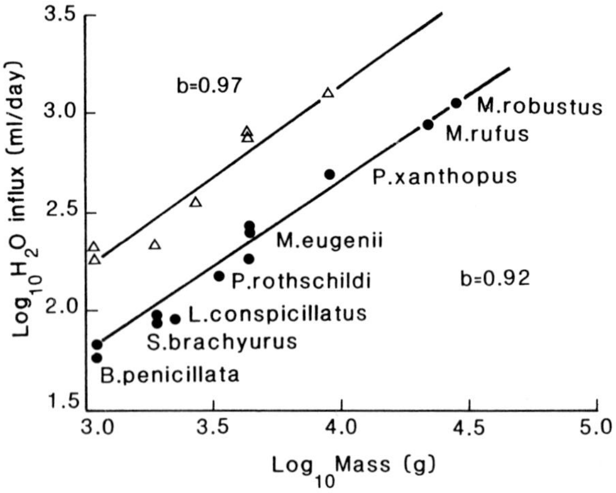

Fig. 3. The relationship between body mass and water influx in free-living macropodoids under "wet" and "dry" conditions. Log_{10} ml/day ("wet") = log_{10} ($g^{0.97}$) −0.69, (r^2 = 0.94, SE_b = 0.11). Log_{10} ml/day ("dry") = log_{10} ($g^{0.92}$) −1.04, (r^2 = 0.98, SE_b = 0.04).

The field water influx rates in summer are generally much lower than rates predicted for captive macropodoids (Denny and Dawson 1975) and marsupials under "standard" conditions (Nicol 1978); i.e., drinking water available and under thermoneutral ambient temperatures. However, summer field water turnover rates are higher than captive rates in *A. rufescens, T. billardieri* and *M. giganteus* and similar in *P. inornata*. These differences presumably reflect the low water status of the summer diet of species inhabiting hot, dry environments compared with those from milder, wetter areas, but could also indicate important physiological differences between arid adapted and non-adapted species (Nicol 1978). Field water influx rates under winter/"wet" conditions are about double those measured under "standard" conditions ($\bar{x}$field/"standard" ratio= 2.01±0.51).

There are limited data available on the field metabolic rates (FMR) of macropodoids, (Table 1 and Fig. 4). The slope of the regression of summer FMR's against body mass is 0.74 (SE=0.08), a value that is intermediate to that described for four macropodoid species by Nagy and Sanson (in prep.), and marsupial herbivores by Nagy (1987); 0.86 and 0.64 respectively. However, the FMR slope presented here is similar to that describing standard metabolic rates

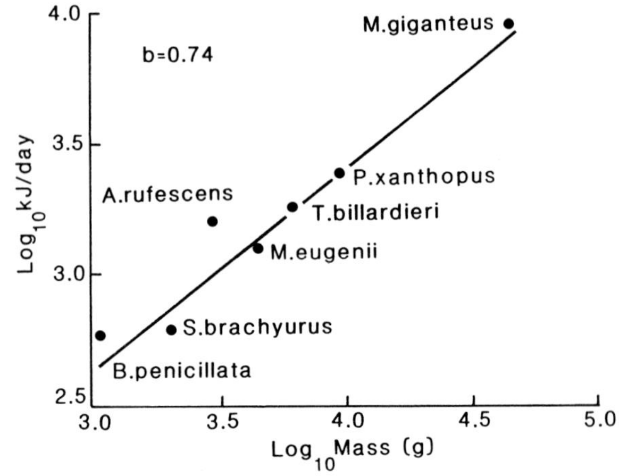

Fig. 4. The relationship between body mass and field metabolic rate in macropodoids during summer. Log_{10} kJ/day = 0.43 + log_{10} ($g^{0.74}$), (r^2 = 0.97, SE_b = 0.08).

(SMR) in marsupials (Dawson and Hulbert 1970) and mammals in general (Kleiber 1975); 0.74 and 0.75 respectively.

The summer FMR's of macropodoids range between 1.7 and 3.3 times the predicted SMR's ($\bar{x}$=2.32±0.52), the lowest value being for *S. brachyurus* (Nagy *et al.* in prep.) and the highest for *A. rufescens* (Green and Wallis, in prep.). There are only two macropodoid species in which FMR's have been studied in seasons other than summer. In the Rufous Rat-kangaroo *(Aepyprymnus rufescens)* winter FMR's are similar to those recorded in summer (Green and Wallis in prep.). However, *Bettongia penicillata* inhabits areas that experience cool winters and FMR's at this time are 33% higher than in summer (Green, Best and Turner, in prep.). It is likely that seasonal influences on FMR's will be more pronounced for smaller species living in cooler climates and that larger macropodoids may not show major seasonal variations in FMR's. These speculations can only be clarified by extensive further research on FMR's in macropodoids, with particular attention being paid to seasonal influences.

The water content of herbage exerts a major influence on the water and energy balance of herbivores. After rainfall the water content of grasses can reach 90% of fresh mass (MacFarlane and Howard 1972). Under these circumstances herbivores are effectively subjected to high obligate water intakes in attempting to achieve an adequate intake of dry matter, and some species lose condition when feeding on such lush vegetation e.g., *M. eugenii* (Bakker *et al.* 1982), *M. agilis* (Bolton 1974).

When high levels of water are present in vegetation there is no requirement for supplementary drinking water; however as plant water contents decline to between 55 and 65% herbivores require access to drinking water (Cooke 1982), or select more succulent dietary items (Ealey *et al.* 1965; MacFarlane 1968).

If either of these options are denied to herbivores, food intake is reduced and animals lose weight, e.g., rabbits (Hayward 1961; Cooke 1982; Wood and Lee 1985). Similar effects have been observed experimentally with macropodoids; Euros *(M. robustus)* require drinking water equivalent to 5% of their body mass when given dry food, and if drinking water is removed food consumption declines by 50% within three days and no food is taken subsequently (Ealey *et al.* 1965). In the field, when natural diets decline to 30% moisture content, non-drinking euros suffered a 25-40% weight loss; a combination of reduced food intake and the catabolism of body reserves (Ealey *et al.* 1965).

The Spectacled Hare-wallaby, *Lagorchestes conspicillatus* also reduces its food consumption when given dry food and deprived of drinking water (Bakker and Bradshaw 1983). Converseley, Bakker *et al.* (1982) noted that Tammar wallabies *(M. eugenii)* that had access to drinking water in the field in summer, were "fatter" than animals inhabiting a beach area without drinking water, presumably because they were able to feed to a greater extent.

Drinking has been observed in a number of macropodoids, particularly during hot, dry periods. Red Kangaroos *(M. rufus)* only drink during severe drought (Newsome 1964) when the availability of green pick is low. A number of workers have utilized drinking requirement as a means of trapping larger macropodoids, such as *M. rufus* and *M. robustus* at watering points (Ealey *et al.* 1965; Dawson *et al.* 1975).

The Yellow-footed Rock-wallaby *(P. xanthopus)* also utilizes drinking water during dry periods, with most individuals visiting water holes every night (Lim *et al.* 1987). Ealey *et al.* (1965) found that few Euros drank every day in summer, but drinking water intake averaged 12% of body mass over a period of a week. There are no other data available on drinking frequencies and volumes ingested by macropodoids in the field.

The relationship between food water content, food intake and drinking is dictated by the water requirement for adequate salivary flow and microbial fermentation in herbivores (MacFarlane 1968). The digestion systems of herbivores require a high water content for efficient functioning, and gut capacity in herbivores scales to body mass $^{0.94}$ (Van Soest 1982). Therefore the water turnover exponent of 0.92 found in the field may result from the superimposition of digestive requirements for water over predicted "standard" water requirements.

Kennedy and Heinsohn (1974) demonstrated that lactating Rock-wallabies *(P. inornata)* had water turnover rates that were 17% higher than non-lactating animals. It is probable that the water turnover rates of lactating macropodoids will be high at peak milk production. There is no published information on the metabolic costs of lactation in free-living macropodoids. The early stages of lactation in marsupials are characterized by the production of small volumes of dilute milk to feed very small pouch young (Green 1984). However, in late lactation when the young are much larger and spend substantial periods outside of the pouch, concentrated milk is produced (Green *et al.* 1980) in large volumes e.g., 86 ml/day in *M. eugenii* (Dove *et al.* 1987; Merchant: this volume). This volume of milk contains about 900 kJ of energy, which is almost equivalent to the summer FMR of *M. eugenii* (Nagy *et al.* in prep.). Consequently the FMR's of late stage lactating Tammar wallabies may be up to double those of non-lactating animals.

Lactation imposes a major drain on maternal water and energy reserves in the later stages of the growth of young. During drought there is an increase in mortality among Red Kangaroo pouch young (Frith and Sharman 1964; Newsome 1965), which is followed by blastocyst reactivation and birth of a new young, as well as a post-partum oestrus/mating. In extreme droughts females may even become anoestrous (Newsome 1964). Other macropodoids also incur increased mortality of young during drought, e.g., *M. robustus* (Ealey 1963; Sadleir 1965), *M. giganteus* (Kirkpatrick and McEvoy 1966) and *M. eugenii* (Inns 1980).

The loss of larger pouch young, which are expensive to nourish, during unfavourable conditions and their subsequent replacement by extremely small neonates, which require very small amounts of milk for nourishment, enables macropodoids to respond quickly to ameliorated environmental conditions and maintain lactation to rear young successfully. While droughts may result in the loss of several cohorts of young this situation is preferable to that whereby females become grossly debilitated while maintaining lactation which cannot be ultimately successful, as occurs in cattle exposed to drought.

It is clear that the water status of herbage exerts a major impact on water turnover, drinking requirement food consumption and energy use in macropodoids. Further research is warranted on a broader range of species to determine the energy costs of lactation, seasonal variations in water and energy turnover and the extent to which species are arid-adapted.

ACKNOWLEDGEMENTS

I am most grateful for the generous access to unpublished data afforded by Ken Nagy and Don

Bradshaw and their research associates, and also Bob Inns. I also thank my own research colleagues for allowing me to use unpublished data in this review.

REFERENCES

BAKKER, H. R. AND BRADSHAW, S. D., 1983. Renal function in the Spectacled Hare-wallaby, *Lagorchestes conspicillatus:* effects of dehydration and protein deficiency. *Aust. J. Zool.* **31:** 101-08.

BAKKER, H. R., BRADSHAW, S. D. AND MAIN, A. R., 1982. Water and electrolyte metabolism of the tammar wallaby *Macropus eugenii. Physiol. Zool.* **55:** 209-19.

BARTHOLOMAI, A., 1972. Aspects of the evolution of the Australian Marsupials. *Proc. R. Soc. Qld.* **82:** V-XVIII.

BOLTON, B. L., 1974. An ecological study of the agile wallaby on the coastal plains of the Northern Territory with a comparison between improved and unimproved areas. M.V. Sc. Thesis, University of Queensland.

COOKE, B. D., 1982. Reduction of food intake and other physiological responses to a restriction of drinking water in captive wild rabbits, *Oryctolagus cuniculus* (L). *Aust. Wildl. Res.* **9:** 247-52.

DAWSON, T. J. AND HULBERT, A. J., 1970. Standard metabolism, body temperature, and surface areas of Australian marsupials. *Am. J. Physiol.* **218:** 1233-38.

DAWSON, T. J., DENNY, M. J. S., RUSSELL, E. M. AND ELLIS, B., 1975. Water usage and diet preferences of free-ranging kangaroos, sheep and feral goats in the Australian arid zone during summer. *J. Zool. Lond.* **177:** 1-23.

DENNY, M. J. S. AND DAWSON, T. J., 1975. Comparative metabolism of tritiated water by macropodid marsupials. *Am. J. Physiol.* **228:** 1794-99.

DOVE, H., CORK, S. J. AND CHRISTIAN, K. R., 1987. Defining the pattern of milk intake in a marsupial herbivore. Pp. 101-02 *in* Herbivore Nutrition Research ed. by M. Rose. Aust. Soc. Anim. Prod.

EALEY, E. H. M., 1963. The ecological significance of delayed implantation in a population of the hill kangaroo *(Macropus robustus)* Pp. 33-48 *in* Delayed Implantation ed. by A. C. Enders. University Press: Chicago.

EALEY, E. H. M., BENTLEY, P. J. AND MAIN, A. R., 1965. Studies on water metabolism of the hill kangaroo, *Macropus robustus* (Gould), in Northwest Australia. *Ecology.* **46:** 473-79.

FRITH, H. J. AND SHARMAN, G. B., 1964. Breeding in wild populations of the red kangaroo, *Megaleia rufa. C.S.I.R.O. Wildl. Res.* **9:** 86-114.

GREEN, B., 1984. Composition of milk and energetics of growth in Marsupials. *Symp. Zool. Soc. Lond.* **51:** 369-87.

GREEN, B., MERCHANT, J. AND NEWGRAIN, K., 1988. Milk consumption and energetics of growth in pouch young of the tammar wallaby, *Macropus eugenii. Aust. J. Zool.* **36:** 217-27.

GREEN, B., NEWGRAIN, K. AND MERCHANT, J., 1980. Changes in milk composition during lactation in the tammar wallaby *(Macropus eugenii). Aust. J. Biol. Sci.* **33:** 35-42.

HORWITZ, P., 1980. Condition, total body water and aspects of water metabolism of quokkas *(Setonix brachyurus)* and tammars *(Macropus eugenii). In* Electrolyte metabolism in the quokka (*Setonix brachyurus,* Quoy and Gaimard) and the tammar (*Macropus eugenii,* Desmarest) ed. by S. Ambrose, S. Bunn, J. Dolra, B. Fergusson, P. Horwitz and A. Kelly. Zoology Department, University of Western Australia.

HUME, I. D., 1982. Digestive physiology and nutrition of marsupials. Cambridge University Press: Cambridge.

HUME, I. D. AND WARNER, A. C. I., 1979. Evolution of microbial digestion in mammals. Pp. 665-84 *in* Digestive physiology and metabolism in ruminants ed. by Y. Ruckebusch and P. Thivend. MTP Press Ltd: England.

INNS, R. W., 1980. Ecology of the Kangaroo Island wallaby, *Macropus eugenii* (Desmarest) in Flinders Chase National Park, Kangaroo Island. Ph.D. Thesis, Zoology Department, University of Adelaide.

KENNEDY, P. M. AND HEINSOHN, G. E., 1974. Water metabolism of two marsupials — the brush-tailed possum, *Trichosurus vulpecula* and the rock-wallaby, *Petrogale inornata* in the wild. *Comp. Biochem. Physiol.* **47**A: 829-34.

KIRKPATRICK, T. H. AND MCEVOY, J. S., 1966. Studies of Macropodidae in Queensland. 5. Effects of drought on reproduction in the grey kangaroo *(Macropus giganteus) Qld. Agric. Anim. Sci.* **23:** 439-42.

KLEIBER, M., 1975. The fire of life. Robert E. Krieger Publishing Co. Inc.: New York.

LIFSON, N. AND MCCLINTOCK, R., 1966. Theory of use of the turnover rates of body water for measuring energy and material balance. *J. Theoret. Biol.* **12:** 46-74.

LIM, L., ROBINSON, A. C., COPLEY, P. B., GORDON, G., CANTY, P. D. AND REIMER, D., 1987. The conservation and management of the yellow-footed rock-wallaby *Petrogale xanthopus* Gray, 1854. Dept. Env. Plan. S. Aust. Spec. Publ. 1987, No. 4, 1-94.

MACFARLANE, W. V., 1968. Adaptations of ruminants to tropics and deserts. Pp. 164-182 *in* The adaptation of domestic animals ed. by E. S. E. Hafex. Lea and Febiger: Philadelphia.

MACFARLANE, W. V. AND HOWARD, B., 1972. Comparative water and energy economy of wild and domestic mammals. *Symp. Zool. Soc. Lond.* **31:** 261-96.

MACFARLANE, W. V., HOWARD, B., HAINES, H., KENNEDY, P. M. AND SHARPE, P. M., 1971. Hierarchy of water turnover of desert mammals. *Nature, Lond.* **234:** 483-84.

NAGY, K. A., 1980. CO_2 production in animals: analysis of potential errors in the doubly labelled water method. *Am. J. Physiol.* **238:** R466-73.

NAGY, K. A. AND COSTA, D. P., 1980. Water flux in animals: analysis of potential errors in the tritiated water method. *Am. J. Physiol.* **238:** R454-65.

NEWSOME, A. E., 1964. Anoestrus in the red kangaroo *Megaleia rufa* (Desmarest). *Aust. J. Zool.* **12:** 9-17.

NEWSOME, A. E., 1965. Reproduction in natural populations of the red kangaroo *Megaleia rufa* (Desmarest), in Central Australia. *Aust. J. Zool.* **13:** 735-59.

NICOL, S. C., 1978. Rates of water turnover in Marsupials and Eutherians: comparative review with new data on the Tasmanian devil. *Aust. J. Zool.* **26:** 465-73.

PETERS, R. H., 1983. The ecological implications of body size. Cambridge University Press: Cambridge.

PHILIP, J. R., 1977. Water on the Earth. Pp. 35-59 *in* Water; Planets, plants and people ed. by A. K. McIntyre. Australian Academy of Science: Canberra.

RICE, G. E., 1976. Water metabolism of the quokka. *In* Salt appetite and condition in the quokka ed. by R. Homes, P. Prideaux, S. Halse, J. Trendall, D. Harrington, O. Nichols and G. Rice. Zoology Department, University of Western Australia.

Recycling of water during lactation and its effects on the estimation of milk intake in a developing macropodid, the Tammar Wallaby *(Macropus eugenii)*

H. DOVE[1], S. J. CORK[2] and K. R. CHRISTIAN[1]

Isotopically-labelled water has been used in a range of species to determine the inflow of water to either the mother or young, or the transfer of water between them as milk. Using data obtained with the Tammar Wallaby *(Macropus eugenii)*, computer simulation was used to examine the effects of water recycling between mother and young on the errors of estimation of water flows, based on either single- or double-isotope procedures.

Simulations indicated that recycling via the mother's consumption of the offspring's excreta resulted, in this species, in only small under-estimates of water flows. In offspring weighing 460 g, estimated errors did not exceed 4%, while in 100 g offspring they were less than 0.4%.

Simulation of the effects of larger litter sizes indicated that the small errors found with one offspring were related to the small body water pool size and milk intake of the offspring, compared with the pool size of the mother.

By contrast water vapour exchange between mother and young resulted in large over-estimates of water flows, especially with the smaller (100 g) offspring. It is suggested that vapour exchange would not be a problem beyond about 200 d of lactation (450-500 g body weight), when the young are furred and increasingly independent, but before this stage, would be likely to lead to unacceptable errors in procedures based on isotopically-labelled water.

INTRODUCTION

SINGLE-ISOTOPE dilution procedures based on tritiated water (TOH) or deuterium oxide (D_2O) have been used to estimate the milk intake of the young of several mammals (e.g., Dove and Freer 1979; Oftedal 1984a), and give accurate estimates provided the young consume only milk. When water is also consumed in other forms, the milk intake of the offspring can then be calculated from double-isotope methods involving the introduction of TOH into the mother and D_2O into the offspring (Holleman *et al*. 1975; Dove and Cork 1988; Cork and Dove 1988). However, several authors have pointed out that, in species such as the mouse or dog, in which the mother consumes the excreta produced by the suckling young, the consequent recycling of isotope through the mother and back into the offspring is a potential source of error in the estimation of isotope dilution and thus of milk intake (Baverstock and Green 1975; Baverstock and Elhay 1981; Oftedal 1984a). Pulmocutaneous exchange of isotope between littermates may also contribute to error (Friedman and Bruno 1976). It has been suggested that, in multiparous species, these effects of recycling and exchange can be accommodated by leaving one littermate uninjected (Baverstock and Green 1975; Oftedal 1984a,b).

From these studies, it has been inferred that isotope recycling would similarly be a problem in early- to mid-lactation in suckling marsupials in which the mother may consume the excreta of the offspring for an extended period (Baverstock and Green 1975; Green and Newgrain 1979). This has led to the development of isotope dilution procedures based on other pools, such as the sodium pool (Green and Newgrain 1979; Green *et al*. 1988), though this latter procedure is itself applicable only up till the time the young begin to consume sodium in forms other than milk. However, the original studies of the effects of recycling were conducted with multiparous species and the effect of the considerably smaller litter weight in macropodids does not appear to have been evaluated.

In a study of the milk intake of the Tammar Wallaby undergoing the transition between pouch life and independent existence, we showed that isotope recycling was present between 210 and 240 d of lactation (Dove and Cork 1988). In this paper, we use our earlier data and computer simulation to assess:

(a) the errors arising from various levels of water recycling via excreta, at about 210 d or 130 d of lactation, with one or more offspring;

(b) the relative sizes of the errors when milk water intake is calculated from single-isotope (e.g., Dove and Freer 1979) or double-isotope procedures (Holleman *et al*. 1975; Dove 1988); and

[1]Division of Plant Industry, C.S.I.R.O., G.P.O. Box 1600, Canberra, Australian Capital Territory 2601.

[2]Division of Wildlife and Ecology, C.S.I.R.O., P.O. Box 84, Lyneham, Australian Capital Territory 2601.

Pages 231-37 in KANGAROOS, WALLABIES AND RAT-KANGAROOS ed. by G. Grigg, P. Jarman and I. Hume. Surrey Beatty & Sons Pty Limited, New South Wales, Australia, 1989.

(c) the likely effects of recycling via exchange of water vapour between mother and offspring, independent of the effects of recycling via excreta.

METHODS

The simulations reported here are based on a study of the milk intake of Tammar Wallaby offspring in mid to late lactation; full details of the experimental procedures used can be found elsewhere (Dove *et al.* 1987; Dove and Cork 1988). Briefly, the total water inflows of eight Tammar mothers (mean weight 4.7±0.74 kg) were estimated from the dilution of injected TOH. Similarly, the total water inflows of their single offspring were determined from the dilution of injected D_2O, using the expression:

$$\text{Inflow} = \frac{-(V_1-V_0).(1n(C_1/C_0))}{t(1n(V_1/V_0))}$$

where C_1 and C_0 are the concentrations of D_2O at the start and finish of time t, over which the body water pool size of the offspring has increased from V_0 to V_1 (Dove and Freer 1979).

The inflow of water as milk was calculated from the fractional turnover rates of TOH and D_2O in mother and offspring respectively and the accumulation, in the offspring, of TOH from the mother, using expression 5 of Holleman *et al.* (1975).

Simulation model

The model used is shown in Fig. 1, and is essentially a simplification of that used by Baverstock and Elhay (1981) in that, where multiple offspring are considered, all are assumed to have been given isotope, so that there are only two pools, the mother and the young.

In setting up the model, it was assumed that:

(a) the pool size of the offspring increased but that of the mother remained constant,

(b) milk water intake increased linearly with time and was the only oral source of water to the offspring,

(c) the rate of recycling, defined as the proportion of the offspring's outflow returned to the mother via her consumption of the offspring's excreta, was constant over the simulation period.

The results presented by Baverstock and Elhay (1981) and by Dove and Cork (1988) suggest that these assumptions are valid or reasonable. Over longer periods, milk intake is exponentially related to time in the Tammar (Dove and Cork 1988), but is essentially linear over the short periods used in this study.

The inputs used initially in the simulation procedures are shown in Table 1. Two cases were studied. The first involved 460 g offspring typical of the older animals used by Green *et al.* (1988) and the youngest animals used by Dove and Cork (1988), which were furred and were spending extended periods outside the pouch. The second case considered much smaller animals (100 g) which would be unfurred and attached to the teat within the pouch.

Pool sizes were calculated assuming that 75.3% of body weight was water (Dove and Cork 1988). Growth rates and thus rates of change of pool size were calculated from the regression equations given by Dove and Cork (1988 — larger offspring) or by Green *et al.* (1988 — smaller offspring). Milk water intakes were based on

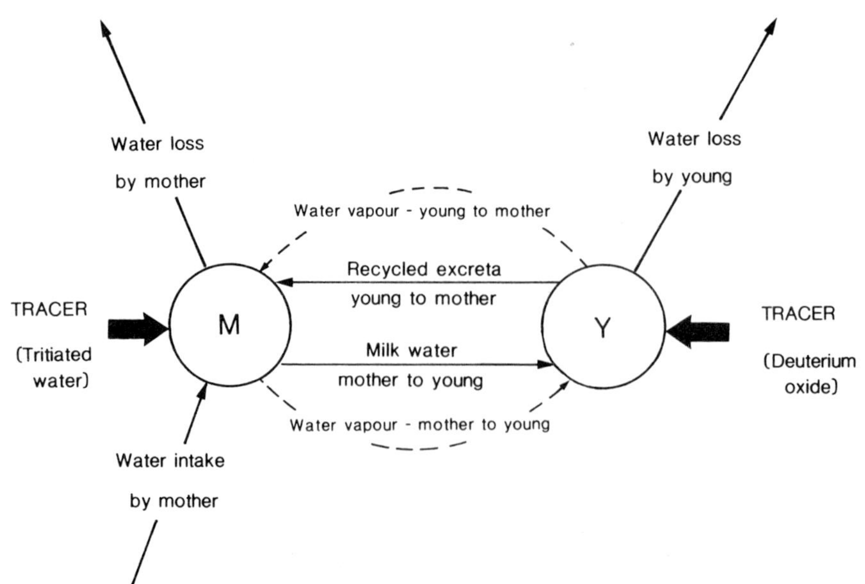

Fig. 1. Two-pool model showing routes of water flow between mother and young of the Tammar Wallaby. (Values for flows and pools are given in Table 1). The tracers introduced into each pool are shown.

Table 1. Inputs used in the simulation of isotope recycling in Tammar Wallabies.

Body weight of offspring (g)	460[A]	100[A]
Offspring		
Pool size (ml)[B]	346	75
Rate of change of pool size (ml/d)	6.0	1.6
Milk water inflow (ml/d)[C]	41.4	5.4
Rate of change of milk water inflow (ml/d.d)	0.9	0.15
Mother		
Body weight (g)	4 700	4 700
Pool size (ml)[B]	3 500	3 500
Rate of outflow of water (ml/d)[D]	505	505

[A]An offspring body weight of 460 g corresponds to approximately 210 d of age and to the commencement of emergence from the pouch. A body weight of 100 g corresponds to approximately 130 d of age.
[B]Pool size is regarded as the isotope dilution space (75.3% of body weight), which, for the offspring, varies at the rate shown. It is held constant for the mother.
[C]At the start of the period of simulation.
[D]For the mother: rate of water inflow = (rate of water outflow) + (milk water flow to offspring). Water recycled as excreta regarded as part of the inflow.

equations relating milk intake to body weight (Dove and Cork 1988; Green *et al.* 1988), while the daily rate of change of milk water intake was calculated by differentiating the equation relating milk intake to age in this species (Dove and Cork 1988). The rate of water outflow from the mother was based on our earlier data (Dove and Cork 1988); the same value was used for the two cases studied.

Based on these inputs, a FORTRAN programme using an iterative procedure and time intervals of 0.1 d was used to generate the changes in isotope concentrations and specific activities over a period of 10 d. Estimates of flows were then calculated from these isotope levels and compared with the input flows. Three situations were examined.

1. Rates of water recycling via excreta, equivalent to 0, 20, 40, 60, 80 or 100% of the offspring's outflow, were used to explore the effect of this source of recycling on the errors of estimation of water flows in both the large and the small offspring.

2. To explore the relationship between litter size, water recycling via excreta and the errors of estimation of water flows, the simulations were repeated with hypothetical litter sizes of two, four or eight offspring. In these cases, it was assumed that all littermates had the same initial size, growth rate, milk water intake and rates of water loss or recycling as each other and as the single offspring. Multiple offspring thus functioned as a single pool. Baverstock and Elhay (1981) presented evidence in support of this assumption.

3. Varying levels of water vapour exchange between the mother and a single offspring were introduced, assuming that recycling via excreta was complete, i.e., there was no external loss from the offspring. It was further assumed that the vapour exchanges between mother and offspring were equal and opposite. In the absence of data on actual rates of vapour transfer in the species, vapour flows of 1, 2, 4 or 8 ml/d were used for the smaller Tammar (100 g) plus flows of 10, 20 and 40 ml/d for the larger offspring (460 g).

RESULTS AND DISCUSSION

Water recycling via excreta

The most marked effect of the simulated recycling of water via excreta, in the case of the larger Tammar, was on the level of D_2O which accumulated in the mother, as can be seen in Fig. 2a. With 40% recycling, which is about the level observed in 210 d old offspring (Dove and Cork 1988), about 11-12% of the dose of D_2O given to the offspring can be found in the mother after 8 d. However, it should be noted that this level of accumulation is principally a result of the large pool size of the mother. The concentration of D_2O in the mother, which determines the extent of recycling to the offspring in milk, is still only 3.6% of the concentration in the offspring.

The effects of recycling of water on the level of TOH in the mother and the level of D_2O in the offspring were very small (Fig. 2b). The effects on the level of TOH in the offspring were negligible and are therefore not shown on the figure. For example, with 0% recycling, 3.63% of the mother's dose of TOH could be detected in the offspring after 10 d; with 100% recycling, this increased to only 3.71%. It follows that when these data were used to calculate water flows in either the mother or the offspring, the net effect of recycling was small. For example, 40% recycling resulted in under-estimates of 1.5% in the mother's inflow (based on TOH dilution) and 1.6% in the offspring's total inflow (based on D_2O dilution). Even with 100% recycling, these under-estimates increased to only 3.8% and 4.1% respectively.

There has been no previous estimation of the effects of recycling on the estimated inflow of water as milk, calculated using the TOH levels in the mother and offspring and the D_2O level in the offspring (Holleman *et al.* 1975). Our simulations suggest that the errors would be slightly smaller than those found with the single-isotope calculations above. With 40% recycling, the estimated error in milk water inflow was 1.2%, increasing to 2.9% with 100% recycling.

These findings differ from those of previous authors (Baverstock and Green 1975; Friedman and Bruno 1976; Baverstock and Elhay 1981), who suggested that the recycling of water via excreta would introduce large errors into the

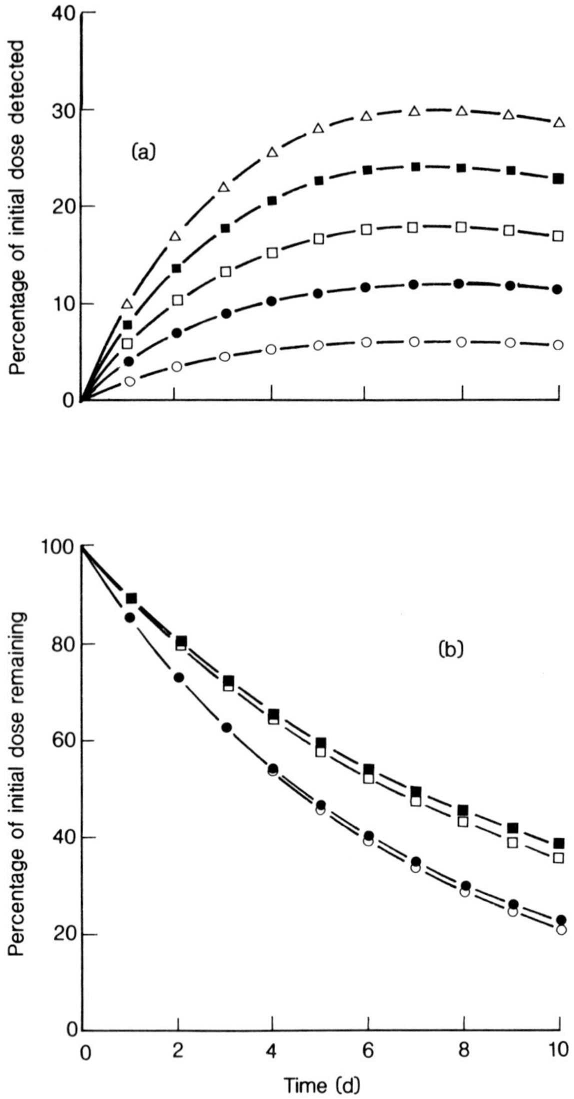

Fig. 2. (a) Simulated effects of the recycling of 20% (○), 40% (●), 60% (□), 80% (■) or 100% (△) of the excreta of the larger offspring (460 g) on the percentage of the initial dose of isotope (D_2O), given to the young, which is ultimately detected in the mother. (b) Simulation of the effects of no recycling (open symbols) or 100% recycling (closed symbols) of the offspring's excreta on the percentage of the initial dose of tritiated water remaining in the mother (○, ●) or the initial dose of deuterium oxide remaining in the offspring (□, ■). The level of tritiated water accumulating in the offspring was barely affected by recycling (see text), and is not shown.

estimates of water-flow in mother and offspring. The small errors cited above, for both single- and double-isotope procedures, can be attributed to the much smaller pool size and milk water intake of the single Tammar offspring, compared with the total offspring pool size and milk water intake in multiparous species and even some uniparous species, as can be seen in the comparisons in Table 2, for young at comparable stages of development.

Such considerations imply that, with the 100 g Tammar, where the offspring's pool size and milk water inflow are even smaller proportions of the mother's pool size (2.1% and 0.2% respectively), the errors associated with recycling via excreta would be very small. This was the outcome of our simulations, with 100% recycling resulting in under-estimates of only 0.3%, 0.4% and 0.2% respectively in the mother's inflow, offspring's total inflow and the offspring's water inflow as milk.

Effects of litter size

The effect of litter size can be emphasized by raising the hypothetical litter size of the Tammar Wallaby in the simulations, to levels similar to those of the multiparous species. In the 460 g offspring, we examined litter sizes of one, two, four and eight, at which point the total offspring pool size is 79% and the total offspring milk water inflow 10.3% of the mother's pool size. The effects on the errors of the estimated milk water inflow are shown in Fig. 3. They support the conclusions of the earlier studies cited, in that as the litter size increases to levels similar to those of the multiparous species in these studies, the estimated errors in any one offspring increase rapidly, from only 2.9% with 100% recycling and one offspring, to a 19.6% under-estimate when there are eight offspring. For the same conditions, the equivalent errors for the total inflows of the mother and offspring would be 21.7% and 24.7% respectively, which are comparable with the errors earlier reported (Baverstock and Green 1975; Baverstock and Elhay 1981).

With a 100 g offspring, 100% recycling with four offspring resulted in under-estimates of only 1.2%, 1.7% and 0.9% respectively in the mother's flow, offspring's total flow and offspring's milk water flow. This would be expected, since four offspring of this size amount to a similar proportion of the mother's pool size as with the single, larger offspring. Even with eight offspring, the equivalent under-estimates were only 2.4, 3.3 and 1.7% respectively.

Recycling as water vapour

Although the small errors cited above indicate that recycling via excreta consumption is not important in estimating water flows in this species, this route is not the only means of recycling in the young Tammar. In early pouch life, where recycling via excreta is likely to be complete, the warm and humid environment of the pouch would also be expected to favour the exchange of water vapour between mother and offspring, by pulmocutaneous and possibly cutaneous-cutaneous routes. The former mode of transfer has been shown to occur in the mouse *(Mus musculus)*, both between mother and offspring and between littermates (Friedman and Bruno 1976; Baverstock and Green 1976;

Table 2. Comparison of the body water pool size and milk water intake at peak lactation of some multiparous and uniparious species with the maternal body water pool size.

Species	Maternal pool size[A] (g)	No. of young	Total offspring pool size[C]	Total offspring /maternal	Milk water inflow (g/d)	Milk water inflow/ maternal pool
Spinifex hopping mouse *(Notomys alexis)*	14.3	3	16.8	1.18	3.33	0.23[a]
Mouse *(Mus musculus)*	28.1	6	28.8	1.02	6.60	0.23[b]
Brown rat *(Rattus norvegicus)*	150	8	102.4	0.68	31.9[B]	0.21[c]
Mink *(Mustela vison)*	720	4.8	33.1	0.46	93.2[B]	0.13[c]
Rabbit *(Oryctolagus cuniculus)*	3 300	6.8	1 380.4	0.42	185.8[B]	0.06[c]
Dog *(Canis familiaris)*	9 525	5	4 500	0.47	811.7[B]	0.09[c]
Sheep *(Ovis aries)*	33 500	1	7 200	0.21	1 536[B]	0.05[d]
Tammar Wallaby *(Macropus eugenii)*	3 500	1	346	0.10	45	0.01[e]

[A]Calculated from published information on body weight, assuming pool size = body weight × 0.75, except for the domestic sheep, where pool sizes were 0.67 and 0.72 of body weight in mother and offspring.
[B]Calculated from published information on peak milk yield and milk water content.
[C]Litter size × individual pool size calculated as in A.
[a]Baverstock and Watts (1975); [b]Baverstock *et al.* (1979); [c]Oftedal (1984b); [d]Dove and Freer (1979); [e]Present study, larger offspring.

Baverstock and Elhay 1981). Under the conditions studied, it appeared that vapour exchange could account for about half of the isotope accumulation in uninjected offspring (Friedman and Bruno 1976; Baverstock and Green 1976). Moreover, observations of a very rapid decline in the level of TOH after injection into the young of the Red- necked Wallaby *(Macropus rufogriseus)* have also been ascribed to these modes of recycling (B. Green and K. Newgrain, pers. comm.), though the effects of vapour exchange in macropodids have not been quantified.

We simulated this effect in both small and large single offspring by invoking vapour flows between mother and offspring, as shown in Fig. 1 (broken lines). Here, it was assumed that vapour flowing into the young was quantitatively returned to the mother and that recycling via excreta was complete. The calculated errors are then those attributable to the vapour transfer, as it increases to levels comparable with the milk water flows (i.e., vapour flows of 40 ml/d in the larger offspring and 8 ml/d in the smaller) as presented in Fig. 4. In both cases, these amount to vapour flows of about 80 mg/g body weight. While there seem to be no data to judge whether these are realistic values, Baverstock *et al.* (1979) obtained values of 48-63 mg/g for the evaporative water loss from the bodies of a range of rodent offspring.

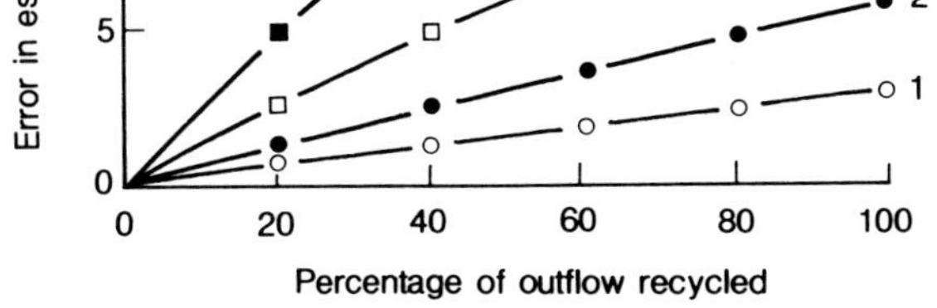

Fig. 3. Effect of litter size and the extent of recycling of the offspring's outflow (as excreta) on the predicted size of the under-estimate of milk water flow. ○, ●, □, ■ show 1, 2, 4, and 8 offspring respectively, each of 460 g body weight.

As Fig. 1 suggests, the calculations of water flows using the dilution of isotopically-labelled water are unable to distinguish water vapour inflow from inflows as water, milk or other food. As a consequence, and as our simulations demonstrated, increasing vapour flows between a mother and offspring led to rapidly increasing over-estimates of water inflows and thus milk intake (Fig. 4). The effect of vapour exchange is thus opposite to and much larger than the effect of recycling via excreta. In effect, as vapour exchange increases, the offspring increasingly behaves as part of the mother's pool.

Clearly, the presence of even a small amount of vapour exchange in the Tammar results in unacceptable errors in the estimates of water flow, especially in smaller offspring. In the larger offspring, the effect is not as marked, with a postulated vapour exchange of 2 ml/d, for

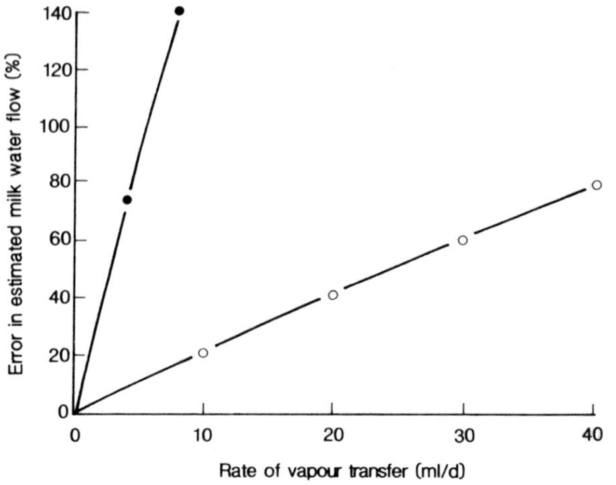

Fig. 4. Effect of water vapour transfer on the predicted extent of over-estimate of milk water flow in the Tammar offspring. ○ 460 g offspring; ● 100 g offspring.

example, resulting in an error of less than 5%. There appear to be no data on the real extent of vapour exchange at this later stage of lactation, though indirect evidence suggests that, if present at all, it would be small. When Tammar offspring near 217 d of lactation were injected with TOH, a mean of 14.3% of the dose could be detected in the mothers after 8 d (Dove and Cork 1988). Simulations in which the extent of excreta recycling or the rate of vapour exchange were varied independently, indicated that for this to occur would require either approximately 45% recycling of water via excreta (Fig. 2) or a vapour exchange of about 27 ml/d (i.e., about 60% of the offspring's milk water inflow). Since, at this stage of lactation the offspring is furred and spends extended periods outside the pouch and since there is no sweat production (Dawson *et al.* 1969), appreciable vapour exchange seems unlikely. By contrast, consumption of some of the offspring's excreta is known still to occur at this stage of lactation. Beyond about 240 d, almost no isotope injected into the offspring can be detected in the mother (Dove and Cork 1988), implying that neither excreta consumption nor vapour exchange occurs.

CONCLUSIONS

From our simulation studies based on data collected with the Tammar Wallaby (Dove and Cork 1988; Green *et al.* 1988), we draw four main conclusions.

1. At all stages until it becomes independent of the pouch, the body water pool size and daily milk water intake of the Tammar offspring are only small fractions of the mother's pool size. As a result, although recycling of water as excreta is known to occur up to about 240 d of age or about 900 g body weight (Dove and Cork 1988), this is unimportant in terms of effects on the errors of estimating water or milk water flows from the dilution of isotopically-labelled water.

2. As the total litter pool size and milk intake increases, the under-estimates of flow attributable to excreta consumption increase to levels similar to those reported for multiparous species (Baverstock and Green 1975; Baverstock and Elhay 1981).

3. In the young Tammar, naked of fur and spending all its time in the pouch, even a small amount of water vapour exchange between mother and offspring causes a more rapid loss of isotope from the pool into which it was injected and an increased rate of accumulation in the other pool. This results in large over-estimates of water flows assumed to come from other sources. In this situation, other approaches, such as that based on ^{22}Na turnover (Green and Newgrain 1979; Green *et al.* 1988), must be used to estimate milk intake.

4. The double-isotope procedure involving TOH and D_2O (Holleman *et al.* 1975; Dove 1988), when used to estimate milk intake in the Tammar undergoing the transition from the pouch to a separate existence, is not affected by recycling of excreta and is unlikely to be affected by vapour exchange at this stage. It has the major advantage of distinguishing between total water intake and that portion of it which is derived from milk.

REFERENCES

Baverstock, P. R. and Elhay, S., 1981. Water-balance of small lactating rodents. III. Estimates of milk production and water recycling in lactating *Mus musculus* under various water regimes. *J. Math. Biol.* **13**: 1-22.

Baverstock, P. R. and Green, B., 1975. Water recycling in lactation. *Science, Wash.* **187**: 657-58.

Baverstock, P. R. and Green, B., 1976. *Science, Wash.* **191**: 410.

Baverstock, P. R. and Watts, C. H. S., 1975. Water-balance of small lactating rodents. I. *Ad libitum* water intakes and effects of water restriction on growth of young. *Comp. Biochem. Physiol.* **50A**: 819-25.

Baverstock, P. R., Watts, C. H. S. and Spencer, L., 1979. Water-balance of small lactating rodents. V. The total water-balance picture of the mother-young unit. *Comp. Biochem. Physiol.* **63A**: 247-52.

Cork, S. J. and Dove, H., 1988. Lactation in the tammar wallaby *(Macropus eugenii)*. II. Intake of milk components and maternal energetic investment. *J. Zool. Lond.* (in press).

Dawson, T. J., Denny, M. J. S. and Hulbert, A. J., 1969. Thermal balance of the macropodid marsupial *Macropus eugenii* (Desmarest). *Comp. Biochem. Physiol.* **31**: 645-53.

Dove, H., 1988. Estimation of the milk intake of lambs, from the turnover of deuterium- or tritium-labelled water. *Brit. J. Nutr.* **60**: 375-87.

DOVE, H. AND CORK, S. J., 1988. Lactation in the tammar wallaby *(Macropus eugenii)*. I. Milk consumption and the algebraic description of the lactation curve. *J. Zool. Lond.* (in press).

DOVE, H. AND FREER, M., 1979. The accuracy of tritiated water turnover rate as an estimate of milk intake in lambs. *Aust. J. Agric. Res.* **30:** 725-39.

DOVE, H., CORK, S. J. AND CHRISTIAN, K. R., 1987. Defining the pattern of milk intake in a marsupial herbivore. Pp. 102-03 *in* Herbivore Nutrition Research ed. by M. Rose. Proc. Second Int. Symp. on Herbivore Nutr. Aust. Soc. Anim. Prod.: Brisbane.

FRIEDMAN, M. I. AND BRUNO, J. P., 1976. Exchange of water during lactation. *Science, Wash.* **191:** 409-10.

GREEN, B. AND NEWGRAIN, K., 1979. Estimation of the milk intake of sucklings by means of ^{22}Na. *J. Mamm.* **60:** 556-59.

GREEN, B., MERCHANT, J. AND NEWGRAIN, K., 1988. Milk consumption and energetics of growth in pouch young of the tammar wallaby *(Macropus eugenii)*. *Aust. J. Zool.* **36:** 217-27.

HOLLEMAN, D. F., WHITE, R. G. AND LUICK, J. R., 1975. New isotope methods for estimating milk intake and yield. *J. Dairy Sci.* **58:** 1814-21.

OFTEDAL, O. F., 1984a. Lactation in the dog: milk consumption and intake by puppies. *J. Nutr.* **114:** 803-12.

OFTEDAL, O. F., 1984b. Milk consumption, milk yield and energy output at peak lactation: a comparative review. Pp. 33-85 *in* Physiological Strategies in Lactation. Symp. Zool. Soc. Lond. ed. by M. Peaker, R. G. Vernon and C. H. Knight. Academic Press: London.

The use of heart rate telemetry in the measurement of energy expenditure in free-ranging Red Kangaroos

F. DOMINIC FANNING[1] and TERENCE J. DAWSON[1]

Field Metabolic Rates (FMR) and the energetic costs of various behaviours were obtained from free-ranging Red Kangaroos using an individual relationship between heart rate and oxygen consumption determined for each animal in the laboratory. Measurement of heart rate from the free-ranging animals by implanted radio-telemeters yielded FMR estimates of 1.5 to 2.3 times the animals' predicted SMR. These ratios are lower than those of similarly-sized placental mammals (2.8 times SMR). Furthermore, the FMR of a placental herbivore is 2.2 times that of a marsupial herbivore of the same size, whilst the ratio between their SMRs is approximately 1.5. Survival in "the wild" consequently appears both relatively and in absolute terms a more expensive process for placentals than for their marsupial equivalents.

INTRODUCTION

OF the several techniques which have been employed to measure the energy expenditure of animals, two currently appear particularly well-suited to the measurement of the energetics of free-ranging animals in their natural habitats. The first of these uses doubly-labelled water to follow the flux of CO_2 and H_2O through the animal, enabling an estimate of oxygen consumption to be made from the rate of production of CO_2 (Nagy 1975). This technique has been used successfully on a number of species (see Nagy 1987 for a comprehensive list) to obtain a broad appreciation of field metabolic rates. Because values are integrated over the period between subsequent samples, this technique provides "average" metabolic rates for free-ranging animals. It is therefore useful for investigating the overall energy budgets of animals in their natural habitats, the different seasonal levels of energy expenditure and the overall cost of processes such as breeding and migration.

However, the energetic costs of individual activities, such as feeding and social behaviour patterns, cannot be elucidated using the doubly-labelled water technique. To investigate these phenomena, the continuous measurement of the energy expenditure of free-ranging animals is necessary. This has been achieved successfully with a number of species using the second of the techniques alluded to above, which exploits the relationship between heart rate and oxygen consumption in animals fitted with heart rate telemeters (Owen 1969; Morhardt and Morhardt 1971; Warnold *et al.* 1977; McNairn and Fairall 1979; Kautz *et al.* 1981; Freddy 1984). Whilst some potential problems exist with this technique as applied to some species (Pauls 1980; Fisher *et al.* 1987), its careful use enables the investigation of aspects of an animal's life which are otherwise not currently quantifiable.

This paper presents an initial analysis of the HR/O_2 technique as applied to Red Kangaroos in the semi-arid zone of western New South Wales, and provides initial data on the energetics of free-ranging animals during winter days. A detailed critique of the potential of this technique and its application is currently being prepared, along with the results of a study into the energy budgets of kangaroos and sheep under various seasonal circumstances at our study area on Fowlers Gap Arid Zone Research Station.

METHODS

Relationship between heart rate and oxygen consumption

Heart rate and oxygen consumption were measured simultaneously in three wild-caught Red Kangaroos at Fowlers Gap Research Station in western New South Wales. Animals were captured using the "stunning technique" (Robertson and Gepp 1982) and given an injection of diazepam (Valium, Roche) at 0.067 ml/kg to prevent capture myopathy. They were transported to the laboratory and physiological measurements were taken 4-12 hours after capture.

Oxygen consumption was determined using open-circuit respirometry. An electric pump was used to suck air through a mask which covered the animal's head, and then pass it through a Servomex model OA272 oxygen analyzer. Air flow rate, barometric pressure and gas meter temperature were recorded on a Leeds and Northrup Speedomax-W chart recorder, along with the oxygen content of the air. The oxygen measurement circuit was calibrated using the Fe-burn technique described by Young *et al.* (1984). In this technique, the amount of oxygen "consumed" by the burning of steel wool is determined by weight gain and compared to the

[1]School of Biological Science, University of New South Wales, P.O. Box 1, Kensington, New South Wales 2033.

Pages 239-44 in KANGAROOS, WALLABIES AND RAT-KANGAROOS ed. by G. Grigg, P. Jarman and I. Hume. Surrey Beatty & Sons Pty Limited, New South Wales, Australia, 1989.

oxygen consumption calculated from the chart recording, generating a correction factor for the whole circuit. The flow and washout characteristics of the circuit were also determined, by the washout of a "bolus" of nitrogen, to enable calculation of "instantaneous" oxygen consumption (Bartholomew, Vleck and Vleck 1981), which is essential when dealing with activity and changing levels of energy expenditure.

Heart rates in the laboratory were obtained by externally attaching a J. Stuart Enterprises transmitter (model ECG-1) to the animal, with stainless steel electrodes (23 gauge hypodermic needles soldered to the transmitter leads) passing through the skin. Heart rate was monitored using a Telonics TR-2 receiver and counting the number of pulses over 30 seconds, although during bouts of activity, shorter periods were sometimes used. A Telonics TDP-2 Digital Processor was used to check the heart rates being recorded and to obtain values during periods of activity. This method of obtaining heart rates in the laboratory was employed because of the ability of the operators to distinguish invalid signals and to adapt to changing circumstances. Animals were blindfolded and then lay quietly while oxygen consumption and heart rate were being measured. Higher rates were recorded during spontaneous struggles by the animals and also when they were allowed to stand up.

Instantaneous oxygen consumption values were determined using custom-written software on an Apple-IIe computer, and these were correlated precisely with the recorded heart rates. A relationship between these variables was obtained and a separate equation derived for each animal.

Data acquisition in the field

Heart rates were obtained from "free-ranging" animals by radio-telemetry. Heart rate transmitters were surgically implanted between the skin and underlying tissues, and were stitched in place with the electrodes facing the skin. The animals were released into a 8-ha enclosure containing natural vegetation and a water trough. No supplementary feed was supplied.

Signals from the implanted transmitters were intercepted by a Telonics TR-2 multi-channel receiver coupled with a Telonics TS-1 automatic scanner, on which the frequencies of all deployed transmitters had been programmed. The pulse periods (intervals between subsequent signals) were determined by the Telonics TDP-2 digital processor. This equipment was connected to a custom-designed Hewlett-Packard Integrated Loop data acquisition system controlled by a HP-41CV pocket calculator. An HP digital multimeter (model 3468-A) monitored the output from the TDP-2 and provided the calculator with pulse period data. Programmes in the calculator analyzed the incoming data and stored the mean values, standard deviations and numbers of readings on micro-cassette tapes. The calculator also controlled the scanning of the transmitter frequencies stored on the Telonics scanner. By this means, 40 seconds of data from each of the transmitters deployed in the enclosure was collected sequentially, with the cycle repeating continuously. A detailed description of these methods is being prepared for publication.

During daylight hours, observations were also made of the behaviour of each animal, over the SAME time period in which its heart rate was being monitored. Observers were stationed in a 7 m high tower in the centre of the enclosure (which also housed the data acquisition equipment) and used high-powered binoculars to watch the animals. Ten different behaviours were recognized and their frequencies of occurrence were grouped into 10% categories during each observation period. This procedure enabled a precise correlation between an animal's heart rate, integrated over the 40-second period, and its simultaneous behaviour. The ten behaviours recorded for this study were:

Lying down	resting, undisturbed
Sitting	prone but observant
Standing quietly	undisturbed
Standing alert	alert, disturbed, observant
Feeding	includes moving slowly along while obtaining food, as well as chewing
Walking	while not feeding
Hopping	
Drinking	
Grooming	
Social interactions	all intraspecific interactions

For this report behaviours have been amalgamated into the following categories:

I	Resting	lying down, standing quietly
II	Feeding	feeding, walking, drinking
III	Restless	standing alert, social interactions, sitting, grooming
IV	Very active	hopping

These behaviour groups were determined for convenience in this paper and appear in Figs 2 and 3 with the animals' derived metabolic rates.

Measurements of relevant environmental variables were recorded using a radiometer (Solar Radiation Instruments, Melbourne) and a copper-constantan thermocouple, both attached to the Leeds and Northrup chart recorder. Air temperature and total incident radiation loads have been calculated for each half-hourly interval for the days presented in this paper.

Data conversion and manipulation

Data analysis was facilitated by a number of programmes developed for the Apple IIe computer in the course of this project. Oxygen consumption was obtained from chart data entered using an Apple Graphics Tablet, as was information on environmental variables. The heart rate stored on HP micro-cassettes were transferred to the Apple IIe for editing and behaviours were entered using a separate programme. Subsequently, all data were transferred to a MacIntosh computer for final manipulation and presentation.

RESULTS

Relationship between heart rate and oxygen consumption

A highly significant correlation was apparent between heart rate and oxygen consumption in each of the three Red Kangaroos (Fig. 1). Data for each animal were subjected to an analysis which permitted comparison between linear, quadratic, exponential and log-log regressions of the two variables. Of these, exponential regressions provided the best fit when both the correlation co-efficient and the standard error of the estimate are included. Data for each animal are displayed in Fig. 1. The equations obtained for the three Red Kangaroos are:

RED #1 $MR=0.406\ e^{0.0287\ HR}$ $(r=0.969)$

RED #3 $MR=0.535\ e^{0.0166\ HR}$ $(r=0.944)$

RED #4 $MR=0.360\ e^{0.0233\ HR}$ $(r=0.937)$

where MR = metabolic rate in $W.kg^{-1}$
HR = heart rate in beats per minute
r = correlation co-efficient

It is clear from these relationships that considerable variation exists, even between similarly sized animals (see Table 1) of the same species and sex, although the form of the relationship appears constant.

Energetics of free-ranging Red Kangaroos

The predicted standard metabolic rates (SMR) and resting heart rates (RHR) of these animals were calculated from the following allometric equations, derived from Dawson and Hulbert (1970) and Dawson and Needham (1981):

$$SMR = 2.35 \times M^{-0.25} \qquad RHR = 116 \times M^{-0.24}$$

where M = mass in kilograms
SMR is in $W.kg^{-1}$
RHR is in beats per minute

Table 1 contains the predicted values for these three animals, as well as the minimum mean heart rates recorded from them in the field, and their calculated metabolic rates.

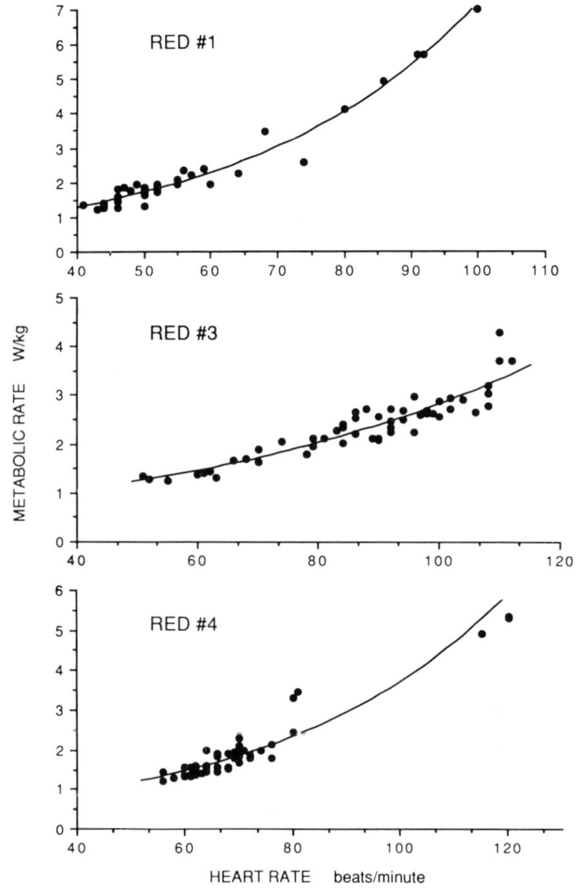

Fig. 1. The relationships between heart rate and metabolic rate for the three Red Kangaroos in this study. Metabolic rate was determined from the animal's oxygen consumption.

The minimal field heart rates of all animals were lower than the predicted RHR values, by up to 23%. Metabolic rates derived from these minimal field heart rates were within 11% of the predicted SMR for each individual.

Data are presented for one of the Red Kangaroos over two consecutive 24-h periods. Figs 2 and 3 illustrate the daily patterns of energy expenditure for RED#3 and the corresponding levels of behaviour (I, II, III, IV as above) during daylight hours. Also included are the air temperatures and nett radiation loads for those days. The negative radiation values at night indicate that the animals would have been radiating heat to their

Table 1. Comparison of the minimal Metabolic and Heart Rates of free-ranging Red Kangaroos with their predicted Resting Metabolic and Heart Rates from allometric relationships. pSMr = predicted Standard Metabolic Rate (from Dawson and Hulbert 1970); pRHR = predicted Resting Heart Rate (from Dawson and Needham 1981).

animal	weight kg	pSMR W/kg	min field MR W/kg	pRHR bpm	min field HR bpm
RED #1	30.85	1.00	1.28	51	40.0
RED #3	24.35	1.06	1.11	54	43.8
RED #4	28.10	1.02	0.91	52	40.0

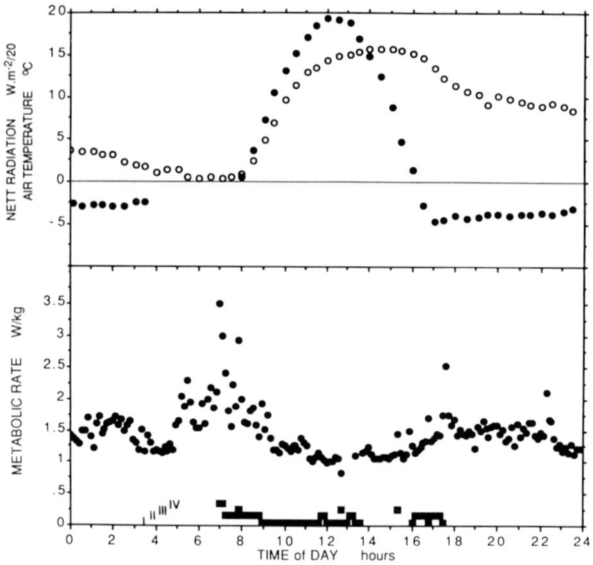

Fig. 2. Metabolic rate from a Red Kangaroo (#3) in the field, and environmental conditions, over one 24-hour period. Metabolic rates were derived from telemetered heart rates using the equation in the text for this animal. Observed behaviours (I-IV) are indicated in the solid bars. Air temperature – open circles; nett radiation – closed circles.

local environment, i.e., that there would have been a nett radiation heat loss rather than a radiation heat load.

Minimum rates of energy expenditure were 0.98-1.05 $W.kg^{-1}$ for this animal. These low rates occurred in discrete segments of the day and, at least during the period of behavioural observations, were associated with the lowest energetic level of activity — lying or standing quietly at rest. The highest rates of energy expenditure observed were correlated with such activities as hopping

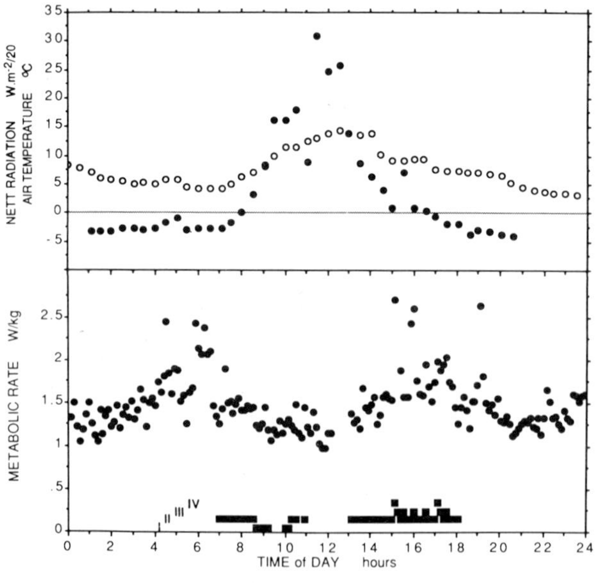

Fig. 3. Metabolic rate from the same Red Kangaroo (#3) in the field, and environmental conditions, over a second 24-hour period.

and standing alert when the animal was alarmed. These rates were up to 3.5 times the minimal values recorded, although we have other data which indicate that rates up to 15 times the minimal may occur during strenuous activity.

The field metabolic rate (FMR) of an animal is the total energetic cost of living in an environment, including the costs of basal metabolic processes, feeding and digestion, locomotion, thermoregulation, reproduction, sociality and all the other aspects of existence. Field metabolic rates for the three Red Kangaroos were obtained by integrating the metabolic rates derived from telemetered heart rates over each day, and then obtaining the mean value for six days. These are presented in Table 2, and compared with the predicted standard metabolic rates (SMR) for each animal.

Table 2. Relationships between the Standard Metabolic Rates, Field Metabolic Rates and minimal Field Metabolic Rates of the three Red Kangaroos. SMR = Standard Metabolic Rate*; FMR (fd) = Field Metabolic Rate by integration, from the data of this study (fd); F/S (fd) = ratio between FMR (fd) and SMR; F/M (fd) = ratio between FMR (fd) and minimal field MR for each animal; FMR (n) = Field Metabolic Rate from Nagy's allometric equation; F/S (n) = ratio between FMR (n) and SMR. *SMR obtained from allometric equation of Dawson and Hulbert (1970).

animal	SMR W/kg	FMR (fd) W/kg	F/S (fd)	F/M (fd)	FMR (n) W/kg	F/S (n)
RED # 1	1.00	2.21	2.3	1.73	1.86	2.0
RED # 3	1.06	1.48	1.5	1.33	2.02	2.0
RED # 4	1.02	1.55	1.6	1.70	1.92	2.0

Post-mortems were executed on the animals at the termination of the study, to evaluate their health status during the period of data acquisition. Specific searches for muscle necrosis in the hind-quarters, neck and back were made, as muscle necrosis is a symptom of the condition known as capture myopathy (Shepherd, pers. comm.). Also, the implantation sites of the heart-rate transmitters were investigated for infection or damage. No evidence of either muscle necrosis or infection was found in these animals.

DISCUSSION

The HR/O_2 technique appears to be a valuable tool for the measurement of energy use in free-ranging Red Kangaroos. We were able to monitor the energy use of free-ranging animals continuously, without interfering with their normal pattern of daily activity, and to assess the energetic costs of a variety of specific behaviours. Integration of the energy expenditures of our animals over 24-h periods provided values for field metabolic rate (FMR) which agreed tolerably well with those calculated from the allometric equations derived by Nagy (1987) using the doubly-labelled water technique.

The HR/O_2 technique enables estimation of the enegetic cost of individual behaviours or patterns of behaviour in free-ranging animals under various environmental conditions, provided that the behaviours can be monitored simultaneously with the animal's heart rate. Consequently, the energetic costs of various thermoregulatory behaviours or of males fighting for mates or defending territories, for instance, can be determined directly from undisturbed animals. Additionally, the effects of manipulating the environment, for example by restricting water supply, can be readily determined on animals in their natural habitats. As the HR/O_2 technique measures the level of energy expenditure by free-ranging animals, it can also be used, in conjunction with feeding and digestion studies, to evaluate the level and pattern of energy extraction by different species from the environment.

The HR/O_2 technique requires close adherence to certain criteria for its valid application. To obtain data of even reasonable accuracy, the specific relationship between heart rate and oxygen consumption (or metabolic rate) must be determined for each individual animal. The use of generalized allometric equations for this relationship, on either a species or any other basis, will yield data of doubtful value (Fisher *et al.* 1987). The relationship obtained for each animal should also be used close to the time of its determination. This is intended to avoid the changes in metabolic rates which may arise at various times of the year, associated with different seasonal conditions and with factors such as breeding. A more detailed analysis of the theory behind this technique and the details of its application is currently being prepared for publication.

Using the allometric equations for herbivorous marsupials derived by Nagy (1987) we estimated the FMR for the three Red Kangaroos in this study. These estimated FMR values were 2.0 times their predicted SMR. Our values for FMR, obtained using the HR/O_2 technique in this study, ranged from 1.5 to 2.3 times the animals' predicted SMR, encompassing those estimated from Nagy's equations. However, the contributions of environmental conditions and different behaviour patterns to the data used by Nagy are not known and cannot be evaluated. The precise costs of various behaviours and their role in the energetics of free-ranging animals can only be accurately determined using the HR/O_2 technique.

Estimations using allometric equations are approximate only, because individual variation has been removed in generating the relationships. Even our estimations for the ratio between FMR and SMR use a value derived from an allometric equation (SMR) and therefore are still only approximations. Consequently, we calculated the minimal metabolic rates of the animals in the field using their minimal daily heart rates, and compared their daily FMR to these. These ratios (Table 2) suggest that one of our animals (RED #1) had an SMR substantially above that predicted from the allometric equation for marsupials. This is also indicated by its relatively high minimal FMR (1.28 times SMR), compared with 1.05 and 0.89 for the other two animals. Also of note are the low minimal heart rates of the animals in the field, these being 77 to 81% of their predicted RHR. Clearly, the animals were entirely relaxed under the experimental conditions, with the lowest heart rates probably occurring during periods of sleep.

Estimations for identically-sized placental mammals, using Nagy's allometric equations, yield FMR values 2.8 times their predicted SMR, compared to the ratio of 2.0 for the marsupials. The energetic costs associated with survival under "natural" conditions thus appear relatively greater for placentals than for marsupials of this body size. Furthermore, the FMR for a placental herbivore using these data is 2.2 times the FMR for a marsupial of the same size, whilst the ratio between their SMRs is approximately 1.5 (Dawson and Needham 1981). Thus, the marsupial way of life may confer significant energetic benefits for animals in the wild. The reasons behind this apparent anomaly are currently under review.

ACKNOWLEDGEMENTS

The authors wish to acknowledge the assistance of many undergraduate and postgraduate students throughout the course of this study. The Earthwatch organization and many of its excellent volunteers also played a significant role in the execution of this project, and are gratefully acknowledged. Dr David Croft provided advice on the behavioural aspects of the study and a number of technical staff, most notable Shane Maloney, assisted with data analysis and presentation. This work has been supported by ARGS grants over several years.

REFERENCES

BARTHOLOMEW, G. A., VLECK, D. AND VLECK, C. M., 1981. Instantaneous measurements of oxygen consumption during pre-flight warm-up and post-flight cooling in Sphingid and Saturnid moths. *J. exp. Biol.* **90**: 17-32.

DAWSON, T. J. AND HULBERT, A. J., 1970. Standard metabolism, body temperature and surface areas of Australian marsupials. *Am. J. Physiol.* **218**: 1233-38.

DAWSON, T. J. AND NEEDHAM, A. D., 1981. Cardiovascular characteristics of two resting marsupials: An insight into the cardio-respirometry allometry of marsupials. *J. Comp. Physiol.* **145B**: 95-100.

FISHER, R., GILBERT, F. F. AND ROBINETTE, J. D., 1987. Heart rate as an indicator of oxygen consumption in the pine marten *(Martes americana)*. *Can. J. Zool.* **65**: 2085-89.

FREDDY, D. J., 1984. Heart rates for activities of mule deer at pasture. *J. Wildl. Manage.* **48**(3): 962-69.

KAUTZ, M. A., MAUTZ, W. W. AND CARPENTER, L. H., 1981. Heart rate as a predictor of energy expenditure of mule deer. *J. Wildl. Manage.* **45**(3): 715-20.

MCNAIRN, I. S. AND FAIRALL, N., 1979. Relationship between heart rate and metabolism in the hyrax *(Procavia capensis)* and guinea pig *(Cavia porcellus)*. *S. Afr. J. Zool.* **14**: 230-32.

MORHARDT, J. E. AND MORHARDT, S. S., 1971. Correlations between heart rate and oxygen consumption in rodents. *Am. J. Physiol.* **221**(6): 1580-86.

NAGY, K. A., 1975. Water and energy budgets of free-living animals: measurement using isotopically labelled water. Pp. 227-45 *in* Environmental Physiology of Desert Organisms ed. by N. F. Hadley. Dowden, Hutchinson and Ross Inc.: Stroudsberg, Pennsylvania.

NAGY, K. A., 1987. Field metabolic rate and food requirement scaling in mammals and birds. *Ecol. Mono.* **57**(2): 111-28.

OWEN, R. B. Jr., 1969. Heart rate, a measure of metabolism in blue-winged teal. *Comp. Biochem. Physiol.* **31**: 431-34.

PAULS, R. W., 1980. Heart rate as an index of energy expenditure in red squirrels *(Tamiasciurus hudsonicus)*. *Comp. Biochem. Physiol.* **67**(A): 409-18.

ROBERTSON, G. G. AND GEPP, B., 1982. Capture of kangaroos by "stunning". *Aust. Wildl. Res.* **9**(3): 393-96.

WARNOLD, I. AND LENNER, R. A., 1977. Evaluation of the heart rate method to determine the daily energy expenditure in disease. A study in juvenile diabetics. *Am. J. Clin. Nutr.* **30**: 304-15.

YOUNG, B. A., FENTON, T. W. AND MCLEAN, J. A., 1984. Calibration methods in respiratory calorimetry. *J. Appl. Physiol.* **56**(4): 1120-25.

The biomechanics and energetics of locomotion in Macropodoidea

R. V. BAUDINETTE[1]

This review examines gait patterns and energy requirements for locomotion in macropodid marsupials. Particular attention is paid to members of the group which hop, and comparisons are made with quadrupedal eutherians of equivalent body mass. Larger, rapidly moving macropodids have the ability to "uncouple" $\mathring{V}O_2$ from the speed of locomotion, presumably due to the storage of elastic strain energy in muscles and tendons. Within the mass range 1.5 to 5 kg this advantage is lost and power requirements for locomotion increase linearly with speed. In all macropodids examined, gait and respiratory frequencies are phase-locked in a 1:1 ratio, and in the Tammar Wallaby, it has been suggested that a "visceral piston" associated with a compliant diaphragm is responsible for lung ventilation during hopping. Entrainment between cardiac and hopping frequencies in not evident in the Tammar, and the consequences of this for arterial pressures are discussed. The review concludes by speculating why bipedalism, with its attendant energy savings at high speed, is only found among large macropodids.

INTRODUCTION

THE marsupial superfamily Macropodoidea comprises species of body mass between 0.5 and 70 kg and covers a wider range of habitats and ecological niches than most equivalent mammalian taxa; it provides an interesting group for comparisons of the mechanics and energetics of locomotion. It is not surprising that most work has been done on the larger wallabies and kangaroos, the only group of large mammals to have developed a bipedal hop as their characteristic gait. Kangaroo hopping was included in the early studies of animal locomotion by Muybridge (1899) and Magne de la Croix (1929) and high speed photographs included in the former are amazing examples of early scientific technology. The Italian zoologist Tamino published results from a study of kangaroo locomotion in 1952, and in 1965 Badoux commented on vertebral dimensions, the anatomy of the hindlimbs and the ballistics of hopping in the Eastern Grey Kangaroo *Macropus giganteus*.

This review will concentrate on descriptions of gait patterns in Macropods, an analysis of the energetic demands or power input during locomotion and some recent work from our laboratory on the physiological correlates of locomotion seen in the cardiorespiratory system. It concludes with a brief comparison of the comparative locomotory performance of marsupials and eutherians, including their respective maximum speeds, and the impact of locomotory patterns on overall daily energy use.

Gait patterns

The most complete survey of the gaits of the Macropodoidea remains that of Windsor and Dagg (1971). Hildebrand (1976) has incorporated this study into a general review of gait patterns and discusses categories of gaits and their derivation. In the former study, gaits were determined in 19 species of macropods from film records taken of zoo animals. Four patterns were determined: the walk, a quadrupedal bound, a "slow progression" in which there is no period of suspension and the tail and forelimbs support the animal during hind-limb movement, and a bipedal hop.

The first of these patterns is found only in the tree kangaroos of the genus *Dendrolagus* and shows nothing distinctive from that seen in other animals which use the gait (Hildebrand 1976). Similarly with macropods which use the quadrupedal bound as a locomotory gait, the Quokka, *Setonix brachyurus*, the tree kangaroos and the potoroo, *Potorous tridactylus* (Buchmann and Guiler 1974). In the latter the authors refer to the observed gait as a "quadrupedal crawl", although their description fits that of a bound (see Dagg 1973). The "slow progression" described by Windsor and Dagg and later referred to as "pentapedal locomotion" by Dawson and Taylor (1973) appears unique to macropods. It involves support by the tail and forelimbs while the hind-limbs are brought forward, and is most pronounced in larger species. It appears that all macropodid species have the ability to hop, including the Potoroo, Quokka and Tree Kangaroos (see Windsor and Dagg 1971; Buchmann and Guiler 1974; Baudinette 1973). The only other mammals which show this pattern of locomotion are some desert rodents, although some high latitude lagamorphs and some prosimians occasionally hop on their hind legs (see Dagg 1973).

Since the first descriptions of wallabies and kangaroos by early European explorers, people have been fascinated by the hopping gait of

[1]School of Biological Sciences, The Flinders University of South Australia, Bedford Park, South Australia 5042.

Pages 245-53 in KANGAROOS, WALLABIES AND RAT-KANGAROOS ed. by G. Grigg, P. Jarman and I. Hume. Surrey Beatty & Sons Pty Limited, New South Wales, Australia, 1989.

these animals. It is the most striking feature of kangaroos. James Cook, in his journals, describes the movement; ". . . its progression is by hopping or jumping seven or eight feet at each hop upon its hind legs only . . .". Captain Watkin Tench, in his account of the settlement at Port Jackson, is probably the first to quote measurements of the stride length of a kangaroo which he claims ". . . has been measured, and found to exceed 20 feet". This fascination is still evident among anatomists and physiologists today because macropods are the only group of large mammals that have ever developed a hopping gait. The question as to why kangaroos alone have adopted such a manner of locomotion will be addressed, somewhat speculatively, below.

Energy costs of running and hopping

The energy costs of locomotion have been measured as rates of oxygen consumption in animals exercising on motor driven treadmills. The first systematic study of this type which considered body size was that of Taylor, Schmidt-Nielsen and Raab (1972). They found that steady-state oxygen consumption ($\dot{V}O_2$) increased linearly with speed, and the relationship between these variables varied as a regular function of body mass. As a comparative index of the relative cost of locomotion they defined a term "cost of transport" as the slope of the regression line relating mass specific metabolic energy and running speed.

The derivation of this convenient relationship for eutherians was repeated for quadrupedal marsupials (Baudinette, Nagle and Scott 1976) and for birds (Fedak, Pinshow and Schmidt-Nielsen 1974). Both studies suggested divergences from the eutherian "model", but this speculation ended when Fedak and Seeherman (1979) published an analysis of data from 66 vertebrate species which showed no difference in the scaling of metabolic energy for locomotion for any taxonomic group, or between bipeds and quadrupeds. More recently Taylor, Heglund and Maloiy (1982) have added data from large wild animals and calculated a revised allometric equation for the energy cost of locomotion. This equation:

$$\dot{V}_{O_2}/M = 0.533M^{-0.316} \cdot \upsilon + 0.300M^{-0.303}$$

(where M is body mass in kg and v is speed in m/s), encompasses all mammalian groups including marsupials, within the 95% level of confidence. There were 11 marsupial species included in this analysis, among them the macropods *Bettongia penicillata, Setonix brachyurus* and *Macropus rufus*.

Provided that animals are permitted to run at a natural gait, metabolic rate increases linearly with speed in all quadrupeds. It has been shown, in fact, that gait changes occur in such a way as to minimize energy usage (Hoyt and Taylor 1981). This pattern of increasing energy consumption with speed does not occur in large macropods. Dawson and Taylor (1973) trained two Red Kangaroos, *M. rufa*, to hop on a treadmill while wearing a ventilated mask to measure rates of oxygen consumption. At slow speeds when the animals were moving pentapedally, energy usage was greater than that expected for a quadruped of equivalent body mass. However, when the animals began to hop at speeds around 1.8 m/s, $\dot{V}O_2$ was independent of speed and even fell slightly. The speed at which a hopping kangaroo expends more energy than a quadruped of the same mass is about 5 m/s. The same pattern has recently been shown in the Tammar, *M. eugenii*, by Baudinette *et al.* (1987). In this species, locomotion at speeds greater than two m/s

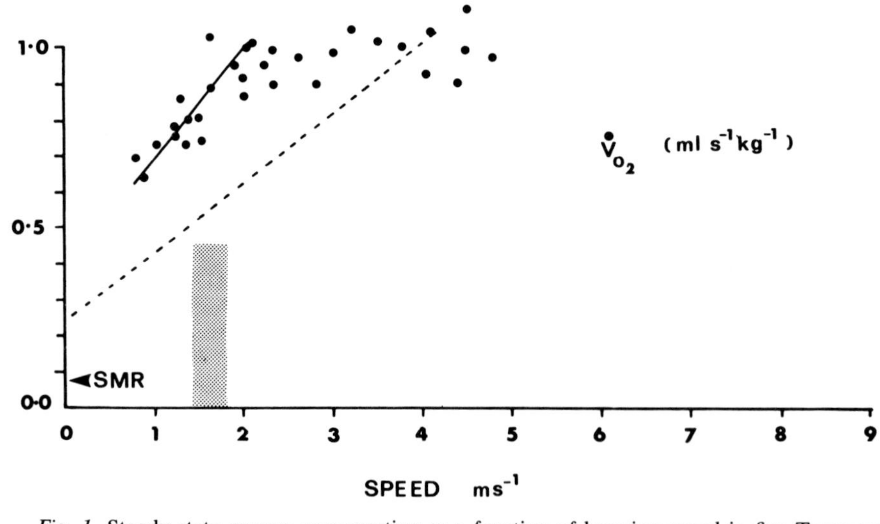

Fig. 1. Steady state oxygen consumption as a function of hopping speed in five Tammars. The shaded area indicates the range of speeds at which the animals became bipedal. The dashed line indicates the expected levels of oxygen consumption in a quadrupedal mammal of similar body mass (after Baudinette *et al.* 1987).

resulted in metabolic power consumption becoming independent of speed, but at lower speeds the slope of the relationship between $\dot{V}O_2$ and speed was greater than expected for a quadruped of similar mass (Fig. 1). The independence of metabolism and speed in this species was not due to maximal levels of $\dot{V}O_2$ being reached, as blood lactate levels were well below those seen in animals where anaerobic metabolism contributes to the available energy. Hence hopping appears to be an energetically cheap way of moving, at least at high speeds. The trade-off is that slow pentapedal locomotion is more costly than movement at the same speed for a quadruped.

Is metabolic power consumption independent of speed for smaller macropods which move with a bipedal gait? Thompson, MacMillen, Burke and Taylor (1980) trained two Brush-tailed Bettongs, *Bettongia penicillata,* along with animals of three other species of bipedal rodents, to hop on a treadmill. Rates of oxygen consumption increased linearly in all four species over a wide range of speeds. In the bettongs, transition speeds from quadrupedal to bipedal locomotion were between three and five km/h, but there was no evidence of $\dot{V}O_2$ becoming independent of speed at any point (Fig. 2). We are left with four species which show a sigmoidal relationship of $\dot{V}O_2$ with speed; the Red Kangaroo, the Tammar, the Red-bellied Pademelon, and a rodent *Notomys cervinus* (Dawson 1976). The results from the latter study have been suggested to result from an artefact in poorly trained animals (Thompson *et al.* 1980), especially since a related species *N. alexis* exhibits a typical linear relation between $\dot{V}O_2$ and speed when moving bipedally (Baudinette *et al.* 1976).

The results from the bettongs have been confirmed in the Long-nosed Potoroo, *P. tridactylus,*

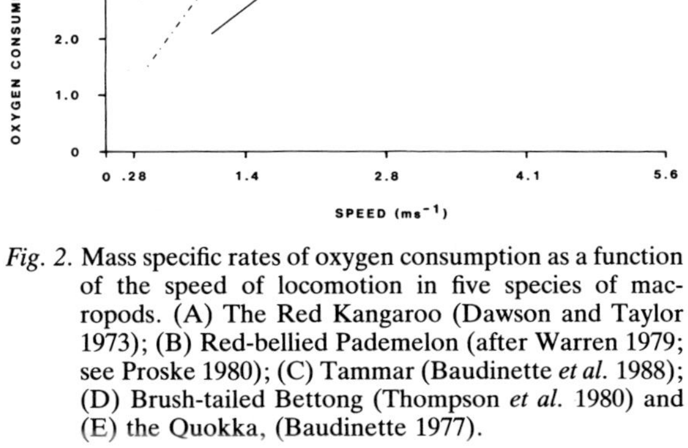

Fig. 2. Mass specific rates of oxygen consumption as a function of the speed of locomotion in five species of macropods. (A) The Red Kangaroo (Dawson and Taylor 1973); (B) Red-bellied Pademelon (after Warren 1979; see Proske 1980); (C) Tammar (Baudinette *et al.* 1988); (D) Brush-tailed Bettong (Thompson *et al.* 1980) and (E) the Quokka, (Baudinette 1977).

which, when hopping, exhibited a linear relationship between $\dot{V}O_2$ and speed (Hinds, Baudinette and Halpern, unpubl.). These animals weigh around 1.5 kg.

Summarized results from four bipedal macropods (Fig. 2) show that only the largest have the peculiar energetic pattern of being able to "uncouple" $\dot{V}O_2$ from the speed of locomotion. Presumably somewhere between 1.5 and 5 kg this advantage in energy saving becomes operative. Furthermore, the results suggest that bipedalism in small macropods (and rodents) has not evolved as an energy saving advantage, but perhaps as a means of predator avoidance or ease of movement in vegetation. Within this mass range between 1.5 and 5 kg, where we predict that an energetic advantage to bipedalism may become evident, is the Quokka. However, at least during treadmill exercise, this species moves quadrupedally. Baudinette (1977) trained three animals to exercise on a treadmill and demonstrated a linear relation between $\dot{V}O_2$ and speed with a gradient similar to that expected from a quadrupedal eutherian (Fig. 2). The gait during treadmill exercise was a bound, punctuated at speeds above 1.8 m/s with six or seven bipedal hops.

The stabilization of stride frequency during locomotion

Increases in the frequency of limb movement or in the length of a stride are the only variables which ultimately contribute to an increase in an animal's running speed. The former variable tends to stabilize in many animals running at high speeds and increases in speed are dependent on changes in stride length (see e.g., Heglund, Taylor and McMahon 1974). This pattern also holds for hopping macropods. During treadmill exercise, the Red Kangaroo maintains hopping frequency relatively constant over a three-fold change in speed (Dawson and Taylor 1973). Similar results have been found in the Tammar, in which hopping frequency was independent of speed up to 9 m/s (Baudinette *et al.* 1987). There is a single field observation based on seismic recordings which also supports this finding. Stewart and Setchell (1974) recorded traces due to Western Grey Kangaroos, *Macropus fuliginosus,* which showed a mean frequency over 466 samples of 2.04±0.13 Hz. This precise mean is remarkable considering the size and speed ranges expected to pass a given seismic station over the one-year period.

A means for comparing speed (u) between animals of different body mass is by the use of a dimensionless Froude number, u^2/gh, in which g is the acceleration due to gravity and h is a characteristic length, usually the height of the hip joint above the ground (see Alexander 1976, 1984a; Alexander, Langman and Jayes 1977). Froude numbers have been used in nautical engineering in which the length of a vessel's hull defines the term. Using an analysis based on Froud numbers, Alexander (1976) formulated an hypothesis relating animal size, speed and stride length. It has subsequently been applied to the gaits of many quadrupedal (Alexander and Jayes 1983), and bipedal vertebrates (Hayes and Alexander 1983). Filmed treadmill locomotion of the Red Kangaroo has been included in the analysis and suggests that stride lengths and the fraction of the duration of the stride for which the foot is on the ground, the duty factor, are related to the Froude number in much the same way as for other vertebrates, whether quadrupedal or bipedal. Furthermore, the kangaroo changes from a pentapedal to a bipedal gait at a Froude number around 0.5, a value similar to that for the initiation of galloping in tetrapods. Alexander treats these results as evidence for his dynamic similarity hypothesis, which predicts that animals of different sizes will adopt the same relative stride lengths (stride length/h) when moving at the same Froude number; i.e., their bodies are geometrically similar. This argument does not hold total acceptance and deviations are apparent in some groups (see e.g., McMahon 1973).

Two other matters relating to hopping should be mentioned briefly, "in-phase hopping" and the role of the tail in gait stability. The two species for which data are available, the Red Kangaroo and the Tammar, set their hind feet simultaneously, a gait described as "in-phase hopping" by Hayes and Alexander (1983). Crows (four species of the family Corvidae) on the other hand set one foot down after the other; the offset or "stagger" lying between 0.5 and 1.1 of the hip height. There is some suggestion that Kangaroo Rats may use a similar gait (see Hayes and Alexander 1983). The energy costs for these two forms of hopping have been estimated from a theoretical model and it is plausible that out-of-phase hopping is more economical in energy use than in-phase hopping. (Alexander, Jayes and Ker 1980; Hayes and Alexander 1983). No information is available on the details of the gait of small bipedal macropods, but such material would be interesting in the light of this analysis.

Alexander's group has also analysed the role of the tail in the hopping kangaroo. In bipedal or quadrupedal running, the changes of angular momentum of the legs cancel out and there is no net torsion on the body which would result in pitching. In hopping, the in-phase motion of the legs creates a moment around the centre of mass which would tend to make the body pitch with each hop. However, in hopping macropods, as the hindlimbs are swinging backwards (or anticlockwise), the tail is swinging downwards (clockwise) and hence tending to cancel the moment established by the limbs. Alexander and Vernon (1975) calculate that, for a Red Kangaroo, the balancing of these moments of inertia would result in the head and trunk pitching through six to 11 degrees, depending on the speed. Actual observations show the head moving through angles of about 10 degrees.

The respiratory and cardiovascular systems during hopping

Throughout the range of hopping speeds, Tammar Wallabies show gait and respiratory frequencies which are phase-locked in a 1:1 ratio (Fig. 3); that is, they take one breath per hop (Baudinette *et al.* 1987). This phenomenon has been confirmed in other species of macropods, including the quadrupedal Tree Kangaroo (Fig. 4; Baudinette and Wells unpubl.). A synchrony in locomotory and respiratory frequencies has previously been observed in jackrabbits, dogs and horses; in the latter a ratio of 1:1 occurs in both the trot and the gallop (Carrier and Bramble 1983). The latter authors coined the term "visceral piston" to describe the movements of the abdominal mass in altering intra-abdominal pressures and therefore inducing respiratory gas flow. Figure 5 shows how such a "piston" may operate in macropods.

Switches in the locomotory-respiratory ratio occur in rabbits, horses and humans; in the latter a ratio of 2:1 is most common, however five others are potentially available (Bramble and Carrier 1983). It appears that bipedalism in

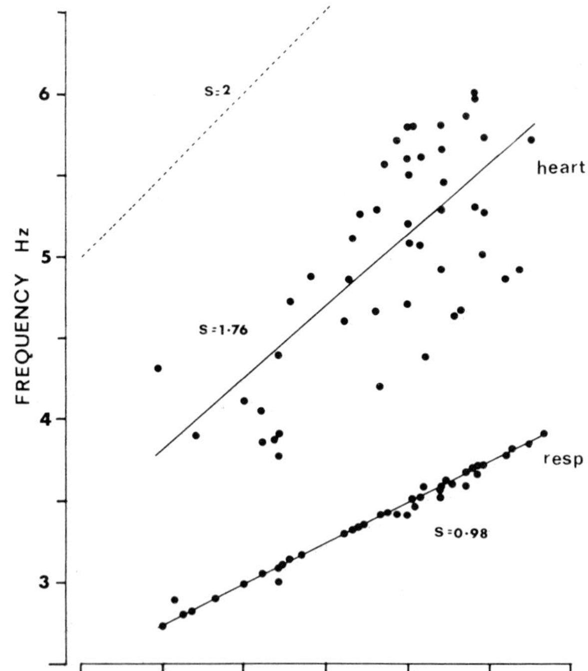

Fig. 3. Respiratory and cardiac frequencies as a function of locomotory speed in the Tammar. The dotted line indicates a theoretical coupling ratio of 1:2; s denotes the slopes of the other lines. (after Baudinette *et al.* 1987):2.

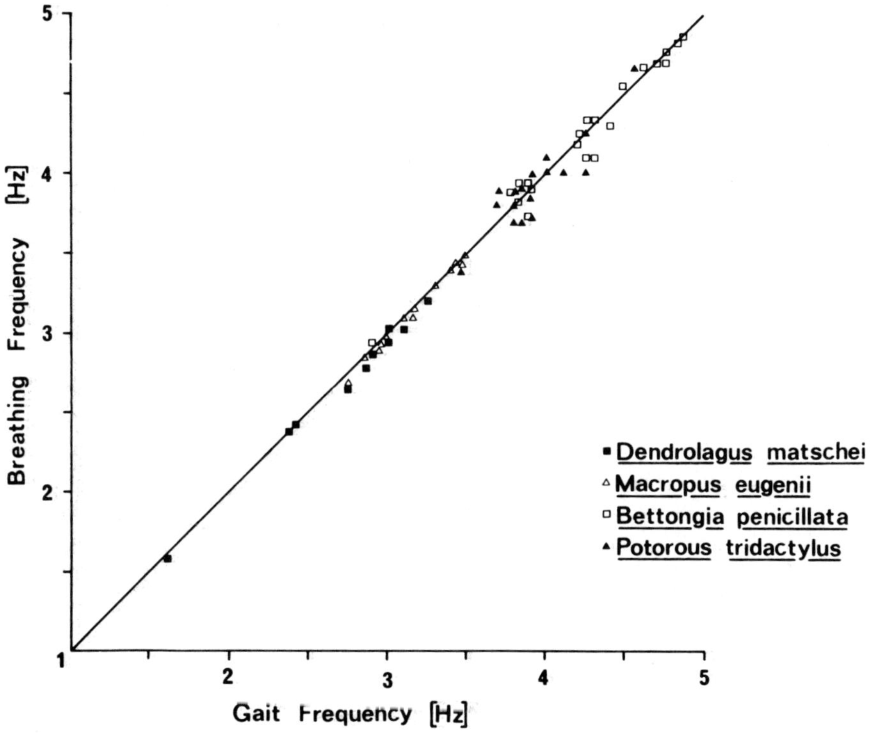

Fig. 4. Correlation of gait and breathing frequencies in four macropod species.

humans has relaxed the mechanical restraints of locomotory-respiratory coupling and a "gear box" analogy has developed. In macropods, bipedalism results in an obligatory 1:1 coupling. Consistent with this difference are the patterns in $\mathring{V}O_2$ with speed in these two species. Humans show a linear increase in metabolic rate with speed, and presumably changes in gait, breathing frequencies and tidal volume all contribute to this increased oxygen supply. Bramble (1983) has suggested that if a single coupling ratio only were available in humans, changes in running speed would have to be met by large increases in tidal volume. In contrast, large macropods change hopping speeds largely independent of $\mathring{V}O_2$, thus negating any required change in respiratory patterns to match an increased oxygen supply. Consistent with this is the finding that tidal volume, and therefore minute volume, in the Tammar remains constant over a doubling of hopping speeds (Baudinette *et al.* 1987). The respiratory diaphragm in macropods also appears adapted to a piston function in that the area of the central tendon is enlarged at the expense of the crural and costal muscles, resulting in a compliant structure on which the visceral mass acts (Baudinette *et al.* 1987). Further discussion on breathing frequencies in wallabies, including some interesting speculations on the natural frequency of the visceral mass can be found in Alexander (1987).

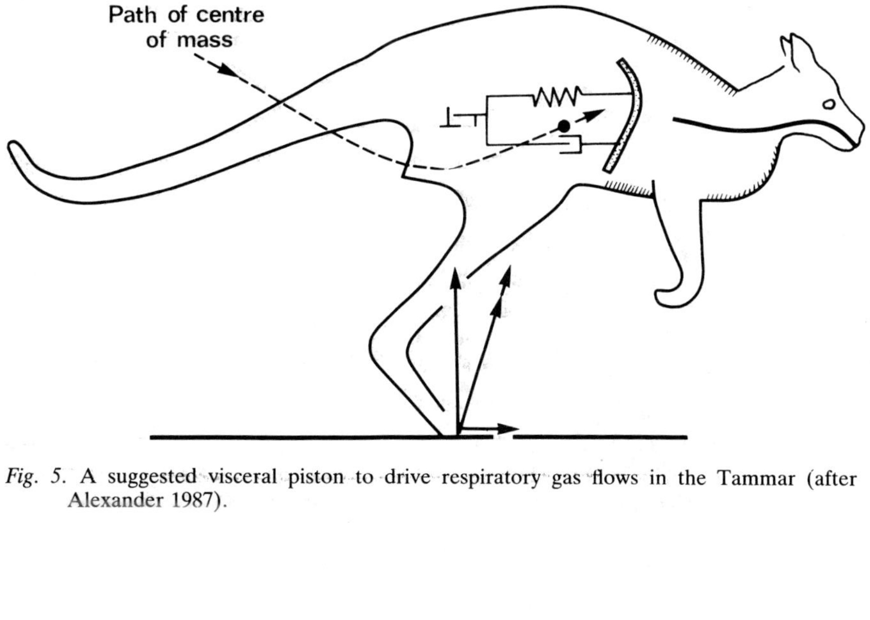

Fig. 5. A suggested visceral piston to drive respiratory gas flows in the Tammar (after Alexander 1987).

The consequences of a synchronous bipedal gait also extend to the cardiovascular system. The triangular profile of kangaroos caused by a large proportion of muscle mass in the hind limbs results in vascular pressure wave reflections which have a marked effect on the aortic pressure pulse (Avolio, Nichols and O'Rourke 1984, 1985). This effect should be even more accentuated during hopping, due to the synchronous movements of the hind limbs. When a kangaroo lands, the muscles in both hind limbs are in an active state and presumably impedance to blood flow is very high. In order to reduce the potentially damaging lateral forces on the wall of the abdominal aorta, it would seem reasonable that cardiac and gait frequencies be phase-locked; i.e., a peak pressure pulse coincides with minimal vascular impedance. This does not occur in the Tammar in which heart rate increases linearly with speed and shows no phase locking with limb frequency (Fig. 3). The site of resolution of the resultant large pressure excursions is unknown, but perhaps the compliance of kangaroo arteries is greater than that found in other mammals.

Structural correlates of hopping

In their original report of the energetic cost of locomotion in kangaroos, Dawson and Taylor (1973) said that hopping was reminiscent of a bouncing ball or a pogo stick, and suggested that analogous elements in the animals store elastic energy. This elastic recoil hypothesis finds general agreement as the mechanism behind the independence of $\dot{V}O_2$ and speed in kangaroo hopping. When the animal hits the substrate, kinetic and gravitational potential energy are stored as elastic strain energy. As this anatomical "spring" recoils at take-off, its energy is then manifest as kinetic and potential energy once again. Thus, after the initial hop, all subsequent hops reguire less energy and a plateau in the relationship between $\dot{V}O_2$ and speed is predicted. Evidence for this hypothesis comes from comparisons of $\dot{V}O_2$ and mechanical energy changes measured on a Red Kangaroo hopping on a force plate (Cavagna, Heglund and Taylor 1977) and from mechanical analyses of muscles and tendons (Bennett *et al.* 1986; Ker, Dimery and Alexander 1986).

Dawson and Taylor's finding that oxygen consumption actually decreases a little at higher speeds is explicable if the role of elastic storage increases at these speeds. Alexander and Vernon (1975) have shown from force plate experiments that the forces involved are greater at high speeds. Since the energy stored in a mechanical spring varies with the square of the force acting on it, it follows that "anatomical" springs would also play a greater role in energy storage at higher hopping speeds.

While the storage of strain energy in anatomical structures finds general agreement, the identity of these structures has not. The main argument centres on whether muscle or tendon is the primary storage site. Alexander and Vernon (1975) used recordings from force platforms and film to calculate the maximum stresses in muscles and tendons of hopping wallabies, *Protemnodon rufogrisea,* and Red Kangaroos. The results showed that the gastrocnemius and plantaris tendons are the main structures for elastic storage of energy but the associated muscles were not important. Alexander and Bennet-Clark (1977) supported this experimental finding with a theoretical treatment and suggested that up to 10 times as much energy could be stored in tendon as in muscle. A similar conclusion was reached by Morgan, Proske and Warren (1978), who performed isolated muscle and tendon experiments on *Thylogale billardieri.* A stretch of the maximally contracting gastrocnemius muscle in these animals produces eight times as much movement in the associated tendon as in muscle fibres themselves. This view has been challenged by Cavgna, Citterio and Jacini (1980), who believe that a significant amount of mechanical strain energy can be stored in the contractile elements of striated muscle, especially when the muscle is stretched at high velocity. A better recovery of this energy would be expected at speeds expected of muscle contraction during hopping, again consistent with the original data of Dawson and Taylor. Further comment on the difficulties with the ideas of muscles as simple springs can be found in Proske (1980).

There are discrepancies in the estimated amounts of energy stored by elastic structures in large hopping macropods. However, it is safe to say that at least one third the metabolic energy the animals would otherwise need for locomotion is saved by using elastic energy stores.

Why not more large bipedal species?

A form of gait which, in large animals, may save about one-third of the energy which would otherwise be used for locomotion would seem an attractive candidate for evolutionary selection. Why is it then that macropods are the only large mammals which have ever developed this specialization? Cavagna, Gitterio and Jacini (1980) suggest that the storage of mechanical energy in the contractile elements of muscle is a property of the tissue which does not appear to have changed during vertebrate evolution. Similarly, Bennett *et al.* (1986) have recently done tensile tests on tendons from the legs of a variety of mammals including the Red-necked Wallaby, *Macropus rufogriseus*, and conclude that there are no consistent differences between different species. In comparing the dimensions of tendons between wallabies and Kangaroo Rats, the only difference seems to be a proportionally greater cross sectional area in the latter (Biewener,

Alexander and Heglund 1981). Thus the two potential sites of elastic storage, tendon and muscle, seem in no way special in macropods compared to quadrupedal animals or smaller bipedal forms which do not rely on elastic storage to a marked degree.

However, it is not just muscle and tendon which correlate with a hopping gait. Discussion earlier in this chapter centred on specializations in the respiratory and cardiovascular systems. Anatomical changes to accommodate the piston-like motion of the viscera, adaptations in the diaphragm, and perhaps changes in the compliance characteristics of arteries are all features which come within a suite of hopping specializations. In the largely stochastic process of evolution, perhaps macropods were the only group to develop these necessary characteristics; hopping is not just dependent on the specialization of the hind limbs.

Marshall (1974) has suggested that anatomical specializations in macropodid ancestor pre-adapted the kangaroos and wallabies to a bipedal cursorial gait. He points out that, unlike in other mammals, the axis of the hind foot in macropods is through digit IV, a legacy of peramelid ancestors which had already developed syndactyly. This permitted the calcaneum to be incorporated into the primary mass transfer vector, an arrangement common in other hopping mammals but not in ungulates. Marshall would argue that the type of cursorial specialization in macropods was therefore predetermined.

THE IMPACT OF BIPEDALISM ON MACROPOD FORAGING PATTERNS

The selection for bipedalism in large macropods and, as suggested from laboratory experiments, a resultant economy of energy for locomotion should have ecological parallels. Are these apparent from what is known of the ecology of the animals? Garland (1983a) and Altman (1987) have both addressed the problem of estimating the daily energy expenditure for locomotion and comparing it to the total energy expenditure of the animal. Both approaches are similar and rely on allometric equations for daily distance moved, incremental energy required for locomotion and daily energy expenditure. Garland then defines "the ecological cost of transport" (ECT) as the percentage of an animal's daily energy expenditure devoted to transport costs, and derives an equation relating this parameter to body mass, M (kg):

$$\text{ECT}\ (\%) = 1.39\ \text{M}^{0.24}$$

If the carnivores are removed from the analysis, the remaining mammals show the relationship:

$$\text{ECT}\ (\%) = 1.17\ \text{M}^{0.21}$$

These equations predict that large animals spend a greater proportion of their daily energy expenditure on locomotion than do small animals. Somewhat surprising are the derived values, ranging from one half of 1% for a 10 g mammal up to 7% for an elephant of mass 6 000 kg. Altman (1987) suggests that these calculated percentages suffer from a systematic underestimation in measurements of path lengths per day and the non-inclusion of the energy associated with changing speed and direction. Selected data from African primates and ungulates seem to support this view. However, even with this uncertainty, for small or reasonably sedentary animals, daily movement has little consequence to the overall energy usage.

A desert eutherian of body mass 62 kg expends 18 764 kJ/day (estimated from equation [28] of Nagy 1987) and has an incremental cost of locomotion 192 kJ/km (from Taylor 1980; assuming 1 ml of oxygen equates to 20.1J). Therefore in moving seven km per day (a typical path length for a large macropod: R. T. Wells, pers. comm.) the animal has expended 7.2% of its daily energy expenditure in energy increments related to locomotion. A macropod of equal body mass, but with an incremental cost of locomotion one-sixth of the value for our hypothetical eutherian (a conservative figure considering the values at low hopping speeds measured by Dawson and Taylor 1973), would expend only 1.2% of its daily energy in this way. This simple calculation involves a circular argument because a lower investment in locomotion will of course feed back to reduce the overall energy cost over the day. There are data available for field energy use in the Eastern Grey Kangaroo, which suggest values of 11 734 kJ/day (Nagy 1987). Assuming the same pattern of energy use during locomotion in this species as in the Red Kangaroo, a recalculation shows that the energy expended in a day on locomotion has only increased to 1.9% of the daily energy expenditure. Is this saving of about 5% in the daily energy budget sufficient for the apparent selection for bipedalism, or are other factors operating? Dawson and Taylor (1973) and Garland (1983b) suggest that the capacity for endurance may be such a factor but unfortunately no information is available to provide support for this idea.

LOCOMOTORY PATTERNS AND THE ECOLOGY OF MACROPODS

Recent analyses deal with the energetic constraints of home range size (e.g., Mace and Harvey 1983), the relation of locomotory energetics to foraging patterns (Altmann 1987) and the overall implications of body size for the ecology of a species (Peters 1983). There is little information from macropods included in these syntheses, since the data base for the group is poor compared to those available for rodents, passerine birds and even African ungulates. This

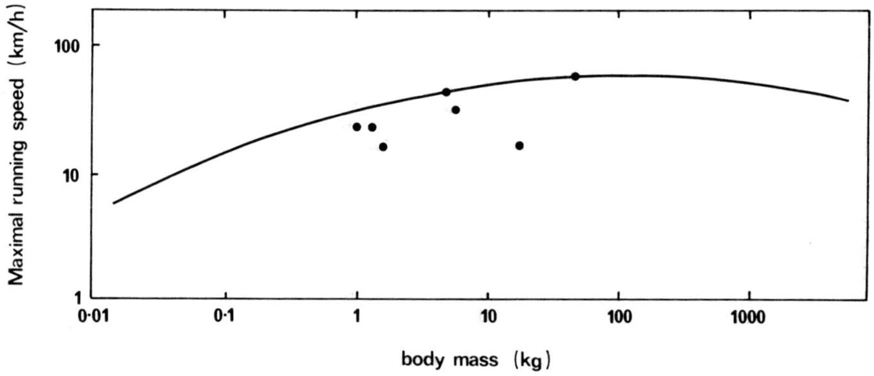

Fig. 6. Maximal running speeds as a function of body mass in 107 species of mammals (solid line) compared with the published estimates from macropods indicated on the figure by the dots (after Garland 1983; Bennett 1987; and Garland, Geiser and Baudinette *et al.* 1988).

lack of information is regrettable since such studies should provide a good theoretical framework on which to base studies on macropods, a taxon which includes many species which lend themselves to ready observation.

Included in the above syntheses are arguments on the relative advantages of energy and time economies in the selection for large body size in arid areas. Altmann (1987) argues that the ability of larger animals to more easily reach areas remote from water, and to thereby use new food sources, has been a source of selection for large body size. Since locomotory energy expenditure may be a small proportion of the total daily energy expenditure in all mammals, the selection of large size would therefore save time, rather than energy *per se,* and result in better food supply. Large desert macropods would seem an attractive vehicle with which to test ideas such as these; what is needed are good measures of daily movement patterns on a seasonal basis.

Of relevance to this argument are comparative locomotory performances of macropods and other mammals. Maximal speeds of seven macropod species given by Bennett (1987) and Garland, Geiser and Baudinette (1988), range from 21.24 km/h in *Potorous apicalis* to 48.6 km/h in *Macropus giganteus* (Fig. 6). In the former of these studies, hopping speeds correlated with body mass and scaled to the power of 0.36. It is interesting that, if this allometric relationship is extended to the extinct macropod *Procoptodon,* an animal about three metres in height, it would predict a maximum speed of about 108 m/s, so presumably the relationship breaks down at large body mass. Garland, Geiser and Baudinette (1988) compared marsupials and eutherians for maximal locomotory speeds scaled for body mass and found no overall significant differences in this regard. However Figure 6, in which the maximum speeds for macropods are compared with the derived polynomial regression equation for all mammals, suggests that the marsupials fall in the lower range of running speeds used by mammals and that bipedalism may result in a compromise in maximal speed. Further analysis using other bipedal species would be of interest.

ACKNOWLEDGEMENTS

R. E. MacMillen, D. S. Hinds, P. B. Frappell and B. J. Gannon provided useful discussions during the preparation of this review. Laura O'Carroll assisted in manuscript preparation. My own work reviewed here has been supported by The Australian Research Committee and The Flinders University Research Committee.

REFERENCES

Alexander, R. McN., 1976. Estimates of speeds of dinosaurs. *Nature* **261**: 129-30.

Alexander, R. McN., 1980. Hopping and running on two legs or four. *Nature* **287**: 187.

Alexander, R. McN., 1984a. The Gaits of Bipedal and Quadrupedal Animals. *Int. J. Robotics Res.* **3**(2): 49-59.

Alexander, R. McN., 1984b. Elastic Energy Stores in Running Vertebrates. *Amer. Zool.* **24**: 85-94.

Alexander, R. McN., 1987. Wallabies vibrate to breathe. *Nature* **328**: 477.

Alexander, R. McN. and Bennet-Clark, H. C., 1977. Storage of elastic strain enegry in muscle and other tissues. *Nature* **265**: 114-17.

Alexander, R. McN. and Jayes, A. S., 1983. A dynamic similarity hypothesis for the gaits of quadrupedal mammals. *J. Zool. Lond.* **201**: 135-52.

Alexander, R. McN., Jayes, A. S. and Ker, R. F., 1980. Estimates of energy cost for quadrupedal running gaits. *J. Zool. Lond.* **190**: 155-92.

Alexander, R. McN., Jayes, A. S., Maloiy, G. M. O. and Wathuta, E. M., 1981. Allometry of the leg muscles of mammals. *J. Zool. Lond.* **194**: 539-52.

Alexander, R. McN., Langman, V. A. and Jayes, A. S., 1977. Fast locomotion of some African ungulates. *J. Zool. Lond.* **183**: 291-300.

Alexander, R. McN. and Vernon, A., 1975. The mechanics of hopping by kangaroos (Macropodidae). *J. Zool. Lond.* **177**: 265-303.

ALTMANN, S. A., 1987. The impact of locomotor energetics on mammalian foraging. *J. Zool. Lond.* **211**: 215-25.

AVOLIO, A. P., NICHOLS, W. W. AND O'ROURKE, M. F., 1984. Exaggerated wave reflection in the kangaroo stimulates arterial counterpulsation. *Am. J. Physiol.* **246**: R267-70.

AVOLIO, A. P., NICHOLS, W. W. AND O'ROURKE, M. F., 1985. Propagation of pressure pulse in kangaroo arterial system. *Am. J. Physiol.* **249**: R335-40.

BADOUX, D. M., 1965. Some notes on the functional anatomy of *Macropus giganteus Zimm.* with general remarks on the mechanics of bipedal leaping. *Acta anat.* **62**: 418-33.

BAUDINETTE, R. V., GANNON, B. J., RUNCIMAN, W. B., WELLS, S. AND LOVE, J. B., 1987. Do cardiorespiratory frequencies show entrainment with hopping in the tammar wallaby? *J. exp. Biol.* **129**: 251-63.

BAUDINETTE, R. V., 1977. Locomotory Energetics in a Marsupial, *Setonix brachyurus. Aust. J. Zool.* **25**: 423-28.

BAUDINETTE, R. V., NAGLE, K. A. AND SCOTT, R. A. D., 1976. Locomotory energetics in a marsupial *(Antechinonomys spencerii)* and a rodent *(Notomys alexis) Experientia* **32**: 583-84.

BENNETT, M. B., 1987. Fast locomotion of some kangaroos. *J. Zool. Lond.* **212**: 457-64.

BENNETT, M. B., KER, R. F., DIMERY, N. J. AND ALEXANDER, R. McN., 1986. Mechanical properties of various mammalian tendons. *J. Zool. Lond.* **209**: 537-48.

BIEWENER, A. A., ALEXANDER, R. McN. AND HEGLUND, N. C., 1981. Elastic energy storage in the hopping of kangaroo rats *Dipodomys spectabilis. J. Zool. Lond.* **195**: 369-83.

BRAMBLE, P. M., 1983. Respiratory patterns and control during unrestrained human running. Pp. 213-20 *in* Modelling and Control of Breathing ed. by B. J. Whipp and D. M. Wiberg. Elsevier.

BRAMBLE, D. M. AND CARRIER, D. R., 1983. Running and Breathing in Mammals. *Science* **219**: 251-56.

BUCHMANN, O. L. K. AND GUILER, E. R., 1974. Locomotion in the potoroo. *J. Mammal.* **55**(1): 203-06.

CAVANGA, G. A., CITTERIO, G. AND JACINI, P., 1980. Elastic storage: role of tendons and muscles. Pp. 231-42 *in* Comparative Physiology: Primitive Mammals ed. by K. Schmidt-Nielsen, L. Bolis and C. R. Taylor. Cambridge University Press.

CAVAGNA, G. A., HEGLUND, N. C. AND TAYLOR, C. R., 1976. Mechanical work in terrestrial locomotion: Two basic mechanisms for minimizing energy expenditure. *Amer. J. Physiol.* **233**:(5): R243-61.

DAGG, A. I., 1973. Gaits in mammals. *Mammal. Rev.* **3**(4): 135-54.

DAWSON, T. J., 1976. Energetic cost of locomotion in Australian hopping mice. *Nature* **259**: 305-07.

DAWSON, T. J. AND TAYLOR, C. R., 1973. Energetic cost of locomotion in kangaroos. *Nature* **246**: 313-14.

FEDAK, M. A., PINSHOW, B. AND SCHMIDT-NIELSEN, K., 1974. Energy cost of bipedal running. *Am. J. Physiol.* **227**: 1038-44.

FEDAK, M. A. AND SEEHERMAN, H. J., 1979. Reappraisal of energetics of locomotion shows identical cost in bipeds and quadrupeds including ostrich and horse. *Nature* **282**: 713-16.

GARLAND, T., GEISER, F. AND BAUDINETTE, R. V., 1988. Comparative locomotor performance of marsupial and placental mammals. *J. Zool. Lond.* **215**: 505-22.

HAYES, G. AND ALEXANDER, R. McN., 1983. The hopping gaits of crows *Corvidae* and other bipeds. *J. Z'ool. Lond.* **200**: 205-13.

HEGLUND, N. C. AND TAYLOR, C. R., 1974. Scaling stride frequency and gait to animal Size: Mice to horses. *Science* **186**: 1112-13.

HILDERBRAND, M., 1976. Analysis of tetrapod gaits: General considerations and symmetrical gaits. *Advances in Behavioural Biology* **18**: 203-36.

HOYT, D. F. AND TAYLOR, C. R., 1981. Gait and the energetics of locomotion in horses. *Nature* **292**: 239-40.

KER, R. F., DIMERY, N. J. AND ALEXANDER, R. McN., 1986. The role of tendon elasticity in hopping in a wallaby *Macropus rufogriseus. J. Zool. Lond.* **208**: 417-28.

MCMAHON, T. A., 1973. Size and shape in biology. *Science* **179**: 1201-04.

MACE, G. M. AND HARVEY, P. H., 1983. Energetic constraints on home-range size. *Am. Nat.* **121**: 120-32.

MAGNE DE LA CROIX, P., 1929. Filogenia de las locomociones cuadrupedal y bipedal en los vertebrados. *An. Soc. cient. Argent.* **108**: 383-406.

MARSHALL, L. G., 1974. Why kangaroos hop. *Nature* **248**: 174-75.

MORGAN, D. L., PROSKE, U. AND WARREN, D., 1977. Measurement of muscle stiffness and the mechanism of elastic storage of energy in hopping kangaroos. *J. Physiol.* **282**: 253-61.

NAGY, K. A., 1987. Field metabolic rate and food requirement scaling in mammals and birds. *Ecol. Monographs.* **52**(2): 111-28.

PETERS, R., 1983. The Ecological Implications of Body Size. Cambridge University Press.

PROSKE, U., 1980. Energy conservation by elastic storage in kangaroos. *Endeavour* **4**(4): 148-53.

RAVEN, H. C. AND GREGORY, W. K., 1946. Adaptive branching in the kangaroo family in relation to habitat. *Am. Mus. Novit.* **309**: 1-33.

STEWART, I. C. F. AND SETCHELL, P. J., 1974. Seismic recording of kangaroo activity. *Search* **5**(3): 107-08.

TAMINO, G., 1952. Sulla locomozione dei canguri. *Archo. zool. Ital.* **37**: 359-81.

TAYLOR, C. R., 1980. Mechanical efficiency of terrestrial locomotion: a useful concept? *In* Aspects of animal movement ed. by H. Y. Elder and E. R. Trueman. Cambridge University Press: Cambridge.

TAYLOR, C. R., HEGLUND, N. C. AND MALOIY, G. M. O., 1982. Energetics and mechanics of terrestrial locomotion. *J. exp. Biol.* **97**: 1-21.

TAYLOR, C. R., SCHMIDT-NEILSEN, K. AND RAAB, J. L., 1970. Scaling of energetic cost of running to body size in mammals. *Am. J. Physiol.* **219**(4): 1104-07.

THOMPSON, S. D., MACMILLEN, R. E., BURKE, E. M. AND TAYLOR, C. R., 1980. The energetic cost of bipedal hopping in small mammals. *Nature* **287**: 223-24.

WINDSOR, D. E. AND DAGG, A. I., 1971. The gaits of the Macropodinae (Marsupialia). *J. Zool. Lond.* **163**: 165-75.

Thermal relations of Macropodoidea in hot environments

H. C. K. McCARRON[1] and T. J. DAWSON[1]

Many members of the Macropodoidea exhibit a complex of inter-related behavioural, physiological and morphological adaptations that enable survival in hot and dry environments. These adaptations are described and compared to those of other mammals. Species from arid and non-arid environments may differ in their level and in certain patterns of response to high temperatures, but further comparative studies with species from mesic environments are required before this can be determined.

INTRODUCTION

SINCE the turn of the century there has been controversy over the thermoregulatory abilities of marsupials. The finding by Martin in 1902 that marsupials had lower body temperatures and metabolic rates than placentals gave rise to the long held view that the marsupials were physiologically primitive, possessing inferior thermoregulatory abilities. While the low standard or basal metabolism of marsupials may indeed be a primitive or plesiomorphic characteristic (Lillegraven *et al.* 1987), the notion that they have inferior thermoregulatory abilities was challenged in the 1950's. At this time studies by Bartholomew (1956) and Robinson and Morrison (1957) showed that some marsupials had temperature regulating abilities at least equal to those of placentals. The issue was still clouded however because licking, a mechanism considered to be inefficient, was reported to be the principal form of heat dissipation at high temperatures. Later studies, however, cast doubts on the thermoregulatory importance of licking in marsupials, and demonstrated that, as in many placentals, panting was the major avenue of heat loss (e.g., Bentley 1960; Dawson 1969; Wallis and Ealey 1974). This lent credence to the emerging view that marsupials, despite their relatively low body temperature and metabolic rate, had a range of thermoregulatory responses which were as efficient as those of advanced placentals (Dawson 1973a). Indeed, more recent work suggests that marsupials have thermoregulatory abilities which are, in many respects, superior to those of many placentals (Dawson 1983).

The macropods, especially wallabies and kangaroos, exhibit some highly developed temperature regulating abilities, which in some cases have unique features. This is most evident in the suite of behavioural and physiological adaptations that enable survival in hot and dry environments. It is this aspect of the biology of the Macropodoidea that will be the subject of this review. The responses of kangaroos and wallabies to cold have been reviewed recently, as part of a general examination of the responses of marsupials to cold (Dawson 1989).

THE THERMAL ENVIRONMENT AND BEHAVIOUR

Almost one-third of macropod species are found in the arid and semi-arid zones. They range in size from the small (1-2 kg) bettongs, up to the larger kangaroos, the males of which may reach 90 kg (Strahan 1983). The arid and semi-arid environment is characterized by high summer air temperatures, and high solar and longwave radiation heatloads which combine to produce a very high integrated radiation temperature. The problem of maintaining body temperature under these conditions is further compounded by the low and erratic rainfall which results in a scarcity of water for evaporative heat loss.

Most desert mammals employ behavioural strategies that help avoid the environmental heat load, thereby effectively "fitting" the environment to their physiological capabilities. The most obvious of these is the adoption of a nocturnal activity pattern coupled with the utilization of daytime resting shelters. While this strategy is a characteristic of marsupials in general, its thermoregulatory significance to macropods living in hot environments has been highlighted in several studies (Ealey 1967; Dawson and Denny 1969; Russell 1969; Kitchener 1972; Russell and Harrop 1976; Dawson and Bennett 1978).

The type of shelter used depends primarily on body size, with smaller mammals generally having access to a wider variety of more sheltered microhabitats. One of the smallest macropod species, the Burrowing Bettong *(Bettongia lesueur)*, escapes the high air temperatures and intense solar radiation by retreating to underground burrows during the day (Stodart 1966). The temperature of most small mammal burrows rarely exceeds 30°C and consequently does not impose a thermal stress (Schmidt-Nielsen 1964). Most small macropods (< 5 kgs), however, seek refuge under bushes or within clumps of vegetation.

[1]School of Biological Science, University of New South Wales, Kensington, New South Wales 2033.

Pages 255-63 in KANGAROOS, WALLABIES AND RAT-KANGAROOS ed. by G. Grigg, P. Jarman and I. Hume. Surrey Beatty & Sons Pty Limited, New South Wales, Australia, 1989.

For example, Spectacled Hare Wallabies *(Lagorchestes conspicillatus)*, on Barrow Island, use "forms" within clumps of porcupine grass as their diurnal shelters. The floors and walls of these shelters only vary between 32° and 34°C during the hottest part of the day, even though unsheltered ground temperatures may reach 58°C (Dawson and Bennett 1978). Kitchener (1972) demonstrated the importance of shelter to the Quokka *(Setonix brachyurus)* on Rottnest Island. On hot days this species sheltered under *Arthrocnemum* and *Gahnia* bushes, where the radiant heat load was only about 5% of the level in the open. The experimental addition of shelter during summer led to an 18% increase in the number of Quokkas in the study area, leading Kitchener to conclude that shelter was a limiting resource.

Information on the importance of shelter to the larger species of macropod has been gained mainly from studies of the microhabitats of the Red Kangaroo *(Macropus rufus)* and the Euro *(Macropus robustus)*. Ealey (1967) found that Euros in the Pilbara district of Western Australia inhabited hilly areas and avoided much of the heat load on hot summer days by "lying up" in caves or under rock ledges. Because of the temperature of the surrounding rock mass, air temperatures inside deep caves were found to never exceed 32°C, even though outside temperatures reached 43°C. Dawson and Denny (1969) compared the microenvironments of the daytime shelters of Euros and Red Kangaroos at Fowlers Gap in western New South Wales. As in the Pilbara district, Euros were mainly found in small caves and under rock ledges on very hot days. Red Kangaroos, which favour the open plains, were typically found in the shelter of small desert trees, e.g., *Casuarina* sp. While only providing sparse shade, these trees were still found to screen out about 80% of solar radiation. However, the radiation temperature in this microenvironment was found to exceed body temperature by 20-30°C at high air temperatures. Consequently, Red Kangaroos were usually faced with a much higher heat load than Euros. Eastern Grey Kangaroos *(Macropus giganteus)* are also present in small numbers in this habitat, and shelter in timbered creeklines and associated stands of Prickly Wattle *(Acacia victoriae)*. At midday, on a hot summer day, the mean radiation temperature of these microenvironments may be more than 50°C lower than that in the open (McCarron, Fanning and Dawson unpublished data).

Although kangaroos spend the heat of the day "lying up" under shelter, they are not totally inactive during these periods. Studies of the diurnal behaviour of the Euro, and Red and Eastern Grey Kangaroos (Caughley 1964; Russell 1969, 1971; Russell and Harrop 1976) have shown these species to be generally restless, especially during the hottest part of the day. This restlessness is largely due to changes in behaviour associated with thermoregulation, such as "licking", and changes in position and posture.

Russell (1969, 1971) showed that for the Euro and Red Kangaroo, the amount of time spent licking increased with temperature. At higher temperatures animals also changed from lying down to standing in a crouched posture (Fig. 1). The reason for this postural change is unclear, since Dawson (1972) showed that the effective radiation temperature is about 6°C higher for a standing animal. However, wind velocity at the level of a standing kangaroo is about twice that at ground level (Bradley 1968). This posture may thus facilitate the convective heat loss of radiation that has been absorbed at the fur surface. Evaporative heat loss from areas of the body that have been licked may likewise be increased. While standing in this posture during periods of peak heat load, Red Kangaroos often bring their tail forward so that it rests between the hind legs. Russell and Harrop (1976) suggested that this behaviour may significantly reduce the absorption of radiation since the tail has a surface area that is approximately 10% of the total surface area, and it also has a low level of fur insulation.

Fig. 1. Posture of the Red Kangaroo in sparse shade on a hot summer day. Tail is pulled between the legs to reduce the surface area for radiant heat inflow. Licking of the vascularized region of the forearms helps heat dissipation, but panting is still the major form of heat loss.

Orientation of the body to the direction of the sun has been suggested as a means of reducing the influx of solar radiation at the time when the sun is not directly overhead (Schmidt-Nielsen 1964). Indeed, both posture and orientation to the sun have been found to have marked effects on the solar profile areas of the Tammar Wallaby *(Macropus eugenii)* (Brown in prep.). Russell and Harrop (1976), however, found no evidence of such orientation in the Red Kangaroo. They suggested that this was because a large part of the radiation influx for an animal in the shade is from the ground, and hence is largely non-directional.

METABOLIC RATE AND BODY TEMPERATURE

Macropods, like most marsupials, have a basal metabolic rate (BMR) that is, on average, about 30% lower than that found for placentals of the same size (Dawson and Hulbert 1970; Kinnear and Shield 1975; Hayssen and Lacy 1985). This low level of metabolism could be viewed as a pre-adaptation to desert life. This is especially the case with the larger species, which operate for much of the time in their thermoneutral zones. Apart from the energy saving, one advantage a reduced BMR confers is that the lower endogenous heat production means that less water has to be expended for evaporative cooling at high temperatures. The adaptive significance of a relatively low BMR is indicated by the fact that some desert placentals, mainly rodents, have a reduced BMR (Schmidt-Nielsen *et al.* 1967; Hart 1971; Shkolnik and Schmidt-Nielsen 1976; McNab 1979). However, McNab (1986) has suggested that the low BMRs of small desert placentals may be a consequence of diet rather than habitat. One marsupial that lives in the arid zone, the small (1-2 kg) Bilby *(Macrotis lagotis)*, has been shown to have a BMR that is reduced by 20% below the general marsupial level (Hulbert and Dawson 1974). Reductions in BMR have not been found in any desert macropods (Dawson and Hulbert 1970; McNab 1978).

In conjunction with a low BMR, macropods also share with other marsupials the characteristic of having a body temperature in thermoneutral environments that is 2-3°C below that generally found for placentals (Dawson and Hulbert 1970). In a hot environment a lower body temperature would be disadvantageous because of the larger gradient for heat flow into the body. Laboratory studies have shown, however, that macropods are able to offset this disadvantage to some degree by allowing body temperature to rise somewhat at high environmental temperatures (e.g., Bartholomew 1956; Dawson 1973b; Kinnear and Shield 1975). The Spectacled Hare Wallaby, for example, has a body temperature of 39°C at air temperatures between 38° and 43°C. This represents a 3°C increase over its level in a thermoneutral environment (Dawson and Bennett 1978).

Macropods do not seem capable of letting their body temperature increase to the marked degree seen in some desert placentals. Ungulates such as the Camel *(Camelus dromedarius)*, Oryx *(Oryx beisa)* and Grant's Gazelle *(Gazella granti)* can allow their deep body temperature to increase by over 6°C during the day (Schmidt-Nielsen *et al.* 1957; Taylor 1970), and in the latter two species, temperatures as high as 46°C have been recorded during dehydration. Increases of this magnitude can result in a considerable water saving, because the gradient for heat inflow is much reduced and stored heat can be subsequently dissipated during the cool of the night by non-evaporative means (Schmidt-Nielsen *et al.* 1957). The tolerance to high body temperature exhibited by these species is facilitated by the presence of a carotid rete which enables the brain to be kept cooler than the rest of the body (Taylor 1970). This vascular adaptation allows the hot arterial blood bound for the brain to be cooled by venous blood returning from the nasal mucosa (Baker and Hayward 1968). It is now well established that many other ungulates, and carnivores, also possess a carotid rete, it being best developed in those species that primarily rely on panting for evaporative cooling (Baker 1982; Kamau *et al.* 1984).

Although panting is the principal mechanism of heat loss in marsupials (Dawson 1983), only one species, the Tasmanian Devil *(Sarcophilus harrisii)*, has thus far been reported to possess a carotid rete (Shah *et al.* 1986). Examination of latex casts of the head vasculature has failed to find a similar structure in kangaroos (Dawson 1973b). Studies of the Red Kangaroo in a field situation have also shown that its brain temperature tracks general body temperature closely on hot summer days and that it does not allow body temperature to reach the very high levels that are seen in some placentals (Fanning and Dawson, unpublished data).

Studies of arid zone macropods suggest that they may extend their capacity for heat storage via another strategy. Brown and Dawson (1977) found that the Red Kangaroo, during summer, dropped its body temperature shortly after sunrise by as much as 3°C below the normal resting level. This was seen as a water conserving strategy since starting the day at a lower body temperature presumably delays the onset of evaporative cooling. A similar pattern has also been reported for the Euro (Dawson 1973b) and Camel (Schmidt-Nielsen *et al.* 1957; Schroter, R. C. *et al.* 1987) during dehydration.

Comparison of the Red and Eastern Grey Kangaroos suggest that macropods from arid and non-arid environments may differ in their strategy of regulating body temperature in a hot environment (McCarron and Dawson, in prep.). While the arid-adapted Red Kangaroo allows body temperature to drift higher during the hottest part of the day, the Eastern Grey Kangaroo, which is largely a forest or woodland species (Tyndale-Biscoe 1973), maintains body temperature at a relatively low level during this period (Fig. 2). The Eastern Grey Kangaroo is thus faced with a larger gradient for heat flow into the body at high air temperatures, and this would necessitate a greater expenditure of water for evaporative cooling. This may partially account for the finding that, while held in outdoor yards in summer, Eastern Grey Kangaroos had a 40% higher tritiated water turnover rate than Red Kangaroos (McCarron and Dawson, in prep.). Also, Caughley (1964) estimated from faecal pellet counts, that in semi-arid country in south-west Queensland, Eastern Grey Kangaroos visit water about three times as often as Red Kangaroos.

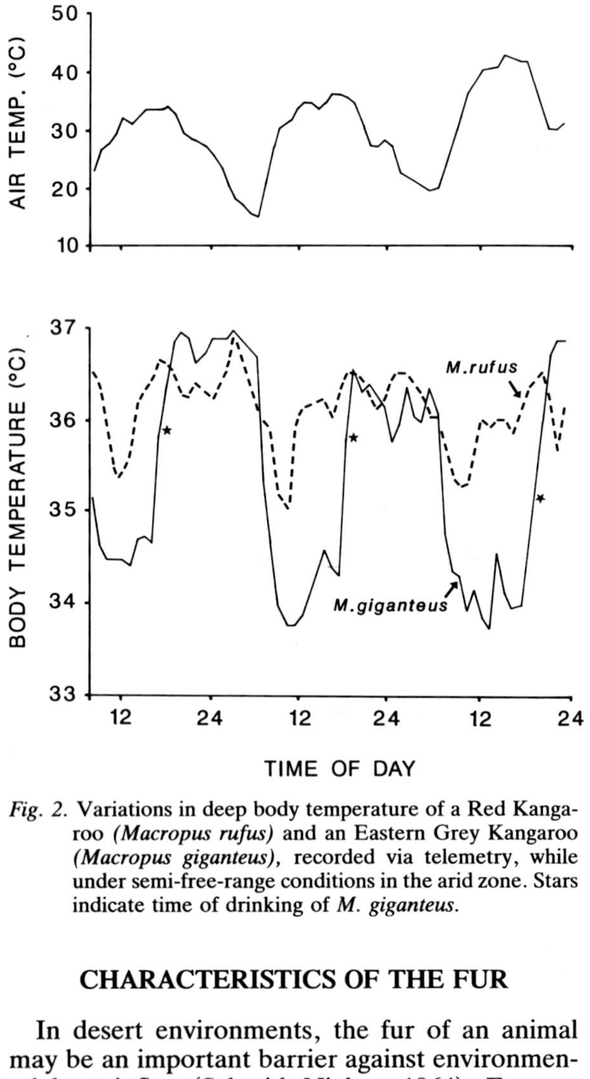

Fig. 2. Variations in deep body temperature of a Red Kangaroo *(Macropus rufus)* and an Eastern Grey Kangaroo *(Macropus giganteus),* recorded via telemetry, while under semi-free-range conditions in the arid zone. Stars indicate time of drinking of *M. giganteus.*

CHARACTERISTICS OF THE FUR

In desert environments, the fur of an animal may be an important barrier against environmental heat influx (Schmidt-Nielsen 1964). Fur can provide protection by:

(a) – reflecting solar radiation; and

(b) – insulating the body against solar radiation that has been absorbed and converted to heat at or near the surface of the fur.

Dawson and Brown (1970) investigated the reflective and insulative properties of the fur of the Euro and Red Kangaroo and found that their summer coats were relatively shallow, having an insulation similar to that found in tropical mammals. The protection against solar radiation afforded by these coats would be low as work on ungulates has indicated that protection generally increases with coat depth (e.g., Schmidt-Nielsen *et al.* 1957; MacFarlane *et al.* 1966; Finch 1972). Brown and Dawson (1977) suggested that, for animals that stay in the shade during the day, a low insulation would be beneficial as it would facilitate heat dissipation during the late afternoon and night. This would be important for an animal that stores heat during the day in order to conserve water. The Eastern Grey Kangaroo, which appears not to rely on heat storage to the same degree as these species, has a fur insulation in summer that is 60-80% greater than that found for the Red Kangaroo (McCarron and Dawson, in prep.). Kelly and Wallis (1986) also noted that the ears of the Eastern Grey Kangaroo were significantly thicker and more furred than those of the Western Grey Kangaroo *(Macropus fuliginosus)*. While the distributions of these two closely related species overlap in some areas (Caughley *et al.* 1984), the Western Grey Kangaroo generally inhabits hotter and more arid regions, and it was suggested that the ears of this species are better adapted for heat dissipation. Another advantage of a low fur insulation is that it would help dissipate the high levels of heat generated during exercise (Dawson *et al.* 1974). In this respect, it is noteworthy that the insulation of the fur on the hip and hind leg of the Red Kangaroo is particularly low (Dawson and Brown 1970).

An important determinant of the protection provided by a shallow coat is its reflectance to solar radiation. The pale fur of the Red Kangaroo is near ideal in this respect, having both a high density (62 fibres/mm^2) and a moderate to high reflectivity (Dawson and Brown 1970). A high density is important because it reduces the level to which unreflected solar radiation penetrates into the coat; penetrance is generally high in very reflective coats (Kovarik 1964; Hutchinson and Brown 1969; Cena and Monteith 1975; Walsberg 1988). The shaggy, grey-brown coat of the Euro has a much lower density (19.6 fibres/mm^2) and reflectance than that of the Red Kangaroo, therefore providing less protection. However, the microhabitat selected by the Euro allows it to avoid high intensities of solar radiation, and in this species, the fur may more importantly serve a cryptic function (Dawson and Brown 1970).

EVAPORATIVE HEAT LOSS

When environmental temperature rises above body temperature, heat can only be dissipated by the evaporation of water. There are three mechanisms for actively increasing evaporative water loss: panting, licking and sweating. Of the three, panting is usually considered to be the most effective (Richards 1970).

In macropods, panting appears to be the principal mechanism of evaporative heat loss, although licking and sweating are also utilized (Dawson *et al.* 1969; Dawson 1973b; Wallis and Ealey 1974; Nicol and Maskrey 1977; Dawson and Bennett 1978). As in the case of many placentals, panting takes the form of a graded response in respiratory rate, with the nasal mucosa being the primary site of evaporation (Dawson and Rose 1970; Needham and Dawson 1984). Air is brought close to the nasal mucosa by an elaborate arrangement of nasal turbinates, presumably to facilitate evaporation (Dawson 1983). In the Red Kangaroo and Euro, the respiratory rate at 45°C increases to approximately fifteen fold over its level in a thermoneutral environment (Fig. 3; Dawson 1973b). Of note, the Red kangaroo is able to evaporate more water for a given rise in respiratory rate than the Euro. A similar pattern of response is seen in the Red-necked Wallaby *(Macropus rufogriseus)*, where at 45°C, the respiratory rate reaches 350-450 breaths/min., and accounts for 84% of the total evaporative heat loss (Gubbins and Johnson 1972).

Under severe heat stress, macropods show a similar response to that of many placentals by shifting to "second phase" or deep open-mouth panting where the pharyngeal region becomes an additional site of evaporation. In these circumstances, the respiratory drive for evaporative heat loss is given priority over acid-base regulation and there is a tendency to develop respiratory alkalosis due to the increased alveolar ventilation (Dawson and Rose 1970).

Licking is the most obvious response of macropods to high temperature. During a typical bout of licking, saliva that has dripped from the mouth is wiped or licked on the forearms; the tongue may not necessarily be involved. Although there have been doubts about the usefulness of licking by macropods (e.g., Bentley 1960; Dawson *et al.* 1969; Wallis and Ealey 1974), studies on the Red Kangaroo suggest that it may play a significant role in temperature regulation. Needham *et al.* (1974) found that in the region of the forearm licked by the Red Kangaroo, there is a dense, superficial network of blood vessels, and during heat stress, blood flow to this region is greatly increased. When wetted, the forearm region can thus be a site of significant heat transfer. Indeed, spreading saliva on the forearms may be making the most efficient use possible of an overflow of fluid from the respiratory system. It has also been shown that the tail has an extensive vascular

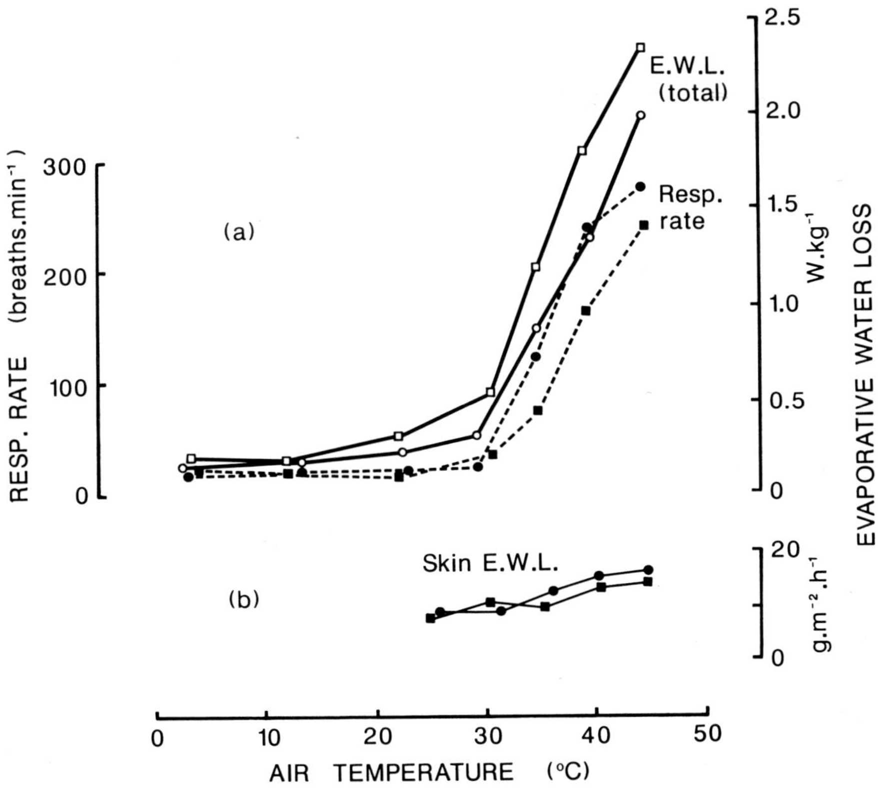

Fig. 3. (a) Total evaporative water loss and respiratory rates in summer of Red Kangaroos and Euros at various air temperatures. Evaporative water loss was expressed in W/kg assuming that the vaporization of 1 gram of water per hour requires 0.674 Watts. Red Kangaroos E.W.L., □—□; Euro E.W.L., ○—○; Red Kangaroo respiratory rate, ■ - - - ■; Euro respiratory rate, ● - - - ●. (b) Cutaneous water loss from the back region of Red Kangaroos and Euros. Red Kangaroo skin E.W.L., ■—■; Euro skin E.W.L., ●—● (modified from Dawson 1973b).

supply similar to that of the forearm (Needham *et al.* 1974; Young and Webster 1974). In the Tammar Wallaby, dry heat loss from the tail under thermoneutral conditions accounts for about 7% of total heat production (Krattenmacher and Rubsamen 1987). However, Needham and Dawson (1984) demonstrated in the Red kangaroo, that this vascular system can be closed down at very high temperatures to help minimize heat inflow to the body (Table 1).

Table 1. Influence of environmental temperature on tissue blood flows ($\mu l.g^{-1}.mins^{-1}$) of Red Kangaroos[a].

Tissue	Air Temperature (°C) 9	24	35	44
Body skin	32	18	24	9
Forelimb skin	137	180	159	341
Nasal mucosa	26	143	589	1 726
Tail tip skin	3	51	133	29

[a]Values are means for two animals (modified from Needham and Dawson 1984).

Sweating in response to an external heat load has only been demonstrated in one species of macropod, the Long-nosed Potoroo *(Potorous tridactylus)* (Hudson and Dawson 1975). This species, which inhabits coastal heath and forest, sweats profusely from its naked tail in a hot environment. At a body weight of about 1 kg, it is the smallest mammal known to sweat. The rate of sweating from the tail is very high, about twice the maximum rate reported for ungulates. As in many small mammals, the body fur of the Potoroo is long and provides important insulation in cool conditions. This insulation evidently necessitates that this active little marsupial have an extra route of heat loss during exercise and hot conditions, in addition to respiratory evaporation (Dawson 1983). The role of sweating has not been investigated in other potoroids, and it would be interesting to find out if potoroids from arid environments respond similarly to an external heat load.

In contrast to the situation seen at rest (Fig. 3), several species of macropod have been demonstrated to sweat in response to exercise (Fig. 4; Gubbins and Johnson 1972; Dawson 1973b; Dawson *et al.* 1974; Hudson and Dawson 1975; Baudinette 1977), and it seems likely that this may occur in most macropods. Hopping generates high levels of heat production and sweating brings into play an additional surface for evaporative heat loss. This may be important since Baudinette *et al.* (1987) have suggested that respiratory ventilation and gait are phase-locked in a 1:1 mode, so that exercise places a mechanical restraint on respiratory evaporation.

Dawson *et al.* (1974) found that of the total evaporative water loss in the exercising Red Kangaroo, at an air temperature of 24°C, 60% was accounted for by sweating and only 40% by increased respiratory evaporation. It was also found that total heat loss fell considerably short of heat production, and a large proportion (43%) of the heat generated during a 20 minute bout of hopping was stored. A similar pattern of heat storage during exercise has been found for the Quokka (Baudinette 1977), and it has been suggested to be characteristic of marsupials in

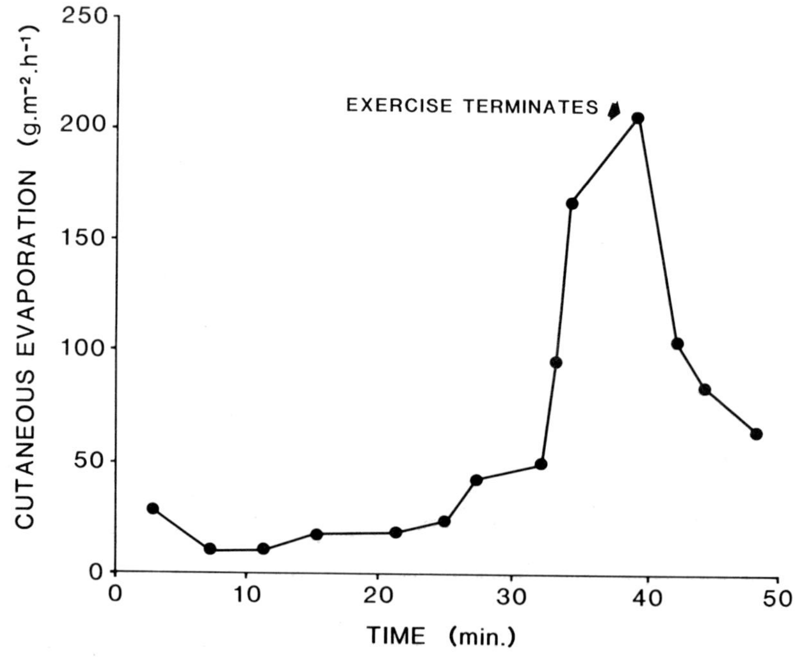

Fig. 4. Sweating of a Red Kangaroo in response to exercise on a treadmill moving at a rate of 4 k.p.h. Some 30 minutes after starting to exercise, the kangaroo began to sweat copiously. When exercise ended, sweating diminished, virtually ceasing within 10 minutes (modified from Dawson *et al.* 1974).

general (Baudinette 1984). Among placentals, species that sprint, rather than engage in prolonged periods of endurance running, also rely heavily on heat storage (Taylor and Rowntree 1973). However, the time course of heat storage during exercise has yet to be examined, especially in kangaroos that can travel long distances. The bipedal hopping gait of macropods allows heat production to be uncoupled from increases in speed, so that at moderate to high speeds they generate a lower heat load than quadrupedal mammals of the same size (Dawson and Taylor 1973; Baudinette 1984; Baudinette *et al.* 1987). Certainly this must be of considerable advantage in a hot environment. This phenomenon results from the storage and subsequent recovery of elastic energy in the tendon and muscle during hopping (Taylor and Heglund 1982). However, this does not appear to be of significance in some of the smaller species (Thompson *et al.* 1980; Baudinette, this volume).

One aspect of the thermoregulatory response to exercise that sets some macropods apart from other mammals is that sweating stops as soon as exercise ceases (Fig. 4; Dawson *et al.* 1974). This is true even when body temperature is still elevated and the animal is panting rapidly. The advantage of switching from sweating to panting once exercise stops lies in water economy. Sweating lowers the skin surface temperature, which in turn reduces the gradient for heat loss in cool and moderate conditions and in hot conditions leads to a greater inflow of heat from the environment. In these circumstances, the utilization of additional water is needed to eliminate this additional heat. Panting does not lower the general body surface temperature. This ability to alternate between sweating and panting is perhaps the most sophisticated heat loss mechanism found in mammals (Dawson 1973b).

In conclusion, it appears that macropods are well able to cope with hot environments, exhibiting a complex of inter-related behavioural, physiological and morphological adaptations. In some instances, these adaptations are even unique among mammals. There are some indications that some of the strategies employed by macropods may be specific to arid zone species, but more comparative studies with species from mesic environments are required before this can be determined.

REFERENCES

BAKER, M. A., 1982. Brain cooling in endotherms in heat and exercise. *Annu. Rev. Physiol.* **44**: 85-96.

BAKER, M. A. AND HAYWARD, J. N., 1968. The influence of nasal mucosa and carotid rete upon hypothalamic temperature in sheep. *J. Physiol. Lond.* **198**: 561-79.

BARTHOLOMEW, G. A., 1956. Temperature regulation in the macropod marsupial, *Setonix brachyurus. Physiol. Zool.* **29**: 26-40.

BAUDINETTE, R. V., 1977. Locomotory energetics in a marsupial, *Setonix brachyurus. Aust. J. Zool.* **25**: 423-28.

BAUDINETTE, R. V., 1984. Thermoregulation in marsupials during exercise. Pp. 423-28 *in* Thermal Physiology ed. by J. R. S. Hales. Raven Press: New York.

BAUDINETTE, R. V., GANNON, B. J., RUNCIMAN, W. B., WELLS, S. AND LOVE, J. B., 1987. Do cardiorespiratory frequencies show entrainment with hopping in the Tammar Wallaby? *J. Exp. Biol.* **129**: 251-63.

BENTLEY, P. J., 1960. Evaporative water loss and temperature regulation in the marsupial *Setonyx brachyurus. Aust. J. Exp. Biol.* **38**: 301-06.

BRADLEY, E. F., 1968. A micrometerorological study of velocity profiles and surface drag in the region modified by a change in surface roughness. *Quart. J. Roy. Met. Soc.* **94**: 361-79.

BROWN, G. D. Solar profile area of the Tammar Wallaby *Macropus eugenii:* the effect of posture and orientation to the sun. In preparation.

BROWN, G. D. AND DAWSON, T. J., 1977. Seasonal variations in the body temperatures of unrestrained kangaroos (Macropodidae: Marsupialia). *Comp. Biochem. Physiol.* **56A**: 59-67.

CAUGHLEY, G., 1964. Social organization and daily activity of the Red Kangaroo and the Grey Kangaroo. *J. Mammal.* **45**: 429-36.

CAUGHLEY, G., BROWN, B., DOSTINE, P. AND GRICE, D., 1984. The Grey Kangaroo overlap zone. *Aust. Wildl. Res.* **11**: 1-10.

CENA, K. AND MONTEITH, J. L., 1975. Transfer processes in animal coats. I. Radiative transfer. *Proc. Roy. Soc. Lond. B.* **188**: 377-94.

DAWSON, T. J., 1969. Temperature regulation and evaporative water loss in the Brush-tailed Possum *Trichosurus vulpecula. Comp. Biochem. Physiol.* **28**: 401-07.

DAWSON, T. J., 1972. Likely effects of standing and lying on the radiation heat load experienced by a resting kangaroo on a summer day. *Aust. J. Zool.* **20**: 17-22.

DAWSON, T. J., 1973a. Primitive Mammals. *In* Comparative Physiology of Thermoregulation ed. by G. C. Whittow. Vol. 3. Academic Press: New York.

DAWSON, T. J., 1973b. Thermoregulatory responses of the arid zone kangaroos, *Megaleia rufa* and *Macropus robustus. Comp. Biochem. Physiol.* **46A**: 153-69.

DAWSON, T. J., 1983. Monotremes and Marsupials: the other mammals. The Institute of Biology Studies in Biology, No. 150. Edward Arnold: London.

DAWSON, T. J., 1989. Responses to cold of monotremes and marsupials. *In* Adaptation of Vertebrates to Cold ed. by L. C. Wang. Springer, Berlin. In press.

DAWSON, T. J. AND BROWN, G. D., 1970. A comparison of the insulative and reflective properties of the fur of desert kangaroos. *Comp. Biochem. Physiol.* **37**: 23-38.

DAWSON, T. J. AND DENNY, M. J. S., 1969. A bioclimatological comparison of the summer day microenvironments of two species of arid zone kangaroo. *Ecology.* **50**: 328-32.

DAWSON, T. J., DENNY, M. J. S. AND HULBERT, A. J., 1969. Thermal balance of the macropod marsupial *Macropus eugenii* Desmarest. *Comp. Biochem. Physiol.* **31**: 645-53.

DAWSON, T. J. AND HULBERT, A. J., 1970. Standard metabolism, body temperature, and surface areas of Australian marsupials. *Am. J. Physiol.* **218**: 1233-38.

DAWSON, T. J., ROBERTSHAW, D. AND TAYLOR, C. R., 1974. Sweating in the kangaroo: a cooling mechanism during exercise, but not in the heat. *Amer. J. Physiol.* **227**: 494-98.

DAWSON, T. J. AND ROSE, R. W., 1970. Influence of the respiratory response to moderate and severe heat on the blood gas values of a macropodid marsupial *(Macropus eugenii)*. *Comp. Biochem. Physiol.* **37**: 55-66.

DAWSON, T. J. AND TAYLOR, C. R., 1973. Energetic cost of locomotion in kangaroos. *Nature.* **246**: 313-14.

DAWSON, W. R. AND BENNETT, A. F., 1978. Energy metabolism and thermoregulation of the Spectacled Hare Wallaby *(Lagorchestes conspicillatus)*. *Physiol. Zool.* **51**: 114-30.

EALEY, E. H. M., 1967. Ecology of the Euro, *Macropus robustus* (Gould), in northwestern Australia. II. Behaviour, movements and drinking patterns. *C.S.I.R.O. Wildl. Res.* **12**: 27-51.

FINCH, V. A., 1972. Energy exchanges with the environment of two East African antelopes, the Eland and the Hartebeest. *Symp. Zool. Soc. Lond.* **31**: 315-26.

GUBBINS, S. AND JOHNSON, K. G., 1972. Evaporative water loss in the wallaby *(Protemnodon rufogrisea)*. *J. Physiol. Lond.* **226**: 110-11.

HART, J. S., 1971. Rodents. *In* Comparative Physiology of Thermoregulation ed. by G. C. Whittow. Vol. 3. Academic Press: New York.

HAYSSEN, V. AND LACY, R. C., 1985. Basal metabolic rates in mammals: taxonomic differences in the allometry of BMR and body mass. *Comp. Biochem. Physiol.* **81A**: 741-54.

HUDSON, J. W. AND DAWSON, T. J., 1975. Role of sweating from the tail in the thermal balance of the rat-kangaroo *Potorous tridactylus*. *Aust. J. Zool.* **23**: 453-61.

HULBERT, A. J. AND DAWSON, T. J., 1974. Standard metabolism and body temperature of perameloid marsupials from different environments. *Comp. Biochem. Physiol.* **47A**: 583-90.

HUTCHINSON, J. C. D. AND BROWN, G. D., 1969. Penetrance of cattle coats by radiation. *J. Appl. Physiol.* **26**: 454-64.

KAMAU, J. M. Z., MAINA, J. N. AND MALOIY, G. M. O., 1984. The design and the role of the nasal passages in temperature regulation in the Dik-dik Antelope *(Rhychotragus kirkii)* with observation on the carotid rete. *Resp. Physiol.* **56**: 183-94.

KELLY, L. C. AND WALLIS, R. L., 1985. Structural and possible thermoregulatory differences in the ears of Grey Kangaroos. *J. Therm. Biol.* **10**: 249.

KINNEAR, A. AND SHIELD, J. W., 1975. Metabolism and temperature regulation in marsupials. *Comp. Biochem. Physiol.* **52A**: 235-45.

KITCHENER, D. J., 1972. The importance of shelter to the Quokka, *Setonix brachyurus* (Marsupialia), on Rottnest Island. *Aust. J. Zool.* **20**: 281-99.

KOVARIK, M., 1964. Flow of heat in an irradiated protective cover. *Nature.* **201**: 1085-87.

KRATTENMACHER, R. AND RUBSAMEN, K., 1987. Thermoregulatory significance of non-evaporative heat loss from the tail of the Coypu *(Myocastor coypus)* and the Tammar Wallaby *(Macropus eugenii)*. *J. Therm. Biol.* **12**: 15-18.

LILLEGRAVEN, J. A., THOMPSON, S. D., MCNAB, B. K. AND PATTON, J. L., 1987. The origin of eutherian mammals. *Biol. J. Linne. Soc.* **32**: 281-336.

MARTIN, C. J., 1902. Thermal adjustments and respiratory exchanges in monotremes and marsupials. *Phil. Trans. Roy. Soc. Lond. B.* **195**: 1-37.

MCCARRON, H. C. K. AND DAWSON, T. J. Energy expenditure and body temperature of kangaroos from desert and forest environments. In preparation.

MCFARLANE, W. V., HOWARD, B. AND MORRIS, R. J. H., 1966. Water metabolism of Merino sheep shorn during summer. *Aust. J. Agric. Res.* **17**: 219-25.

MCNAB, B. K., 1978. The comparative energetics of neotropical marsupials. *J. Comp. Physiol.* **125**: 115-28.

MCNAB, B. K., 1979. Climatic adaptation in the energetrics of heteromyid rodents. *Comp. Biochem. Physiol.* **62A**: 813-20.

MCNAB, B. K., 1986. The influence of food habits on the energetics of eutherian mammals. *Ecol. Mono.* **56**: 1-19.

NEEDHAM, A. D. AND DAWSON, T. J., 1984. Influence of environmental temperature on the blood flow distribution in the Red Kangaroo *Macropus rufus*. Pp. 275-78 *in* Thermal Physiology ed. by J. R. S. Hales. Raven Press: New York.

NEEDHAM, A. D., DAWSON, T. J. AND HALES, J. R. S., 1974. Forelimb blood flow and saliva spreading in the thermoregulation of the Red Kangaroo, *Megaleia rufa*. *Comp. Biochem. Physiol.* **49A**: 555-65.

NICHOL, S. R. AND MASKREY, M., 1977. Panting in small mammals: a comparison of two marsupials and the laboratory rabbit. *J. Appl. Physiol. Respirat. Environ. Exercise Physiol.* **42**: 537-44.

RICHARDS, S. A., 1970. The biology and comparative physiology of thermal panting. *Biol. Rev.* **45**: 223-64.

ROBINSON, K. W. AND MORRISON, P. R., 1957. The reaction to hot atmospheres of various species of Australian marsupial and placental animals. *J. Cell. Comp. Physiol.* **49**: 455-78.

RUSSELL, E. M., 1969. Summer and winter observations of the behaviour of the Euro *Macropus robustus* (Gould). *Aust. J. Zool.* **17**: 655-64.

RUSSELL, E. M., 1971. Changes in behaviour with temperature in the Red Kangaroo, *Megaleia rufa*. *Aust. J. Zool.* **19**: 207-13.

RUSSELL, E. M. AND HARROP, C. J. F., 1976. The behaviour of Red Kangaroos *(Megaleia rufa)* on hot summer days. *Z. Tierpsychol.* **40**: 396-426.

SCHMIDT-NIELSEN, K., 1964. Desert animals: physiological problems of heat and water. Oxford University Press: London.

SCHMIDT-NIELSEN, K., CRAWFORD, E. C., NEWSOME, A. E., RAWSON, K. S. AND HAMMEL, H. T., 1967. Metabolic rate of camels: effect of body temperature and dehydration. *Amer. J. Physiol.* **21**: 341-46.

SCHMIDT-NIELSEN, K., SCHMIDT-NIELSEN, B., JARNUM, S. A. AND HOUPT, T. R., 1957. Body temperature of the camel and its relation to water economy. *Am. J. Physiol.* **188**: 103-13.

SCHROTER, R. C., ROBERTSHAW, D., BAKER, M. A., SHOEMAKER, V. H., HOLMES, R. AND SCHMIDT-NIELSEN, K., 1987. Respiration in heat stressed camels. *Resp. Physiol.* **70**: 97-112.

SHAH, S. K. H., NICHOL, S. C. AND SWAIN, R., 1986. Functional morphology of the cranial vasculature and the nasal passage in the Tasmanian Devil, *Sarcophilus harrisii* (Marsupialia: Dasyuridae): a marsupial carotid rete? *Aust. J. Zool.* **34**: 125-33.

SHKOLNIK, A. AND SCHMIDT-NIELSEN, K., 1976. Temperature regulation in hedgehogs from temperate and desert environments. *Physiol. Zool.* **49**: 56-64.

STODART, E., 1966. Observations on the behaviour of the marsupial *Bettongia lesueuri* (Quoy and Gaimard) in an enclosure. *C.S.I.R.O. Wildl. Res.* **11**: 91-99.

STRAHAN, R. (ed.), 1983. The Australian Museum Complete Book of Australian Mammals. Angus & Robertson: London.

TAYLOR, C. R., 1970. Dehydration and heat: effects on temperature regulation of East African ungulates. *Am. J. Physiol.* **219**: 1136-39.

TAYLOR, C. R. AND HEGLUND, N. C., 1982. Energetics and mechanics of terrestrial locomotion. *Ann. Rev. Physiol.* **44**: 97-107.

TAYLOR, C. R. AND ROWNTREE, V. J., 1973. Temperature regulation and heat balance in running cheetahs: a strategy for sprinters. *Am. J. Phsyiol.* **224**: 848-51.

THOMPSON, S. D., MACMILLEN, R. E., BURKE, E. M. AND TAYLOR, C. R., 1980. The energetic cost of bipedal hopping in small mammals. *Nature* **287**: 223-24.

TYNDALE-BISCOE, H., 1973. Life of Marsupials. Edward Arnold: London.

WALLIS, R. L. AND EALEY, E. H. M., 1974. Thermoregulation in the Parma Wallaby, *Macropus parma* (Marsupialia: Macropodidae). *Aust. Mammal.* **1**: 243-49.

WALSBERG, G. E., 1988. Consequences of skin colour and fur properties for solar heat gain and ultraviolet irradiance in two mammals. *J. Comp. Physiol. B.* **158**: 213-21.

YOUNG, R. J. AND WEBSTER, M. E. D., 1974. Skin temperatures and vascular arrangements in the kangaroo tail *(Macropus giganteus)*. *Proc. Aust. Physiol. Pharmacol. Soc.* **5**: 126-27.

Adrenocortical functions in macropodid marsupials

I. R. McDONALD[1] and I. K. MARTIN[2]

The structure of the adrenal gland in macropodid marsupials resembles that of eutherian mammals. The major glucocorticoid secreted is cortisol and the major mineralocorticoid is aldosterone. Kanagroos, in common with other marsupials, also secrete significant amounts of 21-deoxycortisol, found in man only in cases of adrenogenital syndrome. Cortisol is normally bound to a transcortin-like protein in blood plasma so that only 10-15% is in the free, biologically active form. Cortisol secretion is stimulated by stress and by injection of synthetic ACTH, but the sensitivity to exogenous synthetic ACTH is low. The adrenocortical responses of Red and Eastern Grey Kangaroos to stress appear to be minor in comparison with other species. Red Kangaroos can survive adrenalectomy for long periods without supplementary corticosteroids by voluntarily increasing salt intake, while Quokkas and Tammar Wallabies show no increase in salt appetite and require corticosteroids. With the possible exception of tammar wallabies, macropods are resistant to the diabetogenic and nitrogen mobilizing actions of cortisol, which has no effect on insulin sensitivity, urinary nitrogen excretion or, in the short term, plasma glucose. However, in the long term, high doses of exogenous cortisol increase liver glycogen and plasma glucose concentrations in the Quokka and Red-bellied Pademelon. The lack of effect of cortisol on urinary nitrogen excretion is probably due to recycling of mobilized tissue nitrogen. Cortisol also causes lymphopenia in macropods but its effects on the immune and inflamatory mechnaisms are yet to be explored.

INTRODUCTION

ALTHOUGH the nature of the adrenocortical steroids secreted by all species of mammals is much the same, there may be significant species differences in the major secretory products (Sandor, Sonia and Mehdi 1975) and sensitivity to their actions (Harris and Baxter 1979). This is particularly evident among the different species of marsupials so far investigated. In particular, the macropodid marsupials appear to comprise a distinct group in which some of the accepted metabolic actions of the adrenocortical steroids, readily demonstrable in representatives of other marsupial families, are either not found or are demonstrable only by injection of exogenous corticosteroids at very high dose rates. In this review we assemble the observations published by different groups and attempt to define the role of adrenocortical secretion in macropodid marsupials, both in normal homeostasis and in their responses to environmental disturbances.

Structure of the adrenal gland

As in other marsupials, the structure of the macropodid adrenal gland conforms to the basic mammalian pattern of cords of steroid secreting cells arranged radially around a medulla of chromaffin cells (Fig. 1). At least three distinct zones are recognizable, an outer glomerulosa, where the cells tend to form loops, a middle fasciculata, where the cells form radial columns, and an inner reticularis, where the cells are more randomly dispersed. The blood supply enters through a supcapsular capillary plexus surrounding the glomerulosa cells, which is rearranged into radial sinuses separating the cords of fasciculata cells and rearranged again into a more diffuse plexus surrounding the reticularis cells, before draining into the venous sinuses of the medulla (see Jones 1957). In addition, the medulla receives a more direct arterial supply from branches of the capsular arterioles, which pass directly through the cortex without branching (Fig. 1c). The venous blood drains by way of a short adrenal vein into the renal vein on the left side and the inferior vena cava on the right. The arterial blood is derived mainly from branches arising from the abdominal aorta and renal artery.

When compared with the adrenal cortices of marsupials in other families, such as the Brush-tailed Possum *Trichosurus vulpecula* (Phalangeridae), the macropodid adrenal cortex has a much more distinct zona glomerulosa (Fig. 1), the major source of the salt-retaining hormone, aldosterone. The glomerulosa, inner fasciculata and reticularis are rich in lipid which, under polarised light, shows the characteristic birefringence of cholesterol. The adrenal gland of the Red Kangaroo *Macropus rufus* resembles that of man, having a pyriform instead of the oval shape as in other mammals; and exhibiting a distinct yellowish coloration on the cut face of the cortex. The capsule may be pigmented with melanin, especially in the Eastern Grey Kangaroo *M. giganteus*.

Hearn (1975) investigated the effects of hypophysectomy in the Tammar Wallaby *(Macropus eugenii)*, and noted that this caused a swift decline in adrenal weight. This was due to total disappearance of the zona reticularis and a marked reduction in size of the fasciculata, considered to be the major site of glucocorticoid synthesis. The outer glomerulosa and medulla were little affected.

[1]Department of Physiology, Monash University, Victoria 3168.

[2]Department of Chemistry and Biology, Footscray Institute of Technology, Victoria 3011.

Pages 265-75 in KANGAROOS, WALLABIES AND RAT-KANGAROOS ed. by G. Grigg, P. Jarman and I. Hume. Surrey Beatty & Sons Pty Limited, New South Wales, Australia, 1989.

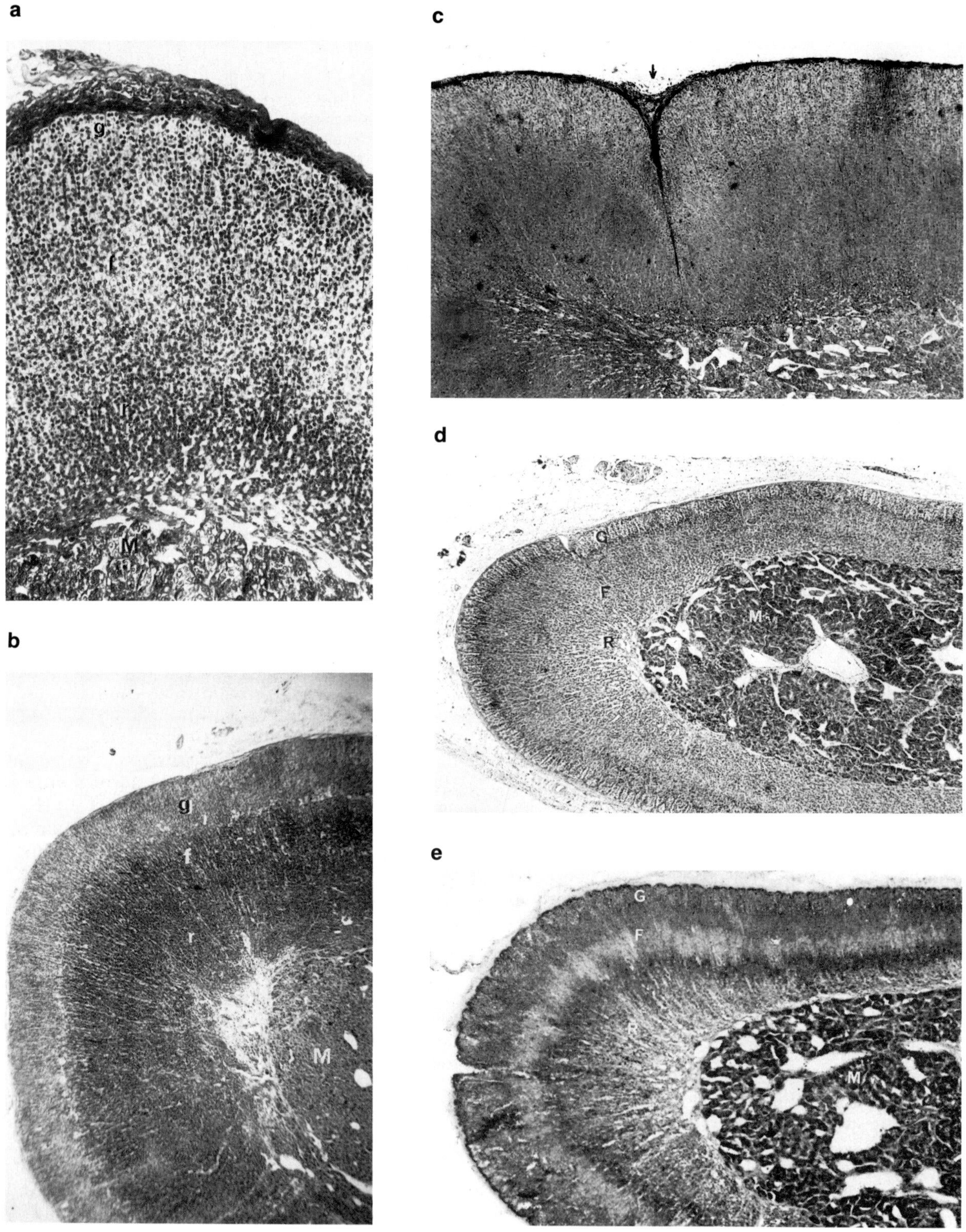

Fig. 1. Histological appearance of adrenal glands in Brushtail Possum *Trichosurus vulpecula* (a), Red Kangaroo *Macropus rufus* (b, c) and Quokka *Setonix brachyurus* (d, e). With the exception of (e), sections were cut from formalin-fixed tissue embedded in paraffin and stained with Masson's Trichrome stain. Note indefinite zonation in the adrenal cortex of *Trichosurus,* compared with the very clear definition of the zona glomerulosa in adrenal cortices of *Macropus* and *Setonix.* An example of the direct arterial blood supply to the medulla in *Macropus* is shown in (c). The distribution of lipid in the adrenal cortex of *Setonix* is shown in a frozen section stained with Oil-red "O" (e). Note the lack of staining in the outer fasciculata. Magnifications: (a) and (c) × 125, remainder × 40.

Nature of the adrenocortical hormones

The mammalian adrenal cortex secretes a wide variety of steroids derived from cholesterol, many of these being precursors or products of the quantitatively major corticosteroids (Fig. 2). Only a few are known to be significant, in terms of rates of secretion and potency of action, to normal physiological functions. Their actions range from highly potent effects on active transport of sodium (mineralocorticoids), through effects on energy metabolism (glucocorticoids) to effects on reproduction (progestagens, androgens and oestrogens).

In marsupials generally and macropods in particular, most attention has been paid to the corticosteroids that are known to influence salt and energy metabolism in eutherian mammals. These are considered to be important to survival in adverse environments. In the adrenal venous or peripheral blood of all the marsupials so far investigated, quantitatively the major corticosteroid in this group is cortisol (Fig. 2). This is the major glucocorticoid in eutherian mammals with the exception of the rodents, in which the major glucocorticoid is corticosterone and cortisol is not secreted. In Kangaroos and Wallabies, significant but lesser concentrations of corticosterone and 11-deoxycortisol are also found, the last being the immediate precursor of cortisol in the normal eutherian biosynthetic pathway. Curiously, kangaroos, in common with some other species of marsupials (Weiss 1968), secrete significant amounts of another earlier precursor. 21-deoxycortisol (Weiss and McDonald 1967). This is secreted in man only by individuals with a deficiency of adrenal 21-hydroxylase, causing excessive secretion of androgens with early virilization, in a disease known as the adrenogenital syndrome (Wieland *et al.* 1965; MacArthur and O'Donnell 1967).

Weiss and Vardalov (1977) investigated corticosteroid biosynthesis by homogenates and slices of kangaroo adrenocortical tissue and concluded that it differs in some ways from that of the eutherian mammals. They concluded that the 11β-hydroxylase system behaved as a single enzyme capable of hydroxylating a variety of substrates at C11, in contrast to the multicomponent, substrate-specific system of eutherian mammals (Hudson *et al.* 1976). The capacity of this system appears to be low relative to the 17α-hydroxylase so that, when adrenocortical secretion is stimulated, cortisol secretion reaches a maximum while the rate of secretion of its immediate precursor, 11-deoxy cortisol, begins to rise (Weiss and McDonald 1967). This also occurs in the Quokka *Setonix brachyurus* (McDonald and Bradshaw 1977) and the Red-bellied Pademelon *Thylogale billardierii* (Martin and McDonald 1986) (Fig. 2). The preferential pathway for cortisol biosynthesis

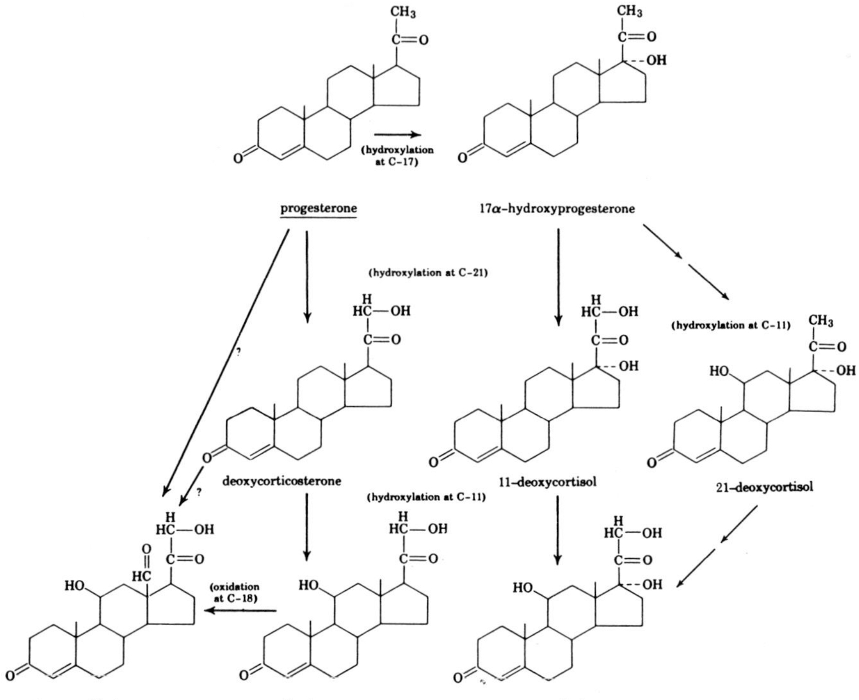

Fig. 2. Pathways of biosynthesis of the major corticosteriods from progesterone. Heavy arrows indicate the preferred pathways determined for eutherian mammals. Lighter interrupted arrows indicate the postulated preferred pathway to cortisol in kangaroos (Weiss 1980). ACTH stimulates corticosteriod biosynthesis from cholesterol before the formation of progesterone. Other stimuli affect aldosterone secretion at the C-18 oxidation step.

in kangaroos appears to be from 11-hydroxy progesterone via 21-deoxy cortisol (Fig. 2) — i.e., 17-hydroxylation precedes 21-hydroxylation (Weiss 1980). This would explain the presence of relatively large amounts of 21-deoxy cortisol in adrenal venous blood of these marsupials. The possible functional significance of these peculiarities is unknown.

In addition to cortisol and its precursors, another significant corticosteroid, though quantitatively minor, is the mineralocorticoid aldosterone (Fig. 2), secreted by all species of mammals so far investigated. Another mineralocorticoid, present in peripheral blood at approximately the same concentration but much less biologically potent than aldosterone, is 11-deoxycorticosterone (Oddie *et al.* 1976).

As in other mammals, the rates of corticosteroid secretion are influenced by anterior pituitary adrenocorticotrophin (ACTH) which stimulates biosynthesis of early precursors in the pathways from cholesterol (Urquhart 1974). In eutherian mammals, secretion of aldosterone is influenced in addition, at the stage of 18-hydroxylation (Fig. 2) by a variety of agents in blood plasma including angiotensin II, potassium, sodium, serotonin and natriuretic peptide (Muller 1988). There is evidence for an effect of the renin-angiotensin system on aldosterone secretion in marsupials (Simpson and Blair-West 1971).

Rates of corticosteroid secretion

Estimates of the rates of corticosteroid secretion have been made in anaesthetized Red and Eastern Grey Kangaroos and Quokkas by cannulation of one adrenal vein and analysis of timed collections of adrenal venous blood (Weiss and McDonald 1967; Illet 1969). These presumably represent the maximum stressed rate of secretion for that gland and the total secretion rate has been estimated from the combined weight of both adrenal glands, assuming equal rates of secretion per unit of gland mass. The rates of cortisol secretion fell within the range 0.6-6.0 nmol/kg/min (Weiss and McDonald 1967). Martin and MacDonald (1986) used a constant-rate isotopic tracer infusion method to estimate cortisol secretion rate in the conscious unrestrained Red-bellied Pademelon before and during stimulation by exogenous ACTH. The calculated mean rates of secretion were 1.15 nmol/kg/min before, and 8.7 nmol/kg/min during ACTH infusion. The resting rate of 1.65 umol/kg/day (0.6 mg/kg/day) falls within the range reported for eutherian herbivores such as the sheep (McDonald and Reich 1959; Blair-West *et al.* 1962; Harrison *et al.* 1964).

The rate of aldosterone secretion has been measured only in anaesthetized Red and Eastern Grey Kangaroos and falls within the range 8-110 pmol/min/kg (4-60 ug/kg/day), similar to that in ACTH-stimulated or Na-depleted sheep (Reich and McDonald 1961; Blair-West *et al.* 1962).

Concentrations in peripheral blood

A wide range of concentrations of the several corticosteroids in whole blood and blood plasma obtained from different species has been reported. Part of the variation can be accounted for by the circumstances in which the samples were taken and part is due to species differences.

Weiss and McDonald (1967) reported plasma cortisol concentrations in the range 55-220 nmol/l in anaesthetized Red and Eastern Grey Kangaroos and found approximately equal amounts in erythrocytes. Coghlan and Scoggins (1967) analysed whole blood of Eastern Grey Kangaroos, obtained by cardiac puncture 1-5 min after the animals had been shot while in their natural environment. The major corticosteroid was cortisol at concentrations in the range 16-66 nmol/l in males and 77-116 nmol/l in females. One group was in an alpine habitat, where the Na content of the food was low and the other was near the coast, where the Na content of the food was high. There was no significant difference between blood cortisol concentration in the two groups. However, the concentration of aldosterone in the blood of the alpine animals was 943-1887 pmol/l, compared with 0-375 pmol/l in the coastal animals. It is interesting that there was little difference between the ranges of plasma cortisol concentrations in animals that were anaesthetized and those that were shot while in their natural environment. This suggests that the adrenocortical response to stress in these macropods is minor when compared with other marsupials.

Oddie *et al.* (1976) made a survey of the concentrations of several corticosteroids in whole blood taken from five different species of macropodid marsupials soon after they had been shot while in their natural environment. These were the Eastern Grey Kangaroo, Swamp Wallaby *(Wallabia bicolor)*, Bennetts Wallaby *(Macropus rufogriseus)*, Red-bellied Pademelon and Quokka. They detected cortisol, corticosterone and aldosterone in the blood of all five species: in addition, 11-deoxy cortisol and 11-deoxy corticosterone were detected in the blood of Eastern Grey Kangaroos and Swamp Wallabies. Cortisol was the major glucocorticoid in all samples with mean concentrations of 30-60 nmol/l in all except the Red-bellied Pademelon, in which the mean value was near 300 nmol/l. There was a similarly wide range of 30-220 pmol/1 in the mean concentrations of aldosterone, the higher values being found in the Swamp Wallaby and the Quokka. These values all fall within the range reported for eutherian mammnls. The concentrations of the other corticosteroids were minor, relative to cortisol.

Table 1. Mean plasma cortisol concentrations and concentration of corticosteroid-binding globulin binding sites in blood plasma of several species of macropodid marsupial.

Species	Mean plasma cortisol conc. (nmol/1) Unstim.	ACTH	Stress	Corticosteroid-binding globulin Conc. (nmol/l)	K_a ($\times 10^{-7}M^{-1}$)
M. rufus	(M) 95*	–	105[1]	110 (268*)	5.0
	(F) 76*	–		284*	
	(J) 60*	–		146*	–
M. giganteus	220[2]	–	165[1]	P	–
M. eugenii[3]	185[a]	1 765	650	370	6.5
	430[b]				
T. billardierii[4]	220	620	1 100	530	–
S. brachyurus[5]	22	400	100	120	4.8
M. rufogriseus[2]	47	–	–	P	
W. bicolor[2]	30	–	–	P	

K_a = high-affinity binding constant; P = high affinity binding present but not characterized; M = adult male; F = adult female; J = juvenile (M & F); a = in metabolism cage; b = in hessian sack; 1 = Weiss and McDonald (1967); 2 = Oddie et al. (1976); 3 = Janssens and Tyndale-Biscoe (1982); 4 = Martin and McDonald (1986); 5 = McDonald and Bradshaw (1977). With the exception of (*) and *T. billardierii* (4), measurements of CBG were made by Sernia, Bradley and McDonald (1979). *= unpublished data from samples obtained by Dr J. Arundel, Department of Veterinary Pathlogy, University of Melbourne, during a culling operation at Lake Menindie, New South Wales.

McDonald and Bradshaw (1977) and Miller and Bradshaw (1979) made more detailed observations on captive and free-ranging Quokkas. The former found that the plasma total corticosteroid concentration in undisturbed captive Quokkas with implanted jugular venous catheters was 20-25 nmol/l. increasing to 400 nmol/l during maximal ACTH stimulation. They also noted that the stress of heart puncture and being chased for 20 min followed by recapture caused plasma total corticosteriod concentration to increase to about 100 nmol/l (Table 1). As in other marsupials, the sensitivity to injected synthetic ACTH was low and even lower than in the Brush-tailed Possum (Than and McDonald 1973), so that approximately 160-fold the human dose i.v. was needed to achieve maximum plasma cortisol concentration (Fig. 3). There was no difference between the sensitivities to synthetic and porcine ACTH. Cortisol was the major glucocorticoid in all circumstances, although stress and ACTH stimulation caused a minor increase in both corticosterone and 11-deoxycortisol concentrations. Miller and Bradshaw (1979) obtained blood from free-ranging Quokkas by cardiac puncture within 2 min of sighting the animal. They reported plasma total corticosteroid concentrations of 30±3 nmol/l during the summer dry season, and 40±2 nmol/l ($P<0.05$) during the colder, wetter winter. They also reported plasma aldosterone concentrations of 200-250 pmol/l with no significant difference between the seasons.

Cooley and Janssens (1977) measured plasma cortisol in tail-vein blood of conscious Tammar Wallabies held in hessian sacks. They found a

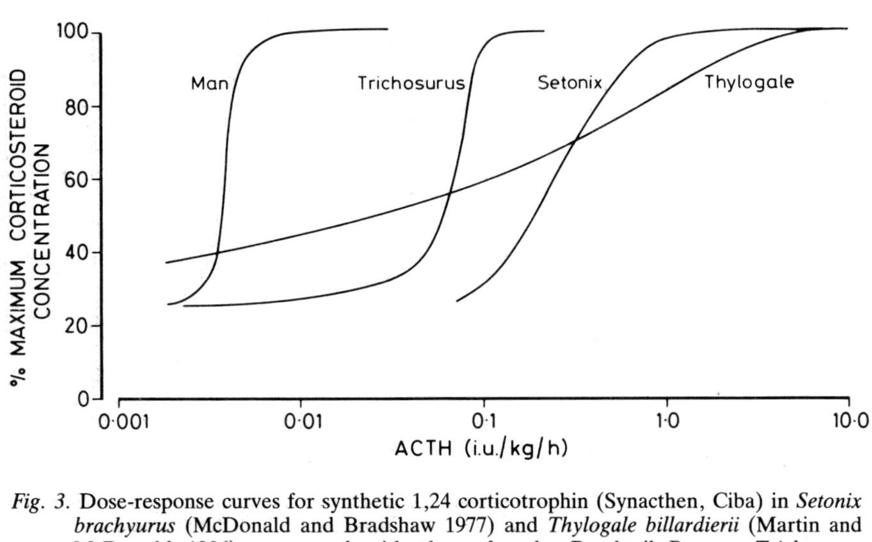

Fig. 3. Dose-response curves for synthetic 1,24 corticotrophin (Synacthen, Ciba) in *Setonix brachyurus* (McDonald and Bradshaw 1977) and *Thylogale billardierii* (Martin and McDonald 1986), compared with those for the Brushtail Possum *Trichosurus vulpecula* (Than and McDonald 1973) and man (Landon *et al.* 1964).

mean value of 430 nmol/l in animals that were quiet, and 600-700 nmol/l in excited animals. Infusion of porcine ACTH i.v. at 0.45 U/kg/h caused plasma cortisol concentration to rise from a mean value of 386 to 1765 nmol/l. Subsequently, Janssens and Hinds (1981) reported a much lower mean concentration of 185 nmol/l in otherwise undisturbed Tammar Wallabies held in metabolism cages (Table 1).

Martin and McDonald (1986) measured peripheral plasma glucocorticoid in captive Red-bellied Pademelons with indwelling jugular venous catheters. They confirmed the observation of Oddie *et al.* (1976) that cortisol was quantitatively the major circulating glucocorticoid and also found that the mean concentration in undisturbed animals was high, in the range 200-250 nmol/1, with irregular episodic variations within the range 100-500 nmol/l. Maximal stimulation with intravenous synthetic ACTH caused a 5-6 fold increase, reaching a peak 2 hours after commencing ACTH infusion. Again, the sensitivity to synthetic ACTH was extremely low, with a very flat dose-response curve (Fig. 3). The stress of ether anaesthesia also caused a rapid 3-7 fold increase in plasma cortisol concentration, reaching maximum values 2-4 hours later (Table 1). It was also noted that administration of the potent synthetic glucocorticoid, dexamethasone, caused a marked fall in plasma cortisol concentration of undisturbed wallabies to almost undetectable levels. This indicates feed-back control of ACTH secretion by changes in plasma glucocorticoid concentration and a significant level of ACTH stimulation of the adrenal cortex in the apparently undisturbed state.

From these observations and the findings of Hearn (1975), it can be concluded that the control of glucocorticoid secretion is similar to that of other mammals, although the reason for the low sensitivity to exogenous ACTH stimulation remains to be determined.

Plasma corticosteroid-binding proteins

In addition to albumin, which has a high capacity for binding corticosteroids but with low affinity, all the macropodid marsupials so far examined have a low capacity, high-affinity glucocorticoid-binding protein in their plasma, resembling eutherian transcortin both in binding affinity and in electrophoretic mobility (Sernia *et al.* 1979). The concentration of binding sites (binding capacity) ranges from 100-500 nmol/l, the high values being found in the Tammar Wallaby (370 nmol/l) and the Red-bellied Pademelon (530 nmol/l, Table 1). In general, the lower the concentration of glucocorticoid binding sites, the lower the unstimulated glucocorticoid concentration. Also the concentration of binding sites is a little lower than the maximally stimulated plasma cortisol concentration, so that the free, biologically active cortisol concentration rises more slowly than the total cortisol concentration until the maximum binding capacity is exceeded. This would provide a useful buffer against excessive glucocorticoid action during moderate stress. Nothing is known about factors that may affect the concentration of binding sites and biological effectiveness of circulating glucocorticoids in these marsupials. Such factors are of significance to interpretation of plasma glucocorticoid concentrations in field observations (McDonald and Waring 1979).

FUNCTIONAL ROLE OF THE CORTICOSTERIODS

Effect of adrenalectomy

The first observations on a macropodid marsupial were made by Buttle, Kirk and Waring (1952) who reported that the Quokka could survive bilateral adrenalectomy for no more than 48 h, in spite of the post-operative administration of saline or adrenocortical extract. The animals showed the classic signs of adrenal insufficiency — anorexia, muscular weakness and lassitude, associated with a decline in plasma Na and rise in plasma K concentration, as in eutherian mammals. Subsequently, McDonald and Bradshaw (unpublished observations) found that adrenalectomized Quokkas can survive bilateral adrenalectomy for at least three months, provided they are given supplementary injections of cortisol, 11-deoxy corticosterone or the synthetic glucocorticoid dexamethasone. However, withdrawal of the hormone supplement resulted in signs of adrenal insufficiency 2-6 days later, depending on the level of the maintenance dose, even though they had access to 1% saline to drink.

McDonald (1974) found that Red Kangaroos also survive bilateral adrenalectomy for long periods if given supplementary injections of cortisol and 11-deoxycorticosterone. Withdrawal of the steriod supplement was followed by reduction in food intake and noticeable weakness 2-3 days later. One animal survived five days without steriod supplements but ceased to eat and was extremely weak. In addition to anorexia and muscle weakness, adrenal insufficiency was characterized by a short-lived increase in urinary Na excretion, fall in plasma Na and rise in plasma K concentration. Restoration of the steroid supplement was followed by rapid increase in strength and appetite, rise in plasma Na and fall in plasma K concentration. During adrenal insufficiency the animals developed an appetite for salt and it was found that they survived withdrawal of the hormone supplement, maintaining food intake and general appearance of well-being, simply by increasing their salt intake through self-selection of isotonic saline. Withdrawal of the saline supplement in one animal was followed by death 24 hours later.

Janssens and Tyndale-Biscoe (1982) found that the Tammar Wallaby could survive bilateral adrenalectomy for at least six weeks if given supplementary injections of 11-deoxycorticosterone but, like the quokka, they survived only 36-48 hours if no steroid supplements were administered even though they had access to extra salt.

It is clear, therefore, that adrenocortical secretions are essential for survival in macropodid marsupials. However, there appear to be species differences in the degree of dependence on the secretions.

Role of mineralocorticoids

The findings of Buttle *et al.* (1952), McDonald (1974) and Janssens and Tyndale-Biscoe (1982) clearly indicate that a major function of adrenocortical secretions is maintenance of salt balance. This is presumably the role of aldosterone and is in keeping with the high degree of development of the zona glomerulosa of the adrenal cortex, the major site of aldosterone biosynthesis in eutherian mammals (Muller 1988).

The findings of Coghlan and Scoggins (1967) are evidence that the secretion of aldosterone is responsive to the dietary salt intake, the low salt intake by the alpine animals being associated with the higher concentrations of aldosterone in the blood. Similarly, Miller and Bradshaw (1979) found that, in Quokkas, plasma aldosterone concentration varied inversely with salt intake. Simpson and Blair-West (1971) found a renin-angiotensin system resembling that of eutherian mammals in Red and Eastern Grey, Bennett's Wallaby and the Red-bellied Pademelon. Scoggins *et al.* (1970) found that the high plasma aldosterone concentration of alpine Eastern Grey Kangaroos was associated with elevated plasma renin concentration, compared with those in a coastal habitat.

Bradshaw and McDonald (cited in Bradshaw 1983) found that adrenal insufficiency in the Quokka was associated with a decrease in plasma Na concentration and a marked reduction in glomerular filtration rate, renal plasma flow and tubular fractional Na reabsorption. These defects could be prevented by administration of aldosterone or cortisol.

It can be concluded that the major function of aldosterone secretion in macropodid marsupials is in the control of Na balance through its influence on renal Na active transport, by responding to changes to dietary Na intake. The rapid decline and death after adrenalectomy or withdrawal of hormone supplements can be attributed mainly to loss of this function. It seems likely that a renin-angiotensin system participates in the control of aldosterone secretion, as in other mammals. Glucocorticoids may also play a role in the control of salt and water balance.

The long-term survival of adrenalectomized Red Kangaroos in apparently good condition while maintained by increased supplementary salt intake alone raises the question of the functional significance of the quantitatively major glucocorticoid, cortisol.

Role of glucocorticoids

It is generally accepted that, in eutherian mammals, glucocorticoids affect carbohydrate metabolism by antagonizing the peripheral actions of insulin and enhancing hepatic gluconeogenesis by induction of gluconeogenic enzymes. In addition, they mobilize tissue alpha-amino nitrogen, providing substrate for new glucose production. At high dose rates, glucocorticoids cause muscle wasting and reduction in the mass of lymphoid tissue, with increase in urinary nitrogen excretion and negative nitrogen balance. In this sense, glucocorticoids are considered to be catabolic. Other effects of high doses are impairment of the immune and inflammatory mechanisms and enhancement of catecholamine actions.

Griffiths, McIntosh and Leckie (1968) found that, in contrast to eutherian herbivores such as the sheep and rabbit, administration of large i.m. doses of cortisone (14 mg/kg) to Red Kangaroos for seven days had no effects on the sensitivity to the hypoglycaemic effects of insulin or on urinary nitrogen excretion. McDonald (1977) subsequently confirmed that the same dose of cortisol for the same period had no effect on the hypoglycaemic response to insulin in normal or adrenalectomized Red Kangaroos (Fig. 4). McDonald and Bradshaw (1981) investigated in the Quokka the metabolic actions of i.m. injections of 1-5 mg cortisol acetate/kg, sufficient to significantly raise plasma cortisol concentration for 3-5 days. They found no effect on plasma glucose concentration, hypoglycaemic action of insulin (Fig. 4) or urinary nitrogen excretion. Subsequently, Martin (unpublished observation) found that, in a single Quokka, daily administration of 14 mg cortisol acetate/kg for seven days caused an increase in plasma glucose and fasting liver glycogen concentrations.

Cooley and Janssens (1977) and Janssens and Hinds (1981) investigated the effect of cortisol administration in the Tammar Wallaby. They found that doses sufficient to increase plasma cortisol concentration to the maximally ACTH-stimulated level caused an increase in plasma glucose concentration both in the short-term (2 hours) and long term (7 days). They found an

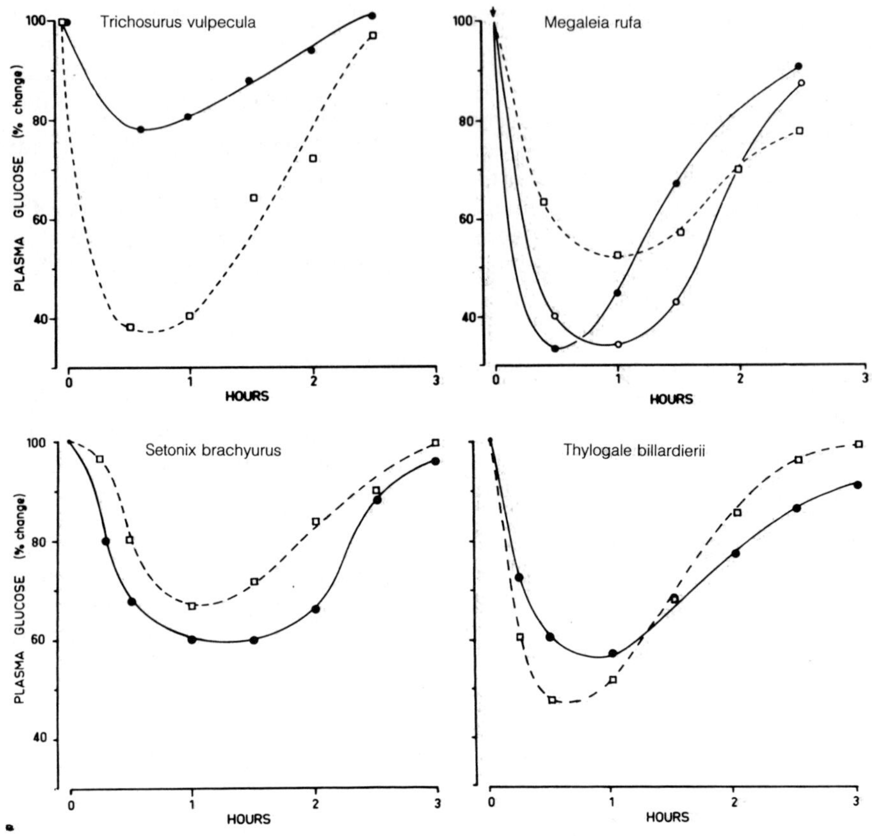

Fig. 4. Effect of long-term cortisol treatment on hypoglycaemic response to insulin in the flucocorticoid-sensitive Brush-tail Possum *Trichosurus vulpecula* and the glucocorticoid-resistant macropodid marsupials, *Macropus rufus, Setonix brachyurus* and *Thylogale billardierii.* Data from Than and McDonald (1973), McDonald (1977), McDonald and Bradshaw (1981), Martin and McDonald (unpublished).
○ – – – ○ Control. ●——● At the end of seven days cortisol treatment.

increase in nitrogen turnover in the long term, although their data show no change in nitrogen balance and do not distinguish between urinary and faecal nitrogen excretion.

Martin (1984) and Martin and McDonald (1986, 1988) made a more extensive investigation of cortisol actions in the Red-bellied Pademelon. They found that, as in the Quokka and in contrast

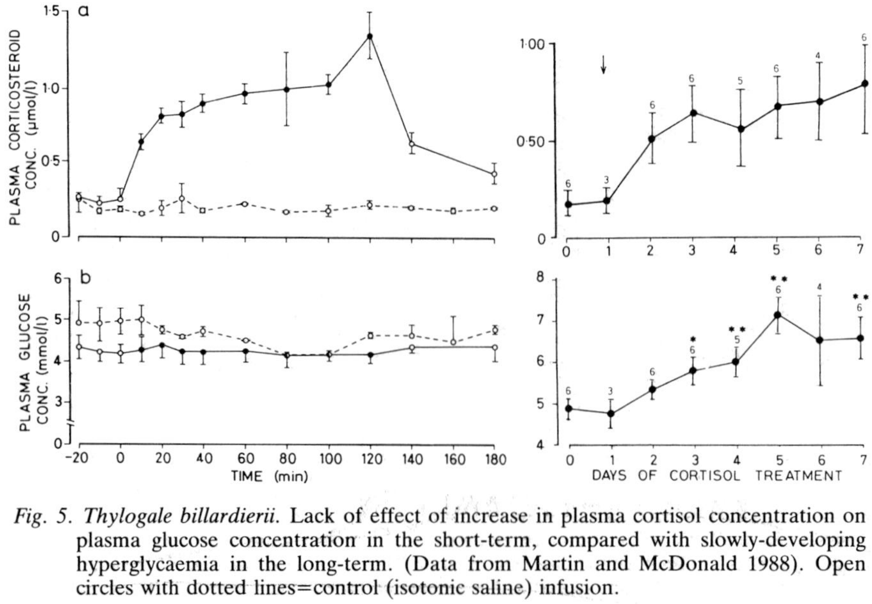

Fig. 5. Thylogale billardierii. Lack of effect of increase in plasma cortisol concentration on plasma glucose concentration in the short-term, compared with slowly-developing hyperglycaemia in the long-term. (Data from Martin and McDonald 1988). Open circles with dotted lines=control (isotonic saline) infusion.

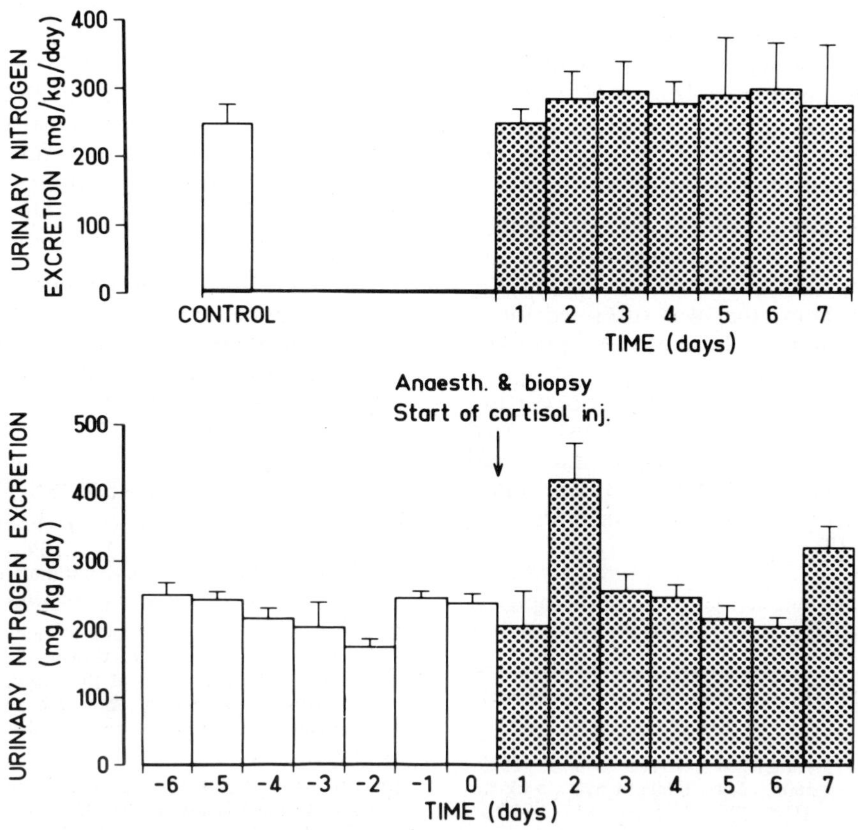

Fig. 6. Urinary nitrogen excretion by *Thylogale billardierii* before and during daily intramuscular administration of 14 mg cortisol/kg. Note lack of change except for the transient increase following liver biopsy under anaesthesia. (Data from Martin and McDonald 1988).

to the Tammar Wallaby, short term (2 hour) infusion of cortisol had no effect on plasma glucose concentration. Also, there was no effect on new glucose production or metabolic clearance of glucose. In the long term, i.m. cortisol administration for seven days caused a slow and progressive increase in plasma glucose concentration with development of glycosuria (Fig. 5). Some animals were more sensitive than others to this effect. This treatment also caused a marked increase in fasting liver glycogen concentration; but there was an increase in new glucose production in only two of four animals. In spite of the increase in plasma glucose concentration, the hypoglycaemic effect of a single i.v. dose of insulin was not changed by cortisol administration (Fig. 4). It was also noted that the stress of ether anaesthesia caused a marked rise in plasma glucose concentration before plasma cortisol concentration began to rise significantly.

In addition to the increase in plasma glucose concentration in the Red-bellied Pademelons, there was an increase in both plasma urea and α-amino nitrogen concentration. While the increase in urea concentration was only transient, that of α-amino nitrogen was sustained for the period of cortisol administration. Although these effects indicate mobilization of tissue nitrogen, there was no change in urinary excretion, except immediately after liver biopsy under ketamine-xylazine anaesthesia. Because it is known that macropodid marsupials are capable of recycling urea to the foregut, where it is converted into microbial nitrogen and re-utilized as a food source (Moir 1968; Kennedy and Milligan 1980), the possibility that this might account for the lack of effect of cortisol on urinary excretion was investigated. It was found that a mean of only 26% of urea, infused i.v. over a period of 6 h, was recovered in the urine during the following 48 hours. Further evidence for urea recycling is a low ratio of urinary urea:total nitrogen (Hume 1982). This was only 0.56 in Red-bellied Pademelon compared with values of 0.72 and 0.78 in Tammar Wallabies and Red Kangaroos under conditions in which urea recycling was considered to be minimal (Kinnear and Main 1975), and the ratio was not affected by cortisol administration. It seems possible therefore, that the lack of effect of cortisol administration on urinary nitrogen excretion can be accounted for by urea recycling. Moderate stress, as in surgical liver biopsy appears to interfere with this mechanism (Fig. 6).

It can be concluded that cortisol has the expected diabetogenic and nitrogen-mobilizing effects found in eutherian mammals. However, with the exception of a short-term effect on plasma glucose in the Tammar Wallaby, this effect is evident only when plamsa cortisol concentration is sustained at high levels for several days. The expected effect on urinary nitrogen excretion is probably masked by recycling the urea to the forestomach.

Martin (1984) found that, in the Red-bellied Pademelon a single i.m. injection of 7 mg cortisol acetate/kg caused a 35% fall in blood lymphocyte count two hours later, that was sustained for at least a further ten hours. There was also a marked fall in eosinophil count and a moderate polymorphonuclear leucocytosis. Long-term cortisol injection (14 mg/kg/day for seven days) caused similar degrees of lymphopaenia and eosinopaenia at the end of the treatment and a more pronounced rise in polymorphonuclear leucocyte count (Table 2).

Table 2. Effect of long-term cortisol treatment (14 mg/kg/day i.m. for seven days) on blood leucocyte counts in three Red-bellied Pademelons.

	Control (No. per mm^3; mean±S.E.M.)	Cortisol	Change
Lymphocytes	5 122±1 280	3 345± 596	−35%
Polymorphs	2 560± 244	6 006±1 611	+135%
Eosinophils	331± 117	81± 81	−76%
Total	8 013±1 048	9 433±2 127	+18%

CONCLUSIONS

The structure of the adrenal cortex, as well as the nature and control of adrenocortical secretions in macropodid marsupials are essentially similar to the classic eutherian pattern. Although there appear to be some quantitative differences in the sensitivity of tissues to some of the metabolic effects of adrenocortical hormones, the same basic mechanisms appear to be present. There is little doubt that a major role of adrenocortical secretion in macropodid marsupials is in the control of salt balance via the secretion of aldosterone. However, the role of the adrenocortical secretions in the control of energy exchange is unclear. It would seem that, with the possible exception of the Tammar Wallaby (however, see Bradley, this Symposium), the major glucocorticoid, cortisol, plays little part in the normal control of carbohydrate metabolism, even though a diabetogenic effect can be demonstrated at high doses over long periods. It has been suggested (Martin and McDonald 1986) that the normal role of glucocorticoids in these marsupials may be to permit key metabolic processes to achieve maximum rates, as conceived by Ingle (1952), and to synergise with other hormones to achieve metabolic control. It also seems likely that the long-term effects of high rates of secretion, as in a stressful environment, could play a part in mobilization and replenishment of energy reserves by gluconeogenesis from mobilized energy reserves. In such a case, the slow rate of response and ability to recycle the urea nitrogen would seem to be a most valuable adaptation in those animals inhabiting mostly arid environments with low food nitrogen levels. In this context, the apparently minor adrenocortical responses of Red and Eastern Grey Kangaroos to stress warrants further investigation.

The influence of the adrenocortical hormones on the immune and inflamatory systems remains to be determined, although the evidence for a lymphopaenic and neutrophilic effect of cortisol suggests similarity with the effects observed in eutherian mammals.

REFERENCES

BLAIR-WEST, J. R., COGHLAN, J. P., DENTON, D. A., GODING, J. R., MUNRO, J. A., PETERSON, R. E. AND WINTOUR, M., 1962. Humoral stimulation of adrenocortical secretion. *J. Clin. Invest.* **41**: 1606-16.

BRADLEY, A. J., 1988. Resistance to the nitrogen-mobilizing, diabetogenic action of cortisol in the Spectacled Hare-wallaby, the Potoroo and the Garden Island Tammar Wallaby. Symposium on Kangaroos, Wallabies and Rat-kangaroos. Australian Mammal Society. (Abstract in press).

BRADSHAW, S. D., 1983. Recent endocrinological research on the Rottnest Island quokka *(Setonix bracyurus) J. Roy. Soc. West. Aust.* **66**: 55-61.

BUTTLE, J. M., KIRK, R. L. AND WARING, H., 1952. Effects of complete adrenalectomy on the wallaby *Setonix brachyurus. J. Endocrin.* **8**: 281-90.

COGHLAN, J. P. AND SCOGGINS, B. A., 1967. The measurement of aldosterone, cortisol and corticosterone in the blood of the wombat (*Vombatus hirsutus,* Perry), and the kangaroo *(Macropus giganteus). J. Endocrin.* **39**: 445-50.

COOLEY, H. AND JANSSENS, P. A., 1977. Metabolic effects of infusion of cortisol and adrenocorticotropin in the tammar wallaby *Macropus eugenii* Desmarest. *Gen. Com. Endocrin.* **33**: 352-58.

GRIFFITHS, M., MCINTOSH, D. L. AND LECKIE, R. M. C., 1968. The effects of cortisone on nitrogen balance and glucose metabolism in diabetic and normal kangaroos, sheep and rabbits. *J. Endocrin.* **44**: 1-12.

HARRIS, A. W. AND BAXTER, J. D., 1979. Variations in cellular sensitivity to glucorticoids: observations and mechanisms. Pp. 423-48 *in* Glucocorticoid Hormone Action ed. by J. D. Baxter and G. G. Rousseau. Berlin: New York, Springer Verlag.

HARRISON, F. A., MCDONALD, I. R. AND PATERSON, J. Y. F., 1964. Comparison of the apparent rate of cortisol turnover determined by isotope dilution and cortisol secretion rate. *J. Endocrin.* **28**: 173-81.

HEARN, J. P., 1975. Hypophysectomy of the tammar wallaby *Macropus eugenii:* Surgical approach and general effects. *J. Endocrin.* **64**: 403-16.

HUDSON, R. W., SCHACTER, H. AND KINNINGER, K. W., 1976. Studies of 11-beta hydroxylation by beef adrenal mitochondria. *J. Steriod Biochem.* **7**: 255-62.

HUME, I. D., 1982. Digestive Physiology and Nutrition of Marsupials. Cambridge University Press: Cambridge.

ILLET, K. F., 1969. Corticosteriods in the adrenal venous and heart blood of the quokka *Setonix brachyurus* (Marsupialia: macropodidae). *Gen. Comp. Endocrin.* **13**: 218-21.

INGLE, D. J., 1952. The role of the adrenal cortex in homeostasis. *J. Endocrin.* **8**: xxiii-xxxvii.

JANSSENS, P. A. AND HINDS, L. A., 1981. Long-term effects of corticosteroid administration in the tammar wallaby *Macropus eugenii. Gen. Comp. Endocrin.* **47**: 221-25.

JANSSENS, P. A. AND TYNDALE-BISCOE, H., 1982. Long-term effects of ACTH in the tammar wallaby *Macropus eugenii:* the role of the adrenal medulla. *Gen. Comp. Endocrin.* **45**: 56-80.

JONES, I. C., 1957. The adrenal cortex. Cambridge University Press: London.

KENNEDY, P. M. AND MILLIGAN, L. P., 1980. The degradation and utilization of endogenous urea in the gastrointestinal tract of ruminants: a review. *Can. J. Anim. Sci.* **60**: 205-21.

KINNEAR, J. E. AND MAIN, A. R., 1975. The recycling of urea nitrogen by the wild tammar wallaby *(Macropus eugenii)* — "ruminant-like" marsupial. *Comp. Physiol. Biochem.* **51A**: 793-810.

LANDON, J., JAMES, V. H. T., CRYER, R. J., WYNN, V. AND FRANKLAND, A. W., 1964. Adrenocorticotrophic effects of a syntheric polypeptide — 1-24 corticotrophin — in man. *J. Clin. Endocrin. Metab.* **24**: 1206-13.

MACARTHUR, E. AND O'DONNELL, V. J., 1967. The formation of 21-deoxycortisol by human adrenal tissue. *J. Endocrin.* **37**: 235-36.

MCDONALD, I. R., 1974. Adrenal insufficiency in the red kangaroo *(Megaelia rufa* Desm). *J. Endocrin.* **62**: 689-90.

MCDONALD, I. R., 1977. Adrenocortical functions in marsupials. Pp. 345-75 *in* The Biology of Marsupials ed. by B. Stonehouse and D. Gilmore. Macmillan Press Ltd: London.

MCDONALD, I. R. AND BRADSHAW, S. D., 1977. Plasma corticosteriods and the effects of adrenocorticotrophin in a macropodid marsupial (*Setonix brachyurus,* Quoy and Gaimard). *J. Endocrin.* **75**: 409-18.

MCDONALD, I. R. AND BRADSHAW, S. D., 1981. Resistance to the nitrogen-mobilizing, diabetogenic actions of cortisol in a small wallaby (*Setonix brachyurus*). *Gen. Comp. Endocrin.* **43**: 39-41.

MCDONALD, I. R. AND REICH, M., 1959. Corticosteriod secretion by the autotransplanted adrenal gland of the conscious sheep. *J. Physiol. (London).* **147**: 33-50.

MCDONALD, I. R. AND WARING, H., 1979. Hormones of marsupials and monotremes. Pp. 873-924 *in* Hormones and Evolution, Vol. 2 ed. by E. J. W. Barrington. Academic Press: London.

MARTIN, I. K., 1984. Glucocorticoid actions in *Thylogale billardierii.* Ph.D. Thesis, Monash University, Victoria, Australia.

MARTIN, I. K. AND MCDONALD, I. R., 1986. Adrenocortical functions in a macropodid marsupial *Thylogale billarderii. J. Endocrin.* **110**: 471-80.

MARTIN, I. K. AND MCDONALD, I. R., 1988. Metabolic actions of cortisol in a macropodid marsupial *Thylogale billardierii. J. Endocrin.* **116**: 71-79.

MILLER, T. AND BRADSHAW, S. D., 1979. Adrenocortical function in a field population of a macropodid marsupial (*Setonix brachyurus,* Quoy and Gaimard). *J. Endocrin.* **82**: 159-70.

MOIR, R. J., 1968. Ruminant digestion and evolution. Pp. 2673-94 *in* Handbook of Physiology, section 6, Vol. 5, Alimentary Canal. American Physiological Society: Washington D.C.

MULLER, J., 1988. Regulation of aldosterone biosynthesis. Physiological and clinical aspects. Berlin: Springer-Verlag.

ODDIE, C. J., BLAINE, E. H., BRADSHAW, S. D., COGHLAN, J. P., DENTON, D. A., NELSON, J. F. AND SCOGGINS, B. A., 1976. Blood corticosteriods in Australian marsupial and placental mammals and one monotreme. *J. Endocrin.* **69**: 341-48.

REICH, M. AND MCDONALD, I. R., 1961. The effect of progressive sodium depletion on corticosteriod secretion by the autotransplanted adrenal gland of conscious sheep. *J. Endocrin.* **21**: 387-400.

SANDOR, T., SONEA, S. AND MEHDI, A. Z., 1975. The possible role of steriods in evolution. *Am. Zoologist.* **15** (suppl. 1): 227-53.

SCOGGINS, B. A., BLAIR-WEST, J. R., COGHLAN, J. P., DENTON, D. A., MYERS, K., NELSON, J. F., ORCHARD, E. AND WRIGHT, R. D., 1970. The physiological and morphological response of mammals to changes in their sodium status. Pp. 576-602 *in* Hormones and the Environment ed. by G. K. Benson and K. G. Phillips. Cambridge University Press: Cambridge.

SERNIA, C., BRADLEY, A. J. AND MCDONALD, I. R., 1979. High affinity binding of adrenocortical and gonadal steriods by plasma proteins of Australian marsupials. *Gen. Comp. Endocrin.* **38**: 469-503.

SIMPSON, P. A. AND BLAIR-WEST, J. R., 1971. Renin levels in the kangaroo, the wombat and other marsupial species. *J. Endocrin.* **51**: 79-90.

THAN, K. A. AND MCDONALD, I. R., 1973. Adrenocortical function in the Australian brush-tailed possum *Trichosurus vulpecula* (Kerr). *J. Endocrin.* **58**: 97-109.

URQUART, J., 1974. Physiological actions of adrenocorticotrophic hormone. Pp. 133-57 *in* The Pituitary Gland and its Neuroendocrine Control. Handbook of Physiology, Section 7, Vol. IV ed. by R. O. Greep and E. B. Astwood. American Physiological Society: Washington D.C.

WEISS, M., 1968. 21-deoxycortisol and other minor steriods in adrenal venous blood of the phalanger *Trichosurus vulpecula. J. Endocrin.* **42**: 293-94.

WEISS, M., 1980. Adrenocorticosteroids in prototherian, metatherian and eutherian mammals. Pp. 285-96 *in* Comparative Physiology of Primitive Mammals ed. by Schmidt-Nielsen, L. Bolis and C. R. Taylor. Cambridge University Press: Cambridge.

WEISS, M. AND MCDONALD, I. R., 1967. Corticosteroid secretion in kangaroos (*Macropus canguru major* and *M. (Megaleia) rufa*). *J. Endocrin.* **39**: 251-61.

WEISS, M. AND VARDALOV, L., 1977. A study of steroid 11-beta hydroxylation by adrenal mitochondria of marsupials. Part 1. A comparison of 11-beta hydroxylase activity and specificity for different steriod substrates by possum *(Trichosurus vulpecula),* kangaroo *(Macropus major)* and beef. *J. Steriod Biochem.* **8**: 1233-42.

WIELAND, R. G., MAYNARD, D. E., RILEY, T. R. AND HAMWI, G. J., 1965. Detection of 21-deoxycortisol in blood from a patient with congential adrenal hyperplasia. *Metabolism.* **14**: 1276-81.

The adaptiveness of reproductive processes

C. H. TYNDALE-BISCOE[1]

The macropodoid pattern of reproduction is common to potoroine, macropodine and sthenurine species, which suggests that it evolved in the Miocene or earlier from the basic marsupial pattern, exemplified by *Trichosurus*. The features that distinguish this from the basic pattern are: gestation occupies 80% to 110% of the oestrous cycle, the profile of plasma progesterone in the reproductive cycle is bimodal, the corpus luteum is sensitive to inhibition by prolactin, and the embryo is capable of entering diapause at the unilaminar blastocyst stage. Species of macropodoids have exploited the ancestral feature of pituitary control of gonadal function, which determines when mature gametes will be produced and fertilization take place, and the more recently evolved features of corpus luteum quiescence and diapause, in different ways to become either obligatory or facultatively seasonal breeders on the one hand, or to become opportunistic breeders on the other. It is of interest that the ancestral features are more amenable to individual response than the more recently evolved features of quiescence and diapause.

INTRODUCTION

LIVING species of macropodoids represent four sub-families, two from each of the Potoroidae and the Macropodidae. These each have fossil representation in the Pliocene or late Miocene (Flannery 1984; this volume), so it is reasonable to infer that any features that they hold in common were possessed by the ancestral macropodoids. Reproductive processes have been investigated in five species of rat kangaroos, representing the Potoroinae, in five species of kangaroo and in about eight species of wallaby, representing the Macropodinae, and in the Banded Hare-wallaby, *Lagostrophus fasciatus*, representing the Sthenurinae (Flannery 1983); the reproduction of *Hypsiprymnodon moschatus*, the sole representative of the Hypsiprymnodontinae, remains almost entirely unknown. The best known reproductive feature that, with one exception, all these species have in common is embryonic diapause.

This feature of macropodoid reproduction was first suspected by Carson (1912), but was not established until Sharman's studies on the Quokka, *Setonix brachyurus* (Sharman 1955a,b), and the Tammar, *Macropus eugenii* (Sharman 1955c). These studies established the relationship between the presence during lactation of an embryo at the unilaminar stage of development in the uterus and an associated, undeveloped, corpus luteum in the adjacent ovary. He showed that, unlike other marsupials, such as *Didelphis virginiana*, and later, *Trichosurus vulpecula*, gestation in the Quokka was approximately the same length as the infertile oestrous cycle and that the subsequent post-partum oestrus and ovulation are not suppressed, as in the former species. The resultant corpus luteum and embryo, however, remained in a dormant state during the subsequent lactation but could be reactivated if the young was removed from the pouch (RPY). In the 30 years that have elapsed since Sharman's work many species of macropodoids have been studied in greater or lesser degree and in all, except the Western Grey Kangaroo, *Macropus fuliginosus,* some aspect of this unusual phenomenon has been observed; in many species dormant blastocysts associated with small corpora lutea have been recovered from lactating females, and in others a female, in the absence of a male, has given birth to another offspring some weeks after the loss or removal of her previous pouch young. In a few species detailed studies on the length of the oestrous cycle and gestation have been reported and in three species these have been supplemented by measurements of progesterone and oestradiol, which reflect the activity of the ovaries and the corpus luteum.

This phenomenon is so pervasive within the super family that it must have evolved early and have been retained because it confers selective advantage. However, to date no clear advantage has been recognized, although several ideas have been put forward (Sharman 1965; Waring 1956; Sadleir 1965; Marshall 1967; Lee and Cockburn 1985). We now have detailed information on the endocrine control of reproduction in one species, the Tammar (see Tyndale-Biscoe, Hinds and McConnell 1986; Tyndale-Biscoe and Renfree 1987), and a considerable amount of information on the ecological aspects of reproduction in many species from a range of different habitats and climates (see Russell 1974; Tyndale-Biscoe 1984; Lee and Cockburn 1985), so that it should now be possible to examine the question of the importance of embryonic diapause in the reproductive strategies of the Macropodoidea. In this chapter I propose to attempt three things; first, from the detailed studies on the Tammar establish the main endocrine features of reproduction in the macropodoids; second, from this understanding interpret the breeding strategies of selected species

[1]C.S.I.R.O. Division of Wildlife and Ecology, P.O. Box 84, Lyneham, Australian Capital Territory 2602.

Pages [illegible] KANGAROOS, WALLABIES AND RAT KANGAROOS ed. by G. Grigg, P. Jarman and I. Hume. Surrey Beatty & Sons Pty Limited, New South Wales, Australia, 1989.

that represent the range of macropodoid adaptations; and third, to speculate on why and how this unusual pattern of reproduction evolved in this group of marsupials.

Main endocrine features of reproduction in the Macropodoidea

When we compare the profile of progesterone through the reproductive cycle of a typical marsupial, such as the Brushtail Possum and any one of the macropodoids (Fig. 1) two features stand out. Unlike in the Brushtail Possum the period of elevated progesterone is prolonged past the middle of the cycle to the extreme end and, if it is a pregnant cycle, declines abruptly at the time of parturition. This abrupt decline is induced by the fetus (Tyndale-Biscoe, Hinds and Horn 1988). The second and more significant feature is the transient pulse of progesterone that occurs during the first week after oestrus. This has now been established to be a consistent feature of the reproductive cycle, both pregnant and non-pregnant, in several species and is closely associated with the development of the embryo past the unilaminar blastocyst stage. In female Tammars that lactate the early pulse is suppressed and, if the young is removed from the pouch the pulse in progesterone occurs 5-6 days later (Hinds and Tyndale-Biscoe 1982). If the corpus luteum is surgically removed before day 6 the pulse does not occur (Fletcher and Renfree 1988), and reactivation and subsequent development of the embryo is impaired (Berger 1970; Tyndale-Biscoe 1970). The first changes in the quiescent blastocyst of the Tammar occur between day 3 and day 4 (Shaw and Renfree 1986), so the pulse is not directly involved in reactivation but it is essential for the further development of the embryo to full term (for reviews see Tyndale-Biscoe 1986; Tyndale-Biscoe and Renfree 1987). We feel confident therefore in concluding that the early pulse of progesterone is an integral component of the macropodoid pattern of reproduction and it is probable that it will be found to occur in the early stage of the cycle of all species.

To conclude, the distinctive features of reproduction in the Macropodoidea are the extension of gestation to occupy 80% to 110% of the oestrous cycle, with a concommitant extension of the secretory phase of the corpus luteum and its bimodal, rather than unimodal, profile, the ability of the corpus luteum to become quiescent and for the associated embryo to enter diapause at the stage of the unilaminar blastocyst.

It must be emphasized that this pattern of a quiescent corpus luteum and an associated embryo in diapause has been grafted on to, rather than replacing, the pre-existent reproductive pattern as seen in the Brushtail Possum; some macropodoids retain the inhibition of follicle

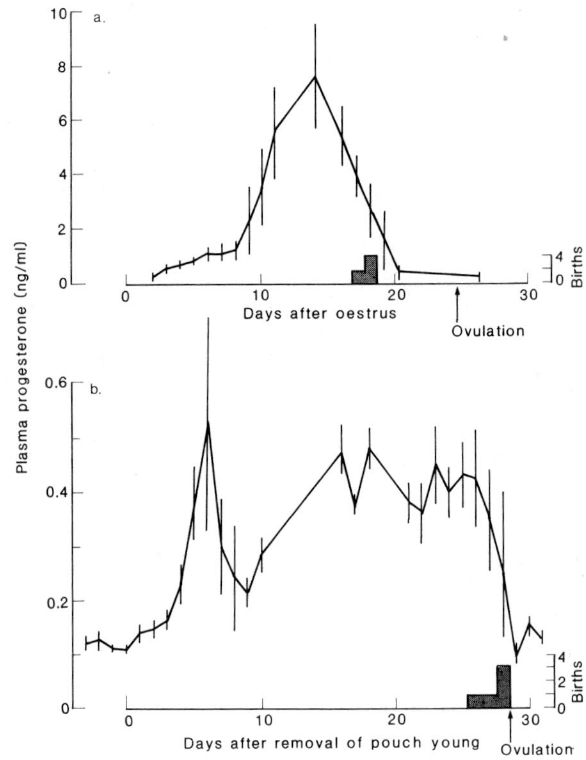

Fig. 1. A comparison of the plasma progesterone profiles through pregnancy in (a) the Brushtail Possum and (b) the Tammar, to illustrate the salient features of the macropodoid cycle: the early peak at about day 6, the extension of the second peak, coinciding with follicular growth, and the precipitate fall at parturition. From Hinds (1983) and Hinds and Tyndale-Biscoe (1982).

growth and ovulation during lactation, which is termed lactational anoestrus, to distinguish it from lactational quiescence, as seen in the Tammar and Quokka. Some also may spend part of the year neither lactating nor undergoing oestrous cycles, being then in seasonal anoestrus, while two species have extended quiescence beyond the duration of lactation and then enter seasonal quiescence. Thus, apart from the period when a macropodoid is undergoing pregnancy or an infertile oestrous cycle it is in one of four reproductive states, lactational or seasonal quiescence, or lactational or seasonal anoestrus. In order to understand how these four states differ from one another and, more importantly, what the endocrine control of each is, it was recognized about 20 years ago that it would be necessary to select one species as a laboratory testbed with which to do properly controlled experiments with adequate replicates. It was hoped that findings from this species could then be used to interpret the others. The first choice was the burrowing bettong *Bettongia lesueur,* which proved to be intractable in captivity although excellent in other respects (Stodart 1966; Tyndale-Biscoe 1968). The second choice was the Tammar or Kangaroo Island Wallaby, which proved to be easy to keep and breed in captivity and which was docile and of a convenient size for experimental study and blood sampling. It had other qualities that were not suspected at first, the most important being seasonal quiescence (Berger 1966); its exquisite sensitivity to change in photoperiod has made it one of the most suitable mammals in which to explore the role of the pineal in the measurement of time and the reproductive responses that follow therefrom.

Hormonal control of reproduction in the Tammar

There are three basic endocrine interactions involved in Tammar reproduction (Fig. 2). (1) The hypothalamus-pituitary-gonadal axis, which is common to males and females of all mammals investigated and is therefore also, probably, the basic interaction in all marsupials. (2) The pituitary-corpus luteum interaction, which is only kncwn to occur in the macropodoids. (3) The corpus luteum-Graafian follicle interaction, which operates between the two ovaries of monovular species and within the ovary and the corpus luteum during the cycle; it is a local and a systemic effect of the corpus luteum upon all follicles and is also probably common to all marsupials. The first and the second of these interactions can be influenced through the hypothalamus by outside factors, such as the sucking stimulus of the pouch young, the nutritional state of the mother and environmental factors such as photoperiod and social stress.

Hypothalamo-hypophyseal-gonadal axis

From studies on eutherian mammals, particularly the sheep (see Karsch *et al.* 1984), it is known that centres in the hypothalamus synthesize the polypeptide Gonadotrophin Releasing Hormone (GnRH), which passes to the anterior pituitary via the hypothalamo-hypophyseal portal system, and there causes the release of the gonadotrophic hormones, Follicle Stimulating Hormone (FSH) and Luteinising Hormone (LH) into the systemic circulation (Fig. 2). At the gonad these hormones bind selectively to specific receptors on the surface of the cells of the developing ovarian follicles and induce the synthesis of steroid hormones, especially oestradiol. Oestradiol itself influences the hypothalamic centre; at low concentrations it inhibits GnRH and thus causes a reduction in the secretion of the gonadotrophins, while at higher concentrations it has a stimulatory effect on the hypothalamus and causes increasingly rapid releases of GnRH, that eventually lead to a large release of LH from the pituitary, which induces ovulation from the follicle or follicles in the ovary. Oestradiol also has a local effect on the cells of the follicle to induce in them the further synthesis of receptors for LH. Thus during follicular growth the most advanced follicles capture more of the available LH in circulation and at the same time inhibit secretion of FSH and LH so that less advanced follicles are not stimulated and so become atretic. As they grow the large Graafian follicles secrete increasing amounts of oestradiol into the circulation, which induces oestrous behaviour in the female and simultaneously induces the pre-ovulatory pulse of LH.

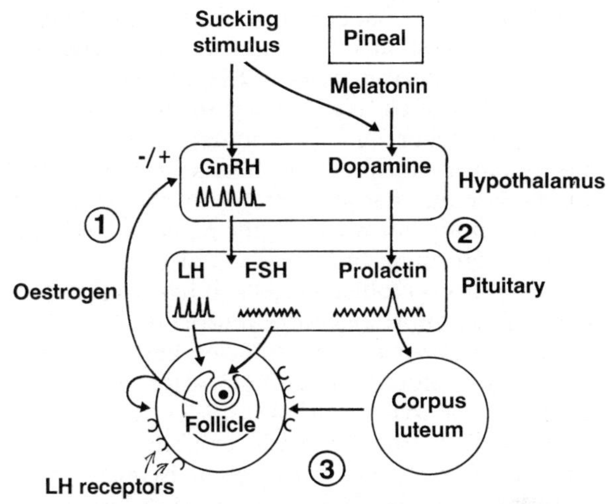

Fig. 2. A diagram of the three endocrine axes controlling reproduction in the Tammar. (1) the hypothalamo-hypophyseal-follicular axis, (2) the pineal-hypothalamo-hypophyseal-corpus luteum axis, and (3) the corpus luteum-ovarian axis. FSH, follicle stimulating hormone; GnRH, gonadotrophin releasing hormone; LH luteinising hormone.

Several aspects of this axis have been investigated in the Tammar. Passive immunization of female Tammars against GnRH during pregnancy totally prevented the growth of follicles in the ovaries, although the pituitary was still present in these animals (Short, Flint and Renfree 1985). In appearance the ovaries of these Tammars resembled those of anoestrous Quokkas (Tyndale-Biscoe 1963). Surgical removal of the pituitary from female Tammars also abolished follicular growth beyond the stage of small antral follicles (Hearn 1974; Panyaniti, Carpenter and Tyndale-Biscoe 1985) and, after 60 days the uteri of such animals resembled those of anoestrous Quokkas (Sharman and Berger 1969), Red Kangaroos (Newsome 1964a) and Eastern Grey Kangaroos (Clark and Poole 1967). Conversely, female Tammars injected with GnRH responded by secreting a pulse of LH within 25 min of the injection (Evans, Tyndale-Biscoe and Sutherland 1980).

The converse influence of the ovary on pituitary function has also been investigated. The levels of LH and FSH in intact female Tammars are very low or undetectable by current assay techniques and no change was observed after removal of the one ovary bearing the corpus luteum. However, after removal of the second ovary the levels of both hormones rose steadily over a period of two weeks, indicating the removal of an inhibitory influence from the ovaries (Evans *et al.* 1980;

Tyndale-Biscoe and Hearn 1981). In another study (C. H. Tyndale-Biscoe, L. A. Hinds and P. Diggle unpublished) the short-term pattern also changed, with the appearance of marked pulsatile releases of LH at about hourly intervals. In ovariectomized Tammars oestradiol was shown to have an inhibitory effect on the elevated levels of both FSH and LH at concentrations equivalent to the basal levels in circulation of intact females and to induce a large pulse of LH at doses equivalent to the pre-ovulatory level in intact females (Horn, Fletcher and Carpenter 1985). Conversely, progesterone, which is secreted by the corpus luteum, had no effect on the levels of either pituitary hormone. While no work has been done to determine the influence of oestradiol on LH receptors on follicular cells, it has been shown that LH receptor concentration on these cells is highest on pre-ovulatory follicles and that they have disappeared altogether after ovulation (Stewart and Tyndale-Biscoe 1982). Thus in all these respects the Tammar exhibits the components of the basic mammalian hypothalamo-hypopyseal-ovarian axis and the only aspect in which it appears to differ from the sheep is in the lack of an influence of season on the axis.

Corpus luteum-follicular axis

In the Tammar removal of the corpus luteum or the ovary bearing the corpus luteum during quiescence or during the first half of the reproductive cycle leads to the premature growth of another follicle and ovulation; oestrus may also occur but is not invariable, particularly during quiescence (Tyndale-Biscoe and Hawkins 1977; Renfree, Wallace and Young 1982). Similar results to these have been obtained in other macropodoids and in the Brushtail Possum (see Tyndale-Biscoe and Renfree 1987). These results indicate a role of the corpus luteum in inhibiting the growth and maturation of follicles during the early period of its activity; an important aspect common to all these results is that it is the undeveloped corpus luteum that exerts the most effective inhibition upon the follicles, during the period when the level of progesterone, secreted by the corpus luteum, is low; the period when progesterone is elevated is precisely the time when the inhibitory effect of the corpus luteum has waned (Fig. 1). Attempts have been made to identify the agent of follicular inhibition by the corpus luteum but so far without conclusion. After removal of the corpus luteum injections of progesterone and androstenedione were without effect (Evans *et al.* 1980; Renfree *et al.* 1982) but oestradiol would inhibit follicle growth and ovulation. However, to date no good evidence has been found for production or secretion of oestradiol by the corpus luteum (Shaw and Renfree 1984; Renfree *et al.* 1984; Fletcher and Renfree 1988) that could account for this being the agent. For oestradiol to be the agent of corpus luteum inhibition of follicular growth it would be expected to act at the hypothalamus to reduce gonadotrophin secretion but, as already mentioned, there is no detectable change in the level of either LH or FSH after removal of the corpus luteum, unlike the effect after total ovariectomy. In all non-macropodoid marsupials the sucking stimulus during lactation is the agent of follicular inhibition, taking over from the inhibition imposed by the early-formed corpus luteum. In macropodoids that display lactational anoestrus this continues to be the agent of inhibition, but in those in lactational quiescence the quiescent corpus luteum exercises follicular inhibition and it has generally been assumed that the means of inhibition is the same, that is to say hypothalamic. An alternative possibility, as yet untested, is that the corpus luteum may act directly on the follicles to inhibit the concentration of receptors for FSH and LH, thus making them less sensitive to the low levels of available hormones; when the corpus luteum is removed receptor concentration could increase and the follicles become responsive to the unchanging level of gonadotrophins in circulation.

Pituitary-corpus luteum axis

Unlike the corpora lutea of eutherian mammals, that of the Tammar does not require any stimulation by the pituitary for its maintenance, growth and development (Hearn 1974; Tyndale-Biscoe and Hawkins 1977) or for its capacity to secrete progesterone (Hinds 1983), nor does it have any receptors for LH (Stewart and Tyndale-Biscoe 1982); once formed after ovulation it appears to be autonomous. From limited data on other species this appears to be true of all marsupials and has been thought to represent the simplest and most primitive form of mammalian corpus luteum (Rothchild 1981). The macropodoids have interposed an inhibitory control by the pituitary, which must be exercised continuously if the corpus luteum is not to escape and continue its autonomous life (Fig. 2). This inhibition is normally induced by the stimulus of the sucking of the young in the pouch on the teat. This in turn causes the pituitary to secrete prolactin for the maintenance of lactation and at the same time to exert a tonic inhibition on the cells of the corpus luteum. The evidence for the latter role comes from the demonstration of abundant receptors for prolactin on the cells of the corpus luteum (Sernia and Tyndale-Biscoe 1979; Stewart and Tyndale-Biscoe 1982), on the demonstration that injections of prolactin, given three times a day to Tammars from which the young had been removed, delayed the onset of reactivation by an equal time (Tyndale-Biscoe and Hawkins 1977), and on the demonstration that the drug bromoiptine, which blocks the secretion of prolactin in eutherian mammals, induced reactivation in lactating Tammars (Tyndale-Biscoe and Hinds 1984) and Bennetts wallabies (Curlewis *et al.* 1986).

In Bennetts Wallaby and the Tammar the pituitary control of the corpus luteum has been extended for several months beyond the end of lactation, so that the species have become secondarily seasonal breeders. This particular control is not known for any other macropodoid and does not even occur in the mainland subspecies of Bennetts Wallaby, the Red-necked Wallaby (Merchant and Calaby 1981), from which we may conclude that it has evolved much later than the lactation-controlled quiescence. In both species seasonal quiescence occurs during that half of the year when daylength is increasing after the winter solstice and, in the Tammar it is associated with a pulse of prolactin that occurs at about dawn (McConnell, Hinds and Tyndale-Biscoe 1986). In recently published experiments Hinds (1989) has found that seasonal quiescence can be maintained in Tammars subjected to decreasing daylength by means of a single injection of prolactin delivered at dawn each day. A tentative conclusion from these results is that in these two species the normal lactational quiescence that occurs in almost all macropodoids has been extended by making use of a pre-existing pattern of prolactin secretion, which itself may be common to all species. This would mean that a change of profound importance to the breeding success of the species was achieved by a relatively small change in the sensitivity of the corpus luteum to prolactin.

Having briefly summarized the three endocrine axes that control reproduction in female Tammars it is now possible to attempt to interpret the conditions described in other species of macropodoid that exhibit different patterns of breeding.

Table 1.

	Quiescence	Anoestrus
Lactational	Potoroinae *Macropus rufus* – – – –>(drought) *M. robustus* *M. agilis* *M. rufogriseus banksianus* <– – – *Setonix brachyurus* (mainland population)	 *M. giganteus* *M. antelopinus* *M. parryi* *M. parma*
Facultative Seasonal	none	*S. brachyurus* (island pop.) *M. irma* *Lagostrophus fasciatus* *Thylogale billardierii*
Obligate Seasonal	*M. eugenii* *M. rufogriseus rufogriseus*	*M. fuliginosus*

Patterns of breeding in the Macropodoidea

We can recognize three breeding stategies among the Macropodoidea — continuous breeders, primarily seasonal breeders and secondarily seasonal breeders (Table 1).

Continuous breeders. The commonest and most widespread pattern among the Macropodoidea is that of continuous production of successive young, resulting from post-partum ovulation, lactational quiescence and reactivation of the corpus luteum and embryo at the end of the pouch life of the previous young. This pattern is seen in all species of *Bettongia*, in *Potorous tridactylus* and in *Aepyprymnus rufescens*, in which females can produce two or three successive young in a year. Births can occur at all times of the year and pouch young of all ages can be found at anytime. The pattern is also seen in the desert-living kangaroos, *Macropus rufus*, *M. robustus* and *M. agilis* and the Quokka on the mainland of Western Australia. This pattern allows the species to exploit a continuous food resource, in the case of the smaller species, and an unpredictable resource in the case of the large desert kangaroos. The selective advantage of this strategy has not been fully appreciated because it has been thought to represent so short a part of lactation (Ealey 1963; Sadleir 1965). However, for the desert-dwelling kangaroos the critical period is when conditions are deteriorating but may improve rapidly if rain falls. In these conditions the pouch young survive until about two months old (Newsome 1965; Frith and Sharman 1964), and then are replaced by a new offspring resulting from the reactivation of a quiescent corpus luteum and embryo. A comparison of this strategy with the alternate strategy of returning to oestrus after the two-months old pouch young is lost and undergoing a pregnancy before recommencing lactation shows that the probability of having a 2-month old pouch young ready to exploit breaking rains is 1 in 60 for the female with lactational quiescence compared to 1 in 90 or 1 in 100 for a female without; this is a very substantial difference, which could be the prime selective advantage for the retention of embryonic diapause in these species. The selective advantage of lactational quiescence for the small potoroids is less clear.

Under very unfavourable conditions in the desert kangaroos the pituitary-corpus luteum axis continues to function but the pituitary-follicular axis shuts down, so that follicular growth and ovulation do not occur at parturition and the kangaroo enters lactational anoestrus (Newsome 1964a). As mentioned earlier, this sequence has been exactly mimicked in the Tammar by immunizing females against GnRH and hence suppressing FSH and LH from the pituitary. Presumably drought conditions have the same effect on the pituitary of the Red Kangaroo. Furthermore, after breaking rains oestrus and ovulation occur within 11-14 d (Newsome 1964b; Frith and Sharman 1964; Sharman and Clark 1967), suggesting that the pituitary-follicular axis is rapidly reactivated if the drought breaks. Thus the Red Kangaroo exploits features of quiescence and anoestrus in its reproductive strategy.

The Eastern Grey Kangaroo, *Macropus giganteus*, could be seen to have extended this strategy beyond the exigencies of extreme drought, in that follicular growth always fails to occur at parturition but can resume later in lactation if nutritional conditions are favourable. It then enters lactational quiescence in the same way as the Red Kangaroo does after the drought breaks. However, in the southern parts of its range the incidence of late lactational quiescence is rare (Poole 1973). It would be particularly interesting to have information from Tasmania, the most southerly part of its range, but that is not available.

The Western Grey Kangaroo, *Macropus fuliginosis*, has taken this strategy of lactational anoestrus even further, so that it has become seasonally anoestrus. In doing so it has lost the capacity to enter lactational quiescence under any circumstances; in the rare instances when females in captivity were observed to show oestrus in late lactation the corpus luteum was not inhibited and birth occurred prematurely and the young entered the already occupied pouch and died (Poole 1975). Maynes (1973) argued that this represents the ancestral condition of macropodoid reproduction, from which the continuous breeding pattern of the Red Kangaroo has been derived. It seems improbable that one of the most advanced species, on dental, skeletal and size criteria, should have retained the ancestral condition when all other species, including the more generalized Potoroinae, have a more advanced pattern of reproduction. To the contrary, it suggests that under selection pressure for a highly seasonal pattern of reproduction the capacity for quiescence has been lost and anoestrus has become obligatory and irreversible in the species.

A second strategy for deriving seasonal breeding from continuous breeding is seen in some of the smaller species from southern Australia. In *Setonix brachyurus* the mainland populations breed continuously but the populations on Bald Island and Rottnest Island have a very restricted breeding season. In the Rottnest population it is known that this is due to supervention of true anoestrus in August and simultaneous disappearance of the corpus luteum and embryo without reactivation (Shield and Woolley 1963; Wallace 1981). However, this seasonal anoestrus can be reversed when nutritional conditions improve, as occurs at the Rottnest rubbish tip and when Quokkas are brought into captivity. Thus this pattern resembles the facultative anoestrus of the Red Kangaroo except for the phenomenon of corpus luteum decline during lactation. *Lagostrophus fasciatus* on Bernier and Dorre Islands and *Macropus irma* on the mainland of Western Australia may have the same strategy, since both species are apparently seasonal breeders, which do not carry quiescent blastocysts through the non-breeding season (Tyndale-Biscoe 1965; Prince 1983; A. Baynes and M. B. Renfree, pers. comm.). *Thylogale billardieri* in Tasmania has a restricted breeding season (Rose and McCartney 1982) but it is not clear which strategy it has adopted in this.

The third strategy by which seasonal breeding has been derived from the macropodoid pattern of continuous breeding is that of *Macropus rufogriseus* in Tasmania and the Tammar. The former is more instructive for the argument proposed here. On the mainland the subspecies is a continuous breeder and only displays lactational quiescence (Merchant and Calaby 1981), whereas the Tasmanian subspecies has evolved the same strategy as the Tammar, with lactational quiescence during the months of shortening daylength from the summer to the winter solstice, and seasonal quiescence during the months of increasing daylength from the winter to the summer solstice. As mentioned above, this has been achieved as a result of the corpus luteum becoming sensitive to prolactin at this time of the year. Neither this subspecies nor the Tammar ever display true anoestrus but they are both, secondarily, highly seasonal breeders, adapted to climates with a highly predictable annual climate. An important feature of this strategy, which appears to have evolved since the eustatic change in the sea level in the Pleistocene, is that it is irreversible: neither species has lost its highly seasonal pattern of reproduction in over 100 years in different climates, or during many years in captivity, where nutritional conditions are likely to favour continuous breeding. The potential to enter seasonal quiescence may exist in the mainland subspecies of *M. rufogriseus* since, in our captive colony at Canberra, some females from which young were removed in October failed subsequently to show elevated plasma progesterone profiles, typical of reactivation and did not give birth to new young until after the summer solstice (L. A. Hinds and J. Wright, unpublished results).

Evolution of reproduction in the Macropodoidea

The treatment of the variations in reproductive strategies of macropodoids given above has followed what appears to be a logical order from the commonest pattern of continuous breeding to the extreme of highly seasonal breeding; both the extreme of irreversible seasonal anoestrus and the extreme of irreversible seasonal quiescence occur at the southern limits of the range of the Macropodoidea on Tasmania and Kangaroo Island. The more variable and responsive patterns of lactational quiescence and lactational anoestrus predominate on the mainland of Australia and New Guinea and allow the species concerned to respond to less predictable climates. The crucial question is whether this regional distribution of reproductive strategies has any bearing on the evolution of macropodoid reproduction.

The question of the evolution of macropodoid reproduction can never be fully answered but, from the knowledge we have of living representatives, a few clues are presented. The phenomenon of embryonic diapause and the associated features are synapomorphic. The original selection may have been prolongation of gestation, by extending the luteal phase of the cycle into the follicular phase, so as to produce more advanced young at birth. It is now generally acknowledged that the neonates of macropodoids are more advanced than those of any other marsupials (Hughes and Hall 1988 and this book). Diapause, as I have argued elsewhere (Tyndale-Biscoe 1968, 1979) may be an intrinsic property of all mammalian blastocysts and is linked to a signal from the corpus luteum. Thus the second adaptation of the ancestral macropodoid was for the luteal cells of the corpus luteum to develop receptors for prolactin and become sensitive thereby to inhibition by prolactin. For small forest-dwelling marsupials this would have facilitated continuous breeding and maximized fecundity, while also providing a rapid response to the return of favourable condition after a dearth

As the decendants occupied other climates and habitats this basic pattern, incorporating the three endocrine axes described above for the Tammar, could be modified to emphasize lactational quiescence or lactational anoestrus or seasonal anoestrus. Seasonal quiescence may have only evolved in the Pleistocene in essentially desert-adapted species during cooling climatic conditions, when plant growth became confined largely to spring and early summer. However, since it clearly evolved independently in these two species, it may, in the distant past, have occurred in some other species now extinct.

REFERENCES

BERGER, P., 1966. Eleven month "Embryonic Diapause" in a marsupial. *Nature.* **211:** 435-36.

BERGER, P., 1970. Reproductive biology of the tammar wallaby *Macropus eugenii.* Ph.D. Thesis, Tulane University. *Diss. Abstr.* **31B:** 3760-61.

CARSON, R. D., 1912. Retarded development in a red kangaroo. *Proc. Zool. Soc. Lond.* **1912:** 234.

CLARK, M. J. AND POOLE, W. E., 1967. The reproductive system and embryonic diapause in the female grey kangaroo, *Macropus giganteus. Aust. J. Zool.* **15:** 441-59.

CURLEWIS, J. D., WHITE, A. S., LOUDON, A. S. I. AND MCNEILLY, A. S., 1986. Effects of lactation and season on plasma prolactin concentrations and response to bromocriptine during lactation in the Bennetts Wallaby *Macropus rufogriseus rufogriseus. J. Endocrinol.* **110:** 59-66.

EALEY, E. H. M., 1963. The ecological significance of delayed implantation in a population of the hill kangaroo *(Macropus robustus). In* Delayed Implantation ed. by A. C. Enders. Chicago University Press.

EVANS, S. M., TYNDALE-BISCOE, C. H. AND SUTHERLAND, R. L., 1980. Control of gonadotrophin secretion in the female tammar wallaby *(Macropus eugenii). J. Endocr.* **86:** 13-23.

FLANNERY, T. F., 1983. Revision of the macropodoid subfamily Sthenurinae (Marsupialia: Macropodoidea) and the relationships of the species of *Troposodon* and *Lagostrophus. Aust. Mammal.* **6:** 15-28.

FLANNERY, T., 1984. Kangaroos: 15 million years of Australian bounders. *In* Vertebrate Zoogeography and Evolution in Australasia ed. by M. Archer and G. Clayton. Hesperian Press: Carlisle, Western Australia.

FLETCHER, T. P. AND RENFREE, M. B., 1988. Effects of corpus luteum removal on progesterone, oestradiol-17β and LH in early pregnancy of the tammar wallaby, *Macropus eugenii. J. Reprod. Fert.* **83:** 89-95.

FRITH, H. J. AND SHARMAN, G. B., 1964. Breeding in wild populations of the red kangaroo, *Megaleia rufa. C.S.I.R.O. Wildl. Res.* **9:** 86-114.

HEARN, J. P., 1974. The pituitary gland and implantation in the tammar wallaby, *Macropus eugenii. J. Reprod. Fert.* **39:** 235-41.

HINDS, L. A., 1983. Progesterone and prolactin in marsupial reproduction. Ph.D. Thesis, Australian National University, Canberra.

HINDS, L. A., 1989. Morning pulse of prolactin maintains seasonal quiescence in the tammar, *Macropus eugenii. J. Reprod. Fert.* (in press).

HINDS, L. A. AND TYNDALE-BISCOE, C. H., 1982. Plasma progesterone levels in the pregnant and non-pregnant tammar, *Macropus eugenii. J. Endocr.* **93:** 99-107.

HORN, C. A. FLETCHER, T. P. AND CARPENTER, S. M., 1985. Effects of oestradiol-17β on peripheral plasma concentrations of LH and FSH in ovariectomised tammars *(Macropus eugenii). J. Reprod. Fert.* **73:** 585-92.

HUGHES, R. L. AND HALL, L. A., 1988. Structural adaptations of the newborn marsupial. *In* The Developing Marsupial. Models for Biomedical Research ed. by C. H. Tyndale-Biscoe and P. A. Janssens. Springer-Verlag: Berlin.

KARSCH, F. J., BITTMAN, E. L., FOSTER, D. L., GOODMAN, R. L., LEGAN, S. J. AND ROBINSON, J. E., 1984. Neuroendocrine basis of seasonal reproduction. *Recent. Prog. Horm. Res.* **40:** 185-225.

LEE, A. K. AND COCKBURN, A., 1985. The Evolutionary Ecology of Marsupials. Cambridge University Press.

MARSHALL, A. J., 1967. Origin of delayed implantation in marsupials. *Nature, Lond.* **216:** 192-93.

MAYNES, G. M., 1973. Reproduction in the parma wallaby, *Macropus parma* Waterhouse. *Aust. J. Zool.* **21:** 331-51.

MCCONNELL, S. J., HINDS, L. A. AND TYNDALE-BISCOE, C. H., 1986. Change in duration of elevated concentrations of melatonin is the major factor in photoperiod response of the tammar, *Macropus eugenii. J. Reprod. Fert.* **77:** 623-32.

MERCHANT, J. C. AND CALABY, J. H., 1981. Reproductive biology of the red-necked wallaby *(Macropus rufogriseus banksianus)* and Bennett's Wallaby *(M. r. rufogriseus)* in captivity. *J. Zool. Lond.* **194:** 203-17.

NEWSOME, A. E., 1964a. Anoestrus in the red kangaroo *Megaleia rufa* (Desmarest). *Aust. J. Zool.* **12:** 9-17.

NEWSOME, A. E., 1964b. Oestrus in the lactating red kangaroo, *Megaleia rufa* (Desmarest). *Aust. J. Zool.* **12:** 315-21.

NEWSOME, A. E., 1965. Reproduction in natural populations of the red kangaroo, *Megaleia rufa* (Desmarest), in Central Australia. *Aust. J. Zool.* **13:** 735-59.

PANYANITI, W., CARPENTER, S. M. AND TYNDALE-BISCOE, C. H., 1985. Effects of hypophysectomy on folliculogenesis in the tammar *Macropus eugenii* (Marsupialia: Macropodidae). *Aust. J. Zool.* **33:** 303-11.

POOLE, W. E., 1973. A study of breeding in grey kangaroos, *Macropus giganteus* Shaw and *M. fuliginosus* (Desmarest), in central New South Wales. *Aust. J. Zool.* **21:** 183-212.

POOLE, W. E., 1975. Reproduction in the two species of grey kangaroos, *Macropus giganteus* Shaw, and *M. fuliginosus* (Desmarest). II. Gestation, parturition and pouch life. *Aust. J. Zool.* **23:** 333-53.

PRINCE, R. I. T., 1983. Banded hare wallaby, *Lagostrophus fasciatus*. Pp. 201-02 *in* The Australian Museum Complete Book of Australian Mammals ed. by R. Strahan. Angus & Robertson: Sydney.

RENFREE, M. B., FLINT, A. P. F., GREEN, S. W. AND HEAP, R. B., 1984. Ovarian steriod metabolism and luteal oestrogens in the corpus luteum of the tammar wallaby. *J. Endocr.* **101:** 231-40.

RENFREE, M. B., WALLACE, G. I. AND YOUNG, I. R., 1982. Effects of progesterone, oestradiol-17β and androstenedione on follicular growth after removal of the corpus luteum during lactational and seasonal quiescence in the tammar wallaby. *J. Endocr.* **92:** 397-403.

ROSE, R. W. AND MCCARTNEY, D. J., 1982. Reproduction of the red-bellied pademelon, *Thylogale billardierii* (Marsupialia). *Aust. Wildl. Res.* **9:** 27-32.

ROTHCHILD, I., 1981. The regulation of the mammalian corpus luteum. *Rec. Progr. Hormone Res.* **37:** 183-298.

RUSSELL, E. M., 1974. The biology of kangaroos (Marsupialia-Macropodidae) *Mammal. Rev.* **4:** 1-59.

SADLEIR, R. M. F. S., 1965. Reproduction in two species of kangaroo *Macropus robustus* and *Megaleia rufa* in the arid Pilbara region of Western Australia. *Proc. Zool. Soc. Lond.* **145:** 239-61.

SERNIA, C. AND TYNDALE-BISCOE, C. H., 1979. Prolactin receptors in the mammary gland, corpus luteum and other tissues of the tammar wallaby, *Macropus eugenii. J. Endocr.* **83:** 79-89.

SHARMAN, G. B., 1955a. Studies on marsupial reproduction. II. The oestrous cycle of *Setonix brachyurus. Aust. J. Zool.* **3:** 44-55.

SHARMAN, G. B., 1955b. Studies on marsupial reproduction. III. Normal and delayed pregnacy in *Setonix brachyurus. Aust. J. Zool.* **3:** 56-70.

SHARMAN, G. B., 1955c. Studies on marsupial reproduction. IV. Delayed birth in *Protemnodon eugenii. Aust. J. Zool.* **3:** 156-61.

SHARMAN, G. B., 1965. Marsupials and the evolution of viviparity. *Viewpoints in Biology* **4:** 1-28.

SHARMAN, G. B. AND BERGER, P. J., 1969. Embryonic diapause in marsupials. *Adv. Reprod. Physiol.* **4:** 211-40.

SHARMAN, G. B. AND CLARK, M. J., 1967. Inhibition of ovulation by the corpus luteum in the red kangaroo, *Megaleia rufa. J. Reprod. Fertil.* **14:** 129-37.

SHAW, G. AND RENFREE, M. B., 1984. Concentrations of oestradiol-17β in plasma and corpora lutea throughout pregnancy in the tammar, *Macropus eugenii. J. Reprod. Fertil.* **72:** 29-37.

SHAW, G. AND RENFREE, M. B., 1986. Uterine and embryonic metabolism after diapause in the tammar wallaby, *Macropus eugenii. J. Reprod. Fertil.* **76:** 339-47.

SHIELD, J. W. AND WOOLLEY, P., 1963. Population aspects of delayed birth in the quokka *(Setonix brachyurus). Proc. Zool. Soc. Lond.* **141:** 783-90.

SHORT, R. V., FLINT, A. P. F. AND RENFREE, M. B., 1985. Influence of passive immunization against GnRH on pregnancy and parturition in the tammar wallaby, *Macropus eugenii. J. Reprod. Fertil.* **75:** 567-75.

STEWART, F. AND TYNDALE-BISCOE, C. H., 1982. Prolactin and luteinizing hormone receptors in marsupial corpora lutea: relationship to control of luteal function. *J. Endocr.* **92:** 63-72.

STODART, E., 1966. Observations on the behaviour of the marsupial *Bettongia lesueur.* (Quoy and Gaimard) in an enclosure. *C.S.I.R.O. Wildl. Res.* **11:** 91-99.

TYNDALE-BISCOE, C. H., 1963. The role of the corpus luteum in the delayed implantation of marsupials. *In* Delayed Implantation ed. by A. C. Enders. Chicago University Press.

TYNDALE-BISCOE, C. H., 1965. The female urogenital system and reproduction of the marsupial *Lagostrophus fasciatus. Aust. J. Zool.* **13:** 255-67.

TYNDALE-BISCOE, C. H., 1968. Reproduction and post-natal development in the marsupial *Bettongia lesueur* (Quoy and Gaimard). *Aust. J. Zool.* **16:** 577-602.

TYNDALE-BISCOE, C. H., 1970. Resumption of development by quiescent blastocysts transferred to primed, ovariectomised recipients in the marsupial, *Macropus eugenii. J. Reprod. Fertil.* **23:** 25-32.

TYNDALE-BISCOE, C. H., 1979. Hormonal control of embryonic diapause and reactivation in the tammar wallaby. Pp. 173-90 *in* Maternal Recognition of Pregnancy. *Ciba Foundation Symp.* 64 (new ser.). Excerpta Medica: Amsterdam.

TYNDALE-BISCOE, C. H., 1984. Mammal — Marsupials. *In* Marshall's Physiology of Reproduction, 4th Ed., Vol. 1., Chapter 5. ed. by G. E. Lamming. Churchill Livingstone: Edinburgh.

TYNDALE-BISCOE, C. H., 1986. Embryonic diapause in a marsupial: roles of the corpus luteum and pituitary in its control. *In* Comparative Endocrinology: Developments and Directions ed. by C. L. Ralph. Alan R. Liss Inc.: New York.

TYNDALE-BISCOE, C. H. AND HAWKINS, J., 1977. The corpora lutea of marsupials, aspects of function and control. *In* Reproduction and Evolution ed. by J. H. Calaby and C. H. Tyndale-Biscoe. Australian Academy of Science: Canberra.

TYNDALE-BISCOE, C. H. AND HEARN, J. P., 1981. Pituitary and ovarian factors associated with seasonal quiescence of the tammar wallaby, *Macropus eugenii. J. Reprod. Fertil.* **65:** 225-30.

TYNDALE-BISCOE, C. H. AND HINDS, L. A., 1984. Seasonal patterns of circulating progesterone and prolactin and response to bromocriptine in the female tammar, *Macropus eugenii. Gen. comp. Endocr.* **53:** 58-68.

TYNDALE-BISCOE, C. H. AND RENFREE, M. B., 1987. Reproductive Physiology of Marsupials. Cambridge University Press: Cambridge.

TYNDALE-BISCOE, C. H., HINDS, L. A. AND HORN, C. A., 1988. Fetal role in the control of parturition in the tammar, *Macropus eugenii*. *J. Reprod. Fertil.* **82**: 419-28.

TYNDALE-BISCOE, C. H., HINDS, L. A. AND MCCONNELL, S. J., 1986. Seasonal breeding in a marsupial: opportunities of a new species for an old problem. *Recent Prog. Horm. Res.* **42**: 471-512.

WALLACE, G. I., 1981. Uterine factors and delayed implantation in macropodoid marsupials. *J. Reprod. Fert., Suppl.* **29**: 173-81.

WARING, H., 1956. Marsupial studies in Western Australia. *Aust. J. Sci.* **18**: 66-73.

Reproduction in male Macropodidae

RUSSELL C. JONES[1]

Reproduction in male Macropodidae is reviewed and compared to other marsupials and eutherians which have been studied. All of the adults which have been studied produce spermatozoa throughout the year except when spermatogenesis is impaired by environmental stress. However, *M. eugenii* shows some seasonality in testosterone levels and mass of the accessory glands of reproduction with peaks coincident with the highest incidences of females coming into oestrus.

Estimates of parameters related to sperm production in *M. eugenii* are of the same order of magnitude as in eutherians. The cytology of spermiogenesis exhibits many of the unique marsupial characteristics. As in eutherians spermiogenesis occurs in an adluminal environment of the seminiferous tubule which is different in composition from blood.

There is an extratesticular component of sperm maturation in the Macropodidae as in other mammals. The epididymis of *M. eugenii* has been studied in detail. Throughout its length it maintains an electrochemical gradient against blood for all inorganic diffusable ions. The anatomical differentiation into a caput, corpus and cauda epididymidis corresponds to the structural and functional differentiation of the epithelium lining the duct. The ductuli efferentes in the caput epididymidis reabsorb most of the fluid leaving the testis. The caput and corpus regions of the ductus epididymidis are involved in sperm maturation; the cytoplasmic droplet which is shed during sperm maturation is phagocytosed by the epithelium lining the corpus epididymidis. Proteins are secreted into the lumen of the caput epididymidis and one of the proteins becomes associated with the sperm membranes. Sperm are stored in the cauda epididymidis in an environment which has a lower osmotic pressure and concentration of organic compounds than in the eutherians.

The prostrate gland and epididymides are the main sources of semen in the Macropodidae as in all marsupials. The semen contains particulate matter from the prostatic epithelium and forms a true copulatory plug as in rodents. The main sugar in the semen is N-acetylglucosamine as in many other marsupials.

The accessory organs of reproduction are dependent on systemic testosterone and in addition the caput and corpus epididymidis, which have high levels of 5α — reductase activity, are also dependent on a luminal connection with the testis.

INTRODUCTION

THE male reproductive system is highly specialized in mammals. The testes descend into a scrotum in most species, spermatozoa are not capable of fertilizing an ovum when they leave the testis and the spermatozoa are considerably diluted during ejaculation by secretions of accessory glands of reproduction. Nevertheless, there is considerable diversity in the structure of spermatozoa and the reproductive system, and the biochemistry of semen in mammals, so that it is pertinent to assess the occurrence and significance of adaptations in particular groups. This review examines reproduction in male Macropodidae in relation to our knowledge of other marsupials and the eutherians. There is particular emphasis on the Tammar, *Macropus eugenii*, as it has received more attention than the other Macropodidae and it is a useful model animal for studying sperm production in mammals.

Fig. 1 illustrates the characteristic arrangement of the reproductive system of Macropodidae. The testes are scrotal in a prepenial position which is characteristic of all marsupials except *Notoryctes typhlops*, in which the testes are inguinal, but only occurs in a few eutherians (Eckstein and Zuckerman 1956). Spermatozoa are transported from the testis via the ductuli efferentes to the ductus epididymidis which is involved in their maturation and storage awaiting ejaculation. Like other marsupials the Macropodidae have a prostate gland and bulbourethral (Cowper's) glands but lack the other accessory glands of reproduction (ampulla, seminal vesicles and coagulating gland) which may be present in the eutheria (Djakiew 1978; Eckstein and Zuckerman 1956; Hann 1964). The glans penis is a single elongate cone which correlates with the presence of a permanent median vagina (as well as lateral vaginae) in parous females, and is unlike the divided structure characteristic of most marsupials (Rodger and Hughes 1973).

SEXUAL DIFFERENTIATION AND EARLY DEVELOPMENT

Gonadal development in the Macropodidae and other marsupials follows the general pattern for mammals. Recent work on the mouse (Byskov and Lintern-Moore 1973; Byskov 1978) and sheep (Zamboni and Upadhyay 1982) confirms the early work on eutherians (Witschi 1951, 1956, 1962) which indicated that the somatic elements of the sex cords originate from the mesonephros. These somatic elements ultimately form the Sertoli cells of the testes and granulosa cells of the ovary. The primordial germ cells migrate into the sex cords from an extragonadal origin. In *M. eugenii* they migrate from the dorsal mesentery surrounding the hind gut at about day 22 of the 27-day gestation and undergo mitotic divisions in the sex cords to establish a population of about 12 000 cells by day 25 of pregnancy (Alcorn 1975; Alcorn and

[1]Department of Biological Sciences, University of Newcastle, New South Wales, Australia 2308.

Pages 287-305 in KANGAROOS, WALLABIES AND RAT-KANGAROOS ed. by G. Grigg, P. Jarman and I. Hume. Surrey Beatty & Sons Pty Limited, New South Wales, Australia, 1989.

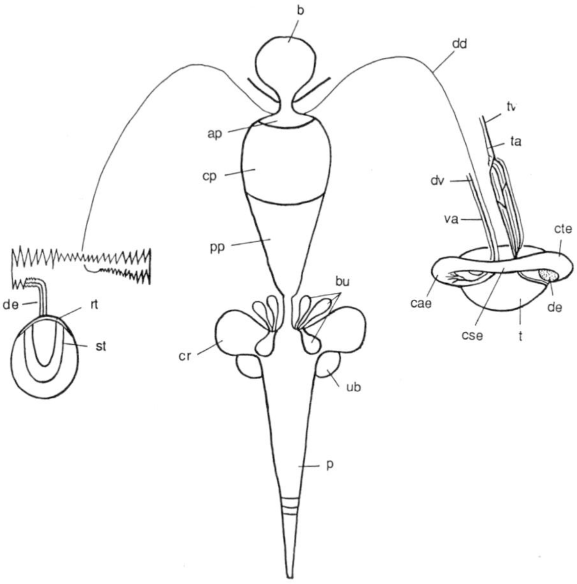

Fig. 1. Diagram of a ventral view of the male reproductive tract of *M. eugenii* showing the gross anatomy of the testis and epididymis on the right of the page and a diagrammatic representation of the arrangement of the sperm ducts on the left side. ap, anterior prostate; b, bladder; bu, bulbourethral (Cowper's) glands; cae, cauda epididymidis; cp, central prostate; cr, crus penis; cse, corpus epididymidis; cte, caput epididymidis; dd, ductus deferens; de, ductuli efferentes; dv, deferential vein; p, penis; pp, posterior prostate; rt, rete testis; st, seminiferous tubule (tubules form loops between the two arms of the rete testis); t, testis; ta, internal spermatic artery; tv, testicular vein; ub, urethral bulb; va, vasal artery.

Robinson 1983). The interstitial tissue of the gonad, from which the Leydig cells of the testis ultimately develop, is formed from primitive mesonephric tissue in eutherians and *Dasyurus viverrinus* (Ullmann 1984). However, Alcorn (1975) considered that it is formed from the rete cords in *M. eugenii*.

The gonads are at the sexually indifferent stage of development at birth in *M. eugenii*. However, the scrotum, mammary gland, gubernaculum and processus vaginalis are sexually dimorphic from five days pre-partum indicating that, unlike development of these accessory sex organs in eutherian mammals, phenotypic differentiation has occurred before gonadal differentiation (Wai-Sum *et al.* 1988). Differentiation of the testis occurs between three and seven days after birth when the primitive mesonephric tissue becomes organized into a central cortex of stroma intimately associated with the sex cords, and a medulla consisting of the medullary cords and an outer layer of fibrous cells under the coelomic epithelium. The Leydig cells probably differentiate at about this time as Catling and Vinson (1976) assayed testosterone in the blood plasma of *M. eugenii* which were one day old.

Work on the development of the genital tract of marsupials (see review by Tyndale-Biscoe and Renfree 1987) indicates that it is homologous with the eutherians. In this respect work on the ram by Zamboni and Upadhyay (1982) concluded that the male genital tract proximal to the ductus epididymidis differentiates from a single "giant" mesonephric nephron. The mesonephric duct (or Wolffian duct) forms the ductus epididymidis and ductus deferens (Witschi 1956). The duct is patent at birth in *M. eugenii* and functions as a urinary duct to convey urine from the mesonephros to the urogenital sinus until about day 14 after birth (Alcorn 1975).

Early workers did not observe macroscopic signs of the scrotum until one to three weeks after birth in the Macropodidae (see Table 24.1 in Setchell 1977 for list of species). However, the microscopic studies of Alcorn (1975) and Wai-Sum *et al.* (1988) indicate that the anlagen develops earlier. As in the scrotal eutherians the testes migrate to the scrotum from a site medial to the kidneys. In *Macropus parma* the descent of the testes is complete by 91 days after birth (Haynes 1973). In *M. eugenii* the testes have commenced to descend by day 56 and are located in the scrotum by day 60 (Alcorn 1975). As in other marsupials the ventral migration of the uretic bud to the neck of the urinary bladder is mesial in the Macropodidae and not lateral as in eutherians (Tyndale-Biscoe and Renfree 1987). Consequently, in adult marsupials the ductus deferens does not loop over the ureters before entering the urethra as it does in eutherians. Presumably as in the eutherians the prostate and Cowper's glands of the Macropodidae develop from the proximal and distal urethra respectively.

PUBERTY AND SEASONALITY

Male kangaroos are capable of achieving fertile matings at 22 to 26 months of age (Sadleir 1965; Sharman and Calaby 1964; Poole 1973) whilst wallabies achieve the capacity about four months younger (Merchant 1976; Bolton, Newsome and Merchant 1982; Johnson 1979; Merchant and Calaby 1981). In general this stage of puberty in the male is achieved two to four months after their female contemporaries have their first oestrus. The age at puberty may vary in arid zones, for example from 20 to 72 months in *Macropus giganteus* and *M. fuliginosus* (Poole 1973, 1976), depending on the year and time of the year of birth. In *Macropus agilis* it varies by two to four months depending upon habitat (Bolton *et al.* 1982).

Whereas there is variation between species of female Macropodidae from being continuous breeders, or continuous breeders exhibiting drought-induced anoestrus, to seasonal breeders, the males of all species which have been studied seem to produce spermatozoa continuously after reaching puberty. The only exceptions are the animals which inhabit arid regions, such as *Macropus rufus*, in which spermatogenesis may be arrested during periods of drought or high temperatures (Newsome 1973). Further, there is an interesting expression of seasonality in male *M. eugenii* in which the females are seasonal breeders. The males show peaks in systemic testosterone concentrations and masses of accessory glands of reproduction during January-February when most females exhibit oestrus, and minor peaks in October when young females enter first oestrus (Fig. 2; Inns 1982). Interestingly there are also small increases in the masses of the testes and epididymides during the period of high testosterone levels which are probably due to an increased volume of Leydig cells in the testes, and an increased number of spermatozoa in the epididymides due to sperm transport through the duct being slowed by the high testosterone levels (Meistrich, Hughes and Bruce 1975). The increases in testosterone levels have been implicated with the presence of females (Catling and Sutherland 1980).

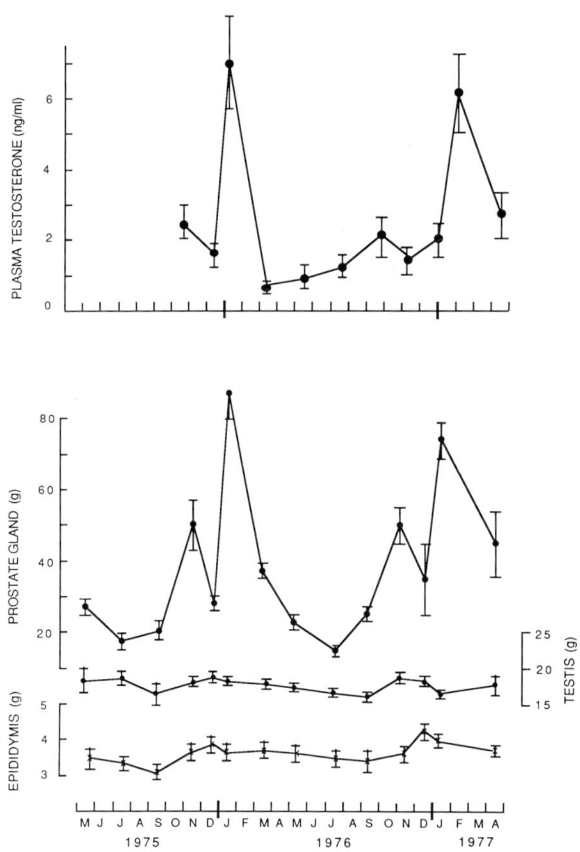

Fig. 2. Seasonal changes (mean±SE) in the mass of the testis, epididymis and prostate, and concentration of plasma testosterone of *M. eugenii* at Kangaroo Island, South Australia, during the years 1975 to 1977. The maximum testosterone concentrations and prostate masses coincide during January-February when the maximum number of females exhibit oestrus, and there are minor peaks in October when young females enter first oestrus. Redrawn from Inns (1982).

THE SCROTUM AND SPERMATIC CORD

The testes of Macropodidae are held in a pendulous scrotum with a long, narrow neck. The skin forms a tight-fitting tunic over the underlying viscera. There is some smooth muscle under the skin (Setchell 1977) which is analagous to the tunica dartos in eutherians, but the skin does not loosen as much during hot weather as in eutherians like the bull, ram and man. The hair covering the scrotum is sparser than over the rest of the body. In *M. eugenii* it varies in density and length with season and is bald over the cauda epididymidis during summer in Canberra. The tunica vaginalis which surrounds each testis is pigmented black in *Macropus canguru* (Heddle and Guiler 1970), *M. eugenii* and *M. rufogriseus* and some other marsupials, but not in *M. rufus* (Setchell 1977). Several explanations for the pigmentation have been proposed (Biggers 1966; Temple-Smith 1984b), but none are convincing.

The neck of the scrotum contains a well developed cremaster muscle which may shorten to hold the testes against the body wall during cold conditions and when the animal hops, or relax allowing the neck to lengthen to about 30 mm during warm conditions. Two spermatic cords each carry blood and lymphatic vessels, nerves and a ductus deferens through the neck of the scrotum (see Fig. 2 in Setchell and Thorburn 1969). Unlike the scrotal eutherians in which only the testicular vein branches to form a network of vessels (the pampiniform plexus; Harrison 1949) in the spermatic cord, the testicular artery and vein of the Macropodidae and other marsupials (except *Notoryctes typhlops*; Setchell 1977) each form a rete mirabile with the branches of the vessels running roughly parallel and in close aposition to one another. The number of arterial and venous vessels may not be equal in an individual but they tend to be most numerous in large animals (Table 1). The only known functional difference between the double retia in the Macropodidae and the presence of only a venous rete in eutherians is that heating the testes of the marsupial increases blood flow by a factor of 1.3 to 2.8 times the resting value (Setchell and Thorburn

Table 1. Number of vessels in the testicular rete mirabile of some Macropodoidea[a].

Species	Arteries	Veins	Lymphatics
Kangaroos			
Macropus rufus	95	74	4
Macropus robustus	120	100	4
Wallabies			
Macropus eugenii	240	170	5
Macropus rufogriseus	97	122	5
Rat-kangaroos			
Caloprymnus campestris	46	38	—
Potorous tridactylus	6	6	—

[a]Selected from tables published by Heddle and Guiler (1970) and Setchell (1970a).

1969) whereas heating has no effect on blood flow in eutherians (Setchell 1970a). The different effect of heating between the two groups is probably due to differences in the cross-sectional area and arrangement of the testicular artery.

It is unlikely that the main purpose of the scrotum and retia mirabilia has been resolved. Setchell and Waites (1969) showed that as in the spermatic cord of scrotal eutherians the retia in *Macropus* and *Macropus robustus* attenuate pulse pressure in the testicular artery with little change in blood pressure, and they also act as a counter current heat exchange organ maintaining the testes at a uniform temperature of 4-5°C below body temperature. Similar temperature gradients have also been recorded for *M. eugenii* and *M. rufogriseus* (Setchell 1977). Heat loss from the scrotum may be assisted by the evaporation of water from sweat (Heddle and Guiler 1970) or licking (Carrick 1969; quoted by Setchell 1977). As in scrotal eutherians and *Trichosurus vulpecula* it has been shown that local heating of the testes disrupts spermatogenesis in *M. eugenii* (Carrick 1969; quoted by Setchell 1977). Also, Newsome (1973) has shown that spermatogenesis was impaired in *Macropus rufus* by prolonged summer temperatures exceeding 32°C, and that the effect was related to the duration of the heat load (and poor nutrition). However, the environmental conditions in both these studies (Carrick 1969 heated the testes to 40°C for 3 hours) were extreme so that there may be some validity in Sadleir's (1965) suggestion that male *Macropus rufus* and *M. robustus* can remain fertile at temperatures which cause infertility of rams in the same environment.

It has also been suggested that the purpose of the retia mirabilia may be to transfer substances from the veins to the arteries in the spermatic cord in order to establish an environment in the testis which is different to the rest of the body. Jacks and Setchell (1973) used radioactive tracers to demonstrate that some exchange does occur in *Macropus rufogriseus* and the ram, but considering other data (e.g., Setchell and Hinks 1969) Setchell (1977) concluded that the effect is "probably minimal".

THE TESTES AND ANDROGEN PRODUCTION

The testes of Macropodidae are oblong with the longest axis parallel with the anterior-posterior axis of the body. The rete testis is a horseshoe-shaped area under the anterior surface of the testis and the seminiferous tubules form loops from one arm to the other (Fig. 1). The seminiferous tubules are highly convoluted except at their junction with the rete where, as in eutherians, they form a transitional region in which the duct narrows and is lined only by apparently modified Sertoli cells (Setchell 1977) which seem to act as one-way valves only allowing the flow of fluid out of the seminiferous tubules. The seminiferous tubules are supported by considerable loose connective tissue containing blood and lymphatic vessels, and Leydig cells. Blood flow through the testes of *M. eugenii* (9.4±0.88 ml/100 g/min) and *Macropus rufa* (3.5 ml/100 g/min; Setchell and Thorburn 1969) is about the same as in eutherians (e.g., 8.4±0.74 ml/100 g/min for the ram; Setchell *et al.* 1966).

The Leydig cells occupy only 5% of the total volume of the testis in Macropodidae as in the ram and bull (Fawcett, Neaves and Flores 1973) and so differ from Didelphidae and Peramelidae which have abundant Leydig cells and only small lymphatic spaces as in the pig (Setchell 1977). According to ultrastructural studies on *Didelphis virginiana* and *Trichosurus vulpecula* it may be presumed that the Leydig cells are the main source of androgens in the Macropodidae (Christensen and Fawcett 1961; Temple-Smith 1984b). Testosterone is the main androgen secreted (Carrick and Cox 1977). It is secreted at a rate which is much the same as for eutherians like the ram (6.8 and 2.6 ng per min per g testis for *Macropus robustus* and *Macropus rufus* respectively; Setchell 1977). The regression of the testes and accessory organs of reproduction which follows hypophysectomy of *M. eugenii* (Hearn 1975) indicates that the pituitary hormones regulate secretion of testosterone and as judged by the rise in systemic LH levels associated with the reduction in testosterone levels following castration of *M. eugenii* (Catling and Sutherland 1980) testosterone has a negative feed-back effect on the pituitary. The rise in FSH levels following castration probably indicates that the testes also produce inhibin (Catling and Sutherland, loc. cite.).

Caution is required in interpreting determinations of blood testosterone concentrations in marsupials as the stress involved in handling animals may cause a drop in concentration (Lincoln 1978). Consequently, the studies by Inns (1982) on *M. eugenii* provide the most reliable data that is available as he collected blood in the field after shooting undisturbed animals (Fig. 2).

SEMINIFEROUS TUBULES AND SPERMATOGENESIS

The structure of the seminiferous tubules follows the general pattern in mammals (Setchell 1977) Spermatogenesis involves a series of mitotic divisions of the diploid spermatogonia which then undergo DNA synthesis as they form the larger primary spermatocytes and enter meiosis. The secondary spermatocytes produced by the first meiotic division are diploid and the spermatids produced by the second division are haploid.

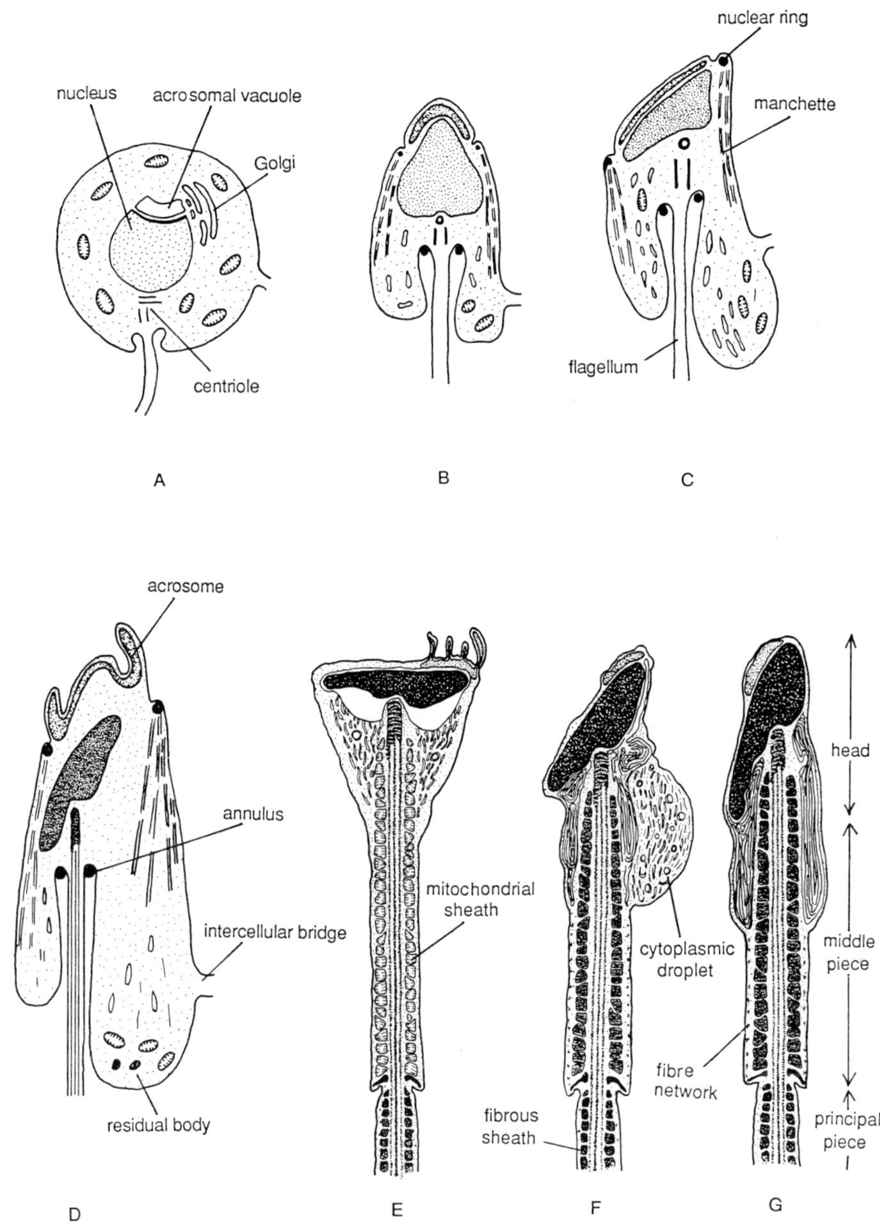

Fig. 3. Diagrammatic representation of the successive stages of spermiogenesis (A-E) and the maturation of spermatozoa in the epididymis (E-G) of *Macropus eugenii.* A, *Early spermatid stage* when the Golgi forms the acrosomal vacuole on the luminal side of the nucleus and the tail originates from the centriole near the Golgi then migrates to the opposite pole. B, *Nuclear protrusion stage* involving reorientation and projection of the nucleus and acrosome from the surface of the spermatid facing the basement membrane of the seminiferous tubule. C, *Nuclear flattening and condensation stage* involving condensation of the nucleus and acrosome and caudal displacement of cytoplasm. D, *Nuclear rotation stage* involving condensation and rotation of the nucleus around the point of insertion of the tail, acrosomal retraction, and lengthening of the tail and formation of the fibrous sheath around it. E, The *post-nuclear rotation stage;* involves movement of the annulus down the flagellum, helical arrangement of the mitochondria around the middle piece and release from the Sertoli cell. E, "testicular" type spermatozoon: found in rete testis, ductuli efferentes and proximal caput epididymidis. F, nuclear rotation stage found in proximal corpus epididymidis. G, "mature" spermatozoon: found in distal corpus epididymidis and cauda epididymidis. Diagrams based on electron micrographs of *M. eugenii* and other species; and drawings by Harding *et al.* (1984), Temple-Smith and Bedford (1976), Sapsford *et al.* (1969) and Fawcett, Anderson and Phillips (1971).

Subsequently, during the process of spermiogenesis the spermatids differentiate to form spermatozoa (Fig. 3).

There have been three studies on the cytology of spermiogenesis in the Macropodidae (Harding 1977; Binder 1927; Kim, Harding and Shorey 1987). Fig. 3 summarizes my interpretation of the structural changes which occur in *M. eugenii* based on the work mentioned above, and on work on *M. eugenii* (Jones, Hinds and Tyndale-Biscoe 1984 and unpublished data), *Trichosurus vulpecula* (Harding, Carrick and Shorey 1975, 1976a, 1976b, 1979) and *Perameles nasuta* (Sapsford and Rae 1969; Sapsford *et al.* 1969). The changes are similar to the eutherian pattern with some exceptions (see Setchell 1977) as in other marsupials, except perhaps the Koala (Harding *et al.* 1979). During nuclear condensation in marsupials the nucleus flattens so that its long axis is perpendicular to the tail and it remains free to rotate, changing its angle with the tail at the point of insertion until completion of the phase of sperm maturation which occurs in the epididymis. Also, as in some other marsupials (Harding *et al.* 1979) the acrosome of sperm of Macropodidae retracts to cover only one third of the rostral end of the nucleus.

Harding and co-workers (Harding 1977; Harding *et al.* 1979; Harding *et al.* 1987) have studied spermatozoa from more than five species of Macropodidae and concluded (Harding *et al.* 1979) that there is "a close resemblance" in their structure and that of spermatozoa from *Trichosurus vulpecula*. In general, spermatozoa from the Macropodidae, except the *Lagorchestes*, show some variation in dimensions and shape of the sperm head but no more than, for example, the eutherian ungulates (Cleland 1969). The head shape of the two *Lagorchestes* studied by Cleland (1969) were quite different from the other Macropodidae, a finding which is consistent with serological data (Kirsch 1977).

The kinetics of spermatogenesis has been studied in *M. eugenii* and *M. rufogriseus* (Carrick and Setchell 1973), and in a possum and two species of bandicoot (Carrick and Setchell 1973; Mason and Blackshaw 1973). The studies confirmed the observations of earlier workers which indicated that different parts of a seminiferous tubule contain definite associations of various types of germ cell. The associations were classified into eight stages of spermatogenesis according to the nuclear morphology of the germ cells (see Fig. 3; method of Roosen-Rungen and Giesel 1950; and Courot *et al.* 1970). Stage VIII corresponds to the early spermatid stage, stage I and II to the nuclear protrusion and flattening stages, stage III to the nuclear rotation and early post-rotational stages and stages IV to VIII to the late post-rotational stage of spermatid differentiation. Setchell and Carrick (1973) determined that the length of one cycle of the seminiferous epithelium was 16 and 17 days for *M. eugenii* and *M. rufogriseus* respectively which is about the same as for *Trichosurus vulpecula* (15 days) and towards the upper end of the range for eutherians. Using these estimates the total duration of spermatogenesis from the first spermatogonial division involving stem cell renewal to release of the spermatozoa from the Sertoli cells (4.5 cycles) is 72 and 71.4 days respectively. Setchell (1977) estimated that the meiotic prophase lasted 21 and 24 days respectively for *M. eugenii* and *M. rufogriseus* and spermiogenesis lasted 25 days in both species.

There have been no studies on the regulation of spermatogenesis in Macropodidae. Hearn (1975) noted that the seminiferous epithelium regresses following hypophysectomy. However, this may be a secondary effect of testosterone withdrawal rather than a direct effect of withdrawing the gonadotrophins.

Sertoli cells extend from the wall to the lumen of the seminiferous tubules and act as "supportive cells" for the germ cells. They probably form intercellular junctions as in other mammals (Fawcett 1975) to establish a barrier (the so-called blood-testis barrier) between pre- and post-meiotic germ cells. The efficacy of the barrier and function of the Sertoli cells in forming a secretion in the lumen of the seminiferous tubules of *M. eugenii* can be judged by the composition of fluid collected from the rete testis (Table 3) or close to it (Setchell 1970b). Setchell *(loc. cite)* found that the fluid flows from the testis at a rate of 0.25 ml/h or 0.0125 mg/g testes/h which is within the range reported for eutherians (Setchell 1974). The fluid is also similar in composition to rete testis fluid from eutherians in containing less sodium and more potassium and chloride than blood plasma, little glucose or protein, and relatively high concentrations of some amino acids. As in eutherians (Koskimes and Kormano 1975) some proteins which are present in rete testis fluid from *M. eugenii* are not present in blood plasma (apparent molecular weight, MW, of 14 700, 22 800, 24 100, 43 200, and 44 800; Jones 1987). The rete testis fluid differs from that in eutherian mammals in that it does not contain the sugar inositol (Setchell 1970b).

Little is known of the functions of the Sertoli cell secretions. It is generally recognized that the fluid may be secreted to maintain a constant milieu for spermiogenesis and to help transport spermatozoa from the testis as most of the fluid is reabsorbed in the ductuli efferentes when it leaves the testis (see below). Further, rete testis fluid is known to contain specific proteins such as transferrin, androgen binding protein and a growth factor in eutherians such as the rat. Growth factor activity has been identified in rete testis fluid from *M. eugenii* (Jones *et al.* 1987) at about the same level as in rete testis fluid from the rat and ram, and it is probably involved in regulating the mitotic divisions during the production of spermatozoa in the testis (Feig *et al.* 1980).

STRUCTURE AND REGULATION OF THE EPIDIDYMIS

The epididymides have been studied in only one Macropodidae, *M. eugenii* (Jones, Hinds and Tyndale-Biscoe 1984; Jones 1987; Jones and Clulow 1987a; Jones *et al.* 1987, 1988). They are anatomically differentiated into a caput, corpus and cauda epididymidis (Fig. 1) and lie medial to and extend anteriorly and posteriorly beyond the testis. The caudal protrusion is quite distinct and is enclosed within a separate pouch formed by the tunica vaginalis to which it is attached at the distal end.

The ductuli efferentes join the rete testis to the ductus epididymidis and form part of the caput epididymidis. Twelve to 16 ductuli leave the anterior pole of the testis adjacent to the vascular bundle and course distally forming a cylindrical cord, the initial zone. The cord widens, due to the ductuli developing a sinuous course (the coni vasculosa), becomes pigmented brown and folds through an angle of about 90°. Some pairs of ductuli join together but most join the common ductulus efferens which then continues on as the ductus epididymidis.

The ductuli are narrow (140±9.1 μm diameter), contain few spermatozoa and each is invested in a tunic of circularly arranged smooth muscle. They are lined by a low, columnar epithelium consisting of principal and ciliated cells, and intraepithelial leucocytes indicating that it is a multifunctional epithelium (Fig. 4). The ciliated cells aid the muscular tunic to move spermatozoa along the ductuli. The principal cells have the structure of absorptive and digestive epithelium with numerous apical vesicles and vacuoles, dense bodies (tertiary lysosomes) and a small Golgi apparatus.

There is an abrupt junction between the common ductulus efferens and the ductus epididymidis which in histological sections is noticeable where the epithelium changes from the low ciliated to a tall non-ciliated type. The ductus epididymidis is about 35 metres long in *M. eugenii* (G. Chaturapanich, unpublished data). Its diameter varies little from the caput epididymidis to the flexure of the cauda epididymidis (diameter ranges from 208±22.5 to 249±19.3 μm). Beyond the flexure (distal cauda epididymidis) it widens (diameter 552±40.0 μm) and its muscular tunic, which increases in thickness 2- to 3-fold in the proximal cauda epididymidis, increases in thickness by a factor of another 2- to 3-fold at the

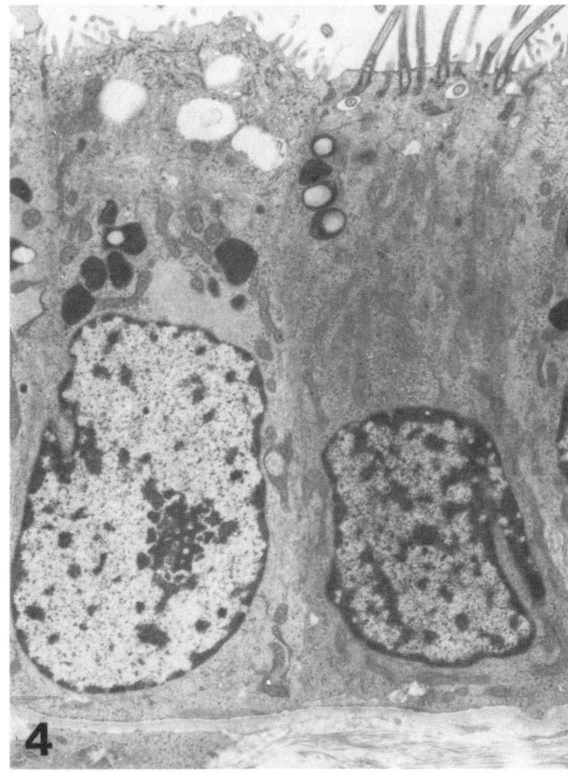

Fig. 4. Electron micrograph of the epithelium lining the ductuli efferentes of *M. eugenii* showing a ciliated cell on the right and principal cell on the left, 5 000×.

flexure between the proximal and distal cauda epididymidis. The ductus is lined by a pseudostratified, columnar epithelium (Figs 5-7) consisting of principal, apical and basal cells, mitochondria rich cells (caput and corpus epididymidis only) and intraepithelial leucocytes, indicating that the epithelium carries out a number of functions. The principal cells contain ribosomes, endoplasmic reticulum and a supranuclear Golgi apparatus indicating that they may be secretory. They also contain apical vesicles, small supranuclear vacuoles, and dense bodies, suggesting that they are involved in absorption and heterophagic digestion.

Based on the structure of the epithelium the ductus epididymidis is differentiated into three structurally distinct segments (Figs 5-7) which correspond to the anatomical differentiation of the epididymis into the caput, corpus and cauda epididymidis, a correspondence which is not so close in the eutherians. Spermatozoa develop the capacity for motility in the caput and corpus epididymidis (Fig. 8). In these regions the epithelium is taller (35.0±1.3 μm and 31.2±1.5 μm respectively) than in the cauda epididymidis (27.4±2.0 to 17.5±2.1 μm) and the epithelium has longer stereocilia. The cytology of the principal cells and occurrence of mitochondria rich cells in the caput epididymidis indicates that it is homologous with the initial segment of eutherians (Benoit 1926; Nicander and Glover 1973). However, the epithelium is lower and the stereocilia

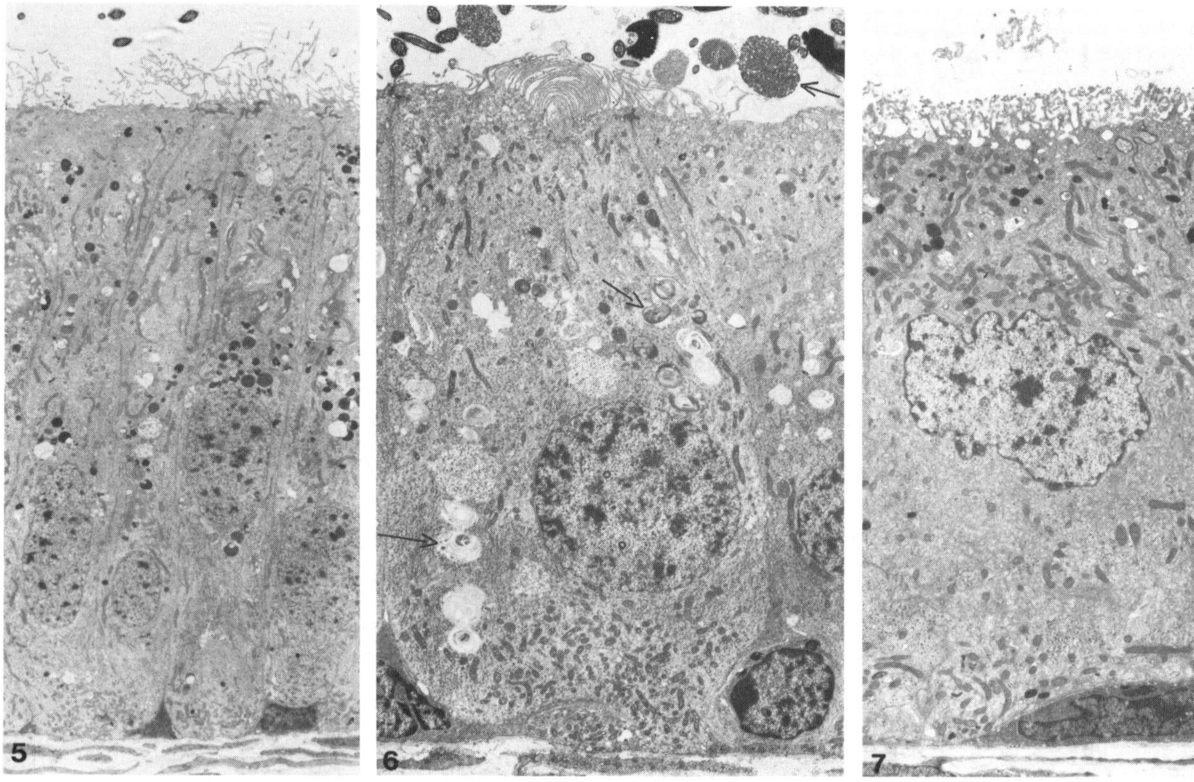

Figs 5-7. Electron micrographs of the epithelia lining the epididymis of *M. eugenii* (from Jones *et al.* 1984, with permission). Note the (dark) basal cell in the bottom right hand corner of Fig. 7 and the intraepithelial leucocyte in the bottom right hand corner of Fig. 6. Fig. 5, caput epididymidis 3 200×. Fig. 6, corpus epididymidis: arrows show the cytoplasmic droplets in the lumen and partly digested droplets (see Jones *et al.* 1984, Figs 5 and 6 for more details) in the epithelium 3 800×. Fig. 7, cauda epididymidis, 6 600×.

shorter than in the eutherians which have been studied except the common shrew, *Sorex araneus*, in which the epithelium is low (14.2±0.44 μm) and the stereocilia very long (Suzuki and Racey 1984).

Studies on *M. eugenii* of the effects of orchidectomy and subsequent replacement therapy with testosterone (Jones *et al.* 1987) showed that whilst orchidectomy caused regression of all segments of the ductus epididymidis, testosterone therapy caused extensive redevelopment only in the cauda epididymidis (Fig. 9). This dependence of the caput and corpus epididymidis on a luminal connection with the testis is similar to the regulation of the initial segments of eutherian mammals and is the basis of a classification (Fawcett and Hoffer 1979) into the proximal "initial segment proper" which has the structural characteristics described by Benoit (1926) and Nicander and Glover (1973) and "distal initial segments". On this basis the caput epididymidis of *M. eugenii* is the initial segment proper and the corpus epididymidis contains the distal initial segments. According to work on the mouse (Abe, Takano and Ito 1984) there is a dense network of fenestrated subepithelial blood capillaries in the initial segment proper which becomes unfenestrated after efferent duct ligation resulting in apoptosis of the epithelium. Abe *et al.* (1984) suggested that a

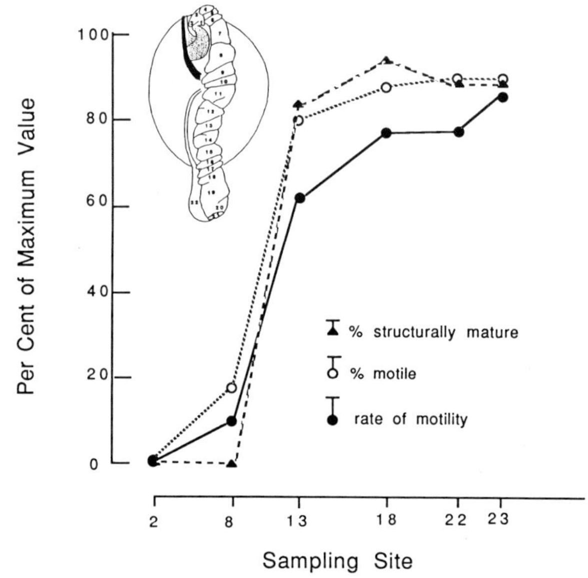

Fig. 8. Graph showing the increase in mean scores of rate of motility (relative to the maximum speed seen for sperm from the cauda epididymidis) and percentage of motile, and counts of structurally mature (as seen in the cauda epididymidis) spermatozoa in different parts of the epididymis of *M. eugenii*. Sampling sites 1 to 22 are shown in the diagram of the testis and epididymis; site 23 was the ductus deferens opposite site 14. Mean from five animals; standard errors shown with symbols. Data from Jones *et al.* (1984).

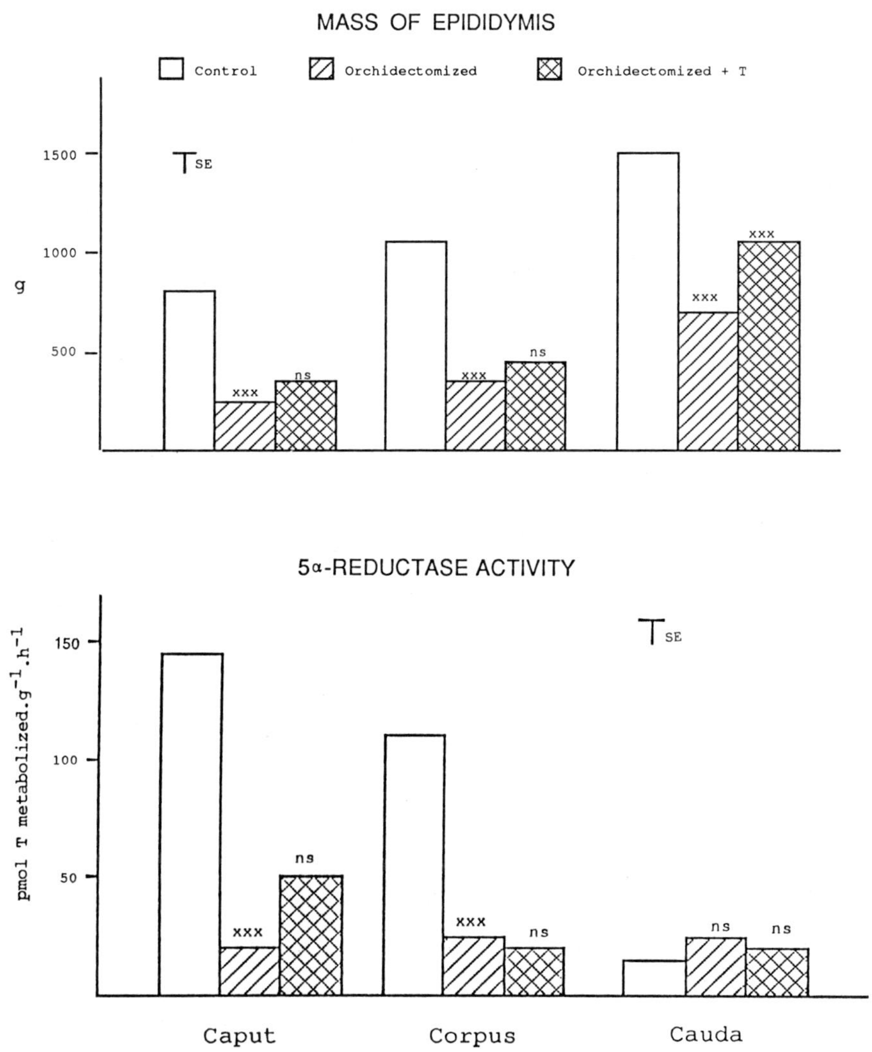

Fig. 9. Histograms showing the effects on mass and 5α-reductase activity of the caput, corpus and cauda epididymidis of *M. eugenii* of castration for six weeks and castration for four weeks followed by administration of a Silastic implant of testosterone propionate two weeks before slaughter. The implant maintained a mean (±SE) blood level of testosterone of 6±0.9 ng/ml; the corresponding level for the untreated (control) group was 1.03±0.5 ng/ml. Mean responses were from five, four and three animals for the mass, testosterone concentration and 5α-reductase activity. The standard errors (SE) shown in the diagrams were calculated from the pooled errors in the analyses of variance. Data from Jones*et al.* (1988).

substance is transported from the testes through the lumen of the extratesticular ducts to support the fenestration. However, this process has not been investigated in the Macropodidae.

A study on *M. eugenii* of the activity of the enzyme 5α-reductase, which converts testosterone to 5α-dihydrotestosterone, indicates that testosterone is probably not the active androgen in the epididymis, at least in the proximal regions (Jones *et al.* 1988). As in most of the eutherians which have been studied (Robaire *et al.* 1988) the enzyme activity is much higher (7.5 times) in the initial segments than the cauda epididymidis. The activity in the initial segments reduces to the level in the cauda epididymidis following orchidectomy, and subsequent testosterone therapy has little effect (Fig. 9) indicating that the activity of the enzyme is probably dependent on a direct luminal connection with the testis.

The structural differentiation and differential regulation of the epididymis which is described above indicates that there is a division of labour (Fig. 10) along the organ into three regions. They are involved mainly in the reabsorption of fluid

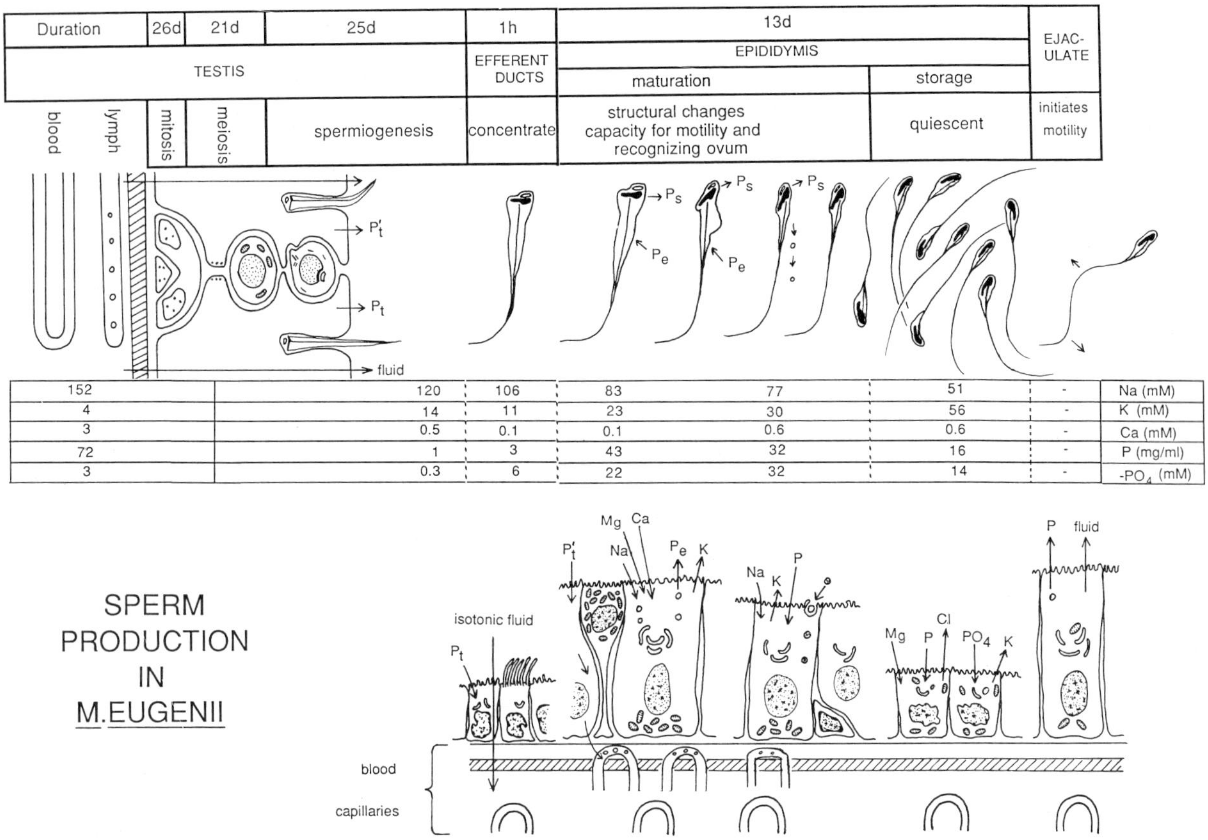

152	120	106	83	77	51	-	Na (mM)
4	14	11	23	30	56	-	K (mM)
3	0.5	0.1	0.1	0.6	0.6	-	Ca (mM)
72	1	3	43	32	16	-	P (mg/ml)
3	0.3	6	22	32	14	-	$-PO_4$ (mM)

Fig. 10. Diagrammatic representation of sperm production in *M. eugenii* showing how the luminal milieu in the sperm ducts is regulated, and the duration of each stage of sperm production. Spermatozoa are produced in the seminiferous tubules in close association with the Sertoli cells which establish an anatomical and physiological barrier between blood and the lumen of the duct and secrete a fluid and some protein into the lumen. Most of the fluid is reabsorbed in the ductuli efferentes and the concentrate is modified by considerable transport of ions, and secretion and absorption of proteins (Pe and P respectively) in the maturation (caput and corpus) and storage (cauda) regions of the epididymis. Sperm maturation includes functional and structural changes, including loss of the cytoplasmic droplet which is phagocytosed by the duct epithelium. Spermatozoa also lose some protein (Ps). Ejaculation involves dilution of spermatozoa mainly with prostatic fluids (indicated by cell in bottom right corner), and initiation of motility. The concentrations of some of the main elements and compounds are shown for blood and the luminal fluids where P refers to protein and PO_4 refers to total phosphates (Jones 1987; Jones and Clulow 1987a).

(the ductuli efferentes), the maturation of spermatozoa (caput and corpus epididymidis) and storage of spermatozoa (cauda epididymidis). Consequently, further discussion on the functions of the epididymis are described below under those three headings. They are based on work on *M. eugenii* (Jones 1987; Jones and Clulow 1987a; Jones *et al.* 1987; Chaturapanich and Jones 1987) and show that the inorganic composition of the luminal millieu along the ducts is maintained against electrochemical gradients and that the ducts' epithelium is actively synthesising, secreting and absorbing proteins.

FLUID REABSORPTION BY THE DUCTULI EFFERENTES

The micropuncture studies on *M. eugenii* (Jones 1987; Jones and Clulow 1987a) have clarified the role of the ductuli efferentes in mammals. Spermatozoa leave the testes of *M. eugenii* and other mammals as a very dilute suspension (Table 2) and the ductuli efferentes reabsorb about 87% of the fluid and about half the protein present. The fluid reabsorption involves the removal of one protein (MW of 14 700) and the isotonic transport of the monovalent cations and possibly calcium. Judging from the structure of the ductuli efferentes (Fig. 4) the protein is reabsorbed by endocytosis as in the rat (Hermo *et al.* 1985, Hermo and Morales 1984).

The fluid transporting role of the ductuli is consistent with their structural arrangement in which numerous narrow ducts are arranged in parallel providing a favourable ratio of surface area of epithelium to luminal volume for reabsorption. Interestingly it is a similar role to their homologue, the proximal convoluted tubules of the kidney, and it seems to be shared by all the other mammals which have been studied (bull and boar, Crabo 1965; African elephant, Jones 1980; echidna, Djakiew and Jones 1983; rat, Jones and Jurd 1987). The significance of the reabsorptive role of the ductuli is that they remove proteins which may inhibit further

Table 2. Changes in the concentration of spermatozoa (spermatocrit) and estimates of percentage of fluid reabsorbed in the epididymis of *M. eugenii*. Estimates (means±SE) from samples collected by micropuncture (Jones 1987).

Sampling site	Spermatocrits (%)	Apsorption+ of fluid as % leaving: testis	adjacent proximal site
Rete testis	1.5±0.2	—	—
Epididymidis			
Proximal caput (1)	11.3±2.5***	87	87
Distal caput (8)	50.8±3.4***	97	78
Distal corpus	69.9±4.5**	98	27
Middle cauda	66.9±1.6	98	0
Distal cauda	61.3±2.6	98	−9

Asterisks indicate that value is statistically different from mean for nearest proximal site ***P<0.001, **P<0.01.
+Positive values indicate absorption from duct lumen and negative values indicate secretion into the lumen.

maturation of spermatozoa. They also deliver a considerably reduced volume of fluid to the ductus epididymidis which modifies its composition appropriate for sperm maturation and storage. Considering the role of the ductuli efferentes it is suggested that the fluid transport which occurs across the mucosa of the ductus epididymidis (Table 2) is related mainly to secretion or absorption of solute, in order to modify the composition of the luminal milieu, rather than to changing the concentration of spermatozoa.

MATURATION OF SPERMATOZOA IN THE EPIDIDYMIS

The maturation of spermatozoa in the epididymis of Macropodidae has been studied in *Macropus eugenii, M. robustus, M. parma, M. rufus* and *Potorous tridactylus* (Harding *et al.* 1979; Jones *et al.* 1984). Based on the immature structure and lack of motility of spermatozoa taken from the rete testis it is assumed that, like eutherian spermatozoa, they are not capable of fertilizing an ovum when they leave the testis.

Figure 10 relates the maturation of spermatozoa to the structural differentiation of the epididymis of *M. eugenii*. Spermatozoa develop some capacity for motility, but there are no structural signs of maturation, in the caput epididymidis (Fig. 8). The structural changes and the development of motility occur soon after spermatozoa enter the corpus epididymidis. Most of the structural changes (Fig. 3) are similar to the changes in other marsupials, but are quite different and more extensive than has been described for the eutherians. They involve compaction of the acrosome from a cup-like structure to an ovate form, reorganization of the membranous material around the neck to form membranous whorls, eccentric displacement then shedding of the cytoplasmic droplet from the proximal middle piece, rotation of the head so that the neck and proximal middle piece lie within a longitudinal groove in the nucleus, and formation of a fibre network under the plasmalemma of the posterior middle piece. No invaginations of the plasmalemma of the middle piece have been described in Macropodidae as in *Trichosurus vulpecula* and some other marsupials (Harding *et al.* 1979).

The structure of the epithelium in the corpus epididymidis indicates that it is more absorptive and less secretory than the caput epididymidis. Further, there is some variation in the structure of the epithelium in the corpus epididymidis which is due to a middle region being involved in the hetrophagic digestion of the cytoplasmic droplets released from spermatozoa in the proximal corpus epididymidis (Fig. 6). Phagocytosis of the cytoplasmic droplets occurs in some other, but not all marsupials (Temple-Smith 1984a; Cummins, Temple-Smith and Renfree 1986). However, it has not been reported in eutherians since, except possibly in some rodents (Amann 1987), the cytoplasmic droplets are retained by spermatozoa until ejaculation.

There are major changes in the composition of the luminal fluids in the caput and corpus epididymidis involving reabsorption of fluid, sodium, potassium, chlorine and magnesium, and sulphur containing compounds with the result that the luminal concentrations of sodium decrease and potassium and phosphorus containing compounds increase. Two proteins which are present in fluid from the rete testis (MW of 43 200 and 44 800) are also reabsorbed (Jones 1987).

There is a net secretion of protein into the caput epididymidis and reabsorption of proteins in the corpus epididymidis of *M. eugenii*. The protein secretion may be important for sperm maturation as *in vitro* studies on the rabbit (Orgebin-Crist and Jahad 1979) have implicated epididymal proteins in the maturation process, and in a number of eutherians it has been shown that epididymal proteins become incorporated into the sperm membranes (Cameo and Blaquier 1976; Oliphant and Singhas 1979; Voglmayr *et al.* 1980). The incorporation is correlated with the development by spermatozoa of the capacity to bind to the zona pellucida (Saling 1982) and to fertilize ova (Gonzales-Echeverria *et al.* 1984). Denatured polyacrylamide gel electrophoresis of micropuncture samples of the luminal fluids from *M. eugenii* (Jones 1987) has shown that six proteins appear in the lumen of the caput epididymidis (MW of 30 000, 31 000, 32 300, 21 400, 18 700 and 17 400) and three in the corpus epididymidis (MW of 12 800, 39 800 and 90 000) which are not present in rete testis fluid or blood. One of those proteins (MW of 30 400) becomes associated with the sperm membrane (Chaturapanich and Jones 1987). Spermatozoa also lose some proteins (MW 28 000, 29 400 and 47 200), but as they do not appear to change the electrophoretic pattern of the epididymal plasma the proteins may be part of the cytoplasmic droplet.

Other organic substances, such as carnitine, inositol and glycerylphosphorylcholine, are secreted into the proximal part of the epididymis of eutherians (Brooks 1979), but their role is uncertain. Apart from the chromatographic demonstration of inositol in semen from *M. eugenii* (Rodger and White 1974b) their occurrence has not been investigated in a marsupial. The increase in phosphorus containing compounds in the caput and corpus epididymidis of *M. eugenii* (Jones and Clulow 1987) probably indicates that phosphoryl compounds are secreted into the duct (Lloyd and Jones unpublished data). Nevertheless, judging from the osmotic difference between the total osmolality of the inorganic electrolytes and the measured osmotic pressure of epididymal plasma (e.g., see Table 3) there is much less non-proteinaceous organic material secreted into the epididymis of *M. eugenii* than in eutherians (Jones 1978).

STORAGE OF SPERMATOZOA IN THE EPIDIDYMIS

The cauda epididymidis of *M. eugenii* can maintain some viable spermatozoa for 28 days between ligatures around the duct (G. Chaturapanich, J. Clulow and R. C. Jones, unpublished data). It has the structural features of the terminal segment of the eutherian epididymis which is considered to store spermatozoa available for ejaculation (Glover and Nicander 1971; Jones and Clulow 1987b). The duct is wide, densely packed with "mature" spermatozoa and lined by an epithelium which appears to be adapted for heterophagic digestion (Golgi and apical vesicles, vacuoles and dense bodies are present) and solute transport (there are dense areas of mitochondria in the apical and basal cytoplasm). Also, the extra layers of smooth muscle around the duct, which are mentioned above, are reminiscent of layers described in man (Baumgarten *et al.* 1971) which, unlike the inner layers of periductal muscle which line all the ductus, are innervated to effect ejaculation.

There is no fat pad between the testis and cauda epididymidis of *M. eugenii* as in eutherians like the rat. However, the anatomical isolation and temporal variation in thermal conductivity of the scrotal skin around the cauda which are described above probably achieve the effect observed in the scrotal eutherians of maintaining the temperature of the cauda epididymidis several degrees Celcius below testicular temperature (Bedford 1977). Unfortunately, there have been no studies on the temperature of the cauda epididymidis of a marsupial.

The inorganic cation composition of the luminal milieu in the cauda epididymidis of *M. eugenii* is similar to eutherians (Table 3 and see Jones 1978). Compared to blood plasma the fluid is characterized by low levels of calcium and sodium and high levels of potassium which are all maintained against electrochemical gradients. Maintaining these concentrations of cations is probably important for sperm storage as calcium from the extracellular fluid is a regulator of the activity of mammalian spermatozoa and the low

Table 3. The composition of blood plasma and luminal fluids from the rete testis (RTF) and cauda epididymidis (CEF) of the laboratory rat and *M. eugenii*. Mean±SE concentrations in mM/l except for protein (mg/ml) and spermatocrits (%).

	Rattus norvegicus a-e			*M. eugenii* f-g		
Substance	Blood	RTF	CEF	Blood	RTF	CEF
Na	139±4	143±4	21±3.0	152±3	120±6.7	51±4.4
K	6±0.3	14±1	55±2.3	4±0.2	14±1.0	56±3.4
Cl	122±4	140±2	24±2.8	114±2.3	130±6.8	129±1.4
Ca	0.5±0.07	0.8±0.09	0.3±0.06	2.7±0.08	0.5±0.25	0.6±0.23
Mg	0.4±0.02	0.4±0.09	2.6±0.21	1.1±0.26	0.4±0.09	3.0±0.27
Total P	2.25±0.09	1.72±0.53	79.4±3.69	2.8±0.15	0.3±0.20	13.9±1.15
GPC	—	nd	41.4±1.97	—	—	—
PC	—	nd	19.8±1.25	—	—	—
PO4	—	nd	16.0±1.94	—	—	—
Inositol	0.08±0.01	2.5±0.12	31.4±0.28	—	—	—
Carnitine	0.03±0.002	<1	53.7±1.26	—	—	—
Total S	3.7±0.67	4.3±1.30	7.0±0.67	20.6±1.80	0.6±0.19	8.2±1.50
Protein	82.2±2.4	1.1±0.1	36.8±3.2	71.6±3.7	0.9±0.4	15.6±2.2
Bicarbonate	30.1	21±5	6.7	29±3.5	10±0.6	2±0.3
pH	7.5±0.02		6.9±0.03	7.5±0.05	7.0±0.03	6.3±0.06
Osmolality (mOs/Kg):						
measured	311±3.3	328±23	329±4.6	288±3.2	296±2.6	299±3.0
difference*	67	57	235	39	32	63
Spermatocrit (%)	—	3.3±0.25	55.5±1.73	—	1.5±0.2	61.3±2.6

a, Hinton (1979); b, Levine and Marsh (1971); c, Wong, Tsang and Lee (1981); d, Jenkins *et al.* (1980); e, Tuck, Setchell, Waites and Young (1970); f, Jones (1987); g, Jones and Clulow (1987a). GPC, glycerylphosphoryl choline; PC phosphoryl choline; PO_4, inorganic phosphate. nd, not detected with assay system used. *measured osmolality — 0.91 (sum concentrations of inorganic electrolytes).

sodium:potassium ratio in the cauda epididymidis is characteristic of the intracellular environment of spermatozoa (O'Donnell 1969) and so would minimize the energy requirements of spermatozoa by reducing the need for active transport of the ions across the sperm plasmalemma.

There is a net reabsorption of protein in the cauda epididymidis. However, polyacrylamide gel electrophoresis of the luminal fluids indicates that three proteins which are not present in blood or rete testis fluid appear in the fluids (MW's of 10 900, 56 300 and 63 000).

ACCESSORY GLANDS AND SEMEN

Semen originates from the epididymis, and prostate and bulbourethral glands of the Macropodidae. The prostate is the disseminate type characteristic of marsupials, with branched tubular glands radiating from the urethra (Rodger and Hughes 1973). It is encapsulated by a tunic of smooth muscle forming a carrot-shaped structure which consists of three segments which are visible on surface examination. All segments secrete a viscous fluid containing bodies (see Fig. 4 in Rodger and White 1975) which originate from the cytoplasm of the glandular epithelium (Rodger and Hughes 1973), and which give semen and the secretion from the central and posterior prostate a milky appearance.

Rodger and Hughes (1973) identified three pairs of Cowper's glands in the seven species of Macropodidae which they studied. The glands are lobular structures which consist of branched tubular glands within a tunic of skeletal muscle and connected to the urethra by ducts surrounded by smooth muscle. They are small and do not contain much epithelium so would appear to contribute little to semen.

The Macropodidae are the only marsupials known to produce semen which coagulates to form a copulatory plug (Poole and Pilton 1964; Rodger and White 1975) like the plug produced by rodents (Walton 1960; Hann 1964). In recently mated females the coagulum extends throughout the vaginae and urogenital sinus (Tyndale-Biscoe and Rodger 1978; J. C. Rodger, pers. comm.) Rodger (1978) correlated the development of the plug in Macropodidae with the parallel evolution of a median vagina, indicating a difference in sperm transport through this canal compared to the lateral vaginae in other marsupials.

N-acetylglucosamine is the major sugar in semen (Rodger and White 1974b) of Macropodidae (336, 339 and 508 mg % for *Macropus giganteus, M. rufus* and *M. eugenii* respectively), Phalangeridae and Peramelidae (Rodger and White 1975). This is unlike semen from Dasyuridae and Didelphidae in which glycogen is the main sugar and semen from eutherians in which fructose is the main sugar (Rodger 1976; Mann 1964). The N-acetylglucosamine is secreted by all segments of the prostate of *M. giganteus, M. rufus* and *M. eugenii* but is most concentrated in the central segment (Rodger and White 1976). There is little glucose in semen from *M. eugenii* (20 mg %) as in semen from most other mammals, but in semen from *M. giganteus* and *M. rufus* its concentration is much higher (82 and 111 mg % respectively) than in eutherians. Glucose and very small amounts of fructose are present in all segments of the prostate of *M. giganteus, M. rufus* and *M. eugenii*, but glucose is most concentrated in the posterior segments in *M. giganteus* and *M. rufus* (Rodger and White 1974a, 1976).

Unfortunately there have been no studies on the metabolism of spermatozoa from Macropodidae. The only report on the metabolism of marsupial spermatozoa is a study on *Trichosurus vulpecula* (Rodger and Suter 1978) which showed that they utilized glucose and N-acetylglucosamine (3μmol and 2μmol per 10^8 spermatozoa per 3 hours respectively). The oxygen consumption was very high (134μl per 10^8 spermatozoa per 3 hour) being twice that of ram spermatozoa and four to six times the rate of rabbit and rat spermatozoa. However, some of the measured metabolism may be due to the prostatic bodies which would not separate from spermatozoa by the washing prior to incubation in the Warburg flasks.

The accessory glands of *M. eugenii* are dependent on systemic androgens (Jones *et al.* 1988). However, the prostate gland, and presumably Cowper's glands, secrete thoughout the year (Jones unpublished data) even when systemic androgen levels are low (Fig. 2, Inns 1982). Consequently, as proposed by Cook *et al.* (1978) the prostate may be supported by androgen transported along the excurrent ducts of the testis into the urethra. The prostate of *M. eugenii*, like the prostate of *Trichosurus vulpecula* (Cook *et al.* 1978; Curlewis and Stone 1985) contains little activity of the enzyme 5α-reductase to convert testosterone to 5α-dihydrotestosterone (G. M. Stone and R. C. Jones, unpublished data). However, as there is a high concentration of 5α-dihydrotestosterone in the prostate of *T. vulpecula*, Curlewis and Stone (1985) consider that it must be formed elsewhere and transported to the prostate where it is the active androgen.

STRATEGY OF SPERM PRODUCTION IN MACROPODIDAE

Table 4 compares estimates of reproductive parameters associated with sperm production for *M. eugenii* and some eutherian mammals representing a range in body size. Values for a bird,

Table 4. Reproductive parameters for *M. eugenii* compared to some eutherian mammals and a non-passerine bird, *Coturnix coturnix japonica.*

	M. eugenii (c,k,i,n,o)	*rat* (d,e,m,q,r)	*rabbit* (h,l,r,s)	*ram* (a,f,j,n)	*bull* (a,b,g,p)	*C. coturnix* (d)*
Mass (g):						
combined testes, t	31.4	3.7	5.6	318.8	359.2	3.34
body, m	5 800	557	3 900	35 000	445 000	149
ratio t:m (%)	0.54	0.66	0.14	0.91	0.08	2.26
Duration of spermatogenesis (d)	72	54-58	51	47	61	14-16
Daily sperm production (10^6/g testis/d)	24.3	23.7	26.5	21	12-17	92.5
Sperm in epididymis:						
region of maturation	I	I,M	I,M	I,M	I,M	PDE
duration of transit (d)	13	8.4	9-10	16.4	8	1
survival in ligated duct (d)	>28	42	28	—	—	5-9
no. (10^6/Kg animal)	1.71	1.32	0.49	1.54	0.11	2.06

*References a, Amann *et al.* (1976); b, Amann *et al.* (1974); c, G. Chaturapanich, J. Clulow and R. C. Jones (unpublished data); d. Clulow and Jones (1982); e, Courot *et al.* (1970); f, Dott and Skinner (1967); g, Hochereau *et al.* (1964); h, Jones *et al.* (1984); i, Jones *et al.* (1988); j, Lincoln and Short (1980); k, Orgebin-Crist (1965); l, Orgebin-Crist (1968); m, Robb *et al.* (1978); n, Setchell (1970b); o, Setchell and Carrick (1973); p, Swierstra (1966); q, White (1932); r, Glover and Nicander (1971); s, Paufler and Foote (1968). The number of spermatozoa in the epididymides of *M. eugenii* and the bull were calculated from the daily sperm production and duration of transit through the epdidymis.

the Japanese quail, *Coturnix coturnix*, are also shown to illustrate the relative magnitude of the variation among mammals compared to that between mammals and birds. Sperm production in the latter is much faster than in mammals, but spermatozoa cannot survive as long in the sperm ducts of *C. coturnix* as in the mammals which have been studied.

Apart from the parameters shown in Table 4 for *M. eugenii* the only other data relating to sperm production in Macropodidae are estimates of gonadal indices (ratio of testes:body mass expressed as a percentage) for five species quoted by Tyndale-Biscoe and Renfree (1987), estimates of the number of spermatozoa per gram of testis and per epididymis published by Bedford, Rodger and Breed (1984) and the estimates of the duration of spermatogenesis for *Macropus rufogriseus* stated above. The data on gonadal indices indicates a difference between the three wallabies (0.23% to 0.53%) and two kangaroos (0.10% and 0.15%) which were examined, but all values lie within the range of values which have been reported for eutherians (Table 4). The estimates of number of spermatozoa in the epididymis which were tabulated by Bedford *et al.* (1984) are an order of magnitude lower than the estimates for *M. eugenii* shown in Table 4. The low values given by Bedford *et al.* (1984) may be partly due to technical problems associated with determining the number of spermatozoa in the epididymis. With *M. eugenii* I found that spermatozoa bind to intact pieces of tissue and so are not counted when the traditional homogenization methods are used for preparing spermatozoa for counting with a haemocytometer.

It is concluded from Table 4 that there is nothing unusual relative to other mammals about the parameters relating to sperm production in *M. eugenii*. The constant factor associated with sperm production in mammmals is the daily sperm production per gram of testis and for this *M. eugenii* falls within the general range for eutherians which varies from 12 millions of spermatozoa/g/d for the dairy bull (Amann *et al.* 1976) to 26.5 millions/g/d for the rabbit (Amann 1981) (except for man who produces 4.4 millions/g/d, but it is not possible to know how much of this is due to diet and clothing). Consequently, as the mass of the testes in mammals is directly proportional to body mass (Kenagy and Trombulak 1986), a rough comparison of sperm production in different species may be made by comparing the mass of the testes weighted according to body mass as shown in Table 4. Sperm transit through the epididymis of *M. eugenii* is at the long end of the range for mammals but this is probably only a reflection of the complexity of sperm maturation in marsupials compared to the eutherians.

CONCLUSIONS

As the main biological role of male Macropodidae is to provide the male pronucleus to seed the next generation, the emphasis in this review has been on the physiology of sperm production. The behavioural aspects determining which males contribute to the next generation are discussed elsewhere in the monograph. It is concluded that the limited data that has been reported on the Macropodidae indicates that compared to our knowledge of other mammals there is nothing unusual about the quantitative aspects of their sperm production. However, our knowledge of the subject, and other aspects of reproduction in the male, is virtually limited to work on *M. eugenii*, so that there is a need for work on other species as well as more work on *M. eugenii*.

It is noteworthy that the considerable contribution by work on *M. eugenii* to our knowledge of the Macropodidae and other marsupials is due to the establishment of breeding colonies which have provided an ensured supply of animals which are familiar to handling. As well as being useful for studying reproduction in male marsupials *M. eugenii* has proved a good animal model for studying the process of sexual differentiation (Wai-Sum *et al.* 1988) and the function of the epididymis (Jones *et al.* 1987) in mammals. The work on sexual differentiation questions the generally accepted role of the gonads in phenotypic differentiation. The work on the epididymis helps clarify the role of the initial segments of the epididymis in sperm maturation and questions whether the proposed classification of the mammalian ductus epididymidis into initial, middle and terminal segments (Glover and Nicander 1971) is appropriate for marsupials. Our earlier work indicated that the proposal is not appropriate for the monotremes in which the ductus is only differentiated into an initial and terminal segment (Djakiew and Jones 1981, 1982, 1983).

M. eugenii is also a useful animal for studying the transport of androgens from the testis to the accessory organs of reproduction. Like other marsupials, the prostate gland is the diffuse type. Consequently, as the animals are spermatorrhoeal (Bolliger 1942) androgens from the epididymis would often be present in the urethral lumen where they would be much closer to the secretory epithelium than in the lobulate prostate of eutherians. A luminal route of androgen transport would also explain why high levels of 5α-dihydrotesterone have been measured in the prostate of *Trichosurus vulpecula* (Curlewis and Stone 1985). Consequently, further work on the subject is required. Work is also required on *M. eugenii* to determine whether there is local transport of androgens from the testis to the epididymis as in the rat, ram and boar (Setchell 1986), and how the epididymis, which seems to be less sensitive to androgens than the prostate (Jones *et al.* 1988), remains fully functional when systemic androgen levels are low and the prostate regresses. Also, there is a need to determine the role of the female and intermale relationships on androgen production by the testes during the year (Fig. 2).

The high metabolic rate of semen from *Trichosurus vulpecula* (Rodger and Suter 1978) compared to the eutherian mammals is interesting and needs to be examined further in an animal like *M. eugenii* for which the techniques are established to collect spermatozoa directly from the epididymis and so avoid confounding the effects on respiration, of spermatozoa and secretions from the prostate gland. More work on the prostate is also needed to determine the source and mode of production of the cytoplasmic bodies in semen and the mechanism of semen coagulation. Further, it is necessary in order to assess which males may contribute to the next generation, to determine whether the role of the copulatory plug is to help in sperm transport in the female or to prevent spermatozoa from subsequent matings to achieve fertilization.

ACKNOWLEDGEMENTS

This work was supported by a grant from the Australian Research Grants Scheme. I am very grateful to Mrs L. Conroy for preparing the drawings and Mr R. Taylor for the graphs, to Mr and Mrs R. M. Jacombs for typing the manuscript and Drs H. R. Harding and J. C. Rodger for comments on the manuscript.

REFERENCES

ABE, K., TAKANO, H. AND ITO, T., 1984. Microvasculature of the mouse epididymis with special reference to fenestrated capillaries localized in the initial segment. *Anat. Rec.* **209**: 209-18.

ALCORN, G. T., 1975. Development of the ovary and urinogenital ducts in the tammar wallaby *Macropus eugenii* (Desmarest, 1817). Ph.D. Thesis, Macquarie University.

ALCORN, G. T. AND ROBINSON, E. S., 1983. Germ cell development in female pouch young of the tammar wallaby *(Macropus eugenii)*. *J. Reprod. Fertil.* **67**: 319-25.

AMANN, R. P., 1981. A critical review of methods for evaluation of spermatogenesis from seminal characteristics. *J. Androl.* **2**: 37-58.

AMANN, R. P., 1987. Functions of the epididymis in bulls and rams. *J. Reprod. Fertil. Suppl.* **34**: 115-31.

AMANN, R. P., JOHNSON, L., THOMPSON, D. L. AND PICKETT, B. W., 1976. Daily spermatozoal production, epididymal spermatozoal reserves and transit time of spermatozoa through the epididymis of the Rhesus monkey. *Biol. Reprod.* **15**: 586-92.

AMANN, R. P., KAVANAUGH, J. F., GRIEL, L. C. AND VOGLMAYR, J. K., 1974. Sperm production of Holstein bulls determined from testicular spermatid reserves, after cannulation of rete testis or vas deferens, and by daily ejaculation. *J. Dairy Sci.* **57**: 93-99.

BAUMGARTEN, H. G., HOLSTEIN, A. F. AND ROSENGREN, E., 1971. Arrangement, ultrastructure, and adrenergic innervation of smooth musculature of the ductuli efferentes, ductus epididymidis and ductus deferens of man. *Cell. Tiss. Res.* **120**: 37-79.

BEDFORD, J. M., 1977. Evolution of the scrotum: the epididymis as the prime mover? Pp. 171-82 *in* Reproduction and Evolution ed. by J. H. Calaby and C. H. Tyndale-Biscoe. Australian Academy of Science: Canberra.

BEDFORD, J. M., RODGER, J. C. AND BREED, W. G., 1984. Why so many mammalian spermatozoa — a clue from marsupials? *Proc. R. Soc. Lond.* **B221**: 221-33.

BENOIT, J., 1926. Recherches anatomiques, cytologiques et histophysiologiques sur les voies excretices du testicule, chez les mammiferes. *Arch. Anat. Histol. Embryol.* **5**: 173-412.

BIGGERS, J. D., 1966. Reproduction in male marsupials. *Symp. Zool. Soc. London* **15**: 251-80.

BINDER, S., 1927. Spermatogenese von *Macropus giganteus*. *Z. Zellforsch.* **5**: 293-346.

BOLLIGER, A., 1942. Spermatorrhea in marsupials, with special reference to the action of sex hormones on spermatogenesis of *Trichosurus vulpecula*. *J. Proc. R. Soc. NSW.* **76:** 86-92.

BOLTON, B. L., NEWSOME, A. E. AND MERCHANT, J. C., 1982. Reproduction in the agile wallaby, *Macropus agilis* (Gould) in the tropical lowlands of the Northern Territory: opportunism in a seasonal environment. *Aust. J. Ecol.* **1:** 261-77.

BROOKS, D. E., 1979. Biochemical environment of sperm maturation. Pp. 23-34 *in* The Spermatozoon. Maturation, Motility, Surface Properties and Comparative Aspects ed. by D. W. Fawcett and J. M. Bedford. Urban and Schwarzenberg Inc.: Baltimore-Munich.

BYSKOV, A. G., 1978. The anatomy and ultrastructure of the rete system in the fetal mouse ovary. *Biol. Reprod.* **19:** 720-35.

BYSKOV, A. G. AND LINTERN-MOORE, S. M., 1973. Follicle formation in the immature mouse ovary. The role of rete ovarii. *J. Anat.* **116:** 207-17.

CAMEO, M. S. AND BLAQUIER, J. A., 1976. Androgen-controlled specific proteins in rat epididymis. *J. Endocr.* **69:** 47-55.

CARRICK, F. N. AND COX, R. I., 1979. Testicular endocrinology of marsupials and monotremes. Pp. 137-42 *in* Reproduction and Evolution ed. by J. H. Calaby and C. H. Tyndale-Biscoe. Australian Academy of Science: Canberra.

CATLING, P. C. AND SUTHERLAND, R. L., 1980. Effect of gonadectomy, season and the presence of females on plasma testosterone, luteinizing hormone and follicle stimulating hormone levels in male tammar wallabies *(Macropus eugenii)*. *J. Endocrin.* **86:** 25-33.

CATLING, P. C. AND VINSON, G. P., 1976. Adrenocortical hormones in the neonate and pouch young of the tammar wallaby, *Macropus eugenii*. *J. Endocrin.* **69:** 447-48.

CHATURAPANICH, G. AND JONES, R. C., 1987. Protein secretion by the epididymis of the tammar, *Macropus eugenii,* and changes in the protein content of sperm. Proc. Aust. Soc. Reprod. Biol. 19th Ann. Conf., Sydney, August. p. 75.

CHRISTENSEN, A. K. AND FAWCETT, D. W., 1961. The normal fine structure of opossum interstitial cells. *J. Biophys. Biochem. Cytol.* **9:** 653-70.

CLELAND, K. W., 1969. Comparative cytology of macropod marsupial spermatozoa. *J. Reprod. Fertil.* **19:** 392. See Fig. 3 by Rodger (1978).

CLULOW, J. AND JONES, R. C., 1982. Production, transport, maturation, storage and survival of spermatozoa in the male Japanese quail, *Coturnix coturnix*. *J. Reprod. Fertil.* **64:** 259-66.

COOK, B., MCDONALD, I. R. AND GIBSON, W. R., 1978. Prostatic function in the brush-tailed possum, *Trichosurus vulpecula*. *J. Reprod. Fertil.* **53:** 369-75.

COUROT, M., HOCHEREAU-DE REVIERS, M. T. AND ORTAVANT, R., 1970. Spermatogenesis. Pp. 339-452 *in* The Testis ed. by A. D. Johnson, W. R. Gomes and N. L. Van Demark. Vol. 1. Academic Press, Inc.: New York.

CRABO, B. O., 1965. Studies on the composition of epididymal content in bulls and boars. *Acta. vet. scand.* **6:** Suppl. **5:** 1-94.

CUMMINS, J. M., TEMPLE-SMITH, P. D. AND RENFREE, M. B., 1986. Reproduction in the male honey possum (*Tarsipes rostratus:* Marsupialia) : the epididymis. *Amer. J. Anat.* **177:** 385-401.

CURLEWIS, J. AND STONE, G. M., 1985. Some effects of breeding season and castration on the prostate and epididymis of the brush-tail possum, *Trichosurus vulpecula*. *Aust. J. Biol. Sci.* **38:** 313-26.

DJAKIEW, D., 1978. Studies of the male reproductive tract of the echidna *(Tachyglossus aculeatus)* with particular emphasis on the epididymis. B.Sc. Hons. Thesis, University of Newcastle.

DJAKIEW, D. AND JONES, R. C., 1981. Structural differentiation of the male genital ducts of the echidna *(Tachyglossus aculeatus)*. *J. Anat.* **132:** 187-202.

DJAKIEW, D. AND JONES, R. C., 1982. Ultrastructure of the ductus epididymidis of the echidna, *Tachyglossus aculeatus*. *J. Anat.* **135:** 625-34.

DJAKIEW, D. AND JONES, R. C., 1983. Sperm maturation, fluid transport and secretion and absorption of protein on the epididymis of the echinda, *Tachyglossus aculeatus*. *J. Reprod. Fert.* **68:** 445-56.

DOTT, H. M. AND SKINNER, J. D., 1967. A reassessment of extragonadal sperm reserves in Suffolk rams. *J. agric. sci.* **69:** 293-95.

ECKSTEIN, P. AND ZUCKERMAN, S., 1956. Morphology of the reproductive tract. Pp. 43-155 *in* Marshall's Physiology of Reproduction ed. by A. S. Parkes. Vol. 1, Part 1. Longmans Green: London.

FAWCETT, D. W., 1975. Ultrastructure and function of the Sertoli cells. Pp. 21-55 *in* Handbook of Physiology Section 7, Endocrinology ed. by R. O. Greep and E. B. Astwood. Am. Physiol. Soc.: Washington D.C.

FAWCETT, D. W., ANDERSON, W. A. AND PHILLIPS, D. M., 1971. Morphogenetic factors influencing the shape of the sperm head. *Devel. Biol.* **26:** 220-51.

FAWCETT, D. W. AND HOFFER, A., 1979. Failure of exogenous androgen to prevent regression of the inital segments of the rat epididymis after efferent duct ligation or orchidectomy. *Biol. Reprod.* **20:** 162-81.

FAWCETT, D. W., NEAVES, W. B. AND FLORES, M. N., 1973. Comparative observations on intertubular lymphatics and the organization of the interstitial tissue of the mammalian testis. *Biol. Reprod.* **9:** 500-32.

FEIG, L. A., BELLVE, A. R., ERICKSON, N. H. AND KLAGSBRUN, M., 1980. Sertoli cells contain a mitogenic polypeptide. *Proc. Natl. Acad. Sci. USA* **77:** 4774-78.

GLOVER, T. D. AND NICANDER, L., 1971. Some aspects of structure and function in the mammalian epididymis. *J. Reprod. Fertil. (Suppl.)* **13:** 39-50.

GONZALES-ECHEVERRIA, F., CUASNICU, P. S., PIAZZA, A., PINEIRO, L. AND BLAQUIER, J. A., 1984. Addition of an androgen-free epididymal protein extract increases the ability of immature hamster spermatozoa to fertilize *in vivo* and *in vitro*. *J. Reprod. Fertil.* **71:** 433-37.

HARDING, H. R., 1977. Reproduction in male marsupials: a critique with additional observations on sperm development and structure. Ph.D. Thesis presented to the University of New South Wales, Australia.

HARDING, H. R., CARRICK, F. N. AND SHOREY, C. D., 1975. Ultrastructural changes in spermatozoa of the brush-tailed possum, *Trichosurus vulpecula* (Marsupialia), during epididymal transit. Part 1: The flagellum. *Cell and Tissue Research* **164:** 121-32.

HARDING, H. R., CARRICK, F. N. AND SHOREY, C. D., 1976a. Ultrastructural changes in spermatozoa of the brush-tailed possum, *Trichosurus vulpecula* (Marsupialia), during epididymal transit. Part II: The acrosome. *Cell and Tissue Research* **171:** 61-73.

HARDING, H. R. CARRICK, F. N. AND SHOREY, C. D., 1976b. Spermiogenesis in the brush-tailed possum, *Trichosurus vulpecula* (Marsupialia): development of the acrosome. *Cell Tissue Res.* **171:** 75-90.

HARDING, H. R., CARRICK, F. N. AND SHOREY, C. D., 1979. Special features of sperm structure and function in marsupials. Pp. 289-303 *in* The Spermatozoon. Maturation, Motility, Surface Properties and Comparative Aspects ed. by D. W. Fawcett and J. M. Bedford. Urban and Schwarzenberg, Inc.: Baltimore-Munich.

HARDING, H. R., CARRICK, F. N. AND SHOREY, C. D., 1984. Acrosome development during spermiogenesis and epididymal sperm maturation in Australian marsupials. Pp. 411-14 *in* The Sperm Cell. Fertilizing Power, Surface Properties, Motility, Nucleus and Acrosome, Evolutionary Aspects ed. by J. Andre. Martinus Hijhoff Pubs.: The Hague, Boston, London.

HARRISON, R. G., 1949. The comparative anatomy of the blood-supply of the mammalian testis. *Zool. Soc. Lond. (Proc)*. **4**: 325-44.

HEARN, J. P., 1975. The role of the pituitary in the reproduction of the male tammar wallaby, *Macropus eugenii*. *J. Reprod. Fertil.* **42**: 399-402.

HEDDLE, R. W. L. AND GUILER, E. R., 1970. The form and function of the testicular rete mirabile of marsupials. *Comp. Biochem. Physiol.* **35**: 415-25.

HERMO, L., CLERMONT, Y. AND MORALES, C., 1985. Fluid-phase and adsorptive endocytosis in ciliated epithelial cells of the rat ductuli efferentes. *Anat. Rec.* **211**: 285-94.

HERMO, L. AND MORALES, C., 1984. Endocytosis in non-ciliated epithelial cells of the ductuli efferentes of the rat. *Amer. J. Anat.* **171**: 59-74.

HINTON, B. T., 1979. The composition and movement of luminal fluids in the testis and epididymis. Ph.D. Thesis presented to the Coun. Nat. Acad. Aw., Cambridge.

HOCHEREAU, M. T., COUROT, M. AND ORTVANT, R., 1964. Duree de la spermatogenese chez la taureau etude par autoradiographic testiculaire. *5th Intern. Congr. Anim. Reprod. A.I.* (Trento) **3**: 541-46.

INNS, R. W., 1982. Seasonal changes in the accessory reproductive system and plasma testosterone levels in the male tammar wallaby, *Macropus eugenii*, in the wild. *J. Reprod. Fertil.* **66**: 675-80.

JACKS, F. AND SETCHELL, B. P., 1973. A technique for studying the transfer of substances from venous to arterial blood in the spermatic cord of wallabies and rams. *J. Physiol. Lond.* **233**: 17-18.

JENKINS, A. D., LECHENE, C. P. AND HOWARDS, S. S., 1980. Concentrations of seven elements in the intraluminal fluids of the rat seminiferous tubules, rete testis and epididymis. *Biol. Reprod.* **23**: 981-87.

JOHNSON, P. M., 1979. Reproduction in the plain rock-wallaby *Petrogale penicillata inornata* Gould, in captivity, with age estimation of the pouch young. *Aust. Wildl. Res.* **6**: 1-4.

JONES, R., 1978. Comparative biochemistry of mammalian epididymal plamsa. *Comp. Biochem. Physiol.* **61B**: 365-70.

JONES, R. C., 1980. Luminal composition and maturation of spermatozoa in the genital duct of the African elephant *Loxodonta africana*. *J. Reprod. Fertil.* **60**: 87-93.

JONES, R. C., 1987. Changes in protein composition of the luminal fluids along the epididymis of the tammar, *Macropus eugenii*. *J. Reprod. Fertil.* **80**: 193-99.

JONES, R. C. AND CLULOW, J., 1987a. Regulation of the elemental composition of the epididymal fluids in the tammar, *Macropus eugenii*. *J. Reprod. Fertil.* **81**: 583-90.

JONES, R. C. AND CLULOW, J., 1987b. Fluid absorption and sperm maturation and storage in the epididymis. Pp. 229-40 *in* Proceedings of the First Congress of the Asian and Oceanian Physiological Societies ed. by C. Pholpramool and R. Sudsuang. The Physiological Society Thailand.

JONES, R. C., CLULOW, J., STONE, G. M. AND SETCHELL, B. P., 1987. The role of the initial segments of the epididymis in sperm maturation in mammals. Pp. 63-71 *in* New Horizons in Sperm Cell Research ed. by H. Mohri. Jap. Sci. Soc. Press: Tokyo. Gordon and Breach Sci. Publ.: New York.

JONES, R. C. HINDS, L. A. AND TYNDALE-BISCOE, C. H., 1984. Ultrastructure of the epididymis of the tammar, *Macropus eugenii*, and its relationship to sperm maturation. *Cell Tiss. Res.* **237**: 525-35.

JONES, R. C. AND JURD, K. M., 1987. Structural differentiation and fluid reabsorption in the ductuli efferentes testis of the rat. *Aust. J. Biol. Sci.* **40**: 79-90.

JONES, R. C., STONE, G. M., HINDS, L. A. AND SETCHELL, B. P., 1988. Distribution of 5α-reductase in the epididymis of the tammar wallaby *(Macropus eugenii)* and dependence of the epididymis on systemic testosterone and luminal fluids from the testis. *J. Reprod. Fertil.* **83**: in press.

KENAGY, G. J. AND TROMBULAK, S. C., 1986. Size and function of mammalian testes in relation to body size. *J. Mammal.* **67**: 1-22.

KIM, J.-W., 1987. Electron-microscopic studies on the spermiogenesis and spermatozoa of the Allied Rock Wallaby *(Petrogale assimilis) Korean. J. Elect. Micros.* **17**: 1-15.

KIRSCH, J. A. W., 1977. The comparative serology of Marsupialia, and a classification of marsupials. *Aust. J. Zool. Suppl.* **52**: 1-152.

KOSKIMIES, A. I. AND KORMANO, M., 1975. Proteins in fluids from different segments of the rat epididymis. *J. Reprod. Fertil.* **43**: 345-48.

LEVINE, N. AND MARSH, D. J., 1971. Micropuncture studies of the electrochemical aspects of fluid and electrolyte transport in individual seminiferous tubules, the epididymis and the vas deferens in rats. *J. Physiol.* **213**: 557-70.

LINCOLN, G. A., 1978. Plasma testosterone profiles in male macropodid marsupials. *J. Endocrin.* **77**: 347-51.

LINCOLN, G. A. AND SHORT, R. V., 1980. Seasonal breeding: nature's contraceptive. *Rec. Progs. Hormone Res.* **36**: 1-52.

MANN, T., 1964. The biochemistry of semen and of the male reproductive tract. Methuen: London.

MASON, K. AND BLACKSHAW, A. W., 1973. The spermatogenic cycle of the bandicoot, *Isoodon macrourus*. *J. Reprod. Fertil.* **32**: 307-08.

MAYNES, G. M., 1973. Reproduction in the parma wallaby, *Macropus parma* Waterhouse. *Aust. J. Zool.* **21**: 331-51.

MEISTRICH, M. L., HUGHES, T. J. AND BRUCE, W. R., 1975. Alteration of epididymal sperm transport and maturation in mice by oestrogen and testosterone. *Nature (Lond.)* **258**: 145-47.

MERCHANT, J. C., 1976. Breeding biology of the agile wallaby, *Macropus agilis* (Gould) (Marsupialia: Macropodidae), in captivity. *Aust. Wildl. Res.* **3**: 93-103.

MERCHANT, J. C. AND CALABY, J. H., 1981. Reproductive biology of the red-necked wallaby *(Macropus rufogriseus banksianus)* and Bennett's wallaby *(Macropus rufogriseus rufogriseus)* in captivity. *J. Zool. (Lond.)* **194**: 203-17.

NEWSOME, A. E., 1973. Cellular degeneration in the testis of red kangaroos during hot weather and drought in Central Australia. *J. Reprod. Fertil. Suppl.* **19**: 191-201.

NICANDER, L. AND GLOVER, T. D., 1973.. Regional history and fine structure of the epididymal duct in the golden hamster *(Mesocricetus auratus)*. *J. Anat.* **114**: 347-64.

O'DONNELL, J. M., 1969. intracellular levels of sodium and potassium in bull spermatozoa in relation to cell metabolism. *J. Reprod. Fertil.* **19:** 207-09.

OLIPHANT, G. AND SINGHAS, C. A., 1979. Iodination of rabbit sperm plasma membrane: relationships of specific surface proteins to epididymal function and sperm capacitation. *Biol. Reprod.* **21:** 937-44.

ORGEBIN-CRIST, M.-C., 1965. passage of spermatozoa labelled with thymidine-^{3}H through the ductus epididymidis of the rabbit. *J. Reprod. Fertil.* **10:** 241-51.

ORGEBIN-CRIST, M.-C., 1968. Gonadal and epididymal sperm reserves in the rabbit: estimation of the daily sperm production. *J. Reprod. Fertil.* **15:** 15-25.

ORGEBIN-CRIST, M.-C. AND JAHAD, N., 1979. Maturation of rabbit epididymal spermatozoa in organ culture: stimulation by epididymal cytoplasmic factors. *Biol. Reprod.* **21:** 511-15.

PAUFLER, S. K. AND FOOTE, R. H., 1968. Morphology, motility and fertility of spermatozoa recovered from different areas of ligated rabbit epididymis. *J. Reprod. Fertil.* **17:** 125-37.

POOLE, W. E., 1973. A study of breeding in grey kangaroos, *Macropus giganteus* (Shaw) and *Macropus fuliginosus* (Desmarest), in Central New South Wales. *Aust. J. Zool.* **21:** 183-212.

POOLE, W. E., 1976. Breeding biology and current status of the grey kangaroo, *Macropus fuliginosus fuliginosus,* of Kangaroo Island, South Australia. *Aust. J. Zool.* **24:** 169-87.

POOLE, W. E. AND PILTON, P. E., 1964. Reproduction in the grey kangaroos, *Macropus canguru,* in captivity. *C.S.I.R.O. Wildl. Res.* **9:** 218-34.

ROBAIRE, B., SCHEER, H. AND HACHEY, C., 1981. Regulation of epididymal steriod metabolizing enzymes. Pp. 487-98 *in* Bioregulators of Reproduction ed. by G. Jagiello and H. J. Vogel. Academic Press: New York.

ROBB, G. W., AMANN, R. P. AND KILLIAN, G. J., 1978. Daily sperm production and epididymal sperm reserves of pubertal and adult rats. *J. Reprod. Fertil.* **54:** 103-07.

RODGER, J. C., 1976. Comparative aspects of the accessory sex glands and seminal biochemistry of mammals. *Comp. Biochem. Physiol.* **55B:** 1-8.

RODGER, J. C., 1978. Male reproduction — its usefulness in discussions of Macropodidae evolution. *Aust. Mammal.* **2:** 73-80.

RODGER, J. C. AND HUGHES, R. L., 1973. Studies of the accessory glands of male marsupials. *Aust. J. Zool.* **21:** 203-20.

RODGER, J. C. AND WHITE, I. G., 1974a. Carbohydrates of the prostate of two Australian marsupials, *Trichosurus vulpecular* and *Megaleia rufa. J. Reprod. Fertil.* **38:** 267-73.

RODGER, J. C. AND WHITE, I. G., 1974b. Free N-acetylglucosamine in marsupial semen. *J. Reprod. Fertil.* **39:** 383-86.

RODGER, J. C. AND WHITE, I. G., 1975. Electroejaculation of Australian marsupials and analyses of the sugars of the seminal plasma from three macropod species. *J. Reprod. Fertil.* **43:** 233-39.

RODGER, J. C. AND WHITE, I. G., 1976. Source of seminal N-acetylglucosamine in Australian Marsupials and further studies of free sugars of the marsupial prostate gland. *J. Reprod. Fertil.* **46:** 467-69.

RODGER, J. C. AND SUTER, D. A. I., 1978. Respiration rates and sugar utilization by marsupial spermatozoa. *Gamete Res.* **1:** 111-16.

ROOSEN-RUNGEN, E. C. AND GIESEL, L. O., 1950. Quantitative studies on spermatogenesis in the albino rat. *Amer. J. Anat.* **17:** 1-30.

SADLEIR, R. M. F. S., 1965. Reproduction in two species of kangaroo (*Macropus robustus* and *Megaleia rufa*) in the arid Pilbara region of Western Australia. *Proc. Zool. Soc. Lond.* **145:** 239-61.

SALING, P. M., 1982. Development of the ability to bind to zonae pellucidae during epididymal maturation: reversible immobilization of mouse spermatozoa by lanthanum. *Biol. Reprod.* **26:** 429-36.

SAPSFORD, C. S. AND RAE, C. A., 1969. Ultrastructural studies on Sertoli cells and spermatids in the bandicoot and ram during the movement of mature spermatids into the lumen of the seminiferous tubule. *Aust. J. Zool.* **17:** 415-45.

SAPSFORD, C. S., RAE, C. A. AND CLELAND, K. W., 1969. Ultrastructural studies on maturing spermatids and on Sertoli cells in the bandicoot *Perameles nasuta* Geoffroy (Marsupialia). *Aust. J. Zool.* **17:** 195-292.

SETCHELL, B. P., 1970a. Testicular blood supply, lymphatic drainage and secretion of fluid. Vol. 1. Pp. 101-239 *in* The Testis ed. by A. D. Johnson, W. R. Gomes and N. L. Van Demark. Academic Press Inc.: New York.

SETCHELL, B. P., 1970b. Fluid secretion by the testes of an Australian marsupial, *Macropus eugenii. Comp. Biochem. Physiol.* **36:** 411-14.

SETCHELL, B. P., 1974. Secretions of the testis and epididymis. *J. Reprod. Fertil.* **37:** 165-77.

SETCHELL, B. P., 1977. Reproduction in male marsupials. Pp. 411-57 *in* The Biology of Marsupials ed. by B. Stonehouse and D. Gilmore. Macmillan: London.

SETCHELL, B. P., 1986. The movement of fluids and substances in the testis. *Aust. J. Biol. Sci.* **39:** 193-207.

SETCHELL, B. P. AND CARRICK, F. N., 1973. Spermatogenesis in some Australian marsupials. *Aust. J. Zool.* **21:** 491-99.

SETCHELL, B. P. AND HINKS, N. T., 1969. Absence of countercurrent exchange of oxygen, carbon dioxide, hydrogen ions or glucose between the arterial and venous blood in the spermatic cords of rams and two marsupials (*Macropus eugenii* and *Megaleia rufa*). *J. Reprod. Fertil.* **20:** 179-81.

SETCHELL, B. P. AND THORBURN, G. D., 1969. The effect of local heating on blood flow through the testes of some Australian marsupials. *Comp. Biochem. Physiol.* **31:** 675-77.

SETCHELL, B. P. AND WAITES, G. M. H., 1969. Pulse attenuation and countercurrent heat exchange in the internal spermatic artery of some Australian marsupials. *J. Reprod. Fertil.* **20:** 165-69.

SETCHELL, B. P., WAITES, G. M. H. AND THORBURN, G. D., 1966. Blood flow in the testis of the concious ram measured with krypton^{85}: effect of heat, catecholamines and acetylcholine. *Circulation Res.* **18:** 755-65.

SHARMAN, G. B. AND CALABY, J. H., 1964. Reproductive behaviour in the red kangaroo, *Megaleia rufa,* in captivity. *C.S.I.R.O. Wildl. Res.* **9:** 58-85.

SUZUKI, F. AND RACEY, P. A., 1984. Light and electron microscopical observations on the male excurrent duct system of the common shrew. *(Sorex araneus). J. Reprod. Fertil.* **70:** 419-28.

SWIERSTRA, E. E., 1966. Structural composition of Shorthorn bull testes and daily spermatozoa production as determined by quantitative testicular histology. *Can. J. Animal Sci.* **46:** 107-19.

TEMPLE-SMITH, P. D., 1984a. Phagocytosis of sperm cytoplasmic droplets by a specialized region in the epididymis of the brushtailed possum, *Trichosurus vulpecula. Biol. Reprod.* **30**: 707-20.

TEMPLE-SMITH, P. D., 1984b. Reproductive structures and strategies in male possums and gliders. Pp. 89-106 *in* Possums and Gliders ed. by A. P. Smith and I. D. Hume. Australian Mammals Society: Sydney.

TEMPLE-SMITH, P. D. AND BEDFORD, J. M., 1976. The features of sperm maturation in the epididymis of a marsupial, the brushtailed possum *Trichosurus vulpecula. Amer. J. Anat.* **147**: 471-500.

TYNDALE-BISCOE, C. H. AND RENFREE, M., 1987. Reproductive Physiology of Marsupials. Cambridge University Press: Cambridge.

TYNDALE-BISCOE, C. H. AND RODGER, J. C., 1978. Differential transport of spermatozoa into the two sides of the genital tract of a monovular marsupial, the tammar wallaby *(Macropus eugenii). J. Reprod. Fertil.* **52**: 37-43.

TUCK, R. R., SETCHELL, B. P., WAITES, G. M. H. AND YOUNG, J. A., 1970. The composition of fluid collected by micropuncture and catheterization from the seminiferous tubules and rete testis of rats. *Pflugers Archiv.* **318**: 225-43.

ULLMAN, S. L., 1984. Early differentiation of the testis in the native cat, *Dasyurus viverrinus* (Marsupialia). *J. Anat.* **138**: 675-88.

VOGLMAYR, J. K., FAIRBANKS, G., JACKOWITZ, M. A. AND COLELLA, J. R., 1980. Post-testicular development changes in the ram sperm cell surface and their relationship to luminal fluid proteins of the reproductive tract. *Biol. Reprod.* **22**: 655-67.

WAI-SUM, O., SHORT, R. V., RENFREE, M. B. AND SHAW, G., 1988. Primary genetic control of somatic sexual differentiation in a mammal. *Nature (London)* **331**: 716-17.

WALTON, A., 1960. Copulation and natural insemination. Pp. 130-60 *in* Marshall's Physiology of Reproduction Vol. 1: part 2 ed. by A. S. Parkes. Longmans Green: London.

WHITE, W. W., 1932. The effect of hypophysectomy on the survival of spermatozoa in the male rat. *Anat. Rec.* **54**: 253-73.

WITSCHI, E., 1951. Embryogenesis of the adrenal and the reproductive glands. Pp. 1-27 *in* Recent Progress in Hormone Research ed. by G. Pincus. Academy Press: New York.

WITSCHI, E., 1956. Development of Vertebrates. Saunders: Philadelphia.

WITSCHI, E., 1962. Embryology of the ovary. Pp. 1-10 *in* The Ovary. Vol. III. Williams and Wilkins: Baltimore.

WONG, P. Y. D., TSANG, A. Y. F. AND LEE, W. M., 1981. Origin of the luminal fluid proteins of the rat epididymis. *Int. J. Androl.* **4**: 331-41.

ZAMBONI, L. AND UPADHYAY, S., 1982. The contribution of the mesonephros to the development of the sheep fetal testis. *Amer. J. Anat.* **165**: 339-56.

Reproductive biology of Rat-kangaroos

R. W. ROSE[1]

The reproductive biology of the Potoroidae, though similar to the related larger members of the superfamily Macropodoidea, differs in several ways. Pouch life is considerably shorter and this allows more than one litter per year in most species. Reproductive anatomy varies little between species and the anterior vaginal expansion is a valid taxonomic characteristic. The median vagina is transient in Rat-kangaroos and must reform before each birth under the influence of the corpus luteum. Lactation is similar in the two families though interspecific differences may account for differential growth rates.

INTRODUCTION

THE reproductive biology of the Potoroidae is generally considered to be remarkedly similar to that of the larger kangaroos (Macropodidae) consequently they are rarely treated as a separate group. However, the rat-kangaroos differ from the macropodids in two major ways. Pouch life duration is relatively short even when considering body size (Rose, this volume) and they generally produce more than one young per year. The Tasmanian Bettong *Bettongia gaimardi,* is able to produce and rear up to six young over a two year period. Despite their fecundity, potoroids are reduced in numbers over most of their range, due to environmental and predation pressures (Rose 1986a). Recent reviews of the reproduction of this group are by Rose (1978), Tyndale-Biscoe and Renfree (1987) and Seebeck and Rose (1988). This paper includes additional data, particularly on lactation and energetics of reproduction as well as unpublished data on male reproduction.

Sex determination

The chromosome number and sex chromosomes are presented in Table 1. *Potorous tridactylus* differs from almost all marsupials in having a sex determining mechanism that differs from XX/XY. The other extant Potoroo, *P. longipes* has a different chromosome number and sex determination despite its similar external appearance to *P. tridactylus*. However, the conclusion of Johnston *et al.* (1984) was that in all likelihood the ancestral *Potorous* karyotype was 2n=24 and that the usual chromosome number of *P. tridactylus* could be derived by evolutionary fusion events. Most macropodoids have a chromosome number of 22, *Aepyprymnus rufescens* is most unusual in having the largest number of chromosomes of all marsupials (32).

Reproductive anatomy

This is reviewed in Rodger (1978) and Rose (1978). On the basis of the presence of an anterior vaginal expansion (vaginal caecum) and other aspects of female reproductive anatomy, Pearson (1944, 1945, 1946, 1950) suggested that rat-kangaroos were a separate family to the larger kangaroos and wallabies. Although there has been some controversy in the literature in regard to the anterior vaginal expansion, it does appear to be a valid taxonomic characteristic of rat-kangaroos (Rose 1978). An additional characteristic is found in *B. lesueur* (Tyndale-Biscoe 1968) and *A. rufescens* (Moors 1975), the anterior cul-de-sac. All potoroids give birth via the transient median vagina (Shaw and Rose 1979) despite the earlier suggestion (Flynn 1922; Pearson 1946) that the fetus of *P. tridactylus* traverses the lateral vagina.

The macropodoid prostate gland is carrot-shaped. The tip of the penis is not bifid (Rodger 1978) though Seebeck and Rose (1988) describe a slight cleft in the tip of *B. gaimardi.*

Reproductive patterns

The reproductive life histories of all rat-kangaroos are summarized in Table 2 based on numerous studies summarized in Lee and Cockburn (1985) and Seebeck and Rose (1988). As with the majority of all macropodoids, the gestation length is only just shorter than the length of the oestrous cycle. Those species with short gestation (e.g., *Bettongia* spp.) correspondingly have short oestrous cycles whereas those with a longer gestation (e.g., *P. tridactylus*) have a long oestrous cycle. Surprisingly amongst the rat-kangaroos one finds both the shortest (*Bettongia* spp.) and longest (*P. tridactylus*) gestation lengths of the Macropodoidea. The close similarity in all reproductive parameters of the genus *Bettongia*, no doubt indicates a close phylogenetic relationship and perhaps recent speciation (Rose 1978).

The close similarity in oestrus cycle and gestation length results in the association of birth with a post-partum oestrus when the female can mate and produce a fertilized egg. Should the pouch

[1]Zoology Department, University of Tasmania, G.P.O. Box 252C,, Hobart, Tasmania 7001.

Pages 307-15 in KANGAROOS, WALLABIES AND RAT-KANGAROOS ed. by G. Grigg, P. Jarman and I. Hume. Surrey Beatty & Sons Pty Limited, New South Wales, Australia, 1989.

Table 1. Chromosome number and sex determination in the Potoroidae (Seebeck and Rose 1988).

Species	2N	Sex Chromosomes
Hypsiprymnodon moschatus	22	XX/XX
Potorous tridactylus	13/12	XY^1-Y^2/XX
Potorous longipes	24	XY/XX
Bettongia lesueur	22	XY/XX
Bettongia gaimardi	22	XY/XX
Bettongia penicillata	22	XY/XX
Aepyprymnus rufescens	32	XY/XX

be occupied at this time, the fertilized egg develops only to the blastocyst stage (approx. 100 cells and 250 μm in diameter, Kerr 1934; Smith 1981) before entering embryonic diapause. The blastocyst remains in this state until near the end of pouch life or after the pouch incumbent is removed or lost. The subsequent pregnancy is termed "delayed gestation" (Table 2). In most species, the delayed gestation length is shorter by 2-3 days than the normal gestation length, however in *P. tridactylus* the difference extends to nine days (Shaw and Rose 1979).

In some species e.g., *B. gaimardi* and *P. tridactylus* the end of pouch life is accompanied, often on the same night, by a new birth and subsequent post-partum oestrus (Rose 1986b, 1987b).

Variations in the female reproductive tract

Cyclic variation in the cellular constituents of the vaginal smear (Table 3) occur in those potoroids so far examined (Hughes 1962; Tyndale-Biscoe 1968; Rose 1987a). For much of the cycle, there appears to be an inverse relationship between cornified cells and leucocytes. The abrupt disappearance of leucocytes, accompanied by a rapid increase in cornified cells usually heralds oestrus. Sperm were found in the smear on only one day of the cycle in *B. gaimardi* though Hughes (1962) found sperm over a longer period in the vaginal smear of *P. tridactylus*.

Shaw and Rose (1979) and Bryant and Rose (1986) describe in detail the histological changes that occur in the ovaries and uteri and vagina of *P. tridactylus* during delayed pregnancy and non-pregnancy after removal of a pouch young. Flynn (1930) and Rose (1985) provide similar details for *B. gaimardi* (Fig. 1). Apart from the time sequence, the changes that occur are similar in both species and to those found in *B. lesueur* by Tyndale-Biscoe (1968).

In the Bettong and Potoroo as in most macropodoids, a corpus luteum is formed after shedding the single egg. This remains small and quiescent during embryonic diapause. Removal of the pouch young, results in reactivation of the corpus luteum and its subsequent stimulation of both uteri. But after 4-5 days in both species, the uterine glands on the same side as the ovary containing the corpus luteum are significantly larger than their opposite side (Fig. 1). This may be mediated via a facilitated transport across the localized vascular complex as concluded by Shaw and Rose (1979). The size of both uterii have doubled by this stage. This uterine stimulation occurs whether or not the mother is pregnant. However, around mid-pregnancy, there is a subtle changeover that leads to the conclusion that the fetus/placenta at this time have a stimulatory effect upon the pregnant uterus as this effect is absent in non-pregnant animals. This local effect continues throughout pregnancy in the Bettong and Potoroo and no doubt contributes to the relatively lengthy gestations of potoroids relative to their oestrous cycles (Shaw and Rose 1979; Rose 1985).

Physiological factors that result in differences between the state of pregnancy and non-pregnancy are usually termed "maternal recognition of pregnancy" and Tyndale-Biscoe (1979) believes that term can be applied to the above phenomenon.

Table 2. Reproductive patterns in Rat-kangaroos.

	Potorous[a] *longipes*	*Potorous*[b] *tridactylus*	*Bettongia*[c] *lesueur*	*Bettongia*[d] *penicillata*	*Bettongia*[e] *gaimardi*	*Aepyprymnus*[f] *rufescens*	*Hypsiprymnodon*[g] *moschatus*
Adult weight (kg)	1.7	1.2	1.1	1.3	1.8	3.5	0.57
Oestrus cycle length (days)	—	42	22.2	22.5	22.6	21-25	—
Gestation length (days)	—	38	21.3	21	21.3	21-23.6	—
Delayed gestation length (days)	—	29	20	21	18	18.7	—
Pouch life (days)	145(?)	125	115	100	106	114	147
Litter size	1	1	1	1	1	1	2
Annual fecundity	1-2	2.5	3	3	3.5	3	—
Age at maturity of female (months)	26	12	9	9	9	10	12
Age at weaning (weeks)	28	24	25	19	24	22	—

(a) Seebeck and Rose 1988; Hughes 1962, Shaw and Rose 1979; (c) Tyndale-Biscoe 1988; (d) Sampson 1971, Parker 1977; (e) Rose 1988a; (f) Johnson 1978; (g) Johnson and Strahan 1982.

Table 3. Variations in the vaginal smear of *Bettongia gaimardi* (from Rose 1987a).

Day of cycle	Cornified cells	Leucocytes
−10	+	±
−8	±	++
−6	±	+
−4	±	++
−2	+	++
Oestrus	++	−
+2	+++	−
+4	+++	−
+6	+++	±
+8	+	±
+10	±	±

Corpus luteum

The corpus luteum in *P. tridactylus* has an important role to play not only in pregnancy, but also in the preparation of the transient birth canal; as in most if not all potoroids, the median vagina is only patent (open along its length) for a short period near birth (Rose 1978). The corpus luteum is also necessary for successful parturition and the onset of lactation (Bryant and Rose 1986). After RPY in the Potoroo, excision of the corpus luteum before day 3 of pregnancy inhibited embryo activation and/or development as embryos were not recovered from the uterus. Removal of the gland (CLX) between 6-21 days of the 29 day delayed gestation, allowed normal development of the fetus but interfered with the birth process and the post-term fetus remained within the uterus. CLX on day 25 resulted in birth but the neonates died shortly after, possibly due to impaired lactation. By day 27 the corpus luteum was no longer essential as young were born and reared.

The requirement of the corpus luteum for such an extended period (93% of gestation) in the Potoroo is greater than in other kangaroos and may be because in this rat-kangaroo and probably most others, the birth canal must be prepared under control of the CL shortly before each birth (Bryant and Rose 1986).

"Maternal recognition of pregnancy" extends to effect other than the unilateral stimulation, resulting in an increased uterine gland diameter described earlier. Oestrous cycles containing a pregnancy are shorter by 1.5 days in *B. gairmardi* than cycles from non-pregnant animals (Rose 1987a). This "Merchant effect" (Merchant 1976, 1979) has only been detected in the

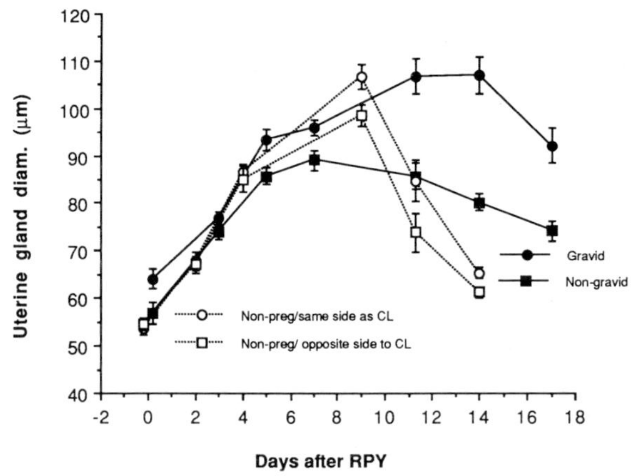

Fig. 1. Change in the size of basal uterine glands after removal of pouch young in *Bettongia gaimardi* (Rose unpubl. data).

Macropodoidea, and results in synchrony between birth and post-partum oestrus, thus ensuring fertilization.

Pouch vacation

The influence of the fetus may be even more profound than that mentioned above. Almost certainly it plays a role in pouch vacation in some species (Rose 1986b). In continuously breeding marsupials, it is possible for successive young to occupy the pouch simultaneously. Although this rarely occurs in nature, it is found in some macropodoids including rat-kangaroos. In most potoroids, the large furred young, soon to vacate the pouch is often replaced by the new neonate born as a result of the activation of the delayed blastocyst. Alternatively, however, the development of an egg fertilized at a post-partum oestrus may not be sufficiently inhibited and can result in two pouch young separated in age by one gestation length as described by Inns (1980) for *M. eugenii* and in *P. tridactylus* (S. Johnson, unpub. data). Support for some sort of genetic control of the duration of pouch life comes from early experiments by Merchant and Sharman (1966) who demonstrated that the length of pouch life of several kangaroo species did not vary when fostered into the pouches of other species. By contrast, others have found that pouch life duration in a number of macropodids is more dependent on size and weight (Johnson 1981). Regardless of species differences, most authors (e.g., Merchant and Sharman 1966) have assumed that it is the young itself that has the major role

Table 4. Age and weight at pouch vacation in *Bettongia gaimardi*. Values are means±standard deviation. (from Rose 1986b).

Group	Associated events	N	Age (days) at pouch vacation Mean	Age (days) at pouch vacation Range	Weight (g) at pouch vacation Mean	Weight (g) at pouch vacation Range
1	Birth+oestrus	10	106.8±1.7	104−109	333.5±27.7	291−371
2	Oestrus only	7	105.9±3.2	101-111	379.5±68.0	293−460
3	Neither	3	112.7±3.5	109-116	430.7± 6.4	427−438

in determining pouch life duration. However, there is no doubt that in *B. gaimardi* and other rat-kangaroos e.g., *B. penicillata* and perhaps *P. tridactylus*, the mother plays an important role in determining pouch life (Rose 1986b; M. Smith, this volume).

During the final two weeks of pouch life, the young Tasmanian Bettong spends an increasing amount of time outside the pouch, returning in the last few days only when startled. On the night of final pouch vacation a precise sequence of events occur. Firstly, there is a dramatic tightening of the pouch sphincter muscle, sufficient to prevent re-entry by the large furred young, then usually on the same night as pouch vacation, both birth and a post-partum oestrus occur. The results from a series of observations are presented in Table 4. The pouch life and weight at pouch vacation of animals in group 3 were significantly elevated. Although there was little difference between groups 1 and 2, the data in group 2 were significantly more variable. It was hypothesized from this data that imminent birth and/or oestrus precipitate pouch vacation. Further experiments to test this hypothesis included replacing pouch young by others of different ages. In these fostering experiment, the replacements remained in the pouch until the night of birth and/or oestrus regardless of their age (Rose 1986b; M. Smith this volume). In order to precipitate birth while a young resided in the pouch, injections were given of bromocriptine, a dopamine agonist, known to result in the initiation of blastocyst development during lactation in Tammar Wallabies (Tyndale-Biscoe and Hinds 1984). However, the bromocriptine injections failed to result in birth or oestrus in the Bettong (Rose 1985; MacFadyen and Rose unpublished data.). Bromocriptine injections near the end of pouch life in the Tasmanian Pademelon, *Thylogale villardierii* did result in premature vacation of the pouch (Rose 1985). The above results all indicate a maternal role in final pouch vacation. The term "maternal recognition of pregnancy" could be applied to this phenomenon as the fetus is also involved, in that when birth occurs a night before oestrus, pouch vacation always occurs on the night of birth (Rose 1986b).

Progesterone and body temperature

Very few measurements have been made of reproductive hormones in potoroids. Hughes *et al.* (1978) report on a few measurements of progesterone in *P. tridactylus,* but these values are somewhat higher than found in other marsupials and were not collected in a systematic way. Figure 2 shows measurement of progesterone in two female Bettongs after their young were removed (Rose and Gemmell unpub. data). The shape of the curve including a small peak on day four are similar to that obtained for the Tammar Wallaby (Hinds and Tyndale-Biscoe 1982) though the actual values are somewhat higher in the Bettong. Peters and Rose (1979) have shown changes in basal body temperature associated with the oestrous cycle in the Common Wombat *Vombatus ursinus*. Similar changes occur in *Bettongia gaimardi* (Rose 1985) and to some extent they mirror the changes in progesterone.

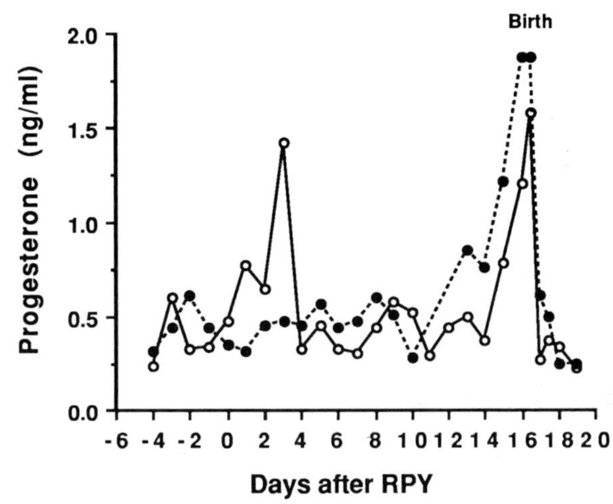

Fig. 2. Progesterone profile for two Bettongs (*B. gaimardi*, Rose and Gemmell unpub. data).

Development

Detailed embryological studies have yet to be performed on any potoroid. However, Kerr (1934, 1936), Flynn (1930), Hughes *et al.* (1978), Shaw and Rose (1979), Rose (1985) and Seebeck and Rose (1988) provide some general details on *P. tridactylus* and *B. gaimardi.* More recently Rose (1988a and this volume) discusses and gives comparative embryonic growth rates. A clear indication of the interval from primitive streak to birth is a valuable index of development. Embryonic growth of *B. gaimardi* proceeds more rapidly than in *P. tridactylus* e.g., it takes seven days to reach the medullary plate stage in *B. gaimardi* but 14 days in *P. tridactylus* (Seebeck and Rose 1988). In fact the embryonic growth rate of *B. gaimardi* is as fast or faster than all other macropodids for which we have information (Rose 1989a). After birth, development continues in the pouch for 4-5 months in most potoroids (Table 2). During this period growth rates vary but are far higher than in the larger kangaroos and wallabies, *B. gaimardi* has a faster rate of relative growth than any other macropodid (Rose 1985, this volume) and this has been related to the high protein content in the milk (Smolenski and Rose 1988).

Details of age estimation during pouch life and beyond are available for most species, *B. lesueur* (Tyndale-Biscoe 1968), *B. penicillata* (Sampson 1971; Christensen 1980), *B. gaimardi* Rose 1989b), *P. tridactylus* (Guiler 1960; Hughes 1962;

Bryant this volume) and *A. rufescens* (Johnson 1978). Rose (1985) and Taylor and Rose (1987) show how *B. gaimardi* grows more slowly in the wild than in captivity on an enriched diet. This indicates that some caution should be used in applying laboratory obtained growth curves to wild populations. It also suggests differences in lactation between wild and captive animals

Energetics of pouch life

At birth all marsupials are ectothermic, but within the pouch they are maintained as passive homeotherms. The development of endothermy in *B. gaimardi* and its ability to maintain body temperature (T_b) has been reported by Rose (1987b). Young removed from the pouch cool less slowly as they become older. In part this is due to changes in physical parameters such as decreased surface area to volume ratio and increased insulation, however, metabolic rate changes gradually become of major importance. The Bettong's metabolism increases with age as does its ability to alter the metabolic rate. However, despite gradual increases in basal metabolism up to week 8-9 of pouch life, the Bettong is unable to increase its metabolic rate above basal levels in response to cold stress. Subsequently, from the age of 10 weeks (pouch life is 15 weeks), the young is able to increase its metabolic rate and maintain its T_b substantially above room temperature (20°C). The Bettong is able to reduce heat loss, prior to the development of an increased metabolic capacity, by behavioural mechanisms such as curling up to minimize surface area and physiological means, initially by peripheral vasoconstriction, which reduces conductance but subsequently by shivering. The Bettong achieves "homeothermy" at about week 12 when fur growth becomes substantial. The first excursions outside the pouch occur one week later. However, the young huddles with its mother for some time before and after final pouch vacation (Rose 1987b). Growth, physiological maturation and maintenance of endothermy requires an increasing amount of energy that can only be provided by the mother's milk.

Lactation

Recent studies of lactation in *P. tridactylus* and *B. gaimardi* (Fig. 3) are by Smolenski and Rose (1988) and Crowley, Woodward and Rose (1988). Comparison of rat-kangaroo milk with other macropodoids is given by Merchant (this volume). The differences in growth rates between the Bettong and Potoroo described elsewhere (Rose 1987b, this volume) may be contributed to by a protein enriched milk early in lactation and later by a greater milk intake (Smolenski and Rose 1988). The substantial changes in milk composition and increases in energy content (Fig. 3) are comparable to those found in the Tammar Wallaby (Green 1984). Despite energy increases in the milk, relative growth rates diminish after pouch vacation (Rose 1987b, this volume). This may be due to partitioning of milk energy away from absolute growth towards other activities of the young such as fueling the increased metabolic rate, general temperature regulation and providing for locomotion, nest construction and food gathering activities etc.

Male reproduction

There are few studies on the reproduction of male potoroids. Hughes (1964) provides details on Potoroo sexual development and sperm morphology and Rodger (1978) reviews this and other details of male reproductive structure. The sperm of potoroids are indistinguishable from other kangaroos (Cleland pers. comm. quoted by Rodger 1978).

Although male marsupials possess a rete mirabile as opposed to the pampiniform plexus of eutherians, both vascular arrangements perform a similar function in cooling the testes. It is well documented that cadmium has a destructive and seemingly selective effect on the testes of numerous eutherian species (Parizek 1960), acting, at least in part, on the blood vessels supplying the testes. Rose, Swain and Whitten (1981) suggested that the marsupial rete may result in a more effective blood-tissue barrier to cadmium than the spirally arranged eutherian pampiniform plexus. Barnett and Brazenor (1958) stated that the Potoroo had the lowest rete vessel number (11 vessels) of all marsupial species. If the marsupial rete does afford some protection against cadmium induced injury, this protection should be minimal in the Potoroo. Bryant and Rose (1985) did find greater testicular damage in the potoroo than previously reported for two other marsupial species. However, this could not be ascribed totally to a small rete as they determined the rete to be much more extensive (56 vessels) than described by Barnett and Brazenor (1958).

Bryant and Rose (unpubl. data) in another study have shown that the degenerative changes that occur in testes surgically transposed to the abdomen in the potoroo are similar to those reported for other marsupials. e.g., *M. eugenii* (Setchell and Thorburn 1970)

Hughes (1964) found no epididymal sperm in a male Potoroo weighing 561 g (testis length 1.2 cm) and very few in an animal a little larger (639 g, 1.5 cm), all animals of 900 g and over (testis length 1.8-2.6 cm) were mature and had numerous sperm. Rose (1985) and Rose and Bradley (unpublished data) report on the maturation of *B. gaimardi* and on testosterone levels from pouch vacation to maturation and subsequently over a twelve month period. They show how testis

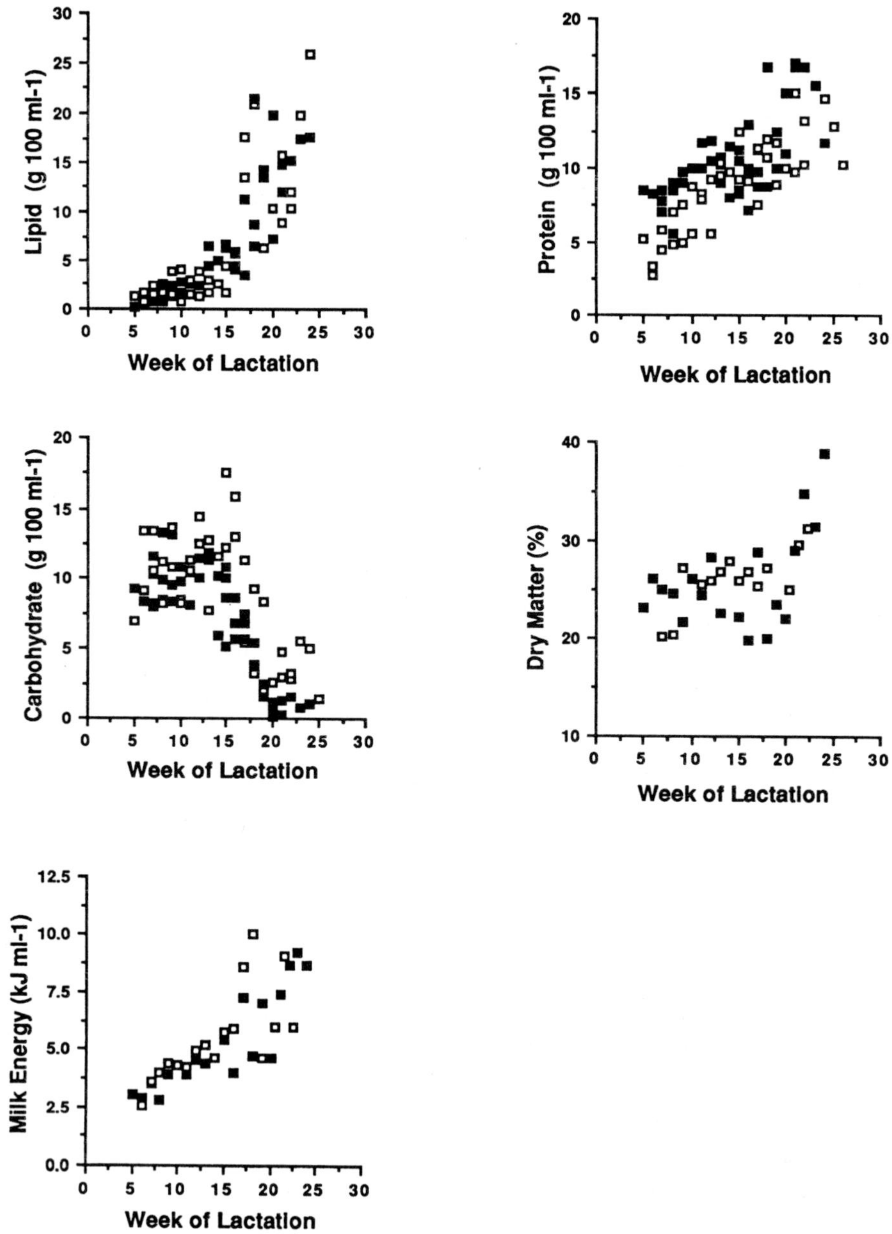

Fig. 3. Changes in lipid, protein, carbohydrate, solids and energy content during lactation in *Potorous tridactylus* (open symbols) and *Bettongia gaimardi* (closed symbols) (from Smolenski and Rose 1988).

size increases sigmoidally with age, increasing most rapidly near puberty. This period is accompanied by peak levels in testosterone that are not repeated in later life (Fig. 4). After maturation there are few changes in testis size and body weight though there are peaks and troughs in female births that coincide to some extent with testosterone levels (Rose and Bradley unpubl. data). Other authors have noted the absence of seasonal changes in testes size and body weight in potoroids (Sampson 1971, Christensen 1980). However, to date accessory glands have not been examined. In some other species accessory glands are known to vary with the breeding season despite only minor variation in testes size e.g., *Macropus eugenii* (Inns 1980).

It has been shown that the male accessory glands in the Potoroo decrease in size after castration. One control animal *P. tridactylus)* had a prostate that weighed 2.7 g and measured 4.2 cm. while an animal that had been castrated 14 days previously had corresponding measurements of 1.7 g. and 3.4 cm. Replacement therapy of testosterone propionate (5 mg every second day for 14 days) was able to maintain the prostate at near normal weight (Rose unpubl. data).

Reproductive behaviour

Certain reproductive behaviour is similar in all species examined, *B. lesueur* (Stodart 1966). *A. rufescens* (Johnson 1980; Jarman 1983) and

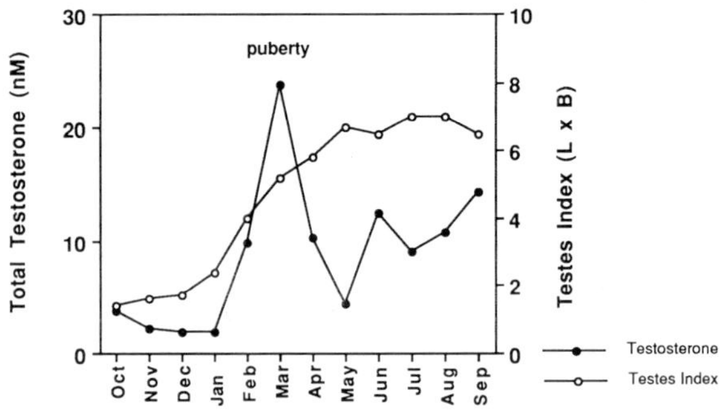

Fig. 4. Maturation of a male bettong *(B. gaimardi)* (Rose and Bradley unpub. data).

B. gaimardi (Rose 1982, 1985). Males regularly inspect the pouch and cloacal region of females in order to detect oestrus. Prolongation of this behaviour often results in non-responsive females laying on their sides and kicking strongly with their large hind limbs. On the day of oestrus, males and females hop to and fro in unison. A male mounts the female and holds onto her flanks with his forelimbs and often bites her neck. The disruption to her fur may be tell-tale evidence of a mating. Mounting and thrusting may occur more than once on a night (e.g., *B. gaimardi* Rose 1985) though it is not known if the male ejaculates more than once. In *B. gaimardi,* females only mate on the night of oestrus but Hughes (1962) reports that mating may occur over many days in *P. tridactylus.*

Birth behavior in *B. gaimardi*, involves the female crouching as she shuffles backwards, usually into a corner for support. The tail is forward and it is licked for most of its length. The region from the uro-genital opening to the lip of the pouch is licked by the female as is the inside of the pouch (MacFadyen and Rose unpubl. obs.). An orange encrustment some times seen on the fur after birth may be remnants of the yolk-sac (Rose 1985).

CONCLUSION

The differences between potoroid and macropodid reproduction are not great, though they exist in conservative areas such as reproductive anatomy. The differences in reproductive biology that do occur include a shorter pouch life during which young grow more rapidly than their larger macropodid relatives. Most potoroids are able to breed continuously unlike most macropodids and as a consequence of this and the shorter pouch life they are able to produce more young per year.

The corpus luteum plays a pivotal role in gestation, birth and the initiation of lactation; it may be required to act over a greater proportion of pregnancy in order to allow the development of the median vagina before each birth.

In view of the antiquity of their undoubted common ancestor and the differences in life style and diet between rat-kangaroos and their relatives the kangaroos and wallabies, it seems that the reproductive specializations of the whole group (e.g., gestation extended past the luteal phase and embryonic diapause) are very successful.

ACKNOWLEDGEMENTS

I wish to thank Sally Bryant and Sherrin Bowden for their assistance in the preparation of this manuscript.

REFERENCES

BARNETT, C. M. AND BRAZENOR, W. C., 1958. The testicular rete mirabile of marsupials. *Aust. J. Zool.* **6:** 23-32.

BRYANT, S. L. (this volume). Age estimation and breeding pattern of the Potoroo *Potorous tridactylus* in Tasmania.

BRYANT, S. L. AND ROSE, R. W., 1985. The effect of cadmium on the reproductive organs of the marsupial *Potorous tridactylus. Aust. J. Biol. Sci.* **38:** 305-11.

BRYANT, S. L. AND ROSE, R. W., 1986. The growth and role of the corpus luteum throughout delayed gestation in the potoroo, *Potorous tridactylus. J. Reprod. Fert.* **76:** 409-14.

CHRISTENSEN, P., 1980. The biology of *Bettongia penicillata* Gray, 1837 and *Macropus eugenii* Desmarest, 1834 in relation to fire. Forest Department of Western Australia Bulletin 91.

CROWLEY, H. M., WOODWARD, D. R. AND ROSE, R. W., 1988. Changes in milk composition during lactation in the potoroo *(Potorous tridactylus). Aust. J. Biol. Sci.* **41:** 289-96.

FLYNN, T. T., 1922. Notes on certain reproductive phenomena in some Tasmanian marsupials. Ann. Mag. N. Hist. Ser. X, Pp. 225-31.

FLYNN, T. T., 1930. The uterine cycle of pregnancy and pseudo-pregnancy as it is in the diprotodont marsupial *Bettongia cuniculus. Proc. Linn. Soc. NSW* **55:** 506-31.

GREEN, B., 1984. Composition of milk and energetics of growth in marsupials. *Symp. Zool. Soc. Lond.* **51:** 369-87.

GUILER, E. R., 1960. The pouch young of the potoroo. *J. Mammal.* **41:** 441-51.

HINDS, L. A. AND TYNDALE-BISCOE, C. H., 1982. Plasma progesterone levels in the pregnant and non-pregnant tammar, *Macropus eugenii. J. Endocr.* **93:** 99-107.

HUGHES, R. L., 1962. Reproduction in the macropod marsupial *Potorous tridactylus. Aust. J. Zool.* **10:** 193-224.

HUGHES, R. L., 1964. Sexual development and spermatozoan morphology in the male macropod marsupial *Potorous tridactylus* (Kerr). *Aust. J. Zool.* **12:** 42-51.

HUGHES, R. L., GEMMELL, R. T., ROSE, R. W. AND SHAW, G., 1978. The major features of embryonic development in the marsupial *Potorous tridactylus. J. Anatomy* **128:** 656.

INNS, R. W., 1980. Occurrence of twins in macropod marsupials. *Search.* **11:** 118-19.

INNS, R. W., 1982. Seasonal changes in the accessory reproductive system and plasma testosterone levels of the male tammar wallaby, *Macropus eugenii,* in the wild. *J. Reprod. Fert.* **66:** 675-80.

JARMAN, P., 1983. Mating system and sexual dimorphism in large terrestrial mammalian herbivores. *Biol. Rev.* **58**: 485-520.

JOHNSON, P. M., 1978. Reproduction in the rufous rat-kangaroo *(Aepyprymnus rufescens)* (Gray) in captivity with age estimation of pouch young. *Qld. J. Agric. Anim. Sci.* **35**: 69-72.

JOHNSON, P. M., 1980. Observations on the behaviour of the Rufous Rat-kangaroo *Aepyprymnus rufescens* (Gray) in captivity. *Aust. Wildl. Res.* **7**: 347-57.

JOHNSON, P. M., 1981. The rearing of marsupial pouch young by foster mothers of different species. *Int. Zoo. Yb.* **21**: 173-76.

JOHNSON, P. M. AND STRAHAN, R., 1982. A further description of the musky rat-kangaroo, *Hypsiprymnodon moschatus.* Ramsay 1876, (Marsupialia, Potoroidae), with notes on its biology. *Aust. Zool.* **21**: 27-46.

JOHNSTON, P. G., DAVEY, R. J. AND SEEBECK, J. M., 1984. Chromosome homologies in *Potorous tridactylus* and *Potorous longipes* (Marsupialia, Macropodidae) based on G-banding patterns. *Aust. J. Zool.* **32**: 319-24.

KERR, T., 1934. Notes on the development of the germ-layers in diprotodont marsupials. *Quart. J. Microsc. Sci.* **77**: 305-15.

KERR, T., 1936. On the primitive streak and associated structures in the marsupial *Bettongia cuniculus. Quart. J. Microsc. Sci.* **78**: 687-98.

LEE, A. K. AND COCKBURN, A., 1985. Evolutionary ecology of marsupials. Cambridge University Press.

MERCHANT, J. C., 1976. Breeding biology of the agile wallaby *Macropus agilis* (Gould); in captivity. *Aust. Wildl. Res.* **3**: 93-103.

MERCHANT, J. C., 1979. The effect of pregnancy on the interval between one oestrus and the next in the tammar wallaby *Macropus eugenii. J. Reprod. Fertil.* **56**: 459-64.

MERCHANT, J. C. AND SHARMAN, G. B., 1966. Observations on the attachment of marsupial pouch-young to the teats and on the rearing of pouch young by foster-mothers of the same or different species. *Aust. J. Zool.* **14**: 592-609.

MERCHANT, J. C. (this volume). Lactation in macropodoid marsupials.

MOORS, P. J., 1975. The urogenital system and notes on the reproductive biology of the female rufous rat-kangaroo *Aepyprymnus rufescens* (Gray) Macropodidae). *Aust. J. Zool.* **23**: 355-61.

PARKER, P., 1977. An ecological comparison of marsupial and placental patterns. *In* The Biology of Marsupials ed. by B. Stonehouse and D. Gilmore. MacMillan: London.

PARIZEK, J., 1960. The sterilization of the male by cadmium salt. *J. Reprod. Fert.* **1**: 294-309.

PEARSON, J., 1944. The vaginal complex of the rat-kangaroos. *Aust. J. Sci.* **7**: 80-83.

PEARSON, J., 1945. The female urogenital systems of the Marsupialia with special reference to the vaginal complex. *Pap. Proc. R. Soc. Tasm.* **1944**: 71-98.

PEARSON, J., 1946. The affinities of the rat-kangaroos (Marsupialia) as revealed by a comparative study of the female urogenital system. *Pap. Proc. R. Soc. Tasm.* **1945**: 13-25.

PEARSON, J., 1950. The relationships of the Potoroidae to the Macropodidae (Marsupialia). *Pap. Proc. R. Soc. Tasm.* **1949**: 211-27.

PETERS, D. G. AND ROSE, R. W., 1979. The oestrous cycle and basal body temperature in the common wombat *(Vombatus ursinus). J. Reprod. Fert.* **57**: 453-60.

RODGER, J. C., 1978. Male reproduction — its usefulness in discussion of Macropodidae evolution. *Aust. Mammal.* **2**: 73-80.

ROSE, R. W., 1978. Reproduction and evolution in female Macropodidae. *Aust. Mamm.* **2**: 65-72.

ROSE, R. W., 1982. Tasmanian bettong *Bettongia gaimardi,* maintenance and breeding in captivity. *In* The Management of Australian Mammals in Captivity ed. by D. D. Evans. Zoo. Board Vic., Melbourne.

ROSE, R. W., 1985. The reproductive biology of the Tasmanian bettong, *Bettongia gaimardi.* Ph.D. Thesis, University of Tasmania.

ROSE, R. W., 1986a. The habitat, distribution and conservation status of the Tasmanian bettong, *Bettongia gaimardi* (Desmarest). *Aust. Wildl. Res.* **13**: 1-6.

ROSE, R. W., 1986b. Control of pouch vacation in the Tasmanian bettong, *Bettongia gaimardi. Aust. J. Zool.* **34**: 485-91.

ROSE, R. W., 1987a. Reproductive biology of the Tasmanian bettong (*Bettongia gaimardi,* Macropodidae). *J. Zool. Lond.* **212**: 59-67.

ROSE, R. W., 1987b. Reproductive energetics of two Tasmanian rat-kangaroos (Potoroinae, Marsupialia). *Symp. Zool. Soc. Lond.* **57**: 149-65.

ROSE, R. W., 1989a. Embryonic growth rates of marsupials with a note on monotremes. *J. Zool. Lond.* **217** (in press).

ROSE, R. W., 1989b. Age estimation of the Tasmanian bettong *(Bettongia gaimardi)* (Macropodoidea, Marsupialia). *Aust. Wildl. Res.* **16** (in press).

ROSE, R. W. (this volume). Comparative growth rates in the Macropodoidea.

ROSE, R. W., SWAIN, R. AND WHITTEN, W. K., 1981. Cadmium toxicity and its effects on the testes of an Australian marsupial *(Trichosurus vulpecula) Bull. Environ. Contam. Toxicol.* **26**: 223-36.

SAMPSON, J. C., 1971. The Biology of *Bettongia pencillata* Gray, 1837. Ph.D. Thesis, University of Western Australia.

SEEBECK, J. H. AND ROSE, R. W., 1988. *Potoroidae.* Pp. 2-24 *in* Mammals of Australia ed. by D. Walton. Canberra Bureau of Flora and Fauna 1B.

SETCHELL, B. P. AND THORBURN, G. D., 1970. The effect of cryptorchidism on the testis of an Australian marsupials *M. eugenii. Comp. Biochem. Physiol.* **38**: 705-08.

SHAW, G. AND ROSE, R. W., 1979. Delayed gestation in the potoroo *Potorous tridactylus* (Kerr). *Aust. J. Zool.* **27**: 901-12.

SMITH, M. J., 1981. Morphological observations on the diapausing blastocyst of some macropodid marsupials. *J. Reprod. Fert.* **61**: 483-86.

SMITH, M. J. (this volume). Release of embryonic diapause in the Brush-tailed Bettong, *Bettongia penicillata.*

SMOLENSKI, A. J. AND ROSE, R. W., 1988. Comparative lactation in two species of Rat-kangaroo (Marsupialia). *Comp. Biochem. Physiol.* **90A**: 459-64.

STODART, E., 1966. Observations on the behaviour of the marsupial *Bettongia lesueur* (Quoy and Gaimard) in an enclosure. *C.S.I.R.O. Wildl. Res.* **11**: 91-99.

TAYLOR, R. J. AND ROSE, R. W., 1987. Comparison of growth of pouch young of the Tasmanian bettong, *Bettongia gaimardi,* in captivity and in the wild. *Aust. Wildl. Res.* **14:** 257-62.

TYNDALE-BISCOE, C. H., 1968. Reproduction and post-natal development in the marsupial *Bettongia lesueur* (Quoy and Gaimard). *Aust. J. Zool.* **16:** 577-602.

TYNDALE-BISCOE, C. H., 1979. Hormonal control of diapause and reactivation in the tammar wallaby. Pp. 173-90 *in* Maternal Recognition of Pregnancy. Ciba Found. Ser. 64.

TYNDALE-BISCOE, C. H. AND HINDS, L. A., 1984. Seasonal patterns of circulating progesterone and prolactin and response to bromocriptine in the female tammar, *Macropus eugenii. Gen. Comp. Endocr.* **53:** 58-68.

TYNDALE-BISCOE, C. H. AND RENFREE, M. B., 1987. Reproductive Physiology of Marsupials. Cambridge University Press: Melbourne. 476 Pp.

Release of embryonic diapause in the Brush-tailed Bettong, *Bettongia penicillata*

MEREDITH J. SMITH[1]

The diapausing embryo of continuously breeding macropodid marsupials is released from diapause while the large young continues to occupy the pouch. If this release is permitted because the large young suckles infrequently, then the age of the young in the pouch should determine the time of release. Pouch young of *Bettongia penicillata* were transferred to a more advanced pouch. The diapause embryo developed at the time expected if the original young had occupied the pouch. Similarly, young transferred to a less advanced pouch did not affect the time of release from diapause.

INTRODUCTION

EMBRYONIC diapause occurs in many macropodid marsupials and in most of these it is associated with lactation, the embryo remaining in diapause while a young suckles within the pouch. However, in several species the embryo is released from diapause before the young leaves the pouch permanently so that the resulting neonate can enter the pouch within a day of the older young leaving it for the last time (Tyndale-Biscoe and Renfree 1987). This termination of diapause is enigmatical. How can the embryo be released from diapause in the continued presence of the stimulus that caused it to enter diapause and, till the time of release, to remain in diapause?

The pouch young of the Red Kangaroo, *Macropus rufus,* remains continuously attached to the teat for about the first five months. Thereafter it releases the teat occasionally, and for longer intervals as it grows older. By the time it is about 190 days old, it starts to make short excursions away from the pouch (Sharman and Calaby 1964). Sharman (1965a) believed that it was during this time of intermittent suckling within the pouch that the embryo was released from diapause.

In contrast with the average 235 days from birth to permanent emergence from the pouch in the Red Kangaroo (Sharman and Calaby 1964), the young of the Brush-tailed Bettong *Bettongia penicillata,* emerges permanently at an average of 97.9 (n=25, s.d. 3.28) days old (Smith, unpublished) the shortest interval known for any macropodid (Tyndale-Biscoe and Renfree 1987). The mean interval from removal of pouch young (RPY) to birth in the Brush-tailed Bettong is 17.75 days (n=20, s.d. 1.12) days (Smith, unpublished) and so the release from diapause must occur when the pouch young is about 80 days old. At this age the young has little fur, its eyes are closed and if removed from the pouch it is unable to stand up (Smith, unpublished). Hence in the Brush-tailed Bettong, release of embryonic diapause occurs while the young remains within the pouch. Almost certainly, the young does release the teat at times, and these occasional interruptions of the sucking stimulus might result in the termination of embryonic diapause. If that were the case, then the time of release of the embryo from diapause would be determined by the age of the pouch young, because its stage of development would determine the amount of time the young did not suckle. In this paper the effect of the age of the young in the pouch on the time of release of embryonic diapause is investigated.

MATERIALS AND METHODS

Animals

The colony of *Bettongia penicillata* at the Evolutionary Biology Unit has been derived from five females and several males received from the stock maintained by the South Australian National Parks and Wildlife Service (Delroy *et al.* 1986). The animals were housed singly or in small groups in indoor pens with natural lighting. The diet consisted of apples, bananas, wholemeal bread, parrot seed mix, dog biscuits, boiled eggs and kangaroo pellets (Milling Industries).

Females nearing the expected time of parturition were caught daily and restrained in a cloth bag while the pouch was examined for presence of a newborn young. Routinely, females were caged with a male at the time of parturition and they mated at the postpartum oestrus. In some females, oestrus and mating were confirmed by the presence of abundant cornified cells and sperm in the vaginal smear.

Experimental design

Experiment 1. Eight females with pouch young aged between 32 and 58 days had their pouch

[1]Evolutionary Biology Unit, South Australian Museum, North Terrace, Adelaide 5000.

Pages 317-21 in KANGAROOS, WALLABIES AND RAT-KANGAROOS ed. by G. Grigg, P. Jarman and I. Hume. Surrey Beatty & Sons Pty Limited, New South Wales, Australia, 1989.

young removed and replaced by one 16 to 19 days younger. Four other females, the controls, had their pouch young replaced by one 1-3 days younger. Transferred young older than 30 days were simply placed in the pouch of the foster mother; they located the lactating teat and attached to it without assistance. A young 30 days old or less was assisted by having the teat gently pushed into its mouth by means of a tiny wooden spatula. After each transferred young attained 70 days of age its foster mother was observed daily until the foster young left the pouch and the mother gave birth.

Experiment 2. Pouch young were exchanged between five pairs of females whose young differed in age by 16-19 days. Pouch young transferred up, i.e., to a more advanced pouch were examined daily after they reached 70 days and down transfers i.e., to a less advanced pouch, after they reached 90 days, the daily observations being continued until the young left the pouch and the foster mother gave birth or mated.

Body measurements of young

At seven day intervals some young in Experiment 1 and all young in Experiment 2 were removed from the pouch of their foster mother, weighed, measured and returned to the pouch. The length of head, right arm, right leg and right foot were measured as described in Sharman, Frith and Calaby (1964). Brief observations were made of developmental features e.g., appearance of pigment, fur and the opening of the eyes.

RESULTS

Experiment 1. Of the 12 young transferred, five experimental and three control young survived to weaning (Table 1). Three experimental young (U3, U4 and U6) died a minimum of 36, 24 and 21 days respectively after transfer and one control (C2) died a minimum of 44 days after transfer. All the transferred young grew at first after transfer, and most of the experimental young grew faster than expected if they had not been transferred. However, four of the experimental young later ceased to gain weight and three of these died. The fourth (U8) survived to permanent vacation of the pouch and was hand reared successfully.

The young left the pouch permanently within a very few hours before the foster mother gave birth to a neonate derived from the embryo that had been in diapause during early lactation. Permanent vacation of the pouch and the foster mother's parturition occurred when the average age of experimental transferred young was 80 days (s.d. 4.3), the average for the control young being 95.3 (s.d. 1.5). The difference was significant (T-test $P<.002$).

However, the interval between parturitions for experimental foster mothers was an average of 97.4 (s.d. 3.36) and for control foster mothers 97.7 (s.d. 2.08), a difference that was not significant statistically (T-test).

Experimental transferred young weighed on the day of permanent vacation of the pouch registered weights of 186.5 (U1), 128.3 (U5), 208.4 (U7) and 60.6 (U8) gm, while the only control weighed at that stage registered 239.6 gm (C1).

Experiment 2. Of the ten young, nine survived to permanent vacation of the pouch (Table 2); the other died after increasing in weight from 5.2 gm at transfer to 13.4 gm three weeks later.

Young that were transferred to a pouch of more advanced lactation grew rapidly (Fig. 1) and developed a plump appearance; subcutaneous fat could be seen clearly. Young transferred to a pouch of less advanced lactation lost weight at first but then increased in weight and at permanent emergence from the pouch, all were within or close to the range of weights of nontransferred young at permanent emergence. Not only were the weight changes in transferred young affected

Table 1. Outcome of transfer of pouch young to a pouch 16-19 days more advanced in lactation (experimental transfers) or of transfer to a pouch 1-3 days less advanced (control transfers).

Transfer number	Pouch young age at transfer (days)	Age of replaced young (days)	Interval between foster mother's parturitions (days)	Suckling young's age (days) at foster mother's parturition
U1	16	32	102	86
U2	17	34	97	80
U3	18	37	P.y. died	—
U4	21	40	P.y. died	—
U5	32	51	96	77
U6	37	55	P.y. died	—
U7	40	57	99	82
U8	40	58	93	75
C1	19	20	96	95
C2	20	22	P.y. died	—
C3	34	37	97	94
C4	37	40	100	97

Table 2. Outcome of exchanging pouch young between pairs of females whose young differed in age by 16 to 19 days.

Transfer number	Pouch young age at transfer (days)	Difference in age from replaced young	Interval (days) between foster mother's parturition and the following parturition and/or oestrus	Suckling young's age (days) at foster mother's parturition and/or oestrus
E1U	30	−16	96	80
E2U	30	−19	99	80
E3U	32	−19	—	Died
E4U	30	−17	96	77
E5U	35	−18	100	82
E1D	46	+16	95	111
E2D	49	+19	94	113
E3D	51	+19	93	112
E4D	47	+17	95	112
E5D	53	+18	98	116

by the transfer (Fig. 1) but also the rate of growth of head (Fig. 2), forearm, leg (Fig. 3) and foot were affected.

Young transferred to a more advanced pouch were last seen with eyes closed at an average of 67.5 days old (s.d 3.87) and first seen with eyes open at an average age of 73.75 days (s.d. 2.5). Young transferred to a less advanced pouch were last seen with eyes closed at an average of 87.0 days old (s.d. 3.16) and first seen with eyes open at an average age of 94.0 days (s.d. 3.16).

Four of the nine foster mothers of surviving young produced a neonate from a delayed blastocyst. The other five all came into oestrus, determined by abundant cornified cells in the vaginal smear. In *B. penicillata* the post-partum oestrus occurs within a few hours after parturition (Smith, unpublished) and so in the absence of parturition, the day of oestrus was used as an estimate of the day parturition would have occurred if a viable delayed blastocyst had been present. The four young that survived to permanent emergence after transfer to a more advanced pouch had an average age of 79.8 days (s.d. 2.06) at the time their foster mothers gave birth and/or came into oestrus. The five transferred to less advanced

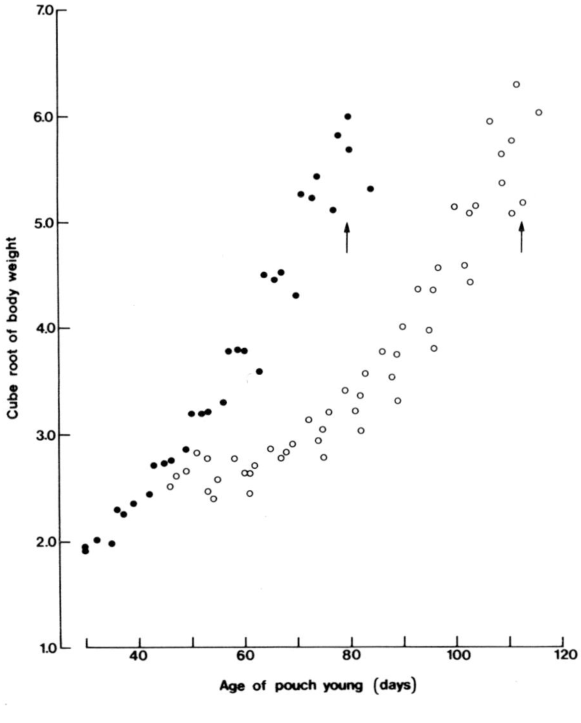

Fig. 1. Relationship between age and body weight in pouch young transferred to a more advanced mammary gland (solid circles) or a less advanced mammary gland (open circles). Arrows indicate average day when foster mothers gave birth and/or mated. Each circle represents a single measurement.

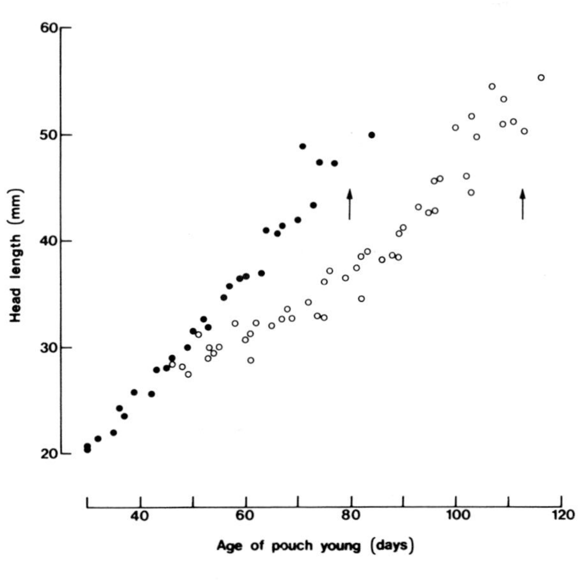

Fig. 2. Relationship between age and headlength in pouch young transferred to a more advanced mammary gland (solid circles) or a less advanced mammary gland (open circles). Arrows indicate average day when foster mothers gave birth and/or mated. Each circle represents a single measurement.

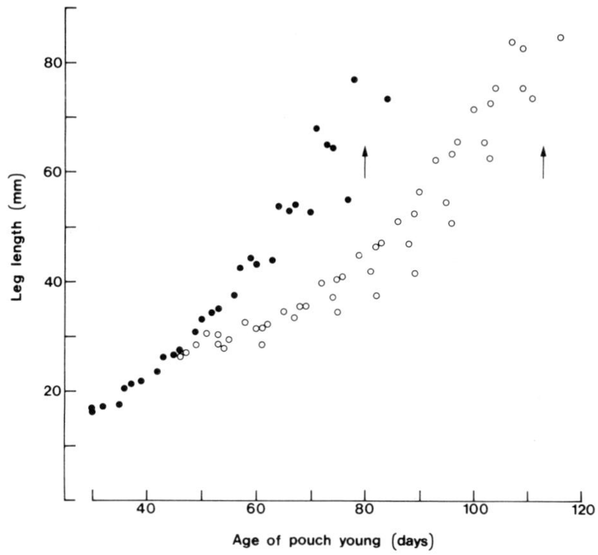

Fig. 3. Relationship between age and leg length in transferred pouch young. Symbols as for Fig. 2.

pouches had an average age of 112.8 days (s.d. 1.92) when their foster mothers gave birth and/or came into oestrus. These means are statistically significantly different (T-test P<.001).

The interval between the foster mother's parturitions was an average of 97.75 days (s.d. 2.06) for the mothers receiving a less advanced young and 95.00 (s.d. 1.87) for mothers receiving a more advanced young. This difference is nonsignificant at the 5% level (T-test).

DISCUSSION

The rate of growth was changed markedly in pouch young of Red Kangaroo *(Macropus rufus)*, Tammar Wallaby *(Macropus eugenii)* and Tasmanian Bettong *(Bettongia gaimardi)* which replaced a young of the same species but different age. In Red Kangaroo and Tammar Wallaby, perturbations to growth were so severe that some young died (Clark 1968; Findlay and Renfree 1984; Rose 1986). The similar result of asynchronous transfers of Brush-tailed Bettong young confirms the suggestion of Findlay and Renfree (1984) that the normal changes in milk composition which occur during lactation appear to be an intrinsic characteristic of the gland and occur regardless of alterations in the local stimuli produced by the pouch young.

Several approaches have been taken by different workers to determine the sequence of physiological responses leading from the young in the pouch to the diapause embryo in the uterus. Sharman (1965b) injected Red Kangaroos with oxytocin for several days after he had removed their pouch young and found that the diapause blastocyst remained in diapause while oxytocin was being administered. Tyndale-Biscoe and Hawkins (1977) repeated these findings in the Tammar Wallaby but found that injection of reserpine or prolactin also caused the embryo to remain in diapause after removal of pouch young. Significantly, only prolactin caused the embryo to remain in diapause after hypophysectomy (Tyndale-Biscoe and Hawkins 1977). It was known from Hearn's (1974) work that hypophysectomy alone caused an existing corpus luteum to develop. Renfree (1979) found that in Tammar Wallabies suckling small pouch young, denervation of the teat allowed milk synthesis and secretion to continue, but led to the embryo resuming development. Sharman (1965a) had previously found that when a Red Kangaroo suckled two young out of the pouch (both of them using the same teat, sequentially) a newly formed blastocyst entered diapause as if one young was suckling continuously within the pouch. Overall, these observations suggest that prolactin, released in response to neural signals from the sucked teat, acts directly on the corpus luteum which remains quiescent and fails to secrete the substance(s), possibly progesterone, necessary for the release of the embryo from diapause.

The concentration of prolactin in the peripheral blood was measured throughout pregnancy and lactation in the Tammar Wallaby (Hinds and Tyndale-Biscoe 1985). The concentration of prolactin was actually very low during the first half of lactation, rising abruptly at about 140 days to high levels that were maintained till the young left the pouch for the last time at about 250 days. At that time environmental factors had taken over control of corpus luteum and embryo, for the Tammar Wallaby breeds seasonally. The observations of falling prolactin levels in the Tammar Wallaby at the time of permanent emergence of the pouch young perhaps could be extrapolated to the continuously-breeding Red Kangaroo, for which blood prolactin concentrations are not known, and used to confirm Sharman's theory (1965b) that intermittent suckling and consequent lower peripheral concentrations of pituitary hormone lead to corpus luteum development and release of the embryo from diapause.

The Brush-tailed Bettong resembles the Red Kangaroo in breeding continuously. However, its rapid rate of growth in the last three weeks of pouch suckling makes unlikely the possibility of falling prolactin levels at the time the embryo resumes development. The present observations that the age of the young in the pouch did not affect the time of release from diapause suggest that it is not simply the frequency of suckling that determines the time of release. It could perhaps be argued that because young transferred to a more advanced pouch grow faster, they might also release the teat more frequently, as would an older young, but this viewpoint is negated by the observations of U5 and U8 which left the pouch at a lower weight viz. 128.3 gm and 60.6 gm

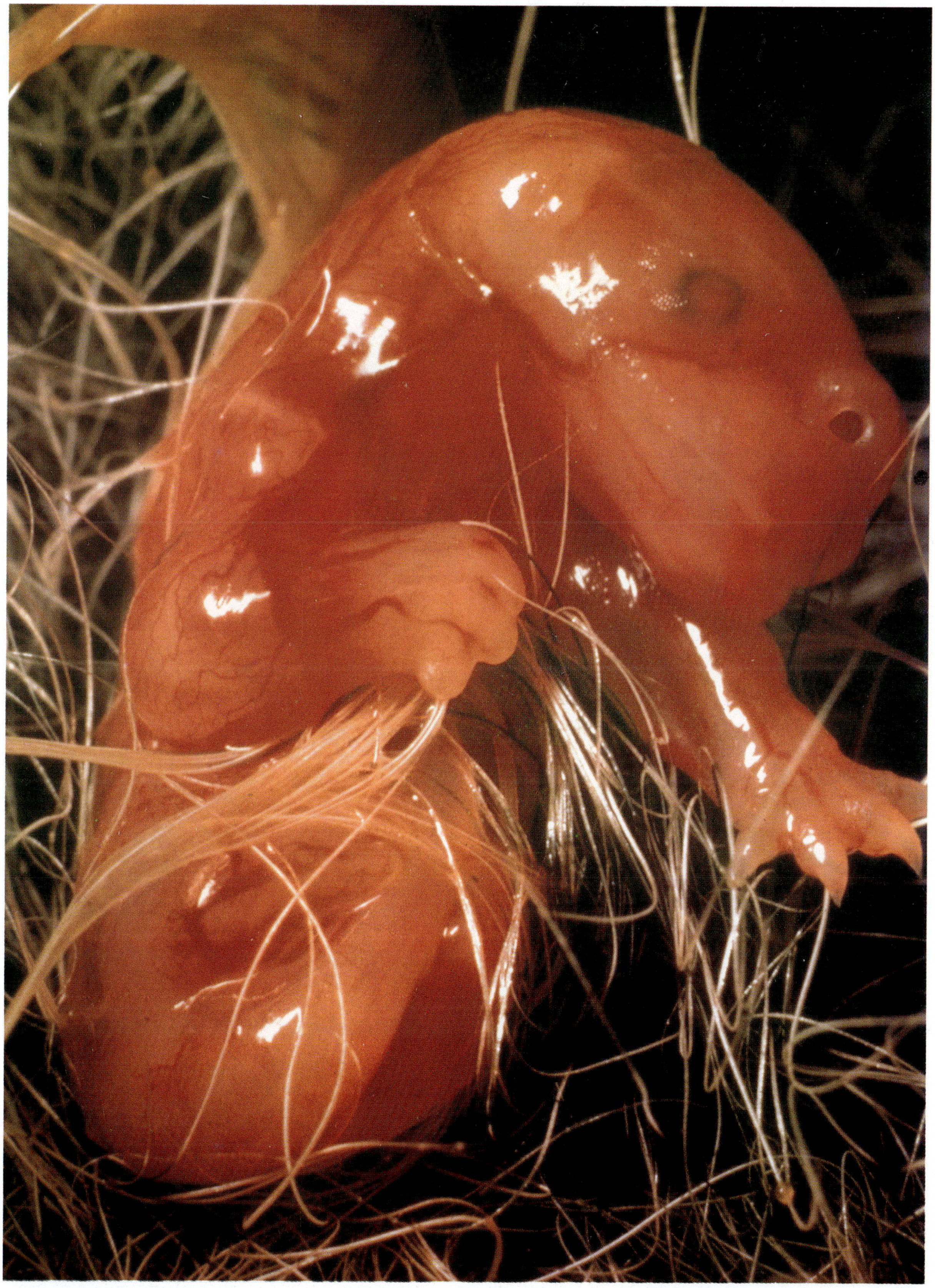

Fig. 1. A neonatal Tammar (16.7 mm crown rump length (CRL) on the mother's abdominal fur, less than a minute after birth. Note the flushed red skin and prominent vasculature. The powerful forelimbs and well-developed claws are used in a swimming motion during the climb to the pouch. The hindlimbs are rudimentary, but the rear half of the body may help direct the upward climb through a pendulum effect (see text). The large nostrils are open at this stage, and olfaction may aid young in finding a teat once it has climbed to the lip of the pouch. Photograph by David Parer.

Fig. 2. (a) An adult female Tammar in its normal posture; (b) As birth approaches the female becomes engrossed in pouch cleaning in a standing position; (c) At the time of birth the female adopts this characteristic posture, sitting far back, with tail forward and hips thrust upward, and intently licks the urogenital opening, pouch and the path between; (d) After birth, males show great interest in the female, persistently following her and sniffing at her urogenital and pouch areas; (e) Females will allow the males to mount repeatedly from about 8-24 h, but occasionally sooner, after birth. Photographs by Elizabeth Parer-Cook.

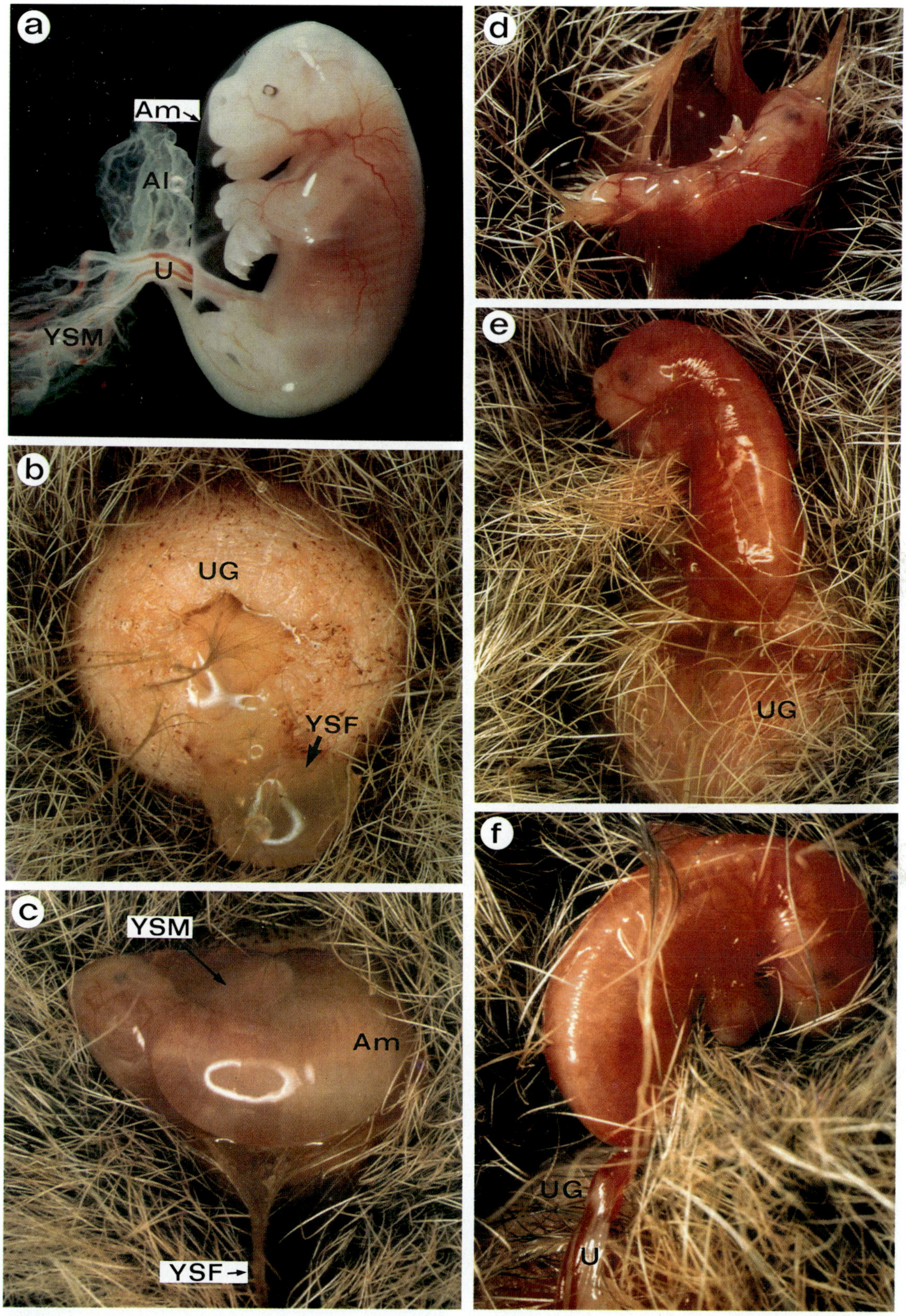

Fig. 3. Birth of the Tammar Wallaby. (a) The full term fetus (CRL 16 mm) at day 25 RPY surrounded by the amnion (AM). The collapsed allantois (Al), part of the yolk sac membrane (YSM), and the umbilical cord (U) can be seen. Note the prominent forelimbs and large head with the mouth and nostrils open. Photograph by I. A. Fox and M. B. Renfree from Renfree (1972). (b) At birth the urogenital opening (UG) "winks". Here a drop of the yellow sac fluid (YSF) has been expelled. (c) The Tammar is born enclosed in fetal membranes. The back of the fetus enclosed in amnion is protruding through a tear in the yolk sac membrane, from which yolk sac fluid is dripping. (d) Within 10-15 seconds of birth the young becomes active, orienting itself to face upwards, and tearing through the fetal membranes with its sharp claws. Note the fiery red colour now compared with (c). (e) Freed of the membranes the young begins to climb with a swimming motion. (f) At this stage the young is still attached by the umbilical cord. Photographs (b-f) by David Parer.

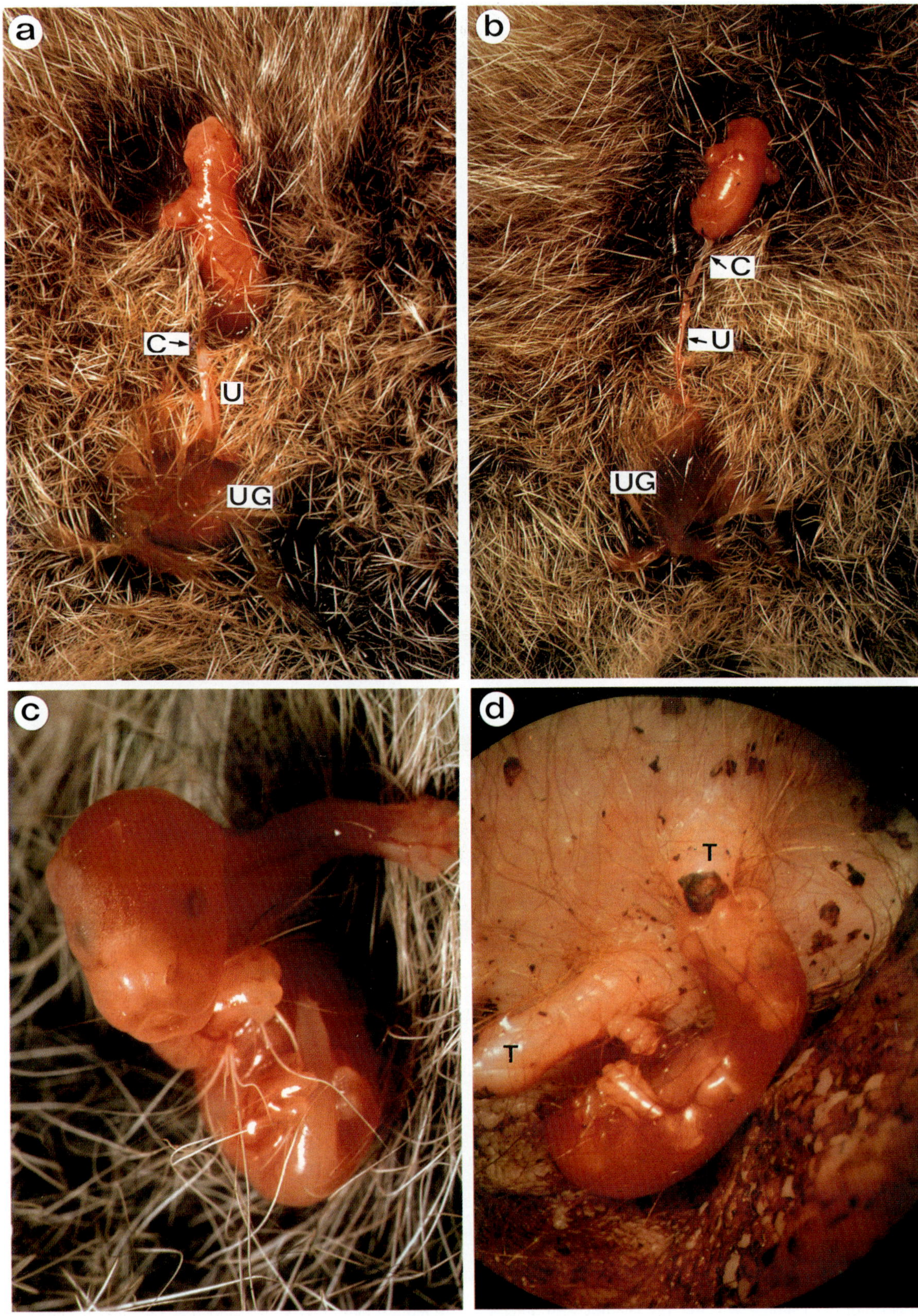

Fig. 4. The journey to the pouch. (a) The young has climbed about 1 cm above the urogenital opening (UG) to which it is still connected by the umbilicus. A constriction (C) can be seen, at which point the cord will soon break; (b) As the ascent proceeds the umbilicus stretches to 2 or 3 cm in length; (c) The umbilicus has broken by the time the young nears the lip of the pouch; (d) Inside the pouch the young takes only a few minutes to locate a teat (T) and attach. Note the remnants of pouch scale on the teat and particularly on the ventral pouch wall. Photographs by David Parer.

Table 2. Sequence of events at the time of birth in Tammars. Sources of hormonal data given in Fig. 7.

	Time before birth	Hormonal profile
Cleaning pouch	0-2 days	Progesterone high
Buds on teats	0-2 days	
Expression of colostrum	0-2 days	
Adoption of birth position and licking	0-4 hours	Prolactin increasing
Regular vigorous licking	0-1 hours	Prostaglandin rapidly increasing
Intense non-stop licking	5 minutes	Prostaglandin high Prolactin high Progesterone decreasing
Adoption of exaggerated birth position	5 minutes	
Female oblivious to disturbance	0-3 minutes	
Winking of urogenital opening	0-2 minutes	
Oozing of yellow yolk sac fluid ± blood	10-60 seconds	
Birth of neonate within membranes	0	Peak prostaglandin
	Time after birth	
Neonate immobile within membranes	10-15 seconds	Progesterone decreasing Prostaglandin decreasing rapidly Prolactin decreasing
Delivery of allantoic sac	5-25 seconds*	
Neonate clawing free of membranes	20-60 seconds	
Start of climbing by neonate	20-100 seconds	
Fluid and blood	1-5 minutes	
Arrival of neonate at pouch opening	1-3 minutes	
Attachment to teat	2-5 minutes	
Female licking and cleaning	0-40 minutes	
Traces of fluid and blood at urogenital opening	5-40 minutes	
Resumption of normal activities, by female but continued, less frequent licking and pouch cleaning	1-4 hours	Prostaglandin low Prolactin low Progesterone low Oestradiol increasing
Oestrus and mating	8-10 hours 12-15 hours	Peak oestradiol Peak LH
Neonate loses fiery red colour	≈ 24 hours	Oestradiol low
Ovulation	≈ 36 hours	LH low

*Note that Sharman and Calaby (1964) record delivery of the allantois before birth of the head.

relaxin; it is assumed the contractions are caused by the actions of oxytocin and/or prostaglandin (see Tyndale-Biscoe and Renfree 1987). It is also the time when progesterone withdrawal is occurring in response to the peripartum pulses of prostaglandin and prolactin (see below). Sharman and Calaby (1964) do not comment on the orientation of the young, but in only one of our eight closely observed births was the baby's head initially pointed up (i.e., towards the mother's head). The fetuses remained immobile for 10-15 seconds after delivery through the urogenital opening before beginning clawing and twisting motions which broke the amniotic membrane and led to their orientation changing so they were head towards the pouch opening, with their ventral surfaces towards the mother. The young moved actively within the amnion using their epitrichial claws to break it free themselves from the membranes (Fig. 3d-f), though some licking by the mother at the urogenital opening may have assisted in this. Fluid and blood were expelled from the urogenital opening after the birth of the baby but we never observed delivery of the placenta as Sharman and Calaby (1964) did.

Post-partum activities

The neonate, still attached by the umbilical cord, began to climb with a swimming motion (Fig. 3e-f), moving the head from side to side and using the forearms in alternate strokes to make progress through the fur. The umbilicus became stretched until it finally broke at a natural constriction when the neonate had climbed about 3 cm upwards (Fig. 4a,b).

The times taken for eight Red Kangaroo young from delivery to disappearance over the edge of the pouch averaged 2.92 minutes (range 1.25-4.33 min) (Sharman and Calaby 1964). In six Tammars in which the distance from urogenital opening to pouch is much less the range of time was 1-3 minutes (mean 2.25 + S.D. 0.75 min).

The female licked continuously throughout the birth process (see Table 2), but she did not lick a track in the fur for the young to follow. During birth and climbing the neonate is largely ignored by the female, who licked most actively for up to 40 minutes after birth around the urogenital opening where fluids and blood ooze. However, the

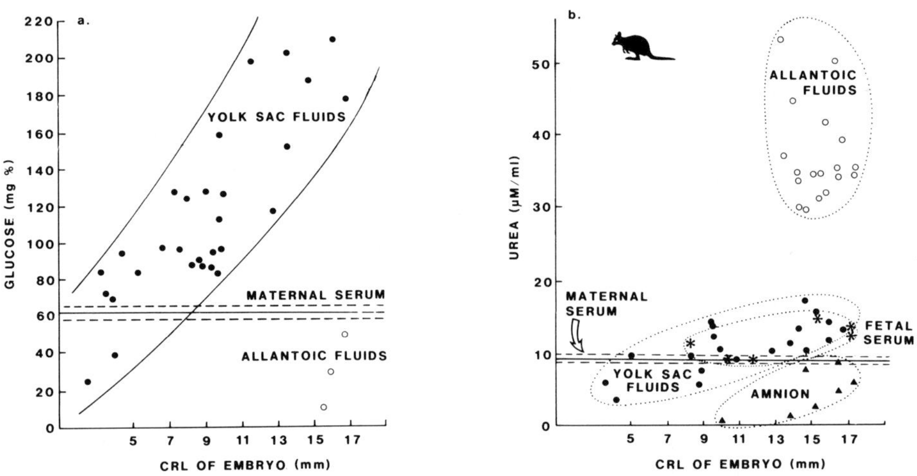

Fig. 5. Concentrations of glucose and urea in the fetal fluids of the Tammar. (a) Very high concentrations of glucose are found in yolk sac fluids towards the end of gestation; (b) In the last few days of gestation the allantois fills with clear fluid rich in urea. Urea concentration in yolk sac and amniotic fluid remain low. Redrawn from Renfree (1973).

mother's licking may facilitate ascent of the neonate since in both our study and that of Sharman and Calaby (1964) the young of anaesthetized animals took longer to reach the pouch. Sharman and Calaby (1964) suggest that this licking is important to break the umbilical cord. Sharman and Pilton (1961) also suggest that the mother occasionally licks over and ahead of the neonate as well as behind it, and that wetting of the young by the mother's tongue may be important to maintain cutaneous respiration. Hartman (1920, 1952) believed that wetting of the fur was important to prevent the young sticking to the fur. We have observed the mother licking over and ahead of the young in four of six closely observed "climbs". In one case the mother knocked the baby upside down. It immediately righted itself. The licking activity was continuous, with about 40% of the attention being given to the region between the cloaca and the pouch, 50% of the licking around the urogenital opening and 10% resting without licking. This licking may indeed be of importance for respiration in all marsupials, as *Monodelphis* fetuses, which are delivered directly into the pouch area at birth, are very roughly and vigorously licked by the mother, during which time they change from a sickly grey colour to a bright pink (J. L. VandenBerg, pers. comm.). The Tammar neonates also changed colour from a light pink when they appeared at the urogenital opening to the fiery red by the time they were attached to a nipple (Fig. 1, 3c,d; Fig. 4). This fiery red colour fades within 24 hours of birth.

Sharman and Calaby (1964) observed that the yolk sac membrane was the last of the embryonic membranes to appear at the urogenital opening, and they presumed these to be eaten by the mother (Sharman and Calaby 1964). This is in contrast to observations made on the Bandicoot, which retains its placental membranes (Lyne 1974), and our observations in which we have found remnants of the yolk sac within the birth canal in Tammars (M. B. Renfree and D. R. Blanden, unpub. obs.) and Potoroos (G. Shaw, unpub. obs.) autopsied within a day of birth.

There is still no agreement on how the newborn young navigates to the pouch. Numerous authors have removed noenates from the teat and placed them on the mother's fur which was inclined at various angles and they found that the young invariably climb upwards, even if this means that they move away from the pouch and against the "lie" of the fur (Hartman 1920, 1952; Sharman and Calaby 1964; Cannon *et al.* 1976). Because opossum neonates have no functional inner ear, Hartman (1920, 1952) suggested that this "negative geotropism" resulted from the fact that since only the foreclaws are used in climbing, the rear of the young remains passive, and acts like a pivot to keep the fore end pointed upwards. Other authors have assumed that the balance organs in the inner ear play a role. We have observed that young presenting at the vulva in a "head down" orientation reorient themselves before rupture of the fetal membranes, so they are clearly negatively geotropic independently of Hartman's pivot effect. Recently, Gemmell and Nelson (1988a) have observed that although young dasyurids have no cochlea and no semi-circular canals, they do have otoliths, cilia and microcilia, which may serve as a balance system, since these structures are neurologically connected to the brain (Gemmell and Rose (this volume) and Hughes *et al.* (this volume)). These authors have also described sensitive touch receptors (Merkel cells) surrounding the mouth of *Dasyurus hallactus, Trichosurus vulpecula,* and *Isoodon macrourus,* and have shown that there are sensory cells within the olfactory epithelium, and conclude that the senses of balance, touch and smell are developed sufficiently to aid the newborn marsupial in reaching the pouch (Gemmell, Peters and Nelson 1988; Gemmell and Nelson 1988a and b). Similar development of the sense organs is noted in macropodids (Gemmell and Rose, this symposium). However, when externally stimulated neonatal Tammars did not respond to touch or smell (Hughes *et al.,* this volume), so these authors conclude that reflex locomotor activity is the primary drive to climb to the pouch, that smell and touch are not of central importance, and that the Merkel cells function to stop the drive to climb and search and initiate teat attachment behaviour. Clearly this is an area for further study.

Once the neonate reaches the rim of the pouch (Fig. 4c) the baby climbs over the edge and down inside towards the nipples. Because the nostrils are open and the olfactory region of the brain is differentiated, it is assumed that this reversal of direction must be an olfactory or pheromonal response. Certainly the young are capable of complex co-ordinated behaviour in response to unknown external cues (see Hughes *et al.,* this symposium).

As observed by Sharman and Calaby (1964) for the Red Kangaroo, Tammar neonates need the ventral wall of the pouch to give them support while drawing the teat into the mouth (Fig. 4d). In three Tammars, attachment occurred within five minutes of entry into the pouch, and in one Red Kangaroo it took less than 17 minutes. Tammar young filmed within the pouch before attachment moved their heads from side to side seeking a teat, whilst firmly grasping the mammary hairs.

Despite our early impressions to the contrary, Tammars do not appear to show any circadian rhythm in the timing of birth (Fig. 6). Births were only slightly more frequent between 7 p.m. and 7 a.m. inclusive (15/26) compared to 8 a.m. to 6 p.m. (11/26). In 24 animals timed from removal of pouch young (RPY), (recorded to within five minutes) to birth (recorded to the minute), the duration of pregnancy was 26.44 ± 0.21 (SEM) days. Since seven previous observations were made only once a day, usually in the morning, (Berger 1970; Renfree and Tyndale-Biscoe 1973; Tyndale-Biscoe and Rodger 1978; Merchant 1979; Young and Renfree 1979; Hinds and Tyndale-Biscoe 1982a; Ward and Renfree 1984), the majority of newborns would not be seen until the morning of day 27 (day of RPY = day 0), and yet such observations based on widely different sample sizes have given remarkably good estimations of the true gestation period (see Tyndale-Biscoe and Renfree 1987). More frequent observations (12-hourly: Shaw and Renfree 1984; Harder *et al.* 1985; 8-hourly: Tyndale-Biscoe *et al.* 1983) provide values very close to those of this study.

INFLUENCE OF HORMONES ON FEMALE BEHAVIOUR IN THE PERI-PARTUM PERIOD

The behaviour which accompanies parturition is presumably induced by the concurrent endocrine changes. From the available evidence it appears that the endocrine control of parturition

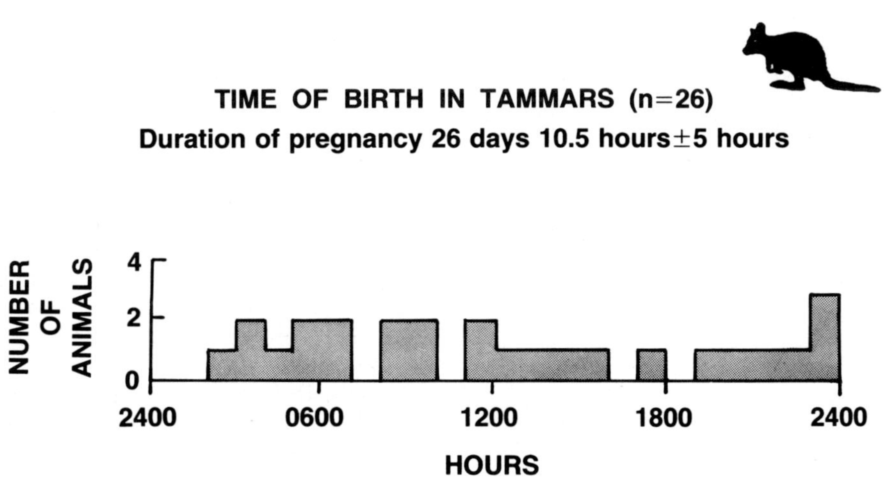

Fig. 6. Time of day at which Tammars give birth. Births can occur at any time of day or night.

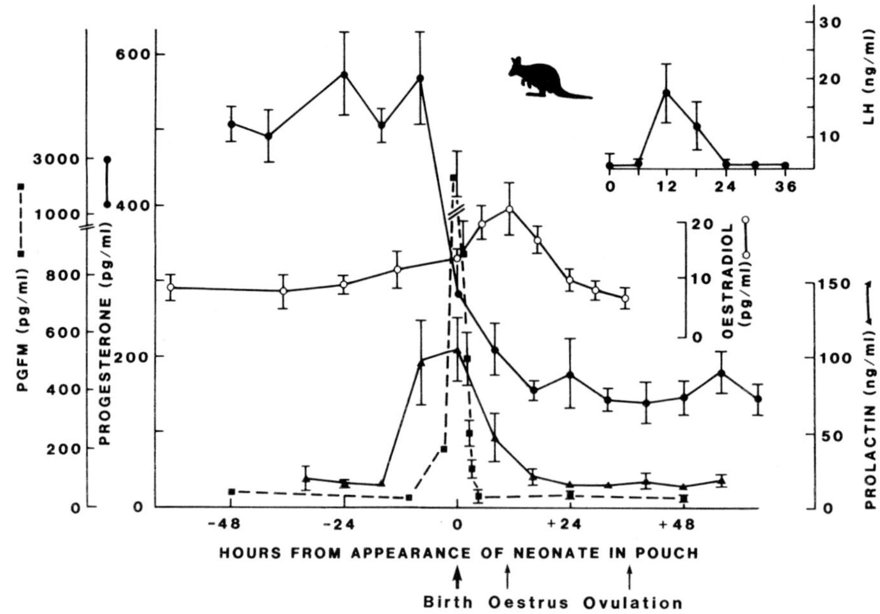

Fig. 7. Changes in the main reproductive hormones during the peri-partum period of the Tammar Wallaby. Data from Hinds and Tyndale-Biscoe 1982a and b; Tyndale-Biscoe *et al.* 1983; Shaw 1983; Shaw and Renfree 1984; Harder *et al.* 1984; Lewis *et al.* 1986.

is broadly similar in marsupials and eutherians. The corpus luteum plays an essential role in parturition, but there is growing evidence that for some species at least, the fetus and placenta may also be involved (Tyndale-Biscoe and Renfree 1987). Marsupial neonates are very small, and the anatomy of the female reproductive system is such that the young has to traverse a long distance, many times its own length, between the uterine cervix and the urogenital opening. Adequate hormonal preparation of the median vaginal birth canal, a transient structure in most non-macropodid species, is essential for successful delivery.

Whilst a great deal is now known about the hormonal changes in the peripartum period, much less is known about the role of hormones in the control of maternal parturient behaviour, and yet the maternal behaviour preceding parturition summarized above is highly specialized and represents one facet of the maternal recognition of preganacy. It is unfortunate that some authors (e.g., Keverne 1985) state so dogmatically that marsupials show little change in behaviour at parturition.

Steroid hormones

Progesterone concentrations are elevated in the last half of pregnancy in Tammars, but fall prior to parturition in all macropodid species studied (Tyndale-Biscoe and Renfree 1987). The fall is very rapid in the Tammar (Fig. 7), taking less than 8 hours (Fig. 8). Since progesterone withdrawal is essential for birth in several eutherian species, it was thought that Tammars (and presumably most other macropodids) required this signal. However exogenous progesterone at the end of pregnancy does not prevent birth, whether administered in physiological or pharmacological amounts, and it does not prevent oestrus (Ward and Renfree 1984). Similarly premature progesterone withdrawal after lutectomy between day 23 and day 25 does not affect the timing of birth or the occurrence of normal maternal parturient behaviour in the Tammar (Young and Renfree 1979; Harder *et al.* 1985), or the Potoroo (Bryant and Rose 1986). Progesterone, perhaps acting in concert with oestradiol and relaxin, may be important in softening the connective tissue of the median vagina before birth (Tyndale-Biscoe 1966). Lutectomy early in pregnancy or during embryonic diapause releases the hypothalamo-pituitary axis from inhibition, so the follicular growth, oestrus and ovulation occur 9-16 days later (Tyndale-Biscoe and Hawkins 1977; Evans, Tyndale-Biscoe and Sutherland 1980; Renfree, Wallace and Young 1982). In the Swamp Wallaby *(Wallabia bicolor)* oestrus and ovulation occur before birth (Tyndale-Biscoe and Renfree 1987) and therefore presumably before a progesterone withdrawal expected to be associated with birth. The conclusion from all these studies is that progesterone withdrawal is not essential for birth or oestrous behaviour.

Oestradiol is apparently not essential for normal parturition but is essential for post-partum oestrous behaviour. In the Tammar, the progesterone fall usually is followed within 8 h by a rise in oestradiol-17β, (Fig. 7) but oestrous behaviour is only seen in those females with an oestrogen rise (Shaw and Renfree 1984; Harder *et al.* 1984, 1985). Removal of the Graafian follicles on day 23 RPY abolishes this rise in oestradiol and the post-partum oestrus, the preovulatory luteinising hormone (LH) surge and ovulation, although birth occurs normally (Harder *et al.* 1985). Similarly, animals immunized against gonadotrophin releasing hormone (GnRH) failed to grow follicles and had no post-partum oestrus, although birth took place normally (Short, Flint and Renfree 1985). Two species, *M. giganteus* and *M. fuliginosis,* do not have a post partum oestrus (Poole 1975), and in many other macropodids oestrus does not invariably follow parturition and therefore progesterone withdrawal (Tyndale-Biscoe, Hearn and Renfree 1974). Conversely, in *M. rufus,* oestrus can occur within 14 days of drought-breaking rain in the absence of an active corpus luteum and, presumably, an exposure to high progesterone (Frith and Sharman 1964; Newsome 1964). Harder *et al.* (1985) suggest that it is the oestrogen:progesterone ratio which signals oestrous behaviour and the LH surge because animals which had Graafian follicles removed at d23 failed to show behavioural oestrous although they had ovulated at d35-36, 13 days after the surgery and six days out of phase with the control animals. However, in the same study mating and oestrus failed to occur at the normal time in five sham-operated Tammars which suffered from heat stress, although 4/5 had a post-partum oestradiol rise with an oestradiol:progesterone ratio similar in magnitude to control animals, and two had ovulated. From the results summarized above it is clear that the progesterone levels have little influence on oestrous behaviour, and that rising oestradiol levels are essential for oestrous behaviour, but that external factors can override even a permissive oestrogen:progesterone ratio.

Pituitary hormones and prostaglandins

It is clear that gonadotrophins are not essential for sustained pregnancy (Hearn 1974) or birth (Short *et al.* 1985). However, a well-defined LH surge, lasting about 12 h, is always associated with oestrus. Mating occurs about 8 h before and ovulation approximately 24 h after the maximal concentration of LH (Fig. 8, Sutherland, Evans and Tyndale-Biscoe 1980; Tyndale-Biscoe, Hinds, Horn and Jenkin 1983). This LH-surge is dependent on the post partum oestradiol rise which it follows by about 7 h (Harder *et al.* 1985).

The fetus or placenta apparently provides an essential stimulus for initiating parturition. Clark (1968) terminated embryonic diapause prematurely by steroid injections up to 18 days before the CL was reactivated by RPY. Birth always occurred at the expected time after the onset of steroid treatment (and thus embryonic reactivation) and was unrelated to the time of RPY,

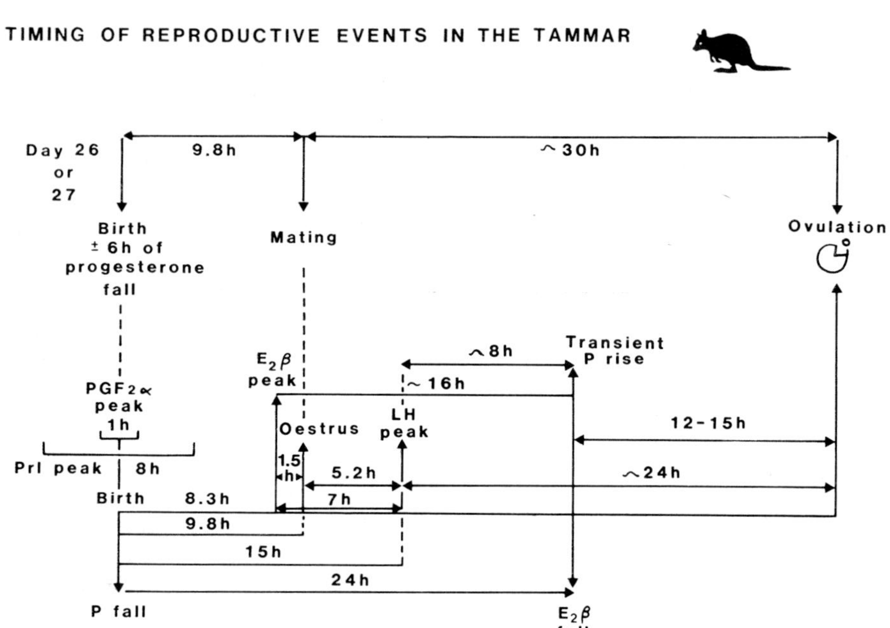

Fig. 8. Schematic diagram illustrating the timing of the major endocrinological and behavioural events in the peri-partum period of the Tammar. Vertical lines and arrows indicate significant events, such as birth, mating, oestrus, and ovulation and peak or trough values of hormones; horizontal lines and arrows indicate the duration of, or the interval between, each event. Data from Tyndale-Biscoe and Rodger 1978; Sutherland *et al.* 1980; Tyndale-Biscoe *et al.* 1983; Shaw and Renfree 1984; Harder *et al.* 1984, 1985; Lewis *et al.* 1986.

whereas oestrus invariably occurred about 33 days after RPY regardless of the timing of steroid treatment. How the full term fetus signals that it is ready for birth is unknown, but there is a pronounced but transient peak of prolactin (Figs 7, 8) from about 4 h before birth which does not occur at a comparable stage of the non-pregnant cycle (Tyndale-Biscoe *et al.* 1983; Hinds and Tyndale-Biscoe 1982b; Tyndale-Biscoe, Hinds and Horn 1988). There is also a marked but extremely short-lived increase in prostaglandin metabolite (PGFM) concentrations in the plasma at the time of birth (Fig. 7; Lewis, Fletcher and Renfree 1986). At present it is not known how this surge of prostaglandin secretion is regulated, or whether it is of uterine, placental or fetal origin, but the timing of the peaks of both prolactin and prostaglandin secretion make them likely candidates for at least some aspects of birth behaviour.

The hormonal control of parturient behaviour in eutherian mammals is still incompletely understood, although prolactin seems to control certain aspects of maternal behaviour (and nesting in particular) in the rabbit (Keverne 1985). Vaginal distension also appears to be important in the ewe for triggering parturient behaviour, since it initiates a strong maternal desire to lick the fetal fluids and the newborn; this reflex can be elicited in non-pregnant nulliparous animals given appropriate oestrogenic priming (Keverne, Levy, Poindron and Lindsay 1983). Prostaglandin apparently has an important role in birth behaviour, because injection of $PGF_{2\alpha}$ into non-pregnant cycling female Tammars caused all of them to adopt the birth position within a few minutes of the injection at either day 23 or day 26 RPY (Hinds, Tyndale-Biscoe, Shaw, Fletcher and Renfree 1988). This behavioural change in response to prostaglandin may be a response to the uterine and vaginal contractions it induces, but there is evidence that it may represent a direct central action of prostaglandin (Shaw, unpublished results).

Fig. 9. Males fighting. (a) The males lock in combat, clawing at each other's faces; (b) Heads are held well back to avoid their opponent's claw in their eyes; (c) The tail is used for rearward support, whilst they rise on to their toes. If the other male retreats even slightly, the hind legs will be drawn up and used in a double footed kick against the opponent's body. Photographs by M. B. Renfree.

MALE BEHAVIOUR AND ENDOCRINOLOGY

A number of marsupials have a seasonal breeding pattern, but in male macropodids spermatogenesis is continuous once they are sexually mature (Setchell 1977). There is no seasonal decline in spermatogenesis, testis size, epididymal size or seminiferous tubule diameter, even in Bennett's and Tammar Wallabies in which the female has a highly seasonal breeding pattern, although in both these species plasma testosterone concentrations and prostate weight increased during the main periods of copulatory activity (Catt 1977; Inns 1982). Hearn (1975) found no seasonal changes in gonadotrophin levels in males, but hypophysectomy immediately resulted in a decline in testis and prostate weights. Similarly, Catling and Sutherland (1980) demonstrated that testosterone, LH and follicle-stimulating hormone (FSH) levels in males *isolated* from females did not change seasonally, but marked seasonal changes in testosterone and LH occurred in males associating with pregnant and oestrous females. The seasonal rise began about two weeks before mating, and so they suggest a pheromone produced by the female may be responsible. Certainly during the 1978-79, 1981 and 1983 seasons on Kangaroo Island we have observed male Tammars closely associating with females to form semi-isolated pairs from late December, which is a month before the first births are expected (M. B. Renfree, A. P. F. Flint, G. Shaw and R. V. Short, unpub. obs.). This is also suggestive of pheromonal effect. By contrast, animals we observed in the wild during the non-breeding season in August and September were in large mixed sex groups.

Once the breeding season has begun, the males begin sniffing around the genital region of the female, for up to two days before oestrus (Fig. 2d). Whether progesterone and/or oestrogen induce a pouch pheromone attractive to males remains to be established. Sharman and Calaby (1964), and Maynes (1973) have observed in the Red Kangaroo and the Parma Wallaby males tasting urine, while Ganslosser (1979), Coulson and Croft (1981) and T. P. Fletcher (unpub. obs.) have observed male Tree Kangaroos, Grey and Red Kangaroos and Tammars displaying flehmen, so pheromones are clearly important in mating behaviour. The males also inspect the female's pouch, and female Tammars examined at the time of birth have swollen pouches, with oedema of the subcutaneous connective tissue of the ventral pouch wall. It is interesting to note that a similar pouch swelling occurs in mid to late pregnancy in dasyurids when no oestradiol is detected but when progesterone is elevated (T. P. Fletcher, unpublished results), so that this change in Tammars may also be progesterone-dependent. By contrast, Rose (1986) has observed a significant tightening of the pouch muscle of the Tasmanian Bettong on the night the young vacates the pouch, immediately before birth and/or oestrus and suggests that oestrogen is the hormone controlling pouch smooth muscle contraction and the maternal response. However, in the light of the effects of prostaglandin in the Tammar, it may be that prostaglandin provides the hormonal signal for pouch exit in this species by both a behavioural influence and its effects on smooth muscle.

In most macropodid species the presence of oestrous females induces conflict between the males, and in the wild this may result in sparring. Fighting was first described in the Red Kangaroo (Sharman and Calaby 1964), and much more extensive details are given by Ganslosser (1989: this volume). Briefly, agnostic behaviour is essentially identical in Tammars. In threat behaviour, the males gain extra height by adopting an upright stance, supported on their tails and the extremities of their hind feet, they may lick or bite the gular presternal glandular area, grasp tufts of grass in their hands and rub their chest on the ground. When fighting, the males clasp each other with their forearms, clawing at each other's eyes, and using the tail for balance they kick towards the opponent's abdomen (Fig. 9). The fights are usually brief but intense, although in several Tammars we have observed bouts that last for 10-15 minutes. The fight concludes with the winning buck chasing the loser, and then pawing and grabbing the ground, chinning the disturbed earth. The male may then retire to lick his wounds, or resume mating.

Copulation has been described in detail in a dozen or so macropodid species (Hughes 1962; Sharman and Calaby 1964; Sharman *et al.* 1966; Stodart 1966; Maynes 1973 and Tyndale-Biscoe and Rodger 1978). Females are followed by the male, who grasps at their tail and inspects the pouch and genital regions. The male occasionally makes a "clucking" noise, and in the smaller macropodids moves the females tail from side to side, occasionally cuffing her. The females usually move away, sometimes making a hissing or coughing sound, for some time before they submit to mounting (Fig. 2e). The male grasps the female around the lower abdomen, while the female has both forefeet on the ground and raises her hindquarters, at times lifting the male clear off the ground. Copulation lasts 2 minutes in the Potoroo and Parma Wallaby, 5 minutes in Bennett's Wallaby, 8 minutes in the Tammar, 15-20 minutes in the Red Kangaroo and 40-50 minutes in the Eastern Grey Kangaroo (Tyndale-Biscoe and Renfree 1987). Most males copulate with the same female only once, but several males may copulate with a female during her oestrus, which lasts about 12-15 hours in the Tammar, (see Fig. 8). During oestrus a female is often persued by six to eight males at once. It must be remembered that at least in the Tammar, all this frenetic mating activity is taking place within a few hours of delivery of the newborn young and its attachment to the teat.

CONCLUSIONS

Successful birth in macropodids is dependent on a co-ordinated series of endocrine and behavioural events. Whilst much is now known of the hormonal changes at this time in one species, the Tammar, our knowledge of how these hormones control maternal parturient behaviour is much more scanty and in many places, conjectural. Up to two days before birth, when there are no obvious changes in circulating hormones, pouch cleaning begins. Four hours before birth, prolactin rises. Pouch cleaning activity also becomes more frequent 4 h before birth but the main period of pouch cleaning and adoption of the birth position is clearly related to rising prostaglandin levels which also cause uterine contractions. Progesterone falls under the influence of the prostaglandin-induced prolactin peak, but progesterone withdrawal is not essential for birth. By contrast, the post-partum oestradiol rise is essential for oestrus and mating, ovulation following much later as a result of the LH peak. We still have little understanding of the remarkable anatomical, physiological and behavioural adaptations that enable the precocious marsupial noenate, to find its way unaided from the urogenital opening to the teat, but the several studies reported in this symposium illustrate its anatomical and physiological complexity. To watch the birth of a marsupial is to witness one of Nature's most amazing events, and we feel privileged to have been onlookers. We hope that the film we have made will enable others to share in this experience.

ACKNOWLEDGEMENTS

We thank Ainslie Brown, Dione Gilmore, Amy Jetton, Martin Stone, David Taggart and Jill Shaw for assistance with various aspects of this research, and Monash University and the Australian Research Grants Scheme for financial support. The film was produced by David Parer and the ABC Natural History Unit, Carre Street, Elsternwick, Melbourne, and is available for borrowing for non-commercial use from M. B. Renfree.

REFERENCES

Berger, P. J., 1970. Reproductive biology of the tammar wallaby, *Macropus eugenii*. Ph.D. Thesis, Tulane University. *Diss. Abst. Int.* **31B**: 3760-61.

Bryant, S. L. and Rose, R. W., 1986. The growth and role of the corpus luteum throughout delayed gestation in the marsupial, *Potorous tridactylus*. *J. Reprod. Fert.* **76**: 409-14.

Cannon, J. R., Bakker, H. R., Bradshaw, S. D. and McDonald, I. R., 1976. Gravity as the sole navigational aid to the newborn quokka. *Nature. Lond.* **259**: 42.

Catling, P. C. and Sutherland, R. L., 1980. Effect of gonadectomy, season, and the presence of females on plasma testosterone, luteinizing hormone, and follicle stimulating hormone levels in male tammar wallabies *(Macropus eugenii)*. *J. Endocrinol.* **86**: 25-33.

Catt, D. C., 1977. The breeding biology of Bennett's wallaby *Macropus rufogriseus fruticus)* in South Canterbury, New Zealand. *J. Zool.* **4**: 401-11.

Clark, M. J., 1968. Termination of embryonic diapause in the red kangaroo, *Megaleia rufa,* by injection of progesterone or oestrogen *J. Reprod. Fert.* **15**: 347-55.

Collie, A., 1830. On some particulars connected with the natural history of the kangaroo. *J. Zool.* **5**: 238-41.

Coulson, G. M. and Croft, D. B., 1981. Flehmen in kangaroos. *Aust. Mammal.* **4**: 139-40.

Evans, S. M., Tyndale-Biscoe, C. H. and Sutherland, R. L., 1980. Control of gonadotrophin secretion in the female tammar wallaby *(Macropus eugenii)*. *J. Endocrinol* **86**: 13-23.

Frith, H. J. and Sharman, G. B., 1964. Breeding in wild populations of the red kangaroo, *Megaleia rufa*. *C.S.I.R.O. Wildl. Res.* **9**: 86-114.

Ganslosser, U., 1979. Soziale Interaktionen des Doriabaumkanguruhs, *Dendrolagus dorianus* Ramsay, 1833 (Marsupialia, Macropodidae). *Saugertierkd.* **44**: 1-18.

Ganslosser, U. (this volume). Agonistic behaviour in Macropodoids – A Review.

Gemmell, R. T. and Nelson, J., 1988a. The ultrastructure of the olfactory system of three newborn marsupial species. *Anat. Rec.* (in press).

Gemmell, R. T. and Nelson, J., 1988b. The vestibular system of the newborn marsupial cat, *Dasyurus hallucatus*. *Anat. and Embryol.* (in press).

Gemmell, R. T., Peters, B. and Nelson, J., 1988. Ultrastructure identification of Merkel cells around the mouth of the newborn marsupial. *Anat. and Embryol.* **177**: 403-08.

Gemmell, R. T. and Rose, R. W. (this volume). The senses involved in movement of some newborn Macropodoidea and other marsupials from cloaca to pouch.

Harder, J. D., Hinds, L. A., Horn, C. A. and Tyndale-Biscoe, C. H., 1984. Oestradiol in follicular fluid and in utero-ovarian venous and peripheral plasma during parturition and post-partum oestrus in the tammar, *Macropus eugenii*. *J. Reprod. Fert.* **72**: 551-58.

Harder, J. D., Hinds, L. A., Horn, C. A. and Tyndale-Biscoe, C. H., 1985. Effects of removal in late pregnancy of the corpus luteum, Graafian follicle or ovaries on plasma progestorone, oestradiol, LH, parturition and post-partum oestrus in the tammar, *Macropus eugenii*. *J. Reprod. Fert.* **75**: 449-59.

Hartman, C. G., 1920. Studies in *Didelphis*. I. The phenomena of parturition. *Anat. Rec.* **19**: 251-62.

Hartman, C. G., 1952. *Possums,* University of Texas Press: Austin, Pp. 174.

Hearn, J. P., 1974. The pituitary gland and implantation in the tammar wallaby, *Macropus eugenii*. *J. Reprod. Fert.* **39**: 235-41.

Hearn, J. P., 1975. The role of the pituitary in the reproduction of the male tammar wallaby, *Macropus eugenii*. *J. Reprod. Fert.* **42**: 399-402.

Hinds, L. A. and Tyndale-Biscoe, C. H., 1982a. Plasma progesterone level in the pregnant and non-pregnant tammar, *Macropus eugenii*. *J. Endocrinol.* **93**: 99-107.

Hinds, L. A. and Tyndale-Biscoe, C. H., 1982b. Prolactin in the marsupial, *Macropus eugenii,* during the estrous cycle, pregnancy and lactation. *Biol. Reprod.* **26**: 391-98.

Hinds, L., Tyndale-Biscoe, C. H., Shaw, G., Fletcher, T. P. and Renfree, M. B., 1988. Role of prostaglandin in parturition in the tammar. *Proc. Aust. Soc. Reprod. Biol.* p. 59.

Hughes, R. L., 1962. Reproduction in the macropod marsupial *Potorous tridactylus* (Kerr). *Aust. J. Zool.* **13**: 533-43.

Hughes, R. L., Hall, L. S., Tyndale-Biscoe, C. H. and Hinds, L. A. (this volume). Evolutionary implications of macropodid organogenesis.

Inns, R. W., 1982. Seasonal changes in the accessory reproductive system and plasma testosterone levels of the male tammar wallaby, *Macropus eugenii,* in the wild. *J. Reprod. Fert.* **66**: 675-80.

Keverne, E. B., 1985. Reproductive Behaviour. Pp. 133-75 *in* Reproductive Fitness, Reproduction in Mammals, Book 4, 2nd edition ed. by C. R. Austin and R. V. Short, Cambridge University Press.

KEVERNE, E. B., LEVY, F., POINDRON, P. AND LINDSAY, D., 1983. Vaginal stimulation: an important determinant of maternal bonding in sheep. *Science* **219**: 81-83.

LEWIS, P. R., FLETCHER, T. P. AND RENFREE, M. B., 1986. Prostaglandin in the peripheral plasma of tammar wallabies during parturition. *J. Endocrinol.* **111**: 103-09.

LYNE, A. G., 1974. Gestation period and birth in the marsupial *Isoodon macrourus*. *Aust. J. Zool.* **22**: 303-09.

MCCRADY, E., 1938. The embryology of the opossum. *Am. Anat. Mem.* **16**: 1-233.

MAYNES, G. M., 1973. Reproduction in the parma wallaby, *Macropus parma* Waterhouse. *Aust. J. Zool.* **21**: 331-51.

MERCHANT, J. C., 1979. The effect of pregnancy on the interval between one oestrus and the next in the tammar wallaby, *Macropus eugenii*. *J. Reprod. Fert.* **56**: 459-63.

MERCHANT, J. C. AND CALABY, J. H., 1981. Reproduction biology of the red-necked wallaby *(Macropus rufogriseus banksianus)* and Bennett's wallaby *(M. r. rufogriseus)* in captivity. *J. Zool. (Lond.)* **194**: 203-07.

MIEGS, C. D., 1847. Memoirs on the reproduction of the opossum *Didelphys virginiana*. *Proc. Am. Philos. Soc.* **4**: 327-30.

NEWSOME, A. E., 1964. Anoestrus in the red kangaroo *Megaleia rufa* (Desmarest). *Aust. J. Zool.* **12**: 9-17.

POOLE, W. E., 1975. Reproduction in the two species of grey kangaroos, *Macropus giganteus* Shaw, and *M. fuliginosus* (Desmarest). II. Gestation, parturition and pouch life. *Aust. J. Zool.* **23**: 333-53.

POOLE, W. E. AND CATLING, P. C., 1974. Reproduction in the two species of grey kangaroo, *Macropus giganteus* Shaw and *M. fuliginosus* (Desmarest). I. Sexual maturity and oestrus. *Aust. J. Zool.* **22**: 277-302.

RENFREE, M. B., 1972. Embryo-maternal relationships in the tammar wallaby, *Macropus eugenii* (Desmarest). Ph.D. Thesis, Australian National University, Canberra.

RENFREE, M. B., 1973. The composition of fetal fluids of the marsupial *Macropus eugenii*. *Dev. Biol.* **33**: 62-79.

RENFREE, M. B., 1983. Marsupial reproduction: the choice between placentation and lactation. *In* Oxford Reviews of Reproductive Biology 5 ed. by C. A. Finn. Oxford University Press.

RENFREE, M. B. AND TYNDALE-BISCOE, C. H., 1973. Intrauterine development after diapause in the marsupial *Macropus eugenii*. *Dev. Biol.* **32**: 28-40.

RENFREE, M. B., WALLACE, G. I. AND YOUNG, I. R., 1982. Effects of progesterone, oestradiol 17β and androstenedione on follicular growth after removal of the corpus luteum during lactational and seasonal quiescence in the tammar wallaby. *J. Endocrinol.* **92**: 397-403.

REYNOLDS, H. C., 1952. Studies on reproduction in the opossum *(Didelphis virginiana virginiana)*. *Univ. Calif. Publ. Zool.* **52**: 223-84.

ROSE, R. W., 1986. The control of pouch vacation in the Tasmanian bettong, *Bettongia gaimardi*. *Aust. J. Zool.* **34**: 485-91.

RUSSELL, E. M. (this volume). Maternal behaviour in the Macropodoidea.

SETCHELL, B. P., 1977. Reproduction in male marsupials. Pp. 411-57 *in* The Biology of Marsupials ed. by B. Stonehouse and D. Gilmore. Macmillan: London.

SHARMAN, G. B. AND CALABY, J. H., 1964. Reproductive behaviour in the red kangaroo, *Megaleia rufa,* in captivity. *C.S.I.R.O. Wildl. Res.* **9**: 58-85.

SHARMAN, G. B., CALABY, J. H. AND POOLE, W. E., 1966. Patterns of reproduction in female diprotodont marsupials. *Symp. Zool. Soc. Lond.* **15**: 205-32.

SHARMAN, G. B. AND PILTON, P. E., 1964. The life history and reproduction of the red kangaroo *(Magaleia rufa)*. *Proc. Zool. Soc. Lond.* **142**: 29-48.

SHAW, G., 1983. Effect of PGF-2α on uterine activity, and concentrations of 13, 14-dihydro-15-keto-PGF-2α in peripheral plasma during parturition in the tammar wallaby *(Macropus eugenii)*. *J. Reprod. Fert.* **69**: 429-36.

SHAW, G. AND RENFREE, M. B., 1984. Concentrations of oestradiol-17β in plasma and corpora lutea throughout pregnancy in the tammar, *Macropus eugenii*. *J. Reprod. Fert.* **72**: 29-37.

SHORT, R. V., FLINT, A. P. F. AND RENFREE, M. B., 1985. Influence of passive immunization against GnRH on pregnancy and parturition in the tammar wallaby, *Macropus eugenii*. *J. Reprod. Fert.* **75**: 567-75.

STODART, E., 1966. Observations on the behaviour of the marsupial *Bettongia lesueuri* (Quoy and Gaimard) in an enclosure. *C.S.I.R.O. Wildl. Res.* **11**: 91-99.

STIRLING, E. C., 1889. On some points in the anatomy of the female organs of generation of the kangaroo, especially in relation to the acts of impregnation and parturition. *Proc. Zool. Soc. Lond.* **1889**: 433-40.

SUTHERLAND, R. L., EVANS, S. M. AND TYNDALE-BISCOE, C. H., 1980. Macropodid marsupial luteinising hormone: validation of assay procedures and changes in plasma levels during the oestrous cycle in the female tammar wallaby *(Macropus eugenii)*. *J. Endocrinol.* **86**: 1-12.

TYNDALE-BISCOE, C. H., 1966. The marsupial birth canal. *Symp. Zool. Soc. Lond.* **15**: 233-50.

TYNDALE-BISCOE, C. H. AND HAWKINS, J., 1977. The corpora lutea of marsupials, aspects of function and control. Pp. 245-252. *In* Reproduction and Evolution ed. by J. H. Calaby and C. H. Tyndale-Biscoe. Australian Academy of Science: Canberra.

TYNDALE-BISCOE, C. H. AND RODGER, J. C., 1978. Differential transport of spermatozoa into the two sides of the genital tract of a monovular marsupial, the tammar wallaby *Macropus eugenii*. *J. Reprod. Fert.* **52**: 37-43.

TYNDALE-BISCOE, C. H., HEARN, J. P. AND RENFREE, M. B., 1974. Control of reproduction in macropodid marsupials. *J. Endocrinol.* **63**: 589-614.

TYNDALE-BISCOE, C. H., HINDS, L. A., HORN, C. A. AND JENKIN, G., 1983. Hormonal changes at oestrus, parturition and post-partum oestrus in the tammar wallaby *(Macropus eugenii)*. *J. Endocrinol.* **96**: 155-61.

TYNDALE-BISCOE, C. H., HINDS, L. A. AND HORN, C. A., 1988. Fetal role in the control of parturition in the tammar, *Macropus eugenii*. *J. Reprod. Fert.* **82**: 419-28.

TYNDALE-BISCOE, C. H. AND RENFREE, M. B., 1987. *Reproductive physiology of marsupials*. Cambridge University Press. 476 pages.

WARD, K. L. AND RENFREE, M. B., 1984. Effects of progesterone on parturition in the tammar, *Macropus eugenii*. *J. Reprod. Fert.* **72**: 21-28.

YOUNG, I. R. AND RENFREE, M. B., 1979. The effects of corpus luteum removal during gestation on parturition in the tammar wallaby *(Macropus eugenii)*. *J. Reprod. Fert.* **56**: 249-54.

The senses involved in movement of some newborn Macropodoidea and other marsupials from cloaca to pouch

R. T. GEMMELL[1] and R. W. ROSE[2]

Evidence is presented that newborn marsupials have a sense of direction, determined in part by the inner ear, and are able to detect odours, and distinguish certain structures by touch. These senses seem to be important in attaining directed movement from the cloaca to the pouch and attaching to the teat immediately after birth.

INTRODUCTION

PRIOR to Hartman's (1920) account of birth in the Virginia Opossum, *Didelphis virginiana*, a plethora of hypotheses were advanced to explain how marsupial pouch young become attached to the teat (see historical introduction to Reproductive Physiology in Marsupials, by Tyndale-Biscoe and Renfree 1987). The observations of Carl Hartman and his wife (Hartman 1920) of the birth of a litter of *D. virginiana* and the subsequent migration of the newborn from the cloaca to the pouch confirmed that the young travelled to the pouch and attached to the teat without obvious assistance from the mother, although the mother may position herself in a manner such that her pouch is easily accessible and in some species is directly above the urogenital sinus. Although she may also lick the region around the pouch, the young make this journey under their own power and direction. Birth has now been observed and described in a number of marsupials (Russell 1982).

The senses used by the newborn marsupial in moving from the urogenital sinus to the pouch and locating the teat are still not understood fully. We postulate a functional role for some organs in the movement to the pouch and attachment to the teat in several marsupial species in particular the rat-kangaroos based on morphological (including ultrastructural) evidence.

NEGATIVE GEOTROPISM AND THE VESTIBULAR SYSTEM

The newborn Virginia Opossum, *D. virginiana* was reported to be "negative geotropic" by Hartman (1920). He stated that "If the skin be tilted, the embryos can be made to travel upward and even away from the pouch, for they are negatively geotropic". Presumably the newborn opossum can sense gravity and crawl upwards. Larsell, McCrady and Zimmermann (1935) and McCrady (1938) in subsequent papers cast doubt on this theory. Larsell *et al.* (1935) suggested that newborn marsupials orientated themselves away from gravity without the assistance of a righting reflex or any consciousness of "up" and "down". The orientation of the young is due to the structure of the body at the time, that it has functional forelegs and non-functional hind legs. Thus, as the young crawls upwards, its body and hindlegs swing downwards in obedience to the pull of gravity. McCrady (1938) stated that Langworthy (1928) had reported that the vestibular system of the opossum does not begin to function until about 41 days after birth and he proposed that if the young can sense gravity, it does so without the assistance of any vestibular mechanism. The senses used by the young to attach to the teat also interested McCrady (1938), who stated that possibly smell and possibly cutaneous sensations of warmth and moisture and touch were involved, but certainly not sight.

Reynolds (1952) in his description of birth in the opossum, noticed that although the young crawled in an upward direction, some young would have missed the pouch by crawling past the pouch entrance. Upon reaching the level of the pouch, young turn at an angle of approximately 90 degrees in order to enter the pouch. Reynolds deduced that there are factors other than negative geotropism which enable the young to reach the pouch. Later reports that some newborn young move other than vertically from the cloaca to the pouch and that the route is horizontal in some species (Lyne *et al.* 1959; Sharman and Calaby 1964; Lyne 1974; Poole 1975) further weakened the gravity hypothesis. However, evidence in support of it was presented by Cannon *et al.* (1976). Negative geotropism was postulated to be the sole mechanism in directing movement of the newborn Quokka, *Setonix brachyurus* towards the pouch.

Larsell, McCrady and Zimmermann (1935) concluded that "no part of the organ, either vestibular or acoustic, is ready to function at

[1]Department of Anatomy, University of Queensland, St. Lucia, Brisbane, Australia 4067.
[2]Department of Zoology, University of Tasmania, Sandy Bay, Hobart, Tasmania 7005.

Pages 339-47 in KANGAROOS, WALLABIES AND RAT-KANGAROOS ed. by G. Grigg, P. Jarman and I. Hume. Surrey Beatty & Sons Pty Limited, New South Wales, Australia, 1989.

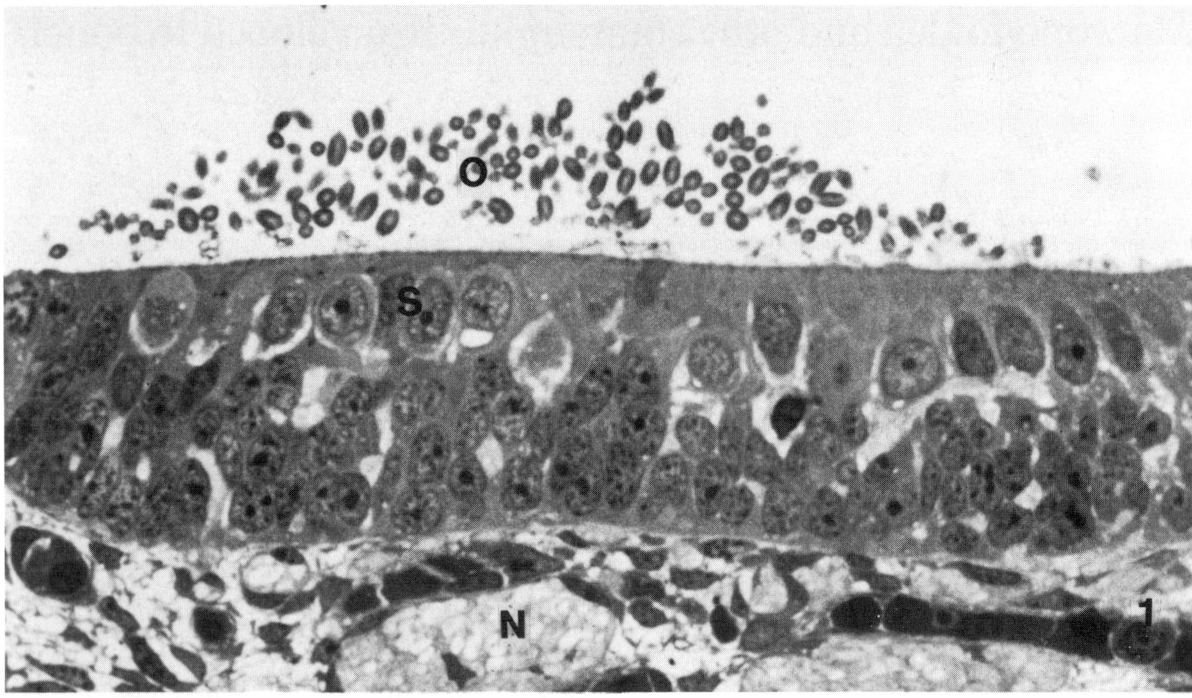

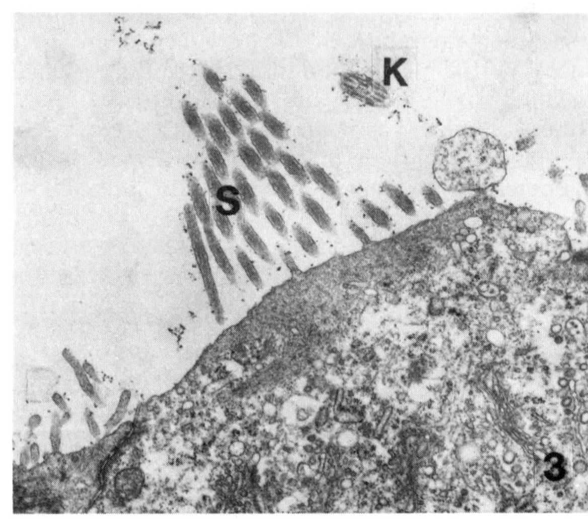

Fig. 1. Otoliths (O) positioned over the sensory cells (S) of the utricle in a newborn brushtail possum. Nerves (N are located in the dermis below the sensory epithelium. X1 000. *Fig. 2.* Scanning electron micrographs of the otoliths present in the ear of the newborn *Trichosurus vulpecula*. X6 000. *Fig. 3.* Stereocilia (S) and kinocilium (K) which project from the luminal surface of a sensory cell in the newborn *Trichosurus vulpecula*. X12 000.

birth", based on light microscopical examination of the ear of the newborn opossum. Recently Gemmell and Nelson (1989) examined the ultra-structure of the vestibular apparatus of the Northern Quoll *Dasyurus hallucatus* and a sensory region, presumably the utricle, was identified in each ear. A discrete population of otoliths were observed overlying a sensory epithelium which possessed cilia and microvilli. The remainder of the vestibular system was not developed anatomically.

The newborn Brushtail Possum *Trichosurus vulpecula* and the rat-kangaroos *Potorous tridactylus* and *Bettongia gaimardi* also have a utricle similar in structure to that of *D. hallucatus*. A layer of sensory cells is positioned below a discrete band of otoliths (Fig. 1), which have a crystalline appearance (Fig. 2). The sensory cell possesses several stereocilia and one kinocillium (Fig. 3). A sensory region has also been observed in the ear of the newborn *Monodelphis domesticus* (B. L. Munger, pers. comm.). Although there is

---

Fig. 4A-I. One micron frontal sections through the head of a newborn *Trichosurus vulpecula* in the region of the developing right ear. Fig. 4A is anterior with respect to Fig. 4I. These micrographs illustrate the structure of the utricle (U) of the developing cochlea (C), and semi-circular canals (SC). X180.

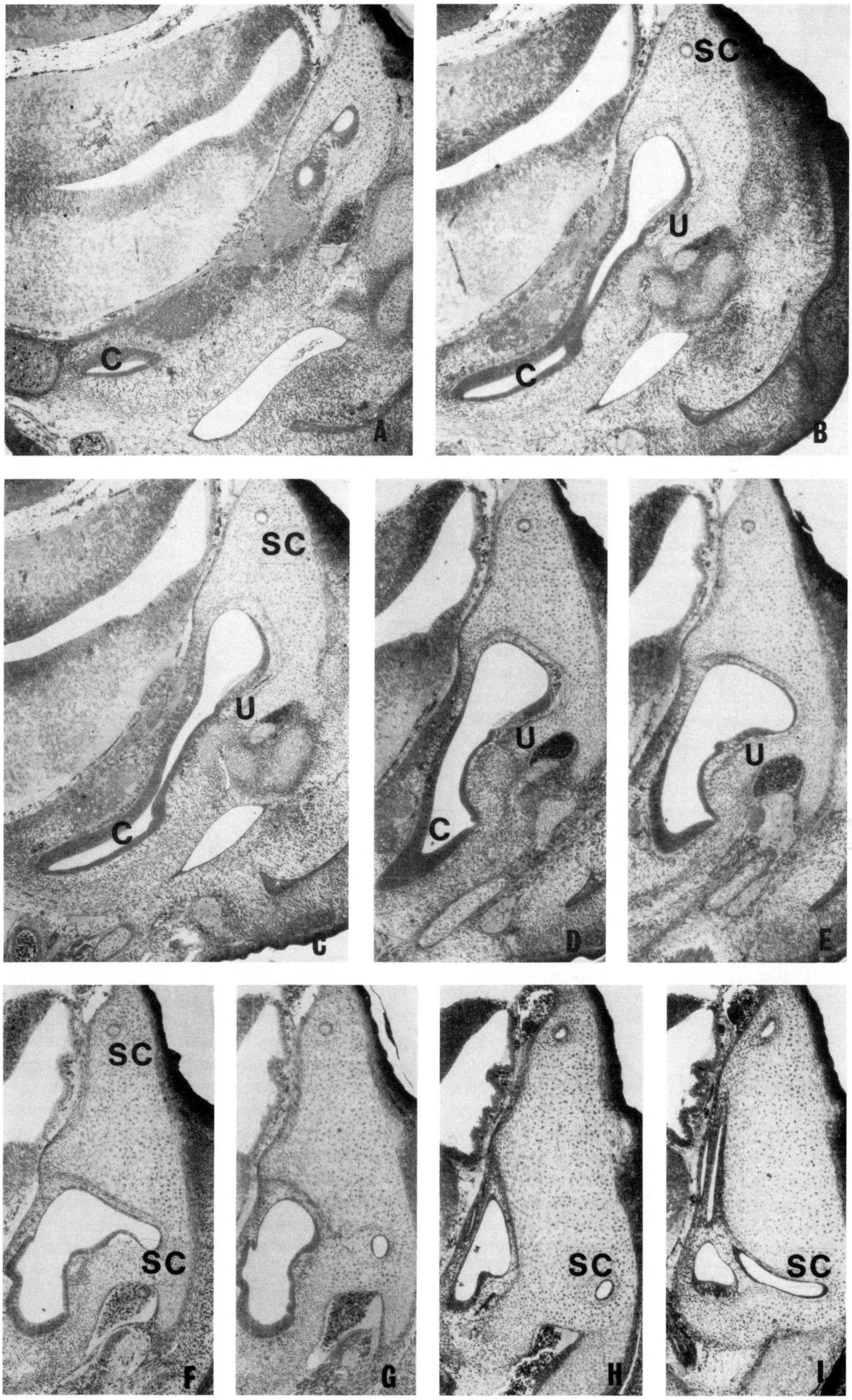
SC
U
C
A
B
C
D
E
F
G
H
I

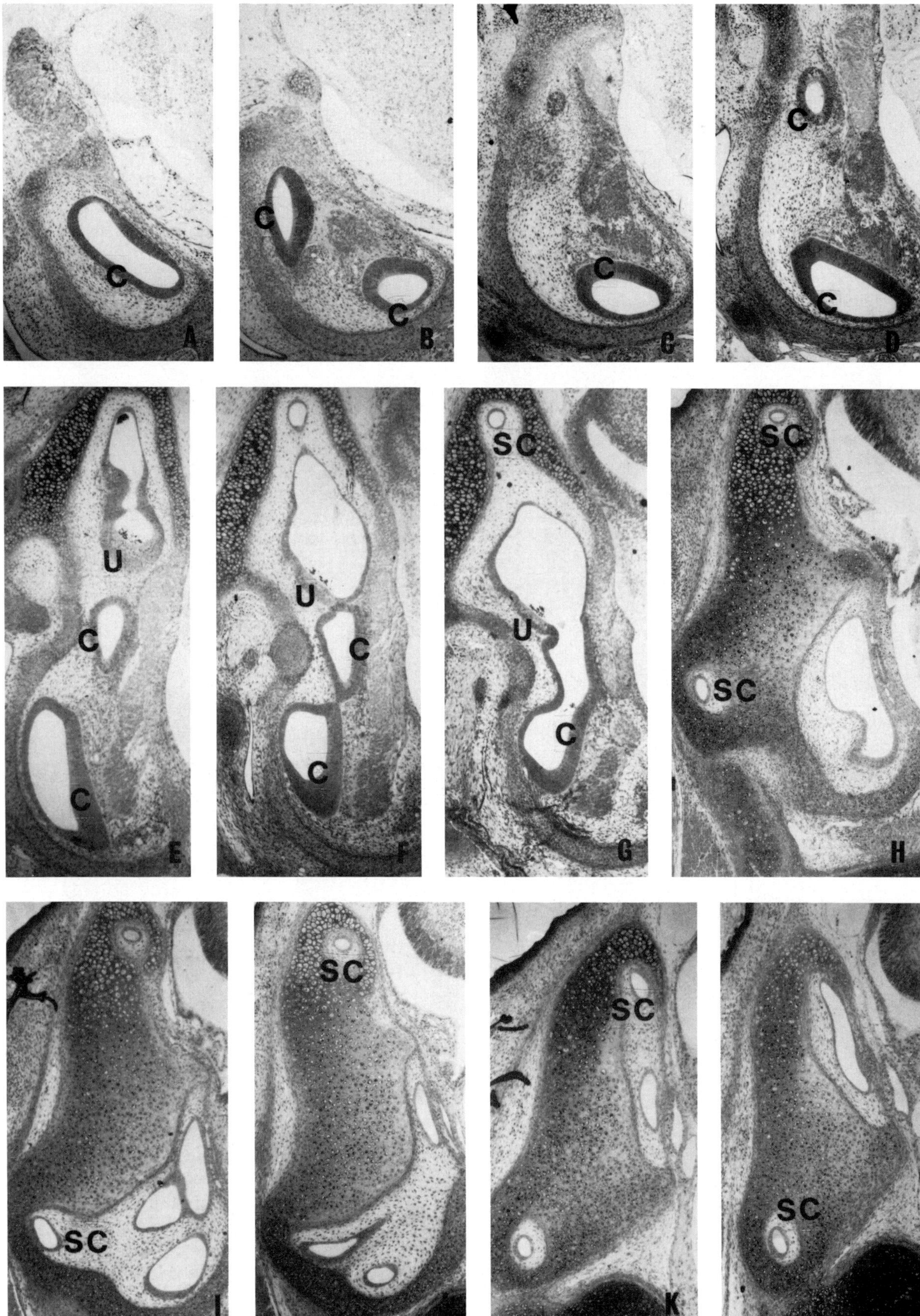
C
C
C
C
C
C
A
B
C
D
U
C
C
E
U
C
C
F
SC
U
C
G
SC
SC
H
SC
SC
I
SC
SC
K
SC

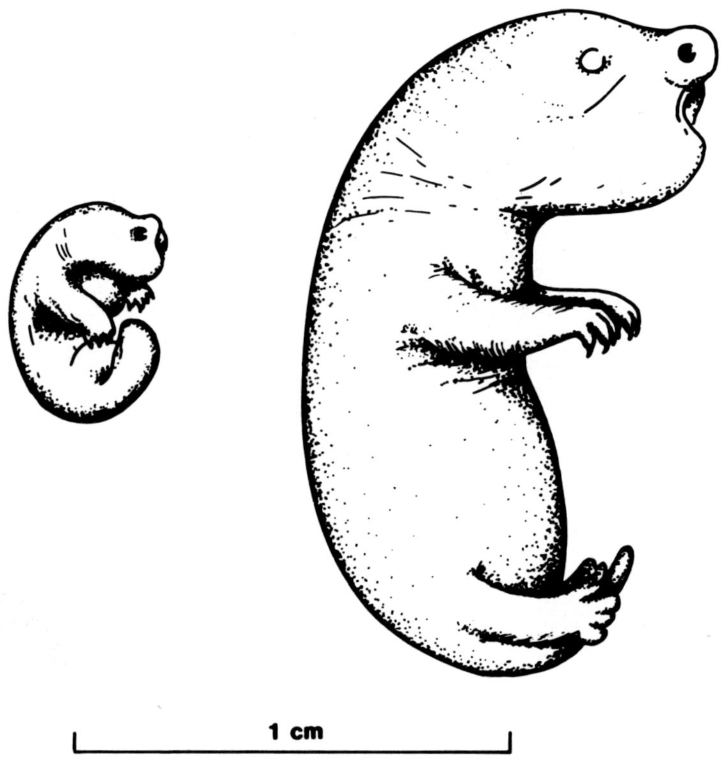

Fig. 6. A newborn *Dasyurus hallucatus,* body weight 18 mgm, and a newborn *Trichorsurus vulpecula,* body weight 198 mgm.

no apparent difference in the morphology of the utricles, the development of the cochlea and semi-circular canals at birth is variable among marsupial species. In the newborn Northern Quoll the semi-circular canals are still forming and the cochlea is rudimentary (Hill and Hill 1955; Gemmell and Nelson 1989). The cochlea has a rudimentary structure and the semi-circular canals are formed in *T. vulpecula* (Fig. 4) and the rat-kangaroo *P. tridactylus* (Fig. 5). The cochlea is more coiled in the rat-kangaroos (Fig. 5) than in the Brushtail Possum (Fig. 4) which in turn appears further developed than that of the Northern Quoll.

OLFACTION

The structural features characteristic of the newborn Eastern Quoll *Dasyurus viverrinus* were described by Hill and Hill (1955). The development of the forelimbs are well advanced compared to the hindlimbs (Fig. 6) as is the shape of the mouth and size of the tongue which aids the young in remaining attached to the teat. Olfactory sense cells are present in the olfactory epithelum and the olfactory bulbs are relatively large. Hill and Hill (1955) considered that the newborn Eastern Quoll probably possesses, to some degree, a sense of smell, and the distribution of the terminal branches of the divisions of the fifth nerve suggested that it is probably capable for responding to tactile stimuli. In a later study, Hughes and Hall (1984) reported that the olfactory epithelium of the Brushtail Possum was also well developed at birth and stated "olfaction was of significance to newborn *Trichosurus vulpecula* in locating the pouch or nipple". However, Kratzing (1986) concluded that the general morphological immaturity of the olfactory system of the bandicoot at birth, particularly the olfactory bulb, indicates that olfaction is not likely to guide the newborn to the pouch and teat. On the basis of ultrastructural examination of the olfactory system of newborn *D. hallucatus, T. vulpecula* and *Isoodon macrourus*, Gemmell and Nelson (1988) concluded that although the olfactory system underwent further development after the young entered the pouch, there was no reason to believe that the sensory cells present in the

Fig. 5A-L. One micron frontal sections through the head of a newborn *Potorous tridactylus* in the region of the developing left ear. Fig. 5A is anterior with respect to Fig. 5L. These micrographs illustrate the structure of the utricle (U) of the developing cochlea (C), and semi-circular canals (SC). X180.

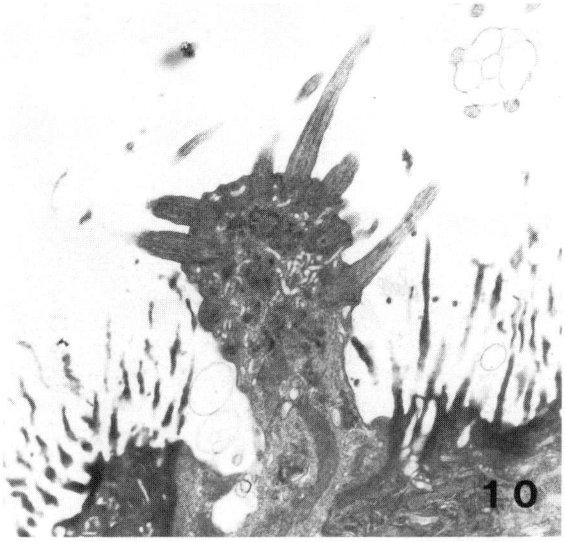

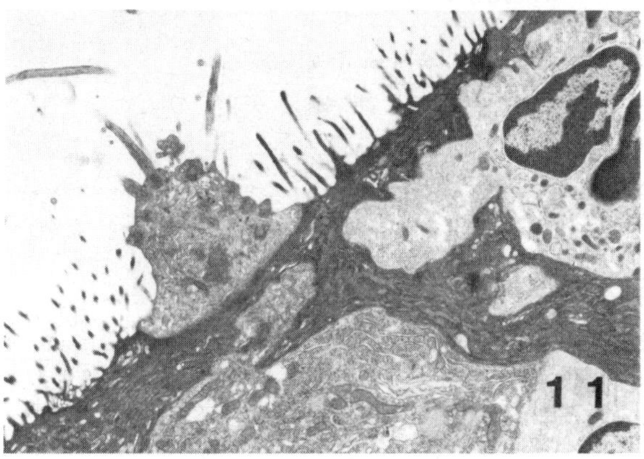

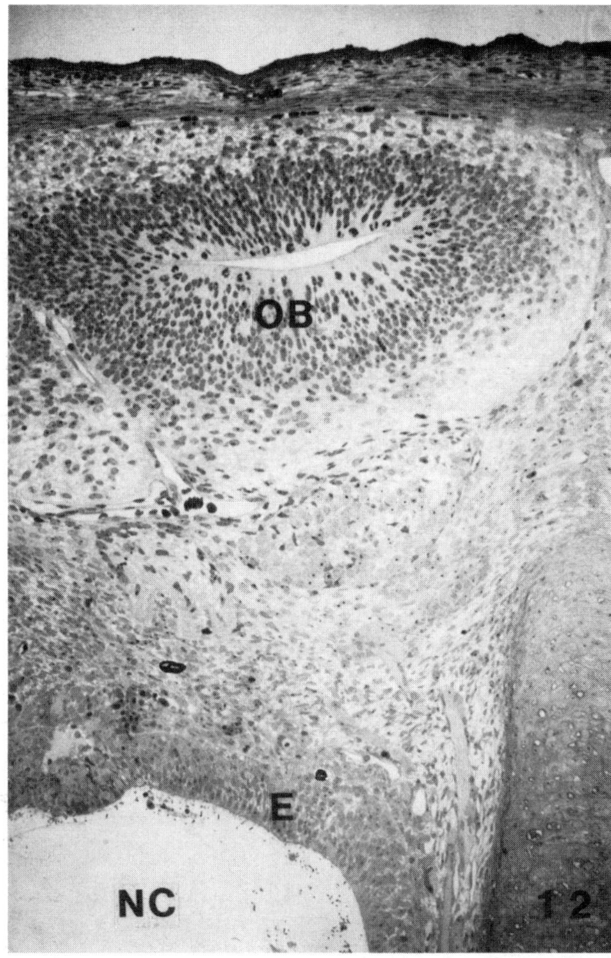

Fig. 10. Cilia projecting from olfactory knobs in newborn *Potorous tridactylus.* X10 000. *Fig. 11.* Olfactory knob in newborn *Bettongia gaimardi.* X6 000. *Fig. 12.* A frontal section of the left side of the head of a newborn *Bettongia gaimardi.* The olfactory bulb (OB), nasal cavity (NC) and olfactory epithelium (E) have been identified. X200.

olfactory epithelium of the newborn were not capable of detecting odours. The olfactory system of *P. tridactylus* and *B. gaimardi* would also appear to be functional. Olfactory knobs are present (Figs 7, 8, 10 and 11), the distribution of which is well illustrated by scanning electron microscopy (Fig. 9). Both rat kangaroos had defined olfactory bulbs (Fig. 12).

MECHANORECEPTORS

Jones and Munger (1985) provided evidence supporting the suggestion by Hill and Hill (1955) that the newborn was capable of detecting tactile stimuli. Merkel cells, highly sensitive touch receptors, were found in the snout skin of the newborn opossum *M. domesticus*. In a subsequent study, these cells were also found around the mouth of the newborn Northern Quoll, *D. hallucatus*, the Brushtail Possum *T. vulpecula* and the Northern Brown Bandicoot *I. Macrourus* (Gemmell, Peters and Nelson 1988). Merkel cells were also observed in the skin around the mouth of the rat-kangaroos *P. tridactylus* and *B. gaimardi*. These cells can be identified just above the basal layer of the epidermis (Fig. 13), by the presence of dense staining granules and an attendant sensory nerve (Figs 14 and 15). The Merkel cells of all the marsupial species examined have a similar morphology.

The presence of Merkel cells suggests that the sense of touch aids the newborn in moving from the urogenital sinus to the pouch and in locating the teat.

In conclusion, morphological evidence suggests that the newborn marsupial in its migration from

Fig. 7. An area of the olfactory epithelium of a newborn *Potorous tridactylus.* The olfactory knobs on the luminal surface of the epithelium are indicated (arrows). X1 000. *Fig. 8.* Two olfactory knobs (arrows) in the epithelium of a newborn *Potorous tridactylus.* X5 000. *Fig. 9.* Cilia (arrows) from the olfactory knobs projecting into the basal cavity of the newborn *Trichosurus vulpecula.* The smaller projections are microvilli. X13 000.

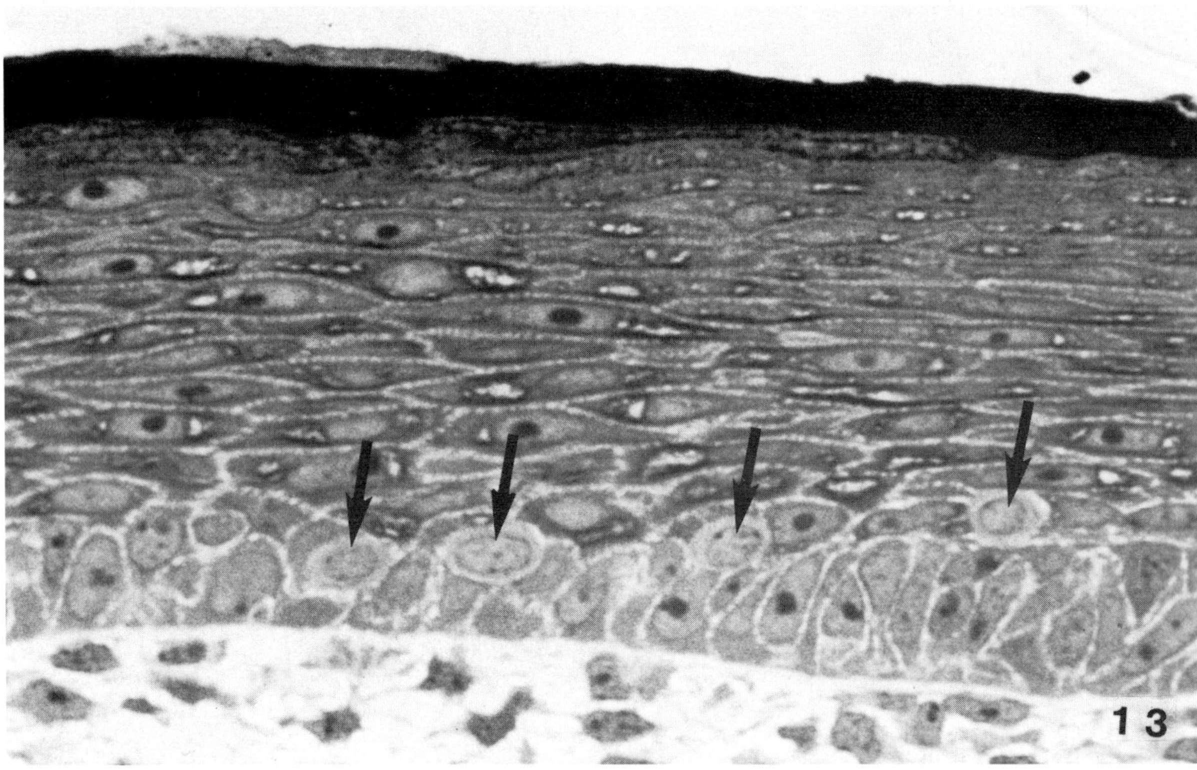

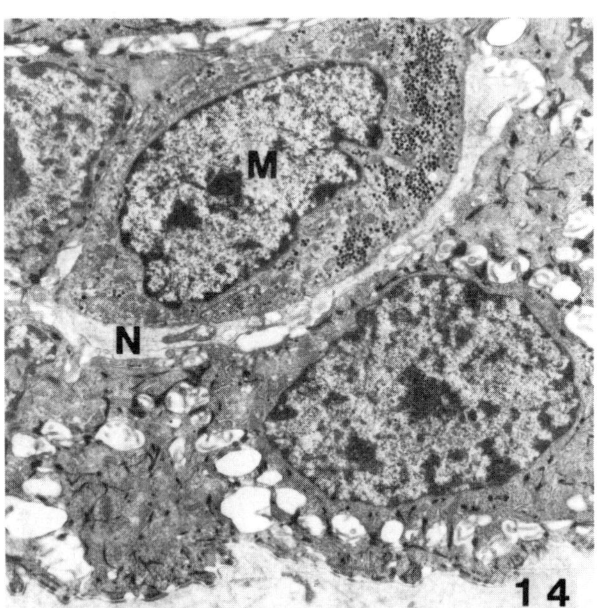

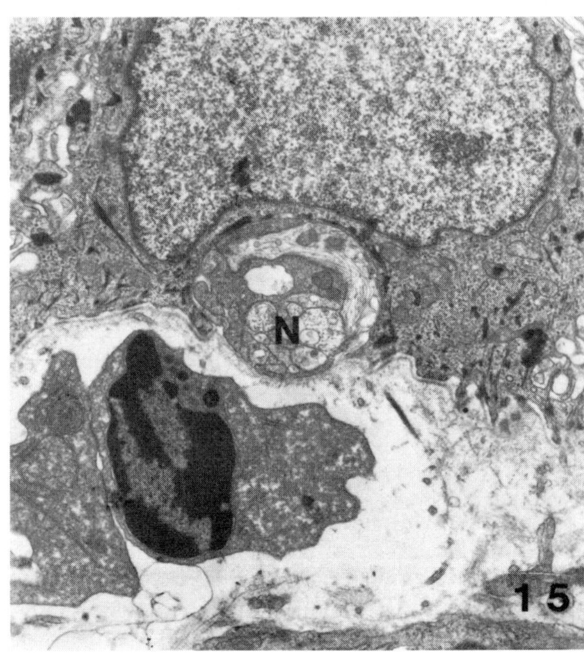

Fig. 13. A section of skin, close to the mouth from a newborn *Potorous tridactylus*. Merkel cells (arrows) are located just above the basal layer of the epithelium. X1 000. *Fig. 14.* A Merkel cell (M) and attendant sensory nerve (N) within the epidermis around the mouth of the newborn *Potorous tridactylus*. X4 200. *Fig. 15.* A nerve bundle (N) at the epidermal junction in skin from close to the mouth of the newborn *Potorous tridactylus*. X4 600.

the urogenital sinus to the pouch would appear to use the senses of gravity, odour and touch.

ACKNOWLEDGEMENT

The authors thank the Australian Research Grants Scheme for financial assistance.

REFERENCES

CANNON, J. R., BAKKER, H. R., BRADSHAW, S. D. AND McDONALD, I. R., 1976. Gravity as the sole navigational aid to the newborn Quokka. *Nature.* **259**: 42.

GEMMELL, R. T. AND NELSON, J., 1988. The ultrastructure of the olfactory system of three newborn marsupial species. *Anat. Rec.* **221**: 655-62.

GEMMELL, R. T. AND NELSON, J., 1989. The vestibular system of the newborn marsupial cat, *Dasyurus hallucatus. Anat. Rec.* (in press).

GEMMELL, R. T., PETERS, B. AND NELSON, J., 1988. Ultrastructural identification of Merkel cells around the mouth of the newborn marsupial. *Anat. Embryol.* **177**: 403-08.

HARTMANN, C. G., 1920. Studies in the development of the opossum *Didelphys virginiana.* V. The phenomena of parturition. *Anat. Rec.* **19**: 1-11.

HILL, J. P. AND HILL, W. C. O., 1955. The growth stages of pouch young of the native cat *(Dasyurus viverrinus)* together with observations on the anatomy of the newborn young. *Trans. Zool. Soc. Lond.* **28**: 349-453.

HUGHES, R. L. AND HALL, L. S., 1984. Embryonic development in the Common Brushtail Possum *(Trichosurus vulpecula).* Pp. 197-212 *in* Possums and Gliders ed. by A. P. Smith and I. D. Hume, Australian Mammal Society, Sydney.

JONES, T. E. AND MUNGER, B. L., 1985. Early differentiation of the afferent system in glabrous snout skin of the opossum *(Monodelphis domesticus). Somatosensory Res.* **3**: 169-84.

KRATZING, J. E., 1986. Morphological maturation of the olfactory epithelia of Australian marsupials. Pp. 57-70 *in* Ontogeny of Olfaction ed. by W. Breipohl, Springer-Verlag: Berlin.

LANGWORTHY, O. R., 1928. The behaviour of pouch young possums correlated with the myelinization of tracts in the nervous system. *J. Comp. Neurol.* **46**: 201-40.

LARSELL, O., McGRADY, E. Jr. AND ZIMMERMANN, A. A., 1935. Morphological and functional development of the membranous labyrinth in the opossum. *J. Comp. Neurol.* **63**: 95-118.

LYNE, A. G., 1974. Gestation period and birth in the marsupial *Isoodon macrourus. Aust. J. Zool.* **22**: 303-09.

LYNE, A. G., PILTON, P. E. AND SHARMAN, G. B., 1959. Oestrous cycle, gestation period and parturition in the marsupial *Trichosurus vulpecula. Nature.* **183**: 622-23.

McGRADY, E. Jr., 1938. The embryology of the opossum. American Anatomical Memoirs Number 16, Wistar Institute: Philadelphia.

POOLE, W. E., 1975. Reproduction in two species of Grey Kangaroos, *Macropus giganteus* Shaw and *M. fuliginosus* (Desmarest). II Gestation, parturition and pouch life. *Aust. J. Zool.* **23**: 333-53.

REYNOLDS, H. C., 1952. Studies on reproduction in the opossum *(Didelphis virginiana). Univ. California Public. Zool.* **52**: 223-84.

RUSSELL, E. M., 1982. Patterns of parental care and parental investment in marsupials. *Biological Rev.* **57**: 423-86.

SHARMAN, G. B. AND CALABY, J. H., 1964. Reproductive behaviour in the Red Kangaroo, *Megaleia rufa,* in captivity. *C.S.I.R.O. Wildlife Res.* **9**: 58-85.

TYNDALE-BISCOE, G. AND RENFREE, M., 1987. Reproductive physiology of Marsupials. Cambridge University Press: Cambridge.

Organ development in some newborn marsupials with particular reference to the rat-kangaroos

R. T. GEMMELL[1] and R. W. ROSE[2]

Morphological evidence suggests that newborn marsupials possess functional pituitary and adrenal glands. The activity of these glands may determine the length of gestation in marsupials as they appear to do in most eutherian mammals. Although the maturation of the various organ systems in both groups of mammals, eutherians and marsupials, follows similar time patterns, the main differences are prompted by the requirement of the young marsupial to leave the uterus and move to the pouch at an early stage of development. Certain organs, such as those involved with respiration, digestion and the formation of urine, are advanced to allow the young marsupial to survive in the pouch. The major part of development of other less essential structures, such as the skeleton, skin, eyes and ears occurs when the young marsupial is in the pouch. Although the sequence of organ development in eutherians and marsupials differs, once they are capable of maintaining a steady body temperature at weaning, the juveniles of both groups are usually at a similar stage of development.

INTRODUCTION

BIRTH is a climactic event in the development of every mammal. This amazing phenomenon allows the young mammal to leave the internal environs of the mother and enter the larger external environment. Although this major event is of importance in the development of all mammalian species, the stage of development at which parturition takes place varies from species to species particularly when marsupials are compared with eutherians. As further understanding of the control of birth is obtained, it is apparent that parturition is a singular event in the life of the mammal and is not a "bench mark" by which species can be compared accurately. The stage of organ development of a newborn lamb and that of a newborn rat differ and these in turn are at a different stage of development from the newborn macropodid marsupial. Nevertheless, the structural variation observed in the eutherian mammals at birth is not seen in marsupial species. Although newborn marsupials may vary in body weight from 4 to 800 mg as a consequence of variation in gestation length and embryonic growth rate (Rose 1989), organ development is similar at birth. In this paper a functional role for various organs will be postulated, whereby these structures initiate birth and allow the newborn marsupial to survive within the pouch.

ENDOCRINE CONTROL OF BIRTH

The extensive studies of the hormonal control of birth in eutherian mammals, especially the sheep, have allowed a greater understanding of the functional role of the various organs in initiating birth. There would appear to be a strong link between fetal maturity and parturition in birds, monotremes, marsupials and eutherians (Thorburn and Rice 1987).

In this present manuscript the hormonal changes which occur around the time of birth in the marsupial mother will not be examined. The reader is referred to Tyndale-Biscoe and Renfree (1987) and Thorburn and Rice (1987) for recent reviews of the relevant literature. This review will report on the structure and possible function of the fetal organs which may play a role in the initiation of birth. The fetal pituitary- adrenal axis is involved in the initiation of birth in several eutherian mammals (Liggins *et al.* 1973; Challis *et al.* 1977; Heap *et al.* 1977; Thorburn and Challis 1979). Parturition fails following hypophysectomy of the fetus in the sheep (Liggins *et al.* 1967), goat (Rawlings and Ward 1978) and pig (Bosc *et al.* 1974) and administration of adrenocorticotrophic hormone (ACTH) to the fetus induces birth in various eutherian mammals (Liggins *et al.* 1973). Therefore the maternal hormonal changes which initiate birth in the eutherian mammal are thought to be influenced not only by the maturation of the fetal adrenal glands but also by the developing pituitary gland.

Ultrastructural evidence of cellular activity, mainly granule synthesis, has been shown in *Macropus eugenii* (Leatherland and Renfree 1983), *M. rufogriseus* (Walker and Gemmell 1983) and *Potorous tridactylus* and *Bettongia gaimardi* (Figs 1, 2 and 3). It has been suggested that active pituitary and adrenal glands play an important

[1]Department of Anatomy, University of Queensland, St. Lucia, Brisbane, Australia 4067.
[2]Department of Zoology, University of Tasmania, Sandy Bay, Hobart, Tasmania 7005.

Pages 349-54 in KANGAROOS, WALLABIES AND RAT-KANGAROOS ed. by G. Grigg, P. Jarman and I. Hume. Surrey Beatty & Sons Pty Limited, New South Wales, Australia, 1989.

role in the gestation length and/or birth of the Red-necked Wallaby (Walker and Gemmell 1983). A subsequent study of the Northern Brown Bandicoot, *Isoodon macrourus*, led Hall and Hughes (1985) to suggest that the pituitary gland may not be active in all newborn marsupial species. Wheeler (1943) and Shorey (1968), who examined the pituitary of the newborn Virginia Opossum, *Didelphis virginiana*, and the Brushtail Possum, *Trichosurus vulpecula*, respectively, both reported the absence of cellular differentiation of the pituitary. Hall and Hughes (1985) reported the lack of capillaries in the pars distalis of *I. macrourus* at birth and suggested that the pituitary may have little influence on parturition. They further suggested that early postnatal cytodifferentiation of the pars distalis may occur in macropodid marsupials with a relatively longer gestation but takes longer in non-macropodid, short gestation marsupials. Presumably, this would account for the more mature appearance of the pituitary of the newborn Tammar and Red-necked Wallabies when compared with that of the Virginia Opossum, Brushtail Possum and Northern Brown Bandicoot. However, the anterior pituitary of each of the newborn Northern Quoll, *Dasyurus hallucatus*, *T. vulpecula* and *I. macrourus* was found to be well vascularised with many cells containing electron-dense, membrane bound granules (Gemmell and Nelson 1988). The pituitary of these species was similar in appearance to that in the newborn wallaby pituitary.

Liggins, Kennedy and Holm (1967) stated that the adrenal cortex of many eutherian mammals was functional at birth and in some species may be involved in parturition. Sharman (1976) noted that Poole's (1975) observations with two Grey Kangaroos suggested that gestation length in the Grey Kangaroo was controlled by the genetic constitution of the fetus. When male Western Grey Kangaroos, *Macropus fuliginosus*, (gestation 30.56±2.55 days) bred with female Eastern Grey kangaroos, *M. giganteus*, (gestation 36.41 ±1.63 days), the offspring had a gestation period of 34.09±0.81 days. Thus the genotype of the embryo appeared to influence its gestation length, as had been previously demonstrated in cattle (Holm and Short 1962) and different breeds of sheep (Smith 1967). These results, together with the considerable development of the fetal adrenal cortex in late gestation in a number of marsupials, led Sharman (1976) to suggest that the adrenal, as in some eutherians, may be responsible for initiating birth in the marsupial.

Endocrinological evidence has been obtained which demonstrates that the adrenal actively produces steroid hormones at birth. Corticosteroid concentrations of 0.9 and 1.5μg/ml were measured in two fetal plasma samples obtained from *M. eugenii*, one or two days before birth (Catling and Vinson 1976). A concentration of 0.094 ng and 0.58 ng of cortisol per adrenal gland was obtained for *I. macrourus* and *M. rufogriseus* respectively (Gemmell *et al.* 1982; Walker and Gemmell 1983). The presence of active steroid dehydrogenase within the adrenal cortex of the tammar approximately four days before birth (Call, Catling and Janssens 1980) and ultrastructural evidence of organelles involved in steroid hormone synthesis in *I. macrourus* (Gemmell *et al.* 1982), *M. rufogriseus* (Walker and Gemmell 1983), *D. hallucatus, T. vulpecula* (Gemmell and Nelson 1988) and the rat-kangaroos *P. tridactylus* and *B. gaimardi* (this study) further substantiates the activity of the adrenal gland at birth in the marsupial.

ORGAN DEVELOPMENT

The newborn marsupial has moved from the aqueous environment of the uterus into the aerobic environment of the pouch. The developing pouch young must be able to absorb oxygen. Selenka (1887) stated that "the construction of the lung is precocious in the sense that alveoli capable of respiratory function are developed in an extraordinarily short time; but it is retarded in the sense that the multiplication of the alveoli is deferred to a later time". Bremer (1904) and later McCrady (1938) confirmed Selenka's original findings and McCrady noted that the "alveoli", later termed respiratory sacs, were lined with a respiratory epithelium. This respiratory epithelium was later shown to contain type II surfactant secreting cells (Krause, Cutts and Leeson 1976). The lungs of the newborn *Dasyurus viverrinus* were described as simple organs, structurally at the lowest level, compatible with some degree of functional efficiency (Hill and Hill 1955). The respiratory chambers were much less developed than that of the opossum. Gemmell and Nelson (1988) described the lungs of another Quoll *D. hallucatus*, as two large respiratory sacs (Fig. 4) with a respiratory lining of squamous cells and surfactant-secreting cells. The lungs of the newborn native cats appear to be more primitive than those of other marsupials. The newborn *I. macrourus* has a lung of branching airways which terminate in large blind sacs; alveolar ducts and alveoli are not present (Gemmell and Little 1982). The lungs of newborn *M. rufogriseus* (Walker and Gemmell 1983), *T. vulpecula* (Hughes and Hall 1984; Gemmell and Nelson 1988) and the rat-kangaroos *P. tridactylus* and *B. gaimardi* (Figs 5 and 6) have a similar structure at birth. The lungs of all the newborn marsupials examined contain type II surfactant secreting cells and are functional. The native cats appear to have less developed lungs compared with those of the opossum, bandicoot, wallaby and possum. The terminal sacs present in the newborn are transformed into alveoli during

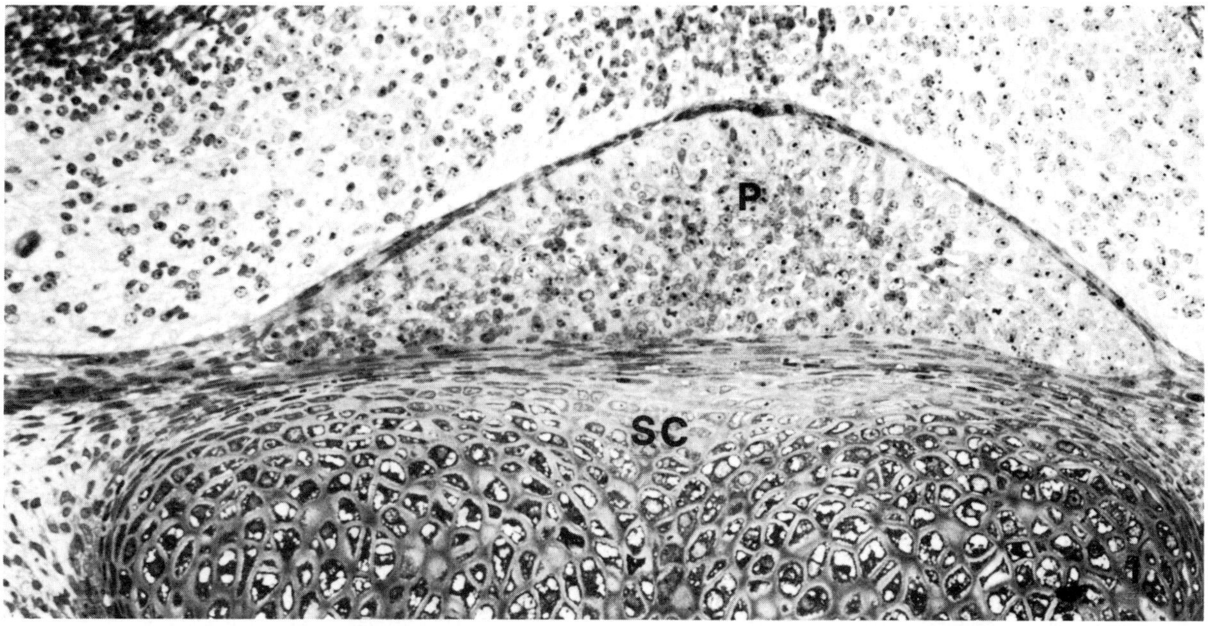

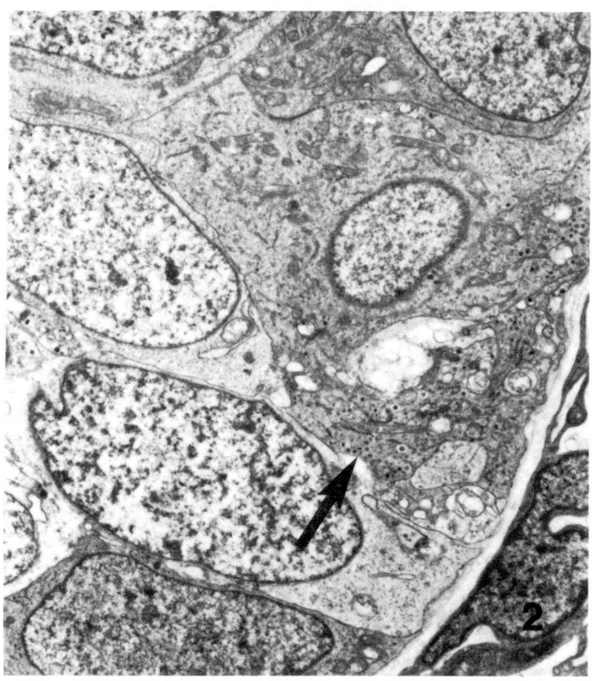

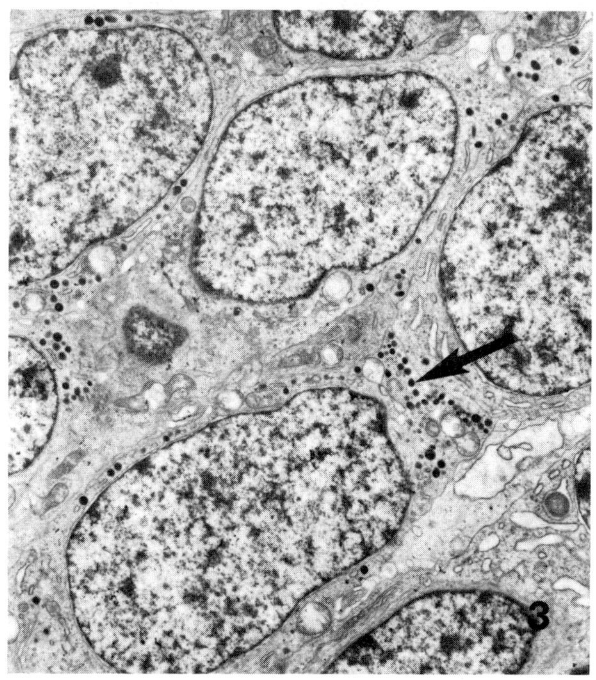

Fig. 1. A light micrograph of a frontal section of the pituitary (P) of a newborn *Bettongia gaimardii.* The pituitary gland is situated dorsal with respect to the sphenoid cartilage (SC). X300. *Fig. 2.* An area of the anterior pituitary of the newborn *Potorous tridactylus.* A few granules (arrow) are discernible. X2 500. *Fig. 3.* Densely staining granules (arrow) within the pituitary cells of the newborn *Potorous tridactylus.* X4 500.

pouch life and the marsupial has an adult form of lung when weaned (Krause and Leeson 1973, 1975; Gemmell 1986).

Other structures required for the existence of the young in the pouch are the urinary and digestive systems. Again these systems, like the lungs, do not have an adult morphology, nevertheless the developing form of the organs are functional. Hill and Hill (1955) described the alimentary tract of the newborn *D. viverrinus* as having no functional glands, no muscular coat, a lining epithelium and very little vascular mesenchyme. Bensley (1902) described the stomach of the newborn opossum as a reservoir and that all milk ingested was absorbed in the upper part of the tract, little material reaching the colon. The structure of the digestive tract of the newborn *M. rufogriseus* (Walker and Rose 1981), *D. viverrinus* (Hill and Hill 1955) and *T. vulpecula* (Hughes and Hall 1984) is similar.

The kidney of the newborn marsupial is still developing at birth. The newborn marsupial voids

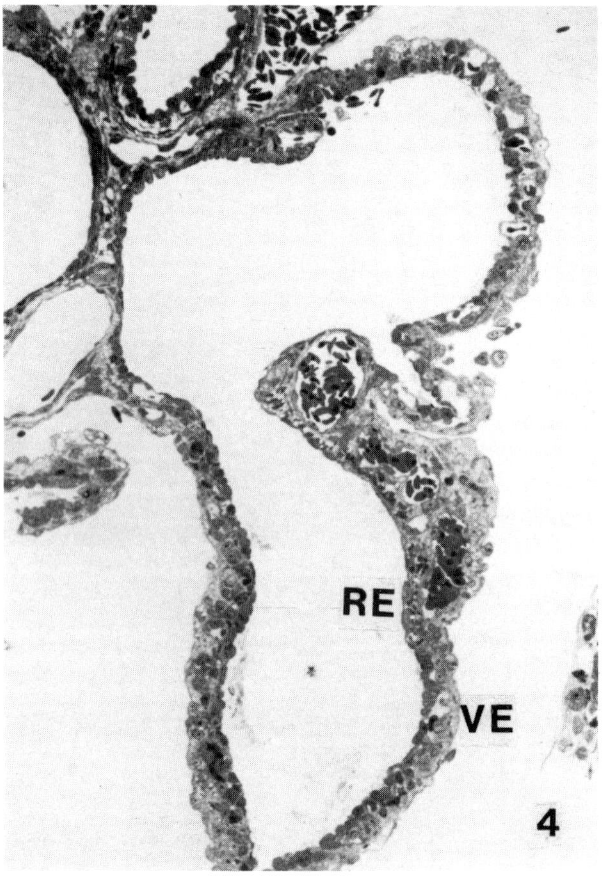

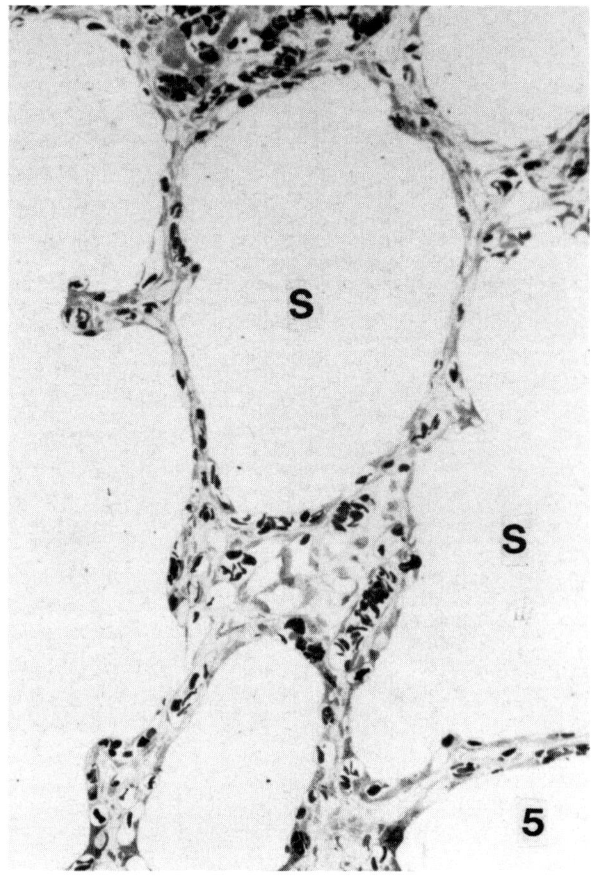

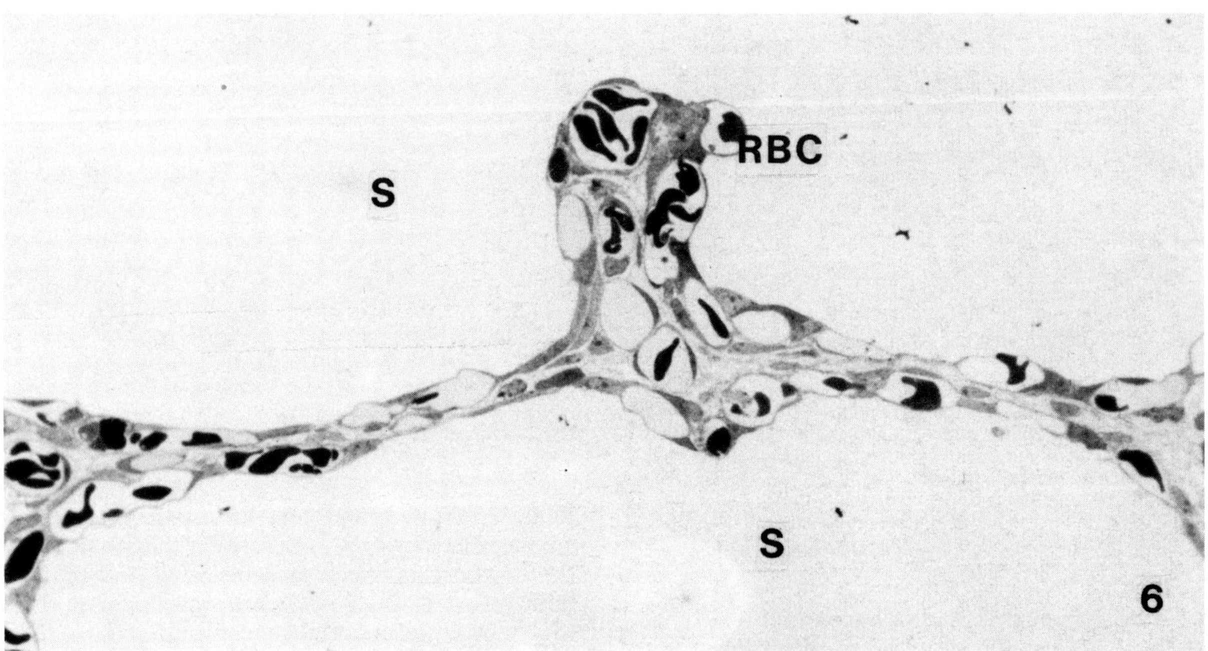

Fig. 4. The right lung of a newborn *Dasyurus hallucatus*. The respiratory epithelium (RE) is located on the opposite side of the wall of the lung from the visceral epithelium (VE). X250. *Fig. 5.* An area of the lung of *Potorous tridactylus*. The respiratory exchange takes place in the large air sacs (S) of the lungs. X250. *Fig. 6.* A photomicrograph of the cellular structure of the sac wall of the newborn *Potorous tridactylus*. The sac (S) is lined with an extensive blood supply, capillaries containing red blood cells (RBC) can be observed. X600.

urine via a mesonephric kidney (Buchanan and Fraser 1918; Gersh 1937; McCrady 1938; Hill and Hill 1955; Walker and Rose 1981; Hughes and Hall 1984). This foetal form of kidney is replaced by the adult form or metanephric kidney during pouch life.

Other organ systems not essential for birth develop while the young are in the pouch. As stated previously the newborn marsupial has the morphological requirements for a sense of gravity, olfaction and touch. The newborn cannot see, hear, taste or balance. All these senses develop while the young is in the pouch. Presumably these senses are not required at birth. It is of interest to note that the stage of eye development varies within marsupial species. Nelson (pers. comm.) has reported four different stages of development in seven species of marsupial, *D. hallucatus, I. macrourus, Perameles nasuta, Antechinus stuartii, T. vulpecula, M. eugenii* and *Macropus giganteus*, the stage of maturity increasing from *D. hallucatus* to *M. giganteus*. Eye development would not appear to have a primary role in parturition.

Marsupials possess a cartilagenous skeleton at birth and ossification takes place immediately after birth. The young marsupial has a fully ossified skeletal system before leaving the pouch (Gemmell, Johnston and Bryden 1988). The general pattern of ossification in the marsupial is similar to that in eutherians, however the long bones of the forelimbs, and the thoracic vertebrae, have a precocious development when compared with that of eutherian mammals. This difference may be due to the requirement of the marsupial forearms for the passage of the young from the cloaca to the pouch. Accelerated rib development occurs, and is possibly related to respiratory movements which are initiated at a very much earlier stage than in eutherian mammals. Thus a fully developed skeletal system is not required at birth, however the sequence of ossification has been modified to accommodate the pouch form of development.

The thyroid gland is not functional at birth in the marsupial. Studies of thyroid development in the Brushtail Possum (Hopkins 1972), Tammar Wallaby (Setchell 1974), opossum (Krause and Cutts 1983) and a bandicoot (Johnston and Gemmell 1987) would suggest that thyroid activation in the marsupial occurs during the latter half of pouch life. There would appear to be a correlation between thyroid activity and thermoregulatory capacities. Marsupials begin to develop thermogenic mechanisms during the latter third of pouch occupancy, and the ability to maintain a constant body temperature is acquired over the last few days before weaning in the opossum (Reynolds 1952), the wallaby *Macropus parma* (Wallis and Maynes 1974), the bandicoot (Gemmell and Johnston 1985), and the rat-kangaroos *P. tridactylus* and *B. gaimardi* (Gemmell, Cepon and Barnes 1987; Rose 1987).

The pituitary and adrenal glands are probably active in the initiation of birth in marsupials as in some, if not all, Eutheria. The glands not active in the birth process such as the thyroid are not fully differentiated and are inactive at birth. When the juvenile marsupial and eutherian achieve the ability to control a steady body temperature, they are at a similar state of maturity. The general structure of their respective respiratory, endocrine, nervous and digestive systems is similar. Therefore the difference in the sequence of organ development between the two groups of mammals is due to the occurrence of birth.

ACKNOWLEDGEMENT

The authors thank the Australian Research Grants Scheme for financial assistance.

REFERENCES

BENSLEY, R. R., 1902. The cardiac glands of the opossum. *Amer. J. Anat.* **2**: 105-56.

BOSC, M. J., DU MENSILDU BOISSON, F. AND LOCATELLI, A., 1974. Mise en evidence d'un controle foetal de la parturition chez la truie. Interaction avec la function luteale. *CR Academie de Science* **278**: 1507-10.

BREMER, J. L., 1904. On the lung of the opossum. *Amer. J. Anat.* **3**: 67-73.

BUCHANAN, G. AND FRASER, E. A., 1918. The development of the urogenital system in the Marsupialia, with special reference to *Trichosurus vulpecula*. Part 1. *J. Anat.* **53**: 35-95.

CALL, R. N., CATLING, P. C. AND JANSSENS, P. A., 1980. Development of the adrenal gland in the Tammar Wallaby, *Macropus eugenii* (Desmarest) (Marsupialia: Macropodidae). *Aust. J. Zool.* **28**: 249-59.

CATLING, P. C. AND VINSON, G. P., 1976. Adrenocortical hormones in the neonate and pouch young of the Tammar Wallaby, *Macropus eugenii. J. Endocr.* **69**: 447-48.

CHALLIS, J. R. G., JONES, C. T., ROBINSON, J. S. AND THORBURN, G. D., 1977. Development of fetal pituitary-adrenal function. *J. Steriod. Biochem.* **8**: 471-78.

GEMMELL, R. T., 1986. Lung development in the marsupial bandicoot, *Isoodon macrourus. J. Anat.* **148**: 193-204.

GEMMELL, R. T. AND JOHNSTON, G., 1985. The development of thermoregulation and the emergence from the pouch of the marsupial bandicoot, *Isoodon macrourus. Physiol. Zool.* **58**: 299-302.

GEMMELL, R. T. AND LITTLE, G. J., 1982. The structure of the lung of the newborn marsupial bandicoot, *Isoodon macrourus. Cell and Tissue Res.* **223**: 445-53.

GEMMELL, R. T. AND NELSON, J., 1988. The ultrastructure of the pituitary and the adrenal gland of three newborn marsupials (*Dasyurus hallucatus, Trichosurus vulpecula* and *Isoodon macrourus*). *Anat. Embryol.* **177**: 395-402.

GEMMELL, R. T. AND NELSON, J., 1988. The ultrastructure of the lung of two newborn marsupial species (*Dasyurus hallucatus* and *Trichosurus vulpecula*). *Cell and Tissue Res.* **252**: 683-85.

GEMMELL, R. T., CEPON, G. AND BARNES, A., 1987. The development of thermoregulation in the rat kangaroo, *Potorous tridactylus*. *Comp. Biochem. Physiol.* **88**: 257-63.

GEMMELL, R. T., JOHNSTON, G. AND BRYDEN, M. M., 1988. Osteogenesis in two marsupial species, the bandicoot *Isoodon macrourus,* and the possum *Trichosurus vulpecula*. *J. Anat.* **159**: 155-64.

GEMMELL, R. T., SINGH-ASA, P., JENKIN, G. AND THORBURN, G. D., 1982. Ultrastructural evidence for steriod hormone production in the adrenal of the marsupial, *Isoodon macrourus,* at birth. *Anat. Rec.* **203**: 505-12.

GERSH, I., 1937. The correlation of structure and function in the developing mesonephros and metanephros. *Cont. Embryol. Carnegie Institute* **153**: 35-58.

HALL, L. S. AND HUGHES, R. L., 1985. The embryological development and cytodifferentiation of the anterior pituitary in the marsupial *Isoodon macrourus*. *Anat. Embryol.* **172**: 353-63.

HEAP, R. B., GALIL, A. K. A., HARRISON, F. A., JENKIN, G. AND PERRY, J. S., 1977. Progesterone and oestrogen in pregnancy and parturition: Comparative aspects and hierarchial control. Pp. 127-57 *in* The Fetus and Birth ed. by J. Knight and M. O'Connor. Ciba Foundation Symposium 47. Excerpta Medica, Amsterdam.

HILL, J. P. AND HILL, W. C. O., 1955. The growth stages of pouch young of the native cat *(Dasyurus viverrinus)* together with observation of the anatomy of the newborn young. *Trans. Zool. Soc. Lond.* **28**: 349-453.

HOLM, L. W. AND SHORT, R. V., 1962. Progesterone in the peripheral blood of Guernsey and Freisian cows during prolonged gestation. *J. Reprod. Fert.* **4**: 137-41.

HOPKINS, P. S., 1972. The role of the thyroid gland in fetal development. Ph.D. Thesis, Macquarie University, Sydney, Australia.

HUGHES, R. L. AND HALL, L. S., 1984. Embryonic development in the Common Brushtail Possum *(Trichosurus vulpecula)*. Pp. 197-212 *in* Possums and Gliders ed. by A. P. Smith and I. D. Hume. Australian Mammal Society: Sydney.

JOHNSTON, G. M. AND GEMMELL, R. T., 1987. Thyroid development in the marsupial bandicoot, *Isoodon macrourus*. *Anat. Rec.* **217**: 178-87.

KRAUSE, W. J. AND CUTTS, J. H., 1983. Postnatal development of the thyroid gland in the opossum *(Didelphis virginiana)*. *Acta Anatomica* **116**: 322-38.

KRAUSE, W. J. AND LEESON, C. R., 1973. The postnatal development of the respiratory system of the opossum. I. Light and scanning microscopy. *Amer. J. Anat.* **137**: 337-56.

KRAUSE, W. J. AND LEESON, C. R., 1975. Postnatal development of the respiratory system of the opossum. II. Electron microscopy of the epithelium and the pleura. *Acta Anat.* **92**: 28-44.

KRAUSE, W. J., CUTTS, J. H. AND LEESON, C. R., 1976. Type II pulmonary epithelial cells of the newborn opossum lung. *Amer. J. Anat.* **146**: 181-88.

LEATHERLAND, J. F. AND RENFREE, M. B., 1983. Structure of the pars distalis in pouch-young Tammar wallabies *(Macropus eugenii)*. *Cell Tiss. Res.* **230**: 587-603.

LIGGINS, G. C., FAIRCLOUGH, R. J., GRIEVES, S. A., KENDALL, J. Z. AND KNOX, B. S., 1973. The mechanisms of initiation of parturition in the ewe. *Rec. Prog. Hor. Res.* **29**: 111-59.

LIGGINS, G. C., KENNEDY, P. C. AND HOLM, L. W., 1967. Failure of initiation of parturition after electrocoagulation of the pituitary of the fetal lamb. *Amer. J. Obstet. Gynec.* **98**: 1080-86.

MCGRADY, E. Jr., 1983. The embryology of the opossum. American Anatomical Memoirs Number 16, Wistar Institute: Philadelphia.

POOLE, W. E., 1975. Reproduction in two species of Grey Kangaroos, *Macropus giganteus* Shaw and *M. fuliginosus* (Desmarest). II. Gestation, parturition and pouch life. *Aust. J. Zool.* **23**: 333-53.

RAWLINGS, N. C. AND WARD, W. R., 1978. Effect of fetal hypophysectomy on the initiation of parturition in the goat. *J. Reprod. Fert.* **52**: 249-54.

REYNOLDS, H. C., 1952. Studies on reproduction in the opossum *(Didelphis virginiana)*. *Univ. California Public. Zool.* **52**: 223-84.

ROSE, R. W., 1987. Reproductive energetics of two Tasmanian rat-kangaroos (Potorinae: Marsupialia). *Symp. Zool. Soc. Lond. No.* **57**: 149-65.

ROSE, R. W., 1989. Embryonic growth rates in marsupials with a note on Monotremes. *J. Zool. (Lond.)* **217**: in press.

SELENKA, E., 1887. Studien uber Entwickelungsgeschichte. Viertes Heft. Das Opossum. Wiesbaden, C. W. Kriedels.

SETCHELL, P. J., 1974. The development of thermoregulation and thyroid function in the marsupial *Macropus eugenii* (Desmarest). *Comp. Biochem. Physiol.* **47A**: 1115-21.

SHARMAN, G. B., 1976. Evolution of viviparity in marsupials. Pp. 32-70 *in* Reproduction in Mammals Book 6 ed. by C. R. Austin and R. V. Short. The Evolution of Reproduction. Cambridge University Press.

SHOREY, C. D., 1968. An electron microscope study of pituitary-ovarian changes during oogenesis, and embryonic development in *Trichosurus vulpecula*. M.Sc. Thesis, University of New South Wales, Sydney, Australia.

SMITH, J. D., 1967. Breed differences in the duration of gestation in sheep. *Aust. Vet. J.* **43**: 63-64.

THORBURN, G. D. AND RICE, G. E., 1987. Fetal maturation and the timing of parturition: a comparative analysis. *Proc. Aust. Physiol. Pharmacol. Soc.* **18**(2): 51-68.

THORBURN, G. D. AND CHALLIS, J. R. G., 1979. Endocrine control of parturition. *Physiol. Rev.* **59**: 863-918.

TYNDALE-BISCOE, G. AND RENFREE, M., 1987. Reproductive Physiology of Marsupials. Cambridge University Press: Cambridge.

WALKER, M. T. AND GEMMELL, R. T., 1983. Organogenesis of the pituitary, adrenal and lung at birth in the wallaby *Macropus rufogriseus*. *Am. J. Anat.* **168**: 331-44.

WALKER, M. T. AND ROSE, R., 1981. Prenatal development after diapause in the marsupial *Macropus rufogriseus*. *Aust. J. Zool.* **29**: 167-87.

WALLIS, R. L. AND MAYNES, G. M., 1972. Ontogeny of thermoregulation in *Macropus parma* (Marsupialia: Macropodidae). *J. Mammal.* **54**: 278-81.

WHEELER, R. S., 1943. Normal development of the pituitary in the opossum and its responses to hormonal treatment. *J. Morph.* **73**: 43-87.

Lactation in macropodoid marsupials

J. C. MERCHANT[1]

Lactation in the macropodoid marsupials is characterized by its generally long duration and the considerable changes that take place in milk composition and production compared to eutherians. Simultaneous lactation in two mammary glands, supporting young of different ages and supplying milk of different compositions, is a uniquely macropodoid feature.

INTRODUCTION

In contrast to the eutherians, lactation has not been widely studied in the marsupials. The macropodoids, however, perhaps because of the relative ease with which large colonies of the Tammar Wallaby, *Macropus eugenii*, can be maintained in captivity, are probably best known because of studies of this species. The Tammar has been the subject of studies into the hormonal control of the initiation and maintenance of lactation, milk composition and milk production. This species therefore provides us with the most comprehensive data for a macropodoid and will be used in this chapter as a model against which comparisons can be made. Numerous references to milk composition exist for a variety of marsupial species but these are often of limited value because many are based on only a few samples or are from unknown stages of lactation.

There is a substantial literature on lactation in the eutherians, particularly domestic species. Their milk is relatively uniform in composition throughout lactation once full lactation is established. There are major differences between eutherian species, however, in the concentration of the milk they produce and the duration of lactation. These differences result from an interplay between lifestyle, suckling regimes and the production of altricial or precocial young (Green 1984).

In the marsupials, on the other hand, milk composition is relatively uniform between species but changes dramatically during the course of lactation (Green 1984). The gestation period of the marsupial is short and the newborn young is extremely small, weighing from 3-6 mg in *Tarsipes rostratus* (Renfree *et al.* 1984) to about 900 mg in the larger kangaroos. Birth is followed by an extended period of lactation during which the young marsupial undergoes the major part of its development in the pouch before leaving to suck from outside over the weaning period until it reaches independence.

The macropodoids usually give birth to only one young which after birth passes through a period when it is permanently attached to the teat, followed by a period during which it may release the teat but remains in the pouch and a period of weaning which may begin before it leaves the pouch but takes place predominantly after permanent pouch vacation. All young macropodoids undergo this sequence of events which differs between species only in the timing of the different phases.

This chapter is concerned mainly with the changes that take place in milk composition and production during lactation once it is initiated and with the energetics of growth in the young. The hormonal control of lactation has been reviewed in detail by Findlay and Renfree (1984); Tyndale-Biscoe and Renfree (1987)and Hinds (1988) and is outlined only briefly.

LACTATIONAL PHASES

Marsupials, in contrast to eutherians, have three, rather than two, phases of lactation (see Tyndale-Biscoe and Janssens 1988). The first is a period during pregnancy when the mammary gland is prepared for lactogenesis and is similar to Stage 1 of the eutherians (Cowie *et al.* 1980). The second phase, which may be considered in two parts (early and mid-lactation), is characterized initially by low milk production and permanent attachment to the teat by a slow growing and fully dependent young. In the latter part of the second phase the young may release the teat and during this phase many changes take place in both the development of the young and the composition of the milk. The second phase is absent in the eutherians. In the third phase (late lactation) the milk is most concentrated, milk production reaches peak values and the young undergoes a period of rapid growth and is eventually weaned. It is in this phase that lactation in the macropodoids is most similar to that in eutherians.

During the first phase of lactation considerable development and differentiation of the mammary gland occurs (Hinds 1988) and by the time parturition is imminent milk-like secretions are

[1]C.S.I.R.O., Division of Wildlife and Ecology, P.O. Box 84, Lyneham, Australian Capital Territory 2062.

Pages 355-66 in KANGAROOS, WALLABIES AND RAT-KANGAROOS ed. by G. Grigg, P. Jarman and I. Hume. Surrey Beatty & Sons Pty Limited, New South Wales, Australia, 1989.

present in the alveoli. Identical changes occur in the mammary glands of non-pregnant females of several marsupial species, including *Macropus rufus (=Megaleia rufa),* and this has been demonstrated by transferring neonates to their teats at comparable times post-oestrus (Sharman 1962; Sharman and Calaby 1964; Merchant and Sharman 1966; Clark 1968; Tyndale-Biscoe *et al.* 1984). The transferred young survive as successfully as young reared by their own mothers.

In several marsupials including *M. rufus* (Sharman and Calaby 1964) and *M. eugenii* (Stewart 1984) the growth and differentiation of the mammary gland is due to the secretion of progesterone by an active corpus luteum. During pregnancy and the oestrous cycle in *M. eugenii* (26 and 28 days respectively; Merchant 1979) there is a 4 to 5 fold increase in the weight of the mammary glands (Stewart 1984). Hinds (1988) notes that an active corpus luteum is essential in the development of a mammary gland which is ready to produce milk if a sucking stimulus is applied; however, she suggests, on the basis of experiments with *M. eugenii*, that adrenal and thyroid hormones cannot be excluded from a possible role in the preparation of the mammary gland.

In the second phase, milk production is initiated in the macropodoid mother in response to the sucking stimulus of the newborn young. The mammary gland is relatively small early in this period but continues to increase in size by hypertrophy and hyperplasia (Stewart 1984; Lincoln and Renfree 1981). It is only the gland associated with the teat to which the newborn young attaches that continues to enlarge (Sharman 1962; Stewart 1984). In *M. eugenii* the gland grows slowly up to about Day 110 by which time it has reached 2.5 g (Fig. 1) (Tyndale-Biscoe *et al.* 1984).

In contrast to the first phase, neither the corpus luteum of pregnancy nor the corpus luteum formed

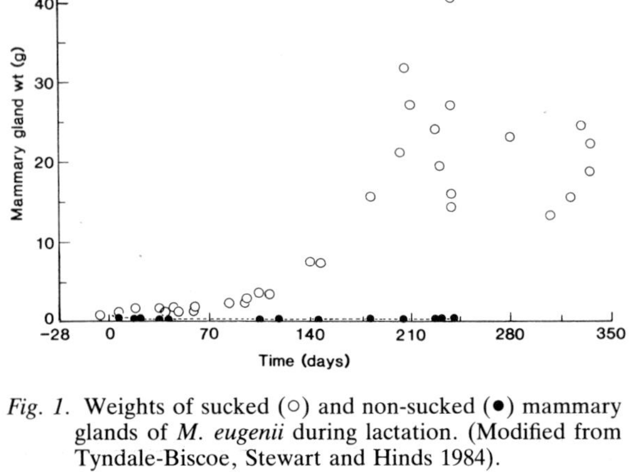

Fig. 1. Weights of sucked (○) and non-sucked (●) mammary glands of *M. eugenii* during lactation. (Modified from Tyndale-Biscoe, Stewart and Hinds 1984).

at post-partum ovulation are necessary for the maintenance of lactation. In both the Quokka, *Setonix brachyurus* (Tyndale-Biscoe 1963) and *M. eugenii* lactation continues after ovariectomy. The pituitary, however, is essential throughout lactation in *M. eugenii* (Hearn 1972). The concentrations of plasma prolactin in this phase are less than 40 ng/ml and are unaffected by either removal of pouch young or injection of bromocriptine, a drug which depresses plasma prolactin in eutherians (Hinds and Tyndale-Biscoe 1982). Despite this apparent lack of response to bromocriptine, Tyndale-Biscoe and Hinds (1984) showed that lactating animals given a single dose of bromocriptine might lose their pouch young within one to three weeks. A similar response was reported for *M. rufogriseus rufogriseus* by Curlewis *et al.* (1986). Hinds (1988) suggests that a low level of prolactin is therefore essential for maintenance of galactopoiesis and that the sucking stimulus does not elicit a release of prolactin in this early stage of lactation.

The third phase in *M. eugenii* commences when the pouch young is about 190 days old and is characterized by increased milk production, a period of rapid growth by the young and independence following a period of weaning from a milk to a fully herbivorous diet.

The mammary gland continues to increase in size as lactation proceeds (Green 1984) and between 110 days and 250 days increases from 2.5 g to 35 g (Fig. 1) (Tyndale-Biscoe *et al.* 1984). Plasma prolactin concentrations increase markedly over the period 140 to 200 days in *M. eugenii* which coincides with the increase in mammary gland size and the period of rapid growth in the pouch young (Hinds 1988). Prolactin levels remain elevated from Days 200 to 250 when young begin to vacate the pouch permanently and then decline rapidly (Hinds and Tyndale-Biscoe 1985). Temporary removal of pouch young between Days 200 and 240 result also in a rapid decline in plasma prolactin which is reversed when the young is replaced (Hinds and Tyndale-Biscoe 1985). Similar changes in plasma prolactin concentrations are shown by *M. r. rufogriseus* (Curlewis *et al.* 1986). Hinds (1988) suggests that the changing prolactin levels are a response to the sucking regime of the young and that the decrease in plasma prolactin as the young vacates the pouch probably reflects a decline in the frequency of sucking as the young is weaned. This is similar to the decline in prolactin seen in the eutherians (Cowie *et al.* 1980).

The macropodoid mother may suckle simultaneously two young of different ages; one which has recently vacated the pouch and a second produced either from a reactivated blastocyst or a recent mating (Fig. 2). This may occur in *M. rufus, M. agilis* and *M. r. banksianus*

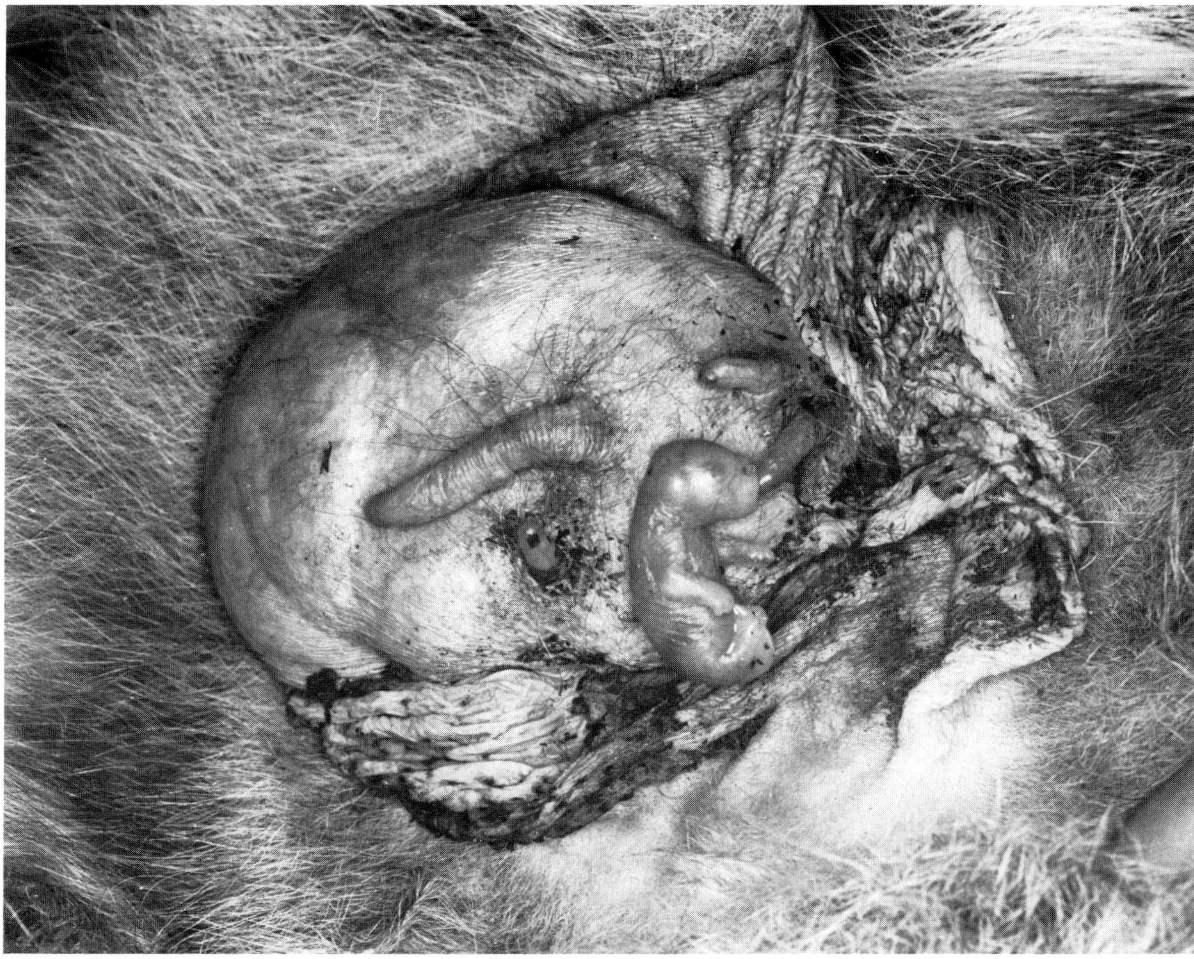

Fig. 2. Newborn young of *M. eugenii* attached to its small teat adjacent to an elongated teat supporting a young-at-heel. Photograph by G. Chapman.

(Griffiths *et al.* 1972; Lincoln and Renfree 1981; Merchant and Calaby 1981) and other continuously breeding species. Other species such as *S. brachyurus, M. eugenii* and *M. r. rufogriseus* which normally produce only one young per season also may exhibit this phenomenon (Yadav and Eadie 1973; Messer and Green 1979 and Merchant and Calaby 1981). It appears probable that all macropodoids have this capacity described by Lincoln and Renfree (1981) as concurrent asynchronous lactation.

MILK COMPOSITION

Total solids

At birth the concentration of solids in the milk of the Tammar, *M. eugenii*, and of the Red-necked Wallaby, *M. r. banksianus* is about 10% (w/w) (Green *et al.* 1980; Merchant, Green, Messer and Newgrain, in press) (Fig. 3a). In the Potoroo, *Potorous tridactylus* (Crowley *et al.* 1988), and the Brush-tailed Bettong, *Bettongia penicillata* (Merchant, unpubl. obs.) milk solids concentrations at about 28 days post-parturition are around 15 to 17% (w/w). In *B. gaimardii* at about 35 days they are slightly higher at about 25% (Smolenski and Rose 1988). These levels are similar to those observed in other marsupial species early in lactation (Green and Merchant 1988).

As lactation proceeds the concentration of solid material in the milk increases so that in the Tammar the solids fraction is about 35% (w/w) when the young first leaves the pouch for short periods at 210 days of age. The young of the larger Red-necked Wallaby begins to leave the pouch when about 230 days old (Merchant and Calaby 1981) and the milk has increased in solids content to about 25%. In the Potoroo at this time, 120-130 days of age (Crowley *et al.* 1988), the solids concentration is around 30%. The Brush-tailed Bettong, of similar size to the Potoroo, leaves the pouch at about 100 days of age and milk solids concentrations are similar to those seen in the Potoroo. The milk of the Yellow-footed Rock-wallaby, *Petrogale xanthopus,* reaches a solids concentration of 34% (Green 1984). Lactation in *P. tridactylus* (about 1 kg) lasts about 180 days, in *M. eugenii* (5 kg) about 320 days, in *M. r. banksianus* (12 kg) about 400 days and in *M. rufus* (24 kg) about 360 days. The duration of lactation in *P. xanthopus* (7 kg) is not accurately known (Poole *et al.* 1985) but is around 300 to 400 days. The data suggest a trend towards higher peak milk solids concentrations in the smaller species.

Proteins

The total protein content increases significantly during lactation in *M. eugenii* (Green *et al.* 1980), *P. tridactylus* (Crowley *et al.* 1988) and *B. gaimardi* (Smolenski and Rose 1988) (Fig. 3b). In *M. eugenii* there is a distinct increase in total protein in later stages of lactation that coincides with the appearance, during electrophoresis, of additional protein bands, one of which has been named late lactation protein (LLP) by Nicholas *et al.* (1987). A function for LLP has not yet been identified (Nicholas 1988).

In *M. eugenii* the milk proteins are predominantly in the whey fraction (Green and Renfree 1982) whereas in most eutherians they are predominantly in the caseins (Thompson and Farrell 1974). Electrophoretic patterns of whey proteins show a general increase in the intensity of protein bands as lactation proceeds. Late lactation protein (LLP) seems to be identical to a protein described as being associated with blastocyst reactivation in *M. rufus* (Bailey and Lemon 1966; Lemon and Barker 1967) and "zone C" proteins of milk from *M. rufus* and *M. giganteus* (McKenzie *et al.* 1983). The proteins once related to blastocyst development are now considered to be a function of lactational stage (Lemon and Poole 1969; McKenzie *et al.* 1983).

The relative concentrations of some amino acids in *M. eugenii* are quite constant during lactation, but others predominate at different times. This is presumably related to the developmental stage of the young as appears to be the case with the presence of higher levels of sulphur-rich amino acids during hair growth (Renfree *et al.* 1981).

Lipids

In *M. eugenii* during early lactation the concentration of lipids in the milk is low and represents only 2% (w/w) of whole milk or about 15% of the total solids fraction (Fig.3c). As lactation proceeds there is a dramatic increase in total

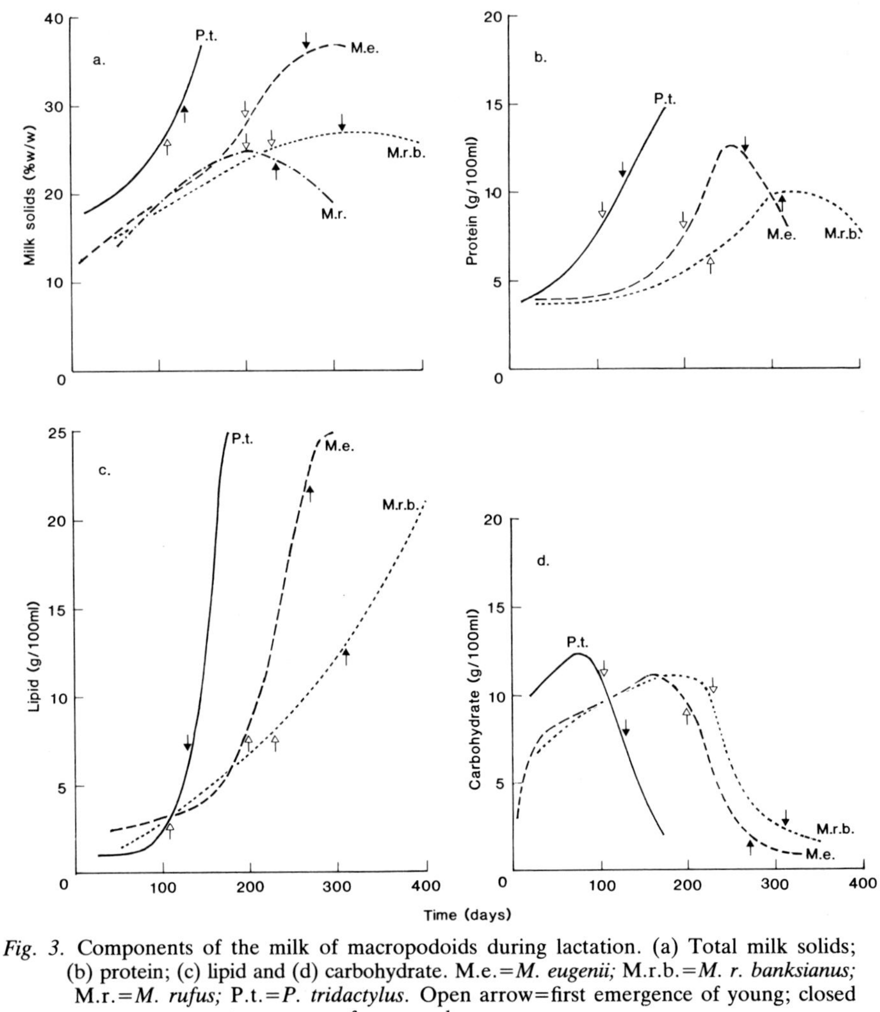

Fig. 3. Components of the milk of macropodoids during lactation. (a) Total milk solids; (b) protein; (c) lipid and (d) carbohydrate. M.e.=*M. eugenii;* M.r.b.=*M. r. banksianus;* M.r.=*M. rufus;* P.t.=*P. tridactylus.* Open arrow=first emergence of young; closed arrow=permanent emergence from pouch.

lipid that coincides with the decline in carbohydrate (see below). This pattern is seen in *M. r. banksianus* (Merchant, Green, Messer and Newgrain, in press), *M. rufus* (Griffiths *et al.* 1972), *P. tridactylus* (Crowley *et al.* 1988) and *B. gaimardi* (Smolenski and Rose 1988). Preliminary results for *B. penicillata* (Merchant unpubl. obs.) suggest that the pattern in this species is similar. This marked increase in milk fats coincides with the emergence of young from the pouch of macropodoids or the deposition of young in a nest as in the dasyurid, *Dasyurus viverrinus* (Green *et al.* 1987). At this time the young have an increased energy requirement for thermoregulation and increased activity associated with independence.

The relative composition of milk lipids changes extensively throughout lactation. On day 10 in *M. eugenii*, 97% of the milk fats are in the form of triglycerides, 2.5% are phospholipids and there are minute amounts of free fatty acids and cholesteryl esters. At 35 to 50 days the triglyceride fraction is 90% and phospholipids have risen to 9%. By 240 days triglycerides are 98.5% while the remainder consists of traces of phospholipid, free fatty acids and cholesteryl esters (M. Griffiths and R. A. Gibson, pers. comm.). Chain lengths of both the saturated and unsaturated fatty acids generally range from C12 to C20 (see Green and Merchant 1988). Macropodid milk is therefore relatively uniform with respect to chain length of fatty acids, in contrast to eutherian milk which contains both short and long chain fatty acids.

The early milk of *M. rufus* (Griffiths *et al.* 1972) and *M. eugenii* (Green *et al.* 1983) contains palmitic acid (C16:0) as the predominant fatty acid; about 60% of the fatty acids of the total triglyceride fraction is palmitic acid and about 20% is oleic acid in *M. eugenii* at 10 days (M. Griffiths and R. A. Gibson, pers. comm.). After about 50 to 70 days, however, the concentration of palmitic acid decreases and it is replaced by oleic acid (C18:1) so that in late lactation oleic acid represents about 43% of the milk triglyceride fatty acids and palmitic only 27% (M. Griffiths and R. A. Gibson, pers. comm.).

Palmitic acid is known to be important in the synthesis of surfactants in the lungs of newborn eutherians (Masoro 1968). Green *et al.* (1983) conclude that in *M. eugenii* and *M. rufus,* however, on the basis of a study of *M. rufus* by Johnson and Nicholas (1979), palmitic acid is not implicated in this role. Green and Merchant (1988) again suggest, however, that high levels of palmitic acid in early milk may be correlated with the synthesis of surfactant in the air sacs of young macropodids. It is obvious that more work needs to be done before there are sufficient data to draw firm conclusions about the role of the various fatty acids or indeed, many of the constituents of macropodoid milk.

Carbohydrates

Like the other constituents of macropodoid milk the carbohydrate content changes markedly during lactation as it does in other marsupials (Green and Merchant 1988). The carbohydrate concentration in the milk of the Tammar at seven days post-partum, measured as total hexose, is 3% (w/w) (Fig. 3d, Messer *et al.* 1984) and may increase to as high as 13% (w/w) at 180 days. The concentration of milk carbohydrate then declines rapidly to about 1% (w/w) and remains at this low level until the end of lactation (Messer and Green 1979). This pattern of declining carbohydrate concentration is seen also in the milks of the Red-necked Wallaby (Merchant, Green, Messer and Newgrain, in press), the Potoroo (Crowley *et al.* 1988), the Red Kangaroo (Lemon and Barker 1967) and the Eastern and Western Grey Kangaroos (Poole *et al.* 1982). The milks of the two Bettong species show a similar decline in carbohydrate (Smolenski and Rose 1988; Merchant, unpubl. obs.). This decline occurs over the period of temporary emergence of the young from the pouch and is correlated with the change in its diet from milk to solids. Messer and Green (1979) and Janssens and Ternouth (1987) have noted and discussed this correlation in *M. eugenii.* Its significance to the digestive physiology of young marsupials generally has been discussed by Janssens and Messer (1988) who suggest that changes in milk carbohydrate concentrations are related to the establishment of conditions in the gut suitable for the microorganism populations necessary for the digestion of herbage.

The predominant carbohydrate in the milk of eutherian mammals is the disaccharide, lactose. The carbohydrate of marsupial milk, however, consists of higher oligosaccharides for the major part of lactation (Messer and Mossop 1977; Messer and Green 1979; Messer *et al.* 1987). Lactose occurs in Tammar milk until about seven days post-partum but is then replaced by tri- to penta-saccharides (Messer *et al.* 1984) and by about 180 days the milk sugars are predominantly higher oligosaccharides (Messer and Green 1979). The sugar molecules then become progressively smaller so that by 280 days all milk sugars are present as monosaccharides (Messer and Green 1979). A similar pattern of increasing size of milk oligosaccharides followed by a decrease towards the end of lactation has been found in *M. r. banksianus* (Merchant, Green, Messer and Newgrain, in press).

The oligosaccharides are mostly composed of a single glucose unit linked to varying numbers of galactose units. The glucose-galactose linkage is joined by a β (1→4) bond as in lactose whereas the galactose-galactose linkages are β (1→3) (Messer *et al.* 1980; Collins *et al.* 1981). Some of the oligosaccharides contain N-acetylglucosamine and sialic acid. The relative proportions of these sugar units have been calculated for *M. eugenii* (Messer and Green 1979). Throughout lactation galactose (about 70-80% w/w of the total) and glucose (about 10-20% w/w) predominate. At 300 days, however, glucosamine (14% w/w), galactosamine (5% w/w) and sialic acid (9% w/w) have increased at the expense of galactose (50% w/w).

Minerals

The concentrations of minerals in the milks of *M. eugenii* and *M. r. banksianus* do not differ greatly from those of other marsupial groups and the range of concentrations for most is within those found in eutherian mammals (Green 1984; Green and Merchant 1988). Copper and iron concentrations, however, are substantially higher in the milks of marsupials generally, particularly in early, as opposed to late stage milk. Copper concentrations in milks of the Cow, the Horse and Man are 0.6, 0.1 and 0.3 mg/l respectively (Johnson 1974; Schryver *et al.* 1986; Macy 1949). In both *M. eugenii* and *M. r. banksianus* copper in early milk is 5 mg/l but falls to 1 and 0.1 mg/l respectively. In a study examining copper deficiency as a possible cause of seasonal anaemia in the Quokka, *Setonix brachyurus*, Barker (1962) showed that a significant and positive correlation existed between copper levels in maternal blood and in milk during late lactation and that the concentrations in milk were usually higher.

Iron levels in early milk of *M. eugenii* and *M. r. banksianus* are similar at 23 and 20 mg/l respectively but decrease to 7 and 5 mg/l later in lactation. The respective values for iron concentrations in milk of the Cow and Man are much lower at 0.5 and 0.3 mg/l. The Quokka mother can concentrate iron in the milk (Kaldor and Ezekiel 1962; Kaldor and Morgan 1986). Milk of this species contains iron at concentrations of about 20 mg/l, about seven to nine times the maternal plasma level, while the young are permanently in the pouch. At the start of weaning the concentration falls abruptly to 4 mg/l, still almost twice that of maternal plasma. Kaldor and Morgan (1986) suggest that, in relation to iron, the period of pouch occupancy resembles the intra-uterine phase of the eutherians. The newborn eutherian has a relatively large body iron content associated with a large red cell mass and non-haem iron stored in the liver (Widdowson 1950). The marsupial, however, is so small at birth that its iron status cannot satisfy the post-natal iron requirement. Most of its growth takes place outside the uterus and it is therefore entirely dependent on milk, not only for iron, but for all its nutritional requirements.

Sodium and potassium concentrations change during lactation in all macropodoids examined so far. These include *M. eugenii* (Green *et al.* 1980), *M. r. banksianus* (Merchant, Green, Messer and Newgrain, in press), *S. brachyurus* (Bentley and Shield 1962) and *P. tridactylus* (Crowley *et al.* 1988). Preliminary observations for *B. penicillata* (Merchant, unpubl. obs.) suggest that this species also conforms to this pattern.

In general sodium levels are elevated during early lactation (Messer *et al.* 1984) but then gradually decline. Potassium, in contrast, tends to commence at levels lower than those of sodium and rise to higher concentrations in late lactation. In *M. eugenii* (Green *et al.* 1980) there is a second crossing over of the two curves as sodium and potassium show a tendency to approach plasma concentrations towards the end of lactation. This trend, although more obscure, probably exists in *P. tridactylus* (Crowley *et al.* 1988).

Linzell and Peaker (1971) demonstrated an inverse relationship between lactose and sodium in the milk of eutherians and this maintains milk at concentrations that are isosmotic to maternal plasma. Messer and Green (1979) pointed out that if lactose was the predominant carbohydrate of Tammar milk, the osmotic pressure of the milk would exceed that of plasma. However, they note that most of the carbohydrate is composed of other oligosaccharides which, weight for weight, exert a lower osmotic pressure than lactose so that this problem does not arise. The osmotic pressure exerted by carbohydrate depends also on the amount present in the milk. In the following discussion comparisons between species are made at mid-lactation which coincides with maximal concentrations of carbohydrate. Messer *et al.* (1987) observed that the oligosaccharides of mid-lactation milk of the small dasyurid, *D. viverrinus*, are of larger molecular size than those of the Tammar and therefore exert lower osmotic pressures. They postulated that the concentration of sodium plus potassium in *D. viverrinus* milk should therefore be higher than that in Tammar milk in order to maintain isosmolarity. They found that the combined electrolyte concentration of Tammar milk was 37 mM (Green *et al.* 1980) whereas that of *D. viverrinus* milk was 83 mM (Green *et al.* 1987). The milk of the Bandicoot, *Isoodon macrourus,* has an electrolyte concentration around 95 mM (Merchant and Libke 1988), making it similar to that of *D. viverrinus*. The concentration of sodium plus potassium in mid-lactation (140 days) milk of the Quokka, *Setonix brachyurus*, is around 40 mM (Bentley and Shield 1962) while in *M. r. banksianus* (200 days) it is 45-50 mM (Merchant, Green, Messer and Newgrain, in press). The combined sodium and potassium concentrations of mid-lactation milk of the Potoroo (about 90 days) are around 80 mM (Crowley *et al.* 1987). The Potoroo, in this respect appears to be similar to the dasyurid, *D. viverrinus* and the Bandicoot, *Isoodon macrourus*. Since the concentrations of electrolytes in the milks of *M. r. banksianus* and *S. brachyurus* are equivalent to that of *M. eugenii* it seems likely that similar osmotic pressures are being exerted by the milk oligosaccharides of all three species, and that the molecular sizes of the oligosaccharides are also similar. Using the same argument it is tempting to speculate that the milk of the Potoroo at a comparable stage post partum contains oligosaccharides of a molecular size similar to those found in *D. viverrinus* milk. Until further research is done, however, the possible reasons for such an association between milk electrolytes and carbohydrates in disparate marsupial groups remain unclear.

Calcium and phosphorus concentrations in *M. eugenii* milk increase throughout lactation presumably to satisfy the requirements of bone formation and growth of the pouch young (Green 1984). In late lactation in *M. eugenii* (Green 1984) and *M. r. banksianus* (Green and Merchant 1988) the concentrations of these minerals are 4 and 5.5 g/l respectively. These levels are therefore similar to those in other non-macropodid marsupials but higher than reported for most eutherian milks with the exception of the Rabbit, *Oryctolagus cuniculus* (Lebas *et al.* 1971) which ranges from 4.1 to 6.8 g/l.

Vitamins

The vitamin content of the milks of marsupials in general has not been extensively examined. The available information is contained in Poole *et al.* (1982) who show that vitamin concentrations in the milks of *M. rufus, M. giganteus* and *M. fuliginosus* are within the ranges shown by eutherian species. Retinol concentrations in the macropodid species were 640 to 1 590 ug/l milk and although higher than in most eutherian species were exceeded by the Blue Whale, *Balaenoptera musculus*, and the Rabbit, *O. cuniculus* at 2 500 and 2 710 ug/l respectively. The carotenoids, absent in many eutherians, are present in *M. giganteus, B. taurus* and *H. sapiens* milks at similar concentrations. Biotin and vitamin B6 milk concentrations vary widely in the eutherians. Vitamin B6, for example, ranges from 67 ug/l in the Goat, *Capra hircus*, to 2 900 ug/l in the Rabbit so that the macropodid species, ranging from 360 to 1 600 ug/l, are well within these limits.

The concentrations of vitamins varied widely between samples, leading Poole *et al.* (1982) to observe that the effects of lactational stage were obscure. However, certain trends were evident. The concentration of riboflavin in the milk of *M. rufus* with young in the pouch averaged 1.4 mg/l while later, when young had left the pouch permanently, it averaged 3.8 mg/l. Values for *M. giganteus* milk were 2.7 and 3.6 mg/l respectively for similar stages. Nicotinic acid, however, decreased during lactation from about 50 to 15 mg/l while pantothenic acid, averaging about 10 to 11 mg/l, showed no clear correlation with stage of lactation.

The data suggest that vitamin concentrations, like the other components of the milk, probably undergo changes as lactation proceeds. These changes are presumably related to the developmental requirements of the young but this must remain speculative until such time as further work on this aspect of milk composition has been done.

Transfer of immunity

Neonatal marsupials are immunologically incompetent (Yadav 1972) and yet inhabit an environment, the pouch, which contains a variety of micro-organisms (Yadav *et al.* 1972). Maternal immunoglobulins (Yadav 1971; Yadav and Eadie 1973; Deane and Cooper 1984) and milk macrophages (Cockson and McNeice 1980) therefore play a critical role in the protection of pouch young against infection during lactation. Around the time of birth a clear secretion is produced by the mammary glands of marsupials (Deane and Cooper 1984) which is similar to the colostrum of eutherians and which is low in fat and contains many free-floating cells (Griffiths *et al.* 1972). Proteins identified as IgG and IgA-like antibodies have been found in the colostrum and neonatal serum of quokkas but not in unsuckled neonates. This suggests that immunoglobulin transfer occurs only via the milk (Yadav and Eadie 1973). Renfree (1973), however, provided evidence that some transfer of immunity occurs *in utero* in *M. eugenii*.

MILK PRODUCTION AND ENERGETICS

Information on milk consumption, milk energy content and the relative efficiencies of growth in marsupials is confined predominantly to *M. eugenii* (Green 1984; Cork and Dove 1986; Dove *et al.* 1987). The energy content of *P. tridactylus* milk is described by Crowley *et al.* (1988) and of *P. tridactylus* and *B. gaimardii* milks by Smolenski and Rose (1988) who include some data on milk consumption.

The rates of milk consumption of Tammar young increase from about 1 ml/day at Day 30 to

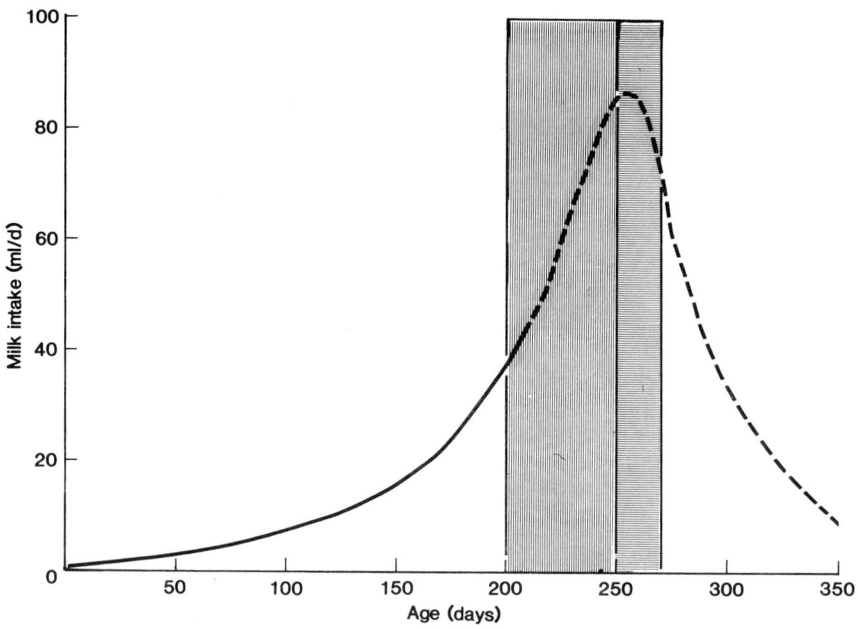

Fig. 4. Milk intake during lactation from birth to independence in *M. eugenii.* Vertical shading=temporary pouch emergence; horizontal shading=permanent emergence. Solid line modified from Green *et al.* (1988); dashed line modified from Dove *et al.* (1987).

9 ml/day at Day 130 (Fig. 4). This is followed by a substantial increase to about 43 ml/day by Day 210 at which time the young commence weaning (Green *et al.* 1988). The isotopic turnover techniques of Green and Newgrain (1979) cannot be used to determine milk intake once sodium is ingested in food other than milk. Cork and Dove (1986), however, using a two isotope technique in *M. eugenii*, were able to estimate the amount of milk consumed after weaning commenced and calculated peak values of 56 ml/day when the young were about 240 days old. In a subsequent paper a correction was applied and peak values were determined to be 86 ml/day (Dove *et al.* 1987).

Green *et al.* (1988) note that increasing variability in milk consumption rates with age is associated with variable growth rates in young Tammars. The body weights of young at Day 210 ranged from 350 to 960 g. These authors also demonstrate that heavier mothers have larger young at a given age and suggest that milk intake probably is influenced predominantly by maternal weight, with these females producing more milk (Fig. 5) and hence faster growing young. The correlation between size of mother and young has been shown also for the Agile Wallaby, *M. agilis* (Dudzinski *et al.* 1978).

The energy content of *M. eugenii* milk increases as lactation proceeds (Green *et al.* 1988). At Day 7 it is 250 kJ/100 g, by Day 180 it is about 500 kJ/100 g and this is followed by a rapid increase to a peak value of 1150 kJ/100 g between Days 250 to 280. The major increase in milk energy content is due to the large increase in milk lipid in late lactation. Crowley *et al.* (1988) and Smolenski and Rose (1988) show that the Potoroo and Bettong also follow this pattern of change in milk energy content and that peak values, about 12 kJ/ml and 9 kJ/ml respectively are similar to those of the Tammar. It is only the timing of these changes that differs, so that while the Tammar reaches peak values at about 210 days the Bettong and the Potoroo take around 170 and 180 days respectively.

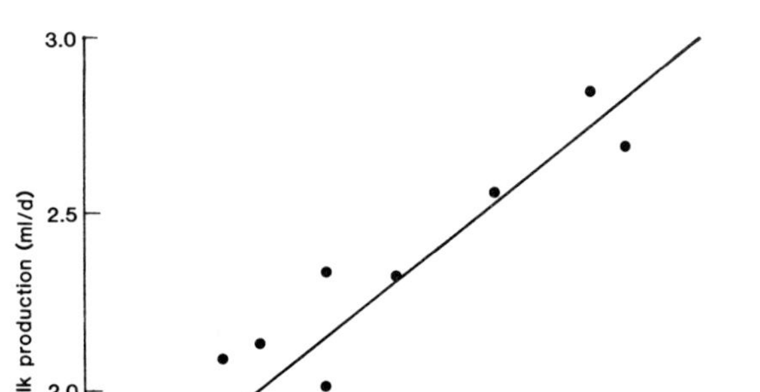

Fig. 5. Maternal weight and milk production at 110 days post-partum in *M. eugenii.* (From Green *et al.* 1988).

The mass of body tissue that is laid down for each ml of milk consumed, the crude growth efficiency (CGE), increases from 0.21 to 0.25 g/ml at Day 150 in the Tammar (Green *et al.* 1988). As the solids content of the milk increases, however, particularly lipid, CGE increases so that by Day 220 it is 0.37 g/ml (Green *et al.* 1988). The estimated proportions of stored energy and milk protein that are converted into body energy and protein increase from 23% and 55% respectively in early lactation to 36% and 87% in late lactation (Green *et al.* 1988). There are comparatively few data on energy and protein conversions in mammals. Energy conversion in the Black-tailed Deer, *Odocoileus hemionus*, is 25%, the Grey Seal, *Halocheirus grypus*, 62%, and the Weddell Seal, *Leptonychotes weddelli* 53% (Sadleir 1980; Fedak and Anderson 1982; Tedman and Green 1986). The values for protein conversion in the Horse, *Equus* sp., range from 51%-70%, *O. hemianus*, 77%, the Dog, *Canis familiaris,* 37% and *L. weddelli,* 79% (Oftedal *et al.* 1983; Sadleir 1980; Oftedal 1984 and Tedman and Green 1986). The Tammar therefore tends to be intermediate in its range of values.

Green *et al.* (1988) argue that the relatively long period of lactation and comparatively slow growth of the young is therefore not due to an inefficient conversion of milk constituents. They suggest that the only alternative is that milk production and therefore growth of the young is limited by the mother. This hypothesis is supported by the observations of Findlay (1982) for *M. eugenii*, and Merchant and Sharman (1966) for several macropodid species, who show that pouch young transferred to the pouches of either larger species or females with larger young may increase in weight at an accelerated rate.

CONCLUSIONS

It is apparent that during lactation the milk of macropodoids undergoes significant quantitative and qualitative changes — a fact which needs to be considered in studies of this type. The macropodoids are not unique in this respect as marsupials in general show dramatic compositional changes in their milks (Green 1984; Green and Merchant 1988). Green *et al.* (1980) suggest that the changes in marsupial milk conform to the hypothesis first proposed by Ben Shaul (1962) that milk composition in many species is influenced by the frequency of suckling.

The macropodoids appear to be unique, however, in their ability to produce simultaneously milks of two distinct qualities and quantities for two young of different ages. The control of this phenomenon is not precisely understood but Lincoln and Renfree (1981) have shown a differential response to oxytocin, the milk let-down hormone, in the two glands of the Agile Wallaby, *M. agilis.* Although the basic dose-response curve is similar to eutherian mammals the macropodid threshold level is lower in early lactation. The significance of this observation lies in the different sucking ability of the small and permanently attached young which can exert presumably only a slight sucking pressure compared to its large sibling. While the small young will evoke a response only in the gland to which it is attached the more vigorous sucking of the young-at-heel will cause milk let-down in the large gland.

It is perhaps surprising that lactation in the macropodoids has not been more closely examined when it is considered that, in contrast to the eutherians, it represents their major reproductive investment. As well as this the pouch young is easily accessible for experimental manipulation, compared to the eutherian foetus, and provides an ideal model in which to study the development of a variety of physiological systems (see Tyndale-Biscoe and Janssens 1988). The relationships between changing milk components and development of the young are unclear. On the basis of evidence such as the presence of high concentrations of sulphur-containing amino acids in the milk around the time of hair follicle development (Renfree *et al.* 1981) or highest sodium and potassium concentrations before the kidneys of the pouch young are fully developed it is tempting to suggest that the milk of marsupials has co-evolved with the changing requirements of the young as they grow and develop. Further physiological or biochemical studies are required, however, before this hypothesis can be accepted.

The macropodoids, ranging in size from the rat-kangaroos to the large kangaroos, therefore present opportunities for elucidating lactational strategies that may have evolved in different environments.

REFERENCES

BAILEY, L. F. AND LEMON, M., 1966. Specific milk proteins associated with resumption of development of the quiescent blastocyst of the lactating red kangaroo. *J. Reprod. Fert.* **11**: 473-75.

BARKER, S., 1962. Copper levels in the milk of a marsupial. *Nature.* **193**: 292.

BEN SHAUL, D. M., 1962. The composition of the milk of wild animals. *Int. Zoo. Yb.* **4**: 333-42.

BENTLEY, P. J. AND SHIELD, J. W., 1962. Metabolism and kidney function in the pouch young of the macropodid marsupial, *Setonix brachyurus. J. Physiol.* **164**: 127-37.

CLARK, M. J., 1968. Termination of embryonic diapause in the red kangaroo, *Megaleia rufa,* by injection of progesterone or oestrogen. *J. Reprod. Fert.* **15**: 347-55.

COCKSON, A. AND MCNEICE, R., 1980. Survival in the pouch: the role of macrophage and maternal milk cells. *Comp. Biochem. Physiol.* **66**: 221-25.

COLLINS, J. G., BRADBURY, J. H., TRIFONOFF, E. AND MESSER, M., 1981. Structures of four new oligosaccharides from marsupial milk, determined mainly by ^{13}C-n.m.r. spectroscopy. *Carbohyd. Res.* **92:** 136-40.

CORK, S. J. AND DOVE, H., 1986. Milk consumption in late lactation in a marsupial, the tammar wallaby *(Macropus eugenii). Proc. Nutr. Soc. Aust.* **11:** 93.

COWIE, A. T., 1969. Variations in the yield and composition of the milk during lactation in the rabbit and the galactopoietic effect of prolactin. *J. Endocr.* **44:** 437-50.

COWIE, A. T., FORSYTH, I. A. AND HART, I. C., 1980. Hormonal control of lactation. *In* Monographs on Endocrinology. Vol. 15 ed. by F. Gross, M. M. Grumback, A. Labhart, M. B. Lipsett, T. Mann, L. T. Samuels and J. Zander. Springer-Verlag: Berlin.

CROWLEY, H., WOODWARD, D. AND ROSE, R., 1988. Changes in milk composition during lactation in the potoroo *Potorous tridactylus* (Marsupialia: Potoroinae). *Aust. J. Biol. Sci.* **41:** 289-96.

CURLEWIS, J. D., WHITE, A. S., LOUDON, A. S. I. AND MCNEILLY, A. S., 1986. Effects of lactation and season on plasma prolactin concentrations and response to bromocriptine during lactation in the Bennetts Wallaby *Macropus rufogriseus rufogriseus. J. Endocr.* **110:** 59-66.

DEANE, E. M. AND COOPER, D. W., 1984. Immunology of pouch young marsupials. I. Levels of immunoglobulin, transferrin and albumin in the blood and milk of euros and wallaroos (Hill Kangaroos: *Macropus robustus*; Marsupialia). *Dev. Comp. Immun.* **8:** 863-76.

DOVE, H., CORK, S. J. AND CRISTIAN, K. R., 1987. Defining the pattern of milk intake in a marsupial herbivore. Pp. 101-02 *in* Herbivore Nutrition Research ed. by M. Rose. *Aust. Soc. Anim. Prod.*

DUDZINSKI, M. L., NEWSOME, A. E. AND MERCHANT, J. C., 1978. Growth rhythms in pouch young of the agile wallaby in Australia. *Acta. Theriol.* **23:** 401-11.

FEDAK, M. A. AND ANDERSON, S. S., 1982. The energetics of lactation: accurate measurements from a large wild mammal, the grey seal. *J. Zool. Lond.* **198:** 473-80.

FINDLAY, L., 1984. The mammary glands of tammar wallaby *(Macropus eugenii)* during pregnancy and lactation. *J. Reprod. Fert.* **65:** 59-66.

FINDLAY, L. AND RENFREE, M. B., 1984. Growth development and secretion of the mammary gland of macropodid marsupials. *Symp. Zool. Soc. Lond.* **51:** 403-32.

GREEN, B., 1984. Composition of milk and energetics of growth in marsupials. *Symp. Zool. Soc. Lond.* **51:** 369-87.

GREEN, B., GRIFFITHS, M. AND LECKIE, R. M. C., 1983. Qualitative and quantitative changes in milk fat during lactation in the tammar wallaby *(Macropus eugenii). Aust. J. Biol. Sci.* **36:** 455-61.

GREEN, B. AND MERCHANT, J. C., 1988. The composition of marsupials milk. *In* The Developing Marsupial. Models for Biomedical research ed. by C. H. Tyndale-Biscoe and P. A. Janssens. Springer-Verlag: Berlin.

GREEN, B., MERCHANT, J. C. AND NEWGRAIN, K., 1987. Milk composition in the eastern quoll, *Dasyurus viverrinus* (Marsupialia: Dasyuridae). *Aust. J. Biol. Sci.* **40:** 379-87.

GREEN, B., MERCHANT, J. C. AND NEWGRAIN, K., 1988. Milk consumption and energetics of growth in pouch young of the tammar wallaby *Macropus eugenii. Aust. J. Zool.* **36:** 217-27.

GREEN, B. AND NEWGRAIN, K., 1979. Estimation of the milk intake of sucklings by means of ^{22}Na. *J. Mammal.* **60:** 556-59.

GREEN, B., NEWGRAIN, K. AND MERCHANT, J. C., 1980. Changes in milk composition during lactation in the tammar wallaby *(Macropus eugenii). Aust. J. Biol. Sci.* **33:** 35-42.

GREEN, S. W. AND RENFREE, M. B., 1982. Changes in the milk proteins during lactation in the tammar wallaby, *Macropus eugenii. Aust. J. Biol. Sci.* **35:** 145-52.

GRIFFITHS, M., MCINTOSH, D. L. AND LECKIE, R. M. C., 1972. The mammary glands of the red kangaroo with observations on the fatty acid components of the milk triglycerides. *J. Zool.* **166:** 265-75.

HEARN, J. P., 1972. The pituitary gland and reproduction in the marsupial *Macropus eugenii* (Desmarest). Ph.D. Thesis. Australian National University A.C.T.

HINDS, L. A., 1988. Hormonal control of lactation. *In* The developing marsupial. Models for biochemical research ed. by C. H. Tyndale-Biscoe and P. A. Janssens. Springer-Verlag: Berlin.

HINDS, L. A. AND TYNDALE-BISCOE, C. H., 1982. Prolactin in the marsupial, *Macropus eugenii* during the estrous cycle, pregnancy and lactation. *Biol. Reprod.* **26:** 391-98.

HINDS, L. A. AND TYNDALE-BISCOE, C. H., 1985. Seasonal and circadian patterns of circulating prolactin during lactation and seasonal quiescence in the tammar *Macropus eugenii. J. Reprod. Fert.* **74:** 173-83.

JANSSENS, P. A. AND TERNOUTH, J. H., 1987. The transition from milk to forage diets. Pp. 281-305 *in* The nutrition of herbivores ed. by J. B. Hacker and J. H. Ternouth. Academic Press: Sydney.

JANSSENS, P. A. AND MESSER, M., 1988. Changes in nutritional metabolism during weaning. *In* The developing marsupial. Models for biochemical research ed. by C. H. Tyndale-Biscoe and P. A. Janssens. Springer-Verlag: Berlin.

JOHNSON, A. H., 1974. The composition of milk. Pp. 1-57 *in* Fundamentals of dairy chemistry ed. by B. H. Webb, A. H. Johnson and J. A. Alford. The AVI Publishing Co. Inc.: Westport.

JOHNSON, R. G. AND NICHOLAS, T. E., 1979. The pattern of pulmonary morphological development and surfactant phospholipid composition in the marsupial *Megaleia rufa. Proc. Aust. Phys. Pharm. Soc.* **10**.

KALDOR, I. AND EZEKIEL, E., 1962. Iron content of mammalian breast milk: measurements in the rat and in a marsupial. *Nature.* **196:** 175.

KALDOR, I. AND MORGAN, E. H., 1986. Iron metabolism during lactation and suckling in a marsupial, the quokka *(Setonix brachyurus). Comp. Biochem. Physiol.* **84:** 691-94.

LEBAS, F., BESANCON, P. AND ABOUYOUB, A., 1971. Composition minerale du lait de lapine. Variations en fonction du stade de lactation. *Am. Zootech.* **20:** 487-95.

LEMON, M. AND BARKER, S., 1967. Changes in milk composition of the red kangaroo, *Megaleia rufa* (Desmarest) during lactation. *Aust. J. Exp. Biol. Med. Sci.* **45:** 213-19.

LEMON, M. AND POOLE, W. E., 1969. Specific proteins in the whey from milk of the Grey Kangaroo. *Aust. J. Exp. Biol. Med. Sci.* **47:** 283-85.

LINCOLN, D. W. AND RENFREE, M. B., 1981. Mammary gland growth and milk ejection in the agile wallaby *Macropus agilis* displaying concurrent asynchronous lactation. *J. Reprod. Fert.* **63:** 193-302.

LINZELL, J. L. AND PEAKER, M., 1971. Mechanism of milk secretion. *Physiol. Rev.* **51:** 564-97.

MACY, I. G., 1949. The composition of human colostrum and milk. *Am. J. Dis. Child.* **78:** 589-603.

MASORO, E. J., 1968. Physiological Chemistry of Lipids in Mammals. W. B. Saunders Company: London.

MCKENZIE, H. A., MULLER, V. J. AND TREACY, G. B., 1983. "Whey" proteins of milk of the red kangaroo *(Macropus rufus)* and eastern grey *(Macropus giganteus)* kangaroo. *Comp. Biochem. Physiol.* **74B:** 259-71.

MERCHANT, J. C., 1979. The effect of pregnancy on the interval between one oestrus and the next in the tammar wallaby *Macropus eugenii. J. Reprod. Fert.* **56:** 459-63.

MERCHANT, J. C. AND CALABY, J. H., 1981. Reproductive biology of the red-necked wallaby *(Macropus rufogriseus banksianus)* and Bennett's wallaby *(M. r. rufogriseus)* in captivity. *J. Zool.* **194:** 203-17.

MERCHANT, J., GREEN, B., MESSER, M. AND NEWGRAIN, K., 1989. Milk composition in the red-necked wallaby, *Macropus rufogriseus banksianus* (Marsupialia). *Comp. Biochem. Physiol.* (in press).

MERCHANT, J. C. AND LIBKE, J. A., 1988. Milk composition in the northern brown bandicoot, *Isoodon macrourus* (Peramelidae, Marsupialia). *Aust. J. Biol. Sci.* **41:** 495-505.

MERCHANT, J. C. AND SHARMAN, G. B., 1966. Observations on the attachment of marsupial young to the teats and on the rearing of pouch young by foster-mothers of the same or different species. *Aust. J. Zool.* **14:** 593-609.

MESSER, M., FITZGERALD, P. A., MERCHANT, J. C. AND GREEN, B., 1987. Changes in milk carbohydrates during lactation in the eastern quoll, *Dasyurus viverrinus* (Marsupialia). *Comp. Biochem. Physiol.* 1083-86.

MESSER, M. AND GREEN, B., 1979. Milk carbohydrates of marsupials II. Quantitative and qualitative changes in milk carbohydrates during lactation in the tammar wallaby *(Macropus eugenii). Aust. J. Biol. Sci.* **32:** 519-631.

MESSER, M., GRIFFITHS, M. AND GREEN, B., 1984. Changes in milk carbohydrates and electrolytes during early lactation in the Tammar Wallaby, *Macropus eugenii. Aust. J. Biol. Sci.* **37:** 1-6.

MESSER, M. AND MOSSOP, G. S., 1977. Milk carbohydrates of marsupials I. Partial separation and characterization of neutral milk oligosaccharides of the eastern grey kangaroo. *Aust. J. Biol. Sci.* **30:** 379-88.

MESSER, M., TRIFONOFF, E., STERN, W., COLLINS, J. G. AND BRADBURY, J. H., 1980. Structure of a marsupial-milk trisaccharide. *Carbohydrate Research* **83:** 327-34.

NICHOLAS, K. R., 1988. Control of milk protein synthesis in the marsupial *Macropus eugenii;* a model system to study prolactin-dependent development. *In* The developing marsupial. Models for biochemical research ed. by C. H. Tyndale-Biscoe and P. A. Janssens. Springer-Verlag: Berlin.

NICHOLAS, K. R., MESSER, M., ELLIOTT, C., MAHER, F. AND SHAW, D. C., 1987. A novel whey protein synthesised only in late lactation by the mammary gland from the Tammar *(Macropus eugenii). Biochem. J.* **241:** 899-904.

OFTEDAL, O. T., 1984. Milk composition, milk yield and energy output at peak lactation: a comparative review. *In* Physiological strategies in lactation ed. by M. Peaker, R. G. Vernon and C. H. Knight. Academic Press: London.

OFTEDAL, O. T., HINTZ, H. F. AND SCHRYVER, H. F., 1983. Lactation in the horse: Milk composition and intake by foals. *J. Nutr.* **113:** 2096-106.

POOLE, W. E., MERCHANT, J. C., CARPENTER, S. M. AND CALABY, J. H., 1985. Reproduction, growth and age determination in the yellow-footed rock wallaby *Petrogale xanthopus* Gray, in captivity. *Aust. Wildl. Res.* **12:** 127-36.

POOLE, W. E., SHARMAN, G. B., SCOTT, K. J. AND THOMPSON, S. Y., 1982. Composition of milk from red and grey kangaroos with particular reference to vitamins. *Aust. J. Biol. Sci.* **35:** 607-15.

RENFREE, M. B., 1973. The composition of fetal fluids of the marsupial *Macropus eugenii. Dev. Biol.* **33:** 62-79.

RENFREE, M. B., MEIER, P., TENG, C. AND BATTAGLIA, F. C., 1981. Relationship between amino acid intake and accretion in a marsupial, *Macropus eugenii.* I. Total amino acid composition of the milk throughout pouch life. *Biol. Neonate* **40:** 29-37.

RENFREE, M. B., RUSSELL, E. M. AND WOOLLER, R. D., 1984. Reproduction and life history of the honey possum, *Tarsipes rostratus, In* Possums and gliders ed. by A. P. Smith and I. D. Hume. Australian Mammal Society: Sydney.

SADLEIR, R. M. F. S., 1980. Energy and protein intake in relation to growth of suckling black-tailed deer fawns. *Can. J. Zool.* **58:** 1347-54.

SCHRYVER, H. F., OFTEDAL, O. T., WILLIAMS, J., CYMBALUK, N. F., ANTCZAK, D. AND HINTZ, H. F., 1986. A comparison of the mineral composition of milk of domestic and captive wild equids *(Equus przewalski, E. zebra, E. burchelli, E. caballus, E. assinus). Comp. Biochem. Physiol.* **85:** 233-35.

SHARMAN, G. B., 1962. The initiation and maintenance of lactation in the marsupial, *Trichosurus vulpecula. J. Endocr.* **25:** 375-85.

SHARMAN, G. B. AND CALABY, J. H., 1964. Reproductive behaviour in the red kangaroo *Megaleia rufa,* in captivity. *C.S.I.R.O. Wildlife Res.* **9:** 58-85.

SMOLENSKI, A. J. AND ROSE, R. W., 1988. Comparative lactation in two species of rat-kangaroo (Marsupialia). *Comp. Biochem. Physiol.* **90A:** 459-63.

STEWART, F., 1984. Mammogenesis and changing prolactin receptor concentrations in the mammary glands of the tammar wallaby *(Macropus eugenii). J. Reprod. Fert.* **71:** 131-48.

TEDMAN, R. AND GREEN, B., 1986. Water and sodium fluxes and lactational energetics in suckling pups of Weddell seals *(Leptonychotes weddelli). J. Zool.* **212:** 29-42.

THOMPSON, M. P. AND FARRELL, H. M., 1974. Genetic variants of the milk proteins. *In* Lactation, a comprehensive treatise ed. by B. L. Larson and V. R. Smith. Academic Press: New York, London.

TYNDALE-BISCOE, C. H., 1963. The role of the corpus luteum in the delayed implantation of marsupials. *In* Delayed Implantation ed. by A. C. Enders. University Press: Chicago, Illinois.

TYNDALE-BISCOE, C. H. AND HINDS, L. A., 1984. Seasonal patterns of circulating progesterone and prolactin and response to bromocriptine in the female tammar *Macropus eugenii. Gen. Comp. Endocrin.* **53:** 58-68.

TYNDALE-BISCOE, C. H. AND JANSSENS, P. A., 1988. *In* The developing marsupial. Models for biochemical research ed. by C. H. Tyndale-Biscoe and P. A. Janssens. Springer-Verlag: Berlin.

TYNDALE-BISCOE, C. H. AND RENFREE, M. B., 1987. *In* Reproductive physiology of marsupials ed. by C. H. Tyndale-Biscoe and M. B. Renfree. Cambridge University Press: Cambridge.

TYNDALE-BISCOE, C. H., STEWART, F. AND HINDS, L. A., 1984. Some factors in the initiation and control of lactation in the tammar wallaby. *Symp. Zool. Soc. Lond.* **51**: 389-401.

WIDDOWSON, E. M., 1950. Chemical composition of newly born mammals. *Nature* **166**: 626-28.

YADAV, M., 1971. The transmission of antibodies across the gut of pouch-young marsupials. *Immunology* **21**: 839-51.

YADAV, M., 1972. Characteristics of blood in the pouch young of a marsupial, *Setonix brachyurus*. *Aust. J. Zool.* **20**: 249-63.

YADAV, M. AND EADIE, M., 1973. Passage of maternal immunoglobulins to the pouch young of a marsupial, *Setonix brachyurus*. *Aust. J. Zool.* **21**: 171-81.

YADAV, M., STANLEY, N. F. AND WARING, H., 1972. The microbial flora of the gut of the pouch-young and pouch of a marsupial, *Setonix brachyurus*. *J. Gen. Microbiol.* **70**: 437-42.

YOUNG, I. R. AND RENFREE, M. B., 1979. The effects of corpus luteum removal during gestation on parturition in the tammar wallaby *(Macropus eugenii)*. *J. Reprod. Fert.* **56**: 249-54.

Metabolic changes during pouch vacation and weaning in macropodoids

P. A. JANSSENS[1] and A. M. T. ROGERS[1]

The development of nutritional independence of macropodoid young from their mother is associated with a change from a carbohydrate-rich to a lipid-rich diet, with the appearance of a population of microorganisms in the fore-stomach, and with changes in the patterns of enzymes present in the alimentary canal, liver and kidney, which accommodate these changes in diet.

In the tammar, *Macropus eugenii*, it is suggested, on the basis of relative timing, that the changes in metabolic enzyme pattern are a consequence of the change in diet, which, in turn, follows the first exits of the young from the pouch. It is argued that as the young increases its heat production to a level which will permit it to maintain its body temperature when it is outside the pouch, the mother will have increasing difficulty in dissipating the excess heat produced while the young remains within the pouch. It is suggested that this increasing loss of thermal compatability between mother and young may play an important role in pouch vacation in macropodoid marsupials.

INTRODUCTION

THE Tammar Wallaby, *Macropus eugenii,* is a medium-sized macropodoid marsupial which has, in the past twenty years or so, become the macropodoid most used in research. There are a number of reasons for this. First, *M. eugenii* can easily be handled by one person, readily adapts to captive colonies, and breeds well in such colonies. In the Canberra colony, most females breed every year and do so for at least five or six years. The colony is thus self-sustaining. A second reason for the popularity of *M. eugenii* as research animals is the seasonal breeding. Female *M. eugenii* carry a dormant blastocyst through the year and this is reactivated at the summer solstice so that the animals give birth towards the end of January. After a post-partum oestrus and mating, the newly fertilized egg develops to the blastocyst stage and then becomes dormant. It is therefore relatively easy to have large groups of individuals which are developing synchronously; this facilitates studies of development. *M. eugenii* is, however, not identical with all other macropodoids, let alone all marsupials, and this must be borne in mind, especially when considering events such as those occurring at the time of pouch vacation, the topic of this chapter. Most of the developmental changes which occur at the time when the young is becoming independent from the mother are nutritional, and very much depend on the diet of the young. At this time, the developmental changes seen in *M. eugenii* may be similar to those of other macropodoids, but are certainly not typical of those seen in other marsupials which have very different diets. The fact that *M. eugenii* is a seasonal breeder, while many other macropodoids are not, also affects the events at pouch vacation; in *M. eugenii*, the exit of one young is not usually followed by the birth of another, whereas this often occurs in non-seasonal breeders such as *Macropus rufus* and *Bettongia gaimardi*.

This chapter will concentrate on *M. eugenii* since little is known of the metabolic changes which occur at pouch exit in any other marsupial. We shall first discuss development in *M. eugenii*, concentrating on the time of pouch vacation, and we shall then consider, in some detail, the stimuli which cause the young to venture out of the security of the pouch.

DEVELOPMENT IN *M. EUGENII*

Development in *M. eugenii* can readily be divided into three fairly distinct periods (Fig. 1). The first, which lasts from birth until about Day 100 of pouch life, is typified by the permanent attachment of the young to the teat. During this period, the young grows rather slowly, and the most noticeable event is the rapid development of the brain which grows faster than the rest of the body between birth and about 130 days (the allometric constant, α, equals 1.1; Renfree *et al.* 1982). At birth, the neocortex is very undeveloped, having only two cell layers, a ventricular zone and a plexiform layer (Reynolds *et al.* 1985; Reynolds and Saunders 1988). The cortical plate is not complete until Day 10-15 and the formation of the six layers of the adult cerebral cortex takes a further 50-60 days. There is further rapid maturation of the brain after Day 70 and it is tempting to think that the development of the brain to the particular level of organization achieved in *M. eugenii* at about Day 100, must precede the maturation of physiological functions.

[1]Department of Zoology, Australian National University, P.O. Box 4, Canberra, Australian Capital Territory 2601.

Pages 367-76 in KANGAROOS, WALLABIES AND RAT-KANGAROOS ed. by G. Grigg, P. Jarman and I. Hume. Surrey Beatty & Sons Pty Limited, New South Wales, Australia, 1989.

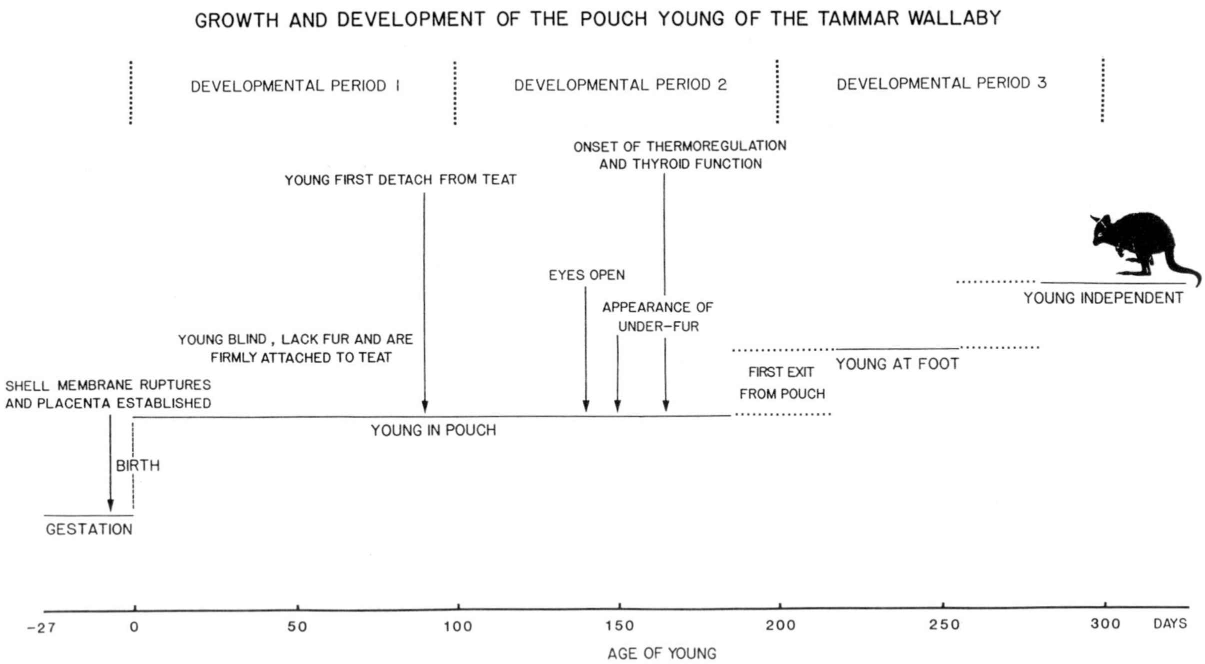

Fig. 1. Growth and development of pouch young of *M. eugenii.* Modified from Fig. 2.2 of Wilkes (1984).

The second period of development lasts from about Day 100 to about Day 200. It begins when the young first voluntarily relinquishes the teat and lasts until it makes its first exit from the pouch. During this period, which is characterized by the development of endothermy (Hulbert 1988) and the maturation of the cardio-respiratory (Baudinette *et al.* 1988a and b), renal (Wilkes and Janssens 1988) and other physiological systems, the young becomes physiologically independent of the mother and therefore has the physiological capacity to exist independently outside the pouch. At the end of this second period of development, the young first gets out of the pouch, albeit for short periods of time initially. In *M. eugenii*, this first separation of mother and young is likely to be the trigger for the third period of development which lasts until the completion of weaning and during which the young becomes nutritionally independent of the mother.

DEVELOPMENT OF NUTRITIONAL INDEPENDENCE

After its first emergence, the young *M. eugenii* spends increasing amounts of time out of the pouch until permanent pouch vacation at about 250 days. During this time it consumes increasing amounts of herbage and there are major changes in the enzyme complement of a number of organs, particularly the alimentary canal and the liver. Within one or two weeks of first pouch exit, the composition of the milk changes dramatically, with the carbohydrate content falling from about 120-140 g/l to less than 20 g/l and total lipid rising from about 50 g/l to about 230 g/l. These changes occur in a 6-8 week period and are reviewed in detail by Green (1984), Green and Merchant (1988) and Merchant (this volume). Contemporaneously with the change in milk composition, a population of microorganisms, essential for the metabolism of herbage, is developing in the fore-stomach, and so the energy supply of the young *M. eugenii* changes from being primarily carbohydrate to being primarily lipid during a period of a few weeks. The changes in the enzyme complement of the liver and alimentary canal reflect this dietary switch.

The milk carbohydrates of *M. eugenii* are in the form of oligosaccharides of which the major hexose component is galactose. Minor components are N-acetylglucosamine and sialic acid. The enzyme profile in the alimentary canal and liver of the sucking young reflect this. β-galactosidase, β-N-acetylglucosaminidase and neuraminidase, enzymes which participate in the breakdown of milk oligosaccharides, are high in activity while milk is the major dietary component, but decrease in activity as herbage becomes increasingly important, falling to very low levels in the adult (Walcott and Messer 1980; Table 1). Fucosidase activity follows the same pattern (Walcott and Messer 1980; Table 1), but its function in the young is not clear. Galactokinase, hexose-1-phosphate uridylyl transferase and UDP glucose 4-epimerase, hepatic enzymes which convert galactose to glucose, are also high in activity in the milk-dependent young and fall in activity during weaning (Vernon *et al.* 1981; Table 1). Thus the young *M. eugenii* whose diet is high in milk carbohydrate, has the enzymes to digest oligosaccharides and convert them to

Table 1. Metabolic changes during weaning in *M. eugenii.*

Age (days)	126-140	161-168	189-196	203-210	224	238	252	280	294	Adult	Units	Source
Liver-enzymes of glycolysis/gluconeogenesis												
Hexokinase	0.09±0.01	0.09±0.03	0.30±0.03	0.47±0.04	0.42±0.05	0.33±0.02	0.49±0.12	0.57±0.06	0.44±0.04	0.16±0.02	units/g liver	Wilkes & Janssens 1986a
Fructose-bis-phosphatase	7.82±0.24	7.03±0.40	6.47±0.20	5.73±0.40	6.76±0.35	7.41±0.34	7.44±0.37	6.02±0.33	4.70±0.25	5.32±0.42	units/g liver	Wilkes & Janssens 1986a
PEPCK (particulate)	5.41±0.15	4.44±0.20	3.02±0.12	2.62±0.24	6.51±0.54	4.06±0.46	5.46±0.29	5.96±0.96	4.96±0.26	5.54±0.62	units/g liver	Wilkes & Janssens 1986a
PEPCK (soluble)	1.07±0.22	0.98±0.19	0.81±0.10	1.05±0.10	2.63±0.31	2.62±0.14	4.16±0.54	3.83±0.34	4.49±0.27	3.82±0.21	units/g liver	Wilkes & Janssens 1986a
Pyruvate kinase	15.9±2.9	18.9±2.7	32.1±2.4	30.5±2.0	26.6±2.0	23.9±0.7	17.5±0.7	15.5±1.0	16.5±1.5	5.7±0.3	units/g liver	Wilkes & Janssens 1986a
Liver glycogen	30.8±3.2	60.2±11.9	79.2±11.4	70.5±2.3	17.9±2.8	15.1±2.2	7.5±0.9	34.4±6.4	—	30.4±2.1	mg/g liver	Wilkes & Janssens 1986a
Liver protein	64.0±1.8	78.5±3.5	65.0±1.8	76.5±4.0	101±4	97±5	118±9	119±3	156±2	144±5	mg/g liver	Wilkes & Janssens 1986b
Blood glucose	—	11.3±1.2	10.2±0.78	10.6±1.3	8.94±0.83	9.78±0.56	10.1±1.56	6.54±0.51	8.08±1.54	3.96±0.18	mM	Wilkes & Janssens 1986a
Liver-ornithine urea cycle enzymes												
Carbamoylphosphate synthetase	1.92±0.48	2.27±0.22	2.00±0.12	2.94±0.36	3.19±0.32	3.88±0.50	4.56±1.00	8.45±0.90	5.95±0.59	6.52±0.68	units/g liver	Wilkes & Janssens 1986b
Ornithine carbamoyltransferase	47.0±1.5	62.9±8.1	67.6±5.7	83.2±9.0	115±10.7	137±10.2	196±14.5	174±13.0	196±20.4	141±18.5	units/g liver	Wilkes & Janssens 1986b
Argininosuccinate synthetase	0.65±0.09	1.14±0.16	0.95±0.19	2.29±0.17	1.69±0.15	2.65±0.31	2.23±0.27	2.30±0.29	2.57±0.21	3.00±0.30	units/g liver	Wilkes & Janssens 1986b
Argininosuccinate lyase	0.58±0.05	0.80±0.30	1.10±0.16	2.75±0.16	4.20±0.70	4.62±0.42	4.70±0.42	4.57±0.50	4.30±0.25	5.00±0.48	units/g liver	Wilkes & Janssens 1986b
Arginase	192±8	238±10	216±11	596±60	564±34	700±44	760±54	562±96	540±38	680±114	units/g liver	Wilkes & Janssens 1986b
Liver-galactose metabolising enzymes												
Galactokinase	30	31	21	20	—	16	9	—	8	5	units/g protein	Vernon *et al.* 1981
Hexose-1-P uridylyl transferase	31	28	32	21	—	—	15	7	6	8	units/g protein	Vernon *et al.* 1981
UDPglucose 4-epimerase	27	24	12	10	—	9	5	1	—	1	units/g protein	Vernon *et al.* 1981
Intestine — milk carbohydrate digesting enzymes												
β-galactosidase	60	140	150	200	—	130	—	—	0	0	units/g protein	Walcott & Messer 1980
β-N-acetylglucosaminidase	450	960	560	1 320	—	920	—	—	30	30	units/g protein	Walcott & Messer 1980
alpha-L-fucosidase	10	10	8	16	—	8	—	—	1	1	units/g protein	Walcott & Messer 1980
Neuraminidase	1	2.4	5.6	8.5	—	7.6	—	—	0	0	units/g protein	Walcott & Messer 1980
Maltase	6	8	14	13	—	13	—	—	42	94	units/g protein	Walcott & Messer 1980
Sucrase	0	0	0	0	—	0	—	—	5.5	14.5	units/g protein	Walcott & Messer 1980
Isomaltase	0	1	2.5	0	—	0	—	—	3	14	units/g protein	Walcott & Messer 1980
Trehalase	0	0	0	0	—	0	—	—	1	9.5	units/g protein	Walcott & Messer 1980
Intestine — protein digesting enzymes												
Cathepsin B	11.2	3.8	4.6	4.4	13.4	15.5	—	5.2	—	3.0	units/g protein	Davies 1981
Cathepsin D	23	18.3	31	12.3	8.3	20.7	—	5.7	—	7.7	units/g protein	Davies 1981
Trypsin	0.01	0.015	0.03	0.1	0.025	0.03	—	0.092	—	2.37	units/g tissue	Davies 1981
Chymotrypsin	0.0021	0.01	0.0086	0.0107	0.0067	0.007	—	0.016	—	0.106	units/g tissue	Davies 1981
Stomach — protein digesting enzymes												
Forestomach pepsinogen	120	160	220	430	500	80	—	0	—	0	units/g protein	Davies 1981
Hindstomach pepsinogen	200	280	300	560	420	580	—	760	—	1 640	units/g protein	Davies 1981

glucose, and this ability is lost as milk disappears from the diet.

Towards the end of weaning, the disaccharidases maltase, sucrase, isomaltase and trehalase appear in the small intestine (Walcott and Messer 1980; Table 1). The significance of the appearance of these enzymes is not so apparent but it is unlikely that enzyme synthesis would occur without good reason. Some disaccharides may escape fermentation in the fore-stomach and be available for digestion in the intestine. Certainly the concentration of glucose in the lumen of the alimentary canal increases progressively from near zero in the fore-stomach to 0.87 mM in the distal portion of the small intestine (Janssens 1984), which suggests that breakdown of either di- or oligosaccharides is occurring.

The proteases of the intestine show comparable changes in activity. Before the start of the weaning period, the casein-digesting enzymes, cathepsins B and D, are at high activity and subsequently fall, whereas more general proteases such as trypsin and chymotrypsin, which are low in activity before the start of weaning, increase thereafter (Davies 1981; Table 1). In the stomach, pepsinogen is present throughout the wall of both fore- and hind-stomachs before weaning but is restricted to the hind-stomach soon after the start of weaning (Davies 1981; Table 1).

This change in diet is a considerable hurdle for the young *M. eugenii* to overcome. As can be seen from Table 1, the liver glycogen content, which is 60 to 80 mg/g between Day 160 and Day 210, falls to less than 20 mg/g during the next few weeks, rising to the adult level of about 30 mg/g by Day 280. These low liver glycogen levels, especially around Day 250, when milk intake is beginning to fall (Dove, Cork and Christian 1987), carry the implication that the young may be short of carbohydrate. Young *M. eugenii* do indeed appear to die more frequently at this time than at other stages of development, and Johnson (1988) has reported a 50% mortality of Red-necked Wallabies, *Macropus rufogriseus banksianus*, at the time when the young are leaving the pouch and being weaned. There is, however, little indication of a fall in blood glucose levels in surviving young (Table 1) and, since adult *M. eugenii* will exhibit symptoms typical of hypoglycaemia in eutherians if blood glucose falls too low, it is likely that young must maintain normal blood glucose levels throughout the weaning period if they are to survive.

This requirement for glucose at a time when the amount of glucose being absorbed from the alimentary canal is falling is met by an increased rate of gluconeogenesis (Janssens *et al.* 1977; Paton and Janssens 1981) and there is a change in the profile of the rate-limiting enzymes of glycolysis and gluconeogenesis in the liver. The most significant change is a 3-5 fold increase in soluble phosphoenolpyruvate carboxykinase (PEPCK) activity between Day 210 and Day 252 (Wilkes and Janssens 1986a; Table 1) and an associated loss of inducibility of this enzyme. Whereas both soluble PEPCK activity and the rate of gluconeogenesis increase after fasting in *M. eugenii* pouch young, there are no such increases after fasting in adults (Paton and Janssens 1981), and it is likely that this increased capacity for gluconeogenesis, together with a loss of inducibility, is associated with the development at weaning of a continuous requirement for glucose synthesis at a relatively constant rate. Between Day 210 and Day 252, there is a halving of hepatic pyruvate kinase activity (Wilkes and Janssens 1986a; Table 1), and the ratio of soluble pyruvate kinase activity to PEPCK activity in *M. eugenii* liver falls by about an order of magnitude during this period. There is thus a fall in the capacity for hepatic glycolysis during the period of weaning which might well be seen as a mechanism for conserving glucose residues and be driven by the reduced availability of glucose from the alimentary canal.

There are also major changes in the pattern of nitrogen and acid-base metabolism during the weaning period (Wilkes and Janssens 1986b). During pouch life, the ammonia content of the urine increases progressively to a maximum of about 100 mM and there is a slight increase in urinary urea concentration. Between Day 175 and Day 224 urinary ammonia content falls rapidly towards the adult level of about 10-25 mM while there is a 2-4 fold increase in urinary urea content. Urine pH, which is consistently less than 6.0 and averages about 5.0 until about Day 224, thereafter becomes very variable but tends to increase towards the adult value of about 8.1. Coincident with the increase in urinary urea content, all five enzymes of the ornithine-urea cycle increase several-fold in activity (Table 1).

These changes in the pattern of nitrogen excretion and urine pH are also probably a direct consequence of the change in diet during the period of weaning. As glucose uptake decreases, both because of the fall in glucose content of the diet and because the developing population of microorganisms in the fore-stomach will ferment any simple sugars eaten, the requirement for gluconeogenesis will increase. Although propionate, produced by fermentative metabolism in the fore-stomach, is probably an increasingly important gluconeogenic substrate, some protein will also be utilized and it is an intriguing thought that the appearance of new proteins in the milk at the time of weaning, such as the late-lactation protein described by Nicholas (1988), may provide substrate for gluconeogenesis. Such protein may also, of course, provide amino acids both to support growth of the young and for the synthesis of the new enzymes which are appearing at this time. Protein which is utilized as gluconeogenic substrate must first be deaminated and this will increase the release of ammonia into the blood. It is interesting that at the time when urinary ammonia concentration is falling rapidly, plasma ammonia content is rising (Wilkes and Janssens 1986b).

At the same time as increasing amounts of ammonia are being released as by-products of gluconeogenesis, herbage is forming an increasing proportion of the diet. Herbivores characteristically produce urine with a high pH, thought to be associated with the high levels of potassium in the diet, and, as the amount of herbage in the diet increases, excess protons, which are presumably being produced as a result of metabolism of dietary milk, will progressively be taken up by the products of herbage metabolism. The pH of the urine thus rises and urinary ammonia concentration, which is positively correlated with urine pH, falls (Wilkes and Janssens 1986b). This change in acid-base balance therefore results in a large decrease in the amount of ammonia excreted as such and this, together with the increase in production of ammonia from gluconeogenesis provides a stimulus for the induction of the enzymes of the ornithine-urea cycle, with a consequent increase in excretion of urea.

All the metabolic changes which occur at about the time of pouch vacation in *M. eugenii* are therefore probably consequent upon the change in diet which occurs at this time (Fig. 2) and which also forms a logical sequence. The establishment of a mixed population of bacteria and protozoa in the fore-stomach is essential for the development of herbivory. Mackie *et al.* (1978) have shown that when sheep are fed diets high in simple carbohydrates, bacteria ferment the carbohydrates to lactic acid causing a fall in fore-stomach pH which is not well tolerated by resident protozoa. It is probable that the fall in carbohydrate content of the milk is an essential precursor for the development of a balanced population of bacteria and protozoa in the fore-stomach (Janssens and Ternouth 1987).

It is not known what causes the major change in composition of the milk which is seen at about Day 200 in *M. eugenii.* The fall in hexose content of the milk always follows the first vacation of the pouch (B. Green and K. Newgrain unpublished findings), but there is no evidence that these two events are causally linked. If the first exit(s) from the pouch do indeed initiate the change in milk composition, there is no suggestion as to what the signal might be. No specific changes in maternal hormone profile have been associated with the first pouch exit. There is a decrease in prolactin towards the end of lactation (Hinds and

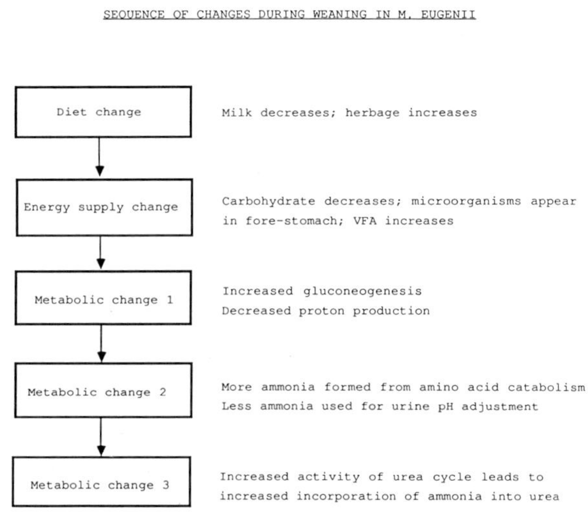

Fig. 2. Model suggesting that the series of metabolic changes which occur during weaning are initiated by the changes in diet.

Tyndale-Biscoe 1985) but this occurs at about Day 250, too late to be regulating the change in milk composition; there is no information on the levels of neurohypophysial peptides, such as oxytocin, at this time. Alternatively, the change in milk composition occurs only about 20 days after first pouch exit and these two associated events may be triggered by a single stimulus rather than one being the cause of the other.

THE STIMULUS FOR FIRST POUCH EXIT

There is considerable circumstantial evidence that thermal intolerance between mother and young may be the stimulus for the first vacation of the pouch in some macropodoids. In *Setonix brachyurus*, Bartholomew (1956) stated that "the temperatures of the fully haired young about ready to leave the pouch closely approximate pouch temperature and sometimes exceed it; those of the lightly haired or naked young are slightly lower than pouch temperature", and Shield (1966) reported that all *S. brachyurus* young of 117 days and older had body temperatures higher than that of the pouch, whereas the reverse applied in younger animals. *S. brachyurus* young leave the pouch at about 180-190 days (Shield 1968). Using oxygen consumption as an indicator of metabolic rate, Shield (1966) reported a thermo-neutral zone between 31 and 32°C in a 178 day young, with oxygen consumption increasing outside these temperatures, and concluded that "Body heat generated by joeys older than 100 days of age is sufficient to raise the body temperature above that of the pouch but the mother, acting as a thermal sink, evidently limits any rise in pouch temperature which is kept at about 36.5°C by her thermoregulatory mechanisms. This is higher than the thermal neutral temperature and may provide a stimulus for older joeys to leave the pouch for short periods before becoming completely independent of the mother".

Loh and Shield (1977), also working with *S. brachyurus*, showed that there was a sudden increase in the capacity for energy production in young between 154 and 169 days, when the young became homeothermic. A 186 day old, *S. brachyurus* young maintained a relatively constant oxygen consumption for 76 hours at an ambient temperature of 20°C. Furthermore, young older than about 80 days, kept in a humid atmosphere at 37.5°C, produced sufficient heat to elevate their body temperature; one 105 day young kept in these conditions had a body temperature of over 39°C and subsequently died. Young older than 162 days kept in a humid atmosphere at only 34°C had body temperatures of over 39°C. Despite this, Loh and Shield (1977) found that the rectal temperature of young in the pouch was little different from the rectal temperature of their mothers towards the end of pouch life but that the pouch temperature was almost a degree lower. They suggested that young in the pouch lose heat both by conductance to the pouch and by evaporative cooling associated with respiration, and concluded that "respiratory discomfort and finally mutual thermal intolerance between mother and joey may then lead to the joey's first ventures out of the pouch". Wallis and Maynes (1973) also showed that homeothermy developed shortly before the first exit from the pouch in the Parma Wallaby *(Macropus parma)*.

Loudon, Rothwell and Stock (1985) described the appearance of brown adipose tissue in Bennett's Wallaby *(Macropus rufogriseus rufogriseus)* about one month before the first emergence from the pouch and reported that, although thyroid hormones can modulate β-adrenergic receptor density and might be involved in the increased metabolic rate and response to noradrenaline associated with the appearance of brown adipose tissue, there was no change in the concentration of triiodothyronine (T_3) during pouch life. However, the level of T_3 which they reported (440 μg/ml or 0.65 mM) is about five orders of magnitude higher than that found in mammals in general and compares with our findings that T_3 levels in *M. eugenii* young increase between day 150 and day 200 of pouch life to the adult level of about 2 nM (P. A. Janssens and A. J. Hulbert unpublished findings). The presence of brown adipose tissue in marsupial young and the role of thyroid hormones in its development deserves further study.

Rogers (1979) has also argued that, towards the end of pouch life, *M. eugenii* young might be thermally incompatible with their mother. The weight specific standard metabolic rate of young changes little during pouch life in both *M. eugenii* (Wright 1975; Baudinette *et al.* 1988a and b) and *S. brachyurus* (Shield 1966) and the allometric constant relating standard metabolic rate and body weight is close to 1 in these species (Hulbert 1988). There is, therefore, a progressive increase in total oxygen consumption at 37°C as pouch young age (Shield 1966; Wright 1975) and, with increasing age, young increase their relative heat production at all ambient temperatures. This is reflected in the rates of heat loss for *M. eugenii* young kept at an ambient temperature of 20°C which demonstrate an increasing ability to regulate their body temperature (Rogers 1979; Figs 3 and 4). The youngest animals able to maintain a fairly constant body temperature were about 180 days old; younger animals were unable to do so, despite shivering (Fig. 3). Further, when continuous recordings of pouch temperature and young rectal temperature were made, in naked pouch young the rectal temperature of the young was usually below the pouch temperature (Rogers 1979; Fig. 5) which implies that heat moves from mother to young at this age, while the rectal temperatures of older young usually exceeded that of the pouch (Rogers 1979; Fig. 6) suggesting a reversal of heat flow. Finally, when young *M. eugenii* were kept at a constant temperature of 37.5°C in a dry atmosphere, young of about 150 days equilibrated to a body temperature of 37.5°C whereas the body temperatures of older young increased and could not be reduced to ambient by panting or licking of the extremities (Rogers 1979; Fig. 7). These findings are consistent with the postulate that, towards the end of pouch life in *M. eugenii*, as the young develops the thermal capacity to survive independently, the mother becomes less able to remove the excess heat produced by the young while it remains in the pouch. By about 200 days of age, when the young is making its first exit from the pouch, it has reached a metabolic rate similar to that of an adult marsupial of similar weight (Baudinette 1988; Hulbert 1988) and will have a body weight of about 450g. Using the formula of Dawson and Hulbert (1970), one gram of adult tissue (body weight 5.0 kg) will produce heat at an average rate of 5.56 J/h whereas one gram of pouch young tissue (body weight 450 g) will produce 10.49 J/h. As the young increases in body weight, the total amount of heat it produces will increase, and increasing thermal intolerance between mother and young may well stimulate the first emergence of the young from the pouch.

The findings reviewed in this section are consistent with the postulate that thermal intolerance between mother and young is the stimulus for pouch vacation, but it should be remembered that all the animals discussed so far are seasonal breeders, and in no case has a causal relationship between pouch exit and thermal distress of either

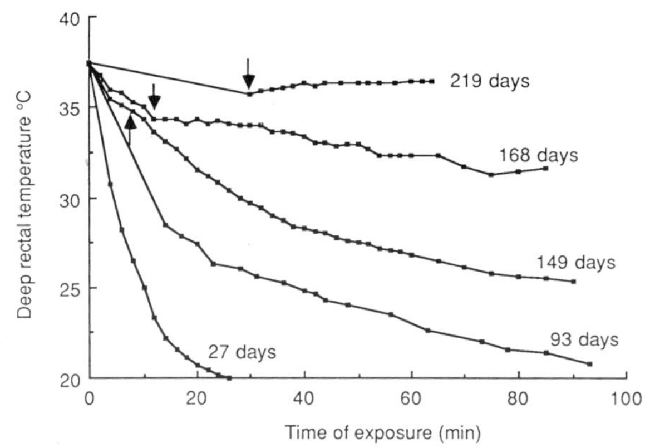

Fig. 3. Representative cooling curves of variously aged *M. eugenii* pouch young at an ambient temperature of 20±1.5°C. Arrows indicate the first appearance of shivering; the age of each young is given on the figure. Rectal temperatures were recorded from copper/constantan thermocouples, calibrated to 0.1°C, inserted in the rectum to a depth of 0.5-4.5 cm depending upon the size of the young. Limbs of older young were taped to limit struggling and prevent removal of the thermocouples. Ambient temperature was recorded from a thermocouple fixed 25 cm above the animal. (From Rogers 1979).

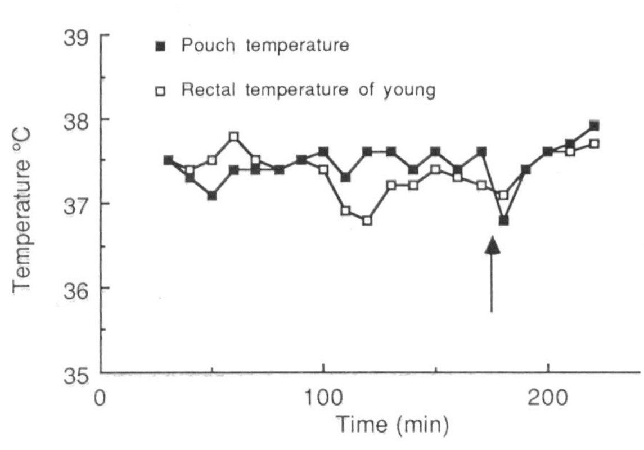

Fig. 5. Continuous simultaneous recordings of maternal *M. eugenii* pouch temperature and the rectal temperature of her 144 day old young. Points are at 10 min intervals. Measurements were taken from copper-constantan thermocouples, one located in the rectum of the young (2.5-5.0 cm deep, depending upon the size of the young) and the other close to the dorsal wall of the pouch, just anterior to the teats. Wires led from the thermocouples beneath the skin of the mother to a small incision on the chest and then passed through vinyl tubing to a potentiometric recorder. A fall of 0.8°C in pouch temperature (arrowed) was associated with the mother's licking of the inner pouch. (From Rogers 1979).

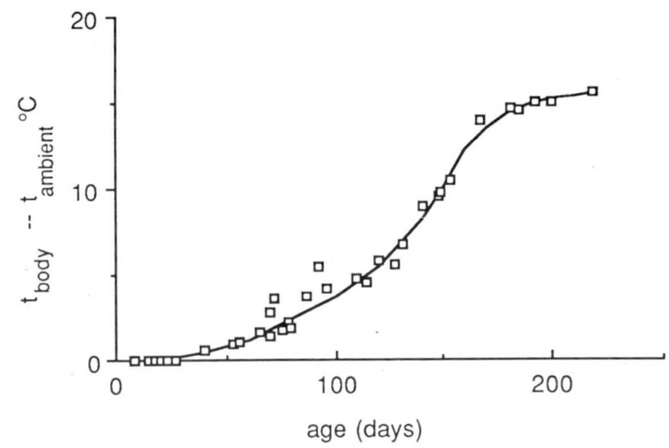

Fig. 4. Temperature difference between rectal and ambient temperatures in *M. eugenii* pouch young of various ages maintained at an ambient temperature of 20±1.5°C. Recordings were made as in the legend to Fig. 3. (From Rogers 1979).

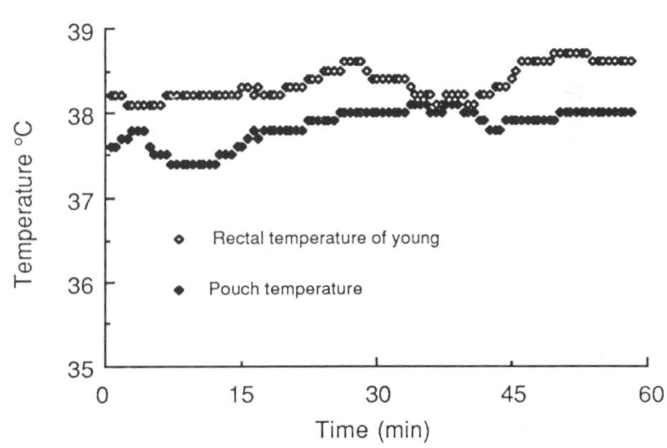

Fig. 6. Continuous, simultaneous recordings of maternal *M. eugenii* pouch temperature and the rectal temperature of her 175 day old young. Points are at 36 sec intervals. Measurements were taken as described in the legend to Fig. 5. (From Rogers 1979).

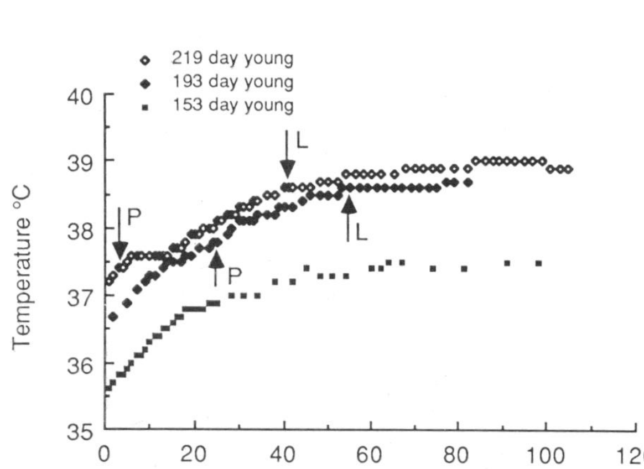

Fig. 7. Representative heating curves of variously aged *M. eugenii* pouch young at an ambient temperature of 37.5±1.0°C. P, indicates the first appearance of panting; L, of licking the extremities. Temperature recordings were made as in the legend to Fig. 3. (From Rogers 1979).

young or mother been established. A clearly contrasting situation exists in continuously breeding macropodoids, such as the rat-kangaroos. Both Rose (1986), working with *Bettongia gaimardi*, and Meredith Smith (this volume) with *B. penicillata*, have demonstrated that the age of the young in the pouch has no discernible effect upon the length of pouch life, which appears to be determined solely by physiological events in the mother. The normal time of pouch vacation is about 109 days in *B. gaimardi* and about 100 days in *B. penicillata* (Rose 1978); birth of a new young follows one day later. Clearly, the older young must be excluded from the pouch immediately before the neonate enters because of the danger of physical damage to the neonate, and this is effected by a tightening of the pouch opening as a result of which the older young, despite repeated attempts, is unable to re-enter (Rose 1986). The date of exclusion is, therefore, set by the date at which the dormant blastocyst is reactivated, which is 18 and 21 days before birth in *B. gaimardi* and *B. penicillata* respectively (Rose 1978). Rose (1986) exchanged three young at 77-85 days with three at 96-98 days and determined their subsequent rate of growth, the time when they emerged from the pouch and their body weight at that time. The smaller young, transferred to a pouch 96-98 days "old" emerged between 8 and 14 days later, on the night of parturition or oestrus, when they were aged 91-95 days. When older young were transferred to a "younger" pouch, pouch vacation occurred between 19 and 26 days later, at the time expected for the mother but when the young were 116-124 days old, 10-18 days older than expected. In both cases, birth of the new young occurred at about the expected time for the mother, and the length of pouch life was unaffected by the exchange of young (Rose 1986). The results of these experiments in which the exchange of young took place at about, or after, the time of blastocyst reactivation were confirmed by Smith (this volume) who exchanged much earlier young of *B. penicillata*, well before the time of blastocyst reactivation, and again showed that such exchanges had no effect on either the length of pouch life or the birth of the new young. It is clear, then, that in both *B. gaimardi* and *B. penicillata*, both the time at which the blastocyst is reactivated and the length of pouch life are determined by maternal factors, and are uninfluenced by the age of the young in the pouch.

In another rat kangaroo, *Potorous tridactylus*, in which the end of pouch life is also determined by a new birth, Gemmell, Cepon and Barnes (1987) demonstrated that, as in *S. brachyurus* (Loh and Shield 1977), the rectal temperatures of mother and young were similar towards the end of pouch life, but that the temperature of the pouch was nearly 1°C lower. They interpreted this finding as indicating that thermal factors do not influence the time at which the young first leaves the pouch and argued that "curiosity and the need for solid food are the primary reasons for young leaving the pouch". We have questioned this view (Janssens and Messer 1988) and have argued that, although permanent pouch vacation in *P. tridactylus* is clearly determined by the birth of the next young, the lowering of pouch temperature towards the end of pouch life may permit the young to remain in the pouch for longer than would be possible if pouch temperature remained high, and so may increase the chances of survival of the young. Similar factors may operate in *S. brachyurus*. The mechanism by which pouch temperature is reduced below core temperature in these two species is not clear. Presumably there must either be heat exchange with a radiating surface, such as the ventral skin on the outside of the pouch wall, or heat loss by convection or evaporation following gaping of the pouch opening. In either case, there may well be thermal gradients within the pouch which probably should not be considered a single thermal unit. In all studies discussed here, pouch temperature was measured with a single sensor and often the precise placing of the sensor in the pouch was not given. The assumption is made that pouch temperature is constant throughout, an assumption which may well be unwarranted. A study of pouch vasculature and of the detailed thermal anatomy of the pouch would be most valuable.

We would argue, then, that much evidence is consistent with thermal incompatibility between mother and young being a major factor driving the first emergence of the young from the pouch in *M. eugenii* and other seasonally breeding macropodoids, although a causal relationship has not been established. Such thermal factors may act through either mother or young; the studies of Russell (1973; this volume) suggest that they might act through the mother. Clearly further studies of this area are warranted. With continuous breeders such as *B. gaimardi*, it would be interesting to know what signals the reactivation of the blastocyst, which is followed by birth and a compulsory pouch vacation 18-29 days later (Rose 1978). In all macropodoids it would be valuable to know how heat is removed from the pouch, whether for example, there is countercurrent exchange between the pouch and the ventral body wall, or whether the mother automatically reduces the temperature of the pouch during gaseous exchange which must be occurring at increasing rates during the second half of pouch life (Baudinette *et al.* 1988a and b). Much remains to be done!

ACKNOWLEDGEMENTS

We thank several persons listed in the text for permission to quote their unpublished work. We also thank our colleagues, Richard Barwick,

Lyn Hinds, Michael Messer and Hugh Tyndale-Biscoe, for encouragement and discussion during the course of these studies. The work was partly supported by a grant from the National Health and Medical Research Council.

REFERENCES

BARTHOLOMEW, G. A., 1956. Temperature regulation in the macropod marsupial, *Setonix brachyurus. Physiol. Zool.* **29**: 26-40.

BAUDINETTE, R. V., GANNON, B. J., RYALL, R. G. AND FRAPPELL, P. B., 1988a. Changes in metabolic rate and blood respiratory characteristics during pouch development of a marsupial, *Macropus eugenii. Respir. Physiol.* **72**: 219-28.

BAUDINETTE, R. V., RUNCIMAN, S. I. C., FRAPPELL, P. B. AND GANNON, B. J., 1988b. Development of the marsupial cardiorespiratory system. Pp. 132-47 *in* The Developing Marsupial: Models for Biomedical Research ed. by C. H. Tyndale-Biscoe and P. A. Janssens. Springer: Heidelberg.

DAVIES, P. H., 1981. The post-natal development of digestive proteinases in the rat *(Rattus rattus)* and the tammar wallaby *(Macropus eugenii)*. Unpublished Honours Thesis, University of Sydney.

DOVE, H., CORK, S. J. AND CHRISTIAN, K. R., 1987. Defining the pattern of milk intake in a marsupial herbivore. *Proc. 2nd Int. Symp. on Nutrition of Herbivores.* Pp. 101-02.

GEMMELL, R. T., CEPON, G. AND BARNES, A., 1987. The development of thermoregulation in the rat-kangaroo, *Potorous tridactylus. Comp. Biochem. Physiol.* **88A**: 257-61.

GREEN, B., 1984. Composition of milk and energetics of growth in marsupials. *Symp. Zool. Soc. Lond.* **51**: 369-87.

GREEN, B. AND MERCHANT, J. C., 1988. The composition of marsupial milk. Pp. 41-54 *in* The Developing Marsupial: Models for Biomedical Research ed. by C. H. Tyndale-Biscoe and P. A. Janssens. Springer: Heidelberg.

HINDS, L. A. AND TYNDALE-BISCOE, C. H., 1985. Seasonal and circadian patterns of circulating prolactin during lactation and seasonal quiescence in the tammar, *Macropus eugenii. Reprod. Fertil.* **74**: 173-83.

HULBERT, A. J., 1988. Metabolism and the development of endothermy. Pp. 148-61 *in* The Developing Marsupial: Models for Biomedical Research ed. by C. H. Tyndale-Biscoe and P. A. Janssens. Springer: Heidelberg.

JANSSENS, P. A., 1984. Nutrition and metabolism in marsupial pouch young with special reference to the tammar wallaby. *Proc. Nutr. Soc. Aust.* **9**: 68-75.

JANSSENS, P. A. AND TERNOUTH, J. H., 1987. The transition from milk to forage diets. Pp. 281-305 *in* The Nutrition of Herbivores ed. by J. B. Hacker and J. H. Ternouth. Academic: Sydney.

JANSSENS, P. A. AND MESSER, M., 1988. Changes in nutritional metabolism during weaning. Pp. 162-75 *in* The Developing Marsupial: Models for Biomedical Research ed. by C. H. Tyndale-Biscoe and P. A. Janssens. Springer: Heidelberg.

JANSSENS, P. A., JENKINSON, L. A., PATON, B. C. AND WHITELAW, E., 1977. The regulation of gluconeogenesis in pouch young of the tammar wallaby, *Macropus eugenii* Desmarest. *Aust. J. Biol. Sci.* **30**: 183-95.

JOHNSON, C. N., 1988. Mortality of immature red-necked wallabies. *J. Mammal.* (in press).

LOH, T. T. AND SHIELD, J., 1977. Temperature regulation and oxygen consumption in the developing macropod marsupial, *Setonix brachyurus. J. Physiol.* **269**: 677-86.

LOUDON, A., ROTHWELL, N. AND STOCK, M., 1985. Brown fat, thermogenesis and physiological birth in a marsupial. *Comp. Biochem. Physiol.* **81A**: 815-19.

MACKIE, R. I., GILCHRIST, F. M. C., ROBBERTS, A. M., HANNAH, P. E. AND SCHWARTZ, H. M., 1978. Microbiological and chemical changes in the rumen during the stepwise adaptation of sheep to high concentrate diets. *J. Agric. Sci.* **90**: 241-42.

NICHOLAS, K. R., 1988. Control of milk protein synthesis in the marsupial *Macropus eugenii;* a model system to study prolactin-dependent development. Pp. 68-85 *in* The Developing Marsupial: Models for Biomedical Research ed. by C. H. Tyndale-Biscoe and P. A. Janssens. Springer: Heidelberg.

PATON, B. C. AND JANSSENS, P. A., 1981. Metabolic changes associated with the switch from a milk to a vegetable diet in the tammar wallaby, *Macropus eugenii. Comp. Biochem. Physiol.* **70B**: 105-13.

RENFREE, M. B., HOLT, A. B., GREEN, S. W., CARR, J. P. AND CHEEK, D. B., 1982. Ontogeny of the brain in a marsupial *(Macropus eugenii)* throughout pouch life. I. Brain Growth. *Brain Behav. Evol.* **20**: 57-71.

REYNOLDS, M. L., CAVANAGH, M. E., DZIEGIELEWSKA, K. M., HINDS, L. A., SAUNDERS, N. R. AND TYNDALE-BISCOE, C. H., 1985. Postnatal development of the telencephalon of the tammar wallaby *(Macropus eugenii)*. An accessible model of neocortical differentiation. *Anat. Embryol.* **173**: 81-94.

REYNOLDS, M. L. AND SAUNDERS, N. R., 1988. Differentiation of the neocortex. Pp. 101-16 *in* The Developing Marsupial: Models for Biomedical Research ed. by C. H. Tyndale-Biscoe and P. A. Janssens. Springer: Hiedelberg.

ROGERS, A. M. T., 1979. Maternal pouch young thermal relationships in the tammar wallaby, *Macropus eugenii* (Desmarest, 1817). Unpublished Honours Thesis, Australian National University, Canberra.

ROSE, R. W., 1978. Reproduction and evolution in female Macropodidae. *Aust. Mammal.* **2**: 65-72.

ROSE, R. W., 1986. The control of pouch vacation in the Tasmanian bettong, *Bettongia gaimardi. Aust. J. Zool.* **34**: 485-91.

RUSSELL, E. M., 1973. Mother-young relationships and early behavioural development in the marsupials *Macropus eugenii* and *Megaleia rufa. Z. Tierphsych.* **33**: 163-203.

SHIELD, J., 1966. Oxygen consumption during pouch development of the macropod marsupial, *Setonix brachyurus. J. Physiol.* **187**: 257-70.

SHIELD, J., 1968. Reproduction of the quokka, *Setonix brachyurus,* in captivity. *J. Zool. Lond.* **155**: 427-44.

VERNON, J. K., MESSER, M. AND GREEN, B., 1981. Enzymes of galactose metabolism in livers of suckling and adult tammar wallabies *(Macropus eugenii)* and other marsupials. *Aust. J. Biol. Sci.* **34**: 401-09.

WALCOTT, P. J. AND MESSER, M., 1980. Intestinal lactase (β-galactosidase) and other glycosidase activities in suckling and adult tammar wallabies. *Aust. J. Biol. Sci.* **33**: 521-31.

WALLIS, R. L. AND MAYNES, G. M., 1973. Ontogeny of thermoregulation in *Macropus parma* (Marsupialia, Macropodidae). *J. Mammal.* **54**: 278-81.

WILKES, G. E., 1984. The development of structure and function of the kidney in the tammar wallaby, *Macropus eugenii.* Unpublished Ph.D. Thesis, Australian National University.

WILKES, G. E. AND JANSSENS, P. A., 1986a. The development of urine concentrating ability in pouch young of a marsupial, the tammar wallaby *(Macropus eugenii). J. Comp. Physiol. B.* **156:** 573-82.

WILKES, G. E. AND JANSSENS, P. A., 1986b. Physiological and metabolic changes associated with weaning in the tammar wallaby *(Macropus eugenii). J. Comp. Physiol. B* **156:** 829-37.

WILKES, G. E. AND JANSSENS, P. A., 1988. The development of renal function Pp.176-89 *in* The Developing Marsupial: Models for Biomedical Research ed. by C. H. Tyndale-Biscoe and P. A. Janssens. Springer: Heidelberg.

WRIGHT, B., 1975. The development of thermoregulation in the pouch young of the tammar wallaby, *Macropus eugenii* (Desmarest). Unpublished Honours Thesis. Australian National University, Canberra.

Evolutionary implication of macropodid organogenesis

R. L. HUGHES[1], L. S. HALL[1], C. H. TYNDALE-BISCOE[2] and L. A. HINDS[2]

Organogenetic profiles within *Macropus eugenii* and other macropodids are characterized by a much slower rate of organogenesis than that found within non-macropodid marsupials so far studied. This development, by marsupial standards, extends over a uniquely long interval of about 12 days from primitive streak to birth. The ultimate structural complexity of the major life support systems of macropodids at birth is sufficiently in advance of that found in other marsupials to justify the recognition of a separate developmental grade (G3) of both taxonomic and evolutionary significance. The locomotor and teat attachment behaviour of new born macropodids, for so long a mystery, is shown to be co-ordinated by the organogenesis of a neuromuscular locomotor generator. Neurogenesis precedes myogenesis. The functional kindling of the locomotor generator as evidenced by both early movements and morphogenetic criteria is attained in *Macropus eugenii* three days before birth. Behavioural and structural evidence suggests that the locomotor generator is switched on developmentally before birth and is internally automated within the neonate until a concentrated perioral field of Merkel cell touch receptors is stimulated. The locomotor generator is then trigeminally switched off as long as stimulation continues and teat attachment behaviour is initiated. Evidence is presented to show that the trigeminal-hind brain-spinal neuronal connections are of major importance in these events.

INTRODUCTION

For more than 200 years kangaroo-like marsupials have been a focus of interest to biologists as well as within the wider community. An almost universally held view has been that kangaroo-like marsupials typify the marsupial grade of organization. A relatively free extrapolation of findings from one marsupial group to another, across the entire marsupial grade, is a common practice in much of the literature on marsupial reproductive physiology and embryology (e.g., Klima 1987).

There are few comprehensive descriptive accounts of the development of macropodid marsupials over the entire organogenetic period (primitive streak to birth) based on accurately aged specimens (Berger 1970; Renfree 1972; Renfree and Tyndale-Biscoe 1973; Walker and Rose 1981). Prior to 1950, studies on the embryology of macropodids were based on specimens of unknown age and it was not until Sharman (1955 a,b,c) described embryonic diapause in macropodids that the technique of removing pouch young (RPY) to re-activate a blastocyst enabled the easy acquisition of known aged embryos. Despite the technique being extensively used in studies on the physiology of reproduction as reviewed by Tyndale-Biscoe and Renfree (1987), few embryologists have exploited the facility to obtain and describe known aged macropodid embryos. As yet no comprehensive embryological study has been undertaken for any large macropodid such as *Macropus robustus, Macropus rufus, Macropus giganteus* or *Macropus fuliginosus*. The few papers dealing with known age macropodid embryos are reviewed by Walker (1983) and Tyndale-Biscoe and Renfree (1987).

In the Tammar Wallaby, as in other macropodids, there is a variation of up to three days in the length of gestation, determined either from an observed mating (29.3 days) or from the removal of the pouch young (26.2 days) and activation of the diapausing embryo (Merchant 1979; Tyndale-Biscoe and Renfree 1987, Table 6.4), with a consequent variability in the stage of intra-uterine development. A better index of expected birth and hence of intra-uterine development is the early pulse of plasma progesterone that occurs four to 10 days after oestrus or removal of pouch young (RPY). The interval from the pulse of progesterone to birth in the Tammar Wallaby ranges from 21.1 to 22.3 days (Tyndale-Biscoe *et al.* 1986), whereas the interval that precedes the pulse accounts for most of the variability in gestation length. Thus, by monitoring the day of the early pulse of plasma progesterone much more accurately timed embryos can be obtained.

In a recent publication Hughes and Hall (1988) showed that a grade in the state of development at birth exists between marsupial groups. This grade (Gl, G2 and G3) places macropodids in the most advanced category. Their paper (Hughes and Hall 1988) also proposed a locomotor generator and modulator model which is involved in the birth process and teat attachments of marsupials.

In this paper we will investigate the locomotor model proposed by Hughes and Hall (1988) by observing the behaviour of newborn young of the Tammar Wallaby, *Macropus eugenii*. The locomotor model will be correlated with internal histological features. Known aged embryos of

[1]Department of Anatomy, University of Queensland, Brisbane, Queensland 4000.
[2]C.S.I.R.O., Division of Wildlife and Ecology, Canberra, Australian Capital Territory 2600.

Pages 377-405 in KANGAROOS, WALLABIES AND RAT-KANGAROOS ed. by G. Grigg, P. Jarman and I. Hume. Surrey Beatty & Sons Pty Limited, New South Wales, Australia, 1989.

M. eugenii will be used to describe the development of external features and the internal development of select life support systems necessary at birth. These features will be briefly compared with equivalent data from other marsupials to test the grades proposed by Hughes and Hall (1988). Where possible a functional interpretation based on morphological features will be exploited. A detailed description of the embryogenesis of all organ systems in *M. eugenii* is well beyond the scope of this article.

METHODS

In this study lactating tammars carrying embryos in diapause were used. Pregnancy was reactivated by RPY and daily blood samples were collected from day 3 to 13. Plasma was assayed for progesterone by the method of Sernia *et al.* (1980) in order to determine the day of the early pulse in each animal. Embryos at stages 9, 6, 3, and 0 days before expected birth were obtained from females on days 13, 16, 19 and 22 respectively after the early progesterone pulse.

RESULTS

The results of timing the development of embryos from the progesterone pulse are shown in Table 1. It is clear that the stage of development of the embryos six days before birth as well as at nine days before birth were not uniform. However, those scheduled for day 22 after the peak had all undergone parturition, within a period of 24 hours, as predicted. Despite this variation in development it is evident that much less uniformity would have been achieved if the embryos had been dated from RPY, as has been done in previous studies (see Renfree and Tyndale-Biscoe 1973; Denker and Tyndale-Biscoe 1986).

EXTERNAL FEATURES

Macropus eugenii nine days before birth (Fig. 1).

The embryo is essentially a flat mid-somite neurula stage. A head fold, pharyngeal arches, and forelimb buds are present. The early primordia of major organ systems have developed, however no neonatal external characters are recognizable. The definitive embryonal area has a preserved weight of 9-10 mg. In *Macropus rufogriseus* at 18 days RPY Walker and Rose (1981) showed a developmental complexity very slightly in advance of our *Macropus eugenii* stage at nine days before birth. In six *Macropus rufogriseus* Walker (1983) reported birth at a mean interval of 26.8 days RPY.

Macropus eugenii six days before birth (Fig. 2).

The definitive embryonic tissues are now moulded away from the surface of the yolk-sac and all of the extra-embryonic membranes have been established. The digit rays of the forelimbs are at an advanced stage of separation, however, as yet the primordia of the epitrichial claws are at a cellular non-keratinized stage. The forelimbs are in an unpronated position. The hindlimbs are paddles. The pharyngeal arches are in their terminal stages of facial and neck modelling. The primordia of all external body structures exhibit an early neonatal appearance. The eye is open and in two specimens early retinal pigmentation could be recognized. The hillocks of the

Table 1. Experimental protocols.

Animal number	Days before expected birth	Days after Prog. peak	Days after RPY	Overall embryonic length mm	Body weight mg	Stage of Development
5927	9.5	12.5	20	8.5	10	Flat somite stage, apposed neural folds, forelimb buds and early amniotic folds.
6201	9.0	13.0	21	8.5	10	Flat somite stage, apposed neural folds, forelimb buds, and early amniotic folds.
6409	9.0	13.0	21	8.5	9	Flat somite stage, apposed neural folds, forelimb buds but no amniotic folds.
6603	9.0	13.0	20	8.3	34	Early branchial clefts, allantois 0.8 mm, forelimb paddles, amnion near completed.
5657	6.0	16.0	21	8.5	50	Late branchial cleft, allantois dia. 2.2 mm.
6250	6.0	16.0	21	10.9	194	Post branchial cleft, allantois dia. 8.3 mm.
6379	6.0	16.0	21	9.5	110	Late branchial cleft, allantois dia. 4.7 mm.
6389	6.0	16.0	20	8.2	100	Late branchial cleft, allantois dia. 1.5 mm.
5585	3.0	19.0	25	15.2	367	Clawed digits, mouth gapes, allantois dia. 11.3 mm.
5630	3.0	19.0	25	14.7	345	Clawed digits, mouth gapes, allantois dia. 11.0 mm.
6220	3.0	19.0	25	15.0	364	Clawed digits, mouth gapes, allantois dia. 10.0 mm.
6253	3.0	19.0	25	15.0	313	Clawed digits, mouth gapes, allantois dia. 9.0 mm.
6277	0.0	21.0	27	15.8	373	New born, less than 10 hours.
6285	0.0	22.0	28	15.0	454	New born, 10 to 18 hours post birth.
6538	0.0	21.0	27	15.0	373	New born, less than 8 hours.

primordia of the external ear are located on the rostral and caudal margins of the first pharyngeal cleft. The medially facing plantar surfaces of the hindlimb paddles do not, as yet, clasp the twisted tail. The four embryos had a modal preserved body weight of about 100 mg (range 50 mg to 194 mg).

Macropus eugenii three days before birth (Fig. 3).

The mouth gapes and the tongue protrudes. The eyes are not covered by lids and the strongly pigmented retina exhibits an annular quadrangular structure. The primordia of the external ear point rostrally. Neither the eye nor the ear is plugged with epitrichium. The neck region is poorly differentiated. The now keratinized epitrichial claws of the digits of the forelimbs are completely separated and are pronated. The hindlimb shafts now exhibit foot plates in which the digit primordia give evidence of syndactyly. The preserved body weight ranged in four fetuses from 313 mg to 367 mg. The modal weight is approximately 350 mg.

Macropus eugenii at birth (Fig. 4).

The tongue protrudes very slightly and has a dorsal trough. The lips are fused so that the mouth no longer gapes. The eye and external ear pinna are covered by a thick layer of epitrichium. A well defined neck region is present. The forelimbs are highly mobile so that the manus is not always in a strictly pronated position. The epitrichial claws are recurved with sharp tips and margins. The inner curvature of the claw is concave. The digits of the hindlimb show early stages of separation with the smaller second and third syndactylous digits easily recognizable. The tail is recurved against the abdomen and exhibits a markedly narrowed terminal papilla. The preserved body weights of two of the three new born young were 373 mg the other weighed 454 mg.

CARDIOVASCULAR SYSTEM

The recently developed functional primordia of the cardiovascular system at nine days before birth is centred on a strongly flexed bulboventricular loop, constricted but without obvious valvular elements, and situated within a pericardial cavity (Fig. 5). The wall of the heart consists of a thin inner layer of endocardium. The outer epimyocardial wall of several layers invests an intermediate layer of cardiac jelly that is an extremely open meshwork of connective tissue (Fig. 5). At nine days before birth *Macropus rufogriseus* exhibits a similar degree of cardiac development (Walker and Rose 1981).

The venous circulation reaches the heart by way of paired vitelline veins that empty into the sinus horns situated within the lateral wall of the anterior intestinal portal of the foregut. Arterial blood leaving the heart passes by way of aortic arches to enter the paired dorsal aortae. The arteries of the lst and 2nd arch are particularly prominent. Intersegmental arteries occur between each somite and are connected with the lateral cardinal veins which return venous blood to the horns of the sinus venosus.

The heart has attained the primordia of the four-chambered condition by six days before birth. The paired thin walled atria are in wide communication. Slightly thickened mesenchymal ridges form the primordia of atrio-ventricular valves and are situated between both the left and right atrioventricular openings.

The walls of the ventricles are prominent and consist of a thickened layer of embryonic cardiac muscle. The interventricular septum is poorly developed so that the ventricles are in wide communication. The primordia of the major organ systems have formed by six days before birth and are served by conspicuous primary vascular beds. The inferior vena cava (IVC), extending from the liver to the right sinus venosus, is an extremely prominent vessel. The arteries of the lst and 2nd aortic arches are no longer the major outlet for arterial blood. Arterial blood has now been channelled via the truncus arteriosus which flows into the aorta.

The walls of the ventricles thicken considerably by three days before birth and the atrioventricular valves exhibit greater development. The truncus arteriosus and pulmonary trunk spiral around each other and are served by thickened valvular ridges. The interatrial and interventricular septa become more defined but these chambers are each in wide lateral communication. A developing sino-atrial valve is located where the sinus venousus enters the right atrium. The post-hepatic segment of the IVC is a thin-walled prominent vessel and enters the right atrium via a rudimentary sino-atrial valve (Fig. 6). The two atria are separated by a highly perforated septum primum (Fig. 6).

At birth the atria are separated by a perforated septum primum. There is no septum secundum. The atrioventricular valves on each side have thickened and are contoured. The spiral valvular ridges within the truncus arteriosus have separated the pulmonary and the aortic trunks. There is a small interventricular foramen near the atrioventricular junction. At the ultrastructural level the branched cardiac muscle exhibits prominent A and I bands as well as intercalated disks (Figs 7, 8).

EXCRETORY SYSTEM

The excretory system at nine days before birth consists of a long nephrogenic cord with a luminal primordium. This structure extends within the

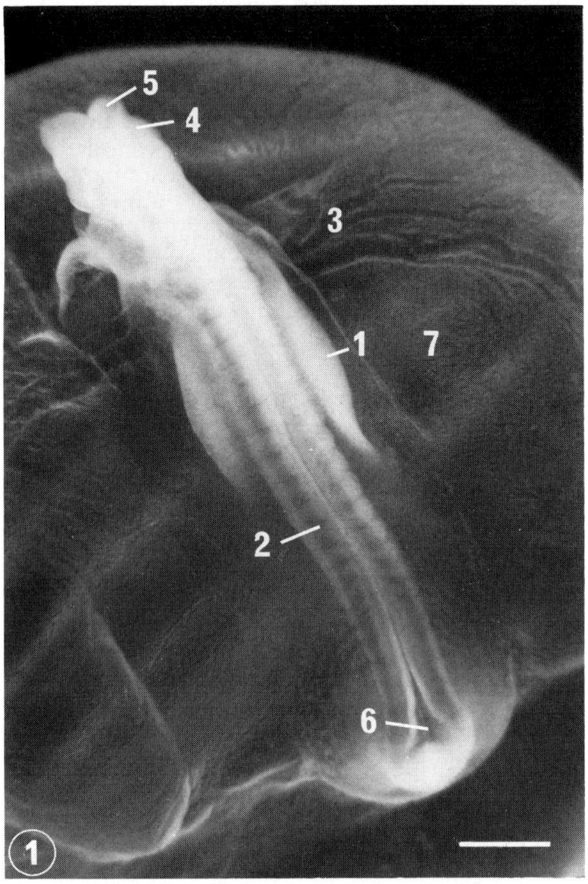
5
4
3
1
7
2
6
1

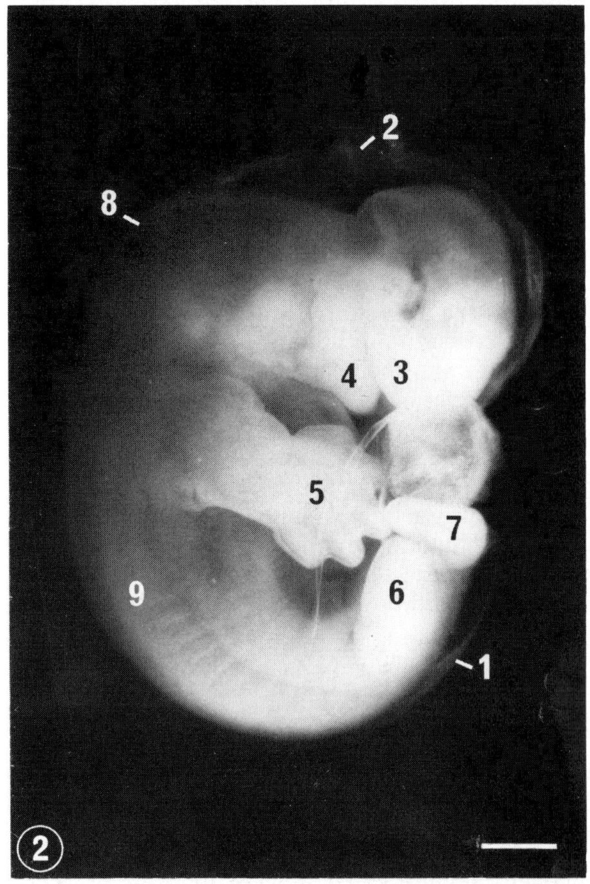
2
8
4
3
5
7
9
6
1
2

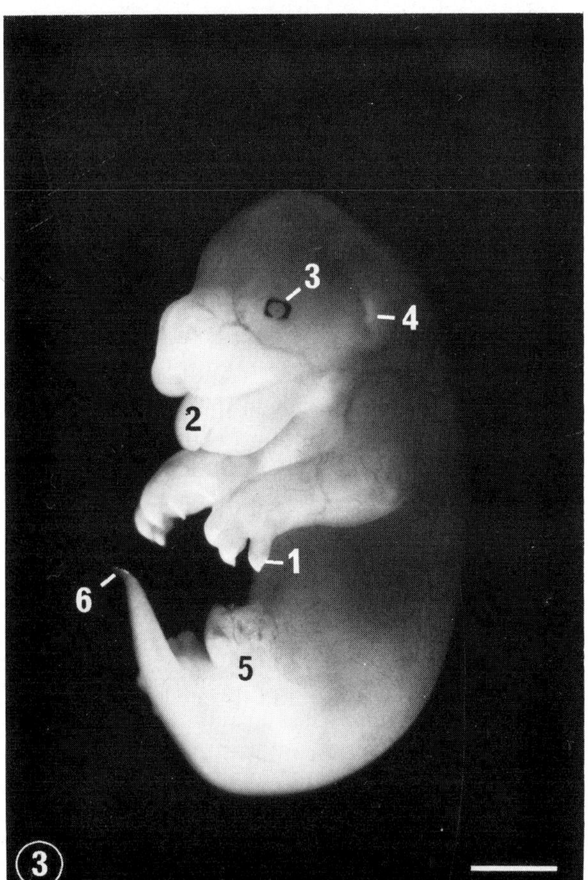
3
4
2
1
6
5
3

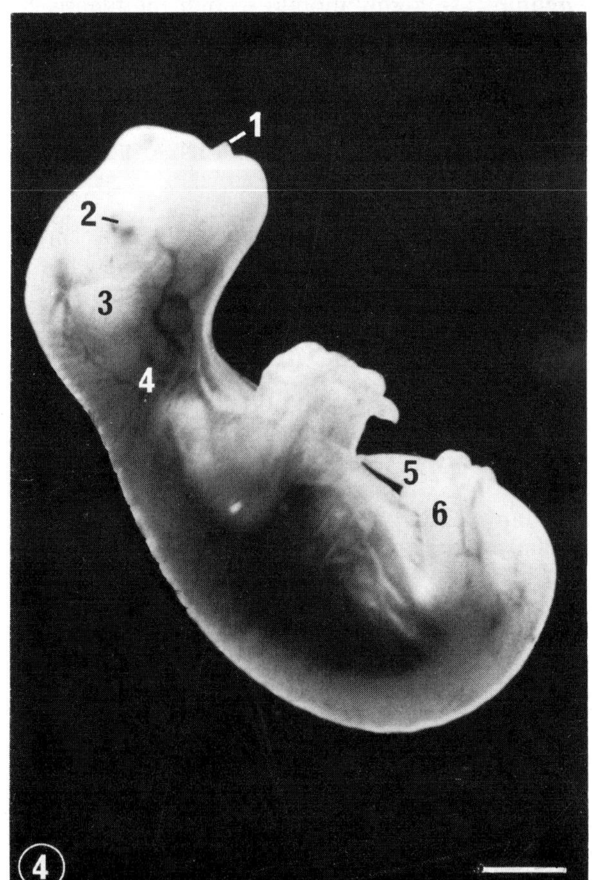
1
2
3
4
5
6
4

lateral nephrogenic intermediate mesoderm from the level of the lst caphalic somite to the 17th somite. The cephalic end of this cord is closed and is associated with at least several mesodermal condensations. These are presumed to represent the pronephric segment. At the caudal end of the nephrogenic cord about 15 non-vascularized vesicles represent the early primordia of the mesonephros (Fig. 9). In *Macropus rufogriseus* Walker and Rose (1981) describe degenerating pronephric tubules and mesonephric primordia nine days before birth. Renfree (1972) reported the appearance of mesonephric tubules in *Macropus eugenii* at about nine days before birth. At six days before birth the mesonephros is recently functional with vascularized glomeruli, associated with mesonephric tubules that enter patent mesonephric ducts. The genital ridge, with prominent germ cells, bulges from the medial aspect of the mesonephros. The mesonephric ducts carry urine into the small vascularized allantoic vesicle by way of the urogenital sinus. The allantoic vesicle has a mean diameter of 4.4 mm (2.2-8.3 mm). An allantoic primordium was reported to be present in *Macropus rufogriseus* at nine days before birth by Walker and Rose (1981). The primordia of the metanephros is present in *Macropus eugenii* at six days before birth as a single unbranched ureteric bud invested by condensed, but undifferentiated metanephrogenic mesenchyme. The state of development of the metanephros now attained is equivalent to that seen in *Sarcophilus harrisii* at birth (Hughes and Hall 1988).

The mesonephros at three days before birth has increased markedly in size (see Fig. 25). The numerous well vascularized glomeruli and collecting tubules now form a prominent thickened and elongated abdominal organ. The allantoic vesicle distended with embryonic urine now has a mean diameter of 10.0 mm (9.0-11.3 mm). The cephalic end of the mesonephros has commenced to degenerate. The gonadal ridge remains attached to the medial aspect of the mesonephros and now contains gonadal sex cords with their associated primordal germ cells. The ureteric bud has undergone several generations of branching in the formation of the metanephric kidney. The metanephrogenic mesenchyme exhibits some flattened cell clusters which represent the first primordia of nephrons. It should be noted that at this stage the metanephros is more advanced than that found in *Trichosurus vulpecula* and *Isoodon macrourus* at birth. These marsupials represent the G2 grade of organogenesis of Hughes and Hall (1988).

At birth the mesonephros of *M. eugenii* is a highly active organ. The extensive vasculature, collecting tubules and well developed glomeruli have resulted in the allantoic vesicle being even more distended (Tyndale-Biscoe and Renfree 1987). At the ultrastructural level the mesonephros at birth contains a glomerular filtration system similar to the mammalian metanephric kidney (Fig. 10). Capillaries of the glomerulus are covered by pedicel processes from podocytes. Filtration slit pores are present and collecting tubules are lined with cells containing basal infoldings and surface microvilli (Figs 11, 12).

The metanephros of *M. eugenii* and other macropodids is rapidly developing at birth. The ureteric bud has undergone complex branching to produce the principal units of the collecting system of the definitive kidney. The metanephrogenic mesoderm has formed vesicular primordia of the nephron that includes several vascularized glomeruli. The metanephros of newborn *M. eugenii* is in a pre-functional condition

---

Fig. 1. M. eugenii nine days before birth is at a flat neurula stage containing approximately 17 somites. This specimen was subsequently serially sectioned for light microscopy. 1=forelimb folds, 2=somite, 3=vitelline veins, 4=head fold, 5=unfused folds of diencephalon, 6=unfused neurocaudal folds, 7=vascularized yolk sac. Scale bar = 1 mm.

Fig. 2. M. eugenii six days before birth is at a late branchial cleft/prefetal stage and is invested by the amnion (1) and proamnion (2). 3=maxillary swelling, 4=mandibular swelling, 5=manus plate with partly separated digits, 6=hindlimb paddle, 7=prominent distal curvature of tail to right hand side, 8=cervical flexure, 9=segmental muscle primordia. Scale bar = 1 mm.

Fig. 3. M. eugenii three days before birth has pronated and clawed digits (1), tongue protruding from gaping mouth (2), and annular quadrangular retinal pigmentation (3), 4=pinna primordium (projects rostrally), 5=pes plate with early modelling of digit rays. The eye, ear and mouth (lips) are not as yet covered by epitrichium and the distal tip of the tail (6) is tapered and is slightly turned to the right. Scale bar = 2 mm.

Fig. 4. M. eugenii at birth has fused lips and the tongue (1) protrudes slightly. The eye promordia (2) and the pinna of the ear (3) are now covered by the thick layer of epitrichium. A neck region (4) is now discernable and the tail (5) is strongly recurved against the ventral abdominal wall. The tail is clasped laterally by well defined pes plates (6) which have partially separated digits that exhibit differential development. Scale bar = 2 mm.

and its development exceeds that of the newborn of the G1 and G2 grades of other marsupials reviewed by Hughes and Hall (1988).

RESPIRATORY SYSTEM

The respiratory diverticulum was not a feature of the embryonic gut in *Macropus eugenii* nine days before birth. However, Walker and Rose (1981) reported the presence of the primordia of a trachea, primary bronchi and two lung buds at 18 days RPY in *Macropus rufogriseus*. McCrady (1938) suggested an unusual but unconfirmed mode of formation of the respiratory primordia from lateral paired pharyngeal primordia in *Didelphis virginiana*. Renfree (1972), apparently following the thoughts of McCrady (1938), considers that paired respiratory primordia were formed in 17 day RPY embryos of *M. eugenii* from paired pharyngeal primordia. A separate study is needed to establish whether the respiratory primordium is a single diverticulum in marsupials as occurs in eutherians.

The respiratory system at six days before birth consists of a long branched trachea with mesodermal condensations indicating the models of future cartilage rings. The lobes of the lung are thick walled and contain branches of the vagus nerve (CNX) as well as major blood vessels. The terminal air sacs exhibited approximately three to four generations of branching of respiratory tree formation. The walls of the terminal air sacs consist of a thick non-vascularized columnar epithelium. The order of branching, but not the level of vascularization, is comparable to that of *Trichosurus vulpecula* and *Isoodon macrourus* at birth (i.e., G2 of Hughes and Hall 1988). However, the order of branching slightly exceeds that seen in the lung of *Sarcophilus harrisii* (G1 of Hughes and Hall 1988). In *M. eugenii* at six days before birth the abdominal and respiratory cavities are only partly separated by pleuroperitoneal membranes.

In *M. eugenii* at three days before birth the terminal air sacs exhibit a marked proliferation (Fig. 13) and the abdominal and pleural cavities are now separated by a thin diaphragm.

At birth the barrel-shaped pleural cavity contains a highly proliferated respiratory tree with its thin walled terminal air sacs richly vascularized by superficial blood vessels that frequently bulge into the lumen. The well-formed precartilaginous rings of the bronchi are a conspicuous feature of the lung. The vascular supply and vagal nerve fibres were prominent in the lung tissue.

At the ultrastructural level the epithelium lining the trachea is richly supplied by mucus cells (Figs 14, 15). The respiratory epithelium of terminal air sacs contains Type 1 and 2 pneumocytes. Lamellated surfactant inclusions were numerous in the cytoplasm of Type 2 pneumocytes (Fig. 16). Neuroepithelial bodies (NEBs) with their small membrane bound granules are found within the wall of the trachea (Figs 14, 15). Small projections from the NEBs reached the surface between mucus cells. Ganglion type cells were found in association with blood vessels in deep layers of the trachea. These cells also contained membrane bound granules. Walker (1983) should be consulted for an ultrastructural account of the neonatal maturation of the terminal air sacs in *Macropus rufogriseus* which closely parallels *Macropus eugenii*.

DIGESTIVE SYSTEM

The primordia of the digestive tract at nine days before birth consists of a mid-gut region continuous with the lumen of the yolk-sac (see also Walker and Rose 1981 for equivalent development in *Macropus rufogriseus*). A well differentiated foregut diverticulum extends into the head. The proximal opening of the anterior intestinal portal is well marked by lateral folds that contain on each side the sinus horns which

---

Fig. 5. Bulboventricular loop of *M. eugenii* nine days before birth. 1=endocardium, 2=cardiac jelly, 3=primitive epimyocardium, arrows=atrioventricular constriction, 4=pericardial cavity, 5=pericardium. Magnification=142.

Fig. 6. Post-hepatic segment of inferior vena cava of *M. eugenii* three days before birth. 1=liver, 2=IVC, 3=primordium of sino-atrial valve, 4=right atrium, 5=left atrium, 6=septum primum, arrows=perforations of septum primum, 7=terminal airsacs of lung. Magnification=38.

Fig. 7. Ultrastructure of cardiac muscle of *M. eugenii* at birth. 1=I band, 2=A band, 3=Z line, 4=mitochondrion, 5=glycogen, 6=nucleus of cardiac muscle fibre. Magnification=18 750.

Fig. 8. Ultrastructure of branching cardiac muscle of *M. eugenii* at birth. 1=intercalated disk, 2=mitochondrion, 3=glycogen. Magnification=18 750.

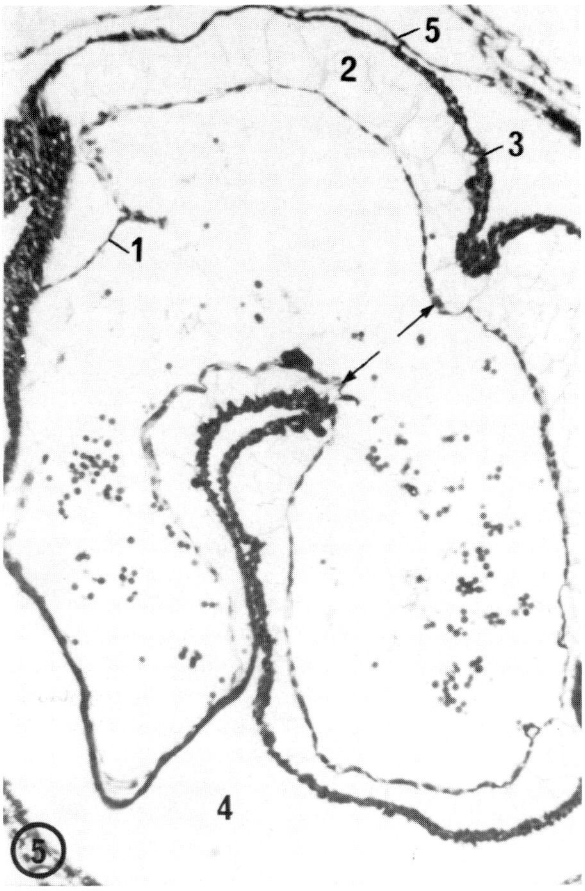
5
2
3
1
4
5

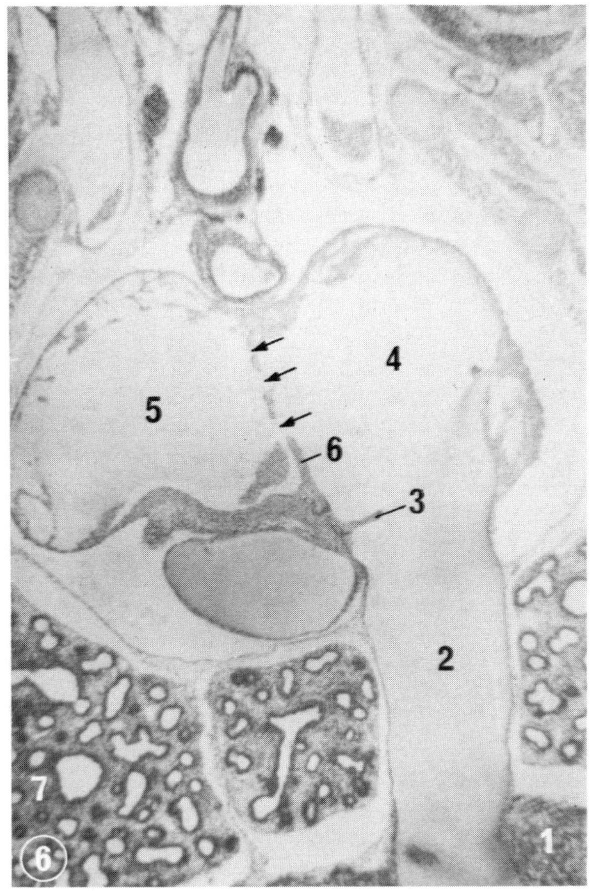
4
5
6
3
2
7
1
6

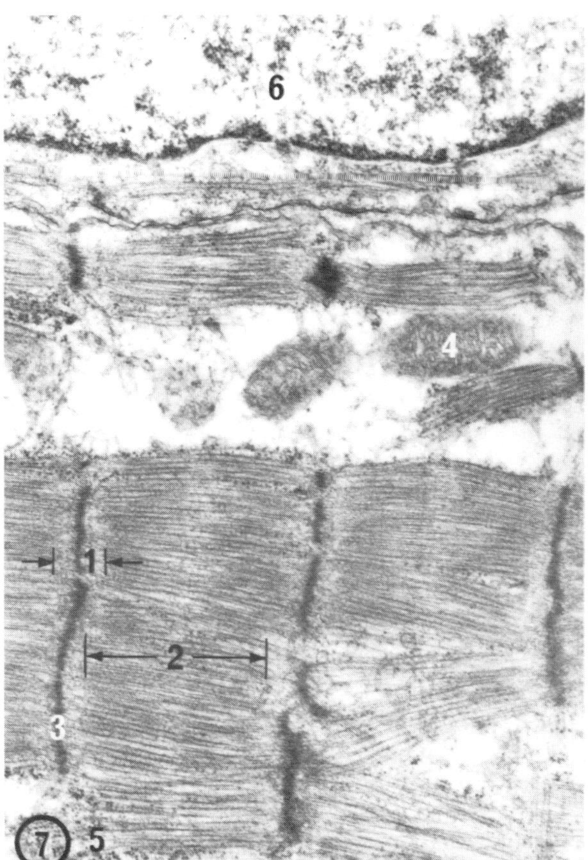
6
4
1
2
3
7
5

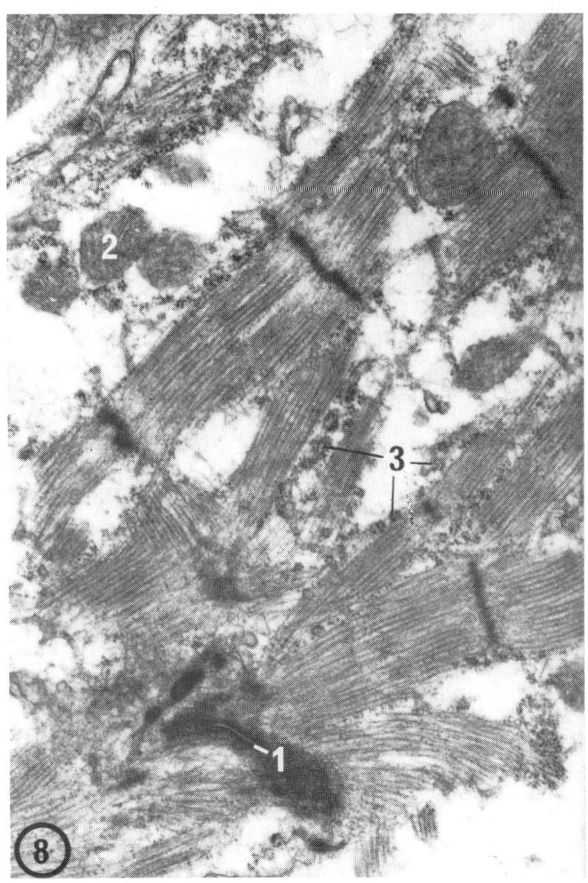
2
3
1
8

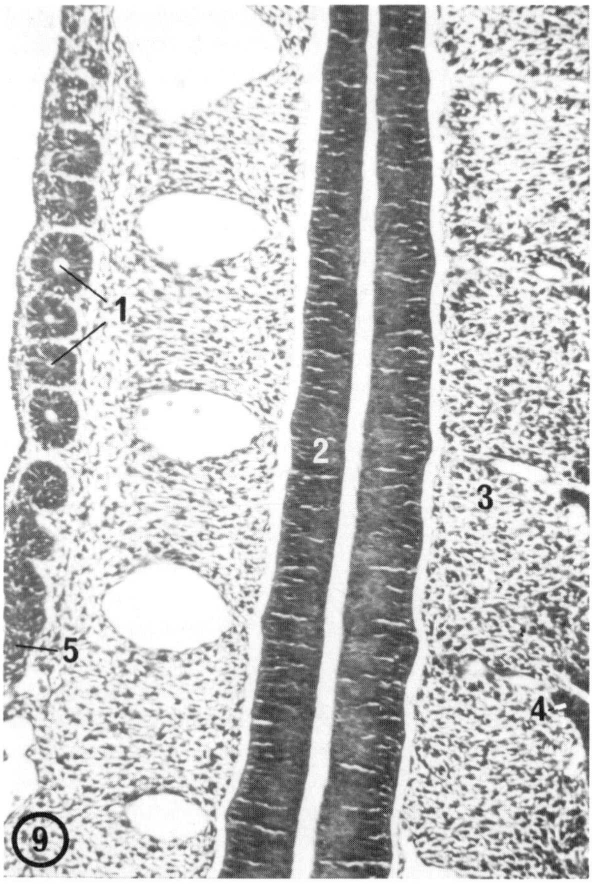
1
2
3
4
5
9

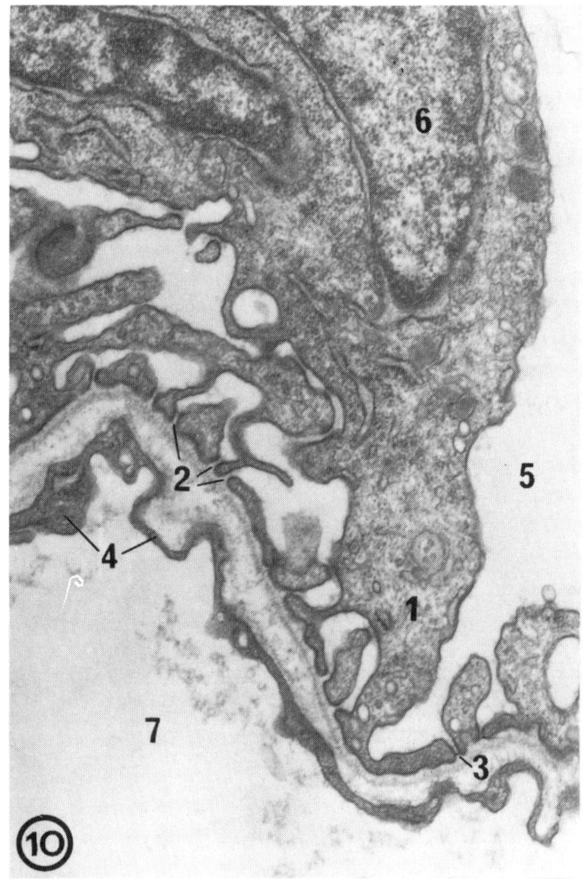
6
2
5
4
1
7
3
10

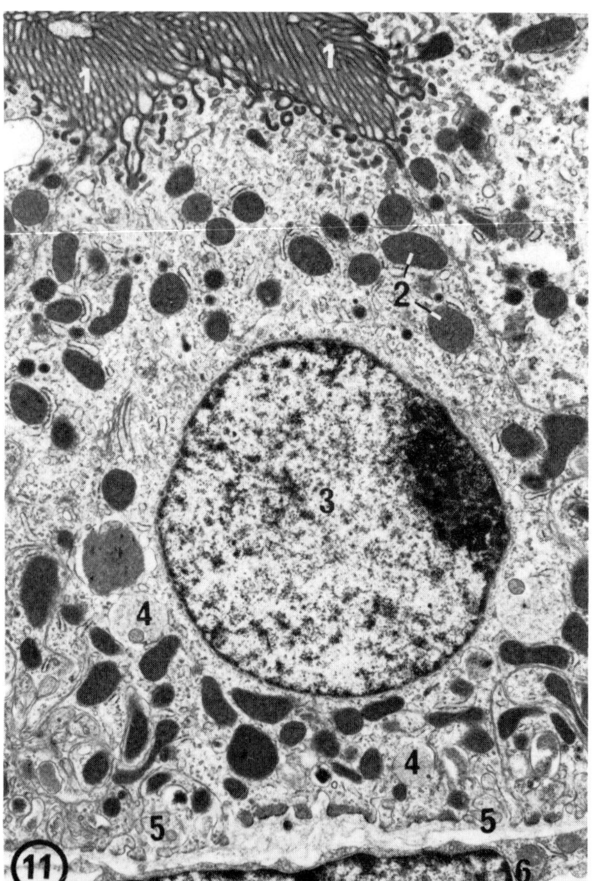
1
1
2
3
4
4
5
5
6
11

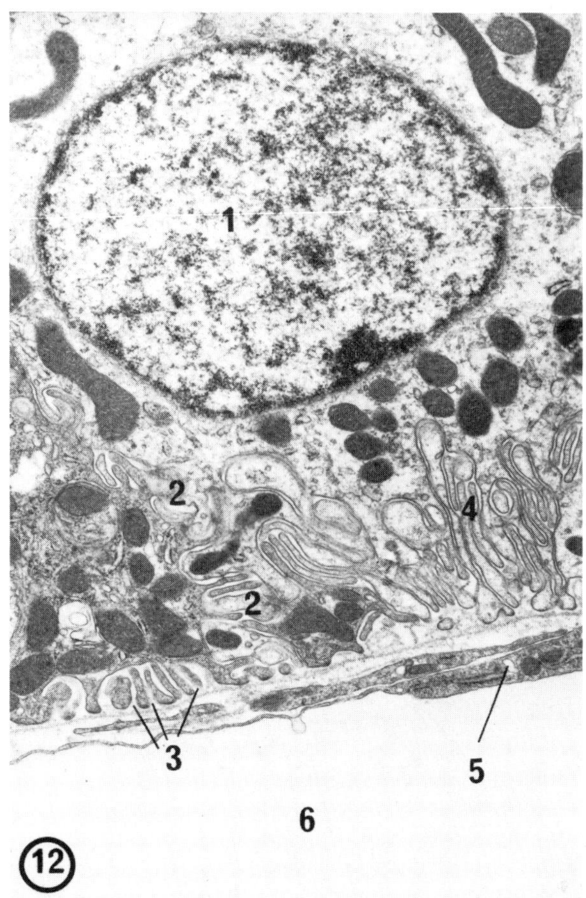
1
2
2
4
3
5
6
12

receive the proportionally large vitelline veins from the yolk-sac. The pharyngeal region of the foregut exhibits well marked lateral pharyngeal pouch extensions with pouches 1 and 2 exhibiting particular prominence. Pouches 3 and 4 are of reduced size. The roof of the pharynx contains a dorsal pocket with a flat distal plate (see pituitary formation). The mid region of this plate is the future site of Rathke's pouch. The slit-like oral opening is located between well differentiated maxillary and mandibular swellings of the first pharyngeal arch. Neither the tongue nor the lip primordia have yet appeared.

The embryonic gut derivatives at six days before birth are easily regionalized. The oral cavity is bounded by facial primordia, maxillary and mandibular swellings. Lip primordia have appeared on the maxillary swelling. Both primary and secondary palate have not yet formed and the oral and nasal cavities are in wide communication. The tongue primordia contains mesenchymal condensations a few of which have formed myoblasts of the intrinsic and extrinsic muscles. The auditory tube has formed from the first pharyngeal pouch and the second pharyngeal pouch contains two deep pockets on each side. The thymus is a prominent dense body of tissue in a medial ventral location to the developing epiglottis. Parathyroids could be seen lateral and slightly caudal to the thymus.

The primordia of the epiglottis projects into the pharynx and is associated with richly innervated mesenchymal condensations that differentiate into muscle and cartilage. Rathke's pouch is well developed, highly folded into an "H" shape and is connected to the pharangeal area by a narrow stalk. The oesophagus is lined by a dense layer of stratified epithelium and is surrounded by a band of early myoblasts representing the muscularis externa. The oesophagus is accompanied by prominent fibres of the vagus (CNX) nerve and it also penetrates the diaphragm to enter the simple saccular stomach. The stomach is lined with a dense stratified epithelium and contains a pale submucosa and a dense outer layer of developing myoblasts. The outer two layers of the stomach are richly vascularized by the celiac artery. The liver is composed of cords of hepatocytes separated by prominent sinuoids. The cystic duct is patent and lined with a dense cuboidal epithelium and lies in the medial caudal edge of the liver. The pancreas and the patent pancreatic duct are located in the loose tissue of the gastrohepatic mesentery. The duodenal component of the gut has differentiated with outer muscle layers and a dense columnar epithelium lining a patent lumen. The short hind gut is less developed and consists of dense epithelium which occludes the lumen and is surrounded by a dense but undifferentiated mesenchyme. The cloacal plate is intact.

Regional differentiation and histogenesis of the gut at three days before birth has progressed to a point where many features are more advanced than those seen in a new born G2 marsupial (e.g., *T. vulpecula, I. macrourus*) (Hughes and Hall 1988). The gaping mouth is a result of incomplete fusion of the facial area. Internally the nasal septum has fused with the lateral palatine shelves (Fig. 17). The tongue protrudes but the striated myotubules of the intrinsic and extrinsic muscles as well as their innervation and vascular supply indicate a near functional level of differentiation (Fig. 18). Innervated taste buds are present on the upper medial and lateral surface areas of the tongue (Figs 19a, 19b). The maxilla and mandible have commenced to ossify (Fig. 17). The rostral portion of the oral cavity is expanded and contoured for teat reception. The epiglottis is well developed and is intranarial. The myotubules of the pharyngeal muscles associated with the epiglottis are much less developed than those of the tongue. The throat area is innervated by prominent fibres of the vagus and glossopharyngeal nerves. The dorsal aspect of the distal portion of the epiglottis is deeply grooved and

---

Fig. 9. Mesonephric primordium of *M. eugenii* nine days before birth. 1=non-segmentally arranged mesonephric vesicles, 2=neural tube, 3=somatic sclerotome, 4=dermo-myotome, 5=pronephric duct. Magnification=142.

Fig. 10. Ultrastructure of mesonephric glomerulus of *M. eugenii* at birth. 1=foot process, 2=slit pores, 3=filtration membrane, 4=endothelial cell of glomerular capillary, 5=urinary space, 6=nucleus of podocyte, 7=capillary lumen. Magnification=18 750.

Fig. 11. Ultrastructure of mesonephric collecting tubule of *M. eugenii* at birth. 1=surface microvilli, 2=mitochondria, 3=nucleus, 4=cytosomes, 5=basal infoldings, 6=endothelial cell of capillary. Magnification=9 450.

Fig. 12. Ultrastructure of mesonephric collecting tubule of *M. eugenii* at birth. 1=nucleus, 2=microvilli in intercellular space, 3=basal infolding, 4=basal interdigitation of cells, 5=endothelial cell of capillary, 6=capillary lumen. Magnification=9 450.

channels air to the respiratory opening. Paired lateral grooves are shaped to direct milk into the oesophagus around the lateral margins of the epiglottis. The pharyngeal area of the oesophagus consists of a dense stratified epithelium surrounded by a submucosa and a ring of differentiating myoblasts which represent the muscularis externa. The level of neuromuscular and cartilaginous development suggest the possibility that the mechanism for swallowing and breathing is developed three days before birth. The stomach is large and saccular and its wall contains a well developed muscularis externa and the luminal epithelium is differentially specialized (Fig. 20). Rugae are appearing at the pyloric end. The liver is the largest abdominal organ and is well vascularized. The gall bladder is well developed and its patent duct enters the highly folded duodenal segment of the gut. Pancreatic acini are prominent. A mid-gut loop enters the umbilical cord. The intestines exhibit marked regional specializations with developing villi and are well vascularized by the superior and inferior mesenteric blood vessels. The terminal portion of the hind-gut is much reduced in diameter and enters the cloaca at the urogenital sinus. The cloacal membrane is intact.

At birth the gut is much more differentiated than that of any other marsupial group (Hughes and Hall 1988). Structures related to the tissue digestive interface that are particularly well developed include the presumptively functional submandibular gland which has a well developed innervation by secretory motor fibres (CNVII) and associated submandibular ganglion cells that have a secondary origin from the mandibular branch of CNV. The lateral and medial upper surfaces of the tongue exhibit innervated taste buds. Presumptive pressure receptors are located within the epithelium of the secondary palate and are strategically located for activation during suckling. Associated with the deep vascular supply of the tongue are large nerve fibre tracts containing small neurons as satellite ganglia of CNVII and seen as early as three days before birth (Figs 18, 21). These deep lingual paraneuronic complexes (DLPCs) (Fig. 22) arise from the CNVII network that supplies the submandibular salivary gland (Fig. 21). Thus the neuro-anatomical basis for oral lubrication and salivary secretion is established. At the ultrastructural level tongue muscles exhibit prominent banding (Fig. 23) and are served by neural motor end-plates (Fig. 24). The separation of food and airways is achieved by the anatomical specializations of the epiglottis which is now only slightly intra-narial. The pharyngeal cartilages and their associated tubular muscle fibres suggest that a swallowing mechanism is present. This mechanism is further indicated by the state of the musculature of the oesophagus. The advanced development of the intrinsic and extrinsic muscles of the tongue and the shape of the tongue are of major importance in the transfer of milk from the teat to the digestive tract. The relatively poor development of the muscularis externa of the stomach indicates that at this stage the stomach primarily acts as a reservoir for milk and is not significantly involved in food transport. The lining of the stomach shows growing specialization for milk absorption. Epithelial cells lining the stomach exhibit a wide range of cell organelles and surface microvilli. The pancreatic acini are vascularized and are more extensive than those in other newborn marsupials (Hughes and Hall 1984, 1988). The duodenal segment of the gut appears to be one of the major areas of absorption. In comparison to three days before birth the duodenum has increased greatly in length and diameter and is lined by a complex folded columnar epithelium. The terminal segment of the hind-gut now contains a patent lumen, the epithelium is unspecialized, and it empties into the cloaca. The midgut loop is confined within the abdominal cavity.

Fig. 13. Respiratory tree of *M. eugenii* three days before birth. 1=trachea with mesenchymal primordia of cartilage rings, 2=primary bronchus, 3=secondary bronchi, 4=terminal air sac, 5=pleural cavity, 6=oesophagus, 7=dorsal aorta, 8=rib cartilage. Magnification=38.

Fig. 14. Ultrastructure of trachea of *M. eugenii* at birth. 1=lumen, 2=mucus cell, 3=neuroepithelial cell, 4=fibroblast, 5=capillary containing nucleated erythrocyte. Magnification=3 000.

Fig. 15. Ultrastructure of neuroepithelial cell in trachea of *M. eugenii* at birth. 1=cytoplasm containing membrane bound granules, 2=nucleus, 3=apical luminal protrusion, 4=cytoplasm of adjacent mucus cell, 5=lumen, 6=basal lamina. Magnification=6 000.

Fig. 16. Ultrastructure of type II pneumocyte of *M. eugenii* at birth. 1=lumen of terminal air sac, 2=microvilli, 3=lamellated vesicle, 4=mitochondria, 5=nucleus. Magnification=30 000.

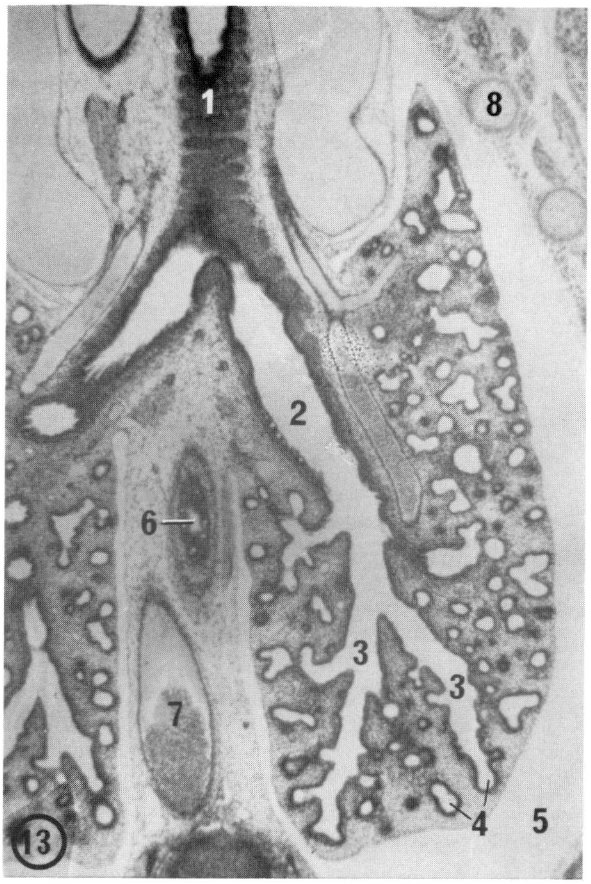
1
8
2
6
3
3
7
4
5
13

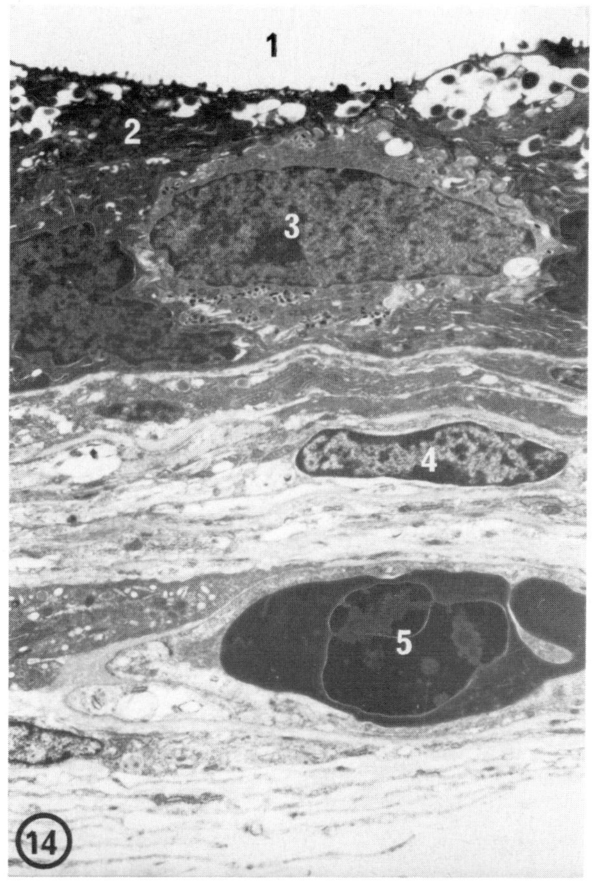
1
2
3
4
5
14

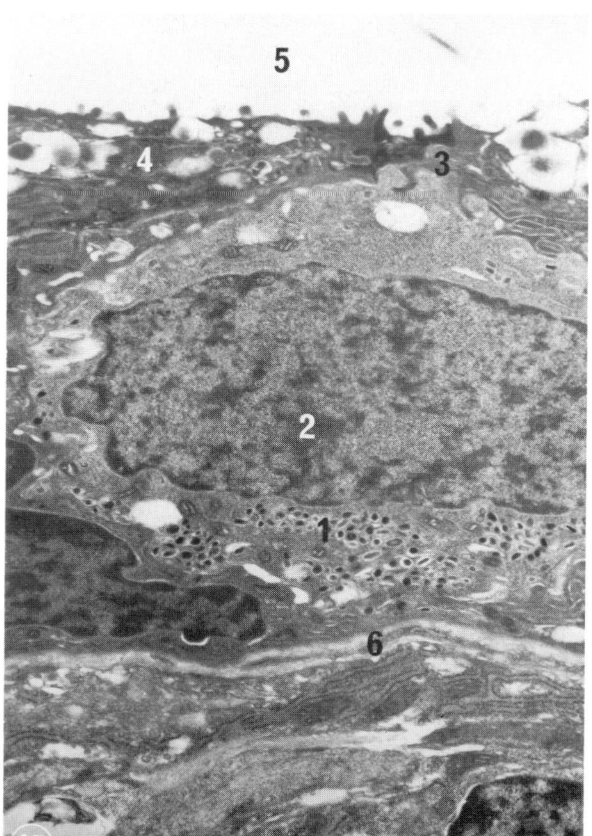
5
4
3
2
1
6
15

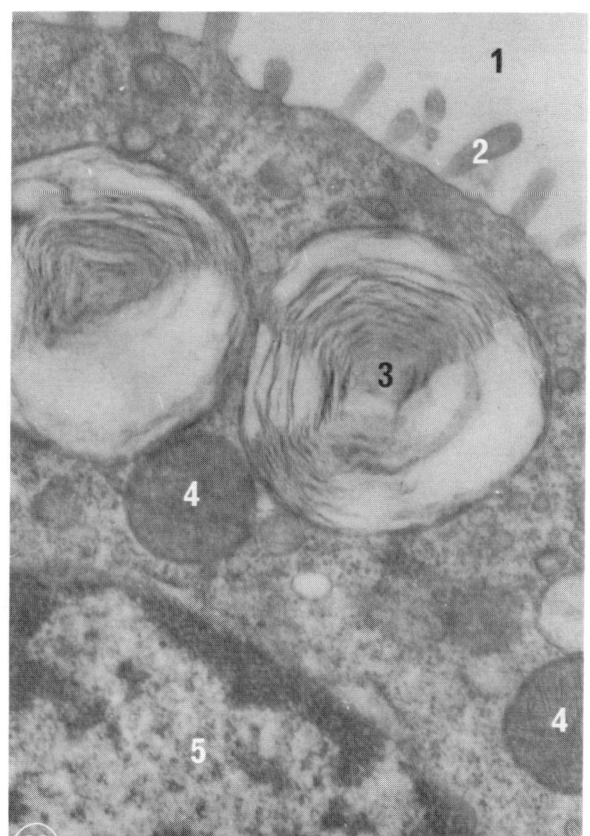
1
2
3
4
4
5
16

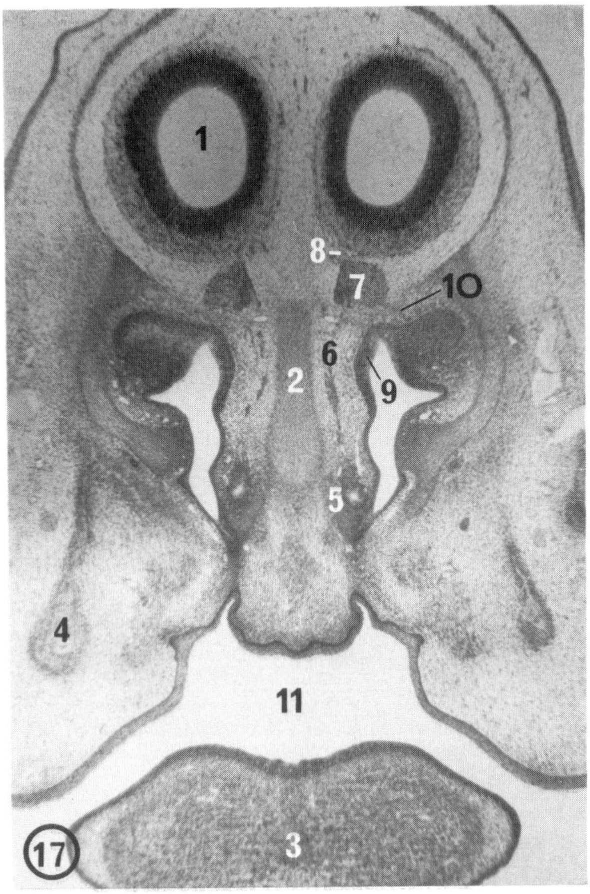
1
8
7
10
6
2
9
5
4
11
3
17

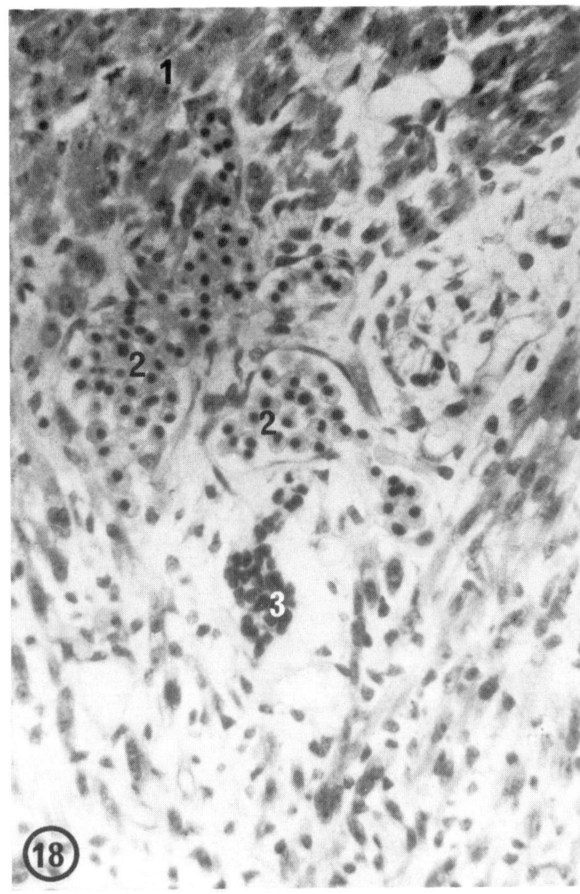
1
2
2
3
18

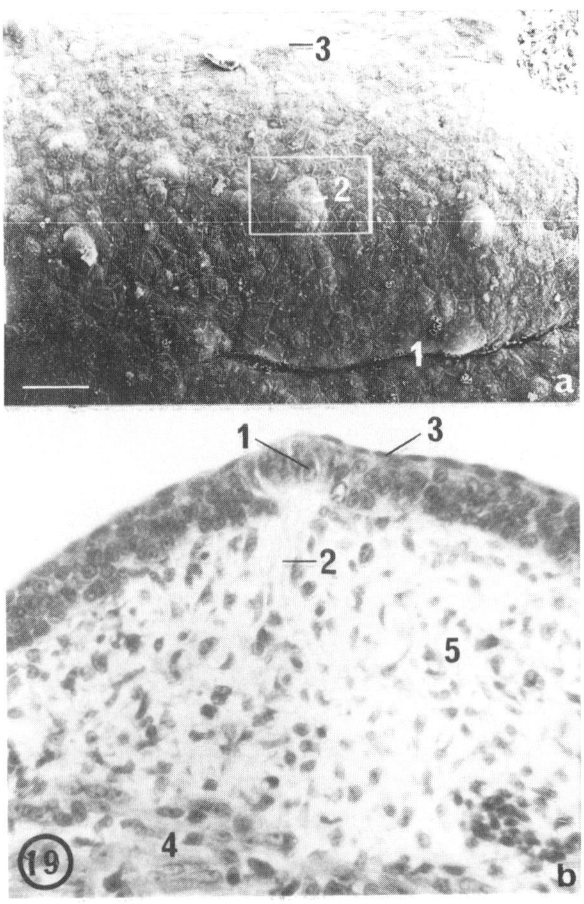
3
2
1
a
1
3
2
5
4
19
b

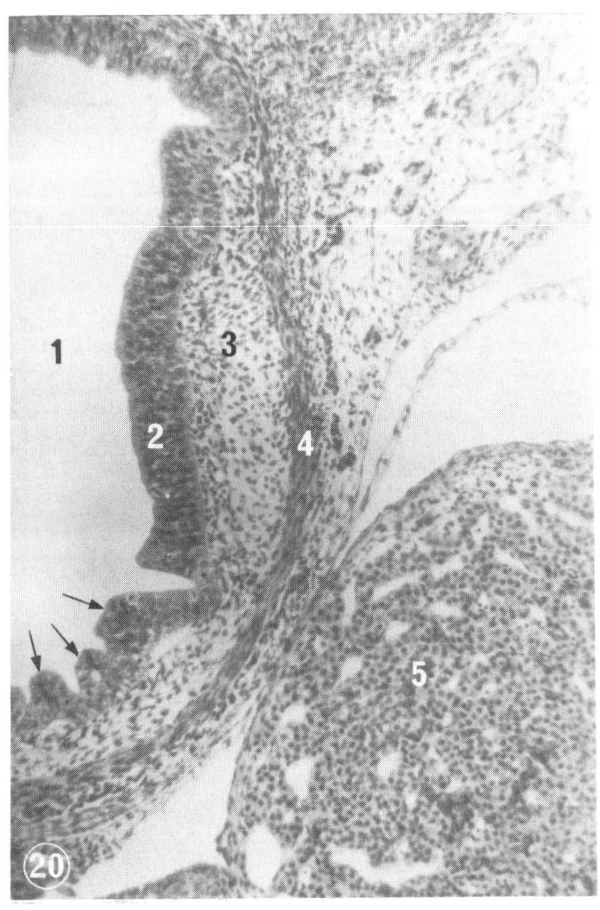
1
3
2
4
5
20

ENDOCRINE SYSTEM

Thyroid gland

The primordia of the thyroid gland in *Macropus eugenii* first appeared six days before birth. The subsequent development of the thyroid gland at three days before birth or at birth requires further investigation. In *Macropus rufogriseus* Walker and Rose (1981) reported that the primordial thyroid gland was connected to the pharynx by a thyroglossal duct at 25 days RPY.

Gonads

Gonadal primordia are absent at nine days before birth. Genital ridges containing a number of enlarged cells believed to be germ cells, are located on the medial aspects of the mesonephroi at six days before birth. The peak of migration of germ cells into the gonad of *Macropus eugenii* occurs at about five days before birth (Alcorn 1975; Alcorn and Robinson 1983). The gonadal ridge of *Macropus rufogriseus* contains germ cells at 20 days RPY. Primary sex cords containing large numbers of germ cells are present in our specimens of *Macropus eugenii* three days before birth. The genital ridges although elongated have proportionately shortened and commenced to separate as gonadal primordia. The surface gonadal mesothelium is cuboidal. At birth the gonads are discrete oval structures still attached to the medial aspect of the mesonephroi. The peripheral gonadal mesothelium is cuboidal and the gonadal stroma contains many loops of sex cords in which germ cells can be distinguished. At the ultrastructural level the germ cells contain a few membrane bound cytoplasmic vesicles. Supportive cells that extend from the basement lamina to the small lumen are also a feature of the sex cords. Gonadal cords of the hilus region are of reduced diameter and both the basement lamina and lumen are poorly defined. We are undecided as to whether these hilus cords should be interpreted as rete cords or degenerating sex cords. The gonadal stromal tissue between the sex cords is poorly vascularized and contains interstitial-like cells and at the ultrastructural level small membrane bound cytoplasmic granules are associated with a high level of organelles involved in protein synthesis and packaging.

Adrenal glands

The primordia of the adrenal glands are absent nine days before birth and are first manifested as poorly vascularized mesenchymal primordia situated at the cranial poles of the mesonephroi near the liver six days before birth. The adrenal glands consist of well vascularized cords three days before birth (Fig. 25). The adrenal glands at birth appear creamy in colour and at the light microscope level the epithelial cords are well separated by an extremely well developed vascular bed. At the ultrastructural level the capillary walls are richly fenestrated (Fig. 26), and the cells of the cords consist mostly of fetal cortical cells packed with smooth endoplasmic reticulum as well as by numerous mitochondria with tubular cristae (Fig. 27). The cytoplasm of these cells contains numerous large lipid droplets. The adrenal cords at birth also contain many presumptive medullary cells that exhibit numerous small membrane bound cytoplasmic granules that are a feature of the catecholamine-secreting cells of the functional adrenal medulla (Fig. 28). The adrenal gland at the ultrastructural level is richly innervated at birth (Fig. 27). Nerve axons and filaments are also a prominent feature. The swollen axonal terminals of many axons contain both clear and membrane bound vesicles in large numbers. The

---

Fig. 17. Nasal and oral structures of *M. eugenii* three days before birth. 1=telencephalon, 2=nasal septum, 3=tongue, 4=primordia of maxilla, 5=vomeronasal organ, 6=vomeronasal nerve, 7=nervus terminals, 8=olfactory nerve, 9=olfactory epithelium, 10=filum olfactorium, 11=oral cavity. Magnification=38.

Fig. 18. Vascularization and innervation of deep tongue muscles of *M. eugenii* three days before birth. 1=deep lingual muscles, 2=capillary bed, 3=cell bodies of deep lingual paraneuronic complex. Magnification=370.

Fig. 19(a). Scanning electron micrograph of dorsal surface of tongue of *M. eugenii* three days before birth. 1=medial sulcus, 2=surface projection of underlying medial taste bud, 3=surface projection of lateral taste bud. Scale bar = 50 μm.

Fig. 19(b). Light micrograph of innervated lateral taste bud of *M. eugenii* three days before birth. 1=distal receptor complex, 2=peripheral nerve fibres, 3=stratified squamous epithelium, 4=developing intrinsic tongue muscles, 5=sub-epithelial mesenchyme. Magnification=370.

Fig. 20. Stomach wall of *M. eugenii* three days before birth. 1=lumen, 2=mucosal epithelium, 3=lamina propria, 4=developing muscularis externa, arrows=developing rugae, 5=vascularized chords of liver. Magnification=142.

ultrastructure is consistent with reports in the literature that the macropodid adrenal is functional at birth (Catling and Vinson 1976; Call *et al.* 1980; Walker 1983; Walker and Gemmell 1983).

Pituitary gland

The site of the future Rathke's pouch is indicated at nine days before birth by a medial dorsal pharyngeal pocket with a plate-like roof (Fig. 29). A small middle section of this dorsal pharyngeal plate, has separated from the pharynx at six days before birth and is linked by a cord of cells to the more deeply located but symmetrically folded Rathke's pouch confirming the observations of Tyndale-Biscoe and Renfree (1987). The lumen of Rathke's pouch in transverse sections is "H" shaped and is lined by a thickened columnar epithelium. The pituitary gland at three days before birth is separated from the pharynx by cartilage and the luminal remnants of Rathke's pouch. The anterior lobe of the developing pituitary in places exhibits a "M" shaped symmetrical lumen (Fig. 30). The infundibular process from the diencephalon is partly invested by Rathke's pouch and forms the primordium of the posterior lobe of the pituitary gland.

At birth the anterior lobe of the pituitary gland is a well vascularized structure. At the ultrastructural level the luminal cells of Rathke's pouch exhibit junctional complexes and some have an occasional cilium. Few of the luminal cells exhibit membrane bound granules within their cytoplasm. The cells of the deeper stroma of the anterior pituitary gland contain a variety of membrane bound granules (Fig. 31). The mean maximum diameter of the granules varies as has been described by Leatherland and Renfree (1983). The distribution of these granular cells is not even throughout the anterior lobe, however the proportion of granular cells of 60% to 70% are quite common (Fig. 31). At the ultrastructural level nerve axonal terminals within the posterior lobe are commonly expanded and contain membrane bound granules (Fig. 32). The level of ultrastructural differentiation of the macropodid pituitary gland at birth greatly exceeds that of other marsupials. The extensive vascular sinusoids as well as the high proportion of cells containing membrane bound granules in the pars distalis as figured for *Macropus rufogriseus* by Walker (1983) and Walker and Gemmell (1983) are much reduced in *Isoodon macrourus* (Hall and Hughes 1985). This differential is particularly striking when comparison is made with the Gl marsupial grade as seen in *Sarcophilus harrisii* (Hughes, unpublished).

Motor generator

The motor generator consists of the functionally innervated skeletomuscular system responsible for movement. Four types of movement can be resolved. First; muscles of the forelimbs and the pectoral girdle together with the paraxial muscles of the trunk are responsible for forward progression. Second; random teat search is the domain of the muscles of the head and neck. Third; the tongue and associated pharyngeal muscles are dominant in sucking. Fourth; intercostal muscles are almost totally responsible for respiratory movement. The function of the heart muscle and the smooth muscle of the gut are outside the domain of the motor generator.

The morphogenesis of the motor generator can conveniently be subdivided into three integrated components that consist of: Central nervous system (CNS) component; neural crest component; and muscular component.

CNS component of motor generator

The central nervous system at nine days before birth consists of well differentiated neural folds

---

Fig. 21. Transverse section in posterior of oral cavity of *M. eugenii* three days before birth. 1=presumptive pons, 2=trigeminal ganglion (CNV), 3=mandibular branch of trigeminal nerve, 4=lingual branch of mandibular nerve, 5=submandibular ganglion cells investing lingual nerve, 6=duct of submandibular gland, 7=deep lingual paraneuronic complex, 8=oral cavity, 9=fusing palatine shelves, 10=nasopharynx, 11=extrinsic tongue muscles, 12=intrinsic tongue muscles, 13=Meckle's cartilage, 14=cartilage of basisphenoid. Magnification=38.

Fig. 22. Ultrastructure of a deep lingual paraneuronic complex in the tongue of *M. eugenii* at birth. 1=nuclei of paraneurones, 2=axonal complex, 3=contractile fibres of tongue muscle. Magnification=2 400.

Fig. 23. Ultrastructure of tongue muscle of *M. eugenii* at birth. 1=I band, 2=A band, 3=Z line, 4=glycogen. Magnification=30 000.

Fig. 24. Ultrastructure of motor end plate in tongue of *M. eugenii* at birth. 1=motor end plate containing synaptic vesicles, 2=nucleus of muscle fibre, 3=Z line of contractile fibres. Magnification=15 000.

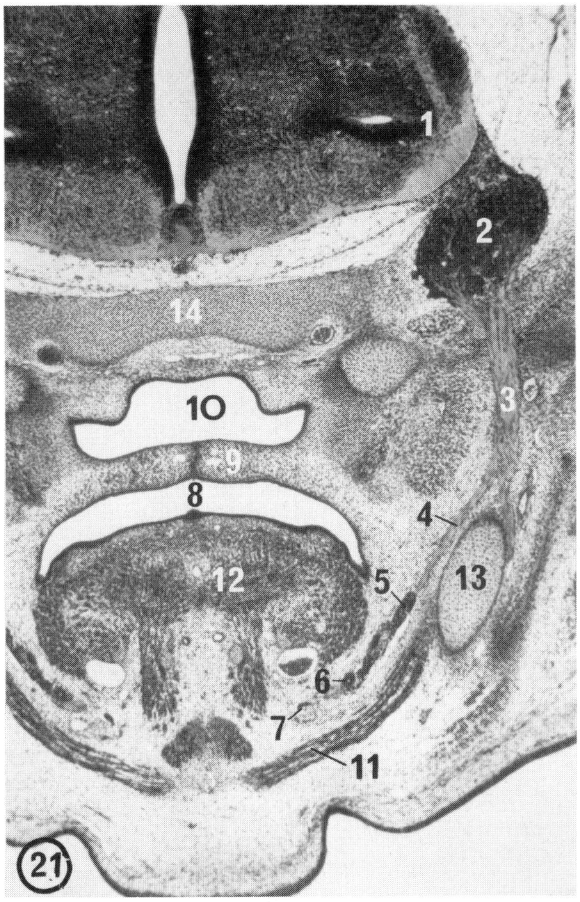
1
2
14
3
10
9
8
4
13
12
5
6
7
11
21

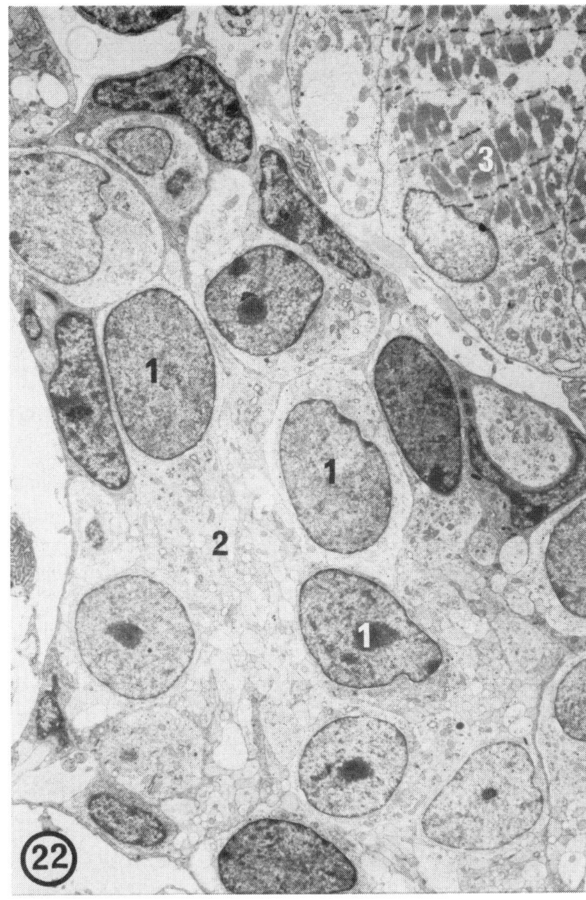
3
1
1
2
1
22

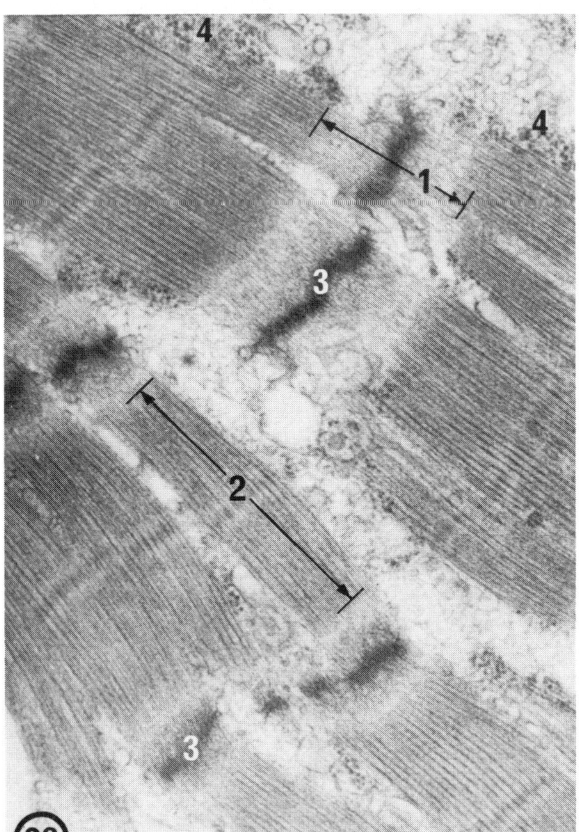
4
4
1
3
2
3
23

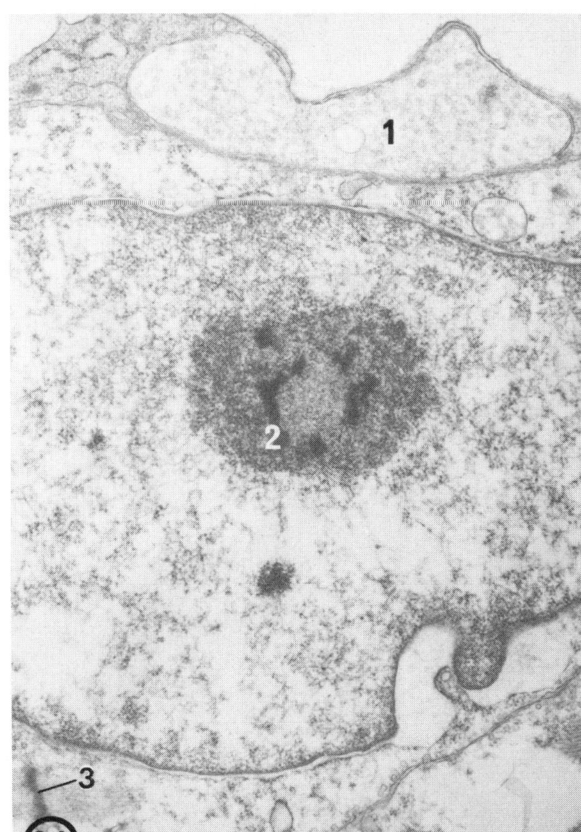
1
2
3
24

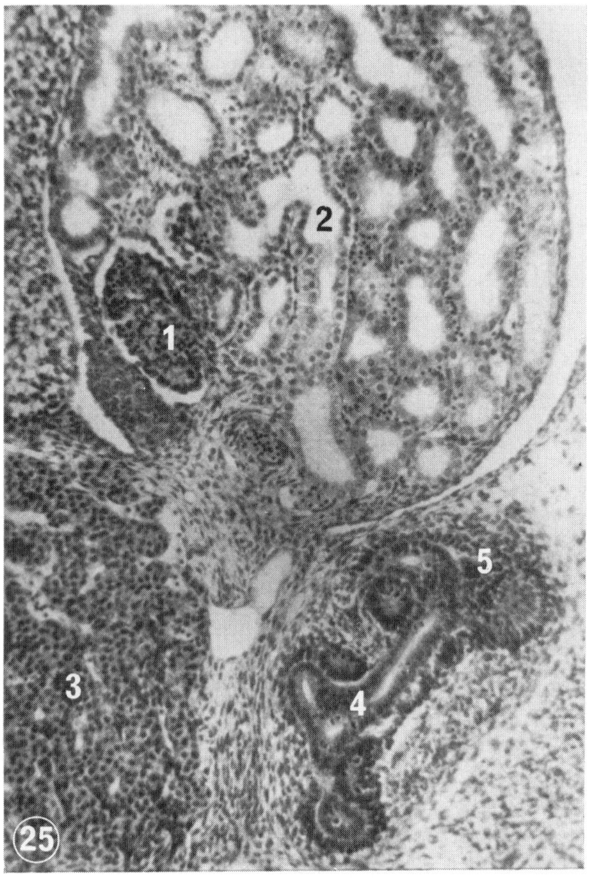

2
1
5
3
4
25

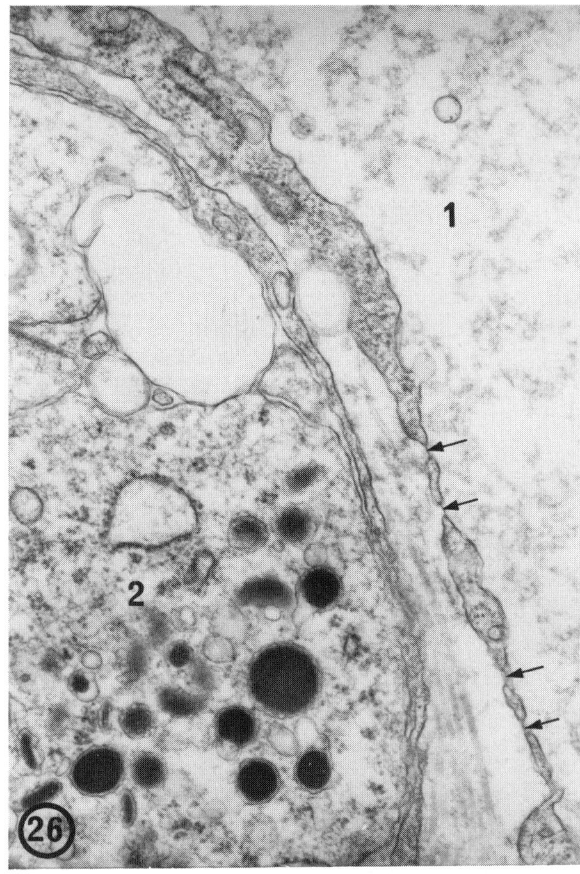

1
2
26

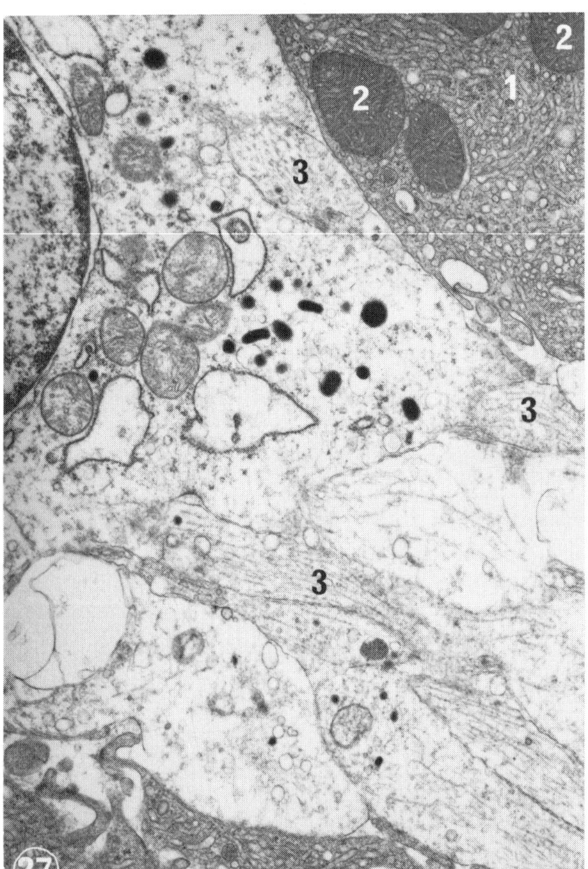

2
2
1
3
3
3
27

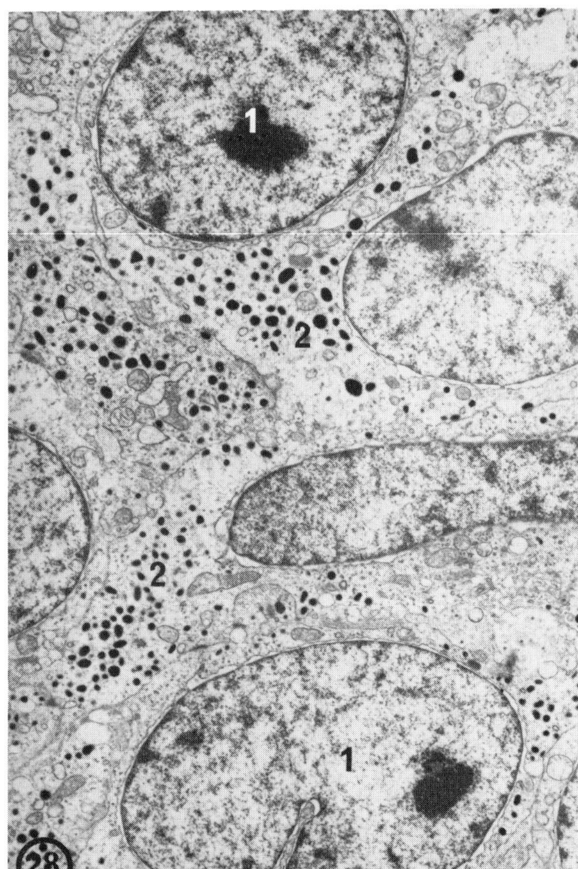

1
2
2
1
28

that have commenced to fuse in the future hindbrain region. The walls of the CNS consist at this stage of a thickened epithelium of unspecialized neural cells (Figs 9, 33). Axonal primordia are not evident at the light microscope level. The primordia of the spinal cord thickens towards the cephalic end and is continuous with four slightly bulbous regions (neuromere primordia) within the area of the future hindbrain (Fig. 33). The forebrain primordia consists of expanded but widely open folds (Fig. 33). Deep optic vesicles arise from, and open onto, the unfused medial surfaces of the presumptive diencephalon (Fig. 29). The central nervous system exhibits marked differentiation by six days before birth. The brain then consists of the five closed secondary vesicles: telencephalon; diencephalon; mesencephalon; metencephalon; and myelencephalon. These regions all exhibit three dimensional complexity by virtue of their own contouring as well as due to gross brain flexures; cervical, midbrain and pontine flexures. Within each brain region, and particularly the ventral motor area of the hindbrain, neuronal specialization includes subdivision of regions into the primordia of brain nuclei and brain axonal tracts. The forebrain and the midbrain regions are poorly differentiated particularly in their dorsal regions. Anatomical continuity of the hindbrain with the spinal cord is particularly evident with longitudinal continuity in presumptive dorsal (sensory) and presumptive ventral (motor) elements. A complexity of interconnecting axonal tracts that deserves special investigation is easily recognizable. This applies particularly to the connections of the hind brain in general and in particular to those relating to the trigeminal ganglion. Thus at six days before birth the primordia of the CNS components of the motor generator have been established and exhibit rapidly increasing complexity over the next six days. In general the ventral motor elements maintain more structural complexity than the dorsal sensory areas.

Neural crest component of the motor generator

The primordia of the trigeminal ganglia (CN V) are evident at nine days before birth as conspicuous undifferentiated cellular condensations located immediately paraxial to the most rostral (first pair) of four slightly bulbous regions of the hindbrain primordium. A similar pair of slightly less conspicuous condensations occur laterally to the third pair of bulbous hindbrain swellings. As these are immediately rostral to the prominent otic placodes they represent the undifferentiated ganglionic primordia of the acousticofacial complex (CN VII-VIII) (Fig. 33). These are the only ganglionic elements present in relation to the entire CNS. With the light microscope no axonal differentiation is evident despite the fact that these ganglionic neural crest primordia immediately abut the developing CNS. Ganglionic development at six days before birth is a spectacular feature of neurogenesis (Fig. 34). The large trigeminal ganglion (CN V) exhibits a wide connection with the presumptive pons region of the CNS. Complex trigeminal tracts are easily visualized and extend towards neuronal nuclear condensations within the hindbrain as well as extensive axonal tracts widely distributed throughout the pons. The ophthalmic, maxillary, and mandibular rami of the trigeminal ganglion are very thick in relation to other axonal tracts and these ramify deeply within their target tissues. Cranial nerve ganglia with strong connections with the hind brain also include the well developed acousticofacial complex (CN VII-CN VIII). Axonal connections from CN VIII immediately abut a thickened portion of the auditory capsule. This arrangement remains conspicuous until at least three days before birth (Fig. 35). The glossopharyngeal (CN IX) ganglia and complex vagal ganglia (CN X) are also centrally connected and exhibit extensive axonal proliferations towards their target tissues. At six days before birth the dorsal root ganglia of the spinal cord are formed with dorsal sensory root

---

Fig. 25. Adrenal gland of *M. eugenii* three days before birth. 1=vascularized mesonephric glomerulus, 2=mesonephric collecting tubules, 3=vascularized cords of adrenal gland, 4=branched ureteric bud (metanephros), 5=early nephron primordia of metanephros. Magnification=142.

Fig. 26. Ultrastructure of fenestrated capillary in the adrenal gland of *M. eugenii* at birth. Arrows=fenestrations in endothelial cell, 1=capillary lumen, 2=cytoplasm of catecholamine cell with membrane bound granules. Magnification=30 000.

Fig. 27. Ultrastructure of adrenal gland of *M. eugenii* at birth. 1=cytoplasm containing dense smooth endoplasmic reticulum (SER) of steriod producing cell in cortex, 2=mitochondrion with tubular cristae, 3=nerve fibres with neurofilaments. Magnification=12 000.

Fig. 28. Ultrastructure of adrenal gland of *M. eugenii* at birth showing a group of catecholamine producing cells. 1=nucleus, 2=membrane bound cytoplasmic granules. Magnification=4 725.

connections. The ventral motor roots of the spinal column form a common nerve trunk with axons emerging from the dorsal root ganglia. This common trunk after a short distance breaks up to ramify towards the target muscle primordia. The shafts of primordia of both hindlimb and forelimb are at this stage well innervated. The brachial plexus of the forelimbs is particularly well formed with confluence of spinal ganglia by well formed axonal interconnections. The size of the dorsal root ganglia diminishes in a craniocaudal direction, however, ganglionic development is much reduced within the sacral and tail regions of the embryo.

The complexity of the CNS and peripheral axonal connections in association with the cranial nerves and the spinal dorsal root ganglia exhibit a marked increase at both three days before birth (Figs 35, 36) and at birth. Parasympathetic ganglia at the light microscope level are conspicuous in association with the abdominal aorta, the lungs, the submandibular salivary ducts as well as the palatine and deep lingual paraneuronic complexes (DLPCs) of the tongue, (Figs 18, 21) at three days before birth. A brief account of the development of the nervous system of *Macropus rufogriseus* at 18, 20 and 25 days RPY is given by Walker and Rose (1981).

Muscular component of motor generator

The primordia of the skeletomuscular system are recognized at nine days before birth as the sclerotomal and dermomyotomal segments of a seventeen somite embryo (Fig. 9). Paraxial somites were reported to be present in *Macropus rufogriseus* at nine days before birth by Walker and Rose (1981). The muscular primordia of the motor generator in *Macropus eugenii* at six days before birth are mostly undifferentiated spindle shaped myoblasts. Early tubular stages of myogenesis are present in those fibres that form dominant elements of muscular contraction at birth, such as the major muscles of the shoulder, forelimbs and the tongue.

A spectacular level of myogenesis occurs during the next three days within those elements functional at birth. A range in the stages of tubular muscle myogenesis parallels the priority of functional usage. The tubular muscle fibres that activate flexion and extension of the forelimbs and also dorsoventral flexion of the thorax and abdomen are thickened and are grouped in well organized muscular elements by three days before birth. Primordial tendons now insert into major skeletal elements. In the case of the long bones of the forelimbs peripheral ossification in their central shaft region is also evident.

Small gross forelimb movements at three days before birth as well as dorsoventral flexion of the trunk of three of the four embryos examined at this stage by stereo microscope, is evidence of the functional maturation of the skeletal and muscular elements involved. Functional kindling of the muscular generator consequently begins at about three days before birth, and is served by a well developed brachial plexus, already developed at six days before birth (Fig. 37). At birth the muscle fibres associated with the pectoral girdle that activate the forelimbs consist of long thickened well innervated tubular elements. At the ultrastructural level these fibres exhibit contractile elements (Fig. 38). The fibres have retained their central vacuolar elements but the fibre nuclei are now not so frequently centrally located having in many cases migrated to the periphery during myogenic maturation. Nerve fibres pass between and near to myofibrils. The shoulder and elbow as well as interphalangeal articulations, at birth, are served by joint capsules. The terminal phalangeal cartilaginous

---

Fig. 29. Transverse section of the pharyngeal area of *M. eugenii* nine days before birth. 1=mid-dorsal pharyngeal plate (site of future Rathke's pouch), 2=unfused neural tube in the region of the diencephalon, 3=optic vesicle, 4=mandibular primordia of stomodeum, 5=pharyngeal lumen (foregut primordium). Magnification=142.

Fig. 30. Pituitary gland primordia in *M. eugenii* three days before birth. This is a transverse section in which the right hand margin of the micrograph represents the dorsal aspect. 1=lumen of infundibulum, 2=primordia of pars intermedia, 3=lumen of Rathke's pouch, 4=primordia of pars distalis, 5=remnants of Rathke's pouch, 6=cartilage of basisphenoid, 7=developing hypophyseal blood vessels. Magnification=142.

Fig. 31. Ultrastructure of pars distalis of *M. eugenii* at birth. 1=nucleus of deep pars distalis cells, 2=membrane bound secretory granules. Magnification=2 400.

Fig. 32. Ultrastructure of pars nervosa of *M. eugenii* at birth. 1=neurosecretory granules in axoplasm of a fibre of the hypothalamo-hypophyseal tract, 2=clear vesicles of axonal terminal, 3=transverse section of nerve fibre containing clear vesicles. Magnification=12 000.

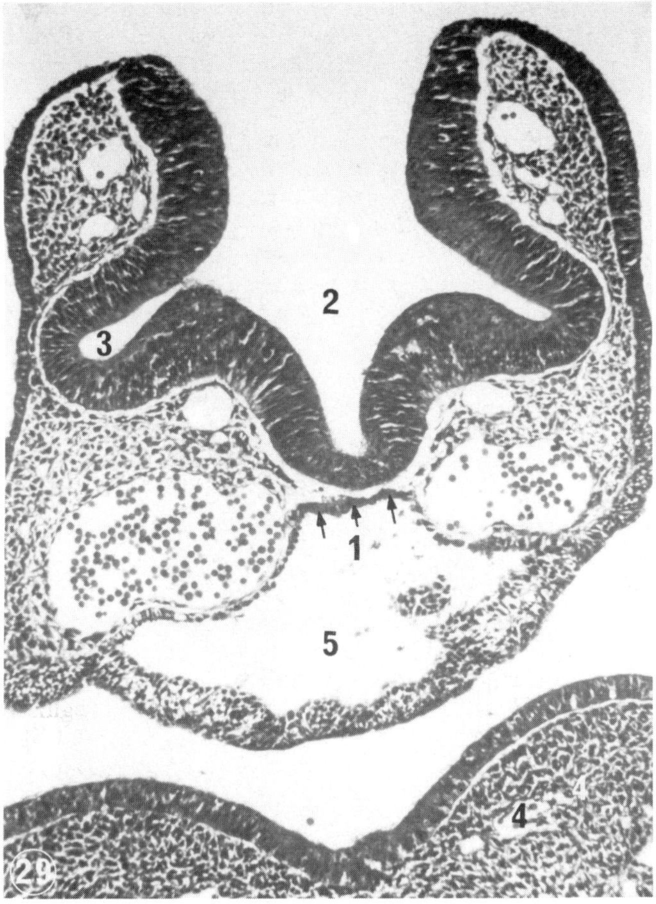
2
3
1
5
4
29

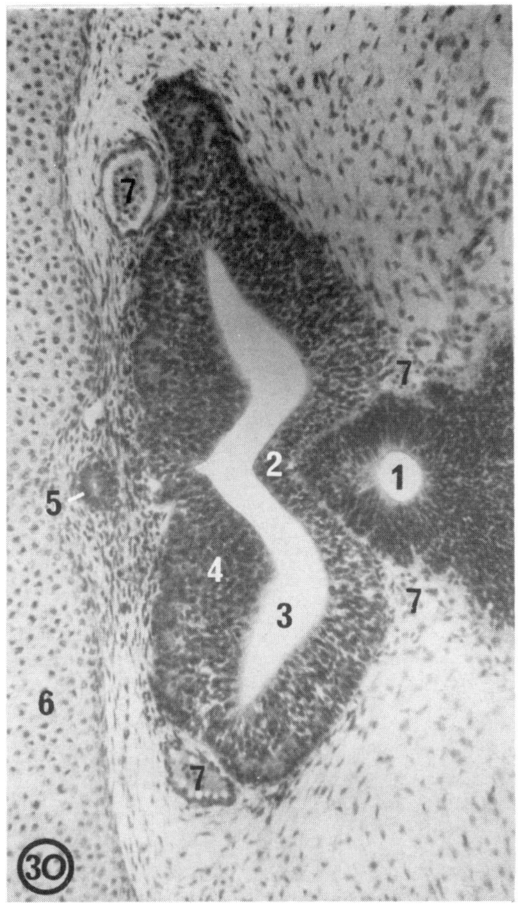
7
7
2
1
5
4
3
7
6
7
30

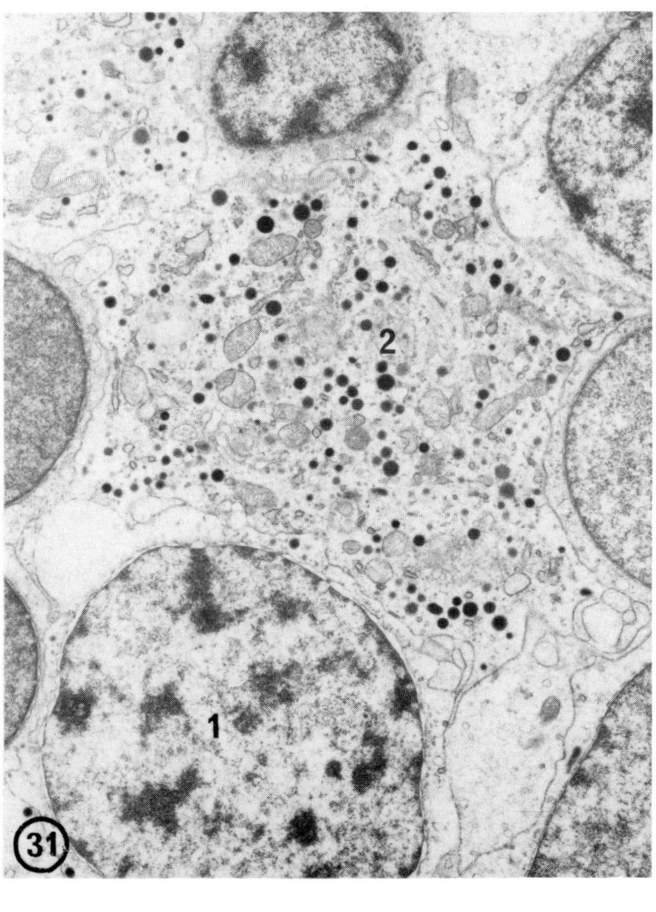
2
1
31

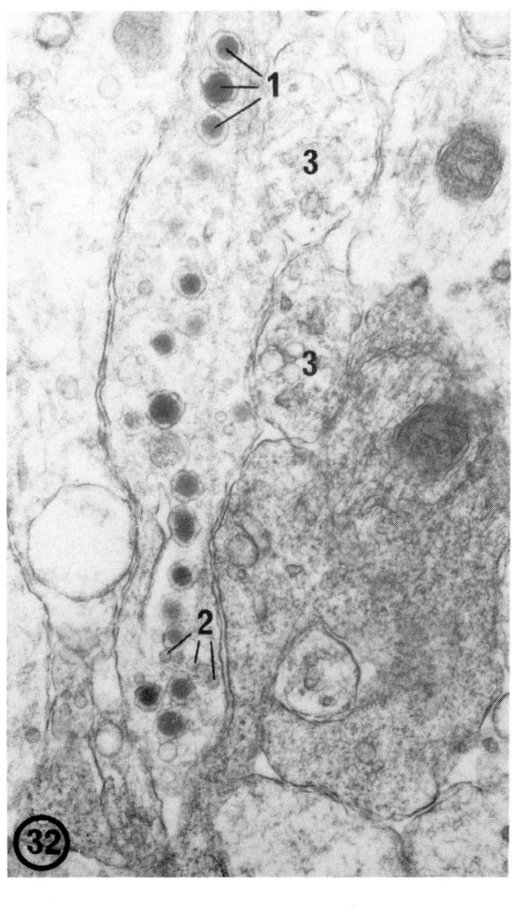
1
3
3
2
32

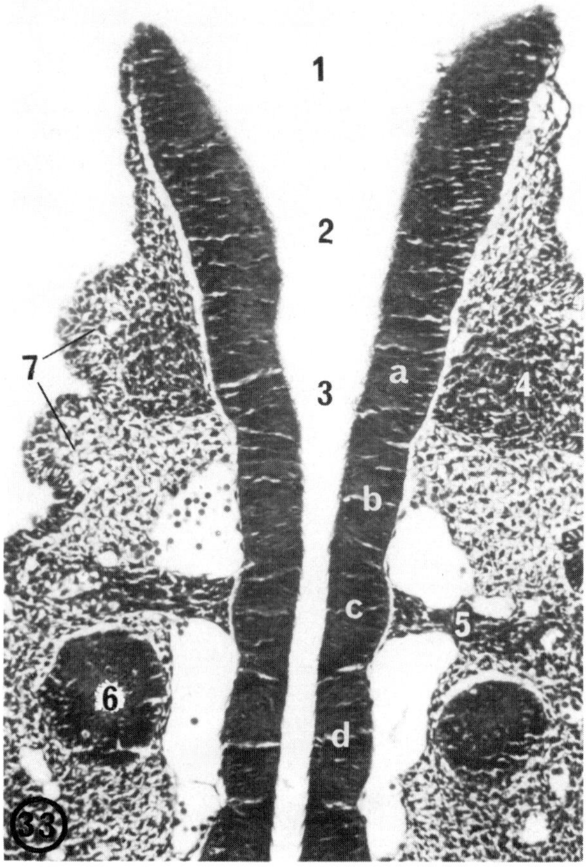
1
2
7
3
a
4
b
c
5
6
d
33

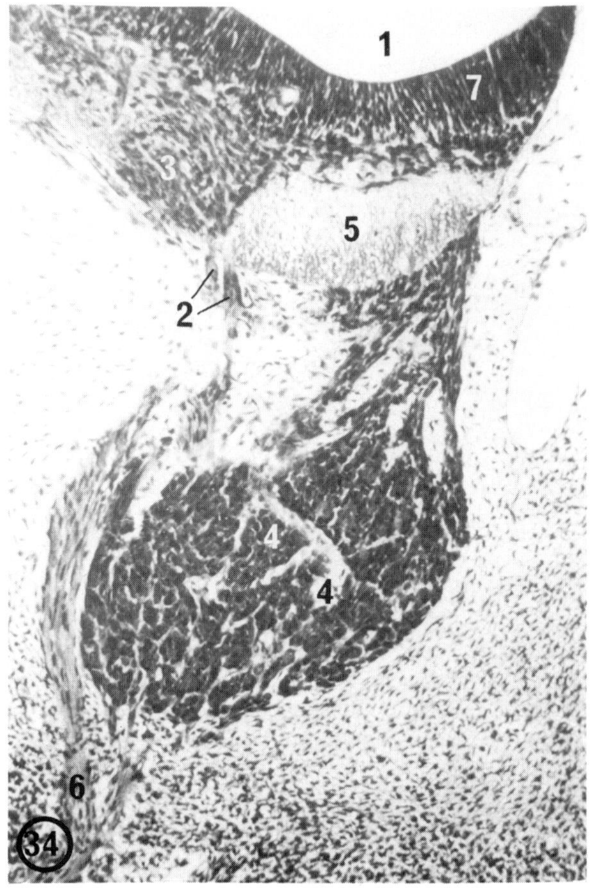
1
7
3
5
2
4
4
6
34

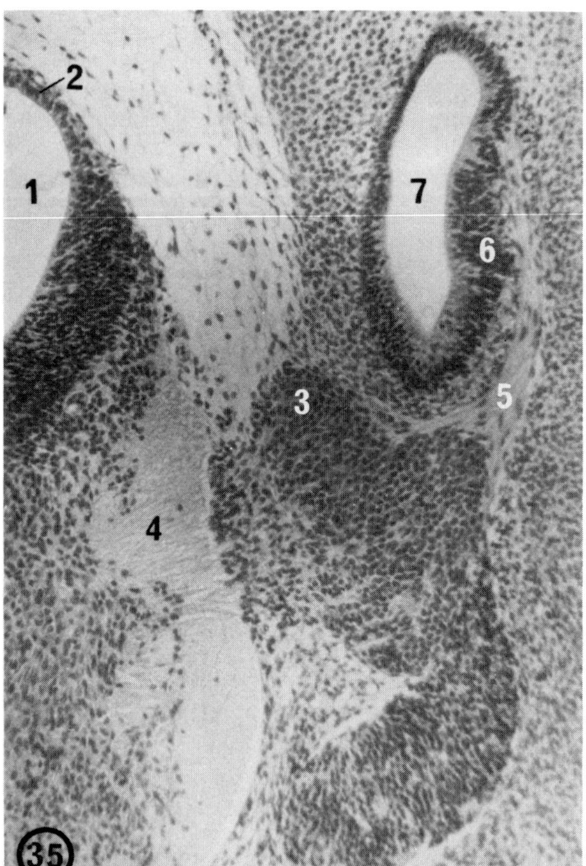
2
1
7
6
3
5
4
35

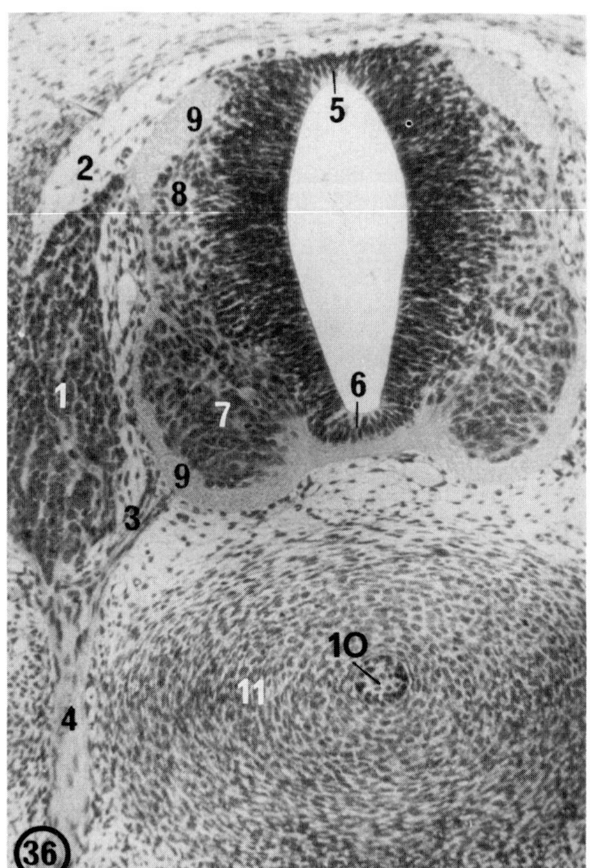
5
9
2
8
1
7
6
9
3
10
11
4
36

models of the forelimb digits are distally ossified and the digitopalmar flexor tendons are inserted at the inner palmar surface of the digits and are obviously strengthened by the terminal ossification centres (See Hughes and Hall 1988). The sharp recurved bill-like epitrichial claws cap the outer surface of the digits (Fig. 39) and are homologous with the epitrichial false nails of human development (Arey 1974). The cellular primordia of the epitrichial claw is first evident six days before birth (Fig. 40). The extensor tendons are attached to the outer surface of the digital cartilaginous skeletal models.

At birth the paraxial musculature of the trunk responsible for dorsoventral flexion and sinuous body movement are variably developed. The length as well as the thickness of their striated muscle tubules diminish in a craniocaudal progression that is particularly evident in the lower abdominal regions.

The well innervated tubular elements of the muscles of the head and neck are among the most well developed in the newborn *M. eugenii*. These are the muscular elements responsible for random teat search behaviour that is characterized by alternating lateral head movements.

The tongue and pharyngeal muscles responsible for sucking and swallowing at birth exhibit a wide range of myotubular differentiation. The extrinsic tongue muscles consist of thick, near mature, striated voluntary muscle. Some of the major intrinsic muscles of the tongue are only slightly less developed with muscle nuclei peripherally arranged and the central vacuolar component of the fibres replaced by contractile elements. The oblique elements of the intrinsic muscle fibres of the tongue range in size with relatively small poorly striated elements still common. The central nuclei of these less developed fibres alternate with vacuoles. Small immature fibres at this early stage of formation are interspersed between the more advanced elements. Oesophageal muscles concerned with swallowing are moderately developed at birth while smaller earlier stages of contractile muscle tubules are associated with the pharyngeal cartilages that might be expected to regulate epiglottal movements. The functional competence of these minor elements deserves special investigation.

Close inspection with a ×5 jeweller's loop of the oral cavity of all three newborn *M. eugenii* revealed sucking movements of the tongue and parallels an earlier report by Hughes (1962) of a similar phenomenon in two newborn *Potorous tridactylus*. McCrady (1938) reported that this was also a feature of newborn *Didelphis virginiana*. Detailed accounts of birth in the Red Kangaroo *Megaleia rufa* were given by Sharman and Pilton (1964) and Sharman and Calaby (1964) including sucking movements at birth.

The intercostal muscles of *Macropus eugenii* at birth are well developed and innervated, while those of the diaphragm are weak and consist of a very thin sheet. It is evident that the respiratory pump is chiefly driven intercostally.

In general the neuromuscular and skeletal development of newborn macropodid marsupials is much more advanced than that found in non-macropodid species. This not only compensates for the marked increase in body size of macropodids in relation to other marsupials but also represents the end point of a much longer and more complex developmental profile. This confers on macropodids a much greater repertoire of

---

Fig. 33. Coronal section through the head region of *M. eugenii* nine days before birth. Unfused neural folds showing presumptive regions of: 1=forebrain, 2=midbrain, 3=hindbrain, 4=neural crest primordia of trigeminal ganglion (CN V), 5=neural crest primordium of acousticofacial (CN VII and CN VIII) ganglion, 6=lumen of otic vesicle, 7=maxillary and mandibular components of 1st pharyngeal arch. a, b, c, d=neuromeres (more apparent in deeper sections). Magnification=142.

Fig. 34. Transverse section of trigeminal ganglion of *M. eugenii* six days before birth. 1=lumen of 4th ventricle, 2=fibre tracts connecting trigeminal ganglion to nucleus in presumptive pons (3), 4=body of trigeminal ganglion, 5=marginal axonal tracts associated with trigeminal connections to pons, 6=first arch nerves originating in trigeminal ganglion, 7=undifferentiated ventricular cells of pons. Magnification=142.

Fig. 35. Coronal section through the acousticofacial ganglion of *M. eugenii* three days before birth. 1=lumen of 4th ventricle, 2=undifferentiated ventricular cells, 3=acoustic ganglia (CN VIII), 4=fibre tracts from acoustic ganglion to hindbrain, 5=fibres innervating auditory epithelium (6), 7=lumen of auditory primordia. Magnification=142.

Fig. 36. Transverse section of spinal cord in lumbar region of *M. eugenii* three days before birth. 1=dorsal root ganglion, 2=dorsal root connection, 3=ventral root connection, 4=nerve fibres supplying hind limb, 5=roof plate, 6=floor plate, 7=motor neuroblasts of the basal plate, 8=alar plate with sensory connections, 9=marginal zone of spinal cord, 10=notochord, 11=developing centrum of lumbar vertebra. Magnification=142.

locomotor movements. This is particularly a feature of the forelimbs where the manus exhibits a much greater range of mobility than that of *Sarcophilus harrisii, Perameles nasuta* or *Trichosurus vulpecula* at birth (Hughes, unpublished data). Macropodids, including *Macropus eugenii*, also have the ability at birth to engage in asynchronous forelimb movement. In *Sarcophilus harrisii* where the entire period of organogenesis from primitive streak to birth occupies only two and two thirds days, the manus at birth has undifferentiated joints that lack capsules (Hughes and Hall 1988) and is consequently at a much earlier stage of morphogenesis than in macropodids. This is reflected in a reduced range of mobility of the forelimbs at birth and consequently a relatively more stereotyped forelimb movement.

Modulators of the motor generator

The previous section has shown that the motor generator develops as a result of several interacting organogenetic profiles of relatively long duration by marsupial standards. The generator primordia present at nine days before birth, first exhibit a period of marked neurogenesis which is well established at six days before birth. This is followed by extensive myogenesis culminating in motor kindling at three days before birth. The development of skeletal elements in *Macropus eugenii* is also of relatively long duration before birth so that its morphological components at birth are more structurally complex than those of other non-macropodid marsupials. On this basis an expected increase in the complexity of the repertoire of movements in newborn macropodids deserves appropriate investigation. In this connection a detailed description of the early development of the shoulder girdle and sternum in marsupials by Klima (1987) is of particular interest. An account of the locomotor and teat attachment movements of two newly born *Potorous tridactylus* is given by Hughes (1962) and parallels observations on the three new born *Macropus eugenii* in the present study. Observations on locomotion in new born *Megaleia rufa* have been reported by Sharman and Pilton (1964) and Sharman and Calaby (1964). Sinuous movements of the trunk in the neonate of *Macropus eugenii* in combination with about 50 alternate forearm strokes per minute result in forward progression. The necessary mechanical advantage is obtained by digitopalmar prehension with the blade-like margins of the epitrichial claws locking the hair against the palmar surface of the manus (Fig. 39). Each forearm stroke thus has potential to add an increment of progression. The brushed hairs of the human forearm permits close inspection of this action. Forward progression is accompanied by alternate lateral scanning movements of the rostral perioral portion of the head. This action possibly also serves to part the hair during forward progression, however its essential function is random search behaviour for the teat. Close inspection of the oral area of newborn *Macropus eugenii* reveals minute sucking movements of the lower jaw and tongue. The whole repertoire of motor movements continues irrespective of the orientation of the newborn *Macropus eugenii* or *Potorous tridactylus*. This includes minimal restraint by free holding with a minute area of dorsal surface epidermis (in mid-air). Likewise all movements persist during temporary immersion within a fluid medium. It is quite apparent that these movements persist in the absence of epidermal stimulation. Certainly the epidermis of the head or forearms could not be involved. With the elapse of time the alternate forearm movements in one animal were observed to slow at room temperature from about 50 strokes per minute on removal from the pouch to about 20 strokes per minute after some 25 minutes had elapsed. This slowing can be attributed to factors such as hypothermia, desiccation, fatigue or depletion of energy reserves. In a previous section it was seen that the locomotor generator began to be developmentally switched on three days

Fig. 37. Limb nerve from brachial plexus in *M. eugenii* six days before birth. 1=neural trunk from brachial plexus, 2=myoblasts of upper limb. Magnification=142.

Fig. 38. Ultrastructure of forelimb muscle of *M. eugenii* at birth. 1=I band, 2=A band, 3=Z line. Magnification=18 750.

Fig. 39. Scanning electron micrograph of epitrichial forelimb claws of *M. eugenii* at birth. 1=blade-like edge of epitrichial claw, 2=finely tapered tip, 3=point of separation of deciduous claw. Scale=0.5 mm.

Fig. 40. Oblique section of forelimb epitrichial claw primordium of *M. eugenii* six days before birth. 1=cellular claw primordium, 2=epitrichial layer, 3=basal germinal layer, 4=marginal capillary plexus, 5=mesenchymal skeletal elements. Magnification=370.

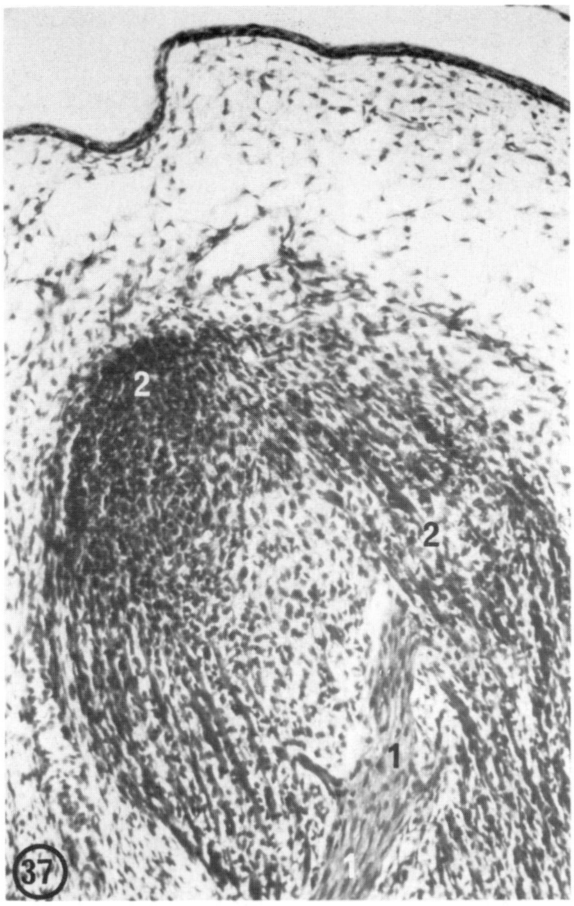
2
2
1
37

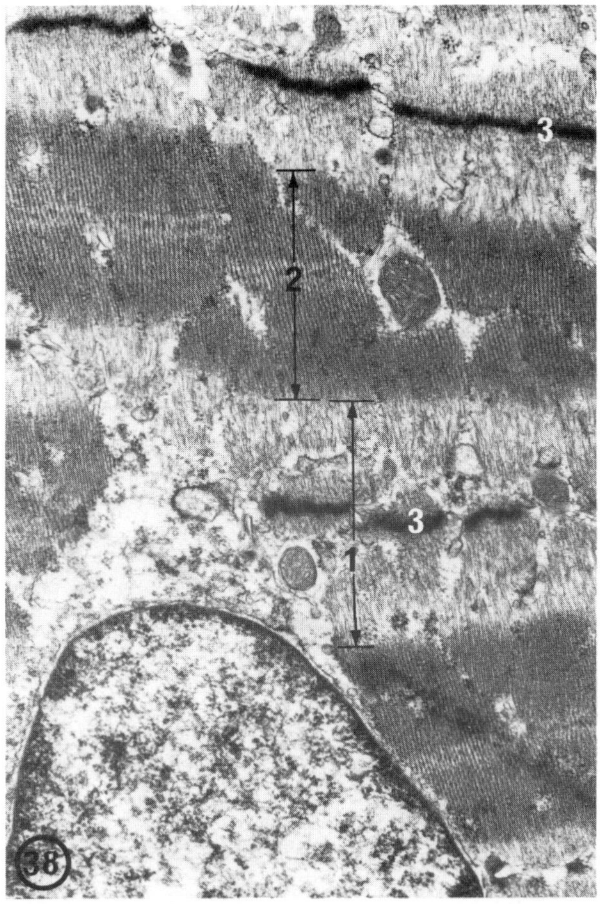
3
2
3
1
38

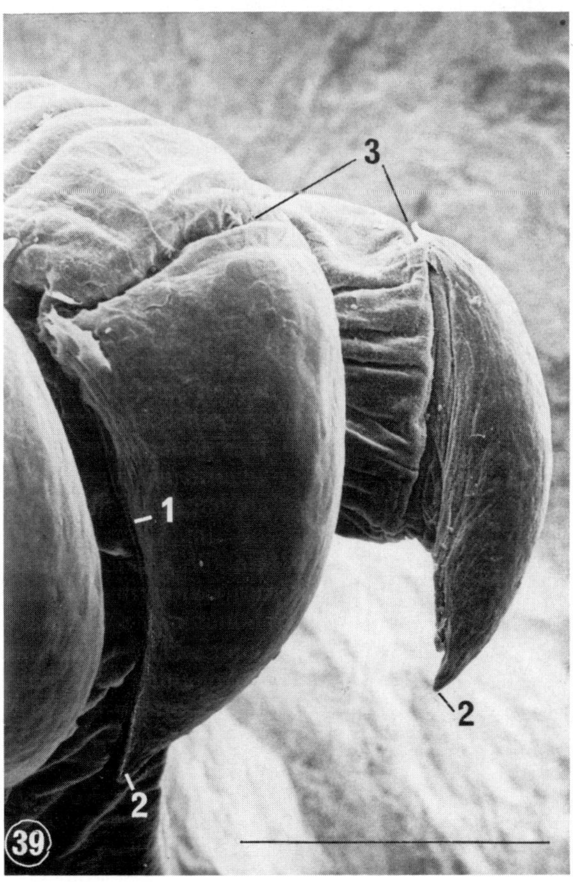
3
1
2
2
39

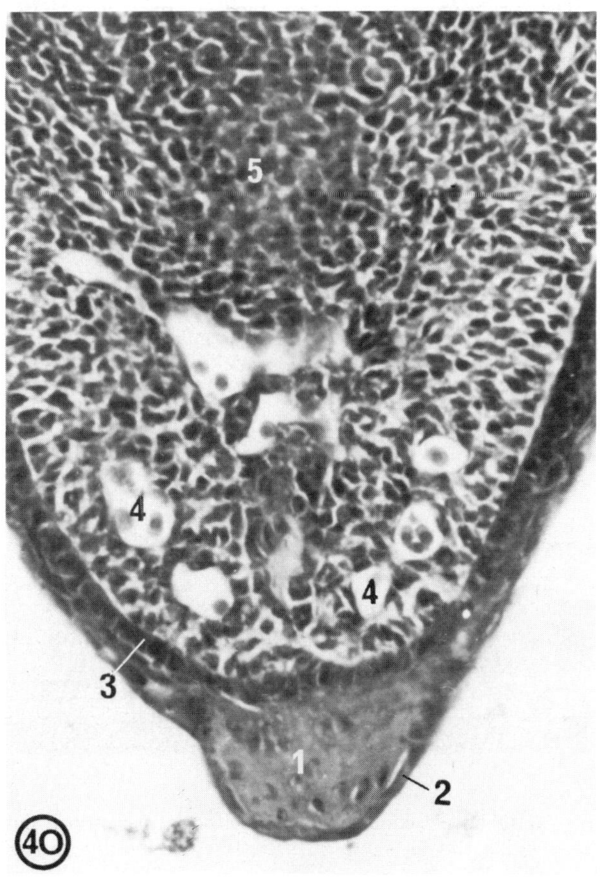
5
4
4
3
1
2
40

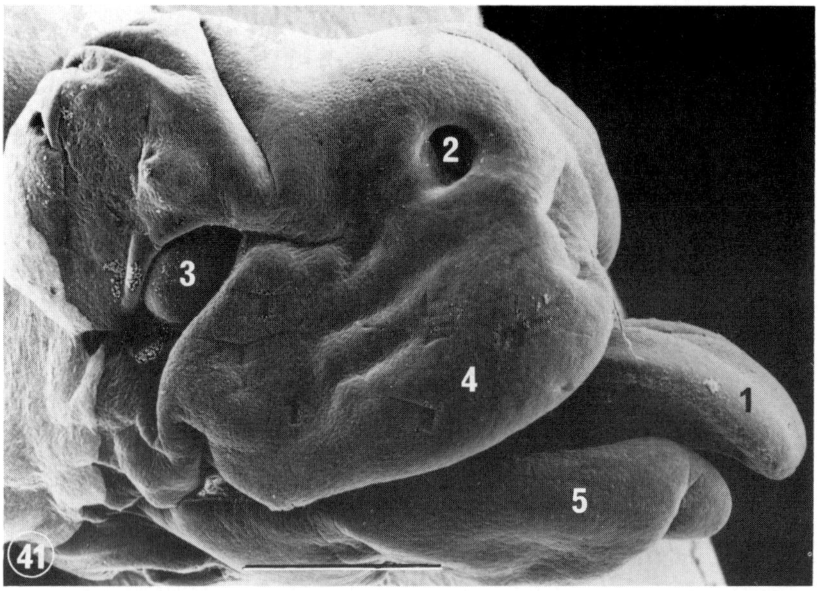
2
3
4
1
5
41

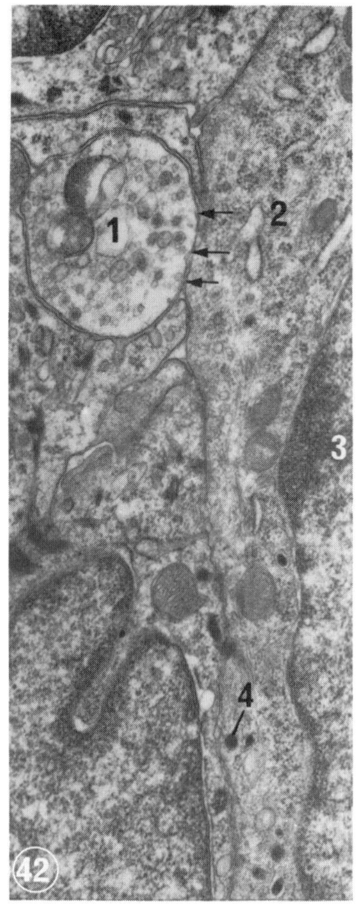
1
2
3
4
42

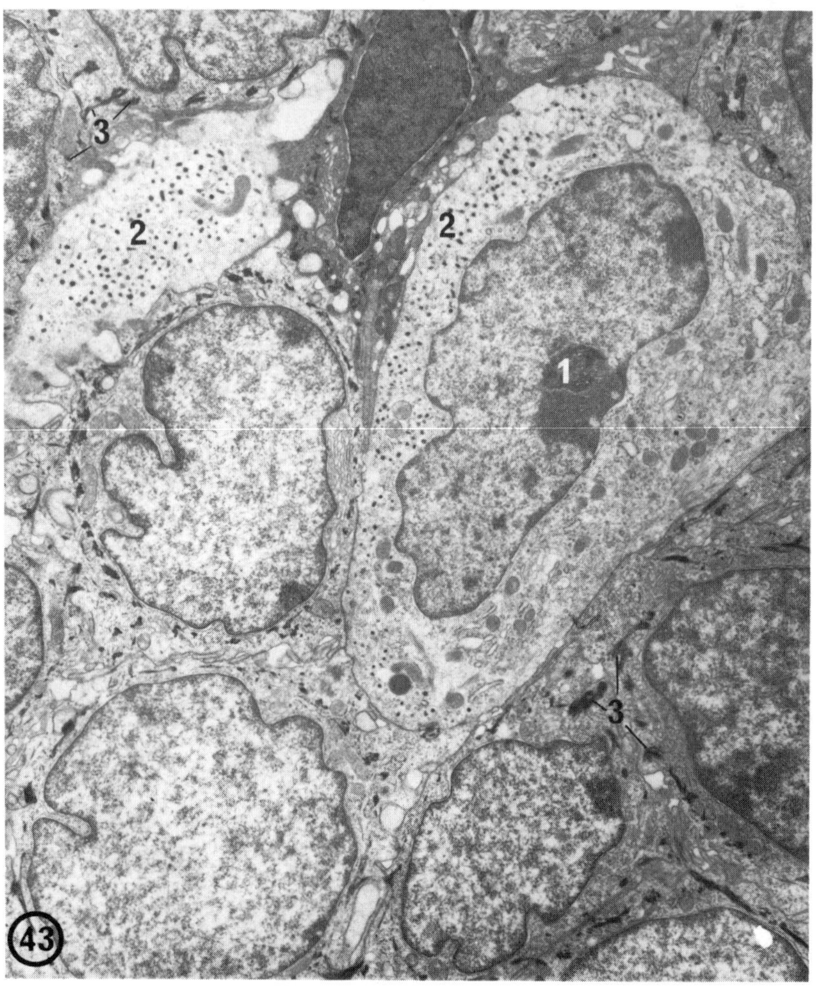
3
2
2
1
3
43

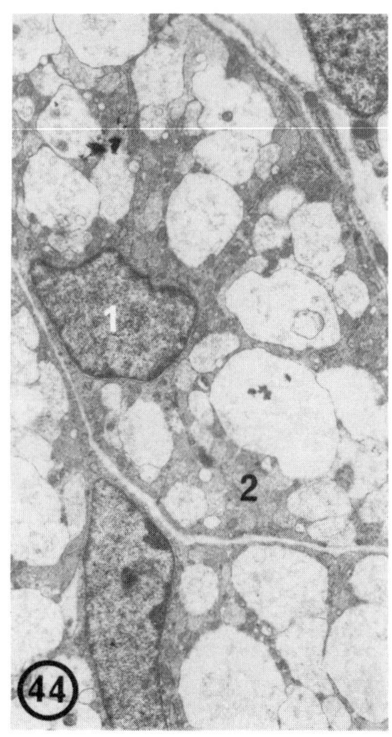
1
2
44

before birth. It is now postulated that movement is internally automated, not requiring epidermal receptors of the type one would expect to find involved in reflex arcs of more mature pouch young or adults. An extensive ultrastructural search of the epidermis of the manus of *Macropus eugenii* at birth revealed the absence of neural receptors. It is our contention that the locomotor generator at birth is developmentally switched on and internally maintained. On arrival at the teat the locomotor generator is switched off and teat attachment behaviour initiated. The structural and behavioural basis for this is now to be considered. The structural integrity of the hindbrain with the spinal cord is essential for the locomotor generator to operate. Cervical decapitation in one of three newborn *Macropus eugenii* immediately eliminated digitopalmar prehension as well as alternate forearm movements, sinuous movements of the trunk and random nipple search movements of the head. However, a minute level of dorsoventral flexion of the trunk of less than 1 mm remained. A similar range of total flexion of the forelimbs (approx. 1 mm) was also evident. The remainder of this section will be concerned with structural and behaviour appraisal of five neonatal receptor sites with a potential for modulation of the locomotor generator.

Olfactory and vomeronasal epithelium

The deep portion of the olfactory chambers near the telecephalon consist of a thickened layer of columnar cells at six days before birth and constitute the primordia of the olfactory epithelium. Axons of the bipolar olfactory neurons of the olfactory epithelium three days before birth leave the olfactory basal lamina as separate fibres of the fila olfactorum but soon aggregate into more condensed olfactory axons that enter a preolfactory lobular condensation, part of which forms a small ganglion, the nervus terminalis (Fig. 17). Dispersed nerve fibres connect the preolfactory bulb condensation with the definitive and slightly thickened olfactory bulb of the telencephalon (Fig. 17).

The olfactory epithelium at three days before birth is served by well developed Bowman's glands, and a branch from the trigeminal ganglion lies within the immediately adjacent subepithelial tissues. At birth the olfactory epithelium constitutes a more extensive area than previously and at the ultrastructural level consists of microvillous supporting cells interspersed between ciliated olfactory bipolar neurons.

The vomeronasal organs consist of a pair of pouches arising from the epithelium adjacent to the lower part of the nasal septum (Fig. 17). The epithelium of the vomeronasal organ is tall columnar and is associated with vomeronasal glands. The vomeronasal complex is innervated by a branch from the nervus terminalis with connections to the olfactory lobe. The undifferentiated primordia are present at six days before birth and at birth this organ is prominent.

The olfactory epithelium as well as the vomeronasal organs of marsupials have been suggested as olfactory guidance receptors at birth (McCrady 1938; Hill and Hill 1955; Sharman and Pilton 1964) and of possible assistance in the journey to the teat (see Chapter by Gemmell and Rose). Two newborn *Macropus eugenii* removed from the pouch and held freely failed to show any lateral olfactory recognition (either attraction or repulsion) to "cotton bud" swabs of their respective mother's pouch. The small "cotton bud" swabs were held at a distance of from 1 mm to 3 mm from the external nasal openings. This was repeated at an interval of several minutes, five tests per nostril in each animal. Likewise there was no recognition (attraction or repulsion) of pouch smears on human forearms where neonates were permitted to crawl freely. An

Fig. 41. Scanning electron micrograph of perioral area of *M. eugenii* three days before birth. 1=protruding tongue with median sulcus, 2=external nares, 3=uncovered eye, 4=maxillary swelling, 5=mandibular swelling. Scale bar = 0.1 mm.

Fig. 42. Ultrastructure of perioral area of *M. eugenii* at birth. 1=terminal maxillary trigeminal fibre containing both membrane bound and clear vesicles, 2=Merkel cell fused to adjacent nerve fibre (arrows), 3=nucleus of Merkel cell, 4=membrane bound cytoplasmic granules in Merkel cell. Magnification=18 750.

Fig. 43. Ultrastructure of Merkel cells in densley innervated perioral region of *M. eugenii* at birth. 1=nucleus of Merkel cell, 2=cytoplasm containing dense membrane bound granules, 3=desmosomes between adjacent keratinocytes. Magnification=6 000.

Fig. 44. Ultrastructure of trigeminal ganglion of *M. eugenii* at birth. 1=nucleus of ganglion neurone, 2=axonal tracts. Magnification=4 725.

incidental observation was that the neonates did not exhibit a typical righting reflex but that the grasping of the manus in its alternate movement achieved this when hair was contacted.

Vibrissal primordia

Vibrissal primordia were absent at six days before birth. Epithelial buds and associated innervated mesodermal papillae of the primordia of facial vibrissae were present at three days before birth. At birth the mystacial vibrissal papillae were well developed and arranged in an area consisting of several externally visible rows.

The vibrissal areas have been suggested as potential areas of tactile recognition by neonates (Hall and Hughes 1987; Hughes and Hall 1988). Three newly born *Macropus eugenii* freely held showed no obvious lateral tactile recognition (attraction or repulsion) when the lateral vibrissal mystacial areas were stimulated by repeated gentle touching with finely pointed jeweller's forceps. An initial notion, that the lateral search movement of the head showed a minimal attractive response could not be confirmed. The final conclusion was that both the lateral teat search movement of the head and the alternate forelimb movements were not detectably modified. Five tests were performed on the right and the left vibrissal areas in each of three newborn young.

Taste buds

Innervated taste buds are present on the upper lateral and particularly the medical concave upper surface of the tongue at three days before birth (Fig. 19). At birth these continue to be prominent features of the tongue. The effectiveness of taste buds as receptors at birth remains to be investigated.

Merkel cells

Trigeminal innervation of the perioral area by the maxillary and madibular rami of the trigeminal ganglion is a predominant feature of the head six days before birth. The trigeminal axonal terminals fan out beneath the basal lamina of the perioral area to form a relatively dense neural arborisation at three days before birth. A scanning EM of the perioral area at three days before birth is shown in Fig. 41. Specimens for ultrastructural observation were prepared at birth from microdissected segments of the medial and lateral margins of the immediate perioral rim. Similar material was obtained from slightly separated adjacent facial segments. The presence of Merkel cell receptors was detected with the transmission electron microscope within the two to three basal layers of the epithelium. In many cases Merkel cells at birth exhibited cell membrane contact zones with terminal trigeminal axons (Fig. 42). In the immediate perioral rim of the single newborn young so far examined the proportion of Merkel cells in the maxillary oral region was up to 20-25% of the basal two cell layers of the epithelium (Fig. 43). High densities of about 18% were also found in the immediate perioral rim of the mandibular components of the lips. In the separated adjacent facial areas Merkel cells contributed less than 3% of the basal two epithelial layers. This suggests a tactile gradient with increasing tactile sensitivity towards the immediate perioral rim. Well developed axonal profiles in the tyrigeminal ganglion are shown in Fig. 44.

Three newborn *Macropus eugenii* were each freely held and tested some five to six times each for evidence of perioral tactile sensitivity by gentle touching with pointed jeweller's forceps. In every instance gentle touching or stroking of either the maxillary or mandibular segments of the inner perioral margins resulted in the cessation of both the alternate forelimb movements and the lateral random head search behaviour as long as the tactile stimulation continued. Consequently locomotor behaviour was switched off and teat attachment behaviour elicited. The head of the neonates then moved so as to rotate the oral cavity about the tips of the forceps. The tongue was seen to be depressed slightly within the oral cavity and expansion of the rostral oral margins also occurred. The head also moved slightly towards the forceps. An increase in the amount of oral tactile stimulus around the inner perimeter of the oral margins increased the magnitude of oral dilation and teat attachment behaviour. In these observations with newborn *Macropus eugenii* the degree of suppression of locomotor activity as well as the teat attachment behaviour was similar to that described during two relatively undisturbed births in another macropodid *(Potorous tridactylus)* reported by Hughes (1962).

These observations support the locomotor model proposed by Hughes and Hall (1988) where it was suggested that locomotor activity was developmentally initiated and internally maintained. In this model cessation of locomotor activity was required to be switched off by receptors including tactile trigeminal elements.

Tongue receptors

A rudimentary tongue is present six days before birth and is grossly innervated, vascularized, and contains myoblasts of some of the extrinsic and intrinsic muscle primordia. Prominent separated branches of the mandibular rami of the trigeminal nerve enter the lateral root of the tongue (one trunk on each side) at three days before birth. Areas adjacent to these trunks are at least in part innervated by parasympathetic ganglion cells

(CNVII) near the base of the tongue (Fig. 21). Isolated satellites of these submandibular ganglion cells accompany the lateral main lingual trunks and occur in the immediately adjacent deep lingual musculature (Figs 18, 22). These deep lingual ganglionic elements we propose to designate as "deep lingual paraneuronic complexes" (DLPCs). The DLPCs remain a prominent feature of the tongue of newborn *Macropus eugenii* and are presumably secreto-motor in function. Merkel cells were not ultrastructurally detected as a feature of the basal layers of the surface epithelium of the tongue at birth. Striated tubular muscle fibres were interspersed by neuronal axons (Figs 23, 24). Motor end plates were associated with muscle fibres and these are presumably the terminals of the hypoglossal nerve, CN XII (Fig. 24). The DLPCs, presumably of CNVII parasympathetic origin via CNV, at the ultrastructural level consisted of bundles of ganglionic cells with cytoplasm containing both clear and membrane bound vesicles. In the nerves of the tongue axonal synaptic vesicles were common. Bulbous axonal regions contain clear vesicles, membrane bound vesicles, or both.

In two of the three newborn *Macropus eugenii* freely held the tip of the tongue was lightly touched (in four tests for each neonate) with the pointed tips of the jeweller's forceps. In each test this resulted in ventral depression of the tongue associated with slight expansion of the oral cavity. It was not possible to determine whether the response lay in tongue receptors or was a secondary effect of indirect stimulation of the Merkel cells of the mandibular oral margin.

General comments on motor generator modulators

It is of interest that all the effective behavioural modulators in our observations of the motor generator are trigeminally connected. The level of tactile trigeminally mediated involvement in abolishing locomotor movement as well as the random nipple search and the initiation of teat attachment behaviour accounts for virtually all the recorded locomotor and teat attachment behaviour of newborn macropodid marsupials (Hughes 1962; Sharman and Pilton 1964; Sharman and Calaby 1964). It would be tempting to conclude that if other factors are involved these are likely to be quite trivial. It is very obvious that trigeminal modulation of the motor generator is of focal importance and at the structural level this is reflected by both the early appearance of its primordium at nine days before birth as well as the extensive area of its neural connections with the pons and other head structures (Fig. 45). Other modulators, if functional, obviously play a very subservient role to the perioral trigeminally based tactile receptors. The original concept of the present motor generator model arose in a less refined form from the studies

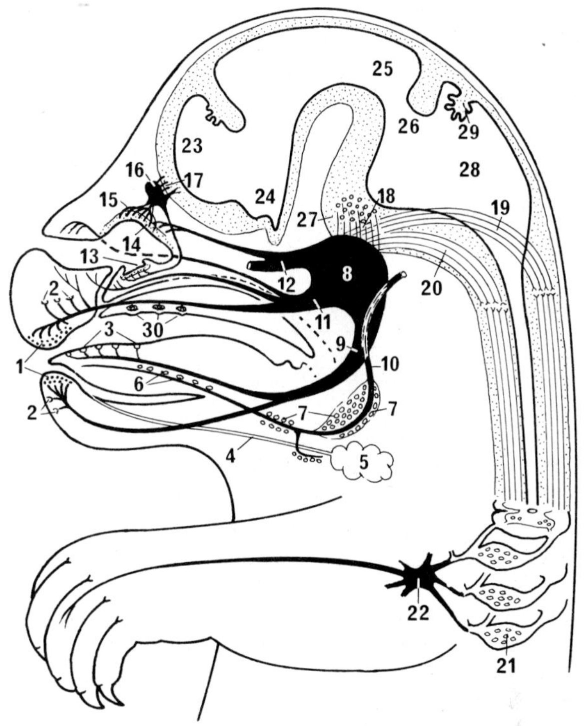

Fig. 45. Schematic diagram constructed from serial section of trigeminal and CNS connections in *M. eugenii* at birth. 1=zone of Merkel cell concentration in perioral area, 2=vibrissal primordia, 3=dorsal lingual taste buds, 4=patent submandibular gland duct, 5=submandibular gland, 6=deep lingual paraneuronic complex (DLPC), 7=paraneurones of submandibular ganglion concentrated around lingual branch of the mandibular ramus, 8=trigeminal ganglion, 9=mandibular ramus, 10=lingual branch (CN VII) from mandibular ramus (CN V) innervating duct of submandibular gland and DLPCs (para sympathetic), 11=maxillary ramus of trigeminal, 12=ophthalmic ramus, 13=vomeronasal organ, 14=olfactory epithelium, 15=fila olfactoria, 16=preolfactory lobe complex including cell bodies of nervus terminalis ganglion, 17=olfactory nerve connections (CNI), 18=complex trigeminal connections with pons, 19=sensory trigeminal/spinal connections, 20=motor trigeminal/spinal connections, 21=dorsal root ganglion, 22=brachial plexus, 23=telencephalon, 24=diencephalon, 25=mesencephalon, 26=metencephalon, 27=pons, 28=myelencephalon, 29=developing choroid plexus, 30=presumptive palatine pressure receptors. Note that neural elements 15, 16 and 17 are separate from the trigeminal system.

of Hughes and Hall (1988). On the basis of our new information we suggest a motor generator model for *Macropus eugenii* and also appropriate to other newborn Macropodidae (Fig. 46).

CONCLUDING REMARKS

Notions concerning the journey of the newborn marsupial to the pouch and of teat attachment for so long a mystery can now be resolved in terms of integrated organogenetic profiles (Fig. 47). The locomotor generator and its modulation is one of the many organogenetic profiles necessary for life support and survival. In macropodid marsupials so far investigated these events occupy about 12 days from primitive streak to birth (Renfree 1972,

MOTOR GENERATOR 1.

DEVELOPMENTALLY *SWITCHED ON*

Neurogenesis	By 6 days prepartum
Myogenesis	By 3 days prepartum
Motor kindling	From 3 days prepartum to birth

MOTOR GENERATOR 2.

AT BIRTH: PERPETUALLY PROGRAMMED INTERNALLY

MOTOR GENERATOR 3.

AT TEAT: TRIGEMINALLY *SWITCHED OFF*
AND
TRIGEMINALLY REPROGRAMMED FOR TEAT ATTACHMENT

Fig. 46. Flow diagram summarizing functional development of motor generator in *M. eugenii.*

1973; Shaw and Rose 1979; Walker and Rose 1981; Rose 1985) and characterize the G3 grade of marsupial development of Hughes and Hall (1988). This period in macropodids is twice the length of G2 type marsupials and more than four times that of G1 type marsupials such as *Sarcophilus harrisii* in which the interval from primitive streak to birth is only about 64 hours (Hughes and Hall 1988). The much longer duration of organogenesis in macropodids although at first supporting a more leisurely rate of development than that of other marsupials results in the sub-terminal days of gestation in a very much greater level of structural specialization that constitutes the highest grade of marsupial organization at birth and characterizes the G3 type of neonatal structural adaptation for survival. The distinctive macropodid organogenetic profiles can be considered to consist of multifaceted structural units. The use of these structural characters on a comparative basis for taxonomic and evolutionary considerations between marsupial groups is an exciting new prospect. Our contribution to an understanding of the journey to the pouch and the subsequent teat attachment provides the basis for a functional neuroembryological approach at a level not previously articulated.

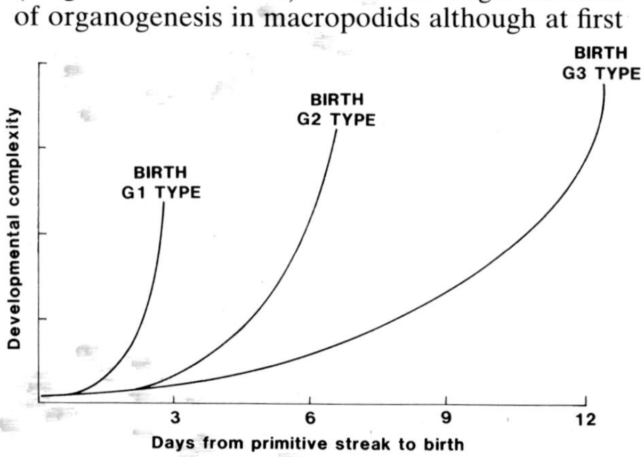

Fig. 47. Organogenetic profiles for marsupials. These discrete intervals from primitive streak stage to birth of about 3, 6 and 12 days in marsupials were shown by Hughes and Hall (1988) to be accompanied by a corresponding separation in both the cytological differentiation as well as in the dimensional complexity of organ systems attained at birth. The elements for establishing profiles of developmental complexity have been described in this paper for the G3 grade as represented by *Macropus eugenii* and a number of other macropodids. The supporting observations and literature for the G1 grade as seen in *Sarcophilus harrisii* and *Dasyurus viverrinus* together with the G2 grade as found in *Trichosurus vulpecula, Isoodon macrourus* and *didelphis virginiana* have been reviewed by Hughes and Hall (1988). The order of magnitude of the developmental complexity as indicated in Fig. 47 have been qualatitively assessed in delineating the overall changing slope of the organogenetic profile for G1, G2, G3. The detailed quantification of these profiles awaits separate investigation.

ACKNOWLEDGEMENTS

We wish to thank the following members of the technical staff of the University of Queensland for their valued assistance: Mr R. Grimmer for scanning electron microscopy; Mr C. Haller for his assistance and interpretation of nervous tissue, including para neurones and neuroepithelial cells used for transmission electron microscopy; Miss E. Rousseau and Miss L. K. Tolley for assistance with electron microscopy; Mr D. Lincoln for serial sections of light microscope preparations of embryos, L. K. Banh and T. R. Burgess for photography.

We also wish to acknowledge the help of J. Wright and G. Richards of C.S.I.R.O. Division of Wildlife and Ecology.

REFERENCES

ALCORN, G. T., 1975. Development of the ovary and urinogenital ducts in the tammar wallaby *Macropus eugenii* (Desmarest, 1817). Ph.D. Thesis. Macquarie University, Sydney.

ALCORN, G. T. AND ROBINSON, E. S., 1983. Germ cell development in female pouch young of the tammar wallaby *(Macropus eugenii)*. *J. Reprod. Fert.* **67**: 319-25.

AREY, L. B., 1974. Developmental Anatomy. 7th Edn. Saunders: Philadelphia.

BERGER, P. J., 1970. Reproductive biology of the tammar wallaby *Macropus eugenii*. Ph.D. Thesis. Tulane University, U.S.A.

CALL, R. N., CATLING, P. C. AND JANSSENS, P. A., 1980. Development of the adrenal gland in the tammar wallaby, *Macropus eugenii* (Desmarest) (Marsupialia: Macropodidae). *Aust. J. Zool.* **28**: 249-59.

CATLING, P. C. AND VINSON, G. P., 1976. Adrenocortical hormones in the neonate and pouch young of the tammar wallaby, *Macropus eugenii*. *J. Endocrin.* **69**: 447-48.

DENKER, H. W. AND TYNDALE-BISCOE, C. H., 1986. Embryo implantation and proteinase activities in a marsupial *(Macropus eugenii)*. *Cell. Tiss. Res.* **246**: 279-91.

HALL, L. S. AND HUGHES, R. L., 1985. The embryological development and cytodifferentiation of the anterior pituitary in the marsupial, *Isoodon macrourus*. *Anat. Embryol.* **172**: 353-63.

HALL, L. S. AND HUGHES, R. L., 1987. An evolutionary perspective of structural adaptations for environmental perception and utilization by the neonatal marsupials *Trichorsurus vulpecula* (Phalangeridae) and *Didelphis virginiana* (Didelphidae). Pp. 257-71 *in* Possums and Opossums: Studies in Evolution ed. by M. Archer. Surrey Beatty & Sons: New South Wales.

HILL, J. P. AND HILL, W. C. O., 1955. The growth stages of the pouch-young of the native cat *(Dasyurus viverrinus)*, together with observations on the anatomy of the newborn young. *Trans. Zool. Soc. Lond.* **28**: 349-52.

HUGHES, R. L., 1962. Reproduction in the macropod marsupial *Potorous tridactylus* (Kerr). *Aust. J. Zool.* **10**: 193-224.

HUGHES, R. L. AND HALL, L. S., 1984. Embryonic development in the common brushtail possum *Trichosurus vulpecula*. Pp. 197-212 *in* Possums and Gliders ed. by A. Smith and I. Hume. Surrey Beatty & Sons: New South Wales.

HUGHES, R. L. AND HALL, L. S., 1988. Structural adaptations of the newborn marsupial. Pp. 8-27 *in* The Developing Marsupial. Models for Biomedical Research ed. by C. H. Tyndale-Biscoe and P. A. Janssens. Springer-Verlag: Berlin.

KLIMA, M., 1987. Early development of the shoulder girdle and sternum in marsupials (Mammalia: Metatheria). *Advances in Anatomy Embryology and Cell Biology* **109**: 1-91.

LEATHERLAND, J. F. AND RENFREE, M. B., 1983. Structure of the pars distalis of adult and pouch-young tammar wallabies *(Macropus eugenii. Cell. Tiss. Res.* **227**: 439-50.

MCCRADY, E., 1938. Embryology of the opossum. *Amer. Anat. Mem.* **16**: 1-233.

MERCHANT, J. C., 1979. The effect of pregnancy on the interval between one oestrus and the next in the tammar wallaby, *Macropus eugenii. J. Reprod. Fert.* **56**: 459-63.

RENFREE, M. B., 1972. Embryo-maternal relationships in the tammar wallaby, *Macropus eugenii*. Ph.D. Thesis. Australian National University: Canberra.

RENFREE, M. B., 1973. The composition of fetal fluids of the marsupial *Macropus eugenii. Devel. Biol.* **33**: 62-79.

RENFREE, M. B. AND TYNDALE-BISCOE, C. H., 1973. Intrauterine development after diapause in the marsupial *Macropus eugenii. Devel. Biol.* **32**: 28-40.

ROSE, R. W., 1985. Reproductive biology of the Tasmanian bettong, *Bettongia gaimardi*. Ph.D. Thesis. University of Tasmania, Hobart.

SERNIA, C., HINDS, L. A. AND TYNDALE-BISCOE, C. H., 1980. Progesterone metabolism during embryonic diapause in the tammar wallaby, *Macropus eugenii. J. Reprod. Fert.* **60**: 139-47.

SHARMAN, G. B., 1955a. Studies on marsupial reproduction. II. The oestrous cycle of *Setonix brachyurus. Aust. J. Zool.* **3**: 44-55.

SHARMAN, G. B., 1955b. Studies on marsupial reproduction. III. Normal and delayed pregnancy in *Setonix brachyurus. Aust. J. Zool.* **3**: 56-70.

SHARMAN, G. B., 1955c. Studies on marsupial reproduction. IV. Delayed birth in *Protemnodon eugenii. Aust. J. Zool.* **3**: 156-61.

SHARMAN, G. B. AND PILTON, P. E., 1964. The life history and reproduction of the red kangaroo *(Megaleia rufa). Proc. Zool. Soc. Lond.* **142**: 29-48.

SHARMAN, G. B. AND CALABY, J. H., 1964. Reproductive behaviour in the red kangaroo, *Megaleia rufa,* in captivity. *C.S.I.R.O. Wildl. Res.* **9**: 58-85.

SHAW, G. AND ROSE, R. W., 1979. Delayed gestation in the potoroo *Potorous tridactylus* (Kerr). *Aust. J. Zool.* **27**: 901-12.

TYNDALE-BISCOE, C. H., HINDS, L. A. AND MCCONNELL, S. J., 1986. Seasonal breeding in a marsupial: opportunities of a new species for an old problem. *Rec. Progr. Hormone Res.* **42**: 471-512.

TYNDALE-BISCOE, C. H. AND RENFREE, M. B., 1987. Reproductive Physiology of Marsupials ed. by C. H. Tyndale-Biscoe and M. B. Renfree. Cambridge University Press: Cambridge.

WALKER, M. T., 1983. Gestational hormones and embryonic maintenence in the marsupial, *Macropus rufogriseus*. Ph.D. Thesis. University of Queensland, Brisbane.

WALKER, M. T. AND ROSE, R. W., 1981. Prenatal development after diapause in the marsupial *Macropus rufogriseus. Aust. J. Zool.* **29**: 167-87.

WALKER, M. T. AND GEMMELL, R. T., 1983. Organogenesis of the pituitary, adrenal and lung at birth in the wallaby *Macropus rufogriseus. Amer. J. Anat.* **168**: 331-44.

Cardiorespiratory structure and function in macropodid marsupials

B. J. GANNON[1] R. V. BAUDINETTE[2] S. I. C. RUNCIMAN[1] and P. B. FRAPPELL[2]

There are unusual features of the cardiovascular and respiratory systems of macropod marsupials as compared to "typical" eutherian mammals; these differences are in adult form and function, and also in the development of these organ systems. In the adult cardiovascular system, differences include persistence of a left anterior vena cava, more prominent atrial appendages, a more crescentric right ventricular lumen, a common origin of anterior and posterior mesenteric arteries, and separate aortic origins of internal and external iliac arteries. The gonadal arteries develop vascular rete with their corresponding vein, and the forelimb veins are densely developed; both these developments are thought to be thermoregulatory adaptations. Developmental division of the primitive atrium is by a single fenestrated septum, there being no equivalent of the valve of the foramen ovale and adult hearts consequently lack a fossa ovalis; closure of the interatrial communication is not complete until several days after birth. Capillaries within the central nervous system of all marsupials are present as hairpin-like loops, unique among mammals, which lack the cross anastamotic network pattern found in eutheria and monotremes.

Heart rates are lower, and stroke volumes higher than for eutheria; increases in exercise are principally by increasing rate, with maximal rates being similar to eutheria. Macropod body shape results in exaggerated systolic pressure wave reflections in the aorta, which may present significant vascular impedence problems during hopping. There is no entrainment of cardiac and locomotor frequencies during hopping, whereas respiratory and locomotor frequencies are phase locked 1:1.

While adult blood respiratory characteristics are similar to eutheria, for new born joeys, haemoglobin-oxygen affinity is markedly lower than in the adult. This affinity rises during early pouch life, and exceeds adult values during the mid-period of pouch development when pouch oxygen tensions are lowest; observed changes in blood 2-3 diphosphoglycerate levels are likely to be involved in regulating this affinity.

Septation of the developing lungs is poorly developed at birth, but the pulmonary gas exchange surface is well vascularized, and surfactant production is established; the lungs accommodate at least 96% of oxygen uptake with transcutaneous O_2 uptake being negligible. Oxygen uptake rates are lower than for newborn eutheria, reflecting the lower newborn marsupial metabolic rate which is possibly due to a thermal dependence of the joey on the mother. Respiratory frequencies for adult macropods are lower, but tidal volumes are higher than in eutheria.

It appears that macropods have adopted some different strategies from eutheria for optimizing their cardiovascular and respiratory systems to provide adequate connective and diffusive gas exchange and nutrient transport. While some of these differences are shared by marsupials in general, others may be restricted to macropods.

INTRODUCTION

CARDIOVASCULAR and respiratory function in marsupials has attracted little scientific interest compared to the reproductive, neurological, digestive and skeletal aspects of these animals. Some comparative anatomical studies of the heart and blood vessels were conducted in the nineteenth century, but such studies were often undertaken from the perspective that marsupials were phylogenetically inferior animals, which provided useful comparative material but little else. The respiratory system appears to have been largely ignored by early workers.

It is not our purpose in this short review to consider the history of this area of macropod biology, but rather to point out areas where significant differences occur in cardiorespiratory anatomy and function between macropods and the general eutherian pattern. We also attempt to suggest where the adaptiveness of these differences may lie. Where information is available, both the adult and the pouch-young, especially neonates, are discussed. Because of the early dependence on aerial breathing in marsupials, many of the adaptive differences in macropods are seen in animals still in the pouch.

CARDIOVASCULAR FUNCTION

The eutherian heart and great vessels

The full-term fetus has vascular right to left shunts, including the ductus arteriosus and the foramen ovale, which permit bypass of the pulmonary circulation before birth; the closure of these shunts effects rapid separation of the pulmonary and systemic circulations after birth, for the first time. The ductus arteriosus connects the pulmonary trunk to the aortic arch, allowing two thirds of right ventricular output to flow into the aorta pre-natally; it constricts at birth and subsequently closes anatomically, directing right ventricular flow totally through the neonatal pulmonary vascular bed (Folkow and Neil 1971).

[1]School of Medicine, Flinders University of South Australia, Bedford Park, South Australia 5042.

[2]School of Biological Sciences, Flinders University of South Australia, Bedford Park, South Australia 5042.

Pages 407-18 in KANGAROOS, WALLABIES AND RAT-KANGAROOS ed. by G. Grigg, P. Jarman and I. Hume. Surrey Beatty & Sons Pty Limited, New South Wales, Australia, 1989.

In prenatal eutherian hearts, the foramen ovale is a large opening in the stout interatrial muscular septum which separates the right from the left atrium (Fig. 1). This opening actually represents an area of the *secondary interatrial septum* which fails to develop embryologically. To the left side of the interatrial septum, the opening is overlapped by the thin flap valve of the foramen ovale; this valve developed earlier as the *primary interatrial septum,* which grows down from the "roof" (most cephalad part) of the primitive single atrium, towards the atrio-ventricular (A.V.) opening. However, prior to its reaching the A.V. opening, the most cephalad part of this growing *primary septum* becomes fenestrated and then disintegrates, forming a new *secondary (interatrial) opening* which lies above (cephalad to) the level of the foramen ovale (Moore 1982).

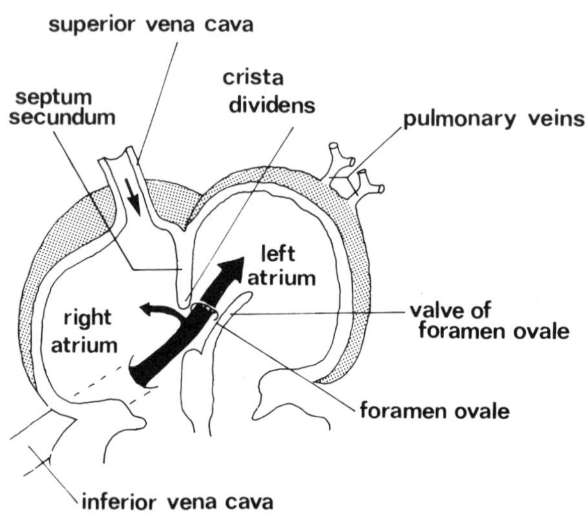

Fig. 1. Interatrial foramen (foramen ovale) in a fetal eutherian. Note the unidirectional flow of blood from right to left atrium determined by the valve of the foramen ovale. (After Moore 1982).

The foramen ovale is located adjacent to and in line with the flow direction of blood in the posterior (inferior) vena cava (PVC), as it enters the right atrium. Thus, most of the PVC blood, which is chiefly comprised of oxygenated umbilical venous blood, can bypass the right ventricle, since it passes from the right atrium through the foramen ovale and into the left atrium. In so doing, this blood deflects the flap valve of the foramen ovale (the primary interatrial septum) into the left atrium and then passes up over its free edge and through the secondary interatrial opening into the left atrium. This blood thus bypasses the fetal lungs and will be pumped by the left ventricle directly into the aorta.

Blood returning via the anterior (superior) vena cava (AVC), which is low in oxygen content, is "aimed" at the right atrio-ventricular valve. This flow passes to the right ventricle, and is pumped out the pulmonary trunk. However, most of this blood also bypasses the fetal lungs, which have a high vascular resistance. This bypass occurs through the ductus arteriosus, which connects the pulmonary trunk to the aortic arch. Thus, prenatally, less than 10% of the combined output of the two ventricles passes through the lungs (see review by Rudolph 1979).

At birth in Eutheria, placental venous return falls rapidly, markedly lowering PVC pressure and hence right atrial pressure. Simultaneously, the pulmonary vascular bed dilates following lung inflation, and there is a profound increase in pulmonary perfusion. Hence, there is increased pulmonary venous return to, and thus pressure in, the left atrium. In consequence, left atrial pressure exceeds that in the right atrium for the first time, and the flap valve of the foramen ovale is slammed closed physiologically by the reversal in the interatrial pressure gradient. Anatomical closure of the foramen ovale occurs by fusion of the endocardium of the valve with the margins of the foramen over several days to weeks; nevertheless, a noticeable depression remains in the right side of the interatrial septum — the *fossa ovalis*, which is readily apparent in adult eutheria.

In macropods, as in all other full-term uterine marsupial young which we have examined, there are analogous, but not identical cardiovascular adaptations to intra-uterine life to those found in eutheria (Runciman 1989; Baudinette, Runciman, Frappell and Gannon 1988b). In each case, there are several fenestrae in the single interatrial septum of the full-term marsupial, which allow blood to flow directly between the right and left atria (or vice versa[1]); these foramina do not possess any valve like structures (Fig. 2).

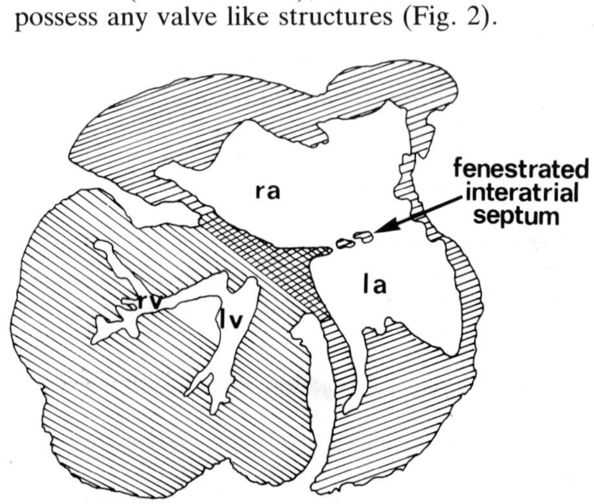

Fig. 2. Interatrial communication between right (ra) and left atria (la) in full term uterine Tammar Wallaby (26 days after removal of pouch young). Camera lucida drawing from one of a serial transverse section series. (rv and lv denote the right and left ventricles respectively).

A widely patent ductus arteriosus connects the pulmonary trunk to the aortic arch in full-term uterine macropods (Baudinette *et al.* 1988b). We presume that this connection allows most of the

Table 1. Conditions of structures permitting left-right (or right-left) vascular shunting in perinatal *M. eugenii* and laboratory mice.

Species	Age	Ductus	Inter-atrial septum	Inter-ventricular septum	Aorto-pulmonary septum
Macropus eugenii	Term embryo	Open	Incomplete	Incomplete	Incomplete
	Newborn 0-12 h	Almost closed	Incomplete	Complete	Complete
	24 hours	Closed	Incomplete	Complete	Complete
	3 days	Closed	Complete (but with fenestrations)	Complete	Complete
	14 days	Closed	Complete	Complete	Complete
Mus musculus	20 days gestation	Open	Incomplete	Incomplete	Incomplete
	Newborn	Open	Valve closed	Complete	Complete
	1 h	Open	Valve closed	Complete	Complete
	1 day	Closed	Complete	Complete	Complete
	3 days	Closed	Complete	Complete	Complete

output from the right ventricle to bypass the immature prenatal lungs and enter the systemic circuit, as in eutheria; however there is no experimental proof of the *direction* of pre-natal blood flow through either the inter-atrial foramina[1] or the ductus arteriosus in marsupials.

In Table 1, the condition of structures permitting L-R vascular shunting in the perinatal mouse, *(Mus musculus)* is compared with those in the Tammar wallaby *(Macropus eugenii)*.

While the posterior vena cava in the uterine joey must carry returning placental venous (oxygenated) blood to the right atrium, there does not appear to be any particular spatial proximity of its entrance to the atrium with the positions of the inter-atrial foramena, which might allow selective shunting of oxygenated PVC blood to the left atrium, as is found in eutheria. Furthermore, because of the persistence in marsupials of a left anterior vena cava (which joins the coronary sinus prior to its opening into the right atrium, adjacent to the PVC opening), a significant admixture of left AVC blood with PVC blood in the atrium must be anticipated before any PVC shunt flow through the inter-atrial formanena could occur. In eutheria (which lack a left AVC), the more highly oxygenated PVC blood (it is comprised chiefly of placental venous blood) is mainly channelled through the right atrium via the foramen ovale to the left atrium and thence to the left ventricle and to the aortic root, to supply the most highly oxygenated blood to the carotid and subclavian arteries. Thus eutherian fetuses are able to achieve a higher degree of oxygen saturation in blood flow to the head, brain and upper limbs, compared to the rest of the developing body; this situation is unlikely to be able to be achieved in intra-uterine macropods because of the persistence of the left AVC, admixture of this blood with PVC blood as it enters the right atrium, and the lack of a specific spatial proximity of the PVC opening to the interatrial openings.

The heart in neonatal macropods

In early macropod pouch-young, the interatrial connections seems to close gradually as a result of proliferation of tissue around the margins of the interatrial fenestrae, over a period of approximately four days after birth. Perhaps this slow closure of interatrial openings and delayed formation of physiologically discreet pulmonary and systemic circuits is important in that the delay allows similar filling pressures to be maintained in both ventricles for some time after birth in the immature joey, until adequate separate control of pulmonary and systemic vascular volumes and resistances is developed. In mice, with gestation periods similar in length to that of the Tammar, there are no functional shunts 24 hours after birth.

The closure of the ductus arteriosus after birth isolates the right from the left ventricular pressures for the first time. Prenatally, right and left ventricular pressures (RVP and LVP, respectively) are identical; however LVP rises, whereas RVP falls postnatally. Thus, the postnatal growth of the heart thickens the left ventricular wall by comparison to the right.

In typical eutheria, postnatal total body mass increases by approximately 20 fold from newborn to adult (e.g., human neonate 3.5 kg $\rightarrow$ adult 70 kg), and increases in cardiac mass are of a similar order; i.e., the linear dimensions ($\sim\sqrt[3]{\text{mass}}$) of the heart increase by a factor of approximately 2.7 times. In macropods, the increase in body

[1]It is not possible to determine in which direction blood might flow through these foramena in marsupials; in contrast due to the valve of the foramen ovale, blood can flow *only* from R$\rightarrow$L atria in eutherian fetuses.

mass is of the order of 13 000 fold between the neonate and adult (e.g., Tammar 450 mg → 6 kg) and thus the increase in cardiac linear dimensions are of the order of 24 times. This approximately 10 fold greater linear dimensional growth of the heart during the postnatal period — i.e., the period when left and right ventricles have to cope with different systolic pressures — may account for the difference between eutheria and marsupials in the anatomical relations between the left and right ventricles; in marsupials the right ventricular lumen is reduced to a more cresentic slit, which extends around the left ventricle to a noticeably greater extent than is the case in eutherians (Fig. 3). This more cresentic right ventricular lumenal shape may also account for the quadracuspid nature of the right atrio-ventricular valve in marsupials and the lack of cordae tendinae attached to the free (non septal) wall of the right ventricle (see below).

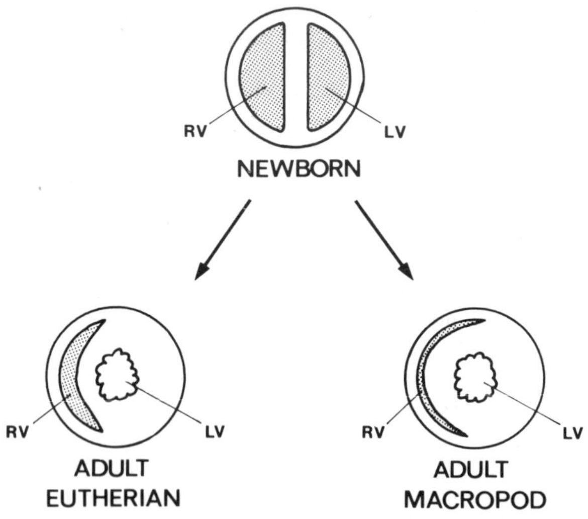

Fig. 3. Diagrammatic changes in relative positions and wall thickness of ventricles between newborn and adult in eutherians and macropodids.

In most mammals, heart rates at birth generally exceed those of the adult and during growth follow expected allometric relations. There is one report of heart rate in a neonatal marsupial macropod, the tammar, in which newborns had lower heart rates than adults and showed marked variation on postnatal day three (Baudinette, Runciman, Frappell and Gannon 1988b). Whether this initially lower rate correlates with the late separation of pulmonary and systemic circuits is not known. This phenomenon has previously been observed only in one strain of laboratory rats (Marcuse and Moore 1943).

The adult macropod heart

Several features of marsupial hearts, including those of macropods, show marked differences from the typical eutherian heart. The pericardium, for example, is suspended from the dorsal aspect of the sternum, a feature macropods share with other marsupials except the Koala (Dowd 1969) and there is little pericardial contact with the diaphragm (Owen 1868). Characteristic of the marsupial heart is a bifurcated right atrium, with the two parts surrounding the base of the aorta. This arrangement is present in members of the genus *Macropus* (Owen 1868) but not in *M. robustus* (Boardman 1941). A third difference is the absence of papillary muscles arising from the non septal wall of the right ventricle (Dowd 1969; Parsons 1868). In addition the moderator band[1] is absent in *Petrogale xanthopus* (Parsons 1868) and is described as small in *Macropus rufus* (Windle and Parsons 1898). The right atrio-ventricular valve in *Petrogale xanthopus* is quadricuspid (Parsons 1868). The right ventricle in marsupials resembles that of birds and monotremes, being much more crescentic in cross section than that in eutherians (Wade and Neely 1949). The left ventricle, the coronary circulation and the conducting system of the heart have been described for the genus *Macropus* and closely resemble the arrangement in eutherians (Blair, Davies and Francis 1942). In *Petrogale* the left atrio-ventricular or mitral valve has only two cusps (Parsons 1868) as in eutheria. The absence of a fossa ovalis or any remnant of fetal structures in the adult interatrial septum has been described in a number of species (Windle and Parsons 1898; Owen 1868; Parsons 1868; Dowd 1969). By comparison to eutheria, the marsupial right atrium has a small smooth walled portion or sinus venarum, derived embryologically from the sinus venosus, and a large atrial appendage derived from the primitive atrium. Two pulmonary veins open close together into the dorsal wall of the left atrium in *Petrogale*, as opposed to four in eutherians (Parsons 1896). This presumably represents the fact that the extent of incorporation of the proximal portions of the pulmonary veins into the primitive left atrium, thus forming the smooth walled definitive atrium, is less extensive in marsupials than in eutheria. Hence on the left side, as on the right, a greater proportion of the total atrial volume is bounded by the relatively larger rough walled auricular appendages than is the case for Eutheria.

Arterial and venous systems

In reviewing the arterial system in marsupials, Pearson (1940) commented on the marked variation

[1]Also termed the "septo-marginal trabeculum": a band of cardiac muscle traversing the right ventricular lumen from the interventricular septum to the base of the anterior papillary muscle in the lower part of the R ventricle; it contains a branch of the right bundle of His (conducting tissue) and a small coronary arterial branch.

in the manner of origin of vessels from the aortic arch, but concluded that the pattern did not reflect on the origins of the species and therefore held no taxonomic value. There has been some difference in the interpretation of the branching of the brachial artery. Parsons (1896) maintains that there are no radial or ulnar arteries, but that the brachial gives off a small superior profunda artery which supplies the humerus. In the middle of the forearm, a superficial branch arises, but this does not accompany the ulnar nerve. The brachial artery then divides into a posterior interosseous and a median artery. Owen (1868) describes this median artery as the ulnar. The thoracic aorta is similar to that in eutheria, but the coeliac, single mesenteric and two renal arteries orginate close together from the abdominal aorta, just posterior to the opening in the diaphragm (Parsons 1896). The mesenteric artery divides almost immediately into an anterior and a posterior branch. In eutheria, these latter two arteries arise independently, directly from the aorta, widely separated from each other as anterior and posterior mesenteric arteries (Owen 1868; Pearson 1940).

Another marked difference between macropods and eutheria is the different origins of the internal and external iliac arteries in macropods. In *Petrogale xanthopus* the aorta trifurcates at the level of the disc between the fifth and sixth lumbar vertebrae into the caudal and two external iliacs, There are no common iliacs as in eutheria, and the internal iliacs originate from the caudal artery at the level of the sixth lumbar vertebra (Parsons 1896; Owen 1868). In *M. giganteus*, the caudal artery branches segmentally in the distal portion of the tail and forms complex arterial clusters which interconnect on the dorsal and ventral surfaces (Young and Webster 1974).

The venous system of marsupials differs from that of eutherians in that the external jugular is particularly large in *Petrogale xanthopus* (Parsons 1896) and there are two anterior vena cavae in all species that have been investigated (e.g., Love *et al.* 1986) with the exception of *Petaurus breviceps* (Pearson 1940). The number of azygous veins varies from one, as in *Macropus rufus* to two, as in *Macropus robustus* (Windle and Parsons 1898; Boardman 1941). This variation has no taxonomic significance. Detailed descriptions of the veins of the head and neck in *M. robustus* are given by Boardman (1941). The caudal veins in macropods are paired along their complete length and may anastomose with the paired lateral tail veins (Young and Webster 1974).

Specialized regions of cardiovascular function

Several areas of the vasculature in marsupials have attracted attention because of their marked difference from the general mammalian pattern, notably vessels of the testis, the ovary, the forelimb, and the central nervous system.

1. *Reproductive Vessels*

Harrison (1948, 1949) examined the testicular artery in *M. eugenii, M. rugogriseus* and *Wallabia bicolor* and described the marked ramification of both arterial and venous vessels. This arrangement forms a vascular rete mirabile between testicular artery and vein. Setchell (1970) has suggested two possible functions for this rete, heat exchange and as a means of reducing pulse pressure in the testicular artery. Heddle and Guiler (1970) suggest that the structure is a vascular pump which, by the association of the arterial vessels with the venous and lymphatic vessels, aids the return of both blood and lymph from the pendulous testes of larger marsupial species.

In females the ovarian artery also shows an arterial plexus which intertwines with a similar venous structure draining the ovary. The function of this apparent counter-current arrangement remains obscure (see Lee and O'Shea 1977).

2. *Forelimb Vessels*

When red kangaroos are heat stressed, they spread saliva by licking the ventral surface of the body and the forelimbs. Needham, Dawson and Hales (1974) described a dense subcutaneous venous network in the latter site, which presents a large surface area for cooling of blood returning from the metacarpal veins in the paw as a result of evaporative cooling of the saliva wetted fur. Forelimb blood flow was shown to increase markedly at high ambient temperatures.

3. *Cranial Vessels*

The vessels supplying and draining the brain, and other head structures, are apparently unexceptional in macropods. Macropods have well developed nasal turbinates, and blood flow to the nasal mucosa of *Macropus rufus* is dramatically increased in environmental heat stress. There are reports in marsupials of lower brain temperatures compared to rectal temperatures in e.g., exercising possum (Baudinette 1984) and Tasmanian Devil (Shah *et al.* 1986). There is, however, no report of the presence of a carotid rete in any macropod (de Boulay and Veritz 1973), which might be utilized in brain cooling. Shah *et al.* (1986) reported a tortuous course of the intracranial internal carotid artery, which they inappropriately termed a "Marsupial carotid rete", arguing that this was analagous to the carotid *rete mirabile* found in e.g., artiodactylia and felidae. The (eutherian) rete mirabile in fact consists of a multiplicity of parallel channels, rather than the tortuous single carotid pathway within the cavernous sinus as reported by Shah *et al.* (1986).

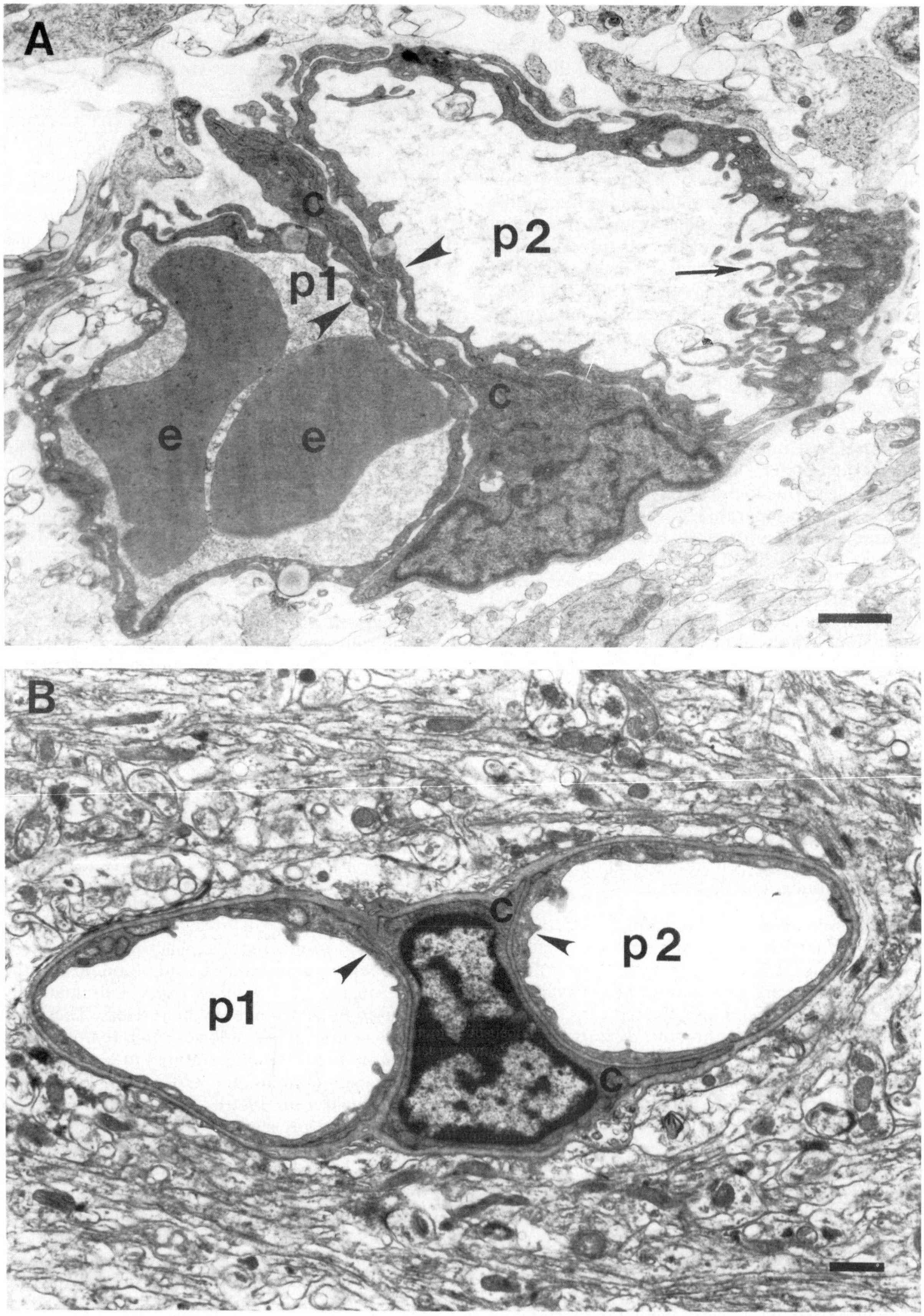

Fig. 4. Transmission electron micrographs of capillary loops in macropodid marsupial central nervous system. A: transverse section of the pair of capillary profiles (p1, p2) of a capillary loop from immersion fixed CNS tissue of *Bettongia penicillata*. Note intercalated cell process (c) intervening between the endothelial cells (arrowheads) lining the paired capillary profiles of the loop; erythrocytes (e) are evident in one profile, and endothelian protruberances into the lumen are evident in the other. B: Paired capillary profiles of a capillary loop from cerebellum of *Macropus eugenii*; perfusion fixed specimen. Note larger separation of the endothelial cells (arrowheads) of the adjacent capillary profiles (p1, p2) of the loop by the intercalated cell (c) than is evident in A above. Calibration bars = 1.0 μm.

At the level of circulation *within* the brain, however, the vascular arrangements in macropods, as in other marsupials (but not monotremes) is quite exceptional, and is totally different to that found in eutherians. In eutherians, as in most other vertebrates, the capillaries of the central nervous system (CNS) are arranged in anastamotic meshworks, such that local blood flow might arise, and exit, via almost any pathway through the brain (Scharrer 1939). Consequently, there are no *end arteries* or *end arterioles*, i.e., feeder vessels without more peripheral anastamoses, in the CNS in "typical" vertebrates. However, as first described over 50 years ago by Wislocki and Campbell (1937), the intra CNS capillaries of *all* marsupials (as is also found in certain primitive fishes, tailed amphibia, and particular lizards — Schöbl 1872; Sterzi 1907) consist entirely of hairpin-like capillary loops, which totally lack cross-anastamoses with adjacent vessels. Thus all arteries and arterioles in the marsupial CNS are end arteries. This arrangement would appear to have potentially catastrophic consequences for macropods in the case of their developing small intra arterial thrombi, which would embolize areas of brain. Since the capillaries are present as hairpin loops, the afferent and efferent capillary limbs are closely approximated, and thus appear in tissue sections as paired capillary profiles (Fig. 4). Given that capillary profile densities in e.g., native cat are similar to those in rat (Snyder *et al.* 1989), mean diffusion distances between capillary lumen and cerebral tissue in marsupials are substantially larger (Table 2) and maximal diffusion distances almost exactly twice those encountered in equivalent areas of eutherian brain, since the capillary profiles in marsupial brains are invariably arranged as adjacent pairs. The precise reasons why capillary loops should exist in marsupial CNS, but not in the peripheral nervous system (or the cranial nerves except the ophthalmic) in the same animals is not known. Possible advantages of this capillary loop arrangement include buffering the brain against changing plasma concentrations of diffusible substances, including lipid soluble substances, which are able to transverse the blood-brain barrier, but which would short circuit from the afferent limb to the efferent limb of the capillary pair in a capillary loop arrangement. Ultra-structural investigations of the marsupial CNS capillary loops reveals continuous, unfenestrated endothelia, with diffusion distances between afferent and efferent limbs typically of less than 1.0 μm. There is a fine intercalated cell process between the basal lamina of the afferent and efferent limb endothelia of brain capillary loops (Fig. 4; Gannon, White and Baudinette, unpublished). The ultra-structure and development of marsupial brain capillary loops remains to be studied in detail; of particular interest will be the microanatomical relationships between capillary loops, glial cell processes and neurones.

Table 2. Capillary profile density in the central nervous system of dasyurid marsupial and rat (from Snyder *et al.* 1989).

	Dasyurid marsupial	Rat
Capillary Profile Density (caps mm^{-2})	428.2±26.4	428.4±19.4
Total Capillary Area (% Tissue Area)	3.5±0.6	3.3±0.4
Diffusion Distance (μm)	22.2±1.7	15.5±0.7*

Numbers are means ±SEM.
*indicates values significantly different at $P<0.05$.

The retina of the eye is derived embryologically as a part of forebrain. Hence its capillary blood vessels and in particular whether these are present as loops in macropods are of interest. Vascular supply to the retina in mammals is, in general derived from two sources — (a) the coroidal circulation which lies behind the outer segments of the rods and cones (between these and the sclera of the eye), and (b) the retinal circulation which travels to the retina with the ophthalmic nerve, and ramifies on the inner surface of the retina, between it and the vitreous humour. In marsupials in general, while there is universal presence of a choroidal circulation, the degree of development of a retinal circulation varies from its being absent, except at the optic disc (the entry of the optic nerve to the retina) (e.g., in diprotodonts — e.g., wombats), to being well developed throughout the retina, (as in Dasyuridae); in macropods, the retinal circulation is restricted to a few clumps of small vessels in the immediate vicinity of the optic disc (Johnson 1901). It will be of interest to determine whether, in macropods in particular and in marsupials in general, the capillary vessels of the retina and indeed also of the choroid are present as capillary loops; such an arrangement of retinal capillaries as capillary loops is not reported for other mammals.

Detailed studies of the blood brain barrier in macropods, as in marsupials in general, are lacking. Our own studies (White, Gannon and Baudinette unpublished) indicate that in general the cerebral microvessels are impervious to Trypan (or Evans) Blue-Albumin, as is the case in eutheria. In eutheria, there are specific brain regions, known collectively as the *circumventricular organs* (CVOs) in which this barrier is lacking; capillaries in these regions in eutheria are fenestrated and relatively permeable to macromolecular permeability tracers (e.g., trypan blue-albumin), in contrast to the continuous capillaries of all the other eutherian brain regions, which possess a blood brain barrier (Gross and Weindl 1987). In Macropods, our preliminary studies (White, Gannon and Baudinette unpublished) indicate that, in the brain regions equivalent to the CVOs in eutheria (e.g., area postrema, pineal) which we have examined, the capillaries are present as capillary loops; furthermore these capillaries are virtually impervious to macromolecular tracers, in contrast to the highly permeable fenestrated capillaries found in CVOs of Eutheria. Whether capillaries in marsupial "CVOs" are fenestrated awaits investigation.

Minimum heart rates and stroke volumes in marsupials

Heart rates and cardiac outputs were first measured in macropodids by Maxwell, Elliot and Kneebone (1964). Their results from anaesthetized animals suggested few differences in these variables from eutherian values. Apart from the effects of the anaesthetic, the impact of this study is somewhat reduced by the pooling of results from three species into a generalized "kangaroo". Using conscious animals of 14 marsupial species, including six macropodids, Kinnear and Brown (1967) demonstrated resting heart rates less than half the values expected from eutherians of equivalent body mass (Fig. 5). At this time these results supported earlier findings from nutritional studies which suggested that marsupials had low rates of metabolism, a fact later confirmed by measures of oxygen consumption by MacMillen and Nelson (1969) and Dawson and Hulbert (1970).

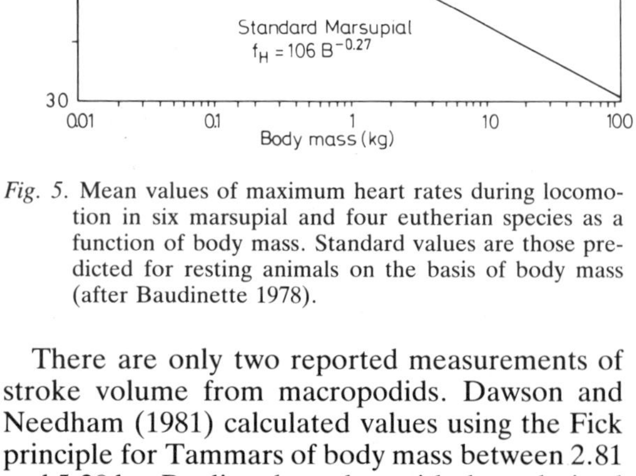

Fig. 5. Mean values of maximum heart rates during locomotion in six marsupial and four eutherian species as a function of body mass. Standard values are those predicted for resting animals on the basis of body mass (after Baudinette 1978).

There are only two reported measurements of stroke volume from macropodids. Dawson and Needham (1981) calculated values using the Fick principle for Tammars of body mass between 2.81 and 5.28 kg. Pooling these data with those derived in an identical manner from Brushtail Possums *(Trichosurus vulpecula)*, enabled the derivation of an allometric eguation;

$$y = 1.22M^{1.07}$$

where y is the stroke volume in ml and M the body mass in kg. Values predicted for marsupials from this equation are about 1.5 times the expected eutherian value. The only other reported value for stroke volume in a macropodid is also from the Tammar (Baudinette *et al.* 1987). Using a dye injected into the pulmonary artery and measuring its subsequent dilution, they reported mean stroke volumes from five resting animals of 6.4 mls, again a value greater than that expected for an eutherian of equivalent body mass.

So in one species of macropodid, the Tammar, a lower resting heart rate than eutherians is accompanied by a higher stroke volume. It will be of interest to determine whether this different cardiac pumping "strategy" holds for other macropodids.

Cardiac function during exercise

In the Tammar, cardiac output measured from tracer dilution showed a significant increase with hopping speed, but this was a consequence of increased heart rate rather than of stroke volume (Baudinette *et al.* 1987). A similar finding had earlier been reported for the Brushtail possum (Baudinette, Seymour and Orbach 1978). Maximum heart rates during exercise have been shown to be similar in marsupials and eutherians (Fig. 4; Baudinette 1978); the analysis included two macropodids, the Tammar and the Rufous Bettong, *Aepyprymnus rufescens.* Since resting heart rates of marsupials are lower than in eutherians, the implication from this study is that the former have a greater capacity to increase heart rates than the latter.

Heart rates in the Rufous Bettong (Baudinette 1978) and the Tammar (Baudinette 1978; Baudinette *et al.* 1987) increase linearly with hopping speed and show no differences in this regard to established eutherian patterns. There is no evidence of entrainment of cardiac and gait frequencies in the latter species, a, perhaps, surprising finding discussed further in the following section.

Propagation of the pressure pulse in the kangaroo arterial system

Body size and shape are major determinants of arterial pressure pulse contours (see review by O'Rourke 1982), and variation between species has been attributed to differences in wave reflection from muscle mass. In extreme cases such as the Wombat, a discrete wave reflection results in a prominent secondary aortic pressure wave, whereas in snakes the wave reflection is minimal.

Kangaroos *(M. robustus)* have more marked secondary aortic pressure wave reflections than any other species examined (Fig. 6). The upper

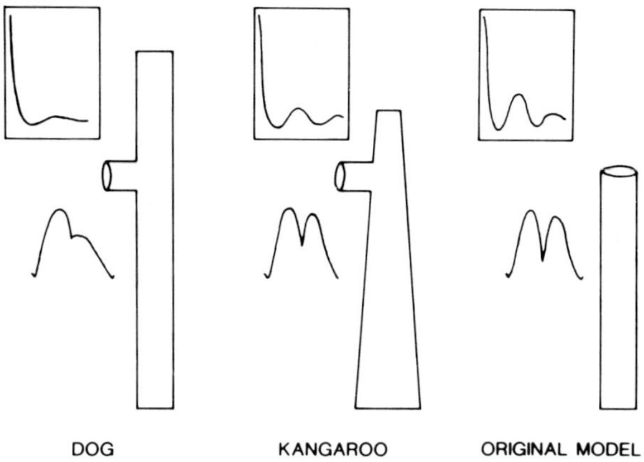

Fig. 6. A model of the arterial system of the kangaroo (centre) compared with two models proposed for eutherians. The insets show the impedance modulus patterns (above) and the pressure wave contour (below) for each model. (From Avolio, Nichols and O'Rourke 1984).

body with its lack of musculature and small brachiocephalic and subclavian arteries provides reflections which are insignificant compared to those from the lower body mass. The effect of this conical body shape is so marked that peak diastolic pressure may exceed peak systolic pressure (Avolio, Nichols and O'Rourke 1984, 1985).

Baudinette *et al.* (1987) have commented on the accentuation of this effect during exercise. In contrast to quadrupedal locomotion, hopping results in opposing sets of muscles acting synchronously. When the animal lands, most of the muscles of the hind limbs will be in an active state and vascular impedance will be very high. This effect should be additive to the effects of shape mentioned above and produce an even more marked reflected pressure wave. It therefore seemed a reasonable hypothesis that locomotory and cardiac frequencies should be phase-locked in order to reduce the potentially large, damaging excursions in arterial pressure. The results of the study failed to support this hypothesis; in the hopping Tammar there is no entrainment of gait and heart frequencies. Unpublished results from our laboratory in which locomotion was simulated in anesthetized animals by stimulating muscles, showed that peak arterial pressures occurred about the origin of the renal arteries. The compliance of the abdominal aorta at this point is greater than that reported for the rabbit and the dog, but a full understanding of the mechanisms responsible for the dissipation of potentially damaging reflected waves is not yet available.

Blood respiratory charactersitics during development in macropodids

There are two published reports of the oxygen carrying properties of marsupial blood in the early postnatal period, both of them on the Tammar. Baudinette *et al.* (1988a) determined haemoglobin-oxygen equilibrium curves and levels of 2,3-diphosphoglycerate for Tammar joeys and related these determinations to the oxygen levels in the pouch and to the metabolic rate of the animals during development over 200 days of pouch life. The oxygen equilibrium curves show a high oxygen tension at which the haemoglobin is 50% saturated (P50) in the neonate, indicating a low affinity of haemoglobin for oxygen. However by three weeks, the P50 levels have reduced to levels typical of the adult (Fig. 7). The intracellular organic phosphate, 2,3-diphosphoglycerate, a molecule which effectively reduces the O2-Haemoglobin affinity, is present in high concentrations in the neonates, but then, consistent with a decrease in P50, the levels reduce to those found in the adult. Holland *et al.* (1988) found similar results for haemoglobin properties and commented on the unusual shape of the Hb-O2 curves in pouch-young less than four days of age, indicating a high degree of interaction between subunits of the molecule. The result of this effect is a steeper equilibrium curve above 50% saturation, compensating in part for the rightward shift of the curve as a whole.

In eutherians, haemoglobin-oxygen affinity is generally high in utero and in the immediate postnatal period. Consistent with this is a low level of 2,3-DPG at birth. It would seem then that macropodids and eutheriens have adopted different strategies for haemoglobin function. The former can accommodate a low affinity system in order to supply rapidly growing tissues because of the relatively oxygen rich atmosphere of the pouch, which in the Tammar rarely falls below 100 torr (Baudinette *et al.* 1988a). The eutherian system is geared towards a high affinity haemoglobin system in the perinatal period, dictated by the low (50 torr) levels of oxygen in the intervillous space of the placenta.

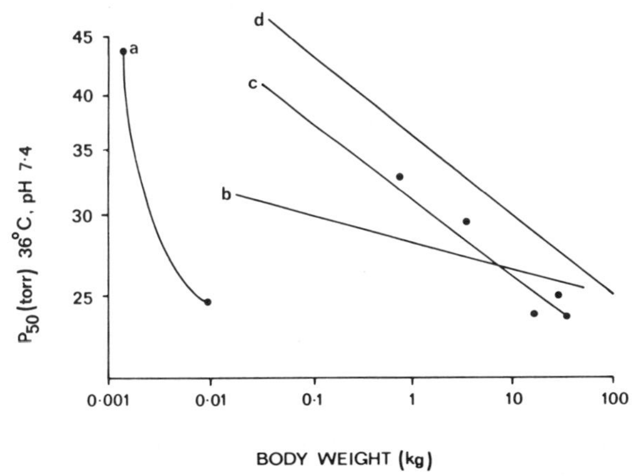

Fig. 7. Haemoglobin P50 values as a function of body mass for (a) Tammar joeys (Baudinette *et al.* 1988), (b) Eutherians (Bartels 1964), (c) marsupials (Bland and Holland 1977) and (d) Eutherians (Schmidt-Nielsen and Larimer 1958). The points indicate data from macropodids included in the latter study.

Richardson and Russell (1969) and Holland *et al.* (1988) have investigated haemoglobin types and red cell morphology in macropodids. Nucleated erythocytes are present in the days following birth in *M. rufus, M. giganteus, M. robstus* and *M. eugenii*. Associated with these cells are embryonic haemoglobins, four in *M. eugenii* and at least one in the larger species. Replacement of these haemoglobins by the adult form of the molecule is complete by day 13 in the Tammar and day 40 in the three larger species.

Blood respiratory characteristics in the adult

Studies of the respiratory properties of adult macropodid blood began with those of Bartels *et al.* (1966) on *M. giganteus.* Bland and Holland (1977) extended the study to seven marsupial species, including five macropodids, and derived an allometric relationship relating haemoglobin P50 to body mass (Fig. 7). Two derivations are available from eutherians with which to compare the line for marsupials. That of Bartels (1964), indicated as (b) in Figure 6, shows an apparent difference in the slope of the line relating P50 and body mass. An alternative relationship, that of Schmidt-Nielsen and Larimer (1958), shows a much closer correlation with the marsupial line and suggests that in this regard marsupials and eutherians do not differ.

Bland and Holland (1977) found 2,3 DPG present in the cells of all the macropodid species and showed that it is a regulator of haemoglobin-O2 affinity, just as in eutherians.

RESPIRATORY FUNCTION

Pouch gas levels

Oxygen levels have been measured in the pouches of Tammar wallabies containing young over a period of 200 days (Baudinette *et al.* 1988a). The concentration of oxygen was close to atmospheric levels for pouches which contained joeys up to 20 days of age. Around 120 to 160 days the oxygen levels reached their lowest levels, equivalent to about 18 per cent, then returned to values close to atmospheric levels. In this latter phase, diffusion through the pouch wall and opening is augmented by ventilation associated with the movement of the joey and licking by the mother.

This pattern is similar to that reported for the Brushtail Possum by Bailey and Dunnet (1980), but differs from the situation in the Virginia Opossum in which oxygen and carbon dioxide levels remain steady after 20 days of pouch life (Farber and Tenney 1971).

Respiratory function in neonatal marsupials

The lung in most new-born mammals is structurally immature, yet functions as a gas exchanger. The septation process, which results in the formation of alveoli, occurs primarily in the first few days after birth, for example between day 5 and 10 in the rat (see review by Burri 1984). In the newborn Tammar, the lung parenchyma is still at the terminal air sac stage and alveoli formation is not complete until around day 75 (Randall *et al.* 1984; Baudinette *et al.* 1988b). In the first few days after birth, the lung is characterised by large air sacs up to 350 μm diameter, with no stainable collagen or elastin. By 33 days some cartilage has begun to form around the largest conducting airways and the development of smaller air chambers has begun. This pattern of development is slower than that seen in even altricial eutherians, but it is not known if the Tammar is typical of the sequence seen in other macropodids.

Despite the slow development of alveolar tissue, pulmonary surfactant is present in the Tammar from soon after birth. Ribbons *et al.* (1989) have demonstrated type II cells which are responsible for the synthesis and secretion of surfactant as well as the lamellar bodies which characterize surfactant release. Furthermore these authors have shown that the composition of the surfactant in this species has the same basic fatty acid profile as in the rat, but shows less change over the first thirty days in phospholipid distribution.

Skin erythema in joeys has lead to suggestions that cutaneous respiration may be important in the early days of pouch life. The relative infrequency of skin capillaries and their distance from the surface (Randall *et al.* 1984), as well as direct measurements of gas exchange across skin and lungs (Baudinette *et al.* 1987), has shown this assumption to be wrong. Even in Tammar joeys of mass less than one gram, Baudinette *et al.* (1988a) have shown that cutaneous gas exchange contributes only about four per cent of the total oxygen exchange.

Little is known of respiratory variables in neonatal marsupials or the development of the chemical control of breathing. In contrast, the dynamics of breathing in newborn eutherians is well documented and allometric relationships have been derived relating these variables to body mass (see review by Mortola 1987). For macropodids the only published data are from the Tammar (Baudinette *et al.* 1988b). In seven day old young of this species, the tidal volume was around seven μl, but by 50 days had increased to around 250 μl. Over this time period respiration rate was between 30 and 60 breaths per minute and was independent of body mass.

Respiratory rates and tidal volumes in adult macropodids

In light of lower metabolic and heart rates in adult macropodids, perhaps we would expect lower minute volumes than in eutherians of similar body mass. Maxwell, Elliot and Kneebone (1964) found this to be so in their lumped sample of kangaroo species and in the Tammar. However these measurements were conducted on anaesthetized animals. Dawson and Needham (1981) used the Fick principle and measurements of oxygen consumption to derive values for tidal volume and respiratory rates in marsupials. Tidal volumes were estimated to be greater than in eutherians of similar body mass; however respiration rates were less. These authors suggested that eutherians and marsupials differ in cardiorespiratory function in that stroke and tidal volumes are greater in the latter, whereas heart rate and respiratory frequencies are lower.

These suggestions from indirect measurements are generally consistent with measurements of tidal and stroke volumes in the Tammar (Baudinette *et al.* 1987). In this study stroke volume was measured by a thermal dilution technique, using a catheter in the right ventricle. Tidal volume was assessed from changes in electrical impedance across the chest wall during a respiratory cycle. In resting animals both measurements were greater than those expected from allometric predictions in eutherians. Support for this conclusion also comes from studies of other marsupials (Nicol and Maskrey 1977; and Nagle and Dawson [unpubl.] in Dawson and Needham 1981).

There is only one published report of respiratory function during locomotion in a macropodid. Baudinette *et al.* (1987) measured tidal volume and respiratory rate in Tammars hopping on a motor-driven treadmill. As discussed in the chapter by Baudinette in this volume, the latter variable was phase and frequency locked with the stride frequency. From rest to a hopping speed of 2.5 ms^{-1}, tidal volume increased by about 50%; however this value showed no further increase with speed, a finding consistent with the independence of oxygen consumption and speed over the same range.

SUMMARY

It is apparent that significant differences exist in cardiorespiratory structure and function between macropodid marsupials and eutherians. Unfortunately many of the reported differences are based on findings from few species and their extension to the group as a whole remains hazardous. Much of the variation relates to developmental differences, such as a short period of gestation and life in the pouch, or to bipedal locomotion. Some differences relate to the lower oxygen demands of macropodids at rest, compared to eutherians. For others, the significance is unknown and may simply reflect evolutionary alternatives in the convective and diffusive complexities of providing efficient respiratory gas exchange to all peripheral cells.

ACKNOWLEDGEMENTS

Our own work summarized in this review has been funded by The Australian Research Grants Scheme and The Flinders University Research Committee.

REFERENCES

Avolio, A. P., Nichols, W. W. and O'Rourke, M. F., 1984. Exaggerated wave reflection in the kangaroo simulates arterial counterpulsation. *Am. J. Physiol.* **246:** R267-70.

Avolio, A. P., Nichols, W. W. and O'Rourke, M. F., 1985. Propagation of pressure pulse in kangaroo arterial system. *J. Physiol.* **249:** R335-40.

Bailey, S. W. and Dunnett, G. M., 1960. The gaseous environment of the pouch young of the brush-tailed possum, *Trichosurus vulpecula* Kerr. *C.S.I.R.O. Wildl. Res.* **5:** 149-51.

Bartles, H., 1964. Comparative physiology of gas transport in mammals. *Lancet (Lond).* **II:** 599-604.

Bartels, H., Riegel, K., Kleihauer, E. and Lang, E. M., 1965. Comparative studies of the respiratory functions of mammalian blood. 2. Marsupialia: Great grey kangaroos and Tasmanian devil. *Respir. Physiol.* **1:** 145-50.

Baudinette, R. V., 1978. Scaling of heart rate during locomotion in mammals. *J. Comp. Physiol.* **127:** 337-42.

Baudinette, R. V., 1984. Thermoregulation in marsupials during exercise. *In* Thermal Physiology ed. by J. R. S. Hales. Raven Press: New York.

Baudinette, R. V., Gannon, B. J., Runciman, W. B., Wells, S. and Love, B. J., 1987. Do cardiorespiratory frequencies show entrainment with hopping in the tammar wallaby? **129:** 251-63.

Baudinette, R. V., Gannon, B. J., Ryall, R. G. and Frappell, P. B., 1988a. Changes in metabolic rates and blood repiratory characteristics during pouch development of a marsupial, *Macropus eugenii. Respir. Physiol.* **72:** 219-28.

Baudinette, R. V., Runciman, S. I. C., Frappell, P. B. and Gannon, B. J., 1988b. Development of the Marsupial Cardiorespiratory system. *In* The Developing Marsupial. Models for Biomedical Research ed. by C. H. Tyndale-Biscoe and P. Janssens. Springer: Berlin.

Baudinette, R. V., Seymour, R. S. and Orbach, J., 1978. Cardiovascular responses to exercise in the brush-tailed possum. *J. Comp. Physiol.* **124:** 143-47.

Blair, D. M., Davis, F. and Francis, E. T. B., 1942. The conducting system of the marsupial heart. *Trans. Roy. Soc. Edin.* **60:** 629-37.

Bland, D. K. and Holland, R. A. B., 1977. Oxygen affinity and 2,3-diphosphoglycerate in blood of Australian marsupials of differing body size. *Repir. Physiol.* **31:** 270-90.

Boardman, W., 1941. On the anatomy and functional adaptation of the thorax and pectoral girdle in the wallaroo, *Macropus robustus. Proc. Linn. Soc. N.S.W* **66:** 349-87.

Burri, P. H., 1984. Foetal and postnatal development of the lung. *Ann. Rev. Physiol.* **46:** 617-28.

Dawson, T. J. and Hulbert, A. J., 1970. Standard metabolism, body temperature and surface areas of Australian marsupials. *Am. J. Physiol.* **218:** 1233-38.

DAWSON, T. J. AND NEEDHAM, A. D., 1981. Cardiovascular characteristics of two resting marsupials: An insight into the cardio-respiratory allometry of marsupials. *J. Comp. Physiol.* **145:** 95-100.

DOWD, D. A., 1969. Gross features of the heart of a marsupial, *Trichosurus vulpecula. Acta. Anat.* **74:** 454-71.

DU BOULAY, G. H. AND VERITY, P. M., 1973. The cranial arteries of mammals. Whitefriars Press Ltd: London.

FARBER, J. P. AND TENNEY, S. M., 1971. The pouch gas of the Virginia opossum *Didelphis virginiana. Respir. Physiol.* **11:** 334-45.

FOLKOW F. B. AND NEIL, E., 1971. Circulation. Oxford University Press.

GROSS AND WIENDEL, 1987. Peering through the windows of the brain. *J. Cereb. blood flow Metab.* **7:** 663-72.

HARRISON, R. G., 1948. Vascular patterns in the testis, with particular reference to *Macropus. Nature.* **161:** 399.

HARRISON, R. G., 1949. The comparative anatomy of the blood supply of the mammalian testis. *Proc. Zool. Soc. Lond.* **119:** 325-44.

HEDDLE, R. W. L. AND GUILER, E. R., 1970. The form and function of the testicular rete mirabile of marsupials. *Comp. Biochem. Physiol.* **25:** 415-25.

HOLLAND, R. A. B. AND ROSS, J. A., 1978. Carbon monoxide:oxygen affinity ratio in blood of Australian marsupials and monotremes. *Comp. Biochem. Physiol.* **59A:** 343-46.

HOLLAND, R. A. B., RIMES, A. F., COMIS, A. AND TYNDALE-BISCOE, C. H., 1988. Oxygen carriage and carbonic anhydrase activity in the blood of a marsupial, the tammar wallaby *Macropus eugenii* during early development. *Respir. Physiol.* **73:** 69-86.

JOHNSON, G. L., 1901. Contribution to the comparative anatomy of the mammalian eye, chiefly based on opthalmoscopic examination. *Proc. Roy. Soc. B.* **194:** 1-82.

KINNEAR, J. E. AND BROWN, G. D., 1967. Minimum heart rates of marsupials. *Nature.* **215:** 1501.

LEE, C. S. AND O'SHEA, J. D., 1977. Observations on the vasculature of the reproductive tract in some Australian marsupials. *J. Morpol.* **154:** 95-114.

LOVE, J. B., GANNON, B., RUNCIMAN, W. B. AND SKOWRONSKI, G. A., 1986. Caught on the hop? *Chest.* **87:** 843-44.

MACMILLEN, R. E. AND NELSON, J. E., 1969. Bioenergetics and body size in dasyurid marsupials. *Am. J. Physiol.* **217:** 1246-51.

MARCUSE, F. L. AND MOORE, A. U., 1943. Low heart rate in the newborn rat. *Amer. J. Physiol.* **139:** 49-51.

MAXWELL, G. M., ELLIOTT, R. B. AND KNEEBONE, G. B., 1964. Hemodynamics of kangaroos and wallabies. *Am. J. Physiol.* **206**(5): 967-70.

MOORE, K. L., 1982. The Developing Human: Clinically Oriented Embryology. 3rd. Ed. W. B. Saunders.: Philadelphia.

MORTOLA, J. P., 1987. Dynamics of breathing in newborn mammals. *Physiol. Rev.* **67**(1): 187-243.

NEEDHAM, A. D., DAWSON, T. J. AND HALES, J. R. S., 1974. Forelimb blood flow and saliva spreading in the thermoregulation of the red kangaroo, *Megaleia rufa. Comp. Biochem. Physiol.* **49A:** 555-65.

NICOL, S. C. AND MASKREY, M., 1977. Panting in small mammals: a comparison of two marsupials and the laboratory rabbit. *J. Appl. Physiol.* **42:** 537-44.

O'ROURKE, M. F., 1982. Vascular impedance in studies of arterial and cardiac function. *Physiol. Rev.* **62**(2): 570-622.

OWEN, R., 1868. On the Anatomy of Vertebrates: Vol. III Mammals. Longmans: London.

PARSONS, F. G., 1896. On the anatomy of *Petrogale xanthopus,* compared with that of other kangaroos. *Proc. Zool. Soc. Lond.* **1896:** 683-714.

PEARSON, J., 1940. Notes on the blood system of the Marsupialia. *Proc. Roy. Soc. Tas.* **1939:** 77-94.

RANDALL, D., GANNON, B. J., RUNCIMAN, S. I. C. AND BAUDINETTE, R. V., 1984. Gas transfer by the neonate in the pouch of the tammar wallaby, *Macropus eugenii. In* Respiration and Metabolism of Embryonic Vertebrates ed. by R. S. Seymour. Dr. W. Junk Publishers.

RIBBONS, K. A., BAUDINETTE, R. V. AND MCMURCHIE, E. J. The development of pulmonary surfactant lipids in a neonatal marsupial and the rat. In Review.

RICHARDSON, B. J. AND RUSSELL, E. M., 1969. Changes with age in the proportion of nucleated red blood cell types and in the type of haemoglobin in kangaroo pouch young. *Aust. J. Exp. Biol. med. Sci.* **47:** 573-80.

RUDOLPH, A. N., 1979. Fetal and neonatal pulmonary circulation. *Ann. Rev. Physiol.* **41:** 383-95.

RUNCIMAN, S., 1989. Central Cardiovascular Shunts in Perinatal Marsupials. *Proc. A.P.P.S.* **20:** 58-65.

SCHARRER, E., 1939. The functional significance of the capillary bed in the brain of the opossum. *Anat. Rec.* **75:** 319-40.

SCHMIDT-NIELSEN, K. AND LARIMER, J. L., 1958. Oxygen dissociation curves of mammalian blood in relation to body size and oxygen consumption. *Am. J. Physiol.* **200:** 746-50.

SCHÖBL, J., 1878. Über eine eingenthumliche Schleifendildung der blutgefasse im Gehirn und Ruckenmark der Saurier. *Arch. Mikrosk. Anat. Entwich.* **15:** 60-64.

SETCHELL, B. P., 1970. Testicular blood supply, lymphatic drainage, and secretion of fluid. *In* The Testis ed. by A. D. Johnson. Vol. 1. Academic Press: New York.

SHAH, S. K. H., NICOL, S. C. AND SWAIN, R., 1986. Functional morphology of the crainal vasculature and the nasal passage in the Tasmanian Devil, *Sarcophilus harrisii* (Marsupialia and Dasyuridae): a marsupial carotid rete? *Aust. J. Zool.* **34:** 125-33.

SNYDER, G. K., GANNON, B. J., BAUDINETTE, R. V. AND NELSON, J. E., 1989. Cerebral vasculature in a dasyurid marsupial. *J. Exp. Zool.* (in press).

STERZI, G., 1907. Il Systema Nervosa Dei Vertebratai. A. Draghi, Padova.

WADE, O. AND NEELEY, P., 1949. Heart and attached vessels of the opossum, a marsupial. *J. Mammal.* **30:** 111-16.

WINDLE, B. C. A. AND PARSONS, F. G., 1898. On the anatomy of *Macropus rufus. J. Anat.* **32:** 119-34.

WISLOCKI, G. B. AND CAMPBELL, A. C. P., 1939. The unusual manner of vascularization of the brain of the opossum *Didelphys virginiana. Anat. Rec.* **67:** 177-91.

YOUNG, R. J. AND WEBSTER, M. E. D., 1974. Skin temperatures and vascular arrangements in the kangaroo tail, *Macropus giganteus. Proc. Aust. Physiol. Pharmacol. Soc.* **5:** 126-27.

Testicular development in the Allied Rock-wallaby *Petrogale assimilis*

J. N. BELL[1], R. L. CLOSE[2] and P. M. JOHNSON[3]

Petrogale of known age were used to establish a relationship between molar eruption stages and age, so that ages of wild-caught animals could be estimated. Eruption of the final molar was not complete before the age of 6.5 years. Full spermatogenic development and increased testis size commonly occur at 23-27 months of age in wild populations of *P. assimilis*. However, captive males of this species have sired young at 20 months of age. Two wild-caught males were sterile: one aged more than 6.5 years, the other, estimated at 4-4.5 years of age, was captured at the contact zone between two chromosomal races.

INTRODUCTION

COLONIES of the Allied Rock-wallaby (*Petrogale assimilis*) are found on rocky outcrops and gorges in northeastern Queensland. Three parapatric chromosome races are currently recognized within the species, the assimilis, Mt Claro and Mareeba races (Eldridge *et al.* 1988). Each race differs from the others by at least one chromosome rearrangement but all are morphologically similar and have cross bred readily in captivity at Macquarie University (Sharman *et al.* 1989).

Testicular development in chromosomally pure adults from the three races of *P. assimilis* was studied and the age of each individual examined was estimated from data of captive *Petrogale* of known age.

METHODS AND MATERIALS

Testes from 16 *P. assimilis* (10 assimilis race, four Mt Claro race and two Mareeba race) were collected between 1981 and 1985, during April to July. Animals were shot from 12 localities (see Fig. 1) and the tissues fixed in 10% formalin or Bouin's solution. The karyotypes of all animals were determined from standard bone-marrow preparations obtained from the sternum or femur.

Tissues were prepared by standard histological techniques, sectioned at 7 μm and stained in Ehrlich's haematoxylin and eosin. Diameters of transverse sections of seminiferous tubules were measured and the degree of spermatogenic development in the tubules was assessed using a modification of the categories of Poole and Catling (1974).

Field records of 18 animals collected from 1976 to 1986 provided information of testicular measurements and the corresponding molar eruption stage (MES) of animals for which testes were not preserved. These 18 comprised 10 assimilis race, six Mt Claro and two Mareeba race animals (see Table 1).

Captive animals of eastern *Petrogale* species were used to estimate age and provide data on the onset of reproductive maturity. These animals were either collected from the wild or bred in captivity at Macquarie University Fauna Park or QNPWS Townsville, and included *P. godmani*, *P. assimilis* (three races), *P. purpureicollis* and *P. penicillata*.

Age determination was calculated using the method of Sharman *et al.* (1964). The molar eruption stages of 15 animals of known age were recorded. Ten of these were born in captivity, the remainder were collected from the wild at the end of pouch life and their ages estimated from head length (unpublished data).

Testis size was measured regularly in immature captive males using vernier calipers. To record the age at which captive males can sire offspring, young males approaching maturity were housed alone with adult females.

RESULTS

Figure 2 shows the curve fitted by regression analysis to the plot of MES against age for 15 captive-raised animals ($y=0.75+0.291A$ [$r^2=96.9\%$], where y=log of age in months and A=MES). Eruption of the final molar (M_4) was not completed before the age of 6.5 years. An older animal of MES IV, aged 11 years and 2 months was not included in this figure.

This relationship was used to assess the age at sexual maturity in natural populations of *P. assimilis*. Full spermatogenic development (Fig. 3) and increased testis size (Table 1) occur at MES II.2 (23-27 months). Only one of five animals of MES<II.2 had shown testis

[1]School of Biological Sciences, Macquarie University, New South Wales 2109.
[2]Division of Science and Mathematics, Macarthur Institute of Higher Education, Campbelltown, New South Wales 2560.
[3]National Parks and Wildlife Service, Northern Regional Centre, Pallarenda, Townsville, Queensland 4810.

Pages 419-22 in KANGAROOS, WALLABIES AND RAT-KANGAROOS ed. by G. Grigg, P. Jarman and I. Hume. Surrey Beatty & Sons Pty Limited, New South Wales, Australia, 1989.

Allied Rock-wallaby *Petrogale assimilis* (race), Mt Stuart, Townsville. Photograph taken by R. L. Close.

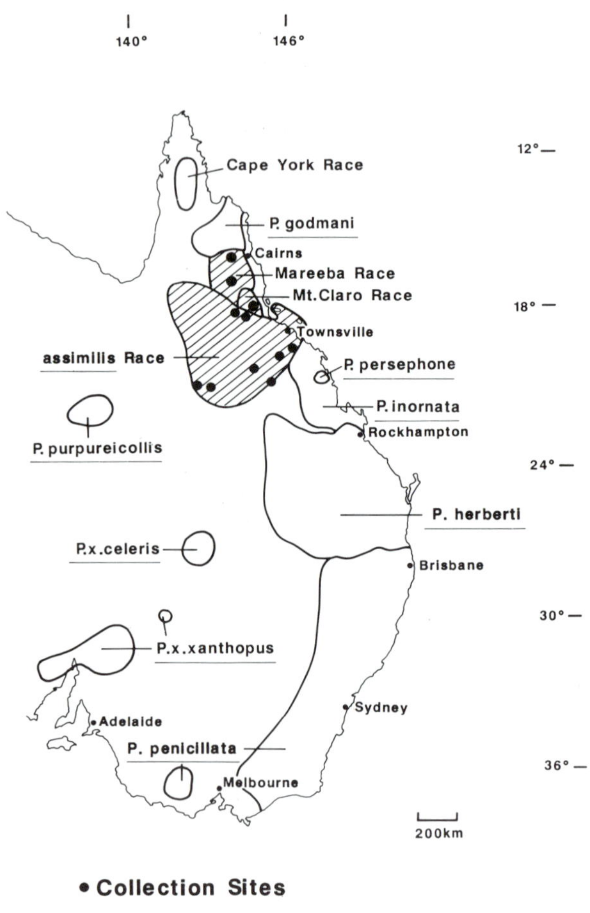

Fig. 1 (left). Distribution of *Petrogale* chromosomal races in eastern Australia. Three races of *P. assimilis* (assimilis, Mt Claro and Mareeba races) are designated by shading.

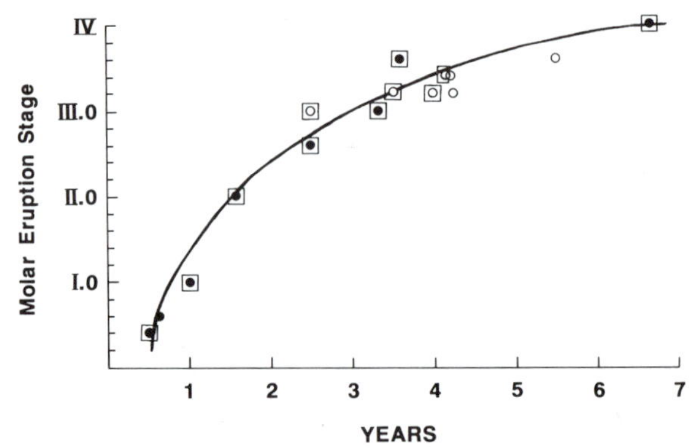

Fig. 2 (above). Regression analysis of molar eruption stages of captive *Petrogale* of known ages. ○ male; ● female; □ *P. assimilis.*

enlargement. The average testis length for 15 adults (MES III.1-IV) was 27.3 mm.

Spermatogenesis in two wild-caught animals was obviously defective (Fig. 3). The elder animal (S472, estimated age greater than 6.5 years) had small testes; tubules were discoloured brown, interstitial cells were abundant but no spermatogenesis was evident. In the younger animal (S853, estimated age between 4 and 4.5 years) spermatogenesis was arrested at spermatid development; spermatids were often clumped together and giant, multinucleate cells were present. This animal was collected from within 6 km of a colony of the neighbouring Mt Claro race, but carried the chromosomes of the assimilis race.

In four captive males, testes began to increase in size at the following ages: Mareeba race — 17 months (MES I.4); *P. penicillata* — 16 months; *P. purpureicollis* — 16 months; *P. xanthopus* — 15 months.

The earliest age at fertile matings in captivity, for an assimilis and a Mt Claro race male, was 20 months. The later age of 24 months for a Mareeba race male, reflected the fact that he was not placed with an oestrous female prior to this age.

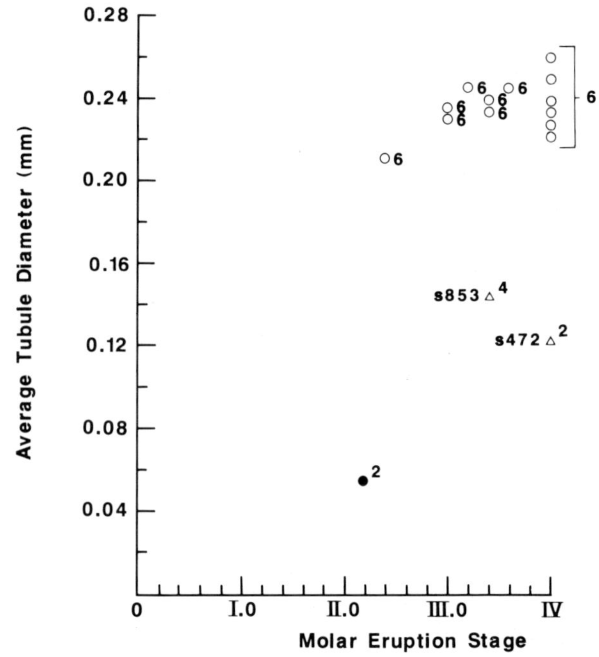

Fig. 3. Tubule size and condition in relation to molar eruption stage. ○ sperm present, open tubules. △ no sperm, open tubules. ● no sperm, closed tubules. Degree of spermatogenic development in tubules as follows: (1) spermatogonia only; (2) few spermatocytes (primary); (3) many spermatocytes (secondary); (4) spermatids present; (5) few sperm; (6) many sperm on edge of lumen.

DISCUSSION

Male and female *P. assimilis* of known age were used to construct MES/age growth curves so that the ages of wild-caught animals could be estimated. Eruption rates for the sexes and the individual races appeared similar. So too were rates obtained from the few available individuals of known age of other *Petrogale* species (as specified in the Methods and Materials).

Molar eruption rates for wild and captive animals may differ. Sharman *et al.* (1964) showed variation in the time of eruption of individual teeth in captive *Macropus rufus*. However, the similarity of curves for *Petrogale* of different sex and taxon indicates considerable uniformity in the eruption process. While some inaccuracy in age estimates of wild individuals may be inevitable, the difference in eruption rates between wild populations with different diets and substrates may be as great as that between captive and some wild populations.

Testis enlargement commenced at MES I.4 (16-18 months) for captive *P. assimilis*. However, in wild animals all testes were immature and, the histological criteria for recognizing sexual maturity were not seen till MES II.2 (23-27 months). This contrasts with results from captive *P. assimilis,* in which the earliest age at fertile matings was 20 months. Johnson (1979) reported the youngest age for copulation in captive *P. inornata* to be 19.5 months. Full spermatogenic development was not seen in captive *P. xanthopus* until MES II.3 (30 months) (Poole *et al.* 1985).

Earlier maturation of captive males compared with wild males was also noted for *M. agilis* (Merchant 1976; Bolton *et al.* 1982) and *M. rufus* (Frith and Sharman 1964; Sharman and Calaby 1964; Sadleir 1965). These authors considered nutrition and individual variation contributed to the differences.

Two of the *P. assimilis* shown in Figure 3 were obviously sterile. The teeth of one of these indicated considerable age, and its sterility may have resulted from senescence. Newsome (1973) has reported a 13% incidence of senescence in aged,

Table 1. Size of testes from field records (*P. assimilis* races). A=testis not mature: length < 10 mm; B=partially developed testis: length between 10-15 mm; C=large testis: length >15 mm.

Estimated Age (months)	14-16	16-18	18-20	20-23	23-27	27-31	31-36	36-41
MES	I.3	I.4	II.0	II.1	II.2	II.3	II.4	III.0
n	—	2A	1B	2A	3C	1C	3C	6C
Bodyweight (kg)	—	3.2	—	2.5-3.0	3.1-3.9	4.0	3.3-4.3	3.9-4.4

wild *M. rufus*. The other individual (S853) is the only case of sterility we have found in a young, wild-caught animal of reproductive age. Apart from the two above, sterile males have only been found by cross-breeding chromosomal races in captivity or in an animal from the *P. godmani*/Mareeba race hybrid zone, identified as a backcross hybrid (Briscoe *et al.* 1982; Sharman *et al.* 1989). Interestingly, S853 was collected within 6 km of known Mt Claro race colonies, with no intervening assimilis race colonies detected.

Although there are other possible explanations for the infertility of S853 e.g., physical damage, the Mt. Claro/assimilis contact zone should be investigated more closely in case this animal was a backcross hybrid. At Macquarie University, a hybrid male (assimilis race × Mt Claro race) was produced which had reduced testes and was sterile (Sharman *et al.* 1989). However, female hybrids have a greater chance of being fertile than equivalent male hybrids (Sharman *et al.* 1989). Backcrosses may therefore be possible and we are currently arranging the appropriate crosses.

ACKNOWLEDGEMENTS

Material and animals used in this study were obtained during the course of a detailed study of rock-wallabies headed by Professor G. B. Sharman and funded by the Australian Department of Environment, Housing and Development and later, the Australian Research Grant Committee and Macquarie University Research Grants. We thank the National Parks and Wildlife Services of Queensland and New South Wales for permits to work on the rock-wallabies. We acknowledge the efforts of many field workers particularly Gerry Maynes who collected much of the material and also, Steve Barker, Geoff Sharman, John Stevens, Terry Bush and Andrew Gooley. We are indebted to Ron Moore and Ron Claassens for their help in the husbandry of the captive animals, to Gillian Rankin and Ian Faulkener for preparation of figures and to the many landholders who allowed us to work on their properties.

REFERENCES

Bolton, L., Newsome, A. E. and Merchant, J. C., 1982. Reproduction in the agile wallaby *Macropus agilis* (Gould) in the tropical lowlands of the Northern Territory: opportunism in a seasonal environment. *Aust. J. Ecol.* **7**: 261-77.

Briscoe, D. A., Calaby, J., Close, R. L., Maynes, G. M., Murtagh, C. E. and Sharman, G. B., 1982. Isolation, introgression and genetic variation in rock-wallabies. Pp. 73-87 *in* Species at Risk: Research in Australia ed. by R. H. Groves and W. D. L. Ride. Australian Academy of Science: Canberra.

Eldridge, M. D. B., Dollin, A. E., Johnston, P. G., Close, R. L. and Murray, J. D., 1988. Chromosomal rearrangements in rock wallabies. *Petrogale* (Marsupialia: Macropodidae). I. The *Petrogale assimilis* species complex. G-banding and synaptonemal complex analysis. *Cytogenet. Cell Genet.* **48**: 228-32.

Frith, H. J. and Sharman, G. B., 1964. Breeding in wild populations of the red kangaroo, *Megaleia rufa. C.S.I.R.O. Wildl. Res.* **9**: 86-114.

Johnson, P. M., 1979. Reproduction in the plain rock-wallaby, *Petrogale penicillata inornata* Gould, in captivity, with age estimation of the pouch young. *Aust. Wildl. Res.* **6**: 1-4.

Merchant, J. C., 1976. Breeding biology of the agile wallaby, *Macropus agilis* (Gould) (Marsupialia: Macropodidae), in captivity. *Aust. Wildl. Res.* **3**: 93-103.

Newsome, A. E., 1973. Cellular degeneration in the testis of red kangaroos during hot weather and drought in Central Australia. *J. Reprod. Fert. Suppl.* **9**: 191-201.

Poole, W. E. and Catling, P. C., 1974. Reproduction in the two species of grey kangaroos, *Macropus giganteus* (Shaw) and *M. fuliginosus* (Desmarest). I. Sexual maturity and oestrus. *Aust. J. Zool.* **22**: 277-302.

Poole, W. E., Merchant, J. C., Carpenter, S. M. and Calaby, J. H., 1985. Reproduction, growth and age determination in the yellow-footed rock-wallaby *Petrogale xanthopus* Gray, in captivity. *Aust. Wildl. Res.* **12**: 127-36.

Sadleir, R. M. F. S., 1965. Reproduction in two species of kangaroo *(Macropus robustus* and *Megaleia rufa)* in the arid pilbara region of Western Australia. *Proc. Zool. Soc. Lond.* **145**: 239-61.

Sharman, G. B. and Calaby, J. H., 1964. Reproductive behaviour in the red kangaroo, *Megaleia rufa*, in captivity. *C.S.I.R.O. Wildl. Res.* **9**: 58-85.

Sharman, G. B., Close, R. L. and Maynes, G. M., 1989. Chromosome evolution, phylogeny and speciation of rock-wallabies (*Petrogale:* Macropodidae). *Aust. J. Zool.* (in press).

Sharman, G. B., Frith, H. J. and Calaby, J. H., 1964. Growth of the pouch young, tooth eruption and age determination in the red kangaroo *Megaleia rufa. C.S.I.R.O. Wildl. Res.* **9**: 20-49.

Comparative growth in the Macropodoidea with particular reference to *Bettongia gaimardi*

R. W. ROSE[1]

Examined in this paper are various methods of measuring and comparing growth rates in the Macropodoidea. These include the Gompertz coefficient, linear equivalent and the relative growth rate. Growth is examined during gestation, pouch life and beyond. New data are presented for the Tasmanian Bettong.

Most kangaroos, wallabies and rat-kangaroos finally leave the pouch at a time when they are approximately 20% of their maternal weight. However, potoroids reach this stage far more rapidly than macropodids because their relative growth rates are almost twice as fast. The Tasmanian Bettong has the fastest relative growth rate during pouch life of all macropodids/potoroids. This may be related to a high protein milk perhaps accompanied by increased milk intake. Five phases of constant relative growth can be delineated and the rates of growth of various body parameters vary. These are adapted to stages of the Bettongs' life cycle.

INTRODUCTION

COMPARATIVE analysis of growth of marsupials have rarely been attempted. Lyne and Verhagen (1957) were perhaps the first to study marsupial growth in any depth. They established the concept of the linear equivalent (Weight$^{1/3}$) in their detailed study of the growth of *Trichosurus vulpecula*. Since then, there have been numerous studies of the changes in body parameters and weight with age in marsupials though few of these addressed the topic of comparative growth. Maynes (1976) in his study of growth of the Parma Wallaby *Macropus parma* derived the value of "k" (relative growth rate) based on earlier work by Huxley (1932). This term, if used correctly, has the value of allowing across the board comparison between species regardless of their size. Other studies have attempted to compare the growth of a larger number of marsupial species, sometimes with eutherian species e.g., Case (1978) while Russell (1982) has completed the most detailed study to date. However, these latter studies suffer in that they are not directly comparable with each other due to the variety of methodologies.

Most recently, Lee and Cockburn (1985) have introduced the use of the Gompertz constant as a measure of marsupial growth. This value has the great advantage of allowing the comparison of the overall growth from birth to maturity but perhaps does not make sufficient allowance for the fact that much marsupial growth occurs in the confines of a pouch within which growth rates may be quite different from that outside — the situation being not entirely analogous to that in eutherians.

This paper, by the use of various methods, provides details on comparative growth rates of macropodoids (kangaroos, wallabies and rat-kangaroos) during embryonic development and pouch life. It includes new data on the Tasmanian Bettong *Bettongia gaimardi*.

METHODS

Embryonic growth rates

Growth rates of macropodid embryos were calculated as in Huggett and Widdas (1951) and Rose (1989a) from data on birth weight and the duration of development from the primitive streak stage to birth. The following equation was used:

$$\text{Growth rate} = \frac{\text{Birth wt}}{t_g - t_o}\ ^{1/3}\ \text{(grams)}$$

where t_g=gestation length (days) and t_o=time of formation of primitive streak.

Pouch life

Data has been compiled on the relationship between maternal weight of macropodoids, the duration of pouch life and the weight of pouch young at the time of final vacation of the pouch.

Growth rates

One way of looking at growth is by use of differential equations or more simply by plotting weight increase logarithmically (Fig. 3). This type of representation relates the amount of weight increase to the actual weight of the animal and allows comparisons between the growth of different sized animals or within the same species at different stages of the life cycle.

Linear segments represent periods during which relative growth occurs at a constant rate (Table 2). The slope of these various linear

[1]Zoology Department, University of Tasmania, G.P.O. Box 252C, Hobart, Tasmania 7001.

Pages 423-31 in KANGAROOS, WALLABIES AND RAT-KANGAROOS ed. by G. Grigg, P. Jarman and I. Hume. Surrey Beatty & Sons Pty Limited, New South Wales, Australia, 1989.

Table 1. Embryonic growth rates in macropodoid marsupials.

Species	Approx. birth weight (mg)	Gestation (days)	Relative growth rate	Source
Potorous tridactylus	300	38	0.06	Shaw and Rose (1979)
Bettongia gaimardi	330	21	0.07	Rose (1985)
Setonix brachyurus	350	27	0.08	Sharman (1961)
Thylogale billardierii	400	30	0.07	Rose and McCartney (1982, unpubl.)
Macropus eugenii	444	29	0.07	Renfree and Tyndale-Biscoe (1973); Renfree (pers. comm.)
Macropus rufogriseus	500	30	0.07	Walker and Rose (1981)

segments represent the mean "instantaneous relative growth rate" (k) Maynes (1976).

It is calculated as:

$$\frac{\log_e W_2 - \log_e W_1}{day_2 - day_1}$$

W=weight in grams.

Weekly measurements of the weight of twelve Bettongs were obtained throughout pouch life and subsequently up until maturity. At the same time measurements of head, tail and pes length were taken as in Rose (1989b). Milk samples were collected from some of these Bettongs and a number of Potoroos as described by Smolenski and Rose (1988) and Crowley, Woodward and Rose (1988).

Linear equivalent

This general measure of body size was calculated as Weight$^{1/3}$ (g) as in Lyne and Verhagen (1957).

Growth in the wild

Measurements of pes length were plotted against weight as an index of condition in the Tasmanian Bettong for animals caught in the wild and reared in captivity. The data is from Rose (1985) and Taylor and Rose (1987).

RESULTS

Embryonic growth rates

Table 1 gives details of embryonic growth rates from a number of macropodoids (Rose 1989a). Most macropodids develop at similar rates to each other. There is a direct but non linear relationship between birth weight and maternal weight (Russell 1982) as smaller species have young that are relatively larger (as a percentage of maternal weight) than do larger kangaroo species.

Pouch life

When one compares maternal weight on a log/log scale against the weight of the young at final pouch vacation there is a very close correlation (Fig. 1). All macropodids leave the pouch at a constant weight, relative to their mothers (approx. 20%). However, plotted similarly, the duration of pouch life is less closely related to maternal weight (Fig. 2). Potoroids remain in the pouch for a shorter period than would be predicted.

Growth rates

The usual representation of growth of marsupial young, that is age plotted against weight, seems to demonstrate that initial growth is slow followed by a progressive increase coincident with rises in milk energy content and milk intake (Rose 1987a).

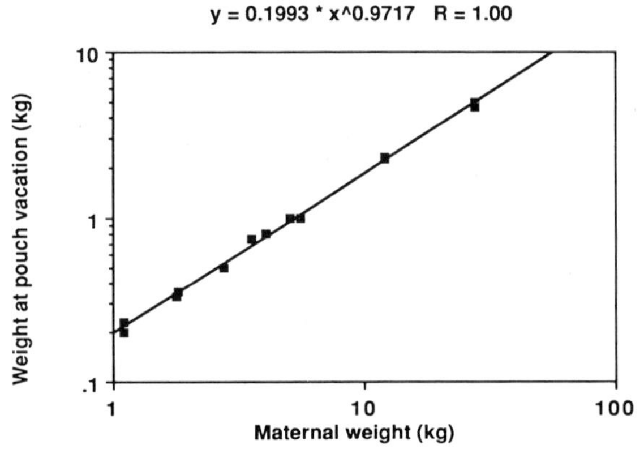

Fig. 1. Log/log plot of weight at pouch vacation in Macropodoids.

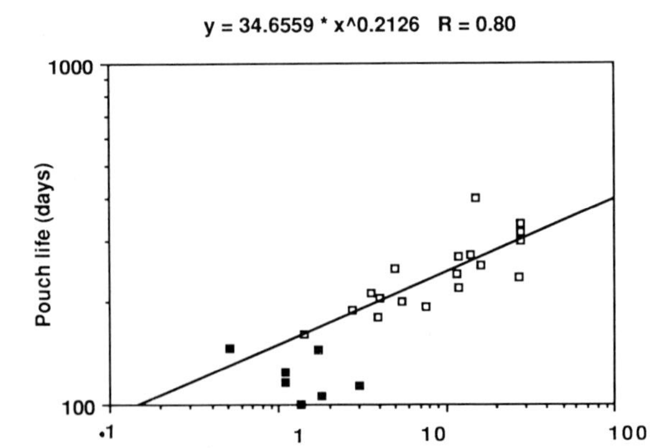

Fig. 2. Log/log plot of duration of pouch life in Macropodoids. □ Macropodids; ■ Potoroids.

Table 2. Mean instantaneous relative growth rate (k) in *B. gaimardi* during periods of constant growth (from Fig. 3) (N=17).

Growth curve segment	Age range (days)	Growth rate (k) $\log_e$.g.day^{-1}	Stage of development
1	1-7	1.86×10^{-1}	New born young, first week of life
2	7-35	8.49×10^{-2}	Young firmly attached to teat
3	49-105	4.56×10^{-2}	Pouch life until final emergence, young able to release teat
4	119-147	2.27×10^{-2}	Young at foot, sucking
5	168-224	4.75×10^{-3}	Young no longer sucking, until near maturity

Relative growth rates are constant for most of pouch life up until final pouch vacation (Fig. 3, Table 2) when growth rates slow. This holds true for *Macropus parma* (Maynes 1976), the Bettong and for many other macropodoids (Rose 1985). The values for k during this "stable" period have been calculated for all available species and are presented in Table 3.

During the major portion of pouch life in the Bettong when the young is growing at a constant rate, the value of k obtained (4.56×10^{-2} Table 2) is quite high when compared with those of other macropodids (Table 3). Although there are some anomalies, it can be seen that k values of 2.2 to 2.5×10^{-2} predominate within the Macropodidae but that all of the Potoroidae have elevated k values. *B. gaimardi* has a larger k value than that found in *B. lesueur* thus the Tasmanian Bettong is considerably larger than the Boodie at pouch vacation even though the duration of pouch life is similar. In a similar manner amongst species of kangaroos, differences in k values to some extent reflect the duration of pouch life, e.g., *Thylogale billardierii* and *Macropus eugenii* are similar sized animals with similar sized young at pouch vacation (Rose and McCartney 1982) but the pouch life of Tammar (250 days) is 50 days longer than in the Tasmanian Pademelon.

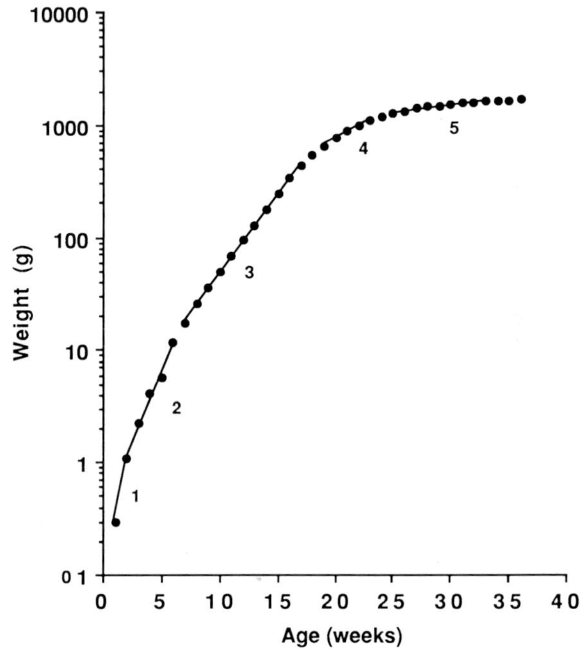

Fig. 3. Log plot of weight against age in *B. gaimardi* (Rose 1985).

Table 4 summarizes the composition of Potoroo and Bettong milk during lactation. There are some dramatic changes as the young leave the pouch; the carbohydrate levels drop and lipid levels rise. During much of pouch life milk carbohydrates are primarily oligosaccharides but

Table 3. Relative growth rates (k) during pouch life in the Macropodoidea.

Species	Maternal weight (approx. kg)	$k\times10^{-2}$ ($\log_e$ g.day^{-1})	Source
Bettongia gaimardi	1.8	4.56	Rose (1985)
Bettongia lesueur	1.15	4.15	Tyndale-Biscoe (1968)
Potorous tridactylus	1.2	3.7	Bryant and Rose (unpubl.)
Aepyprymnus rufescens	3.0	4.38	Jarman (pers. comm.)
Peradorcas concinna	1.6	2.31	Nelson and Goldstone (1986)
Setonix brachyurus	2.6	2.52	Shield and Woolley (1961)
Thylogale billardierii	4.0	2.37	Rose and McCartney (1982)
Macropus parma	3.3	2.35	Maynes (1972)
Macropus eugenii	4.5	1.98	Murphy and Smith (1970)
Macropus rufus	26.0	2.45	Sharman *et al.* (1964)
Macropus robustus	15.0	1.84	Sadleir (1963); Ealey (1967)
Macropus giganteus	32.0	1.91	Poole *et al.* (1982a)
Macropus fuliginosus	28.0	1.84	Poole *et al.* (1982b)
Wallabia bicolor	10.1	2.24	Merchant and Sharman (1966)

Table 4. Milk composition at three stages of lactation in the *B. gaimardi* and *P. tridactylus* (Rose 1987; Smolenski and Rose 1988 and Rose, this volume).

Species	Stages of lactation (week)	Lipid ($g.100 ml^{-1}$)	Protein ($g.100 ml^{-1}$)	Carbohydrate ($g.100 ml^{-1}$)
	Early (5)	0.3	8.5	9.3
Bettong	Mid (12)	2.4	10.5	11.5
	Late (24)	17.6	11.7	0.9
	Early (6)	1.3	3.4	9.4
Potoroo	Mid (14)	2.3	9.9	12.3
	Late (25)	20.4	14.5	1.8

at pouch emergence these virtually disappear, to be replaced by mono- and disaccharides (Crowley, Woodward and Rose 1988). The composition of Bettong milk is generally similar to that of the Potoroo, as are the sequence of changes (Smolenski and Rose 1988). However, Bettong milk, for most of pouch life and particularly during early and mid-lactation has significantly more protein ml^{-1} than does the potoroo.

Growth of the Tasmanian Bettong

A description of the change in body parameters (head, foot length, weight etc.) and molar eruptions with age can be found in Rose (1989b). The weight increment ($g.week^{-1}$) of growing animals rises until the nineteenth week after birth; weight increase peaks during this week (3-4 weeks after pouch vacation) with an average of 133 $g.week^{-1}$ (Fig. 4). Subsequently weekly increments slowly decrease as maturity approaches, although they remain at about 50 $g.week^{-1}$ for some time.

If the weekly mean weight increment is expressed as a percentage of an animal's actual weight at the start of the week (Fig. 5) it can be seen that after a dramatic weight increase in the first week of life (266%), over the next three to four weeks the percentage weight increase is approximately 100% per week. For most of the subsequent pouch life, weekly weight increase stabilizes at about 35-40%. Following pouch vacation there is a gradual decrease in weekly weight increase to below 5% near maturity.

Linear equivalence

Prior to the end of pouch life, the head length curve (Fig. 6) is parallel to the linear equivalence ($Weight^{1/3}$) curve but after pouch vacation the increase in head length as a percentage of adult size decreases in relation to the linear equivalence. During the first few weeks of life, the foot is small and well below the linear equivalence. However, the growth of the foot is above the linear equivalence for approximately the second half of pouch life. Relative foot length shows its greatest divergence from the linear equivalence at the time of pouch emergence. Tail length remains below the linear equivalence for the

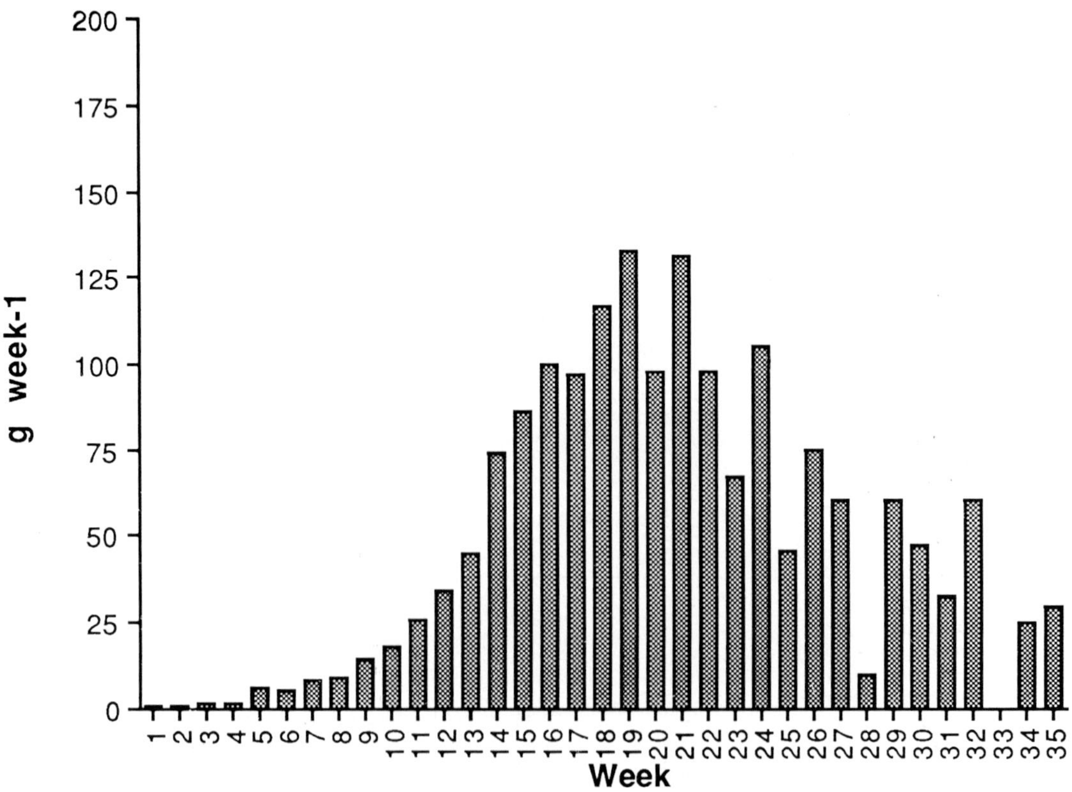

Fig. 4. Absolute weight increase in *B. gaimardi* ($g.week^{-1}$) (Rose 1985).

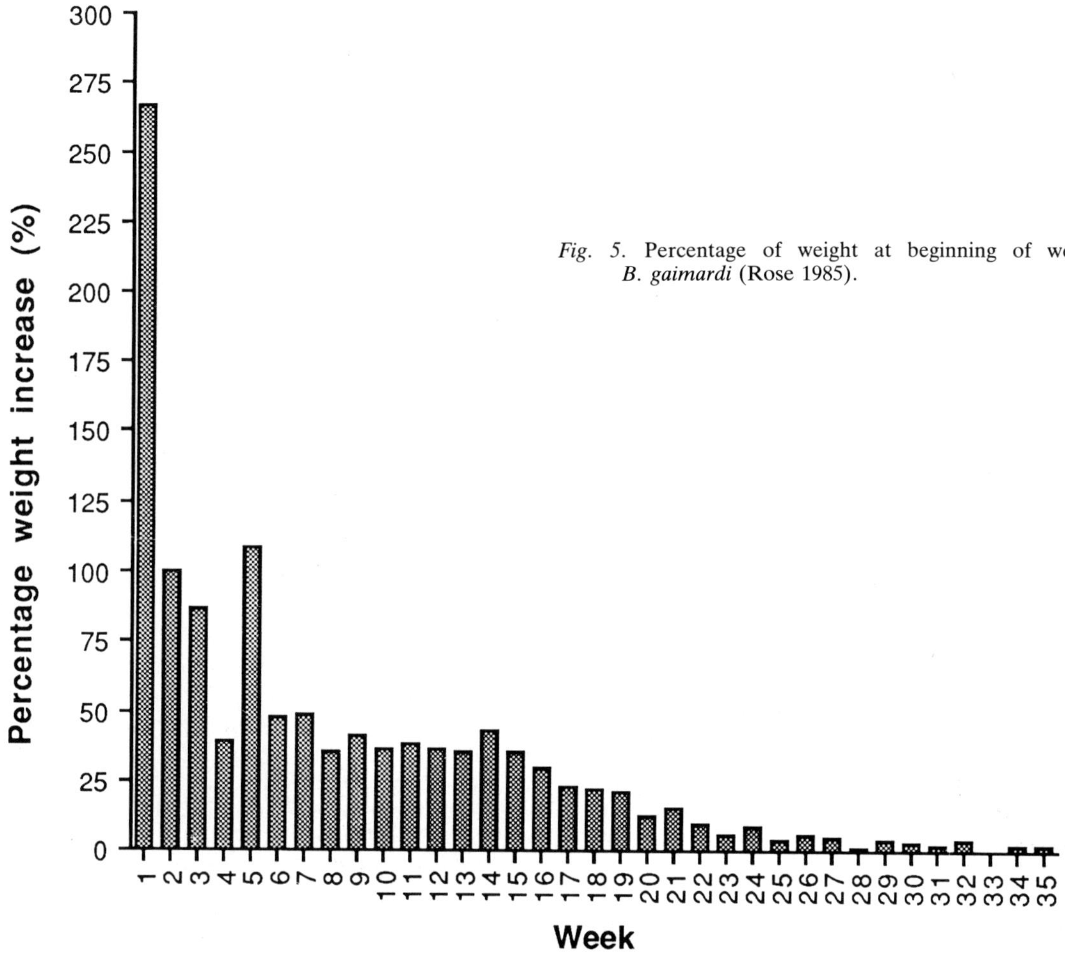

Fig. 5. Percentage of weight at beginning of week in *B. gaimardi* (Rose 1985).

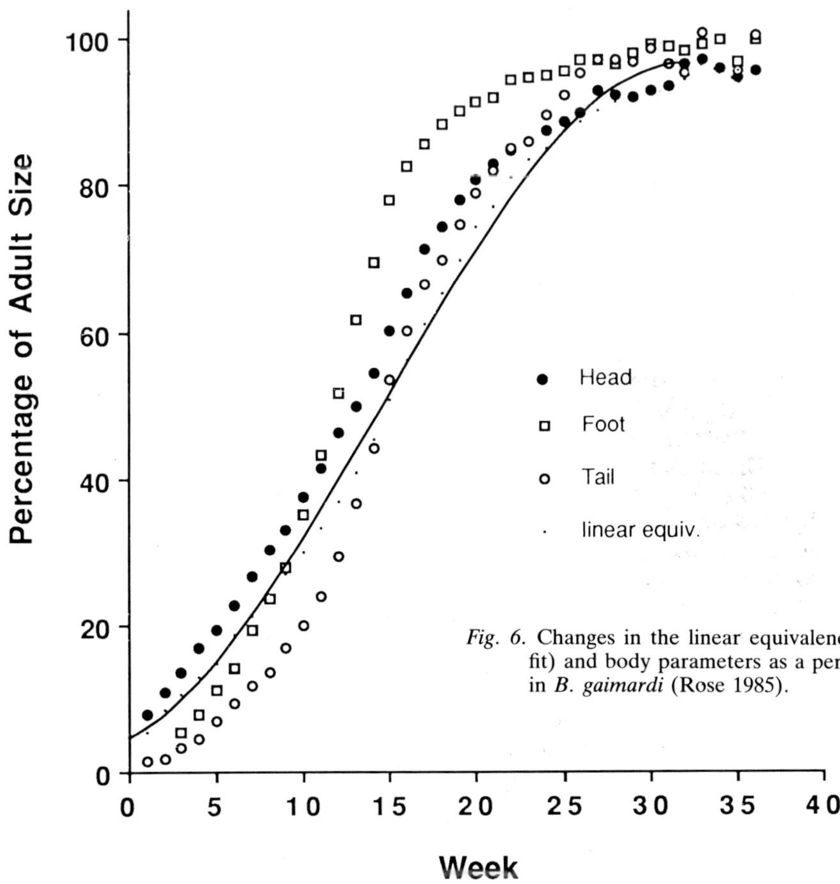

Fig. 6. Changes in the linear equivalence ($W^{1/3}$, line of best fit) and body parameters as a percentage of adult size in *B. gaimardi* (Rose 1985).

majority of pouch life but approaches it just prior to pouch vacation.

Growth in the wild

In the wild, older Bettong pouch young gain weight more slowly than in captivity and in addition the body condition of young is higher in captivity (Rose 1985; Taylor and Rose 1987). It is not known at what age wild young vacate the pouch but it seems likely that they weigh a similar percentage of their mothers weight (which is lighter in the wild) as in captivity.

The weights of field and laboratory animals are similar for individuals with foot length in the range of 2-8 cm. For individuals with larger feet, field animals weigh considerably less than captive animals, with this difference increasing as foot size increases (Taylor and Rose 1987). Animals with a foot length of 8 cm are approximately 13 weeks of age (Rose 1985) and this is the age at which young first leave the pouch.

DISCUSSION

The Tasmanian Bettong and the Potoroo are similar both in neonatal size and embryonic growth rates. However, the gestation length of the Potoroo (38 days) is almost twice that of the Bettong (21 days). The reason for this variation in growth rates is unclear, although Tyndale-Biscoe (1968), Shaw and Rose (1979) and Rose (1989a) have suggested it is caused by variations in the duration of the "slow phase" in embryonic development. Even before the primitive streak has formed, the bettong blastocyst expands more rapidly than in the Potoroo (Rose 1985). The Potoroo has a slow cell cycle (Bick and Brown 1969) and as a consequence, cell division and multiplication are slower than in other mammals. Subsequently it is possible that this provides a partial explanation of its slow growth. Also, after birth the Potoroo has a slower growth rate than other potoroids.

That all macropodoids finally leave the pouch at a similar weight relative to their mother suggests perhaps that the mode of locomotion (hopping) places similar constraints on the size that young can reach before becoming an unacceptable physical burden upon the mother. From that relationship one might expect that there would also be a good correlation between maternal weight and duration of pouch life. Figure 2 shows that larger kangaroos have young that remain in the pouch for lengthy periods, after all these young must reach 20% of a larger weight. However, the rat-kangaroos, with the exception of *Hypsiprymnodon* (which has two young) all have a pouch life duration considerably less than expected, in other words they all have faster growth rates than kangaroos and wallabies. The tree kangaroo *Dendrolagus matchiei* lies well above the line, i.e., this species has a relatively long pouch life. If, as seems probable, its relative weight at pouch vacation is similar to that of the other macropodids, the tree kangaroo must have slow development within the pouch. The diet is probably quite different from other macropodids and this postulated slow growth rate is paralleled by other marsupial arboreal folivores (Russell 1982)

Lee and Cockburn (1985) show that interspecific differences in growth may be interpreted as the result of trade-offs between selected factors which favour longer or shorter periods of growth. Ricklefs (1979) discusses how growth is dependent on the partitioning of maternal resources between actual growth and maintenance activities (e.g., metabolic rate increases and thermoregulation) as well as being dependent on the rate of proliferation and growth of cells, the latter being used earlier to explain differences in embryonic growth rates between the *Potorous tridactylus* and *Bettongia gaimardi*. The former argument (partitioning of resources) was used by Rose (1987) to explain why the relative growth rate of Bettong young declines. In that study it was demonstrated that as homeothermy/endothermy was gradually achieved, growth rates correspondingly decreased as milk energy was diverted towards the maintenance of the pouch young's increased metabolic rate.

Although the maternal weights of the Quokka *Setonix brachyurus* (2.6 kg) and Rock Wallaby *Peradorcas concinna* (1.6 kg) are not greatly different from that of the rat-kangaroos, their k values are similar to those of the other larger kangaroos. It seems likely therefore that there are family differences in the general value of k that may reflect phylogenetic relationships. Also, the Gompertz constant (0.015) when calculated for the Bettong as in Lee and Cockburn (1985) is higher relative to its body weight than that for other macropodoids including *B. lesueur*.

The fact that many kangaroos of widely different size have similar k values (i.e., relative growth rates) is interesting. It may indicate similarities in both the quantity and quality of the milk at comparable phases of pouch life as well as similar patterns of sucking. Green *et al.* (1980) in a study of lactation in the Tammar Wallaby showed how the constituents of milk change through pouch life and they suggested that similar changes were likely to occur in other macropodids. They also considered that variations in body weight, growth rate and duration of lactation may be more closely correlated with changing milk compositions than with the actual age of the young. This, combined with a substantial intake of milk over and above that found in the Potoroo (Smolenski and Rose 1988), may explain the more rapid growth in the Bettong.

It is perhaps unfortunate that the linear equivalent has not been more frequently measured as it is a useful parameter that allows one to compare measurements of body parameters e.g., the head, foot and tail against some generalized standard. Lyne and Verhagen (1957) used it to compare growth rates of *Trichosurus vulpecula* with non-marsupial species (man, mouse, cow and sheep) and within different (Tasmanian and mainland) Possum populations. In the Bettong the fact that the value of the head length expressed as a percentage of the adult head size remains above that of the linear equivalence for the whole of pouch life no doubt reflects the overall importance of the head with its associated sense organs and brain in allowing the young initially to reach the pouch and then to suck (Gemmell and Rose, this volume). Foot length on the same basis remains below the linear equivalence until after the mid-point of pouch life when rapid growth of this parameter occurs. By the time the young emerge from the pouch, they have the characteristically large feet of macropodids. The functional adaptation of this mode of development is towards swift and agile movements when the young has to "keep up" with the pace of its mother and to evade predators; this is not a requirement of early pouch life and consequently, hindlimbs can remain relatively small during that period.

For much of pouch life the tail remains smaller than the corresponding linear equivalence. The Bettong tail is used in locomotion and balance and in the construction of a nest. Maynes (1976) showed that in the Parma Wallaby, tail length was above the linear equivalence only during the period when the young was leaving the pouch. After this period it fell and remained below the linear equivalence. The situation is somewhat different in the Bettong; once the tail length exceeds the linear equivalence (approximately one week before final pouch vacation) it remains so until maturity, prior to which time the young will have separated from its mother and constructed a nest of its own. Sexual dimorphism does not occur in the Tasmanian Bettong or other Bettongs. It does occur in most larger kangaroos (Maynes 1976; Jarman 1983) but usually does not develop until some time after the young have left the pouch.

Darwinian theory suggests that the larger size of males in those species with sexual dimorphism, is the result of competition among males for mates. Thus the more polygamous the species (males mate with several females) the larger the males should be relative to the females (Ralls 1978). Bettong males are certainly aggressive towards each other (Rose 1982). Bigger mothers may be better mothers particularly in the latter stages of pouch life when increased metabolic demands must be met so that the large pouch young may be nourished and transported. However, it would be more than coincidental if females and males reached a similar size but for evolutionary reasons that differed widely. Jarman (1983) discusses sexual dimorphism within the Macropodoidea in detail. He presents arguments to explain the lack of sexual dimorphism in the mono-(or what he terms homo)-morphic rat-kangaroos. There is a compromise between ranging ability (related to size?) of the male with perhaps a reduced ability to constantly monitor any individual female. On the other hand, females are sufficiently large to fight off over-eager males and to defend access to nest sites and food items (Jarman 1983).

The variation in growth rate in Bettongs in the wild and captivity is interesting. Food in captivity consisted of bread, apples and solid dog food. The diet of these animals in the wild consists mainly of underground fungi (Rose 1986a). Kinnear *et al.* (1979), who worked with a related West Australian species, *Bettongia penicillata* state that as a source of amino acids, fungi are both deficient and unbalanced. However, those workers showed how microbial intervention in the foregut was able to correct this deficiency. Frappell and Rose (1986) have suggested that both the Potoroo and Tasmanian Bettong are similar to *B. penicillata* in their ability to manufacture essential amino acids in the foregut. Previous studies have documented a reduction in pouch young condition only during time of severe environmental stress (*M. rufus*: Newsome 1965; *M. robustus:* Ealey 1967). As environmental conditions during this study were not severe it was considered that the poorer body condition of Bettongs in the wild was the norm (Taylor and Rose 1987). The explanation for the differential growth may lie in a greater partitioning of energy in the field towards maintenance. In addition, it could be an indication of poor nutrition of the mother near the end of pouch life of her young when absolute growth of the young is high. The Bettong in the wild may not be able to assimilate sufficient high-quality food, or manufacture sufficient amino acids in her rumen-like foregut near the time that the young leaves the pouch. In some mammals, females often drain their own resources before any reduction in milk production occurs (Sadleir 1969). The Bettong appears to be no exception, as the mother may lose weight near the end of pouch occupancy (Rose 1987).

In conclusion, it appears obvious that rat-kangaroos grow more rapidly than do the larger kangaroos and wallabies. It is not clear how this growth is achieved, although it seems likely that an increased milk intake may be involved as milk composition in potoroids is similar to the Tammar. During pouch life the relative growth rate (k) is a useful comparative measure allowing comparisons over a wide range of body weights.

By use of the linear equivalence it is possible to compare the growth of individual body parameters with some generalized measure of growth allowing intraspecific comparison.

That the feeding of relatively high protein diets in captivity result in pouch young that are in better condition is not surprising, though it has rarely been reported in the marsupial literature.

REFERENCES

BICK, Y. A. E. AND BROWN, J. K., 1969. Variations in radiosensitivity during the cell cycle in marsupial cell lines. *Mutation Res.* **8**: 613-22.

CASE, T. J., 1978. On the evolution and adaptive significance of postnatal growth rates in terrestrial vertebrates. *Quart. Rev. Biol.* **53**: 243-81.

CROWLEY, H. M., WOODWARD, D. R. AND ROSE, R. W., 1988. Changes in milk composition during lactation in the Potoroo *(Potorous tridactylus). Aust. J. Biol. Sci.* **41**: 289-96.

EALEY, E. H. M., 1967. Ecology of the euro, *Macropus robustus* (Gould) in north-western Australia. IV. Age and Growth. *C.S.I.R.O. Wildl. Res.* **12**: 67-80.

FRAPPELL, P. B. AND ROSE, R. W., 1986. Function of the gastro-intestinal tract in the potoroo *(Potorous tridactylus). Aust. J. Zool.* **34**: 463-71.

GEMMELL, R. T. AND ROSE, R. W. (this volume). The senses involved in movement of some marsupials from cloaca to pouch.

GREEN, B., NEWGRAIN, K. AND MERCHANT, J., 1980. Changes in milk composition during lactation in the tammar wallaby *(Macropus eugenii). Aust. J. Biol. Sci.* **33**: 35-42.

HUGGETT, A. ST. G. AND WIDDAS, W. F., 1951. The relationship between feotal weight and conception age. *J. Physiol. Lond.* **57**: 149-65.

HUXLEY, J. S., 1932. Problems of Relative Growth. Metheun: London.

JARMAN, P., 1983. Mating system and sexual dimorphism in larger terrestrial, mammalian herbivores. *Bio. Rev.* **58**: 485-520.

KINNEAR, J. E., COCKSON, A., CRISTENSEN, P. AND MAIN, A. R., 1979. The nutritional biology of the ruminants and ruminant-like mammals — a new concept. *Comp. Biochem. Physiol.* **64A**: 357-65.

LEE, A. K. AND COCKBURN, A., 1985. Evolutionary ecology of marsupials. Cambridge University Press.

LYNE, A. G. AND VERHAGEN, A. M., 1957. Growth of the marsupial *Trichosurus vulpecula* and a comparison with some higher mammals. *Growth* **21**: 167-95.

MAYNES, G. M., 1972. Age estimation in the parma wallaby, *Macropus parma* Waterhouse. *Aust. J. Zool.* **20**: 107-18.

MAYNES, G. M., 1976. Growth of the parma wallaby, *Macropus parma* Waterhouse. *Aust. J. Zool.* **24**: 217-35.

MERCHANT, J. C. AND SHARMAN, G. B., 1966. Observations on the attachment of marsupial pouch young to the teats and on the rearing of pouch young by foster-mothers of the same or different species. *Aust. J. Zool.* **14**: 593-609.

MURPHY, C. R. AND SMITH, J. R., 1970. Age determination of pouch young and juvenile kangaroo island wallabies. *Trans R. Soc. S. Aust.* **94**: 15-20.

NELSON, J. E. AND GOLDSTONE, A., 1986. Reproduction in *Perdorcas concinna* (Marsupialia: Macropodidae). *Aust. Wildl. Res.* **13**: 501-06.

NEWSOME, A. E., 1965. Reproduction in natural populations of the red kangaroos *Megaleia rufus* (Desmarest) in Central Australia. *Aust. J. Zool.* **13**: 735-59.

POOLE, W. E., CARPENTER, S. M. AND WOOD, J. T., 1982a. Growth of grey kangaroos and the reliability of age determination from body measurements. I. The eastern grey kangaroo *Macropus giganteus. Aust. Wildl. Res.* **9**: 9-20.

POOLE, W. E., CARPENTER, S. M. AND WOOD, J. T., 1982b. Growth of grey kangaroos and the reliability of age determination from body measurements II. The western grey kangaroos, *Macropus fuliginosus fuliginosus, M. f. melanops* and *M. f. ocydromus. Aust. Wildl. Res.* **9**: 203-12.

RALLS, K., 1978. When bigger is best. *New Scientist* 9th Feb. 360-62.

RENFREE, M. B. AND TYNDALE-BISCOE, C. H., 1973. Intrauterine development after diapause in the marsuipal *Macropus eugenii. Devel. Biol.* **32**: 28-40.

RICKLEFS, R. W., 1979. Adaptations, constraint and compromise in avian development. *Biol. Rev.* **54**: 269-90.

ROSE, R. W., 1982. Tasmanian bettong *Bettongia gaimardi:* maintenance and breeding in captivity. *In* The Management of Australian Mammals in Captivity ed. by D. D. Evans. Zoological Board Victoria: Melbourne.

ROSE, R. W., 1985. The reproductive biology of the Tasmanian bettong, *Bettongia gaimardi.* Ph.D. Thesis, University of Tasmania.

ROSE, R. W., 1986a. The habitat, distribution and conservation status of the Tasmanian bettong, *Bettongia gaimardi* (Desmarest). *Aust. Wildl. Res.* **13**: 1-6.

ROSE, R. W., 1986b. Control of pouch vacation in the Tasmanian Bettong, *Bettongia gaimardi. Aust. J. Zool.* **34**: 485-91.

ROSE, R. W., 1987. Reproductive energetics of two Tasmanian rat-kangaroos (Potoroinae: Marsupialia). *Symp. Zool. Soc. Lond.* **57**: 149-65.

ROSE, R. W., 1989a. Embryonic growth rates of marsupials with a note on monotremes. *J. Zool. Lond.* **217**: in press.

ROSE, R. W., 1989b. Age estimation of the Tasmanian Bettong *(Bettongia gaimardi)* (Macropodoidea: Marsupialia). *Aust. Wildl. Res.* **16**: in press.

ROSE, R. W. (this volume). Reproductive biology of Rat Kangaroos.

ROSE, R. W. AND MCCARTNEY, D. J., 1982. Reproduction of the red-bellied Pademelon *Thylogale billardierii* **9**: 27-32.

RUSSELL, E., 1982. Patterns of parental care and parental investment in marsupials. *Biol. Rev.* **57**: 423-86.

SADLEIR, R. H. F. S., 1963. Age estimation by measurements of joeys of the euro *Macropus robustus* Gould in Western Australia. *Aust. J. Zool.* **11**: 241-49.

SADLEIR, R. H. F. S., 1969. The Ecology of Reproduction in Wild and Domestic Mammals. Methuen and Co. Ltd.

SHARMAN, G. B., 1961. Embryonic membranes and placentation in five genera of diprotodont marsupials. *Proc. Zool. Soc. Lond.* **137**: 197-200.

SHARMAN, G. B., FRITH, H. J. AND CALABY, J. H., 1964. Growth of the pouch young, tooth eruption and age determination in the red kangaroo *Megaleia rufa. C.S.I.R.O. Wildl. Res.* **9**: 20-49.

SHAW, G. AND ROSE, R. W., 1979. Delayed gestation in the potoroo *Potorous tridactylus* (Kerr). *Aust. J. Zool.* **27**: 901-12.

SHIELD, J. W. AND WOOLLEY, P., 1961. Age estimation by measurement of pouch young of quokka *(Setonix brachyurus)*. *J. Physiol. Lond.* **9**: 14-23.

SMOLENSKI, A. J. AND ROSE, R. W., 1988. Comparative lactation in two species of Rat-kangaroo (Marsupialia). *Comp. Biochem. Physiol.* **90A**: 459-64.

TAYLOR, R. J. AND ROSE, R. W., 1987. Comparison of growth of pouch young of the Tasmanian Bettong *Bettongia gaimardi* in captivity and in the wild. *Aust. Wildl. Res.* **14**: 257-62.

TYNDALE-BISCOE, C. H., 1968. Reproduction and post-natal development in the marsupial *Bettongia lesueur* (Quoy and Gaimard). *Aust. J. Zool.* **6**: 577-602.

WALKER, M. T. AND ROSE, R. W., 1981. Prenatal development after diapause in the marsupial *Macropus rufogriseus*. *Aust. J. Zool.* **29**: 167-87.

Sexual dimorphism in Macropodoidea

PETER J. JARMAN[1]

The forms of sexual dimorphism found in the Macropodoidea (Marsupialia; kangaroos, wallabies and rat-kangaroos) are described and their distribution within the superfamily reviewed. All extant species of the Potoroidae and the smallest of the Macropodidae (below about 4 kg) are homomorphic; the medium and large Macropodidae are heteromorphic. In the homomorphic species both sexes grow at the same rate to a plateau weight, achieved early in adult life, and remain at that weight thereafter. In the heteromorphic species the sexes grow at different rates. In some medium-sized macropod species, both sexes continue to grow quite strongly throughout life, males and females showing significant increases in size during adult life. In the larger macropodid species, female growth decelerates much more than does male growth. As a consequence, males grow to be very much larger than females, and, more importantly, younger males. The degree of dimorphism in size is thus not randomly distributed across the superfamily: there is a significant trend in degree of heteromorphism with increasing body size. Males of heteromorphic species are more than just a larger version of females. Their forelimbs and pectoral girdles become exaggerated, with increases in muscularity, in length of arm (and other) bones, and in robustness of claws. The relative exaggeration in forearm muscularity is greatest in some of the medium-sized species. All these aspects of forelimb development reflect the role of the arms as weapons in fighting. In exaggerating the forearms, macropods have parallelled the bovids' or cervids' exaggeration of cranial weaponry. Like heteromorphic species of those families, heteromorphic male macropods display their weapons by demonstrating their arm length and musculature.

INTRODUCTION

THE ecological and behavioural adaptations of the Macropodoidea (Marsupialia; kangaroos, wallabies and rat-kangaroos) are commonly compared to those of the other great Pliocene-Pleistocene (Quaternary) radiations of mammalian terrestrial herbivores, the Bovidae and Cervidae, in that their species cover a similar range of sizes; range from solitary to social in their organizations; feed upon a range of diets selected from plants mainly in the ground layer of vegetation; and occupy habitats ranging from rainforest to desert communities. However, they differ profoundly in aspects of their metabolic and reproductive physiology, and their morphology. In exploring forms of social organization among macropods, it became clear to me that the forms of sexual dimorphism in the superfamily were poorly documented and largely unexplained. Since many arguments about the adaptiveness of social behaviour have in part leaned upon interpretations of dimorphism (see Jarman 1983), the dearth of published information needs to be rectified.

This paper sets out to display accumulated data on sexual dimorphism in macropods, hoping in part to stimulate further research, or at least the acquisition of more, pertinent data. The existing data are used to investigate trends in dimorphism within the superfamily. For those species that are sexually heteromorphic, the paper considers how the sexes differ and how those differences develop. In particular it considers the parts played by prolonged allometric growth, on the one hand, and by exaggeration of males' fore-limbs as weapons, on the other.

MATERIALS

Despite the large numbers of macropods, of most species, which have passed through scientists' hands in the last two centuries, published data on measurements suitable for describing dimorphism are scarce. Moreover, because of the relationship between age and size in many Macropodidae (see below), individual data which lack associated data on age are of limited use. Thus many of the data used in this paper have been collected specifically with this study in mind. They come from animals that were live-caught, shot, or found dead. Some museum series are sufficiently large, and carry sufficient standardized body measurements, to be useful; those made by Dr T. H. Kirkpatrick and deposited in the Queensland Museum are particularly noteworthy. Other useful collections are those held in the Australian National Wildlife Collection (often collected by Dr J. H. Calaby), and the live animals studied by Dr W. E. Poole, C.S.I.R.O. Division of Wildlife and Ecology (Poole *et al.* 1982a,b, 1984), which provide measurements of animals of known age. The measurements which I routinely took and found to be most useful in my analyses were:

1. weight (live or dead),
2. pes (without claw),
3. forearm length (elbow to bent wrist),
4. forearm girth (maximum circumference of the extended arm, measured over skin and fur, between the elbow and the wrist),
5. length (as a straight-line measurement) of skull (of dead animal) or head (dead or alive),
6. relative muscle bulk of the forearm or hindleg. This was measured as the ratio of the weight

[1]Department of Ecosystem Management, University of New England, Armidale, New South Wales 2351.

Pages 433-47 in KANGAROOS, WALLABIES AND RAT-KANGAROOS ed. by G. Grigg, P. Jarman and I. Hume. Surrey Beatty & Sons Pty Limited, New South Wales, Australia, 1989.

of all the freshly dissected muscle from the forearm (between elbow and wrist) to the fresh weight of the ulnar plus radius, or the same for the muscles of the hindleg between knee and ankle and the tibia plus fibula,

7. relative age. This was taken to be shown by the stage of molar eruption and progression, the assumption being (despite Newsome *et al.* 1977) that these changed at the same rate in the two sexes. Criteria of absolute age are not yet available for most species. It should be remembered that, where molar progression has been compared with known age, the molar index bears some sort of logarithmic relationship to real age. In this paper, "molar index" refers to the molar progression index developed by Kirkpatrick (1964, 1965).

Dimorphism in general appearance

Unlike most species of Bovidae and Cervidae, male macropods do not carry sex-specific cranial weapons such as horns or antlers. Even in those species which retain canines (mainly the potoroids), those of males are only slightly larger than those of females and are not exposed as tusks. With two exceptions, the sexes do not differ strikingly in colouration, although slight differences are quite widespread. Areas of red fur are more emphatically coloured in males than in females of *Macropus rufogriseus, Thylogale thetis, T. stigmatica, Petrogale penicillata* and *Wallabia bicolor.* In *M. giganteus*, fur of the throat, chest and belly of males may be darker than that of females. In this species and in *M. rufus, M. robustus* and *M. fuliginosus* there may be extensive staining of the skin and fur of the throat and chest of large males by secretions from dispersed skin glands. Such secretions may be present in large males of many other species, without affecting their appearance. For instance, copious brownish secretions on the chests and necks of large male *W. bicolor* do not show up against their dark fur. Males of most macropod species carry paracloacal glands which are larger than those of females. Those, the males' skin glands, and the common male practice of enurination, suggest that males differ strongly from females in smell, even though this is apparent to humans only at close range.

Unlike all other extant macropod species, *M. rufus* and *M. robustus* are generally dimorphically coloured. The degree of the differences varies between races of *M. robustus*. In *M. rufus* the sex-linkage of colouration is imperfect, with distinct regional variation in the proportion of males that are female-coloured and vice versa (Oliver 1986). In both these species, the sex-specific colouration is present from the time the juvenile is first furred; it is not gained only on attainment of sexual maturity. In this respect, and in the imperfection of sex-linkage in colouration, these macropods differ from such typically dimorphic ungulates as *Hippotragus niger* or *Antilope cervicapra*.

Differences in weight between the sexes

Tables 1 and 2 illustrate differences in weight between the sexes in potoroids and macropodids respectively. Maximum known weights have been used because, as will be discussed below, males and sometimes females of many macropod species grow continuously; thus a "mean" weight for adults is meaningless in this context. To avoid problems of regional variation in body size, comparisons have been made between minimal weights for the sexes sampled from the same population or region.

Males and females of all potoroids, and of all the smallest macropodids, grow to similar sizes. Males of the larger species grow to a relatively far greater size than females. Between these extremes are many macropodid species in which the largest males are only moderately larger (10-40% larger) than the largest females. Thus the extant macropods (Fig. 1) range in dimorphism from homomorphic, through mildly to strongly heteromorphic (*sensu* Jarman 1983).

Table 1. Maximum recorded weights for males and females in populations of five potoroid species, and the ratios of male to female weight.

	Max. recorded weights Male	Max. recorded weights Female	Ratio Male:Female weight	Authority for measurements
Aepyprymnus rufescens (NSW)	3.22	3.08	1.05	Schlager 1981
Aepyprymnus rufescens (Qld)	3.00	3.50	0.86	Johnson, P. M. 1978
Bettongia gaimardi	1.89	1.76	1.07	Johnson K. A., pers. comm.
Potorous tridactylus (Tas)	1.74	1.60	1.09	Johnston and Sharman 1976
Bettongia lesueur	1.05	1.05	1.00	Tyndale-Biscoe 1968
Hypsiprymnodon moschatus	0.68	0.55	1.24	Johnson and Strahan 1982

Table 2. Maximum recorded weights for males and females in populations of twenty macropodid species, and the ratios of male to female weight. Access to unpublished records was provided by those authorities listed with no dates; Queensland Museum refers to my abstraction of data from collections made by Kirkpatrick and deposited in the Queensland Museum; A.N.W.C. refers to materials held by C.S.I.R.O. Wildlife in the Australian National Wildlife Collection; C.C.N.T. data refers to measurements, provided by Johnson and Lundie-Jenkins, of animals held by the Conservation Commission of the Northern Territory.

Species	Max. recorded weights Male	Female	Ratio Male:Female weight	Authority for measurements
Macropus rufus	93.0	37.0	2.51	Denny, M. J. S
M. giganteus giganteus	75.5	40.0	1.89	Poole, W. E
M. giganteus tasmaniensis	74.0	38.0	1.95	Pearse, 1981
M. fuliginosus	72.0	39.0	1.85	Poole, W. E.
M. robustus robustus	49.0	23.5	2.09	Jarman, P. J.
M. robustus robustus	46.0	23.0	2.00	Poole, W. E.
M. robustus erubescens	58.0	25.0	2.32	Denny, M. J. S.
M. agilis	29.25	14.65	2.00	Merchant, J. C.
M. agilis	23.8	12.7	1.87	Kirkpatrick, T. H. Queensland Museum
M. rufogriseus banksianus	27.0	14.5	1.86	Merchant, J. C. and Calaby, J. H.
M. rufogriseus banksianus	20.4	12.2	1.67	Kirkpatrick, T. H. Queensland Museum
M. rufogriseus fruticus	26.75	16.0	1.67	Merchant, J. C. and Calaby, J. H. (field, Tasmania)
M. dorsalis	19.0	8.5	2.24	A.N.W.C.
M. eugenii	9.1	6.9	1.32	Poole, W. E. and Brown, G. D. (field, Kangaroo Is.)
M. parma	5.3	4.1	1.29	Maynes, 1976
Wallabia bicolor	22.0	15.0	1.47	Jarman, P. J.
Wallabia bicolor	20.5	16.5	1.24	Merchant, J. C.
Thylogale thetis	9.2	4.8	1.92	Johnson, K. A., 1977
Setonix brachyurus	3.3	2.7	1.22	Dunnet, 1962
Petrogale brachyotis (W.A.)	5.60	4.65	1.20	A.N.W.C. and Maynes, G.M.
Petrogale brachyotis (Arnhem Land)	4.50	3.20	1.41	A.N.W.C.
P. penicillata NSW/Qld	7.60	6.60	1.15	A.N.W.C.
P. xanthopus	8.00	7.00	1.14	Poole *et al.* 1985
P. xanthopus (asymptoic values)	10.4	8.3	1.25	Poole *et al.* 1985
P. burbidgei	1.43	1.40	1.02	Kitchener and Sanson, 1978
Peradorcas concinna	1.50	1.70	0.88	Sanson, G.
Lagostrophus fasciatus	1.90	1.90	1.00	Hughes, 1965
Lagorchestes conspicillatus	4.30	4.20	1.02	A.N.W.C. (Calaby, J. H.)
L. hirsutus	1.64	1.64	1.00	C.C.N.T. data (Johnson, K. A. and Lundie-Jenkins, G.)

Development of dimorphism with age

Figures 2 to 7 illustrate growth in weight with age (or molar index) in several macropod species. The data usually come from large samples of free-living individuals from one population or region, thus avoiding the confounding effects of geographical variation. Both sexes of most species show much variation in size within each age class, and this individual variability seems to typify macropods.

Growth of individuals against real time (rather than the indexed ages used to produce Figs 2 to 7) of species held in captivity by C.S.I.R.O. Division of Wildlife and Ecology (data kindly made available by W. E. Poole on behalf of J. Calaby, J. Merchant and himself) confirm the form of growth of these wild animals. Unfortunately few captive animals, especially males of the larger species, were weighed and measured beyond about eight years of age, so data from captive animals do little to confirm that growth persists in the all-important later years. Nevertheless, the studies by Poole *et al.* (1982a,b, 1984) of grey kangaroos remain the most thorough published descriptions of growth in any macropodid species. The data from captive *M. fuliginosus* show variations in weight between individuals of the same

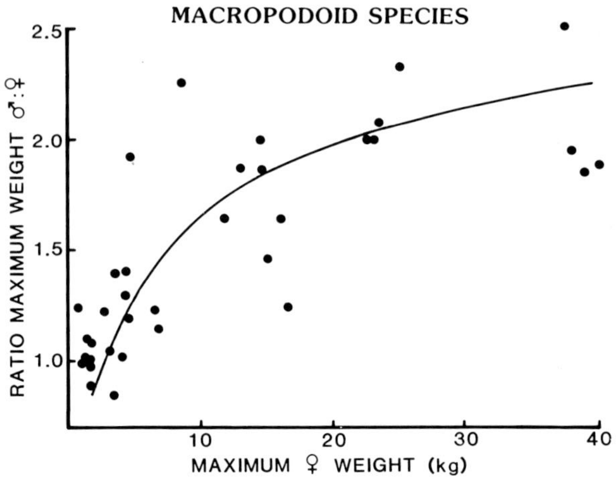

Fig. 1. The ratios of male to female weights (using the largest values known for the species or population) plotted against the maximum female weight for that species or population, for potoroids and macropodids.

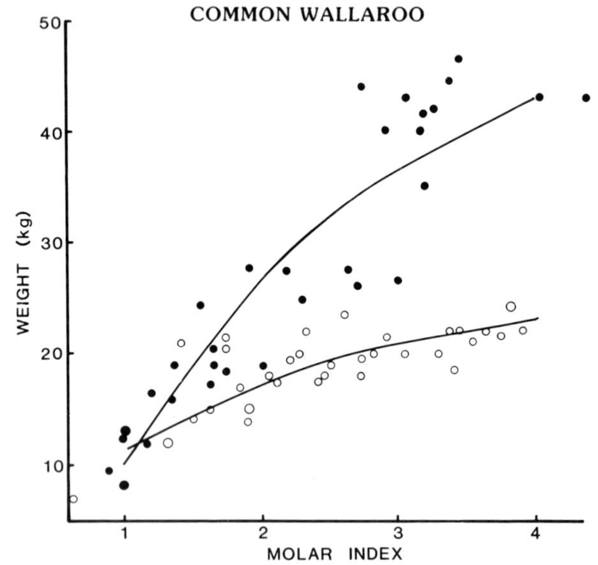

Fig. 3. Individual weight plotted against molar index for male (black dot) and female (open circle) *Macropus robustus robustus*. Larger symbols indicate two coincident records. The sample was taken from a population on the New England Tablelands of New South Wales.

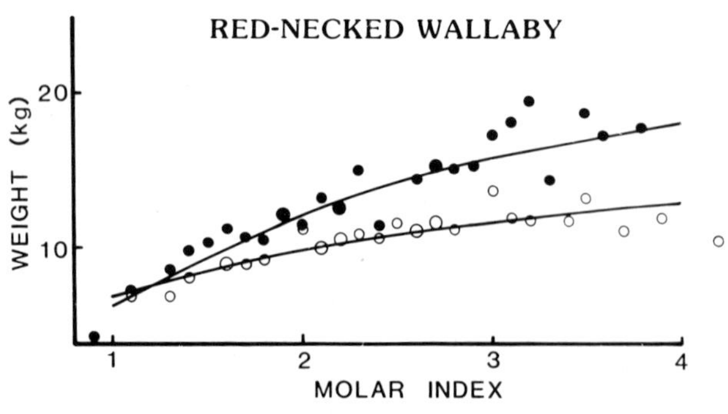

Fig. 5. Mean weights for age-classes, plotted against molar index, for male (black dots) and female (open circles) *Macropus rufogriseus*. Larger symbols indicate means of more than three individuals. Data were taken from the Kirkpatrick collection in the Queensland Museum.

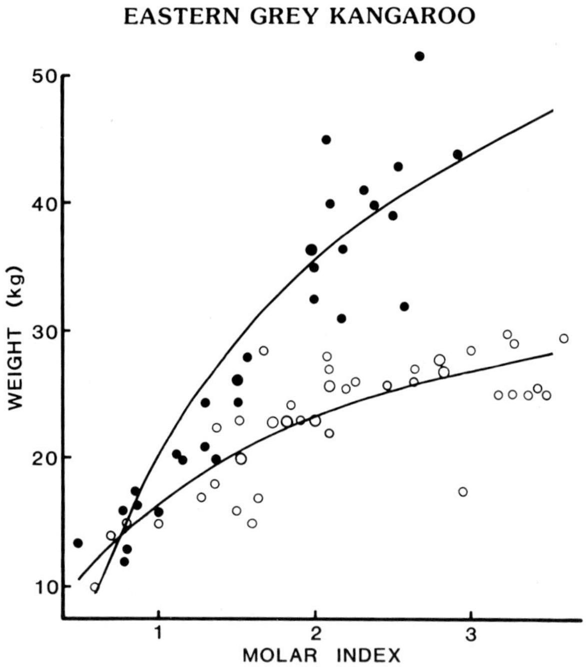

Fig. 2. Individual weight plotted against molar index for male (black dot) and female (open circle) *Macropus giganteus*. Larger symbols indicate two coincident records. The sample was taken from a population on the New England Tablelands of New South Wales.

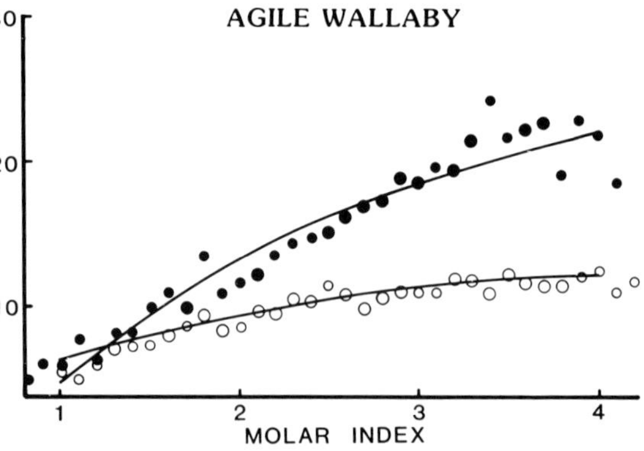

Fig. 4. Mean weights for age-classes, plotted against molar index, for male (black circles) and female (open circles) *Macropus agilis*. Larger symbols indicate means of more than three individuals. Data were taken from the Kirkpatrick collection in the Queensland Museum.

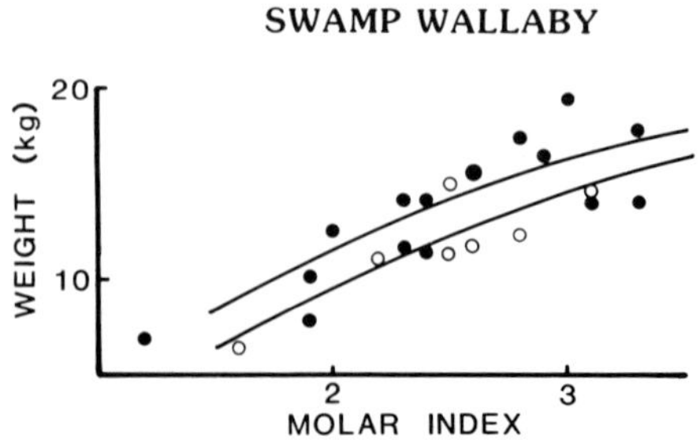

Fig. 6. Individual weight plotted against molar index for male (black dots) and female (open circles) *Wallabia bicolor*. The sample was taken from northeastern New South Wales.

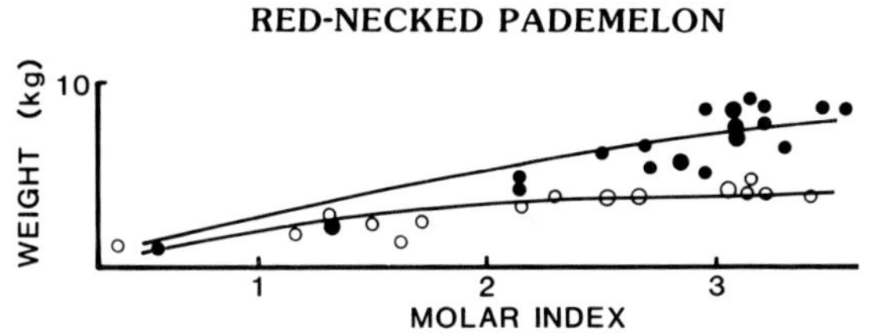

Fig. 7. Individual weight plotted against molar index for male (black dots) and female (open circles) *Thylogale thetis*. Larger symbols indicate two coincident records. The sample was taken from the Dorrigo plateau, north-eastern New South Wales, from animals collected by K. A. Johnson.

sex and age, and between successive measurements on the same individual, similar to the variance in the field data.

Measurements of known-age captive *Macropus parma* have been used to describe growth in weight with age in that species (Maynes 1976). Dunnett (1962) has published growth in weight of wild *Setonix brachyurus*.

Data on individual growth in a potoroid, *Aepyprymnus rufescens* (Fig. 8), and on the comparably small macropodids *Peradorcas concinna* (data from G. Sanson) and *Lagorchestes hirsutus* (data from G. Lundie-Jenkins), show that such species in captivity apparently cease growing early in "adult" life. This is confirmed by field observations of *Aepyprymnus rufescens*, which showed that animals attained an "adult" weight about which their weight then fluctuated (data from I. Wallis, and my own unpublished records).

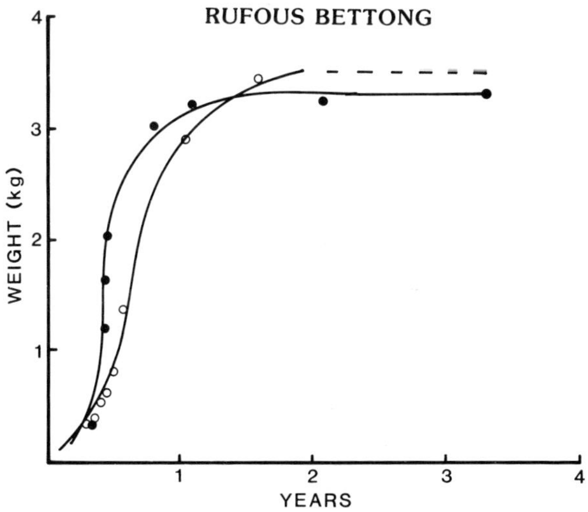

Fig. 8. Plots of individual growth of a male (black dots) and a female (open circles) *Aepyprymnus rufescens* in captivity, showing growth to a plateau weight.

The macropods thus show two broadly differing forms of growth. The extant potoroids and at least some small macropodids (probably all those under 3 or 4 kg) grow to a plateau weight which is retained, allowing for fluctuations, for the rest of their lives. In the medium and large macropodids, by contrast, at least one and usually both sexes grow throughout most of their lives. These two forms of growth are further distinguished by the former characterizing homomorphic species, the latter heteromorphic ones in which the sexes differ in their rates of growth and sometimes in the degree of persistence of growth. Among the heteromorphic species, the larger and most heteromorphic show the greatest differences between males and females in rates of growth. This suggests that at least part of their dimorphism arises simply from differences in rate of general growth. This point is investigated further in the rest of this paper.

Growth in skull length

There are few macropod species for which samples of weights of individuals of known ages are available. For more species there are good collections of skulls. These can be used as indicators of both relative age (by molar eruption and progression) and body size. Examples of growth in skull length with age appear in Figs 9 to 14. Significant correlations between body weight and skull length for all those species in which both were recorded indicate that persistent skull growth implies persistent growth in body weight. Records of skull length against relative age therefore expand the sample of species whose form of growth can be inferred. From these figures it appears that the sexes of *Macropus agilis, M. dorsalis, M. parryi* and *M. rufogriseus* differ considerably in their rates of growth in skull size, the females' rates of growth decelerating faster than the males'. The sexes in *Wallabia bicolor* and possibly *Petrogale penicillata* differ much less in their rates of skull growth.

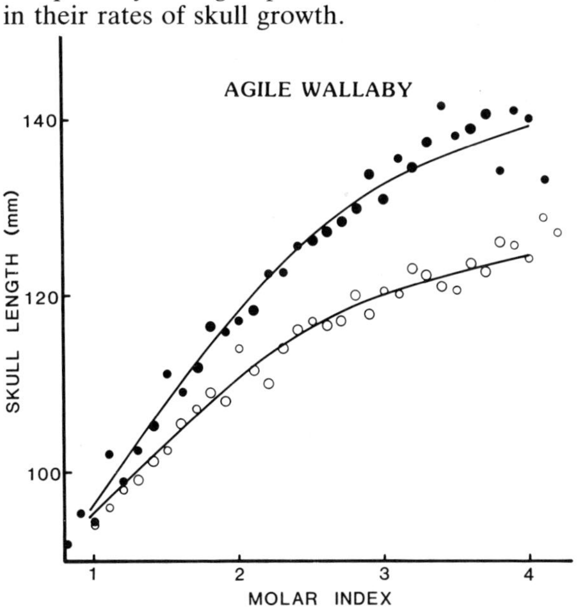

Fig. 9. Mean skull length for age-classes, plotted against molar index, for male (black dots) and female (open circles) *macropus agilis*. Larger symbols indicate means of more than three individuals. Data were taken from the Kirkpatrick collection in the Queensland Museum.

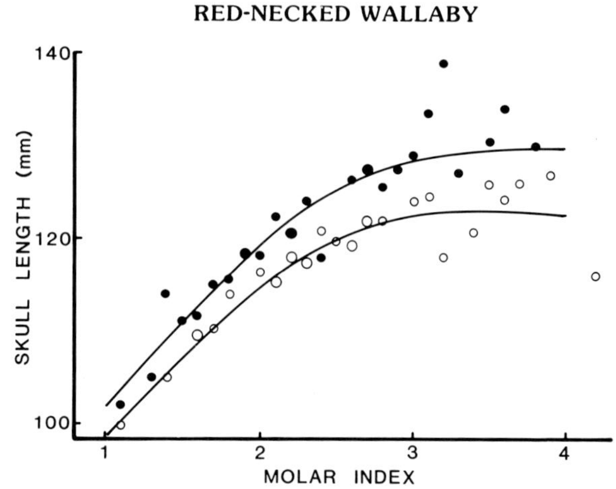

Fig. 10. Mean skull length for age-classes, plotted against molar index, for male (black dots) and female (open circles) *Macropus rufogriseus*. Larger symbols indicate means of more than three individuals. Data were taken from the Kirkpatrick collection in the Queensland Museum.

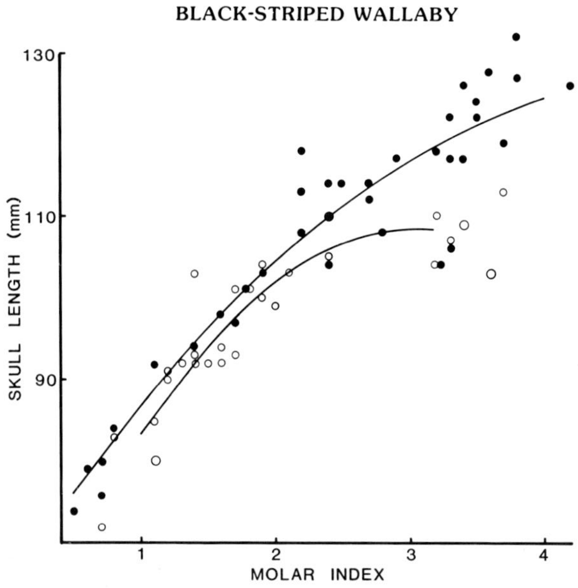

Fig. 12. Individual skull lengths for male (black dots) and female (open circles) *Macropus dorsalis*, plotted against molar index. Sampled animals came from southern Queensland and northern New South Wales.

Absolute and relative growth of the forelimbs

Other body parts of at least the larger kangaroos grow so consistently that they, particularly head, pes and tail, have been used as indices of age (Sharman *et al.* 1964; Poole *et al.* 1982a,b, 1984, for example), although it was recognized that their usefulness declined with age as their rates of growth fell. Growth in the lengths of pes, tail, head and body, or ear are not explored further in this paper. However, I will explore relative growth of the forelimbs. In doing this, I am looking for departures from simple isometric growth in body parts, that is, for changes in shape. These changes may or may not be allometric. I am also looking for differences in shape between males and females.

It has been noted by several authors (e.g., Maynes 1976; Johnson 1977; Jarman 1983) that the arms of large males of the larger macropod species are relatively longer, and the forearm girdles in general carry relatively more muscle, than those of females. This is exemplified in Johnson's (1977) study of rates of growth in length of head, pes and arms of male and female

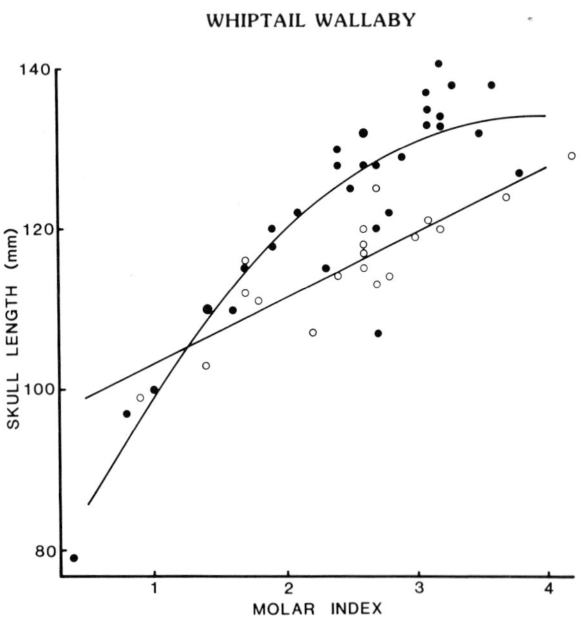

Fig. 11. Individual skull lengths for male (black dots) and female (open circles) *Macropus parryi*, plotted against molar index. Sampled animals came from southern Queensland and northern New South Wales.

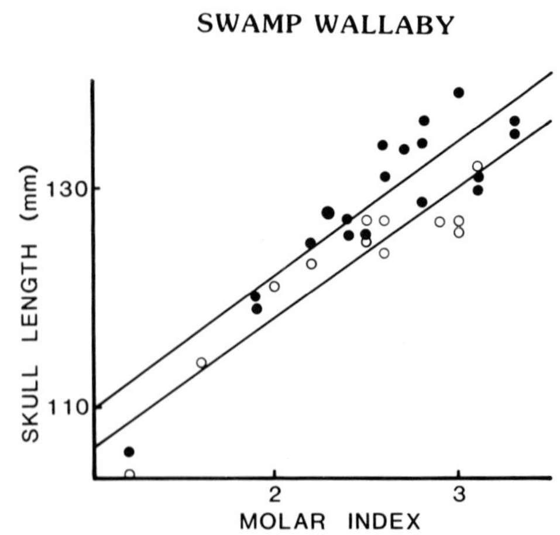

Fig. 13. Individual skull lengths for male (black dots) and female (open circles) *Wallabia bicolor*, plotted against molar index. Sampled animals came from northern New South Wales.

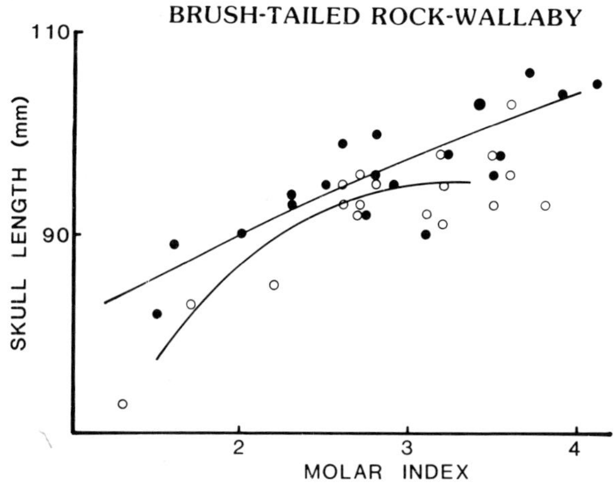

Fig. 14. Individual skull lengths for male (black dots) and female (open circles) *Petrogale penicillata,* plotted against molar index. Sampled animals came from the C.S.I.R.O. Australian National Wildlife Collection, mainly from New South Wales, and from the New England Tablelands.

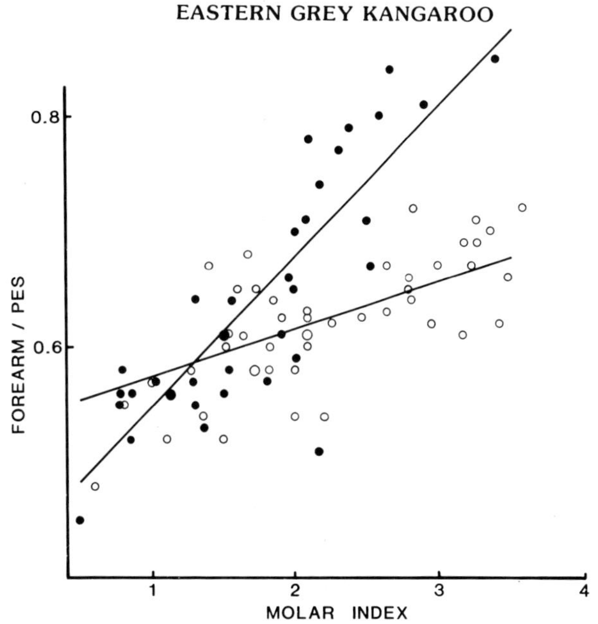

Fig. 16. The ratio of the length of forearm to length of pes, plotted against molar index, for male (black dots) and female (open circles) *Macropus giganteus.* Larger symbols indicate two coincident records. The animals were sampled on the New England Tablelands of northern New South Wales.

Thylogale thetis. The ratio of male to female growth constants for these were: pes 1.32:1, head 1.39:1 and arm 2.00:1, demonstrating that growth of the arms of males is accelerated, relative to growth of other parts of the body, compared to that of females. The growth of arm length against molar index for this species is shown in Fig. 15.

Similar evidence for the changes in bodily proportions with age, produced by exaggerated growth of the forelimbs in males, compared to females, is given by plots of the ratios of lengths of arm to pes against age or an age index, as in Figs 16 and 17. These figures show that, in both *M. giganteus* and *M. robustus*, the arm length:foot length ratios of both males and females increase with age, but that those of the males do so more rapidly. These results are comparable to those of Poole *et al.* (1982a,b) for *M. giganteus* and *M. fuliginosus*.

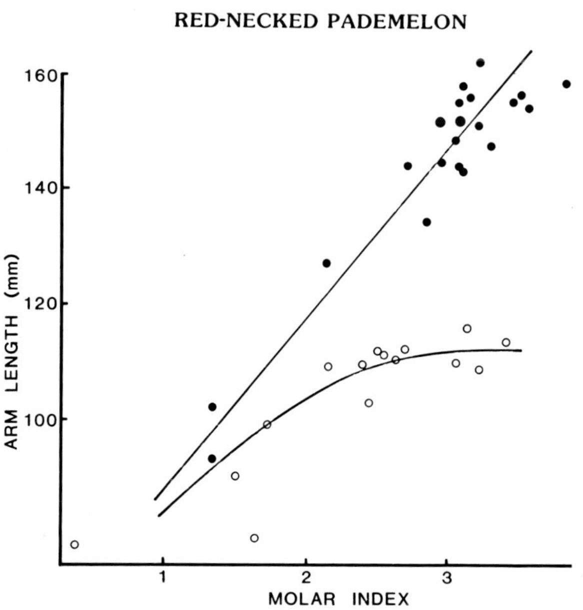

Fig. 15. Individual arm lengths plotted against molar index for male (black dots) and female (open circles) *Thylogale thetis.* Larger symbols indicate two coincident records. The animals were sampled (mainly by K. A. Johnson) on the Dorrigo Plateau, northern New South Wales.

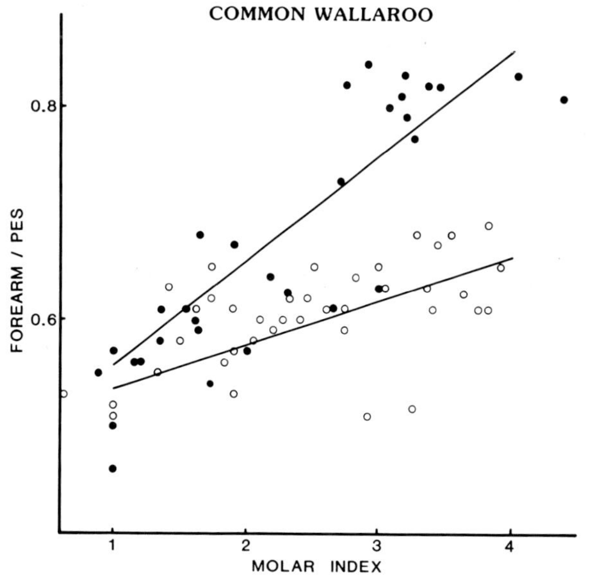

Fig. 17. The ratio of the length of forearm to length of pes, plotted against molar index, for male (black dots) and female (open circles) *Macropus robustus.* The animals were sampled on the New England Tablelands of northern New South Wales.

Muscularity is reflected in the circumference of the forearm. Unlike bone measurements, this measurement is likely to vary with the animal's physical condition. Nevertheless, it is a useful reflection of any dimorphic development of the musculature of the forelimb girdle. Figures 18 and 19 show that the ratio of forearm circumference to length increases with age in males but not females of *M. giganteus* and *M. robustus*, suggesting that in these species not only the relative length, but also the relative muscularity, of males' arms increases with age. Females' arms do not appear to become relatively more muscular with age. The increase in forearm relative muscle bulk in *M. robustus* is further illustrated in Figure 20, which plots the ratio of the weight of dissected muscle from the forearm to the weight of the bones, against the body weight of the individual. This shows that the larger males do, indeed, have relatively more muscular forearms.

Dimorphism in the forelimbs of macropod species

The outcome of these differences in growth of arms relative to growth of other body parts in a range of macropod species is recorded in Table 3, which lists the ratios of arm length to pes length for the largest specimens of males and females (unless otherwise indicated) for sampled populations of 15 macropodid and three potoroid species. The Table also gives the ratio of male to female values for the arm:pes ratio. This is intended to illustrate interspecific variation in the extent of this aspect of dimorphism. It shows that, at their maximum, male relative arm length (relative to his own foot length) is much greater than

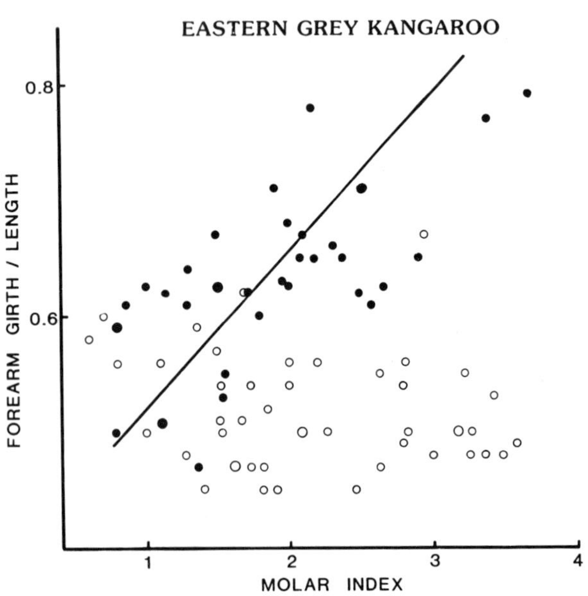

Fig. 18. The ratio of the girth (maximal circumference) to the length of the forearm, plotted against molar index, for male (black dots) and female (open circles) *Macropus giganteus*. Larger symbols indicate two coincident records. The animals were sampled on the New England Tablelands of northern New South Wales.

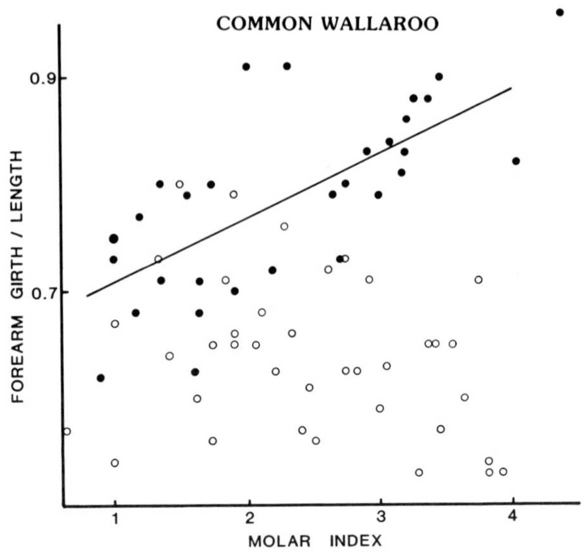

Fig. 19. The ratio of the girth (maximal circumference) to the length of the forearm, plotted against molar index, for male (black dots) and female (open circles) *Macropus robustus robustus*. Larger symbols indicate two coincident records. The animals were sampled on the New England Tablelands of northern New South Wales.

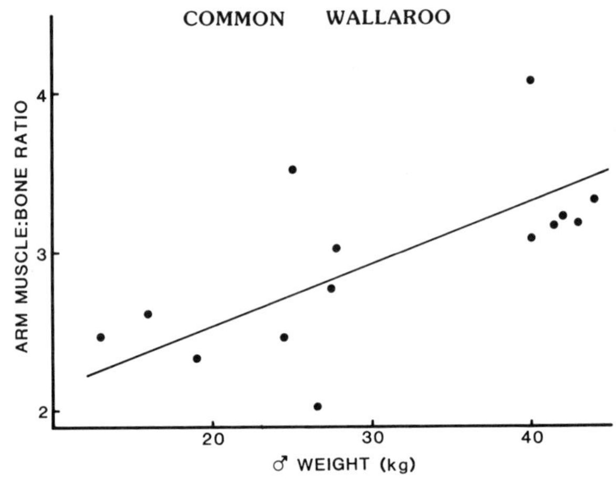

Fig. 20. The ratio of the weight of fresh muscle to fresh bone in the forearms of male *Macropus robustus robustus*, plotted against the individual's weight. Animals were sampled on the New England Tablelands of northern New South Wales.

female relative arm length in the largest *Macropus* species, little greater in the smaller *Macropus* species and other medium-sized wallabies, and no greater in the smallest macropodid and potoroid species. These ratios of male to female arm:pes ratios are plotted against each species' (or populations') maximal female weight in Figure 21, which confirms that the largest species have males which eventually develop the most exaggeratedly long forearms.

Table 3. The ratio of forearm length to pes length in fifteen macropodid and three potoroid species. Unless indicated, the ratios are from the measurement of the largest available individuals of each sex.

Species	Ratio of forearm length to pes length Male	Female	Ratio Male:Female weight	Authority for Measurements
M. rufus	1.171	0.806	1.453	Jarman, P. J.
M. giganteus	1.000	0.682	1.466	Jarman, P. J.
M. giganteus	0.937	0.697	1.344	Poole, W. E.
M. fuliginosus melanops	0.953	0.700	1.361	Poole, W. E.
M. fuliginosus ocydromus	0.948	0.711	1.333	Poole, W. E.
M. robustus robustus	0.906	0.685	1.323	Jarman, P. J.
M. rufogriseus banksianus	0.780	0.640	1.219	Jarman, P. J.
M. rufogriseus banksianus	0.795	0.663	1.199	Merchant, J. C. and Calaby, J. H.
M. rufogriseus fruticus	0.820	0.686	1.195	Merchant, J. C. and Calaby, J. H. (Tasmania)
M. antilopinus	0.764	0.630	1.213	Poole, W. E.
M. parma (maximum)	0.761	0.686	1.100	Maynes, G. M. and Jarman, P. J.
M. parma (mean)	0.755	0.683	1.105	Maynes, G. M.
M. agilis (N.T.)	0.838	0.687	1.202	Merchant, J. C.
Wallabia bicolor	0.771	0.667	1.156	Jarman, P. J.
Thylogale thetis	1.143	0.941	1.215	Johnson, K. A.
T. stigmatica	0.827	—	—	Jarman, P. J.
Petrogale xanthopus	0.700	0.656	1.067	Poole, W. E.
Petrogale xanthopus (asymptotic values)	0.676	0.630	1.073	Poole *et al.* 1985
Petrogale burbidgei (mean)	0.639 (N=4)	0.601 (N=4)	1.063	Kitchener and Sanson, 1978
Peradorcas concinna (mean)	0.572 (N=4)	0.577 (N=4)	0.991	Sanson, G.
Lagorchestes hirsutus	0.643	0.553	1.163	CCNT (Johnson and Lundie-Jenkins)
Aepyprymnus rufescens	0.532	0.515	1.033	Jarman, P. J.
Bettongia gaimardi	0.524	0.480	1.092	Johnson, K. A.
Potorous tridactylus	0.762	0.743	1.026	Johnson, K. A.

Table 4 lists the ratios of forearm circumference to length for males and females of two potoroid and eight macropodid species, and presents values only for males of two other macropodid species. These ratios vary between species, for both sexes. Some species have forearms which are very stocky for their length; others have slender arms. The two potoroids (which use the forearms to dig up food) are among the stocky-armed species, with girth-to-length ratios close to 1. Among the eight macropodid species for which both male and female values of the ratio are available, there is a weak correlation between the male and female values in each species (Spearman rank coefficient $r_s=0.7143$, $N=8$, $p<0.05$). Species with stocky-armed males tend to have stocky-armed females: ones with slender-armed males have slender-armed females. Yet over these eight macropodid species and the two potoroids there is an interesting relationship between the ratio of male to female values for the arm girth:length ratio and the species' weight (maximum female weight), shown in Figure 22. The species which are most heteromorphic in their muscularity are not the largest, but are actually some of the medium-sized wallabies.

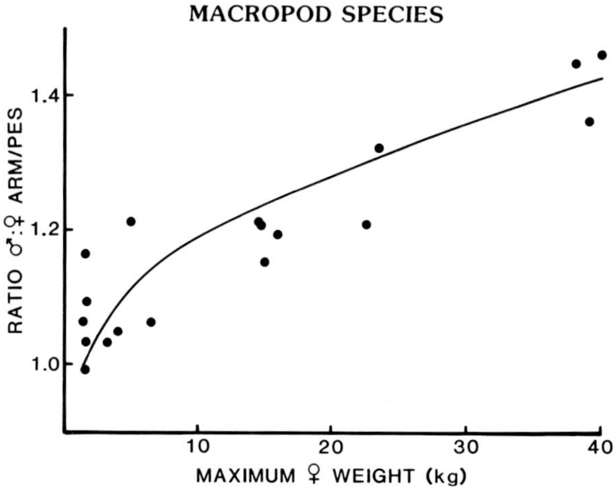

Fig. 21. The distribution, across a range of sizes of species, of the degree of relative elongation of the arms of male macropods. This is shown by the ratio of male to female values of the ratios of arm length to pes length, plotted against maximum known female weight, for a range of macropod species.

One consequence of the exaggerated growth of the arms in many male macropods is that old or large males have forepaws that are very much larger than those of females (Plate 1). They also have relatively larger and stouter claws on the manus (but not noticeably so on the pes). Table 5 presents measurements of the third digit of the manus of the males and females of one potoroid

Table 4. The ratios of forearm circumference to length for two potoroid and ten macropodid species. Measurements for *Aepyprymnus rufescens* and *Lagorchestes hirsutus* are means of the indicated numbers of samples; all others are values from the largest animals I have inspected.

Species	Ratio of forearm circumference to length: Male	Female	Ratio of Male:Female weight	Authority for measurements
Bettongia gaimardi	1.136	1.053	1.079	Johnson, K. A.
Aepyprymnus rufescens	1.045 (N=4)	1.048 (N=3)	0.9975	Jarman, P. J.
Lagorchestes hirsutus	0.625 (N=7)	0.541 (N=11)	1.155	Lundie-Jenkins, G. and Johnson, K. A.
Lagorchestes hirsutus (maximum)	0.676	0.599	1.129	Lundie-Jenkins, G. and Johnson, K. A.
Thylogale stigmatica	0.922	—	—	Jarman, P. J.
T. thetis	0.875	—	—	Jarman, P. J.
Macropus parma	0.747	0.558	1.339	Jarman, P. J.
M. rufogriseus	0.564	0.375	1.504	Jarman, P. J.
W. bicolor	0.768	0.500	1.536	Jarman, P. J.
M. robustus	1.231	0.787	1.564	Jarman, P. J.
M. giganteus	0.790	0.600	1.317	Jarman, P. J.
M. rufus	0.588	0.526	1.118	Jarman, P. J.
M. fuliginosus	0.574	0.546	1.051	Jarman, P. J.

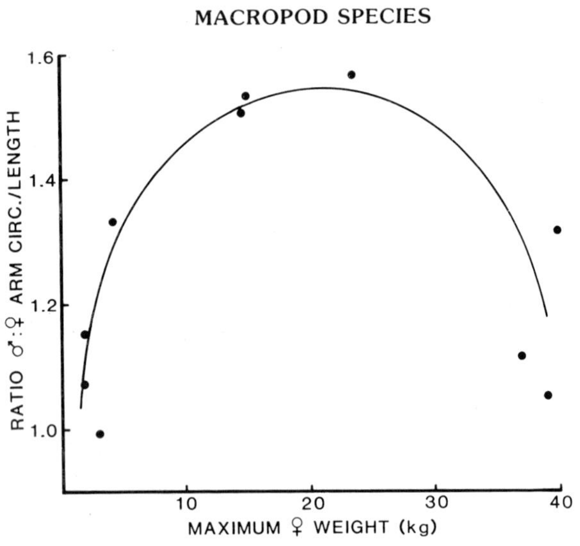

Fig. 22. The distribution, across a range of sizes of species, of the degree of relative exaggeration of forearm musculature in male macropods. This is shown by the ratio of male to female values of the ratio of arm girth to length, plotted against maximum female weight, for two potoroid and eight macropodid species.

and six macropodid species, as well as the male:female weight ratios for the largest specimens of each species. In the potoroid, males and females have claws that are equally long and slender (and are used for digging up food). Among the macropodids, the claws of females were always relatively more slender than those of males.

APPENDIX

Appendix 1 lists calculated values for the sampled species in some equations describing growth in weight or body parts.

DISCUSSION

The macropods reviewed in this paper show at least two forms of growth. The potoroids and smallest macropodids grow rapidly to a plateau weight, retaining that for the rest of their lives. The medium and large macropodids show continuing growth in at least males, and often both sexes, throughout life. There is some further differentiation among the latter species between

Table 5. Measurements of the claw of the third (central) digit of the manus in males and females of some macropod species. The measurements for *Aepyprymnus rufescens* are means for those of six adults of each sex. For other species the measurements are taken from the largest specimen of each sex that I have been able to inspect.

Species	Weight ratio Male:Female	Claw on third digit on manus: Length Male	Length Female	Width Male	Width Female	Ratio of length:width Male	Ratio of length:width Female
A. rufescens (Male, N=6; Female, N=6)	1.05	19.32	19.05	2.8	2.9	6.78	6.50
T. thetis	1.92	16.9	13.5	6.6	3.6	2.56	3.75
W. bicolor	1.47	21.6	14.7	6.6	3.4	3.27	4.32
M. parma	1.29	14.7	13.8	—	—	—	—
M. rufogriseus	1.86	26.5	18.3	6.8	3.8	3.90	4.82
M. robustus	2.09	39.2	22.5	11.7	6.3	3.35	3.57
M. giganteus	1.89	51.9	29.0	11.4	5.3	4.55	5.47

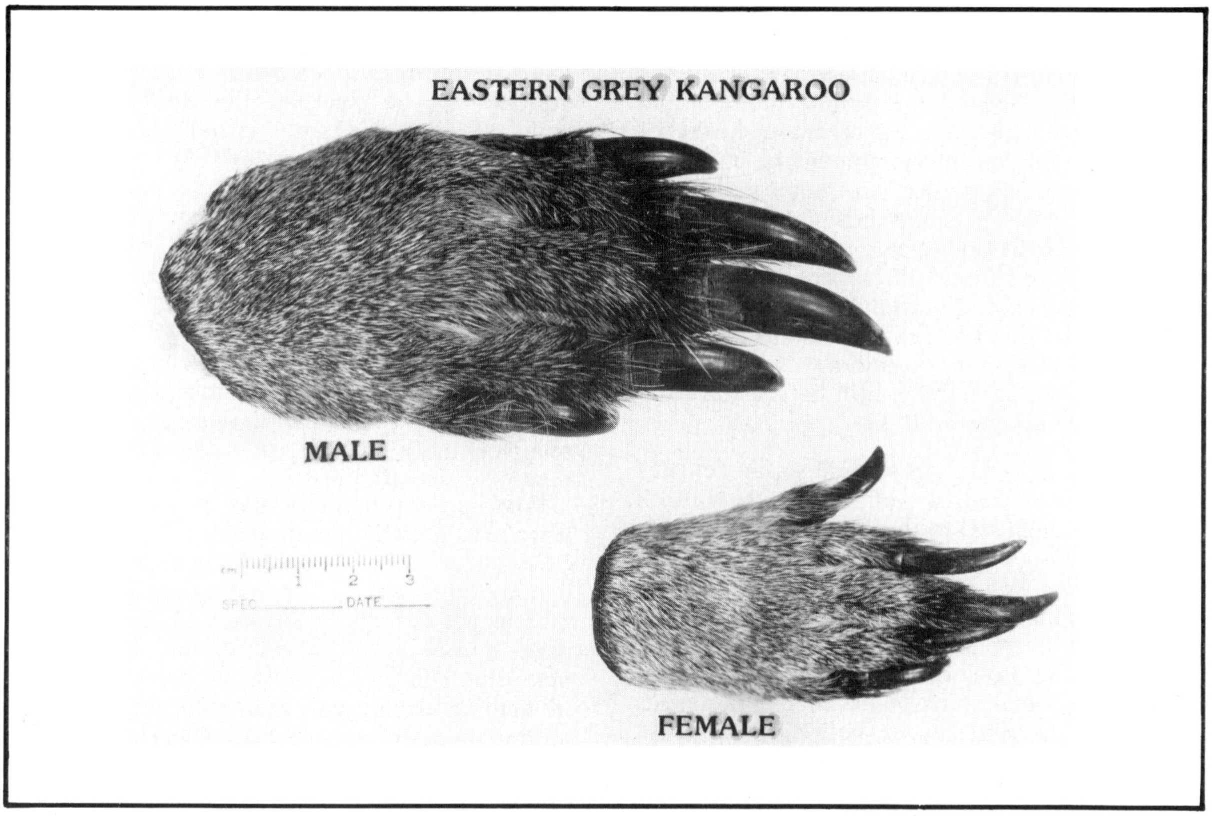

Plate 1. The forepaws of large male and female *Macropus giganteus*, illustrating the relatively stouter claws of males (which have the upper paw in this plate).

Plate 2. A large male *Macropus giganteus*, illustrating the exaggeration of forelimb girdle development, which is being displayed by this male.

those whose females decelerate their rate of growth strongly, growing only slowly past middle age, and those (such as *Petrogale* and *Wallabia* species) whose females continue quite rapid growth. The consequences of these forms of growth are homomorphism among the potoroids and smallest macropodids, and heteromorphism among the other macropodids (Jarman 1983). Among the medium and large macropodids there may again be functional differences. In the larger species, males exceed maximal female weight for much of their adult life; in the medium-sized macropodid species, there may always be some females larger than a male until he is a considerable way through his adult life.

In all macropod species studied so far, females first experience oestrus well before they reach maximum weight. Even in homomorphic species such as *Aepyprymnus rufescens* a female may conceive at about half her eventual plateau weight. In the heteromorphic species females may still be growing quite rapidly at the time of the first oestrus. There are several consequences of this. Primiparous females are smaller than multiparous ones. The size differences between them and males courting them is exaggeratedly large. This may be unimportant, unless active rejection of importunate males is a part of the female's reproductive strategy.

Considerable variability in weight between individuals of the same sex and age seems to be a common feature of macropod species. This, coupled with field observations that males of *M. rufogriseus* (Johnson. C. N., pers. comm.) and *M. giganteus* (my data), may show spurts of more rapid growth which are not obviously triggered by improvements in the season or resources, and are not synchronized with their peers, suggest that growth is plastic in macropods.

Heteromorphism in macropods has two components: an increase in overall body size of males relative to females, brought about by faster and more prolonged fast growth in males than in females; and a change in male shape, relative to females, brought about by exaggerated development of elements of the forelimb girdle. In this paper I have concentrated on measurements of the forearm. In fact, all parts of the forelimb girdle are involved, from the scapula to the claws, and all the associated musculature. So obvious is the development of this part of the body in males that it is the easiest way to sex and age males of the medium and large macropods in the field (see Plate 2). Moreover, males use postures and actions which reveal their arm length and muscularity when displaying to other males.

It is interesting that there are different ways among macropod species of exaggerating the forelimb development. Some species have concentrated on elongation of the forelimbs; others have exaggerated the muscularity of the forelimbs. Poole (pers. comm.) points out that there are even subspecific differences in forelimb morphology; he contrasts the stocky, thick-set build of the Kangaroo Island subspecies of Western Grey Kangaroo *M. fuliginosus fuliginosus*, and the more gracile subspecies *M. f. melanops* on the adjacent mainland (Poole *et al.* 1984). I believe, although I have not measured this, that those species with exaggeratedly elongate arms may have exaggerated the size of the manus more than have those with exaggeratedly muscular arms. The two modes are not distinct; males of all medium and large species develop arms which are both relatively longer and relatively more muscular than those of females.

Forelimb development may be seen as equivalent to the growth of cranial weapons in bovids or cervids (Jarman 1983), providing organs for both fighting and display. The timing of the development makes it highly likely that it is controlled by male hormones produced at and after sexual maturity. The forelimbs are used by macropods in fighting to grapple and hold the opponent around the head, neck and shoulders, and holding him close to the chest. Wallaroos are also notorious biters, both of each other (shown by the torn ears of large wallaroo males) and of humans who handle them (Poole, pers. comm.; and painful pers. obs.); this suits their "close-in" style of fighting. Other species, typfied by *M. giganteus,* hold the opponent at arm's length, clawing at the face, positioning him for a kick with the hind feet. The former tend towards muscular arms, the latter towards long ones.

Although all macropod species kick with the hind feet when fighting, I do not think that the hindlimbs develop dimorphically. Development of the forelimbs as weapons may be possible only because they are relatively lightly committed for other functions such as locomotion, while the hindlimbs are unalterably dedicated to locomotion. However, it would be interesting to test for dimorphism in the hindlimbs, since species vary in the form and frequency of kicking in their fighting styles.

The adaptiveness of dimorphism and the forms that it takes in the macropods has been argued elsewhere (Jarman 1983). Briefly, in heteromorphic species, persistent body growth and exaggeration of the forelimbs as weapons is reflected behaviourally in males establishing a dominance hierarchy based upon fighting ability, which is itself the result of size and weaponry (arm length or muscularity). Position in the hierarchy partly determines a male's probability of gaining undisputed access to females in or approaching oestrus, in order to court and mate them. The male that survives longest, or organizes its growth so as to become the biggest, of those in its cohort is able to gain a disproportionately great share of the matings. In the absence of territoriality, harems, or any other pre-emptive exclusion of other males from females, age, size and fighting weaponry alone decide inter-male competition for access to oestrous females. The implied plasticity in growth suggests that a male might grow tactically as well as strategically.

This emphasis on persistent growth in all heteromorphic macropods distinguishes the Macropodidae from the Bovidae and Cervidae, in which few heteromorphic species show persistent growth (Jarman 1983), but many more species are territorial. These differences in social and spatial organization have yet to be convincingly explained. Differences in food resources, feeding styles, or type of fighting and weaponry might all be involved.

The phenomenon of persistent growth in female macropods deserves more attention. If it were homologous to extended growth in males, I would expect it to be adaptive in inter-female competition of some kind, for resources or the oportunity to breed or both. There is anecdotal evidence (K. P. Joblin, pers. comm.) that breeding opportunity among females of *Petrogale penicillata*, when sharing a refuge, is determined by size, the largest female raising young most often. This appears to be a species in which females show strongly persistent growth. Whether the two phenomena are regularly linked deserves further research.

ACKNOWLEDGEMENTS

It will be seen from Tables 1 to 4 that I owe a debt of gratitude to many colleagues who have given me access to their accumulated data; these include John Calaby, Martin Denny, Ken Johnson, Geoff Lundie-Jenkins, Bill Poole, Gordon Sanson and Robert Taylor. I am also indebted to those collectors, especially Tom Kirkpatrick and John Calaby, whose carefully measured and documented field specimens are available in museums. Many associates have, over the past 14 years, helped me measure live or dead macropods; in particular I thank Gary Martin, Stuart Green and Robyn Liddle. My most long-standing debt is to Ken Johnson, with whom I initially developed my interest in macropod dimorphism and who enthusiastically collected measurements on animals in New South Wales, Tasmania and the Northern Territory. To these, and the many students who have autopsied macropods with me, I am very grateful.

REFERENCES

DUNNET, G. M., 1962. A population study of the quokka, *Setonix brachyurus* (Quoy and Gaimard) (Marsupialia): II. Habitat, movements, breeding and growth. *C.S.I.R.O. Wildlife Research.* **7**: 13-32.

HUGHES, R. D., 1965. On the age composition of a small sample of individuals from a population of the banded hare wallaby, *Lagostrophus fasciatus* (Péron and Lesueur). *Aust. J. Zool.* **13**: 75-95.

JARMAN, P. J., 1983. Mating systems and sexual dimorphism in large, terrestrial, mammalian herbivores. *Biological Reviews.* **58**: 485-520.

JOHNSON, K. A., 1977. Ecology and management of the red-necked pademelon, *Thylogale thetis,* on the Dorrigo Plateau of northern New South Wales. Ph.D. Thesis, University of New England.

JOHNSON, P. M., 1978. Reproduction in the rufous rat-kangaroo *Aepyprymnus rufescens* (Gray) in captivity with age estimation of pouch young. *Qld. J. Agric. Anim. Sci.* **35**: 69-72.

JOHNSON, P. M. AND STRAHAN, R., 1982. A further description of the musky rat-kangaroo, *Hypsiprymnodon moschatus* Ramsay, 1876 (Marsupialia, Potoroidae), with notes on its biology. *Aust. J. Zool.* **21**: 27-46.

JOHNSTON, P. G. AND SHARMAN, G. B., 1976. Studies on populations of *Potorous* Marsupialia. Part 1. Morphological variation. *Aust. J. Zool.* **24**: 573-88.

KIRKPATRICK, T. H., 1964. Molar progression and macropod age. *Qld. J. Agric. Sci.* **21**: 163-65.

KIRKPATRICK, T. H., 1965. Studies of Macropodidae in Queensland. 2. Age estimation in the grey kangaroo, the red kangaroo, the eastern wallaroo and the red-necked wallaby, with notes on dental abnormalities. *Qld. J. Agric. Anim. Sci.* **22**: 301-17.

KITCHENER, D. J. AND SANSON, G., 1978. *Petrogale burbidgei* (Marsupialia, Macropodidae), a new rock wallaby from Kimberley, Western Australia. *Rec. W.A. Mus.* **6**: 269-85.

MAYNES, G. M., 1976. Growth of the parma wallaby, *Macropus parma* Waterhouse. *Aust. J. Zool.* **24**: 217-36.

NEWSOME, A. E., MERCHANT, J. C., BOLTON, J. C. AND DUDZINSKI, M. I., 1977. Sexual dimorphism in molar progression and eruption in the agile wallaby. *Aust. Wildl. Res.* **4**: 1-5.

OLIVER, A. J., 1986. Social organization and dispersal in the red kangaroo. Ph.D. Thesis, Murdoch University.

PEARSE, R. J., 1981. Notes on breeding, growth and longevity of the forester or eastern grey kangaroo, *Macropus giganteus* Shaw, in Tasmania. *Aust. Wildl. Res.* **8**: 229-35.

POOLE, W. E., CARPENTER, S. M. AND WOOD, J. T., 1982a. Growth of grey kangaroos and the reliability of age determination from body measurements. I. The Eastern Grey Kangaroo, *Macropus giganteus. Aust. Wildl. Res.* **9**: 9-20.

POOLE, W. E., CARPENTER, S. M. AND WOOD, J. T., 1982b. Growth of grey kangaroos and the reliability of age determination from body measurements. II. The Western Grey Kangaroos, *Macropus fuliginosus fuliginosus, M. f. melanops* and *M. f. ocydromus. Aust. Wildl. Res.* **9**: 203-12.

POOLE, W. E., CARPENTER, S. M. AND WOOD, J. T., 1984. Growth of grey kangaroos and the reliability of age determination from body measurements. III. Interspecific comparisons between Eastern and Western Grey Kangaroos, *Macropus giganteus* and *M. fuliginosus. Aust. Wildl. Res.* **11**: 11-19.

POOLE, W. E., MERCHANT, J. C., CARPENTER, S. M. AND CALABY, J. H., 1985. Reproduction, growth and age determination in the yellow-footed rock-wallaby, *Petrogale xanthopus* Gray, in captivity. *Aust. Wildl. Res.* **12**: 127-36.

SCHLAGER, F. E., 1981. The distribution, status and ecology of the rufous rat-kangaroo, *Aepyprymnus rufescens,* in northern New South Wales. M.Nat.Res. Thesis, University of New England.

SHARMAN, G. B., FRITH, H. J. AND CALABY, J. H., 1964. Growth of the pouch young, tooth eruption and age determination in the red kangaroo, *Megaleia rufa. C.S.I.R.O. Wildl. Res.* **9**: 20-49.

TYNDALE-BISCOE, C. H., 1968. Reproduction and post-natal development in the marsupial *Bettongia lesueur* (Quoy and Gaimard). *Aust. J. Zool.* **16**: 577-602.

APPENDIX I

The following values are for constants in equations used to describe the relationship among measured attributes of samples of macropods. The correlations described by the curves are significant ($p \leq 0.05$) unless otherwise indicated ("ns" by the R^2 value).

Units of measurement are: MI, molar index as described by Kirkpatrick's (1964, 1965) method; weight in kilograms; skull or head length in millimetres; logarithms are natural.

Computation was carried out by Ms. P. Greet, to whom I am most grateful.

(i) Values in the equation: Weight = a + b. $\log_n$MI.

Species	Sex	N.	a.	b.	R^2
M. agilis	M	128	10.51	27.12	0.725
M. agilis	F	157	14.02	9.07	0.553
M. giganteus	M	30	20.29	21.53	0.827
M. giganteus	F	48	16.77	9.14	0.666
M. robustus	M	31	10.23	23.82	0.849
M. robustus	F	38	11.44	8.45	0.704
M. rufogriseus	M	60	13.38	19.19	0.619
M. rufogriseus	F	69	15.15	9.37	0.474
P. brachyotis	M	14	0.69	2.38	0.752
P. brachyotis	F	11	1.40	1.10	0.616
T. thetis	M	25	2.01	4.83	0.779
T. thetis	F	20	2.05	1.70	0.753
W. bicolor	M	14	3.74	11.35	0.709
W. bicolor	F	6	1.02	12.36	0.794

(ii) Values in the equation: Weight = a + b.MI.

Species	Sex	N.	a.	b.	R^2
T. thetis	M	25	−1.06	2.77	0.866
T. thetis	F	20	0.68	1.13	0.828
W. bicolor	M	14	0.85	5.16	0.695
W. bicolor	F	6	−1.36	5.40	0.756

(iii) Values in the equation: Weight = a + b. $\log_n$MI +c. $[\log_n \text{MI}]^2$.

Species	Sex	N.	a.	b.	c.	R^2
M. agilis	M	128	13.56	14.56	8.86	0.739
M. agilis	F	157	11.68	15.81	−4.00	0.566
M. giganteus	M	30	19.15	16.83	8.06	0.852
M. giganteus	F	48	16.61	11.70	−2.45	0.682
M. robustus	M	31	11.97	13.75	7.56	0.862
M. robustus	F	38	11.05	11.20	−2.20	0.719
M. rufogriseus	M	60	13.07	20.11	−0.59	0.619
M. rufogriseus	F	69	10.54	22.03	−7.72	0.520

(iv) Values in the equation: Skull length (or* Head length) = a + b.MI.

Species	Sex	N.	a.	b.	R^2
M. agilis	M	128	86.93	14.76	0.859
M. agilis	F	157	91.81	8.87	0.809
M. dorsalis	M	37	74.23	13.73	0.868
M. dorsalis	F	30	78.82	8.93	0.704
M. parryi	M	32	87.16	14.44	0.775
M. parryi	F	19	95.18	8.11	0.757
M. rufogriseus	M	60	98.81	9.78	0.703
M. rufogriseus	F	69	100.35	6.96	0.536
P. brachyotis	M	14	57.45	7.86	0.762
P. brachyotis	F	11	62.05	4.76	0.835
P. penicillata	M	20	75.57	7.21	0.715
P. penicillata	F	18	69.05	8.10	0.671
S. brachyurus	M	31	52.25	10.89	0.845
S. brachyurus	F	79	49.15	11.31	0.851
*T. thetis**	M	25	76.04	14.51	0.890
*T. thetis**	F	20	73.68	11.40	0.808
W. bicolor	M	21	85.79	12.73	0.787
W. bicolor	F	13	84.06	12.02	0.850

(v) Values in the equation: Skull length (or* Head length) = a + b.MI +c.MI^2.

Species	Sex	N.	a.	b.	c.	R^2
M. agilis	M	128	66.53	33.49	−3.83	0.900
M. agilis	F	157	75.59	22.38	−2.53	0.850
M. dorsalis	M	37	64.65	24.51	−2.39	0.894
M. dorsalis	F	30	53.16	36.09	−5.90	0.857
M. parryi	M	32	70.50	32.66	−4.19	0.835
M. parryi	F	19	90.96	11.81	−0.74	0.764 (ns)
M. rufogriseus	M	60	76.58	29.32	−4.00	0.761
M. rufogriseus	F	69	74.07	28.71	−4.20	0.679
P. brachyotis	M	14	50.43	13.24	−0.96	0.768 (ns)
P. brachyotis	F	11	64.04	3.20	0.29	0.837 (ns)
P. penicillata	M	20	72.11	9.86	−0.47	0.717
P. penicillata	F	18	39.14	32.67	−4.69	0.811
S. brachyurus	M	31	42.92	26.08	−4.72	0.886
S. brachyurus	F	79	41.95	21.51	−3.12	0.875
*T. thetis**	M	25	65.31	26.06	−2.58	0.903
*T. thetis**	F	20	66.06	20.54	−2.24	0.831
W. bicolor	M	21	56.25	38.99	−5.54	0.862
W. bicolor	F	13	52.71	42.78	−6.99	0.948

(vi) Values in the equation: Weight = a + b.Y. Where Y = Skull (or* Head) length.

Species	Sex	N.	a.	b.	R^2
M. agilis	M	128	−71.75	0.852	0.868
M. agilis	F	157	−22.84	0.392	0.602
M. rufogriseus	M	60	−69.98	0.813	0.711
M. rufogriseus	F	69	−36.09	0.505	0.788
P. brachyotis	M	14	−4.36	0.094	0.656
P. brachyotis	F	11	−4.48	0.092	0.773
*T. thetis**	M	25	−14.03	0.178	0.850
*T. thetis**	F	20	−6.13	0.094	0.935
W. bicolor	M	14	−31.16	0.382	0.822
W. bicolor	F	6	−42.89	0.476	0.877

(vii) Values in the equation: (FL:P) = a + b.MI. Where (FL:P) is the ratio of forearm to pes lengths.

Species	Sex	N.	a.	b.	R^2
M. giganteus	M	30	0.419	0.130	0.716
M. giganteus	F	48	0.534	0.041	0.332
M. robustus	M	31	0.431	0.105	0.776
M. robustus	F	38	0.523	0.034	0.411

Growth, development and breeding pattern of the Long-nosed Potoroo, *Potorous tridactylus* (Kerr, 1792) in Tasmania

SALLY L. BRYANT[1]

In Tasmania, past studies on the growth, development and age estimation of *Potorous tridactylus* have been poorly defined or hampered by small sample sizes. In this study, growth curves for body weight and several body parts have been constructed using both wild-caught and known-age pouch young. Morphological development from birth through pouch vacation are described and some seasonality of births has been determined. Comparisons are made with other Potoroidae and Macropodidae.

INTRODUCTION

THE Long-nosed Potoroo *Potorous tridactylus* (Kerr, 1792) is a member of the family Potoroidae or rat-kangaroos and is the smallest macropodoid in Tasmania (Plate 1). It is found in a wide range of habitats including coastal heath and dry and wet sclerophyll forests (Johnson 1983).

The potoroo has an oestrous cycle of 42 days and a gestation period of 38 days, making it one of the longer cycle lengths of any macropodoid species (Hughes 1962). Delayed gestation occupies 29 days (Shaw and Rose 1979). Long-nosed Potoroos are sexually mature by the end of their first year, and in captivity individuals have been reported to live and continue breeding for seven to 10 years (Ullmann and Brown 1983).

A characteristic of macropodoid growth is the uniformity between small pouch young during the period of permanent teat attachment. Pes and head length constitute the two most easily measured body parts for young while still attached to the teat. Sampson (1971) used sequential measures of known-age pouch young to construct growth curves for *Bettongia penicillata*. To validate this curve he constructed a second plot using simultaneous head and pes measurements. By assuming no error in the head curve, which approximated a straight line, the variation about the pes measures reflected the error and accuracy of both curves. Techniques have since been developed to account for between- and within-animal variation in growth (Wood, Carpenter and Poole 1981).

Growth rates for two *Potorous* pouch young were presented by Hughes (1962) and for five individuals by Heinsohn (1968) based on Hughes' curves. Both workers vary by up to eight days in their estimation of the age of small young. Guiler (1960a) sampled a significantly larger number of females with pouch young and drafted a series of crude growth curves and a "nomogram". Guiler's results differed in age estimation from Heinsohn by up to 12 days, Hughes by up to 30 days and this study by up to 20 days of age. While some of these differences may be due to season or trapping location, the curves for age estimation are in need of revision.

Guiler (1960a,b) trapped females with young at most times of the year, and suggested that births occurred in every month and that breeding was continuous. Earlier, Guiler (1958) had described two periods of breeding with a maximum number of births occurring in late winter to mid-spring (August to October) and again in summer (January and February). From a very small sample size, Hughes (1962) also found that there were periods of reduced breeding, but that the pattern was complicated by delayed pregnancy and a failure of some females to undergo post-partum oestrus.

The aims of this paper are to present additional information on the age estimation and age-related developmental stages of the long-nosed Potoroo based on a large sample size and to provide accurate growth curves from birth to adulthood. Also to determine whether the potoroo has any defined breeding season and to associate this information with life history traits of other species of Potoroidae and Macropodidae.

MATERIALS AND METHODS

Long-nosed Potoroos were trapped at sites 20 km south of Hobart from July 1982 to March 1983. Measurements were obtained for young over a range of ages from birth to pouch vacation. These comprised single sets of observations on 32 pouch young from the wild, a series of sequential

[1]Department of Zoology, University of Tasmania, G.P.O. Box 252C, Hobart, Tasmania 7001.

Pages 449-56 in KANGAROOS, WALLABIES AND RAT-KANGAROOS ed. by G. Grigg, P. Jarman and I. Hume. Surrey Beatty & Sons Pty Limited, New South Wales, Australia, 1989.

Plate 1. Captive adult Long-nosed Potoroos feeding during the day.

measurements and weights from eight young born and raised in captivity until weaning, and from 13 additional young, some of known age, maintained in captivity over a period of time. Head, crown-rump, ear, pes, manus and tail measurements were made using vernier calipers and were based on those of Sharman, Frith and Calaby (1964). Body weights were obtained with Salter spring balances. Observations on the eruption of vibrissae were made using a dissecting microscope and followed the notation by Lyne (1959). Molar eruption, skin and nail pigmentation, pelage and other features were classified according to Sharman, Frith and Calaby's (1964) description for the Red Kangaroo, *Macropus rufus.*

Body measurements and body weights were also obtained from 45 female and 34 male adult potoroos in a similar manner.

Growth curves

A growth curve from birth to pouch vacation was constructed using head and pes lengths from eight pouch young born in captivity and from 13 young for which sequential measures were available. Using the technique of Sampson (1971), a check on the accuracy of the curve was made by fitting the head measurements of wild caught young onto this curve and analysing the mean degree of scatter between the two pes lines. The two pes curves were superimposed with no significant difference between the scatter for pes (known age) and pes check (unknown age) ($t=0.45$, $df=71$, $p>0.375$). The correlation coefficient for known age and pes was 0.986, $R=0.973$, and for unknown age and pes was 0.984, $R=0.967$, indicating very little difference between the curves and a high degree of accuracy in estimating age using this technique.

The instantaneous relative growth rate (k) for pouch young was calculated by a log weight transformation using the equation:

$$k = \frac{\text{Log}_e \text{ weight}_2 - \log_e \text{ weight}_1}{\text{day 2} - \text{day 1}} \times 10^{-2}$$

Linear sections on the graph represent periods when growth is occurring at a constant relative rate. This method has been described by Maynes (1976) and used by Rose (1985 and this volume) for *Bettongia gaimardi.*

RESULTS

Growth curves

The growth of a number of parameters from newborn to adult are shown in Figures 1 to 5. The growth of head (Fig. 1) and crown rump (Fig. 4) followed a straight line (age = 2.145 × head − 13.885, R = 1.00; age = 0.912 × crown rump − 10.045, R = 0.98). The growth of other body parameters approximated Gompertz curves which were difficult to fit by the measures of least squares. A mathematical formula has been adopted for marsupials with this growth (Lee and Cockburn 1985); however, in this study a visual line of best fit was drawn. The accuracy of the curves were estimated at ± three days for small young and approximately ± eight days for young nearing pouch vacation.

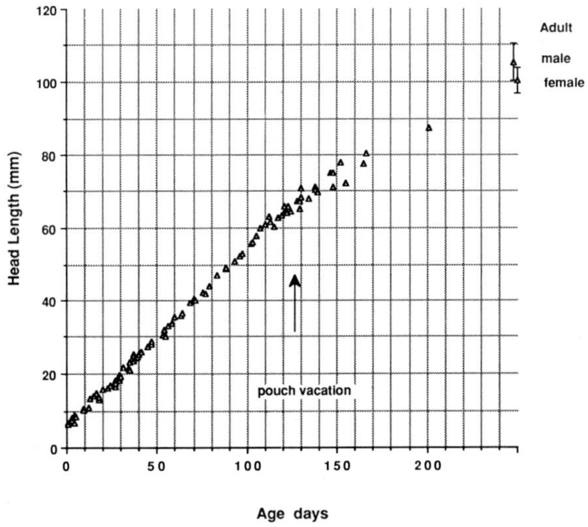

Fig. 1. Increase in head length from birth to adult in *P. tridactylus* (n=50 pouch young; 21 female, 12 male adults mean ± s.d.).

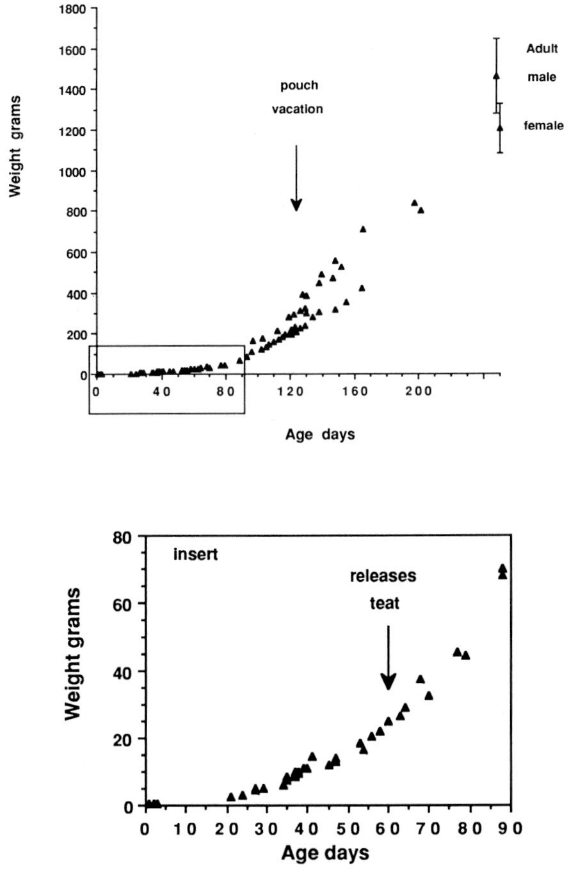

Fig. 2. Weight from birth to maturity in *P. tridactylus* (n=49 pouch young; 46 female, 34 male adults). Insert shows weights of small pouch young s.d. ± mean.

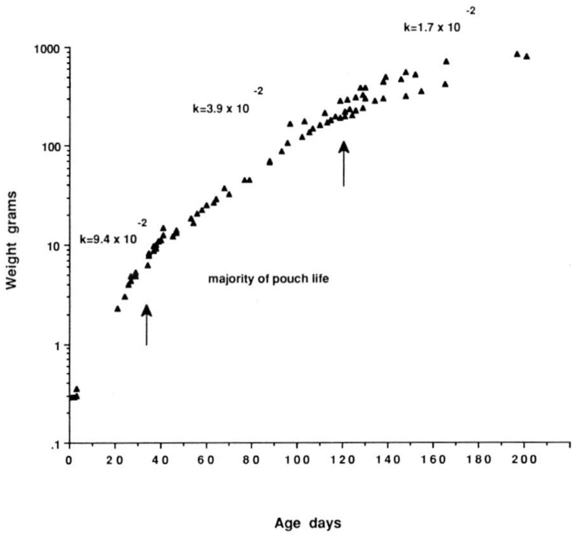

Fig. 3. Relative growth rates of *P. tridactylus*. For the majority of pouch life (shown between arrows) the growth rate was 3.9×10^{-2}.

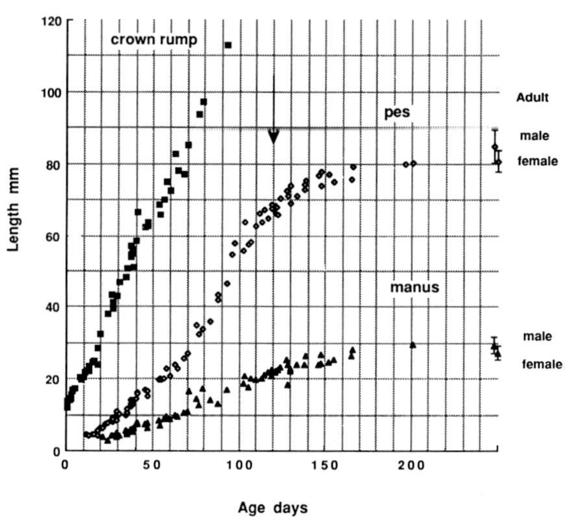

Fig. 4. Growth of crown-rump, pes and manus in *P. tridactylus* (n=48 young; 21 female, 12 male adults). The arrow indicates pouch vacation s.d. ± mean.

Pouch activity and the female cycle

In the wild, the oldest young in the pouch were between 102 and 107 days of age. In captivity, young were periodically out of the pouch between 110 and 125 days of age. Between 120 and 125 days (weight ranging from 225 to 291 grams, n=5) young permanently left the pouch. Young at foot continued feeding from an elongated teat protruded through the contracted pouch opening for five to six weeks later. In captivity, young were weaned at approximately 170 days of age (24 weeks) and weighed approximately 500 grams.

In this and a previous study (Bryant and Rose 1986), the uteri of pregnant females contained quiescent blastocysts while suckling young were between 20 and 95 days of age. This is 15 days shorter than the interval reported by Shaw and Rose (1979) and may be a direct consequence of different aging techniques.

Weight and growth rate

At birth, neonates weighed 0.292 ± 0.003 grams (±s.d., n =5). This comprised 0.025% of

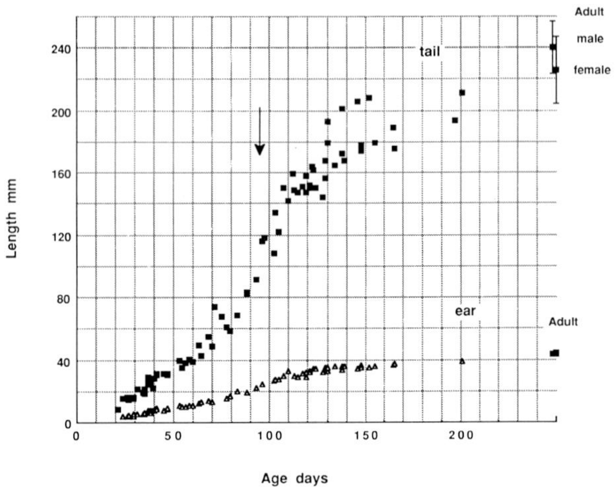

Fig. 5. Growth of tail and ear length in *P. tridactylus.* (n=48 young; 21 female, 12 male adults). Arrow shows start of locomotory activity s.d. ± mean.

the female's body weight. Weight increases were evident after the young relinquished the teat near 60 days of age (Fig. 2). However, there was actually a very rapid rate of growth from birth until 35 days of age ($k=9.4 \times 10^{-2}$, Fig. 3) although this value may be inflated by few data points between day 3 and day 20. The growth rate throughout the majority of pouch life (i.e., day 35 to 125) had a k factor of 3.9×10^{-2}. After weaning and until adult weight, the growth rate decreased to a k factor of 1.7×10^{-2}.

Adult males range in body weight from 1.08 to 1.83 kg with a mean of 1.47 ± 0.18 kg (s.d., n=34). Adult females show a smaller weight range between 0.88 to 1.47 kg with a mean of 1.21 ± 0.12 kg (s.d., n=46). There were no significant seasonal variations in female body weight during the months of July to March ($F_{[3, 39]}=1.35$, $p>0.25$).

Eyes

The bulbous eye areas on the newborn are darkly pigmented, and subsequent development is visible beneath a covering of translucent skin. Differentiation of the eyelid was first noticeable in young aged between 16 and 24 days and was clearly visible by 35 days of age. After 35 days, the eyelashes appeared from small papillae. The eyes were still closed in two young aged 79 days but open in three young aged 84 ± 4 days. Guiler (1960a) estimated that one or both of the eyes opened between 90 and 98 days.

Mouth development, vocalization and tooth eruption

The mouth of the newborn has fused lips which form an opening to secure the mother's teat. Differentiation of the mouth by lip grooves was apparent near day 27, and the lips commenced separation by day 37. By 56 days of age, the lips were fully separated, and from 60 days onwards the young relinquished the teat. This is earlier than the 70 to 77 days reported for *P. longipes* (Seebeck 1988).

In this study, the earliest vocalization was noted from one young aged 34 days but high pitched hisses were frequently heard from 37 days onwards. Hughes (1962) noted a weak squeal from one animal immediately after birth.

The lower incisors were first visible beneath the gums by 50 ± 7 days. They began erupting by 57 ± 5 days of age and had fully erupted by 70 ± 9 days. Guiler (1960a) estimated these teeth erupted at 90 ± 4 days of age. The points of the lower premolars and then the upper premolars were visible above the gum line between 70 ± 9 days and emerged over the next 20 days. One loph on an anterior molar was observed in an animal aged 90 days and it is assumed the molar teeth gradually erupt after the young have left the pouch. The dental formula of the adult potoroo is I^3_1, C^1_0, P^1_1, M^4_4.

Vibrissae, skin pigmentation and pelgae

The young have pink, translucent skin at birth. The papillae of the facial vibrissae were evident around four days after birth. Rows of mystacial vibrissae had erupted by 20 ± 4 days, followed by the supra-orbital and genal- (two single) hairs. The medial antebrachial, ulnar carpal and sub-mental hairs had all emerged by 37 ± 3 days. The inter-ramal hairs appeared after 40 days of age, and the calcaneal and anconeal vibrissae had appeared by approximately 90 days of age. Rhinal vibrissae were not obvious on pouch young. Guiler (1960a) reported that the facial papillae did not appear until 25 days after birth and that the vibrissae erupt some 25 days later.

The perimeter of the ear acquired a light brown pigmentation by 25 ± 2 days. This continued to darken and cover the entire ear by approximately 60 days. From 60 days onwards, the body assumed a light grey colour which gradually darkened to a charcoal grey by 90 days. Potoroo young were lightly furred by 93 days, at which time their eyes had opened and they were venturing outside the pouch. Animals aged 105 days and older were fully furred.

Pes and manus

The potoroo is typical of all marsupial newborn and has paddle-like back limbs but relatively well developed forelimbs bearing deciduous claws. Twelve days after birth, the pes measured approximately 4.5 mm in length (Fig. 4) but was still paddle shaped. Pes and manus claws had formed by 25 days and a light pigmentation at

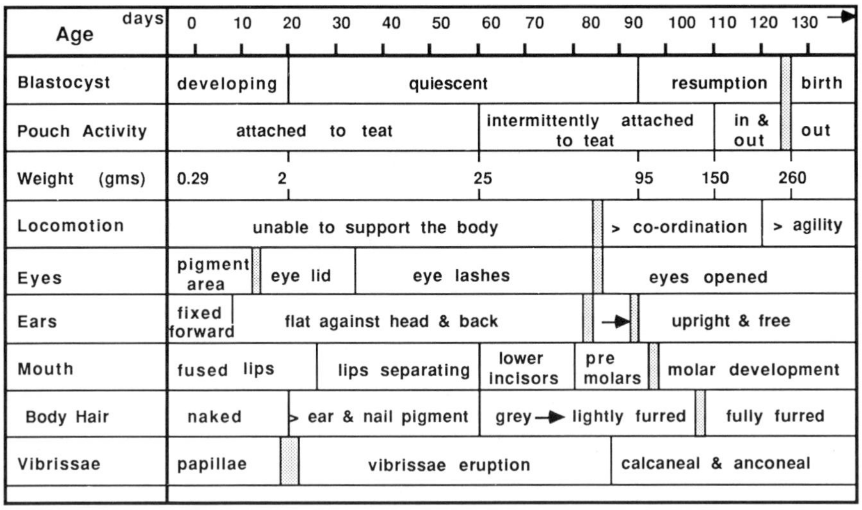

Fig. 6. Stages of development from birth to pouch vacation in the Long-nosed Potoroo, *Potorous tridactylus*. Stippling represent variation in days.

their tips was visible by 40 ± 5 days. Adult males had a mean pes length of 84.8 ± 4.6 mm (range 76.5 to 93.3 mm) and manus length of 29.3 ± 2.3 mm (range 24.6 to 32.8 mm). Adult females had a mean pes length of 80.7 ± 3.0 mm (range 74.0 to 85.3 mm) and manus length of 27.2 ± 1.9 (range 23.5 to 30.9 mm).

Locomotion

Young could support their own body weight when aged between 86 and 110 days (n=5) and weighing 88.0 to 160.0 grams. Between 105 and 110 days of age (weight 136.4 to 183.2 grams) they could crawl with poorly co-ordinated movements (Fig. 5). Competent locomotion was demonstrated by young aged 119 days or older (250 grams or greater) and from 119 days onwards the young spent increasing time out of the pouch.

A summary of the emergence of these characteristics is presented in Figure 6.

Sexual dimorphism

The male scrotum and female pouch anlagen were distinguishable two to three days after birth which is similar in timing to *P. longipes* (Seebeck

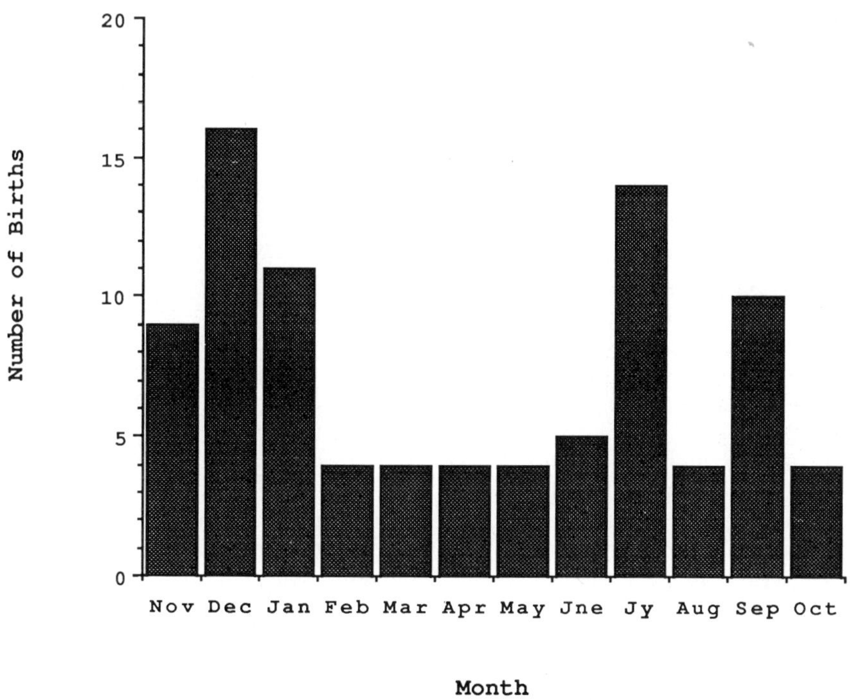

Fig. 7. Estimated month of birth for the Long-nosed Potoroo (includes Shaw's (1977) observations).

1988). The adult male potoroo was significantly heavier ($t = 7.53$, d.f $= 78$, $p<0.0005$) and larger than the female for most body parts except ear length (t-test, d.f 31, head, pes and manus length, $0.0005<p<0.005$, tail length, $0.025<p>0.05$). Forty-eight pouch young were sexed in this study; 30 were male and 18 were female (ratio 1.7 males:1 female).

Breeding season

Of the 34 adult females trapped between December and March, seven had young that were born in November (20%), 14 that were born in December (41%) and six in January (17%) (Fig. 7). Six of nine females trapped in January carried young aged between 30 and 45 days. The remaining births appeared to be evenly distributed between months. Immature animals were trapped in the field only during December, January and February (n=14).

In an earlier study by Shaw (1977), of 43 females trapped in a similar area but in opposite months of the year, 13 females had young born in July (30.2%) and eight had young born in September (19.0%). Thirteen females trapped in September all had young that were older than 100 days (n=5) or newborn (=8). From Shaw's study, young born between July and September, with a pouch life of approximately 4.5 months, would vacate the pouch from mid-November to January. This would be consistent with the timing of births in this study.

Sixty-three percent of young aged in this study were born between November and January and 53.0% of young aged in Shaw's study were born between July and September. As neither study trapped over one full year this may be the reason why two breeding peaks were not detected in the one set of data.

DISCUSSION

Growth and development of *Potorous tridactylus* closely parallels that of *P. longipes* (Seebeck 1988) and is similar in pattern to that of other Macropodidae and Potoroidae (Sharman, Frith and Calaby 1964; Murphey and Smith 1970; Sampson 1971; Maynes 1976; Rose and McCartney 1982; Poole *et al.* 1985; Rose 1988). Development during the first half of pouch life includes differentiation of the mouth and eyes and emergence of vibrissae and claws. During the second half of pouch life, the eruption of the incisors coincides with the time at which the young are no longer permanently attached to the teat. At around 90 days of age, just prior to the first exploratory activity outside the pouch, the eyes are open, the ears are upright and the body is furred. Pes and tail growth-rates are also maximal at this time as these parts are soon required for locomotion and stability.

Adult *Potorous tridactylus* are sexually dimorphic in terms of weight and most body dimensions. This agrees with other studies which have found significant differences in the weight (Heinsohn 1968) and skull measurements (Johnston and Sharman 1976) of male and female potoroos. Adult *P. longipes* are also sexually dimorphic in weight (Seebeck 1988) as are the majority of marsupials (reviewed by Tyndale-Biscoe and Renfree 1987). In contrast, *Bettongia gaimardi* is not sexually dimorphic for any of these features (Rose 1988).

In comparison to the larger Macropodidae, the Long-nosed Potoroo has a short pouch life and a fast rate of growth. It gives birth to a young that comprises approximately 0.025% of the female body weight. This is consistent with other members of the Potoroidae, such as *Potorous longipes; Bettongia gaimardi, B. lesueur* and *B. penicillata,* but is considerably greater than found in Macropodidae (*Settonix brachyurus* 0.009%, *Macropus rufus* 0.003% and *M. eugenii* 0.008%) (reviewed by Maynes 1976; Rose 1978; Tyndale-Biscoe and Renfree 1987; Rose, this volume). For the first 35 days of pouch life, the Long-nosed Potoroo has a very rapid rate of growth represented by a k factor of 9.4×10^{-2}. During the majority of pouch life, the growth rate slows but still remains much faster than that of macropodids. Some of the reasons for this have been attributed to the high carbohydrate levels in *P. tridactylus* milk (Crowley, Woodward and Rose 1988) and to the efficient digestion of hypogean fungi compared to the herbivorous diet of macropodids (Hume 1982; Frappell and Rose 1986).

In comparison to other Potoroidae, *P. tridactylus* has the longest gestation period and, apart from *Hypsiprymnodon moschatus,* the longest pouch life (Lee and Cockburn 1985). *P. tridactylus* produce young of similar size and development to *Bettongia gaimardi* and at pouch vacation both these young represent 21.0% of the female body weight. The young of the potoroo, however, remain in the pouch for approximately three weeks longer than the bettong (Rose 1987). Smolenski and Rose (1988) suggest that one reason for the difference in growth between the two species may be because *Bettongia* utilizes food of higher nutrient content and the pouch young have a greater milk intake.

While breeding may be continuous, *P. tridactylus* show at least two breeding peaks per year, July to September and December to January. This latter period may not be accompanied by post-partum copulation (Hughes 1962) resulting in a short interval of anoestrous prior to July. The annual breeding cycle therefore comprises the birth of two young culminating in nine months of continuous lactation (2×4.5 months). This is followed by a short anoestrous period before the oestrous cycle resumes. Heinsohn (1968) recorded one female in the field giving birth twice in succession and in a study by Bryant and Rose (1986) the ovaries of captive females contained corpora lutea resulting from a recent pregnancy. As the quiescent blastocyst in this species resumes development prior to the young emerging from the pouch, a new young is born coincidental with pouch vacation. This breeding pattern would ensure that, in the majority of cases, potoroo young were not weaned during the winter months, although to date there is no evidence of differing mortality rates during this time.

Although some Potoroidae do not breed seasonally (Sampson 1971; Rose 1985), *Bettongia lesueur* shows peaks in breeding during autumn (April, May) and late spring (November, December) (Tyndale-Biscoe 1968). Moors (1975) found female *Aepyrymnus rufescens* carried young of two age classes at certain times in the year and suggested that this species was not a continuous breeder but instead had a long breeding season. While the breeding pattern of *P. tridactylus* is not well defined, as shown by the fact that females give birth to young in every month of the year (Fig. 7), some seasonal component is evident. Long term trapping studies are required to define this pattern in the wild in greater detail. A large degree of size and weight variation is common throughout the geographic range of *Potorous tridactylus* (Johnston and Sharman 1976) and this variation together with seasonal differences may be important when considering the breeding patterns and estimating ages of young.

ACKNOWLEDGEMENTS

Sincere thanks to R. Rose for encouragement and J. Seebeck and G. Shaw for allowing me to quote from unpublished material.

REFERENCES

BRYANT, S. L. AND ROSE, R. W., 1986. The growth and role of the corpus luteum throughout delayed gestation in the potoroo *Potorous tridactylus*. *J. Reprod. Fert.* **76**: 409-14.

CROWLEY, H. M., WOODWARD, D. R. AND ROSE, R. W., 1988. Changes in milk composition during lactation in the potoroo, *Potorous tridactylus* (Marsupialia: Potoroinae). *Aust. J. Biol. Sci.* **41**: 289-96.

FRAPPELL, P. B. AND ROSE, R. W., 1986. A radiographic study of the gastrointestinal tract of *Potorous tridactylus*, with a suggestion as to the role of the foregut and hindgut in potoroine marsupials. *Aust. J. Zool.* **34**: 463-71.

GUILER, E. R., 1958. Observations on a population of small marsupials in Tasmania. *J. Mammal.* **39**(1): 44-58.

GUILER, E. R., 1960a. The pouch young of the potoroo. *J. Mammal.* **41**(4): 441-51.

GUILER, E. R., 1960b. The Breeding season of *Potorous tridactylus* (Kerr). *Aust. J. Sci.* **23**: 126-27.

HEINSOHN, G. E., 1968. Habitat requirements and reproductive potential of the macropod marsupial *Potorous tridactylus* in Tasmania. *Mammalia* **32**(1): 30-43.

HUGHES, R. L., 1962. Reproduction in the macropod marsupial *Potorous tridactylus* (Kerr). *Aust. J. Zool.* **10**(2): 193-224.

HUME, I. D., 1982. Digestive Physiology and Nutrition of Marsupials. Cambridge University Press: Cambridge.

JOHNSTON, P. G. AND SHARMAN, G. B., 1976. Studies on populations of *Potorous* Desmarest (Marsupialia). 1. Morphological variation. *Aust. J. Zool.* **24**: 573-88.

LEE, A. K. AND COCKBURN, A., 1985. Evolutionary Ecology of Marsupials. Monographs on Marsupial Biology. Cambridge University Press: Cambridge.

LYNE, A. G., 1959. The systematic and adaptive significance of the vibrissae in the marsupialia. *Proc. Zool. Soc. Lond.* **133**(1): 79-133.

MAYNES, G. M., 1976. Growth of the Parma wallaby *Macropus parma*. *Aust. J. Zool.* **24**: 217-35.

MOORS, P. J., 1975. The urogenital system and notes on the reproductive biology of the female Rufous Rat-kangaroo, *Aepyprymnus rufescens* (Gray) (Macropodidae). *Aust. J. Zool.* **23**: 355-61.

MURPHEY, C. R. AND SMITH, J. R., 1970. Age determination of pouch young and juvenile Kangaroo Island wallabies. *Trans. R. Soc. S. Aust.* **94**: 15-20.

POOLE, W. E., MERCHANT, J. C., CARPENTER, S. M. AND CALABY, J. H., 1985. Reproduction, growth and age determination in the Yellow-Footed Rock-wallaby *Petrogale xanthopus* (Gray) in captivity. *Aust. Wildl. Res.* **12**(2): 127-37.

ROSE, R. W., 1978. Reproduction and evolution in female Macropodidae. *Aust. Mammal.* **2**: 65-72.

ROSE, R. W., 1985. Reproductive biology of the Tasmanian Bettong, *Bettongia gaimardi*. Ph.D. Thesis: University of Tasmania.

ROSE, R. W., 1987. Reproductive energetics of two Tasmanian rat-kangaroos (Potoroinae: Marsupialia). *Symp. Zool. Soc. Lond.* **57**: 149-65.

ROSE, R. W., 1988. Age estimation of the Tasmanian bettong *(Bettongia gaimardi)* (Macropodidae, Marsupialia). *Aust. Wildl. Res.* in press.

ROSE, R. W. AND MCCARTNEY, D. J., 1982. Growth of the red-bellied pademelon, *Thylogale billardierii*, and age estimation of pouch young. *Aust. Wildl. Res.* **9**: 33-38.

SAMPSON , J. C., 1971. The biology of *Bettongia penicillata* Gray, 1837. Ph.D. Thesis, University of Western Australia, Nedlands.

SEEBECK, J. H., 1988. Breeding, growth and development of captive *Potorous longipes* (Marsupialia: Potoroidae) submitted manuscript.

SHARMAN, G. B., FRITH, H. J. AND CALABY, J. H., 1964. Growth of the pouch young, tooth eruption and age determination in the red kangaroo, *Megaleia rufa*. *C.S.I.R.O. Wildl. Res.* **9**: 20-49.

SHAW, G., 1977. Reproduction in the Tasmanian potoroo *Potorous tridactylus apicalis*. Hons. Thesis. University of Tasmania.

SHAW, G. AND ROSE, R. W., 1979. Delayed gestation in the potoroo, *Potorous tridactylus* (Kerr). *Aust. J. Zool.* **27**: 901-12.

SMOLENSKI, A. J. AND ROSE, R. W., 1988. Comparative lactation in two species of Rat-kangaroo (Marsupialia). *Comp. Biochem. Physiol.* **90A**: 459-64.

TYNDALE-BISCOE, C. H., 1968. Reproduction and post-natal development in the marsupial *Bettongia lesueur* (Quoy and Gaimard). *Aust. J. Zool.* **16:** 577-602.

TYNDALE-BISCOE, C. H. AND RENFREE, M. B., 1987. Reproductive Physiology of Marsupials. Monographs on Marsupial Biology. Cambridge University Pres: Cambridge.

ULLMANN, S. L. AND BROWN, R., 1983. Further observations on the potoroo *(Potorous tridactylus)* in captivity. *Lab. Animals* **17:** 133-37.

WOOD, J. T., CARPENTER, S. M. AND POOLE, W. E., 1981. Confidence intervals for ages of marsupials determined from body measurements. *Aust. Wildl. Res.* **8:** 269-74.

KANGAROOS, WALLABIES and RAT-KANGAROOS

KANGAROOS, WALLABIES and RAT-KANGAROOS

Volume 2

Edited by

Gordon Grigg, Peter Jarman and Ian Hume

Published by

Surrey Beatty & Sons Pty Limited

The National Library of Australia
Cataloguing-in-Publication entry:

Kangaroos, Wallabies and Rat-kangaroos.

Bibliography.
Includes index.
ISBN 0 949324 20 5.
ISBN 0 949324 22 1 (v. 1).
ISBN 0 949324 23 X (v. 2).

1. Marsupialia — Australia. 2. Marsupialia — Papua New Guinea. 3. Kangaroos. 4. Wallabies. 5. Potoroidae. I. Grigg, Gordon C. (Gordon Clifford). II. Jarman, P. (Peter). III. Hume, Ian D. (Ian Douglas).

599.20994

Published November, 1989

PRINTED AND PUBLISHED IN AUSTRALIA BY
SURREY BEATTY & SONS PTY LIMITED
43 Rickard Road, Chipping Norton, NSW 2170

CONTENTS

(VOLUME 2)

SOCIAL ORGANIZATION AND BEHAVIOURAL ECOLOGY

POPULATION ECOLOGY AND MANAGEMENT

(VOLUME 1)

EVOLUTION, ZOOGEOGRAPHY AND COMMUNITY ECOLOGY

DIET, FEEDING AND DIGESTION

ENERGY, WATER AND TEMPERATURE

REPRODUCTION

DEVELOPMENT, GROWTH AND SEXUAL DIMORPHISM

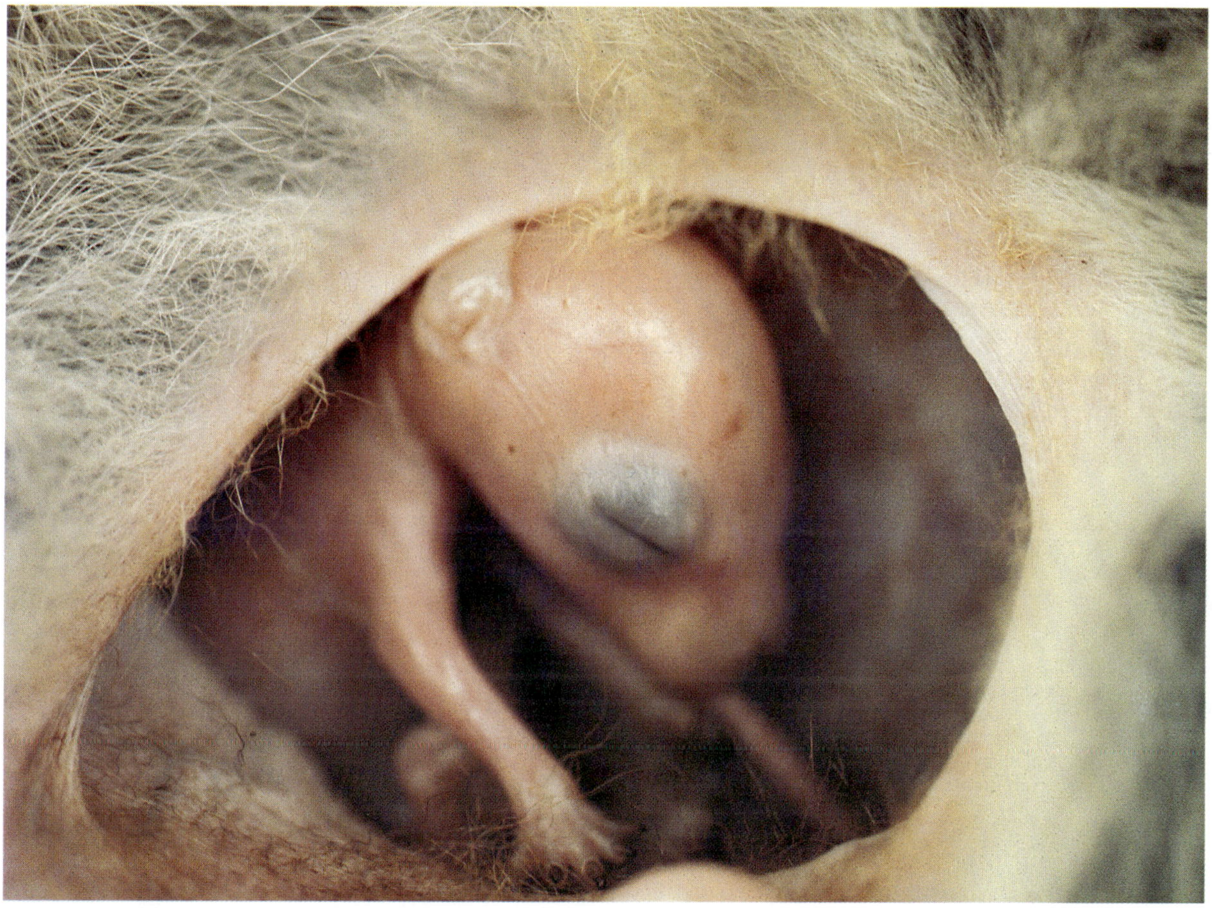

The suckling pouch-young of a Spectacled Hare Wallaby *(Lagorchestes conspicillatus)*. Photo by Gordon Grigg.

LIST OF COLOUR ILLUSTRATIONS

Repertoires of social behaviour in the Macropodoidea

GRAEME COULSON[1]

Repertoires of social behaviour have been studied extensively in some members of the Macropodoidea (e.g., *Macropus rufus*), while others (e.g., *M. irma, Dorcopsis, Onychogalea*) have been largely neglected. These studies differ in their sampling effort, sample size and composition, and use of captive and free-ranging animals. A logarithmic relationship between sampling effort and the number of novel behavioural acts that are detected suggests that the behavioural catalogues of most species are likely to be incomplete. A generalized classification system was used to code the acts occurring in four social contexts: general, agonistic, sexual and maternal. The distribution of these acts within the Macropodoidea reveals a number of acts that are apparently universal and primitive. The disjunct distribution of other acts can be partly accounted for in terms of morphological and ecological constraints on behaviour; some are capable of differentiating between taxa at generic and subgeneric levels

INTRODUCTION

THE range of ecological variation displayed by the superfamily Macropodoidea presents valuable research opportunities in behavioural ecology. At one extreme are small omnivorous species such as the Long-nosed Potoroo, *Potorous tridactylus*, which occupies closed habitat and is nocturnal, cryptic and solitary. The other extreme is exemplified by large grazing species such as the Red Kangaroo, *Macropus rufus*, which is partly diurnal, highly visible and gregarious.

As Eisenberg and Golani (1977) point out, macropods can serve as a "control" group for testing hypotheses about the evolution of behaviour in eutherian taxa. The behaviour of macropods has most often been compared with that of ungulates (e.g., Kaufmann 1974; Jarman 1983). These comparisons have also drawn attention to the evolution of behaviour in terms of the distinctive characteristics of the Macropodoidea, notably their characteristic hopping form of locomotion which has permitted the evolution of the forelimbs of males as weapons (Jarman 1983), and their marsupial mode of reproduction which has resulted in unique patterns of maternal care (Russell 1982; Johnson 1987).

Comparative studies require a sound knowledge of the species' behavioural repertoires. The difficulty encountered in attempting comparisons is that the diversity of the Macropodoidea has engendered a patchy research effort. Some species are readily obtained, while the study of others presents considerable logistic problems. The outcome is that a core of species has been studied extensively, but most have received considerably less attention and for some there are no behavioural data available.

The aim of this paper is to review the studies of social behaviour in the Macropodoidea. At the intraspecific level, a number of parameters that are likely to influence the completeness of a repertoire are examined. Acknowledging the limitations of the data available for each species, the patterns of social behaviour within the superfamily are then outlined, are compared in general terms to the patterns shown by ungulates, and are related to the phylogeny of the Macropodoidea as it is presently understood.

CLASSIFICATION OF BEHAVIOUR

The behavioural data used in this analysis have been drawn from many sources, including a number of unpublished theses, listed in Table 1. The list is not exhaustive; other studies by Immelmann (1965), Merchant and Sharman (1966), Sharman *et al.* (1966), Veselovsky (1969) and Coulson and Croft (1981) have recorded specific observations on a range of species. Many of thc carlicr studies (e.g., Hughes 1962; Sharman and Calaby 1964) were primarily investigations of reproductive biology, of which descriptions of sexual behaviour formed a minor part. Stodart (1966) was the first to provide a comprehensive report on the behaviour of one species, the Burrowing Bettong, *Bettongia lesueur*. Studies of a number of other species then followed, and more recently there has been an increasing number of studies of specific aspects of behaviour, particularly maternal and agonistic interactions.

The diversity in studies has resulted in a lack of conformity in behavioural classification and terminology. The behaviour categories that were identified by Stodart (1966) and Russell (1970a,b) have been modified and extended in subsequent studies on an *ad hoc* basis, and various systems for coding behaviour have also been produced (e.g., Southwell 1976; Batchelor 1980; Jaremovic 1984). In order to make comparisons between

[1]Department of Zoology, University of Melbourne, Parkville, Victoria 3052.

Pages 457-73 in KANGAROOS, WALLABIES AND RAT-KANGAROOS ed. by G. Grigg, P. Jarman and I. Hume. Surrey Beatty & Sons Pty Limited, New South Wales, Australia, 1989.

Table 1. Characteristics of studies of social behaviour in the Macropodoidea. Blank spaces indicate a lack of information. N is the maximum number of individuals available; C, captive; F, field; N, night; D, day. Subgenera of *Macropus* are in parentheses.

Genus	Species	Study	Focus	Site	Time	N	Hours	Anim.h
Hypsiprymnodon	*moschatus*	Johnson & Strahan 1982	general					
Aepyprymnus	*rufescens*	P. M. Johnson 1980	general	C	N	7	90	540
Bettongia	*lesueur*	Stodart (1966)	general	C	N	8	>61	
	penicillata	Sampson (1971)	general	C				
	gaimardi	Lukschanderl & Lukschanderl (1969)	general	C				
		Coulson (unpublished)	social	C	N	4	19	50
Potorous	*longipes*	Seebeck & Johnston (1980)	general	C		2		
	tridactylus	Hughes (1962)	sexual	C				
		Kitchener (1967)	agonistic	F	N			
		S. D. Johnson (1987)	maternal	C	N	28	93	
Dorcopsis	*luctuosa*[1]	Ganslosser & Holz (1984)[2]	agonistic	C	D	8	~200	
		Ganslosser & Schappert (1985)[2]	play	C	D+N	26	~400	
		Schappert (1984)[2]	social	C	D	26	~400	
		Bourke (this volume)	social	C	D	4	60	300
Dendrolagus	*dorianus*	Ganslosser (1977)[2]	general	C	D+N	10	~100	
		Ganslosser (1979a, b)[2]	social	C	D+N		~200	
		Ganslosser (1983)[2]	social	C	D+N	7	>450	>2 360
		Ganslosser (1984)[2]	social	C	D	11	>450	
		Ganslosser & Schappert (1985)	play	C	D+N			
	goodfellowi	Ganslosser (1984)[2]	social	C	D+N	3	~250	
	inustus	Schneider (1954)	general	C	D	7		
		Ganslosser (1977)	general	C	D	3		
	lumholtzi	Procter-Gray (1985)	general	F	D+N			292
Peradorcas	*concinna*	Sanson *et al.* (1985)	general	F+C	D			
		Goldstone & Nelson (1986)	agonistic	C	N[3]	2		
Petrogale	*assimilis*[4]	Dwyer (1972)	social	F		80		
		Davies (1976, 1979); Russell (1985)	general	F		150		
		Barker (1982)	general	F	D+N	>49	183	
		Horsup (1986)	general	F	D+N	~72	334	
	lateralis	Gepp (1973)	general	F	D	>12	320	
		Hornsby (1978)	maternal	C	D			
	penicillata	Batchelor (1980)	general	C	D	9	19	<171
		Joblin (1981)	general	F				
	xanthopus	Nicholls (1972)	general	C	D	26	>136	
		Nicholls (1972)	general	F	D	12	92	
		Hornsby (1978)	maternal	C	D			
		Lim (1987)	general	C	D+N	6	151	
		Lim (1987)	general	F	D			
Thylogale	*billardierii*	Morton & Burton (1973)	general	C	D	14	81	~891
		Rose & McCartney (1982)	sexual	C		19		
		Clancy (1982)	general	C	D+N	11	106	~1 097
		Ganslosser & Werner (1984)[2]	social	C	D+N	6	~220	
	brunii	Ganslosser (1978)[2]	agonistic	C	D			
	stigmatica	Cooke (1979)	general	F	N			
	thetis	K. A. Johnson (1977)	general	F+C	D		>94	
Setonix	*brachyurus*	Packer (1969)	general	C	N	10	50	~375
Wallabia	*bicolor*	Crebbin (1982)	general	C	D+N	12	224	
Macropus								
(Prionotemnus)	*agilis*	Russell (1985)	olfactory					
	dorsalis	Heathcote (this volume)	social	C	D	7	35	385
	eugenii	Russell (1973)	maternal	C	D	6	450	450
		Russell & Giles (1974)	maternal	C		19	200	
		Vujcich (1979)	general	F	D	~108		
		Tyndale-Biscoe & Renfree (1987)	sexual	C				
	parma	Maynes (1973)	sexual	C		>47		
		Vujcich (1979)	general	F	D	~84		
		Herter *et al.* (1985)	social	C		19	>215	
	parryi	Kaufmann (1974)	general	F	D	150	~3 024	~18 144
	rufogriseus	LaFollette (1971)	agonistic	C	D	7	>76	>505
		Murböck (1975)	agonistic	C	D	2		
		Southwell (1976)	general	F	D+N			~98
		Murböck (1979)	general	C	D+N	26		
		Fleming *et al.* (1983)	sexual	F				
		C. N. Johnson (1987)	maternal	F	D			
(Macropus)	spp.	Poole & Pilton (1964)	sexual					
	fuliginosus	Herrmann (1967)[5]	maternal	C	D	5		
		Germann-Meyer & Schenkel (1970)[5]	agonistic	C				
		Germann-Meyer (1974)[5]	general	C	D	19		
		Coulson (unpublished)	social	C	D	15	250	
	giganteus	Herrmann (1971)	general	F	D	~45		

Table 1 — continued

Genus	Species	Study	Focus	Site	Time	N	Hours	Anim.h
(Macropus)	*giganteus*	Grant (1973, 1974)	general	C	D+N	9	360	2 520
		Grant (1973, 1974)	general	F	D+N	23	137	
		Kaufmann (1975)	general	F	D	37		
		Southwell (1976)	general	F	D+N			~98
		Southwell (1981)	general	F	D+N		~720	
		Jaremovic (1984)	general	F	N	191		3 691
		Coulson (unpublished)	social	C	D	48	250	
(Osphranter)	*antilopinus*	Croft (1982)	general	F		50		
		Coulson (unpublished)	social	C	D	5	100	500
	robustus	Osazuwa (1978)	general	F+C	D+N	20	846	>7 771
		Croft (1981a)	general	F	D	62	1 600	
		Croft (1985)	agonistic	F	D		120	
(Megaleia)	*rufus*	Sharman & Pilton (1964)	sexual	C				
		Russell (1970a, b)	general	C	D+N	29	126	~1 827
		Cicala *et al.* (1970)	general	C		5	18	90
		Russell (1973)	maternal	C	D	6	110	110
		Russell & Nicholls (1974)	auditory	C		4		
		Croft (1981b)	general	F	D		1 500	
		Croft (1985)	agonistic	F	D		120	
		Ganslosser & Wilhelm (1986)	social	C	D	9	422	

[1]Previously referred to as *D. muelleri* (Groves and Flannery, this volume).
[2]Additional methodological data supplied by U. Ganslosser (pers. comm.).
[3]Reversed light cycle.
[4]Previously referred to as *P. inornata* (R. Close, pers. comm.).
[5]Misidentified as *M. giganteus* (Poole 1982).

these studies, it was necessary to devise a universal system for categorizing and coding behaviour. The categories (represented by codes) are referred to here as acts. The system of acts has been designed to meet four criteria:

1. The resolving power of an act should be appropriate to as many studies as possible without excessive loss of information. The acts used here are generally broader than the categories used by Ganslosser and colleagues (e.g., Ganslosser and Wilhelm 1986), but offer higher resolution of behavioural components than do vague terms such as "play" or "fight".

2. An act should reflect the social context in which it occurs. As an illustration, allogrooming of a female by a male during courtship is treated as distinct from allogrooming of a young-at-foot by its mother.

3. Acts should be named descriptively, and should not include any interpretation of possible function. Use of terms such as "threat" and "submission" was avoided.

4. Acts are recorded simply as present or absent in the behavioural descriptions given by each study; few studies provide data on the relative frequency, intensity or duration of acts.

Acts were identified from the original descriptions of behaviour given in each study, then were coded using a system that was designed to be flexible enough for expansion to incorporate acts that have yet to be described, and for modification to permit finer resolution of behaviour in more detailed studies. The coding system uses letter codes which, as far as possible, are the initial letters of the acts. The major code (upper case) indicates the general type of act, and may have additional qualifying letters where necessary: a prefix (lower case) identifies the age-sex class of the actor in contexts where ambiguity arises; a suffix (also lower case) indicates the site of the act on the body of the recipient. Less often an upper case suffix is added to clarify the major code. These combinations are illustrated by the following examples:

C cuff
yC young cuff (in maternal contexts)
Nc nose cloaca
KL kick lying

Table 2 lists the codes used and gives a brief clarification of each act, including some non-social acts for completeness. In total 72 social acts were identified, and 61 of these are considered here.

Vocalizations proved difficult to classify. Eisenberg *et al.* (1975) cautioned against the use of verbal descriptions of vocalizations since they are subjective and may obscure differences in the structure of calls. Only two studies (Russell and Nicholls 1974; Clancy 1982) have presented sonogram analyses of the physical properties of calls, so no attempt has been made to establish homologies between calls other than for the unmistakable buccal clicks referred to by Eisenberg *et al.* (1975). Other calls are classified according to the context in which they are given, a procedure which inevitably underestimates the richness of vocal repertoires in the Macropodoidea.

Table 2. A list of coded behavioural acts identified in the Macropodoidea. The clarification of acts is from the standpoint of the actor.

Code	Act	Clarification
AD	Adjacent	Remain beside conspecific
AV	Avoid	Move away from conspecific
B	Bite	Bite conspecific
BC	Bite chest	Bite own sternal region
C	Cuff	Strike or grab conspecific with paws
DC	Drag cloaca	Rub cloaca on ground
EN	Enter pouch	Young enters pouch
EX	Exit pouch	Young leaves pouch
F	Feed	
FD	Feed (displacement)	Feed warily during fight
FS	Feed (suckling)	Suckle from mother
G	Groom	Groom self
GA	Allogroom	Groom conspecific
GD	Groom (displacement)	Exaggerated self-grooming in fight
GP	Groom pouch	Clean own pouch
H	Hop	
HA	Hop after	Chase conspecific
HC	Hop in circles	Hop away from and return to mother
I	Intromission	
KG	Kick ground	Thump or drum feet on ground
KH	Kick (hopping)	Kick conspecific while hopping
KL	Kick (lying)	Kick conspecific while lying
KT	Kick (tail)	Kick conspecific supported on tail
L	Lie	
M	Mount	Mount conspecific
N	Nose	Nose conspecific
NJ	Nose jab	Thrust muzzle at conspecific
P	Paw	Touch or push conspecific with paws
PG	Paw ground	Scratch or pull at grass with paws
PW	Paw wrestle	Grip conspecific tightly
Q	Quiver	Shake or vibrate head
R	Rub	Rub chest on conspecific
RB	Rub branch	Rub chest on branch or tree
RG	Rub ground	Rub chest on ground or tussock
S	Stand	Stand in front of conspecific
SH	Stand high	Stand fully erect
TL	Tail lash	Move tail sinuously
U	Urinate	
W	Walk	"Pentapedal" gait
WS	Walk stiff-legged	Walk on tips of feet and tail
X	Erection	Erection of penis
Vocalizations:		
VA	Alarm	When predator detected
VC	Cluck	Buccal clicking sound
VL	Lost	When separated from mother
VD	Dominant	When dominant over conspecific
VS	Subordinate	When subordinate to conspecific

Prefixes		*Suffixes*	
m	male adult	b	back
f	female adult	c	cloaca
s	subadult	f	flank
y	young in pouch	h	head
		n	nose
		p	pouch
		t	tail
		u	urine

COVERAGE OF SPECIES

Some species are well-represented in behavioural studies while others have been neglected (Table 1). The Red Kangaroo has received the most attention, being the object of thousands of hours of cumulative observations in at least eight separate studies. With three exceptions, other species of *Macropus* have been covered by at least one comprehensive study. A surprising exception is the Agile Wallaby, *Macropus agilis*; it is commonly held in captivity and other aspects of its biology have been studied. The apparent neglect of the Western Brush (Black-gloved) Wallaby, *Macropus irma,* is also puzzling although most aspects of its biology are poorly known. The third exception, the Black Wallaroo, *Macropus bernardus,* is rarely held in captivity and is likely to remain unresearched unless the difficulties of observing it in the field can be overcome. Of the other extant genera, *Aepyprymnus, Bettongia, Petrogale, Thylogale, Setonix* and *Wallabia* have been moderately well covered, whereas *Hypsiprymnodon, Potorous, Dorcopsis* (except *luctuosa*) and *Peradorcas* need further attention. Behavioural data are completely lacking for four genera *Lagorchestes, Lagostrophus, Onychogalea* and *Dorcopsulus,* although studies on the first two are under way (J. Courtenay, pers. comm.; G. Lundie-Jenkins, pers. comm.).

It cannot be assumed that a species' behavioural repertoire will be fully understood after one study, however comprehensive, of its behaviour. Repertoire completeness is influenced by a number of methodological parameters.

Sampling effort

A common observation in biological sampling is that the occurrence of novel events is related to the log of sample size (Fagen and Goldman 1977). In practice this means that new acts will accumulate rapidly at first, but become decreasingly frequent as a study progresses. In the case of behavioural repertoires, which are presumably finite, a logarithmic relationship will not hold indefinitely and the curve will eventually become asymptotic for very large samples. Fagen and Goldman (1977) presented analytical techniques for estimating the maximum size of a repertoire using data on the number of acts observed (types) and the total occurrence of all acts (tokens). Unfortunately, very few studies of macropod behaviour present data which are appropriate to a type-token analysis. Instead, two measures of sampling effort are used here as an approximation of the number of tokens: total hours of observation, and the product of the number of individuals and the hours for which they were observed (animal-hours). The latter measure should more closely approximate the number of tokens, but it was available from fewer studies. In studies where observation hours or animal-hours were not specified it was often possible to estimate one or both from other data given. These estimates necessarily relied on a number of assumptions, and are distinguished (by~signs) in Table 1 from the values which were stated. Some other studies give minimum or maximum values (indicated by $>$ and $<$); these were taken to be the actual value.

As predicted, a plot of distinct acts against observation hours from all studies (Fig. 1) fits a logarithmic regression ($y=5.819\ x^{0.238}$, $r=0.598$, d.f.$=31$, $P<0.001$). A similar equation applied to the regression of acts against animal-hours ($y=3.446\ x^{0.257}$, $r=0.632$, d.f.$=17$, $P<0.01$). There is no obvious explanation for the outlier (Clancy 1982), which was excluded from the regressions, but it suggests that under favourable conditions a skilled observer may accumulate novel acts more rapidly than would be expected on average. It would be unwise to pursue this analysis in an attempt to estimate the completeness of repertoires because, in the absence of type-token data for individual species, it would be necessary to make the improbable assumption that all species have repertoires of equal finite size. However, at a general level these relationships suggest that roughly 1 000 hours (or 5 000 animal-hours) yield only about 75% of acts in the full repertoire in an average study of macropod behaviour. Since few studies have sampling intensities that exceed 500 hours (or 2 500 animal-hours), most repertoires that have so far been recorded are likely to be incomplete.

Sample size and composition

There is considerable variation in repertoire size at low sampling intensities (Fig. 1). This is certain to be due, at least in part, to differences in sample size and sample composition. The use of animal-hours in preference to hours alone as a measure of sampling intensity can correct mathematically for the effects of sample size, and the practical implication is that more time must be devoted to observations in order to compensate for a small sample. However, it is not possible to allow for the effects of gross differences in the age-sex composition of the sample. This problem is best illustrated by the high proportion of studies (e.g., Russell 1970a; Ganslosser and Werner 1984; Bourke, this volume) which have only one adult male in each study group. Clearly it would have been impossible for these studies to have recorded acts that are restricted to male-male interactions. Less frequently, a gap in the repertoire is caused by an absence of young (e.g., Packer 1969). More subtle effects of varied sample composition are also possible even when all age-sex classes are adequately represented. The balance of age-sex classes, and hence the mixture of dyad-specific acts, may differ between studies, with the result that the probability of observing a given act in one study is not the same as in another. In addition, it has been shown in studies of the Red-necked Wallaby, *Macropus rufogriseus,* by Murböck (1979) and of the Swamp Wallaby, *Wallabia bicolor,* by Crebbin (1982) that enclosure size can influence the rate of interactions in captivity by altering the effective population density for a given sample size. Taken together, the limitations of sample size and composition are further reasons for incompleteness of behavioural repertoires in the Macropodoidea.

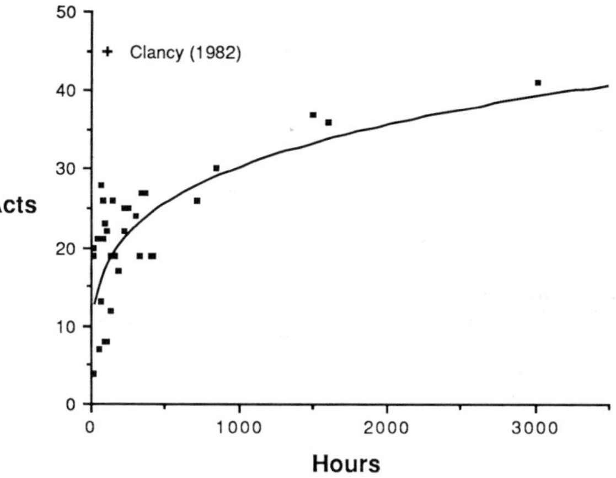

Fig. 1. The relationship between hours of observation and novel behavioural acts observed in studies of general social behaviour in the Macropodoidea. See comments in text on the outlier (Clancy 1982).

Sampling time and location

A large proportion of macropods are cryptic and are strictly or predominantly nocturnal (Kaufmann 1974). Some studies have therefore made observations at night with the aid of auxiliary lighting (e.g., Clancy 1982) or night-vision devices (e.g., Southwell 1981). Although the larger species are more visible and tend to be crepuscular, with few exceptions (e.g., Jarman *et al.* 1987) they remain difficult to approach and observe. To overcome these difficulties a number of studies have been conducted on captive colonies; a few (e.g., Nicholls 1972; Grant 1973, 1974) have followed captive observations with a field study.

The value of captive studies has often been questioned on two grounds: the constraints of an artificial captive environment may limit full expression of the behavioural repertoire (e.g., Martin and Bateson 1986), or the captive environment may elicit aberrant behaviour that is not part of the natural repertoire, reaching its most extreme manifestations in the primates (e.g., Erwin and Deni 1979). Both effects are likely to occur in macropods. In the Eastern Grey Kangaroo, *Macropus giganteus*, for example, my experience suggested that the repertoire was limited in captivity, since elements of "grass pulling" (see Table 2) were either subdued or absent in a captive group by contrast with the elaborate and intense sequences observed in the field; Grant (1974) argued that aberrant behaviour also occurred in this species, as shown by the frequent mutual grooming he observed between two captive females.

The data generated by captive studies of macropods should properly be interpreted with caution, but where captive and field studies have been conducted on the same species there is a close correspondence in the repertoires they have yielded. Captive studies also give the advantage of proximity to the animals, thus enhancing the probability of detecting subtle actions or close-range vocalizations. This is illustrated by "Flehmen", a fleeting and unobtrusive behaviour in macropods, which was first detected in captive animals (Coulson and Croft 1981). Similarly, Croft (1981b) was unable to detect soft "clucking" vocalizations made by Red Kangaroos in the field although he suspected that they did occur in the same contexts as had been reported in captivity by Russell (1970a). Captive studies may thus complement studies conducted on free-ranging macropods. Significantly, for some species, captive studies provide the only behavioural data.

PATTERNS OF BEHAVIOUR

The approach taken in this analysis is to compile a composite repertoire for each species. Where more than one study (captive or field) has been conducted on one species, the data on acts have been pooled to broaden the sample size, sampling effort and sampling conditions, thereby maximizing the size of the recorded behavioural repertoire.

General acts

Acts classified as general are those which apparently do not occur within specific sexual, agonistic or maternal contexts (see definitions below), or acts which are common to all social contexts. As defined here, general acts embrace categories which have been created in other studies either by inclusion of specific acts, such as Kaufmann's (1975) category of "friendly" behaviour, or by exclusion, such as Croft's (1981a) category of non-agonistic, non-sexual and non-maternal interactions. The general context also includes acts which have been identified as "transitional" or "neutral" elements in sequential analyses by Ganslosser and colleagues (e.g., Herter *et al.* 1985).

The occurrence of general acts in the Macropodoidea is summarized in Table 3. The act that occurs at the highest frequency is nose-to-nose (Nn). Russell (1985) also noted the high incidence of this act, which is probably universal in the Macropodoidea. Nosing (N) of other parts of the body was also common in general contexts. The frequency of both forms of nosing has led to the assumption that olfactory cues are important in identification of age-sex classes or possibly of individuals (e.g., Kaufmann 1974; Croft 1981b), but with the exception of Johnson's (1980) observation of an increase in nosing following the introduction of a new individual, there is little direct evidence for this hypothesis. The common observation that individuals nose each other after feeding side by side for some time suggests that recognition may not be the sole function of nosing.

Allogrooming in this context was classified separately from its occurrence in other contexts. It has been reported in about half of the species surveyed, and also in the Rufous Hare-wallaby, *Lagorchestes hirsutus* (G. Lundie-Jenkins, pers. comm.). In most studies it has been noted as a rare event, but has been claimed to be more common in the Antilopine Wallaroo *Macropus antilopinus* (Croft 1982) and the Black-striped Wallaby *Macropus dorsalis* (Heathcote, this volume).

Dragging the cloaca (DC) on the ground, recorded in only two species, is included as a general act. Russell (1985) noted that macropods possess paracloacal glands and the "marking" behaviour observed could serve to deposit secretions on the substrate, but the function of this marking was not clear in either species.

Kicking the ground (KG) to produce an audible thump is a common response to disturbance by a possible predator. Thumps appear to be produced by hopping and striking the ground heavily with both feet simultaneously in all species except the Grey Kangaroos, *Macropus (Macropus),* which strike with one foot slightly sooner than the other to produce a distinct double thump. With the probable exception of the Tree-kangaroos, *Dendrolagus,* thumping is likely to be universal in the Macropodoidea. In addition to the species listed in Table 3, thumping has been recorded in the Quokka, *Setonix brachyurus* (W. Packer, pers. comm.; W. Holsworth, pers. comm.), the Rufous Hare-wallaby (G. Lundie-Jenkins, pers. comm.) and the Red-necked Wallaby (Coulson, pers. obs.). Alarm vocalizations (VA) have been recorded in only four species, two potoroids and two wallaroos, *Macropus (Osphranter),* and in three of these species the known repertoire also includes thumping. Thumps may also occur in other contexts: Barker (1982) reported aggressive foot-thumping by male Allied Rock-wallabies, *Petrogale assimilis,* and a form of thumping by males of several species occurs in sexual contexts (see below).

Table 3. The occurrence of acts in the Macropodoidea, in general and maternal contexts, based on composite repertoires for 33 species.

		General acts							Maternal acts									
Genus	Species	Nn	N	P	GA	DC	KG	VA	Gp	fGA	yGA	FS	yP	yC	yK	HC	VL	VC
Hypsiprymnodon	*moschatus*																	
Aepyprymnus	*rufescens*	•	•		•	•	•	•	•	•							•	
Bettongia	*lesueur*	•						•										
	penicillata																	
	gaimardi	•	•								•	•						
Potorous	*longipes*																•	
	tridactylus						•		•	•	•	•					•	
Dorcopsis	*luctuosa*	•	•		•				•	•	•	•		•				
Dendrolagus	*dorianu*	•	•	•	•				•	•		•						
	goodfellowi	•		•	•													
	lumholtzi									•	•	•	•					
Peradorcas	*concinna*						•			•						•		
Petrogale	*assimilis*	•			•		•			•	•	•					•	
	lateralis	•								•	•	•					•	
	penicillata	•			•		•			•							•	
	xanthopus	•	•		•		•		•	•		•				•		
Thylogale	*billardierii*	•			•		•		•	•	•	•	•	•		•	•	•
	brunii																	
	stimatica	•	•			•	•			•	•	•					•	•
	thetis						•							•			•	•
Setonix	*brachyurus*	•																
Wallabia	*bicolor*	•	•	•	•		•					•				•	•	
Macropus																		
(Prionotemnus)	*agilis*																	
	dorsalis	•	•		•													
	eugenii	•	•				•		•	•		•	•	•	•	•	•	
	parma	•	•	•	•		•					•					•	
	parryi	•	•				•			•	•	•	•	•	•	•		
	rufogriseus	•		•	•					•		•		•	•	•		
(Macropus)	*fuliginosus*	•	•		•		•		•	•	•	•	•	•	•	•	•	•
	giganteus	•	•	•	•		•		•	•	•	•	•	•	•	•	•	•
(Osphranter)	*antilopinus*	•	•		•		•	•	•	•	•	•	•	•				
	robustus	•	•	•			•	•	•	•	•	•	•	•	•	•		
(Megaleia)	*rufus*	•	•	•	•		•		•	•		•	•	•	•	•	•	•

Maternal acts

The maternal context is considered here to include all acts observed in mother-young interactions, performed by both the female and the young during the period of dependence. Behaviour associated with birth has not been included here because it has received considerable attention in the past and has been reviewed by Tyndale-Biscoe and Renfree (1987). Overall, there are fewer maternal acts reported for potoroids, in which female and young typically share a nest (Russell 1982) and thus are difficult to observe, and it is likely that many acts have been missed. The general remarks on maternal behaviour which follow are thus based predominantly on macropodids.

Females groom the pouch (GP) as the young develop, removing urine and faeces and probably also grooming the young, although this is difficult to detect until the young are large. When the young leaves the pouch (either temporarily or permanently), it is groomed by the female (fG) in most species, and less commonly grooms the female (yG). The young also feeds by suckling (FS) from outside the pouch, and often remains adjacent (AD) to the female. Two forms of play have been observed: pawing (yP), cuffing (yC) and sometimes kicking (yK) of the female by the young, which generally elicit subdued retaliation by the female, and solitary play by the young which hops (HC) rapidly away from the female then returns to her. The two forms are often interspersed.

Two functionally distinct vocalizations have been reported. The young of many species vocalize when lost or disturbed (VL), giving calls variously described as hisses, squeaks and coughs. In a few species, the female has been heard to give a clicking vocalization (VC) in reply to the young's call or when disturbance is detected. Although this vocalization tends to be soft, it has been heard in only a small proportion of those species in which a structurally similar call given by males has been reported (see Sexual contexts). Females are known to click only in Pademelons, *Thylogale,* Grey Kangaroos and the Red Kangaroo. Russell (1973) noted that the click was absent from the repertoire of female Tammar Wallabies, *Macropus eugenii*, which instead call their young with a deeper version of the young's lost call. Unusually, young of the Yellow-footed Rock-wallaby, *Petrogale xanthopus,* are reported to give a clicking call, commonly directed at the female (Lim 1987).

Table 4. The occurrence of acts in the Macropodoidea in agonistic contexts, based on composite repertiores for 33 species.

		Agonistic acts																					
Genus	Species	AV	HA	B	NJ	P	C	VD	VS	Q	KL	KH	KT	SH	PW	PG	RG	RB	BC	GA	GD	FD	WS
Hypsiprymnodon	*moschatus*																						
Aepyprymnus	*rufescens*	•	•	•		•		•			•												
Bettongia	*lesueur*		•				•		•		•												
	penicillata			•			•																
	gaimardi	•	•	•	•	•	•	•			•												
Potorous	*longipes*																						
	tridactylus	•	•				•	•															
Dorcopsis	*luctuosa*		•	•	•		•	•	•					•	•			•			•		
Dendrolagus	*dorianus*	•	•	•		•	•	•	•	•					•			•	•				
	goodfellowi																						
	lumholtzi																						
Peradorcas	*concinna*		•	•		•	•		•			•											
Petrogale	*assimilis*	•	•	•	•	•	•	•	•	•		•									•		•
	lateralis	•	•	•		•	•					•											
	penicillata		•	•		•	•		•			•		•	•	•							
	xanthopus	•	•	•		•		•	•										•				
Thylogale	*billardierii*	•	•			•	•	•		•		•		•	•	•	•			•			
	brunii						•			•		•											
	stigmatica	•					•	•				•		•	•	•		•					
	thetis		•	•		•	•	•				•		•	•	•							
Setonix	*brachyurus*		•	•			•					•											
Wallabia	*bicolor*		•			•	•	•				•		•	•								
Macropus																							
(Prionotemnus)	*agilis*															•							
	dorsalis		•			•	•	•						•	•	•				•			
	eugenii	•	•			•	•	•	•	•		•			•	•	•			•	•		
	parma	•	•	•	•	•	•	•		•		•		•	•								
	parryi	•		•		•	•	•	•			•		•	•	•				•		•	•
	rufogriseus	•	•		•	•	•	•				•		•	•					•	•		
(Macropus)	*fuliginosus*	•				•	•	•	•	•			•	•	•					•	•	•	•
	giganteus	•	•			•	•	•	•				•	•	•	•	•			•	•	•	•
(Osphranter)	*antilopinus*	•				•	•	•	•				•	•	•	•				•			
	robustus	•	•	•		•	•	•	•				•	•	•	•	•				•	•	
(Megaleia)	*rufus*	•	•	•		•	•	•	•	•			•	•	•		•		•	•	•		•

Agonistic acts

Acts which occur in agonistic contexts include both aggressive and defensive acts (Table 4). The simplest form of agonistic interaction is avoidance (AV), which probably occurs throughout the Macropodoidea. Aggressive chasing (HA) is also performed by most species. The nature of combat differs between potoroids and macropodids.

In potoroids, interactions tend to be rapid, direct and violent, and a number of authors have commented on the difficulty of keeping males together in captivity. There is no apparent difference between the sexes in fighting techniques (see also Ganslosser, this volume). The data in Table 4 together with additional data on the Brush-tailed Bettong, *Bettongia penicillata* (U. Ganslosser, pers. comm.) indicate that Bettongs, *Aepyprymnus* and *Bettongia,* bite (B), paw (P) and cuff (C) their opponents, and kick at them while lying down (KL). Similar behaviour occurs in the Long-nosed Potoroo, *Potorous tridactylus,* except that kicks are delivered downwards or backwards when hopping over or past the opponent (O. Buchmann, pers. comm.). Vocalizations by dominant (VD) or submissive (VS) individuals have been described in some species.

In macropodids, agonistic interactions tend to be more prolonged, stylized and complex (see also Ganslosser, this volume). While male and female macropodids also bite, paw, cuff and vocalize (VD and VS), their kicks differ from those of potoroids in that they are generally delivered from an upright posture. In the larger macropodids, the tail serves as a prop while kicking (KT) as shown in Fig. 2, but the smaller species merely hop vertically before kicking (KH). In the rock-wallabies, *Peradorcas* and *Petrogale,* the initial hop may be quite high and the two combatants often leap towards each other simultaneously. The "turnaway kick" described in the Red-necked Wallaby by LaFollette (1971) and Southwell (1976), is regarded as a variant of KH. Another variant, hopping over and kicking down on the opponent, has been reported in the Red-necked Wallaby (LaFollette 1971) and the Brush-tailed Rock-wallaby, *Petrogale penicillata* (Batchelor 1980); it is not exclusive to the Nabarlek, *Peradorcas concinna,* as claimed by Goldstone and Nelson (1986). Tree-kangaroos probably do not kick at all during fights. Two other acts are exhibited by both sexes of macropodids: standing high (SH) often occurs as a prelude to fighting and wrestling (PW) in an

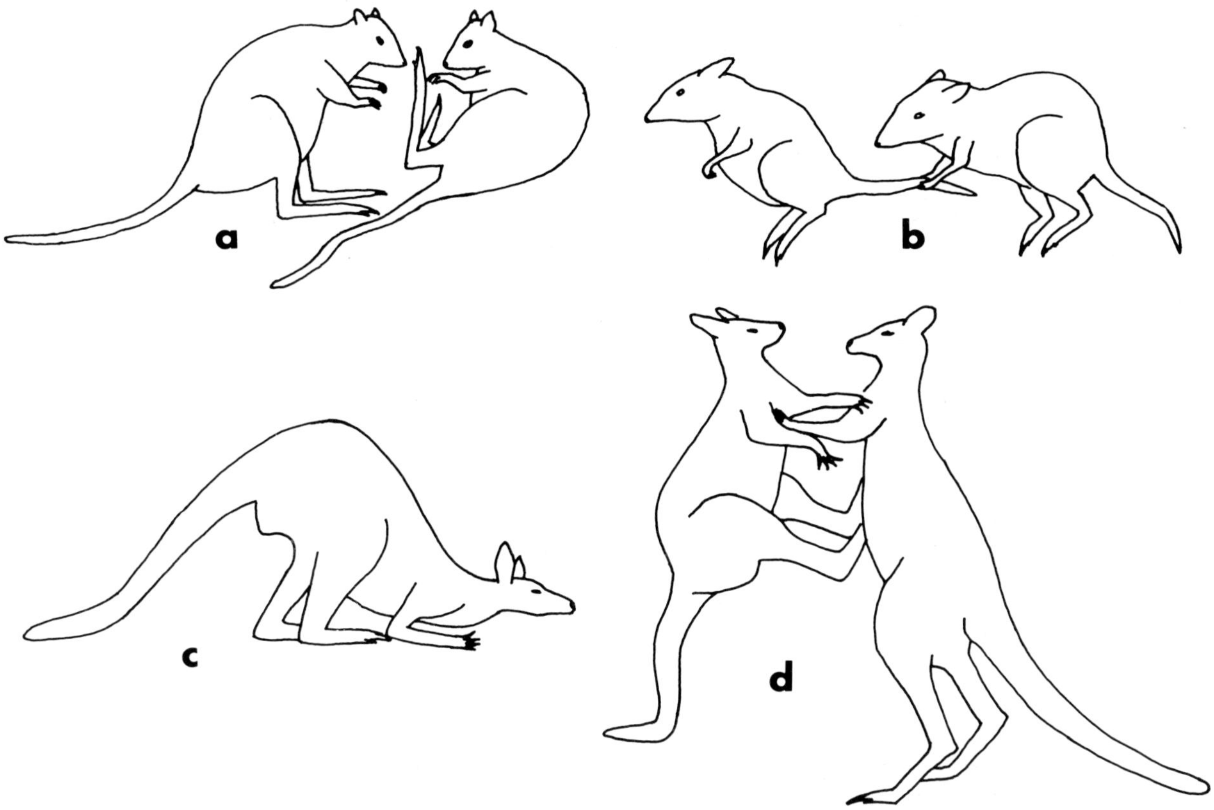

Fig. 2. Behavioural acts in the Macropodoidea. Sexual interactions: (a) a female Tasmanian Bettong, *Bettongia gaimardi*, lies kicking at a male as he attempts to nose her cloaca; (b) male Long-nosed Potoroo, *Potorous tridactylus*, paws at the tail of a female (redrawn from Veselovsky 1969). Agonistic interactions: (c) a male Antilopine Wallaroo, *Macropus antilopinus*, rubs his chest on the ground after pawing at a clump of vegetation; (d) a male Eastern Grey Kangaroo, *Macropus giganteus*, uses his tail for support as he kicks at an opponent.

attempt to push the opponent down or off balance.

A suite of additional, more stylized acts occurs almost exclusively in male-male agonistic interactions. Males of many species perform grass-pulling (PG) and less commonly rub their chest on branches (RB) or on the ground (RG) as shown in Fig. 2. Grass-pulling, chest-rubbing and standing-high may alternate in complex sequences, such as the "bush display" reported by Croft (1981a) in the Euro, *Macropus robustus*. These sequences are sometimes accompanied by erection of the penis. There are some reports of males spraying urine, directed at an opponent in the Red Kangaroo (Russell 1985), at spinifex clumps in the Rufous Hare-wallaby (G. Lundie-Jenkins, pers. comm.), and at the ventral surface of the displaying individual in the Western Grey Kangaroo, *Macropus fuliginosus* (W. Poole, pers. comm. to Russell 1985). Arching of the head often occurs during high-standing, and Croft (1982) reported a distinctive backwards tossing of the head in male Antilopine Wallaroos, *Macropus anilopinus*. Biting at the chest (BC) has been reported in three species, but seems to be most developed in the Red Kangaroo. In the larger species, males also perform exaggerated grooming (GD) as shown in Fig. 3, or feed warily (FD) between bouts of fighting. The striking stiff-legged walk (WS), shown in Fig. 4, is known to occur in five species.

Males use this array of acts, together with the acts common to both sexes, in prolonged bouts of fighting. Many authors (e.g., Kaufmann 1974; Croft 1981a) have commented that such bouts are often relaxed and perfunctory. When motivational levels are apparently higher, as happens in the presence of an oestrous female, the stylised acts are dispensed with and fights are reduced to intense and brief exchanges limited to the basic acts such as cuffing, kicking and wrestling.

Sexual acts

Sexual interactions include routine sexual "checking" (Kaufmann 1974) as well as the more complex aspects of courtship leading to copulation (Table 5). Checking involves the male nosing the cloaca (Nc) of the female, an act which is probably universal in the Macropodoidea. The male also noses the pouch (Np) in most species, although this tends to occur less frequently. Nosing of the cloaca frequently elicits urination by the female, and males of many species have

Fig. 3. Exaggerated grooming during a fight between male Grey Kangaroos, *Macropus fuliginosus* (left) and *M. giganteus* (right).

been seen to nose urine (Nu), sometimes putting their muzzles in the stream of urine (Fig. 5), but also nosing the substrate where urine has fallen. While "Flehmen" has been reported in this context in a few species, Coulson and Croft (1981) suggested that this unobtrusive behaviour may be widespread in the Macropodoidea and has simply been overlooked.

Checking differs markedly between at least some potoroids and the rest of the Macropodoidea. In *Aepyprymnus* and *Bettongia,*

Fig. 4. The stiff-legged walk performed by an adult male Eastern Grey Kangaroo, *Macropus giganteus.*

Table 5. The occurrence of acts in the Macropodoidea in sexual contexts, based on composite repertoires for 33 species.

		Sexual acts																					
Genus	Species	Nc	Nu	Np	HA	KG	mS	Ph	Rh	Pt	Pb	Pf	Pc	Pp	mGA	X	VC	TL	XM	mM	I	fM	MH
Hypsiprymnodon	*moschatus*						•	•															
Aepyprymnus	*rufescens*	•		•	•	•												•		•	•	•	
Bettongia	*lesueur*	•		•												•	•	•		•	•		
	penicillata																						
	gaimardi	•				•										•		•		•		•	
Potorous	*longipes*																						
	tridactylus	•	•		•					•								•		•	•		
Dorcopsis	*luctuosa*	•	•	•	•	•		•			•				•	•	•			•	•		
Dendrolagus	*dorianus*	•	•	•								•								•	•		
	goodfellowi																						
	lumholtzi																						
Peradorcas	*concinna*																						
Petrogale	*assimilis*	•		•					•	•					•	•	•			•	•		
	lateralis	•	•	•	•				•	•										•			•
	penicillata	•		•	•				•											•			
	xanthopus	•	•	•	•												•	•		•		•	
Thylogale	*billardierii*	•	•	•	•		•	•				•		•	•	•	•	•	•	•	•		
	brunii																						
	stigmatica	•		•	•			•								•	•		•	•	•		
	thetis	•	•	•			•	•				•				•	•	•		•			
Setonix	*brachyurus*																•			•	•		
Wallabia	*bicolor*	•	•	•			•		•	•	•					•		•		•	•		•
Macropus																							
(Prionotemnus)	*agilis*	•	•							•								•					
	dorsalis	•	•	•				•	•						•	•		•					
	eugenii	•	•	•	•		•	•		•						•	•	•		•	•		
	parma	•	•	•			•	•	•	•						•	•	•		•			
	parryi	•	•	•	•					•	•	•	•			•	•	•	•	•	•		•
(Macropus)	*rufogriseus*	•	•	•	•			•		•			•			•	•	•		•	•		
	fuliginosus	•	•	•	•		•	•		•						•	•			•	•		•
	giganteus	•	•	•	•		•	•	•	•			•	•	•	•	•		•	•	•	•	•
(Osphranter)	*antilopinus*	•	•				•	•								•							
	robustus	•	•	•	•		•		•	•						•				•	•		
(Megaleia)	*rufus*	•	•	•	•		•	•		•						•	•	•	•	•	•		•

including the Brush-tailed Bettong (U. Ganslosser, pers. comm.) the female typically falls on her side at the approach of the male (Fig. 2) and kicks out at him while lying down (KL); the male advances slowly while thumping the ground repeatedly with his feet (KG) and making jerky movements of the abdomen, and attempts to nose the cloaca or pouch. This repetitive thumping action appears to be quite distinct from the alarm thump (see general acts) and is more appropriately termed "drumming" to distinguish the two acts. Both sexual and alarm thumping occur at least in the Rufous Bettong, *Aepyprymnus rufescens*. Little information is available for other potoroids. J. Seebeck (pers. comm.) has not observed kicking while lying down by the female or drumming by the male in the Long-nosed Potoroo or the Long-footed Potoroo, *Potorous longipes*. Following a frontal approach, male Long-footed Potoroos have instead been observed to nose the head of the female then climb over her back, leaving a trail of urine from shoulder to rump. In macropodids, males often approach and stand in front of the female (mS), apparently blocking her path in some species. The male generally paws at the head or shoulder (Ph) of the female, grasping her head and rubbing it (Rh) against his chest in many species. Sternal rubbing of the female's head appears to be most frequent and persistent in the Rock-wallabies, *Petrogale*. The Allied Rock-wallaby is also unusual in that males have been observed to mark the base of the female's tail with urine (Horsup 1986). Approaches by Rufous Hare-wallabies resemble more the potoroid pattern, since the female may lie on her side and kick at the male (G. Lundie-Jenkins, pers. comm.). Males of some species also paw at the female's cloaca (Pc) or pouch (Pp). After (or as an alternative to) a frontal approach male macropodids may move behind the female and paw at her tail (Pt) or high on her back (Pb). Pawing at both sites has been observed in the Swamp Wallaby (Crebbin 1982) and the Whiptail Wallaby, *Macropus parryi* (Kaufmann 1974), while pawing at either site seemingly does not occur in Tree-kangaroos, *Dendrolagus*, or Pademelons, *Thylogale*. Pawing at the flanks (Pf) of the female has been reported in Pademelons and the Whiptail Wallaby. In Doria's Tree-kangaroo, *Dendrolgaus dorianus*, however, the male rubs the female's flanks with his hind feet (Ganslosser 1979a). Allogrooming by the male also occurs in sexual contexts in a diverse minority of species.

Sexual arousal of males in the Macropodoidea is typically accompanied by erection of the penis

Fig. 5. Flehmen in the Eastern Grey Kangaroo, *Macropus giganteus*. The male's incisors are visible as he noses the stream of urine produced by the female.

(X), a clucking vocalization (VC) and a sinuous lashing of the tail (TL). Tail-lashing has also been noted in some species in non-sexual contexts, typically when an individual is alerted to a potential predator (e.g., Cooke 1979; Vujcich 1979), and Lim (1987) recorded "wagging" of the Yellow-footed Rock-wallaby's tail in agonistic contexts. Tail-lashing does not usually occur in the largest species, but Sharman and Calaby (1969) reported it during masturbation by solitary captive male Red Kangaroos. Masturbation has also been reported in four other species, and is not merely an artifact of captivity since three of these species were observed in the field.

Copulation, involving mounting by the male (mM) and intromission (I) is similar in all members of the Macropodoidea, although the frequency and duration of mountings varies considerably. The failure to observe such a mandatory act as copulation in many species is clear evidence of the deficiency of behavioural repertoires. Homosexual mounting by males (Mm) and females (fM) has been observed in a range of species, and is relatively common in mammals (Dagg 1984).

EVOLUTION OF BEHAVIOUR

The Macropodoidea evolved from a phalangerid ancestor in the Miocene, then underwent an explosive radiation in the Pliocene and Pleistocene (Flannery 1984). The ancestral macropod was a small and presumably nocturnal inhabitant of the rainforest floor. Extant macropods range from about 500 g to over 50 kg in body weight; they occupy habitats ranging from rainforest trees to desert plains, and their activity patterns range from nocturnal to diurnal (Pople, this volume). This review has revealed that there is also a diversity of behavioural repertoires in the Macropodoidea. A number of acts are shared by all taxa, and could be considered to comprise the primitive behavioural repertoire of the superfamily. Many presumably primitive acts, such as nosing the head and cloaca, pawing at the tail, and buccal click vocalizations are shared with the Phalangeridae (Eisenberg and Golani 1977; Russell 1984), and are likely to have been part of the repertoire of ancestral macropods. Subsequent modifications and additions to this basic repertoire have produced the range of variation in behaviour that is now evident.

Comparisons with the ungulates

Comparisons have often been drawn between the Macropodoidea and the ungulates, particularly the Cervidae and Bovidae, in terms of morphology and ecology (e.g., Hume 1982; Jarman 1983). Behavioural repertoires of ungulates, as reviewed by Leuthold (1977) and Eisenberg (1981), also show convergence with those of the macropods in a number of aspects.

In general contexts, acts such as nosing (N) occur commonly in both taxa. Thumping (KG) in response to disturbance has no direct equivalent in the ungulates, although foot-stamping when alarmed has been recorded in some species such as the White-tailed Deer, *Odocoileus virginianus* (LaGory 1987). It is, however, common in another large-footed taxon, the lagomorphs (Ewer 1968). Most studies have concluded that thumps function as an alarm signal in macropods, alerting conspecifics (and less commonly other macropod species) to the presence of a predator. By analogy with "stotting" in Thomson's Gazelles, *Gazella thomsoni* (Caro 1986), it is equally probable that the primary function of thumps is to inform the predator that it has been detected.

Many aspects of maternal behaviour show general similarities between the ungulates and the macropods. C. Johnson (1987) has pointed out differences between these taxa in the prevalence of "hiding" and "following" strategies in dependent young, arguing that the differences relate to the characteristics of their eutherian and marsupial modes of reproduction. Sexual behaviour also shows numerous instances of convergence: induced urination by females and testing of urine (NU) by males occurs throughout both groups, although "Flehmen" by males is much less spectacular in macropods; sexual following or driving by males is equally common and the "Laufschlag" (leg beat) of ungulates is reminiscent of tail-clasping (PT) in macropods.

Perhaps the most fundamental difference between ungulates and macropods, in terms of its implications for behavioural repertoires, is related to quadrupedal *versus* predominantly bipedal locomotion. As Jarman has argued, this has led to differentiation of the limbs in the Macropodoidea, allowing the forelimbs of macropods to become more specialized for the location and manipulation of food, and, particularly in males, to have evolved under sexual selection as weapons. By contrast, antlers and horns are the predominant weapons in the ungulates. Some agonistic acts have obvious analogues, such as the broadside displays in ungulates and the stiff-legged walk in macropods which both exaggerate the height of the animal. Many other acts reflect the differences in morphology of weapons: macropods predominantly use and emphasize the forelimbs in fighting techniques and agonistic displays.

Morphological and ecological constraints

Some variation in behavioural repertoires within the Macropodoidea can be accounted for by constraints imposed by differences in morphology. The occurrence of tail-lashing (TL) by sexually-aroused males, and kicking using the tail as a support (KT) provide a clear example of morphological constraints. The tails of small and medium-sized macropods are essentially cylindrical and fairly flexible, whereas the tails of Wallaroos, Grey Kangaroos and Red Kangaroos taper from a wide muscular butt and are relatively inflexible. It is likely that only the large species have sufficient strength in the tail to use it as a support, while the smaller species have tails that apparently lack sufficient rigidity for support and are flexible enough to be lashed. Similarly, the absence of wrestling and the preference for biting, pawing and cuffing in potoroids could be interpreted in terms of their lack of sexual dimorphism in forelimb size but possession of well-developed (and sharp) teeth and claws.

Ecological factors, particularly habitat structure, can also account for variation in behavioural repertoires. The influence of habitat is exemplified by the Tree-kangaroos: unlike other macropods, they apparently do not kick when fighting, or give an alarm thump. Their arboreal habitat would make these acts difficult and dangerous to perform. Other hazardous acts, such as hopping in circles by newly-emerged pouch young, are absent in Tree-kangaroos but widespread in the terrestrial species.

Eisenberg and Golani (1977) and Russell (1984) have pointed out that habitat preference and activity pattern will also impose limitations on the sensory modalities used in communication between macropods. Russell (1984) concluded that the major channels are tactile and olfactory, while visual channels become significant in the large macropods, which are fairly diurnal occupants of open habitats. In this context, it is worth reconsidering the evolution of the grass-pulling and chest-rubbing "displays" performed by most macropodids. Noting Mykytowycz and Nay's (1964) report of apocrine sternal glands in these species, many authors (Kaufmann 1974; Croft 1981a; Russell 1984, 1985) have suggested that the display originated as sternal marking of the substrate and may still partly serve that function. However, grass-pulling also has elements in common with the feeding technique of potoroids which dig for food. This interpretation was strengthened by my observation of an exaggerated digging action performed by a male Tasmanian Bettong, *Bettongia gaimardi;* it was accompanied by the rhythmic abdominal contractions often seen during sexual checking, and followed a rebuff by a female. The intriguing marking sequence observed by Bourke (this volume) in the Grey Dorcopsis, *Dorcopsis muelleri,* could also be interpreted as a modification of a potoroid behavioural pattern since it includes foot-drumming, a characteristically potoroid act.

Distinctiveness of taxa

The variations in behavioural repertoires within the Macropodoidea should reflect the phylogeny of the superfamily as it is currently understood. The ordering of taxa in Tables 3-5 incorporates a crude scale from the most primitive *(Hypsiprymnodon)* to the most derived *Macropus (Megaleia).* A number of trends in the distribution of acts along this scale are evident; some acts can be identified as shared primitive characteristics and many others have discontinuous distributions. It should therefore be possible to identify behavioural acts that are useful taxonomic characters. This analysis is necessarily a preliminary attempt and is limited by the resolving power of the acts and the coverage of repertoires. To permit a closer examination of *Macropus* the subgenera proposed by Bartholomai (1975) have been adopted, except that *Megaleia* has been excluded from *M. (Osphranter)*.

In the Potoroidae, *Hypsiprymnodon* has not been studied sufficiently to identify distinctive social acts, but it can be readily separated from all other macropods on one behavioural criterion: it is the only diurnal species (Johnson and Strahan 1982). The Bettongs, *Aepyprymnus* and *Bettongia,* are indistinguishable in their behaviour according to this analysis, but the Rufous Rat-kangaroo, *Aepyprymnus rufescens,* produces a distinctive "chainsaw" vocalization when disturbed (P. M. Johnson 1980). Bettongs are distinct from *Potorous* in sexual interactions, since male Bettongs do not paw the female's tail.

In the Macropodoidea, more research is needed on the New Guinea Wallabies, although the Grey Dorcopsis apparently has distinctive marking behaviour. *Dendrolagus* is distinguished by the absence of kicking and alarm-thumping, and *Thylogale* males apparently do not paw the tail or back of females during sexual interactions. At present the Rock-wallabies (*Peradorcas* and *Petrogale*) and *Setonix* form an amorphous group with no apparent characteristics which could distinguish them from each other or from other taxa. *Wallabia* is also difficult to differentiate, but it is one of the few taxa in which males routinely paw high on the back of the female during sexual interactions and can thus be separated from other large wallabies which rarely or never paw the back. *Wallabia* also has a distinctive crouching posture used in threat (U. Ganslosser, pers. comm.).

Within *Macropus, M. (Prionotemnus)* is distinguished by the presence of tail-lashing and the absence of kicking with the tail as a prop. *M. (Macropus)* has the unique double alarm thump, and *M. (Osphranter)* has an additional alarm hiss. On behavioural grounds, the Red Kangaroo apparently does not belong in *M. (Osphranter)*. The clicks made by male and female Red Kangaroos are not shared by the wallaroos, and the Red Kangaroo lacks the alarm hiss. Chest-biting by male Red Kangaroos is also distinctive.

This review has attempted to bring a more rigorous approach to the study of behavioural repertoires in the Macropodoidea. The deficiencies in the data have been highlighted and the known distribution of behavioural acts has been analysed. Further examination of the phylogeny and ecology of behavioural repertoires in the Macropodoidea must await development in two main areas: investigations of poorly studied species and more detailed studies of others, including the validation of captive studies in the field, and the use of more sophisticated techniques of recording and analysis of data, including measurement of rates and temporal patterns of behaviour.

ACKNOWLEDGEMENTS

I am grateful to Angus Martin for his critical comments on the manuscript and his patience over the years, to Udo Ganslosser for his valuable comments on the manuscript, and to the many researchers who responded promptly to my requests for unpublished material. Much of my work in this area could not have been carried out without the co-operation of the Royal Melbourne Zoological Gardens (where I am particularly indebted to Ernst Weiher and Bev Drake), and Randy Rose generously gave access to the Bettong colony at the University of Tasmania. Finally, Ruth Coulson has been extremely supportive and tolerant of my macropodophilia.

REFERENCES

Barker, S. C., 1982. Ecology, behaviour and social organization of the plain rock-wallaby, *Petrogale inornata assimilis* (Macropodinae). Hons. Thesis, James Cook University of North Queensland.

Bartholomai, A., 1975. The genus *Macropus* Shaw (Marsupialia: Macropodidae) in the Upper Cainozoic deposits of Queensland. *Mem. Qd. Mus.* **17**: 195-235.

Batchelor, T. A., 1980. the social organization of the brush-tailed rock-wallaby *(Petrogale penicillata penicillata)* on Motutapa Island. M.Sc. Thesis, University of Auckland.

Bourke, D. W. (this volume). Observations on the behaviour of the Grey Dorcopsis Wallaby, *Dorcopsis luctuosa,* in captivity.

Caro, T. M., 1986. The functions of stotting in Thomson's gazelles: some tests of the predictions. *Anim. Behav.* **34:** 663-84.

Cicala, G. A., Albert, I. A. and Ulmer, F. A., 1970. Sleep and other behaviours of the red kangaroo *(Megaleia rufa)*. *Anim. Behav.* **18:** 787-90.

Clancy, T. F., 1982. Aspects of the behaviour of *Thylogale billardierii* in captivity. Hons. Thesis, University of Tasmania.

Cooke, B. N., 1979. Field observations of the macropod marsupial *Thylogale stigmatica* (Gould). M.Sc. Thesis, University of Queensland.

Coulson, G. M. and Croft, D. B., 1981. Flehmen in Kangaroos. *Aust. Mammal.* **4:** 139-40.

Crebbin, A., 1982. Social organization and behaviour of the swamp wallaby, *Wallabia bicolor* (Desmarest) (Marsupialia: Macropodidae) in captivity. Hons. Thesis, University of New South Wales.

Croft, D. B., 1981a. Social behaviour of the euro, *Macropus robustus,* in the Australian arid zone. *Aust. Wildl. Res.* **8:** 13-49.

Croft, D. B., 1981b. Behaviour of Red Kangaroos, *Macropus rufus* (Desmarest 1922) in northwestern New South Wales, Australia. *Aust. Mammal.* **4:** 5-58.

CROFT, D. B., 1982. Some observations of the behaviour of the Antilopine Wallaroo *Macropus antilopinus* (Marsupialia: Macropodidae). *Aust. Mammal.* **5**: 5-13.

CROFT, D. B., 1985. Inter- and intraspecific conflict between arid-zone kangaroos at watering points. *Aust. Wildl. Res.* **12**: 337-48.

DAGG, A. I., 1984. Homosexual behaviour and female-male mountings in mammals — a first survey. *Mammal. Rev.* **14**: 155-85.

DAVIES, W. G., 1976. Some observations on the ontogeny of social behaviour in the rock wallaby. *Aust. Mamm. Soc. Bull.* **3**(1): 43.

DAVIES, W. G., 1979. Social organization of the rock wallaby *(Petrogale puella)* in North Queensland. *Aust. Mamm. Soc. Bull.* **6**(1): 18-19.

DWYER, P. D., 1972. Social organization of a population of rock-wallabies, *Petrogale inornata. Aust. Mammal.* **1**: 72.

EISENBERG, J. F., 1981. The Mammalian Radiations. An Analysis of Trends in Evolution, Adaptation, and Behaviour. University of Chicago Press: Chicago.

EISENBERG, J. F. COLLINS, L. R. AND WEMMER, C., 1975. Communication in the Tasmanian Devil *(Sarcophilus harrisii)* and a survey of auditory communication in the Marsupalia. *Z. Tierpsychol.* **37**: 379-99.

EISENBERG, J. F. AND GOLANI, I., 1977. Communication in Metatheria. Pp. 575-99 *in* How Animals Communicate ed. by T. A. Seebeok. Indiana University Press: Bloomington.

ERWIN, J. AND DENI, R., 1979. Strangers in a strange land: abnormal behaviour or abnormal environments? Pp. 1-28 *in* Captivity and Behaviour ed. by J. Erwin, T. L. Maple and G. Mitchell. Van Nostrand Reinhold: New York.

EWER, R. F., 1968. Ethology of Mammals. Logos Press: London.

FAGEN, R. M. AND GOLDMAN, R. N., 1977. Behavioural catalogue analysis methods. *Anim. Behav.* **25**: 261-74.

FLANNERY, T., 1984. Kangaroos: 15 million years of Australian bounders. Pp. 817-35 *in* Vertebrate Zoogeography and Evolution in Australia ed. by M. Archer and G. Clayton. Hesperian Press. Carlisle, Western Australia.

FLEMING, D., CINDEREY, R. N. AND HEARN, J. P., 1983. The reproductive biology of Bennett's wallaby *(Macropus rufogriseus rufogriseus)* ranging free at Whipsnade Park. *J. Zool. Lond.* **201**: 283-91.

GANSLOSSER, U., 1977. Beobachtungen an Doria-Baumkänguruhs *(Dendrolagus dorianus)* und Grauen Baumkänguruhs *(Dendrolagus inustus)* in Zoologischen Gärten. *Zool. Anz.* **198**: 303-412.

GANSLOSSER, U., 1978. Zum Kampfverhalten des Neuguinea-Filanders, *Thylogale brunii* (Schreber, 1778). *Säugertierkd. Mitt.* **26**: 195-98.

GANSLOSSER, U., 1979a. Soziale Interaktionen des Doria-Baumkänguruhs *Dendrolagus dorianus* Ramsay, 1833 (Marsupialia, Macropodidae). *Z. Säugertierkunde* **44**: 1-18.

GANSLOSSER, U., 1979b. Soziale Kommunikation, Gruppenleben, Spiel- und Jugendverhalten des Doria-Baumkänguruhs *(Dendrolagus dorianus)* Ramsay, 1833. *Z. Säugertierkunde* **44**: 137-53.

GANSLOSSER, U., 1983. Quantitative Untersuchungen zum Sozialverhalten des Doria-Baumkänguruhs (*Dendrolagus dorianus* Ramsay, 1833) in Gefangenschaft I. Struktur und Entwicklung einer Familiengruppe. *Zool. Anz.* **211**: 1-29.

GANSLOSSER, U., 1984. On the occurrence of female coalitions in tree-kangaroos (Marsupialia: Macropodidae, *Dendrolagus*). *Aust. Mammal.* **7**: 219-21.

GANSLOSSER, U. (this volume). Agonistic behaviour in macropodoids. A review.

GANSLOSSER, U. AND HOLZ, C., 1984. Zum Kampfverhalten verschiedener sozialer Kategorien beim Streifenbuschkänguruh *(Dorcopsis muelleri). Verh. Dtsch. Zool. Ges.* **77**: 281.

GANSLOSSER, U. AND SCHAPPERT, I., 1985. Zum Sozialspiel bei Doria-Baumkänguruhs *(Dendrolagus dorianus)* und Streifenbuschkänguruhs *(Dorcopsis muelleri* (Macropodidae). *Verh. Dtsch. Zool. Ges.* **78**: 206.

GANSLOSSER, U. AND WERNER, I., 1984. Zur Häufigkeit freundlicher und agonistischer Verhaltensmuster in einer Gruppe von Rotbauchfilandern (*Thylogale billardieri* Desmarest) (Marsupialia, Macropodidae). *Zool. Anz.* **213**: 51-67.

GANSLOSSER, U. AND WILHELM, P., 1986. Sex- and age-dependant differences in frequencies of social interactions between captive Red Kangaroos, *Macropus rufus* (Desmarest, 1822). *Zool. Anz.* **216**: 58-71.

GEPP, B., 1973. The social organization of the Pearson Island rock-wallaby, *(Petrogale penicillata pearsoni).* Hons. Thesis, University of Adelaide.

GERMANN-MEYER, V., 1974. Verhaltensstudien am Grauen Riesenkänguruh, *Macropus (Macropus) giganteus.* Juris: Zurich.

GERMANN-MEYER, V. AND SCHENKEL, R., 1970. Über das Kampfverhalten des grauen Riesenkänguruhs, *Macropus giganteus. Rev. Suisse. Zool.* **77**: 938-42.

GOLDSTONE, A. D. AND NELSON, J. E., 1986. Aggressive behaviour in two female *Peradorcas concinna* (Macropodidae) and its relation to oestrus. *Aust. Wildl. Res.* **13**: 375-85.

GRANT, T. R., 1973. Dominance and association among members of a captive and a free-ranging group of grey kangaroos *(Macropus giganteus). Anim. Behav.* **21**: 449-56.

GRANT, T. R., 1974. Observations of enclosed and free-ranging grey kangaroos *Macropus giganteus. Z. Säugetierkunde* **39**: 65-78.

GROVES, G. P. AND FLANNERY, T. (this volume). Revision of the genus *Dorcopsis* (Macropodoidae).

HEATHCOTE, C. F. (this volume). Social behaviour of the Black-striped Wallaby, *Macropus dorsalis,* in captivity.

HERRMANN, D., 1967. Beutelreinigung beim grauen Reisenkänguruh, *Macropus giganteus. Zool. Garten.* **33**: 265-78.

HERRMANN, D., 1971. Beobachtungen des Gruppenlebens ostaustralischer Graugrosskänguruhs, *Macropus giganteus* (Zimmermann, 1777) und Bennetkänguruhs, *Protemnodon rufogrisea* (Desmarest, 1817). *Säugetierk. Mitt.* **19**: 352-62.

HERTER, U., SCHNEIDER, I. AND GANSLOSSER, U., 1985. Some remarks on the behavioural repertoire of the Parma-Wallaby, *Macropus parma* Waterhouse. *Zool. Anz.* **215**: 287-97.

HORNSBY, P. E., 1978. A note on the pouch life of rock wallabies. *Victorian Nat.* **95**: 108-11.

HORSUP, A. B., 1986. The behaviour of the allied rock wallaby *Petrogale assimilis* (Macropodinae). Hons. Thesis, James Cook University of North Queensland.

HUGHES, R. L., 1962. Reproduction in the macropod marsupial *Potorous tridactylus* (Kerr). *Aust. J. Zool.* **10**: 193-224.

HUME, I. D., 1982. Digestive Physiology and Nutrition of Marsupials. Cambridge University Press: Cambridge.

IMMELMANN, K., 1965. Verhaltensstudien an Känguruhs in Australien. *Zool. Garten.* **31**: 144-57.

JAREMOVIC, R., 1984. Space and time related behaviour in eastern grey kangaroos (*Macropus giganteus* Shaw). Ph.D. Thesis, University of New South Wales.

JARMAN, P., 1983. Mating systems and sexual dimorphism in large, terrestrial, mammalian herbivores. *Biol. Rev.* **58**: 485-520.

JARMAN, P. J., JOHNSON, C. N., SOUTHWELL, C. J. AND STUART-DICK, R., 1987. Macropod studies at Wallaby Creek. I. The area and the animals. *Aust. Wildl. Res.* **14**: 1-14.

JOBLIN, K., 1981. Social organization and spatial distribution in the brush-tailed rock-wallaby, *Petrogale penicillata. Aust. Mamm. Soc. Bull.* 7(1): 43-44.

JOHNSON, K. A., 1977. Ecology and management of the red-necked pademelon, *Thylogale thetis,* on the Dorrigo Plateau of northern New South Wales. Ph.D. Thesis, University of New England.

JOHNSON, C. N., 1987. Relationships between mother and infant red-necked wallabies *(Macropus rufogriseus banksianus). Ethology* **74**: 1-20.

JOHNSON, P. M., 1980. Observation of the behaviour of the rufous rat-kangaroo, *Aepyprymnus rufescens* (Gray), in captivity. *Aust. Wildl. Res.* **7**: 347-57.

JOHNSON, P. M. AND STRAHAN, R., 1982. A further description of the Musky Rat-kangaroo, *Hypsiprymnodon moschatus* Ramsay, 1876 (Marsupialia, Potoroidae), with notes on its biology. *Aust. Zool.* **21**: 27-46.

JOHNSON, S. N., 1987. Aspects of the behaviour and ecology of the potoroo, *Potorous tridactylus* (Marsupialia: Potoridae). Hons. Thesis, University of Tasmania.

KAUFMANN, J. H., 1974. Social ethology of the whiptail wallaby, *Macropus parryi,* in northeastern New South Wales. *Anim. Behav.* **22**: 281-369.

KAUFMANN, J. H., 1975. Field observations of the social behaviour of the eastern grey kangaroo, *Macropus giganteus. Anim. Behav.* **23**: 214-21.

LAGORY, K. E., 1987. The influence of habitat and group characteristics on the alarm and flight response of white-tailed deer. *Anim. Behav.* **35**: 20-25.

LEUTHOLD, W., 1977. African Ungulates. A Comparative Review of their Ethology and Behavioural Ecology. Springer-Verlag: Berlin.

LIM, L. T., 1987. Conservation and management of the Yellow-footed Rock-wallaby *Petrogale xanthopus* Gray 1854 (Macropodidae). Ph.D. Thesis, Macquarie University.

LAFOLLETTE, R. M., 1971. Agonistic behaviour and dominance in confined wallabies, *Wallabia rufogrisea frutica. Anim. Behav.* **19**: 93-101.

LUKSCHANDERL, L. AND LUKSCHANDERL, K., 1969. Einige Beobachtungen zur Jugendentwicklung von *Bettongia cuniculus. Zool. Garten* **37**: 117-26.

MARTIN, P. AND BATESON, P., 1986. Measuring Behaviour. Cambridge University Press: Cambridge.

MAYNES, G. M., 1973. Reproduction in the parma wallaby, *Macropus parma* Waterhouse. *Aust. J. Zool.* **21**: 332-51.

MAYNES, G. M., 1977. Distribution and aspects of the biology of the parma wallaby, *Macropus parma,* in New South Wales. *Aust. Wildl. Res.* **4**: 109-25.

MERCHANT, J. C. AND SHARMAN, G. B., 1966. Observations on the attachment of marsupial pouch young to the teats and on the rearing of pouch young by foster-mothers of the same or different species. *Aust. J. Zool.* **14**: 593-609.

MORTON, S. R. AND BURTON, T. C., 1973. Observations on the behaviour of the macropodid marsupial *Thylogale billardieri* (Desmarest) in captivity. *Aust. Zool.* **18**: 1-14.

MURBÖCK, A., 1979. Welchen Einfluss haben Gehegegrösse und Tierdichte auf das Verhalten des Bennettwallabis, *Macropus rufogriseus fruticus* Ogilby, 1838? *Säugetierk. Mitt.* **27**: 216-40.

MURBÖCK, T., 1975. Zum Kampfverhalten des Bennettwallabis, *Protemnodon rufogrisea frutica* (Ogilby, 1838). *Säugetierk. Mitt.* **23**: 79-80.

MYKYTOWYCZ, R. AND NAY, T., 1964. Studies on the cutaneous glands and hair follicles of some species of Macropodidae. *C.S.I.R.O. Wildl. Res.* **9**: 200-17.

NICHOLLS, C., 1972. A study of the behaviour and social organization of the Yellow-footed Rock-wallaby *(Petrogale xanthopus).* Hons. Thesis, University of Adelaide.

OSAZUWA, J. E., 1978. Activity and behaviour of the wallaroo, *Macropus robustus* Gould (Marsupialia: Macropodidae). M.Nat.Res. Thesis, University of New England.

PACKER, W. C., 1969. Observations on the behaviour of the marsupial *Setonix brachyurus* (Quoy and Gaimard) in an enclosure. *J. Mammal.* **50**: 8-20.

POOLE, W. E., 1982. Management of captive Macropodidae. Pp. 91-99 *in* The Management of Australian Mammals in Captivity ed. by D. D. Evans. The Zoological Board of Victoria: Melbourne.

POOLE, W. E. AND PILTON, P. E., 1964. Reproduction in the grey kangaroo, *Macropus canguru,* in captivity. *C.S.I.R.O. Wildl. Res.* **9**: 218-34.

POPLE, A. (this volume). Habitat associations of macropods.

PROCTER-GRAY, E., 1985. The behaviour and ecology of Lumholtz's Tree-kangaroo, *Dendrolagus lumholtzi* (Marsupialia: Macropodidae). Ph.D. Thesis, Harvard University.

ROSE, R. W. AND MCCARTNEY, D. J., 1982. Reproduction of the red-bellied pademelon, *Thylogale billardierii* (Marsupialia). *Aust. Wildl. Res.* **9**: 27-32.

RUSSELL, E. M., 1970. Agonistic interactions in the red kangaroo *(Megaleia rufa). J. Mammal.* **51**: 80-88.

RUSSELL, E. M., 1970. Observations on the behaviour of the red kangaroo *(Megaleia rufa)* in captivity. *Z. Tierpsychol.* **27**: 385-404.

RUSSELL, E. M., 1973. Mother-young relations and early behavioural development in the marsupials *Macropus eugenii* and *Megaleia rufa. Z. Tierpsychol.* **33**: 163-203.

RUSSELL, E. M., 1982. Patterns of parental care and parental investment in marsupials. *Biol. Rev.* **57**: 423-86.

RUSSELL, E. M., 1984. Social behaviour and social organization of marsupials. *Mammal. Rev.* **14**: 101-54.

RUSSELL, E. M., 1985. The metatherians: order Marsupialia. Pp. 45-104 *in* Social Odours in Mammals ed. by R. E. Brown and D. W. MacDonald. Oxford University Press: Oxford.

RUSSELL, E. M. (this volume). Maternal behaviour in the Macropodoidea.

RUSSELL, E. M. AND GILES, D. C., 1974. The effects of young in the pouch on pouch cleaning in the tammar wallaby, *Macropus eugenii* Desmarest (Marsupialia). *Behaviour* **51**: 19-37.

RUSSELL, E. M. AND NICHOLLS, D. G., 1974. Distress vocalization of young of the red kangaroo, *Megaleia rufa* (Marsupialia: Macropodidae). *Aust. Mammal.* **1**: 251-54.

SAMPSON, J. C., 1971. The biology of *Bettongia penicillata* Gray, 1837. Ph.D. Thesis, University of Western Australia.

SANSON, G. D., NELSON, J. E. AND FELL, P., 1985. Ecology of *Peradorcas concinna* in Arnhemland in a wet and dry season. *Proc. Ecol. Soc. Aust.* **13**: 65-72.

SCHAPPERT, I., 1984. Verhaltensbeobachtungen beim Buschkänguruh, *Dorcopsis muelleri*, in Zoologischen Gärten. *Vivarium Darmstadt* **1**: 2-12.

SCHNEIDER, K. M., 1954. Von Baumkänguruh *(Dendrolagus leucogenys)*. *Zool. Garten* **21**: 63-106.

SEEBECK, J. H. AND JOHNSTON, P. G., 1980. *Potorous longipes* (Marsupialia: Macropodidae); a new species from eastern Victoria. *Aust. J. Zool.* **28**: 119-34.

SHARMAN, G. B. AND CALABY, J. H., 1964. Reproductive behaviour in the red kangaroo, *Megaleia rufa,* in captivity. *C.S.I.R.O. Wildl. Res.* **9**: 58-85.

SHARMAN, G. B., CALABY, J. H. AND POOLE, W. E., 1966. Patterns of reproduction in female diprotodont marsupials. Pp. 205-32 *in* Comparative Biology of Reproduction in Mammals ed. by I. W. Rowlands. 15th Symposium Zoological Society London.

SHARMAN, G. B. AND PILTON, P. E., 1964. The life history and reproduction of the red kangaroo *(Megaleia rufa)*. *Proc. Zool. Soc. Lond.* **142**: 29-48.

SOUTHWELL, C., 1976. Behaviour and social organization of three species of macropod at Diamond Flat, New South Wales. Hons. Thesis, University of New England.

SOUTHWELL, C. J., 1981. Sociobiology of the eastern grey kangaroo, *Macropus giganteus*. Ph.D. Thesis, University of New England.

STODART, E., 1966. Observation on the behaviour of the marsupial *Bettongia lesueuri* (Quoy and Gaimard) in an enclosure. *C.S.I.R.O. Wildl. Res.* **11**: 91-101.

TYNDALE-BISCOE, H. AND RENFREE, M., 1987. Reproductive Physiology of Marsupials. Cambridge University Press: Cambridge.

VESELOVSKY, Z., 1969. Beitrag zur Kenntnis des Fortpflanzungsverhaltens der Känguruhs. *Zool. Garten* **37**: 93-107.

VUJCICH, M. V., 1979. Aspects of the biology of the parma (*Macropus parma* Waterhouse) and Darma (*Macropus eugenii* Desmarest) wallabies with particular emphasis on social organization. M.Sc. Thesis, University of Auckland.

Agonistic behaviour in Macropodoids — a review

U. GANSLOSSER[1]

A review of agonistic behaviour in macropodoids, at the levels of interactions and relationships, shows that taxon-specific patterns of behaviour occur. The general evolutionary trend in the superfamily is for the potoroids to display a set of more plesiomorphic behaviour patterns, especially in fighting, with more advanced patterns in especially the larger macropodids. Agonistic interactions between males have become strongly ritualized in the larger and strongly heteromorphic macropodids. Submissive displays, hitherto usually overlooked, do exist, especially in potoroids. Supplants are the most frequent agonistic interactions in macropodoids, usually through displays in larger species with well differentiated, ritualized patterns. Ritualized fights, the most frequent form of agonistic interactions between males of larger species, are characterized by highly stereotyped behaviour of both opponents, long duration and separation of the fight into bouts. Unidirectional transition in forms of interaction is usual, from scrapping, through wrestling, to boxing and kicking (in the large macropodids), with predictable chains of acts in each form. In the overall evolutionary trend in macropodoid fighting, the number of male-typical elements increases, females sharing some elements in smaller but not larger species; fights become organized in bouts, and elements typifying winners and losers emerge, in larger species. Dominance hierarchies, often size- or age-based, can be present in males or females; the larger sex dominates in intersexual encounters. Fights initiated by the potential winner are important in maintaining and stabilizing dominance relationships. Dominance relationships are shown to be dynamic in some species.

1. INTRODUCTION

AGONISTIC behaviour, like all components of social behaviour, can be studied on three different levels (*sensu* Hinde 1977), the second and third level each consisting of the totality of elements of the next-lower one. The three levels are (a) interactions, (b) relationships and (c) structures. This paper intends to review agonistic behaviour on the levels of interactions and relationships. The influence of agonistic relationships on social structure will become evident from the papers of Croft, Jarman and others (this volume).

Agonistic behaviour, a category introduced into ethology by Scott and Fredericson (1951), is all social behaviour related to attack, fight, defence and flight or submission. This behavioural category is thus broader in scope than pure aggression, and, as can be demonstrated by motivational analyses of displays, often involves a conflict between attack and flight. Motivational conflicts are indeed one of the most important proximate causes for ritualization (Baerends 1975; McFarland 1981). They may result in displacement activities (such as auto-grooming), redirected attacks, intention movements or so-called ambivalent behaviour (e.g., rapidly moving to and fro, thus alternately showing intentions to attack and escape, respectively). These behaviour patterns are often made more conspicuous by the evolution of bright colours, manes or other morphological features. These structures are demonstrated in more-or-less stereotyped behaviour patterns in communication. Displays often are performed with a so-called typical intensity, e.g., limbs are moved with a certain, characteristic velocity or in a characteristic rhythm or frequency, or the animal even "freezes" in a conspicuous, static position. This evolutionary process is called ritualization, and the resulting behaviour, mostly taxon-specific, is a display. In his famous studies of ungulate communication, Walther (e.g., 1983) distinguished threat displays, superiority displays and submissive display in agonistic behaviour (see below). Ritualization can take place not only in the evolution of communicatory behaviour but also in actual fighting behaviour. Here it is normally connected with the development of weapons that might become too dangerous to use in an uncontrolled way; thus, by evolving a standardized fighting technique, animals are able to test their strength in combat without risking serious injuries.

Agonistic behaviour in animals normally occurs in competition for some limited resource (Parker 1978; Maynard-Smith 1978, 1984; Hammerstien 1985). These limited resources, be they food, water holes, shelters or mates, tend to be non-shareable or, at least, a few individuals tend to gain access to a large proportion of these resources. The ability to gain, exploit and defend a resource is characterized by the resource-holding-potential, or RHP, of a given individual. The strength of the RHP depends on body size, weapons, fighting ability, and other traits of the individual. Normally the individual with the larger RHP should win a conflict. It has been demonstrated for a large array of animal species

[1]Zoology Institute I, Staudtstr 5, University of Erlangen-Nürnberg, 8520 Erlangen, West Germany.

Pages 475-503 in KANGAROOS, WALLABIES AND RAT-KANGAROOS ed. by G. Grigg, P. Jarman and I. Hume. Surrey Beatty & Sons Pty Limited, New South Wales, Australia, 1989.

that individuals are able to assess the RHP of of their opponents, and normally challenge only those with an RHP similar to their own (see below). However, differences in RHP are not the only reason for asymmetric contests; e.g., another reason may be the fact that one of the two opponents is the owner of the resource. In that case a convention to settle the conflict often helps to avoid injurious fighting: "Resouce holder wins", i.e., ownership between roughly equal-sized opponents is repected.

To avoid fighting that is potentailly dangerous to both opponents, animals should announce their status (e.g., as resource-holder) and RHP to their opponents. This is achieved by displays, be they territorial marking, superiority displays to "show off" strength and weapon size, or threat displays which signal the opponent that the sender, even if frightened, is able and ready to attack.

If a conflict cannot be solved by displays, actual fighting may follow. The fight can either be a ritualized test of strength, thus continuing the comparison of RHPs, or an escalated, potentially dangerous fight. The game – theory analysis of conflicts e.g., Maynard-Smith (1978, 1984) leads to several conclusions with regard to escalation into potentially dangerous fighting. One is that "retaliate" is an evolutionarily stable strategy, i.e., if one individual tries to damage his opponent, the latter should respond in kind. Another conclusion is that escalation should be more likely either if the resource is extremely valuable for at least one opponent, or if the risk of serious injury even at unrestricted fighting is low. From this last-mentioned fact predictions can be derived concerning the escalation of fights in smaller versus larger species. Novellie *et al.* (1984) mention a further consideration for the likelihood of escalations, namely the ability of injured animals to hide themselves from predators while healing their wounds; this ability should be better for animals in closed habitats vs open ones. In all these considerations it should be kept in mind that it is in each animal's interest not to get injured, and thus, if possible, to avoid any act that might elicit unnecessary escalation on the part of the rival. No group, or species, selection levels are necessary for this explanation.

Based on game-theoretical analyses a dispute arose as to whether animals should exchange "honest" information (about their motivations, status, fighting ability etc.) or whether they should "cheat" by pretending to be less frightened and more determined to fight. Maynard-Smith (1984), among others, has stressed the fact that displays concerning the RHP should be proof against cheating if they are connected with some real advantage (e.g., size), and that in other cases the contestants might be "bargaining". Displays and also aggressive acts of increasing intensity might lead to a compromise between the two opponents, especially if a resource (like a territory) is divisible. However, there are few studies of mammals with regard to these theories as yet. On the other hand Hinde (1981) emphasized that the traditional view of displays rarely, if ever, collides with the game-theoretical approach; e.g., one of the most important aspects of threat displays is their ambiguity, the uncertainty which behaviour is most likely to follow.

Studies of agonistic conflicts and fighting must be based on careful analyses of each of the contestants' acts, and must be accompanied by monitoring the status of respective animals before and after the trials. Often, frame-by-frame video or movie analyses are necessary, as well as experimental tests of dominance, etc. Such studies can be very profitable using captive individuals. Studies in captivity often reveal patterns of behaviour that are rarely noticed in field situations, and only in a few cases have free-ranging populations been monitored as closely and completely as is possible is a captive study. Field and captive situations thus should be regarded as complementary settings to refine and develop reseach projects, and not as an alternative. Walther (1963), Hendrichs (1978), Eisenberg (1981) and Kummer (1984) have discussed the value of studies in captivity and their relevance for field studies from different points of view.

2. THE LEVEL OF INTERACTIONS

Hinde (*in* McFarland 1981) described interactions thus: "In a social encounter one individual may elicit a response from another, which in its turn elicits another response from the first individual. Such a case where each individual influences the behaviour of the other, can be described as an interaction.

In the following review, behavioural elements used in social interaction shall be termed patterns, which more or less corresponds to the term "fixed action pattern" in ethological textbooks, however, with variability not only as to orientation but also in intensity (e.g., graded signals).

Table 1 reviews the literature of agonistic behaviour patterns among those macropodoid genera and species for which published information was available. When no species or genera are cited as a specific example in the following discussion, "large species" means the kangaroos, wallaroos and large wallabies (e.g., *M. parryi, M. agilis,* to some extent *M. rufogriseus*), while "smaller species" refers to rock-wallabies, pademelons, Parma Wallaby *M. parma* and similar sized macropodines.

A. Agonistic behaviour patterns

Even before analysing the two modes of interactions several remarkable points with regard to

Table 1a. Patterns of agonistic behaviour in macropodoids. Abbreviations used throughout this paper, listed alphabetically with short definitions of each element.

Abbreviation	Definition
A	Attack: one animal jumps at another one.
AF	Frontal attack
AL	Lateral attack
AH	Attack from behind
AV	Avoiding contact or interaction with another animal by changing direction or turning away.
B	Biting
BLO	Blocking; one animal overtaking another, barring the latter's way by standing at right angles in front of it.
BK	Break: between bouts of ritualized fighting, initiated by one animal lowering itself, autogrooming or feeding.
BX	Boxing: a prolonged fight, mostly consisting of rapid, rigid sequences of wrestling, pushing and kicking.
C	Chase: in rapid, bipedal hops.
CF	Crossing forearms: the animal holds its forearms crossed in front of its chest or abdomen.
CL	Clinging: one animal preventing another one from escape by clinging to it with its forepaws.
D	Defence, raising hand: one animal raises one or both hands, sometimes performing hitting movements without touching opponent.
DP	Passive supplant by approaching another animal which gets out of the way or flees.
DU	Ducking down: one animal pressing itself to the ground in a low, quadrupedal crouch.
EB	Embrace: one animal putting its forearms around another's body, pulling or pressing the recipient vigorously against itself.
EBF	Embrace from frontal position.
EBL	Embrace from lateral position or around neck.
EBH	Embrace from behind.
F	Fleeing by rapidly moving away, normally in bipedal hopping gait.
FBA	Forcing another animal backwards by approaching rapidly.
GP	Grass-pulling, picking up plant material, raising it or pressing it against one's chest area.
GR	Grasping another animal's fur or any body region.
GRT	Grasping the other's tail.
GZ	Gazing at another animal.
H	Hitting another animal violently with one (H 1) or both (H 2) forepaws.
H air	The animal performs hitting movements with its forepaws without actually hitting the opponent; intention movement to H.
HI	Hiding from a conspecific under bushes or behind obstacles
HT	Head-turning: turning snout away from partner.
HTW	Rapidly hopping towards another animal.
IF	Interfering: one animal interferes in an encounter between two conspecifics, terminating or taking over one part of it.
JOB	Jumping upon another animal from behind, clinging onto its back, often losing contact with the ground.
K	Kicking with hindlegs.
KLi	Kicking while lying on one's side.
MO	Manipulation of objects, standing in a fully upright position under a tree or shrub, performing fighting and chest-rubbing movements towards the branches.
NoN	Nose over nose position, the snout being held (for a certain time) over the conspecific's nose.
OM	Open mouth, lips are drawn back, teeth exposed.
PD	Pressing another animal down into a low, quadrupedal position.
PHB	Pushing head backwards: one animal pushing the other's head back into a backwardly stretched position.
PU	Pushing another animal away with one or both forepaws.
RU	Raised up position, standing almost fully erect on hindlegs and tail.
RUM	Maximally raised up, standing on tiptoe.
SAP	Snapping, shooting one's head towards another animal as if biting.
SAU	Stretching arms upright, one or both arms being raised over the head.
SC	Scrapping, a sequence of repeated mutual hits with both forepaws.
SCD	Scratching in displacement with hasty, nervous movements, preferrably chest, flank, or ventral areas.
SF	Slapping face and ears of opponent with alternating hits of both forepaws.
SHF	Stretching head foreward, with neck extended, muzzle slightly upward.
SHB	Stretching the head backward, sticking the muzzle into the air.
SK	Skipping: hopping to and fro in small steps, in full upright posture, in front of an opponent.
STO	Stooping: lowering into a low posture.
T+	One animal holding its head above the other's neck, remaining in this position.
TD	Throwing down, causing another animal to lose its balance.
TH	Thump with hindfeet.
TL	Tail-lashing.
WO	Wriggling out from another's embrace.
WR	Wrestling: two animals standing upright, trying to make each other fall by vigorously pushing and pressing with their forelimbs.

Some non-agonistic elements had to be included in order to get continuous, uninterrupted sequences:

LOK	Locomotion.
SM	Staying motionless.
SPT	Hissing, coughing or spitting.
TA	Turning away.
HT	Head turning.
GO	Social grooming.
GRO	Autogrooming.

Table 1b-e. Occurrence of agonistic patterns in taxonomic sub-groups. For each species available literature has been used; where no literature is cited the data come from my research. As far as possible, occurrence in male/male, male/female and female/female dyads has been differentiated; + in male/female column means that this element is performed by either male or female. Where no information was available, a blank space remains. − in any column means that I am sure the species does not perform this element (either from literature or prolonged observation). (In the fourth column, beneath the references, a + means that the element occurs in the species, but data on sex-typical occurrence were not available.

Table 1b: Potoroidae

	Bettongia penicillata				*Bettongia lesueur*				*Aepyprymnus rufescens*			
	♂♂	♂♀	♀♀	Lit.	♂♂	♂♀	♀♀	Lit.	♂♂	♂♀	♀♀	Lit.
A_F						+		1				
A_H	+	+	+						+	+	+	
A_L												
AV	+	+	+						+	+	+	2
B	+	+	+						+	+	+	2,3
BK	−	−	−		−	−	−		−	−	−	
BLO												
BX	−	−	−		−	−	−		−	−	−	
C	+	+	+		+	+	+		+	+	+	2,3
CF	−	−	−						−	−	−	
CL												
D	+	+	+						+	+	+	
DP												
DU		−	+						+	−	+	2
EBF	−	−	−						−	−	−	
EBH	−	−	−						−	−	−	
EBL	−	−	−						−	−	−	
F	+	+	+						+	+	+	
FBA	−	−	−						−	−	−	
GP	−	−	−						−	−	−	
GR	+	+	+						+	+	+	
GRT	−	−	−						−	−	−	
GZ	+	+	+						+	+	+	
H	+	+	+						+	+	+	2,3
HI	+	+	+						+	+	+	
HTW	+	+	+						+	+	+	2
IF	−	−	−						−	+	+	2
JOP	−	−	−						−	−	−	
K	+	+	+						+	+	+	3
MO	−	−	−						−	−	−	
NON	−	−	−						−	−	−	
OM	−	−	−						−	−	−	
PD	−	−	−						−	−	−	
PHB	−	−	−						−	−	−	
RU	+	+	+						+	+	+	
RUM												
SAP	−	−	−						−	−	−	
SAU	−	−	−						−	−	−	3
SC	−	−	−						−	−	−	
SF	−	−	−						−	−	−	
SHB	−	−	−						−	−	−	
SHF	−	−	−						−	−	−	
SK	−	−	−						−	−	−	
STL												
STO	−	−	−						−	−	−	
T+	−	−	−						−	−	−	
TD	+	+	+						+	+	+	
TH	−	−	−						+	+	+	
TL	+	+	+						+	+	−	
UP	−	−	−						−	−	±	
VP												
WO	+	+	+						+	+	+	
WR	−	−	−						−	−	−	

Table 1c: Dorcopsis/Dendrolagus

	Dorcopsis luctuosa				*Dengrolagus inustus*				*Dendrolagus goodfellowi*				*Dendrolagus ursinus*				*Dendrolagus matschiei*				*Dendrolagus dorianus*			
	♂♂	♂♀	♀♀	Lit.	♂♂	♂♀	♀♀	Lit.	♂♂	♂♀	♀♀	Lit.	♂♂	♂♀	♀♀	22	♂♂	♂♀	♀♀	Lit.	♂♂	♂♀	♀♀	Lit.
A_F	+	+	+		±	±	±		+	+	+						+	+	+		+	+	+	
A_H	+	+	+		+	+	+		+	+	+						+				+	+	+	
A_L																								
AV	+	+	+		+	−	+		+	+	+		+	−	+						+	−	+	
B	+	+	+		+	+	+		+	+	+		+	+	+		+	+	+		+	+	+	
BK	−	−	−		−	−	−		−	−	−		−	−	−		−	−	−		−	−	−	
BLO																								
BX	−	−	−		−	−	−		−	−	−		−	−	−		−	−	−		−	−	−	
C	+	−	+		+	+	+		+	−	+		+	+	+			+	+		+	−	+	
CF	−	−	−		−	−	−		−	−	−		−	−	−		−	−	−		−	−	−	
CL																								
D	+	+	+		+	+	+		+	+	+							+	+		+	+	+	
DP	−	−	−		−	−	−		−	−	−										+	+	+	
DU	−	+	+		−	−	−		−	−	−						−	−	−					
EBF	+	+	+		+	+	+		+	+	+		+	+	+			+	+		+	+	+	
EBH	+	+	+		+	+	+		+	+	+		+	+	+			+	+		+	+	+	
EBL	+	+	+		+	+	+		+	+	+		+	+	+			+	+		+	+	+	
F	+	+	+		+	−	+		+	−	+						+				+	−	+	
FBA	−	−	−		−	−	−		−	−	−										−	−	−	
GP	−	−	−		−	−	−		−	−	−										+	±	−	
GR	+	+	+		+	+	+		+	+	+						+				+	+	+	
GRT	−	+	−		−	+	+		−	+	−										−	+	+	
GZ	−	−	−		+	+	+		+	+	+										+	+	−	
H	+	+	+		+	+	+		+	+	+		+	+	+						+	+	+	
HI	−	+	+		−	−	−		−	−	−										−	−	−	
HTW	+	+	+		+	+	+		+	+	+										+	+	+	
IF	−	−	−		−	−	−		−	−	−										−	−	−	
JOP	−	−	−		−	−	−		−	−	−										−	−	−	
K	+	−	+		−	−	−		−	−	−										−	−	−	
MO	−	−	−		+	+	+		+	−	−	17									+	−	−	
NOI	+	+	+		−	−	−		−	−	−										+	−	−	
OM	+	+	+		−	−	−		−	−	−			+	+						−	−	−	
PD	+	+	−		−	−	−		−	−	−										−	−	−	
PHB	−	−	−		−	−	−		−	−	−										−	−	−	
RU	+	+	+		+	+	+		+	+	+						+				+	+	+	
RUM																								
SAP	+	+	+		+	+	+		+	+	+										+	+	+	
SAU	−	−	−		−	−	−		−	−	−										−	−	−	
SC	−	−	−		−	−	−		−	−	−						[illegible]				−	−	−	
SCD	−	−	−						±	±	±										±	±	−	
SHB	−	−	−		−	−	−		−	−	−										−	−	−	
SHF	+	+	+		−	−	−	20	−	−	−	17									+	+	+	17
SK	+	−	−						−	−	−										−	−	−	
SF	−	−	−		−	−	−		−	−	−										−	−	−	
STO	+	+	+		−	−	−		−	−	−										±	−	±	
T+	+	+	±		−	−	−		−	−	−										−	−	−	
TD	−	−	−		−	−	−		−	−	−										−	−	−	
TH	+	+	+		−	−	−		−	−	−										−	−	−	
TL	−	−	−		+	+	−		+	+	−										+	−	−	
UP	+	+	+		±	±	±		+	+	+										+	+	+	
VP	+	+	+		−	−	−		−	−	−										+	+	−	
WO	−	−	−		+	+	+		+	−	+						+				+	+	+	
WR	+	+	+		+	+	+		+	+	+		+	+	+		+	+	+		+	+	+	

behaviour patterns can be deduced from the comparative overview:

A.1. TAXON-SPECIFIC PATTERNS

Potoroidae: The rat-kangaroos possess a unique set of behaviour patterns, which for the most part seem to be more primitive than the corresponding patterns in Macropodidae. The most obvious of these patterns is the kick, which is normally delivered while the animal is lying on its side. This certainly has to be regarded as a plesiomorphic mode of delivering hindlimb kicks (compare fighting in e.g., dasyurids). A remarkable and rat-kangaroo-specific threat display also has evolved from this primitive type of kicking. The animal lies on its side, raising one hindleg, ready to kick, and (typical for ambivalence in ritualization) sometimes repeatedly jerks the foot without actually hitting the opponent. Wrestling

Table 1d: Thylogale, Setonix, Petrogale

	Thylogale billardierii			*(T. stigmatica)*	*Setonix brachyurus*				*Petrogale penicillata*				*Petrogale lateralis*			*(P. inornata)*
	♂♂	♂♀	♀♀	Lit.	♂♂	♂♀	♀♀	Lit.	♂♂	♂♀	♀♀	Lit.	♂♂	♂♀	♀♀	Lit.
A_F	+								+				+			
A_H	+	+	+						+	+	+		−	+	−	
A_L	+	+	+						+				+	−	−	
AV									+	+	+		+	+	+	
B	+	+	+		+		−	10	+	+	+		+	+	+	(12)
BK	−	−	−						+	−	−		+	−	−	
BLO	+	−	−													
BX	+			4	−			10	+				+	−	−	
C	+	+	±		+		−	10	+	−	+					(13) (14)
CF	−	−	−						−	−	−		(−)	(−)	(−)	
CL	+	+	+						+	+	+		+	+	+	
D	+	+	+	(5)					+	+	+					
DP	+	+	+	6					−	−	−		−	−	−	
DU	+	+	+							+	+		+	+	+	
EBF	+								+	+			+	+	−	
EBH	+								+	+	+		+	−	−	
EBL	+								+	+			+	−	−	
F	+	+	+						+	+			+	+	−	
FBA	−	−	−						−	−	−					
GP																
GR	+	(+)	+						+	+	+		+	+		(12)
GRT		(+)											−	+	−	
GZ									+	+	+		+	+	+	(13)
H	+	+	+		+		+		+	+	+		+	+	−	(12)
HI		+	+							+	+		−	+	−	(12)
HTW	+	+	+			+			+	+	+		+	+	+	
IF	−	−	−						−	−	−		−	−	−	
JOP	−	−	−						−	−	−		−	−	−	
K	+	(+)	±		+		−	10	+	±	+		+	−	−	
MO	−	−	−						−	−	−		−	−	−	
NON	−	−	−						−	−	−		−	−	−	
OM	−	−	−						−	−	−		−	−	−	
PD	+												+	−	−	
PHB	+	−	−						−	−	−		−	−	−	
RU	+	+	+						+	+	+		+	+	+	
RUM	+	+	+						+				+	+		
SAP	+	+	+						+	+	+		+	+	+	
SAU	−	−	−						−	−	−		−	−	−	
SC	(+)	(+)	(+)	(5)												
SF									−	−	−		−	−	−	
SHB	−	−	−						+	+		−	+	+	−	
SHF																
SK	−	−	−										+	+	−	
STL													+	−	−	
STO									+	+	−					
T+	−	−	−						−	−	−		−	−	−	
TD													−	−	−	
TH	+	+	+										−	+	−	(13) (14) (16)
TL																
UP																
VP																
WO	−	+	−										−	+	−	
WR	+	+	+						+	+	+		+	−	−	

and biting often are also performed while both opponents are lying on their sides. Other primitive features in rat-kangaroo agonistic behaviour are the absence of almost all ritualized displays, and the fact that attacks and bites normally are delivered by jumping onto the opponent's back from behind (AH). A very limited degree of ritualization can be found in sexual interactions, however.

Dorcopsis: Within the macropodids the genera *Dorcopsis* and *Dorcopsulus* seem to have an isolated position (Groves, pers. comm.; Flannery this volume). A unique set of behavioural

Table 1e: Wallabia/Macropus.

	Wallabia bicolor				*Macropus parma*				*Macropus eugenii*			
	♂♂	♂♀	♀♀	Lit.	♂♂	♂♀	♀♀	Lit.	♂♂	♂♀	♀♀	Lit.
A_F				+	+	±	±					
A_H				+	+	−	±	23				
A_L				+	+	−	±					
AV				+	+	+	±					
B				+	+	+	+					
BK	−	−	−		−	−	−					
BLO				+	±	±	−					
BX												
C				+	+	−	±					
CF	−	−	−		−	−	−					
CL				+	+	+	+					
D				−	+	+	+	23				
DP				−								
DU				+								
EBF				+								
EBH				+	+	+	+					
EBL				+								
F				+	+	−	+					
FBA				−	−	−	−					
GP					+			23				
GR				+	+	+	+					
GRT				+								
GZ				+	−	+	+					
H					+	+	+	23				
HI					+			23				
HTW				−								
IF					−	−	−					
JOP	+	+	+		−	−	−					
K				+	+	+	+	23				
MO				−	−	−	−					
NON				−	−	−	−					
OM				−	−	−	−					
PD				+								
PHB				−								
RU				+	+	+	−					
RUM				−	±	−	−					
SAP				+	±	+	+					
SAU				−	−	−	−					
SC					−	−	−					
SCD												
SHB				+	±	±	+					
SHF				+	+	+	+					
SK				−	−	−	−					
STL												
STO				+								
T+				−	−	−	−					
TD				−	+	±	±	23				
TH				+	+	+	+					
TL				+		+		23				
UP												
VP												
WO												
WR				+	+	+	+	23				
SF				+								
PU					+	+	+					

	Macropus rufogriseus				*Macropus agilis*				*Macropus parryi*			
	♂♂	♂♀	♀♀	Lit.	♂♂	♂♀	♀♀	Lit.	♂♂	♂♀	♀♀	Lit.
A_F	+	+	−		+	−	−		+	−	±	
A_H	+	−	−		+	±	±		±	−	−	24
A_L	+	±	−									
AV	+	+	+		+	+	+		+	+	+	
B	+	−	−		−	±	+					
BK	−	−	−		+	−	−		+			24
BLO	+	+	−						+	+	±	
BX	+	−	−		+	−	−		+			16
C	+	−	−		+	−	±		+	−	+	
CF	−	−	−		−	−	−		+	−	−	25
CL	+	±	−		+	+	+		+	+	+	
D	+	±	+		+		+		±	±	+	
DP	+				+				+	±	+	24
DU	+	−	−		+	−	−					
EBF	+	+	−		+	+	−		+	+	+	
EBH	+	+	−		+	+	−		±	+	+	
EBL	+	−	−		+	+	+		+			
F	+	−	−		+	−	+		+	+	+	24
FBA												
GP	+	±	−		+				+			16
GR	+	+	−		+	+	+		+	+	+	
GRT									−			
GZ	+	+	+	24								
H	+	+	−		+	+	+		+	+	+	
HI												
HTW	+	+	+		+	±	−					
IF									+	−	−	24
JOP												
K	+	−	−		+	+	+		+		−	24
MO									+			
NON												
OM												
PD	+	−	−		+	+			+			
PHB	+	−	−		+	+	−		+			
RU	+	+	±		+	+	+		+	±	±	
RUM	+	±	±		+	±	−		+	−	−	
SAP	−	+	−		±	+	±					
SAU	−	−	−		−	−	−					
SC					+							
SCD					+	+						
SHB					+	+			+			
SHF												
SK					+	−	−		+			24
STL					+				+			24
STO	+	−	−		+	−	−					
T+	−	−	−						−	−	−	
TD	+	−	−		+				+			24
TH	+	+	+		+	+	+		+	+	+	
TL	±	+	−		+				+	+	−	
UP												
VP												
WO	+	−	−		+	±	±					
WR	+	±	±		+	+	+		+	+	±	24
SF	−	−	−		+	+	±					
PU	+	+	−		+	+	+		−	+	+	

	Macropus giganteus				*Macropus robustus*				*Macropus antilopinus*				*Macropus rufus*			
	♂♂	♂♀	♀♀	Lit.	♂♂	♂♀	♀♀	Lit.	♂♂	♂♀	♀♀	Lit.	♂♂	♂♀	♀♀	Lit.
A_F					+			30					+	−	−	34
A_H				−	+	±	−						−	−	−	30
A_L					+	±	−						+	−	−	
AV					+	+	+						+	+	±	34,35
B					+	+	+	30					+	−	−	30,34
BK	+	−	−	7,26	+	−	−	31					+	−	−	
BLO	+				+								+	±	−	
BX	+			27,28									+	−	−	30,32
C	+	−	−		+	−	+	30					+	−	−	30
CF	+	−	−		−	−	−						+	−	−	25,30,36
CL				+									+	±	−	
D				+	+	+	+						±	+	−	30
DP				+	+			30					+			30
DU	+	−	−										+	−	−	
EBF				+	+	±	−						+	+	−	34
EBH				+	+	−	−						+	+	−	
EBL				+	+	±	−						+	−	−	
F				+	+	+	±						+	+	±	34
FBA				+	−	−	−						+	−	−	
GP	+			27,28	+			31					+	−	−	
GR				+	+	+	±	30					+	+	−	37
GRT						+							+		−	25,30,35
GZ					+	+	+						+			25,30
H	+			26	+	+	+	30					+	±	±	30,39
HI													−	−	−	
HTW	+			26	+	+	+	30					+		+	34
IF					+								+	−	−	38
JOP	+				+	−	−						+	−	−	34
K	+				+	±	−	30					+	−	−	30
MO					+								+	±	−	30
NON													−	−	−	
OM													−	−	−	
PD	+				+	−	−						+	−	−	30
PHB	+			7	+	−	−						+	±	−	
RU				+	+	+	+						+	±	±	30
RUM					+								+		−	
SAP					+	+	±						+			34
SAU	+	−	−		+		−						+	−	−	40,30
SC													+	±	−	39
SCD	+			7	+								+	+	±	30
SHB	+			7,26	+	±	−						+	±	±	30
SHF	+															
SK					+	−	−						+	−	−	30,34
STL					+	±							+			41
STO				+	+								+	±	−	34
T+				−									−	−	−	
TD				−	+	−	−						+	−	−	34
TH				+	+	+	+						+	+	+	
TL				28,29									+			
UP																
VP																
WO	+			+	+	−	−						+	−	−	39
WR				+7	+	−	−	30					+	±	−	30,34
SF				+									+	±	−	30,34
PU					+	+	−	30								+30,39

Literature cited (Tables 1b-1e): 1 = Stoddart 1966; 2 = Ganslosser and Fuchs 1988; 3 = P. Johnson 1978; 4 = Clancy 1982; 5 = Cooke 1979; 6 = Leuthold 1977; 7 = Croft 1981a; 8 Murböck 1975; 9 Morton and Burton 1973; 10 = Packer 1969; 11 = Cicala *et al.* 1970; 12 = Barker 1982; 13 = Batchelor 1980; 14 = Nicholls 1979; 15 = Bell 1972; 16 = Kaufmann 1974a; 17 = Russell 1985; 18 = Ganslosser 1979a,b; 19 = Dmoch 1975; 20 = Dmoch, pers. comm.; 21 = Gewalt 1965; 22 = Schneider 1955; 23 = Vujcich 1979; 24 = Kaufmann 1974; 25 = Sharman and Calaby 1964; 26 = German-Meyer and Schenkel 1970; 27 = Grant 1974; 28 = Kaufmann 1975; 29 = Grant 1973; 30 = Croft 1981b; 31 = Croft 1982; 32 = Coulson 1980 and pers. comm.; 33 = Ganslosser and Wilhelm 1985; 34 = Russell 1970a; 35 = Russell 1973; 36 = Murböck 1977; 38 = Russell 1970; 39 = Frith and Calaby 1969.

patterns correspondends to this phylogenetically isolated position, mostly positions or movements of head and snout in a typical intensity and fashion (T+, T-, OM). SHB seems to have been ritualized independently in *Dorcopsis* (as well as *Thylogale*) and occurs not only while standing bipedally in front of an opponent, but also in a quadrupedal position, at a greater distance from the other animal. Like MA, this display might have evolved by ambivalence between biting (or SHF) and RU, that is between a more unritualized and a more ritualized form of attack. Conflicting tendencies to shoot the head forward from a quadrupedal position (SHF, SAP, B, in that order of intensity) and to rise onto the hindlegs with head raised may result in this position of stretching the head extremely backwards while crouching quadrupedally.

Dendrolagus: The patterns of social behaviour in tree-kangaroos differ in many ways from those of the ground-dwelling species. For most of these differences a direct or indirect link exists to the modified locomotor anatomy. Tree-kangaroos, with their less powerful tails, and (in most species) shorter hindlegs, are obviously no longer able to deliver kicks, at least not while standing bipedally. Instead, fights are performed by wrestling, hitting with the forepaw, and biting. In high-intensity fighting, the primitive mode of wrestling, scratching and biting while lying down occurs. The fighting technique thus seems to be quite primitive. However, again some genus-typical ritualizations have evolved. A superiority display in male *D. dorianus* consists of rising onto the hindlegs, stretching the back, lifting both arms with the powerful claws stretched (until they are almost level with the snout) and raising the snout into the air, standing not face to face, but at a more broadside angle to the opponent. Within the genus, the threat display of "intention to bite" seems to have been further standardized (from SAP, e.g., in *D. goodfellowi,* to simple SHF in *D. dorianus*). D also has a typical intensity only in *D. dorianus*. In general, however, the agonistic behaviour of *Dendrolagus* also lacks the sophisticated displays of large, ground-living species (compare A.3, B.1), a fact which has to be kept in mind because of their comparatively strong sexual dimorphism (see below).

A.2. CHANGES IN INTENSITY, ORIENTATION AND CONTEXT

Attacks in potoroids (see above) and in smaller genera of macropodids *(Dorcopsis, Thylogale, Petrogale, Dendrolagus)* are often performed by jumping onto the opponent from behind or from lateral positions, often immediately following a rapid hopping towards the other individual without any preliminary signals. In these species such an attack usually is followed by grabbing the other's fur, biting or scratching. In larger species (e.g., the larger wallabies) these attacks are mostly performed from a frontal position and the opponent is hit with the hindlegs, which then leads to the kick (see below). In the smaller species again, AF may also lead to biting and scratching.

Changes in posture, intensity and orientation towards the opponent are evident when comparing kicks (K) between macropodoids. Again the unique form in potoroids, together with its own ritualization, has been discussed. Within the macropodids kicks become more and more standardized with regard to the orientation towards the opponent (in larger species, kicks are delivered directly from a frontal position), the use of the tail as a prop (in smaller species, e.g., pademelons, the tail completely loses contact with the ground; see Fig. 2), the behaviour immediately following the blow (in larger species, contact with the opponent is mostly held with the arms; in smaller ones again, contact is broken and the animal either drops back or even turns in mid air and jumps away momentarily), and area of opponent hit (as a result of the frontal position with locked forearms, in large species of *Macropus* the kicks normally hit only chest or abdomen). In the latter case, kicks can be delivered repeatedly in rapid succession, and without the risk of being hit unguarded by the opponent while in mid air.

Some elements of sexual behaviour can occur in agonistic contexts in certain species. This is illustrated by anal sniffing which, as Clancy's (1982) and my own observations reveal, has connections to agonistic behaviour in *Thylogale billardierii, Petrogale, Wallabia* and perhaps some others as well. Even more evident is the occurrence of EBH/Mount in an agonistic context. Whereas EBH is a common element of unritualized fighting (in connection with AH, biting and scratching) in smaller macropods, tree-kangaroos and potoroids, it is restricted to male/male competition in specific, unritualized situations in larger species (see below).

Aggressive mounting, however, seems to occur only in a few species (e.g., among females of Doria's Tree-kangaroo *D. dorianus*, Whiptail-wallabies *M. parryii*), in dominance/subordinance situations. Mention should be made here (though it will be discussed as an interaction) of the use of BO and PHC by dominant, and SCH by subordinate Parma-wallabies, Red-bellied Pademelons, and perhaps other species (Clancy 1982; Herter *et al.* 1985).

A.3. RITUALIZED PATTERNS AS DISPLAYS

In connection with ritualized fighting (see below) several displays evolved as more or less stereotyped patterns of behaviour. These include elements that can be classed as superiority displays. For example RUM, CF simply exaggerate the animals' strength and size; others like SAU, SHB and perhaps FBA seem to originate in a conflict, either between attack vs not attack (FBA), or between displaying vs defensive threat (SAU may be a result of RUM vs D). The peculiar head-bob display in *M. antilopinus* (Croft 1982) reveals its root in a motivational conflict even more clearly than simple SHB. Again, these displays exist as stereotyped, static positions in the larger species of *Macropus;* whereas if they occur in smaller species of macropods at all, they tend to be short, and directly connected to non-ritualized aggressive acts (e.g., Clancy 1982; and my own data for *T. billardierii;* Barker 1982; and our own data for *Petrogale* sp.).

At present, a motivational interpretation for blocking (BLO), connected with STL (stiff-legged walk), seems to be uncertain. Whereas its occurrence in courting as well as agonistic contexts seems to categorize it as a superiority display, observations (Kaufmann 1974) indicate that it might be more frequent in situations of unclear dominance, though in these situations Kaufmann found it to be more associated with the prospective winner of the encounter. According to Kaufmann, at least in *M. parryi,* STL rarely is connected with fighting, which might point to its resulting from a motivational conflict, thus being a threat display. With regard to stereotypy and "typical intensity" again, BLO occurs connected with STL, or even as a static position, standing quadrupedally, stiff-legged, at right angles to the opponent, in larger species; whereas in *Dorcopsis, Dendrolagus, Thylogale* and *M. parma* normally no fixed position or stance is achieved.

Two threat displays, of different intensity, can be derived from intention to bite: SAP, during which the snout actually shoots towards the opponent, and SHF, when the head is held still with the snout raised and pointed towards the other animal. These displays, especially SAP, are often accompanied by a low growl. In both cases, normally, no actual bite follows, which is in accordance with the nature of an intention movement originating in a motivational conflict.

In some species (e.g., *W. bicolor,* and *M. parma*), SHF is connected with a characteristic, quadrupedal crouched position which in itself resembles the crouched position immediately before an attack in potoroids and small macropodids. In these two species SHF can be followed by actual attacks (ambiguity).

Most authors (Kaufmann 1974) mention that almost no submissive displays could be found in macropodoids. However, there are several behaviour patterns which can be observed in agonistic contexts, are characteristic for submissive animals and normally stop the other's attack. The most obvious submissive patterns can be seen in potoroids; whenever one animal, be it a juvenile, or a female, is harrassed by another one, e.g., adult, or male, the submissive one throws itself onto its side or even back, spreads its hindlegs and allows the other to sniff anally. After this, the other one turns away without further harrassment. This is clearly a submissive display. In smaller macropodids from closed habitats, a stooped posture (DU), which seems to be derived from hiding behind an obstacle, with lowered head and crouched back, is adopted whenever one animal is chased, or repeatedly attacked, by another. This normally also leads to a cessation of the aggression. In *Dorcopsis luctuosa,* this posture is combined with rapid tapping of forelegs on the ground, which seems to indicate a tendency to flee that is inhibited.

In all species which show RUM, or even RU, as a superiority display, the "antithetical pattern", i.e., lowering one's body into a bipedal position with hindfeet flat on the ground and rounded back, results in a cessation of the opponent's attacks. Thus, technically, this return to "normal posture" functions as a submissive display as well.

A.4. SEX-TYPICAL AND SEX-SPECIFIC PATTERNS

A distinction is often drawn between sex-typical behaviour (which occurs in both sexes, though with significantly different frequencies) and sex-specific patterns, which are shown only by one sex, at least in adulthood.

However (see Coulson this volume), a differentiation between sex-specific patterns and those that are sex-typical with an extremely low frequency in one sex can be very difficult. As can be seen from the tables, sex-specific agonistic behaviour patterns are lacking in potoroids, as well as smaller macropodids. The occurrence of sex-specific elements is restricted to the large wallabies and kangaroos. In these species, most displays connected with fighting are male-specific. Most remarkable is the fact that several elements which are regularly used by females in smaller species (especially kicks) from a frontal position are restricted more and more to males in the large species. Thus not only newly evolved, but also "old" patterns become connected with the male/male fight.

The only display in agonistic contexts which seems to be male-specific even in smaller macropods and tree-kangaroos may be chest-rubbing and manipulating objects (GP, MO).

Sex-typical agonistic patterns evolved earlier in macropodoid phylogeny. Even in *A. rufescens* some (unritualized) patterns are more frequently performed by females, others more by males (P. Johnson 1980, Ganslosser and Fuchs 1988). In

Dorcopsis luctuosa female-typical agonistic patterns include SHF, OM, B and aggressive approach whereas male-typical patterns are, among others, GR, GRT, D and HH.

Among tree-kangaroos, unfortunately, data exist only for *D. dorianus*. In this species, nearly all aggressive elements are male-typical and some are shown equally by both sexes (e.g., B, PU); only F is typical for females.

In rock-wallabies and pademelons elements connected with fights are shown more by males, although females still perform kicks in both genera. However, in *Petrogale,* these female kicks are extremely rare, and normally the first kick settles the dispute for good. Among the pademelons at least female *T. brunii* regularly deliver kicks in female/female conflicts (Ganslosser 1978), whereas in female *T. billardierii* these elements seem to be rare.

M. parma, the only small species of *Macropus* analysed so far, reveals the following sex-typical elements: D, AV, SHF, VIB are female-typical, RU and H male-typical. Attacks by jumping at the opponent, A, wrestling, WR and other elements are shown by both sexes without any significant differences in frequency. In *M. rufogriseus,* a larger wallaby, most agonistic elements are male-specific; of the remaining, only H and D in varying intensity are female-typical.

M. parryi, the largest of the wallabies analysed here, is almost identical to kangaroos and wallaroos in sex-typical distributions. In *M. robustus,* avoiding or terminating contact is the only female-typical element; all other agonistic elements are performed by males more than expected. In *M. rufus* similarly, nearly all agonistic elements are male-typical, with the exception of D, which is shown by both sexes with non-significant differences.

B. Agonistic interactions

Patterns of agonistic behaviour as have been described above normally are combined into interactions. Distribution and context of these interactions again differ among species and between sexes.

B.1. SUPPLANTING (RUSSELL 1970; CROFT 1981; AND OTHERS)

Supplants are the most frequent agonistic interactions in macropodoids. A supplanting interaction may either be a passive one, with one animal approaching another (HTW), and the approached moving away without any further behaviour, or an active supplant, which consists of a display or an overt aggressive act, which then, or after some more aggressive behaviour patterns, leads to one animal leaving the site. Animals may supplant another when coming into too close proximity; subadult or small males are supplanted by females when courting (Southwell 1981, and see below); or, as in most cases of supplants, a disputed resource (shelter in rock-wallabies; lying-up place in the shade in *M. rufus*; water (Croft 1985); females; nest in rat-kangaroos; but rarely food) can be recognized as the reason.

Not surprisingly, active supplanting is performed more and more by displays in those species possessing well differentiated ritualized patterns. In most cases threat displays (SHF, SAP, D, accompanied by growls, hisses or stares toward the opponent) occur during active supplants. However, Croft (1981a,b) and others also describe the use of RUM, SAU and SHB, sometimes even CF, as a "threat" in supplanting interactions. Most interesting, and in accordance with the interpretation of CF as a superiority display to demonstrate size, and of SAU resulting more from conflict (see above), Croft mentions that CF is rarer in supplants then AHW.

The occurrence of stiff-legged walks in conflict situations has already been discussed. Southwell's (1981) finding that supplanting in *M. giganteus* is more frequent than fighting among large males stands in perfect accordance with theoretical expectations regarding risk of fighting in "heavily armed" individuals.

B.2. FIGHTS

Fighting here means a repeated exchange of aggressive patterns between at least two individuals, with at least part of the behaviour patterns involving physical force against the opponent. Fights in macropodoids normally are distinguished into what most authors describe as ritualized verses non-ritualized fights. Before discussing ritualization in macropodoid fighting, the so-called non-ritualized fights shall be discussed briefly.

Non-ritualized fighting includes behaviour patterns that are not stereotyped in form or intensity, and which can lead to serious injury in the opponent e.g., biting, scratching, attacks from beside or behind which are followed by B, GR or kicks delivered at whatever region of the opponent's body may be hit. Within a fight there are no apparent temporal features, behavioural sequences are variable and, once physical aggression has commenced, displays are only rarely interspersed. These unritualized fights are the only form of fighting to be observed in rat-kangaroos, tree-kangaroos (perhaps with the exception of *D. dorianus,* see below) and when females of smaller and medium-sized wallabies are involved. They are normally short (exceptions being male/male fights e.g., in potoroids, and some special cases of male/female interactions in pademelons and tree-kangaroos, see below), and they often lead to the winner chasing its fleeing

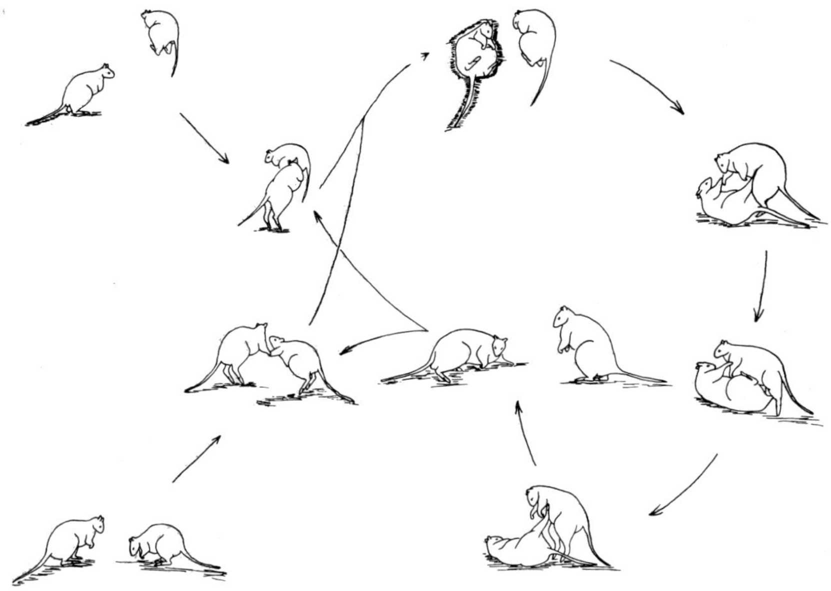

Fig. 1. Typical sequences of fighting in rat-kangaroos. The interaction may start with one animal jumping upon (AL, AH) a hopping (e.g., already fleeing) opponent (upper left) or from a crouched position with head stretched toward the opponent (lower left). After the attack, biting (B) and wrestling (WR) occur, normally at least one animal throwing itself on to its side (above centre), and one or both hitting (H) and lashing with fore- and hindlimbs (K_{LY})

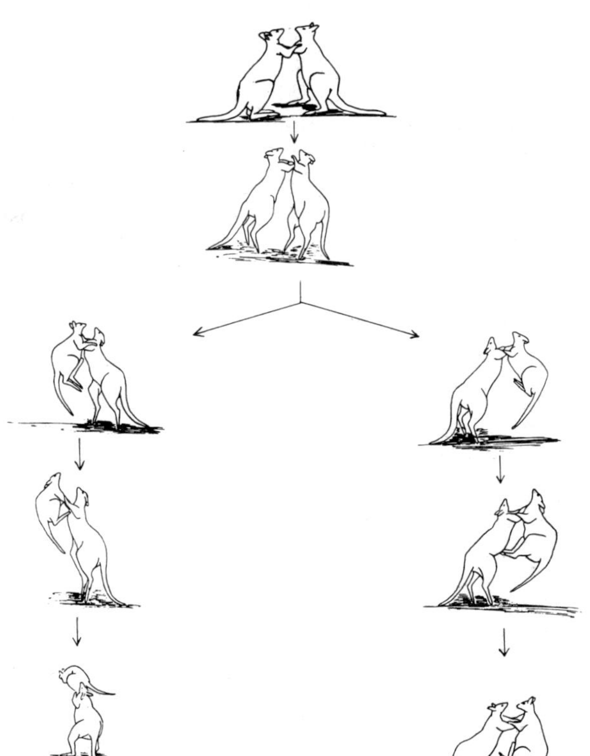

opponent, with the loser often adopting submissive postures after fleeing, at least in captivity.

Non-ritualized fighting in larger species, especially kangaroos, is restricted to situations which tend to be highly competitive. In *M. rufogriseus,* our own studies as well as Murböck (1975, 1977) connect unritualized fighting with the proximity of oestrous females. Murböck described these fights as the only ones during which he saw AL and AH. Contrary to this suggestion, our study of captive Bennett's Wallabies revealed an eight-fold increase in agonistic acts during the mating season, but most of this increase was due to a rising frequency of "weak agonistic acts" (e.g., D, SAP, aggressive vocalizations) (see below). Thus, if a stable hierarchy existed obviously it did not break down. However, there were more unritualized fights, apparently especially among males that are equal in size. Croft (1983) studied unritualized fights in male Red Kangaroos and used frame by frame analysis of films for a detailed comparison with ritualized fights. His data, and mine for *M. r. rufogriseus* during the mating season, reveal the

Fig. 2. Possible sequences of fighting in small and medium-sized wallabies. They face each other in RU or RUM, and one jumps up, delivering a kick (K) without losing contact with the opponent (right). Note that the tail is not used as a prop. In a second, less ritualized form, the kicking animal breaks contact during or after the kick (AF-K) and turns in midair; sometimes a second kick (AL) is delivered while jumping away. Opponent normally chases, attacks from behind (AH) and bites (B).

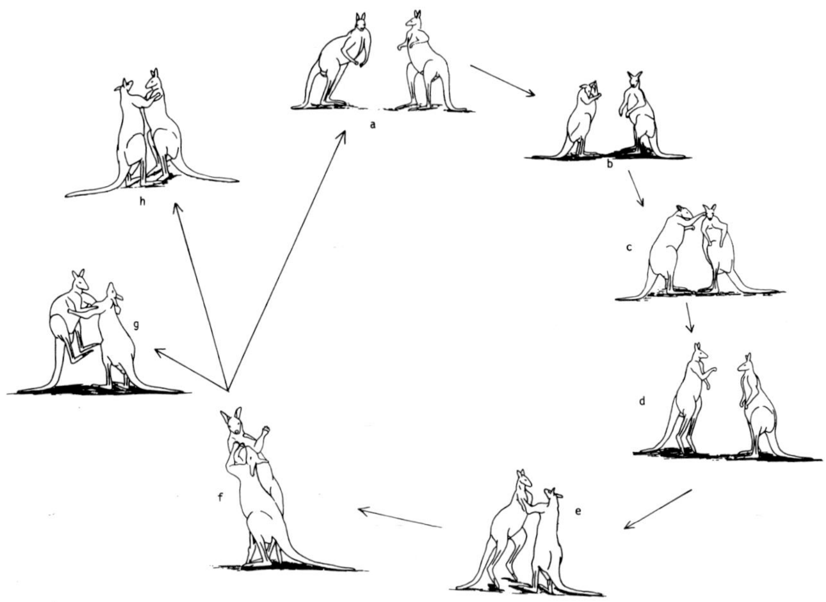

Fig. 3. Ritualized fight in kangaroos and wallabies. Both stand in RU (a), at least one facing away (TA); the left one grooms himself in displacement (b), hitting (H) while the opponent scratches himself (c), left one RUM, starts with SAU (d), both opponents in scrapping stage (e), smaller one SHB and SAU (f), (wrestling stage), then either left one kicks (note tail as a prop), boxing stage, (g) or wrestling continues (h) till left one initiates a break (BK, a) again.

following characteristics for unritualized fights in males of larger species: short duration (15.5 sec mean duration in *M. r. r.*), a change in winner-loser typical distributions (kicks, hitting against the opponent's body, clasping from behind, lateral attack and lateral aggressive approach become winner-typical elements, approach and attack from behind are loser-typical), occurrence of unritualized elements such as AL and even kicks from a lateral direction. Sequential analyses of the non-ritualized fights in both species demonstrate less fixed sequences of behavioural elements, especially in intra-individual sequences, which means that the predictability of the animal's acts is reduced and transitions between elements are more random (see below); there is a lack of longer wrestling and mutual pushing phases and only short duration of AFM (which, again, becomes winner-typical). The most frequent display in our *M. rufogriseus* study was SHF in quadrupedal stance, which again is a display more connected with unritualized attacks. Observations by Southwell (1976, 1981) also describe the lack of long fighting when opponents are chased away from an oestrous female (in *M. rufogriseus* and *M. giganteus*). Germann-Meyer and Schenkel (1970) observed an unritualized fight with AG and EBH among two male Grey Kangaroos unfamiliar with each other, again in the presence of an oestrous female.

Ritualized fights (Croft 1981 and others) are the most frequent form of agonistic interactions between males of larger species. The ritualized fights are characterized by a highly stereotyped behaviour of both opponents, longer duration, separation of the total time of a fight into bouts, interspersed with breaks, during which both opponents perform individual behaviour like feeding and self-grooming, and the occurence of display (SAU, CF, RUM, SHB, GRO_{ch}, MO). Another obvious characteristic is the initiation of a fight (or the reinitiation after a break): in most cases the two opponents face each other, adopt RU posture, scratch and bite their chest and flank area, with heads turned away from each other (see Fig. 3). After these initial displays, the actual fight starts by one animal embracing and pawing the opponent. Once the fight has started, its intensity rises again, through several stages. Our sequential analyses (see below) show a distinction between the scrapping fight (mutual rapid pawing, alternating with both hands, of the opponent), leading to the wrestling stage, and finally to boxing and kicking with forearms locked in RUM. The transitions between these complexes are unidirectional i.e., boxing follows, but does not precede wrestling. The number of intra-individual, non-random transitions between behavioural elements is very high, which means that animals perform more acts immediately following each other, without interruption by the opponent, in a non-random, predictable chain. Again, these chains often are unidirectional, which further increases the predictability and non-randomness. (For a discussion of methods and interpretation of the sequential analysis see below: Appendix).

The characteristic elements of ritualized fights are most evident in a comparison of fighting techniques between several species from different levels of social evolution. We analyzed fighting scenes on movie and video films, frame by frame, of *Dorcopsis luctuosa, Thylogale brunii, Macropus agilis, M. rufogriseus* and *M. rufus*. The results, together with Croft's (1983) data on *M. rufus,* depict the evolutionary path in macropod fighting.

In *T. brunii,* females and males perform similar fights, but male fights tend to be longer (mean value of more than 7s versus 5s in females) and contain more RU, WR and K. Bouts of fighting, interrupted by breaks, occur in both sexes, as do chases of one animal by the other. Male-typical elements in *T. brunii* fights are H, EBF, and C, whereas females show AV, WR, GRT, D, K and JO. Initiation of fighting happens by a simple sequence of RUM – D. Inter-individual sequences tend to be short and variable, and longer chains of behaviour in fixed, non-random ways are absent in both sexes; the mode of fighting thus is unpredictable for the opponents.

In *Dorcopsis luctuosa* (Ganslosser and Holz 1984 and in prep) sequences again are not very different in male verses female fights. Fights tend to escalate from more ritualized to more unritualized stages (the latter being characterized by AH, B, etc.). However, a distinction into stages (c.f. *M. rufus*) is not possible, because ritualized and non-ritualized patterns occur during the whole fight, though with different frequencies. In fights between male and female, the male performs more ritualized patterns, including the *Dorcopsis*-typical head and snout positions. Females tend to escalate earlier than males during each single fight, and perform only some rare, single ritualized patterns between unritualized ones, while the reverse occurs in males. Female/female fights in *D. luctuosa* are characterized by unritualized elements with increasing frequency in both opponents, by the almost total absence of intra-individual fixed sequences, and the predominance of biting over other aggressive patterns. SAP is restricted to female/female fights. Female/female fights mostly start from a lateral orientation, male/male fights from frontal positions or from AS. The duration of female fights is considerably shorter (a mean value of about 2.5's) than fights with at least one male participant (more than 10s).

In *Macropus agilis* female/female fights were too rare for statistical sequential analysis. However, a trend can be stated: there seems to be no fixed initiation; fights tend to start with lateral approach, AL, AH, EBH and kicks (also lateral), and are concluded by one animal chasing the other. Within the short fights (less than 10s vs mean duration of 35s in male/male fights), some fixed sequences occur at both inter- and intra-individual levels. This is more obvious, however, in male/male fights. These conflicts are characterized by fixed sequences of highly non-random combinations, especially at the inter-individual levels; one animal responds in a non-random, and thus predictable, way to the others act. A further differentiation now has to be made: fights between two similar-sized males are characterized by kicks (but, contrary to female fights, from frontal positions), SK, H, WR and PU, whereas fights between animals of different size include long breaks between bouts, the break being initiated by the smaller animal by STO and AH, terminated by the larger one with RU, D, rapid approach or AH. The breaks tend to increase in length during the course of a fight.

In *M. rufus* fights including females are virtually absent (Croft 1981; Russell 1970; and our own data). Thus, only male/male fights were analysed. The typical sequences, consisting of long, fixed, unidirectional intra-individual chains, have already been mentioned. Elements and overall patterns of the fights are highly differentiated among prospective winners and losers (see below), with prospective winners being characterized by EBA, GR, H, pushing, SAP, HD and TD, losers preferring K, A_F, initiating breaks and terminating fights. The mean duration of the fights analysed by us was 2.5 min. In the course of a fight, the tendency to rise into AFM increases: the more bouts, the earlier AFM occurs.

Thus, the overall trend in the evolution of macropodoid fighting can be summarized as follows. The number of male-typical elements increases, with females performing K, A_F, WR and SI in smaller species, but not in larger ones. The fights are organized in bouts, interspersed with breaks, in the larger species. Elements can be recognized as typical for winners and losers at least in larger species.

Sequences of non-randomly associated behavioural elements occur more in larger species, with intra-individual sequences being more variable than inter-individual ones within every species so far studied except *M. rufus*. The fighting of small species tends to change rapidly beween ritualized and non-ritualized elements, whereas the fights of larger species escalate in unidirectional way from more ritualized to more unritualized patterns.

Females in small species tend to fight, although fights are shorter and less ritualized than in males (up to 40% non-ritualized elements), whereas females in larger species hardly fight at all. In small species, differences in frequencies between sexes can be detected, especially for those elements becoming sex-specific in the more highly-evolved taxa. Thus, a sort of preadaptation to sex-specific fighting seems to be evident in the smaller species with a less highly evolved social behaviour.

In trying to separate effects of phylogeny from those of size, ecology, mating systems and resulting sexual dimorphism, data from small species of *Macropus* are of special interest. *M. parma* for example, when compared to species of *Thylogale* and *Petrogale*, has a more complex behavioural repertoire, more stereotyped displays and a higher degree of ritualization (Herter *et al.* 1985).

B.3. DOMINANCE/SUBORDINATE INTERACTIONS

Though dominant/subordinate relationships will be discussed below, some interactions between animals that indicate this particular relationship are mentioned here. In most cases, as described above, dominance is expressed in displays by one animal with the other retreating. However, dominance relationships should be regarded also as an active act on the part of the subordinate animal, which has to "acknowledge" its lower rank and some interactions occur in this context.

A ritualized form of "dominance mounting" seems to be present in males of *M. rufus, M. robustus, M. giganteus* and in both sexes of *W. bicolor:* JOP, which consists of the higher-ranking animal's jumping onto its subordinate's back, clinging to it in a way similar to mounting, though in a less powerful grasp, and sometimes losing ground contact with its feet.

Another interaction also derived from sexual contexts, and thus deserving to be called ritualized even in the strictest sense, was described by Herter *et al.* (1985) in *M. parma,* and by Clancy (1982) in *T. billardierii:* the higher-ranking individual leans over the subordinate (BO), pressing the latter's head firmly towards the former's chest (PHC) or in *M. parma,* performing PTF, the ritualized form of PHC, with the subordinate sniffing the dominant's chest (NCR). Among males of large species, an interaction derived from the same context may be dominant's standing RUM, PRS, and the subordinate's grooming the former's chest (GO_{ch}).

C. Overall structure of agonistic behaviour — Sequential analysis

In order to understand the results of sequential analyses of macropodoid agonistic behaviour, the method is described briefly in an appendix.

Sequential analyses of behavioural repertoires can answer several questions. Patterns of ritualized behaviour can be detected and, using criteria of homology, can be related to the non-ritualized elements from which they originate (e.g., if in one species individuals always raise their heads and point their snouts towards an opponent where in another species in the same context, they really bite, and if the opponent in the first species regularly flees after this act of snout-pointing, we may assume that it is a ritualized bite). A second possible use of sequential analysis lies in the field of motivational analysis. Displays (e.g., in threatening or in courtship) can be related to the motivational conflict from which they may derive, and superiority displays can be distinguished from them, by comparing their transitional probabilities to behaviour of escape, aggression, etc.

Both applications of sequential analysis have already been used in this paper. A further strength of sequencing with regard to motivation can be demonstrated in briefly discussing my own results on overall sequences of social behaviour. In several species (e.g., *Dorcopsis luctuosa, Dendrolagus dorianus,* rock-wallabies and *M. parryi*) the so-called socio-positive behaviour (social grooming and other amicable activities) can be distinguished as a separate complex of behavioural elements, which are connected mostly with each other, and with transitional elements. In other species (e.g., *Thylogale, M. rufogriseus, M. robustus*) socio-positive behaviour is directly connected with sexual behaviour, whereas in *M. giganteus* and *M. rufus* connections to agonistic behaviour exist. This result is remarkable with regard to the fact that in *M. parryi*, rock-wallabies and Doria's Tree-kangaroo, stronger, long-lasting individual bonds (which involved more contact than sharing a home range) among adults (especially related females) seem to exist. The connections of agonistic behaviour with social grooming have already been discussed with regard to P_{Br} and should be kept in mind for dominance relationships (see 3).

The most important contributions of sequential analysis in the context of this paper, however, concern the non-random transitions within agonistic behaviour itself. The concepts of game-theory discussed in the introduction predict that animals should avoid escalation in fighting more if they themselves might suffer serious injuries from retaliation. Escalation of fighting can for example be avoided by both opponents being able, with at least a reasonable margin of certainty, to predict that the other will not immediately escalate either. The more non-random transitions between elements of agonistic behaviour exist, the more certainty as to what will follow next. This concerns X-Y transitions (if X does that, Y will probably do the same) and X-X transitions (if X does A and B consecutively, the probability of his doing C next is high, the probability of doing Z, e.g., escalate, is low). The danger of accidental escalation is lowest when agonistic behaviour is structured into

Figs 4-13. Results of sequential analysis of agonistic behaviour in adult macropodoids. Methods see appendix. Elements of two subcomplexes of ritualized fighting (see Fig. 13a,b) have been shaded or hatched as in the other species, to demonstrate their occurrence in connection with a variety of other elements (see text). Abbreviations see Table 1a. Non-agonistic elements that had to be included because of their connections with agonistic behaviour have been surrounded with rectangles instead of circles.

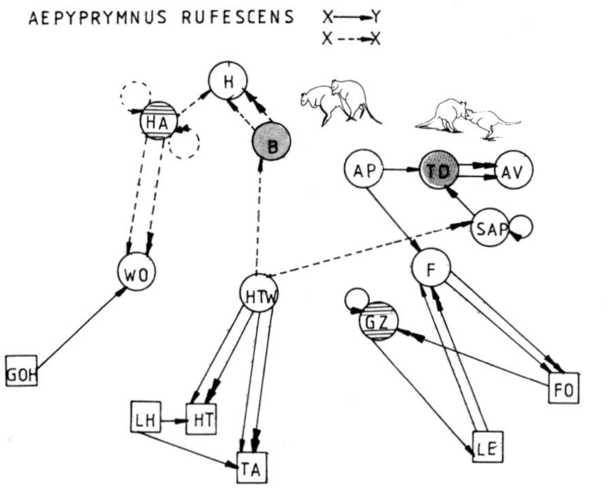

Fig. 4. Aepyprymnus rufescens. Intra-individual and inter-individual connections in one diagram. Reading example: Rapid, aggressive approach (HTW) is followed by biting (B) in intra-individual transition more frequently than expected; the transition is significant on C_2-level. Thus HTW ⇢ B. TD is followed inter-individually by AV, significant on the higher, C_3-level of significance; thus TD ↠ AV.

sub-complexes that clearly indicate the instantaneous level of intensity at which each contestant operates at the moment.

The results of sequential analysis from several species are shown in Figs 4-14. In *A. rufescens,* agonistic behaviour is structured only to a very limited extent, with X-X connections being even rarer than X-Y. Almost no ritualization can be detected; even HA and SAP are still directly connected with physical attacks. In *Dorcopsis luctuosa,* agonistic behaviour has more non-random transitions, but still no subcomplexes can be distinguished based on C_3-values, AH/K and RU/SK/HH being the only exceptions.

Both species of tree-kangaroos, in contrast, allow a subdivision of agonistic behaviour. However, in *D. goodfellowi,* the two subcomplexes of agonsitic behaviour can be separated only by means of C2-values, which means the probability of changing from one to the other is still higher. Further in *D. goodfellowi,* both subcomplexes still contain physical acts of direct aggression.

In *D. dorianus* more X-X connections are significant on the C_3-level, especially the ritualized complex of SHF, DP, AV, and the distinction between intense fighting (AL, C, EBH, WR; and low level fights (H, D, WR in bipedal stance) is obvious.

In *Petrogale lateralis* longer lasting fights (WR, AH, H, WO) are clearly distinguished. Ritualized elements in aggressive behaviour (e.g., RUM) have central positions in aggressive behaviour and fightings are in bouts with breaks (BK) in between. The transitions are mostly significant on the C_3- levels, both in X-X and X-Y. A similar structure can be described for *W. bicolor*, where two subcomplexes of agonistic behaviour are clearly separated on the C_3-level in X-X and X-Y.

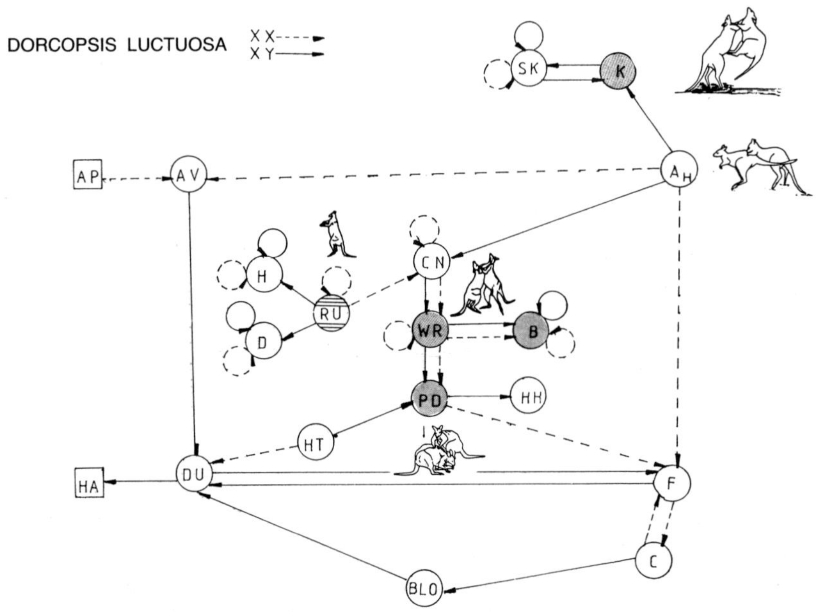

Fig. 5. Dorcopsis luctuosa (=muelleri). Note that more elements are included, i.e., more elements of agonistic behaviour are connected by non-random transitions. Still, most elements are connected with each other, and only RU, H and D can be distinguished as a group frequently occurring together.

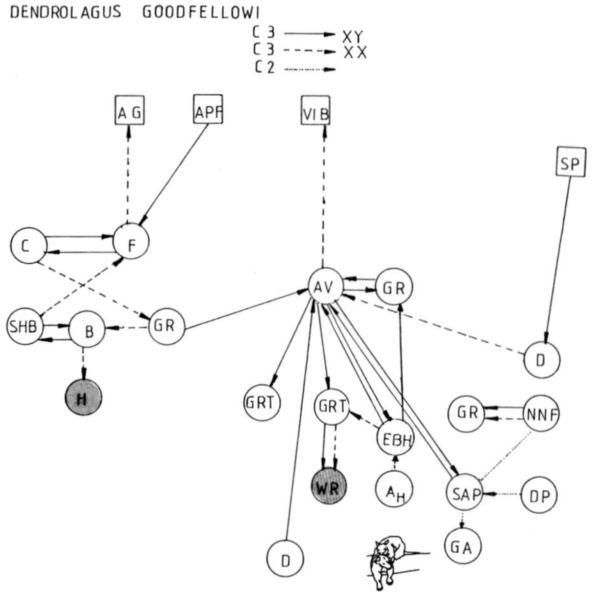

Fig. 6. Dendrolagus goodfellowi. X-X transitions and X-Y transitions in one diagram.

Within the genus *Macropus,* the repertoire of *M. parma* once again represents the most primitive stage: no subdivisions in agonistic behaviour, and a moderate amount of fixed intra-individual sequences (mostly on C_2-level). In larger wallabies, two complexes of weak aggression (mostly elements used in active supplanting) and heavy fighting (with K, AH, C and F), are separated on the C_3-level. *M. parryi* has even more X-X connections in this context than *M. rufogriseus* or *M. agilis*.

In kangaroos and wallaroos, the subdivision reaches its peak. A high amount of C_3-connections in X-X transitions points to long, fixed behavioural chains, most of which are unidirectional.

Purely displaying behaviour, connected to grass-pulling and object manipulation, is distinguished as a third subcomplex. As was already mentioned in discussing ritualized fights, the subcomplexes of fighting follow each other in a predictable, intensity-dependent, unidirectional way.

Summarizing these results, the trend within the macropodoid agonistic behaviour is obvious: small species are less predictable in their agonistic behaviour than larger ones. This is closely related to the amount of sexual dimorphism (the more sexually dimorphic the species, the higher the risk of injury) and to social organization in general, with more gregarious species *(M. parryi, D. dorianus)* having more structured behavioural repertoires. However, phylogenetic influences ("phylogenetic inertia") must be considered as well, as can be seen in comparing *M. parma* with *Petrogale* and *Thylogale* or *Dorcopsis.* These influences are discussed below (section 6).

3. AGONISTIC RELATIONSHIPS

Social relationships "involve a series of social interactions over time. Whereas each interaction occupies a strictly limited span of time, relationships can be discussed only in the context of a much longer period. To describe a relationship we must describe not only the content and quality of the interactions . . . but also how these interactions are patterned, that is, their absolute and relative frequencies, and how interactions of one type are related to . . . other types" (Hinde in McFarland 1981). Hinde lists dimensions that are important for understanding relationships, among others the content of interactions used in a particular relationship, the reciprocity or nonreciprocity of the interaction (e.g., in a dominance relationship, all interactions are non-reciprocal

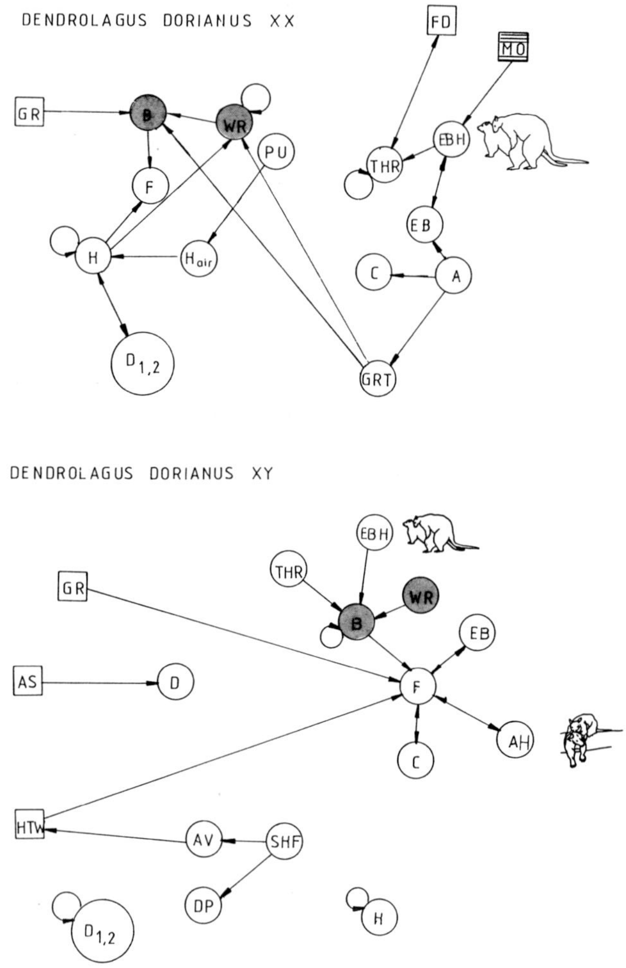

Fig. 7. Dendrolagus dorianus. Note that in tree-kangaroos, at least two subcomplexes can be distinguished, though in *D. goodfellowi* the distinction has been made on the C_2-level. Arrows of C_2-transitions are omitted for clarity. In *D. dorianus,* one complex of mostly ritualized slightly agonistic acts (H, H air as intention, $D_{1,2}$ with one or both hands) occurs in XX, another one of heavy fights (THR, B, WR, AH) in XY. A purely ritualized displacement (SHF → AV) also exists. *D. goodfellowi,* on the contrary, has unritualized élements such as WR, EBF also in the "weak agonistic subcomplex".

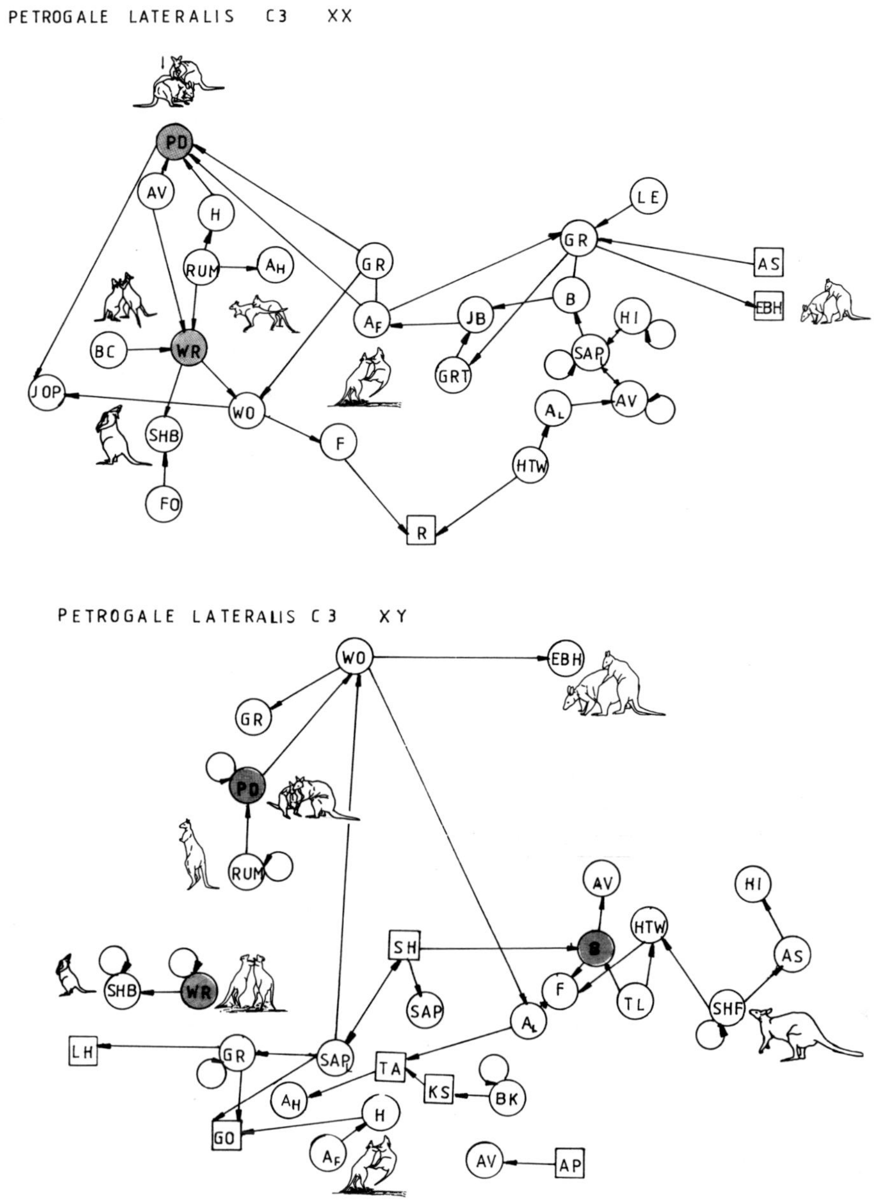

Fig. 8a,b. Petrogale lateralis. A subcomplex of (male/male) fighting can be distinguished around RUM, AH, H, PD and AF in XX; note that elements of these fights are connected more in X-X than in X-Y.

but complementary), the question of who controls whose behaviour (which need not be similar to who is being dominant e.g., in male/female relationship), the quality of the interactions, and developmental aspects.

In the following agonistic relationships shall be discussed first with regard to dominance, then agonistic dyadic relationships between age/sex classes in general. Some aspects which still await further study, like triadic effects, and development of relationships are briefly mentioned as a stimulation for future research.

A. Dominance/Subordinate relationships

Dominance "... relates to a situation in which individuals can be ranked so that a higher-ranking individual may attack or show aggressive threat, while the subordinate avoids conflict ... A dominance hierarchy is commonly established through fighting ... or through differences in age" (Poole 1985). Dominance normally results in preferential access to limited resources. An important point which has to be stressed is that dominance is a relationship, and not a characteristic feature of the dominant one!

A.1. DOMINANCE AMONG MALES

Dominance in male macropodoids is closely connected with sexual dimorphism (see K. Johnson 1977; Jarman 1983; Ganslosser and Jarman 1984; for reviews, and nearly all studies of single species for examples). Dominance in males also is a size-related phenomenon, thus

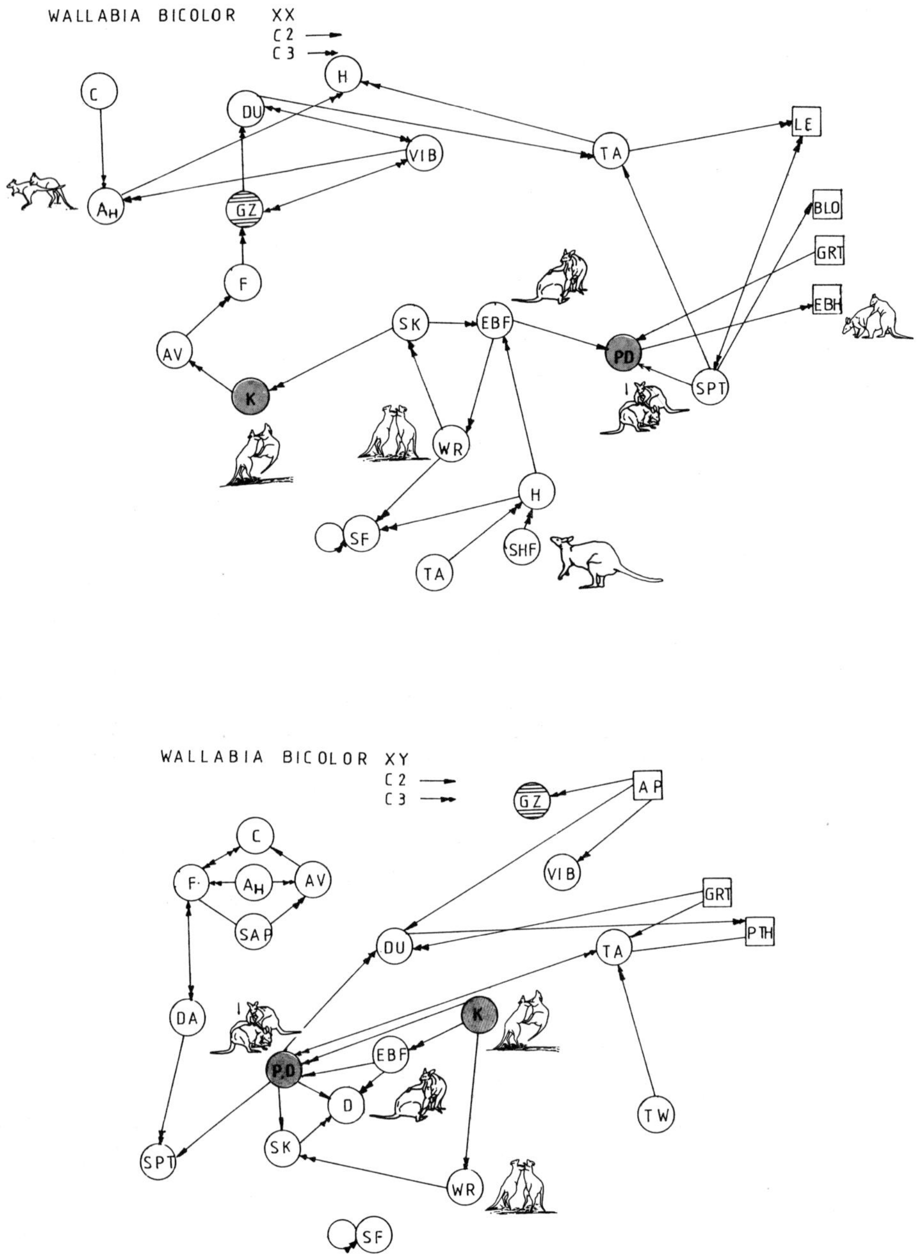

Fig. 9. Wallabia bicolor. Two subcomplexes can be distinguished, both in XX and XY; however, both subcomplexes still include non-ritualized elements of "physical violence".

being an aspect not only of individual, but also age-class relationships. Dominance is exerted by active or even passive supplanting, with passive supplants becoming more frequent when the size difference between the opponents increases (e.g., Croft 1981a for *M. rufus;* Southwell 1981 for *M. giganteus*). It has already been mentioned that dominance is exerted by superiority displays such as CF if the hierarchial position of both opponent is clear, in contrast to occurence of threat displays (e.g., SHB, SAU) in more uncertain situations. Grass-pulling, and object-manipulation are also behaviour patterns associated with dominant individuals (e.g., Kaufmann 1974; Croft 1981a,b).

Under certain circumstances, dominance relations may become weaker in the proximity of oestrous females, leading to more frequent active displacements or even brief, unritualized fights, than at other times. This is to be expected, since a female in oestrous is a more valuable resource than a non-oestrous one, and thus may lead to escalation more readily. However, most authors describe intense, unritualized fighting even in the presence of oestrous females only when the males were unknown to each other (e.g., Germann-Meyer and Schenkel 1970 for *M. giganteus*; K. Johnson 1977 for *T. thetis*) or a formerly dominant male was re-introduced after longer

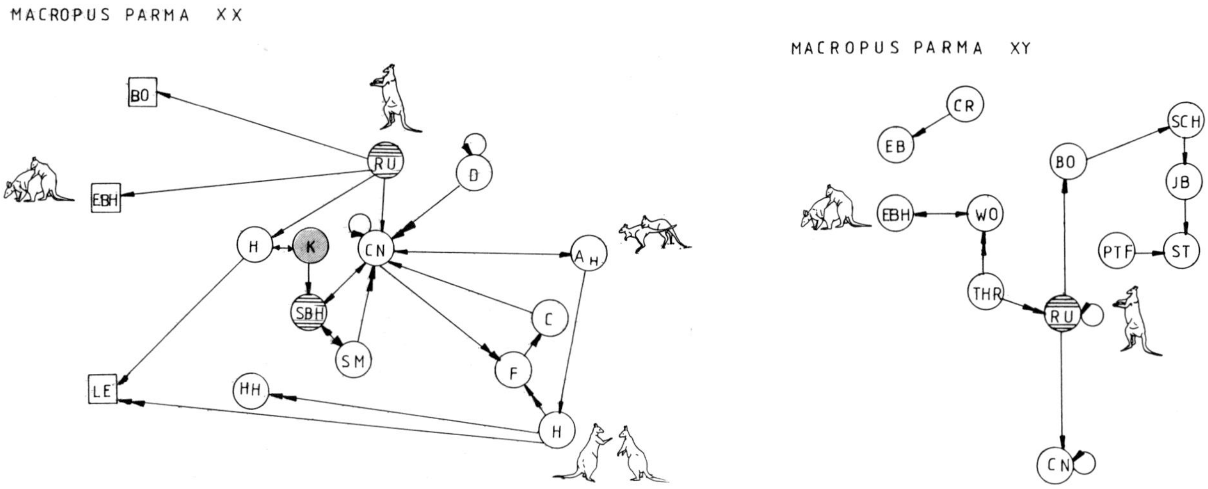

Fig. 10. Macropus parma. Only a few C_3-connections are present, and the agonistic repertoire seems not to be ordered into subcomplexes.

separation (La Follette 1974 for *M. rufogriseus*). In *M. r. rufogriseus,* Murböck (1976) describes a "break-down of the hierarchy" during the mating season. Apparently, the dominance relations in his group were not very fixed. In our study, males performed more weak agonistic acts against each other during the mating season (almost a fourteen-fold increase compared to a seven-fold increase of strong agonistic behaviour; the latter category still includes, for technical reasons, H and some other acts of supplanting as well as heavy fighting). When percentage values are compared, this becomes even more obvious. Strong agonistic acts among adult males in the mating season contribute 17.3%, to the animal's average total sum of behaviour shown versus 12.9%, during the rest of the year. Weak agonistic acts contribute 8.2% vs 3.2%. Thus, a breakdown of the hierarchy was not observed.

In smaller species without holobiotic heteromorphism (Jarman 1983), age, though not always recognizable from size, still seems to be important for dominance (e.g., Kitchener 1972 for *Setonix*). Kaufmann (1974) also found that dominance in *M. parryi* seemed to depend on individual recognition of the opponent, not his size. Our observations in male *Dendrolagus dorianus* and several potoroids suggest that somehow age seems to make the animals more "secure" and "self-confident" (perhaps based on hormone levels?) and this results in their approaching their younger conspecifics in a bolder way (e.g., faster or with higher raised snouts). This then seems to lead to the younger ones yielding even without a remarkable size difference; however, the exact nature of these relationships still has to be analysed in detail.

A.2. DOMINANCE AMONG FEMALES

Female macropodoids develop dominance hierarchies as well as males.

In captive studies, a size or age related dominance was found at least in some potoroids (*A. rufescens* and *B. penicillata,* my data), in *T. billardieri* (Clancy 1982 and my data; Ganslosser and Werner 1984), in *D. dorianus* (Ganslosser 1983) and in *M. parryi* (Schneider and Ganslosser

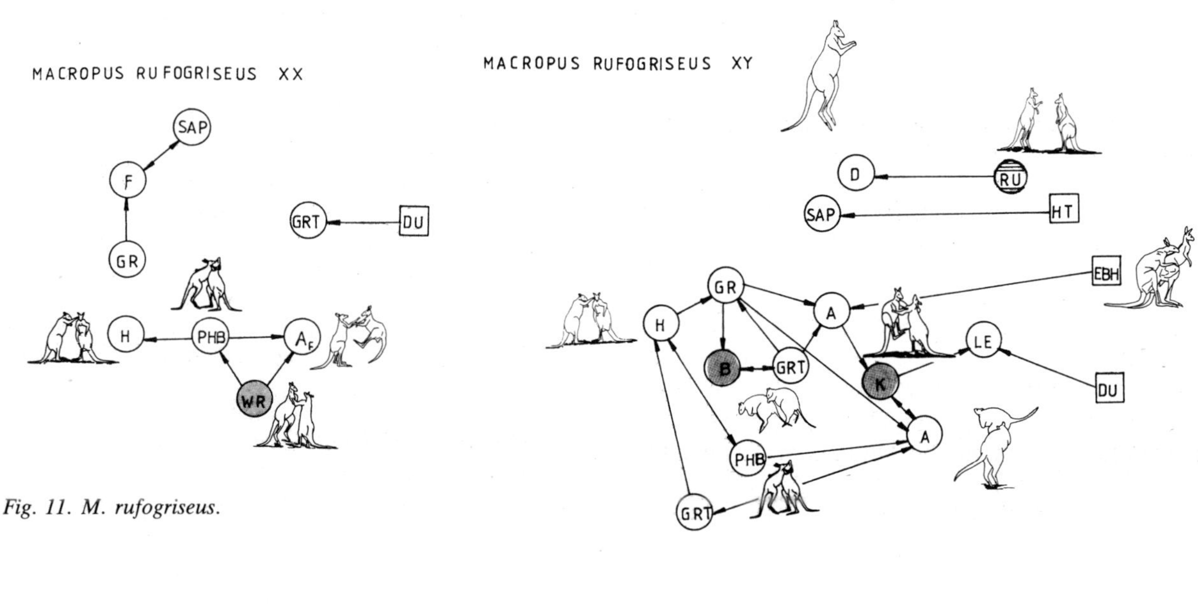

Fig. 11. M. rufogriseus.

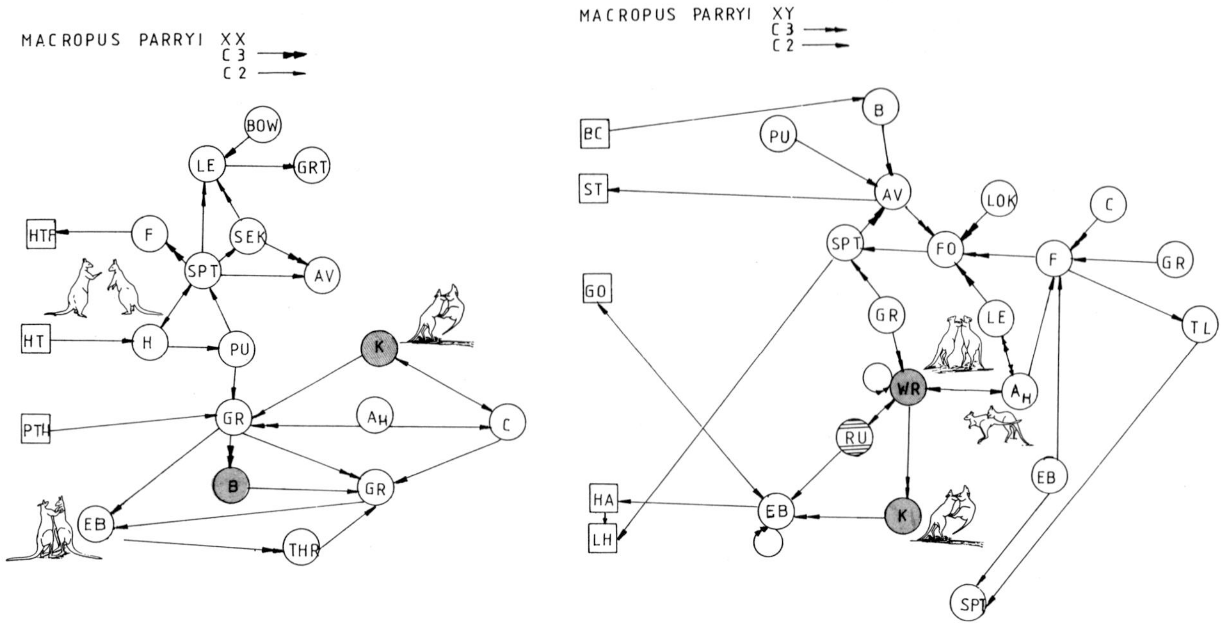

Fig. 12. M. parma. Subcomplexes in XX can be distinguished, e.g., centred around the coughing vocalization (SPT).

in prep.). Dominance hierarchies have been found among free-ranging females of *M. giganteus* (Jarman, pers. comm.) and *Petrogale lateralis* (Engel, pers. comm.) as well. In the case of the *Petrogale* colony a group of five dominant females at the top of the female hierarchy could be identified, with the rest of the adult females below them, perhaps without a linear hierarchy. Attempts to find dominance hierarchies among several other species in captivity have failed (e.g., in *Dorcopsis luctuosa*, Ganslosser and Holz 1984). Female hierarchies seem to be related to reproductive success (Jarman, pers. comm.; Schneider and Ganslosser in prep. for *M. parryi,*

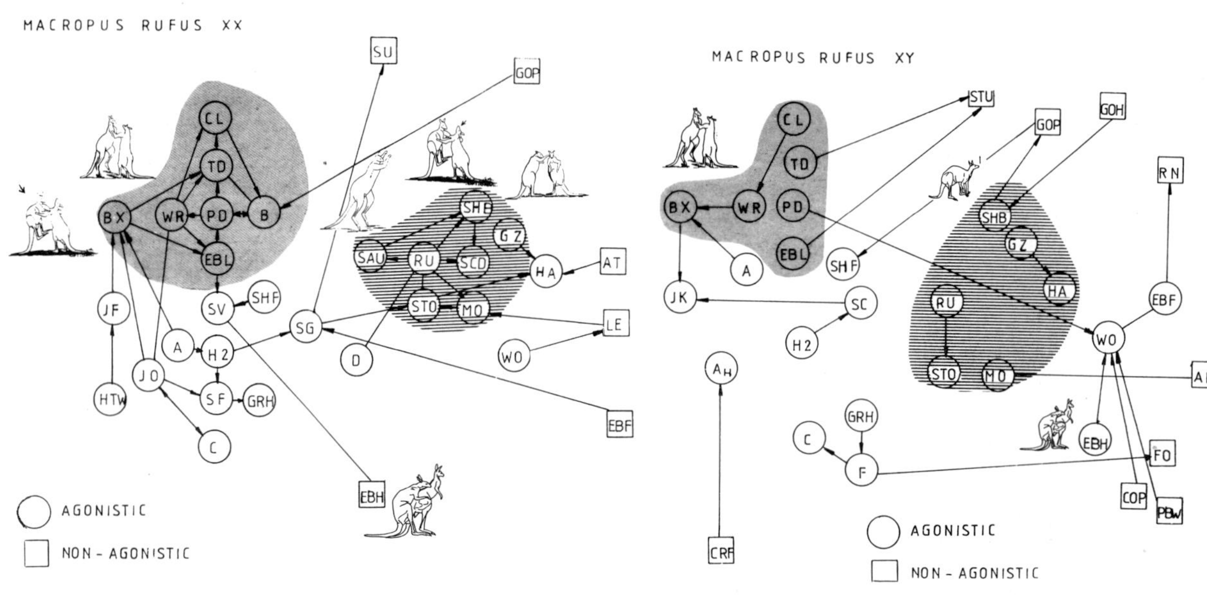

Fig. 13. M. rufus. Fighting has reached the highest level of ritualization, several sub-complexes are shaded to demonstrate them more clearly. In X-X more significant transitions occur than in X-Y. Elements of supplanting (H2, SC, SF), which have a low probability of escalation are not connected directly to fights.

own unpublished observations for *B. penicillata*). Again, in species where females do not grow continually during their life-span, dominance seems to result from "self-confidence" in for example approaching conspecifics. It should be mentioned, however, that none of the linear hierarchies among females has been tested statistically so far.

A.3. INTERSEXUIAL HIERARCHIES

These are currently not studied enough to draw many firm conclusions. In sexually dimorphic species, the larger sex (female rat-kangaroos, male macropodids) seems to dominate.

A.4. WINNER/LOSER PECIFICITY IN FIGHTS

A special consequence of dominance in macropodoids is the widespread occurrence of winner/loser specific behaviour in prolonged fights. This has been observed by many authors both in the field (e.g., Kaufmann 1974 for *M. parryi;* Cook 1979 for *T. stigmatica;* Croft 1981a,b for *M. robustus* and *M. rufus*), as well as in captivity (e.g., Clancy 1982 for *T. billardierii* and my data on several species). Even without frame-by-frame analyses, the following observations can be made. Fights are normally initiated by the prospective winner, who also ends a break between bouts by a new challenge. Losers tend to terminate fights, tend to end bouts and thus initiate breaks, and tend to use kicks either earlier, or more often. To elucidate the winner/loser typical elements further we again performed video analyses on several species. Croft's (1981a, 1983) results on *M. rufus* and Clancy's (1982) on *T. billardierii,* the only detailed studies of winner/loser specific behaviour so far, have been included.

In general, thus, the winner/loser typical elements tend to increase in the evolution of macropod fighting. Losers tend to wrestle, attack and embrace from lateral positions, winners from frontal positions. Interestingly, in fights between similar sized opponents, winner-typical elements occur earlier than in fights between animals of different size. RUM tends to become more loser-typical in males and in larger species, whereas in females and in small species it often characterizes the winner (also observed by Cook 1982 in free-ranging *T. stigmatica*).

The reversal of winner/loser typical behaviour in female versus male fights is in accordance with the interpretation of female fighting as less ritualized. In female fights, winner/loser typical distributions resemble the one found by Croft (1983) for *M. rufus* and by me in *M. rufogriseus* during brief, non-ritualized fights or other highly competitive situations.

Table 2. Winner/loser typical distribution of elements in fights of *Dorcopsis luctuosa, Thylogale brunii, Macropus agilis* and *M. rufus,* data from video analyses. Data by Croft (1983) for *M. rufus* and Clancy (1982) for *T. billardierii,* have been included together with results from our study.

	♂/♂ fight		♂/♀ fight		♀/♀ fight	
Species	Winner	Loser	♂ (normally winner)	♀ (normally loser)	Winner	Loser
Dorcopsis luctuosa	B	H, RU, NoN			B, H, SAP	
T. brunii	C, FO, H, GR and H	H against rump			AL, AF	
T. billardierii (Clancy 1982)	Embrace in head/neck area Clancy states that RUM is more frequent in dominant animals but frequency may not be the same as duration	DP, DU, AL, AF longer in RUM			Approach in bipedal hop GRT, HA, H, GR towards head RUM	K SK, PHB AV
M. agilis	H, GR towards head, RUM, AF, EB	AL, WR, SK, K, attempts to kick, AV	Combined winner- and loser-typical elements		AL, EBL, GR, K from behind	AV, F
M. rufus	FBA, GR and H against head SAP, PD, TD	K, A combined with K, AV Initiation of breaks, earlier and longer in RUM, escalate earlier		No fights including ♀ ♀ observed		

Table 3. Agonistic relationships between age/sex-classes of macropodids, as revealed by differences in frequency distribution between them. Typical elements for each dyad (those that are performed more frequently than expected, at least with p. < 0,05, χ^2 test) are listed.

Species*	♂/♂	♂/♀	♂/♂ juv.	♂/♀ juv.	♀/♂	♀/♀	♀/♂ juv.	♀/♀ juv.	♂ juv./♂	♂juv./♀	♂juv./♂juv	♂juv./♀juv	♀juv./♂	♀juv./♀	♀juv./♂juv.	♀juv./♀juv.
Dorcopsis luctuosa	AF, AH, D B, K, SK GRT	EBF, B, D	SK, D	—	B, D, AH	B, D, AH K, AF	AF, AH, D B, K, SK GRT	AF, AH, D B, K, SK GRT, H	AF, K, SK RU	AF, K, SK RU	AF, K, SK RU	AF, K, SK RU	—	AF, K	AF, K, SK RU	AF, K, SK RU
Dendrolagus dorianus	D	AH, D, B DP, C, WR H, PU, GRT	DP, FH, WR PU, EBH	DP, FH, WR PU	AH, D, B GR, F, WR H, PU	AH, B, DP C, H	AH, D, B DP, FH, WR PU, EBH, EBF	DP, FH, WR PU, EBH	AL, B, DP WR, H, PU	AL, B, DP WR, H, PU GRT	B, DP, WR H, PU	B, DP, WR H, PU	B	B, DP, WR H, PU	B, DP, WR H	B, DP, H
Thylogale billardierii	Fight	B, AH, GR SHF, WR, H K	AP, WR, H K, F, C	—	B, AH, H DU, WR, HA	AH, D	AH, WR, H	H	B, DU, WR	B, WR	H, K	AL, GR, H	B, HJ, F WR, HA	—	HJ, F, H HA	B, HJ, F C, H, K
Petrogale lateralis (free-ranging colony)	AL, AV, B WP, DP, AF	B, H	WR, DP, H	H	AV, SAP	AV, SAP, H	H	—	WR, DP, H	H	AH, Fight	WR, B	WR, DP, H	WR, H	WR, DP, H	H, WR
Wallabia bicolor (Crebbin 1982)	C (large) WR (medium)	H, Frequent agonistic acts (medium)	All agonistic acts (large)	All agonistic acts (large)	H	H										
Macropus parma	FIGHT	H, PU, D RU	D, AH	D	SAP, RU, AV, HA, SHF	D, GR, EBH F, C	D, SAP	D, SAP, K	AH, AV, F HA, H, K	K, HA			AH, AV, F HA	SHF, K	AH	AH
M. rufogriseus	C, WR, AF B, K, GRT RUM	GR			RU, SAP, PU D, B	B										
M. parryi (Kaufmann 1974) free-ranging (frequencies)	Frequent agonistic interactions		Frequent agonsitic interactions		Frequent	Frequent agonsitic acts	Frequent	Frequent	Highest frequency		High frequency	Lowest frequency		High frequency	Lowest frequency	
Our data in captivity	Ritualized fight	PRS	PRS, WR, H		H, D, AV	AH, AV, D C, H,	H, RU, AV B, D, HA WR	H, C, AH	WR, K, RU	WR, RU	K, WR, RU			RU, D, PRS WR, AV (other clans)	—	AH
M. robustus (Croft 1961a, free-ranging)	Fight, C	Nonagonistic	Supplant	—	Frequent agonistic acts mostly to medium males	H	H	—	Fight		Fight					
M. antilopinus (Croft 1982) similar to *M. robustus*	Allogroom															
M. giganteus (Southwell 1981, free-ranging)	Supplant	—	Aggressive acts	—	Supplant (small males)	Random	Random	Random	Frequent aggressive acts		Fight		Random	Random	Random	Random
M. fuliginosus	AP (large); SAU (medium), RUM, K, H (medium)	—	AP (large); WR, SAU, RUM, K, H (medium)	—	D, H, (medium)	PU, WR	D, H	WR	WR, RUM K	WR, K, H	Fight		D, H	H	D, H, K	PU, WR
M. rufus (see Croft 1981b, free-ranging, Russell 1970a,b; Ganslosser and Wilhelm 1986 in captivity for details)	Most frequent agonistic	Rare (medium males)	H, supplant		H (medium)	—	Frequent aggressive acts	K, SK	A, AF, AL	AF, K	Fight	Fight	A, AF, AL	AF, K	Fight	Fight

Ritualized fights are often initiated by the dominant animal, the prospective winner. They seem to be an important element in maintaining and stabilizing dominance relationships.

The initiation of a fight at one's own "convenience" seems to allow dominant individuals to demonstrate their continuous fighting ability by constantly forcing the subordinates to lose fights.

B. Agonistic relationships within and between age/sex classes

So far, most studies of agonistic relationships have been restricted to the frequency of agonistic acts directed by members of one age/sex class towards another, without studying quality or other dimensions mentioned by Hinde. Thus discussion of agonistic relationships between age/sex classes in macropodoids means discussing sociograms. This shall be attempted by taking a few examples.

In captive groups of *A. rufescens* (P. Johnson 1980; Ganslosser and Fuchs 1988 and unpublished observations), longer-lasting agonistic relationships between males do not exist. In the presence of females they fight until the death of one. Agonistic relationships of females are characterized by strong aggressive acts towards subadult females, including EBH, B and C. Subadult females tend to avoid adult females. Their relationships towards adult males seem to be more aggressive, with more H and B, towards a male, in connection with sexual behaviour. Subadult males, until they are attacked heavily by adult males, seem to have no characteristic agonistic relationships towards other age/sex classes.

To summarize, the age/sex classes in macropodoids tend to differentiate more among recipients of agonistic acts in the larger, more sexually dimorphic species, whereas the smaller species distribute more agonistic patterns more evenly among all age/sex classes available. Females tend to act agonistically towards subadult and medium-sized males, but less towards large, adult males in dimorphic species. Subadult females often perform agonistic acts in a frequency distribution more like males, which may include a certain amount of aggressive play and play fighting as well.

C. Triadic and polyadic relationships

Polyadic relationships, i.e., social relationships involving more than two individuals, have not been studied extensively in macropodoids so far. There are however, several effects of a third party on agonistic relationships that are described regularly in literature. The most important of these triadic or polyadic effects is the interference of a dominant individual in a fight between two subordinates. This interference may either take the form of approaching and supplanting both participants, or the dominant animal, normally the largest male present, challenges one of the contestants and takes over the fight himself. Interference and taking over of fights has been described from field and captive studies in most large wallabies and kangaroos (e.g., Kaufmann 1974 in *M. parryi;* Osazuwa 1978 in *M. robustus;* Croft 1981a, Ganslosser and Wilhelm 1986 and unpublished in *M. rufus*). The dominant animal normally challenges the larger of the contestants.

A second form of agonistic interference is related to sexual behaviour. Packer (1969) describes for *S. brachyurus* that females squeak when harrassed by subordinate males; the dominant male then approaches and chases off the smaller one. The same observations were made for other species, for example Jarman and Southwell (1986) in *M. giganteus;* Coulson (pers. comm.) in *M. fuliginosus*. Similarly in *A. rufescens* (Ganslosser and Fuchs 1988) the male regularly approaches, sniffs and sometimes attacks females which fight vigorously, and with male-typical elements. This also might be an attempt to detect a seemingly male intruder.

In *Dendrolagus dorianus* (Ganslosser 1984) coalitions between mother and daughter occur in agonistic contexts: whenever the daughter fled from the male (who was newly introduced at that time) she approached her mother, both exchanged some friendly behaviour patterns (e.g., allogrooming), and then mother and daughter approached the male together and tried to displace him from a resting site or other place. In *D. goodfellowi,* a sort of precursor to this seems to exist. Here, the daughter only exchanges friendly behaviour with the mother, without joint action afterwards.

Females of *P. penicillata* (e.g., Batchelor 1980) frequently share, and defend, dens. These female pairs groom each other regularly, and at least some cases have been mother and daughter.

D. Dynamics of agonistic relationships

Again, few data on dynamics of relationships exist so far. La Follette (1971) separated, and re-introduced after varying periods of time, the dominant male from a group of *M. rufogriseus*. When the dominant male was separated for more than 3-5 days, a dominance reversal took place, with the former β taking (and holding after re-introduction) the α position. After a separation for only 48 hours a serious fight arose, which might however have resulted from the fact that a female was oestrous at the time of reintroduction.

In a colony of free-ranging *P. lateralis,* Engel (pers. comm.) observed a similar reversal of dominance, with the β male taking over for about three weeks, while the former α looked slightly ill. After three weeks, a vigorous fight occurred (the only one observed among adult males in 400 h of observation during five months). This fight resulted in the original α male regaining his top status.

During our current study of the development of social relationships (Kappes *et al.* 1987 and Ganslosser in prep.) a similarity to formation of dyadic and polyadic relationships in primates (e.g., Kummer 1975, Abbegglen 1984) was found: the developing relationships follow a fixed sequence of stages, each one characterized by a certain group of behavioural elements. The velocity of the process, and the progression from one stage to the next, depend upon factors such as status, size and age of the participants, status differences, and the influence of other conspecifics.

The first stage, avoidance, and the third, aggression and often vigorous fighting, consist of agonistic elements. In certain species and dyads, these agonistic stages can be skipped, whereas in some species certain dyads never seem to be able to advance to higher stages (one-sided social grooming being the next one). The stages reached by animals in a given dyad might become a useful tool for describing and perhaps quantifying the quality of relationships e.g., for comparison between species, habitats or age/sex classes.

4. AGONISTIC BEHAVIOUR IN SEXUAL CONTEXTS

Mention should be made of the occurrence of agonistic behaviour in sexual contexts. As has already been noted above, females tend to reject males, especially small and subadult males, when courted. In fact, most female/male agonistic acts result from a male courting either a female in anoestrus or the male being rejected because of his belonging to the "wrong" age/size class.

In potoroids (P. Johnson 1980 and my unpublished observations) this rejection can be quite vigorous and the male can be injured by the female's vigorous kicks. Due to the "reversed sexual dimorphism" most females are able to hold their ground against males. However, normally the female allows a brief anal sniffing after a certain time, and then is no longer harrassed by the male when in anoestrus.

In smaller wallabies and tree-kangaroos (our data) females perform a certain amount of mate choice by reacting aggressively against males, perhaps even fighting with them briefly, and selectively allowing courting of older, larger males, even sometimes approaching these older males on their own. The male normally performs behaviour that is neither typical for pure male fighting nor female-typical but something in between.

In large species (e.g., Croft 1981a, for *M. rufus;* Southwell 1981 for *M. giganteus;* my data for several species in captivity) females react even more selectively towards the male size classes; whereas subadult males are supplanted actively when courting, medium sized males are rejected by hitting and SAP, and females simply move away from large males. Medium sized males also tend to react more aggressively in sexual contexts, whereas large males rarely do so. This might point to a difference in the males' mating strategies at different ages.

5. ONTOGENETIC ASPECTS

A. Play

Agonistic play in macropodoids is considered to occur but rarely. Most authors from field studies (e.g., Kaufmann 1974, for *M. parryi;* Osazuwa 1978 for *M. robustus;* Croft 1981a,b for *M. robustus* and *M. rufus*) and captive studies (e.g., Russell 1970a,b for *M. rufus*) report playfighting to be restricted almost totally to the mother/young dyad.

In captive *T. billardierii, Dorcopsis luctuosa* and *Dendrolagus dorianus* we observed play fighting, initiated normally by AL, also directed towards other adults, especially adult males (Ganslosser and Schappert 1985). Playfighting between juveniles has also been observed in captive animals of these species as well as free-ranging rock-wallabies (Engel, pers. comm.). Playfighting in these species is characterized by: a lack of vocalizations, biting and other elements of serious fights; high frequencies of SK and circling movements; a combination of elements from different contexts (frequent transitions between attack and flight, social grooming and HT connected with aggressive acts); a reduction of significant, non-random transitions between all behavioural elements (thus high variability); and generally slow, "inhibited" performance of acts that might otherwise lead to injury.

In captive *M. rufus* social play also consists of playfighting and, frequently in juveniles, of simultaneous bouts of running play, chasing each other and engaging in playfights when meeting for example in a corner (Wilhelm and Ganslosser 1989). Playfighting in itself was not restricted to the mother-young dyad in our study, contrary to Croft (1981). Playfighting frequently occurred, e.g., between a subadult male and an adult female. A playfight frequently is initiated by mutual grooming (similar to Croft's observations of ritualized fights in males). Normally adult females remain on the flat of their hindfeet during playfighting with subadults.

Elements regularly occurring in playfights are upright postures (RU), neck-grasping (EBF), rubbing noses (RN), pawing at head (SF),

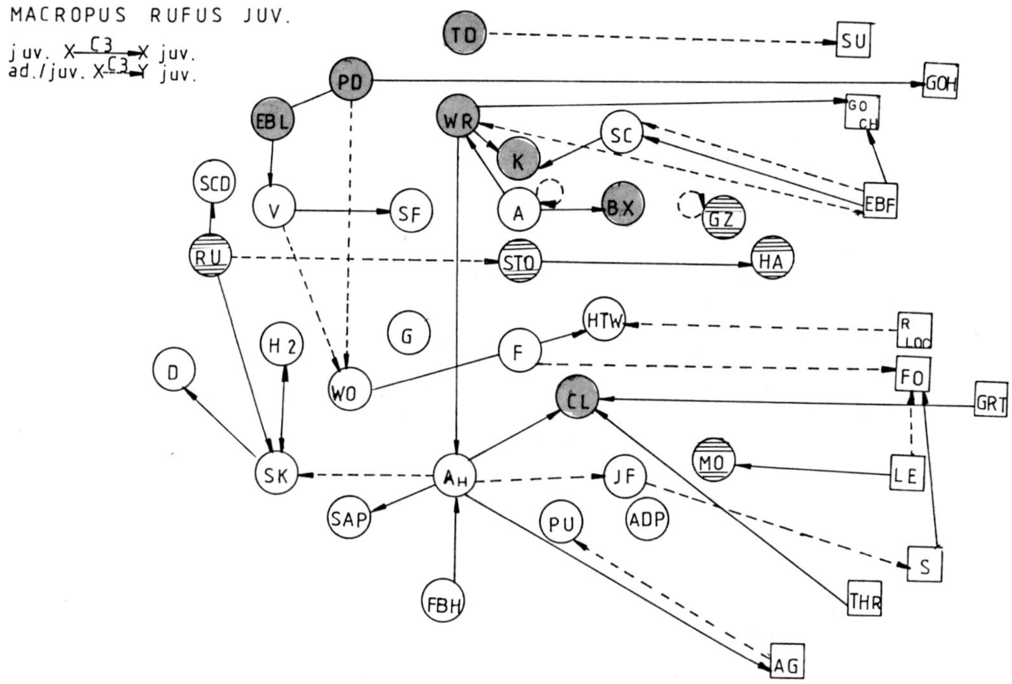

Fig. 14. Sequences of juvenile *M. rufus*. The transitions between elements are more variable, no subcomplexes can be distinguished, and nearly all elements are connected with each other, so the predictability of the next element is low.

shoulders and neck (H), kicks with hindlegs (K) (normally shown by the young only; the forearms remained locked while kicking in a playfight). Frequently these playfights were interrupted by mutual licking and grooming (GO) or skipping (SK)*.

B. Sequences of behaviour in juveniles

Fig. 14 depicts the result of a sequential analysis in juvenile *M. rufus* (Wilhelm and Ganslosser 1989). Though all elements of adult social behaviour (except those shown in high-intensity sexual behaviour) are present in the young too, the sequential organization is not as fixed as in adults. It is not possible to distinguish functional categories, since transitions between nearly all elements of the behavioural repertoire are possible. This is especially the case in fighting behaviour where the sequences in adults are largely fixed and predictable, whereas in juveniles, all elements of agonistic behaviour have frequent connections with transitional elements or those of grooming (e.g., compare the few arrows entering agonistic behaviour, especially in X-X in Fig. 13 with the many different connections in Fig. 14). Some elements occurring only in sexual behaviour of adults are shown totally out of context by young. Certain sequences, however, that are fixed in adults also occur in the same way in the young, e.g., RU-STO, C-F, A-BX. Similar studies in other species, especially smaller ones, are currently under way.

C. Differences in frequencies

Frequency distributions of agonistic behaviour have already been described in section 3B. Thus, only a few general remarks will be made concerning ontogeny (see also Ganslosser and Wilhelm 1986; Ganslosser and Fuchs 1988; Ganslosser and Schappert (in press).

The most striking is the behaviour of subadult females. These are more comparable to males not only in occurrence but also in distribution of agonistic behaviour patterns. This is illustrated best by the different forms of clasping and embracing in *M. rufus*. The more frequently a form of clasping is used in non-sexual contexts (EBF, EBL vs EBH) the more it is performed by subadult females. Subadult females tend to direct agonistic behaviour more randomly towards more age/sex classes, again comparable to subadult males, who also tend to differentiate less between classes of recipients than do adults.

Ontogeny of behaviour tends to recapitulate earlier stages of the phylogeny of behavioural differentiation: (a) young-at-foot in *T. billardierii,* and probably other species as well, show the "potoroid-type" submissive display (throwing themselves onto their back) when approached and sniffed at by large males (Lissowsky, pers. comm.). (b) the sequential organization of behaviour in juveniles (compare Fig. 14 with diagrams of any other of the adults of smaller species) is more random, like in more primitive stages of behavioural differentiation. (c) juveniles

*In playfights of free-ranging Wallaroos *(M. robustus)* Kappes and Krettinger (pers. comm.) even describe kicks while lying down "potoroid-like".

direct their social behaviour towards other conspecifics more randomly, again like adults of smaller species (compare juveniles of e.g., *M. rufus* with adults of smaller species in Table 3, or subadult females *M. rufus* with adult females of their own species, and of smaller species).

6. CONCLUSIONS

1. Behaviour patterns and interactions become more ritualized and stereotyped in macropodoid phylogeny. Ritualization on this level, and occurrence of certain behaviour patterns, depends on phylogenetic position.
2. Agonistic behaviour in total becomes more ritualized and stereotyped, thus more predictable. This evolutionary trend runs parallel to sexual dimorphism and the risk of injury.
3. Dyadic agonistic relationships between age/sex classes become more complex, and triadic and polyadic effects occur, with differentiation in social organization. Social relationships thus are influenced by habitat and other aspects of ecology.
4. In ontogeny agonistic behaviour is modified in structure (sequences become more rigid, ritualized patterns occur) as well as direction (elements being performed more randomly, against more age/sex classes, by juveniles). Female subadults behave more like males than do adult females.

ACKNOWLEDGEMENTS

First of all, I have to thank about two dozen students, who during the last eight years, worked for me as student helpers, research assistants, or doing their Diploma and Staatsexamens theses. They provided a lot of the data cited here, and discussions in our research group. Though for technical reasons the list of references cites only those papers that have been published, and are thus obtainable by regular library processes, the work of the others was equally important. Special thanks go to those who, in various stages, helped in editing and cross-referencing the data: Martina Jordan, Ilona Schappert-Sattler and Regina Walther; to Ursula Herter and Regina Muschketat for the computations; and to my mother for typing endless pages of manuscripts. Mascha Lissowsky created nice kangaroos from videotapes and crude sketches. The directors and staff of the Zoos in which a good deal of the observations was made, especially Dr R. Faust of Frankfurt, Prof. W. Neugebauer of Stuttgart and Prof. H. Wiesner of Munich, are to be thanked for their co-operation and help.

The study was financially supported by Deutsche Forschungs-Gemeinschaft (Ga 283/1), Universitätsbund Erlangen and Nixon-Griffis Fund for Zoological Research, New York.

Last but not least, the discussions with my wife Reinhild Etter-Ganslosser, and with the friends and colleagues at the symposium and afterwards greatly improved the manuscript.

REFERENCES

Abbegglen, J.-J., 1984. On Socialization in Hamadryas Baboons. Associated University Presses: London.

Baerends, G., 1975. An Evaluation of the Conflict Hypothesis. Pp. 187-228 *in* Function and Evolution in Behaviour ed. by G. Baerends, C. Beer and A. Manning. Oxford University Press: Oxford.

Barker, S. C., 1982. Ecology, Behaviour and Social Organization of the Plain Rock-Wallaby *Petrogale inornata assimilis* (Macropodinae). Hons. Thesis, James Cook University, Townsville.

Batchelor, T. A., 1980. The Social Organization of the Brushtailed Wallaby *(Petrogale penicillata penicillata)* on Motutapu Island. M.Sc. Thesis, University of Auckland.

Bell, H. M., 1973. The Ecology of three macropod marsupial species in an area of open forest and savannah woodland in North Queensland. *Mammalia* **37**: 527-44.

Cicala, G. A., Albert, I. B. and Ulmer, F. A., 1970. Sleep and other behaviour of the Red Kangaroo *(Megaleia rufa). Animal Behaviour* **18**: 787-90.

Clancy, T., 1982. Aspects of the behaviour of *Thylogale billardieri* in captivity. B.Sc. Hons. Thesis University of Tasmania, Hobart.

Cook, P., 1973. Some observations on bettongs. *Discovery* **2**: 93-96.

Cooke, B. N., 1979. Field Observations of the Behaviour of the macropod marsupial *Thylogale stigmatica* (Gould). M.Sc. Thesis, University of Queensland.

Coulson, G., 1980. Behaviour of *Macropus antilopinus* in captivity. *Bull. Aust. Mamm. Soc.* **6**: 29.

Crebbin, A., 1982. Social Organization and Behaviour of the Swamp Wallaby, *Wallabia bicolor* in Captivity. B.Sc. Hons. Thesis, University of New South Wales, Kensington.

Croft, D. B., 1981a. Behaviour of the Red Kangaroos *Macropus rufus* (Desmarest, 1822) in north-western New South Wales, Australia. *Aust. Mammal.* **4**: 5-58.

Croft, D. B., 1981b. Social Behaviour of the Euro, *Macropus robustus* (Gould), in the Australian arid zone. *Aust. Wildl. Res.* **8**: 13-49.

Croft, D. B., 1981. Some observations on the behaviour of the Antilopine Wallaroo *Macropus antilopinus* (Marsupialia: Macropodidae). *Aust. Mammal.* **5**: 5-13.

Croft, D. B., 1983. Fighting behaviour in *M. rufus*. Abstracts IEC **18**: 70.

Croft, D. B., 1985. Inter- and intraspecific conflict between arid-zone kangaroos at watering points. *Aust. Wildl. Res.* **12**: 337-48.

Croft, D. B., 1987. Socio-ecology of the Antilopine Wallaroo, *Macropus antilopinus* in the Northern Territory with observations on sympatric, *M. robustus woodwardi* and *M. agilis. Aust. Wildl. Res.* **14**: 243-55.

Dmoch, R., 1975. Die Evolution schlägt Haken – Baumkänguruhs *Zoofreund, Hannover* **14**: 2-3.

Eisenberg, J. F., 1981. The Mammalian Radiations. Chicago University Press: Chicago.

FRITH, H. J. AND CALABY, J. H., 1969. Kangaroos. Hurst: London.

GANSLOSSER, U., 1978. Zum Kampfverhalten des Neuguinea-Filanders *Thylogale brunii* (Schreber, 1778). *Säugetierk. Mitt.* **26**: 195-98.

GANSLOSSER, U., 1979a. Soziale Interaktionen des Doria-Baumkänguruhs, *Dendrolagus dorianus,* Ramsay 1883, *Z. Säugetierkunde* **44**: 1-18.

GANSLOSSER, U., 1979b. Soziale Kommunikation, Gruppenleben, Spiel–und Jugendverhalten des Doria-Baumkänguruhs. (*Dendrolagus dorianus,* Ramsay 1883). *Z. Säugetierkunde* **44**: 137-53.

GANSLOSSER, U., 1983. Quantitative Untersuchungen zum Sozialverhalten des Doria-Baumkänguruhs (*Dendrolagus dorianus* Ramsay 1883) in Gefangenschaft, *Zool. Anz.* **211**: 1-29.

GANSLOSSER, U., 1984. On the occurrence of female coalitions in Tree-kangaroos (Marsupialia, Macropodidae, *Dendrolagus*). *Aust. Mammal.* **7**: 219-21.

GANSLOSSER, U. AND FUCHS, C., 1988. Some remarks on social behaviour in captive Rufous Rat-kangaroos, *Zool. Anz.* **220**: 300-12.

GANSLOSSER, U. AND HOLZ, C., 1984. Zum Kampferhalten verschiedener sozialer Kategorien beim Streifenbuschkänguruh. *Dorcopsis muelleri. Verh. Dt. Zool. Ges.* **77**: 281.

GANSLOSSER, U. AND JARMAN, P. J., 1984. Mating strategies in Macropodoidea. Abstracts Ethol. treffen Göttingen, 23.

GANSLOSSER , U. AND SCHAPPERT, I., 1985. Zum Sozialspiel bei Doria-Baumkänguruhs *(Dendrolagus dorianus)* und Streifenbuschkänguruhs *(Dorcopsis muelleri)* (Macropodidae). *Verh. Dt. Zool. Ges.* **78**: 206.

GANSLOSSER , U. AND SCHAPPERT, I., (in press). Sex-typical social behaviour in *Dorcopsis muelleri* (Lesson, 1827). *Zool. Garten.*

GANSLOSSER, U. AND WERNER, I., 1984. Zur Häufigkeit freundlicher und agonistischer Verhaltensmuster in einer Gruppe von Rotbauchfilandern (*Thylogale billardieri* Desmarest (Marsupialia, Macropodidae)). *Zool. Anz.* **213**: 51-67.

GANSLOSSER, U. AND WILHELM, P., 1986. Sex- and age-dependant differences in frequences of social interactions between captive Red Kangaroos, *Macropus rufus* (Desmarest, 1822). *Zool. Anz.* **216**: 58-71.

GERBER, U., 1976. Ethogramm und Sequenzanalyse bei handaufgezogenen Rehkitzen *(Capreolus capreolus).* Diplomarbeit, Zürich.

GERMANN-MEYER, V., 1974. Verhaltensstudien am Grauen Riesenkänguruh *Macropus (Macropus) giganteus.* Ph.D.Diss. Basel.

GERMANN-MEYER, V. AND SCHENKEL, R., 1970. Über das Kampfverhalten des Grauen Riesenkänguruhs, *Macropus giganteus. Rev. Suisse Zool.* **77**: 938-42.

GEWALT, W., 1965. Kleine Beobachtungen an seltenen Beuteltieren im Berliner Zoo. III. Rotes Baumkänguruh (*Dendrolagus matschiei.* Förster and Rothschild 1907). *Zool. Garten* **31**: 240-49.

GRANT, T. R., 1973. Dominance and association among members of a captive and free-ranging group of Grey Kangaroos *(Macropus giganteus). Anim. Behav.* **21**: 449-56.

GRANT, T. R., 1974. Observation of enclosed and free-ranging Grey kangaroos *Macropus giganteus. Z. Säugetierkunde.* **39**: 65-78.

HAMMERSTEIN, P., 1985. Spieltheorie und Kampfverhalten von Tieren. *Verh. Dt. Zool. Ges.* **78**: 83-99.

HENDRICHS, H., 1978. Die soziale Organisation von Säugetier-populationen. *Säugetierk. Mitt.* **26**: 81-116.

HERTER, U., SCHNEIDER, I. AND GANSLOSSER, U., 1985. Some remarks on the behavioural repertoire of the Parma Wallaby *Macropus parma,* Waterhouse. *Zool. Anz.* **215**: 287-97.

HINDE, R., 1981. Animal Signals: Ethological and games-theory approaches are not incompatible. *Anim. Behav.* **29**: 535-42.

HINDE, R. A. (ed.), 1983. Primate Social Relationships. Blackwell: Oxford.

JARMAN, P. J., 1983. Mating system and sexual dimorphism in large, terrestrial, mammalian herbivores. *Biol. Rev.* **58**: 485-520.

JOHNSON, K. A., 1977. Ecology and Management of the Red-necked Pademelon, *Thylogale thetis,* on the Dorrigo Plateau of Northern New South Wales, Ph.D. Thesis, University of New England, Armidale.

JOHNSON, P. M., 1980. Observations on the behaviour of the Rufous Rat-kangaroo, *Aepyprymnus rufescens* (Gray), in captivity. *Aust. Wild. Res.* **7**: 347-57.

KAPPES, G., KRETTINGER, B. AND GANSLOSSER, U., 1987. Ausbildung von Sozialbeziehungen bei Macropodoidea – Ergebnisse dyadischer und triadischer Konfrontationen. *Z. Säugetierkunde.* **52**: Suppl. 24/25.

KAUFMANN, J. H., 1974. Social ethology of the Whiptail Wallaby, *Macropus parryi* in north-eastern New South Wales. *Anim. Behav.* **22**: 281-369.

KAUFMANN, J. H., 1975. Field observations of the social behaviour of the Eastern Grey Kangaroo, *Macropus giganteus. Anim. Behav.* **23**: 214-21.

KELLER, R., 1982. Das Markierungsverhalen des Kleinen Pandas. *Zool. Garten.* **52**: 369-92.

KITCHENER, D. J., 1972. The importance of shelter to the Quokka, *Setonix brachyurus* (Marsupialia), on Rottnest Island. *Aust. J. Zool.* **20**: 281-99.

KUMMER, H., 1975. Rules of dyad and group formation among captive Gelada Baboons. *(Theropithecus gelada).* Pp. 129-60 *in* Symposia of the 5th Congress of the International Primatological Society ed. by S. Kondo, M. Kawai, A. Ehara, S. Kawamura. Japan Science Press: Tokyo.

KUMMER, H., 1984. From laboratory to desert and back: A social system of Hamadryas Baboons. *Anim. Behav.* **32**: 965-71.

LA FOLLETTE, R. M., 1971. Agonistic behaviour and dominance in confined wallabies, *Wallabia rufogrisea frutica. Anim. Behav.* **19**: 93-101.

LEHNER, P., 1979. A Handbook of Ethological Methods. Garland: New York.

LEUTHOLD, W., 1977. African Ungulates. Springer: Berlin-Heidelberg.

MCFARLAND, D. (ed.), 1981. The Oxford Companion to Animal Behaviour. Oxford University Press: Oxford.

MAYNARD-SMITH, J., 1978. Evolution and the theory of games. Pp. 258-70 *in* Readings in Sociobiology ed. by T. H. Clutton-Brock and P. H. Harvey. Freeman: Reading.

MAYNARD-SMITH, J., 1984. Game theory and the evolution of behaviour. *Behav. Brain Sci.* **7**: 95-125.

MORTON, S. R. AND BURTON, T. C., 1973. Observations on the behaviour of the macropodid marsupial *Thylogale billardieri* (Desmarest) in captivity. *Aust. Zool.* **18**: 1-14.

MURBÖCK, T., 1975. Zum Kampfverhalten des Bennett-wallabies. *Säugetierkund. Mitt.* **23**: 79-80.

MURBÖCK, T., 1977. Komment- und Beschädigungskampf beim Bennett-Wallaby, *Macropus rufogriseus fruticus. Verh. Dt. Zool. Ges.*

NICHOLLS, C., 1972. A study of the behaviour and social organization of the Yellow-footed Rock-wallaby *(Petrogale xanthopus)*. B.Sc. Hons. Thesis, University of Adelaide, Adelaide.

NOVELLIE, P. A., MANSON, J. AND BIGALKE, R. C., 1984. Behavioural ecology and communication in the Cape Grysbok. *S. Afr. J. Zool.* **19**: 22-30.

OSAZUWA, J. E., 1978. Activity and behaviour of the Wallaroo, *Macropus robustus* Gould (Marsupialia: Macropodidae). M.Nat.Res. Thesis, University of New England, Armidale.

PACKER, W. C., 1969. Observations on the behaviour of the marsupial *Setonix brachyurus* (Quoy and Gaimard) in an enclosure. *J. Mammal.* **50**: 8-20.

PARKER, G. A., 1978. Assessment strategy and the evolution of fighting behaviour. Pp. 271-93 *in* Readings in Sociobiology ed. by T. H. Clutton-Brook and P. H. Harvey. Freeman: Reading.

POOLE, T. B., 1985. Social Behaviour of Mammals. Blackie: Oxford, London.

RUSSELL, E. M., 1970a. Observations on the behaviour of the Red Kangaroo *(Megaleia rufa)* in captivity. *Z. Tierpsychol.* **27**: 385-404.

RUSSELL, E. M., 1970b. Agonistic interactions in the Red Kangaroo *(Megaleia rufa)*. *J. Mammal.* **51**: 80-88.

RUSSELL, E. M., 1973. Mother-young relations and early behavioural development in the marsupials. *Macropus eugenii* and *Megaleia rufa*. *Z. Tierpsychol.* **33**: 163-203.

SCHNEIDER, I., 1985. Sozialbeziehungen beim Hübschgesicht-Känguruh *(Macropus parryi,* Bennett 1835). *Z. Säugetierk.* **50**: Suppl. 65-66.

SCHNEIDER, I. AND GANSLOSSER, G. (in prep). Female dominance in *Macropus parryi*.

SCHNEIDER, K. M., 1953. Vom Baumkänguruh *(Dendrolagus leucogenys)*. *Zool. Garten.* **21**: 63-106.

SCOTT, J. P. AND FREDERICSON, E., 1951. The causes of fighting in mice and rats. *Physiol. Zool.* **24**: 273-309.

SHARMAN, G. B. AND CALABY, J. H., 1964. Reproductive behaviour in the Red Kangaroo, *Megaleia rufa,* in captivity. *C.S.I.R.O. Wildl. Res.* **9**: 58-85.

SOUTHWELL, C. J., 1976. Behaviour and social organization of three species of macropods at Diamond Flat, New South Wales. Hons. Thesis, University of New England, Armidale.

SOUTHWELL, C. J., 1981. Sociobiology of the Eastern Grey Kangaroo, *Macropus giganteus*. Ph.D. Diss. University of New England, Armidale.

STODART, E., 1966. Observation on the behaviour of the marsupial *Bettongia lesueuri* (Quoy and Gaimard) in an enclosure. *C.S.I.R.O. Wildl. Res.* **11**: 91-101.

VUJCICH, M. V., 1979. Aspects of the biology of the Parma (*Macropus parma* Waterhouse) and Dama (*Macropus eugenii,* Desmarest) wallabies with particular emphasis on social organization. M.Sc. Thesis, University of Auckland.

WALTHER, F. R., 1963. Über die Möglichkeit der Verhaltensforschung in Tiergärten. *Säugetierk. Mitt.* **11**: 62-68.

WALTHER, F. R., 1984. Communication and Expression in Hoofed Mammals. Indiana University Press: Blomington.

APPENDIX

Sequential Analysis

The elements of a behavioural repertoire, which are performed by animals in a constant flow over time, normally are not distributed at random. They occur in certain patterns, in more or less fixed sequences, e.g., courtship behaviour.

Sequential analysis is a mathematical tool to discover these patterns. Basically, probabilities of transitions of all elements of the animal's behavioural repertoire are calculated under the hypothesis that the elements are distributed at random, i.e., each element can follow each other with the same probability. These expected values then are compared to the observed frequencies of transition between elements, and the difference between observed and expected values can be tested for significance. The larger the deviations between observed and expected values the more rigid and stereotyped is a sequence of behavioural elements. Data for sequential analysis are collected by focal-animal sampling, i.e., every act performed by the animal under observation is noted. Sequential analysis of the data is performed first by calculating expected frequencies of transitions between behavioural elements and then dividing the values for observed and expected frequencies (for details see Lehner 1979 p.274).

Sequences of behavioural elements performed by one animal without interruptions are termed X-X, and calculated separately from sequences of two animals shown as a reaction to each other (X-Y sequences, e.g., A bites B, B hits A, A leaves B). In order to get a clearer picture of significant transitions between elements of the behavioural repertoire, two additional levels of testing are introduced using a method developed by Gerber (1976) in Zürich and cited by Keller (1982) and Ganslosser (1984).

The highest level of significance is reached when:

$$C_3 = \frac{\text{Obs.}>1.0}{E+10\sqrt{E}}$$

with Obs = observed, E = expected frequency.

The next level is:

$$C_2 = \frac{\text{Obs.}>1.0}{E+2\sqrt{E}}$$

The elements are then arranged into functional categories on the basis of C_3 values >1.0.

Each pair of elements having a transitional value at the highest level of deviation (i.e., $C_3>1.0$) in linked with an arrow pointing from the preceding to the following element. In the next step, so-called central elements of each functional category (social grooming for amicable behaviour, biting and kicking for agonsitic behaviour, etc.) are picked out, and all elements connected with them are arranged in their vicinity. Elements having most of the connections on the C_3-level to these central elements are regarded as belonging to the same functional category. In the next step, all elements having most of their connections on C_3-level to elements already grouped into this category also are included into the same category. Thus, elements within one functional category are those that have more connections on $C_3>1.0$ to each other than to elements of other functional categories. For some elements that could not be arranged into one single category on C_3-level, the connections on the level of $C_2>1.0$ were considered as well.

Certain behavioural elements cannot be attributed to single functional categories, because they regularly occur in several contexts, e.g., nose-sniffing, approaching or leaving. These elements connect patterns of different function or motivation in the actual flow of behaviour observed as well as in the diagrams of sequential analysis. They are called "transitional elements".

In the last step of constructing the diagrams, elements are regrouped several times in order to avoid as much crossing-over of arrows as possible, making them easier to read and understand.

Social organization of the Macropodoidea

D. B. CROFT[1]

Four grades of group size, group structure and individual home-range overlap pattern during the species-typical active and inactive periods for 21 macropodoid species were used to define three social Types by cluster and factor analyses. Type 1 species are solitary except for reproductive associations; Type 2 species are often solitary but aggregate on favoured resource patches; Type 3 species are gregarious. Factor analyses using variables for shelter and feeding habitats, activity pattern, size, sexual dimorphism, annual fecundity, annual birth distribution and diet were used to discriminate the social Types of subsets of these 21 species. Type 1 and 2 species are small to medium-sized, homomorphic or moderately heteromorphic, use closed shelter sites, are nocturnal, and forage in or close to cover. Their diets predominantly contain items that are more discrete, limited and require longer search and, often, handling times than Type 3 species. The gregarious Type 3 species tend to be large, strongly heteromorphic, partially diurnally active, occupy open shelter and foraging habitats, and eat predominantly grass. Both Type 2 and 3 species show spacing behaviour where the feeding ranges of both sexes overlap and territorial behaviour is absent. Male home-ranges are usually larger than female ranges and more so in Type 2 species where females are not as gregarious. Male home-range size tends to increase with male size rank, and high ranked-males centre their home range on the densest segments of the female population. There these high-ranked males hold dominion and have high mating success for tenures of about a year. Intrasexual conflict in males is more frequent than in females. Predation as a factor in social evolution of macropodoids has been underestimated in the past. The review finds that further research on potoroids and smaller macropodids is essential to define the social organization of the Macropodoidea. In comparison with cervid and bovid counterparts, territoriality and longer-term female defence or female resource defence polygyny are absent in macropodoids. This difference between these placental and marsupial herbivore assemblages deserves further research.

INTRODUCTION

THE social organization of a species encompasses behavioural systems for resource exploitation, predator avoidance, mating and rearing of the young; and refers to an abstraction of the totality of social relationships among all members of sample populations for a species (Hinde 1976). The social organization of a species is built from certain structural features (McBride 1976) which can be empirically determined (e.g., the nature and patterning of social interactions and the constraints under which they occur). The social structure formed depends on at least the interaction between external environmental (ecological) variables and conservative species parameters (Crook *et al.* 1976). The supposition is that there is some selection from amongst all possible structures in a given environment to produce the one that maximizes reproductive success for some or all of the individuals who engage in that structure. Socio-ecologists attempt to identify population and/or species-typical social organizations while recognizing that behavioural characters are often labile and the organization is dynamic at both the individual level (e.g., transition through age classes) and the population level (e.g., in response to high variance in the environment). Inter- and intraspecific comparisons of social organization have both been informative in delimiting selective forces which have led to the evolution of various underlying structures (Rubenstein and Wrangham 1986).

Interspecific comparisons of social organization in primates (Eisenberg *et al.* 1972; Clutton-Brock and Harvey 1977) and ungulates (Geist 1974; Jarman 1974; Leuthold 1977) have been made several times. Two principal factors, resource distribution and predation, are identified as responsible for variation in social structure. Wrangham and Rubenstein (1986) add a third, intraspecific competition. Species which feed on items that are discrete, scarce and require relatively long search and processing times are less likely to aggregate in groups than those where items are locally abundant and easily obtained (Jarman 1982). Foraging behaviour and the dietary niche relate to environmental variation and the constraints of such species parameters as body size, dental morpholoy and digestive physiology. High predation risk favours either large group formation or solitariness depending on whether an individual relies on cryptic behaviour to avoid predation (Pulliam and Caraco 1984); the latter is usually the strategy for individuals of small body size whose mobility potential is insufficient to outrun a predator. Intraspecific competition shapes the social structure, particularly where the distribution and behaviour of one sex influences these in the other sex. Thus females, for example, may act as a classical ecological resource in shaping the behaviour of males as discussed in analyses of mating systems in mammals (Wittenberger 1979; Dunbar 1982; Gosling 1986).

Interspecific comparisons of the social organization of the Macropodoidea have been made in whole (Kaufmann 1974a; Russell 1974) or part (Jarman 1983; Russell 1984) several times. The relationships between ecological variables, species parameters and social structure found in other herbivorous mammalian taxa generally hold. Grazers of large body size and open habitat form larger groups than browsers or selective

[1]School of Biological Science, University of New South Wales, P.O. Box 1, Kensington, New South Wales 2003, Australia.

Pages 505-25 in KANGAROOS, WALLABIES AND RAT-KANGAROOS ed. by G. Grigg, P. Jarman and I. Hume. Surrey Beatty & Sons Pty Limited, New South Wales, Australia, 1989.

feeders of small body size and closed habitat. This review incorporates new information on the social organization of macropodoids and takes a multivariate statistical approach to the analysis of relationships between social structures and the variables identified above. Three social types are identified in the Macropodoidea and their relationships with some ecological variables and species parameters are examined. Spatial relationships and patterns of association and interaction in various species are discussed in relation to intraspecific competition as a shaping force in social structure. Predation and its effect on macropodoid social organization are briefly considered.

Sociality in the Macropodoidea

The social structures of species in a taxon are likely to be distributed along a continuum with potentially considerable variation at any point, especially for those species with broad geographic ranges. The sociality of a species identifies its position on this continuum, where the range is normally from solitary individuals to individuals forming highly-structured social relationships in closed-membership groups. For interspecific comparison one ideally requires similarly continuous variables which reflect species-typical behaviour in relation to species-typical values of variables which are extrinsic to the animals (ecological), intrinsic to the animals (species parameters) or interactions between these (Jarman 1982). This ideal is seldom realized (Clutton-Brock and Harvey 1984). In the macropodoids, studies of social organization range from brief comments as part of a wider study (e.g., Ride and Tyndale-Biscoe 1962), data on a limited number of social structure variables (e.g., P. M. Johnson 1980a), to detailed studies at the individual level (e.g., C. N. Johnson 1986). Mean group size is known from several species but in some species (e.g., *Macropus giganteus,* Southwell 1984a) it varies in a predictable relationship with density but not in others (e.g., *Macropus rufogriseus,* C. N. Johnson 1985). Its use as a continuous variable across all studies is limited by a lack of information on density and density-dependence. Field research is unevenly distributed across macropodoid species. The genus *Petrogale* is second in species diversity to *Macropus* but the social organization and even basic species parameters of few of its species have been described. Field studies are strongly biased towards the large, open habitat and more diurnally active macropodid species. For these reasons discrete graded variables are used to describe the sociality of species and analyses are based on a subset of extant species.

Some species are dispersed in individual rest sites and aggregate on patches while actively feeding (e.g., *Thylogale thetis*, K. A. Johnson 1980); others potentially share rest sites and are more dispersed when feeding (e.g., *Setonix brachyurus*, Kitchener 1972). Therefore the periods when a species is normally inactive or active are distinguished to describe dispersion and group structure. The values of three variables in each of the active and inactive periods were determined (Appendix 1). Each variable was scaled from one to four. Group size took the values: (1) singleton, (2) male-female pair counting the mother and young-at-foot as a single unit, (3) occasional aggregations of 3-5 or mean group size in the range >1-2, (4) species usually seen in groups or mean group size >2. Group structure took the values: (1) singleton, (2) male-female (± young-at-foot) pair, (3) uni-male group, (4) open-membership mixed-sex group. Spacing pattern took the values: (1) individual home ranges with minimal overlap, (2) single male and female (± young-at-foot) share common but relatively exclusive home range, (3) female home ranges overlap but male home ranges segregate, (4) males and females share large proportion of their home ranges.

All variables will usually correlate for the singleton but there could be temporal segregation on a partly shared home range. Similarly for a pair, but a male whose home range overlapped with several females might usually be in consort with one or other of them but females do not aggregate. These social structure variables may correlate in the active and inactive periods in some species (e.g., *Macropus rufus*, Croft 1981a) but in others, such as *Petrogale penicillata* (Batchelor 1980), individuals intermingle while feeding but defend against or supplant others at rest sites. Evidence in macropodoids for stable pair formation across the population (e.g., *Setonix brachyurus*, Kitchener 1972) or uni-male groups (harems) (e.g., *Petrogale inornata*, Dwyer 1972) are either inconclusive or poorly documented and so I have tended to be conservative in assigning species to these classes. Captive studies which might lead to conclusions contrary to field evidence were rejected in favour of the latter. For example, *Bettongia lesueur* may form uni-male groups in captivity (Stodart 1966) but Dahl (1897) observed colonial burrow systems with interconnecting tunnels and individuals intermingling. Similarly P. M. Johnson (1980b) observed a high degree of male intolerance in *Aepyprymnus rufescens* in captivity but Jarman (1983) states that home ranges of both sexes overlap in the field. Across species, most data are available on group size and least on spacing pattern. In some cases the latter could reasonably be inferred. Conclusive field data for no New Guinea species were found. Thus the analysis was restricted to 21 species of Australian macropodoids. For *Macropus robustus*, two subspecies, *M. r. robustus* and *M. r. erubescens*, were included since they differ in reproductive variables. (Poole and Merchant 1987).

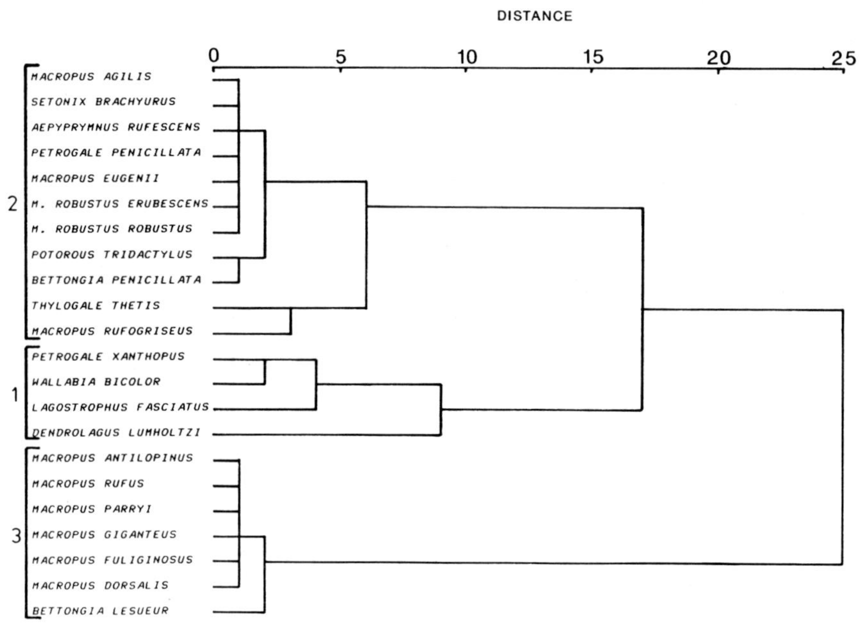

Fig. 1. Dendrogram of UPGMA cluster analysis of social system variable for 21 species (Appendix 1). Distance scaled to 25 units.

A UPGMA (unweighted pair-group method using arithmetic averages) cluster analysis (SPSSX, Norusis 1985) was performed using the values of the six variables for the sample of 22. Three clusters were distinguished (Fig. 1). The cluster labelled "1" was formed from an heterogeneous group of macropodids. The cluster labelled "2" linked a homogeneous group of macropodids and one potoroid to a pair of potoroids, with *Thylogale thetis* and *Macropus rufogriseus* as linked but outlying species. The inactive home ranges of individuals in the latter two species overlap more than those of individuals in other species in this cluster. The cluster labelled "3" contained a homogeneous group of macropodids linked to *Bettongia lesueur*. Factor (principal components or pc) analysis (SPSSX, Norusis 1985) was used to define these clusters in terms of the variables examined (Fig. 2). The first axis separated species in terms of their social structure during the inactive period and the second during the active period. Combined they explained 89% of the variance in the data set. Cluster 1 species are usually solitary but may be found in pairs or tend towards uni-male groups. Cluster 2 species are relatively solitary during inactivity and occasionally gregarious when active with sexes intermingling. Cluster 3 species are usually gregarious whether active or inactive and sexes mix relatively freely.

Three Types of sociality are respectively defined for these clusters, ranked in order of increasing sociality. Type 1 species are solitary, a male associating with one or more females for reproductive behaviour. Type 2 species are often solitary but aggregate on favoured resource patches. Type 3 species are gregarious; one factor influencing group size may be population density.

Habitat, activity, species parameters and sociality

The relationships between the previously determined Types of sociality and a combination of discrete and continuous variables encompassing habitat, activity and certain species parameters were investigated. Both discriminant analysis using maximization of Wilk's lambda as criterion and Factor (pc) analysis (SPSSX, Norusis 1985) were used.

Habitat was divided into that used during the inactive and active periods. At rest, all species use some form of shelter relating to anti-predator or thermoregulatory behaviour, or both. For most species the type of shelter selected is perennial, although some vary exposure to sun with diurnal temperature (e.g., *Petrogale penicillata*, Batchelor 1980). In others there may be seasonal variation and so I have selected the site the species is most commonly found in over the annual cycle. For example, *M. r. erubescens* uses cave shelters in mid-summer (Ealey 1967; Croft 1981b) but only at daytime temperatures exceeding 37°C (Croft unpubl.) and not all individuals use such shelters. Thus this subspecies is not assigned to the "rock cavity" shelter category given below. Shelter site took the values: (1) burrow, (2) nest, (3) rock cavity, (4) dense shrub or tussock grass, (5) large canopy tree, (6) small canopy tree/shrub. The common foraging (active) habitat took the values: (1) closed (e.g., rainforest, wet sclerophyl forest with dense understorey, dense scrub or heath), (2) dense vegetation with small open

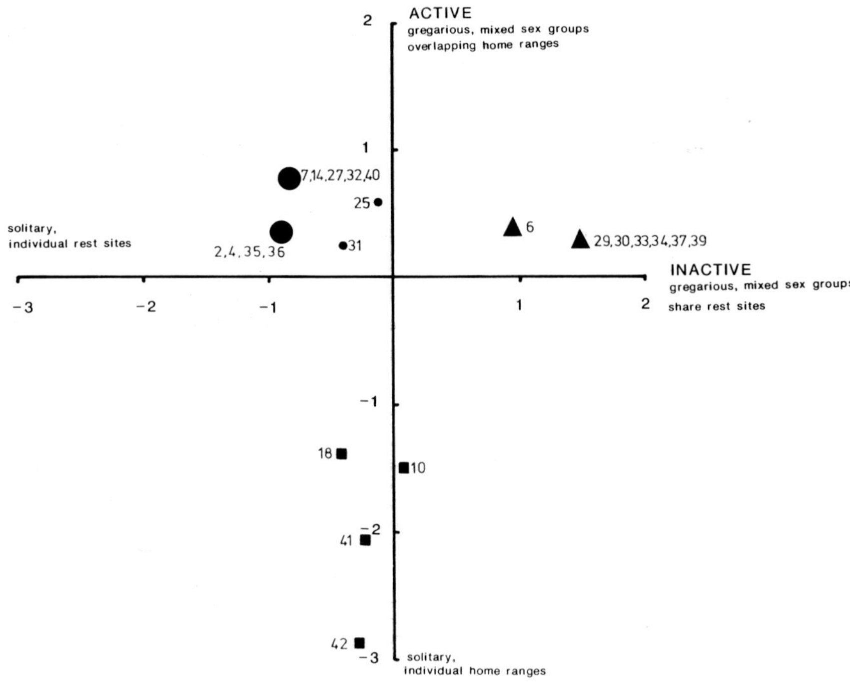

Fig. 2. Plot of factor scores from a Factor analysis of social system variables for 21 macropodoid species (Appendix 1). ■ social Type 1 species, ● social Type 2 species, ▲ social Type 3 species. KEY: 2, *Potorous tridactylus;* 4, *Bettongia penicillata;* 6, *B. lesueur;* 7, *Aepyprymnus rufescens;* 10, *Lagostrophus fasciatus;* 14, *Petrogale penicillata;* 18, *P. xanthopus;* 25, *Thylogale thetis;* 27, *Macropus eugenii;* 29, *M. parryi;* 30, *M. dorsalis;* 31, *M. rufogriseus;* 32, *M. agilis;* 33, *M. giganteus;* 34, *M. fuliginosus;* 35, *M. robustus erubescens;* 36, *M. robustus robustus;* 37, *M. antilopinus;* 39, *M. rufus;* 40, *Setonix brachyurus;* 41, *Wallabia bicolor;* 42, *Dendrolagus lumholtzi.*

spaces relative to the species' size (e.g., tussock grass), (3) forest edge (i.e., open short vegetation bordering relatively tall dense vegetation), (4) woodland (e.g., dry sclerophyl forest with sparse understorey), (5) open (i.e., no or very sparse lateral cover across a large vegetation zone). In classifying habitats, I have attempted to relate them to the animal's perspective. For example, tussock grassland and shrub steppe would be open habitats for *Macropus rufus* but moderately closed for *Lagorchestes hirsutus*.

The activity pattern of species was divided into three classes: (1) diurnal, (2) nocturnal, (3) crepuscular/nocturnal. With the exception of *Hypsiprymnodon moschatus*, all macropodoids are nocturnally active. Most species in *Macropus* are active and feed in the periods of a few hours before dusk and following dawn and so fall into class 3. Potoroids generally emerge from their nests at dusk and return before dawn and so are nocturnal. Some species, such as *Petrogale penicillata* (Batchelor 1980) and *Thylogale thetis* (K. A. Johnson 1980), may be active and feed within their shelter habitat but emerge on to their foraging habitat after dark. Thus I have emphasized the activity pattern in the species-typical foraging ("active") habitat.

The species parameters selected reflect size, dimorphism and reproduction. Except for birth distribution these are continuous variables. Macropodoids grow continuously through life at a decreasing rate. Small species are usually homomorphic and large species show holobiotic heteromorphism (Jarman 1983). Heteromorphism is expressed in body weight and morphology (e.g., colour, size of forelimb). Skeletal dimensions are better predictors of age in juveniles (e.g., Poole *et al.* 1982a; 1982b) and a field population of *Macropus rufus* (Edwards, unpubl.). Body weight may vary within the annual cycle reflecting variation in the plane of nutrition (e.g., Christensen 1980). Even so, body weight and the ratio of male:female body weight are the most readily available measures of body size and dimorphism, respectively, across the macropodoids and are adequate for interspecific comparisons. Data were generally taken from Strahan (1983) and original data sets were sought for confirmation of these values. In some cases, such as *Petrogale penicillata* (Batchelor 1980), a field data set was used. Female body weight was taken as a measure of the species' size and the ratio of male:female body weight as a measure of dimorphism.

Two reproductive variables were examined. Annual fecundity was calculated from the division of mean pouch life (d) + mean gestation length (d) into 365 d. All macropodoids are capable of producing at least one young per year. For those species having a distinct single birth season in any year (e.g., *Macropus eugenii*), annual fecundity was set to one. The time from conception to permanent exit from the pouch in marsupials has generally been equated to the length of gestation in placentals (Russell 1982). Time from conception to weaning is usually more variable and known for fewer species. The data set used generally agrees with values on gestation length and mean pouch life in Tables 2.1 and 2.2 in Tyndale-Biscoe and Renfree (1987). Birth season was divided into three classes: (1) seasonal, (2) clumped, and (3) continuous. Published data on the monthly distribution of births was tested against a model of an equal frequency of births across months using the G-statistic (Sokal and Rohlf 1969). The data set used was ordered in decreasing preference from field data from a population at or nearest the locality where sociality data were collected (e.g., Rottnest Island population of *Setonix brachyurus*), field data with a testable sample, captive data. Seasonal species show birth distributions with highly significant G values ($P<<0.001$), a distinct single birth peak across a few months, and months when no births occur. Clumped species show birth distributions with significant G values ($P<0.05$), some births in most months, but a distinct peak in a few months. Continuous species give birth in all months and the G statistic is not significant ($P>0.05$).

Examples of these classes are given in Fig. 3. The distribution of female oestrus is more likely to affect social structure across the population

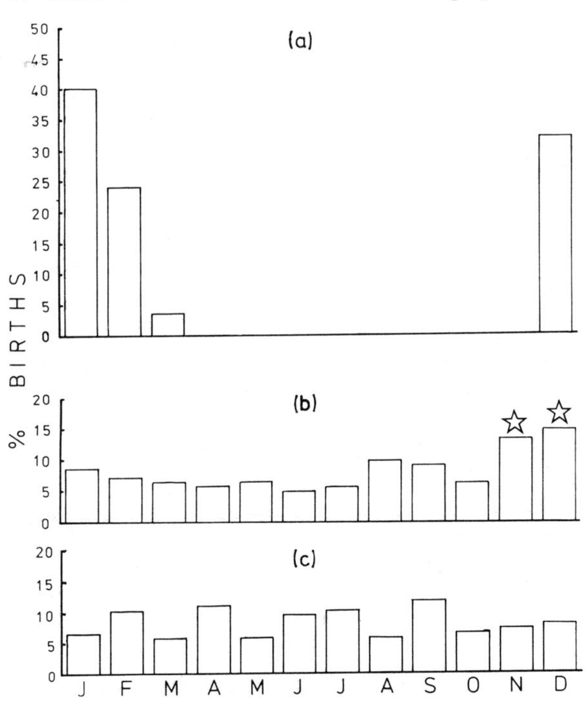

Fig. 3. Percentage of monthly births showing three patterns of macropodoid birth distribution: (a) seasonal (data for *Macropus eugenii* from Christensen 1980); (b) clumped (data for *Bettongia penicillata* from Christensen 1980); (c) continuous (data for *Macropus agilis* from Kirkpatrick and Johnson 1969).

than births. However, in reproductive Type 3 species (Tyndale-Biscoe and Renfree 1987), the majority of Macropodoids, the two will be strongly correlated given the incidence of post-partum oestrus. Tyndale-Biscoe and Renfrees' (1987) reproductive Type 3i species may be at variance with this trend but distributions of oestruses are not available. Other studies of the relationship between reproductive season and social organization (e.g., Ridley 1986) have used birth distributions because of the reliability of these data.

Some other species parameters were examined but not included in this analysis. Data on male and female age at sexual maturity are available but these ages are more variable in field than captive populations (e.g., Poole and Catling 1974). Across species, data are usually available only for captive populations. In heteromorphic species, sexual maturity is unlikely to indicate mating activity. Males achieve high mating success at an age well beyond sexual maturity (e.g., Jarman and Southwell 1986). That age is not well-known across species. Dental morphology (Sanson 1978) or digestive anatomy and physiology (Hume 1982) could be used to reflect dietary niche. Data on actual diets are available for fewer species but contain more information and so this limited data set has been used in preference (see next section).

The discriminant analysis provided reasonable separation of the three social types and correct classifications to 75%. The three discriminating variables and their canonical coefficients were habitat inactive (−2.48), habitat active (1.46) and dimorphism (1.74). The opposing signs of the habitat variables tended to counteract habitat effects and a key gregarious species, *Macropus parryi*, was classified as Type 2. Previous attempts at discriminating degrees of sociality from activity and habitat (Kaufmann 1974a; Russell 1974) revealed only trends rather than clear separations so an ambivalent result is expected. The Factor analysis is more informative about why this may be so (Fig. 4). Factor 1 explained 56.3% of the variance from a combination of the habitat, activity and body-size variables. Species loading positively on this axis use open shelter and foraging habitats, are active both crespuscularly and nocturnally, are large and heteromorphic. *Macropus rufus* exemplifies these characters. All Type 3 species are found at the positive end of this axis with the exception of *Bettongia lesueur*. *Bettongia lesueur* is small, homomorphic, nocturnal and uses a burrow for shelter. It was ranked as gregarious because of its communal burrow use but its behaviour at night is not well documented. For these reasons it is distinctive and a special case.

Of greater import is the mix of Type 2 and Type 3 species positively associated with Factor 1. The

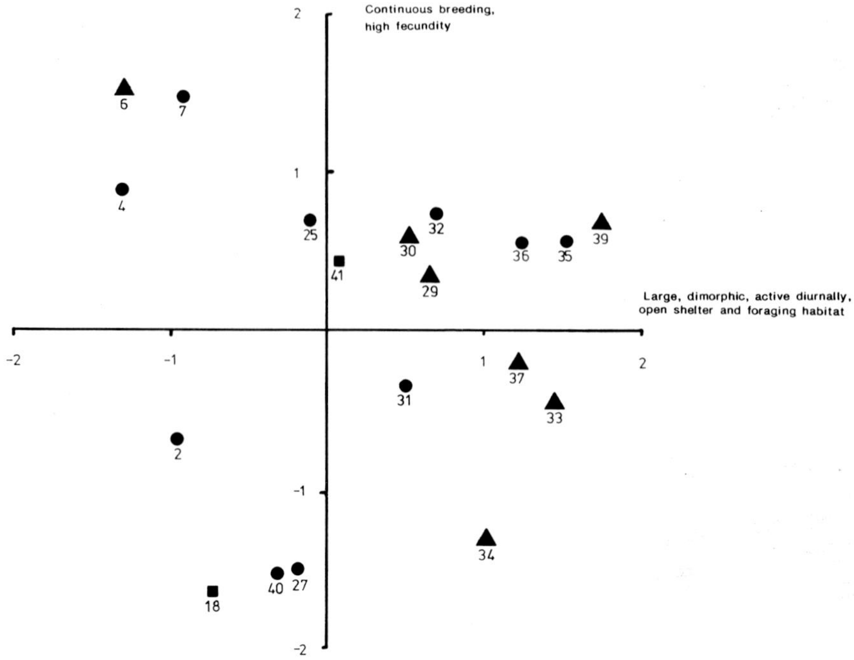

Fig. 4. Plot of factor scores from a Factor analysis of habitat, activity, size, sexual dimorphism and reproductive variables for 19 macropodoid species (Appendix 2). ■ social Type 1 species, ● social Type 2 species, ▲ social Type 3 species. KEY: 2, *Potorous tridactylus;* 4, *Bettongia penicillata;* 6, *B. lesueur;* 7, *Aepyprymnus rufescens;* 18, *P. xanthopus;* 25, *Thylogale thetis;* 27, *Macropus eugenii;* 29, *M. parryi;* 30, *M. dorsalis;* 31, *M. rufogriseus;* 32, *M. agilis;* 33, *M. giganteus;* 34, *M. fuliginosus;* 35, *M. robustus erubescens;* 36, *M. robustus robustus;* 37, *M. antilopinus;* 39, *M. rufus;* 40, *Setonix brachyurus;* 41, *Wallabia bicolor.*

two subspecies of *Macropus robustus* share similar characters to the other large gregarious *Macropus* species. However, they occupy elevated areas of high relief relative to sympatric Type 3 species (Croft 1981b; Taylor 1985). Croft (1987) argued that the velocity of a cursorial predator's attack would be less on steep, rocky and uneven slopes than flat ground, thus potentially reducing predation pressure on individuals of this species. *Macropus robustus robustus* may be a more selective feeder than *Macropus giganteus* in terms of the quality of grasses eaten (Taylor 1983a). Likewise *M. r. erubescens* may be more selective of grasses relative to their abundance than *Macropus rufus* (T. J. Dawson, pers. comm.). Favoured grasses are probably more dispersed and require longer search times on slopes than flood-plains or creek flats. Thus individual predation risk and diet characters not considered in this analysis may explain the anomalous position of the relatively asocial *Macropus robustus*. The medium-sized wallabies tend to group together on this axis. *Macropus parryi*, as Kaufmann (1974a) has argued, shares similar activity and habitat characters to the larger *Macropus* species and differentiates by being smaller and more often in open woodland. The other three species share similar characters; they use moderate cover as daytime rest sites and feed at night in more open habitats close to cover (C. N. Johnson 1985; Croft 1987). The anomalous species is therefore *Macropus dorsalis* which apparently form cohesive groups (Calaby 1966; C. N. Johnson 1985; T. J. Dawson, pers. comm.). Diet characters possibly resolve this anomaly (see next section). Dietary niche also probably discriminates the marginally positive *Wallabia bicolor*.

The remaining species spread along the negative end of Factor 1. They are small, nocturnal and occupy varying degrees of closed habitats. The latter as well as the degree of heteromorphism account for most of the separation between species. There are too few data on Type 1 species to resolve their peculiar characteristics relative to the other two social Types.

The second axis, Factor 2, explained a further 19.9% of the variance and related to reproductive characters. Seasonal breeders fall at the negative end and continuous breeders at the positive end. Small continuous breeders like *Aepyprymnus rufescens* are more fecund than large continuous breeders and exemplify the characters loading on this axis. This is a common relationship in mammals (Lee and Cockburn 1985). Reproductive characters do not obviously discriminate social Types.

Diet and sociality

Data on the average proportion of various plant categories, fungi, and insects were available for a subset of 12 species used in the previous analyses. The average proportions of the following

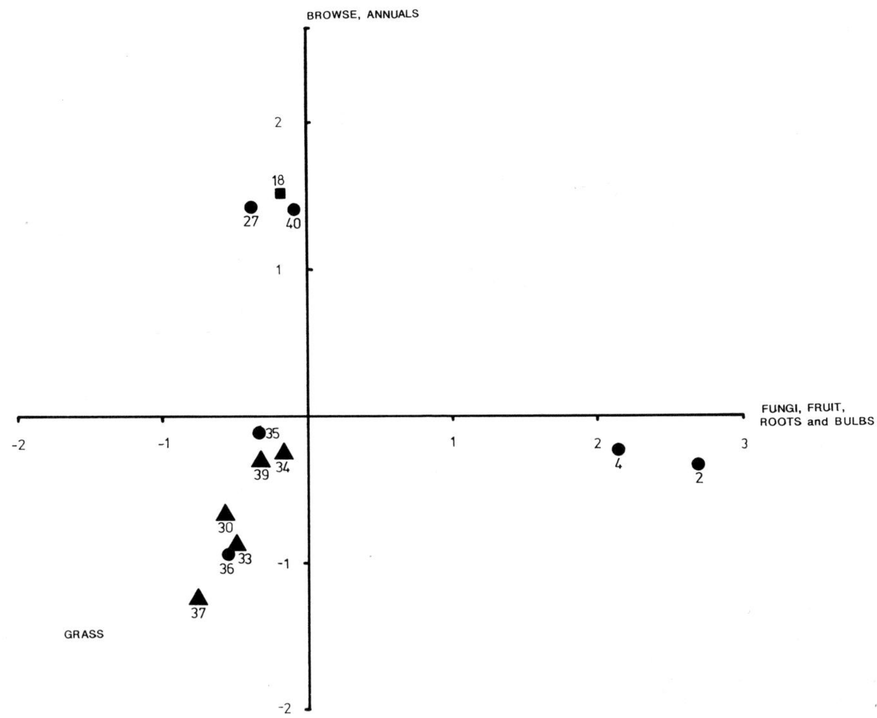

Fig. 5. Plot factor scores on factors 1 and 2 from a Factor analysis of diets of 12 macropodoid species (Appendix 3). ■ social Type 1 species, ● social Type 2 species, ▲ social Type 3 species. KEY: 2, *Potorous tridactylus;* 4, *Bettongia penicillata;* 18, *P. xanthopus;* 27, *Macropus eugenii;* 30, *M. dorsalis;* 33, *M. giganteus;* 34, *M. fuliginosus;* 35, *M. robustus erubescens;* 36, *M. robustus robustus;* 37, *M. antilopinus;* 39, *M. rufus;* 40, *Setonix brachyurus.*

categories in the diets of species were calculated from published data. The categories were: fungi, insects, fruit and seeds, roots and bulbs, forbs, annuals, shrubs, browse, grasses. For arid zone species, "good" times data were used rather then "bad" times (e.g., Barker 1987) as more representative of diet selection. Likewise, Taylor's (1983a) data from unimproved rather than improved pastures were used. Categories are mutually exclusive for single studies of sympatric species at the same site. Different classifications between studies means that this is not necessarily so for allopatric species or species in different sites. The categories annuals, shrubs and browse are least well differentiated. For example, the chenopodiaceous shrubs have both annual and perennial forms whose spatio-temporal distributions differ. Thus part of this shrub component is "annual", part "browsed" from perennial shrubs. Browse may include fallen leaves as well as classically "browsed" vegetation (T. J. Dawson, pers. comm.). "Succulents" found in some diets (e.g., *Setonix*) were classified as "annuals".

Proportions were arcsine transformed to approximate multivariate normality and a Factor (pc) analysis performed. The insect category had no variance and was dropped. Four factors were extracted explaining 36.4%, 18.2%, 14.2% and 12.7% of the variance, respectively. Factor 1 represents a high proportion of fungi, fruit and seeds, and bulbs and roots in the diet. Factor 2 represents high proportions of browse and annuals. Grasses loaded strongly negatively on both factors; thus the lower left-hand quadrant represents grazers. All social Type 3 species eat high proportions of grass (Fig. 5). The limited data on Type 1 and Type 2 species suggests that they separated on dietary characters. The potoroids fall into a distinctive diet class but they eat food which is discrete and requires relatively long search times and large effort to obtain (e.g., hypogeous fungi, roots and bulbs). All species apparently have similar diets (Strahan 1983) and will fall into a common class. The other Type 1 and 2 species separate from Type 3 by a higher proportion of browse and annuals in the diet. The general statements made about other species such as *Macropus agilis, M. rufogriseus* and *Wallabia bicolor,* and *Lagorchestes* in Strahan (1983) suggest that they would fall into this group. Likewise other *Petrogale* species probably have similar diets to *P. xanthopus*. Herbivores feeding on, at least, browse are argued to be more selective and have longer search times (Jarman 1974). *Macropus robustus* subspecies again fall amongst the Type 3 species. Reasons for their anomalous position have been discussed.

Plots of Factors 3 and 4 which related to shrubs and forbs, respectively, provided no obvious separation of species by social Type.

This analysis includes an insufficient diversity of species but points to diet as potentially an important correlate of sociality in the Macropodoids as argued by Kaufmann (1974a) and C. N. Johnson (1985). This aspect of their socio-ecology deserves further research but a common scale of measurement should be aimed for if interspecific comparisons are to be made (Jarman 1982). The actual proportion of the item selected in the diet is less important than its spatio-temporal distribution, search and handling times. This information will be difficult to gather for predominantly nocturnal species.

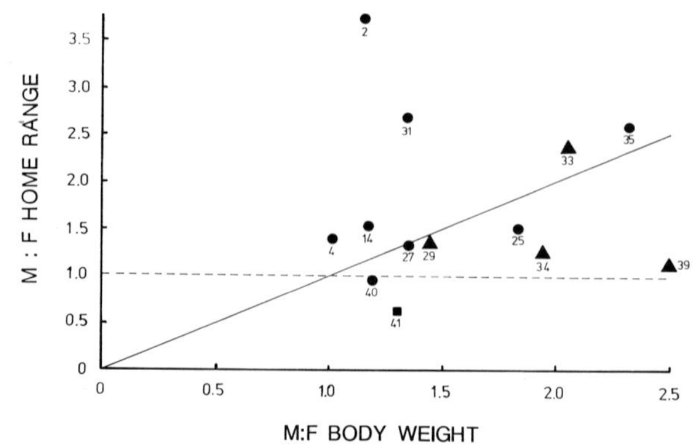

Fig. 6. The relationship between the ratio of male:female home-range size and male:female body weight for 13 macropodoid species. Solid line represents the relationship Y=X, broken line represents home-range ratio equal to unity. ■ social Type 1 species, ● social Type 2 species, ▲ social Type 3 species. Data from sources listed in Appendix 4. KEY: 2, *Potorous tridactylus;* 4, *Bettongia penicillata;* 14, *Petrogale penicillata;* 25, *Thylogale thetis;* 27, *Macropus eugenii;* 29, *M. parryi;* 31, *M. rufogriseus;* 33, *M. giganteus;* 34, *M. fuliginosus;* 35, *M. robustus erubescens;* 39, *M. rufus;* 40, *Setonix brachyurus;* 41, *Wallabia bicolor.*

Home range and spacing behaviour

Wrangham and Rubenstein (1986) cite intraspecific competition as an important factor in the evolution of social structures, at least in the sense of shaping such structures. An analysis of spatial relationships, associations and interactions between individuals provides some insight into this organizing factor, particularly in relation to mating systems. Home ranges of individuals in a number of macropodoid species have been determined and this information was used in the determination of social Types. The purpose of this section is to explore in more detail the social organization of Type 2 and 3 species. Interspecific comparisons of mammalian home ranges have focussed on relationships with body mass, food habits and group size (Harestad and Bunnell 1979; McNab 1983). Home-range size is not proportional to metabolic rate. Large mammals have even larger home ranges than the difference in mass from small mammals. These interesting relationships tend to ignore social factors as determinants of home range.

The disparate methods used in the calculation of home-range size in macropodoids make a regression on body mass dubious. Jaremovic and Croft (1987) discuss biases in the calculation of home-range size by various methods for a macropodid example. Ratios of male:female home-range size can be used without this methodological bias. The relationship with body mass was examined by regression on the ratio of male:female body weight with a range from homomorphic to markedly heteromorphic species (Fig. 6). No linear or non-linear model provided a slope significantly different from zero. No simple 1:1 scaling of male:female body weight and male:female home-range size was proven, in contrast to North American mammals (Harestad and Bunnell 1979). There are various possible reasons why no relationship was established, not the least of which is the quality of the data set (e.g., small samples, bias in representation of age/sex classes, bias towards the sedentary segment of the population). The one relevent to this paper is differences in social organization.

Leuthold (1977) categorized the social classes of African bovids by the pattern of home-range use by various age/sex classes. Gosling (1986) explored this further in relation to the evolution of mating strategies in male antelope. Male antelope that follow females will have home-range sizes similar to females. Males that move between territorial female groups will have much larger ranges than females. Males that defend female groups will have slightly larger home ranges than females as they repulse intruders or advertise territorial boundaries. Males that defend resources attractive to female groups will have smaller home ranges than the females who move through their territories. With the exception of *Wallabia bicolor* (the samples of locations and individuals were very small), male macropodoids studied have similar sized or more usually larger home ranges than females (Fig. 6). Most of these species have a mating system described as hierarchical promiscuity (Russell 1984). This is similar to male antelope who move between female groups but there is no conclusive evidence that female macropodoids are territorial. At most, C. N. Johnson (1986) describes matrilineal groups of *Macropus rufogriseus* whose home ranges overlap more within the group than between groups but group ranges are not overtly defended.

Males of social Type 2 species tend to have larger home ranges than their females (median=1.52, range=0.96-3.73, N=8) as compared to males of Type 3 species (median=1.29, range=1.12-2.36, N=4). This difference is not significant (U=9, P>0.05, Mann-Whitney

U test). Both social Types show a pattern of overlapping home ranges shared by both sexes.

In Type 2 Potoroids, nest sites are usually exclusive to an individual but feeding ranges overlap. *Potorous tridactylus* shows this pattern. The site studied contained a mixture of sedentary residents and individuals which used the site transiently (Kitchener 1973). *Bettongia penicillata* day ranges (these include nests) overlap by an average of 19.5% while night feeding ranges overlap by an average of 63.9% (Christensen 1980). Nest areas of males do not overlap. In Type 2 Macropodids, day rest sites are usually exclusive but night feeding areas overlap and may lead to aggregations on profitable patches. *Thylogale thetis* are unusual in having larger day than night ranges (K. A. Johnson 1980). Individuals disperse in the forest during the day but aggregate at night on pasture close to the forest edge. *Setonix brachyurus* on Rottnest Island usually defend individual rest sites although a male and female may share (Kitchener 1972). Holsworth (1967) determined a pattern of overlapping home ranges for groups of 25-150 individuals. *Petrogale penicillata* on Motutapu Island in New Zealand, aggregate on ledges of a rocky outcrop during the day and feed out from the rocks in overlapping ranges at night (Batchelor 1980). Rest sites were exclusive in 95.4% of 2 368 records and never shared by adult males. In *Macropus rufogriseus* in northeastern New South Wales, a similar pattern of overlapping home ranges occurs (C. N. Johnson 1987). Female offspring are philopatric and matrilineal clusters form in the overall pattern of dispersion of the population (C. N. Johnson 1986). Centres of activity may shift in relation to female reproductive class and the distribution of food. High size-ranked males centre their home ranges amongst the dense segments of the female population; low ranked males are more peripheral. *Macropus robustus* in northwestern New South Wales likewise has a pattern of overlapping home ranges (Fig. 7a). High ranking male ranges are more central to the female population than low ranking ones (Clancy unpubl.). Rest sites are usually exclusive and not shared by large adult males (Croft 1981b).

Type 3 macropodoids differ from Type 2 in that females are more likely to be aggregated. A male may therefore encompass more females within a smaller relative home range than Type 2 macropodoids. Type 3 species are also more likely to share rest sites although an individual distance is maintained (Croft 1981a). Kaufmann (1974b) showed considerable overlap in the home ranges of *Macropus parryi* although there was individual variation in use of the common area amongst adult females. Jaremovic (1984) found substantial overlap in the diurnal feeding ranges of female *Macropus giganteus*. Some males used the area similarly; others ranged more widely associating

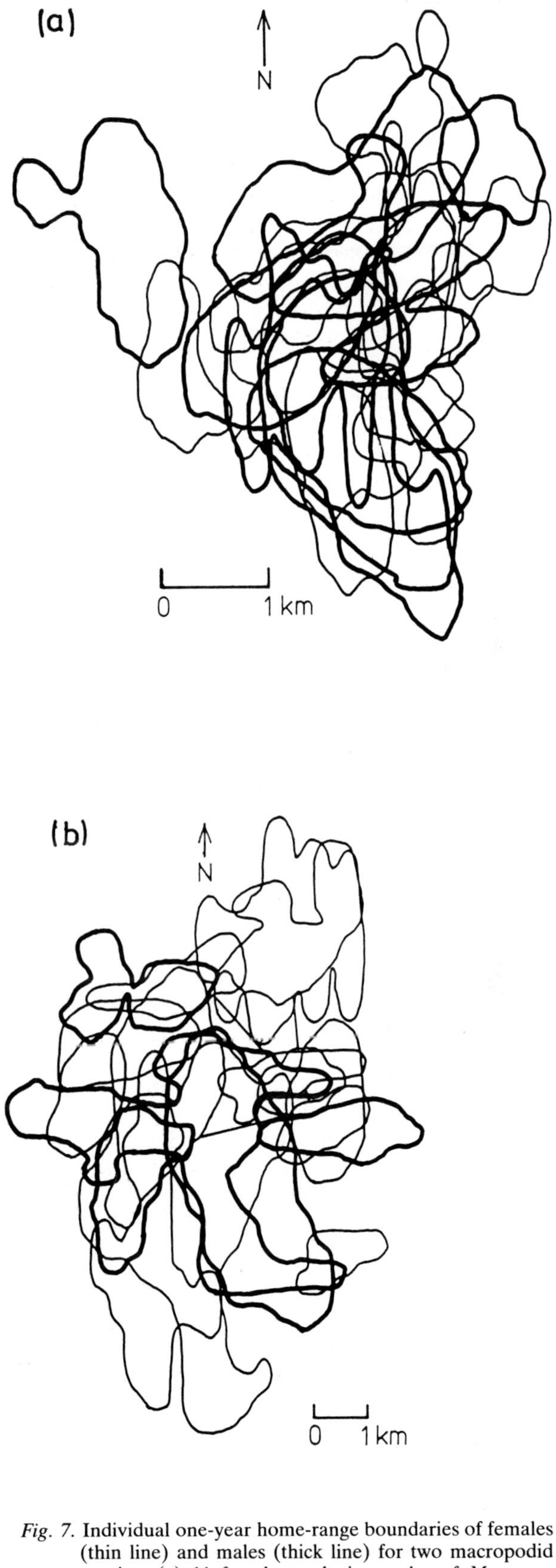

Fig. 7. Individual one-year home-range boundaries of females (thin line) and males (thick line) for two macropodid species: (a) 11 females and nine males of *Macropus robustus erubescens;* (b) nine females and five males of *M. rufus*. MAP (95) boundaries calculated as in Jaremovic and Croft (1987) from weekly sightings of radio-collared individuals captured at the same location for each species at Fowlers Gap.

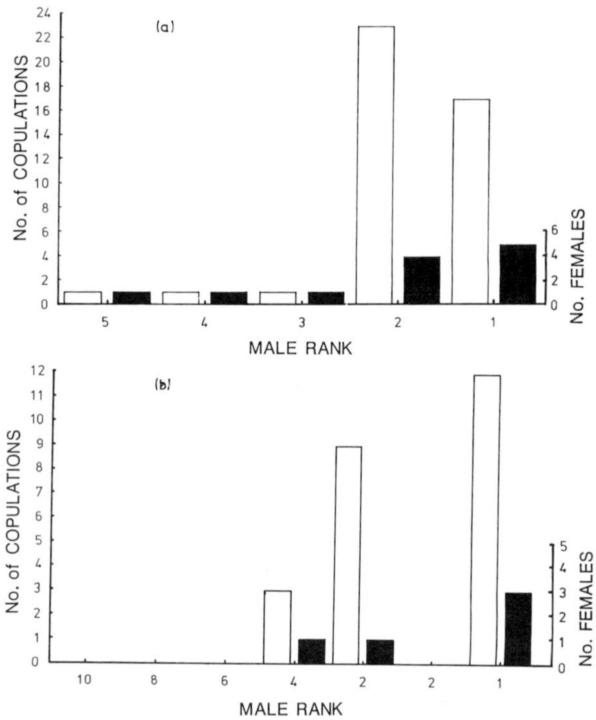

Fig. 8. Relationship between male dominance rank and either number of copulations (open bars) or number of females copulated with (solid bars) for two macropodid species: (a) *Petrogale penicillata* (data from Batchelor 1980), (b) *Macropus parryi* (data from Kaufmann 1974b).

with various female groups at different times. In *Macropus rufus*, neither males nor females maintain exclusive areas and home ranges overlap considerably (Fig. 7b).

In most mammals, the distribution of males is organized by the pattern of dispersion of females or the resources of highest quality to females (Wrangham and Rubenstein 1986). Macropodoids are no exception. Type 2 macropodoids feed on resources that are more discrete, scattered and incur higher costs in search and, possibly, handling time than Type 3 species. Type 2 females are more likely to be dispersed than aggregated. In the moderately heteromorphic Type 2 macropodids, males are organized hierarchically (Kitchener 1972; Batchelor 1980; Croft 1981b; C. N. Johnson 1985) with high-ranking individuals maintaining dominion over oestrous females encompassed by their home range. These high-ranked individuals copulate more and with more females then lower ranks (Fig. 8a). The dispersion of females probably makes it uneconomical to defend female "groups" continuously; although the constraints of limited shelter sites in *Petrogale* species might provide males with a defendable female "group" (Davies, pers. comm. to Russell 1984). The scattered dispersion of food resources probably makes it uneconomical to defend "female-preferred" areas particularly since oestrus is brief and breeding is usually clumped or continuous. Instead males compete throughout life to establish a reference for dominance and those of high status (correlated with body size) range more widely (C. N. Johnson 1987 and Fig. 9) and encompass more females than lower ranks. The behaviour of the homomorphic potoroids is not well-documented in the field. Jarman (1983) states that male *Aepyprymnus rufescens* emerge earlier in the evening than females and check the few females encompassed by their home range at their nest site. Male-male interactions in the field are not documented but males are relatively intolerant of each other in captivity (P. M. Johnson 1980b).

Type 3 macropodoid males potentially have defendable female aggregations. Females are usually quite sedentary (Kaufmann 1974b; Jaremovic 1984; Jarman and Southwell 1986;

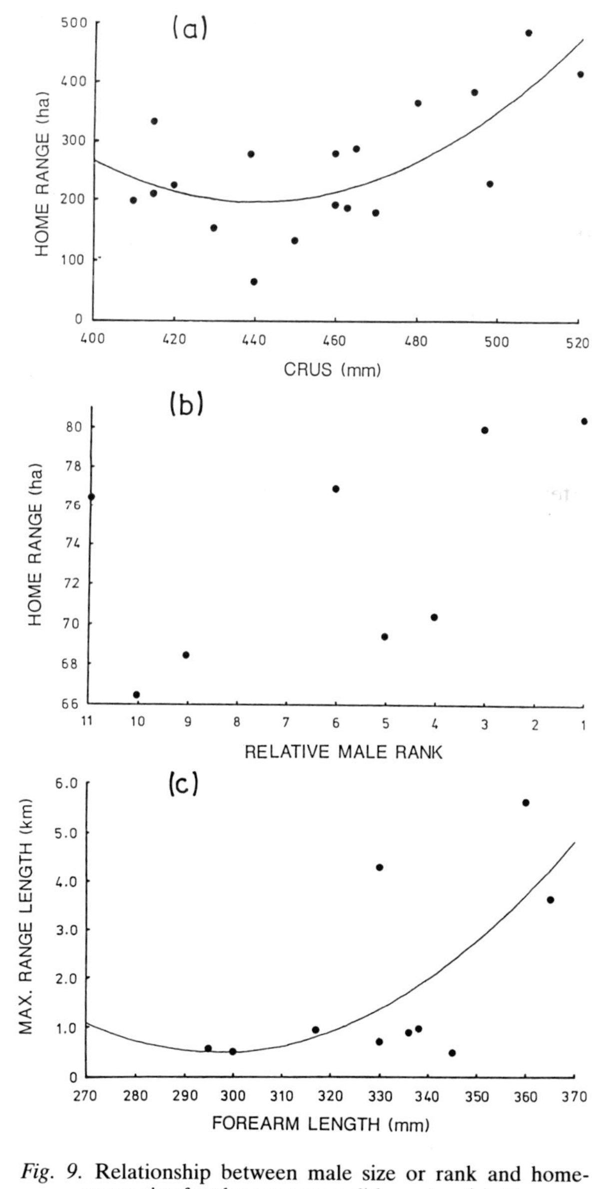

Fig. 9. Relationship between male size or rank and home-range size for three macropodid species: (a) *Macropus robustus erubescens* (data for annual home-range size and crus length measured at capture from Clancy unpubl.), (b) *M. parryi* (data from Kaufmann 1974b). (c) *M. giganteus* (data from Jaremovic 1984).

Priddel 1987) although associations between females are labile (Kaufmann 1974b; Croft 1981a; Jarman and Southwell 1986). Perhaps *Bettongia lesueur* males exploit the aggregation of females in burrows to form harems as Stodart (1966) suggests, but field evidence is lacking. Instead of expressing female-defence polygyny (Emlen and Oring 1977) as a mating system, these species display hierarchical promiscuity of a similar kind to Type 2 species. Male *Macropus parryi* of high rank copulate more and with more females than lower ranks (Fig. 8b) but in this population rank and home range size appear to be uncorrelated (Fig. 9b). Males of all classes intermingle in a mob (Kaufmann l974b). Jarman and Southwell (1986) found a similar structure in *Macropus giganteus*. An alpha male held tenure for about a year, ranged more frequently between females then subordinates, and enjoyed disproportionately higher mating success. In Jaremovic's (1984) study, very stable aggregations of female *M. giganteus* formed in discrete areas. The birth season in this region of southeastern New South Wales peaks strongly in spring-summer (Poole 1983) and so a male had a potentially defendable resource of oestrous females in this period. Although males of large size were disproportionately associated with female groups at this time relative to their abundance in the population, no such exclusive defense was observed. Rather, very large males consorted exclusively with oestrous females amidst other males of lesser size. The population density was locally very high (up to 200 individuals km^2), possibly swamping any attempt at female defence. As Jarman and Southwell (1986) note at their more northerly site, the probability of synchronous oestrus in two females is slight and so an alpha male loses little in consorting with and defending oestrous females individually rather than defending a group of potentially oestrous females. Male range length tends to increase with body size in *M. giganteus* (Fig. 9c). *Macropus rufus* males of high size-status associate disproportionately more with females likely to enter oestrus than lesser ranks (Johnson and Bayliss 1981). The mating system (Croft 1981a) is similar to that described for *M. giganteus* and more data are currently being gathered on this subject. The interesting aspect of *M. rufus* is that the male:female home-range ratio is lower than the other large macropodids. Both sexes in a population contain a mix of sedentary and transient individuals. Adult males and females can move distances in excess of 200 km (Bailey 1971; Priddel 1987; Croft, unpubl.). Populations in a single 812-ha paddock at Fowlers Gap have fluctuated between 7.4 to 35.5 individuals km^{-2} in three years (Edwards, unpubl.). The fluctuating intrusion of transient females into a large male's home range may compensate for matings that could be gained by ranging more widely across the home ranges of the sedentary females.

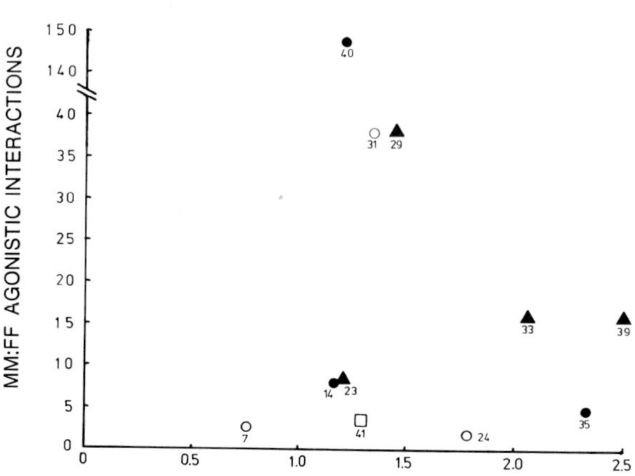

Fig. 10. Relationship between ratio of frequencies of male-male:female-female agonistic interactions and male:female body weight for 10 macropodoid species. ■ social Type 1 species, ● social Type 2 species, ▲ social Type 3 species. Open symbols, captive studies; solid symbols, field studies. Data from sources listed in Appendix 5. KEY: 7, *Aepyprymnus rufescens;* 14, *Petrogale penicillata;* 23, *Thylogale stigmatica;* 24, *T. billardierii;* 29, *M. parryi;* 31, *M. rufogriseus;* 33, *M. giganteus;* 35, *M. robustus erubescens;* 39, *M. rufus;* 40, *Setonix brachyurus;* 41, *Wallabia bicolor.*

Intraspecific competition shapes the social organization of macropodoids in two ways. There is probably a degree of feeding interference between individuals which is expressed in the female population by their dispersion. Type 2 species because of the distribution of their food are more dispersed than Type 3 except where highly profitable patches occur (e.g., the improved pasture in K. A. Johnson's 1980 study). Type 3 females maintain individual spacing, aggressively so on sparse winter pastures (Jaremovic 1984). High female density may lead to breeding suppression (C. N. Johnson 1986). Intense intrasexual competition occurs in males. For those species where agonistic interaction rates have been recorded the ratio of MM:FF interactions always exceeds 1 (Fig. 10). If the degree of heteromorphism is an expression of the intensity of male competition, we would expect this ratio to correlate positively with the ratio of M:F body weight. With the available mix of captive and field data this is not so. The data for *Setonix brachyurus* are probably anomalous because interactions were recorded at artificial feeding sites. Data on homomorphic species are also relatively lacking. The ratio was lower in Type 2 species (median=6.47, range=1.64-145, $\underline{N}$=5) than Type 3 species (median=16.1, range=16.1-38.3, $\underline{N}$=3) but the difference was not significant even if the outlying *S. brachyurus* value is removed (Mann-Whitney U tests). This trend is expected if Type 2 females maintain greater spacing through aggressive means in comparison to the more gregarious Type 3 females.

Males establish a reference for dominance in, usually undamaging, sparring contests (Croft 1981a; 1981b). This behaviour continues through a male's life, peaking in frequency when of intermediate size in captive *Macropus rufogriseus* (Watson, unpubl.). In relatively closed populations or "mobs" an assessment hierarchy in males may form (Kitchener 1972; Kaufmann 1974b; Batchelor 1980); in others, a confidence hierarchy probably forms (see Barnard and Burk (1979) for classification of dominance hierarchies). Males of high status centre their home ranges in the most dense segments of the female population and maintain a dominion for a year or more consorting and mating with most oestrous females and enjoying potentially high reproductive success. Females may exploit the relationships between males by increasing ranging on the periphery of their home range (C. N. Johnson 1985; Jarman and Southwell 1986; Clancy, unpubl.) during pro-oestrus, thereby increasing the likelihood of attracting an alpha male.

Some species are organized into relatively stable, closed-membership groups or "mobs" (Holsworth 1967; Kaufmann 1974b; Batchelor 1980; Jaremovic 1984; C. N. Johnson 1985). Holsworth argued that *Setonix brachyurus* groups occupied group territories but Nicholls (1971) concluded that group home-ranges were constrained by resource distributions not social factors. In other species with group home-ranges similar conclusions have been drawn and there is no evidence for defence of a group territory.

Sociality and predation

Predation risk has been cited as promoting aggregation in large open-habitat species (Kaufmann 1974a) and crypsis in small species (e.g., Heinsohn 1968). Predation pressure on macropodoids has generally been considered to be low relative to assemblages of mammalian herbivores in other zoogeographic regions (Calaby 1971). The relative isolation of the mother and young-at-foot unit in large macropodids has been cited as evidence that predation risk is low (Croft 1981a). However, other reasons related to maintenance of the mother-young bond have been proposed (Croft 1981a; 1981b) and supported in various species (Southwell 1984b; C. N. Johnson 1985). Jarman (1987) has shown in the gregarious *Macropus giganteus* that aggregation benefits the individual by increased total vigilance of the group. The young-at-foot is probably a vigilant companion since it feeds less (Croft 1981a) and is easily startled relative to the mother. Johnson (1987) has shown a shift in habitat selection by *Macropus rufogriseus* mothers to provide cover for their young-at-foot; this is not found in *M. giganteus* (Jarman and Southwell 1986).

Macropodoids universally show some behavioural adaptations advantageous in predator avoidance. Potoroids clearly demonstrate crypsis, especially at rest, to a potentially broad suite of reptilian, avian and mammalian predators. All species show some form of alarm behaviour, probably universally the foot thump, but also vocalizations in some species. The large open-habitat species flee in scattered, potentially confusing, patterns; with some co-ordination in the direction of group movements. Medium-sized species grazing in open habitats flee to nearby cover (Croft 1987).

The brief association of the macropodoid radiation with Man (40 000 y) and the Dingo, *Canis familiaris* (5 000 y), has been cited as too short to lead to significant evolutionary change in the behaviour of macropodoids in response to predation by these species (Russell 1984). However, behaviour is a labile character and, within the constraints of conservative morphological and physiological characters, such adaptations could occur. Wrangham and Rubenstein (1986) cite research by Watanabe (1981) on a leaf-eating monkey, *Nasalis concolor*, whose social organization has changed in populations hunted by Man in the last 2 000 y. Predation by Aboriginal man on Macropodoids has provided and continues to provide a significant protein source in the hunter-gatherer diet (e.g., Altman 1984). Various communal and solitary techniques were adapted to hunting macropodoids (Finlayson 1945; Meagher and Ride 1978). Current hunting for commercial utilization and pest control removes up to 12% of the population of the largest macropodids (Shepherd and Giles 1982). Dingoes preferentially prey on larger mammals, and if macropodoids are the most abundant, they form a significant part of the diet (Robertshaw and Harden 1986). The percent occurrence in dingo stomachs range from 54% (Robertshaw and Harden 1985) to 69% (Whitehouse 1977). Thus recent predation pressure on macropodoids is probably substantial.

Predation must be considered as an important factor shaping the current social organization of macropodoids (Croft 1987). Species which have experienced little predation by Man within a few decades may habituate to Man's presence (Jarman *et al.* 1987). Robertshaw and Harden (1986) compare the dispersion of *Macropus giganteus* populations of similar density but different predation pressure from Dingoes and find a smaller group size where pressure is lower. Wrangham and Rubenstein (1986) emphasize the lability of variation in intraspecific social organization with different solutions to subtly different pressures. The murky question of the scaling of predation risk in the evolution of macropodoid social organization deserves further study at intra-

and interspecific levels. This topic is developed further by Jarman and Coulson in this volume.

CONCLUSION

The striking feature of the social organizations of many species of macropodoids is their apparent disorder. Viewed at a distance, age/sex classes intermingle haphazardly, feeding ranges are not defended, mating is promiscuous. The broad array of patterns of dispersion and mating systems of like-sized bovids and cervids is comparatively narrow in the macropodoids. Even so these three herbivore assemblages share a common trend for the small, selective feeders to be dispersed and solitary, and the large unselective grazers to be aggregated and gregarious. Detailed analysis of the patterns of association of age/sex classes and individuals of various species (Croft 1981a, 1981b; C. N. Johnson 1983; Taylor 1983b; Southwell 1984b; Jarman and Southwell 1986; Croft 1987) reveal a subtle structure confirmed across some species in different study sites. Females relate to the distribution of their resources in ways which, we presume, maximize their reproductive success. Males overlap the distribution of females such that high-status individuals gain high reproductive success. Individuals of many species appear to prefer the company of familiars, constraining their movements within a common home range. Kinship between females may play a role in the local organizetion of female clusters (C. N. Johnson 1986).

The absence of any strong territoriality in macropodoids suggests that resources, with the exception of specific rest and shelter sites, are rarely economically defendable. We clearly need to know more about the spatio-temporal distribution and variation of macropodoid resources in order to contrast this behavioural difference with the frequently territorial bovids and cervids. For males, oestrous females also appear to be an unpredictable resource which is only economically defendable in the very short term and possible after a lengthy apprenticeship in the lower ranks. Perhaps the pattern of replacement of lost young and a subsequent oestrus found in most species (Tyndale-Biscoe and Renfree 1987) leads to this unpredictability. *Macropus rufogriseus rufogriseus* and *M. eugenii*, where quiescence results in strict birth seasons, provide a control for this effect but their mating system in the field is poorly documented. The comprehensive studies of the reproductive physiology of macropodoids have rarely considered reproductive behaviour. Proceptive behaviour in females is indicated in field observations (e.g., C. N. Johnson 1985; Jarman and Southwell 1986) but our knowledge of the relationships between proceptive behaviour, attractiveness and receptivity and the reproductive physiology of the female is scant. Clearly we need to know whether the peculiarities of macropodoid reproductive physiology lead to peculiarities in reproductive behaviour and hence social organization relative to placental counterparts.

Behavioural ecologists have focussed their effort on the medium-large sized species which are most readily observed. For these species there are still many mysteries but to complete our knowledge of the social organization of the Macropodoidea we clearly need innovative and dedicated field workers to discover much more about the potoroids and smaller macropodids.

ACKNOWLEDGEMENTS

I thank Peter Gillespie for drawing the figures. Tim Clancy, Duncan Watson, Glenn Edwards and Prof. Terry Dawson from the School of Biological Science, University of New South Wales, provided unpublished data used in some analyses and were valuable discussants of the ideas presented in this paper. I thank the referees for useful sugestions on improvements to this paper.

REFERENCES

ALTMAN, J. C., 1984. The dietary utilization of flora and fauna by contemporary hunter-gatherers at Momega Outstation, north-central Arnhem Land. *Aust. Aboriginal Studies* **1**: 35-46.

ANDREWARTHA, H. G. AND BARKER, S., 1969. Introduction to a study of the ecology of the Kangaroo Island wallaby, *Protemnodon eugenii* (Desmarest) within Flinders Chase, Kangaroo Island, South Australia. *Trans. Roy. Soc. S.A.* **93**: 127-32.

BAILEY, P. T., 1971. The red kangaroo, *Megaleia rufa* (Desmarest) in north-western New South Wales. I. Movements. *C.S.I.R.O. Wildl. Res.* **16**: 11-28.

BARKER, R. D., 1987. The diet of herbivores in the sheep rangelands. Pp. 69-83 *in* Kangaroos: their ecology and management in the sheep rangelands of Australia ed. by G. Caughley, N. Shepherd and J. Short. Cambridge University Press: Cambridge.

BARNARD, C. J. AND BURK, T., 1979. Dominance hierarchies and the evolution of "individual recognition". *J. Theor. Biol.* **81**: 65-72.

BATCHELOR, T. A., 1980. The social organization of the brush-tailed rock-wallaby *(Petrogale penicillata penicillata)* on Motutapu Island. M.Sc. Thesis, University of Auckland, Auckland.

CALABY, J. H., 1966. Mammals of the Upper Richmond and Clarence Rivers, New South Wales. *C.S.I.R.O. Div. Wildl. Res.* Technical Paper No. 10.

CALABY, J. H., 1971. Man, fauna and climate in aboriginal Australia. Pp. 80-93 *in* Aboriginal Man and Environment in Australia ed. by J. D. Mulvaney and J. Golson. A.N.U. Press: Canberra.

CALABY, J. H. AND POOLE, W. E., 1971. Keeping kangaroos in captivity. *Intern. Zoo. Yrbk.* **11**: 5-12.

CHRISTENSEN, P. E. S., 1980. The biology of *Bettongia penicillata* Gray, 1837, and *Macropus eugenii* Desmarest, 1804, in relation to fire. *Forests Department of Western Australia,* Bulletin No. 91.

CLANCY, T. F., 1982. Aspects of the behaviour of the red-bellied pademelon *Thylogale billardierii* (Desmarest 1822) in captivity. B.Sc. (Hons.) Thesis, University of Tasmania, Hobart.

CLUTTON-BROCK, T. H. AND HARVEY, P. H., 1977. Primate ecology and social organization. *J. Zool. Lond.* **183:** 1-39.

CLUTTON-BROCK, T. H. AND HARVEY, P. H., 1984. Comparative approaches to investigating adaptations. Pp.7-29 *in* Behavioural Ecology: An Evolutionary Approach, 2nd Edition ed. by J. R. Krebs and N. B. Davies. Blackwell: Oxford.

COOKE, B. N., 1979. Field observations of the behaviour of the macropod marsupial *Thylogale stigmatica* (Gould). M.Sc. Thesis, University of Queensland, Brisbane.

COPLEY, P. B. AND ROBINSON, A. C., 1983. Studies on the yellow-footed rock-wallaby, *Petrogale xanthopus* Gray (Marsupialia: Macropodidae). II. Diet. *Aust. Wildl. Res.* **10:** 63-76.

CREBBIN, A., 1982. Social organization and behaviour of the swamp wallaby, *Wallabia bicolor* (Desmarest) (Marsupialia: Macropodidae) in captivity. B.Sc. (Hons.) Thesis, University of New South Wales, Sydney.

CROFT, D. B., 1981a. Behaviour of red kangaroos, *Macropus rufus* (Desmarest, 1822), in north-western New South Wales, Australia. *Aust. Mammal.* **4:** 5-58.

CROFT, D. B., 1981b. Social behaviour of the euro, *Macropus robustus* (Gould), in the Australian arid zone. *Aust. Wildl. Res.* **8:** 13-49.

CROFT, D. B., 1982. Some observations on the behaviour of the antilopine wallaroo, *Macropus antilopinus*. *Aust. Mammal.* **5:** 5-15.

CROFT, D. B., 1987. Socio-ecology of the antilopine wallaroo, *Macropus antilopinus,* in the Northern Territory, with observations on sympatric *M. robustus woodwardii* and *M. agilis*. *Aust. Wildl. Res.* **14:** 243-55.

CROOK, J. H., ELLIS, J. E. AND GOSS-CUSTARD, J. D., 1976. Mammalian social systems: structure and function. *Anim. Behav.* **24:** 261-74.

DAHL, K., 1897. Biological notes on north Australian mammalia. *The Zoologist* **671:** 189-216.

DIXON, J. M. AND HUXLEY, L. (eds) 1985. Donald Thompson's Mammals and Fishes of Northern Australia. Nelson: Melbourne.

DUNBAR, R. I. M., 1982. Intraspecific variation in mating strategy. Pp. 385-431 *in* Perspectives in Ethology Vol. 5 ed. by P. P. G. Bateson and P. H. Klopfer. Plenum Press: New York.

DWYER, P. D., 1972. Social organization of a population of rock-wallabies, *Petrogale inornata. Aust. Mammal.* **1:** 72.

EALEY, E. M. H., 1967. Ecology of the euro, *Macropus robustus* (Gould), in north-western Australia. II. Behaviour, movements and drinking patterns. *C.S.I.R.O. Wildl. Res.* **12:** 27-51.

EDWARDS, G. P. AND EALEY, E. M. H., 1974. Aspects of the ecology of the swamp wallaby *Wallabia bicolor* (Marsupialia: Macropodidae). *Aust. Mammal.* **1:** 307-17.

EISENBERG, J. F., MUCKENHIRN, N. A. AND RUDRAN, R., 1972. The relation between ecology and social structure in primates. *Science* **176:** 863-74.

ELLIS, B. A., RUSSELL, E. M., DAWSON, T. J. AND HARROP, C. J. F., 1977. Seasonal changes in diet preferences of free-ranging red kangaroos, euros and sheep in western New South Wales. *Aust. Wildl. Res.* **4:** 127-44.

EMLEN, S. T. AND ORING, L. W., 1977. Ecology, sexual selection, and the evolution of mating systems. *Science* **197:** 215-23.

FINLAYSON, H. H., 1945. The Red Centre: Man and beast in the heart of Australia. Halstead Press: Sydney.

GEIST, V., 1974. On the relationships of social evolution and ecology in ungulates. *Amer. Zool.* **14:** 205-20.

GOSLING, I. M., 1986. The evolution of mating strategies in male antelopes. Pp. 244-81 *in* Ecological Aspects of Social Evolution ed. by D. I. Rubenstein and R. W. Wrangham. Princeton University Press: Princeton.

GUILER, E. R., 1971. Food of the potoroo (Marsupialia, Macropodidae). *J. Mammal.* **52:** 232-34.

HARESTAD, A. S. AND BUNNELL, F. L., 1979. Home range and body weight — a re-evaluation. *Ecology* **60:** 389-402.

HEINSOHN, G. E., 1968. Habitat requirements and reproductive potential of the macropod marsupial *Potorous tridactylus* in Tasmania. *Mammalia* **32:** 30-43.

HINDE, R. A., 1976. Interactions, relationships and social structure. *Man* **11:** 295-311.

HOLSWORTH, W. N., 1967. Population dynamics of the quokka, *Setonix brachyurus,* on the west end of Rottnest Island, Western Australia. I. Habitat and the distribution of the quokka. *Aust. J. Zool.* **15:** 29-46.

HUGHES, R. L., 1962. Reproduction in the macropod marsupial *Potorous tridactylus* (Kerr). *Aust. J. Zool.* **13:** 533-43.

HUME, I. D., 1982. Digestive Physiology and Nutrition of Marsupials. Cambridge University Press: Cambridge.

INNS, R. W., 1980. Ecology of the Kangaroo Island wallaby, *Macropus eugenii* (Desmarest), in Flinders Chase National Park, Kangaroo Island. Ph.D. Thesis, University of Adelaide, Adelaide.

JAREMOVIC, R. V., 1984. Space and time related behaviour in eastern grey kangaroos (*Macropus giganteus* Shaw). Ph.D. Thesis, University of New South Wales, Sydney.

JAREMOVIC, R. V. AND CROFT, D. B., 1987. Comparison of techniques to determine eastern grey kangaroo home range. *J. Wildl. Manage.* **51:** 921-30.

JARMAN, P. J., 1974. The social organization of antelope in relation to their ecology. *Behaviour* **58:** 215-67.

JARMAN, P. J., 1982. Prospects for interspecific comparison in sociobiology. Pp. 323-42 *in* Current Problems in Sociobiology ed. by King's College Sociobiology Group. Cambridge University Press: Cambridge.

JARMAN, P. J., 1983. Mating system and sexual dimorphism in large, terrestrial, mammalian herbivores. *Biol. Rev.* **58:** 485-520.

JARMAN, P. J., 1987. Group size and activity in eastern grey kangaroos. *Anim. Behav.* **35:** 1044-50.

JARMAN, P. J., JOHNSON, C. N., SOUTHWELL, C. J. AND STUART-DICK, R., 1987. Macropod studies at Wallaby Creek. I. The area and animals. *Aust. Wildl. Res.* **14:** 1-14.

JARMAN, P. J. AND SOUTHWELL, C. J., 1986. Grouping, associations, and reproductive strategies in eastern grey kangaroos. Pp. 399-428 *in* Ecological Aspects of Social Evolution ed. by D. I. Rubenstein and R. W. Wrangham. Princeton University Press: Princeton.

JOHNSON, C. N., 1983. Variation in group size and composition in red and western grey kangaroos, *Macropus rufus* (Desmarest) and *M. fuliginosus* (Desmarest). *Aust. Wildl. Res.* **10**: 25-31.

JOHNSON, C. N., 1985. Ecology, social behaviour and reproductive success in a population of red-necked wallabies. Ph.D. Thesis, University of New England, Armidale.

JOHNSON, C. N., 1986. Philopatry, reproductive success of females and maternal investment in the red-necked wallaby. *Behav. Sociobiol. Ecol.* **19**: 143-50.

JOHNSON, C. N., 1987. Macropod studies at Wallaby Creek. IV. Home range and movements of the red-necked wallaby. *Aust. Wildl. Res.* **14**: 125-32.

JOHNSON, C. N. AND BAYLISS, P. G., 1981. Habitat selection by sex, age and reproductive class in the red kangaroo, *Macropus rufus* in western New South Wales. *Aust. Wildl. Res.* **8**: 465-74.

JOHNSON, K. A., 1977. Ecology and management of the red-necked pademelon, *Thylogale thetis,* on the Dorrigo plateau of northern New South Wales. Ph.D. Thesis, University of New England, Armidale.

JOHNSON, K. A., 1980. Spatial and temporal use of habitat by the red-necked pademelon, *Thylogale thetis* (Marsupialia: Macropodidae). *Aust. Wildl. Res.* **7**: 157-66.

JOHNSON, P. M., 1978. Studies of Macropodidae in Queensland. 9. Reproduction of the rufous rat-kangaroo (*Aepyprymnus rufescens* (Gray)) in captivity with age estimation of pouch young. *Qld. J. Agric. Anim. Sci.* **35**: 69-72.

JOHNSON, P. M., 1980a. Field observations on group compositions in the agile wallaby, *Macropus agilis* (Gould) (Marsupialia: Macropodidae). *Aust. Wildl. Res.* **7**: 327-31.

JOHNSON, P. M., 1980b. Observations of the behaviour of the rufous rat-kangaroo, *Aepyprymnus rufescens. Aust. Wildl. Res.* **7**: 347-67.

KAUFMANN, J. H., 1974a. The ecology and evolution of social organization in the kangaroo family (Macropodidae). *Amer. Zool.* **14**: 51-62.

KAUFMANN, J. H., 1974b. Social ethology of the whiptail wallaby, *Macropus parryi,* in north-eastern New South Wales. *Anim. Behav.* **22**: 281-369.

KIRKPATRICK, T. H. AND JOHNSON, P. M., 1969. Studies on Macropodidae in Queensland. 7. Age estimation and reproduction in the agile wallaby (*Wallabia agilis* (Gould)). *Qld. J. Agric. Anim. Sci.* **26**: 691-98.

KITCHENER, D. J., 1972. The importance of shelter to the quokka, *Setonix brachyurus* (Marsupialia) on Rottnest Island. *Aust. J. Zool.* **20**: 281-99.

KITCHENER, D. J., 1973. Notes on home range and movement in two small macropods, the potoroo *(Potorous apicalis)* and the quokka *(Setonix brachyurus). Mammalia* **37**: 231-40.

LEE, A. K. AND COCKBURN, A., 1985. Evolutionary Ecology of Marsupials. Cambridge University Press: Cambridge.

LEUTHOLD, W., 1977. African Ungulates: A Comparative Review of Their Ethology and Behavioural Ecology. Springer-Verlag: Berlin.

MAYNES, G. M., 1973. Aspects of reproduction in the whiptail wallaby *Macropus parryi. Aust. Zool.* **18**: 43-46.

MCBRIDE, G., 1976. The study of social organizations. *Behaviour* **59**: 96-115.

MCNAB, B. K., 1983. Ecological and behavioural consequences of adaptation to various food resources. Pp. 664-97 *in* Advances in the Study of Mammalian Behaviour ed. by J. F. Eisenberg and D. G. Kleiman. American Society of Mammalogists: Washington DC.

MEAGHER, S. J. AND RIDE, W. D. L., 1978. Use of natural resources by the aborigines of south-western Australia. Pp. 66-80 *in* Aborigines of the West. Their Past and Their Present ed. by R. M. Berndt and C. H. Berndt. University of Western Australia Press: Perth.

MERCHANT, J. C. AND CALABY, J. H., 1981. Reproductive biology of the red-necked wallaby *(Macropus rufogriseus banksianus)* and Bennett's wallaby *(M. r. rufogriseus)* in captivity. *J. Zool. Lond.* **194**: 203-17.

MURPHY, C. R. AND SMITH, J. R., 1970. Age determination of pouch young and juvenile Kangaroo Island wallabies. *Trans. Roy. Soc. S.A.* **94**: 15-20.

NICHOLS, D. G., 1971. Daily and seasonal movements of the quokka, *Setonix brachyurus* (Marsupialia) on Rottnest Island. *Aust. J. Zool.* **19**: 215-26.

NORUSIS, M. J., 1985. SPSSX: Advanced Statistics Guide. McGraw-Hill: New York.

PARKER, P., 1977. An ecological comparison of marsupial and placental patterns of reproduction. Pp. 273-86 *in* Biology of Marsupials ed. by B. Stonehouse and D. Gilmore. Macmillan: London.

POOLE, W. E., 1975. Reproduction in the two species of grey kangaroos, *Macropus giganteus* Shaw, and *M. fuliginosus* (Desmarest). II. Gestation, parturition and pouch life. *Aust. J. Zool.* **23**: 333-53.

POOLE, W. E., 1976. Breeding biology and current status of the grey kangaroos, *Macropus fuliginosus fuliginosus,* of Kangaroo Island, South Australia. *Aust. J. Zool.* **24**: 169-87.

POOLE, W. E., 1983. Breeding in the grey kangaroo, *Macropus giganteus,* from widespread locations in eastern Australia. *Aust. Wildl. Res.* **10**: 453-66.

POOLE, W. E. AND CATLING, P. C., 1974. Reproduction in two species of grey kangaroo, *Macropus giganteus* Shaw and *M. fuliginosus* (Desmarest). I. Sexual maturity and oestrus. *Aust. J. Zool.* **22**: 277-302.

POOLE, W. E., CARPENTER, S. M. AND WOOD, J. T., 1982a. Growth of grey kangaroos and the reliability of age determination from body measurements. I. The eastern grey kangaroo, *Macropus giganteus. Aust. Wildl. Res.* **9**: 9-20.

POOLE, W. E., CARPENTER, S. M. AND WOOD, J. T., 1982b. Growth of grey kangaroos and the reliability of age determination from body measurements. I. The western grey kangaroos, *Macropus fuliginosus fuliginosus, M. f. melanops* and *M. f. ocydromus. Aust. Wildl. Res.* **9** 203-12.

POOLE, W. E. AND MERCHANT, J. C., 1987. Reproduction in captive wallaroos: the eastern wallaroo, *Macropus robustus robustus,* the euro, *M. r. erubescens* and the antilopine wallaroo, *M. antilopinus. Aust. Wildl. Res.* **14**: 225-42.

POOLE, W. E., MERCHANT, J. C., CARPENTER, S. M. AND CALABY, J. H., 1985. Reproduction, growth and age determination in the yellow-footed rock-wallaby *Petrogale xanthopus* Gray, in captivity. *Aust. Wildl. Res.* **12**: 127-36.

PRIDDEL, D., 1987. The mobility and habitat utilization of kangaroos. Pp. 100-118 *in* Kangaroos: their ecology and management in the sheep rangelands of Australia ed. by G. Caughley, N. Shepherd and J. Short. Cambridge University Press: Cambridge.

PROCTOR-GRAY, E. AND GANSLOSSER, U., 1986. The individual behaviour of Lumholtz's tree-kangaroo: repertoire and taxonomic implications. *J. Mammal.* **67**: 343-52.

PULLIAM, H. R. AND CARACO, T., 1984. Living in groups: Is there an optimal group size? Pp. 122-47 *in* Behavioural Ecology: An Evolutionary Approach 2nd Edition ed. by J. R. Krebs and N. B. Davies. Blackwell: Oxford.

RIDE, W. D. L. AND TYNDALE-BISCOE, C. H., 1962. Mammals. Pp. 54-97 *in* The Results of an Expedition to Bernier and Dorre Islands, Shark Bay, Western Australia ed. by A. J. Fraser. Fauna Bulletin No. 2: Fisheries Department of Western Australia.

RIDLEY, M., 1986. The number of males in a primate troop. *Anim. Behav.* **34**: 1848-58.

ROBERTSHAW, J. D. AND HARDEN, R. H., 1985. The ecology of the dingo in north-eastern New South Wales. II. Diet. *Aust. Wildl. Res.* **12**: 39-50.

ROBERTSHAW, J. D. AND HARDEN, R. H., 1986. The ecology of the dingo in north-eastern New South Wales. IV. Prey selection by dingoes, and its effect on the major prey species, the swamp wallaby *Wallabia bicolor* (Desmarest). *Aust. Wildl. Res.* **13**: 141-64.

RUBENSTEIN, D. I. AND WRANGHAM, R. W., 1986. Ecological Aspects of Social Evolution. Princeton University Press: Princeton.

RUSSELL, E. M., 1974. The biology of kangaroos (Marsupialia-Macropodidae). *Mammal Rev.* **4**: 1-59.

RUSSELL, E. M., 1982. Patterns of parental care and parental investment in marsupials. *Biol. Rev.* **47**: 423-86.

RUSSELL, E. M., 1984. Social behaviour and social organization of marsupials. *Mammal Rev.* **14**: 101-54.

RUSSELL, E. M. AND RICHARDSON, B. J., 1971. Some observations on the breeding, age structure, dispersion and habitat of populations of *Macropus robustus* and *Macropus antilopinus* (Marsupialia). *J. Zool. Lond.* **165**: 131-42.

SADLEIR, R. M. F. S., 1965. Reproduction in two species of kangaroo *Macropus robustus* and *Megaleia rufa* in the arid Pilbara region of Western Australia. *Proc. Zool. Soc. Lond.* **145**: 239-61.

SANSON, G. D., 1978. The evolution and significance of mastication in the Macropodidae. *Aust. Mammal.* **2**: 23-28.

SHARMAN, G. B., 1955a. Studies on marsupial reproduction. II. The oestrous cycle of *Setonix brachyurus. Aust. J. Zool.* **3**: 44-55.

SHARMAN, G. B., 1955b. Studies on marsupial reproduction. III. Normal and delayed pregnancy in *Setonix brachyurus. Aust. J. Zool.* 3: 56-70.

SHARMAN, G. B. AND CALABY, J. H., 1964. Reproductive behaviour in the red kangaroo, *Megaleia rufa,* in captivity. *C.S.I.R.O. Wildl. Res.* **9**: 58-85.

SHARMAN, G. B. AND PILTON, P. E., 1964. The life history and reproduction of the red kangaroo *(Megaleia rufa). Proc. Zool. Soc. Lond.* **142**: 29-48.

SHAW, G. AND ROSE, R. W., 1979. Delayed gestation in the potoroo *Potorous tridactylus* (Kerr). *Aust. J. Zool.* **27**: 901-12.

SHEPHERD, N. AND GILES, J., 1982. Kangaroo management in new South Wales. *Parks and Wildlife* "Kangaroos". Pp. 50-55.

SHIELD, J., 1964. A breeding season difference in two populations of the Australian macropod marsupial *(Setonix brachyurus). J. Mammal.* **45**: 616-25.

SHORT, J. C., 1980. Ecology of the brush-tail rock-wallaby, *Petrogale penicillata* (Griffith, Smith and Pidgeon). M.Sc. Thesis, University of Sydney, Sydney.

SOKAL, R. R. AND ROHLF, F. J., 1969. "Biometry". W. H. Freeman: San Francisco.

SOUTHWELL, C. J., 1984a. Variability in grouping in the eastern grey kangaroo, *Macropus giganteus.* I. Group density and group size. *Aust. Wildl. Res.* **11**: 423-35.

SOUTHWELL, C. J., 1984b. Variability in grouping in the eastern grey kangaroo, *Macropus giganteus.* II. Dynamics of group formation. *Aust. Wildl. Res.* **11**: 437-49.

STODART, E., 1966. Observations on the behaviour of the marsupial *Bettongia lesueuri* (Quoy and Gaimard) in an enclosure. *C.S.I.R.O. Wildl. Res.* **11**: 91-101.

STORR, G. M., 1964. Studies on marsupial nutrition. IV. Diet of the quokka, *Setonix brachyurus* (Quoy and Gaimard), on Rottnest Island, Western Australia. *Aust. J. Biol. Sci.* **17**: 469-81.

STRAHAN, R. (ed.), 1983. The Australian Museum Complete Book of Australian Mammals. Angus & Robertson: Sydney.

TAYLOR, R. J., 1983a. The diet of the eastern grey kangaroo and wallaroo in areas of improved and native pasture in the New England Tablelands. *Aust. Wildl. Res.* **10**: 203-11.

TAYLOR, R. J., 1983b. Association of social classes of the wallaroo, *Macropus robustus* (Marsupialia: Macropodidae). *Aust. Wildl. Res.* **10**: 229-37.

TAYLOR, R. J., 1985. Habitat use by the eastern grey kangaroo and wallaroo in an area of sympatry. *Mammalia* **49**: 173-86.

TYNDALE-BISCOE, C. H., 1968. Reproduction and post-natal development in the marsupial *Bettongia lesueur* (Quoy and Gaimard). *Aust. J. Zool.* **16**: 577-602.

TYNDALE-BISCOE, C. H. AND RENFREE, M., 1987. Reproductive Physiology of Mammals. Cambridge University Press: Cambridge.

WITTENBERGER, J. F., 1979. The evolution of mating systems in birds and mammals. Pp. 271-349 *in* Handbook of Behavioural Neurobiology, Vol. 3 ed. by P. Marler and J. G. Vandenbergh. Plenum Press: New York.

WHITEHOUSE, S. J. O., 1977. The diet of the dingo in Western Australia. *Aust. Wildl. Res.* **4**: 145-50.

WRANGHAM, R. W. AND RUBENSTEIN, D. I., 1986. Social evolution of birds and mammals. Pp. 452-70 *in* Ecological Aspects of Social Evolution ed. by D. I. Rubenstein and R. W. Wrangham. Princeton University Press: Princeton.

APPENDIX 1

Social system and habitat variables used in the analysis of macropodoid social types with references to data sources. ID, species identity number; G, group size; S, group structure; R, home range overlap pattern; H, habitat type. Variables and their values are defined in text.

Species		Inactive				Active				Reference
	ID	G	S	R	H	G	S	R	H	
Potorous tridactylus	2	1	1	1	2	3	4	4	1	Heinsohn (1968), Kitchener (1973)
Bettongia penicillata	4	1	1	1	2	3	4	4	2	Christensen (1980)
Bettongia lesueur	6	4	3	3	1	4	4	4	2	Dahl (1897), Ride & Tyndale-Biscoe (1962), Stodard (1966)
Aepyprymnus rufescens	7	1	1	1	2	3	4	4	4	Jarman (1983), Strahan (1983)
Lagostrophus fasciatus	10	1	1	3	4	3	3	3	3	Strahan (1983)
Petrogale penicillata	14	1	1	1	3	4	4	4	5	Batchelor (1980), Short (1980)
Petrogale xanthopus	18	1	1	1	3	3	3	3	5	Hornsby (Unpubl.)
Thylogale thetis	25	1	1	4	4	4	4	4	3	Johnson K. A. (1980)
Macropus eugenii	27	1	1	1	4	4	4	4	5	Christensen (1980, Inns (1980)
Macropus parryi	29	4	4	4	5	4	4	4	4	Kaufmann (1974b)
Macropus dorsalis	30	4	4	4	5	4	4	4	3	Strahan (1983), Johnson C. N. (1985)
Macropus rufogriseus	31	1	1	3	4	3	4	4	5	Johnson C. N. (1985)
Macropus agilis	32	1	1	1	4	4	4	4	5	Johnson P. M. (1980a) Croft (1987)
Macropus giganteus	33	4	4	4	5	4	4	4	5	Jaremovic (1984)
Macropus fuliginosus	34	4	4	4	5	4	4	4	4	Johnson C. N. (1983), Priddel (1987)
M. robustus erubescens	35	1	1	1	6	3	4	4	5	Croft (1981b, unpubl.)
M. robustus robustus	36	1	1	1	5	3	4	4	4	Taylor (1983b)
Macropus antilopinus	37	4	4	4	5	4	4	4	5	Croft (1982, 1987)
Macropus rufus	39	4	4	4	6	4	4	4	5	Croft (1981a, unpubl.)
Setonix brachyurus	40	1	1	1	4	4	4	4	5	Holsworth (1967), Kitchener (1972)
Wallabia bicolor	41	1	1	1	4	3	2	3	1	Edwards and Ealey (1974), Robertshaw and Harden (1986)
Dendrolagus lumholtzi	42	1	1	1	5	1	2	3	1	Dixon and Huxley (1985), Proctor-Gray and Ganslosser (1986)

APPENDIX 2

Activity (Act), mean female body weight (F.B. Wt) and male:female body-weight ratio (M:F ratio) variables used for analysis in Fig. 4 with references to sources. Values for activity variable are defined in the text.

Species	ID	Act[1]	F.B. Wt[2] (kg)	M:F ratio[2]	Reference
Potorous tridactylus	2	2	1.02	1.16	Strahan (1983)
Bettongia pencillata	4	2	1.33	1.02	Christensen (1980)
Bettongia lesueur	6	2	1.1	1	Strahan (1983)
Aepyprymnus rufescens	7	2	3.6	0.76	Strahan (1983)
Lagostrophus fasciatus	10	2	3.0	0.83	Strahan (1983)
Petrogale penicillata	14	2	5.8	1.18	Batchelor (1980)
Thylogale thetis	25	2	3.8	1.84	Strahan (1983)
Macropus eugenii	27	2	5.5	1.36	Strahan (1983)
Macropus parryi	29	3	11.0	1.45	Strahan (1983)
Macropus dorsalis	30	2	6.5	2.46	Strahan (1983)
Macropus rufogriseus[3]	31	3	13.8	1.35	Strahan (1983)
Macropus agilis	32	3	11.0	1.73	Strahan (1983)
Macropus giganteus	33	3	32.0	2.06	Strahan (1983)
Macropus fuliginosus	34	3	27.5	1.95	Strahan (1983)
M. robustus erubescens	35	3	20.0	2.33	Strahan (1983)
M. robustus robustus	36	3	20.0	2.33	Strahan (1983)
Macropus antilopinus	37	3	20.0	2.45	Strahan (1983)
Macropus rufus	39	3	26.5	2.5	Strahan (1983)
Setonix brachyurus	40	2	2.9	1.2	Strahan (1983)
Wallabia bicolor	41	3	13.0	1.31	Strahan (1983)
Dendrolagus lumholtzi	42	3	5.9	1.25	Strahan (1983)

[1]Primary references for activity listed in appendix 1.

[2]Where a range is given in Strahan (1983) and no mean, the maximum was used.

[3]*Macropus rufogriseus banksianus.*

APPENDIX 3

Reproductive variables used in analysis in Fig. 4 with references to sources. GL, mean gestation length (d); PL, mean pouch life (d); FE, annual fecundity; BS, birth season type (values defined in text); G, G statistic value; P, probability of G (ns, not significant; $*P<0.05$, $**P<0.01$, $***P<0.001$).

Species	GL	PL	FE	BS[1]	G	P	Reference
Potorous tridactylus	38	130	1	2	—	—	Hughes (1962), Heinsohn (1968), Shaw & Rose (1979)
Bettongia penicillata	21	100	3.02	2	36.09	**	Parker (1977), Christensen (1980)
Bettongia lesueur	21	115	2.68	3	—	—	Tyndale-Biscoe (1968)
Aepyprymnus rufescens	22.8	112	2.71	3	6.28	ns	Johnson P. M. (1978) Strahan (1983)
Lagostrophus fasciatus	—	180	1	1	—	—	Strahan (1983)
Petrogale penicillata	—	—	—	3	—	—	Batchelor (1980)
Petrogale xanthopus	31.8	195	1.61	3	—	—	Poole *et al.* (1985)
Thylogale thetis	—	181	1.72	3	—	—	Johnson K. A. (1977)
Macropus eugenii	29.3	250	1.31	1	144	***	Andrewartha & Barker (1969) Murphy & Smith (1970) Christensen (1980)
Macropus parryi	36	281	1.15	3	—	—	Maynes (1973) Calaby & Poole (1971)
Macropus dorsalis	34	210	1.50	3	—	—	Strahan (1983)
Macropus rufogriseus[2]	30	285	1.16	2	11.36	*	Merchant & Calaby (1981) Johnson C. N. (1985)
Macropus agilis	30.5	209	1.53	3	8.66	ns	Kirkpatrick & Johnson (1969)
Macropus giganteus[3]	36.4	319	1.03	3	46.02	***	Poole (1975), Poole (1983)
Macropus fuliginosus	30.8	302	1	1	26.24	**	Poole (1976) Poole — Catling (1974) Poole (1975)
M. robustus erubescens	33.3	243	1.32	3	10.9	ns	Poole & Merchant (1987) Russell & Richardson (1971)
M. robustus robustus	32.6	260	1.25	3	—	—	Poole & Merchant (1987)
Macropus antilopinus	33.9	269	1.21	2	—	—	Poole & Merchant (1987) Russell & Richardson (1971)
Macropus rufus	33.2	235	1.36	3	18.86	ns	Sadleir (1965), Sharman & Calaby (1964), Sharman & Pilton (1964), Calaby & Poole (1971)
Setonix brachyurus	27	190	1	1	—	—	Sharman (1955a, 1955b), Shield (1964)
Wallabia bicolor	35.5	256	1.25	3	—	—	Calaby & Poole (1971)

[1]In absence of testable birth distribution data, Strahan (1983) and Tyndale-Biscoe and Renfree (1987) were used to define BS.

[2]*Macropus rufogriseus banksianus.*

[3]Cootamundra, New South Wales sample.

APPENDIX 4

Mean %fungi (FU), %insects (IN), %fruit and seeds (FR), %roots and bulbs (RO), %forbs (FO), %annuals (AN), %shrubs (SH), %browse (BR), and %grasses (GR) in diets of Macropodoids used in analysis in Fig. 5 with references to sources.

Species	Mean Percentage									Reference
	FU	IN	FR	RO	FO	AN	SH	BR	GR	
Potorous tridactylus	80	0	6	12	2	0	0	0	0	Guiler (1971)
Bettongia penicillata	62	0	1	33	4	0	0	0	0	Christensen (1980)
Petrogale xanthopus	0	0	0	0	28	13	0	30	29	Copley & Robertson (1983)
Macropus eugenii	0	0	0	0	0	0	0	40	60	Christensen (1980)
Macropus dorsalis	0	0	0	0	0	1	1	5	93	Dawson (this volume)
Macropus giganteus	0	0	0	0	9	0	0	0	91	Taylor (1983a)
Macropus fuliginosus	0	0	0	0	45	6	7	0	42	Barker (1987)
M. robustus erubescens	0	0	0	0	9	16	4	1	70	Ellis *et al.* (1977)
M. robustus robustus	0	0	0	0	8	0	0	0	92	Taylor (1983a)
Macropus antilopinus	0	0	0	0	0	0	0	0	100	Croft (1987)
Macropus rufus	0	0	0	0	19	11	1	0	69	Ellis *et al.* (1977)
Setonix brachyurus	0	0	0	0	6	55	13	0	27	Storr (1964)

APPENDIX 5

Ratios of male:female home ranges (M:F HR) and male-male:female-female agonistic interactions (MM:FF AG) for various macropodoids in analyses in Fig. 6 and Fig. 10 with references to sources.

Species	ID	M:F HR	MM:FF AG	Reference
Potorous tridactylus	2	3.73	—	Kitchener (1973)
Bettongia penicillata	4	1.39	—	Christensen (1980)
Aepyprymnus rufescens	7	—	2.73	Johnson P. M. (1980b)
Petrogale penicillata	14	1.53	8.15	Batchelor (1980)
Thylogale stigmatica	23	—	8.39	Cooke (1979)
Thylogale billardierii	24	—	1.64	Clancy (1982)
Thylogale thetis	25	1.51	—	Johnson K. A. (1980)
Macropus eugenii	27	1.32	—	Inns (1980)
Macropus parryi	29	1.35	38.3	Kaufmann (1974b)
Macropus rufogriseus	31	2.68	38.1	Johnson C. N. (1987), Watson (Unpubl.)
Macropus giganteus	33	2.36	16.1	Jaremovic (1984)
Macropus fuliginosus	34	1.25	—	Priddel (1987)
M. robustus erubescens	35	2.59	4.8	Clancy (Unpubl.), Croft (1981b)
Macropus rufus	39	1.12	16.1	Croft (Unpubl.), Croft (1981a)
Setonix brachyurus	40	0.96	145	Holsworth (1967), Kitchener (1972)
Wallabia bicolor	41	0.62	3.4	Edwards & Ealey (1974), Crebbin (1982)

Dynamics and adaptiveness of grouping in macropods

PETER J. JARMAN[1] and GRAEME COULSON[2]

Macropod groups are distinguished from aggregations, mobs or colonies, and ways of describing group size are reviewed. With few possible exceptions, gregarious macropods form open-membership groups; these can adjust their size and composition in response to their members' perceptions of the environment. In the better-studied species, groups show a small degree of organization, social classes differing slightly in their propensity for grouping. The macropods have, throughout their evolution, been preyed upon by enough terrestrial and aerial predators to have evolved avoidance of the risk of predation.

There are broad interspecific correlations between species' group size and body size, and group size and feeding style. The use of daytime refuges such as nests or squats is associated with solitariness; the use of rocky refuges or burrow systems promotes gregariousness (at least while resting). The largest macropods show intraspecific positive correlations between group size and population density (which smaller species do not show), and negative correlations between group size and lateral cover in their habitats. Group formation, persistence and disintegration are affected by the separation of individuals in groups and their rates of movement, which both relate to pasture conditions or the dispersion of acceptable food items.

The benefits of grouping relate to either reproduction or the avoidance of the risk of predation. In one large gregarious species, individual feeding time increases, and time spent surveying decreases, with increasing group size. Several costs of grouping are suggested; their existence may be inferred from situational variation in attributes such as group spread or organization of group composition. We suggest a simple dynamic model of grouping in macropods.

INTRODUCTION

THE extant Macropodoidea form one of the world's great arrays of large, terrestrial, herbivorous mammals, comparable in diversity of ecological and behavioural adaptations to the Cervidae or Bovidae. The parallels between these eutherian and metatherian radiations in the role that grouping plays in each species' response to its environment have long been evident (compare Kaufmann 1974a and Jarman 1974, for example). In each radiation gregariousness typifies larger, sexually heteromorphic species which eat a relatively unselected diet (often grasses) and live in open habitats, while small species, which feed highly selectively, are homomorphic, live in dense cover, and tend to live more solitarily. Yet the forms of social organization which accompany gregariousness or solitariness differ between (and within) these radiations. They thus present a mixture of similar and contrasting adaptive responses to their environments. Before the similarities and contrasts can be understood, we need to consider carefully why species in each radiation do or do not form groups.

This paper reviews what is known of the occurrence and dynamics of grouping within the Macropodidae and Potoroidae. We relate interspecific variations in group size or type to the variations in intrinsic or extrinsic ecological factors which characterize the species, especially body size, diet, the need for a structured shelter, and lateral cover in the habitat. Intra-specific variations are also considered. In particular, differences between classes of animals in their assumed vulnerability to predation are related to their tendency to take part in grouping. These reviews are used to argue for the adaptiveness of grouping as a response to a complex set of selection pressures which include minimization of the risk of predation, the acquisition of food and shelter, and the need for social interactions. We argue that the dynamics of grouping so far described in studies of macropods reflect the processes whereby individuals adjust their social environment to balance the costs and benefits of being with conspecifics in groups.

Definitions

By a *group* we mean a number of animals observed in substantially closer proximity to one another than to other members of the population, and potentially able to communicate and interact cohesively with, and respond to the actions of, other members of the group. This definition has spatial and behavioural components, but does not imply that a group persists through time. In some studies of macropods, arbitrary maximal separations have been used, so that, for example, a group may be defined as consisting of animals each of which is no more than 30 m from its nearest neighbour in the group (e.g., Johnson 1985). Other studies have emphasized behavioural criteria (e.g., Southwell 1981, who

[1]Department of Ecosystem Management, University of New England, Armidale, New South Wales 2351.

[2]Department of Zoology, University of Melbourne, Victoria 3052.

Pages 527-47 in KANGAROOS, WALLABIES AND RAT-KANGAROOS ed. by G. Grigg, P. Jarman and I. Hume. Surrey Beatty & Sons Pty Limited, New South Wales, Australia, 1989.

defined a group as consisting of "animals apparently communicating visually"). In practice there is usually less difficulty in deciding in the field what constitutes a group than there is in constructing a rigid set of criteria for its definition. For convenience in this paper we describe a lone animal as being in a group of one, although we recognize that this strains our definition.

The behavioural component of the definition distinguishes a group from an *aggregation*. The latter arises when animals are drawn to a spatially concentrated resource (a patch of fallen fruit; a grassy clearing in a forest; a water hole; a shady tree), are brought into proximity, but do not interact cohesively with each other. Johnson (1977) observed an aggregation of 57 Red-necked Pademelons *Thylogale thetis* simultaneously grazing on a 0.5 ha patch of oats, at a density of one animal to 90 m^2. He was so struck by the species' lack of response to conspecifics that he coined the term *monadic* to describe such species, in which individuals may come into close proximity while using resources, without being secondarily either drawn to or repelled from others. However, in other species, individuals brought into proximity by common attraction to a spatially limited resource may react cohesively; they would then be counted by us as a group, not an aggregation, the distinction being one of current behaviour, not initial cause of the coming together. An aggregation may contain several separate groups.

Some observers distinguish *subgroups*, sets of animals which, while clearly behaving as part of the group, nevertheless are closer to, or more cohesively interacting with, each other than the rest of the group. It may require quite long observation of a group before a subgroup can be discerned. Some workers (e.g., Stuart-Dick 1987) have found the subgroup concept to be useful in considering regularity of association between individuals.

Groups of animals differ in the persistence of their membership through time. Those whose component individuals remain the same for months or years, varying only as birth, death, dispersal or similarly rare life-events add or subtract individuals, we call *closed-membership groups*. Those contrast with groups whose composition can change from hour to hour or day to day through readily reversible joining or leaving by individuals. These we call *open-membership groups*. They are sometimes equated with what has been called "*fission-fusion society*", although that term may also cover social organizations in which small, permanently associating (i.e., closed-membership) groups come together to form larger groups which may then split apart into the same smaller groups.

A level of organization above the group, first described for Whiptail Wallabies *Macropus parryi* by Kaufmann (1974b) is the *mob*, a set of individuals whose home ranges overlap, which interact commonly, and all of whom may associate (although it would be rare for all members of a mob to be found together in one group). While a group specifies the instantaneous social neighbourhood of an individual (the number and density of conspecifics with which it is interacting cohesively), and a subgroup defines a closer (and often more persistent) set of immediate social partners, the mob is the long-term pool of associates amongst whom the individual leads its social life. One mob is discernibly separated from neighbouring mobs by lack of overlapping ranges or lack of associations between respective members. The mob is still a loosely defined concept but its major implication, that a population may be broken into sub-populations (mobs) within which most associating and interacting occurs, is useful. It has been reported for Tammar Wallabies *M. eugenii* (Inns 1980) and Eastern Grey Kangaroos *M. giganteus* in several studies (Herrmann 1971; Kaufmann 1975; Jaremovic 1984; Jarman and Southwell 1986). The observations of Eastern Grey Kangaroos by Bell (1973) and Grant (1973) are also consistent with this structure. So, too, are some observations of "colonies" of Brush-tailed Rock wallabies *Petrogale penicillata* (Joblin, pers. comm.). Each colony is restricted in the daytime to a section of cliff containing suitable refuges, and is separated from the next colony by less suitable habitat. Young adult males may disperse between colonies or mobs. Whether or not mobs are induced by ecological hetereogeneity does not matter as far as the behavioural consequences are concerned.

It should be noted that Kaufmann (1974b) used "sub-group" for the groups in which mob members were observed, not in the way we have defined subgroup above; and Jaremovic (1984) sometimes used "subgroup" in a similar way to Kaufmann, while referring to sets of regularly (but not invariably) associating individuals as "groups", whereas we have integrated them as mobs.

This paper focuses mainly on the group since that is the unit which mediates the individual's moment-by-moment experience of the world.

Measures of group size

Many studies have reported group sizes for species or populations (Table 1). Most have meant the same thing by "group" as we do, but a few may have included aggregations with "groups", which will have inflated slightly some values of mean and typical group-size in Table 1 and Fig. 2.

Variation in group size within a population is most completely displayed as a frequency distribution of group size. The distribution shows

Table 1. Group size measures reported in the Macropodoidea. A range of values indicates the extent of seasonal variation; two entries for one study indicate separate study sites, or density strata. Studies may have differed in their criteria for "groups".

Species	Study	Groups	Indiv.	Group size Mean	Group size Typical	Group size Mode
Aepyprymnus rufescens	Southgate (1980)	319	360	1.1†	1.2	1
Dendrolagus lumholzi	Proctor-Gray (1985)*	206	221	1.1	1.1	1
Macropus agilis	Bell (1972)	199	397	2.0†	3.3†	1-2
	P. M. Johnson (1980)	583	>873	> 1.5†	>2.0†	1
	Croft (1987)			1.1- 1.4	1.2- 1.8	1
M. antilopinus	Russell and Richardson (1971)	53	131	2.5	4.6	–
	Croft (1987)	449		2.2- 5.4	3.3-11.7	1
M. dorsalis	C. N. Johnson (1985)			4.3		–
	Rabbidge (1987)	35	146	4.2†	10.4†	1
M. eugenii	Vujcich (1979)	228	521	2.3†	3.0†	2
M. fuliginosus	C. N. Johnson (1983)	642		2.0- 3.1	1.8- 5.0	–
	Coulson (unpublished)	1195	4432	2.5- 6.3	3.8-16.1	1-3
M. giganteus giganteus	Caughley (1964a)	174	548	3.3	4.5†	2
	Kirkpatrick (1966)	272				1
	Bell (1972)	262	1022	3.9†	5.6†	2
	Grant (1973)					1
	Kaufmann (1975)	189	840	3.7- 6.4	4.9- 8.0	2-3
	Southwell (1976)	162	392	2.5	3.2†	2
	Taylor (1982)		162	2.7	4.4†	2
			564	4.5	8.8†	2
	Southwell (1984a)			2.8-10.5	3.3-20.1	2
				1.8- 2.1	2.3- 2.9	1
	Coulson and Raines (1985)	48	153	3.2		–
	Jarman and Southwell (1986)			3.3- 5.9		–
	Heathcote (1987)			5.4-12.5	10.0-22.9	–
M. giganteus tasmaniensis	Clancy (1982)	1412		3.6	5.4	–
M. parma	Maynes (1977)	71	95	1.3	1.6†	1
	Vujcich (1979)	93	97	1.0†	1.1†	1
M. parryi	Bell (1972)	39	121	3.1†		–
	Kaufmann (1974b)		2185	6.0		2
M. rufogriseus banksianus	Kaufmann (1974a)	739	1120	1.5	2.1†	2
	Southwell (1976)	287	489	1.7	2.3†	1
	C. N. Johnson (1985)			1.6	2.3	1
M. rufogriseus rufogriseus	Clancy (1982)	599		1.6	2.5	–
M. robustus erubescens	Russell and Richardson (1971)	135	189	1.4†	2.2†	1
	C. N. Johnson (1979)	25	57	2.3	3.6	–
	Croft (1981a)	366		2.1	2.6	1
M. robustus robustus	Kaufmann (1974a)	153	323	2.1	3.2†	1
	Joblin (1980)	100	182	1.8†	2.6†	1
	Taylor (1982)		233	1.8	2.6†	1
			1610	2.2	3.6†	1
M. robustus woodwardi	Russell and Richardson (1971)	58	65	1.1†	1.2†	1
	Croft (1987)			1.0- 1.1	1.0- 1.3	–
M. rufus	Caughley (1964a)	106	266	2.5	3.4†	2
	Frith (1964)	1190				2
	Jarman (1977)	375		1.9	3.9	–
		30		1.4	1.7	–
	Russell (1979)	760	1935	2.6	3.7	2
	Croft (1981b)	613		2.2	3.7	2
	C. N. Johnson (1983)	1252		2.6	3.7	–
Petrogale assimilis	Horsup (1986)	3483	4424	1.3	1.4†	1
P. penicillata	Joblin (1980)	177	237	1.3†	1.7†	1
Thylogale stigmatica	C. N. Johnson (1985)			1.5		–
Wallabia bicolor	Kaufmann (1974a)	58	66	1.1	1.3†	1
	Southwell (1976)	125	142	1.1	1.3†	1
	Wyre (1981)	108	134	1.2	1.4†	1

*excludes young-at-foot
†calculated by us from the presented data.

the frequency (or % frequency) of groups that fall within each class interval, ranging from groups of one to the largest recorded (Fig. 1). Many studies have presented group-size frequency distributions for populations of macropodoids. These distributions can be tested for goodness-of-fit to a theoretical distribution such as a Poisson series, but only Caughley (1964a), Taylor (1982) and Croft (1987) have done so. Some studies have used group-size frequency distributions to draw comparisons within a species (e.g., Southwell 1984a) and between species (e.g., Kaufmann 1974a). However, these distributions are unwieldy tools for comparisons. Instead, many workers have

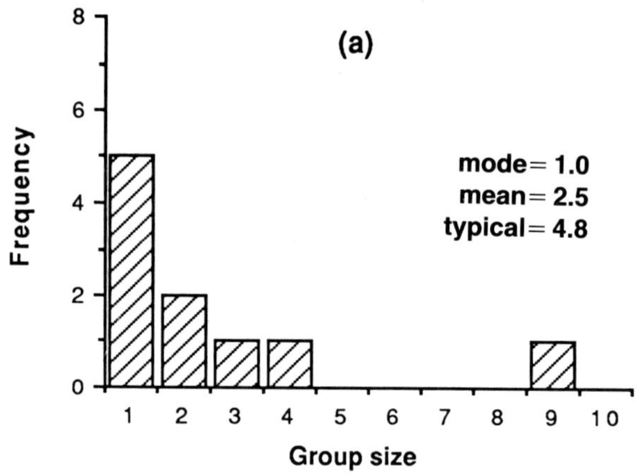

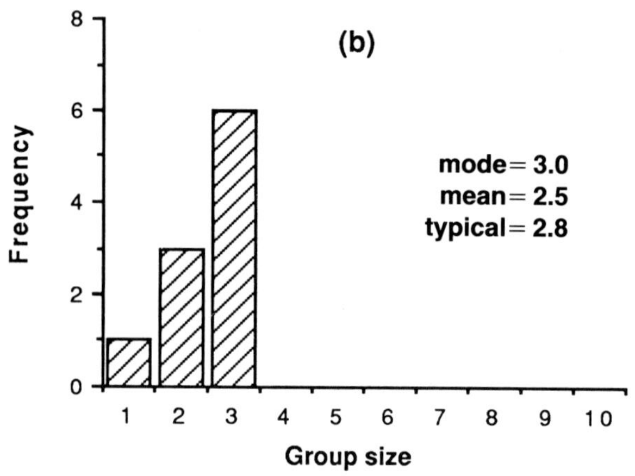

Fig. 1. Distributions of frequency of group sizes, and measures describing group size for the population, for two hypothetical populations, illustrating differences between the measures in their ability to reflect the distributions.

opted for a single measure of central tendency. Three measures have been used, and are compared in Fig. 1.

1. MODAL GROUP-SIZE

Modal group-size is simply the group-size class in which the highest frequency was observed. The concept of modal group-size is inherent in many non-statistical descriptions of sociality that are expressed in such terms as "mostly found alone" (e.g., Kirkpatrick 1968; McEvoy 1970). Many statistically-based studies have also presented modal group-size data (Table 1). The limitation of modal group-size is that it conveys no information about the less frequent size classes in the frequency distribution. This is illustrated in Fig. 1 where population (a) has a modal group-size of one, but the majority of individuals occur in groups.

2. MEAN GROUP-SIZE

The arithmetic mean is a stable measure of central tendency which is little affected by the presence of outliers in the frequency distribution. Fig. 1 shows that two quite different frequency distributions can yield identical mean group-sizes. Many studies have presented data on the arithmetic mean of group sizes (Table 1). None has used harmonic and geometric means which offer different degrees of sensitivity to the shape of the frequency distribution.

3. TYPICAL GROUP-SIZE

Jarman (1974) introduced typical group-size as a measure of the group in which the average individual occurs, intending it to reflect the social experience of the individual. The method of calculation of typical group-size may be compared with arithmetic mean group-size by examination of the formulae

$$\text{Mean} = \sum_{i=1}^{N} X_i \Big/ N \qquad \text{Typical} = \sum_{i=1}^{N} X_i^2 \Big/ \sum_{i=1}^{N} X_i$$

where X_i is the number of individuals in the ith group and N is the number of groups. Typical group size is more sensitive to variations in the group-size frequency distribution than is arithmetic mean group-size as shown in Fig. 1, where the population with the greater spread of size classes has a markedly higher typical group-size. In practice, typical group-size is approximately equal to mean group-size plus one standard deviation. A number of studies have presented data on typical group-size, often in parallel with mean group-size (Table 1). Confusingly, Croft (1987) used typical group-size as a measure but was forced to refer to it as modal group-size, which it clearly is not.

Other grouping parameters

Group size alone is insufficient to describe fully the phenomenon of grouping, or to test adequately all hypotheses about the processes involved in group formation. Several additional parameters have been measured in macropodid groups, and their applications are illustrated later in this paper.

(i) GROUP COMPOSITION

A number of studies have examined the composition of groups in terms of broad age-sex categories; most (e.g., Croft 1981a; Taylor 1983) have also categorized the reproductive status of females by reference to the presence of young-at-foot and pouch-young of different sizes. In a minority of studies (e.g., Jaremovic 1984; Jarman and Southwell 1986) it has been possible to record the presence of known individuals in groups.

(ii) GROUP DENSITY

The density of the "population" of Eastern Grey Kangaroo groups was estimated by Southwell (1984a), as well as mean group-size.

(iii) GROUP FLUX

The lability of groups, or group flux, has been measured directly by Southwell (1984b) in terms of individuals joining and leaving groups of Eastern Grey Kangaroos.

(iv) GROUP SPREAD

The spacing of animals within a group could be calculated from the area covered by the group divided by the number in it, but the lack of a clear boundary to the group normally precludes that. More usually it is reflected in a measure of separation between individuals, such as mean nearest-neighbour distance. Southwell (1981) used a measure which he called group spread = s/n-1 where s = standard deviation of distances between all pair combinations of group members and n = group size (for groups greater than two). He explained that this measure, unlike the mean of distances between all pair combinations, is independent of group size.

Types of grouping

The above distinction between species whose groups were of closed or open membership is important for our theme because the two types of species differ profoundly in their ability to vary group size. By definition, members of closed-membership groups do not leave or join the group in response to their immediate perception of the environment; members of open-membership groups might do so. This distinction in group type clearly does not apply to any species whose members are habitually solitary.

Prolonged observation of populations containing a high proportion of known individuals can be the only source of evidence for the existence of closed-membership groups, for which the observations must show that the same individuals, and no others, were always to be found together in a group. Open-membership grouping, on the other hand, may be accepted if animals can be seen leaving or joining groups.

Apart from essentially solitary species (see below), all studied macropods exhibit open-membership grouping, with the possible exceptions of Black-striped Wallaby *M. dorsalis,* and rock-wallabies *Petrogale* species. Tree-kangaroos *Dendrolagus* species, although forming groups, are too little studied in the wild for comments on their type of grouping to be made. Rock-wallabies are well studied on their daytime refuges and their many species form groups which consist of persistent associations between females, dependent young, and, generally, one adult male. However, until more is known of the composition of their groups while foraging, we hesitate to classify them as having closed-membership groups. Indeed, observations of individuals moving away from refuges separately from other refuge-users suggest an openness of grouping. In contrast, Black-striped Wallabies are noted for being seen moving in tight groups, Indian-file and close together. Limited field observations do not indicate open-membership grouping; Johnson (pers. comm.) observed two marked females regularly together in a group. However, the detailed studies needed to confirm closed-membership grouping have yet to be made on this interesting species. Thus for the rest of this paper we will be discussing macropod species as though their individuals were either solitary or formed open-membership groups and were thus free to adjust the size or composition of group in which they occurred.

That freedom to associate is not constrained by any territoriality or defence of groups either. Although defence of nests, rock refuges or sheltering sites occurs (see below for discussion), this does not appear to extend to defence of whole or partial feeding ranges.

GROUP COMPOSITION

If, as we suppose, nearly all the gregarious species of macropods form open-membership groups, it is of interest to know whether social classes differ in their freedom to join or withdraw from groups. Is any group likely to be a simple cross-section of the population's composition, or are there social forces at work determining the group's composition?

Relevant data come mainly from the large, grazing, open-country species: the Red *M. rufus,* Eastern and Western *M. fuliginosus* Grey Kangaroos, and Common *M. robustus* and Antilopine *M. antilopinus* Wallaroos (Russell and Richardson 1971; Russell 1979; Southwell 1981; Croft 1981a and b, 1982, 1987; Johnson 1983; Taylor 1983; Jarman and Southwell 1986). There are some common features in the group composition of all these species.

The most striking feature is that at some time in their reproductive cycle, usually when the young is at the stage of permanently leaving the pouch or is a small young-at-foot, the female tends to isolate herself and her offspring from other kangaroos. Thus for all these species a group of two animals commonly consisted of a female with her young-at-foot. The proportion of females in this social class which were found without company of other adults contrasts with females at all other stages (Table 2). The proportion of females with young-at-foot which are found "alone" varies greatly between samples, and may be affected by season,

Table 2. The proportions of females found without adult company ("alone") in the two reproductive classes, with and without young-at-foot, in five species of large macropods.

Species	Female Classes: With young-at-foot %	N	Without young-at-foot %	N	References
Red Kangaroo	38.3	97	18.7	56	Croft 1981a
Red Kangaroo	53.6	240	16.0	699	Johnson 1983
Eastern Grey Kangaroo	22.0	177	7.9	480	Jarman and Southwell 1986
Western Grey Kangaroo					
spring and summer	81.5	98	28.6	248	Johnson 1983
autumn and winter	56.2	57	13.7	117	Johnson 1983
Common Wallaroo	54.0	389	26.4	1005	Taylor 1983
Antilopine Wallaroo	21.5	28	0	31	Croft 1982

population density and habitat. It is worth noting that this solitariness of females with young-at-foot is not absolute; generally a fifth to a half of the females in this category were alone, but the others were with other kangaroos. Thus even this most obvious non-randomness in the composition of groups reveals that grouping is still facultative in this class.

The solitariness of females with young-at-foot may be, as Stuart-Dick (1987) has suggested, an adaptive response by females to the behavioural problems associated with expelling the young from the pouch but allowing it still to suckle. It may also be compared to the new-born phase of many ungulates, when the mother keeps the obviously small and weak young away from the herd in which it would be the outstandingly most vulnerable target for a cursorial predator. Croft (1981a) suggested that it might reflect the poor "locomotor ability" of the young, restricting the pair's movements, or alternatively be a way of avoiding the confusion that such young can suffer when in a group during alarm and flight.

The other social class which is abnormally solitary in these species (although less so than females with young-at-foot) is the large adult males. Table 3 shows that large males may be two or three times as solitary as are other males. However, like that of females with young-at-foot, their solitariness is facultative, not obligatory. It appears to arise from a greater tendency among large males to move far and frequently between groups (see Clarke, Jones and Jarman, this volume). The most dominant large males in a population of Eastern Grey Kangaroos were found by Jarman and Southwell (1986) to be even more often solitary than other large males, so there appear to be even finer social distinctions, below the size class, involved.

Another non-randomness in group composition apparently common to these species is the clustering of males. This results in more groups of females without any males, and more males in those groups which contain any males, than would be expected by chance. This has been demonstrated for Eastern Grey Kangaroos (Jaremovic 1984; Jarman and Southwell 1986) and can be inferred for Red Kangaroos (Croft 1981a; Johnson 1983) and Western Grey Kangaroos (Johnson 1983). Adult males appear to move in groups, or are attracted to other males, or aggregate because they are attracted to a common resource (such as a class of female). The evidence from known individuals in a population of Eastern Grey Kangaroos shows that males certainly do not associate permanently in fixed-membership sub-groups. Males in that population do to some extent share a preference for association with a particular class of females (those with very small pouch young, the class most likely to enter oestrus), but the tendency is not strong enough to produce the clustering of males (Jarman, unpubl. records). The explanation seems to be that males join other males, either to interact with them or because those other males indicate the presence of an oestrous female. The latter circumstance produces the common macropod phenomenon of a female being followed by a string of males in descending order of size, which can produce some of the largest groups in such relatively ungregarious species as Red-necked Wallaby *M. rufogriseus* or Red-necked Pademelon. However, groupings of males without any females occur in all these species, sometimes at quite high rates. For instance, in autumn and winter samples of Western Grey Kangaroos, Johnson (1983) recorded 45% of adult males in male-only groups. The extent of the phenomenon varied seasonally in this population.

Non-random associations between some classes are apparent in more detailed analyses. For example, in small groups of Common Wallaroos (Taylor 1981, 1983) medium males were significantly more often with other medium males, females (without young-at-foot) with females, and females (without young-at-foot) with subadult females, than would be expected if those animals were apportioned between groups randomly. Females with young-at-foot were more often accompanied by large males or subadult males than expected. In large groups of Common Wallaroos, the strongest departures from

Table 3. The proportion of large males, and all other adult males, found alone in studies of four large species of macropods.

Species	Male Classes: Large adults %	N	Other adults %	N	Reference
Red Kangaroo	21.7	34	11.7	44	Croft 1981a
Red Kangaroo	25.2	340	20.4	437	Johnson 1983
Eastern Grey Kangaroo	11.0	414	3.4	991	Jarman and Southwell 1986
Common Wallaroo	36.7	267	33.6	532	Taylor 1983
Antilopine Wallaroo	30.0	20	10.7	28	Croft 1982

randomness were shown by medium males, which were unexpectedly often accompanied by other medium males, large males, and subadult males. In Eastern Grey Kangaroos (Jarman and Southwell 1986; Jarman, unpubl. records) there is similarly a tendency for females with no young, with young-at-foot and with small pouch-young to be unexpectedly frequently accompanied by others of their classes. Classes of males, other than large males, also tended to associate with their own class. However, as with solitariness, these associative tendencies are far from obligatory. In fact, very large samples of group compositions are needed before most of them can be confidently discerned.

Potential predators

The influence of predation on macropods has been a matter of debate in the literature. In particular, there has been disagreement about predation as a selective force in the evolution of social organization. Caughley (1964b) first suggested that predators may have exerted selective pressure on behaviour when he attempted to explain an observed difference in habitat preference between Eastern Grey Kangaroos and Red Kangaroos in terms of past predation, proposing that Eastern Grey Kangaroos exhibited a preference for habitat that offered greater lateral cover because they had been exposed to predators from the Pleistocene to the present, whereas Red Kangaroos had not. Kaufmann (1974a) formalized this argument and applied it to account for the large groups that he observed in the Whiptail Wallaby. He argued that their preference for open feeding areas exposed them to cursorial predators, so there would be selection for individuals to form groups and benefit from the vigilance and alarm signals of other individuals. Kaufmann suggested that the same principle could apply to Eastern Grey Kangaroos. Low (1979) also considered predation in open areas to have been significant in the past. She found that Red Kangaroos exhibited a general preference for wooded areas, and that the size of groups was larger in the open regardless of the population density there.

Other studies (e.g., Croft 1981b; Taylor 1982) have also discussed the implications of predation for habitat preferences and grouping in kangaroos. However, in a recent review of social organization in the Macropodoidea, Russell (1984) drew a cautious conclusion: "Predation pressure as a factor promoting sociality in large marsupials has not been convincingly substantiated." Her view was echoed more emphatically in another review by Lee and Cockburn (1985): "Predation avoidance may have little influence on grouping in marsupials".

As Oliver (1986) has noted, there has been a tendency to discount predation as a selective force in the evolutionary history of the Macropodoidea. In part this apparently stems from Calaby's (1971) observation that the ratio of carnivores to herbivores was comparatively low in the historic and prehistoric mammalian fauna of Australia. This comment has been cited, for example, by Croft (1981b) and Russell (1984), as evidence of low levels of predation on kangaroos. We consider that Calaby's observation has been misinterpreted in three ways. Firstly, Calaby was referring to species diversity (or richness) of carnivores and herbivores, not to their relative population densities; a few species of generalist carnivores could exert much the same predation pressure as a higher diversity of more specialized species. Secondly, as Low (1979) has pointed out, predation pressure that is insignificant in terms of population control of the prey species could nonetheless be sufficient to influence their behavioural traits. Thirdly, the predators of macropods were not necessarily mammals.

We believe that it is time to reassess the role of predation in the evolution of social behaviour in the Macropodoidea. What follows is an analysis of the potential predators and the niches they occupy or may have occupied.

1. THYLACOLEONIDAE

Three genera of "marsupial lions" (Thylacoleonidae) are recognized (Archer and Dawson 1982; Rauscher 1987). *Priscileo pitikantensis,* the most primitive and oldest, dates to the middle Miocene. *Wakaleo* is represented by at least three species from the middle to late

Miocene. *Thylacoleo* comprised at least three species that spanned the Pliocene and Pleistocene. The largest and most recent of these, *Thylacoleo carnifex,* has received most attention. Despite earlier confusion about its diet, recent studies of dental morphology and jaw mechanics have concluded that this leopard-sized thylacoleonid was a specialized carnivore. It probably captured its prey with the aid of the powerful grasping mechanism of the manus, and killed it by a stabbing motion of the procumbent lower incisors (Finch 1982; Wells *et al.* 1982). Opinions differ, however, on its hunting techniques. Finch's (1982) analysis of limb proportions led her to conclude that it was a cursorial hunter capable of only short rushes at prey, whereas Wells *et al.* (1982) argued that a scansorial habit was indicated by the grasping ability of the manus and pes. Which species it preyed on is also unclear. Wells *et al.* (1982) noted its frequent association in fossil deposits with the browsing short-faced kangaroos, *Sthenurus.* Case (1985) examined species associations at a number of sites and found that, in general, it occurred most frequently with a number of macropods (*Sthenurus, Macropus, Bettongia* and *Protemnodon*) and with *Diprotodon.*

2. THYLACINIDAE

The dingo-sized Thylacine *Thylacinus cynocephalus* has been recorded in Pliocene, Pleistocene and Recent faunas, and an earlier species, *Thylacinus potens,* has been recorded from the Miocene (Dawson 1982a). The Thylacine was widely distributed in mainland Australia, but by the time of European settlement it had become confined to Tasmania and is probably now extinct (M. Smith 1982). The Thylacine was unquestionably carnivorous. In fossil deposits it is most frequently associated with smaller genera of mammalian prey species, including *Bettongia, Potorous* and the smaller *Macropus* (Case 1985). Many authors have remarked on the high degree of morphological convergence between the Thylacine and the placental canids, particularly the Wolf *Canis lupus.* On closer examination of some of these convergent characteristics, Keast (1982) concluded that the limb proportions of the Thylacine were indicative of a pursuit carnivore, albeit not one as specialized as the Wolf. However, it should be noted that the Thylacine emerged from Keast's analysis as more specialized for this role than the Cape Hunting-dog *Lycaon pictus,* a pursuit carnivore *par excellence.* In a similar analysis of skull shape, Werdelin (1986) concluded that the Thylacine was more like the Red Fox *Vulpes vulpes* than the Wolf, and by analogy may have hunted more by ambush than by running after the prey. Eyewitness accounts of the Thylacine hunting in Tasmania have been collated by S. Smith (1980), M. Smith (1982) and Guiler (1985). Together they provide a picture of a predator which actively hunted macropods by dogged (THYLACINID?) pursuit, but which undoubtedly also utilized other techniques for a variety of prey.

3. DASYURIDAE

The Tasmanian Devil *Sarcophilus harrissii,* the largest of the extant Dasyuridae, was widely distributed on the mainland in Pleistocene and Recent deposits but now occurs only in Tasmania (Dawson 1982b). Two other *Sarcophilus* species from the Pleistocene have been recognized, differing little from the Tasmanian Devil except in size (Crabb 1982; Dawson 1982b). The impressive dental array and robust form of the Tasmanian Devil attest to a diet consisting mainly of carrion; it is slow-moving and deals awkwardly with live prey (Buchmann and Guiler 1977). Quolls (*Dasyurus*) are the only other dasyurids likely to be predators of macropods. The largest extant species is the Spotted-tailed Quoll *Dasyurus maculatus,* which feeds on a range of invertebrate and vertebrate prey (Edgar 1983). The smaller species feed primarily on invertebrates (e.g., Blackhall 1980).

4. REPTILES

Two genera of monitors (Varanidae) are potential predators of macropods. *Megalania* is a genus of very large monitors from the Pliocene and Pleistocene (Molnar 1984). The largest and most recent was *Megalania prisca* which was almost twice the body length of the largest extant varanid, the Komodo Dragon *Varanus komodoensis.* By analogy with the ambush techniques of the Komodo Dragon, Hecht (1975) suggested that *Megalania prisca* may have preyed on the Pleistocene megafauna, which included the larger macropods such as *Sthenurus* and *Macropus titan.* Smaller extant varanids, notably the Lace Monitor *Varanus varius* and the Sand Monitor *Varanus gouldi,* date back to the Pleistocene (Molnar 1984). Both are active carnivores that feed on small vertebrates (Cogger 1975), and so could prey on small macropods. Two other reptiles from the Pleistocene were regarded by Molnar (1984) as significant predators: *Wonambi naracoortensis,* a boid snake 5 m in length, and *Quinkana fortirostrum,* a terrestrial crocodilian.

5. RAPTORS

There were apparently more species of large birds of prey (Accipitridae) in Australia during the Pleistocene than at present (Rich and van Tets 1985). They included species similar to the extant Wedge-tailed Eagle *Aquila audax,* and perhaps also to the White-bellied Sea-Eagle *Haliaeetus leucogaster,* as well as at least three other species referable to genera from New Zealand and New Guinea. Studies of the diet of the Wedge-tailed

Eagle by Leopold and Wolfe (1970) and Brooker and Ridpath (1980) have found remains of a wide range of vertebrates, including Red Kangaroos, Western Grey Kangaroos, Euros and Agile Wallabies. Of the macropod remains which could be aged by Brooker and Ridpath (1980), 83% were from early young-at-foot. Remains of smaller species such as the Burrowing Bettong (Glauert 1929), Rufous Bettong (Jarman, pers. obs.) and the Quokka (Storr 1965), have also been recorded. Although it is not possible to distinguish the relative contributions of carrion and live prey to data of this sort, the Wedge-tailed Eagle will attack smaller species such as the Quokka (Glauert 1929) and the Yellow-footed Rock-wallaby (Hornsby 1978), and would thus be capable of killing young of the larger macropodids. Eagles have even been observed to attack adult kangaroos and wallaroos (Frith and Calaby 1969; Brooker and Ridpath 1980; Croft 1981a; Jaremovic 1984). The only other raptor likely to be a predator of macropods is the White-bellied Sea-Eagle; Sanson *et al.* (1985) reported a kill of a Narbalek *Peradorcas concinna,* and also noted that the eagles disturbed Narbaleks and Short-eared Rock-wallabies *Petrogale brachyotis.*

6. DINGOES

The Dingo *Canis familiaris dingo* has been in Australia for no more than 4 000 years (Gollan 1984). There is abundant evidence of Dingo predation on macropods, ranging from indirect measures to direct observation of attacks. One indirect measure is provided by the marked difference in population density of kangaroos in north-western New South Wales, where kangaroos are common and Dingoes are rare, and in adjacent areas of South Australia and Queensland where the relative abundance of kangaroos and Dingoes are reversed (Jarman and Denny 1976; Caughley *et al.* 1980). Analyses by Whitehouse (1977), Newsome *et al.* (1983a), and Robertshaw and Harden (1985, 1986) of prey remains in stomachs and scats have revealed a wide range of vertebrate prey including numerous macropod species: Red Kangaroo, Eastern Grey Kangaroo, Common Wallaroo, Euro, Red-necked Wallaby, Swamp Wallaby *Wallabia bicolor,* Parma Wallaby *M. parma,* Red-necked Pademelon, Spectacled Hare-wallaby *Lagorchestes conspicillatus,* Brush-tailed Rock-wallaby, and Long-nosed Potoroo *Potorous tridactylus.* In each of these studies, kangaroos or larger wallabies made up the greatest part of the diet. Direct observations have been made of successful and unsuccessful attacks on a number of species, including the Red Kangaroo (Shepherd 1981; Oliver 1986), Eastern Grey Kangaroo (Newsome *et al.* 1983b; Jarman *et al.* 1987), Whiptail Wallaby (Kaufmann 1974b) and Swamp Wallaby (Newsome *et al.* 1983b). Descriptions of these attacks indicate that a chase is usually involved; 17% of the concerted chases witnessed by Oliver (1986) were successful.

7. INTRODUCED PREDATORS

The Red Fox *Vulpes vulpes* and the Cat *Felis catus* are recent introductions to the Australian fauna. Foxes, according to analyses of stomach contents and scats, consume an array of vertebrates, invertebrates and plants (e.g., Coman 1973; Croft and Hone 1978; Triggs *et al.* 1984). Macropods comprise a generally low proportion of the diet of Foxes, and although there have been reports of attacks on a Swamp Wallaby (Wakefield 1954), a Red-necked Pademelon (Johnson 1977) and a Euro (Hornsby 1982) it is usually assumed that the remains of larger species originated as carrion. Dietary studies of feral Cats (e.g., Coman and Brunner 1973; Jones and Coman 1981; Triggs *et al.* 1984) indicate that their diet is broadly similar to that of Foxes, but their smaller body size would mean that they were less likely to consume macropods. Many species of small macropods have undergone a dramatic reduction in range since European settlement (Poole 1979). Foxes, and to a lesser extent cats, have been implicated in these declines, as illustrated by the Bettongs (see King *et al.* 1981). Rock-wallabies also appear to be prone to predation by Foxes (e.g., Kinnear *et al.* 1984), as are Rufous Hare-wallabies (Lundie-Jenkins, pers. comm.; Johnson and Burbidge, this volume).

8. HUMANS

Aborigines arrived in Australia approximately 50 000 years before present (White and O'Connell 1979). They routinely hunted macropods of all sizes using a variety of techniques (e.g., Dixon and Huxley 1985). Europeans have continued hunting (mainly the larger species) in many areas (Poole 1984).

The Macropodoidea first evolved from a phalangerid ancestor in the Miocene, then underwent an explosive radiation in the Pliocene which continued into the Pleistocene (Flannery 1984, this volume). They co-evolved with a suite of predators which included thylacoleonids, thylacinids, varanids and raptors. Together these predators were capable of preying on the full range of body sizes in the Macropodoidea, and must have done so. In particular, the Macropodoidea have shared a long evolutionary history with the Thylacine. As Archer (1979) has noted, the Thylacine was apparently replaced by the Dingo on the Australian mainland, suggesting a high degree of niche overlap between the two species. The predatory load on macropods would thus have been maintained.

Our contention is that predation on the Macropodoidea has been significant and sustained. It has certainly been sufficient to select for a range of morphological and behavioural characteristics that enable individuals to detect and avoid predators (Oliver 1986) and could therefore also

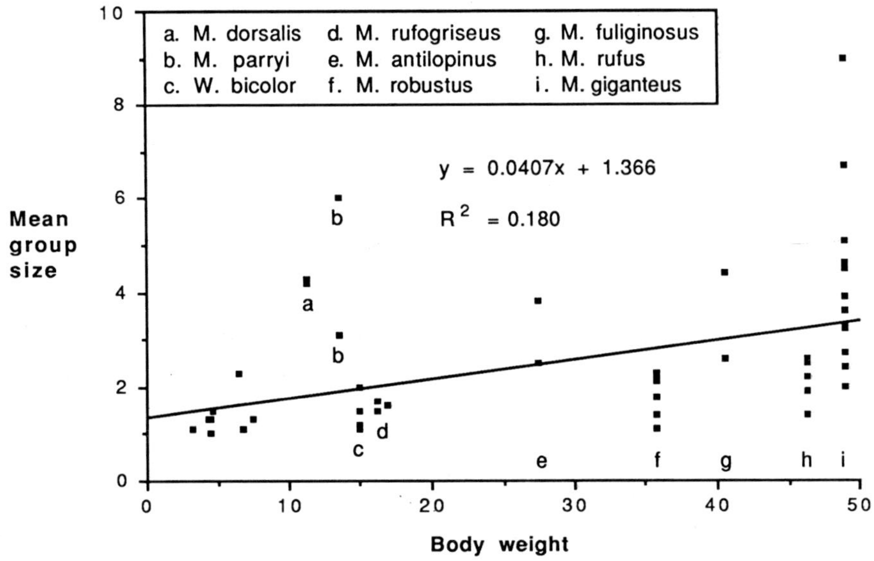

Fig. 2. Regression of the typical group-size on body weight for species in the Macropodoidea. Group-size values are from Table 1, taking the mid-point of the range where appropriate. Body weight values are from Strahan (1983), taking the mid-point of ranges and mean of male and female weights.

have influenced the social behaviour of the Macropodoidea.

Correlates with group size

Group-size variables have been measured in a number of studies of social behaviour in the Macropodoidea (Table 1). Although some species have been the subject of a number of studies, the range of taxa is not extensive and data are lacking for many species.

BODY SIZE AND GROUP SIZE

A linear regression of mean group-size on body weight in the Macropodoidea shows a general trend of larger groups with increased body weight (Fig. 2). However, some species deviate from the line. Of the smaller species, the Black-striped Wallaby and the Whiptail Wallaby have consistently higher group-sizes than would be predicted from the regression. Kaufmann (1974a) regarded both as highly gregarious species, and saw the Whiptail Wallaby in particular as the acme of sociality in the Macropodoidea. Of the larger species, the Antilopine Wallaroo also has a higher mean group-size than predicted, and Croft (1987) described it as one of the most gregarious species. By contrast, the Common Wallaroo (including the Euro) and the Red Kangaroo have smaller groups than predicted, confirming the commonly held view of the former as fairly solitary (e.g., Taylor 1982; Croft 1987) but at variance with the view of the latter as one of the more gregarious species (e.g., Russell 1984; Southwell 1984a). Considerable intraspecific variation is also evident, particularly in the well-studied Eastern Grey Kangaroo.

POPULATION DENSITY AND GROUP SIZE

Group size has been found to be positively related to population density in the Eastern Grey Kangaroo (Taylor 1982; Southwell 1984a; Jarman unpubl. data), the Common Wallaroo (Taylor 1982) and the Red Kangaroo (Johnson 1983). This led Southwell (1984a) to argue that group-size data for a macropod species ought not to be interpreted without reference to appropriate data on its population density. Using data for the Eastern Grey Kangaroo, he constructed a log-linear regression of mean and typical group-size against population density to describe the grouping response of a "model" gregarious macropod. Comparing group-size and population density data from other species with this regression, Southwell concluded that the Whiptail Wallaby, the Western Grey Kangaroo and the Red Kangaroo showed a similar response to that of the Eastern Grey Kangaroo, while the Common Wallaroo and Euro formed significantly smaller groups than predicted. Croft (1987) showed that the Antilopine Wallaroo fitted the regression and could be included amongst the social species, whereas the Agile Wallaby fell below the line and could be regarded as essentially solitary. Johnson (1985) found that group size of Red-necked Wallabies did not increase with density. Much of the intraspecific variation in group size, at least for the larger and more social species, arises from differences in population density.

HABITAT STRUCTURE AND GROUP SIZE

Some deviations from Southwell's (1984a) regression line are evident within the Eastern Grey Kangaroo. Southwell's data from this species in dense habitat (closed shrubland) fell below the model line which was derived from populations in more open habitat (woodland and open forest), a difference he interpreted as being due to the cessation of breeding in the drought conditions that prevailed in the closed shrubland study site. Additionally, the mean group-size of

Differences between group members in their rates of satisfaction of requirements could create problems in maintaining group co-ordination and cohesion. This can lead to the fragmentation of groups, when some satisfied individuals lie down to rest while others move on, still foraging; or it may mean that some individuals, in order to remain with a group, must restrict their foraging to the neighbourhood of the resting group, or forego foraging.

Grouping potentially allows other animals to parasitise the food-finding ability of one individual. If a species' food-items take little finding (i.e., they are numerous, well dispersed and obvious), this matters very little. However, for species seeking scarce, individually valuable items, each of which must be sought and discovered, this is a good reason for not foraging in a group. Bettongs, which must seek and dig up fruiting bodies of hypogean fungi or roots of plants, are appropriately solitary.

The increase in visual detectability of a group with increasing size may not matter to most individuals provided that increasing group-size improves the probability of detecting predators and that the individuals are almost invulnerable to a detected predator. However, if the individuals would be vulnerable, then being in a group would incur heavy costs and no benefits. This applies not just to small species but also to the sick adults or newly emerged young-at-foot of larger species. This may be one factor contributing to the withdrawal by females with young-at-foot from groups.

DISCUSSION

We have argued in this review that grouping in macropods, in particular the size of foraging groups, is an adaptive response by the individuals to several interacting factors, including body size (as a determinant of metabolic requirements per unit body weight and hence of quality of food item; of running speed and hence potential to escape cursorial predators; and of height in relation to obscuring vegetation), the quality and dispersion of plant parts which might be used for food, the structure of the habitat, and the risk of predation. The arguments are essentially those used elsewhere (see Jarman 1974; Jarman and Sinclair 1979), and the studies discussed here show macropods responding to these factors much as do bovids or cervids. However, it is particularly noticeable that foraging-group size of some gregarious macropod species correlates tightly with population density, a relationship that has been far less well revealed by studies of eutherian herbivores. In general, the relationship between specific body size and group size in these metatherian herbivores is similar to that for eutherian herbivores.

Risk of predation has tended to be ignored in considering the adaptiveness of macropod behaviour for reasons given above. There is another, general reason, we believe, for its being regarded as unimportant: that is, that the effects of predation are confused with the risk. The effects on populations are now amply demonstrated for several species (see above, and Robertshaw and Harden, this volume); the effects on individuals are less appreciated. Stuart-Dick (1987) has shown the impact that dingo predation can have on the individual reproductive success of female Eastern Grey Kangaroos, simply in terms of survival of juveniles. Predation is unlike almost all the other ecological factors affecting an individual's success: it does not lessen or enhance growth rate, or age at maturity, or interbirth interval, or hierarchical status; it either kills the animal or it does not. So it is inappropriate for the individual to respond as though a greater or lesser "dose" of predation could be traded off against another cost or benefit. In evolutionary terms the successful animal must behave at all times as though there were a reasonable risk of being killed by a predator. At any moment it must act to minimize that risk before indulging in any other activity. There is virtually no feed-back, informing the individual of the level of that risk (unless a predator is actually in sight), which would allow the animal to adjust its level of response to the perceived threat. An animal that is detected by a predator twice daily can not afford to be half as vigilant as one that is detected four times daily, unless it has a way of knowing when the daily detection has passed. All that the individual can do is to schedule its activities, choose its habitats and companions, and fine-tune its behaviour, so as to minimize either the risk of being detected by a predator or the cost of avoiding detection, or to maximize the probability of detecting a predator before it is attacked, if it would then be able to take evasive action successfully.

We believe that the individual's response to the risk of detection and attack by predators is crucial to the adaptiveness of its participation (or not) in grouping and the size of group it occupies. Risk of predation denies the smallest macropod species the opportunity to form groups, since their best way of avoiding predation is to hide, and a group would be more obvious than a single animal, even when hiding. Risk of predation makes it highly beneficial for individuals of larger species, which are capable of escape from a detected predator, to be in a group. But to benefit from membership of that group, they must be able to co-ordinate activities with the rest of the group and remain in communication with all other members. If a group becomes too large for co-ordination and communication, membership is a liability, since the group will be more readily detected by a predator than would solitary animals.

The choice is not necessarily to group or not to group. Individuals will vary in their ability to coordinate and communicate, and the extent to which they would be the outstandingly vulnerable members of a group. These relative factors may make individuals in the population differ in the size of group which would be optimum for them. Some associations are obligatory or nearly so: those between a mother and her dependent young, a male and the female he is courting, or between males which need to interact to establish relative dominance. Any individual needs to indulge in such associations in order to be successful. Thus these associations, which persist for differing lengths of time, seem to us to be of a basically different kind to the association (or dissociation) for avoidance of predation. Nevertheless, they inevitably impinge upon that second kind of grouping and are usually indistinguishable in records of group size.

We have tried to make the point that the appropriateness of grouping, or of group size, will vary with activity as the spatial nature of the resources required for that activity, and of the behaviours involved in securing them optimally, vary. Thus individuals resting in shade, or sunbathing, in which each animal requires only a small space around it (in order to retain control over its next possible move), are usually more closely spaced than when feeding. The patch-size of the subhabitat (patch of shade or sunshine) providing these physical resources limits the size of a resting group more precisely than the spatial limits of the habitat providing food resources, although food-rich patches within the habitat may still limit group size. Similarly, the size of a secure physical refuge may limit the number of animals that can occupy it. So far, most studies of grouping in macropods have been of species using well dispersed foods, and the size of food-rich patches has not been considered.

Activity may have a sequential effect upon grouping, when the group size appropriate to one activity (such as resting) carries on into the next activity (such as feeding). This may account for Taylor's (1981) finding that mean feeding-group size of Eastern Grey Kangaroo was slightly greater in late afternoon than in the morning, after resting-group size had been greater still during the middle of the day. His data suggested that feeding-group size correlated more closely with density than did resting-group size for both Eastern Grey Kangaroos and Common Wallaroos.

A simple model of grouping in macropods emerges from our interpretation of the relationship between population density and the slight organization found within and between groups of large macropod species. Taylor (1983) noted for Common Wallaroo that the proportion of animals that were solitary was greater in a low-density population than in a high-density one. Johnson (1983) commented that male-only groups of Red Kangaroos were more common at low than at high densities. In an Eastern Grey Kangaroo population, the proportions of females found alone in each of six classes were always greater in a low-density stratum than in a high-density one (Jarman, unpubl. records). Indeed, all of the manifestations of "organization" in grouping tended to be exaggerated at low density. These reports suggest that social associations in macropod grouping are quite loosely organized around pairs or small numbers of individuals that are interacting cohesively. At low density, groups are small (see below) and may, indeed, consist just of the interacting animals; consequently, associative or dissociative tendencies stand out clearly. But at higher densities, as groups coalesce to form larger groups, other interactors are present, and those tendencies are far less apparent, at least at the coarse level of group composition. That they still exist is suggested by analyses of nearest-neighbour preferences (Jarman, unpubl. records), and the presence of persistent sub-groups (Stuart-Dick 1987), in Eastern Grey Kangaroos.

This view of the organization of group composition is borne out by grouping in smaller and less gregarious macropod species. In their groups, an immediate, functional relationship between group members can often be inferred: a male consorting with an oestrous female; a female with a young-at-foot; a female and her subadult offspring; and so on. What is missing in nearly all the studied macropods is evidence for strong and persistent organization imposed by one class excluding others from the group, as happens in so many bovid and cervid species. Macropod groups seem to be very open indeed, so that the influence of extrinsic factors upon grouping of at least the larger species may be considered without confounding problems of socially determined organization.

Our model proposes these steps in the process of group formation in a macropod species: the association of individuals for essentially reproductive reasons; the meeting of units (individuals or associating animals) with possible formation of groups; and adjustment of group size. Mating systems in part reflect the species' typical grouping.

If an individual is a female with a dependent juvenile, she will associate with that juvenile. If an individual is an adult male, he will tend to try to inspect any adult female he meets and to check his status against any male he meets. If he meets a female in or approaching oestrus he will attempt to consort with her. These reasons for associating to some extent cut across other determinants of group size. Thus in the smallest species, individuals will associate for these reasons, but are otherwise strongly solitary. In the larger species, such associations form a subset of the units from which groups can form, the other units being members of the population not in those defined categories.

Units in the population seek resources appropriate to their activity state. If the resource is at a single location, or clustered, use of it may lead to aggregation, clustering of the units. If the resource is well dispersed, units will meet less predictably during their ranging. Frequency of meeting will be related to population density, group density, rate of movement, and characteristics of the habitat such as visibility.

Upon meeting, units may coalesce to form cohesive groups. The observed typical group-sizes result from units coalescing and splitting around a group-size which gives the unit (individual or associating animals) enough reduction in surveillance or increase in foraging time to balance the cost of being in that group. Coalescing may not happen if the probability of agonistic interactions would be so high that increased within-group vigilance would outweight reduced individual surveillance for threats from outside the group. Larger species are gregarious because their individual food-items are abundant, easily found, and therefore cheap, so that the costs of being in a group should rise only slowly with increasing group-size. Because the value of food-items varies between habitats and seasons, group sizes also vary. Characteristics within groups are little affected by population density, and the rate at which animals leave groups is much more affected by group size than by population density. However, the rate at which animals join groups is directly related to density.

The process of grouping described here affects the mating system of macropod species, which are all very similar in the basic male mating strategy, albeit with species-typical tactical variations. Males, perhaps with the exception of rock-wallabies on daytime refuges, do not defend access to an area containing females. Instead they use their size-derived dominance to defend their consorting and mating with an oestrous female. The degree to which females in the population are grouped will affect the proportion of the population that each male can be with at any moment, and thus the extent to which a dominant male can thwart attempts at courtship by subordinate males. This is one of the factors influencing the adaptiveness of heteromorphism such that, in general, the degree of dimorphism should relate to the typical size of groups of females, expressed as a proportion of the population in the male's home-range (Jarman 1983).

Groups will persist if the benefits of doing so outweight the disadvantages. The principal benefit, we have argued, would be the reduction in individual time spent in surveying the surroundings for predators, and hence an increase in potential foraging time. There may be other benefits that have yet to be revealed. The disadvantages are manifold. They include the greater detectability of a group, possible interference with the individual's foraging track or freedom of use of other resources, the need to remain co-ordinated with the group's activities, and the possibility of being an outstandingly vulnerable target for a predator compared with other members. Sick or injured animals, or juveniles that are small but have been excluded from the pouch, would appear relatively vulnerable, and are the most likely to suffer predation by being in a group. To remain co-ordinated with the group's activities, an individual should be with others having similar requirements; this may entail associating with animals of a similar size or reproductive class. The potential for detrimental interference in free use of a resource will increase with the value of each unit of the resource. Rare, highly nutritious food items, which must be individually sought and processed, may be so valuable that no interference can be tolerated or risked; animals depending on such resources, such as most potoroids, may forage alone. In the larger species, increasing value will correlate with greater scatter of items, and result in animals either interfering more with each other's feeding paths, or keeping further apart, or following divergent paths. These will lead to splitting of the group.

Improvements can still be made in the comparative study of grouping in macropods. Southwell (1981) suggested that comparisons between species should be made on the basis of the slope of the regression of their typical group-size on density. In the smallest species, the slope is zero, while in Eastern Grey Kangaroos it rises steeply. Within-species comparisons show that habitat factors (e.g., lateral cover) further influence the relationship between group-size and population density. So, presumably, do the availability and dispersion of food items; yet their effect is relatively slight, judged by the small seasonal variation in group-size in many macropod species. Two areas of study seem particularly worth pursuing. One is the foraging behaviour of animals in relation to habitat, food availability and dispersion, and group-size. The other is the sequential behaviour of individuals before and after leaving feeding groups. These should provide further insight into the dynamics of adjustment of group-size.

ACKNOWLEDGEMENTS

Coulson thanks Mike Clarke, Margie Beilharz, Cathy Heathcote, Angus Martin and other members of the (Melbourne) Animal Behaviour Discussion Group for encouragement to follow up the question of predators past and present. He acknowledges financial assistance from a University of Melbourne Postgraduate Scholarship, the Department of Conservation, Forests and Lands (Victoria), and Melbourne College of Advanced Education. Jarman thanks many colleagues who have contributed to the facts and ideas in this review; they include Ken Johnson, John Osazuwa, Felix Schlager, Richard Southgate, Gordon Wyre, Keiran McNamara, Cath Handel, Rob Taylor, Ken Joblin, Chris Johnson, Robyn Stuart-Dick, Col Southwell, Menna Jones, Jeff Rabbidge, Karen Higginbottom, Jo Clarke, and Geoff Lundie-Jenkins. He acknowledges financial support from the Australian Research Grants Scheme and the University of New England. We both thank those who read drafts of the manuscripts, especially David Croft, Angus Martin and the Wallaby Creek crew.

REFERENCES

ARCHER, M., 1979. *Wabularoo naughtoni* gen. et. sp. nov., an enigmatic kangaroo (Marsupialia) from the Middle Tertiary Carl Creek limestone of northwestern Queensland. Results of the Ray E. Lemley Expedition, Part 4. *Mem. Qld Mus.* **19:** 299-307.

ARCHER, M. AND DAWSON, L., 1982. Revision of marsupial lions of the genus *Thylacoleo* Gervais (Thylacoleonidae, Marsupialia) and thylacoleonid evolution in the late Cainozoic. Pp. 477-94 *in* Carnivorous Marsupials ed. by M. Archer. Royal Zoological Society of New South Wales, Mosman, Australia.

BELL, H. M., 1972. The behaviour and ecology of three Macropodid species in an area of open forest and savannah woodland in north Queensland. B.Sc. (Hons.) Thesis, James Cook University of North Queensland, Australia.

BELL, H. M., 1973. The ecology of three macropod marsupial species in an area of open forest and savannah woodland in North Queensland, Australia. *Mammalia* **37:** 527-44.

BLACKHALL, S., 1980. Diet of the eastern native-cat, *Dasyurus viverrinus* (Shaw), in southern Tasmania. *Aust. Wildl. Res.* **7:** 191-97.

BROOKER, M. G. AND RIDPATH, M. G., 1980. The diet of the wedge-tailed eagle, *Aquila audax,* in Western Australia. *Aust. Wildl. Res.* **7:** 433-52.

BUCHMANN, O. L. K. AND GUILER, E. R., 1977. Behaviour and ecology of the Tasmanian Devil *Sarcophilus harrisii.* Pp. 155-68 *in* Biology of Marsupials ed. by B. Stonehouse and E. Gilmore. Macmillan: London.

BURBIDGE, A. A., 1983. Crescent Nailtail Wallaby *Onychogalea lunata.* Pp. 206 *in* The Complete Book of Australian Mammals ed. by R. Strahan. Australian Museum, Sydney, Australia.

CALABY, J. H., 1971. The current status of Australian Macropodidae. *Aust. Zool.* **16:** 17-31.

CASE, J. A., 1985. Differences in prey utilization by Pleistocene marsupial carnivores, *Thylacoleo carnifex* (Thylacoleonidae) and *Thylacinus cynocephalus* (Thylacinidae). *Aust. Mammal.* **8:** 45-52.

CAUGHLEY, G., 1964a. Social organization and daily activity of the Red Kangaroo and the Grey Kangaroo. *J. Mammal.* **45:** 429-36.

CAUGHLEY, G., 1964b. Density and dispersion of two species of kangaroo in relation to habitat. *Aust. J. Zool.* **12:** 238-49.

CAUGHLEY, G., GRIGG, G. C., CAUGHLEY, J. AND HILL, G. J. E., 1980. Does dingo predation control the densities of kangaroos and emus? *Aust. Wildl. Res.* **7:** 1-12.

CLANCY, T. F., 1982. Aspects of the behaviour of *Thylogale billardieri* in captivity. B.Sc. (Hons.) Thesis, University of Tasmania, Australia.

CLARKE, J. L., JONES, M. E. AND JARMAN, P. J. (this volume). A day in the life of a kangaroo: Activities and movements of Eastern Grey Kangaroos *Macropus giganteus* at Wallaby Creek.

COGGER, H. G., 1975. Reptiles and Amphibians of Australia. Reed: Sydney.

COMAN, B. J., 1973. The diet of red foxes, *Vulpes vulpes* L., in Victoria. *Aust. J. Zool.* **21:** 391-401.

COMAN, B. J. AND BRUNNER, H., 1972. Food habits of the feral house cat in Victoria. *J. Wildl. Manage.* **36:** 848-53.

COULSON, G. M. (this volume). Repertoires of social behaviour in the Macropodoidea.

COULSON, G. M. AND RAINES, J. A., 1985. Methods for small-scale surveys of grey kangaroo populations. *Aust. Wildl. Res.* **12:** 119-25.

CLUTTON-BROCK, T. H. AND HARVEY, P. H., 1977. Primate ecology and social organization. *J. Zool. Lond.* **183:** 1-39.

CRABB, P. L., 1982. Pleistocene dasyurids (Marsupialia) from southwestern New South Wales. Pp. 511-6 *in* Carnivorous Marsupials ed. by M. Archer. Royal Zoological Society of New South Wales, Mosman, Australia.

CROFT, D. B., 1981a. Social behaviour of the euro, *Macropus robustus,* in the Australian arid zone. *Aust. Wildl. Res.* **8:** 13-49.

CROFT, D. B., 1981b. Behaviour of Red Kangaroos, *Macropus rufus* (Desmarest, 1822) in northwestern New South Wales, Australia. *Aust. Mammal.* **4:** 5-58.

CROFT, D. B., 1982. Some observations on the behaviour of the Antilopine Wallaroo *Macropus antilopinus* (Marsupialia: Macropodidae). *Aust. Mammal.* **5:** 5-13.

CROFT, D. B., 1987. Socio-ecology of the Antilopine Wallaroo, *Macropus antilopinus,* in the Northern Territory, with observations on sympatric *M. robustus woodwardii* and *M. agilis. Aust. Wildl. Res.* **14:** 243-55.

CROFT, J. D. AND HONE, L. J., 1978. The stomach contents of foxes, *Vulpes vulpes,* collected in New South Wales. *Aust. Wildl. Res.* **5:** 85-93.

DAWSON, L., 1982a. Taxonomic status of fossil thylacines (*Thylacinus,* Thylacinidae, Marsupialia) from late Quaternary deposits in eastern Australia. Pp. 527-36 *in* Carnivorous Marsupials. ed. by M. Archer. Royal Zoological Society of New South Wales, Mosman, Australia.

DAWSON, L., 1982b. Taxonomic status of fossil devils (*Sarcophilus,* Dasyuridae, Marsupialia) from late Quaternay eastern Australian localities. Pp. 517-25 *in* Carnivorous Marsupials ed. by M. Archer. Royal Zoological Society of New South Wales, Mosman, Australia.

DIXON, J. M. AND HUXLEY, L. (eds.) 1985. Donald Thomson's Mammals and Fishes of Northern Australia. Nelson: Melbourne.

EDGAR, R., 1983. Spotted-tailed Quoll. Pp. 244-7 *in* The Australian Museum Complete Book of Mammals ed. by R. Strahan. Australian Museum, Sydney.

FINCH, M. E., 1982. The discovery and interpretation of *Thylacoleo carnifex* (Thylacoleonidae, Marsupialia). Pp. 537-51 *in* Carnivorous Marsupials ed. by M. Archer. Royal Zoological Society of New South Wales, Mosman, Australia.

FLANNERY, T., 1984. Kangaroos: 15 million years of Australian bounders. Pp. 817-35 *in* Vertebrate Zoogeography and Evolution in Australasia ed. by M. Archer and G. Clayton. Hesperian Press: Western Australia.

FLANNERY, T. (this volume). Phylogeny of the Macropodoidea; a study in convergence.

FRITH, H. J., 1964. Mobility of the Red Kangaroo, *Megaleia rufa*. *C.S.I.R.O. Wildl. Res.* **9**: 1-19.

FRITH, H. J. AND CALABY, J. H., 1969. Kangaroos. Cheshire: Melbourne.

GLAUERT, L., 1929. Eagles and wallabies. *Victorian Nat.* **45**: 243.

GOLLAN, K., 1984. The Australian dingo: in the shadow of Man. Pp. 921-7 *in* Vertebrate Zoogeography and Evolution in Australasia ed. by M. Archer and G. Clayton. Hesperian Press: Carlisle, Western Australia.

GORDON, G., 1983. Bridled Nailtail Wallaby *Onychogalea fraenata*. Pp. 205 *in* The Complete Book of Australian Mammals ed. by R. Strahan. Australian Museum, Sydney, Australia.

GRANT, T. R., 1973. Dominance and association among members of a captive and a free-ranging group of Grey Kangaroos (*Macropus giganteus*). *Anim. Behav.* **21**: 449-56.

GUILER, E. R., 1985. Thylacine: The Tragedy of the Tasmanian Tiger. Oxford University Press: Melbourne.

HANDEL, K. J., 1987. Aspects of spacing and social behaviour in a captive population of Swamp Wallabies (*Wallabia bicolor*). B.Nat.Res. Thesis, University of New England, Armidale, Australia.

HEATHCOTE, C. F., 1987. Grouping of eastern grey kangaroos in open habitat. *Aust. Wildl. Res.* **14**: 343-48.

HECHT, M., 1975. The morphology and relationships of the largest known terrestrial lizard, *Megalania prisca* Owen, from the Pleistocene of Australia. *Proc. Roy. Soc. Vic.* **87**: 239-50.

HERRMANN, D., 1971. Beobachtungen des Gruppenlebens ostaustralischer Graugrosskanguruhs, *Macropus giganteus* (Zimmerman, 1777) und Bennetkanguruhs, *Protemnodon rufogrisea* (Desmarest, 1817). *Saugetierk. Mitt.* **19**: 352-62.

HORNSBY, P., 1978. Some features of later pouch life and early juvenile behaviour in the rock wallaby, *Petrogale xanthopus* and *P. penicillata pearsoni*. *Aust. Mammal. Soc. Bull.* **5**: 29.

HORNSBY, P., 1982. Social organization, seasonal movements and breeding behaviour in the yellow-footed rock wallaby. *Aust. Mammal. Soc. Bull.* **7**: 35.

HORSUP, A. B., 1986. The behaviour of the allied rock wallaby *Petrogale assimilis* (Macropodinae). B.Sc. (Hons.) Thesis, James Cook University of North Queensland, Australia.

INNS, R. W., 1980. Ecology of the Kangaroo Island Wallaby, *Macropus eugenii* (Desmarest), in Flinders Chase National Park, Kangaroo Island. Ph.D. Thesis, University of Adelaide, Australia.

JAREMOVIC, R., 1984. Space and time related behaviour in Eastern Grey Kangaroos (*Macropus giganteus* Shaw). Ph.D. Thesis, University of New South Wales, Kensington, Australia.

JARMAN, P. J., 1974. The social organization of antelope in relation to their ecology. *Behaviour* **48**: 215-67.

JARMAN, P. J., 1977. Data on large animals, from aerial surveys. Pp. 41-57 *in* Flooding In The Ecosystems of Inland Eastern Australia. A report to the Department of the Environment, by M. J. S. Denny, P. J. Jarman, R. Longmore and E. Wyndham.

JARMAN, P. J., 1983. Mating system and sexual dimorphism in large, terrestrial, mammalian herbivores. *Biol. Rev.* **58**: 485-520

JARMAN, P. J., 1984. The dietary ecology of macropod marsupials. *Proc. Nutr. Soc. Aust.* **9**: 82-87.

JARMAN, P. J., 1987. Group size and activity in eastern grey kangaroos. *Anim. Behav.* **35**: 1044-50.

JARMAN, P. J. AND DENNY, M. J. S., 1976. Red Kangaroos and land use along the New South Wales, Queensland, and South Australian borders. Pp. 55-67 *in* Agriculture, Forestry or Wildlife: Conflict or Co-existence? ed. by P. J. Jarman. University of New England.

JARMAN, P. J. AND SINCLAIR, A. R. E., 1979. Feeding strategy and the pattern of resource-partitioning in ungulates. Pp. 130-63 *in* Serengeti: Dynamics of an Ecosystem ed. by A. R. E. Sinclair and M. Norton-Griffiths. University of Chicago Press: Chicago.

JARMAN, P. J. AND SOUTHWELL, C. J., 1986. Grouping, associations, and reproductive strategies in Eastern Grey Kangaroos. Pp. 399-428 *in* Ecological Aspects of Social Evolution ed. by D. I. Rubenstein and R. W. Wrangham. Princeton University Press: Princeton, USA.

JARMAN, P. J., JOHNSON, C. N., SOUTHWELL, C. J. AND STUART-DICK, R., 1987. Macropod studies at Wallaby Creek. I. The area and animals. *Aust. Wildl. Res.* **14**: 1-14.

JOBLIN, K. P., 1980. A preliminary report on the behaviour and ecology of the wallaroo *M. r. robustus* and the brushtailed rock-wallaby *P. penicillata*. M.Nat.Res. (prelim.) Report, University of New England, Armidale, Australia.

JOHNSON, C. N., 1979. Social structures of red and western grey kangaroos in Kinchega National Park, western New South Wales. B.Nat.Res. Thesis, University of New England, Armidale, Australia.

JOHNSON, C. N., 1983. Variations in group size and composition in Red and Western Grey Kangaroos, *Macropus rufus* (Desmarest) and *M. fuliginosus* (Desmarest). *Aust. Wildl. Res.* **10**: 25-31.

JOHNSON, C. N., 1985. Ecology, social behaviour and reproductive success in a population of Red-necked Wallabies. Ph.D. Thesis, University of New England, Armidale, Australia.

JOHNSON, K. A., 1977. Ecology and management of the Red-necked Pademelon, *Thylogale thetis*, on the Dorrigo Plateau of northern New South Wales. Ph.D. Thesis, University of New England, Armidale, Australia.

JOHNSON, K. A., BURBIDGE, A. AND MCKENZIE, N. L. (this volume). Australian Macropodoidea: status, causes of decline and future research and management.

JOHNSON, P. M., 1980. Field observations on group compositions in the Agile Wallaby, *Macropus agilis* (Gould) (Marsupialia: Macropodidae). *Aust. Wildl. Res.* **7**: 327-31.

JONES, E. AND COMAN, B. J., 1981. Ecology of the feral cat, *Felis catus* (L.), in south-eastern Australia. I. Diet. *Aust. Wildl. Res.* **8**: 537-47.

KAUFMANN, J. H., 1974a. The ecology and evolution of social organization in the kangaroo family (Macropodidae). *Amer. Zool.* **14**: 51-62.

KAUFMANN, J. H., 1974b. Social ethology of the Whiptail Wallaby, *Macropus parryi,* in northeastern New South Wales. *Anim. Behav.* **22**: 281-369.

KAUFMANN, J. H., 1975. Field observations of the social behaviour of the Eastern Grey Kangaroo, *Macropus giganteus. Anim. Behav.* **23**: 214-21.

KEAST, A., 1982. The Thylacine (Thylacinidae, Marsupialia): how good a pursuit carnivore? Pp. 675-84 *in* Carnivorous Marsupials ed. by M. Archer. Royal Zoological Society of New South Wales, Mosman, Australia.

KING, D. R., OLIVER, A. J. AND MEAD, R. J., 1981. *Bettongia* and fluoroacetate: a role for 1080 in fauna management. *Aust. Wildl. Res.* **8**: 529-36.

KINNEAR, J., ONUS, M. AND BROMILOW, A., 1984. Foxes, feral cats and rock wallabys. S.W.A.N.S. **14**: 3-8.

KIRKPATRICK, T. H., 1966. Studies of the Macropodidae in Queensland. 4. Social organization of the Grey Kangaroo (*Macropus giganteus*). *Qld. J. Agric. Anim. Sci.* **23**: 318-22.

KIRKPATRICK, T. H., 1968. Studies on the Wallaroo. *Qld. Agric. J.* **94**: 362-65.

LEE, A. K. AND COCKBURN, A., 1985. Evolutionary Ecology of Marsupials. Cambridge University Press: Cambridge.

LEOPOLD, A. S. AND WOLFE, T. O., 1970. Food habits of nesting wedge-tailed eagles, *Aquila audax,* in south-eastern Australia. *C.S.I.R.O. Wildlife Res.* **15**: 1-17.

LOW, B. S., 1979. The predictability of rain and the foraging patterns of the Red Kangaroo (*Megaleia rufa*) in central Australia. *J. Arid Env.* **2**: 61-76.

MAYNES, G. M., 1977. Distribution and aspects of the biology of the Parma Wallaby, *Macropus parma,* in New South Wales. *Aust. Wildl. Res.* **4**: 109-25.

MCEVOY, J. S., 1970. Red-necked Wallaby in Queensland. *Qld. Agric. J.* **96**: 114-16

MOLNAR, R., 1984. Cainozoic reptiles from Australia (and some amphibians). Pp. 337-41 *in* Vertebrate Zoogeography and Evolution in Australia ed. by M. Archer and G. Clayton. Hesperian Press: Carlisle, Western Australia.

NEWSOME, A. E., CATLING, P. C. AND CORBETT, L. K., 1983a. The feeding ecology of the dingo. 2. Dietary and numerical relationships with fluctuating prey populations in south-eastern Australia. *Aust. J. Ecol.* **8**: 345-66.

NEWSOME, A. E., CORBETT, L. K., CATLING, P. C. AND BURT, R. J., 1983b. The feeding ecology of the dingo. I. Stomach contents from trapping in south-eastern Australia, and the non-target wildlife also caught in dingo traps. *Aust. Wildl. Res.* **10**: 477-86.

OLIVER, A., 1986. Social organization and dispersal in the Red Kangaroo. Ph.D. Thesis, Murdoch University, Australia.

POOLE, W. E., 1979. The status of the Australian Macropodidae. Pp. 13-27 *in* The Status of Endangered Australasian Wildlife ed. by M. J. Tyler. Royal Zoological Society of South Australia, Adelaide.

POOLE, W. E., 1984. Management of kangaroo harvesting in Australia (2nd edition). Australian National Parks and Wildlife Service, Canberra, Australia.

PROCTOR-GRAY, E., 1985. The behavior and ecology of Lumholtz's tree-kangaroo, *Dendrolagus lumholtzi* (Marsupialia: Macropodidae). Ph.D. Thesis, Harvard University, Cambridge, Massachusetts, USA.

RABBIDGE, J. J., 1987. The Black-striped Wallaby in the south-west Narrabri region. B.Nat.Res. Thesis, University of New England, Armidale, Australia.

RAUSCHER, B., 1987. *Prisciléo pitikantensis,* a new genus and species of thylacoleonid marsupial (Marsupialia, Thylacoleonidae) from the Miocene Etadunna formation, South Australia. Pp. 423-32 *in* Possums and Opossums: Studies in Evolution ed. by M. Archer. Surrey Beatty & Sons Pty Ltd: New South Wales.

RICH, P. V. AND VAN TETS, G. F., 1985. Kadimakara: Extinct Vertebrates of Australia. Pioneer Design Studio, Lilydale, Australia. 284 pp.

ROBERTSHAW, J. D. AND HARDEN, R. H., 1985. The ecology of the dingo in north-eastern New South Wales. II. Diet. *Aust. Wildl. Res.* **12**: 39-50.

ROBERTSHAW, J. D. AND HARDEN, R. H., 1986. The ecology of the dingo in north-eastern New South Wales. IV. Prey selection by dingoes and its effect on the major prey species, the swamp wallaby, *Wallabia bicolor* (Desmarest). *Aust. Wildl. Res.* **13**: 141-63.

ROBERTSHAW, J. D. AND HARDEN, R. H. (this volume). Predation on Macropodoidea: a review.

RUSSELL, E. M., 1979. The size and composition of groups in the Red Kangaroo, *Macropus rufus. Aust. Wildl. Res.* **6**: 237-44.

RUSSELL, E. M., 1984. Social behaviour and social organization of marsupials. *Mammal Rev.* **14**: 101-54.

RUSSELL, E. M. AND RICHARDSON, B. J., 1971. Some observations on the breeding, age structure, dispersion and habitat of populations of *Macropus robustus* and *Macropus antilopinus* (Marsupialia). *J. Zool. Lond.* **165**: 131-42.

SANSON, G. D., NELSON, J. E. AND FELL, P., 1985. Ecology of *Peradorcas concinna* in Arnhem Land in a wet and dry season. *Proc. Ecol. Soc. Aust.* **13**: 65-72.

SCHLAGER, F. E., 1981. The distribution, status and ecology of the rufous rat-kangaroo, *Aepyprymnus rufescens,* in northern New South Wales. M.Nat.Res. Thesis, University of New England, Armidale, Australia.

SEEBECK, J. H., BENNETT, A. F. AND SCOTTS, D. J. (this volume). Ecology of the Potoroidae — A review.

SHEPHERD, N. C., 1981. Predation of red kangaroos, *Macropus rufus,* by the dingo, *Canis familiaris dingo* (Blumenbach), in north-western New South Wales. *Aust. Wildl. Res.* **8**: 255-62.

SMITH, M., 1982. Review of the Thylacine (Marsupialia, Thylacinidae). Pp. 237-53 *in* Carnivorous Marsupials ed. by M. Archer. Royal Zoological Society of New South Wales, Mosman, Australia.

SMITH, S. J., 1980. The Tasmanian Tiger — 1980. Hobart: Tasmanian National Parks and Wildlife Service.

SOUTHGATE, R., 1980. Factors affecting the spatial and temporal use of habitat by the rufous rat-kangaroo *Aepyprymnus rufescens.* B.Nat.Res. Thesis, University of New England, Armidale, Australia.

SOUTHWELL, C. J., 1976. Behaviour and social organization of three species of macropod at Diamond Flat, New South Wales. B.Sc. (Hons.) Thesis, University of New England, Armidale, Australia.

SOUTHWELL, C. J., 1981. Sociobiology of the Eastern Grey Kangaroo, *Macropus giganteus.* Ph.D. Thesis, University of New England, Armidale, Australia.

SOUTHWELL, C. J., 1984a. Variability in grouping in the Eastern Grey Kangaroo, *Macropus giganteus*. I. Group density and group size. *Aust. Wildl. Res.* **11**: 423-35.

SOUTHWELL, C. J., 1984b. Variability in grouping in the Eastern Grey Kangaroo, *Macropus giganteus*. II. Dynamics of group formation. *Aust. Wild. Res.* **11**: 437-49.

STORR, G. M., 1965. Notes on Bald Island and the adjacent mainland. *West. Aust. Nat.* **9**: 187-96.

STUART-DICK, R. I., 1987. Parental investment and rearing schedules in the Eastern Grey Kangaroo. Ph.D. Thesis, University of New England, Armidale, Australia.

TAYLOR, R. J., 1981. The comparative ecology of the eastern grey kangaroo and wallaroo in the New England tablelands of New South Wales. Ph.D. Thesis, University of New England, Armidale, Australia.

TAYLOR, R. J., 1982. Group size in the Eastern Grey Kangaroo, *Macropus giganteus,* and the Wallaroo, *Macropus robustus*. *Aust. Wildl. Res.* **9**: 229-37.

TAYLOR, R. J., 1983. Association of social classes of the Wallaroo, *Macropus robustus* (Marsupialia: Macropodidae). *Aust. Wildl. Res.* **10**: 39-45.

TRIGGS, B., BRUNNER, H. AND CULLEN, J. M., 1984. The food of the fox, dog and cat in Croajingalong National Park, south eastern Victoria. *Aust. Wildl. Res.* **11**: 491-99.

VUJCICH, M. V., 1979. Aspects of the biology of the parma (*Macropus parma* Waterhouse) and darma (*Macropus eugenii* (Desmarest) wallabies with particular emphasis on social organization. M.Sc. Thesis, University of Auckland, New Zealand.

WAKEFIELD, N. A., 1954. The rediscovery of the rock-wallaby in Victoria. *Victorian Nat.* **70**: 202-06.

WELLS, R. T., HORTON, D. R. AND ROGERS, P., 1982. *Thylacoleo carnifex* Owen (Thylacoleonidae, Marsupialia): marsupial carnivore? Pp. 573-86 *in* Carnivorous Marsupials ed. by M. Archer. Royal Zoological Society of New South Wales, Mosman, Australia.

WERDELIN, L., 1986. Comparison of skull shape in marsupial and placental carnivores. *Aust. J. Zool.* **34**: 109-17.

WHITE, J. P. AND O'CONNELL, J. F., 1979. Australian prehistory: new aspects of antiquity. *Science* **203**: 21-28.

WHITEHOUSE, S. J. O., 1977. The diet of the dingo in Western Australia. *Aust. Wildl. Res.* **4**: 145-50.

WYRE, G., 1981. Aspects of the activity and behaviour of the Swamp Wallaby *Wallabia bicolor* (Desmarest) at Hillgrove, New South Wales. B.Nat.Res. Thesis, University of New England, Armidale, Australia.

Maternal behaviour in the Macropodoidea

ELEANOR M. RUSSELL[1]

Maternal behaviour is a basic character that must have occurred in early mammals along with altricial young nourished by lactation. The consequences of marsupial reproduction for maternal behaviour are most evident in the Macropodoidea, where a single pouch-young (PY) is carried in a deep pouch until it reaches nearly 2% of maternal body mass. Maternal behaviour is described briefly in five stages and significant aspects emphasized: 1: *Parturition*. The mother's behaviour must remain within the limits of what the rudimentary nervous system of the young is capable of doing; her behaviour at parturition is expected to be stereotyped. 2. *Birth to first emergence from the pouch (FEP).* The pattern of growth in marsupial PY means that for almost the first half of pouch life the young is very small and the most obvious maternal behaviour is an increased level of pouch cleaning. Nevertheless, the physiological effect of sucking even by a newborn PY is profound. 3. *FEP to Permanent emergence (PEP)*. Although this period lasts 1-2 months, rapid change occurs only in the last 1-2 weeks; periods out of the pouch become longer and attempts to return fewer. The final transition is almost imperceptible, with little overt aggression by the mother.

Communication and recognition between mother and young and the control of maternal behaviour, by hormones and by stimulation from the young, are discussed and identified as important areas for further research.

INTRODUCTION

PARENTAL behaviour is behaviour of a parent directed towards an offspring which has the function of increasing the latter's chances of survival. Parental behaviour has a cost and the corresponding benefit can be measured in terms of fitness, the proportion of an individual's genes that is represented in future generations. Such fitness can be improved by increasing the number of surviving offspring. If some form of care by the parent can better the chance of survival by the offspring, then we have conditions favouring the evolution of parental care. In the case of mammals, with lactation, this parental care is of necessity provided significantly by the mother, which is why this chapter is about maternal behaviour.

Early mammals were small, perhaps 10-15 cm long, weighing 20-30 g; if, as supposed, they filled the niche of nocturnal insectivore left vacant by the dinosaurs, they were probably homeothermic (Crompton 1980). Hopson (1973) has argued that endothermy and small size necessitated the production of immature, altricial, ectothermic young, with the development of elaborate parental care; small endothermic young would have relatively far greater energy requirements, and Case (1978) suggested that the presence of a large volume of non-thermogenic yolk would have severely compromised the homeothermy of the female herself. Altricial young depend on the parental provision of food and shelter and so lactation and parental care must have evolved along with homeothermy. The pattern of tooth eruption and replacement, with premolars replacing milk molars and molars added sequentially and not replaced is taken to indicate the occurrence of lactation; if young do not require functional teeth until after weaning, a single replacement of milk teeth and the addition of a few molar teeth is all that is needed to accommodate the remaining growth after weaning. In reptiles (or mammal-like reptiles) the hatchlings needed teeth; since teeth do not grow, numerous replacements were necessary to accommodate growth from hatchling to adult. Pond (1977) suggested that the evolution of lactation and associated brood care were more fundamental adaptations in mammalian reproduction than viviparity. The existence of the monotremes is evidence that lactation and viviparity did not necessarily evolve together. Thus maternal behaviour of maintaining contact with young, keeping ectothermic young warm, providing milk to them and promoting their independence as they grow, must be a basic mammalian attribute.

From their comprehensive review of marsupial reproduction, Tyndale-Biscoe and Renfree (1987) concluded that viviparity with the delivery of highly altricial young had probably evolved before the major dichotomy of the Metatheria and the Eutheria. At the time of the metatherian-eutherian dichotomy, the differences between the groups would have been small, and indeed, today, differences between the Didelphidae and Dasyuridae on the one hand and the Insectivora on the other are not great (Eisenberg 1981); all produce extremely altricial young after a short gestation. Insectivoran young are left in a nest from birth while marsupial young are at first

[1]C.S.I.R.O. Division of Wildlife and Ecology, LMB No. 4, P.O. Midland, Western Australia 6056.

Pages 549-69 in KANGAROOS, WALLABIES AND RAT-KANGAROOS ed. by G. Grigg, P. Jarman and I. Hume. Surrey Beatty & Sons Pty Limited, New South Wales, Australia, 1989.

carried on the abdomen of the female, and later left in a nest (Russell 1982). As in eutherians, the marsupial radiation exhibits a trend towards a life-history strategy with increased body size, and a number of correlated characteristics; smaller litter size, increased longevity, longer development, increased period of dependence, increased age at sexual maturity, slower growth rate and less investment in individual young (Western and Ssemakula 1982). Radiation occurred from the original small insectivore stock into various other ecological niches including a branch which gave rise to the large grazing and browsing hopping herbivores, the Macropodoidea.

Where pouches fit into this story is less easy to determine. Clearly, for a form such as the monotremes which produced milk without benefit of teats and produced highly altricial young from eggs, the pouch is necessary, not so much for the egg as for the newly hatched young, since there are no teats for it to attach to. Whether the pouch was a basic mammalian character shared by ancestors of monotremes and the metatherian-eutherian ancestor, or a derived character in both groups is related to the question of when teats arose; some modern marsupials rear their young adequately with no pouch because the young are attached firmly to teats. If female mammals at the time of the marsupial-eutherian dichotomy were, as seems likely, polytocous, then it is most likely that the newborn young after birth became attached to the teats and remained there until they were large enough to be left on their own. The physical problems of nurturing several tiny "embryos" seem sufficient argument in favour of their remaining attached. Whether there was a pouch or not at this stage is a matter of speculation (Kirsch 1977; Archer 1984). Further pouch development, as an improvement on this process, came later and in several groups of marsupials, including the Macropodoidea, the pouch has increased in size, depth and ability to contain the young to a later stage in development. Many functions of maternal behaviour that in eutherian mammals are performed by the mother in other ways are in macropodoids performed by the pouch. Thus there are some elements of maternal behaviour common to all mammals and others relating to the pouch that are unique to marsupials.

Maternal behaviour in the Macropodoidea

The kangaroos and rat-kangaroos have litters of one, with the sole exception of *Hypsiprymnodon* with two; the pouch is no longer just a ridge of skin, but a deep bag large enough to contain the single young until it has reached a size relative to its mother that is equivalent to, and generally larger than, that of many similar sized eutherians at birth (especially ungulates) (Maynes 1976; Russell 1982; see below).

In broad outline, maternal behaviour in the Macropodoidea needs little introduction. After birth, the neonate climbs to the pouch, attaches to a teat, and remains in the pouch for a period known as pouch life, after which it may be left in a refuge or follows its mother, putting its head into the pouch from time to time to suckle. Eventually it is weaned and sooner or later becomes independent. But there is much more to it than that.

Maternal behaviour is inseparable from the behavioural development of the young and I consider five stages:

1. Parturition: just before and at the time of birth.
2. From birth to First Emergence from the Pouch (FEP).
3. From FEP to Permanent Emergence from the Pouch (PEP).
4. From PEP to weaning.
5. Post-weaning.

The duration of these stages depends largely on the body size of a particular species. As with other mammals (Western 1979; Western and Ssemakula 1982; Calder 1984), such life-history attributes as age at FEP, PEP and weaning are strongly correlated with body size (Russell 1982). As the young progresses from birth to weaning, so the balance of its mother's behaviour changes from predominantly caring for the young to promoting its independence.

Parturition

Although eutherian young at birth are relatively much larger, and birth may take longer and be more difficult for the mother, birth in marsupials is not the trivial event it is sometimes assumed to be. Even though the young is small, the mechanisms have to exist to expel it from the uterus. And because it is so small, undeveloped and fragile, it is important that the mother should behave in such a way as to give the young the best possible chance of reaching the pouch and attaching to a teat. At birth, the precocially developed anatomical features of the neonate necessary for crawling from the vulva to the pouch are controlled by a rudimentary nervous system (Hill and Hill 1955; Sharman 1973; Walker and Rose 1981; Klima and Bangma 1987). The reflex control of the young's behaviour must be matched by behaviour of the mother which must remain within the limits of what the young can do. Not surprisingly there is precise control of both the timing and the nature of the female's behaviour at the time of birth (Renfree *et al.*, this volume).

The first detailed descriptions of birth in macropodids were for *Macropus rufus* by Sharman and Pilton (1964) and Sharman and Calaby (1964). These accounts were amplified by

Tyndale-Biscoe and Renfree (1987), and Renfree *et al.* (this volume) give a detailed description of parturition in *M. eugenii*.

In those macropodids where birth has been observed, very similar behaviour has been described, with stereotyped birth position and increased frequency of pouch licking before birth. In most species so far described, the birth position is the Tail-Forward position (Fig. 1a) as described for *M. rufus* by Sharman and Calaby (1964) (*M. eugenii*: Renfree *et al.*, this volume; *M. rufogriseus*: Merchant and Calaby 1981; Fleming *et al.* 1983; *Potorous tridactylus*: Hughes 1962; *Bettongia lesueur*: Tyndale-Biscoe 1968; *Petrogale inornata*: Johnson 1979; *M. robustus*: Poole and Merchant 1987; *Setonix brachyurus*: Sharman *et al.* 1966; Cannon *et al.* 1976). This "birth position" is one commonly adopted for grooming the tail and pouch; it is also used in hot weather, shading the tail (Russell and Harrop 1976), and is a commonly adopted resting position (*M. eugenii*, Russell 1973; Fig. 8b). Poole and Pilton (1964, p.227) describe a female *M. giganteus* "squatting over a saucer-shaped depression in the ground with her tail in the normal position behind her, but with her legs thrust forward and toes in the air as if taking part of her weight on her heels which were forced against the rim of the depression." Poole (1975) again observed this position. Poole and Pilton (1964) also quote descriptions of this same birth position in *M. fuliginosus*. There is as yet no information on the variation in preliminary pouch-cleaning behaviour between species with post-partum oestrus (e.g., *M. rufus*) and without (*M. giganteus, M. fuliginosus*).

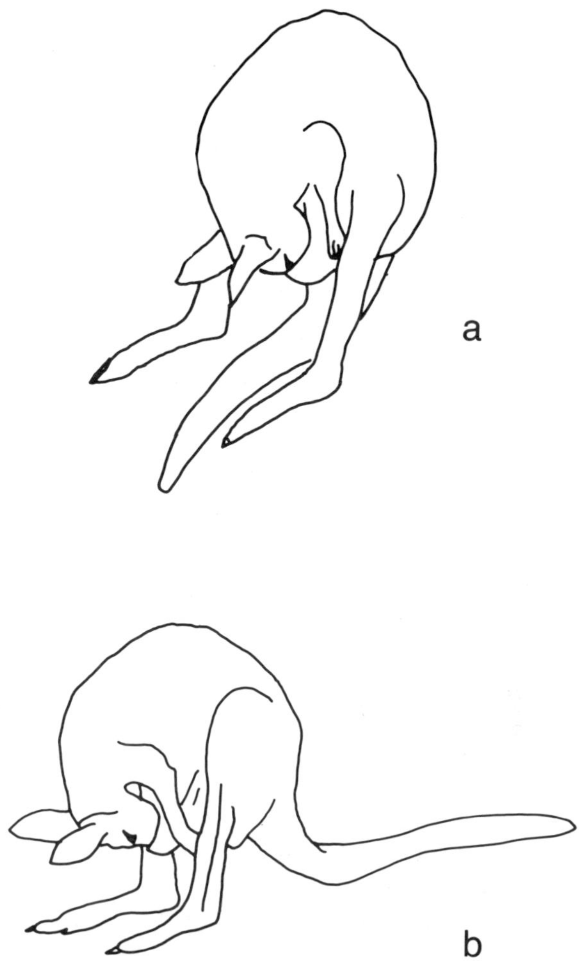

Fig. 1. (a) Female *Macropus rufus* cleaning her pouch just before giving birth. Some of her weight is still being taken by the hind legs; at later stages, the legs are extended further forwards and all the weight taken on the butt of her tail. (b) Female *Macropus rufus* pouch cleaning when a small pouch young is present.

From birth to FEP

The development of young marsupials, especially macropodoids, is distinguished by a long early period when growth is slow. For example, in *M. eugenii*, although the young leaves the pouch for the first time after 190 d, weighing *ca* 500 g (and PEP occurs after *ca* 250 d, at *ca* 1 000 g), in the first 18 weeks after birth, the young reaches only 100 g (Green 1984). For the first 100 d, it is continuously attached to the teat and for most of the time between birth and FEP, the young imposes relatively little burden on its mother, either as a weight to be carried or as a consumer of energy (Green 1984). It is only during the last third of this period that the absolute growth rate increases, and the body mass of the young reaches a level which might be considered a significant burden to the female (Fig. 2).

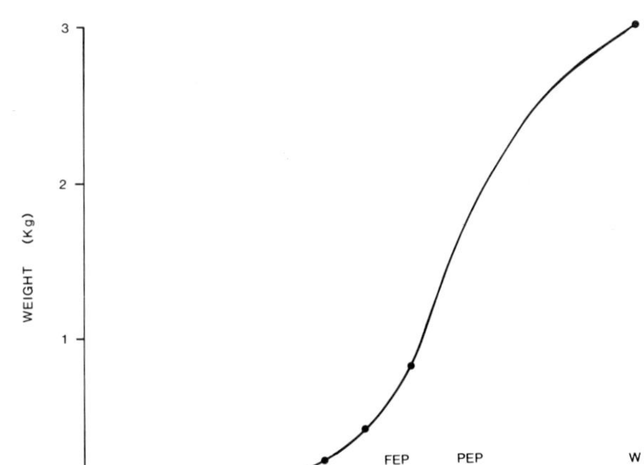

Fig. 2. Growth of young *Macropus eugenii* (body mass) from birth to weaning, drawn from data in Green (1984).

The description of maternal behaviour between birth and FEP is based primarily on intensive observations of *Macropus eugenii* and *M. rufus* (Russell 1973). Until the young first puts its head out of the pouch (in *M. eugenii* at about 160 d), the only obvious maternal behaviour was increased grooming of the pouch, and grooming of the young in the pouch (Russell and Giles 1974), commonly called Pouch Cleaning (Fig. 1b). Soon after the young was born, pouch cleaning increased above background levels for females with no PY and continued to increase as the PY grew, but still did not occupy more than 1% of the time. The female consumed urine and faeces produced by the young, recycling about one-third of the water used in milk production (Baverstock and Green 1975). Observations of older young indicate that the licking by the mother stimulated micturition and defaecation.

The activity of the young increased as it grew, until eventually it could be seen wriggling in the pouch before any part of it protruded. The mother rarely showed any response to these movements in the period before FEP, although clearly she must have received considerable tactile stimulation in the abdominal area. During the last third of this period, when the PY had begun to grow rapidly, bouts of wriggling in the pouch sometimes led to the protrusion of an arm or a leg. Again the mother remained apparently oblivious. Eventually, at about 160 d, the head of the young first appeared out of the pouch during one of these bouts of wriggling. The duration of periods with head out were at first very short. The head hung passively, and was retracted as the PY continued to wriggle. Gradually the PY began actively to put its head out of the pouch, to look around, to groom and to sniff at its mother; by week 29 after birth, more than 20% of the time was spent with its head out of the pouch. Mothers were seen to sniff at the head of the PY and to groom it occasionally, but levels of interaction were low compared with that just before the time of PEP (Russell 1973).

First emergence to permanent emergence: the "In-Out" period

During the period from first emergence from the pouch (FEP) to permanent emergence (PEP), the important aspects of maternal behaviour continue from the previous stage: pouch cleaning and grooming the young, at first in the pouch and later, also, outside. In addition, maternal behaviour develops in relation to controlling when the young leaves the pouch and when it returns, general surveillance of the young when out of the pouch, and "education" of the young towards the time of PEP. The presence of the growing young in the pouch also has modifying effects on the behaviour of the mother, reducing the area over which she forages. The interesting question during this in-out period is to what extent does the mother control the comings and goings of her PY? Two aspects of this are frequently confused. First, how the PY gets in and out of the pouch, and second, more significant early in this period, is whether the female by her behaviour allows it to happen at all. The pouch is simply a cutaneous invagination through the superficial ventral musculature (*panniculus carnosus* in Langworthy 1932; Barbour 1977; *pars marsupialis* of the *thoracoabdominalis* part of the *cutaneus maximus*, after Johnson-Murray 1987). The muscle fibres, running fore and aft, form a semi-circle on each side around the pouch opening, and thus by contracting, both close the opening and pull the ventral wall of the pouch closer to the body. In addition, Rose (1986) described the occurrence of smooth muscle fibres in the wall of the pouch. The female has voluntary control of contractions of the striated muscle. When the muscle is relaxed, the pouch hangs down gaping wide and the young easily falls out. When the muscle is contracted, the young can either be held in the pouch or tipped out if the contraction is used to push the young through the pouch opening.

The early excursions from the pouch are very brief. The young may leave the pouch in a number of ways, summarized below, but this can only occur if the female is behaving appropriately:

(i) *The young falls out*

(a) The mother is pouch cleaning, with pouch relaxed, and the mouth of the pouch held open; the young has its head out, wriggles, overbalances and falls out or may just gradually slide out onto the ground ("oozes out"). The first excursions by young *M. rufus* and *M. eugenii* that I have seen occurred in this way, apparently by accident.

(b) The mother is lying down, with the pouch and its opening very relaxed; the young is relaxed too, or just wriggles, and "oozes" out onto the ground. When the mother stands up, the young is out of the pouch.

(c) The young falls out by overbalancing, stretching out too far to explore at a distance, or to pick up food.

(ii) *Young ejected by mother*

An obvious contraction of the ventral wall of the pouch or a movement of the mother's whole body tips the young from the pouch. Seen chiefly when the female is pouch cleaning; clearly distinguishable from 1(a). First excursions by young *M. giganteus* described by Stuart-Dick (1987) occurred in this way.

(iii) *Young gets out by its own efforts*

Most frequently, when the mother is lying down, the young just stands up out of the pouch. This also occurs by the young stretching its forepaws out onto the ground and pulling the rest of the body out of the pouch.

Young may be almost entirely outside the pouch many times before they actually have to stand up and climb back in. This may happen when the mother is lying down or she may let the pouch slacken and gape open so that the young is almost out (Stuart-Dick 1987). The earliest excursions from the pouch include many which occur during bouts of pouch cleaning (Russell 1973; Stuart-Dick 1987). In studies in captivity, obvious tipping by the mother was less frequent than excursions when the young just appeared to fall out of the pouch (Russell 1973). Later, as the young spent more time out of the pouch, it also began to get out by its own efforts. From observations of *M. eugenii* young in the three weeks before PEP I found that for 108 cases where the mode of leaving was known, in 33% the young itself got out of the pouch by standing up when the female was lying down with the pouch relaxed so that the young slipped out; in 30% the young "oozed" out when the female was pouch cleaning, standing with the pouch very relaxed; in 23% the young fell out, and in 14% the female was clearly seen to eject the young by contracting the ventral wall of the pouch.

At first it was only when the mother was standing and pouch cleaning, bent over feeding or just generally standing relaxed and looking about that the young fell or was tipped out of the pouch. Clearly when the mother was alert or alarmed, standing erect with the pouch pulled up fairly tight, the young did not get out of the pouch; it would be physically difficult for it to do so. Later, the young could or did get out when the female was lying down or feeding with the pouch relaxed; again in circumstances when she was not alert. When she was alert or alarmed, the pouch was pulled up tighter, and may have prevented the young from leaving. As the young gets older and larger, the mother appears to have less "control" over its leaving the pouch. We talk about the female controlling what the young does; but whether she really does so in any deliberate sense, or whether it just happens that way has not been demonstrated.

Field observations on young entering and leaving the pouch suggest that the mother does exercise some control, directly or indirectly. Johnson (1987) found that young *M. rufogriseus* were rarely seen to leave the pouch if their mothers were close to other wallabies; he suggested that this avoided the confusion which arose when a young did not recognize its mother and either followed or tried to get into the pouch of a strange female, often leading to aggression towards the young. He described (1987, p. 5) older PY as eager to go on excursions from the pouch, but they could only re-enter at their mother's discretion.

For *M. parryi*, Kaufmann (1974) considered that although a PY usually left the pouch only when it wanted to, occasionally it was dumped out unceremoniously by its mother in the manner described above. Stuart-Dick (1987) concluded that in *M. giganteus* the female was primarily responsible for the timing of her infant's excursions from the pouch, in that the young were more likely to leave the pouch when the female was grooming, interacting with her PY or looking around.

My own observations of *M. eugenii* found a higher incidence of older young clearly leaving the pouch by their own efforts. Although these were captive animals, they were observed in large quiet outdoor yards from a permanent hide which did not disturb the mothers. Certainly for the first few weeks after FEP, excursions from the pouch occurred rarely during the middle of the day when the female was inactive, and more often early in the morning or late afternoon when the end or start of activity was preceded by a bout of grooming, including pouch cleaning. Older young, just before PEP, are likely to be out of the pouch at any time of day (Stuart-Dick 1987; author unpublished observations). Seasonal differences in periods out of the pouch and in pouch life are not well documented.

When a young *M. eugenii* was first out of the pouch, it stood just where it fell, and, usually within 30 s, got straight back into the pouch again. The mother made no attempt to leave her young. She might sniff at it, and as soon as it attempted to enter the pouch it was allowed to do so. This is reflected by the short duration of periods out of the pouch (Russell 1973). For the first 4-5 weeks after FEP, there were frequent short excursions, lasting generally less than one and not more than ten minutes. If the mother went away from her young, or it went away from her, no more than 1-2 metres at this stage, it immediately turned and went back to her and put its head into her pouch. Generally, the young just stood in front of its mother with its head in the pouch and perhaps groomed itself briefly. The mother might groom its back, tail or urogenital opening. Generally, the first attempt by the young to return to the pouch was successful; the young stood with its head in the pouch and kicked off with its hind legs and the mother co-operated fully.

For a young to get into the pouch successfully in any species of macropodoid, the mother has to co-operate in a synchronized operation. Slow-motion analysis of videotape (author unpublished) allowed this to be observed in detail in *M. eugenii*. The young put its head and forelimbs into the pouch, then kicked off with the hind-legs, at which point the female bent forward. The angle between the tibia and the foot did not change, but the upper part of the body bent forward at the hips. This brought the pouch opening closer to the ground and to the young. If this did not happen, the young was left with its head and forelimbs in the pouch, kicking with the hind-legs but unable to get over the lip of the pouch. The whole process happened quickly and smoothly, and the bending forward was imperceptible without slow-motion analysis. The head and thorax may lean forward, but if the mother did not bend forward at the hip, the young could not get in. Obviously, it is more important for the mother to bend lower and co-operate more in the first week or two after FEP. There are slight differences in the same process as described and illustrated for *M. rufus* by Sharman and Calaby (1964), in which the female clearly leans backwards, increasing the angle between tibia and foot, and lowering the pouch opening.

Late in pouch life, mothers have a variety of ways of preventing or discouraging young from entering the pouch. They may move away as the young puts its head in the pouch. A mother may push her young away more or less gently; she may hold it away from the pouch, or she may just not co-operate in its attempts to get in, by standing at the wrong angle. At this time it is not uncommon to see a young which has got so far as to have its head and forelimbs in the pouch, and is kicking off with its hindlegs, while its mother resolutely stands just a little higher than the angle which will let the young get in. Eventually after much vigorous kicking, either the young gives up or the mother bends a few more degrees and the young succeeds. In other circumstances, the mother may simply stay lying down in response to vigorous pushing at the pouch by the young with its head and forelimbs (Pestering, Fig. 8c, d). She may eventually stand up and let the young in, or she may not until a disturbance occurs — bird noises, thunder, another animal going past — when she quickly stands up and the young is in the pouch within three seconds. If the pouch is completely relaxed, the young cannot get in, because it has nothing to push against.

Stuart-Dick (1987) determined for *M. giganteus* that when the young was out of the pouch for longer periods and the distance between the mother and the young had increased, the mother spent more time alert, looking round "surveying". In response to a perceived threat she gave a soft clucking vocalization and generally approached her young (as it approached her), and effectively retrieved it. This was particularly important if other animals were about, since when alarmed the young might approach the wrong female, try to get into her pouch and be chased away, becoming thoroughly alarmed and confused. The behaviour of the mother generally ensured that she was there for her young to approach, since visual recognition, at this stage at least, is not very well developed (see below).

In the period of transition towards the end of pouch life, the behaviour of the mother and her young changes markedly. I shall illustrate this with some data on *M. eugenii* in captivity (Russell, unpub. obs.).

Females with young of known age (aged to within one week from pouch checking, and later aged by measurements) were observed from a hide. They were watched for 3 h a day on alternate days for three weeks before the predicted date of PEP and for 2-3 weeks after actual PEP. In the early stages, all periods out of the pouch were scored for all individuals; later, focal sampling was used to accumulate at least 3 h detailed observation on each mother-young pair per week. Behaviour was sampled at 30 s intervals and the frequency of all interactions was scored. Individuals varied in age at PEP, and records are standardized to weeks before and after PEP, which was determined as the day after the last day on which the young was seen in the pouch. This was tested in each case by approaching the young when it was out of the pouch; before PEP, young returned to the pouch at this disturbance.

A large increase in time spent out of the pouch did not occur until 1-2 weeks before PEP (Table 1). Even two weeks before PEP, although the young was out of the pouch more often most periods were less than 10 min, with an occasional longer period. It was only in the week before PEP that young spent periods of more than one hour out of the pouch, and most excursions were for more than 10 min.

Even three weeks before PEP, most of the time was spent less than one metre from the mother; the difference from 100% is accounted for by a few brief excursions away by the young and periods when mother and young lay close together but less than one metre apart (Fig. 3). The distance between mother and young remained generally less than one metre until the week before PEP, when the young was spending longer periods out of the pouch. During these periods, the female frequently moved away from her young, as far as the yard would allow (>20 m), generally grazing, and the young remained, standing or lying down, until eventually she came back.

While the young was out of the pouch during this time, play with the mother began to occur; generally play-fighting or sparring (Russell 1973; Croft 1981a), with some fast-running play as well, where the young dashed away from the mother and back to her, immediately going to the pouch and pushing its head in briefly. Such play was at its highest level 2-3 weeks before PEP and declined thereafter (Fig. 4a).

It was very noticeable during these three weeks before PEP that, whenever the young came close to its mother after having been separated by more than about a metre, it immediately put its head

Table 1. Number and duration of periods out of the pouch during the three weeks preceeding permanent emergence from the pouch for two individual young of *Macropus eugenii* (based on unpublished observations by the author).

Weeks before PEP	Animal number	Hours observed	Number of periods out of pouch Total	/hour	Median duration (mins) (range)
3	14	11.25	12	1.1	1 min 24 s (20 s – 10 min 25 s)
3	21	11.0	13	1.2	4 min 24 s (8 s – 37 min)
2	14	7.25	9	1.2	6 min 10 s (15 s – >14 min)
2	21	7.5	11	1.5	7 min 9 s (9 s – >20 min)
1	14	11.75	9	0.7	29 min (1 min – >2.75 h)
1	21	10	8	0.8	40 min (45 s – >50 min)

in the pouch, often only briefly (Fig. 4b). This was nothing like the behaviour related to suckling, and had all the appearance of the young seeking contact with its mother and the pouch. It was comparable to the clinging behaviour seen in young primates, where there is also much early mother-young contact from which the young has to become independent.

A significant feature of behaviour during this time was a sudden increase in the frequency of the mother leaving her young and the young approaching its mother, when the distance between them changed from <1 m to >1 m and vice versa (Figs 5a,b). In weeks 3 and 2 before PEP, mother and young were <1 m apart more than 95% of the time. During weeks 3, 2, and 1 before PEP, median frequencies of mother-leaves young and young-approaches-mother were

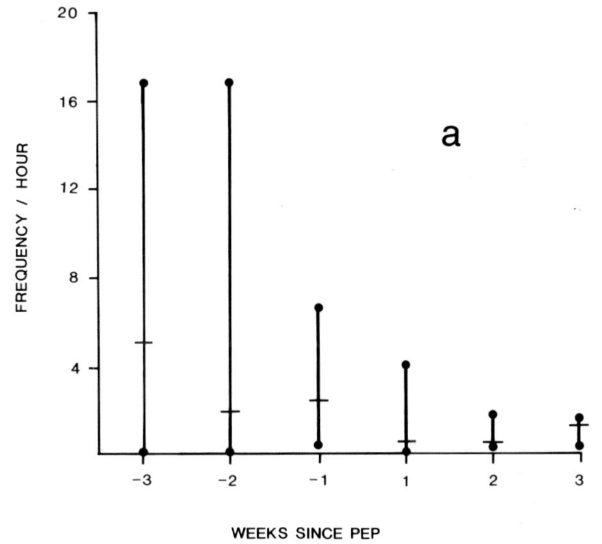

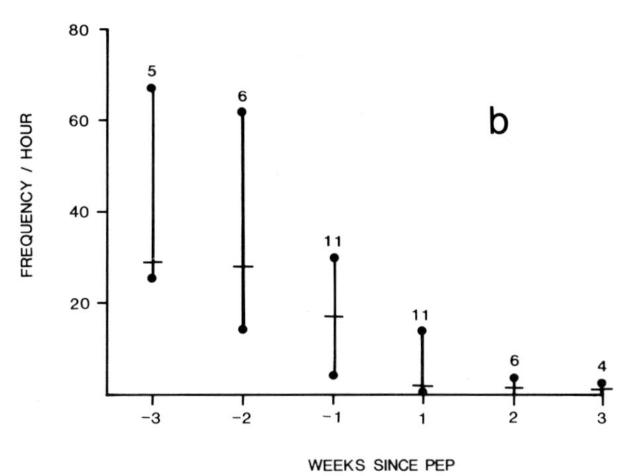

Fig. 4. (a) Play behaviour between female *Macropus eugenii* and their offspring. Median and range of frequency per hour of play interactions, from three weeks before permanent emergence to three weeks after. (b) Frequency of head-to-pouch by young *Macropus eugenii* when temporarily out of the pouch and by YAF. Each insertion of head to pouch was scored, regardless of the length of time the young remained with head in pouch. Numbers are number of mother-young pairs sampled.

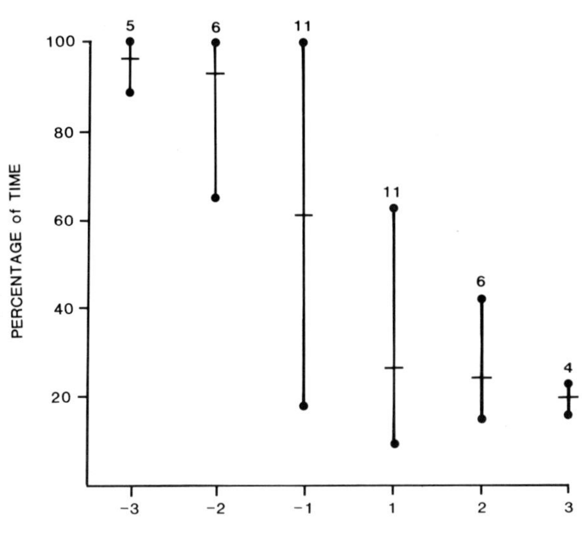

Fig. 3. Distance of young *Macropus eugenii* from its mother: median and range of proportion of time spent <1 m apart, from three weeks before permanent emergence to three weeks after. Numbers are numbers of mother-young pairs sampled.

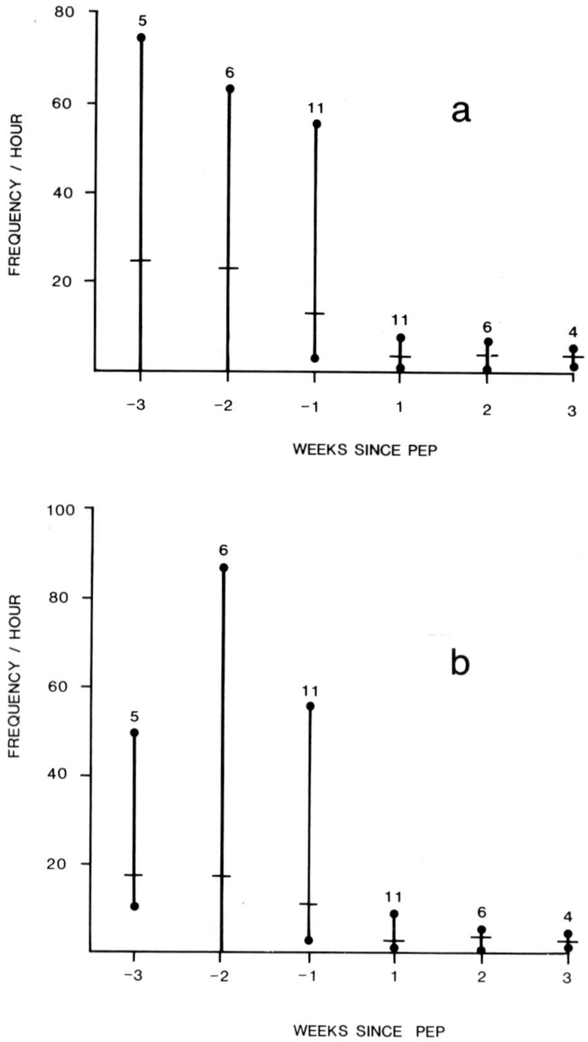

Fig. 5. Female *Macropus eugenii* and their young. Frequency of (a) mother leaving young; (b) young approaching mother. From three weeks before to three weeks after permanent emergence from the pouch. Numbers across the top of the figure are the number of mother-young pairs sampled. Approach or leave involves a change in distance apart from >1 m to <1 m or vice versa.

very high, with an extreme value for one pair of more than 50 times an hour. What this measured is behaviour which started 2-3 weeks before PEP; it involved the mother walking slowly a few steps (1-2 m) away from her young. The young immediately walked to her, frequently putting its head into the pouch; the mother then walked away again, the young went to her and so on. At first it happened four or five times during a short period (1-2 min) out of the pouch, later 2-3 times in 1-2 h. After PEP it stopped, suggesting that it was not just abnormal behaviour or a captive stereotypy. It had the appearance of a "training" process, of the "when I go, you follow" kind; at some stage, too, the young must improve its recognition of its mother. It will take more detailed data, specifically collected, to elucidate this. Later, just before and after PEP, the mother went away and the young did *not* follow. I am not certain what signalled "stay" or "follow"; possibly the rate at which the mother moved is important, and whether she walked or hopped.

I have described how mothers may keep young out of the pouch and also how periods out of the pouch increase in frequency and duration as the young advances towards PEP. To what extent was the mother responsible for this increase by keeping her young out? One might expect that just before PEP there would be an increase in refusals by the mother to let her young into the pouch. What actually happened was a fall in the success rate of attempts to re-enter the pouch. Initially, just after FEP, all attempts succeeded; by three weeks before PEP, many attempts were made,

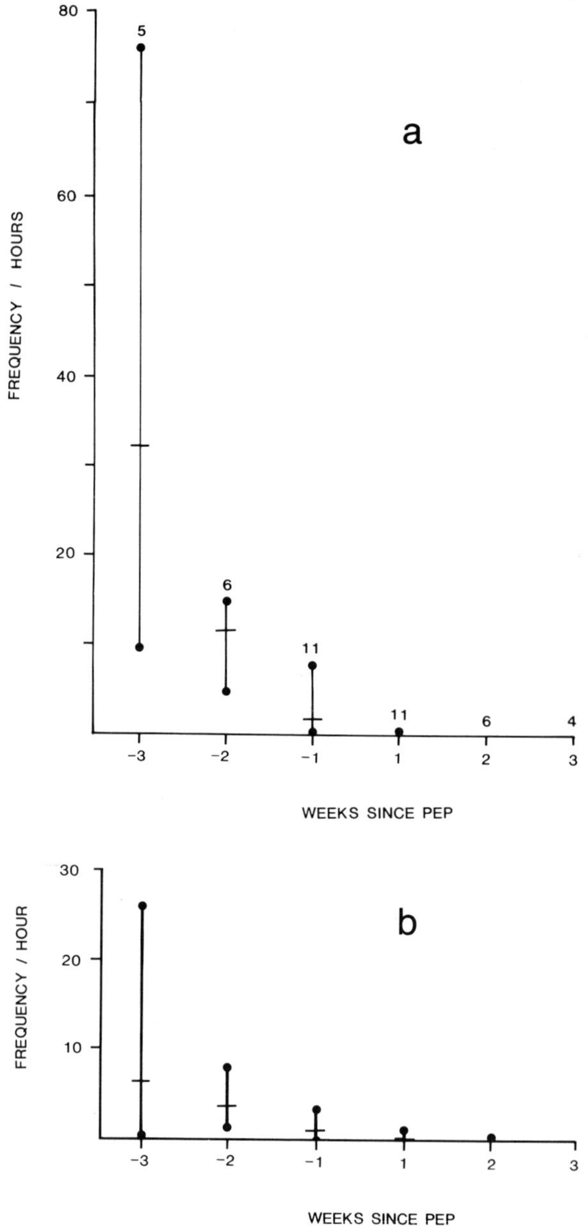

Fig. 6. Attemps by young *Macropus eugenii* to return to the pouch which (a) succeeded or (b) failed. Median frequency/hour and range. Numbers at the top of the figure are the number of mother-young pairs sampled.

but some did not succeed (Fig. 6a,b); two weeks before PEP, there were fewer attempts, with a median success rate of 2.8 successes per failure (range 1.3-10, N=6). In the last week before PEP, very few attempts were made, and the median success rate was 1.4 per failure (N=11), ranging from 100% failure to 100% success from about two attempts per hour. In the week after PEP, there were none of the frantic attempts to get in that might be expected; one or two attempts, but that was all.

There was never at any stage a high level of obvious aggression by the mother related to keeping her young out of the pouch. In Fig. 7, all possible types of aggressive behaviour towards the young are scored, together with behaviour where the female tried to move away from her young which had its head in the pouch, dragging the young with her. The main agonistic behaviour towards young at this stage involved pushing the young away from the pouch. Most unsuccessful tries were due to lack of co-operation. Thus by the time of PEP, the young has learned that most of its attempts to enter the pouch will not succeed, and that only in cases of extreme alarm will it be allowed in; the final transition is very quickly made.

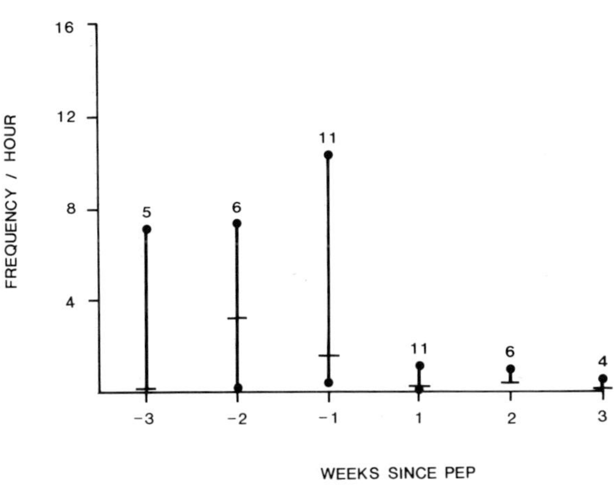

Fig. 7. Obvious attempts by female *Macropus eugenii* to keep young out of the pouch, during the three weeks before and after permanent emergence. Numbers at the top of the figure are the number of mother-young pairs sampled.

It was also clear that during this stage, the mother, by refusing to allow the young back into the pouch on every occasion when it tried, was controlling the amount of time that her young spent out of the pouch. If something alarmed her, she allowed the young to re-enter immediately. Field observations confirm this. Johnson (1987, p. 5) says "if the mother consistently prevented the young from returning to the pouch for more than a few days, it stopped attempting to do so". It was not possible to designate the exact instant of permanent emergence.

I have not made any distinction here between behaviour of mother towards male and female young; there is some indication that there are differences, possibly due, at least in part, to differences in the behaviour of male and female young (Johnson 1987; Stuart-Dick 1987; Stuart-Dick and Higginbottom (this volume).

PEP-Weaning

After PEP, lactation continues for a variable time, depending on species, body size and reproductive pattern. The time from birth to weaning is related to maternal body size (Russell 1982; Lee and Cockburn 1985). After PEP, the young is generally referred to as young-at-foot (YAF).

Suckling occurs in discrete bouts, when the young stands with its head in the pouch of its mother. It is impossible to tell by observation whether the young is really being suckled, but there are occasions when the young goes to its mother and puts its head in the pouch very briefly, especially after it has been away. After PEP, the YAF puts its head into the pouch only infrequently, and the occasional longer bouts are undoubtedly bouts of suckling. Field observation on *M. rufogriseus* (Johnson 1987) found that bouts of suckling lasted about 8-11 min, and occurred at a rate of about 0.2 bouts/h or 5/d, with a tendency for larger YAF to suck longer. For *M. giganteus*, Stuart-Dick (1987) found that bouts were about 11-15 min long (increasing with age of the YAF) and she estimated that the YAF were probably being suckled 8-12 times a day. Most bouts of suckling were initiated or terminated by the young (Russell 1973; Johnson 1987; Stuart-Dick 1987). To initiate a bout of suckling, a YAF of *M. eugenii* frequently pestered its mother if she was lying down by pushing at her pouch, at her head and shoulders, or by climbing on her back until she stood and allowed it to get its head into the pouch (Fig. 8c,d). While suckling her YAF, the female stood relaxed; eventually the young took its head out of the pouch, and frequently some further interaction with the mother followed; sniffing, sparring or play-fighting (Fig. 8e,f).

The typical view of the relationship between mother and YAF after PEP is that the young follows its mother everywhere and is suckled from time to time. This is more or less true for the larger kangaroos. Kaufmann (1974; *M. parryi*) and Croft (1981a,b; *M. rufus* and *M. robustus*) observed that females with YAF had a more restricted ranging area than other females. From following known female *M. giganteus*, Stuart-Dick (1987) confirmed this, showing that at the time of PEP and when they had a small YAF, females had smaller home ranges and were more likely to be isolated from the centres of activity of other females.

Fig. 8. Interactions between female *Macropus eugenii* and their young. (a) Female licks young in pouch (b) Female sits tail forward (c) Young pestering mother lying down; push at head (d) Young pestering mother lying down; climb on back (e) Nose-to-nose sniffing interaction (f) Play-fighting (sparring).

Information has been accumulating that the smaller kangaroos and rat-kangaroos are rather different, and there is now circumstantial evidence from a number of species of "hiding young", reviewed by Johnson (1985, 1987), whose evidence of hiding in *M. rufogriseus* is the most significant to date. An early report by Hornsby (1978) described young rock-wallabies (*Petrogale penicillata* and *P. xanthopus*) left behind in safe places while their mothers travelled long distances to feed or drink. Johnson (1987) observed YAF of *M. rufogriseus*, especially in the first month after PEP, alone in dense cover and apart from their mothers which were grazing at some distance in the open. I have observed similar behaviour in *M. eugenii* in captive animals at the time of PEP (Fig. 9). During daytime periods of inactivity, mother and YAF rested together, generally lying or sitting 1-2 m apart. Towards evening when the mother began to graze, it was obvious that the YAF remained alone for long periods, as far distant from its mother as the yard allowed (30 m), and made no attempt to follow her. In this situation, the YAF generally stood or lay in a small shed in the corner of the yard. After some time, the mother came back to the shed, the YAF came out to meet her, and a bout of suckling generally ensued. By the second week

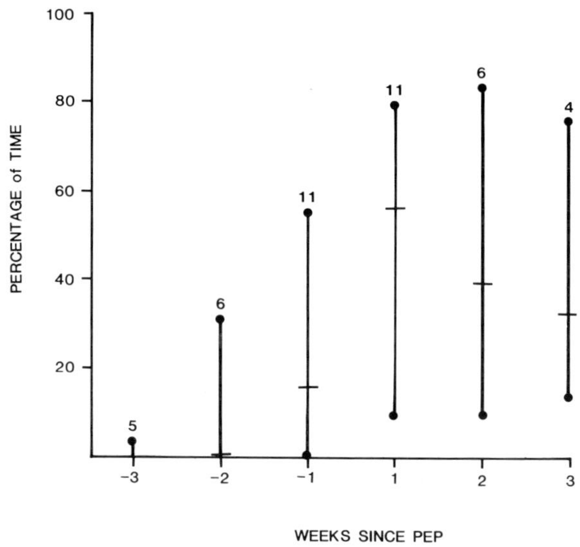

Fig. 9. Distance of young *Macropus eugenii* from their mothers. Median per cent of time spent >5 m apart, and range; from three weeks before to three weeks after permanent emergence. Numbers at the top of the figures are the number of mother-young pairs sampled.

after PEP, the YAF was following its mother more and spending less time alone (Fig. 9). It was not clear what determined whether the YAF stayed or followed.

Johnson observed a number of separations, where the YAF entered cover while the mother went to feed in the open. The young in cover stood, crouched or lay down, but did not freeze if alarmed, as is typical in ungulates. In other respects, hiding behaviour in smaller macropodids is similar to that of ungulates. The difference between large and small macropodids may be in part related to the more open habitat of the larger species and the larger absolute size of the YAF (and therefore greater speed).

It is clearly established that in general, when mother and young *are* in proximity, the YAF is responsible for staying with its mother. Observations on a number of species in captivity and in the field have shown that the young moves to maintain its position relative to its mother much more often than the mother maintains her position relative to her YAF. She moves away and it is up to her young to follow (Russell 1973; Croft 1981a,b; Johnson 1987; Stuart-Dick 1987). This is not to say that, as with young out of the pouch, she is not aware of her young and its position, and ready to move towards it in the case of an alarm or if it becomes lost or disoriented and gives the loud coughing distress call of macropodoids. Nevertheless, as the YAF gets older, it does not stay so close to its mother (Fig. 10), the frequency of social interactions declines and those that do occur are most often during the contact associated with suckling bouts. Excursions away from the mother become more extensive and last for longer.

Trivers (1974) demonstrated that an individual offspring will always tend to favour a longer period of parental investment (in this case, lactation) than it is in a mother's best interests to provide (considering her possible future production). The expected "weaning conflict" (Trivers 1974) is not very obvious in macropodids. Johnson (1987, p. 5) "at no time (saw) a female display impatience while her offspring suckled". Kaufmann (1974) and Croft (1981a,b) made no comment at all. Stuart-Dick (1987) predicted an increase in the termination by the mother of bouts of suckling at the time of weaning and the occurrence of some agonistic behaviour towards the young. In *M. eugenii*, the size of the suckled mammary gland declines after PEP (Stewart 1984) and there is a correlated decline in the level of prolactin (Hinds and Tyndale-Biscoe 1985). As the young begins to eat more solid food when it leaves the pouch, so milk production and milk ingestion decline (Cork and Dove 1986). It is not possible to know how much of the time that a YAF has its head in the pouch is actually spent sucking and how much it is just standing "in contact". The way in which, at an earlier age, young returned to their mother after each brief excursion away and put their head in her pouch briefly, not to suck but possibly seeking contact, suggests that the social function of sucking may last longer than its nutritive one.

The clearly established quantitative differences in the pattern of interaction of mothers with male and female YAF are discussed elsewhere (Stuart-Dick and Higginbottom, this volume).

Post weaning

In larger macropodids which have been studied in the field there is now clear evidence of continued association of mothers and offspring

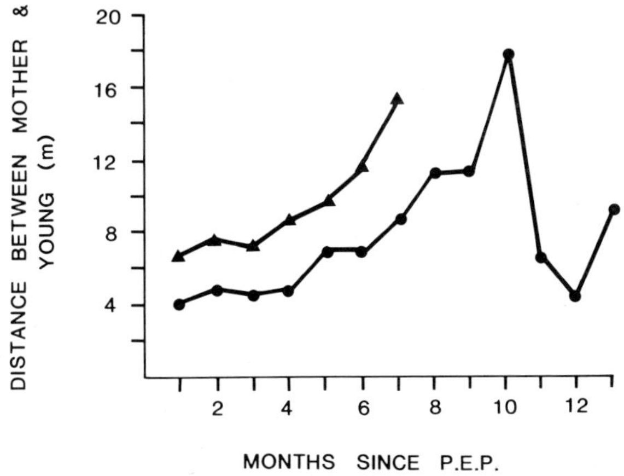

Fig. 10. Median distance recorded between male (filled circles) and female (triangles) *Macropus rufogriseus* young and their mothers when in association with them, in relation to age of the young (from Johnson 1987).

(usually daughters) after weaning. Sub-adult males appear to disperse from their mother's home range 1-2 years after weaning, but sub-adult females remain on their mother's home range area to sexual maturity and beyond (Kaufmann 1974; Johnson 1987; Stuart-Dick 1987), and rarely disperse very far, giving rise to kin-groups of females (termed matrilines). The aspects of this behaviour which relate to continued investment in offspring are discussed elsewhere (Stuart-Dick and Higginbottom, this volume). The behaviour on the part of the mother is continued tolerance of her offspring, continued association in the same area, and interactions between mother and daughter which are generally non-agonistic.

Smaller macropodids and potoroids are less well known in the field. Sexual maturity occurs earlier in smaller species (see Tyndale-Biscoe and Renfree 1987, Table 1). The more solitary social organization of most species would suggest that such association does not occur to the same extent, although overlap of home ranges of mothers and daughters may still occur.

Pouch life

The period from birth to PEP, generally called Pouch Life, is commonly used as a summary of life history pattern, maternal behaviour and reproductive strategy, and quoted as characteristic of a species. However, times quoted for FEP and PEP should be treated with caution. First excursions are brief, and any disturbance sends the young back into the pouch. A brief daily inspection will not always pick up FEP; unless longer observation is possible, the ability of the PY to stand and inspection of its feet for signs of soil are perhaps the best guide (e.g., Poole and Merchant 1987). PEP is also difficult to determine precisely without continuous observation. Close to PEP, casual inspection may suggest that the young has left the pouch, but it may still return to the pouch for brief periods. Perhaps the best indication is whether the young returns to the pouch when alarmed.

Thus the measurement "pouch life" is rarely determined to the exact day, although it is generally given that way. Variability in Pouch Life is rarely considered, but there is evidence that pouch life can vary with season (Hinds and Tyndale-Biscoe 1985; Johnson 1985; Stuart-Dick 1987). If thermal intolerance between mother and PY is an important cause of pouch vacation (Janssens and Rogers, this volume), seasonal differences in pouch life are a likely consequence. Green *et al.* (1988) found that PY of *M. eugenii* of the same age varied considerably in body weight; whether this is expressed as lower weight at PEP or longer pouch life is not known. Nevertheless, Pouch Life is the best available measure for comparisons between species of the duration of intensive maternal care.

Pouch Life, like many other life history attributes (Western 1979; Russell 1982; Lee and Cockburn

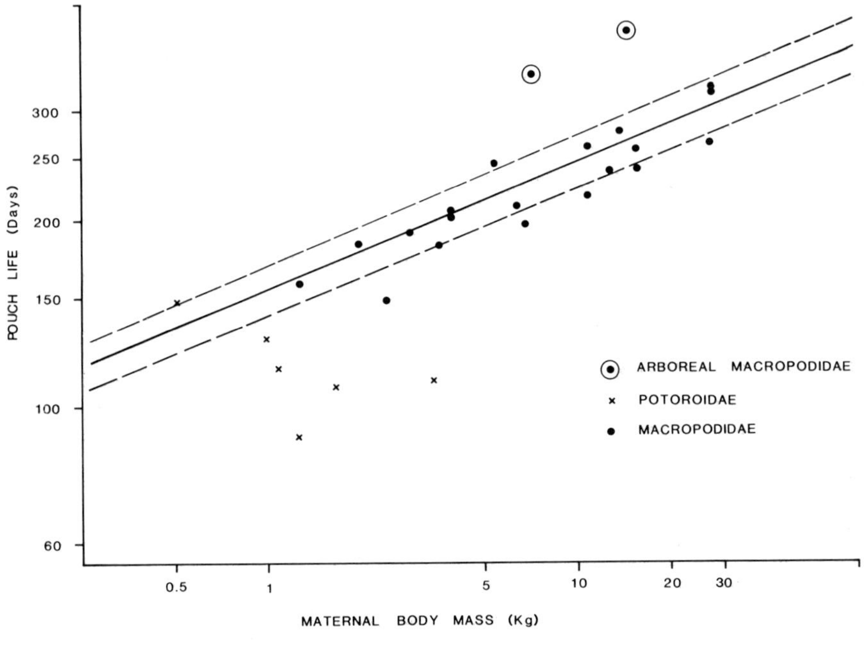

Fig. 11. Relationship of pouch life to maternal body mass. Potoroidae (N=6): crosses; Macropodidae (N=23): filled circles, including two tree kangaroos: *Dendrolagus*: dot in circle). The regression line is that calculated for Macropodidae (N=21) without *Dendrolagus* spp. (for equation, see text). The dotted lines parallel to the regression line indicate pouch life 10% above and below, allowing for errors in estimating pouch life.

1985; see Calder 1984 for a comprehensive review) is to a large extent a function of the body size of the species. This was reviewed in detail for all marsupial families by Russell (1982) and is also discussed by Smith and Lee (1984) and Lee and Cockburn (1985). Data for the families Macropodidae and Potoroidae are presented in Fig. 11. Pouch Life increases with body size; the scaling is described by the allometric equation

$$PL = 19.99\ M^{0.27}$$

where PL is pouch life in days, and M is maternal body mass in grams (N=29 species). The regression coefficient (S.E.=0.04) is significantly different from zero ($t=7.3$, $df=27$, $P<0.001$).

This is similar to the equation given by Russell (1982, and quoted by Lee and Cockburn 1985) based on 22 species. However, inspection of Fig. 11 shows that the larger potoroid species do not fit the line very well, and that in fact, for the six species, there is no correlation between pouch life and body size. If the relationship between pouch life and body size is calculated for the 23 macropodids, the equation is

$$PL=35.22\ M^{0.21} \quad (r^2=0.59).$$

Much of the variability is due to the two species of *Dendrolagus* which have very long pouch lives for their size. Slow growth rates appear to be characteristic of arboreal mammals (McNab 1986). If these two species are removed, the regression equation changes little to

$PL=39.15\ M^{0.20}$, but r^2 increases to 0.77.

For the terrestrial Macropodidae, there is a close relationship between pouch life and body size. For most species pouch life lies within 10% of the predicted value. Why then should there be apparently no relationship between pouch life and body size in the Potoroidae? The trend to increase in life history attribute with size is so universal (Calder 1984) that it suggests either selection for shorter pouch life or some limitation to increase. Thompson (1987) suggests that the increased litter frequency which follows from shorter pouch life may be important for the Potoroidae.

Communication, recognition

Auditory communication is important in the maintenance of contact if mother and young become visually isolated. In this situation, the young gives a loud harsh coughing vocalization which alerts the mother, and usually brings her towards her young and re-establishes contact. There are differences between individuals in the nature of this call, and obviously the possibility of individual recognition exists, although in captivity, a distress call attracts all adults within range.

Females produce a soft clucking vocalization and Stuart-Dick (1987) describes the context of this "chkk" vocalization in *M. giganteus* as when a female is inviting pouch entry by a young temporarily out of the pouch, especially in response to an alarming situation. Kaufmann (1974) said he did not hear any such sound, but was possibly not close enough to do so, as the animals he studied were less habituated than those observed by Stuart-Dick. Russell (1973) observed *M. rufus* in captivity at very close range, and heard this soft clucking vocalization only, when after a sudden disturbance, females approached their young as they approached her; in normal occurrences of pouch entry it did not occur. Sharman and Calaby (1964) also describe this vocalization as occurring in *M. rufus* females when young were approaching and entering the pouch. Female *M. rufus* also make a syllabic sound similar to the distress call of young but louder, harsher and more cough-like. Sharman and Calaby (1964) heard it given by females which had temporarily lost their young and hopped round actively looking for them, and Russell (1973) described it as occurring in a female approaching a distress-calling young. It also appears to occur in other stressful situations, such as an animal fleeing from an attacker (Croft 1981b; Author, pers. obs.), and during mating chases of *M. giganteus* (Stuart-Dick 1987).

The significance of visual displays in macropodoid social organization and our own ability to recognize individual animals by sight suggests that some visual information also may be important in recognition, but as yet we know nothing about this.

Olfactory information is important in individual recognition. Approach to another individual is always followed by nose-to-nose sniffing (Russell 1985); a lost young which approaches a strange female is sniffed before agonistic rejection (Croft 1981b), and a mother sniffs her own young when it approaches her. Contact between mother and PY would allow olfactory recognition to become established before the young leaves the pouch for the first time (Fig. 8a). Olfaction is an important sense in marsupials (Russell 1985), and the frequency of "sniffing" interactions suggests that it is important in maternal recognition of young, as it is in noenate lambs (Alexander and Stevens 1982), but experimental confirmation is lacking.

Some females will accept transfers of PY of their own or other species, and in captivity in large outdoor yards female *M. eugenii* have been known to accept a second young into their pouch, or to exchange young (Author, unpublished observations). It is impossible to say whether this indicates lack of recognition or merely that recognition of a PY as strange does not impede its acceptance; in the pouch it would soon acquire the "right" smell. Successful transfers of older PY (after FEP) have been made (Merchant and Sharman 1966; Rose 1986); it may be that complete recognition by the mother does not occur until after PEP. Even so, some females are particularly tolerant of strange YAF. Sharman (1965) was able to foster a second YAF to a number of female *M. rufus*, although other females would not accept foster young.

Young do not appear to recognize their mothers when they first leave the pouch. By the time of PEP, one would expect that they would have learned to do so, but this has not been investigated. One possible function for the more solitary behaviour of females of more social species when they have large PY may be the need for recognition to develop without too much possibility of confusion.

Control of maternal behaviour

Two principal phases are distinguished in the regulation of maternal behaviour in eutherian mammals (Rosenblatt and Siegel 1981): the onset of maternal behaviour, governed by endocrine processes, and its maintenance, governed mainly by sensory stimuli and motivational processes. In marsupials, the same steroid hormones are present (in general) as in eutherian mammals, although not always with the same function. Understanding the function of various hormones in regulating the processes of reproduction in marsupials has progressed far ahead of understanding their role in the initiation and maintenance of maternal behaviour (Tyndale-Biscoe and Renfree 1987). Renfree *et al.* (this volume) delineate the hormonal background to the behaviour of macropodoids at the time of parturition and compare this with information available for eutherians.

Among the closely related members of the Families Macropodidae and Potoroidae, considerable variety exists in the annual pattern of reproduction and in the stage of the pattern at which the female gives birth (Table 2). In non-seasonally breeding species, an active pregnancy may be concurrent with the later stages of pouch life of a previous offspring, so that when the female gives birth she is also suckling a YAF which has just left the pouch. In species where the length of gestation is close to the length of the oestrous cycle, birth may occur in association with a post-partum (as in *M. rufus* and *B. gaimardi*), or a pre-partum oestrus (as in *Wallabia bicolor*). Where oestrous cycle length is several days longer than gestation, post-partum oestrus is usually suppressed until later in pouch life, when the female may return to oestrus (as in *M. giganteus*) and conception may occur, but the resulting blastocyst is subject to diapause. Such variations in the possible hormonal status of female kangaroos at the time of birth and during pouch life suggest that the elaborations of the reprodoctive pattern and the hormonal changes associated with oestrus itself have had little effect on the initiation of maternal behaviour at the time of birth. In most other marsupial families, birth occurs divorced from oestrus, and without the complications of embryonic diapause.

It is important to make a distinction between the behaviour of the mother at the time of parturition, and subsequent behaviour when the young is in the pouch. The process of birth and the hormonal processes associated with it are unnecessary for the initiation of subsequent maternal behaviour towards the PY, at least in kangaroos and possums. Sharman and Calaby (1964) reported the successful transfer of new-born young less than one day old to the pouch of un-mated foster mothers in *M. rufus*. The transfers were done at about the time the foster mothers were in oestrus. The foster-young were successfully reared until they left the pouch, and grew at the same rate as PY reared by their own mothers. Even a virgin female successfully reared

Table 2. Examples of variation in possible reproductive status of female macropodoid at time of birth and during pouch life (sources as for Table 3).

Species	Pregnant[2] during last weeks of pouch life	Suckling YAF + PY concurrent	Post-partum oestrus	Pre-partum oestrus
Macropus giganteus[1]	±	+	–	–
Macropus fuliginosus[1]	±	+	–	–
Macropus rufus	±	+	+	–
Wallabia bicolor	±	+	–	+
Macropus eugenii[3]	±	±	+	–
Setonix brachyurus	±	±	+	–
Bettongia gaimardi	±	+	+	–

[1]Females of both species of grey kangaroo may undergo oestrus during pouch life. Conception is unlikely but possible, depending on nutritional state. Embryonic diapause does not occur in *M. fuliginosus*.

[2]A female which did not mate at a previous post-partum oestrus would have no diapaused blastocyst to resume development.

[3]It is possible for a female *M. eugenii* to give birth in June if a previous PY is lost, and then to mate post-partum. This young would still be in the pouch at the time of resumption of development of the diapaused blastocyst in December.

a fostered young. Successful transfers of neonates to virgin or unmated females have been made in two other species of kangaroo, *M. giganteus* (Merchant and Sharman 1966) and *M. eugenii* (Tyndale-Biscoe (unpublished), quoted by Findlay, Ward and Renfree 1982). Part of the explanation of this phenomenon lies in the similarity of the pregnant and non-pregnant cycles. The hormonal changes of the oestrous cycle are sufficient to prepare the mammary glands for a normal lactation which will ensue provided a sucking stimulus is applied to the gland at the correct stage of the cycle. Even the prolactin surge before birth is not necessary, since only rarely does it occur during the non-pregnant oestrous cycle in *M. eugenii*.

In eutherians, the mammary glands develop during pregnancy, and lactation begins at a high level with maximum yields reached soon after birth. In marsupials, however, neonates weigh less than 1 g at birth, and at first grow slowly so that maximum milk production is not reached for some time, several months in the larger kangaroos. Related to the intial small size of the PY, levels of prolactin are very low in the first half of lactation, and in *M. eugenii* are not distinguishable from pre-partum concentrations until after 100 d of lactation (Hinds and Tyndale-Biscoe 1982b). Nevertheless, it is clear that the sucking stimulus has a significant effect on the hormonal status of the female, and probably also on aspects of maternal behaviour. If the PY dies or is removed from the teat, the mammary gland involutes (Findlay 1982) and during this time the corpus luteum resumes growth. Renfree (1979) observed that denervation of the suckled mammary gland during early lactation did not impair milk secretion although inhibition of the corpus luteum was removed. Clearly some information from the pouch, perhaps from the presence of the PY, in addition to information from the suckled gland, was still getting through to the pituitary gland, and was enough for the maintenance of lactation, although not enough to maintain inhibition of the corpus luteum (Hinds and Tyndale-Biscoe 1982a).

Obviously at this early stage in lactation, the whole system has to be exquisitely sensitive for the sucking stimulation from the neonate to have the effect it does on maternal physiology. Lincoln and Renfree (1981) demonstrated that the suckled mammary gland is extremely sensitive to oxytocin early in lactation and becomes less sensitive as lactation progresses. They suggested that the activities of a small PY could result in a small but sustained release of oxytocin sufficient to influence the highly sensitive mammary gland. In addition, the pressures generated at milk ejection early in lactation are sufficently high to force milk from the nipple, suggesting that as long as the PY is permanently attached to the teat, response of the mammary gland to any activity of the PY, not just to sucking, is possible.

This sensitivity to oxytocin early in lactation does not explain how lactation can continue in the presence of the low levels of prolactin which are found at that time (no different from prepartum levels) and how lactation can be established even in the absence of the potentially priming surge of prolactin which occurs at parturition. Although prolactin concentrations do not change, the sensitivity of the tissues to prolactin does change. Prolactin receptor concentrations in non-lactating mammary glands in *M. eugenii* are highest at about the time of parturition, then decline rapidly in the unsucked glands (Stewart 1984). In a gland sucked by a newborn young, prolactin receptors increase markedly within 7 d of birth, and receptor concentrations per cell continued to increase up to 12 weeks of lactation, and then remained constant, although numbers of cells (and of receptors) increased (Stewart 1984).

Once the PY is in the pouch, the most significant aspects of maternal behaviour during this first phase when the PY is attached to the teat are simply acceptance of the young, pouch cleaning and grooming of the PY. Acceptance of the PY is taken for granted and has never been investigated, but just as one may ask how it is that a female rat allows her pups to suckle rather than eating them, so one may ask what it is that prevents a female kangaroo from removing the young from the pouch during pouch cleaning. Despite the small size of the PY (in *M. eugenii*, mass at birth 0.4 g; after 40 d, 10 g), its presence in the pouch is sufficient for the level of pouch cleaning in females with small PY (<40 d) to be significantly higher than in females with no PY. The level of pouch cleaning increases during pouch life, but has already reached a high level before the young begins its period of rapid absolute growth at about 100 d (Russell and Giles 1974), which is when higher levels of prolactin are found. The consumption of urine and faeces by the female during pouch cleaning provides a changing olfactory and chemical stimulus to the mother with a possible but unknown influence on behaviour.

Once the young grows and becomes more active there is more interaction between mother and young, and co-ordination of behaviour becomes important. As the young begins to eat solid food and becomes less dependent on its mother for milk, so the suckling stimulus declines. Finally the young no longer returns to the pouch (PEP). What determines the time of first emergence and the end of pouch life is not yet clear. The behavioural aspects of the change have already been discussed. Janssens and Rogers (this volume) describe the physiological changes in the

Table 3. The variability in duration of pouch life in young kangaroos and wallabies.

Species	N	Age of young kangaroos at PEP (±SD)	CV (%)	Reproductive state of female at PEP	Author
Bettongia gaimardi	10	106.8+1.7	1.6	birth, oestrus after <1 d	Rose 1986
	7	105.93.2	3.0	oestrus, no birth	
	3	112.7+3.5	3.1	anoestrous	
Macropus agilis	15	219.3+9.7	4.4	birth, oestrus 1-3 d after PEP	Merchant 1976
Macropus robustus					Poole and Merchant 1987
M. r. robustus	26[1]	260+5	1.9	birth, oestrus	
M. r. erubescens	25[2]	243+7	2.9	1-2 d after PEP	
Macropus rufogriseus					
M. r. rufogriseus	15	274.9+7.2	2.6	birth, oestrus 27 d after PEP or seasonal anoestrus	Merchant and Calaby 1981
M. r. banksianus	9	284.9+12.3	4.3	birth, oestrus 22 d after PEP	
Macropus rufus	16	235+8.5	3.6[3]	birth, oestrus <1 d after PEP	Sharman and Calaby 1964
	8	235.4+4.4	1.9[3]	oestrus, no birth	
Petrogale xanthopus	7	194.9+3.9	2.0	birth, oestrus 1-4 d after PEP	Poole, Merchant and Calaby 1985
Macropus parma	10	211.9+1.0	0.5	17% oestrus	Maynes 1973
Macropus giganteus	22	319.2+18.4	5.8	birth, no oestrus in some or female cycling does not conceive	Poole 1975
Macropus fuliginosus	24	323.1+22.9	7.1	Lactational anoestrus	Poole 1975

[1]26 cycles for six females

25 cycles for seven females

[3]Assuming Sharman and Calaby gave the S.E. of the mean.

young which appear to be important in triggering FEP. Non-seasonally breeding species with embryonic diapause may have an embryo developing in the uterus while a large young is still in the pouch; that PY has to leave or be ejected before the new one is born. In such species as *M. rufus*, *M. agilis* and *Bettongia gaimardi*, this sets a limit to the end of pouch life, and suggests that approaching birth and oestrus may produce changes in the mother that signal the end of pouch life.

In species such as *M. eugenii, S. brachyurus*, and some *M. r. rufogriseus*, the female does not give birth again until well after one young has left the pouch. There is no oestrus to signal the end of pouch life. There is some indication that pouch life is more variable in species without regular diapause, but the end of pouch life is not easy to determine exactly without constant observation, and some of the variation may be due to observer error (Table 3).

The situation is complicated by those species with no post-partum oestrus which may undergo oestrus during the course of pouch life (*M. giganteus, M. fuliginosus, M. parma* and *M. parryi*). In *M. giganteus*, a female may return to oestrus 110 d or more after the birth of a PY. If she conceives, the ensuing blastocyst undergoes diapause, and resumes development late in pouch life and the young is born after PEP of the large PY (Clark and Poole 1967). Although diapause can occur, in the animals studied by Poole and Catling (1974) it was relatively infrequently that females conceived when they had a young in the pouch and most young were born well after PEP of the previous PY. In *M. fuliginosus*, diapause does not occur, although females may mate during periods of oestrus when they have a young in the pouch (Poole and Catling 1974). Thus the variable pouch life (Table 3) came from animals with a low incidence of diapause, and few females gave birth close to the time of PEP. In these four species, females may undergo oestrus while still carrying a PY. In some cases, the oestrus occurs when the PY is quite small, before it has begun to leave the pouch at all, so it would have no effect on re-entry. Later oestrus may affect re-entry to the pouch, and it may be that the proximate cause of variation in the pouch life of grey kangaroos is the very variable timing of oestrus. It is clearly a fascinating field for further investigation.

Sharman and Calaby (1964, p.77) noted that the post-partum oestrus which occurred after PEP in *M. rufus* was usually accompanied, or preceded, by a dramatic tightening of the pouch musculature and shrinkage of the pouch to less than half the size it was when occupied by the large young, suggesting that hormonal changes were responsible for the changes in the pouch which made it impossible for the young to enter. Rose (1986) found that in *Bettongia gaimardi*, on the night of birth and/or oestrus, there is a tightening of the pouch muscle which prevents re-entry of the young. If a large young was replaced by a smaller, the replacements were rejected on the night of birth, regardless of their age. Similarly, when small young were replaced by larger, the replacements stayed in the pouch for 9-17 d longer than the usual pouch life, not leaving the pouch until the female gave birth or returned to oestrus. Similar changes in the pouch also occur in *Thylogale billardierii* (Rose and McCartney 1982).

In various cross-fostering experiments, PY have been reared in the pouch of a female of another species. From such experiments, it appears that the time of PEP is appropriate for the species of the PY, not to the species of the foster mother, suggesting that it is the rate of development of the PY which determines the time of PEP, provided that the transfers are not made after a quiescent blastocyst has reactivated (Merchant and Sharman 1966; Clark 1968; Johnson 1981). Changes in the pouch of the female may signal the final PEP, but the rate of development of the PY determines how long it takes to initiate these changes.

The transfers made by Rose (1986) appear to contradict this, in that the PY stayed in the pouch longer, but in these transfers, all the PY were close to the end of pouch life (*ca* 98 d), and the older young transferred were not ejected from the pouch until the foster mothers had gone through the period of gestation after the quiescent blastocyst resumed development. Exchange of PY of *B. penicillata* at an earlier age (Smith, this volume) had no effect on the length of pouch life or the birth of a new PY, which was determined by maternal factors. All transfers so far described have been into the pouches of non-seasonally breeding females in which embryonic diapause can occur. Thus once the young reaches a certain size and suckling declines, the corpus luteum reactivates and the female will come into oestrus/give birth, which will signal the end of pouch life. Transfers of young between females of different sized species, even if the PY are evenly matched for age, introduces the problem of mammary glands with different levels of production, and therefore differences in growth. This is an area where many fascinating experiments are waiting to be done, including fostering PY onto females which are in seasonal quiescence when the PY is due to leave the pouch, as in *M. eugenii*, so that neither birth nor oestrus can affect PEP.

This is not the whole story, however. The amount of time spent in the pouch and the success of attempts to return to it change rapidly in a few weeks before PEP. It is unlikely to be due to approaching oestrus or birth, since it has been observed to occur in a range of species (*M. rufus,* Russell 1973; *M. parryi,* Kaufmann 1974; *M. rufogriseus,* Johnson 1987 and *M. giganteus,* Stuart-Dick 1987), including those (e.g., *M. eugenii*, Russell 1973 and this paper) where no birth or oestrus occurs following PEP. In those *M. giganteus* where the female is cycling but has no diapaused blastocyst, the next oestrus may signal the end of pouch life. The end of pouch life may be due to developmental changes in the PY, which induce corresponding changes in the female. Birth and oestrus may signal a definite end to a process already started in response to other physiological changes. Thermal intolerance by the mother of increased heat production by the PY seems the most likely candidate (Janssens and Rogers, this volume).

In eutherians, the hormonal changes near the time of birth provide the means for switching on maternal behaviour (Rosenblatt and Siegel 1981) which, once initiated, can be maintained largely by stimulation from the young. Similarly, there are probably two phases in the control of maternal behaviour in marsupials. The first phase relates to the behaviour of the female around the time of parturition. The stereotyped birth positions provide exactly the stimuli and route to the pouch that the barely functional embryonic nervous system of the newborn requires. Although no specific investigations of the hormonal bases of maternal behaviour at parturition have been done, the hormonal changes are there which could have the necessary effect on behaviour.

The second phase of regulation of maternal behaviour starts once the young is born and in the pouch. Little is required of the mother for some time until the young has grown and starts to do more than just wriggle from time to time and suck. The stimulation and hormonal changes of birth are not necessary for the performance of maternal behaviour in this phase. What is necessary is stimulation from the young, and the control mechanism is extremely sensitive to stimulation from the tiny PY early in lactation. A completely inexperienced female is thus an adequate mother, which suggests that the more important aspects of maternal behaviour are "built-in". To what extent experience contributes to a female's success in rearing young is not known. One might expect that it would be more important in such species as kangaroos which live longer, have one young at a time, and a long association with that young.

Much of what I have said in this section is speculative; I have tried to outline a fascinating but neglected problem, best exemplified by the following description from Merchant and Sharman (1966, p. 64) of a female *M. rufus* with two PY, her own and a *M. giganteus* foster young transferred at the age of 250 d:

> "Both young shared the single lactating teat and mammary gland available and both left and returned to the pouch on occasions for a period of 34 days, after which they vacated the pouch completely. At this time the red kangaroo young was 236 days old (i.e., due to leave the pouch) and the grey kangaroo 284 days old (i.e., 28 days younger than the average age at which grey kangaroo young leave the pouch permanently). The foster-young re-entered the pouch 13 days after vacating it, but the red kangaroo young never entered the pouch again. During the 13-day period when the pouch was empty it tightened up considerably, so that it would have been impossible for either young to enter, and the female came into oestrus. Both young were observed inserting their heads, in turn, into the pouch to suck from the lactating teat during this period. After re-entering the pouch the foster-young occupied it intermittently until it reached the age of 302 days, when it vacated the pouch entirely."

CONCLUSION

The functions of maternal behaviour are to protect, keep warm, and feed the young, especially if it is altricial. The aspects of this which relate to carrying the young attached to the abdomen (whether in a pouch or not) are specialized in marsupials and not shared with other mammals. For other maternal behaviour, such as the mother keeping in contact with her young, feeding her young when it is no longer carried, and the gradual process of weaning, there is no functional reason for macropodoid behaviour to be very different from that of eutherians, and it is not, provided comparisons are made with ecologically equivalent terrestrial herbivore species.

Parental care cannot be divorced from parental investment. The iteroparous life-history strategy of macropodoids means that although in one reproductive attempt they produce one young with relatively less total investment measured crudely as mass of young/maternal body mass) than in a litter of six *Antechinus* which at weaning weighs three times maternal body mass, they do it more than once, and over a lifetime they may invest the same by increased investment for longer in fewer young (Russell 1982; Lee and Cockburn 1985). Part of the increase in investment involves carrying the young in the pouch until a later stage in its development. The basic marsupial pattern of maternal behaviour is probably that in which, as soon as the litter of >1 release the teat, the young are left in a nest, with little fur, eyes still closed and unable to thermoregulate, a pattern typical of the Dasyuridae and Didelphidae (Pattern A, Russell 1982). The patterns of maternal behaviour that we see in the Macropodoidea are different and are amongst the most specialized to be found in living marsupials; in them the single young is carried by the mother until it is well furred, homeothermic and capable of independent locomotion (Pattern C of Russell 1982).

Pouch life in macropodoids is also long relative to the period of dependence of an equivalent sized eutherian; the time from conception to pouch vacation and weaning are longer and the period of maternal care relatively long (Russell 1982; Lee and Cockburn 1985; Thompson 1987), so that costs of maintenance and transport while the young is carried are high. Maynes (1976) first quoted figures suggesting that macropodoids at PEP were heavier and more advanced in development than corresponding eutherian (ungulate) young at birth. There is a close relationship between body mass of the young at PEP and maternal body mass (MBM). The equation is

$$M_{PEP}=0.34\ M^{0.93} \qquad (r^2=0.96,\ N=17 \text{ species}).$$

The corresponding equation calculated by Blueweiss *et al.* (1978) for a range of eutherians is

$$M_{neonate}=0.97\ M^{0.92}$$

The slopes are similar but the intercept is different. Thus for both groups, there is not much change with body size in the relative size at birth or PEP, but for marsupials, the young at PEP is considerably larger relative to MBM than are eutherian young at birth. The most appropriate comparison is with ungulates; from Figure 4 in Robbins and Robbins (1974), birth mass for ungulates of less than 100 kg MBM is not above 15% of MBM. For macropodoids, mass of the young at PEP has a mean of 19.4% (S.D. = 2.8, N=17; the allometric equation quoted by Robbins and Robbins contains an error in the value for the intercept, and cannot be used as a basis for the comparison).

The increased proportion of the period of dependence spent in the pouch has significance for the evolution of social organization (Russell 1982). When the young is in the pouch, there is little contribution that a male can make to parental care, and he is more likely to increase his fitness by going off to mate again and leaving the female to raise the offspring. Not surprisingly, promiscuity is the most frequent mating system found in the Macropodidae, with no long-term association between male and female; as in most other marsupials, males generally make little or no contribution to the care of their offspring. If the nature of the food resource was such as to favour a territorial system, then male assistance in defence could be important, but the terrestrial herbivore niche of most macropodoids does not appear to favour a social organization based on territory.

Thus the pouch does a lot of the work that is done inside mothers of (even precocial) newborn eutherian young. If, as several authors have argued (Maynes 1976; Russell 1982; Johnson 1987), FEP is more nearly equivalent to eutherian birth than is PEP, then the time after FEP is "extra" time in the pouch when the young grows larger (to beyond birth weight in comparable eutherians). It is only in the last weeks before PEP that marsupial young spend much time out of the pouch. There are not yet the data on sufficient species to allow comparison of immediate post-natal mortality rates of eutherians with mortality after FEP and soon after PEP. We might expect that the increased costs of transport are balanced by improved survival.

The period from FEP to PEP is most comparable with the period just after birth of a precocial eutherian, with the difference that the pouch is there for the young to go back to. While it is in and out of the pouch, the young has the chance to learn quite a lot about its environment. The macropodoid response to predators is to hide or run, not to stand and fight, which has important consequences for maternal behaviour. The important thing for the mother to do is to keep track of her young and to be alert for predators. What is specific to marsupials is the use of the pouch with behaviour and signals related to getting in and out of it, finally leading to the end of pouch life.

The complex interplay of the mother's behaviour and the developing behaviour of the young involves a highly specialized system of signalling that is not yet obvious or well understood. It enables recognition of each other to develop. The mother is very aware of her young and what is going on when it is out of the pouch; in fact, a lot of her behaviour is a matter of being in the right place at the right time, ready to do the right thing. If she is not, then the young becomes separated, and when lost may be attacked by a strange male or female, or taken by a predator. That the system works well is shown by the low mortality during pouch life (reviewed by Russell 1982; also see Johnson 1987 and Stuart-Dick 1987). Much of the changing pattern of maternal behaviour is driven by the development of the young, either in direct response to changing stimuli from the young, or mediated by complex hormonal mechanisms. Either way, we do not know much about it yet and there is plenty left to do in the next two hundred years.

ACKNOWLEDGEMENTS

I thank Robin Stuart-Dick for allowing me to quote unpublished data from her Ph.D. Thesis and for comments on the manuscript; Ian Rowley and Hugh Tyndale-Biscoe for constructive comments on the manuscript, Jeanette Coffey for drawing the figures, and C.S.I.R.O. Division of Wildlife and Ecology for office and library facilities. Unpublished data on Tammar wallabies were collected at the University of New South Wales with support provided by ARGS.

REFERENCES

ALEXANDER, G. AND STEVENS, D., 1982. Failure to mask lamb odour with odoriferous substances. *Appl. Anim. Ethol.* **8**: 253-60.

ARCHER, M., 1984. Origins and early radiation of marsupials. Pp. 585-625 *in* Vertebrate Zoogeography and Evolution in Australia ed. by M. Archer and G. Clayton. Hesperian Press: Perth.

BARBOUR, R. A., 1977. Anatomy of Marsupials. Pp. 237-72 *in* The Biology of Marsupials ed. by B. Stonehouse and D. Gilmore. Macmillian: London.

BAVERSTOCK, P. AND GREEN, B., 1975. Water recycling in Lactation. *Science, Wash.* **87**: 657-58.

BLUEWEISS, L., FOX, H., KUDZMA, V., NAKASHIMA, D., PETERS, R. AND SAMS , S., 1978. Relationships between body size and some life history parameters. *Oecologia (Berl.)* **37**: 257-72.

CALDER, W. A., 1984. Size, Function and Life History. Harvard University Press: Cambridge.

CANNON, J. R., BAKKER, H. R., BRADSHAW, S. D. AND MCDONNALD, I. R., 1976. Gravity as the sole navigational aid to the newborn quokka. *Nature, Lond.* **259**: 42.

CASE, T. J., 1978. Endothermy and parental care in the terrestrial vertebrates. *Amer. Nat.* **112**: 861-74.

CLARK, M. J., 1968. Growth of pouch young of the red kangaroo in the pouches of foster mothers of the same species. *Int. Zoo Yearb.* **8**: 102-6.

CLARK, M. J. AND POOLE, W. E., 1967. The reproductive system and embryonic diapause in the female grey kangaroo, *Macropus giganteus*. *Aust. J. Zool.* **15**: 441-59.

CORK, S. J. AND DOVE, H., 1986. Milk consumption in late lactation in a marsupial, the tammar wallaby *(Macropus eugenii)*. *Proc. Nutr. Soc. Aust.* **11**: 93.

CROFT, D. B., 1981a. Behaviour of red kangaroos, *Macropus rufus* (Desmarest, 1822), in northwestern New South Wales. *Aust. Mammal.* **4**: 5-58.

CROFT, D. B., 1981b. Social behaviour of the euro *Macropus robustus* (Gould), in the Australian Arid Zone. *Aust. Wildl. Res.* **8**: 13-49.

CROMPTON, A. W., 1980. Biology of the earliest mammals. Pp. 1-12 *in* Comparative Physiology: Primitive Mammals ed. by K. Schmidt-Nielsen, L. Bolis and C. R. Taylor. Cambridge University Press: London.

EISENBERG, J. F., 1981. The Mammalian Radiations. University of Chicago Press: Chicago.

FINDLAY, L., 1982. The mammary gland of the tammar wallaby *(Macropus eugenii)* during pregnancy and lactation. *J. Reprod. Fert.* **65**: 59-66.

FINDLAY, L., WARD, K. L. AND RENFREE, M. B., 1983. Mammary gland lactose, plasma progesterone and lactogenesis in the marsupial *Macropus eugenii*. *J. Endocr.* **97**: 425-36.

FLEMING, D., CINDEREY, R. N. AND HEARN, J. P., 1983. The reproductive biology of Bennett's wallaby *(Macropus rufogriseus rufogriseus)* ranging free at Whipsnade Park. *J. Zool., Lond.* **201**: 283-91.

GREEN, B., 1984. Composition of milk and energetics of growth in marsupials. *Symp. Zool. Soc. Lond.* **51**: 369-87.

GREEN, B., MERCHANT, J. C. AND NEWGRAIN, K., 1988. Milk consumption and energetics of growth in pouch young of the tammar wallaby, *Macropus eugenii*. *Aust. J. Zool.* **36**: 217-27.

HILL, J. P. AND HILL, W. C. O., 1955. The growth stages of the pouch young of the native cat *(Dasyurus viverrinus)* together with observations on the anatomy of the new born young. *Trans. Zool. Soc. Lond.* **28**: 349-453.

HINDS, L. A. AND TYNDALE-BISCOE, C. H., 1982a. Prolactin in the marsupial *Macropus eugenii*, during the oestrus cycle, pregnancy and lactation. *Biol. Reprod.* **26**: 391-98.

HINDS, L. A. AND TYNDALE-BISCOE, C. H., 1982b. Plasma progesterone levels in the pregnant and non-pregnant tammar, *Macropus eugenii*. *J. Endocr.* **93**: 99-107.

HINDS, L. A. AND TYNDALE-BISCOE, C. H., 1985. Seasonal and circadian patterns of circulating prolactin during lactation and seasonal quiescence in the tammar, *Macropus eugenii*. *J. Reprod. Fert.* **74**: 173-83.

HOPSON, J. A., 1973. Endothermy, small size, and the origin of mammalian reproduction. *Amer. Natur.* **107**: 446-52.

HORNSBY, P. A., 1978. A note on the pouch life of rock-wallabies. *Vict. Nat.* **95**: 108-11.

HUGHES, R. L., 1962. Reproduction in the macropod marsupial *Potorous tridactylus* (Kerr). *Aust. J. Zool.* **10**: 193-224.

JANSSENS, P. A. AND ROGERS, A. M. T. (this volume). Metabolic changes during pouch vacation and weaning in macropodids.

JOHNSON, C. N., 1985. Ecology, social behaviour and reproductive success in a population of red-necked wallabies. Ph.D. Thesis, University of New England, Armidale.

JOHNSON, C. N., 1987. Relationships between mother and infant red-necked wallabies *(Macropus rufogriseus banksianus)*. *Ethology* **74**: 1-20.

JOHNSON, P. M., 1979. Reproduction in the plain rock-wallaby, *Petrogale penicillata inornata* Gould, in captivity, with age estimation of the pouch young. *Aust. Wildl. Res.* **6**: 1-4.

JOHNSON, P. M., 1981. The rearing of marsupial pouch young by foster mothers of different species. *Int. Zoo. Yearb.* **8**: 173-76.

JOHNSON-MURRAY, J. L., 1987. The comparative myology of the gliding membranes of *Acrobates, Petauroides* and *Petaurus* contrasted with the cuntaneous myology of *Hemibelideus* and *Pseudocheirus* (Marsupialia: Phalangeridae) and with selected gliding Rodentia (Sciruridae and Anamoluridae). *Aust. J. Zool.* **35**: 101-13.

KAUFMANN, J. H., 1974. Social ethology of the whiptail wallaby, *Macropus parryi*, in north-eastern New South Wales. *Anim. Beh.* **22**: 281-369.

KIRSCH, J. A. W., 1977. The six-percent solution; second thoughts on the adaptedness of the Marsupialia. *Amer. Sci.* **65**: 276-88.

KLIMA, M. AND BANGMA, G. C., 1987. Unpublished drawings of marsupial embryos from the Hill collection and some problems of marsupial ontogeny. *Z. Saugetierk.* **52**: 201-11.

LANGWORTHY, O. R., 1932. The *panniculus carnosus* and pouch masculature of the opossum, a marsupial. *J. Mammal.* **13**: 241-51.

LEE, A. K. AND COCKBURN, A., 1985. Evolutionary Ecology of Marsupials. Cambridge University Press: Cambridge.

LEWIS, P. R., FLETCHER, T. P. AND RENFREE, M. B., 1986. Prostaglandin in the peripheral plasma of tammar wallabies during parturition. *J. Endocrinol.* **111**: 103-9.

LINCOLN, D. W. AND RENFREE, M. B., 1981. Mammary gland growth and milk ejection in the agile wallaby, *Macropus agilis*, displaying asynchronous lactation. *J. Reprod. Fert.* **63**: 193-203.

MCNAB, B. K., 1986. Food habits, energetics, and the reproduction of marsupials. *J. Zool., Lond. (A)* **208**: 595-614.

MAYNES, G., 1973. Reproduction in the parma wallaby, *Macropus parma* Waterhouse. *Aust. J. Zool.* **21**: 331-51.

MAYNES, G., 1976. Growth of the parma wallaby *Macropus parma* Waterhouse. *Aust. J. Zool.* **24**: 217-36.

MERCHANT, J. C., 1976. Breeding biology of the agile wallaby, *Macropus agilis* (Gould) (Marsupialia: Macropodidae), in captivity. *Aust. Wildl. Res.* **3**: 93-103.

MERCHANT, J. C. AND CALABY, J. H., 1981. Reproductive biology of the red-necked wallaby *(Macropus rufogriseus banksianus)* and Bennett's wallaby *(M. r. rufogriseus)* in captivity. *J. Zool., Lond.* **194**: 203-17.

MERCHANT, J. C. AND SHARMAN, G. B., 1966. Observations on the attachment of marsupial pouch young to the teats and on the rearing of pouch young by foster-mothers of the same or different species. *Aust. J. Zool.* **14**: 593-609.

POND, C. M., 1977. The significance of lactation in the evolution of mammals. *Evolution* **31**: 177-99.

POOLE, W. E., 1975. Reproduction in the two species of grey kangaroos, *Macropus giganteus* Shaw and *M. fuliginosus* (Desmarest) II. Gestation, parturition and pouch life. *Aust. J. Zool.* **23**: 333-53.

POOLE, W. E. AND CATLING, P. C., 1974. Reproduction in the two species of grey kangaroos, *Macropus giganteus* Shaw and *M. fuliginosus* (Desmarest). I. Sexual maturity and oestrus. *Aust. J. Zool.* **22**: 277-302.

POOLE, W. E. AND MERCHANT, J. C., 1987. Reproduction in captive wallaroos: the eastern wallaroo, *Macropus robustus robustus*, the euro, *M. r. erubescens* and the antilopine wallaroo, *M. antilopinus*. *Aust. Wildl. Res.* **14**: 225-42.

POOLE, W. E. AND PILTON, P. E., 1964. Reproduction in the grey kangaroo, *Macropus canguru*, in captivity. *C.S.I.R.O. Wildl. Res.* **9**: 218-34.

POOLE, W. E., MERCHANT, J. C. AND CALABY, J. H., 1985. Reproduction, growth and age determination in the yellow-footed rock-wallaby *Petrogale xanthopus* Gray in captivity. *Aust. Wildl. Res.* **12**: 127-36.

RENFREE, M. B., 1979. Initiation of development of diapausing embryo by mammary denervation during lactation in a marsupial. *Nature, Lond.* **287**: 549-51.

RENFREE, M. B., FLETCHER, T. P., BLANDEN, D. R., LEWIS, P. R., SHAW, G., GORDON, K., SHORT, R. V., PARER-COOK, E. AND PARER, D. (this volume). Physiological and behavioural events around the time of birth in macropodid marsupials.

ROBBINS, C. T. AND ROBBINS, B. L., 1979. Fetal and neonatal growth patterns and maternal reproductive effort in ungulates and subungulates. *Amer. Natur.* **114**: 101-16.

ROSE, R. W., 1986. The control of pouch vacation in the Tasmanian bettong *Bettongia gaimardi*. *Aust. J. Zool.* **34**: 485-91.

ROSE, R. W. AND MCCARTNEY, D. J., 1982. Reproduction of the red-bellied pademelon, *Thylogale billardierii* (Marsupialia). *Aust. Wildl. Res.* **9**: 27-32.

ROSENBLATT, J. S. AND SIEGEL, H. I., 1981. Factors governing the onset and maintenance of maternal behaviour among non-primate mammals. Pp. 13-76 *in* Parental Care in Mammals ed. by D. J. Gubernick and P. H. Klopfer. Plenum Press: New York.

RUSSELL, E. M., 1973. Mother-young relations and early behavioural development in the marsupials *Macropus eugenii* and *Megaleia rufa*. *Z. Tierpsychol.* **33**: 163-203.

RUSSELL, E. M., 1982. Patterns of care and parental investment in marsupials. *Biol. Rev.* **57**: 423-86.

RUSSELL, E. M., 1985. The metatherians: Order Marsupialia. Pp. 45-104 *in* Social Odours in Mammals, Volume 1 ed. by R. E. Brown and D. W. Macdonald. Clarendon Press: Oxford.

RUSSELL, E. M. AND GILES, D. C., 1974. The effects of young in the pouch on pouch cleaning in the tammar wallaby, *Macropus eugenii* Desmarest (Marsupialia). *Behaviour* **51**: 19-37.

RUSSELL, E. M. AND HARROP, C. J. F., 1976. The behaviour of red kangaroos *(Megaleia rufa)* on hot summer days. *Z. Tierpsychol.* **40**: 396-426.

SHARMAN, G. B., 1965. The effects of suckling on normal and delayed cycles of reproduction in the red kangaroo. *Z. Saugetierk.* **30**: 10-20.

SHARMAN, G. B., 1973. Adaptations of marsupial pouch young for extra-uterine existence. Pp. 67-90 *in* The Mammalian Fetus *in vitro* ed. by C. R. Austin. Chapman and Hall: London.

SHARMAN, G. B. AND CALABY, J. H., 1964. Reproductive behaviour in the red kangaroo, *Megaleia rufa* in captivity. *C.S.I.R.O. Wildl. Res.* **9**: 58-85.

SHARMAN, G. B. AND PILTON, P. E., 1964. Life history and reproduction of the red kangaroo *(Megaleia rufa)*. *Proc. Zool. Soc. Lond.* **142**: 29-48.

SHARMAN, G. B., CALABY, J. H. AND POOLE, W. E., 1966. Patterns of reproduction in female diprotodont marsupials. *Symp. Zool. Soc. Lond.* **15**: 205-32.

SMITH, A. P. AND LEE, A. K., 1984. The evolution of strategies for survival and reproduction in possums and gliders. Pp. 17-33 *in* Possums and Gliders ed. by A. Smith and I. Hume. Surrey Beatty & Sons: New South Wales.

SMITH, M. J. (this volume). Release of embryonic diapause in the Brush-tailed Bettong, *Bettongia penicillata*.

STEWART, F., 1984. Mammogenesis and changing prolactin receptor concentrations in the mammary gland of the tammar wallaby *(Macropus eugenii)*. *J. Reprod. Fert.* **71**: 141-48.

STUART-DICK, R. I., 1987. Parental Investment and Rearing Schedules in the Eastern Grey Kangaroo. Ph.D. Thesis, University of New England, Armidale.

STUART-DICK, R. I. AND HIGGINBOTTOM, K. (this volume). Strategies of parental investment in macropodoids.

THOMPSON, S. D., 1987. Body size, duration of parental care, and the intrinsic rate of natual increase in eutherian and metatherian mammals. *Oecologica (Berl.)* **71**: 201-19.

TRIVERS, R. L., 1974. Parent-offspring conflict. *Amer. Zool.* **14**: 249-64.

TYNDALE-BISCOE, C. H., 1968. Reproduction and post-natal development in the marsupial *Bettongia lesueur* (Quoy and Gaimard). *Aust. J. Zool.* **16**: 577-602.

TYNDALE-BISCOE, C. H. AND RENFREE, M. B., 1987. Reproductive Physiology of Marsupials. Cambridge University Press: Cambridge.

WALKER, M. T. AND ROSE, R. W., 1981. Prenatal development after diapause in the marsupial *Macropus rufogriseus*. *Aust. J. Zool.* **29**: 167-87.

WESTERN, D., 1979. Size, life history and ecology in mammals. *Afr. J. Ecol.* **17**: 185-204.

WESTERN, D. AND SSEMAKULA, J., 1982. Life history patterns in birds and mammals and their evolutionary interpretation. *Oecologia (Berl.)* **54**: 281-90.

Strategies of parental investment in Macropodoids

R. I. STUART-DICK[1] and K. B. HIGGINBOTTOM[1]

Research into strategies of parental investment in mammals has recently found that females may vary their reproductive tactics according to their own rank, age, physical condition, environmental conditions, or sex of their offspring. We await a unifying theory which explains the variety of parental-investment patterns observed. This paper focuses on within-species variation in parental-investment tactics of two macropod species, and briefly examines whether other macropod species might exhibit similar degrees of variation.

Delineating female reproductive tactics requires data from relative long-term studies of individually known females. Two such studies, of the Red-necked Wallaby, *Macropus rufogriseus,* and the Eastern Grey Kangaroo, *Macropus giganteus,* have found that mothers appear to employ a complex combination of tactics. In both species preweaning investment in sons did not appear to exceed that in daughters, while post-weaning investment was probably greater in daughters. This was "balanced" by a greater overall production of sons. Both kangaroo and wallaby mothers produced sons to take advantage of temporarily good environmental conditions, and there was evidence that wallaby mothers who were able to give "extra parental-care" tended to produce sons. The two species exhibited contrasting age-related tactics in relative production of the sexes which we interpret in terms of the different ratios between costs and benefits to mothers of sharing their home ranges with daughters.

INTRODUCTION

THE manner in which females rear and invest in offspring is an integral part of their life-history strategy, and as such is assumed to be part of a set of interdependently adapted traits (Stearns 1976), which have been selected to maximize reproduction. Much of the recent research on female life-history strategies has examined interspecific differences in levels of parental investment in relation to morphometric principles and the general ecology of the species (Cole 1954; Millar 1977; Russell 1982; Stearns 1976). However, some populations of mammals show considerable individual variation in investment schedules (Altmann 1980; Clark *et al.* 1986; Silk *et al.* 1981; Simpson and Simpson 1985). Each offspring which a mammalian female produces and rears represents a considerable proportion of her potential contribution to the next generation's gene pool; it follows that females should be under strong selection to optimize their investment schedules facultatively (Wasser 1983; Werren and Charnov 1978; Werren and Taylor 1984). The ratio between benefit gained by the offspring from maternal care and cost to the mother of producing and rearing that offspring should be maximized (over all offspring). An increasing body of theory supports the idea that females' rearing tactics may be related to:

1. inherent differences between the mothers, such as their rank-related access to local resources, their age, or their physical condition;
2. the offspring's expected life-history, which may differ according to characteristics such as its sex;
3. the season or local environmental fluctuations such as the local weather conditions, food availability, or probability of predation (see Dunbar 1984; Gadgil and Bossert 1970; Goodman 1979; Rubenstein 1982; Trivers and Willard 1973).

However, no unifying theory yet describes or adequately predicts the observed variation in parental-investment tactics either between or within species. Rather, *post hoc* explanations, which assume that the observed variation is adaptive, abound in the literature. Until a more unified theory is developed, that enables us to test predictions of how females should be adapted, we should remain particularly sceptical of assuming adaptive significance of any observed variation in parental investment.

The present paper examines individual variation in the manner in which female macropods rear offspring. Macropodoid males do not directly provide care to their offspring, and therefore we are concerned only with maternal investment. Russell (1982) has provided an excellent review of the levels of parental investment in macropods, and we do not intend to conduct another such review. Instead, the focus will be on the description of parental-investment tactics within species. To date only two studies, one of the Eastern Grey Kangaroo, *Macropus giganteus,* and one of the Red-necked Wallaby, *Macropus rufogriseus banksianus,* have been comprehensive enough to enable interpretation of individual females' tactics. We use these studies to highlight the complexity found in parental-investment tactics, and to make preliminary attempts to unravel the principles determining the tactics of macropodid females.

[1]Department of Ecosystem Management, University of New England, Armidale, New South Wales 2351.

Pages 571-92 in KANGAROOS, WALLABIES AND RAT-KANGAROOS ed. by G. Grigg, P. Jarman and I. Hume. Surrey Beatty & Sons Pty Limited, New South Wales, Australia, 1989.

The degree of variation in parental-investment tactics might differ between species in relation to the extent of differences between mothers or between offspring. Thus variation in parental investment tactics might be related to, for example, the degree of sexual dimorphism in the species or to allometric principles affecting the morphology and development of the individuals of a species. This paper also makes a preliminary investigation of the potential for variation in parental-investment tactics for a range of macropod species.

Where we have referred to females' tactical decisions in a manner that implies a decision process on their part, we have done so to reduce the complexity of expression which the reader is otherwise asked to digest. We emphasize that we do not suggest that females necessarily make "cognitive decisions". The tactics employed by the females are presumably those which have been selected over the previous generations.

The range of tactics employed and the circumstances in which these tactics are employed are taken to constitute the females' strategy.

Theoretical aspects of tactical variation within species

Parental investment is usually defined as "investment by the parent in an individual offspring that increases the offspring's chance of survival (and hence reproductive success) at the cost of the parent's ability to invest in other offspring" (Trivers 1978, p.55). Parental investment may be costly in terms of the time involved, the incurred risk of mortality, or metabolic expenditure (Bell 1980; Calow 1981; Knapton 1984). According to Fisher (1930), the genetic contribution of males to the next generation will always match that of females, since each sex provides one half of the gene pool, so total parental-investment in sons should be equal to that in daughters. In practice the different types of investment are not comparably quantifiable. Fisher's theory is usually assumed to be valid and statements to the contrary require rigorous supporting evidence. A corollary of Fisher's hypothesis is that if individuals of one sex are "cheaper" to produce than those of the other sex, then mothers will produce a greater number of the cheaper sex.

Trivers and Willard (1973) proposed that under certain conditions the sex ratio of offspring would vary within the population and/or the average level of parental care given to each sex would differ. They predicted that this pattern would occur if male and female life-histories and the variances in their expected lifetime reproductive success were quite different. This is the case, for example, in large sexually dimorphic polygynous mammals where only a few males, but the majority of females, in a population breed. In these populations a mother can expect, on average, that a son will be less likely to reproduce than a daughter. However, if a son becomes an "alpha male" he is likely to contribute genes to a larger number of weaned offspring than a daughter could. Thus, according to Trivers and Willard (1973), if the following apply:

1. the condition of the offspring after the parental-investment period is correlated with the condition of the mother during the parental-investment period;
2. differences in the physical condition of the offspring at the end of the parental-investment period endure into its adult life; and
3. a son's expected lifetime reproductive success is positively correlated to his physical condition;

then a mother who is in good condition should preferentially conceive sons and provide them with an above-average level of parental care, and a mother who is in poor condition should preferentially conceive daughters and provide them with a below-average level of parental care (see also Maynard Smith 1980). By "parental care" we refer directly to activities of the mother which are quantified in terms of potential benefit gained by her offspring. We will use the phrase "extra parental-care" when referring to the *difference* between parental care received (or benefit gained) by offspring of "above-average" and offspring of "below-average" mothers. It is important to note that Trivers and Willard do not predict differences in levels of parental investment between the sexes (in terms of cost to the mother); they assume that this remains constant and it is the level of parental care which varies.

Trivers and Willard's theory implies that individual parents inherently differ in their ability to raise offspring throughout their lives, so that the total proportion of each sex produced by individual females should vary accordingly. However, the ability to give extra parental care may often be a temporary phenomenon. If changes in environmental conditions through time significantly affect the mother's condition then preferential production of the sexes may be correlated to the environmental fluctuations (Werren and Charnov 1978; Werren and Taylor 1984). The degree of parental care given to offspring may also increase as a mother ages; if so, mothers may tend to produce daughters preferentially when young and sons when old (Charlesworth and Leon 1976; Gadgil and Bossert 1970).

Altmann (1980) has proposed that under certain conditions it might be adaptive for mothers preferentially to produce and rear daughters. She suggested that in populations where there are closed-membership social groups and the rank of a daughter is directly related to her mother's rank, high-ranking mothers would positively influence their daughters' lifetime reproductive expectations to a greater extent than would low-ranking mothers. Hence high-ranking mothers should preferentially produce and rear daughters. This assumes that high-ranking females are more successful in rearing offspring than are low-ranking females; and that, since sons disperse, their expected lifetime reproductive success is less directly influenced by their mother's rank than that of daughters.

As a special case of Altmann's hypothesis, Dunbar (1984) has suggested that if a mother can expect to influence positively her daughters' reproductive-success *by her presence* to a greater extent than can other mothers, then she should preferentially produce daughters at an early age, rather than at a late age, so that she may be present when many of her daughters commence reproduction. This could occur when mother-daughter pairs become coalitions which either increase the probability of the individuals obtaining the local resources which are necessary to breed, or reduce the potential competition from non-relatives by, for example, feeding in the company of tolerant relatives (Clutton-Brock *et al.* 1986).

Altmann's and Dunbar's hypotheses focus on the variable degree of benefit a daughter receives from a mother and the consequent influence of the mother on the daughter's lifetime reproductive success. Another hypothesis, the Local Resource Competition hypothesis (Clarke 1978), focuses rather on the variable cost of daughters to mothers in relation to the degree to which the mother shares her resources with her daughters. This hypothesis states that if a daughter remains and breeds in the mother's home range and a son disperses, a mother should produce fewer daughters than sons because each daughter represents a greater cost to the mother than each son.

Both Altmann's and Trivers and Willard's hypotheses propose that mothers who can significantly boost the expected lifetime reproductive success of one sex to a greater extent than that of the other sex should preferentially produce the former sex. However, Altmann does not require that one sex's lifetime reproductive success (that of daughters) be significantly more variable than it is for the other sex. The Local Resource Competition hypothesis is concerned, not with the breeding expectations of offspring, but with the different effects of sons and daughters on their mothers' future breeding potential.

Overall, there is much debate as to exactly what influences the high-ranking mothers or the mothers in better-than-average condition to produce one sex preferentially (Armitage 1987; Clutton-Brock *et al.* 1986; Meikle *et al.* 1984; Silk *et al.* 1981). Research is required which investigates the following variables and their interactions within populations: the variances in the sexes' lifetime reproductive success, parental-investment schedules for each sex, and the offspring sex ratios of both the population and the individual mothers. Both the studies selected have attempted such investigations.

To compare predictions from each of the above hypotheses and to look at patterns of parental investment in the sexes in general, the wallaby and kangaroo data will be discussed in relation to three main questions:

1. Does the amount or type of parental investment vary between sons and daughters?
2. How does variation between individuals and environmental conditions affect the relative production of sons and daughters?
3. What are the age-related differences in mothers' parental-investment tactics?

THE STUDIES

Red-necked Wallabies and Eastern Grey Kangaroos

Both the studies of parental investment in Red-necked Wallabies and in Eastern Grey Kangaroos were conducted in a valley known as "Wallaby Creek" (152°45′E 28°45′S). The study area, described in detail by Jarman *et al.* (1987), is partially cleared land surrounded by forest, and receives a relatively high rainfall (1 011 mm). Rain occurs throughout the year but more heavily and consistently in summer. Years of drought are not uncommon; we use the Pasture Protection Board's assessment of which were drought years for this area.

The macropods in the valley are unmanaged. Dingoes *Canis familiaris dingo* and Wedge-tailed Eagles *Aquila audax* prey on both the wallabies and the kangaroos.

The Red-necked Wallabies have been intensively monitored since March 1981 and the Eastern Grey Kangaroos since April 1982. Data used in this paper are derived from the first seven and four years respectively. All animals in the local populations were individually recognized by 18 and 13 months respectively from the commencement of the studies (see Johnson 1985; and Stuart-Dick 1987, for greater detail of the studies). Three classes of offspring were described.

1. pouch-young (PY), which spend either all or part of their time inside the mother's pouch;
2. pouch-infants (PI), which spend part of their time inside and part outside the mother's pouch.
3. young-at-foot (YAF), which have permanently emerged from the pouch (PEP), but still suck

from their mothers. A "small" YAF permanently emerged from the pouch less than three months previously. A "large YAF" emerged more than three months previously.

The "age" index (Stuart-Dick 1987) used in the study of the kangaroos was subjective. It assumed that animals are no different from humans in showing signs of physical deterioration with age (Comfort 1979; Austad 1988), and is supported by the finding that significantly more of the "older" than "younger" females alive in 1984 had disappeared by 1988 (G=6.457, df=1, $p<0.02$). The age categories used for Red-necked Wallabies were either known from long-term records, or were derived by assuming a maximum female lifespan of eleven years (Higginbottom, unpubl. data). The physical-condition indices for both species were based on subjective assessments of how well covered an individual was with muscle and fat (for details see Stuart-Dick 1987).

In this paper we have decided to discuss results with a 10% probability of falsely rejecting the null hypothesis because so many lay between the 10% and 5% probability levels. We believe these results are worth discussing as they appear to follow consistent patterns which are biologically feasible. Misleading conclusions about maternal tactics are equally possible from accepting a false null hypothesis as from accepting a false alternative hypothesis. We distinguish those results with probability values between 0.1 and 0.05 by following them with a †; results which have probability values of less than 0.05 are distinguished by an *. Except for chi-squared tests, all probability values are two-tailed unless otherwise specified. All G tests and chi-squared tests had one degree of freedom unless otherwise specified.

WITHIN-SPECIES VARIATION IN INVESTMENT: RESULTS AND DISCUSSION

Investment in individual sons and daughters

We can not quantify a mother's total parental investment in or care of an offspring. However, we can determine for each of a range of variables, such as suckling, pouch-life duration, weaning age, mother-offspring interactions, and post-weaning association levels, whether these measures of parental investment in sons and daughters differ. The often-observed reduction in probability of survival to weaning of an offspring following one who was reared to weaning may be related to either:

a) a cost attached to the presence of kin in an individual's home range which affects the survival of subsequently produced offspring, or

b) mothers who have just reared an offspring being in poorer condition than those who have not, and consequently being less able to invest in the next offspring (Clutton-Brock *et al.* 1982a and 1982b; Dittus 1979; Johnson 1985; van Schaik and van Noordwijk 1985; Silk 1983; Silk and Boyd 1983; Simpson and Simpson 1985).

Thus if mothers invest more in sons than daughters, an offspring following a weaned son may be less likely to survive than an offspring following a daughter, or there may be a greater delay in its production.

The Eastern Grey Kangaroo

In this kangaroo population significantly more sons than daughters were produced and weaned (96 males; 65 females, sex ratio at permanent emergence=1.48, $\chi^2=5.969$, $p<0.05$). However, even if all known unsexed pouch-young were females, the offspring sex-ratio at birth would still have favoured males. There was no difference between the sexes of offspring in the proportion surviving to weaning (44% males, 49% females; $\chi^2=0.467$).

Although maternal investment through the duration of pouch-life and the YAF stage for some sons was particularly long, for other sons it was quite short (Table 1). The variation in the duration of these stages was related to the timing of the sons' permanent emergence (PEP) (pouch duration (in weeks)=34.6+4.07 (month of PEP) −0.26 (month of PEP)^2; $R^2=0.101$; linear probability=0.020 and quadratic probability=0.041; YAF duration (in weeks)=20.49+2.94 (month of PEP)−0.22 (month of PEP)^2; $R^2=0.238$; linear probability=0.008 and quadratic probability=0.012). Neither of these relationships between month of the year of permanent emergence and the duration of pouch life or of YAF stage, nor the great range in duration of pouch-life was found for daughters. However, the mean durations were the same for the sexes (Table 1).

Table 1. Eastern Grey Kangaroo: the durations of pouch-life, the pouch-infant phase, and the YAF phase for sons and daughters. The tests used were t-tests (t) and F ratio tests (F). The t-tests used pooled (t_p) or unpooled (t_{up}) standard deviations, depending on whether the variances were significantly different. N represents sample size, SE represents standard error, and $\bar{X}$ represents mean.

duration (weeks)	sons	daughters	test	probability
pouch-life:				
N	62	45		
$\bar{X}$	47.9	46.8	$t_{up}=0.67$	=0.502ns
SE	1.22	0.98	F=2.13	<0.01*
pouch-infant:				
N	36	33		
$\bar{X}$	17.1	17.4		
SE	2.71	2.4	F=1.29	ns
YAF phase:				
N	30	25		
$\bar{X}$	28.8	29.4	$t_p=0.61$	=0.542ns
SE	0.73	0.72	F=1.22	=0.63 ns

The frequency of suckling (a commonly used indicator of maternal investment) also was not significantly different for YAF sons and daughters (Table 2). Pouch-infant sons did suck significantly more than daughters while out of the pouch. However, we did not measure what they did inside the pouch.

Table 2. Eastern Grey Kangaroo: percent of activity budget spent in sucking by son and daughter pouch-infants, small YAF, and large YAF. Note that for pouch-infants the out-of-pouch behaviour only was used. The tests used were G tests (G).

offspring stage	sons	daughters	test	probability
pouch-infants:	50%	32%	G=15.5	<0.05*
small YAF:				
winter	9.7%	9.1%	G=1.003	ns
summer	9.3%	11.3%	G=0.252	ns
large YAF				
winter	7.8%	10.3%	G=1.709	ns
summer	5.9%	6.1%	G=0.910	ns

Other preweaning differences in the manner sons and daughters were reared were:

1. mothers were more active in maintaining association with daughters than sons (mothers moved toward rather than away from small YAF daughters more than they do for sons, G=8.162, df=1, total moves=192, p<0.05,*; mothers moved toward rather than away from large YAF daughters more than they do for sons, G=11.605, df=1, total moves=52, p<0.05,*);
2. mothers allogroomed their small YAF daughters more than they did small YAF sons (Table 3);
3. mothers tended to be more active in making physical contact with pouch-infant sons than daughters (Table 3); and
4. indirectly mothers may have spent more time surveying if they had a pouch-infant daughter than son. This is suggested because mothers tended to survey more than usual when their

Table 3. Eastern Grey Kangaroo: mother-offspring contact rates with, types of contact with, and distances from sons and daughters separately. The tests used were G tests (G), and Mann-Whitney U tests (MWU). N represents sample size, SE represents standard error, and $\bar{X}$ represents mean. As the analyses for the types of contact states with sons and daughters involved lumping individuals results are more likely to lead to rejecting the null hypothesis inappropriately. Thus we do not highlight the analyses with probabilities between 0.1 and 0.05.

variable	sons	daughters	test	probability
mothers' contacts/contact minute				
pouch-infant:				
N	7	7		
$\bar{X}$	1.195	0.729	MWU=11	p<0.1†
SE	0.247	0.372		
YAF:				
N	6	5	(i.e., months post PEP)	
$\bar{X}$	0.858	1.705	(G tests by month)	
range	(0.375-1.895)	(0.003-3.568)	(post PEP, ns)	
mother's types of contact states				
pouch-infant:				
allogroom	79%	68%	for the 3 states	
passive contact	18%	28%	G=3.959	p<0.05ns
"play-fight"	3%	4%	df=2	
small YAF:				
allogroom	45%	70%	for the 3 states	
passive contact	45%	23%	G=9.862	p<0.01*
"play-fight"	10%	7%	df=2	
large YAF:				
allogroom	59%	48%	for the 3 states	
passive contact	24%	33%	G=3.668	p>0.05ns
"play-fight"	17%	19%	df=2	
percent time mother and pouch-infant spent in various distance categories				
	N=7	N=7		
contact time	65%	50%	MWU=15	ns
≤2 metres	30%	36%	MWU=17	ns
>2 *to* ≤10 metres	4.4%	11.7%	MWU=10	p<0.1†
>10 metres	0.2%	2.1%	MWU= 3	p<0.05*

pouch-infant was at a distance (G=3.661, df=1, p<0.1, †) from them and pouch-infant daughters spent more time at a greater distance from their mothers (Table 3).

If, by remaining in their mothers' home range after weaning, offspring gain benefit at the mothers' expense, then daughters were likely to receive more postweaning investment than sons (see Table 4 for statistics). Fewer daughters than sons were known to disperse, and subadult and adult daughters tended to associate with their mothers to a greater extent than did sons. Further, a small sample of daughters with mothers still present in their home range first conceived at an earlier age than daughters whose mothers had disappeared shortly after they were weaned (the difference was not significant). Significantly greater number of offspring survived to weaning of those daughters with mothers present than of those without. We do not yet have enough data on matrilines in this population to test for a depressive effect on the reproductive success of mothers who share a home range with their daughters.

Preweaning parental-investment in sons and daughters did not significantly affect the next offspring's survival or mean duration of pouch-life (see Table 5). There was a slightly greater tendency for offspring following a son, rather than a daughter, to be lost. This is likely to be related to the more variable level of preweaning care given to sons. Sons' pouch-life durations tended to be more affected by whether the previous son was weaned or lost. A similar tendency was not found for daughters. Sons who disappeared before weaning tended to have shorter pouch-life durations than those who survived to weaning.

Overall, levels of parental care and parental investment in the early life of sons and daughters are similar, although given to sons in a different manner and in more variable quantities than to daughters. Daughters may receive greater investment postweaning than sons. However, in accordance with Fisher's hypothesis of equal investment in the sexes, this extra parental-investment in individual daughters might have been equivalent to the extra parental-investment in sons arising from the production of more sons than daughters.

Table 4. Eastern Grey Kangaroo: results related to potential postweaning investment in sons and daughters: the percent of offspring who remain in their natal home-range when adults; the distance between mother and subadult home-range centres; the percent of subgroups of all of a mother's subgroups in which she was with her subadult, the percent of subgroups of all of a mother's subgroups in which she was with her adult offspring; the daughter's age at first conception when her mother was still alive (present) or not (absent), and the percent of a daughter's offspring who survived to weaning when the daughter's mother was still present in the home range or absent. The tests used were t-tests (t), G-tests (G), and Mann Whitney U tests (MWU). The t-tests used pooled (t_p) or unpooled (t_{up}) standard deviations, depending on whether the variances were significantly different. N represents sample size, SE represents standard error, and $\bar{X}$ represents mean.

variable	sons	daughters	test	probability
% offspring as adults remain in the natal home range				
	4/29	15/26	G=12.210	p<0.05*
	14%	58%		
distance (metres) between mother and subadult home range centres				
N	12	9		
$\bar{X}$	96.8	54.8	t_{up}=2.08	<0.1†
SE	1.87	0.78		
% of mother's subgroups she was with subadult				
N	19	11		
$\bar{X}$	39	51	t_p=1.39	<0.1†
SE	5.7	5.5		
% of mother's subgroups she was with adult offspring				
N	10	6		
$\bar{X}$	17	26	t_p=1.66	<0.1†
SE	3.5	5.5		
	mother:			
	present	absent		
daughter's age at first conception (months)				
N	16	8		
$\bar{X}$	32.7	35.4	MWU=43	ns
SE	1.5	1.7		
daughter's young				
lost	8	6		
weaned	5	0	G=4.58	p<0.05*

The Red-necked Wallaby

There was no significant difference between the total numbers of male and female offspring produced at PEP, although there was a preponderance of males (77 males; 57 females, sex ratio=1.35; χ^2=2.7, probability=0.11). The relative proportions of each sex did not differ significantly from those of the kangaroos (χ^2=0.07, ns). As the mortality rate of pouch-young was very low, this biased production of males would probably have been present from birth. Most offspring which died, disappeared between their permanent emergence and weaning; males were not found to be significantly more likely to disappear in this age group (7/73 males sexed and 7/52 females sexed disappeared, χ^2=0.16, ns). Despite the small numbers of YAF who were sexed before disappearing, it seems unlikely that the biased sex ratio resulted from a differential mortality of the sexes as YAF.

The duration of pouch-life was longer for daughters than sons, but their YAF stages were of very similar duration (Table 6).

The mean durations of sucking bouts and the observed frequency of sucking were very similar for sons and daughters (Table 7). However, YAF daughters had significantly more variable durations of sucking bouts than sons.

Table 5. Eastern Grey Kangaroo: indirect measures of parental investment in sons and daughters: percent of offspring who survived following a son or a daughter; the pouch-life duration of sons and daughters following a previous weaned or lost son or daughter; and the pouch-life duration of lost and weaned sons and daughters. The tests used were t-tests (t), G-tests (G), and Mann-Whitney U tests (MWU). The t-tests used pooled (t_p) or unpooled (t_{up}) standard deviations, depending on whether the variances were significantly different.

variable	sons	daughters	test	probability
% offspring surviving following previous son or daughter				
previous lost	43%	54%		
previous weaned	23%	44%	G=1.885	ns
previous son:lost vs weaned			G=2.485	ns
previous daughter:lost vs weaned			G=0.394	ns
previous sibling:son vs daughter			G=2.273	ns
pouch-life duration of son/daughter following previous sibling: (weeks)				
previous lost	50.1	47.4		
previous weaned	43.9	45.9		
brother lost	52.1	48.5		
brother weaned	43.1	48.0		
sister lost	47.3	47.4		
sister weaned	46.0	44.2		
MWU tests previous sibling effect on current son/daughter				
lost vs weaned	z=−3.127*	z=−1.442ns		
brother lost/wean	z=−3.039*	z=−1.501ns		
sister lost/wean	z=−0.880ns	z=−1.141ns		
other comparisons not significant				
pouch-life duration (weeks) of lost and weaned offspring				
lost	45.8	47.3		
weaned	51.8	46.3		
t test of means for lost vs weaned sons			t_{up}=1.94	p=0.064†
t test of means for lost vs weaned daughters			t_p= 0.53	ns

Mother-offspring interaction rates were similar for both sexes in the month following permanent emergence from the pouch. However, Johnson (1987) found that over the whole YAF life the interaction rates tended to be greater for sons than daughters (see Table 8).

The degree of association between mother and YAF offspring as measured by Johnson (1987) was greater for sons than for daughters (Table 8). However, once again there was no significant difference in the month following permanent emergence (our data). Johnson (1987) also notes that YAF sons were significantly closer to their mothers than were daughters (Table 8).

Sub-adult sons spend more time associating with their mothers than did sub-adult daughters (Table 9). However, on average postweaning parental-investment is probably greater in daughters than sons, as more daughters breed in their mothers' home range, and thus may compete with their mothers for a longer period of time than sons (Table 9). Further, there may be a depressive effect on a female's reproductive success if she shares her home range with female relatives (Johnson 1985). On the other hand, there was no evidence that daughters benefitted by staying in their mothers' home ranges; those who dispersed tended on the contrary to have a (non-significantly) higher offspring survival rate than those that stayed and to conceive for the first time at similar ages to them (Table 9).

The indirect measures of parental investment, that is the effect of a previous son or daughter on a mother Red-necked Wallaby's next reproductive attempt, did not indicate that individual sons were

Table 6. Red-necked Wallaby: comparisons between sons and daughters of duration of pouch-life and of YAF stages. The tests used were Mann-Whitney U tests (MWU).

duration (months)	sons	daughters	test	probability
pouch-life:				
sample size	44	29		
mean	9.2	9.7	MWU:Z=−2.7	p<0.01*
YAF phase:				
sample size	22	10		
mean	4.5	4.4	MWU:Z=0.95	ns

Table 7. Red-necked Wallaby: comparisons between sons and daughters of YAF sucking. The tests used were F-ratio test (F), and Mann-Whitney U tests (MWU).

sucking	sons	daughters	test	probability
duration (minutes)				
mean	9.0	9.5	MWU=113.5	ns
variances (between bouts)	2.0	8.2	F-ratio=4.0	p<0.05*
frequency (bouts per hour)				
mean	0.15	0.17	MWU=75.5	ns

Table 8. Red-necked Wallaby: comparisons between mother-son and mother-daughter pairs of mother-offspring interaction rates, association levels and distances. The tests used were Mann-Whitney U tests (MWU).

variable	sons	daughters	test	probability
mother-offspring interaction rates (per hour)				
month after PEP:	0.15	0.13	MWU=73	ns
all YAF-life:	0.38	0.17		
(taken from Johnson 1987)				
mother-offspring association (medians)				
month after PEP:	57%	37%	MWU=66.0	ns
all YAF-life:	56%	38%	MWU= 2.0	p<0.01*
(taken from Johnson 1987)				
mother-offspring distance (medians, metres)				
all YAF-life	5.41	8.27	MWU= 2.0	p<0.005*
(taken from Johnson 1987)				

more costly to rear than daughters (Table 10). In contrast to the kangaroos, the daughter wallabies had a more variable effect than the sons on their mothers' next reproductive attempt. The next offspring was significantly less likely to survive if it followed a daughter reared during good conditions (41% of next offspring survived) than if it followed a daughter reared during drought conditions (74% of next offspring survived) (χ^2=3.1, probability <0.1†). This relationship was not found for offspring following sons (χ^2=1.2, probability >0.1, ns).

Overall the measures of preweaning parental investment did not indicate that mothers consistently invested more in sons than daughters. Postweaning investment was likely to be greater in daughters than sons. However, as wallaby daughters were much more likely to disperse than kangaroo daughters, the level of postweaning investment in individual daughters is likely to be more variable for wallaby than kangaroo mothers.

Table 9. Red-necked Wallaby: potential postweaning investment in sons and daughters: percent of a mother's time spent associating with subadult sons and daughters; percent of sons and daughters who remain as adults in their natal home range; daughter's age at first conception when her mother was present in the same home range, or not (absent); and the proportion of a daughter's offspring who survived to weaning when the daughter's mother was present in the same home range or not (absent). The tests used were chi squared tests, F-ratio test (F), and Mann-Whitney U tests (MWU). N represents sample size, and $\bar{X}$ represents mean.

variable	sons	daughters	test	probability
% time spent associating with sub-adults: (taken from Johnson 1987)				
	33%	9%	MWU=4.5	p<0.01*
% offspring as adults remain in natal home range:				
	0/47(0%)	12/26(46%)	χ^2=22.5	p<0.01,*
Mother:	**present**	**absent**		
daughters' age at first conception (months)				
(N=21)	26.3	22.6	MWU=34.5	ns
proportion of daughter's young surviving:				
(N=17)	0.61	0.94	MWU=19.5	ns

Individual and environmental variation in sex ratios

In this section we examine whether sons and daughters were preferentially produced by different mothers or at different times. According to Werren and Charnov (1978) and Werren and Taylor (1984), if one sex's physical condition at the end of the parental investment period is more positively correlated with their lifetime reproductive success than the other sex's physical condition, then that first sex should be preferentially produced during periods when offspring are more likely to be weaned in good condition. We have assumed that the YAF stage rather than the pouch-infant phase would be timed to the season in which condition is most readily gained. Pouch-young are believed to grow to an all-or-none schedule and YAF show some variation in growth rate (Johnson 1977; Poole *et al.* 1982; Sharman *et al.* 1964). Pouch-young obtain all of their nutrition from the mother while YAF suck and spend a considerable proportion of the day grazing (Green 1984; Janssens and Ternouth 1987; Stuart-Dick 1987).

The Eastern Grey Kangaroo

The biased production of males was particularly strong when:

1. the offspring's permanent emergence was timed to late winter or early spring (Table 11) (when the pasture conditions during the YAF phase are probably at their optimum, unpublished data from Higginbottom; the difference in sex ratios between seasons was not significant);
2. the offspring's conception occurred in autumn (Table 11) (which is the season when females were in their best physical condition (Stuart-Dick 1987); an offspring is typically conceived 15 months prior to that offspring's permanent emergence from the pouch; note again that this difference in sex ratios between seasons is not significant);

Table 10. Red-necked Wallaby: indirect measures of parental investment in sons and daughters: percent of offspring which survive to weaning following weaned and lost, sons and daughters. The tests used were chi squared tests. N represents sample size.

variable	sons	daughters	test	probability
% offspring surviving following previous offspring				
	current offspring of both sexes			
previous lost	53%(N=64)		χ^2=0.4	ns
previous weaned	59%(N=89)		df=1	
current offspring	previous son weaned	previous daughter weaned		
% weaned	59%(N=49)	60%(N=40)	χ^2=0.1	ns

3. high rainfall had been received in the two to three months prior to their conception (R^2=0.618, sex ratio=0.386+1.6×10^{-6}× (rain)^2, probability=0.0003; observations of J. Clarke, K. Higginbottom, and R. Stuart-Dick support the view that high rainfall is most likely to be followed by good pasture conditions in this area);
4. the mother's home-range site was located in the north of the study area (Table 12) (with generally poorer food quality according to unpublished data from Higginbottom);
5. the mother had a relatively low rate of weaning offspring (Table 12); and
6. the mother was an "older" female (Table 12).

Further, mothers tended to produce offspring of the same sex as their previous offspring (χ^2=6.242, p=0.013).

Since the mother's age was subjectively assessed, we checked to see if the offspring who were known to be the first and last that a mother produced were more frequently daughters and sons respectively (Table 13). There was an insignificant bias towards production of daughters by the small sample of primiparae. The last offspring a female produced before disappearing was significantly more often a male than were non-last offspring.

The proportion of daughters among the offspring a female produced was not significantly related to her average condition relative to other females. Nor was it related to whether or not she was known to produce only offspring who permanently emerged in the peak season (Table 12).

As predicted by Trivers and Willard (1973), Werren and Charnov (1978), and Werren and Taylor (1984), mothers appeared more likely to conceive sons during that period of the year when they were in good condition or when pasture was good, and there was some indication that sons rather than daughters tended to have their YAF phase timed to span the spring-summer period of pasture growth. Sons' expected lifetime reproductive success might well be positively correlated to their early growth rate (there were other indirect relationships supporting this suggestion, but see Stuart-Dick 1987). However, a mother's strategy for determining whether to invest in an individual son or a daughter did not appear related by a simple function to her condition as predicted by Trivers and Willard. The mothers' offspring sex ratio was not related to the condition index used in the study. Further, a mother's preferential production of daughters was significantly correlated to a high rate of weaning offspring. It is possible that mothers who were likely to rear an offspring to weaning preferentially produced

Table 11. Eastern Grey Kangaroo: season offspring sex ratios for permanent emergence from the pouch (PEP) and for conception. The conception date is estimated using changes in the females' reproductive classes. The tests are chi-squared tests comparing obtained frequencies with the expected of parity and then comparing between seasons. The peak season is that half of the year with the greater number of offspring permanently emerging.

season		males	ratio females	m/f	χ^2	probability
	PEP date					
offpeak	April-June	16	9	1.78	1.96	ns
peak	July-Sept.	36	21	1.71	3.95	<0.05*
peak	Oct.-Dec.	32	25	1.28	0.86	ns
offpeak	Jan.-March	12	10	1.20	0.18	ns
sex by season:		April to Sept. vs Oct. to March			0.995	ns
	conception date					
	Jan.-March	15	11	1.36	0.31	ns
	April-June	33	14	2.36	7.68	<0.05*
	July-Sept.	23	20	1.15	0.21	ns
	Oct.-Dec.	19	17	1.12	0.111	ns
	sex by season:		df=3		3.616	ns

Table 12. Eastern Grey Kangaroo: variables relating to which mothers exhibit a preferential production of sons or of daughters. The abbreviations used and the full terms are as follows: FSR, Female offspring sex ratio index; RWY, rate of weaning young; ARCond, average relative condition. The average relative condition index was calculated by averaging over the months a female was assessed, her score for condition minus the average female's score (for that month). The tests used were F-ratio tests (F), t-tests (t), Mann-Whitney U tests (MWU), and regressions (R^2). The female offspring sex ratio index indicates the proportion of a female's offspring which are daughters. The values range between 1.0 (sons only) and 2.0 (daughters only). Thus a positive R^2 indicates females produce proportionally more daughters with an increase in the value of the related variable.

variable related to **FSR**	equation	R^2	probability
RWY	$100RWY=44.62-35.29(FSR)+7.42(FSR)^2$	0.127	lin.0.023* quad.0.016*
'age'		−0.295	0.032*
ARcond	$FSR=-3.60+7.12(ARcond)-3.00(ARcond)^2$	0.101	lin.0.478ns quad.0.560ns

	FSR: mean	**SE**	**(N)**	**test**	**probability**
home-range site:					
North	1.14	0.062	(13)	F=3.09	0.043*
South	1.51	0.073	(29)	$t_{up}=3.82$ df=38	<0.001*
				MWU,z=2.803	0.005*
type of female: whether or not she has offspring PEP in peak season only					
peak	1.45	0.070	(23)	F=1.42	0.405ns
aseasonal	1.36	0.070	(29)	$t_p=0.81$ df=50	0.422ns

daughters. Alternatively, this relationship might have arisen because a slightly greater proportion of daughters than sons survived to weaning. Female offspring have a less variable pouch-life duration; and offspring with short pouch-lives are less likely to survive than those with longer pouch-lives. A mother might have a higher rate of rearing offspring to weaning because she had produced more daughters (fortuitously?). Nevertheless, there was also a significant tendency for the "younger" mothers and the southern mothers of the study population to produce more daughters than other mothers. These former two groups were in better-than-average or at least not in poorer condition than others (Table 14). Therefore mothers were not simply preferentially producing daughters when in poor condition and sons when in good condition.

The Red-necked Wallaby

Mother Red-necked Wallabies were more likely to conceive males:

1. in Good (nondrought) years (Table 15) (the survival rate of offspring at permanent emergence was high in Good than in Drought years; Mann-Whitney U, z=2.29, p<0.05).
2. if the offspring was conceived in winter during a Good year (Table 16).
3. if the offspring emerged permanently in spring during good conditions (Table 17) (during Good conditions the survival rate of offspring

Table 13. Eastern Grey Kangaroo: The numbers of male and female offspring produced by primiparae and multiparae mothers, and the sex of the last offspring of a mother versus non-last offspring. The tests used were G-tests.

	males	females	ratio	test	probability
offspring sex ratio of primiparae vs multiparae					
primiparae	6	9	0.67	G=2.008	ns
multiparae	123	86	1.43	between mothers	
offspring sex ratio for known last vs others					
last	18	5	3.60	G=4.683	p<0.05*
others	110	88	1.25	between mothers	

Table 14. Eastern Grey Kangaroo: Females' physical condition related to "age" and home range site. The tests are Mann-Whitney U tests. The condition index values range between 0 very poor and 1.0 very good.

	October 1984			March 1985		
	sample size	mean	SE	sample size	mean	SE
"age"						
young	8	0.63	0.036	6	0.58	0.041
mature	18	0.53	0.025	17	0.54	0.027
old	12	0.38	0.027	9	0.41	0.073
home range site						
North	10	0.43	0.038	9	0.42	0.036
South	20	0.54	0.031	15	0.57	0.031
Mann-Whitney U tests						
"age"	*z*	*p*		*z*	*p*	
young vs mature	−2.531	0.006*		−0.808	>0.1	
mature vs old	−3.156	0.001*		−1.593	0.055†	
young vs old	−3.397	0.0004*		−1.421	0.077†	
home-range site						
North vs South	−2.135	0.033*		−3.201	0.001*	

Table 15. Red-necked Wallaby: Relationships between individual and environmental variables and the sex ratio of offspring. The tests used are chi squared tests. Good and drought climates refer to the years when the Pastoral Protection Board did not and did respectively declare the area as drought-stricken.

variable	males	females	sex ratio m/f	test	probability
mother's home range site					
South	40	24	1.67	$\chi^2=3.5$	p<0.1†
North and Slopes	36	32	1.13	$\chi^2=0.1$	ns
	between South and North:			$\chi^2=0.9$	ns
climate at conception					
Drought	31	33	0.94	$\chi^2=0.02$	ns
Good	46	24	1.92	$\chi^2=6.3$	p<0.05*
	between Drought and Good:			$\chi^2=3.4$	p<0.1,†
age of mother at conception					
in Good climate					
young	22	6	3.67	$\chi^2=9.1$	p≤0.05*
old	20	18	1.11	$\chi^2=0.03$	ns
	between young & old:			$\chi^2=3.7$	p<0.1†
total					
young	28	15	1.87	$\chi^2=3.3$	p<0.1†
old	21	26	0.81	$\chi^2=0.5$	ns
	between young and old:			$\chi^2=3.0$	p<0.1†

who permanently emerged in spring was higher than for the other seasons; $\chi^2=7.8$, $p<0.05$).

4. when the mother's home range was situated in the south of the study site (Table 15) (with generally higher food quality according to unpublished data Higginbottom). (However, the sex ratios for mothers in the south and north were not significantly different).
5. if the mothers when young, particularly during Good climatic conditions (Table 15) (old mothers tended to produce as many daughters as sons).
6. if the mothers had a relatively high rate of weaning offspring (weaning rate was correlated with the proportion of males among a mother's offspring; Spearman's r=0.43, probability <0.05*).

There was, however, no evidence that individuals tended to specialize in one sex or the other, although the data were derived from only a part of each female's lifespan (for all mothers with the same numbers of offspring of known sex the frequencies of mothers that produced the various proportion of sons and daughters were tested against the expected frequencies; the χ^2 values

Table 16. Red-necked Wallaby: The numbers of male and female offspring estimated to be conceived in each season, for years of Good, Drought and all conditions. The tests used were chi squared tests. Good and Drought years have been defined according to the Pastoral Protection Board's assessment.

season	males	females	sex ratio m/f	condition ranked (good→poor)
Drought				
spring	9	8	1.12	
summer	10	8	1.25	
autumn	5	7	0.71	
winter	7	8	0.88	
Test for homogeneity between seasons:			$\chi^2=0.7$, df=3	ns
Good				
spring	15	7	2.14	
summer	12	12	1.0	
autumn	9	4	2.25	
winter	9	1	9.00	
Test for homogeneity between seasons:			$\chi^2=5.5$, df=3	ns
total				
spring	24	15	1.60	4
summer	22	20	1.10	3
autumn	14	11	1.27	1
winter	16	9	1.78	2
Test for homogeneity between seasons:			$\chi^2=1.2$, df=3	ns

All tests comparing above frequencies with 1:1 not significant except: Good winter, $\chi^2=4.9$, $p<0.05$, *;(Good spring, $\chi^2=2.2$).

Table 17. Red-necked Wallaby: the numbers of male and female offspring estimated to permanently emerge from the pouch (PEP) in each season. The tests used were chi squared tests.

season	males	females	sex ratio m/f	condition ranked (good→poor)	% offspring surviving
spring	27	17	1.59	4	64%
summer	9	11	0.82	3	46%
autumn	15	11	1.36	1	53%
winter	26	16	1.63	2	61%

Test for homogeneity between seasons: χ^2=1.9, df=3,ns
(also non-significant within Drought and Good)
tests comparing above frequencies with 1:1 not significant
except: spring. χ^2=3.4,p<0.05,*

were not significant). The proportion of sons among the offspring a mother produced was not significantly related to her average condition relative to other females (Spearman r=0.10, p=0.3, sample size=20).

In sum, it is possible that male rather than female wallabies tend to be produced to take advantage of good conditions as Trivers and Willard (1973), Werren and Charnov (1978) and Werren and Taylor (1984) predicted. Further, sons tended to disperse at a younger age if their YAF phase was timed to a good rather than a drought year (Mann-Whitney U: z=2.1, p<0.05 *). However, young females rather than old females preferentially produced sons. This is contrary to predictions about age-related tactics based on Trivers and Willard's arguments if sons, on average, received more parental care than daughters, or if as predicted older mothers rather than younger mothers give their offspring more parental care.

Since the relationships between the sex of offspring and both season and weaning rate may have been masking effects of the age of the mother, we can not interpret these results without examining age-related patterns of investment in more detail.

Age-related patterns of parental investment

Although there were no differences in parental-investment patterns indicating a greater investment in individual sons than daughters, the mothers who preferentially produce sons, the older kangaroos and younger wallabies, might be those who invest more intensely in offspring. Thus variation in investment between the sexes would be indirectly given through age-related investment tactics of the mothers. If mothers do differ in investment patterns then sons and daughters of different aged mothers might have different ages of dispersal or of first conception.

Table 18. Age-related differences in mothers' parental investment tactics in the Eastern Grey Kangaroo. The tests used were G-tests, Pearsons correlation, and regresion analyses. The latter two types of tests have used the age index score, which ranges from 1 (very young) to 6 (very old) in value. Sample size is abbreviated to N. The sex ratio is males/females.

variable (age score:)	young (1&2)	mature (3&4)	old (5&6)	test	probability
offspring					df=2
sex ratio:	0.833	1.333	2.857	G=4.841	p<0.1†
rate of weaning offspring (offspring/year)					
mean	0.52	0.42	0.44	R^2=0.166	$P_{quadratic}$=0.032*
% offspring surviving					
%	48%	33%	39%	G=1.464	ns
N	21	69	46	df=2	
rate of producing PEP offspring (offspring/year)					
mean	1.01	0.99	0.96		
pouch duration (weeks)					
means:	46	46.3	52.2	Pearson's r:	p=0.024*
N	5	18	13	0.333	
frequencies of females who produced only peak offspring or bred aseasonally					
peak only	8	6	9	$G_{young/mature}$=6.462	p<0.05*
seasonal	2	13	6	$G_{mature/old}$=2.176	ns
sons' age at dispersal: (months)					
dispersal age=21.5+2.41(mother's age)				R^2=0.210	p=0.074†
daughters' age at first conception: (months)					
	33.2	32.6	34		

Table 19. Age-related differences in mothers' parental-investment tactics in the Red-necked Wallaby. The tests used were chi-squared tests, analysis of variance F-tests, and Mann-Whitney U tests (MWU). Sample size is abbreviated to N. The sex ratio is males/females.

variable	young	old	test	probability
offspring sex ratio:	1.87	0.81	χ^2=3.0	p<0.1†
pouch duration: (months)				
means	9.1	9.5	AOV:F=4.63 df=2	p<0.05*
YAF duration: (months)				
medians	4.17	4.67	χ^2=0.26	ns
birth to wean: (frequencies)				
≤14 months	6	11	χ^2=3.2	p<0.1†
>14 months	0	11	(between mothers)	
mean % of monitored females with offspring PEP in each season				
	N=64	N=78		
peak season	35%	35%		
offpeak season	17%	25%		
peak vs offpeak for young			χ^2=15.6	p<0.001,*
peak vs offpeak for old:			χ^2=2.3	ns
% survival of offspring in respective seasons (in good only):				
overall	66%	67%	χ^2=0.03	ns
peak	75%(N=28)	69%(N=26)	χ^2=0.24	ns
offpeak	68%(N=37)	62%(N=47)	χ^2=0.31	ns
rate of producing PEP offspring: (offspring/year)				
mean	1.08	1.2	MWU=21 N=9,9	p=0.1†
rate of weaning offspring (offspring/year)				
mean	0.56	0.72	MWU=29 N=9,9	ns
sons' age at dispersal: (months)				
mean	24.4	20.0	MWU=3 N=7,5	p<0.05
daughters' age at first conception: (months)				
mean	22.9	22.0	MWU=14 N=4,8	ns
contact rates for YAFS in month after PEP: (per hour)				
means	0.10	0.07	MWU=106	ns N=10,21
suckle rates for YAFS in months after PEP((per hour)				
means	0.07	0.18	MWU=56	p<0.1,† N=10,19
proportion of time associating with mothers for YAFS in month after PEP:				
means	54%	54%	MWU=104	ns N=10,21
Survival of daughters offspring:				
	60%	85%	χ^2=1.18	ns N=20,13

The Eastern Grey Kangaroo

The results discussed in this section are presented in Table 18.

Young Eastern Grey Kangaroo mothers had a significantly higher rate of weaning offspring than older mothers. However, the percent of offspring weaned of those produced did not differ significantly between old and young mothers. Nor did the rate of production of offspring significantly vary between them. Probably a weak trend in both factors contributed to the significant difference in weaning rate.

All primiparae produced only offspring which permanently emerged in the peak season. In general both the very young and very old mothers tended to produce only peak-season offspring, while the middle-aged mothers were more likely to reproduce to an aseasonal schedule. The offspring of the mothers who produced only peak offspring were significantly less likely to survive (33%) than were the peak offspring of mothers who bred aseasonally (61%) (G=8.825, df=1, $p<0.01$ *). Thus it is unlikely that the offspring of seasonal breeders (typically the very young and old mothers) were benefitting more from maternal investment than were the offspring of aseasonal breeders.

The old mothers were more likely to give their offspring long pouch durations. But again this may not necessarily indicate a greater level of care by the mother nor benefit gained by the offspring. Older mothers' sons tended to disperse at an older age than those of younger mothers. From subjective field observations it seemed that these sons were maturing at a slower rate. Objective assessments are required to substantiate this. However, the daughters of younger or mature mothers averaged a nonsignificantly earlier first conception age than those of older mothers.

Overall, if the offspring of different aged mothers receive different benefit, it is more likely that those of young or mature mothers gain more than those of old mothers. It was the old mothers who preferentially produced sons.

The Red-necked Wallaby

The different parental-investment patterns for the younger and older Red-necked Wallaby mothers are presented in Table 19.

The offspring of the young wallaby mothers (those who preferentially produced sons) were more likely to have had a short pouch-life and as a consequence be weaned at an earlier age. They were not significantly more likely to survive than the offspring of old mothers. These young mothers were particularly likely to have their offspring's permanent emergence timed to spring during good climatic conditions.

Older mothers, despite their longer period of preweaning investment in each offspring, tended to have a higher rate of weaning offspring; this was a result of the younger mothers having a higher rate of reproductive failure. The older mothers, on average, may have given more parental care to each offspring than younger mothers (reflected in a later age at weaning and higher suckling rates). This may have enabled sons to disperse earlier, as a result of a boosted growth or maturation rate. In addition, although daughters of older mothers did not conceive earlier, there was a tendency for their offspring to have a higher rate of survival than did the daughters of younger mothers.

The greater relative production of sons by mothers with higher weaning rates is not a confounding effect of age; instead, mothers who had higher weaning rates tended to produce males even though the young females who produced more males tended to have relatively low weaning rates. Thus there may be a tendency for females who are able to give extra parental-care to produce sons preferentially, regardless of their age.

CONCLUSIONS

Strategies of parental investment

Before discussing the strategies that we believe the Eastern Grey Kangaroo and the Red-necked Wallaby mothers may be following in relation to the various hypothesis, we need to establish whether the conditions necessary for each of the hypotheses apply in each species.

According to Trivers and Willard (1973) selection for investing extra parental-care in male rather than female offspring is related to a greater variance in expected lifetime reproductive success for males than females.

Red-necked Wallaby and Eastern Grey Kangaroo males are much larger than females (Jarman 1983; and Strahan 1983); the sexual dimorphism is more extreme in the kangaroos (Jarman, this volume). Males have the potential to grow throughout their lives. In the local population of kangaroos (including approximately 50 adult females and 24 adult males) the largest male obtains about 75% of the matings; this male's tenure of the "alpha" position typically lasts one year. Six at most contribute genes to the annual cohort of offspring produced by the females. In contrast greater proportion of the adult male wallabies probably contribute genes to an annual cohort of offspring. There are typically two to four alpha male wallabies in the study area at any one time (and 30 to 40 adult female wallabies); an alpha male wallaby's tenure lasts a mean of 1.7 years which is longer than that of kangaroo males.

The means and variances in the lifetime-reproductive-success of the males and the females were estimated for both species (Table 20). Wallabies weaned offspring at a higher rate than did kangaroos, but the breeding life of females in both species was estimated to be about eight years. So the average female wallaby could expect to wean more offspring in its life than could the average female kangaroo; this allows greater variance in the lifetime-reproductive-success for female wallabies than kangaroos.

For each species, we used the following formula, which uses the standard error (SE) and mean ($\bar{X}$) of the values of estimates of individuals' lifetime reproductive success (values of 0 were included in the calculations for the offspring estimated to have been produced but which did not live to breed), as a measure of the difference between the sexes in the extent of variation in lifetime reproductive success.

$$\frac{SE/\bar{X}\ males}{SE/\bar{X}\ females}$$

This ratio is similar for the two species and in both cases the expectations of males are much more variable than are those of females (see Table 20).

Table 20. General features of the reproductive ecology of the Red-necked Wallaby (RNW) and the Eastern Grey Kangaroo (EGK) at Wallaby Creek. N refers to sample size, SE refers to standard error, and $\bar{X}$ refers to the mean.

variable	RNW	EGK
estimated lifetime reproductive success		
females:	N=72	N=198
mean (offspring in lifetime)	3.12	0.94
standard error	0.45	0.14
males:	N=131	N=296
mean (offspring in lifetime)	3.5	0.6
standard error	1.06	0.21
SE/$\bar{X}$males/SE/$\bar{X}$females:	2.2	2.3
female reproductive rate:		
(offspring/months)	1/10.4	1/10.8
age at first conception: (months)	24	33
typical breeding life: (years)	8	8
total offspring	9	9
maximum number of daughter's offspring overlapping		
mothers' offspring:	7	6
dispersal age of sons: (months)	23	31
weaning rate:		
(offspring/month)	0.054	0.037
offspring weaned/total born:	134/245 (54.7%)	73/190 (38.4%)
number of females		
sharing a home range	4	26

Thus Trivers and Willard would propose that female wallabies and kangaroos who can give more parental care than the average female should preferentially produce sons. This could occur for females who are in better-than-average condition, or when better-than-average environmental conditions prevail. In addition, according to the age hypothesis, old females should preferentially produce sons and invest in each one heavily. There should be no difference between the wallaby and kangaroo mothers' parental investment strategies. Note that the above predictions depend on sons' lifetime-reproductive-success being positively correlated to their condition at the end of the period of parental investment.

Altmann's hypothesis relies on a mother's ability to influence her daughters' expected lifetime reproductive success to a greater extent than her sons' expected lifetime reproductive success, by postweaning care or inheritance rank.

Dunbar's (1983) age-related extension to this hypothesis depends on mothers being able to influence the reproductive success of their adult daughters by their presence, through, for example, reduced risk of predation or formation of coalitions for defending resources.

Male-biased dispersal occurs in both species; but female philopatry is probably stronger in the kangaroos than in wallabies (Table 20). Maternal influence on sons' expected lifetime reproductive success for both species is likely to be restricted to the short-term pre-dispersal forms of parental investment which influence a son's growth rate or age and physical condition at dispersal. In contrast, maternal influence on a daughter's expected lifetime-reproductive-success might extend post-weaning to more than half of the kangaroo or wallaby daughter's breeding life (see Table 20). As kangaroo females have a much larger home range and share it with a greater number of other females (approximately 25 other females) than do wallaby females (approximately three other females), fewer wallaby than kangaroo daughters can share a mother's home range. Further, compared to wallaby females, kangaroo females are found more often in groups and these groups tend to be larger than those of wallabies; and mother and adult daughter are found more frequently (25% kangaroo, 5% wallaby).

This means that in kangaroos there may be a greater potential for coalitions or other positive social effects which increase the daughters' reproductive success. Because more wallaby daughters than kangaroo daughters disperse, the average wallaby daughter is likely to receive less investment and to be influenced less by her mother's presence than are kangaroo daughters. Thus we might expect that there would be stronger selection for daughters to be produced at an early age in kangaroos than in wallabies.

The Local Resource Competition hypothesis rests on sons' greater level of dispersal and daughters' greater tendency to remain and breed in their natal home range with a consequent greater cost of daughters than sons to the mother.

We have just established that sons have a higher level of dispersal than daughters in both species. However, the costs to a mother of sharing her home range with a daughter may be greater for the wallabies than kangaroos, as proportionally more of the mother's local resources would be shared by an individual daughter. Although, on average, daughter wallabies may not be any more costly to a mother than are daughter kangaroos (since more wallaby daughters disperse), if for any one daughter the expected costs of competition for resources greatly exceed those of sons then a mother would do better to produce those daughters late in her breeding life, all else being equal.

Having examined the extent to which the prerequisite conditions for each hypothesis apply to each species, we will now attempt to interpret our results in relation to these hypotheses.

For both Eastern Grey Kangaroos and Red-necked Wallabies individual male offspring did not consistently receive a greater quantity of preweaning parental-care than individual female offspring: daughters received more of some forms of preweaning parental care while sons received more of others. Postweaning care was probably greater in daughters. However, it is interesting that in the kangaroos, levels of parental care were more variable for males than females, while the opposite was found in the wallabies. In both species more male than female offspring were reared and weaned (although the ratio was not statistically significantly different from parity for wallabies).

Males of both species were preferentially produced during Good environmental conditions. According to Werren and Taylor (1984) and Werren and Charnov (1978), this could indicate that the expected lifetime-reproductive-success of males is affected by the environmental conditions in which they are reared more than that of females. The male offspring in these populations may have benefitted from being opportunistically produced to take advantage of Good conditions, perhaps via a rapid growth rate or particularly good physical condition at the end of the parental investment period. However, as predicted by Trivers and Willard (1973), there should not and did not appear to be any "extra" cost to the mothers.

The kangaroo mothers who tended to produce sons did not appear to be those with the highest ability to invest as proposed by Trivers and Willard. Sons were not preferentially produced by mothers in above-average condition, and the older mothers, who preferentially produced sons, were in poorer condition than the younger mothers. These older mothers appeared to be giving less parental care than younger mothers, and mothers with high weaning rates tended to produce daughters, not sons. Thus there is no evidence that differences in condition of the Eastern Grey Kangaroo mothers are related to the preferential production of the sexes in the manner proposed by Trivers and Willard's hypothesis.

Some of the wallaby mothers who tended to produce sons may have been those able to give extra parental-care, since those with higher weaning rates (due chiefly to higher survival rates) and those in the higher quality home-ranges preferentially produced sons. However, sons were not preferentially produced by mothers in above-average condition, despite the younger mothers, who tended to produce more sons, being in better condition than older mothers. This may simply mean that our measure of condition was not related to the ability of mothers to give extra parental-care and was not independent of age. Although Trivers and Willard's hypothesis may account for some of the individual variation in relative production of the sexes, it does not explain the age-related pattern of production of the sexes since younger mothers appeared to give less parental care than older mothers.

So why were there age-related tactical changes in mothers' investment in sons and daughters? In the case of kangaroos, mothers may have preferentially produced daughters according to their relative ability to influence daughters' expected lifetime-reproductive-success as Altmann proposed. The studies have not been of long enough duration to measure offspring lifetime-reproductive-success. However, there was a non-significant tendency for kangaroos to conceive at an earlier age if their mothers were young, and a stronger tendency for kangaroos to conceive at a young age if their mother was still present in the study area when the daughters became adults. Further, the offspring of daughters with mothers present were more likely to survive than offspring of daughters without mothers present. Thus it is possible that a kangaroo mother's presence can significantly influence her daughter's lifetime-reproductive-success. There was no such tendency for wallabies to be more likely to commence reproduction at an early age if their mothers were young. Although the daughters tended (not significantly) to be less likely to disperse if their mothers were young, their age at first conception was not significantly related to the degree to which they shifted their home range from their natal site. It appears that wallaby mothers do not influence their daughter's lifetime-reproductive-success as consistently as kangaroo mothers, but as yet the data are few.

Both the kangaroo and wallaby mothers' ages were related to their sons' age at first dispersal. However, the mothers who produced early-dispersing sons were not those who preferentially produced sons. Early dispersal was related to a good climate for wallabies and also probably for kangaroos.

Tentatively, Altmann's hypothesis may apply to the young kangaroo mothers because they can, by their presence, influence the reproductive success of their daughters more than that of their sons. Young kangaroo mothers preferentially produced daughters and probably invested heavily in them postweaning; as the kangaroo mother aged she might have tended preferentially to produce sons who do not require as much postweaning investment to influence their reproductive expectations. In this manner a kangaroo mother might most influence her daughters' breeding by maximizing the overlap of her own and her daughters' lifespan. Overall more sons than daughters are produced by a mother, but she invests in her daughters over a longer period. In this manner Fisher's hypothesis of equal investment in the sexes might be upheld.

However, mother wallabies appear to follow a different strategy. We suggest that as mother wallabies do not share their home range with more than three other females, some daughters must disperse, but those who do not may be costly to a mother wallaby's future reproduction (Johnson 1986). Compared to kangaroos, wallaby mothers and daughters may gain a smaller amount of benefit from sharing their home range (wallabies being less social). As old mothers' future reproductive expectations are reduced, the cost to them of daughters remaining to breed in their home range is less than it would be for young mothers. Consequently, individual daughters and sons might be equally costly to an old wallaby mother, and accordingly she produces as many daughters as sons. As mother wallabies' age-related tactics are expected to be a function not just of the relative cost of sons and daughters, but also a function of the relative benefit gained by sons and daughters, it would be of interest to gather more long-term data to check whether daughters' expected lifetime-reproductive-success does vary according to their mothers' age or presence.

In sum, we believe that in both species, an excess of males is produced which is balanced by a greater post-weaning investment in daughters, such that Fisher's hypothesis is upheld. Further, males are produced to take advantage of Good environmental conditions since males' expected lifetime-reproductive-success is likely to be more strongly influenced by early growth rate and physical condition than is that of females. Preferentially producing males during times when they can gain condition and grow rapidly may be readily selected for as there would be little extra cost to the mothers compared to the benefit they gain.

Wallabies seem to be combining two extra tactics in their overal parental investment strategy. Mothers who can give extra parental-care tend to produce males since the boost they can give to their sons' lifetime-reproductive-success exceeds that which they can give to their daughters, as proposed by Trivers and Willard. The second tactic is related to age. Individual mothers tend to produce the "extra" sons when young rather than old. In this manner wallaby mothers reduce the costs of competition from breeding daughters (the Local Resource Competition hypothesis); these costs are not greatly offset by benefits accruing to mother and daughter from sharing their home range.

Kangaroos also appear to combine two extra tactics in their parental investment strategy. Individual females who can give extra parental-care appear to produce daughters preferentially. These females might be better able to invest over the long term in the more expensive daughters than sons. Daughters' lifetime-reproductive-success is less variable than sons' and thus represents a safer investment for these mothers. Kangaroo males begin active breeding much later in life than do wallabies. Thus kangaroo sons' lifetime-reproductive-success might not be as *strongly* correlated to their condition at the end of the parental investment period (a condition of Trivers and Willards' hypothesis) as it is in wallabies. This suggestion requires substantiation from long-term data. The second tactic of kangaroo mothers is to produce daughters preferentially when young and sons when old. In this manner mother and daughter might gain benefit from a social coalition (Dunbar's hypothesis).

Overall, if we have made a valid interpretation of the parental-investment strategies of Eastern Grey Kangaroos and Red-necked Wallabies, it appears that Altmann's hypothesis can be extended to include social organizations with open-membership grouping. However, a mother's influence on her daughters' lifetime-reproductive-success may be limited to the number of daughters who can share her home range or to whether or not social affiliations between matrilineal relatives affect the individuals' reproductive expectations. Altmann's hypothesis probably works in conjunction with the Local Resource Competition hypothesis: mothers will preferentially produce daughters if the benefits of sharing their home ranges with daughters outweigh the costs. However, Trivers and Willards' hypothesis of preferentially conceiving sons if a mother can expect her sons to have a greater than average chance of becoming an "alpha" male also appears to explain some of the wallaby and kangaroo mothers' tactics.

A better way of conceptualizing the importance of the various hypotheses used to explain the data is to view the mothers strategy as employing tactics which weight the benefits and costs of producing a son against the benefits and costs of producing a daughter in order to optimize the inclusive fitness of the mother (including the reproductive expectations of the mother and all her offspring). There should be selection for a strategy which uses variables such as a mother's age or local environmental conditions as cues to when to produce sons or daughters preferentially when there is a strong relationship between these variables and the mother's inclusive fitness.

We stress that much more long-term research is required to test these preliminary findings and interpretations. We hope these studies stimulate interest in this area.

OTHER SPECIES

The degree and type of variation in parental-investment tactics may differ between species of macropods. For example, if Trivers and Willard (1973) were correct in their predictions, homomorphic species should show less variation in parental investment tactics than hetereomorphic ones. Determining which species exhibit much and which exhibit little tactical variation and why the differences exist would increase our understanding of the general potential for facultative adjustment of parental investment by macropods and mammals in general. This knowledge in turn could aid long-term management of macropod populations.

In both the Eastern Grey Kangaroos and the Red-necked Wallaby the duration of pouch-life varied considerably. The degree of variation and the mean duration were related to factors such as the sex of the offspring, season of permanent emergence, age of the mothers, and the mothers' previous reproductive history. Thus pouch-life duration would seem an excellent variable for using as a tool to propose whether and which other species of macropods might exhibit variation in parental-investment tactics.

In Figure 1A we present a scattergram of the standard error of pouch-life duration for various macropod species. The sources of the data are presented in Appendix A. It was not possible to fit a significant curve to the values. This lack of a relationship between the mean and standard error of pouch-life duration for macropod species might be indicative of a differing tendency between the species for within-species variation in parental-investment tactics. Alternatively, it might just result from "random" variation in sampling the variation in pouch-life duration.

In order to discover characteristics which might distinguish between species with highly variable and those with fairly invariable pouch-life durations, we looked at these species' relative positions on the following curves:

1. age at first emergence against pouch-life duration (Figure 1B);
2. YAF phase duration against pouch-life duration (Figure 1C); and
3. pouch-life duration against maternal weight (Figure 1D);

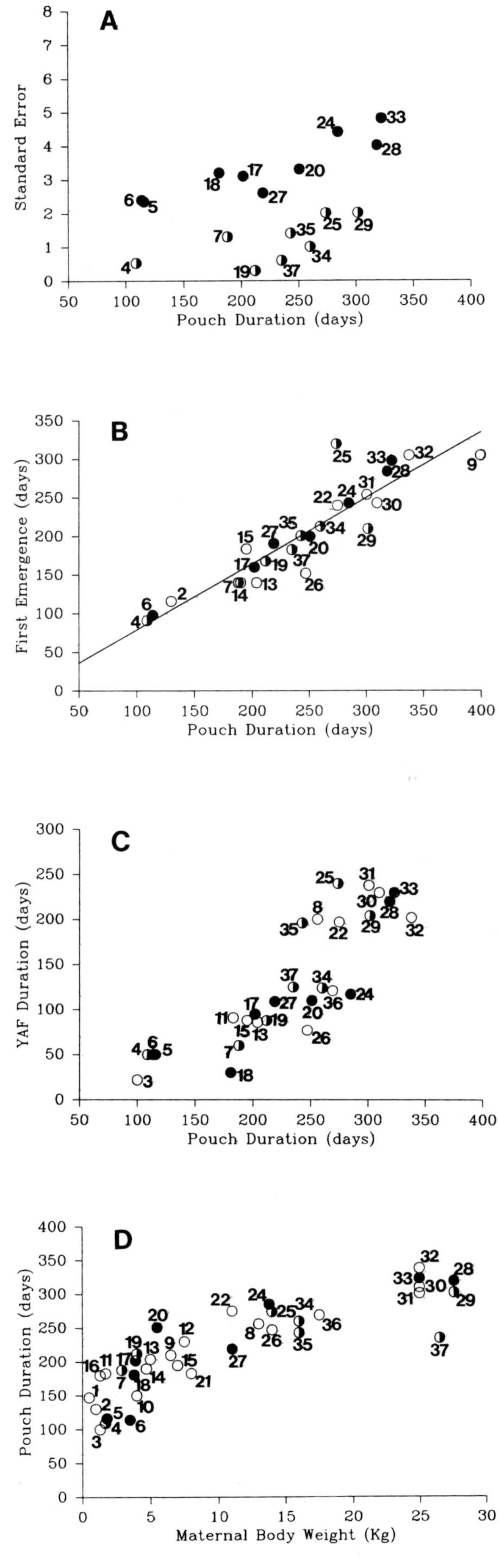

Fig. 1. Graphs depicting pouch-life duration for a variety of macropod species against values of A) the standard error of pouch-life duration, B) the age of first emergence from the pouch, C) the duration of YAF life, and D) maternal body weight. Species are identified by number; for species code number see Appendix A. The species with standard errors greater than 2.0 are represented by fully shaded circles and those with standard errors less than 2.0 are represented by half shaded circles. The age at first emergence is significantly related to pouch duration: first emergence=−6.152+0.854 (pouch duration), R^2=0.828, and probability<0.001 *. There is a significant relationship between pouch duration and YAF duration: pouch duration=−60.5+2.1 (YAF duration) −0.005 (YAF duration)^2, R^2=0.801, quadratic probability <0.01, *.

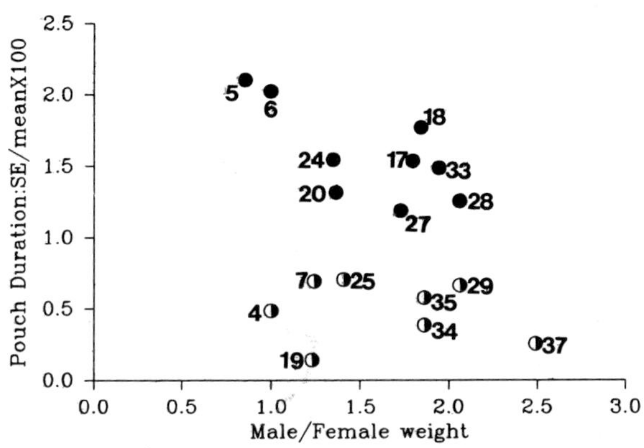

Fig. 2. The standard error over mean of pouch-life duration plotted against the male over female weight for a variety of macropod species. Species are identified by number; for species code number see Appendix A. The species with standard errors greater than 2.0 are represented by fully shaded circles and those with standard errors less than 2.0 are represented by half shaded circles.

4. male/female weight against standard error/ mean of pouch-life duration (Fig. 2).

Unfortunately we could not distinguish any characteristics which consistently differentiated between the species with more and those with less variable pouch-life durations. Thus if the standard errors of pouch-life duration do reflect the species' propensity for variation in parental-investment tactics, then neither maternal weight nor sexual dimorphism are related to this propensity in an obvious manner. The latter is a particularly interesting non-relationship, as according to Trivers and Willards hypothesis a biased investment in males is in pouch-life duration does not appear to be associated in a simple manner with particularly short or long YAF or pouch-infant phases.

Age at first emergence and the duration of the YAF phase may also indicate variation in tactics; however, the standard error or related variable is too infrequently given in the literature to allow us to generate a scattergram similar to that for pouch-life duration.

ACKNOWLEDGEMENTS

We thank Peter Jarman for his considerable help and advice, and in sharing data. Chris Johnson has contributed in many ways, by sharing data he gathered on the wallabies and in readily discussing wallaby and kangaroo tactics with both of us. These studies would not have been possible without the very generous nature of the land-holders of the study area: Iris and Jack Hayes, Margaret and Ned Hayes, Anne and Rusty Bell, Victor Mulcahy, Glen and Rosemary Kennedy, and Len Grayson. These owners, in paticular the Hayes, are responsible for allowing the native wildlife to flourish and have made field-life thoroughly enjoyable by their seemingly endless hospitality and even help in collecting data. The research was supported by Commonwealth Post-graduate Research Awards and an Australian Research Grant Scheme grant D18115308 to Peter Jarman.

REFERENCES

ALTMANN, J., 1980. Baboon mothers and Infants. Harvard University Press: Cambridge.

ANDREWARTHA, H. G. AND BARKER, S., 1969. Introduction to a study of the ecology of the Kangaroo Island wallaby, *Protemnodon eugenii* (Desmarest) within Flinders Chase, Kangaroo Island, South Australia. *Transactions of the Royal Society of South Australia* **93**: 127-32.

ARMITAGE, K. B., 1987. Do female yellow-bellied marmots adjust the sex ratios of their offspring? *Amer. Nat.* **129**: 501-19.

AUSTAD, S. N., 1988. The adaptable opossum. *Scien. Amer.* **258**: 54-59.

BELL, G., 1980. The costs of reproduction and their consequences. *Amer. Nat.* **116**: 45-76.

CALABY, J. H. AND POOLE, W. E., 1971. Keeping kangaroos in captivity. *Internat. Zoo Yearbk.* **11**: 5-12.

CALOW, P., 1981. Resource utilization and reproduction. *In* Physiological Ecology: an evolutionary approach to resource use ed. by C. R. Townsend and P. Calow. Blackwell Scientific Publ.: Oxford.

CATT, D. C., 1977. The breeding biology of Bennett's wallaby (*Macropus rufogriseus fruticus*) in South Canterbury, New Zealand. *NZ Jour. of Zool.* **4**: 401-11.

CHARLESWORTH, B. AND LEON, J. A., 1976. The relation of reproductive effort to age. *Amer. Nat.* **110**: 449-59

CLARK, A. B., 1978. Sex ratio and local resource competition in a prosimian primate. *Science, N.Y.* **201**: 163-65.

CLARK, M. M., SPENCER, C. A. AND GALEF, B. G. Jnr., 1986. Reproductive life history correlates of early and late sexual maturation in female Mongolian gerbils (*Meriones unguiculatus*). *Anim. Behav.* **34**: 551-60.

CLUTTON-BROCK, T. H., ALBON, S. D. AND GUINNESS, F. E., 1986. Great expectations: dominance, breeding success and offspring sex ratios in red deer. *Anim. Behav.* **34**: 460-79.

CLUTTON-BROCK, T. H., ALBON, S. P. AND GUINNESS, F. E., 1982a. Competition between female relatives in a matrilocal mammal. *Nature* **300**: 178-80.

CLUTTON-BROCK, T. H., GUINNESS, F. E. AND ALBON, S. D., 1982b. Red Deer: Behaviour and ecology of two sexes. University of Chicago Press: Chicago.

COLE, L. C., 1954. The population consequences of life history phenomena. *Quart. Rev. Biol.* **29**: 103-37.

COMFORT, A., 1979. The Biology of Scenescence. (3rd ed.). Churchill Livingstone: Edinburgh.

DITTUS, W. P. J., 1979. The evolution of behaviours regulating density and age-specific sex ratios in a primate population. *Behaviour* **69**: 265-301.

DUNBAR, R. I. M., 1984. Reproductive Decisions: An economic analysis of gelada baboon social strategies. Princeton University Press: Princeton, New Jersey.

FISHER, R. A., 1930. The Genetical Theory of Natural Selection. Clarendon Press: Oxford.

FLEMING, D., CINDEREY, R. N. AND HEARN, J. P., 1983. The reproductive biology of Bennett's Wallaby (*Macropus rufogriseus rufogriseus*) ranging free at Whipsnade Park. *Jour. of Zool. Lond.* **201**: 283-91.

GADGIL, M. AND BOSSERT, W. H., 1970. Life history consequences of natural selection. *Amer. Nat.* **104**: 1-24.

GOODMAN, D., 1979. Regulating reproductive effort in a changing environment. *Amer. Nat.* **113**: 735-48.

GREEN, B., 1984. Composition of milk and energetics of growth in marsupials. *Symp. Zool. Soc. Lond.* **51**: 369-87.

GUILER, E. R., 1960. The pouch young of the potoroo. *Jour. Mammal.* **41**: 441-51.

HEINSOHN, G. E., 1968. Habitat requirements and reproductive potential of the macropod marsupial *Potorous tridactylus* in Tasmania. *Mammalia* **32**: 30-43.

HORNSBY, P., 1978. A note on the pouch life of rock wallabies. *Vic. Nat.* **95**: 108-11.

HUGHES, R. L., 1962. Reproduction in the macropod marsupial *Potorous tridactylus* (Kerr). *Aust. Jour. Zool.* **10**: 193-224.

JANSSENS, P. A. AND TERNOUTH, J. H., 1987. The transition from milk to forage diets. *In* Nutrition of herbivores ed. by J. B. Hacker and J. H. Ternouth. Academic Press: Australia.

JARMAN, P. J., 1983. Mating system and sexual dimorphism in large terrestrial mammalian herbivores. *Biol. Rev.* **58**: 485-520.

JARMAN, P. J. (this volume). Sexual dimorphism in macropods.

JARMAN, P. J., JOHNSON, C. N., SOUTHWELL, C. J. AND STUART-DICK, R., 1987. Macropod studies at Wallaby Creek. I. The area and animals. *Aust. Wildl. Res.* **14**: 1-14.

JOHNSON, C. N., 1985. Ecology, Social Behaviour and Reproductive Success in a Population of Red-necked Wallabies. Ph.D. Thesis, University of New England, Armidale.

JOHNSON, C. N., 1986. Philopatry, reproductive success of females, and maternal investment in the red-necked wallaby. *Behav. Ecol. Sociobiol.* **19**: 143-50.

JOHNSON, C. N., 1987. Relationships between mother and infant red-necked wallabies (*Macropus rufogriseus banksianus*). *Ethology* **740**: 1-20.

JOHNSON, K. A., 1977. Ecology and management of the red-necked pademelon, *Thylogale thetis,* on the Dorrigo plateau of northern New South Wales. Ph.D. Thesis, University of New England, Armidale, Australia.

JOHNSON, P. M., 1978. Husbandry of the rufous rat-kangaroo *Aepyprymnus rufescens* and brush-tailed rock-wallaby *Petrogale penicillata* in captivity. *Internat. Zoo Yearbk.* **18**: 156-57.

JOHNSON, P. M., 1979. Reproduction in the plain rock-wallaby *Petrogale penicillata inornata* Gould, in captivity, with age estimation of the pouch young. *Aust. Wildl. Res.* **6**: 1-14.

JOHNSON, P. M. AND STRAHAN, R., 1982. A further description of the Musky Rat-kangaroo, *Hypsiprymnodon moschatus* Ramsay, 1876 (Marsupialia, Potoroidae), with notes on its biology. *Aust. Zool.* **21**: 27-46.

KNAPTON, R. W., 1984. Parental investment: the problem of currency. *Can. Journ. Zool.* **62**: 2673-674.

LEE, A. K. AND COCKBURN, A., 1985. Evolutionary Ecology of Marsupials. Monographs on Marsupial Biology. Cambridge University Press: Cambridge.

MAYNARD SMITH, J., 1980. A new theory of sexual investment. *Behav. Ecol. Sociobiol.* **7**: 247-51.

MAYNES, G., 1973a. Reproduction in the parma wallaby, *Macropus parma* Waterhouse. *Aust. Zool.* **21**: 331-51.

MAYNES, G., 1973b. Aspects of reproduction in the whiptail wallaby, *Macropus parryi. Aust. Zool.* **18**: 43-46.

MCCARTNEY, D. J., 1978. Reproductive biology of the Tasmanian pademelon, *Thylogale billardierii.* Honours Thesis, University of Tasmania, Hobart.

MEIKLE, D. B., TILFORD, B. L. AND VASSEY, S. H., 1984. Dominance rank, secondary sex ratio, and reproduction of offspring in polygonous primates. *Amer. Nat.* **124**: 173-88.

MERCHANT, J. C., 1976. Breeding biology of the agile wallaby *Macropus agilis* (Gould) (Marsupialia: Macropodidae), in captivity. *Aust. Wildl. Res.* **3**: 93-103.

MERCHANT, J. C. AND CALABY, J. H., 1981. Reproductive biology of the red-necked wallaby (*Macropus refogriseus banksianus*) and Bennett's wallaby (*M. r. rufogriseus*) in captivity. *Journ. Zool. Lond.* **194**: 203-17.

MILLAR, J. S., 1977. Adaptive features of mammalian reproduction. *Evolution* **31**: 370-86.

MURPHY, C. R. AND SMITH, J. R., 1970. Age determination of pouch young and juvenile Kangaroo Island wallabies. *Transac. Roy. Soc. Sth. Aust.* **94**: 15-20.

OLDS, T. J. AND COLLINS, L. R., 1972. Breeding Matschie's tree kangaroo in captivity. *Internat. Zoo Yearbk.* **13**: 123-26.

PARKER, P., 1977. An ecological comparison of marsupial and placental patterns of reproduction. *In* The Biology of Marsupials ed. by B. Stonehouse and D. Gilmore. Macmillan: London.

POOLE, W. E., 1973. A study of breeding in grey kangaroos, *Macropus giganteus* (Shaw) and *M. fuliginosus* (Desmarest) in central New South Wales. *Aust. Zool.* **21**: 183-212.

POOLE, W. E., 1975. Reproduction in the two species Grey Kangaroos, *Macropus giganteus* Shaw and *M. fuliginosus* (Desmarest). II. Gestation, Parturition and pouch life. *Aust. Wildl. Res.* **23**: 333-54.

POOLE, W. E. AND MARCHANT, J. C., 1987. Reproduction in captive wallaroos: the Eastern Wallaroo, *Macropus robustus robustus,* the Euro, *M. r. erubescens* and the Antilopine Wallaroo, *M. antilopinus. Aust. Wildl. Res.* **14**: 225-42.

POOLE, W. E., CARPENTER, S. M. AND WOOD, J. T., 1982. Growth of grey kangaroos and the reliability of age determination from body measurements. I. The eastern grey kangaroo, *Macropus giganteus. Aust. Wildl. Res.* **9**: 9-20.

POOLE, W. E., MERCHANT, J. C. AND CALABY, J. H., 1985. Reproduction, growth and age determination in the yellow-footed rock-wallaby *Petrogale xanthopus* Gray in captivity. *Aust. Wildl. Res.* **12**: 127-36.

ROSE, R., 1986. The control of pouch vacation in the Tasmanian Bettong *Bettongia gaimardi. Aust. Zool.* **43**: 485-91.

ROSE, R. W., 1978. Reproduction and evolution in female Macropodidae. *Aust. Mammal.* **2**: 65-72.

ROSE, R. W., 1986. The control of pouch vacation in the Tasmanian bettong *Bettongia gaimardi. Aust. Zool.* **34**: 485-91.

ROSE, R. W. AND MCCARTNEY, D. J., 1982. Reproduction of the red-bellied pademelon, *Tylogale billardierii* (Marsupialia). *Aust. Wildl. Res.* **9**: 27-32.

RUBENSTEIN, D. I., 1982. Risk, uncertainty and evolutionary strategies. *In* Current Problems in Sociobiology ed. by King's College Sociobiology Group. Cambridge University Press: Cambridge.

RUSSELL, E. M., 1973. Mother-young relations and early behavioural development in the marsupials *Macropus eugenii* and *Megaleia rufa. Zeitschrift fur Tierpsycholgie* **33**: 163-203.

RUSSELL, E. M., 1982. Patterns of parental investment in marsupials. *Biol. Rev.* **57**: 423-86.

SAMPSON, J. C., 1971. The biology of *Bettongia penicillata* Gray 1837. Ph.D. Thesis, University of Western Australia, Perth.

VAN SCHAIK, C. P. AND NOORDWIJK, M. A., 1985. Interannual variability in fruit abundance and the reproductive seasonality in Sumatran long-tailed macaques (*macaca fascicularis*). *Journ. Zool. Lond.* **206**: 533-49.

SHARMAN, G. B. AND CALABY, J. H., 1964. Reproductive behaviour in the Red Kangaroo, *Megaleia rufa,* in captivity. *C.S.I.R.O. Wildl. Res.* **9**: 58-85.

SHARMAN, G. B. AND PILTON, P. E., 1964. The life history and reproduction of the red kangaroo (*Megaleia rufa*). *Proc. Zool. Soc. Lond.* **142**: 29-48.

SHARMAN, G. B., FRITH, H. J. AND CALABY, J. H., 1964. Growth of the pouch young, tooth eruption, and age determination in the red kangaroo, *Megaleia rufa. C.S.I.R.O. Wildl. Res.* **9**: 20-49.

SHIELD, J., 1968. Reproduction of the quokka, *Setonix brachyurus,* in captivity. *Journ. Zool. Lond.* **155**: 427-44.

SILK, J. B., 1983. Local resource competition and facultative adjustment of sex ratios in relation to competitive abilities. *Amer. Nat.* **121**: 56-66.

SILK, J. B. AND BOYD, R., 1983. Co-operation, Competition, and mate choice in matrilineal macaque groups. *In* Social Behaviour of Female Vertebrates ed. by S. K. Wasser. Academic Press: New York.

SILK, J. B., CLARK-WHEATLEY, C. B., RODMAN, P. S. AND SAMUELS, A., 1981. Differential reproductive success and facultative adjustment of sex ratios among captive female bonnet macaques (*Macaca radiata*). *Anim. Behav.* **29**: 1106-120.

SIMPSON, A. E. AND SIMPSON, M. J. A., 1985. Short-term consequences of different breeding histories for captive rhesus macaque mothers and young. *Behav. Ecol. Sociobiol.* **18**: 83-89.

STEARNS, S. C., 1976. Life-history tactics: a review of the ideas. *Quart. Rev. Biol.* **51**: 3-47.

STRAHAN, R. (ed.), 1983. The Australian Museum Complete Book of Australian Mammals. Angus & Robertson: London.

STUART-DICK, R. I., 1987. Parental investment in the Eastern Grey Kangaroo. Ph.D. Thesis, University of New England, New South Wales.

TRIVERS, R. L., 1978. Parental investment and sexual selection. *In* Readings in Sociobiology ed. by T. H. Clutton-Brock and P. H. Harvey. W. H. Freeman and Company: Reading.

TRIVERS, R. L. AND WILLARD, D. E., 1973. Natural selection of parental ability to vary the sex ratio of offspring. *Science* **179**: 90-92.

TYNDALE-BISCOE, C. H., 1968. Reproduction and post-natal development in the marsupial *Bettongia lesueur* (Quoy and Gaimard). *Aust. Zool.* **16**: 577-602.

WARING, H., SHARMAN, G. B., LOVAT, D. AND KAHAN, M., 1955. Studies on marsupial reproduction: I. General features and techniques. *Aust. Zool.* **3**: 34-43.

WASSER, S. K., 1983. Reproductive competition and co-operation among female yellow baboons. *In* Social Behaviour of Female Vertebrates ed. by S. K. Wasser. Academic Press: New York.

WERREN, J. H. AND CHARNOV, E. L., 1978. Facultative sex ratios and population dynamics. *Nature* **272**: 349-50.

WERREN, J. H. AND TAYLOR, P. D., 1984. The effect of population recruitment on sex ratio selection. *Amer. Nat.* **124**: 143-48.

APPENDIX A

Species' code numbers and references for data in Figures 1 and 2.

Code number	Species	References
1	*Hypsiprymnodon moschatus*	Johnson and Strahan 1982
2	*Potorous tridactylus*	Guiler 1960; Heinsohn 1968; Hughes 1962
3	*Bettongia penicillata*	Parker 1977; Sampson 1971
4	*Bettongia gaimardi*	Rose 1986
5	*Bettongia lesueur*	Tyndale-Biscoe 1968
6	*Aepyprymnus rufescens*	Johnson 1978; Strahan 1983
7	*Setonix brachyurus*	McCartney 1978; Sheild 1968; Waring *et al.* 1955
8	*Wallabia bicolor*	Calaby and Poole 1971; Strahan 1983
9	*Dendrolagus matschiei*	Olds and Collins 1972
10	*Lagorchestes conspicillatus*	Johnson 1981; Strahan 1983
11	*Lagorchestes hirsutus*	Russell 1982
12	*Petrogale penicillata*	Batchelor unpubl.; Joblin unpubl.
13	*Petrogale inornata*	Johnson 1979
14	*Petrogale inornata*	Strahan 1983
15	*Petrogale xanthopus*	Calaby and Poole 1971; Hornsby 1978; Poole *et al.* 1985
16	*Peradorcas concinna*	Sanson in Lee and Cockburn 1985
17	*Thylogale billardierii*	McCartney 1978; Rose and McCartney 1982b
18	*Thylogale thetis*	Johnson 1977
19	*Macropus parma*	Maynes 1973a
20	*Macropus eugenii*	Andrewartha and Barker 1969; Murphy and Smith 1970; Russell 1973
21	*Macropus irma*	Strahan 1983
22	*Macropus parryi*	Maynes 1973b
23	*Macropus dorsalis*	Strahan 1983
24	*Macropus rufogriseus banksianus*	this study
25	*Macropus rufogriseus rufogriseus* (overseas)	Catt 1977; Fleming *et al.* 1983
26	*Macropus rufogriseus rufogriseus* (Australia)	Merchant and Calaby 1981
27	*Macropus agilis*	Merchant 1976
28	*Macropus giganteus*	Poole 1973; 1975
29	*Macropus giganteus*	this study
30	*Macropus fuliginosus melanops*	Poole 1973; 1975
31	*Macropus fuliginosus fuliginosus*	Poole 1973; 1975
32	*Macropus fuliginosus ocydromus*	Poole 1973; 1975
33	*Macropus fuliginosus* (combined)	Poole 1973; 1975
34	*Macropus robustus robustus*	Poole and Merchant 1987
35	*Macropus robustus erubescens*	Poole and Merchant 1987
36	*Macropus antilopinus*	Poole 1975
37	*Macropus rufus*	Sharman and Calaby 1964; Sharman and Pilton 1964

A: The suckling pouch-young of a Spectacled Hare Wallaby *(Lagorchestes conspicillatus)*. Photo by Gordon Grigg.

C: A resting group of mother Eastern Grey Kangaroos *(Macropus giganteus)* with young-at-foot, illustrating the close mother-young association and the association of females at similar stages in reproduction.

PLATE I

B: Large male Eastern Grey Kangaroo *(Macropus giganteus)*, showing extensive development of the forelimb musculature. Photo by Peter Jarman.

D: Sub-adult male Eastern Grey Kangaroos *(Macropus giganteus)* in a group, play-fighting: wrestling and kicking. Northeastern New South Wales. Photo by R. I. Stuart-Dick.

E: Eastern Grey Kangaroos *(Macropus giganteus)* in northeastern New South Wales. Male, female and young-at-foot. Note the contrast in build between males and females. Photo by Peter Jarman.

F: Female Red-necked Wallaby *(Macropus rufogriseus)* with a large pouch young, close to stage of permanent emergence; the young is out of the pouch and is sitting typically close to the mother. Northeastern New South Wales. Photo by C. N. Johnson.

A: A female Eastern Grey Kangaroo *(Macropus giganteus)* with her not long emerged pouch-young crouched typically by her tail while she feeds. Northeastern New South Wales. Photo by R. I. Stuart-Dick.

PLATE II

D: Female Red-necked Wallaby *(Macropus rufogriseus)* with a large pouch young, close to stage of permanent emergence, in a daytime sheltering site in tall grass. Northeastern New South Wales. Photo by J. L. Clarke.

B: Female Eastern Grey Kangaroo *(Macropus giganteus)* with juvenile at the stage of permanent emergence from the pouch. Northeastern New South Wales. Photo by R. I. Stuart-Dick.

E: *Petrogale xanthopus*, Yellow-footed Rock-wallaby. Photo N.P.I.A.W.

C: *Onychogalea fraenata*, Bridled Nailtail Wallaby. Photo N.P.I.A.W.

Dispersal and philopatry in the Macropodoids

C. N. JOHNSON[1]

Information on dispersal and philopatry is available for nine species of macropods. In most of these species dispersal is predominantly by juvenile males, and females are philopatric; this pattern may be most characteristic of species in the genus *Macropus*. Dispersal by juveniles does not appear to be initiated by aggression from older residents or by other social or environmental factors, but is probably innate; adults may disperse in response to changed environmental conditions. The philopatry of females in the *Macropus* kangaroos and wallabies produces life-long associations among related females, and there is some evidence of competition for resources among related females. Preliminary comparisons of sex ratios of pouch young suggest that male-biased dispersal may be accompanied by male-biased sex ratios.

INTRODUCTION

THE word "dispersal" takes two quite distinct meanings in the ecological literature (Baker 1982). It is sometimes used to refer to the scattering of groups (as when populations of Red Kangaroos *Macropus rufus* disperse from refuge habitats after drought-breaking rains), and at other times it refers to the movement of individuals from birth places to breeding sites (as when young Red Kangaroos leave their natal home-ranges and resettle elsewhere).

This review is about the second kind of dispersal. I follow Greenwood (1980) in dividing it into two categories: movement from the birth-place to the site at which breeding commences is "natal dispersal", and movement from one breeding site to another is "breeding dispersal". In Greenwood's terminology, "philopatry" refers to continued residence close to the birth-place. The criterion adopted in this review to distinguish dispersal from philopatry is that dispersal must involve a complete departure from a previously-held home-range, whereas philopatry implies the retention of some part of an original home-range.

These definitions assume that individuals normally concentrate their activities within stable home-ranges. This assumption appears to be well-founded for a wide variety of macropods (e.g., the Long-nosed Potoroo *Potorous tridactylus* Heinsohn 1968; Bennett 1987a; the Red-necked Pademelon *Thylogale thetis* K. A. Johnson 1980; the Red-necked Wallaby *Macropus rufogriseus* Johnson 1987; the Eastern Grey Kangaroo *M. giganteus* Jaremovic and Croft 1987). Individuals of some species are remarkably faithful to their home-ranges. Brush-tailed Bettongs *Bettongia penicillata,* for example, refuse to move even during severe fires (Christensen 1980). The only species of macropod which has been considered to be in any sense nomadic is the Red Kangaroo *M. rufus*, a species subject to violent environmental fluctuations and capable of rapid movement over long distances (Bailey 1971). Denny (1985) has shown, however, that changes in local abundance of red kangaroos, which may be initiated by rainstorms, fires or even by strong winds, are consistent with the retention by individuals of large home-ranges (see also Frith 1964; Caughley 1977; Oliver 1986; Priddel 1987).

There have been few studies of dispersal in macropods, and very few which have looked beyond the genus *Macropus*. This is unfortunate, and somewhat ironical. The smaller macropods should be excellent subjects for studies of dispersal, because most of them are reasonably trappable and, since young can be marked in the pouch when their mothers are caught, it is possible to record their life histories with more completeness than is attainable for most eutherians. The value of this approach has been demonstrated in other marsupials by Cockburn *et al.* (1985), and it is reasonable to hope that studies of dispersal in the macropods will proliferate over the next few years. For the moment, however, our knowledge of dispersal in macropods is sketchy and unbalanced.

This review is divided into three sections. I begin by summarizing the little that is currently known of patterns of dispersal and philopatry in macropods, then discuss the proximate causes and ultimate origins of sex-biases in dispersal, and comment on some consequences of male-biased dispersal in the macropods.

PATTERNS OF DISPERSAL AND PHILOPATRY IN MACROPODS

This section describes those studies of macropods which have been sufficiently detailed and extensive to reveal patterns of dispersal and philopatry. General criteria for inclusion of studies are that they have extended over at least a year, have included at least 10 adults and 10 subadults of each sex, and have used marking, or

[1]Queensland National Parks and Wildlife Service, P.O. Box 1395, Rockhampton, Queensland, Australia 4700.

Pages 593-601 in KANGAROOS, WALLABIES AND RAT-KANGAROOS ed. by G. Grigg, P. Jarman and I. Hume. Surrey Beatty & Sons Pty Limited, New South Wales, Australia, 1989.

individual recognition by natural features, to record the movements of these individuals.

Natal dispersal

The nine species in which natal dispersal has been studied can be divided into three categories: those in which both sexes disperse (one species), those in which natal dispersal is absent (two species), and those in which males disperse and females are philopatric (six species).

Natal dispersal by both sexes (the Long-nosed Potoroo). Natal dispersal by male Long-nosed Potoroos coincides approximately with the onset of spermatogenesis; females disperse before their first parturition (Bennett 1987a,b). Males entering Bennett's study population tended to be heavier than incoming females, although ages at first dispersal were similar for both sexes (of 11 immigrants weighing less than 650 g, only one was male, whereas six of 10 heavier immigrants were male; X^2=4.03, $P<0.05$, data from Bennett 1987b). This observation suggests that males may take longer to settle after dispersing than females.

Natal dispersal by neither sex (the Quokka and Tammar Wallaby). Three mark-and-recapture studies of the Quokka *Setonix brachyurus* have failed to find evidence of regular dispersal by any class of animals (Dunnett 1962; Holsworth 1964, 1967; Kitchener 1970 cited in Kaufmann 1974). Holsworth showed that adult Quokkas were sedentary over long periods, and of 73 animals first caught when less than a year old and subsequently recaptured as adults, only seven had dispersed (over an average distance of about 750 m). Holsworth gave the sexes of five dispersing sub-adults: all but one of these were female. Dispersal in Holsworth's study was judged to be a shift from one "group-territory" to another; group territories covered, on average, 10 ha and contained in the order of 50 to 100 Quokkas, who used home-ranges of 5 to 10 ha each.

The three studies of the Quokka cited above were conducted on Rottnest Island, and it is possible that their failure to record dispersal was due to the limitations of space on a small island. This seems unlikely. Holsworth's study area, for example, covered 142 ha, and included 15 group-territories. Individuals in this study population should have had space available to them to shift their home-ranges had they been inclined to disperse.

Christensen (1980) could find little evidence of dispersal by subadults in undisturbed populations of the Tammar Wallaby *M. eugenii;* his observations on Tammars are described in the section on breeding dispersal (below).

Natal dispersal predominantly by males (the Brush-tailed Bettong, and Macropus *spp.).* Christensen noted that a high proportion of Brush-tailed Bettongs recolonizing areas depopulated by fire were subadult males (Christensen 1980; Christensen and Maisey 1987). This suggests that subadult males may be more inclined to disperse than other classes of bettongs, but because very little dispersal was observed in undisturbed populations, the general significance of this inclination is unclear. Christensen (1980) suggested that young male bettongs typically disperse cautiously over short distances, and only when local vacancies arise.

Male Red-necked Wallabies in Johnson's (1986a) study dispersed when about two years old; this is approximately the age at which they reached sexual maturity. No males of this age failed to disperse. Four young males who were observed in their new home-ranges had moved a mean distance of 670 m from the centres of their natal ranges; this distance was just over three times the diameter of the average adult female's home range. Departures of young males from the study population were balanced by arrivals from outside it. Females in Johnson's study settled, apparently for life, in home-ranges overlapping those of their mothers, but subsequent work on the same population has revealed that some females disperse after having their first offspring (Stuart-Dick and Higginbottom, this volume; Higginbottom, pers. comm.). Solitary wallabies have been seen up to 12 km beyond the normal range of the population of feral Bennett's wallabies *M. r. rufogriseus* in the Peak District of England. These wallabies were probably emigrants, but unfortunately no information on their sex and age is available (Yalden 1988).

Natal dispersal is also male-biased in the Eastern Grey Kangaroo, in which males disperse between the ages of two and four years while females remain within their natal home-ranges (Stuart-Dick 1987; Jarman, Jones, Southwell and Stuart-Dick, pers. comm.), and in the Euro *M. robustus* (Clancy, pers. comm.).

Kaufmann (1974) made three sets of observations which suggested dispersal by subadult males in the Whiptail Wallaby *M. parryi.* Two young males disappeared from the study population over the year it was kept under observation; another five made transient appearances in the population. A further four males were, at the beginning of the study, using large home-ranges which overlapped two separate wallaby communities, but by the end of the study all had settled in one community or the other. All 11 of these males appeared to be about 20 months old and must have been almost sexually mature. Adult males and females occupied stable home-ranges.

Information on dispersal by Red Kangaroos comes from broad-scale mark/recovery studies, particularly those of Oliver (1986) and Denny (1982, 1985). In both cases, individuals found far from their points of capture were likely to be males who had been subadult when marked. Oliver (1986) showed that female Red Kangaroos were more sedentary than males (see also Bailey 1971); this appeared to be true of subadult as well as adult females, but samples of subadult females were small. Dispersers moved varying distances, some in excess of 100 km (see also Bailey 1971). The data on Red Kangaroos are not clear enough to show whether and to what extent females inherit the home ranges of their mothers. Croft (this volume), citing the unpublished observations of Edwards from Fowler's Gap in western New South Wales, comments that many Red Kangaroos of both sexes are no more than transient members of local populations. It is not yet clear whether this transience represents dispersal or regular movement within very large home-ranges.

All of the subadult male Brush-tailed Rock-wallabies *Petrogale penicillata,* and most of the subadult females, observed by K. P. Joblin (pers. comm.) disappeared from their natal refuges. Joblin concluded that these disappearances represented dispersal, but mortality may also have been involved. The data are not conclusive, but they are suggestive of a male-biased dispersal of subadults in this species.

Breeding dispersal

Breeding dispersal has been documented in three species of macropods, the Long-nosed Potoroo, the Tammar Wallaby and the Red Kangaroo.

Two of the adult males in Bennett's (1987b) population of Long-nosed Potoroos shifted their home-ranges, by more than one kilometre. Most adult males, and all adult females, were sedentary.

Christensen (1980) recorded 10 home-range shifts in a two-year study of an undisturbed population of Tammar Wallabies in southwestern Western Australia. Two of these were by subadult males, one was by a subadult female, the remaining seven were by adult males. Rates of dispersal were higher from recently-burnt areas: of 16 migrants from such areas, three were subadult males, one was a subadult female, seven were adult males and five were adult females. These data suggest that adult male Tammar Wallabies are more likely to disperse than adult females. A radio-telemetry study of adult Tammar Wallabies (eight males and eight females) on Kangaroo Island did not record home-range shifts in either sex (Inns 1980). It is therefore uncertain how complete or consistent is breeding dispersal by males in this species.

Oliver (1986) recorded dispersal by some adult Red Kangaroos under conditions of declining food availability during drought, and noted that dispersal distances increased as the drought deepened.

Seasonal shifts in home-range, with a return to the original range, are known in the Quokka (Nicholls 1971), in the Euro (Ealey 1967), and in male Eastern Grey Kangaroos in some populations (Jaremovic and Croft 1987). Because such patterns of movement involve re-use of the same breeding ranges, they do not fall within the definition of breeding dispersal used in this review.

Exploration and settlement

Practically nothing is known of the criteria which determine where dispersing macropods settle. Denny (1985) noted that dispersing Red Kangaroos moved at variable rates, and most were resighted in areas where freshly-growing grass had become locally abundant. These observations are consistent with the idea that dispersing individuals explore, assess habitats, and settle in the most favourable sites (Baker 1982). It is not clear from Denny's account whether systematic exploration might be more typical of one class of disperser than of others. In Oliver's (1986) study, dispersal by adults favoured certain directions, whereas subadults dispersed to all points of the compass.

Subadult male Eastern Grey Kangaroos expand their home-ranges in what appears to be an exploratory phase preceeding dispersal (Jarman, Jones, Southwell and Stuart-Dick, pers. comm.). Johnson's (1987) data on Red-necked Wallabies and Kaufmann's (1974) on Whiptail Wallabies also show some variability in home-range size of sub-adult males which might indicate pre-dispersal exploration.

Generalizations and comparisons

Dispersal in macropods appears to be predominantly by sub-adult males: of the nine species studied to date, dispersal is male-biased in seven, and in six of these species subadult males are more likely to disperse than are adults. Where natal dispersal has been carefully studied, it involves all young or all members of one sex of young. Breeding dispersal has been properly documented in three species but in two of these it occurs less regularly than natal dispersal. Exceptions to these trends are found only among small-bodied species (Table 1).

The sample of species is at present heavily biased towards the genus *Macropus,* and any generalizations drawn from it should be regarded as preliminary until more is known about dispersal in other genera.

Male-biased dispersal has been reported in other marsupials (e.g., *Antechinus* spp. Cockburn *et al.* 1985; Common Brush-tailed Possum *Trichosurus vulpecula* Clout and Efford 1984) and

Table 1. Observations of dispersal in the macropods (xx: most individuals disperse; x: a minority of individuals disperse; —: very few, if any individuals disperse). Species ranked in approximate order of increasing body weight. Sources cited in text.

	Natal dispersal		Breeding dispersal	
Species	Males	Females	Males	Females
Potorous tridactylus	xx	xx	x	—
Bettongia penicillata	x	—	—	—
Setonix brachyurus	—	—	—	—
Macropus eugenii	x	—	xx	—
M. rufogriseus	xx	—	—	x
M. parryi	xx	—	—	—
M. giganteus	xx	—	—	—
M. robustus	xx	—	—	—
M. rufus	xx	x	x	x

appears to be quite a general phenomenon among mammals. Greenwood (1980) collated records of natal dispersal in 65 species of mammals and breeding dispersal in 25 species; dispersal was male-biased in 69% of species with natal dispersal and 84% of species with breeding dispersal (Greenwood's review included only one macropod, the Whiptail Wallaby). In many species with male-biased natal dispersal, all males disperse whereas females do so only under exceptional circumstances (Pusey and Packer 1987).

Moehlman (1986) has suggested that in the Canidae dispersal is generally male-biased in small species, is unbiased in species of intermediate body-size and is female-biased in large species. Similarly, data currently available on primates suggest that dispersal is more consistently male-biased in small than in large species: in the 34 species in which dispersal has been studied, it is male-biased in 88% of those below the median weight for the sample, but is male-biased in only 47% of those above the median body-weight (X^2=4.84, $P<0.05$; Charles-Dominique 1977; Dawson 1978; Neyman 1978; Struhsaker and Leland 1979; Terbogh and Goldzein 1985; Harcourt and Nash 1986; Bearder 1987; McFarland Symington 1987; Pusey and Packer 1987; Rodman and Mitani 1987; Smuts *et al.* 1987. This comparison does not include monogamous, territorial species in which breeding is not co-operative; dispersal of both sexes of offspring is inevitable in such species). Dispersal is unvaryingly male-biased in ground-squirrels (Michener 1983). If it is really true that dispersal is less consistently male-biased in small species of macropods than in large ones, this could provide an interesting contrast to the data on canids, primates and ground-squirrels.

PROXIMATE CAUSES OF DISPERSAL

The incidence of breeding dispersal in macropods is generally low. Data on breeding dispersal in Long-nosed Potoroos are too sketchy to suggest causes, but in Tammar Wallabies and Red Kangaroos the incidence of breeding dispersal increases as environmental conditions deteriorate (Oliver 1986). This suggests a simple model of breeding dispersal, in which adults tend to remain within their home-ranges unless forced to move by severe local shortages of resources. The level of the threshold above which adults disperse is presumably determined by the advantages of remaining within a familiar area (and perhaps by the results of exploratory movements undertaken during an initial phase of home-range expansion). The rarity of breeding dispersal suggest that this threshold must generally be high in macropods.

The causes of natal dispersal are not easy to discern. Its regular coincidence with the attainment of sexual maturity suggest that it is triggered by some aspect of the maturation process itself, rather than by independent stimuli from outside the individual. External influences would presumably be somewhat variable in their action but, at least in male Red-necked Wallabies and Whiptail Wallabies and in male and female Long-nosed Potoroos, all dispersers leave at approximately the same age (Kaufmann 1974; Johnson 1985, 1986a; Bennett 1987b).

There is, in any case, little direct evidence that dispersal by subadult macropods is initiated by extrinsic stimuli. Two external factors that could be imagined as triggering dispersal in subadults are aggression from older residents, and attractiveness of mates outside the natal home-range.

Aggression by dominant males towards subadults is not conspicuous in those species of macropods whose subadult males disperse. In the Red-necked Wallaby, for example, young males may occasionally be displaced by older males, but are not actively pursued or persecuted by them. Most aggressive interactions in this species involve males who are several years post-dispersal (Johnson 1989a; see also Kaufmann 1974 on Whiptail Wallabies and Croft 1981 on Red Kangaroos). Persistent aggression by mothers towards their subadult sons is likewise absent. Bennett (1987b) commented that dispersal by sub-adult Long-nosed Potoroos was not associated with the emergence from the pouch of the mother's subsequent offspring; it therefore cannot be construed as resulting from an onset of maternal aggression at that time.

Similarly, there are no observations to suggest a close association of dispersal with increased sexual activiy. In the Red Kangaroo, the most characteristic pattern of grouping of young adult males in their first few years after dispersal is their occurrence in all-male groups (Croft 1981), and in all studies of *Macropus* species, young males are prevented by dominant males from having close contact with oestrous females (Kaufmann

1974; Croft 1981a; Jarman and Southwell 1986; Johnson 1989a).

EVOLUTION OF MALE-BIASED DISPERSAL

Most proposed explanations of the evolution of male-biased dispersal in mammals assume it to confer some advantage on dispersing individuals. Several hypotheses which do not make this assumption are reviewed and discarded by Greenwood (1980). The hypothesis that dispersal by juvenile males is caused by aggression from adult males, for example, is an attractive one in species whose adult males harass their young colleagues, and has been advocated by several authors (e.g., Moore and Ali 1984; Liberg and von Schantz 1985) but it is clearly inappropriate in species (like the macropods) in which such aggression is absent, but in which subadult males nevertheless disperse (Packer 1985).

Adaptive explanations of sex-biased dispersal take the general approach of comparing the costs and benefits of dispersal, in terms of expected reproductive outcomes, for individual dispersers of either sex. The major costs of dispersal are generally assumed to be the risks and energetic demands incurred through the abandonment of a familiar area. Such costs may be outweighed in either sex by the benefits of avoiding inbreeding (Greenwood *et al.* 1978; Packer 1979; Ralls *et al.* 1986) or of finding vacant living spaces when these are scarce close to the natal site (Waser 1985). Of these two postulated benefits of dispersal, the avoidance of inbreeding is most consistent with the evolution of sex-biased dispersal, because it is only necessary that the members of one sex disperse for inbreeding to be eliminated (Pusey 1987; Pusey and Packer 1987).

If it is accepted that sex-biased dispersal has evolved in response to selection for inbreeding avoidance, further arguments must still be offered to account for the directions of the sex-biases. A number of specific hypotheses have been raised to account for male-biased dispersal in mammals. Waser and Jones (1983) have suggested that because female mammals are generally responsible for the care of offspring and because this is likely to be carried out most effectively by females who are familiar with the distribution of food, dens and other resources in their home-ranges, they will benefit more from philopatry than males (see also Baker 1978, 1982). A complementary suggestion is that males should be more predisposed to disperse because their reproductive success is generally limited by their ability to find mates and to deal with competition from other males (Clutton-Brock and Harvey 1976; Packer 1979); dispersing males might increase their reproductive success by settling in areas not occupied by more dominant males (Dobson 1982), or by selecting areas in which females concentrate (Greenwood 1980). It can also be argued that where females typically begin breeding at younger ages than males (as is commonly the case in mammals in which males compete for access to mates), juvenile dispersal will cause a greater delay in age at first breeding for females than for males (Johnson 1986b).

All of these hypotheses predict a general tendency for male-biased dispersal in polygynous mammals, and any of them could be used to "explain" male-biased dispersal in the genus *Macropus* with reasonable plausability. They do make different predictions on sex-differences in dispersal in potoroos: there is no evidence of male parental care in potoroos so, according to Waser and Jones' hypothesis, female potoroos should be philopatric and males should disperse. However, because inter-male competition may be less intense among male potoroos than in larger macropods (at least insofar as inter-male competition is indicated by sexual dimorphism), the hypotheses of Greenwood, Dobson and Johnson accurately predict that sex-differences in dispersal in potoroos should be slight. Potoroos can reasonably be compared with larger macropods because their social organization — home-range overlap among all sex and age classes, with apparently transient association between males and females — is fundamentally similar to that of the larger species (Bennett 1978a).

The arguments reviewed above are also consistent with the absence of dispersal in the Quokka which, as Baker (1978) remarks, is quite exceptional among mammals. The home-ranges of Quokkas on Rottnest Island, where the species has been studied extensively, overlap so broadly (50 to 100 individuals may occupy approximately congruent home-ranges) that random mating in the absence of dispersal could not be expected to lead to much close inbreeding; there may therefore be little general selection for regular dispersal in this population.

CONSEQUENCES OF SEX-BIASED DISPERSAL IN THE MACROPODS

Any tendency for individual males and females to disperse over different distances will produce sex-differences in the distribution of kin within populations. In mammals generally, and in the *Macropus* kangaroos and wallabies in particular, the philopatry of females means that they tend to live out their lives close to female relatives, whereas male kin are more nearly randomly distributed within populations.

The clustering of related females appears to have had a profound influence on social evolution in mammals. It is more common for female than for male mammals to live in stable social groups and these groups are most often matrilineal families, attended with varying degrees of fidelity by unrelated adult males (see for example Michener 1983 on ground squirrels and Pusey and Packer 1987 on primates).

Female macropods do not live in groups of fixed membership. The short-term instability of groups and the absence of any strong patterns in group compositions even in the more gregarious species, are probably the most striking attributes of social behaviour in the macropods (Croft, this volume; Jarman and Coulson, this volume). Individual whiptail wallabies (Kaufmann 1974) and Eastern Grey Kangaroos (Kaufmann 1975; Jarman and Taylor 1983; Jarman and Southwell 1986) do live in stable communities, however, and the philopatry of females means that they will often be closely related to other female community-members. It is also possible that the degree to which any female associates with other community members could vary with relatedness.

The Red-necked Wallaby provides an example of such variation. Female Red-necked Wallabies in the population studied by Johnson (1989b) live in home-ranges which typically overlap those of a number of other females of varying degrees of relatedness. The amount of time that any two females spend in one another's company is loosely correlated with the extent to which their home-ranges overlap but, given this correlation, closely related females spend significantly more time together than do females known or suspected not to be close relatives. This observation suggests that closely-related females (in this case mothers and their daughters) exercise some preference for one another's company or, what is very difficult to distinguish from preferential association, tend to make coincident selections of feeding and resting sites and paths of travel, and therefore encounter one another frequently. Such fine-grained matching of habitat selection within matrilines is to be expected if daughters inherit the movement patterns of their mothers, and if mothers are tolerant of close association with female relatives.

Social organization of females is a little-studied aspect of macropod behavioural ecology. The data on Red-necked Wallabies suggest that the analysis of the links between relatedness of females, their patterns of association and their degree of overlap in fine-grained habitat utilization may be a productive line of approach. This would be most appropriate in species like the rock-wallabies, in which habitat utilization is very tightly structured around refuges and pathways.

Life-long association of related females obviously raises the potential for the evolution, by kin selection, of co-operation within matrilines. This has not so far been recorded in macropods, except in the rather diffuse sense that individuals spend less time alert and more time feeding when in groups than when alone (Jarman and Coulson, this volume) and, to the extent that groups are composed of relatives, this benefit is shared among kin. Further work in this area might usefully ask whether females derive more benefit from the company of relatives than from non-relatives; this could happen if related females form well-coordinated sub-groups within large feeding groups, or direct alarm signals at one another.

Although outright co-operation among related females seems not to have evolved in the macropods, there is evidence of competition for resources within matrilines in one species. Individual female Red-necked Wallabies who associate regularly with other females (who, from the forgoing, are likely to be relatives), have lower reproductive success than do more solitary females (Johnson 1986a). This suggests an indirect form of competition within matrilines, in which the availability of resources to an individual female is diminished in proportion to the number of relatives using the same set of resources (Clutton-Brock *et al.* 1982; van Schaick 1983).

As Hamilton (1967) has pointed out, sex-biased competition between relatives has an evolutionary significance not shared by other forms of competition. Hamilton argued that where siblings of one sex compete with one another, equal investment by parents in the two sexes ceases to be evolutionarily stable, as it is under conditions of the random distribution of kin within populations. Hamilton's argument accounts for female-biased sex ratios in certain insects in which brothers compete with one another to mate with their sisters (Charnov 1982). It has also been used to account for the production of male-biased sex ratios at birth in the Muskrat *Ondata zibethicus*: in this species females are philopatric while males disperse; sisters are therefore likely to compete with one another for the use of living sites made available by the deaths of their mothers (Caley *et al.* 1988).

Sex ratios at birth may also be expected to vary when daughters compete with their mothers. Such competition, which must be common in species, like many macropods, whose females are philopatric and long-lived, increases the cost to the mother of producing daughters, and is expected therefore to set up selection either for a relative increase in the numbers of males produced or for some other form of balancing investment in males. This argument has been used to explain male-biased sex ratio at birth in tree squirrels *Tamiasciurus* spp. (Smith 1968), Red Deer *Cervus elaphus* (Clutton-Brock *et al.* 1982), *Antechinus* spp. (Cockburn *et al.* 1985) and primates (Clark 1978; Johnson 1988).

Sex ratios of pouch young in seven of the macropod species discussed in this review are listed in Table 2. Sex ratios are generally slightly (but

Table 2. Sex ratios of pouch young in species of macropods in which dispersal is not sex-biased (the Quokka), and in which dispersal is male-biased (the remainder).

Species	Sex ratio (prop^n. male)	Source
Setonix brachyurus	0.44 (n= 398), P<0.05	Dunnett (1963)
Bettongia penicillata	0.60 (n= 105), P=0.05	Christensen (1980)
Macropus eugenii	0.54 (n= 429), n.s.	Inns (1980)
M. rufogriseus	0.55 (n= 615), P<0.05	Kirkpatrick (pers. comm.); Walker (pers. comm.); Merchant (pers. comm.); Fleming *et al.* (1983); Catt (1977)
M. giganteus	0.54 (n=1115), P<0.05	Refs. in Johnson and Jarman (1983); Stuart-Dick (pers. comm.)
M. robustus	0.47 (n= 596), n.s.	Ealey (1963)
M. rufus	0.52 (n=3805), P<0.05	Refs. in Johnson and Jarman (1983)

often significantly) male-biased, or close to parity, in species with male-biased dispersal, but are significantly female-biased in the Quokka. The direction of this bias is as predicted by the hypothesis of local resource competition, although it is not clear why the Quokka's sex ratio should be positively female-biased. The most extreme sex ratio of pouch young known for a macropod is the value of 0.67 (n=61) recorded in a captive population of the Yellow-footed Rock-Wallaby *P. xanthopus* (Poole *et al.* 1985). Although data on dispersal and philopatry in this species are not available, dispersal appears to be male-biased in the Brush-tailed Rock-wallaby (see above). Competition between female rock-wallabies is likely to be particularly close because of the species' strict reliance on scattered rocky refuges.

CONCLUDING REMARKS

The most obvious conclusion to be drawn from this review is that we still have very little good information on patterns of dispersal and philopatry in the macropods. Further studies should concentrate on small-bodied species outside the genus *Macropus,* and on the factors which determine dispersal distances and govern choices of settlement sites, as our ignorance of these subjects is almost complete. Rock-wallabies may be particularly appropriate subjects for studies of the relations between dispersal, social behaviour and sex allocation. The data at hand do not show up any major differences between the macropods and eutherians. There is, however, sufficient variation in dispersal patterns within the macropods to suggest that further study will ultimately be rewarded with clear insights into the evolution of dispersal in the group.

ACKNOWLEDGEMENTS

I am grateful to Tim Clancy, Karen Higginbottom, Peter Jarman, Ken Joblin, Menna Jones, Tom Kirkpatrick, Jim Merchant, Colin Southwell, Robyn Stuart-Dick and Marcus Walker for allowing me to cite their unpublished observations.

REFERENCES

BAILEY, P. T., 1971. The red kangaroo, *Megaleia rufa* (Desmarest) in north-western New South Wales 1. Movements. *C.S.I.R.O. Wildl. Res.* **16**: 11-28.

BAKER, R. R., 1978. The Evolutionary Ecology of Animal Migration. Hodder and Stoughton: London.

BAKER, R. R., 1982. Migration: Paths Through Space and Time. Hodder and Stoughton: London.

BEARDER, S. K., 1987. Lorises, bushbabies and tarsiers: diverse societies in solitary foragers. *In* Primate Societies ed. by B. B. Smuts, D. L. Cheney, R. M. Seyfarth, R. W. Wrangham and T. T. Struhsaker. University of Chicago Press.

BENNETT, A. F., 1987a. Conservation of mammals within a fragmented forest environment: the contributions of insular biogeography and autecology. *In* Nature Conservation: the Role of Remnants of Native Vegetation ed. by D. A. Saunders, G. W. Arnold, A. A. Burbidge and A. J. M. Hopkins. Surrey Beatty & Sons: New South Wales.

BENNETT, A. F., 1987b. Biogeography and conservation of mammals in a fragmented forested environment in south-eastern Victoria. Ph.D. Thesis, University of Melbourne.

CALEY, M. J., BOUTIN, S. AND MOSES, R. A., 1988. Male-biased reproduction and sex-ratio adjustment in muskrats. *Oecologia,* **74**: 501-06.

CATT, D. C., 1977. The breeding biology of Bennett's wallaby *(Macropus rufogriseus fruticus)* in South Canterbury, New Zealand. *N.Z. J. Zool.* **4**: 401-11.

CAUGHLEY, G., 1977. Analysis of Vertebrate Populations. John Wiley and Sons: Chichester.

CHARLES-DOMINIQUE, P., 1977. Ecology and Behaviour of Nocturnal Prosimians. Duckworth: London.

CHARNOV, E., 1982. The Theory of Sex Allocation. Princeton University Press.

CHRISTENSEN, P. E. S., 1980. The biology of *Bettongia penicillata* (Gray, 1837) and *Macropus eugenii* (Desmarest 1917) in relation to fire. Forests Dept. Western Australia Bulletin No. 91.

CHRISTENSEN, P. E. S. AND MAISEY, K., 1987. The use of fire as a management tool in fauna conservation reserves. *In* Nature Conservation: the Role of Remnants of Native Vegetation ed. by D. A. Saunders, G. W. Arnold, A. A. Burbidge and A. J. Hopkins. Surrey Beatty & Sons: New South Wales.

CLARK, A. B., 1978. Sex ratio and local resource competition in a prosimian primate. *Science* **201**: 163-65.

CLOUT, M. N. AND EFFORD, M. G., 1984. Sex differences in dispersal and settlement of brushtailed possums *(Trichosurus vulpecula)*. *J. Anim. Ecol.* **53**: 737-49.

CLUTTON-BROCK, T. H., ALBON, S. D. AND GUINESS, F. E., 1982. Competition between female relatives in a matrilocal mammal. *Nature* **300**: 178-80.

CLUTTON-BROCK, T. H. AND HARVEY, P. H., 1976. Evolutionary rules and primate societies. *In* Growing Points in Ethology ed. by P. G. Bateson and R. A. Hinde. Cambridge University Press.

COCKBURN, A., SCOTT, M. P. AND DICKMAN, C. R., 1985. Sex ratio and intrasexual competition in mammals. *Oecologia* **66**: 427-29.

COCKBURN, A., SCOTT, M. P. AND SCOTTS, D., 1985. Inbreeding avoidance and male-biased dispersal in *Antechinus* spp. (Marsupialia: Dasyuridae). *Anim. Behav.* **33**: 908-15.

CROFT, D. B., 1981. Behaviour of the red kangaroo *Macropus rufus* (Desmarest) in north western New South Wales. *Aust. Mammal.* **4**: 5-58.

DAWSON, G. A., 1978. Composition and stability of social groups of the tamarin, *Saguinus oedipus geoffroyi,* in Panama: Ecological and behavioural implications. *In* The Biology and Conservation of the Callitrichidae ed. by D. G. Kleiman. Smithsonian Institution Press: Washington DC.

DENNY, M. J. S., 1982. Adaptions of the red kangaroo and euro (Macropodidae) to aridity. *In* Evolution of the Flora and Fauna of Arid Australia ed. by R. W. Barker and P. J. M. Greenslade. Peacock Publications: Frewville, South Australia.

DENNY, M. J. S., 1985. The red kangaroo and the arid environment. *In* The Kangaroo Keepers ed. by T. H. Lavery. University of Queensland Press: St. Lucia.

DOBSON, F. S., 1982. Competition for mates and predominant juvenile male dispersal in mammals. *Anim. Behav.* **30**: 1183-93.

DUNNETT, G. M., 1962. A population study of the quokka, *Setonix brachyurus* Quoy and Gaimard (Marsupialia) III. The estimation of population parameters by means of the recapture technique. *C.S.I.R.O. Wildl. Res.* **8**: 78-117.

EALEY, E. H. M., 1963. The ecological significance of delayed implantation in a population of the hill kangaroo *(Macropus robustus)*. *In* Delayed Implantation ed. by A. C. Enders. University of Chicago Press.

EALEY, E. H. M., 1967. Ecology of the euro, *Macropus robustus* (Gould) in north western Australia II. Behaviour, movements and drinking patterns. *C.S.I.R.O. Wildl. Res.* **12**: 27-51.

FLEMING, D., CINDEREY, R. W. AND HEARN, J. P., 1983. The reproductive biology of Bennett's wallaby *(Macropus rufogriseus rufogriseus)* ranging free at Whipsnade Park. *J. Zool. (Lond.)*. **201**: 283-91.

FRITH, H. J., 1964. Mobility of the red kangaroo, *Megaleia rufa*. *C.S.I.R.O. Wildl. Res.* **9**: 1-19.

GREENWOOD, P. J., 1982. Mating systems, philopatry and dispersal in birds and mammals. *Anim. Behav.* **28**: 1140-62.

GREENWOOD, P. J., HARVEY, P. H. AND PERRINS, C. M., 1978. Inbreeding and dispersal in the great tit. *Nature* **271**: 52-54.

HAMILTON, W. D., 1967. Extraordinary sex ratios. *Science.* **156**: 477-88.

HARCOURT, C. S. AND NASH, L. T., 1986. Social organization of galagos in Kenyan coastal forests 1. *Galago zanzibaricus*. *Am. J. Primatol.* **10**: 339-55.

HEINSOHN, G. E., 1968. Habitat requirements and reproductive potential of the macropod marsupial *Potorous tridactylus* in Tasmania. *Mammalia* **32**: 30-43.

HOLSWORTH, W. N., 1964. Marsupial behaviour with special reference to population homeostasis in the quokka on the west end of Rottnest Island. Ph.D. Thesis, University of Western Australia.

HOLSWORTH, W. N., 1967. Population dynamics of the quokka, *Setonix brachyurus* on the west end of Rottnest Island, Western Australia. 1. Habitat and distribution of the quokka. *Aust. J. Zool.* **15**: 29-46.

INNS, R. W., 1980. Ecology of the Kangaroo Island wallaby *Macropus eugenii* (Desmarest) in Flinders Chase National Park, Kangaroo Island. Ph.D. Thesis, University of Adelaide.

JAREMOVIC, R. V. AND CROFT, D. B., 1987. Comparison of techniques to determine eastern grey kangaroo home range. *J. Wildl. Manage.* **51**: 921-30.

JARMAN, P. J. AND TAYLOR, R. J., 1983. Ranging of eastern grey kangaroos and wallaroos on a New England pastoral property. *Aust. Wildl. Res.* **10**: 33-38.

JARMAN, P. J. AND SOUTHWELL, C. J., 1986. Grouping, association and reproductive strategies in the eastern grey kangaroo. *In* Ecological Aspects of Social Evolution ed. by D. I. Rubenstein and R. W. Wrangham. Princeton University Press: Princeton, New Jersey.

JOHNSON, C. N., 1985. Ecology, social behaviour and reproductive success in a population of red-necked wallabies. Ph.D. Thesis, University of New England, Armidale, New South Wales.

JOHNSON, C. N., 1986a. Philopatry, maternal investment and the reproductive success of females in the red-necked wallaby. *Behav. Ecol. Sociobiol.* **19**: 143-50.

JOHNSON, C. N., 1986b. Sex-biased philopatry and dispersal in mammals. *Oecologia* **69**: 626-27.

JOHNSON, C. N., 1987. Macropod studies at Wallaby Creek IV. Home range and movements of the red-necked wallaby. *Aust. Wildl. Res.* **14**: 125-32.

JOHNSON, C. N., 1988. Dispersal and the sex ratio at birth in primates. *Nature.* **332**: 726-28.

JOHNSON, C. N., 1989a. Dominance and reproductive tactics among adult red-necked wallabies. *J. Zool. (Lond.)*. **217**: 267-80.

JOHNSON, C. N., 1989b. Grouping and the structure of association in the red-necked wallaby. *J. Mammal.* **70**: 18-26.

JOHNSON, C. N. AND JARMAN, P. J., 1983. Geographical variation in offspring sex ratios in kangaroos. *Search.* **14**: 152-53.

KAUFMANN, J. H., 1974. Social ethology of the whiptail wallaby *Macropus parryi* in northeastern New South Wales. *Anim. Behav.* **22**: 281-369.

KAUFMANN, J. H., 1975. Field observations on the social behaviour of the eastern grey kangaroo, *Macropus giganteus*. *Anim. Behav.* **23**: 214-21.

KITCHENER, D. J., 1970. Aspects of the response of the quokka to environmental stress. Ph.D. Thesis, University of Western Australia.

LIBERG, O. AND VON SCHANTZ, T., 1985. Sex-biased philopatry and dispersal in birds and mammals: the Oedipus hypothesis. *Am. Nat.* **126**: 129-35.

MACFARLAND SYMINGTON, M., 1987. Sex ratio and maternal rank in wild spider monkeys: when daughters disperse. *Behav. Ecol. Sociobiol.* **20**: 421-25.

MICHENER, G. R., 1983. Kin identification, matriarchies and the evolution of sociality in ground-dwelling sciurids. *In* Advances in the Study of Mammalian Behaviour ed. by J. F. Eisenberg and D. G. Kleiman. The American Society of Mammalogists, Special Publication No. 7.

MOEHLMAN, P. D., 1986. Ecology of cooperation in canids. *In* Ecological Aspects of Social Evolution: Birds and Mammals ed. by R. I. Rubenstein and R. W. Wrangham. Princeton University Press: Princeton New Jersey.

MOORE, J. AND ALI, R., 1984. Are dispersal and inbreeding avoidance related? *Anim. Behav.* **32:** 94-112.

NEYMAN, P. F., 1978. Aspects of the ecology and social organization of free-ranging cotton-top tamarins *(Saguinus oedipus)* and the conservation status of the species. *In* The Biology and Conservation of the Callitrichidae ed. by D. G. Kleiman. Smithsonian Institution Press: Washington DC.

NICHOLLS, D. G., 1971. Daily and seasonal movements of the quokka *(Setonix brachyurus). Aust. J. Zool.* **19:** 215-26.

OLIVER, A. J., 1986. Social organization and dispersal in the red kangaroo. Ph.D. Thesis, Murdoch University, Western Australia.

PACKER, C., 1979. Inter-troop transfer and inbreeding avoidance in *Papio anubis. Anim. Behav.* **27:** 1-36.

PACKER, C., 1985. Dispersal and inbreeding avoidance. *Anim. Behav.* **33:** 676-78.

POOLE, W. E., MERCHANT, J. C., CARPENTER, S. M. AND CALABY, J. H., 1985. Reproduction, growth and age determination in the yellow-footed rock-wallaby *Petrogale xanthopus* Gray, in captivity. *Aust. Wild. Res.* **12:** 127-36.

PRIDDEL, D., 1987. The mobility and habitat utilization of kangaroos. *In* Kangaroos: Their Ecology and Management in the Sheep Rangelands of Australia ed. by G. Caughley, N. Shepherd and J. Short. Cambridge University Press.

PUSEY, A. E., 1987. Sex-biased dispersal and inbreeding avoidance in birds and mammals. *Trends Ecol. Evol.* **2:** 295-99.

PUSEY, A. E. AND PACKER, C., 1987. Dispersal and philopatry. *In* Primate Societies ed. by B. B. Smuts, D. L. Cheney, R. M. Seyfarth, R. W. Wrangham and T. T. Struhsaker. University of Chicago Press.

RALLS, K., HARVEY, P. H. AND LYLES, A. M., 1986. Inbreeding in natural populations of birds and mammals. *In* Conservation Biology: the Science of Scarcity and Diversity ed. by M. E. Soule. Sinauer Associates Inc.: Sunderland, Massachusetts.

RODMAN, P. S. AND MITANI, J. C., 1987. Orangutans: sexual dimorphism in a solitary species. *In* Primate Societies ed. by B. B. Smuts, D. L. Cheney, R. M. Seyfarth, R. W. Wrangham and T. T. Struhsaker. University of Chicago Press.

VAN SCHAICK, C. P., 1983. Why are diurnal primates living in groups? *Behav.* **87:** 120-44.

SMITH, C. C., 1968. The adaptive nature of social organization in the genus of tree squirrels, *Tamiasciurus. Ecol. Monog.* **38:** 31-64.

SMUTS, B. B., CHENEY, D. L., SEYFARTH, R. M., WRANGHAM, R. W. AND STRUHSAKER, T. T. (eds), 1987. Primate Societies. University of Chicago Press.

STUART-DICK, R., 1987. Parental investment and rearing schedules in the eastern grey kangaroo *(Macropus giganteus).* Ph.D. Thesis, University of New England, Armidale, New South Wales.

STRUHSAKER, T. T. AND LELAND, L., 1979. Socioecology of five sympatric monkey species in the Kibale Forest, Uganda. *In* Advances in the Study of Behaviour ed. by J. S. Rosenblatt, R. A. Hinde, C. Beer and M. C. Busnel. Academic Press: New York.

TERBOGH, J. AND GOLDIZEN, A., 1985. On the mating system of the cooperatively breeding saddle-backed tamarin *(Saguinus fuscicollis). Behav. Ecol. Sociobiol.* **16:** 293-99.

WASER, P. M., 1985. Does competition drive dispersal? *Ecology* **66:** 1170-75.

WASER, P. M. AND JONES, W. J., 1983. Natal philopatry among solitary mammals. *Q. Rev. Biol.* **58:** 355-90.

YALDEN, D. W., 1988. Feral wallabies in the Peak District, 1971-1985. *J. Zool.* **215:** 369-74.

Space-use patterns of the Common Wallaroo *Macropus robustus erubescens* in the arid zone

T. F. CLANCY[1,2] and D. B. CROFT[1]

Inter- and intrasexual comparisons were made using indices of space use, based on the radio-telemetry data collected on 19 male and 16 female Common Wallaroos, *Macropus robustus erubescens*. Results indicate that individuals of different age/sex classes employ different strategies in their movement patterns. Males shifted their activity centres more extensively than females on a short-term basis, and therefore increased the effective area of their home-range. The positions of the majority of male and female home-ranges were stable from year to year; however, a small proportion of both males and females progressively shifted their centres of activity. The implications of variation in shifts of activity centres and its effect on estimates of home-range size are discussed.

INTRODUCTION

INVESTIGATIONS of the ways in which animals distribute their activities in time and space have played an important role in ecological studies since the beginning of this century (Dixon and Chapman 1980). While studies on movement patterns of mammals in general have been numerous (Sanderson 1966), this has not been the case with macropodids (Jaremovic and Croft 1987). Ealey (1967) reported on the ranging behaviour of the Common Wallaroo (=Euro), *Macropus robustus erubescens,* based on repeated sightings of marked animals. Since his pioneer study there have been numerous advances in the field of radio telemetry (both in data acquisition and analysis) that permit more detailed information to be obtained on animal's space-use patterns.

Two parameters are generally considered to be the most useful in the investigation of animals' spatial and temporal movements: home-range area and activity centre (Sanderson 1966). This study concentrated on variation in shifts in activity centres and the effect such shifts have on home-range size. The interaction between shifts in activity centre and size of home-range has rarely been investigated previously. The two phenomena are clearly closely related, an animal's home-range area being dependent on both the spread of activities around its activity centres and the relative positions of these centres in space. Therefore, the area encompassed by a home-range can be increased by either shifting the location of the home-range in space or by spreading activities over a larger area. The strategy chosen will be related to the reasons for the increase.

METHODS

Data collection

This study was part of a wider investigation into the behaviour and movement patterns of *M. r. erubescens* conducted at Fowlers Gap Research Station in northwestern New South Wales, from January 1984 to February 1987 (Clancy 1989). The study site was located nearby and similar to that described by Croft (1981); however, there were no significant cave shelters present within our site. Thirty five individual Wallaroos (19 males, 16 females) were captured by either cannon-netting (Day *et al.* 1980) or stunning (Robertson and Gepp 1982) and fitted with collars carrying radio transmitters (Telonics Inc., Arizona, U.S.A.). Radio-collared individuals were tracked on foot on an approximately weekly basis over the three-year period of field studies. A hand-held directional antenna (Telonics, model RA-2A) and a scanner/receiver (Telonics, model Nos. TS-1, TR-2E) were used, and the procedure involved walking toward the signal source until the animal was sighted. Animal locations were recorded by determining their position when first encountered using "photo-maps" created by 2 × photographic enlargements of 1:25000 topographic maps (Central Mapping Authority, Dept. of Lands, map Nos. 70-417, 70-418) divided into 50 × 50 m grid squares. Location data were generally collected in either the early morning or the late afternoon. Individual animals were tracked at intervals of four to seven days, where possible, for the life of their transmitter. A total of 54 full-year samples (28 males and 26 females) of hand-tracking data were collected on radio-collared wallaroos. Fifteen animals were monitored for two years or more and six of these (three males, three females) for the full three years.

[1]School of Biological Science, University of New South Wlales, P.O. Box 1, Kensington, New South Wales 2033.

[2]Present address: Conservation, Parks and Wildlife, P.O. Box 149, Charleville, Queensland 4470.

Pages 603-09 in KANGAROOS, WALLABIES AND RAT-KANGAROOS ed. by G. Grigg, P. Jarman and I. Hume. Surrey Beatty & Sons Pty Limited, New South Wales, Australia, 1989.

Intensive tracking sessions from two fixed-point stations were conducted in winter and summer of each year. Tracking sessions consisted of two 5×24 h "weeks" with a rest day in between. In all future discussion, weekly home-range corresponds to one set of 120 h tracking data. Bearings were taken, from each station, on one fixed-point transmitter (placed at a surveyed location) and on nine or ten radio-collared wallaroos every 15 minutes around the clock. The beginning times of each set of readings were synchronized between stations by a CB radio link. A different set of animals was generally followed in each "week"; however, in some cases animals were tracked for the full 10 days. Each tracking station was fitted with two 4-m, 8-element "Yagi" antennae (Telonics, model RA-4B, gain = 11.8 dBd) mounted in parallel with elements vertical on the ends of a 2-m (1λ) cross-beam. The centre of the cross-beam was attached to a 4-m mast coupled to a rotatable axle assembly. A 0.5-m diameter compass rose marked in 1° divisions was attached to and rotated with the axle assembly across a fixed pointer. Signals from the two antennae were combined 180° out of phase using a two-port null converter (Telonics, model no. TAC-5). Animals were located by finding the null point between the two loudest peaks of signal amplitude and recording the compass rose position of the antennae. The readings of both stations were corrected to true north against those of the fixed-point transmitter, and the animals' position calculated trigonometrically and resolved into cartesian co-ordinates using one tracking station as the origin. These co-ordinates were translated to the national grid co-ordinates of the Fowlers Gap topographic map. Readings were taken of the fixed-point transmitter during each 15-min run to guard against system errors (Heezen and Tester 1967) caused by shifts in antenna position, temperature changes, etc. The locations of the two tracking stations were chosen, based on initial hand-tracking data, to minimize the error polygon (Heezen and Tester 1967) and maximize signal quality for the majority of animals. Stations were therefore positioned so that most of the radio-collared animals were between the two stations and not directly in line with them. The minimum error polygon caused by angular error is found at the intersection of the bearings 45° and 315° from the respective stations (Heezen and Tester 1967). To optimize signal quality the highest practical locations were chosen. Altitudes of the stations were 250 m and 275 m, respectively, which meant that in practice the receiving antennae height above the signal source varied between five and 100 m. The average error of radio-tracking readings inherent in the system was assessed to be 0.3 ha based on the variation of readings taken on stationary animals located at varying distances and directions from the tracking stations (Clancy 1989). The accuracy of the system was also verified by comparing radio-tracking locations with hand-tracking locations collected simultaneously.

Data analyses

All locations of animals gathered by hand tracking were initially plotted out and compared with a topographic map to ensure no errors had been made when transcribing grid-reference co-ordinates. The inherent error of radio-tracking data collected from stations is much greater than hand tracking (Lee *et al.* 1985), therefore more rigorous editing procedures were used. Firstly data sets where null reading extended over a range of several degrees were discarded because this indicates very poor signal quality. Samuel and Garton's (1985) bivariate normal UD test was applied to all station-tracking data to identify outliers. As it was not the intent to force normal distributions on these data, editing of these outliers was performed subjectively based upon comparisons with the original data set and with what was known about the animals' movements. For example, if the data set showed an animal moving two kilometres from one point and back again to its original position in three successive readings (a movement that is possible but highly improbable) the errant point was discarded; however, if the data showed the animal moving progressively from one point to another and both stations had obtained good signals (nulls of 1° or 2°) these points were included. Generally points with a weighted Mahalanobis distance value <0.6 were excluded, as recommended (Samuel and Garton 1985). A minimum sample size of 30 was used. A detailed description of procedures for the creation of data sets for analyses and for the editing of errant points from the station tracking data is given in Clancy (1989).

The harmonic mean centre (Dixon and Chapman 1980) was chosen to estimate the activity centre of an individual's home-range and to locate the home-range's position in space. The harmonic mean centre of an areal distribution is located in the area of the most dense packing of points (i.e., where location points exhibit the highest clustering) and gives a close approximation to the true activity centre of an animal's home-range. Jaremovic and Croft's (1987) method of calculating isopleth values was used for calculating harmonic mean centres (HMC). A grid size of 50 m was superimposed on each data set and the lowest inverse first areal moment was used as an estimate of HMC. The overall distance between successive daily activity centres and the maximum distance between daily activity centres over a week's tracking are good indices of relative animal mobility for that time period. Two different measures were used to investigate daily shifts in HMC over weekly station tracking periods: cumulative shift and maximum shift.

Cumulative shift was defined as the cumulative distance between successive daily HMCs and maximum shift as the distance between the two most widely separated HMCs over the week's tracking. Distances between HMC of weekly data sets were used in measurements of shifts on six monthly and longer bases. Where hand-tracking data had been collected on individuals for two years or longer, the positions of their yearly HMC from consecutive years were also compared.

For estimates of home-range area, the Fourier transform method (Anderson 1982) was used. This technique was chosen because it is non-parametric and is independent of sample size above a certain sample size. The Fourier transform method allows for the estimation of minimum areas used at different probability levels (MAP values) for the description of animals' home-ranges. In accordance with other studies, MAP (95) and MAP (50) areas were calculated and used to represent total home-range areas and core areas, respectively, for the time period under consideration. The programme to generate the 3-dimensional co-ordinates of the Fourier transform utilization distribution was obtained from J. Bradbury and J. Wilkinson of the University of California, San Diego. Contour and 3-dimensional plots were generated by the Surface II graphics system (Sampson 1975).

The home-ranges of the six animals that were monitored for the full three years were investigated to ascertain the effect of increasing time-span upon them. Total home-range areas and core areas were calculated by the Fourier-transform method for seven different time-spans: day, week, year, 18 months, two years, 30 months, three years. Daily and weekly home-ranges were calculated for each of the animals using data from the total station-tracking data sets of each animal. All longer duration home-ranges were calculated by subdivision of their complete hand-tracking data sets. Four regression models were fitted to test the relationship between home-range size and time, viz. linear, exponential, geometric and quadratic. For all animals tracked for >2 years, their average yearly MAP (50) area was compared with their MAP (50) area over two years. If the location of an individual's home-range is stable in time, the ratio of 1:2 year home-range area will approach unity whereas if it shifts the location over time the ratio will be much greater than 1. MAP (50) values are much less affected by sample-size variations than MAP (95) (Anderson 1982) and are therefore better suited to comparisons between the different time intervals.

RESULTS

The mean cumulative shifts and mean maximum shifts in activity centres for the six tracking sessions are shown in Fig. 1. For both these measures of shifts in daily HMC there was a significant difference between the sexes (cumulative shifts $F_{1,77}=13.48$, $P<0.01$; maximum shifts $F_{1,77}=11.35$, $P<0.01$). There were no significant year, session or interaction effects. Males' cumulative-shift and maximum-shift distances were both approximately 60% longer than females', indicating that they were more mobile on a day-to-day basis. In the case of males, cumulative-shift and maximum-shift distances were slightly larger in winter than summer but this difference was not significant. Females cumulative shifts were also higher on average in winter than summer but not significantly so.

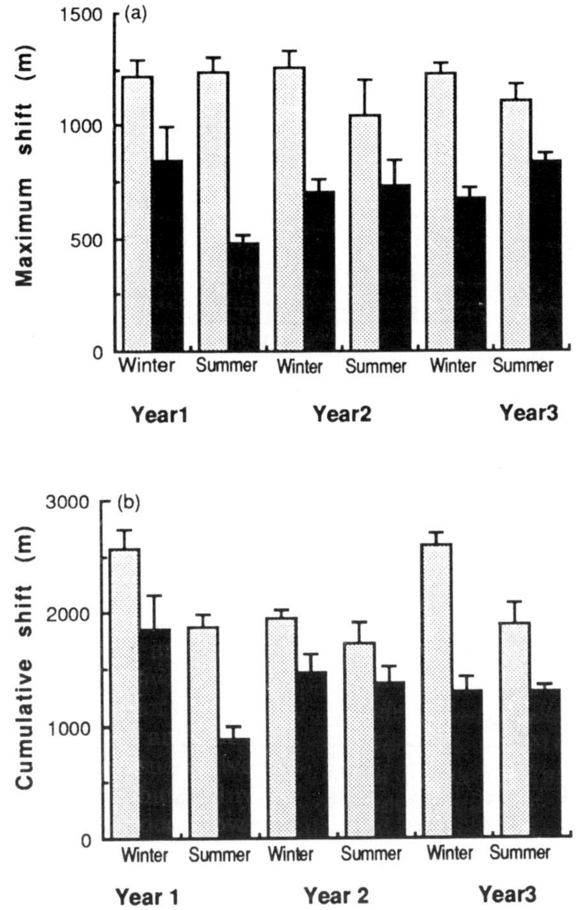

Fig. 1. Mean differences (+standard error) in indices of shifts in daily HMCs over the six station tracking sessions: (a) maximum shift; (b) cumulative shift. Light stipple, males; dark stipple, females.

Mean distances between weekly HMC over six month, 12 month and "long term" (18-30 months) time-spans for individual animals are shown in Fig. 2. The average distance between daily HMCs is included for comparison. There is a significant difference between sexes with the females' weekly HMC shifting to a much smaller degree than males' (Table 1). Also, the distances between males' weekly HMCs increased with time, whereas the distances between females' weekly

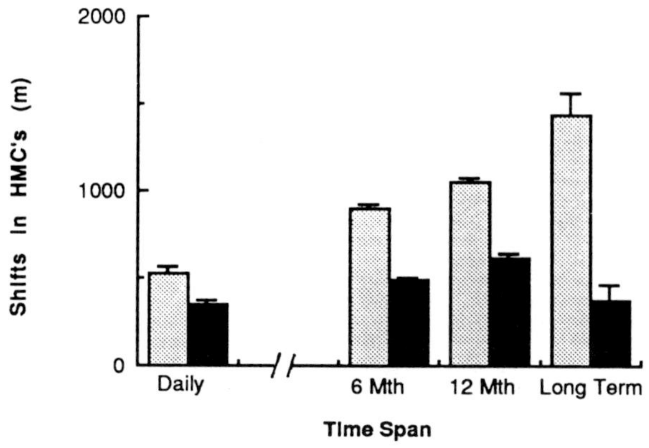

Fig. 2. Mean shifts (+standard error) in weekly HMCs over the three time spans considered and mean shifts in daily HMCs. Light stipple, males; dark stipple, females.

Table 1. Analyses of sex differences in shifts of weekly HMCs over different time spans. *P<0.05.

Time Span	D.F.	t-statistic	P
Six months	54	2.63	*
Year	35	2.43	*
Long Term	9	2.23	*

HMCs remained relatively constant. For long-term time-spans, the distances between females' weekly HMCs were similar in magnitude to their average daily shifts (Fig. 2).

Mean differences (±standard error) between HMC of yearly home-ranges for males (n=6) was 642.8±58.1 and for females (n=7) was 495.1±52.1. Where animals had two measures of HMC shift (i.e., year 1 → year 2 + year 2 → year 3) the mean was used. There was no significant difference in the magnitude of HMC shifts between the sexes (t=0.74, df=11, P=0.52). There is variation in the amount that individual animals' HMC shifted and therefore differences in positional fidelity of home-ranges. Examples of the HMC shifts of animal's yearly home-ranges are shown in Fig. 3. Three different patterns are apparent:

(1) Animals whose yearly HMCs' locations did not shift, or shifted only slightly, from year to year and therefore the position of their home-ranges was stable with time, e.g., females C3WT130 and T144C43 (Fig. 3a,b).

(2) Animals whose yearly HMCs' locations shifted from year to year; however, the direction of shift reversed so that the position of their long-term home-range was relatively stable with time, e.g., female T187165 and male T192 (Fig. 3c,d).

(3) Animals whose yearly HMCs' locations shifted from year to year and the directions of shift were similar and therefore the position of their long-term home-range changed progressively, e.g., female T151164 and male T180169 (Fig. 3e,f).

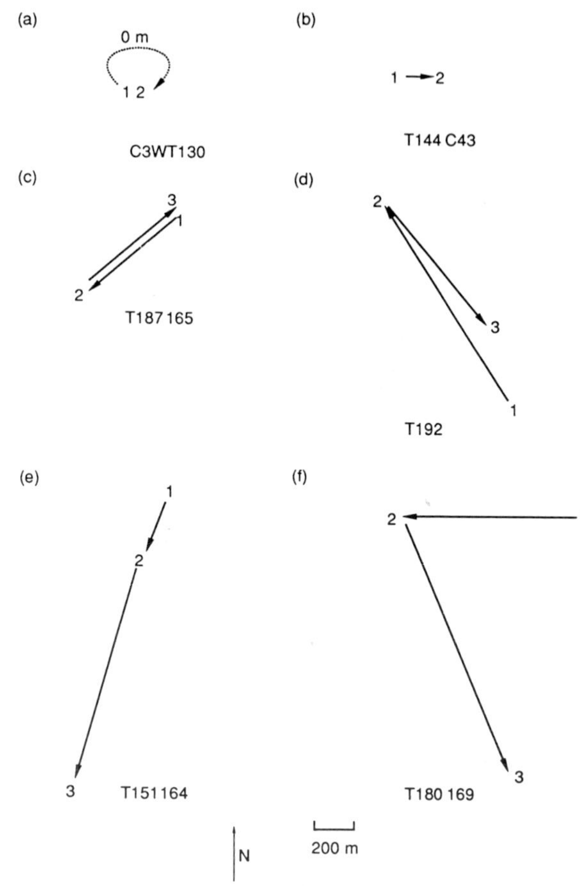

Fig. 3. Examples of shifts in yearly HMCs from year to year by individual radio-collared wallaroos: (a) C3WT130; (b) T144C43); (c) T187165; (d) T192; (e) T151164; (f) T180169. See text for explanation.

The average size of home-ranges, for each time-span considered, of animals that were monitored for the full three years are shown in Fig. 4. For all of the animals a geometric (power) regression best explained the distribution of home-range sizes, except in the case of female T151164 for which a linear regression gave the best fit (R^2 linear=0.95 cf. R^2=0.91 for the geometric). A summary of the regression results is given in Table 2. Because two different methods for collecting data were used, the short-term and long-term home-ranges are not strictly comparable. Nevertheless, the general pattern of home-range area increasing with increasing time-span, but at an ever decreasing rate, is probably realistic; weekly home-ranges were much larger than daily and much smaller than yearly ones. The slopes of linear regression lines calculated for hand-tracked data on time-spans one year to three year were not significantly different from zero for any of the six animals except T151164 (Table 2). The ratios of average yearly MAP (50) areas and two-yearly MAP (50) areas for 15 animals are shown in Table 3. There is no significant difference between the sexes (Mann-Whitney $U_{7,8}$=35, P>0.10). The majority of animals' average yearly home-ranges were greater

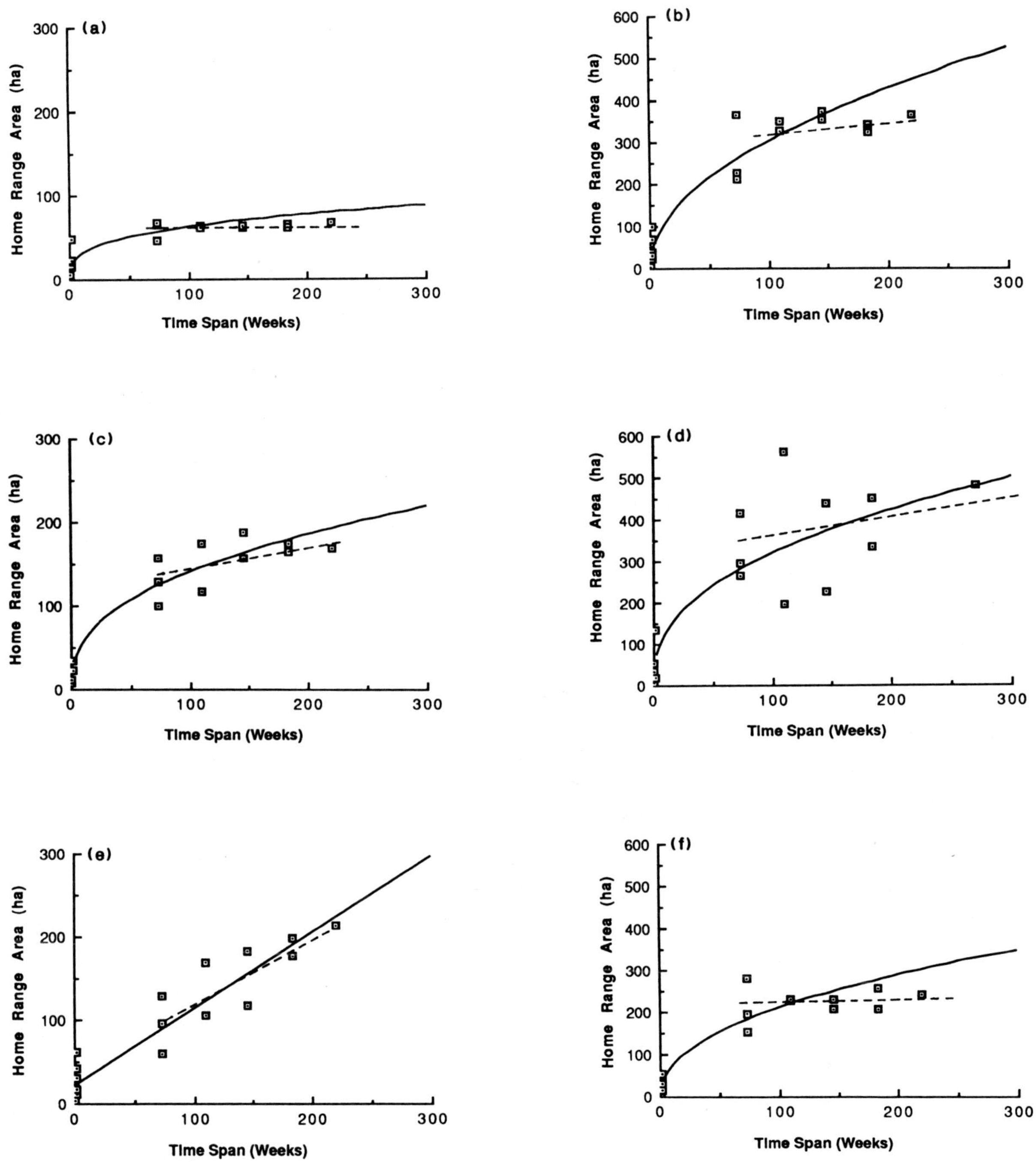

Fig. 4. Relationship between home-range size and time for the six animals monitored for a full three years: (a) T153151; (b) T192; (c) T187165; (d) T193; (e) T151164; (f) T180169. Females are on the left side, males on the right. Solid line, regression line for time spans from one day to three years; dashed lines, one year to three years.

Table 2. Regression results of home-range size (HS) vs time (t) for the six wallaroos tracked for three years. Home-range sizes were calculated for MAP (95)s. Time spans were from one day to three years and from one year to three years. See text for details. **P<0.01, ***P<0.001, n.s. not significant. M = male; F = female.

Animal	Sex	Daily → 3 Year R^2	Daily → 3 Year $F_{1,n}$	Daily → 3 Year P	D.F.$_n$	1 Year → 3 year R^2	1 Year → 3 year $F_{1,9}$	1 Year → 3 year P
T180169	M	0.91	212.3	***	2	0.31	3.5	n.s.
T192	M	0.80	52.9	***	4	0.11	0.9	n.s.
T193	M	0.91	204.9	***	22	0.08	0.7	n.s.
T151164	F	0.90	190.9	***	20	0.69	18.0	**
T153151	F	0.86	129.9	***	22	0.14	1.3	n.s.
T187165	F	0.96	328.6	***	13	0.26	2.8	n.s.

Table 3. Ratios of average yearly MAP (50) to two-yearly MAP (50). [a]Animals whose home-range area increased progressively from year to year. See text for explanation.

FEMALES	RATIO	MALES	RATIO
T1431	0.82	T147	1.23
T144C43	1.19	T148	3.15[a]
T151164	1.53[a]	T1582	1.01
T166C3H	1.12	T180169	1.24
T173180	1.06	T181	1.87[a]
T187165	1.30	T192	0.89
		T193	1.04
Median	1.12		1.24

than 75% of their two-year home-range. In general adult Common Wallaroos apparently range over most of their long-term home-range over a period of about one year. There are, however, exceptions; for example, males T147 and T181 and female T151164 all ranged over much larger areas as time increased.

DISCUSSION

There are two possible explanations for the higher mobility of male wallaroos compared with females on a short-term basis. The difference may be related to the high degree of sexual dimorphism causing differences in feeding patterns. Difference in body size has been postulated as a mechanism in ecological separation (Illius and Gordon 1987). Alternatively, the difference in short-term mobility may be a reflection of the different reproductive strategies. This seems the more likely explanation. Sex differences in feeding patterns are more likely to be reflected in differences in the spread of activities around daily activity centres and in the relative positions of the sexes' activity centres in space. Female wallaroos have been shown to have relatively small, discrete home-ranges (Clancy 1989). The most efficient way for a male to increase his number of encounters with females is to shift the position of his short-term home-range with time. Croft (1981) found that male wallaroos frequently checked the reproductive condition of females and suggested that they moved from female to female on a regular basis. Wallaroos in arid areas are non-seasonal breeders (Tyndale-Biscoe and Renfree 1987) therefore no seasonal differences would be expected in sexual checking activity. Jaremovic and Croft (1987) found that medium and large male Eastern Grey Kangaroos, *Macropus giganteus,* shifted their activity centres seasonally and that this was probably related to higher levels of reproductive activity in spring.

Male Common Wallaroos have been shown to have significantly larger mean (±1 SD) weekly home-ranges in winter than females (males=77.2±47.5 ha, females=30.5±16.5 ha); however, their summer home-range areas are similar in size to those of females (males=30.2±20.4 ha, females=27.6±15.0 ha) (Clancy 1989). As there was no significant seasonal effect for either of the measures of shift in daily activity centre, the differences between winter and summer home-ranges are not due to them shifting the positions of their daily home-ranges to a greater extent in winter but to their increasing their movement about their daily centres of activity. Jaremovic and Croft (1987) found that seasonal differences in sheltering behaviour caused changes in the location of activity centres in Eastern Grey Kangaroos. The slightly longer distances between daily activity centres in winter compared with summer recorded for both sexes of Common Wallaroos may reflect seasonally related behaviour changes. There were, however, no caves or other areas of localised shelter points within the study site which might have more strongly influenced summer movement patterns.

Male and female wallaroos differed markedly in their measured shifts in weekly home-range position. These results suggest that the larger year and long-term home-ranges of males compared to females (Clancy 1989) are due in part to their shifting their centres of activity more extensively. Differences between the sexes in the distances between weekly activity centres over the various time-spans are probably related not only to differences in reproductive strategies but also to differences in feeding patterns. There is a tendency for males to move to lower-lying areas during dry periods where the effective rainfall is higher (Clancy 1989). Such differences in feeding behaviour and habitat use are common in sexually dimorphic mammals (Clutton-Brock *et al.* 1987).

The small average shift in activity centres from year to year indicates that wallaroos are highly sedentary. The sedentary nature of the majority of wallaroos is also reflected in the relative stability of home-range area over periods of a year or more. These results concur with those of Ealey (1967). Positional fidelity of home-ranges has been documented also in Red Kangaroos *M. rufus* (Croft 1982; Priddel 1987). Males on average shifted their activity centres slightly more from year to year than females; however, it is probable that this is more a reflection on their higher mobility and larger overall home-ranges than a less sedentary nature. Female T151164 exemplifies the more nomadic nature of some wallaroos, her centre of activity shifting markedly and her home-range area increasing linearly with time (Fig. 3e, Fig. 4e). Whether there exists in wallaroo populations a small percentage of individuals that are continuously shifting their home-ranges can be ascertained only by following individuals for much longer periods than the three-year period of this study. It is possible that such shifts in home-range are only spasmodic rather than continuous.

There have been several studies relating home-range area to metabolic needs (McNab 1963; Harestad and Bunnell 1979; Clutton-Brock and Harvey 1981; Gittleman and Harvey 1982; Mace *et al.* 1982). Clearly, in both inter and intraspecific comparisons of home-range sizes it is necessary to distinguish between changes in home-range area due to the shifting of short-term home-ranges and changes due to increased activity about activity centres, as the functional reasons for such changes are likely to be different. The measurement of shifts in HMC is therefore a powerful tool in the investigation of space-use patterns.

ACKNOWLEDGEMENTS

We thank the many volunteers who aided in the capture of wallaroos. We especially thank the Centre for Field Research (Boston U.S.A.) for support and the many Earthwatch volunteers who assisted with the station tracking in all six sessions. Peter Gillespie helped in the preparation of data. Nicola Marlow assisted in the preparation of figures. This study was supported by a Commonwealth Postgraduate award to Tim Clancy and an A.R.G.S. award (A18315987) to David Croft.

REFERENCES

ANDERSON, D. J., 1982. The Home Range: A New Non-parametric Estimation Technique. *Ecology* **63**: 103-12.

CLANCY, T. F., 1989. Factors influencing movement patterns of the Euro *(Macropus robustus erubescens)* in the arid zone. Unpubl. Ph.D. Thesis. University of New South Wales.

CLUTTON-BROCK, T. H. AND HARVEY, P. H., 1981. The functional significance of variation in body size among mammals. *In* Mammalian Behaviour ed. by J. F. Einsenberg. Special publication of the American Society of Mammalogists.

CLUTTON-BROCK, T. H., IASON, G. R. AND GUINNESS, F. E., 1987. Sexual segregation and density-related changes in habitat use in male and female Red deer *(Cervus elaphus)*. *J. Zool., Lond.* **211**: 275-89.

CROFT, D. B., 1981. Social behaviour of the Euro, *Macropus robustus* (Gould), in the Australian arid zone. *Aust. Wildl. Res.* **8**: 13-49.

CROFT, D. B., 1982. Radio-telemetry studies on arid zone kangaroos. C.S.I.R.O. Arid Zone Newsletter.

DAY, G. I., SCHEMNITZ, S. D. AND TABER, R. D., 1980. Capturing and marking wild animals. *In* Wildlife management techniques manual ed. by S. D. Schemnitz. The Wildlife Society: Washington, D.C.

DIXON, K. R. AND CHAPMAN, J. A., 1980. Harmonic mean measure of animal activity areas. *Ecology* **61**: 1040-44.

EALEY, E. H., 1967. Ecology of the Euro, *Macropus robustus* (Gould) in North-Western Australia. II. Behaviour, movement and drinking patterns. *C.S.I.R.O. Wildl. Res.* **12**: 27-51.

GITTLEMAN, J. L. AND HARVEY, P. H., 1982. Carnivore home-range size, metabolic needs and ecology. *Behav. Ecol. Sociobiol.* **10**: 57-63.

HARESTAD, A. S. AND BUNNELL, F. L., 1979. Home range and body weight — a re-evaluation. *Ecology* **60**: 389-402.

HEEZEN, K. L. AND TESTER, J. R., 1967. Evaluation of Radio-Tracking by triangulation with special reference to deer movements. *J. Wildl. Manage.* **31**: 124-41.

ILLIUS, A. W. AND GORDON, I. J., 1987. The allometry of food intake in grazing ruminants. *In* Herbivore Nutrition Research. Research papers presented at the second international symposium of the nutrition of herbivores. Australian Society of Animal Production.

JAREMOVIC, R. V. AND CROFT, D. B., 1987. Comparison of techniques to determine Eastern Grey Kangaroo home range. *J. Wildl. Manage.* **51**: 921-30.

LEE, J. E., WHITE, G. C., BARTRAM, R. M. AND ALLDREGE, A. W., 1985. Assessing accuracy of a radiotelemetry system for estimating animal locations. *J. Wildl. Manage.* **49**: 658-63.

MACE, G. M., HARVEY, P. H. AND CLUTTON-BROCK, T. H., 1983. Vertebrate home range size and metabolic requirements. *In* The Ecology of animal movement ed. by P. J. Swingland and P. J. Greenwood. Oxford University Press: Oxford.

MCNAB, B. K., 1963. Bioenergetics and the determination of home range size. *Am. Nat.* **97**: 133-40.

PRIDDEL, D., 1987. The mobility and habitat utilization of kangaroos. Pp. 100-18 *in* Kangaroos, their ecology and management in the sheep rangelands of Australia. Cambridge University Press: Cambridge.

ROBERTSON, G. G. AND GEPP, B., 1982. Capture of Kangaroos by "stunning". *Aust. Wildl. Res.* **9**: 393-96.

SAMPSON, R. J., 1975. Surface II graphics system. Computer Serv. Sect. Kans. Geol. Surv., Lawrence.

SAMUEL, M. D. AND GARTON, E. O., 1985. Home range: a weighted normal estimate and tests of underlying assumptions. *J. Wildl. Manage.* **49**: 513-19.

SANDERSON, G. C., 1966. The study of mammal movements — a review. *J. Wildl. Manage.* **30**: 215-35.

TYNDALE-BISCOE, H. AND RENFREE, M., 1987. Reproductive physiology of marsupials. Cambridge University Press: Cambridge.

A day in the life of a Kangaroo: Activities and movements of Eastern Grey Kangaroos *Macropus giganteus* at Wallaby Creek

J. L. CLARKE[1], M. E. JONES[1] and P. J. JARMAN[1]

To obtain sequential data on activity scheduling and distance moved, 24 Eastern Grey Kangaroos were each followed on foot for a 24-hour period. The daily schedule of activities was variable, with almost any activity state being possible at any time of day or night. However, there was a regular pattern of decreased feeding and increased resting in the middle of the day. The time that individuals spent feeding ranged from 10.5 h to 18.5 h. Males and females did not differ in time spent feeding but both sexes fed for longer in winter than in summer. Rates of biting were lower in winter than in summer and, although kangaroos compensated for lower biting rates by decreasing the proportion of time spent surveying whilst feeding, low total intakes were probably the cause of long feeding times in winter. Increased rates of movement within a few hours of dawn and dusk reflected a strong tendency in summer for animals to move uphill or into the forest in the evening, and to emerge in the morning. Total distance moved in 24 hours varied greatly, the extreme values of 400 m and 5 600 m both being recorded from large adult males. Both males and females moved further and faster in summer than in winter, these differences being reflected in both non-feeding and feeding movements. Males moved further and faster than females in summer, particularly in the period between midnight and dawn. Within the same season, the sexes moved at similar rates during feeding bouts.

INTRODUCTION

SOME species of animals are tightly constrained in their daily activities, being strictly limited by extrinsic or intrinsic factors to specific quantities, schedules or sequences of activities. As an example among the macropods, the Rufous Bettong *Aepyprymnus rufescens* spends the hours of daylight crouched in a nest, emerging at dusk to forage. Its length of foraging time is limited by the length of the night, and its day-time activities cannot include foraging and social interactions. Other species are less constrained, and their individuals may facultatively vary the quantities of time devoted to activities, or their scheduling or sequencing. In such species information on variability between classes or individuals in their daily quantities or sequencing of activities may help us to recognize individual and class strategies of apportionment of time and energy amongst activities.

The large macropod species, in the few studies that have been made of their activity (e.g., Caughley 1964; Bell 1972; Osazuwa 1978; Southwell 1981; Pridell 1986), differ in the degree to which the daily quantities and schedules of their activities are constrained. Some species, especially those from the semi-arid zone, seem to be highly constrained by high day-time temperatures and follow structured schedules, while others are more flexible and distribute their activities more evenly through the day. Despite those differences, there is some commonality in these species' schedules of activity. All the studied species forage least frequently in the middle of the day, most moving and feeding occurring at night. All these species vary their daily schedules seasonally as the clock-times of dawn and dusk shift and as day or night temperatures fluctuate. The Eastern Grey Kangaroo *Macropus giganteus* is one of the less constrained species (Bell 1972; Southwell 1981). As a part of our studies of the behavioural ecology of the species at Wallaby Creek, we wished to investigate daily activity, to see whether social classes differed in their quantities, scheduling or sequencing of activities.

STUDY AREA, ANIMALS, AND METHODS

The kangaroos were studied at Wallaby Creek, in northeastern New South Wales. The study site and kangaroo population are fully described by Jarman *et al.* (1987). The site consists of roughly cleared, cattle-grazed pasture on the slope and bottom of the eastern side of a forested valley. All the kangaroos in the immediate study area (between 85 and 100 in the period covered by this aspect of the work) were individually known to Clarke and Jones, who carried out the field work.

The techniques used were simple, but depended upon the observers' familiarity with the animals and upon the animals' habituation to the presence of observers. A focal animal was chosen and followed, by one observer at a time, on foot, for 24 hours. Followings began three hours after first light. This starting time was chosen because by then most kangaroos had moved to their day-time resting spot. Locations and activities of the focal animal were recorded at 5-minute intervals.

[1]Department of Ecosystem Management, University of New England, Armidale, New South Wales, Australia 2350.

Pages 611-18 in KANGAROOS, WALLABIES AND RAT-KANGAROOS ed. by G. Grigg, P. Jarman and I. Hume. Surrey Beatty & Sons Pty Limited, New South Wales, Australia, 1989.

Night observations were made by moonlight or using image-intensifying binoculars; spotlights were used to confirm identification of animals after dark.

Activities were: Feeding, when animals were either biting, selecting bites, chewing or regurgitating; Alert, when they had their ears and eyes focussed on a source of disturbance; Moving, when animals were moving; Standing, when animals were not Feeding or Alert; and Lying, when they were lying down. The last two categories were both considered to be resting activities. Movements were generally recorded on a scan when kangaroos were hopping large distances, rather than when they were taking steps whilst feeding. Bite-rates were recorded at half-hour intervals during feeding bouts by watching the focal animal for three minutes and recording onto a tape recorder each incidence of a bite, step and raising of the head to survey. Two bite-rate measures were calculated: the first we have termed the interrupted bite-rate and is based on the number of bites taken in the three minutes; the second (the uninterrupted bite-rate) is based on the time taken for 10 bites without the animal raising its head from the sward (see Clarke and Loudon 1985). The interrupted bite-rates were used to calculate the total bite number, whilst the uninterrupted bite rates were used as a measure of the animal's biting response to sward conditions. The ratio of the two measures (the Biting Ratio) is an indication of the proportion of feeding time that kangaroos were actually biting or selecting bites.

Focal animals were chosen to represent the adult population classes as evenly as possible, with animals being followed in all areas of the study site. Males were allocated to a population class according to their size and degree of development of musculature. Females were classed on the basis of the presence or absence (as we could determine from observation at a short distance), and relative size, of a pouch-young or young-at-foot. All the animals followed were adults; the small males were of known age, as were young females.

Despite some successful studies of macropod activities which have used remote techniques such as radio-telemetry (Johnson 1977; Oliver 1986; Priddel 1986), we chose to follow animals on foot for several reasons. We were studying an excellently habituated population, and did not want to risk the possible destruction of that habituation through catching, handling and collaring. We would not have been able, technically or monetarily, to put transmitters on 100 animals; nor did we want to differentiate among animals physically by collaring some and not others. Thus we kept open our options for choosing which animals to follow by having them all equally available. Moreover, there were many data which we could gather by close following and observation, which would not be obtained remotely. These included details of feeding (e.g., bite-rates, feeding site and sub-habitat, and type of food), interactions, identity of nearest neighbours, sequences of behaviour, and contexts for behaviour (especially alertness). A drawback to the technique is its inevitable slowness (but this is no worse than the use of telemetry); it is also very tiring for the observers, especially on winter nights or in constant rain.

RESULTS

Between December 1985 and December 1987, 31 followings were attempted. Of these, 24 were completed (i.e., the animal was not out of sight for more than 5% of the time and its general activity during that time could be safely inferred). The animals that were "lost" tended to disappear at night, during a phase of movement. The results may therefore under-represent nocturnal movement. Followings were repeated on "lost" animals so as not to bias the sample towards animals that were easy to follow. The successful 24 followings were all of different individuals and were evenly divided between adult males and females; nine could be assigned to summer conditions (December, January or February) and 10 to winter (June or July), with nearly equal representation of males and females in each season.

Because of the small numbers of each sex and class studied in each season, and the variability found in most measures of activity, we have reported differences which were significant at the 0.1 level but not at the 0.05 level. Where we have stated that two samples differ, this indicates significance at the 0.05 level, while statements that two samples "tended" to differ indicate a result with a significance level between 0.05 and 0.1 but where other results in the same direction gave us grounds for believing that a difference existed.

General schedule of activities

Fig. 1 shows the activity schedules for kangaroos in summer and winter. The major differences between the seasons are the earlier daily onset, and greater amount, of resting in summer. In both seasons lying was least common in the four hours before dawn, and just before and after dusk. Moving (i.e., moving other than as a part of feeding), which formed only a small proportion of the daily activity, tended to occur at dawn and (particularly) dusk in summer, and in the first two hours after dawn and in the middle of the night in winter.

There was some consistency in location of the kangaroos at different times of the day. Using a pair-wise comparison of easterly locations at the start and end of each time period (one-sample

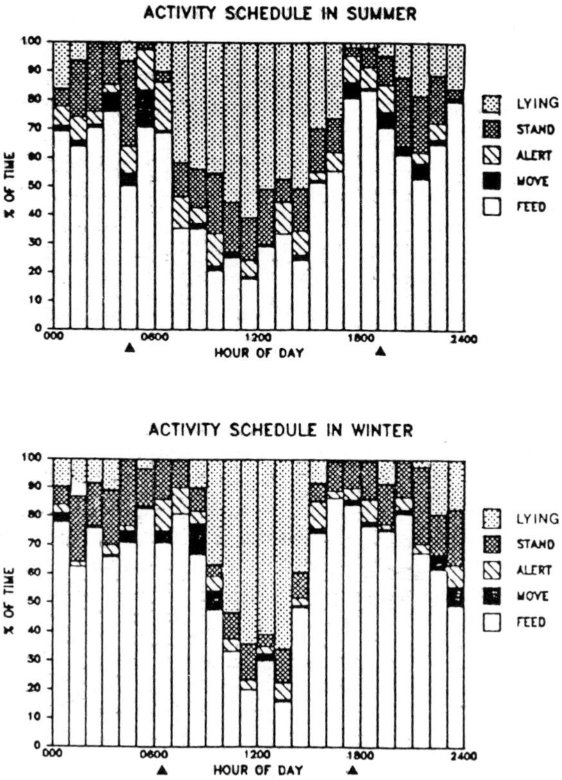

Fig. 1. Summer (a) and winter (b) schedules of activity for Eastern Grey Kangaroos, displayed as the percentage of each hour spent in one of the following activities. Feeding, including biting, selecting bites, moving while selecting or biting and chewing; Moving, while not feeding; Standing, when animals were crouched or standing but not alert; Alert, when animals were standing with their ears and eyes focussed on a source of disturbance; and Lying, when they were lying down.

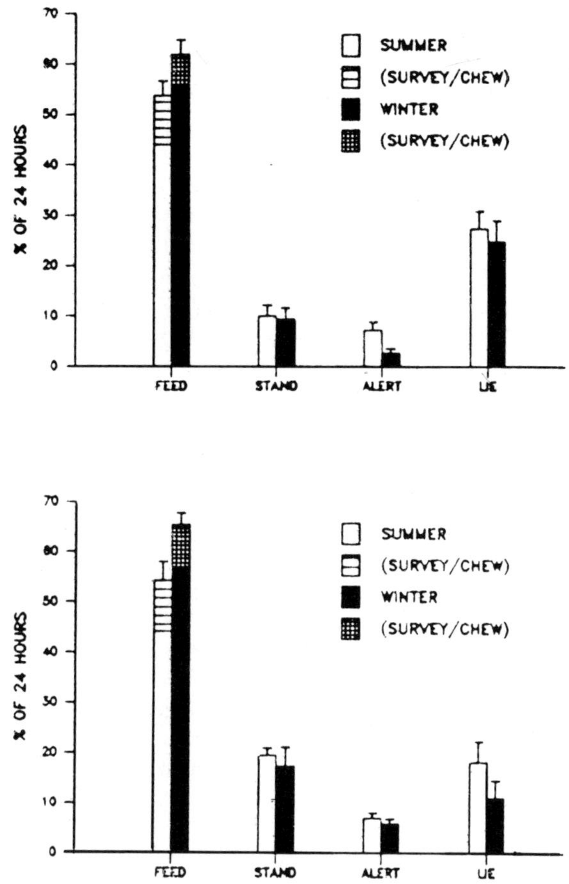

Fig. 2. A comparison of the proportion of time spent by males (a) and females (b), in summer and winter, Feeding, Standing, Alert and Lying as means ± 1 standard error. Note that surveying while chewing is distinguished within feeding.

Wilcoxon signed-rank test), kangaroos showed a significant movement uphill in the three-hour period before dusk ($p<0.02$) and downhill in the three-hour period after dawn ($p<0.05$) in summer. This movement applied equally to males and females; it was not seen in winter. Four out of the eight "northern" animals followed moved, not uphill but north into the open forest at night, and two of the remaining four northern animals stayed in the forest throughout the 24-hour period. However, there were equal numbers of northern animals followed in summer and winter so this movement into the forest cannot explain the observed movement of kangaroos uphill in summer and not in winter. The difference may be explained by the difference in distribution of kangaroos prior to dusk. In winter kangaroos spent the day further upslope than they did in

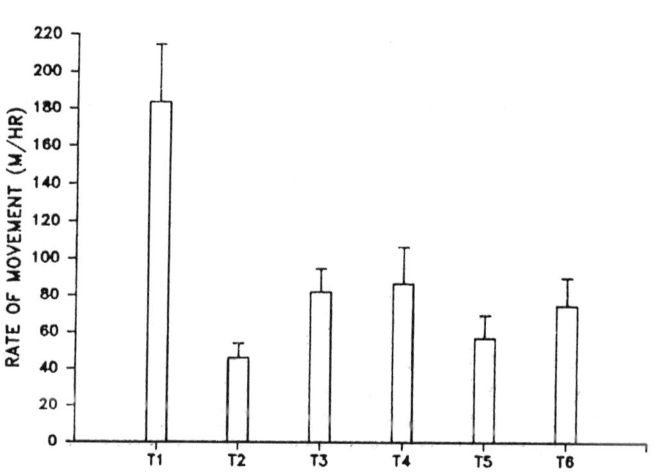

Fig. 3. Rates of movement (both sexes, all seasons, calculated from each individual's distance moved in a whole time period) as means ± 1 standard error, for the six time periods listed in Table 4.

summer (difference of means 240 m at 3 hours before dusk; Mann-Whitney U test, $p<0.05$), so that they were as far upslope at dusk in winter as they were at dusk in summer.

In summer and winter there were no differences between males and females in their daily time spent feeding or in the proportion of feeding time that they were chewing/surveying (Figs 2 and 3). However, in a two-way analysis of variance, with sex and season as grouping factors, the proportion of time spent feeding, the proportion of feeding time spent biting or selecting bites, and the proportion of time spent alert differed between summer and winter (Table 1). Kangaroos fed for longer in winter than in summer and spent more of that feeding time biting or selecting bites. They also spent more time alert in summer than in winter. A higher proportion of feeding time spent biting or selecting bites may reflect a difficulty in selecting food items, or a need to increase rates of food intake, or both. The proportions of time spent standing and lying differed between the sexes, reflecting differences in their preferred postures for resting with females standing more and lying less than males. There were no significant interactions between sex and season for any factor.

Table 1. Significant results of a two-way analysis of variance comparing the proportion of the time spent in different activity states, with sex and season as grouping factors.

Source	d.f.	F	p
feed × season	1	10.49	0.006
bite × season	1	12.18	0.003
stand × sex	1	10.62	0.005
alert × season	1	5.45	0.034
lie × sex	1	9.02	0.009

Table 2. Biting rates and total number of bites in 24 h, for kangaroos in summer and winter. Probability values based on two-tailed Student's t tests.

	Summer	Winter	p
Interrupted Bite-Rate	30.9 ± 1.7	21.1 ± 1.5	0.001
Uninterrupted Bite-Rate	47.4 ± 3.3	28.6 ± 2.5	0.001
Biting Ratio	0.69 ± 0.03	0.79 ± 0.03	0.05
Total Bite number	24 200 ± 1 700	19 400 ± 1 600	0.06
Sample size	9	10	

There were no differences between males and females in their rates of biting or their total bite number for 24 hours, so data were combined for the sexes in Table 2. Both the interrupted and uninterrupted bite rates were higher in summer than winter and the Biting Ratio was lower in summer. Low uninterrupted bite rates in winter suggest that in this season kangaroos were spending more time selecting each bite. The poor availability of green grass in winter, when live leaf comprised only 30-50% of the sward surface (compared to 60-80% in summer; Clarke, unpubl. data), made the selection of green leaf more difficult, leading to low biting rates. Low biting rates combined with a poor quality diet led to high feeding time being recorded in winter (mean of 15.3 ± 0.5 hours per day) and little time spent surveying whilst feeding, as the kangaroos attempted to maintain intake.

Movement

Total distance moved in 24 hours ranged from 400 to 5 600 m, both these values being recorded from large adult males. The overall daily mean distance moved was 1 873 m ± 272 m (s.e.); the male mean was 2 280 m ± 423 m and the female mean 1467 m ± 145 m. Kangaroos moved further in summer than in winter ($F_{3,15}=9.25$, $p<0.0001$); males moved more than twice as far in summer as in winter but the difference for females was not as great (Table 3). Males tended to move further than females in summer but not in winter.

Table 3. A comparison of the mean daily distances moved by male and female Eastern Grey Kangaroos in summer and winter. Probability values based on a two-tailed Student's t test.

	Mean Distances ± s.e. : Metres		
	Males	Females	p
Summer	3 090 ± 630	1 800 ± 330	0.13
(n)	(5)	(4)	
Winter	1 200 ± 430	1 050 ± 100	0.75
(n)	(5)	(5)	
p	0.04	0.16	

Two-way analysis of variance source	df	F	p
sex	1	3.16	0.096
season	1	9.25	0.008
sex × season	1	2.02	0.176

There was much variation in the daily distance moved by males, some of which appeared to relate to the male's age or home-range size, or to the presence or absence of oestrous females. For example, a small adult male who was in the same group as his mother for all the 24 hours moved 970 m, a typical daily distance for females. By contrast, an older male, closer to the age of dispersal, spent only 30 minutes in the same group as his mother, and moved a total of 3 900 m. However, having studied this population for several years, we feel that some males are consistently conservative in their movements, while others regularly move long distances. These differences have important implications for the individuals' energy budgets and their abilities to find oestrous females, and they deserve further investigation.

Mean rates of movement differed between time periods (two-way analysis of variance, with repeated measures: $F_{5,15}=8.18$ $p<0.005$, $n=24$)

Table 4. Average rates of movement, in metres per hour, for the whole of each time period, and for the full 24 hours, of male and female Eastern Grey Kangaroos in summer (December to February) and winter (June and July). The time periods were: T1, from dawn to 3 hours after dawn; T2, from then until the start of T3; T3, from 3 hours before dusk until dusk; T4, from dusk until 3 hours after dusk; T5, from then until T6; T6, from 3 hours before dawn until dawn. Superscripts indicate significant differences between matched pairs (Student's t tests: p<0.05).

Records	(n)	T1	T2	T3	T4	T5	T6	ALL
Males								
Summer	5	353.0[1]	56.4	132.0	139.0	116.0[2]	109.6[3]	128.8[4]
(s.e.)		(91)	(15)	(35)	(77)	(38)	(17)	(26)
Winter	5	74.0[1]	35.2	50.8	42.8	31.2	103.0	50.0[4]
(s.e.)		(21)	(28)	(20)	(16)	(10)	(6)	(18)
Females								
Summer	4	222.0	48.5	106.7	58.0	24.0[2]	26.2[3]	72.2
(s.e.)		(61)	(11)	(28)	(15)	(8)	(7)	(10)
Winter	5	127.6	28.6	48.8	37.4	28.6	22.6	43.6
(s.e.)		(30)	(5)	(15)	(11)	(10)	(8)	(4)

Two-way analysis of variance source	df	F	p
season	1	5.22	0.04
sex	1	8.27	0.01
time	5	8.18	0.0001
time × sex	5	2.88	0.02

and were faster in the morning (T1) than in other time periods, as shown in Figure 3. This time period encompasses the movements downhill or out of the forest in summer and the movement to shade at the end of the morning feeding period. There was a significant interaction between sex and time period ($F_{5,15}=2.88$, $p<0.05$). Table 4 shows that females' rates of movement in both winter and summer were greatest in Tl and lowest for periods T5 and T6 (i.e., from 3 h after dusk until dawn). Males were more variable. In summer they moved very fast in T1, and slowest around midday (T2), but in winter their peak rate of movement was in T6 (before dawn). Rates of movement differed between the sexes ($F_{1,15}=5.22$, $p<0.05$), with males moving faster than females.

These patterns in rates of movement are reflected in the proportional distribution of the daily movement. Table 5 gives values for the proportional distribution of the daily movement per hour. This has been calculated to reveal the scheduling of movement, regardless of the extent of each animal's daily movement. Again the distribution of movement differed significantly between time periods (two-way analysis of variance with repeated measures: $F_{5,15}=7.90$, $p<0.0001$). Averaging values for all animals, proportional movement is greatest in T1 and least in T2 and T5. Scheduling differed between the sexes, females showing lower proportional movement in T6 than males. This difference between the sexes in T6 was significant in summer but not

Table 5. The percentage of a day's total movement executed per hour by Eastern Grey Kangaroos, in the six time periods defined in Table 4. Superscripts indicate significant differences between matched pairs (Student's t tests: p<0.05).

		Time Periods					
Records	(n)	T1 %	T2 %	T3 %	T4 %	T5 %	T6 %
All	24	9.49	2.40	4.96	4.37	2.96	3.96
All Males	12	8.66	1.94	5.03	4.16	3.45	5.44[2]
All Females	12	10.31	2.86	4.89	4.57	2.47	2.47[2]
Summer Dec-Feb	9	11.82	2.35	5.07	3.74	2.85	2.85
Winter Jun-Jly	10	9.23	2.27	5.67	3.43	2.94	4.40
Summer Males	5	11.13[1]	1.98	4.40	3.73	4.06	3.80[3]
Summer Females	4	12.67	2.82	5.91	3.75	1.33	1.67[3]
Winter Males	5	6.93[1]	1.76	6.80	3.53	2.79	6.73
Winter Females	5	11.53	2.78	4.53	3.33	3.09	2.07

Two-way analysis of variance source	df	F	p
sex	1	3.95	0.07
time	5	7.90	0.0001
time × sex	5	2.69	0.03

Table 6. The percentage of 5-minute periods, in the six time periods defined in Table 4, during which a change in location was recorded for the Eastern Grey Kangaroo being watched. Superscripts indicate significant differences between means (Student's t tests: $p<0.05$).

		Time Periods						
		T1	T2	T3	T4	T5	T6	ALL
Records used	(n)	%	%	%	%	%	%	%
All	24	27.2	11.0	22.4	18.9	15.3	17.9	16.2
All Males	12	30.1	9.5	22.4	22.5	18.7	24.0^{1}	17.3
All Females	12	24.3	12.6	22.4	15.4	12.0	11.8^{1}	15.2
Summer	9	38.1^{2}	12.7	30.4^{3}	23.3	20.5^{4}	21.4	22.5^{5}
Winter	10	19.6^{2}	8.2	18.6^{3}	13.0	9.2^{4}	12.3	12.3^{5}
Male, Sum.	5	42.4	12.0	33.2	29.4	27.2^{6}	29.6^{7}	25.0^{8}
Male, Win.	5	14.8	4.8	17.8	12.8	10.0^{6}	18.0	11.8^{8}
Female, Sum.	4	32.7	13.5	27.0	15.7	12.2	11.2^{7}	17.2^{9}
Female, Win.	5	24.4	11.6	19.4	13.2	8.4	6.6	12.8^{9}

Two-way analysis of variance source	df	F	p
season	1	10.98	0.006
time period	5	7.54	0.0004
time period × sex	5	3.88	0.02

(log transformation and Greenhouse-Geisser correction used)

in winter. Proportional movement in T1 was greater for males in summer than in winter.

Movement is a product of the time spent moving and the rate of movement. Table 6 gives the proportions of the 5-minute periods in which any change of location was recorded, for the six time periods. These changes of location include movement during feeding bouts as well as the non-feeding movement shown in Figure 1. Overall, these proportions were high for T1 and T3 (the two daylight feeding periods), low for T2, and fairly even through the night (T4 to T6). Kangaroos changed location in more 5-minute time periods in summer than in winter ($F_{1,15}=10.98$, $p<0.01$), although the pattern of distribution of the values between time periods remained very similar. However, as shown in Tables 4 and 5, the pattern of distribution of values between time periods did differ between the sexes ($F_{5,15}=7.54$, $p<0.001$).

Some of the differences between males and females in rates of movement (Table 4), distributions of proportional movement (Table 5), and the frequency of movement (Table 6) are reflected in Table 7. This table shows the distance moved per 5 minutes, when any change of location was recorded (so this differs from Table 3 in that 5-minute periods in which the animal did not move are ignored). Males move faster than females in T6 in summer and move faster in T1 in summer than in winter because they move more frequently and they move further per movement. Similarly, kangaroos move further in summer than in winter (Table 2) because they move more frequently and they move further per movement. Movement was particuarly fast in T1 but in other time periods animals displayed two fairly consistent speeds of movement: a summer speed of 35.5 ± 2.4 m/5-minutes; and a winter speed of 28.4 ± 3.3 m/5-minutes.

DISCUSSION

Our rate of success, in following individual kangaroos for 24 hours each, was high, with 24 out of 31 attempted followings being completed. The data obtained, a full record of each individual's activity behaviour, associates and

Table 7. Distance moved (in metres) per 5-minute period in which the studied Eastern Grey Kangaroo was recorded as changing location, for the six time periods defined in Table 4. Superscripts indicate significant differences between means (Student's t tests: $p<0.05$).

		Time Periods						
Records Used	(n)	T1	T2	T3	T4	T5	T6	ALL
Males, Sum.	5	70.3^{1}	38.7	31.6	32.6	32.1^{2}	31.9^{3}	41.9
Males, Win.	5	37.7^{1}	31.2	21.6	24.4	24.3	37.8	35.3
Female, Sum.	4	56.5	31.6	31.8	28.2	15.3^{2}	19.7^{3}	30.5
Female, Win.	5	44.0	22.2	19.5	29.0	26.6	22.6	28.7

Two-way analysis of variance source	df	F	p
time period	5	6.26	0.0001

caught, collared, and relocated on each of these occasions, their weights (if caught during that field trip), and the numbers of nests used by each during that field trip. Only four animals were located, at five nests, in April, and those are included in the July tally in the table.

Attributes of nests and nest sites

Table 2 presents an analysis of the attributes of the 129 nest sites discovered during the study. Over half the nests were in open forest with medium undergrowth, but a quarter were in open country; dense undergrowth was not often used. In general it appears that nest sites are related to the structure of the vegetation community at ground level rather than to the canopy cover. Bettongs appear to choose a structure that provides cover for the nest without obstructions to rapid movement. In forested country they preferred medium undergrowth; in tree-cleared pasture they nested near shrubs. Nests on sloping land were almost always on a flat site on the slope. Nest sites were often surrounded by fallen timber, amongst which bettongs navigate with precision when flushed from their nests.

Table 2. The type of country, and features of the locations, in which 129 Rufous Bettong nests were found.

(a) Type of Country		
Open (tree-cleared)	31	24%
Open forest with:		
dense understorey	12	9%
medium understorey	71	55%
no understorey	15	12%
(b) Features of the Nest Location		
Watercourse <20 m	31	24%
Fallen Timber <10 m	66	51%
Blady grass bed <10 m	45	35%
Large tree (>300 mm dbh) <5 m	44	34%
Rocky area	9	7%
N.B.: A nest site may fall into more than one of these categories.		
(c) Features of Nest Locations in Open Country		
Shrubs within 1 m	25	76%
No shrub within 1 m	8	24%

A quarter of the nests were close to watercourses, which is significantly ($\chi^2=135.2$, $p<0.001$) more than would have occurred if nests were randomly distributed, since country within 20 m of a watercourse accounted for no more than 4% of the study area. Over half were near fallen timber (which is similar to Southgate's (1980) findings at Wallaby Creek), a third near beds of blady grass, and a third near large trees. The probabilities of these occurring by chance are untested but these proportions strike us as being unexpectedly high. By contrast, the small proportion sited in rocky ground is not unexpectedly high. Some nests were, of course, near more than one of these features.

For 94 of the nests, details were recorded of their placement relative to surrounding cover. Nests were most commonly placed under a tussock of grass (27%), against a log (27%), under a fallen branch (21%), or among rocks (14%). Two nests were among flood debris beside a watercourse, two under bushes, four under low-hanging grasstrees, and three among bracken. One was in an embankment and four were in the open in flat ground; these last five used a natural depression or were dug so far into the ground that nesting material did not protrude above ground (Table 3).

Table 3. Placement of 94 studied nests of Rufous Bettongs.

Under grass tussocks	28	30%
Amongst rocks	15	16%
Under or against a log	25	27%
Under a fallen branch	20	21%
Amongst flood debris	2	2%
Under a bush	2	2%
Under a grasstree	5	5%
Amongst bracken	3	3%
Dug into flat ground in the open	5	5%
Dug into an open embankment	1	1%

N.B.: A nest might qualify for more than one of these categories.

Nests did not face in a consistent direction (e.g., they did not all have the entrance facing north to avoid incident solar radiation). Instead, each entrance, and thus long axis of the nest, was situated to take full advantage of the available cover, but still leave the animal free passage if it had to run from the nest. Where nests were on slopes, they rarely faced uphill.

Nearly all nests were longer than they were wide, shaped generally like a blunt Rugby or Australian Rules football. The dimensions of nests were like those reported by Schlager (1981). None had more than one entrance, and this was 100 to 200 mm in diameter. The entrance's outer circumference ranged from 270 to 500 mm, depending on how far the nest was dug into the ground. The inner length varied from 210 to 400 mm, and the outer length (over the curve) from 450 to 650 mm. Most nests were dug into the ground, sometimes only shallowly, but occasionally nests in completely open habitat were almost fully below ground level. Fresh nests usually had some earth turned up at the entrance.

We confirmed Schlager's (1981) finding that bettongs at Cheviot Hills use a wide variety of nesting materials. The inner lining of the nest was always of fine material, usually dry grasses. The outer layer was often made from the vegetation sheltering the nest, such as grass-tree leaves or the leaves of a fallen branch. Often these had not been detached from their source, so that the outer layer of the nest blended into the sheltering vegetation, visually and physically.

Distribution of individuals' nests

When the nests used by each individual were mapped, it was evident that each individual for which we had enough records used nests clustered in one or two nesting areas (Figs 1 and 2). These nesting areas might be closely adjacent to, or even overlap, those of neighbours. The extent of overlap in nesting areas within the population cannot be quantified from our data, since not all animals were tracked simultaneously.

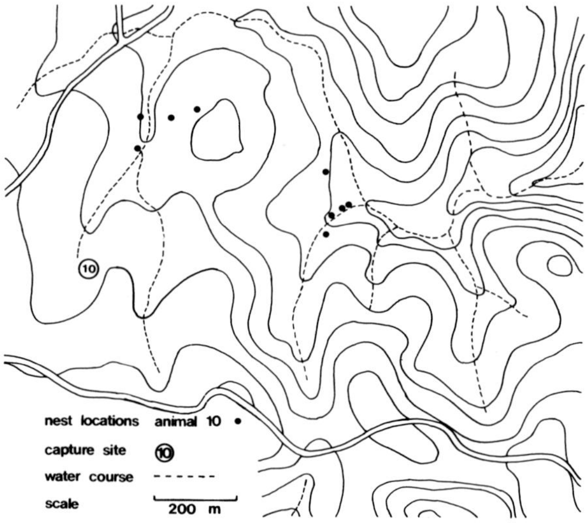

Fig. 1. Point of capture while feeding (a circle with the animal's number in it) and subsequently located nest sites of animal number 10, an adult male. The two clusters of nest sites are in pasture (the cluster to the left) and forest.

The feeding areas at which individuals were caught are indicated in Figs 1 and 2. Some nest clusters were close to feeding areas (<50 m), while others were over one km away. An individual's two nesting areas might be up to 500 m apart and adjacent to different feeding areas, or on different sides of the same one.

Four bettongs were tracked in both summer and winter. In each case they used the same nesting areas, but not necessarily the same feeding areas, in each season. Within a season an individual used several different nest sites (Table 1). Unfortunately, nests were located because animals had to be recaptured. Consequently our records of daily nest use show only that disturbed animals changed nest site, and that daily disturbance led to daily change of site. However, on a few occasions animals were tracked on successive days without being flushed, and on some of these occasions were known to have used the same nest two days running. Thus, daily change of nest site is not obligatory, and nests can be used more than once. Nests can last several months, and one was known to be used on occasions six months apart.

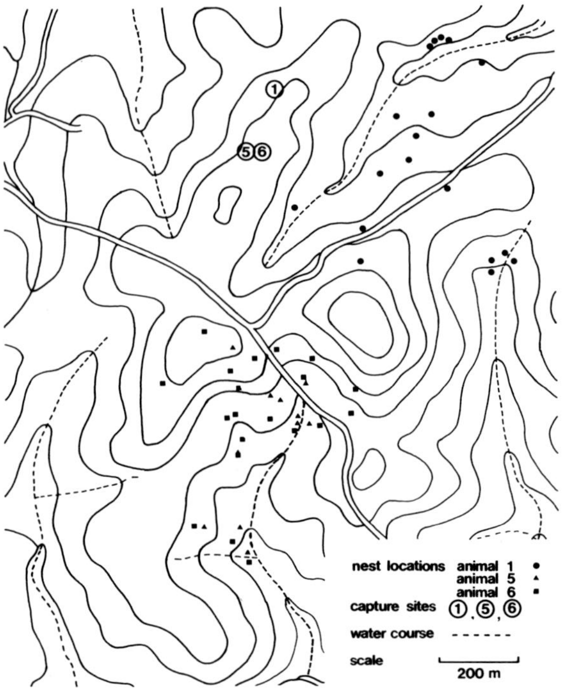

Fig. 2. Points of capture while feeding (each a circle with the animal's number in it) and subsequently located nest sites of animals numbers 1, a female; 5, a male; and 6, a female. Numbers 5 and 6 often nested close to each other and had fully overlapping nesting ranges.

No individual that was flushed from several nests used or constructed nests in only one type of site or under only one type of cover. Individuals thus appear flexible in their choice of site and cover. Rarely was an individual flushed from a nest that was known to have been used by another animal, but one nest was used, at separate times, by a female with her young-at-foot and a male.

When an individual was flushed from a nest without being caught by an encircling net, it would bound out and sprint rapidly away. Within 30 minutes (often much less) it could be tracked to another nest, which could not have been built in that time. The implication is that the animal knows the location of another usable nest and goes straight to it. Animals flushed from nests were either alone or were a mother with dependent young. Mothers and young-at-foot might escape in different directions, but be reunited within a day. On a few occasions a mother was flushed from a nest and left behind a juvenile which was old enough to be out of the pouch for part of the day while in the nest. This stage of juvenile development lasts less than a week in captivity (Jarman, unpubl. recs.). The mother later returned and retrieved the young.

The function of nests

We believe that our findings contribute to an understanding of the function of the numerous nests so carefully positioned and constructed by Rufous Bettongs.

Although adults were alone (except for dependent young) when flushed from nests, there was a tantalising but unprovable indication that two particular individuals consistently nested near to each other and fed together. A. Dennis (pers. comm.) reports that the species is less solitary when feeding at his Black Rock study site, in northern inland Queensland, than it has been reported to be in northern New South Wales (Southgate 1980; Schlager 1981). For short periods, a high proportion of adult bettongs at Wallaby Creek (northeastern New South Wales) have been seen in pairs, although for most of the year they forage solitarily (J. L. Clarke, pers. comm. and Wallaby Creek field records). At Drake, a large proportion of the night-time sightings (as many as half on some field trips) were of pairs of animals (Wallis, pers. obs.). Thus, solitariness appears to be facultative in Rufous Bettongs when feeding; by contrast, it seems to be obligatory when nesting. The use by an individual of one or two clusters of nest sites, the rarity of asynchronous use by two individuals of the same nest, and an individual's obvious knowledge of the location of alternative, usable nests, all suggest that each nest is a resource which the individual must be able to find and use dependably in an emergency. This justifies the time and effort put into nest construction and emphasizes the need for their careful location.

Our findings suggest that nests do more than shade animals from direct solar radiation; mere overhead cover would achieve that. The extent to which they ameliorate ambient temperature (Rübsamen *et al.* 1983) could be achieved without building an elaborate nest. The built nest, usually under cover or against an object like a log, sometimes dug into the ground or between rocks, gives its occupant protection against detection by both aerial and terrestrial predators hunting by sight in daylight. Most nests which were built under vegetation incorporated some of the vegetation into their outer layer, contributing to their inconspicuousness as well as providing a water-shedding outer cover. Each nest had its entrance in such a way that the bettong could burst from the nest at full speed. The clustering of an individual's nests ensured that the escaping bettong would have another nest nearby to "disappear" into. Even with radio-tracking equipment, it was difficult to pinpoint the site of many occupied nests exactly; there were often potential alternative sites within a few metres of the actual nest. Even if a predator detected by scent that a bettong was in the area, it might be unable with any degree of certainty to locate the real nest by its appearance; attacking grass tussocks randomly in the hope that one contained an occupied nest would not be a profitable predatory strategy. However, some nests were more obvious than others, and bettongs seemed to leave such nests readily when approached.

Although these findings must be considered cautiously because they derive from bettongs being flushed from nests daily, they extend our knowledge of the diversity of types of sites used by Rufous Bettongs for nesting. They also show that each individual may use a number of nests sequentially and has knowledge of their locations. An individual's nests are clustered. We believe that these findings make best sense if it is supposed that nesting throughout the daylight hours primarily has an anti-predator function.

ACKNOWLEDGEMENTS

We owe a debt of gratitude to the Ramsays of Cheviot Hills, who housed us and allowed us to range freely over their property. They helped us with spotlighting and capturing of animals, and enthusiastically shared their knowledge of the fauna and flora. The volunteer bettong-catchers, without whom the study would not have succeeded, were Leah Brien, Joanne Loughlin, John Friend, Tim Harrold and Ian Hume. We thank all of these, and John Seebeck, Chris Johnson and Denis Wallis for their constructive comments on a draft of this paper.

REFERENCES

RÜBSAMEN, U., HUME, I. D. AND RÜBSAMEN, K., 1983. Effect of ambient temperature on autonomic thermoregulation and activity patterns in the rufous rat-kangaroo (*Aepyprymnus rufescens:* Marsupialia). *J. Comp. Physiology B.* **153:** 175-79.

SCHLAGER, F. E., 1981. The distribution, status and ecology of the rufous rat-kangaroo, *Aepyprymnus rufescens,* in northern New South Wales. M.Nat.Res. Thesis, University of New England, Armidale.

SEEBECK, J. H., BENNETT, A. F. AND SCOTTS, D. J. (this volume). Ecology of the Potoridae — A Review.

SOUTHGATE, R., 1980. Factors affecting the spatial and temporal use of habitat by the rufous rat-kangaroo *Aepyprymnus rufescens.* B.Nat.Res. Thesis, University of New England, Armidale.

Social behaviour of the Black-striped Wallaby *Macropus dorsalis*, in captivity

C. F. HEATHCOTE[1]

A behavioural study of a captive group of *M. dorsalis* was conducted at the Royal Melbourne Zoo. A group of seven individuals (six females and one male) was observed for thirty five hours. Neutral and agonistic interactions between females and neutral and sexual interactions between the male and females are described. The observed repertoire of social acts is similar to those of other *Macropus* species. High frequencies of allogrooming acts and low levels of aggression suggest that *M. dorsalis* is a moderately social species.

INTRODUCTION

THE Black-striped Wallaby, *Macropus dorsalis*, is distributed throughout northeastern New South Wales and central eastern Queensland; forest with a dense shrub layer is the preferred habitat (Kirkpatrick 1983). Little work has been undertaken on the social organization of this species. Marlow (1962) and Calaby (1966) note that the species is gregarious, with the latter author also mentioning that individuals run off together when alarmed. Kaufmann (1974a) reports that *M. dorsalis* occurs in small, discrete groups, that there is a continuing bond between female and young, and that males tolerate young.

The types and frequencies of social acts may be useful indicators of a species' level of sociality (Russell 1984). This study investigates the social behaviour of a captive group of *M. dorsalis* with the aim of obtaining an insight into the level of sociality in *M. dorsalis*, in comparison with other species of *Macropus*.

METHODS

This study was conducted at the Royal Melbourne Zoological Gardens in June and July 1983. The *M. dorsalis* group was observed over eighteen days for a total of thirty-five hours, usually between 0900 and 1400 hours. Individuals were recognizable (e.g., by coat colour) and social behaviour of all could be monitored simultaneously; ad libitum sampling was used (Martin and Bateson 1986). The group was composed of six adult females and one adult male, precluding the study of male-male and female-young interactions.

A social act was defined as any behaviour of an individual that was directed towards, or involved, another individual. A social interaction between two individuals was defined as a sequence of social acts, and ended when individuals moved away, when no social act occurred for two minutes, or when the initiator interacted with another individual or commenced feeding. Social interactions were defined as being agonistic or sexual if they included an agonistic or sexual act respectively, and neutral if they included neither. Quantitative data on agonistic acts are presented only for female-female interactions; male-female agonistic interactions were observed on only four occasions.

RESULTS

1. Neutral acts

(a) Mutual nose-touching: the extension by two individuals of the neck and head towards each other, with nose contact being made, or at least the nares being brought very close together. Mutual nose-touching was common in female-female and male-female neutral interactions (Table 1).

(b) Nosing of the body: sniffing of another individual's body, excluding the cloacal region but including the pouch area (nosing of the pouch could not always be distinguished from general body-nosing). This act was common in all interactions (Table 1).

(c) Allogrooming: licking, and possibly nibbling with the teeth, the fur of another individual, predominately the neck and back regions. Touching the other individual with the forepaws sometimes occurred during allogrooming and was also observed between females on two occasions in the absence of allogrooming. Allogrooming was defined such that an individual could be recorded as allogrooming only once during an interaction. Allogrooming was common in male-female interactions, with the male allogrooming more than the female (116 occasions and 42 occasions respectively), and was also common in female-female interactions (Table 1).

[1]Department of Zoology, University of Melbourne, Parkville, Victoria 3052.

Pages 625-28 in KANGAROOS, WALLABIES AND RAT-KANGAROOS ed. by G. Grigg, P. Jarman and I. Hume. Surrey Beatty & Sons Pty Limited, New South Wales, Australia, 1989.

Table 1. Frequency of neutral acts in neutral, agonistic and sexual interactions between captive *M. dorsalis*. f-f=female-female and m-f=male-female interactions. Figures in parentheses refer to the number of interactions observed. Mutual nose-touching and nosing of the body can occur more than once in an interaction.

Act	f-f neutral (165)	f-f agonistic (49*)	m-f neutral (193)	m-f sexual (180)
Mutual nose touching	40	6	61	2
Nosing of body	70	23	174	115
Allogrooming	40	6	90	70

*This figure includes only interactions where contact occurred (total f-f agonistic interactions equals 103).

2. *Agonistic acts between females*

A passive displacement occurred when an individual moved toward another, causing it to move away, or when it nosed an individual, which then moved. Agonistic acts were predominantly passive (70 passive acts, 33 overt acts). The following overt acts were observed:

(a) Cuffing: an individual hitting or clawing another individual with its forepaws. This accounted for 60.6% of overt acts observed. Mutual cuffing was observed on only one occasion and was of brief duration.

(b) Chase: an individual moving after a retreating animal for a short distance. The individual initiating the chase usually started off in an upright stance with the back vertical and the head raised. This accounted for 9.1% of overt acts observed.

(c) Push: an individual pushing with the forepaws on the back region of another individual. This accounted for 6.1% of overt acts observed.

(d) Kick: an aggressor kicking out with its hindlimb. This was observed on one occasion (3.0% of overt acts).

(e) Wrestling: one or both individuals grasping each other about the neck, usually followed by both individuals standing erect and pushing at each other; cuffing or kicking by an individual sometimes occurred. Wrestling accounted for 21.2% of overt acts observed; allogrooming occurred during three wrestling acts.

The number of agonistic acts initiated by females towards other females is shown in Table 2. Wrestling acts are excluded as the initiator was not always easy to determine. A dominance hierarchy existed between females, closely correlated with age. Using the method of Appleby (1983) the hierarchy was near-linear (k=0.75) but the degree of linearity was not significant (p=0.12). The two youngest females were obviously dominated; interactions between the four oldest females were infrequent, but a trend according to age is apparent. The male dominated all the females; females were not seen to act aggressively towards the male and the male was occasionally seen to supplant females from resting areas.

Table 2. Age of female *M. dorsalis* and the frequency of agonistic acts between them.

Initiator	Age (months)	Recipient 1448	1212	1200	1686	1204	1207
1448	42	*	2	3	—	6	15
1212	29		*	2	2	9	11
1200	26			*	1	4	5
1686	20	2			*	13	24
1204	4					*	12
1207	3						*

3. *Sexual acts*

(a) Nosing of the cloaca: the male reaching in under the female's body and nosing the cloacal area; this occurred in 28.9% of male-female interactions. Nosing of a male's cloacal area by a female was observed on two occasions, and nosing of a female's cloaca by another female was observed on three occasions.

(b) Flehmen: the male placing his muzzle into a stream of female urine (usually following his nosing of the cloacal area) and, after withdrawing his head, moving his lips up and down. This behaviour was observed during four male-female interactions, and probably occurred more often than noted.

(c) Cuffing of the face and thigh: the male cuffing, lightly or quite forcibly, with his forepaws, the face and thigh (or rump) of the female. Cuffing of the face occurred in 18.7% of male-female interactions, and the female often ducked her head or raised her forepaws as if in defence. Cuffing of the thigh was observed in 14.1% of male-female interactions.

(d) Hugging: the male, usually while "standing crouched" (*sensu* Russell 1970a) in front of the female, pushing his chest up against her chest; occasionally the male also grasped the female about the neck. This act occurred in 5.6% of male-female interactions, but was observed only between the male and one female, occurring in 15% of interactions between the two.

(e) Grass-pulling: the male pawing or grasping a tussock with the forepaws. This was observed on two occasions, both times occurring after the male had nosed a female's cloaca. Grass-pulling was of brief duration.

(f) Tail-lashing: the male moving the tail along the ground sinuously. This was observed only on one occasion, occurring while the male cuffed the face of a female.

(g) Hissing. A hissing vocalization was given by the female in 11.2% of male-female interactions, occurring when the male nosed the females cloaca and pouch area, cuffed her face or thigh, hugged her, or pursued her. The only occasion on which a female was heard to vocalize in a female-female interaction was when a female hissed as she cuffed an individual that had disturbed her from resting.

Hissing by a female was also heard during capture for ear marking. The male was not heard to vocalize, but was not captured during the study.

DISCUSSION

1. Neutral Acts

Odours of the mouth and nose area are important in social interactions of the Macropodidae, with most studies describing nasonasal touching and sniffing (Russell 1985). The incidence of sniffing behaviour observed in *M. dorsalis*, of both the mouth and general body regions, indicates that olfactory stimuli are also important in this species.

Allogrooming in the Macropodidae is seen commonly between females and their young, and occasionally between adult females, and between male and female (Russell 1985). *M. dorsalis* exhibited a high rate of allogrooming (0.84 occurences per animal hour), occurring in both the female- female and male-female dyads. Possible functions of allogrooming are the passive transfer of odour from one individual to another (Russell, 1985), removal of parasites and/or maintenance of social cohesion (Sparks 1967).

2. Agonistic acts

The overt agonistic acts observed in *M. dorsalis*, such as cuffing, chasing, pushing and kicking, are similar to those of other species of *Macropus* (e.g., Ganslosser, this volume). Wrestling or fighting behaviour has been described in a number of species of *Macropus*, including the Whiptail Wallaby, *M. parryi* (Kaufmann 1974b) and the Red-necked Wallaby, *M. rufogriseus* (LaFollette 1971). Allogrooming during pauses in fights has also been described in these studies. Croft (1981) suggested that sparring encounters between males may be a form of play behaviour rather than a competitive interaction for some immediate resource, and this may apply to the wrestling behaviour of *M. dorsalis* females.

The high incidence of passive agonistic acts between female *M. dorsalis* suggests the existence of a dominance hierarchy; there is evidence of an age-related linear hierarchy among females. Similar relationships have been found in other studies (mostly captive) of Macropodidae; for example, Russell (1970b) found that the rank of females was correlated with age and weight in the Red Kangaroo, *M. rufus*. The existence of such hierarchies suggest that in the field older or larger individuals may gain improved access to limited resources, such as favoured resting positions.

3. Sexual acts

In the Macropodidae, olfactory stimuli are important in the recognition of the sexual status of females, with most studies reporting sniffing of the cloaca and pouch area of females by males (Russell 1985). Flehmen has been reported in a number of species of *Macropus* (Coulson and Croft 1981).

Cuffing of a female's face and thigh area by a male during sexual interactions has been described in *M. parma* (Maynes 1973; Herter *et al.* 1985) and in *M. parryi* (Kaufmann 1974b), with cuffing being more gentle in the latter species than in *M. dorsalis*.

Hugging behaviour of the male may achieve sternal gland marking of the female, which has been reported for a number of species of Macropodidae (Coulson, this volume). In particular, hugging in *M. dorsalis* is similar to behaviour of *M. giganteus*, where the male stands in front of the female and rubs his chest on her head, or grasps her head and forces it against his chest (Grant 1974).

Grass-pulling has been observed in a number of species of *Macropus* (Coulson, this volume). Russell (1985) suggests that this display originated in marking behaviour with the sternal gland, even if it does not now serve that function.

Tail-lashing is a component of the courtship sequence of many smaller and medium-sized macropodids (Sharman and Calaby 1964). Kaufmann (1974b) reported that tail-lashing sometimes was accompanied by an aggressive grass-pulling display in *M. parryi*; in *M. dorsalis,* tail-lashing occurred in the same interaction as a grass-pulling act, but was not vigorous.

4. Vocalizations

Macropods are not very vocal animals, and their repertoire of auditory signals is limited (Kaufmann 1974b). The vocalization given by *M. dorsalis* females may be a threat, indicating an intention to act aggressively. Kaufmann (1974b) reported hiss-growls by females in sexual interactions in *M. parryi.*

5. Conclusion

Behavioural acts of *M. dorsalis* (involving female-female and male-female interactions), of a neutral, agonistic and sexual nature are similar to those exhibited by other species of *Macropus*. The low frequency of agonistic encounters (0.44 agonistic interactions per animal hour) is comparable with that of other macropodids (e.g., 0.22

encounters per animal hour in *M. rufus,* Russell 1970b), and may be even lower in non-captive animals (Russell 1970b). The relatively high frequency of allogrooming in *M. dorsalis* along with the low level of aggression might be indicative of moderate sociality.

A captive study such as this has limitations; the small size and composition of the group, kin relationships in the group and the constraints of the captive environment will undoubtedly produce artifacts in behaviour. The general inference that *M. dorsalis* is a social species, by comparison with other species in the Macropodidae, can be confirmed only by field studies.

ACKNOWLEDGEMENTS

I would like to thank the Royal Melbourne Zoo management and staff for their cooperation with this study. I am grateful for critical comments on the manuscript by Graeme Coulson and Angus Martin.

REFERENCES

APPLEBY, M. C., 1983. The probability of linearity in hierarchies. *Anim. Behav.* **31:** 600-8.

CALABY, J. H., 1966. Mammals of the upper Richmond and Clarence rivers, New south Wales. *C.S.I.R.O. Div. Wildl. Res. Tech. Pap.* **10:** 1-55.

COULSON, G. (this volume). Repertoires of social behaviour in the Macropodoidea.

COULSON, G. M. AND CROFT, D. B., 1981. Flehmen in kangaroos. *Aust. Mammal.* **4:** 139-40.

CROFT, D. B., 1981. Social behaviour of the euro, *Macropus robustus,* in the Australian arid zone. *Aust. Wildl. Res.* **8:** 13-49.

GANSLOSSER, U. (this volume). Agonistic behaviour in macropodoids – a review.

GRANT, T. R., 1974. Observations of enclosed and free-ranging grey kangaroos *Macropus giganteus. Z. Säugetierkunde.* **39:** 65-78.

HERTER, U., SCHNEIDER, I. AND GANSLOSSER, U., 1985. Some remarks on the behavioural repertoire of the parma-wallaby, *Macropus parma* Waterhouse. *Zool. Anz.* **215:** 287-97.

KAUFMANN, J. H., 1974a. Habitat use and social organisation of nine sympatric species of macropodid marsupials. *J. Mammal.* **55:** 66-80.

KAUFMANN, J. H., 1974b. Social ethology of the whiptail wallaby, *Macropus parryi,* in northeastern New South Wales. *Anim. Behav.* **22:** 281-369.

KIRKPATRICK, T. H., 1983. Black-striped wallaby. P. 238 *in* The Australian Museum Complete Book of Australian Mammals ed. by R. Strahan. Angus & Robertson: London.

LAFOLLETTE, R. M., 1971. Agonistic behaviour and dominance in confined wallabies, *Wallabia rufogrisea frutica. Anim. Behav.* **19:** 93-101.

MARLOW, B., 1962. Marsupials of Australia. Jacaranda Press: Brisbane.

MARTIN, P. AND BATESON, P., 1986. Measuring behaviour. Cambridge University Press: Cambridge.

MAYNES, G. M., 1973. Reproduction in the parma wallaby, *Macropus parma* Waterhouse. *Aust. J. Zool.* **21:** 331-51.

RUSSELL, E. M., 1970a. Observations on the behaviour of the red kangaroo *(Megaleia rufa)* in captivity. *Z. Tierpsychol.* **27:** 385-404.

RUSSELL, E. M., 1970b. Agonistic interactions in the red kangaroo *(Megaleia rufa). J. Mammal.* **51:** 80-88.

RUSSELL, E. M., 1984. Social behaviour and social organization of marsupials. *Mammal. Rev.* **14:** 101-54.

RUSSELL, E. M., 1985. The metatherians: order Marsupialia. Pp. 45-104 *in* Social odours in Mammals ed. by R. E. Brown and D. W. MacDonald. Oxford University Press: Oxford.

SHARMAN, G. B. AND CALABY, J. H., 1964. Reproductive behaviour in the red kangaroo, *Megaleia rufa,* in captivity. *C.S.I.R.O. Wildl. Res.* **9:** 58-85.

SPARKS, J., 1967. Allogrooming in primates: a review. Pp. 190-225 *in* Primate ethology ed. by D. Morris. Weidenfield and Nicholson: London.

Observations on a captive forest wallaby *(Dorcopsis luctuosa)* colony

J. I. MENZIES[1]

Dorcopsis spp. are distinguished from other wallabies by their posture; they rest and move with their tail held in an arc so that only the tip touches the ground. The premolar and four permanent molars are functional throughout life. When feeding, *Dorcopsis* pick up food with the incisors, transfer it to the hands, then to the side of the mouth to be processed by the premolars. Total pouch life is 180-190 days, birth is followed by a post-partum oestrus and embryonic diapause, but there seems to be no seasonality in breeding, although these observations need confirmation since they are based on a small captive colony.

FOREST wallabies (*Dorcopsis* species) are now confined to the island of New Guinea and some satellite islands, although fossils are known from Australia. There are four or five species of which *muelleri* (formerly known as *veterum*), *hageni* and *luctuosa* (formerly *muelleri luctuosa)* are the most common. *Dorcopsis luctuosa* occurs in lowland forests on the south-east side of the island, *D. hageni* (which is bigger and has a white dorsal stripe), occur in lowland forests on the north side, while *D. muelleri* is found in western lowlands.

Breeding colonies of *Dorcopsis luctuosa* have been maintained successfully in various Australian and German zoos but little has been published on their biology. Uka (1980) reported on husbandry at Melbourne Zoo, and Brewer and Dickson (1987) recorded breeding at Taronga Zoo. Schurer (1978) studied the resting posture of various macropodids including *Dorcopsis*.

Dorcopsis species are distinguished from other wallabies by their posture and dentition. They rest and move with the tail held in an arc (Figs 1 and 2) so that only the tip touches the ground. The stance is less upright than in wallabies of other genera. In captivity in this museum they are crepuscular and spend the time between 0800 and 1500 hours resting in a squatting position with the tail projecting forwards between the legs and the head depressed. *Dorcopsis* never lie on their sides as wallabies of other genera may do.

The permanent premolars are long, narrow teeth forming a pair of shears more than half the length of the four permanent molars (Figs 3 and 4). The last molar erupts shortly after the premolar and all five teeth are functional throughout life. The molars are comparatively low-crowned.

Dorcopsis wallabies are browsers and much use is made of the sectorial premolars in dealing with their food. Although fed on a common mixed diet in this institution, *Dorcopsis, Thylogale* and *Macropus agilis* tackle their food in different ways. *Dorcopsis* pick up food with the incisors, transfer it to the hands and then introduce it into the side of the mouth to be bitten into pieces with the premolars. The other two wallabies use their incisors only (Figs 5 and 6).

Few observations of wild *Dorcopsis* have been made but it is probable that they are crepuscular or nocturnal, resting during the heat of the day. They seem remarkably unaggressive to one another in captivity, with only mild aggression, such as shoving with the hands and hissing at each other, over food. Vocalization includes noisy expiration of air in intra- or inter-specific aggression, clicking denoting sexual interest and a sneeze-like noise accompanying curiosity or excitement, such as the arrival of food. Alarm is denoted by stamping of the feet.

Reproductive performance of two related (mother and daughter) females is graphed in Fig. 7. Total pouch life is 180-190 days, which is comparable to other macropodids of similar size. Birth is followed by a post-partum oestrus and embryonic diapause, as was first reported by

Fig. 1. Dorcopsis luctuosa from southeastern Papua.

[1]University of Papua New Guinea, Box 320, University P.O.

Pages 629-31 in KANGAROOS, WALLABIES AND RAT-KANGAROOS ed. by G. Grigg, P. Jarman and I. Hume. Surrey Beatty & Sons Pty Limited, New South Wales, Australia, 1989.

Brewer and Dickson (1987). Females become reproductively mature at 15 months, males a little later. There seems to be no seasonality in breeding, births have occurred in February, May and August.

These observations are based on a small colony (maximum of seven individuals) over five years and need confirmation. In view of the numbers being kept in captivity elsewhere, it is hoped that those zoos keeping *Dorcopsis* colonies will publish some of their observations.

Fig. 2. Dorcopsis hageni from north coast New Guinea has a white dorsal stripe.

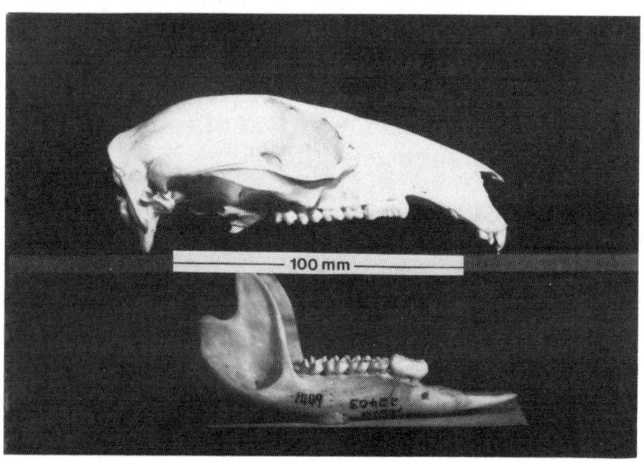

Fig. 3. Skull of *Dorcopsis luctuosa* showing upper canines, sectorial premolars and low-crowned molars.

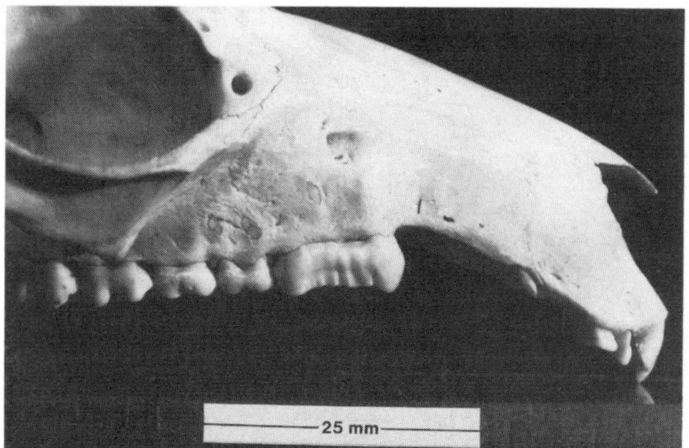

Fig. 4. Enlarged view of right upper premolar of *Dorcopsis luctuosa.*

REFERENCES

BREWER, C. AND DICKSON, J., 1987. Possible embryonic diapause in Grey Dorcopsis Wallabies, *Dorcopsis muellerii luctuosa. Thylacinus.* **12:** 1 page.

SCHURER, U., 1978. Vergleichende Untersuchung des Ruheverhaltens von Kanguruhs (Macropodidae). Unpublished dissertation of the Freien Universitat, Berlin.

UKA, D., 1980. The husbandry of the Grey Dorcopsis Wallaby *(Dorcopsis muelleri luctuosa)* at the Royal Melbourne Zoological Gardens. Unpublished report from an Australian conference of zookeepers.

Fig. 5. Macropus agilis bites hard food with the incisors.

Fig. 6. Dorcopsis luctuosa uses the premolars.

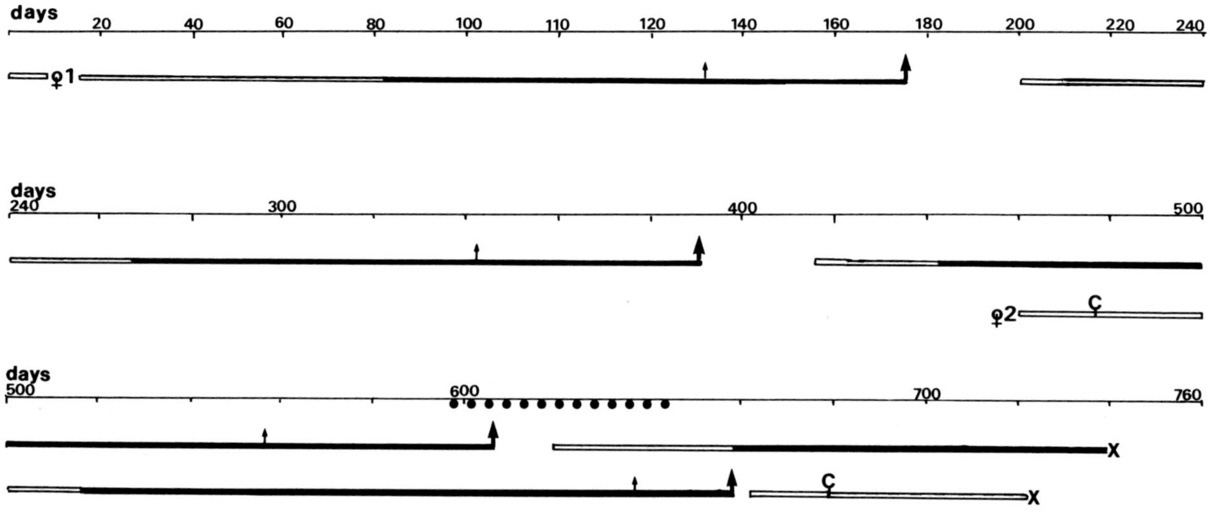

Fig. 7. Reproductive performance of two female *Dorcopsis luctuosa.* Solid bars – joey known to be in pouch; open bars, presumed time from birth; small arrows – joey first looking out; large arrows – first emergence; C – observed copulation. ●●●●●●●●● – time when no male was present in the group; X – did not complete development.

Observations on the behaviour of the Grey Dorcopsis Wallaby, *Dorcopsis luctuosa* (Marsupialia: Macropodidae), in captivity

DARREN W. BOURKE[1]

Observations were made on the diurnal behaviour of a small captive group of Grey Dorcopsis Wallabies, *Dorcopsis luctuosa*. A large proportion of time was spent resting, usually as a group, and mostly during the middle of the day. Non-agonistic interactions between individuals, such as allogrooming, occurred most often during these periods. The wallabies were most active during the morning and late afternoon and were more dispersed throughout the enclosure than when resting. The incidence of agonistic interactions was low and any agonistic encounters were brief. Although *Dorcopsis* is regarded as a primitive macropodine, the Grey Dorcopsis Wallaby shows the tendency for social grouping and exhibits some behaviours consistent with an advanced level of social organization.

INTRODUCTION

THE Grey Dorcopsis Wallaby, *Dorcopsis luctuosa,* is a medium sized wallaby with a long face, broad naked nose and small rounded ears (Fig. 1). The animal is short-furred, almost naked on the inner aspect of the limbs; the pelage is dark smokey grey with the underparts grey-based. The short heavy tail is evenly haired except for the tip which lacks hair and is heavily cornified. An adult male weighs approximately 11.6 kg, while the female is considerably smaller weighing only 3.6 kg (Archer *et al.* 1985).

No dietary studies of the Grey Dorcopsis Wallaby have been published but Sanson (1980) cited two personal communications suggesting that *Dorcopsis* is a browser, selecting softer vegetation, flowers and fruits. The Grey Dorcopsis Wallaby is primarily an animal of dense lowland forests, but may also inhabit secondary-growth bush and man-made grasslands (Ziegler 1977). It occurs in southeastern and eastern New Guinea at altitudes less than 400 m (George 1984).

The genus *Dorcopsis* has been the subject of much taxonomic confusion. However, in a recent review, Groves and Flannery (this volume) confirmed the use of *D. luctuosa* as the specific name for the Grey Dorcopsis Wallaby.

The present study of a small captive group of Grey Dorcopsis Wallabies at the Royal Melbourne Zoological Gardens is aimed at providing a preliminary assessment of aspects of this species' behaviour. While a zoo environment is unnatural in many ways, macropod behaviour seems not to be seriously affected by captivity, and a captive study may make a significant contribution to research into macropod behaviour (Coulson, this volume). This study provides the basis for further investigation into the habits and ecology of the New Guinea forest wallabies (*Dorcopsis* and *Dorcopsulus* spp.), of which little is known.

MATERIALS AND METHODS

The composition of the wallaby group during the study is shown in Table 1. A display enclosure, in the macropod exhibit at the Royal Melbourne Zoological Gardens, housed the wallabies during the study. The enclosure was approximately 200 m^2 in area. It contained two sheds, a small water trough, and was heavily vegetated with tree-ferns, scattered clumps of tussocks and several large trees. Both sheds were lined with hay and contained a feeding tray for "macropod pellets" and carrots. Other supplementary food included lucerne and soft dicotyledonous vegetation.

Observations totalling 60 hours were made between April and September, 1987. Initially, a 12-hour sampling programme was carried out to determine the time budgeting of certain behaviours during daylight hours. This consisted of instantaneous sampling periods at five minute intervals, during which each individual's behaviour at that time was recorded.

Behaviour was divided into five categories:

1. Feeding — When animals ate grass, leaves and supplementary foods.
2. Moving — When animals either walked or hopped.
3. Stationary — When animals either sat or stood (i.e., did not move).
4. Grooming — When animals cleaned themselves.

[1]Department of Zoology, University of Melbourne, Parkville, Victoria 3052.

Pages 633-40 in KANGAROOS, WALLABIES AND RAT-KANGAROOS ed. by G. Grigg, P. Jarman and I. Hume. Surrey Beatty & Sons Pty Limited, New South Wales, Australia, 1989.

Fig. 1. A male Grey Dorcopsis Wallaby. Note the long narrow face, broad naked nose and short rounded ears. (Photo: P. Dilorenzo).

5. Social Behaviour — When two or more animals interacted, including sexual, agonistic and non-agonistic (e.g., allogrooming) interactions.

All observations were made from outside the enclosure during daylight hours, with the aid of 7 ×50 binoculars when required. Individuals were recognized by ear markings, representing their zoo registration number, and by size and coat colour. In this paper, individuals are identified by one letter names (Table 1). The duration of observation periods ranged from 30 minutes to 180 minutes with the majority being about 90 minutes. They were made at various times of the day from 1000 to 1600 hours. Cold and wet

Table 1. Composition of *D. luctuosa* group. Data drawn from zoo records. All animals originated in the zoo.

Zoo Number /(Name)	Sex	Date of Leaving Pouch	Offspring	Sire	Dam	Removal From Group
1272/4 (R)	male	23.v.82	267/3 male 215/11 male 2149/12 female	105/-	1255/-	
102/- (M)	female	30.xi.81	215/11 male 106/19 female 2149/21 female	105/-	481/y	
268/24 (G)	female	25.vii.79	127/- male 267 male 214/12 female	386/10	25/18	22.iv.87 (died)
121/18 (C)	female	26.vi.86	—	267/3	214/12	
106/19 (K)	female	16.xi.86	—	267/3	102/-	
2149/21 (Y)	female	24.vi.87	—	1272/4	102/-	16.vii.87 (died)

weather often caused the animals to retreat inside the sheds, which made observation impossible.

RESULTS

Individual behaviour

DAILY ACTIVITY PATTERN

The percentage of time spent by each individual in six categories of behaviour during daylight hours is shown in Fig. 2. A large proportion of the wallabies' time was spent resting, usually as a group, and such periods of inactivity occurred mostly during the middle of the day. These periods of rest were often interrupted by periods of grooming, which generally occupied a small proportion of the day's activity, except for female M. This female had a pouch young during sampling and spent more time grooming her pouch than other females. While resting or grooming, the wallabies usually adopted a sitting posture with the tail thrust forward between the hind legs and the back against a fence or tree.

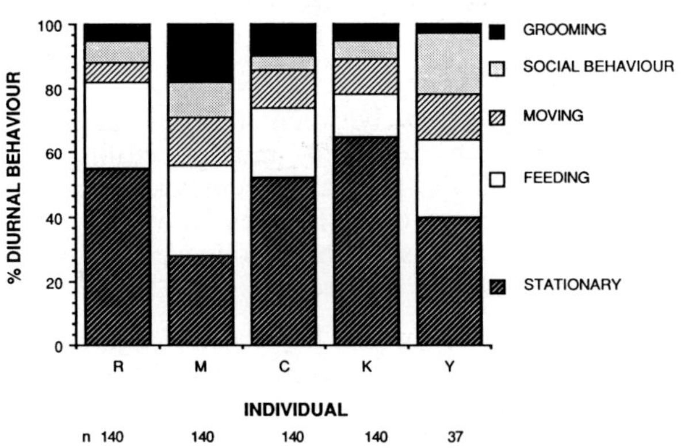

Fig. 2. Diurnal activity budget of each individual. (n=number of sampling periods).

For most individuals, feeding occupied a large proportion of the day's activities. The wallabies showed a strong preference for leaves and occasionally small twigs or bark, but some areas of grass were also grazed. The Grey Dorcopsis Wallaby has a large P3 tooth with a long sectorial cutting edge, which was used effectively for feeding on leaves and grass. Most movement was associated with feeding and occurred during the mornings and late afternoon, when most members of the group fed. The hopping gait of the Grey Dorcopsis Wallaby is similar to that of other wallabies, though the body is usually kept more horizontal to the ground. In the enclosure, rapid locomotion was rare unless the animals were disturbed.

All animals usually showed a preference for resting or grooming in an area of direct sunlight. However, the wallabies' movement and activities were greatly reduced during periods of cold or rainy weather, when they retreated to the sheds and remained there for periods lasting several hours. If the sheds were closed, the animals would find protection in the enclosure under tussocks or against the side of a fence.

BODY POSTURE

When stationary, the wallabies adopted a distinct standing crouched posture, where the short tail was held off the ground with the proximal portion being bent upward and the distal portion being bent downwards, so that only the tip was in contact with the ground (Fig. 3). This posture is typical of the genus *Dorcopsis* (Archer *et al.* 1985). This posture was adopted when the animals were feeding, moving slowly or during short rest periods. However, when sitting, the tail was not held off the ground but lay flat along the ground, behind the animal.

Fig. 3. Standing crouched: the short tail is held off the ground so that only the tip of the tail is in contact with the ground.

Social behaviour

In general, with the exception of female M and her young (Y), social behaviour occupied a small proportion of the diurnal time budget (Fig. 2). Social behaviour was classified into four broad categories of social interaction — spatial relationships, agonistic, non-agonistic and sexual interactions.

SPATIAL RELATIONSHIPS

It was very rare for all individuals within the group to be by themselves (i.e., not within five metres of another animal) for long periods. Close proximity between individuals occurred mostly during the middle of the day, with the wallabies grooming and resting together for long periods. Schappert (1984) also observed that during the day, captive Grey Dorcopsis Wallabies rested in groups. However, in the morning and late afternoons, when activity was at its greatest, individuals usually spaced themselves throughout the enclosure.

NON-AGONISTIC INTERACTIONS

Non-agonistic social behaviour includes social behaviour not directly related to aggression or reproduction. Non-agonistic behaviour is likely to promote cohesion of a social group by encouraging the close proximity of interactees (Croft 1981a). Non-agonistic interactions performed within the context of sexual behaviour are discussed under Sexual Behaviour.

(i) Sniffing and Touching

This behaviour involved only one of the interactees sniffing the other's head or body. This behaviour was rare between females and involved one female briefly sniffing the head and neck region of another female. Mutual nose touching and sniffing, a behaviour common among macropodids (Russell 1985), was observed on only a few occasions, between a male and a female.

(ii) Allogrooming

Allogrooming is grooming of one individual by another. Allogrooming involved both male-female and female-female interactions, and was usually unidirectional (Table 2). Allogrooming was frequent in the mother-young dyads (M/K, M/Y). In the M/Y dyad, allogrooming was almost always associated with the period when the young

Table 2. Frequency of allogrooming interactions between individuals.

		RECIPIENT				
		R	M	C	K	Y[A]
	R	—	2	4	5	0
	M	1	—	5	14	9
INITIATOR	C	2	3	—	2	0
	K	1	12	1	—	0
	Y[A]	0	0	0	0	—

[A]Observed for only five hours.

(Y) was suckling or just after suckling. Female K, the previous young of female M, frequently associated with her mother, and mutual allogrooming between the two individuals was observed on several occasions. Most allogrooming between adults was brief (less than one minute in duration) and usually occurred between animals sitting close together (Fig. 4).

AGONISTIC INTERACTIONS

The frequency of agonistic interactions between individuals is shown in Table 3. Overall, the level of aggression between individuals was low.

(i) Male-Male

Only one male was present in the enclosure during the study and so no observations were able to be made on male-male aggressive interactions. Surplus males are held in bachelor groups to eliminate fighting in the breeding group (D.Uka, pers. comm.).

(ii) Female-Female

Most aggression between females included fighting, which involved mutual cuffing or wrestling with the forepaws. Females were also observed to chase other females for varying distances. Such agonistic altercations were of short duration (less than 30 seconds).

Although there was a low level of aggression between females, there appeared to be a dominance hierarchy among the three females, which may be related to size and age. A comparison of size and age of the three females can be seen in

Fig. 4. Allogrooming: a female licks the back of another female. From the left are females K, M and C. Female M was the largest and oldest of the females, while female C was five months older than female K and slightly larger. (Photo: Author).

Figure 4. Female M was the dominant female of the group, while female C was usually dominant over female K (Table 3).

(iii) Male-Female

Aggressive behaviour between male and females was usually associated with sexual interactions. Females were observed to occasionally cuff, poke with their heads or attempt to bite the head region of the male when he attempted to sniff the female's pouch or cloacal region. The male usually retreated after such an interaction. On several occasions when the male was holding a female around the neck, she retaliated by poking her head into his abdominal region, causing him to let go. Ganslosser and Holz (1984) observed that in male-female fighting, female Grey Dorcopsis Wallabies preferred to bite, while the males preferred to hit. Further elements of male-female aggression are described under Sexual Behaviour.

Table 3. Frequency of agonistic interactions between individuals.

		RECIPIENT				
		R	M	C	K	Y[A]
INITIATOR	R	—	5	2	3	1
	M	1	—	3	9	0
	C	10	1	—	4	6
	K	4	1	0	—	0
	Y[A]	0	0	0	0	—

[A]Observed for only five hours.

Sexual behaviour

The most common male-female sexual interaction in the Grey Dorcopsis Wallaby was sexual "checking" of a female by a male. The male approached a female from the front or side and closely sniffed the cloacal region and less frequently the pouch opening, while occasionally placing his forepaws on the female's neck or body. The male was also observed to sniff fresh urine and faeces of females after sniffing their cloacal regions. Flehmen, the lateral and upward retraction of the upper lips exposing the incisors, was not observed during sexual checking. This phenomenon is regarded as a mechanism to aspirate the vomeronasal organ and facilitate the detection of the oestrus state of the female. Failure to observe this behaviour, if it occurs, may be due to its fleeting and unspectactular nature in macropodines (Coulson and Croft 1981).

A soft clucking vocalization by the male accompanied nearly all sexual interactions. The female generally terminated sexual interactions by moving away, although a female sometimes aggressively rejected a sexual approach by poking or cuffing around the head region of the male. The male also displayed aggression during some sexual interactions and was observed on several occasions to cuff the female around the head and neck while trying to sniff her cloacal region. Most non-agonistic interactions observed in a sexual context were performed by the male and included nosing the female's body, licking the female's face and mouth, and allogrooming.

Copulation was observed only once; between male R and female M on June 22. Initially, the male followed the female around the enclosure and into the sheds for approximately 20 minutes. During this time, the male vocalized with soft clucking while trying to sniff the cloacal region of the female. The penis of the male was sometimes erect during these encounters.

The male then approached the sitting female, sniffed her cloacal region and gripped her flanks, causing the female to adopt a standing crouched posture. He then mounted the female and copulation occurred. The copulation sequence consisted of eight intromissions over 30 minutes. These intromissions, which included short periods of thrusting by the male, averaged two minutes but ranged from 40 seconds to 3 minutes. All mountings were terminated by the female suddenly moving away. Both the male and female remained adjacent and groomed themselves between intromissions.

Display behaviour

On three occasions, male R was observed to perform a display which involved standing erect against a tree and rubbing his chest briefly up and down against it. He then clawed at the ground at the base of the tree, while thumping his hind legs up and down. The same tree was used on each occasion and the total display lasted less than one minute in each case. On two occasions, less than 20 minutes after the display, the male then approached a large clump of grass and vigorously bit and pulled at the grass with the forelimbs, while thumping the ground with his hind feet. No response by the females to the displaying male was observed and no individual was observed to sniff at the tree or grass used in the display.

On several occasions, while walking around the enclosure, male R was observed to rub his cloaca on the ground several times by squatting on his hind feet and moving his hindquarters from side to side. This was interpreted as marking behaviour. On one such occasion, the male marked in three of the corners of the enclosure.

Mother-Young interactions

The pouch life of one young was estimated from zoo records to be approximately 34 weeks. Post-partum oestrus and embryonic diapause have been reported in the Grey Dorcopsis Wallaby (Brewer and Dickson 1987). Interactions between female M and her two offspring, K and

Y, were observed during the study. However, the young Y died three weeks after leaving the pouch and only five hours of observations were made of it's behaviour. When the young was in the pouch all the time, the mother (M) was observed to frequently clean the pouch and young. One week after first leaving the pouch, the young was observed to have a large patch of fur missing from the back of the neck and one ear was also chewed. These wounds were probably caused by the other individuals in the group who were frequently seen to aggressively chase away the young if it ventured too close. On several occasions, female C was observed to bite the young on the neck when it ventured too close. Schappert (1987) also observed that some female Grey Dorcopsis Wallabies behaved aggressively towards juveniles in captivity.

During the three weeks after first leaving the pouch, the young was not observed to re-enter the pouch but spent all of its time out of the pouch. The young spent much of its time out of the pouch away from its mother. Suckling was observed only once and lasted approximately 15 minutes. While suckling, the mother (M) groomed the back and neck of the young. Allo-grooming between mother and young was frequent, while mutual grooming between mother and young-at-foot (K) was also observed. Sparring between mother and young-at-foot was observed several times and consisted of brief wrestling while up on the hind legs, for up to one minute; but most interactions were less than 20 seconds in duration.

DISCUSSION

General behaviour

In captivity, the Grey Dorcopsis Wallaby is crepuscular and its diurnal activities appear to be similar to those observed in the Red-bellied Pademelon, *Thylogale billardieri* (Morton and Burton 1973). In the wild, Grey Dorcopsis Wallabies are presumably nocturnal, although there is some evidence that they do move about during the day, in dense forests (Nowak and Paradiso 1983).

The low level of active hopping displayed by the wallabies may be an adaptation to a closed habitat, or it may be the result of the small, heavily vegetated enclosure. Grzimek (1972) suggested that these wallabies do not appear as adapted for hopping as most other wallabies. However, despite short hind limbs and feet, when disturbed the wallabies were able to move briskly, easily hopping over logs in the enclosure.

Holding the short tail off the ground when the animals are standing crouched is a unique feature amongst the macropodids and is only found in the New Guinea forest wallabies (*Dorcopsis* and *Dorcopsulus* spp.). The reasons for this unusual habit and morphology are unknown.

While the opportunity for feeding on the grass is limited in the undisturbed forests in which the Grey Dorcopsis Wallaby lives, in captivity they were observed to graze on grass but preferred softer vegetation, such as leaves. This is in contrast to the genus *Macropus,* which comprises predominantly grazers. Sanson (1980) suggested that the masticatory features of *Dorcopsis* are correlated with a diet of low-fibre unabrasive vegetation.

Social organization

The importance of olfactory information in the social behaviour of macropods is generally not recognized (Russell 1984). Olfaction in the Grey Dorcopsis Wallaby was used primarily for sexual checking, which is presumably important in the recognition of the sexual state of females. The male was observed to sniff both the pouch and perineal region of all females almost daily, and this behaviour has been reported in many species of macropods (Russell 1985).

The display performed by the male appears to be both an olfactory and visual form of communication. Since Grey Dorcopsis Wallabies possess sternal and chin scent glands (Schaffer 1940), the display may serve to spread glandular secretions of the chin and sternum, which may be used in olfactory communication. The thumping of the ground may provide a visual reinforcement of the marking behaviour.

Specialized visual displays do occur in other macropodids such as the Red Kangaroo, *Macropus rufus* (Sharman and Calaby 1964), and the Whiptail Wallaby, *Macropus parryi* (Kaufmann 1974a). Croft (1981b) described a "bush display" in the Euro, *Macropus robustus erubescens,* in which the male hugged a low shrub or tussock to his chest and wiped his neck back and forth across it. This display was performed by males in the company of an oestrous female.

Marking the ground by cloacal dragging may release chemical secretions to mark the boundary of a territory. Although many of the Macropodidae are known to possess paracloacal glands, active cloacal marking does not appear to be widespread (Russell 1985). The only definite report of active cloacal marking is in the Rufous Bettong, *Aepyprymnus rufescens,* which has been observed to mark the substrate and introduced objects, in an enclosure (Johnson 1980).

The Grey Dorcopsis Wallaby is not a very vocal animal and its repertoire of auditory signals is limited. The main vocalizations emitted were the soft clucking by the male in sexual interactions and these vocalizations have been described, in a similar context, in many macropodids including the Red Kangaroo (Russell 1970a) and the Red-bellied Pademelon (Morton and Burton 1973).

In the group of Grey Dorcopsis Wallabies, the incidence of agonistic responses was low and agonistic encounters were brief. Ganslosser and Schappert (1985) found that the play behaviour of the Grey Dorcopsis Wallaby is marked by the absence of very aggressive behavioural patterns. Low levels of aggression have also been described for captive groups of Burrowing Bettongs, *Bettongia lesueur* (Stodart 1966), Quokkas, *Setonix brachyurus* (Packer 1969) and Red Kangaroos (Russell 1970b). In all these animals, some form of dominance between females was observed, as well as in Red-bellied Pademelons (Morton and Burton 1973), which display a high level of aggression between females. The dominance hierarchy observed between the three females in the study group may be a mechanism that reduces intraspecific aggression in the wild. This may suggest a tendency for social grouping in the Grey Dorcopsis Wallaby as group cohesion would be enhanced by low levels of intraspecific aggression.

Allogrooming between adult Grey Dorcopsis Wallabies appears to have more significance as a social interaction than the functional, co-operative removal of ectoparasites. Thus allogrooming may serve to maintain group cohesion. Similarly, Ganslosser (1979) considered allogrooming between adult Doria's Tree-kangaroos, *Dendrolagus dorianus,* to serve a bonding function within the group. Croft (1982) considered allogrooming between male Antilopine Wallaroos, *Macropus antilopinus,* to promote and maintain close social contact.

Ecological factors may be important determinants of a species' social organization. Kaufmann (1974b) has identified a number of factors that may facilitate group formation. In comparison of 23 species of macropodids, Kaufmann found that social organization is more closely associated with diurnal tendencies, and that most social species lived in open forests and plains and were exclusively grazers under normal conditions; while the smallest macropodids were generally solitary. These ecological restrictions on social organization suggest that *D. luctuosa* may be a predominantly solitary species because of its closed habitat, browsing mode of feeding and predominantly nocturnal activities. However, it seems that the Grey Dorcopsis Wallaby, despite potential ecological restrictions, shows the tendency to aggregate in small groups in captivity. Schappert (1984) also concluded that captive Grey Dorcopsis Wallabies have a "moderate group life".

The ancestors of the Macropodidae were small omnivorous to herbivorous animals of closed habitat and probably of nocturnal habits (Kaufmann 1974b). The principle evolutionary trends in the Macropodinae have been to an increased size and a corresponding greater mobility, a herbivorous diet with a late specialization to grazing, more crepuscular to diurnal activity patterns and increasing utilization of open habitats (Bartholomai 1978). In terms of habits and ecology, the above descriptions indicate that *D. luctuosa* may represent a primitive grade of macropodid.

Flannery (1984) suggested that the species of *Dorcopsis* and, the more primitive species of *Dendrolagus*, show the primitive macropodine dental morphology and that these species may represent the slightly modified basal stock from which other macropodines evolved. Therefore, in phylogenetic terms *D. luctuosa* is regarded as a primitive macropodine. However, *D. luctuosa* shows some behavioural characteristics that occur in more evolutionarily advanced species such as those in the genus *Macropus*.

Further study, in the field, needs to be undertaken to provide more information about the behaviour and ecology of Grey Dorcopsis Wallaby. Information obtained on *Dorcopsis* may prove significant in the comparative study of macropod social organization.

ACKNOWLEDGEMENTS

I am most grateful to all those who helped in various ways during the study: The Zoological Board of Victoria for giving permission for this study at the Royal Melbourne Zoological Gardens; many members of the zoo staff, particularly Bev Drake; Dr A. A. Martin for suggesting and supervising the study; Mr G. M. Coulson who provided invaluable assistance during the study and in the preparation of this paper; Dr M. F. Clarke for his generous assistance during the study and in the preparation of this paper, Mr J. I. Menzies for discussions on the captive Grey Dorcopsis Wallabies in New Guinea and S. Atzinger for the translation of German articles. Mr G. M. Coulson, Dr M. F. Clarke, Dr A. A. Martin and Mr J. I. Menzies provided useful criticisms of earlier drafts of this manuscript.

REFERENCES

ARCHER, M., FLANNERY, T. F. AND GRIGG, G. C., 1985. The Kangaroo. Weldons: McMahons Point.

BARTHOLOMAI, A., 1978. The fossil kangaroos. *Aust. Mammal.* **2**: 15-22.

BREWER, C. AND DICKSON, J., 1987. Possible embryonic diapause in Grey Dorcopsis Wallabies, *Dorcopsis mullerii luctuosa. Thylacinus* **12**: 6.

COULSON, G. M. AND CROFT, D. B., 1981. Flehmen in kangaroos. *Aust. Mammal.* **4**: 139-40.

COULSON, G. M. (this volume). Repertoires of social behaviour in the Macropodoidea.

CROFT, D. B., 1981a. Behaviour of Red Kangaroos, *Macropus rufus* (Desmarest, 1822), in north-western New South Wales, Australia. *Aust. Mammal.* **4**: 5-58.

CROFT, D. B., 1981b. Social behaviour of the Euro, *Macropus robustus* (Gould), in the Australian arid zone. *Aust. Wildl. Res.* **8**: 13-49.

CROFT, D. B., 1982. Some observations on the behaviour of the Antilopine Wallaroo, *Macropus antilopinus* (Marsupialia: Macropodidae). *Aust. Mammal.* **5**: 5-13.

FLANNERY, T., 1984. Kangaroos: 15 million years of Australian bounders. Pp. 817-35 *in* Veterbrate Zoogeography and Evolution in Australasia ed. by M. Archer and G. Clayton. Hesperian Press: Carlisle.

GANSLOSSER, U., 1979. Soziale Kommunikation Gruppenleben, Spiel-und Jugendverhalten des Doria Baumkänguruhs (*Dendrolagus dorianus* Ramsay 1833). *Z. Saugetierk.* **44**: 137-53.

GANSLOSSER, U. AND HOLTZ, C., 1984. Zum Kampfverhalten verschiedener sozialer Kategorien beim Streifenbusckänguruh *(Dorcopsis muelleri). Verh. Dtsch. Zool. Ges.* **77**: 281.

GANSLOSSER, U. AND SCHAPPERT, I., 1985. Zum Sozialspiel bei Doria-Baumkänguruhs *(Dendrolagus dorianus)* und Streifenbuschkänguruhs *(Dorcopsis muelleri)* (Macropodidae). *Verh. Dtch. Zool. Ges.* **78**: 206.

GEORGE, G. G., 1984. The present day land mammal fauna of Papua New Guinea. Pp. 1137-41 *in* Vertebrate Zoogeography and Evolution in Australasia ed. by M. Archer and G. Clayton. Hesperian Press: Carlisle.

GROVES, C. P. AND FLANNERY, T. (this volume). Revision of the genus *Dorcopsis* (Macropodidae).

GRZIMEK, B., 1972. Grzimek's Animal Life Encyclopeadia, Vol. 10. Van Nostrand Reinhold: New York.

JOHNSON, P. M., 1980. Observations on the behaviour of the rufous rat kangaroo, *Aepyprymnus rufescens* (Gray), in captivity. *Aust. Wildl. Res.* **7**: 347-57.

KAUFMANN, J. H., 1974a. Social ethology of the whiptail wallaby, *Macropus parryi,* in northeastern New South Wales. *Anim. Behav.* **22**: 281-369.

KAUFMANN, J. H., 1974b. The ecology and evolution of the social organization in the kangaroo family (Macropodidae). *Am. Zool.* **14**: 51-62.

MORTON, S. R. AND BURTON, T. C., 1973. Observations on the behaviour of the macropodid marsupial *Thylogale billardieri* (Desmarest) in captivity. *Aust. Zool.* **18**: 1-14.

NOWAK, R. M. AND PARADISO, J. L., 1983. Walker's Mammals of the World. 4th edition, Volume 1. John Hopkins Press: Baltimore.

PACKER, W. C., 1969. Observations on the behaviour of the marsupial, *Setonix brachyurus* (Quoy and Gaimard) in an enclosure. *J. Mammal.* **50**: 8-20.

RUSSELL, E. M., 1970a. Observations on the behaviour of the red kangaroo, *Megaleia rufa,* in captivity. *Z. f. Tierpsychol.* **27**: 385-404.

RUSSELL, E. M., 1970b. Agonistic interactions in the red kangaroo, *Megaleia rufa. J. Mammal.* **51**: 80-88.

RUSSELL, E. M., 1984. Social behaviour and social organization of marsupials. *Mammal. Rev.* **14**: 101-54.

RUSSELL, E. M., 1985. The Metatherians: order Marsupialia. Pp. 45-104 *in* Social Odours in Mammals, Vol. 1 ed. by R. E. Brown and D. A. Macdonald. Oxford University Press.

SANSON, G. D., 1980. The morphology and occlusion of the molariform cheek teeth in some Macropodinae (Marsupialia: Macropodidae). *Aust. J. Zool.* **28**: 341-65.

SCHAFFER, J., 1940. Die Hautdrusenorgane der Saugetiere. Urban and Schwarzenberg: Berlin.

SCHAPPERT, I., 1984. Verhaltensbeobachtungen beim Buschkänguruh, *Dorcopsis muelleri,* in Zoologischen Garten. *Vivarium Darmstadt Informationen.* **1**: 2-12.

SHARMAN, G. B. AND CALABY, J. H., 1964. Reproductive behaviour in the red kangaroo, *Megaleia rufa,* in captivity. *C.S.I.R.O. Wildl. Res.* **9**: 58-85.

STODDART, D. M., 1966. Observations on the behaviour of the marsupial, *Bettongia lesueuri* (Quoy and Gaimard) in an enclosure. *C.S.I.R.O. Wildl. Res.* **11**: 91-102.

ZIEGLER, A. C., 1977. Evolution of New Guinea's Marsupial fauna in response to a forested environment. Pp. 123-78 *in* The biology of marsupials ed. by B. Stonehouse and D. Gilmore. Macmillian: London.

PLATE III

A: A Brush-tailed Rock Wallaby *(Petrogale penicillata)* in the New England area of New South Wales. This species is found on rocky outcrops in sclerophyll forest from Central Queensland to Victoria. It is common except in Victoria where its range is greatly reduced. Photo by K. P. Joblin.

B: A Whiptail Wallaby *(Macropus parryi)* from northern New South Wales. These are common in eastern Queensland and northeastern New South Wales. They may be a pest to agriculture and some are taken commercially in Queensland for the hide. This individual is using its paw to assist with feeding. Photo by P. J. Jarman.

C: A Swamp Wallaby *(Wallabia bicolor)* in the New England area of New South Wales. This species is common in thick undergrowth in forest, woodland or heath, in the eastern parts of Queensland, New South Wales and Victoria. Photo by K. P. Joblin.

D: A Red-necked Wallaby *(Macropus rufogriseus)* in the New England area of New South Wales. This species is a common large wallaby found in forests of southeastern Australia, including Tasmania. Photo by K. P. Joblin.

E and F: (female and male, respectively). The Eastern sub-species *(Macropus robustus robustus)* of the Common Wallaroo. It is usually found associated with steep escarpments. Photo by K. P. Joblin.

A: The most widespread sub-species of the Common Wallaroo *(Macropus robustus erubescens)* is usually known as the Euro. This pair (female at right, with radio collar) are in typical habitat, at Fowler's Gap in northwestern New South Wales. Photo by Tim Clancy.

PLATE IV

B: A large male Red Kangaroo *(Macropus rufus)*. Males may grow to more than 80 kg and are the largest extant marsupials. This individual is in poor condition during dry times. Drought is a very common cause of mortality in Red Kangaroos and populations may decline markedly. Photo by P. J. Jarman.

C: Red Kangaroos *(Macropus rufus)* drinking at a man-made dam. The advent of pastoralism, with its provision of watering points, has led to increased numbers of Red Kangaroos and both species of Grey Kangaroos, which has in turn led to them being recognized as a pest to graziers. Making superb leather, having excellent meat and being less injurous than sheep to pasture, it is surprising that more attention has not been paid to them for their potential economic benefit to graziers. Photo by P. J. Jarman.

D: Distributed in the tropical grasslands across the northern third of Australia, the Spectacled Hare Wallaby *(Lagorchestes conspicillatus)* is the only one of five species of Hare Wallaby which is anywhere common. It is widespread and common in parts of Queensland. The other species are either rare or extinct, thought to be the consequence of changed land management practices associated with grazing. Photo by Gordon Grigg.

E: A cannon net is set at a water trough to catch Euros *(Macropus robustus erubescens)*, in a study at Fowler's Gap, northwestern New South Wales. Photo by Tim Clancy.

F: Releasing a Euro *(Macropus robustus erubescens)* at Fowler's Gap, New South Wales, after a radio collar has been fitted. Photo by J. Kealey.

uncommon. Burbidge and McKenzie (1989) concluded that steep rocky slopes and rock piles provided a buffer against environmental stress because these habitats:

- are difficult for most exotic herbivores to penetrate.
- have a patchy distribution of fuel which breaks up fires,
- provide effective shelter from most cursorial predators, and;
- provide energy cheap, thermally buffered shelters.

Vegetation near to rock surfaces also receives shading, seepage and runoff, during storms which encourages greater plant species diversity, extended periods of growth, and occasional flushes during drought.

Fossorial behaviour *per se* is no protection against decline or extinction since the only truly fossorial macropod, *Bettongia lesueur,* disappeared rapidly from an immense area of continental Australia during the first half of this century.

Overkill

There is considerable debate about the impact and ethics of harvesting of macropods but the species which are currently being harvested are widespread and abundant (Burbidge 1977).

Jarman and Johnson (1977) documented the considerable numbers of large, medium and small macropods harvested in New South Wales for the bonus payable on their scalps during the period 1881 to 1920. In the Tamworth district almost 261 000 "kangaroo" (probably *M. giganteus*) returns were made in 1884. 243 814 "wallabies" in 1897, and 45 755 "rat kangaroos" (*Aepyprymnus* or *Bettongia*) in 1895. Despite these substantial harvests there was no evidence that overkill was involved in the change in species' abundance.

Robinson and Young (1983) considered the effects of hunting on *Macropus greyi* to be difficult to determine but noted that the hunting of *M. fuliginosus* for scalp bonuses most probably had a severe effect. *M. fuliginosus* apparently increased in numbers within the range of *M. greyi* in much the same way as *M. giganteus* did in the Tamworth district. The "wholesale slaughter" was officially stopped in 1892 when the South Australian Government closed the season for kangaroos, but Robinson and Young (1983) note that the smaller wallabies were still regarded as fair game in the district.

Burbidge and McKenzie (1989) reviewed the evidence for overkill of species by Europeans as a primary cause of extinction but concluded that no land mammals had been so affected. Reduction in numbers or fragmentation of a species range could nevertheless make that species more susceptible to other forces.

Habitat destruction

Land clearing, introduced grazing and browsing animals, drainage, erosion and changed fire regimes can cause significant changes in the botanical composition and structure of vegetation which in turn may cause profound changes in the status of native fauna.

(a) LAND CLEARING

Much of the Bassian region of southern Australia and the Torresian region of the east coast has been cleared of native vegetation to enable cropping and/or intensive grazing. Where this process is complete and no sizeable relict patches of native vegetation remain the macropods have vanished. Whether land clearing was the primary cause of extinction or whether some other additional factor was responsible depends on regional history.

Potorous platyops inhabited the wheatbelt region of southwestern Western Australia in historical times. This area has been 65% cleared for cereal production and much of the uncleared land is at the driest periphery of the district (Burbidge and McKenzie 1989). *P. platyops* is now extinct but it was last recorded in 1875, well before the major push in land clearing commenced.

Bettongia penicillata has vanished from the agricultural and intensive grazing areas of southern Australia and survives in timbered areas in southwestern Western Australia. However, it has also vanished from the Eyrean region of central Australia (Burbidge *et al.* 1988) where there has been no land clearing.

Macropus greyi appears to have vanished from the south-east of South Australia and the southwest of Victoria where there has been extensive modification to the landscape. Swamps formed a significant portion of the habitat once occupied by the species but most of them have now been drained in a series of schemes that commenced as early as 1862 (Robinson and Young 1983). These authors conclude that destruction of habitat, including clearing of vegetation and drainage, was the major cause of the demise of the species.

Burbidge and McKenzie (1989) note that *S. brachyurus* inhabits swampy vegetation on the mainland, but although it has been eliminated from drained or altered (e.g., salinized) areas it has not become extinct.

No extinct species other than *P. platyops* and *M. greyi* had a range restricted to areas now extensively cleared or drained and with the probable exception of *M. greyi,* it is plain that land clearing was not a primary cause of the historical declines and extinctions of Australian macropods. It is important, however, in the continuing change in status of some species that are now reduced to remnant populations (e.g., *M. eugenii)*.

(b) FIRE

The propensity for Aborigines to burn the country was the subject of comment by some of the earliest Europeans to visit Australia (see Nicholson 1981 for a review) and continued to be recorded wherever explorers visited new country (e.g., Giles 1889, p83). Some, such as Mitchell (1848, p412) recognized the biological impact of fire on the environment. He was drawn to comment "Fire, grass, kangaroos and human inhabitants, seem all dependent on each other for existence in Australia; for any one of these being wanting, the others could no longer continue". He (Mitchell 1848, p143) also noted that "The omission of the annual periodical burning by natives, of the grass and young saplings, has already produced in the open forest lands nearest to Sydney, thick forests of young trees, where, formerly a man might gallop without impediment, and see whole miles before him". Howitt (1890) observed that the river valleys of eastern Victoria were originally ". . . clothed with grass and a few large scattered trees . . ." maintained by the . . . . . annual bush fires of the Aborigines". He reported that the first settlers discouraged burning and soon young saplings appeared and after some years of occupation whole tracts of land became covered with forest. Some settlers in the Cypress Pine country of New South Wales abandoned their properties when lack of fire enabled dense regrowth to become established (Jeans 1972). Rolls (1981, p402) concluded that the abundance of trees had increased since European settlement despite the ringbarking and elimination of rainforest.

The change in the tree layer was a manifestation of changes that were also taking place below among the grasses and herbs. Fire facilitates the germination and growth of many understorey species in forests (Ashton 1981), woodland (Fox and Fox 1986), heathlands (Specht 1981) and hummock grasslands (Burbidge 1944). Many of these species contribute to the diet (Bolton and Latz 1978; Christensen 1980) and provide shelter (Christensen *et al.* 1981) for macropods.

In much of the non-pastoral Eyrean region the movement of Aborigines to settlements, missions and pastoral stations has occurred only in the last 25-60 years and it is possible to crudely reconstruct the pre-white Australian situation. Displacement of Aborigines by Europeans has in most places resulted in a change in the fire regime from one where fires occurred throughout the year to one dominated by infrequent, large, intense summer fires that have an homogenising effect on the vegetation. The vegetation has changed from one of small patches in different stages of recovery following fire to larger less diverse patches. Burbidge *et al.* (1988) have shown that the disappearance of mammals from these desert regions coincided with the depopulation of the area by Aborigines and the subsequent change in fire regime. Bolton and Latz (1978) concluded that the persistence of habitat for *Lagorchestes hirsutus* had depended on deliberate burning by Aborigines.

Fire-induced patchiness in the vegetation is necessary for the persistence of suitable habitat for *B. penicillata* and *M. eugenii* in the south-west of Western Australia (Christensen 1977; Christensen and Maisey 1987). Fire, and more particularly the manner in which it occurs, thus appears to be a key element in the decline of at least three Australian macropods.

Precisely what the Australian landscape looked like and how the macropods were distributed over the continent at the time Aborigines arrived remains unclear but it is certain that significant elements of the megafauna existed (Hope 1982) and that the *Eucalyptus* and other fire-adapted species were much less common over large areas (Singh *et al.* 1981). The interaction between climatic swings and land management practised by Aborigines induced a new set of conditions in which the megafauna were notable failures.

Competitors

Europeans brought with them a suite of 30 herbivorous mammals of which 19 have established wild populations (Frith 1973, p142).

(a) DOMESTIC LIVESTOCK

Kangaroos initially prospered in the first few decades wherever sheep and cattle were introduced. While sheep numbers were rising from 600 000 to 6 000 000 between 1860 and 1890 in New South Wales the numbers of "Kangaroos" were increasing at a considerable rate also. The clamour of landholders in this state led eventually to the passing of the Stock and Pastures Protection Act in 1880 and the paying of a bounty on kangaroo scalps by the various regional districts. Before this the landholders were using whatever means were available to them to control kangaroo numbers including the organization of "drives" such as one at Trinkey Station near Gunnedah in New South Wales where 80 000 were killed in one exercise (Rolls 1969, p391). Some landholders blamed the earlier elimination of Dingoes (Rolls 1969, p391) while others regarded the situation as being far more complex and involving the diversifying effects of grazing by sheep and cattle (Newsome 1971), the establishment of stock watering points, and the introduction of exotic pasture and weed species.

Small macropods prospered also. In the Tamworth Pastures Protection Board district between 1880 and 1920, rat kangaroo (probably *Aepyprymnus rufescens* and/or *Bettongia gaimardi* scalp returns rose to a peak of 78 938 in 1892, about the same year as the peak in sheep numbers, and then oscillated in a general downward trend until 1918 when no bounties were paid. Wallaby (probably *Macropus rufogriseus, M. dorsalis* and/or *Wallabia bicolor)* numbers peaked after that of sheep and oscillated widely until about 1910 when a general downward trend occurred and few bounties were paid during World War I and thereafter. Curiously there was a brief surge in "pademelon" (identified confidently as *Onychogales fraenata*) numbers in the same district around 1910 when some 70 000 bounties were paid about 30 years and several droughts after the sheep numbers peaked (Jarman and Johnson 1977). There was no clear relationship between the changing abundance of sheep and the rise or decline in native herbivores. There was a clearer relationship with the arrival and eruption of Rabbits and Foxes.

Burbidge and McKenzie (1989) compared change in status of native mammals in two districts in the Eyrean region of Western Australia: one pastoral that had been grazed by sheep, goats and cattle over more than 90% of its area for more than 60 years, and the other largely unoccupied desert. Both districts have suffered a great reduction in CWR species and the pastoral area had a lower, but not significantly lower, proportion of D and E species than the desert district. When the rockpile inhabiting species were excluded from the analysis the two districts produced similar figures. Burbidge and McKenzie (1989) could find no relationship between a history of pastoral land use and extinction or decline in their critical weight range of 35 g to 5 500 g, but pointed out that the decline could have been caused by different factors in the two districts.

(b) FERAL HERBIVORES

There is no evidence to support the conclusion that feral livestock were a primary factor in the decline of mammals (Burbidge and McKenzie 1989). This conclusion was reinforced with data from islands such as Bernier Island which has suffered significant reduction in density of vegetation due to goats (Ride and Tyndale-Biscoe 1962; Burbidge unpubl. data) without loss of the four CWR mammals.

Feral goats have been linked with the suppression of *Petrogale xanthopus* (Lim *et al.* 1980, 1981) but there is no evidence that competition between the two has led to local extinctions of the latter (Lim *et al.* 1987). In examining the history of *P. xanthopus* in the Flinders Ranges of South Australia, Lim *et al.* (1981) found that its major decline occurred in the 1880s and 1890s whereas goats were not common in the region until the early 1960s.

Rabbits *(Oryctolagus cuniculus)* have been implicated in the decline of several Australian mammals (Rolls 1969) but macropods have vanished from the northern Gibson, Great Sandy, and Tanami Deserts (Gibson 1986; Burbidge *et al.* 1988) where Rabbits have never reached or have never been common. Rabbits may have been significant in the decline of *Bettongia lesueur* in some districts (Jones 1924, p211) but their effect has no general application.

Introduced predators

There are three introduced predators now widespread in Australia. The Dingo, *Canis familiaris,* was introduced at least 3 000 years B.P. and had a pan-continental distribution at the time of European settlement. The remaining two species, the Feral Cat, *Felis catus,* and the Red Fox, *Vulpes vulpes,* were introduced after European settlement and might therefore be implicated in historical changes in status of macropods. *F. catus* inhabits the entire mainland region as well as several offshore islands including Tasmania (Jones 1983). The Red Fox, *Vulpes vulpes,* does not occur in Tasmania, and it is largely absent from the northern wet-dry tropics (Coman 1983). It is rare in the principal habitats of the Great Sandy, Gibson, Great Victoria and Tanami Deserts, but is locally common along salt lakes of the Great Sandy Desert (King and Smith 1985; Gibson 1986).

a) FERAL CAT

Feral Cats have been present in Australia for a very long time. Burbidge *et al.* (1988) report statements by western desert Aborigines that Feral Cats had "always been present".

Aborigines have shown one of us (K.A.J.) tracks indicating that a Feral Cat had stalked and killed a young *L. hirsutus* in the Tanami Desert. Triggs *et al.* (1984) found that 56% of scats from Feral Cats in Croajingalong National Park in southeastern Victoria contained Ringtail Possum *(Pseudocheirus peregrinus)*, a species that weighs between 700 g and 1 100 g (McKay 1983). *Potorous platyops* lies well within this weight range and Feral Cats might well explain its apparent demise by the late 1870s. However, *Caloprymnus campestris* (MABW 890 g, Smith 1983), *Lagorchestes asomatus* (MABW *ca* 1 500 g), many bandicoots, and the Western Quoll *(Dasyurus geoffroii)* survived until the 1930s and 1950s before declining suggesting that Feral Cats have not been the primary cause of their extinction or decline.

(b) FOX

The Fox arrived much later than the Feral Cat. The earliest release may have been in 1845 (Rolls

1969, p318) but Coman (1983) places the time of establishment in the 1860s. It had colonized most of South Australia by 1923 (Jones 1925, p357), having established in that State by 1888 (Rolls 1969, p323). It arrived in Western Australia about 1911 or 1912 and had colonized most of that State by the early 1930s (Crawford 1912 and Long 1972 in King and Smith 1985). By 1893, 13 years after the Rabbit, it had reached the Murray River and by 1900, at about the same time as the Rabbit, it reached Armidale in northern New South Wales (Frith 1973; Jarman and Johnson 1977).

The absence of D and E species in the wet-dry tropics of northern Australia is consistent with the absence or low numbers of Foxes in the area. Foxes have been linked with the decline and extinction of many native Australian mammals (Finlayson 1927, 1961; Poole 1979; Southgate unpubl.) but the relationship is not absolute. Jones (1925, p358) noted that in many parts of South Australia the marsupials had already begun to decline before the arrival of the Fox. Calaby (1971) noted that whereas the decline of some species such as *Setonix brachyurus* was coincident with the arrival of the Fox, there were other areas such as Western New South Wales where CWR mammals disappeared before it invaded. Gibson (1986) found that many species vanished from the Tanami Desert before the Fox had arrived.

Bettongia penicillata survives in the south-west of Western Australia where Foxes are common but, has vanished from the remaining part of its range that is less densely occupied by Foxes (King and Smith 1985). Calaby (1966) described the remarkable richness of macropod fauna in the Wallaby Creek region of northeastern New South Wales where Jarman *et al.* (1987) reported that Dingoes were common but Foxes and Feral Cats were very rare. The extinction of *Bettongia gaimardi* on mainland Australia and its survival in Tasmania is attributed by some to the absence of the Fox in Tasmania.

Kinnear *et al.* (1988) have demonstrated a Fox induced decline of *Petrogale lateralis* in south-western Western Australia and showed that removal of Foxes led to the recovery of populations.

We conclude that while the Fox may not have been the primary cause of decline and extinction in all areas it is clearly a factor in the local extinction of remnant populations.

(c) IMPACT OF ABORIGINES ON CURSORIAL PREDATORS

The former impact of Aborigines on predators should not be underestimated. Foxes were eaten (R. Kimber, pers. comm.) and Feral Cats are still a highly prized food item of Aborigines in the desert regions of Western Australia and the Northern Territory (Burbidge *et al.* 1988). They are relatively easy to catch in open country because they have little stamina, especially in hot weather, and tend to seek cover after a relatively short period of flight (K. A. Johnson, pers. obs.).

Dingoes form an integral part of Aboriginal culture but they were formerly eaten in times of need. Such a time was during the large gatherings of people for ceremonial purposes (P. Latz, pers. comm.). The movement of Aborigines to settlements and other centres may have lifted a restraining force on Feral Cats, Foxes and Dingoes that were then able to have a greater impact on prey species.

The evidence for an exotic predator (other than *Homo sapiens*) causing the extinction of native species on continents anywhere in the world is unconvincing (Frankel and Soulé 1981). One must be mindful, however, that in the Australian example, the effect is of eutherian predators on marsupial and eutherian prey that in the last 20 000 to 30 000 years have not been challenged by a Fox-sized predator.

Island macropods

A large number of islands were isolated from the Australian mainland as the seas rose towards the end of the last ice age. The timing varied but severence most probably occurred between 12 000 to 13 500 years B.P. when the relatively deep Bass Strait waters were inundated, and 8 000 to 6 500 B.P. when the shallower Torres Strait region was flooded (Galloway and Kemp 1981 p59). The islands inhabited by macropods vary in size from the quite tiny Westall Island of 0.7 km^2 to large land masses like Tasmania (68 000 km^2) and New Guinea (890 000 km^2).

There are at least 22 Australian macropods represented on islands around the Australian coast. Some of these have been artificially established (A. A. Burbidge and I. J. Abbott, unpubl.). In addition *M. parma* was introduced onto Kawau Island off the west coast of New Zealand in the 1870s (Wodzicki and Flux 1967) and *B. penicillata* has been established and re-established on islands off South Australia (Delroy *et al.* 1986). Islands are a critical buffer to the process of decline and extinction that has occurred on the mainland. Three E species (*B. lesueur, B. gaimardi* and *L. fasciatus*) were formerly widespread on the continent but are now confined to islands. *Lagorchestes hirsutus* is represented by about 20 individuals in the wild but reasonable numbers of the sub-species *L. h. dorreae* and *L. h. bernieri* persist on the respective islands in Shark Bay, Western Australia.

Not all island macropods have persisted and there is a strong correlation of decline in status with the presence of exotic predators. Foxes have been blamed for the extinction of *Petrogale lateralis* on Depuch Island (Kinnear *et al.* 1984) and Feral Cats for the elimination of *Lagostrophus fasciatus* and *Bettongia lesueur* from Dirk Hartog Island (Burbidge and George 1978), and *Lagorchestes conspicillatus* from Hermite Island (Burbidge 1971). Jones (1924, p215) attributed the extinction of *Bettongia penicillata* on Saint Francis Island to the introduction of Feral Cats. All the above are CWR species.

The taxonomic relationship between mainland and island macropods is not well studied and probably varies with population size and period of isolation. There is some evidence for behavioural differences, e.g., Jones (1924 p215) reported the lack of fear of humans and diurnal activity of *Bettongia penicillata* on Saint Francis Island. This characteristic would make them particularly susceptible to the sudden arrival of an effective predator to their island.

The importance of Feral Cats is less clear in Tasmania where three CWR species, *Potorous tridactylus, Bettongia gaimardi* and *Thylogale billardierii* have survived Feral Cat predation and substantial modification to habitat through forestry and agricultural practices (K. A. Johnson, unpubl. data). *Potorous tridactylus* persists in reasonable numbers on Flinders Island in Bass Strait despite the presence of Feral Cats, the clearing and pasture improvement of about 50% of the island, and the grazing by domestic livestock (G. Hocking, pers. comm.).

The apparent inconsistency of the role of Feral Cats might be related to rainfall. The Western Australian islands referred to lie in the 250-500 mm mean annual rainfall belt whereas Tasmania and Flinders Island are cooler with 2 to 6 times the rainfall (Fitzpatrick and Nix 1973 p7). Prey populations on large islands might also reach an equilibrium with the predator whereas they might be readily extinguished on small islands.

A large population of Feral Goats (now removed) on Bernier Island appears not to have affected the populations of *Bettongia lesueur, Lagorchestes hirsutus* and *Lagostrophus fasciatus* (Ride and Tyndale-Biscoe 1962). A vegetation change on Rottnest Island from woodlands and scrub to low heath as a result of changed fire regimes (Pen and Green 1983) has not affected the status of *Setonix brachyurus*. Burbidge and McKenzie (1989) concluded that habitat change in the absence of exotic predators had not resulted in any declines on Western Australian islands. Moreover they could detect no relationship between the size of disturbed islands (880 ha to 54 360 ha) and the status of macropods and other native species.

DISCUSSION

Our appraisal of the patterns of decline and extinction among Australian macropods reveals no single cause or consistent combination of causes relating to European settlement. In contrast, the natural features of climate and body weight are generally implicated (see Burbidge and McKenzie 1989). Species within the CWR from low rainfall areas are more vulnerable to decline and extinction than larger species or those from high rainfall areas.

There are exceptions, inconsistencies and enigmas when attempting to apply possible causal factors in a general sense and plainly their impact differed in relation to both the macropod species and the areas involved. Drawing together the various forces that appear to have affected macropods, two general themes emerge.

1. Modification of vegetation (structure and botanical composition) by changed fire regimes, grazing by introduced herbivores and introduction of exotic plant species.
2. Direct reduction in numbers due to predation and destruction of habitat.

These two themes and their discrete components must be viewed in their ecological perspective: the components and themes are all highly interactive. Their collective impacts on a species varies in relation to time and geographic location and we believe particularly in relation to seasonality in rainfall. Rainfall is a significant factor in all regions but more so in the Eyrean region where it has a 1:1 relationship with primary productivity (Pianka 1986) and variation from year to year is greatest. Rainfall records over a period of 69 years across a gradient between Alice Springs and Darwin show 14.5 drought years in the northern Torresian region, 25.3 such years on the interzone between the Torresian and Eyrean regions, and a massive 30.0 years in the central Eyrean region (Foley 1957).

Friend (1987) examined some examples of population collapse and recovery following drought and reviewed the genetic consequences of recovery by immigration and reproductive increase. Recovery limited to reproductive increase is likely to cause loss of genetic diversity and loss of fitness unless population increase following collapse is fast.

Desert populations expand and collapse in relation to seasonal conditions. During the high rainfall years of 1980 to 1984 in the Tanami Desert *Lagorchestes conspicillatus* and the Greater Bilby *(Macrotis lagotis)* were widely distributed over most habitats. In the following two years of drought both were reduced to small scattered pockets which, for the Bilby, were centred on patches of high food availability.

Species under stress and confined to refugia due to drought will be further stressed if they must also cope with habitats that lack diversity because of partial clearing, single event large and intense wildfire, grazing by introduced herbivores, homogenisation by introduced plant species, and reduced presence of diversifying agents such as patchy and seasonally distributed fire. Exposure to exotic cursorial predators would further depress population size and possibly prevent its recovery altogether. Depending on the severity of the circumstances any one or combination of the above factors might tip an isolated population into extinction. The Eyrean region suffers more frequent, longer and less predictable droughts than the more mesic areas and we believe this has significantly influenced its loss of macropods.

Little can be done now about most introduced plant species but feral and domestic livestock are large enough to be readily controlled and should present straight forward management options. One should be mindful however, that reduction of the standing crop of grasses and herbs by grazing directly affects the potential fire regime through its impact on the amount and distribution of fuel. Removal of domestic and/or feral livestock from an area set aside for wildlife conservation may create problems through increased fuel loads and higher potential for widespread, intense, and destructive wildfires.

Fire is one management tool that has general application to macropod management but it is a tool that has subtle effects which are only beginning to be untangled.

Predation by Dingoes on macropods in Nadgee Nature Reserve in southeastern New South Wales increased substantially following wildfire when numbers of small mammals decreased and Dingoes switched their prey preferences (Newsome *et al.* 1975). Recher and Christensen (1981) reported greatly increased predation of *B. penicillata* by the Fox and *Dasyurus geoffroii* following fire which removed substantial portions of shelter habitat. Predators are able to suppress macropod populations (Kinnear *et al.* 1988; Caughley *et al.* 1980) and they are implicated in the extirpation of island populations (Burbidge and George 1978; Jones 1923). They were responsible for the rapid extinction of one small isolated mainland population of *L. hirsutus* (K. A. Johnson, unpubl. data). Manipulation of habitat for CWR macropods through prescriptive fire management will therefore require careful assessment of the potential feedback on the population by predators.

In the Perup Nature Reserve in the south-west of Western Australia, *Macropus eugenii* occupies *Melaleuca viminea* and heartleaf *(Gastrolobium bilobum)* thickets which are patchily distributed and small in total area (Christensen and Maisey 1987). Hot fires consuming all fuel are required for regeneration of *Melaleuca* and heartleaf thickets which are then occupied by *M. eugenii* 5-6 years after burning when the canopy is high enough to give freedom of movement underneath. The thicket structure is maintained for 25-30 years when it again needs fire treatment. In the absence of fire the thickets open up, other shrub species invade, conditions become favourable for *Macropus fuliginosus,* and *M. eugenii* disappear.

The snapshot response to fire is not necessarily the same for all CWR species. *M. eugenii* on recently burned country lost weight rapidly whereas *Bettongia penicillata* in the same area eating hypogean fungi gained weight (Christensen 1977).

Dingoes had a pan-continental distribution at the time of European settlement but have been greatly reduced in numbers or eliminated from that part now used for sheep grazing and intensive agriculture. The Fox is generally common in these low-Dingo areas and the Feral Cat is distributed over the entire continent: both occur on islands. The diets of the three predators have been extensively studied but there have been few investigations where two or more are in sympatry. Only one study has been carried out where all three were co-existing (Triggs *et al.* 1984) but here the authors avoided the term Dingo and referred only to *Canis familiaris.* Further studies are required before the relative impact of these predators can be properly assessed.

Results of the single species studies indicate that where Rabbits are common they form a major part of the diet of Dingoes (Corbett and Newsome 1987), Foxes (Brunner *et al.* 1976; Seebeck 1978) and Feral Cats (Jones and Coman 1981). Where Rabbits are uncommon, Dingoes are able to take larger prey items, principally macropods (Robertshaw and Harden 1985; Triggs *et al.* 1984) than Foxes or Feral Cats, and will switch to them when fire (Catling and Newsome 1981) or drought (Corbett and Newsome 1987) suppress the numbers of smaller mammalian species. A programme to reduce competition for a CWR macropod by controlling Rabbits alone is likely to increase predation on those macropods by carnivores looking for alternative food. The size and distribution of the macropods will determine how significant this predation will be and whether coincident control of carnivores is indicated.

Macropods have coexisted with Dingoes for several thousand years so it could be an initial premise that if there was a choice between the three main cursorial predators, a wildlife manager would prefer the presence of the Dingo. There is circumstantial evidence for antagonism between the three predator species with the Dingo being dominant where it is left unmolested by *Homo sapiens* (A. Newsome and P. Catling, pers. comm.; K. A. Johnson, pers. obs.). The role of Dingoes in suppressing Fox and Feral Cat numbers needs thorough investigation and the specific control of Dingoes in the presence of other predators should be viewed with some caution, especially where CWR macropods exist.

Species priorities for research and management

The employment of the scarce human and financial resources in conservation of the remaining macropod species needs careful direction in order to prevent more species joining the six already extinct.

Our appraisal of the patterns of decline and extinction indicates that the CWR, rainfall, presence or absence on islands, and present and former geographic range are useful indicators in priority assessment. We have used two categories — High and Low Priority — and the first approximation of the High species is given below.

(a) CWR, Declined and not established on islands.

Bettongia penicillata
Bettongia tropica
Onychogalea fraenata

(b) CWR, Declined and restricted to islands.

Bettongia gaimardi
Bettongia lesueur
Thylogale billardierii
Lagostrophus fasciatus

(c) Species with very restricted geographic ranges.

Potorous longipes
Petrogale persephone

(d) CWR, occurs on islands but mainland populations declined.

Lagorchestes hirsutus
Petrogale lateralis
Macropus eugenii
Setonix brachyurus

In the second and final categorization (Table 4) we have excluded *Thylogale billardierii* because it occurs in mesic areas, which our review has shown to be occupied by species remarkably resilient to decline providing their habitat is not eliminated. It also occurs on many islands and is in numbers high enough to warrant management aimed at reducing its abundance for crop protection purposes. *Bettongia gaimardi* also occurs in the relatively mesic, large island of Tasmania and we have down-graded its priority accordingly.

Bettongia penicillata, *B. tropica* and *Onychogalea fraenata* stand out as being of high priority because of their history of alarming decline and absence from islands. The order for the remaining species is less clear. *Potorous longipes* and *Petrogale persephone* both occur in high rainfall areas and it would appear that habitat destruction or modification would be their major threat. This is possibly a problem with *P. longipes,* given its proximity to wood chip areas.

The remaining seven species are all represented on islands and have all experienced substantial declines in geographic range. We recognize that our list provides a global view for Australia and that State/Territory, regional and local government and non-government agencies may apply additional criteria. Our list provides a general guideline. Because eight of the 12 priority species are Eyrean and/or Bassian it is clear that emphasis should now be applied to research and management of species in the arid and semi-arid lands.

Current indications of research priority

Of the 44 extant Australian macropods, 27 (61%) are CWR species and 17 (39%) are larger non-CWR species. Research into the Australian macropods in the past 30 years has been strongly biased towards the large species and this trend has continued into recent times. In the past five years (1983-87), *Australian Wildlife Research,* the principal vehicle for publication of applied wildlife science in Australia, has published 35 papers directly related to macropods, and a further 13 that are indirectly related (concerning survey results, predator diet and 1080 use). Of the 35 macropod papers only five (14.3%) concern CWR species. Only three of these papers

Table 4. Priority order for research and management.

Species	% of historical mainland range still occupied	No. islands now inhabited
Bettongia penicillata	<1	0*
Onychogalea fraenata	<1	0
Potorous longipes	?	0
Petrogale persephone	100?	0
Bettongia tropica	<5	0
Lagostrophus fasciatus	0	2
Bettongia lesueur	0	3
Lagorchestes hirsutus	<1	2
Setonix brachyurus	10-15	2
Macropus eugenii	5	13
Petrogale lateralis	5	9
Bettongia gaimardi	0	2

* recently re-introduced to islands but establishment not yet assured.

concern any of the 11 species we consider to be of highest priority.

Management

Management of CWR macropods is in its infancy and resources committed to them are minor compared with the resources invested in managing the larger species. Captive breeding populations of D species (*Potorous longipes, Bettongia penicillata, Onychogalea fraenata, Macropus eugenii* and *Lagorchestes hirsutus*) have been established and could be potentially used in efforts to rehabilitate the species in the wild. Fox control has been experimentally applied to colonies of *P. lateralis* (Kinnear *et al.* 1988) and habitat management using fire has been experimentally applied to *Lagorchestes hirsutus* (Johnson 1988), *B. penicillata* and *M. eugenii* (Christensen and Maisey 1987). *O. fraenata, M. eugenii, P. lateralis* and *B. penicillata* are the only CWR macropods for which land has been reserved for their specific conservation on the mainland.

Establishment on islands has commenced for *Bettongia penicillata* (Delroy *et al.* 1986) and *L. fasciatus* (R. Prince, pers. comm.), and there is potential for other species to be established in the relatively uncomplicated security that islands provide. Such programmes must however be viewed as holding operations to secure a genetic aliquot of particular species. Aspects relevant to the CWR species management are reviewed below.

a. LEGISLATION

All States and Territories have legislation in place which either prohibits or controls the utilization of macropods. These controls are important in limiting/preventing the critical reduction in numbers in local areas. We have nevertheless shown that overkill on a national scale has not been responsible for the decline or extinction of any species with the possible exception of *Macropus greyi.*

Legislation that controls the removal of natural vegetation (e.g., the Vegetation Retention Scheme of South Australia) can potentially protect the habitat of species at a local level. The CWR species are dependent to varying degrees on the presence of dense vegetation in which to shelter, but they are generally unattractive as species for either commercial harvest or sport shooting. Legislation controlling killing is therefore unlikely to assist the survival or prosperity of our critical species, but in some circumstances legislation protecting habitat may be useful.

b. LAND RESERVATION

The creation of national parks, nature reserves and equivalent areas has historically been considered one of the key steps in securing the status of a species. Barrow Island Nature Reserve off the north-west coast of Western Australia was declared as early as 1908 for the conservation of *Lagorchestes conspicillatus* among other native species. Bernier and Dorre Islands were also declared Nature Reserves for the conservation of CWR macropods (among other species). These island reserves are, as we have argued, special cases that escape the full range of forces which now stress the ecosystems of the mainland.

Comprehensive and on-going management will be required on (preferably large) mainland reserves if long term conservation of CWR macropods is to be achieved. There are indications also of subtle balances that will require resolution before effective management can be achieved. For Example the Wallaby Creek area of northeastern New South Wales is inhabited by 10 species of macropods that are relatively abundant (Calaby 1966). No macropod species (including four CWR species) known to have occurred in this uncommonly rich area has been lost in historical times (Jarman *et al.* 1987). The area is not reserved and is currently managed for low intensity cattle and timber production, which incorporates infrequent logging, tree clearing, cattle grazing, a scatter of improved pasture, acceptance of a low number of Dingoes (maybe to the detriment of foxes), prohibition of shooting, and use of fire, all of which Jarman *et al.* (1987) and Southwell and Jarman (1987) argue has fostered the macropods of the valley. A change in the existing management (e.g., the removal of stock, cessation of logging and a modification to the fire regime) might set in motion a chain of events that would lead to the long term degradation of the valley for macropods. Acquisition of the land specifically for the conservation of macropods would need to be contemporary with management developed on the basis of well documented research.

c. ZOOS

Zoos are often regarded as a last resort in the conservation of a species (Conway 1980). They might also be considered as a last ditch stand: a Tobruk. But like Tobruk they are potentially able to provide the reserves to break back into and re-establish in former territory. *Potorous longipes, Lagorchestes hirsutus* and *Bettongia penicillata* are the only CWR species currently maintained by Australian zoos (G. George, pers. comm.) and there is obvious potential for greater involvement. Maintenance of genetic diversity through exchange of animals between zoos and wild populations would be beneficial. Frankel and Soulé (1981) recommend the transfer of one to five individuals per generation for most zoo populations.

d. RE-INTRODUCTION OR TRANSLOCATION

Given the sensitivity of remnant populations to environmental stress, solutions for securing any of the priority CWR species in the long term must include programmes of range extension. This may be achieved by natural extension at the boundaries of existing populations or re-introduction to parts of the species' former range. Re-introduction offers the greatest chance of greatly extending a species range and several attempts are currently being made to re-establish wild populations using captive bred stock. These include *B. penicillata* in South Australia (Delroy *et al.* 1986), *Lagostrophus fasciatus* in Western Australia (Prince unpubl.) and *Lagorchestes hirsutus* in the Northern Territory (Johnson 1988).

An adjunct of range extension is re-introduction to parts of the former range of a species and this has shown promise for *L. hirsutus* (Johnson 1988), *B. penicillata* (Christensen and Maisey 1987), and non-macropod species such as *Myrmecobius fasciatus* (J. A. Friend, pers. comm.) and *Macrotis lagotis* (R. I. Southgate, pers. comm.). The premise for such programmes is that the major forces that effected the decline of the species concerned can be countered. Some important considerations for translocation programmes are as follows.

Choice of release site

Choice of release site is simplified if it is to be to a predator free island, or a place near an existing population and in similar habitat. In the second case, studies of the natural population should reveal the habitat requirements of the species and conditions and that should favour re-establishment. Where species are to be moved well beyond their existing range to a very different habitat or from islands to the mainland the process is more difficult (e.g., the translocation of *Bettongia penicillata* from Perup Nature Reserve in the south-west of Western Australia to where it formerly occurred in the spinifex desert of central Australia, or *O. fraenata* from brigalow *(Acacia harpophylla)* shrubland in Queensland to the open grassy woodlands of north central New South Wales). Judgments of which habitat will be suitable must involve a degree of subjectivity. This may be an expensive and frustrating process for a researcher if mistakes are made.

Number to be released

The number of founder individuals to establish a viable population has been reviewed by Friend (1987) who noted a "... severe shortage of directly useful experimental evidence." Foose (1983) considered five to ten pairs could be adequate but that the number was not nearly as important as the subsequent management. In the context of macropods this includes whether individuals disperse from the release area and make no contribution to the breeding group, the presence or absence of predators, and whether there are favourable seasons that foster rapid recruitment to the population.

Main and Yadav (1971) nominated a minimum population of 200-300 for long term (10 000 years) survival of macropods on an island after relating occurrence on islands off the Western Australian coast to island size and available evidence on population density. Franklin (1980) proposed a minimum population size of 50 for short term viability (based on theoretical consideration of inbreeding), and 500 for long term viability based from work on *Drosophila* bristle numbers and "common sense reasoning" (Frankel 1983). A range of 50 to 500 was accepted by Kinnear (1987).

Theoretical considerations of founder populations and minimum viable population sizes are at best a guide: initial and subsequent management will be the significant determining factor. From our experience we suggest a founder population of 10 to 20 pairs, or, a founder and minimum viable population size of "as large as can be managed".

Predator control

Predator control appears to be essential and can be achieved by direct reduction (e.g., poisoning) or exclusion (e.g., fencing). Dingoes, Foxes and Feral Cats utilize tracks and roads (Corbett and Newsome 1987; Robertshaw and Harden 1985; and K. A. Johnson, pers. obs.) so a reduction in predator presence can be assisted by minimizing road works and fire breaks in a release area and ensuring that tracks do not form a natural circuit route.

Electric fencing has excluded Dingoes and Feral Cats from a 1 km^2 area in the Tanami Desert (D. Clarke, pers. comm.) for a six month period despite the presence of *L. hirsutus* inside. Kinnear *et al.* (1988) have used poison to reduce Fox numbers at various *P. lateralis* populations in Western Australia but found that their study area was constantly reinvaded by Foxes from surrounding areas. Baiting was required on a regular basis.

Baiting in areas where Dingoes, Foxes and Feral Cats co-exist may have the potential disadvantage of favouring one predator at the expense of another. Until more is known about predator-predator interactions and the relative importance of each species to CWR macropods the use of baits in such situations would require careful review.

Rabbits are important as prey to the three predators and could sustain them while they eliminated a less numerous but more easily caught species (Jarman and Johnson 1977). Eradication of Rabbits in and around the release area would seem prudent.

Dingoes in the southern Northern Territory switched from their preferred smaller prey (Rabbits and rodents) to the larger *M. rufus* and cattle during drought but switched back again as smaller species increased upon the return to favourable seasons (Corbett and Newsome 1987). Cattle were seen to be one of the agents buffering Dingo numbers which could then suppress natural population increases of native species. Removal of stock from a release area and for some distance around it may aid the success of a re-establishment programme. Removal of Rabbits and cattle could initially increase predation on CWR macropods and predator control might be indicated.

Habitat manipulation

Various studies have shown fire to be an almost universal modifier of habitat. Its application in desert areas is simpler than in southern forested areas because people are few, vulnerable property developments are sparse, and the fuel load lighter and more predictable. It is possible to have controlled fires throughout the year in desert areas under conditions of patchy fuel but prescribed burning in forests containing valuable timber or near urban developments can be worrisome.

Notwithstanding these difficulties, a re-introduction programme should be wary of restricting fire management effort to the more controllable cool season burns. The long-term result may be a modification of the botanical and structural diversity of vegetation that disadvantages the target macropod. Where possible, fires should be patchy, so that clumps of shelter habitat remain, and seasonally well spread so that advantage can be taken of varying seasonal conditions.

CONCLUSIONS

Areas for future research

In this paper we began by examining the forces potentially responsible for the decline and extinction of macropods and then discussed aspects of those we regard as most relevant. In most cases there is sketchy understanding of processes affecting habitat change, of the interaction between predators, of environmental variables such as drought, of predators and changing abundances of potential prey, and of genetic variation in populations under stress. In addition to autecological studies we regard the following areas of research to be important for the successful future management of CWR macropods.

1. Habitat

a. The role of different fire regimes with varying fuel loads, weather, time of year and subsequent seasonal conditions in shifting botanical composition and physical structure of vegetation in the long term.

b. Habitat use with emphasis on resources critical to the persistence of populations in times of stress (e.g., drought and fire).

2. Genetics

a. Genetic variation and shift in captive and wild populations in the long term and the effect this has on fitness.

b. Genetic variability of populations throughout their ranges and clarification of the genetic implications of translocations (e.g., the re-establishment of species on the mainland from island populations).

c. Minimum numbers needed for successful translocations.

3. Cursorial Predators

a. Movements of predators and hunting effectiveness in relation to habitat structure and the effect of human developments such as roads, firebreaks, artificial watering points, and livestock enterprises.

b. Antagonism between predator species.

c. Biological control of Foxes: a monospecific genus in Australia that has no commercial value as a domestic species.

d. Selective control of Foxes, Dingoes and Feral Cats by poisoning or other methods.

4. Re-introduction

a. Minimum numbers required in relation to environmental conditions and the probability of effecting successful re-introduction in the long term.

b. Long-term monitoring of changing distribution, abundance and genetic variation.

c. Development of techniques, including fencing, that enclose founder populations but exclude predators, selection of release sites, short term cossetting (food, water, etc.) to ensure short term survival and reproduction, and methods for gradual release to the natural environment.

If, as we suspect, the forces that caused the decline and extinction of CWR macropods are still operating, the decline of D species on mainland Australia will probably continue and some S species may start to be affected. Management is required and there will be no sunset clause to any management programme.

Finally, there are indications that CWR macropods in northern Australia are starting to decline. These include *Petrogale brachyotis* in the Gulf region of the Northern Territory (Calaby 1976; K. A. Johnson, unpubl. data) and several species in the Kimberley region of Western Australia (Burbidge and McKenzie 1989).

The decline of CWR macropods may be entering a new phase! Mammalogists and nature conservation agencies must switch emphasis from abundant macropods to CWR species before it is too late.

REFERENCES

ASHTON, D. H., 1981. Fire in tall open-forests (wet sclerophyll forests). Pp. 339-66 *in* Fire and the Australian Biota ed. by A. M. Gill, R. H. Groves and I. R. Noble. Australian Academy of Science: Canberra.

BOLTON, B. L. AND LATZ, P. K., 1978. The Western Hare-Wallaby *Lagorchestes hirsutus* (Gould) (Macropodidae) in the Tanami Desert. *Aust. Wildl. Res.* **5**: 285-93.

BRUNNER, H., AMOR, R. L. AND STEVENS, P. L., 1976. The use of predator scat analysis in a mammal survey at Dartmouth in north-eastern Victoria. *Aust. Wildl. Res.* **3**: 85-90.

BURBIDGE, A. A., 1971. The fauna and flora of the Monte Bello Islands. Report No. 9. Perth, Department of Fisheries and Fauna.

BURBIDGE, A. A. (ed.), 1977. The status of kangaroos and wallabies in Australia: Report of the Working Group on Macropod Habitat of the Standing Committee of the Council of Nature Conservation Ministers. Australian Government Publishing Service: Canberra, 20 pp.

BURBIDGE, A. A. AND FULLER, P. F., 1978. The mammals of the Warburton region, Western Australia. *Rec. West. Aust. Mus.* **8**: 57-73.

BURBIDGE, A. A. AND GEORGE, A. S., 1978. The flora and fauna of Dirk Hartog Island, Shark Bay, Western Australia. *J. Roy. Soc. W. Aust.* **60**: 71-90.

BURBIDGE, A. A. AND MCKENZIE, N. L. (1989). Patterns in the modern decline of Western Australia's vertebrate fauna: causes and conservation implications. *Biol. Cons.* **50**: 143-98.

BURBIDGE, N. T., 1944. Ecological succession observed during regeneration of *Triodia pungens* R.Br. after burning. *Proc. Roy. Soc. W. Aust.* **28**: 149-56.

CALABY, J. H., 1966. Mammals of the Upper Richmond and Clarence Rivers, New South Wales. Tech. Pap. Div. Wildl. Res. C.S.I.R.O. No. 10.

CALABY, J. H., 1971. The current status of Australian macropodidae. *Aust. Zool.* **16**: 17-29.

CALABY, J. H., 1976. Mammals. *In* A Survey of the Fauna of the Lower McArthur River Region, Northern Territory ed. by J. Calaby. Unpubl. report to MIMETS Development Pty Ltd by C.S.I.R.O. Div. Wildl. Res.

CALABY, J. H. AND RICHARDSON, B. J., 1988a. Potoroidae. Pp. 53-59 *in* Zoological catalogue of Australia. Vol. 5. Mammalia ed. by D. W. Walton. Australian Government Publishing Service: Canberra.

CALABY, J. H. AND RICHARDSON, B. J., 1988b. Macropodidae. Pp. 60-80 *in* Zoological catalogue of Australia. Vol. 5. Mammalia ed. by D. W. Walton. Australian Government Publishing Service: Canberra.

CATLING, P. C. AND NEWSOME, A. E., 1981. Responses of the Australian vertebrate fauna to fire: an evolutionary approach. Pp. 273-310 *in* Fire and the Australian Biota ed. by A. M. Gill, R. H. Groves and I. R. Noble. Australian Academy of Science: Canberra.

CAUGHLEY, C., GRIGG, G. C., CAUGHLEY, J. AND HILL, G. J. E., 1980. Does dingo predation control the densities of kangaroos and emus? *Aust. Wildl. Res.* **7**: 1-12.

CAUGHLEY, J., 1981. Native quolls and tiger quolls *Dasyurus viverrinus, Dasyurus geoffroii, Dasyurus maculatus.* Pp. 45-48 *in* Parks and Wildlife: Endangered Animals of New South Wales ed. by C. Haige. New South Wales Government Printer.

CHRISTENSEN, P., 1977. The biology of *Bettongia penicillata* (Gray, 1837) and *Macropus eugenii* (Demarest. 1804) in relation to fire. Ph.D. Thesis, University of Western Australia.

CHRISTENSEN, P., 1980. The biology of *Bettongia penicillata.* (Gray, 1837) and *Macropus eugenii* (Desmarest, 1817) in relation to fire. Forests Department of Western Australia. Bull. 91. Pp. 90.

CHRISTENSEN, P. AND MAISEY, K., 1987. The use of fire as a management tool in fauna conservation reserves. Pp. 323-29 *in* Nature Conservation: The Role of Remnants of Native Vegetation ed. by D. A. Saunders, G. W. Arnold, A. A. Burbidge and A. J. M. Hopkins. Surrey Beatty & Sons: New South Wales.

CHRISTENSEN, P., RECHER, H. AND HOARE, J., 1981. Responses of open forests (dry sclerophyll forests) to fire regimes. *In* Fire and the Australian Biota ed. by A. M. Gill, R. H. Groves and I. R. Noble. Australian Academy of Science: Canberra.

COMAN, B. J., 1983. Fox *Vulpes vulpes* P. 486 *in* The Australian Museum Complete Book of Australian Mammals ed. by R. Strahan. Angus & Robertson.

CONWAY, G. W., 1980. Where we go from here. *Int. Zoo. Yearbk.* **20**: 184-89.

CORBETT, L. K. AND NEWSOME, A. E., 1987. The feeding ecology of the dingo. III. Dietary relationships with widely fluctuating prey populations in arid Australia: an hypothesis of alternation of predation. *Oecologia* Berlin **74**: 215-27.

DELROY, L. B., EARL, J., RADBONE, I., ROBINSON, A. C. AND HEWETT, M., 1986. The breeding and re-establishment of the brush-tailed Bettong, *Bettongia penicillata,* in South Australia. *Aust. Wildl. Res.* **13**: 387-96.

FINLAYSON, H. H., 1927. Observations on the South Australian members of the subgenus '*Wallabia*'. *Trans. R. Soc. S. Aust.* **51**: 363-77.

FINLAYSON, H. H., 1961. On central Australian mammals. Part IV — The distribution and status of central Australian species. *Rec. S. Aust. Mus.* **14**: 141-91.

FITZPATRICK, E. A. AND NIX, H. A., 1973. The climatic factor in Australian grassland ecology. Pp.3-26 *in* Australian Grasslands ed. by R. M. Moore. Australian National University Press: Canberra.

FOLEY, J. C., 1957. Droughts in Australia from earliest years of settlement to 1955. Bureau Meteorology Bull. No. 43 Pp. 281.

FOOSE, T. J., 1983. The relevance of captive populations to the conservation of biotic diversity. *In* Conservation Biology: An Evolutionary-Ecological Perspective ed. by M. Soule and B. Wilcox. Sinauer Associates: Sunderland, Massachusetts.

FOX, M. D. AND FOX, B. J., 1986. The effect of fire frequency on the structure and floristic composition of a woodland understorey. *Aust. J. Ecol.* **11**: 77-85.

FRANKEL, O. H., 1983. The place of management in conservation. *In* Genetics and Conservation: A reference for managing wild animal and plant populations ed. by C. M. Schonewald-Cox, S. M. Chambers, B. MacBryde and W. L. Thomas.

FRANKEL, O. H. AND SOULÉ, M. E., 1981. Conservation and Evolution. Cambridge University Press: New York.

FRANKLIN, I. R., 1980. Evolutionary change in small populations. Pp. 135-49 *in* Conservation Biology: An Evolutionary-Ecological Perspective ed. by M. Soulé and B. Wilcox. Sinauer Associates: Sunderland, Massachusetts.

FRIEND, J. A., 1987. Local decline, extinction and recovery: relevance to mammal populations in vegetation remnants. Pp. 53-64 *in* Nature Conservation: The Role of Remnants of Native Vegetation ed. by D. A. Saunders, G. W. Arnold, A. A. Burbidge and A. J. M. Hopkins. Surrey Beatty & Sons: New South Wales.

FRITH, H. J., 1973. Wildlife Conservation. Pp. 414. Angus & Robertson: Sydney.

GALLOWAY, R. W. AND KEMP, M., 1981. Late cainozoic environments in Australia. Pp. 51-80 *in* Ecological Biogeography of Australia ed. by A. Keast. Publ. Junk: The Hague.

GIBSON, D. F., 1986. A biological survey of the Tanami Desert in the Northern Territory. Conservation Commission of the Northern Territory. Tech. Rep. No. 30 Pp. 1-258.

GILES, E., 1889. Australia Twice Traversed: The Romance of Exploration, being a Narrative Compiled from the Journals of Five Exploring Expeditions into and through Central South Australia, and Western Australia, from 1872 to 1876. Sampson Low, Marston, Searle and Rivington: London.

GREEN, R. H., 1973. The Mammals of Tasmania. Pp. 76. Mary Fisher Bookshop: Launceston.

HOWITT, A. W., 1890. The eucalypts of Gippsland. *Trans. Roy. Soc. Vic.* **22:** 81-120.

JARMAN, P. J. AND JOHNSON, K. A., 1977. Exotic mammals, indigenous mammals, and land-use. *Proc. Ecol. Soc. Aust.* **10:** 146-66.

JARMAN, P. J., JOHNSON, C. N., SOUTHWELL, C. J. AND STUART-DICK, R., 1987. Macropod studies at Wallaby Creek. 1. The area and animals. *Aust. Wildl. Res.* **14:** 1-14.

JEANS, D. N., 1972. An historical geography of New South Wales to 1901. Reed Education: Sydney.

JOHNSON, K. A., 1988. Rare and Endangered: Rufous Hare-Wallaby. *Aust. Nat. Hist.* **22:** 406-07.

JONES, E., 1983. Feral Cat *Felis catus*. P. 489 *in* The Australian Museum Complete Book of Australian Mammals ed. by R. Strahan. Angus & Robertson.

JONES, E. AND COMAN, B. J., 1981. Ecology of the feral cat *Felis catus* (L) in south-eastern Australia. *Aust. Wildl. Res.* **8:** 537-47.

JONES, F. W., 1923. The Mammals of South Australia. Part I. The monotremes and the carnivorous marsupials. (The Ornithodelphia and the didactylous Didelphia). Pp. 1-131. Government Printer: Adelaide.

JONES, F. W., 1924. The Mammals of South Australia. Part II. The Bandicoots and the Herbivorous Marsupials. (The syndactylous Didelphia). Pp. 132-270. Government Printer: Adelaide.

JONES, F. W., 1925. The Mammals of South Australia. Part III. The Monodelphia. Pp. 271-458. Government Printer: Adelaide.

KING, D. R. AND SMITH, L. A., 1985. The distribution of the European Red Fox (*Vulpes vulpes*) in Western Australia. *Rec. West. Aust. Mus.* **12:** 197-205.

KINNEAR, J. , 1907. Viability of small populations of plants and animals and the role of introductions and translocations. P. 365 *in* Nature Conservation: The Role of Remnants of Native Vegetation ed. by D. A. Saunders, G. W. Arnold, A. A. Burbidge and A. J. M. Hopkins. Surrey Beatty & Sons: New South Wales.

KINNEAR, J., ONUS, M. AND BROMILOW, R. N., 1984. Foxes, feral cats and rock wallabies. *Swans* **4:** 3-8.

KINNEAR, J., ONUS, M. AND BROMILOW, R. N., 1988. Fox control and rock wallaby population dynamics. *Aust. Wildl. Res.* **15:** in press.

LEEPER, G. W., 1970. Climates. Pp. 12-20 *in* The Australian Environment ed. by G. W. Leeper. C.S.I.R.O. and Melbourne University Press.

LIM, L., MAHOOD, I. AND SIEPEN, G., 1980. Yellow-footed Rock Wallaby *Petrogale xanthopus*. Pp. 27-32 *in* Parks and Wildlife ed. by C. Haig. Publication New South Wales National Parks and Wildlife Service.

LIM, L., ROBINSON, A. AND COPLEY, P., 1981. Rock wallabies Genus Petrogale. Pp. 21-26 *in* Parks and Wildlife ed. by C. Haig. Publication of New South Wales National Parks and Wildlife Service.

LIM, L., ROBINSON, A. C., COPLEY, P. B., GORDON, G., CANTY, P. D. AND REIMER, D., 1987. The Conservation and Management of the Yellow-footed Rock-wallaby *Petrogale xanthopus* Gray, 1854. *Dept. Env. Plan. S. Aust. Spec. Publ.* No. **4:** 1-94.

MAIN, A. R. AND YADAV, M., 1971. Conservation of macropods in reserves in Western Australia. *Biol. Conserv.* **3:** 123-33.

MCKAY, G. M., 1983. Common Ringtail Possum *Pseudocheirus peregrinus*. Pp. 126-27 *in* The Australian Museum Complete Book of Australian Mammals ed. by R. Strahan. Angus & Robertson.

MCKENZIE, N. L., 1981. Mammals of the Phanerozoic South-west Kimberley, Western Australia: biogeography and recent changes. *J. Biogeog.* **8:** 263-80.

MITCHELL, T. L., 1848. Journal of an Expedition into the Interior of Tropical Australia in Search of a Route from Sydney to the Gulf of Carpentaria. Pp. 437. Longman, Green and Longmans: London (Reprinted 1969 by Greenwood: USA).

NEWSOME, A. E., 1971. Competition between wildlife and domestic livestock. *Aust. Vet. J.* **47:** 577-86.

NEWSOME, A. E., MCILROY, J. AND CATLING, P., 1975. Effects of an extensive wildfire on populations of twenty ground vertebrates in south-east Australia. *Proc. Ecol. Soc. Aust.* **9:** 107-24.

NICHOLSON, P. H., 1981. Fire and the Australian Aborigine — an enigma. Pp. 55-76 *in* Fire and the Australian Biota ed. by A. M. McGill, R. H. Groves and I. R. Noble. Australian Academy of Science: Canberra.

PIANKA, E. R., 1986. Ecology and Natural History of Desert Lizards. New Jersey, Princeton University Press.

PEN, L. J. AND GREEN, J. W., 1983. Botanical exploration and vegetational changes on Rottnest Island. *J. Roy. Soc. W. Aust.* **66:** 20-24.

POOLE, W. E., 1979. The status of endangered Australian Macropodidae. Pp. 13-27 *in* The Status of Endangered Australian Wildlife ed. by M. J. Tyler. Royal Zoological Society of South Australia.

RECHER, H. F. AND CHRISTENSEN, P. E., 1981. Fire and the evolution of the Australian biota. *In* Ecological Biogeography of Australia ed. by A. Keast. W. Junk: The Hague.

RIDE, W. D. L. AND TYNDALE-BISCOE, C. H., 1962. Mammals. Pp. 54-97 *in* The Results of an expedition to Bernier and Dorre Islands, Shark Bay, Western Australia in July, 1959, Fauna Bulletin No. 2 ed. by A. J. Fraser. Perth Fisheries Department.

ROBERTSHAW, H. D. AND HARDEN, R. H., 1985. The ecology of the Dingo in north-eastern New South Wales. 2. Diet. *Aust. Wildl. Res.* **12:** 39-50.

ROBINSON, A. C. AND YOUNG, M. C., 1983. The Toolache Wallaby (*Macropus greyi* Waterhouse). *Dept. Env. Plan. S. Aust. Spec. Publ. 1983. No. 2,* 1-54.

ROLLS, E. C., 1969. They All Ran Wild. The story of pests on the land in Australia. Angus & Robertson: Australia.

ROLLS, E. C., 1981. A Million Wild Acres: 200 Years of Man and an Australian Forest. Pp. 465. Thomas Nelson: Melbourne.

SEEBECK, J. H., 1978. Diet of the fox *Vulpes* in a western Victorian forest. *Aust. J. Ecol.* **3:** 105-8.

SERVENTY, D. L. AND WHITTELL, H. M., 1967. Birds of Western Australia. Pp. 440. Lamb Publications: Perth.

SMITH, M. J., 1983. Desert Rat-kangaroo *Caloprymnus campestris*. P. 192 *in* The Australian Museum Complete Book of Australian Mammals ed. by R. Strahan. Angus & Robertson.

SOUTHWELL, C. J. AND JARMAN, P. J., 1987. Macropod studies at Wallaby Creek III. The effect of fire on pasture utilization by macropodids and cattle. *Aust. Wildl. Res.* **14:** 117-24.

SPECHT, R. L., 1981. Responses to fires in heathlands and related shrublands. Pp. 395-415 *in* Fire and the Australian Biota ed. by A. M. Gill, R. H. Groves and I. R. Noble. Australian Academy of Science: Canberra.

SPENCER, W. B., 1896. Report on the Horn Expedition to central Australia. Part I. Introduction, Narrative, Summary of Results, Supplement to Zoological Report, Map ed. by B. Spencer. Pp. 220. Melville, Mullen and Slade: Melbourne.

STRAHAN, R. (ed.), 1983. The Australian Museum Complete Book of Australian Mammals. Angus & Robertson. Pp. 530.

THORNBACK, J. AND JENKINS, M., 1982. The I.U.C.N. Mammal Red Data Book: Part 1 Publ. International Union for Conservation of Nature and Natural Resources. Pp. 516.

TRIGGS, B., BRUNNER, H. AND CULLEN, J. M., 1984. The food of the fox, dog, and cat in Croajingalong National Park, south-eastern Victoria. *Aust. Wildl. Res.* **11:** 491-99.

WHITE, S. R., 1952. The occurrence of the Quokka in the south-west. *West. Aust. Nat.* **3:** 101-03.

WHITEHOUSE, S. J. O., 1977. The diet of the dingo in Western Australia. *Aust. Wildl. Res.* **4:** 145-50.

WODZICKI, K. A. AND FLUX, J. E. C., 1967. Rediscovery of the white-throated wallaby, *Macropus parma* Waterhouse 1846, on Kawau Island, New Zealand. *Aust. J. Sci.* **29:** 429-30.

Techniques for monitoring the abundance of kangaroo and wallaby populations

COLIN SOUTHWELL[1]

The application of total count, mark-recapture, addition-removal, catch-effort and sample count methods to the monitoring of kangaroo and wallaby populations is reviewed. Application of total count and addition-removal methods has been restricted to the assessment of the accuracy of sample count methods rather than for monitoring *per se*. Mark-recapture methods are not suitable for the medium to large sized macropods and have received little use. Catch-effort methods have been used extensively to monitor populations of those species subject to commercial culling, but there has been little empirical research into the assumptions and requirements of the method. The great majority of monitoring studies have utilized direct sample counts. Aerial survey has been applied extensively across inland Australia to monitor populations of the large kangaroos. Ground survey by vehicle and foot has been applied to a wide range of species and locations but has generally been limited in extent. Indirect counts of faecal pellets have similarly been limited in extent. Several aspects of sample count methods are discussed, including repeatability, precision, accuracy and survey design. Deficiencies in current knowledge and areas for further research are identified.

INTRODUCTION

THE superfamily Macropodoidea contains a diverse array of species differing greatly in biology and ecology. No single monitoring technique could be used effectively for all of these species; one must instead select from a suite of available techniques one or a combination of methods that are most suited to a particular species. In this paper the techniques available for population monitoring in general and their application to some members of the Macropodoidea are reviewed. The macropodoid species to which this review is aimed are those of medium to large size (>4 kg), those which are relatively unspecialized in their habitat requirements and those which are relatively abundant. Approximately 20% of the 50 macropodoid species extant in Australia fall into these categories, and it is to these species that most monitoring has been directed. Table 1 lists the species for which monitoring work has been reviewed. For convenience these kangaroos, wallabies and pademelons will hereafter be referred to generally as macropods.

Over the past three decades biologists have expended much effort in monitoring the abundance of macropods. The body of information available on this subject is now quite large but is scattered widely over many sources in the literature. This review attempts to bring this information together to provide an overview of the development and application of monitoring techniques to macropod populations, and to assist in the further development of techniques by identifying areas where knowledge is most lacking and where the greatest scope for improvement exists. The review attempts to be comprehensive in scope and as up-to-date as possible; not only work published in the scientific literature, but also work presented in unpublished reports and theses, and in some cases work still in preparation or in progress, is covered.

Monitoring wildlife abundance

Wildlife abundance may be monitored across either space or time. The change in abundance across space is the species distribution and the change in abundance across time is the rate of increase. The rate of increase statistic is a direct and concise summary of all population processes (fecundity, mortality, immigration and emigration) and is arguably the most powerful and useful statistic available to the wildlife biologist.

The reasons for monitoring wildlife abundance vary. The practical wildlife manager may wish to monitor the response of a population to a particular management action; for example, habitat alteration aimed at increasing the abundance of a rare species, or control aimed at reducing the abundance of a common or pest species. At the other extreme, the theoretical population biologist may monitor population abundance in order to understand the cause of changing abundance.

Whether the question is the practical "what happens" or the more theoretical "how or why does it happen", monitoring requires the repeated measurement of abundance. In this review the strict definition of monitoring (i.e., repeated measurement) has been extended to include studies where abundance has been measured at only one point in space or time. These "one-off" efforts, while not strictly aimed at population monitoring, are useful in considering monitoring techniques.

[1]National Kangaroo Monitoring Unit, Australian National Parks and Wildlife Service, G.P.O. Box 636, Canberra, Australian Capital Territory 2601.

Pages 659-93 in KANGAROOS, WALLABIES AND RAT-KANGAROOS ed. by G. Grigg, P. Jarman and I. Hume. Surrey Beatty & Sons Pty Limited, New South Wales, Australia, 1989.

Table 1. Macropod species for which monitoring work is reviewed.

Common name	Scientific name	Abbreviation used in subsequent tables
Red Kangaroo	*Macropus rufus*	RDK
Eastern Grey Kangaroo	*Macropus giganteus*	EGK/GK*
Western Grey Kangaroo	*Macropus fuliginosus*	WGK/GK*
Wallaroo	*Macropus robustus*	WLR
Antilopine Wallaroo	*Macropus antilopinus*	AWL
Whiptail Wallaby	*Macropus parryi*	WTW
Red-necked Wallaby	*Macropus rufogriseus*	RNW
Swamp Wallaby	*Wallabia bicolor*	SPW
Red-necked Pademelon	*Thylogale thetis*	RNP
Red-bellied Pademelon	*Thylogale billardierii*	RBP

*abbreviation GK used when two Grey Kangaroo species have not been separated.

Measures of abundance

Caughley (1977) listed three ways in which abundance can be measured:

1. as population size. This measure has a biological meaning only when the population has a distinct boundary. When no such boundary exists, density rather than size provides the most realistic measure of abundance;
2. as absolute density, the number of animals per unit area; and
3. as a density index i.e., some measurable correlate of absolute density.

Figure 1 shows a number of possible relationships between a density index and absolute density. The most useful density indices have a relationship with absolute density that is linear through the origin (A in Figure 1). Any proportional change in abundance will then be reflected in the same proportional change in the index being measured. Indices that are non-linear and/or do not pass through the origin (B, C and D in Figure 1) can still be useful, but unless the form of the relationship between the index and absolute density is known their use is restricted to the ranking of densities (Caughley 1977).

In practice most biologists rely on common sense judgments and theoretical reasoning rather than empirical testing when choosing an index for its qualities of linearity and passing through the origin. Empirical testing is more desirable, but this requires control populations of known size which are rarely available.

The need for a linear index can depend on the purpose of the monitoring programmes. The aims of the theoretical population biologist are such that density indices would need to be linear across the entire abundance range. The practical wildlife manager may not always require strict adherence to linearity. If, for example, a manager undertakes a monitoring programme to ensure that a species' abundance does not fall below a critically low level, an index which departs from linearity only at high abundance levels may be quite adequate.

Provided an appropriate index is used, relative indices of density can be just as effective as measurements of absolute density in monitoring abundance. Since relative indices are often far cheaper and easier to obtain, the use of absolute estimates for monitoring may be wasteful of time and resources. Sometimes the measurement of absolute density may be desirable or necessary, but the benefits obtained must be weighed against the costs.

Accuracy, repeatability and precision

From the foregoing discussion it should be obvious that accuracy is not an essential quality of a monitoring technique. If, for example, a biologist consistently misses half of the animals in a survey area, the counts, even though inaccurate, will provide a useful density index because the relationship between the number counted and the number actually present is linear through the origin.

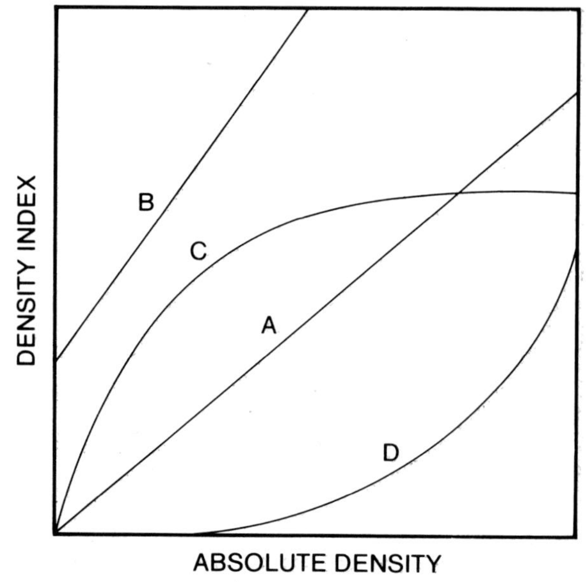

Fig. 1. Some possible relationships between a density index and absolute density. A: linear through the origin; B: linear, not through the origin; C: curvilinear through the origin; D: curvilinear, not through the origin.

Of greater importance are the qualities of repeatability and precision. A method is repeatable if it is unaffected by variations in the conditions of counting, catching, etc. For example, if at one time half of the animals present in an area are counted but at another time only one quarter are counted (because lighting conditions had worsened, vegetation cover had increased, etc.), the counts would not constitute a repeatable index of abundance and would be of no use for monitoring unless they could be adjusted for variations in countability. The second important quality, precision, is related to random errors associated with sampling. If the sampling effort is inadequate then abundance estimates will be imprecise and inferences about the population cannot be made with confidence.

Causes of changing abundance

As stated previously, the rate of increase statistic summarizes the population processes of fecundity, mortality, immigration and emigration. The contribution of these processes to changing abundance will depend on the size of the area over which abundance is measured and the frequency at which monitoring occurs.

If the survey area is small (for example, 10's to 100's of square kilometres) then fecundity and mortality are unlikely to contribute to short-term changes in abundance. Instead changes would be due to movement of macropods across the survey area boundaries (i.e., immigration and emigration). On the other hand, when both the survey area and time-scale are large then fecundity and mortality are the most likely causes of changing abundance. When the survey area is small and time-scale large then all four processes may be important.

The above inferences depend of course on the abundance measure being repeatable.

Techniques for measuring abundance

Caughley (1977) and Eberhardt (1978) have reviewed the many techniques available for measuring animal abundance. These can be broadly classified into five main groups:

1. *Total counts.* In a total count the biologist attempts to locate and count every animal in the survey erea. This requires that the survey area is thoroughly and totally (i.e., a 100% sample) searched, that all animals in the area are counted and that no animals are counted twice.

2. *Sample counts.* Often the large size of the survey area rules out the possibility of a total count and some form of sampling is required. Sample counts have the advantages over total counts of requiring less work, reducing the chances of double-counted or missed animals, and being less disruptive to the population (Caughley 1977; Norton-Griffiths 1978). Counts may be of the animals themselves (direct counts) or of signs left by them, such as faeces or tracks (indirect counts).

3. *Mark-recapture methods.* These methods provide an estimate of population size. Absolute density can be easily inferred if the population is circumscribed by a distinct boundary, but if not the inference can be more difficult and dependent on survey design.

In its simplest form, a mark-recapture estimate of numbers requires the marking of a number of animals (M) on one occasion and the recording of the proportion of marked animals (m) in a sample (n) captured on a second occasion. Population size (N) is then estimated as $N=Mn/m$, the Petersen estimate (Caughley 1977). The general assumptions required of mark-recapture methods are:

(a) the probability of capturing an individual is the same for all individuals in the population,

(b) no animal is born or immigrates to the study area between marking and recapturing,

(c) marked and unmarked individuals die or leave the area at the same rate, and

(d) no marks are lost.

In reality, most of these assumptions will be violated to some extent. The Petersen estimate is not very robust to such violations, but there are numerous more complex and robust models available [Otis *et al.* 1978 present models that are robust to violation of assumption (a)].

4. *Addition-removal methods.* In these methods population size is estimated by surveying the population before and after a known number of animals (C) are added to or removed from it. If the manipulation is non-selective (animals of any type are added or removed) and a linear index of density is obtained before (I_1) and after (I_2) the manipulation, the pre-manipulation size (N) can be estimated as $N=I_1C/(I_2-I_1)$ (Caughley 1977).

If the population can be classified into two or more classes — for example, by sex, age or colour — then the manipulation may be selective in being confined to only one of the classes. The methods assume that the population is closed during the manipulation period i.e., there are no births, deaths, immigrations or emigrations.

The main advantage of the index-manipulation-index method lies in calibrating a relative index to an absolute measure. It has no inherent value in monitoring because this is already achieved by the before and after indices. Also, if the manipulation is an artificial one then naturally occurring temporal variations in abundance cannot be monitored because of the disruptive effect of the manipulation. The method is most appropriate when a game population is supplemented by individuals raised at a game farm or hatchery; then the addition is rapid and the assumption that the population is closed throughout the experiment is unlikely to be violated. If animals are removed the manipulation period is longer and the chance of violating this assumption greater.

5. *Catch-effort methods*. When a population is being exploited the trend in abundance may be monitored using catch per unit effort methods. These methods are very cost-effective when the population is being exploited because the data are collected during the course of exploitation, and little extra work is required to calculate the catch-effort index. In the absence of exploitation the method has the disadvantages of requiring more labour and causing more disturbance to the population than alternative methods (Caughley 1977). Catch-effort methods require the following assumptions:

(a) conditions of catching are standardized,
(b) catching efficiency is standardized,
(c) catching gear is standardized,
(d) the catching of one animal does not affect the catching of another, and
(e) animals must not learn to court or avoid capture.

When all of these conditions are met, the regression of absolute density on catch per unit effort is linear through the origin (Ricker 1940).

THE APPLICATION OF MONITORING TECHNIQUES TO KANGAROO AND WALLABY POPULATIONS

Total count, mark-recapture and addition-removal methods

These three techniques have been considered together because of several characteristics they share generally and with regard to the monitoring of macropod abundance, *viz.* they are relevant to the estimation of population size or absolute density only, they have received only infrequent use, they have been applied over small areas only, and they have usually been applied to assess the accuracy of other monitoring techniques of more general utility than for population monitoring *per se* (Table 2).

A total count can be achieved through a variety of means. In Africa, Europe and North America total counts of large mammals have employed aerial photography of groups or herds (Watson and Turner 1965; Sinclair 1973), the census of individually recognized animals (Clutton-Brock *et al.* 1982) and drive counts (Ryel 1959). These counts have been dependent on the animals being conspicuous or on certain unusual characteristics of the population being surveyed or the survey

Table 2. Total count, mark-recapture and addition-removal studies of macropod abundance. Abbreviations are as follows: Species, see Table 1. Method used: C, census of individually recognized animals; DC, drive count; PE, Petersen estimate; RD, removal by drought.

Reference	Size of survey area (km^2)	Species	Method used
TOTAL COUNT			
Kaufmann (1974)	3	WTW	C
Coulson and Raines (1985)	3	EGK	DC
Short and Hone (in press)	12	RDK,WGK WLR	DC
Johnson and Jarman (1987) Southwell *et al.* (in prep)	2	EKG,RNW	C
MARK-RECAPTURE			
Denny (1977, 1978)	91	RDK	PE
Johnson (1977)	—	RNW	PE
Arnold and Maller (1987)	3	WGK	PE
ADDITION-REMOVAL			
Morgan (1983a)	169	WGK	RD
Bayliss (1987)	484	RDK,WGK	RD

area. In Australia only a few attempts have been made at total counting of macropod populations (Table 2). Kaufmann (1974), Johnson and Jarman (1987) and Southwell *et al.* (in prep.) relied on habituation of the population to observers to obtain a total count by censusing individually recognized animals. These studies required close and continual observation over long periods of time to ensure that all animals were individually recognized and none were missed. Counts were obtained at the very high cost of hundreds of person-days effort concentrated into areas of a few square kilometres. Coulson and Raines (1985) and Short and Hone (in press) used a drive count to obtain a total count. Both of their survey areas consisted of long, narrow strips of land surrounded by kangaroo-proof barriers, either large tracts of water or a kangaroo-proof fence. This arrangement allowed the areas to be searched with a line of beaters aligned across the areas' narrow axis. In these studies an effort of 30-40 person-days was required to count over areas of 12 km^2 or less.

Total counts often appeal intuitively as an accurate method of estimating population size, perhaps because they are not subject to sampling error. However, they are subject to other sources of error, notably the obvious one of failing to count all animals accurately. Norton-Griffiths (1978) relates a study where a supposed total count of elephants from the air increased by 40% from the first to the fourth count. The large increase was well beyond the reproductive capacity of the species and was most likely related to improved searching efficiency of the survey team. Similarly, if observational time is insufficient individual censuses can suffer from misidentification and the missing of individuals, or if the organization of a drive count is tardy (there are too few beaters or they are bunched rather than spread evenly), then animals may escape unseen. Even with many beaters and excellent organization, animals that use stealth or crypsis as avoidance strategies can be missed in a drive count. The larger kangaroos rarely hide and are easily seen when moving due to their hopping gait, but the smaller wallabies are more cryptic and the drive count technique is less likely to be successful for them. While these problems are unlikely to have arisen in the above macropod studies they should be borne in mind in future counts.

The infrequent use of mark-recapture methods for monitoring of the larger macropods undoubtedly relates primarily to the difficulties of capturing and marking large animals. All workers who have attempted mark-recapture studies of kangaroo populations (Table 2) have expressed the opinion that the method is too labour intensive to be of general use. Arnold and Maller (1987) report than an average of 15 person-days effort was required to obtain a mark-recapture estimate of Western Grey Kangaroo abundance with a precision of 15% over an area of three km^2.

The sensitivity of the method to assumption violations, and evidence of violation in kangaroo populations, may be other reasons for infrequent use of the method. Mark-recapture assumptions have been shown to be violated in several ways by kangaroo populations. The tendency of kangaroos to suffer from post-capture myopathy (Shepherd 1981) can lead to violations of assumption c (see introduction). Arnold *et al.* (1986) found use of the drug alpha-chloralose to capture Western Grey Kangaroos resulted in the animals developing a taste aversion to the drug and in a differential capture rate for males and females (violations of assumption a). Jarman and Taylor (1983) reported a significant shedding of collars by Eastern Grey Kangaroos and Wallaroos (violation of assumption d). No attempt has been made to assess the degree to which such violations bias estimates of kangaroo population size.

Addition-removal methods have only limited application to monitoring of macropod abundance. Like total counts and mark-recapture methods, the greatest potential for use of addition-removal methods lies in the evaluation of other techniques of more general utility. Morgan (1983a) and Bayliss (1987) have taken advantage of drought-induced mortality as a removal manipulation to assess the accuracy of ground line transect counts and aerial strip transect counts respectively (Table 2). Similarly, total counts and mark-recapture estimates of population size have been used for comparison with sample count estimates of density. These comparisons will be reviewed in more detail in a later section on sample count accuracy.

Catch-effort methods

The control of macropods by professional shooters to alleviate crop and pasture damage in the Australian rangelands is a well-known aspect of kangaroo management. This kind of management practice provides the opportunity for catch-effort monitoring of abundance, and the opportunity has been utilized on most of the seven species subject to commercial culling (Table 3).

One of the most frequently used catch-effort indices of wildlife abundance is number caught per unit time spent hunting. Wildlife authorities in New South Wales, South Australia, Western Australia and Tasmania collect such data for the monitoring of macropod abundance (Table 3). In Tasmania a number of randomly chosen shooters are asked to provide information on number taken and number of days spent hunting via regular questionnaire surveys carried out by the Tasmanian National Parks and Wildlife Service (Anon. 1984a). In New South Wales, South Australia and Western Australia shooters are required to provide wildlife authorities information on number taken and time spent hunting, either as number per night (New South Wales and South

Table 3. Catch-effort monitoring of macropod abundance. Abbreviations are as follows: Species, see Table 1. Catch-effort index: SR, Harvest sex ratio; CW, harvest carcass weight; CS, harvest carcass size; CT, catch per unit time (unit of time); HA, harvest age structure.

Reference	Area (approx. km^2)	State	Species	Catch-effort index
Anon. (1984a)	50 000	TAS	RNW,RBP	CT (hunting day)
Anon. (1987a)	550 000	NSW	RDK,GK	SR,CW,CT (hunting night)
Anon. (1984b), Alexander and Best (1987)	282 000	SA	RDK,WGK, WLR	SR,CW,CT (hunting night)
Anon. (1984c, 1988)	800 000	QLD	RDK,EGK, WLR	SR,CW,CS,HA
Crook and Prince (1984), McNamara and Prince (1986), Prince (1984)	1 000 000	WA	RDK,WGK	SR,CW,CT (hunting hour)

Australia), or more precisely, number taken per hour spent searching and shooting (Western Australia) (Anon. 1987a; Anon. 1984b; Prince 1984).

A variation on this usual index applied to kangaroo populations in the mainland states is monitoring of the sex ratio, age structure or size/ weight of kangaroos harvested. The basis for this index lies in the marked sexual dimorphism of adult kangaroo size, and the difference in size between adults and subadults. Professional kangaroo shooters are paid by weight of carcass or size of skin, and so prefer to take the larger animals wherever possible. The selectivity of shooters is assumed to be determined primarily by the status of the population being hunted, with high abundance allowing high selectivity and low abundance low selectivity. Consequently, parameters such as percent of males, average age, average carcass weight and average skin size have been used to monitor population status of the larger kangaroos (Red Kangaroo, Eastern and Western Grey Kangaroos and Wallaroo) in the mainland states (especially Queensland) where commercial culling occurs.

The situations where catch-effort monitoring has been used on macropod populations provide an insight into the advantages of the method over other techniques. In Tasmania state-wide monitoring of wallaby and pademelon populations is very difficult because of the dense forest habitats and large survey area. Populations can only be monitored directly by roadside counts, but this method itself has disadvantages (see section on direct sample counts). Catch-effort data provide additional information that is interpreted in conjunction with direct counts. The immense size of the two largest mainland states, Queensland and Western Australia, makes state-wide monitoring of kangaroo populations logistically very difficult and expensive, even with sampling. Wildlife authorities in these states have consequently placed considerable reliance on the more cost-effective catch-effort methods. In the smaller states of New South Wales and South Australia broad-scale monitoring is principally based on direct sample counts, but catch-effort data are also collected as a secondary source of monitoring data.

The advantages of cost-effectiveness, however, must be weighed against the potential disadvantages and pitfalls of catch-effort monitoring. For kangaroos (and large mammals in general) there are strong theoretical reasons for the relationship between number shot per unit time spent hunting and abundance to be non-linear. The reason is time saturation (Caughley 1977), whereby the number shot is ultimately related to the time available to a hunter for shooting than to the abundance of animals (this is a violation of the assumption that the catching of one animal does not affect the catching of another). Thus the form of the relationship is more likely to be curvilinear than linear with slope decreasing as abundance increases until, at a point of total saturation, the slope is zero and catch per unit effort is thereafter constant regardless of abundance (C in Figure 1). A shooter operating in the zone of saturation would return the same catch per unit effort whether populations were increasing, stable or decreasing. This effect is most likely to occur when the catch-effort index is relatively coarse (for example, number shot per week), but even for more precise indices (such as number shot per hour spent searching and shooting) some departures from linearity at high densities would be expected. It is therefore important to know the form of the relationship between catch-effort indices and abundance and the level of absolute density at which any significant departure from linearity occurs. The problem of curvilinearity may also apply to sex, age and size-related catch-effort indices.

The success of a kangaroo shooter is known to vary with weather conditions, being lowest in windy or wet weather, or following periods of heavy rain when travel is difficult and ground cover high (Prince 1984). This is a violation of the assumption requiring standardized conditions of catching. A more subtle violation of this assumption could result from broad-scale and long-term changes in habitat; for example, widespread clearance of tree cover, which could allow a gradual increase in areas accessible to hunting, or widespread regeneration of tree cover, which could have the opposite effect.

Technical improvements in gear used for shooting kangaroos could also confound catch-effort indices. Unlike fisheries, where technical improvement in catching gear has been very rapid, the equipment used for kangaroo shooting has changed little over many years. Nevertheless, one needs to monitor equipment to ensure that any change is detected.

Another problem is that catch-effort data alone will not effectively monitor abundance if the size of the area occupied by the population changes but density within this area does not, because areas of low abundance are usually not sampled by shooters. Additional information on the area over which shooting occurs and/or the number of shooters operating would be required for interpretation of catch-effort data in this situation.

Of course, these are only potential problems and would not of necessity invalidate catch-effort monitoring of macropod populations within the context of the monitoring aims. For example, if the aim of monitoring is to ensure that a population does not decline to a dangerously low level, it may not matter that a saturation effect occurs at high (=safe) levels of abundance. Unfortunately, it is presently difficult to assess the extent to which the above problems are realized and their influence on the effectiveness of catch-effort monitoring because developmental research on the method has been limited.

One of the few studies of catch-effort monitoring of macropod populations is the development of a simulation model by Nance (1985) and T. H. Kirkpatrick (pers. comm.) which aims to determine whether harvest parameters provide a reliable index of population status. This model has been developed specifically for the Eastern Grey Kangaroo but the investigators are confident it can be extended to other macropod species by incorporation of relevant data on reproduction and mortality. Nance (1985) concluded that the harvest sex ratio is a reliable index of Eastern Grey Kangaroo population status. It is important to note that the Nance-Kirkpatrick model aims to monitor population status rather than abundance *per se*, population status being defined as a state of over- or under-harvest, or the probability of a population's long-term survival when subject to a particular harvest level. Hence the existence of a saturation effect would be unimportant to this model provided harvest levels within the zone of saturation do not threaten the population's long-term survival.

A second study of catch-effort monitoring is that by Prince (1984) for the Red Kangaroo in Western Australia. He adopted an empirical approach by examining spatial and temporal variations in historical harvest data and relating these variations to environmental variables, particularly rainfall. Prince (1984) concluded that, despite expected biases due to effects of short and long-term weather patterns on the behaviour of both kangaro populations and kangaroo shooters, the overall patterns of variation supported the ability of catch-effort parameters to monitor kangaroo abundance or status.

Neither of these studies has addressed directly the relationship between catch-effort parameters and abundance; in the first case this was deliberate as the model was concerned with population status rather than abundance *per se,* and in the second case a lack of abundance data forced the drawing of indirect inferences only. Knowledge on the effectiveness of catch-effort monitoring of macropod abundance is currently limited by the lack of any direct investigations.

A final problem of catch-effort monitoring is that data quality is not under the control of the biologist. The reliability of the method may therefore be impaired by tardy data recording or deliberate falsification of data by hunters. This problem can be reduced by designing elaborate systems of data collection which incorporate cross-checks on data validity. Prince (1984) did not consider deliberate data falsification to be a major problem in Western Australia, arguing that the allowance for professional shooting in the kangaroo management programme encouraged shooters to act responsibly. Prince did, however, refer to persistent problems in obtaining full supporting detail additional to the basic data on numbers and weight.

SAMPLE COUNTS

Direct sample counts

Direct counts can be made from fixed points, within quadrats or along transects. These methods differ in the shape of the sampling unit and the means by which they are searched. In fixed-point counts observers stationed at a number of points throughout the survey area count the number of animals seen or heard within a defined time. Quadrats may be any shape but are characterized by having similar dimensions of length and breadth. They are searched by traversing the area several times in any desired pattern (zig-zag, systematic, random) until one is satisfied that all animals within the quadrat have been counted. Transects are long, narrow sampling units that are searched by a single traverse of the long axis (the transect line).

(i) FIXED-POINT COUNTS

Fixed-point counts have been used frequently to monitor bird populations (for example, De Sante 1981; Pyke and Recher 1985), but have received infrequent use on large mammals. Likely factors responsible for this differential use between animal groups are a lack of response to observers and high densities of many bird populations compared with mammal populations. The few fixed-point counts of macropods have been conducted at only one or a few stations in habitats unrepresentative of the species preferred habitat. Hence, the counts have been of no use in monitoring overall abundance. Instead they have monitored utilization by macropods of a particular habitat or resource, for example, paddocks of improved pasture (Johnson 1977; Naarding 1979) and drinking sites (Ealey 1967). In Johnson's study counts were made visually, but in the other two studies counts were made with simple mechanical devices.

(ii) QUADRAT COUNTS

Quadrat counts have been popular in aerial survey of large mammals in North America (for example, Siniff and Skoog 1964; Evans *et al.* 1966; Goddard 1969), but have received little use in Africa and Australia. Quadrat counts have several disadvantages compared to transect counts. They are relatively inefficient because the sampling effort is concentrated into a small locality, the boundaries of the quadrat can be difficult to locate, and travel time between quadrats can be high. A major disadvantage is that animals may leave the quadrat before the surveyor has been able to search all of it. These problems are lessened by using transects. To my knowledge there has been no attempt to use quadrat counts for the monitoring of macropod abundance.

(iii) TRANSECT COUNTS

There are two basic types of transects: strip transects and line transects. In the strip transect method all animals observed within a specified distance (the strip half-width) from the transect line are counted, and animals beyond that distance are ignored. The strip half-width may be fixed for the entire length of the transect, or it may be varied along the transect according to sightability (Figure 2). In the line transect method, all animals seen are counted regardless of their distance from the transect line, and information on the distance and angle from observer to the animal when first sighted are recorded (Figure 2). Density from a strip transect count is obtained simply by dividing the number of animals seen by the area of the strip. Density from a line transect count is obtained by modelling the decline in sightings away from the transect line or observer with an appropriate line transect model.

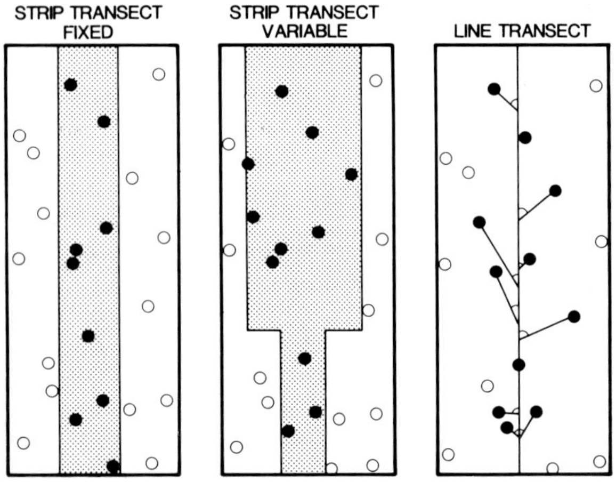

Fig. 2. Strip and line transects. Closed circles represent animals that are seen and recorded, open circles represent animals that are either not seen (line transect) or not recorded (strip transect). Shaded region is the area on which strip transect density estimation is based.

Both methods require satisfactory fulfillment of several assumptions for accurate density estimation. For strip transects animals must be accurately classified as in or out of the strip. For line transects distances and angles must be measured accurately. For both methods animals must not move large distances away from or toward the transect line before being sighted and animals must not be counted twice. For strip transects, all animals within the strip must be seen, but for line transects only all animals directly on the line must be seen.

Failure of these assumptions would invalidate accurate density estimation but would not necessarily invalidate estimation of a density index. For example, density will be underestimated if some animals within a strip transect are missed, but if the proportion missed from survey to survey is constant the counts will still provide a valid density index for population monitoring.

If animals are counted on transects but no information on distance or area surveyed is collected, estimation of absolute density is not possible. Expressed in the form of number counted per unit length of transect, the information may still be useful as a relative index provided observing conditions remain constant from survey to survey.

Details of macropod transect surveys are summarized in Tables 4-6. The surveys have been categorized according to the mode of travel along transects (aircraft, vehicle and foot) and to the scale of the survey: (broad-scale: >50 000 km^2, medium scale: 1 000-50 000 km^2 and small-scale: <1 000 km^2).

Aerial survey was first applied to kangaroo populations in the 1960s to investigate short-term changes in abundance over medium-sized areas (Frith 1964; Newsome 1965; Bailey 1971). It was not until a decade later, following the developmental work of Caughley and his colleagues (Caughley *et al.* 1976) that aerial survey was used on a regular basis over broad-scale areas. After initial surveys of the pastoral areas of the four mainland states where commercial culling of kangaroos occurs (New South Wales in 1975 — Caughley *et al.* 1977; South Australia in 1978 — Caughley and Grigg 1981; Queensland in 1980 — Caughley and Grigg 1982, and Western Australia in 1981 — Short *et al.* 1983), broad-scale aerial survey has been widely adopted as a regular monitoring procedure by state and federal wildlife authorities. Surveys have since been maintained on an annual basis in New South Wales, South Australia and Queensland (since 1984), and on a tri-annual basis in Western Australia. In 1980-81 and 1984 additional areas in these states beyond the pastoral zones, as well as parts of Victoria and the Northern Territory (1980-81 only), were surveyed with the aim of obtaining an Australia-wide population estimate for the three large kangaroo species (Caughley *et al.* 1983; Grigg *et al.* 1985).

With annual surveys now spanning over a decade in New South Wales and South Australia, and several years in Western Australia and Queensland, a solid data-base for the examination of kangaroo population dynamics exists. These surveys have also provided valuable information on broad-scale distribution patterns (Caughey *et al.* 1987).

The application of aerial survey to small and medium-sized areas has generally been on a one-off or short-term basis (Table 4). The work of Dudzinski *et al.* 1982) and Bayliss (1985), with 82 and 23 repeated surveys respectively, are exceptions. Even in small-scale aerial surveys few workers have attempted to obtain fine-grained information about the population, such as habitat utilization or population composition, as aerial survey is not suited to this purpose.

For reasons of safety and visibility aerial surveys have concentrated in the flat and sparsely vegetated interior of the continent and have been used only on the larger species of kangaroo (Red, Eastern Grey and Western Grey Kangaroo — Table 4). Because of cost almost all aerial surveys have employed fixed wing aircraft (usually a Cessna 182) in preference to helicopters.

All aerial surveys have used strip transects in preference to line transects. Line transects have several practical and theoretical disadvantages compared to strip transects for aerial survey (see section on accuracy). The earliest aerial surveys used strips of variable width (Frith 1964; Newsome 1965; Dudzinski *et al.* 1982), but all subsequent surveys have used strips of fixed width with correction factors applied to counts where necessary (correction factor work is reviewed in the section on accuracy).

The majority of vehicle surveys have been undertaken in small to medium-sized areas on a one-off basis (Table 5). A few workers have carried out repeated surveys (McKenzie 1976; Bradley 1978), but the interval between surveys and size of the survey area were so small that any variations in counts were more likely to reflect movement of animals across survey area boundaries or poor repeatability of the method than overall change in abundance. The only application of vehicle surveys over broad-scale areas occurs in Tasmania where wallaby populations are monitored on an annual basis. Vehicle surveys have been undertaken in both the arid interior and wetter coastal areas. Both strip and line transect techniques have been used, but the former has been used most frequently, and in all cases strip transects have been of variable width. No distance measurements are taken on roadside counts of wallabies in Tasmania, so these counts are used only as relative indices. Unlike aerial survey, vehicle surveys have been undertaken with a variety of aims additional to population estimation, including determination of habitat utilization (McKenzie 1976; Bradley 1978; Johnson and Bayliss 1981), the assessment of population composition (Johnson and Bayliss 1981) and calibration of aerial survey (Short and Bayliss 1985).

Predictably, most foot surveys have covered only small areas (Table 6). Two exceptions are regular surveys of Wallaroos in South Australia (Anon. 1984b; Alexander and Best 1987) and a recent survey of Whiptail Wallabies in south-east Queensland (Southwell and Fletcher 1988). Small-scale surveys have generally been one-off studies or have involved repeated surveys separated by short intervals. Contrasting with this generalization are the annual surveys of kangaroo populations at Wyperfeld National Park in Victoria by Morgan (1983a, 1984a) which, with a total of 16 repetitions, form the longest running monitoring study of kangaroo abundance. There has been a strong bias in the distribution of foot surveys toward the wet, forested areas of the continent and toward the use of line transect methodology (Table 6). Like vehicle survey, many foot surveys have been carried out for reasons other than population estimation *per se*.

(a) *Survey design*

One of the most important aspects in the design of a transect survey is the placement of transects within the survey area. For unbiased inferences about abundance transects should ideally be randomly located. However, strict random placement can often lead to logistical difficulties in locating and accessing transects. The overriding practical principle for unbiased sampling is that the distribution of transects is independent of the distribution of animals being surveyed.

Vehicle surveys suffer from a major disadvantage over foot and aerial surveys; one is usually restricted to counting from roads or tracks. All but one of the vehicle surveys listed in Table 5 were conducted along roads or tracks. Since roads are invariably non-random in their distribution (avoiding where possible rough and difficult terrain and tending along rather than across contours), the principle of random or independent location of transects is violated and the potential for biased counts is high. For this reason estimates of absolute density from vehicle surveys conducted along roads should be considered cautiously. However, provided the density of animals along roads bears a constant relationship to the density of animals in the survey area, counts along roads, even if biased, can be useful in monitoring population trends.

Transects conducted on foot are rarely constrained in their location by vegetation or topography as are vehicle counts, and provided the topography is flat, aerial transects are similarly unconstrained. However, even for the foot and aerial surveys listed in Tables 4 and 6 strict random placement of transects has rarely been employed. For the practical reasons already mentioned, transects have more often been placed systematically (usually parallel and equally

Table 4. Details of macropod aerial surveys. Abbreviations are as follows: Species: See Table 1. Type of transect: FS, fixed strip; VS, variable strip. Orientation of Transects: P, parallel; EW, east-west; NS, north-south; Rd, radiating from central point; B, relative to survey area boundaries; V, relative to pattern within survey area, to maximize variance within and minimize variance between transects; N, subjective location between known points, but independent of population; R, random. Spacing between transects: S, systematic; R, random; N, relative to known points. Regularity: First number indicates number of repetitions at time of writing, number in brackets indicates interval in months between repetitions.

Reference	Size of survey area (km^2)	State	Species	Type of transect	Speed (km hr^{-1})	Height (m)	Strip half-width (m)	Use of stratification	Orientation of transects	Spacing between transects	Regularity
Broad-scale (>50 000 km^2)											
Caughley *et al.* (1977, 1979, 1984), Anon. (1987a)	496 000	NSW	RDK,GK	FS	185	76	200	No	P,EW	R	12(12)
Caughley and Grigg (1981), Anon. (1984b) Alexander and Best (1987), Grigg *et al.* (1987)	207 000	SA	RDK,WGK	FS	185	76	200	No	P,EW	S	10(12)
Caughley and Grigg (1982), Caughley (unpub.), Southwell and Fletcher (unpub.)	537 000	QLD	RDK,GK	FS	185	76	200	No	P,EW	S	5(12)
Short and Grigg (1982)	133 000	VIC,SA	RDK,GK	FS	185	76	200	No	P,EW	S	1(–)
Short *et al.* (1983), Fletcher and Southwell (1984, 1988)	1 528 000	WA	RDK,WGK	FS	185	76	200	No	P,EW	S	3(36)
Caughley *et al.* (1983), Grigg *et al.* (1985)	5 800 000	NSW,QLD NT,WA SA,VIC	RDK,GK	FS	185	76	200	No	P,EW	S	2(36)
Medium-scale (1 000-50 000 km^2)											
Frith (1964)	2 300	NSW	RDK	VS	145	91	800-1 300	Yes	P,NS	S	8(1-24)
Frith (1964)	8 200	NSW	RDK	VS	145	91	800-1 300	Yes	N	N	4(4-32)
Newsome (1965)	2 642	NT	RDK	VS	128	91	116-434	Yes	V,R	R	2(7)
Bailey (1971)	25 600	NSW	RDK	FS	145	–	256	No	Rd	N	6(3-17)
Denny (1979)	6 900	NSW	RDK	FS	162	100	100	No	Single transect		1(–)
Hill (1981)	47 100	QLD	EGK	FS	160	76	200	Yes	P,EW	R	1(–)
Short and Grigg (1982)	1 107	VIC	RDK,GK	FS	185	76	200	No	P,EW	S	1(–)
Small-scale (<1 000 km^2)											
Dudzinski *et al.* (1982)	170	NT	RDK	VS	–	30	200-800	Yes	–	–	82(0.5)
Short and Grigg (1982)	480	VIC	RDK,WGK	FS	185	76	200	No	P,EW	S	1(–)
Short and Grigg (1982)	500	VIC	RDK,GK	FS	185	76	200	No	P,EW	S	1(–)
Short and Grigg (1982)	350	VIC	RDK,GK	FS	185	76	200	No	P,EW	S	1(–)
Caughley *et al.* (1985)	116	NSW	RDK,WGK	FS	185	76	200	No	P,EW,NS	R	2(8)
Short and Bayliss (1985)	226	NSW	RDK,GK	FS	185	76	200	No	P,B	R	1(–)
Short and Bayliss (1985)	130	NSW	RDK,WGK	FS	185	76	200	No	P,B	R	1(–)
Bayliss (1985), Bayliss and Giles (1985)	484	NSW	RDK,WGK	FS	167	90	100	No	P,B	S	23(3)
Southwell *et al.* (1986)	300	QLD	RDK,WLR	FS	185	76	200	No	P,EW	S	1(–)
Short and Hone (in press)	12	NSW	RDK,WGK,WLR	FS	185	76	200	No	P,B	S	2(12)

Table 5. Details of macropod vehicle surveys. Abbreviations are as follows: Species: See Table 1. Type of transect: VS, variable strip; LT, line transect; I, index. Orientation of transects: T, along roads or tracks; R, random. Spacing between transects: R, random. Regularity: First number indicates the number of repetitions at time of writing, number in brackets indicates interval in months between repetitions.

Reference	Size of survey area (km^2)	State	Species	Type of transect	Speed (km hr^{-1})	Strip half-width (m)	Use of stratification	Orientation of transects	Spacing between transects	Regularity
Broad-scale (>50 000 km^2)										
Anon. (1984a), Hocking (unpub.)	50 000	TAS	RNW,RBP	I	20	–	No	T	–	12(12)
Medium-scale (1 000-50 000 km^2)										
McNamara (1976)	3 400	NSW	RDK	LT	–	–	No	T	–	1(–)
Denny (1978)	3 890	NSW	RDK	VS	16	–	No	T	–	1(–)
Denny (1978)	9 000	QLD	RDK	VS	16	–	No	T	–	1(–)
Denny (1978)	1 154	QLD	RDK	VS	16	–	No	T	–	1(–)
Denny (1979)	6 900	NSW	RDK	VS	–	–	No	T	–	1(–)
Anon. (1984b), Alexander and Best (1987)	>2 500	SA	RDK,WGK	VS	25-30	50-100	No	T	–	6(12)
Small-scale (<1 000 km^2)										
McKenzie (1976)	60	NSW	EGK,RNW	LT	20	–	Yes	T	–	12(0.3)
Bradley (1978)	30	NSW	EGK,RNW, SPW,RNP	LT	10	–	Yes	T	–	6(1)
Denny (1978)	543	NSW	RDK	VS	16	–	No	T	–	1(–)
Denny (1978)	555	NSW	RDK	VS	16	–	No	T	–	1(-)
Denny (1978)	280	NSW	RDK	VS	16	–	No	T	–	1(–)
Johnson and Bayliss (1981)	250	NSW	RDK,WGK	VS	–	125-295	Yes	T	–	4(4-12)
Short and Bayliss (1985)	226	NSW	RDK,WGK	VS	–	30-100	Yes	R	R	1(–)
Short and Bayliss (1985)	130	NSW	RDK,WGK	VS	–	30-100	No	T	–	1(–)
Southwell and Fletcher (1985b)	300	QLD	WTW	LT	15	–	No	T	–	1(–)

Table 6. Details of macropod surveys by foot. Abbreviations are as follows: Species: See Table 1. Type of transect: FS, Fixed strip; VS, variable strip; LT, line transect. Orientation of transects: P, parallel; R, random; N, non-random but independent of population; T, along roads or tracks. Spacing between transects: S, systematic; R, random; N, non-random but independent of population. Regularity: First number indicates number of repetitions at time of writing, number in brackets indicates interval in months between repetitions.

Reference	Size of survey area (km^2)	State	Species	Type of transect	Speed ($km\ hr^{-1}$)	Strip half-width (m)	Use of stratification	Orientation of transects	Spacing between transects	Regularity
Broad-scale (>50 000 km²)										
Anon. (1984b), Alexander and Best (1987)	67 000	SA	WLR	FS	–	50	Yes	P,R	S,N	3(12-24)
Southwell and Fletcher (1988)	146 000	QLD	WTW	LT	1	–	Yes	R	R	1(–)
Small-scale (<1 000 km²)										
Gibson (1975)	20	NSW	EGK,WLR	VS	–	50-1 500	Yes	N	N	5(1)
Robertshaw and Harden (1976)	70	NSW	SPW,RNW,EGK	LT	3	–	No	T	–	18(1)
Robertshaw and Harden (1976)	40	NSW	SPW,RNW,EGK	LT	3	–	No	T	–	18(1)
Robertshaw and Harden (1976)	25	NSW	SPW,RNW,EGK	LT	3	–	No	T	–	18(1)
McKenzie (1976)	60	NSW	EGK,RNW,SPW, RNP	LT	–	–	Yes	T	–	7(0.5)
Johnson (1977)	4	TAS	RNW,RBP	LT	–	–	No	N	N	
Bradley (1978)	30	NSW	EGK,RNW,SPW, RNP	LT	–	–	Yes	T	–	6(1)
Taylor (1982)	10	NSW	EGK,WLR	VS	–	–	Yes	N	N	12(1)
Taylor (1982)	10	NSW	EGK,WLR	VS	–	–	Yes	N	N	12(1)
Morgan (1983a, 1984a)	169	VIC	WGK	LT	–	–	Yes	N	N	16(12)
Morgan (1983b, 1984b)	495	VIC	RDK,WGK	LT	–	–	Yes	P	S	6(12)
Morgan (1983c)	424	VIC	EGK	LT	–	–	Yes	P	S	3(12-24)
Southwell (1984)	3	NSW	EGK	LT	2	–	No	N	N	8(3)
Southwell (1984)	3	NSW	EGK	LT	2	–	No	N	N	4(3)
Southwell (1984)	4	NSW	EGK	LT	2	–	No	N	N	4(3)
Southwell (1984)	6	NSW	EGK	LT	2	–	No	N	N	1(–)
Southwell (1984)	8	NSW	EGK	LT	2	–	No	N	N	1(–)
Coulson and Raines (1985)	3	VIC	EGK	LT	1	–	No	P	S	1(–)
Southwell (1986)	160	QLD	EGK	LT	2	–	No	R	R	1(–)
Arnold and Maller (1987)	3	WA	WGK	LT	2	–	No	P	S	25(0.2)
Croft (1987)	6	NT	AWL	FS	–	50	No	N	–	–
Southwell *et al.* (in prep.)	2	NSW	EGK,RNW	LT	2	–	No	P	S	18(1)
Southwell and Fletcher (unpub.)	12	NSW	RDK,WGK,WLR	LT	2	–	No	P	S	1(–)

spaced). This design will be unbiased provided the transects do not match some systematic pattern in the survey area. Systematic placement has the theoretical disadvantage of providing an invalid basis for estimation of precision. Caughley (1977) argues that robustness is a more important and useful concept to the biologist that statistical validity, and on the basis that estimation of precision is generally robust to departures from random sampling, recommended the more practical option of systematic sampling for aerial survey. As is evident from Table 4, most workers have adopted Caughley's recommendation.

In aerial survey parallel transects have generally been oriented in an east-west direction. Since aerial surveys are usually conducted in the early morning and late afternoon when visibility is best and the sun relatively low in the sky, east-west transects provide more even sighting conditions for observers on each side of an aircraft than would north-south transects. In broad-scale surveys an east-west orientation also offers the advantage of transects being easily allocated to a degree block or map sheet, the usual framework for the survey design. No such consistent directional orientation of transects has occurred for foot surveys. In most foot surveys transects have been oriented in relation to the direction of survey area boundaries or located at points that are easily recognized on the ground.

Another important aspect of survey design is stratification. If the distribution of animals in a survey area is clumped the most precise estimate of density is obtained by dividing the area into strata within which distribution is homogeneous and combining densities measured separately in each strata. From another viewpoint, stratification allows a given precision to be achieved with the most efficient use of resources.

Few broad-scale macropod surveys have incorporated stratification into their survey design. The main reason is that surveyors often count more than one species in a single survey, and the species have differing and sometimes opposing distributions. Red and Grey Kangaroos occur widely across inland Australia and usually both are counted on aerial surveys (Table 4), but the proportion of Reds counted to Greys increases with increasing aridity. Similarly, Red-necked Wallabies and Red-bellied Pademelons occur widely across Tasmania and are both counted during annual vehicle surveys (Table 5), but the proportion of wallabies counted to pademelons is highest in dry scleropyll forest and lowest in wet scleropyll forest. In these cases stratification would improve precision for one species but lessen it for the other. If equal importance is placed on both species the most efficient overall design may not incorporate stratification.

A recent broad-scale ground survey of Whiptail Wallaby populations (Southwell and Fletcher 1988) provides an example of where stratification can be useful. Whiptail Wallabies have very specific habitat requirements, preferring hilly country with a grassy open forest vegetation (Southwell 1987; Calaby 1966). Such habitat occurs as discrete units within a background of unsuitable habitat. The survey area of 146 000 km^2 was divided into three habitat strata; optimal (steep, open forest, 32 000 km^2), sub-optimal (undulating, open forest, 32 000 km^2) and unsuitable (flat and/or vegetation either cleared or dense forest, 82 000 km^2), with optimal habitat being sampled more intensively that sub-optimal habitat, and unsuitable habitat receiving no sampling. Stratification thus substantially reduced the size of the area sampled, and provided a more precise estimate of density than would have unstratified sampling. Although data on all species of macropod observed on transects were recorded, these species were of secondary importance to the Whiptail Wallaby, and it was accepted that density estimates for them would be less precise than for the target species.

In contrast to broad-scale survey, stratification has been employed in many medium to small-scale macropod surveys. However, usually this has been done to obtain information on density within each strata (=habitat utilization) rather than to improve the efficiency of density estimation over the entire survey area.

(b) *Precision and sampling intensity*

All sample counts are subject to sampling error. Sampling error or precision is most conveniently expressed as a coefficient of variation (C.V.), the standard error of counts expressed as a percentage of the mean. Ideally one attempts to obtain a very precise estimate of abundance (very low C.V.), so that real trends are not masked by large sampling errors. However, this usually requires a very large effort in counting, and a reasonable compromise between precision and effort must be met. The effort required to obtain an estimate of specified precision is the sampling intensity.

Sampling intensity is best determined from a small pilot study undertaken prior to a full-scale survey. The small amount of effort expended is usually more than compensated for by the efficient use of resources in the survey proper. In the absence of a pilot study the surveyor must either guess or be guided by the efforts of others. As is obvious from Tables 4-6, there are now numerous studies to which surveyors may refer for guidance.

Sampling intensity is influenced by several factors, Robinette *et al.* (1958) listing the more important ones as the density of target objects, survey area size, survey design and the desired precision. I have examined data from previous macropod surveys to determine whether any general relationships between these variables are evident for use as guidelines in future studies. Only data from studies where transects were replicated over space (rather than time) were considered. Most survey designs were unstratified, but for the few stratified designs data were considered separately for each strata. Animal density was divided into two broad categories of low (1-10 km^{-2}) and high (>10 km^{-2}). The efforts expended in each survey or strata (as kilometre of transect) were standardized to a common precision of 15% (C.V.) according to the equation

$$K_R = KxCV^2/225$$

where K_R= kilometre of transect required for 15% C.V., K=kilometre of transect undertaken for the survey and C.V.=coefficient of variation obtained in the survey. In Figures 3 and 4 the standardized efforts are plotted against survey area size.

For high density populations (Fig. 3) a reasonable relationship exists, effort increasing logarithmically with survey area size. Thus, for areas of around 10 km^2 approximately 50 km of transect have been required to obtain a precision of 15%. Successive order-of-magnitude increases in survey area size have each required approximately an additonal 260 km of transect, so that for areas of 50 000 km^2 around 1 000 km of transect have been required. The form of this relationship probably reflects varying degrees of contagion in macropod populations; reasonably uniform in very small (<10 km^2) homogeneous areas, much more clumped in larger areas (1 000s of km^2) containing a diversity of habitats, and only marginally more clumped in very large areas (10s of 1 000s of km^2 or greater) containing a slightly more diverse array of habitats. Of note is the fact that the speed of the surveyor has little or no effect on precision of counts; aerial, vehicle and foot surveys differ from each other in speed by one or two orders of magnitude, but the precision returned per length of transect is similar for the three means of travel (Fig. 3).

For low density populations the relationship between effort and survey area size is less clear (Fig. 4). Again a general logarithmic increase in effort with survey area size is evident, but there

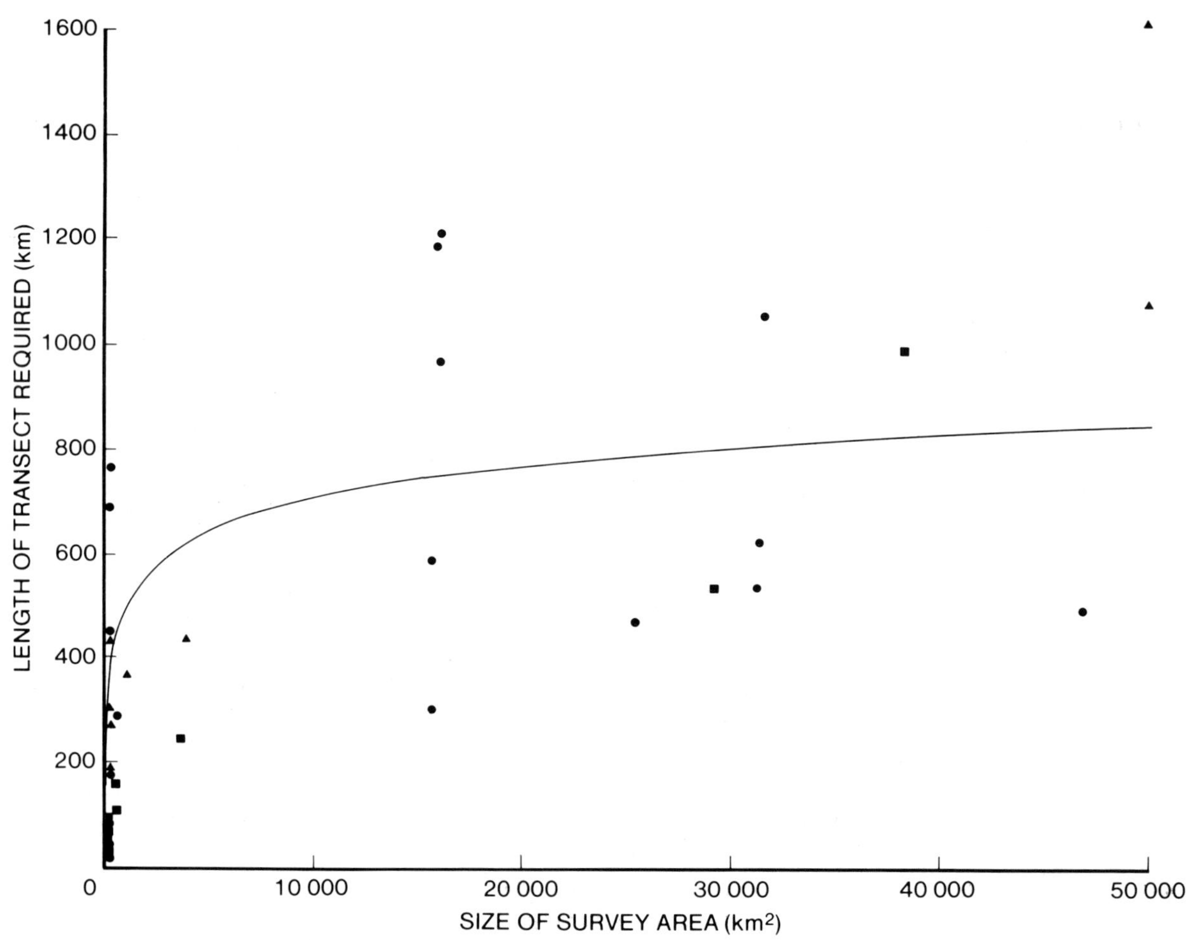

Fig. 3. Length of transect required for a 15% precision against survey area for high density macropod populations (>10 km^{-2}). Circles: aerial survey; Triangles: vehicle survey; Squares: foot survey. Curve of best fit: $y=205\ log_{10}x-118$, $r^2=0.58$, $F_{1,40}=56.14$, P=0.0000.

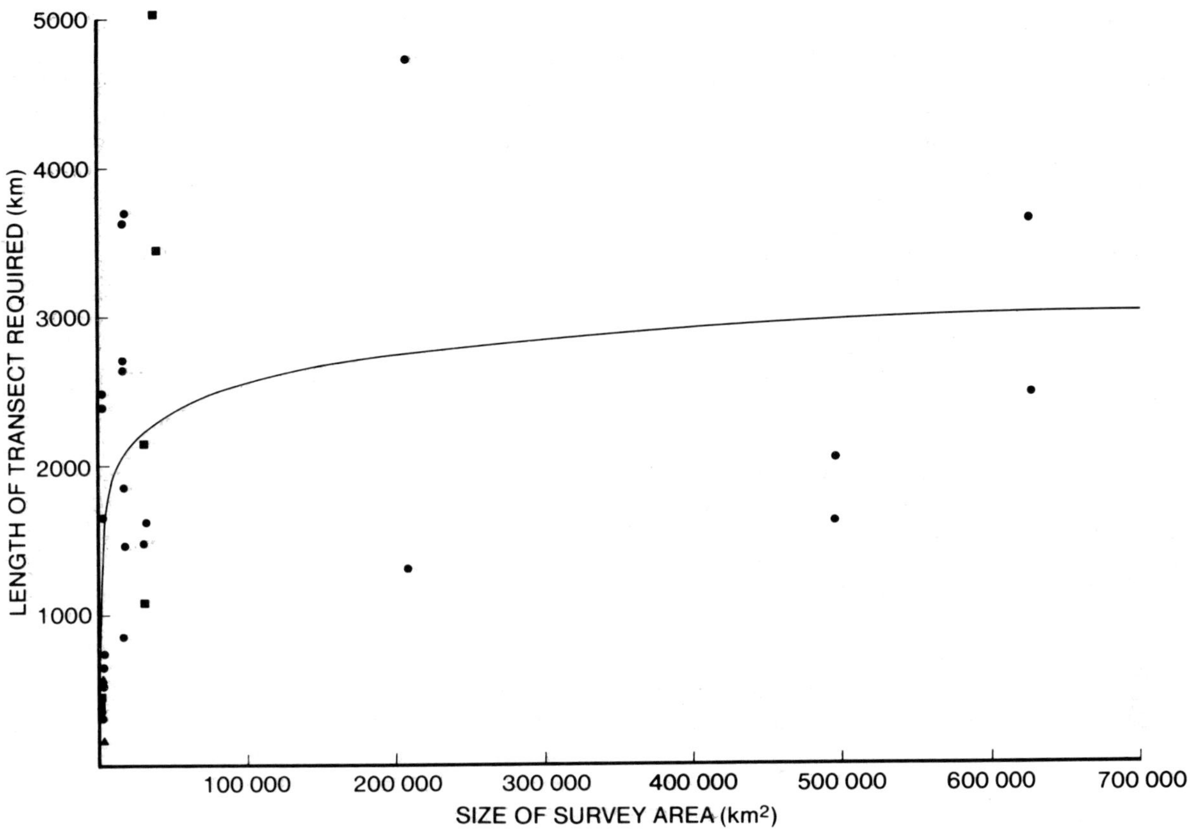

Fig. 4. Length of transect required for a 15% precision against survey area size for low density macropod populations (<10 km^{-2}). Circles: aerial survey; Triangles: vehicle survey; Squares: foot survey. Curve of best fit: $y=586 \log_{10}x-390$, $r^2=0.34$, $F_{1,30}=15.49$, $P=0.0005$.

is more variability about this trend. As would be theoretically expected, low density populations have required more effort in sampling than high density populations.

(c) Repeatability

Many factors are known to affect counts of macropods. These include time of day (Hill *et al.* 1985; Short and Hone, in press) temperature (Bayliss and Giles 1985), cloud cover (Short and Bayliss 1985), vegetation density (Caughley *et al.* 1976; Southwell and Fletcher 1985a), strip width (Caughley *et al.* 1976) and height above ground (Caughley *et al.* 1976). With so many factors affecting countability, the question arises of how repeatable are transect counts of macropods.

The effects of many factors on countability can be negated by strict standardization of counting procedures. Broad-scale aerial surveys of kangaroos are standardized for speed, height, strip width, time of day, season and observer ability. Since regular broad-scale surveys began these variables have generally been applied consistently by kangaroo aerial survey practitioners (Table 4), enabling raw counts to be directly compared across space and time. Similarly, roadside counts of wallabies in Tasmania are standardized for vehicle speed, observer position, time of night, season and observer ability (Anon. 1984a).

However, some factors that affect countability are beyond the control of the observer. Lighting and weather conditions can change within days and between days, and both are known to affect countability of kangaroos from the air (Bayliss and Giles 1985; Short and Bayliss 1985). Use of roadside verges by macropods may vary within and between seasons according to rain-induced variability in grass and herb growth. The ability to standardize for such variables is limited. In theory surveys can be postponed or re-organized so that counting occurs only within a narrow range of environmental conditions, but in practice this is often not feasible. One is forced to standardize within limits and then either ignore or correct for variation beyond these limits.

Broad-scale aerial surveys of kangaroos are routinely undertaken in the winter months when temperatures are low and skies clear. Nevertheless, occasional unseasonal weather can affect counts. Bayliss and Giles (1985) have developed temperature correction factors for aerial counts of Red and Western Grey Kangaroos enabling the effects of unseasonally warm weather to be accounted for. They found that countability was constant in temperatures below 15°C, but declined with increasing temperature beyond that threshold. Aerial counts are now corrected for temperature when necessary according to Bayliss and Giles' correction factors. G. Caughley (pers. comm.) has recently undertaken experiments to assess the effects of environmental variables additional to temperature on aerial counts of kangaroos, allowing further improvement in repeatability.

Effects on repeatability of aerial strip transects due to characteristics of the population being surveyed have not yet been investigated for the macropods. A number of effects relating to variable dispersion are possible. Kangaroos are gregarious animals, but group size is known to vary with absolute density (Southwell 1984). Since group size affects countability for Elk, *Cervus elephus*, (Samuel *et al.* 1987) and cattle (Newsome *et al.* 1981), countability of kangaroos may vary with density. A similar effect may occur in relation to climatic conditions, as kangaroos are frequently reported to aggregate around watering points during drought and disperse more evenly in good conditions. Variable habitat utilization may also affect repeatability. To give a simple and extreme example, suppose a survey area consists of two habitats, grassland and dense mulga, which are identical in area but differ in sightability (1 and 0.1 respectively). Suppose also that the area is occupied by a constant number of kangaroos, and that the distribution of kangaroos varies over time (90% in grassland and 10% in shrubland at one time, the opposite at a second time). An index derived from an uncorrected strip transect count under these conditions would be almost five times higher at time one than at time two, even though population size was constant. This confounding would not occur if counts are adjusted with correction factors that are accurate and specific to each vegetation type. Studies by Hill (1982) on the Eastern Grey Kangaroo and Newsome (1965) and Dudzinski *et al.* (1982) on the Red Kangaroo indicate a tendency for kangaroos to utilize open areas more frequently in poor conditions and wooded areas more frequently in good conditions, suggesting that a potential for some confounding of uncorrected aerial counts may be realized.

Repeatability of roadside counts may be confounded by variable use of the "roadside verge" habitat in a similar way to that described above. Bradley (1978) found monthly counts of Red-necked Wallabies and Eastern Grey Kangaroos along forestry tracks in north-east New South Wales to steadily decline over a six-month period of study. Progulske and Duerre (1964) and McCullough (1982) found roadside counts of deer to also vary between months. This variability in counts is more likely the result of differential use of roadside habitats than of actual change in abundance. Research into factors affecting the use of roadside habitats by macropods is needed to improve this method, which is one of the few techniques available for monitoring macropod populations in dense forest habitats.

Other things being equal, line transect counts should be more repeatable than strip transect and index counts. In line transect methodology variations in countability are automatically accounted for; the only variation that must not occur is the critical assumption that all animals on the line are counted. Data from Southwell *et al.* (in prep.) illustrate this point (Table 7). They conducted foot line transect counts on a population of Eastern Grey Kangaroos at four times of day when kangaroo activity and lighting conditions varied but density did not. The number of kangaroos counted per kilometre was not a repeatable index over these conditions, with extremes differing by 240%. Repeatability for a strip transect of 410 m width was little better, but improved as strip width decreased until, at the most narrow strip width of 30 m, repeatability began to worsen. This would be due to small sample size and resultant random errors caused by the narrow strip. Line transect density estimates varied by a maximum of 20% between times and were more repeatable than strip transects of any width.

(d) *Accuracy*

Tests of the accuracy of transect counts require information on actual density obtained by some

Table 7. Repeatability of line transect, strip transect and index counts for a population of Eastern Grey Kangaroos. The index of repeatability is calculated as the ratio of maximum and minimum values from four times of day. A low index indicates high repeatability. Line and strip transect count values are in kangaroos km^{-2}, and the index count value is number of kangaroos seen per km of transect. Actual density was 62.7/km^2. Data from Southwell *et al.* (in prep.). Line transect density estimates were obtained using the Exponential Power Series model.

		Morning	Midday	Afternoon	Night	Index of repeatability
Line transect count		64.4	57.1	66.3	56.8	1.2
Strip transect count	(15 m half-width)	79.9	51.4	73.1	61.3	1.6
	(25 m half-width)	68.4	50.9	68.3	51.8	1.3
	(55 m half-width)	59.0	49.1	62.3	38.2	1.6
	(105 m half-width)	55.4	51.8	56.7	25.7	2.2
	(205 m half-width)	41.7	35.8	40.1	13.2	3.2
Index count		18.5	15.5	18.1	5.4	3.4

independent and accurate means. Since this information often can only be obtained from transect counts, the ability to undertake such tests is frequently invalidated by circular reasoning. Some direct tests of accuracy have been possible in the rare situations where a total count has been made. In the absence of a total count, others have attempted to test accuracy by indirect means or by assuming that estimates based on ground transect counts are accurate. Each of these tests is outlined below.

The line transect count method with observers walking along randomly or independently located transects is potentially the most accurate transecting method for estimating macropod population density. Line transect density estimation is based on several assumptions, *viz:* (1) all animals on the line are seen, (2) animals do not move before being seen and none are counted twice, (3) distance and angle to animals are measured accurately and (4) sightings are independent events (Burnham *et al.* 1980). In addition an appropriate line transect model must be used. Macropods have the advantage of being very visible when moving because of their hopping gait, so animals are unlikely to move far before being seen if within the observer's visual field. However, macropods are highly mobile and tend to avoid approaching humans, hence the possibility of undetected movement beyond the observer's visual field, with subsequent violation of the first two assumptions, cannot be ignored.

Southwell *et al.* (in prep.) tested the accuracy of foot transect counts using line transect methodology against total counts of Eastern Grey Kangaroo, Red-necked Wallaby and Red Kangaroo populations. These populations were well habituated to observers and so not typical with regard to movement. They did, however, allow an examination of the accuracy of line transect methodology in relation to variations in visibility conditions, because counts were made at different times of day and night when animal activity and light conditions varied and on three species differing in conspicuousness. From a total of over 20 different line transect models examined, less than half returned relatively low bias for the range of sightability conditions encountered. These included several nonparametric and generalized parametric perpendicular distance models and two radial distance models proposed by Morgan (1986). The other models, which included simple parametric perpendicular distance models, Hayne radial distance models, the Eberhardt-Cox and Kelker methods and several *ad hoc* models (i.e., lacking a sound theoretical basis) were rejected on the basis of high bias or subjectivity in the estimation procedure.

Further tests on the accuracy of line transect counts in the more realistic situation where kangaroos are not habituated to the observer are required before the general utility of the method can be accepted. If those models which returned acceptably low bias in Southwell *et al.*'s work on habituated populations remain acceptable when applied to unhabituated populations, then it can be inferred that the method is robust to violations of the assumptions concerning movement. Three studies undertaken to date on unhabituated populations support that inference. Two of these compared line transect density estimates with a drive count (Coulson and Raines 1985; and Southwell and Fletcher unpubl. data using the drive count of Short and Hone in press), and the other made use of a severe drought as an index-manipulation-index experiment (Morgan 1983a). In the latter study the number of animals removed by drought was estimated by line transect counts of skeletons less than one year old, and line transect counts of live animals were carried out before and after the drought. In all three cases, models judged as acceptable in Southwell *et al.*'s work returned estimates with relatively low bias (<15%).

These studies clearly indicate that line transect counts can return accurate density estimates in realistic situations, but it should not be assumed that the method is appropriate to all situations. Line transect counts are best suited to species which inhabit sparsely to moderately wooded country (for example, Red Kangaroos, Eastern and Western Grey Kangaroos, Wallaroos and Whiptail Wallabies), and to populations of low or moderate density. Dense, shrubby forest hinders inconspicuous movement by the observer, possibly forewarning animals of his or her approach and allowing them to move away unseen and unheard. Smaller species which prefer dense habitats, such as the Red-necked Wallaby and Swamp Wallaby, may therefore be suitable for line transecting only in some particular and unrepresentative situations. In dense populations animals fleeing from the observer may cause others further along the line to flee unseen (chain flushings) or they may be confused with uncounted animals. This is unlikely to be a problem in low density populations. The method therefore should not be used without due thought to the species, habitat and population being surveyed.

Many surveys of macropod populations have been conducted from vehicles along roads (Table 5). Roads offer the advantage of rapid coverage of transects and in densely forested habitats may be the only route that can be traversed without causing major disturbance to the target animals. Regardless of whether line or strip methods are used, the potential for bias in transect counts along roads is high because of the selective location of roads and possible effect of roads on animal distribution. Rost and Bailey (1979) for example found both Mule Deer, *Odocoileus hemionus,* and Elk, *Cervus canadensis,* avoided roads, particularly areas within 200 m of a road. They attributed this effect to a number of factors, including disturbance by traffic and hunting and the distribution of forage and shelter relative to roads. Despite the high potential for bias, little effort has been made to determine whether the potential is realized, and if so, its magnitude for macropod populations. Southwell and Fletcher (1985b) found that line transect counts of Whiptail Wallabies from tracks gave extremely negatively biased density estimates compared to random foot line transect counts because roads generally avoided the species' preferred habitat of hilly, wooded country. A similar effect is likely for the Wallaroo due to its preference for steep, rocky habitat. The effect should be less pronounced for Red and Grey Kangaroos in plains country, but no data are available to test this assumption. In densely forested areas where roads form the only convenient routes for travel potential for bias is very high, but there is no certainty on the direction of bias; macropods may either avoid the disturbance caused by roads or be attracted to the growth of grass that often occurs on roadside verges.

It is generally accepted that observers counting animals within strip transects from an aircraft will miss a substantial proportion of those actually present (Jolly 1969; Caughley and Goddard 1972; Caughley 1974; Pollock and Kendall 1987). Aerial strip transect counts of macropods will therefore be negatively biased unless the counts are corrected. Several studies have been undertaken to estimate correction factors so that aerial counts can be converted from density indices to estimates of absolute density, and these are summarized in Table 8.

The first such study used an indirect regression technique whereby counts were regressed with speed, height and strip width and then extrapolated back to zero values of these variables to estimate actual density (Caughley *et al.* 1976). Experiments for Red Kangaroos in open and medium-wooded country gave correction factor estimates of 2.29 and 2.43 for these habitats respectively for the routinely-used survey variables of 200 m strip-width, 76 m height and 185 km hr^{-1} speed. In subsequent aerial surveys these correction factors were applied to Grey Kangaroos on the assumption that sightability for Red and Grey Kangaroos was similar.

These experiments have since been criticized on several grounds, including low precision of density estimates (Jolly 1981; Barnes *et al.* 1986), extrapolation beyond the range of observations (Jolly 1981; Bayliss and Giles 1985; Barnes *et al.*

Table 8. Correction factor studies of aerial strip transects of kangaroos. Abbreviations are as follows: Species: See Table 1. Method of estimation: R, regression method; AG, comparison of aerial and ground surveys; ST, strip transect ground survey; LT, line transect ground survey; V, ground survey by vehicle; F, ground survey by foot; DC, ground survey by drive count; N, ground survey at night; D, ground survey during day; I, index-manipulation-index. Weather conditions: S, sunny; O, overcast. Time of day: M, morning; A, afternoon; number indicates the length of time in hours from sunrise or sunset that aerial survey counts were made. Habitat: G, open grassland; W, medium woodland; S, open shrubland; M, mulga. *obtained by extrapolating to height of zero and survey temperature of 15°C. Estimate likely to be underestimates because effect of strip width and speed not included. **estimate considered by authors to lack generality because habitat was not typical for Wallaroos. ***based on limited ground survey data.

Reference	Method of estimation	Speed (km hr^{-1})	Height (m)	Strip half-width (m)	Time of day	Weather	Species	Habitat	Estimate
Bailey (1971)	AG,ST,V,N	145	91	256	–	–	RDK	G,S	1.75
Caughley *et al.* (1976)	R	185	76	200	M and A,3	–	RDK	G	2.29
	R	185	76	200	M and A,3	–	RDK	W	2.43
Short and Bayliss (1985)	AG,ST,V,N	185	76	200	M,1	S	RDK	S	1.34
	AG,ST,V,N	185	76	200	M,1	S	RDK	W	2.02
	AG,ST,V,N	185	76	200	M,1	O	RDK	S	2.46
	AG,ST,V,N	185	76	200	M,1	O	RDK	W	2.75
	AG,ST,V,N	185	76	200	M,1	S	WGK	S	3.71
	AG,ST,V,N	185	76	200	M,1	S	WGK	W	14.15
	AG,ST,V,N	185	76	200	M,1	O	WGK	S	2.69
	AG,ST,V,N	185	76	200	M,1	O	WGK	W	8.76
Southwell *et al.* (1986)***	AG,LT,F,D	185	76	200	M and A,2.5	S	RDK	G	1.80
	AG,LT,F,D	185	76	200	M and A,2.5	S	RDK	M	4.20
	AG,LT,F,D	185	76	200	M and A,2.5	S	WLR	G	3.90
	AG,LT,F,D	185	76	200	M and A,2.5	S	WLR	M	23.30
Bayliss (1987)	R*	167	91	100	M,2	S	RDK	S,W	1.84
	R*	167	91	100	M,2	S	WGK	S,W	2.30
Bayliss (1987)	I	167	91	100	M,2	S	RDK	S,W	1.83
	I	167	91	100	M,2	S	WGK	S,W	2.53
Bayliss (1987)	AG,ST,V	167	91	100	M,2	S	RDK	S,W	1.02
	AG,ST,V	167	91	100	M,2	S	WGK	S,W	2.88
Short and Horne (in press)	AG,DC	185	76	200	M,3.5	S	RDK	S	2.63
	AG,DC	185	76	200	M,3.5	S	WGK	S	6.67
	AG,DC	185	76	200	M,3.5	S	WLR**	S	11.10

1986), the existence of a crossover between results for open and wooded habitats (Bayliss and Giles 1985), the accuracy of density estimation (Hone 1986) and the assumption of equal visibility for Red and Grey Kangaroos (Short and Bayliss 1985). Bayliss and Giles (1985) applied the regression technique to helicopter counts of Red and Western Grey Kangaroos and concluded that the exercise was of dubious value. Caughley *et al.*'s (1976) experiments were of considerable value, however, in showing that the survey variables of speed, height and strip width must be standardized between surveys and that correction factors are specific for particular values of the survey variables. As a result of criticisms of the regression method, efforts were made to obtain correction factors by alternative means.

Short and Bayliss (1985) and Bayliss (1987) attempted to estimate correction factors for Red and Western Grey Kangaroos by comparing aerial strip counts with ground strip counts conducted from a vehicle at night. Similar work had been conducted earlier by Bailey (1971) on Red Kangaroos. Only Short and Bayliss employed the survey variable levels of 76 m height, 200 m strip width and 185 km hr^{-1} speed that have generally been used for broad-scale kangaroo aerial surveys (compare Tables 4 and 8) so the results of the other two studies cannot be generally applied. Short and Bayliss found significant differences in sightability between species, tree cover, weather and observers, with the weather effect acting differentially between species. Estimated correction factors suggested that those developed by Caughley *et al.* (1976) for the Red Kangaroo may overestimate populations and that those assumed for Grey Kangaroos by Caughley *et al.* (1976) would markedly underestimate Western Grey Kangaroo populations (Table 8). All three studies relied on the untested assumption that the ground count was accurate. Since counts were made from roads in Bailey's study and in the wooded habitat of Short and Bayliss' study, the assumption is in these cases questionable. Also untested in all studies was the assumption that all kangaroos within the strip were counted. This assumption is generally less robust than the equivalent assumption for line transecting whereby only all kangaroos directly on the line need be counted (Burnham and Anderson 1984). Another potential confounding factor in Short and Bayliss' study was that aerial counts were undertaken over short one-hour periods in the early morning and late afternoon (J. Short, pers. comm.), whereas Caughley *et al.*'s (1976) work and most broad-scale aerial surveys of kangaroos have been carried out over three-hour periods in the morning and afternoon. Hill *et al.* (1985) found that time from sunrise and sunset has a large effect on counts of Eastern Grey Kangaroos.

Short and Hone (in press) attempted to eliminate those potential flaws by comparing aerial counts, for which survey variables and times of day were similar to those employed for broad-scale surveys, against a total count obtained from a drive count. This work confirmed the marked underestimation of Western Grey Kangaroo populations using correction factors from Caughley *et al.*'s experiments but supported the accuracy of Caughley *et al.*'s correction factors for Red Kangaroos. Amalgamation of results from this and Short and Bayliss' study indicates that correction factors of 4.8-5.9 and 16.7 for Western Grey Kangaroos in open and wooded country respectively, and 1.8-2.5 and 2.4-2.8 for Red Kangaroos in open and wooded country respectively, are required for estimation of absolute density using standard survey variables and times.

Further information on correction factors for Red Kangaroos in extreme vegetation types of open grassland and dense mulga shrubland is available from Southwell *et al.* (1986). They conducted concurrent aerial strip transect and foot line transect counts. Although the ground survey data were limited in extent, the estimates of 1.8 for grassland and 4.2 for shrubland are generally consistent with those given above (the higher value for shrubland may be due to a denser vegetation cover between this and other studies).

Another means of obtaining correction factor estimates is the index-manipulation-index method. Bayliss (1987) employed this technique, taking advantage of a severe drought as the manipulating tool, to determine correction factors for Red and Western Grey Kangaroos in a mixture of open and wooded habitats. The estimates however are not generally applicable because the survey variables differed from those used in broad-scale surveys.

Eastern and Western Grey Kangaroos occur sympatrically over large areas of eastern Australia and no attempt is made to separate the species from the air because of their similar appearance. All correction factor work to date has concentrated on Western Greys. One would expect correction factors for Eastern and Western Greys to be far more similar than for Reds and Greys, but further work on Eastern Greys is required to test this expectation.

Wallaroos are seen and noted by many aerial survey teams even though the target species are Red and Grey Kangaroos. However, until recently no information on correction factors for Wallaroos was available. Short and Hone (in press) counted Wallaroos in addition to Red and Western Grey Kangaroos in their study, and estimated a correction factor of 11.1 for their study site, considerably higher than that for the other two species. The researchers attributed the high correction factor to the Wallaroo's preference for a narrow strip of wooded vegetation along one side of the site, and to the species' reluctance to flush from a passing aircraft. They cautioned that the estimate may have little generality because the study site was not typical of Wallaroo habitat. However, data from Southwell *et al.* (1986) supports the generalization that Wallaroos are much harder to see from the air than Red or Western Grey Kangaroos. In their study, Wallaroo density estimates from foot line transect counts were higher than aerial strip transect density estimates by factors of 3.9 for open vegetation and 23.3 for mulga shrubland (overall correction factor for both habitats combined of 15.3). Again a preference for densely wooded habitats and a reluctance to flush were thought to be responsible for these high correction factors.

The very high correction factors for some species in some habitats brings into question the unrestricted use of aerial survey. Extreme visibility bias will increase random error in counts. It may therefore be necessary to exclude some habitats from aerial survey where sightability is very low, increase the sampling intensity in those habitats, or simply acknowledge that in these situations aerial survey can provide only a rough guide to population levels.

Several workers have adapted the principles of mark-recapture techniques to estimate aerial survey correction factors for large mammals and birds, but this approach has not been applied to kangaroos. Adaptations which involve actual capture of animals include marking of animals with conspicuous collars that can be seen from the air (Rice and Harder 1977) and use of radiotelemetry (Packard *et al.* 1985). Both techniques could be applied to kangaroos, but would be very costly and labour intensive. Another adaptation which does not require physical capture of animals involves the use of multiple observers, whereby animals or groups of animals are counted independently by two observers simultaneously scanning the same transect, and sightings are partitioned into those seen by one observer but missed by the other and those seen by both observers (Caughley and Grice 1982). This method is most appropriate for animals that are either solitary or form tight groups, and for populations that are sparse, because under these conditions the decision as to which animals or groups were seen by both observers is least difficult. Kangaroos tend to form loose groups of variable size and are often highly variable in their local density, so they are not ideal animals for this method. However, the method may be useful in some circumstances.

A final means of improving the accuracy of aerial counts involves modification of line transect methodology normally used on ground surveys. The main difference between line transecting from the air and the ground is that frequently the area directly beneath a plane (including the transect line) cannot be searched. The procedure suggested here is that perpendicular distance from the line is estimated for animals sighted from the aircraft and the resultant detection histogram, minus sightings from the strip that cannot be searched, is used to estimate density. Leatherwood *et al.* (1982) and D. G. Morgan (pers. comm.) have modified line transect models to cope with missing data in the innermost strip. From a practical viewpoint, perpendicular distances would have to be grouped and the number of distance-groupings small (probably about five). From a theoretical viewpoint, one must assume that the detection function follows the same form in the searched and unsearched areas (Leatherwood *et al.* 1982) and that all animals on the line would be seen if it could be searched (Pollock and Kendall 1987). This latter assumption may be particularly prone to failure. D. G. Morgan (pers. comm.) is currently investigating the accuracy of aerial line transect counts for kangaroo populations.

The main advantage of multiple observer and line transect methods in correcting aerial survey counts is that corrections can be specific to each survey or to each transect within a survey, and the combined effects of all factors affecting countability may be accounted for. This contrasts with the "correction factor" approach, whereby a separate correction factor is developed for each factor that affects countability. If there are many such factors, this approach can become somewhat messy.

Indirect sample counts

All animals leave a record of their presence in the form of signs such as faecal pellets, tracks, dens, scent posts, mounds and nests. An index based on indirect counts of such signs has several advantages over direct counts of the animals themselves. Caughley (1977) lists these as: its accuracy is less dependent on the skill of the observer, its measurement is easier to standardize between observers, it is less affected by viewing conditions, and the act of observing does not influence what is observed. The disadvantages compared with direct counts are the less direct relationship with actual density and the presence of a time lag between the sign being laid down and observed.

The type of sign most commonly left by macropods is faecal pellets. In some situations other signs such as tracks or diggings may be of use.

(i) FAECAL PELLET COUNTS

The faecal pellet count method requires measurement of pellet accumulation per unit area and time. Provided the rate at which pellets are voided is constant in space and time, pellet accumulation can be used as an index for the monitoring of abundance. Furthermore, if pellet defecation rate is known, the index can be converted to an estimate of absolute density.

The measurement of pellet standing crop does not provide an adequate assessment of accumulation rate, because pellets decay or disappear with time and this parameter itself varies in time and space. Ideally then, sampling units or plots need to be cleared of the pellet standing crop and the number of pellets deposited by the species of interest over a defined time counted, with care taken to ensure that counts are not biased by loss of pellets through decay in the counting period.

Table 9 summarizes details of pellet count surveys of macropod populations. The great majority of surveys have been conducted over very small areas of only a few square kilometres. The extensive surveys of Caughley (1964) and Sinclair (1980) aimed to assess broad patterns of habitat utilization rather than obtain a precise index or estimate of abundance. Several of the small-scale surveys were undertaken to assess habitat utilization at a finer-scale (Gibson 1975; McKenzie 1976; Bradley 1978; Taylor 1980; Floyd 1980; Andrew and Lange 1986). Pellet count surveys have been applied to a wide range of species, but there has been a strong bias in the location of surveys toward the wet coastal areas of the continent, reflecting the difficulty of monitoring populations by direct counts in these areas.

In most surveys counts have been repeated at regular intervals (Table 9), but the interval between repetitions has generally been small (one or two months) and the surveys therefore relatively short-term in duration. As a consequence, variations in counts are more likely the result of local change in abundance due to movement across survey area boundaries, or of non-repeatable counting methods, than of changing abundance due to variations in fecundity and mortality.

(a) Survey design

Pellet count surveys are subject to similar decisions concerning survey design as are direct counts. One must decide upon the type of sampling unit to be used, the placement of these units in the survey area, whether to stratify, etc. However, because the target objects differ from the animals themselves in being immobile, the investigator's response to some of these decisions (such as whether to use quadrats or transects) may differ from those for direct counts. In addition, decisions specific to the pellet count method, such as whether to count pellets or pellet-groups, must be made. Pellet count studies of macropod populations have utilized a range of survey designs (Table 9).

Most pellet count studies of ungulates have used pellet-groups rather than individual pellets as the basic counting unit. This is a logical decision when the number of pellets per defecation is large and all pellets within a defecation are deposited as a discrete group. White-tailed Deer, *Odocoileus virginianus,* for example, deposit approximately 70 pellets per defecation (Rogers 1987). However, because macropods void a small but variable number of pellets per defecation (Table 13), the decision of whether to count pellets or pellet-groups is not so clear-cut.

Several workers have addressed the question of pellets versus pellet-groups for macropod surveys. The collective conclusion of these workers is in favour of pellets. Johnson (1977), Taylor (1980), Perry and Braysher (1986), Arnold and Maller (1987) and M. Statham (pers. comm.) opted for pellets because of the difficulty of defining pellet-groups. This difficulty arises from the highly variable number of pellets per defecation, the tendency of macropods to produce several defecations in quick succession (increasing the chance of two groups being counted as one), and the tendency for pellets within groups to be scattered by animals walking or grazing through them (increasing the chance of one group being counted as two or more). Johnson and Jarman (1987) concluded that these latter two effects caused no bias in their study, because the number of pellets observed per defecation was no different from the number of pellets per group found on the ground. They point out, however, that it may have been for one of these reasons that counts of pellets estimated the very dense population studied by Coulson and Raines (1985) more accurately than did counts of pellet-groups. Hill (1978) opted for pellets instead of pellet-groups because he found the number of pellets per group to vary markedly between sites.

While the general consensus is for counting pellets rather than pellet-groups, Johnson and Jarman (1987) recognize there are some disadvantages to this choice. Pellets are likely to decay more quickly and thus be harder to detect than pellet-groups. Pellets are also more likely to be obliterated if stepped on while fresh and soft than are pellet-groups. The assumption of pellet-groups being independent (which is necessary if one is to assign confidence limits to estimates of density) should cause less distortion than the assumption of pellets being independent (since pellets within each defecation are clearly not independently sampled). As is often the case one has to make a decision between alternatives based on which method has the fewest disadvantages.

Sinclair (1980) and G. W. Arnold (pers. comm.) have found a high correlation between number of pellets per plot and the frequency of plots containing pellets. This finding allows inferences on relative density of pellets to be made from presence-absence data. If pellet density is

Table 9. Details of macropod pellet count studies. Abbreviations are as follows: Species: See Table 1. Shape of basic plot: C, circular; B, belt; P, plotless. Placement of plots: S, systematically along transects; TS, transects systematically located; TV, transects located to maximize variance within and minimize variance between; TR, transects randomly located; TN, method of locating transects not stated; T, adjacent to roads or tracks; R, random. Type of plot: P, permanent; T, temporary. Attribute counted: P, pellets; G, pellet-groups; W, weight of pellets; PA, presence-absence of pellets; D, distance, V, pellet volume. Regularity: First number indicates number of repetitions, number in brackets indicates interval in months between repetitions.

Reference	Size of survey area (km^2)	Species	Shape of basic plot	Area of basic plot (m^2)	Dimensions of basic plot (m)	Use of stratification	Placement of plots	Type of plot	Basic unit for estimation of mean and variance (m^2)	Attribute counted	Regularity
Caughley (1964)	Sheep property	RDK,GK	C	4.0	1.14	Yes	S,TS	T	20.3	PA	3(5)
Warren (1971)	4.8	RDK,WLR	B	64.0	100×0.64	No	S,TN	T	64.0	W	2(28)
Warren (1971)	52.0	RDK,WLR	B	64.0	100×0.64	No	S,TN	T	64.0	W	2(28)
Gibson (1975)	20.0	EGK,WLR	C	10.0	1.8	Yes	S,TS	P	40.0	P	5(1)
McKenzie (1976)	20.0	EGK,RNW, SPW,RNP	B	40.0	40×1	Yes	S,TS	P	40.0	P	3(1)
Johnson (1977)	0.3	RNW,RBP	C	3.14	1	No	S,TN	P	3.14	P,G	4(1-2)
Johnson (1977)	0.3	RNW,RBP	B	10.0	10×1	No	S,TN	P	10.0	P,G	4(1-2)
Johnson (1977)	0.3	RNW,RBP	P	–	–	No	S,TN	–	–	D	1(–)
Bradley (1978)	1.8	EGK,RNW, SPW,RNP	B	50.0	50×1	Yes	R,TN	P	50.0	P	5(1)
Floyd (1980)	2.5	SPW	B	50.0	50×1	Yes	R	P	50.0	P	1(–)
Sinclair (1980)	10 000	RDK,GK	C	1.0	0.56	No	S,TR	T	–	PA	1(–)
Taylor (1980)	6.0	EGK	C	4.9	1.25	Yes	S,TN	T	–	P,G	1(–)
Hill (1981)	33.0	EGK	C	10.0	1.8	No	S,TR	T	250.0	P,G	2(2)
Hill (1982)	1.3	EGK	C	10.0	1.8	No	S,TR,TV	T	50.0	P	14(2)
Stewart (1982)	–	EGK	C	50.0	4.0	–	–	P	50.0	G	1(–)
Coulson and Raines (1985)	2.6	EGK	C	20.0	2.5	No	S,TS	P	20.0	P,G	1(–)
Andrew and Lange (1986)	5.1	RDK	B	6.1	20.4×0.3	No	R	–	–	V	1(–)
Andrew and Lange (1986)	13.4	RDK,WGK	B	10.0	30×0.3	No	S,TS	T	–	PA	2(12)
Andrew and Lange (1986)	21.7	RDK,WGK	B	240.0	120×2	No	S,TS	P	–	W	17(1)
Perry and Braysher (1986)	3.3	EGK	C	3.14	1	No	S,TV	T	78.6	G	1(–)
Arnold and Maller (1987)	3.1	WGK	B	100.0	100×1	Yes	R	P	100.0	P	60(1)
Arnold and Maller (1987)	2.7	WGK	B	100.0	100×1	Yes	R	P	100.0	P	60(1)
Arnold and Maller (1987)	3.3	WGK	B	100.0	100×1	Yes	R	P	100.0	P	60(1)
Johnson and Jarman (1987)	1.6	EGK,RNW	C	3.0	1	No	S,TS	P	9.0	P,G	2(1)
Stratham (pers. comm.)	1.0	RNW	B	20.0	20×1	No	S,TV	P	20.0	P,G	–

high and ground cover sparse, presence-absence data can be obtained much faster than counts of pellets.

Two basic sampling unit shapes have been used in macropod pellet count surveys; circular units and belts (analogous to quadrats and transects respectively). The two types of units have received similar usage (Table 9). The only worker to apply the plotless nearest-neighbour method developed by Batchelor (1975) to macropod pellet counts found measurement of nearest neighbour distances in dense vegetation to be very difficult and impractical (Johnson 1977).

Gibson (1975) and Johnson (1977) have attempted to assess the relative efficiencies of circular and belt plots for macropod pellet counts. Gibson found counts were more variable for circular plots than for belts and on that basis concluded that belt plots were more efficient. He reasoned that this was probably due to belt plots being more spread out (and hence containing more variation within the plot) than circular plots, which were more concentrated in the area they covered. Gibson's experiment however was not conclusive, because it confounded the effect of plot size. With respect to counting efficiency the most important aspect of plot size is not total area but the size of the "searching edge", for example, the radius of a circle, or the width of a rectangle (circles are usually searched by rotating a straight object, for example a metre stick, fixed to the circle's centre and counting pellets as they pass the object, and rectangles are usually searched by moving a straight object along the belt and counting pellets as they cross the object). In Gibson's experiment the "searching edge" was 1 m for belt plots but 1.8 m for circular plots. Since Batchelor (1975) found circular plots of 1.1 m radius were five times more efficient than plots of 3.6 m radius for counting Red Deer pellets, Gibson's experiment clearly confounds shape and size.

Johnson's (1977) experiments provide a better assessment of the efficiencies of the two plot shapes. He also compared circular and belt plots, but circle-radius and belt-width were both 1 m. Data given in Johnson (1977) have been reanalysed and are shown in Table 10. Belt plots sampled three times as much area and returned a lower co-efficient of variation than the same number of circular plots. At face-value belt plots appear more efficient. However, the most realistic assessment of efficiency is the time taken to achieve a particular level of precision. Belt plots took twice as long to count as circular plots. When time taken to achieve a specified level of precision is calculated, the previous conclusion is contradicted, and circular plots emerge as the slightly more efficient plot shape.

The size of the "searching edge" employed in macropod pellet surveys has varied considerably. Belt plots have varied in width from 0.3 m to 2 m (most commonly 1 m), and plot radii have varied from 1 m to 4 m. There has apparently been no serious attempt to determine the most efficient "searching edge" of sampling units for use in macropod pellet counts. For Red Deer pellet surveys Batchelor (1975) found circular plots of intermediate size (0.8 m to 1.1 m radius) were more efficient in time required for a specified precision than larger (3.6 m radius) or smaller (0.4 m radius) plots. His work should serve as a reasonable guide for macropod studies, but similar experimental work on macropods is desirable given the very different defecation patterns of these animals.

The placement of plots in the survey area is another important aspect of experimental design. Circular plots have most commonly been systematically spaced along transects which themselves are either systematically or randomly placed within the survey area (Table 9). This effectively combines the advantages of quadrats and transects. Circular quadrats have the advantages over transects of having a lower perimeter to area ratio, thus reducing the chance of incorrectly judging a pellet or pellet-group as in or out of the sampling unit, and being easier to set up and count (Robinette *et al.* 1958). Transects, on the other hand, have the advantage over quadrats of increasing the dispersion of effort and hence improving sampling efficiency. Pechanec and Stewart (1940) found line-plot sampling with sub-units spaced at systematic intervals along transects was more efficient than either rectangular or square units in sampling vegetation. In contrast to circular plots, belt plots have more often been located randomly than systematically.

Table 10. Re-analysis of Johnson's (1977) data on relative efficiencies of circular and rectangular plots.

	Johnson's data		Re-calculated values for 15% C.V.	
	Circular	Belt	Circular	Belt
Number of plots	160	160	473	268
Area sampled (m^2)	503	1 600	1 488	2 676
Average coefficient of variation (%)	25.8	19.4	15.0	15.0
Time (arbitrary units)	1	2	2.96	3.35

When line-plot designs have been used in macropod pellet count surveys data have usually been pooled over several adjacent plots, or over all plots in a transect, and these pooled data used as the basic unit for estimating a mean and variance. Johnson and Jarman (1987) achieved the same effect by arranging plots in clusters of three along transects. Pooling and clustering increases the area of the basic estimation unit and hence reduces the number of zero values for estimation. Areas of the basic estimation unit for line-plot designs have ranged from 3 to 250 m^2, with the most common size around 50 m^2 (Table 9). In contrast to circular plots, no clustering or pooling of data has been attempted for belt plots. As for circular plots, the size of the basic estimation unit for belt plots has been highly variable with the most common size again around 50 m^2.

Many pellet count surveys have employed stratification in their design (Table 9). The aim has not been to improve the efficiency of estimating abundance in the entire survey area, but to examine small-scale habitat utilization (McKenzie 1976; Bradley 1978; Floyd 1980; Taylor 1980). Use of pellet counting for this purpose depends on the assumption that defecation rate does not vary with time or activity. This assumption is known to be false for several species of macropod and ungulate, hence the method is of dubious value for this purpose (see section on defecation rate).

A final aspect of survey design is the use of permanent and temporary plots. Permanent plots involve the identification and marking of each plot and its boundaries, initial clearing of all pellets from plots, and subsequent counting and removal of pellets from the plots at specified times. Temporary plots are visited only once and no clearing of pellets from plots is required. Instead only those pellets considered to be of a certain age or less are counted and pellets considered to be older are ignored. Age is usually assessed by reference to a set of standard pellets of known age, and is based on visual characteristics such as colour and degree of disintegration. In many areas of North America biologists make use of the seasonal phenomena of autumn leaf-fall and winter snowfall to assess age. Apparently these phenomena can be fixed accurately in time and hide most or all pellets voided before their occurrence.

Temporary plots have the obvious advantage of being quick and cost-effective and the potential disadvantage of being inaccurate. Examination of these two considerations by Freddy and Bowden (1983) for deer pellet-group surveys confirmed the cost-effectiveness of temporary plots but gave no indications of inaccuracies. Permanent plots were twice as expensive to establish in the initial year, 16% more expensive in subsequent years, and gave similar estimates of pellet-group density as temporary plots.

The two types of plots have received similar frequency of use in macropod pellet count surveys (Table 9). In Australia, biologists cannot rely on convenient seasonal phenomena as an aid in ageing pellets on temporary plots, so all ageing must be done by reference to "standard" pellets. No attempt has been made to assess the relative efficiency and accuracy of temporary and permanent plots for macropod pellet counts.

(b) Pellet decay

Studies of pellet decay are necessary for an accurate assessment of pellet accumulation. A considerable body of information on pellet decay now exists for a wide range of species and situations (Table 11). The species studied range from the large Eastern and Western Grey Kangaroo down to medium-sized wallabies and pademelons. Most trials have been conducted over relatively short periods of one to two months, this being the most common period allowed for accumulation

Table 11. Macropod pellet decay studies. Abbreviations are as follows: Species: See Table 1. Habitat: P, pasture (L) long and (S) short; G, native grassland; DS, dry sclerophyll forest; WS, wet sclerophyll forest; OW, open woodland. Season: S, summer (Nov-Feb); A, autumn (Mar-May); W, winter (Jun-Aug); Sp, spring (Sept-Nov).

Reference	Species	Habitat	Season when trial commenced	Length of trial in months
Gibson (1975)	EGK/WLR	G,OW	A	3
McKenzie (1976)	EGK,RNW, SPW,RNP	P,DS,WS	W	1
Johnson (1977)	RNW,RBP	P,DS	W	4
Hill (1978)	EGK	–	A,W	1
Bradley (1978)	EGK,RNW, SPW,RNP	DS,WS	A	5
Hill (1982)	EGK	P	S,A,W,Sp	2
Stewart (1982)	EGK	–	–	5
Johnson and Jarman (1987)	EGK,RNW	P(L),P(S)	A,W	1-2
Perry and Braysher (1986)	EGK	G	Sp	6
Arnold (unpubl.)	WGK	OW	S,A,W,Sp	1-6
Statham (unpubl.)	RNW,RBP	P(S)	W,Sp	1

of pellets. Some trials, however, have run over longer periods of 4-6 months (Johnson 1977; Bradley 1978; Stewart 1982; Perry and Braysher 1986 and G. W. Arnold unpub. data).

The form of decline for the longer trials varies from linear declines for the Red-necked Wallaby and Red-bellied Pademelon in Tasmania (Johnson 1977) to curvilinear (concave-downward) for the Eastern Grey Kangaroo in the A.C.T. (Perry and Braysher 1986). Rates of decay also vary considerably between trials. Johnson and Jarman (1987) found decay of Eastern Grey Kangaroo and Red-necked Wallaby pellets was more rapid in short grass than in long grass and in warm, moist weather than cool, dry weather. Rain caused fresh pellets to disappear quickly but had no apparent effect on older, dry pellets. Hill (1982) recorded lower decay rates for Eastern Grey Kangaroo pellets in winter than in summer in south-east Queensland, but Arnold (unpub. data) observed the opposite for Western Grey Kangaroo pellets in south-west Western Australia. Johnson (1977) found decay rates of Red-necked Wallaby and Red-bellied Pademelon pellets were higher in pasture than in dry sclerophyll forest.

In theory, pellet counts can extend over long periods and be corrected for decay. However, large corrections should be avoided to circumvent possible errors in the decay estimate being introduced into the accumulation estimate. It would therefore be unwise to conduct pellet accumulation counts over periods of several months or more, and in some cases very short periods of a few weeks or even days may be necessary to avoid significant decay. Unfortunately, shortening the period for accumulation has the disadvantage of increasing the number of plots that are required to achieve desirable precision, so one must seek a trade-off between these two counter-acting forces.

Macropod pellet decay trials have been aimed directly at ensuring accurate pellet accumulation estimation rather than studying the process *per se*. In contrast many North American deer pellet count studies have focused on decay as the parameter of interest. After measuring decay in relation to a number of parameters including precipitation, wind speed, evaporation, ambient temperature, ground cover, canopy cover and slope, several workers identified rainfall as the single most important decaying agent of deer pellets, acting through its washing effect and its effect on insect activity (Wallmo *et al.* 1962; Caughley and Johnson 1981; Harested and Bunnell 1987). Cochran and Staines (1961) found temperature correlated best with decay in cottontail rabbit pellets; decay was high at temperatures >25°C but negligible at very low temperatures. In this case temperature acted through its effect on insect activity. Studies of this kind improve the surveyor's ability to predict pellet decay and hence improve the efficiency and accuracy of pellet count surveys.

No similar studies of macropod pellet decay have been undertaken, but data from the many decay trials summarized in Table 11 provide a basis for preliminary examination. A series of step-wise multiple regression analyses were run on these data with pellet decay as the dependent variable and rainfall, average maximum temperature, grass density, shrub density, tree density and diet as independent variables. Climatic data were obtained from meteorological records at the trial site or at nearby meteorological stations, vegetation density was estimated broadly on a five-point scale from site descriptions, and diet was categorized broadly into three groups (1 for grass, 2 for mixed and 3 for browse diets).

Unfortunately, no clear results emerged from the analysis. Decay in the first month was associated positively with diet and negatively with grass density and shrub density (indicating that decay has least for grazing macropods in areas with high grass and shrub density), but these three variables explained only 16% of the variability in decay rates (Table 12). A second analysis on decay over a two month period gave similarly unconvincing results.

Table 12. Results of step-wise multiple regression analyses of macropod pellet decay data. NS: P>0.05; *: P<0.05; **: P<0.01.

	N	r^2	Variable	B	SE	Sign
1st month	46	0.1580	Diet	14.82	3.73	**
			Grass density	−7.35	2.27	**
			Shrub density	−5.24	2.42	*
			Constant	8.01	8.04	NS
2nd month	38	0.1808	Diet	13.72	3.83	**
			Tree density	−3.25	1.58	*
			Constant	−2.38	5.31	NS

No doubt a major reason for the poor results lies in the coarse nature of data for the independent variables. Also important would be the exclusion of relevant variables; insect activity, for example, is particulary relevant but was not quantified. It is also likely that the effects of variables are complex and interacting. Johnson and Jarman (1987), for example, note that rain causes rapid decay of very fresh pellets but has little effect if high temperatures have previously "baked" the pellets' surface. It is presently very difficult to provide general predictions of macropod pellet decay rates.

(c) Defecation rate

Knowledge of defecation rate is essential if relative density indices based on pellet counts are to be converted to estimates of absolute density. Even monitoring of relative density over space or time by pellet-counting should be based on a

Table 13. Available information on defecation rate in macropods. Abbreviations are as follows: Species: See Table 1. Situation: W, wild; C, captive; CG, held in cage; E, held in an enclosure; number in brackets indicates area of enclosure in hectares. Conditions: A, average; D, drought; G, good. Diet: N, naturally occurring pasture or browse; L, lucerne chaff; S, supplementary feed provided. *adjusted for body weight of animals.

Reference	Species	Situation	Month	Conditions	Diet	Pellets per day	Pellet-groups per day	Pellets per group: mean	Pellets per group: range
Johnson *et al.* (1987)	EGK	W	June/July	A	N	493	89.5	5.5	1-14
Hill (1978)	EGK	W	June	–	N	–	–	3.7	1-30
Coulson and Raines (1985)	EGK	W	September	–	N	521	113.9	3.9	–
Perry and Braysher (1986)	EGK	W	–	D	N	–	–	5.2	–
Perry and Braysher (1986)	EGK	W	–	D	N	–	–	5.0	–
Perry and Braysher (1986)	EGK	W	–	G	N	–	–	7.0	–
Perry and Braysher (1986)	EGK	W	–	G	N	–	–	6.9	–
Stewart (1982)	EGK	C	–	–	–	–	24.2	–	–
Bradley (1978)	EGK	C,CG*	–	–	L	185	–	–	–
McKenzie (1976)	EGK	C,CG*	–	–	L	114	–	–	–
Gibson (1975)	EGK	C,CG*	–	–	–	490	–	–	–
Perry and Braysher (1986)	EGK	C,E,(25)	–	D	N	–	–	3.9	–
Perry and Braysher (1986)	EGK	C,E,(25)	–	G	N	–	–	6.4	–
Hill (1978)	EGK	C,E,0.7	June	–	N,S	412	68.4	6.0	1-30
Johnson *et al.* (1987)	RNW	W	June,July	A	N	311	81.4	3.8	1-11
Johnson (1977)	RNW	W	–	–	N	–	–	2.8	1-11
Bradley (1978)	RNW	C,CG*	–	–	L	230	–	–	–
McKenzie (1976)	RNW	C,CG*	–	–	L	214	–	–	–
Johnson (1977)	RNW	C,E	July	–	N,S	128	–	–	–
Statham (pers. comm.)	RNW	C,E(1)	May	–	N	151	–	–	–
Statham (pers. comm.)	RNW	C,E(1)	June	–	N	121	–	–	–
Statham (pers. comm.)	RNW	C,E(1)	July	–	N	94	–	–	–
Statham (pers. comm.)	RNW	C,E(1)	August	–	N	95	–	–	–
Statham (pers. comm.)	RNW	C,E(1)	September	–	N	77	–	–	–
Statham (pers. comm.)	RNW	C,E(1)	October	–	N	66	–	–	–
Statham (pers. comm.)	RNW	C,E(1)	November	–	N	77	–	–	–
Bradley (1978)	SPW	C,CG*	–	–	L	150	–	–	–
McKenzie (1976)	SPW	C,CG*	–	–	L	108	–	–	–
Johnson (1977)	RBP	W	–	–	N	–	–	3.2	1-8
McKenzie (1976)	RNP	C,CG*	–	–	L	107	–	–	–

knowledge of defecation rate as significant variations in defecation rate will invalidate such monitoring.

Knowledge on this aspect of macropod biology is currently very poor. Table 13 summarizes the available information, most of which comes from captive animals (either in laboratory cages or in small enclosures). Only two studies have been conducted on wild macropods. Information is available for four species, but only for the Eastern Grey Kangaroo and Red-necked Wallaby can this information be considered as reliable.

One obvious feature in Table 13 is the large discrepancy in defecation rates for captive and wild animals; for both the Eastern Grey Kangaroo and Red-necked Wallaby defecation rates are much lower for captive animals than for wild animals (Gibson's data for captive Eastern Grey Kangaroos are an exception). Detailed studies of defecation rate in North American deer species have shown that defecation patterns of captive animals are not representative of wild animals. Rogers (1987) observed the defecation rate of tame, free-ranging White-tailed Deer and found much higher rates than for previous studies on captive animals. Furthermore, when the free-ranging animals were confined to small pens their defecation rates declined. The decline in defecation rate with captivity could operate through several factors, including diet, stress and activity levels (Rogers *et al.* 1958; Rollins *et al.* 1984; Mitchell *et al.* 1985; Rogers 1987).

The use of captive animals to obtain defecation rates is of doubtful value, but often this is the only way in which this parameter can be measured. Accurate determination of defecation rate for wild animals is probably the most difficult aspect of the pellet count technique to overcome. There are two means by which measurements can be made; by direct observation of wild animals, and by combining knowledge of population size with measurement of pellet accumulation rate. Johnson *et al.* (1987) used observation to obtain data on defecation rates for Eastern Grey Kangaroos and Red-necked Wallabies by taking advantage of uniquely habituated but free-ranging populations (Jarman *et al.* 1987). Even in this ideal situation they still encountered problems, and reliable observations for Red-necked Wallabies at night and during the middle of the day could not be obtained. Observation of defecation rate is considerably more difficult for macropods than for quadrupedal animals of similar size because the cloaca is often close to the ground and hidden by vegetation. The estimate of defecation rate attributed to Coulson and Raines (1985) in Table 13 has been calculated from their data on actual population size and pellet accumulation rate. This study also took advantage of a unique situation whereby the population was counted accurately with a drive count. Neither of the situations utilized in the above two studies are likely to occur frequently, so further improvement in knowledge of defecation rates is likely to be slow.

There is currently no information on seasonal variations in defecation rate for wild macropods. The studies of Johnson *et al.* (1987) and Coulson and Raines (1985) on the Eastern Grey Kangaroo were both conducted in winter and gave very similar defecation rate estimates. Only a single estimate is available for wild Red-necked Wallabies (Johnson *et al.* 1987). The only information on seasonal variation comes from studies of captive Red-necked Wallabies by M. Statham (unpub. data). He found a steady decline in defecation rate from early winter through to early summer, with the extremes differing by over 100%. Since the conditions of the experiment were reasonably realistic (animals in a one ha enclosure feeding predominantly on naturally occurring pasture), similar seasonal variation for wild animals is strongly suggested. Improved knowledge of seasonal variation of defecation rate in wild populations is essential if pellet counting is to be used for monitoring macropod abundance over seasons or years.

Both Hill (1978) and Johnson *et al.* (1987) found pronounced variation in defecation rate with activity, macropods defecating rarely while resting in the middle of the day, most frequently in the hour after resting ceased, and with intermediate frequency during the night and morning. This finding has significant consequences for pellet counting in the assessment of small-scale patterns of habitat utilization, which requires the assumption that defecation rate does not vary with activity. A study by Collins and Urness (1978) on Elk provides an example of how this effect can lead to incorrect conclusions. They found that while Elk generally defecate when grazing, most defecations (40%) occur when walking from one area to another (4% of their time). As a result most pellets were found in habitats where food was sparse and scattered, because Elk were forced to spend more time walking between food sources. Pellet counting does not appear to be a reliable method for the assessment of small-scale habitat utilization by macropods. Johnson *et al.* (1987) point out that activity-related variations in defecation rate could also bias estimates of absolute density unless all areas used by the population over the entire diurnal cycle of activity are sampled.

Two further considerations with regard to defecation rate are age and sex. Only the study of Johnson *et al.* (1987) has considered these variables for macropods. They found adult female Eastern Grey Kangaroos to defecate twice as often as adult males, but found little difference between adults and subadults. If this finding has general relevance to the macropods it introduces the additional assumption that population sex ratio must not differ significantly in time or space for pellet counting to be a reliable monitoring technique for macropod populations.

(d) Precision and sampling intensity

Given the analysis of transect count data in relation to sampling intensity produced some useable generalizations, it is worthwhile considering a similar analysis for pellet counts. Such an analysis however is restricted by the small number of pellet count surveys undertaken, the small range of survey area size and by differing survey designs.

In Figure 5, effort (as total area surveyed) is plotted against survey area size, with effort required standardized to a desired precision of 15% (C.V.). Few data were available for low density populations, so only data for populations of density >10 km^{-2} were considered. The scatter of points is very wide and no tight relationship is evident. For small areas of less than 10 km^2 most surveys have required around 0.2 ha to 0.3 ha of ground sampled for a 15% precision. The equivalent effort in person-days for counting such an area (not including time for clearing plots) has been around 2-3 person-days (Coulson and Raines 1985; Johnson and Jarman 1987; G. J. E. Hill, pers. comm.; R. Saillard, pers. comm.). However, some exceptional sites have required up to 4 ha of sampling. For areas of 20 km^2 or greater effort has varied from 1.5 ha to 7.5 ha of sampling (five to 14 person-days effort — G. J. E. Hill, pers. comm.; G. W. Arnold, pers. comm.). The expected increase in effort with survey area size is therefore evident, but only weakly.

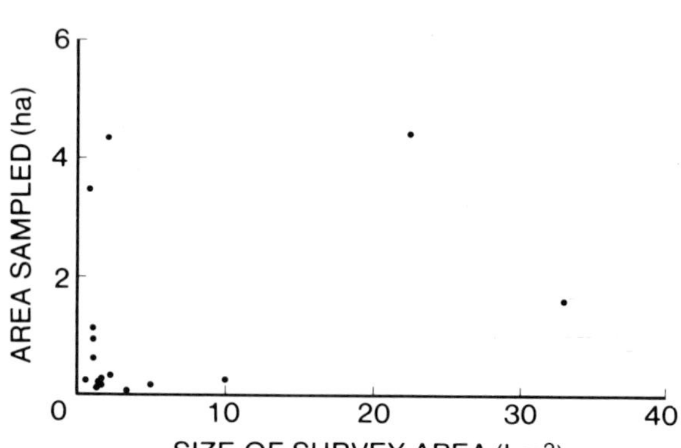

Fig. 5. Area of faecal pellet counts required for a 15% precision against survey area size for high density macropod populations (>10 km^{-2}).

(e) Repeatability

Repeatability of pellet counting has been assessed in two studies (Johnson and Jarman 1987; Hill 1981). In both cases estimates of pellet accumulation rate were obtained with reasonable precision (C.V. <20%) at two points in time when population density was either known or very likely to be the same. In neither study was a significant change in pellet accumulation rate detected.

These tests provide some support for the repeatability of the method, but the strength of the tests is limited by the short periods between counts (one to two months). In such short times it is unlikely that any of the parameters of pellet counting, such as defecation rate and decay rate, will vary as much as they would between seasons or years, the more likely interval between counts for monitoring studies. Further tests of repeatability over seasons or years are therefore desirable.

(f) Accuracy

Tests of the accuracy of pellet counting in estimating absolute density require, in addition to information on pellet accumulation rate, data on defecation rate and actual density, the latter obtained by some independent and accurate means. Several such tests have been attempted, with varying success.

Johnson (1977) acquired the necessary information to test accuracy by estimating the accumulation rate of Red-necked Wallaby pellets in a paddock of improved pasture, measuring the defecation rate of captive Red-necked Wallabies, and determining actual density from a fixed-point count of the paddock. Estimated density exceeded the actual by 400%, and Johnson attributed the overestimate to a bias in defecation rate. Another attempt by Coulson and Raines (1985) on an Eastern Grey Kangaroo population censused by a drive count overestimated density variously by 130% to 550% depending on the estimate of defecation rate used. Defecation rates were again derived from captive animals (Hill 1978; Stewart 1982) and were the probable source of error. A third test by Johnson and Jarman (1987) addressed the likely error in the previous two tests by estimating defecation rate directly for the free-ranging populations of Eastern Grey Kangaroos and Red-necked Wallabies under study. This time density estimates, compared to the actual population known from individual recognition of animals, were relatively unbiased (generally within 15% of actual). Furthermore, when Johnson and Jarman applied the defecation rates obtained in their study to density estimates from the previous two tests, the biases were dramatically reduced.

It appears then that accurate density estimation can be obtained from pellet counts, but the main limitation is the difficulty of obtaining reliable information on defecation rate.

(ii) OTHER SIGNS

Few studies have attempted to use signs other than faecal pellets to monitor macropod populations. Other possible signs include trails (Coulson 1979), holes in fences (Coulson 1979) and spoor (Denny 1977; Statham 1983). Counts of these signs have been used to assess small-scale habitat utilization or the use of particular resources by macropods.

DISCUSSION AND SUMMARY

Emphasis of past monitoring effort

Much of the monitoring effort to date has been undertaken in response to the culling of macropods where they are thought to conflict with rural enterprises. Because the perceived conflict has centred predominantly on the large Red and Grey Kangaroos in the arid and semi-arid rangelands, the pattern of development and application of monitoring techniques for the macropods has been heavily biased in some particular areas:

(a) species — the large kangaroos are among the most intensively studied wildlife species in the world, but relatively little is known of the population dynamics of the many smaller macropod species.

(b) location and extent of monitoring — monitoring has focussed on the arid and semi-arid rangelands which cover much of inland Australia. As a result, knowledge of kangaroo population dynamics is much better for the unpredictable xeric environments than for the more stable mesic environments. Because of the huge area of rangelands in Australia, techniques appropriate to broad-scale monitoring have been most fully developed.

(c) techniques — catch-effort and direct sample count techniques have been applied. Total counts, indirect sample counts, and mark-recapture methods are not appropriate to broad-scale monitoring of macropods, although the former two can be useful in assessing the repeatability and accuracy of other methods.

(d) aerial and ground survey — to obtain a precise sample of abundance over broad-scale areas a very large sampling effort is required. Consequently, aerial survey has often been used in preference to ground survey because large inaccessible areas can be sampled most efficiently and cheaply from the air.

(e) strip and line transects — strip transects have received more use than line transects even though they are less repeatable and accurate. There are several reasons for this pattern. Line transect aerial survey has both theoretical and practical difficulties, as discussed earlier. These difficulties are less evident for ground survey, but even so use of line transect methodology for ground survey has been limited. This may be due to both the timing of theoretical developments in line transect methodology (many of the macropod surveys cited in this review were undertaken prior to important and very recent developments in line transect methodology) and the logistics of line transect surveys (because more measurements must be taken, line transect surveys are more time-consuming than strip transect surveys).

(f) accuracy and repeatability — more effort has been directed toward improving accuracy of sample counts than repeatability, even though repeatability is more important for monitoring population trends. Aerial survey is a prime case in point. When broad-scale kangaroo aerial surveys were first commenced it was assumed that rigid standardization of survey variables such as height, speed and strip width would ensure repeatable counts. With repeatability thus "accounted for", effort was directed toward the development of vegetation-specific correction factors to account for the more obvious problem of bias. In doing so, researchers underestimated the sensitivity of strip transect counts to factors beyond their control. Only recently, after a decade of regular aerial surveying and numerous correction-factor studies, has attention been directed to improving repeatability.

Deficiencies in current knowledge and areas for further research

With the exception of point (f), all the above emphases are perfectly justified and reasonable with regard to the primary reason for monitoring. However, there are many other situations in which monitoring of macropod abundance may be necessary. For example, it is likely that increased fragmentation and intensive management of forest habitats in the future will require use of techniques which are as yet poorly developed. It is in the situations where catch-effort and aerial survey methods are not appropriate that most deficiencies in current knowledge occur. At the same time, catch-effort monitoring, although widely applied, is based on limited developmental work, and aerial survey, although well researched, could be further refined.

Catch-effort monitoring

In the context of the population being hunted for reasons other than monitoring, catch-effort monitoring has the advantages of being very cost-effective and easily applied to situations where other methods are least appropriate. However, the method is subject to several potential deficiencies, and despite its widespread use, there has been little research on the technique. Further research into catch-effort monitoring of macropod populations is therefore highly desirable: improved knowledge would either vindicate and refine further use of the method, with possible savings in unnecessary and expensive direct counts, or reject the method, again with savings in unnecessary and unreliable collection and analysis of catch-effort data.

The modelling approach adopted by Nance (1985) and Kirkpatrick is a significant contribution, but the current model is a "first approximation" and further improvements are possible. De la Mare (1988) argues that the Nance-Kirkpatrick model should incorporate the size of animals as well as their age and provides an alternative model on this basis. The relaxation of a constant environment assumption in the Nance-Kirkpatrick model may improve predictions for populations in variable xeric environments. Regardless of further developments and improvements in this approach, models need to be tested empirically before they are accepted as reliable [the Nance-Kirkpatrick model is now being field-tested by T. H. Kirkpatrick (pers. comm.)]

Direct empirical investigation would complement the modelling approach. Most important would be an examination of the form of the relationship between catch-effort indices and abundance. Since several indices of varying precision and form have been used in the past, relationships should be developed for them all so that comparisons between indices could be made. Some indices may be more sensitive to change in abundance than others. It would also be desirable to determine the effects of such factors as weather, vegetation cover and degree of access on catch-effort indices. It should be possible to examine the effects of all these factors (including abundance) in a single experiment with a multivariate design. The concurrent measurement of catch-effort indices, abundance, weather, vegetation cover and degree of access at a number of sites covering a representative range of factor levels would enable an empirical catch-effort model to be developed. The reliability of catch-effort monitoring could then be gauged from the amount of variation in catch-effort indices that is attributable to abundance. It would of course be necessary to measure abundance with a density index that is linear through the origin or, more preferably, with an absolute measure.

Aerial survey

Aerial survey has received more developmental research than any other monitoring technique. This work has addressed the major factors likely to affect both repeatability and accuracy and has substantially improved the technique since its first application to kangaroo populations in the 1960's. However, improvements in some areas can still be made.

The greatest need is to ensure that repeatability is satisfactory. Studies of repeatability are currently limited in number and location. Most work has concentrated on the effects of various survey variables on counts in an attempt to improve accuracy. In retrospect, this work was probably more important in standardizing aerial survey procedures on a nation-wide basis, and thus contributing to improved repeatability, than in improving accuracy. Recent work has concentrated on the effects of environmental factors such as temperature and cloud cover on repeatability (G. Caughley, pers. comm.). It may be necessary to duplicate this work over several areas that vary climatically to ensure that results can be applied generally. Yet to be investigated is the effect of animal dispersion (both within groups and within habitats) on repeatability.

Much of the research on aerial survey has been concerned with the development of correction factors to improve accuracy. Following criticism of Caughley *et al.*'s (1976) initial correction factor estimates using an indirect regression technique, several studies have recently been undertaken using total counts or ground transect counts as the basis for correction factor estimation. Further work of this nature would be worthwhile, with some qualifications. Firstly, experiments must match closely the conditions of broad-scale surveys to ensure that results are directly applicable. Secondly, one must be reasonably confident that the ground count is accurate. Since the circumstances that allow total counts occur only rarely, air-ground comparisons will need to rely primarily on ground transect counts. The accuracy of ground transect counting is itself poorly studied, and in the past methods with high potential for inaccuracy (such as road surveys) have been used. A more complete assessment of ground transect counts should therefore be undertaken before further air-ground comparisons are made. The greatest deficiency in current knowledge of correction factors for the large kangaroos concerns the Eastern Grey Kangaroo: no work has yet been conducted on this species. Less importantly, there is a need to investigate sightability over a wider range of vegetation conditions for all three large species.

Researchers should also consider alternative means for improving the repeatability and accuracy of aerial surveys. Multiple observers and line transect methodology offer the potential for improvements without the need for multiple correction factors. However, there may be theoretical and practical problems with these methods. The use of multiple observers has not yet been attempted on kangaroo populations, and line transect methodology has received only preliminary investigation.

Improvements in aerial survey have been rapid in recent years and this is likely to continue into the near future. For improvements to be of any use, however, they must be able to be incorporated into the analysis of data from future, and hopefully past, surveys. This requires the data to be collected in a form amenable to such analysis. Of particular importance is the need to collect auxiliary data on environmental variables which may affect countability, and to record any deviations from the generally accepted survey design (height, speed, strip width, etc.). It is also important to recognize that incorporation of new knowledge will require a re-assessment of population estimates or trends from earlier studies.

Ground survey

Ground survey techniques are often assumed to be accurate and repeatable despite there having been few tests of this assumption. In fact there are many ways in which the assumptions of ground survey techniques can be violated.

A major deficiency in current knowledge is the usefulness of roads or tracks as transect routes in ground survey. Vehicle surveys on roads offer the advantage of rapid coverage of an area (compared to walking), and in densely forested habitats may be the only means of travel without causing major disturbance to the animals being surveyed. However, there has been virtually no research to determine the use of roadside habitats by macropods. Differential use of roadside and non-roadside habitats, and variation in the use of these two habitats over time, could affect both repeatability and accuracy of vehicle surveys.

The more frequent use of line transect than strip transect methodology for foot surveys is justified on both theoretical and empirical grounds. Recent research has shown that line transect foot surveys of macropod populations are both more repeatable and more accurate than strip transects. In fact with appropriate survey design line transect ground survey is probably the most repeatable and accurate method generally available for survey of macropod populations. However, it is not suitable to all situations, and further research to determine the limits of its use is necessary.

The greatest difficulty in testing the accuracy of any technique lies in finding a population of known density for experimentation. Total counts of any wildlife species are unfortunately very difficult to obtain and are usually dependent on certain special characteristics of the population or the survey area. In the few cases where total counts of macropod populations have been made, significant improvements in survey techniques have been achieved. Thus any future opportunities for total counts of macropod populations should be thoroughly utilized. Addition-removal methods could also be used to assess the accuracy of ground (and aerial) survey techniques. Severe drought has been successfully used as a removal manipulation to assess the accuracy of both ground and aerial survey. The main problem is that droughts occur unpredictably in time and space, hence planning of pre- and post-drought surveys can be difficult. It should also be possible to utilize man-induced manipulations, for example, the removal of large numbers of macropods from areas to allow pasture regeneration, or the re-introduction of macropods to parts of their former range.

Improved knowledge of line transect methods would in turn allow the potential for improvements in most other survey techniques. Line transect ground survey could be used to determine correction factors for aerial survey, to assess the repeatability and accuracy of vehicle surveys, and to investigate catch-effort methods. This could lead to improvements in methods appropriate to broad-scale survey. As a consequence, I would place the further assessment of line transecting as a very high priority in future research because of its potential for general improvement in monitoring techniques.

Pellet counts

The utility of pellet counting as a monitoring technique is restricted by the paucity of information on defecation rates of macropods, the highly variable and sometimes very rapid decay of macropod pellets, and the great expense of applying the method to all but very small areas.

The lack of information on defecation rate is probably the greatest restriction to general use of the method. We need much more information on temporal variation in defecation rates to assess repeatability and accuracy. Since information on defecation rate is very difficult to obtain, improvement in this area is likely to be slow. The problem of pellet decay is an area that needs to be adderessed more specifically than in the past. At present it is very difficult to predict the rate at which pellets will decay from general climatic data alone. From a climatic viewpoint Australia is not as well suited as some northern hemisphere countries to general use of the pellet count method, lacking the cold, dry conditions which ensure minimal pellet decay and the convenient seasonal phenomena of autumn leaf-fall and winter snowfall which can be used to reduce counting effort.

Even with improved knowledge in these areas, pellet counting is not likely to be useful as a general utility monitoring technique. It is simply too time-consuming and expensive to apply over anything but very small areas. Its main use is therefore for specific problems in small areas, for example, monitoring the use of small areas of pasture by macropods, or monitoring the abundance of small localized populations of macropods.

The application described by Perry and Braysher (1986), where pellet counting is used to estimate the number of kangaroos grazing on individual properties in the A.C.T. is an example of the appropriate use of the method. For some forest-dwelling species pellet counting may be the only potential technique for monitoring. The very poor knowledge of population fluctuations in these species compared to that known for the large kangaroos may alone be justification for further efforts at improving the method.

ACKNOWLEDGEMENTS

This review benefited greatly from inclusion of much unpublished information. To the authors of many unpublished reports I express my thanks for permission to cite their work. Graeme Arnold, Graeme Coulson, Greg Hill, David Morgan, John Robertshaw, Jeff Short and Mick Statham responded generously to my requests for unpublished data. For comments on an earlier draft of the manuscript I thank Derek Ovington, Malcolm Forbes, Tom Kirkpatrick and two anonymous referees. Pat Protas, Sue Hazel and Olga Mangos patiently typed and edited several drafts of the manuscript. Peter Tieman and Rob Scougall drew the diagrams.

REFERENCES

ANON., 1984a. The status and management of Bennetts Wallaby *(Macropod rufogriseus)* and Rufous Wallaby *(Thylogale billardierii)* in Tasmania. Pp. 33-59 *in* Kangaroo Management Programmes of the Australian States. Commonwealth of Australia.

ANON., 1984b. The macropod conservation programme in South Australia Part A. Management of the large kangaroos. Pp. 81-119 *in* Kangaroo Management programmes of the Australian States. Commonwealth of Australia.

ANON., 1984c. Kangaroo conservation and management in Queensland. Pp. 1-31 *in* Kangaroo Management Programmes of the Australian States. Commonwealth of Australia.

ANON., 1987a. Kangaroo management in New South Wales. Unpublished programme prepared by the New South Wales National Parks and Wildlife Service.

ANON., 1988. 1988 Queensland kangaroo management programme. Unpublished programme prepared by the Queensland National Parks and Wildlife Service.

ALEXANDER, P. AND BEST, L., 1987. The macropod conservation programme in South Australia. Part A. Management of the large kangaroos. Unpublished programme prepared by the South Australian National Parks and Wildlife Service, Department of Environment and Planning.

ANDREW, M. H. AND LANGE, R. T., 1986. The spatial distributions of sympatric populations of kangaroos and sheep: examples of dissociation between these species. *Aust. Wildl. Res.* **13**: 367-73.

ARNOLD, G. W. AND MALLER, R. A., 1987. Monitoring population densities of Western Grey Kangaroos in remnants of native vegetation. Pp. 219-25 *in* Nature Conservation: The Role of Remnants of Native Vegetation ed. by D. A. Saunders, G. W. Arnold, A. A. Burbidge and A. J. Hopkins. Surrey Beatty & Sons: New South Wales.

ARNOLD, G. W., STEVEN, D., WEELDENBURG, S. J. AND BROWN, O. E., 1986. The use of alpha-chloralose for the repeated capture of Western Grey Kangaroos, *Macropus fuliginosus. Aust. Wildl. Res.* **13**: 527-33.

BAILEY, P. T., 1971. The Red Kangaroo, *Megaleia rufa* Desmarest, in north-western New South Wales. 1. Movements. *C.S.I.R.O. Wildl. Res.* **16**: 11-28.

BARNES, A., HILL, G. J. E. AND WILSON, G. R., 1986. Correcting for incomplete sighting in aerial surveys of kangaroos. *Aust. Wildl. Res.* **13**: 339-48.

BATCHELOR, C. L., 1975. Development of a distance method for deer census from pellet groups. *J. Wildl. Manage.* **37**: 641-52.

BAYLISS, P., 1985. The population dynamics of Red and Western Grey Kangaroos in arid New South Wales, Australia. I. Population trends and rainfall. *J. Anim. Ecol.* **54**: 111-25.

BAYLISS, P., 1987. In Appendix 8.1. *In* Kangaroos: their Ecology and Management in the Sheep Rangelands of Australia ed. by G. Caughley, N. Shepherd and J. Short. Cambridge University Press.

BAYLISS, P. AND GILES, J., 1985. Factors affecting the visibility of kangaroos counted during aerial surveys. *J. Wildl. Manage.* **49**: 686-92.

BRADLEY, L. R., 1978. A comparison of methods for censusing macropods in the Styx River State Forest, New South Wales. Hons. Thesis, University of New England.

BURNHAM, K. P. AND ANDERSON, D. R., 1984. The need for distance data in transect counts. *J. Wildl. Manage.* **48**: 1248-54.

BURNHAM, K. P., ANDERSON, D. R. AND LAAKE, J. L., 1980. Estimation of density from line transect sampling of biological populations. *Wildl. Monog.* No. 72.

CALABY, J. H., 1966. Mammals of the upper Richmond and Clarence Rivers, New South Wales. *C.S.I.R.O. Div. Wildl. Res. Tech. Pap.* No. 1.

CAUGHLEY, G., 1964. Density and dispersion of two species of kangaroos in relation to habitat. *Aust. J. Zool.* **12**: 238-49.

CAUGHLEY, G., 1974. Bias in aerial survey. *J. Wildl. Manage.* **38**: 921-33.

CAUGHLEY, G., 1977. Analysis of Vertebrate Populations. John Wiley and Sons.

CAUGHLEY, G. AND GODDARD, J., 1972. Improving the estimates from inaccurate censuses. *J. Wildl. Manage.* **36**: 135-40.

CAUGHLEY, G. AND GRICE, D., 1982. A correction factor for counting Emus from the air, and its application to counts in Western Australia. *Aust. Wildl. Res.* **9**: 253-59.

CAUGHLEY, G. AND GRIGG, G. C., 1981. Survey of the distribution and density of kangaroos in the pastoral zone of South Australia, and their bearing on the feasibility of aerial survey in large and remote areas. *Aust. Wildl. Res.* **8**: 1-11.

CAUGHLEY, G. AND GRIGG, G. C., 1982. Numbers and distribution of kangaroos in the Queensland pastoral zone. *Aust. Wildl. Res.* **9**: 365-71.

CAUGHLEY, G., GRIGG, G. AND SHORT, J., 1983. How many kangaroos? *Search* **14**: 151-52.

CAUGHLEY, G., SHORT, J., GRIGG, G. C. AND NIX, H., 1987. Kangaroos and climate: and analysis of distribution. *J. Anim. Ecol.* **56**: 751-61.

CAUGHLEY, G., SINCLAIR, R. G. AND GRIGG, G., 1979. Trend of kangaroo populations in New South Wales, Australia. *J. Wildl. Manage.* **43**: 775-77.

CAUGHLEY, G., SINCLAIR, R. G. AND SCOTT-KEMMIS, D., 1976. Experiments in aerial survey. *J. Wildl. Manage.* **40**: 290-300.

CAUGHLEY, G., SINCLAIR, R. G. AND WILSON, G. R., 1977. Numbers, distribution and harvesting rate of kangaroos on the inland plains of New South Wales. *Aust. Wildl. Res.* **4**: 99-108.

CAUGHLEY, G., BROWN, G. AND NOBLE, J., 1985. Movement of kangaroos after a fire in mallee woodland. *Aust. Wildl. Res.* **12**: 349-53.

CAUGHLEY, J., BAYLISS, P. AND GILES, J., 1984. Trends in kangaroo numbers in western New South Wales and their relation to rainfall. *Aust. Wildl. Res.* **11**: 415-22.

CLUTTON-BROCK, T. H., GUINESS, F. E. AND ALBON, S. D., 1982. Red Deer. Behaviour and Ecology of Two Sexes. University of Chicago Press.

COCHRAN, G. A. AND STAINES, W. J., 1961. Deposition and decomposition of faecal pellets by cottontails. *J. Wildl. Manage.* **25**: 432-35.

COLLINS, W. B. AND URNESS, P., 1978. Elk pellet group distribution and rates of deposition in aspen and lodgepole pine habitats. Pp. 140-44 *in* North American Elk: Ecology, Behaviour and Management ed. by M. Boyce and I. Hayden-Wing. University of Wyoming, Laramie.

COULSON, G. M., 1979. Some methods for surveying kangaroo populations. *Vict. Nat.* **96**: 184-89.

COULSON, G. M. AND RAINES, J., 1985. Methods for small-scale surveys of Grey Kangaroo populations. *Aust. Wildl. Res.* **12**: 119-25.

CROFT, D. B., 1987. Socio-ecology of the Antilopine Wallaroo, *Macropus antilopinus,* in the Northern Territory, with observations on sympatric *M. robustus woodwardii* and *M. agilis. Aust. Wildl. Res.* **14**: 243-55.

CROOK, I. G. AND PRINCE, R. I. T., 1984. Kangaroo management in Western Australia. Pp. 121-39 *in* Kangaroo Management Programs of Australian States. Commonwealth of Australia.

DE LA MARE, W. K., 1988. Simulation studies on the use of the sex ratio for monitoring the effects of exploitation on kangaroo populations. Unpublished paper presented to the National Advisory Committee on Kangaroos, Canberra (mimeo).

DENNY, M. J. S., 1977. Red Kangaroo arid zone investigations. Unpublished report prepared for the Australian National Parks and Wildlife Service.

DENNY, M. J. S., 1978. Red Kangaroo arid zone investigations. Unpublished report prepared for the Australian National Parks and Wildlife Service.

DENNY, M. J. S., 1979. Ground versus aerial surveys. *In* Aerial Surveys of Fauna Populations. Australian National Parks and Wildlife Service Special Publication No. 1. Australian Government Publishing Service: Canberra.

DE SANTE, D. F., 1981. A field test of the variable circular-plot censusing technique in a California coastal scrub breeding bird community. *Stud. Avian. Biol.* **6**: 177-85.

DUDZINSKI, M. L., LOW, W. A., MULLER, W. J. AND LOW, B. S., 1982. Joint use of habitat by Red Kangaroos and Shorthorn Cattle in arid central Australia. *Aust. J. Ecol.* **7**: 69-74.

EALEY, E. H. M., 1967. Ecology of the Euro, *Macropus robustus* (Gould), in northwestern Australia II. Behaviour, movements and drinking patterns. *C.S.I.R.O. Wildl. Res.* **12**: 27-51.

EBERHARDT, L. L., 1978. Appraising variability in population studies. *J. Wildl. Manage.* **42**: 207-38.

EVANS, C. D., TROYER, W. A. AND LENSINK, C. J., 1966. Aerial census of Moose by quadrat sampling. *J. Wildl. Manage.* **30**: 767-76.

FLETCHER, M. AND SOUTHWELL, C., 1984. 1984 survey of the distribution and abundance of the Red Kangaroo in Western Australia. Unpublished report prepared for the West Australian Government.

FLETCHER, M. AND SOUTHWELL, C., 1988. Report on the 1987 aerial survey in Western Australia. Unpublished report prepared for the Department of Conservation and Land Management.

FREDDY, D. J. AND BOWDEN, D. C., 1983. Efficacy of permenent and temporary pellet plots in juniper-pinyan woodland. *J. Wildl. Manage.* **47**: 512-16.

FLOYD, R. B., 1980. Density of *Wallabia bicolor* (Desmarest) (Marsupialia: Macropodidae) in eucalypt plantations of different ages. *Aust. Wildl. Res.* **7**: 333-37.

FRITH, H. J., 1964. Mobility of the Red Kangaroo, *Megaleia rufa. C.S.I.R.O. Wildl. Res.* **9**: 1-19.

GIBSON, D. F., 1975. A population study of five mammal species on Newholme with emphasis on counting techniques. B. Nat. Res. Thesis, University of New England.

GODDARD, J., 1969. Aerial census of Black Rhinoceros using stratified random sampling. *E. Afr. Wildl. J.* **7**: 105-14.

GRIGG, G., BEARD, L. A., CAUGHLEY, G., GRICE, D., CAUGHLEY, J. A., SHEPPARD, N., FLETCHER, M. AND SOUTHWELL, C., 1985. The Australian kangaroo populations, 1984. *Search* **16**: 277-79.

GRIGG, G., SMITH, L., ALEXANDER, P. AND POPLE, T., 1987. The 1987 survey of kangaroos in the pastoral area of South Australia. Unpublished report prepared for the South Australian National Parks and Wildlife Service.

HARESTED, A. S. AND BUNNELL, F. L., 1987. Persistence of Black-tailed Deer fecal pellets in coastal habitats. *J. Wildl. Manage.* **51**: 33-37.

HILL, G. J. E., 1978. Preliminary assessment of defecation patterns for the Eastern Grey Kangaroo *(Macropus giganteus) Aust. Zool.* **19**: 291-300.

HILL, G. J. E., 1981. A study of Grey Kangaroo density using pellet counts. *Aust. Wildl. Res.* **8**: 237-43.

HILL, G. J. E., 1982. Seasonal movement patterns of the Eastern Grey Kangaroo in southern Queensland. *Aust. Wildl. Res.* **9**: 373-87.

HILL, G. J. E., BARNES, A. AND WILSON, G. R., 1985. Time of day and aerial counts of Grey Kangaroos. *J. Wildl. Manage.* **49**: 843-49.

HONE, J., 1986. Accuracy of the multiple regression method for estimating density in strip transects. *Aust. Wildl. Res.* **13**: 121-26.

JARMAN, P. J. AND TAYLOR, R. J., 1983. Ranging of Eastern Grey Kangaroos and Wallaroos on a New England pastoral property. *Aust. Wildl. Res.* **10**: 33-38.

JOHNSON, C. N. AND BAYLISS, P., 1981. Habitat selection by sex, age and reproductive class in the Red Kangaroo, *Macropus rufus,* in western New South Wales. *Aust. Wildl. Res.* **8**: 465-74.

JOHNSON, C. N. AND JARMAN, P. J., 1987. Macropod studies at Wallaby Creek. VI. A validation of the use of dung-pellet counts for measuring absolute densities of populations of macropodids. *Aust. Wildl. Res.* **14**: 139-45.

JOHNSON, C. N., JARMAN, P. J. AND SOUTHWELL, C. J., 1987. Macropod studies at Wallaby Creek. V. Patterns of defecation by Eastern Grey Kangaroos and Red-necked Wallabies. *Aust. Wildl. Res.* **14**: 133-38.

JOHNSON, K., 1977. Methods for census of wallaby and possum in Tasmania. Tasmanian National Parks and Wildlife Service Technical Report.

JOLLY, G. M., 1969. The treatment of errors in aerial counts of wildlife populations. *E. Afr. Agric. For. J.* **34**: 50-55.

JOLLY, G. M., 1981. Methods of bias correction. Monog. 4. ILCA, Addis Ababa. Pp. 181-84.

KAUFMANN, J. H., 1974. Social ethology of the Whiptail Wallaby, *Macropus parryi,* in northeastern New South Wales. *Anim. Behav.* **22**: 281-369.

LEATHERWOOD, S., SHOW, I. T., REEVES, R. R. AND WRIGHT, M. B., 1982. Proposed modification of transect models to estimate population size from aircraft with obstructed downward visibility. *Rep. Int. Whal. Comm.* **32**: 577-80.

MCCULLOUGH, D. R., 1982. Evaluation of night spotlighting as a deer study technique. *J. Wildl. Manage.* **46**: 963-73.

MCKENZIE, D. C., 1976. The distribution and abundance of macropod species in rainforest, eucalypt forest and exotic pine plantation at Clouds Creek, New South Wales. B. Nat. Res. Thesis, University of New England.

MCNAMARA, K., 1976. An evaluation of line transect methods for estimating wildlife populations. B. Nat. Res. Thesis, University of New England.

MCNAMARA, K. J. AND PRINCE, R. I. T., 1986. Kangaroo management in Western Australia. Wildlife Management Programme No. 3. Department of Conservation and Land Management.

MITCHELL, B., ROWE, J. R., RATCLIFFE, P. R. AND HINGE, M. D. C., 1985. Defecation frequency in Roe Deer *(Capreolus capreolus)* in relation to accumulation rates of faecal deposits. *J. Zool. Lond.* **207**: 1-7.

MORGAN, D. G., 1983a. Wyperfeld National Park. Kangaroo populations, August 1983. Unpublished report prepared for the National Parks Service, Ministry for Conservation, Victoria.

MORGAN, D. G., 1983b. Hattah-Kulkyne National Park and Murray-Kulkyne State Park. Kangaroo populations, June 1983. Unpublished report prepared for the National Parks Service, Ministry for Conservation, Victoria.

MORGAN, D. G., 1983c. Puckapunyal Military Reserve: Kangaroo populations, December 1983. Unpublished report prepared for the Fisheries and Wildlife Division, Department of Conservation, Forests and Lands, Victoria.

MORGAN, D. G., 1984a. Wyperfeld National Park. Kangaroo populations, August 1984. Unpublished report prepared for the National Parks Service, Ministry for Conservation, Victoria.

MORGAN, D. G., 1984b. Hattah-Kulkyne National Park and Murray-Kulkyne State Park. Kangaroo populations, July 1984. Unpublished report prepared for the National Parks Service, Ministry for Conservation, Victoria.

MORGAN, D. G., 1986. Estimating vertebrate population densities by line transect methods. Melbourne College of Advanced Education Occasional Paper No. 11.

NAARDING, J. A., 1979. A simple mechanical device for counting animals. *Aust. Wildl. Res.* **6**: 63-68.

NANCE, C. A., 1985. Macropodid management of cereal and pasture regions of Queensland. Unpublished report prepared for the Reserve Bank of Australia.

NEWSOME, A. E., 1965. The abundance of Red Kangaroos, *Megaleia rufa* (Desmarest), in central Australia. *Aust. J. Zool.* **13**: 269-87.

NEWSOME, A. E., DUDZINSKI, R. C. AND LOW, W. A., 1981. Measuring bias in aerial surveys due to dispersion of animals. ILCA Monograph No. 4. Addis Ababa. Pp. 185-94.

NORTON-GRIFFITHS, M., 1978. Counting Animals. Handbook No. 1. African Wildlife Leadership Foundation, Nairobi, Kenya.

OTIS, D. L., BURNHAM, K. P., WHITE, G. C. AND ANDERSON, D. R., 1978. Statistical inference from capture data on closed animal populations. *Wildl. Monog.* No. 62.

PACKARD, J. M., SUMMERS, R. C. AND BARNES, L. B., 1985. Variation of visibility bias during aerial surveys of Manatees. *J. Wildl. Manage.* **49**: 347-51.

PECHANEC, J. F. AND STEWART, G., 1940. Sagebrush-grass range sampling studies: size and structure of sampling unit. *Amer. Soc. Agron. J.* **32**: 669-82.

PERRY, R. J. AND BRAYSHER, M. L., 1986. A technique for estimating the numbers of Eastern Grey Kangaroos, *Macropus giganteus,* grazing a given area of pasture. *Aust. Wildl. Res.* **13**: 335-38.

POLLOCK, K. W. AND KENDALL, W. L., 1987. Visibility bias in aerial surveys: a review of estimation procedures. *J. Wildl. Manage.* **51**: 502-10.

PRINCE, R. I. T., 1984. Exploitation of kangaroos and wallabies in Western Australia. II. Exploitation and management of the Red Kangaroo: 1970-1979. Wildlife Research Bulletin Western Australia No. 14.

PROGULSKE, D. R. AND DUERRE, D. C., 1964. Factors influencing spotlight counts of deer. *J. Wildl. Manage.* **28**: 27-34.

PYKE, G. H. AND RECHER, H. F., 1985. Estimating forest bird densities by variable distance point counts. *Aust. Wildl. Res.* **12**: 307-19.

RICE, W. R. AND HARDER, J. D., 1977. Application of multiple aerial sampling to a mark-recapture census of White-tailed Deer. *J. Wildl. Manage.* **41**: 197-206.

RICKER, W. E., 1940. Relation of "catch per unit effort" to abundance and rate of exploitation. *J. Fish. Res. Bd. Canada.* **5**: 43-70.

ROBERTSHAW, J. D. AND HARDEN, R. H., 1986. The ecology of the Dingo in north-eastern New South Wales. IV. Prey selection by Dingoes and its effect on the major prey species, the Swamp Wallaby, *Wallabia bicolor* (Desmarest). *Aust. Wildl. Res.* **13**: 141-63.

ROBINETTE, W. L., FERGUSON, R. B. AND GASHWILER, J. S., 1958. Problems involved in the use of deer pellet group counts. *Trans. N. Am. Wildl. Conf.* **23**: 411-25.

ROGERS, G., JULANDER, O. AND ROBINETTE, W. L., 1958. Pellet group counts for deer census and range use index. *J. Wildl. Manage.* **22**: 193-99.

ROGERS, L. L., 1987. Seasonal changes in defecation rates in free-ranging White-tailed Deer. *J. Wildl. Manage.* **51**: 330-33.

ROLLINS, D., BRYANT, F. C. AND MONTANDON, R., 1984. Faecal pH and defecation rates of eight ruminants fed known diets. *J. Wildl. Manage.* **48**: 857-913.

ROST, G. R. AND BAILEY, J. A., 1979. Distribution of Mule Deer and Elk in relation to roads. *J. Wildl. Manage.* **43**: 634-41.

RYEL, L. A., 1959. Deer pellet group surveys on an area of known herd size. *Michigan Dept. Conserv. Game Div. Rept.* 2252.

to a sudden shortage of readily usable carbohydrate, leading to a drop in volatile fatty acid production, a rise in gastric pH and a subsequent proliferation of *Salmonella* in the stomach.

Salmonella can cause gastroenteritis and septicaemia in captive adult macropodids and in orphaned joeys (Finnie 1978; Speare 1988) and in theory salmonellosis could occur in free-ranging macropodoids, particularly when they are stressed and environmental contamination is high. However, salmonellosis has not been reported from free-ranging populations. In the comprehensive study of Hart *et al.*(1985) clinical cases of salmonellosis in Quokkas were not found.

Most studies on Salmonellae in macropodids have been concerned with the zoonotic implications of commercial harvesting. Kangaroo meat has been surveyed in Australia and overseas and prevalences of contaminated meat have ranged from 58% to 11% (Mayer and Haug 1962; Anderson *et al.* 1964; Suzuki *et al.* 1967). Three serotypes of *Salmonella* at an overall prevalence of 11.1% were found in 81 kangaroo carcasses at chillers in south-east Queensland (Ekaputra 1985). High coliform and total bacterial counts indicated that heavy contamination had occurred. Ekaputra (1985) suggested that the presence of *Salmonella* and high coliform counts indicated that evisceration procedures in the field operations were "suspect" and needed further improvements to reduce intestinal contamination on the carcasses. With minimal contamination during evisceration and adequate refrigeration of carcasses to prevent the growth of *Salmonella,* human consumption of kangaroo meat should not pose a great problem with regard to public health (Ekaputra 1985). Meat workers are at a risk of acquiring salmonellas from the products they handle (Howie 1979). Obviously this risk increases the greater the degree of contamination of the product. The potential zoonotic hazards of orphaned macropodid joeys has been noted (Speare and Thomas 1988) and all workers with macropodids and their products should maintain good hygenic practices.

Leptospirosis

Leptospira, fragile spirochaetal bacteria, can infect an extremely wide range of hosts, but clinical disease in most hosts is uncommon. Macropodoidea are no exception. Although serological surveys have shown that infection with various serovars of *Leptospira* occur (Table 3), no clinical illness or significant postmortem lesions have been reported in wild or captive macropodids. Durfee and Presidente (1979) found focal interstitial nephritis of mild degree in two *M. rufus* and four *M. rufogriseus*, but could not correlate these pathologic changes with serology. Focal interstitial nephritis was reported in wild *Macropus agilis* (Speare *et al.* 1983) and in *M. rufus* (Arundel *et al.* 1979). Leptospiral titres were not estimated.

DISEASES CAUSED BY VIRUSES

Macropod Herpesvirus

Macropod Herpesvirus (MHV) causes a disease manifested by conjunctivitis, muco-cutaneous blisters and ulcerations with varying degrees of lethargy and may progress to death. Focal necrosis occurs in mucous membranes, liver, heart, lung, spleen and kidney (Acland 1981; Callinan and Kefford 1981). Intranuclear viral inclusion bodies also occur in these organs. Nine macropod herpesvirus strains have been isolated. By the use of restriction endonucleases, dot-blot hybridizations and analytical ultracentrifugation in caesium chloride, Johnson *et al.* (1985) classified the strains into two groups, the MHV 1, which includes the Parma Wallaby herpes virus (PWHV), the virus originally isolated by Finnie *et al.* (1976), and MHV 2, which contains the other eight isolates.

MHV has been reported as a cause of disease in captive *Aepyprymnus rufescens, Bettongia penicillata, Dendrolagus ursinus, Dorcopsis luctuosa, M. eugenii, Macropus parma, M. rufus, Macropus fuliginosus, Potorous tridactylus* and *S. brachyurus* (Finnie *et al.* 1976; Finnie 1978; Spence 1979; Dickson *et al.* 1980; Callinan and Kefford 1981; Wilks *et al.* 1981; Schoon *et al.* 1982), but clinical disease has not been reported in free-ranging Macropodidea. Antibody to MHV is widespread (Webber and Whalley 1978; Kerr *et al.* 1981). In a study of 192 wild macropodids Webber and Whalley (1978) reported serum neutralizing antibody to PWHV from all states of Australia and Kawau Island in New Zealand, a site of an introduced population of *M. parma.*

Table 3. Prevalences of positive antibody titres to *Leptospira* serovars in free-ranging Macropodidae (only surveys with positive titres are included).

Leptospira	Hosts	Location	Status	Prevalence No. (%)	Reference
L. ballum	*W. bicolor*	NSW and Victoria	W	2-12 (16.5%)	Milner *et al.* (1981)
L. grippotyphosa	*T. stigmatica*	North Queensland	W	1/2 (50%)	Emmanuel *et al.* (1964)
L. hyos	*P. tridactylus*	Tasmania	W	3/6 (50%)	Munday (1972a)
L. pomona	*M. eugenii*	NSW and Victoria	C	1/26 (3.9%)	Milner *et al.* (1981)
	M. rufogriseus rufogriseus	Tasmania	W	1/7 (14.3%)	Durfee and Presidente (1979)

The overall prevalence was 22.9%. *Lagorchestes conspicillatus, M. giganteus, M. parma, Macropus robustus, Petrogale* spp., *P. tridactylus, Thylogale stigmatica* and *Thylogale thetis* had positive antibody titres, while limited numbers of *M. rufus* were negative (Webber and Whalley 1978). However, in the same study captive *M. rufus* had positive titres. Kerr *et al.* (1981) found 69% of 218 *M. giganteus* from Victoria and *M. fuliginosus* from Kangaroo Island, South Australia had antibody titres to PWHV. Geometric mean titre was 105. Mature animals of both species had greater prevalences (almost 90%) compared with young. This pattern is similar to that found for herpesviruses in other hosts and indicates that animals are exposed to recurring infection during their lifetime or carry a latent infection. Captive macropodids with clinical disease usually had titres at higher levels (≥512) than clinically normal animals (Webber and Whalley 1978).

Althouh MHV is the most pathogenic viral disease of macropodids, little is known of its epidemiology. The virus is probably capable of infecting all species of macropodids. The half life of the strain from Parma Wallabies at 37°C is less than 1 hr (Whalley and Webber 1979) and explains why the virus appears to require close contact for transmission. Like herpesviruses of other species stress may allow latent macropod herpesvirus to be expressed as clinical disease.

Macropod Pox Virus

The causative agent of macropod pox has been seen by electron microscopy, but has not been isolated in culture. The ultrastructure is typical of a pox virus (Papadimitriou and Ashman 1972; McKenzie *et al.* 1979; Rothwell *et al.* 1984) and, as is characteristic of *Poxviridae,* Macropod Pox Virus re-activates Vaccinia Virus (Stanley and McKenzie 1983). A pox virus isolated from the Quokka showed cross-reactivity with Fowlpox in a complement fixation test (Stanley and McKenzie 1983).

Pox virus lesions have been described in *M. agilis* (Speare 1988), *M. fuliginosus* (Presidente 1978; Rothwell *et al.* 1984), *M. giganteus* (McKenzie *et al.* 1979; Speare 1988), *M. rufus* (Bagnall and Wilson 1974; Presidente 1978; Arundel *et al.* 1979) and *S. brachyurus* (Papadimitriou and Ashman, 1972). Host specificity is indicated by clinical observations (Speare 1988). Two types of gross lesions have been described, an umbilicated papule with smooth edges and a slightly depressed, pale centre (Speare 1988) and a larger, pigmented hyperkeratotic wart-like lesion (Papadimitriou and Ashman 1972; McKenzie *et al.* 1979; Rothwell *et al.* 1984; Speare 1988). The gross appearance of the umbilicated papule is typical of human molluscum contagiosum, but this form is uncommon and has been observed to progress to a papilloma (Speare 1988). Most lesions are first seen as papillomata (Speare 1988). On histopathology the diagnostic finding is the presence of large, rounded eosinophilic intracytoplasmic inclusions which displace the nucleus of epidermal cells above the basal layer. The epidermis in papillomata is thickened, hyperkeratosis is marked with eosinophilic inclusions apparent in the keratinized material, acanthosis may be present and a mononuclear infiltration may be present in the dermis below the lesions (Speare 1988).

Macropod pox has been seen in wild *M. agilis, M. giganteus, M. rufus* and *S. brachyurus.* It has been reported in furred joeys and juveniles (Bagnall and Wilson 1974; Presidente 1978; McKenzie *et al.* 1979; Speare 1988) and in adult macropodids (Papadimitriou and Ashman 1972; Rothwell *et al.* 1984). In some joeys in captivity individual lesions have persisted for about two months, but crops of new lesions appeared over a period of about three months (Speare 1988). Captive joeys appeared to lose all remaining lesions over a short period before maturity was reached (Speare 1988). The natural history of macropod pox in free-ranging macropodids is unknown.

Arboviruses

Serological surveys of wild kangaroos and wallabies have shown antibodies to a number of Group A arboviruses (Doherty *et al.* 1964, 1970; see Potkay 1977). Experimental infection with Murray Valley Encephalitis Virus (MVE) was established in *M. agilis* and *M. giganteus* by allowing feeding of orally infected *Culex annulirostris* (Kay *et al.* 1982, 1986). Although viraemia was detected in the Grey Kangaroos, but not the Agile Wallabies, both were able to reinfect *C. annulirostris.* Grey Kangaroos were considered to be a very effective amplifier of MVE (Kay *et al.* 1986). Both species of macropodids developed viraemia from infection with Ross River Virus (Kay *et al.* 1986) and are good amplifiers. Macropodids are probably the primary host of Ross River Virus (Kay *et al.* 1986). No ill effects were noted after infection with either virus.

Adenovirus

Munday and Obendorf (1983) noted lesions in the renal collecting tubules of wild *B. gaimardi.* Lesions consisted of hypertrophy of tubule cells, margination of chromatin and eosinophilic intranuclear inclusions. Viral particles seen on electron microscopy were approximately 70 nm in diameter, roughly hexagonal to spherical and contained a single large electron-dense core. Morphology of the viral particles was typical of an adenovirus (Munday and Obendorf 1983).

Viruses of introduced ruminants

Experimental infections of macropodids with Foot and Mouth Disease Virus showed that *Dendrolagus matschiei, M. giganteus, M. rufogriseus, M. rufus* and *P. tridactylus* could be infected by parenteral inoculation (Snowdon 1968, 1988). Red Kangaroos could also be infected by close confinement with cattle with clinical foot and mouth disease. Clinical signs were noted in Red Kangaroos and Tree Kangaroos. Vesicles up to 4 cm in diameter developed on the tongue and the pads of the feet 3-7 days after intramuscular inoculation. Although experimentally infected Red Kangaroos could transmit FMD to cattle, Snowdon (1968, 1988) considered that macropodids would be unlikely to play a significant role if FMD was introduced to Australia.

No clinical disease was noted in a kangaroo drenched with blood from cattle infected with rinderpest (Snowdon 1988).

Antibody against Mucosal Disease Virus was detected in 4.5% of 44 wild Bennett's Wallabies in Tasmania (Munday 1972a). The epidemiological significance of this finding apparently has not been investigated further.

DISEASES CAUSED BY RICKETTSIA

Coxiella burneti

Coxiella burneti is capable of infecting a wide range of hosts, including marsupials and man (Beveridge 1981). Infected animals show no disease, while man develops Q Fever, an important and potentially fatal zoonosis. Cattle and sheep are usually implicated as the sources of Q Fever in man.

Pope *et al.* (1960) found *M. rufus* and *M. major* seropositive to *C. burneti* over considerable areas of western Queensland. The Grey Kangaroos surveyed were close to the limits of the north-eastern extension of *M. fuliginosus* into Queensland, but essentially outside the range given by Caughley *et al.* (1984). Consequently, Pope *et al.* (1960) were probably dealing predominantly with *M. giganteus* under their taxon, *M. major*. Smaller numbers of *M. dorsalis, M. robustus* and *W. bicolor* were seronegative. *C. burneti* was isolated from the blood of a Grey Kangaroo, thus confirming that patent infections were occurring in kangaroos. *C. burneti* was also isolated from the kangaroo tick *Amblyomma triguttatum* Koch, 1844, collected from kangaroos and sheep (Pope *et al.* 1960). A kangaroo-tick cycle was postulated (Dane and Beech 1955; Pope *et al.* 1960). *Ornithodoros gurneyi* Warburton, 1926, also known as the kangaroo tick, fed on guinea pigs infected with *C. burneti* became infected although the rickettsia could not be transmitted by the tick to other guinea pigs (Smith 1942). *O. gurneyi* was not found in the survey by Pope *et al.* (1960). A titre of 1:20 to *C. burneti* was found in Tasmania in a *M. r. rufogriseus* (Munday 1972a). Two *A. rufescens* were experimentally infected by *C. burneti* (Derrick *et al.* 1940). In one Rat-kangaroo the infection resulted in splenomegaly with multiple foci of necrosis in the liver. *C. burneti* was recovered from the blood of both animals and the spleen of the one with splenomegaly. Neither Rat-kangaroo had a febrile response.

C. burneti is of little significance to free-ranging macropods, but may be of zoonotic significance to persons coming into close contact with wild macropods or their products. However, Pope *et al.* (1960) noted no cases reported in kangaroo shooters. *C. burneti* is an extremely resistant organism and can survive for considerable periods of time in dust. Inhalation of aerosols or infected dust is thought to be the most common route of infection in humans, although infection can be acquired through abrasions (Beveridge 1981). A vaccine for use in humans is currently undergoing clinical trials in South Australia (Anon. 1988), and if epidemiological data indicate that workers in the kangaroo industry are at risk, vaccination for Q Fever could be considered.

Rickettsia rickettsii

R. rickettsii is the causative agent of north Queensland tick typhus. Two of 11 wild *A. rufescens* in north Queensland were found to have antibodies against *R. rickettsii* (Fenner 1946).

DISEASES CAUSED BY FUNGI

Dermatophytosis

Ringworm or dermatophytosis, the skin disease caused by keratinophilic fungi, has not been reported from free-ranging macropods. A survey of wild macropods found 67% of 18 Grey Kangaroos and the sole *A. rufescens* to carry keratinophilic fungi, but only one isolate, *Microsporum cookei*, from a Grey Kangaroo was a potential pathogen (Rees 1967). In captivity *M. eugenii, M. rufus* and *S. brachyurus* were found to carry *M. cookei* and *Trichophyton ajelloi*, both potential pathogens (McAleer 1980). The prevalence of positive isolations was high at 89%, although no comment was made on the presence or absence of clinical lesions. In soils from the yards in which these animals were held *M. cookei* was commonly isolated, with occasional isolations of *T. ajelloi* and another potential pathogenic fungus, *Microsporum gypseum* (McAleer 1980). Unspecified kangaroos kept as pets were implicated in eight cases of human dermatophytoses caused by *Trichophyton mentagrophytes* (Donald 1960). *M. rufus* was implicated as the source in other cases of human ringworm due to *T. mentagrophytes* (Barker *et al.* 1963). Other species of dermatophytes isolated from captive macropodids are *M. canis* (Speare 1988) and *M. gypseum* from a *Dendrolagus lumholtzi* (pers. obs.).

Clinically obvious dermatophytosis occurs occasionally in captive macropods (Munday 1978; Presidente 1978; Speare 1988), but has not been reported in free-ranging Macropodoidea.

DISEASES CAUSED BY PROTOZOA

Coccidia use Macropodoidea either as difinitive hosts, for example, *Eimeria* and *Klossellia,* or as intermediate hosts (*Sarcocystis* and *Toxoplasma*). Infections with other parasitic protozoa are uncommon.

Intestinal coccidia

Intestinal coccidia in macropodids predominantly infect the cells of the mucosa of the small intestine, although some species of coccidia have stages in gastric mucosa and non-enteric locations. Transmission is by environmentally resistant oocysts passed in the faeces. Although only 12 species of coccidia have been named, many others exist with most belonging to *Eimeria* and some to *Isospora* (I. Barker, pers. comm.). Coccidia occur in wild macropodids of all ages. Surveys of wild *M. rufus* found prevalences of 13.5%, 10% and 5.9% in three locations and an increased prevalence at about 12 months of age (Mykytowycz 1964). In wild *M. agilis* endogenous stages of *Eimeria* were found in 4% of histological sections of ileum (Speare *et al.* 1983). Pathological changes were not observed in either survey.

Coccidian oocysts are commonly found in captivity in the faeces of all species of macropodid, but coccidiosis, the disease associated with the schizonts is generally uncommon and occurs predominantly in juveniles of several species only. *M. giganteus* is most at risk (Gilruth and Bull 1911; Winter 1959; Mykytowycz 1964; Calaby and Poole 1971; Finnie 1974; Munday 1978; Speare 1988). Other species in which clinical coccidiosis have been reported are *M. dorsalis, M. eugenii, M. fuliginosus, M. parryi, M. rufogriseus,* and *M. rufus* (Gilruth and Bull 1911; Keymer 1972; Wilson *et al.* 1984; Munday 1978, 1988; Presidente 1978; Speare *et al.* 1982; Speare 1988). Pathologic change due to coccidia in free-ranging macropodids has been reported in *M. giganteus* (Barker *et al.* 1972; Arundel *et al.* 1977), *M. eugenii* (Presidente 1978; Hartley in Munday 1988) and *M. fuliginosus* (Arundel *et al.* 1977; Presidente 1978). In *M. eugenii* scattered granulomatous lesions were found in hepatic portal triads associated with large multilocular coccidian schizonts (Presidente 1978; Hartley in Munday 1988). Juveniles which died in captivity showed the same lesions as well as a haemorrhagic enteritis (Presidente 1978). Focal necrosis in the pyloric antrum of *M. fuliginosus* was attributed to large multilocular schizonts (Arundel *et al.* 1977).

A localized epidemic of coccidiosis caused deaths of wild *M. giganteus* trapped for several weeks on "islands" by rising flood water in Barmah Forest in Victoria (Barker *et al.* 1972). Only juveniles were affected by coccidiosis. Density of *M. giganteus* was high, and resulted in depletion of fodder and heavy contamination of soil by faeces. Pathologic findings were haemorrhagic enteritis, most intense in the jejunum, associated with villous atrophy and oedema probably secondary to hypoproteinaemia due to protein loss through the small intestine mucosa. The species of coccidia responsible for the pathological changes in the trapped *M. giganteus* could not be identified, but oocysts of *E. cunnamullensis* and *E. kogoni* were found in faeces. Barker *et al.* (1972) hypothesized that young, susceptible animals had been stressed by flooding, malnutrition and overcrowding, and had been massively infected by exposure to large numbers of oocysts over a short period of time. These authors suggested that the frequency and periodicity of flooding of the Barmah Forest was such that conditions suitable for similar outbreaks of coccidiosis would be expected to occur with a probability of over 50% in a given year.

Klossiella

Klossiella is a genus of coccidia which has endogenous stages in the epithelial cells of renal tubules. Wild Macropodoidea are commonly infected and a number of species have been described and named (Barker *et al.* 1975; Presidente 1978). These protozoans are not pathogenic and cause no damage apart from enlargement of the parasitised cell.

Toxoplasma gondii *(Nicholle and Manceaux, 1908)*

Toxoplasma gondii is a coccidian parasite with sexual stages in the small intestine of Felidae and asexual stages in organs of mammals and birds (Speare 1985). Infected cats excrete resistant oocysts and herbivores are infected by ingestion of oocysts. In some areas of Australia a cat-sheep cycle is important with pastures being contaminated by oocysts excreted by domestic and feral cats (Munday 1970, 1975). The prevalence of clinical disease is much less common than the prevalence of infection, since intermediate hosts infected by *T. gondii* usually show no disease upon initial infection and the organism becomes dormant in tissues. Hosts with toxoplasmosis may have damage caused by an initial infection or due to a reactivation of dormant organisms. Depression of immunity and stress appear to be critical events in the development of most cases of toxoplasmosis. Toxoplasmosis is a common disease in some colonies of captive macropodids (Barker *et al.* 1963; Arundel *et al.* 1977).

Wild Macropodoidea in most areas have not been extensively surveyed. The prevalence of serological evidence of infection with *Toxoplasma* was low in some regions (Cook and Pope 1959; Smith and Munday 1965; Jakob-Hoff and Dunsmore 1983), but quite high in others. Higher prevalences may be related to mean annual rainfall of greater than 61 mm (Smith and Munday 1965). A survey of *S. brachyurus* from Rottnest Island using serologic, histologic and cultural methods, found 32 (21%) of 150 to have evidence of infection with *T. gondii* (Gibb *et al.* 1966). Only animals close to the settlement on the island were found to be infected. No clinical cases of toxoplasmosis or pathological response to the tissue cysts were found. Forty-one percent of 17 *T. billardierii* in Tasmania were serologically positive (Munday 1970).

A histological survey of wild Macropodidae from many areas of Australia found six species with cysts consistent with *T. gondii* (Munday *et al.* 1978). Reported free-ranging macropodid hosts of *T. gondii* include *M. agilis, M. giganteus, M. rufus, M. rufogriseus, S. brachyurus, T. billardierii* and *W. bicolor.*

Clinical toxoplasmosis in free-ranging *T. billardierii* and *M. rufogriseus* was reported by Obendorf and Munday (1983). About 50 macropodids of both species were affected on a farm in Tasmania. Animals were uncoordinated, apparently blind and some had died. Severe non-suppurative myocarditis, encephalitis and hepatitis were associated with cysts of *T. gondii.* Similar signs and deaths of wallabies had occurred on the same farm in late autumn and early winter of the previous two years. Obendorf and Munday (1983) considered that nutritional stresses due to scarcity of food had caused latent infection to be manifest as acute disease.

Sarcocystis

Sarcocystis is a genera of coccidia with a life cycle similar to that of *Toxoplasma,* but with a greater specificity for intermediate hosts (Fayer 1980). Cysts typical of *Sarcocystis* have been found in muscle of *M. rufus, Peradorcas concinna,* and *Petrogale venustula* (Munday *et al.* 1978; Arundel *et al.* 1979). Thin walled sarcocysts with small zoited (Type B of Munday *et al.* 1978) were found in the cardiac and skeletal muscle of one of two *W. bicolor* (Beveridge *et al.* 1985). These may have been *T. gondii.* Sarcocysts were found in 2% of skeletal muscles of wild *M. agilis* (Speare *et al.* 1983). Multiple Type A cysts were noted in the skeletal muscle of a wild *D. lumholtzi* in north Queensland (pers. obs.).

Haemoprotozoa

Haemoprotozoa have been reported from two species of Macropodoidea and none have been described and named. A parasite identified as an *Eperythrozoon* was implicated as a possible cause of seasonal anaemia in *M. eugenii* on Kangaroo Island, South Australia (Murphy 1970). *Eperythrozoon* is now considered to be a rickettsial genus. Poor nutrition was postulated to stress animals sufficiently to allow the parasite to increase from a latent infection to numbers sufficient to cause a hypochromic normocytic anaemia (Murphy 1970). *Theileria* sp. was noted in *P. tridactylus* (Mackerras 1958).

DISEASES CAUSED BY METAZOAN PARASITES

Trematodes

Trematodes are uncommon parasites in Macropodoidea with only three species reported from free-ranging hosts and none from captive macropodids. Two species, *Gemellicotyle wallabicola* and *Macropotrema pertinax,* are found only in *M. agilis,* the former in the stomach and the latter in the caecum and proximal colon (Prudhoe 1975; Blair *et al.* 1979; Speare *et al.* 1983). Both parasites are paramphistomes. Intermediate mollusc hosts are expected for the paramphistomes, but there is no information on their life cycles.

The other trematode is *Fasciola hepatica,* a large parasite found in the biliary system and not native to Australia, but introduced with sheep and cattle. This trematode is common and widespread in temperate, wet, agricultural areas in southeastern Australia (Boray 1969). Without doubt 1988 is the bicentenary of the arrival of *F. hepatica* in Australia and the parasite has successfully colonized the bile ducts of six species of marsupial, including four species of Macropodidae, as well as *Trichosurus vulpecula* and *Vombatus ursinus* (Johnston 1909, 1911; Howkins 1966a; Arundel *et al.* 1977; McManus 1979; Spratt and Presidente 1981). A native rodent, *Rattus fuscipes* can also act as a host (Spratt and Presidente 1981). *F. hepatica* has as intermediate host, the snail *Lymnea tomentosa.* The distribution of the parasite is determined by the distribution of this snail which is dependent on suitable habitat (Boray 1969). Infection with *F. hepatica* is by ingestion of metacercariae encysted on pasture.

Paramphistomes

The paramphistomes of *M. agilis* cause pathologic changes, but of a localized nature and are of little consequence to the health of the host. Each attaches to the mucosa of their respective sites by means of a large, muscular sucker. *G. wallabicola* causes an irregular hypertrophy of the mucosa with adjacent ulceration, microhaemorrhage and infiltration of the lamina propria by mononuclear cells and small numbers of eosinophils (Speare *et al.* 1983). *M. pertinax* can cause atrophy of mucosal cells at its attachment sites (Blair *et al.* 1979) or a hyperplastic response in heavier infections (Speare *et al.* 1983). Pathological changes with both parasites are minor.

Fasciolia hepatica *Linnaeus, 1758*

In contrast to the paramphistomes *F. hepatica* can cause severe changes. Infections with *F. hepatica* have been reported in *M. giganteus* and *M. rufogriseus* in southeastern New South Wales, Victoria and Tasmania (Johnston 1909, 1911; McManus 1979; Spratt and Presidente 1981), in *T. billardierii* in Tasmania (McManus 1979) and in *W. bicolor* in southeastern New South Wales and Victoria (Spratt and Presidente 1981). Mean prevalences of *F. hepatica* in southeastern Australia were 59% in *M. giganteus,* 15% in *M. rufogriseus banksianus* and 5% in *W. bicolor* (Spratt and Presidente 1981). Prevalences were higher in macropodids collected on agricultural pastures grazed by sheep and cattle. However, at Tidbinbilla Nature Reserve, an area of 5 500 hectares in the A.C.T., the life cycle of *F. hepatica* appeared to be maintained by a macropodid-snail cycle, in the absence of introduced ruminants (Spratt and Presidente 1981). *M. giganteus, M. rufogriseus banksiana, M. rufogriseus rufogriseus* and *T. billardierii* appear to have infections with egg outputs capable of significantly contaminating pasture (McManus 1979; Spratt and Presidente 1985). Infections in *W. bicolor* appear also to be patent (Spratt and Presidente 1981), but data on egg production has not been reported.

As in domestic ruminants, *F. hepatica* in macropodids causes fibrosis and thickening of bile ducts with bile duct hyperplasia and an eosinophilic cholangiohepatitis (Beveridge 1978; Presidente 1978; McManus 1979). Migrating immature flukes cause parenchymal fibrosis. Infected macropodids may show no ill effects or severe cachexia and anaemia. Possible differences exist between species in the pathologic response to infection with *F. hepatica* (Presidente in Beveridge 1978). *M. giganteus* tolerates moderate infections of up to 80 flukes without clinical disease. Severe fibrosis of the major bile ducts is the principal lesion, with bile duct hyperplasia and focal parenchymal fibrosis. In *M. rufogriseus* the bile ducts become thickened and cystic and an anaemia may be present. Bile ducts of *T. billardierii* are thickened, the liver shows marked fibrosis, anaemia may be severe and infection can be fatal. McManus (1979) considered that Tasmanian macropodids were less susceptible to the pathogenic effects of *F. hepatica* than sheep and in particular acute fascioliasis, a widespread focal necrosis of the hepatic parenchyma due to migrating immature flukes, had not been reported. Spratt and Presidente (1981) commented that *F. hepatica* is one of the most pathogenic parasites encountered in marsupials.

Nematodes

Wild kangaroos have an impressive number and array of nematodes in their stomachs. Both Grey Kangaroos may have a total of more than 30 000 nematodes from 20 species of strongyles (Beveridge and Arundel 1979). Minor pathological changes are noted in the stomach with larvae of some of the strongyles, but the adult nematodes appear to be non-pathogenic. The most pathogenic nematode in the macropodid stomach is *Strongyloides* sp. In the small intestine nematodes are generally few in number and species. *Globocephaloides* is a significant pathogen. Nematodes are slightly more abundant and diverse in the large intestine. *Hypodontus* is potentially pathogenic. In the large kangaroos high numbers of oxyurids are sometimes found, but these appear to have no effect. Pathogenic nematodes found in sites outside the gastrointestinal tract include lungworms, filarioids (Beveridge 1978) and *Durikainema macropi* (Spratt and Speare 1982; Speare and Spratt 1982).

Globocephaloides

Deaths of free-ranging juvenile *M. giganteus* were shown to be associated with a severe anaemia and hypoproteinaemia (Arundel *et al.* 1977). The number of *G. trifidospicularis* recovered from the small intestine of the juveniles was much greater than that seen in adult Eastern Grey Kangaroos (Beveridge and Arundel 1979). A prospective study in the following year noted a rise in the numbers of *Globocephaloides* and a corresponding fall in haemoglobin, haematocrit and plasma protein of juveniles (Arundel *et al.* 1977). Death of a captive *M. giganteus* due to anaemia and hypoproteinaemia was caused by another species, *G. macropodis* Yorke and Maplestone, 1926 (Beveridge *et al.* 1984). Although no histo-pathological studies are available, it appears that *Globocephaloides* has a hookworm-like pathogenicity and causes loss of red blood cells as well as protein.

Strongyloides

Although strongyloidiasis has been responsible for deaths of captive macropodids (Winter 1958; Arundel *et al.* 1977; Speare *et al.* 1982), and lesions are seen in free-ranging Macropodoidea (Arundel *et al.* 1977; Speare *et al.* 1983), deaths have not been reported. *Strongyloides* live in tunnels in the mucosal epithelium of the areas of the stomach with cardiac glands and, in some species, in the mucosa of the caecum and upper colon (Arundel *et al.* 1977; Speare *et al.* 1982; Speare *et al.* 1983). They are normally well tolerated by the host. Occasionally in free-ranging *M. agilis* a hyperplastic response occurs with the formation of isolated nodules of hyperplastic cardiac glands, sometimes with the nodule penetrating through the lamina propria into the sub-mucosa (pers. obs.). A localized inflammatory response may occur. Arundel *et al.* (1977) noted a focal hyperaemia associated with *Strongyloides* in juvenile Eastern Grey Kangaroos. The species of *Strongyloides* in macropodids has yet to be described.

Hypodontus Macropi Monnig, 1929

H. macropi is found in the ileum and large intestine of *M. agilis, M. fuliginosus, M. robustus, M. rufus, W. bicolor* (Arundel *et al.* 1979; Beveridge 1979; Speare *et al.* 1983). It is a strongyle which attaches to the intestinal wall by grasping a plug of mucosa and submucosa in its globular buccal caspule (Beveridge 1979). The mucous membrane is lysed, an inflammatory reaction is elicited and with infections of more than 50 worms small foci of haemorrhage from prior feeding sites can be seen. *H. macropi* does not appear to feed selectively on blood, but on the tissue of the intestine (Beveridge 1979). Deaths of wild *M. rufus* were thought to be due to *H. macropi* (Copeman in Arundel *et al.* 1977).

Lungworms

Nematodes from several families occur in the lungs of Macropodoidea (Spratt 1979, 1984). There is little data on the prevalences of nematodes in the lungs of Macropodoidea. *Marsupostrongylus* sp. in *W. bicolor* occurred at a prevalence of 46%, and was more common in southern than northern Australia (Beveridge *et al.* 1985). Nematodes in the lungs of macropodids may be associated with an inflammatory reaction. The severity of the reaction in terms of intensity and volume of lung involved determines the significance of the parasite to the host. Most reactions are localized to the region of the parasites and severe reactions occur only with larger numbers of nematodes. Beveridge *et al.* (1985) in a survey of *W. bicolor* are the only authors to describe pathological lesions which they considered severe enough to compromise the host.

Metastrongyloid lungworms are often associated with a mild to severe interstitial pneumonia with the most prominent reaction in and around the small airways (McColl and Spratt 1982). Verminous pneumonia has been reported for several host-parasite combinations, but apart from the reaction in *W. bicolor* the pathology has not been well described (Beveridge *et al.* 1985). Extensive, severe bronchopneumonia was found in two *W. bicolor*. Peribronchiolitis with hyperplasia and inflammation of bronchiolar epithelium and infiltration with eosinophils was noted (Beveridge *et al.* 1985). Dead worms were surrounded by granulomas of eosinophils and foreign body giant cells. In another *W. bicolor* there were scattered focal granulomas of eosinophils and mononuclear cells, as well as smooth muscle hyperplasia and mild bronchiolitis (Beveridge *et al.* 1985).

Capillaria sp. has caused severe pneumonia and death in a colony of captive *A. rufescens* (D. M. Spratt, pers. comm., 1988). The condition has yet to be described. There are no reports of *Capillaria* in the lungs of free-ranging *A. rufescens*.

Pelecitus Roemeri (von Linstow, 1905)

P. roemeri is a large filarioid nematode found in the subcutaneous and intermuscular connective tissue planes of many species of macropodid (Spratt and Varughese 1975). The adult parasites occur in the pelvic region and hind limbs, with the majority in the region of the knee joint (Spratt 1972a, 1975; Beveridge 1978). The taxon was known previously as *Dirofilaria roemeri,* but has now been transferred to *Pelecitus* and a hypothesis proposed that macropodids originally acquired this parasite from birds (Bartlett and Greiner 1986). Microfilariae occur in the blood of suitable hosts and transmission is by haemaphagous tabanid flies (Spratt 1972a, 1972b, 1972c, 1974). Macropodids can be classified into normal, abnormal or unsuitable hosts (Table 4). In the normal host, *M. robustus,* the adult parasites provoke a minimal host reaction and microfilariae occur in significant numbers in the blood (Spratt 1972a, 1974, 1975). Some adult *P. roemeri* in *M. robustus* are not encapsulated (Spratt 1972a). The abnormal host reacts to the adult parasite by forming a fibrous capsule around it. Microfilaremia, if it occurs, is short lived and of low intensity (Spratt 1972a, 1974). In the unsuitable host the reaction to the parasite is marked, often with complete degeneration of the nematode. In *M. agilis,* for example, calcified irregular nodules are sometimes found in the tissues surrounding the knee joint. These nodules are most likely degenerated *P. roemeri.* Johnston and Mawson (1949) had reported *P. roemeri* from *M. agilis,* but this observation was not confirmed in the survey by Speare *et al.* (1983). However, subsequently a *M. agilis* was found with encapsulated *P. roemeri. M. agilis* is obviously an unsuitable host.

Capsules surrounding the adult *P. roemeri* can be divided into three types. Thin walled capsules with a minimal host response in which the wall is composed of scattered fibroblasts among loose strands of collagen with some lymphocytes, neutrophils, plasma cells and mast cells occur in normal hosts (Spratt 1972a). Viable microfilariae occur in the lumen and in the wall of the capsule (Spratt 1972a, 1974). Thin walled capsules with a moderate host response in which the wall consists of scattered fibroblasts among loose strands of

Table 4. Hosts of *Pelicetus roemeri* and their suitability.

Host	Prevalence	Host status	Reference
Macropus agilis	0.0%	Unsuitable	Speare *et al.* (1983)
	rare		this paper
Macropus giganteus	37.9%	Abnormal	Spratt (1972a)
	25.0%		Spratt (1974)
	18.0%		Spratt (1975)
Macropus rufogriseus	0.0%	Abnormal	Spratt (1972a)
	22.2%		Spratt (1974)
Macropus rufus	6.0%	Abnormal	Spratt (1975)
Macropus robustus	7.9%	Normal	Spratt (1972a)
	19.4%		Spratt (1974)
	22.0%		Spratt (1975)
Wallabia bicolor	11.0%	Abnormal	Beveridge *et al.* (1985)

collagen were noted in *M. rufus* (Spratt 1974). Lymphocytes, eosinophils, neutrophils, plasma cells and mast cells, many degranulated, are scattered through the wall of the capsule and microfilariae are not seen in the lumen or wall of the capsule. Thick walled capsules with numerous eosinophils, plasma cells, macrophages and fibroblasts in well developed, laminated fibrous connective tissue are typical of the response in abnormal hosts (Spratt 1972a; Spratt 1974). Some neutrophils, lymphocytes, mast cells and foreign body giant cells are also present. Many microfilariae with pyknotic nuclei occur in the lumen and wall of the capsule. Microfilariae in the lumen are surrounded by eosinophils, nuetrophils, plasma cells and macrophages. These capsules often contain pale coloured, hard pellets consisting of concentric layers of dense fibrous tissue with no cellular elements or of degenerate microfilariae surrounded by eosinophils and macrophages inside whorls of fibrous tissue. Spratt (1974) also noted an increase of eosinophils in the spleen and the interstitial areas of the lung, but no microfilariae were seen. In abnormal hosts microfilariae are destroyed by the local host response and microfilaremia is either transient or does not occur (Spratt 1972a, 1972c, 1975).

The prevalence of infection in abnormal hosts may be quite high (Spratt 1972a, 1974, 1975). Unsuitable hosts have very low prevalences of infection and *P. roemeri* has not been found in the Potoroinae.

P. roemeri may pose a problem as kangaroo meat gains acceptence as a high quality meat for human consumption. In abnormal hosts, such as *M. giganteus* and *M. rufus,* nodules can be located and trimmed off meat if infection is not extensive. However, in *M. robustus,* the parasites occur in tissue planes between muscles and may not be encapsulated. The minimal host reaction makes dissection of individual worms difficult. In some areas of southern Queensland the prevalence of infection in commercially harvested kangaroos was high, 18% for *M. giganteus,* 6% for *M. rufus* and 22% for *M. robustus* (Spratt 1975). This would represent a considerable loss in meat wasted in trimming and in labour costs. The adult parasite is of no danger to humans, raw or cooked, but a forceful advertising campaign will be required to convince the consumer of the unique health benefits of *P. roemeri.*

Durikainema macropi Spratt and Speare, 1982

An unusual nematode, *D. macropi,* occurs in the portal veins, pulmonary arterioles, intracardiac veins and epicardial lymphatics of *L. conspicillatus, M. agilis, M. giganteus, M. fuliginosus, M. robustus* (Spratt and Speare 1982; Speare and Spratt 1982; Spratt 1984) and in *M. rufus* (pers. obs.). *D. macropi* causes an eosinophilic phlebitis of the terminal radicals of the portal venules and an epicardial lymphangitis (Speare and Spratt 1982). The reactions can become so severe that vessels are obliterated. No reaction has been noted in the lungs even when inflammation is present in other sites.

Cestodes

Cestodes in Macropodidea belong to the family Anoplocephalidae, Davaineidae, Linstowiidae and Taeniidae (Beveridge 1978, 1986). Macropodids act as intermediate hosts for several species, *Echinococcus granulosus,* whose definitive host is the dog and dingo, *Anoplotaenia dasyuri,* whose definitive hosts are the larger dasyurids and *Dasyurotaenia dasyuri,* with *Satanellus maculatus* as definitive host. There is a single report of a macropodid infected with the larval stage of *Taenia serialis* (Dunsmore and Howkins 1967). Pathological change occurs with cestodes in bile ducts and the metacestode stages, but does not appear to result from the cestodes in the intestine even when they are present in large numbers.

Progamotaenia festiva (Rudolphi, 1819)

The anoplocephalid cestode, *P. festiva,* occurs in the bile ducts and gall bladder of a number of macropodids (Beveridge 1976). A similar parasite, *P. effigia,* is found in the biliary system of *M. fuliginosus.* The cestodes elicit a hypertrophic and hyperplastic response with inflammation leading to fibrosis (Presidente and Beveridge 1978). The main hepatic ducts are prominent and dilated with slightly thickened walls. The mucos glands of the mucosa of the major ducts proliferate and inflammation is evidenced by infiltration with mononuclear cells and smaller numbers of eosinophils with varying degrees of fibrosis in major ducts and in portal triads. Hyperplasia of bile ductules in portal triads can occur. The pathologic effects varies with the species of host (Presidente and Beveridge 1978; Speare *et al.* 1983; Beveridge *et al.* 1985). *M. rufus* and *M. agilis* are least affected (Presidente and Beveridge 1978; Speare *et al.* 1983). The response in *O. unguifera* is the most severe of all the macropodids with marked dilatation of major ducts and a severe inflammatory response (pers. obs.).

Anoplotaenia dasyuri Beddard, 1911

A. dasyuri was originally placed in the family Taeniidae, but metacestode morphogenesis indicates that it more accurately belongs to the family Linstowiidae (Beveridge 1986). The characters which originally led to the placement of *A. dasyuri* in the Taeniidae are a remarkable example of convergent evolution (Beveridge 1986). Tasmanian macropodids, *M. rufogriseus, T. billardierii* and *P. tridactylus,* act as intermediate hosts for this cestode. The definitive hosts are *D. vivverrinus, S. harrisii* and *S. maculatus*. The adult cestode is found in the caecum and colon. The metacestode stage of *A. dasyuri* occurs in the heart, lung and skeletal muscles of macropodids and are seen as small, white foci 1-2 mm in diameter (Beveridge 1978). A fibrous capsule occurs around the metacestode larvae with varying degrees of cellular infiltration, predomiantly mononuclear, but including plasma cells, eosinophils and occasionally nuetrophils and giant cells (Gregory *et al.* 1975; Beveridge *et al.* 1975). Macropodids become infected by ingestion of eggs with pasture. *M. fuliginosus, M. giganteus, M. rufogriseus* and *Trichosurus vulpecula* were experimentally infected (Gregory *et al.* 1975; Beveridge *et al.* 1975).

Dasyurotaenia dasyuri Beveridge, 1984

Metacestodes originally identified as *D. robusta* Beddard, 1912 were reported from the peritoneal cavity of *P. tridactylus* (Gregory in Beveridge 1978). Subsequent studies showed them to belong to a new species, *D. dasyuri*, whose definitive host is *S. maculatus* (Beveridge 1984). The response of the macropodid host has not been described. The intermediate hosts of *D. robusta* from *S. harrisi* have not been identified.

Echinococcus granulosus Rudolphi, 1801

The adult *E. granulosus* is a small cestode with four or five segments which attaches to the mucosa of the small intestine of canids. Eggs in canid faeces contaminate pasture and are ingested by herbivores. The cestode becomes a metacestode larvae (a hydatid cyst) which, in order to complete the life cycle, must be eaten by the definitive host. A number of strains of *E. granulosus* have been identified in Australia and overseas (Thompson 1986). Three distinct strains which differ in the morphology of the adult and metacestode stages, the developmental characteristics and the isoelectric protein profiles have been identified in Australia (Thompson and Kumaratilake 1982; Thompson 1986). The strain in Tasmania is perpetuated in a sheep-dog cycle (Thompson 1986). In a comprehensive survey of the three species of Macropodidae in Tasmania, Howkins (1966a) found no hydatid cysts. Macropodids are apparently exposed to infective *E. granulosus* eggs of this strain, but hydatidosis has not been reported (Kumaratilake and Thompson 1982). The sheep-dog strain on the mainland is identical to that found in the United Kingdom and was probably introduced with the early settlers (Thompson 1986). This strain may be poorly infective to macropodids (Sweatman and Williams 1962). The third strain is a sylvatic strain which utilizes a dingo-macropodid cycle. The existence of this cycle was first proposed by Durie and Reik (1952) for Queensland and appears to occur in northern Queensland (Banks 1984), New South Wales (Gemmell 1959; Howkins 1966b) and Victoria (Coman 1972). The cycle occurs in geographic foci where macropodids and infected dinoes share a common environment (Banks 1984). The domestic strain has been found occasionally in dingoes and macropodids and the sylvatic strain in dogs (Kumaratilake and Thompson 1984). The strains are adapted to their particular hosts, but are obviously not confined totally to them.

Metacestode larvae of *E. granulosus* have been noted in ten species of macropodids and one hybrid in Australia (Table 5). Cysts are usually found in the lungs, and less commonly in the liver and other sites (Beveridge 1978). Cysts disseminated throughout the peritoneal cavity have been reported (Beveridge 1978; Beveridge *et al.* 1985). The reaction of Macropodidae to *Echinococcus* is minimal, and effects on the host appear to be solely related to the physical consequences of a space occupying lesion. Reports of disability in macropodids are rare. Durie and Riek (1952) described a wild *M. rufogriseus*, with a large cyst in the left lung, which could be captured by hand. They suggested that, by decreasing fitness, hydatids in the lungs of wallabies may predispose infected hosts to capture by dinoes.

Taenia serialis (Gervais, 1847)

A *M. giganteus,* with progressive muscular weakness, was reported to have multiple cysts of

Table 5. Macropodid marsupials in which the hydatid cysts have been noted.

Species	References
M. dorsalis	Bancroft (1889, 1893); Young (1939); Durie and Riek (1952); Presidente (1978); Banks (1984); this paper.
M. eugenii	Johnston (1911); Young (1939); Beveridge (1978)
M. giganteus	Johnston (1909; Howkins (1966b); Herd and Coman (1975); Presidente (1978); Kumaratilake and Thompson (1982); Banks (1984); this paper
M. giganteus×M. fuliginosus	Kumaratilake and Thompson (1982)
M. parryi	Durie and Reik (1952)
M. robustus	Johnston (1911); Banks (1984)
M. rufogriseus	Durie and Reik (1952); Sweatman and Williams (1962); Munday and Green (1972)
P. assimilis	This paper
T. stigmatica	Durie and Reik (1952)
T. thetis	Johnston (1909); Young (1939); Presidente (1978)
W. bicolor	Johnston (1909); Durie and Reik (1952); Coman (1972); Beveridge (1978); Presidente (1978); Kumaratilake and Thompson (1982); Beveridge *et al.* (1985)

Taenia serialis (Coenurus serialis) subcutaneously and in the musculature (Dunsmore and Howkins 1967). The animal had been acquired as an orphaned joey and raised in captivity, but the minder was reported to have noted subcutaneous cysts at the time of acquisition. The definitive host of *T. serialis* is the domestic dog and cat.

Ectoparasites

Diptera

Diptera affect macropodids directly by their feeding activities, notable midges and mosquitoes (Muller *et al.* 1981), and indirectly as vectors of microorganisms, in particular arboviruses, and as vectors of helminth parasites. The role of tabanids as vectors of the filarioid nematode, *P. roemeri,* has been discussed previously.

Hippoboscid flies

Some macropodids are hosts to hippoboscid flies (Beveridge *et al.* 1985). The flies are not pathogenic.

Black flies

In central Queensland the black fly, *Austrosimulium pestilens* Mackerras and Mackerras, was reported to cause swelling of eyelids and muzzle in *M. major, M. robustus, M. rufus* and a wallaby (*Wallabia* sp.) (McCarthy 1961a). In some cases the regions of the eyelids and the muzzle were markedly inflamed and denuded of hair. The ears were not affected. Severely affected animals moved in small circles of only a metre or so in diameter and their sight appeared to be impaired. *A. pestilens* is a biting fly which occurs in heavy infestations after rains and floods of the summer months in the rivers of the Gulf Country and in western and southern Queensland and in the Fitzroy and Burnett watersheds (Seddon 1951; Roberts 1952). Dense clouds of flies attack all livestock as well as man and the flies tend to concentrate around the head. Domestic livestock crowd together when attacked (Seddon 1951), but infested macropodids are found singly and are characteristically seen shaking their heads (McCarthy 1961a). McCarthy (1961a) commented that their general health did not seem to be impaired by the fly attack itself. During an infestation macropodids may die (Seddon 1951; McCarthy 1961a), but usually in small numbers and carcasses are usually found away from water (McCarthy 1961a). Death may be due to dehydration (McCarthy 1961a) secondary to visual impairment or disorientation.

Bot flies

Larval stages of a bot fly, *Tracheomyia macropi* (Froggatt, 1913), occurs commonly in the trachea of *M. rufus,* particularly the upper part, but also in the bronchi and bronchioles (McCarthy 1961b; Mykytowycz 1963; Arundel *et al.* 1979). The parasite has also been reported from the Euro, *M. robustus* (Ealey in Mykytowycz 1963), but from no other macropodids. The larvae move freely over the mucosal surface, but if an attempt is made to dislodge a living specimen, the firmness of the attachment is surprising. The larvae can produce pathologic change (Mykytowycz 1963; Beveridge 1978). Gross changes in the trachea are usually minor, with reddening of the mucosa and slight ulceration. In heavy infections with third stage larvae extensive superficial erosions were noted, but histological changes were not described (Mykytowycz 1963). In infections with less than 10 larvae histological changes are obvious. The mucosa is ulcerated, the submucosa is infiltrated by mononuclear cells and the mucosa on the edges of the ulceration show squamous metaplasia (pers. obs.). The general health of the animals appear not to be affected.

Dung beetles

A surprising, but consistent finding, particularly noted during the wet season in north Queensland, was the presence of a dung beetle, *Orthophagus parvus* Blanchard, 1853, on the fur of the pericloacal region of wild Agile Wallabies (Speare *et al.* 1983). No ill effects were noted.

Lice

Free-ranging macropodids usually carry biting lice of the family Boopidae (Keler 1971). They do not appear to cause any pathological changes.

Ticks

Ticks have been reported from most macropodids (Roberts 1970). However, the only description of the pathologic effects of ticks is the brief comment by Beveridge *et al.* (1985) that *Amblyomma triguttatum* Koch, 1844 and *Haemaphysalis bancrofti* Nuttall and Warburton, 1915 elicited a mild, focal inflammation at their attachment sites. *Ornithodoros gurneyi* is known as the Kangaroo Tick although apparently no specimens have been collected from kangaroos (Roberts 1970). It is usually found in inland Australia in soil under trees where kangaroos rest. The ticks emerge to feed when a host is available and become fully engorged in about 5-10 min (Roberts 1936). On humans, the bite causes both local and systemic effects (Henry 1938). The effect on macropodids has not been studied, but a wild kangaroo was noted to have neurological signs presumably soon after an *O. gurneyi* had fed on it (Henry 1938). The flight reaction was greatly diminished and ataxia was present. Partial recovery took 15 minutes.

Mites

Although 13 species of mesostigmatic mite have been reported from Macropodoidea (Domrow 1987), there are no reports of the responses of the hosts. Trombiculid mites have been reported from Macropodoidea (Domrow and Lester 1985). Adult and nymphs of trombiculid mites live free in the soil while the larvae are parasitic on vertebrates. Trombiculid mites were commonly found embedded in the skin of the medial aspects of the thighs of wild *M. agilis* in north Queensland (Speare *et al.* 1983). Surrounding the attachment site the skin was inflamed for up to 5 mm, with a central depression containing encrustations. *Eutrombicula hirsti* (Sambon 1927) was recovered by dissection from the centre of these lesions. Large numbers of mites caused an obvious dermatitis, most severe on the inner aspects of the thighs.

TOXIC DISEASES

Fluoroacetate poisoning

Sodium fluoroacetate (compound 1080) is a poison used to kill vertebrate pests. It is highly toxic to mammals due to its conversion within the body to fluorocitrate which interferes with carbohydrate metabolism to cause severe dysfunction. In Australia it has been used extensively to control rabbits and dingoes, and has also been used in Australia and New Zealand against marsupials, *Trichosurus vulpecula, Macropus rufogriseus*, and *Thylogale billardierii* (McIlroy 1981). Macropodids are susceptible to poisoning and may die after variable periods of time depending on the dose. At lower doses *M. giganteus* may take up to three days to die and many eastern Australian macropodids are more susceptible than the rabbit (McIlroy 1981, 1986). Pouch young and young-at-heel of Tammar Wallabies are more sensitive to 1080 poison than older animals (McIlroy 1981). *M. fuliginosus* in Western Australia can be poisoned by 1080, but has a dose related tolerance to 1080 mediated through defluorination of fluoroacetate or a metabolite (King *et al.* 1978). *M. fuliginosus* from the West were able to safely ingest doses of 1080 which killed *M. rufus* (King *et al.* 1978). This tolerance is thought to have evolved to enable *M. fuliginosus* to utilize fluoroacetate containing plants such as *Gastrolobium* and *Oxylobium* (King *et al.* 1978).

Deaths in wild macropodids have been noted in poisoning campaigns against rabbits and pigs (McIlroy 1986). In a capture-release study on *M. eugenii* in south-west Western Australia. Christensen (1980) observed a "shock syndrome" in captured wallabies possibly related to subclinical toxicity with fluoroacetate from heartleaf (*Gastrolobium* spp.) becoming exacerbated by stress. Some trapped wallabies collapsed, appeared to be paralysed, pupils were dilated and death occurred. No specific lesions were detected at autopsy. Mortality induced by acute stress, such as droving is commonly seen with introduced ruminants grazing fluoroacetate-containing plants (Everist 1974).

Phalaris staggers

Typical phalaris staggers was observed in *M. fuliginosus* on Kangaroo Island (Phillips in Munday 1988). On histology pigmentation of neurones was similar to that seen in domestic herbivores (Hartley in Munday 1988).

Pyrollizadine alkaloids

Hepatic changes typical of pyrollizadine alkaloidosis were reported in wild *M. rufus* in an area where plants containing pyrollizadine alkaloids were known to occur (Hartley in Munday 1988). The plants which are noted for these alkaloids are *Crotalaria* sp., *Senecio* sp. and several less important genera (Everist 1974).

METABOLIC DISEASE

Sodium deficiency

M. giganteus in the Snowy Mountains, an alpine region with low sodium concentrations in soil and vegetation, were shown to be deficient in sodium (Blair-West *et al.* 1968). The mean plasma concentration of sodium was lower, the urinary sodium was markedly decreased and the aldosterone concentration in blood was elevated in Grey Kangaroos from the Snowy Mountains compared with animals from coastal Victoria (Coghlan and Scoggins 1967; Blair-West *et al.* 1967). Glucocorticoid concentrations were similar. Adrenal glands of the alpine kangaroos were almost double the weight of the coastal animals and the zona glomerulosa in alpine *M. giganteus* represented twice the percentage of total area compared with that of the coastal kangaroos (Blair-West *et al.* 1968). The salivary glands of the sodium deficient kangaroos had a more extensive system of ducts, cells lining the ducts were increased in height, and blood vessels were increased around ducts (Blair-West *et al.* 1968). The changes in the salivary glands are a morphological sign of chronic hyperactivity in active sodium resorption (Blair-West *et al.* 1968). This is probably mediated by increased concentrations of aldosterone associated with the hyperplasia of the zona glomerulosa. Administration of deoxycorticosterone acetate to *M. rufus* in an experimental situation resulted in a marked reduction in the concentration of sodium in parotid saliva (Beal 1987).

Salt-hunger has been noted in these alpine kangaroos (Blair-West *et al.* 1968). Sodium deficient rabbits from the same area were less able to tolerate stress than sodium replete rabbits, but, since the kangaroos were not handled, the authors were unable to comment on their tolerance to stress.

Heat stress

Newsome (1973) noted impaired spermatogenesis in *M. rufus* in central Australia and showed a significant relationship to heat load. Temperatures in excess of 32°C for 15 days were sufficient to affect spermatogenesis in some kangaroos. Spermatozoa were greatly reduced and production of spermatids declined. The amount of interstitium also declined in some kangaroos. Repair occurred after temperatures fell and was accelerated by good rains, but probably took about six weeks (Newsome 1973, 1977). The fertility of males recovered later than the fertility of females when the season became more favourable. Some large males had small discoloured testes with no evidence of spermatogenesis and were considered senescent (Newsome 1973).

Drought

The effect of drought has been evaluated for the Grey Kangaroos, the Red Kangaroo and the Euro *(Macropus robustus erubescens)* (Frith and Sharman 1964; Newsome 1964a, 1964b, 1965, 1973, 1977; Ealey and Main 1967; Newsome *et al.* 1967; Frith and Calaby 1969; Bailey 1971; Myers *et al.* 1976; Caughley, J. *et al.* 1984; Caughley *et al.* 1985). The sizes of populations are reduced during drought, but recover after the return of good conditions (Grigg *et al.* 1985). Deaths of kangaroos in the summer of 1982-83 occurred over four months (Caughley *et al.* 1985). The decline in Red Kangroos is due initially to mortality in pouch young and young-at-foot, reduction of breeding due to anoestrus and decreased male fertility, and subsequently to deaths of adults (Frith and Sharman 1964; Newsome 1964a, 1964b, 1965, 1973, 1977; Frith and Calaby 1969; Myers *et al.* 1976). Nutrition, heat stress and later, dehydration appear to be significant factors.

However, few studies have examined the pathological findings in animals in drought. The cause of death of pouch young is not known, but is thought to be related to reduced production of milk (Frith and Sharman 1964; Newsome 1964a, b; Frith and Calaby 1969). Green (1984) has shown that the growth rate of the joey is determined by the volume of milk consumed. In orphaned joeys maintained in captivity inadequate nutrition is usually associated with lymphoid involution and predisposes the joey to infection with its enteric micro-flora (Speare 1988). Decreased growth rate in malnourished orphaned joeys usually precedes clinical disease. A reduction in growth rate occurs in pouch young of *M. rufus* during drought (Myers *et al.* 1976). Young are rarely found dead in the pouch and so the nature of the terminal event in drought stricken pouch young has not been recorded. Similarly, the pathological findings in young-at-foot have not been reported.

The total body weight, hepatic and splenic weights, and weight of the adrenal glands of adult Red Kangaroos declined during drought (Myers *et al.* 1976). All zones of the adrenal gland decreased in size and the cortex became folded and nodular (Myers *et al.* 1976). Arundel *et al.* (1979) suggested that these atrophic changes in the adrenal cortex were preceded by hyperplasia in response to environmental stress with progressive atrophy occurring subsequently. The pathological findings in adult kangaroos dying during drought have not been comprehensively studied. Tomlinson and Gooding (1954) reported deaths from lumpy jaw in drought stricken Red Kangaroos. Newsome (1977) noted one instance of apparent death from dehydration. Cachexia was noted in Euros during drought (Ealey and Main 1967) and a seasonal anaemia was associated with reduced protein content of vegetation. Unfortunately, no concurrent disease studies were undertaken.

Signs of ill-health (unspecified) in Red Kangaroos were noted after drought breaking rains (Myers *et al.* 1976) and were thought to be related to leaching of soluble carbohydrate and minerals from the dry grass.

Floods

Macropodids can be isolated on "islands" by flood waters and suffer nutritional stress and exposure to high concentrations of infectious agents (Barker *et al.* 1972). Near drowning of adult Agile Wallabies and drowning of their pouch young in the wet season was reported by Bolton *et al.* (1982). Fertility of *M. agilis* declined and mortality of pouch young was increased during the wet season (Bolton *et al.* 1982).

NEOPLASTIC DISEASES

Apart from papillomas associated with the Macropod Pox Virus, neoplasms are uncommon in free-ranging macropodidae.

Trichoepithelioma

A trichoepithelioma, a benign neoplasm of the epithelial cells of the wall of the hair follicle, was reported from a mature, wild *M. giganteus* (Speare and Banks 1987). The lesion was solitary, measured 3.5×3.0×1.3 cm and consisted of multiple acini, many cystic, with abundant melanin. Although the age of the host was not given in the original report, the Grey Kangaroo had a molar index of 3.5. According to Kirkpatrick (1965) this is consistent with an age of 8.3 years.

Squamous cell carcinoma of stomach

A well differentiated squamous cell carcinoma was found in the stomach of a wild *T. billardierii* in Tasmania (Munday 1971). Neoplastic cells had infiltrated deeply into the submucosa, with epithelial pearls surrounded by less mature epithelial elements.

Large numbers of fourth stage *Labiostrongylus* sp. larvae were present in the mucosal surface of the carcinoma. Larval stages of gastric strongyles are commonly found in Macropodidae with their anterior ends buried in the mucosa and submucosa of the cardiac glandular regions of the stomach. These larvae produce small elevated nodules due to a localized fibroplastic and inflammatory response (Beveridge and Presidente 1978; Speare *et al.* 1983) and at least one species causing these isolated nodules has been identified (Beveridge and Presidente 1978). However, the larvae of *Labiostrongylus* spp. show a tendency to crowd together at a single site with their anterior ends embedded in the stomach mucosa and submucosa (Smales 1977). The host develops a prominent hyperplastic response with formation of mucosal nodules (Smales 1977; Speare *et al.* 1983; Beveridge *et al.* 1985). As suggested by Munday (1971) it is possible that neoplasia could be initiated by a chronic irritative response to these larvae. Apart from this case, neoplastic transformation has not been noted.

Papillomas

Papillomas and papules due to the Macropod Pox Virus are the neoplasm most frequently encountered on the skin of free-ranging macropodids. They have been discussed previously.

DEGENERATIVE DISEASES

There appear to be no reports of degenerative diseases in wild Macropodoidea. This may be a reflection of environmental selection for "fitness". Maximal longevity in the wild has been measured and estimated for several species of Macropodidae (Table 6). However, for studies on the effects of disease on populations the most important information is the age composition of the whole population (Frith and Calaby 1969).

Table 6. Maximum longevity in the wild for Macropodoidea (E=estimated; M=measured).

Species	Max Age (years)	Tech.	References
M. agilis	10	E	Kirkpatrick and Johnson (1969)
M. eugenii	15	M	Inns (1982)
M. giganteus	19.9±3	M	Kirkpatrick (1965)
M. parma	9-10	M	Maynes (1977)
M. robustus	18.6±4.5	M	Kirkpatrick (1965)
M. rufogriseus	18.6±3.4	M	Kirkpatrick (1965)
M. rufus	22	M	Kirkpatrick (1965)
P. tridactylus	4-5	E	Guiler (1957)
S. brachyurus	11	E	Niven (1970)

TRAUMA

Predation

The major predator of macropodids is man. Hunting is larely confined to the larger species of macropodids and was shown to have a significant effect on some populations of *M. rufus* and Grey Kangaroos (Frith and Calaby 1969; Sinclair 1977). The effect of illegal hunting on the smaller species of Macropodoidea is probably negligeable compared to the indirect effects of habitat destruction and degradation. The effect of road traffic accidents on macropodid populatious has not been determined, but it could prove significant in isolated populations, such as "pocket-hankie" national parks traversed by busy roads. Dingoes and the introduced fox, *Vulpes vulpes,* prey on macropodids particularly the smaller species or juveniles of the larger ones (Frith and Calaby 1969; Coman 1972; Harden 1978; Christensen 1980; Hornsby 1981; Shepherd 1981; Southwell 1981; Banks 1984). The behaviour of dingoes and *M. rufus* in predator-prey interactions was described by Shepherd (1981). Predation of *B. penicillata* by the native cat, presumably *Dasyurus geoffroii,* was reported by Christensen (1980). Although wedge-tailed eagles, *Aquila audax*, largely eat macropodids which have already died, there are reports of predation on *B. penicillata* (Christensen 1980) and attacks on *M. robustus* and *M. rufus* (Frith and Calaby 1969).

Postmortem findings in *M. rufus* killed by dingoes were described by Shepherd (1981). Carcasses were usually found in lateral recumbancy with the head thrown back and unless they were eaten, there was usually no visible external signs. Internally extensive trauma was always present. Bite wounds were deep and penetrating and caused subcutaneous and intramuscular haemorrhage. Body cavities and major organs were often penetrated. Only 53% of carcasses had been eaten and none totally consumed. The most commonly eaten parts were the abdominal contents, hindquarter muscle and thoracic contents. Speare (1988) reported similar external and internal findings in captive juvenile macropodids attacked by dogs.

Direct effects of trauma

There are few reports of the direct effects of trauma on wild Macropodoidea. A fracture of radius and ulna caused by a shot gun blast in a nine-year-old male Eastern Grey Kangaroo was reported by Keep (1972). Osteomyelitis was present and *Streptococcus equisimilis* isolated. Multiple injuries due to shot gun pellets in a pouch young of *T. stigmatica* was reported by Speare (1988). Osteomyelitis occurred in the forearm of a free-ranging *M. giganteus* after a bite from a dog (pers. obs.). A dislocated hock joint in a wild Red Kangaroo was associated with moderate hyperplasia of the adrenal cortex (Arundel *et al.* 1979).

Indirect effects of trauma

Myonecrosis was observed in a free-living Bennett's Wallaby (Munday 1972b). The wallaby was unable to move and at postmortem symmetrical paleness of many muscle masses were noted. On histological examination hyaline degeneration and marked interstitial oedema was found in the muscles. Similar changes have been observed in wild *M. agilis* pursued by dogs (pers. obs.). Myopathy in macropodids may be associated with a deficiency of Vitamin E or related to physical exertion (Munday 1988). In wild macropodids the stresses of capture commonly result in clinical or subclinical capture myopathy (Munday 1988). The aetioloy of capture myopathy may be multifactorial, but both physical and psychological stress appear to play a role in capture myopathy in macropodids (Shepherd 1982; Munday 1988).

EPIDEMIC MORTALITY

Ealey and Main (1967) noted two types of mortality in Euros in the Pilbara district of Western Australia. One is a regular seasonal mortality and the other a spectacular "crash" type. With the first type deaths occur in a predictable manner and seem to be related to poor nutrition, alone or aggravated by pathogens. The seasonal mortality in Euros in some areas of the Pilbara (Ealey and Main 1967), in Quokkas on Rottnest Island (Hart *et al.* 1985) and possibly Tammar Wallabies on Flinders Island (Murphy 1970) appear to follow this pattern. Degradation of the natural vegetation, overcrowding and poor rainfall seem to be factors involved. The sudden "die-off" is an epidemic mortality that is not predictable on a seasonal basis. Ealey and Main (1967) considered that for Euros in some areas in the Pilbara inadequate vegetation allowed a population to expand beyond the capacity of the area to support it and then deaths occurred in large numbers. Newsome (1977) also noted this pattern in Red Kangaroos and proposed that marked increases in populations in an area are indicative of a degenerating environment, usually secondary to the effects of grazing of ruminants, and will often be followed by spectacular declines. A spectacular population "crash" of unknown aetiology affected Red Kangaroos, Grey Kangaroos and Eastern Wallaroos in June-August 1983 in the Boulia-Bedourie-Windora area of western Queensland (pers. obs.). Kangaroos were reported to be in good condition and to have no gross pathological changes. Deaths were reported to begin in some areas before drought-breaking rains and several months after in others. The primary aetiological agents have been identified in four localized epidemics (Table 7). Caughley *et al.* (1985) considered that droughts and the consequent mortality were intrinsic to the ecology

Table 7. Population "crashes" reported in free-ranging Macropodoidea.

Species	Disease	Aetiology	Stresses	Location	Date	Reference
M. eugenii	?Anaemia	*Eperythrozoon*	Nutritional Drought	S.A.	7/1968	Murphy (1970)
M. giganteus	Coccidiosis	*Eimeria* sp.	Flooding	Vic.	9-10/1970	Barker *et al.* (1972)
M. giganteus	Anaemia	*Globocephaloides*	Unknown	Vic.	1971-72	Arundel *et al.* (1977)
M. giganteus	Unknown	Unknown	Unknown	Qld.	6-8/1983	This paper
M. robustus	?Anemia	Unknown	Nutritional	W.A.	1954	Ealey and Main (1967)
M. robustus	Unknown	Unknown	Unknown	Qld.	6-8/1983	This paper
M. rufogriseus	Toxoplasmosis	*T. gondii*	Nutritional	Tas.	6/1982	Obendorf and Munday (1983)
M. rufus	Lumpy jaw	*F. necrophorum*	Drought Overcrowding	W.A.	1954	Tomlinson and Gooding (1954)
M. rufus	Unknown	Unknown	Unknown	Qld.	6-8/1983	This paper
S. brachyurus	Unknown	Unknown	Unknown	W.A. (mainland)	1930's	Arundel *et al.* (1977); Hart *et al.* (1985)
T. billardierii	Toxoplasmosis	*T. gondii*	Nutritional	Tas.	6/1982	Obendorf and Munday (1983)

Table 8. Diseases and pathologic changes reported from free-ranging Macropodoidea (A=Arthropoda; B=Bacteria; C=Cestoda; I=Insecta; N=Nematoda; P=Protozoa; T=Trematoda).

Species	Disease	Aetiology	Reference
MACROPODIDAE			
L. conspicillatus	Cholangitis	*P. festiva* (C)	Presidente and Beveridge (1978)
M. agilis	Cholangitis	*P. festiva* (C)	Speare *et al.* (1983)
	Dermatitis	*E. hirsti* (A)	Speare *et al.* (1983)
	Gastritis — focal	*G. wallabicola* (T)	Speare *et al.* (1983)
	Gastric nodules	*Strongyloides*	This paper
	Myopathy	Pursuit by dogs	This paper
	Nodules — s/c	*P. roemeri* (N)	This paper
	Nephritis — interstitial	Unknown	Speare *et al.* (1983)
	Papillomata	Macropod Pox Virus	Speare (1988)
	Phlebitis — eosinophilic	*D. macropi* (N)	Speare and Spratt (1982)
	Typhylitis — focal	*M. pertinax* (T)	Blair *et al.* (1979); Speare *et al.* (1983)
M. eugenii	Anaemia	*Eperythrozoon* sp.	Munday (1970)
	Coccidiosis – hepatic	*Eimeria* sp. (P)	Presidente (1978); Munday (1988)
	Lumpy jaw	Bacteria	Arundel *et al.* (1977)
	Predation	European fox	Christensen (1980)
	Shock syndrome	Fluoroacetate	Christensen (1980)
M. fuliginosus	Cholangitis	*P. effugia* (C)	Presidente and Beveridge (1978)
	Coccidiosis – gastric	*Eimeria* sp. (P)	Arundel *et al.* (1977); Presidente (1978)
	Papillomata	Macropod pox virus	Presidente (1978); Rothwell *et al.* (1984)
	Phalaris staggers		Munday (1988)
	Phlebitis – eosinophilic	*D. macropi* (N)	Speare and Spratt (1982)
M. giganteus	Anaemia	*G. trifidospicularis* (N)	Arundel *et al.* (1977)
	Cholangitis	*F. hepatica* (T)	Beveridge (1978)
	Cholangitis	*P. festiva* (C)	McCarthy (1961b); Presidente and Beveridge (1978)
	Coccidiosis – intestinal	*Eimeria* sp. (P)	Barker *et al.* (1972)
	Gastritis – focal	*Strongyloides* (N)	This paper
	Nodules – s/c and i/m	*P. roemeri* (N)	Spratt (1972a, 1975)
	Papillomata	Macropod Pox Virus	McKenzie *et al.* (1979); Speare (1988)
	Phlebitis – eosinophilic	*D. macropi* (N)	Speare and Spratt (1982)
	Osteomyelitis	Gun shot	Keep (1972)
	Osteomyelitis	Dog bite	This paper
	Trichoepithelioma	Unknown	Speare and Banks (1987)
'*M. major*'	Death	*A. pestilens* (I)	McCarthy (1961a)
	Facial swelling	*A. pestilens* (I)	McCarthy (1961a)
M. parryi	Tetanus	*C. tetani* (B)	This paper
M. robustus	Cholangitis	*P. festiva* (C)	McCarthy (1961b)
M. rufogriseus	Carcinoma – gastric SCC	*Labiostrongylus* sp.	Munday (1971)
	Cholangitis	*F. hepatica* (T)	Beveridge (1978)
	Hydatidosis	*E. granulosa* (C)	Durie and Reik (1952)
	Lumpy jaw	Bacteria	Arundel *et al.* (1977); Horton and Samuel (1978)
	Metacestode nodules	*A. dasyuri* (C)	Gregory *et al.* (1975); Beveridge *et al.* (1985)
	Myopathy	Unknown	Munday (1972b)
	Phlebitis – eosinophilic	*D. macropi* (N)	Speare and Spratt (1982)
	Nephritis – interstitial	Unknown	Durfee and Presidente (1979)
	Nodules – s/c and i/m	*P. roemeri* (N)	Spratt (1972a, 1975)
	Toxoplasmosis	*T. gondii* (P)	Obendorf and Munday (1983)
M. rufus	Adrenocortical atrophy	Drought	Myers *et al.* (1976)
	Adrenocortical hyperplasia	Stress	Arundel *et al.* (1979)
	Alopecia – focal	Pyogranuloma	Arundel *et al.*(1979)
	Anaemia	*H. macropi* (N)	Arundel *et al.* (1977)
	Bronchiolitis	Unknown	Arundel *et al.* (1979)
	Cholangitis	*P. festiva* (C)	McCarthy (1961b); Presidente and Beveridge (1978)
	Death	*A. pestilens* (I)	McCarthy (1961a)
	Dislocated hock	Trauma	Arundel *et al.* (1979)
	Enteritis	*H. macropi* (N)	Arundel *et al.* (1979)
	Epicarditis	Unknown	Arundel *et al.* (1979)
	Facial swelling	*A. pestilens* (I)	McCarthy (1961a)
	Glomerulosclerosis	Unknown	Arundel *et al.* (1979)

Table 8 — continued.

Species	Disease	Aetiology	Reference
M. rufus — cont.	Hepatic atrophy	Unknown	Arundel *et al.* (1979)
	Hepatitis	Unknown	Arundel *et al.* (1979)
	Infertility	Heat stress	Newsome (1973)
	Lumpy jaw	*F. necrophorum* (B)	Tomlinson and Gooding (1954)
	Lymphoid involution	Stress	Arundel *et al.* (1979)
	Myocarditis	Unknown	Arundel *et al.* (1979)
	Nephritis – interstitial	Unknown	Arundel *et al.* (1979); Durfee and Presidente (1979)
	Nodules – gastric	*F. flagrifer* (N)	Arundel *et al.* (1979)
	Nodules – s/c and i/m	*P. roemeri* (N)	Spratt (1972a, 1975)
	Oesophageal erosions	Unknown	Arundel *et al.* (1979)
	Papillomata	Macropod Pox Virus	Bagnall and Wilson (1974); Presidente (1978); Arundel *et al.* (1979)
	Pneumonia	Unknown	Arundel *et al.* (1979)
	Predation	Dingo	Shepherd (1981)
	Pyrollizadine alkaloidosis	Plants	Munday (1988)
	Tracheitis	*T. macropi* (I)	Mykytowycz (1963); This paper
	Urinary calculi	Unknown	Arundel *et al.* (1979)
O. unguifera	Cholangitis	*P. festiva* (C)	This paper
S. brachyurus	Papillomata	Macropod Pox Virus	Papadimitriou and Ashman (1972)
T. billardierii	Cholangitis	*F. hepatica* (T)	Beveridge (1978)
	Metacestode nodules	*A. dasyuri* (C)	Gregory *et al.* (1975); Beveridge *et al.* (1985)
	Metacestode nodules	*D. dasyuri* (C)	Beveridge (1984)
	Toxoplasmosis	*T. gondii*	Obendorf and Munday (1983)
	Squamous cell Carcinoma	*Labiostrongylus* sp. (N)	Munday (1971)
T. stigmatica	Multiple trauma	Shot gun	Speare (1988)
W. bicolor	Dermatitis	Ticks (A)	Beveridge *et al.* (1985)
	Lumpy jaw	Bacteria	Arundel *et al.* (1977)
	Nodules – s/c and i/m	*P. roemeri* (N)	Beveridge *et al.* (1985)
	Peritonitis	*B. mundayi* (N)	Beveridge *et al.* (1985)
	Pneumonia – interstitial	Lungworms (N)	Beveridge *et al.* (1985)
	Tetanus	*C. tetani* (B)	Speare (1988)
POTOROIDAE			
B. penicillata	Predation	European fox Native cat Wedge Tailed eagle	Christensen (1980)
	Shock	Post-fire	Christensen (1980)
P. tridactylus	Metacestode nodules	*A. dasyuri* (C)	Gregory *et al.* (1975); Beveridge *et al.* (1985)

of kangaroos. They noted that localized droughts occurred almost every year and regional droughts about every 10 years, and that populations were able to recover between droughts. Unfortunately, no disease studies have been reported on animals dying during the more widespread mortalities.

MISCELLANEOUS CONDITIONS

Miscellaneous conditions reported in wild macropodids are listed in Table 8. An examination by host of the diseases or pathologic changes reported from free-ranging Macropodoidea reveal that the Macropodidea have a greater number of reports than the Potoroidae (Table 8). Within the Macropodoidea the small species are poorly represented. This probably reflects a bias in sampling rather than a resistance to disease.

REFERENCES

ACLAND, H. M., 1981. Parma wallaby herpesvirus infection. *J. Wild. Dis.* **17**: 471-77.

ANDERSON, K., CROWDER, E. F. AND WOODRUFF, P., 1964. The isolation of Salmonellae from kangaroo meat sold as pet food. *Med. J. Aust.* **1964**(2): 668-69.

ANON (1988). Viruses, chlamydias, coxiellas, rickettsias and mycoplasmas reporting scheme. Communicable Diseases Intelligence, Bulletin No. 88/7.

ARUNDEL, J. H., BARKER, I. K. AND BEVERIDGE, I., 1977. Diseases of marsupials. Pp. 141-54 *in* The Biology of Marsupials ed. by B. Stonehouse and D. Gilmore. Macmillan: London.

ARUNDEL, J. H., BEVERIDGE, I. AND PRESIDENTE, P. J., 1979. Parasites and pathological findings in enclosed and free-ranging populations of *Macropus rufus* (Desmarest) (Marsupialia) at Menindee, New South Wales. *Aust. Wildl. Res.* **6**: 361-79.

ASHTON, D. G., 1982. Whipsnade Park. *In* Scientific Report: 1979-1981. Zoological Society. *J. Zool.* **197**: 1-123.

ATKINSON, N., CARTER, M. C., WOLLASTON, J. M. AND WALL, M., 1953. The occurrence of Salmonellae types in Australia. *Aust. J. Exp. Biol. Med. Sci.* **31**: 465-72.

BAGNALL, B. G. AND WILSON, G. R., 1974. Molluscum contagiosum in a red kangaroo. *Aust. J. Dermatol.* **15**: 115-20.

BAILEY, P. T., 1971. The red kangaroo, *Megaleia rufa* (Desmarest), in north-western New South Wales. *C.S.I.R.O. Wildl. Res.* **16**: 11-28.

BANCROFT, T. L., 1889. On echinococcus in a wallaby. *Proc. R. Soc. Qld.* **7**: 31.

BANCROFT, T., 1893. Entozoal parasites (from Kumaratilake and Thompson 1982). *Aust. Med. Gaz.* **12**: 258-60.

BANKS, D. J. D., 1984. The epidemiology of echinococcosis in tropical Queensland. Ph.D. Thesis, James Cook University of North Queensland.

BARKER, I. K., HARRIGAN, K. E. AND DEMPSTER, J. K., 1972. Coccidiosis in wild grey kangaroos. *Int. J. Parasitol.* **2**: 187-92.

BARKER, I. K., MUNDAY, B. L. AND HARRIGAN, K. E., 1975. *Klossiella* spp. in the kidneys of peramelid, petaurid and macropodid marsupials. *Z. Parasitenk.* **46**: 35-41.

BARKER, S., CALABY, J. H. AND SHARMAN, G. B., 1963. Diseases of Australian laboratory marsupials. *Vet. Bull.* **33**: 539-44.

BARTLETT, C. M. AND GREINER, E. C., 1986. A revision of *Pelecitus* Railliet and Henry, 1910 (Filaroidea, Dirofilarinae) and evidence for the "capture" by mammals of filaroids from birds. *Bull. Mus. Natn. Hist. Nat., Paris* **8**: 47-99.

BEAL, A. M., 1987. Effect of diet and mineralocorticoid administration on the concentration of anions in parotid saliva from the red kangaroo, *Macropus rufus*. *Aust. J. Zool.* **35**: 133-45.

BEVERIDGE, I., 1976. A taxonomic revision of the Anoplocephalidae (Cestoda: Cyclophyllidea) of Australian marsupials. *Aust. J. Zool. Suppl. Ser. No.* **44**: 1-110.

BEVERIDGE, I., 1978. Helminth parasites of Australian marsupials. Pp. 273-85 *in* Proc. No. 36 Fauna — Part B. Post Graduate Committee in Veterinary Science, University of Sydney: Sydney.

BEVERIDGE, I., 1979. *Hypodontus macropi* Mönning, 1929, a hookworm-like parasite of macropodid marsupials. *J. Helminthol.* **53**: 229-44.

BEVERIDGE, I., 1984. *Dasyurotaenia robusta* Beddard, 1912, and *D. dasyuri* sp. nov., from carnivorous Australian marsupials. *Trans. R. Soc. S. Aust.* **108**: 185-95.

BEVERIDGE, I., 1986. Coevolutionary relationships of the helminth parasites of Australian marsupials. Pp. 93-117 *in* Coevolution and Systematics ed. by A. R. Stone and D. L. Hawksworth. Clarendon Press: Oxford.

BEVERIDGE, I. AND ARUNDEL, J. H., 1979. Helminth parasites of Grey Kangaroos, *Macropus giganteus* Shaw and *M. fuliginosus* (Desmarest), in eastern Australia. *Aust. Wildl. Res.* **6**: 69-77.

BEVERIDGE, I. AND PRESIDENTE, P. J. A., 1978. *Rugopharynx rosmariae* new species (Nematoda: Pharyngostrongylidae) from grey kangaroos (*Macropus giganteus* and *M. fuliginosus*) with life cycle stages and associated pathology. *Int. J. Parasitol.* **8**: 379-87.

BEVERIDGE, I., PRESIDENTE, P. J. A. AND SPEARE, R., 1985. Parasites and associated pathology of the swamp wallaby, *Wallabia bicolor* (Marsupialia). *J. Wildl. Dis.* **21**: 377-85.

BEVERIDGE, I., RICKARD, M. D., GREGORY, G. G. AND MUNDAY, B. L., 1975. Studies on *Anoplotaenia dasyuri* Beddard, 1911 (Cestoda: Taenidae), a parasite of the Tasmanian devil: observations on the egg and metacestode. *Int. J. Parasitol.* **5**: 257-67.

BEVERIDGE, I., SPEARE, R. AND JOHNSON, P. M., 1984. New records of Globocephaloidinae (Nematoda: Trichostrongyloidea) from Macropodidae in north Queensland. *Trans. R. Soc. S. Aust.* **108**: 197-201.

BEVERIDGE, W. I. B., 1981. Animal health in Australia. Vol. I. Viral diseases of farm livestock. Australian Bureau of Animal Health. Australian Government Publishing Service: Canberra.

BLAIR, D., BEVERIDGE, I. AND SPEARE, R., 1979. *Macropotrema pertinax* gen. et sp. nov. (Digenea: Paramphistomidae) from a wallaby, *Macropus agilis,* in northern Australia, and associated pathology. *Ann. Parasitol.* **54**: 585-92.

BLAIR-WEST, J. R., COGHLAN, J. P., DENTON, D. A., NELSON, J. F., ORCHARD, E., SCOGGINS, B. A., WRIGHT, R. D., MYERS, K. AND JUNQUIERA, C. L., 1968. Physiological, morphological and behavioural adaptation to a sodium deficient environment by wild native Australian and introduced species of animals. *Nature (Lond.)* **217**: 922-28.

BOLTON, B. L., NEWSOME, A. E. AND MERCHANT, J. C., 1982. Reproduction in the Agile Wallaby *Macropus agilis* (Gould) in the tropical lowlands of the Northern Territory: opportunism in a seasonal environment. *Aust. J. Ecol.* **7**: 261-77.

BORAY, J. C., 1969. Experimental fascioliasis in Australia. *Adv. Parasitol.* **7**: 95-210.

BUTLER, R. AND BURTON, J. D., 1980. Necrobacillosis of macropods — control and therapy. Pp. 137-40. Ann. Proc. Amer. Assoc. Zoo Vet. American Association of Zoo Veterinarians: Washington D.C.

CALABY, J. H. AND POOLE, W. E., 1971. Keeping kangaroos in captivity. *Int. Zoo Yearbk.* **11**: 5-20.

CALLINAN, R. B. AND KEFFORD, B., 1981. Mortalities associated with herpesvirus infection in captive macropods. *J. Wildl. Dis.* **17**: 311-17.

CAUGHLEY, G., BROWN, B., DOSTINE, P. AND GRICE, D., 1984. The grey kangaroo overlap zone. *Aust. Wildl. Res.* **11**: 1-10.

CAUGHLEY, G., GRIGG, G. C. AND SMITH, L., 1985. The effect of drought on kangaroo populations. *J. Wildl. Manag.* **49**: 679-85.

CAUGHLEY, J., BAYLISS, P. AND GILES, J., 1984. Trends in kangaroo numbers in western New South Wales and their relation to rainfall. *Aust. Wildl. Res.* **11**: 415-22.

CHRISTENSEN, P. E. S., 1980. The biology of *Bettongia penicillata* Gray, 1837, and *Macropus eugenii* (Desmarest, 1817) in relation to fire. Forests Department of Western Australia, Bulletin 91.

COGHLAN, J. P. AND SCOGGINS, B. A., 1967. The measurement of aldosterone, cortisol and corticosterone in the blood of the wombat (*Vombatus hirsutus* Perry) and the kangaroo *(Macropus giganteus)*. *J. Endocr.* **39**: 445-48.

COMAN, B. J., 1972. A sylvatic cycle for the hydatid tapeworm *(Echinococcus granulosus)* in remote areas of eastern Victoria. *Aust. Vet. J.* **48**: 522-23.

COOK, I. AND POPE, J. H., 1959. *Toxoplasma* in Queensland. III. A preliminary survey of animal hosts. *Aust. J. Exp. Biol. Med. Sci.* **37**: 253-62.

DANE, D. S. AND BEECH, M. D., 1955. Institute of Medical Veterinary Science, 17th Annual Report P. 37 (from Pope *et al.* 1960).

DERRICK, E. H., SMITH, D. J. W. AND BROWN, H. E., 1940. Studies on the epidemiology of Q Fever. 6. The susceptibility of various animals. *Aust. J. Exp. Biol. Med. Sci.* **18**: 409-13.

DICKSON, J., HOPKINSON, W. I. AND COAKLEY, W., 1980. Herpesvirus hepatitis in rat kangaroos. *Aust. Vet. J.* **56**: 463.

DOHERTY, R. L., CARLEY, J. G. AND GORMAN, B. M., 1964. Studies of arthropod-borne virus infections in Queensland. IV. Further serological investigations of antibodies to group B arboviruses in man and animals. *Aust. J. Exp. Biol. Med. Sci.* **42**: 149-64.

DOHERTY, R. L., WHITEHEAD, R. H., WETTERS, E. J., GORMAN, B. M. AND CARLEY, J. G., 1970. A survey of antibody to 10 arboviruses (Koongol group, Mapputta group and ungrouped) isolated in Queensland. *Trans. R. Soc. Trop. Med. Hyg.* **64:** 748-53.

DOMROW, R., 1987. Acari Mesostigmata parasitic on Australian vertebrates: an annotated checklist, keys and bibliography. *Invertebr. Taxon.* **1:** 817-948.

DOMROW, R. AND LESTER, L. N., 1985. Chiggers of Australia (Acari: Trombiculidae): an annotated checklist, keys and bibliography. *Aust. J. Zool., Suppl. Ser.* **114:** 1-111.

DONALD, G. F., 1960. Tinea contracted from kangaroos. *Med. J. Aust.* **47:** 852-53.

DUNSMORE, J. D. AND HOWKINS, A. B., 1967. *Coenurus serialis* in a grey kangaroo. *Aust. J. Sci.* **30:** 465.

DURFEE, P. T. AND PRESIDENTE, P. J. A., 1979. A serological survey of Australian wildlife for antibodies to leptospires of the Hebdomadis serogroup. *Aust. J. Exp. Biol. Med. Sci.* **57:** 177-89.

DURIE, P. H. AND RIEK, R. F., 1952. The role of the dingo and wallaby in the infestation of cattle with hydatids (*Echinococcus granulosus* (Batsch, 1786) Rudolphi, 1805) in Queensland. *Aust. Vet. J.* **28:** 249-54.

EALEY, E. H. M. AND MAIN, A. R., 1967. Ecology of the euro, *Macropus robustus* (Gould), in north-western Australia. III. Seasonal changes in nutrition. *C.S.I.R.O. Wildl. Res.* **12:** 53-65.

EKAPUTRA, I. G. M. A., 1985. *Salmonella* in game meat. M.Vet. Studies Thesis, University of Queensland.

EMMANUEL, M. L., MACKERRAS, I. M. AND SMITH, D. J. W., 1964. The epidemiology of leptospirosis in north Queensland. 1. General survey of animal hosts. *J. Hyg.* **62:** 451-84.

EVERIST, S. L., 1974. Poisonous Plants of Australia. Angus & Robertson Publishers: London.

FAYER, R., 1980. Epidemiology of protozoan infections: the coccidia. *Vet. Parasitol.* **6:** 75-103.

FENNER, F., 1946. The epidemiology of North Queensland Tick Typhus: natural mammalian hosts. *Med. J. Aust.* **1946**(2): 666-68.

FINNIE, E. P., 1974. Prophaylaxis against coccidiosis in kangaroos. *Aust. Vet. J.* **50:** 276.

FINNIE, E. P., 1978. Disease description. Pp. 415-20 *in* Zoo and Wild Animal Medicine ed. by M. E. Fowler. W. B. Saunders Company: Philadelphia.

FINNIE, E. P., LITTLEJOHNS, I. R. AND ACLAND, H. M., 1976. Mortalities in parma wallabies *(Macropus parma)* associated with probable herpesvirus. *Aust. Vet. J.* **52:** 293.

FOWLER, M. E., 1982. Wildlife diseases of the Pacific Basin. Pp. 12-14 *in* Wildlife Diseases of Pacific Basin and Other Countries ed. by M. E. Fowler. Wildlife Diseases Assoc. Proc. 4th Inter. Conf. Wild. Dis. Assoc., Sydney, August 25-28, 1981.

FRITH, H. J. AND CALABY, J. H., 1969. Kangaroos. F. W. Cheshire: Melbourne.

FRITH, H. J. AND SHARMAN, G. B., 1964. Breeding in wild populations of the red kangaroo, *Megaleia rufa. C.S.I.R.O. Wild. Res.* **9:** 86-114.

GEMMELL, M. A., 1959. Hydatid disease in Australia. IV. Observations on the carnivora of New South Wales as definitive hosts of *Echinococcus granulosus* (Batsch, 1786) (Rudolphi, 1801), and their role in the spread of hydatidosis in domestic animals. *Aust. Vet. J.* **35:** 450-55.

GIBB, D. G. A., KAKULAS, B. A., PERRET, D. H. AND JENKYN, D. J., 1966. Toxoplasmosis in the Rottnest Quokka *(Setonix brachyurus). Aust. J. Exp. Biol. Med. Sci.* **44:** 665-72.

GILRUTH, J. A. AND BULL, L. B., 1911. Enteritis associated with infection of the intestinal wall by cyst-forming protozoa (Neosporidia), occurring in certain native animals (wallaby, kangaroo, wombat). *Proc. R. Soc. Vict.* **24:** 432-50.

GREEN, B., 1984. Composition of milk and energetics of growth in marsupials. *Symp. Zool. Soc. Lond.* **51:** 369-87.

GREENWOOD, A. G. AND TAYLOR, D. C., 1980. Metronidazole in the treatment of lumpy jaw. *Vet. Rec.* **106:** 161.

GREGORY, G. G., MUNDAY, B. L., BEVERIDGE, I. AND RICKARD, M. D., 1975. Studies on *Anoplotaenia dasyuri* Beddard, 1911 (Cestoda: Taeniidae), a parasite of the Tasmanian Devil: life cycle and epidemiology. *Int. J. Parasitol.* **5:** 187-97.

GRIGG, G. C., BEARD, L. A., CAUGHLEY, G., GRICE, D., CAUGHLEY, J. A., SHEPHERD, N., FLETCHER, M. AND SOUTHWELL, C., 1985. The Australian kangaroo populations, 1984. *Search* **16:** 277-79.

GRINER, L. A., 1983. Pathology of Zoo Animals. Zoological Society of San Diego: San Diego.

GUILER, E. R., 1957. Longevity in the wild Potoroo, *Potorous tridactylus* (Kerr). *Aust. J. Sci.* **20:** 26.

HARDEN, R. H., 1978. The biology of the dingo. Pp. 451-55 *in* Proceedings: No. 36, Fauna — Part B. Post Gradute Committee in Veterinary Science, University of Sydney: Sydney.

HART, R. P., BRADSHAW, S. D. AND IVESON, J. B., 1985. *Salmonella* infections in a marsupial, the Quokka *(Setonix brachyurus),* in relation to seasonal changes in condition and environmental stress. *Appl. Environmental Microbiol.* **49:** 1276-81.

HART, R. P., IVESON, J. B., BRADSHAW, S. D. AND SPEED, T. P., 1982. A study of isolation procedures for multiple infections of *Salmonella* and *Arizona* in a wild marsupial, the Quokka *(Setonix brachyurus). J. Appl. Bacteriol.* **53:** 395-406.

HENRY, M., 1938. The kangaroo tick. *Aust. Vet. J.* **14:** 69-71.

HERD, R. P. AND COMAN, B. J., 1975. Transmission of *Encinococcus granulosus granulosus* from kangaroos to domestic dogs. *Aust. Vet. J.* **51:** 591-92.

HORNSBY, P. E., 1981. Predation of the euro *Macropus robustus* (Marsupialia: Macropodidae) by the European fox *Vulpes vulpes* (Placentalia: Canidae). *Aust. Mammal.* **5:** 225-27.

HORTON, D. L. AND SAMUEL, J. L., 1978. Paleopathology of a fossil macropod population. *Aust. J. Zool.* **26:** 279-92.

HOWIE, J., 1979. Policies in the UK to ensure that a food factory does not distribute food-poisoning micro-organisms — a personal view. *J. Appl. Bacteriol.* **47:** 233-35.

HOWKINS, A. B., 1966a. The role of Macropodidae in Tasmania as intermediate hosts in hydatid disease. *Aust. Vet. J.* **42:** 240-41.

HOWKINS, A. B., 1966b. Epidemiology of hydatid disease in eastern New South Wales 2. Incidence in Macropodidae and other possible intermediate hosts. *Aust. Vet. J.* **42:** 313-34.

INNS, R. W., 1982. Life-spans of the Tammar Wallaby *Macropus eugenii* (Marsupialia: Macropodidae) in wild populations. *Aust. Mammal.* **5:** 283-84.

IVESON, J. B. AND BRADSHAW, S. D., 1973. *Salmonella javiana* infection in an infant associated with a marsupial, the quokka, *Setonix brachyurus,* in Western Austalia. *J. Hyg.* **71:** 423-32.

IVESON, J. B. AND MACKAY-SCOLLEY, E. M., 1972. An evaluation of strontium chloride, Rappaport and strontium selenite enrichment for the isolation of salmonellas from man, animals, meat products and abattoir effluents. *J. Hyg.* **70:** 367-84.

JAKOB-HOFF, R. M. AND DUNSMORE, J. D., 1983. Epidemiological aspects of toxoplasmosis in southern Western Australia. *Aust. Vet. J.* **60:** 217-18.

JOHNSON, M. A., WHALLEY, J. M., LITTLEJOHNS, I. R., DICKSON, J., SMITH, V. W., WILKS, C. R. AND REISNER, A. H., 1985. Macropodid herpes viruses 1 and 2: Two herpes viruses from Australian marsupials differentiated by restriction endonucleases, DNA composition and hydridization. *Arch. Virol.* **85:** 313-19.

JOHNSTON, T. H., 1909. The entozoa of Monotremata and Australian Marsupialia. No. 1. *Proc. Linn. Soc. NSW* **34:** 514-23.

JOHNSTON, T. H., 1911. The entozoa of Monotremata and Australian Marsupialia. *No. 2. Proc. Linn. Soc. NSW* **36:** 47-57.

JOHNSTON, T. H. AND MAWSON, P. M., 1949. Some nematodes from Australian hosts, together with a note on *Rhabditis allgeni. Trans. R. Soc. S. Aust.* **73:** 63-71.

JONES, D. M., 1980. Department of Veterinary Science. *In* Scientific Report: Zoological Society. *J. Zool.* **190:** 477-590.

KAY, B. H., YOUNG, P. L., FANNING, I. D. AND HALL, R. A., 1982. Which vertebrates amplify Murray Valley Encephalitis virus in southern Australia? Pp. 36-39 *in* Wildlife Diseases of the Pacific Basin and Other Countries ed. by M. E. Fowler. Wildlife Disease Association. Proc. 4th Int. Conf. Wild. Dis. Assoc., Sydney, 25-28 August, 1981.

KAY, B. H., HALL, R. A., FANNING, I. D., MOTTRAM, P., YOUNG, P. L. AND POLLITT, C. C., 1986. Experimental infection of vertebrates with Murray Valley Encephalitis and Ross River Viruses. Pp. 71-75 *in* Arbovirus Research in Australia ed. by T. D. St. George, B. H. Kay and J. Blok. Q.I.M.R.: Brisbane. Proc. 4th Symp. May 6-9, 1986.

KEANE, C., TAYLOR, M. R. H., WILSON, P., SMITH, L., CUNNINGHAM, B., DEVINE, P. AND ENGLISH, L. F., 1977. *Bacteriodes ruminicola* as a possible cause of "lumpy-jaw" in Bennett's wallabies. *Vet. Microbiol.* **2:** 179-83.

KEEP, J. M., 1972. Streptococcal osteomyrelitis in a Grey Kangaroo. *Aust. Vet. J.* **48:** 119-20.

KELER, S. VON., 1971. A revision of the Australian Boopidae (Insecta: Phthiraptera). *Aust. J. Zool. Suppl. Ser. No.* **6:** 1-126.

KERR, A., WHALLEY, J. M. AND POOLE, W. E., 1981. Herpes virus neutralizing antibody in Grey Kangaroos. *Aust. Vet. J.* **57:** 347-48.

KEYMER, I. F., 1972. Report of the pathologist, 1969 and 1970. *In* Scientific Report: Zoological Society. *J. Zool.* **166:** 515-40.

KEYMER, I. F., 1974. Report of the pathologist, 1971 and 1972. *In* Scientific Report: Zoological Society. *J. Zool.* **173:** 51-83.

KING, D. R., OLIVER, A. J. AND MEAD, R. J., 1978. The adaptation of some Western Australian mammals to food plants containing fluroacetate. *Aust. J. Zool.* **26:** 699-712.

KIRKPATRICK, T. H., 1965. Studies of Macropodidae in Queensland 2. Age estimation in the Grey Kangaroo, the Red Kangaroo, the Eastern Wallaroo and the Red-necked Wallaby, with notes on dental abnormalities. *Qld. J. Agric. Anim. Sci.* **22:** 301-17.

KIRKPATRICK, T. H. AND JOHNSON, P. M., 1969. Studies of Macropodidae in Queensland 7. Age estimation and reproduction in the Agile Wallaby (*Wallabia agilis* (Gould)). *Qld. J. Agric. Anim. Sci.* **26:** 691-98.

KUMARATILAKE, L. M. AND THOMPSON, R. C. A., 1982. Hydatidosis/echinococcosis in Australia. *Helminth. Abst.* **51:** 233-52.

LAURIE, W. AND KOVAKS, N., 1960. Report of the Commissioner of Public Health, Western Australia. Appendix 2: 20.

LEE, P. E. AND MACKERRAS, I. M., 1955. Salmonella infections of Australian native animals. *Aust. J. Exp. Biol. Med. Sci.* **33:** 117-26.

LINDAU, K. H., 1964. Uber todesfalle von kangeruhs im Zoologischen Garten Koln. *Tijdschr. Diergeneesk., Suppl. 1,* **89:** 143-48.

MACKERRAS, M. J., 1958. Catalogue of Australian mammals and their recorded internal parasites. I-IV. Part 1. Monotremes and Marsupials. *Proc. Linn. Soc. NSW.* **83:** 101-25.

MAYER, H. AND HAUG, E., 1962. Salmonellen in kanguruhfleisch. *Archiv. Lebensmittelhyg.* **13:** 34-36.

MAYNES, G. M., 1977. Breeding and age structure of the population of *Macropus parma* on Kawau Island, New Zealand. *Aust. J. Ecol.* **2:** 207-14.

MCALEER, R., 1980. Keratinophilic fungi on four animal groups. *Aust. Vet. J.* **56:** 387-90.

MCCARTHY, P. H., 1961a. *Austrosimulium pestilens:* a source of severe irritation to marsupials. *Aust. Vet. J.* **37:** 334.

MCCARTHY, P. H., 1961b. Two parasites of marsupials in central Queensland. *Aust. Vet. J.* **37:** 405.

MCCOLL, K. A. AND SPRATT, D. M., 1982. Parasitic pneumonia in a koala *(Phascolarctos cinereus)* from Victoria, Australia. *J. Wild. Dis.* **18:** 511-12.

MCILROY, J. C., 1981. The sensitivity of Australian animals to 1080 poison I. Intraspecific variation and factors affecting acute toxicity. *Aust. Wild. Res.* **8:** 369-83.

MCILROY, J. C., 1986. The sensitivity of Australian animals to 1080 poison IX. Comparisons between the major groups of animals, and the potential danger non-target species face from 1080 poisoning campaigns. *Aust. Wild. Res.* **13:** 39-48.

MCKENZIE, R. A., FAY, F. R. AND PRIOR, H. C., 1979. Pox-virus infection of the skin of an eastern grey kangaroo. *Aust. Vet. J.* **55:** 188-90.

MCMANUS, T. J., 1979. Native fauna and liver fluke. *J. Agric., Tas.* **50:** 43-45.

MILLER, W. A. AND BEIGHTON, D., 1979. Bone abnormalities in two groups of macropod skulls: a clue to the origin of Lumpy Jaw. *Aust. J. Zool.* **27:** 681-89.

MILLER, W. A., BEIGHTON, D. AND BUTLER, R., 1980. Histological and osteological observations on the early stages of lumpy jaw. Pp. 231-39 *in* The Comparative Pathology of Zoo Animals ed. by R. J. Montali and G. Migaki. Smithsonian Institution Press: Washington.

MILNER, A. R., WILKS, C. R., SPRATT, D. M. AND PRESIDENTE, P. J. A., 1981. The prevalence of anti-leptospiral agglutinins in sera of wildlife in southeastern Australia. *J. Wild. Dis.* **17:** 197-202.

MULLER, M. J., MURRAY, M. D. AND EDWARDS, J. A., 1981. Blood-sucking midges and mosquitoes feeding on mammals at Beatrice Hill, Northern Territory. *Aust. J. Zool.* **29:** 573-88.

MUNDAY, B. L., 1966. Diseases of Tasmania's free-living animals. Tasmanian Department of Agriculture Research Bulletin No. 5. Hobart. Pp. 29.

MUNDAY, B. L., 1970. The epidemiology of toxoplasmosis with particular reference to the Tasmanian environment. M.V.Sc. Thesis, University of Melbourne.

MUNDAY, B. L., 1971. Gastric squamous tumor in the stomach of a pademelon *(Thylogale billardieri)* associated with an infestation with *Labiostrongylus* sp. larvae. *J. Wild. Dis.* **7**: 125.

MUNDAY, B. L., 1972a. A serological study of some infectious diseases of Tasmanian wildlife. *J. Wild. Dis.* **8**: 169-75.

MUNDAY, B. L., 1972b. Myonecrosis in free-living and recently captured macropods. *J. Wild. Dis.* **8**: 191-92.

MUNDAY, B. L., 1975. Prevalence of toxoplasmosis in Tasmanian meat animals. *Aust. Vet. J.* **51**: 315-16.

MUNDAY, B. L., 1978. Marsupial disease. Pp. 355-85 *in* Fauna — Part B, Proc. No. 36. Post Graduate Committee in Veterinary Science, University of Sydney: Sydney.

MUNDAY, B. L., 1988. Marsupial diseases. Pp. 299-365 *in* Australian Wildlife, Proc. No.104. Post Graduate Committee in Veterinary Science, University of Sydney: Sydney.

MUNDAY, B. L. AND GREEN, R. H., 1972. Parasites of Tasmanian native and feral fauna. Part ii, Helminthes. *Rec. Queen Vic. Mus. No.* **44**: 15.

MUNDAY, B. L., MASON, R. W., HARTLEY, W. J., PRESIDENTE, P. J. A. AND OBENDORF, D., 1978. *Sarcocystis* and related organisms in Australian wildlife: I. Survey findings in mammals. *J. Wild. Dis.* **14**: 417-33.

MUNDAY, B. L. AND OBENDORF, D. L., 1983. Cytomegalic lesions in Australian marsupials. *J. Wild. Dis.* **19**: 132-35.

MURPHY, C. R., 1970. Haematological studies on the Kangaroo Island Wallaby, *Protemnodon eugenii* (Desmarest). M.Sc. Thesis, University of Adelaide.

MUSHIN, R. AND ASHBURNER, F. M., 1964. Studies on Coli-aerogenes and other gram-negative intestinal bacteria in various animals and birds. *J. Appl. Bacter.* **27**: 392-407.

MYERS, K., BAILEY, P. J. AND DUDZINSKI, M. L., 1976. The effects of a severe drought on the morphology of the adrenal glands and other indices of health in the Red Kangaroo *Megaleia rufa* (Desmarest) in north-western New South Wales. *Aust. J. Ecol.* **1**: 289-302.

MYKYTOWYCZ, R., 1963. Occurrence of bot-fly larvae *Tracheomyia macropi* Froggatt (Diptera: Oestridae) in wild Red Kangaroos, *Megaleia rufa* (Desmarest). *Proc. Linn. Soc. NSW.* **88**: 307-12.

MYKYTOWYCZ, R., 1964. Coccidia in wild populations of the Red Kangaroo, *Megaleia rufa* (Desmarest), and the Grey Kangaroo, *Macropus canguru* (Muller). *Parasit.* **54**: 105-15.

NEWSOME, A. E., 1964a. Anoestrus in the Red Kangaroo, *Megaleia rufa* (Desmarest). *Aust. J. Zool.* **12**: 9-17.

NEWSOME, A. E., 1964b. Oestrus in the lactating Red Kangaroo, *Megaleia rufa* (Desmarest). *Aust. J. Zool.* **12**: 315-21.

NEWSOME, A. E., 1965. Reproduction in natural populations of the Red Kangaroo, *Megaleia rufa* (Desmarest). *Aust. J. Zool.* **13**: 735-59.

NEWSOME, A. E., 1973. Cellular degeneration in the testis of Red Kangaroos during hot weather and drought in central Australia. *J. Reprod. Fert. Suppl.* **19**: 191-201.

NEWSOME, A. E., 1977. The Red Kangaroo — an example of biological indicators of environmental change. Pp. 27-48 *in* Australian Animals and Their Environment ed. by H. Messel and S. T. Butler. Shakespeare Head Press: Sydney.

NEWSOME, A. E., STEPHENS, D. R. AND SHIPWAY, A. K., 1967. Effect of a long drought on the abundance of Red Kangaroos in central Australia. *C.S.I.R.O. Wild. Res.* **12**: 1-8.

NIVEN, B. S., 1970. Mathematics of populations of the Quokka, *Setonix brachyurus* (Marsupialia) I. A simple deterministic model for quokka populations. *Aust. J. Zool.* **18**: 209-14.

OBENDORF, D. L. AND MUNDAY, B. L., 1983. Toxoplasmosis in wild Tasmanian wallabies. *Aust. Vet. J.* **60**: 62.

OKOH, A. E. J. AND ONAZI, M., 1980. Notes of salmonellae isolated from wildlife in Kano Zoological Gardens. *J. Wild. Dis.* **16**: 7-10.

OSMAN HILL, W. C., 1957. Report of the Society's Prosector for the years 1955 and 1956. *Proc. Zool. Soc. London* **129**: 431-46.

PAPADIMITRIOU, J. M. AND ASHMAN, R. B., 1972. A poxvirus in a marsupial papilloma. *J. Gen. Virol.* **16**: 87-89.

POOLE, W. E., 1971. The status of the Australian Macropodidae. Pp. 13-27 *in* The Status of Endangered Australasian Wildlife ed. by M. J. Tyler. Royal Zoological Society South Australia: Adelaide Proceedings Centenary Symposium Royal Zoological Society South Australia, Adelaide, 21-23 September, 1978.

POPE, J. H., SCOTT, W. AND DWYER, R., 1960. *Coxiella burneti* in kangaroos and kangaroo ticks in western Queensland. *Aust. J. Exp. Biol.* **38**: 17-28.

POTKAY, S., 1977. Diseases of marsupials. Pp. 456-506 *in* The Biology of Marsupials ed. by D. Hunsaker. Academic Press Inc.: New York.

PRESIDENTE, P. J. A., 1978. Diseases seen in free-ranging marsupials and those held in captivity. Pp. 457-71 *in* Fauna — Part B, Proceedings: No. 36. Post Graduate Committee in Veterinary Science, University of Sydney: Sydney.

PRESIDENTE, P. J. A. AND BEVERIDGE, I., 1978. Cholangitis associated with species of *Progamotaenia* (Cestoda: Anoplocephalidae) in the bile ducts of marsupials. *J. Wild. Dis.* **14**: 371-77.

PRUDHOE, S., 1975. A new genus of paramphistome trematodes from a wallaby. Dr B. S. Chauhan Comm. Vol., 63-68.

REES, R. G., 1967. Keratinophilic fungi from Queensland — I. Isolations from animal hair and scales. *Sabouraudia* **5**: 165-72.

ROBERTS, D., 1982. Bacteria of public health significance. Pp. 319-86 *in* Meat Microbiology ed. by M. H. Brown. Applied Science Publishers Ltd: London.

ROBERTS, F. H. S., 1936. A note on *Argas (Ornithodoros) gurneyi* Warburton. *Aust. Vet. J.* **12**: 239-40.

ROBERTS, F. H. S., 1952. Insects Affecting Livestock: with Special Reference to Important Species Occurring in Australia. Angus & Robertson: Sydney.

ROBERTS, F. H. S., 1970. Australian Ticks. Commonwealth Scientific and Industrial Research Organization: Melbourne.

ROTHWELL, T. L. W., KEEP, J. M., FU-NAN XU AND MIDDLETON, D. J., 1984. Poxvirus in marsupial skin lesions. *Aust. Vet. J.* **61**: 409-10.

SALISBURY, R. M., 1958. Salmonella infections in animals and birds in New Zealand. *New Zeal. Vet. J.* **6**: 76-86.

SAMUEL, J. L., 1982. *Salmonella* in macropods. Pp. 61-63 *in* Wildlife Diseases of the Pacific Basin and Other Countries ed. by M. E. Fowler. Wildlife Disease Association, Proceedings 4th International Conference Wildlife Diseases Association, Sydney, 25-28 August 1981.

SAMUEL, J. L., 1983. Jaw disease in macropod marsupials: bacterial flora isolated from lesions and from the mouths of affected animals. *Vet. Microbiol.* **8**: 373-87.

SCHOON, H.-A., VEBERSCHAR, S. AND BRANDT, H.-P., 1982. Herpesvirus hepatitis bei baren barunkanguruhs *(Dendrolagus ursinus)*. Pp. 273-78 *in* Erkrankungen der Zootiere ed. by R. Ippen and H. D. Schroder. Verhandlungsbericht des XXIV Internationalen Symposiums uber dis Erkankungen der Zootiere in Vespzpiem.

SEDDON, H. R., 1951. Diseases of domestic animals in Australia. Serv. Publ. No. 6. Division Vet. Hygiene: Canberra.

SHEPHERD, N. C., 1981. Predation of red kangaroos, *Macropus rufus,* by the dingo, *Canis familiaris dingo* (Blumenbach), in north-western New South Wales. *Aust. Wild. Res.* **8**: 255-62.

SHEPHERD, N. C., 1982. Capture myopthay in the Red Kangaroo. *(Macropus rufus)* Pp. 239-47 *in* Wildlife Diseases of the Pacific Basin and Other Countries ed. by M. E. Fowler. Wildlife Diseases Association Proceedings 4th International Conference Wildlife Diseases Association, Sydney, 25-28 August, 1981.

SINCLAIR, R. G., 1977. Harvesting kangaroos in New South Wales. *Aust. Wild. Res.* **4**: 207-18.

SMALES, L. R., 1977. The life history of *Labiostrongylus eugenii,* a nematode parasite of the Kangaroo Island wallaby *(Macropus eugenii)*: the parasitic stages. *Int. J. Parasit.* **7**: 457-61.

SMITH, D. J. W., 1942. (from Pope *et al.* 1960). *Aust. J. Exp. Biol.* **20**: 295.

SMITH, I. D. AND MUNDAY, B. L., 1965. Observations on the incidence of *Toxoplasma gondii* in native and introduced feral fauna in Eastern Australia. *Aust. Vet. J.* **41**: 285-86.

SNOWDON, W. A., 1968. The susceptibility of some Australian fauna to infection with foot-and-mouth disease virus. *Aust. J. Exp. Biol. Med. Sci.* **46**: 667-87.

SNOWDON, W. A., 1988. Feral mammals/exotic diseases. Proc.: No. 104 Australian Wildlife. Pp. 931-35. Post Graduate Committee in Veterinary Science, University of Sydney: Sydney.

SOUTHWELL, C. J., 1981. Sociobiology of the Eastern Grey Kangaroo, *Macropus giganteus.* Ph.D. Thesis, University of New England.

SPEARE, R., 1985. The veterinary aspects of man as an intermediate host for *Toxoplasma gondii. Aust. Vet. Practit.* **15**: 11-18.

SPEARE, R., 1988. Clinical evaluation, diseases and management of the orphaned macropod joey. Pp. 211-96 *in* Proc.: No. 104, Australian Wildlife. Post Graduate Committee in Veterinary Science, University of Sydney: Sydney.

SPEARE, R. AND BANKS, D., 1987. A trichoepithelioma in a wild eastern grey kangaroo *(Macropus giganteus). J. Wild. Dis.* **23**: 508-9.

SPEARE, R., BEVERIDGE, I., JOHNSON, P. M. AND CORNER, L. A., 1983. Parasites of the agile wallaby, *Macropus agilis* (Marsupialia). *Aust. Wild. Res.* **10**: 89-96.

SPEARE, R., JOHNSON, P. M. AND HAFFENDEN, A., 1982. Management of disease in captive macropods in north Queensland. Pp. 117-24 *in* The Management of Australian Mammals in Captivity ed. by D. D. Evans. Zoological Parks Board of Victoria: Melbourne.

SPEARE, R. AND SPRATT, D. M., 1982. Pathology due to a remarkable nematode parasitic in the portal veins of Macropodidae. Pp. 113-17 *in* Wildlife Diseases of the Pacific Basin and Other Countries ed. by M. E. Fowler. Wildlife Diseases Association Proceedings 4th International Conference Wildlife Diseases Association, Sydney, 25-28 August, 1981.

SPEARE, R. AND THOMAS, A. D., 1988. Orphaned kangaroo and wallaby joeys as potential zoonotic source of *Salmonella. Med. J. Aust.* **148**: 619-23.

SPENCE, T., 1979. Annual report. P. 16. Zoological Gardens Board: Perth.

SPRATT, D. M., 1972a. Aspects of the life history of *Dirofilaria roemeri* in naturally and experimentally infected kangaroos, wallaroos and wallabies. *Int. J. Parasitol.* **2**: 139-56.

SPRATT, D. M., 1972b. Natual occurrence, histopathology and developmental stages of *Dirofilaria roemeri* in the intermediate host. *Int. J. Parasitol.* **2**: 201-8.

SPRATT, D. M., 1972c. Transplantation of adult *Dirofilaria roemeri* to grey kangaroos and laboratory rats. *J. Helminthol.* **16**: 81-89.

SPRATT, D. M., 1974. Comparative epidemiology of *Dirofilaria roemeri* infection in two regions of Queensland. *Int. J. Parasitol.* **4**: 481-88.

SPRATT, D. M., 1975. Further studies of *Dirofilaria roemeri* (Nematoda: Filarioidea) in naturally and experimentally infected Macropodidae. *Int. J. Parasitol.* **5**: 561-64.

SPRATT, D. M., 1979. A taxonomic revision of the lungworms (Nematoda: Metastrongyloidea from Australian marsupials) *Aust. J. Zool. Suppl. Ser. No.* **67**: 1-45.

SPRATT, D. M., 1984. Further studies on lung parasites (Nematoda) from Australian marsupials. *Aust. J. Zool.* **32**: 283-310.

SPRATT, D. M. AND PRESIDENTE, P. J. A., 1981. Prevalence of *Fasciola hepatica* infection in native animals in south-eastern Australia. *Aust. J. Exp. Biol. Med. Sci.* **59**: 713-21.

SPRATT, D. M. AND SPEARE, R., 1982. *Durikainema macropi* gen. et sp. nov. (Muspiceoidea: Robertdollfusidae), a remarkable nematode from Macropodidae (Marsupialia). *Ann. Parasitol.* **57**: 53-62.

SPRATT, D. M. AND VARUGHESE, G., 1975. A taxonomic revision of filarioid nematodes from Australiasian marsupials. *Aust. J. Zool. Suppl. Ser. No.* **35**: 1-99.

STANLEY, N. F. AND MACKENZIE, J. S., 1983. Microbiological studies on Rottnest Island. *J. R. Soc. West. Aust.* **66**: 31-34.

STORR, G. M., 1964. Studies on marsupial nutrition. IV. Diet of the quokka, *Setonix brachyurus* (Quoy and Gaimard) on Rottnest Island, Western Australia. *Aust. J. Biol. Sci.* **17**: 469-81.

SUZUKI, A., KAWANISHI, T., SUZUKI, K., IWATA, N. AND TAKAYAMA, S., 1967. A survey on the *Salmonella* contamination of imported meats during 1965-1966. *Bull. Nat. Inst. Hyg. Sci., Tokyo* **85**: 188-95.

SWEATMAN, G. K. AND WILLIAMS, R. J., 1962. Wild animals in New Zealand as hosts of *Echinococcus granulosus* and other taeniid tapeworms. *Trans. R. Soc. N.Z., Zool.* **2**: 221-50.

THOMPSON, R. C. A., 1986. Biology and systematics of *Echinococcus*. Pp. 5-43 *in* The biology of *Echinococcus* and Hydatid Disease ed. by R. C. A. Thompson. George Allen and Unwin: Boston.

THOMPSON, R. C. A. AND KUMARATILAKE, L. M., 1982. Intraspecific variation in *Echinococcus granulosus:* the Australian situation and perspectives for the future. *Trans. R. Soc. Trop. Med. Hyg.* **76:** 13-16.

TOMLINSON, A. R. AND GOODING, C. G., 1954. A kangaroo disease — Investigations into "Lumpy Jaw" on the Murchinson, 1954. *J. Agric. West. Aust.* **3:** 715-18.

TONGEREN, H. A. E. VAN, VERLINDE, J. D. AND PETERS, J. C., 1950. Salmonellosis bij de kangeoroe. *Tijdschrift voor Diergeneeskunde* **75:** 480-81.

WEBBER, C. E. AND WHALLEY, J. M., 1978. Widespread occurrence in Australian marsupials of neutralizing antibodies to a herpesvirus from a parma wallaby. *Aust. J. Exp. Biol. Med. Sci.* **56:** 351-57.

WHALLEY, J. M. AND WEBBER, C. E., 1979. Characteristics of parma wallaby herpes virus grown in marsupial cells. *J. Gen. Virol.* **45:** 423-30.

WIESNER, H., 1976. Marsupials. Pp. 291-300 *in* Handbook of Zoo Medicine ed. by R. Goltenboth and D. Jarofke. Van Nostrand Reinhold Co.: New York.

WILKS, C. R., KEFFORD, B. AND CALLINAN, R. B., 1981. Herpesvirus as a cause of fatal disease in Australian wallabies. *J. Comp. Path.* **91:** 461-65.

WILSON, P., WEAVERS, E. AND KAVANAGH, J., 1984. An outbreak of coccidiosis in kangaroos in a zoological collection. *Irish Vet. J.* **38:** 33-35.

WINTER, H., 1958. Gastric strongyloidiosis in kangaroos. *Aust. Vet. J.* **35:** 301-3.

WINTER, H., 1959. Coccidiosis in kangaroos. *Aust. Vet. J.* **35:** 301-3.

WINTER, H. AND O'CONNOR, R. F., 1957. *Salmonella birkenhead* from a kangaroo. *Aust. Vet. J.* **33:** 83.

YADAV, M., STANLEY, N. F. AND WARING, H., 1972. The microbial flora of the gut of the pouch-young and the pouch of a marsupial, *Setonix brachyurus*. *J. Gen Microb.* **70:** 437-42.

YOUNG, M. R., 1939. Helminth parasites of Australia. Imp. Bur. Agric. Parasit. Helmonthology: England.

Predation on Macropodoidea: A review

J. D. ROBERTSHAW[1] and R. H. HARDEN[2]

Historically, Macropodoidea have been prey for the Marsupial Lion *(Thylacoleo carnifex),* the Thylacine *(Thylacinus cynocephalus)* and Aboriginal man. Today there are three groups of predators which prey on macropods. Minor predators include the larger dasyurids, the Spotted-tailed Quoll *(Dasyurus maculatus)* and the Western Quoll *(D. geoffroii)*; the larger eagles, the Wedge-tailed Eagle *(Aquila audax)* and the White-breasted Sea-eagle *(Haliaeetus leucogaster);* and the Carpet and Amethystine Pythons (*Morelia spilotes* and *Liasis* spp.). Locally significant predators, foxes *(Vulpes vulpes)* and cats *(Felis catus),* are predators of the smaller macropods but further study is required to quantify their impact. In some unique situations they have devastated isolated populations of the smaller species of macropod. The major predators, man and dingo *(Canis familiaris dingo),* kill a wide range of macropods of all sizes over a large geographic area. Direct predation by man over the last 200 years has had little effect on reducing the numbers or distribution of macropods except for the Forester *(Macropus giganteus tasmaniensis)* in Tasmania and the Toolache Wallaby *(Macropus greyi)* in South Australia. However, minor local and temporal effects have occurred. The dingo has been able to locally and temporally alter the number and distribution of macropods but predation has not been shown to have any regulatory effects on macropod populations in the studies undertaken.

INTRODUCTION

SCIENTIFIC accounts in the 1960s stated that predation was of little importance to macropod populations. As a convenient summary of these, the Australian House of Representatives Select Committee on Wildlife Conservation (1971, pt30) concluded that "the kangaroo has no predators of significance other than man, but seasonal conditions, particularly drought, have a very marked effect on kangaroo populations . . . ". This conclusion was based on the submissions of many scientists and naturalists (see Australian Parliament 1971, their Appendix IV).

The purpose of the present paper is to review information on predation on the Macropodoidea in the light of information available in the late 1980s. The first part involves collating information on predation. The second part assesses the effect of such predation on macropod populations and their distributions. The latter task is made more difficult by the dearth of published information. In this review, we will consider predation to have had an effect if the predation has, or is likely to have, changed either the size or distribution of a macropod population. We will use the term "impact" under these circumstances.

PLEISTOCENE PREDATION

During the Pleistocene the Marsupial Lion *(Thylacoleo carnifex)* was widespread throughout Australia. There was an initial controversy as to whether this predator was actually carnivorous (Wells *et al.* 1982). *T. carnifex* may have been able to hold large prey enabling it to pierce and tear the throat of its victim (Finch 1982; Archer 1984). Case (1985) found the remains of *T. carnifex* from South Australia and Victoria in assocation with the remains of megafauna including *Sthenurus,* large species of *Macropus, Osphranter* and *Procoptodon* and concluded it was largely a predator of these species. However, it is probable that *T. carnifex* was markedly sexually dimorphic (Archer and Dawson 1982) and geographically variable in size. Some individuals in eastern Australia were much larger than specimens from Western Australia (Archer 1984). The smaller individuals may not have been able to prey upon the larger megafauna but utilized the smaller species of *Macropus* and wallabies.

This potential range of prey species for *T. carnifex* may indicate that the only other large marsupial predator, the Thylacine *(Thylacinus cynocephalus),* shared an overlapping prey base (cf. Case 1985). The Thylacine was considered a pursuit carnivore which preyed upon smaller wallabies and had little specialized structural adaptation to the role of predator (Keast 1982).

T. carnifex became extinct by 17 000 years BP (Flood 1983), at the same time as the megafauna, while the Thylacine was still present in southern mainland Australia 3 300 years ago (Flood 1983), in northwestern Australia 0±80 years BP (Archer 1984a) and in Tasmania until the 1930s (Smith 1982). The effect of their predation on the distribution or numbers of early macropod species is not known.

Another predator of the Pleistocene was Aboriginal man. Direct evidence of Aboriginal occupation in the far south-west and south-east by 40 000 years BP indicates their arrival on the continent during a period of low sea level around 53 000 years ago (Flood 1983). While man and

[1]Department of Zoology, University of New England, Armidale, New South Wales 2351.

[2]Environmental Survey and Research Branch, New South Wales National Parks and Wildlife Service, c/- Department of Zoology, University of New England, Armidale, New South Wales 2351.

Pages 735-53 in KANGAROOS, WALLABIES AND RAT-KANGAROOS ed. by G. Grigg, P. Jarman and I. Hume. Surrey Beatty & Sons Pty Limited, New South Wales, Australia, 1989.

megafauna were contemporary, there is little evidence that Aboriginal man preyed on macropods in the megafauna. However, Dortch (1979) believed that bones of *Protemnodon* and *Sthenurus* found in the lowest levels of Devil's Lair aged at 33 000 years BP, were probably fractured by man. Generally early Aboriginal man was thought to hunt the smaller macropods. Rednecked Wallabies (Bennett's Wallabies) were the main prey of Aborigines in south-west Tasmania during the last ice-age (Jones 1981 cited by Flood 1983). This predation may have been seasonal and resulted from annual small scale migrations from areas of depleted food resources. Such cases have been revealed at many Recent sites on the mainland (White and O'Connell 1982, Pp. 133-70). As an extension of the "coastal colonization hypothesis" (Bowdler 1977), techniques allowing the intensive hunting of large kangaroos may not have been developed until 12 000 years ago.

Indirectly, Aborigines have affected the number and distribution of macropods. Widespread use of fire to facilitate movement and hunting is one of the reasons given for the extinction of the megafauna (Horton 1980). However, this fauna may have become extinct in central Australia before human penetration (Main 1978; Archer 1984b) and it is uncertain when widespread use of fire started (Luke Godwin, pers. comm.). Either way, Australian vegetation and fauna are, to a certain extent, artefacts created by Aborigines with their fire-sticks. By providing a localized diversity of habitat types Aboriginal impact was probably more beneficial than harmful.

EXISTING PREDATORS

With the arrival of the dingo approximately 5 000 years BP (Corbett 1985) and, much more recently, European man and other predators (foxes and cats), the predator-prey interactions have been changing constantly. The effect of the dingo on the composition of the Australian fauna is not known but no macropod species became extinct in Recent times (Flannery 1984) corresponding to the time of arrival of the dingo. Since European settlement six macropod species and three sub-species have become extinct, one (the Toolache Wallaby *Macropus greyi*) mainly as a result of "sport". Also seven species are considered endangered although two have only recently been discovered and described (see Strahan 1983 and Australian National Parks and Wildlife Service 1988).

In this latter period the predators can be broadly separated into three groups based on their impact on macropod populations. The

Table 1. Minor predators of macropods.

Predator		Prey	Source
Larger Dasyurids	*Dasyurus geoffroii*	*Bettongia penicillata*	Christensen (1980)
	Dasyurus maculatus	*Bettongia gaimardi*	Rose (1986)
	Sarcophilus harrisii	Small wallabies, Rat-kangaroos	Le Souef and Burrell (1926)
Birds	*Haliaeetus leucogaster* (White-breasted Sea-eagle)	*Peradorcas concinna* (Narbarlek or Little Rock-wallaby)	Sanson (1983)
	Aquila audax (Wedge-tailed Eagle)	*Aepyprymnus rufescens*	Jarman *et al.* (1987)
		Bettongia penicillata	Christensen (1980)
		Thylogale thetis	Johnson (1977), Kirkpatrick (1985)
		Thylogale stigmatica	Kirkpatrick (1985)
		Macropus eugenii	Christensen (1980)
		Wallabia bicolor	Leopold and Wolfe (1970), Hull (1985)
		Petrogale penicillata	Lim *et al.* (1981)
		Macropus rufogriseus	Leopold and Wolfe (1970), Jarman *et al.* (1987)
		Macropus giganteus	Leopold and Wolfe (1970), Hull (1985)
		Macropus robustus	Leopold and Wolfe (1970), Ealey (1962), Brooker and Ridpath (1980)
		Macropus rufus	Leopold and Wolfe (1970), Brooker and Ridpath (1980)
		Macropus fuliginosus	Leopold and Wolfe (1970), Brooker and Ridpath (1980)
Reptiles	*Morelia spilotes* (Carpet Python)	Young Potoroids	Jarman *et al.* (1987)
		Young of small macropods including Pademelons and Rock-wallabies	Kirkpatrick (1985)
	Liasis amethystinus and *Liasis olivaceus* Amethystine Pythons	Young of small macropods including Pademelons and Rock-wallabies	Kirkpatrick (1985)
	Varanus varanus (Lace Monitor)	Young Potoroids	Jarman *et al.* (1987)

from 4.2 million (1881) to 3.1 million (1889). Comparable figures are apparently not available again until 1960 when commercial harvesting figures were recorded (Appendix 1) and in 1975 when population numbers were again estimated (Appendix 2). In 1981, the number of kangaroos in the commercial zone was estimated at between 6.7 and 7 million (Caughley *et al.* 1983 and National Parks and Wildlife Service 1988). With an estimated additional 1.7 million grey kangaroos from the remainder of the state (Caughley *et al.* 1983) the population estimate was 8 or 9 million. The harvesting figures for this period are much lower than that of last century (mean 360 000 per year) while the population estimates of the larger species derived from only the commercial zone are similar. While it has been suggested that both the 1880s population estimates (Denny 1981) collected by stock inspectors and the recent aerial survey estimates (Caughley *et al.* 1977; Bayliss and Giles 1985; Short and Bayliss 1985) may be under-estimates, their comparison suggests a similarity between the 1881 and 1981 populations of large macropods even considering the different methodologies by which the data were collected.

In Queensland, a minimum of 66 million kangaroos and wallabies have been harvested since 1877 (Table 7), an average of 600 000 p.a. for the 110 year period. Between 1877 and 1906 20 million bounty payments were made on kangaroos. In 1906 bounty payments ceased on kangaroos but between 1906 and 1917 there were 6.4 million bounties paid on wallabies before the bounty was discontinued on these species. Harvesting was highest between 1897 and 1902 and from 1979 to 1987 with the number taken between 1964 and 1966 also high. While the numbers harvested have varied markedly between years as a result of climatic changes (Kirkpatrick and Amos 1985) there has been a rough similarity in the numbers harvested over larger time intervals (see Tables 7 and 8). This indicates a degree of stability in the harvesting level. Provided that the species composition has not altered greatly, which seems likely since their distributions have not (Poole 1984), these figures suggest not only the harvesting level but also the mortality as a result of road-deaths, fence kills and illegal shooting has not had a marked impact on macropods in Queensland. For greater detail see Kirkpatrick and Amos (1985, Pp. 95-98).

Table 5. The number of kangaroos killed commercially (C) and non-commercially (N) since 1975 for the different Australian states. (Source primarily Australian National Parks and Wildlife Service but additional data from NSW NPWS and Kirkpatrick (pers. comm.). Not all figures are available for all years. In New South Wales and Victoria non-commercial figures indicate the number authorized to be killed under permit and this allocation is rarely completely taken. In New South Wales these figures are also in two parts, those from within the commercial zone (#) and those from the non-commercial zone (*)).

Year	Type	NSW	Qld	SA	WA	Tas.	Vic.	NT	A.C.T.	Total
1975	C	123 000	497 138	43 677	122 000	6 601				792 416
	N						8 908			
1976	C	96 700	620 555	34 524	162 000	21 426				935 205
	N						8 695	2 000		
1977	C	167 200	588 404	60 698	166 000	45 471				1 027 773
	N						8 691			
1978	C	220 000	600 382	135 996	157 500	100 803				1 214 681
	N						10 091			
1979	C	520 000	1 314 260	93 234	195 500	244 918				2 367 912
	N					693 000	14 451			
1980	C	619 023	943 099	148 329	162 930	220 226				2 093 607
	N	*73 185		21 700		599 000	26 535		782	
1981	C	488 647	679 384	193 326	175 000	108 999	<1 000			c.1 646 357
	N	*47 217		25 583		534 000	24 083	5 726	144	
1982	C	664 342	1 136 798	229 453	177 536	77 000	<1 000			c.2 286 129
	N	*38 077		44 901		661 000	31 748	18	180	
1983	C	400 477	877 350	155 785	202 402	110 000				1 746 014
	N	*27 172				780 000	16 540		286	
1984	C	222 015	860 670	99 274	178 500	118 000				1 478 459
	N	*10 140	1 380	19 155		806 000	8 024		426	845 125
1985	C	326 028	1 126 116	114 643	204 325	92 000				1 863 112
	N	*16 422	14 868	27 287		546 525	9 937	6	332	615 377
1986	C	444 509	1 343 511	128 923	194 629	60 000				2 171 572
	N	*33 235	100 633	22 217		701 000				823 850
		#17 758								
1987	C	473 454	1 232 889							
	N	*24 222								
		#18 448								
C mean		357 662	882 306	118 994	174 819	100 370				
std. dev.		193 777	294 124	61 243	23 284	71 355				
coef. var.		54.2%	33.3%	51.5%	13.3%	71.1%				

Table 6. A comparison of the mean number of licensed trappers in the different states. (Not all figures are available for the two types of trappers, commercial (C) and non-commercial (N) in the different periods. (a) Annual reports submitted to Parliament, (b) Kirkpatrick and Amos 1985, (c) licensed to take Red Kangaroo for sale, (d) NSW NPWS, (e) sporting licences, (f) licences in the non-commercial zone, (g) licences in the commercial zone for non-commercial culling; *2 376 in 1979-80 and # 3 years only).

Years	Tasmania (a) C	Tasmania (a) N	Queensland C(b)	Queensland N(a)	New South Wales C(a)	New South Wales N(d)	Western Australia (a) (c) C	Western Australia (a) (c) N
1971-75	153	3 533	1 387	c.280(e)	c.225	—	53	—
1976-80	789*	5 620	1 714	c.150(e)	c.529	—	80	—
1981-85	535	5 437	1 482#		c.428	—	144	—
>1985					c.285	c.190(f) c.220(g)	—	—

In Tasmania, the Forester Kangaroo *(M. giganteus tasmaniensis)* was almost extinct by the 1850s due to shooting (Rolls 1983), but recent strict management has resulted in a substantial increase in its numbers (Poole 1984). Macropod harvest figures have been kept since 1922 (Australian National Parks and Wildlife Service 1984 and Appendix 1). The harvest in Tasmania is unique because it consists almost entirely of two species of wallaby, because the major proportion of these species is taken non-commercially (see Table 5) and because the basis for the killing is not solely to alleviate the damage caused by these pest species. Historically, trapping, snaring and shooting have been a pastime of many Tasmanians who, more than the peoples from other states, use macropod products locally. This is reflected by the large number of "trappers" in Tasmania (Table 6) which is more than three times the number in either Queensland or New South Wales. However, their impact on the wallaby population does not appear to be causing a decline in numbers (Poole 1984) or a reduction in distribution.

In Western Australia, limited export of kangaroo skins began in 1843 (Garden 1977 cited by Prince 1984). The number of skins exported between 1843 and 1860 averaged less than 6 000 but by the mid 1880's export value (and hence the number of skins exported) had increased significantly. Prince (1984) estimated that between 1885 and 1913 the annual harvest may have averaged 200 000. Records from 1914 to 1986 (Prince 1984 and Appendix 1) suggest a similar mean figure. This consistency in the level of harvest despite climatic variation and the harvesting pressure *per se,* is unique to this state. There were only two periods which are notable for their deviation from this norm; 1919-1920 and 1935-1936 when an average of 650 000 macropods were taken commerically each year. This increase was largely due to a much greater number of Red Kangaroos being harvested.

Except for the past 20 years, we were unable to find estimates of either population size or destruction of macropods in the remaining states. Today the only other state where commerical shooting occurs is South Australia where over the last 20 years the number harvested has ranged from 35 000 to 230 000. This is considerably less than in New South Wales or Queensland but the proportion of the kangaroo population harvested is similar. In the remaining three states relatively few macropods are culled (see Table 5). For Victoria these figures over-estimate the actual number killed as they represent the number authorized to be killed only. Commercial culling in Victoria has not occurred in the past 35 years except for 1981 and 1982 when quotas of about 30 000 were set, but less than 1 000 were actually taken (Mark Fletcher, pers. comm.). In the Northern Territory the figures in Table 5 are for Red Kangaroos only and there are no figures available for the number of Agile Wallabies killed in open seasons north of latitude 15°S.

The following conclusions can be drawn from the historical evidence. In New South Wales, populations of the larger macropod species appear to be of a similar order to those 100 years ago, and in Queensland and Western Australia it

Table 7. The average number of kangaroos and wallabies killed in Queensland between 1877 and 1987 for particular time periods. (Data excludes Pademelons and Rat Kangaroos.) Sources: 1877 to 1918 Reports of the Department of Agriculture and Stock for the years 1917-1918 and 1918-1919; 1917 to 1960 Roff and Kirkpatrick 1962; 1954 to 1979 Kirkpatrick and Amos 1985; 1980 to 1986 Australian National Parks and Wildlife Service; 1987 figure was provisional and supplied by Kirkpatrick. N.B. Bounties on kangaroos and wallaroos were discontinued in 1906 and for the other marsupials on the 1/1/1919. (* No returns).

Period	Kangaroos and Wallaroos	Wallabies	Combined	Total numbers
1877-1879	390 476	198 510	588 987	1 766 958
1880-1881*				
1882-1891	309 615	470 442	792 816	7 020 516
1891-1895*				
1896-1906	361 565	602 978	1 012 280	10 609 963
1906-1918		531 333	531 333	6 375 991
1917-1950			350 000	11 550 000
1950-1960			450 000	4 500 000
1954-1959	432 102	49 740	481 837	
1960-1969	837 140	12 370	849 510	8 495 097
1970-1979	655 150	39 986	695 136	6 951 359
1980-1987	1 000 853	23 988	1 024 842	8 198 735
				65 468 619

Table 8. The average number of macropods harvested commercially in five year blocks. (The figures are rounded to the nearest 1 000. Periods of incomplete data are indicated: – not available, * data for one year only, ** two years, *** three years. Sources include Prince (1984), Kirkpatrick and Amos (1985), Kirkpatrick (pers. comm.), McNamara and Prince (1986), Caughley *et al.* (1987), Annual Reports by the different faunal authorities, and Australian National Parks and Wildlife Service).

Period	Queensland '000s	New South Wales '000s	Western Australia '000s	South Australia '000s	Tasmania '000s
1945-1949	–	–	161	–	137
1950-1954	218*	–	116	–	96
1955-1959	520	–	>52	–	22
1960-1964	710	110	>60	–	41
1965-1969	989	162	229	110***	50
1970-1974	666	104	212	61	32
1975-1979	724	225	161	74	84
1980-1984	899	479	179	136	127
1985-1987	1 234***	415***	199***	122**	76**

appears to have been possible to take similar numbers of macropods each year for the past 110 years. Thus it is hard to visualize that human predation has had an impact on the macropod populations of these states in the long term. However, both short term and localized impacts are indicated by the temporary closure of individual kangaroo management zones to harvesting. The difference between the long and short term effects is probably a result of the actual level of harvesting at different times and places, over which must be superimposed widely fluctuating seasonal conditions.

This is quite different to the Tasmanian experience, where the populations of the Forester Kangaroo was brought to very low numbers as early as 1850, supposedly as a result of human predation. It is curious that the Tasmanian result should be so very different to that in both New South Wales and Queensland, particularly when wallaby populations in Tasmania have been subject to very high predation without any apparent impact.

We now turn to examine recent research which may be useful in assessing the impact of the present human predation (harvesting). The most instructive is that research which has been aimed broadly at determining the effects of various levels of harvesting on these large macropods.

Recent studies indicate that kangaroos have a high rate of increase. Caughley (J) *et al.* (1984) found in average rainfall years that kangaroos increased despite commercial culling; a 30-35% increase in Red Kangaroos and 25% in Grey Kangaroos. In above average rainfall this rate can be higher. On properties subject to commercial harvesting it may reach 55% (Bayliss 1985a) or up to 70-80% in a national park subject to no culling (Bayliss 1985b). In western New South Wales, Caughley (J) *et al.* (1984) found that kangaroo numbers did not decline until rainfall was 50-100 mm below average. These figures suggest that kangaroo populations can sustain a cull of 10-20% in average rainfall or better years. During drought with substantial mortality (43% of Reds, Caughley (J) *et al.* 1984; 38% Reds and 44% Western Greys, Grigg 1984; 30% Reds and 67% of Greys, Robertson 1986) this level of cull would help partially alleviate "competition" by more quickly taking those that may have died anyway as a consequence of the drought. During drought, harvest rates are often higher (Kirkpatrick and Amos 1985; Caughley *et al.* 1987) but the effect of this higher rate cannot be separated from the severe environmental factors. With the return of better conditions the high rate of population growth in excess of the culling rate soon would bring about a marked increase in kangaroo numbers.

These studies have been used to develop a more rigorous approach to setting harvesting levels, although in Queensland greater emphasis is placed on population parameters such as species distribution, the age, weight and sex as well as the numbers previously harvested. It is reasonable to expect that harvesting 10% to 20% of the macropod population, which is increasing in an average year by more than 30%, should not result in an impact on the population. In fact, in recent years while the harvest quotas have ranged between 10% and 30%, the actual harvest has generally only been between 10 and 15% of the estimated population (Table 9, Appendix 2 and 3).

Estimation of population size and distribution are also necessary, both to detect trends in the population and to allow the percentage harvest offtake to be converted to an actual number of macropods. The latter has the additional benefit of providing a feedback loop between the population number and the number to be harvested through varying the percentage offtake. On the mainland, population estimates are usually made by aerial surveys which are usually underestimates (particularly in the case of Grey Kangaroos), thus further reducing potential impact. (Correction factors for the estimate of Grey Kangaroo numbers are being currently modified, Dr Jack Giles, pers. comm.). In Tasmania, population trends are monitored through surveys of

Table 9. The relationship between population estimates, quota and harvest figures for four states in 1981 and 1984 for Red and Grey Kangaroos. (Numbers are rounded to the nearest '000, figures in parentheses are percentages of population estimates).

		Year							
		1981				1984			
State	Species	Pop. Est.	Quota	Harvest	% quota harvested	Pop. Est.	Quota	Harvest	% quota harvested
New South Wales	Red	3 836	333 (8.7)	317 (8.3)	95.2	1 663	399 (24.0)	158 (9.5)	39.5
	Grey	2 812	357 (12.7)	171 (6.1)	48.0	1 128	100 (8.9)	64 (5.7)	63.8
Queensland	Red	2 156	450 (20.9)	220 (10.2)	48.9	1 760	250 (14.2)	226 (12.9)	90.5
	Grey	3 132	900 (28.7)	399 (12.7)	44.3	2 372	500 (21.1)	543 (22.9)	108.7
South Australia	Red	1 085	198 (18.2)	164 (15.1)	83.2	710	115 (16.2)	83 (11.8)	72.6
	Grey	256	44 (17.2)	16 (6.4)	37.3	126	22 (17.5)	10 (7.9)	45.0
Western Australia	Red	1 010	160 (15.8)	136 (13.5)	85.0	2 001	140 (7.0)	136 (6.8)	97.1
	Grey	426	88 (20.7)	34 (8.0)	38.6	666	50 (7.5)	37 (5.6)	74.0
Total	Red	8 087	1 141 (14.1)	838 (10.4)	73.4	6 134	904 (14.7)	603 (9.8)	66.7
	Grey	6 626	1 389 (21.0)	620 (9.4)	44.7	4 292	672 (15.7)	654 (15.2)	97.3
		14 713	2 530 (17.2)	1 458 (9.9)	57.6	10 426	1 576 (15.1)	1 257 (12.1)	79.8

shooters aimed at indicating the status of the wallaby populations.

It is too early to be able to assess the effect of this new approach (since 1975) to macropod harvesting, particularly when widely fluctuating seasonal conditions probably have much more effect naturally than does the harvest. To achieve the potential of the system, it is probably important that it be applied within small geographic units, certainly smaller than whole states. However, there is every reason to expect that it will result in minimum impact on the populations generally but provide some mitigation of pasture/crop losses to graziers and farmers. Certainly it is a more rigorous and conservative approach than has been applied in the past 100 years in which the larger macropod species have survived with little or no long term impact.

(ii) *Dingo:* Since 1971 the diet of the dingo *(Canis familiaris dingo)* has been studied across Australia with major studies in most mainland states. The diet has been reported to contain macropods as their most frequently occurring prey item except for some drier inland studies where rabbits are an alternative prey source (Coman 1972; Newsome *et al.* 1973; Corbett 1974; Whitehouse 1977; Shepherd 1981; Newsome *et al.* 1983; Robertshaw and Harden 1985a and 1986; Thomson 1986; Corbett and Newsome 1987).

Thus macropods are a major source of prey for dingoes (see Table 3) although there is some debate about what consitiutes "staple" prey for dingoes. Newsome *et al.* (1983) and Corbett and Newsome (1987) believed that dingoes preferred small and medium-sized prey, probably because they were easily caught. This view was not shared by Robertshaw and Harden (1985a, 1986) whose studies spanned the years from 1969 to 1974 and 1979 to 1980 and concluded that macropods were important prey for dingoes. Combined with unpublished data, a total of 6 581 scats were examined between 1969 and 1985, and macropods were found to be the most frequently occurring prey in each year. The Swamp Wallaby *(Wallabia bicolor)* was the most frequently occurring prey species in 10 of these years. This dominance by the larger native prey in the diet, in particular the macropods, was found in nine out of 10 sites examined in an unpublished study between Barrington Tops and the Queensland border.

However, the impact of this predation on the macropods is not clearly understood. Shepherd (1981) found that 80 of 83 Red Kangaroos killed by dingoes were juvenile, with a weight range (mean 15 kg) that would include the adult weight of wallabies from eastern Australia. He postulated that if this selectivity for juveniles and younger adult females was widespread the impact of dingo predation would primarily be on the recruitment into the adult population. Robertshaw and Harden (1985b) also suggested that dingo predation could cause a high mortality in juvenile macropods reducing the recruitment rate. This was further examined in two sites (Robertshaw and Harden 1986) differing in relative macropod prey availability. In both sites there was a high mortality of pouch young and young at foot which resulted in an age structure suggesting a declining wallaby population (see Fig. 1), and wallaby numbers were recorded as declining. One outcome of the higher predation level was a breakdown of synchronization in a seasonal peak of births as females ejected pouch young which were heavy enough to become a burden when pursued (>c. 800 g for *W. bicolor*). These offspring were soon replaced, as 96% of sexually mature females carried a blastocyst. With a high incidence of this behaviour the output of young would become continuous instead of occurring in a spring-summer peak. The changes in breeding patterns were reflected in differences in the number of ovulations per female and differences in testicular and epididymi weights between wallaby populations.

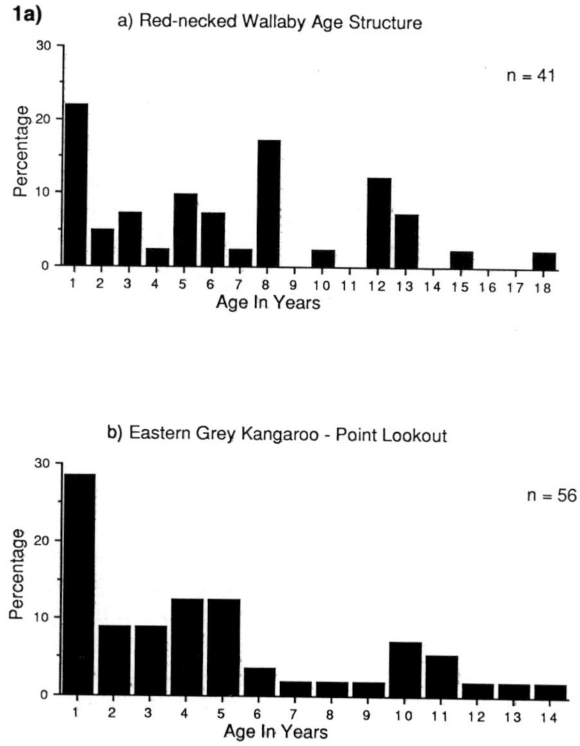

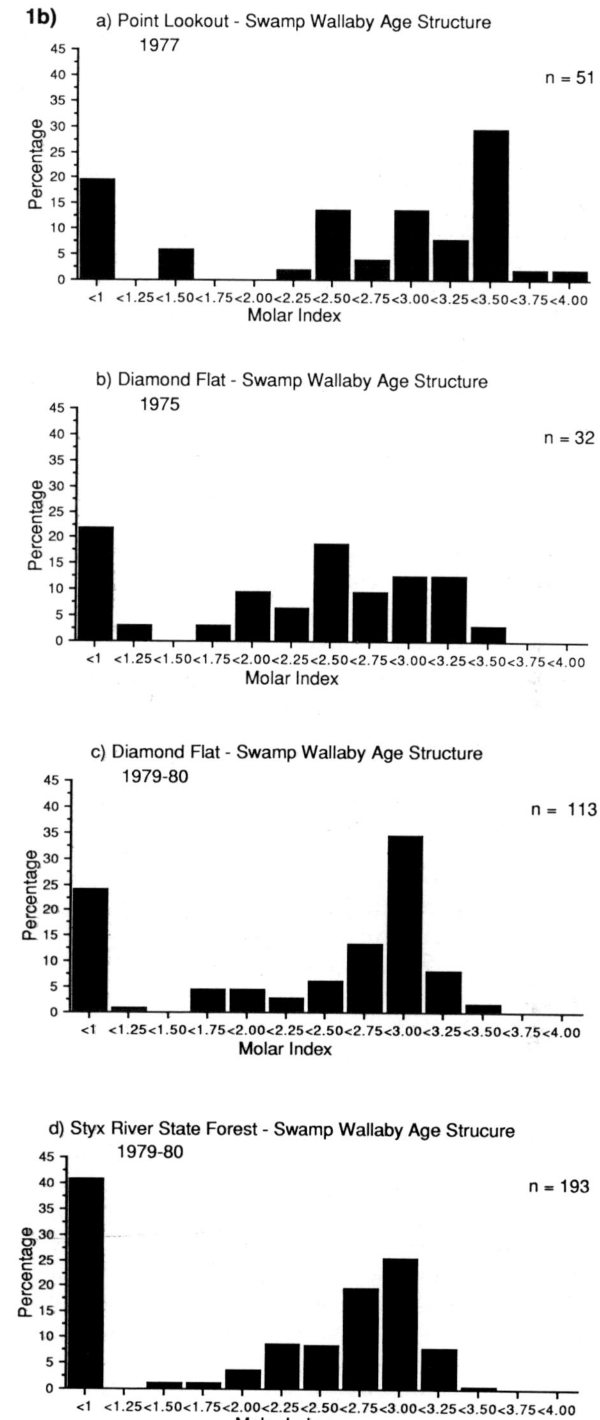

Fig. 1a. The age structures of Red-necked Wallabies and Eastern Grey Kangaroos shot at Point Lookout, New South Wales, 1977. (Year 1 includes pouch young.) *Fig. 1b.* The age structure of Swamp Wallaby populations. (Year 1 <1 contains pouch young.)

While it is likely that dingoes do affect the recruitment rate of macropods, no study has demonstrated that dingoes control macropod populations. Such control has been implied to explain the contrast in density of Red Kangaroos on the two sides of the New South Wales and Queensland border (dingo) fence (Jarman and Denny 1976; Caughley *et al.* 1980). In northeastern New South Wales the Swamp Wallaby is the main macropod prey species and its occurrence in the diet is proportionally higher than would be if based on the relative number of observed macropods (Robertshaw and Harden 1986). Whether this results from a marked preference for Swamp Wallaby or is a result of the dingo's ability to catch this species, the dingo has had a marked impact on Swamp Wallaby numbers. Between 1972 and 1975 wallaby numbers remained similar but dingo numbers increased. There were increases in the proportion of older macropods consumed, in the utilization of their carcasses, and in the percentage occurrence of macropod in the diet (52% to 73%). From 1976 to 1985 further increases in the utilization of older macropods took place. Although there was an initial decline in dingo numbers from 1978 to 1980 macropod prey availability declined further as a result of a marked decline in Swamp Wallaby numbers after 1978. The percentage occurrence of Swamp Wallaby fell by more than half during this period (48% to 17%) but the occurrence of macropod only fell from 64% to 47% as Parma Wallaby (4% to 20%) became a more important prey item.

However, while dingo predation had an impact on the Swamp Wallaby population, it was not possible to demonstrate that dingo predation controlled or regulated their numbers.

We also have other unpublished information in the same study area that suggests an impact by dingoes on other macropod species. Between 1982 and 1985 the small and isolated populations of Grey Kangaroo and Red-necked Wallaby in the south-east of the study area appeared to have been eliminated by dingoes. The localized impact on these populations may be an example of how dingo populations could affect the distribution of macropods in circumstances where the prey population was small. The age structures of shot samples of Grey Kangaroos, Red-necked Wallabies and Swamp Wallabies taken in 1977-78 from the north of the study area were also all "abnormal" or not stable (see Fig. 1). As drought in this area has not been observed to affect macropod numbers, we suspect the perturbations in these age structures may have been caused by changing predation levels. The dingo was the main predator as little shooting occurred in this site which was adjacent to a national park.

Corbett and Newsome (1987) found that Red Kangaroos were more vulnerable to predation during droughts as a result of aggregation around watering holes and the remaining feed and that the post-drought populations remained low relative to the pre-drought levels. At one site eight years after the drought, the population had not recovered as dingoes had become more common as rabbits became abundant. This partial control required the presence of alternative prey before the dingo could reduce the density of the prey species without a negative feedback on its own numbers (see Caughley *et al.* 1980). Further, it may be necessary for the numbers of prey to be lowered relative to the number of predators before control (or a major impact) of the prey population could result, especially if the predator numbers are modified by variables not affecting its prey. These variables may include a stable prey base and not just a temporary alternative or man's attempts at controlling dingo numbers. Thus drought may have reduced the kangaroo numbers sufficiently for dingoes to be able to keep Red Kangaroo numbers lowered while using rabbits as an alternative prey.

In conclusion, there seems reasonable evidence that dingo predation does have an impact on macropod populations. However, its long term effect and the possibility of a regulatory effect is far from clear and requires further work. Such work would need to manipulate dingo and macropod numbers experimentally and, possibly, the availability of alternative prey species. Extension of the work and ideas of Jarman and Denny (1976), Caughley *et al.* (1980) and Shepherd (1981) using the South Australian, New South Wales and Queensland border fence area may prove successful as some experimental blocks exist already and the prey base is relatively simple in comparison to the eastern seaboard.

CONCLUSIONS

While the impact of seasonal conditions is demonstrably greater than that of predation, a number of predators, both native and introduced, take macropods as prey. However, for the great majority of predators, macropods are only occasional prey and the predation has either no or at the most only local impact on macropod populations. Today, while the impact of fox predation is undetermined, only two species, man and the dingo, can be considered important predators of macropods. Both take a wide range of species continually over a large geographic area. Predation by man is largely confined to the common species when they are abundant and at odds with agricultural practices.

In the case of man, the historical evidence is that predation may have resulted in both local and temporal impacts but has not had a significant impact on the viability of macropod populations (the Tasmanian Forester Kangaroo was almost an exception). Recent changes in the methods of determining the level of predation (cull) have resulted in very conservative levels which can be expected to minimize future impact. Importantly, these changes also provide a means of monitoring population size and have an inbuilt feedback loop.

Like man, dingo predation also appears to result in local and temporal impact on macropod populations. The degree to which these impacts are long term or regulatory is unclear.

ACKNOWLEDGEMENTS

We wish to thank the various state authorities who initially collected and collated much of the "kangaroo management" data. In particular we would like to thank Mark Fletcher, Malcolm Forbes and Karen Upton (Australian National Parks and Wildlife Service); Dr Tom Kirkpatrick (Queensland National Parks and Wildlife Service); and Leighton Llewellyn, Nick Sheppard and Kath Everleigh (New South Wales National Parks and Wildlife Service).

For typing and word-processing we would like to thank Adriana Ferraro. The Department of Zoology of the University of New England provided many facilities for which we are most grateful. We also express out gratitude for critical comments on this paper provided by Dr Jack Giles and an anonymous referee.

REFERENCES

ARCHER, M., 1984a. The Australian marsupial radiation. Pp. 633-808 *in* Vertebrate zoogeography and evolution in Australasia ed. by M. Archer and G. Clayton. Hesperian Press: Perth.

ARCHER, M., 1984b. Effects of humans on the Australian vertebrate fauna. Pp. 151-61 *in* Vertebrate zoogeography and evolution in Australasia ed. by M. Archer and G. Clayton.

ARCHER, M. AND DAWSON, L., 1982. Revision of marsupial lions of the genus *Thylacoleo* Gerais (Thylacoleonidae, Marsupialia) and Thylacoleonid evolution in the late cainozoic. Pp. 477-94 *in* Carnivorous marsupials ed. by M. Archer. Royal Zoological Society, New South Wales: Sydney.

AUSTRALIAN NATIONAL PARKS AND WILDLIFE SERVICE, 1984. Kangaroo Management programmes of Australian States, Canberra.

AUSTRALIAN NATIONAL PARKS AND WILDLIFE SERVICE, 1988. Kangaroos in Australia: Conservation Status and Management. Occasional Paper No. 14, Pp. 1-45.

AUSTRALIAN PARLIAMENT HOUSE OF REPRESENTATIVES SELECT COMMITTEE ON WILDLIFE CONSERVATION, 1971. Interim Report Conservation and Commercial Exploitation of Kangaroos. Commonwealth Government Printing Office: Canberra.

BAYLISS, P., 1985a. The population dynamics of red and western grey kangaroos in arid New South Wales, Australia. I. Population trends and rainfall. *J. Animal Ecol.* **54:** 111-25.

BAYLISS, P., 1985b. The population dynamics of red and western grey kangaroos in arid New South Wales, Australia. II. The numerical response function. *J. Animal. Ecol.* **54:** 127-35.

BAYLISS, P. AND GILES, J., 1985. Factors affecting the visibility of kangaroos counted during aerial surveys. *J. Wildl. Manage.* **49:** 686-92.

BOWDLER, S., 1977. The coastal colonization of Australia. Pp. 205-46 *in* Sunda and Sahul: Prehistoric Studies in southeast Asia, Melanesia and Australia ed. by J. Allen, J. Golson and R. Jones. Academic Press: London.

BROOKER, M. G. AND RIDPATH, M. G., 1980. The diet of the wedge-tailed eagle, *Aquila audax*, in Western Australia. *Aust. Wildl. Res.* **7:** 433-52.

BUCHMANN, O. L. K. AND GUILER, E. R., 1977. Behaviour and ecology of the Tasmanian devil, *Sarcophilus harrisii.* Pp. 155-68 *in* The Biology of Marsupials ed. by B. Stonehouse and D. Gilmore. Macmillan: London.

BURBIDGE, A. A., 1983. Burrowing Bettong, *Bettongia lesueur*. Pp. 187-88 *in* The Complete Book of Australian Mammals ed. by R. Strahan. Angus & Robertson Publishers: London.

BURBIDGE, A. A. AND JOHNSON, P. M., 1983. Spectacled Hare-wallaby, *Lagorchestes conspicillatus.* Pp. 197-98 *in* The Complete Book of Australian Mammals ed. by R. Strahan. Angus & Robertson Publishers: London.

CASE, J. A., 1985. Differences in prey utilization by Pleistocene marsupial carnivores, *Thylacoleo carnifex* (Thylacoleonidae) and *Thylacinus cyancephalus* (Thylacinidae). *Aust. Mammal.* **8:** 45-52.

CAUGHLEY, G., GRIGG, G. C., CAUGHLEY, J. AND HILL, G. J. E., 1980. Does dingo predation control the densities of kangaroos and emus? *Aust. Wildl. Res.* **7:** 1-12.

CAUGHLEY, G., SHEPHERD, N. AND SHORT, J., 1987. Kangaroos: their ecology and management in the sheep rangelands of Australia. Pp. 254. Cambridge University Press: Sydney.

CAUGHLEY, G., BROWN, B., DOSTING, N. AND GRICE, D., 1984. The grey kangaroo overlap zone. *Aust. Wildl. Res.* **11:** 151-52.

CAUGHLEY, G., GRIGG, G. C. AND SHORT, J., 1983. How many kangaroos? *Search* **14:** 151-52.

CAUGHLEY, G., SINCLAIR, R. G. AND WILSON, G. R., 1977. Numbers, distribution and harvesting rate of kangaroos on the inland plains of New South Wales. *Aust. Wildl. Res.* **4:** 99-108.

CAUGHLEY, J., BAYLISS, P. AND GILES, J., 1984. Trends in kangaoo numbers in western New South Wales and their relation to rainfall. *Aust. Wildl. Res.* **11:** 415-22.

CHRISTENSEN, P. E. S., 1980. The biology of *Bettongia penicillata* Gray, 1837, and *Macropus eugenii* (Desmarest, 1817) in relation to fire. *Forests Department of Western Australia Bull.* **91:** 1-90.

CHRISTENSEN, P. E. S., 1983. Western Brush Wallaby, *Macropus irma.* P. 235 *in* The Complete Book of Australian Mammals ed. by R. Strahan. Angus & Robertson Publishers: London.

COMAN, B. J., 1972. Helminth parasites of the dingo and feral dog in Victoria with some notes on the diet of the host. *Aust. Vet. J.* **48:** 456-61.

COMAN, B. J., 1973. The diet of red foxes, *Vulpes vulpes* L., in Victoria. *Aust. J. Zool.* **21:** 391-401.

COMAN, B. J. AND BRUNNER, H., 1972. Food habits of the feral house cat. In Victoria. *J. Wildl. Manage.* **36:** 348-53.

COPLEY, P. B., 1983. Studies on the Yellow-footed Rock-wallaby, *Petrogale xanthopus* Gray (Marsupialia: Macropodidae). I. Distribution in South Australia. *Aust. Wildl. Res.* **10:** 47-61.

CORBETT, L. K., 1985. Morphological comparisons of Australian and Thai dingoes: a reappraisal of dingo status, distribution and ancestry. *Proc. Ecol. Soc. Aust.* **13:** 277-91.

CORBETT, L. K. AND NEWSOME, A. E., 1987. The feeding ecology of the dingo. III. Dietary relationships with widely fluctuating prey populations in arid Australia: an hypothesis of alternation of predation. *Oecologia* **74:** 215-27.

CROFT, D. B., 1987. Socio-ecology of the Antilopine Wallaroo, *Macropus antilopinus,* in Northern Territory with observations on sympatric *M. robustus woodwardii* and *M. agilis. Aust. Wildl. Res.* **14:** 243-55.

CROFT, J. D. AND HONE, L. J., 1978. The stomach contents of foxes, *Vulpes vulpes,* collected in New South Wales. *Aust. Wildl. Res.* **5:** 85-92.

DENNY, M., 1981. Kangaroos: An historical perspective. Pp. 36-45 *in* Parks and Wildlife: Kangaroos ed. by C. Haig. New South Wales National Parks and Wildlife Service: Sydney.

DORTCH, C., 1979. 33 000 year old stone and bone artefacts from Devil's Lair, Western Australia. *Records of the Western Australian Museum* **7:** 329-67.

EALEY, E. H. M., 1962. Biology of the euro, *Macropus robustus corvinus* Thomas with emphasis on its interaction with sheep in the Pilbera District of Western Australia. Ph.D. Thesis, University of Western Australia: Perth.

FINCH, M. E., 1982. The discovery and interpretation of *Thylacoleo carnifex*. (Thylacoleonidae, Marsupialia). Pp. 537-51 *in* Carnivorous marsupials ed. by M. Archer. Royal Zoological Society, New South Wales: Sydney.

FLANNERY, T. F., 1984. Kangaroos: 15 million years of Australian bounders. Pp. 817-35 *in* Vertebrate Zoogeography and Evolution in Australasia ed. by M. Archer and G. Clayton. Hesperian Press: Perth.

FLOOD, J., 1983. Archaeology of the Dreamtime. Collins: Sydney. Pp. 146-59.

FRITH, H. J. AND CALABY, J. H., 1969. Kangaroos. F. W. Cheshire: Melbourne.

GREEN, K. AND OSBORNE, W. S., 1981. The diet of foxes, *Vulpes vulpes* (L.), in relation to abundance to prey above the winter snowline in New South Wales. *Aust. Wildl. Res.* **8:** 349-60.

GRIGG, G. C., 1984. Roo harvesting — are kangaroos really under threat? *Aust. Nat. Hist.* **21:** 123-29.

GRIGG, G. C., BEARD, L. A., CAUGHLEY, G., GRICE, D., CAUGHLEY, J. A., SHEPPARD, N., FLETCHER, M. AND SOUTHWELL, C., 1985. The Australian kangaroo populations, 1984. *Search* **16:** 277-79.

HORNSBY, P. E., 1982. Predation of the Euro *Macropus robustus* (Marsupialia Macropodidae) by the European Fox, *Vulpes vulpes* (Placentalia: Canidae). *Aust. Mammal.* **5:** 225-27.

HORTON, D. R., 1980. A review of the extinction question. Man, climate and megafauna. *Archaeology and Physical Anthropology in Oceania* **15:** 86-97.

HULL, C., 1985. The diet of the wedge-tailed eagle, *Aquila audax*, breeding near Melbourne. *Corella* **10:** 21-24.

JARMAN, P. J. AND DENNY, M. J. S., 1976. Red kangaroos and landuse along the New South Wales, Queensland and South Australian borders. Pp. 55-67 *in* Proceedings of a workshop on agriculture, forestry and wildlife: Conflict or Co-existence. University of New England: Armidale.

JARMAN, P. J., JOHNSON, C. N., SOUTHWELL, C. J. AND STUART-DICK, R., 1987. Macropod studies at Wallaby Creek. I. The area and animals. *Aust. Wildl. Res.* **14:** 1-14.

JOHNSON, K. A., 1977. Ecology and management of the red-necked pademelon, *Thylogale thesis,* on the Dorrigo Plateau of north-eastern New South Wales. Ph.D. Thesis, University of New England, Armidale, New South Wales.

JOHNSON, K. A. AND BURBIDGE, A. A., 1988. Australian macropods: Status, causes of decline and future research. Abstract, *Aust. Mammal Soc. Newsletter (Winter)* P. 58.

JONES, E. AND COMAN, B. J., 1981. Ecology of the feral cat, *Felis catus* (L.), in south-eastern Australia. I. Diet. *Aust. Wildl. Res.* **8:** 537-47.

KAUFMANN, J. H., 1974. Social ethology of the whiptail wallaby, *Macropus parryi* in north-eastern New South Wales, Australia. *Animal Behaviour* **22:** 281-369.

KEAST, A., 1982. The Thylacine (Thylacinidae Marsupialia): how good a pursuit carnivore? Pp. 675-84 *in* Carnivorous marsupials ed. by M. Archer. Royal Zoological Society, New South Wales: Sydney.

KINNEAR, J., ONUS, M. AND BROWILOW, B., 1984. Foxes, feral cats and rock wallabies. *Swans* **14:** 3-8.

KIRKPATRICK, T. H., 1985. Biology for management. Pp. 135-60 *in* The kangaroo keepers ed. by H. J. Lavery. University of Queensland Press: St. Lucia.

KIRKPATRICK, T. H. AND AMOS, P. J., 1985. Kangaroo industry. Pp. 75-102 *in* The kangaroo keepers ed. by H. J. Lavery. University of Queensland Press: St Lucia.

LAVERY, H. J. AND TIERNEY, P. J., 1985. Scarcity and extinction. Pp. 25-52 *in* The kangaroo keepers ed. by H. J. Lavery. University of Queensland Press: St. Lucia.

LEOPOLD, A. S. AND WOLFE, T. O., 1970. Food habits of nesting wedge-tailed eagles, *Aquila audax,* in south-eastern Australia. *C.S.I.R.O. Wildl. Res.* **5:** 1-17.

LE SOUEF, A. S. AND BURRELL, H., 1926. The wild animals of Australasia, with a chapter on bats by Ellis LeG. Troughton. George Harrap: London.

LIM, L., ROBINSON, A. AND COPLEY, P., 1981. Rock wallabies, Genus *Petrogale.* Pp. 21-26 *in* Parks and Wildlife: Kangaroos ed. by C. Haig. New South Wales National Parks and Wildlife Service: Sydney.

MCINTOSH, D. L., 1963. Food of the fox in the Canberra district. *C.S.I.R.O. Wildl. Res.* **8:** 1-20.

MCNAMARA, K. J. AND PRINCE, R. I. T., 1986. Kangaroo Management in Western Australia. Wildlife Management Programme No. 3. Department of Conservation and Land Management.

MAINS, A. R., 1978. Ecophysiology: Towards an understanding of Late Pleistocene marsupial extinction. Pp. 169-84 *in* Biology and Quaternary Environments ed. by D. Walker and J. C. Guppy. Australian Academy of Sciences: Canberra.

MARTENSZ, P. N., 1971. Observations on the food of the fox, *Vulpes vulpes* (L.), in an arid environment. *C.S.I.R.O. Wildl. Res.* **16:** 73-75.

MAYNES, G. M. AND SHARMAN, G. B., 1983a. Brush-tailed rock-wallaby, *Petrogale penicillata.* Pp. 211-12 *in* The Complete Book of Australian Mammals ed. by R. Strahan. Angus & Robertson Publishers: London.

MAYNES, G. M. AND SHARMAN, G. B., 1983b. Yellow-footed rock-wallaby, *Petrogale xanthopus.* Pp. 217-18 *in* The Complete Book of Australian Mammals ed. by R. Strahan. Angus & Robertson Publishers: London.

NEWSOME, A. E., CORBETT, L. K., BEST, L. W. AND GREEN, B., 1973. The dingo. *A.M.R.C. Review No.* **14:** 1-11.

NEWSOME, A. E., CATLING, P. C. AND CORBETT, L. K., 1983. The feeding ecology of the dingo II. Dietary and numerical relationships with fluctuating prey populations in south-eastern Australia. *Aust. J. Ecol.* **8:** 345-66.

NEW SOUTH WALES NATIONAL PARKS AND WILDLIFE SERVICE, 1988. Kangaroo management in New South Wales including a summary of information relevent to kangaroo management in New South Wales, Sydney.

POOLE, W. E., 1984. Management of kangaroo harvesting in Australia (1984). *Aust. Nat. Parks and Wildl. Service. Occ. Pap. No.* **9:** 1-25.

PRINCE, R. I. T., 1984. Exploitation of kangaroos and wallabies in Western Australia. I. A review to 1970, with special emphasis on the Red and Grey Kangaroos. *Wildl. Res. Bull. WA* **13:** 1-78.

ROBERTSHAW, J. D., 1983. Mammal survey of the Apsley pump storage project areas. Report to Electricity Commission of New South Wales, 1-25.

ROBERTSHAW, J. D. AND HARDEN, R. H., 1985a. The ecology of the dingo in north-eastern New South Wales. II. Diet. *Aust. Wildl. Res.* **12:** 39-50.

ROBERTSHAW, J. D. AND HARDEN, R. H., 1985b. The ecology of the dingo in north-eastern New South Wales III. Analysis of macropod bone fragments in dingo scats. *Aust. Wildl. Res.* **12**: 163-71.

ROBERTSHAW, J. D. AND HARDEN, R. H., 1986. The ecology of the dingo in north-eastern New South Wales IV. Prey selection by dingoes and its effect on the major prey species, the Swamp wallaby, *Wallabia bicolor* (Desmarest). *Aust. Wildl. Res.* **13**: 141-63.

ROBERTSON, G., 1986. The mortality of kangaroos in drought. *Aust. Wildl. Res.* **13**: 349-54.

ROFF, C. AND KIRKPATRICK, T., 1962. The kangaroo industry in Queensland. *Qld. J. Agric. Sci.* **19**: 385-401.

ROLLS, E. C., 1983. They all ran wild: the animals and plants that plague Australia. (Revised Ed.) Angus & Robertson: Sydney.

ROSE, R. W., 1986. The habitat, distribution and conservation status of the Tasmanian Bettong, *Bettongia gaimardii* (Desmarest). *Aust. Wildl. Res.* **13**: 1-6.

RYAN, G. E. AND CROFT, J. D., 1974. Observations on the food of the fox, *Vulpes vulpes* (L.) in Kinchega National Park, Menindee, New South Wales. *Aust. Wildl. Res.* **1**: 89-94.

SANSON, G. D., 1983. Nabarlek, *Peradorcas concinna*. P. 223 *in* The Complete Book of Australian Mammals ed. by R. Strahan. Angus & Robertson Publishers: London.

SHEPHERD, N., 1981. Predation of Red Kangaroos, *Macropus rufus,* by the dingo, *Canis familiaris dingo* (Blumenbach), in north-western New South Wales. *Aust. Wildl. Res.* **8**: 255-62.

SHORT, J. AND BAYLISS, P., 1985. Bias in aerial survey estimates of kangaroos. *J. Appl. Ecol.* **22**: 435-47.

SMITH, M., 1982. Review of the Thylacine (Marsupialia Thylacinidae) Pp. 237-53 *in* Carnivorous marsupials ed. by M. Archer. Royal Zoological Society, New South Wales: Sydney.

SMITH, M. J., 1983. Tammar Wallaby, *Macropus eugenii.* Pp. 232-33 *in* The Complete Book of Australian Mammals ed. by R. Strahan. Angus & Robertson Publishers: London.

STRAHAN, R., 1983. The Complete Book of Australian Mammals. Angus & Robertson: Sydney. Pp. 175-267.

THOMSON, P., 1986. The effectiveness of aerial baiting for the control of dingoes in north-western Australia. *Aust. Wildl. Res.* **13**: 165-76.

TRIGGS, B., BRUNNER, H. AND CULLEN, J. M., 1984. The food of fox, dog and cat in Croajingalong National Park, south-eastern Victoria. *Aust. Wildl. Res.* **11**: 491-99.

WAKEFIELD, N. A., 1963. Notes on rock-wallabies. *Vic. Nat.* **77**: 322-32.

WALLIS, R. L. AND BRUNNER, H., 1987. Changes in mammalian prey of foxes, *Vulpes vulpes* (Carnivora Lanidae), over 12 years in a forest park near Melbourne, Victoria. *Aust. Mammal.* **10**: 43-44.

WELLS, R. T., HORTON, D. R. AND ROGERS, P., 1982. *Thylacoleo carnifex* Owen (Thylacoleonidae): Marsupial carnivore? Pp. 573-86 *in* Carnivorous marsupials ed. by M. Archer. Royal Zoological Society, New South Wales: Sydney.

WHITE, J. P. AND O'CONNELL, J. F., 1982. A Prehistory of Australia, New Guinea and Sahul. Academic Press: Sydney.

WHITEHOUSE, S. J. O., 1977. The diet of the dingo in Western Australia. *Aust. Wildl. Res.* **4**: 145-50.

APPENDIX 1

The number of macropods commercially harvested in the various Australian states. The figures are derived from various sources. For the years prior to 1975: Queensland, Kirkpatrick and Amos (1985); New South Wales, Caughley *et al.* (1987), Annual Reports of N.P.W.S.; Western Australia, Prince (1984), McNamara and Prince (1986), Annual Reports to Parliament; South Australia, Annual Reports to Parliament; Tasmania, Australian National Parks and Wildlife Service (1984). After 1975 the figures are derived from a table "kangaroos killed by state and species for the period 1975-1986" supplied by Australian National Parks and Wildlife Service and 1987 figures supplied by the various state authorities. (* closed season except for King and Flinders Islands, † change to calendar year).

Year	Queensland	New South Wales	Western Australia	South Australia	Tasmania	Victoria	Total
1944			92 249		311 026		
1945			133 796		150 705		
1946			168 183		311 280		
1947			152 178		48 952*		
1948			182 618		46 741*		
1949			168 153		126 580		
1950			158 517		61 618		
1951			154 517		121 206		
1952			106 630		99 079		
1953			90 879		153 203		
1954	218 459		>70 700		43 253		
1955	305 616		>51 800		18 856		
1956	361 462		>68 000		36 449		
1957	631 034		>46 600		7 808		
1958	295 820		>29 000		5 385		
1959	1 006 919		>62 500		39 753		
1960	769 948	76 000	>27 400		65 604		
1961	471 640	86 000	>29 300		34 203		
1962	548 771	90 000	>32 100		23 941		
1963	642 063	120 000	>64 500		51 931		
1964	1 118 813	177 000	146 800		28 065		
1965	1 168 887	270 000	142 500		32 856		
1966	924 629	252 000	187 000	146 529	48 080		1 558 238
1967	871 390	170 000	192 500	129 050	56 430		1 419 870
1968	862 981	71 497	199 000	80 999	61 916		1 276 393
1969	1 116 065	44 692	425 000	82 966	50 982		1 691 705
1970	900 202	77 746	292 000	98 103	56 805		1 335 856
1971	581 679	80 156	184 000	80 395	31 035		957 265
1972	1 073 732	137 000	214 000	18 500†	37 247		1 480 479
1973	457 720	132 394	132 500	67 610	29 442		819 666
1974	317 479	94 815	140 000	40 597	7 620		601 011
1975	497 138	122 868	122 000	43 677	6 601		792 416
1976	620 555	96 738	162 000	34 524	21 426		935 205
1977	588 404	167 208	166 000	60 698	45 471		1 027 773
1978	600 382	220 000	157 500	135 996	100 803		1 214 681
1979	1 314 260	520 000	195 500	93 234	244 918		2 367 912
1980	943 099	619 023	162 930	148 329	220 226		2 093 607
1981	679 384	488 647	175 000	193 326	108 000	<1 000	1 645 357
1982	1 136 798	664 342	177 536	229 453	77 000	<1 000	2 286 129
1983	877 350	400 477	202 402	155 785	110 000		1 746 014
1984	860 670	222 015	178 500	99 274	118 000		1 478 459
1985	1 126 116	326 028	204 325	114 643	92 000		1 863 112
1986	1 343 511	444 509	194 629	128 923	60 000		2 171 572
1987	1 231 889	473 454					

APPENDIX 2

Population estimates, quotas and harvest figures for New South Wales between 1975 and 1987. (Table based on "Table 2" in New South Wales Kangaroo Management Programme 1988. Estimates from 1975 to 1983 are based on surveys of blocks of approximately 16 000 sq. km. selected at random in 1977. Estimates for 1984 to 1987 are based on surveys conducted across virtually (90%) the whole of the commercial harvesting area).

		Maximum harvest (quota) based on previous years winter population estimate		Actual commercial cull	
Year	Population Estimate	Number	% of previous years population	Number	% of previous years population
1975	3 653 000	212 900		123 700	
1976		319 000	8.74	96 000	2.65
1977	4 699 000	321 500		167 200	
1978	4 383 000	345 000	7.34	220 000	4.68
1979	4 288 000	645 000	14.71	520 000	11.86
1980	6 174 000	645 000	15.04	619 023	14.44
1981	7 046 000	694 500	11.24	488 647	7.91
1982	9 400 000	843 000	11.96	664 342	9.43
1983	5 500 000	843 000	8.97	400 477	4.26
1984	2 738 000	500 000	9.09	229 484	4.17
1985	4 155 000	300 000	10.96	326 028	11.91
1986	4 662 000	577 000	13.89	444 509	10.70
1987	5 425 000	577 000	12.38	473 454	10.16

APPENDIX 3

Population estimates, quotas and harvest figures for Red and Western Grey Kangaroos in the pastoral zone of South Australia between 1978 and 1984. Sources were for population estimates Grigg (1984) and for quotas and harvest figures Australian National Parks and Wildlife Service.

		Quota		Harvest		
Year	Pop. Est.	Nos	% previous estimate	Nos	% previous estimate	% Quota harvested
1978	1 202 600	80 000		135 996		
1979	1 216 700	104 000	8.6	93 234	7.8	89.6
1980	1 236 900	200 000	16.4	148 329	12.2	74.2
1981	2 275 600	250 000	20.2	193 326	15.6	77.3
1982	1 496 300	400 000	17.6	229 453	10.1	57.4
1983	907 000	300 000	20.0	155 785	10.4	51.9
1984		143 000	15.8	99 275	10.9	69.4

APPENDIX 4

DINGO ATTACKS

Descriptions of attacks on macropods by dingoes have not been widely reported. Newsome *et al.* (1983) gave accounts of macropods avoiding dingoes by taking refuge in water. Red Kangaroos and Euros have been reported anecdotally to drown cattle dogs by standing waist-deep in dams and holding the swimming dog under the water with the forearms. Newsome *et al.* (1983) state that a single dingo was seen to chase, catch and kill a Swamp Wallaby.

Personal observations also suggest that single dingoes are quite capable of killing large adult Swamp Wallabies and adult female Eastern Grey Kangaroos. On three occasions Swamp Wallabies were observed being chased and killed. The dingo would run to one side to kill the wallaby with a bite to the neck. There was little evidence of external damage but most of the neck vertebrae were crushed. On one of these occasions a Swamp Wallaby was seen moving rapidly down a ridge through a mob of Grey Kangaroos. Within seconds, but with the Swamp Wallaby out of sight, a dingo ran head to the ground along the same course taken by the Swamp Wallaby. Unperturbed by the kangaroos, the dingo went after the wallaby into a swampy region. Five minutes later the dingo returned back up the ridge to be seen 10 minutes later coming down a transverse ridge straight for the exhausted wallaby which was initially looking in the opposite direction. The pair then went back up the ridge only to come back down within a minute with the dingo gaining on the wallaby. The dingo killed the wallaby before they reached the bottom. This pursuit probably lasted over half an hour and included the dingo breaking off the direct chase to circle around and come at the wallaby from the other side. We have not seen a dingo harrass a female wallaby with a large pouch young and observe the female to eject the young but we have elicited the response on three occasions ourselves.

A few sightings were made where dingoes tried unsuccessfully to ambush Eastern Grey Kangaroos. One dingo was seen to stalk the feeding Greys and once close to them would make a short dash to drive the Greys for cover where another dingo was seen waiting. On one occasion a fortuitous alarm from a Masked Lapwing (Spur-winged Plover) saved the would-be victims before the ambushers were set. On a successful occasion, a Grey Kangaroo was seen in flight only to be attacked suddenly by a dingo from in front and slightly to one side. The impact of the two brought the kangaroo down with the dingo biting its neck. Before the kangaroo was still, another dingo appeared on the track from behind.

Habitat associations of Australian Macropodoidea

A. POPLE[1]

Habitat associations of the Macropodoidea can be viewed at two scales, a biogeographic scale and the scale of a local population. In most studies, habitat relationships have been studied at the latter scale. Time scales have included year to year variation in distribution and abundance as well as changes in habitat use during the day and between seasons. In this paper, habitat associations of macropods are briefly compared at both spatial scales with reference to their status of abundance. Four key factors – food, shelter, water and predation – appear to determine the distribution and abundance of most macropods. The relative importance of each varies between species and also within the range of a species. Heterogeneity has been found to be an important attribute of the habitat of many species.

INTRODUCTION

WITHIN Australia, macropods occur in most major terrestrial habitats (Frith and Calaby 1969). As a result, making generalizations about habitat associations of nearly 50 species appears quite difficult. Descriptions of these habitats and how they are utilized have been made at various temporal and spatial scales. At a broad, biogeographic scale, one may be interested simply in a species' range or in its distribution within that range. These distributions are not necessarily static and may change from year to year. At the finer scale of a local population, some individuals may use different habitats in different seasons and even at different times of the day. Other members of the same population may or may not show the same habitat associations. The scale studied will depend ultimately on the questions being asked.

This paper is a general discussion of habitat associations in Australian macropods, describing relationships at various spatial scales and how they vary temporally. The term "habitat associations" is used to incorporate both broad and fine scale studies of macropod habitat. Using this term also avoids misinterpretation of words like "preference" or "selection". Though macropods are also found in New Guinea, and have been introduced to areas outside Australia and New Guinea, there is little data on their habitat associations there. These areas have therefore been excluded from this discussion.

BROAD SCALE: THE BIOGEOGRAPHIC LEVEL

Zoogeography

At the broadest scale, that of the whole continent, one may simply wish to know the range of a particular species. As an extension, the range of a species can be related to various aspects of the environment. This range may be set by the availability of suitable habitat or the lack of dispersal to suitable areas outside this range.

Caughley *et al.* (1987) have related the range of the three largest species of macropods, the Red Kangaroo, *Macropus rufus,* the Western Grey Kangaroo, *M. fuliginosus,* and the Eastern Grey Kangaroo, *M. giganteus,* to the climate in Australia. *Macropus fuliginosus* is found in areas of uniform or winter rainfall, but *M. giganteus* occurs where rainfall either shows no seasonal trend or falls mainly in summer. In contrast, the distribution of *M. rufus* is best described in terms of mean annual temperature as well as mean annual rainfall. Caughley *et al.* (1987) also suggested that the absence of these three species from much of northern Australia can be explained in terms of climate. They were also able to identify areas that were climatically suitable but unoccupied and considered the value of these species as paleoclimatic indicators of fossil faunas. They also point out that the distributions of these species are probably more directly related to habitat than climate itself.

Australia can be subdivided into various floristic or phytogeographic zones. Burbidge (1960) identified three major zones and these are shown in Figure 1. The delineation of each zone is linked closely with recent climates and the general pattern is very similar to other biogeographic schemes that have been proposed for Australia (Archer and Fox 1984). As well as these three major zones, Burbidge (1960) recognized several areas of special interest. One is the Macpherson-Macleay Overlap, where many tropical elements of the flora are found intermixed with temperate elements. Three interzone areas were recognized where temperate or tropical elements have merged into the Eremaean zone. Tasmania, southwestern Western Australia and north-east Queensland were also singled out due to highly endemic or relict flora. The major floristic zones in which each macropod species occurs are shown in Table 1. All regions are represented and the greatest species diversity occurs in both the tropical and temperate zones with a notable

[1]School of Biological Sciences, University of Sydney, New South Wales 2006.

Pages 755-66 in KANGAROOS, WALLABIES AND RAT-KANGAROOS ed. by G. Grigg, P. Jarman and I. Hume. Surrey Beatty & Sons Pty Limited, New South Wales, Australia, 1989.

Table 1. Relationship between size, habitat, phytogeographic region, diet, activity, habitat utilization and status of knowledge in Australian macropods. The bulk of the table has been adapted from Russell (1974). Species are grouped according to their status of distribution and abundance (Fletcher and Maynes 1988).

Species	Size	Phytogeographic region	Closed forest	Closed scrub	Tall open forest	Open forest	Open scrub	Woodland	Grassland	Open Woodland	Low shrubland	Rocky	Diet	Activity	Resting Habitat	Movements	Use of disturbed habitat	Status of knowledge		References
A. Widespread and abundant																				
Macropus agilis	3	1				1,4		1,4	1,4					1	3	2	3	1	C	4,72
M. dorsalis	3	1,4,6	2	3		3								1	3	3	2	1	C	11,67,97
M. fuliginosus	4	2,3,5				1,2	1,2	1,2		1,2				1	3	2	4	1	B	10,13,16,17,19,20,37,77,78,90,94,95
M. giganteus	4	1. . .6				1,2	1,2	1,2						1	3	2	4	1	A	4,11,12,16,17,19,37,39,40,41,42,45 52,90,94,97,98,101,103,104
M. parryi	3	1,4,6				1,2		1,2						1	2	2	2	2	A	4,11,52,97,98
M. robustus	4	1. . .6	x	x		x	x	x	x	x		x	1	1	3	3	2	2	A	11,21,25,29,52,65,72,83,85,102, 103,104
M. rufogriseus	3	1,3,4,6			3,4	2,3	2,3						1	2	2	3	2	1	B	11,19,46,52,97,98
M. rufus	4	2,5,6						1	1	1		1		1	3	1	4	2	A	3,10,12,15,16,17,21,25,26,27,33,37, 47,62,63,64,68-72,74,77,78,94,95, 96
Wallabia bicolor	3	1,3,4,6			3	3	3							3	3	3	2	1	B	11,13,30,32,52
B. Moderate range and population																				
Aepyprymnus rufescens	1	1,3,4,6				3,4	3,4	3,4						3	4	4	2	2	B	11,52,89,97
Lagorchestes conspicillatus	1	1,2,6				3		3	3	3				3	4	4	2	2	C	65,72
M. antilopinus	4	1						1,4		1,4				1	3	2	3	2	C	22,72,85
M. bernadus	3	1	x					x					1	1	3	3	2	3	C	72
M. irma	2	3				1,2	1,2	1,2						1	2	2	2	2	C	
Onychogalea unguifera	2	1,2,6						2,3	2,3	2,3				2	3	2	2	2	D	72
Petrogale brachyotis	2	1						x		x			1	3	3	3	2	3	D	72
P. godmani	2	1	x			x							1	2	3	3	2	3	C	
P. inornata	2	1,6	x			x		x					1	2	3	3	2	3	B	
P. penicillata	2	1,3,4,6	x			x		x					1	1	3	3	2	3	C	11,52,93
P. rothschildi	2	1,2							x			x	1	2	3	3	2	3	C	65
P. xanthopus	2	2								x			1	2	3	3	2	3	A	18,36,60,61,81,105
Potorous tridactylus	1	3,4		3,4	3,4									4	4	4	2	2	B	5,11,38,56,91,97
Thylogale billardierii	2	3	2	2,3	2,3	2,3								2	3	3	2	1	B	
T. stigmatica	2	1,3,4	2	2,3	2,3									3	2	3	1	3	D	11,52
T. thetis	2	3,4	2	2,3	2,3									2	2	3	2	1	D	11,48,52
C. Restricted range and low populations																				
Bettongia gaimardi	1	3				2								4	4	4	2	3	B	82
Dendrolagus bennettianus	3	1	2											4	2	2	1	2	C	
D. lumholtzi	2	1	2											4	2	2	1	2	B	79
Hypsiprymnodon moschatus	1	1	2											4	1	4	1	3	B	76
M. eugenii	2	3		3,4		3,4	3,4							1	3	3	3	2	A	44,65

Table 1. — Continued

Species	Size	Phytogeographic region	Closed forest	Closed scrub	Tall open forest	Open forest	Open scrub	Woodland	Grassland	Open Woodland	Low shrubland	Rocky	Diet	Activity	Resting Habitat	Movements	Use of disturbed habitat	Status of knowledge	References
C. Restricted range and low populations — *Continued*																			
M. parma	2	3,4		3,4	3,4								1	3	3	2	2	B	66
Peradorcas concinna	1	1						x		x		1	2	3	3	2	3	C	
Petrogale burbidgei	1	1								x		1	2	3	3	2	3	D	72,87
P. lateralis	2	1,3,5						x	x	x		1	2	3	3	2	3	C	65
Setonix brachyurus	2	3		3,4							3,4		2	4	4	3	2	A	28,43,55,56,57,65,73,75
D. Endangered or extinct																			
B. lesueur	1	2									3,4		3	4	5	1	2	B	65
B. penicillata	1	1,3				3	3	3					4	4	4	2	2	B	24,86
Lagorchestes hirsutus	1	2							1,2				2	4	5	2	2	B	6,49,65,88
Lagostrophus fasciatus	1	3									2,3		2	4	4	2	3	C	65
O. fraenata	2	6		3				2,3					3	3	3	2	2	C	35,59
Petrogale persephone	2	1	x							x		1	1	3	3	2	3	C	
Potorous longipes	1	3				3							4	4	4	2	2	C	
Caloprymnus campestris	1	2							1		1		?	4	4	?	3	C	
Lagorchestes asomatus	1	2							1,2				?	?	?	?	3	D	
L. leporides	1	3,5						1,2	1,2				?	4	4	?	3	D	
M. greyi	3	3					2						1	3	2	2	3	C	
O. lunata	1	2,3								2			?	?	3	?	3	D	
Potorous platyops	1	3								3			?	?	?	?	3	D	

Phytogeographic Region (see Figure 1): 1. Tropical; 2. Eremaean; 3. Temperate; 4. Macpherson/Macleay overlap; 5. Interzone 1, Interzone 2; 6. Interzone 3.

Activity: 1. Diurnal; 2,3. Crepuscular/Nocturnal; 4. Nocturnal.

Resting Areas: 1. Open; 2. Moderate shelter; 3. Dense shelter; 4. Thickets, nests; 5. Burrows.

Movements: 1. Feed in approximately the same area as resting; 2. Distinct feeding and resting areas; 3. As for 2., but with seasonal shifts; 4. As for 2., but with marked seasonal shifts.

Diet: 1. Predominantly grasses; 2. Mixture of grasses and browse; 3. Predominantly browse; 4. Other (e.g., fungi, tubers, insects).

Use of Disturbed Habitat: 1. Use of cultural areas (e.g., agricultural pest); 2. Use of degraded areas (e.g., pastoral areas); 3. Only partly disturbed/undisturbed areas.

Status of Knowledge (from Fletcher and Maynes 1988): A. Very good knowledge; B. Adequate knowledge, more research desirable; C. Some knowledge, but much more research desirable; D. Little or no knowledge.

Size: 1. Very small (<3 kg); 2. Small (3-10 kg); 3. Medium (10-25 kg); 4. Large (>25 kg).

Habitat: Classification of major habitat type according to Specht (1970) with subjective estimate of cover actually available at the level of the animal. 1. <10%; 2. 10-30%; 3. 30-70%; 4. >70%; x. Habitat around rocky area.

References: Strahan (1983) was used as a general reference. Numbers refer to references dealing specifically with habitat association of that species.

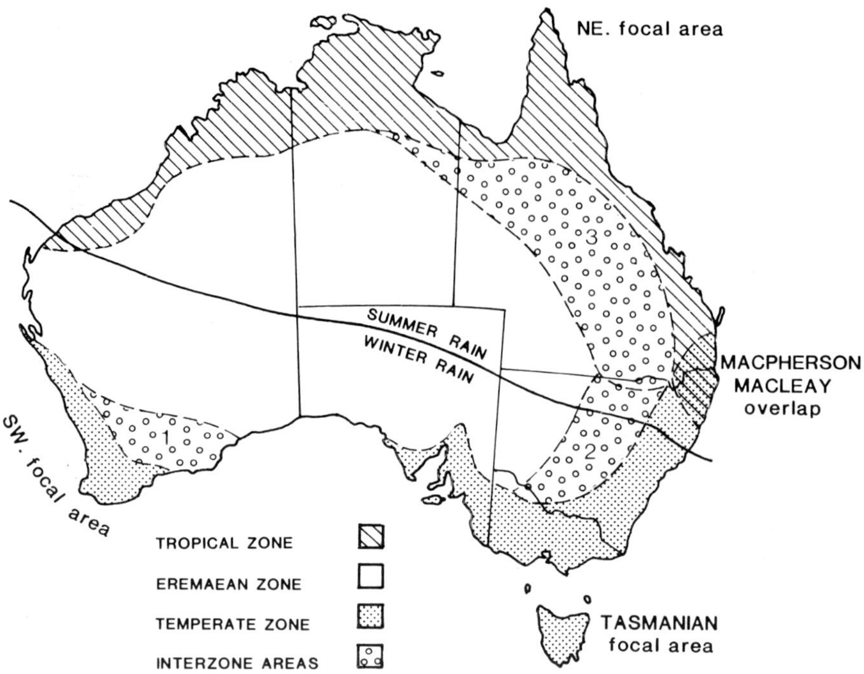

Fig. 1. Floristic regions of Australia. After Burbidge (1960).

concentration of 10 species in the Macleay-Macpherson Overlap.

Distribution within a range

The range of a species may be described by noting its presence or absence in an area. However, within that range there are usually variations in density and these may be linked to various characteristics of the environment. Knowing the range of a species only suggests a range of habitats that it may be found in, so more detailed analyses are possible knowing the variations in density within that range.

The major habitat types in which macropods occur are shown in Table 1. These are only a general guide and many exceptions can be found. The Common Wallaroo, *M. robustus,* for example, has been reported occurring along the heavily timbered banks of the Darling River in New South Wales (Frith and Calaby 1969) and in thick mulga scrub in central Queensland (Russell and Richardson 1971), both locations being far from its more typical rocky, hilly country. However, these areas contain the more important elements of the wallaroo's more typical habitat, namely food, water, and probably most importantly, shelter.

Aerial surveys since the mid 1970s have shown the larger macropods to be distributed patchily within their respective ranges. Aerial survey, though restricted to the less heavily forested areas of Australia and to the larger species, is the only feasible technique for this scale of study.

Australia-wide aerial surveys of Red Kangaroos between 1980 and 1982 showed that the greatest numbers occurred in the eastern states (Caughley *et al.* 1983). Short *et al.* (1983) remarked on the close correspondence between Red Kangaroo densities and sheep densities in New South Wales, South Australia and Western Australia, suggesting similar habitat requirements for both species. Within the areas surveyed in each state, highest densities of Red Kangaroos were found in areas where sheep grazing was the dominant land use. However, within both Western Australia (Short *et al.* 1983) and New South Wales (Sinclair 1980), Red Kangaroo density was negatively correlated with stock density, which was predominantly sheep. In contrast, South Australian Red Kangaroos were positively associated with more intensive sheep grazing within the pastoral zone. This shows the importance of the scale and the environmental range examined to the relationships that are observed.

Australia-wide aerial surveys were again carried out in 1984 (Grigg *et al.* 1985). Both Red and Western Grey Kangaroo numbers had dropped dramatically in the eastern states, while in Western Australia an increase had occurred. The differences between states was probably the result of the breaking of a drought in Western Australia, while the eastern states were still in one. This highlights the temporal as well as the spatial variability in macropod distributions which can occur on a large scale.

Both the environment and densities of species may vary temporally as well as spatially. Examination of data from nine consecutive years of aerial survey in the pastoral zone of South Australia (Grigg *et al.* 1987) has enabled Cairns *et al.* (1988) to examine the distribution of *M. rufus* and *M. fuliginosus* in relation to components of the environment. This period covered both drought and non-drought years. The distribution of *M. rufus* on half-degree square blocks for drought and non-drought years is shown in Figure 2. The highest densities were always found south of Lake Frome. During drought years the highest densities were further south than during non-drought years. Also, densities in the northern blocks were lowest during the drought years, while the western blocks showed the opposite trend. This appears to be due to uneven rainfall during the drought, though it is possible that uniformly low rainfall may have had a similar effect. In support of this latter possibility, Cairns and Grigg (1988) showed that numerical responses of Red Kangaroos to rainfall differed between the western and the central and eastern areas of the South Australian pastoral zone. *Macropus fuliginosus* also showed year-to-year changes in its distribution within the pastoral zone, but these were not as marked.

Sinclair (1980) and Short *et al.* (1983) also examined relationships between Red Kangaroos and the environment in New South Wales and Western Australia respectively and some common patterns emerged. Highest densities of Red Kangaroos were generally found in more arid parts of the pastoral zones in each state and, more specifically, in the mulga woodlands. In New South Wales, Sinclair (1980) considered a mosaic of land systems to be important as these areas supported relatively high densities of Red Kangaroos. The importance of these areas has been examined at a finer scale and will be discussed later. Cairns *et al.* (1988) found aridity, but not mulga, to be an important determinant of Red Kangaroo density in the South Australian pastoral zone. However, where high densities did occur, mulga was still present and these areas tended not to be dominated by one particular soil type, again indicating the importance of habitat heterogeneity. In all three studies, low densities of Red Kangaroos were found in areas of mallee and cultivated land. These areas were considered suboptimal for kangaroos by Short and Grigg (1982). They point out though, that as these areas are close to what has been a static boundary of distribution of Red Kangaroos, other factors such as climate may be determining the low densities found in these areas. However, they also point out that relatively high densities of Red Kangaroos have been found close to their boundary of distribution in different habitat (Caughley *et al.* 1977).

In southern inland Queensland, Hill (1981a) used LANDSAT imagery to identify broadscale habitat zones for the Eastern Grey Kangaroo. Given the importance attached to cover for shelter and open areas for forage from studies at

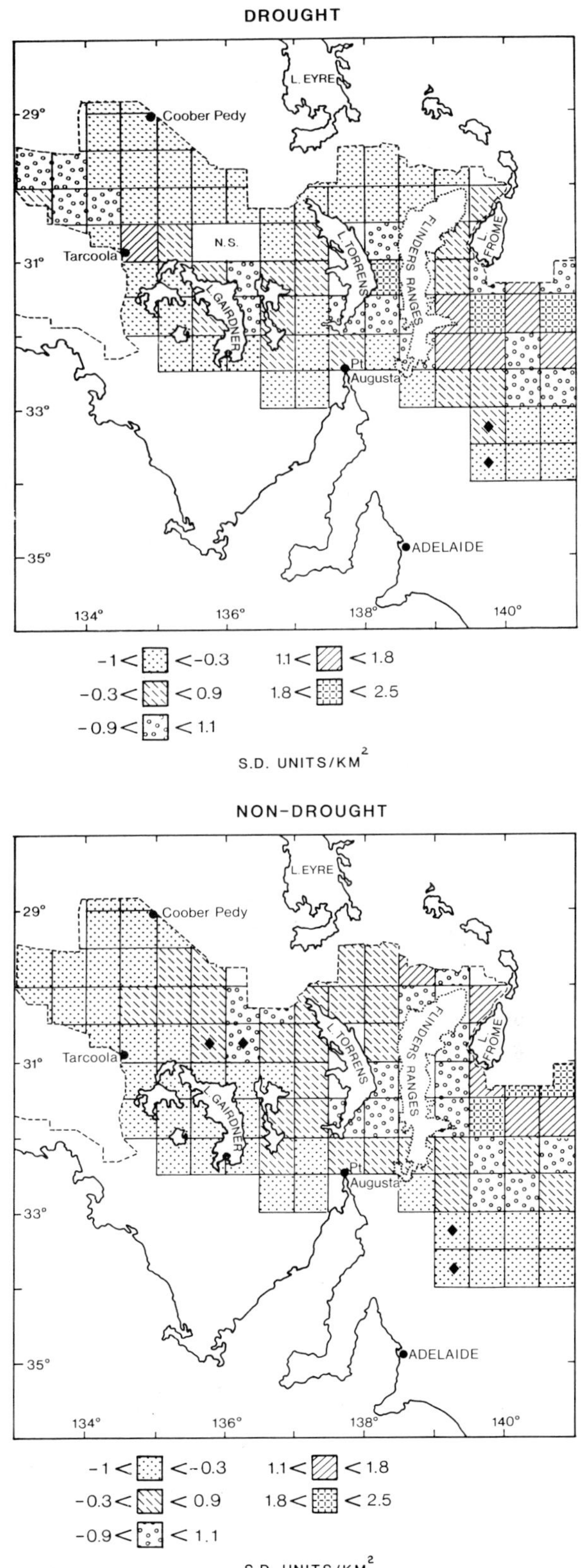

Fig. 2. Standardized densities of Red Kangaroos on half-degree square blocks in the pastoral zone of South Australia. Drought years (1982, 1983) and non-drought years (1978-1981, 1984-1986) are shown. Densities have been standardized to a mean of zero and standard deviation of one. The vermin proof fence (dotted line) bounds the pastoral zone in the north and in the west. N.S. not surveyed. (♦) surveyed infrequently. After Cairns *et al.* (1988).

a local scale (see below), zones were defined on the basis of a balance between open and wooded country. Density of Eastern Grey Kangaroos, derived from aerial survey, was highest in areas where there was a balance between cover and open areas rather than predominantly wooded or predominantly open country. The results reflected those found at a finer scale and are in agreement with regional studies of habitat associations of Grey Kangaroos in New South Wales (Scott-Kemmis 1979), Western Australia (Short *et al.* 1983) and South Australia (Cairns *et al.* 1988). As was found for Red Kangaroos, these studies also suggested that a heterogeneous habitat is important for Grey Kangaroos.

LANDSAT imagery has also been used to identify suitable habitat for species with limited distributions. Saxon (1983) was able to identify locations of possible habitat for the endangered Rufous Hare-wallaby, *Lagorchestes hirsutus,* in the Tanami Desert in the Northern Territory. Of three sites that were identified by LANDSAT, one appeared suitable on closer inspection. It was, however, considered too small to sustain a population of Rufous Hare-wallabies. Nevertheless, the value of the technique was demonstrated as it allowed rapid identification of small patches of habitat over a vast area. Additionally, the technique allows for regular monitoring in an environment where the location and size of suitable habitat may vary through time. This would clearly be important for assessing conservation areas.

Predation appears to play a major role in determining the abundance of many macropods. Caughley *et al.* (1980) found a marked disparity in the density of kangaroos on either side of the vermin proof fence in northwestern New South Wales. This pattern did not seem to reflect any environmental gradient, but was suggested to be the direct result of dingo predation. Shepherd's (1981) direct observations of dingo predation on kangaroos support this hypothesis. Kinnear *et al.* (1981) commented on the reduction in numbers of populations of rock wallabies in Western Australia and their low numbers or absence in some areas. They suggested that this is due to predation by foxes and feral cats. Other authors have also suggested that the disappearance of several species of macropods from the Australian mainland has been the result of fox predation (e.g., Johnson and Rose 1983; Archer *et al.* 1985; Johnson *et al.*, this volume). In addition to the exclusion of predators, artificial watering points have been provided in pastoral areas. This has been suggested as the reason for a possible extension in the range of the Eastern Grey Kangaroo (Kirkpatrick 1967; Caughley *et al.* 1984) and changes in the distribution of the Red Kangaroo (Newsome 1975; Oliver 1986).

FINE SCALE: THE SCALE OF A LOCAL POPULATION

At this level of resolution, not only can smaller patches of habitat be examined in greater detail, but an animal's behaviour can become a component of the study. A more detailed examination of a habitat may explain why a species is concentrated in a particular area, it may explain how several species can coexist or, conversely, what aspects of the habitat should be used in describing broad scale patterns of distribution and abundance. Whereas broad scale studies have relied upon aerial surveys or extrapolation from ground surveys, studies at a finer scale have employed a variety of techniques to examine habitat use and local distribution of macropods. While aerial surveys have been used at this scale, ground transects that are either walked or driven have also been used. Animals may be either observed directly, or indirectly by pellet counts. Individuals may be identified by tagging or monitored by radio telemetry, while smaller, more cryptic species may be studied in mark-recapture programmes. Movements of Quokkas have also been studied using a cotton reel technique (Kitchener 1973) and by monitoring the use of gates in fences (Packer 1965). Each technique has its limitations, but the techniques are clearly complementary. Knowledge of the local distribution and habitat utilization in the better studied species (Table 1) has usually been gained from the use of several of these techniques.

Generally, it has been the larger, widespread and abundant species that have been best studied. Habitat utilization in the Red Kangaroo has been studied in New South Wales (Frith 1964; Bailey 1971; Denny 1985; Priddel 1987), Queensland (Caughley 1964), central Australia (Newsome 1965a, 1965b; Low *et al.* 1981) and Western Australia (Oliver 1986). At a broad scale, Red Kangaroos were found to reach high densities in arid areas offering a mosaic of land systems. Finer scale studies have demonstrated the value to Red Kangaroos of having access to a variety of areas between which they can move, taking advantage of the best pasture and most appropriate shelter. During dry times in Kinchega National Park, western New South Wales, pastures tend to be more persistent and possibly more palatable on heavier textured soils on alluvial plains. However, lighter textured soils require less water to initiate growth (Robertson 1987) and Red Kangaroos moved onto these soil types following rain (Priddel 1987). Western Grey Kangaroos showed similar seasonal changes in habitat use (Priddel 1987), but their response to changing pasture conditions was not as rapid. Red Kangaroos have also been shown to seek different forms of shelter depending on prevailing environmental conditions. Summer resting areas offered more shade than winter sites which provided shelter from cold winds while permitting basking (Denny 1985).

Hill (1979, 1982) found that the use of open country by Eastern Grey Kangaroos increased during dry times as forage reserves declined in both quality and quantity. With this knowledge, Hill (1979) suggested that LANDSAT imagery could be used on a broad scale to monitor forage status and therefore Grey Kangaroo vulnerability to shooting. Usually, the biomass of food in open areas is greater than that in more wooded areas and, as a result, Eastern Grey Kangaroos make greater use of open areas during grazing periods as opposed to resting periods which are spent in more sheltered areas (Taylor 1985). In open areas, kangaroos are more abundant where pasture biomass and quality is greatest (Taylor 1984). Hill (1981b) found that Eastern Grey Kangaroo densities were higher in partly cleared forest than in either open woodland or climax forest. Appropriate shelter has been defined as areas offering good lateral cover (Caughley 1964). These associations between Eastern Grey Kangaroo density and the environment parallel those found at the broader scale. Mallee was considered a "biological desert" for macropods by Caughley *et al.* (1977) and this judgment was confirmed by Short and Grigg (1982). The lack of kangaroos in mallee appears to be due to lack of available food and water. Heavily cleared areas were also described as suboptimal for kangaroos. Here, the limiting factors are probably shelter and the intolerance of landowners.

Fine scale habitat associations for these large macropods, particularly in arid areas, are quite labile. The marked seasonal movements between land systems represent one end of a spectrum of movements displayed by macropods, and these are shown in Table 1. The majority of species have distinct feeding and resting areas. Feeding areas are generally open areas, particularly for grazers. In contrast, a resting area usually incorporates some form of shelter, ranging from an isolated tree to thick scrub and even burrows. Exceptions are generally those species solely restricted to the rainforest. An interesting contrast is between the pademelons. At night, *Thylogale thetis* (Johnson 1980) and *T. billardierii* (Johnson and Rose 1983) both move from their daytime refuges of dense cover to feed on adjoining open areas which can include agricultural crops and pasture. In contrast, *T. stigmatica* does not move as far from cover as its congeners (Calaby 1966) and this is probably due to the smaller proportion of grass in its diet (Jarman and Phillips, this volume). There is, however, limited movement data for this species. Perhaps one non-rainforest species that feeds and rests in the same area is *Bettongia lesueur,* whose shelter requirements are met by burrowing.

Most macropods are largely nocturnal, many with peaks in activity around sunrise and sunset. Many species show varying degrees of activity through the day (Table 1). Though most active at night, *T. thetis* also moves widely within the rainforest throughout the day, feeding and basking (Johnson 1980). Lumholtz's Tree Kangaroo, *Dendrolagus lumholtzi*, browses intermittently throughout the day (Proctor-Gray 1984), while another rainforest species, the Musky Rat-kangaroo, *Hypsiprymnodon moschatus,* is strictly diurnal.

In addition to the Red and Grey Kangaroos, many other macropods show seasonal changes in their habitat use. The use of crops and pastures by many pest species increases during dry times (Lavery and Kirkpatrick 1985). Seasonal differences in home range have been found in both the Tammar Wallaby, *M. eugenii* (Inns 1982) and the Quokka, *Setonix brachyurus* (Nicholls 1971) in seasonally harsh environments. In both cases, it was suggested that animals moved over larger areas in dry times in search of food. For the Quokka, water was also sought (Dunnett 1962). Not all arid zone species show marked seasonal movements. Compared with the Red Kangaroo, the Common Wallaroo is relatively sedentary (Newsome 1975) and has not been found to shift its home range in response to rainfall (Croft 1982). As with rock wallabies, Common Wallaroos show close attachment to rocky areas which offer a distinct microenvironment in terms of food, shelter and water. In tropical northern Australia, species such as the Agile Wallaby, *M. agilis,* show a seasonal shift in local distribution, presumably avoiding the seasonal inundation of lowland areas (Bolton *et al.* 1982).

Use of habitat is clearly linked to diet, but there are also physiological considerations and predators to worry about. In summer, the Quokka (Kitchener 1981) and the Common Wallaroo (Russell 1969) choose shelters that offer the lowest heat load, while the sparse shelter used by Red Kangaroos still affords them substantial relief in their arid environment (Dawson and Denny 1969). Choice of shelter has been shown to vary seasonally and with temperature in these three species. Predation pressure appears to influence the abundance of many species at a broad scale, but it also appears to have a major effect on habitat utilization at finer scales. Numerous authors have suggested that the use of shelter during resting periods by some species is the result of past or present predation pressure rather than the preferable microclimate of the shelter (e.g., Caughley 1964; Johnson 1980). This is supported by the use of open areas only at night, associations with areas of high lateral cover as opposed to high overhead cover and the flight to cover of animals when disturbed.

Disturbed habitat is used to different extents by various species (Table 1). Generally, widespread land clearances have been detrimental to most species, though some species appear more tolerant than others. Calaby (1966) recorded only Red-necked Wallabies, *M. rufogriseus,* in the most heavily disturbed areas in a survey in north-eastern New South Wales. Here, the wallabies sheltered in small strips of uncleared forest adjacent to large tracts of cleared land. Where areas have only been partly cleared, many species may persist, often reaching high densities. In these areas, habitat requirements of both feeding and resting areas are still being met. In central Australia, Newsome (1975) suggested that the subclimax grassland created by introduced ruminants has greatly improved the food supply of the Red Kangaroo and this has resulted in greater numbers. However, this changed environment has been detrimental to the small arid zone macropods whose shelter has been removed.

Fire is another form of disturbance that can affect the distribution of a species. Though obviously devastating when widespread, it can produce regrowth areas that are attractive to kangaroos. Several species may congregate on recently burnt areas (e.g., Southwell and Jarman 1987), though recently burnt areas do not always attract kangaroos. Caughley *et al.* (1985) provide an example of the latter. Here, the burnt area was mallee, generally considered poor kangaroo habitat, and surrounding areas had received recent rain. For some small species, fire can also produce a mosaic of areas in which suitable shelter and food can be found. In Victoria, relatively high densities of the Long-nosed Potoroo, *Potorous tridactylus,* were found in seral stages offering a high density of cover (Seebeck 1981). Bolton and Latz (1978) describe the Rufous Hare-wallaby as occurring only in areas where cool winter fires have resulted in a tight moasic of different regenerating vegetation, providing both food and shelter.

Many studies have examined the habitat relationships of sympatric macropods (e.g., Bell 1973; Kaufman 1974; Southwell 1987). In each case, habitat was deemed a major factor in the ecological separation of the species (see also Fox, this volume). Southwell (1987), for example, found that the four most abundant macropods on his study were spatially separated and this was related to canopy cover, pasture and topography. In Kinchega National Park, Priddel (1987) found that although Red and Grey Kangaroos showed similar responses to changing pasture conditions, they were generally found in different areas. A greater proportion of the Red Kangaroo population was found on sandplains and in open areas, whereas densities of Western Grey Kangaroos were highest on floodplains and in woodlands. Western Grey Kangaroos were also more evenly dispersed than Red Kangaroos (Johnson and Bayliss 1981).

Population classes may segregate in different areas. Differentially distributed population classes have been found in Red Kangaroos (Johnson and Bayliss 1981), Common Wallaroos (Taylor 1983), Eastern Grey Kangaroos (Jaremovic 1984) and Antilopine Kangaroos, *M. antilopinus* (Croft 1987). Johnson and Bayliss (1981) found that lactating females as well as large males predominated in areas of better pasture. The population as a whole may not show the same seasonal changes in habitat use and this may be partially the result of population segregation. To account for the observed seasonal changes in the dispersion of Red and Western Grey Kangaroos, Priddel (1987) explains that only just under half the population need have moved. Johnson and Bayliss (1981) also found the dispersion of Red Kangaroo population classes to be stable through varying environmental conditions and movements of population classes. They suggested that large males were denying subordinate males access to females, while lactating females were selecting better pasture. Though not finding a larger proportion of lactating female *M. antilopinus* in areas of better pasture, Croft (1987) found the sex ratio to be biased towards females and that large rather than medium males predominated in these areas. Another explanation for sexual segregation comes from work on dimorphic species of ungulates, where Illius and Gordon (1987) argue that it is the result of "scramble competition for limited resources" with the larger sex being excluded. Female Red Kangaroos (Oliver 1986) and female Red-necked Wallabies (Johnson 1987) with young at foot have also been found to utilize areas closer to cover than other population classes. This is probably the result of the high vulnerability of young at foot to predation.

SYNTHESIS

As Wiens *et al.* (1986) point out, we must be aware of the constraints imposed on our studies by their temporal and spatial scales. Comparisons of studies conducted at different spatial scales are likely to be invalid, as are comparisons of studies carried out under a different set of environmental conditions (Wiens *et al.* 1987). Broad scale studies relating kangaroo distributions to the environment within Australian states, are examining relationships under different environmental conditions. Comparisons between studies conducted in different states must take this into account. Species responses to aspects of the environment may also vary over a large area. One must therefore be careful when seeking linear relationships between species and the environment over this range (Wiens *et al.* 1987). Hill (1981b) also cautions us against taking fine scale associations of species with habitat out of their regional context. Aerial surveys of Eastern Grey Kangaroos in southern Queensland can be stratified by combining observations in open country, medium woodland and dense woodland. However, the survey crosses several habitat zones, and kangaroo density is greater in open country in the east than open country in the more arid western zones making such stratifications confounding unless considered within each habitat zone.

It is of interest to note the close parallels and complementary nature of studies of macropod habitat associations at the two spatial scales. As a family, their greatest diversity occurs where there is a great variety of habitats, while for many species their greatest densities occur where the habitat is heterogeneous, offering suitable food and shelter. The distribution of a species, both within its range and throughout the continent, can be explained through fine scale studies examining habitat use as well as by correlations with broad scale environmental patterns. While several species show wide ecological tolerances, specific habitat requirements can be identified for most and this appears to have allowed many to be sympatric. Macropods occur in virtually every major Australian habitat, and their use of these habitats is flexible enough to cope with an environment that can be both seasonally harsh and unpredictable.

ACKNOWLEDGEMENTS

This work was part of a study funded by the Australian National Parks and Wildlife Service to whom I am extremely grateful.

I would like to thank Gordon Grigg and Stuart Cairns for many useful suggestions for this review and criticisms of the manuscript. I am also grateful to the referees for their comments. Finally, I would like to thank Anita Gordon for typing much of this manuscript.

REFERENCES

ARCHER, M. AND FOX, B., 1984. 1. Background to vertebrate zoogeography in Australia. Pp. 1-15 *in* Vertebrate Zoogeography and Evolution in Australasia ed. by M. Archer and G. Clayton. Hesperian Press: Carlisle, Western Australia.

ARCHER, M., FLANNERY, T. F. AND GRIGG, G. C., 1985. 2. The Kangaroo. Weldons: Sydney.

BAILEY, P. T., 1971. 3. The Red Kangaroo, *Megaleia rufa* (Desmarest), in northwestern New South Wales. I. Movements. *C.S.I.R.O. Wildl. Res.* **16**: 11-28.

BELL, H. M., 1973. 4. The ecology of three macropod marsupial species in an area of open forest and savannah woodland in north Queensland, Australia. *Mammalia* **37**: 527-44.

BENNETT, A., 1984. 5. Ecology of the Long-Nosed Potoroo *Potorous tridactylus* within a fragmented forest environment in southwestern Victoria. *Bull. Aust. Mamm. Soc.* **8**: 95.

BOLTON, B. L. AND LATZ, P. K., 1978. 6. The Western Hare-wallaby, *Lagorchestes hirsutus* (Gould) (Macropodidae), in the Tanami Desert. *Aust. Wildl. Res.* **5**: 285-93.

BOLTON, B. L., NEWSOME, A. E. AND MERCHANT, J. C., 1982. 7. Reproduction in the Agile Wallaby, *Macropus agilis*, in tropical lowlands of north Australia. *Aust. J. Ecol.* **7**: 261-77.

BURBIDGE, N. T., 1960. 8. The phytogeography of the Australian region. *Aust. J. Bot.* **8**: 75-211.

CAIRNS, S. C. AND GRIGG, G. C., 1988. 9. The numerical responses of Red Kangaroo *(Macropus rufus)* and Western Grey Kangaroo *(M. fuliginosus)* populations in the pastoral zone of South Australia. Australian National Parks and Wildlife Service unpublished report.

CAIRNS, S. C., POPLE, A. R. AND GRIGG, G. C., 1988. 10. Habitat associations of kangaroos in the South Australian pastoral zone. Australian National Parks and Wildlife Service unpublished report.

CALABY, J. H., 1966. 11. Mammals of the upper Richmond and Clarence Rivers, New South Wales. *C.S.I.R.O. Div. Wildl. Res. Tech. Paper.* **10**: 1-55.

CAUGHLEY, G. J., 1964. 12. Density and dispersion of two species of kangaroo in relation to habitat. *Aust. J. Zool.* **12**: 238-49.

CAUGHLEY, G., BROWN, B. AND NOBLE, J., 1985. 13. Movement of kangaroos after a fire in mallee woodland. *Aust. Wildl. Res.* **12**: 349-53.

CAUGHLEY, G., BROWN, B., DOSTINE, P. AND GRICE, D., 1984. The grey kangaroo overlap zone. *Aust. Wildl. Res.* **11**: 1-10.

CAUGHLEY, G., GRIGG, G. C. AND SHORT, J., 1983. 14. How many kangaroos? *Search* **14**: 151-52.

CAUGHLEY, G., SINCLAIR, R. G. AND WILSON, G. R., 1977. 15. Numbers, distribution and harvesting rate of kangaroos on the inland plains of New South Wales. *Aust. Wildl. Res.* **4**: 99-108.

CAUGHLEY, G., GRIGG, G. C., CAUGHLEY, J. AND HILL, G. J. E., 1980. 16. Does dingo predation control the densities of kangaroos and emus? *Aust. Wildl. Res.* **7**: 1-12.

CAUGHLEY, G., SHORT, J., GRIGG, G. C. AND NIX, H., 1987. 17. Kangaroos and climate: an analysis of distribution. *J. Anim. Ecol.* **56**: 751-61.

COPLEY, P. B., 1983. 18. Studies on the Yellow-Footed Rock Wallaby, *Petrogale xanthopus* Gray (Marsupialia: Macropodidae). I. Distribution in South Australia. *Aust. Wildl. Res.* **10**: 47-61.

COULSON, G. M., 1979. 19. Habitat preferences in three sympatric species of *Macropus*. *Bull. Aust. Mamm. Soc.* **5**: 23.

COULSON, G. AND NORBURY, G., 1984. 20. Preliminary observations on habitat utilization and preferred diet of the Western Grey Kangaroo, *Macropus fuliginosus melanops* (Desmarest), in northwestern Victoria. *Bull. Aust. Mamm. Soc.* **8**: 107.

CROFT, D. B., 1982. 21. Movement patterns of kangaroos on Fowler's Gap Station, northwestern New South Wales. *Bull. Aust. Mamm. Soc.* **7**: 26.

CROFT, D. B., 1987. 22. Socio-ecology of the Antilopine Wallaroo, *Macropus antilopinus*, in the Northern Territory, with observations on sympatric *M. robustus woodwardii* and *M. agilis*. *Aust. Wildl. Res.* **14**: 243-55.

DAWSON, T. J. AND DENNY, M. J. S., 1969. 23. A bioclimatological comparison of the summer day microenvironments of two species of arid-zone kangaroos. *Ecology* **50**: 328-32.

DELROY, L. B., EARL, J., RADBONE, I., ROBINSON, A. C. AND HEWETT, M., 1986. 24. The breeding and re-establishment of the Brush-tailed Bettong, *Bettongia penicillata*, in South Australia. *Aust. Wildl. Res.* **13**: 387-96.

DENNY, M., 1982. 25. Adaptations of the Red Kangaroo and Euro (Macropodidae) to aridity. Pp. 179-83 *in* Evolution of the flora and fauna of arid Australia ed. by W. R. Barker and P. M. J. Greenslade. Peacock Publications: Adelaide.

DENNY, M., 1985. 26. The Red Kangaroo and the arid environment. Pp. 55-72 *in* The Kangaroo Keepers ed. by H. J. Lavery. University of Queensland Press: St. Lucia.

DUDZINSKI, M. L., LOW, W. A., MULLER, W. J. AND LOW, B. S., 1982. 27. Joint use of habitat by Red Kangaroos and shorthorn cattle in arid central Australia. *Aust. J. Ecol.* 7: 69-74.

DUNNET, G. M., 1962. 28. A population study of the Quokka, *Setonix brachyurus* (Quoy and Gaimard) (Marsupialia). II. Habitat, movements, breeding and growth. *C.S.I.R.O. Wildl. Res.* 7: 13-32.

EALEY, E. H. M., 1967. 29. Ecology of the euro, *Macropus robustus* (Gould), in northwestern Australia. II. Behaviour, movements and drinking patterns. *C.S.I.R.O. Wildl. Res.* **12**: 27-51.

EDWARDS, G. P. AND EALEY, E. H. M., 1975. 30. Aspects of the ecology of the Swamp Wallaby *Wallabia bicolor* (Marsupialia: Macropodidae). *Aust. Mammal.* **1**: 307-17.

FLETCHER, M. AND MAYNES, G., 1988. 31. Macropod research needs and priorities. Unpublished discussion paper, Australian National Parks and Wildlife Service, Canberra.

FLOYD, R. B., 1980. 32. Density of *Wallabia bicolor* (Desmarest) (Marsupialia: Macropodidae) in eucalypt plantations of different ages. *Aust. Wildl. Res.* **7**: 333-37.

FOX, B. J. (this volume). 1989. Community ecology of macropodoids.

FRITH, H. J., 1964. 33. Mobility of the Red Kangaroo, *Megaleia rufa. C.S.I.R.O. Wildl. Res.* **9**: 1-19.

FRITH, H. J. AND CALABY, J. H., 1969. 34. Kangaroos. F. W. Cheshire: Melbourne.

GORDON, G. AND LAWRIE, B. C., 1980. 35. The rediscovery of the Bridled Nailtail Wallaby, *Onychogalea fraenata* (Gould) (Marsupialia: Macropodidae) in Queensland. *Aust. Wildl. Res.* **7**: 339-45.

GORDON, G., MCGREEVY, D. G. AND LAWRIE, B. C., 1978. 36. The Yellow-footed Rock-wallaby, *Petrogale xanthopus* Gray (Macropodidae), in Queensland. *Aust. Wildl. Res.* **5**: 295-97.

GRIGG, G. C., SMITH, L., ALEXANDER, P. AND POPLE, T., 1987. Report of the 1987 survey of kangaroos in the pastoral area of South Australia. South Australian National Parks and Wildlife Service unpublished report.

GRIGG, G. C., BEARD, L. A., CAUGHLEY, G., GRICE, D., CAUGHLEY, J. A., SHEPHERD, N., FLETCHER, M. AND SOUTHWELL, C., 1985. 37. The Australian kangaroo populations, 1984. *Search* **16**: 277-79.

HEINSOHN, G. E., 1968. 38. Habitat requirements and reproductive potential of the macropod marsupial *Potorous tridactylus* in Tasmania. *Mammalia* **32**: 30-43.

HILL, G. J. E., 1979. 39. Application of LANDSAT to habitat analysis for the grey kangaroo. *Proc. N.Z. Geog. Conf.* (49th ANZAAS). **10**: 293-96.

HILL, G. J. E., 1981a. 40. A study of habitat preferences in the grey kangaroo. *Aust. Wildl. Res.* **8**: 245-54.

HILL, G. J. E., 1981b. 41. Distribution of grey kangaroos in south, inland Queensland. *Aust. Rangel. J.* **3**: 58-66.

HILL, G. J. E., 1982. 42. Seasonal movement patterns of the Eastern Grey Kangaroo in southern Queensland. *Aust. Wildl. Res.* **9**: 373-87.

HOLSWORTH, W. N., 1967. 43. Population dynamics of the Quokka, *Setonix brachyurus,* on the west end of Rottnest Is., Western Australia. *Aust. J. Zool.* **15**: 29-46.

ILLIUS, A. W. AND GORDON, I. J., 1987. The allometry of food intake in grazing ruminants. *J. Anim. Ecol.* **56**: 988-99.

INNS, R. W., 1980. 44. Ecology of the Kangaroo Island wallaby, *Macropus eugenii* (Desmarest), in Flinders Chase National Park, Kangaroo Island. Ph.D. Thesis, University of Adelaide, Adelaide.

JAREMOVIC, R. V., 1984. 45. Space and time related behaviour in Eastern Grey Kangaroos (*Macropus giganteus* Shaw). Ph.D. Thesis, University of New South Wales, Sydney.

JARMAN, P. J. AND PHILLIPS, C. M. (this volume). Diets in a community of macropod species.

JOHNSON, C. N., 1987. 46. Macropod studies at Wallaby Creek. IV. Home range and movements of the Red-necked Wallaby. *Aust. Wildl. Res.* **14**: 125-32.

JOHNSON, C. N. AND BAYLISS, P. G., 1981. 47. Habitat selection by sex, age and reproductive class in the Red Kangaroo, *Macropus rufus,* in western New South Wales. *Aust. Wildl. Res.* **8**: 465-74.

JOHNSON, K. A., 1980. 48. Spatial and temporal use of habitat by the Red-necked Pademelon, *Thylogale thetis* (Marsupialia: Macropodidae). *Aust. Wildl. Res.* **7**: 157-66.

JOHNSON, K., 1982. 49. The Rufous Hare-Wallaby in central Australia. *Bull. Aust. Mamm. Soc.* **7**: 26.

JOHNSON, K. A., 1983. 50. Tasmanian Bettong. Pp. 186 *in* Complete Book of Australian Mammals ed. by R. Strahan. Angus & Robertson: Sydney.

JOHNSON, K. A., BURBIDGE, A. AND MCKENZIE, N L. (this volume). Australian macropodoidea: status, causes of decline and future research and management.

JOHNSON, P. M., 1983. 51. Red-legged Pademelon. Pp. 225 *in* Complete Book of Australian Mammals ed. by R. Strahan. Angus & Robertson: Sydney.

KAUFMANN, J. H., 1974. 52. Habitat use and social organization of nine sympatric species of macropodid marsupials. *J. Mammal.* **55**: 66-80.

KINNEAR, J., ONUS, M. AND BROMILOW, B., 1984. 53. Foxes, feral cats and Rock Wallabies. *Swans* **14**: 3-8.

KIRKPATRICK, T. H., 1967. 54. The grey kangaroo in Queensland. *Queensl. Agric. J.* **93**: 550-52.

KITCHENER, D. J., 1972. 55. The importance of shelter to the Quokka, *Setonix brachyurus* (Marsupialia), on Rottnest Island. *Aust. J. Zool.* **20**: 281-99.

KITCHENER, D. J., 1973. 56. Notes on home range and movement in two small macropods, the Potoroo *(Potorous apicalis)* and the Quokka *(Setonix brachyurus). Mammalia* **37**: 231-40.

KITCHENER, D. J., 1981. 57. Factors influencing selection of shelter by individual Quokkas, *Setonix brachyurus* (Marsupialia), during hot summer days on Rottnest Island. *Aust. J. Zool.* **29**: 875-84.

LAVERY, H. J. AND KIRKPATRICK, T. H., 1985. 58. Kangaroos as pests. Pp. 103-32 *in* The Kangaroo Keepers ed. by H. J. Lavery. University of Queensland Press: Brisbane.

LAVERY, H. J. AND TIERNEY, P. J., 1985. 59. Scarcity and extinction. Pp. 25-52 *in* The Kangaroo Keepers ed. by H. J. Lavery. University of Queensland Press: Brisbane.

LIM, L., 1980. 60. A review of the distribution and status of the Yellow-footed Rock-wallaby *(Petrogale xanthopus)* with particular reference to New South Wales. *Bull. Aust. Mamm. Soc.* **6:** 49.

LIM, T. L. AND GILES, J. R., 1987. 61. Studies on the Yellow-footed Rock-wallaby, *Petrogale xanthopus* Gray (Marsupialia: Macropodidae). III. Distribution and management in western New South Wales. *Aust. Wildl. Res.* **14:** 147-61.

LOW, B. S., 1979. 62. The predictability of rain and the foraging patterns of the Red Kangaroo *(Megaleia rufa)* in central Australia. *Journal of Arid Environments.* **2:** 61-76.

LOW, B. S., BIRK, E., LENDON, C. AND LOW, W. A., 1973. 63. Community utilization by cattle and kangaroos in mulga near Alice Springs, N.T. *Tropical Grasslands* **7:** 149-56.

LOW, W. A., MULLER, W. J., DUDZINSKI, M. L. AND LOW, B. S., 1981. 64. Population fluctuations and range community preference of Red Kangaroos in central Australia. *J. Appl. Ecol.* **18:** 27-36.

MAIN, A. R. AND YADAV, M., 1971. 65. Conservation of macropods in reserves in Western Australia. *Biol. Conserv.* **3:** 123-33.

MAYNES, G. M., 1977. 66. Distribution and aspects of the biology of the Parma Wallaby, *Macropus parma,* in New South Wales. *Aust. Wildl. Res.* **4:** 109-25.

MCCANN, J., 1976. 67. Distribution and habitat of *Wallabia dorsalis* in the Upper Clarence Region. *Bull. Aust. Mamm. Soc.* **3:** 44.

NEWSOME, A. E., 1965a. 68. The distribution of Red Kangaroos, *Megaleia rufa* (Desmarest), about sources of persistent food and water in central Australia. *Aust. J. Zool.* **13:** 288-99.

NEWSOME, A. E., 1965b. 69. The abundance of Red Kangaroos, *Megaleia rufa* (Desmarest), in central Australia. *Aust. J. Zool.* **13:** 269-87.

NEWSOME, A. E., 1971. 70. The ecology of Red Kangaroos. *Aust. Zoologist* **16:** 32-50.

NEWSOME, A. E., 1975. 71. An ecological comparison of two arid-zone kangaroos of Australia, and their anomalous prosperity since the introduction of ruminant stock to their environment. *Q. Rev. Biol.* **50:** 389-424.

NEWSOME, A. E., 1983. 72. The grazing marsupials. Pp. 441-61 *in* Tropical Savannahs ed. by F. Bourliere. Elsevier: Amsterdam.

NICHOLLS, D. G., 1971. 73. Daily and seasonal movements of the Quokka, *Setonix brachyurus* (Marsupialia), on Rottnest Island. *Aust. J. Zool.* **19:** 215-26.

OLIVER, A. J., 1986. 74. Social organization and dispersal in the Red Kangaroo. Ph.D. Thesis, Murdoch University, Perth.

PACKER, W. C., 1965. 75. Environmental influences on daily and seasonal activity in *Setonix brachyurus* (Quoy and Gaimard (Marsupialia). *Anim. Behav.* **13:** 270-83.

PREEN, T., 1984. 76. The response of mammals to selective logging of a tropical rainforest in N.E. Queensland. *Bull. Aust. Mamm. Soc.* **8:** 79.

PRIDDEL, D., 1983. 77. The movements of Red and Western Grey Kangaroos. Ph.D. Thesis, University of Sydney, Sydney.

PRIDDEL, D., 1987. 78. The mobility and habitat utilization of kangaroos. Pp. 100-18 *in* Kangaroos: their ecology and management in the sheep rangelands of Australia ed. by G. Caughley, N. Shepherd and J. Short. Cambridge University Press: Cambridge.

PROCTER-GRAY, E., 1984. 79. Dietary ecology of the Coppery Brushtail Possum, Green Ringtail Possum and Lumholtz's Tree-Kangaroo in north Queensland. Pp. 129-35 *in* Possums and Gliders ed. by A. P. Smith and I. D. Hume. Surrey Beatty & Sons: New South Wales.

ROBERTSON, G., 1987. 80. Plant dynamics. Pp. 50-68 *in* Kangaroos: their ecology and management in the sheep rangelands of Australia ed. by G. Caughley, N. Shepherd and J. Short. Cambridge University Press: Cambridge.

ROBINSON, A. C., 1980. 81. The Yellow-footed Rock-wallaby *(Petrogale xanthopus)* in South Australia. *Bull. Aust. Mamm. Soc.* **6:** 59.

ROSE, R. W., 1986. 82. The habitat, distribution and conservation status of the Tasmanian Bettong, *Bettongia gaimardi* (Desmarest). *Aust. Wildl. Res.* **13:** 1-6.

RUSSELL, E. M., 1969. 83. Summer and winter observations of the behaviour of the euro *Macropus robustus* (Gould). *Aust. J. Zool.* **17:** 655-64.

RUSSELL, E. M., 1974. 84. The biology of kangaroos (Marsupialia: Macropodidae). *Mammal Review.* **4:** 1-59.

RUSSELL, E. M. AND RICHARDSON, B. J., 1971. 85. Some observations on the breeding, age structure, dispersion and habitat of populations of *Macropus robustus* and *Macropus antilopinus* (Marsupialia). *J. Zool.* (Lond.). **165:** 143-62.

SAMPSON, J., 1971. 86. The ecology of the Rat-Kangaroo, *Bettongia penicillata.* Ph.D. Thesis, University of Western Australia, Perth.

SANSON, G. D., NELSON, J. E. AND FELL, P., 1985. 87. Ecology of *Peradorcas concinna* in Arnhemland in a wet and a dry season. *Proc. Ecol. Soc. Aust.* **13:** 65-72.

SAXON, E. C., 1983. 88. Mapping the habitats of rare animals in the Tanami Wildlife Sanctuary (central Australia): an application of satellite imagery. *Biol. Conserv.* **27:** 243-57.

SCHLAGER, F. E., 1981. 89. Evaluation of factors influencing status and distribution of the Rufous Bettong, *Aepyprymnus rufescens,* in New South Wales. *Bull. Aust. Mamm. Soc.* **7:** 58.

SCOTT-KEMMIS, D., 1979. 90. The distribution and abundance of grey kangaroos in relation to environment in arid and semi-arid areas of New South Wales. M.Sc. Thesis, University of Sydney, Sydney.

SEEBECK, J. H., 1981. 91. *Potorous tridactylus* (Kerr) (Marsupialia: Macropodidae): its distribution, status and habitat preferences in Victoria. *Aust. Wildl. Res.* **8:** 285-306.

SHEPHERD, N. C., 1981. 92. Predation of Red Kangaroos, *Macropus rufus,* by the Dingo, *Canis familiaris dingo* (Blumenbach), in northwestern New South Wales. *Aust. Wildl. Res.* **8:** 255-62.

SHORT, J., 1982. 93. Habitat requirements of the Brush-tailed Rock-wallaby, *Petrogale penicillata,* in New South Wales. *Aust. Wildl. Res.* **9:** 239-46.

SHORT, J. AND GRIGG, G. C., 1982. 94. The abundance of kangaroos in suboptimal habitats: wheat, intensive pastoral, and mallee. *Aust. Wildl. Res.* **9:** 221-27.

SHORT, J., CAUGHLEY, G., GRICE, D. AND BROWN, B., 1983. 95. The distribution and abundance of kangaroos in relation to environment in Western Australia. *Aust. Wildl. Res.* **10:** 435-51.

SINCLAIR, R. G., 1980. 96. Red Kangaroos in New South Wales. Ph.D. Thesis, University of Sydney, Sydney.

SOUTHWELL, C., 1987. 97. Macropod studies at Wallaby Creek. II. Density and distribution of macropod species in relation to environmental variables. *Aust. Wildl. Res.* **14**: 15-33.

SOUTHWELL, C. J. AND JARMAN, P. J., 1987. 98. Macropod studies at Wallaby Creek. III. The effect of fire on pasture utilization by macropodids and cattle. *Aust. Wildl. Res.* **14**: 117-24.

SPECHT, R. L., 1970. 99. Vegetation. Pp. 44-67 *in* The Australian Environment ed. by G. W. Leeper. Melbourne University Press: Melbourne.

STRAHAN, R. (ed.), 1983. 100. "Complete Book of Australian Mammals". Angus & Robertson: Sydney.

TAYLOR, R. J., 1980. 101. Distribution of feeding activity of the Eastern Grey Kangaroo, *Macropus giganteus,* in coastal lowland of south-east Queensland. *Aust. Wildl. Res.* **7**: 317-25.

TAYLOR, R. J., 1983. 102. Association of social classes of the Wallaroo, *Macropus robustus* (Marsupialia: Macropodidae). *Aust. Wildl. Res.* **10**: 39-45.

TAYLOR, R. J., 1984. 103. Foraging in the Eastern Grey Kangaroo and the wallaroo. *J. Anim. Ecol.* **53**: 65-74.

TAYLOR, R. J., 1985. 104. Habitat use by the Eastern Grey Kangaroo and Wallaroo in an area of sympatry. *Mammalia* **49**: 173-86.

WILSON, G. R., GERRITSEN, J. AND MILTHORPE, P. L., 1976. 105. The Yellow-footed Rock-wallaby, *Petrogale xanthopus* (Macropodidae), in western New South Wales. *Aust. Wildl. Res.* **3**: 73-78.

WIENS, J. A., ROTENBERRY, J. T. AND VAN HORNE, B., 1987. Habitat occupancy patterns of North American shrubsteppe birds: the effects of spatial scale. *Oikos* **48**: 132-47.

WIENS, J. A., ADDICOTT, J. F., CASE, T. J. AND DIAMOND, J., 1986. Overview: the importance of spatial and temporal scale in ecological investigations. Pp. 145-53 *in* Community Ecology ed. by J. Diamond and T. J. Case. Harper and Row: New York.

Habitat requirements of the Northern Nailtail Wallaby *Onychogalea unguifera* (Marsupialia: Macropodidae) in the Northern Territory and Western Australia

S. INGLEBY[1], M. WESTOBY[1] and P. K. LATZ[2]

Percentage cover of herbs, grasses and shrubs with foliage below one metre was measured at 21 sites occupied by *Onychogalea unguifera* in the Northern Territory from July-September, 1986. Also recorded was the abundance of potential shelters including grass tussocks, spinifex hummocks (*Triodia* and *Plectrachne* spp.) and low shrubs and trees. All sites possessed trees and shrubs together with a percentage cover of either herbs >1% and green grass foliage >0.5% or green grass shoots >9%.

The 1986 results were tested within the species' range in Western Australia and eastern Northern Territory in 1987. The abundance of *O. unguifera* was scored at control sites (sites containing all features considered essential on the basis of 1986 data) and test sites (sites lacking 1-2 "essential" features). These data confirmed the requirement for >1% cover of herbs but indicated that 4% cover of new green grass shoots could substitute. Some feeding areas were in open plains country more than 2-3 kilometres from trees and shrubs.

INTRODUCTION

THE Northern Nailtail Wallaby *Onychogalea unguifera* (Gould) inhabits tropical and sub-tropical regions of northern Australia extending south to around 20°S (Finlayson 1961). It is the only member of its genus remaining widespread and relatively common and its status seems secure at present. However, information concerning its habitat requirements and dietary preferences is extremely limited (Burbidge 1977; Gordon 1983).

O. unguifera is known to inhabit large open coastal plains interspersed with *Melaleuca* thickets near Broome in Western Australia (Dahl 1897). In the Northern Territory it occurs in lightly wooded floodplains in the north and long grass savannah in the south and there are occasional records from rocky slopes in the Tanami Range (Parker 1973). In northern Queensland *O. unguifera* habitats vary from open woodlands with a grassy understorey to grasslands with scattered trees (Gordon 1983).

There is no detailed information on the dietary requirements of *O. unguifera*. They are reported to eat green grass in the wet season and *Melaleuca* leaves and grass roots during the dry season and their dental characteristics suggest they are grazers rather than browsers (Dahl 1897; Sanson 1978).

O. unguifera is nocturnal and the literature suggests that they feed close to shelter. Dahl (1897) found *O. unguifera* sleeping in "tussocks of coarse spear grass" and in *Melaleuca* thickets. He observed them feeding in "open patches" amongst these thickets in the evening. Le Souef and Burrell (1926) also commented that *O. unguifera* sheltered during the day in "dense thickets" and fed at night in the "cleared spaces".

This study examines the habitat requirements of *O. unguifera* within the Northern Territory and Western Australia. Dietary preferences were examined to help expose particular characteristics which make a location acceptable habitat for *O. unguifera*.

A key enabling the identification of habitat suitable for *O. unguifera* is provided together with remarks on the species' conservation.

MATERIALS AND METHODS

The study was divided into two periods. Year 1 (1986) assessed habitat parameters on occupied sites and developed a model for suitable habitat. The results are presented in the form of a key to suitable habitat (Appendix 1). Year 2 (1987) tested this model in new localities by random selection of sites based on habitat qualities and by comparing occupied sites with nearby sites having low degrees of fit to the model. A modified key was produced in light of these tests (Appendix 2).

(a) Selection of study areas in 1986

Locality records obtained from Museums and the Biological Records Scheme (Johnson and Fleming 1985) together with information from colleagues and local inhabitants were used to create a distribution map. Using this map 21 sites were selected in seven ecological domains in the Northern Territory (Newsome 1983) (Fig. 1, Appendix 3). Sampling took place during the dry season from July to September 1986.

[1]Biological Sciences Department, Macquarie University, New South Wales, 2109 Australia.
[2]Wildlife Research Section, Conservation Commission of the Northern Territory, Box 1046, Alice Springs, Northern Territory 0871.

Pages 767-82 in KANGAROOS, WALLABIES AND RAT-KANGAROOS ed. by G. Grigg, P. Jarman and I. Hume. Surrey Beatty & Sons Pty Limited, New South Wales, Australia, 1989.

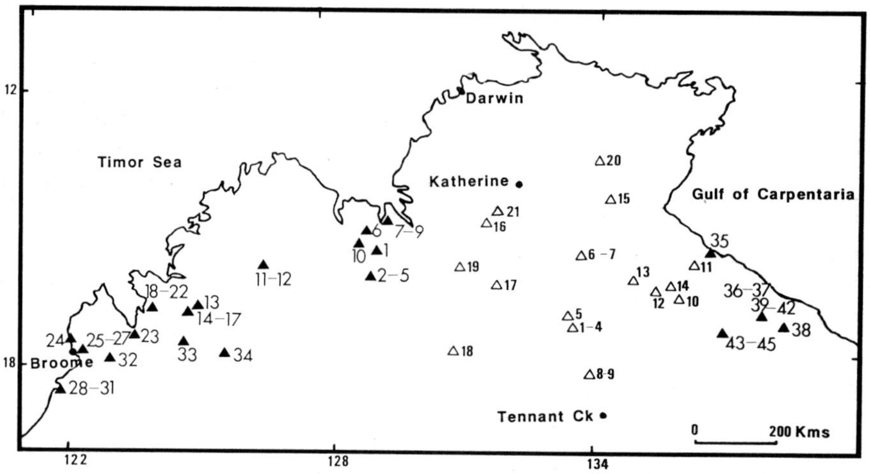

Fig. 1. Sites examined during the 1986 and 1987 dry season surveys. Open triangles represent 1986 sites and closed triangles represent 1987 sites. Numbers refer to site location given in Appendices 3 and 4.

Sites occupied by *O. unguifera* were identified by spotlighting or by daytime searches for characteristic tracks and droppings. A "site" varied in size from one to a few hectares. Places where *O. unguifera* were seen feeding were classed as "feeding areas".

Macropods are known to defaecate mostly while feeding. Consequently, areas used for feeding have high numbers of pellets and areas used for resting have few (Johnson *et al.* 1987). For this reason places with a high density of *O. unguifera* scats and tracks were also classed as feeding areas. Feeding areas were smaller regions within a site.

(b) Sites chosen for evaluation of the model in 1987

During 1987 sites were sought in areas which had not been visited during 1986 and in which locality records indicated *O. unguifera* should occur. Thirty-four sites in the Kimberleys area of Western Australia were examined during July and August 1987 (Fig. 1, Appendix 4). Of these sites 22 were located in the west Kimberleys, two in the north Kimberleys and 10 in the east Kimberleys. In addition, 11 sites were examined in eastern Northern Territory during September and October 1987.

Three hypotheses generated by the 1986 survey were selected for testing in 1987 (see results). Two different experimental designs were used to test these hypotheses.

DESIGN 1: COMPARISONS BASED ON OCCUPIED SITES

The importance of various habitat characteristics in determining *O. unguifera* abundance was examined by paired control and test sites. Control sites were occupied by *O. unguifera* and had optimal levels of the particular habitat characteristic being assessed. Test sites were sub-optimal in one or more characteristics and each was arbitrarily selected to be between 400 m and 4 000 m from its respective control site. It was not known in advance whether test sites would be occupied by *O. unguifera*. Relative *O. unguifera* abundance at control and test sites was compared using methods described later.

DESIGN 2: COMPARISONS BASED ON KEY-SATISFYING SITES

The aim of these experiments was to test whether sites fitting the optimum habitat model were more likely to be occupied than sites with one or more sub-optimum characteristics. Sites based on closeness of fit to the model were selected without reference to the presence or absence of *O. unguifera*. Abundance and presence/absence of *O. unguifera* were then assessed.

(c) Assessment of habitat attributes

At each site a sketch map was drawn to record the distribution of various landscape components such as black soil plains, floodplains, hills and swamps in relation to the feeding area. Surface soil texture was recorded using terminology described by Northcote (1979). Estimated distance to permanent water, landuse activities in the area, and the presence of other macropods or cattle (potential competitors) and dingoes or cats (potential predators) were also recorded.

A list of common plant species in the feeding and shelter areas was compiled. Terminology follows Jessop (1981).

PERCENTAGE COVER ESTIMATES

Aerial cover due to various plant categories in the feeding area was recorded using a wheelpoint (Griffin 1985). The feeding area was criss-crossed with the wheelpoint until at least 200 points one metre apart had been scored. Only plants which could potentially provide food for *O. unguifera* were scored and those trees or shrubs with no leaves below one metre (i.e., above wallaby feeding-height) were ignored.

The following six types of cover were distinguished in 1986:

(1) Hummocks: spinifex hummocks of *Triodia* or *Plectrachne* spp.
(2) Tussocks: grasses growing in bunches as distinct from those sending runners along the ground (includes annual grasses).
(3) Other grasses: including runner grasses.
(4) Herbs: dicotyledonous plants without perennial woody stems; includes members of the Malvaceae, Solanaceae, Portulacaceae, Euphorbiaceae, Papilionaceae and Amaranthaceae.
(5) Woody: woody dicotyledonous plants with foliage below one metre (includes tree saplings).
(6) Open ground: bare soil or litter.

PERCENTAGE GREEN OF GRASSES

The proportion of green to brown foliage for each grass genus was estimated. The data for each genus were pooled to provide an estimate of the percentage cover of green grass in the feeding area.

In 1987 the greenness category was further subdivided into the following categories.

(1) New shoots: young tender leaf which was still emerging.
(2) New leaf: recently emerged leaf which was still actively growing.
(3) Mature leaf: leaf which remained green but was not actively growing.
(4) Mature leaf becoming brown: including mature green leaves which retained some green pigment but had started haying off.

OBSERVATIONS ON FOOD TYPES IN THE FEEDING AREA OF 1987 SITES

The presence or absence of plants having edible rhizomes, bulbs or corms was recorded in the feeding area of each 1987 site. The number of ripe fruits encountered in a 10×10 m portion of the 1987 feed areas was also recorded. Plant species which showed signs of being grazed by *O. unguifera* and the plant part grazed were noted.

Relative *O. unguifera* abundance was scored by recording the number of faecal pellets collected in 10 minutes of searching. The technique of using dung pellet counts to estimate macropod densities is most reliable in cool, dry conditions in areas with little ground vegetation cover (Johnson and Jarman 1987). These conditions applied at most sites examined in this study.

SCAT ANALYSES

O. unguifera pellets were distinguished in the field from those of *Macropus rufus, M. agilis, M. antilopinus, M. robustus* and *Lagorchestes conspicillatus* by their small size and cylindrical shape. The presence of distinctive tracks made by *O. unguifera* also helped us to distinguish their scats from those of other macropods.

Five to 50 fresh faecal pellets were collected at each site. They would have contained food materials eaten not more than 38-58 hours before the time when the site was visited by wallabies (Hume 1982).

Five to 10 pellets from each site were macerated and examined under a microscope using the low-power objective. The proportions of grass and non-grass material in each pellet were scored as either "minor" (<20%), "moderate" (21-70%) or "major" (>70%). Also scored was the proportion of fruit remains such as skin and seed.

Seeds and seed cases enabled some plant species in the scats to be identified, and these were recorded.

SHELTER MEASUREMENTS

Information on the types of shelters used by *O. unguifera* was obtained from the literature (Dahl 1897; Le Souef and Burrell 1926), discussions with colleagues and local inhabitants, and personal observations. The following shelter units were defined.

(1) Spinifex hummock at least 40 cm high and 80 cm wide.
(2) Woody shrub or coppicing tree at least 40 cm high and 80 cm wide with foliage dense between 0 and 1 m.
(3) Grass tussocks greater than 30 cm high and 20 cm wide providing shelter on one or two sides but little overhead protection (e.g., patches of *Dichanthium fecundum, Eulalia fulva, Heteropodon contortus* and *Aristida inaequiglumis*).
(4) Woody plants providing greater than 50% shade at ground level, with foliage dense above two metres but sparse at lower levels. These units could be single or multistemmed trees or tall shrubs (i.e., *Eucalyptus, Acacia, Bauhinia* and *Maytenus* species).
(5) Other units such as fallen logs which could provide concealment in at least one or two directions.

During the 1986 survey we identified structures which might provide shelter for *O. unguifera*, and estimated the number of each structure-type available within 50 m of a random point within the feeding area.

Conclusions about shelter requirements of *O. unguifera* determined from 1986 data were tested in 1987.

Some sites examined during the 1987 survey and found to lack more than one factor in the key were not assessed for habitat attributes because we felt that sites lacking only one factor provided less ambiguous data. However, the presence or absence of *O. unguifera* at these sites was noted.

RESULTS

A description of the general vegetation characteristics at occupied sites examined in 1986 and 1987 is given in Appendix 5.

Characteristics of feeding areas

O. unguifera feeding areas examined in 1986 and 1987 were generally open areas such as the edges of black soil plains (11 sites) or other run-on areas such as claypans, swamps, floodplains, gilgais or road-verges (24 sites). The remaining sites had either been heavily grazed by cattle (7 sites), recently burnt (3 sites) or were paddocks with introduced pasture species (2 sites).

Soil textures in the feed areas ranged from loamy sands to cracking clays but sandy loams were the most common.

General characteristics of the diet in 1986 and 1987

We found no evidence that shrub foliage was eaten. We therefore directed our investigation into the importance of dicot herbs versus grasses. The relevant data are in Appendices 6 and 7.

During the 1986 and 1987 dry seasons *O. unguifera* normally concentrated its feeding on dicot herbs. At all levels of herb cover above 1%, herbs contributed 70-100% of the scat material irrespective of herb cover except where there was substantial fruit in the scats (Fig. 2). Grass consumption was high only at sites with <1% herb cover.

Sites with relatively high grass consumption and low herb consumption did not have distinctively high levels of grass cover (Fig. 3) or of green grass cover (Fig. 4).

Plant species in the diet during 1986 and 1987

The following species were eaten by *O. unguifera*: *Aeschynomene* sp., *Alternanthera nodiflora, Amaranthus* sp., *Borreria* sp., *Portulaca pilosa, Portulaca oleracea, Portulaca oligosperma, Stylosanthes humilis, Sida cordifolia, Neptunia gracilis, Phyllanthus* sp., *Corchorus aestuans, Ipomoea* sp., *Sida acuta, Sida rohlenae, Cleome viscosa, Trianthema triquetra, Indigofera linifolia, Euphorbia* sp., *Gomphrena* sp., *Dactyloctenium radulans, Enneapogon polyphyllus, Digitaria ciliaris, Panicum* sp., *Chrysopogon* sp., *Digitaria bicornis, Bulbostylis* sp., *Yakirra majuscula, Cenchrus* sp., *Fimbristylis dichotoma, Cynodon dactylon* and *Scleria* sp.

Fruit remains of *Solanum esuriale, Carissa lanceolata, Cucumis melo, Cassytha filiformis* and *Capparis* sp. were found in scats collected at 14 sites examined in 1987 (Appendix 7).

Observations on the position of tracks and scats in the feeding area indicated that *O. unguifera*

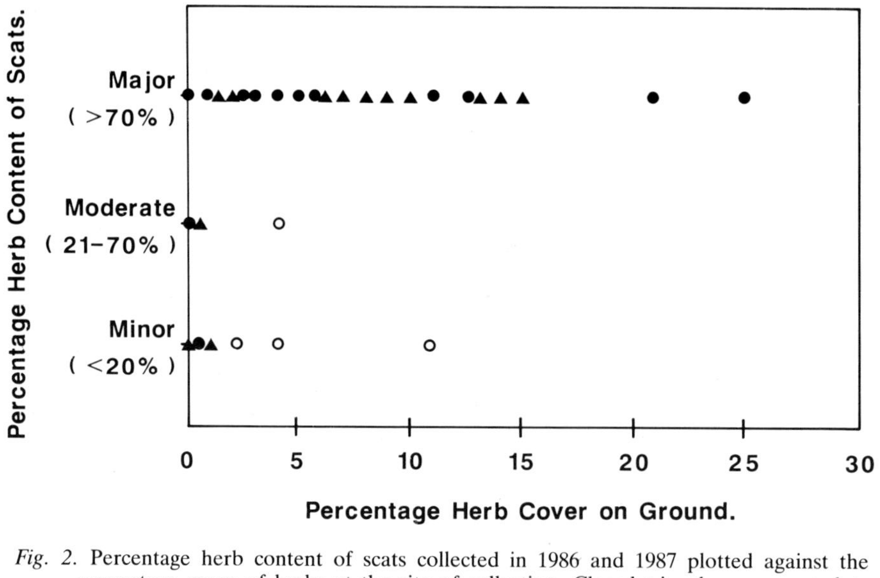

Fig. 2. Percentage herb content of scats collected in 1986 and 1987 plotted against the percentage cover of herbs at the site of collection. Closed triangles represent data from the 1986 survey, closed circles data from 1987 and open circles represent scats collected in 1987 which contained a high proportion of fruit. Percentage cover estimates are of canopy cover from wheelpoint transects. Herbs refers to herbaceous dicots. Division of scat contents into percentage grass and percentage dicot was made by inspection of at least five scats.

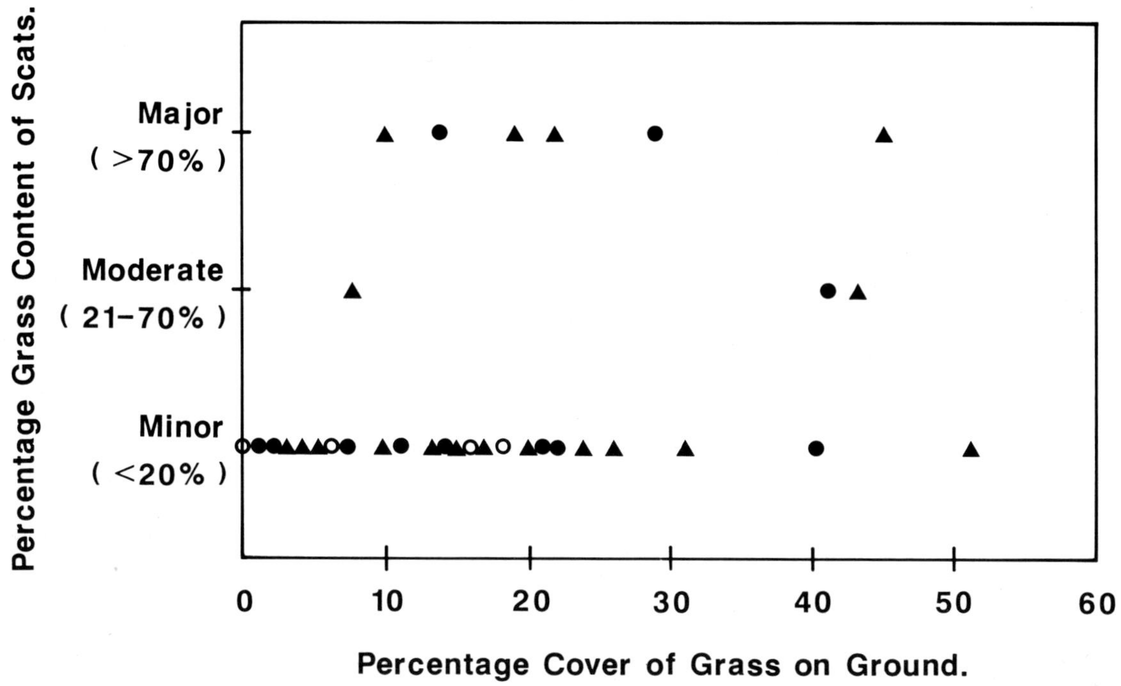

Fig. 3. Percentage grass content of scats collected in 1986 and 1987 plotted against the percentage cover of grass at the site of collection. Closed triangles represent data from the 1986 survey, closed circles data from 1987 and open circles represents scats collected in 1987 which contained a high proportion of fruit. Percentage cover estimates are of canopy cover from wheelpoint transects. Grass refers to grasses other than spinifex. Division of scat contents into percentage grass and percentage dicot was made by inspection of at least five scats.

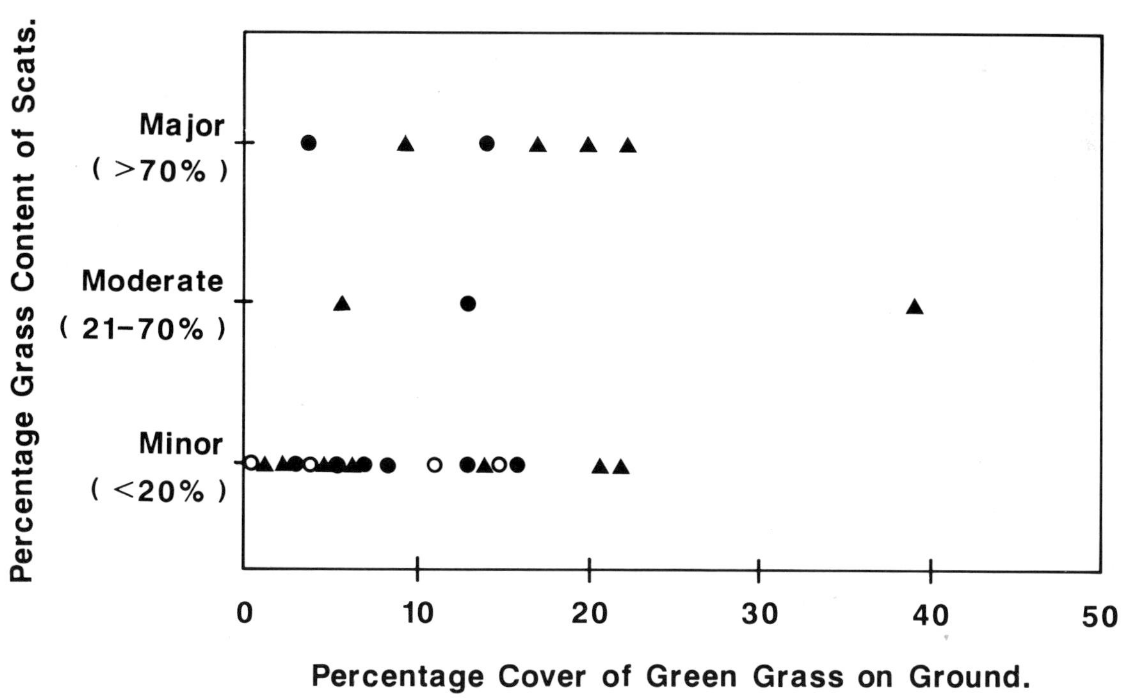

Fig. 4. Percentage grass content of scats collected in 1986 and 1987 plotted against the percentage cover of green grass at the site of collection. Closed triangles represent data from the 1986 survey, closed circles data from 1987 and open circles represent scats collected in 1987 which contained a high proportion of fruit. Percentage cover of green grass foliage was obtained by multiplying percentage grass cover estimates by the corresponding estimates of the proportion of green to brown foliage on the grasses (see methods). Division of scat contents into percentage grass and percentage dicot was made by inspection of at least five scats.

also ate the leaves of *Dichanthium fecundum, Xerochloa imberbis, Ectrosia* sp., *Eriachne glauca* and *Eriachne obtusa*. Species of *Chrysopogon* with buried fibrous butts (stem bases) or rhizomes (underground stems) were recorded at 22 sites occupied by *O. unguifera* during 1987 but only two sites showed signs of digging at the base of tussocks.

Shelter characteristics of sites examined in 1986

All sites had areas with more than 10 shade trees per hectare within 200 m of the feed area. *O. unguifera* generally fled to these areas when disturbed at night and often crouched in the grass understorey to avoid detection.

There were at least 3-4 low shrubs per hectare at each site. At fifteen sites this shrub was *Carissa lanceolata* which was seen to be used as a shelter near Elliott, Northern Territory.

Numerous Nailtail Wallaby squats (>10/ha) were found at the base of *Macropteranthes keckwickii* in a tall shrubland near Elliott. These structures provided good shade and the multi-stemmed base provided concealment in one or two directions.

There were more than 10 tussock grasses per hectare at 18 sites and none at three. Sites lacking tussock grasses also lacked spinifex hummocks. Spinifex hummocks were present at only three sites.

Hypotheses generated by the 1986 survey

The set of sites studied in 1986 were characterized by >1% herb cover and >0.5% cover of green grass foliage (relevant genera are listed in Appendix 1). The few sites with <1% herb cover always had substantial availability of green grass (Appendix 6). One hypothesis to be generated was that sites which had little cover of either herbs or green grass would not be acceptable to *O. unguifera*. Other interpretations would be that at such sites *O. unguifera* eats some third food type or that sites with low herb-cover always have some green grass cover.

All the sites at which *O. unguifera* were found had both shade trees and shrubs (most often *C. lanceolata* thickets) within 200 m. Grass tussocks or hummocks were absent from some occupied sites so it appeared they were not essential providing other types of shelter were available.

We chose three hypotheses to test during 1987. These were:

(1) Sites with <1% cover of "herbs" are unacceptable to *O. unguifera*.

(2) Sites with <2-3 "shrubs" per hectare are unacceptable.

(3) Sites with <10 trees per hectare within 200 m of the feed area are unacceptable.

It should be noted that the term "herbs" refers only to species listed in Appendix 1 and "shrubs" to woody shrubs or coppicing trees at least 40 cm high and 80 cm wide with foliage dense between zero and one metre.

Limitations of the draft key

Conclusions about habitat requirements based on the 1986 survey were regarded as hypotheses to be tested by future field work for the following reasons. Firstly, as occupied sites only were examined, we had no evidence that habitats which did not meet the key's criteria in one or two respects were not habitable. Secondly, the evidence was collected only during the dry season and only from the Northern Territory part of the geographical range of *O. unguifera*. In addition, we did not know to what extent utilized sites were a selection of habitats present in the region.

A key based on dry-season sampling should indicate preferred habitat because habitat requirements are likely to be more restrictive during the dry season than during the wet. Shelter would be available in the same places in both seasons but satisfactory food would probably be available in a much larger proportion of the landscape during the wet season. Dry-season feeding areas may sometimes be flooded during the wet, but in these cases adjacent unflooded land should provide satisfactory feeding areas. Limited wet-season surveys near Elliott, Northern Territory indicated that the dry-season sites continued to be occupied during the wet-season.

Work in 1987 was directed at testing the working hypotheses derived from the 1986 survey and addressing the other limitations.

Evaluation of the model in 1987

HYPOTHESIS 1: REQUIREMENT FOR HERBS

Comparisons based on occupied sites showed herbs to be a primary factor influencing *O. unguifera* abundance. Scat numbers were high at control sites with >1% herb cover and dropped markedly when moving to test sites with <1% herb cover (Appendix 7A).

Fruit was identified in scats collected at four control sites but no fruit was recorded at the respective test sites. It is possible that either fruit or herbs or both affected *O. unguifera* abundance at these sites. We believe that herbs were the main factor involved because our other data indicated that *O. unguifera* were often present at sites which lacked fruit (Appendix 8).

Comparisons based on key-satisfying sites (Appendix 7B) showed that sites lacking herbs were acceptable to *O. unguifera* provided new green shoots on the grasses were present. Four sites with green grass shoots but without herbs were found and these had been either recently burnt or heavily grazed by cattle. *O. unguifera* were present at all four sites. Thirteen additional sites had neither herbs nor green grass (Appendix 9) and none were occupied by *O. unguifera*.

Four sites (at which habitat attributes were not assessed) had sub-optimal levels of herbs, shrubs and green grass. *O. unguifera* were absent from all four sites. We do not know if fruits were available at these sites.

O. unguifera were abundant (30-50 scats collected in 10 mins) at three sites with >1% herb cover but where no green grass was found (Appendix 7). This suggests that green grass is not a requirement if herbs are available.

HYPOTHESIS 2: REQUIREMENT FOR SHRUBS

Evidence collected in 1987 indicated that shrubs (with foliage dense below one metre) were not an essential requirement for *O. unguifera* as they were present in both moderate to high numbers (>5 scats in 10 mins) at six sites and in reduced numbers (1-5 scats in 10 mins) at five sites without shrubs (Appendix 7).

Only one pair of sites was found at which to test the importance of shrubs using comparisons based on occupied sites. In all other cases, sites some distance away from the control sites and lacking shrubs also lacked herbs. Similarly, four sites visited (at which habitat attributes were not assessed) which lacked shrubs also lacked herbs (Appendix 7). These data suggest that shrubs may be associated with some other essential requirement of *O. unguifera*, probably herbs.

HYPOTHESIS 3: REQUIREMENT FOR TREES

Only one site was found at which to test whether sites lacking trees are acceptable using comparisons based on key-satisfying sites (Appendix 7). This was because sites lacking trees generally lacked one to three other key factors. For the same reason no comparisons based on occupied sites could be made.

In total, eleven sites were found where *O. unguifera* were present and no shrubs were recorded. Eight sites had a woodland with a long grass understorey near to the feed area and two were woodlands with open understoreys. At one site near Legune, Northern Territory, *O. unguifera* were observed feeding on an open plain at least 4 km from the closest tree or shrub shelter. The nearest potential cover was a belt of long grass about 3 km away.

DISCUSSION AND CONCLUSIONS

Observations in relation to the literature

"Large generalist herbivores" tend to choose diets not by taking a particular food with a given probability (whether high or low) each time it is encountered, but rather by taking a diet with appropriate overall characteristics (Westoby 1974). To the extent that this is true, important foods should appear in the diet at a level which does not respond to availability of the food in the environment except at very low levels of availability. Such a pattern seems to hold for dry-season consumption of herb foliage by *O. unguifera*. Grass shoots apparently can be substituted for herb foliage, but they are substituted because of shortage of herbs rather than because of abundance of green grass.

O. unguifera also ate some fruits and seeds. There appears to be no preference between herbs or fruit when both are available. At some sites *O. unguifera* ate herb foliage and at others they ate fruit. In three of the four cases where the latter occurred there is evidence that the particular fruit present *(Cassytha filiformis)* was preferred to the particular herb *(Trianthema triquestra)*. The question as to whether fruit can be substituted for herbs hardly ever arises since virtually all sites with fruit also had herbs present.

O. unguifera is reported to eat grass roots and *Melaleuca* leaves (Dahl 1897). Our data suggest that rhizomes or fibrous butts may be eaten on occasion but they do not form a major part of the diet during the dry season. Similarly we do not think that shrubs are an important part of the diet at that time.

O. unguifera are considered to use a variety of shelter types, and the literature suggests that food and shelter are available in roughly the same place (Dahl 1897; Le Souef and Burrell 1926). Most sites examined during this study had some form of shelter, either trees, shrubs or long grass, within 200 m of the feeding areas. However, observations in 1987 suggested that they may move more than 3 km from daytime shelter areas to night-time feed areas when the two are widely separated.

Our conclusions based on data collected during 1987 are:

(1) Sites without herbs and without green grass are generally unacceptable as feeding areas for *O. unguifera*.

(2) Sites without herbs are acceptable providing new green grass shoots are available.

(3) Green grass is not essential, providing herbs are available.

(4) Factors in Key 1 are partially linked i.e., most sites which have green grass towards the end of the dry season also have >1% cover of herbs.

It is not clear at this stage whether sites lacking herbs and grasses would be acceptable if sufficient fruit was available.

Habitat requirements in relation to human landuse

Cattle provide the main economic use of land in regions of Western Australia and Northern Territory occupied by *O. unguifera* and fire is a widespread modifier of habitat in these areas. We do not think that cattle grazing at current levels in non-drought years would make sites unacceptable to *O. unguifera*.

Firstly, most of the range of *O. unguifera* coincides with areas grazed by cattle and there is little evidence of a decline in the numbers or range of *O. unguifera* since the advent of pastoralism (Gordon 1983).

Secondly, cattle are unlikely to graze to the soil surface and usually leave prostrate herbs ungrazed. Moderate grazing pressure probably benefits *O. unguifera*, at least in the short term, by removing tall rank grass thereby making herbs more accessible. In addition cattle grazing often removes mature leaves from grass tussocks and induces the plant to produce new shoots (Newsome 1975). However, grazing or trampling severe enough to remove most plant cover in the understorey would disadvantage *O. unguifera*.

One possible area of dietary overlap between cattle and *O. unguifera* is the use of green grass shoots. We do not consider this to be a serious threat to *O. unguifera* because grass shoots are eaten under limited circumstances. In addition, *O. unguifera* would probably compete favourably with cattle for shoots since macropods are able to select items on a finer scale than cattle (Short 1986; Southwell and Jarman 1987).

Cattle grazing may remove grass tussocks used as shelter by *O. unguifera* but it is unlikely to remove tree or shrub shelters. The removal of grass tussocks would probably disadvantage *O. unguifera* only if alternative shelters were not available.

Fire is a common tool used in land management by Aboriginal people and pastoralists in northern Australia. Pastures are often burnt annually to encourage grasses to use residual soil moisture to produce new shoots known as "green pick" (Shaw and Norman 1973; Stocker and Mott 1981). Fires are generally lit in the early dry season leading to a mosaic pattern of burnt and unburnt patches (Lacey *et al.* 1982). This practice could provide temporarily acceptable feeding areas for *O. unguifera* providing shelter was still available.

In summary, *O. unguifera* is not likely to be much disadvantaged by judicious cattle grazing or pastoral burning.

Percentage of "suitable" habitat occupied during the 1987 dry season survey

Of the fourteen sites examined during 1987 and classed as acceptable by the key, eleven (78%) proved to be occupied by *O. unguifera*. Two of these exceptions could be credibly explained. One site where the species was absent was at the southern limits of the range in Western Australia. Another contained herb species which have not been recorded in the diet. This suggests that when the definition of "suitable" habitat in the key has been refined, nearly all suitable habitat will prove to be occupied.

Key 1 has been modified in light of tests carried out during 1987. Future work in Queensland is planned to test the rate of occupancy of sites described in Key 2 (Appendix 2).

ACKNOWLEDGEMENTS

The research was carried out with funding and assistance from the Australian National Parks and Wildlife Service under the States Co-operative Assistance Scheme. We are grateful to Dr Ken Johnson and Dr Michael Fleming for their helpful advise throughout the project and for revising the manuscript. We also thank Greg Leach and Bruce Thompson for identifying large numbers of plant specimens. For their help with field work at various times we thank Powder O'Keith, Lasse Jorgensen, Blair McFarland, Marion James, Julie Nugent, Miguel Heatwole, Gordon Wyre, Brad Lyndenmayer and Phillip Hasall. We greatly appreciate the logistic support given by the Conservation Commission of the Northern Territory.

REFERENCES

Burbidge, A. A., 1977. The status of Kangaroos and Wallabies in Australia; Report of the working group on Macropod habitat of the Standing Committee of the Council of Nature Conservation Ministers. Canberra. Aust. Govt. Publ. Service. 20p.

Dahl, K., 1897. Biological notes on Australian Mammalia. *The Zoologist.* 189-216.

Finlayson, H. H., 1961. On Central Australian Mammals. Part 4. The distribution and status of Central Australian species. *Rec. S. Aust. Mus.* **14**(1): 141-91.

Gordon, G., 1983. The Northern Nailtail Wallaby. *In* The Complete Book of Australian Mammals ed. by R. Strahan. Angus & Robertson: Sydney.

Griffin, G. F., 1985. Manual for collection and analysis of data for fire behaviour predictions: applications of a fire management strategy at Uluru National Park. *C.S.I.R.O. Division of Wildlife and Rangelands Research. Technical Memorandum* No. 22.

Hume, I. D., 1982. Digestive Physiology and Nutrition of Marsupials. Cambridge University Press. 256 p.

Jessop, J. P., 1981. Flora of Central Australia. The Australian Systematic Botany Society. Reed Pty Ltd: Sydney.

Johnson, C. N. and Jarman, P. J., 1987. Macropod studies at Wallaby Creek VI. Validation of the use of dung-pellet counts for measuring absolute densities of populations of macropodids. *Aust. Wildl. Res.* **14**: 139-45.

JOHNSON, C. N., JARMAN, P. J. AND SOUTHWELL, C. J., 1987. Macropod studies at Wallaby Creek V. Patterns of Defaecation by Eastern Grey Kangaroos and Red-Necked Wallabies. *Aust. Wildl. Res.* **14:** 133-38.

JOHNSON, K. A. AND FLEMING, M. R., 1985. A biological record scheme for the Northern Territory. Survey methods for Nature Conservation ed. by K. Myers and C. R. Margules. C.S.I.R.O.: A.C.T.

LACEY, C. J., WALKER, J. AND NOBLE, I. R., 1982. Fire in Australian tropical savannahs. *In* Ecology of Tropical Savannahs ed. by B. J. Huntley and B. H. Walker. Springer-Verlag: Berlin.

LE SOUEF, A. S. AND BURRELL, H., 1926. Wild Animals of Australasia. G. G. Harrap. London. 381p.

NEWSOME, A. E., 1975. An ecological comparison of two arid zone kangaroos of Australia, and their anomalous prosperity since the introduction of ruminant stock to their environment. *Quart. Rev. Biol.* **50:** 389-424.

NEWSOME, A. E., 1983. The Grazing Australian Marsupials. *In* Tropical Savannahs ed. by F. Bourliere. Elsevier Scientific Publ. Co. Amsterdam.

NORTHCOTE, K. H., 1979. A factual key for the recognition of Australian soils. Rellim Tech. Publ. Adelaide, South Australia.

PARKER, S. A., 1973. An annotated checklist of the native land Mammals of the Northern Territory. *Rec. S. A. Mus.* **16:** 1-57.

SANSON, G. D., 1978. The evolution and significance of mastication in the Macropodidae. *Aust. Mammal.* **2:** 23-28.

SHAW, N. H. AND NORMAN, M. J. T., 1973. Tropical and subtropical woodlands and grasslands. *In* Australian Grasslands ed. by R. M. Moore. Australian National University Press: Canberra.

SHORT, J., 1986. The effect of pasture availability on food intake, species selection and grazing behaviour of kangaroos. *J. Appl. Ecol.* **23:** 559-71.

SOUTHWELL, C. J. AND JARMAN, P. J., 1987. Macropod studies at Wallaby Creek 3. The effect of fire on pasture utilization by Macropodids and cattle. *Aust. Wildl. Res.* **14:** 117-24.

STOCKER, G. C. AND MOTT, J. J., 1981. Fire in the tropical forests and woodlands of Northern Australia. *In* Fire and the Australian Biota ed. by A. M. Gill, R. H. Groves and I. R. Noble. Australian Academy of Science: Canberra, A.C.T.

WESTOBY, M., 1974. An analysis of diet selection by large generaiist herbivores. *Amer. Nat.* 108. No. **961:** 290-304.

APPENDIX 1

Key 1: Draft key used to assess suitable habitat for *O. unguifera* based on the 1986 dry season survey within the Northern Territory.

(1) Is there at least 1-2% herb cover* (green at the end of the dry season) plus at least 0.5% cover of green grass foliage† (excluding spinifex)?

no 2

yes 3

(2) Is there at least 9% cover of green grass shoots toward the end of the dry season?

no .... not acceptable habitat

yes 3

(3) Are there shade trees (single or multistemmed, with foliage dense above 2 m) with a density of at least 10/ha over 2-3 ha?

no 4

yes 5

(4) Is there a woodland or shrubland covering more than 2-3 ha within 200 m?

no .... not acceptable habitat

yes 5

(5) Are there low shrubs (foliage dense somewhere between 0 and 1 m i.e., *Carissa lanceolata*) with a density of at least 3-5/ha?

no .... not acceptable habitat

yes .... acceptable habitat

*Relevant Herb genera include; *Sida, Evolvulus, Neptunia, Alternanthera, Bonamia, Boerhavia, Phyllanthus, Indigofera, Trephrosia, Heliotropium, Portulaca, Euphorbia.*

†Relevant grass genera include; *Dichanthium, Eulalia, Aristida, Heteropogon, Enneapogon, Digitaria, Panicum, Dactyloctenium, Eriachne. Eragrostis.*

APPENDIX 2

Key 2: Key used to assess suitable *O. unguifera* habitat during the 1987 dry season survey in the Northern Territory and Western Australia.

(1) Is there at least 1-2% herb* cover (with green foliage at the end of the dry season)?

no 2

yes 3

(2) Is there at least 4% cover of new green grass shoots (excluding spinifex) towards the end of the dry season?

no .... not acceptable habitat

yes 3

(3) Is there some form of shelter within 3 km, either long grass or spinifex hummocks (>40 cm high), shrubs (2-3/ha with foliage dense somewhere below 1 m) or a woodland (with a density of at least 5-7 trees per hectare)?

no .... not acceptable habitat

yes .... acceptable habitat

* Relevant Herb genera include; *Sida, Evolvulus, Neptunia, Alternanthera, Bonamia, Boerhavia, Phyllanthus, Indigofera, Trephrosia, Heliotropium, Corchorus, Stylosanthes, Polymeria, Trianthema, Ipomoea, Aeschynomene, Borreria, Portulaca, Euphorbia* and *Cleome.*

APPENDIX 3

Sites examined during the 1986 dry season survey.

Site	Location	Latitude	Longitude
(1)	5 km N Elliott	17 30	133 31
(2)	40 km SE Elliott	17 51	133 44
(3)	8 km W Elliott	17 36	133 29
(4)	9 km W Elliott	17 35	133 28
(5)	10 km NW Newcastle Waters	17 18	133 20
(6)	17 km NE Daly Waters	16 07	133 26
(7)	27 km NE Daly Waters	16 10	133 29
(8)	25 km S Banka Banka	19 01	134 05
(9)	24 km S Banka Banka	19 00	134 03
(10)	4 km W Mallapunya	16 58	135 46
(11)	4 km W Borroloola	16 05	136 16
(12)	34 km SE Broadmere	16 38	135 23
(13)	3 km S Tanumbirini	16 28	134 37
(14)	6 km E Cape Crawford	16 37	135 46
(15)	35 km W Roper Bar	14 45	134 11
(16)	5 km NW Willeroo	15 14	131 33
(17)	7 km S Top Springs	16 36	130 48
(18)	2 km NW Lajamanu	18 19	130 37
(19)	4 km NE Victoria R. Downs	16 23	131 01
(20)	2 km SW Mainoru	14 03	134 04
(21)	2 km W Scott Creek	14 52	131 52

APPENDIX 4

Sites examined during the 1987 dry season survey.

Site	Location	Latitude	Longitude
1	35 km NW Newry Stn	15 46	129 37
2-3	15 km N Rosewood Stn	16 18	129 04
4	1 km N Rosewood Stn	16 27	129 02
5	20 km W Newry Stn	16 02	129 05
6	40 km NNE Kununurra	15 28	128 54
7	1 km S Legune Stn	15 12	129 27
8	9 km SW Legune Stn	15 16	129 24
9	10 km SW Legune Stn	15 18	129 24
10	20 km E Kununurra	15 51	128 56
11	12 km S Mt Barnett Stn	16 49	125 58
12	17 km S Mt Barnett Stn	16 47	125 52
13	13 km NE Napier Downs Stn	17 12	124 52
14	2 km W Windjana Gorge	17 25	124 58
15	Napier Downs	17 11	124 41
16	23 km W Windjana Gorge	17 23	124 45
17	6 km W Napier Downs Stn	17 22	124 47
18	7 km SE Meda Stn	17 24	124 03
19-21	1 km W Meda Stn	17 21	124 00
22	5 km SE Meda Stn	17 24	124 02
23	7 km SW Willare Bridge	17 46	123 06
24	22 km N Quandong Pt	17 24	122 09
25	1 km N Roebuck Stn	17 55	122 28
26	3 km E Roebuck Stn	17 55	122 30
27	7 km N Roebuck Stn	17 52	122 28
28	9 km SE Frazier Downs Stn	18 52	121 46
29	8 km E Cape Jaubert	18 58	121 39
30	9 km E Cape Jaubert	18 58	121 39
31	11 km E Cape Jaubert	18 59	121 40
32	20 km WNW Dampier Downs	18 28	122 18
33	14 km WNW Blina Stn	17 41	124 25
34	10 km E Fitzroy Crossing	18 06	125 39
35	King Ash Bay	15 54	136 27
36	20 km Wollogorang	17 11	137 46
37	2 km W Wollogorang	17 12	137 57
38	5 km E Hells Gate	17 30	138 24
39-40	10 km W Wollogorang	17 15	137 52
41-42	1 km W Wollogorang	17 12	137 57
43-44	60 km SW Calvert Hills	17 32	136 54
45	60 km E Creswell Downs	17 54	136 19

APPENDIX 5

General Vegetation Characteristics at Occupied Sites.

Dry season survey in 1986

Most occupied sites examined in 1986 were open woodlands with a tussock grass understorey. In the drier southern areas (sites 1-4, 8-9, 16-19) tree species were commonly *Eucalyptus microtheca, E. pruinosa, Macropteranthes keckwickii, Bauhinia cunninghamii* and *E. terminalis.* The understorey comprised a variety of grass species such as *Aristida inaequiglumis, A. pruinosa, A. browniana, Eulalia fulva, Dichanthium fecundum, Enneapogon polyphyllus* and *Eriachne obtusa. Carissa lanceolata* was a common shrub at most sites. In northern areas (sites 6-7, 10-12, 14-15, 20-21) common tree species were *E. terminalis, E. pruinosa, E. polycarpa, E. confertifolia,* and occasionally *Melaleuca viridiflora.* Tussock grasses included *Heteropogon contortus, Chrysopogon fallax, Themeda australis* and *Sorghum australe. Carissa lanceolata* was common at most sites.

Two sites (5 and 13) were tall shrublands of *Acacia shirleyi* and *Macropteranthes keckwickii* with an open understorey. Grass species were similar to those at southern savannah woodlands.

Dry season survey in 1987

Most occupied sites examined in 1987 were open woodlands with a tussock grass understorey (sites 1-2, 4-6, 12, 16, 19, 21-23, 25, 33-34, 47-51, 53, 54). Characteristic tree species at these sites were *E. microtheca, E. pruinosa, Terminalia arostrata, Adansonia gregorii, Atalaya hemiglauca, Erythrophleum chlorostachys* and *Melaleuca viridiflora.* The understorey comprised various mid-height or short grasses including *Dichanthium fecundum, D. sericeum, D. tenuiculum, Heteropogon contortus, Chrysopogon fallax, Themeda australis, Eulalia fulva, Aristida inaequiglumis, A. pruinosa, A. holathera, A. hygrometrica, Yakirra majuscula, Enneapogon polyphyllus, E. planifolius, Eriachne obtusa, E. glauca, Sporobolus australasicus, Brachyachne convergens* and *Dactyloctenium radulans.* The sedge *Bulbostylis* was occasionally present. A variety of herb species grew in the open areas between the tussocks, among them *Bonamia media, B. pannoosa, Indigofera linifolia, I. linnaei, Neptunia* sp., *Corchorus aestuans, Boerhavia* sp., *Stylosanthes humilis, Sida fibulifera, S. acuta, S. cordifolia, Waltheria indica, Cleome viscosa* and *Polymeria ambigua. Cucumis melo* plants bearing fruits were recorded at 5 sites. Shrubs, most often *Carissa lanceolata,* were present at most sites.

One site was a tall *Acacia* shrubland with scattered *Adansonia gregorii* and *Atalaya hemiglauca.* The understorey comprised the grasses *Aristida hygrometrica, Paspalidium rarum, Brachiaria piligera, Sorghum australiense* and *Chrysopogon* sp. Herbs including *Portulaca* sp., *Sida* sp., and *Waltheria indica* grew in open areas.

The remaining sites occupied by *O. unguifera* were in habitats not encountered during the 1986 survey. Three sites (11, 13, 45) were grasslands with scattered trees. Characteristic grass species at these sites were *Chrysopogon latifolius, C. pallidus, Dichanthium fecundum, Panicum decompositum. Sorghum plumosum, Heteropogon contortus, Aristida inaequiglumis, Sehima nervosum, Eriachne glauca* and *E. tenelula. Melaleuca, Grevillea* or *Eucalyptus* species generally formed a sparse tree layer.

Three sites (29-31) near Cape Jaubert were located in a narrow belt of *Melaleuca* shrubland with a fairly open understorey. The main understorey plants were *Trianthema triquetra* covered with *Cassytha filiformis,* Spinifex and short grasses including *Eriachne obtusa, E. falcata, Aristida holathera, Fimbristylis microcarya, Sporobolus actinocladus. S. virginicus, Cynoden dactylon* and *Whitechloa cymbiformis.* The *Melaleuca* belt formed an interzone between more extensive areas of *Acacia* shrubland interspersed with Spinifex.

Three sites near Legune (7-9) were saline short grass plains interspersed with *Pandanus* or *Melaleuca* thickets. Characteristic grass and sedge species were *Xerochloa imberbis, Cyperus* sp., *Fimbristylis trachycarya, Echinochloa colona, Ectrosia* sp., *Eulalia fulva* and *Perotis rara.* One site had herbs, namely *Solanum, Triantherma triquetra, Cressa cretica* and *Heliotropium,* growing among the tussocks. The shrub layer was absent at all sites.

APPENDIX 6

Characteristics of the feeding area at sites examined during the 1986 dry season and of scat material collected at those sites.

Site numbers refer to site locations given in Appendix 3. Percentage cover estimates are of canopy cover from wheelpoint transects (see Materials and methods). Hummock refers to grasses belonging to the genera *Plectrachne* and *Triodia;* tussock to other grasses; herb, to herbaceous dicots; woody to shrub and tree cover below in 1 m height; open, to bare or litter covered ground; and green, to percentage cover of green grass foliage obtained by multiplying percentage grass cover estimated by the corresponding estimates of the proportion of green to brown foliage on the grasses (see Material and methods). Division of scat contents into percentage grass and percentage dicot was made by inspection of at least five scats.

Site	% cover estimates in the feed area						Scat contents	
	Hummock	Tussock	Herb	Woody	Open	Green	% grass	% dicot
1	—	10	2	—	88	0.5	<10	>90
2	—	51	15	1	33		<20	>80
3	—	20	9	2	70	1	<20	>80
4	—	14	10	0	75	1	<20	>80
5	1	18	3	6	73	16	—	—
6	—	18	4	2	23	16	—	—
7	—	31	1	1	66	1	<20	>80
8	—	22	0	0	77	20	>80	<20
9	6	17	2	8	67	1	<20	>80
10	2	24	7	1	66	22	<20	>80
11	—	5	6	5	83	5	<20	>80
12	—	45	—	10	49	22	>90	<10
13	—	19	1	5	76	17	>80	<20
14	—	4	8	0	87	3	<20	>80
15	—	43	—	2	56	39	50	50
16	—	10	1	0	88	9	>90	<10
17	—	26	13	1	61	21	<20	>80
18	—	20	7	0	73	2	<20	>80
19	0	9	13	5	75	6	<30	>70
20	—	3	2	5	89	3	<20	>80
21	—	15	14	—	71	14	<20	>80

APPENDIX 7

Characteristics of the feed area at sites examined during 1987 and of scat material collected at those sites.

Appendix 7A presents data for comparisons based on occupied sites. Control sites (upper line) are compared to paired test sites (lower line) with <1% herb cover (part 1) or without shrubs (part 2). Data are also presented for control sites which lack a corresponding test site (part 3). Appendix 7B shows data from comparisons based on key-satisfying sites. Site numbers refer to site locations given in Appendix 4. Percentage cover estimates are of canopy cover from wheelpoint transects (see methods). Grass refers to species of Poaceae, herb to herbaceous dicots, green to percentage cover of green grass foliage obtained by multiplying percentage grass cover estimates by the corresponding estimates of the proportion of green to brown foliage on the grasses (see Materials and Methods). Fruit refers to the estimated number of ripe fruits per hectare in the feed area. Foliage refers to the type of foliage present on grasses in the feed area: NS=New shoots, NL=New leaf, ML=Mature leaf. Division of scat contents into percentage of grass, dicot or fruit was made by inspection of at least five scats. P designates sites where *O. unguifera* was known to be present (from sightings or from tracks) but where no scats were found; asterisks mark sites without shrubs where *O. unguifera* was present; and † marks sites where the main constituent of scats was fruit.

APPENDIX 7A

Comparisons based on occupied sites.

	% cover estimates in the feed area					Scat contents			*O. unguifera*
Site	% herb	% grass	% green	foliage	fruit	% grass	% herb	% fruit	Abundance (scats/10 mins)
(1) Comparisons between control sites and corresponding test sites with <1% herb cover.									
2	11	11	0	NL,ML	10-100	<10	>80	10	40
3	1	58	9	ML	—	—	—	—	<5
8*	25	15	15	NS,NL	—	<10	>90	—	36
9*	1	22	11	ML	—	<10	>90	—	4
19†	4	6	—	—	—	<10	>80	<10	30
20	1	37	9	NS,NL,ML	—	—	—	—	—
25	21	6	4	NL,ML	—	<10	>90	—	34
27	1	30	22	NL,ML	—	—	—	—	—
39*	13	2	1	NS,NL,ML	—	<10	>90	—	25
40*	1	1	—	—	—	—	—	—	2
41*	3	5	3	NS,NL,ML	—	not clear		<10	12
42*	—	19	9	NS,NL,ML	—	—	—	—	—
43†	3	1	—	—	—	<10	>70	<20	50
44	—	13	—	—	—	—	—	—	3
(2) Comparisons between control sites (upper line) and their paired test site lacking shrubs.									
25	21	6	4	NL,ML	—	<10	>90		34
26	16	7	1	NS,NL	—	—	—	—	—
(3) Control sites for which no test site was located.									
5	5	11	0	NS,NL,ML	—	not clear		10	3
13	6	22	2	NL,ML	—	<10	>80	<10	8
18	9	18	1	—	—	—	—	—	P
34	1	21	7	NL,ML	—	<20	>80	—	30

APPENDIX 7 — *Continued*

APPENDIX 7B

Comparisons based on key-satisfying sites.

	% cover estimates in the feed area					Scat contents			*O. unguifera*
Site	herb	grass	green	foliage	fruit/ha	% grass	% herb	% fruit	Abundance
Control sites.									
1	5	40	14	NL,ML	10-100	<10	>80	<10	11
4	11	10	4	NL,ML	—	—	—	—	5
10	1	33	3	NL	—	—	—	—	—
17	14	2	1	NL,ML	—	—	—	—	—
21	2	22	2	NS	—	—	—	—	P
22	3	7	3	NS,NL,ML	—	<20	>80	—	30
23	16	14	5	NL	—	—	—	—	2
28	11	14	4	NL,ML	—	—	—	—	—
29†	4	18	16	NL,ML	10-100	<10	0-40	50-90	10
30†	2	6	5	NS,NL,ML	10-100	<10	0-30	60-90	>50
31†	4	16	11	NS,NL,ML	—	<10	40-50	40-50	20
33	2	23	0	NS,NL	—	<30	>70	—	2
35	1	5	3	NS,NL	—	<10	>90	—	25
45†	11	—	—	—	—	<20	0-30	50-80	50
Test sites lacking herbs.									
7	—	16	8	NS,NL	—	<10	>90	—	10
11*	0	41	13	NS,NL,ML	—	50	50	—	20
12*	—	14	14	NS	—	>70	<20	<10	12
38	—	29	4	NS,NL,ML	—	>90	<10	—	5
Test sites lacking shrubs.									
6*	16	13	5	NL,ML	—	<10	>80	10	4
14	7	58	3	NL,ML	—	—	—	—	—
15	1	14	11	NS	—	—	—	—	—
16*	7	18	5	NL,ML	—	<10	>80	10	4
32*	3	24	0	NL,ML	—	—	—	—	3
36	2	18	2	NS,NL	—	—	—	—	—
37*	2	6	1	NS,NL,ML	—	<10	>90	—	12
Test sites lacking trees.									
24	8	32	6	NL,ML	—	—	—	—	0

APPENDIX 8

Number of sites occupied by *O. unguifera* in relation to the presence or absence of herbs and fruit.

Appendix 8A presents data from comparisons based on occupied sites and Appendix 8B shows data from comparisons based on key-satisfying sites. Fruit present means there was evidence of fruit in scats or ripe fruits were found in the feed area at that site. Fruit absent means no such evidence was found. Herb cover refers to canopy cover due to herbaceous dicots in the feed area (measured during wheelpoint transects). *O. unguifera* present means that characteristic tracks or scats were found or animals were seen at a site.

(A) Comparisons based on occupied sites.

		Herb cover >1%		Herb cover <1%	
Fruit present	*O. unguifera* present	4	*O. unguifera* present	1
	absent	0	absent	0
Fruit absent	*O. unguifera* present	2	*O. unguifera* present	3
	absent	0	absent	2

(B) Comparisons based on key-satisfying sites.

		Herb cover >1%		Herb cover <1%
Fruit present	*O. unguifera* present	7	*O. unguifera* present	1
	absent	0	absent	0
Fruit absent	*O. unguifera* present	4	*O. unguifera* present	3
	absent	3	absent	0

APPENDIX 9

Number of control sites and test sites (used in comparisons based on key-satisfying sites) which were occupied by *O. unguifera* during the 1987 dry season survey. *O. unguifera* abundance is given as the number of scats collected in 10 minutes searching. When no scats or tracks were found in 30 minutes searching, the species was considered to be absent. Control sites are those which satisfy key 1. Test sites lack the minimum requirement for herb cover specified by the key but have >4% cover of new green grass shoots in the feed area. Sites visited but habitat attributes not assessed are those failing to satisfy the key with respect to more than one factor (i.e., they have <1% herb cover and <0.5% cover of green grass and in some cases lack shrubs with foliage dense below one metre).

O. unguifera abundance (scats/10 min)	Control sites	Test sites (<1% herb cover with new green grass shoots)	Sites examined, habitat attributes not assessed (<1% herb cover without new green grass shoots)	Sites examined, habitat attributes not assessed (<1% herb cover without green shoots and without shrubs)
Present (>5)	8	3	0	0
Present (1-5)	3	1	0	0
Absent (0)	3	0	13	4

The abundance and distribution of Black Wallaroos *Macropus bernardus* and Common Wallaroos *Macropus robustus* on the Arnhem Land Plateau Australia

A. J. PRESS[1]

Macropus bernardus (Black Wallaroo) and *Macropus robustus* (Common Wallaroo) are sympatric on the Arnhem Land Plateau in northern Australia. A helicopter was used to survey the species in this remote and inaccessible landscape and to obtain indices of relative abundance in different habitats and sites within Kakadu National Park. Over all areas surveyed, *Macropus bernardus* was three times more abundant than *Macropus robustus.* Both species occurred in monsoon forest, woodland and spinifex habitats, but *Macropus robustus* was spread sparsely through these habitats, while *Macropus bernardus* was found in significantly higher observed densities in some sites. There were no significant differences in the observed density of *Macropus robustus* between all the sites surveyed. *Macropus bernardus* was most abundant in two monsoon forest sites although there were significant differences in observed densities among all the monsoon forest sites. There were significant differences in the observed density of *M. bernardus* among the spinifex sites but not among the woodland sites. The observed population densities were probably most negatively biased in the monsoon forest areas and least in the spinifex sites.

INTRODUCTION

THE Black Wallaroo, *Macropus bernardus* Rothschild, is a little known kangaroo restricted to the sandstone rocklands of western Arnhem Land and its outliers. Sympatric with *M. bernardus* in this region is the widely distributed Common Wallaroo, *M. robustus* Gould (Russell and Richardson 1971; Calaby 1983; Archer *et al.* 1985; Press 1988). In terms of status, both *M. bernardus* and *M. robustus* are listed by Burbidge (1977) as common species which have had little change to their distribution since European settlement in Australia. Much more is known about the ecology of *M. robustus* than *M. bernardus.* It occurs over most of Australia in suitable habitat — rocky ranges, hills, scarps, ridges and plateaux (Finlayson 1961; Russell and Richardson 1971; Caughley and Bayliss 1982; Archer *et al.* 1985). It is found in a wide range of vegetation associations from grasslands to rainforest (Archer *et al.* 1985). The habitat requirements of *M. bernardus* are superficially very similar to those of *M. robustus:* "steep slopes and gorges of the escarpment" (Bolton 1970, p36); "on boulder-strewn hillsides" (Parker 1971, p41); and "steep scarps and hills of Arnhem Land and nearby areas" (Russell and Richardson 1971, p132). Calaby (1983, p245) describes *M. bernardus* as inhabiting the escarpments and plateaux which are "variously covered with eucalypt woodland with a sparse understorey of grass and shrubs, a sparse community of spinifex and shrubs . . . or small patches of rainforest (and) . . . much bare rock".

Press (1988) recorded the presence of *M. bernardus* on the sandstone escarpment and plateau areas in Kakadu National Park and on the outliers such as Mt Brockman and Nourlangie Rock. It was found in many different vegetation associations and appeared to be most common in heavily dissected sandstone. *Macropus robustus* was recorded on rocky scree slopes at the foot of the escarpment, hills, and on heavily eroded sandstone remnants (Press 1988) and was later recorded on the Arnhem Land plateau and on top of Nourlangie Rock (Press, unpubl. data). It also occurred in a number of vegetation associations similar to *M. bernardus* and was syntopic with *M. bernardus* in parts of its range in Kakadu National Park. Far fewer *M. robustus* were recorded than *M. bernardus.*

The present study was undertaken to assess the relative abundances of *M. bernardus* and *M. robustus* within the dominant habitats of the Arnhem Land plateau.

METHODS

Study area

The study area comprised the sandstone plateau of western Arnhem Land within Kakadu National Park (Stages 1, 2 and 3; Fig. 1). The plateau dominates the southern and eastern portions of Kakadu National Park and extends south to the Katherine area, north to Nabarlek and east to the Mann River. Inside Kakadu National Park the plateau varies in elevation from about 200 m to 300 m with residual peaks to 500 m.

[1]Australian National Parks and Wildlife Service, Kakadu National Park, Jabiru, Northern Territory, Australia 0886.

Pages 783-86 in KANGAROOS, WALLABIES AND RAT-KANGAROOS ed. by G. Grigg, P. Jarman and I. Hume. Surrey Beatty & Sons Pty Limited, New South Wales, Australia, 1989.

The sandstone plateau supports a great many different vegetation associations, including *Eucalyptus* dominated woodland and open forest, *Allosyncarpia* dominated monsoon forest (closed forest or rainforest), spinifex grasslands and mixed vegetation associations (Schodde *et al.* 1987). There are also many areas of bare rock.

Survey sites

Three vegetation types were chosen for survey as these represented the majority of the sandstone vegetation mapped by Schodde *et al.* (1987). These were: sandstone monsoon forest (sandstone rainforest of Schodde *et al.* 1987, but also including elements of their "broadleaf shrubbery"); sandstone woodland and forest; and sandstone spinifex. The woodland and forest habitats cover the most area of the plateau in Kakadu National Park and the monsoon forests the least. Areas were selected for survey using the "Vegetation Habitats" maps of Schodde *et al.* (1987). Areas were chosen on the basis of uniformity of the dominant mapped vegetation types. The survey areas are shown in Figure 1. One spinifex site, site S1 (Fig. 1) was situated on a sandstone sub-plateau shelf.

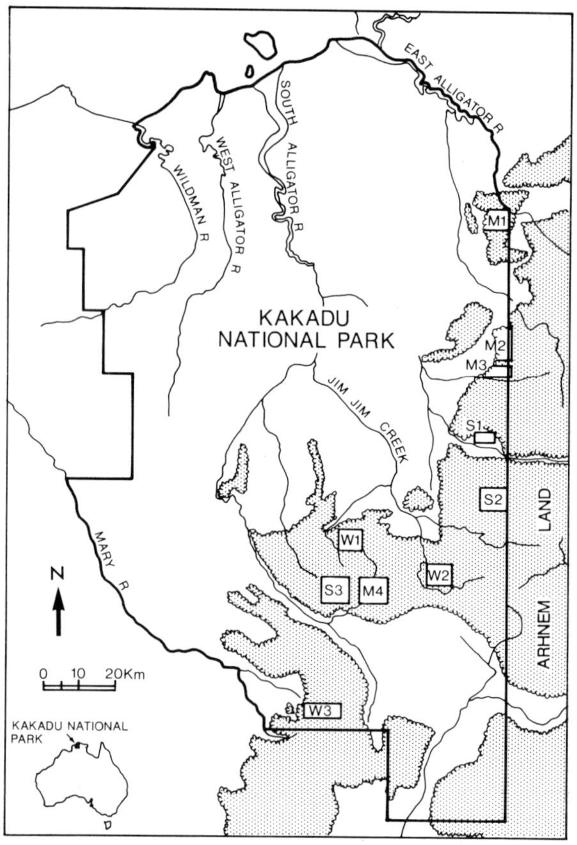

Fig. 1. Survey sites in Kakadu National Park. The shaded area represents the sandstone plateau. Site S1 is on a sandstone sub-plateau shelf. Monsoon forest sites: M1, M2, M3 and M4; Woodland sites: W1, W2 and W3; Spinifex sites: S1, S2 and S3.

Survey methods

East-west transects were marked on 1:100 000 topographic maps of each survey site. The length of the transects varied from site to site depending on the size of the chosen survey area. In some areas, where the escarpment formed one edge of the survey area (sites M1, M2 and M3, Fig. 1) the length of the transects varied within sites. The distance between transects was either 500 m (site M1, Fig. 1), 1 000 m (sites M2, M3, W1, W2, S1, S2, S3, Fig. 1), or 2 000 m (sites M4, W3, Fig. 1).

A Bell 206 "Jetranger" helicopter was used to survey the macropods. The helicopter carried a pilot and an observer (the author) in the front and a navigator in the rear seat. The helicopter flew at a constant speed of 30 knots, and as far as was possible in the rugged terrain, at a height of 20 m above ground level. A line was marked on the bubble of the helicopter which defined a transect of 100 m on the ground from the centre of the helicopter to a point to the left of the observer when the helicopter was at 20 m above ground level. Only animals seen within this transect were used in analyses.

Surveys were undertaken between first light and 1030 hours during January 1987 (sites M1, M2, M3, S1, S2, W1, W2) and February 1988 (sites S3, M4, W3). This was done to take advantage of overcast and cool wet season conditions, when it was expected that the animals would be more active in the daylight hours.

Analyses

The number of animals seen on each transect was divided by the length of each transect in kilometres and multiplied by 10 to give an index of the number of animals seen per square kilometre. These figures were used in analyses of variance with the transects representing the replicates in the analyses.

Analysis of variance of the observed density of animals recorded on each transect was undertaken with the factors "Species", "Habitat" and "Site" (nested within Habitat) included in the analysis (Zar 1984; Pp. 253-60). The analysis is given in Table 1. There were two species, *M. bernardus* and *M. robustus*; three habitats, Monsoon, Woodland and Spinifex; and three sites within each habitat (one monsoon site, M4 Fig. 1, was randomly discarded from this analysis). There were four replicates in each cell of the analysis with the transects chosen randomly as replicates in those sites with more than four replicates. One way analyses of variance were undertaken after the above analysis revealed a significant third order interaction. This was done in order to compare data from all transects in all sites sampled.

Table 1. Analyses of variance of the observed density of *M. bernardus* and *M. robustus*. Sp, Species; H, Habitats; S, Sites. Brackets denote that one factor is nested within the bracketed factor. *, P<0.05.

Source of variation	Sums of Squares x 10^2	Degrees of freedom	Mean Squares x 10^2	F ratio
Species	5.7235	1	5.7235	3.41
Habitats	8.0175	2	4.0088	2.43
Sites (Hab)	9.9025	6	1.6504	2.94*
Sp × H	7.2419	2	3.6210	2.16
Sp × S(H)	10.0708	6	1.6785	2.99*
Residual	30.3220	54	0.5616	
TOTAL	71.2790	71		

Colour aerial photographs (scale 1:25 000) were used to derive habitat indices. The transects flown were marked on the photographs and the following indices derived:

(a) *Creek* — the number of water courses marked on the topographic maps which were crossed by the transects were divided by the length of the transect; (b) *Jungle* — the transect was divided into 0.25 kilometre sections and the number of these selections containing patches of monsoon forest was divided by the length of the transect. (c) *Ratio* — the amount of rock (i.e., non-forested areas) and the amount of forested vegetation was calculated as in (c). The ratio of *rock:forest* was calculated. These habitat indices were used in regression analyses of the numbers seen per transect. The habitat indices are summarized in Table 2. When regression analyses of the observed densities (of *M. bernardus* and *M. robustus*) against the habitat indices were undertaken, all transects flown were used in the analyses (this included one transect between sites M2 and M3).

RESULTS

The observed densities of *M. bernardus* and *M. robustus* were not distributed evenly among sites and habitats, as shown by the significant third order interaction in the analysis of variance in Table 1 (Species X Sites nested within Habitats). The mean observed density for each species

Table 2. Habitat indices. Under "Site", the site code is given and the number of transects and average transect length are shown in brackets; the mean index per transect and standard deviation (in brackets) are given for the habitat variables.

Site	Creek	Jungle	Ratio
M1(9,5.8)	0.23(0.09)	0.58(0.27)	2.05(1.56)
M2(5,3.4)	0.12(0.04)	0.32(0.16)	5.47(4.26)
M3(4,5.5)	0.12(0.08)	0.22(0.16)	3.69(0.85)
M4(4,10.0)	0.22(0.04)	0.41(0.21)	0.48(0.23)
W1(6,7.0)	0.25(0.07)	0.02(0.04)	0.30(0.33)
W2(7,9.0)	0.21(0.05)	0.01(0.02)	0.07(0.14)
W3(4,5.0)	0.00(0.00)	0.00(0.00)	0.00(0.00)
S1(5,5.0)	0.14(0.08)	0.09(0.09)	9.02(9.22)
S2(6,5.0)	0.30(0.05)	0.07(0.05)	14.00(7.75)
S3(4,10.0)	0.25(0.03)	0.08(0.10)	1.99(0.76)

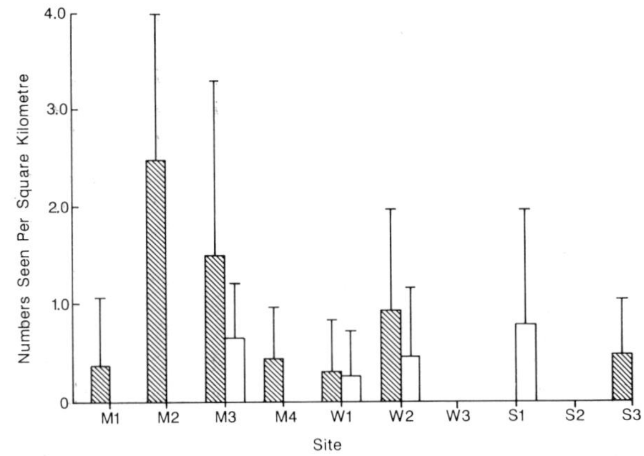

Fig. 2. The observed density of Black Wallaroos and Common Wallaroos at each site. Black Wallaroos, solid bars; Common Wallaroos, open bars. The mean observed density and standard deviation is given for each site. Sites labelled as in Table 1.

recorded at each survey site are shown in Figure 2. Averaged over all sites, 0.62 ± 1.07 *M. bernardus* and 0.21 ± 0.59 *M. robustus* were seen per square kilometre.

One way analyses of variance using data from all the sites and transects showed that: (a) there were significant differences in the observed densities of *M. bernardus* among all sites ($F_{9,44}$ = 4.22, P<0.001); (b) there were significant differences in the observed densities of *M. bernardus* among the monsoon forest sites ($F_{3,18}$ = 4.34, P<0.05); (c) there were significant differences in the observed densities of *M. bernardus* among spinifex sites ($F_{2,12}$ = 4.40, P<0.05); (d) there were no significant differences in the observed densities of *M. bernardus* among woodland sites ($F_{2,14}$ = 1.97, P>0.10); (e) there were no significant differences in the observed densities of *M. robustus* among all sites ($F_{9,44}$ = 1.54, P>0.10).

Multiple regression analyses of the observed densities of *M. bernardus* and *M. robustus* against the variables Creek, Jungle and Ratio showed no significant relationships with any or all of the variables (unweighted least squares regression of the counts of *M. bernardus*: overall $F_{2,51}$ = 1.06, P>0.30 and R^2= 0.06; and with *M. robustus:* $F_{2,51}$ <1, P>0.50 and R^2 = 0.03).

DISCUSSION

The counts of *M. bernardus* and *M. robustus* in this study are indices of the true population densities and not estimates (see Caughley 1979 and Barnes *et al.* 1986). In the present study no corrections were made for sightability. It is probable that *M. bernardus* and *M. robustus* were equally visible during the surveys — they were both flushed by the low flying helicopter and although *M. bernardus* is smaller than *M. robustus* it was easily seen. Differences in sightability between the different habitats were likely, though. Short and Bayliss (1985) and Bayliss and Yeomans (in press) have shown that canopy cover affects sightability during aerial survey. These studies found that the greater the density of canopy cover the more the population was underestimated by the aerial count. In the present study, the spinifex sites had the least canopy cover and the greatest average rock:forest ratio (Table 2) and therefore the greatest sightability. The woodland sites had the next densest canopy and the least average rock:forest ratio (Table 2). The monsoon forest patches had the densest canopy cover but because they occurred in discrete patches, the monsoon forest sites had intermediate rock:forest ratios. It is likely that the true population density was most greatly underestimated in the monsoon forest sites and least in the spinifex sites. Thus the real differences in density between the monsoon forest sites and other sites was probably greater than this study showed.

Macropus bernardus was more common in this survey than the sympatric *M. robustus*. The greatest observed densities of *M. bernardus* were in monsoon forest (sites M2 and M3, Fig. 2), but there were significant differences among all monsoon forest sites. There were no significant differences in the observed density of *M. bernardus* among the woodland sites, although none were observed at one site (W3, Fig. 2). No *M. bernardus* were counted in the transects in two out of the three spinifex sites. *Macropus robustus* was recorded in fewer sites than *M. bernardus* and generally had lower observed densities at each site (Fig. 2). There were no significant differences in the observed densities of *M. robustus* among all the sites surveyed.

The observed densities of *M. bernardus* and *M. robustus* were not significantly related to any of the habitat indices derived. The fact that *M. bernardus* had significantly different observed densities in different sites indicates that *M. bernardus* had habitat requirements which were not measured in this study. Russell (1969) states that at high temperatures, *M. robustus* used shelters which gave the greatest protection from solar radiation. Qualitative observations made during the present study and by Press (1988) have shown that *M. bernardus* was often seen in areas of deeply dissected sandstone. This may have been associated with requirements for shelter, but the data were not adequate to draw firm conclusions. Further investigations using more intensive sampling techniques such as on ground counts, are needed to determine differences in the habitat requirements of these two species in this region.

ACKNOWLEDGEMENTS

This project was undertaken as part of the research and monitoring programmes conducted by the Australian National Parks and Wildlife Service in Kakadu National Park. I wish to thank the Director of A.N.P.W.S., Prof. J. D. Ovington for the opportunity to carry out this study. I thank the staff of Kakadu National Park, especially A. Skeat, P. Godfrey and M. Sandford who navigated. A. Skeat gave critical advice on the manuscript.

REFERENCES

ARCHER, M., FLANNERY, T. F. AND GRIGG, G. C., 1985. The Kangaroo. Weldons: Sydney.

BARNES, A., HILL, G. J. E. AND WILSON, G. R., 1986. Correcting for incomplete sighting in aerial survey of kangaroos. *Aust. Wildl. Res.* **13**: 339-48.

BAYLISS, P. AND YEOMANS, K. (in press). Factors affecting the visibility of feral livestock counted during aerial surveys in the Northern Territory. *J. Anim. Ecol.*

BOLTON, B. L., 1970. Northern Territory kangaroos — a unique group. *Turnoff* **2**: 32-37.

BURBIDGE, A. A., 1977. The Status of Kangaroos and Wallabies in Australia. Australian Government Publishing Service: Canberra.

CALABY, J. H., 1983. Black Wallaroo. P. 254 *in* The Complete Book of Australian Mammals ed. by R. Strahan. Angus & Robertson: Sydney.

CAUGHLEY, G., 1974. Bias in aerial survey. *J. Wildl. Manage.* **38**: 921-33.

CAUGHLEY, J. AND BAYLISS, P., 1982. Kangaroo populations in inland New South Wales. Pp.46-49 *in* Parks and Wildlife — Kangaroos and Other Wildlife of New South Wales ed. by C. Haigh. New South Wales National Parks and Wildlife Service: Sydney.

FINLAYSON, H. H., 1961. On central Australian mammals. Part IV — The distribution and status of central Australian species. *Rec. South Aust. Mus.* **14**: 141-96.

PARKER, S. A., 1971. Notes on the small black wallaroo *Macropus bernardus* (Rothschild, 1904) of Arnhem Land. *Vic. Nat.* **88**: 41-43.

PRESS, A. J., 1988. The distribution and status of macropods in Kakadu National Park, Northern Territory, Australia. *Aust. Mammal.* **11**: 103-08.

RUSSELL, E. M., 1969. Summer and winter observations of the Euro *Macropus robustus* (Gould). *Aust. J. Zool.* **17**: 655-64.

RUSSELL, E. M. AND RICHARDSON, B. J., 1971. Some observations on the breeding, age structure, dispersion, and habitat of populations of *Macropus robustus* and *Macropus antilopinus* (Marsupialia). *J. Zool. Lond.* **165**: 131-42.

SCHODDE, R., HEDLEY, A. B., MAISON, I. J. AND MARTENSZ, P. N., 1987. Vegetation Habitats, Kakadu National Park, Alligator Rivers Region, Northern Territory, Australia. Unpublished report to the Australian National Parks and Wildlife Service, Canberra.

SHORT, J. AND BAYLISS, P., 1985. Bias in aerial survey estimates of kangaroo density. *J. Applied Ecol.* **22**: 415-22.

ZAR, J. H., 1984. Biostatistical Analysis. Prentice-Hall: New Jersey.

Age structures, reproduction and mortality of the Eastern Grey Kangaroo (*Macropus giganteus* Shaw) from Yan Yean, Victoria

DARREN G. QUIN[1]

Information is presented on 240 Eastern Grey Kangaroos shot in the Yan Yean catchment area of Victoria, and a further 324 carcasses (mortality resulting from natural causes) found on this site. The shot samples (from 1962-1964) and the mortality samples (from 1960, 1962 and 1963) were examined to ascertain patterns of reproduction, mortality and age structure. A majority of females matured sexually between their second and third year, and most adult females appeared to raise one young per year. Young were born throughout the year with a marked summer peak. The sex ratio of 47 pouch young did not differ significantly from parity. The overall sex ratio of the adult population was slightly female-biased and was significantly different from equality. Age structures of the shot sample display a high proportion of kangaroos younger than three years (59%) and very few animals (7%) older than ten years. The number of males found dead always exceeded that of females; however, the difference was significant only for the 1960 sample. Almost all known mortality occurred during winter and this mortality appeared to be age-specific in the one to two year class during 1962, but apparently had little impact on the age composition of the population.

INTRODUCTION

THE Eastern Grey Kangaroo, *Macropus giganteus* Shaw, has a continuous distribution throughout eastern Australia including southeastern South Australia, and central and northeastern Tasmania; it occupies various habitats throughout a wide geographic and climatic range (Poole 1982; Caughley *et al.* 1984). In certain areas, Eastern Grey Kangaroos are culled as a pest control measure and this action has provided information on their biology. Age structures of populations were presented by Wilson (1975) for Eastern Grey Kangaroos shot at various localities in New South Wales.

Field studies of reproduction in Eastern Grey Kangaroos have been conducted in southern Queensland (Kirkpatrick 1965b; Kirkpatrick and McEvoy 1966), central New South Wales (Poole 1973) and Tasmania (Pearse 1981). Poole (1983) also provided information on the breeding biology of this species from widespread locations in eastern Australia. However, little is known of the population biology of Eastern Grey Kangaroos in Victoria.

Presented in this paper are seasonal age structures, sex ratios and female breeding information of a population of Eastern Grey Kangaroos shot during 1962-1964 from the Yan Yean catchment area of Victoria. Kangaroos were shot in order to prevent overstocking in the catchment, and to investigate the heavy mortality observed in previous years, particularly during winter. Age structures and sex ratios of culled kangaroos are compared with those of kangaroos found dead in 1960, 1962 and 1963 on this site.

Collection site

The Yan Yean catchment area (37°34′S, 145°09′E) comprises approximately 527 ha of improved pasture now in a state of regeneration (MMBW 1984). It is situated 34 km north-east of Melbourne. The vegetation is savannah-like, covered in parts with eucalypt and pine stands (MMBW 1984). The kangaroo population at Yan Yean is unnaturally confined by the reservoir water body and a 2 m-high fence located along approximately half the catchment boundaries, and as a result there is the potential for overstocking. The population density has been estimated as being between 3.8 and 5.7 kangaroos ha^{-1} in the catchment (MMBW 1984). Rainfall averages 666 mm per year with at least 45 mm falling each month on average. Mean maximum and minimum temperatures in the hottest and coldest months respectively are: February 26.8°-16.3°C; July 12.9°-6.7°C (Bureau of Meteorology).

MATERIALS AND METHODS

(i) *Shot sample*

A representative sample of 245 kangaroos with 47 pouch young were shot by wildlife officers. The number shot in each season (excluding five animals whose molar index and basal length of the skull did not conform to regression curves; see later) are shown in Table 1. The wildlife officers also recorded the sex and various body measurements for each kangaroo and the heads were removed and numbered, the skulls cleaned and lodged at the Museum of Victoria. In 1984 their age was determined by the anterior progression of the molars (Kirkpatrick 1964, 1965a).

[1]Department of Zoology, University of Melbourne, Parkville, Victoria 3052.
Present address: Department of Ecosystem Management, University of New England, Armidale, New South Wales 2351.

Pages 787-94 in KANGAROOS, WALLABIES AND RAT-KANGAROOS ed. by G. Grigg, P. Jarman and I. Hume. Surrey Beatty & Sons Pty Limited, New South Wales, Australia, 1989.

The 1962 and 1963 autumn and winter data, and the 1962 and 1964 spring data were pooled in analyses of seasonal patterns of reproduction of females, and age-sex composition.

(ii) *Mortality sample*

Skulls were collected by wildlife officers whenever carcasses of kangaroos which had died from natural causes were found within the catchment area. The numbers collected during 1960, 1962 and 1963 were 127, 135 and 62 respectively. Where possible, the sex of each carcass was recorded and an estimate was made of the animal's approximate time of death. The age of each animal was determined by molar progression. Where carcasses could not be sexed measurements of molar index and basal length (anterior tips of premaxillae to anterior rim of foramen magnum) of the skull were used to establish the sex by reference to regressions relating these variables derived from a sample of animals that were shot or found dead of known sex (Quin unpublished). However, the sex of many smaller skulls could not be determined because the regressions for each sex do not clearly diverge until approximately three years of age; some skulls were damaged, preventing basal length measurements. Kirkpatrick's (1967) method of sexing unknown animals by using discriminant values of molar index for skulls of a particular basal length could not be used in this study because the population at Yan Yean was found to be composed of larger animals for a given age. Hence, the basal length of known females from Yan Yean frequently exceeded Kirkpatrick's discriminant values.

(iii) *Female reproductive status, birth distribution and age of pouch young*

Wildlife officers examined the pouch and uteri of shot females, and allocated each to one of seven categories (see legend, Figure 1).

Ages of pouch young were estimated using Kirkpatrick's (1965a) regressions for hind-foot length and tail length measurements, averaging the age obtained in days. Date of birth of pouch young was estimated by subtracting the age of the pouch young from the date its mother was shot.

RESULTS

Age distribution and sex ratios

(i) SHOT SAMPLE

A higher proportion of females than males were shot in each season of each year although the sex ratio was never significantly different from unity. However, the overall sex ratio of 71 males per 100 females was significantly different from parity ($P<0.05$).

Table 1. The numbers of male and female Eastern Grey Kangaroos shot during various seasons from 1962-1964 at Yan Yean, Victoria. The chi-square values (with Yates correction for 1 degree of freedom) for disparity between the number of males and females shot are also presented. (N.S.=not significant, N.T.=not testable).

Year and Season	Number of Males	Number of Females	Chi-square value for Disparity Males : Females
Autumn 1962	24	32	1.16 N.S.
Winter 1962	14	17	0.32 N.S.
Spring 1962	16	22	0.98 N.S.
Summer 1962/63	19	28	1.75 N.S.
Autumn 1963	9	10	0.11 N.S.
Winter 1963	3	5	N.T.
Spring 1964	15	26	2.98 N.S.
TOTAL	100	140	6.67 ($P<0.05$)

Figure 1 shows seasonal age distributions pooled for autumn 1962 and 1963, winter 1962 and 1963, spring 1962 and 1964 and the summer sample of 1962/63. Age distributions indicate a very young population at Yan Yean and the pattern is similar to the pyramidal structure of declining frequency with age recorded elsewhere for kangaroo species (Wilson 1975; Norbury *et al.* 1988). In all but the combined winter sample, the <3 years age group was represented in greater than 60% of the sample. From 1962 to 1964, only 16 animals (7%) >10 years were shot, while 59% were <3 years old. Significantly more subadult (<3 years) males were taken than subadult females from 1962-1964 (70% vs. 51%, X_1^2=9.63, $P<0.01$) while significantly more females were taken in the 3-10 years (40% vs. 27%, X_1^2=4.43, $P<0.05$) and >10 years (9% vs. 3%, X_1^2=3.93, $P<0.05$) age classes. The overall ratio of subadults to adults was 1.7:1.

(ii) MORTALITY SAMPLE

Approximately 300, 200 and 140 animals were thought to have died during the winters of 1960, 1962 and 1963 respectively (Dempster 1964). However, not all the skulls were apparently collected. Of the 324 skulls collected from carcasses found on the catchment, an approximate time of death had been ascribed to 72.5%. The 1960 sample contained 65 (51%) skulls

Fig. 1 (opposite). Age distributions of male and female Eastern Grey Kangaroos shot at Yan Yean, Victoria: A autumn 1962 and 1963 combined; B winter 1962 and 1963 combined; C spring 1962 and 1964 combined; D summer 1962/63. Open bars — males. Legend for reproductive status of females: Light stipple – immature pouch clean, small, tight and undeveloped: Heavy stipple — mature; not breeding: pouch large and developed. Diagonal hatching — mature; breeding: with pouch young. Vertical hatching — mature; lactating: presumably suckling young at foot, but no pouch young. Horizontal hatching — mature; with both pouch young and young at foot. Cross hatching — mature; embryo in uterus, but no pouch young. Solid bars — mature; young at foot and embryo in uterus, but no pouch young.

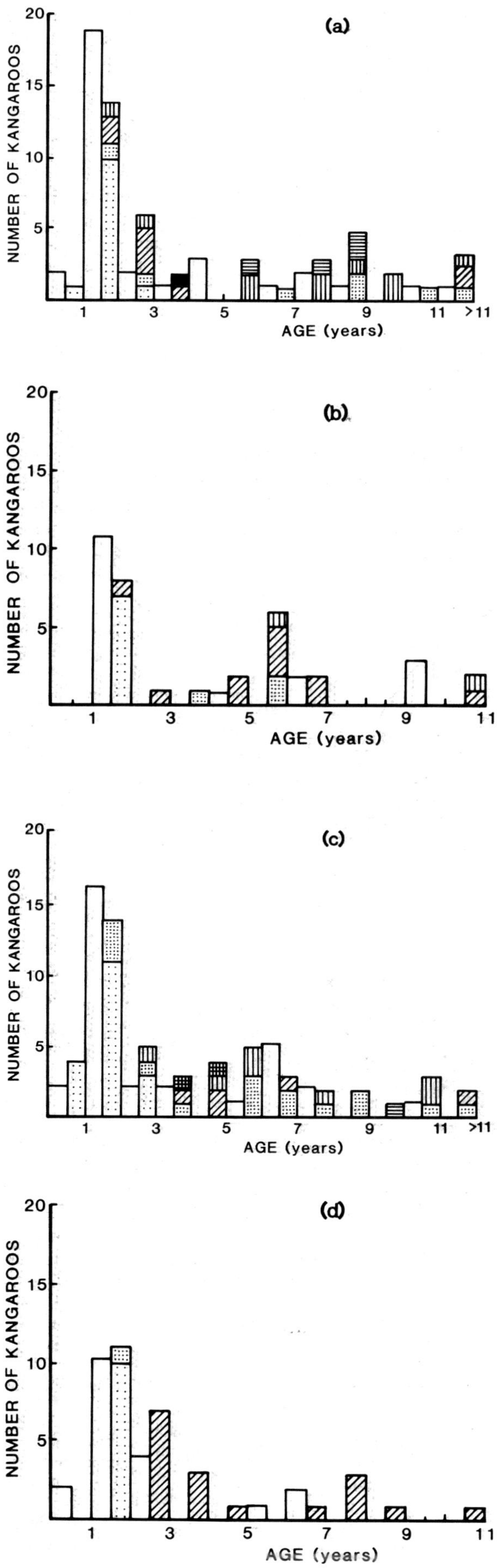

whose time of death could not be estimated. The 1962 and 1963 samples contained 15% and 2% of skulls respectively with no estimate of the time of death. Nearly all known mortality (227 of the 238 skulls whose time of death was estimated) was thought to have occurred during winter months during 1960, 1962 and 1963. When the age at death was estimated, the 1960 and 1963 winter samples were similar, with 97% (59 of 61) and 94% (58 of 62) of animals being <2 years old. The pattern of mortality during winter and early spring in 1962 was spread more evenly across the age classes with 71% (82 of 115) of the animals aged <2 years old.

Table 2 presents the numbers of males, females and unsexed skulls collected from carcasses for 1960, 1962 amd 1963. Where the sex of the animal was determined, males always outnumbered females but the difference was significant only for the 1960 sample. The overall sex ratio of the mortality sample (1.6:1) was also significantly different from parity. However, a large number of skulls, particularly during 1960 and 1962, could not be sexed.

Table 2. The numbers of male and female Eastern Grey Kangaroos and the number of unsexed kangaroos found dead at Yan Yean, Victoria during 1960, 1962 and 1963. The chi-square values (with Yates correction for 1 degree of freedom) for disparity between the number of males and females (i.e., excluding unsexed animals) found dead are also presented. (N.S.=not significant).

Year	Number of Males	Number of Females	Number Unsexed	Chi-square value for Disparity Males : Females
1960	51	29	47	6.06 (P<0.02)
1962	32	24	79	1.16 N.S.
1963	30	19	13	2.49 N.S.
Total	113	72	139	9.09 (P<0.01)

Age distributions of animals found dead during each year, including kangaroos whose time of death was not known, are shown in Figure 2. All skulls collected in 1963 were from animals aged <2 years, and close to 75% of the skulls collected in 1960 and 1962 were <3 years. Over the three years, 80% of the skulls were aged at <3 years while 18% were between three and ten years old, and 2% were >10 years. The sex ratios of three age categories in the population (subadults aged <3 years; adults aged 3-10, and >10 years) did not differ from parity over the three years in which skulls were collected at Yan Yean (P>0.6, all tests). Similarly, individual chi-square analysis of the 1-2 year age class of animals found dead of known sex suggest that there was no evidence of male-biased mortality in this age class for each year (P>0.05, all tests).

(iii) COMPARISON OF CULL AND MORTALITY SAMPLES

On the assumption that the shot and mortality samples represent the age structures of the living

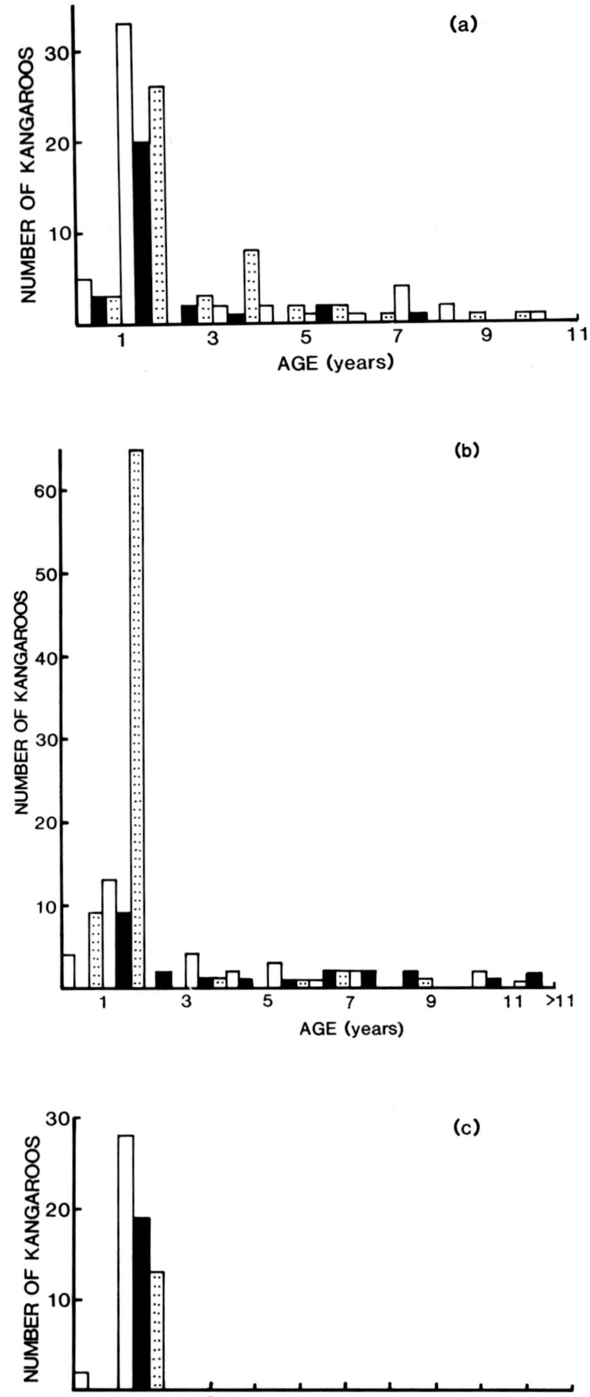

Fig. 2. Age distributions of Eastern Grey Kangaroos found dead during (a) 1960, (b) 1962 and (c) 1963 at Yan Yean, Victoria. (Males — open bars, females — solid bars, unsexed animals — stippled bars).

population and the dying portion of the population respectively, an analysis of the two can be undertaken. The 1962 shot samples of autumn (N=56) and winter (N=31) are assumed to represent the "living population" prior to, and during the period of heavy mortality. The mortality samples of 1962 is assumed to consist of those individuals from the "living population" that died during winter. Hence, only those individuals that were estimated to have died during winter or very early spring (N=115) are considered. Six age classes were used: 1-2 yr., 2-4 yr., 4-6 yr., 6-8 yr., 8-10 yr. and >10 yr. The individuals in the 0-1 yr. age group are excluded from this analysis because individuals in this age class will be under-represented relative to other age classes in the mortality sample due to faster decomposition rates and under-represented in the "living population" as pouch young of shot mothers were not included in the samples of age structures.

The age class distribution of the mortality sample of 1962 differs significantly from that of the live population (pooled autumn-winter sample) of 1962 ($X^2=24.72$, 5 d.f., $P<0.001$). A significantly greater number of kangaroos died in the 1-2 year age class (71% vs. 37%, $X_1^2=21.77$, $P<0.001$) than would be expected if the probability of dying was similar over the age classes. Animals aged 1-2 years were subsequently excluded from further analyses of the remaining age classes and tests showed no significant differences between the animals dying and the "living population" in their representation of age classes.

If the spring 1962 and summer 1963 cull samples represent the "living population" following winter mortality, and, assuming culling during autumn and winter had no effect on age structures, the effects of mortality on age composition can be examined. Considering the same six age classes, the age structure of the post-mortality population (spring and summer pooled sample) differs slightly from that of the "living population" prior to and during the period of heavy mortality (autumn and winter 1962 pooled sample) ($X^2=16.69$, d.f.=5, $P<0.01$). Following the peak mortality period, significantly fewer animals aged 4-6 years (6% vs. 18%, $X_1^2=5.70$, $P<0.02$) and 8-10 years (2% vs. 12%, $X_1^2=5.57$, $P<0.02$) were shot, while significantly more animals in the 2-4 years (27% vs. 13%, $X_1^2=5.70$, $P<0.02$) were shot.

Breeding season and juvenile sex ratios

Figure 3 shows the estimated month of birth of 47 pouch young at Yan Yean from 1962 to 1964. Births were recorded in every month except May and July, and only one birth was recorded in June. Overall, 64% of young were born in summer, 11% in autumn, 4% during winter and 21% in spring. The sex ratio of pouch young (0.88:1) was not significantly different from parity ($X_1^2=0.21$; $P<0.7$).

Age at sexual maturity and breeding success of females

The reproductive status of female kangaroos is shown in Figure 1. In summary, females matured

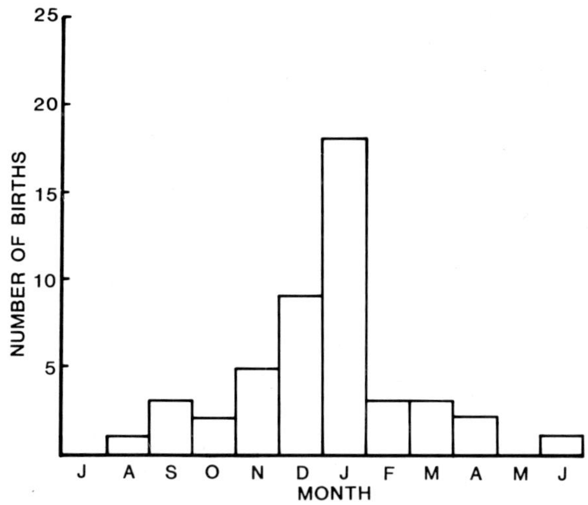

Fig. 3. Estimated month of birth of 47 pouch young at Yan Yean, Victoria, from 1962-1964.

sexually in their second to third year. Approximately 9% of 1-2 year olds gave birth in their second year, while 68% of the 2-3 year olds gave birth. Few females were still classed as immature in their third year. Older females continue breeding until the end of their life.

DISCUSSION

Age estimation of kangaroos

Results presented for age structures depend on a technique for estimating age based on a study of captive Eastern Grey Kangaroos in southern Queensland (Kirkpatrick 1965a). Its relationship to wild Victorian animals is unknown. Sanson (1982) has shown that molar progression is a direct function of chewing and, since this will almost certainly differ between captive and free-ranging populations, data derived from the technique must be conservatively used (see Norbury *et al.* 1988). Furthermore, as there is no current information available on growth rates of captive or wild Victorian Eastern Grey Kangaroos, similar comments apply to the use of foot and tail length measurements in estimating the age of pouch young as regressions used in this study were derived for captive Queensland animals (Kirkpatrick 1965a).

Reproduction

The breeding season of Eastern Grey Kangaroos at Yan Yean (64% of births in summer months) closely conforms to that recorded for this species in southern Queensland (Kirkpatrick 1965b), central New South Wales (Poole 1983), and for a mixed sample of Eastern and Western *(M. fuliginosus)* Grey Kangaroos from central New South Wales (Poole 1973). The consistent pattern of a spring-summer birth peak over the species' wide geographic range was noted by Poole (1983). However, in Tasmania, a pronounced birth peak is slightly later than that at Yan Yean, with 75% of births between January and March (Pearse 1981). Tasmanian young vacate the pouch permanently 10 months later during the following spring, when the weather is warmer and grasses most abundant (Pearse 1981). The timing of reproduction thus appears to ensure optimal food resources for weaning young and lactating females as argued by Poole (1973), Short *et al.* (1983) and Norbury *et al.* (1988).

Some females began breeding in their second year, and older females bred each year. This was also the case in Queensland (Kirkpatrick 1965b) and Tasmania (Pearse 1981). The youngest five females to give birth were thought to be aged at least between 19-21 months. Similarly Poole's (1983) field studies indicated that females matured sexually between 16 and 20 months. Four animals from this study had not reached sexual maturity by their third year. Hence, even when environmental conditions appear favourable, not all females mature sexually at the same time. However, delayed sexual maturity could have resulted from a long-term environmental effect occurring earlier in these females' lives. Stuart-Dick (pers. comm.) also observed that some females on a site in northern New South Wales had not conceived by three to four years of age. The oldest breeding female in this study was a 16-17 year old whose pouch young was 112 days old.

The autumn samples contained five mature but non-breeding females from a sample of 18 females aged between five and twelve years, four of which were from autumn 1962. Examination of rainfall data showed below-average rainfall for summer 1961/1962 and, in particular, during March and April of 1962, which may have caused some females to enter anoestrus (Kirkpatrick and McEvoy 1966; Poole 1983).

Thirteen females of various ages between three and twelve years were shot in spring (October and early November) 1964. Of these, eleven were mature but non-breeding, the reasons for such a high proportion of non-breeders being uncertain. Rainfall had been well above average during September and October 1964, and on average for winter months. Pouch young should have been vacating the pouch for the first time in spring (see above); hence, shooters should have encountered lactating females, but only a small number were apparently breeding. However, only 4% of average rainfall had fallen during January 1964 when a majority of births would have occurred; hence, dry conditions prevailing at that time may have caused many females to lose pouch young, and below-average rain in early autumn may have delayed further breeding.

The alternative hypothesis proposed to account for the high proportion of non-breeding females in spring 1964 relates to the "r" and "K" selection premise. Richardson (1975) ascribed the life history patterns, characteristic of species subject to "r" and "K" selection, to Red *(M. rufus)* and Eastern Grey Kangaroos, respectively. His views were supported by Poole (1983). They suggest that the Eastern Grey Kangaroo is adapted to producing young best able to enter the adult population in the face of competition for limited resources, rather than the efficiency of the mother to convert the limited resources obtained into young. An additional correlate of K-selection is that the population is at or near the carrying capacity, and in this confined situation the population could well be near its carrying capacity.

The density of Eastern Grey Kangaroos at Yan Yean (3.8-5.7 ha^{-1}) is extraordinarily high. Furthermore, Eastern Grey Kangaroos can breed when good conditions occur at the beginning of the first breeding season after reaching 18 months of age, but may wait one, two or more years (Richardson 1975). Delayed sexual maturity and reduced birth rates (unexpected apparent anoestrus in this study) may be attributed to the high density resulting from confined conditions. Reduced birth rates and delayed sexual maturity in various large mammal populations are thought to result from density-dependent effects (see Fowler and Smith 1981).

Juvenile and overall sex ratios

Although it is generally assumed that offspring sex ratios will be at parity, Johnson and Jarman (1983) found that pouch-young sex ratios for Eastern and Western Grey Kangaroos and the Red Kangaroo were closely correlated with short and long-term rainfall. Hence, offspring sex ratios may vary in association with local rainfall conditions. Given a long-term median rainfall of 641 mm at Yan Yean, the slight female-biased sex ratio (88 males:100 females) supports their prediction. In contrast, Norbury *et al.* (1988) recorded a male-biased sex ratio for the Western Grey Kangaroo from a drier locality in northwestern Victoria which also conformed to their suggestions.

Interpretation of adult sex ratios may be confounded by sampling bias (Caughley 1966) but the evidence for shooting bias in kangaroos is conflicting (see Norbury *et al.* 1988). Females may have been easier to shoot than males and their proportion subsequently over-represented in the samples. However, the slightly female-biased sex ratio was never significant within each seasonal sample but is consistent with evidence from studies on Western Grey and Red Kangaroos (Newsome 1977; Johnson and Bayliss 1981; Norbury *et al.* 1988).

Mortality patterns suggest that deaths at Yan Yean are male-biased which has also been recorded in Western Grey Kangaroos during drought (Robertson 1986; Norbury *et al.* 1988). However, only Norbury's *et al.* (1988) study attempted to consider the sex ratio of the pre-drought population which was markedly female-biased. Sharman *et al.* (1964) also predicted higher survival rates in female Red Kangaroos, but without supporting evidence. Male-biased mortality at Yan Yean is consistent with the declining proportion of males with age, in the shot samples.

The three hypotheses proposed to account for male-biased mortality in Western Grey Kangaroos (see Norbury *et al.* 1988) appear equally likely to apply to Eastern Grey Kangaroos. Higginbottom (pers. comm.) has also suggested that male Red-necked Wallabies *(Macropus rufogriseus)* tend to display higher dispersal rates than females which may also be associated with higher male mortality. Eastern Grey Kangaroo males disperse as young adults (P. Jarman, pers. comm.); hence, male-biased mortality associated with dispersion may also apply to this species.

Age structures of shot and mortality samples

The most overwhelming feature of the comparison between age structures of the cull sample prior to and during the period of heavy mortality of 1962, and the age composition of the mortality sample (winter and early spring) at Yan Yean, is the high proportion of kangaroos dying in the 1-2 year age class relative to their proportion in the cull samples. Jaremovic (1984) observed high winter mortality in unconfined subadult Eastern Grey Kangaroos in southern New South Wales and noted the age of deaths corresponded to the age at which Grey Kangaroos are weaned (18 months), a time of great susceptibility for young kangaroos, particularly if the biomass of winter forage is low. However, in her particular population, high subadult mortality was observed only during an abnormally wet winter preceeding a dry autumn. Norbury *et al.* (1988) similarly speculated that the impact of mortality on Western Grey Kangaroos in northwestern Victoria was in the 1-2 year age class. In contrast to speculations by both Jaremovic (1984) and Norbury *et al.* (1988) was the finding that, although mortality appeared to be somewhat male-biased, male-biased mortality did not appear to predominate in the 1-2 year age class but spread evenly across the age classes. The three-year mortality samples suggested that heavy winter subadult mortality in both sexes is not an unusual phenomenon at Yan Yean. This mortality was evident in 1962 and again in 1963 when cull and mortality samples were obtained. It is likely that the situation was similar in 1960 despite there being no cull sample for comparison. It must be remembered that a large number of skulls could not be sexed and that skulls were apparently not collected from all carcasses.

The slight differences in age composition of shot samples preceding the winter mortality, relative to shot samples prior to and during the winter, are not interpreted as reflecting the effects of mortality on the population. These may only reflect sampling effects and larger sample sizes might eliminate the discrepancies. Hence, age-specific winter mortality in 1962 appeared to have little impact on age composition of the population.

The causative organism thought responsible for the winter mortality of subadult Grey Kangaroos at Yan Yean has been identified as an intestinal parasite, *Globocephaloides trifidospicularis* (Nematoda; Amidostomatidae) (Arundel *et al.* 1977; Beveridge 1979). Dempster (1964) suggested that death was most prominent in animals aged between 14-20 months of age, and was associated with general loss of condition, severe anaemia, and depletion of body fat and blood protein. Older animals exhibit only slight losses of condition (see also Arundel *et al.* 1977). It is thought that age-specific (1-2 year age class) mortality at Yan Yean is a natural process aggravated by high densities and confined conditions (Dempster 1964; pers. obser.). Higher densities in confined situations may result in a higher concentration/density of parasite eggs being passed to the soil relative to that in unconfined conditions. Hence, subadult kangaroos have a higher probability of contacting the eggs. It appears that the eggs passed by subadult kangaroos in late winter and spring survive over summer and infect the juveniles of the following year (Arundel *et al.* 1977).

The demographic processes exhibited by the confined Eastern Grey Kangaroo population at Yan Yean reveal some striking similarities to unconfined Eastern Grey Kangaroos. The breeding season, age at sexual maturity of females and response of adult females to below-average rainfall appear similar across this species' wide distribution. However, delayed sexual maturity in females and unexpected non-breeding in mature females, and the repeatability of high subadult winter mortality may be a characteristic of a dense, confined kangaroo population which infrequently occurs in unconfined populations and only during abnormal environmental conditions.

ACKNOWLEDGEMENTS

I am indebted to Graeme Coulson and Angus Martin for advice and helpful suggestions while supervising this project and for criticising an earlier version of this manuscript. Graeme also assisted with measuring of the skulls and provided unpublished data. I thank Joanna Clarke, Karen Higginbottom, Peter Jarman and Robyn Stuart-Dick for their helpful comments on the manuscript. Keith Dempster (Fisheries and Wildlife Division) initiated the project and allowed access to the skulls for study whilst collection of skulls was undertaken by various wildlife officers. Julie Gould measured a large number of the skulls. Joan Dixon and Linda Huxley provided assistance in the Museum of Victoria where skulls were housed. Peter Jarman, Rosemary Lott and Ian Davies provided helpful advice on analyzing age structures. Bill Poole gave informative discussion and valuable data which improved this paper. Finally, I am grateful to Robyn Liddle for her patience and expertise in typing this manuscript and drawing of graphs.

REFERENCES

ARUNDEL, J. H., BARKER, I. K. AND BEVERIDGE, I., 1977. Diseases of marsupials. Pp. 141-54 *in* The Biology of Marsupials ed. by B. Stonehouse and D. Gilmore. MacMillan Press: London.

BEVERIDGE, I., 1979. A review of the Globocephaloidinae Inglis (Nematoda: Amidostomatidae) from macropodid marsupials. *Aust. J. Zool.* **27**: 151-75.

CAUGHLEY, G., 1966. Mortality patterns in mammals. *Ecology* **47**: 906-18.

CAUGHLEY, G., BROWN, B., DOSTINE, P. AND GRICE, D., 1984. The Grey Kangaroo overlap zone. *Aust. Wildl. Res.* **11**: 1-10.

DEMPSTER, J. K., 1964. Control and conservation of kangaroos in Victoria. Pp. 1-2 *in* Australian Vermin Control Conference, Working Papers. C.S.I.R.O.: Melbourne.

FOWLER, C. N. AND SMITH, T. D., 1981. Dynamics of Large Mammal Populations. John Wiley and Sons: New York.

JAREMOVIC, R., 1984. Space and time related behaviour in Eastern Grey Kangaroos (*Macropus giganteus* Shaw). Ph.D. Thesis, University of New South Wales, Sydney.

JOHNSON, C. N. AND BAYLISS, P. G., 1981. Habitat selection by sex, age and reproductive class in the Red Kangaroo, *Macropus rufus,* in western New South Wales. *Aust. Wildl. Res.* **8**: 465-74.

JOHNSON, C. N. AND JARMAN, P. J., 1983. Geographical variation in offspring sex ratios in kangaroos. *Search* **14**: 152-54.

KIRKPATRICK, T. H., 1964. Molar progression and macropod age. *Queensl. J. Agric. Anim. Sci.* **21**: 163-65.

KIRKPATRICK, T. H., 1965a. Studies of Macropodidae in Queensland. 2. Age estimation in the Grey Kangaroo, the Red Kangaroo, the Eastern Wallaroo and the Red-necked Wallaby, with notes on dental abnormalities. *Queensl. J. Agric. Anim. Sci.* **22**: 301-17.

KIRKPATRICK, T. H., 1965b. Studies of Macropodidae in Queensland. 3. Reproduction in the Grey Kangaroo *(Macropus major)* in southern Queensland. *Queensl. J. Agric. Anim. Sci.* **22**: 319-28.

KIRKPATRICK, T. H., 1967. Studies of Macropodidae in Queensland. 6. Sex determination of adult skulls of the Grey Kangaroo and the Red Kangaroo. *Queensl. J. Agric. Anim. Sci.* **34**: 131-33.

KIRKPATRICK, T. H. AND MCEVOY, J. S., 1966. Studies of Macropodidae in Queensland. 5. Effects of drought on reproduction in the Grey Kangaroo *(Macropus giganteus). Queensl. J. Agric. Amin. Sci.* **23**: 439-42.

MELBOURNE METROPOLITAN BOARD OF WORKS, 1984. Yan Yean References Areas. Victoria Reference Area System Management Plan.

NEWSOME, A. E., 1977. Imbalance in the sex ratio and age structure of the Red Kangaroo, *Macropus rufus,* in central Australia. Pp. 221-33 *in* The Biology of Marsupials ed. by B. Stonehouse and D. Gilmore. MacMillan Press: London.

NORBURY, G. L., COULSON, G. M. AND WALTERS, B. L., 1988. Aspects of the demography of the Western Grey Kangaroo, *Macropus fuliginosus melanops,* in semiarid north-west Victoria. *Aust. Wildl. Res.* **15**: 257-66.

PEARSE, R. J., 1981. Notes on breeding, growth and longevity of the Forester or Eastern Grey Kangaroo, *Macropus giganteus* Shaw, in Tasmania. *Aust. Wildl. Res.* **8**: 229-35.

POOLE, W. E., 1973. A study of breeding in Grey Kangaroos *Macropus giganteus* Shaw and *M. fuliginosus* (Desmarest), in central New South Wales. *Aust. J. Zool.* **21**: 183-212.

POOLE, W. E., 1982. *Macropus giganteus. Mamm. Spec.* **187**: 1-8.

POOLE, W. E., 1983. Breeding in the Grey Kangaroo, *Macropus giganteus,* from widespread locations in eastern Australia. *Aust. Wildl. Res.* **10**: 453-66.

RICHARDSON, B. J., 1975. r and K selection in kangaroos. *Nature* **225**: 323-24.

ROBERTSON, G., 1986. The mortality of kangaroos in drought. *Aust. Wildl. Res.* **13**: 349-54.

SANSON, G. D., 1982. Evolution of feeding adaptations in fossil and recent macropodids. Pp. 489-506 *in* The Fossil Vertebrate Record of Australasia ed. by P. V. Rich and E. M. Thompson. Monash University, Melbourne.

SHARMAN, G. B., FRITH, H. J. AND CALABY, J. H., 1964. Growth of the pouch young, tooth eruption, and age determination, in the Red Kangaroo, *Megaleia rufa. C.S.I.R.O. Wildl. Res.* **9**: 20-49.

SHORT, J., CAUGHLEY, G., GRICE, D. AND BROWN, B., 1983. The distribution and abundance of kangaroos in relation to environment in Western Australia. *Aust. Wildl. Res.* **10**: 435-51.

WILSON, G. R., 1975. Age structures of populations of kangaroos (Macropodidae) taken by professional shooters in New South Wales. *Aust. Wildl. Res.* **2**: 1-9.

The interaction between macropodids and sheep: a review

GLENN P. EDWARDS[1]

Domestic sheep, and four species of large macropodid, the Red Kangaroo, the Euro, the Eastern Grey Kangaroo and the Western Grey Kangaroo, represent the dominant large herbivores across most of southern Australia. Much of our knowledge of sheep and macropodid ecology, particularly the nature of the interaction between them, is limited and based upon circumstantial evidence. Available data suggest that a separation of diets is the general pattern. Spacing behaviour and differential activity periods seem to ensure that behavioural interactions are isolated occurrences. Patterns of habitat and forage utilization are complex and largely entwined with climatic factors which determine pasture availability. Recent data appear to confirm the existence of competition between macropodids and sheep in certain situations.

INTRODUCTION

THIS paper reviews various aspects of the ecologies of sheep and macropodids in an effort to evaluate the extent of the interaction between them.

The spread of pastoralism

Between 1850 and 1900 most of arid Australia was settled. Sheep ranged across much of the southern part of the continent (Fig. 1a). Widespread clearing and the construction of artificial watering points were concomitant with the spread of this industry. Also, the major predator, the Dingo, was persecuted to the point of exclusion in many areas. Grazing intensity increased dramatically and food was seen to replace water as the limiting resource for herbivore populations across the more arid areas (Newsome 1971a, 1975; Caughley 1987a; Robertson *et al.* 1987). Additionally, with the ranging of large herds of hooved animals, the land was subject to a suite of new pressures (Moore 1962).

Geographical ranges

The geographical distribution of sheep *(Ovis aries)* across southern Australia is coincident with that of several large macropodid species: the Red Kangaroo (*Macropus rufus*, Fig. 1b), the Eastern Grey (*M. giganteus*, Fig. 1c) and the Western Grey Kangaroo (*M. fuliginosus*, Fig. 1c), and the Euro (*M. robustus*, Fig. 1d).

Several species of smaller macropodids also overlapped geographically with sheep (Table 1). Many of these smaller macropodids are now rare or extinct, and although the spread of pastoralism has been implicated in their demise (Newsome 1971a; Newsome and Corbett 1977; Caughley 1987a), the impact of the European rabbit *(Oryctolagus cuniculus)* and predation by the red fox *(Vulpes vulpes)* are also considered to have played a role (Newsome 1971a).

The four large species of *Macropus* are still abundant. Since most research concerning sheep-macropodid interactions has been confined to these I shall similarly constrain my treatment of the subject.

Population dynamics

Population sizes for both sheep and macropodid species have been estimated recently (Table 2). The population dynamics of large macropodids, formerly dictated by the availability of drinking water, are now largely governed by pasture abundance which is in turn governed by rainfall (Bayliss 1985). Populations increase during years of pasture abundance (Caughley *et al.* 1985) and crash dramatically during prolonged droughts (Newsome *et al.* 1967; Caughley *et al.* 1985). Population crashes during drought result from a cessation of breeding, death of pouch young and adult mortality due to starvation (Newsome *et al.* 1967; Shepherd 1987).

Gross sheep numbers have not changed markedly over the past 80 years (Fig. 2). Short-term changes in sheep population dynamics are largely controlled through grazing management practices (Robinson 1982; Condon 1983). During times of drought destocking occurs, together with natural losses. Numbers are increased to "usual" levels when pasture conditions improve. In this regard sheep population dynamics essentially mimic those of large macropodids.

Feeding ecology, digestive physiology and metabolic rate

There are many behavioural and physiological similarities between kangaroos and sheep, and also some major differences. Typical adult body weights for each species are listed (Table 3). Macropodids and sheep have analogous digestive systems (Hume 1982) which comprise about

[1]School of Biological Science, University of New South Wales, P.O. Box 1, Kensington, New South Wales 2033.

Pages 795-803 in KANGAROOS, WALLABIES AND RAT-KANGAROOS ed. by G. Grigg, P. Jarman and I. Hume. Surrey Beatty & Sons Pty Limited, New South Wales, Australia, 1989.

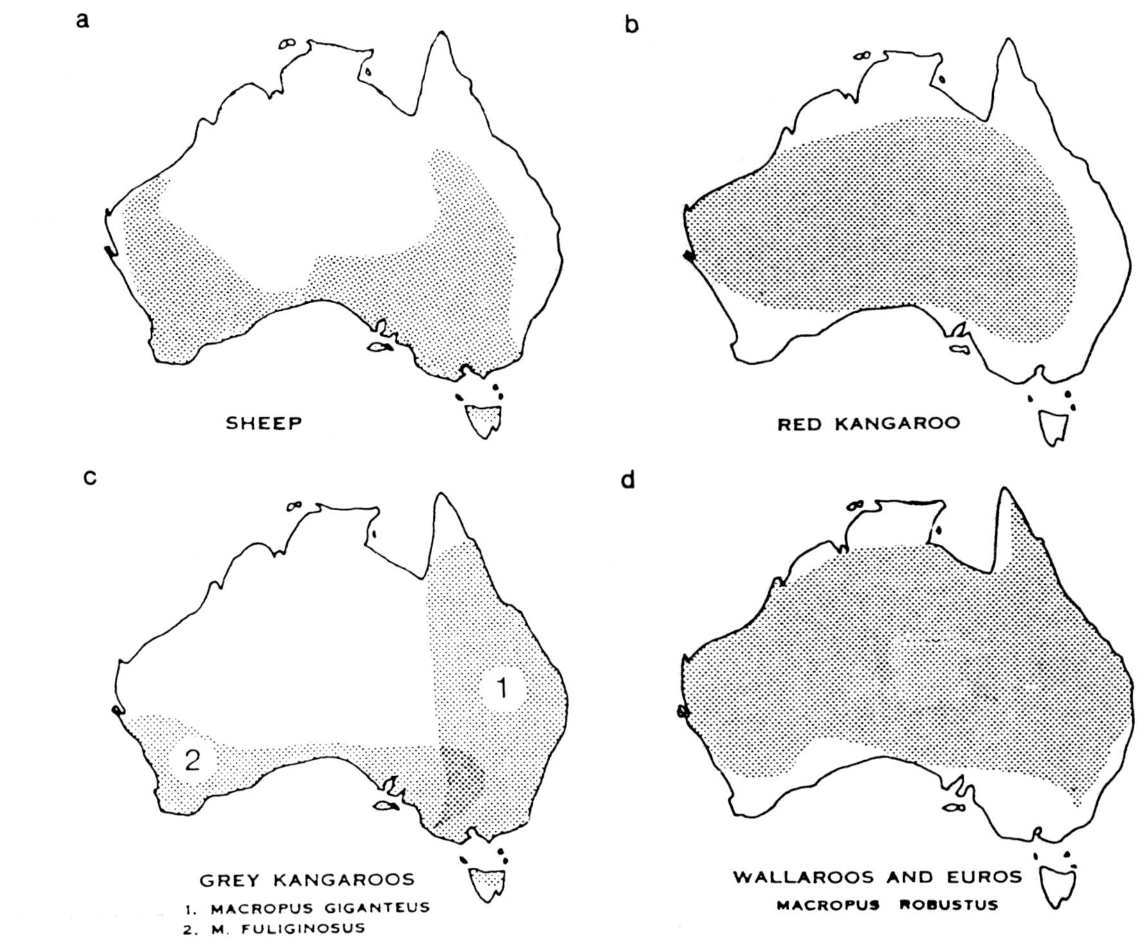

Fig. 1. Distribution of sheep and large *Macropus* species in Australia. (From Fennessy 1970).

Table 1. Small macropodid species whose distribution (or former distribution) overlaps with that of sheep. (Adapted from Strahan (ed.) 1983).

Common name	Scientific name	Status
Broad Faced Potoroo	*Potorous platyops*	extinct
Brush Tailed Bettong	*Bettongia penicillata*	rare
Tasmanian Bettong	*B. gaimardi*	extinct on mainland
Burrowing Bettong	*B. lesueur*	extinct on mainland
Rufous Bettong	*Aepyprymnus rufescens*	common
Eastern Hare Wallaby	*Lagorchestes leporides*	extinct
Rufous Hare Wallaby	*L. hirsutus*	rare
Banded Hare Wallaby	*Lagostrophus fasciatus*	extinct on mainland
Crescent Nail Tail Wallaby	*Onychogalea lunata*	extinct
Bridled Nail Tail Wallaby	*O. fraenata*	rare
Black Footed Rock Wallaby	*Petrogale lateralis*	rare
Yellow Footed Rock Wallaby	*P. xanthopus*	common
Brush Tailed Rock Wallaby	*P. penicillata*	common
Rothschild's Rock Wallaby	*P. rothschildi*	common
Toolache Wallaby	*Macropus greyi*	extinct

Table 2. Population estimates in millions for sheep and macropodid species for southern Australia in 1981. (Adapted from Caughley 1987a). Euro numbers are not shown but are probably comparable to that of Reds.

Species	Estimated population
Sheep	120.9
Red Kangaroo	6.2
Eastern Grey	1.9
Western Grey	1.6

10%-12% and 14% of total body mass, respectively (Denny and Dawson 1975; Dawson in prep.). While sheep are true ruminants, kangaroos also utilize foregut fermentation to digest plant fibre (structural carbohydrate). Relative digestive efficiencies are similar although macropodids on high fibre diets have lower fibre digestion than sheep (Hume 1982). This is probably the result of a faster rate of passage of digesta in macropodids (Hume 1982). Despite the lower retention time of the macropodid gut, sheep and macropodids have similar field food intakes (60-80 g dry matter/$kg^{0.75}$ day; Short 1986). This probably reflects the lower basal

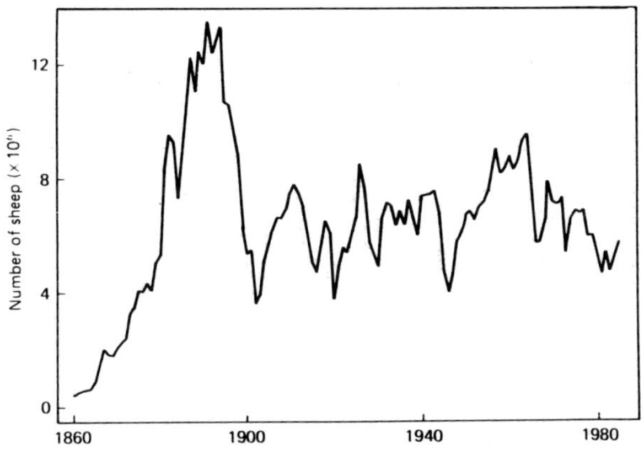

Fig. 2. Sheep numbers in New South Wales' rangelands. (From Caughley 1987a).

Table 3. Typical adult body weights in kilograms for sheep and macropodid species.

Species	Body weight	
	female	male
Sheep	60-70	70-90
Red Kangaroo	30-35	70-90
Eastern Grey	30-35	60-80
Western Grey	30-35	60-80
Euro	17-21	50-60

metabolic rate of marsupials relative to eutherians (Dawson and Hulbert 1970) which results in lower energy maintenance requirements (Hume 1982).

Sheep and kangaroos have different grazing efficiencies yet both are able to graze pastures to similar heights (Short 1987). When feeding, the biting rate of sheep is similar to that of kangaroos (50-60 bites/min; Short 1986). However, sheep have a much larger bite size (Short 1986) and are therefore able to harvest vegetation more quickly than kangaroos. Differences in bite size are related to adult jaw width measurements: 3.3 cm for sheep, 1.7 cm for Red Kangaroos and 1.9 cm for Western Grey Kangaroos (Short 1986).

THE INTERACTION BETWEEN MACROPODIDS AND SHEEP

Given that sheep and the large macropodids are the dominant large herbivores across much of southern Australia and that they are physiologically analogous in many respects, an understanding of their ecology and the extent of the interaction between them is essential to our management of grazing ecosystems within Australia, particularly those of the more fragile arid environments.

Forage selection and diet overlap

Macropodid and sheep diets are discussed in detail elsewhere in this volume (Dawson) and here a brief summary will suffice. The large macropodids are essentially grass eaters whereas sheep accept a much wider variety of plants. The sheep can therefore be described as a generalist herbivore with a larger dietary niche breadth than the more specialized macropodids. Differences in mouth morphology appear to support this assertion, the narrower jaw in macropodids facilitating greater selectivity.

In arid and semi-arid Australia, diet selection by herbivores and hence the extent of interspecific dietary overlap, hinges upon environmental variability. The Australian arid ecosystem is driven by rainfall which is the least predictable in the world (Perry 1973; Low 1979; Shepherd and Caughley 1987). As a result, pasture conditions vary greatly from year to year and season to season. Ellis *et al.* (1977) for example, found that total vegetation cover of the flood-plains at Fowler's Gap in western New South Wales varied from about 30% in the dry summer of 1973 to over 80% during the winter of 1974. In addition, arid environments are generally patchy in terms of landform, soil and vegetation even on a scale of a few metres (Wiens 1976), and patches commonly respond differently to environmental stimuli. However, despite these vagaries, the fundamental pattern is a separation in the diets of sheep and large macropodids (Dawson, this volume) with instantaneous overlap being greatest in times of pasture abundance, and least during drought (Barker 1987; Dawson, this volume). Temporal overlap in diets is variable but one generalization seems to apply; grass is always the major component of large macropodid diets but is only occasionally prominent in the diet of sheep.

It should be remembered that there are usually some components common to the diets of both macropodids and sheep, even when dietary overlap is small. Unfortunately, diet studies have not been sufficiently detailed to address the question of whether sheep and kangaroos eat different species or parts of plants within a vegetation class common to the diets of both. Nor do we have sufficient data to determine whether dietary divergence is forced upon either the large macropodids or sheep through interspecific competition. This possibility will be explored later.

Behavioural interactions

The grazing strategies of sheep and large macropodids largely dictate the extent of behavioural interactions between them. Sheep graze predominantly during the early morning and late afternoon (Arnold and Dudzinski 1978; Squires 1981) and usually camp in a mob at night to ruminate. Kangaroos tend to graze at night (Priddel 1986; Short 1986; Southwell 1987) and over 78% of feeding activity is confined to this period (Priddel 1986). Sheep graze in a flock (Squires 1981) and flock size is inversely related to food availability and positively related to lateral visibility which is restricted in undulating and shrubby environments (Lynch 1974; Arnold and Dudzinski 1978). Large macropodids are, generally speaking, less gregarious than sheep and, although they tend to occupy fixed home ranges (Oliver 1986; Priddel 1987; Clancy and Croft, this volume), they do not defend territories (Jarman and Southwell 1986). Eastern Grey Kangaroos may form large feeding aggregations in more mesic environments (Jaremovic 1984), however grouping behaviour has been poorly studied in the drier parts of its range where its distribution tends to overlap significantly with that of sheep. Red Kangaroos rarely form large groups, mean group size ranging from 1.54 to 2.2 animals (Croft 1980; Johnson 1983). Western Grey Kangaroos in western New South Wales form groups with mean size ranging from 1.25 to 1.55 animals (Johnson 1983). Euros tend to be even more solitary than either Red or Grey Kangaroos (Croft 1981). The general effect of this is that large macropodids tend to be more evenly distributed throughout a given area at any one time than do sheep (Caughley 1964) and the likelihood of large scale behavioural interactions is therefore reduced.

Very few behavioural interactions between kangaroos and sheep have been described. Sheep have been observed to disrupt the activities of Red Kangaroos, dispersing feeding groups and displacing individuals from water troughs and shade trees (Croft 1980). However, spacing behaviour and differential activity periods would seem to ensure that behavioural interactions are isolated occurrences and usually of minor consequence, except perhaps in cases where sheep and/or kangaroos occur at high densities or when sheep flocks split into smaller units in response to decreasing forage availability.

Grazing impact and differential habitat use

It is widely accepted that sheep grazing has had a marked effect upon vegetation communities throughout Australia (Moore 1962; Ealey 1967; Newman and Condon 1969; Leigh 1974; Wilson and Graetz 1979; Robertson *et al.* 1987). Measurable change in the vegetation and soil has occurred over large areas, particularly in the more arid regions with the low shrublands (shrub steppe) being hardest hit (Newman and Condon 1969). A common result of continued heavy sheep grazing is the creation of sub-climax communities dominated by ephemeral species (Moore 1962) and grasses (Newman and Condon 1969). Other effects include a reduction in abundance of palatable perennial grasses such as kangaroo grass (*Themeda australis*: Moore 1962), destruction and replacement of perennial *Atriplex* species by copperburrs (*Sclerolaena* spp.) and other unpalatable shrubs (Moore 1962; Leigh 1977), and an increase in "woody weeds" in semi-arid areas (Harrington *et al.* 1979). These large scale and dramatic vegetation changes have been implicated in the decline and subsequent extinction of some of the smaller native fauna mentioned above.

However, not all native fauna have suffered as a result of vegetational changes wrought by heavy grazing. It is generally agreed that the large macropodids have been favoured by the creation of suitable sub-climax communities and have increased greatly in numbers in response to this as well as the proliferation of water points and widespread eradication of the Dingo (Newsome 1971b; Frith 1973; Leigh 1974; Denny 1980). Newsome (1971a,b; 1975) maintains that an important factor favouring the large kangaroos is that the grazing of domestic stock (sheep and cattle) facilitates that of kangaroos by reducing the standing biomass of dry perennial grasses causing a flush of green growth under suitable conditions. This results in the creation of widespread "marsupial lawns" which are preferred feeding areas for the Red Kangaroo (Newsome 1971a,b). In the more mesic environments, widespread clearing of trees and shrubs by pastoralists created similar "marsupial lawns" (Moore 1962; Kaufmann 1974) favoured by both species of Grey Kangaroo (Poole 1983).

A much quoted example of sheep induced vegetation change favouring kangaroos was documented by Ealey (1967). Heavy sheep grazing in the Pilbara region of Western Australia resulted in replacement of the more palatable and nutritious grasses by the hardy spinifex, *Triodia pungens*. Euros and Red Kangaroos increased in numbers in response to the ready availability of stock water, but ultimately during drought, only the Euro was able to subsist on the new sub-climax vegetation and was thereby able to greatly extend its range at the expense of both Red Kangaroos and sheep. Perhaps this is not so much a case of sheep modifying the vegetation to facilitate the Euro, but rather one of them modifying the vegetation so as to exclude themselves and the Red Kangaroo.

Large macropodids have contributed to the dramatic increase in herbivore pressure on rangelands over the last century. Population densities in many areas are higher than in pre-European times and there is increased kangaroo survival into drought which places added stress on dwindling pasture reserves. However, it is during more "normal" years that the impact of kangaroos on pasture is greatest (Gardiner 1986a,b). Grazing kangaroos may suppress recruitment and survival of some desirable perennial grasses and chenopod sub-shrubs while promoting the spread of so called woody weeds (Gardiner 1986a,b). While the impact of kangaroos on vegetation has not been as severe as that of sheep, they may play a role in maintaining sub-climax vegetation regimes created by domestic stock (Newsome 1975). There are no available data to indicate whether the effect of kangaroo grazing facilitates the grazing of sheep.

The intricacies of habitat use by kangaroos and sheep have been poorly studied. An important factor governing patterns of habitat use by sheep is that they graze into the wind (Arnold and Dudzinski 1978). This phenomenon has also been reported for Red Kangaroos although not all individuals appear to adopt this behaviour (Denny 1980). Sheep are also constrained by fences and thus restricted in their choice of available habitats. Other factors which play a role in determining habitat choice by sheep include the availability of water in summer, topography and the location of camp sites and shade (Arnold and Dudzinski 1978). Unhindered by fences, kangaroos are more easily able to seek and use preferred habitats. Red Kangaroos are reported to favour areas with green herbage (Newsome 1965a,b; Bailey 1971; Johnson and Bayliss 1981), and Grey Kangaroos areas affording cover and shelter (Caughley 1964). Euros prefer habitats containing rocky hills and ridges (Croft 1981). Recent evidence has implicated the grazing of sheep as a factor affecting the choice of habitat by kangaroos.

Andrew and Lange (1986) provided data based on dung counts illustrating the apparent dissociation of both Red and Western Grey Kangaroos from sheep when given the opportunity. In the first case, the density of Red Kangaroo dung was about 10 times higher within a large regeneration paddock exclosed from sheep for 50 years. In the second case, Red and Western Grey Kangaroos usually preferred to graze inside sheep-proof exclosures following depletion of grass reserves outside these exclosures. The dominant grasses in the study area were annuals which do not re-shoot in response to heavy grazing, rather than the perennial grasses involved in the formation of Newsome's "marsupial lawns". Andrew and Lange proposed that the kangaroos moved into the exclosures to utilize "better pasture" and that direct behavioural antagonism appeared to play no role in this regard. The most heavily utilized exclosures were closest to a watering point where sheep impact upon the pasture was most severe. Newsome (1971b) also noted that Red Kangaroos shunned heavily degraded areas around watering points, preferring to graze farther out where pasture was of better quality. These results allude to the existence of interspecific competition between macropodids and sheep, yet do not unequivocally demonstrate that it occurs.

In a third case study, Andrew and Lange (1986) reported that Western Grey Kangaroos preferred areas of a paddock not frequented by sheep. However, in this last case, kangaroo distribution was biased towards an area of dense woodland at one end of the paddock which provided cover and shelter.

In a similar study to that of Andrew and Lange (1986), Watson *et al.* (in prep.) found that Red Kangaroos favoured paddocks stocked lightly with sheep (8.4-12.7 ha/sheep) in preference to those stocked more heavily (2.5-5.6 ha/sheep). This effect became most evident when pasture conditions deteriorated suggesting that kangaroos sought the "better" (more abundant) pastures in the lightly stocked paddocks. The authors reported no behavioural antagonism between kangaroos and sheep and their results appear to confirm those of Andrew and Lange (1986). When interpreting these results it must be remembered that a contributing factor to pasture depletion, particularly at higher effective stocking rates, is that of sheep trampling vegetation. This phenomenon is readily observed around water points (Lange 1969) and is amplified during drought when plant growth ceases and sheep have to cover more ground to obtain enough forage. Thus, although diet overlap may be small, sheep may still deplete the amount of forage available to kangaroos through trampling.

The dynamics of pasture utilization by macropodids and sheep are complex and largely entwined with climatic factors which determine pasture availability. In some instances the grazing of sheep may facilitate that of kangaroos by stimulating perennial grasses to shoot following heavy grazing. However, when conditions become dry, plant growth ceases and heavy grazing no longer stimulates regrowth. Herbivores must subsist on the standing dry plant biomass (Low and Low 1975). When pasture becomes scarce kangaroos appear to seek out areas where preferred pasture is relatively more abundant and perhaps of better quality (Priddel 1987). In most cases sheep are restrained by fences and are unable to do likewise. Interspecific competition may play a role in determining the characteristics of pasture and habitat utilization by kangaroos and sheep and this is discussed in the following section.

DO SHEEP AND LARGE MACROPODIDS COMPETE?

Interspecific competition has traditionally been divided into two basic types depending upon the mechanism of interaction between the competitors (Schoener 1983). The first, interference competition, results from individuals harming one another through direct confrontation e.g., fighting, harassment, etc. The second, exploitative competition, is less direct and involves the harming of a victim through deprivation of a shared resource due to a competitor. By definition, competition occurs between two or more species when harm or deprivation decreases the victim's population size through reduced fitness which may manifest itself through a decrease in body condition, lowered reproductive output or lowered survival rates (Schoener 1983).

Sympatric species may reduce competition by partitioning their use of available resources to enable coexistence (Sale 1974). This partitioning may be achieved through increased specialization through a narrowing of requirements or by a change in the pattern of resource use (Sale 1974). Both mechanisms appear to be important in the ecological separation of some African ungulates which establish a temporal grazing succession. In this system the grazing of one species modifies the vegetation in such a way that the change appears to facilitate the grazing of successive species (Gwynne and Bell 1968; Jarman and Sinclair 1979).

The bulk of available data suggest that the large macropodids have prospered since domestic stock were introduced as indicated by the dramatic increase in population sizes. They appear well suited to the present sub-climax vegetation in many areas. Sheep numbers peaked during the 1890's before crashing dramatically (Fig. 2). They have since stabilized at lower levels. It seems likely that many grazing systems within Australia have now reached a dynamic equilibrium under the new herbivore regime (Caughley 1987a) and because food has become the limiting resource, the possibility of exploitative competition deserves particular consideration.

Since competition for space appears to be minimized by differential activity periods, interference competition between large macropodids and sheep seems unlikely. The fact that behavioural antagonism has not been reported to date supports this assertion. The displacement of macropodids at watering troughs seems inconsequential because water is no longer a limiting resource.

The degree of dietary overlap between large macropodids and sheep has been shown to be small under most circumstances and resource partitioning appears widespread (Dawson, this volume). Even during drought, which Wiens (1977) would predict to be a "crunch time" with heightened competition, diets have been shown to diverge. An important question now arises: is this instantaneous dietary divergence the result of interspecific competition? Most graziers certainly believe this to be the case (Newsome 1971b; Newsome and Corbett 1977). Unfortunately there has been no controlled examination of macropodid and sheep diets in an exclusion situation to help support or invalidate this assertion (Shepherd and Caughley 1987).

Competition induced dietary divergence may occur in two ways. Where species have co-evolved, selective pressures may favour character displacement resulting in resource partitioning. Here a species changes its preference or pattern of use of available resources so as to enable co-existence. The second is "sequential competition" (Low and Low 1975). This concept holds that "the forage grazed by an animal species in a community already grazed by another species is what has been left by the first species" (Low and Low 1975). Such a phenomenon may occur when separate species are sequentially using different habitats and may easily be confused with facilitated grazing. In the case of facilitated grazing, a species utilizes a resource which it would not normally use following modification of that resource through grazing by another species. With sequential competition, species are deprived of a resource they would otherwise utilize. Careful experimentation is obviously needed to distinguish between different interactions between species, competitive or otherwise (Schoener 1983). To date, as mentioned above, dietary studies have been lacking in this regard. In any case, it should be remembered that dietary overlap in itself does not demonstrate the existence of competition. One or both of the competitors must be harmed in some way.

Recently, Bayliss (1985, 1987) postulated that sheep grazing suppressed kangaroo densities (Reds and Western Greys) outside Kinchega National Park by 50% through competition for food. In contrast, Denny (1980) found the population density for the Red Kangaroo on Sturt National Park to be comparable to that of several properties carrying either sheep or cattle. He also found no correlation between kangaroo numbers and stock density. The data presented in Table 4 confirm that in 1977, kangaroo densities outside Kinchega were comparable to those estimated for Sturt and the other properties surveyed by Denny. The resolution of the apparent anomaly between these two studies may lie in the fact that since 1967 Kinchega has been surrounded by a kangaroo-proof fence which limits kangaroo movements (Priddel 1987) and hence dispersal out of the park. Bayliss assumed that the dispersal of individuals outside the park was inconsequential. However, the validity of this assumption may be suspect. Red Kangaroo populations have been reported to comprise a sedentary and a relatively more mobile component (Bailey 1971; Johnson and Bayliss 1981), with long range dispersal, particularly of young males, reported in several studies (Bailey 1971; Denny 1980; Priddel 1987; author's unpublished data). The dispersal of sub-adult males has also been noted for Western Grey Kangaroos (McCarron, pers. comm.). Population levels on Kinchega may therefore be artificially high.

A complicating factor in both these studies is the harvesting of kangaroos outside national parks. Caughley (1987b) has shown that "a harvesting rate between 10% and 15% per year depresses the kangaroo population to between

Table 4. Estimated sheep and kangaroo densities (animals/km^2) for (a) Kinchega National Park and surrounding properties, and (b) Sturt National Park and surrounding properties in 1977.

Area	Sheep density	Total macropodid density	Source
a. Kinchega	0	40	Bayliss 1987
Surrounds	20 (approx.)	16	Bayliss 1987
b. Sturt	0	12	Denny 1980
Boorara	5.2	13	Denny 1980
Pine view	1.9	11.4	Denny 1980

60% and 70% of its unharvested mean density". Harvesting outside Kinchega undoubtedly has an impact and the combined effects of harvesting and dispersal (or the suppression thereof) may account for the observed difference in macropodid densities. Harvesting outside Sturt would similarly reduce macropodid numbers and perhaps attract immigrants from within the park to replace those harvested. Harvesting may therefore indirectly reduce the density of macropodids on Sturt and therefore complicate the comparison of densities of kangaroos on Sturt (no harvesting, no stock) with other properties (stock, harvesting). Comparison with a property with stock and no harvesting would have been preferable.

Based upon the premise of shared food supply, Caughley (1987b) has suggested that sheep and kangaroos compete on semi-arid pastures at pasture biomasses below 300 kg/ha. He argued that this competition was asymmetrical with sheep impacting more upon kangaroos than kangaroos upon sheep. Simulations showed that sheep suppressed long term kangaroo densities by 42%. However, as Caughley points out, these findings are suggestive, and given the degree of dietary divergence, the assumption concerning shared food resources may be tenuous.

To date the possibility of exploitative competition has remained rather clouded as there has been a lack of controlled experiments addressing this problem (Caughley 1987b). However, such experiments are currently in progress. The first is being conducted by the C.S.I.R.O. near Cobar in central western New South Wales. This research involves determination of the optimal sheep stocking rate in semi-arid mulga woodlands and includes examination of the competitive interaction between sheep and kangaroos (Reds and Greys) combined at different densities (0.76-1.25 ha/animal), and associated pasture dynamics. Preliminary results indicate that increasing kangaroo stocking density has a negative effect upon sheep productivity, with sheep body weight being affected more than wool production (Wilson and MacLeod, in prep.).

The second experiment is being conducted by the University of New South Wales at Fowler's Gap located north of Broken Hill. This research aims to examine sheep-Red Kangaroo competitive interactions and associated pasture dynamics in the arid chenopod shrublands and involves a large scale removal experiment. The experiment was designed to mimic conditions experienced by the grazier on a typical arid zone property. There are three grazing treatments: sheep only, kangaroos only and sheep and kangaroos combined. Sheep stocking rates are conservative (7.5 ha/sheep), yet largely typical of the arid zone, being based upon those recommended by the Soil Conservation Service of New South Wales. Kangaroos are excluded from the "sheep only" paddocks but otherwise their numbers are unmanaged and animals are allowed to free-range throughout the study area. Preliminary results indicate that, under pasture conditions ranging from poor to good, free-ranging kangaroos do not impact upon sheep body weights or wool production. Under poor pasture conditions, however, kangaroos in paddocks without sheep were found to have better body condition than those from paddocks containing sheep (author's unpublished data). This suggests that preferred pasture in the "kangaroo only" paddocks was more abundant and perhaps of higher quality. These results seem to support the assertion of Caughley (1987b) concerning the probable asymmetrical nature of sheep-kangaroo competition.

It seems, therefore, that sheep and large macropodids do compete to some degree. This competition is possibly asymmetrical with sheep impacting more upon kangaroos than kangaroos upon sheep, though both effects appear to be significant. It seems likely that such competition is an intermittent process as is often the case (Wiens 1977) and that competition becomes more pronounced at high stocking densities and/or during times of pasture shortage. However, it must be stressed that these results are preliminary and that long-term studies of closely monitored populations are central to the attainment of unequivocal interpretations of the level and overall mechanisms of competition (Wiens 1977). This is especially true in arid and semi-arid environments where climatic and environmental variability confound the overall picture.

CONCLUSION

Despite the fact that sheep and the macropodids are the dominant large herbivores across much of southern Australia, much of our knowledge of their ecology, particularly the nature of the interaction between them, is limited and based upon circumstantial evidence. Sheep have been implicated in the demise and extinction of several smaller macropodid species through habitat modification. The larger macropodids appear to have prospered under similar circumstances and to have increased greatly in numbers.

The extent of the interaction between the large macropodids and sheep appears to be a function of both grazing intensity and pasture availability. Environmental factors impact greatly upon this interaction because they largely govern the amount, type and quality of available forage. Recent data appear to confirm the existence of competition between kangaroos and sheep in certain situations although the intricacies of this competition require further elucidation. More detailed examination of feeding behaviour is warranted as this is undoubtedly an important factor in such competition events. Given the apparent dietary divergence of large macropodids and sheep, two further points are worthy of consideration. Firstly, the possibility that the productivity of rangelands may be better utilized by the grazing of sheep together with macropodids; and secondly, that pastures may be maintained in a more ecologically desirable equilibrium under a mixed grazing regime than when grazed by sheep or macropodids alone. Further research along these lines is fundamental to the proper future management of Australian rangeland ecosystems.

REFERENCES

Andrew, M. H. and Lange, R. T., 1986. The spatial distributions of sympatric populations of kangaroos and sheep: examples of dissociation between these species. *Aust. Wildl. Res.* **13**: 367-73.

Arnold, G. W. and Dudzinski, M. L., 1978. Ethology of free-ranging domestic animals. Elsevier: Amsterdam, Oxford, New York.

Bailey, P. T., 1971. The red kangaroo, *Megaleia rufa* (Desmarest), in north-western New South Wales. I. Movements. *C.S.I.R.O. Wildl. Res.* **16**: 11-28.

Barker, R. D., 1987. The diet of herbivores in the sheep rangelands. Pp. 69-83 *in* Kangaroos: their ecology and management in the sheep rangelands of Australia ed. by G. Caughley *et al.* Cambridge University Press: Cambridge.

Bayliss, P., 1985. The population dynamics of red and western grey kangaroos in arid New South Wales, Australia. I. Population trends and rainfall. *J. Anim. Ecol.* **54**: 111-25.

Bayliss, P., 1987. Kangaroo dynamics. Pp. 119-34 *in* Kangaroos: their ecology and management in the sheep rangelands of Australia ed. by G. Caughley *et al.* Cambridge University Press: Cambridge.

Caughley, G., 1964. Density and dispersion of two species of kangaroo in relation to habitat. *Aust. J. Zool.* **12**: 238-49.

Caughley, G., 1987a. Introduction to sheep rangelands. Pp. 1-14 *in* Kangaroos: their ecology and management in the sheep rangelands of Australia ed. by G. Caughley *et al.* Cambridge University Press: Cambridge.

Caughley, G., 1987b. Ecological relationships. Pp.159-88 *in* Kangaroos: their ecology and management in the sheep rangelands of Australia ed. by G. Caughley *et al.* Cambridge University Press: Cambridge.

Caughley, G., Grigg, G. C. and Smith, L., 1985. The effect of drought on kangaroo populations. *J. Wildl. Manage.* **49**: 679-85.

Clancy, T. and Croft, D. B. (this volume). Space use patterns of the common wallaroo, *Macropus robustus erubescens,* in the arid zone.

Condon, R. W., 1983. Pastoralism. Pp. 54-60 *in* What future for Australia's arid lands? ed. by J. Messer and G. Mosley. Australian Conservation Foundation.

Croft, D. B., 1980. Behaviour of red kangaroos, *Macropus rufus* (Desmarest, 1822) in northwestern New South Wales, Australia. *Aust. Mammal.* **4**: 5-58.

Croft, D. B., 1981. Social behaviour of the euro, *Macropus robustus* (Gould), in the Australian arid zone. *Aust. Wildl. Res.* **8**: 13-49.

Dawson, T. J. (this volume). Diets of macropodid marsupials: general patterns and environmental influences.

Dawson, T. J. (in prep.). Kangaroos. University of New South Wales Press: Sydney.

Dawson, T. J. and Hulbert, A. J., 1970. Standard metabolism, body temperature, and surface areas of Australian marsupials. *Am. J. Physiol.* **218**: 1233-38.

Denny, M. J. S., 1980. Red kangaroo arid zone studies: final report. Unpublished.

Denny, M. J. S. and Dawson, T. J., 1975. Effects of dehydration on body-water distribution in desert kangaroos. *Am. J. Physiol.* **229**: 251-54.

Ealey, E. H. M., 1967. Ecology of the euro, *Macropus robustus* (Gould), in north-western Australia I. The environment and changes in euro and sheep populations. *C.S.I.R.O. Wildl. Res.* **12**: 9-25.

Ellis, B. A., Russell, E. M., Dawson, T. J. and Harrop, C. J. F., 1977. Seasonal changes in diet preferences of free-ranging red kangaroos, euros and sheep in western New South Wales. *Aust. Wildl. Res.* **4**: 127-44.

Fennessy, B. V., 1970. Native fauna. Pp. 152-63 *in* The Australian environment 4th Ed. ed. by G. W. Leeper. Melbourne University Press: Melbourne.

Frith, H. J., 1973. The herbivorous wild animals. Pp. 74-83 *in* Australian grasslands ed. by R. M. Moore. Australian National University Press: Canberra.

Gardiner, H. G., 1986a. Dynamics of perennial plants in the mulga (*Acacia aneura* F. Muell.) zone of Western Australia. I. Rates of population change. *Aust. Rangel. J.* **8**: 18-27.

Gardiner, H. G., 1986b. Dynamics of perennial plants in the mulga (*Acacia aneura* F. Muell.) zone of Western Australia. II. Survival of perennial shrubs and grasses. *Aust. Rangel. J.* **8**: 28-35.

Gwynne, M. D. and Bell, R. H. V., 1968. Selection of vegetation components by grazing ungulates in the Serengeti National Park. *Nature* **220**: 390-93.

Harrington, G. N., Oxley, R. E. and Tongway, D. J., 1979. The effects of European settlement and domestic livestock on the biological system in poplar box *(Eucalyptus populnea)* lands. *Aust. Rangel. J.* **1**: 271-79.

Hume, I. D., 1982. Digestive physiology and nutrition of marsupials. Cambridge University Press: Cambridge.

Jaremovic, R., 1984. Space and time related behaviour in eastern grey kangaroos. Ph.D. Thesis, University of New South Wales.

Jarman, P. J. and Sinclair, A. R. E., 1979. Feeding strategy and the pattern of resource partitioning in ungulates. Pp. 130-63 *in* Serengeti: dynamics of an ecosystem ed. by A. R. E. Sinclair and M. Norton-Griffiths. University of Chicago Press: Chicago.

JARMAN, P. J. AND SOUTHWELL, C. J., 1986. Grouping, associations and reproductive strategies in eastern grey kangaroos. Pp. 399-428 *in* Aspects of social evolution ed. by D. I. Rubenstein and R. W. Wrangham. Princeton University Press: Princeton.

JOHNSON, C. N., 1983. Variations in group size and composition in red and western grey kangaroos, *Macropus rufus* (Desmarest) and *M. fuliginosus* (Desmarest). *Aust. Wildl. Res.* **10**: 25-31.

JOHNSON, C. N. AND BAYLISS, P. G., 1981. Habitat selection by sex, age and reproductive class in the red kangaroo, *Macropus rufus,* in western New South Wales. *Aust. Wildl. Res.* **8**: 465-74.

KAUFMANN, J. H., 1974. Habitat use and social organization of nine sympatric species of macropodid marsupials. *J. Mammal.* **55**: 66-80.

LANGE, R. T., 1969. The piosphere: sheep track and dung patterns. *J. Range Manage.* **22**: 396-400.

LEIGH, J. H., 1974. Diet selection and the effects of grazing on the composition and structure of arid and semi-arid vegetation. Pp. 102-26 *in* Studies in the Australian arid zone II. Animal production ed. by A. D. Wilson. C.S.I.R.O.: Australia.

LEIGH, J. H., 1977. Some effects of diet selectivity by vertebrate and invertebrate herbivores on arid and semi-arid rangelands. Pp.125-48 *in* The impact of herbivores on arid and semi-arid rangelands. Australian Rangeland Society: Perth.

LOW, B. S., 1979. The predictability of rain and the foraging patterns of the red kangaroo *(Megaleia rufa)* in central Australia. *J. Arid. Env.* **2**: 61-76.

LOW, B. S. AND LOW, W. A., 1975. Feeding interactions of red kangaroos and cattle in an arid ecosystem. Pp. 87-94 *in* Proceedings of III World Conference on Animal Production ed. by R. L. Reid. Sydney University Press: Sydney.

LYNCH, J. J., 1974. Merino sheep: some factors affecting their distribution in very large paddocks. Pp. 697-707 *in* The behaviour of ungulates and its relation to management ed. by V. Geist and F. Walther. International Union for Conservation of Nature and Natural Resources: Morges, Switzerland.

MOORE, R. M., 1962. Effects of the sheep industry on Australian vegetation. Pp. 170-83 *in* The simple fleece — studies in the Australian wool industry ed. by A. Barnard. Melbourne University Press: Melbourne.

NEWMAN, J. C. AND CONDON, R. W., 1969. Land use and present condition. Pp. 105-32 *in* Arid lands of Australia ed. by R. O. Slatyer and R. A. Perry. Australian National University Press: Canberra.

NEWSOME, A. E., 1965a. The abundance of red kangaroos, *Megaleia rufa* (Desmarest), in central Australia. *Aust. J. Zool.* **13**: 269-87.

NEWSOME, A. E., 1965b. The distribution of red kangaroos, *Megaleia rufa* (Desmarest), about sources of persistent food and water in central Australia. *Aust. J. Zool.* **13**: 289-99.

NEWSOME, A. E., 1971a. Competition between wildlife and domestic stock. *Aust. Vet. J.* **47**: 577-86.

NEWSOME, A. E., 1971b. The ecology of red kangaroos. *Aust. Zool.* **16**: 32-50.

NEWSOME, A. E., 1975. An ecological comparison of the two arid-zone kangaroos of Australia, and their anomalous prosperity since the introduction of ruminant stock to their environment. *Q. Rev. Biol.* **50**: 389-424.

NEWSOME, A. E. AND CORBETT, L. K., 1977. The effects of native, feral and domestic animals on the productivity of the Australian rangelands. Pp. 331-56 *in* The impact of herbivores on arid and semi-arid rangelands. Australian Rangeland Society: Perth.

NEWSOME, A. E., STEPHENS, D. R. AND SHIPWAY, A. K., 1967. Effect of a long drought on the abundance of red kangaroos in central Australia. *C.S.I.R.O. Wildl. Res.* **12**: 1-8.

OLIVER, A. J., 1986. Social organization and dispersal in the red kangaroo. Ph.D. Thesis, Murdoch University: Western Australia.

PERRY, R. A., 1973. Arid shrublands and grasslands. Pp. 246-59 *in* Australian Grasslands ed. by R. M. Moore. Australian National University Press: Canberra.

POOLE, W. E., 1983. Pp. 244-49 *in* The Australian Museum Complete Book of Australian Mammals ed. by R. Strahan. Angus & Robertson: London.

PRIDDEL, D., 1986. The diurnal and seasonal patterns of grazing of the red kangaroo, *Macropus rufus,* and the western grey kangaroo, *M. fuliginosus. Aust. Wildl. Res.* **13**: 113-20.

PRIDDEL, D., 1987. The mobility and habitat utilization of kangaroos. Pp. 100-18 *in* Kangaroos: their ecology and management in the sheep rangelands of Australia ed. by G. Caughley *et al.* Cambridge University Press: Cambridge.

ROBERTSON, G., SHORT, J. AND WELLARD, G., 1987. The environment of the Australian sheep rangelands. Pp. 14-34 *in* Kangaroos: their ecology and management in the sheep rangelands of Australia ed. by G. Caughley *et al.* Cambridge University Press: Cambridge.

ROBINSON, I. B., 1982. Drought relief schemes for the "pastoral zone". *Aust. Rangel. J.* **4**: 67-77.

SALE, P. F., 1974. Overlap in resource use, and interspecific competition. *Oecologia (Berl.)* **17**: 245-56.

SCHOENER, T. W., 1983. Field experiments on interspecific competition. *Am. Nat.* **122**: 240-85.

SHEPHERD, N., 1987. Condition and recruitment of kangaroos. Pp. 135-58 *in* Kangaroos: their ecology and management in the sheep rangelands of Australia ed. by G. Caughley *et al.* Cambridge University Press: Cambridge.

SHEPHERD, N. AND CAUGHLEY, G., 1987. Options for management of kangaroos. Pp. 188-219 *in* Kangaroos: their ecology and management in the sheep rangelands of Australia ed. by G. Caughley *et al.* Cambridge University Press: Cambridge.

SHORT, J., 1986. The effect of pasture availability on food intake, species selection and grazing behaviour of kangaroos. *J. Appl. Ecol.* **23**: 559-71.

SHORT, J., 1987. Factors affecting food intake of rangeland herbivores. Pp. 84-99 *in* Kangaroos: their ecology and management in the sheep rangelands of Australia ed. by G. Caughley *et al.* Cambridge University Press: Cambridge.

SOUTHWELL, C., 1987. Activity patterns of the eastern grey kangaroo, *Macropus giganteus. Mammalia* **51**: 211-23.

SQUIRES, V., 1981. Livestock management in the arid zone. Inkata Press: Melbourne, Sydney, London.

STRAHAN, R. (ed.), 1983. The Australian Museum Complete Book of Australian Mammals. Angus & Robertson: London.

WATSON, I. W., HOLM, A. MCR., SHACKLETON, K. R. AND ALLEN, R. J., in prep. The effect of sheep stocking rate on red kangaroo grazing pressure.

Wiens, J. A., 1976. Population responses to patchy environments. *Ann. Rev. Ecol. Syst.* **7**: 81-120.

Wiens, J. A., 1977. On competition and variable environments. *Am. Sci.* **65**: 590-97.

Wilson, A. D. and Graetz, R. D., 1979. Management of the semi-arid and arid rangelands of Australia. Pp. 83-111 *in* Management of semi-arid ecosystems ed. by B. H. Walker. Elsevier: Amsterdam.

Wilson, A. D. and MacLeod, N. D., in prep. The effect of kangaroos on the productivity of sheep in the mulga woodlands.

Kangaroos, Wallabies and Rat-kangaroos: Their past, present and future in the Woodchip Agreement Area of the Eden Region, south-east New South Wales

DANIEL LUNNEY[1]

Historical and contemporary studies indicate that the Woodchip Agreement Area within the Eden Region in south-east New South Wales was richer in kangaroo, wallaby and rat-kangaroo species late last century than at present. There were at least nine, and possibly 11, macropodid and potoroid species last century, but within 80 years of initial settlement in 1830 at least three, and possibly five, species had become extinct. A survey in 1986-87 of the Woodchip Agreement Area found that only one species is now common, and the others are uncommon or of uncertain status, with two recognized as endangered species. Recent studies in Mumbulla State Forest and Nadgee Nature Reserve have shown that logging and fire adversely affect the numbers and distribution of macropodids. The conservation of macropodids and potoroids largely depends upon a re-evaluation of land-use options to ensure that large areas are reserved from logging, and upon major changes to the current plan of management for those forests which are being intensively logged for woodchips and sawlogs and are subject to control burning.

INTRODUCTION

In the Woodchip Agreement Area of the Eden Region of south-east New South Wales, the woodchip industry has remained controversial since its inception in 1969 (Lunney and Moon 1987). Lack of information on the fauna has hindered resolution of the conflict. One recent study (Lunney and Leary 1988) found that kangaroos, wallabies and rat-kangaroos constitute an extinction-prone group in the area, and thus one deserving particular attention. This paper reviews evidence of the effect of human activities in the Woodchip Agreement Area on species diversity and population sizes of macropodids and potoroids. It also suggests directions for further research and measures for greater retention of the forest habitat upon which the species depend.

STUDY AREA

The Eden forestry region in south-east New South Wales consists of three forestry districts (Bega, Bombala and Eden) and the Eden Woodchip Agreement Area. With an area of about 278 529 ha, State Forests cover about 46.5% of the Woodchip Agreement Area and are managed under the Eden Native Forest Management Plan (Forestry Commission 1982). About 39.4% of the Woodchip Agreement Area is freehold land, which is mostly farmland with dairy farming as the dominant industry. National Parks, Nature Reserves and a State Recreation Area occupy about 10.5%. About 3% is Vacant Crown Land, and a further 0.7% is Forestry Commission Forest Preserve and Flora Reserve (Fig. 1). The region stretches westwards from the coast, which is mainly rocky with several estuaries and through a low-relief forested range (300 m altitude) with a few high peaks such as Mumbulla Mountain (704 m). Inland is a wide strip (about 20 km) of gently undulating country, including the broad, cleared valleys of the Bega, Brogo and Towamba Rivers. Further west is a steep, forested escarpment running north-south and rising more than 500 m to a forested range which exceeds 1 000 m in places. An undulating forested plateau drops westward to the low-relief country of the southern tableland of the Monaro, most of which has been cleared for grazing and farming. Unlike those on the coast, the inland National Parks and Nature Reserves are predominantly mountainous.

REVIEW OF THE EVIDENCE

1. Survey

(a) Intensive local studies. Two areas have recently been the subject of intensive surveys: Mumbulla State Forest on the coast near Bega (Lunney and Barker 1986, 1987; Lunney *et al.*, in press; O'Connell and Lunney, unpublished); and Nadgee Nature Reserve on the coast near Eden (Fox 1978; Newsome *et al.* 1983). Both carry the same four species: Grey Kangaroo *Macropus giganteus,* Red-necked Wallaby *M. rufogriseus,* Swamp Wallaby *Wallabia bicolor,* and Long-nosed Potoroo *Potorous tridactylus.* The evidence for the presence and status of macropodids and potoroids in Nadgee was based on analysis of scats (the identification of hairs in faeces) of Dogs or Dingoes *Canis familiaris,* identification of footprints or tracks, visual observations and cage trapping. The evidence from Mumbulla was based on scat analysis of *C. familiaris* and Foxes *Vulpes vulpes,* dung counts of macropodids and visual observations. There was a sufficient degree of

[1]National Parks and Wildlife Service, New South Wales, P.O. Box 1967, Hurstville, New South Wales 2220.

Pages 805-12 in KANGAROOS, WALLABIES AND RAT-KANGAROOS ed. by G. Grigg, P. Jarman and I. Hume. Surrey Beatty & Sons Pty Limited, New South Wales, Australia, 1989.

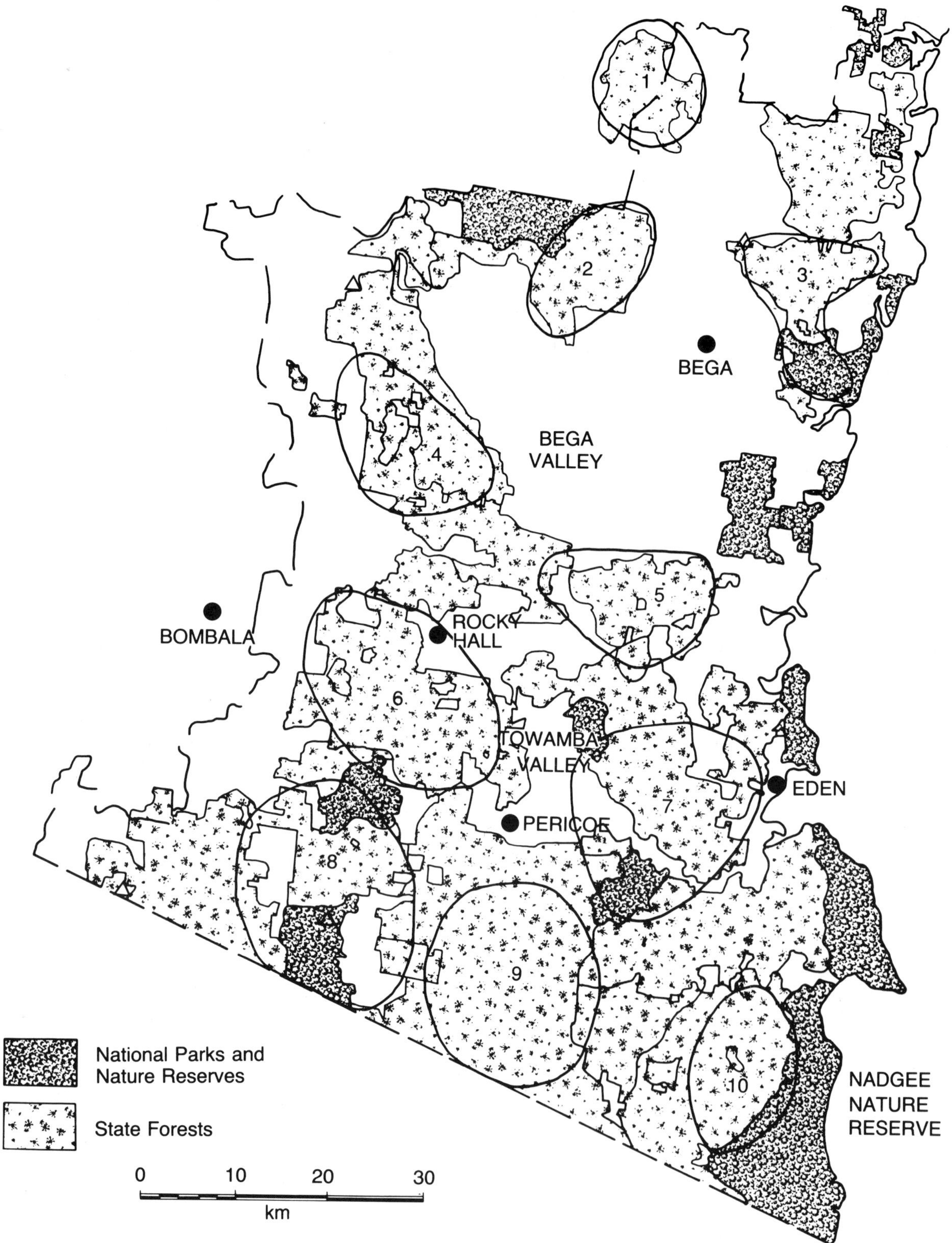

Fig. 1. The location of State Forests, National Parks and Nature Reserves, and places mentioned in the text, in the Eden Woodchip Agreement Area within the Eden forestry region in southeastern New South Wales. The border of the Woodchip Agreement Area is marked by the New South Wales-Victorian border on the southern boundary of the map, and a broken line on the western and northern edge. This line follows parish boundaries and is not necessarily consistent with topographic or land-use features. The white areas within the Woodchip Agreement Area are freehold land, which in the Bega Valley is predominantly for dairy farming, while the western edge near Bombala is predominantly for sheep grazing. The rough circles, numbered 1-10, are the forest areas in which the predator scats were collected and from which the prey or carrion eaten by the dogs and foxes were taken. The boundaries between each State Forest are delineated, but the name only of the forest from which all or most of the scats were collected within the circle is given (Table 1).

consistency between the methods of the Mumbulla and Nadgee studies to allow the general conclusion that in both places the two wallaby species are locally common, with *W. bicolor* more common and *P. tridactylus* locally rare. *M. giganteus* is uncommon in Nadgee and rare in Mumbulla.

(b) A regional survey. A survey of mammals by predator scat analysis was conducted throughout the Woodchip Agreement Area in 1986-87 by the National Parks and Wildlife Service of New South Wales. This survey technique was found to be the most efficient for all mammal species except bats in and near Mumbulla State Forest (Lunney and Barker 1986). Two hundred scats were collected from each of 10 forest areas (Fig. 1). (In the case of Mumbulla, the 200 scats were taken as a random sample from those collected in 1981-82 [Lunney and Barker 1986, and Lunney *et al.*, in press]). The 10 forest areas shown in Figure 1 roughly correspond to State Forests, and include parts of some National Parks or Nature Reserves. The name of the predominant forest within each area is given in Table 1. The scats were collected along the network of tracks and roads well within the circle designating each forest area, which represents the approximate outer limit of the source of prey or carrion consumed. The scat-based survey identified 27 native species. Of these, there were four macropodids (*M. giganteus, M. rufogriseus, M. robustus* and *W. bicolor*) and two potoroids (*P. tridactylus* and *P. longipes*). The most common species was *W. bicolor,* which occurred in all 10 areas and was present in 32% of all scats containing the hair of native species. The next was *M. rufogriseus,* which occurred in low numbers (only 2% of scats) in nine areas. *M. giganteus,* found in 2% of scats in seven areas, was also uncommon. *P. tridactylus* was rare. The survey revealed the presence, in the high country of Bondi State Forest, of the Long-footed Potoroo *P. longipes,* which was previously known only from Victoria (Seebeck and Johnson 1980; Seebeck 1983), as well as what is most likely to be the Wallaroo *M. robustus*, which was found in the Nullica State Forest/Egan Peaks Nature Reserve area near Merimbula. (The element of uncertainty in this is referred to in Table 1).

2. *Historical studies*

A detailed historical study of the Bega district showed that the status of mammals, and macropodids and potoroids in particular, has changed since European settlement in 1830 (Lunney and Leary 1988). The information was obtained from primary sources, including: interviews with local residents; examination of all issues of local papers published between 1860 and 1955 and held by the Mitchell Library; records, maps and photographs held by the State Archives, Bega Historical Society, Australian Museum, Mitchell Library; and portion plans from the 1860s, with surveyor's comments on the vegetation, held by the Department of Lands. At least four species (*M. parma, M. robustus, Thylogale thetis* and a rat-kangaroo, possibly *Bettongia gaimardii*) had become extinct by about 1910. The historical presence of *M. parma* and *T. thetis* confirmed Maynes' (1977) view that both species had a former south coast distribution which had contracted before adequate records were taken. The Rufous Bettong *Aepyprymnus rufescens* may also have occurred in the district. With four of eight species, or five of nine species, becoming extinct, macropodids and potoroids were found to be the most extinction-prone of the mammal groups. The populations of most species did not, as might be thought, decline steadily as land clearing proceeded. Most, if not all, species rose to plague proportions and then dropped sharply, in some cases to extinction, in others to a lower, short-lived peak after a lapse of some decades. That this finding parallels the study by Jarman and Johnson (1977) of the mammals of the western slopes of the Great Dividing Range suggests there is an ecological basis for the spectacular population fluctuations which occurred in the Bega

Table 1. Proportion of scats containing macropodids and potoroids in each of 10 forest areas in the Eden Woodchip Agreement Area. * means that the identification (microscopic examination of the hair) contained an element of uncertainty at the species level. *Potorous* sp. means definitely from that genus when it was not possible to distinguish between the species.

Species	Forest Area (see below) 1	2	3	4	5	6	7	8	9	10	Number of Areas containing each species
Eastern Grey Kangaroo *M. giganteus*	0.02	0.00	0.01	0.01	0.01	0.00	0.04	0.04	0.00	0.07	7
Red-necked Wallaby *M. rufogriseus*	0.02	0.02	0.08	0.02	0.00	0.03	0.01	0.03	0.02	0.02	9
Swamp Wallaby *W. bicolor*	0.31	0.17	0.48	0.49	0.41	0.30	0.31	0.26	0.17	0.32	10
Long-nosed Potoroo *P. tridactylus*	0.00	0.00	0.02	0.00	0.00	0.00	0.00	0.00	0.00	0.00	1
Long-footed Potoroo *P. longipes*	0.00	0.00	0.00	0.00	0.00	0.00	0.00	0.01	0.00	0.00	1
Wallaroo *M. robustus**	0.00	0.00	0.00	0.00	0.00	0.00	0.01	0.00	0.00	0.00	1
Potoroo *Potorous* sp.	0.00	0.00	0.00	0.00	0.00	0.00	0.01	0.00	0.00	0.00	1
Total species/Forest Area	3	2	4	3	2	2	5	4	2	4	

Forest Area. 1. Murrabrine State Forest 2. Bemboka S. F. 3. Mumbulla S. F. 4. Tantawangalo S. F. 5. Yurammie S. F. 6. Coolangubra S. F. 7. Nullica S. F. 8. Bondi S. F. 9. Yambulla S. F. 10. Nadgee S. F.

district. The changes in population sizes and the loss of species are attributed to the changes in the land-use and land-management practices of the latter decades of the nineteenth century and early twentieth century. These included the clearing of the arable land of the Bega Valley, the shooting of kangaroos and wallabies for a thriving fur trade and as pests, and the invasion and population irruptions of the Hare *Lepus capensis,* the Rabbit *Oryctolagus cuniculus,* and the Fox. Both Fox and Dingo populations increased rapidly around the turn of the century, sustained on superabundant hare-rabbit prey. Increased predation from canids and humans, combined with the rapid loss of habitat, resulted in a severe depletion of the macropodid and potoroid fauna of the Bega district.

3. *Habitat selection study*

An intensive study of habitat selection, based on dung counts of the macropodids and estimates of cover provided by the vegetation, was carried out in Mumbulla State Forest near Bega in 1981 and 1982 (Lunney and O'Connell 1988; O'Connell and Lunney, submitted). The study encompassed both ridges and gullies of unlogged forest, recently logged forest and 10-15 year regeneration. The distribution of both *W. bicolor* and *M. rufogriseus* varied with the topography and age-class of the forest, and demonstrated that logging and a recent fire had a marked effect on habitat selection. Refuge cover was shown to be crucial in habitat selection. The habitats least selected were the ridges which has been logged and further exposed by fire.

4. *Fire study*

A long-term study in Nadgee Nature Reserve showed that the numbers of macropodids and potoroids fluctuated markedly following a major wildfire in 1972 (Newsome *et al.* 1983). The fall in numbers in the immediate post-fire years was attributed to Dingo predation. Numbers rose above pre-fire levels as the thick regeneration of the vegetation provided increasingly abundant cover and food.

5. *Other studies*

According to Wakefield (1961), old residents of the Towamba Valley reported that the Brush-tailed Rock-wallaby *Petrogale penicillata* was formerly abundant in the mountains above the eastern escarpment of the Monaro Tableland in the vicinity of Rocky Hall, Pericoe and Nungatta. The date of Wakefield's paper, and the reference to old residents, suggest that the sightings were made last century. There has been no confirmed recent sighting, nor definitive evidence, such as a specimen, that it ever occurred there. However, the potential for its existence, or prior existence, in this area is consistent with its known habitat and geographic range (e.g., Maynes and Sharman 1983).

CONCLUSIONS

The Eden Woodchip Agreement Area was once rich in macropodid and potoroid species with at least nine, if not 11 species. Today, at least five, and probably six, species are present. Of these five, only one *(W. bicolor)* is widespread and common. Recent studies have shown that logging and fire affect macropodid numbers. Since the State Forests, which constitute 46.5% of the Woodchip Agreement Area, are being managed for an intensive integrated woodchip and sawlog operation, with associated pre- and post-logging control burning, it is reasonable to conclude that such a major change to the forest habitat will again have a significant impact on macropodids and potoroids. The evidence to date suggests that all populations will fluctuate, with some falls potentially reaching extinction.

On current plans, logging operations will open up extensive tracts of previously unlogged or selectively logged forest within 40 years of 1969 (Forestry Commission 1982; Lunney and Moon 1989). In this respect, woodchipping resembles nineteenth century clearing in that a large proportion of the forested country is transformed within decades from mature forest to virtually open country. The main difference between the current operation and nineteenth century land clearing lies in the fact that the native forests will be allowed to regrow. However, they will become widespread stands of even-aged regrowth which are scheduled to be logged again after 40 years, at which stage the forest will still be comparatively young and markedly different from mature forest.

Since the current logging operations are occurring in forest surrounding the areas cleared in the nineteenth century, current planners need to recognize that they provide crucial refuge for all forest-dependent species. There also needs to be a recognition of the fact that these forests are generally on soils of lower fertility than those cleared last century. They are located on hilly country and comprise different tree species, or combination of species, than those which occurred on the fertile flat or gently undulating country. As such, the existing forested land is not necessarily the optimum habitat for all the remaining wildlife species, and may constitute marginal habitat for some. Such species are therefore probably less resilient to further major habitat change. There also needs to be a clear acknowledgement that most of the forest set aside from the current woodchip-sawlog operation is the least loggable because of factors such as steepness and soil quality, including susceptibility to erosion.

In conjunction with the Plan of Management for the forests of the Woodchip Agreement Area (Forestry Commission 1982), the Forestry Commission also uses a second major strategy to plan the use of all State Forest areas. Called the

Preferred Management Priority (PMP) Classification, the strategy divides the forest into developed and undeveloped natural forest (Dobbyns 1986). On the PMP maps (1:25 000 topographic maps), currently available only in the relevant district and regional offices of the Forestry Commission, the forest set aside from logging (undeveloped forest) is coloured brown, and that designated to be logged (developed) is coloured light green. A major benefit of the colour-coded topographic maps is that it enables anyone inspecting them to appreciate both the pattern of the forest to be logged and remain unlogged, and the topographic location of both.

What emerges from an inspection of these maps is that, like the nineteenth century land clearing phase the woodchip-sawlog operation is concentrated on the forests of one type, that is, those on the richer soils and on the flatter country. The important exception to this generalization is the provision, within designated developed forest, of special areas retained for one primary purpose such as visual values, research, water catchment, recreation, flora and fauna, historical and Aboriginal sites (Dobbyns 1986). These are coloured dark green on the PMP maps. Of particular importance for fauna are the wildlife corridors, especially those along major creeks and rivers and on river flats. Nevertheless, the area of forest that will remain unlogged will be primarily the least fertile and steepest, so it will be unlikely to contain suitable habitat for all wildlife species.

This review aims not only to point out the losses since European settlement and the effects of logging and the associated policy of both pre- and post-logging control burning, but also to suggest courses of action to conserve the macropodids and potoroids of the region. The conclusions drawn for macropodids and potoroids are similar to conclusions for other species of mammals (e.g., Lunney 1987; Lunney and Ashby 1987; Lunney *et al.* 1987, 1988, 1989). The important difference between ground-dwelling and tree-dwelling species (such as bats, possums and gliders) is that the latter require certain tree species and sizes, with the large old trees providing crucial breeding sites for many species. Reptiles (Lunney and Barker 1986; Lunney *et al.*, in press) and birds (e.g., Smith 1984, 1985, 1986, 1989) are also affected by logging and fire and would benefit from a careful consideration of their specific habitat requirements. Management options for macropodids and potoroids could also provide a widespread benefit for all faunal groups. The options fall into two categories: 1. research and 2. directions for change in forest management.

1. *Research*

(a) Continued survey of the mammals of all the forests is necessary. The 200 scats in each of 10 areas collected and analyzed during the 1986-87 survey is far from being a definitive survey of all the forested land of the Woodchip Agreement Area. In the study of 1 999 scats in Mumbulla State Forest, the last of the native mammals to be recorded was in scat number 475 (Lunney *et al.*, in press). By scat 200, 18 of the 22 native mammals had been recorded. This indicates that more species may exist in each of these forests than was found by 200 scats. As scat analysis cannot identify the habitat occupied by each species, habitat selection needs to be determined by slower, more labour-intensive methods such as trapping.

(b) Of particular importance is the need to determine the range of *P. longipes* and whether populations of *M. robustus* and *P. penicillata* occur in the Woodchip Agreement Area. Consideration also needs to be given to the possibility that small populations of one or more of the species present last century may still occur.

(c) There is an urgent need to undertake specific studies of rare species, particularly both species of *Potorous,* with respect to habitat selection, population dynamics and the impact of logging and fire. This is particularly the case with *P. longipes* because it is very rare and has been found to eat large amounts of hypogeal fungi, a food which the investigators (Seebeck and Scotts 1989) suggest may be adversely affected by logging and burning.

(d) Since the effects of logging and fire on forest regrowth will continue for many decades, there is a strong case for setting up long-term monitoring studies in conjunction with detailed research programmes.

(e) Predators (Dogs and Foxes) affect macropodid and potoroid populations (e.g., Newsome *et al.* 1983; Robertshaw and Harden 1986; Lunney and Leary 1988; Lunney and O'Connell 1988) so it is important to continue to study these predator populations in conjunction with fluctuations in their prey populations.

(f) The examination of macropodid and potoroid remains in Aboriginal middens, such as on the coast near Bega (Hughes and Sullivan 1978; Byrne 1983), could add to our knowledge of the species present prior to European settlement. For example, it could confirm the presence of *A. rufescens* and *P. penicillata.*

2. *Directions for change in forest management practices*

In planning the conservation of macropodids and potoroids in the region, managers of commercially utilized forest could substantially improve the present situation by:

(a) recognizing that macropodids and potoroids constitute a group in need of specific conservation action. Even the still widespread *M. rufogriseus,* which has become a species of concern because of its decline in nearby Victoria (Norris and Mansergh 1981; Norris *et al.* 1983), will require close monitoring.

(b) modification of the logging pattern to prevent the creation of large stands of uniform forest by spreading the logging operation so that coupes (logged areas) of a similar age of regrowth are widely separated. The current procedure of alternate-coupe logging is a step in that direction, but a greater diversity in age classes would offer more scope for dampening population fluctuations and ensuring the survival of all species.

(c) extending the time of each logging cycle so that the age of the forest is not limited to 40 years. For those forests where the first set of coupes has been logged, managers could extend the time before the logging of the second set of coupes. Again, this would have the effect of minimizing the proportion of exposed forest, dampening population fluctuations, and allowing more time for consideration of other management options. This is immediately relevant because logging of the previously unlogged alternate coupes is planned for early in the 1990s.

(d) minimizing control burning because it intensifies the effect of logging by removing ground cover.

(e) setting aside areas, with buffer zones, to retain the known and likely habitats of rare species, which should be protected from the impact of logging, fire, traffic, visitors and dogs. It also needs to be recognized that corridors along roads are unsuitable for ground-dwelling species, such as all the macropodids and potoroids.

(f) setting aside, throughout each forest, representative refuge areas, particularly gullies which have been identified as providing the best available habitat for a number of species, and making them sufficiently large to hold refuge populations.

The current plan of management (Forestry Commission 1982) requires the setting aside of filter strips for hydrological reasons, and these also serve as wildlife refuges. These filter strips are 20 m wide on either side of a gully, creek or river where the catchment exceeds 60 ha, or 40 ha in areas with easily eroded soils. Logging may occur in the outer 10 m, but heavy machinery may not enter. In addition, the plan of management provides for 100 m wildlife corridors to be set aside and to remain unlogged. The position of these corridors may be inspected on maps in the Forestry Commission offices in the Eden region. A map indicating their position but not their size (they were not drawn to scale), was given in the draft environmental impact statement prepared by Harris-Daishowa (1986). Other areas within State Forests will remain unlogged because they are too steep, or contain uncommercial trees, or have been set aside as Flora Reserves or Forest Preserves. The main question for forest planners is whether the areas currently allocated to wildlife represent all habitats and are large enough to maintain all wildlife populations until the forest fully regrows.

The retention of wildlife corridors follows from work done principally in the Eden and Bombala districts by an Australian Museum-Forestry Commission team (Recher *et al.* 1980). Regional Commission staff have devised such a system in recognition of the value of wildlife and the need to retain conservation options (Dobbyns and Ryan 1983; Ryan 1984). Nonetheless, Recher *et al.* (1980) called for a continuing evaluation of their recommendations, while a review by Recher *et al.* (1987), of the value of remnant mature forest, considered there to be much room for improvement and for increasing its reservation. They repeated the call for periodic review of wildlife conservation strategies, for longer rotation times and for the reservation from logging of fauna-rich compartments. The debate about Eden woodchips and proposed national parks, particularly Coolangubra and Tantawangalo, was reviewed by Lunney and Moon (1987). Lunney and Recher (1986) reviewed the theoretical aspects of, and Australian findings on, the optimal size of parks and reserves for effective nature conservation. The primary conclusion drawn was that large areas are essential for long-term wildlife conservation. Thus there is a strong case to suggest that the current areas set aside, or designated to be set aside from woodchipping, are inadequate for the long-term survival of all macropodid and potoroid populations in the Woodchip Agreement Area, particularly away from the coast, and that a major revision of both land-use and land management options is warranted.

ACKNOWLEDGEMENTS

I am indebted to many people for their contributions, particularly E. Ashby and J. Barker for scat collection; B. Triggs for scat analysis; E. Ashby for the map, and E. Ashby, I. Dunn, C. Moon, H. Recher, D. Sheppard and I. Sim for critical comments on the manuscript. The information on the regional survey is from unpublished work with L. Dovey and B. Triggs.

REFERENCES

BYRNE, D., 1983. The Five Forests: An Archaeological and Anthropological Investigation. National Parks and Wildlife Service, New South Wales, Sydney.

DOBBYNS, R., 1986. The forests of south-east New South Wales: Management and Change. *Forest and Timber* **22**: 35-38.

DOBBYNS, R. AND RYAN, D., 1983. Birds, Glider Possums and Monkey Gums: The wildlife reserve system in the Eden district. *Forest and Timber* **19**: 12-15.

FORESTRY COMMISSION OF NEW SOUTH WALES, 1982. Eden Native Forest Management Plan. Forestry Commission of New South Wales, Sydney.

FOX, A. M., 1978. The '72 fire of Nadgee Nature Reserve. *Parks and Wildlife* **2**: 5-24.

HARRIS-DAISHOWA (AUST) PTY LTD, 1986. Draft environmental impact statement for the Eden (New South Wales) export woodchip operation for the period 1989-2009. Harris-Daishowa (Aust) Pty Ltd, Eden.

HUGHES, P. J. AND SULLIVAN, M. E., 1978. A preliminary archaeological report to National Parks and Wildlife Service (New South Wales). National Parks and Wildlife Service, Sydney.

JARMAN, P. J. AND JOHNSON, K. A., 1977. Exotic mammals, indigenous mammals, and land use. *Proc. Ecol. Soc. Aust.* **10**: 146-66.

LUNNEY, D., 1987. Effects of logging, fire and drought on possums and gliders in the coastal forests near Bega, New South Wales. *Aust. Wildl. Res.* **14**: 263-74.

LUNNEY, D. AND ASHBY, E., 1987. Population changes in *Sminthopsis leucopus* (Gray) (Marsupialia: Dasyuridae) and other small mammal species, in forest regenerating from logging and fire near Bega, New South Wales. *Aust. Wildl. Res.* **14**: 275-84.

LUNNEY, D. AND BARKER, J., 1986. Mammals of the coastal forests near Bega, New South Wales I. Survey. *Aust. Zool.* **23**: 19-28.

LUNNEY, D. AND BARKER, J., 1987. Mammals of the coastal forests near Bega, New South Wales. II. Annotated Checklist. *Aust. Zool.* **23**: 41-49.

LUNNEY, D., BARKER, J., PRIDDEL, D. AND O'CONNELL, M., 1988. Roost selection by Gould's Long-eared Bat *Nyctophilus gouldi* Tomes (Chiroptera: Vespertilionidae) in logged forest on the south coast of New South Wales. *Aust. Wildl. Res.* **15**: 375-84.

LUNNEY, D., CULLIS, B. AND EBY, P., 1987. Effects of logging and fire on small mammals in Mumbulla State Forest, near Bega, New South Wales. *Aust. Wildl. Res.* **14**: 163-81.

LUNNEY, D., EBY, P. AND O'CONNELL, M. (in press). Effects of logging, fire and drought on three species of lizards in Mumbulla State Forest on the south coast of New South Wales. *Aust. J. Ecol.*

LUNNEY, D. AND LEARY, T., 1988. The impact on the native mammals of land-use changes and exotic species in the Bega District (New South Wales), since settlement. *Aust. J. Ecol.* **13**: 67-92.

LUNNEY, D. AND MOON, C., 1987. The Eden woodchip debate (1969-1986). *Search* **18**: 15-20.

LUNNEY, D. AND MOON, C., 1989. An ecological view of the history of logging and fire in Mumbulla State Forest on the south coast of New South Wales. Pp. 23-61 *in* Australia's Ever Changing Forests. Proceedings First National Conference on Australian Forest History ed. by K. Frawley and N. Semple. Aust. Defence Force Acad.: Canberra.

LUNNEY, D. AND O'CONNELL, M., 1988. Habitat selection by the Swamp Wallaby *Wallabia bicolor*, the Red-necked Wallaby *Macropus rufogriseus* and the Common Wombat *Vombatus ursinus* in logged, burnt forest near Bega, New South Wales. *Aust. Wildl. Res.* **15**: 695-706.

LUNNEY, D., O'CONNELL, M., SANDERS, J. AND FORBES, S., 1989. Habitat of the White-footed Dunnart *Sminthopsis leucopus* (Dasyuridae: Marsupialia) in a logged, burnt forest near Bega, New South Wales. *Aust. J. Ecol.* **14**: 335-44.

LUNNEY, D. AND RECHER, H., 1986. The living landscape: an ecological view of national parks and nature conservation. Pp. 294-328 *in* A Natural Legacy: Ecology in Australia (2nd ed.) ed. by H. F. Recher, D. Lunney and I. Dunn. Pergamon Press: Sydney.

LUNNEY, D., TRIGGS, B., EBY, P. AND ASHBY, E. (in press). An analysis of the scats of Dogs *Canis familiaris* and Foxes *Vulpes vulpes* (Canidae: Carnivora) in the Coastal Forests near Bega, New South Wales. *Aust. Wildl. Res.*

MAYNES, G. M., 1977. Distribution and aspects of the biology of the Parma Wallaby *Macropus parma* in New South Wales. *Aust. Wildl. Res.* **4**: 109-25.

MAYNES, G. M. AND SHARMAN, G. B., 1983. Brush-tailed Rock-Wallaby *Petrogale penicillata.* Pp. 211 *in* The Australian Museum Complete Book of Australian Mammals ed. by R. Strahan. Angus & Robertson: Sydney.

NEWSOME, A. E., CORBETT, L. K. AND CATLING, P. C., 1983. The feeding ecology of the dingo II. Dietary and numerical relationships with fluctuating prey populations in south-eastern Australia. *Aust. J. Ecol.* **8**: 345-66.

NORRIS, K. C. AND MANSERGH, I. M., 1981. Sites of Zoological Significance in East Gippsland. Ministry for Conservation, Victoria. Environmental Study Report Series No. 321. Recommendations. No. 342.

NORRIS, K. D., MANSERGH, I. M., AHERN, L. D., BELCHER, C. A., TEMBY, I. E. AND WALSH, N. G., 1983. Vertebrate fauna of the Gippsland Lakes Catchment Victoria. Occasional Paper Series No. 1. Fisheries and Wildlife Division, Victoria, Ministry for Conservation, East Melbourne.

RECHER, H. F., ROHAN-JONES, W. AND SMITH, P., 1980. Effects of the Eden woodchip industry on terrestrial vertebrates with recommendations for management. Research Note No. 42. Forestry Commission of New South Wales: Sydney.

RECHER, H. F., SHIELDS, J., KAVANAGH, R. AND WEBB, A., 1987. Retaining remnant mature forest for nature conservation at Eden, New South Wales: A review of theory and practice. Pp. 177-94 *in* Nature Conservation: the role of remnants of native vegetation ed. by D. A. Saunders, G. W. Arnold, A. A. Burbidge and A. J. M. Hopkins. Surrey Beatty & Sons: New South Wales.

ROBERTSHAW, J. D. AND HARDEN, R. H., 1986. The Ecology of the Dingo in north-eastern New South Wales IV. Prey Selection by Dingoes, and its Effect on the Major Prey Species, The Swamp Wallaby, *Wallabia bicolor* (Desmarest). *Aust. Wildl. Res.* **13**: 141-63

RYAN, D., 1984. Keeping the options open: The forestry ethic in practice in Eden. *Aust. Nat. Hist.* **21**: 155-59.

SEEBECK, J., 1983. Long-footed Potoroo *Potorous longipes.* Pp. 183 *in* The Australian Museum Complete Book of Australian Mammals ed. by R. Strahan. Angus & Robertson: Sydney.

SEEBECK, J. AND JOHNSON, P. M., 1980. *Potorous longipes* (Marsupialia: Macropodidae) a new species from eastern Victoria. *Aust. J. Zool.* **28**: 119-34.

SEEBECK, J. H. AND SCOTTS, D. J., 1989. Ecology of *Potorous longipes* (Marsupialia: Potoroidae); and preliminary recommendations for management of its habitat in Victoria. Arthur Rylah Institute for Environmental Research. Technical Report. Series No. 62. Conservation, Forests and Lands, National Parks and Wildlife Division, Victoria (in press).

SMITH, P. J., 1984. The forest avifauna near Bega, New South Wales. I. Differences between forest types. *Emu* **84**: 200-10.

SMITH, P. J., 1985. Effects of intensive logging on birds in Eucalyptus forest near Bega, New South Wales. *Emu* **85**: 15-21.

SMITH, P. J., 1986. Woodchip logging and birds near Bega, New South Wales. Pp. 259-71 *in* Birds of eucalypt forests and woodlands: ecology, conservation and management ed. by A. Keast, H. F. Recher, H. Ford and D. Saunders. R.A.O.U. Surrey Beatty & Sons: New South Wales.

SMITH, P. J., 1989. Changes in a forest bird community during a period of fire and drought near Bega, New South Wales. *Aust. J. Ecol.* **14**: 41-54.

WAKEFIELD, M., 1961. Victoria's Rock-Wallabies. *Vic. Nat.* **77**: 322-33.

Changes in macropodoid communities and populations in the past 200 years, and the future

J. H. CALABY[1] and G. C. GRIGG[2]

This paper presents tables which summarize probable changes in populations of Australian Macropodoidea, and probable causes, since white settlement. Of 50 species, 30 have suffered declines in range and/or numbers and 14 are thought to have increased or remained essentially stable. Six species have declined to extinction and an additional four are extinct on the mainland. Only about half, 28 species, remain common where they now occur. Habitat changes are likely to be the major causative agents of change, whether positive or negative; the most important are the clearing of forests, pastoralism and grazing. Feral animals may have been significant in the decline of five species. Despite the losses, the full range of macropod genera is still well represented, by good luck rather than good management. Tree-kangaroos, Hare-wallabies and Nailtail Wallabies are most in need of active conservation if diversity is to be preserved.

INTRODUCTION

It is common knowledge that there have been changes in the distribution and population status of Macropodoidea, along with all Australian mammals, throughout Quarternary and Holocene times. In historical times at least, the change in the majority of cases has been one of decline, sometimes drastic, and for many of the species there is no general agreement on causes. Few cases of change in status were studied while they were in progress. The species have usually quietly slipped away unnoticed and usually no alarm has been sounded until it suddenly dawned upon country people that one species or another was no longer seen and perhaps had not been seen for years. Because of this, the time when a decline began is always uncertain and the suggested causes are usually based on circumstantial evidence or are no more than speculation. A major problem when considering the early phases of a decline is that there are no measures of relative abundance before European settlement had much effect. Virtually the only sources of information are the results of early museum collecting expeditions, e.g., Blandowski and Krefft in the Murray-Darling junction area 1856-1857 (Wakefield 1966), and Shortridge in south-western Australia 1904-1907 (Shortridge 1910). The latter complained that "up to quite recently — within the last 25 or 30 years — from abundant evidence, many of the Western Australian mammals had a much wider range than at the present time, their disappearance, which is said to have been first noticed about 1880, being most sudden and unaccountable". Disappearance or not, Shortridge was still able to collect in the south-west large numbers of specimens of species now entirely gone from the mainland, e.g., *Onychogalea lunata,* now presumed extinct everywhere — 23 specimens, *Lagostrophus fasciatus* — 22 specimens, *Bettongia lesueur* — 24 specimens.

Many accounts have been published which summarize the present status of macropod species, as far as that is known (e.g., Calaby 1971; Burbidge 1977; Poole 1979; Ride and Wilson 1982). However, there have been fewer systematic reviews of opinions about the likely changes since white settlement and the reasons for these changes. An attempt to provide such an up-to-date review forms the basis of this paper. This review is presented in the form of a table (Table 1), a revised version of one published previously (Grigg 1985), and general conclusions are then presented in summary tables. Much of the review represents opinion rather than hard fact, but solid information about most of the presumed changes and causes will never be available, and it seems useful to have a systematic record of the opinions held and educated guesses made by those closer to the events than people will be in, say, another 200 years.

From Table 1 we have distilled generalizations about the likely changes themselves (Table 2), the reasons for the changes (Table 3) and the current status of groups of species (Table 4). Only Australian species are considered, information about New Guinean species being too sketchy. The groupings are the same as those in Grigg (1985).

CHANGES SINCE EUROPEAN SETTLEMENT (Table 2)

Thirty species (60%) have suffered a large decrease in range and in numbers, seven to the

[1]C.S.I.R.O. Wildlife and Ecology, P.O. Box 84, Lyneham, Australian Capital Territory 2602.

[2]School of Biological Sciences, The University of Sydney, New South Wales 2006.
Present address: Department of Zoology, University of Queensland, St. Lucia, Queensland 4067.

Pages 813-20 in KANGAROOS, WALLABIES AND RAT-KANGAROOS ed. by G. Grigg, P. Jarman and I. Hume. Surrey Beatty & Sons Pty Limited, New South Wales, Australia, 1989.

Table 1. Australian Macropodoidea, their conservation status, likely change since white settlement and suggested reasons for the change.

Scientific name	Common name	Status	Pest destruction sometimes?	Commercial use?	Change since settlement	Suggested reasons
FAMILY POTOROIDAE (POTOROOS, BETTONGS AND RAT-KANGAROOS)						
Hypsiprymnodon moschatus	Musky Rat-kangaroo	Common, limited	—	—	Range reduced	Rainforest clearing in North Qld.
Potorous tridactylus	Long-nosed Potoroo	Common, limited	—	—	Range reduced	Clearing of east coast forests
Potorous platyops	Broad-faced Potoroo	Extinct	—	—	Extinction	Unknown
Potorous longipes	Long-footed Potoroo	Rare, endangered	—	—	Unknown	Habitat possibly threatened by woodpulp industry
Bettongia penicillata	Brush-tailed Bettong	Rare, endangered	—	—	Range reduced severely	Land clearing and impact of grazing animals, foxes
Bettongia tropica	Tropical Bettong	Limited, vulnerable	—	—	Range reduced	?
Bettongia gaimardi	Tasmanian Bettong	Common, limited	—	—	Extinction on mainland, common in Tasmania	Fox predation on mainland, habitat change
Bettongia lesueur	Burrowing Bettong	Limited, vulnerable (WA islands)	—	—	Extinction on mainland	Rabbits, foxes, cats
Aepyprymnus rufescens	Rufous Bettong (or Rat-kangaroo)	Common	—	—	Range reduced	Land clearing, grazing, foxes
Caloprymnus campestris	Desert Rat-kangaroo	Presumed extinct	—	—	Extinction?	Unknown
FAMILY MACROPODOIDEA (KANGAROOS AND WALLABIES)						
HARE-WALLABIES						
Lagorchestes leporides	Eastern Hare-wallaby	Extinct	—	—	Extinction	Grazing
Lagorchestes conspicillatus	Spectacled Hare-wallaby	Common (Qld) elsewhere common to rare, abundant Barrow Island	—	—	Decline on mainland	Grazing
Lagorchestes hirsutus	Rufous Hare-wallaby	Rare, vulnerable	—	—	Range severely reduced	Grazing, changed wildfire regime
Lagorchestes asomatus	Central Hare-wallaby	Extinct	—	—	Extinction	
Lagostrophus fasciatus	Banded Hare-wallaby	Vulnerable, limited to two WA islands	—	—	Extinction on mainland	
NAILTAIL WALLABIES						
Onychogalea unguifera	Northern Nailtail	Common	—	—	Local reductions in numbers	Grazing
Onychogalea fraenata	Bridled Nailtail	Endangered	Yes	—	Range severely reduced	Pastoral industry, clearing of habitat
Onychogalea lunata	Crescent Nailtail	Presumed extinct	—	—	Decline to extinction	?
ROCK-WALLABIES						
Petrogale lateralis	Black-footed Rock-wallaby	Vulnerable, scattered	—	—	Range reduction	Foxes, farming
Petrogale penicillata	Brush-tailed Rock-wallaby	Common	Yes	—	Range reduction	Foxes
Petrogale inornata	Unadorned Rock-wallaby	Common	—	—	Unknown	
Petrogale assimilis	Allied Rock-wallaby	Common	—	—	Probably little change	
Petrogale godmani	Godman's Rock-wallaby	Vulnerable	—	—	Unknown	
Petrogale rothschildi	Rothschild's Rock-wallaby	Common, limited	—	—	Unknown	
Petrogale xanthopus	Yellow-footed Rock-wallaby	Vulnerable, limited	—	Little in 19th century	Range considerably reduced	Land clearing, goats, grazing
Petrogale persephone	Proserpine Rock-Wallaby	Rare, endangered	—	—	Unknown	May be threatened by "natural" circumstances
Petrogale brachyotis	Short-eared Rock-wallaby	Common	—	—	Probably little change	
Petrogale burbidgei	Warabi	Common, limited	—	—	Little or none	
Petrogale concinna	Nabarlek, Little Rock-wallaby	Common in restricted area of occurrence	—	—	Little or none	
PADEMELONS						
Thylogale stigmatica	Red-legged Pademelon	Common	—	—	Range reduction	Clearing of rainforest
Thylogale billardierii	Tasmanian Pademelon	Abundant	Yes	Yes	Extinction on mainland, prob. increase in Tasmania	Unknown, clearing for agriculture adjacent to forest
Thylogale thetis	Red-necked Pademelon	Common	Yes (but not (legally)	No	Local eruptions	Clearing for agriculture adjacent to forest

Table 1. continued.

Scientific name	Common name	Status	Pest destruction sometimes?	Commercial use	Change since settlement	Suggested reasons
TYPICAL WALLABIES AND KANGAROOS						
Macropus parma	Parma Wallaby	Secure, scattered	—	—	Range reduction	Vulnerable to forest clearing
Macropus eugenii	Tammar Wallaby	Common, limited	Sometimes on Kangaroo Is.	Yes*	Range severely reduced	Land clearing for wheat, foxes
Macropus greyi	Toolache Wallaby	Extinct	Yes? Scalp bounty	—	Extinction	Hunting, foxes. Used for coursing
Macropus irma	Western Brush Wallaby	Common	—	—	Population decline	Land clearing, foxes
Macropus parryi	Whiptail Wallaby	Common	Yes	Yes (Qld)	Local eruptions	Grazer, benefitted from grassland increase
Macropus dorsalis	Black-striped Wallaby	Common	Yes (Qld)	Yes (Qld)*	Range reduction, but local eruptions (Qld)	Agriculture, grazing
Macropus rufogriseus	Red-necked or Bennett's Wallaby	Common	Yes	Yes (Tas & Qld*)	Little change, local eruptions	Agriculture
Macropus agilis	Agile Wallaby	Abundant	Yes (Qld)	No	Little change or increase	Pastoral industry
Macropus giganteus	Eastern Grey Kangaroo	Abundant	Yes	Yes	Increase in numbers and possibly range	Pastoral industry, land land clearing and dingo destruction
Macropus fuliginosus	Western Grey Kangaroo	Abundant	Yes	Yes	Increase in numbers and possibly range	Pastoral industry
Macropus robustus	Common Wallaroo	Abundant	Yes	Yes	Little change or increase	Pastoral industry
Macropus antilopinus	Antilopine Wallaroo	Common	—	—	Little change or increase	Pastoral industry
Macropus bernardus	Black Wallaroo	Restricted	—	—	Little change	
Macropus rufus	Red Kangaroo	Abundant	Yes	Yes	Increase in numbers and possibly range	Pastoral industry incl. dingo destruction and provision of water
ANOMALOUS WALLABIES						
Setonix brachyurus	Quokka	Vulnerable, limited	—	—	Range reduced	?
Wallabia bicolor	Swamp Wallaby	Common	Yes (Qld)	Yes (Qld)	Little change, local eruptions	
TREE-KANGAROOS						
Dendrolagus lumholtzi	Lumholtz's Tree-Kangaroo	Vulnerable, limited	—	—	Range severely reduced	Clearing of rainforests
Dendrolagus bennettianus	Bennett's Tree-Kangaroo	Vulnerable, limited	—	—	Range reduced	Clearing of rainforests

Note: extinction of rock-wallabies (?*P. lateralis*) on some offshore islands by hunting and ?foxes.
*Commercial harvesting discontinued.

point of extinction. Note that they are the smaller species.

Fourteen species (28%) are thought to have been unaffected so far or to have increased.

About six (12%) of the species (five of them rock-wallabies) little is known and/or authors have been reluctant to make a guess about the likely changes in the last 200 years.

Most significantly, *only some* of the kangaroos and wallabies in the genus *Macropus* are thought to have shown either no change, or an increase.

REASONS FOR CHANGES (Table 3)

Habitat changes (clearing, pastoralism) are thought to have been implicated in 18 of the 30 extinctions or declines. On the other hand, five species are thought to have benefitted, particularly some of the kangaroos and wallabies in the genus *Macropus*.

Feral predators are less implicated than has often been suggested. There is no real evidence that cats are of concern as predators of macropodoids; however, fox predation may be a serious problem, as discussed later in this paper.

A rare and interesting situation brought to light in recent studies of genetics of rock-wallabies, is that *Petrogale godmani* may be endangered in the long term by introgressive expansion northward of the Mareeba race of *P. assimilis* (Briscoe *et al.* 1982).

PRESENT STATUS (Table 4)

Looking at the eight groups of species which can be identified within the Macropodoidea, it is with relief that we can note the good spread of presently "safe" species. Diversity still exists at the genetic level and we should certainly be taking preservation of diversity into account when identifying conservation priorities.

By that logic, conservation of Tree-kangaroos is an urgent priority, with Hare-wallabies, and then Nailtail Wallabies next on the list.

Pademelons and species of *Macropus* are best off.

Table 2. Summary of likely changes in Australian Macropodoidea, distilled from Table 1. Of 50 species, 29 have declined and 15 have increased or remained stable. The situation in six species is unknown. In parentheses are the numbers of species extinct on the mainland but surviving on islands.

	Number	Extinct	Range reduction	Increase in number	About the the same	Unknown
POTOROIDAE						
Potoroos, Bettongs and Rat Kangaroos	10	2 (2)	7	0		1
MACROPODOIDEA						
Hare-Wallabies	5	2 (1)	3			
Nailtail Wallabies	3	1	2			
Rock-Wallabies	11	0	3		3	5
Pademelons	3	0 (1)	2	1		
Typical Kangaroos and Wallabies	14	1	4	4	5	
'Anomalous' Wallabies	2	0	1		1	
Tree-Kangaroos	2	0	2			
Totals:	50	6 (4)	24	5	9	6

Only 56% of the species may be classified as "common" where they occur and are probably safe for the immediately foreseeable future.

It is worth noting that the premise is often accepted that all 50 macropodoids would have remained safe if Europeans had not arrived. This is not necessarily so; extinction is, after all, a natural process. Three species, *P. platyops, L. fasciatus* and *M. greyi* (discussed below) may have been already in terminal decline and there may have been others.

The Broad-faced Potoroo *(Potorous platyops)* apparently occupied a drier habitat than the other species of potoroos. It was collected as a living animal in only three localities in southwestern Australia, and the last specimens were registered in the Museum of Victoria in 1875, although their dates of collection are unknown. At that date the amount of land cleared for agriculture in Western Australia was small, and no exotic predatory or herbivorous mammals, except the dingo and possibly cats, had become established in the wild in that region. The species has an extensive distribution, however, in Pleistocene to late Holocene cave and Aboriginal occupational deposits in coastal regions of southern Western and South Australia, as far east as southeastern South Australia (localities listed in Lundelius and Turnbull 1984).

The Banded Hare-wallaby *(Lagostrophus fasciatus)* was apparently fairly widespread in shrub woodland and shrubby sand-plain in south-western Australia. The last specimen was collected there in 1906 (one of Shortridge's sample of 22!). It is found today only on Bernier and Dorre Islands. Both Gould (1849) and Sturt (1849) recorded it as a living animal in South Australia but there are no known modern specimens from the state and these records have usually been ignored. However, remains of the species are common throughout a dated Aboriginal occupational deposit on the lower Murray, accumulated over the past approximately 5 000 years (Wakefield 1964).

In historical times the Toolache Wallaby *(Macropus greyi)* occurred in a relatively limited area

Table 3. Summary of likely reasons for changes in Australian Macropodoidea since white settlement, distilled from Table 1.

	Reasons for extinction or decline				Reasons for increase	
	Pastoralism and grazing	Forest clearing	Ferals	Unknown	Pastoralism and grazing	Forest clearing
POTOROIDAE						
Potoroos, Bettongs etc.	2	3	2	2		
MACROPODOIDEA						
Hare-Wallabies	3			2		
Nailtail Wallabies	2			1		
Rock-Wallabies	1		2			
Pademelons		1		1		1
Typical Kangaroos and Wallabies		3	1		3	1
Anomalous W				1		
Tree-Kangaroos		2				
Totals	8	9	5	7	3	2

Table 4. Summary of present status of Australian Macropodoidea, categorized by groups of species. In parentheses are the numbers of species extinct on the mainland but surviving on islands. Tree-kangaroos, hare-wallabies and nailtail wallabies are identified as most in need of attention if diversity is to be preserved.

	Total species	Extinct	Vulnerable, rare or endangered	Common where they remain
POTOROIDAE				
Potoroos, Bettongs and Rat-Kangaroos	10	2 (2)	4	4 (30%)
MACROPODOIDEA				
Hare-wallabies	5	1	2	1 (20%)
Nailtail Wallabies	3	1	1	1 (33%)
Rock-Wallabies	11	0	5	6 (55%)
Pademelons	3	0 (1)		3 (100%)
Typical Kangaroos and Wallabies	14	1	0	13 (93%)
'Anomalous' W	2	0	1	1 (50%)
Tree-kangaroos	2	0	2	0 (0%)
Totals:	50	6 (4)	15	28 (56%)

of southeastern South Australia and a small adjacent area of western Victoria, although Wakefield (1963) has pointed out that there are no modern specimens from Victoria. Its preferred habitat was an almost treeless dense tall sedge and grassland community that was swampy in winter. Agricultural development and human persecution led to an accelerating decline in the species' numbers and it survived for only 85 years of European occupation. The last known specimen was rescued from dogs in 1927 and died in captivity in 1939 (Robinson and Young 1983). The species had a much broader distribution in late Pleistocene to Holocene time, including northwestern Tasmania, Hunter Island, Bass Strait, Kangaroo Island, South Australia, and western Victoria (to about 200 km east of the South Australian border), (Horton and Murray 1980; Wakefield 1963).

GENERALIZATIONS RELATING TO HABITAT

In spite of the historical and ecological uncertainties, some points are clear. The least changes have taken place in the monsoonal tropical woodlands and forest. This part of Australia has always been the most sparsely populated, at least by non-Aborigines. The major European influences have been the introduction of livestock and changed burning regimes. There has been a minor retraction of range in some species, e.g., the eastern fringe of the range of *Petrogale brachyotis* towards the Queensland border. This is a cattle production area (depending on the export price of beef!) and appears to have been subject to over-enthusiastic burning (Calaby 1976). Populations of *Macropus antilopinus* and *M. agilis* are found in largest concentration in managed cattle country which may signify that there has been an increase in population of these species.

In the tropical rainforest there has been a reduction in range of all species, notably the two species of *Dendrolagus*; however, no rainforest species can be said to be endangered. There is surely no argument other than that the cause of the decrease is clearing of the habitat. Although not a rainforest animal, one enigmatic tropical species is *Bettongia tropica* known from a relatively limited area in the Atherton Tableland, and from a specimen collected in 1884 in the Dawson Valley, WSW of Rockhampton (Wakefield 1967; Sharman *et al.* 1980).

The larger macropodids of the forest of eastern, and southeastern Australia, including Tasmania, are still abundant in spite of the fact that the forests have been exploited for more than a century and the region has the highest human density in Australia. There have been range reductions because of the clearing of forest for agriculture and urban development. On the other hand, creation of forest edge and grassed fields have led to increases in numbers of some species. In contrast to the large species, the smaller ones have fared badly. Two species have disappeared from the mainland; *Bettongia gaimardi,* which formerly ranged from southern Victoria into southeastern Queensland, and *Thylogale billardierii,* found originally in southern Victoria. Both species have maintained their status in Tasmania. Suggestions as to why these species became extinct on the mainland are no more than guesses. *Potorous tridactylus* is still common on the mainland in a number of localities but its range has been reduced and it appears to exist now in a number of isolated populations. The species disappeared from its limited range in southwestern Australia during the last century. The recently described *P. longipes* is known from a very limited area of far eastern Victoria and a very small adjacent part of New South Wales.

It is in the vast area of semi-arid woodlands (the rangelands) and arid shrublands that occupy the greater part of Australia, that the greatest calamity has overtaken the smaller Macropodoidea. Paradoxically the large species have increased greatly in numbers in large parts of this enormous area and these species have never been at risk, except in some local areas. Four species (*Caloprymnus campestris, Onychogalea lunata, Lagorchestes leporides, L. asomatus*) which once occupied these habitats are presumed to be extinct everywhere. Likewise *Potorous platyops,* which apparently lived in much drier woodlands or heaths than its congeners. Two formerly very wide-ranging species *(Bettongia lesueur, Lagostrophus fasciatus)* have gone altogether from the mainland and survive only on islands, and three others *(Onychogalea fraenata, Lagorchestes hirsutus, Bettongia penicillata)* are reduced to remnant colonies. A number of reasons have been invoked in attempts to explain this situation but, again, evidence is scant and superficial if it exists at all. It is probable that a number of factors are involved and there is no single cause. A major factor in the exploited rangelands was almost certainly the introduction of livestock.

Bolton and Latz (1978) have produced evidence that *L. hirsutus* in central desert areas may have declined to its present remnant status because of changed burning patterns since Aborigines gave up their traditional land-management practices. This does not explain, however, the species' disappearance about the beginning of the century from the woodlands of southwestern Australia. Kinnear (1987) has shown that foxes are an important factor in the decline of *Petrogale lateralis* in southwestern Australia, and Christensen (1980) has shown that foxes are heavy predators on *Bettongia penicillata* and *Macropus eugenii* in the period following burning of their habitat, before their shelter vegetation is re-established.

THE LARGER SPECIES

It was pointed out long ago (Calaby 1971) that smaller macropodoids seem to be more vulnerable than larger species and, elsewhere in this volume, Johnson *et al.*, refer to a threshold weight of 5.5 kg separating more from less vulnerable species. The extent to which this correlation relates to body weight or to taxonomic affinities is quite uncertain because, of the 17 species with mean body weights greater than 5.5 kg, all but five are in the genus *Macropus.*

Whatever the reason, several of the large "typical" wallabies and kangaroos appear to have prospered with the advent of the pastoral industry (Table 2). This is hardly surprising since three major activities of pastoralists, introduction of watering points, clearing to promote grasslands and the reduction or elimination of dingoes, are each and all likely to favour increase in the rangeland kangaroos. Collectively, the Red *(M. rufus)*, Western Grey *(M. fuliginosus)* and Eastern Grey *(M. giganteus)* Kangaroos number in the tens of millions.

It is these three species of which most is known about their distribution and population dynamics. Extensive and intensive aerial surveys have been undertaken since the mid-1970's, annually in New South Wales (Caughley, Bayliss and Giles 1984) and South Australia (Caughley and Grigg 1981; Grigg, Beard and Cairns, in prep.), in parts of Queensland (Caughley and Grigg 1982), and Western Australia (Short *et al.* 1983). The whole distribution of these three species, except for the Great Dividing Range, has been surveyed in 1981 (Caughley *et al.* 1983) and in 1984 (Grigg *et al.* 1985). The surveys have clarified the distribution of these species and have returned data which provide a fertile field for analysing distribution in relation to climatic variables (e.g., Caughley *et al.* 1987; Short *et al.* 1983), habitat (e.g., Short and Grigg 1982; Cairns, Pople and Grigg 1988) and particularly, the effects of drought (Caughley *et al.* 1985) and construction of rain-driven models (e.g., Caughley *et al.* 1984; Cairns, this volume). No attempt will be made at an extensive review here. In a nutshell, this extensive research has shown that habitat changes towards wheat and intensive pastoral activity reduce kangaroo numbers and that rain or its lack is the main driving force in population dynamics. All of the available evidence suggests that the long term operation of the kangaroo industry has induced no significant widespread reduction in kangaroo populations. Most compelling, even more than the aerial survey data (which are available for only 10-12 years) is the lack of downturn in harvests over the last 100 years or so. In Queensland, for example, 66 million kangaroos are known to have been harvested for commercial use since 1879 (Robertshaw and Harden, this volume), without any apparent trend upwards or downwards. Data for New South Wales allow the same conclusion to be drawn. Recently, the idea has been promoted that kangaroo meat should be the subject of an intensive marketing drive, raising prices to an extent that will encourage graziers in marginal grazing lands to reduce sheep numbers in favour of the endemic kangaroos (Grigg 1987). The proposal was the subject of close scrutiny at a conference of the Royal Zoological Society of New South Wales (Lunney and Grigg 1988). Whether there is long term benefit in this proposal remains to be seen.

THE FUTURE

What of the future for the Macropodoidea? Well, some comfort can be taken from the observation that "now safe" species represent a wide diversity of known genera. At least the genetic material is still there for the long term conservation of this diversity to be at least a possibility.

With the rare or vulnerable species what is needed is a great deal more survey work to determine their present numbers and define their present distribution more precisely, and more particularly a great deal more ecological research into such things as factors involved in changes in habitat over time.

Habitat change appears to present the largest possible threat, and the restriction of so many species to offshore islands is a cause for concern.

The present situation is one rather of good luck than good management, and too much money, comparatively, is being spent on the most abundant species instead of on those whose status is unknown or uncertain. While, in response to public pressure, the focus of most expenditure is the large commercial species, most Australians are at best only vaguely aware of the rich diversity that the macropods represent. Most people overseas seem to think that the large kangaroos are endangered and under threat from commercial harvesting, despite the wealth of evidence to the contrary.

How is it that we have a situation where the large kangaroos, which are among the world's most widespread and abundant large mammals, are or have been on the U.S. Fish and Wildlife Service's list of Endangered Species? While the focus is on the large species, there will be little community pressure for expenditure of funds on active measures to conserve the smaller, apparently more vulnerable species. What has happened of course, is that scientists have lost the initiative in setting the agenda for the conservation of the Macropodoidea. We should be trying to regain that initiative.

REFERENCES

BOLTON, B. L. AND LATZ, P. K., 1978. The western hare-wallaby *Lagorchestes hirsutus* (Gould) (Macropodidae) in the Tanami Desert. *Aust. Wildl. Res.* **5**: 285-93.

BRISCOE, D. A., CALABY, J. H., CLOSE, R. L., MAYNES, G. M., MURTAGH, C. E. AND SHARMAN, G. B., 1982. Isolation, introgression and genetic variation in rock-wallabies. Pp. 73-87 *in* Species at risk: research in Australia ed. by R. H. Groves and W. D. L. Ride. Australian Academy of Science: Canberra.

BURBIDGE, A. A. (ed.), 1977. The status of kangaroos and wallabies in Australia. Report of the Working Group on macropod habitat of the Standing Committee of the Council of Nature Conservation Ministers. Australian Government Publishing Service: Canberra.

CALABY, J. H., 1971. The current status of Australian Macropodidae. *Aust. Zool.* **16**: 17-29.

CALABY, J. H., 1976. Mammals. *In* A survey of the fauna of the lower McArthur River region, Northern Territory. Unpublished report for Mimets Development Pty Ltd.

CAIRNS, S. C., POPLE, A. R. AND GRIGG, G. C., 1988. Habitat associations of kangaroos in the South Australian pastoral zone. Australian National Parks and Wildlife Service unpublished report.

CAUGHLEY, G. J. AND GRIGG, G. C., 1981. Surveys of the distribution and density of kangaroos in the pastoral zone of South Australia and their bearing on the feasibility of aerial survey in large and remote areas. *Aust. Wildl. Res.* **8**: 1-11.

CAUGHLEY, G. J. AND GRIGG, G. C., 1982. Numbers and distribution of kangaroos in the Queensland Pastoral Zone. *Aust. Wildl. Res.* **9**: 365-71.

CAUGHLEY, G. J., GRIGG, G. C. AND SHORT, J., 1983. How many kangaroos? *Search* **14**: 99-108.

CAUGHLEY, G. J., GRIGG, G. C. AND SMITH, L., 1985. The effect of drought on kangaroo populations. *J. Wildl. Manage.* **49**: 679-85.

CAUGHLEY, G. J., SHORT, J., GRIGG, G. C. AND NIX, H., 1987. Kangaroos and climate: an analysis of distribution. *J. Anim. Ecol.* **56**: 751-61.

CAUGHLEY, J., BAYLISS, P. AND GILES, J., 1984. Trends in kangaroo numbers in western New South Wales and their relation to rainfall. *Aust. Wildl. Res.* **11**: 415-22.

CHRISTENSEN, P. E. S., 1980. The biology of *Bettongia penicillata* Gray 1837, and *Macropus eugenii* (Desmarest 1817) in relation to fire. *Forests Dept. W. Aust. Bull.* 11.

GOULD, J., 1849. *Lagorchestes fasciatus*. Text to plate 56. The Mammals of Australia, Vol. 2. The Author, London.

GRIGG, G. C., 1985. Kangaroos and the future. *In* The Kangaroo ed. by M. Archer and T. F. Flannery. Weldon: Sydney.

GRIGG, G. C., 1987. Kangaroos — A better economic alternative in the marginal grazing lands? *Aust. Zool.* **24**: 73-80.

GRIGG, G. C., BEARD, L. A., CAUGHLEY, C. J., GRICE, D., CAUGHLEY, J. A., SHEPHERD, N., FLETCHER, M. AND SOUTHWELL, C., 1985. The Australian kangaroo population 1984. *Search* **16**: 277-79.

GRIGG, G. C., BEARD, L. A. AND CAIRNS, S. C. (in prep.). Trends in the populations of red kangaroos *(Macropus rufus)* and western grey kangaroos *(M. fuliginosus)* in relation to rainfall in the pastoral zone of South Australia.

HORTON, D. R. AND MURRAY, P., 1980. The extinct toolache wallaby *(Macropus greyi)* from a spring mound in north western Tasmania. *Rec. Queen Victoria Mus.* No. **71**: 1-12.

KINNEAR, J. E., 1987. The red fox *(Vulpes vulpes)*: a major threat to the conservation of Australian mammals. Australian Mammalogy Society. Abstracts for 33rd Scientific meeting. P.7.

LUNDELIUS, E. L. AND TURNBULL, W. D., 1984. The mammalian fauna of Madura Cave, Western Australia, Part VI: Macropodidae: Potorinae. *Fieldiana Geology* New Ser. No. 14.

LUNNEY, D. AND GRIGG, G. C. (eds.), 1988. Kangaroo harvesting and the conservation of arid and semi-arid lands. *Aust. Zool.* **24**: 121-93.

POOLE, W. E., 1979. The status of the Australian Macropodidae. Pp. 13-27 *in* The status of endangered Australasian wildlife ed. by M. Tyler. Royal Zoological Society South Australia, Adelaide.

RIDE, W. D. L. AND WILSON, G. R., 1982. The conservation status of Australian animals. Pp. 27-44 and Appendix 1. Pp. 191-203 *in* Species at risk: research in Australia ed. by R. H. Groves and W. D. L. Ride. Australian Academy of Science, Canberra.

ROBINSON, A. C. AND YOUNG, M. C., 1983. The toolache wallaby (*Macropus greyi* Waterhouse). Nat. Parks Wildl. Serv., Dept. Env. Plan. S. Aust. Spec. Publ. No. 2: 1-54.

SHARMAN, G. B., MURTAGH, C. E., JOHNSON, P. M. AND WEAVER, C. M., 1980. The chromosomes of a rat-kangaroo attributable to *Bettongia tropica* (Marsupialia: Macropodidae). *Aust. J. Zool.* **28:** 59-63.

SHORT, J., CAUGHLEY, G. J., GRICE, D. AND BROWN, B., 1983. The distribution and abundance of kangaroos in relation to environment in Western Australia. *Aust. Wildl. Res.* **10:** 435-51.

SHORT, J. AND GRIGG, G. C., 1982. The abundance of kangaroos in suboptimal habitats: wheat, intensive pastoral and mallee. *Aust. Wildl. Res.* **9:** 221-27.

SHORTRIDGE, G. C., 1910. Account of the geographical distribution of the marsupials and monotremes of south-west Australia, having special reference to the specimens collected during the Balston Expedition of 1904-1907. *Proc. Zool. Soc. Lond., 1909:* 803-48.

STURT, C., 1849. Narrative of an expedition into central Australia . . . T. and W. Boone: London.

WAKEFIELD, N. A., 1963. Subfossils from Mount Hamilton, Victoria. *Vict. Nat.* **79:** 323-30.

WAKEFIELD, N. A., 1964. Mammal remains. *Proc. R. Soc. Vict.* **77:** 494-98.

WAKEFIELD, N. A., 1966. Mammals of the Blandowski Expedition to northwestern Victoria, 1856-57. *Proc. R. Soc. Vict.* **79:** 371-91.

WAKEFIELD, N. A., 1967. Some taxonomic revision in the Australian marsupial genus *Bettongia* (Macropodidae), with description of a new species. *Vict. Nat.* **84:** 8-22.

INDEX

Bold figures indicate major reference

Q

R

T

TABLES OF DATA

These tables may contain information on various macropodoids which has not been listed in the index.